Beginning and Intermediate Algebra

with Applications and Visualization

Gary K. Rockswold
Minnesota State University, Mankato

Terry A. Krieger
Rochester Community and Technical College

PEARSON

Boston   Columbus   Indianapolis   New York   San Francisco   Upper Saddle River
Amsterdam   Cape Town   Dubai   London   Madrid   Milan   Munich   Paris   Montreal   Toronto
Delhi   Mexico City   Sao Paulo   Sydney   Hong Kong   Seoul   Singapore   Taipei   Tokyo

Editorial Director Chris Hoag
Executive Editor Cathy Cantin
Executive Content Editor Kari Heen
Senior Content Editor Lauren Morse
Editorial Assistant Kerianne Okie
Executive Director of Development Carol Trueheart
Senior Managing Editor Karen Wernholm
Associate Managing Editor Tamela Ambush
Senior Production Project Manager Sheila Spinney
Digital Assets Manager Marianne Groth
Supplements Production Coordinator Kerri Consalvo
Associate Media Producer Jonathan Wooding
Associate Content Manager Eric Gregg
Senior Content Developer Mary Durnwald
Marketing Manager Rachel Ross
Marketing Assistant Ashley Bryan
Senior Author Support/Technology Specialist Joe Vetere
Rights and Permissions Advisor Michael Joyce
Image Manager Rachel Youdelman
Procurement Manager Evelyn Beaton
Procurement Specialist Debbie Rossi
Media Procurement Specialist Ginny Michaud
Associate Director of Design, USHE North and West Andrea Nix
Senior Designer Barbara T. Atkinson
Text Design, Production Coordination, Composition, and Illustrations Cenveo Publisher
    Services/Nesbitt Graphics, Inc.
Cover Image Water cascading over rocks @ Khomulo Anna/Shutterstock
Cover Design Studio Montage

For permission to use copyrighted material, grateful acknowledgment is made to the copyright
holders on page G-8, which is hereby made part of this copyright page.

Many of the designations used by manufacturers and sellers to distinguish their products are claimed
as trademarks. Where those designations appear in this book, and Pearson was aware of a trademark
claim, the designations have been printed in initial caps or all caps.

Library of Congress Cataloging-in-Publication Data
Rockswold, Gary K.
   Beginning and intermediate algebra with applications and visualization / Gary K.
   Rockswold, Terry A. Krieger. –3rd ed.
     p. cm.
   Includes bibliographical references and index.
  ISBN-13: 978-0-321-75651-0 (hardcover : alk. paper)
  ISBN-10: 0-321-75651-7 (hardcover : alk. paper)
   1. Algebra–Textbooks. I. Krieger, Terry A. II. Title.
QA152.3.R595 2013

512–dc22   2011007441

5 6 7 8 9 10—V057—14

ISBN 13: 978-0-321-75651-0
ISBN 10: 0-321-75651-7

To

Gary, Brian, and their crews for excellence and integrity

Gary Rockswold has been a professor and teacher of mathematics, computer science, astronomy, and physical science for over 35 years. Not only has he taught at the undergraduate and graduate college levels, but he has also taught middle school, high school, vocational school, and adult education. He received his BA degree with majors in mathematics and physics from St. Olaf College and his PhD in applied mathematics from Iowa State University. He has been a principal investigator at the Minnesota Supercomputer Institute, publishing research articles in numerical analysis and parallel processing. He is currently an emeritus professor of mathematics at Minnesota State University–Mankato. He is an author for Pearson Education and has over 10 current textbooks at the developmental and precalculus levels. His developmental co-author and friend is Terry Krieger. They have been working together for over a decade. Making mathematics meaningful for students and professing the power of mathematics are special passions for Gary. In his spare time he enjoys sailing, doing yoga, and spending time with his family.

Terry Krieger has taught mathematics for over 20 years at the middle school, high school, vocational, community college, and university levels. His undergraduate degree in secondary education is from Bemidji State University in Minnesota, where he graduated summa cum laude. He received his MA in mathematics from Minnesota State University–Mankato. In addition to his teaching experience in the United States, Terry has taught mathematics in Tasmania, Australia, and in a rural school in Swaziland, Africa, where he served as a Peace Corps volunteer. Terry is currently teaching at Rochester Community and Technical College in Rochester, Minnesota. He has been involved with various aspects of mathematics textbook publication for more than 15 years and has joined his friend Gary Rockswold as co-author of a developmental math series published by Pearson Education. In his free time, Terry enjoys spending time with his wife and two boys, physical fitness, wilderness camping, and trout fishing.

Contents

1 Introduction to Algebra   1

4 — Systems of Linear Equations in Two Variables — 245

5 — Polynomials and Exponents — 297

6 Factoring Polynomials and Solving Equations   359

9   Systems of Linear Equations   590

10   Radical Expressions and Functions   631

11   Quadratic Functions and Equations   713

Preface

Beginning and Intermediate Algebra with Applications and Visualization, Third Edition, connects the real world to mathematics in ways that are both meaningful and motivational. Students using this textbook will have no shortage of realistic and convincing answers to the question, "Why is math important in my life?" Mathematical concepts are introduced by using relevant applications from students' lives that support their conceptual understanding. This textbook does not present math as a sequence of disconnected procedures to be memorized; rather, it shows that math does make sense.

In addition to presenting symbolic techniques, we present mathematical concepts in both visual and numerical ways whenever possible. Graphs, diagrams, and tables of values are often included to provide students with an opportunity to understand math in more than one way. Multiple representations (symbolic, graphical, and numerical) allow students to solve problems using three different methods. This textbook prepares students for future mathematics courses by teaching them both reasoning and problem-solving skills.

As mathematicians, we understand the profound impact that mathematics has on the world around us; as teachers with a combined experience of over 50 years, we know that our students often have difficulty both learning mathematics and recognizing its relevance. We believe that applications and visualization are essential pathways to empowering students with mathematics. Our goal in *Beginning and Intermediate Algebra with Applications and Visualization* is to help students succeed.

This textbook is one of four textbooks in our series:

- *Prealgebra*
- *Beginning Algebra with Applications and Visualization*, Third Edition
- *Beginning & Intermediate Algebra with Applications and Visualization*, Third Edition
- *Intermediate Algebra with Applications and Visualization*, Fourth Edition.

New to this Edition

- **Updated Data** The data in hundreds of exercises and examples have been updated to keep the applications relevant and fresh.
- **More Visualization** Whenever appropriate, we have made mathematics more visual by using graphs, tables, charts, color, and diagrams to explain important concepts with fewer words.
  - We have added titles to the graphs and also comments on/near the graphs/tables to make them more (immediately) understandable. Now students don't always have to read the text to understand the table or graph.
  - We have added green equation labels, written in standard mathematical notation, to calculator graphs.
- **Teaching Examples** Every example in the text is now paired with a teaching example in the Annotated Instructor's Edition that instructors can use in class to promote further understanding.
- **New Vocabulary Checklist** At the start of every section that contains new mathematical definitions, a checklist of new vocabulary has been added to highlight the math concepts that are defined in the section.
- **Updated Glossary** The glossary has been updated to include all New Vocabulary words.

- **Reading Checks** To be sure that students are grasping important concepts, short questions about the material have been added throughout each section.
- **Study Tips** Throughout the text, study tips are now included to give not only general study advice but also pointers that relate to learning specific mathematical concepts.
- **New Application Topics** Students are more engaged when mathematics is tied to current and relevant topics. Several new examples have been added that discuss the mathematics of the Internet, social networks, tablet computers, and other contemporary topics.
- **Online Exploration** New exercises that invite students to use the Internet to explore real-world mathematics are now included throughout the text.
- **Refinement of Content**

CHAPTER 1: New step diagrams were added to provide students with an organized, visual process for finding the greatest common factor and the least common denominator.

CHAPTER 2: New tables summarizing words and phrases associated with mathematical symbols are now included to give students at-a-glance references to aid in setting up and solving application problems.

CHAPTER 3: At the request of reviewers, exercises that require students to complete tables of values for linear equations in two variables are now given in both vertical and horizontal format. Also, additional exercises asking students to interpret the meaning of the slope and the intercepts in applications modeled by linear equations in two variables are now included.

CHAPTER 4: The discussion and visual representations of the types of systems of equations have been moved to the first section of Chapter 4 to give students immediate exposure to the types of systems that they will encounter throughout the chapter.

CHAPTER 5: New tables summarizing rules for exponents are now included to give students immediate access to these important concepts.

CHAPTER 7: A new visual method for finding the least common denominator for rational expressions has been added to make this important concept more accessible to students.

CHAPTER 8: New visuals for several topics, such as functions, evaluating functions, absolute value equations, and absolute value inequalities, have been added. At the request of reviewers, more explanation is now included for finding the domains of functions. Additional opportunities to represent functions symbolically, numerically, graphically, and verbally have also been included.

CHAPTER 9: A new subsection in Section 9.2 related to social networks and matrices was added to keep topics current for students.

CHAPTER 10: At the request of reviewers, Section 10.1 was split into two sections so that radical expressions and rational exponents can be taught in two different sections. Section 10.1 was reorganized so that the square root and cube root functions can now be taught in Section 10.1 rather than later in the chapter. At the request of reviewers, greater emphasis has been given to converting between radical forms and forms with rational exponents.

CHAPTER 11: At the request of reviewers, Sections 11.1 and 11.2 have been reorganized to put topics in better order for student learning. The topic of increasing and decreasing functions is now optional in Section 11.2. In Section 11.5 the visuals for quadratic inequalities have been increased to make this topic more accessible to students.

CHAPTER 12: Section 12.2 has a new subsection on percent change and exponential functions. Greater emphasis was given to percentages and how they relate to exponential functions. At the request of reviewers, the topic of paying interest more than once a year was added in Section 12.2.

Hallmark Features

Math in the Real World

The Rockswold/Krieger series places an emphasis on teaching algebra in context. Students typically understand best when concepts are tied to the real world or presented visually. We believe that meaningful applications and visualization of important concepts are pathways to success in mathematics.

- **Chapter Openers** Each chapter opens with an application that motivates students by offering insight into the relevance of that chapter's mathematical concepts.

- **A Look into Math** Each section is introduced with a practical application of the math topic students are about to learn.

- **Real World Connection** Where appropriate, we expand on specific math topics and their connections to everyday life.

- **Applications** We integrate applications into both the discussions and the exercises to help students become more effective problem solvers.

- **Modeling Data** We provide opportunities for students to model real and relevant data with their own functions.

- **NEW! Online Exploration** These exercises invite students to find data on the Internet and then use mathematics to analyze the data.

Understanding the Concepts

Conceptual understanding is critical to success in mathematics.

- **Math from Multiple Perspectives** Throughout the text, we present concepts by means of verbal, graphical, numerical, and symbolic representations to support multiple learning styles and problem-solving methods.

- **New Vocabulary** At the beginning of each section, we direct the students' attention to important terms before they are discussed in context.

- **NEW! Reading Check** Reading Check questions appear along with important concepts, ensuring that students understand the material they have just read.

- **Study Tips** Study Tips offer just-in-time suggestions to help students stay organized and focused on the material at hand.

- **Making Connections** Throughout the text, we help students understand the relationship between previously learned concepts and new concepts.

- **Critical Thinking** One or more Critical Thinking exercises are included in most sections. They pose questions that can be used for classroom discussion or homework assignments.

- **Putting It All Together** At the end of each section, we summarize the techniques and reinforce the mathematical concepts presented in the section.

Practice

Multiple types of exercises in this series support the application-based and conceptual nature of the content. These exercise types are designed to reinforce the skills students need to move on to the next concept.

- **Extensive Exercise Sets** The comprehensive exercise sets cover basic concepts, skill-building, writing, applications, and conceptual mastery. These exercise sets are further enhanced by several special types of exercises that appear throughout the text:

  - **Now Try Exercises** Suggested exercises follow every example for immediate reinforcement of skills and concepts.

  - **Checking Basic Concepts** These mixed review exercises appear after every other section and can be used for individual or group review. These exercises require 10–20 minutes to complete and are also appropriate for in-class work.

  - **Thinking Generally** These exercises appear in most exercise sets and offer open-ended conceptual questions that encourage students to synthesize what they have just learned.

  - **Writing About Mathematics** This exercise type is at the end of most sections. Students are asked to explain the concepts behind the mathematics procedures they have just learned.

- **Group Activities: Working with Real Data** This feature occurs once or twice per chapter, and provides an opportunity for students to work collaboratively on a problem that involves real-world data. Most activities can be completed with limited use of class time.

- **Graphing Calculator Exercises** The icon denotes an exercise that requires students to have access to a graphing calculator.

- **Technology Notes** Occurring throughout the book, optional Technology Notes offer students guidance, suggestions, and cautions on the use of a graphing calculator.

Mastery

By reviewing the material and putting their abilities to the test, students will be able to assess their level of mastery as they complete each chapter.

- **Chapter Summary** In a quick and thorough review, we combine key terms, topics, and procedures with illuminating examples to assist in test preparation.

- **Chapter Review Exercises** Students can work these exercises to gain confidence that they have mastered the material.

- **Extended And Discovery Exercises** These capstone projects at the end of every chapter challenge students to synthesize what they have learned.

- **Chapter Test** Students can reduce math anxiety by using these tests as a rehearsal for the real thing.

- **Chapter Test Prep Videos** Students can prepare for the chapter test by viewing videos of an instructor working through the complete solutions for all Chapter Test exercises for each chapter.

- **Cumulative Review Exercises** Starting with Chapter 2 and appearing in all subsequent chapters, Cumulative Review Exercises help students see the big picture of math by reviewing topics and skills they have already learned.

Supplements

STUDENT RESOURCES

Student's Solutions Manual

- Contains solutions for the odd-numbered section-level exercises (excluding Writing About Mathematics and Group Activity exercises), and solutions to all Concepts and Vocabulary exercises, Checking Basic Concepts exercises, Chapter Review Exercises, Chapter Test exercises, and Cumulative Review Exercises.

  ISBNs: 0-321-75655-X, 978-0-321-75655-8

Worksheets

These lab- and classroom-friendly worksheets provide

- Learning objectives and key vocabulary terms for every text section, along with vocabulary practice problems.
- Extra practice exercises for every section of the text with ample space for students to show their work.
- Additional opportunities to explore multiple representation solutions.

  ISBNs: 0-321-75661-4, 978-0-321-75661-9

Video Resources

- A series of lectures correlated directly with the content of each section of the text.
- Material presented in a format that stresses student interaction, often using examples from the text.
- Ideal for distance learning and supplemental instruction.
- Video lectures include English captions.
- The Chapter Test Prep Videos let students watch instructors work through step-by-step solutions to all the Chapter Test exercises from the textbook. Chapter Test Prep videos are available on YouTube™ (search using author name and book title) as well as in MyMathLab®.
- Available in MyMathLab.

INSTRUCTOR RESOURCES

Annotated Instructor's Edition

- Contains Teaching Tips, Teaching Examples, and answers to every exercise in the textbook, excluding the Writing About Mathematics exercises.
- Answers that do not fit on the same page as the exercises themselves are supplied in the Instructor's Answer Appendix at the back of the textbook.

  ISBNs: 0-321-75652-5, 978-0-321-75652-7

Instructor's Resource Manual with Tests and Mini Lectures (Download Only)

Includes resources designed to help both new and adjunct faculty with course preparation and classroom management:

- Teaching tips and additional exercises for selected content.
- Five sets of Cumulative Review Exercises that cover Chapters 1–3, 1–6, 1–9, 1–12, and 1–14.
- Notes for presenting graphing calculator topics as well as supplemental activities.
- Three free-response alternate test forms and one multiple-choice test form per chapter; one free-response and one multiple-choice final exam.
- Available in MyMathLab and on the Instructor's Resource Center.

Instructor's Solutions Manual (Download Only)

- Provides solutions to all section-level exercises (excluding Writing About Mathematics exercises), and solutions to all Checking Basic Concepts exercises, Chapter Review Exercises, Chapter Test exercises, and Cumulative Review Exercises.
- Available in MyMathLab and on the Instructor's Resource Center.

PowerPoint Slides

- Key concepts and definitions from the text.
- Available in MyMathLab or can be downloaded from the Instructor's Resource Center.

TestGen®

TestGen (www.pearsoned.com/testgen) enables instructors to build, edit, print, and administer tests using a computerized bank of questions developed to cover all the objectives of the text. TestGen is algorithmically based, allowing instructors to create multiple but equivalent versions of the same question or test with the click of a button. Instructors can also modify test bank questions or add new questions. The software and testbank are available for download from Pearson Education's online catalog.

MathXL® Online Course (access code required)

MathXL is the homework and assessment engine that runs MyMathLab. (MyMathLab is MathXL plus a learning management system.) With MathXL, instructors can:

- Create, edit, and assign online homework and tests using algorithmically generated exercises correlated at the objective level with the textbook.
- Create and assign their own online exercises and import TestGen tests for added flexibility.
- Maintain records of all student work tracked in MathXL's online gradebook.

With MathXL, students can:

- Take chapter tests in MathXL and receive personalized study plans and/or personalized homework assignments based on their test results.
- Use the study plan and/or the homework to link directly to tutorial exercises for the objectives they need to study.
- Access supplemental animations and video clips directly from selected exercises.

MathXL is available to qualified adopters. For more information, visit **www.mathxl.com** or contact your Pearson representative.

MyMathLab Online Course (access code required)

MyMathLab delivers **proven results** in helping individual students succeed.

- MyMathLab has a consistently positive impact on the quality of learning in higher education math instruction. MyMathLab can be successfully implemented in any environment—lab-based, hybrid, fully online, traditional—and demonstrates the quantifiable effect that integrated usage has on student retention, subsequent success, and overall achievement.
- MyMathLab's comprehensive online gradebook automatically tracks students' results on tests, quizzes, and homework, and in the study plan. Instructors can use the gradebook to quickly intervene if their students have trouble or to provide positive feedback on a job well done. The data within MyMathLab are easily exported to a variety of spreadsheet programs, such as Microsoft Excel. Instructors can determine which points of data they want to export, and then analyze the results to determine success.

MyMathLab provides **engaging experiences** that personalize, stimulate, and measure learning for each student.

- **Exercises:** The homework and practice exercises in MyMathLab are correlated with the exercises in the textbook, and they regenerate algorithmically to give students unlimited opportunity for practice and mastery. The software offers immediate, helpful feedback when students enter incorrect answers.
- **Multimedia Learning Aids:** Exercises include guided solutions, sample problems, animations, videos, and eText clips for extra help at point-of-use.
- **Expert Tutoring:** Although many students describe the whole of MyMathLab as "like having your own personal tutor," students using MyMathLab do have access to live tutoring from qualified Pearson math and statistics instructors.

And, MyMathLab comes from a **trusted partner** with educational expertise and an eye on the future.

Knowing that you are using a Pearson product means knowing that you are using quality content. That means that our eTexts are accurate, that our assessment tools work, and that our questions are error-free. And whether you are just getting started with MyMathLab, or have a question along the way, we're here to help you learn about our technologies and how to incorporate them into your course.

To learn more about how MyMathLab combines proven learning applications with powerful assessment, visit **www.mymathlab.com** or contact your Pearson representative.

MyMathLab Ready to Go Course (access code required)

These new Ready to Go courses provide students with all the same great MyMathLab features that you're used to, but make it easier for instructors to get started. Each course includes pre-assigned homework and quizzes to make creating your course even simpler. Ask your Pearson representative about the details of this particular course or to see a copy of this course.

Acknowledgments

Many individuals contributed to the development of this textbook. We thank the following reviewers, whose comments and suggestions were invaluable in preparing *Beginning and Intermediate Algebra with Applications and Visualization,* Third Edition.

David Atwood, *Rochester Community and Technical College*
Cristina Berisso, *Shasta College*
Michael Bowen, *Oxnard College*
Barbara Burke, *Hawaii Pacific University*
Linda Chan, *California State Polytechnic University*
Patrick Cross, *University of Oklahoma*
Christopher Donnelly, *Macomb Community College*
John Driscoll, *Middlesex Community College—Middletown*
Di Dwan, *Yavapai College*
Lucy Edwards, *Las Positas College*
Jeri Hamilton, *Yavapai College*
Amy Hlavacek, *Saginaw Valley State University*
Brian Hons, *San Antonio College*
Steve Howard, *Rose State College*
Quillie Hunt, *Central Piedmont Community College*
Dr. Philip Kaatz, *Mesalands Community College*
Frederick Lippman, *Shasta College*
Ben Mayo, *Yakima Valley Community College*
Benjamin Moulton, *Utah Valley University*
Charlotte Newsom, *Tidewater Community College—Virginia Beach Campus*
Caleb Olaleye, *South Suburban College*
Gary Parker, *Blue Mountain Community College*
Amy Petty, *South Suburban College*
Pamelyn Reed, *Lone Star College—Cy-Fair*
Matt Shelton, *McLennan Community College*
Eden Thompson, *Utah Valley University*
Alma Wlazlinski, *McLennan Community College*
Bella Zamansky, *University of Cincinnati*
Loris Zucca, *Kingwood College*

We would like to welcome Jessica Rockswold to our team. She has been instrumental in developing, writing, and proofing new applications, graphs, and visualizations for this text.

Paul Lorczak, Mitchell Johnson, Perian Herring, Shannon d'Hemecourt, Lynn Baker, and Mark Rockswold deserve special credit for their help with accuracy checking. Without the excellent cooperation from the professional staff at Pearson Education, this project would have been impossible. Thanks go to Greg Tobin and Maureen O'Connor for giving their support. Particular recognition is due Cathy Cantin, who gave essential advice and assistance. The outstanding contributions of Lauren Morse, Sheila Spinney, Joe Vetere, Michelle Renda, Tracy Rabinowitz, Rachel Ross, Ashley Bryan, and Jonathan Wooding are greatly appreciated. Special thanks go to Kathy Diamond, who was instrumental in the success of this project.

Thanks go to Wendy Rockswold and Carrie Krieger, whose unwavering encouragement and support made this project possible. We also thank the many students and instructors who used the previous editions of this textbook. Their suggestions were insightful and helpful.

Please feel free to send us your comments and questions at either of the following e-mail addresses: gary.rockswold@mnsu.edu or terry.krieger@roch.edu. Your opinion is important to us.

Gary K. Rockswold
Terry A. Krieger

Index of Applications

Geometry

Government and Human Services

1

Introduction to Algebra

> Unless you try to do something beyond what you have already mastered, you will never grow.
>
> —RONALD E. OSBORN

Just over a century ago only about one in ten workers was in a professional, technical, or managerial occupation. Today this proportion is nearly one in three, and the study of mathematics is essential for anyone who wants to keep up with the technological changes that are occurring in nearly every occupation. Mathematics is the *language of technology*.

In the information age, mathematics is being used to describe human behavior in areas such as economics, medicine, advertising, social networks, and Internet use. For example, mathematics can help maximize the impact of advertising by analyzing social networks such as Facebook. Today's business managers need employees who not only understand human behavior but can also describe that behavior using mathematics.

It's just a matter of time before the *majority* of the workforce will need the analytic skills that are taught in mathematics classes every day. No matter what career path you choose, a solid background in mathematics will provide you with opportunities to reach your full potential in your vocation, income level, and lifestyle.

Source: A. Greenspan, "The Economic Importance of Improving Math-Science Education."

1.1   Numbers, Variables, and Expressions

Natural Numbers and Whole Numbers • Prime Numbers and Composite Numbers • Variables, Algebraic Expressions, and Equations • Translating Words to Expressions

A LOOK INTO MATH ▶

Numbers are an important concept in every society. A number system once used in southern Africa consisted of only the numbers from 1 to 20. Numbers larger than 20 were named by counting groups of *twenties*. For example, the number 67 was called *three twenties and seven*. This base-20 number system would not work well in today's technologically advanced world. In this section, we introduce two sets of numbers that are used extensively in the modern world: natural numbers and whole numbers.

Natural Numbers and Whole Numbers

NEW VOCABULARY

☐ Natural numbers
☐ Whole numbers
☐ Product
☐ Factors
☐ Prime number
☐ Composite number
☐ Prime factorization
☐ Variable
☐ Algebraic expression
☐ Equation
☐ Formula

One important set of numbers is the set of **natural numbers**. These numbers comprise the *counting numbers* and may be expressed as follows.

$$1, 2, 3, 4, 5, 6, \ldots$$

Because there are infinitely many natural numbers, three dots are used to show that the list continues without end. A second set of numbers is called the **whole numbers**, and may be expressed as follows.

$$0, 1, 2, 3, 4, 5, \ldots$$

Whole numbers include the natural numbers and the number 0.

CRITICAL THINKING

Give an example from everyday life of natural number or whole number use.

▶ **REAL-WORLD CONNECTION** Natural numbers and whole numbers can be used when data are not broken into fractional parts. For example, the bar graph in Figure 1.1 shows the number of apps on a student's iPad for the first 5 months after buying the device. Note that both natural numbers and whole numbers are appropriate to describe these data because a fraction of an app is not possible.

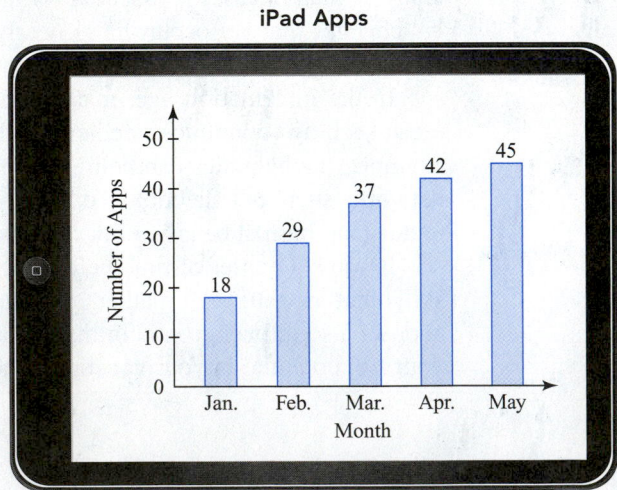

Figure 1.1

Prime Numbers and Composite Numbers

When two natural numbers are multiplied, the result is another natural number. For example, multiplying the natural numbers 3 and 4 results in 12, a natural number. The result 12 is called the **product** and the numbers **3** and **4** are **factors** of **12**.

$$3 \quad \cdot \quad 4 \quad = \quad 12$$
$$\text{factor} \quad \text{factor} \quad \quad \text{product}$$

NOTE: Products can be expressed in several ways. For example, the product $3 \cdot 4 = 12$ can also be written as $3 \times 4 = 12$ and $3(4) = 12$.

A natural number greater than 1 that has *only* itself and 1 as natural number factors is a **prime number**. The number 7 is prime because the only natural number factors of 7 are 1 and 7. The following is a partial list of prime numbers. There are infinitely many prime numbers.

$$2, \quad 3, \quad 5, \quad 7, \quad 11, \quad 13, \quad 17, \quad 19, \quad 23, \quad 29$$

READING CHECK

• How do prime numbers differ from composite numbers?

A natural number greater than 1 that is not prime is a **composite number**. For example, the natural number 15 is a composite number because $3 \cdot 5 = 15$. In other words, 15 has factors other than 1 and itself.

Every composite number can be written as a product of prime numbers. For example, we can use a factor tree such as the one shown in Figure 1.2 to find the prime factors of the composite number 120. Branches of the tree are made by writing each composite number as a product that includes the smallest possible prime factor of the composite number.

CRITICAL THINKING

Suppose that you draw a tree diagram for the prime factorization of a prime number. Describe what the tree will look like.

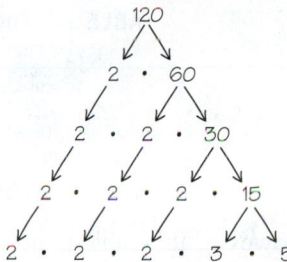

Figure 1.2 Prime Factorization of 120

Figure 1.2 shows that the **prime factorization** of 120 is

$$120 = 2 \cdot 2 \cdot 2 \cdot 3 \cdot 5.$$

Every composite number has a unique prime factorization, and it is customary to write the prime factors in order from smallest to largest.

MAKING CONNECTIONS

Factor Trees and Prime Factorization

The prime factors of 120 can also be found using the following tree. Even though this tree is different than the one used earlier, the prime factors it reveals are the same.

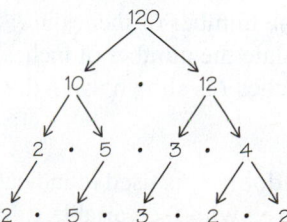

No matter what tree is used, a prime factorization is *always* unique.

| EXAMPLE 1 | **Classifying numbers as prime or composite** |

Classify each number as prime or composite, if possible. If a number is composite, write it as a product of prime numbers.
(a) 31   **(b)** 1   **(c)** 35   **(d)** 200

Solution
(a) The only factors of 31 are itself and 1, so the number 31 is prime.
(b) The number 1 is neither prime nor composite because prime and composite numbers must be greater than 1.
(c) The number 35 is composite because 5 and 7 are factors. It can be written as a product of prime numbers as $35 = 5 \cdot 7$.
(d) The number 200 is composite because 10 and 20 are factors. A factor tree can be used to write 200 as a product of prime numbers, as shown in Figure 1.3. The factor tree reveals that 200 can be factored as $200 = 2 \cdot 2 \cdot 2 \cdot 5 \cdot 5$.

Now Try Exercises 13, 15, 17, 21

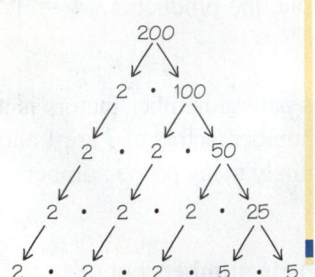

Figure 1.3 Prime Factorization of 200

Variables, Algebraic Expressions, and Equations

There are 12 inches in 1 foot, so 5 feet equal $5 \cdot 12 = 60$ inches. Similarly, in 3 feet there are $3 \cdot 12 = 36$ inches. To convert feet to inches frequently, Table 1.1 might help.

TABLE 1.1 Converting Feet to Inches

Feet	1	2	3	4	5	6	7
Inches	12	24	36	48	60	72	84

However, this table is not helpful in converting 11 feet to inches. We could expand Table 1.1 into Table 1.2 to include $11 \cdot 12 = 132$, but expanding the table to accommodate every possible value for feet would be impossible.

TABLE 1.2 Converting Feet to Inches

Feet	1	2	3	4	5	6	7	11
Inches	12	24	36	48	60	72	84	132

READING CHECK

- What is a variable?
- Give one reason for using variables in mathematics.

Variables are often used in mathematics when tables of numbers are inadequate. A **variable** is a symbol, typically an italic letter such as x, y, z, or F, used to represent an unknown quantity. In the preceding example, the number of feet could be represented by the variable F, and the corresponding number of inches could be represented by the variable I. The number of inches in F feet is given by the *algebraic expression* $12 \cdot F$. That is, to calculate the number of inches in F feet, multiply **F** by **12**. The relationship between feet F and inches I is shown using the *equation* or *formula*

$$I = \mathbf{12} \cdot \mathbf{F}.$$

A dot ($\cdot$) is used to indicate multiplication because a multiplication sign ($\times$) can be confused with the variable x. Many times the multiplication sign is omitted altogether. Thus all three formulas

$$I = 12 \cdot F, \quad I = 12F, \quad \text{and} \quad I = 12(F)$$

represent the same relationship between feet and inches. If we wish to find the number of inches in 10 feet, for example, we can replace F in one of these formulas with the number 10. If we let $F = \textbf{10}$ in the first formula,

$$I = 12 \cdot \textbf{10} = 120.$$

That is, there are 120 inches in 10 feet.

More formally, an **algebraic expression** consists of numbers, variables, operation symbols such as $+$, $-$, $\cdot$, and $\div$, and grouping symbols such as parentheses. An **equation** is a mathematical statement that two algebraic expressions are equal. Equations *always* contain an equals sign. A **formula** is a special type of equation that expresses a relationship between two or more quantities. The formula $I = 12F$ states that to calculate the number of inches in F feet, multiply F by 12.

EXAMPLE 2   **Evaluating algebraic expressions with one variable**

Evaluate each algebraic expression for $x = 4$.

(a) $x + 5$ **(b)** $5x$ **(c)** $15 - x$ **(d)** $\dfrac{x}{(x - 2)}$

Solution
(a) Replace x with 4 in the expression $x + \textbf{5}$ to obtain $\textbf{4} + \textbf{5} = 9$.
(b) The expression $5x$ indicates multiplication of 5 and x. Thus $5x = 5 \cdot \textbf{4} = 20$.
(c) $15 - x = 15 - \textbf{4} = 11$
(d) Perform all arithmetic operations inside parentheses first.

$$\frac{x}{(x - 2)} = \frac{\textbf{4}}{(\textbf{4} - 2)} = \frac{4}{2} = 2$$

▌ **Now Try Exercises 45, 47, 51, 53**

▶ **REAL-WORLD CONNECTION** Some algebraic expressions contain more than one variable. For example, if a car travels 120 miles on 6 gallons of gasoline, then the car's *mileage* is $\frac{120}{6} = 20$ miles per gallon. In general, if a car travels M miles on G gallons of gasoline, then its mileage is given by the expression $\frac{M}{G}$. Note that $\frac{M}{G}$ contains two variables, M and G, whereas the expression $12F$ contains only one variable, F.

EXAMPLE 3   **Evaluating algebraic expressions with two variables**

Evaluate each algebraic expression for $y = 2$ and $z = 8$.

(a) $3yz$ **(b)** $z - y$ **(c)** $\dfrac{z}{y}$

Solution
(a) Replace y with 2 and z with 8 to obtain $3yz = 3 \cdot \textbf{2} \cdot \textbf{8} = 48$.
(b) $z - y = \textbf{8} - \textbf{2} = 6$

(c) $\dfrac{z}{y} = \dfrac{\textbf{8}}{\textbf{2}} = 4$

▌ **Now Try Exercises 55, 57, 59**

EXAMPLE 4 **Evaluating formulas**

Find the value of y for $x = 15$ and $z = 10$.

(a) $y = x - 3$ **(b)** $y = \dfrac{x}{5}$ **(c)** $y = 8xz$

Solution

(a) Substitute 15 for x and then evaluate the right side of the formula to find y.

$$
\begin{aligned}
y &= x - 3 && \text{Given formula} \\
&= 15 - 3 && \text{Replace } x \text{ with 15.} \\
&= 12 && \text{Subtract.}
\end{aligned}
$$

(b) $y = \dfrac{x}{5} = \dfrac{15}{5} = 3$

(c) $y = 8xz = 8 \cdot 15 \cdot 10 = 1200$

Now Try Exercises 61, 67, 69

Translating Words to Expressions

Many times in mathematics, algebraic expressions are not given; rather, we must write our own expressions. To accomplish this task, we often translate words to symbols. The symbols $+$, $-$, $\cdot$, and $\div$ have special mathematical words associated with them. When two numbers are added, the result is called the *sum*. When one number is subtracted from another number, the result is called the *difference*. Similarly, multiplying two numbers results in a *product* and dividing two numbers results in a *quotient*. Table 1.3 lists many of the words commonly associated with these operations.

READING CHECK

• Write three words associated with each of the symbols $+$, $-$, $\cdot$, and $\div$.

TABLE 1.3 Words Associated with Arithmetic Symbols

Symbol	Associated Words
$+$	add, plus, more, sum, total
$-$	subtract, minus, less, difference, fewer
$\cdot$	multiply, times, twice, double, triple, product
$\div$	divide, divided by, quotient

EXAMPLE 5 **Translating words to expressions**

Translate each phrase to an algebraic expression. Specify what each variable represents.
(a) Four more than a number
(b) Ten less than the president's age
(c) A number plus 10, all divided by a different number
(d) The product of 6 and a number

Solution
(a) If the number were 20, then four more than the number would be $20 + 4 = 24$. If we let n represent the number, then four more than the number would be $n + 4$.
(b) If the president's age were 55, then ten less than the president's age would be $55 - 10 = 45$. If we let A represent the president's age, then ten less would be $A - 10$.

(c) Let x be the first number and y be the other number. Then the expression can be written as $(x + 10) \div y$. Note that parentheses are used around $x + 10$ because it is "all" divided by y. This expression could also be written as $\frac{x + 10}{y}$.

(d) If n represents the number, then the product of 6 and n is $6 \cdot n$ or $6n$.

Now Try Exercises 73, 77, 83, 85

▶ **REAL-WORLD CONNECTION** Cities are made up of large amounts of concrete and asphalt that heat up in the daytime from sunlight but do not cool off completely at night. As a result, urban areas tend to be warmer than surrounding rural areas. This effect is called the *urban heat island* and has been documented in cities throughout the world. The next example discusses the impact of this effect in Phoenix, Arizona.

EXAMPLE 6 **Translating words to a formula**

For each year after 1970, the average nighttime temperature in Phoenix has increased by about $0.1°C$.

(a) What was the increase in the nighttime temperature after 20 years, or in 1990?

(b) Write a formula (or equation) that gives the increase T in average nighttime temperature x years after 1970.

(c) Use your formula to estimate the increase in nighttime temperature in 2000.

Solution

(a) The average nighttime temperature has increased by $0.1°C$ per year, so after **20** years the temperature increase would be $0.1 \cdot \mathbf{20} = 2.0°C$.

(b) To calculate the nighttime increase in temperature, multiply the number of years past 1970 by 0.1. Thus $T = 0.1x$, where x represents the number of years after 1970.

(c) Because $2000 - 1970 = 30$, let $x = \mathbf{30}$ in the formula $T = 0.1x$ to get

$$T = 0.1(\mathbf{30}) = 3.$$

The average nighttime temperature increased by $3°C$.

Now Try Exercise 91

Another use of translating words to formulas is in finding the areas of various shapes.

EXAMPLE 7 **Finding the area of a rectangle**

The area A of a rectangle equals its length L times its width W, as illustrated in the accompanying figure.

(a) Write a formula that shows the relationship between these three quantities.

(b) Find the area of a standard sheet of paper that is 8.5 inches wide and 11 inches long.

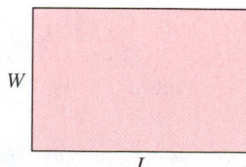

Solution

(a) The word *times* indicates that the length and width should be multiplied. The formula is given by $A = LW$.

(b) $A = 11 \cdot 8.5 = 93.5$ square inches

Now Try Exercise 97

1.1   Putting It All Together

STUDY TIP

Putting It All Together gives a summary of important concepts in each section. Be sure that you have a good understanding of these concepts.

CONCEPT	COMMENTS	EXAMPLES
Natural Numbers	Sometimes referred to as the *counting numbers*	1, 2, 3, 4, 5, …
Whole Numbers	Includes the natural numbers and 0	0, 1, 2, 3, 4, …
Products and Factors	When two numbers are multiplied, the result is called the product. The numbers being multiplied are called factors.	$6 \cdot 7 = 42$ **factor factor product**
Prime Number	A natural number greater than 1 whose only factors are itself and 1; there are infinitely many prime numbers.	2, 3, 5, 7, 11, 13, 17, and 19 are the prime numbers less than 20.
Composite Number	A natural number greater than 1 that is *not* a prime number; there are infinitely many composite numbers.	4, 9, 25, 39, 62, 76, 87, 91, 100
Prime Factorization	Every composite number can be written as a product of prime numbers. This unique product is called the prime factorization.	$60 = 2 \cdot 2 \cdot 3 \cdot 5$, $84 = 2 \cdot 2 \cdot 3 \cdot 7$
Variable	Represents an unknown quantity	x, y, z, A, F, and T
Algebraic Expression	May consist of variables, numbers, operation symbols such as $+, -, \cdot$, and $\div$, and grouping symbols such as parentheses	$x + 3$, $\frac{x}{y}$, $2z + 5$, $12F$, $x + y + z$, $x(y + 5)$
Equation	An equation is a statement that two algebraic expressions are equal. An equation always includes an equals sign.	$2 + 3 = 5$, $x + 5 = 7$, $I = 12F$, $y = 0.1x$
Formula	A formula is a special type of equation that expresses a relationship between two or more quantities.	$I = 12F$, $y = 0.1x$, $A = LW$, $F = 3Y$

1.1   Exercises

CONCEPTS AND VOCABULARY

1. The _____ numbers comprise the counting numbers.

2. A whole number is either a natural number or the number _____.

3. The factors of a prime number are itself and _____.

4. A natural number greater than 1 that is not a prime number is called a(n) _____ number.

5. Because $3 \cdot 6 = 18$, the numbers 3 and 6 are _____ of 18.

6. A(n) _____ is a special type of equation that expresses a relationship between two or more quantities.

7. A symbol or letter used to represent an unknown quantity is called a(n) _____.

8. Equations always contain a(n) _____.

9. When one number is added to another number, the result is called the _____.

10. When one number is multiplied by another number, the result is called the _____.

11. The result of dividing one number by another is called the _____.

12. The result of subtracting one number from another is called the _____.

PRIME NUMBERS AND COMPOSITE NUMBERS

Exercises 13–24: Classify the number as prime, composite, or neither. If the number is composite, write it as a product of prime numbers.

13. 4

14. 36

15. 1

16. 0

17. 29

18. 13

19. 92

20. 69

21. 225

22. 900

23. 149

24. 101

Exercises 25–36: Write the composite number as a product of prime numbers.

25. 6

26. 8

27. 12

28. 20

29. 32

30. 100

31. 39

32. 51

33. 294

34. 175

35. 300

36. 455

Exercises 37–44: State whether the given quantity could accurately be described by the whole numbers.

37. The population of a country

38. The cost of a gallon of gasoline in dollars

39. A student's grade point average

40. The Fahrenheit temperature in Antarctica

41. The number of apps on an iPad

42. The number of students in a class

43. The winning time in a 100-meter sprint

44. The number of bald eagles in the United States

ALGEBRAIC EXPRESSIONS, FORMULAS, AND EQUATIONS

Exercises 45–54: Evaluate the expression for the given value of x.

45. $3x$ $x = 5$

46. $x + 10$ $x = 8$

47. $9 - x$ $x = 4$

48. $13x$ $x = 0$

49. $\dfrac{x}{8}$ $x = 32$

50. $\dfrac{5}{(x - 3)}$ $x = 8$

51. $3(x + 1)$ $x = 5$

52. $7(6 - x)$ $x = 3$

53. $\left(\dfrac{x}{2}\right) + 1$ $x = 6$

54. $3 - \left(\dfrac{6}{x}\right)$ $x = 2$

Exercises 55–60: Evaluate the expression for the given values of x and y.

55. $x + y$ $x = 8,\ y = 14$

56. $5xy$ $x = 2,\ y = 3$

57. $6 \cdot \dfrac{x}{y}$ $x = 8,\ y = 4$

58. $y - x$ $x = 8,\ y = 11$

59. $y(x - 2)$ $x = 5,\ y = 3$

60. $(x + y) - 5$ $x = 6,\ y = 3$

Exercises 61–64: Find the value of y for the given value of x.

61. $y = x + 5$　　$x = 0$

62. $y = x \cdot x$　　$x = 7$

63. $y = 4x$　　$x = 7$

64. $y = 2(x - 3)$　$x = 3$

Exercises 65–68: Find the value of F for the given value of z.

65. $F = z - 5$　　$z = 12$

66. $F = \dfrac{z}{4}$　　$z = 40$

67. $F = \dfrac{30}{z}$　　$z = 6$

68. $F = z \cdot z \cdot z$　$z = 5$

Exercises 69–72: Find the value of y for the given values of x and z.

69. $y = 3xz$　　　$x = 2, \quad z = 0$

70. $y = x + z$　　$x = 3, \quad z = 15$

71. $y = \dfrac{x}{z}$　　　$x = 9, \quad z = 3$

72. $y = x - z$　　$x = 9, \quad z = 1$

TRANSLATING WORDS TO EXPRESSIONS

Exercises 73–86: Translate the phrase to an algebraic expression. State what each variable represents.

73. Five more than a number

74. Four less than a number

75. Three times the cost of a soda

76. Twice the cost of a gallon of gasoline

77. The sum of a number and 5

78. The quotient of two numbers

79. Two hundred less than the population of a town

80. The total number of dogs and cats in a city

81. A number divided by six

82. A number divided by another number

83. The product of a car's speed and traveling time

84. The difference between 220 and a person's heart rate

85. A number plus seven, all divided by a different number

86. One-fourth of a number increased by one-tenth of a different number

APPLICATIONS

87. *Dollars to Pennies* Write a formula that converts D dollars to P pennies.

88. *Quarters to Nickels* Write a formula that converts Q quarters to N nickels.

89. *Yards to Feet* Make a table of values that converts y yards to F feet. Let $y = 1, 2, 3, \ldots, 7$. Write a formula that converts y yards to F feet.

90. *Gallons to Quarts* Make a table of values that converts g gallons to Q quarts. Let $g = 1, 2, 3, \ldots, 6$. Write a formula that converts g gallons to Q quarts.

91. *NASCAR Speeds* On the fastest speedways, some NASCAR drivers reach average speeds of 3 miles per minute. Write a formula that gives the number of miles M that such a driver would travel in x minutes. How far would this driver travel in 36 minutes?

92. *NASCAR Speeds* On slower speedways, some NASCAR drivers reach average speeds of 2 miles per minute. Write a formula that gives the number of miles M that such a driver would travel in x minutes. How far would this driver travel in 42 minutes?

93. **Thinking Generally** If there are 6 blims in every drog, is the formula that relates B blims and D drogs $D = 6B$ or $B = 6D$?

94. *Heart Beat* The resting heart beat of a person is 70 beats per minute. Write a formula that gives the number of beats B that occur in x minutes. How many beats are there in an hour?

95. *Cost of album downloads* The table lists the cost C of downloading x albums. Write an equation that relates C and x.

Albums (x)	1	2	3	4
Cost (C)	$12	$24	$36	$48

96. *Gallons of Water* The table lists the gallons *G* of water coming from a garden hose after *m* minutes. Write an equation that relates *G* and *m*.

Minutes (*m*)	1	2	3	4
Gallons (*G*)	4	8	12	16

97. *Area of a Rectangle* The area of a rectangle equals its length times its width. Find the area of the rectangle shown in the figure.

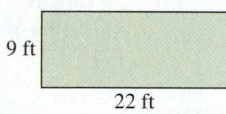

9 ft

22 ft

98. *Area of a Square* A square is a rectangle whose length and width have equal measures. Find the area of a square with length 14 inches.

WRITING ABOUT MATHEMATICS

99. Give an example in which the whole numbers are not sufficient to describe a quantity in real life. Explain your reasoning.

100. Explain what a prime number is. How can you determine whether a number is prime?

101. Explain what a composite number is. How can you determine whether a number is composite?

102. When are variables used? Give an example.

1.2   Fractions

Basic Concepts • Simplifying Fractions to Lowest Terms • Multiplication and Division of Fractions • Addition and Subtraction of Fractions • An Application

A LOOK INTO MATH ▶

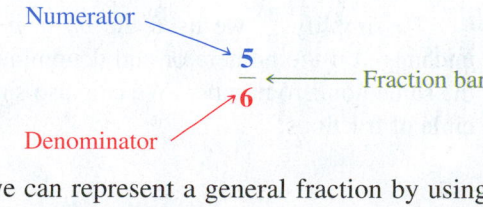

Historically, natural and whole numbers have not been sufficient for most societies. Early on, the concept of splitting a quantity into parts was common, and as a result, fractions were developed. Today, fractions are used in many everyday situations. For example, there are four quarters in one dollar, so each quarter represents a fourth of a dollar. In this section we discuss fractions and how to add, subtract, multiply, and divide them.

NEW VOCABULARY

☐ Lowest terms
☐ Greatest common factor (GCF)
☐ Basic principle of fractions
☐ Multiplicative inverse or reciprocal
☐ Least common denominator (LCD)

Basic Concepts

If we divide a circular pie into 6 equal slices, as shown in Figure 1.4, then each piece represents one-sixth of the pie and can be represented by the fraction $\frac{1}{6}$. Five slices of the pie would represent five-sixths of the pie and can be represented by the fraction $\frac{5}{6}$.

The parts of a fraction are named as follows.

Numerator

$$\frac{5}{6}$$ ← Fraction bar

Denominator

Sometimes we can represent a general fraction by using variables. The fraction $\frac{a}{b}$ can represent *any* fraction with numerator *a* and denominator *b*. However, the value of *b* cannot equal 0, which is denoted $b \neq 0$. (The symbol $\neq$ means "not equal to.")

NOTE: The fraction bar represents division. For example, the fraction $\frac{1}{2}$ represents the result when 1 is divided by 2, which is 0.5. We discuss this concept further in Section 1.4.

Figure 1.4

EXAMPLE 1   **Identifying numerators and denominators**

Give the numerator and denominator of each fraction.

(a) $\dfrac{6}{13}$ **(b)** $\dfrac{ac}{b}$ **(c)** $\dfrac{x-5}{y+z}$

Solution

(a) The numerator is 6, and the denominator is 13.
(b) The numerator is ac, and the denominator is b.
(c) The numerator is $x - 5$, and the denominator is $y + z$.

Now Try Exercise 2

Simplifying Fractions to Lowest Terms

Consider the amount of pizza shown in each of the three pies in Figure 1.5. The first pie is cut into sixths with three pieces remaining, the second pie is cut into fourths with two pieces remaining, and the third pie is cut into only two pieces with one piece remaining. In all three cases half a pizza remains.

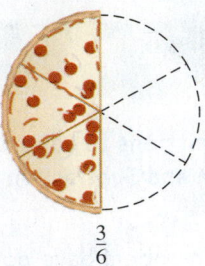 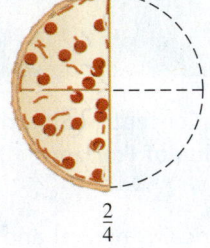 

$\dfrac{3}{6}$ $\dfrac{2}{4}$ $\dfrac{1}{2}$

Figure 1.5

READING CHECK

• How can you tell if a fraction is written in lowest terms?

Figure 1.5 illustrates that the fractions $\frac{3}{6}, \frac{2}{4}$, and $\frac{1}{2}$ are equal. The fraction $\frac{1}{2}$ is in **lowest terms** because its numerator and denominator have no factors in common, whereas the fractions $\frac{3}{6}$ and $\frac{2}{4}$ are not in lowest terms. In the fraction $\frac{3}{6}$, the numerator and denominator have a common factor of 3, so the fraction can be simplified as follows.

$$\frac{3}{6} = \frac{1 \cdot \mathbf{3}}{2 \cdot \mathbf{3}} \quad \text{Factor out 3.}$$

$$= \frac{1}{2} \qquad \frac{a \cdot c}{b \cdot c} = \frac{a}{b}$$

STUDY TIP

A positive attitude is important. The first step to success is believing in yourself.

To simplify $\frac{3}{6}$, we used the *basic principle of fractions*: The value of a fraction is unchanged if the numerator and denominator of the fraction are multiplied (or divided) by the same nonzero number. We can also simplify the fraction $\frac{2}{4}$ to $\frac{1}{2}$ by using the basic principle of fractions.

$$\frac{2}{4} = \frac{1 \cdot \mathbf{2}}{2 \cdot \mathbf{2}} = \frac{1}{2}$$

READING CHECK

• What is the greatest common factor for two numbers?

When simplifying fractions, we usually factor out the *greatest common factor* of the numerator and the denominator. The **greatest common factor (GCF)** of two or more numbers is the largest factor that is common to those numbers. For example, to simplify $\frac{27}{36}$, we first find the greatest common factor of 27 and 36. The numbers **3** and **9** are *common* factors of 27 and 36 because

$$\mathbf{3} \cdot 9 = 27 \text{ and } \mathbf{3} \cdot 12 = 36, \quad \text{and} \quad \mathbf{9} \cdot 3 = 27 \text{ and } \mathbf{9} \cdot 4 = 36.$$

However, the *greatest* common factor is **9** because it is the largest number that divides evenly into both 27 and 36. The fraction $\frac{27}{36}$ simplifies to lowest terms as

$$\frac{27}{36} = \frac{3 \cdot \mathbf{9}}{4 \cdot \mathbf{9}} = \frac{3}{4}.$$

SIMPLIFYING FRACTIONS

To simplify a fraction to lowest terms, factor out the greatest common factor c in the numerator and in the denominator. Then apply the **basic principle of fractions**:

$$\frac{a \cdot c}{b \cdot c} = \frac{a}{b}.$$

NOTE: This principle is true because multiplying a fraction by $\frac{c}{c}$ or 1 does not change the value of the fraction.

The greatest common factor for two numbers is not always obvious. The next example demonstrates two different methods that can be used to find the GCF.

EXAMPLE 2 **Finding the greatest common factor**

Find the greatest common factor (GCF) for each pair of numbers.
(a) 24, 60   **(b)** 36, 54

Solution
(a) One way to determine the greatest common factor is to find the prime factorization of each number using factor trees, as shown in Figure 1.6.

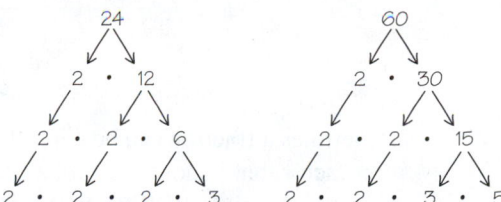

Figure 1.6 Prime Factorizations of 24 and 60

The prime factorizations have two 2s and one 3 in common.

$$24 = \mathbf{2} \cdot \mathbf{2} \cdot 2 \cdot \mathbf{3}$$
$$60 = \mathbf{2} \cdot \mathbf{2} \cdot \mathbf{3} \cdot 5$$

Thus the GCF of 24 and 60 is $\mathbf{2} \cdot \mathbf{2} \cdot \mathbf{3} = 12$.

(b) Another way to find the greatest common factor is to create a factor step diagram. Working downward from the top, the numbers in each step are found by dividing the two numbers in the previous step by their *smallest common prime factor*. The process continues until no common prime factor can be found. A factor step diagram for 36 and 54 is shown in Figure 1.7 on the next page.

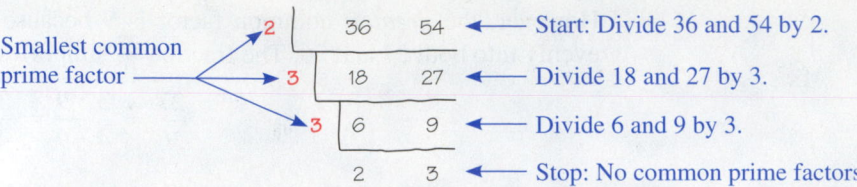

Smallest common prime factor

2 ⟶ 36 54 ⟵ Start: Divide 36 and 54 by 2.
3 ⟶ 18 27 ⟵ Divide 18 and 27 by 3.
3 ⟶ 6 9 ⟵ Divide 6 and 9 by 3.
2 3 ⟵ Stop: No common prime factors

Figure 1.7 Factor Step Diagram for 36 and 54

The greatest common factor is the product of the prime numbers along the side of the diagram. So the GCF of 36 and 54 is **2 · 3 · 3** = 18.

▮ **Now Try Exercises 15, 17**

EXAMPLE 3 **Simplifying fractions to lowest terms**

Simplify each fraction to lowest terms.

(a) $\dfrac{24}{60}$ (b) $\dfrac{42}{105}$

Solution

(a) From Example 2(a), the GCF of 24 and 60 is 12. Thus

$$\frac{24}{60} = \frac{2 \cdot \mathbf{12}}{5 \cdot \mathbf{12}} = \frac{2}{5}.$$

CRITICAL THINKING

Describe a situation from everyday life in which fractions would be needed.

(b) The prime factorizations of 42 and 105 are

$$42 = 2 \cdot \mathbf{3} \cdot \mathbf{7} \quad \text{and} \quad 105 = 5 \cdot \mathbf{3} \cdot \mathbf{7}.$$

The GCF of 42 and 105 is **3 · 7** = 21. Thus

$$\frac{42}{105} = \frac{2 \cdot \mathbf{21}}{5 \cdot \mathbf{21}} = \frac{2}{5}.$$

▮ **Now Try Exercises 29, 33**

MAKING CONNECTIONS

Simplifying Fractions in Steps

Sometimes a fraction can be simplified to lowest terms in multiple steps. By using *any* common factor that is not the GCF, a new fraction in *lower* terms will result. This new fraction may then be simplified using a common factor of its numerator and denominator. If this process is continued, the result will be the given fraction simplified to lowest terms. The fraction in Example 3(b) could be simplified to lowest terms in two steps.

$$\frac{42}{105} = \frac{14 \cdot \mathbf{3}}{35 \cdot \mathbf{3}} = \frac{14}{35} = \frac{2 \cdot \mathbf{7}}{5 \cdot \mathbf{7}} = \frac{2}{5}$$

Multiplication and Division of Fractions

Suppose we cut *half* an apple into *thirds*, as illustrated in Figure 1.8. Then each piece represents one-sixth of the original apple. One-third of one-half is described by the product

$$\frac{1}{3} \cdot \frac{1}{2} = \frac{1}{6}.$$

Figure 1.8

This example demonstrates that the numerator of the product of two fractions is found by multiplying the numerators of the two fractions. Similarly, the denominator of the product of two fractions is found by multiplying the denominators of the two fractions. For example, the product of $\frac{2}{3}$ and $\frac{5}{7}$ is

$$\frac{2}{3} \cdot \frac{5}{7} = \frac{\mathbf{2 \cdot 5}}{\mathbf{3 \cdot 7}} = \frac{10}{21}.$$

Multiply numerators.

Multiply denominators.

NOTE: The word "of" in mathematics often indicates multiplication. For example, the phrases "one-fifth of the cookies," "twenty percent of the price," and "half of the money" all suggest multiplication.

MULTIPLICATION OF FRACTIONS

The product of $\frac{a}{b}$ and $\frac{c}{d}$ is given by

$$\frac{a}{b} \cdot \frac{c}{d} = \frac{ac}{bd},$$

where b and d are not 0.

EXAMPLE 4   **Multiplying fractions**

Multiply. Simplify the result when appropriate.

(a) $\dfrac{4}{5} \cdot \dfrac{6}{7}$ **(b)** $\dfrac{8}{9} \cdot \dfrac{3}{4}$ **(c)** $3 \cdot \dfrac{5}{9}$ **(d)** $\dfrac{x}{y} \cdot \dfrac{z}{3}$

Solution

(a) $\dfrac{4}{5} \cdot \dfrac{6}{7} = \dfrac{4 \cdot 6}{5 \cdot 7} = \dfrac{24}{35}$

(b) $\dfrac{8}{9} \cdot \dfrac{3}{4} = \dfrac{8 \cdot 3}{9 \cdot 4} = \dfrac{24}{36}$; the GCF of 24 and 36 is 12, so

$$\frac{24}{36} = \frac{2 \cdot \mathbf{12}}{3 \cdot \mathbf{12}} = \frac{2}{3}.$$

(c) Start by writing 3 as $\dfrac{3}{1}$.

$$3 \cdot \frac{5}{9} = \frac{3}{1} \cdot \frac{5}{9} = \frac{3 \cdot 5}{1 \cdot 9} = \frac{15}{9}$$

The GCF of 15 and 9 is 3, so

$$\frac{15}{9} = \frac{5 \cdot \mathbf{3}}{3 \cdot \mathbf{3}} = \frac{5}{3}.$$

(d) $\dfrac{x}{y} \cdot \dfrac{z}{3} = \dfrac{x \cdot z}{y \cdot 3} = \dfrac{xz}{3y}$

When we write the product of a variable and a number, such as $y \cdot 3$, we typically write the number first, followed by the variable. That is, $y \cdot 3 = 3y$.

Now Try Exercises 35, 39, 43, 47

MAKING CONNECTIONS

Multiplying and Simplifying Fractions

When multiplying fractions, sometimes it is possible to change the order of the factors to rewrite the product so that it is easier to simplify. In Example 4(b) the product could be written as

$$\frac{8}{9} \cdot \frac{3}{4} = \frac{3 \cdot 8}{9 \cdot 4} = \frac{3}{9} \cdot \frac{8}{4} = \frac{1}{3} \cdot \frac{2}{1} = \frac{2}{3}.$$

Instead of simplifying $\frac{24}{36}$, which contains larger numbers, the fractions $\frac{3}{9}$ and $\frac{8}{4}$ were simplified first.

EXAMPLE 5 **Finding fractional parts**

Find each fractional part.
(a) One-fifth of two-thirds (b) Four-fifths of three-sevenths (c) Three-fifths of ten

Solution
(a) The phrase "one-fifth of" indicates multiplication by one-fifth. The fractional part is

$$\frac{1}{5} \cdot \frac{2}{3} = \frac{1 \cdot 2}{5 \cdot 3} = \frac{2}{15}.$$

(b) $\frac{4}{5} \cdot \frac{3}{7} = \frac{4 \cdot 3}{5 \cdot 7} = \frac{12}{35}$

(c) $\frac{3}{5} \cdot 10 = \frac{3}{5} \cdot \frac{10}{1} = \frac{30}{5} = 6$

Now Try Exercises 49, 51, 53

▶ **REAL-WORLD CONNECTION** Fractions can be used to describe particular parts of the U.S. population. In the next application we use fractions to find the portion of the population that has completed 4 or more years of college.

EXAMPLE 6 **Estimating college completion rates**

About $\frac{17}{20}$ of the U.S. population over the age of 25 has a high school diploma. About $\frac{8}{25}$ of those people have gone on to complete 4 or more years of college. What fraction of the U.S. population over the age of 25 has completed 4 or more years of college? (*Source:* U.S. Census Bureau.)

Solution
We need to find $\frac{8}{25}$ of $\frac{17}{20}$.

$$\frac{8}{25} \cdot \frac{17}{20} = \frac{8 \cdot 17}{25 \cdot 20} = \frac{136}{500} = \frac{34 \cdot 4}{125 \cdot 4} = \frac{34}{125}.$$

About $\frac{34}{125}$ of the U.S. population over the age of 25 has completed 4 or more years of college.

Now Try Exercise 117

The **multiplicative inverse**, or **reciprocal**, of a nonzero number a is $\frac{1}{a}$. Table 1.4 lists several numbers and their reciprocals. Note that the product of a number and its reciprocal is always 1. For example, the reciprocal of 2 is $\frac{1}{2}$, and their product is $2 \cdot \frac{1}{2} = 1$.

TABLE 1.4 Numbers and Their Reciprocals

Number	Reciprocal
3	$\frac{1}{3}$
$\frac{1}{4}$	4
$\frac{3}{2}$	$\frac{2}{3}$
$\frac{21}{37}$	$\frac{37}{21}$

▶ **REAL-WORLD CONNECTION** Suppose that a group of children wants to buy gum from a gum ball machine that costs a half dollar for each gum ball. If the caregiver for the children has 4 dollars, then the number of gum balls that can be bought equals the number of half dollars that there are in 4 dollars. Thus 8 gum balls can be bought.

This calculation is given by $4 \div \frac{1}{2}$. To divide a number by a fraction, multiply the number by the reciprocal of the fraction.

$$4 \div \frac{1}{2} = 4 \cdot \frac{2}{1} \qquad \text{Multiply by the reciprocal of } \frac{1}{2}.$$

$$= \frac{4}{1} \cdot \frac{2}{1} \qquad \text{Write 4 as } \frac{4}{1}.$$

$$= \frac{8}{1} \qquad \text{Multiply the fractions.}$$

$$= 8 \qquad \frac{a}{1} = a \text{ for all values of } a.$$

Justification for multiplying by the reciprocal when dividing two fractions is

$$\frac{a}{b} \div \frac{c}{d} = \frac{\dfrac{a}{b}}{\dfrac{c}{d}} = \frac{\dfrac{a}{b} \cdot \dfrac{d}{c}}{\dfrac{c}{d} \cdot \dfrac{d}{c}} = \frac{\dfrac{a}{b} \cdot \dfrac{d}{c}}{1} = \frac{a}{b} \cdot \frac{d}{c}.$$

These results are summarized as follows.

DIVISION OF FRACTIONS

For real numbers a, b, c, and d, with b, c, and d not equal to 0,

$$\frac{a}{b} \div \frac{c}{d} = \frac{a}{b} \cdot \frac{d}{c}.$$

EXAMPLE 7 **Dividing fractions**

Divide. Simplify the result when appropriate.

(a) $\dfrac{1}{3} \div \dfrac{3}{5}$ **(b)** $\dfrac{4}{5} \div \dfrac{4}{5}$ **(c)** $5 \div \dfrac{10}{3}$ **(d)** $\dfrac{x}{2} \div \dfrac{y}{z}$

Solution
(a) To divide $\frac{1}{3}$ by $\frac{3}{5}$, multiply $\frac{1}{3}$ by $\frac{5}{3}$, which is the reciprocal of $\frac{3}{5}$.

$$\frac{1}{3} \div \frac{3}{5} = \frac{1}{3} \cdot \frac{5}{3} = \frac{1 \cdot 5}{3 \cdot 3} = \frac{5}{9}$$

(b) $\frac{4}{5} \div \frac{4}{5} = \frac{4}{5} \cdot \frac{5}{4} = \frac{4 \cdot 5}{5 \cdot 4} = \frac{20}{20} = 1$. Note that when we divide any nonzero number by itself, the result is 1.

(c) $5 \div \frac{10}{3} = \frac{5}{1} \cdot \frac{3}{10} = \frac{15}{10} = \frac{3}{2}$

(d) $\dfrac{x}{2} \div \dfrac{y}{z} = \dfrac{x}{2} \cdot \dfrac{z}{y} = \dfrac{xz}{2y}$

Now Try Exercises 59, 63, 67, 71

NOTE: In Example 7(c) the answer $\frac{3}{2}$ is an *improper fraction*, which could be written as the mixed number $1\frac{1}{2}$. However, in algebra, fractions are often left as improper fractions.

TECHNOLOGY NOTE

When entering a mixed number into a calculator, it is usually easiest first to convert the mixed number to an improper fraction. For example, enter $2\frac{2}{3}$ as $\frac{8}{3}$. Otherwise, enter $2\frac{2}{3}$ as $2 + \frac{2}{3}$. See the accompanying figure.

```
8/3
        2.666666667
2+2/3
        2.666666667
```

EXAMPLE 8 | **Writing a problem**

Describe a problem for which the solution could be found by dividing 5 by $\frac{1}{6}$.

Solution
One possible problem could be stated as follows. If five pies are each cut into sixths, how many pieces of pie are there? See Figure 1.9.

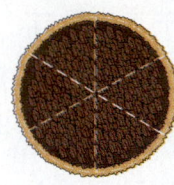

Figure 1.9 Five Pies Cut into Sixths

Now Try Exercise 121

Addition and Subtraction of Fractions

FRACTIONS WITH LIKE DENOMINATORS Suppose that a person cuts a sheet of paper into eighths. If that person picks up two pieces and another person picks up three pieces, then together they have

$$\frac{2}{8} + \frac{3}{8} = \frac{5}{8}$$

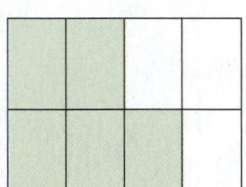

Figure 1.10

of a sheet of paper, as illustrated in Figure 1.10. When the denominator of one fraction is the same as the denominator of a second fraction, the sum of the two fractions can be found by adding their numerators and keeping the common denominator.

Similarly, if someone picks up 5 pieces of paper and gives 2 away, then that person has

$$\frac{5}{8} - \frac{2}{8} = \frac{3}{8}$$

of a sheet of paper. To subtract two fractions with common denominators, subtract their numerators and keep the common denominator.

ADDITION AND SUBTRACTION OF FRACTIONS

To add or subtract fractions with a common denominator d, use the equations

$$\frac{a}{d} + \frac{b}{d} = \frac{a + b}{d} \quad \text{and} \quad \frac{a}{d} - \frac{b}{d} = \frac{a - b}{d},$$

where d is not 0.

EXAMPLE 9   **Adding and subtracting fractions with common denominators**

Add or subtract as indicated. Simplify your answer to lowest terms when appropriate.

(a) $\dfrac{5}{13} + \dfrac{12}{13}$ (b) $\dfrac{11}{8} - \dfrac{5}{8}$

Solution

(a) Because the fractions have a common denominator, add the numerators and keep the common denominator.

$$\frac{5}{13} + \frac{12}{13} = \frac{5 + 12}{13} = \frac{17}{13}$$

(b) Because the fractions have a common denominator, subtract the numerators and keep the common denominator.

$$\frac{11}{8} - \frac{5}{8} = \frac{11 - 5}{8} = \frac{6}{8}$$

The fraction $\frac{6}{8}$ can be simplified to $\frac{3}{4}$.

Now Try Exercise 75

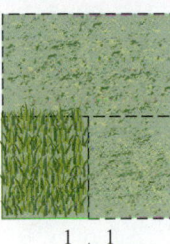

$\frac{1}{2} + \frac{1}{4}$

(a)

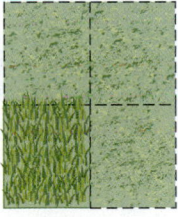

$\frac{2}{4} + \frac{1}{4} = \frac{3}{4}$

(b)

Figure 1.11

FRACTIONS WITH UNLIKE DENOMINATORS Suppose that one person mows half a large lawn while another person mows a fourth of the lawn. To determine how much they mowed together, we need to find the sum $\frac{1}{2} + \frac{1}{4}$. See Figure 1.11(a). Before we can add fractions with unlike denominators, we must write each fraction with a common denominator. The *least common denominator* of 2 and 4 is 4. Thus we need to write $\frac{1}{2}$ as $\frac{?}{4}$ by multiplying the numerator and denominator by the *same nonzero number*.

$$\frac{1}{2} = \frac{1}{2} \cdot \frac{2}{2} \qquad \text{Multiply by 1.}$$

$$= \frac{2}{4} \qquad \text{Multiply fractions.}$$

Now we can find the needed sum.

$$\frac{1}{2} + \frac{1}{4} = \frac{2}{4} + \frac{1}{4} = \frac{3}{4}$$

Together the two people mow three-fourths of the lawn, as illustrated in Figure 1.11(b).

To add or subtract fractions with unlike denominators, we first find the least common denominator for the fractions. The **least common denominator (LCD)** for two or more fractions is the smallest number that is divisible by every denominator.

READING CHECK

• What is the LCD and why is it needed?

FINDING THE LEAST COMMON DENOMINATOR (LCD)

STEP 1: Find the prime factorization for each denominator.

STEP 2: List each factor that appears in one or more of the factorizations. If a factor is repeated in any of the factorizations, list this factor the maximum number of times that it is repeated.

STEP 3: The product of this list of factors is the LCD.

The next example demonstrates two different methods that can be used to find the LCD. In part (a), the LCD is found using the three-step method shown on the previous page, and in part (b), a factor step diagram is used instead.

EXAMPLE 10   **Finding the least common denominator**

Find the LCD for each set of fractions.

(a) $\dfrac{5}{6}, \dfrac{3}{4}$ (b) $\dfrac{7}{36}, \dfrac{5}{54}$

Solution

(a) **STEP 1:** For the fractions $\frac{5}{6}$ and $\frac{3}{4}$ the prime factorizations of the denominators are

$$6 = 2 \cdot 3 \text{ and } 4 = 2 \cdot 2.$$

STEP 2: List the factors: **2, 2, 3**. Note that, because the factor 2 appears a maximum of two times, it is listed twice.

STEP 3: The LCD is the product of this list, or $2 \cdot 2 \cdot 3 = 12$.

NOTE: Finding an LCD is equivalent to finding the smallest number that each denominator divides into evenly. Both 6 and 4 divide into 12 evenly, and 12 is the smallest such number. Thus 12 is the LCD for $\frac{5}{6}$ and $\frac{3}{4}$.

(b) The *same* factor step diagram that was used in Example 2(b) to find the GCF of 36 and 54 can be used to find the LCD for $\frac{7}{36}$ and $\frac{5}{54}$; however, the final step differs slightly. As in Example 2(b), we find the numbers in each step by dividing the two numbers in the previous step by their *smallest common prime factor*. The process continues until no common prime factor can be found, as shown in Figure 1.12.

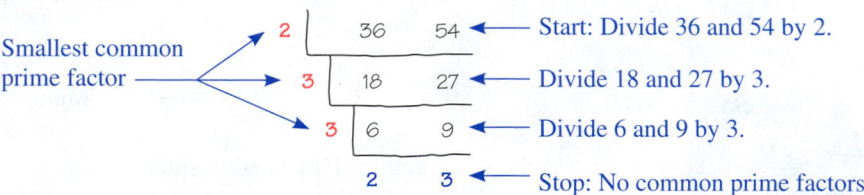

Figure 1.12 Factor Step Diagram for Finding the LCD

The process for finding the LCD differs from that used to find the GCF in that we find the LCD by multiplying not only the numbers along the side of the diagram but also the numbers at the bottom of the diagram. The LCD is $2 \cdot 3 \cdot 3 \cdot 2 \cdot 3 = 108$.

Now Try Exercises 79, 83

Once the LCD has been found, the next step in the process for adding or subtracting fractions with unlike denominators is to rewrite each fraction with the LCD.

EXAMPLE 11   **Rewriting fractions with the LCD**

Rewrite each set of fractions using the LCD.

(a) $\dfrac{5}{6}, \dfrac{3}{4}$ (b) $\dfrac{7}{12}, \dfrac{5}{18}$

Solution

(a) From Example 10(a) the LCD is 12. To write $\frac{5}{6}$ with a denominator of 12, we multiply the fraction by 1 in the form $\frac{2}{2}$.

$$\frac{5}{6} \cdot \frac{2}{2} = \frac{5 \cdot 2}{6 \cdot 2} = \frac{10}{12}$$

To write $\frac{3}{4}$ with a denominator of 12, we multiply the fraction by 1 in the form $\frac{3}{3}$.

$$\frac{3}{4} \cdot \frac{3}{3} = \frac{3 \cdot 3}{4 \cdot 3} = \frac{9}{12}$$

Thus $\frac{5}{6}$ can be rewritten as $\frac{10}{12}$ and $\frac{3}{4}$ can be rewritten as $\frac{9}{12}$.

(b) From Example 10(b) the LCD is 36. To write $\frac{7}{12}$ with a denominator of 36, multiply the fraction by $\frac{3}{3}$. To write $\frac{5}{18}$ with a denominator of 36, multiply the fraction by $\frac{2}{2}$.

$$\frac{7}{12} \cdot \frac{3}{3} = \frac{7 \cdot 3}{12 \cdot 3} = \frac{21}{36} \quad \text{and} \quad \frac{5}{18} \cdot \frac{2}{2} = \frac{5 \cdot 2}{18 \cdot 2} = \frac{10}{36}$$

Thus $\frac{7}{12}$ can be rewritten as $\frac{21}{36}$ and $\frac{5}{18}$ can be rewritten as $\frac{10}{36}$.

▌ **Now Try Exercises 89, 93**

The next example demonstrates how the concepts shown in the last two examples can be used to add or subtract fractions with unlike denominators.

EXAMPLE 12 ▌ **Adding and subtracting fractions with unlike denominators**

Add or subtract as indicated. Simplify your answer to lowest terms when appropriate.

(a) $\dfrac{5}{6} + \dfrac{3}{4}$ (b) $\dfrac{7}{12} - \dfrac{5}{18}$ (c) $\dfrac{1}{4} + \dfrac{2}{5} + \dfrac{7}{10}$

Solution

(a) From Example 10(a) the LCD is 12. Begin by writing each fraction with a denominator of 12, as demonstrated in Example 11(a).

$$\frac{5}{6} + \frac{3}{4} = \frac{5}{6} \cdot \frac{2}{2} + \frac{3}{4} \cdot \frac{3}{3} \qquad \text{Change to LCD of 12.}$$

$$= \frac{10}{12} + \frac{9}{12} \qquad \text{Multiply the fractions.}$$

$$= \frac{10 + 9}{12} \qquad \text{Add the numerators.}$$

$$= \frac{19}{12} \qquad \text{Simplify.}$$

(b) Using Example 10(b) and Example 11(b), we perform the following steps.

$$\frac{7}{12} - \frac{5}{18} = \frac{7}{12} \cdot \frac{3}{3} - \frac{5}{18} \cdot \frac{2}{2} \qquad \text{Change to LCD of 36.}$$

$$= \frac{21}{36} - \frac{10}{36} \qquad \text{Multiply the fractions.}$$

$$= \frac{21 - 10}{36} \qquad \text{Subtract the numerators.}$$

$$= \frac{11}{36} \qquad \text{Simplify.}$$

(c) The LCD for 4, 5, and 10 is 20.

$$\frac{1}{4} + \frac{2}{5} + \frac{7}{10} = \frac{1}{4} \cdot \frac{5}{5} + \frac{2}{5} \cdot \frac{4}{4} + \frac{7}{10} \cdot \frac{2}{2} \qquad \text{Change to LCD of 20.}$$

$$= \frac{5}{20} + \frac{8}{20} + \frac{14}{20} \qquad \text{Multiply the fractions.}$$

$$= \frac{5 + 8 + 14}{20} \qquad \text{Add the numerators.}$$

$$= \frac{27}{20} \qquad \text{Simplify.}$$

Now Try Exercises 97, 103, 107

An Application

The next example illustrates a situation where fractions occur in a real-world application.

EXAMPLE 13   **Applying fractions to carpentry**

A board measuring $35\frac{3}{4}$ inches is cut into four equal pieces, as depicted in Figure 1.13. Find the length of each piece.

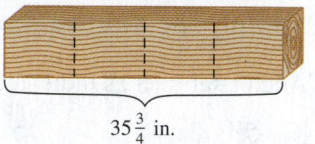

$35\frac{3}{4}$ in.

Figure 1.13

Solution
Begin by writing $35\frac{3}{4}$ as the improper fraction $\frac{143}{4}$ (because $4 \cdot 35 + 3 = 143$). Because the board is to be cut into four equal parts, the length of each piece should be

$$\frac{143}{4} \div 4 = \frac{143}{4} \cdot \frac{1}{4} = \frac{143}{16}, \quad \text{or} \quad 8\frac{15}{16} \text{ inches.}$$

Now Try Exercise 111

1.2   Putting It All Together

CONCEPT	COMMENTS	EXAMPLES
Fraction	The fraction $\frac{a}{b}$ has numerator a and denominator b.	The fraction $\frac{xy}{2}$ has numerator xy and denominator 2.
Lowest Terms	A fraction is in lowest terms if the numerator and denominator have no factors in common.	The fraction $\frac{3}{8}$ is in lowest terms because 3 and 8 have no factors in common.

CONCEPT	COMMENTS	EXAMPLES
Greatest Common Factor (GCF)	The GCF of two numbers equals the largest number that divides into both evenly.	The GCF of 12 and 18 is 6 because 6 is the largest number that divides into 12 and 18 evenly.
Simplifying Fractions	Use the principle $$\frac{a \cdot c}{b \cdot c} = \frac{a}{b}$$ to simplify fractions, where c is the GCF of the numerator and denominator.	The GCF of 24 and 32 is 8, so $$\frac{24}{32} = \frac{3 \cdot 8}{4 \cdot 8} = \frac{3}{4}.$$ The GCF of 20 and 8 is 4, so $$\frac{20}{8} = \frac{5 \cdot 4}{2 \cdot 4} = \frac{5}{2}.$$
Multiplicative Inverse or Reciprocal	The reciprocal of $\frac{a}{b}$ is $\frac{b}{a}$, where a and b are not zero. The product of a number and its reciprocal is 1.	The reciprocals of 5 and $\frac{3}{4}$ are $\frac{1}{5}$ and $\frac{4}{3}$, respectively, because $5 \cdot \frac{1}{5} = 1$ and $\frac{3}{4} \cdot \frac{4}{3} = 1$.
Multiplication and Division of Fractions	$$\frac{a}{b} \cdot \frac{c}{d} = \frac{ac}{bd}$$ $$\frac{a}{b} \div \frac{c}{d} = \frac{a}{b} \cdot \frac{d}{c}$$	$$\frac{3}{5} \cdot \frac{4}{9} = \frac{12}{45} = \frac{4}{15}$$ $$\frac{3}{2} \div \frac{6}{5} = \frac{3}{2} \cdot \frac{5}{6} = \frac{15}{12} = \frac{5}{4}$$
Addition and Subtraction of Fractions with Like Denominators	$$\frac{a}{d} + \frac{c}{d} = \frac{a + c}{d} \quad \text{and}$$ $$\frac{a}{d} - \frac{c}{d} = \frac{a - c}{d}$$	$$\frac{3}{5} + \frac{4}{5} = \frac{3 + 4}{5} = \frac{7}{5} \quad \text{and}$$ $$\frac{17}{12} - \frac{11}{12} = \frac{17 - 11}{12} = \frac{6}{12} = \frac{1}{2}$$
Least Common Denominator (LCD)	The LCD of two fractions equals the smallest number that both denominators divide into evenly.	The LCD of $\frac{5}{12}$ and $\frac{7}{18}$ is 36 because 36 is the smallest number that both 12 and 18 divide into evenly.
Addition and Subtraction of Fractions with Unlike Denominators	First write each fraction with the least common denominator. Then add or subtract the numerators.	The LCD of $\frac{3}{4}$ and $\frac{7}{10}$ is 20. $$\frac{3}{4} + \frac{7}{10} = \frac{3}{4} \cdot \frac{5}{5} + \frac{7}{10} \cdot \frac{2}{2}$$ $$= \frac{15}{20} + \frac{14}{20}$$ $$= \frac{29}{20}$$

1.2 Exercises

MyMathLab

CONCEPTS AND VOCABULARY

1. A small pie is cut into 4 equal pieces. If someone eats 3 of the pieces, what fraction of the pie does the person eat? What fraction of the pie remains?

2. In the fraction $\frac{11}{21}$ the numerator is _____ and the denominator is _____.

3. In the fraction $\frac{a}{b}$, the variable b cannot equal _____.

4. The fraction $\frac{7}{12}$ is in _____ terms because 7 and 12 have no factors in common.

5. (True or False?) The numerator of the product of two fractions is found by multiplying the numerators of the two fractions.

6. (True or False?) The denominator of the sum of two fractions is found by adding the denominators of the two fractions.

7. $\frac{ac}{bc} = $ _____

8. In the phrase "two-fifths of one-third," the word *of* indicates that we _____ the fractions $\frac{2}{5}$ and $\frac{1}{3}$.

9. What is the reciprocal of a, provided $a \neq 0$?

10. To divide $\frac{3}{4}$ by 5, multiply $\frac{3}{4}$ by _____.

11. $\frac{a}{b} \cdot \frac{c}{d} = $ _____

12. $\frac{a}{b} \div \frac{c}{d} = $ _____

13. $\frac{a}{b} + \frac{c}{b} = $ _____

14. $\frac{a}{b} - \frac{c}{b} = $ _____

LOWEST TERMS

Exercises 15–20: Find the greatest common factor.

15. 4, 12

16. 3, 27

17. 50, 75

18. 45, 105

19. 100, 60, 70

20. 36, 48, 72

Exercises 21–24: Use the basic principle of fractions to simplify the expression.

21. $\frac{3 \cdot 4}{5 \cdot 4}$

22. $\frac{2 \cdot 7}{9 \cdot 7}$

23. $\frac{3 \cdot 8}{8 \cdot 5}$

24. $\frac{7 \cdot 16}{16 \cdot 3}$

Exercises 25–34: Simplify the fraction to lowest terms.

25. $\frac{4}{8}$

26. $\frac{4}{12}$

27. $\frac{10}{25}$

28. $\frac{5}{20}$

29. $\frac{12}{36}$

30. $\frac{16}{24}$

31. $\frac{12}{30}$

32. $\frac{60}{105}$

33. $\frac{19}{76}$

34. $\frac{17}{51}$

MULTIPLICATION AND DIVISION OF FRACTIONS

Exercises 35–48: Multiply and simplify to lowest terms when appropriate.

35. $\frac{3}{4} \cdot \frac{1}{5}$

36. $\frac{3}{2} \cdot \frac{5}{8}$

37. $\frac{5}{3} \cdot \frac{3}{5}$

38. $\frac{21}{32} \cdot \frac{32}{21}$

39. $\frac{5}{6} \cdot \frac{18}{25}$

40. $\frac{7}{9} \cdot \frac{3}{14}$

41. $4 \cdot \frac{3}{5}$

42. $5 \cdot \frac{7}{9}$

43. $2 \cdot \frac{3}{8}$

44. $10 \cdot \frac{1}{100}$

45. $\frac{x}{y} \cdot \frac{y}{x}$

46. $\frac{x}{y} \cdot \frac{y}{z}$

47. $\frac{a}{b} \cdot \frac{3}{2}$

48. $\frac{5}{8} \cdot \frac{4x}{5y}$

Exercises 49–54: Find the fractional part.

49. One-fourth of three-fourths

50. Three-sevenths of nine-sixteenths

51. Two-thirds of six

52. Three-fourths of seven

53. One-half of two-thirds

54. Four-elevenths of nine-eighths

Exercises 55–58: Give the reciprocal of each number.

55. (a) 5   (b) 7   (c) $\frac{4}{7}$ (d) $\frac{9}{8}$

56. (a) 3   (b) 2   (c) $\frac{6}{5}$ (d) $\frac{3}{8}$

57. (a) $\frac{1}{2}$ (b) $\frac{1}{9}$ (c) $\frac{12}{101}$ (d) $\frac{31}{17}$

58. (a) $\frac{1}{5}$ (b) $\frac{7}{3}$ (c) $\frac{23}{64}$ (d) $\frac{63}{29}$

Exercises 59–74: Divide and simplify to lowest terms when appropriate.

59. $\frac{1}{2} \div \frac{1}{3}$

60. $\frac{3}{4} \div \frac{1}{5}$

61. $\frac{3}{4} \div \frac{1}{8}$

62. $\frac{6}{7} \div \frac{3}{14}$

63. $\frac{4}{3} \div \frac{4}{3}$

64. $\frac{12}{21} \div \frac{4}{7}$

65. $\frac{32}{27} \div \frac{8}{9}$

66. $\frac{8}{15} \div \frac{2}{25}$

67. $10 \div \frac{5}{6}$

68. $8 \div \frac{4}{3}$

69. $\frac{9}{10} \div 3$

70. $\frac{32}{27} \div 16$

71. $\frac{a}{b} \div \frac{2}{b}$

72. $\frac{3a}{b} \div \frac{3}{c}$

73. $\frac{x}{y} \div \frac{x}{y}$

74. $\frac{x}{3y} \div \frac{x}{3}$

ADDITION AND SUBTRACTION OF FRACTIONS

Exercises 75–78: Add or subtract. Write each answer in lowest terms.

75. (a) $\frac{5}{12} + \frac{1}{12}$ (b) $\frac{5}{12} - \frac{1}{12}$

76. (a) $\frac{3}{2} + \frac{1}{2}$ (b) $\frac{3}{2} - \frac{1}{2}$

77. (a) $\frac{18}{29} + \frac{7}{29}$ (b) $\frac{18}{29} - \frac{7}{29}$

78. (a) $\frac{5}{33} + \frac{2}{33}$ (b) $\frac{5}{33} - \frac{2}{33}$

Exercises 79–88: Find the least common denominator.

79. $\frac{4}{9}, \frac{2}{15}$ **80.** $\frac{1}{11}, \frac{1}{2}$

81. $\frac{2}{5}, \frac{3}{15}$ **82.** $\frac{8}{21}, \frac{3}{7}$

83. $\frac{1}{6}, \frac{5}{8}$ **84.** $\frac{1}{9}, \frac{5}{12}$

85. $\frac{1}{2}, \frac{1}{3}, \frac{1}{4}$ **86.** $\frac{2}{5}, \frac{2}{3}, \frac{1}{6}$

87. $\frac{1}{4}, \frac{3}{8}, \frac{1}{12}$ **88.** $\frac{2}{15}, \frac{7}{20}, \frac{1}{30}$

Exercises 89–96: Rewrite each set of fractions with the least common denominator.

89. $\frac{1}{2}, \frac{2}{3}$ **90.** $\frac{3}{4}, \frac{1}{5}$

91. $\frac{7}{9}, \frac{5}{12}$ **92.** $\frac{5}{13}, \frac{1}{2}$

93. $\frac{1}{16}, \frac{7}{12}$ **94.** $\frac{5}{18}, \frac{1}{24}$

95. $\frac{1}{3}, \frac{3}{4}, \frac{5}{6}$ **96.** $\frac{4}{15}, \frac{2}{9}, \frac{3}{5}$

Exercises 97–108: Add or subtract. Write your answer in lowest terms.

97. $\frac{5}{8} + \frac{3}{16}$ **98.** $\frac{1}{9} + \frac{2}{15}$

99. $\frac{25}{24} - \frac{7}{8}$ **100.** $\frac{4}{5} - \frac{1}{4}$

101. $\frac{11}{14} + \frac{2}{35}$ **102.** $\frac{7}{8} + \frac{4}{15}$

103. $\frac{5}{12} - \frac{1}{18}$ **104.** $\frac{9}{20} - \frac{7}{30}$

105. $\frac{3}{100} + \frac{1}{300} - \frac{1}{200}$ **106.** $\frac{43}{36} + \frac{4}{9} + \frac{1}{4}$

107. $\frac{7}{8} - \frac{1}{6} + \frac{5}{12}$ **108.** $\frac{9}{40} - \frac{3}{50} - \frac{1}{100}$

APPLICATIONS

109. *American Flag* According to Executive Order 10834, the length of an official American flag should be $1\frac{9}{10}$ times the width. If an official flag has a width that measures $2\frac{1}{2}$ feet, find its length.

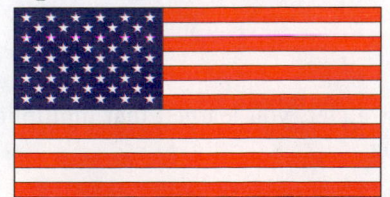

110. *American Flag* The blue rectangle containing the stars on an American flag is called the *union*. On an official American flag, the width of the union should be $\frac{7}{13}$ of the width of the flag. If an official flag has a width of $32\frac{1}{2}$ inches, what is the width of the union?

111. *Carpentry* A board measuring $64\frac{5}{8}$ inches is cut in half. Find the length of each half.

112. *Cutting Rope* A rope measures $15\frac{1}{2}$ feet and needs to be cut in four equal parts. Find the length of each piece.

113. *Geometry* Find the area of the triangle shown with base $1\frac{2}{3}$ yards and height $\frac{3}{4}$ yard. (*Hint:* The area of a triangle equals half the product of its base and height.)

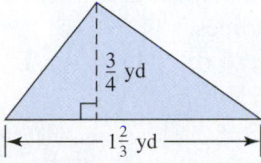

114. *Geometry* Find the area of the rectangle shown.

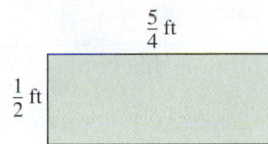

115. *Distance* Use the map to find the distance between Smalltown and Bigtown by traveling through Middletown.

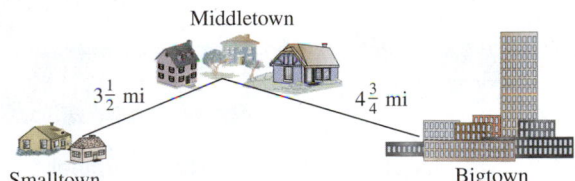

116. *Distance* An athlete jogs $1\frac{3}{8}$ miles, $5\frac{3}{4}$ miles, and $3\frac{5}{8}$ miles. In all, how far does the athlete jog?

117. *Vegetarian Diets* About $\frac{4}{125}$ of U.S. adults follow a vegetarian diet. Within this part of the population, $\frac{1}{200}$ is vegan. What fraction of the adult U.S. population is vegan? (*Source: Vegetarian Times—2010.*)

118. *Women College Students* In 2010, about $\frac{29}{50}$ of all college students were women. Of these women, about $\frac{13}{20}$ were part-time students. What fraction of college students were women who were part-time students? (*Source: National Center for Education Statistics.*)

119. *Accidental Deaths* For the age group 15 to 24, motor vehicle accidents account for $\frac{31}{42}$ of all accidental deaths, and firearms account for $\frac{31}{1260}$ of all accidental deaths. What fraction of all accidental deaths do vehicle accidents *and* firearms account for? (*Source:* National Safety Council.)

120. *Illicit Drug Use* For the age group 18 to 25, the fraction of people who used illicit drugs during their lifetime was $\frac{3}{5}$, whereas the fraction who used illicit drugs during the past year was $\frac{7}{20}$. What fraction of

this population has used illicit drugs but not during the past year? (*Source:* Department of Health and Human Services.)

WRITING ABOUT MATHEMATICS

121. Describe a problem in real life for which the solution could be found by multiplying 30 by $\frac{1}{4}$.

122. Describe a problem in real life for which the solution could be found by dividing $2\frac{1}{2}$ by $\frac{1}{3}$.

SECTIONS 1.1 AND 1.2 ## Checking Basic Concepts

1. Classify each number as prime, composite, or neither. If a number is composite, write it as a product of primes.
   (a) 19   **(b)** 28   **(c)** 1   **(d)** 180

2. Evaluate $\frac{10}{(x+2)}$ for $x = 3$.

3. Find y for $x = 5$ if $y = 6x$.

4. Translate the phrase "a number x plus five" into an algebraic expression.

5. Write a formula that converts F feet to I inches.

6. Find the greatest common factor.
   (a) 3, 18   **(b)** 40, 72

7. Simplify each fraction to lowest terms.
   (a) $\frac{25}{35}$ **(b)** $\frac{26}{39}$

8. Give the reciprocal of $\frac{4}{3}$.

9. Evaluate each expression. Write each answer in lowest terms.
   (a) $\frac{2}{3} \cdot \frac{3}{4}$ **(b)** $\frac{5}{6} \div \frac{10}{3}$
   (c) $\frac{3}{10} + \frac{1}{10}$ **(d)** $\frac{3}{4} - \frac{1}{6}$

10. A recipe needs $1\frac{2}{3}$ cups of flour. How much flour should be used if the recipe is doubled?

1.3   Exponents and Order of Operations

Natural Number Exponents • Order of Operations • Translating Words to Expressions

A LOOK INTO MATH ▶

If there are 15 energy drinks on a store shelf and 3 boxes of 24 drinks each in the storage room, then the total number of energy drinks is given by the expression $15 + 3 \cdot 24$. How would you find the total number of energy drinks? Is the total

$$15 + 3 \cdot 24 \stackrel{?}{=} 18 \cdot 24 \stackrel{?}{=} 432, \quad \text{or} \quad 15 + 3 \cdot 24 \stackrel{?}{=} 15 + 72 \stackrel{?}{=} 87?$$

(add first, then multiply)   (multiply first, then add)

In this section, we discuss the *order of operations agreement*, which can be used to show that the expression $15 + 3 \cdot 24$ evaluates to 87.

NEW VOCABULARY

☐ Exponential expression
☐ Base
☐ Exponent

Natural Number Exponents

In elementary school you learned addition. Later, you learned that multiplication is a fast way to add. For example, rather than adding

$$3 + 3 + 3 + 3, \quad \longleftarrow \text{4 terms}$$

we can multiply $3 \cdot 4$. Similarly, exponents represent a fast way to multiply. Rather than multiplying

$$3 \cdot 3 \cdot 3 \cdot 3, \quad \longleftarrow \text{4 factors}$$

we can evaluate the *exponential expression* 3^4. We begin by discussing natural numbers as exponents.

READING CHECK

- What type of expression represents a fast way to multiply?

The area of a square equals the length of one of its sides times itself. If the square is 5 inches on a side, then its area is

$$5 \cdot 5 = \overset{\text{Exponent}}{5^{2}} = 25 \text{ square inches.}$$
$$\underset{\text{Base}}{}$$

The expression 5^2 is an **exponential expression** with *base* 5 and *exponent* 2. Exponential expressions occur in a variety of applications. For example, suppose an investment doubles 3 times. Then, the calculation

$$\underbrace{2 \cdot 2 \cdot 2}_{\text{Factors}} = 2^{\overset{\text{Exponent 3}}{3}} = 8$$

shows that the final value is 8 times as large as the original investment. For example, if \$10 doubles 3 times, it becomes \$20, \$40, and finally \$80, which is 8 times as large as \$10. Table 1.5 contains examples of exponential expressions.

TABLE 1.5 Exponential Expressions

Repeated Multiplication	Exponential Expression	Base	Exponent
$2 \cdot 2 \cdot 2 \cdot 2$	2^4	2	4
$4 \cdot 4 \cdot 4$	4^3	4	3
$9 \cdot 9$	9^2	9	2
$\frac{1}{2}$	$\left(\frac{1}{2}\right)^1$	$\frac{1}{2}$	1
$b \cdot b \cdot b \cdot b \cdot b$	b^5	b	5

3 Squared

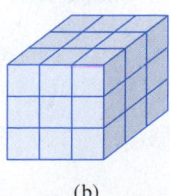

(a)

3 Cubed

(b)

Figure 1.14

Read 9^2 as "9 squared," 4^3 as "4 cubed," and 2^4 as "2 to the fourth power." The terms *squared* and *cubed* come from geometry. If the length of a side of a square is 3, then its area is

$$3 \cdot 3 = 3^2 = 9$$

square units, as illustrated in Figure 1.14(a). Similarly, if the length of an edge of a cube is 3, then its volume is

$$3 \cdot 3 \cdot 3 = 3^3 = 27$$

cubic units, as shown in Figure 1.14(b).

NOTE: The expressions 3^2 and 3^3 can also be read as "3 to the second power" and "3 to the third power," respectively. In general, the expression x^n is read as "x to the nth power" or "the nth power of x."

EXPONENTIAL NOTATION

The expression b^n, where n is a natural number, means

$$b^n = \underbrace{b \cdot b \cdot b \cdots \cdot b}_{n \text{ factors}}.$$

The **base** is b and the **exponent** is n.

EXAMPLE 1 **Writing products in exponential notation**

Write each product as an exponential expression.

(a) $7 \cdot 7 \cdot 7 \cdot 7$ (b) $\dfrac{1}{4} \cdot \dfrac{1}{4} \cdot \dfrac{1}{4}$ (c) $x \cdot x \cdot x \cdot x \cdot x$

Solution
(a) Because there are four factors of 7, the exponent is **4** and the base is **7**.
   Thus $7 \cdot 7 \cdot 7 \cdot 7 = 7^{\mathbf{4}}$.
(b) Because there are three factors of $\frac{1}{4}$, the exponent is 3 and the base is $\frac{1}{4}$.
   Thus $\frac{1}{4} \cdot \frac{1}{4} \cdot \frac{1}{4} = \left(\frac{1}{4}\right)^3$.
(c) Because there are five factors of x, the exponent is 5 and the base is x.
   Thus $x \cdot x \cdot x \cdot x \cdot x = x^5$.

Now Try Exercises 13, 15, 17

EXAMPLE 2 **Evaluating exponential notation**

Evaluate each expression.

(a) 3^4 (b) 10^3 (c) $\left(\dfrac{3}{4}\right)^2$

Solution
(a) The exponential expression 3^4 indicates that 3 is to be multiplied times itself 4 times.

$$3^{\mathbf{4}} = \underbrace{3 \cdot 3 \cdot 3 \cdot 3}_{4 \text{ factors}} = 81$$

CALCULATOR HELP

To evaluate an exponential
expression with a calculator,
see Appendix A (page AP-1).

(b) $10^3 = 10 \cdot 10 \cdot 10 = 1000$
(c) $\left(\frac{3}{4}\right)^2 = \frac{3}{4} \cdot \frac{3}{4} = \frac{9}{16}$

Now Try Exercises 25, 27

EXAMPLE 3 **Writing numbers in exponential notation**

Use the given base to write each number as an exponential expression. Check your results with a calculator, if one is available.
(a) 100 (base 10) (b) 16 (base 2) (c) 27 (base 3)

Solution
(a) $100 = 10 \cdot 10 = 10^2$
(b) $16 = \mathbf{4} \cdot \mathbf{4} = 2 \cdot 2 \cdot 2 \cdot 2 = 2^4$
(c) $27 = \mathbf{3} \cdot \mathbf{9} = 3 \cdot 3 \cdot 3 = 3^3$

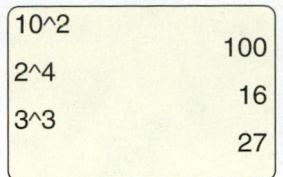

Figure 1.15

These values are supported in Figure 1.15, where exponential expressions are evaluated with a calculator by using the "^" key. (Note that some calculators may have a different key for evaluating exponential expressions.)

■ **Now Try Exercises 29, 33, 35**

▶ **REAL-WORLD CONNECTION** Computer memory is often measured in bytes, with each *byte* capable of storing one letter of the alphabet. For example, it takes four bytes to store the word "math" in a computer. Bytes of computer memory are often manufactured in amounts equal to powers of 2, as illustrated in the next example.

EXAMPLE 4 **Analyzing computer memory**

In computer technology 1 K (kilobyte) of memory equals 2^{10} bytes, and 1 MB (megabyte) of memory equals 2^{20} bytes. Determine whether 1 K of memory equals one thousand bytes and whether 1 MB equals one million bytes.

Solution
Figure 1.16 shows that $2^{10} = 1024$ and $2^{20} = 1{,}048{,}576$. Thus 1 K represents slightly more than one thousand bytes, and 1 MB represents more than one million bytes.

■ **Now Try Exercise 75**

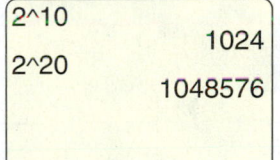

Figure 1.16

CRITICAL THINKING

One gigabyte of memory equals 2^{30} bytes and is often referred to as 1 billion bytes. If you have a calculator available, determine whether 1 gigabyte is exactly 1 billion bytes.

Order of Operations

When the expression $10 - 2 \cdot 3$ is evaluated, is the result

$$8 \cdot 3 = 24 \quad \text{or} \quad 10 - 6 = 4?$$

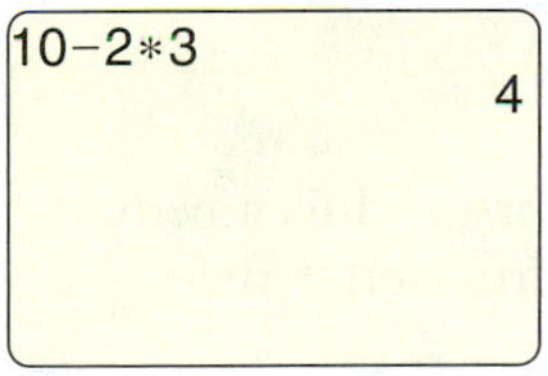

Figure 1.17 shows that a calculator gives a result of 4. The reason is that multiplication is performed before subtraction.

Because arithmetic expressions may contain parentheses, exponents, absolute values, and several operations, it is important to evaluate these expressions consistently. (Absolute value will be discussed in Section 1.4.) To ensure that we all obtain the same result when evaluating an arithmetic expression, the following rules are used.

```
10−2∗3
          4
```

Figure 1.17

ORDER OF OPERATIONS

Use the following order of operations. First perform all calculations within parentheses and absolute values, or above and below the fraction bar.

1. Evaluate all exponential expressions.
2. Do all multiplication and division from *left to right*.
3. Do all addition and subtraction from *left to right*.

EXAMPLE 5 **Evaluating arithmetic expressions**

Evaluate each expression by hand.

(a) $10 - 4 - 3$ **(b)** $10 - (4 - 3)$ **(c)** $5 + \dfrac{12}{3}$ **(d)** $\dfrac{4 + 1}{2 + 8}$

Solution

(a) There are no parentheses, so we evaluate subtraction from *left to right*.

$$\mathbf{10 - 4} - 3 = \mathbf{6} - 3 = 3$$

(b) Note the similarity between this part and part (a). The difference is the parentheses, so subtraction inside the parentheses must be performed first.

$$10 - (\mathbf{4 - 3}) = 10 - \mathbf{1} = 9$$

(c) We perform division before addition.

$$5 + \dfrac{\mathbf{12}}{\mathbf{3}} = 5 + \mathbf{4} = 9$$

(d) Evaluate the expression as though both the numerator and the denominator have parentheses around them.

$$\dfrac{4 + 1}{2 + 8} = \dfrac{\mathbf{(4 + 1)}}{\mathbf{(2 + 8)}} = \dfrac{\mathbf{5}}{\mathbf{10}} = \dfrac{1}{2}$$

Now Try Exercises 43, 45, 57

EXAMPLE 6 **Evaluating arithmetic expressions**

Evaluate each expression by hand.

(a) $25 - 4 \cdot 6$ **(b)** $6 + 7 \cdot 2 - (4 - 1)$ **(c)** $\dfrac{3 + 3^2}{14 - 2}$ **(d)** $5 \cdot 2^3 - (3 + 2)$

Solution

(a) Multiplication is performed before subtraction, so evaluate the expression as follows.

$$
\begin{aligned}
25 - \mathbf{4 \cdot 6} &= 25 - \mathbf{24} && \text{Multiply.} \\
&= 1 && \text{Subtract.}
\end{aligned}
$$

(b) Start by performing the subtraction within the parentheses first and then perform the multiplication. Finally, perform the addition and subtraction from left to right.

$$
\begin{aligned}
6 + 7 \cdot 2 - (4 - 1) &= 6 + 7 \cdot 2 - 3 && \text{Subtract within parentheses.} \\
&= 6 + 14 - 3 && \text{Multiply.} \\
&= 20 - 3 && \text{Add.} \\
&= 17 && \text{Subtract.}
\end{aligned}
$$

(c) First note that parentheses are implied around the numerator and denominator.

$$
\begin{aligned}
\dfrac{3 + 3^2}{14 - 2} &= \dfrac{(3 + 3^2)}{(14 - 2)} && \text{Insert parentheses.} \\
&= \dfrac{(3 + 9)}{(14 - 2)} && \text{Evaluate the exponent first.} \\
&= \dfrac{12}{12} && \text{Add and subtract.} \\
&= 1 && \text{Simplify.}
\end{aligned}
$$

(d) Begin by evaluating the expression inside parentheses.

$$
\begin{aligned}
5 \cdot 2^3 - (3 + 2) &= 5 \cdot 2^3 - 5 && \text{Add within parentheses.} \\
&= 5 \cdot 8 - 5 && \text{Evaluate the exponent.} \\
&= 40 - 5 && \text{Multiply.} \\
&= 35 && \text{Subtract.}
\end{aligned}
$$

▎**Now Try Exercises 59, 61, 63**

Translating Words to Expressions

Sometimes before we can solve a problem we must translate words into mathematical expressions. For example, if a cell phone plan allows for 500 minutes of call time each month and 376 minutes have already been used, then "five hundred minus three hundred seventy-six," or $500 - 376 = 124$, is the number of minutes remaining for the month.

EXAMPLE 7 ▎ **Writing and evaluating expressions**

Translate each phrase into a mathematical expression and then evaluate it.
(a) Two to the fourth power plus ten
(b) Twenty decreased by five times three
(c) Ten cubed divided by five squared
(d) Sixty divided by the quantity ten minus six

Solution
(a) $2^4 + 10 = 2 \cdot 2 \cdot 2 \cdot 2 + 10 = 16 + 10 = 26$
(b) $20 - 5 \cdot 3 = 20 - 15 = 5$
(c) $\dfrac{10^3}{5^2} = \dfrac{1000}{25} = 40$
(d) Here, the word "quantity" indicates that parentheses should be used.

$$60 \div (10 - 6) = 60 \div 4 = 15$$

▎**Now Try Exercises 67, 69, 73**

1.3   Putting It All Together

CONCEPT	COMMENTS	EXAMPLES
Exponential Expression	If n is a natural number, then b^n equals $$\underbrace{b \cdot b \cdot b \cdot \cdots \cdot b}_{n \text{ factors}}$$ and is read "b to the nth power."	$5^1 = 5,$ $7^2 = 7 \cdot 7 = 49,$ $4^3 = 4 \cdot 4 \cdot 4 = 64,$ and $k^4 = k \cdot k \cdot k \cdot k$
Base and Exponent	The base in b^n is b and the exponent is n.	7^4 has base 7 and exponent 4, and x^3 has base x and exponent 3.

continued on next page

continued from previous page

CONCEPT	COMMENTS	EXAMPLES
Order of Operations	First, perform all calculations within parentheses and absolute values, or above and below a fraction bar. 1. Evaluate all exponential expressions. 2. Do all multiplication and division from *left to right*. 3. Do all addition and subtraction from *left to right*.	$10 \div 5 + 3 \cdot 7 = 2 + 21$ $= 23$ $27 \div (4 - 1)^2 = 27 \div 3^2$ $= 27 \div 9$ $= 3$

1.3 Exercises

MyMathLab

 PRACTICE   WATCH   DOWNLOAD   READ   REVIEW

CONCEPTS AND VOCABULARY

1. Exponents represent a fast way to _____.

2. In the expression 2^5, there are five factors of _____ being multiplied.

3. In the expression 5^3, the number 5 is called the _____ and the number 3 is called the _____.

4. Use symbols to write "6 squared."

5. Use symbols to write "8 cubed."

6. When evaluating the expression $5 + 6 \cdot 2$, the result is _____ because _____ is performed before _____.

7. When evaluating the expression $10 - 2^3$, the result is _____ because _____ are evaluated before _____ is performed.

8. The expression $10 - 4 - 2$ equals _____ because subtraction is performed from _____ to _____.

9. (True or False?) The expressions 2^3 and 3^2 are equal.

10. (True or False?) The expression 5^2 equals $5 \cdot 2$.

NATURAL NUMBER EXPONENTS

Exercises 11–20: Write the product as an exponential expression.

11. $3 \cdot 3 \cdot 3 \cdot 3$

12. $10 \cdot 10$

13. $2 \cdot 2 \cdot 2 \cdot 2 \cdot 2$

14. $4 \cdot 4 \cdot 4$

15. $\frac{1}{2} \cdot \frac{1}{2} \cdot \frac{1}{2} \cdot \frac{1}{2}$

16. $\frac{5}{7} \cdot \frac{5}{7} \cdot \frac{5}{7} \cdot \frac{5}{7} \cdot \frac{5}{7}$

17. $a \cdot a \cdot a \cdot a \cdot a$

18. $b \cdot b \cdot b \cdot b$

19. $(x + 3) \cdot (x + 3)$

20. $(x - 4) \cdot (x - 4) \cdot (x - 4)$

Exercises 21–28: Evaluate each expression.

21. (a) 2^4 (b) 4^2 22. (a) 3^2 (b) 5^3

23. (a) 6^1 (b) 1^6 24. (a) 17^1 (b) 1^{17}

25. (a) 2^5 (b) 10^3 26. (a) 10^5 (b) 3^4

27. (a) $\left(\frac{2}{3}\right)^2$ (b) $\left(\frac{1}{2}\right)^5$

28. (a) $\left(\frac{1}{10}\right)^3$ (b) $\left(\frac{4}{3}\right)^1$

Exercises 29–40: (Refer to Example 3.) Use the given base to write the number as an exponential expression. Check your result if you have a calculator available.

29. 8   (base 2)   30. 9   (base 3)

31. 25   (base 5)   32. 32   (base 2)

33. 49   (base 7)   34. 81   (base 3)

35. 1000   (base 10)   36. 256   (base 4)

37. $\frac{1}{16}$ $\left(\text{base } \frac{1}{2}\right)$ 38. $\frac{9}{25}$ $\left(\text{base } \frac{3}{5}\right)$

39. $\frac{32}{243}$ $\left(\text{base } \frac{2}{3}\right)$ 40. $\frac{216}{343}$ $\left(\text{base } \frac{6}{7}\right)$

ORDER OF OPERATIONS

Exercises 41–64: Evaluate the expression by hand.

41. $5 + 4 \cdot 6$ **42.** $6 \cdot 7 - 8$

43. $6 \div 3 + 2$ **44.** $20 - 10 \div 5$

45. $100 - \frac{50}{5}$ **46.** $\frac{200}{100} + 6$

47. $10 - 6 - 1$ **48.** $30 - 9 - 5$

49. $20 \div 5 \div 2$ **50.** $500 \div 100 \div 5$

51. $3 + 2^4$ **52.** $10 - 3^2 + 1$

53. $4 \cdot 2^3$ **54.** $100 - 2 \cdot 3^3$

55. $(3 + 2)^3$ **56.** $5 \cdot (3 - 2)^8 - 5$

57. $\dfrac{4 + 8}{1 + 3}$ **58.** $5 - \dfrac{3 + 1}{3 - 1}$

59. $\dfrac{2^3}{4 - 2}$ **60.** $\dfrac{10 - 3^2}{2 \cdot 4^2}$

61. $10^2 - (30 - 2 \cdot 5)$ **62.** $5^2 + 3 \cdot 5 \div 3 - 1$

63. $\left(\dfrac{1}{2}\right)^4 + \dfrac{5 + 4}{3}$ **64.** $\left(\dfrac{7}{9}\right)^2 - \dfrac{6 - 5}{3}$

TRANSLATING WORDS TO EXPRESSIONS

Exercises 65–74: Translate the phrase into a mathematical expression and then evaluate it.

65. Two cubed minus eight

66. Five squared plus nine

67. Thirty decreased by four times three

68. One hundred plus five times six

69. Four squared divided by two cubed

70. Three cubed times two squared

71. Forty divided by ten, plus two

72. Thirty times ten, minus three

73. One hundred times the quantity two plus three

74. Fifty divided by the quantity eight plus two

APPLICATIONS

75. *Flash Memory* (Refer to Example 4.) Determine the number of bytes on a 512-MB memory stick.

76. *iPod Memory* Determine the number of bytes on a 60-GB video iPod. (*Hint:* One gigabyte equals 2^{30} bytes.)

77. *Population by Gender* One way to measure the gender balance in a given population is to find the number of males for every 100 females in the population. In 1900, the western region of the United States was significantly out of gender balance. In this region, there were 128 males for every 100 females. (*Source: U.S. Census Bureau.*)
   (a) Find an exponent k so that $2^k = 128$.
   (b) During this time, how many males were there for every 25 females?

78. *Solar Eclipse* In early December 2048 there will be a total solar eclipse visible in parts of Botswana. Find an exponent k so that $2^k = 2048$. (*Source: NASA.*)

79. *Rule of 72* Investors sometimes use the *rule of 72* to determine the time required to double an investment. If 72 is divided by the annual interest rate earned on an investment, the result approximates the number of years needed to double the investment. For example, an investment earning 6% annual interest will double in value approximately every $72 \div 6 = 12$ years.
   (a) Approximate the number of years required to double an investment earning 9% annual interest.
   (b) If an investment of $10,000 earns 12% annual interest, approximate the value of the investment after 18 years.

80. *Doubling Effect* Suppose that a savings account containing $1000 doubles its value every 7 years. How much money will be in the account after 28 years?

WRITING ABOUT MATHEMATICS

81. Explain how exponential expressions are related to multiplication. Give an example.

82. Explain why agreement on the order of operations is necessary.

Group Activity   Working with Real Data

Directions: Form a group of 2 to 4 people. Select someone to record the group's responses for this activity. All members of the group should work cooperatively to answer the questions. If your instructor asks for your results, each member of the group should be prepared to respond.

Converting Temperatures To convert Celsius degrees C to Fahrenheit degrees F, use the formula $F = 32 + \frac{9}{5}C$. This exercise illustrates the importance of understanding the order of operations.

(a) Complete the following table by evaluating the formula in the two ways shown.

Celsius	$F = \left(32 + \frac{9}{5}\right)C$	$F = 32 + \left(\frac{9}{5}C\right)$
$-40°\,C$		
$0°\,C$		
$5°\,C$	$169°F$	$41°F$
$20°\,C$		
$30°\,C$		
$100°\,C$		

(b) At what Celsius temperature does water freeze? At what Fahrenheit temperature does water freeze?

(c) Which column gives the correct Fahrenheit temperatures? Why?

(d) Explain why having an agreed order for operations in mathematics is necessary.

1.4   Real Numbers and the Number Line

Signed Numbers • **Integers and Rational Numbers** • **Square Roots** • **Real and Irrational Numbers** • **The Number Line** • **Absolute Value** • **Inequality**

A LOOK INTO MATH ▶   So far in this chapter, we have discussed natural numbers, whole numbers, and fractions. All of these numbers belong to a set of numbers called *real numbers*. In this section, we will see that real numbers also include *integers* and *irrational numbers*. Real-world quantities such as temperature, computer processor speed, height of a building, age of a fossil, and gas mileage are all described with real numbers.

Signed Numbers

The idea that numbers could be negative was a difficult concept for many mathematicians. As late as the eighteenth century, negative numbers were not readily accepted by everyone. After all, how could a person have -5 oranges?

However, negative numbers make more sense to someone working with money. If you owe someone 100 dollars (a debt), this amount can be thought of as -100, whereas if you have a balance of 100 dollars in your checking account (an asset), this amount can be thought of as $+100$. (The positive sign is usually omitted.)

The **opposite**, or **additive inverse**, of a number a is $-a$. For example, the opposite of 25 is -25, the opposite of -5 is $-(-5)$, or 5, and the opposite of 0 is 0 because 0 is neither positive nor negative. The following double negative rule is helpful in simplifying expressions containing negative signs.

DOUBLE NEGATIVE RULE

Let a be any number. Then $-(-a) = a$.

Thus $-(-8) = 8$ and $-\left(\mathbf{-(-10)}\right) = -(\mathbf{10}) = -10$.

READING CHECK

• How is the opposite of a number written?

EXAMPLE 1 **Finding opposites (or additive inverses)**

Find the opposite of each expression.

(a) 13   **(b)** $-\dfrac{4}{7}$ **(c)** $-(-7)$

Solution
(a) The opposite of 13 is -13.
(b) The opposite of $-\frac{4}{7}$ is $\frac{4}{7}$.
(c) $-(-7) = 7$, so the opposite of $-(-7)$ is -7.

Now Try Exercises 17, 19, 21

NOTE: To find the opposite of an exponential expression, evaluate the exponent first. For example, the opposite of 2^4 is

$$-2^4 = -(2 \cdot 2 \cdot 2 \cdot 2) = -16.$$

EXAMPLE 2 **Finding an additive inverse (or opposite)**

Find the additive inverse of $-t$, if $t = -\frac{2}{3}$.

Solution
The additive inverse of $-t$ is $t = -\frac{2}{3}$ because $-(-t) = t$ by the double negative rule.

Now Try Exercise 27

Integers and Rational Numbers

In the opening section of this chapter we discussed natural numbers and whole numbers. Because these sets of numbers do not include negative numbers, fractions, or decimals, other sets of numbers are needed. The **integers** include the natural numbers, zero, and the opposites of the natural numbers. The integers are given by the following.

$$\dots, -3, -2, -1, 0, 1, 2, 3, \dots$$

A **rational number** is any number that can be expressed as the ratio of two integers, $\frac{p}{q}$, where $q \neq 0$. Rational numbers can be written as fractions, and they include all integers. Rational numbers may be positive, negative, or zero. Some examples of rational numbers are

$$\frac{2}{3}, \quad -\frac{3}{5}, \quad \frac{-7}{2}, \quad 1.2, \quad \text{and} \quad 3.$$

The numbers 1.2 and 3 are both rational numbers because they can be written as $\frac{12}{10}$ and $\frac{3}{1}$.

NOTE: The fraction $\frac{-7}{2}$ can also be written as $\frac{7}{-2}$ and $-\frac{7}{2}$. The position of the negative sign does not affect the value of the fraction.

The fraction bar can be thought of as a division symbol. As a result, rational numbers have decimal equivalents. For example, $\frac{1}{2}$ is equivalent to $1 \div 2$. The division

$$\begin{array}{r} 0.5 \\ 2\overline{)1.0} \end{array}$$

shows that $\frac{1}{2} = 0.5$. In general, a rational number may be expressed in a decimal form that either *repeats* or *terminates*. The fraction $\frac{1}{3}$ may be expressed as $0.\overline{3}$, a repeating decimal, and the fraction $\frac{1}{4}$ may be expressed as 0.25, a terminating decimal. The overbar indicates that $0.\overline{3} = 0.3333333\ldots$.

STUDY TIP

The word "NOTE" is used to draw attention to important concepts that may otherwise be overlooked.

READING CHECK

• When a number is written in decimal form, how do we know if the number is a rational number?

▶ **REAL-WORLD CONNECTION** Integers and rational numbers are used to describe quantities such as change in population. Figure 1.18 shows the change in population from 2000 to 2009 for selected U.S. cities. Note that both positive and negative numbers are used to describe these population changes. (*Source:* U.S. Census Bureau.)

Population Change 2000–2009

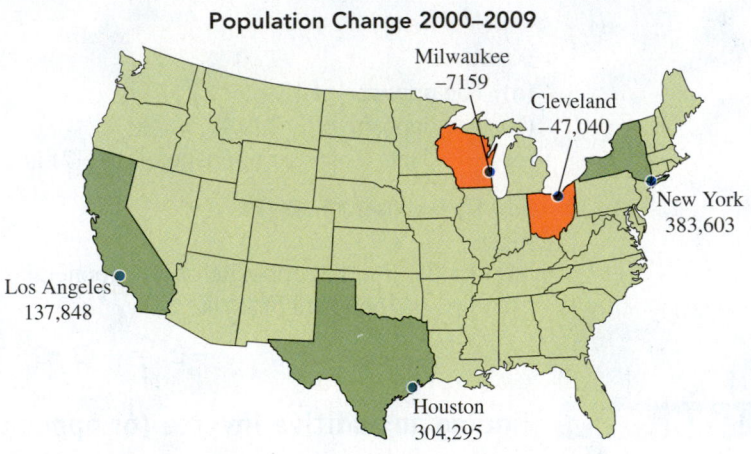

Figure 1.18

EXAMPLE 3 **Classifying numbers**

Classify each number as one or more of the following: natural number, whole number, integer, or rational number.

(a) $\frac{12}{4}$ **(b)** -3 **(c)** 0 **(d)** $-\frac{9}{5}$

Solution
(a) Because $\frac{12}{4} = 3$, the number $\frac{12}{4}$ is a natural number, whole number, integer, and rational number.
(b) The number -3 is an integer and a rational number but it is not a natural number or a whole number.
(c) The number 0 is a whole number, integer, and rational number but not a natural number.
(d) The fraction $-\frac{9}{5}$ is a rational number because it is the ratio of two integers. It is not a natural number, a whole number, or an integer.

▌ **Now Try Exercises 41, 43, 45, 47**

CALCULATOR HELP

To evaluate square roots with a calculator, see Appendix A (page AP-1).

Square Roots

The number b is a **square root** of a number a if $b \cdot b = a$. Every *positive* number has one positive square root and one negative square root. For example, the positive square root of 9 is 3 because $3 \cdot 3 = 9$. The negative square root of 9 is -3. (We will show that $-3 \cdot (-3) = 9$ in Section 1.6.) If a is a positive number, then the **principal square root** of a, denoted $\sqrt{a}$, is the positive square root of a. For example, $\sqrt{25} = 5$ because $5 \cdot 5 = 25$ and the number 5 is positive. Note that the principal square root of 0 is 0. That is, $\sqrt{0} = 0$.

EXAMPLE 4 | **Calculating principal square roots**

Evaluate each square root. Approximate your answer to three decimal places when appropriate.
(a) $\sqrt{36}$ **(b)** $\sqrt{100}$ **(c)** $\sqrt{5}$

Solution
(a) $\sqrt{36} = 6$ because $6 \cdot 6 = 36$ and 6 is positive.
(b) $\sqrt{100} = 10$ because $10 \cdot 10 = 100$ and 10 is positive.
(c) We can estimate the value of $\sqrt{5}$ with a calculator. Figure 1.19 reveals that $\sqrt{5}$ is *approximately* equal to 2.236. However, 2.236 does not exactly equal $\sqrt{5}$ because $2.236 \cdot 2.236 = 4.999696$, which does not equal 5.

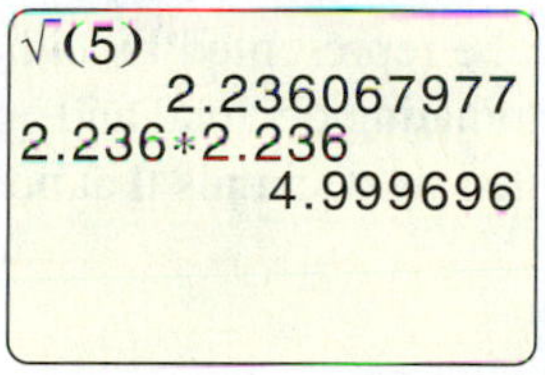

```
√(5)
            2.236067977
2.236*2.236
               4.999696
```

Figure 1.19

Now Try Exercises 49, 51, 53

Real and Irrational Numbers

If a number can be represented by a decimal number, then it is a **real number**. Every fraction has a decimal form, so real numbers include rational numbers. However, some real numbers cannot be expressed by fractions. They are called **irrational numbers**. The numbers $\sqrt{2}$, $\sqrt{15}$, and π are examples of irrational numbers. Every irrational number has a decimal representation that does not terminate or repeat.

NOTE: For any positive integer a, if $\sqrt{a}$ is not an integer then $\sqrt{a}$ is an irrational number.

Examples of real numbers include

$$-17, \quad \frac{4}{5}, \quad -\sqrt{3}, \quad 21\frac{1}{2}, \quad 57.63, \quad \text{and} \quad \sqrt{7}.$$

Any real number may be approximated by a terminating decimal. The symbol $\approx$ is used to represent **approximately equal**. Each of the following real numbers has been approximated to two *decimal places*.

$$\frac{1}{7} \approx 0.14, \quad 2\pi \approx 6.28, \quad \text{and} \quad \sqrt{60} \approx 7.75$$

Figure 1.20 on the next page shows the relationships among the different sets of numbers. Note that each real number is either a rational number or an irrational number but not both. All natural numbers, whole numbers, and integers are rational numbers.

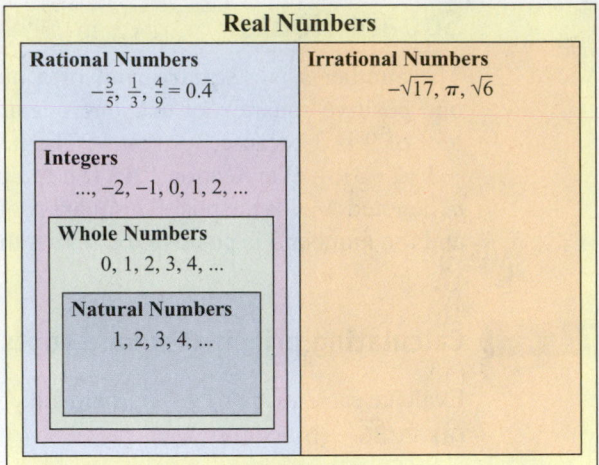

Figure 1.20 The Real Numbers

MAKING CONNECTIONS

Rational and Irrational Numbers

Both rational and irrational numbers can be written as decimals. However, rational numbers can be represented by either terminating or repeating decimals. For example, $\frac{1}{2} = 0.5$ is a terminating decimal and $\frac{1}{3} = 0.333\ldots$ is a repeating decimal. Irrational numbers are represented by decimals that neither terminate nor repeat.

EXAMPLE 5   **Classifying numbers**

Identify the natural numbers, whole numbers, integers, rational numbers, and irrational numbers in the following list.

$$-\sqrt{5},\quad 9,\quad -3.8,\quad \sqrt{49},\quad \frac{11}{4},\quad \text{and}\quad -41$$

Solution

Natural numbers: 9 and $\sqrt{49} = 7$

Whole numbers: 9 and $\sqrt{49} = 7$

Integers: 9, $\sqrt{49} = 7$, and -41

Rational numbers: 9, -3.8, $\sqrt{49} = 7$, $\frac{11}{4}$, and -41

Irrational number: $-\sqrt{5}$

Now Try Exercises 55, 59, 63

▶ **REAL-WORLD CONNECTION** Even though a data set may contain only integers, we often need decimals to describe it. For example, integers are used to represent the number of wireless subscribers in the United States for various years. However, the *average* number of subscribers over a longer time period may be a decimal. Recall that the **average** of a set of numbers is found by adding the numbers and then dividing by how many numbers there are in the set.

EXAMPLE 6 **Analyzing cell phone subscriber data**

Table 1.6 lists the number of wireless subscribers, in millions, in the United States, for various years. Find the average number of subscribers for these years. Is the result a natural number, a rational number, or an irrational number?

TABLE 1.6 U.S. Wireless Subscribers

Year	1995	2000	2005	2010
Subscribers	28	97	194	292

Source: CTIA.

Solution
The average number of subscribers is

$$\frac{28 + 97 + 194 + 292}{4} = \frac{611}{4} = 152.75 \text{ million.}$$

The average of these four natural numbers is an integer divided by an integer, which is a rational number. However, it is neither a natural number nor an irrational number.

Now Try Exercise 105

CRITICAL THINKING

Think of an example in which the sum of two irrational numbers is a rational number.

The Number Line

The real numbers can be represented visually by using a number line, as shown in Figure 1.21. Each real number corresponds to a unique point on the number line. The point associated with the real number 0 is called the **origin**. The positive integers are equally spaced to the right of the origin, and the negative integers are equally spaced to the left of the origin. The number line extends indefinitely both to the left and to the right.

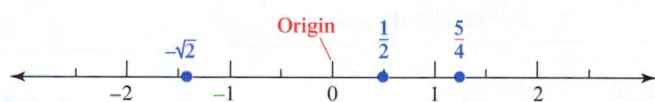

Figure 1.21 A Number Line

Other real numbers can also be located on the number line. For example, the number $\frac{1}{2}$ can be identified by placing a dot halfway between the integers 0 and 1. The numbers $-\sqrt{2} \approx -1.41$ and $\frac{5}{4} = 1.25$ can also be placed (approximately) on this number line. (See Figure 1.21.)

EXAMPLE 7 **Plotting numbers on a number line**

Plot each real number on a number line.

(a) $-\frac{3}{2}$ **(b)** $\sqrt{3}$ **(c)** π

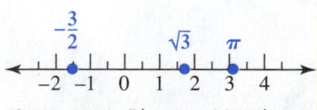

Figure 1.22 Plotting Numbers

Solution
(a) $-\frac{3}{2} = -1.5$. Place a dot halfway between -2 and -1, as shown in Figure 1.22.
(b) A calculator gives $\sqrt{3} \approx 1.73$. Place a dot between 1 and 2 so that it is about three-fourths of the way toward 2, as shown in Figure 1.22.
(c) $\pi \approx 3.14$. Place a dot just past the integer 3, as shown in Figure 1.22.

Now Try Exercises 67, 73

Absolute Value

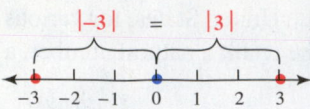

Figure 1.23

The **absolute value** of a real number equals its distance on the number line from the origin. Because distance is never negative, the absolute value of a real number is *never negative*. The absolute value of a real number a is denoted $|a|$ and is read "the absolute value of a." Figure 1.23 shows that the absolute values of -3 and 3 are both equal to 3 because both have distance 3 from the origin. That is, $|-3| = 3$ and $|3| = 3$.

EXAMPLE 8 **Finding the absolute value of a real number**

Evaluate each expression.
(a) $|3.1|$ **(b)** $|-7|$ **(c)** $|0|$

Solution
(a) $|3.1| = 3.1$ because the distance between the origin and 3.1 is 3.1.
(b) $|-7| = 7$ because the distance between the origin and -7 is 7.
(c) $|0| = 0$ because the distance is 0 between the origin and 0.

Now Try Exercises 75, 77, 79

Our results about absolute value can be summarized as follows.

$$|a| = a, \quad \text{if } a \text{ is positive or 0.}$$
$$|a| = -a, \quad \text{if } a \text{ is negative.}$$

Inequality

If a real number a is located to the left of a real number b on the number line, we say that a is **less than** b and write $a < b$. Similarly, if a real number b is located to the right of a real number a, we say that b is **greater than** a and write $b > a$. For example, $-3 < 2$ because -3 is located to the left of 2, and $2 > -3$ because 2 is located to the right of -3. See Figure 1.24.

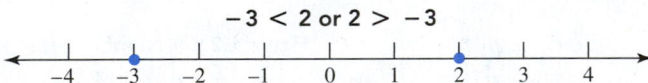

Figure 1.24

NOTE: Any negative number is always less than any positive number, and any positive number is always greater than any negative number.

We say that a is **less than or equal to** b, denoted $a \le b$, if either $a < b$ or $a = b$ is true. Similarly, a is **greater than or equal to** b, denoted $a \ge b$, if either $a > b$ or $a = b$ is true.

Inequalities are often used to compare the relative sizes of two quantities. This can be illustrated visually as shown in the figure below.

Visualizing Inequality by Relative Size

READING CHECK

• Which math symbols are used to express inequality?

EXAMPLE 9 **Ordering real numbers**

List the following numbers from least to greatest. Then plot these numbers on a number line.

$$-2, \quad -\pi, \quad \sqrt{2}, \quad 0, \quad \text{and} \quad 2.5$$

Solution

First note that $-\pi \approx -3.14 < -2$. The two negative numbers are less than 0, and the two positive numbers are greater than 0. Also, $\sqrt{2} \approx 1.41$, so $\sqrt{2} < 2.5$. Listing the numbers from least to greatest results in

$$-\pi, \quad -2, \quad 0, \quad \sqrt{2}, \quad \text{and} \quad 2.5.$$

These numbers are plotted on the number line shown in Figure 1.25. Note that these numbers increase from left to right on the number line.

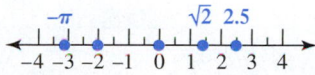

Figure 1.25

Now Try Exercise 103

1.4 | Putting It All Together

CONCEPT	COMMENTS	EXAMPLES
Integers	Include the natural numbers, their opposites, and 0	$\ldots, -2, -1, 0, 1, 2, \ldots$
Rational Numbers	Include integers, all fractions $\frac{p}{q}$, where p and $q \neq 0$ are integers, and all repeating and terminating decimals	$\frac{1}{2}, -3, \frac{128}{6}, -0.335, 0, 0.25 = \frac{1}{4}$, and $0.\overline{3} = \frac{1}{3}$
Irrational Numbers	Any decimal numbers that neither terminate nor repeat; real numbers that are not rational	$\pi, \sqrt{3}$, and $\sqrt{15}$
Real Numbers	Any numbers that can be expressed in decimal form; include the rational and irrational numbers	$\pi, \sqrt{3}, -\frac{4}{7}, 0, -10, 0.\overline{6} = \frac{2}{3}, 1000$, and $\sqrt{15}$
Average	To find the average of a set of numbers, find the sum of the numbers and then divide the sum by how many numbers there are in the set.	The average of 2, 6, and 10 is $$\frac{2 + 6 + 10}{3} = \frac{18}{3} = 6.$$
Number Line	A number line can be used to visualize the real number system. The point associated with the number 0 is called the origin.	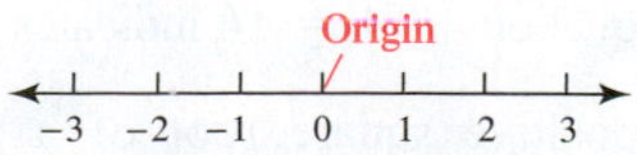

continued on next page

continued from previous page

CONCEPT	COMMENTS	EXAMPLES
Absolute Value	The absolute value of a number a equals its distance on the number line from the origin. If $a \geq 0$, then $\lvert a \rvert = a$. If $a < 0$, then $\lvert a \rvert = -a$. $\lvert a \rvert$ is *never* negative.	$\lvert 17 \rvert = 17$ $\lvert -12 \rvert = 12$ $\lvert 0 \rvert = 0$
Inequality	If a number a is located to the left of a number b on a number line, then a is less than b (written $a < b$). If a is located to the right of b, then a is greater than b (written $a > b$). Symbols of inequality include: $<$, $>$, $\leq$, $\geq$, and $\neq$.	$-3 < -2$ less than $6 > 4$ greater than $-5 \leq -5$ less than or equal $18 \geq 0$ greater than or equal $7 \neq 8$ not equal

1.4 Exercises

MyMathLab | Math XL PRACTICE 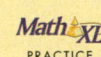   WATCH   DOWNLOAD   READ   REVIEW

CONCEPTS AND VOCABULARY

1. The opposite of the number b is _____.

2. The integers include the natural numbers, zero, and the opposites of the _____ numbers.

3. A number that can be written as $\frac{p}{q}$, where p and q are integers with $q \neq 0$, is a(n) _____ number.

4. If a number can be written in decimal form, then it is a(n) _____ number.

5. If a real number is not a rational number, then it is a(n) _____ number.

6. (True or False?) A rational number can be written as a repeating or terminating decimal.

7. (True or False?) An irrational number cannot be written as a repeating or terminating decimal.

8. Write 0.272727… using an overbar.

9. The decimal equivalent for $\frac{1}{4}$ can be found by dividing _____ by _____.

10. The equation $4 \cdot 4 = 16$ indicates that $\sqrt{16} =$ _____.

11. The positive square root of a positive number is called the _____ square root.

12. The symbol $\neq$ is used to indicate that two numbers are _____.

13. The symbol $\approx$ is used to indicate that two numbers are _____.

14. The origin on the number line corresponds to the number _____.

15. Negative numbers are located to the (left/right) of the origin on the number line.

16. The absolute value of a number a gives its distance on the number line from the _____.

SIGNED NUMBERS

Exercises 17–24: Find the opposite of each expression.

17. (a) 9   (b) -9

18. (a) -6 (b) 6

19. (a) $\frac{2}{3}$ (b) $-\frac{2}{3}$

20. (a) $-\left(\frac{-4}{5}\right)$ (b) $-\left(\frac{-4}{-5}\right)$

21. (a) $-(-8)$ (b) $-\left(-(-8)\right)$

22. (a) $-\left(-(-2)\right)$ (b) $-(-2)$

23. (a) a (b) $-a$

24. (a) $-b$ (b) $-(-b)$

25. Find the additive inverse of t, if $-t = 6$.

26. Find the additive inverse of $-t$, if $t = -\frac{4}{5}$.

27. Find the additive inverse of $-b$, if $b = \frac{1}{2}$.

28. Find the additive inverse of b, if $-b = \frac{5}{-6}$.

NUMBERS AND THE NUMBER LINE

Exercises 29–40: Find the decimal equivalent for the rational number.

29. $\frac{1}{4}$ **30.** $\frac{3}{5}$

31. $\frac{7}{8}$ **32.** $\frac{3}{10}$

33. $\frac{3}{2}$ **34.** $\frac{3}{50}$

35. $\frac{1}{20}$ **36.** $\frac{3}{16}$

37. $\frac{2}{3}$ **38.** $\frac{2}{9}$

39. $\frac{7}{9}$ **40.** $\frac{5}{11}$

Exercises 41–48: Classify the number as one or more of the following: natural number, whole number, integer, or rational number.

41. 8   **42.** -8

43. $\frac{16}{4}$ **44.** $\frac{5}{7}$

45. 0   **46.** $-\frac{15}{31}$

47. $-\frac{7}{6}$ **48.** $-\frac{10}{5}$

Exercises 49–54: Evaluate the square root. Approximate your answer to three decimal places when appropriate.

49. $\sqrt{25}$ **50.** $\sqrt{81}$

51. $\sqrt{49}$ **52.** $\sqrt{64}$

53. $\sqrt{7}$ **54.** $\sqrt{11}$

Exercises 55–66: Classify the number as one or more of the following: natural number, integer, rational number, or irrational number.

55. -4.5 **56.** π

57. $\frac{3}{7}$ **58.** $\sqrt{25}$

59. $\sqrt{11}$ **60.** $-\sqrt{3}$

61. $\frac{8}{4}$ **62.** -5

63. $\sqrt{49}$ **64.** $3.\overline{3}$

65. $1.\overline{8}$ **66.** $\frac{9}{3}$

Exercises 67–74: Plot each number on a number line.

67. (a) 0   (b) -2 (c) 3

68. (a) $-\frac{3}{2}$ (b) $\frac{3}{2}$ (c) 0

69. (a) $\frac{1}{2}$ (b) $-\frac{1}{2}$ (c) 2

70. (a) 1.3   (b) -2.5 (c) 0.7

71. (a) -10 (b) -20 (c) 30

72. (a) 5   (b) 10   (c) -10

73. (a) π (b) $\sqrt{2}$ (c) $-\sqrt{3}$

74. (a) $\sqrt{11}$ (b) $-\sqrt{5}$ (c) $\sqrt{4}$

ABSOLUTE VALUE

Exercises 75–82: Evaluate the expression.

75. $|5.23|$ **76.** $|\pi|$

77. $|-8|$ **78.** $|-\sqrt{2}|$

79. $|2 - 2|$ **80.** $|\frac{2}{3} - \frac{1}{3}|$

81. $|\pi - 3|$ **82.** $|3 - \pi|$

83. Thinking Generally Find $|b|$, if b is negative.

84. Thinking Generally Find $|-b|$, if b is positive.

INEQUALITY

Exercises 85–96: Insert the symbol $>$ or $<$ to make the statement true.

85. 5 _____ 7   **86.** -5 _____ 7

87. -5 _____ -7 **88.** $\frac{3}{5}$ _____ $\frac{2}{5}$

89. $-\frac{1}{3}$ _____ $-\frac{2}{3}$ **90.** $-\frac{1}{10}$ _____ 0

91. -1.9 _____ -1.3 **92.** 5.1 _____ -6.2

93. $|-8|$ _____ 3   **94.** 4 _____ $|-1|$

95. $|-2|$ _____ $|-7|$ **96.** $|-15|$ _____ $|32|$

Exercises 97–100: **Thinking Generally** *Insert the symbol $<$, $=$, or $>$ to make the statement true.*

97. If a number a is located to the right of a number b on the number line, then a _____ b.

98. If $b > 0$ and $a < 0$, then b _____ a.

99. If $a \geq b$, then either $a > b$ or a _____ b.

100. If $a \geq b$ and $b \geq a$, then a _____ b.

Exercises 101–104: List the given numbers from least to greatest.

101. $-3, 0, 1, -9, -2^3$

102. $4, -2^3, \frac{1}{2}, -\frac{3}{2}, \frac{3}{2}$

103. $-2, \pi, \frac{1}{3}, -\frac{3}{2}, \sqrt{5}$

104. $9, 14, -\frac{1}{12}, -\frac{3}{16}, \sqrt{7}$

APPLICATIONS

105. *Marriage Age* The table lists the average age of females at first marriage during selected years.

Year	2001	2003	2005	2007
Age	25.1	25.3	25.5	25.9

Source: U.S. Census Bureau.

(a) What was this age in 2005?

(b) Mentally estimate the average marriage age for these selected years.

(c) Calculate the average marriage age. Is your mental estimate in reasonable agreement with your calculated result?

106. *Music Sales* The table lists the percentage of recorded music purchased through digital downloads during selected years.

Year	2005	2006	2007	2008
Percent	6.0	6.8	12.0	13.5

Source: Recording Industry Association of America.

(a) What was this percentage in 2006?

(b) Mentally estimate the average percentage for this 4-year period.

(c) Calculate the average percentage. Is your mental estimate in reasonable agreement with your calculated result?

WRITING ABOUT MATHEMATICS

107. What is a rational number? Is every integer a rational number? Why or why not?

108. Explain why $\frac{3}{7} > \frac{1}{3}$. Now explain in general how to determine whether $\frac{a}{b} > \frac{c}{d}$. Assume that a, b, c, and d are natural numbers.

SECTIONS 1.3 AND 1.4 — Checking Basic Concepts

1. Write each product as an exponential expression.
 (a) $5 \cdot 5 \cdot 5 \cdot 5$ (b) $7 \cdot 7 \cdot 7 \cdot 7 \cdot 7$

2. Evaluate each expression.
 (a) 2^3 (b) 10^4 (c) $\left(\frac{2}{3}\right)^3$

3. Use the given base to write the number as an exponential expression.
 (a) 64 (base 4)   (b) 64 (base 2)

4. Evaluate each expression without a calculator.
 (a) $6 + 5 \cdot 4$ (b) $6 + 6 \div 2$
 (c) $5 - 2 - 1$ (d) $\frac{6 - 3}{2 + 4}$
 (e) $12 \div (6 \div 2)$ (f) $2^3 - 2\left(2 + \frac{4}{2}\right)$

5. Translate the phrase "five cubed divided by three" to an algebraic expression.

6. Find the opposite of each expression.
 (a) -17 (b) a

7. Find the decimal equivalent for the rational number.
 (a) $\frac{3}{20}$ (b) $\frac{5}{8}$

8. Classify each number as one or more of the following: natural number, integer, rational number, or irrational number.
 (a) $\frac{10}{2}$ (b) -5 (c) $\sqrt{5}$ (d) $-\frac{5}{6}$

9. Plot each number on the same number line.
 (a) 0 (b) -3 (c) 2
 (d) $\frac{3}{4}$ (e) $-\sqrt{2}$

10. Evaluate each expression.
 (a) $|-12|$ (b) $|6 - 6|$

11. Insert the symbol $>$ or $<$ to make the statement true.
 (a) 4 ____ 9 (b) -1.3 ____ -0.5
 (c) $|-3|$ ____ $|-5|$

12. List the following numbers from least to greatest.

$$\sqrt{3}, \quad -7, \quad 0, \quad \frac{1}{3}, \quad -1.6, \quad 3^2$$

1.5 Addition and Subtraction of Real Numbers

Addition of Real Numbers • Subtraction of Real Numbers • Applications

A LOOK INTO MATH ▶

Addition and subtraction of real numbers occur every day at grocery stores, where the prices of various items are added to the total and discounts from coupons are subtracted. Even though prices are usually expressed in decimal form, the rules for adding and subtracting real numbers are the same no matter how the real numbers are expressed.

Addition of Real Numbers

NEW VOCABULARY

☐ Addends
☐ Sum
☐ Difference

In an addition problem the two numbers added are called **addends**, and the answer is called the **sum**. For example, in the following addition problem the numbers **3** and **5** are the **addends** and the number **8** is the **sum**.

$$3 \quad + \quad 5 \quad = \quad 8$$
$$\text{Addend} \quad \text{Addend} \quad \quad \text{Sum}$$

READING CHECK

• What is the answer to an addition problem called?

The *opposite* (or *additive inverse*) of a real number a is $-a$. When we add opposites, the result is 0. For example, $4 + (-4) = 0$. In general, the equation $a + (-a) = 0$ is true for every real number a.

EXAMPLE 1 | **Adding opposites**

Find the opposite of each number and calculate the sum of the number and its opposite.

(a) 45   **(b)** $\sqrt{2}$ **(c)** $-\dfrac{1}{2}$

Solution
(a) The opposite of 45 is -45. Their sum is $45 + (-45) = 0$.
(b) The opposite of $\sqrt{2}$ is $-\sqrt{2}$. Their sum is $\sqrt{2} + (-\sqrt{2}) = 0$.
(c) The opposite of $-\frac{1}{2}$ is $\frac{1}{2}$. Their sum is $-\frac{1}{2} + \frac{1}{2} = 0$.

Now Try Exercises 11, 13, 15

▶ **REAL-WORLD CONNECTION** When adding real numbers, it may be helpful to think of money. A positive number represents income, and a negative number indicates debt. The sum $9 + (-5) = 4$ would represent being paid $9 and owing $5. In this case $4 would be left over. Similarly, the sum $-3 + (-6) = -9$ would represent debts of $3 and $6, resulting in a total debt of $9. To add two real numbers we can use the following rules.

STUDY TIP

Do you know your instructor's name? Do you know the location of his or her office and the hours when he or she is available for help? Make sure that you have the answers to these important questions so that you can get help when needed.

ADDITION OF REAL NUMBERS

To add two real numbers with *like* signs, do the following.

1. Find the sum of the absolute values of the numbers.
2. Keep the common sign of the two numbers as the sign of the sum.

To add two real numbers with *unlike* signs, do the following.

1. Find the absolute values of the numbers.
2. Subtract the smaller absolute value from the larger absolute value.
3. Keep the sign of the number with the larger absolute value as the sign of the sum.

EXAMPLE 2   **Adding real numbers**

Find each sum by hand.

(a) $-2 + (-4)$ (b) $-\dfrac{2}{5} + \dfrac{7}{10}$ (c) $6.2 + (-8.5)$

Solution

(a) The numbers are both negative, so we add the absolute values $|-2|$ and $|-4|$ to obtain 6. The signs of the addends are both negative, so the answer is -6. That is, $-2 + (-4) = -6$. If we owe \$2 and then owe \$4, the total amount owed is \$6.

(b) The numbers have opposite signs, so we subtract their absolute values to obtain

$$\frac{7}{10} - \frac{2}{5} = \frac{7}{10} - \frac{4}{10} = \frac{3}{10}.$$

The sum is positive because $\left|\frac{7}{10}\right|$ is greater than $\left|-\frac{2}{5}\right|$ and $\frac{7}{10}$ is positive. That is, $-\frac{2}{5} + \frac{7}{10} = \frac{3}{10}$. If we spend \$0.40 $\left(-\frac{2}{5} = -0.4\right)$ and receive \$0.70 $\left(\frac{7}{10} = 0.7\right)$, we have \$0.30 $\left(\frac{3}{10} = 0.3\right)$ left.

(c) $6.2 + (-8.5) = -2.3$ because $|-8.5|$ is 2.3 more than $|6.2|$. If we have \$6.20 and we owe \$8.50, we are short \$2.30.

Now Try Exercises 33, 39, 41

ADDING INTEGERS VISUALLY One way to add integers visually is to use the symbol ⌢ to represent a positive unit and to use the symbol ⌣ to represent a negative unit. Now adding opposites visually results in "zero," as shown.

For example, to add $-3 + 5$, we draw three negative units and five positive units.

Because the "zeros" add no value, the sum is two positive units, or 2.

EXAMPLE 3   **Adding integers visually**

Add visually, using the symbols ⌢ and ⌣.

(a) $3 + 2$ (b) $-6 + 4$ (c) $2 + (-3)$ (d) $-5 + (-2)$

Solution

(a) Draw three positive units and then draw two more positive units.

Because no zeros were formed, the sum is five positive units, or 5.

(b) Draw six negative units and then draw four positive units.

Ignoring the zeros that were formed, the sum is two negative units, or -2.

(c) Draw two positive and three negative units.

The sum is -1.

(d) ⌣ ⌣ ⌣ ⌣ ⌣   ⌣ ⌣ $-5 + (-2) = -7$

The sum is -7.

Now Try Exercises 17, 19, 21, 23

Another way to add integers visually is to use a number line. For example, to add $4 + (-3)$ start at 0 (the origin) and draw an arrow to the right 4 units long from 0 to 4. The number -3 is a negative number, so draw an arrow 3 units long to the left, starting at the tip of the first arrow. See Figure 1.26a. The tip of the second arrow is at 1, which equals the sum of 4 and -3.

Adding Integers on a Number Line

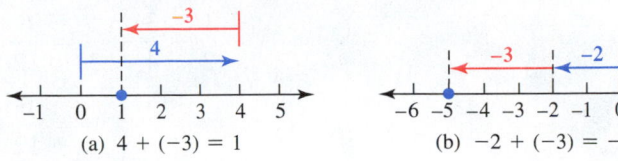

Figure 1.26

To find the sum $-2 + (-3)$, draw an arrow 2 units long to the left, starting at the origin. Then draw an arrow 3 units long to the left, starting at the tip of the first arrow, which is located at -2. See Figure 1.26b. Because the tip of the second arrow coincides with -5 on the number line, the sum of -2 and -3 is -5.

Subtraction of Real Numbers

READING CHECK

• What is the answer to a subtraction problem called?

The answer to a subtraction problem is the **difference**. When subtracting two real numbers, changing the problem to an addition problem may be helpful.

SUBTRACTION OF REAL NUMBERS

For any real numbers a and b,

$$a - b = a + (-b).$$

To subtract b from a, add a and the opposite of b.

EXAMPLE 4   **Subtracting real numbers**

Find each difference by hand.
(a) $10 - 20$ **(b)** $-5 - 2$ **(c)** $-2.1 - (-3.2)$ **(d)** $\frac{1}{2} - \left(-\frac{3}{4}\right)$

Solution
(a) $10 - 20 = 10 + (-20) = -10$
(b) $-5 - 2 = -5 + (-2) = -7$
(c) $-2.1 - (-3.2) = -2.1 + 3.2 = 1.1$
(d) $\frac{1}{2} - \left(-\frac{3}{4}\right) = \frac{1}{2} + \frac{3}{4} = \frac{2}{4} + \frac{3}{4} = \frac{5}{4}$

Now Try Exercises 45, 47, 51, 53

In the next example, we show how to add and subtract groups of numbers.

EXAMPLE 5   **Adding and subtracting real numbers**

Evaluate each expression by hand.
(a) $5 - 4 - (-6) + 1$ **(b)** $\frac{1}{2} - \frac{3}{4} + \frac{1}{3}$ **(c)** $-6.1 + 5.6 - 10.1$

Solution

(a) Rewrite the expression in terms of addition only, and then find the sum.

$$5 - 4 - (-6) + 1 = 5 + (-4) + 6 + 1$$
$$= 1 + 6 + 1$$
$$= 8$$

(b) Begin by rewriting the fractions with the LCD of 12.

$$\frac{1}{2} - \frac{3}{4} + \frac{1}{3} = \frac{6}{12} - \frac{9}{12} + \frac{4}{12}$$
$$= \frac{6}{12} + \left(-\frac{9}{12}\right) + \frac{4}{12}$$
$$= -\frac{3}{12} + \frac{4}{12}$$
$$= \frac{1}{12}$$

CRITICAL THINKING

Explain how subtraction of real numbers can be performed on a number line.

(c) The expression can be evaluated by changing subtraction to addition.

$$-6.1 + 5.6 - 10.1 = -6.1 + 5.6 + (-10.1)$$
$$= -0.5 + (-10.1)$$
$$= -10.6$$

Now Try Exercises 59, 65, 67

Applications

In application problems, some words indicate that we should add, while other words indicate that we should subtract. Tables 1.7 and 1.8 show examples of such words along with sample phrases.

TABLE 1.7 Words Associated with Addition

Words	Sample Phrase
add	add the two temperatures
plus	her age plus his age
more than	5 cents more than the cost
sum	the sum of two measures
total	the total of the four prices
increased by	height increased by 3 inches

TABLE 1.8 Words Associated with Subtraction

Words	Sample Phrase
subtract	subtract dues from the price
minus	his income minus his taxes
fewer than	18 fewer flowers than shrubs
difference	the difference in their heights
less than	his age is 4 less than yours
decreased by	the weight decreased by 7

READING CHECK

- What words are associated with addition?
- What words are associated with subtraction?

▶ **REAL-WORLD CONNECTION** Sometimes temperature differences are found by subtracting positive and negative real numbers. At other times, addition of positive and negative real numbers occurs at banks if positive numbers represent deposits and negative numbers represent withdrawals. The next two examples illustrate these situations.

EXAMPLE 6 **Calculating temperature differences**

The hottest outdoor temperature ever recorded in the shade was 136°F in the Sahara desert, and the coldest outside temperature ever recorded was −129°F in Antarctica. Find the difference between these two temperatures. (*Source: Guinness Book of Records.*)

Solution
The word *difference* indicates subtraction. We must subtract the two temperatures.

$$136 - (-129) = 136 + 129 = 265°F.$$

▪ **Now Try Exercise 87**

EXAMPLE 7 **Balancing a checking account**

The initial balance in a checking account is $285. Find the final balance if the following represents a list of withdrawals and deposits: −$15, −$20, $500, and −$100.

Solution
Find the sum of the five numbers.

$$285 + (-15) + (-20) + 500 + (-100) = 270 + (-20) + 500 + (-100)$$
$$= 250 + 500 + (-100)$$
$$= 750 + (-100)$$
$$= 650$$

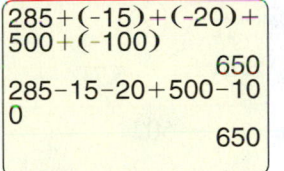

```
285+(-15)+(-20)+
500+(-100)
              650
285-15-20+500-10
0
              650
```

Figure 1.27

The final balance is $650. This result may be supported with a calculator, as illustrated in Figure 1.27. Note that the expression has been evaluated two different ways.

▪ **Now Try Exercise 83**

TECHNOLOGY NOTE

Subtraction and Negation
Calculators typically have *different* keys to represent subtraction and negation. Be sure to use the correct key. Many graphing calculators have the following keys for subtraction and negation.

$$\boxed{-} \qquad \boxed{(-)}$$
Subtraction   Negation

1.5   Putting It All Together

CONCEPT	COMMENTS	EXAMPLES
Addition	To add real numbers, use a number line or follow the rules found in the box on page 45.	$-2 + 8 = 6$ $0.8 + (-0.3) = 0.5$ $-\frac{1}{7} + \left(-\frac{3}{7}\right) = \frac{-1 + (-3)}{7} = -\frac{4}{7}$ $-4 + 4 = 0$ $-3 + 4 + (-2) = -1$
Subtraction	To subtract real numbers, transform the problem to an addition problem by adding the opposite. $a - b = a + (-b)$	$6 - 8 = 6 + (-8) = -2$ $-3 - 4 = -3 + (-4) = -7$ $-\frac{1}{2} - \left(-\frac{3}{2}\right) = -\frac{1}{2} + \frac{3}{2} = \frac{2}{2} = 1$ $-5 - (-5) = -5 + 5 = 0$ $9.4 - (-1.2) = 9.4 + 1.2 = 10.6$

1.5   Exercises

MyMathLab

CONCEPTS AND VOCABULARY

1. The solution to an addition problem is the _____.

2. When you add opposites, the sum is always _____.

3. (True or False?) If two positive numbers are added, the sum is always a positive number.

4. (True or False?) If two negative numbers are added, the sum is always a negative number.

5. If two numbers with opposite signs are added, the sum has the same sign as the number with the larger _____.

6. The solution to a subtraction problem is the _____.

7. When subtracting two real numbers, it may be helpful to change the problem to a(n) _____ problem.

8. To subtract b from a, add the _____ of b to a. That is, $a - b = a +$ _____.

9. The words *sum*, *more*, and *plus* indicate that _____ should be performed.

10. The words *difference*, *less than*, and *minus* indicate that _____ should be performed.

ADDITION AND SUBTRACTION OF REAL NUMBERS

Exercises 11–16: Find the opposite of the number and then calculate the sum of the number and its opposite.

11. 25

12. $-\frac{1}{2}$

13. $-\sqrt{21}$

14. $-\pi$

15. 5.63

16. -6^2

Exercises 17–24: Refer to Example 3. Find the sum visually.

17. $1 + 3$

18. $3 + 1$

19. $4 + (-2)$

20. $-4 + 6$

21. $-1 + (-2)$

22. $-2 + (-2)$

23. $-3 + 7$

24. $5 + (-6)$

Exercises 25–32: Use a number line to find the sum.

25. $-1 + 3$

26. $3 + (-1)$

27. $4 + (-5)$

28. $2 + 6$

29. $-10 + 20$

30. $15 + (-5)$

31. $-50 + (-100)$

32. $-100 + 100$

Exercises 33–44: Find the sum.

33. $5 + (-4)$

34. $-9 + 7$

35. $-1 + (-6)$

36. $-10 + (-23)$

37. $\frac{3}{4} + \left(-\frac{1}{2}\right)$

38. $-\frac{5}{12} + \left(-\frac{1}{6}\right)$

39. $-\frac{6}{7} + \frac{3}{14}$

40. $-\frac{2}{9} + \left(-\frac{1}{12}\right)$

41. $0.6 + (-1.7)$

42. $4.3 + (-2.4)$

43. $-52 + 86$

44. $-103 + (-134)$

Exercises 45–56: Find the difference.

45. $5 - 8$

46. $3 - 5$

47. $-2 - (-9)$

48. $-10 - (-19)$

49. $\frac{6}{7} - \frac{13}{14}$

50. $-\frac{5}{6} - \frac{1}{6}$

51. $-\frac{1}{10} - \left(-\frac{3}{5}\right)$

52. $-\frac{2}{11} - \left(-\frac{5}{11}\right)$

53. $0.8 - (-2.1)$

54. $-9.6 - (-5.7)$

55. $-73 - 91$

56. $201 - 502$

Exercises 57–70: Evaluate the expression.

57. $10 - 19$

58. $5 + (-9)$

59. $19 - (-22) + 1$

60. $53 + (-43) - 10$

61. $-3 + 4 - 6$

62. $-11 + 8 - 10$

63. $100 - 200 + 100 - (-50)$

64. $-50 - (-40) + (-60) + 80$

65. $1.5 - 2.3 + 9.6$

66. $10.5 - (-5.5) + (-1.5)$

67. $-\frac{1}{2} + \frac{1}{4} - \left(-\frac{3}{4}\right)$

68. $\frac{1}{4} - \left(-\frac{2}{5}\right) + \left(-\frac{3}{20}\right)$

69. $|4 - 9| - |1 - 7|$

70. $|-5 - (-3)| - |-6 + 8|$

Exercises 71–80: Write an arithmetic expression for the given phrase and then simplify it.

71. The sum of two and negative five

72. Subtract ten from negative six

73. Negative five increased by seven

74. Negative twenty decreased by eight

75. The additive inverse of the quantity two cubed

76. Five minus the quantity two cubed

77. The difference between negative six and seven (*Hint:* Write the numbers for the subtraction problem in the order given.)

78. The difference between one-half and three-fourths

79. Six plus negative ten minus five

80. Ten minus seven plus negative twenty

81. Online Exploration Use the Internet to find the highest and lowest points in the continental United States, and then find the difference in their heights.

82. Online Exploration In 1972, residents of Loma, Montana, experienced the largest 24-hour temperature swing ever recorded in the United States. Use the Internet to find the starting and ending temperatures, and then write a subtraction equation that shows the difference.

APPLICATIONS

83. *Checking Account* The initial balance in a checking account is $358. Find the final balance resulting from the following withdrawals and deposits: −$45, $37, $120, and −$240.

84. *Savings Account* A savings account has $1245 in it. Find the final balance resulting from the following withdrawals and deposits: −$189, $975, −$226, and −$876.

85. *Football Stats* A running back carries the ball five times. Find his total yardage if the carries were 9, −2, −1, 14, and 5 yards.

86. *Tracking Weight* A person weighs himself every Monday for four weeks. He gains one pound the first week, loses three pounds the second week, gains two pounds the third week, and loses one pound the last week.
   (a) Using positive numbers to represent weight gains and negative numbers to represent weight losses, write a sum that gives the total gain or loss over this four-week period.
   (b) If he weighed 170 pounds at the beginning of this process, what was his weight at the end?

87. *Deepest and Highest* The deepest point in the ocean is the Mariana Trench, which is 35,839 feet below sea level. The highest point on Earth is Mount Everest, which is 29,029 feet above sea level. What is the difference in height between Mount Everest and the Mariana Trench? (*Source: The Guinness Book of Records.*)

88. *Greatest Temperature Ranges* The greatest temperature range on Earth occurs in Siberia, where the temperature can vary between 98°F in the summer and −90°F in the winter. Find the difference between these two temperatures. (*Source: The Guinness Book of Records.*)

WRITING ABOUT MATHEMATICS

89. Explain how to add two negative numbers. Give an example.

90. Explain how to subtract a negative number from a positive number. Give an example.

<table>
<tr><td>**1.6**</td><td>## Multiplication and Division of Real Numbers</td></tr>
</table>

Multiplication of Real Numbers • Division of Real Numbers • Applications

A LOOK INTO MATH ▶ Exam and quiz scores are sometimes displayed in fraction form. For example, a student who answers 17 questions correctly on a 20-question exam may see the fraction $\frac{17}{20}$ written at the top of the exam paper. If the instructor assigns grades based on percents, this fraction must be converted to a percent. Doing so involves division of real numbers. In this section, we discuss multiplication and division of real numbers.

Multiplication of Real Numbers

In a multiplication problem, the numbers multiplied are called the *factors*, and the answer is called the *product*.

READING CHECK

• What is the answer to a multiplication problem called?

$$7 \cdot 4 = 28$$
Factor Factor Product

Multiplication is a fast way to perform addition. For example, $5 \cdot 2 = 10$ is equivalent to finding the sum of five 2s, or

$$2 + 2 + 2 + 2 + 2 = 10.$$

Similarly, the product $5 \cdot (-2)$ is equivalent to finding the sum of five -2s, or

$$(-2) + (-2) + (-2) + (-2) + (-2) = -10.$$

Thus $5 \cdot (-2) = -10$. In general, the product of a positive number and a negative number is a negative number.

What sign should the product of two negative numbers have? To answer this question, consider the following patterns.

What values should replace the question marks to continue the pattern?

$$
\begin{array}{rcl}
-4 \cdot \mathbf{3} &=& -\mathbf{12} \\
-4 \cdot \mathbf{2} &=& -\mathbf{8} \\
-4 \cdot \mathbf{1} &=& -\mathbf{4} \\
-4 \cdot \mathbf{0} &=& \mathbf{0} \\
-4 \cdot (-\mathbf{1}) &=& ? \\
-4 \cdot (-\mathbf{2}) &=& ?
\end{array}
$$

Decrease by 1 — Increase by 4

Continuing the pattern results in $-4 \cdot (-\mathbf{1}) = \mathbf{4}$ and $-4 \cdot (-\mathbf{2}) = \mathbf{8}$. This pattern suggests that if we multiply two negative real numbers, the product is positive.

> **SIGNS OF PRODUCTS**
>
> The product of two numbers with *like* signs is positive. The product of two numbers with *unlike* signs is negative.

NOTE: Multiplying a number by -1 results in the additive inverse (opposite) of the number. For example, $-1 \cdot 4 = -4$, and -4 is the additive inverse of 4. Similarly, $-1 \cdot (-3) = 3$, and 3 is the additive inverse of -3. In general, $-1 \cdot a = -a$.

EXAMPLE 1 **Multiplying real numbers**

Find each product by hand.
(a) $-9 \cdot 7$ **(b)** $\frac{2}{3} \cdot \frac{5}{9}$ **(c)** $-2.1(-40)$ **(d)** $(-2.5)(4)(-9)(-2)$

Solution
(a) The resulting product is negative because the factors -9 and 7 have unlike signs. Thus $-9 \cdot 7 = -63$.
(b) The product is positive because both factors are positive.

$$\frac{2}{3} \cdot \frac{5}{9} = \frac{2 \cdot 5}{3 \cdot 9} = \frac{10}{27}$$

(c) As both factors are negative, the product is positive. Thus $-2.1(-40) = 84$.
(d) $(-2.5)(4)(-9)(-2) = (-10)(-9)(-2) = (90)(-2) = -180$

Now Try Exercises 13, 23, 27, 29

READING CHECK

• How can the number of negative factors be used to decide if a product is positive or negative?

MAKING CONNECTIONS

Multiplying More Than Two Negative Factors

Because the product of two negative numbers is positive, it is possible to determine the sign of a product by counting the number of negative factors. For example, the product $-3 \cdot 4 \cdot (-5) \cdot 7 \cdot (-6)$ is negative because there are an odd number of negative factors and the product $2 \cdot (-1) \cdot (-3) \cdot (-5) \cdot (-4)$ is positive because there are an even number of negative factors.

When evaluating expressions such as

-5^2 (**Square** and then **negate**)   and $(-5)^2$ (**Negate** and then **square**),

it is important to note that the first represents the opposite of five squared, while the second indicates that negative five should be squared. The next example illustrates the difference between the opposite of an exponential expression and a power of a negative number.

EXAMPLE 2   **Evaluating real numbers with exponents**

Evaluate each expression by hand.
(a) $(-4)^2$ (b) -4^2 (c) $(-2)^3$ (d) -2^3

Solution
(a) Because the exponent is outside of parentheses, the base of the exponential expression is -4. The expression is evaluated as
$$(-4)^2 = (-4)(-4) = 16.$$
(b) This is the negation of an exponential expression with base 4. Evaluating the exponent before negating results in $-4^2 = -(4)(4) = -16$.
(c) $(-2)^3 = (-2)(-2)(-2) = -8$
(d) $-2^3 = -(2)(2)(2) = -8$

Now Try Exercises 35, 37, 39, 41

MAKING CONNECTIONS

Negative Square Roots

Because the product of two negative numbers is positive, $(-3) \cdot (-3) = 9$. That is, -3 is a square root of 9. As discussed in Section 1.4, every positive number has one positive square root and one negative square root. For a positive number a, the positive square root is called the *principal square root* and is denoted $\sqrt{a}$. The negative square root is denoted $-\sqrt{a}$. For example, $\sqrt{4} = 2$ and $-\sqrt{4} = -2$.

Division of Real Numbers

READING CHECK

• What is the answer to a division problem called?

In a division problem, the **dividend** is divided by the **divisor**, and the result is the **quotient**.

$$\underset{\text{Dividend}}{20} \div \underset{\text{Divisor}}{4} = \underset{\text{Quotient}}{5}$$

This division problem can also be written in fraction form as $\frac{20}{4} = 5$. Division of real numbers can be defined in terms of multiplication and reciprocals. The **reciprocal**, or **multiplicative inverse**, of a real number a is $\frac{1}{a}$. The number 0 has *no reciprocal*.

TECHNOLOGY NOTE

Try dividing 5 by 0 with a calculator. On most calculators, dividing a number by 0 results in an error message.

DIVIDING REAL NUMBERS

For real numbers a and b with $b \neq 0$,

$$\frac{a}{b} = a \cdot \frac{1}{b}.$$

That is, to divide a by b, multiply a by the reciprocal of b.

NOTE: Division by 0 is undefined because 0 has no reciprocal.

EXAMPLE 3 **Dividing real numbers**

Evaluate each expression by hand.

(a) $-16 \div \frac{1}{4}$ **(b)** $\dfrac{\frac{3}{5}}{-8}$ **(c)** $\frac{-8}{-36}$ **(d)** $9 \div 0$

Solution

(a) $-16 \div \frac{1}{4} = \frac{-16}{1} \cdot \frac{4}{1} = \frac{-64}{1} = -64$ The reciprocal of $\frac{1}{4}$ is $\frac{4}{1}$.

(b) $\dfrac{\frac{3}{5}}{-8} = \frac{3}{5} \div (-8) = \frac{3}{5} \cdot \left(-\frac{1}{8}\right) = -\frac{3}{40}$ The reciprocal of -8 is $-\frac{1}{8}$.

(c) $\frac{-8}{-36} = -8 \cdot \left(-\frac{1}{36}\right) = \frac{8}{36} = \frac{2}{9}$ The reciprocal of -36 is $-\frac{1}{36}$.

(d) $9 \div 0$ is undefined.   The number 0 has no reciprocal.

Now Try Exercises 49, 51, 59, 63

When determining the sign of a quotient, the following rules may be helpful. Note that these rules are similar to those for signs of products.

SIGNS OF QUOTIENTS

The quotient of two numbers with *like* signs is positive. The quotient of two numbers with *unlike* signs is negative.

NOTE: To see why a negative number divided by a negative number is a positive number, remember that division is a fast way to perform subtraction. For example, because

$$20 - 4 - 4 - 4 - 4 - 4 = 0,$$

a total of **five 4**s can be subtracted from **20**, so $\frac{20}{4} = 5$. Similarly, because

$$-20 - (-4) - (-4) - (-4) - (-4) - (-4) = 0,$$

a total of **five** -4s can be subtracted from -20, so $\frac{-20}{-4} = 5$.

▶ **REAL-WORLD CONNECTION** In business, employees may need to convert fractions and mixed numbers to decimal numbers. The next example illustrates this process.

EXAMPLE 4   **Converting fractions to decimals**

Convert each measurement to a decimal number.

(a) $2\frac{3}{8}$-inch bolt   **(b)** $\frac{15}{16}$-inch diameter   **(c)** $1\frac{1}{3}$-cup flour

Solution

(a) First divide 3 by 8.

$$
\begin{array}{r}
0.\textbf{375} \\
8\overline{)3.000} \\
\underline{24} \\
60 \\
\underline{56} \\
40 \\
\underline{40} \\
0
\end{array}
$$

So, $2\frac{3}{8} = \textbf{2.375}$.

(b) Divide 15 by 16.

$$
\begin{array}{r}
0.\textbf{9375} \\
16\overline{)15.0000} \\
\underline{144} \\
60 \\
\underline{48} \\
120 \\
\underline{112} \\
80 \\
\underline{80} \\
0
\end{array}
$$

So, $\frac{15}{16} = \textbf{0.9375}$.

(c) First divide 1 by 3.

$$
\begin{array}{r}
0.\textbf{333}\ldots \\
3\overline{)1.000} \\
\underline{9} \\
10 \\
\underline{9} \\
10 \\
\underline{9} \\
1
\end{array}
$$

So, $1\frac{1}{3} = 1.\overline{3}$.

▮ **Now Try Exercises 69, 71, 75**

In the next example, numbers that are expressed as terminating decimals are converted to fractions.

EXAMPLE 5   **Converting decimals to fractions**

Convert each decimal number to a fraction in lowest terms.

(a) 0.06   **(b)** 0.375   **(c)** 0.0025

Solution

(a) The decimal 0.06 equals six hundredths, or $\frac{6}{100}$. Simplifying this fraction gives

$$\frac{6}{100} = \frac{3 \cdot \textbf{2}}{50 \cdot \textbf{2}} = \frac{3}{50}.$$

(b) The decimal 0.375 equals three hundred seventy-five thousandths, or $\frac{375}{1000}$. Simplifying this fraction gives

$$\frac{375}{1000} = \frac{3 \cdot \textbf{125}}{8 \cdot \textbf{125}} = \frac{3}{8}.$$

(c) The decimal 0.0025 equals twenty-five ten thousandths, or $\frac{25}{10,000}$. Simplifying this fraction gives

$$\frac{25}{10,000} = \frac{1 \cdot \textbf{25}}{400 \cdot \textbf{25}} = \frac{1}{400}.$$

▮ **Now Try Exercises 81, 85, 87**

FRACTIONS AND CALCULATORS (OPTIONAL)   Many calculators have the capability to perform arithmetic on fractions and express the answer as either a decimal or a fraction. The next example illustrates this capability.

| EXAMPLE 6 | **Performing arithmetic operations with technology** |

Use a calculator to evaluate each expression. Express your answer as a decimal and as a fraction.

(a) $\frac{1}{3} + \frac{2}{5} - \frac{4}{9}$ **(b)** $\left(\frac{4}{9} \cdot \frac{3}{8}\right) \div \frac{2}{3}$

Solution

(a) Figure 1.28(a) shows that

$$\frac{1}{3} + \frac{2}{5} - \frac{4}{9} = 0.2\overline{8}, \text{ or } \frac{13}{45}.$$

(b) Figure 1.28(b) shows that $\left(\frac{4}{9} \cdot \frac{3}{8}\right) \div \frac{2}{3} = 0.25$, or $\frac{1}{4}$.

CALCULATOR HELP

To find fraction results with a calculator, see Appendix A (pages AP-1 and AP-2).

```
(1/3)+(2/5)-(4/9
)
          .2888888889
(1/3)+(2/5)-(4/9
)▶Frac
               13/45
```
(a)

```
((4/9)*(3/8))/(2
/3)
                .25
((4/9)*(3/8))/(2
/3)▶Frac
                1/4
```
(b)

Figure 1.28

Now Try Exercises 89, 91

NOTE: Generally it is a good idea to put parentheses around fractions when you are using a calculator.

MAKING CONNECTIONS

The Four Arithmetic Operations

READING CHECK

• How are addition and subtraction related?
• How are multiplication and division related?

If you know how to add real numbers, then you also know how to subtract real numbers because subtraction is defined in terms of addition. That is,

$$a - b = a + (-b).$$

If you know how to multiply real numbers, then you also know how to divide real numbers because division is defined in terms of multiplication. That is,

$$\frac{a}{b} = a \cdot \frac{1}{b}.$$

Applications

There are many instances when we may need to multiply or divide real numbers. Two examples are provided here.

| EXAMPLE 7 | **Comparing top-grossing movies** |

Even though *Avatar* (2009) was an extremely successful film that set domestic box office records, it ranks only fourteenth among top-grossing movies of all time when calculated by using estimated total admissions. The total admissions for *Avatar* are $\frac{17}{25}$ of the total admissions for *The Sound of Music* (1965), which ranks third. (**Source:** Box Office Mojo.)

(a) If *The Sound of Music* had estimated total admissions of 142 million people, find the estimated total admissions for *Avatar*.

(b) The top-grossing movie of all time is *Gone With the Wind* (1939), which had estimated total admissions of 202 million. How many more people saw *Gone With the Wind* than saw *Avatar*?

Solution

(a) To find the total admissions for *Avatar* we multiply the real numbers 142 and $\frac{17}{25}$ to obtain $142 \cdot \frac{17}{25} = \frac{2414}{25} \approx 97$.

  Total admissions for *Avatar* were about 97 million people.

(b) The difference is $202 - 97 = 105$ million people.

■ **Now Try Exercise 97**

EXAMPLE 8 **Analyzing the federal budget**

It is estimated that $\frac{14}{125}$ of the federal budget is used to pay interest on loans. Write this fraction as a decimal. (*Source:* U.S. Office of Management and Budget.)

Solution

One method for writing the fraction as a decimal is to divide 14 by 125, using long division. An alternative method is to multiply the fraction by $\frac{8}{8}$ so that the denominator becomes 1000. Then write the numerator in the thousandths place in the decimal.

$$\frac{14}{125} \cdot \frac{8}{8} = \frac{112}{1000} = 0.112$$

■ **Now Try Exercise 99**

1.6 Putting It All Together

CONCEPT	COMMENTS	EXAMPLES
Multiplication	The product of two numbers with like signs is positive, and the product of two numbers with unlike signs is negative.	$6 \cdot 7 = 42$ Like signs $6 \cdot (-7) = -42$ Unlike signs

continued on next page

continued from previous page

CONCEPT	COMMENTS	EXAMPLES
Division	For real numbers a and b, with $b \neq 0$, $$\frac{a}{b} = a \cdot \frac{1}{b}.$$ The quotient of two numbers with like signs is positive, and the quotient of two numbers with unlike signs is negative.	$\dfrac{42}{6} = 7$ Like signs $\dfrac{-42}{6} = -7$ Unlike signs
Converting Fractions to Decimals	The fraction $\frac{a}{b}$ is equivalent to $a \div b$, where $b \neq 0$.	The fraction $\frac{2}{9}$ is equivalent to $0.\overline{2}$, because $2 \div 9 = 0.222\ldots$.
Converting Terminating Decimals to Fractions	Write the decimal as a fraction with a denominator equal to a power of 10 and then simplify this fraction.	$0.55 = \dfrac{55}{100} = \dfrac{11 \cdot 5}{20 \cdot 5} = \dfrac{11}{20}$

1.6   Exercises

CONCEPTS AND VOCABULARY

1. The solution to a multiplication problem is the _____.

2. The product of a positive number and a negative number is a(n) _____ number.

3. The product of two negative numbers is _____.

4. The solution to a division problem is the _____.

5. The reciprocal of a nonzero number a is _____.

6. Division by zero is undefined because zero has no _____.

7. To divide a by b, multiply a by the _____ of b.

8. In general, $-1 \cdot a =$ _____.

9. $\dfrac{a}{b} = a \cdot$ _____

10. A negative number divided by a negative number is a(n) _____ number.

11. A negative number divided by a positive number is a(n) _____ number.

12. To convert $\frac{5}{8}$ to a decimal, divide _____ by _____.

MULTIPLICATION AND DIVISION OF REAL NUMBERS

Exercises 13–32: Multiply.

13. $-3 \cdot 4$

14. $-5 \cdot 7$

15. $6 \cdot (-3)$

16. $2 \cdot (-1)$

17. $0 \cdot (-2.13)$

18. $-2 \cdot (-7)$

19. $-6 \cdot (-10)$

20. $-3 \cdot (-1.7) \cdot 0$

21. $-\frac{1}{2} \cdot \left(-\frac{2}{4}\right)$

22. $-\frac{3}{4} \cdot \left(-\frac{5}{12}\right)$

23. $-\frac{3}{7} \cdot \frac{7}{3}$

24. $\frac{5}{8} \cdot \left(-\frac{4}{15}\right)$

25. $-10 \cdot (-20)$

26. $1000 \cdot (-70)$

27. $-0.5 \cdot 100$

28. $-0.5 \cdot (-0.3)$

29. $-2 \cdot 3 \cdot (-4) \cdot 5$

30. $-3 \cdot (-5) \cdot (-2) \cdot 10$

31. $-6 \cdot \frac{1}{6} \cdot \frac{7}{9} \cdot \left(-\frac{9}{7}\right) \cdot \left(-\frac{3}{2}\right)$

32. $-\frac{8}{5} \cdot \frac{1}{8} \cdot \left(-\frac{5}{7}\right) \cdot -7$

33. **Thinking Generally** Is the product given by the expression $a \cdot (-a) \cdot (-a) \cdot a \cdot (-a)$ positive or negative if $a > 0$?

34. **Thinking Generally** Is the product given by the expression $a \cdot (-a) \cdot (-a) \cdot a \cdot (-a)$ positive or negative if $a < 0$?

Exercises 35–44: Evaluate the expression.

35. $(-5)^2$

36. -5^2

37. $(-1)^3$

38. $(-6)^2$

39. -2^4

40. $-(-4)^2$

41. $-(-2)^3$

42. $3 \cdot (-3)^2$

43. $5 \cdot (-2)^3$

44. -1^4

Exercises 45–68: Divide.

45. $-10 \div 5$

46. $-8 \div 4$

47. $-20 \div (-2)$

48. $-15 \div (-3)$

49. $\frac{-12}{3}$

50. $\frac{-25}{-5}$

51. $-16 \div \frac{1}{2}$

52. $10 \div \left(-\frac{1}{3}\right)$

53. $0 \div 3$

54. $\frac{0}{-5}$

55. $\frac{-1}{0}$

56. $\frac{0}{-2}$

57. $\frac{1}{2} \div (-11)$

58. $-\frac{3}{4} \div (-6)$

59. $\frac{-\frac{4}{5}}{-3}$

60. $\frac{\frac{7}{8}}{-7}$

61. $\frac{5}{6} \div \left(-\frac{8}{9}\right)$

62. $-\frac{11}{12} \div \left(-\frac{11}{4}\right)$

63. $-\frac{1}{2} \div 0$

64. $-9 \div 0$

65. $-0.5 \div \frac{1}{2}$

66. $-0.25 \div \left(-\frac{3}{4}\right)$

67. $-\frac{2}{3} \div 0.5$

68. $\frac{1}{6} \div 1.5$

CONVERTING BETWEEN FRACTIONS AND DECIMALS

Exercises 69–80: Write the number as a decimal.

69. $\frac{1}{2}$

70. $\frac{3}{4}$

71. $\frac{3}{16}$

72. $\frac{1}{9}$

73. $3\frac{1}{2}$

74. $2\frac{1}{4}$

75. $5\frac{2}{3}$

76. $6\frac{7}{9}$

77. $1\frac{7}{16}$

78. $6\frac{1}{12}$

79. $\frac{7}{8}$

80. $\frac{11}{16}$

Exercises 81–88: Write the decimal number as a fraction in lowest terms.

81. 0.25

82. 0.8

83. 0.16

84. 0.35

85. 0.625

86. 0.0125

87. 0.6875

88. 0.21875

Exercises 89–96: Use a calculator to evaluate each expression. Express your answer as a decimal and as a fraction.

89. $\left(\frac{1}{3} + \frac{5}{6}\right) \div \frac{1}{2}$

90. $\frac{4}{9} - \frac{1}{6} + \frac{2}{3}$

91. $\frac{4}{5} \div \frac{2}{3} \cdot \frac{7}{4}$

92. $4 - \frac{7}{4} \cdot 2$

93. $\frac{15}{2} - 4 \cdot \frac{7}{3}$

94. $\frac{1}{6} - \frac{3}{5} + \frac{7}{8}$

95. $\frac{17}{40} + 3 \div 8$

96. $\frac{3}{4} \cdot \left(6 + \frac{1}{2}\right)$

APPLICATIONS

97. *Top-Grossing Movies* (Refer to Example 7.) *The Ten Commandments* (1956) is the fifth top-grossing movie of all time. Find the total admissions for *The Ten Commandments* if they were $\frac{13}{20}$ of the total admissions for the top-grossing movie of all time, *Gone With the Wind* (1939), which had total admissions of 202 million. (*Source*: Box Office Mojo.)

98. *Planet Climate* Saturn has an average surface temperature of $-220°$F. Neptune has an average surface temperature that is $\frac{3}{2}$ times that of Saturn. Find the average surface temperature on Neptune. (*Source*: NASA.)

99. *Uninsured Americans* In 2010, the fraction of Americans who did not have health insurance coverage was $\frac{21}{125}$. Write this fraction as a decimal. (*Source*: U.S. Census Bureau.)

100. *Uninsured Minnesotans* In 2010, the fraction of Minnesotans who did not have health insurance coverage was $\frac{23}{250}$. Write this fraction as a decimal. (*Source*: U.S. Census Bureau.)

WRITING ABOUT MATHEMATICS

101. Division is a fast way to subtract. Consider the division problem $\frac{-6}{-2}$, whose quotient represents the number of -2s in -6. Using this idea, explain why the answer is a positive number.

102. Explain how to determine whether the product of three integers is positive or negative.

Checking Basic Concepts

1. Find each sum.
   (a) $-4 + 4$ (b) $-10 + (-12) + 3$

2. Evaluate each expression.
   (a) $\frac{2}{3} - \left(-\frac{2}{9}\right)$ (b) $-1.2 - 5.1 + 3.1$

3. Write an arithmetic expression for the given phrase and then simplify it.
   (a) The sum of negative one and five
   (b) The difference between four and negative three

4. The hottest temperature ever recorded at International Falls, Minnesota, was $99°F$, and the coldest temperature ever recorded was $-46°F$. What is the difference between these two temperatures?

5. Find each product.
   (a) $-5 \cdot (-7)$ (b) $-\frac{1}{2} \cdot \frac{2}{3} \cdot \left(-\frac{4}{5}\right)$

6. Evaluate each expression.
   (a) -3^2 (b) $4 \cdot (-2)^3$ (c) $(-5)^2$

7. Evaluate each expression.
   (a) $-5 \div \frac{2}{3}$ (b) $-\frac{5}{8} \div \left(-\frac{4}{3}\right)$

8. What is the reciprocal of $-\frac{7}{6}$?

9. Simplify each expression.
   (a) $\frac{-10}{2}$ (b) $\frac{10}{-2}$ (c) $-\frac{10}{2}$ (d) $\frac{-10}{-2}$

10. Convert each fraction or mixed number to a decimal number.
   (a) $\frac{3}{5}$ (b) $3\frac{7}{8}$

1.7 Properties of Real Numbers

Commutative Properties • Associative Properties • Distributive Properties • Identity and Inverse Properties • Mental Calculations

A LOOK INTO MATH ▶

The order in which you perform actions is often important. For example, putting on your socks and then your shoes is not the same as putting on your shoes and then your socks. In mathematical terms, the action of putting on footwear is not *commutative* with respect to socks and shoes. However, the order in which you tie your shoes and put on a sweatshirt does not matter. So, these two actions are *commutative*. In mathematics some operations are commutative and others are not. In this section we discuss several properties of real numbers.

Commutative Properties

The **commutative property for addition** states that two numbers, a and b, can be added in any order and the result will be the same. That is, $a + b = b + a$. For example, if a person buys 4 DVDs and then buys 2 DVDs or first buys 2 DVDs and then buys 4 DVDs, as shown in Figure 1.29, the result is the same. Either way the person buys a total of $4 + 2 = 2 + 4 = 6$ DVDs.

Figure 1.29 Commutative Property: $4 + 2 = 2 + 4$

There is also a **commutative property for multiplication**. It states that two numbers, a and b, can be multiplied in any order and the result will be the same. That is, $a \cdot b = b \cdot a$.

For example, if one person rolls 3 dice, each resulting in 6, and another person rolls 6 dice, each resulting in 3, as shown in Figure 1.30, then each person has rolled a total of $3 \cdot 6 = 6 \cdot 3 = 18$.

Figure 1.30 Commutative Property: $3 \cdot 6 = 6 \cdot 3$

We can summarize these results as follows.

COMMUTATIVE PROPERTIES

For any real numbers a and b,

$$a + b = b + a \quad \text{and} \quad a \cdot b = b \cdot a.$$

Addition   Multiplication

EXAMPLE 1 **Applying the commutative properties**

Use a commutative property to rewrite each expression.
(a) $15 + 100$ **(b)** $a \cdot 8$

Solution
(a) By the commutative property for addition $15 + 100$ can be written as $100 + 15$.
(b) By the commutative property for multiplication $a \cdot 8$ can be written as $8 \cdot a$ or $8a$.

Now Try Exercises 13, 19

While there are commutative properties for addition and multiplication, the operations of subtraction and division are *not* commutative. Table 1.9 shows each of the four arithmetic operations along with examples illustrating whether or not each operation is commutative.

READING CHECK

• Which of the four arithmetic operations are commutative and which are not?

TABLE 1.9 Commutativity of Operations

Operation	Commutative?	Example
+	Yes	$4 + 9 = 9 + 4$
−	No	$5 - 3 \neq 3 - 5$
·	Yes	$8 \cdot 5 = 5 \cdot 8$
÷	No	$4 \div 2 \neq 2 \div 4$

Associative Properties

The *associative properties* allow us to change how numbers are grouped. For example, if a person buys 1, 2, and 4 energy drinks, as shown in Figure 1.31, then the total number of drinks can be calculated either as

$$(1 + 2) + 4 = 3 + 4 = 7 \quad \text{or as} \quad 1 + (2 + 4) = 1 + 6 = 7.$$

Figure 1.31 Associative Property: $(1 + 2) + 4 = 1 + (2 + 4)$

In either case we obtain the same answer, 7 drinks, which is the result of the **associative property for addition**. We did not change the order of the numbers; we only changed how the numbers were grouped. There is also an **associative property for multiplication**, which can be illustrated by considering the total number of flowers shown in Figure 1.32, where 2 shelves hold 3 pots each, and each pot contains 4 flowers. The total number of flowers can be calculated either as

$$(2 \cdot 3) \cdot 4 = 6 \cdot 4 = 24 \quad \text{or as} \quad 2 \cdot (3 \cdot 4) = 2 \cdot 12 = 24.$$

Figure 1.32

We can summarize these results as follows.

ASSOCIATIVE PROPERTIES

For any real numbers a, b, and c,

$$(a + b) + c = a + (b + c) \quad \text{and} \quad (a \cdot b) \cdot c = a \cdot (b \cdot c).$$

Addition   Multiplication

NOTE: Sometimes we omit the multiplication dot. Thus $a \cdot b = ab$ and $5 \cdot x \cdot y = 5xy$.

EXAMPLE 2 **Applying the associative properties**

Use an associative property to rewrite each expression.
(a) $(5 + 6) + 7$ **(b)** $x(yz)$

Solution
(a) The given expression is equivalent to $5 + (6 + 7)$.
(a) The given expression is equivalent to $(xy)z$.

Now Try Exercises 21, 23

EXAMPLE 3 **Identifying properties of real numbers**

State the property that each equation illustrates.
(a) $5 \cdot (8y) = (5 \cdot 8)y$ **(b)** $3 \cdot 7 = 7 \cdot 3$ **(c)** $x + yz = yz + x$

Solution
(a) This equation illustrates the associative property for multiplication because the grouping of the numbers has been changed.
(b) This equation illustrates the commutative property for multiplication because the order of the numbers 3 and 7 has been changed.
(c) This equation illustrates the commutative property for addition because the order of the terms x and yz has been changed.

Now Try Exercises 53, 55, 63

While there are associative properties for addition and multiplication, the operations of subtraction and division are *not* associative. Table 1.10 shows each of the four arithmetic operations along with examples illustrating whether or not each operation is associative.

TABLE 1.10 Associativity of Operations

Operation	Associative?	Example
+	Yes	$(3 + 6) + 7 = 3 + (6 + 7)$
−	No	$(10 - 3) - 1 \neq 10 - (3 - 1)$
·	Yes	$(4 \cdot 5) \cdot 3 = 4 \cdot (5 \cdot 3)$
÷	No	$(16 \div 8) \div 2 \neq 16 \div (8 \div 2)$

READING CHECK

• Which of the four arithmetic operations are associative and which are not?

STUDY TIP

The information in Making Connections ties the current concepts to those studied earlier. By reviewing your notes often, you can gain a better understanding of mathematics.

MAKING CONNECTIONS

Commutative and Associative Properties

Both the commutative and associative properties work for addition and multiplication. However, neither property works for subtraction or division.

Distributive Properties

The **distributive properties** are used frequently in algebra to simplify expressions. Arrows are often used to indicate that a distributive property is being applied.

$$4(2 + 3) = 4 \cdot 2 + 4 \cdot 3$$

The 4 must be multiplied by *both* the 2 and the 3—not just the 2.

The distributive property remains valid when addition is replaced with subtraction.

$$4(2 - 3) = 4 \cdot 2 - 4 \cdot 3$$

We illustrate a distributive property geometrically in Figure 1.33. Note that the area of one rectangle that is 4 squares by 5 squares is the same as the area of two rectangles: one that is 4 squares by 2 squares and another that is 4 squares by 3 squares. In either case the total area is 20 square units.

Distributive Property

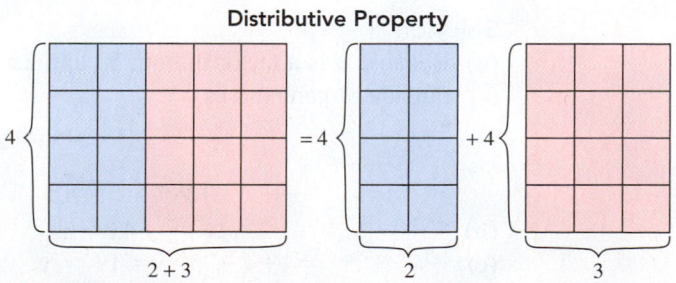

Figure 1.33 $4(2 + 3) = 4 \cdot 2 + 4 \cdot 3$

DISTRIBUTIVE PROPERTIES

For any real numbers a, b, and c,

$$a(b + c) = ab + ac \quad \text{and} \quad a(b - c) = ab - ac.$$

NOTE: Because multiplication is commutative, the distributive properties can also be written as

$$(b + c)a = ba + ca \quad \text{and} \quad (b - c)a = ba - ca.$$

EXAMPLE 4 **Applying the distributive properties**

Apply a distributive property to each expression.
(a) $3(x + 2)$ **(b)** $-6(a - 2)$ **(c)** $-(x + 7)$ **(d)** $15 - (b + 4)$

Solution
(a) Both the x and the 2 must be multiplied by 3.

$$3(x + 2) = 3 \cdot x + 3 \cdot 2 = 3x + 6$$

(b) $-6(a - 2) = -6 \cdot a - (-6) \cdot 2 = -6a + 12$

(c) $-(x + 7) = (-1)(x + 7)$ In general, $-a = -1 \cdot a$.

$\quad\quad\quad\quad = (-1) \cdot x + (-1) \cdot 7$ Distributive property

$\quad\quad\quad\quad = -x - 7$ Multiply.

(d) $15 - (b + 4) = 15 + (-1)(b + 4)$ Change subtraction to addition.

$\quad\quad\quad\quad\quad\quad = 15 + (-1) \cdot b + (-1) \cdot 4$ Distributive property

$\quad\quad\quad\quad\quad\quad = 15 - b - 4$ Multiply.

$\quad\quad\quad\quad\quad\quad = 11 - b$ Simplify.

NOTE: To simplify the expression $15 - (b + 4)$, we subtract *both* the b and the 4. Thus we can quickly simplify the given expression to

$$15 - (b + 4) = 15 - b - 4.$$

Now Try Exercises 31, 35, 37, 41

EXAMPLE 5 **Inserting parentheses using the distributive property**

Use a distributive property to insert parentheses in the expression and then simplify the result.
(a) $5a + 2a$ **(b)** $3x - 7x$ **(c)** $-4y + 5y$

Solution
(a) Because a is a factor in both $5a$ and $2a$, we use the distributive property to write the a outside of parentheses.

$$5a + 2a = (5 + 2)a \quad \text{Distributive property}$$

$$= 7a \quad \text{Simplify.}$$

(b) $3x - 7x = (3 - 7)x = -4x$

(c) $-4y + 5y = (-4 + 5)y = 1y = y$

Now Try Exercises 45, 47, 51

| EXAMPLE 6 | **Identifying properties of real numbers** |

State the property or properties illustrated by each equation.

(a) $4(5 - x) = 20 - 4x$ **(b)** $(4 + x) + 5 = x + 9$

(c) $5z + 7z = 12z$ **(d)** $x(y + z) = zx + yx$

Solution

(a) This equation illustrates the distributive property with subtraction.

$$4(5 - x) = 4 \cdot 5 - 4 \cdot x = 20 - 4x$$

(b) This equation illustrates the commutative and associative properties for addition.

$$(4 + x) + 5 = (x + 4) + 5 \quad \text{Commutative property for addition}$$
$$= x + (4 + 5) \quad \text{Associative property for addition}$$
$$= x + 9 \quad \text{Simplify.}$$

(c) This equation illustrates the distributive property with addition.

$$5z + 7z = (5 + 7)z = 12z$$

(d) This equation illustrates a distributive property with addition and commutative properties for addition and multiplication.

$$x(y + z) = xy + xz \quad \text{Distributive property}$$
$$= xz + xy \quad \text{Commutative property for addition}$$
$$= zx + yx \quad \text{Commutative property for multiplication}$$

Now Try Exercises 55, 57, 59, 61

Identity and Inverse Properties

The **identity property of 0** states that if 0 is added to any real number a, the result is a. The number 0 is called the **additive identity**. Examples include

$$-4 + 0 = -4 \quad \text{and} \quad 0 + 11 = 11.$$

The **identity property of 1** states that if any number a is multiplied by 1, the result is a. The number 1 is called the **multiplicative identity**. Examples include

$$-3 \cdot 1 = -3 \quad \text{and} \quad 1 \cdot 8 = 8.$$

We can summarize these results as follows.

IDENTITY PROPERTIES

For any real number a,

$$a + 0 = 0 + a = a \quad \text{and} \quad a \cdot 1 = 1 \cdot a = a.$$

Additive identity   Multiplicative identity

READING CHECK

- Restate the identity property for 0 in your own words.
- Restate the identity property for 1 in your own words.

The *additive inverse*, or *opposite*, of a number a is $-a$. The number 0 is its own opposite. The sum of a number a and its additive inverse equals the additive identity 0. Thus $-5 + 5 = 0$ and $x + (-x) = 0$.

The *multiplicative inverse*, or *reciprocal*, of a nonzero number a is $\frac{1}{a}$. The number 0 has no multiplicative inverse. The product of a number and its multiplicative inverse equals the multiplicative identity 1. Thus $-\frac{5}{4} \cdot \left(-\frac{4}{5}\right) = 1$.

INVERSE PROPERTIES

For any real number a,

$$a + (-a) = 0 \quad \text{and} \quad -a + a = 0. \quad \text{Additive inverse}$$

For any *nonzero* real number a,

$$a \cdot \frac{1}{a} = 1 \quad \text{and} \quad \frac{1}{a} \cdot a = 1. \quad \text{Multiplicative inverse}$$

EXAMPLE 7 | **Identifying identity and inverse properties**

State the property or properties illustrated by each equation.

(a) $0 + xy = xy$ **(b)** $\frac{36}{30} = \frac{6}{5} \cdot \frac{6}{6} = \frac{6}{5}$

(c) $x + (-x) + 5 = 0 + 5 = 5$ **(d)** $\frac{1}{9} \cdot 9y = 1 \cdot y = y$

Solution

(a) This equation illustrates use of the identity property for 0.
(b) Because $\frac{6}{6} = 1$, these equations illustrate how a fraction can be simplified by using the identity property for 1.
(c) These equations illustrate use of the additive inverse property and the identity property for 0.
(d) These equations illustrate use of the multiplicative inverse property and the identity property for 1.

Now Try Exercises 67, 69, 71, 73

Mental Calculations

Properties of numbers can be used to simplify calculations. For example, to find the sum

$$\textcolor{blue}{4 + 7 + 6 + 3}$$

we might apply the commutative and associative properties for addition to obtain

$$(\textbf{4 + 6}) + (\textbf{7 + 3}) = 10 + 10 = 20.$$

Suppose that we are to add $128 + 19$ mentally. One way is to add 20 to 128 and then subtract 1.

$$128 + \textbf{19} = 128 + (\textbf{20} - \textbf{1}) \quad 19 = 20 - 1.$$
$$= (128 + 20) - 1 \quad \text{Associative property}$$
$$= 148 - 1 \quad \text{Add.}$$
$$= 147 \quad \text{Subtract.}$$

The distributive property can be helpful when we multiply mentally. For example, to determine the number of people in a marching band with 7 columns and 23 rows we need to find the product $7 \cdot 23$. To evaluate the product mentally, think of 23 as $20 + 3$.

$$7 \cdot 23 = 7(20 + 3) \qquad 23 = 20 + 3.$$
$$= 7 \cdot 20 + 7 \cdot 3 \quad \text{Distributive property}$$
$$= 140 + 21 \qquad \text{Multiply.}$$
$$= 161 \qquad \text{Add.}$$

EXAMPLE 8 **Performing calculations mentally**

Use properties of real numbers to calculate each expression mentally.
(a) $21 + 15 + 9 + 5$ **(b)** $\frac{1}{2} \cdot \frac{2}{3} \cdot 2 \cdot \frac{3}{2}$
(c) $523 + 199$ **(d)** $6 \cdot 55$

Solution
(a) Use the commutative and associative properties to group numbers into pairs that sum to a multiple of 10.

$$21 + 15 + 9 + 5 = (21 + 9) + (15 + 5) = 30 + 20 = 50$$

(b) Use the commutative and associative properties to group numbers with their reciprocals.

$$\frac{1}{2} \cdot \frac{2}{3} \cdot 2 \cdot \frac{3}{2} = \left(\frac{1}{2} \cdot 2\right) \cdot \left(\frac{2}{3} \cdot \frac{3}{2}\right) = 1 \cdot 1 = 1$$

(c) Instead of adding **199**, add 200 and then subtract 1.

$$523 + 200 - 1 = 723 - 1 = 722$$

(d) Think of **55** as $50 + 5$ and then apply the distributive property.

$$6 \cdot (50 + 5) = 300 + 30 = 330$$

Now Try Exercises 77, 81, 85, 91

CRITICAL THINKING

How could you quickly calculate $5283 - 198$ without a calculator?

The next example illustrates how the commutative and associative properties for multiplication can be used together to simplify a product.

EXAMPLE 9 **Finding the volume of a swimming pool**

An Olympic swimming pool is 50 meters long, 25 meters wide, and 2 meters deep. The volume V of the pool is found by multiplying 50, 25, and 2. Use the commutative and associative properties for multiplication to calculate the volume of the pool mentally.

Solution
Because $50 \cdot 2 = 100$ and multiplication by 100 is relatively easy, it may be convenient to order and group the multiplication as $(50 \cdot 2) \cdot 25 = 100 \cdot 25 = 2500$. Thus the pool contains 2500 cubic meters of water.

Now Try Exercise 103

1.7   Putting It All Together

PROPERTY	DEFINITION	EXAMPLES
Commutative	For any real numbers a and b, $a + b = b + a$ and $a \cdot b = b \cdot a.$	$4 + 6 = 6 + 4$ $4 \cdot 6 = 6 \cdot 4$
Associative	For any real numbers a, b, and c, $(a + b) + c = a + (b + c)$ and $(a \cdot b) \cdot c = a \cdot (b \cdot c).$	$(3 + 4) + 5 = 3 + (4 + 5)$ $(3 \cdot 4) \cdot 5 = 3 \cdot (4 \cdot 5)$
Distributive	For any real numbers a, b, and c, $a(b + c) = ab + ac$ and $a(b - c) = ab - ac.$	$5(x + 2) = 5x + 10$ $5(x - 2) = 5x - 10$
Identity (0 and 1)	The identity for addition is 0, and the identity for multiplication is 1. For any real number a, $a + 0 = a$ and $a \cdot 1 = a.$	$5 + 0 = 5$ and $5 \cdot 1 = 5$
Inverse	The additive inverse of a is $-a$, and $a + (-a) = 0$. The multiplicative inverse of a nonzero number a is $\frac{1}{a}$, and $a \cdot \frac{1}{a} = 1.$	$8 + (-8) = 0$ and $\frac{2}{3} \cdot \frac{3}{2} = 1$

1.7   Exercises

MyMathLab    PRACTICE   WATCH   DOWNLOAD   READ   REVIEW

CONCEPTS AND VOCABULARY

1. $a + b = b + a$ illustrates the _____ property for _____.

2. $a \cdot b = b \cdot a$ illustrates the _____ property for _____.

3. $(a + b) + c = a + (b + c)$ illustrates the _____ property for _____.

4. $(a \cdot b) \cdot c = a \cdot (b \cdot c)$ illustrates the _____ property for _____.

5. (True or False?) Addition and multiplication are commutative.

6. (True or False?) Subtraction and division are associative.

7. $a(b + c) = ab + ac$ illustrates the _____ property.

8. $a(b - c) = ab - ac$ illustrates the _____ property.

9. The equations $a + 0 = a$ and $0 + a = a$ each illustrate the _____ property for _____.

10. The equations $a \cdot 1 = a$ and $1 \cdot a = a$ each illustrate the _____ property for _____.

11. The additive inverse, or opposite, of a is _____.

12. The multiplicative inverse, or reciprocal, of a nonzero number a is _____.

PROPERTIES OF REAL NUMBERS

Exercises 13–20: Use a commutative property to rewrite the expression. Do not simplify.

13. $-6 + 10$ **14.** $23 + 7$

15. $-5 \cdot 6$ **16.** $25 \cdot (-46)$

17. $a + 10$ **18.** $b + c$

19. $b \cdot 7$ **20.** $a \cdot 23$

Exercises 21–28: Use an associative property to rewrite the expression. Do not simplify.

21. $(1 + 2) + 3$ **22.** $-7 + (5 + 15)$

23. $2 \cdot (3 \cdot 4)$ **24.** $(9 \cdot (-4)) \cdot 5$

25. $(a + 5) + c$ **26.** $(10 + b) + a$

27. $(x \cdot 3) \cdot 4$ **28.** $5 \cdot (x \cdot y)$

29. Thinking Generally Use the commutative and associative properties to show that $a + b + c = c + b + a$.

30. Thinking Generally Use the commutative and associative properties to show that $a \cdot b \cdot c = c \cdot b \cdot a$.

Exercises 31–42: Use a distributive property to rewrite the expression. Then simplify the expression.

31. $4(3 + 2)$ **32.** $5(6 - 9)$

33. $a(b - 8)$ **34.** $3(x + y)$

35. $-4(t - z)$ **36.** $-1(a + 6)$

37. $-(5 - a)$ **38.** $12 - (4u - b)$

39. $(a + 5)3$ **40.** $(x + y)7$

41. $12 - (a - 5)$ **42.** $4x - 2(3y - 5)$

43. Thinking Generally Use properties of real numbers to show that the distributive property can be extended as follows.

$$a \cdot (b + c + d) = ab + ac + ad.$$

44. Thinking Generally Use properties of real numbers to show that the distributive property can be extended as follows.

$$a \cdot (b - c - d) = ab - ac - ad.$$

Exercises 45–52: Use the distributive property to insert parentheses in the expression and then simplify the result.

45. $6x + 5x$ **46.** $4y - y$

47. $-4b + 3b$ **48.** $2b + 8b$

49. $3a - a$ **50.** $-2x + 5x$

51. $13w - 27w$ **52.** $25a - 21a$

Exercises 53–66: State the property or properties that the equation illustrates.

53. $x \cdot 5 = 5x$ **54.** $7 + a = a + 7$

55. $(a + 5) + 7 = a + 12$

56. $(9 + a) + 8 = a + 17$

57. $4(5 + x) = 20 + 4x$ **58.** $3(5 + x) = 3x + 15$

59. $x(3 - y) = 3x - xy$ **60.** $-(u - v) = -u + v$

61. $6x + 9x = 15x$ **62.** $9x - 11x = -2x$

63. $3 \cdot (4 \cdot a) = 12a$ **64.** $(x \cdot 3) \cdot 5 = 15x$

65. $-(t - 7) = -t + 7$

66. $z \cdot 5 - y \cdot 6 = 5z - 6y$

IDENTITY AND INVERSE PROPERTIES

Exercises 67–76: State the property or properties that are illustrated.

67. $0 + x = x$ **68.** $5x + 0 = 5x$

69. $1 \cdot a = a$ **70.** $\frac{1}{7} \cdot (7 \cdot a) = a$

71. $\frac{25}{15} = \frac{5}{3} \cdot \frac{5}{5} = \frac{5}{3}$ **72.** $\frac{50}{40} = \frac{5}{4} \cdot \frac{10}{10} = \frac{5}{4}$

73. $\frac{1}{xy} \cdot xy = 1$ **74.** $\frac{1}{a + b} \cdot (a + b) = 1$

75. $-xyz + xyz = 0$ **76.** $\frac{1}{y} + \left(-\frac{1}{y}\right) = 0$

MENTAL CALCULATIONS

Exercises 77–92: Use properties of real numbers to calculate the expression mentally.

77. $4 + 2 + 9 + 8 + 1 + 6$

78. $21 + 32 + 19 + 8$

79. $45 + 43 + 5 + 7$

80. $5 + 7 + 12 + 13 + 8$

81. $129 + 49$ **82.** $87 + 99$

83. $178 - 99$ **84.** $500 - 101$

85. $6 \cdot 15$ **86.** $4 \cdot 56$

87. $8 \cdot 102$ **88.** $5 \cdot 999$

89. $\frac{1}{2} \cdot \frac{1}{2} \cdot \frac{1}{2} \cdot 2 \cdot 2 \cdot 2$ **90.** $\frac{1}{2} \cdot \frac{4}{5} \cdot \frac{7}{3} \cdot 2 \cdot \frac{5}{4}$

91. $\frac{7}{6} \cdot \frac{1}{2} \cdot \frac{1}{2} \cdot \frac{1}{2} \cdot \frac{8}{7}$ **92.** $\frac{4}{11} \cdot \frac{11}{6} \cdot \frac{6}{7} \cdot \frac{7}{4}$

MULTIPLYING AND DIVIDING BY POWERS OF 10 MENTALLY

93. *Multiplying by 10* To multiply an integer by 10, attach one 0 to the number. For example, $10 \cdot 23 = 230$. Simplify each expression mentally.
  (a) $10 \cdot 41$ **(b)** $10 \cdot 997$
  (c) $-630 \cdot 10$ **(d)** $-14{,}000 \cdot 10$

94. *Multiplying by 10* To multiply a decimal number by 10, move the decimal point one place to the right. For example, $10 \cdot 23.45 = 234.5$. Simplify each expression mentally.
  (a) $10 \cdot 101.68$ **(b)** $10 \cdot (-1.235)$
  (c) $-113.4 \cdot 10$ **(d)** $0.567 \cdot 10$
  (e) $10 \cdot 0.0045$ **(f)** $-0.05 \cdot 10$

95. *Multiplying by Powers of 10* To multiply an integer by a power of 10 in the form

$$10^k = \underbrace{100 \ldots 0,}_{k \text{ zeros}}$$

attach k zeros to the number. Some examples of this are $100 \cdot 45 = 4500$, $1000 \cdot 235 = 235{,}000$, and $10{,}000 \cdot 12 = 120{,}000$. Simplify each expression mentally.
  (a) $1000 \cdot 19$ **(b)** $100 \cdot (-451)$
  (c) $10{,}000 \cdot 6$ **(d)** $-79 \cdot 100{,}000$

96. *Multiplying by Powers of 10* To multiply a decimal number by a power of 10 in the form

$$10^k = \underbrace{100 \ldots 0,}_{k \text{ zeros}}$$

move the decimal point k places to the right. For example, $100 \cdot 1.234 = 123.4$. Simplify each given expression mentally.
  (a) $1000 \cdot 1.2345$ **(b)** $100 \cdot (-5.1)$
  (c) $45.67 \cdot 1000$ **(d)** $0.567 \cdot 10{,}000$
  (e) $100 \cdot 0.0005$ **(f)** $-0.05 \cdot 100{,}000$

97. *Dividing by 10* To divide a number by 10, move the decimal point one place to the left. For example, $78.9 \div 10 = 7.89$. Simplify each expression mentally.
  (a) $12.56 \div 10$ **(b)** $9.6 \div 10$
  (c) $0.987 \div 10$ **(d)** $-0.056 \div 10$
  (e) $1200 \div 10$ **(f)** $4578 \div 10$

98. *Dividing by Powers of 10* To divide a decimal number by a power of 10 in the form

$$10^k = \underbrace{100 \ldots 0,}_{k \text{ zeros}}$$

move the decimal point k places to the left. For example, $123.4 \div 100 = 1.234$. Simplify each expression mentally.

  (a) $78.89 \div 100$ **(b)** $0.05 \div 1000$
  (c) $5678 \div 10{,}000$ **(d)** $-9.8 \div 1000$
  (e) $-101 \div 100{,}000$ **(f)** $7.8 \div 100$

APPLICATIONS

99. *Earnings* Earning $100 one day and $75 the next day is equivalent to earning $75 the first day and $100 the second day. What property of real numbers does this example illustrate?

100. *Leasing a Car* An advertisement for a lease on a new car states that it costs $2480 down and $201 per month for 20 months. Mentally calculate the cost of the lease. Explain your reasoning.

101. *Gasoline Mileage* A car travels 198 miles on 10 gallons of gasoline. Mentally calculate the number of miles that the car travels on 1 gallon of gasoline.

102. *Gallons of Water* A wading pool is 50 feet long, 20 feet wide, and 1 foot deep.
  (a) Mentally determine the number of cubic feet in the pool. (*Hint:* Volume equals length times width times height.)
  (b) One cubic foot equals about 7.5 gallons. Mentally calculate the number of gallons of water in the pool.

103. *Dimensions of a Pool* A small pool of water is 13 feet long, 5 feet wide, and 2 feet deep. The volume V of the pool in cubic feet is found by multiplying 13 by 5 by 2.
  (a) To do this calculation mentally, would you multiply $(13 \cdot 5) \cdot 2$ or $13 \cdot (5 \cdot 2)$? Why?
  (b) What property allows you to do either calculation and still obtain the correct answer?

104. *Digital Images of Io* The accompanying picture of Jupiter's moon Io is a digital picture, created by using a rectangular pattern of small pixels. This image is 500 pixels wide and 400 pixels high, so the total number of pixels in it is $500 \cdot 400 = 200{,}000$ pixels. (*Source:* NASA.)

  (a) Find the total number of pixels in an image 400 pixels wide and 500 pixels high.
  (b) Suppose that a picture is x pixels wide and y pixels high. What property states that it has the same number of pixels as a picture y pixels wide and x pixels high?

105. To determine the cost of tuition, a student tries to compute $16 \cdot \$96$ with a calculator and gets $15,360. How do you know that this computation is not correct?

106. A student performs the following computation by hand.

$$20 - 6 - 2 + 8 \div 4 \div 2 \overset{?}{=} 20 - 4 + 8 \div 2$$
$$\overset{?}{=} 20 - 4 + 4$$
$$\overset{?}{=} 20$$

Find any incorrect steps and explain what is wrong. What is the correct answer?

107. The computation $3 + 10^{20} - 10^{20}$ performed on a calculator gives a result of 0. (Try it.) What is the correct answer? Why is it important to know properties of real numbers even though you have a calculator?

108. Does the "distributive property"

$$a + (b \cdot c) \overset{?}{=} (a + b) \cdot (a + c)$$

hold for all numbers a, b, and c? Explain your reasoning.

Group Activity   Working with Real Data

Directions: Form a group of 2 to 4 people. Select someone to record the group's responses for this activity. All members of the group should work cooperatively to answer the questions. If your instructor asks for your results, each member of the group should be prepared to respond.

Winning the Lottery In the multistate lottery game *Powerball,* there are 120,526,770 possible number combinations, only one of which is the grand prize winner. The cost of a single ticket (one number combination) is $1. (*Source:* Powerball.com.)

Suppose that a *very* wealthy person decides to buy tickets for every possible number combination to be assured of winning a $150 million grand prize.

(a) If this individual could purchase one ticket every second, how many hours would it take to buy all of the tickets? How many years is this?

(b) If there were a way for this individual to buy all possible number combinations quickly, discuss reasons why this strategy would probably lose money.

1.8   Simplifying and Writing Algebraic Expressions

Terms • Combining Like Terms • Simplifying Expressions • Writing Expressions

A LOOK INTO MATH ▶   Filing taxes might be easier if several of the complicated tax formulas could be *combined* into a single formula. Furthermore, some of the large formulas might be easier to evaluate if they were *simplified* to be more concise. In mathematics, we *combine like terms* when we *simplify* expressions. In this section, we simplify algebraic expressions.

Terms

One way to simplify expressions is to combine like terms. A **term** is a number, a variable, or a *product* of numbers and variables raised to natural number powers. Examples include

$$4, \quad z, \quad 5x, \quad \frac{2}{5}z, \quad -4xy, \quad -x^2, \quad \text{and} \quad 6x^3y^4. \quad \text{Terms}$$

NEW VOCABULARY

☐ Term
☐ Coefficient
☐ Like terms

Terms do not contain addition or subtraction signs, but they can contain negative signs. The **coefficient** of a term is the number that appears in the term. If no number appears, then the coefficient is understood to be either -1 or 1. Table 1.11 on the next page shows examples of terms and their corresponding coefficients.

- How do you identify the coefficient in a term?

TABLE 1.11 Terms and Coefficients

Term	Coefficient
$\frac{1}{2}xy$	$\frac{1}{2}$
$-z^2$	-1
12	12
$-5w^4$	-5

EXAMPLE 1 **Identifying terms**

Determine whether each expression is a term. If it is a term, identify its coefficient.
(a) 51 **(b)** $5a$ **(c)** $2x + 3y$ **(d)** $-3x^2$

Solution
(a) A number is a term. This term's coefficient is 51.
(b) The product of a number and a variable is a term. This term's coefficient is 5.
(c) The sum (or difference) of two terms is not a term.
(d) The product of a number and a variable with an exponent is a term. This term's coefficient is -3.

Now Try Exercises 7, 9, 11, 13

STUDY TIP

One of the best ways to prepare for class is to read a section before it is covered by your instructor. Reading ahead gives you the chance to formulate any questions you might have about the concepts in the section.

MAKING CONNECTIONS

Factors and Terms

When variables and numbers are multiplied, they are called *factors*. For example, the expression $4xy$ has factors of 4, x, and y. When variables and numbers are added or subtracted, they are called *terms*. For example, the expression $x - 5xy + 1$ has three terms: x, $-5xy$, and 1.

Combining Like Terms

- Which property of real numbers is used to combine like terms?

Suppose that we have two boards with lengths $2x$ and $3x$, where the value of x could be any length such as 2 feet. See Figure 1.34. Because $2x$ and $3x$ are *like terms* we can find the total length of the two boards by applying the distributive property and *adding like terms*.

$$2x + 3x = (2 + 3)x = 5x$$

The combined length of the two boards is $5x$ units.

Visualizing Like Terms

Figure 1.34 $2x + 3x = 5x$

We can also determine the difference between the lengths of the two boards by *subtracting like terms*.

$$3x - 2x = (3 - 2)x = 1x = x$$

The second board is x units longer than the first board.

If two terms contain the same variables raised to the same powers, we call them **like terms**. We can add or subtract (combine) like terms but cannot combine *unlike* terms. For example, if one board has length $2x$ and the other board has length $3y$, then we cannot determine the total length other than to say that it is $2x + 3y$. See Figure 1.35. The terms $2x$ and $3y$ are unlike terms and *cannot* be combined.

READING CHECK

• How can you tell whether two terms are like terms?

Visualizing Unlike Terms

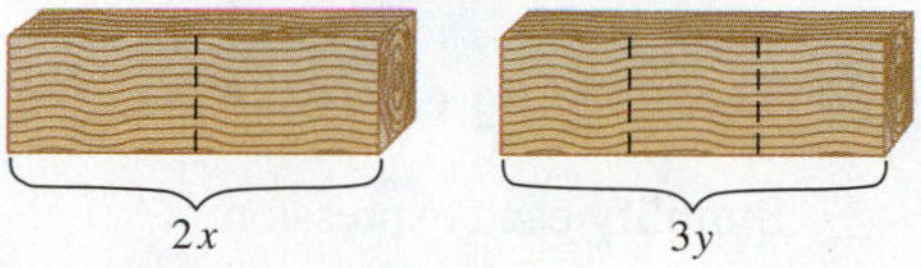

Figure 1.35 $2x + 3y$

EXAMPLE 2 **Identifying like terms**

Determine whether the terms are like or unlike.
(a) $-4m, 7m$ **(b)** $8x^2, 8y^2$ **(c)** $\frac{1}{2}z, -3z^2$ **(d)** $5, -4n$

Solution
(a) The variable in both terms is m (with power 1), so they are like terms.
(b) The variables are different, so they are unlike terms.
(c) The term $-3z^2$ contains $z^2 = z \cdot z$, whereas the term $\frac{1}{2}z$ contains only z. Thus they are unlike terms.
(d) The term 5 has no variable, whereas the term $-4n$ contains the variable n. They are unlike terms.

Now Try Exercises 19, 21, 23, 27

EXAMPLE 3 **Combining like terms**

Combine terms in each expression, if possible.
(a) $-3x + 5x$ **(b)** $-x^2 + 5x^2$ **(c)** $\frac{1}{2}y - 3y^3$

Solution
(a) Combine terms by applying a distributive property.

$$-3x + 5x = (-3 + 5)x = 2x$$

(b) Note that $-x^2$ can be written as $-1x^2$. They are like terms and can be combined.

$$-x^2 + 5x^2 = (-1 + 5)x^2 = 4x^2$$

(c) They are unlike terms, so they cannot be combined.

Now Try Exercises 31, 39, 43

Simplifying Expressions

The area of a rectangle equals length times width. In Figure 1.36 on the next page the area of the first rectangle is $3x$, the area of the second rectangle is $2x$, and the area of the third rectangle is x. The area of the last rectangle equals the total area of the three smaller rectangles. That is,

$$3x + 2x + x = (3 + 2 + 1)x = 6x.$$

The expression $3x + 2x + x$ can be *simplified* to $6x$.

Simplifying an Expression Visually

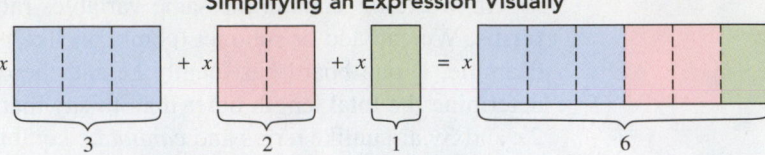

Figure 1.36 $3x + 2x + x = 6x$

EXAMPLE 4 **Simplifying expressions**

Simplify each expression.

(a) $2 + x - 6 + 5x$ **(b)** $2y - (y + 3)$ **(c)** $\dfrac{-1.5x}{-1.5}$

Solution

(a) Combine like terms by applying the properties of real numbers.

$$
\begin{aligned}
2 + x - 6 + 5x &= 2 + (-6) + x + 5x & \text{Commutative property} \\
&= 2 + (-6) + (1 + 5)x & \text{Distributive property} \\
&= -4 + 6x & \text{Add.}
\end{aligned}
$$

(b)
$$
\begin{aligned}
2y - 1(y + 3) &= 2y + (-1)y + (-1) \cdot 3 & \text{Distributive property} \\
&= 2y - 1y - 3 & \text{Definition of subtraction} \\
&= (2 - 1)y - 3 & \text{Distributive property} \\
&= y - 3 & \text{Subtract.}
\end{aligned}
$$

(c)
$$
\begin{aligned}
\frac{-1.5x}{-1.5} &= \frac{-1.5}{-1.5} \cdot \frac{x}{1} & \text{Multiplication of fractions} \\
&= 1 \cdot x & \text{Simplify the fractions.} \\
&= x & \text{Multiplicative identity}
\end{aligned}
$$

Now Try Exercises 47, 59, 75

NOTE: The expression in Example 4(c) can be simplified directly by using the basic principle of fractions: $\frac{ac}{bc} = \frac{a}{b}$.

EXAMPLE 5 **Simplifying expressions**

Simplify each expression.

(a) $7 - 2(5x - 3)$ **(b)** $3x^3 + 2x^3 - x^3$

(c) $3y^2 - z + 4y^2 - 2z$ **(d)** $\dfrac{12x - 8}{4}$

Solution

(a)
$$
\begin{aligned}
7 - 2(5x - 3) &= 7 + (-2)(5x + (-3)) & \text{Change subtraction to addition.} \\
&= 7 + (-2)(5x) + (-2)(-3) & \text{Distributive property} \\
&= 7 - 10x + 6 & \text{Multiply.} \\
&= 13 - 10x & \text{Combine like terms.}
\end{aligned}
$$

(b)
$$
\begin{aligned}
3x^3 + 2x^3 - x^3 &= (3 + 2 - 1)x^3 & \text{Distributive property} \\
&= 4x^3 & \text{Add and subtract.}
\end{aligned}
$$

(c) $3y^2 - z + 4y^2 - 2z = 3y^2 + 4y^2 + (-1z) + (-2z)$ Commutative property

$\qquad\qquad\qquad\quad = (3 + 4)y^2 + (-1 + (-2))z$ Distributive property

$\qquad\qquad\qquad\quad = 7y^2 - 3z$ Add.

(d) $\dfrac{12x - 8}{4} = \dfrac{12x}{4} - \dfrac{8}{4}$ Subtraction of fractions

$\qquad\qquad\quad = 3x - 2$ Simplify fractions.

Now Try Exercises 57, 65, 71, 79

Recall that the commutative and associative properties of addition allow us to rearrange a sum in any order. For example, if we write the expression

$$2x - 3 - 4x + 10 \quad \text{as the sum} \quad 2x + (-3) + (-4x) + 10,$$

the terms can be arranged and grouped as

$$(2x + (-4x)) + (-3 + 10).$$

Applying the distributive property and adding within each grouping results in

$$(2 + (-4))x + (-3 + 10) = -2x + 7.$$

Note that the terms in the result can be found directly by combining like (same color) terms in the given expression, where addition indicates that a term is positive and subtraction indicates that a term is negative. In the next example, we simplify an expression in this way.

EXAMPLE 6   **Simplifying an expression directly**

Simplify the expression $6x + 5 - 2x - 8$.

Solution
The like terms and their indicated signs are shown.

$$6x + 5 - 2x - 8$$

By combining like terms, the expression can be simplified as

$$4x - 3.$$

Now Try Exercise 53

Writing Expressions

▶ **REAL-WORLD CONNECTION** In real-life situations, we often have to translate words to symbols. For example, to calculate federal and state income tax we might have to multiply taxable income by 0.15 and by 0.05 and then find the sum. If we let x represent taxable income, then the total federal and state income tax is $0.15x + 0.05x$. This expression can be simplified with a distributive property.

$$0.15x + 0.05x = (0.15 + 0.05)x$$

$$= 0.20x$$

Thus the total income tax on a taxable income of $x = \$20{,}000$ would be

$$0.20(20{,}000) = \$4000.$$

EXAMPLE 7 **Writing and simplifying an expression**

A sidewalk has a constant width w and comprises several short sections with lengths 12, 6, and 5 feet, as illustrated in Figure 1.37.
(a) Write and simplify an expression that gives the number of square feet of sidewalk.
(b) Find the area of the sidewalk if its width is 3 feet.

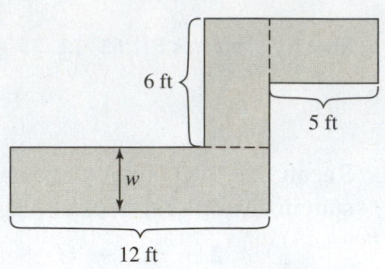

6 ft

5 ft

w

12 ft

Figure 1.37

Solution
(a) The area of each sidewalk section equals its length times its width w. The total area of the sidewalk is

$$12w + 6w + 5w = (12 + 6 + 5)w = 23w.$$

(b) When $w = 3$, the area is $23w = 23 \cdot 3 = 69$ square feet.

Now Try Exercise 89

1.8 Putting It All Together

CONCEPT	COMMENTS	EXAMPLES
Term	A term is a number, variable, or product of numbers and variables raised to natural number powers.	$12, -10, x, -y$ $-3x, 5z, xy, 6x^2$ $10y^3, \frac{3}{4}x, 25xyz^2$
Coefficient of a Term	The coefficient of a term is the number that appears in the term. If no number appears, then the coefficient is either 1 or -1.	12 Coefficient is 12 x Coefficient is 1 $-xy$ Coefficient is -1 $-4x$ Coefficient is -4
Like Terms	Like terms have the same variables raised to the same powers.	$5m$ and $-6m$ x^2 and $-74x^2$ $2xy$ and $-xy$
Combining Like Terms	Like terms can be combined by using a distributive property.	$5x + 2x = (5 + 2)x = 7x$ $y - 3y = (1 - 3)y = -2y$ $8x^2 + x^2 = (8 + 1)x^2 = 9x^2$

MyMathLab

CONCEPTS AND VOCABULARY

1. A(n) _____ is a number, a variable, or a product of numbers and variables raised to natural number powers.

2. The number 7 in the term $7x^2y$ is called the _____ of the term.

3. When variables and numbers are multiplied, they are called _____. When they are added or subtracted, they are called _____.

4. If two terms contain the same variables raised to the same powers, they are (like/unlike) terms.

5. We can add or subtract (like/unlike) terms.

6. We can combine like terms in an expression by applying a(n) _____ property.

LIKE TERMS

Exercises 7–16: Determine whether the expression is a term. If the expression is a term, identify its coefficient.

7. 91

8. -12

9. $-6b$

10. $9z$

11. $x + 10$

12. $20 - 2y$

13. x^2

14. $4x^3$

15. $4x - 5$

16. $5z + 6x$

Exercises 17–28: Determine whether the terms are like or unlike.

17. $6, -8$

18. $2x, 19$

19. $5x, -22x$

20. $19y, -y$

21. $14, 14a$

22. $-33b, -3b$

23. $18x, 18y$

24. $-6a, -6b$

25. $x^2, -15x^2$

26. $y, 19y$

27. $3x^2, \frac{1}{5}x$

28. $12y^2, -y^2$

29. **Thinking Generally** Are the terms $4ab$ and $-3ba$ like or unlike?

30. **Thinking Generally** Are the terms $-xyz^2$ and $3yz^2x$ like or unlike?

Exercises 31–46: Combine terms, if possible.

31. $-4x + 7x$

32. $6x - 8x$

33. $19y - 5y$

34. $22z + z$

35. $28a + 13a$

36. $41b - 17b$

37. $11z - 11z$

38. $4y + 4y$

39. $5x - 7y$

40. $3y + 3z$

41. $5 + 5y$

42. $x + x^2$

43. $5x^2 - 2x^2$

44. $25z^3 - 10z^3$

45. $8y - 10y + y$

46. $4x^2 + x^2 - 5x^2$

SIMPLIFYING AND WRITING EXPRESSIONS

Exercises 47–82: Simplify the expression.

47. $5 + x - 3 + 2x$

48. $x - 5 - 5x + 7$

49. $-\frac{3}{4} + z - 3z + \frac{5}{4}$

50. $\frac{4}{3}z - 100 + 200 - \frac{1}{3}z$

51. $4y - y + 8y$

52. $14z - 15z - z$

53. $-3 + 6z + 2 - 2z$

54. $19a - 12a + 5 - 6$

55. $-2(3z - 6y) - z$

56. $6\left(\frac{1}{2}a - \frac{1}{6}b\right) - 3b$

57. $2 - \frac{3}{4}(4x + 8)$

58. $-5 - (5x - 6)$

59. $-x - (5x + 1)$

60. $2x - 4(x + 2)$

61. $1 - \frac{1}{3}(x + 1)$

62. $-3 - 3(4 - x)$

63. $\frac{3}{5}(x + y) - \frac{1}{5}(x - 1)$

64. $-5(a + b) - (a + b)$

65. $0.2x^2 + 0.3x^2 - 0.1x^2$

66. $32z^3 - 52z^3 + 20z^3$

67. $2x^2 - 3x + 5x^2 - 4x$

68. $\frac{5}{6}y^2 - 4 + \frac{1}{12}y^2 + 3$

69. $a + 3b - a - b$

70. $2z^2 - z - z^2 + 3z$

71. $8x^3 + 7y - x^3 - 5y$

72. $4y - 6z + 2y - 3z$

73. $\dfrac{8x}{8}$

74. $\dfrac{-0.1y}{-0.1}$

75. $\dfrac{-3y}{-y}$

76. $\dfrac{2x}{7x}$

77. $\dfrac{-108z}{-108}$

78. $\dfrac{3xy}{-6xy}$

79. $\dfrac{9x - 6}{3}$

80. $\dfrac{18y + 9}{9}$

81. $\dfrac{14z + 21}{7}$

82. $\dfrac{15x - 20}{5}$

Exercises 83–88: Translate the phrase into a mathematical expression and then simplify. Use the variable x.

83. The sum of five times a number and six times the same number

84. The sum of a number and three times the same number

85. The sum of a number squared and twice the same number squared

86. One-half of a number minus three-fourths of the same number

87. Six times a number minus four times the same number

88. Two cubed times a number minus three squared times the same number

APPLICATIONS

89. *Street Dimensions* (Refer to Example 7.) A street has a constant width w and comprises several straight sections having lengths 600, 400, 350, and 220 feet.

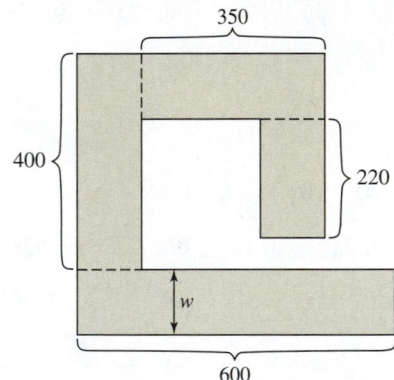

(a) Write a simplified expression that gives the square footage of the street.

(b) Find the area of the street if its width is 42 feet.

90. *Sidewalk Dimensions* A sidewalk has a constant width w and comprises several short sections having lengths 12, 14, and 10 feet.
(a) Write a simplified expression that gives the number of square feet of sidewalk.
(b) Find the area of the sidewalk if its width is 5 feet.

91. *Snowblowers* Two snowblowers are being used to clear a driveway. The first blower can remove 20 cubic feet per minute, and the second blower can remove 30 cubic feet per minute.
(a) Write a simplified expression that gives the number of cubic feet of snow removed in x minutes.
(b) Find the total number of cubic feet of snow removed in 48 minutes.
(c) How many minutes would it take to remove the snow from a driveway that measures 30 feet by 20 feet, if the snow is 2 feet deep?

92. *Winding Rope* Two motors are winding up rope. The first motor can wind up rope at 2 feet per second, and the second motor can wind up rope at 5 feet per second.
(a) Write a simplified expression that gives the length of rope wound by both motors in x seconds.
(b) Find the total length of rope wound up by the two motors in 3 minutes.
(c) How many minutes would it take to wind up 2100 feet of rope by using both motors?

WRITING ABOUT MATHEMATICS

93. The following expression was simplified *incorrectly*.

$$3(x - 5) + 5x \stackrel{?}{=} 3x - 5 + 5x$$
$$\stackrel{?}{=} 3x + 5x - 5$$
$$\stackrel{?}{=} (3 + 5)x - 5$$
$$\stackrel{?}{=} 8x - 5$$

Find the error and explain what went wrong. What should the final answer be?

94. Explain how to add like terms. What property of real numbers is used?

Checking Basic Concepts

1. Use a commutative property to rewrite each expression.
   (a) $y \cdot 18$ (b) $10 + x$

2. Use an associative and a commutative property to simplify $5 \cdot (y \cdot 4)$.

3. Simplify each expression.
   (a) $10 - (5 + x)$ (b) $5(x - 7)$

4. State the property that the equation $5x + 3x = 8x$ illustrates.

5. Simplify $-4xy + 4xy$.

6. Mentally evaluate each expression.
   (a) $32 + 17 + 8 + 3$ (b) $\frac{5}{6} \cdot \frac{7}{8} \cdot \frac{6}{5} \cdot 8$
   (c) $567 - 199$

7. Determine whether the terms are like or unlike.
   (a) $-3x, -3z$ (b) $4x^2, -2x^2$

8. Combine like terms in each expression.
   (a) $5z + 9z$ (b) $5y - 4 - 8y + 7$

9. Simplify each expression.
   (a) $2y - (5y + 3)$
   (b) $-4(x + 3y) + 2(2x - y)$
   (c) $\frac{20x}{20}$ (d) $\frac{35x^2}{x^2}$

10. Write "the sum of three times a number and five times the same number" as a mathematical expression with the variable x and then simplify the expression.

CHAPTER 1 Summary

SECTION 1.1 ■ NUMBERS, VARIABLES, AND EXPRESSIONS

Sets of Numbers

Natural Numbers — $1, 2, 3, 4, \ldots$

Whole Numbers — $0, 1, 2, 3, \ldots$

Products and Factors — Two numbers that are multiplied are called factors and the result is called the product.

Example: $3 \cdot 5 = 15$
Factor   Factor   Product

Prime and Composite Numbers

Prime Number — A natural number greater than 1 whose only natural number factors are itself and 1

Examples: 2, 3, 5, 7, 11, 13, and 17

Composite Number and Prime Factorization — A natural number greater than 1 that is not prime; a composite number can be written as a product of two or more prime numbers.

Examples: $24 = 2 \cdot 2 \cdot 2 \cdot 3$ and $18 = 2 \cdot 3 \cdot 3$

Important Terms

Variable — A symbol or letter that represents an unknown quantity

Examples: $a, b, F, x,$ and y

Algebraic Expression — Consists of numbers, variables, arithmetic symbols, and grouping symbols

Examples: $3x + 1$ and $5(x + 2) - y$

Equation — A statement that two algebraic expressions are equal; an equation always contains an equals sign.

Examples: $1 + 2 = 3$ and $z - 7 = 8$

Formula   A special type of equation that expresses a relationship between two or more quantities.

Examples: $P = 2l + 2w$ and $A = lw$

Fractions and Lowest Terms

Fraction   A fraction has a numerator, a denominator, and a fraction bar.

Example: Numerator $\searrow$ $\dfrac{4}{9}$ $\leftarrow$ Fraction bar
Denominator $\nearrow$

Lowest Terms   A fraction is in lowest terms if the numerator and denominator have no factors in common.

Example: The fraction $\frac{3}{5}$ is in lowest terms because 3 and 5 have no factors in common.

Simplifying Fractions

$$\frac{a \cdot c}{b \cdot c} = \frac{a}{b}$$

Example: $\frac{20}{35} = \frac{4 \cdot 5}{7 \cdot 5} = \frac{4}{7}$

Multiplicative Inverse, or Reciprocal The reciprocal of a nonzero number a is $\frac{1}{a}$.

Examples: The reciprocal of -5 is $-\frac{1}{5}$, and the reciprocal of $\frac{2x}{y}$ is $\frac{y}{2x}$, provided $x \neq 0$ and $y \neq 0$.

Multiplication and Division

$$\frac{a}{b} \cdot \frac{c}{d} = \frac{ac}{bd} \quad \text{and} \quad \frac{a}{b} \div \frac{c}{d} = \frac{a}{b} \cdot \frac{d}{c}$$

b and d are nonzero. b, c, and d are nonzero.

Examples: $\frac{3}{4} \cdot \frac{7}{5} = \frac{21}{20}$ and $\frac{3}{4} \div \frac{7}{5} = \frac{3}{4} \cdot \frac{5}{7} = \frac{3 \cdot 5}{4 \cdot 7} = \frac{15}{28}$

Addition and Subtraction with Like Denominators

$$\frac{a}{b} + \frac{c}{b} = \frac{a + c}{b} \quad \text{and} \quad \frac{a}{b} - \frac{c}{b} = \frac{a - c}{b}$$

Examples: $\frac{3}{5} + \frac{1}{5} = \frac{3 + 1}{5} = \frac{4}{5}$ and $\frac{3}{5} - \frac{1}{5} = \frac{3 - 1}{5} = \frac{2}{5}$

Addition and Subtraction with Unlike Denominators Write the expressions with the LCD. Then add or subtract the numerators, keeping the denominator unchanged.

Examples: $\frac{2}{9} + \frac{1}{6} = \frac{2}{9} \cdot \frac{2}{2} + \frac{1}{6} \cdot \frac{3}{3} = \frac{4}{18} + \frac{3}{18} = \frac{7}{18}$ LCD is 18.

$\frac{2}{9} - \frac{1}{6} = \frac{2}{9} \cdot \frac{2}{2} - \frac{1}{6} \cdot \frac{3}{3} = \frac{4}{18} - \frac{3}{18} = \frac{1}{18}$

Exponential Expression

Base $\rightarrow 5^3 \leftarrow$ Exponent

Example: $3^4 = 3 \cdot 3 \cdot 3 \cdot 3 = 81$

Order of Operations

Use the following order of operations. First perform all calculations within parentheses and absolute values, or above and below the fraction bar.

1. Evaluate all exponential expressions.
2. Do all multiplication and division from *left to right*.
3. Do all addition and subtraction from *left to right*.

NOTE: Negative signs are evaluated after exponents, so $-2^4 = -16$.

Examples:

$$100 - 5^2 \cdot 2 = 100 - 25 \cdot 2 \quad \text{and} \quad \frac{4+2}{5-3} \cdot 4 = \frac{6}{2} \cdot 4$$

$$= 100 - 50 \qquad\qquad\qquad = 3 \cdot 4$$

$$= 50 \qquad\qquad\qquad\qquad = 12$$

SECTION 1.4 ■ REAL NUMBERS AND THE NUMBER LINE

Opposite or Additive Inverse The opposite of the number a is $-a$.

Examples: The opposite of 5 is -5, and the opposite of -8 is $-(-8) = 8$.

Sets of Numbers

Integers	$\ldots, -3, -2, -1, 0, 1, 2, 3, \ldots$
Rational Numbers	$\frac{p}{q}$, where p and $q \neq 0$ are integers; rational numbers can be written as decimal numbers that either repeat or terminate.

Examples: $-\frac{3}{4}, 5, -7.6, \frac{6}{3},$ and $\frac{1}{3}$

Real Numbers   Numbers that can be written as decimal numbers

Examples: $-\frac{3}{4}, 5, -7.6, \frac{6}{3}, \pi, \sqrt{3},$ and $-\sqrt{7}$

Irrational Numbers   Real numbers that are not rational

Examples: $\pi, \sqrt{3},$ and $-\sqrt{7}$

Average To calculate the average of a set of numbers, find the sum of the numbers and then divide the sum by how many numbers there are in the set.

Example: The average of 4, 5, 20, and 11 is

$$\frac{4 + 5 + 20 + 11}{4} = \frac{40}{4} = 10.$$

We divide by 4 because we are finding the average of 4 numbers.

The Number Line

The origin corresponds to the number 0.

Absolute Value If a is positive or 0, then $|a| = a$, and if a is negative, then $|a| = -a$. The absolute value of a number is *never* negative.

Examples: $|5| = 5, \quad |-5| = 5, \quad$ and $\quad |0| = 0$

Inequality If a is located to the left of b on the number line, then $a < b$. If a is located to the right of b on the number line, then $a > b$.

Examples: $-3 < 6$ and $-1 > -2$

SECTION 1.5 ■ ADDITION AND SUBTRACTION OF REAL NUMBERS

Addition of Opposites

$$a + (-a) = 0$$

Examples: $3 + (-3) = 0$ and $-\frac{5}{7} + \frac{5}{7} = 0$

Addition of Real Numbers
To add two real numbers with *like* signs, do the following.

1. Find the sum of the absolute values of the numbers.
2. Keep the common sign of the two numbers as the sign of the sum.

To add two real numbers with *unlike* signs, do the following.

1. Find the absolute values of the numbers.
2. Subtract the smaller absolute value from the larger absolute value.
3. Keep the sign of the number with the larger absolute value as the sign of the sum.

Examples: $-4 + 5 = 1$, $3 + (-7) = -4$, $-4 + (-2) = -6$,   and $8 + 2 = 10$

Subtraction of Real Numbers For any real numbers a and b, $a - b = a + (-b)$.

Examples: $5 - 9 = 5 + (-9) = -4$ and $-4 - (-3) = -4 + 3 = -1$

SECTION 1.6 ■ MULTIPLICATION AND DIVISION OF REAL NUMBERS

Important Terms

Factors Numbers multiplied in a multiplication problem

Example: 5 and 4 are factors of 20 because $5 \cdot 4 = 20$.

Product The answer to a multiplication problem

Example: The product of 5 and 4 is 20.

Dividend, Divisor, and Quotient If $\frac{a}{b} = c$, then a is the dividend, b is the divisor, and c is the quotient.

Example: In the division problems $30 \div 5 = 6$ and $\frac{30}{5} = 6$, 30 is the dividend, 5 is the divisor, and 6 is the quotient.

Dividing Real Numbers For real numbers a and b with $b \neq 0$,

$$\frac{a}{b} = a \cdot \frac{1}{b}.$$

Examples: $\dfrac{8}{\frac{1}{2}} = \dfrac{8}{1} \cdot \dfrac{2}{1} = 16$, $14 \div \dfrac{2}{3} = \dfrac{14}{1} \cdot \dfrac{3}{2} = \dfrac{42}{2} = 21$,   and $5 \div 0$ is undefined.

Signs of Products or Quotients The product or quotient of two numbers with like signs is positive. The product or quotient of two numbers with unlike signs is negative.

Examples: $-4 \cdot 6 = -24$, $-2 \cdot (-5) = 10$, $\frac{-18}{6} = -3$,   and $-4 \div (-2) = 2$

Writing Fractions as Decimals To write the fraction $\frac{a}{b}$ as a decimal, divide b into a.

Example: $\frac{4}{9} = 0.\overline{4}$ because division of 4 by 9 gives the repeating decimal $0.4444\ldots$.

Writing Decimals as Fractions Write the decimal as a fraction with a denominator equal to a power of 10 and then simplify this fraction.

Example: $0.45 = \frac{45}{100} = \frac{9 \cdot 5}{20 \cdot 5} = \frac{9}{20}$

SECTION 1.7 ■ PROPERTIES OF REAL NUMBERS

Important Properties

Commutative

$a + b = b + a$ and $a \cdot b = b \cdot a$

Examples: $3 + 4 = 4 + 3$ and $-6 \cdot 3 = 3 \cdot (-6)$

Associative

$(a + b) + c = a + (b + c)$ and $(a \cdot b) \cdot c = a \cdot (b \cdot c)$

Examples: $(2 + 3) + 4 = 2 + (3 + 4)$
$(2 \cdot 3) \cdot 4 = 2 \cdot (3 \cdot 4)$

Distributive

$a(b + c) = ab + ac$ and $a(b - c) = ab - ac$

Examples: $3(x + 5) = 3 \cdot x + 3 \cdot 5$
$4(5 - 2) = 4 \cdot 5 - 4 \cdot 2$

Identity

$a + 0 = 0 + a = a$ Additive identity is 0.
$a \cdot 1 = 1 \cdot a = a$ Multiplicative identity is 1.

Examples: $5 + 0 = 0 + 5 = 5$ and $1 \cdot (-4) = -4 \cdot 1 = -4$

Inverse

$a + (-a) = 0$ and $-a + a = 0$
$a \cdot \frac{1}{a} = 1$ and $\frac{1}{a} \cdot a = 1$

Examples: $5 + (-5) = 0$ and $\frac{1}{2} \cdot 2 = 1$

NOTE: The commutative and associative properties apply to addition and multiplication but *not* to subtraction and division.

SECTION 1.8 ■ SIMPLIFYING AND WRITING ALGEBRAIC EXPRESSIONS

Important Terms

Term

A term is a number, a variable, or a product of numbers and variables raised to natural number powers.

Examples: $5, -10x, 3xy,$ and x^2

Coefficient

The number portion of a term

Examples: The coefficients for the terms $3xy$, $-x^2$, and -7 are 3, -1, and -7, respectively.

Like Terms

Terms containing the same variables raised to the same powers; their coefficients may be different.

Examples: The following pairs are like terms:

$5x$ and $-x$; $6x^2$ and $-2x^2$; $3y$ and $-\frac{1}{2}y$.

Combining Like Terms To add or subtract like terms, apply a distributive property.

Examples: $4x + 5x = (4 + 5)x = 9x$ and $5y - 7y = (5 - 7)y = -2y$

CHAPTER 1   Review Exercises

SECTION 1.1

Exercises 1–6: Classify the number as prime, composite, or neither. If the number is composite, write it as a product of prime numbers.

1. 29

2. 27

3. 108

4. 91

5. 0

6. 1

Exercises 7–10: Evaluate the expression for the given values of x and y.

7. $2x - 5$ $x = 4$

8. $7 - \dfrac{10}{x}$ $x = 5$

9. $9x - 2y$ $x = 2,\ y = 3$

10. $\dfrac{2x}{x - y}$ $x = 6,\ y = 4$

Exercises 11–14: Find the value of y for the given values of x and z.

11. $y = x - 5$ $x = 12$

12. $y = xz + 1$ $x = 2,\ z = 3$

13. $y = 4(x - z)$ $x = 7,\ z = 5$

14. $y = \dfrac{x + z}{4}$ $x = 14,\ z = 10$

Exercises 15–18: Translate the phrase into an algebraic expression. State precisely what each variable represents when appropriate.

15. Three squared increased by five

16. Two cubed divided by the quantity three plus one

17. The product of three and a number

18. The difference between a number and four

SECTION 1.2

Exercises 19 and 20: Find the greatest common factor.

19. 15, 35

20. 12, 30, 42

21. Use the basic principle of fractions to simplify each expression.
    (a) $\dfrac{5 \cdot 7}{8 \cdot 7}$
    (b) $\dfrac{3a}{4a}$

22. Simplify each fraction to lowest terms.
    (a) $\dfrac{9}{12}$
    (b) $\dfrac{36}{60}$

Exercises 23–30: Multiply and then simplify the result to lowest terms when appropriate.

23. $\dfrac{3}{4} \cdot \dfrac{5}{6}$

24. $\dfrac{1}{2} \cdot \dfrac{4}{9}$

25. $\dfrac{2}{3} \cdot \dfrac{9}{10}$

26. $\dfrac{12}{11} \cdot \dfrac{22}{23}$

27. $4 \cdot \dfrac{5}{8}$

28. $\dfrac{2}{3} \cdot 9$

29. $\dfrac{x}{3} \cdot \dfrac{6}{x}$

30. $\dfrac{2}{3} \cdot \dfrac{9x}{4y}$

31. Find the fractional part: one-fifth of three-sevenths.

32. Find the reciprocal of each number.
    (a) 8   (b) 1   (c) $\dfrac{5}{19}$ (d) $\dfrac{3}{2}$

Exercises 33–38: Divide and then simplify to lowest terms when appropriate.

33. $\dfrac{3}{2} \div \dfrac{1}{6}$

34. $\dfrac{9}{10} \div \dfrac{7}{5}$

35. $8 \div \dfrac{2}{3}$

36. $\dfrac{3}{4} \div 6$

37. $\dfrac{x}{y} \div \dfrac{3}{y}$

38. $\dfrac{4x}{3y} \div \dfrac{9x}{5}$

Exercises 39 and 40: Find the least common denominator for the fractions.

39. $\dfrac{1}{8}, \dfrac{5}{12}$

40. $\dfrac{3}{14}, \dfrac{1}{21}$

Exercises 41–46: Add or subtract and then simplify to lowest terms when appropriate.

41. $\dfrac{2}{15} + \dfrac{3}{15}$

42. $\dfrac{5}{4} - \dfrac{3}{4}$

43. $\dfrac{11}{12} - \dfrac{1}{8}$

44. $\dfrac{6}{11} - \dfrac{3}{22}$

45. $\dfrac{2}{3} - \dfrac{1}{2} + \dfrac{1}{4}$

46. $\dfrac{1}{6} + \dfrac{2}{3} - \dfrac{1}{9}$

SECTION 1.3

Exercises 47–52: Write the expression as an exponential expression.

47. $5 \cdot 5 \cdot 5 \cdot 5 \cdot 5 \cdot 5$

48. $\dfrac{7}{6} \cdot \dfrac{7}{6} \cdot \dfrac{7}{6}$

49. $x \cdot x \cdot x \cdot x \cdot x$

50. $3 \cdot 3 \cdot 3 \cdot 3$

51. $(x + 1) \cdot (x + 1)$

52. $(a - 5) \cdot (a - 5) \cdot (a - 5)$

53. Use multiplication to rewrite each expression, and then evaluate the result.
    (a) 4^3 (b) 7^2 (c) 8^1

54. Find a natural number n such that $2^n = 32$.

Exercises 55–66: Evaluate the expression by hand.

55. $7 + 3 \cdot 6$

56. $15 - 5 - 3$

57. $24 \div 4 \div 2$

58. $30 - 15 \div 3$

59. $18 \div 6 - 2$

60. $\frac{18}{4 + 5}$

61. $9 - 3^2$

62. $2^3 - 8$

63. $2^4 - 8 + \frac{4}{2}$

64. $3^2 - 4(5 - 3)$

65. $7 - \frac{4 + 6}{2 + 3}$

66. $3^3 - 2^3$

SECTION 1.4

Exercises 67 and 68: Find the opposite of each expression.

67. (a) -8 **(b)** $-(-(-3))$

68. (a) $-\left(\frac{-3}{7}\right)$ **(b)** $\frac{-2}{-5}$

Exercises 69 and 70: Find the decimal equivalent for the rational number.

69. (a) $\frac{4}{5}$ **(b)** $\frac{3}{20}$

70. (a) $\frac{5}{9}$ **(b)** $\frac{7}{11}$

Exercises 71–76: Classify the number as one or more of the following: natural number, whole number, integer, rational number, or irrational number.

71. 0

72. $-\frac{5}{6}$

73. -7

74. $\sqrt{17}$

75. π

76. 3.4

77. Plot each number on the same number line.

(a) 0 **(b)** -2 **(c)** $\frac{5}{4}$

78. Evaluate each expression.

(a) $|-5|$ **(b)** $|-\pi|$ **(c)** $|4 - 4|$

79. Insert the symbol $>$ or $<$ to make each statement true.

(a) -5 _____ 4 **(b)** $-\frac{1}{2}$ _____ $-\frac{5}{2}$

(c) -3 _____ $|-9|$ **(d)** $|-8|$ _____ $|-1|$

80. List the numbers $\sqrt{3}$, -3, 3, $-\frac{2}{3}$, and $\pi - 1$ from least to greatest.

SECTIONS 1.5 AND 1.6

Exercises 81 and 82: (Refer to Example 3 in Section 1.5.) Find the sum visually.

81. $-5 + 9$

82. $4 + (-7)$

Exercises 83 and 84: Use a number line to find the sum.

83. $-1 + 2$

84. $-2 + (-3)$

Exercises 85–96: Evaluate the expression.

85. $5 + (-4)$

86. $-9 - (-7)$

87. $11 \cdot (-4)$

88. $-8 \cdot (-5)$

89. $11 \div (-4)$

90. $-4 \div \frac{4}{7}$

91. $-\frac{5}{9} - \left(-\frac{1}{3}\right)$

92. $-\frac{1}{2} + \left(-\frac{3}{4}\right)$

93. $-\frac{1}{3} \cdot \left(-\frac{6}{7}\right)$

94. $\frac{\frac{4}{5}}{-7}$

95. $-\frac{3}{2} \div \left(-\frac{3}{8}\right)$

96. $\frac{3}{8} \div (-0.5)$

Exercises 97 and 98: Write an arithmetic expression for the given phrase and then simplify.

97. Three plus negative five

98. Subtract negative four from two

Exercises 99 and 100: Write the fraction or mixed number as a decimal.

99. $\frac{7}{9}$

100. $2\frac{1}{5}$

Exercises 101 and 102: Write the decimal number as a fraction in lowest terms.

101. 0.6

102. 0.375

SECTION 1.7

Exercises 103 and 104. Use a commutative property to rewrite the expression. Do not simplify.

103. $3 + 16$

104. $14 \cdot (-x)$

Exercises 105 and 106: Use an associative property to rewrite the expression by changing the parentheses. Do not simplify.

105. $-4 + (1 + 3)$

106. $(x \cdot y) \cdot 5$

Exercises 107 and 108: Use a distributive property to rewrite the expression. Then simplify the expression.

107. $5(x + 12)$

108. $-(a - 3)$

Exercises 109–112: State the property or properties that are illustrated.

109. $y + 0 = y$

110. $b \cdot 1 = b$

111. $\frac{1}{4} \cdot 4 = 1$

112. $-3a + 3a = 0$

Exercises 113–122: State the property that the equation illustrates.

113. $z \cdot 3 = 3z$

114. $6 + (7 + 5x) = (6 + 7) + 5x$

115. $2(5x - 2) = 10x - 4$

116. $5 + x + 3 = 5 + 3 + x$

117. $1 \cdot a = a$

118. $3 \cdot (5x) = (3 \cdot 5)x$

119. $12 - (x + 7) = 12 - x - 7$

120. $a + 0 = a$

121. $-5x + 5x = 0$

122. $-5 \cdot \left(-\frac{1}{5}\right) = 1$

Exercises 123–128: Use properties of real numbers to evaluate the expression mentally.

123. $7 + 9 + 12 + 8 + 1 + 3$

124. $500 - 199$

125. $25 \cdot 99$

126. $4581 + 1999$

127. 54.98×10

128. $4356 \div 100$

SECTION 1.8

Exercises 129–132: Determine whether the expression is a term. If the expression is a term, identify its coefficient.

129. $55x$

130. $-xy$

131. $9xy + 2z$

132. $x - 7$

Exercises 133–144: Simplify the expression.

133. $-10x + 4x$

134. $19z - 4z$

135. $3x^2 + x^2$

136. $7 + 2x - 6 + x$

137. $-\frac{1}{2} + \frac{3}{2}z - z + \frac{5}{2}$

138. $5(x - 3) - (4x + 3)$

139. $4x^2 - 3 + 5x^2 - 3$

140. $3x^2 + 4x^2 - 7x^2$

141. $\dfrac{35a}{7a}$

142. $\dfrac{0.5c}{0.5}$

143. $\dfrac{15y + 10}{5}$

144. $\dfrac{24x - 60}{12}$

APPLICATIONS

145. *Painting a Wall* Two people are painting a large wall. The first person paints 3 square feet per minute while the second person paints 4 square feet per minute.
   (a) Write a simplified expression that gives the total number of square feet the two people can paint in x minutes.
   (b) Find the number of square feet painted in 1 hour.
   (c) How many minutes would it take for them to paint a wall 8 feet tall and 21 feet wide?

146. *Area of a Triangle* Find the area of the triangle shown.

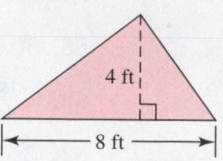

147. *Gallons to Pints* There are 8 pints in 1 gallon. Let $G = 1, 2, 3, \ldots, 6$, and make a table of values that converts G gallons to P pints. Write a formula that converts G gallons to P pints.

148. *Text Messages* The table lists the cost C of sending x text messages after the number of text messages included in a monthly plan is reached. Write an equation that relates C and x.

Texts (x)	1	2	3	4
Cost (C)	\$0.05	\$0.10	\$0.15	\$0.20

149. *Aging in the United States* In 2050, about $\frac{1}{5}$ of the population will be age 65 or over and about $\frac{1}{20}$ of the population will be age 85 or over. Estimate the fraction of the population that will be between the ages of 65 and 85 in 2050. (*Source:* U.S. Census Bureau.)

150. *Rule of 72* (Refer to Exercise 79 in Section 1.3.) If an investment of \$25,000 earns 9% annual interest, approximate the value of the investment after 24 years.

151. *Carpentry* A board measures $5\frac{3}{4}$ feet and needs to be cut in five equal pieces. Find the length of each piece.

152. *Distance* Over four days, an athlete jogs $3\frac{1}{8}$ miles, $4\frac{3}{8}$ miles, $6\frac{1}{4}$ miles, and $1\frac{5}{8}$ miles. How far does the athlete jog in all?

153. *Checking Account* The initial balance in a checking account is \$1652. Find the final balance resulting from the following sequence of withdrawals and deposits: $-\$78$, $-\$91$, \$256, and $-\$638$.

154. *Temperature Range* The highest temperature ever recorded in Amarillo, Texas, was $108°\text{F}$ and the lowest was $-16°\text{F}$. Find the difference between these two temperatures. (*Source:* The Weather Almanac.)

155. *Top-Grossing Movies* Titanic (1997) is the sixth top-grossing movie of all time. Find the total admissions for *Titanic* if they were $\frac{16}{25}$ of the total admissions for the top-grossing movie of all time, *Gone With the Wind* (1939), which had total admissions of 202 million. (*Source:* Box Office Mojo.)

1. Classify the number as prime or composite. If the number is composite, write it as a product of prime numbers.
   (a) 29   (b) 56

2. Evaluate the expression $\dfrac{5x}{2x-1}$ for $x = -3$.

3. Translate the phrase "four squared decreased by three" to an algebraic expression. Then find the value of the expression.

4. Simplify $\frac{24}{32}$ to lowest terms.

5. Find the fractional part: two-tenths of one-sixth.

6. Find the least common denominator for $\frac{3}{8}$ and $\frac{5}{12}$.

7. Evaluate each expression. Write your answer in lowest terms.
   (a) $\frac{5}{8} + \frac{1}{8}$ (b) $\frac{5}{9} - \frac{3}{15}$ (c) $\frac{3}{5} \cdot \frac{10}{21}$
   (d) $6 \div \frac{8}{5}$ (e) $\frac{5}{12} + \frac{4}{9}$ (f) $\frac{10}{13} \div 5$

8. Write $y \cdot y \cdot y \cdot y$ as an exponential expression.

9. Evaluate each expression.
   (a) $6 + 10 \div 5$ (b) $4^3 - (3 - 5 \cdot 2)$
   (c) $-6^2 - 6 + \frac{4}{2}$ (d) $11 - \frac{1+3}{6-4}$

10. Write $\frac{7}{20}$ as a decimal.

11. Classify the number as one or more of the following: natural number, whole number, integer, rational number, or irrational number.
    (a) -1 (b) $\sqrt{5}$

12. Plot each number on the same number line.
    (a) -2 (b) $\frac{1}{3}$ (c) $\sqrt{7}$

13. Insert the symbol $>$ or $<$ to make each statement true.
    (a) $2 \underline{\quad} |-5|$ (b) $|-1| \underline{\quad} |0|$

14. Evaluate the expression.
    (a) $-5 \div \frac{5}{6}$ (b) $-7 \cdot (-3)$

15. Write 0.75 as a fraction in lowest terms.

16. State the property or properties that each equation illustrates.
    (a) $6x - 2x = 4x$ (b) $12 \cdot (3x) = 36x$
    (c) $4 + x + 8 = 12 + x$

17. Use properties of real numbers to evaluate $17 \cdot 102$ mentally.

18. Simplify each expression.
    (a) $5 - 5z + 7 + z$ (b) $12x - (6 - 3x)$
    (c) $5 - 4(x + 6) + \dfrac{15x}{3}$

19. *Mowing a Lawn* Two people are mowing a lawn. The first person has a riding mower and can mow $\frac{4}{3}$ acres per hour; the second person has a push mower and can mow $\frac{1}{4}$ acre per hour.
    (a) Write a simplified expression that gives the total number of acres that the two people mow in x hours.
    (b) Find the total acreage that they can mow in an 8-hour work day.

20. A wire $7\frac{4}{5}$ feet long is to be cut in 3 equal parts. How long should each part be?

21. *Cost Equation* The table lists the cost C of buying x tickets to a hockey game.

Tickets (x)	3	4	5	6
Cost (C)	\$39	\$52	\$65	\$78

   (a) Find an equation that relates C and x.
   (b) What is the cost of 17 tickets?

22. The initial balance for a savings account is \$892. Find the final balance resulting from withdrawals and deposits of $-\$57$, \$150, and $-\$345$.

CHAPTER 1   Extended and Discovery Exercises

1. *Arithmetic Operations* Insert one of the symbols $+, -, \cdot,$ or $\div$ in each blank to obtain the given answer. Do not use any parentheses.

$$2 \_ 2 \_ 2 \_ 2 = 0$$

$$3 \_ 3 \_ 3 \_ 3 = 10$$

$$4 \_ 4 \_ 4 \_ 4 = 1$$

$$6 \_ 6 \_ 6 \_ 6 = 36$$

$$7 \_ 7 \_ 7 \_ 7 = 63$$

2. *Magic Squares* The following square is called a "magic square" because it contains each of the numbers from 1 to 9 and the numbers in each row, column, and diagonal sum to 15.

8	3	4
1	5	9
6	7	2

Complete the following magic square having 4 rows and 4 columns by arranging the numbers 1 through 16 so that each row, column, and diagonal sums to 34. The four corners will also sum to 34.

	2		13
5			
		6	
4			1

2

Linear Equations and Inequalities

Education is not the filling of a pail, but the lighting of a fire.

— WILLIAM BUTLER YEATS

Mathematics is a unique subject that is essential for describing, or modeling, events in the real world. For example, ultraviolet light from the sun is responsible for both tanning and burning exposed skin. Mathematics lets us use numbers to describe the intensity of ultraviolet light. The table shows the maximum ultraviolet intensity measured in milliwatts per square meter for various latitudes and dates.

Latitude	Mar. 21	June 21	Sept. 21	Dec. 21
0°	325	254	325	272
10°	311	275	280	220
20°	249	292	256	143
30°	179	248	182	80
40°	99	199	127	34
50°	57	143	75	13

If a student from Chicago, located at a latitude of 42°, spends spring break in Hawaii with a latitude of 20°, the sun's ultraviolet rays in Hawaii will be approximately $\frac{249}{99} \approx 2.5$ times as intense as they are in Chicago. Equations can be used to describe, or model, the intensity of the sun at various latitudes. In this chapter we will focus on *linear equations* and the related concept of *linear inequalities*.

Source: J. Williams, *The USA Today Weather Almanac.*

2.1 Introduction to Equations

Basic Concepts • Equations and Solutions • The Addition Property of Equality • The Multiplication Property of Equality

A LOOK INTO MATH ▶

The Global Positioning System (GPS) consists of 24 satellites that travel around Earth in nearly circular orbits. GPS can be used to determine locations and velocities of cars, airplanes, and hikers with an amazing degree of accuracy. New cars often come equipped with GPS, and their drivers can determine their cars' locations to within a few feet. To create GPS, thousands of equations were solved, and mathematics was essential in finding their solutions. In this section we discuss many of the basic concepts needed to solve equations. (*Source:* J. Van Sickle, *GPS for Land Surveyors.*)

Basic Concepts

NEW VOCABULARY

☐ Solution
☐ Solution set
☐ Equivalent equations

▶ **REAL-WORLD CONNECTION** Suppose that during a storm it rains 2 inches before noon and 1 inch per hour thereafter until 5 P.M. Table 2.1 lists the total rainfall R after various elapsed times x, where $x = 0$ corresponds to noon.

TABLE 2.1 Rainfall *x* Hours Past Noon

Elapsed Time: x (hours)	0	1	2	3	4	5
Total Rainfall: R (inches)	2	3	4	5	6	7

The data suggest that the total rainfall R in inches is **2** more than the elapsed time **x**. A formula that *models*, or *describes*, the rainfall x hours past noon is given by

$$R = x + 2.$$

For example, **3** hours past noon, or at 3 P.M.,

$$R = 3 + 2 = 5$$

inches of rain have fallen. Even though $x = 4.5$ does not appear in the table, we can calculate the amount of rainfall at 4:30 P.M. with the formula as

$$R = 4.5 + 2 = 6.5 \text{ inches.}$$

The advantage that a formula has over a table of values is that a formula can be used to calculate the rainfall at *any* time x, not just at the times listed in the table.

Equations and Solutions

In the rainfall example above, how can we determine when 6 inches of rain have fallen? From Table 2.1 the *solution* is 4, or 4 P.M. To find this solution without the table, we can *solve* the equation

$$x + 2 = 6.$$

READING CHECK

• What is a solution to an equation?

An equation can be either true or false. For example, the equation $1 + 2 = 3$ is true, whereas the equation $1 + 2 = 4$ is false. When an equation contains a variable, the equation may be true for some values of the variable and false for other values of the variable. Each value of the variable that makes the equation true is called a **solution** to the equation, and the *set of all solutions* is called the **solution set**. *Solving an equation* means finding all of its solutions. Because $4 + 2 = 6$, the solution to the equation

$$x + 2 = 6$$

is 4, and the solution set is $\{4\}$. Note that *braces* $\{\}$ are used to denote a set. Table 2.2 shows examples of equations and their solution sets. Note that some equations, such as the third one in the table, can have more than one solution.

TABLE 2.2 **Equations and Solution Sets**

Equation	Solution Set	True Equation(s)
$3 - x = 1$	$\{2\}$	$3 - 2 = 1$
$10 - 4y = 6$	$\{1\}$	$10 - 4(1) = 6$
$x^2 = 4$	$\{-2, 2\}$	$(-2)^2 = 4$ and $2^2 = 4$

Many times we cannot solve an equation simply by looking at it. In these situations we must use a step-by-step procedure. During each step an equation is transformed into a different but equivalent equation. **Equivalent equations** are equations that have the same solution set. For example, the equations

$$x + 2 = 5 \quad \text{and} \quad x = 3$$

are equivalent equations because the solution set for both equations is $\{3\}$.

MAKING CONNECTIONS

Equations and Expressions

READING CHECK

• What is the difference between an equation and an expression?

Although the words "equation" and "expression" occur frequently in mathematics, they are *not* interchangeable. An equation *always* contains an equals sign but an expression *never* contains an equals sign. We often want to *solve* an equation, whereas an expression can sometimes be *simplified*. Furthermore, the equals sign in an equation separates two expressions. For example, $3x - 5 = x + 1$ is an equation where $3x - 5$ and $x + 1$ are each expressions.

When solving equations, it is often helpful to transform a more complicated equation into an equivalent equation that has an obvious solution, such as $x = 3$. The *addition property of equality* and the *multiplication property of equality* can be used to transform an equation into an equivalent equation that is easier to solve.

The Addition Property of Equality

STUDY TIP

The addition property of equality is used to solve equations throughout the remainder of the text. Be sure that you have a firm understanding of this important property.

When solving an equation, we have to apply the same operation to each side of the equation. For example, one way to solve the equation

$$x + 2 = 5$$

is to add -2 to each side. This step results in isolating the x on one side of the equation.

$x + 2 = 5$	Given equation
$x + 2 + (-2) = 5 + (-2)$	Add -2 to each side.
$x + 0 = 3$	Addition of real numbers
$x = 3$	Additive identity

These four equations (on the bottom of the previous page) are equivalent, but the solution is easiest to see in the last equation. When -2 is added to each side of the given equation, the addition property of equality is used.

ADDITION PROPERTY OF EQUALITY

If a, b, and c are real numbers, then

$$a = b \quad \text{is equivalent to} \quad a + c = b + c.$$

That is, adding the same number to each side of an equation results in an equivalent equation.

NOTE: Because any subtraction problem can be changed to an addition problem, *the addition property of equality also works for subtraction.* That is, if the same number is subtracted from each side of an equation, the result is an equivalent equation.

EXAMPLE 1 **Using the addition property of equality**

Solve each equation.
(a) $x + 10 = 7$ (b) $t - 4 = 3$ (c) $\frac{1}{2} = -\frac{3}{4} + y$

Solution
(a) When solving an equation, we try to isolate the variable on one side of the equation. If we add -10 to (or **subtract 10** from) each side of the equation, we find the value of x.

$x + 10 = 7$	Given equation
$x + 10 - 10 = 7 - 10$	Subtract 10 from each side.
$x + 0 = -3$	Addition of real numbers
$x = -3$	Additive identity

The solution is -3.
(b) To isolate the variable t, **add 4** to each side.

$t - 4 = 3$	Given equation
$t - 4 + 4 = 3 + 4$	Add 4 to each side.
$t + 0 = 7$	Addition of real numbers
$t = 7$	Additive identity

The solution is 7.
(c) To isolate the variable y, **add $\frac{3}{4}$** to each side.

$\dfrac{1}{2} = -\dfrac{3}{4} + y$	Given equation
$\dfrac{1}{2} + \dfrac{3}{4} = -\dfrac{3}{4} + \dfrac{3}{4} + y$	Add $\frac{3}{4}$ to each side.
$\dfrac{5}{4} = 0 + y$	Addition of real numbers
$\dfrac{5}{4} = y$	Additive identity

The solution is $\frac{5}{4}$.

Now Try Exercises 17, 19, 23

MAKING CONNECTIONS

Equations and Scales

Think of an equation as an old-fashioned scale, where two pans must balance, as shown in the figure. If the two identical golden weights balance the pans, then adding identical red weights to each pan results in the pans remaining balanced. Similarly, removing (subtracting) identical weights from each side will also keep the pans balanced.

CHECKING A SOLUTION To check a solution, substitute it in the *given* equation to find out if a true statement results. To check the solution for Example 1(c), substitute $\frac{5}{4}$ for y in the given equation. Note that a question mark is placed over the equals sign when a solution is being checked.

CRITICAL THINKING

When you are checking a solution, why do you substitute your answer in the *given* equation?

$$\frac{1}{2} = -\frac{3}{4} + y \qquad \text{Given equation}$$

$$\frac{1}{2} \stackrel{?}{=} -\frac{3}{4} + \frac{5}{4} \qquad \text{Replace } y \text{ with } \frac{5}{4}.$$

$$\frac{1}{2} \stackrel{?}{=} \frac{2}{4} \qquad \text{Add fractions.}$$

$$\frac{1}{2} = \frac{1}{2} \; ✓ \qquad \text{The answer checks.}$$

The answer of $\frac{5}{4}$ checks because the resulting equation is true.

EXAMPLE 2   **Solving an equation and checking a solution**

Solve the equation $-5 + y = 3$ and then check the solution.

Solution
Isolate y by **adding 5** to each side.

$$-5 + y = 3 \qquad \text{Given equation}$$

$$-5 + 5 + y = 3 + 5 \qquad \text{Add 5 to each side.}$$

$$0 + y = 8 \qquad \text{Addition of real numbers}$$

$$y = 8 \qquad \text{Additive identity}$$

The solution is 8. To check this solution, substitute **8** for **y** in the given equation.

$$-5 + y = 3 \qquad \text{Given equation}$$

$$-5 + 8 \stackrel{?}{=} 3 \qquad \text{Replace } y \text{ with 8.}$$

$$3 = 3 \; ✓ \qquad \text{Add; the answer checks.}$$

Now Try Exercise 21

The Multiplication Property of Equality

We can illustrate the multiplication property of equality by considering a formula that converts yards to feet. Because there are 3 feet in 1 yard, the formula $F = 3Y$ computes F, the number of feet in Y yards. For example, if $Y = 5$ yards, then $F = 3 \cdot 5 = 15$ feet.

Now consider the reverse, converting 27 feet to yards. The answer to this conversion corresponds to the solution to the equation

$$27 = 3Y.$$

To find the solution, **multiply** each side of the equation by the reciprocal of 3, or $\frac{1}{3}$.

$$27 = 3Y \qquad \text{Given equation}$$

$$\frac{1}{3} \cdot 27 = \frac{1}{3} \cdot 3 \cdot Y \qquad \text{Multiply each side by } \tfrac{1}{3}.$$

$$9 = 1 \cdot Y \qquad \text{Multiplication of real numbers}$$

$$9 = Y \qquad \text{Multiplicative identity}$$

Thus 27 feet are equivalent to 9 yards.

MULTIPLICATION PROPERTY OF EQUALITY

If a, b, and c are real numbers with $c \neq 0$, then

$$a = b \quad \text{is equivalent to} \quad ac = bc.$$

That is, multiplying each side of an equation by the same nonzero number results in an equivalent equation.

NOTE: Because any division problem can be changed to a multiplication problem, *the multiplication property of equality also works for division*. That is, if each side of an equation is divided by the same nonzero number, the result is an equivalent equation.

READING CHECK

• Why do we use the multiplication property of equality?

EXAMPLE 3

Using the multiplication property of equality

Solve each equation.

(a) $\frac{1}{3}x = 4$ **(b)** $-4y = 8$ **(c)** $5 = \frac{3}{4}z$

Solution

(a) We start by **multiplying** each side of the equation by **3**, the reciprocal of $\frac{1}{3}$.

$$\frac{1}{3}x = 4 \qquad \text{Given equation}$$

$$3 \cdot \frac{1}{3}x = 3 \cdot 4 \qquad \text{Multiply each side by 3.}$$

$$1 \cdot x = 12 \qquad \text{Multiplication of real numbers}$$

$$x = 12 \qquad \text{Multiplicative identity}$$

The solution is 12.

(b) The coefficient of the y-term is -4, so we can either multiply each side of the equation by $-\frac{1}{4}$ or **divide** each side by -4. This step will make the coefficient of y equal to 1.

$$-4y = 8 \qquad \text{Given equation}$$

$$\frac{-4y}{-4} = \frac{8}{-4} \qquad \text{Divide each side by } -4.$$

$$y = -2 \qquad \text{Simplify fractions.}$$

The solution is -2.

(c) To change the coefficient of z from $\frac{3}{4}$ to 1, **multiply** each side of the equation by $\frac{4}{3}$, the reciprocal of $\frac{3}{4}$.

$$5 = \frac{3}{4}z \qquad \text{Given equation}$$

$$\frac{4}{3} \cdot 5 = \frac{4}{3} \cdot \frac{3}{4}z \qquad \text{Multiply each side by } \frac{4}{3}.$$

$$\frac{20}{3} = 1 \cdot z \qquad \text{Multiplication of real numbers}$$

$$\frac{20}{3} = z \qquad \text{Multiplicative identity}$$

The solution is $\frac{20}{3}$.

Now Try Exercises 37, 43, 45

EXAMPLE 4 **Solving an equation and checking a solution**

Solve the equation $\frac{3}{4} = -\frac{3}{7}t$ and then check the solution.

Solution
Multiply each side of the equation by $-\frac{7}{3}$, the reciprocal of $-\frac{3}{7}$.

$$\frac{3}{4} = -\frac{3}{7}t \qquad \text{Given equation}$$

$$-\frac{7}{3} \cdot \frac{3}{4} = -\frac{7}{3} \cdot \left(-\frac{3}{7}\right)t \qquad \text{Multiply each side by } -\frac{7}{3}.$$

$$-\frac{7}{4} = 1 \cdot t \qquad \text{Multiplication of real numbers}$$

$$-\frac{7}{4} = t \qquad \text{Multiplicative identity}$$

The solution is $-\frac{7}{4}$. To check this answer, substitute $-\frac{7}{4}$ for t in the given equation.

$$\frac{3}{4} = -\frac{3}{7}t \qquad \text{Given equation}$$

$$\frac{3}{4} \stackrel{?}{=} -\frac{3}{7} \cdot \left(-\frac{7}{4}\right) \qquad \text{Replace } t \text{ with } -\frac{7}{4}.$$

$$\frac{3}{4} = \frac{3}{4} \checkmark \qquad \text{Multiply; the answer checks.}$$

Now Try Exercise 47

▶ **REAL-WORLD CONNECTION** Twitter is a microblogging Web site that is used to post short messages called "tweets" on the Internet. In its early years, Twitter's popularity increased dramatically and new accounts were added at an amazing rate. People from around the world began posting millions of tweets every day. (*Source:* Twitter.)

EXAMPLE 5 **Analyzing Twitter account data**

In the early months of 2010, Twitter added 0.3 million new accounts every day.
(a) Write a formula that gives the number of new Twitter accounts T added in x days.
(b) At this rate, how many days would be needed to add 18 million new accounts?

Solution
(a) In 1 day $0.3 \cdot 1 = 0.3$ million new accounts were added, in 2 days $0.3 \cdot 2 = 0.6$ million new accounts were added, and in x days $0.3 \cdot x = 0.3x$ new accounts were added. So the formula is $T = 0.3x$, where x is in days and T is in millions.
(b) To find the number of days needed for Twitter to add 18 million new accounts, replace the variable T in the formula with **18** and solve the resulting equation.

$$T = 0.3x \qquad \text{Formula from part (a)}$$
$$18 = 0.3x \qquad \text{Replace } T \text{ with 18.}$$
$$\frac{18}{0.3} = \frac{0.3x}{0.3} \qquad \text{Divide each side by 0.3.}$$
$$60 = x \qquad \text{Simplify.}$$

At this rate, it takes 60 days to add 18 million accounts.

Now Try Exercise 59

2.1 Putting It All Together

CONCEPT	COMMENTS	EXAMPLES
Equation	An equation is a mathematical statement that two expressions are equal. An equation can be either true or false.	The equation $2 + 3 = 5$ is true. The equation $1 + 3 = 7$ is false.
Solution	A value for a variable that makes an equation a true statement	The solution to $x + 5 = 20$ is 15, and the solutions to $x^2 = 9$ are -3 and 3.
Solution Set	The set of all solutions to an equation	The solution set to $x + 5 = 20$ is $\{15\}$, and the solution set to $x^2 = 9$ is $\{-3, 3\}$.
Equivalent Equations	Two equations are equivalent if they have the same solution set.	The equations $$2x = 14 \quad \text{and} \quad x = 7$$ are equivalent because the solution set to both equations is $\{7\}$.

CONCEPT	COMMENTS	EXAMPLES
Addition Property of Equality	The equations $$a = b \quad \text{and} \quad a + c = b + c$$ are equivalent. This property is used to solve equations.	To solve $x - 3 = 8$, add 3 to each side of the equation. $$x - 3 + 3 = 8 + 3$$ $$x = 11$$ The solution is 11.
Multiplication Property of Equality	When $c \neq 0$, the equations $$a = b \quad \text{and} \quad a \cdot c = b \cdot c$$ are equivalent. This property is used to solve equations.	To solve $\frac{1}{5}x = 10$, multiply each side of the equation by 5. $$5 \cdot \frac{1}{5}x = 5 \cdot 10$$ $$x = 50$$ The solution is 50.
Checking a Solution	Substitute the solution for the variable in the given equation and then simplify each side to see if a true statement results.	To show that 8 is a solution to $$x + 12 = 20,$$ substitute 8 for x. $$8 + 12 \stackrel{?}{=} 20$$ $$20 = 20 \quad \text{True}$$

2.1 Exercises

MyMathLab  PRACTICE   WATCH   DOWNLOAD   READ   REVIEW

CONCEPTS AND VOCABULARY

1. Each value of a variable that makes an equation true is called a(n) _____.

2. The equation $1 + 3 = 4$ is (true/false).

3. The equation $2 + 3 = 6$ is (true/false).

4. The _____ is the set of all solutions to an equation.

5. To solve an equation, find all _____.

6. Equations with the same solution set are called _____ equations.

7. If $a = b$, then $a + c =$ _____.

8. Because any subtraction problem can be changed to an addition problem, the addition property of equality also works for _____.

9. If $a = b$ and $c \neq 0$, then $ac =$ _____.

10. Because any division problem can be changed to a multiplication problem, the multiplication property of equality also works for _____.

11. To solve an equation, transform the equation into a(n) _____ equation that is easier to solve.

12. To check a solution, substitute it for the variable in the _____ equation.

THE ADDITION PROPERTY OF EQUALITY

13. To solve $x - 22 = 4$, add _____ to each side.

14. To solve $\frac{5}{6} = \frac{1}{6} + x$, add _____ to each side.

15. To solve $x + 3 = 13$, subtract _____ from each side.

16. To solve $\frac{3}{4} = \frac{1}{4} + x$, subtract _____ from each side.

Exercises 17–30: Solve the equation. Check your answer.

17. $x + 5 = 0$ | **18.** $x + 3 = 7$

19. $a - 12 = -3$ | **20.** $a - 19 = -11$

21. $9 = y - 8$ | **22.** $97 = -23 + y$

23. $\frac{1}{5} = z - \frac{3}{2}$ | **24.** $\frac{3}{4} + z = -\frac{1}{2}$

25. $t - 0.8 = 4.3$ | **26.** $y - 1.23 = -0.02$

27. $4 + x = 1$ | **28.** $16 + x = -2$

29. $1 = \frac{1}{3} + y$ | **30.** $\frac{7}{2} = -2 + y$

31. Thinking Generally To solve $x - a = b$ for x, add _____ to each side.

32. Thinking Generally To solve $x + a = b$ for x, subtract _____ from each side.

THE MULTIPLICATION PROPERTY OF EQUALITY

33. To solve $5x = 4$, multiply each side by _____.

34. To solve $\frac{4}{3}y = 8$, multiply each side by _____.

35. To solve $6x = 11$, divide each side by _____.

36. To solve $0.2x = 4$, divide each side by _____.

Exercises 37–52: Solve the equation. Check your answer.

37. $5x = 15$ | **38.** $-2x = 8$

39. $-7x = 0$ | **40.** $25x = 0$

41. $-35 = -5a$ | **42.** $-32 = -4a$

43. $-18 = 3a$ | **44.** $-70 = 10a$

45. $\frac{1}{2}x = \frac{3}{2}$ | **46.** $\frac{3}{4}x = \frac{5}{8}$

47. $\frac{1}{2} = \frac{2}{5}z$ | **48.** $-\frac{3}{4} = -\frac{1}{8}z$

49. $0.5t = 3.5$ | **50.** $2.2t = -9.9$

51. $-1.7 = 0.2x$ | **52.** $6.4 = 1.6x$

53. Thinking Generally To solve $\frac{1}{a} \cdot x = b$ for x, multiply each side by _____.

54. Thinking Generally To solve $ax = b$ for x, where $a \neq 0$, divide each side by _____.

APPLICATIONS

55. *Rainfall* On a stormy day it rains 3 inches before noon and $\frac{1}{2}$ inch per hour thereafter until 6 P.M.
  (a) Make a table that shows the total rainfall R in inches, x hours past noon, ending at 6 P.M.
  (b) Write a formula that calculates R.

(c) Use your formula to calculate the total rainfall at 3 P.M. Does the answer agree with the value in your table from part (a)?
  (d) How much rain has fallen by 2:15 P.M.?

56. *Cold Weather* A furnace is turned on at midnight when the temperature inside a cabin is $0°$ F. The cabin warms at a rate of $10°$ F per hour until 7 A.M.
  (a) Make a table that shows the cabin temperature T in degrees Fahrenheit, x hours past midnight, ending at 7 A.M.
  (b) Write a formula that calculates T.
  (c) Use your formula to calculate the temperature at 5 A.M. Does the answer agree with the value in your table from part (a)?
  (d) Find the cabin temperature at 2:45 A.M.

57. *Football Field* A football field is 300 feet long.
  (a) Write a formula that gives the length L in feet of x football fields.
  (b) Use your formula to write an equation whose solution gives the number of football fields in 870 feet.
  (c) Solve your equation from part (b).

58. *Acreage* An acre equals 43,560 square feet.
  (a) Write a formula that converts A acres to S square feet.
  (b) Use your formula to write an equation whose solution gives the number of acres in 871,200 square feet.
  (c) Solve your equation from part (b).

59. *Twitter Accounts* (Refer to Example 5.) In the early months of 2010, Twitter added 0.3 million new accounts every day. At this rate, how many days would be needed to add 15 million new Twitter accounts? (*Source:* Twitter.)

60. *Web Site Visits* If a Web site was increasing its number of visitors by 14,000 every day, how many days would it take for the site to gain a total of 98,000 new visitors?

61. Online Exploration The city of Winnipeg is located in the province of Manitoba in Canada.
  (a) Use the Internet to find the latitude of Winnipeg to the nearest degree.
  (b) Use the table on page 89 to determine how many times as intense the sun's ultraviolet rays are at the equator (latitude $0°$) on March 21 compared to the sun's intensity in Winnipeg. Round your answer to 1 decimal place.

62. Online Exploration Columbus is a city located in the center of Ohio. It is the state's capital city.
  (a) Use the Internet to find the latitude of Columbus to the nearest degree.

(b) Use the table on page 89 to determine how many times as intense the sun's ultraviolet rays are in Limon, Costa Rica, (latitude 10°N) on June 21 compared to the sun's intensity in Columbus. Round your answer to 1 decimal place.

63. *Cost of a Car* When the cost of a car is multiplied by 0.07 the result is $1750. Find the cost of the car.

64. *Raise in Salary* If an employee's salary is multiplied by 1.06, which corresponds to a 6% raise, the result is $58,300. Find the employee's current salary.

WRITING ABOUT MATHEMATICS

65. A student solves an equation as follows.

$$x + 30 = 64$$
$$x \stackrel{?}{=} 64 + 30$$
$$x \stackrel{?}{=} 94$$

Identify the student's mistake. What is the solution?

66. What is a good first step for solving the equation $\frac{a}{b}x = 1$, where a and b are natural numbers? What is the solution? Explain your answers.

2.2   Linear Equations

Basic Concepts • Solving Linear Equations • Applying the Distributive Property • Clearing Fractions and Decimals • Equations with No Solutions or Infinitely Many Solutions

A LOOK INTO MATH ▶

Billions of dollars are spent each year to solve equations that lead to the creation of better products. If our society could not solve equations, we would not have HDTV, high-speed Internet, satellites, fiber optics, CAT scans, smart phones, or accurate weather forecasts. In this section we discuss *linear equations* and some of their applications. Linear equations can always be solved by hand.

Basic Concepts

NEW VOCABULARY

☐ Linear equation
☐ Identity
☐ Contradiction

Suppose that a bicyclist is 5 miles from home, riding *away* from home at 10 miles per hour, as shown in Figure 2.1. The distance between the bicyclist and home for various elapsed times is shown in Table 2.3.

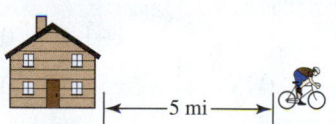

Figure 2.1 Distance from Home

TABLE 2.3 Distance from Home

Elapsed Time (hours)	0	1	2	3
Distance (miles)	5	15	25	35

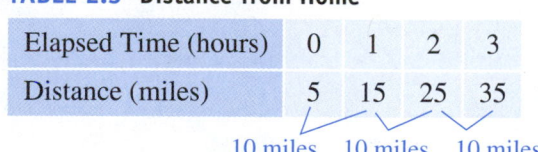

The bicyclist is moving at a constant speed, so the distance increases by 10 miles every hour. The distance D from home after x hours can be calculated by the formula

$$D = 10x + 5.$$

For example, after **2** hours the distance is

$$D = 10(2) + 5 = 25 \text{ miles.}$$

Table 2.3 verifies that the bicyclist is 25 miles from home after 2 hours. However, the table is less helpful if we want to find the elapsed time when the bicyclist is 18 miles from home. To answer this question, we could begin by substituting **18** for D in the formula to obtain the equation

$$\mathbf{18} = 10x + 5.$$

The equation $18 = 10x + 5$ can be written in a different form by applying the addition property of equality. Subtracting 18 from each side gives an equivalent equation.

$$18 - \textbf{18} = 10x + 5 - \textbf{18} \quad \text{Subtract 18 from each side.}$$
$$0 = 10x - 13 \quad \text{Simplify.}$$
$$10x - 13 = 0 \quad \text{Rewrite the equation.}$$

Even though these steps did not result in a solution to the equation $18 = 10x + 5$, applying the addition property of equality allowed us to rewrite the equation as $10x - 13 = 0$, which is an example of a *linear equation*. (See Example 3(a) for a solution to the equation $10x - 13 = 0$.) Linear equations can model applications in which things move or change at a constant rate.

LINEAR EQUATION IN ONE VARIABLE

A **linear equation** in one variable is an equation that can be written in the form

$$ax + b = 0,$$

where a and b are constants with $a \neq 0$.

If an equation is linear, writing it in the form $ax + b = 0$ should not require any properties or processes other than the following.

- using the distributive property to clear any parentheses
- combining like terms
- applying the addition property of equality

For example, the equation $18 = 10x + 5$ is linear because applying the addition property of equality results in $10x - 13 = 0$, as shown above.

Table 2.4 gives examples of linear equations and values for a and b.

TABLE 2.4 Linear Equations

Equation	In $ax + b = 0$ Form	a	b
$x = 1$	$x - 1 = 0$	1	-1
$-5x + 4 = 3$	$-5x + 1 = 0$	-5	1
$2.5x = 0$	$2.5x + 0 = 0$	2.5	0

READING CHECK

- Name three things that tell you that an equation is not a linear equation.

NOTE: An equation *cannot* be written in the form $ax + b = 0$ if after clearing parentheses and combining like terms, any of the following statements are true.

1. The variable has an exponent other than 1.
2. The variable appears in a denominator of a fraction.
3. The variable appears under the symbol $\sqrt{}$ or within an absolute value.

EXAMPLE 1   **Determining whether an equation is linear**

Determine whether the equation is linear. If the equation is linear, give values for a and b that result when the equation is written in the form $ax + b = 0$.

(a) $4x + 5 = 0$ **(b)** $5 = -\frac{3}{4}x$ **(c)** $4x^2 + 6 = 0$ **(d)** $\frac{3}{x} + 5 = 0$

Solution
(a) The equation is linear because it is in the form $ax + b = 0$ with $a = 4$ and $b = 5$.
(b) The equation can be rewritten as follows.

$$5 = -\frac{3}{4}x \qquad \text{Given equation}$$

$$\frac{3}{4}x + 5 = \frac{3}{4}x + \left(-\frac{3}{4}x\right) \qquad \text{Add } \tfrac{3}{4}x \text{ to each side.}$$

$$\frac{3}{4}x + 5 = 0 \qquad \text{Additive inverse}$$

The given equation is linear because it can be written in the form $ax + b = 0$ with $a = \frac{3}{4}$ and $b = 5$.

NOTE: If 5 had been subtracted from each side, the result would be $0 = -\frac{3}{4}x - 5$, which is an equivalent linear equation with $a = -\frac{3}{4}$ and $b = -5$.

(c) The equation is *not* linear because it cannot be written in the form $ax + b = 0$. The variable has exponent 2.
(d) The equation is *not* linear because it cannot be written in the form $ax + b = 0$. The variable appears in the denominator of a fraction.

Now Try Exercises 9, 11, 13, 15

Solving Linear Equations

Every linear equation has *exactly one* solution. Showing that this is true is left as an exercise (see Exercise 59). Solving a linear equation means finding the value of the variable that makes the equation true.

SOLVING LINEAR EQUATIONS NUMERICALLY One way to solve a linear equation is to make a table of values. A table provides an organized way of checking possible values of the variable to see if there is a value that makes the equation true. For example, if we want to solve the equation

$$2x - 5 = -7,$$

we substitute various values for x in the left side of the equation. If one of these values results in -7, then the value makes the equation true and is the solution. In the next example a table of values is used to solve this equation.

EXAMPLE 2   **Using a table to solve an equation**

Complete Table 2.5 for the given values of x. Then solve the equation $2x - 5 = -7$.

TABLE 2.5

x	-3	-2	-1	0	1	2	3
$2x - 5$	-11						

Solution
To complete the table, substitute $x = -2, -1, 0, 1, 2$, and 3 into the expression $2x - 5$. For example, if $x = -2$, then $2x - 5 = 2(-2) - 5 = -9$. The other values shown in Table 2.6 on the next page can be found similarly.

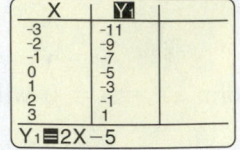
TABLE 2.6

x	-3	-2	-1	0	1	2	3
$2x - 5$	-11	-9	-7	-5	-3	-1	1

From the table, $2x - 5$ equals -7 when $x = -1$. So the solution to $2x - 5 = -7$ is -1.

Now Try Exercise 25

SOLVING LINEAR EQUATIONS SYMBOLICALLY Although tables can be used to solve some linear equations, the process of creating a table that contains the solution can take a significant amount of time. For example, the solution to the equation $9x - 4 = 0$ is $\frac{4}{9}$. However, creating a table that reveals this solution would be quite challenging.

The following strategy, which involves the addition and multiplication properties of equality, is a method for solving linear equations symbolically.

SOLVING A LINEAR EQUATION SYMBOLICALLY

STEP 1: Use the distributive property to clear any parentheses on each side of the equation. Combine any like terms on each side.

STEP 2: Use the addition property of equality to get all of the terms containing the variable on one side of the equation and all other terms on the other side of the equation. Combine any like terms on each side.

STEP 3: Use the multiplication property of equality to isolate the variable by multi- plying each side of the equation by the reciprocal of the number in front of the variable (or divide each side by that number).

STEP 4: Check the solution by substituting it in the given equation.

When a linear equation does not contain parentheses, we can start with the second step in the strategy shown above. This is the case for the equations in the next example.

EXAMPLE 3   **Solving linear equations**

Solve each linear equation. Check the answer for part (b).
(a) $10x - 13 = 0$ **(b)** $\frac{1}{2}x + 3 = 6$ **(c)** $5x + 7 = 2x + 3$

Solution
(a) First, isolate the x-term on the left side of the equation by adding 13 to each side.

$$10x - 13 = 0 \qquad \text{Given equation}$$
$$10x - 13 + 13 = 0 + 13 \qquad \text{Add 13 to each side. (Step 2)}$$
$$10x = 13 \qquad \text{Add the real numbers.}$$

To obtain a coefficient of 1 on the x-term, divide each side by 10.

$$\frac{10x}{10} = \frac{13}{10} \qquad \text{Divide each side by 10. (Step 3)}$$

$$x = \frac{13}{10} \qquad \text{Simplify.}$$

The solution is $\frac{13}{10}$.

(b) Start by subtracting 3 from each side.

$$\frac{1}{2}x + 3 = 6 \qquad \text{Given equation}$$

$$\frac{1}{2}x + 3 - 3 = 6 - 3 \qquad \text{Subtract 3 from each side. (Step 2)}$$

$$\frac{1}{2}x = 3 \qquad \text{Subtract the real numbers.}$$

$$2 \cdot \frac{1}{2}x = 2 \cdot 3 \qquad \text{Multiply each side by 2. (Step 3)}$$

$$x = 6 \qquad \text{Multiply the real numbers.}$$

The solution is 6. To check it, substitute **6** for x in the equation, $\frac{1}{2}x + 3 = 6$.

$$\frac{1}{2} \cdot \mathbf{6} + 3 \overset{?}{=} 6 \qquad \text{Replace } x \text{ with 6. (Step 4)}$$

$$3 + 3 \overset{?}{=} 6 \qquad \text{Multiply.}$$

$$6 = 6 \checkmark \qquad \text{Add; the answer checks.}$$

(c) Since this equation has two x-terms, we need to get all x-terms on one side of the equation and all real numbers on the other side. To do this, begin by subtracting $2x$ from each side.

$$5x + 7 = 2x + 3 \qquad \text{Given equation}$$

$$5x - 2x + 7 = 2x - 2x + 3 \qquad \text{Subtract } 2x \text{ from each side. (Step 2)}$$

$$3x + 7 = 3 \qquad \text{Combine like terms.}$$

$$3x + 7 - 7 = 3 - 7 \qquad \text{Subtract 7 from each side. (Step 2)}$$

$$3x = -4 \qquad \text{Simplify.}$$

$$\frac{3x}{3} = \frac{-4}{3} \qquad \text{Divide each side by 3. (Step 3)}$$

$$x = -\frac{4}{3} \qquad \text{Simplify the fractions.}$$

The solution is $-\frac{4}{3}$.

Now Try Exercises 31, 33, 37

▶ **REAL-WORLD CONNECTION**   In recent years, the number of worldwide Internet users has increased at a constant rate. Recall that linear equations are often used to model situations that exhibit a constant rate of change. In the next example we solve a linear equation that models Internet use.

EXAMPLE 4   **Estimating numbers of worldwide Internet users**

The number of Internet users I in millions during year x can be approximated by the formula

$$I = 241x - 482{,}440,$$

where $x \geq 2007$. Estimate the year when there were 2210 million (2.21 billion) Internet users. (*Source:* Internet World Stats.)

Solution

Let $I = \mathbf{2210}$ in the formula $I = 241x - 482{,}440$ and solve for x.

$$
\begin{aligned}
\mathbf{2210} &= 241x - 482{,}440 && \text{Equation to be solved} \\
484{,}650 &= 241x && \text{Add 482,440 to each side.} \\
\frac{484{,}650}{241} &= \frac{241x}{241} && \text{Divide each side by 241.} \\
\frac{484{,}650}{241} &= x && \text{Simplify.} \\
x &\approx 2011 && \text{Approximate.}
\end{aligned}
$$

During 2011 the number of Internet users reached 2210 million.

Now Try Exercise 73

- In the strategy for solving linear equations symbolically, which step involves the use of the distributive property?

Applying the Distributive Property

Sometimes the distributive property is helpful in solving linear equations. The next example demonstrates how to apply the distributive property in such situations. Use of the distributive property appeared in Step 1 of the strategy for solving linear equations discussed earlier.

EXAMPLE 5   **Applying the distributive property**

Solve each linear equation. Check the answer for part (a).
(a) $4(x - 3) + x = 0$ **(b)** $2(3z - 4) + 1 = 3(z + 1)$

Solution
(a) Begin by applying the distributive property.

$$
\begin{aligned}
4(x - 3) + x &= 0 && \text{Given equation} \\
4x - 12 + x &= 0 && \text{Distributive property (Step 1)} \\
5x - 12 &= 0 && \text{Combine like terms.} \\
5x - 12 \mathbf{+ 12} &= 0 \mathbf{+ 12} && \text{Add 12 to each side. (Step 2)} \\
5x &= 12 && \text{Add the real numbers.} \\
\frac{5x}{\mathbf{5}} &= \frac{12}{\mathbf{5}} && \text{Divide each side by 5. (Step 3)} \\
x &= \frac{12}{5} && \text{Simplify.}
\end{aligned}
$$

To see if $\frac{12}{5}$ is the solution, substitute $\frac{12}{5}$ for x in the equation $4(x - 3) + x = 0$.

$$
\begin{aligned}
4\left(\frac{\mathbf{12}}{\mathbf{5}} - 3\right) + \frac{\mathbf{12}}{\mathbf{5}} &\overset{?}{=} 0 && \text{Replace } x \text{ with } \tfrac{12}{5}. \text{ (Step 4)} \\
4\left(\frac{12}{5} - \frac{15}{5}\right) + \frac{12}{5} &\overset{?}{=} 0 && \text{Common denominator} \\
4\left(-\frac{3}{5}\right) + \frac{12}{5} &\overset{?}{=} 0 && \text{Subtract within parentheses.} \\
-\frac{12}{5} + \frac{12}{5} &\overset{?}{=} 0 && \text{Multiply.} \\
0 &= 0 \checkmark && \text{Add; the answer checks.}
\end{aligned}
$$

(b) Begin by applying the distributive property to each side of the equation. Then get all z-terms on the left side and terms containing only real numbers on the right side.

$$2(3z - 4) + 1 = 3(z + 1) \quad \text{Given equation}$$

$$6z - 8 + 1 = 3z + 3 \quad \text{Distributive property (Step 1)}$$

$$6z - 7 = 3z + 3 \quad \text{Add the real numbers.}$$

$$3z - 7 = 3 \quad \text{Subtract } 3z \text{ from each side. (Step 2)}$$

$$3z = 10 \quad \text{Add 7 to each side. (Step 2)}$$

$$\frac{3z}{3} = \frac{10}{3} \quad \text{Divide each side by 3. (Step 3)}$$

$$z = \frac{10}{3} \quad \text{Simplify.}$$

The solution is $\frac{10}{3}$.

Now Try Exercises 43, 45

READING CHECK

• Why do some people prefer to clear fractions or decimals from an equation?

Clearing Fractions and Decimals

Some people prefer to do calculations without fractions or decimals. For this reason, clearing an equation of fractions or decimals before solving it can be helpful. To clear fractions or decimals, multiply each side of the equation by the least common denominator (LCD).

EXAMPLE 6   **Clearing fractions from linear equations**

Solve each linear equation.
(a) $\frac{1}{7}x - \frac{5}{7} = \frac{3}{7}$ **(b)** $\frac{2}{3}x - \frac{1}{6} = x$

Solution
(a) Multiply each side of the equation by the LCD 7 to clear (remove) fractions.

$$\frac{1}{7}x - \frac{5}{7} = \frac{3}{7} \quad \text{Given equation}$$

$$7\left(\frac{1}{7}x - \frac{5}{7}\right) = 7 \cdot \frac{3}{7} \quad \text{Multiply each side by 7.}$$

$$\frac{7}{1} \cdot \frac{1}{7}x - \frac{7}{1} \cdot \frac{5}{7} = \frac{7}{1} \cdot \frac{3}{7} \quad \text{Distributive property}$$

$$x - 5 = 3 \quad \text{Simplify.}$$

$$x = 8 \quad \text{Add 5 to each side.}$$

The solution is 8.
(b) The LCD for 3 and 6 is 6. Multiply each side of the equation by 6.

$$\frac{2}{3}x - \frac{1}{6} = x \quad \text{Given equation}$$

$$6\left(\frac{2}{3}x - \frac{1}{6}\right) = 6 \cdot x \quad \text{Multiply each side by 6.}$$

$$\frac{6}{1} \cdot \frac{2}{3}x - \frac{6}{1} \cdot \frac{1}{6} = 6 \cdot x \quad \text{Distributive property}$$

$$4x - 1 = 6x \quad \text{Simplify.}$$

$$-1 = 2x \quad \text{Subtract } 4x \text{ from each side.}$$

$$-\frac{1}{2} = x \quad \text{Divide each side by 2.}$$

The solution is $-\frac{1}{2}$.

Now Try Exercises 53, 55

EXAMPLE 7   **Clearing decimals from a linear equation**

Solve each linear equation.
(a) $0.2x - 0.7 = 0.4$ **(b)** $0.01x - 0.42 = -0.2x$

Solution
(a) The least common denominator for 0.2, 0.7, and 0.4 $\left(\text{or } \frac{2}{10}, \frac{7}{10}, \text{ and } \frac{4}{10}\right)$ is 10. Multiply each side by 10. When multiplying by 10, move the decimal point 1 place to the right.

$0.2x - 0.7 = 0.4$	Given equation
$\mathbf{10}(0.2x - 0.7) = \mathbf{10}(0.4)$	Multiply each side by 10.
$\mathbf{10}(0.2x) - \mathbf{10}(0.7) = \mathbf{10}(0.4)$	Distributive property
$2x - 7 = 4$	Simplify.
$2x = 11$	Add 7 to each side.
$x = \dfrac{11}{2}$	Divide each side by 2.

The solution is $\frac{11}{2}$, or 5.5.

(b) The least common denominator for 0.01, 0.42, and 0.2 $\left(\text{or } \frac{1}{100}, \frac{42}{100}, \text{ and } \frac{2}{10}\right)$ is 100. Multiply each side by 100. To do this move the decimal point 2 places to the right.

$0.01x - 0.42 = -0.2x$	Given equation
$\mathbf{100}(0.01x - 0.42) = \mathbf{100}(-0.2x)$	Multiply each side by 100.
$\mathbf{100}(0.01x) - \mathbf{100}(0.42) = \mathbf{100}(-0.2x)$	Distributive property
$x - 42 = -20x$	Simplify.
$x - 42 \; \mathbf{+ \, 20x \, + \, 42} = -20x \; \mathbf{+ \, 20x \, + \, 42}$	Add $20x$ and 42.
$21x = 42$	Combine like terms.
$x = 2$	Divide each side by 21.

The solution is 2.

Now Try Exercises 49, 51

Equations with No Solutions or Infinitely Many Solutions

Some equations that appear to be linear are not because when they are written in the form $ax + b = 0$ the value of a is 0 and no x-term appears. This type of equation can have no solutions or infinitely many solutions. An example of an equation that has no solutions is

$$x = x + 1$$

because a number x cannot equal itself plus 1. If we attempt to solve this equation by subtracting x from each side, we obtain the equation $0 = 1$, which has no x-term and is *always false*. An equation that is always false has no solutions.

An example of an equation with infinitely many solutions is

$$5x = 2x + 3x,$$

because the equation simplifies to

$$5x = 5x,$$

which is true for any real number x. If $5x$ is subtracted from each side the result is $0 = 0$, which has no x-term and is *always true*. When an equation containing a variable is always true, it has infinitely many solutions.

NOTE: An equation that is always true is called an **identity**, and an equation that is always false is called a **contradiction**.

EXAMPLE 8   Determining numbers of solutions

Determine whether the equation has no solutions, one solution, or infinitely many solutions.
(a) $3x = 2(x + 1) + x$ **(b)** $2x - (x + 1) = x - 1$ **(c)** $5x = 2(x - 4)$

Solution
(a) Start by applying the distributive property.

$$3x = 2(x + 1) + x \qquad \text{Given equation}$$
$$3x = 2x + 2 + x \qquad \text{Distributive property}$$
$$3x = 3x + 2 \qquad \text{Combine like terms.}$$
$$0 = 2 \qquad \text{Subtract } 3x \text{ from each side.}$$

The equation $0 = 2$ is always false, so the given equation is a contradiction with no solutions.

(b) Start by applying the distributive property.

$$2x - (x + 1) = x - 1 \qquad \text{Given equation}$$
$$2x - x - 1 = x - 1 \qquad \text{Distributive property}$$
$$x - 1 = x - 1 \qquad \text{Combine like terms.}$$
$$x = x \qquad \text{Add 1 to each side.}$$
$$0 = 0 \qquad \text{Subtract } x \text{ from each side.}$$

The equation $0 = 0$ is always true, so the given equation is an identity that has infinitely many solutions. Note that the solution set contains all real numbers.

(c) Start by applying the distributive property.

$$5x = 2(x - 4) \qquad \text{Given equation}$$
$$5x = 2x - 8 \qquad \text{Distributive property}$$
$$3x = -8 \qquad \text{Subtract } 2x \text{ from each side.}$$
$$x = -\frac{8}{3} \qquad \text{Divide each side by 3.}$$

Thus there is one solution.

Now Try Exercises 61, 63, 67

CRITICAL THINKING

What must be true about b and d for the equation

$$bx - 2 = dx + 7$$

to have no solutions? What must be true about b and d for this equation to have exactly one solution?

READING CHECK

• How can you tell when an equation will have no solutions, one solution, or infinitely many solutions?

MAKING CONNECTIONS

Number of Solutions

When solving the general form $ax + b = 0$, where a and b can be *any* real numbers, the resulting equivalent equation will indicate whether the given equation has no solutions, one solution, or infinitely many solutions.

No Solutions: The result is an equation such as $4 = 0$ or $3 = 2$, which is *always false* for any value of the variable.

One Solution: The result is an equation such as $x = 1$ or $x = -12$, which is true for *only one* value of the variable.

Infinitely Many Solutions: The result is an equation such as $0 = 0$ or $-3 = -3$, which is *always true* for any value of the variable.

2.2   Putting It All Together

CONCEPT	COMMENTS	EXAMPLES
Linear Equation	Can be written as $$ax + b = 0,$$ where $a \neq 0$; has one solution	The equation $5x - 8 = 0$ is linear, with $a = 5$ and $b = -8$. The equation $2x^2 + 4 = 0$ is not linear.
Solving Linear Equations Numerically	To solve a linear equation numerically, complete a table for various values of the variable and then select the solution from the table, if possible.	The solution to $2x - 4 = -2$ is 1. <table><tr><td>x</td><td>-1</td><td>0</td><td>1</td></tr><tr><td>$2x - 4$</td><td>-6</td><td>-4</td><td>-2</td></tr></table>
Solving Linear Equations Symbolically	Use the addition and multiplication properties of equality to isolate the variable. See the four-step approach to solving a linear equation on page 102.	$5x - 8 = 0$ Given equation $5x = 8$ Add 8 to each side. $x = \dfrac{8}{5}$ Divide each side by 5.
Equations with No Solutions	Some equations that appear to be linear have no solutions. Solving will result in an equivalent equation that is always false.	The equation $$x = x + 5$$ has no solutions because a number cannot equal itself plus 5.
Equations with Infinitely Many Solutions	Some equations that appear to be linear have infinitely many solutions. Solving will result in an equivalent equation that is always true.	The equation $$2x = x + x$$ has infinitely many solutions because the equation is true for all values of x.

2.2   Exercises

CONCEPTS AND VOCABULARY

1. Linear equations can model applications in which things move or change at a(n) _____ rate.

2. A linear equation can be written in the form _____ with $a \neq 0$.

3. How many solutions does a linear equation in one variable have?

4. When a table of values is used to solve a linear equation, the equation is being solved _____.

5. What two properties of equality are frequently used to solve linear equations?

6. To clear fractions or decimals from an equation, multiply each side by the _____.

7. If solving an equation results in $0 = 4$, how many solutions does it have?

8. If solving an equation results in $0 = 0$, how many solutions does it have?

IDENTIFYING LINEAR EQUATIONS

Exercises 9–22: (Refer to Example 1.) Is the equation linear? If it is linear, give values for a and b that result when the equation is written in the form ax + b = 0.

9. $3x - 7 = 0$

10. $-2x + 1 = 4$

11. $\frac{1}{2}x = 0$ **12.** $-\frac{3}{4}x = 0$

13. $4x^2 - 6 = 11$ **14.** $-2x^2 + x = 4$

15. $\frac{6}{x} - 4 = 2$ **16.** $2\sqrt{x} - 1 = 0$

17. $1.1x + 0.9 = 1.8$ **18.** $-5.7x - 3.4 = -6.8$

19. $2(x - 3) = 0$ **20.** $\frac{1}{2}(x + 4) = 0$

21. $|3x| + 2 = 1$ **22.** $3x = 4x^3$

SOLVING LINEAR EQUATIONS

Exercises 23–28: Complete the table for the given values of x. Then use the table to solve the given equation numerically.

23. $x - 3 = -1$

x	-1	0	1	2	3
$x - 3$	-4				

24. $-2x = 0$

x	-2	-1	0	1	2
$-2x$	4				

25. $-3x + 7 = 1$

x	0	1	2	3	4
$-3x + 7$	7				

26. $5x - 2 = 3$

x	-1	0	1	2	3
$5x - 2$	-7				

27. $4 - 2x = 6$

x	-2	-1	0	1	2
$4 - 2x$	8				

28. $9 - (x + 3) = 4$

x	-2	-1	0	1	2
$9 - (x + 3)$	8				

Exercises 29–58: Solve the given equation and check the solution.

29. $11x = 3$ **30.** $-5x = 15$

31. $x - 18 = 5$ **32.** $8 = 5 + 3x$

33. $\frac{1}{2}x - 1 = 13$ **34.** $\frac{1}{4}x + 3 = 9$

35. $-6 = 5x + 5$ **36.** $31 = -7x - 4$

37. $3z + 2 = z - 5$ **38.** $z - 5 = 5z - 3$

39. $12y - 6 = 33 - y$ **40.** $-13y + 2 = 22 - 3y$

41. $4(x - 1) = 5$ **42.** $-2(2x + 7) = 1$

43. $1 - (3x + 1) = 5 - x$

44. $6 + 2(x - 7) = 10 - 3(x - 3)$

45. $(5t - 6) = 2(t + 1) + 2$

46. $-2(t - 7) - (t + 5) = 5$

47. $3(4z - 1) - 2(z + 2) = 2(z + 1)$

48. $-(z + 4) + (3z + 1) = -2(z + 1)$

49. $7.3x - 1.7 = 5.6$ **50.** $5.5x + 3x = 51$

51. $-9.5x - 0.05 = 10.5x + 1.05$

52. $0.04x + 0.03 = 0.02x - 0.1$

53. $\frac{1}{2}x - \frac{3}{2} = \frac{5}{2}$ **54.** $-\frac{1}{4}x + \frac{5}{4} = \frac{3}{4}$

55. $-\frac{3}{8}x + \frac{1}{4} = \frac{1}{8}$ **56.** $\frac{1}{3}x + \frac{1}{4} = \frac{1}{6} - x$

57. $4y - 2(y + 1) = 0$

58. $(15y + 20) - 5y = 5 - 10y$

59. **Thinking Generally** A linear equation has exactly one solution. Find the solution to the equation $ax + b = 0$, where $a \neq 0$, by solving for x.

60. **Thinking Generally** Solve the linear equation given by $\frac{1}{a}x - b = 0$ for x.

EQUATIONS WITH NO SOLUTIONS OR INFINITELY MANY SOLUTIONS

Exercises 61–70: Determine whether the equation has no solutions, one solution, or infinitely many solutions.

61. $5x = 5x + 1$ **62.** $2(x - 3) = 2x - 6$

63. $8x = 0$ **64.** $9x = x + 1$

65. $4x = 5(x + 3) - x$ **66.** $5x = 15 - 2(x + 7)$

67. $5(2x + 7) - (10x + 5) = 30$

68. $4(x + 2) - 2(2x + 3) = 10$

69. $x - (3x + 2) = 15 - 2x$

70. $2x - (x + 5) = x - 5$

APPLICATIONS

71. Distance Traveled A bicyclist is 4 miles from home, riding away from home at 8 miles per hour.
   (a) Make a table that shows the bicyclist's distance D from home after 0, 1, 2, 3, and 4 hours.
   (b) Write a formula that calculates D after x hours.
   (c) Use your formula to determine D when $x = 3$ hours. Does your answer agree with the value found in your table?
   (d) Find x when $D = 22$ miles. Interpret the result.

72. Distance Traveled An athlete is 16 miles from home, running *toward* home at 6 miles per hour.
   (a) Write a formula that calculates the distance D that the athlete is from home after x hours.
   (b) Determine D when $x = 1.5$ hours.
   (c) Find x when $D = 5.5$ miles. Interpret the result.

73. Internet Users (Refer to Example 4.) The number of Internet users I in millions during year x, where $x \geq 2007$, can be approximated by the formula

$$I = 241x - 482{,}440.$$

Approximate the year in which there were 1730 million (1.73 billion) Internet users.

74. HIV Infections The cumulative number of HIV infections N in thousands for the United States in year x can be approximated by the formula

$$N = 42x - 83{,}197,$$

where $x \geq 2000$. Approximate the year when this number reached 970 thousand. (*Source:* Centers for Disease Control and Prevention.)

75. State and Federal Inmates The number N of state and federal inmates in millions during year x, where

$x \geq 2002$, can be approximated by the formula $N = 0.03x - 58.62$. Determine the year in which there were 1.5 million inmates. (*Source:* Bureau of Justice.)

76. Government Costs From 1960 to 2000 the cost C (in billions of 1992 dollars) to regulate social and economic programs could be approximated by the formula $C = 0.35x - 684$ during year x. Estimate the year in which the cost reached \$6.6 billion. (*Source:* Center for the Study of American Business.)

77. Hospitals The number of hospitals H with more than 100 beds during year x is estimated by the formula $H = -33x + 69{,}105$, where x is any year from 2002 to 2008. In which year were there 2841 hospitals of this type? (*Source:* AHA Hospital Statistics.)

78. Home Size The average size F in square feet of new U.S. homes built during year x is estimated by the formula $F = 34x - 65{,}734$, where x is any year from 2002 to 2008. In which year was the average home size 2504 square feet? (*Source:* U.S. Census Bureau.)

WRITING ABOUT MATHEMATICS

79. A student says that the equation $4x - 1 = 1 - x$ is not a linear equation because it is not in the form $ax + b = 0$. Is the student correct? Explain.

80. A student solves a linear equation as follows.

$$4(x + 3) = 5 - (x + 3)$$
$$4x + 3 \overset{?}{=} 5 - x + 3$$
$$4x + 3 \overset{?}{=} 8 - x$$
$$5x \overset{?}{=} 5$$
$$x \overset{?}{=} 1$$

Identify and explain the errors that the student made. What is the correct answer?

SECTIONS 2.1 and 2.2

Checking Basic Concepts

1. Determine whether the equation is linear.
   (a) $4x^3 - 2 = 0$ (b) $2(x + 1) = 4$

2. Complete the table for each value of x. Then use the table to solve $4x - 3 = 13$.

x	3	3.5	4	4.5	5
$4x - 3$	9				17

3. Solve each equation and check your answer.
   (a) $x - 12 = 6$
   (b) $\frac{3}{4}z = \frac{1}{8}$
   (c) $0.6t + 0.4 = 2$
   (d) $5 - 2(x - 2) = 3(4 - x)$

4. Determine whether each equation has no solutions, one solution, or infinitely many solutions.
   (a) $x - 5 = 6x$
   (b) $-2(x - 5) = 10 - 2x$
   (c) $-(x - 1) = -x - 1$

5. Distance Traveled A driver is 300 miles from home and is traveling toward home on a freeway at a constant speed of 75 miles per hour.
   (a) Write a formula to calculate the distance D that the driver is from home after x hours.
   (b) Write an equation whose solution gives the hours needed for the driver to reach home.
   (c) Solve the equation from part (b).

2.3   Introduction to Problem Solving

Steps for Solving a Problem • Percent Problems • Distance Problems • Other Types of Problems

A LOOK INTO MATH ▶

Throughout history people have found it quite difficult to predict human behavior. With the development of the Internet and social networks, accurately predicting behavior is becoming possible. For example, using mathematics, it is possible to analyze a person's Facebook friends to predict with "reasonable certainty" the person's hobbies, interests, and even health issues. Solving problems requires problem-solving skills, which we introduce in this section.

Steps for Solving a Problem

Word problems can be challenging because formulas and equations are not usually given. To solve such problems we need a strategy. The following steps are based on George Polya's (1888–1985) four-step process for problem solving.

NEW VOCABULARY

☐ Percent change

STEPS FOR SOLVING A PROBLEM

STEP 1: Read the problem carefully and be sure that you understand it. (You may need to read the problem more than once.) Assign a variable to what you are being asked to find. If necessary, write other quantities in terms of this variable.

STEP 2: Write an equation that relates the quantities described in the problem. You may need to sketch a diagram or refer to known formulas.

STEP 3: Solve the equation. Use the solution to determine the solution(s) to the original problem. Include any necessary units.

STEP 4: Check your solution in the original problem. Does it seem reasonable?

READING CHECK

• Why doesn't the problem-solving strategy stop after the equation is solved in Step 3?

STUDY TIP

Step 4 in this problem-solving strategy provides good advice for working any math problem. Checking your answer, especially when taking an exam, can lead to fewer errors and better scores.

Even if we understand the problem that we are trying to solve, we may not be able to find a solution if we cannot write an appropriate equation. Table 2.7 lists several common words that are associated with the math symbols needed to write equations.

TABLE 2.7 Math Symbols and Associated Words

Symbol	Associated Words
+	add, plus, more, sum, total, increase
−	subtract, minus, less, difference, fewer, decrease
·	multiply, times, twice, double, triple, product
÷	divide, divided by, quotient, per
=	equals, is, gives, results in, is the same as

EXAMPLE 1 | **Translating sentences into equations**

Translate the sentence into an equation using the variable x. Then solve the resulting equation.
(a) Three times a number minus 6 is equal to 18.
(b) The sum of half a number and 5 is zero.
(c) Sixteen is 4 less than twice a number.

Solution

(a) The phrase "Three times a number" indicates that we multiply x by 3 to get $3x$. The word "minus" indicates that we then subtract 6 from $3x$ to get $3x - 6$. This expression "equals" 18, so the equation is $3x - 6 = 18$. The solution is 8 as shown here.

$$3x - 6 = 18 \qquad \text{Equation to be solved}$$
$$3x = 24 \qquad \text{Add 6 to each side.}$$
$$\frac{3x}{3} = \frac{24}{3} \qquad \text{Divide each side by 3.}$$
$$x = 8 \qquad \text{Simplify the fractions.}$$

(b) The word "sum" indicates that we add "half a number" and 5 to get $\frac{1}{2}x + 5$. The word "is" implies equality, so the equation is $\frac{1}{2}x + 5 = 0$. The solution is -10 as shown here.

$$\frac{1}{2}x + 5 = 0 \qquad \text{Equation to be solved}$$
$$\frac{1}{2}x = -5 \qquad \text{Subtract 5 from each side.}$$
$$x = -10 \qquad \text{Multiply each side by 2.}$$

(c) To translate "4 less than twice a number" into a mathematical expression, we write $2x - 4$. If this seems backwards, consider how you would calculate "4 less than your age." The equation is $16 = 2x - 4$ and the solution is 10 as shown here.

$$16 = 2x - 4 \qquad \text{Equation to be solved}$$
$$20 = 2x \qquad \text{Add 4 to each side.}$$
$$\frac{20}{2} = \frac{2x}{2} \qquad \text{Divide each side by 2.}$$
$$x = 10 \qquad \text{Simplify and rewrite.}$$

Now Try Exercises 13, 15, 19

In the next example we apply the four-step process to a word problem that involves three unknown numbers.

EXAMPLE 2 | **Solving a number problem**

The sum of three consecutive natural numbers is 81. Find the three numbers.

Solution

STEP 1: Start by assigning a variable n to an unknown quantity.

$$n: \text{smallest of the three natural numbers}$$

Next, write the other two natural numbers in terms of n.

$$n + 1: \text{next consecutive natural number}$$
$$n + 2: \text{largest of the three consecutive natural numbers}$$

STEP 2: Write an equation that relates these unknown quantities. The sum of the three consecutive natural numbers is **81**, so the needed equation is

$$n + (n + 1) + (n + 2) = 81.$$

STEP 3: Solve the equation in Step 2.

$n + (n + 1) + (n + 2) = 81$	Equation to be solved
$(n + n + n) + (1 + 2) = 81$	Commutative and associative properties
$3n + 3 = 81$	Combine like terms.
$3n = 78$	Subtract 3 from each side.
$n = 26$	Divide each side by 3.

The smallest of the three numbers is 26, so the three numbers are 26, 27, and 28.

STEP 4: To check this solution we can add the three numbers to see if their sum is 81.

$$26 + 27 + 28 = 81$$

The solution checks.

▌ **Now Try Exercise 21**

▶ **REAL-WORLD CONNECTION** The infant mortality rate for a country measures the number of deaths of infants under one year of age per 1000 live births in a given year. A high infant mortality rate may indicate a lack of good quality health care for infants. In the next example, a number problem is solved to find the infant mortality rate in Iceland, which has one of the lowest rates in the world.

EXAMPLE 3	**Finding an infant mortality rate**

Sierra Leone, in West Africa, has one of the highest infant mortality rates in the world at 160. This rate is 10 more than 50 times the rate in Iceland. Find the infant mortality rate in Iceland. (*Source:* World Population Prospectus.)

Solution
STEP 1: Let r represent the unknown infant mortality rate in Iceland.
STEP 2: Since the rate 160 in Sierra Leone is 10 more than 50 times the rate r in Iceland, the equation to solve is $160 = 10 + 50r$.
STEP 3: Rewrite the equation in Step 2 and then solve it.

$50r + 10 = 160$	Equation to be solved (rewritten)
$50r = 150$	Subtract 10 from each side.
$r = 3$	Divide each side by 50.

The infant mortality rate in Iceland is 3 deaths per 1000 live births.
STEP 4: To check this solution, we verify that 10 more than 50 times the Iceland infant mortality rate of 3 gives the Sierra Leone infant mortality rate of 160.

$$50(3) + 10 = 150 + 10 = 160$$

▌ **Now Try Exercise 33**

Percent Problems

▶ **REAL-WORLD CONNECTION** Problems involving percentages occur in everyday life. Wage increases, sales tax, and government data all make use of percentages. For example, it is estimated that 10% of the world's population will be older than 65 in 2025. (*Source:* World Health Organization.)

Percent notation can be changed to either fraction or decimal notation.

> **PERCENT NOTATION**
>
> The expression $x\%$ represents the fraction $\frac{x}{100}$ or the decimal number $x \cdot 0.01$.

NOTE: To write $x\%$ as a decimal number, move the decimal point in the number x two places to the *left* and then remove the % symbol. For example, 13.6% can be written as a decimal number as follows.

$$13.6\% \to 13.6\% \to 0.136$$

READING CHECK

• How do we write $x\%$ as a decimal?

EXAMPLE 4 | **Converting percent notation**

Convert each percentage to fraction and decimal notation.
(a) 23%   **(b)** 5.2%   **(c)** 0.3%

Solution
(a) *Fraction Notation*: To convert to fraction notation, divide 23 by 100. Thus $23\% = \frac{23}{100}$.
Decimal Notation: Move the decimal point two places to the left and then remove the % symbol: $23\% = 0.23$.
(b) *Fraction Notation*: $5.2\% = \frac{5.2}{100} = \frac{52}{1000} = \frac{13 \cdot 4}{250 \cdot 4} = \frac{13}{250}$
Decimal Notation: Move the decimal point two places to the left and then remove the % symbol: $5.2\% = 0.052$.
(c) *Fraction Notation*: $0.3\% = \frac{0.3}{100} = \frac{3}{1000}$
Decimal Notation: Move the decimal point two places to the left and then remove the % symbol: $0.3\% = 0.003$.

Now Try Exercises 49, 53, 55

EXAMPLE 5 | **Converting to percent notation**

Convert each real number to a percentage.
(a) 0.234   **(b)** $\frac{1}{4}$ **(c)** 2.7

Solution
(a) Move the decimal point two places to the *right* and then insert the % symbol to obtain $0.234 = 23.4\%$.
(b) $\frac{1}{4} = 0.25$, so $\frac{1}{4} = 25\%$.
(c) Move the decimal point two places to the right and then insert the % symbol to obtain $2.7 = 270\%$. Note that percentages can be greater than 100%.

Now Try Exercises 57, 59, 63

PERCENT CHANGE When prices increase (or decrease), the actual amount of the increase (or decrease) is often not as significant as the *percent change* in the price. For example, consider the $1000 increases shown in Figure 2.2.

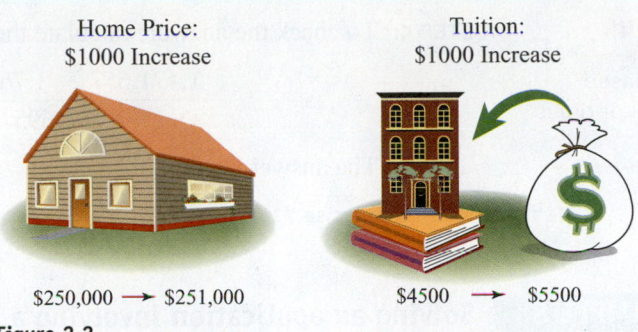

Home Price:
$1000 Increase

Tuition:
$1000 Increase

$250,000 → $251,000 $4500 → $5500

Figure 2.2

Even though both prices increased by $1000, the percent increase in the price of the home is four-tenths of 1% and the percent increase in the cost of tuition is more than 22%. The increase in tuition is much more dramatic than the increase in the price of the home.

If a quantity changes from an **old amount** to a **new amount**, the **percent change** is given by

$$\text{Percent Change} = \frac{\text{new amount} - \text{old amount}}{\text{old amount}} \cdot 100.$$

The reason for multiplying by 100 is to change the decimal representation to a percentage.

NOTE: A positive percent change corresponds to an increase and a negative percent change corresponds to a decrease.

READING CHECK

- What does it mean when a percent change is negative?

EXAMPLE 6 **Calculating percent increase in Skype subscriptions**

From 2007 to 2010 the number of users paying for a Skype subscription increased from 4.6 million to 8.1 million. Calculate the percent increase in paying Skype subscribers from 2007 to 2010. (*Source:* Skype SEC filing.)

Solution
The **old amount** is **4.6** and the **new amount** is **8.1**. The percent increase is

$$\frac{\text{new amount} - \text{old amount}}{\text{old amount}} \cdot 100 = \frac{8.1 - 4.6}{4.6} \cdot 100 \approx 76\%.$$

Now Try Exercise 69

As we previously indicated, percentages frequently occur in applications.

EXAMPLE 7 **Solving an application involving a percent**

In 2008, there were 1,766,695 surgical cosmetic procedures performed. In 2009, this number decreased by 16.7%. Find the number of surgical cosmetic procedures performed in 2009. (*Source:* American Society for Aesthetic Plastic Surgery.)

Solution
STEP 1: Let N represent the number of surgical cosmetic procedures performed in 2009.
STEP 2: N equals 1,766,695 minus 16.7% of 1,766,695, and 16.7% = 0.167, so

$$N = 1{,}766{,}695 - (0.167)1{,}766{,}695.$$

NOTE: The word "of" in percent problems often indicates multiplication.

STEP 3: To find N, evaluate the expression on the right side to get

$$N = 1{,}766{,}695 - (0.167)1{,}766{,}695 \approx 1{,}471{,}657.$$

There were about 1,471,657 surgical cosmetic procedures performed in 2009.

CRITICAL THINKING

If your salary increased 200%, by what factor did it increase?

STEP 4: To check the answer, calculate the percent decrease.

$$\frac{1,471,657 - 1,766,695}{1,766,695} \cdot 100 \approx -16.7\%$$

The answer checks.

Now Try Exercise 71

EXAMPLE 8 **Solving an application involving a percent**

In 2025, about 10% of the world's population, or 800 million people, will be older than 65. Find the estimated population of the world in 2025. (*Source:* World Population Prospectus.)

Solution
STEP 1: Let P represent the world's population in millions in 2025.
STEP 2: Since 10% of P equals 800 million, this information is described by

$$0.10P = 800.$$

STEP 3: To solve the equation in Step 2, divide each side by 0.10.

$$0.10P = 800 \qquad \text{Equation to be solved}$$
$$P = 8000 \qquad \text{Divide by 0.10.}$$

In 2025 the estimated world population is 8000 million, or 8 billion.
STEP 4: To check the answer, determine whether 10% of 8 billion is 800 million.

$$(0.10)8,000,000,000 = 800,000,000$$

The answer checks.

Now Try Exercise 73

Distance Problems

If a person drives on an interstate highway at 70 miles per hour for 3 hours, then the total distance traveled is $70 \cdot 3 = 210$ miles. In general, $d = rt$, where d is the distance traveled, r is the rate (or speed), and t is time. In this example, the distance is in miles, the time is in hours, and the rate is expressed in miles per hour. In general, the rate in a distance problem is expressed in units of distance per unit of time.

EXAMPLE 9 **Solving a distance problem**

A person drives for 2 hours and 30 minutes at a constant speed and travels 180 miles. See Figure 2.3. Find the speed of the car in miles per hour.

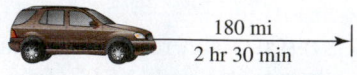

180 mi
2 hr 30 min

Figure 2.3

Solution
STEP 1: Let r represent the car's rate, or speed, in miles per hour.
STEP 2: The rate is to be given in miles per hour, so change 2 hours and 30 minutes to 2.5 or $\frac{5}{2}$ hours. Because $d = 180$ and $t = \frac{5}{2}$, the equation $d = rt$ becomes

$$180 = r \cdot \frac{5}{2}.$$

STEP 3: Solve the equation in Step 2 for r by multiplying each side of the equation by $\frac{2}{5}$, which is the reciprocal of $\frac{5}{2}$.

$$180 = \frac{5}{2} \cdot r \qquad \text{Equation to be solved}$$

$$\frac{2}{5} \cdot 180 = r \cdot \frac{5}{2} \cdot \frac{2}{5} \qquad \text{Multiply each side by } \tfrac{2}{5}.$$

$$72 = r \qquad \text{Simplify.}$$

The speed of the car is 72 miles per hour.

STEP 4: Because 2 hours and 30 minutes is equivalent to $\frac{5}{2}$ hours, traveling for 2 hours and 30 minutes at a constant rate of **72** miles per hour results in a distance of

$$d = rt = 72 \cdot \frac{5}{2} = 180 \text{ miles.}$$

The answer checks.

■ **Now Try Exercise 83**

EXAMPLE 10   **Solving a distance problem**

An athlete jogs at two speeds, covering a distance of 7 miles in $\frac{3}{4}$ hour. If the athlete runs $\frac{1}{4}$ hour at 8 miles per hour, find the second speed.

Solution

STEP 1: Let r represent the second speed of the jogger in miles per hour.

STEP 2: The total time spent jogging is $\frac{3}{4}$ hour, so the time spent jogging at the second speed must be $\frac{3}{4} - \frac{1}{4} = \frac{1}{2}$ hour. The total distance of 7 miles is the result of jogging at 8 miles per hour for $\frac{1}{4}$ hour and at r miles per hour for $\frac{1}{2}$ hour. See Figure 2.4. The **distance** traveled at 8 miles per hour for $\frac{1}{4}$ hour is given by $8 \cdot \frac{1}{4}$ and the **distance** traveled at r miles per hour for $\frac{1}{2}$ hour is given by $r \cdot \frac{1}{2}$. The sum of these distances must equal **7** miles. Thus

$$8 \cdot \frac{1}{4} + r \cdot \frac{1}{2} = 7.$$

STEP 3: Solve the equation in Step 2 for r.

$$8 \cdot \frac{1}{4} + r \cdot \frac{1}{2} = 7 \qquad \text{Equation to be solved}$$

$$2 + \frac{r}{2} = 7 \qquad \text{Simplify.}$$

$$\frac{r}{2} = 5 \qquad \text{Subtract 2 from each side.}$$

$$r = 10 \qquad \text{Multiply each side by 2.}$$

The athlete's second speed is 10 miles per hour.

STEP 4: Jogging at a rate of 8 miles per hour for $\frac{1}{4}$ hour results in a distance of $8 \cdot \frac{1}{4} = 2$ miles. Jogging at a rate of 10 miles per hour for $\frac{1}{2}$ hour results in a distance of $10 \cdot \frac{1}{2} = 5$ miles. The total distance is $2 + 5 = 7$ miles and the total time is $\frac{1}{4} + \frac{1}{2} = \frac{3}{4}$ hour, so the answer checks.

■ **Now Try Exercise 87**

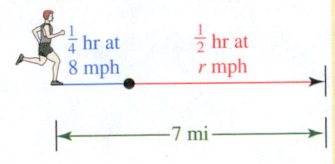

$\frac{1}{4}$ hr at 8 mph $\frac{1}{2}$ hr at r mph

|← 7 mi →|

Figure 2.4

Other Types of Problems

▶ **REAL-WORLD CONNECTION** Many applied problems involve linear equations. For example, right after people have their wisdom teeth pulled, they may need to rinse their mouth with salt water. In the next example, we use linear equations to determine how much water must be added to dilute a concentrated saline solution.

EXAMPLE 11   **Diluting a saline solution**

A solution contains 4% salt. How much pure water should be added to 30 ounces of the solution to dilute it to a 1.5% solution?

Solution

STEP 1: Assign a variable x as follows.

$$x: \text{ounces of pure water (0\% salt solution)}$$

$$30: \text{ounces of 4\% salt solution}$$

$$x + 30: \text{ounces of 1.5\% salt solution}$$

In Figure 2.5 three beakers illustrate this situation.

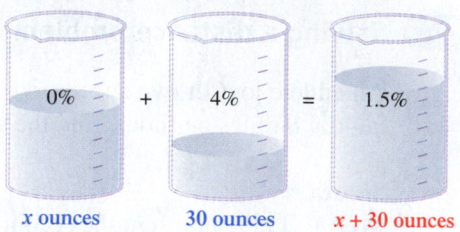

Figure 2.5 Mixing a Saline Solution

STEP 2: Note that the amount of salt in the first two beakers must equal the amount of salt in the third beaker. We use Table 2.8 to organize our calculations. The amount of salt in a solution equals the concentration times the solution amount, as shown in the last column of the table.

TABLE 2.8 Mixing a Saline Solution

Solution Type	Concentration (as a decimal)	Solution Amount (ounces)	Salt (ounces)
Pure Water	0% = 0.00	x	$0.00x$
Initial Solution	4% = 0.04	30	$0.04(30)$
Diluted Solution	1.5% = 0.015	$x + 30$	$0.015(x + 30)$

The amount of salt in the first two beakers is

$$0.00x + 0.04(30) = 0 + 1.2 = \mathbf{1.2} \text{ ounces.}$$

The amount of salt in the final beaker is

$$\mathbf{0.015(x + 30)} \text{ ounces.}$$

Because the amounts of salt in the solutions before and after mixing must be equal, the following equation must hold.

$$\mathbf{0.015(x + 30) = 1.2}$$

STEP 3: Solve the equation in Step 2.

$$0.015(x + 30) = 1.2 \qquad \text{Equation to be solved}$$
$$0.015x + 0.45 = 1.2 \qquad \text{Distributive property}$$
$$0.015x = 0.75 \qquad \text{Subtract 0.45 from each side.}$$
$$\frac{0.015x}{0.015} = \frac{0.75}{0.015} \qquad \text{Divide each side by 0.015.}$$
$$x = 50 \qquad \text{Simplify fractions.}$$

Fifty ounces of water should be added.

STEP 4: Adding 50 ounces of water will yield $50 + 30 = 80$ ounces of water containing $0.04(30) = 1.2$ ounces of salt. The concentration is $\frac{1.2}{80} = 0.015$ or 1.5%, so the answer checks.

Now Try Exercise 91

▶ **REAL-WORLD CONNECTION** Many times interest rates for student loans vary. In the next example, we present a situation in which a student has to borrow money at two different interest rates.

EXAMPLE 12 ## Calculating interest on college loans

A student takes out a loan for a limited amount of money at 5% interest and then must pay 7% for any additional money. If the student borrows $2000 more at 7% than at 5%, then the total interest for one year is $440. How much does the student borrow at each rate?

Solution

STEP 1: Assign a variable x as follows.

$$x: \text{loan amount at 5\% interest}$$
$$x + 2000: \text{loan amount at 7\% interest}$$

STEP 2: The amount of interest paid for the 5% loan is 5% of x, or **$0.05x$**. The amount of interest paid for the 7% loan is 7% of $x + 2000$, or **$0.07(x + 2000)$**. The total interest equals $**440**$, so we solve the equation

$$\mathbf{0.05x + 0.07(x + 2000) = 440}.$$

STEP 3: Solve the equation in Step 2 for x.

$$0.05x + 0.07(x + 2000) = 440 \qquad \text{Equation to be solved}$$
$$0.05x + 0.07x + 140 = 440 \qquad \text{Distributive property}$$
$$0.12x + 140 = 440 \qquad \text{Combine like terms.}$$
$$0.12x = 300 \qquad \text{Subtract 140 from each side.}$$
$$x = 2500 \qquad \text{Divide each side by 0.12.}$$

The student borrows $2500 at 5% and $2500 + 2000 = \$4500$ at 7%.

STEP 4: The amount of interest on $2500 borrowed at 5% is $0.05(2500) = \$125$ and the amount of interest on $4500 borrowed at 7% is $0.07(4500) = \$315$. Thus the total interest is $125 + 315 = \$440$. Furthermore, the amount borrowed at 7% is $2000 more than the amount borrowed at 5%. The answer checks.

Now Try Exercise 93

2.3   Putting It All Together

In this section on page 111 we presented a four-step approach to problem solving. However, because no approach works in every situation, solving mathematical problems takes time, effort, and creativity.

CONCEPT	COMMENTS	EXAMPLES
Percent Notation	$x\%$ represents either the fraction $\frac{x}{100}$ or the decimal $x \cdot 0.01$.	$17\% = \frac{17}{100} = 0.17$ $1.5\% = \frac{1.5}{100} = \frac{15}{1000} = 0.015$ $234\% = \frac{234}{100} = 2.34$
Converting Fractions to Percent Notation	Divide the numerator by the denominator, multiply by 100, and insert the % symbol.	$\frac{1}{2} = 0.5 = 50\%$ $\frac{2}{3} = 0.\overline{6} \approx 66.7\%$ $\frac{8}{5} = 1.6 = 160\%$
Percent Change	If a quantity changes from an old amount to a new amount, then the percent change is $$\frac{\text{new amount} - \text{old amount}}{\text{old amount}} \cdot 100.$$	If a price changes from \$1.50 to \$1.35, the percent change is $$\frac{1.35 - 1.50}{1.50} \times 100 = -10\%.$$ The price decreases by 10%.
Distance Problems	Distance d, rate r, and time t are related as expressed in the equation $$d = rt.$$	If a car travels 60 mph for 3 hours, then the distance d is $$d = rt = 60 \cdot 3 = 180 \text{ miles.}$$

2.3   Exercises

MyMathLab
PRACTICE   WATCH   DOWNLOAD   READ   REVIEW

CONCEPTS AND VOCABULARY

1. When you are solving a word problem, what is the last step?

2. The words *more*, *sum*, and *increase* are associated with the symbol _____.

3. The words *is*, *gives*, and *results in* are associated with the symbol _____.

4. Given an integer n, what are the next two consecutive integers?

5. The expression $x\%$ equals the fraction _____.

6. The expression $x\%$ equals the decimal $x \cdot$ _____.

7. To write 63.2% as a decimal, move the decimal point 2 places (left/right) and then remove the % symbol.

8. To write 0.349 as a percentage, move the decimal point 2 places (left/right) and then insert the % symbol.

9. If a price changes from A to B, then the percent change equals _____.

10. A positive percent change corresponds to a(n) _____ and a negative percent change corresponds to a(n) _____.

11. If a car travels at speed r for time t, then distance d is given by $d =$ _____.

12. In general, the rate in a distance problem is expressed in units of _____ per unit of _____.

NUMBER PROBLEMS

Exercises 13–20: Using the variable x, translate the sentence into an equation. Solve the resulting equation.

13. The sum of 2 and a number is 12.

14. Twice a number plus 7 equals 9.

15. A number divided by 5 equals the number decreased by 24.

16. 25 times a number is 125.

17. If a number is increased by 5 and then divided by 2, the result is 7.

18. A number subtracted from 8 is 5.

19. The quotient of a number and 2 is 17.

20. The product of 5 and a number equals 95.

Exercises 21–28: Find the number or numbers.

21. The sum of three consecutive natural numbers is 96.

22. The sum of three consecutive integers is -123.

23. Three times a number equals 102.

24. A number plus 18 equals twice the number.

25. Five times a number is 24 more than twice the number.

26. Three times a number is 18 less than the number.

27. Six times a number divided by 7 equals 18.

28. Two less than twice a number, divided by 5, equals 4.

29. *Finding Age* In 10 years, a child will be 3 years older than twice her current age. What is the current age of the child?

30. *Finding Age* A mother is 15 years older than twice her daughter's age. If the mother is 49 years old, how old is the daughter?

31. *Weight Loss* After losing 30 pounds, an individual weighs 110 pounds more than one-third his previous weight. Find the previous weight of the individual.

32. *Weight Gain* After gaining 25 pounds, a person is 115 pounds lighter than double his previous weight. How much did the person weigh before gaining 25 pounds?

33. *Hazardous Waste* In 2005, the number of federal hazardous waste sites in California was 2 less than twice the number of sites in Washington. How many hazardous waste sites were there in Washington if there were 24 such sites in California? (*Source:* Environmental Protection Agency.)

34. *Air Quality* In 2007, there were 100 unhealthy air quality days in Los Angeles. This is 1 more than 3 times the number of unhealthy air quality days in San Diego. How many unhealthy air quality days were there in San Diego? (*Source:* Environmental Protection Agency.)

35. *Endangered Species* There were 92 birds on the endangered species list in 2010. This is 12 more than twice the number of reptiles on the list. How many reptiles were on the endangered species list in 2010? (*Source:* U.S. Fish and Wildlife Service.)

36. *Millionaires* In 2009, there were three times as many millionaires in Kentucky as there were in Rhode Island. Together, a total of 84 thousand millionaires lived in the two states. Find the number of millionaires in each of these states. (*Source:* Internal Revenue Service.)

37. *U.S. Oil Production* There were 1562 million fewer barrels of crude oil produced in Alaska in 2009 than there were in the lower 48 states. Find the number of barrels of crude oil (in millions) produced in the lower 48 states in 2009 if there were 250 million barrels produced in Alaska that year. (*Source:* U.S. Energy Information Administration.)

38. *Power Production* In 1995, U.S. hydroelectric power production reached 311 billion kilowatthours. This is 189 billion kilowatt-hours less than twice the 2009 production level. Find the amount of hydroelectric power produced in 2009. (*Source:* U.S. Department of Energy.)

39. *Cosmetic Surgery* If the number of cosmetic surgeries (in thousands) performed in 2009 on persons under the age of 18 is multiplied by 2 and then decreased by 4, the result is the number (in thousands) of cosmetic surgeries performed on persons over age 65 that year. If there were 70 (thousand) such surgeries performed on persons over age 65, find the number of cosmetic surgeries performed on persons under age 18. (*Source:* American Society for Aesthetic Plastic Surgery.)

40. *Open Heart Surgery* In 2002, open heart surgery was performed 368 thousand times in the United States on persons over the age of 65. This is 22 thousand fewer than 13 times the number of this type of surgery performed on persons under the age of 15. How many times was open heart surgery performed on persons under the age of 15 in 2002? (*Source:* U.S. Department of Health and Human Services.)

41. *Geometry* If the perimeter of the following rectangle is 106 inches, find the value of x.

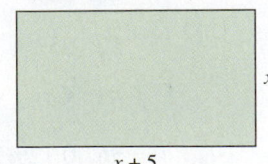

$x + 5$

42. *Geometry* If the perimeter of the following triangle is 24 inches, find the value of x.

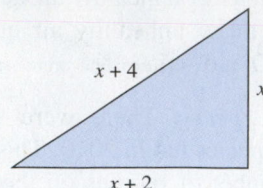

43. *Rectangle Dimensions* The length of a rectangle is 7 inches longer than the width. If the perimeter of the rectangle is 62 inches, find the measures of the length and width.

44. *Triangle Dimensions* The shortest side of a triangle measures 15 feet less than the longest side. If the third side is 6 feet shorter than the longest side and the perimeter is 102 feet, find the measures of the three sides of the triangle.

45. *Social Networking* In January 2008, the number of active MySpace users was 10 million fewer than double the number of active Facebook users. If the two social networking sites had a total of 170 million users, how many people used each of the Web sites? (*Source:* comScore.)

46. *Facebook Users* In February 2010, Facebook had 225 million more active users than it had the same month in 2009. If there were 400 million active users in February 2010, how many were there in February 2009? (*Source:* Facebook.com)

47. *Troops in Iraq* In 2010, there were 52,000 fewer U.S. troops in Iraq than there were in 2003. If the total number of U.S. troops in Iraq for the two years was 248,000, find the number of troops for each year. (*Source:* USA Today.)

48. *Farm Land* In 2009, there were 12 million fewer acres of farm land in Arizona than there were in 1980. If the sum of acreage for these two years is 64 million acres, find the number of farm land acres in Arizona for each of these years. (*Source:* U.S. Department of Agriculture, 2010.)

PERCENT PROBLEMS

Exercises 49–56: Convert the percentage to fraction and decimal notation.

49. 37%   **50.** 52%

51. 148%   **52.** 252%

53. 6.9%   **54.** 8.1%

55. 0.05%   **56.** 0.12%

Exercises 57–68: Convert the number to a percentage.

57. 0.45   **58.** 0.08

59. 1.8   **60.** 2.97

61. 0.006   **62.** 0.0001

63. $\frac{2}{5}$ **64.** $\frac{1}{3}$

65. $\frac{3}{4}$ **66.** $\frac{7}{20}$

67. $\frac{5}{6}$ **68.** $\frac{53}{50}$

69. *Voter Turnout* In the 1980 election for president there were 86.5 million voters, whereas in 2008 there were 132.6 million voters. Find the percent change in the number of voters.

70. *College Degrees* In 2005, about 594,000 people received a master's degree, and by 2010 this number had increased to 659,000. Find the percent change in the number of master's degrees received over this time period. (*Source:* The College Board.)

71. *Wages* A part-time instructor is receiving $950 per credit taught. If the instructor receives a 4% increase, how much will the new per credit compensation be?

72. *Tuition Increase* Tuition is currently $125 per credit. There are plans to raise tuition by 8% for next year. What will the new tuition be per credit?

73. *Rural Forestland* There are about 13.6 million acres of rural forestland in Wisconsin. This is about 38% of the state's total area. Approximate the total area of Wisconsin in millions of acres. (*Source:* U.S. Department of Agriculture.)

74. *Women in Uniform* In 2009 there were 203,375 female active-duty military personal, up 0.38% from 2008. How many female military personnel were there in 2008? (*Source:* U.S. Department of Defense.)

75. **Thinking Generally** Calculate the percent change if the price of an item increases from $1.20 to $1.50. Now calculate the percent change if the price of the item decreases from $1.50 to $1.20.

76. **Thinking Generally** Refer to the previous exercise. Suppose an amount increases from A_1 to A_2 and the percent change is calculated to be 30%. If that amount now decreases from A_2 back to A_1, will the percent change be -30%?

DISTANCE PROBLEMS

Exercises 77–82: Use the formula $d = rt$ to find the value of the missing variable.

77. $r = 4$ mph, $t = 2$ hours

78. $r = 70$ mph, $t = 2.5$ hours

79. $d = 1000$ feet, $t = 50$ seconds

80. $d = 1250$ miles, $t = 5$ days

81. $d = 200$ miles, $r = 40$ mph

82. $d = 1700$ feet, $r = 10$ feet per second

83. *Driving a Car* A person drives a car at a constant speed for 4 hours and 15 minutes, traveling 255 miles. Find the speed of the car in miles per hour.

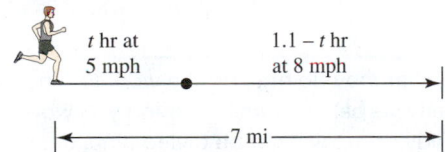

255 mi
4 hr 15 min

84. *Flying an Airplane* A pilot flies a plane at a constant speed for 5 hours and 30 minutes, traveling 715 miles. Find the speed of the plane in miles per hour.

85. *Jogging Speeds* One runner passes another runner traveling in the same direction on a hiking trail. The faster runner is jogging 2 miles per hour faster than the slower runner. Determine how long it will be before the faster runner is $\frac{3}{4}$ mile ahead of the slower runner.

86. *Distance Running* An athlete runs 8 miles, first at a slower speed and then at a faster speed. The total time spent running is 1 hour. If the athlete runs $\frac{1}{3}$ hour at 6 miles per hour, find the second speed.

87. *Jogging Speeds* At first an athlete jogs at 5 miles per hour and then jogs at 8 miles per hour, traveling 7 miles in 1.1 hours. How long does the athlete jog at each speed? (*Hint:* Let t represent the amount of time the athlete jogs at 5 mph. Then $1.1 - t$ represents the amount of time the athlete jogs at 8 mph.)

t hr at 5 mph $1.1 - t$ hr at 8 mph

7 mi

88. *Distance and Time* A bus is 160 miles east of the North Dakota–Montana border and is traveling west at 70 miles per hour. How long will it take for the bus to be 295 miles west of the border?

89. *Distance and Time* A plane is 300 miles west of Chicago, Illinois, and is flying west at 500 miles per hour. How long will it take for the plane to be 2175 miles west of Chicago?

500 mph ── ✈ ← 300 mi ─●
Chicago
2175 mi

90. *Finding Speeds* Two cars pass on a straight highway while traveling in opposite directions. One car is traveling 6 miles per hour faster than the other car. After 1.5 hours the two cars are 171 miles apart. Find the speed of each car.

OTHER TYPES OF PROBLEMS

91. *Saline Solution* (Refer to Example 11.) A solution contains 3% salt. How much water should be added to 20 ounces of this solution to make a 1.2% solution?

92. *Acid Solution* A solution contains 15% hydrochloric acid. How much water should be added to 50 milliliters of this solution to dilute it to a 2% solution?

93. *College Loans* (Refer to Example 12.) A student takes out two loans, one at 5% interest and the other at 6% interest. The 5% loan is $1000 more than the 6% loan, and the total interest for 1 year is $215. How much is each loan?

94. *Bank Loans* Two bank loans, one for $5000 and the other for $3000, cost a total of $550 in interest for one year. The $5000 loan has an interest rate 3% lower than the interest rate for the $3000 loan. Find the interest rate for each loan.

95. *Mixing Antifreeze* How many gallons of 70% antifreeze should be mixed with 10 gallons of 30% antifreeze to obtain a 45% antifreeze mixture?

96. *Mixing Antifreeze* How many gallons of 65% antifreeze and how many gallons of 20% antifreeze should be mixed to obtain 50 gallons of a 56% mixture of antifreeze? (*Hint:* Let x represent the number of gallons of 65% antifreeze. Then $50 - x$ represents the amount of 20% antifreeze.)

97. *Hydrocortisone Cream* A pharmacist needs to make a 1% hydrocortisone cream. How many grams of 2.5% hydrocortisone cream should be added to 15 grams of cream base (0% hydrocortisone) to make the 1% cream?

98. *Credit Card Debt* A person carries a balance on two credit cards, one with a monthly interest rate of 1.5% and the other with a monthly rate of 1.75%. The balance on the 1.5% card is $600 less than the balance on the 1.75% card. If the total interest for the month is $49.50, what is the balance on each card?

WRITING ABOUT MATHEMATICS

99. State the four steps for solving a word problem.

100. The cost of living has increased about 600% during the past 50 years. Does this percent change correspond to a cost of living increase of 6 times? Explain.

Group Activity   Working with Real Data

Directions: Form a group of 2 to 4 people. Select someone to record the group's responses for this activity. All members of the group should work cooperatively to answer the questions. If your instructor asks for your results, each member of the group should be prepared to respond.

Exercises 1–5: In this set of exercises you are to use your mathematical problem-solving skills to find the thickness of a piece of aluminum foil without measuring it directly.

1. *Area of a Rectangle* The area of a rectangle equals length times width. Find the area of a rectangle with length 12 centimeters and width 11 centimeters.

2. *Volume of a Box* The volume of a box equals length times width times height. Find the volume of the box shown, which is 12 centimeters long, 11 centimeters wide, and 5 centimeters high.

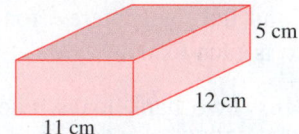

5 cm

12 cm

11 cm

3. *Height of a Box* Suppose that a box has a volume of 100 cubic centimeters and that the area of the bottom of the box is 50 square centimeters. Find the height of the box.

4. *Volume of Aluminum Foil* One cubic centimeter of aluminum weighs 2.7 grams. If a piece of aluminum foil weighs 5.4 grams, find the volume of the aluminum foil.

5. *Thickness of Aluminum Foil* A rectangular sheet of aluminum foil is 50 centimeters long and 20 centimeters wide, and weighs 5.4 grams. Find the thickness of the aluminum foil in centimeters.

2.4   Formulas

Formulas from Geometry • Solving for a Variable • Other Formulas

A LOOK INTO MATH ▶

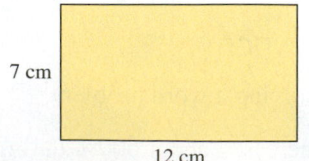

Have you ever wondered how your grade point average (GPA) is calculated? A formula is used that involves the number of credits earned at each possible grade. Once a formula has been derived, it can be used over and over by any number of people. For this reason, formulas provide a convenient and easy way to solve certain types of recurring problems. In this section we discuss several types of formulas.

STUDY TIP

If you are studying with classmates, make sure that they do not "do the work for you." A classmate with the best intentions may give too many verbal hints while helping you work through a problem. Remember that members of your study group will not be giving hints during an exam.

NEW VOCABULARY

☐ Degree
☐ Circumference

Formulas from Geometry

Formulas from geometry are frequently used in various fields, including surveying and construction. In this subsection we discuss several important formulas from geometry.

RECTANGLES If a rectangle has length l and width w, then its perimeter P and its area A are given by the formulas

$$P = 2l + 2w \quad \text{and} \quad A = lw.$$

For example, the rectangle in Figure 2.6 has length **12** centimeters and width **7** centimeters. Its perimeter is

$$P = 2(12) + 2(7) = 24 + 14 = 38 \text{ centimeters,}$$

7 cm

12 cm

Figure 2.6 Rectangle

and its area is

$$A = 12 \cdot 7 = 84 \text{ square centimeters.}$$

READING CHECK

• What do *l* and *w* stand for in formulas for rectangles?

NOTE: All measurements used in a geometry formula must have the same units.

TRIANGLES If a triangle has base *b* and height *h*, as shown in Figure 2.7(a), then its area *A* is given by the formula

$$A = \frac{1}{2}bh.$$

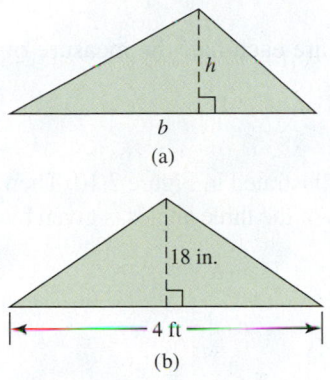

Figure 2.7 Triangles

For example, to find the area of a triangle with base **4** feet and height **18** inches as illustrated in Figure 2.7(b), we begin by making sure that all measurements have the same units. By converting 18 inches to **1.5** feet, the area of the triangle is

$$A = \frac{1}{2}bh = \frac{1}{2}(4)(1.5) = 3 \text{ square feet.}$$

If the base had been converted to **48** inches, the area is

$$A = \frac{1}{2}bh = \frac{1}{2}(48)(18) = 432 \text{ square inches.}$$

READING CHECK

• What do *b* and *h* stand for in the triangle area formula?

EXAMPLE 1   **Calculating area of a region**

A residential lot is shown in Figure 2.8. It comprises a rectangular region and an adjacent triangular region.
(a) Find the area of this lot.
(b) An acre contains 43,560 square feet. How many acres are there in this lot?

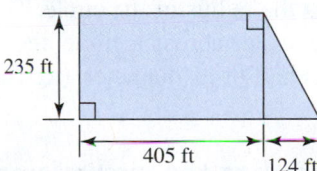

235 ft

405 ft

124 ft

Figure 2.8

Solution
(a) The rectangular portion of the lot has length **405** feet and width **235** feet. Its area A_R is

$$A_R = lw = 405 \cdot 235 = 95{,}175 \text{ square feet.}$$

The triangular region has base **124** feet and height **235** feet. Its area A_T is

$$A_T = \frac{1}{2}bh = \frac{1}{2} \cdot 124 \cdot 235 = 14{,}570 \text{ square feet.}$$

The total area *A* of the lot equals the sum of A_R and A_T.

$$A = 95{,}175 + 14{,}570 = 109{,}745 \text{ square feet}$$

(b) Each acre equals 43,560 square feet, so divide 109,745 by 43,560 to calculate the number of acres.

$$\frac{109{,}745}{43{,}560} \approx 2.5 \text{ acres}$$

Now Try Exercise 27

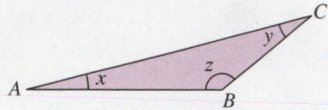

Figure 2.9

ANGLES Angles are often measured in degrees. A **degree** (°) is $\frac{1}{360}$ of a revolution, so there are 360° in one complete revolution. In any triangle, the sum of the measures of the angles equals 180°. In Figure 2.9, triangle ABC has angles with measures x, y, and z. Therefore

$$x + y + z = 180°.$$

EXAMPLE 2   Finding angles in a triangle

In a triangle the two smaller angles are equal in measure and are each half the measure of the largest angle. Find the measure of each angle.

Solution
Let x represent the measure of each of the two smaller angles, as illustrated in Figure 2.10. Then the measure of the largest angle is $2x$, and the sum of the measures of the three angles is given by

$$x + x + 2x = 180°.$$

This equation can be solved as follows.

$$x + x + 2x = 180° \qquad \text{Equation to be solved}$$
$$4x = 180° \qquad \text{Combine like terms.}$$
$$\frac{4x}{4} = \frac{180°}{4} \qquad \text{Divide each side by 4.}$$
$$x = 45° \qquad \text{Divide the real numbers.}$$

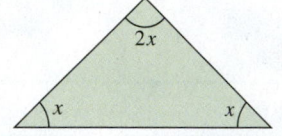

Figure 2.10

The measure of the largest angle is $2x = 2 \cdot 45° = 90°$. Thus the measures of the three angles are 45°, 45°, and 90°.

▌ **Now Try Exercise 33**

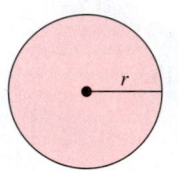

Figure 2.11 Circle

CIRCLES The *radius* of a circle is the distance from its center to the perimeter of the circle. The perimeter of a circle is called its **circumference** C and is given by $C = 2\pi r$, where r is the radius of the circle. The area A of a circle is given by $A = \pi r^2$. See Figure 2.11. The distance across a circle, through its center, is called the *diameter*. Note that a circle's radius is half of its diameter. (Recall that $\pi \approx 3.14$.)

EXAMPLE 3   Finding the circumference and area of a circle

A circle has a diameter of 25 inches. Find its circumference and area.

Solution
The radius is half the diameter, or **12.5** inches.

$$\textit{Circumference: } C = 2\pi r = 2\pi(\mathbf{12.5}) = 25\pi \approx 78.5 \text{ inches.}$$
$$\textit{Area: } A = \pi r^2 = \pi(\mathbf{12.5})^2 = 156.25\pi \approx 491 \text{ square inches.}$$

▌ **Now Try Exercise 37**

READING CHECK

• What is the approximate value of π when rounded to the nearest hundredth?

CALCULATOR HELP

To evaluate π on a calculator, see Appendix A (page AP-1).

CRITICAL THINKING

Write formulas for the circumference and area of a circle having diameter d.

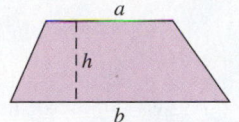

Figure 2.12 Trapezoid

TRAPEZOIDS A trapezoid such as the one shown in Figure 2.12 has bases a and b, and height h. The area A of a trapezoid is given by the formula

$$A = \frac{1}{2}(a + b)h.$$

EXAMPLE 4 **Finding the area of a trapezoid**

Find the area of the trapezoid shown in Figure 2.13.

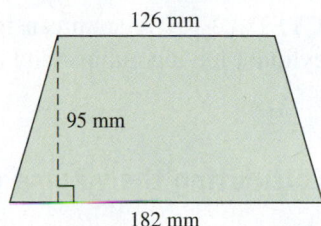

Figure 2.13

Solution

The bases of the trapezoid in Figure 2.13 are **126** millimeters and **182** millimeters, and its height is **95** millimeters. Substituting these values in the area formula gives an area of

$$A = \frac{1}{2}(a + b)h = \frac{1}{2}(\mathbf{126} + \mathbf{182}) \cdot \mathbf{95} = \frac{1}{2}(308) \cdot 95 = 14{,}630 \text{ square millimeters.}$$

Now Try Exercise 19

BOXES The box in Figure 2.14 has length l, width w, and height h. Its volume V is given by

$$V = lwh.$$

The surface of the box comprises six rectangular regions: top and bottom, front and back, and left and right sides. The total surface area S of the box is given by

$$S = lw + \quad lw \quad + wh + wh + \quad lh \quad + \quad lh.$$
$$\text{(top + bottom + front + back + left side + right side)}$$

When we combine like terms, this expression simplifies to

$$S = 2lw + 2wh + 2lh.$$

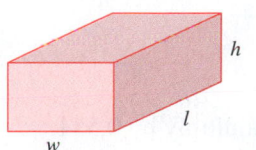

Figure 2.14 Rectangular Box

EXAMPLE 5 **Finding the volume and surface area of a box**

Find the volume and surface area of the box shown in Figure 2.15.

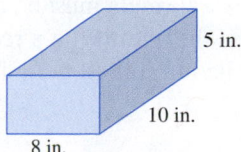

Figure 2.15

Solution

Figure 2.15 shows that the box has length $l = \mathbf{10}$ inches, width $w = \mathbf{8}$ inches, and height $h = \mathbf{5}$ inches. The volume of the box is

$$V = lwh = \mathbf{10} \cdot \mathbf{8} \cdot \mathbf{5} = 400 \text{ cubic inches.}$$

Since $l = \textcolor{red}{10}$, $w = \textcolor{red}{8}$, and $h = \textcolor{red}{5}$, the surface area of the box is

$$S = 2lw + 2wh + 2lh$$
$$= 2(\textcolor{red}{10})(\textcolor{red}{8}) + 2(\textcolor{red}{8})(\textcolor{red}{5}) + 2(\textcolor{red}{10})(\textcolor{red}{5})$$
$$= 160 + 80 + 100$$
$$= 340 \text{ square inches.}$$

Now Try Exercise 41

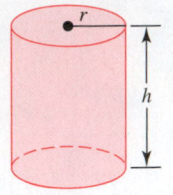

Figure 2.16 Cylinder

CYLINDERS A soup can is usually made in the shape of a cylinder. The volume of a cylinder having radius r and height h is $V = \pi r^2 h$. See Figure 2.16.

EXAMPLE 6   **Calculating the volume of a soda can**

A cylindrical soda can has a radius of $1\frac{1}{4}$ inches and a height of $4\frac{3}{8}$ inches.
(a) Find the volume of the can.
(b) If 1 cubic inch equals 0.554 fluid ounce, find the number of fluid ounces in the can.

Solution
(a) Changing mixed numbers to improper fractions gives the radius as $r = \frac{5}{4}$ inches and the height as $h = \frac{35}{8}$ inches.

$$V = \pi r^2 h \qquad \text{Volume of the soda can}$$

$$= \pi \left(\frac{5}{4}\right)^2 \left(\frac{35}{8}\right) \qquad \text{Substitute.}$$

$$= \pi \left(\frac{875}{128}\right) \qquad \text{Multiply the fractions.}$$

$$\approx 21.48 \text{ cubic inches} \qquad \text{Approximate.}$$

(b) To calculate the number of fluid ounces in 21.48 cubic inches, multiply by 0.554.

$$21.48(0.554) \approx 11.9 \text{ fluid ounces}$$

Note that a typical aluminum soda can holds 12 fluid ounces.

Now Try Exercise 49

Solving for a Variable

A formula establishes a relationship between two or more variables (or quantities). Sometimes a formula must be rewritten to solve for the needed variable. For example, if the area A and the width w of a rectangular region are given, then its length l can be found by solving the formula $A = lw$ for l.

$$A = lw \qquad \text{Area formula}$$

$$\frac{A}{w} = \frac{lw}{w} \qquad \text{Divide each side by } w.$$

$$\frac{A}{w} = l \qquad \text{Simplify the fraction.}$$

$$l = \frac{A}{w} \qquad \text{Rewrite the equation.}$$

If the area A of a rectangle is 400 square inches and its width w is 16 inches, then the rectangle's length l is

$$l = \frac{A}{w} = \frac{400}{16} = 25 \text{ inches.}$$

Once the area formula has been solved for l, the resulting formula can be used to find the length of *any* rectangle whose area and width are known.

EXAMPLE 7 **Finding the base of a trapezoid**

The area of a trapezoid is given by

$$A = \frac{1}{2}(a + b)h,$$

where a and b are the bases of the trapezoid and h is the height.
(a) Solve the formula for b.
(b) A trapezoid has area $A = 36$ square inches, height $h = 4$ inches, and base $a = 8$ inches. Find b.

Solution
(a) To clear the equation of the fraction, multiply each side by 2.

$$A = \frac{1}{2}(a + b)h \qquad \text{Area formula}$$

$$2A = (a + b)h \qquad \text{Multiply each side by 2.}$$

$$\frac{2A}{h} = a + b \qquad \text{Divide each side by } h.$$

$$\frac{2A}{h} - a = b \qquad \text{Subtract } a \text{ from each side.}$$

$$b = \frac{2A}{h} - a \qquad \text{Rewrite the formula.}$$

(b) Let $A = 36$, $h = 4$, and $a = 8$ in $b = \frac{2A}{h} - a$. Then

$$b = \frac{2(36)}{4} - 8 = 18 - 8 = 10 \text{ inches.}$$

Now Try Exercise 59

EXAMPLE 8 **Solving for a variable**

Solve each equation for the indicated variable.
(a) $c = \frac{a + b}{2}$ for b **(b)** $ab - bc = ac$ for c

Solution
(a) To clear the equation of the fraction, multiply each side by 2.

$$c = \frac{a + b}{2} \qquad \text{Given formula}$$

$$2c = a + b \qquad \text{Multiply each side by 2.}$$

$$2c - a = b \qquad \text{Subtract } a.$$

The formula solved for b is $b = 2c - a$.

(b) In $ab - bc = ac$, the variable c appears in two terms. We will combine the terms containing c by using the distributive property. Begin by moving the term on the left side containing c to the right side of the equation.

$$
\begin{aligned}
ab - bc &= ac & &\text{Given formula} \\
ab - bc + \boldsymbol{bc} &= ac + \boldsymbol{bc} & &\text{Add } bc \text{ to each side.} \\
ab &= (a + b)c & &\text{Combine terms; distributive property.} \\
\frac{ab}{a + b} &= \frac{(a + b)c}{(a + b)} & &\text{Divide each side by } (a + b). \\
\frac{ab}{a + b} &= c & &\text{Simplify the fraction.}
\end{aligned}
$$

The formula solved for c is $c = \frac{ab}{a + b}$.

Now Try Exercises 63, 67

CRITICAL THINKING

Are the formulas $c = \frac{1}{a - b}$ and $c = \frac{-1}{b - a}$ equivalent? Why?

Other Formulas

To calculate a student's GPA, the number of credits earned with a grade of A, B, C, D, and F must be known. If a, b, c, d, and f represent these credit counts respectively, then

$$
\text{GPA} = \frac{4a + 3b + 2c + d}{a + b + c + d + f}.
$$

This formula is based on the assumption that a 4.0 GPA is an A, a 3.0 GPA is a B, and so on.

EXAMPLE 9 **Calculating a student's GPA**

A student has earned 16 credits of A, 32 credits of B, 12 credits of C, 2 credits of D, and 5 credits of F. Calculate the student's GPA to the nearest hundredth.

Solution
Let $a = \boldsymbol{16}$, $b = \boldsymbol{32}$, $c = \boldsymbol{12}$, $d = \boldsymbol{2}$, and $f = \boldsymbol{5}$. Then

$$
\text{GPA} = \frac{4 \cdot \boldsymbol{16} + 3 \cdot \boldsymbol{32} + 2 \cdot \boldsymbol{12} + \boldsymbol{2}}{\boldsymbol{16} + \boldsymbol{32} + \boldsymbol{12} + \boldsymbol{2} + \boldsymbol{5}} = \frac{186}{67} \approx 2.78.
$$

The student's GPA is 2.78.

Now Try Exercise 71

EXAMPLE 10 **Converting temperature scales**

In the United States, temperature is measured with either the Fahrenheit or the Celsius temperature scale. To convert Fahrenheit degrees F to Celsius degrees C, the formula $C = \frac{5}{9}(F - 32)$ can be used.
(a) Solve the formula for F to find a formula that converts Celsius degrees to Fahrenheit degrees.
(b) If the outside temperature is $20°C$, find the equivalent Fahrenheit temperature.

Solution

(a) The reciprocal of $\frac{5}{9}$ is $\frac{9}{5}$, so multiply each side by $\frac{9}{5}$.

$$C = \frac{5}{9}(F - 32) \qquad \text{Given equation}$$

$$\frac{9}{5}C = \frac{9}{5} \cdot \frac{5}{9}(F - 32) \qquad \text{Multiply each side by } \frac{9}{5}.$$

$$\frac{9}{5}C = F - 32 \qquad \text{Multiplicative inverses}$$

$$\frac{9}{5}C + 32 = F \qquad \text{Add 32 to each side.}$$

The required formula is $F = \frac{9}{5}C + 32$.

(b) If $C = \mathbf{20}°$C, then $F = \frac{9}{5}(\mathbf{20}) + 32 = 36 + 32 = 68°$F.

▎**Now Try Exercise 75**

2.4	**Putting It All Together**

In this section we discussed formulas and how to solve for a variable. The following table summarizes some of these formulas.

CONCEPT	FORMULA	EXAMPLES
Area and Perimeter of a Rectangle	$A = lw$ and $P = 2l + 2w$, where l is the length and w is the width.	If $l = 10$ feet and $w = 5$ feet, then $A = 10 \cdot 5 = 50$ square feet and $P = 2(10) + 2(5) = 30$ feet. 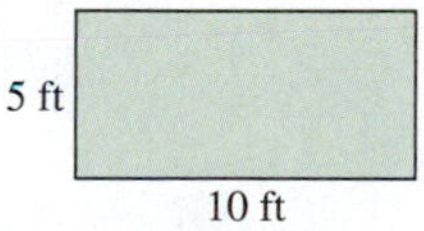 5 ft ▢ 10 ft
Area of a Triangle	$A = \frac{1}{2}bh$, where b is the base and h is the height.	If $b = 5$ inches and $h = 6$ inches, then the area is $A = \frac{1}{2}(5)(6) = 15$ square inches. 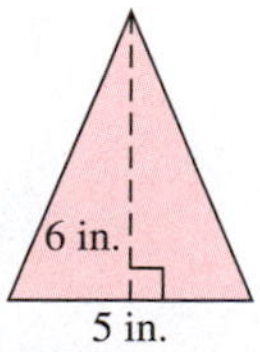 6 in. 5 in.
Angle Measure in a Triangle	$x + y + z = 180°$, where x, y, and z are the angle measures.	If $x = 40°$ and $y = 60°$, then $z = 80°$ because $40° + 60° + 80° = 180°$ 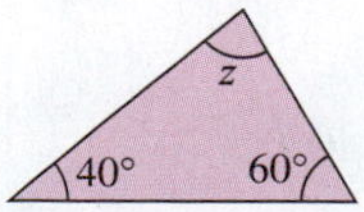 z 40° 60°

continued on next page

continued from previous page

CONCEPT	FORMULA	EXAMPLES
Circumference and Area of a Circle	If a circle has radius r, then its circumference is $$C = 2\pi r$$ and its area is $$A = \pi r^2.$$	If $r = 6$ inches, $C = 2\pi(6) = 12\pi \approx 37.7$ inches and $A = \pi(6)^2 = 36\pi \approx 113.1$ square inches. 6 in.
Area of a Trapezoid	$$A = \tfrac{1}{2}(a + b)h,$$ where a and b are the bases and h is the height.	If $a = 4$, $b = 6$, and $h = 3$, then the area is $$A = \frac{1}{2}(4 + 6)(3) = 15 \text{ square units.}$$ 4 3 6
Volume and Surface Area of a Box	If a box has length l, width w, and height h, then its volume is $$V = lwh$$ and its surface area is $$S = 2lw + 2wh + 2lh.$$	If a box has dimensions $l = 4$ feet, $w = 3$ feet, and $h = 2$ feet, then $$V = 4(3)(2) = 24 \text{ cubic feet}$$ and $$S = 2(4)(3) + 2(3)(2) + 2(4)(2)$$ $$= 52 \text{ square feet.}$$ 2 ft 4 ft 3 ft
Volume of a Cylinder	$$V = \pi r^2 h,$$ where r is the radius and h is the height.	If $r = 5$ inches and $h = 20$ inches, then the volume is $$V = \pi(5^2)(20) = 500\pi \text{ cubic inches.}$$ 5 in. 20 in.
Calculating Grade Point Average (GPA)	GPA is calculated by $$\frac{4a + 3b + 2c + d}{a + b + c + d + f},$$ where a, b, c, d, and f represent the number of credits earned with grades of A, B, C, D, and F, respectively.	10 credits of A, 8 credits of B, 6 credits of C, 12 credits of D, and 8 credits of F results in a GPA of $$\frac{4(10) + 3(8) + 2(6) + 12}{10 + 8 + 6 + 12 + 8} = 2.0.$$

CONCEPT	FORMULA	EXAMPLES
Converting Between Fahrenheit and Celsius Degrees	$C = \dfrac{5}{9}(F - 32)$ $F = \dfrac{9}{5}C + 32$	212°F is equivalent to $C = \dfrac{5}{9}(212 - 32) = 100°C.$ 100°C is equivalent to $F = \dfrac{9}{5}(100) + 32 = 212°F.$

2.4 Exercises

CONCEPTS AND VOCABULARY

1. A(n) _____ can be used to calculate one quantity by using known values of other quantities.

2. The area A of a rectangle with length l and width w is $A = $ _____.

3. The area A of a triangle with base b and height h is $A = $ _____.

4. One degree equals _____ of a revolution.

5. There are _____ degrees in one revolution.

6. The sum of the measures of the angles in a triangle equals _____ degrees.

7. The volume V of a box with length l, width w, and height h is $V = $ _____.

8. The surface area S of a box with length l, width w, and height h is $S = $ _____.

9. The circumference C of a circle with radius r is $C = $ _____.

10. The area A of a circle with radius r is $A = $ _____.

11. The volume V of a cylinder with radius r and height h is $V = $ _____.

12. The area A of a trapezoid with height h and bases a and b is $A = $ _____.

FORMULAS FROM GEOMETRY

Exercises 13–20: Find the area of the region shown.

13.

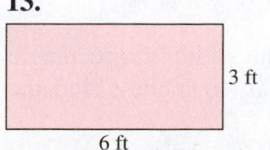

6 ft · 3 ft

14.

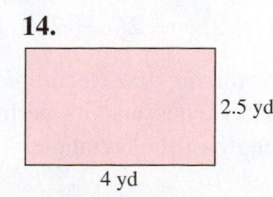

4 yd · 2.5 yd

15.

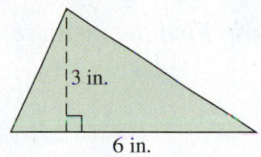

3 in. · 6 in.

16.

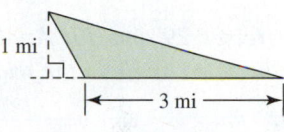

1 mi · 3 mi

17.

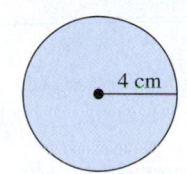

4 cm

18.

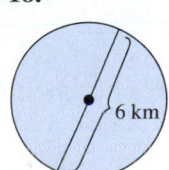

6 km

19.

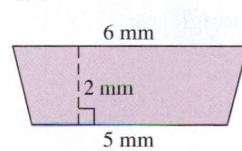

6 mm, 2 mm, 5 mm

20.

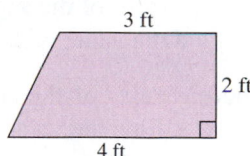

3 ft, 2 ft, 4 ft

21. Find the area of a rectangle having a 7-inch width and a 13-inch length.

22. Find the area of a triangle having a 9-centimeter base and a 72-centimeter height.

23. Find the area of a rectangle having a 5-foot width and a 7-yard length.

24. Find the area of a triangle having a 12-millimeter base and a 6-millimeter height.

25. Find the circumference of a circle having an 8-inch diameter.

26. Find the area of a circle having a 9-foot radius.

27. *Area of a Lot* Find the area of the lot shown, which consists of a square and a triangle.

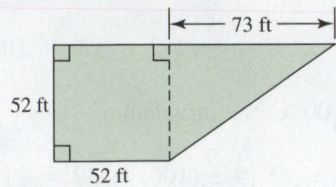

28. *Area of a Lot* Find the area of the lot shown, which consists of a rectangle and two triangles.

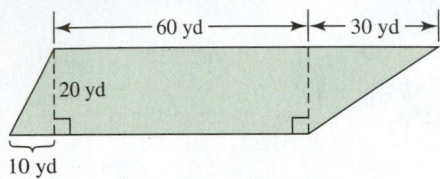

Exercises 29 and 30: Angle Measure Find the measure of the third angle in the triangle.

29.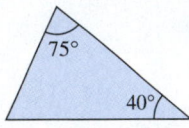

30.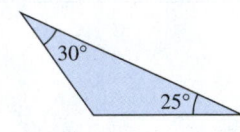

31. A triangle contains two angles having measures of 23° and 76°. Find the measure of the third angle.

32. The measures of the angles in an *equilateral triangle* are equal. Find their measure.

33. The measures of the angles in a triangle are x, $2x$, and $2x$. Find the value of x.

34. The measures of the angles in a triangle are $3x$, $4x$, and $11x$. Find the value of x.

35. In a triangle the two smaller angles are equal in measure and are each one third of the measure of the larger angle. Find the measure of each angle.

36. In a triangle the two larger angles differ by 10°. The smaller angle is 50° less than the largest angle. Find the measure of each angle.

37. The diameter of a circle is 12 inches. Find its circumference and area.

38. The radius of a circle is $\frac{5}{4}$ feet. Find its circumference and area.

39. The circumference of a circle is 2π inches. Find its radius and area.

40. The circumference of a circle is 13π feet. Find its radius and area.

Exercises 41–44: A box with a top has length l, width w, and height h. Find the volume and surface area of the box.

41. $l = 22$ inches, $w = 12$ inches, $h = 10$ inches

42. $l = 5$ feet, $w = 3$ feet, $h = 6$ feet

43. $l = \frac{2}{3}$ yard, $w = \frac{2}{3}$ foot, $h = \frac{3}{2}$ feet

44. $l = 1.2$ meters, $w = 0.8$ meter, $h = 0.6$ meter

Exercises 45–48: Use the formula $V = \pi r^2 h$ to find the volume of a cylindrical container for the given r and h. Leave your answer in terms of π.

45. $r = 2$ inches, $h = 5$ inches

46. $r = \frac{1}{2}$ inch, $h = \frac{3}{2}$ inches

47. $r = 5$ inches, $h = 2$ feet

48. $r = 2.5$ feet, $h = 1.5$ yards

49. *Volume of a Can* (Refer to Example 6.)
 (a) Find the volume of a can with a radius of $\frac{3}{4}$ inch and a height of $2\frac{1}{2}$ inches.
 (b) Find the number of fluid ounces in the can if one cubic inch equals 0.554 fluid ounces.

50. *Volume of a Barrel* (Refer to Example 6.) A cylindrical barrel has a diameter of $1\frac{3}{4}$ feet and a height of 3 feet. Find the volume of the barrel.

SOLVING FOR A VARIABLE

Exercises 51–68: Solve the formula for the given variable.

51. $9x + 3y = 6$ for y **52.** $-2x - 2y = 10$ for y

53. $4x + 3y = 12$ for y **54.** $5x - 2y = 22$ for y

55. $A = lw$ for w **56.** $A = \frac{1}{2}bh$ for b

57. $V = \pi r^2 h$ for h **58.** $V = \frac{1}{3}\pi r^2 h$ for h

59. $A = \frac{1}{2}(a + b)h$ for a **60.** $C = 2\pi r$ for r

61. $V = lwh$ for w **62.** $P = 2l + 2w$ for w

63. $s = \dfrac{a + b + c}{2}$ for b **64.** $t = \dfrac{x - y}{3}$ for x

65. $\dfrac{a}{b} - \dfrac{c}{b} = 1$ for b **66.** $\dfrac{x}{y} + \dfrac{z}{y} = 5$ for z

67. $ab = cd + ad$ for a

68. $S = 2lw + 2lh + 2wh$ for w

69. *Perimeter of a Rectangle* If the width of a rectangle is 5 inches and its perimeter is 40 inches, find the length of the rectangle.

70. *Perimeter of a Triangle* Two sides of a triangle have lengths of 5 feet and 7 feet. If the triangle's perimeter is 21 feet, what is the length of the third side?

OTHER FORMULAS AND APPLICATIONS

Exercises 71–74: (Refer to Example 9.) Let the variable a represent the number of credits with a grade of A, b the number of credits with a grade of B, and so on. Calculate the corresponding grade point average (GPA). Round your answer to the nearest hundredth.

71. $a = 30, b = 45, c = 12, d = 4, f = 4$

72. $a = 70, b = 35, c = 5, d = 0, f = 0$

73. $a = 0, b = 60, c = 80, d = 10, f = 6$

74. $a = 3, b = 5, c = 8, d = 0, f = 22$

Exercises 75–78: (Refer to Example 10.) Convert the Celsius temperature to an equivalent Fahrenheit temperature.

75. $25°C$ **76.** $100°C$ **77.** $-40°C$ **78.** $0°C$

Exercises 79–82: (Refer to Example 10.) Convert the Fahrenheit temperature to an equivalent Celsius temperature.

79. $23°F$ **80.** $98.6°F$ **81.** $-4°F$ **82.** $-31°F$

83. *Gas Mileage* The formula $M = \frac{D}{G}$ can be used to calculate a car's gas mileage M after it has traveled D miles on G gallons of gasoline. Suppose that a truck driver leaves a gas station with a full tank of gas and the odometer showing 87,625 miles. At the next gas stop, it takes 38 gallons to fill the tank and the odometer reads 88,043 miles. Find the gas mileage for the truck.

84. *Gas Mileage* (Refer to Exercise 83.) A car that gets 34 miles per gallon is driven 578 miles. How many gallons of gasoline are used on this trip?

85. *Lightning* If there is an x-second delay between seeing a flash of lightning and hearing the thunder, then the lightning is $D = \frac{x}{5}$ miles away. Suppose that the time delay between a flash of lightning and the sound of thunder is 12 seconds. How far away is the lightning?

86. *Lightning* (Refer to Exercise 85.) Doppler radar shows an electrical storm 2.5 miles away. If you see the lightning from this storm, how long will it be before you hear the thunder?

WRITING ABOUT MATHEMATICS

87. A student solves the formula $A = \frac{1}{2}bh$ for h and obtains the formula $h = \frac{1}{2}bA$. Explain the error that the student is making. What is the correct answer?

88. Give an example of a formula that you have used and explain how you used it.

SECTIONS 2.3 and 2.4   **Checking Basic Concepts**

1. Translate the sentence into an equation containing the variable x. Then solve the resulting equation.
   (a) The product of 3 and a number is 36.
   (b) A number subtracted from 35 is 43.

2. When three consecutive integers are added, the sum is -93. Find the three integers.

3. Convert 9.5% to a decimal.

4. Convert $\frac{5}{4}$ to a percentage.

5. *Violent Crime* In 2009, New York City experienced 46,357 violent crimes. This figure represented a 38.8% decrease from the number in 2000. Find the number of violent crimes in 2000. (*Source: New York State Crime Update.*)

6. *Driving a Car* How many hours does it take the driver of a car to travel 390 miles at 60 miles per hour?

7. *College Loans* A student takes out two loans, one at 6% and the other at 7%. The 6% loan is $2000 more than the 7% loan, and the total interest for one year is $510. Find the amount of each loan.

8. *Gas Mileage* A car that gets 28 miles per gallon is driven 504 miles. How many gallons of gasoline are used on this trip?

9. *Height of a Triangle* The area of a triangle having a base of 6 inches is 36 square inches. Find the height of the triangle.

10. *Area and Circumference* Find the area and circumference of the circle shown.

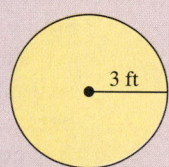

11. *Angles* Find the value of x in the triangle shown.

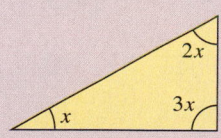

12. *Solving a Formula* Solve $A = \pi r^2 + \pi r l$ for l.

2.5   Linear Inequalities

Solutions and Number Line Graphs • The Addition Property of Inequalities •
The Multiplication Property of Inequalities • Applications

A LOOK INTO MATH ▶ At an amusement park, a particular ride might be restricted to people at least 48 inches tall. A child who is x inches tall may go on the ride if $x \geq 48$ but may not go on the ride if $x < 48$. A height of 48 inches represents the *boundary* between being allowed on the ride and being denied access to the ride.

Solving linear inequalities is closely related to solving linear equations because equality is the boundary between *greater than* and *less than*. In this section we discuss techniques used to solve linear inequalities.

Solutions and Number Line Graphs

A **linear inequality** results whenever the equals sign in a linear equation is replaced with any one of the symbols $<, \leq, >,$ or $\geq$. Examples of linear equations include

$$x = 5, \quad 2x + 1 = 0, \quad 1 - x = 6, \quad \text{and} \quad 5x + 1 = 3 - 2x.$$

Therefore examples of linear inequalities include

$$x > 5, \quad 2x + 1 < 0, \quad 1 - x \geq 6, \quad \text{and} \quad 5x + 1 \leq 3 - 2x.$$

Table 2.9 shows how each of the inequality symbols is read.

NEW VOCABULARY

☐ Linear inequality
☐ Solution
☐ Solution set
☐ Interval notation
☐ Set-builder notation

TABLE 2.9 Inequality Symbols

Symbol	How the Symbol Is Read
$>$	greater than
$<$	less than
$\geq$	greater than or equal to
$\leq$	less than or equal to

A **solution** to an inequality is a value of the variable that makes the statement true. The set of all solutions is called the **solution set**. Two inequalities are *equivalent* if they have the same solution set. Inequalities often have infinitely many solutions. For example, the solution set for the inequality $x > 5$ includes all real numbers greater than 5.

A number line can be used to graph the solution set for an inequality. The graph of all real numbers satisfying $x < 2$ is shown in Figure 2.17(a), and the graph of all real numbers satisfying $x \leq 2$ is shown in Figure 2.17(b). A parenthesis ")" is used to show that 2 is not included in Figure 2.17(a), and a bracket "]" is used to show that 2 is included in Figure 2.17(b).

READING CHECK

• How many solutions do inequalities have?

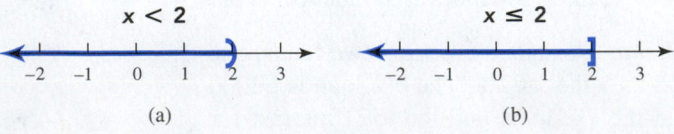

Figure 2.17

EXAMPLE 1 Graphing inequalities on a number line

Use a number line to graph the solution set to each inequality.
(a) $x > 0$ (b) $x \geq 0$ (c) $x \leq -1$ (d) $x < 3$

Solution

(a) First locate $x = 0$ (or the origin) on a number line. Numbers greater than 0 are located to the right of the origin, so shade the number line to the right of the origin. Because the inequality is $x > 0$, the number 0 is not included, so place a parenthesis "(" at 0, as shown in Figure 2.18(a).

(b) Figure 2.18(b) is similar to the graph in part (a) except that a bracket "[" is placed at the origin because the inequality symbol is $\geq$ and 0 is included in the solution set.

(c) First locate $x = -1$ on the number line. Numbers less than -1 are located to the left of -1. Because -1 is included, a bracket "]" is placed at -1, as shown in Figure 2.18(c).

(d) Real numbers less than 3 are graphed in Figure 2.18(d).

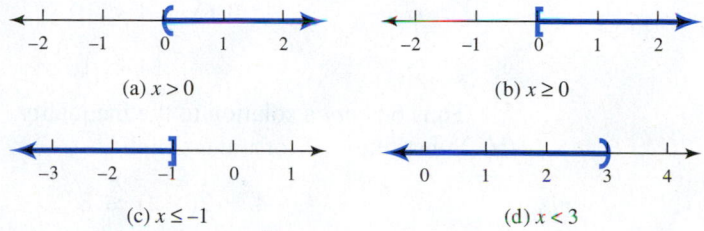

(a) $x > 0$ (b) $x \geq 0$

(c) $x \leq -1$ (d) $x < 3$

Figure 2.18

Now Try Exercises 11, 13, 15, 17

INTERVAL NOTATION (OPTIONAL) The solution sets graphed in Figure 2.18 can also be represented in a convenient notation called **interval notation**. Rather than draw the entire number line, we can use brackets or parentheses to indicate the interval of values that represent the solution set. For example, the solution set shown in Figure 2.18(a) can be represented by the interval $(0, \infty)$, and the solution set shown in Figure 2.18(b) can be represented by the interval $[0, \infty)$. The symbol ∞ refers to infinity and is used to indicate that the values increase without bound. Similarly, $-\infty$ can be used when the values decrease without bound. The solution sets shown in Figure 2.18(c) and (d) can be represented by $(-\infty, -1]$ and $(-\infty, 3)$, respectively.

EXAMPLE 2 Writing solution sets in interval notation

Write the solution set to each inequality in interval notation.
(a) $x > 4$ (b) $y \leq -3$ (c) $z \geq -1$

Solution

(a) Real numbers greater than 4 are represented by the interval $(4, \infty)$.
(b) Real numbers less than or equal to -3 are represented by the interval $(-\infty, -3]$.
(c) The solution set is represented by the interval $[-1, \infty)$.

Now Try Exercises 25, 27, 29

CHECKING SOLUTIONS We can check possible solutions to an inequality in the same way that we checked possible solutions to an equation. For example, to check whether 5 is a solution to the equation $2x + 3 = 13$, we substitute **5** for x in the equation.

$$2(5) + 3 \stackrel{?}{=} 13 \quad \text{Replace } x \text{ with 5.}$$

$$13 = 13 \quad \text{A true statement}$$

Thus 5 is a solution to this equation. Similarly, to check whether 7 is a solution to the inequality $2x + 3 > 13$, we substitute **7** for x in the inequality.

$$2(\mathbf{7}) + 3 \overset{?}{>} 13 \qquad \text{Replace } x \text{ with 7.}$$
$$17 > 13 \checkmark \qquad \text{A true statement}$$

Thus 7 is a solution to the inequality.

EXAMPLE 3 ## Checking possible solutions

Determine whether the given value of x is a solution to the inequality.
(a) $3x - 4 < 10, \quad x = 6$ **(b)** $4 - 2x \leq 8, \quad x = -2$

Solution
(a) Substitute **6** for x and simplify.

$$3(\mathbf{6}) - 4 \overset{?}{<} 10 \qquad \text{Replace } x \text{ with 6.}$$
$$14 < 10 \; \textbf{X} \qquad \text{A false statement}$$

Thus 6 is *not* a solution to the inequality.
(b) Substitute $-\mathbf{2}$ for x and simplify.

$$4 - 2(-\mathbf{2}) \overset{?}{\leq} 8 \qquad \text{Replace } x \text{ with } -2.$$
$$8 \leq 8 \checkmark \qquad \text{A true statement}$$

Thus -2 is a solution to the inequality.

▪ **Now Try Exercises 33, 35**

Just as solving a linear equation means finding the value of the variable that makes the equation true, solving a linear inequality means finding the *values* of the variable that make the inequality true.

Making a table is an organized way of checking possible values of the variable to see if there are values that make an inequality true. In the next example, we use a table to find solutions to an equation and related inequalities.

EXAMPLE 4 ## Finding solutions to equations and inequalities

In Table 2.10 the expression $2x - 6$ has been evaluated for several values of x. Use the table to determine any solutions to each equation or inequality.
(a) $2x - 6 = 0$ **(b)** $2x - 6 > 0$ **(c)** $2x - 6 \geq 0$ **(d)** $2x - 6 < 0$

TABLE 2.10

x	0	1	2	**3**	4	5	6
$2x - 6$	-6	-4	-2	**0**	2	4	6

Solution
(a) From Table 2.10, $2x - 6$ equals **0** when $x = \mathbf{3}$.
(b) The values of x in the table that make the expression $2x - 6$ greater than 0 are 4, 5, and 6. These values are all greater than 3, which is the solution found in part (a). It follows that $2x - 6 > 0$ when $x > 3$.
(c) The values of x in the table that make the expression $2x - 6$ greater than or equal to 0 are 3, 4, 5, and 6. It follows that $2x - 6 \geq 0$ when $x \geq 3$.
(d) The expression $2x - 6$ is less than 0 when $x < 3$.

▪ **Now Try Exercise 39**

The Addition Property of Inequalities

▶ **REAL-WORLD CONNECTION** Suppose that the speed limit on a country road is 55 miles per hour, and this is 25 miles per hour faster than the speed limit in town. If x represents lawful speeds in town, then x satisfies the inequality

$$x + 25 \leq 55.$$

To solve this inequality we **add** -25 to (or subtract 25 from) each side of the inequality.

$$x + 25 + (-25) \leq 55 + (-25) \qquad \text{Add } -25 \text{ to each side.}$$
$$x \leq 30 \qquad \text{Add the real numbers.}$$

Thus drivers are obeying the speed limit in town when they travel at 30 miles per hour or less. To solve this inequality the addition property of inequalities was used.

ADDITION PROPERTY OF INEQUALITIES

Let a, b, and c be expressions that represent real numbers. The inequalities

$$a < b \quad \text{and} \quad a + c < b + c$$

are equivalent. That is, the same number may be added to (or subtracted from) each side of an inequality. Similar properties exist for $>$, $\leq$, and $\geq$.

To solve some inequalities, we apply the addition property of inequalities to obtain a simpler, equivalent inequality.

EXAMPLE 5 **Applying the addition property of inequalities**

Solve each inequality. Then graph the solution set.
(a) $x - 1 > 4$ **(b)** $3 + 2x \leq 5 + x$

Solution
(a) Begin by adding 1 to each side of the inequality.

$$x - 1 > 4 \qquad \text{Given inequality}$$
$$x - 1 + 1 > 4 + 1 \qquad \text{Add 1 to each side.}$$
$$x > 5 \qquad \text{Add the real numbers.}$$

The solution set is given by $x > 5$ and is graphed as follows.

$$\xleftarrow{\hspace{1cm}} \overset{\substack{-2 \quad 0 \quad 2 \quad 4 \quad 6 \quad 8}}{\underset{}{\rule{4cm}{0.4pt}}} \xrightarrow{\hspace{1cm}}$$

(b) Begin by subtracting x from (or adding $-x$ to) each side of the inequality.

$$3 + 2x \leq 5 + x \qquad \text{Given inequality}$$
$$3 + 2x - x \leq 5 + x - x \qquad \text{Subtract } x \text{ from each side.}$$
$$3 + x \leq 5 \qquad \text{Combine like terms.}$$
$$3 + x - 3 \leq 5 - 3 \qquad \text{Subtract 3 from each side.}$$
$$x \leq 2 \qquad \text{Subtract the real numbers.}$$

The solution set is given by $x \leq 2$ and is graphed as follows.

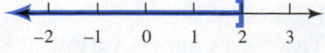

Now Try Exercise 47

MAKING CONNECTIONS

The addition property of inequalities can be illustrated by an old-fashioned pan balance. In the figure on the left, the weight on the right pan is heavier than the weight on the left because the right pan rests lower than the left pan. If we add an equal amount of weight (blue in the figure) to both pans, or if we subtract an equal amount of weight from both pans, the pan on the right still weighs more than the pan on the left by the same amount that it previously did.

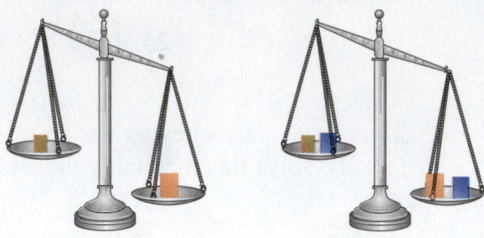

EXAMPLE 6 ▶ **Applying the addition property of inequalities**

Solve $5 + \frac{1}{2}x \le 3 - \frac{1}{2}x$. Then graph the solution set.

Solution
Begin by subtracting 5 from each side of the inequality.

$$5 + \frac{1}{2}x \le 3 - \frac{1}{2}x \qquad \text{Given inequality}$$

$$5 + \frac{1}{2}x - 5 \le 3 - \frac{1}{2}x - 5 \qquad \text{Subtract 5 from each side.}$$

$$\frac{1}{2}x \le -\frac{1}{2}x - 2 \qquad \text{Subtract real numbers.}$$

$$\frac{1}{2}x + \frac{1}{2}x \le -\frac{1}{2}x + \frac{1}{2}x - 2 \qquad \text{Add } \frac{1}{2}x \text{ to each side.}$$

$$x \le -2 \qquad \text{Combine like terms.}$$

The solution set is given by $x \le -2$ and is graphed as follows.

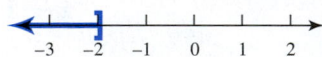

Now Try Exercise 53

STUDY TIP

If you miss something in class, section video lectures provide a short lecture for each section in this text. These lectures, presented by actual math instructors, offer you the opportunity to review topics that you may not have fully understood before. These videos are available as streaming videos and video podcasts within MyMathLab.

The Multiplication Property of Inequalities

The multiplication property of inequalities differs from the multiplication property of *equality*. When we multiply each side of an inequality by the same nonzero number, we may need to reverse the inequality symbol to make sure that the resulting inequality remains true. Table 2.11 shows various results that occur when each side of a true inequality is multiplied by the same nonzero number.

TABLE 2.11 Determining When the Inequality Symbol Should Be Reversed

True Statement	Multiply Each Side By	Resulting Inequality	Is the Result True or False?	Reverse the Inequality Symbol
$-3 < 5$	4	$-12 \overset{?}{<} 20$	True	Not needed
$7 > -1$	-2	$-14 \overset{?}{>} 2$	False	$-14 < 2$
$-2 > -5$	3	$-6 \overset{?}{>} -15$	True	Not needed
$4 < 9$	-11	$-44 \overset{?}{<} -99$	False	$-44 > -99$

Table 2.11 indicates that the inequality symbol must be reversed when each side of the given inequality is *multiplied by a negative number*. This result is summarized below.

MULTIPLICATION PROPERTY OF INEQUALITIES

Let a, b, and c be expressions that represent real numbers with $c \neq 0$.

1. If $c > 0$, then the inequalities $a < b$ and $ac < bc$ are equivalent. That is, each side of an inequality may be multiplied (or divided) by the same positive number.
2. If $c < 0$, then the inequalities $a < b$ and $ac > bc$ are equivalent. That is, each side of an inequality may be multiplied (or divided) by the same negative number, provided the inequality symbol is reversed.

Note that similar properties exist for $\leq$ and $\geq$.

NOTE: Remember to reverse the inequality symbol when either multiplying or *dividing* by a negative number.

READING CHECK

- When solving an inequality, when does it become necessary to reverse the inequality symbol?

EXAMPLE 7 **Applying the multiplication property of inequalities**

Solve each inequality. Then graph the solution set.
(a) $3x < 18$ **(b)** $-7 \leq -\frac{1}{2}x$

Solution
(a) To solve for x, divide each side by 3.

$$3x < 18 \qquad \text{Given inequality}$$

$$\frac{3x}{3} < \frac{18}{3} \qquad \text{Divide each side by 3.}$$

$$x < 6 \qquad \text{Simplify fractions.}$$

The solution set is given by $x < 6$ and is graphed as follows.

(b) To isolate x in $-7 \leq -\frac{1}{2}x$, multiply each side by -2 and reverse the inequality symbol.

$$-7 \leq -\frac{1}{2}x \qquad \text{Given inequality}$$

$$-2(-7) \geq -2\left(-\frac{1}{2}\right)x \qquad \text{Multiply by } -2; \text{ reverse the inequality.}$$

$$14 \geq 1 \cdot x \qquad \text{Multiply the real numbers.}$$

$$x \leq 14 \qquad \text{Rewrite the inequality.}$$

The solution set is given by $x \leq 14$ and is graphed as follows.

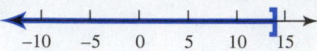

Now Try Exercises 55, 57

SET-BUILDER NOTATION Because $x \leq 14$ is an inequality with infinitely many solutions and is not itself a set of solutions, a notation called **set-builder notation** has been devised for writing the solutions to an inequality as a set. For example, the solution set consisting of "all real numbers x such that x is less than or equal to 14" can be written as $\{x \mid x \leq 14\}$. The vertical line segment "$\mid$" is read "such that."

In the next example, the solution sets are expressed in set-builder notation. However, this notation is not widely used throughout this text.

EXAMPLE 8 Applying both properties of inequalities

Solve each inequality. Write the solution set in set-builder notation.
(a) $4x - 7 \geq -6$ **(b)** $-8 + 4x \leq 5x + 3$ **(c)** $0.4(2x - 5) < 1.1x + 2$

Solution
(a) Start by adding 7 to each side.

$$4x - 7 \geq -6 \qquad \text{Given inequality}$$

$$4x - 7 + 7 \geq -6 + 7 \qquad \text{Add 7 to each side.}$$

$$4x \geq 1 \qquad \text{Add real numbers.}$$

$$\frac{4x}{4} \geq \frac{1}{4} \qquad \text{Divide each side by 4.}$$

$$x \geq \frac{1}{4} \qquad \text{Simplify.}$$

In set-builder notation, the solution set is $\left\{x \mid x \geq \frac{1}{4}\right\}$.
(b) Begin by adding 8 to each side.

$$-8 + 4x \leq 5x + 3 \qquad \text{Given inequality}$$

$$-8 + 4x + 8 \leq 5x + 3 + 8 \qquad \text{Add 8 to each side.}$$

$$4x \leq 5x + 11 \qquad \text{Add real numbers.}$$

$$4x - 5x \leq 5x + 11 - 5x \qquad \text{Subtract } 5x \text{ from each side.}$$

$$-x \leq 11 \qquad \text{Combine like terms.}$$

$$-1 \cdot (-x) \geq -1 \cdot 11 \qquad \text{Multiply by } -1; \text{ reverse the inequality.}$$

$$x \geq -11 \qquad \text{Simplify.}$$

The solution set is $\{x \mid x \geq -11\}$.

CRITICAL THINKING

Solve
$-5 - 3x > -2x + 7$ without having to reverse the inequality symbol.

(c) Begin by applying the distributive property.

$$0.4(2x - 5) < 1.1x + 2 \quad \text{Given inequality}$$
$$0.8x - 2 < 1.1x + 2 \quad \text{Distributive property}$$
$$0.8x < 1.1x + 4 \quad \text{Add 2 to each side.}$$
$$-0.3x < 4 \quad \text{Subtract } 1.1x \text{ from each side.}$$
$$x > -\frac{4}{0.3} \quad \text{Divide by } -0.3; \text{ reverse the inequality.}$$

Since $-\frac{4}{0.3} = -\frac{40}{3}$, the solution set is $\left\{ x \mid x > -\frac{40}{3} \right\}$.

■ **Now Try Exercises** 67, 77, 85

Applications

To solve applications involving inequalities, we often have to translate words or phrases to mathematical statements. Table 2.12 lists words and phrases that are associated with each inequality symbol.

TABLE 2.12 Words and Phrases Associated with Inequality Symbols

Symbol	Associated Words and Phrases
$>$	greater than, more than, exceeds, above, over
$<$	less than, fewer than, below, under
$\geq$	greater than or equal to, at least, is not less than
$\leq$	less than or equal to, at most, does not exceed

READING CHECK

• Give two phrases associated with each inequality symbol.

EXAMPLE 9 **Translating words to inequalities**

Translate each phrase to an inequality. Let the variable be x.
(a) A number that is more than 30
(b) An age that is at least 18
(c) A grade point average that is at most 3.25

Solution
(a) The inequality $x > 30$ represents a number x that is more than 30.
(b) The inequality $x \geq 18$ represents an age x that is at least 18.
(c) The inequality $x \leq 3.25$ represents a grade point average x that is at most 3.25.

■ **Now Try Exercises** 95, 101

▶ **REAL-WORLD CONNECTION** In the atmosphere the air temperature generally becomes colder as the altitude increases. One mile above Earth's surface the temperature is about $19°F$ colder than the ground-level temperature. As the air cools, there is an increased chance of clouds forming. In the next example we estimate the altitudes where clouds may form. (**Source:** A. Miller and R. Anthes, *Meteorology.*)

EXAMPLE 10 **Finding the altitude of clouds**

If the ground temperature is $79°F$, then the temperature T above Earth's surface is given by the formula $T = 79 - 19x$, where x is the altitude in miles. Suppose that clouds form only where the temperature is $3°F$ or colder. Determine the heights at which clouds may form.

Solution

Clouds may form at altitudes at which $T \le 3$ degrees. Since $T = 79 - 19x$, we can substitute the expression $79 - 19x$ for T to obtain the inequality $79 - 19x \le 3$.

$$79 - 19x \le 3 \qquad \text{Inequality to be solved}$$

$$-19x \le -76 \qquad \text{Subtract 79 from each side.}$$

$$\frac{-19x}{-19} \ge \frac{-76}{-19} \qquad \text{Divide by } -19 \text{; reverse the inequality.}$$

$$x \ge 4 \qquad \text{Simplify the fractions.}$$

Clouds may form at 4 miles or higher.

▌ **Now Try Exercise 115**

EXAMPLE 11 **Calculating revenue, cost, and profit**

For a computer company, the cost to produce one laptop computer (variable cost) is $1320 plus a one-time cost (fixed cost) of $200,000 for research and development. The revenue received from selling one laptop computer is $1850.

(a) Write a formula that gives the cost C of producing x laptop computers.
(b) Write a formula that gives the revenue R from selling x laptop computers.
(c) Profit equals revenue minus cost. Write a formula that calculates the profit P from selling x laptop computers.
(d) How many computers need to be sold to yield a positive profit?

Solution

(a) The cost of producing the first laptop is

$$1320 \times 1 + 200{,}000 = \$201{,}320.$$

The cost of producing two laptops is

$$1320 \times 2 + 200{,}000 = \$202{,}640.$$

And, in general, the cost of producing x laptops is

$$1320 \times x + 200{,}000 = 1320x + 200{,}000.$$

Thus $C = 1320x + 200{,}000$.

(b) Because the company receives $1850 for each laptop, the revenue for x laptops is given by $R = 1850x$.

(c) Profit equals revenue minus cost, so

$$P = R - C$$
$$= 1850x - (1320x + 200{,}000)$$
$$= 530x - 200{,}000.$$

Thus $P = 530x - 200{,}000$.

(d) To determine how many laptops need to be sold to yield a positive profit, we must solve the inequality $P > 0$.

$$530x - 200{,}000 > 0 \qquad \text{Inequality to be solved}$$

$$530x > 200{,}000 \qquad \text{Add 200,000 to each side.}$$

$$x > \frac{200{,}000}{530} \qquad \text{Divide each side by 530.}$$

Because $\frac{200{,}000}{530} \approx 377.4$, the company must sell at least 378 laptops. Note that the company cannot sell a fraction of a laptop.

▌ **Now Try Exercise 111**

2.5 Putting It All Together

CONCEPT	COMMENTS	EXAMPLES
Linear Inequality	If the equals sign in a linear equation is replaced with $<$, $>$, $\leq$, or $\geq$, a linear inequality results.	*Linear Equation* *Linear Inequality* $4x - 1 = 0$ $4x - 1 > 0$ $2 - x = 3x$ $2 - x \leq 3x$ $4(x + 3) = 1 - x$ $4(x + 3) < 1 - x$ $-6x + 3 = 5$ $-6x + 3 \geq 5$
Solution to an Inequality	A value for the variable that makes the inequality a true statement	6 is a solution to $2x > 5$ because $2(6) > 5$ is a true statement.
Set-Builder Notation	A notation that can be used to identify the solution set to an inequality	The solution set for $x - 2 < 5$ can be written as $\{x \mid x < 7\}$ and is read "the set of real numbers x such that x is less than 7."
Solution Set to an Inequality	The set of all solutions to an inequality	The solution set to $x + 1 > 5$ is given by $x > 4$ and can be written in set-builder notation as $\{x \mid x > 4\}$.
Number Line Graphs	The solutions to an inequality can be graphed on a number line.	$x < 2$ is graphed as follows. $x \geq -1$ is graphed as follows.
Addition Property of Inequalities	$a < b$ is equivalent to $a + c < b + c,$ where a, b, and c represent real number expressions.	$x - 5 \geq 6$ Given inequality $x \geq 11$ Add 5. $3x > 5 + 2x$ Given inequality $x > 5$ Subtract $2x$.
Multiplication Property of Inequalities	$a < b$ is equivalent to $ac < bc$ when $c > 0$, and is equivalent to $ac > bc$ when $c < 0$.	$\dfrac{1}{2}x \geq 6$ Given inequality $x \geq 12$ Multiply by 2. $-3x > 5$ Given inequality $x < -\dfrac{5}{3}$ Divide by -3; reverse the inequality symbol.

2.5   Exercises

MyMathLab Math XL PRACTICE WATCH DOWNLOAD READ REVIEW

CONCEPTS AND VOCABULARY

1. A linear inequality results whenever the equals sign in a linear equation is replaced by any one of the symbols _____, _____, _____, or _____.

2. Equality is the boundary between _____ and _____.

3. A(n) _____ is a value of the variable that makes an inequality statement true.

4. Two linear inequalities are _____ if they have the same solution set.

5. (True or False?) When a linear equation is solved, the solution set contains one solution.

6. (True or False?) When a linear inequality is solved, the solution set contains infinitely many solutions.

7. The solution set to a linear inequality can be graphed by using a(n) _____.

8. The addition property of inequalities states that if $a > b$, then $a + c$ _____ $b + c$.

9. The multiplication property of inequalities states that if $a < b$ and $c > 0$, then ac _____ bc.

10. The multiplication property of inequalities states that if $a < b$ and $c < 0$, then ac _____ bc.

SOLUTIONS AND NUMBER LINE GRAPHS

Exercises 11–18: Use a number line to graph the solution set to the inequality.

11. $x < 0$

12. $x > -2$

13. $x > 1$

14. $x < -\frac{5}{2}$

15. $x \le 1.5$

16. $x \ge -3$

17. $z \ge -2$

18. $z \le -\pi$

Exercises 19–24: Express the set of real numbers graphed on the number line as an inequality.

19.
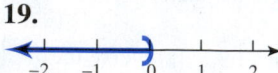

20.

21.
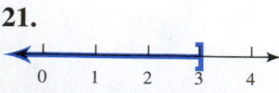

22.
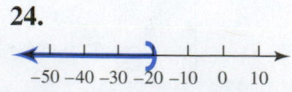

23.
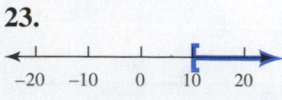

24.

Exercises 25–30: Write the solution set to the inequality in interval notation.

25. $x \ge 6$

26. $x < 3$

27. $y > -2$

28. $y \ge 1$

29. $z \le 7$

30. $z < -5$

Exercises 31–38: Determine whether the given value of the variable is a solution to the inequality.

31. $x + 5 > 5$ $x = 4$

32. $-3x \le -8$ $x = -2$

33. $5x \ge 25$ $x = 3$

34. $4y - 3 \le 5$ $y = -3$

35. $3y + 5 \ge -8$ $y = -3$

36. $-(z + 7) > 3(6 - z)$ $z = 2$

37. $5(z + 1) < 3z - 7$ $z = -4$

38. $\frac{3}{2}t - \frac{1}{2} \ge 1 - t$ $t = \frac{3}{5}$

TABLES AND LINEAR INEQUALITIES

Exercises 39–42: Use the table to solve the inequality.

39. $3x + 6 > 0$

x	-4	-3	-2	-1	0
$3x + 6$	-6	-3	0	3	6

40. $6 - 3x \le 0$

x	1	2	3	4	5
$6 - 3x$	3	0	-3	-6	-9

41. $-2x + 7 > 5$

x	-1	0	1	2	3
$-2x + 7$	9	7	5	3	1

42. $5(x - 3) \le 4$

x	3.2	3.4	3.6	3.8	4
$5(x - 3)$	1	2	3	4	5

Exercises 43–46: Complete the table. Then use the table to solve the inequality.

43. $-2x + 6 \leq 0$

x	1	2	3	4	5
$-2x + 6$	4				-4

44. $3x - 1 < 8$

x	0	1	2	3	4
$3x - 1$	-1				

45. $5 - x > x + 7$

x	-3	-2	-1	0	1
$5 - x$	8				4
$x + 7$	4				8

46. $2(3 - x) \geq -3(x - 2)$

x	-2	-1	0	1	2
$2(3 - x)$					
$-3(x - 2)$					

SOLVING LINEAR INEQUALITIES

Exercises 47–54: Use the addition property of inequalities to solve the inequality. Graph the solution set.

47. $x - 3 > 0$ **48.** $x + 6 < 3$

49. $3 - y \leq 5$ **50.** $8 - y \geq 10$

51. $12 < 4 + z$ **52.** $2z \leq z + 17$

53. $5 - 2t \geq 10 - t$ **54.** $-2t > -3t + 1$

Exercises 55–62: Use the multiplication property of inequalities to solve the inequality. Graph the solution set.

55. $2x < 10$ **56.** $3x > 9$

57. $-\frac{1}{2}t \geq 1$ **58.** $-5t \leq -6$

59. $\frac{3}{4} > -5y$ **60.** $10 \geq -\frac{1}{7}y$

61. $-\frac{2}{3} \leq \frac{1}{7}z$ **62.** $-\frac{3}{10}z < 11$

Exercises 63–68: Solve the linear inequality and write the solution in set-builder notation.

63. $x + 6 > 7$ **64.** $x + 4 < 1$

65. $-3x \leq 21$ **66.** $4x \geq -20$

67. $2x - 3 < 9$ **68.** $-5x + 4 < 44$

Exercises 69–94: Solve the linear inequality.

69. $3x + 1 < 22$ **70.** $4 + 5x \leq 9$

71. $5 - \frac{3}{4}x \geq 6$ **72.** $10 - \frac{2}{5}x > 0$

73. $45 > 6 - 2x$ **74.** $69 \geq 3 - 11x$

75. $5x - 2 \leq 3x + 1$ **76.** $12x + 1 < 25 - 3x$

77. $-x + 24 < x + 23$ **78.** $6 - 4x \leq x + 1$

79. $-(x + 1) \geq 3(x - 2)$

80. $5(x + 2) > -2(x - 3)$

81. $3(2x + 1) > -(5 - 3x)$

82. $4x \geq -3(7 - 2x) + 1$

83. $1.6x + 0.4 \leq 0.4x$

84. $-5.1x + 1.1 < 0.1 - 0.1x$

85. $0.8x - 0.5 < x + 1 - 0.5x$

86. $0.1(x + 1) - 0.1 \leq 0.2x - 0.5$

87. $-\frac{1}{2}\left(\frac{2}{3}x + 4\right) \geq x$

88. $-5x > \frac{4}{5}\left(\frac{10}{3}x + 10\right)$

89. $\frac{3}{7}x + \frac{2}{7} > -\frac{1}{7}x - \frac{5}{14}$

90. $\frac{5}{6} - \frac{1}{3}x \geq -\frac{1}{3}\left(\frac{5}{6}x - 1\right)$

91. $\frac{x}{3} + \frac{5x}{6} \leq \frac{2}{3}$ **92.** $\frac{3x}{4} - \frac{x}{2} < 1$

93. $\frac{6x}{7} < \frac{1}{3}x + 1$ **94.** $\frac{5x}{8} - \frac{3x}{4} \leq 8$

TRANSLATING PHRASES TO INEQUALITIES

Exercises 95–102: Translate each phrase to an inequality. Let x be the variable.

95. A speed that is greater than 60 miles per hour

96. A speed that is at most 60 miles per hour

97. An age that is at least 21 years old

98. An age that is less than 21 years old

99. A salary that is more than $40,000

100. A salary that is less than or equal to $40,000

101. A speed that does not exceed 70 miles per hour

102. A speed that is not less than 70 miles per hour

APPLICATIONS

103. *Geometry* Find all values for x so that the perimeter of the rectangle is less than 50 feet.

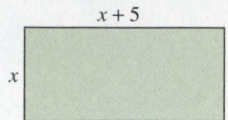

104. *Geometry* A triangle with height 12 inches is to have area less than 120 square inches. What must be true about the base of the triangle?

105. *Grade Average* A student scores 74 out of 100 on a test. If the maximum score on the next test is also 100 points, what score does the student need to maintain at least an average of 80?

106. *Grade Average* A student scores 65 and 82 on two different 100-point tests. If the maximum score on the next test is also 100 points, what score does the student need to maintain at least an average of 70?

107. *Parking Rates* Parking in a student lot costs $2 for the first half hour and $1.25 for each hour thereafter. A partial hour is charged the same as a full hour. What is the longest time that a student can park in this lot for $8?

108. *Parking Rates* Parking in a student lot costs $2.50 for the first hour and $1 for each hour thereafter. A nearby lot costs $1.25 for each hour. In both lots a partial hour is charged as a full hour. In which lot can a student park the longest for $5? For $11?

109. *Car Rental* A rental car costs $25 per day plus $0.20 per mile. If someone has $200 to spend and needs to drive the car 90 miles each day, for how many days can that person rent the car? Assume that the car cannot be rented for part of a day.

110. *Car Rental* One car rental agency charges $20 per day plus $0.25 per mile. A different agency charges $37 per day with unlimited mileage. For what mileages is the second rental agency a better deal?

111. *Revenue and Cost* (Refer to Example 11.) The cost to produce one compact disc is $1.50 plus a one-time fixed cost of $2000. The revenue received from selling one compact disc is $12.
   (a) Write a formula that gives the cost C of producing x compact discs. Be sure to include the fixed cost.
   (b) Write a formula that gives the revenue R from selling x compact discs.
   (c) Profit equals revenue minus cost. Write a formula that calculates the profit P from selling x compact discs.

(d) How many compact discs need to be sold to yield a positive profit?

112. *Revenue and Cost* The cost to produce one laptop computer is $890 plus a one-time fixed cost of $100,000 for research and development. The revenue received from selling one laptop computer is $1520.
   (a) Write a formula that gives the cost C of producing x laptop computers.
   (b) Write a formula that gives the revenue R from selling x laptop computers.
   (c) Profit equals revenue minus cost. Write a formula that calculates the profit P from selling x laptop computers.
   (d) How many computers need to be sold to yield a positive profit?

113. *Distance and Time* Two athletes are jogging in the same direction along an exercise path. After x minutes the first athlete's distance in miles from a parking lot is given by $\frac{1}{6}x$ and the second athlete's distance is given by $\frac{1}{8}x + 2$.
   (a) When are the athletes the same distance from the parking lot?
   (b) When is the first athlete farther from the parking lot than the second?

114. *Altitude and Dew Point* If the dew point on the ground is $65°F$, then the dew point x miles high is given by $D = 65 - 5.8x$. Determine the altitudes at which the dew point is greater than $36°F$. (*Source:* A. Miller.)

115. *Altitude and Temperature* (Refer to Example 10.) If the temperature on the ground is $90°F$, then the air temperature x miles high is given by $T = 90 - 19x$. Determine the altitudes at which the air temperature is less than $4.5°F$. (*Source:* A. Miller.)

116. *Size and Weight of a Fish* If the length of a bass is between 20 and 25 inches, its weight W in pounds can be estimated by the formula $W = 0.96x - 14.4$, where x is the length of the fish. (*Source:* Minnesota Department of Natural Resources.)
   (a) What length of bass is likely to weigh 7.2 pounds?
   (b) What lengths of bass are likely to weigh less than 7.2 pounds?

WRITING ABOUT MATHEMATICS

117. Explain each of the terms and give an example.
   (a) Linear equation   **(b)** Linear inequality

118. Suppose that a student says that a linear equation and a linear inequality can be solved the same way. How would you respond?

SECTION 2.5 — Checking Basic Concepts

1. Use a number line to graph the solution set to the inequality $x + 1 \geq -1$.

2. Express the set of real numbers graphed on the number line by using an inequality.

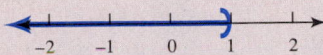

3. Determine whether -3 is a solution to the inequality $4x - 5 \leq -15$.

4. Complete the table. Then use the table to solve the inequality $5 - 2x \leq 7$.

x	-2	-1	0	1	2
$5 - 2x$					1

5. Solve each inequality.
   (a) $x + 5 > 8$
   (b) $-\frac{5}{7}x \leq 25$
   (c) $3x \geq -2(1 - 2x) + 3$

6. Translate the phrase "a price that is not more than $12" to an inequality using the variable x.

7. *Geometry* The length of a rectangle is 5 inches longer than twice its width. If the perimeter of the rectangle is more than 88 inches, find all possible widths for the rectangle.

CHAPTER 2   Summary

SECTION 2.1 ■ INTRODUCTION TO EQUATIONS

Equations An equation is a mathematical statement that two expressions are equal. Every equation contains an equals sign. An equation can be either true or false.

Important Terms

Solution	A value for a variable that makes the equation true
Solution Set	The set of all solutions
Equivalent Equations	Equations that have the same solution set
Checking a Solution	Substituting the solution in the given equation to verify that the solution makes the equation true

Example: 3 is the solution to $4x - 2 = 10$ because $4(3) - 2 = 10$ is a true statement.

Properties of Equality

Addition Property $a = b$ is equivalent to $a + c = b + c$.

Example: $x - 3 = 0$ and $x - 3 + 3 = 0 + 3$ are equivalent equations.

Multiplication Property $a = b$ is equivalent to $ac = bc$, provided $c \neq 0$.

Example: $2x = 5$ and $2x \cdot \frac{1}{2} = 5 \cdot \frac{1}{2}$ are equivalent equations.

SECTION 2.2 ■ LINEAR EQUATIONS

Linear Equation Can be written in the form $ax + b = 0$, where $a \neq 0$

Examples: $3x - 5 = 0$ is linear, whereas $5x^2 + 2x = 0$ is *not* linear.

Solving Linear Equations Numerically To solve a linear equation numerically, complete a table for various values of the variable and then select the solution from the table.

Example: The solution to $3x - 4 = -1$ is 1.

x	-1	0	1
$3x - 4$	-7	-4	-1

Solving Linear Equations Symbolically The following steps can be used as a guide for solving linear equations symbolically.

STEP 1: Use the distributive property to clear any parentheses on each side of the equation. Combine any like terms on each side.

STEP 2: Use the addition property of equality to get all of the terms containing the variable on one side of the equation and all other terms on the other side of the equation. Combine any like terms on each side.

STEP 3: Use the multiplication property of equality to isolate the variable by multiplying each side of the equation by the reciprocal of the number in front of the variable (or divide each side by that number).

STEP 4: Check the solution by substituting it in the given equation.

Distributive Properties $a(b + c) = ab + ac$ or $a(b - c) = ab - ac$

Examples: $5(2x + 3) = 10x + 15$ and $5(2x - 3) = 10x - 15$

Clearing Fractions and Decimals When fractions or decimals appear in an equation, multiplying each side by the least common denominator can be helpful.

Examples: Multiply each side of $\frac{1}{3}x - \frac{1}{6} = \frac{2}{3}$ by 6 to obtain $2x - 1 = 4$.

Multiply each side of $0.04x + 0.1 = 0.07$ by 100 to obtain $4x + 10 = 7$.

Number of Solutions Equations that can be written in the form $ax + b = 0$, where a and b are *any* real number, can have no solutions, one solution, or infinitely many solutions.

Examples: $x + 3 = x$ is equivalent to $3 = 0$. (No solutions)

$2y + 1 = 9$ is equivalent to $y = 4$. (One solution)

$z + z = 2z$ is equivalent to $0 = 0$. (Infinitely many solutions)

SECTION 2.3 ■ INTRODUCTION TO PROBLEM SOLVING

Steps for Solving a Problem The following steps can be used as a guide for solving word problems.

STEP 1: Read the problem carefully and be sure that you understand it. (You may need to read the problem more than once.) Assign a variable to what you are being asked to find. If necessary, write other quantities in terms of this variable.

STEP 2: Write an equation that relates the quantities described in the problem. You may need to sketch a diagram or refer to known formulas.

STEP 3: Solve the equation. Use the solution to determine the solution(s) to the original problem. Include any necessary units.

STEP 4: Check your solution in the original problem. Does it seem reasonable?

Percent Problems

The expression x%

Represents the fraction $\frac{x}{100}$ or the decimal number given by $x \cdot 0.01$.

Examples: $45\% = \dfrac{45}{100} = 0.45$

$7.1\% = \dfrac{7.1}{100} = \dfrac{71}{1000} = 0.071$

Changing Fractions to Percents

Divide the numerator by the denominator, multiply by 100, and insert the % symbol.

Example: Since $3 \div 4 = 0.75$, the fraction $\frac{3}{4}$ is written in percent notation as $0.75 \cdot 100 = 75\%$.

Percent Change

If a quantity changes from an old amount to a new amount, then the percent change equals $\dfrac{(\text{new amount}) - (\text{old amount})}{(\text{old amount})} \cdot 100$.

Example: If a price increases from \$2 to \$3, then the percent change equals $\dfrac{3 - 2}{2} \cdot 100 = 50\%$.

Distance Problems If an object travels at speed (rate) r for time t, then the distance d traveled is calculated by $d = rt$.

Example: A car moving at 65 mph for 2 hours travels

$$d = rt = 65 \cdot 2 = 130 \text{ miles.}$$

SECTION 2.4 ■ FORMULAS

Formula A formula is an equation that can be used to calculate a quantity by using known values of other quantities.

Formulas from Geometry

Area of a Rectangle

$A = lw$, where l is the length and w is the width.

Perimeter of a Rectangle

$P = 2l + 2w$, where l is the length and w is the width.

Area of a Triangle

$A = \frac{1}{2}bh$, where b is the base and h is the height.

Degree Measure

There are $360°$ in one complete revolution.

Angle Measure

The sum of the measures of the angles in a triangle equals $180°$.

Circumference

$C = 2\pi r$, where r is the radius.

Area of a Circle

$A = \pi r^2$, where r is the radius.

Area of a Trapezoid

$A = \frac{1}{2}(a + b)h$, where h is the height and a and b are the bases of the trapezoid.

Volume of a Box

$V = lwh$, where l is the length, w is the width, and h is the height.

Surface Area of a Box

$S = 2lw + 2wh + 2lh$, where l is the length, w is the width, and h is the height.

Volume of a Cylinder

$V = \pi r^2 h$, where r is the radius and h is the height.

Other Formulas

Grade Point Average (GPA)

$\text{GPA} = \frac{4a + 3b + 2c + d}{a + b + c + d + f}$, where a represents the number of A credits earned, b the number of B credits earned, and so on.

Temperature Scales

$F = \frac{9}{5}C + 32$ and $C = \frac{5}{9}(F - 32)$, where F is the Fahrenheit temperature and C is the Celsius temperature.

See Putting It All Together in Section 2.4 for examples.

SECTION 2.5 ■ LINEAR INEQUALITIES

Linear Inequality When the equals sign in a linear equation is replaced with any one of the symbols $<$, $\leq$, $>$, or $\geq$, a linear inequality results.

Examples: $x > 0$, $\quad 6 - \frac{2}{3}x \leq 7$, and $\quad 4(x - 1) < 3x - 1$

Solution to an Inequality Any value for the variable that makes the inequality a true statement

Example: 3 is a solution to $2x < 9$ because $2(3) < 9$ is a true statement.

Number Line Graphs A number line can be used to graph the solution set to a linear inequality.

Example: The graph of $x \leq 1$ is shown in the figure.

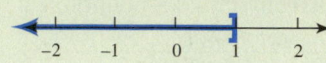

Properties of Inequality

Addition Property $\qquad\qquad\qquad$ $a < b$ is equivalent to $a + c < b + c$.

$\qquad\qquad\qquad\qquad\qquad\qquad$ **Example:** $x - 3 < 0$ and $x - 3 + 3 < 0 + 3$ are equivalent inequalities.

Multiplication Property $\qquad\qquad$ When $c > 0$, $a < b$ is equivalent to $ac < bc$.
$\qquad\qquad\qquad\qquad\qquad\qquad$ When $c < 0$, $a < b$ is equivalent to $ac > bc$.

$\qquad\qquad\qquad\qquad\qquad\qquad$ **Examples:** $2x < 6$ is equivalent to $2x\left(\frac{1}{2}\right) < 6\left(\frac{1}{2}\right)$ or $x < 3$.

$\qquad\qquad\qquad\qquad\qquad\qquad\qquad\qquad$ $-2x < 6$ is equivalent to $-2x\left(-\frac{1}{2}\right) > 6\left(-\frac{1}{2}\right)$ or $x > -3$.

CHAPTER 2 Review Exercises

SECTION 2.1

Exercises 1–8: Solve the equation. Check your solution.

1. $x + 9 = 3$ $\qquad\qquad$ **2.** $x - 4 = -2$

3. $x - \frac{3}{4} = \frac{3}{2}$ $\qquad\qquad$ **4.** $x + 0.5 = 0$

5. $4x = 12$ $\qquad\qquad\qquad$ **6.** $3x = -7$

7. $-0.5x = 1.25$ $\qquad\quad$ **8.** $-\frac{1}{3}x = \frac{7}{6}$

SECTION 2.2

Exercises 9 and 10: Decide whether the equation is linear. If the equation is linear, give values for a and b so that it can be written in the form ax + b = 0.

9. $-4x + 3 = 2$ $\qquad\qquad$ **10.** $\frac{3}{8}x^2 - x = \frac{1}{4}$

Exercises 11–20: Solve the equation. Check the solution.

11. $4x - 5 = 3$ $\qquad\qquad$ **12.** $7 - \frac{1}{2}x = -4$

13. $5(x - 3) = 12$ $\qquad\quad$ **14.** $3 + x = 2x - 4$

15. $2(x - 1) = 4(x + 3)$

16. $1 - (x - 3) = 6 + 2x$

17. $3.4x - 4 = 5 - 0.6x$

18. $-\frac{1}{3}(3 - 6x) = -(x + 2) + 1$

19. $\frac{2}{3}x - \frac{1}{6} = \frac{5}{12}$

20. $2y - 3(2 - y) = 5 + y$

Exercises 21–24: Determine whether the equation has no solutions, one solution, or infinitely many solutions.

21. $4(3x - 2) = 2(6x + 5)$

22. $5(3x - 1) = 15x - 5$

23. $8x = 5x + 3x$

24. $9x - 2 = 8x - 2$

Exercises 25 and 26: Complete the table. Then use the table to solve the given equation.

25. $-2x + 3 = 0$

x	0.5	1.0	1.5	2.0	2.5
$-2x + 3$	2				

26. $-(x + 1) + 3 = 2$

x	-2	-1	0	1	2
$-(x + 1) + 3$					

SECTION 2.3

Exercises 27–30: Using the variable x, translate the sentence into an equation. Solve the resulting equation.

27. The product of a number and 6 is 72.

28. The sum of a number and 18 is -23.

29. Twice a number minus 5 equals the number plus 4.

30. The sum of a number and 4 equals the product of the number and 3.

Exercises 31 and 32: Find the number or numbers.

31. Five times a number divided by 3 equals 15.

32. The sum of three consecutive integers is -153.

Exercises 33–36: Convert the percentage to fraction and decimal notation.

33. 85%   **34.** 5.6%

35. 0.03%   **36.** 342%

Exercises 37–40: Convert the number to a percentage.

37. 0.89   **38.** 0.005

39. 2.3   **40.** 1

Exercises 41–44: Use the formula $d = rt$ to find the value of the missing variable.

41. $r = 8$ miles per hour, $t = 3$ hours

42. $r = 70$ feet per second, $t = 55$ seconds

43. $d = 500$ yards, $t = 20$ seconds

44. $d = 125$ miles, $r = 15$ miles per hour

SECTION 2.4

Exercises 45 and 46: Find the area of the region shown.

45.   **46.**

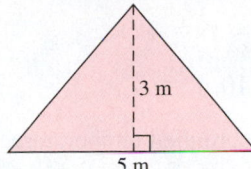   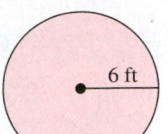

47. Find the area of a rectangle having a 24-inch width and a 3-foot length.

48. Find the perimeter of a rectangle having a 13-inch length and a 7-inch width.

49. Find the circumference of a circle having an 18-foot diameter.

50. Find the area of a circle having a 5-inch radius.

51. Find the measure of the third angle in the triangle.

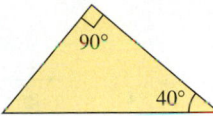

52. The angles in a triangle have measures x, $3x$, and $4x$. Find the value of x.

53. If a cylinder has radius 5 inches and height 25 inches, find its volume. (*Hint:* $V = \pi r^2 h$.)

54. Find the area of a trapezoid with height 5 feet and bases 3 feet and 18 inches. (*Hint:* $A = \frac{1}{2}(a + b)h$.)

Exercises 55 and 56: Find the area of the figure shown.

55.

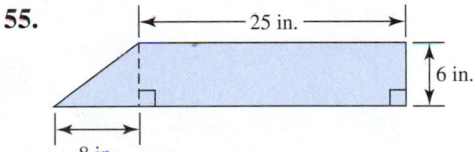

56.

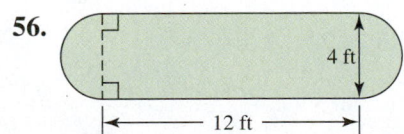

Exercises 57–62: Solve the given formula for the specified variable.

57. $3x = 5 + y$ for y **58.** $16 = 2x + 2y$ for y

59. $z = 2xy$ for y **60.** $S = \dfrac{a + b + c}{3}$ for b

61. $T = \dfrac{a}{3} + \dfrac{b}{4}$ for b **62.** $cd = ab + bc$ for c

Exercises 63 and 64: Let the variable a represent the number of credits with a grade of A, b the number of credits with a grade of B, and so on. Calculate the grade point average (GPA). Round your answer to the nearest hundredth.

63. $a = 20, b = 25, c = 12, d = 4, f = 4$

64. $a = 64, b = 32, c = 20, d = 10, f = 3$

65. Convert 15°C to an equivalent Fahrenheit temperature.

66. Convert 113°F to an equivalent Celsius temperature.

SECTION 2.5

Exercises 67–70: Use a number line to graph the solution set to the inequality.

67. $x < 2$

68. $x > -1$

69. $y \geq -\frac{3}{2}$

70. $y \leq 2.5$

Exercises 71 and 72: Express the set of real numbers graphed on the number line with an inequality.

71.

72.

Exercises 73 and 74: Determine whether the given value of x is a solution to the inequality.

73. $1 - (x + 3) \geq x$ $x = -2$

74. $4(x + 1) < -(5 - x)$ $x = -1$

Exercises 75 and 76: Complete the table and then use the table to solve the inequality.

75. $5 - x > 3$

x	0	1	2	3	4
$5 - x$	5				

76. $2x - 5 \leq 0$

x	1	1.5	2	2.5	3
$2x - 5$	-3				

Exercises 77–82: Solve the inequality.

77. $x - 3 > 0$

78. $-2x \leq 10$

79. $5 - 2x \geq 7$

80. $3(x - 1) < 20$

81. $5x \leq 3 - (4x + 2)$

82. $3x - 2(4 - x) \geq x + 1$

Exercises 83–86: Translate the phrase to an inequality. Let x be the variable.

83. A speed that is less than 50 miles per hour

84. A salary that is at most $45,000

85. An age that is at least 16 years old

86. A year before 1995

APPLICATIONS

87. *Rainfall* On a stormy day 2 inches of rain fall before noon and $\frac{3}{4}$ inch per hour fall thereafter until 5 P.M.
  (a) Make a table that shows the total rainfall at each hour starting at noon and ending at 5 P.M.
  (b) Write a formula that calculates the rainfall R in inches, x hours past noon.
  (c) Use your formula to calculate the total rainfall at 5 P.M. Does your answer agree with the value in your table from part (a)?
  (d) How much rain had fallen at 3:45 P.M.?

88. *Cost of a Laptop* A 5% sales tax on a laptop computer amounted to $106.25. Find the cost of the laptop.

89. *Distance Traveled* At noon a bicyclist is 50 miles from home, riding toward home at 10 miles per hour.
  (a) Make a table that shows the bicyclist's distance D from home after 1, 2, 3, 4, and 5 hours.
  (b) Write a formula that calculates the distance D from home after x hours.
  (c) Use your formula to determine D when $x = 3$ hours. Does your answer agree with the value shown in your table?
  (d) For what times was the bicyclist at least 20 miles from home? Assume that $0 \leq x \leq 5$.

90. *Master's Degree* In 2001, about 468,500 people received a master's degree, and in 2008, about 625,000 did. Find the percent change in the number of master's degrees received between 2001 and 2008.

91. *Car Speeds* One car passes another car on a freeway. The faster car is traveling 12 miles per hour faster than the slower car. Determine how long it will be before the faster car is 2 miles ahead of the slower car.

92. *Dimensions of a Rectangle* The width of a rectangle is 10 inches less than its length. If the perimeter is 112 inches, find the dimensions of the rectangle.

93. *Saline Solution* A saline solution contains 3% salt. How much water should be added to 100 milliliters of this solution to dilute it to a 2% solution?

94. *Investment Money* A student invests two sums of money, $500 and $800, at different interest rates, receiving a total of $55 in interest after one year. The $500 investment receives an interest rate 2% lower than the interest rate for the $800 investment. Find the interest rate for each investment.

95. *Geometry* A triangle with height 8 inches is to have an area that is not more than 100 square inches. What lengths are possible for the base of the triangle?

96. *Grade Average* A student scores 75 and 91 on two different tests of 100 points. If the maximum score on the next test is also 100 points, what score does the student need to maintain an average of at least 80?

97. *Parking Rates* Parking in a lot costs $2.25 for the first hour and $1.25 for each hour thereafter. A partial hour is charged the same as a full hour. What is the longest time that someone can park for $9?

98. *Profit* The cost to produce one DVD player is $85 plus a one-time fixed cost of $150,000. The revenue received from selling one DVD player is $225.
  (a) Write a formula that gives the cost C of producing x DVD players.
  (b) Write a formula that gives the revenue R from selling x DVD players.
  (c) Profit equals revenue minus cost. Write a formula that calculates the profit P from selling x DVD players.
  (d) What numbers of DVD players sold will result in a loss? (*Hint:* A loss corresponds to a negative profit.)

CHAPTER 2   **Test**   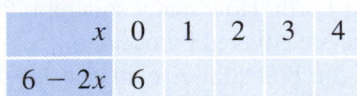 Step-by-step test solutions are found on the Chapter Test Prep Videos available in **MyMathLab** and on **YouTube** (search "RockswoldComboAlg" and click on "Channels").

Exercises 1–4: Solve the equation. Check your solution.

1. $9 = 3 - x$ **2.** $4x - 3 = 7$

3. $4x - (2 - x) = -3(2x + 6)$

4. $\frac{1}{12}x - \frac{2}{3} = \frac{1}{2}\left(\frac{3}{4} - \frac{1}{3}x\right)$

Exercises 5 and 6: Determine the number of solutions to the given equation.

5. $6(2x - 1) = -4(3 - 3x)$

6. $4(2x - 1) = 8x - 4$

7. Complete the table. Then use the table to solve the equation $6 - 2x = 0$.

x	0	1	2	3	4
$6 - 2x$	6				

Exercises 8 and 9: Translate the sentence into an equation, using the variable x. Then solve the resulting equation.

8. The sum of a number and -7 is 6.

9. Twice a number plus 6 equals the number minus 7.

10. The sum of three consecutive natural numbers is 336. Find the three numbers.

11. Convert 3.2% to fraction and decimal notation.

12. Convert 0.345 to a percentage.

13. Use the formula $d = rt$ to find r when $d = 200$ feet and $t = 4$ seconds.

14. *Area* Find the area of the triangle shown.

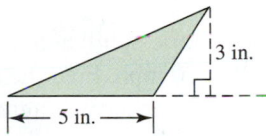

15. Find the circumference and area of a circle with a 30-inch diameter.

16. The measures of the angles in a triangle are x, $2x$, and $3x$. Find the value of x.

Exercises 17 and 18: Solve the formula for x.

17. $z = y - 3xy$ **18.** $R = \dfrac{x}{4} + \dfrac{y}{5}$

19. Use a number line to graph the solution set to $x \le 3$.

20. Express the set of real numbers graphed on the number line with an inequality.

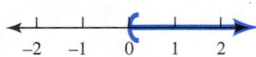

Exercises 21 and 22: Solve the inequality.

21. $-3x + 9 \ge x - 15$

22. $3(6 - 5x) < 20 - x$

23. *Snowfall* Suppose 5 inches of snow fall before noon and 2 inches per hour fall thereafter until 10 P.M.
    (a) Write a formula that calculates the snowfall S in inches, x hours past noon.
    (b) Use your formula to calculate the total snowfall at 8 P.M.
    (c) How much snow had fallen at 6:15 P.M.?

24. *Mixing Acid* A solution is 45% hydrochloric acid. How much water should be added to 1000 milliliters of this solution to dilute it to a 15% solution?

25. *Cable TV* In 2010, the average cable TV subscriber was paying $75 per month. By 2015, this cost is expected to rise to $95 per month. Find the percent increase in cable TV costs over this time period. (*Source:* CNN Money.)

CHAPTER 2   Extended and Discovery Exercises

Exercises 1–4: Average Speed If someone travels a distance d in time t, then the person's average speed is $\frac{d}{t}$. Use this fact to solve the problem.

1. A driver travels at 50 mph for the first hour and then travels at 70 mph for the second hour. What is the average speed of the car?

2. A bicyclist rides 1 mile uphill at 5 mph and then rides 1 mile downhill at 10 mph. Find the average speed of the bicyclist. Does your answer agree with what you expected?

3. At a 3-mile cross-country race an athlete runs 2 miles at 8 mph and 1 mile at 10 mph. What is the athlete's average speed?

4. A pilot flies an airplane between two cities and travels half the distance at 200 mph and the other half at 100 mph. Find the average speed of the airplane.

5. *A Puzzle About Coins* Suppose that seven coins look exactly alike but that one coin weighs less than any of the other six coins. If you have only a balance with two pans, devise a plan to find the lighter coin. What is the minimum number of weighings necessary? Explain your answer.

6. *Global Warming* If the global climate were to warm significantly as a result of the greenhouse effect or other climatic change, the Arctic ice cap would start to melt. This ice cap contains the equivalent of some 680,000 cubic miles of water. More than 200 million people live on land that is less than 3 feet above sea level. In the United States several large cities have low average elevations. Three examples are Boston (14 feet), New Orleans (4 feet), and San Diego (13 feet). In this exercise you are to estimate the rise in sea level if the Arctic ice cap were to melt and to determine whether this event would have a significant impact on people living in coastal areas.
    (a) The surface area of a sphere is given by the formula $4\pi r^2$, where r is its radius. Although the shape of Earth is not exactly spherical, it has an average radius of 3960 miles. Estimate the surface area of Earth.
    (b) Oceans cover approximately 71% of the total surface area of Earth. How many square miles of Earth's surface are covered by oceans?
    (c) Approximate the potential rise in sea level by dividing the total volume of the water from the ice cap by the surface area of the oceans. Convert your answer from miles to feet.
    (d) Discuss the implications of your calculation. How would cities such as Boston, New Orleans, and San Diego be affected?
    (e) The Antarctic ice cap contains some 6,300,000 cubic miles of water. Estimate how much the sea level would rise if this ice cap melted. (*Source:* Department of the Interior, Geological Survey.)

CHAPTERS 1–2   Cumulative Review Exercises

Exercises 1 and 2: Classify the number as prime or composite. If the number is composite, write it as a product of prime numbers.

1. 45

2. 37

Exercises 3 and 4: Multiply or divide and then simplify to lowest terms when appropriate.

3. $\frac{4}{3} \cdot \frac{3}{8}$

4. $\frac{2}{3} \div 6$

Exercises 5 and 6: Add or subtract and then simplify to lowest terms when appropriate.

5. $\frac{11}{12} - \frac{3}{8}$

6. $\frac{2}{3} + \frac{1}{5}$

Exercises 7 and 8: Classify the number as one or more of the following: natural number, whole number, integer, rational number, or irrational number.

7. -1

8. $\sqrt{3}$

Exercises 9–12: Evaluate the expression by hand.

9. $15 - 4 \cdot 3$

10. $30 \div 6 \cdot 2$

11. $23 - 4^2 \div 2$

12. $11 - \frac{3 + 1}{6 - 4}$

Exercises 13 and 14: Simplify the expression.

13. $5x^3 - x^3$

14. $4 + 2x - 1 + 3x$

Exercises 15–18: Solve the equation.

15. $x - 3 = 11$

16. $4x - 6 = -22$

17. $5(6y + 2) = 25$

18. $11 - (y + 2) = 3y + 5$

Exercises 19 and 20: Determine whether the equation has no solutions, one solution, or infinitely many solutions.

19. $6x + 2 = 2(3x + 1)$

20. $2(3x - 4) = 6(x - 1)$

21. Find three consecutive integers whose sum is 90.

22. Convert 4.7% to decimal notation.

23. Convert 0.17 to a percentage.

24. Find the average speed of a car that travels 325 miles in 5 hours.

25. The angles in a triangle have measures $2x$, $3x$, and $4x$. Find the value of x.

26. Find the area of a circle having a 10-inch diameter.

Exercises 27 and 28: Solve the formula for x.

27. $a = 3xy - 4$

28. $A = \dfrac{x + y + z}{3}$

Exercises 29 and 30: Solve the inequality and graph the solution set.

29. $7 - 3x > 4$

30. $6x \le 5 - (x - 9)$

31. *Yards to Inches* There are 36 inches in 1 yard. Write a formula that converts Y yards to I inches.

32. *Checking Account* The initial balance in a checking account is $468. Find the final balance resulting from the following sequence of withdrawals and deposits: $-\$14$, $\$200$, $-\$73$, $-\$21$, and $\$58$.

33. *Acid Solution* How much of a 4% acid solution should be added to 150 milliliters of a 10% acid solution to dilute it to a 6% acid solution?

34. *Energy Production* In 1997, U.S. hydroelectric power production hit an all-time high of 356 billion kilowatt-hours. This is 188 billion kilowatt-hours less than twice the 2007 production level. Find the production level in 2007. (*Source:* U.S. Department of Energy.)

3

Graphing Equations

> We all have ability. The difference is how we use it.
>
> —STEVIE WONDER

During the first decade of the 21st century, the U.S. economy experienced both incredible growth and substantial decline. One way that economists monitor changes in the economy is by tracking median home prices. The *line graph* in the accompanying figure shows median home prices displayed in two-year intervals. Graphs such as this one provide an excellent way to visualize data trends. In this chapter, we discuss line graphs and other types of graphs that can be used to represent mathematical information visually.

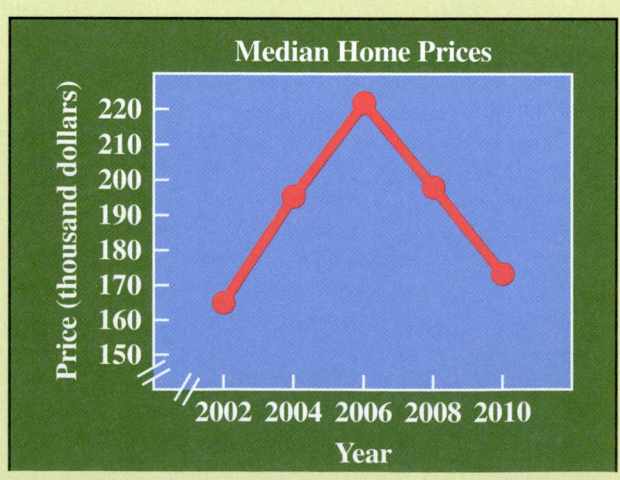

Source: National Association of Realtors.

3.1   Introduction to Graphing

Tables and Graphs • The Rectangular Coordinate System • Scatterplots and Line Graphs

A LOOK INTO MATH ▶
As Internet connection speeds have improved, the amount of visual information displayed on Web pages has increased dramatically. A page of computer graphics contains much more information than a page of printed text. In math, we visualize data by plotting points to make *scatterplots* and *line graphs*. In this section, we discuss the rectangular coordinate system and practice basic point-plotting skills.

Tables and Graphs

When data are displayed in a table, it is often difficult to recognize any trends. In order to visualize information provided by a set of data, it can be helpful to create a graph.

NEW VOCABULARY

☐ Rectangular coordinate system (xy-plane)
☐ x-axis
☐ y-axis
☐ Origin
☐ Quadrants
☐ Ordered pair
☐ Scatterplot
☐ x-coordinate
☐ y-coordinate
☐ Line graph

▶ **REAL-WORLD CONNECTION**   Table 3.1 lists the per capita (per person) income for the United States for selected years. These income amounts have *not* been adjusted for inflation. When a table contains only four data values, as in Table 3.1, we can easily see that income increased from 2002 to 2008. However, if a table contained 1000 data values, determining trends in the data would be extremely difficult. In mathematical problems, there are frequently infinitely many data points!

Rather than always using tables to display data, presenting data on a graph is often more useful. For example, the data in Table 3.1 are graphed in Figure 3.1. This line graph is more visual than the table and shows the trend at a glance. Line graphs will be discussed further later in this section.

TABLE 3.1 U.S. Per Capita Income

Year	Amount
2002	$31,444
2004	$33,857
2006	$37,679
2008	$40,649

Source: Bureau of Economic Analysis.

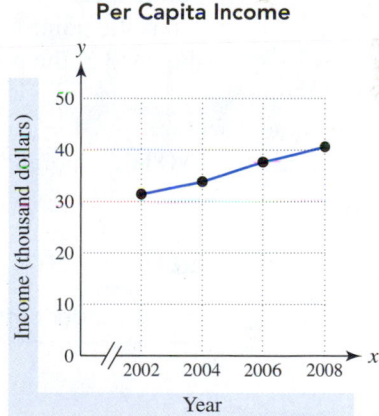

Figure 3.1

NOTE: In Figure 3.1 the double hash marks // on the *x*-axis indicate a break in the *x*-axis data values. The years before 2002 have been skipped.

The Rectangular Coordinate System

One common way to graph data is to use the **rectangular coordinate system**, or **xy-plane**. In the *xy*-plane the horizontal axis is the **x-axis**, and the vertical axis is the **y-axis**. The axes can be thought of as intersecting number lines. The point of intersection is called the **origin** and is associated with zero on each axis. Negative values are located left of the origin on the *x*-axis and below the origin on the *y*-axis. Similarly, positive values are located right of the origin on the *x*-axis and above the origin on the *y*-axis. The axes divide the *xy*-plane into four regions called **quadrants**, which are numbered I, II, III, and IV counterclockwise, as shown in Figure 3.2.

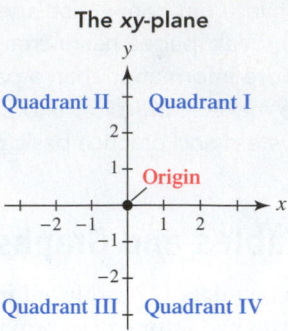

The xy-plane

Figure 3.2

Plotting a Point

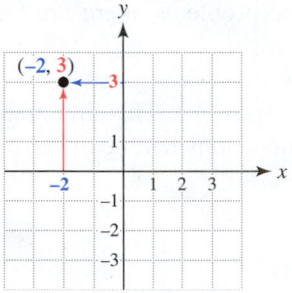

Figure 3.3

Before we can plot data, we must understand the concept of an **ordered pair** (*x*, *y*). In Table 3.1 we can let *x*-values correspond to the year and *y*-values correspond to the per capita income. Then the fact that the per capita income in **2002** was $**31,444** can be summarized by the ordered pair (**2002**, **31444**). Similarly, the ordered pair (2006, 37679) indicates that the per capita income was $37,679 in 2006.

Order is important in an ordered pair. The ordered pairs given by (1950, 2025) and (2025, 1950) are different. The first ordered pair indicates that the per capita income in 1950 was $2025, whereas the second ordered pair indicates that the per capita income in 2025 will be $1950.

To plot an ordered pair such as (−2, 3) in the *xy*-plane, begin at the origin and move left to locate *x* = −2 on the *x*-axis. Then move upward until a height of *y* = 3 is reached. Thus the point (−2, 3) is located 2 units left of the origin and 3 units above the origin. In Figure 3.3, the point (−**2**, **3**) is plotted in quadrant II.

NOTE: A point that lies on an axis is not located in a quadrant.

READING CHECK

• Is the order of the numbers in an ordered pair important? Explain.

EXAMPLE 1 | **Plotting points**

Plot the following ordered pairs on the same *xy*-plane. State the quadrant in which each point is located, if possible.
(a) (3, 2) **(b)** (−2, −3) **(c)** (−3, 0)

Solution
(a) The point (3, 2) is located in quadrant I, 3 units to the right of the origin and 2 units above the origin. See Figure 3.4.
(b) The point (−2, −3) is located 2 units to the left of the origin and 3 units below the origin. Figure 3.4 shows the point (−2, −3) in quadrant III.
(c) The point (−3, 0) is not in any quadrant because it is located 3 units left of the origin on the *x*-axis, as shown in Figure 3.4.

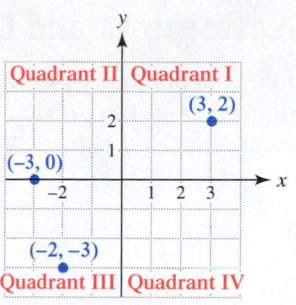

Figure 3.4

■ Now Try Exercise 13

| EXAMPLE 2 | **Reading a graph containing YouTube data** |

Figure 3.5 shows the average number of hours of video posted to YouTube every *minute* during selected months. Use the graph to estimate the number of hours of video posted to YouTube every minute in June 2009 and December 2010. (*Source:* YouTube.)

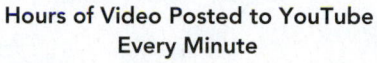

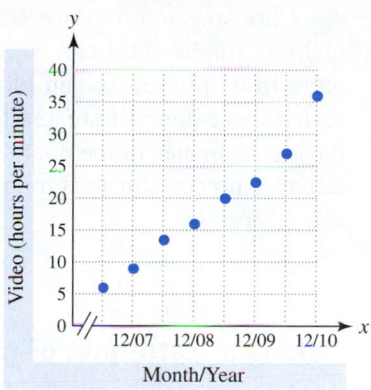

Figure 3.5

Solution

To find the number of hours of video posted to YouTube every minute in June 2009, start by locating Jun. '09 (halfway between Dec. '08 and Dec. '09) on the *x*-axis and then move vertically upward to the data point. From the data point, move horizontally to the *y*-axis. Figure 3.6(a) shows that about 20 hours of video was posted every minute in June 2009. Similarly, about 36 hours of video was posted every minute in December 2010, as shown in Figure 3.6(b).

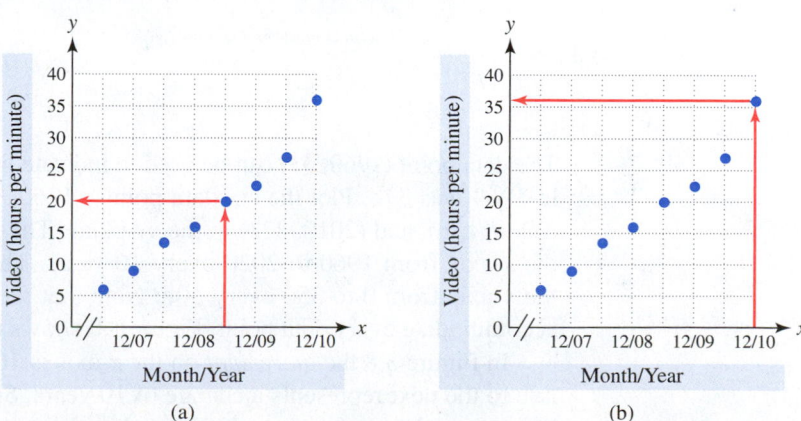

Figure 3.6

■ Now Try Exercise 41

Scatterplots and Line Graphs

If distinct points are plotted in the xy-plane, then the resulting graph is called a **scatterplot**. Figure 3.5 on the previous page is an example of a scatterplot that displays information about YouTube videos. A different scatterplot is shown in Figure 3.7, in which the points $(1, 3)$, $(2, 2)$, $(3, 1)$, $(4, 4)$, and $(5, 1)$ are plotted.

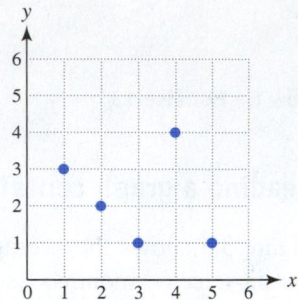

Figure 3.7 A Scatterplot

READING CHECK

• Which value in an ordered pair is the x-coordinate, and which is the y-coordinate?

Choosing appropriate scales for the axes is important when plotting points and making graphs. This can be accomplished by looking at the *coordinates* of the ordered pairs to be plotted. When plotting in the xy-plane, the first value in an ordered pair is called the **x-coordinate** and the second value is called the **y-coordinate**. In Figure 3.7, the x-coordinates of the points are 1, 2, 3, 4, and 5 and the y-coordinates are 1, 2, 3, and 4. Because no coordinate is more than 6 units from the origin, the scale shown in Figure 3.7 is appropriate.

> **EXAMPLE 3** | **Making a scatterplot of gasoline prices**

Table 3.2 lists the average price of a gallon of gasoline for selected years. Make a scatterplot of the data. These prices have *not* been adjusted for inflation.

TABLE 3.2 Average Price of Gasoline

Year	1960	1970	1980	1990	2000	2010
Cost (per gal)	31¢	36¢	119¢	115¢	156¢	273¢

Source: Department of Energy.

Solution

The data point (**1960**, **31**) can be used to indicate that the average cost of a gallon of gasoline in **1960** was **31**¢. Plot the six data points (1960, 31), (1970, 36), (1980, 119), (1990, 115), (2000, 156), and (2010, 273) in the xy-plane. The x-values vary from 1960 to 2010, so label the x-axis from 1960 to 2020 every 10 years. The y-values vary from 31 to 273, so label the y-axis from 0 to 350 every 50¢. Note that the x- and y-scales must be large enough to accommodate every data point. Figure 3.8 shows the scatterplot.

In Figure 3.8 the *increment* on the x-axis is 10 because each step from one vertical grid line to the next represents a change of 10 years. Similarly, the increment on the y-axis is 50 because each step from one horizontal grid line to the next represents a 50-cent change in price. This example demonstrates that the scale and increment on one axis are not always the same as those on the other axis.

Price of Gasoline

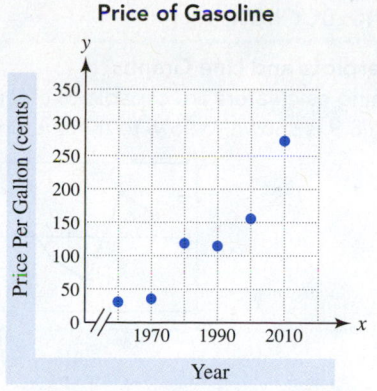

Figure 3.8

❚ **Now Try Exercise 25**

Sometimes it is helpful to connect consecutive data points in a scatterplot with line segments. This type of graph visually emphasizes changes in the data and is called a **line graph**. When making a line graph, be sure to plot all of the given data points *before* connecting the points with line segments. The points should be connected consecutively from left to right on the scatterplot, even if the data are given "out of order" in a table.

NOTE: In Section 3.2 we will discuss the graph of a linear equation. It is important to note that a line graph is **not** the same as the graph of a linear equation. A line graph is a *finite* number of data points connected with line segments, while a graph of a linear equation is a single straight line that represents an *infinite* number of data points.

EXAMPLE 4   **Making a line graph**

Use the data in Table 3.3 to make a line graph.

TABLE 3.3

x	-2	-1	0	1	2
y	1	2	-2	-1	1

Solution
The data in Table 3.3 are represented by the five ordered pairs $(-2, 1)$, $(-1, 2)$, $(0, -2)$, $(1, -1)$, and $(2, 1)$. Plot these points and then connect consecutive points with line segments, as shown in Figure 3.9.

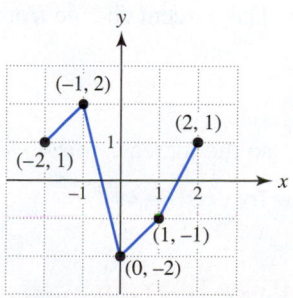

Figure 3.9 A Line Graph

❚ **Now Try Exercise 35**

TECHNOLOGY NOTE

Scatterplots and Line Graphs
Graphing calculators are capable of creating line graphs and scatterplots. The line graph in Figure 3.9 is shown (below) to the left, and the corresponding scatterplot is shown to the right.

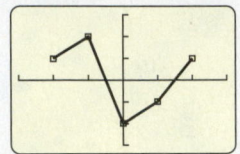 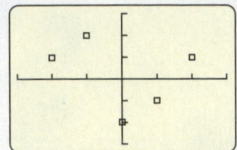

EXAMPLE 5 | **Analyzing a line graph**

The line graph in Figure 3.10 shows the per capita energy consumption in the United States from 1960 to 2010. Units are in millions of Btu, where 1 Btu equals the amount of energy necessary to heat 1 pound of water $1°F$. (*Source:* Department of Energy.)

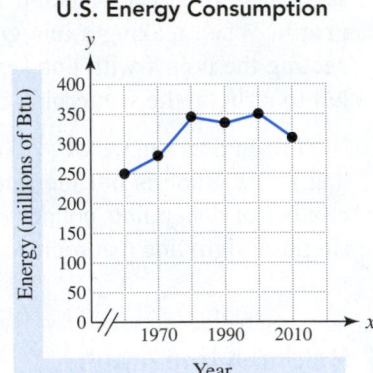

U.S. Energy Consumption

Figure 3.10

(a) Did energy consumption ever decrease during this time period? Explain.
(b) Estimate the energy consumption in 1960 and in 2000.
(c) Estimate the percent change in energy consumption from 1960 to 2000.

Solution
(a) Yes, energy consumption decreased slightly between 1980 and 1990, and again between 2000 and 2010.
(b) From the graph, per capita energy consumption in 1960 was about 250 million Btu. In 2000 it was about 350 million Btu.

  NOTE: When different people read a graph, values they obtain may vary slightly.

(c) The percent change from 1960 to 2000 was

$$\frac{350 - 250}{250} \cdot 100 = 40,$$

so the increase was 40%. (To review the percent change formula, refer to Section 2.3.)

Now Try Exercise 49

READING CHECK

• What is the main difference between a scatterplot and a line graph?

CRITICAL THINKING

When analyzing data, do you prefer a table of values, a scatterplot, or a line graph? Explain your answer.

3.1 Putting It All Together

CONCEPT	EXPLANATION	EXAMPLES
Ordered Pair	Has the form (x, y), where the order of x and y is important	$(1, 2)$, $(-2, 3)$, $(2, 1)$ and $(-4, -2)$ are distinct ordered pairs.
Rectangular Coordinate System, or xy-plane	Consists of a horizontal x-axis and a vertical y-axis that intersect at the origin Has four quadrants, which are numbered counterclockwise as I, II, III, and IV	
Scatterplot	Individual points that are plotted in the xy-plane	The points $(1, 1)$, $(2, 3)$, $(3, 2)$, $(4, 5)$, and $(5, 4)$ are plotted in the graph.
Line Graph	Similar to a scatterplot except that line segments are drawn between consecutive data points	

3.1 Exercises

MyMathLab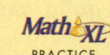

CONCEPTS AND VOCABULARY

1. Another name for the rectangular coordinate system is the _____.

2. The point where the x-axis and y-axis intersect is called the _____.

3. In the xy-plane, the origin corresponds to the ordered pair _____.

4. How many quadrants are there in the xy-plane?

5. (True or False?) Every point in the xy-plane is located in one of the quadrants.

6. In the *xy*-plane, the first value in an ordered pair is called the _____ -coordinate and the second value is called the _____ -coordinate.

7. If distinct points are plotted in the *xy*-plane, the resulting graph is called a(n) _____.

8. If the consecutive points in a scatterplot are connected with line segments, the resulting graph is called a(n) _____ graph.

RECTANGULAR COORDINATE SYSTEM

Exercises 9–12: Identify the coordinates of each point in the graph.

9.
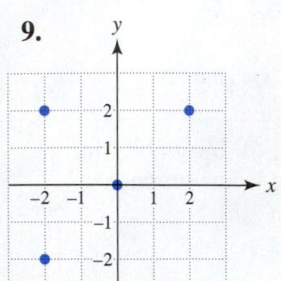

10.
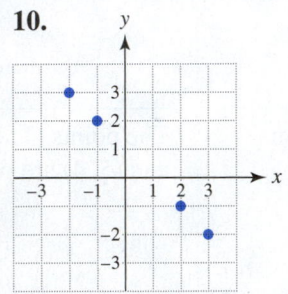

11.
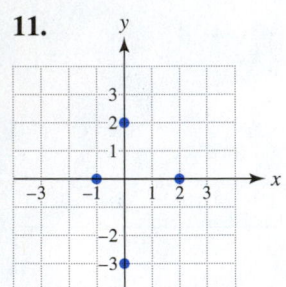

12.
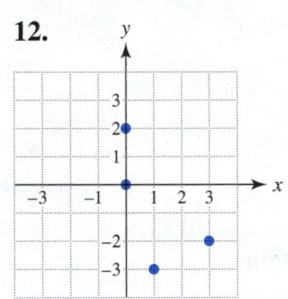

Exercises 13–16: Plot the given ordered pairs in the same xy-plane. If possible, state the quadrant in which each of the points is located.

13. $(1, 3)$, $(0, -3)$, and $(-2, 2)$

14. $(4, 0)$, $(-3, -4)$, and $(2, -3)$

15. $(0, 6)$, $(8, -4)$, and $(-6, -6)$

16. $(-4, 8)$, $(6, 8)$, and $(-8, 0)$

Exercises 17–22: If possible, identify the quadrant in which each point is located.

17. (a) $(1, 4)$ **(b)** $(-1, -4)$

18. (a) $(-2, -3)$ **(b)** $(2, -3)$

19. (a) $(7, 0)$ **(b)** $(0.1, 7)$

20. (a) $(100, -3)$ **(b)** $(-100, 3)$

21. (a) $\left(-\frac{1}{2}, \frac{3}{4}\right)$ **(b)** $\left(\frac{3}{4}, -\frac{1}{2}\right)$

22. (a) $(1.2, 0)$ **(b)** $(0, -1.2)$

23. Thinking Generally Which of the four quadrants contain points whose *x*- and *y*-coordinates have the same sign?

24. Thinking Generally Which of the four quadrants contain points whose *x*- and *y*-coordinates have different signs?

SCATTERPLOTS AND LINE GRAPHS

Exercises 25–34: Make a scatterplot by plotting the given points. Be sure to label each axis.

25. $(0, 0)$, $(1, 2)$, $(-3, 2)$, $(-1, -2)$

26. $(0, -3)$, $(-2, 1)$, $(2, 2)$, $(-4, -4)$

27. $(-1, 0)$, $(4, -3)$, $(0, -1)$, $(3, 4)$

28. $(1, 1)$, $(-2, 2)$, $(-3, -3)$, $(4, -4)$

29. $(2, 4)$, $(-4, 4)$, $(0, -4)$, $(-6, 2)$

30. $(4, 8)$, $(8, 4)$, $(-8, -4)$, $(-4, 0)$

31. $(5, 0)$, $(5, -5)$, $(-10, -20)$, $(10, -10)$

32. $(10, 30)$, $(-20, 10)$, $(40, 0)$, $(-30, -10)$

33. $(0, 0.1)$, $(0.2, -0.3)$, $(-0.1, 0.4)$

34. $(1.5, 2.5)$, $(-1, -1.5)$, $(-2.5, 0)$

Exercises 35–40: Use the table to make a line graph.

35.

x	−2	−1	0	1	2
y	2	1	0	−1	−2

36.

x	−4	−2	0	2	4
y	4	−2	3	−1	2

37.

x	−10	−5	0	5	10
y	20	−10	10	0	−20

38.

x	1	2	3	4	5
y	2	3	1	5	4

39.

x	−5	5	−10	10	0
y	10	20	30	20	40

40.

x	3	-2	2	1	-3
y	4	3	3	-2	-3

Exercises 41 and 42: Identify the coordinates of each point in the graph. Then explain what the coordinates of the first point indicate.

41. Cigarette consumption in the United States (*Source:* U.S. Department of Health and Human Services.)

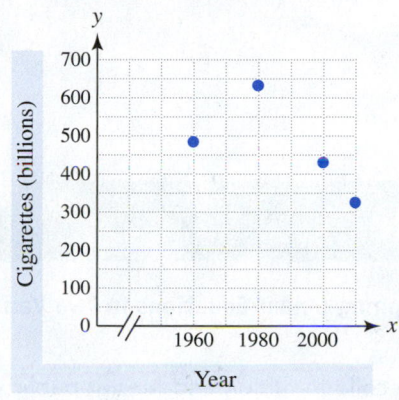

U.S. Cigarette Consumption

42. Number of unhealthy air quality days in Pittsburgh (*Source:* Environmental Protection Agency.)

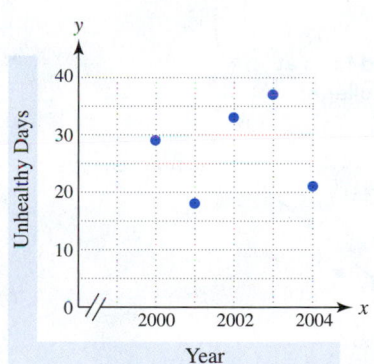

GRAPHING REAL DATA

*Exercises 43–48: **Graphing Real Data** The table contains real data.*
   (a) Make a line graph of the data. Be sure to label the axes.
   (b) Comment on any trends in the data.

43. Percent P of total music sales that were digital downloads during year t

t	2006	2007	2008	2009	2010
P	10%	20%	30%	40%	50%

Source: The NPD Group.

44. Federal income tax receipts I in billions during year t

t	1970	1980	1990	2000	2010
I	90	244	467	1003	2165

Source: Office of Management and Budget.

45. Welfare beneficiaries B in millions during year t

t	1970	1980	1990	2000	2010
B	7	11	12	6	4

Source: Administration for Children and Families.

46. U.S. cotton production C in millions of bales during year t

t	2003	2004	2005	2006	2007
C	18	23	24	22	19

Source: U.S. Department of Agriculture.

47. Projected U.S. Internet users y in millions during year x

x	2010	2011	2012	2013	2014
y	221	229	237	244	251

Source: eMarketer.

48. Number of farms in Iowa F in thousands during year x

x	2001	2003	2005	2007
F	92	90	89	92

Source: U.S. Department of Agriculture.

49. The line graph shows the U.S. infant mortality rate for selected years.

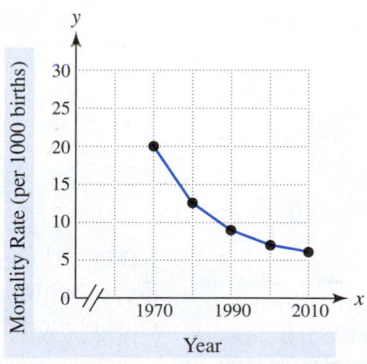

   (a) Comment on any trends in the data.
   (b) Estimate the infant mortality rate in 1990.
   (c) Estimate the percent change in the infant mortality rate from 1970 to 2010.

50. The line graph shows the population of the midwestern states, in millions, for selected years.

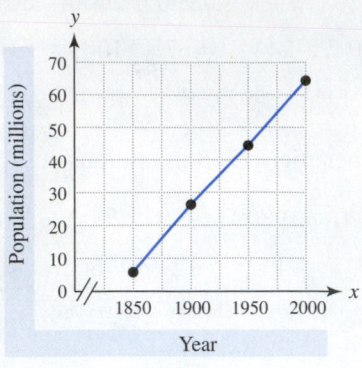

(a) Comment on any trends in the data.
(b) Estimate this population in 1900.
(c) Estimate the percent change in population from 1850 to 2000.

WRITING ABOUT MATHEMATICS

51. Explain how to identify the quadrant that a point lies in if it has coordinates (x, y).

52. Explain the difference between a scatterplot and a line graph. Give an example of each.

3.2   Linear Equations in Two Variables

Basic Concepts • Tables of Solutions • Graphing Linear Equations in Two Variables

A LOOK INTO MATH ▶

Figure 3.11 shows a scatterplot of average college tuition and fees at public colleges and universities (in 2010 dollars), together with a line that models the data. If we could find an equation for this line, then we could use it to estimate tuition and fees for years without data points, such as 2015. In this section we discuss linear equations, whose graphs are lines. Linear equations and lines are often used to approximate data. (*Source:* The College Board.)

NEW VOCABULARY

☐ Standard form (of a linear equation in two variables)
☐ Linear equation in two variables

Tuition and Fees at Public Colleges

Figure 3.11

STUDY TIP

If you have tried to solve a homework problem but need help, ask a question in class. Other students will likely have the same question.

NOTE: The graph shown in Figure 3.11 is *not* a line graph that has consecutive points connected with line segments. Rather, it is a scatterplot together with a single straight line that closely models the data.

Basic Concepts

Equations can have any number of variables. In Chapter 2 we solved equations having one variable. The following are examples of equations with two variables.

$$y = 2x, \quad 3x + 2y = 4, \quad z = t^2, \quad \text{and} \quad a - b = 1$$

Equations with two variables

A solution to an equation with one variable is one number that makes the statement true. For example, the solution to $x - 1 = 3$ is 4 because $4 - 1 = 3$ is a true statement. A solution to an equation with two variables consists of two numbers, one for each variable, which can be expressed as an ordered pair. For example, one solution to the equation $y = 5x$ is given by $x = 1$ and $y = 5$ because $5 = 5(1)$ is a true statement. This solution can be expressed as the ordered pair $(1, 5)$. Table 3.4 lists several ordered pairs and shows whether each ordered pair is a solution to the equation $y = 5x$.

TABLE 3.4 Checking Ordered-Pair Solutions to $y = 5x$

Ordered Pair	Check	Is it a solution?
$(1, 5)$	$5 \stackrel{?}{=} 5(1)$	Yes, because $5 = 5$
$(2, 12)$	$12 \stackrel{?}{=} 5(2)$	No, because $12 \neq 10$
$(-2, -10)$	$-10 \stackrel{?}{=} 5(-2)$	Yes, because $-10 = -10$
$(4, 20)$	$20 \stackrel{?}{=} 5(4)$	Yes, because $20 = 20$

READING CHECK

• How are solutions to an equation in two variables expressed?

EXAMPLE 1 **Testing solutions to equations**

Determine whether the given ordered pair is a solution to the given equation.

(a) $y = x + 3, (1, 4)$ (b) $2x - y = 5, \left(\frac{1}{2}, -4\right)$ (c) $-4x + 5y = 20, (-5, 1)$

Solution
(a) Let $x = 1$ and $y = 4$ in the given equation.

$$y = x + 3 \qquad \text{Given equation}$$
$$4 \stackrel{?}{=} 1 + 3 \qquad \text{Substitute.}$$
$$4 = 4 \ \checkmark \qquad \text{The solution checks.}$$

The ordered pair $(1, 4)$ is a solution.

(b) Let $x = \frac{1}{2}$ and $y = -4$ in the given equation.

$$2x - y = 5 \qquad \text{Given equation}$$
$$2\left(\frac{1}{2}\right) - (-4) \stackrel{?}{=} 5 \qquad \text{Substitute.}$$
$$1 + 4 \stackrel{?}{=} 5 \qquad \text{Simplify the left side.}$$
$$5 = 5 \ \checkmark \qquad \text{The solution checks.}$$

The ordered pair $\left(\frac{1}{2}, -4\right)$ is a solution.

(c) Let $x = -5$ and $y = 1$ in the given equation.

$$-4x + 5y = 20 \qquad \text{Given equation}$$
$$-4(-5) + 5(1) \stackrel{?}{=} 20 \qquad \text{Substitute.}$$
$$20 + 5 \stackrel{?}{=} 20 \qquad \text{Simplify the left side.}$$
$$25 \neq 20 \ \text{✗} \qquad \text{The solution does \textit{not} check.}$$

The ordered pair $(-5, 1)$ is *not* a solution.

Now Try Exercises 11, 15, 17

Tables of Solutions

A table can be used to list solutions to an equation. For example, Table 3.5 lists solutions to $x + y = 5$, where the sum of each xy-pair equals 5.

TABLE 3.5 $x + y = 5$

x	-2	-1	0	1	2
y	7	6	5	4	3

READING CHECK

• When is it helpful to have a table that lists a few solutions to an equation?

Most equations in two variables have infinitely many solutions, so it is impossible to list all solutions in a table. However, when you are graphing an equation, having a table that lists a few solutions to the equation is often helpful. The next two examples demonstrate how to complete a table for a given equation.

EXAMPLE 2 Completing a table of solutions

Complete the table for the equation $y = 2x - 3$.

x	-4	-2	0	2
y				

Solution

Start by determining the corresponding y-value for each x-value in the table. For example, when $x = -2$, the equation $y = 2x - 3$ implies that $y = 2(-2) - 3 = -4 - 3 = -7$. Filling in the y-values results in Table 3.6.

TABLE 3.6 $y = 2x - 3$

x	-4	-2	0	2
y	-11	-7	-3	1

Now Try Exercise 21

EXAMPLE 3 Making a table of solutions

Use $y = 0, 5, 10,$ and 15 to make a table of solutions to $5x + 2y = 10$.

Solution

Begin by listing the required y-values in the table. Next determine the corresponding x-values for each y-value by using the equation $5x + 2y = 10$.

TABLE 3.7 $5x + 2y = 10$

x	2	0	-2	-4
y	0	5	10	15

When $y = 0,$ When $y = 5,$ When $y = 10,$ When $y = 15,$

$5x + 2(0) = 10$ $5x + 2(5) = 10$ $5x + 2(10) = 10$ $5x + 2(15) = 10$

$5x + 0 = 10$ $5x + 10 = 10$ $5x + 20 = 10$ $5x + 30 = 10$

$5x = 10$ $5x = 0$ $5x = -10$ $5x = -20$

$x = 2$ $x = 0$ $x = -2$ $x = -4$

Filling in the x-values results in Table 3.7.

Now Try Exercise 23

▶ **REAL-WORLD CONNECTION** Formulas can sometimes be difficult for people to understand. As a result, newspapers, magazines, and books often list numbers in a table rather than presenting a formula for the reader to use. The next example illustrates a situation in which a table might be preferable to a formula.

EXAMPLE 4   **Calculating appropriate lengths of crutches**

People with leg injuries often need crutches. An appropriate crutch length L in inches for an injured person who is t inches tall is estimated by $L = 0.72t + 2$. (*Source: Journal of the American Physical Therapy Association.*)

(a) Complete the table. Round values to the nearest inch.

t	60	65	70	75	80
L					

(b) Use the table to determine the appropriate crutch length for a person 5 feet 10 inches tall.

Solution

(a) In the formula $L = 0.72t + 2$, if $t = 60$, then $L = 0.72(60) + 2 = 43.2 + 2 = 45.2$, or about 45 inches. If $t = 65$, then $L = 0.72(65) + 2 = 46.8 + 2 = 48.8 \approx 49$. Other values in Table 3.8 are found similarly.

TABLE 3.8 Crutch Lengths

t	60	65	70	75	80
L	45	49	52	56	60

(b) A person who is 5 feet 10 inches tall is $5 \cdot 12 + 10 = 70$ inches tall. Table 3.8 shows that a person 70 inches tall needs crutches that are about 52 inches long.

Now Try Exercise 71

Graphing Linear Equations in Two Variables

Many times graphs are used in mathematics to make concepts easier to visualize and understand. Graphs can be either curved or straight; however, graphs of linear equations in two variables are always straight lines.

EXAMPLE 5   **Graphing a linear equation in two variables**

Make a table of values for the equation $y = 2x$, and then use the table to graph this equation.

Solution

Start by selecting a few convenient values for x, such as $x = -1, 0, 1$, and 2. Then complete the table by doubling each x-value to obtain the corresponding y-value.

NOTE: Tables can be either horizontal or vertical. Table 3.9 is given in a vertical format.

TABLE 3.9 $y = 2x$

x	y
-1	-2
0	0
1	2
2	4

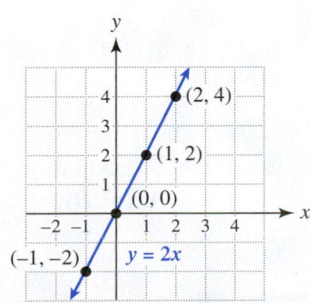

Figure 3.12

To graph the equation $y = 2x$, start by plotting the points $(-1, -2)$, $(0, 0)$, $(1, 2)$, and $(2, 4)$ given in Table 3.9, as shown in Figure 3.12.

Note that all four points in Figure 3.12 on the previous page appear to lie on the same (straight) line. Because there are infinitely many points that satisfy the equation, we can draw a line through these points.

NOTE: Because the graph of a linear equation in two variables is a straight line, you may want to use a ruler or other straight edge when graphing this kind of equation.

❙ **Now Try Exercise 39**

MAKING CONNECTIONS

Line graphs and graphs of linear equations

Do not confuse a line graph with the graph of a linear equation. A *line graph* results when consecutive points from a table of values are connected with line segments. However, the *graph of a linear equation* is a single, straight line that passes through points from a table of values and continues indefinitely in two directions.

READING CHECK

• Is a line graph the same as a graph of a linear equation?

Not every equation with two variables has a graph that is a straight line. Those equations whose graphs are straight lines are called *linear* equations in two variables. Every linear equation in two variables can be written in the following **standard form**.

LINEAR EQUATION IN TWO VARIABLES

A **linear equation in two variables** can be written as

$$Ax + By = C,$$

where A, B, and C are fixed numbers (constants) and A and B are not both equal to 0. The graph of a linear equation in two variables is a line.

NOTE: In mathematics a line is always *straight*.

In Example 5, the equation $y = 2x$ is a linear equation because it can be written in standard form by adding $-2x$ to each side.

$$y = 2x \qquad \text{Given equation}$$
$$-2x + y = -2x + 2x \qquad \text{Add } -2x \text{ to each side.}$$
$$-2x + y = 0 \qquad \text{Simplify.}$$

The equation $-2x + y = 0$ is a linear equation in two variables because it is in the form $Ax + By = C$ with $A = -2$, $B = 1$, and $C = 0$.

EXAMPLE 6 **Graphing linear equations**

Graph each linear equation.
(a) $y = \frac{1}{2}x - 1$ (b) $x + y = 4$

Solution

(a) Because $y = \frac{1}{2}x - 1$ can be written in standard form as $-\frac{1}{2}x + y = -1$, it is a linear equation in two variables and its graph is a line. Two points determine a line. However, it is a good idea to plot three points to be sure that the line is graphed correctly. Start by choosing three values for x and then calculate the corresponding y-values, as shown in Table 3.10. In Figure 3.13, the points $(-2, -2)$, $(0, -1)$, and $(2, 0)$ are plotted and the line passing through these points is drawn.

TABLE 3.10 $y = \frac{1}{2}x - 1$

x	y
-2	-2
0	-1
2	0

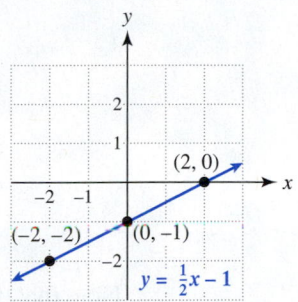

Figure 3.13

(b) The equation $x + y = 4$ is a linear equation in two variables, so its graph is a line. If an ordered pair (x, y) is a solution to the given equation, then the sum of x and y is 4. Table 3.11 shows three examples. In Figure 3.14, the points $(0, 4)$, $(2, 2)$, and $(4, 0)$ are plotted with the line passing through each one.

TABLE 3.11 $x + y = 4$

x	y
0	4
2	2
4	0

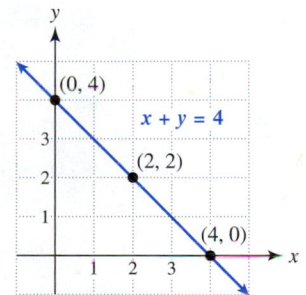

Figure 3.14

Now Try Exercises 43, 57

TECHNOLOGY NOTE

CALCULATOR HELP

To graph an equation, see Appendix A (page AP-5).

Graphing Equations
Graphing calculators can be used to graph equations. Before graphing the equation $x + y = 4$, solve the equation for y to obtain $y = 4 - x$. A calculator graph of Figure 3.14 is shown, except that the three points have not been plotted.

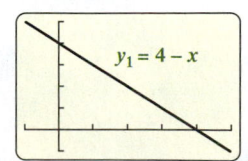

When a linear equation in two variables is given in standard form, it is sometimes difficult to create a table of solutions. Solving an equation for y often makes it easier to select x-values for the table that will make the y-values simpler to calculate. This is demonstrated in the next example.

EXAMPLE 7 | **Solving for *y* and then graphing**

Graph the linear equation $4x - 3y = 12$ by solving for *y* first.

Solution
First solve the given equation for *y*.

$$4x - 3y = 12 \qquad \text{Given equation}$$

$$-3y = -4x + 12 \qquad \text{Subtract } 4x \text{ from each side.}$$

$$\frac{-3y}{-3} = \frac{-4x + 12}{-3} \qquad \text{Divide each side by } -3.$$

$$\frac{-3y}{-3} = \frac{-4x}{-3} + \frac{12}{-3} \qquad \text{Property of fractions, } \frac{a+b}{c} = \frac{a}{c} + \frac{b}{c}$$

$$y = \frac{4}{3}x - 4 \qquad \text{Simplify fractions.}$$

Note that dividing each side of an equation by -3 is equivalent to dividing each term in the equation by -3.

Select multiples of 3 (the denominator of $\frac{4}{3}$) as *x*-values for the table of solutions. For example, if $x = \mathbf{6}$ is chosen, $y = \frac{4}{3}(\mathbf{6}) - 4 = \frac{24}{3} - 4 = 8 - 4 = 4$. Table 3.12 lists the solutions $(0, -4)$, $(3, 0)$, and $(6, 4)$, which are plotted in Figure 3.15 with the line passing through each one.

TABLE 3.12 $y = \frac{4}{3}x - 4$

x	y
0	−4
3	0
6	4

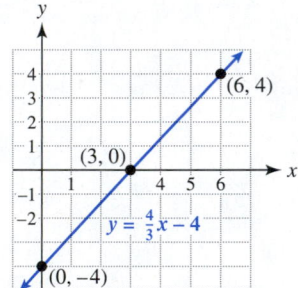

Figure 3.15

Now Try Exercise 59

MAKING CONNECTIONS

Graphs and Solution Sets

A graph visually depicts the set of solutions to an equation. Each point on the graph represents one solution to the equation.

NOTE: In a linear equation, each *x*-value determines a unique *y*-value. Because there are infinitely many *x*-values, there are infinitely many points located on the graph of a linear equation. Thus the graph of a linear equation is a *continuous* line with no breaks.

3.2 Putting It All Together

CONCEPT	EXPLANATION	EXAMPLES
Equation in Two Variables	An equation that has two variables	$y = 4x + 5$, $4x - 5y = 20$, and $u - v = 100$
Solution to an Equation in Two Variables	An ordered pair (x, y) whose x- and y-values satisfy the equation Equations in two variables often have infinitely many solutions.	$(1, 3)$ is a solution to $3x + y = 6$ because $3(1) + 3 = 6$ is a true statement. The equation $y = 2x$ has infinitely many solutions, such as $(1, 2)$, $(2, 4)$, $(3, 6)$, and so on.
Table of Solutions	A table can be used to list solutions to an equation in two variables.	The following table lists solutions to the equation $y = 3x - 1$. $\begin{array}{c\|cccc} x & -1 & 0 & 1 & 2 \\ \hline y & -4 & -1 & 2 & 5 \end{array}$
Linear Equation in Two Variables	Can be written as $$Ax + By = C,$$ where A, B, and C are fixed numbers and A and B are not both equal to 0	$3x + 4y = 5$ $y = 4 - 3x$ (or $3x + y = 4$) $x = 2y + 1$ (or $x - 2y = 1$) The graph of each equation is a line.

3.2 Exercises

MyMathLab   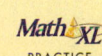 Math XL PRACTICE   WATCH   DOWNLOAD   READ   REVIEW

CONCEPTS AND VOCABULARY

1. (True or False?) The equation $4x + 6y = 24$ is an equation in two variables.

2. (True or False?) The equation $4x - 3x = 10$ is a linear equation in two variables.

3. A solution to an equation with two variables consists of two numbers expressed as a(n) _____.

4. A(n) _____ equation in two variables can be written in the form $Ax + By = C$.

5. (True or False?) A table of solutions lists all of the solutions to an equation in two variables.

6. (True or False?) A linear equation in two variables has infinitely many solutions.

7. An equation's _____ visually depicts its solution set.

8. The graph of a linear equation in two variables is a(n) _____.

9. (True or False?) Every equation in two variables has a graph that is a straight line.

10. (True or False?) A line graph and the graph of a linear equation in two variables are the same thing.

SOLUTIONS TO EQUATIONS

Exercises 11–20: Determine whether the ordered pair is a solution to the given equation.

11. $y = x + 1$, $(5, 6)$

12. $y = 4 - x$, $(6, 2)$

13. $y = 4x + 7$, $(2, 13)$

14. $y = -3x + 2$, $(-2, 8)$

15. $4x - y = -13$, $(-2, 3)$

16. $3y + 2x = 0$, $(-2, 3)$

17. $y - 6x = -1$, $\left(\frac{1}{2}, 2\right)$

18. $\frac{1}{2}x + \frac{3}{2}y = 0$, $\left(-\frac{3}{2}, \frac{1}{2}\right)$

19. $0.31x - 0.42y = -9$, $(100, 100)$

20. $0.5x - 0.6y = 4$, $(20, 10)$

TABLES OF SOLUTIONS

Exercises 21–28: Complete the table for the given equation.

21. $y = 4x$

x	−2	−1	0	1	2
y	−8				

22. $y = \frac{1}{2}x - 1$

x	0	1	2	3	4
y	−1				

23. $3y + 2x = 6$

x					
y	−2	0	2	4	8

24. $3x - 5y = 30$

x					
y	−9	−6	−3	0	3

25. $y = x + 4$

x	y
−8	−4
	0
	4
	8
	12

26. $2x - y = 1$

x	y
−1	−3
	−1
	0
	1
	3

27. $2x - 6y = 12$

x	y
−6	
0	
6	
12	
18	

28. $4x + 3y = 12$

x	y
9	
6	
3	
0	
−3	

Exercises 29–32: Use the given values of the variable to make a horizontal table of solutions for the equation.

29. $y = 3x$ $x = -3, 0, 3, 6$

30. $y = 1 - 2x$ $x = 0, 1, 2, 3$

31. $x + y = 6$ $y = -2, 0, 2, 4$

32. $2x - 3y = 9$ $y = -3, 0, 1, 2$

Exercises 33–36: Use the given values of the variable to make a vertical table of solutions for the equation.

33. $y = \dfrac{x + 4}{2}$ $x = -8, -4, 0, 4$

34. $y = \dfrac{x}{3} - 1$ $x = 0, 2, 4, 6$

35. $y - 4x = 0$ $y = -2, -1, 0, 1$

36. $-4x = 6y - 4$ $y = -1, 0, 1, 2$

37. Thinking Generally If a student wishes to avoid fractional *y*-values when making a table of solutions for $y = \frac{3}{5}x - 7$, what must be true about any selected integer *x*-values?

38. Thinking Generally If a student wishes to avoid fractional *y*-values when making a table of solutions for $y = \frac{a}{b}x$, where *a* and *b* are natural numbers, what must be true about any selected integer *x*-values?

GRAPHING EQUATIONS

Exercises 39–44: Make a table of solutions for the equation, and then use the table to graph the equation.

39. $y = -2x$ **40.** $y = 2x - 1$

41. $x = 3 - y$ **42.** $x = y + 1$

43. $x + 2y = 4$ **44.** $2x - y = 1$

Exercises 45–58: Graph the equation.

45. $y = x$ **46.** $y = \frac{1}{2}x$

47. $y = \frac{1}{3}x$ **48.** $y = -2x$

49. $y = x + 3$ **50.** $y = x - 2$

51. $y = 2x + 1$ **52.** $y = \frac{1}{2}x - 1$

53. $y = 4 - 2x$ **54.** $y = 2 - 3x$

55. $y = 7 + x$ **56.** $y = 2 + 2x$

57. $y = -\frac{1}{2}x + \frac{1}{2}$ **58.** $y = -\frac{3}{4}x + 2$

Exercises 59–70: Graph the linear equation by solving for y first.

59. $2x + 3y = 6$ **60.** $3x + 2y = 6$

61. $x + 4y = 4$ **62.** $4x + y = -4$

63. $-x + 2y = 8$ **64.** $-2x + 6y = 12$

65. $y - 2x = 7$ **66.** $3y - x = 2$

67. $5x - 4y = 20$ **68.** $4x - 5y = -20$

69. $3x + 5y = -9$ **70.** $5x - 3y = 10$

APPLICATIONS

71. *U.S. Population* For the years 2010 to 2050, the projected percentage P of the U.S. population that will be over the age of 65 during year t is estimated by $P = 0.178t - 344.6$. (*Source:* U.S. Census Bureau.)
  (a) Complete the table. Round each resulting value to the nearest tenth.

t	2010	2020	2030	2040	2050
P					

  (b) Use the table to find the year when the percentage of the population over the age of 65 is expected to reach 16.7%.

72. *U.S. Population* For the years 2010 to 2050, the projected percentage P of the U.S. population that will be 18 to 24 years old during year t is estimated by $P = -0.025t + 60.35$. (*Source:* U.S. Census Bureau.)
  (a) Complete the table. Round each resulting value to the nearest tenth.

t	2010	2020	2030	2040	2050
P					

  (b) Use the table to find the year when the percentage of the population that is 18 to 24 years old is expected to be 9.4%.

73. *Solid Waste in the Past* In 1960 the amount A of garbage in pounds produced after t days by the average American is given by $A = 2.7t$. (*Source:* Environmental Protection Agency.)
  (a) Graph the equation for $t \geq 0$.
  (b) How many days did it take for the average American in 1960 to produce 100 pounds of garbage?

74. *Solid Waste Today* Today the amount A of garbage in pounds produced after t days by the average American is given by $A = 4.5t$. (*Source:* Environmental Protection Agency.)
  (a) Graph the equation for $t \geq 0$.
  (b) How many days does it take for the average American to produce 100 pounds of garbage today?

75. *Digital Music Sales* From 2006 to 2010 the percent P of total music sales from digital downloads is modeled by $P = 10t - 20,050$, where t is the year. (*Source:* Recording Industry Association of America.)
  (a) Evaluate P for $t = 2006$ and for $t = 2010$.
  (b) Use your results from part (a) to graph the equation from 2006 to 2010.
  (c) In what year was $P = 30\%$?

76. *U.S. HIV Infections* The cumulative number of HIV infections I in thousands during year t is modeled by the equation $I = 42t - 83,197$ where $t \geq 2000$. (*Source:* Department of Health and Human Services.)
  (a) Evaluate I for $t = 2000$ and for $t = 2005$.
  (b) Use your results from part (a) to graph the equation from 2000 to 2005.
  (c) In what year was $I = 971$?

WRITING ABOUT MATHEMATICS

77. The number of welfare beneficiaries B in millions during year t is shown in the table. Discuss whether a linear equation might work to approximate these data from 1970 to 2010.

t	1970	1980	1990	2000	2010
B	7	11	12	6	4

Source: Administration for Children and Families.

78. The Asian-American population P in millions during year t is shown in the table. Discuss whether a linear equation might model these data from 2002 to 2010.

t	2002	2004	2006	2008	2010
P	12.0	12.8	13.6	14.4	15.2

Source: U.S. Census Bureau.

Checking Basic Concepts

1. Identify the coordinates of the four points in the graph. State the quadrant, if any, in which each point lies.

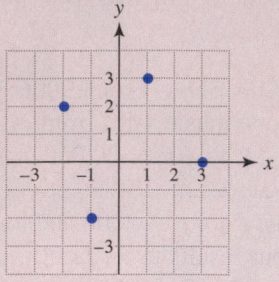

2. Make a scatterplot of the five points $(-2, -2)$, $(-1, -3)$, $(0, 0)$, $(1, 2)$, and $(2, 3)$.

3. *U.S. Population* The table gives the percentage P of the U.S. population that was over the age of 85 during year t. Make a line graph of the data in the table, and then comment on any trends.   (*Source:* U.S. Census Bureau.)

t	1970	1980	1990	2000	2010
P	0.7	1.0	1.2	1.5	2.1

4. Determine whether $(-2, -3)$ is a solution to the equation $-2x - y = 7$.

5. Complete the table for the equation $y = -2x + 1$.

x	-2	-1	0	1	2
y	5				

6. Graph the equation.
   (a) $y = \frac{1}{2}x$ (b) $4x + 6y = 12$

7. *Total Federal Receipts* The total amount of money A in trillions of dollars collected by the federal government in year t from 1990 to 2010 can be approximated by $A = 0.085t - 168.68$. (*Source:* Internal Revenue Service.)
   (a) Find A when $t = 1990$ and $t = 2010$. Interpret each result.
   (b) Use your results from part (a) to graph the equation from 1990 to 2010. Be sure to label the axes.
   (c) In what year were the total receipts equal to $1.83 trillion?

3.3   More Graphing of Lines

Finding Intercepts • Horizontal Lines • Vertical Lines

A LOOK INTO MATH ▶

When things move or change at a *constant* rate, linear equations can often be used to describe or model the situation. For example, when a car moves at a constant speed, a graph of a linear equation can be used to visualize its driver's distance from a particular location. In real-world situations such as this, the x- and y-intercepts of the graph often provide important information. We begin this section by discussing intercepts and their significance in applications.

Finding Intercepts

NEW VOCABULARY

☐ x-intercept
☐ y-intercept

▶ **REAL-WORLD CONNECTION** Suppose that someone leaves a family gathering at a state park and drives home at a constant speed of 50 miles per hour. The graph in Figure 3.16 shows the distance of the driver from home at various times. The graph intersects the y-axis at 200 miles, which is called the *y-intercept*. In this situation the y-intercept represents the initial distance (when $x = 0$) between the driver and home. The graph also intersects the *x*-axis at 4 hours, which is called the *x-intercept*. This intercept represents the elapsed time when the distance of the driver from home is 0 miles (when $y = 0$.) In other words, the driver arrived at home after 4 hours of driving.

Distance from Home

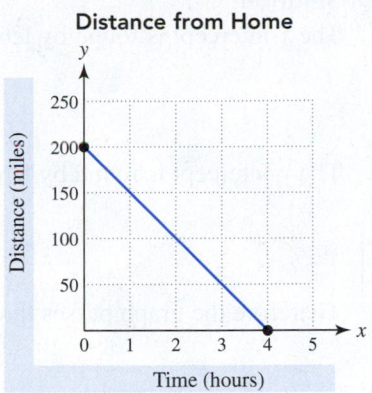

Figure 3.16

FINDING *x*- AND *y*-INTERCEPTS

The *x*-coordinate of a point where a graph intersects the *x*-axis is an **x-intercept**. To find an *x*-intercept, let $y = 0$ in the equation and solve for *x*.

The *y*-coordinate of a point where a graph intersects the *y*-axis is a **y-intercept**. To find a *y*-intercept, let $x = 0$ in the equation and solve for *y*.

READING CHECK

• Where are the intercepts located on a graph?

NOTE: Each intercept is defined as a single number, rather than an ordered pair. The *x*-coordinate of the point where a graph intersects the *x*-axis and the *y*-coordinate of the point where a graph intersects the *y*-axis often have special meaning in applications.

The graph of the linear equation $3x + 2y = 6$ is shown in Figure 3.17. The *x*-intercept is **2** and the *y*-intercept is **3**. These two intercepts can be found without a graph. To find the *x*-intercept, let $y = \mathbf{0}$ in the equation and solve for *x*.

$$3x + 2(\mathbf{0}) = 6 \quad \text{Let } y = 0.$$
$$3x = 6 \quad \text{Simplify.}$$
$$x = \mathbf{2} \quad \text{Divide each side by 3.}$$

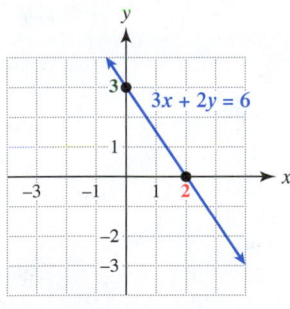

Figure 3.17

To find the *y*-intercept, let $x = \mathbf{0}$ in the equation and solve for *y*.

$$3(\mathbf{0}) + 2y = 6 \quad \text{Let } x = 0.$$
$$2y = 6 \quad \text{Simplify.}$$
$$y = \mathbf{3} \quad \text{Divide each side by 2.}$$

Note that the *x*-intercept **2** corresponds to the point $(\mathbf{2}, \mathbf{0})$ on the graph and the *y*-intercept **3** corresponds to the point $(\mathbf{0}, \mathbf{3})$ on the graph.

CRITICAL THINKING

If a line has no *x*-intercept, what can you say about the line?
If a line has no *y*-intercept, what can you say about the line?

EXAMPLE 1

Using intercepts to graph a line

Use intercepts to graph $2x - 6y = 12$.

Solution

The x-intercept is found by letting $y = \mathbf{0}$.

$$2x - 6(\mathbf{0}) = 12 \qquad \text{Let } y = 0.$$
$$x = \mathbf{6} \qquad \text{Solve for } x.$$

The y-intercept is found by letting $x = \mathbf{0}$.

$$2(\mathbf{0}) - 6y = 12 \qquad \text{Let } x = 0.$$
$$y = \mathbf{-2} \qquad \text{Solve for } y.$$

Therefore the graph passes through the points $(\mathbf{6}, \mathbf{0})$ and $(\mathbf{0}, \mathbf{-2})$, as shown in Figure 3.18.

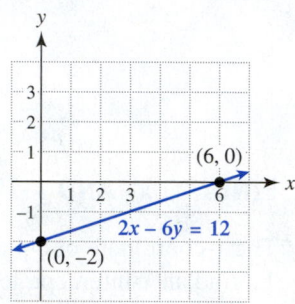

Figure 3.18

Now Try Exercise 21

In the next example a table of solutions is used to determine the x- and y-intercepts.

EXAMPLE 2

Using a table to find intercepts

Complete the table for the equation $x - y = -1$. Then determine the x-intercept and y-intercept for the graph of the equation $x - y = -1$.

x	-2	-1	0	1	2
y					

Solution

Substitute $-\mathbf{2}$ for x in $x - y = -1$ to find the corresponding y-value.

$$-\mathbf{2} - y = -1 \qquad \text{Let } x = -2.$$
$$-y = 1 \qquad \text{Add 2 to each side.}$$
$$y = -1 \qquad \text{Multiply each side by } -1.$$

The other y-values can be found similarly. See Table 3.13.

The x-intercept corresponds to a point on the graph whose y-coordinate is 0. Table 3.13 shows that the y-coordinate is 0 when $x = -1$. So the x-intercept is $\mathbf{-1}$. Similarly, the y-intercept corresponds to a point on the graph whose x-coordinate is 0. Table 3.13 shows that the x-coordinate is 0 when $y = 1$. So the y-intercept is $\mathbf{1}$.

TABLE 3.13 $x - y = -1$

x	-2	-1	0	1	2
y	-1	0	1	2	3

READING CHECK

- How can the intercepts of a graph be found using a table of values?

Now Try Exercise 17

EXAMPLE 3   **Modeling the velocity of a toy rocket**

A toy rocket is shot vertically into the air. Its velocity v in feet per second after t seconds is given by $v = 160 - 32t$. Assume that $t \geq 0$ and $t \leq 5$.

(a) Graph the equation by finding the intercepts. Let t correspond to the horizontal axis (x-axis) and v correspond to the vertical axis (y-axis).

(b) Interpret each intercept.

Solution

(a) To find the t-intercept, let $v = 0$.

$$0 = 160 - 32t \quad \text{Let } v = 0.$$
$$32t = 160 \quad \text{Add } 32t \text{ to each side.}$$
$$t = 5 \quad \text{Divide each side by 32.}$$

To find the v-intercept, let $t = 0$.

$$v = 160 - 32(0) \quad \text{Let } t = 0.$$
$$v = 160 \quad \text{Simplify.}$$

Therefore the graph passes through $(5, 0)$ and $(0, 160)$, as shown in Figure 3.19.

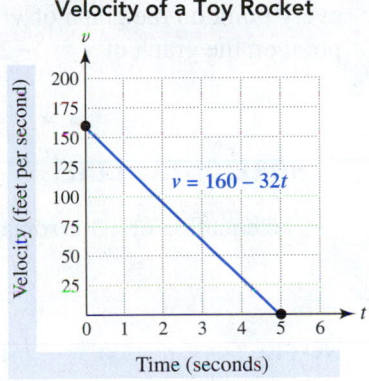

Velocity of a Toy Rocket

$v = 160 - 32t$

Time (seconds)

Figure 3.19

(b) The t-intercept indicates that the rocket had a velocity of 0 feet per second after **5** seconds. The v-intercept indicates that the rocket's initial velocity was **160** feet per second.

Now Try Exercise 77

Horizontal Lines

▶ **REAL-WORLD CONNECTION**   Suppose that someone drives a car on a freeway at a constant speed of 70 miles per hour. Table 3.14 shows the speed y after x hours.

TABLE 3.14 Speed of a Car

x	1	2	3	4	5
y	70	70	70	70	70

We can make a scatterplot of the data by plotting the five points $(1, 70)$, $(2, 70)$, $(3, 70)$, $(4, 70)$, and $(5, 70)$, as shown in Figure 3.20(a) on the next page. The speed is always

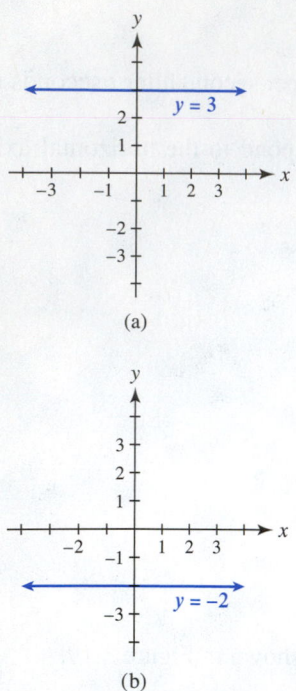

Figure 3.21

70 miles per hour and the graph of the car's speed is a horizontal line, as shown in Figure 3.20(b). The equation of this line is $y = 70$ with y-intercept 70. There are no x-intercepts.

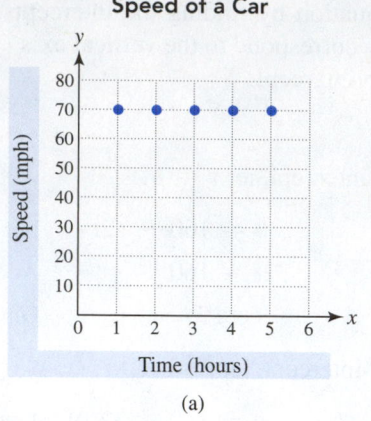

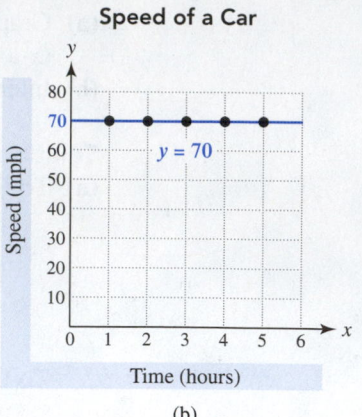

Figure 3.20

In general, the equation of a horizontal line is $y = b$, where b is a constant that corresponds to the y-intercept. Examples of horizontal lines are shown in Figure 3.21. Note that every point on the graph of $y = 3$ in Figure 3.21(a) has a y-coordinate of 3, and that every point on the graph of $y = -2$ in Figure 3.21(b) has a y-coordinate of -2.

READING CHECK

• Which variable, x or y, is *not* present in the equation for a horizontal line?

HORIZONTAL LINE

The equation of a horizontal line with y-intercept b is $y = b$.

NOTE: The equation $y = b$ is a linear equation in the form $Ax + By = C$ with $A = 0$, $B = 1$, and $C = b$. (In general B and b do not represent the same number.)

EXAMPLE 4 **Graphing a horizontal line**

Graph the equation $y = -1$ and identify its y-intercept.

Solution
The graph of $y = -1$ is a horizontal line passing through the point $(0, -1)$, as shown in Figure 3.22. Its y-intercept is -1.

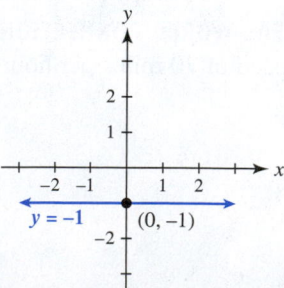

Figure 3.22

Now Try Exercise 45(a)

TABLE 3.15 **Text Messages**

x (months)	y (texts)
1	1631
1	12
1	2314
1	359

Vertical Lines

▶ **REAL-WORLD CONNECTION** A father looks at the number of text messages sent in one month by each of the four people on his wireless family plan. Table 3.15 shows the results of his research, where x represents the number of months and y represents the number of text messages.

We can make a scatterplot of the data by plotting the points $(\mathbf{1}, 1631)$, $(\mathbf{1}, 12)$, $(\mathbf{1}, 2314)$, and $(\mathbf{1}, 359)$, as shown in Figure 3.23(a). In each case the time is always 1 month and each point lies on the graph of a vertical line, as shown in Figure 3.23(b). This vertical line has the equation $x = \mathbf{1}$ because each point on the line has an x-coordinate of 1 and there are no restrictions on the y-coordinate. This line has x-intercept $\mathbf{1}$ but no y-intercept.

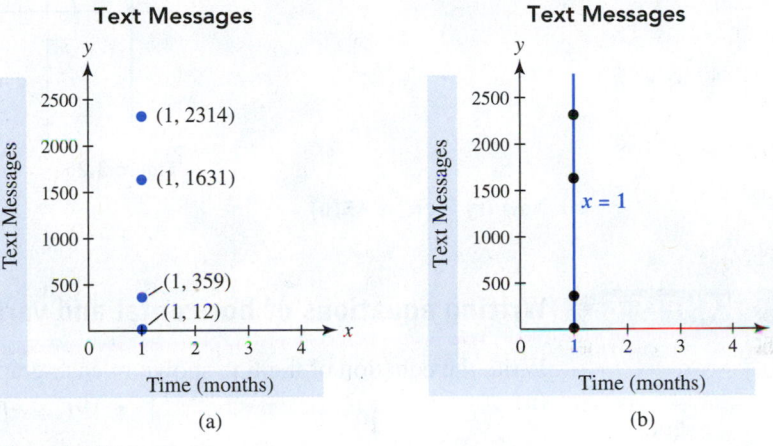

Figure 3.23

In general, the graph of a vertical line is $x = k$, where k is a constant that corresponds to the x-intercept. Examples of vertical lines are shown in Figure 3.24. Note that every point on the graph of $x = 3$ in Figure 3.24(a) has an x-coordinate of $\mathbf{3}$ and that every point on the graph of $x = -2$ shown in Figure 3.24(b) has an x-coordinate of $-\mathbf{2}$.

CALCULATOR HELP

To graph a vertical line, see Appendix A (page AP-5).

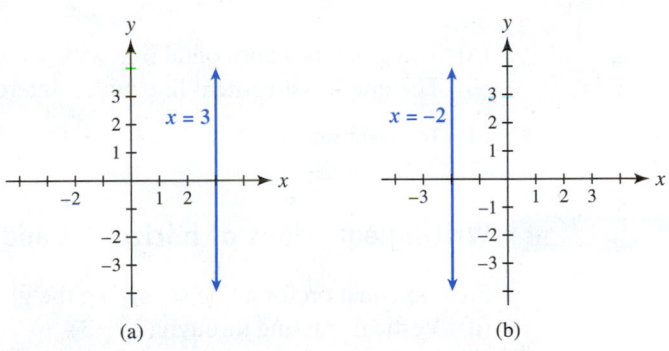

Figure 3.24

READING CHECK

• Which variable, x or y, is *not* present in the equation for a vertical line?

VERTICAL LINE

The equation of a vertical line with x-intercept k is $x = k$.

NOTE: The equation $x = k$ is a linear equation in the form $Ax + By = C$ with $A = 1$, $B = 0$, and $C = k$.

EXAMPLE 5 | **Graphing a vertical line**

Graph the equation $x = -3$, and identify its x-intercept.

Solution

The graph of $x = -3$ is a vertical line passing through the point $(-3, 0)$, as shown in Figure 3.25. Its x-intercept is -3.

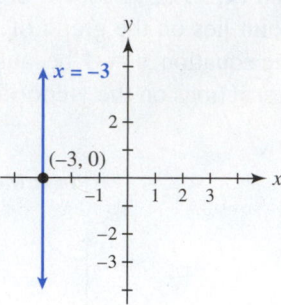

Figure 3.25

Now Try Exercise 45(b)

EXAMPLE 6 | **Writing equations of horizontal and vertical lines**

Write the equation of the line shown in each graph.

(a)

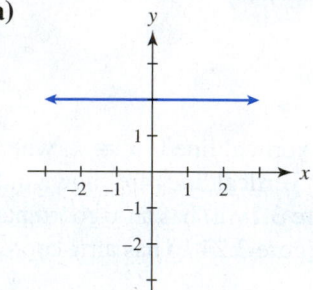

(b)

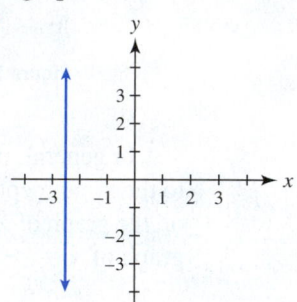

Solution

(a) The graph is a horizontal line with y-intercept 2. Its equation is $y = 2$.

(b) The graph is a vertical line with x-intercept -2.5. Its equation is $x = -2.5$.

Now Try Exercises 51, 53

EXAMPLE 7 | **Writing equations of horizontal and vertical lines**

Find an equation for a line satisfying the given conditions.

(a) Vertical, passing through $(2, -3)$

(b) Horizontal, passing through $(3, 1)$

(c) Perpendicular to $x = 3$, passing through $(-1, 2)$

Solution

(a) A vertical line passing through $(2, -3)$ has x-intercept 2, as shown in Figure 3.26(a). The equation of a vertical line with x-intercept 2 is $x = 2$.

(b) A horizontal line passing through $(3, 1)$ has y-intercept 1, as shown in Figure 3.26(b). The equation of a horizontal line with y-intercept 1 is $y = 1$.

(c) Because the line $x = 3$ is vertical, a line that is perpendicular to this line is horizontal, as shown in Figure 3.26(c). The equation of a horizontal line passing through the point $(-1, 2)$ is $y = 2$.

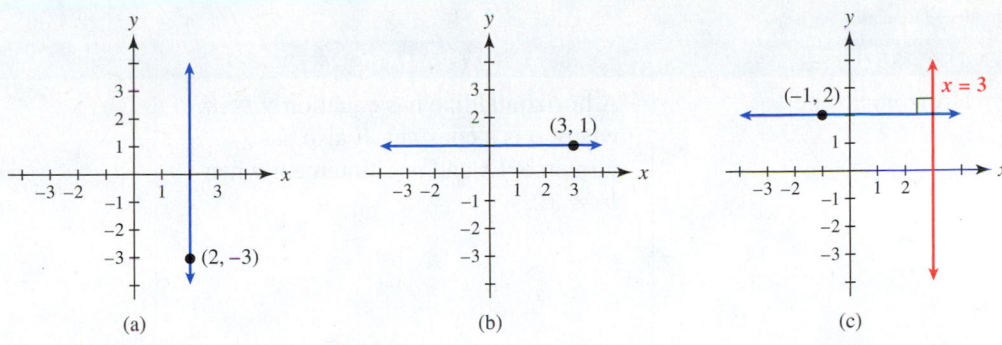

Figure 3.26 Horizontal and Vertical Lines

Now Try Exercises 65, 67, 69

MAKING CONNECTIONS

Lines and Linear Equations

The equation of any line can be written in the standard form $Ax + By = C$.

1. If $A = 0$ and $B \neq 0$, the line is horizontal.
2. If $A \neq 0$ and $B = 0$, the line is vertical.
3. If $A \neq 0$ and $B \neq 0$, the line is neither horizontal nor vertical.

3.3 Putting It All Together

CONCEPT	EXPLANATION	EXAMPLES
x- and y-Intercepts	The x-coordinate of a point at which a graph intersects the x-axis is called an x-intercept. The y-coordinate of a point at which a graph intersects the y-axis is called a y-intercept.	**x-intercept -3 and y-intercept 2**
Finding Intercepts	To find x-intercepts, let $y = 0$ in the equation and solve for x. To find y-intercepts, let $x = 0$ in the equation and solve for y.	Let $4x - 5y = 20$. x-intercept: $4x - 5(0) = 20$ $\phantom{x\text{-intercept: }4x - 5(0)} x = 5$ y-intercept: $4(0) - 5y = 20$ $\phantom{y\text{-intercept: }4(0) - 5} y = -4$ The x-intercept is 5, and the y-intercept is -4.

continued on next page

continued from previous page

CONCEPT	EXPLANATION	EXAMPLES
Horizontal Line	A horizontal line has equation $y = b$, where b is a constant. It also has y-intercept b and no x-intercept when $b \neq 0$.	**y-intercept b** $y = b$
Vertical Line	A vertical line has equation $x = k$, where k is a constant. It also has x-intercept k and no y-intercept when $k \neq 0$.	**x-intercept k** $x = k$

3.3 Exercises

MyMathLab

CONCEPTS AND VOCABULARY

1. The x-coordinate of a point where a graph intersects the x-axis is a(n) _____.

2. To find an x-intercept, let $y = $ _____ and solve for x.

3. The y-coordinate of a point at which a graph intersects the y-axis is a(n) _____.

4. To find a y-intercept, let $x = $ _____ and solve for y.

5. The graph of the linear equation $Ax + By = C$ with $A = 0$ and $B = 1$ is a(n) _____ line.

6. A horizontal line with y-intercept b has equation _____.

7. The graph of the linear equation $Ax + By = C$ with $A = 1$ and $B = 0$ is a(n) _____ line.

8. A vertical line with x-intercept k has equation _____.

FINDING INTERCEPTS

Exercises 9–16: Identify any x-intercepts and y-intercepts in the graph.

9.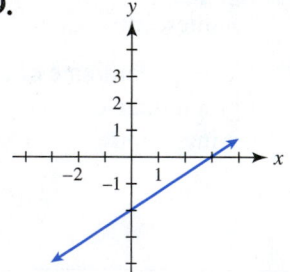

10.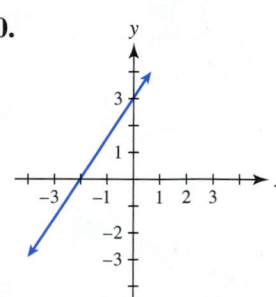

11.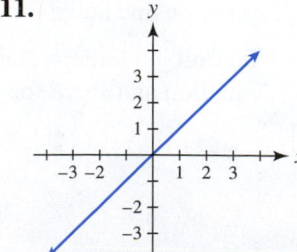

12.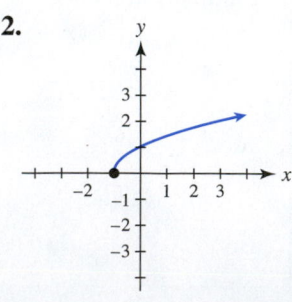

13.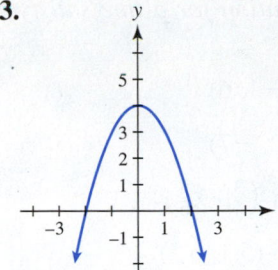

14.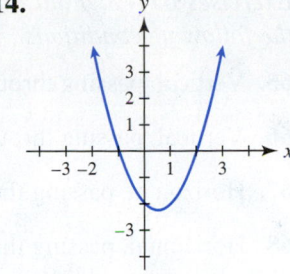

15.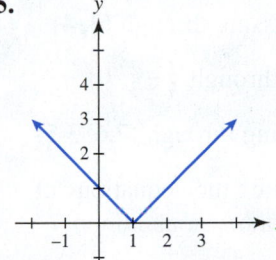

16.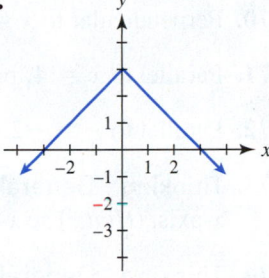

27. $3x + 7y = 21$

28. $-3x + 8y = 24$

29. $40y - 30x = -120$

30. $10y - 20x = 40$

31. $\frac{1}{2}x - y = 2$

32. $x - \frac{1}{2}y = 4$

33. $-\dfrac{x}{4} + \dfrac{y}{3} = 1$

34. $\dfrac{x}{3} - \dfrac{y}{4} = 1$

35. $\dfrac{x}{3} + \dfrac{y}{2} = 1$

36. $\dfrac{x}{5} - \dfrac{y}{4} = 1$

37. $0.6y - 1.5x = 3$

38. $0.5y - 0.4x = 2$

39. Thinking Generally Find any intercepts for the graph of $Ax + By = C$.

40. Thinking Generally Find any intercepts for the graph of $\frac{x}{A} + \frac{y}{B} = 1$.

HORIZONTAL AND VERTICAL LINES

Exercises 41–44: Write an equation for the line that passes through the points shown in the table.

41.

x	−2	−1	0	1	2
y	1	1	1	1	1

42.

x	0	1	2	3	4
y	−10	−10	−10	−10	−10

43.

x	−6	−6	−6	−6	−6
y	5	4	3	2	1

44.

x	20	20	20	20	20
y	−2	−1	0	1	2

Exercises 17–20: Complete the table. Then determine the x-intercept and the y-intercept for the graph of the equation.

17. $y = x + 2$

x	−2	−1	0	1	2
y					

18. $y = 2x - 4$

x	−2	−1	0	1	2
y					

19. $-x + y = -2$

x	−4	−2	0	2	4
y					

20. $x + y = 1$

x	−2	−1	0	1	2
y					

Exercises 21–38: Find any intercepts for the graph of the equation and then graph the linear equation.

21. $-2x + 3y = -6$

22. $4x + 3y = 12$

23. $x - 3y = 6$

24. $5x + y = -5$

25. $6x - y = -6$

26. $5x + 7y = -35$

Exercises 45–50: Graph each equation.

45. (a) $y = 2$ (b) $x = 2$

46. (a) $y = -4$ (b) $x = -4$

47. (a) $y = 0$ (b) $x = 0$

48. (a) $y = -\frac{1}{2}$ (b) $x = -\frac{1}{2}$

49. (a) $y = \frac{3}{2}$ (b) $x = \frac{3}{2}$

50. (a) $y = -1.5$ (b) $x = -1.5$

Exercises 51–58: Write an equation for the line shown in the graph.

51.
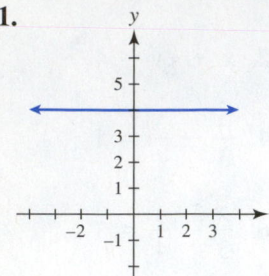

52.
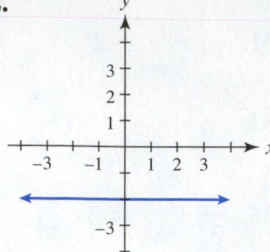

53.
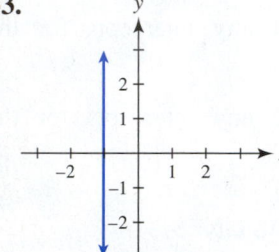

54.
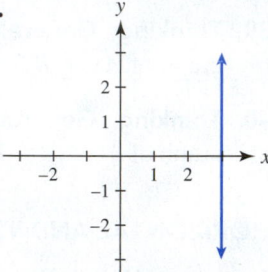

55.
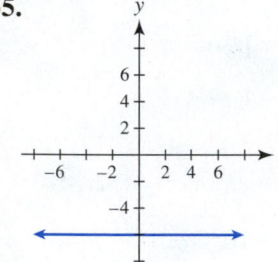

56.
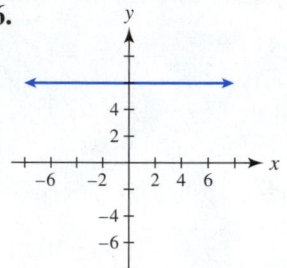

57.
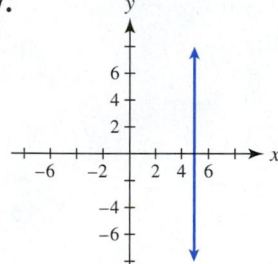

58.
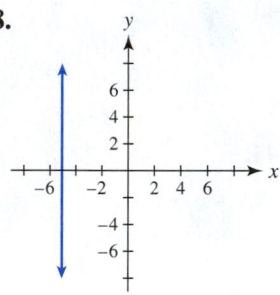

Exercises 59–64: Write the equations of a horizontal line and a vertical line that pass through the given point. (Hint: Make a sketch.)

59. $(1, 2)$

60. $(-3, 4)$

61. $(20, -45)$

62. $(-5, 12)$

63. $(0, 5)$

64. $(-3, 0)$

Exercises 65–72: Find an equation for a line satisfying the following conditions.

65. Vertical, passing through $(-1, 6)$

66. Vertical, passing through $(2, -7)$

67. Horizontal, passing through $\left(\frac{3}{4}, -\frac{5}{6}\right)$

68. Horizontal, passing through $(5.1, 6.2)$

69. Perpendicular to $y = \frac{1}{2}$, passing through $(4, -9)$

70. Perpendicular to $x = 2$, passing through $(3, 4)$

71. Parallel to $x = 4$, passing through $\left(-\frac{2}{3}, \frac{1}{2}\right)$

72. Parallel to $y = -2.1$, passing through $(7.6, 3.5)$

73. **Thinking Generally** Write the equation of the *x*-axis. (*Hint:* The *x*-axis is a horizontal line.)

74. **Thinking Generally** Write the equation of the *y*-axis. (*Hint:* The *y*-axis is a vertical line.)

APPLICATIONS

*Exercises 75 and 76: **Distance** The distance of a driver from home is illustrated in the graph.*

(a) Find the intercepts.

(b) Interpret each intercept.

75.
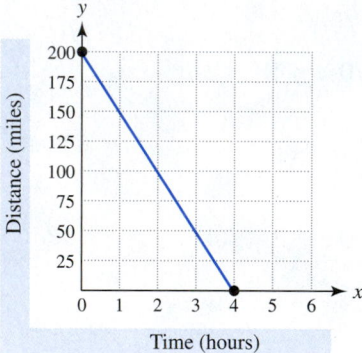

76.
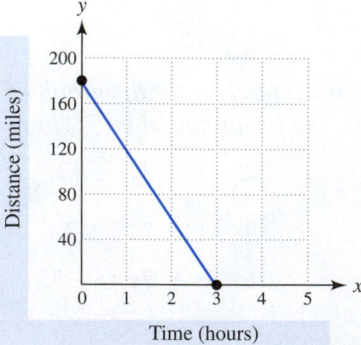

Exercises 77 and 78: **Modeling a Toy Rocket** *(Refer to Example 3.) The velocity v of a toy rocket in feet per second after t seconds of flight is given. Assume that t ≥ 0 and t does not take on values greater than the t-intercept.*

   (a) *Find the intercepts and then graph the equation.*
   (b) *Interpret each intercept.*

77. $v = 128 - 32t$ **78.** $v = 96 - 32t$

Exercises 79 and 80: **Water in a Pool** *The amount of water in a swimming pool is depicted in the graph.*

   (a) *Find the intercepts.*
   (b) *Interpret each intercept.*

79.

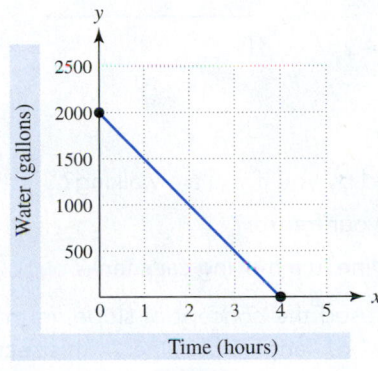

80.

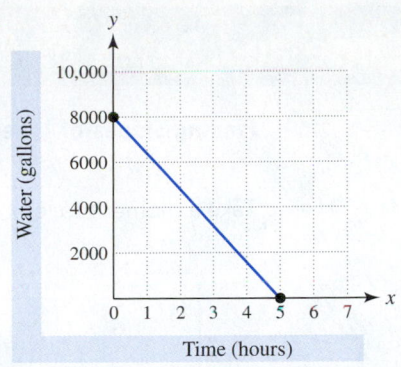

WRITING ABOUT MATHEMATICS

81. Given an equation, explain how to find an *x*-intercept and a *y*-intercept.

82. The form $\frac{x}{a} + \frac{y}{b} = 1$ is called the **intercept form** of a linear equation. Explain how you can use this equation to find the intercepts. (*Hint:* Graph $\frac{x}{2} + \frac{y}{3} = 1$ and find its intercepts.)

Group Activity   Working with Real Data

Directions: Form a group of 2 to 4 people. Select someone to record the group's responses for this activity. All members of the group should work cooperatively to answer the questions. If your instructor asks for the results, each member of the group should be prepared to respond.

1. *Radio Stations* The approximate number of radio stations on the air for selected years from 1960 to 2010 is shown in the table.

x (year)	1960	1970	1980
y (stations)	4100	6800	8600

x (year)	1990	2000	2010
y (stations)	10,800	12,600	14,500

Source: M. Street Corporation.

Make a line graph of the data. Be sure to label both axes.

2. *Estimation* Discuss ways to estimate the number of radio stations on the air in 1975. Compare your

estimates with the actual value of 7700 stations. Repeat this estimate for 1985 and compare it to the actual value of 10,400. Discuss your results.

3. *Modeling Equation* Substitute each *x*-value from the table into the equation $y = 220x - 427{,}100$ and determine the corresponding *y*-value. Do these *y*-values give reasonable approximations to the *y*-values in the table? Explain your answer.

4. *Making Estimates* Use $y = 220x - 427{,}100$ to estimate the number of radio stations on the air in 1975 and 1985. Compare the results to your answer in Exercise 2.

3.4   Slope and Rates of Change

Finding Slopes of Lines • Slope as a Rate of Change

A LOOK INTO MATH ▶

Take a moment to look at the graphs in Figure 3.27, where the horizontal axis represents time.

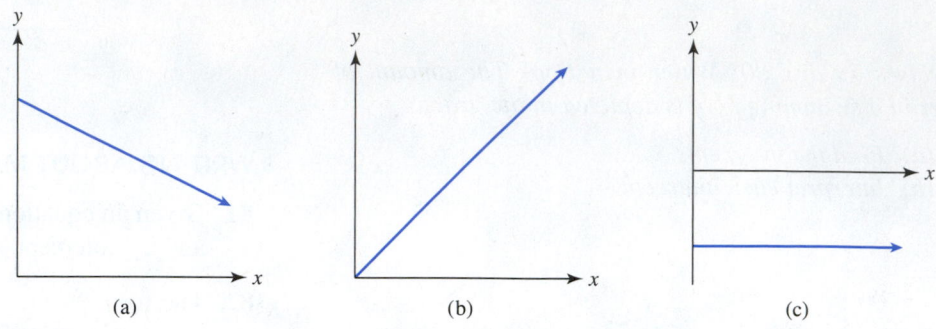

Figure 3.27

NEW VOCABULARY

☐ Rise
☐ Run
☐ Slope
☐ Positive slope
☐ Negative slope
☐ Zero slope
☐ Undefined slope
☐ Rate of change

Which graph might represent the distance traveled by you if you are walking?

Which graph might represent the temperature in your freezer?

Which graph might represent the amount of gasoline in a moving car's tank?

To be able to answer these questions, you probably used the concept of slope. In mathematics, slope is a real number that measures the "tilt" or "angle" of a line. In this section we discuss slope and how it is used in applications.

Finding Slopes of Lines

STUDY TIP

If you have trouble keeping your course materials organized or have difficulty managing your time, you may be able to find help at the student support services office on your campus. Students who learn to manage their time and keep organized find the college experience to be less overwhelming and more enjoyable.

▶ **REAL-WORLD CONNECTION** The graph shown in Figure 3.28 illustrates the cost of parking for x hours. The graph tilts upward from left to right, which indicates that the cost increases as the number of hours increases. Note that, for each hour of parking, the cost increases by \$2.

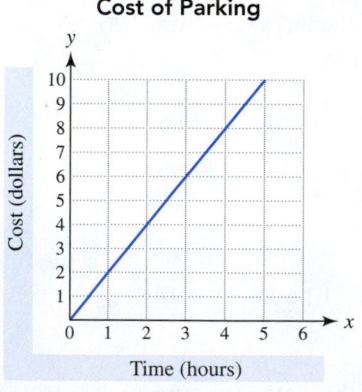

Cost of Parking

Figure 3.28

The graph *rises* 2 units vertically for every horizontal unit of *run*, and the ratio $\frac{\text{rise}}{\text{run}}$ equals the *slope* of the line. The slope m of this line is 2, which indicates that the cost of parking is \$2 per hour. In applications, slope indicates a *rate of change*.

A more general case of the slope of a line is shown in Figure 3.29, where a line passes through the points (x_1, y_1) and (x_2, y_2). The **rise**, or *change in y*, is $y_2 - y_1$, and the **run**, or *change in x*, is $x_2 - x_1$. Slope m is given by $m = \frac{y_2 - y_1}{x_2 - x_1}$.

Slope m of a Line

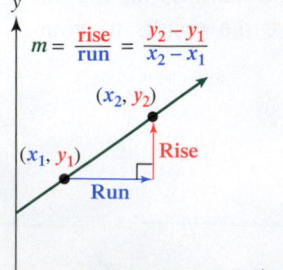

$$m = \frac{\text{rise}}{\text{run}} = \frac{y_2 - y_1}{x_2 - x_1}$$

Figure 3.29

NOTE: The symbol x_1 has a *subscript* of 1 and is read "*x* sub one" or "*x* one". Thus x_1 and x_2 are used to denote two different *x*-values. Similar comments apply to y_1 and y_2.

> **SLOPE**
>
> The **slope** *m* of the line passing through the points (x_1, y_1) and (x_2, y_2) is
>
> $$m = \frac{\text{rise}}{\text{run}} = \frac{y_2 - y_1}{x_2 - x_1},$$
>
> where $x_1 \neq x_2$. That is, slope equals rise over run.

NOTE: If $x_1 = x_2$, the line is vertical and the slope is undefined.

EXAMPLE 1 **Calculating the slope of a line**

Use the two points labeled in Figure 3.30 to find the slope of the line. What are the rise and run between these two points? Interpret the slope in terms of rise and run.

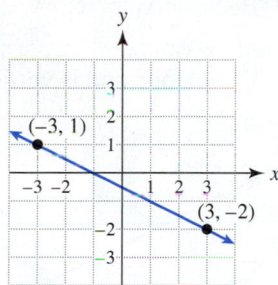

Figure 3.30

Solution

The line passes through the points $(-3, 1)$ and $(3, -2)$, so let $(x_1, y_1) = (\mathbf{-3, 1})$ and $(x_2, y_2) = (\mathbf{3, -2})$. The slope is

$$m = \frac{y_2 - y_1}{x_2 - x_1} = \frac{\mathbf{-2 - 1}}{\mathbf{3 - (-3)}} = \frac{\mathbf{-3}}{\mathbf{6}} = -\frac{1}{2}.$$

Starting at the point $(-3, 1)$, count 3 units downward and then 6 units to the right to return to the graph at the point $(3, -2)$. Thus the "rise" is -3 units and the run is **6** units. See Figure 3.31(a). The ratio $\frac{\text{rise}}{\text{run}}$ is $\frac{-3}{6}$, or $-\frac{1}{2}$.

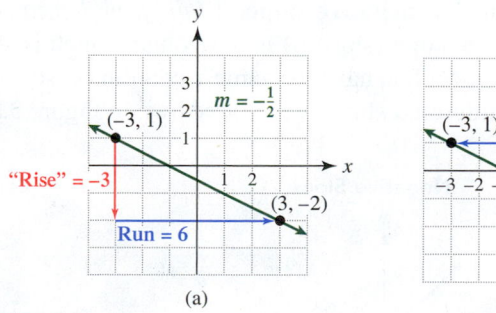

(a)   (b)

Figure 3.31

Figure 3.31(b) shows an alternate way of finding this slope. Starting at the point $(3, -2)$, count 3 units upward and then 6 units to the left to return to the graph at the point $(-3, 1)$. Here, the rise is **3** units and the run is -6 units so that the ratio $\frac{\text{rise}}{\text{run}}$ is $\frac{3}{-6}$, or $-\frac{1}{2}$. In either case, the slope is $-\frac{1}{2}$, which can be written as $\frac{-1}{2}$ indicating that the graph falls 1 unit for every 2 units of run (to the right).

Now Try Exercise 27

NOTE: In Example 1 the same slope would result if we let $(x_1, y_1) = (3, -2)$ and $(x_2, y_2) = (-3, 1)$. In this case the calculation would be

$$m = \frac{y_2 - y_1}{x_2 - x_1} = \frac{1 - (-2)}{-3 - 3} = \frac{3}{-6} = -\frac{1}{2}.$$

A graph is not needed to find slope. Any two points on a line can be used to find the line's slope, as demonstrated in the next example.

EXAMPLE 2 **Calculating the slope of a line**

Calculate the slope of the line passing through each pair of points. Graph the line.
(a) $(-2, 3), (2, 1)$ **(b)** $(-1, 3), (2, 3)$ **(c)** $(-3, 3), (-3, -2)$

Solution
(a) $m = \frac{y_2 - y_1}{x_2 - x_1} = \frac{1 - 3}{2 - (-2)} = \frac{-2}{4} = -\frac{1}{2}$. This slope indicates that the line falls 1 unit for every 2 units of horizontal run, as shown in Figure 3.32(a).

(b) $m = \frac{y_2 - y_1}{x_2 - x_1} = \frac{3 - 3}{2 - (-1)} = \frac{0}{3} = 0$. The line is horizontal, as shown in Figure 3.32(b).

(c) Because $x_1 = x_2 = -3$, the slope formula does not apply. If we try to use it, we obtain $m = \frac{y_2 - y_1}{x_2 - x_1} = \frac{-2 - 3}{-3 - (-3)} = \frac{-5}{0}$, which is an undefined expression. The line has undefined slope and is vertical, as shown in Figure 3.32(c).

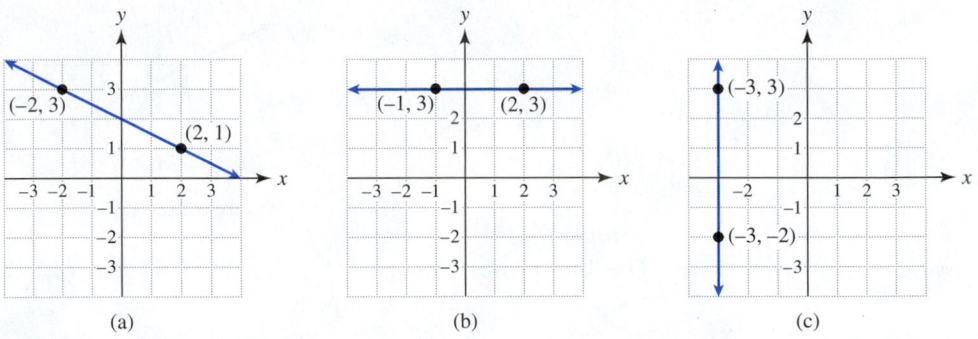

(a)   (b)   (c)

Figure 3.32

Now Try Exercises 41, 43, 45

READING CHECK

• When a line is neither vertical nor horizontal, how can you tell if it has a positive or a negative slope?

If a line has **positive slope**, it *rises from left to right*, as shown in Figure 3.33(a). If a line has **negative slope**, it *falls from left to right*, as shown in Figure 3.33(b). A line with **zero slope** (slope 0) is horizontal, which is shown in Figure 3.33(c). Any two points on a vertical line have the same x-coordinate, so the run always equals 0. Thus a vertical line has **undefined slope**, which is shown in Figure 3.33(d).

Positive Slope

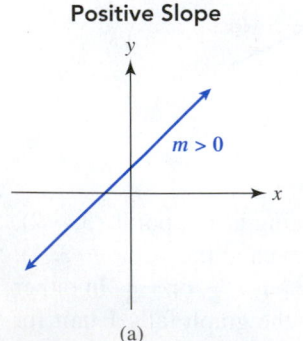

(a)

Negative Slope

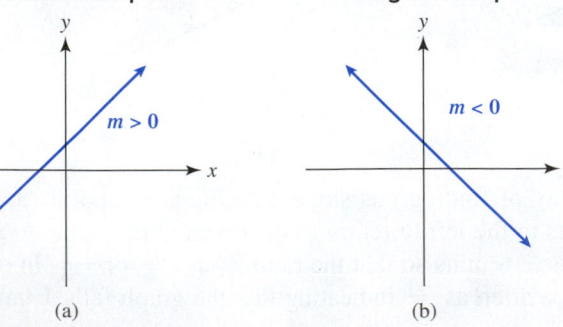

(b)

Zero Slope

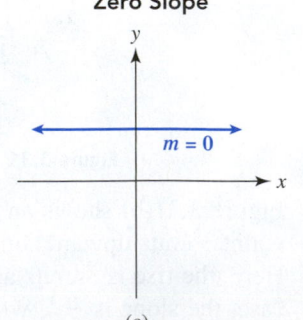

(c)

Undefined Slope

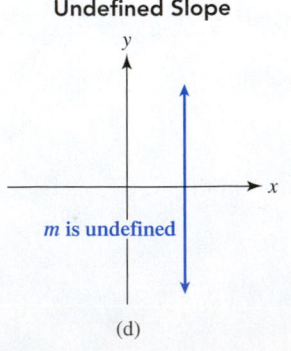

(d)

Figure 3.33

SLOPE OF A LINE

1. A line that rises *from left to right* has positive slope.
2. A line that falls *from left to right* has negative slope.
3. A horizontal line has zero slope.
4. A vertical line has undefined slope.

EXAMPLE 3 **Finding slope from a graph**

Find the slope of each line.

(a)

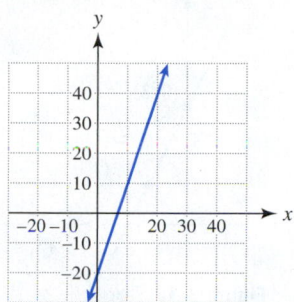

(b)

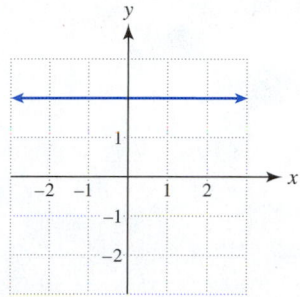

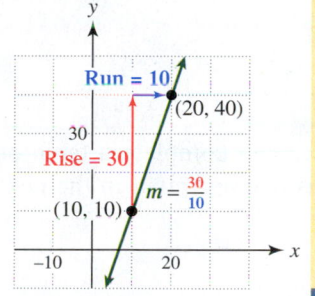

Figure 3.34

Solution

(a) The graph rises 30 units for every 10 units of run. For example, the graph passes through **(10, 10)** and **(20, 40)**, so the line rises **40 − 10 = 30** units with **20 − 10 = 10** units of run, as shown in Figure 3.34. Therefore the slope is

$$m = \frac{\text{rise}}{\text{run}} = \frac{30}{10} = 3.$$

(b) The line is horizontal, so the slope is 0.

Now Try Exercises 19, 29

A point and a slope also determine a line, as illustrated in the next example.

EXAMPLE 4 **Sketching a line with a given slope**

Sketch a line passing through the point (1, 4) and having slope $-\frac{2}{3}$.

Solution

Start by plotting the point (1, 4). A slope of $-\frac{2}{3}$ can be written as $\frac{-2}{3}$, which indicates that the y-values *decrease* 2 units each time the x-values increase by 3 units. That is, the line *falls* 2 units for every 3-unit increase in the run. Because the line passes through (**1, 4**), a **2**-unit decrease in y and a **3**-unit increase in x results in the line passing through the point (**1** + **3**, **4** − **2**) or (**4, 2**). See Figure 3.35.

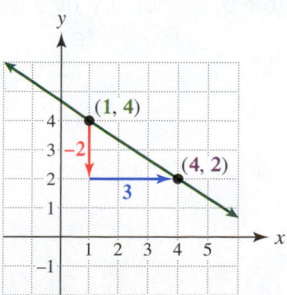

Figure 3.35 Slope $m = -\frac{2}{3}$

Now Try Exercise 59

EXAMPLE 5   **Sketching a line with a given *y*-intercept**

Sketch a line with slope -2 and *y*-intercept 3.

Solution

For the *y*-intercept of 3, plot the point $(0, 3)$. The slope of -2 can be written as $\frac{-2}{1}$, so the *y*-values decrease 2 units for each unit increase in *x*. Increasing the *x*-value in the point $(\mathbf{0}, \mathbf{3})$ by $\mathbf{1}$ and decreasing the *y*-value by $\mathbf{2}$ results in the point $(\mathbf{0 + 1}, \mathbf{3 - 2})$ or $(1, 1)$. Plot $(\mathbf{0}, \mathbf{3})$ and $(\mathbf{1}, \mathbf{1})$ and then sketch the line, as shown in Figure 3.36.

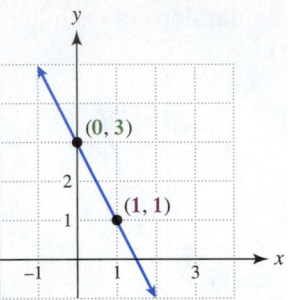

Figure 3.36 Slope: -2; *y*-intercept: 3

■ **Now Try Exercise 55**

If we know the slope of a line and a point on the line, we can complete a table of values that gives the coordinates of other points on the line, as demonstrated in the next example.

EXAMPLE 6   **Completing a table of values**

A line has slope 2 and passes through the first point listed in the table. Complete the table so that each point lies on the line.

x	-2	-1	0	1
y	1			

Solution

Slope 2 indicates that $\frac{\text{rise}}{\text{run}} = 2$. Because consecutive *x*-values in the table increase by 1 unit, the run from one point in the table to the next is 1 unit. Substituting **1** for the **run** in the slope equation results in $\frac{\text{rise}}{1} = 2$. Thus the **rise** is **2** and consecutive *y*-values shown in Table 3.16 increase by 2 units.

TABLE 3.16

	+1	+1	+1	
x	-2	-1	0	1
y	1	3	5	7
	+2	+2	+2	

■ **Now Try Exercise 63**

Slope as a Rate of Change

▶ **REAL-WORLD CONNECTION** When lines are used to model physical quantities in applications, their slopes provide important information. Slope measures the **rate of change** in a quantity. We illustrate this concept in the next four examples.

EXAMPLE 7   **Interpreting slope**

The distance y in miles that an athlete training for a marathon is from home after x hours is shown in Figure 3.37.
(a) Find the y-intercept. What does it represent?
(b) The graph passes through the point $(1, 10)$. Discuss the meaning of this point.
(c) Find the slope of this line. Interpret the slope as a rate of change.

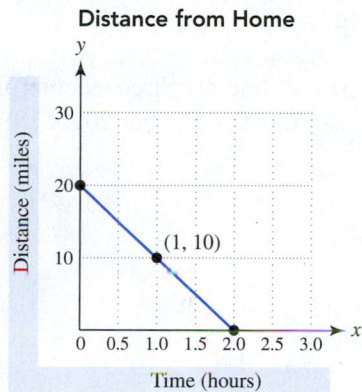

Distance from Home

Figure 3.37

Solution
(a) On the graph the y-intercept is 20, so the athlete is initially 20 miles from home.
(b) The point $(1, 10)$ means that after 1 hour the athlete is 10 miles from home.
(c) The line passes through the points $(0, 20)$ and $(1, 10)$. Its slope is

$$m = \frac{10 - 20}{1 - 0} = -10.$$

Slope -10 means that the athlete is running *toward* home at 10 miles per hour. A *negative* slope indicates that the distance between the runner and home is decreasing.

▌ **Now Try Exercise 87**

NOTE: The units for a rate of change are determined by putting the y-axis units over the x-axis units. In Example 7, the units for the rate of change are $\frac{\text{miles}}{\text{hour}}$, or miles per hour.

EXAMPLE 8   **Interpreting slope**

When a company manufactures 2000 MP3 players, its profit is \$10,000, and when it manufactures 4500 MP3 players, its profit is \$35,000.
(a) Find the slope of the line passing through $(2000, 10000)$ and $(4500, 35000)$.
(b) Interpret the slope as a rate of change.

Solution
(a) $m = \frac{35{,}000 - 10{,}000}{4500 - 2000} = \frac{25{,}000}{2500} = 10$

(b) Profit increases, *on average*, by \$10 for each additional MP3 player made.

▌ **Now Try Exercise 91**

EXAMPLE 9   **Analyzing tetanus cases**

Table 3.17 lists numbers of reported cases of tetanus in the United States for selected years.

TABLE 3.17

Year	1960	1970	1980	1990	2000	2010
Cases of Tetanus	368	148	95	64	45	18

Source: Department of Health and Human Services.

(a) Make a line graph of the data.
(b) Find the slope of each line segment.
(c) Interpret each slope as a rate of change.

Solution

(a) A line graph connecting the points (1960, 368), (1970, 148), (1980, 95), (1990, 64), (2000, 45), and (2010, 18) is shown in Figure 3.38.

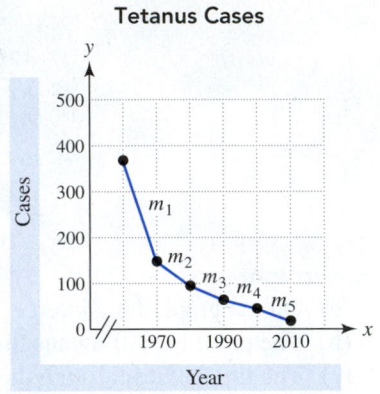

Figure 3.38

(b) The slope of each line segment may be calculated as follows.

$$m_1 = \frac{148 - 368}{1970 - 1960} = -22.0, \qquad m_2 = \frac{95 - 148}{1980 - 1970} = -5.3,$$

$$m_3 = \frac{64 - 95}{1990 - 1980} = -3.1, \qquad m_4 = \frac{45 - 64}{2000 - 1990} = -1.9$$

$$m_5 = \frac{18 - 45}{2010 - 2000} = -2.7$$

(c) Slope $m_1 = -22.0$ indicates that, *on average*, the number of tetanus cases *decreased* by 22.0 cases per year between 1960 and 1970. The other four slopes can be interpreted similarly.

NOTE: The number of tetanus cases did not decrease by *exactly* 22.0 cases each year between 1960 and 1970. However, the yearly *average* decrease was 22.0 cases.

▌ **Now Try Exercise 89**

READING CHECK

• When a line models a real-world situation, how are the units of the slope found?

EXAMPLE 10   **Sketching a model**

During a storm, rain falls at the constant rates of 2 inches per hour from 1 A.M. to 3 A.M., 1 inch per hour from 3 A.M. to 4 A.M., and $\frac{1}{2}$ inch per hour from 4 A.M. to 6 A.M.
(a) Sketch a graph that shows the total accumulation of rainfall from 1 A.M. to 6 A.M.
(b) What does the slope of each line segment represent?

Solution

(a) At 1 A.M. the accumulated rainfall is 0, so place a point at $(1, 0)$. Rain falls at a constant rate of 2 inches per hour for the next 2 hours, so at 3 A.M. the total rainfall is 4 inches. Place a point at $(3, 4)$. Because the rainfall is constant, sketch a line segment from $(1, 0)$ to $(3, 4)$, as shown in Figure 3.39. Similarly, during the next hour 1 inch of rain falls, so draw a line segment from $(3, 4)$ to $(4, 5)$. Finally, 1 inch of rain falls from 4 A.M. to 6 A.M., so draw a line segment from $(4, 5)$ to $(6, 6)$.

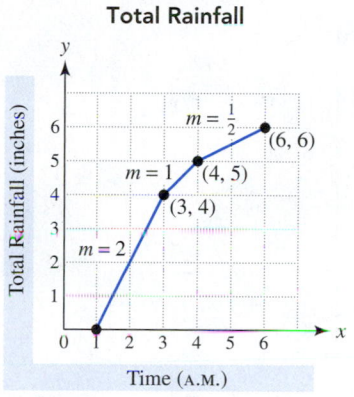

Total Rainfall

Figure 3.39

(b) The slope of each line segment represents the rate at which rain is falling. For example, the first segment has slope 2 because rain falls at a rate of 2 inches per hour during that period of time.

Now Try Exercise 81

3.4   Putting It All Together

CONCEPT	COMMENTS	EXAMPLE
Rise, Run, and Slope	Rise is a vertical change in a line, and run is a horizontal change in a line. The ratio $\frac{\text{rise}}{\text{run}}$ is the slope m when run is nonzero. **1.** A line that rises from left to right has positive slope. **2.** A line that falls from left to right has negative slope. **3.** A horizontal line has zero slope. **4.** A vertical line has undefined slope.	

continued on next page

continued on previous page

CONCEPT	COMMENTS	EXAMPLE
Calculating Slope	For any two points (x_1, y_1) and (x_2, y_2), slope m is $$m = \frac{y_2 - y_1}{x_2 - x_1},$$ where $x_1 \neq x_2$.	The slope of the line passing through $(-2, 3)$ and $(1, 5)$ is $$m = \frac{5 - 3}{1 - (-2)} = \frac{2}{3}.$$ The line rises vertically 2 units for every 3 horizontal units of run.
Slope as a Rate of Change	Slope measures the "tilt" or "angle" of a line. In applications, slope measures the rate of change in a quantity.	Slope $m = -66\frac{2}{3}$ indicates that water is *leaving* the pool at the rate of $66\frac{2}{3}$ gallons per hour. **Water in a Pool** *(graph: Water (gallons) vs. Time (hours), line from (0, 2000) to (30, 0), labeled $m = -66\frac{2}{3}$)*

3.4   Exercises

 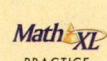

CONCEPTS AND VOCABULARY

1. (True or False?) The change in the horizontal distance along a line is called run.

2. (True or False?) The change in the vertical distance along a line is called rise.

3. Slope m of a line equals _____ over _____.

4. Slope 0 indicates that a line is _____.

5. Undefined slope indicates that a line is _____.

6. If a line passes through (x_1, y_1) and (x_2, y_2) where $x_1 \neq x_2$, then $m =$ _____.

7. A line that rises from left to right has _____ slope.

8. A line that falls from left to right has _____ slope.

9. When a line models a physical quantity in an application, its slope measures the _____ of change in the quantity.

10. If a line that models the distance between a hiker and camp has a negative slope, is the hiker moving away from or toward camp?

Exercises 11–18: State whether the slope of the line is positive, negative, zero, or undefined.

11.

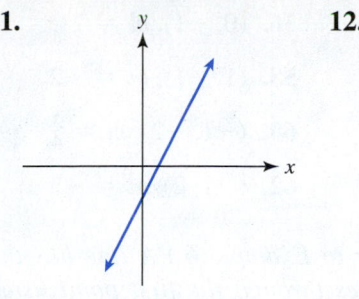

12.

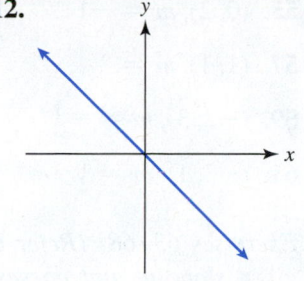

13.

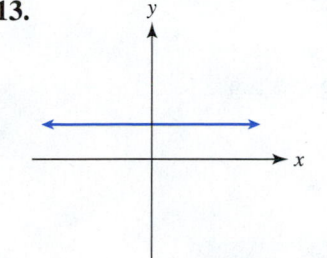

14.

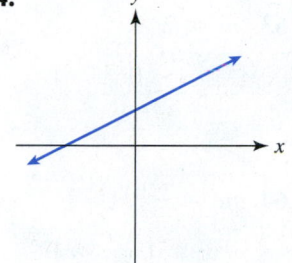

15.

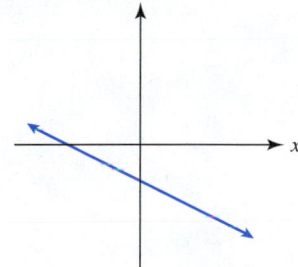

16.

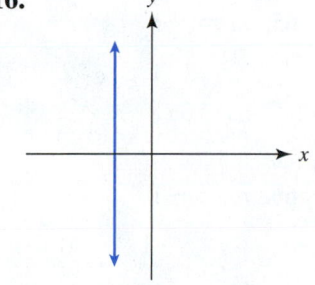

17.

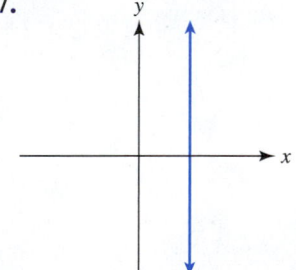

18.

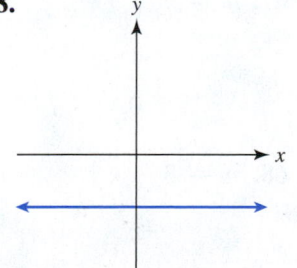

FINDING SLOPES OF LINES

Exercises 19–30: If possible, find the slope of the line. Interpret the slope in terms of rise and run.

19.

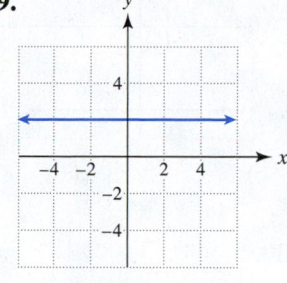

20.

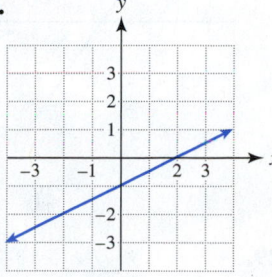

21.

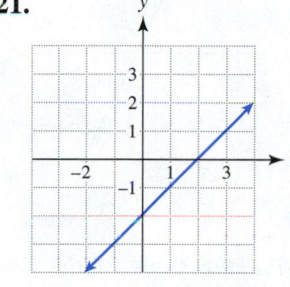

22.

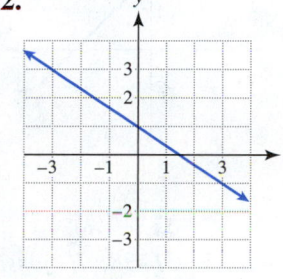

23.

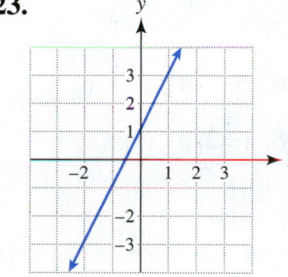

24.

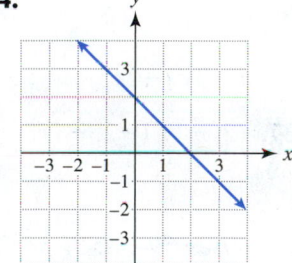

25.

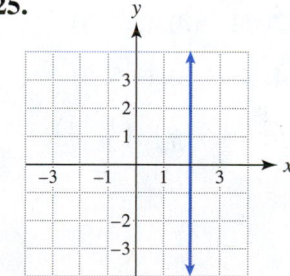

26.

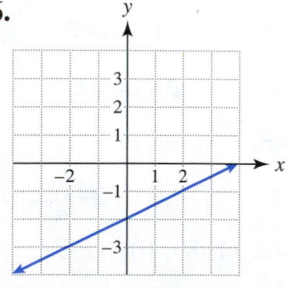

27.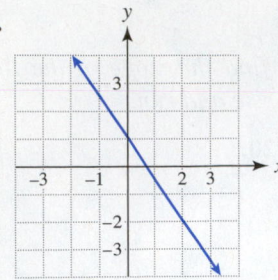

28.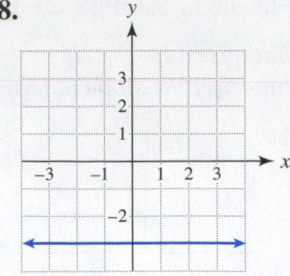

29.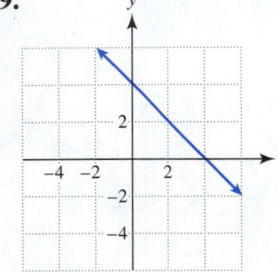

30.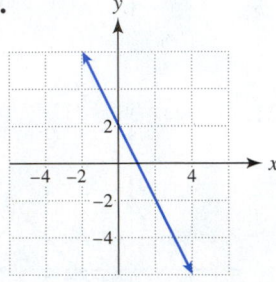

Exercises 31–38: If possible, find the slope of the line passing through the two points. Graph the line.

31. $(1, 2), (2, 4)$ **32.** $(-4, 7), (7, -4)$

33. $(2, 1), (2, 4)$ **34.** $(-1, 3), (-1, -1)$

35. $(1, 3), (-2, 5)$ **36.** $(0, -4), (-4, 6)$

37. $(2, -1), (-2, -1)$ **38.** $(-2, 3), (1, 3)$

Exercises 39–54: If possible, find the slope of the line passing through the two points.

39. $(4, -2), (-3, -9)$ **40.** $(15, -3), (20, 9)$

41. $(-3, 4), (4, -2)$ **42.** $(1, -3), (3, -5)$

43. $(-3, 5), (2, 5)$ **44.** $(-3, 3), (-5, 3)$

45. $(-1, 6), (-1, -4)$ **46.** $\left(\frac{1}{2}, -\frac{2}{7}\right), \left(\frac{1}{2}, \frac{13}{17}\right)$

47. $(1980, 5), (2000, 18)$

48. $(1989, 10), (1999, 16)$

49. $(1950, 6.1), (2000, 10.6)$

50. $(1900, 10), (1950, 35)$

51. $\left(\frac{1}{3}, -\frac{2}{7}\right), \left(-\frac{2}{3}, \frac{3}{7}\right)$

52. $(-1.3, 5.6), (-2.6, -2.5)$

53. $(12, -34), (14, 64)$

54. $(-25, 105), (60, 55)$

Exercises 55–62: (Refer to Example 4.) Sketch a line passing through the point and having slope m.

55. $(0, 2), m = -1$ **56.** $(0, -1), m = 2$

57. $(1, 1), m = 3$ **58.** $(1, -1), m = -2$

59. $(-2, 3), m = -\frac{1}{2}$ **60.** $(-1, -2), m = \frac{3}{4}$

61. $(-3, 1), m = \frac{1}{2}$ **62.** $(-2, 2), m = -3$

Exercises 63–68: (Refer to Example 6.) A line has the given slope m and passes through the first point listed in the table. Complete the table so that each point in the table lies on the line.

63. $m = 2$

x	0	1	2	3
y	-4			

64. $m = -\frac{1}{2}$

x	0	1	2	3
y	2			

65. $m = -3$

x	1	2	3	4
y	4			

66. $m = -1$

x	-1	0	1	2
y	10			

67. $m = \frac{3}{2}$

x	-4	-2	0	2
y	0			

68. $m = 3$

x	-2	0	2	4
y	-4			

SLOPE AS A RATE OF CHANGE

*Exercises 69–72: **Modeling** Choose the graph (a.–d.) in the next column that models the situation best.*

69. Cost of buying x gum balls at a price of 25¢ each

70. Total number of movies in VHS format (videotape) purchased during the past 5 years

71. Average cost of a new car over the past 30 years

72. Height of the Empire State Building after x people have entered it

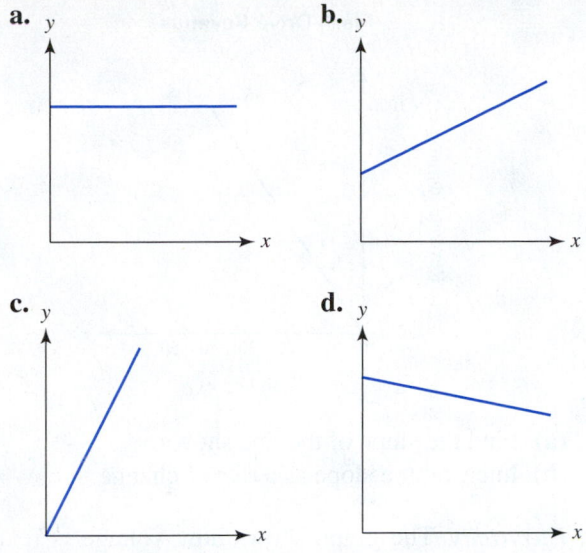

a.   **b.**

c.   **d.**

Exercises 73 and 74: **Modeling** *The line graph represents the gallons of water in a small swimming pool after x hours. Assume that there is a pump that can either add water to or remove water from the pool.*

(a) *Estimate the slope of each line segment.*
(b) *Interpret each slope as a rate of change.*
(c) *Describe what happened to the amount of water in the pool.*

73.   Water in a Pool

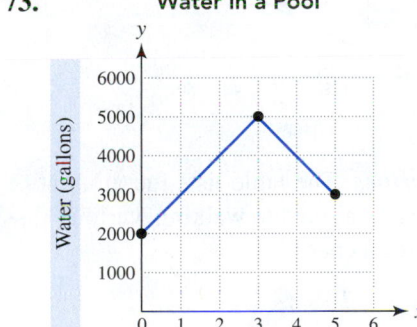

74.   Water in a Pool

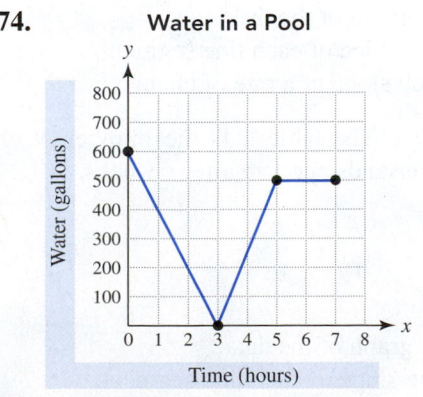

Exercises 75 and 76: **Modeling** *An individual is driving a car along a straight road. The graph shows the distance that the driver is from home after x hours.*

(a) *Find the slope of each line segment in the graph.*
(b) *Interpret each slope as a rate of change.*
(c) *Describe both the motion of the car and its distance from home.*

75.   Distance from Home

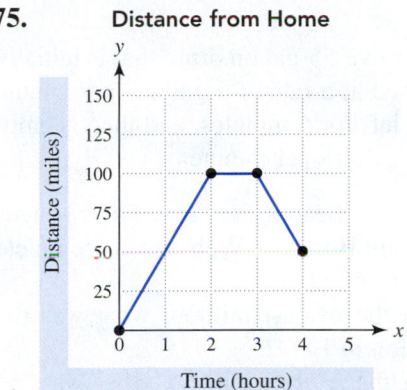

76.   Distance from Home

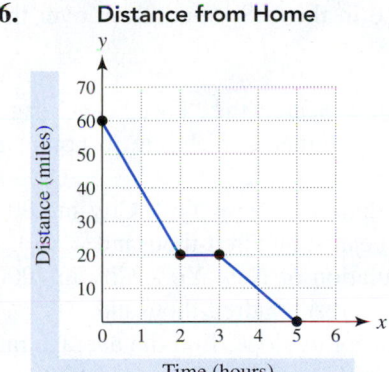

77. Thinking Generally If a line is used to model physical data where the x-axis is labeled "Time (minutes)" and the y-axis is labeled "Volume (cubic meters)," what are the units for the rate of change represented by the slope of the line?

78. Thinking Generally If a line is used to model physical data where the x-axis is labeled "Cookies" and the y-axis is labeled "Chocolate Chips," what are the units for the rate of change represented by the slope of the line?

Exercises 79–82: **Sketching a Model** *Sketch a graph that models the given situation.*

79. The distance that a boat is from a harbor if the boat is initially 6 miles from the harbor and arrives at the harbor after sailing at a constant speed for 3 hours

80. The distance that a person is from home if the person starts at home, walks away from home at 4 miles per hour for 90 minutes, and then walks back home at 3 miles per hour

81. The distance that an athlete is from home if the athlete jogs for 1 hour to a park that is 7 miles away, stays for 30 minutes, and then jogs home at the same pace

82. The amount of oil in a 55-gallon drum that is initially full, then is drained at a rate of 5 gallons per minute for 4 minutes, is left for 6 minutes, and then is emptied at a rate of 7 gallons per minute

83. **Online Exploration** Search the "Fast Facts" pages on the U.S. Census Bureau's Web site to complete the following.
   (a) Rounding to the nearest million, what was the U.S. population in 1900?
   (b) Rounding to the nearest million, what was the U.S. population in 2000?
   (c) Using the concept of slope, find the average rate of change in the U.S. population over this time.

84. **Online Exploration** Search the "Fast Facts" pages on the U.S. Census Bureau's Web site to complete the following.
   (a) Find the population of New York City in 1900. Round to the nearest hundred-thousand.
   (b) Find the population of New York City in 2000. Round to the nearest hundred-thousand.
   (c) Using the concept of slope, find the average rate of change in the New York City population over this time.

APPLICATIONS

85. *Twitter Followers* One year after a celebrity first began posting comments on Twitter, she had 25,600 followers. Four years after this celebrity first began posting comments, she had 148,000 followers.
   (a) Find the slope of the line that passes through the points (1, 25600) and (4, 148000).
   (b) Interpret the slope as a rate of change.

86. *Profit from Tablet Computers* When a company manufactures 500 tablet computers, its profit is $100,000, and when it manufactures 1500 tablet computers, its profit is $400,000.
   (a) Find the slope of the line passing through the points (500, 100000) and (1500, 400000).
   (b) Interpret the slope as a rate of change.

87. *Revenue* The graph shows revenue received from selling x flash drives.

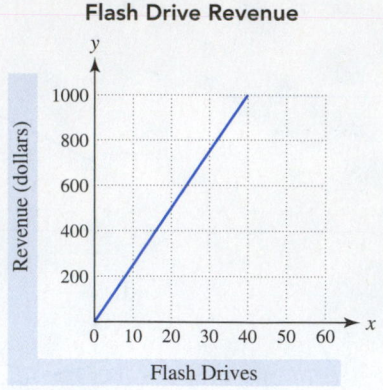

Flash Drive Revenue

   (a) Find the slope of the line shown.
   (b) Interpret the slope as a rate of change.

88. *Electricity* The graph shows how voltage is related to amperage in an electrical circuit. The slope corresponds to the resistance in ohms. Find the resistance in this electrical circuit.

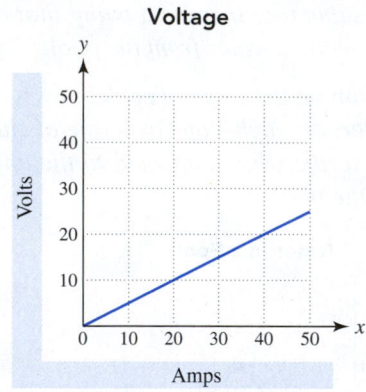

Voltage

89. *Walking for Charities* The table lists the amount of money M in dollars raised for walking various distances x in miles for a charity.

x	0	5	10	15
M	0	100	250	450

   (a) Make a line graph of the data.
   (b) Calculate the slope of each line segment.
   (c) Interpret each slope as a rate of change.

90. *Insect Population* The table lists the number N of black flies in thousands per acre after x weeks.

x	0	2	4	6
N	3	4	10	18

   (a) Make a line graph of the data.
   (b) Calculate the slope of each line segment.
   (c) Interpret each slope as a rate of change.

91. *Median Household Income* In 2000, median family income was about $42,000, and in 2008 it was about $50,000. (*Source:* Department of the Treasury.)
   (a) Find the slope of the line passing through the points (2000, 42000) and (2008, 50000).
   (b) Interpret the slope as a rate of change.
   (c) If this trend continues, estimate the median family income in 2014.

92. *Minimum Wage* In 1995, the minimum wage was $4.25 per hour, and by 2010 it had increased to $7.25 per hour. (*Source:* Department of Labor.)
   (a) Find the slope of the line passing through the points (1995, 4.25) and (2010, 7.25).
   (b) Interpret the slope as a rate of change.
   (c) If this trend continues, estimate the minimum wage in 2015.

93. *Rate of Change* Suppose that $y = -2x + 10$ is graphed in the first quadrant of the xy-plane where the x-axis is labeled "Time (minutes)" and the y-axis is labeled "Distance (feet)." If this graph represents the distance y that an ant is from a stone after x minutes, answer each of the following.
   (a) Is the ant moving toward or away from the stone?
   (b) Initially, how far from the stone is the ant?
   (c) At what rate is the ant moving?
   (d) What is the value of x (time) when the ant reaches the stone?

94. *Rate of Change* Suppose that we graph $y = 15x + 8$ in the first quadrant of the xy-plane where the x-axis is labeled "Time (minutes)" and the y-axis is labeled "Distance (feet)." If this graph represents the distance y that a frog is from a tree after x minutes, answer each of the following.
   (a) Is the frog moving toward or away from the tree?
   (b) Initially, how far from the tree is the frog?
   (c) At what rate is the frog moving?
   (d) What is the value of x when the frog is 53 feet from the tree?

WRITING ABOUT MATHEMATICS

95. If you are given two points and the slope formula $m = \frac{y_2 - y_1}{x_2 - x_1}$, does it matter which point is (x_1, y_1) and which point is (x_2, y_2)? Explain.

96. Suppose that a line approximates the distance y in miles that a person drives in x hours. What does the slope of the line represent? Give an example.

97. Describe the information that the slope m gives about a line. Be as complete as possible.

98. Could one line have two slopes? Explain.

SECTIONS 3.3 and 3.4	**Checking Basic Concepts**

1. Identify the x- and y-intercepts in the graph.

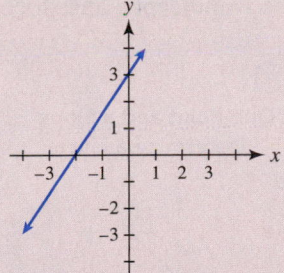

2. Complete the table for the equation $2x - y = 2$. Then determine the x- and y-intercepts.

x	-2	-1	0	1	2
y	-6				

3. Find any intercepts for the graphs of the equations and then graph each linear equation.
   (a) $x - 2y = 6$ **(b)** $y = 2$ **(c)** $x = -1$

4. Write the equations of a horizontal line and a vertical line that pass through the point $(-2, 4)$.

5. If possible, find the slope of the line passing through each pair of points.
   (a) $(-2, 3), (2, 6)$ **(b)** $(-5, 3), (0, 3)$
   (c) $(1, 5), (1, 8)$

6. Find the slope of the line shown.

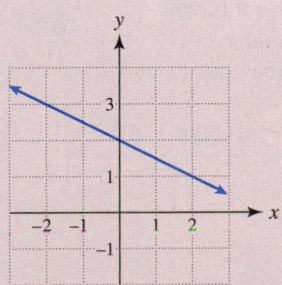

continued on next page

continued from previous page

7. Sketch a line passing through the point $(-3, 1)$ and having slope 2.

8. *Modeling* The line graph to the right shows the depth of water in a small pond before and after a rain storm.
   (a) Estimate the slope of each line segment.
   (b) Interpret each slope as a rate of change.
   (c) Describe what happened to the amount of water in the pond.

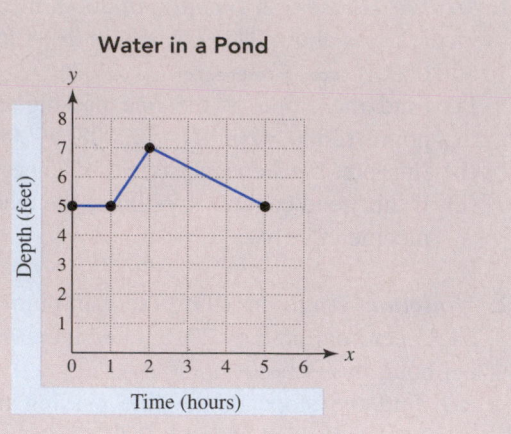

Water in a Pond

3.5   Slope–Intercept Form

Basic Concepts • Finding Slope–Intercept Form • Parallel and Perpendicular Lines

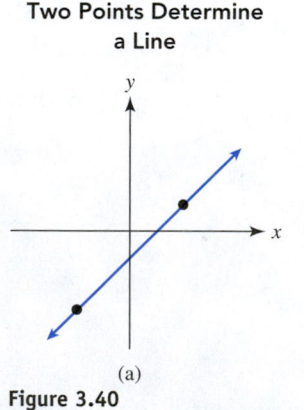

When a line models a real-world situation, slope represents a rate of change and the *y*-intercept often represents an initial value. For example, if there are initially **30** gallons of water in a small wading pool, and the pool is being filled by a garden hose at a constant rate of **2** gallons per minute, a line representing this situation has a slope of **2** and a *y*-intercept of **30**. In this section we discuss how to find the equation of a nonvertical line, given its slope and *y*-intercept.

Basic Concepts

For any two points in the *xy*-plane, we can draw a unique line passing through them, as illustrated in Figure 3.40(a). Another way we can determine a unique line is to know the *y*-intercept and the slope. For example, if a line has *y*-intercept 2 and slope $m = 1$, then the resulting line is shown in Figure 3.40(b).

NEW VOCABULARY

☐ Slope–intercept form
☐ Negative reciprocals

STUDY TIP

A new concept is often easier to learn when we can find a relationship between the concept and our personal experience. Try to list five real-life situations that have an initial value and a constant rate of change.

Two Points Determine a Line

One Point and a Slope Determine a Line

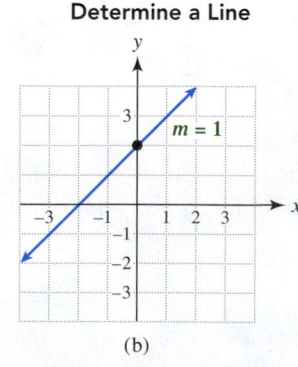

(a)

(b)

Figure 3.40

READING CHECK

• How many points does it take to determine a line?

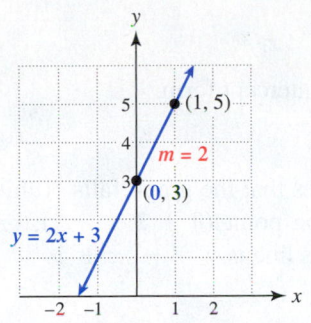

Figure 3.41 Slope 2, y-intercept 3

Finding Slope–Intercept Form

The graph of $y = 2x + 3$ passes through $(0, 3)$ and $(1, 5)$, as shown in Figure 3.41. The slope of this line is

$$m = \frac{5 - 3}{1 - 0} = 2.$$

If $x = 0$ in $y = 2x + 3$, then $y = 2(0) + 3 = 3$. Thus the graph of $y = 2x + 3$ has slope **2** and y-intercept **3**. In general, the graph of $y = mx + b$ has slope m and y-intercept b. The form $y = mx + b$ is called the *slope–intercept form*.

SLOPE–INTERCEPT FORM

The line with slope m and y-intercept b is given by

$$y = mx + b,$$

the **slope–intercept form** of a line.

Table 3.18 shows several equations in the form $y = mx + b$ and lists the corresponding slope and y-intercept for the graph associated with each.

READING CHECK

- In $y = mx + b$, which value is the slope and which is the y-intercept?

TABLE 3.18 $y = mx + b$ Form

Equation	Slope	y-intercept
$y = 4x - 3$	4	-3
$y = 12$	0	12
$y = -x - \frac{5}{8}$	-1	$-\frac{5}{8}$
$y = -\frac{2}{3}x$	$-\frac{2}{3}$	0

EXAMPLE 1 **Using a graph to write the slope–intercept form**

For each graph write the slope–intercept form of the line.

(a)

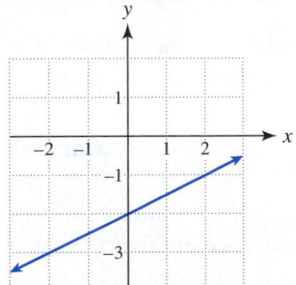

(b)

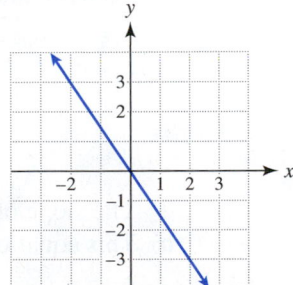

Solution

(a) The graph intersects the y-axis at -2, so the y-intercept is -2. Because the graph rises **1** unit for each **2**-unit increase in x, the slope is $\frac{1}{2}$. The slope–intercept form of the line is $y = \frac{1}{2}x - 2$.

(b) The graph intersects the y-axis at 0, so the y-intercept is 0. Because the graph falls 3 units for each 2-unit increase in x, the slope is $-\frac{3}{2}$. The slope–intercept form of the line is $y = -\frac{3}{2}x + 0$, or $y = -\frac{3}{2}x$.

Now Try Exercises 15, 19

EXAMPLE 2   **Sketching a line**

Sketch a line with slope $-\frac{1}{2}$ and y-intercept 1. Write its slope–intercept form.

Solution

For the y-intercept of 1 plot the point $(0, 1)$. Slope $-\frac{1}{2}$ indicates that the graph falls **1** unit for each **2**-unit increase in x. Thus the line passes through the point $(0 + 2, 1 - 1)$, or $(2, 0)$, as shown in Figure 3.42. The slope–intercept form of this line is $y = -\frac{1}{2}x + 1$.

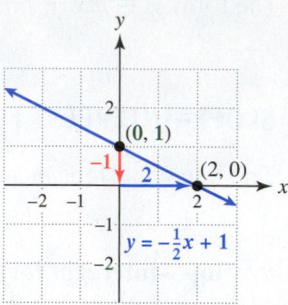

Figure 3.42

Now Try Exercise 23

When a linear equation is not given in slope–intercept form, the coefficient of the x-term may not represent the slope and the constant term may not represent the y-intercept. For example, the graph of $2x + 3y = 12$ does *not* have slope 2 and it does *not* have y-intercept 12. To find the correct slope and y-intercept, the equation can first be written in slope–intercept form. This is demonstrated in the next example.

EXAMPLE 3   **Writing an equation in slope–intercept form**

Write each equation in slope–intercept form. Then give the slope and y-intercept of the line.
(a) $2x + 3y = 12$ **(b)** $x = 2y + 4$

Solution

(a) To write the equation in slope–intercept form, solve for y.

$$2x + 3y = 12 \qquad \text{Given equation}$$

$$3y = -2x + 12 \qquad \text{Subtract } 2x \text{ from each side.}$$

$$y = -\frac{2}{3}x + 4 \qquad \text{Divide each side by 3.}$$

The slope of the line is $-\frac{2}{3}$, and the y-intercept is **4**.

(b) This equation is *not* in slope–intercept form because it is solved for x, not y.

$$x = 2y + 4 \qquad \text{Given equation}$$

$$x - 4 = 2y \qquad \text{Subtract 4 from each side.}$$

$$\frac{1}{2}x - 2 = y \qquad \text{Divide each side by 2.}$$

$$y = \frac{1}{2}x - 2 \qquad \text{Rewrite the equation.}$$

The slope of the line is $\frac{1}{2}$, and the y-intercept is -2.

Now Try Exercises 39, 41

EXAMPLE 4 **Writing an equation in slope–intercept form and graphing it**

Write the equation $2x + y = 3$ in slope–intercept form and then graph it.

Solution
First write the given equation in slope–intercept form.

$$2x + y = 3 \qquad \text{Given equation}$$
$$y = -2x + 3 \qquad \text{Subtract } 2x \text{ from each side.}$$

The slope–intercept form is $y = -2x + 3$, with slope -2 and y-intercept 3. To graph this equation, plot the y-intercept as the point $(0, 3)$. The line falls **2** units for each **1**-unit increase in x, so plot the point $(0 + \textbf{1}, 3 - \textbf{2}) = (1, 1)$. Sketch a line passing through $(0, 3)$ and $(1, 1)$, as shown in Figure 3.43.

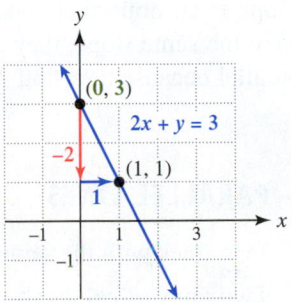

Figure 3.43

Now Try Exercise 51

MAKING CONNECTIONS

Different Ways to Graph the Same Line

In Example 4 the line $2x + y = 3$ was graphed by finding its slope–intercept form. A second way to graph this line is to find the x- and y-intercepts of the line. If $y = 0$, then $x = \frac{3}{2}$ makes this equation true, and if $x = 0$, then $y = 3$ makes this equation true. Thus the x-intercept is $\frac{3}{2}$, and the y-intercept is 3. Note that the line in Figure 3.43 passes through the points $\left(\frac{3}{2}, 0\right)$ and $(0, 3)$, which could be used to graph the line.

EXAMPLE 5 **Modeling cell phone costs**

Roaming with a cell phone costs $5 for the initial connection and $0.50 per minute.
(a) If someone talks for 23 minutes while roaming, what is the charge?
(b) Write the slope–intercept form that gives the cost of talking for x minutes.
(c) If the charge is $8.50, how long did the person talk?

Solution
(a) The charge for 23 minutes at $0.50 per minute plus $5 would be

$$\textbf{0.50} \cdot 23 + \textbf{5} = \$16.50.$$

(b) The rate of increase is $0.50 per minute with an initial cost of $5. Let $y = \textbf{0.5}x + \textbf{5}$, where the slope or rate of change is **0.5** and the y-intercept is **5**.

(c) To determine how long a person can talk for \$8.50, we solve the following equation.

$$0.5x + 5 = 8.5 \quad \text{Equation to solve}$$
$$0.5x = 3.5 \quad \text{Subtract 5 from each side.}$$
$$x = \frac{3.5}{0.5} \quad \text{Divide each side by 0.5.}$$
$$x = 7 \quad \text{Simplify.}$$

The person talked for 7 minutes. Note that this solution is based on the assumption that the phone company did not round up a fraction of a minute.

■ **Now Try Exercise 73**

Parallel and Perpendicular Lines

Slope is an important concept for determining whether two lines are parallel. If two lines have the same slope, they are parallel. For example, the lines $y = 2x$ and $y = 2x - 1$ are parallel because they both have slope **2**, as shown in Figure 3.44.

Parallel Lines

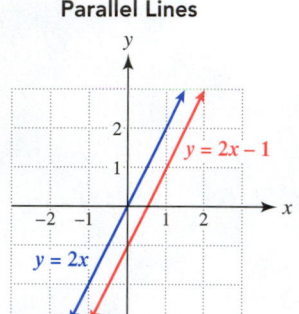

Figure 3.44

PARALLEL LINES

Two lines with the same slope are parallel.

Two nonvertical parallel lines have the same slope.

NOTE: Two vertical lines are parallel and the slope of each is undefined.

EXAMPLE 6   **Finding parallel lines**

Find the slope–intercept form of a line parallel to $y = -2x + 3$ and passing through the point $(-2, 3)$. Sketch each line in the same xy-plane.

Solution
Because the line $y = -2x + 3$ has slope -2, any parallel line also has slope -2 with slope–intercept form $y = -2x + b$ for some value of b. The value of b can be found by substituting the point $(-2, 3)$ in the slope–intercept form.

$$y = -2x + b \quad \text{Slope–intercept form}$$
$$3 = -2(-2) + b \quad \text{Let } x = -2 \text{ and } y = 3.$$
$$3 = 4 + b \quad \text{Multiply.}$$
$$-1 = b \quad \text{Subtract 4 from each side.}$$

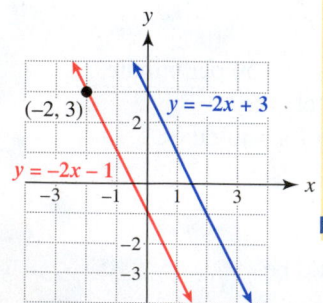

Figure 3.45

The y-intercept is -1, and so the slope–intercept form is $y = -2x - 1$. The graphs of the equations $y = -2x + 3$ and $y = -2x - 1$ are shown in Figure 3.45. Note that they are parallel lines, both with slope -2 but with different y-intercepts, 3 and -1.

■ **Now Try Exercise 63**

The lines shown in Figure 3.46 are perpendicular because they intersect at a 90° angle. Rather than measure the angle between two intersecting lines, we can determine whether two lines are perpendicular from their slopes. The slopes of perpendicular lines satisfy the properties given in the box below Figure 3.46.

Perpendicular Lines

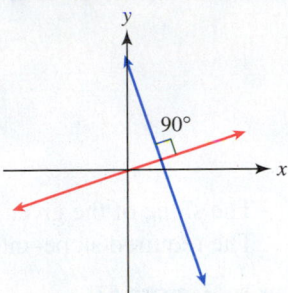

Figure 3.46

PERPENDICULAR LINES

If two perpendicular lines have nonzero slopes m_1 and m_2, then $m_1 \cdot m_2 = -1$.

If two lines have slopes m_1 and m_2 such that $m_1 \cdot m_2 = -1$, then they are perpendicular lines.

NOTE: A vertical line and a horizontal line are perpendicular.

Table 3.19 shows examples of slopes m_1 and m_2 that result in perpendicular lines. Note that $m_1 \cdot m_2 = -1$ and that $m_2 = -\frac{1}{m_1}$. That is, the product of the two slopes is -1, and the two slopes are **negative reciprocals** of each other. Note that in each case $m_1 \cdot m_2 = -1$.

TABLE 3.19 Slopes of Perpendicular Lines

m_1	1	$-\frac{1}{2}$	-4	$\frac{2}{3}$	$\frac{3}{4}$
m_2	-1	2	$\frac{1}{4}$	$-\frac{3}{2}$	$-\frac{4}{3}$

READING CHECK

- What does it mean for one number to be a negative reciprocal of another?

We can use these concepts to find equations of perpendicular lines, as illustrated in the next two examples.

EXAMPLE 7 **Finding equations of perpendicular lines**

For each of the given lines, find the slope–intercept form of a line passing through the origin that is perpendicular to the given line.

(a) $y = 3x$ **(b)** $y = -\frac{2}{5}x + 5$ **(c)** $-3x + 4y = 24$

Solution

(a) If a line passes through the origin, then its y-intercept is 0 with slope–intercept form $y = mx$. The given line $y = 3x$ has slope $m_1 = 3$, so a line perpendicular to it has a slope that is the negative reciprocal of 3.

$$m_2 = -\frac{1}{m_1} = -\frac{1}{3}$$

The required slope–intercept form is $y = -\frac{1}{3}x$.

(b) The given line $y = -\frac{2}{5}x + 5$ has slope $m_1 = -\frac{2}{5}$, so a line perpendicular to it has slope $m_2 = \frac{5}{2}$ because $-\frac{2}{5} \cdot \frac{5}{2} = -1$. The required slope–intercept form is $y = \frac{5}{2}x$.

(c) To find the slope of the given line, write $-3x + 4y = 24$ in slope–intercept form.

$$-3x + 4y = 24 \qquad \text{Given equation}$$
$$4y = 3x + 24 \qquad \text{Add } 3x \text{ to each side.}$$
$$y = \frac{3}{4}x + 6 \qquad \text{Divide each side by 4.}$$

The slope of the given line is $m_1 = \frac{3}{4}$, so a line perpendicular to it has slope $m_2 = -\frac{4}{3}$. The required slope–intercept form is $y = -\frac{4}{3}x$.

Now Try Exercise 67

EXAMPLE 8   **Finding a perpendicular line equation**

Find the slope–intercept form of the line perpendicular to $y = -\frac{1}{2}x + 1$ and passing through the point $(1, -1)$. Sketch each line in the same xy-plane.

Solution
The line $y = -\frac{1}{2}x + 1$ has slope $m_1 = -\frac{1}{2}$. Any line perpendicular to it has slope $m_2 = 2$ (because $-\frac{1}{2} \cdot 2 = -1$) with slope–intercept form $y = 2x + b$ for some value of b. The value of b can be found by substituting the point $(1, -1)$ in the slope–intercept form.

$$y = 2x + b \qquad \text{Slope–intercept form}$$
$$-1 = 2(1) + b \qquad \text{Let } x = 1 \text{ and } y = -1.$$
$$-1 = 2 + b \qquad \text{Multiply.}$$
$$-3 = b \qquad \text{Subtract 2 from each side.}$$

The slope–intercept form is $y = 2x - 3$. The graphs of $y = -\frac{1}{2}x + 1$ and $y = 2x - 3$ are shown in Figure 3.47. Note that the point $(1, -1)$ lies on the graph of $y = 2x - 3$.

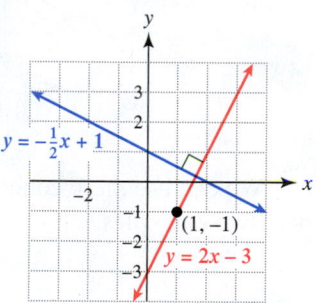

Figure 3.47

Now Try Exercise 69

3.5   Putting It All Together

CONCEPT	COMMENTS	EXAMPLE
Slope–Intercept Form $y = mx + b$	A unique equation for a line, determined by the slope m and the y-intercept b	An equation of the line with slope $m = 3$ and y-intercept $b = -5$ is $y = 3x - 5$.

CONCEPT	COMMENTS	EXAMPLE
Parallel Lines	$y = m_1x + b_1$ and $y = m_2x + b_2$, where $m_1 = m_2$ Nonvertical parallel lines have the same slope. Two vertical lines are parallel.	The lines $y = 2x - 1$ and $y = 2x + 2$ are parallel because they both have slope 2. 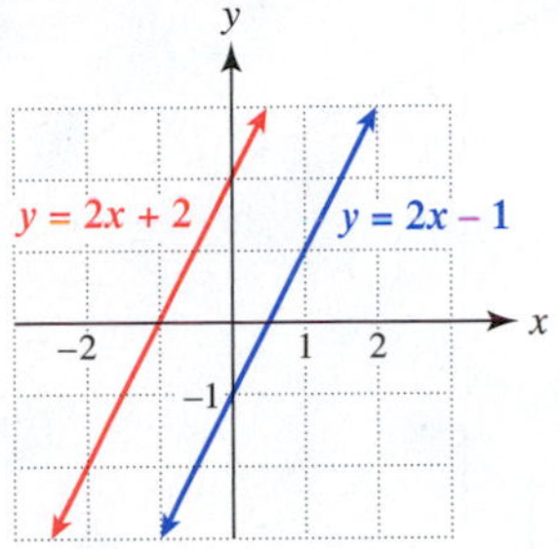 $y = 2x + 2$ $y = 2x - 1$
Perpendicular Lines	$y = m_1x + b_1$ and $y = m_2x + b_2$, where $m_1m_2 = -1$ Perpendicular lines which are neither vertical nor horizontal have slopes whose product equals -1. A vertical line and a horizontal line are perpendicular.	The lines $y = 3x - 1$ and $y = -\frac{1}{3}x + 2$ are perpendicular because $m_1m_2 = 3\left(-\frac{1}{3}\right) = -1.$ 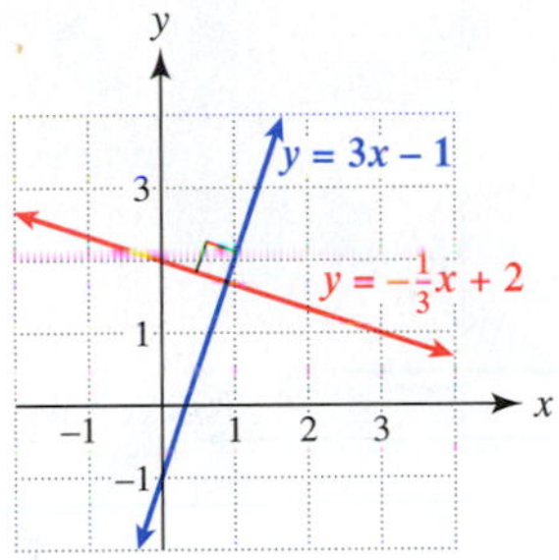 $y = 3x - 1$ $y = -\frac{1}{3}x + 2$

3.5 Exercises

CONCEPTS AND VOCABULARY

1. The slope–intercept form of a line is _____.

2. In the slope–intercept form of a line, m represents the _____ of the line.

3. In the slope–intercept form of a line, b represents the _____ of the line.

4. If $b = 0$ in the slope–intercept form of a line, then its graph passes through the _____.

5. Two lines with the same slope are _____.

6. Two nonvertical parallel lines have the same _____.

7. If m_1 and m_2 are the slopes of two lines where $m_1 \cdot m_2 = -1$, the lines are _____.

8. If two perpendicular lines have nonzero slopes, the slopes are negative _____.

Exercises 9–14: Match the description with its graph (a.–f.) at the top of the next page.

9. A line with positive slope and negative y-intercept

10. A line with positive slope and positive y-intercept

11. A line with negative slope and y-intercept 0

12. A line with negative slope and nonzero y-intercept

13. A line with no x-intercept

14. A line with no y-intercept

a.

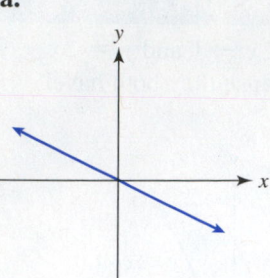

b.

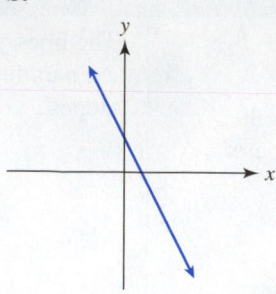

c.

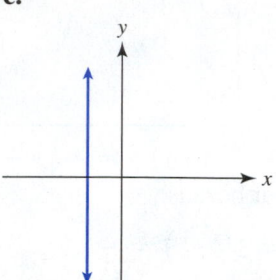

d.

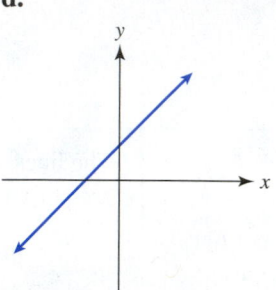

e.

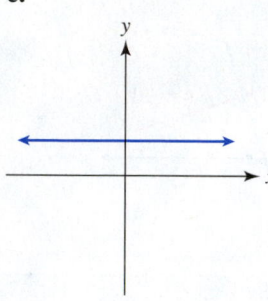

f.

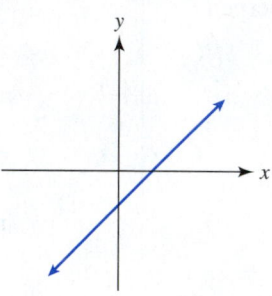

SLOPE–INTERCEPT FORM

Exercises 15–22: Write the slope–intercept form for the line shown in the graph.

15.

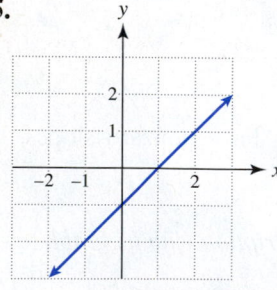

16.

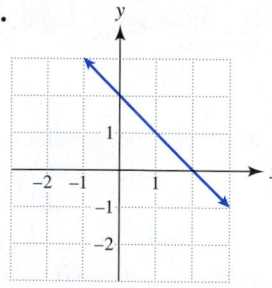

17.

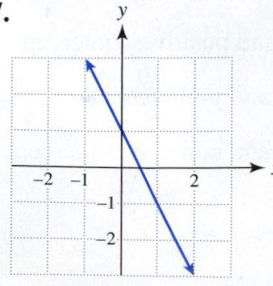

18.

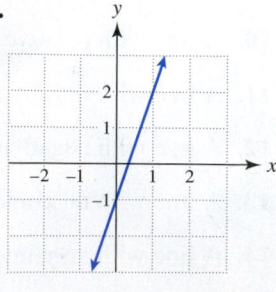

19.

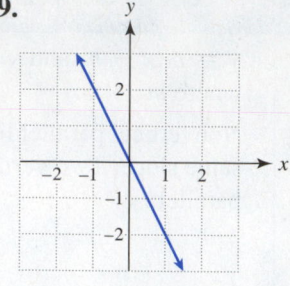

20.

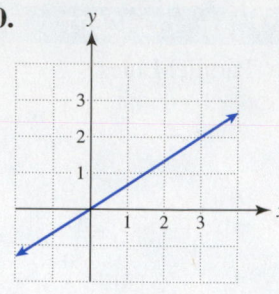

21.

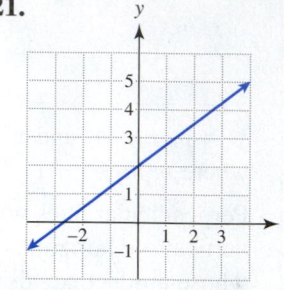

22.

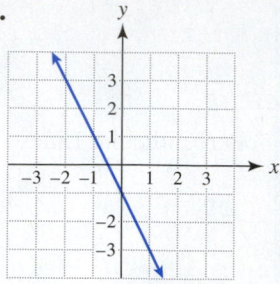

Exercises 23–32: Sketch a line with the given slope m and y-intercept b. Write its slope–intercept form.

23. $m = 1, b = 2$

24. $m = -1, b = 3$

25. $m = 2, b = -1$

26. $m = -3, b = 2$

27. $m = -\frac{1}{2}, b = -2$

28. $m = -\frac{2}{3}, b = 0$

29. $m = \frac{1}{3}, b = 0$

30. $m = 3, b = -3$

31. $m = 0, b = 3$

32. $m = 0, b = -3$

Exercises 33–44: Do the following.

(**a**) *Write the equation in slope–intercept form.*
(**b**) *Give the slope and y-intercept of the line.*

33. $x + y = 4$

34. $x - y = 6$

35. $2x + y = 4$

36. $-4x + y = 8$

37. $x - 2y = -4$

38. $x + 3y = -9$

39. $2x - 3y = 6$

40. $4x + 5y = 20$

41. $x = 4y - 6$

42. $x = -3y + 2$

43. $\frac{1}{2}x + \frac{3}{2}y = 1$

44. $-\frac{3}{4}x + \frac{1}{2}y = \frac{1}{2}$

Exercises 45–56: Graph the equation.

45. $y = -3x + 2$

46. $y = \frac{1}{2}x - 1$

47. $y = \frac{1}{3}x$

48. $y = -2x$

49. $y = 2$

50. $y = -3$

51. $x + y = 3$

52. $-\frac{2}{3}x + y = -2$

53. $x + 2y = 2$

54. $-2x - y = -2$

55. $x = 2 - y$

56. $x = -\frac{1}{3}y + \frac{2}{3}$

Exercises 57–60: The table shows points that all lie on the same line. Find the slope–intercept form for the line.

57.

x	0	1	2
y	2	4	6

58.

x	−1	0	1
y	4	8	12

59.

x	−2	0	2
y	−4	−2	0

60.

x	0	2	4
y	6	3	0

PARALLEL AND PERPENDICULAR LINES

Exercises 61–70: Find the slope–intercept form of the line satisfying the given conditions.

61. Slope $\frac{4}{7}$, y-intercept 3

62. Slope $-\frac{1}{2}$, y-intercept -7

63. Parallel to $y = 3x + 1$, passing through $(0, 0)$

64. Parallel to $y = -2x$, passing through $(0, 1)$

65. Parallel to $2x + 4y = 5$, passing through $(1, 2)$

66. Parallel to $-x - 3y = 9$, passing through $(-3, 1)$

67. Perpendicular to $y = -\frac{1}{2}x - 3$, passing through $(0, 0)$

68. Perpendicular to $y = \frac{3}{4}x - \frac{1}{2}$, passing through $(3, -2)$

69. Perpendicular to $x = -\frac{1}{3}y$, passing through $(-1, 0)$

70. Perpendicular to $6x - 3y = 18$, passing through $(4, -3)$

APPLICATIONS

71. *Rental Cars* Driving a rental car x miles costs $y = 0.25x + 25$ dollars.
  (a) How much would it cost to rent the car but not drive it?
  (b) How much does it cost to drive the car 1 *additional* mile?
  (c) What is the y-intercept of $y = 0.25x + 25$? What does it represent?
  (d) What is the slope of $y = 0.25x + 25$? What does it represent?

72. *Calculating Rainfall* The total rainfall y in inches that fell x hours past noon is given by $y = \frac{1}{2}x + 3$.
  (a) How much rainfall was there at noon?
  (b) At what rate was rain falling in the afternoon?

  (c) What is the y-intercept of $y = \frac{1}{2}x + 3$? What does it represent?
  (d) What is the slope of $y = \frac{1}{2}x + 3$? What does it represent?

73. *Cell Phone Plan* A cell phone plan costs $3.95 per month plus $0.07 per minute. (Assume that a partial minute is not rounded up.)
  (a) During July, a person talks a total of 50 minutes. What is the charge?
  (b) Write an equation in slope–intercept form that gives the monthly cost C of talking on this plan for x minutes.
  (c) If the charge for one month amounts to $8.64, how much time did the person spend talking on the phone?

74. *Electrical Rates* Electrical service costs $8 per month plus $0.10 per kilowatt-hour of electricity used. (Assume that a partial kilowatt-hour is not rounded up.)
  (a) If the resident of an apartment uses 650 kilowatt-hours in 1 month, what is the charge?
  (b) Write an equation in slope–intercept form that gives the cost C of using x kilowatt-hours in 1 month.
  (c) If the monthly electrical bill for the apartment's resident is $43, how many kilowatt-hours were used?

75. *Cost of Driving* The cost of driving a car includes both fixed costs and mileage costs. Assume that it costs $164.30 per month for insurance and car payments and $0.35 per mile for gasoline, oil, and routine maintenance.
  (a) Find values for m and b so that $y = mx + b$ models the monthly cost of driving the car x miles.
  (b) What does the value of b represent?

76. *Antarctic Ozone Layer* The ozone layer occurs in Earth's atmosphere between altitudes of 12 and 18 miles and is an important filter of ultraviolet light from the sun. The thickness of the ozone layer is frequently measured in Dobson units. An average value is 300 Dobson units. In 2007, the reported minimum in the antarctic *ozone hole* was about 83 Dobson units. (*Source:* NASA.)
  (a) The equation $T = 0.01D$ describes the thickness T in millimeters of an ozone layer that is D Dobson units. How many millimeters thick was the ozone layer over the antarctic in 2007?
  (b) What is the average thickness of the ozone layer in millimeters?

WRITING ABOUT MATHEMATICS

77. Explain how the values of m and b can be used to graph the equation $y = mx + b$.

78. Explain how to find the value of b in the equation $y = 2x + b$ if the point $(3, 4)$ lies on the line.

Group Activity   Working with Real Data

Directions: Form a group of 2 to 4 people. Select someone to record the group's responses for this activity. All members of the group should work cooperatively to answer the questions. If your instructor asks for the results, each member of the group should be prepared to respond.

Exercises 1–5: In this set of exercises you are to use your knowledge of equations of lines to model the average annual cost of tuition and fees (in 2010 dollars).

1. *Cost of Tuition* In 2005, the average cost of tuition and fees at *private* four-year colleges was $23,410, and in 2010 it was $27,290. Sketch a line that passes through the points (2005, 23410) and (2010, 27290). (*Source:* The College Board.)

2. *Rate of Change in Tuition* Calculate the slope of the line in your graph. Interpret this slope as a rate of change.

3. *Modeling Tuition* Find the slope–intercept form of the line in your sketch. What is the *y*-intercept and does it have meaning in this situation?

4. *Predicting Tuition* Use your equation to estimate tuition and fees in 2008 and compare it to the known value of $24,950. Estimate tuition and fees in 2015.

5. *Public Tuition* In 2005, the average cost of tuition and fees at *public* four-year colleges was $6130, and in 2010 it was $7610. Repeat Exercises 1–4 for these data. Note that the known value for 2008 is $6530. (*Source:* The College Board.)

3.6   Point–Slope Form

Derivation of Point–Slope Form • Finding Point–Slope Form • Applications

A LOOK INTO MATH ▶

In 1995, there were 690 female officers in the Marine Corps, and by 2010 this number had increased to about 1110. This growth is illustrated in Figure 3.48, where the line passes through the points (1995, 690) and (2010, 1110). Because two points determine a unique line, we can find the equation of this line and use it to *estimate* the number of female officers in other years (see Example 6). In this section we discuss how to find this equation by using the *point–slope form*, rather than the slope–intercept form of the equation of a line. (*Source:* Department of Defense.)

NEW VOCABULARY

☐ Equation of a line
☐ Point–slope form

STUDY TIP

Spend some extra time learning words in the language of mathematics. A strong mathematical vocabulary is one of the keys to success in any math course.

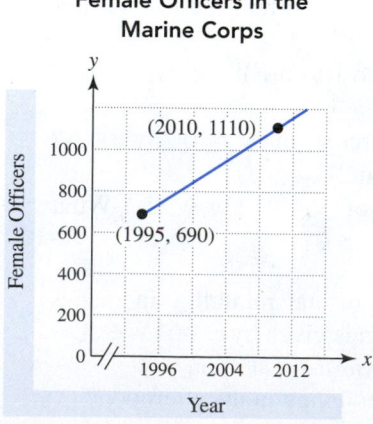

Female Officers in the Marine Corps

Figure 3.48

Derivation of Point–Slope Form

Slope of a Line

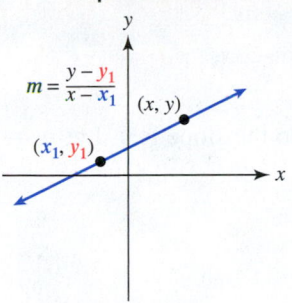

Figure 3.49

If we know the slope and y-intercept of a line, we can write its slope–intercept form, $y = mx + b$, which is an example of an **equation of a line**. The point–slope form is a different type of equation of a line.

Suppose that a (nonvertical) line with slope m passes through the point (x_1, y_1). If (x, y) is a different point on this line, then $m = \frac{y - y_1}{x - x_1}$. See Figure 3.49. Using this slope formula, we can find the point–slope form.

$$m = \frac{y - y_1}{x - x_1} \qquad \text{Slope formula}$$

$$m \cdot (x - x_1) = \frac{y - y_1}{x - x_1} \cdot (x - x_1) \qquad \text{Multiply each side by } (x - x_1).$$

$$m(x - x_1) = y - y_1 \qquad \text{Simplify.}$$

$$y - y_1 = m(x - x_1) \qquad \text{Rewrite the equation.}$$

The equation $y - y_1 = m(x - x_1)$ is traditionally called the *point–slope form*. By adding y_1 to each side of this equation we get $y = m(x - x_1) + y_1$, which is an equivalent form that is helpful when graphing. Both equations are in *point–slope form*.

POINT–SLOPE FORM

The line with slope m passing through the point (x_1, y_1) is given by

$$y - y_1 = m(x - x_1)$$

or, equivalently, $\qquad y = m(x - x_1) + y_1,$

the **point–slope form** of a line.

Finding Point–Slope Form

In the next example we find a point–slope form for a line. Note that *any* point that lies on the line can be used in its point–slope form.

EXAMPLE 1 | **Finding a point–slope form**

Use the labeled point in each figure to write a point–slope form for the line and then simplify it to the slope–intercept form.

(a)

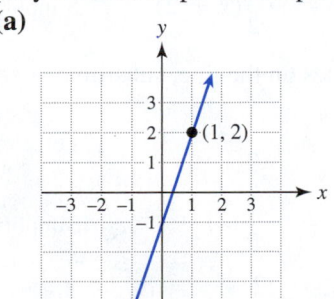

(b)

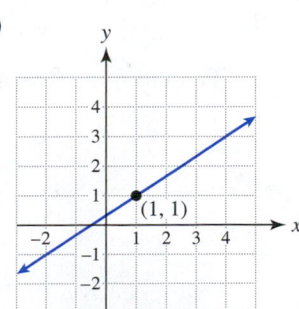

Solution

(a) The graph rises 3 units for each unit of horizontal run, so the slope is $\frac{3}{1}$, or 3. Let $m = 3$ and $(x_1, y_1) = (1, 2)$ in the point–slope form.

$$y - y_1 = m(x - x_1) \qquad \text{Point–slope form}$$

$$y - 2 = 3(x - 1) \qquad \text{Let } m = 3, x_1 = 1, \text{ and } y_1 = 2.$$

A point–slope form for the line is $y - 2 = 3(x - 1)$.

To simplify to the slope–intercept form, apply the distributive property to the equation $y - 2 = 3(x - 1)$.

$$y - 2 = 3x - 3 \qquad \text{Distributive property}$$
$$y = 3x - 1 \qquad \text{Add 2 to each side.}$$

The slope–intercept form is $y = 3x - 1$.

(b) The graph rises 2 units for each 3 units of horizontal run, so the slope is $\frac{2}{3}$. Let $m = \frac{2}{3}$ and $(x_1, y_1) = (\mathbf{1}, \mathbf{1})$ in the point–slope form.

$$y - y_1 = m(x - x_1) \qquad \text{Point–slope form}$$
$$y - \mathbf{1} = \frac{2}{3}(x - \mathbf{1}) \qquad \text{Let } m = \frac{2}{3}, x_1 = 1, \text{ and } y_1 = 1.$$

A point–slope form for the line is $y - 1 = \frac{2}{3}(x - 1)$.

To simplify to the slope–intercept form, apply the distributive property to the equation $y - 1 = \frac{2}{3}(x - 1)$.

$$y - 1 = \frac{2}{3}x - \frac{2}{3} \qquad \text{Distributive property}$$
$$y = \frac{2}{3}x + \frac{1}{3} \qquad \text{Add 1, or } \frac{3}{3}, \text{ to each side.}$$

The slope–intercept form is $y = \frac{2}{3}x + \frac{1}{3}$.

Now Try Exercises 13, 15

EXAMPLE 2 **Finding a point–slope form**

Find a point–slope form for a line passing through the point $(-2, 3)$ with slope $-\frac{1}{2}$. Does the point $(2, 1)$ lie on this line?

Solution
Let $m = -\frac{1}{2}$ and $(x_1, y_1) = (-\mathbf{2}, \mathbf{3})$ in the point–slope form.

$$y - y_1 = m(x - x_1) \qquad \text{Point–slope form}$$
$$y - \mathbf{3} = -\frac{\mathbf{1}}{2}(x - (-\mathbf{2})) \qquad x_1 = -2, y_1 = 3, \text{ and } m = -\frac{1}{2}$$
$$y - 3 = -\frac{1}{2}(x + 2) \qquad \text{Simplify.}$$

To determine whether $(2, 1)$ lies on the line, substitute $x = \mathbf{2}$ and $y = \mathbf{1}$ in the equation.

$$\mathbf{1} - 3 \overset{?}{=} -\frac{1}{2}(\mathbf{2} + 2) \qquad \text{Let } x = 2 \text{ and } y = 1.$$
$$-2 \overset{?}{=} -\frac{1}{2}(4) \qquad \text{Simplify.}$$
$$-2 = -2 \checkmark \qquad \text{A true statement}$$

The point $(2, 1)$ lies on the line because it satisfies the point–slope form.

Now Try Exercises 9, 19

In the next example we use the point–slope form to find the equation of a line passing through two points.

EXAMPLE 3   **Finding an equation of a line passing through two points**

Use the point–slope form to find an equation of the line passing through the points $(1, -4)$ and $(-2, 5)$.

Solution
Before we can apply the point–slope form, we must find the slope.

$$m = \frac{y_2 - y_1}{x_2 - x_1} \quad \text{Slope formula}$$

$$= \frac{5 - (-4)}{-2 - 1} \quad x_1 = 1, y_1 = -4, x_2 = -2, \text{ and } y_2 = 5$$

$$= -3 \quad \text{Simplify.}$$

We can let either the point $(1, -4)$ or the point $(-2, 5)$ be (x_1, y_1) in the point–slope form. If $(x_1, y_1) = (1, -4)$, then the equation of the line becomes the following.

$$y - y_1 = m(x - x_1) \quad \text{Point–slope form}$$

$$y - (-4) = -3(x - 1) \quad x_1 = 1, y_1 = -4, \text{ and } m = -3$$

$$y + 4 = -3(x - 1) \quad \text{Simplify.}$$

If we let $(x_1, y_1) = (-2, 5)$, the point–slope form becomes

$$y - 5 = -3(x + 2).$$

Now Try Exercise 25

NOTE: Although the two point–slope forms in Example 3 might appear to be different, they actually are equivalent because they simplify to the same slope–intercept form.

$y + 4 = -3(x - 1)$	$y - 5 = -3(x + 2)$	Point–slope forms
$y = -3(x - 1) - 4$	$y = -3(x + 2) + 5$	Addition property
$y = -3x + 3 - 4$	$y = -3x - 6 + 5$	Distributive property
$y = -3x - 1$	$y = -3x - 1$	Same slope–intercept form

MAKING CONNECTIONS

Slope–Intercept and Point–Slope Forms

READING CHECK

- Is the point–slope form of the equation of a line unique?

The slope–intercept form, $y = mx + b$, is unique because any nonvertical line has one slope m and one y-intercept b. The point–slope form, $y - y_1 = m(x - x_1)$, is *not* unique because (x_1, y_1) can be any point that lies on the line. However, any point–slope form can be simplified to a unique slope–intercept form.

EXAMPLE 4   **Finding equations of lines**

Find the slope–intercept form for the line that satisfies the conditions.
(a) Slope $\frac{1}{2}$, passing through $(-2, 4)$
(b) x-intercept -3, y-intercept 2
(c) Perpendicular to $y = -\frac{2}{3}x$, passing through $\left(\frac{2}{3}, 3\right)$

Solution

(a) Substitute $m = \frac{1}{2}$, $x_1 = -2$, and $y_1 = 4$ in the point–slope form.

$$y - y_1 = m(x - x_1) \qquad \text{Point–slope form}$$

$$y - 4 = \frac{1}{2}(x + 2) \qquad \text{Substitute and simplify.}$$

$$y - 4 = \frac{1}{2}x + 1 \qquad \text{Distributive property}$$

$$y = \frac{1}{2}x + 5 \qquad \text{Add 4 to each side.}$$

(b) The line passes through the points $(-3, 0)$ and $(0, 2)$. The slope of the line is

$$m = \frac{2 - 0}{0 - (-3)} = \frac{2}{3}.$$

Because the line has slope $\frac{2}{3}$ and y-intercept **2**, the slope–intercept form is

$$y = \frac{2}{3}x + 2.$$

(c) The slope of $y = -\frac{2}{3}x$ is $m_1 = -\frac{2}{3}$, so the slope of a line perpendicular to it is the negative reciprocal of $-\frac{2}{3}$, or $\frac{3}{2}$. Let $m = \frac{3}{2}$, $x_1 = \frac{2}{3}$, and $y_1 = 3$ in the point–slope form.

$$y - y_1 = m(x - x_1) \qquad \text{Point–slope form}$$

$$y - 3 = \frac{3}{2}\left(x - \frac{2}{3}\right) \qquad \text{Substitute.}$$

$$y - 3 = \frac{3}{2}x - 1 \qquad \text{Distributive property}$$

$$y = \frac{3}{2}x + 2 \qquad \text{Add 3 to each side.}$$

▌**Now Try Exercises** 43, 47, 51

In the next example the point–slope form is used to find the slope–intercept form of a line that passes through several points given in a table.

EXAMPLE 5 **Using a table to find slope–intercept form**

The points in the table lie on a line. Find the slope–intercept form of the line.

x	2	4	6	8
y	2	1	0	-1

Solution

The y-values in the table decrease one unit for every two-unit increase in the x-values, so the line has a "rise" of -1 when the run is 2. The slope is $m = \frac{\text{rise}}{\text{run}} = \frac{-1}{2} = -\frac{1}{2}$. *Any* point from the table can be used to obtain a point–slope form of the line, which can then be simplified to slope–intercept form. Letting $(x_1, y_1) = (2, 2)$ and $m = -\frac{1}{2}$ in the point–slope form yields the following result.

$$y - y_1 = m(x - x_1) \qquad \text{Point–slope form}$$

$$y - 2 = -\frac{1}{2}(x - 2) \qquad \text{Substitute.}$$

$$y - 2 = -\frac{1}{2}x + 1 \qquad \text{Distributive property}$$

$$y = -\frac{1}{2}x + 3 \qquad \text{Add 2 to each side.}$$

This result can be checked by substituting each x-value from the table in the equation. For example, when $x = 4$, the corresponding y-value is $y = -\frac{1}{2}(4) + 3 = -2 + 3 = 1$. This agrees with the table.

Now Try Exercise 53

MAKING CONNECTIONS

Finding Slope–Intercept Form Without Using Point–Slope Form

The equation in Example 5 can be obtained without using the point–slope form. The value of b can be found by letting $(x, y) = (2, 2)$ and $m = -\frac{1}{2}$ in the slope–intercept form.

$$y = mx + b \qquad \text{Slope–intercept form}$$

$$2 = -\frac{1}{2}(2) + b \qquad \text{Substitute.}$$

$$2 = -1 + b \qquad \text{Multiply.}$$

$$3 = b \qquad \text{Add 1 to each side.}$$

Because $m = -\frac{1}{2}$ and $b = 3$, the slope–intercept form is $y = -\frac{1}{2}x + 3$.

Applications

In the next example we find the equation of the line that models the Marine Corps data presented in A Look into Math at the beginning of this section.

EXAMPLE 6   **Modeling numbers of female officers**

In 1995, there were 690 female officers in the Marine Corps, and by 2010 this number had increased to about 1110. Refer to Figure 3.48 at the beginning of this section.
(a) Use the point (1995, 690) to find a point–slope form of the line shown in Figure 3.48.
(b) Interpret the slope as a rate of change.
(c) Use Figure 3.48 to estimate the number of female officers in 2006. Then use your equation from part (a) to approximate this number. How do your answers compare?

Solution
(a) The slope of the line passing through (1995, 690) and (2010, 1110) is

$$m = \frac{1110 - 690}{2010 - 1995} = 28.$$

If we let $x_1 = 1995$ and $y_1 = 690$, then the point–slope form becomes

$$y - 690 = 28(x - 1995) \quad \text{or} \quad y = 28(x - 1995) + 690.$$

(b) Slope $m = 28$ indicates that the number of female officers increased, *on average*, by about 28 officers per year.
(c) From Figure 3.48, it appears that the number of female officers in 2006 was about 1000. To estimate this value let $x = 2006$ in the equation found in part (a).

$$y = 28(2006 - 1995) + 690 = 998$$

Although the graphical estimate and calculated answers are not exactly equal, they are approximately equal. Estimations made from a graph usually are not exact.

Now Try Exercise 65

In the next example we review several concepts related to lines.

EXAMPLE 7   **Modeling water in a pool**

A small swimming pool is being emptied by a pump that removes water at a constant rate. After 1 hour the pool contains 5000 gallons, and after 3 hours it contains 3000 gallons.
(a) How fast is the pump removing water?
(b) Find the slope–intercept form of a line that models the amount of water in the pool. Interpret the slope.
(c) Find the y-intercept and the x-intercept. Interpret each.
(d) Sketch a graph of the amount of water in the pool during the first 6 hours.
(e) The point (2, 4000) lies on the graph. Explain its meaning.

Solution
(a) The pump removes $5000 - 3000 = 2000$ gallons of water in 2 hours, or 1000 gallons per hour.
(b) The line passes through the points (1, 5000) and (3, 3000), so the slope is

$$m = \frac{3000 - 5000}{3 - 1} = -1000.$$

One way to find the slope–intercept form is to use the point–slope form.

$$y - y_1 = m(x - x_1) \qquad \text{Point–slope form}$$
$$y - 5000 = -1000(x - 1) \qquad m = -1000, x_1 = 1, \text{ and } y_1 = 5000$$
$$y - 5000 = -1000x + 1000 \qquad \text{Distributive property}$$
$$y = -1000x + 6000 \qquad \text{Add 5000 to each side.}$$

Slope -1000 means that the pump is *removing* 1000 gallons of water per hour.
(c) The y-intercept is 6000 and indicates that the pool initially contained 6000 gallons of water. To find the x-intercept let $y = 0$ in the slope–intercept form.

$$0 = -1000x + 6000 \qquad \text{Let } y = 0.$$
$$1000x = 6000 \qquad \text{Add } 1000x \text{ to each side.}$$
$$x = \frac{6000}{1000} \qquad \text{Divide by 1000.}$$
$$x = 6 \qquad \text{Simplify.}$$

An x-intercept of 6 indicates that the pool is empty after 6 hours.
(d) The x-intercept is **6**, and the y-intercept is **6000**. Sketch a line from (**6**, 0) to (0, **6000**), as shown in Figure 3.50.

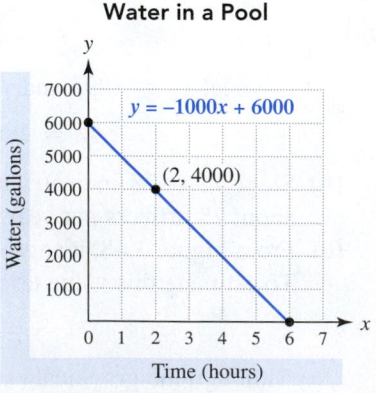

Water in a Pool

Figure 3.50

Now Try Exercise 59

(e) The point (2, 4000) indicates that after 2 hours the pool contains 4000 gallons of water.

CRITICAL THINKING

Suppose that a line models the amount of water in a swimming pool. What does a positive slope indicate? What does a negative slope indicate?

3.6   Putting It All Together

CONCEPT	COMMENTS	EXAMPLE
Point–Slope Form $y - y_1 = m(x - x_1)$ or $\quad y = m(x - x_1) + y_1$	Used to find an equation of a line, given two points or one point and the slope Can always be simplified to slope–intercept form Does *not* provide a unique equation for a line because *any* point on the line can be used	For two points $(1, 2)$ and $(3, 5)$, first compute $m = \frac{5-2}{3-1} = \frac{3}{2}$. An equation of this line is $$y - 2 = \frac{3}{2}(x - 1).$$ 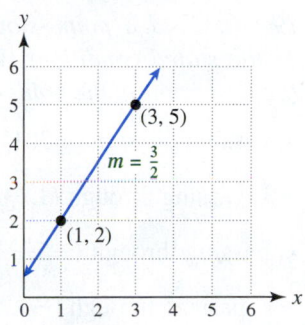

3.6   Exercises

MyMathLab  Math XL PRACTICE   WATCH   DOWNLOAD   READ   REVIEW

CONCEPTS AND VOCABULARY

1. (True or False?) One line is determined by two distinct points.

2. (True or False?) One line is determined by a point and a slope.

3. Give the slope–intercept form of a line.

4. Give the point–slope form of a line.

5. If the point–slope form is written for a line passing through $(1, 3)$, then $x_1 = $ _____ and $y_1 = $ _____.

6. To write a point–slope equation in slope–intercept form, use the _____ property to clear the parentheses.

7. Is the slope–intercept form of a line unique? Explain.

8. Is the point–slope form of a line unique? Explain.

POINT–SLOPE FORM

Exercises 9–12: Determine whether the given point lies on the line.

9. $(3, -11)$ $y + 1 = -2(x + 3)$

10. $(1, 4)$ $y - 3 = -(x - 1)$

11. $(0, 4)$ $y = \frac{1}{2}(x + 4) + 2$

12. $(2, -5)$ $y = -\frac{1}{3}(x - 5) - 6$

Exercises 13–18: Use the labeled point to write a point–slope form for the line.

13.

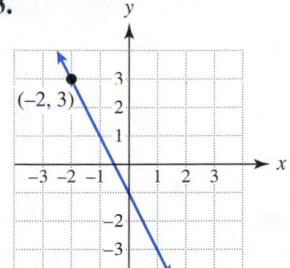

14.

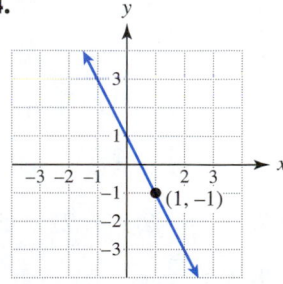

15.

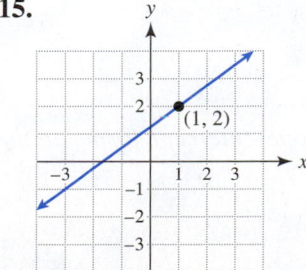

16.

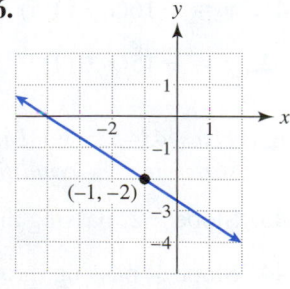

17.

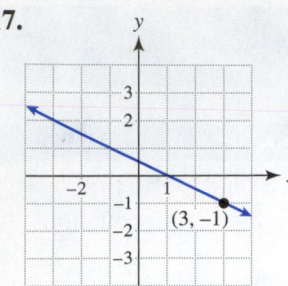

18.

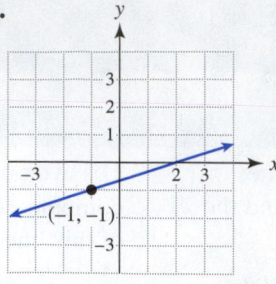

Exercises 19–30: Find a point–slope form for the line that satisfies the stated conditions. When two points are given, use the first point in the point–slope form.

19. Slope 4, passing through $(-3, 1)$

20. Slope -3, passing through $(1, -2)$

21. Slope $\frac{1}{2}$, passing through $(-5, -3)$

22. Slope $-\frac{2}{3}$, passing through $(-3, 6)$

23. Slope 1.5, passing through $(2010, 30)$

24. Slope -10, passing through $(2014, 100)$

25. Passing through $(2, 4)$ and $(-1, -3)$

26. Passing through $(3, -1)$ and $(1, -4)$

27. Passing through $(5, 0)$ and $(0, -3)$

28. Passing through $(-2, 0)$ and $(0, -1)$

29. Passing through $(2003, 15)$ and $(2013, 65)$

30. Passing through $(2009, 5)$ and $(2014, 30)$

Exercises 31–42: Write the point–slope form in slope–intercept form.

31. $y - 4 = 3(x - 2)$ **32.** $y - 3 = -2(x + 1)$

33. $y + 2 = \frac{1}{3}(x + 6)$ **34.** $y - 1 = \frac{2}{5}(x + 10)$

35. $y - \frac{3}{4} = \frac{2}{3}(x - 1)$ **36.** $y + \frac{2}{3} = -\frac{1}{6}(x - 2)$

37. $y = -2(x - 2) + 5$ **38.** $y = 4(x + 3) - 7$

39. $y = \frac{3}{5}(x - 5) + 1$ **40.** $y = -\frac{1}{2}(x + 4) - 6$

41. $y = -16(x + 1.5) + 5$

42. $y = -15(x - 1) + 100$

Exercises 43–52: Find the slope–intercept form for the line satisfying the conditions.

43. Slope -2, passing through $(4, -3)$

44. Slope $\frac{1}{5}$, passing through $(-2, 5)$

45. Passing through $(3, -2)$ and $(2, -1)$

46. Passing through $(8, 3)$ and $(-7, 3)$

47. x-intercept 3, y-intercept $\frac{1}{3}$

48. x-intercept 2, y-intercept -3

49. Parallel to $y = 2x - 1$, passing through $(2, -3)$

50. Parallel to $y = -\frac{3}{2}x$, passing through $(0, 20)$

51. Perpendicular to $y = -\frac{1}{2}x + 3$, passing through the point $(6, -3)$

52. Perpendicular to $y = \frac{3}{5}(x + 1) + 3$, passing through the point $(1, -2)$

Exercises 53–56: The points in the table lie on a line. Find the slope–intercept form of the line.

53.

x	1	2	3	4
y	-3	-5	-7	-9

54.

x	2	3	4	5
y	5	8	11	14

55.

x	-1	1	3	5
y	-3	-2	-1	0

56.

x	-1	5	11	17
y	1	-3	-7	-11

57. Thinking Generally Find the y-intercept of the line given by $y - y_1 = m(x - x_1)$.

58. Thinking Generally Find the x-intercept of the line given by $y - y_1 = m(x - x_1)$.

GRAPHICAL INTERPRETATION

59. *Change in Temperature* The outside temperature was $40°$F at 1 A.M. and $15°$F at 6 A.M. Assume that the temperature changed at a constant rate.
   (a) At what rate did the temperature change?
   (b) Find the slope–intercept form of a line that models the temperature T at x A.M. Interpret the slope as a rate of change.
   (c) Assuming that your equation is valid for times after 6 A.M., find and interpret the x-intercept.
   (d) Sketch a graph that shows the temperature from 1 A.M. to 9 A.M.
   (e) The point $(4, 25)$ lies on the graph. Explain its meaning.

60. *Cost of Fuel* The cost of buying 5 gallons of fuel oil is $12 and the cost of buying 15 gallons of fuel oil is $36.
(a) What is the cost of a gallon of fuel oil?
(b) Find the slope–intercept form of a line that models the cost of buying *x* gallons of fuel oil. Interpret the slope as a rate of change.
(c) Find and interpret the *x*-intercept.
(d) Sketch a graph that shows the cost of buying 20 gallons or less of fuel oil.
(e) The point (11, 26.40) lies on the graph. Explain its meaning.

61. *Water and Flow* The graph shows the amount of water *y* in a 500-gallon tank after *x* minutes have elapsed.
(a) Is water entering or leaving the tank? How much water is in the tank after 4 minutes?
(b) Find the *y*-intercept. Explain its meaning.
(c) Find the slope–intercept form of the line. Interpret the slope as a rate of change.
(d) After how many minutes will the tank be full?

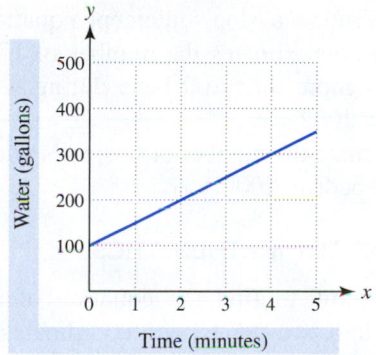

62. *Water and Flow* A hose is used to fill a 100-gallon barrel. If the hose delivers 5 gallons of water per minute, sketch a graph of the amount *A* of water in the barrel during the first 20 minutes.

63. *Distance and Speed* A person is driving a car along a straight road. The graph at the top of the next column shows the distance *y* in miles that the driver is from home after *x* hours.
(a) Is the person traveling toward or away from home?
(b) The graph passes through (1, 250) and (4, 100). Discuss the meaning of these points.
(c) How fast is the driver traveling?
(d) Find the slope–intercept form of the line. Interpret the slope as a rate of change.

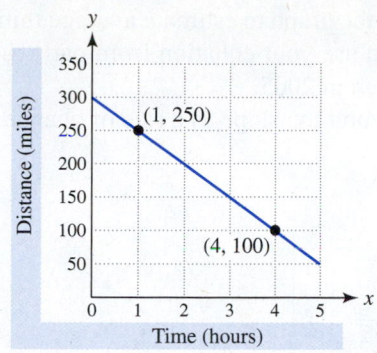

64. *Distance and Speed* A person rides a bicycle at 10 miles per hour, first away from home for 1 hour and then toward home for 1 hour. Sketch a graph that shows the distance *d* between the bicyclist and home after *x* hours.

APPLICATIONS

65. *Home Size* The graph models the average size in square feet of new U.S. homes built from 2002 to 2008. (*Source:* U.S. Census Bureau.)
(a) The line passes through the points (2002, 2334) and (2008, 2538). Explain the meaning of the first point.
(b) Use the first point to write a point–slope form for the equation of this line.
(c) Use the graph to estimate the average size of a new home in 2007. Then use your equation from part (b) to estimate the average size of a new home in 2007.
(d) Interpret the slope as a rate of change.

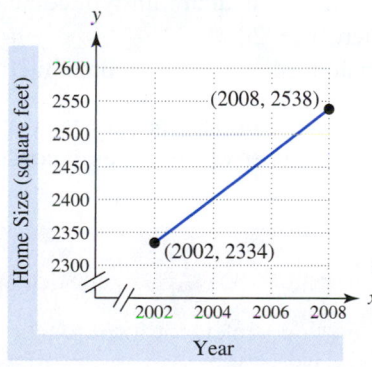

66. *Wyoming Tuition* The graph on the next page models average tuition at Wyoming public 2-year colleges after 2000, where 1 corresponds to 2001, 2 corresponds to 2002, and so on. (*Source:* SSTI *Weekly Digest.*)
(a) The line passes through the points (3, 1557) and (8, 1912). What is the meaning of the first of these two points?
(b) Use the first point to write a point–slope form for the equation of this line.

(c) Use the graph to estimate average tuition in 2005. Then use your equation from part (b) to estimate tuition in 2005.

(d) Interpret the slope as a rate of change.

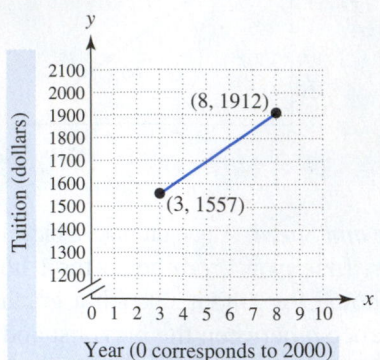

Year (0 corresponds to 2000)

67. *International Adoption* In 2006, there were 731 children adopted into the United States from Ethiopia, and this number was increasing by 515 children per year. (*Source:* U.S. Department of State.)

(a) Determine a point–slope equation of a line that approximates the number of children who were adopted into the United States from Ethiopia during year x, where $x \geq 2006$.

(b) Estimate the number of children adopted into the United States from Ethiopia in 2008.

68. *Median Family Income* In 2000, median income for U.S. families was \$41,950, and this number was increasing at a rate of \$1250 per year. (*Source:* Department of the Treasury.)

(a) Determine a point–slope equation of a line that approximates median family income during year x, where $x \geq 2000$.

(b) Estimate median family income in 2010.

69. *Cigarette Consumption* For any year x from 1975 to 2015, the number of cigarettes y consumed in the

United States is modeled by $y = -10.33x + 21{,}087$, where y is in billions. Interpret the slope as a rate of change. (*Source: The Tobacco Outlook Report.*)

70. *Alcohol Consumption* If x represents a year from 2000 to 2008, then the *average* number of gallons y of pure alcohol consumed annually in the United States by each person age 14 years and older is modeled by $y = 0.015x - 27.825$. Interpret the slope as a rate of change. (*Source: National Institutes of Health.*)

71. *HIV Infection Rates* In 2009, there were an estimated 33 million HIV infections worldwide, with an annual infection rate of 2.7 million. (*Source:* UNAIDS)

(a) Find the slope–intercept form of a line that approximates the cumulative number of HIV infections in millions during year x, where $x \geq 2009$.

(b) Estimate the number of HIV infections in 2012.

72. *Hospitals* In 2002, there were 3039 U.S. hospitals with more than 100 beds, and this number was decreasing at a rate of 33 hospitals per year. (*Source:* AHA Hospital Statistics.)

(a) Determine a slope–intercept equation of a line that approximates the number of U.S. hospitals with more than 100 beds during year x, where $x \geq 2002$.

(b) Estimate the number of hospitals with more than 100 beds in 2009.

WRITING ABOUT MATHEMATICS

73. Explain how to find the equation of a line passing through two points with coordinates (x_1, y_1) and (x_2, y_2). Give an example.

74. Explain how slope is related to rate of change. Give an example.

Checking Basic Concepts

1. Write the slope–intercept form for the line shown in the graph.

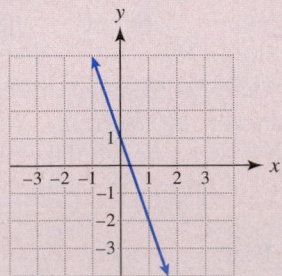

2. Write $4x - 5y = 20$ in slope–intercept form. Give the slope and y-intercept.

3. Graph $y = \frac{1}{2}x - 3$.

4. Write the slope–intercept form of a line that satisfies the given conditions.

(a) Slope 3, passing through $(0, -2)$

(b) Perpendicular to $y = \frac{2}{3}x$, passing through the point $(-2, 3)$

(c) Passing through $(1, -4)$ and $(-2, 3)$

5. Write a point–slope form for a line with slope -2, passing through $(-1, 3)$.

6. Write the equation $y + 3 = -2(x - 2)$ in slope–intercept form.

7. Find the slope–intercept form of the line passing through the points in the table.

x	-3	-1	1	3
y	-3	1	5	9

8. *Distance and Speed* A bicyclist is riding at a constant speed and is 36 miles from home at 1 P.M. Two hours later the bicyclist is 12 miles from home.

(a) Find the slope–intercept form of a line passing through $(1, 36)$ and $(3, 12)$.
(b) How fast is the bicyclist traveling?
(c) When will the bicyclist arrive home?
(d) How far was the bicyclist from home at noon?

9. *Snowfall* The total amount of snowfall S in inches t hours past noon is given by $S = 2t + 5$.

(a) How many inches of snow fell by noon?
(b) At what rate did snow fall in the afternoon?
(c) What is the S-intercept for the graph of this equation? What does it represent?
(d) What is the slope for the graph of this equation? What does it represent?

3.7   Introduction to Modeling

Basic Concepts • Modeling Linear Data

A LOOK INTO MATH ▶ For centuries people have tried to understand the world around them by creating models. For example, a weather forecast is based on a model. Mathematics is used to create these weather models, which often contain thousands of equations.

A model is an *abstraction* of something that people observed. Not only should a good model describe *known* data, but it should also be able to predict *future* data. In this section we discuss linear models, which are used to describe data that have a constant rate of change.

Basic Concepts

▶ **REAL-WORLD CONNECTION** Figure 3.51(a) shows a scatterplot of the number of inmates in the federal prison system from 2003 to 2009. The four points in the graph appear to be "nearly" collinear. That is, they appear almost to lie on the same line. Using mathematical modeling, we can find an equation for such a line. Once we have found it, we can use it to make estimates about the federal inmate population. An example of such a line is shown in Figure 3.51(b). (See Exercise 49 at the end of this section.)

STUDY TIP

There is often more than one way to accurately model data. Don't be afraid to try different methods. Part of the modeling process is trying to justify the reasoning behind a particular modeling choice.

READING CHECK

• What are some uses for a mathematical model?

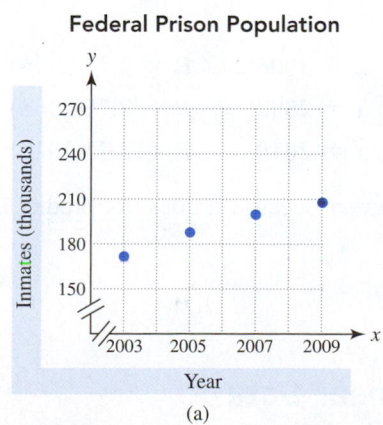

(a)

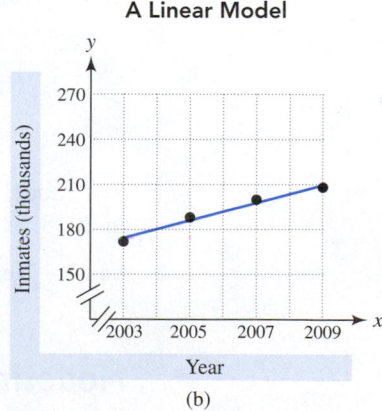

(b)

Figure 3.51

Generally, mathematical models are not *exact* representations of data. Data that might appear to be linear may not be. Although the line in Figure 3.51(b) on the previous page appears to touch every point, it does not pass through all four points exactly. Figure 3.52(a) shows data modeled *exactly* by a line, whereas Figure 3.52(b) shows data modeled *approximately* by a line. In applications a model is more likely to be approximate than exact.

Exact Model

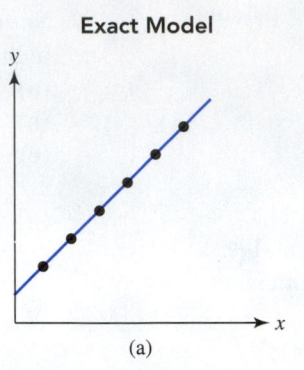

(a)

Approximate Model

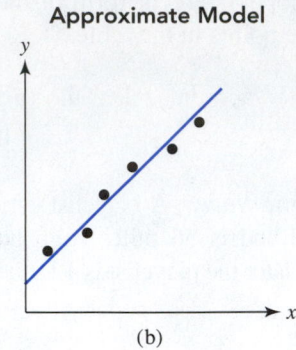

(b)

Figure 3.52 Linear Models

READING CHECK

• Does a mathematical model always touch every point in a scatterplot?

EXAMPLE 1 **Determining whether a model is exact**

A person can vote in the United States at age 18 or over. Table 3.20 shows the voting-age population P in millions for selected years x. Does the equation $P = 2.75x - 5291.5$ model the data exactly? Explain.

TABLE 3.20 **Voting-Age Population**

x	2006	2008	2010
P	225	230	236

Source: U.S. Census Bureau.

Solution

To determine whether the equation models the data exactly, let $x = 2006, 2008,$ and 2010 in the given equation.

$$x = \mathbf{2006}: \quad P = 2.75(\mathbf{2006}) - 5291.5 = 225$$
$$x = \mathbf{2008}: \quad P = 2.75(\mathbf{2008}) - 5291.5 = 230.5$$
$$x = \mathbf{2010}: \quad P = 2.75(\mathbf{2010}) - 5291.5 = 236$$

The model is *not exact* because it does not predict a voting-age population of 230 million in 2008.

Now Try Exercise 13

Modeling Linear Data

A line can model linear data. In the next example we use a line to model gas mileage.

EXAMPLE 2   **Determining gas mileage**

TABLE 3.21

x	2	4	6	8
y	30	60	90	120

Table 3.21 shows the number of miles y traveled by an SUV on x gallons of gasoline.
(a) Plot the data in the xy-plane. Be sure to label each axis.
(b) Sketch a line that models the data. (You may want to use a ruler.)
(c) Find the equation of the line and interpret the slope of the line.
(d) How far could this SUV travel on 11 gallons of gasoline?

Solution
(a) Plot the points (2, 30), (4, 60), (6, 90), and (8, 120), as shown in Figure 3.53(a).

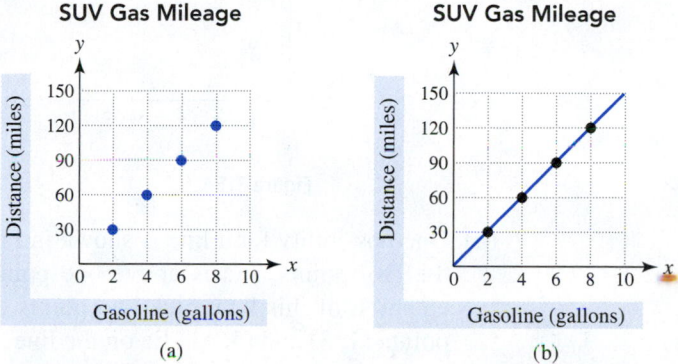

Figure 3.53

(b) Sketch a line similar to the one shown in Figure 3.53(b). This particular line passes through each data point.
(c) First find the slope m of the line by choosing two points that the line passes through, such as (2, 30) and (8, 120).

$$m = \frac{120 - 30}{8 - 2} = \frac{90}{6} = 15$$

Now find the equation of the line passing through (**2, 30**) with slope **15**.

$$y - y_1 = m(x - x_1) \quad \text{Point–slope form}$$
$$y - 30 = 15(x - 2) \quad x_1 = 2, y_1 = 30, \text{ and } m = 15$$
$$y - 30 = 15x - 30 \quad \text{Distributive property}$$
$$y = 15x \quad \text{Add 30 to each side.}$$

The data are modeled by the equation $y = 15x$. Slope 15 indicates that the mileage of this SUV is 15 miles per gallon.
(d) On 11 gallons of gasoline the SUV could go $y = 15(11) = 165$ miles.

Now Try Exercise 51

EXAMPLE 3   **Modeling linear data**

Table 3.22 contains ordered pairs that can be modeled *approximately* by a line.
(a) Plot the data. Could a line pass through all five points?
(b) Sketch a line that models the data and then determine its equation.

TABLE 3.22

x	1	2	3	4	5
y	3	5	6	10	11

Solution

(a) Plot the ordered pairs $(1, 3)$, $(2, 5)$, $(3, 6)$, $(4, 10)$, and $(5, 11)$, as shown in Figure 3.54(a). The points are not collinear, so it is impossible to sketch a line that passes through *all* five points.

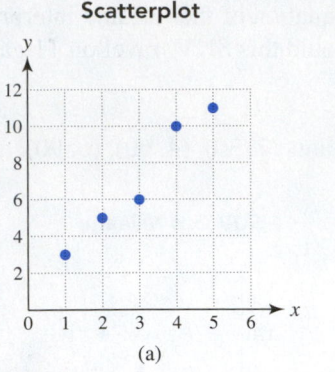

Figure 3.54

(b) One possibility for a line is shown in Figure 3.54(b). This line passes through three of the five points, and is above one point and below another point. To determine the equation of this line, pick two points that the line passes through. For example, the points $(1, 3)$ and $(5, 11)$ lie on the line. The slope of this line is

$$m = \frac{11 - 3}{5 - 1} = \frac{8}{4} = 2.$$

The equation of the line passing through $(\mathbf{1}, \mathbf{3})$ with slope $\mathbf{2}$ can be found as follows.

$$y - y_1 = m(x - x_1) \quad \text{Point–slope form}$$
$$y - \mathbf{3} = \mathbf{2}(x - \mathbf{1}) \quad x_1 = 1, y_1 = 3, \text{ and } m = 2$$
$$y - 3 = 2x - 2 \quad \text{Distributive property}$$
$$y = 2x + 1 \quad \text{Add 3 to each side.}$$

The equation of this line is $y = 2x + 1$.

▌ **Now Try Exercise 33**

NOTE: The equation found in Example 3 represents one possible linear model. Other linear models are possible.

▶ **REAL-WORLD CONNECTION** When a quantity increases at a constant rate, it can be modeled with the linear equation $y = mx + b$. This concept is illustrated in the next example.

EXAMPLE 4 | **Modeling worldwide HIV/AIDS cases in children**

At the beginning of 2009, a total of 2.5 million children (under age 15) were living with HIV/AIDS. The rate of new infections was 0.4 million per year.

(a) Write a linear equation $C = mx + b$ that models the total number of children C in millions that were living with HIV/AIDS, x years after January 1, 2009.

(b) Estimate C at the beginning of 2012.

Solution

(a) In the equation $C = mx + b$, the rate of change in HIV infections corresponds to the slope m, and the initial number of cases at the beginning of 2009 corresponds to b. Therefore the equation $C = 0.4x + 2.5$ models the data.

(b) The beginning of 2012 is 3 years after January 1, 2009, so let $x = \mathbf{3}$.

$$C = 0.4(\mathbf{3}) + 2.5 = 3.7 \text{ million}$$

NOTE: A total of 2.5 million children were living with HIV/AIDS at the beginning of 2009, with an infection rate of 0.4 million per year. After 3 years an additional $0.4(3) = 1.2$ million children would be infected, raising the total number to $2.5 + 1.2 = 3.7$ million.

❚ **Now Try Exercise 47**

EXAMPLE 5   **Modeling with linear equations**

Find a linear equation in the form $y = mx + b$ that models the quantity y after x days.
(a) A quantity y is initially 500 and increases at a rate of 6 per day.
(b) A quantity y is initially 1800 and decreases at a rate of 25 per day.
(c) A quantity y is initially 10,000 and remains constant.

Solution
(a) In the equation $y = mx + b$, the y-intercept b represents the initial amount and the slope m represents the rate of change. Therefore $y = 6x + 500$.
(b) The quantity y is decreasing at the rate of 25 per day with an initial amount of 1800, so $y = -25x + 1800$.
(c) The quantity is constant, so $m = 0$. The equation is $y = 10,000$.

❚ **Now Try Exercises 25, 27, 29**

MODELING WITH A LINEAR EQUATION

To model a quantity y that has a constant rate of change, use the equation

$$y = mx + b,$$

where $m =$ (constant rate of change) and $b =$ (initial amount).

3.7   Putting It All Together

CONCEPT	COMMENTS	EXAMPLE
Linear Model	Used to model a quantity that has a constant rate of change	If a total of 2 inches of rain falls before noon, and if rain falls at the rate of $\frac{1}{2}$ inch per hour, then $y = \frac{1}{2}x + 2$ models the total rainfall x hours past noon.
Modeling Linear Data with a Line	1. Plot the data. 2. Sketch a line that passes either through or nearly through the points. 3. Pick two points on the line and find the equation of the line.	To model $(0, 4)$, $(1, 3)$, and $(2, 2)$, plot the points and sketch a line as shown in the accompanying figure. Many times one line cannot pass through all the points. The equation of the line is $y = -x + 4$.

3.7   Exercises

MyMathLab
PRACTICE   WATCH   DOWNLOAD   READ   REVIEW

CONCEPTS AND VOCABULARY

1. Linear data are modeled by a(n) _____ equation.

2. If a line passes through all the data points, it is a(n) _____ model.

3. If a line passes near but not through each data point, it is a(n) _____ model.

4. Linear models are used to describe data that have a(n) _____ rate of change.

5. If a quantity is modeled by the equation $y = mx + b$, then m represents the _____.

6. If a quantity is modeled by the equation $y = mx + b$, then b represents the _____.

Exercises 7–12: **Modeling** *Match the situation to the graph (a.–f.) that models it best.*

7. College tuition from 2000 to 2012

8. Yearly average temperature in degrees Celsius at the North Pole

9. Profit from selling boxes of candy if it costs $200 to make the candy

10. Height of Mount Hood in Oregon

11. Total amount of water delivered by a garden hose if it flows at a constant rate

12. Sales of music on CDs from 2000 to 2010

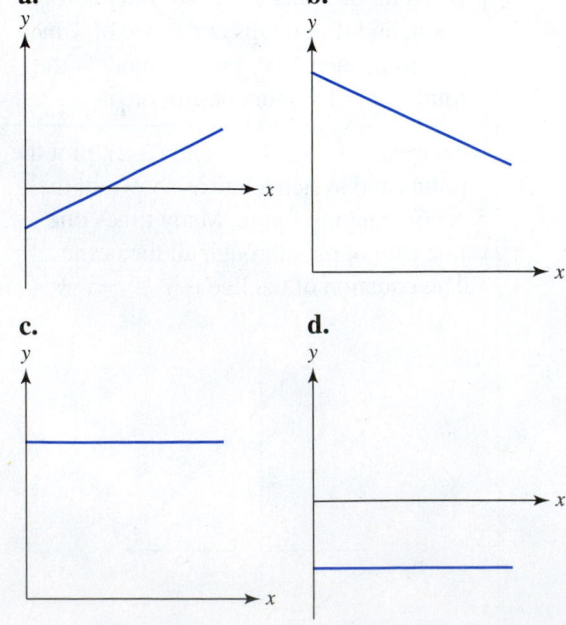

e.   f.

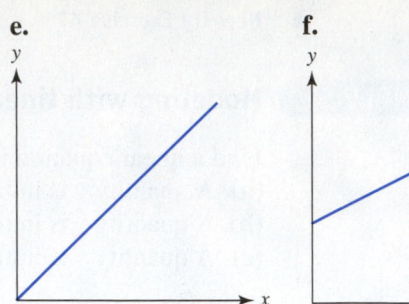

MODELING LINEAR DATA

Exercises 13–18: State whether the ordered pairs in the table are modeled exactly by the linear equation.

13. $y = 2x + 2$

x	0	1	2
y	2	4	6

14. $y = -2x + 5$

x	0	1	2
y	5	3	0

15. $y = -4x$

x	-1	0	1
y	4	0	-8

16. $y = 5 - x$

x	-2	1	4
y	7	4	1

17. $y = 1.4x - 4$

x	0	5	10
y	-4	3	9

18. $y = -\frac{4}{3}x - \frac{13}{3}$

x	-7	-4	-1
y	5	1	-3

Exercises 19–24: State whether the linear model in the graph is exact or approximate. Then find the equation of the line.

19.   20.

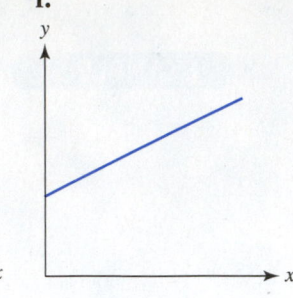

21.   22.

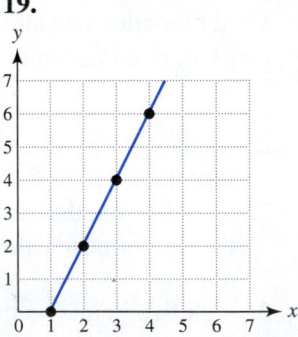

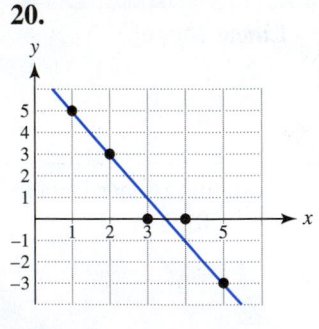

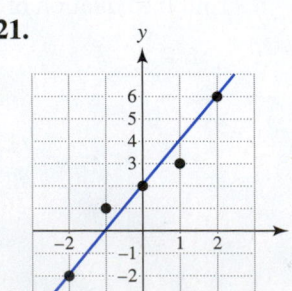

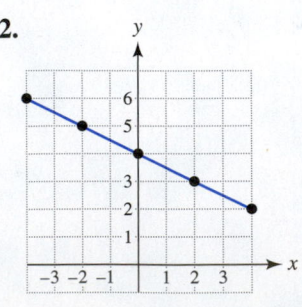

23.

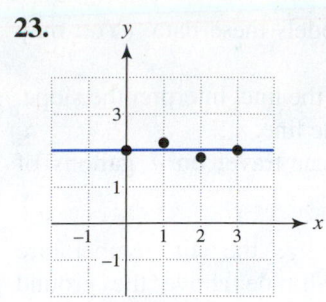

24.

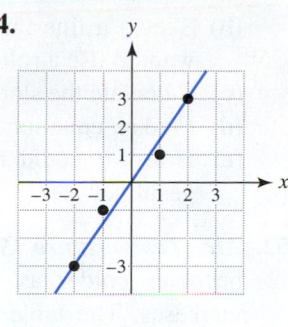

36.

x	−4	−2	0	2	4
y	1	2	3	4	5

37.

x	−6	−3	0	3	6
y	−3	−2	−0.5	0	0.5

38.

x	−3	−2	−1	0	1
y	1	2	3	4	5

Exercises 25–30: Find an equation $y = mx + b$ that models the quantity y after x units of time.

25. A quantity y is initially 40 and increases at a rate of 5 per minute.

26. A quantity y is initially −60 and increases at a rate of 1.7 per minute.

27. A quantity y is initially −5 and decreases at a rate of 20 per day.

28. A quantity y is initially 5000 and decreases at a rate of 35 per day.

29. A quantity y is initially 8 and remains constant.

30. A quantity y is initially −45 and remains constant.

*Exercises 31–38: **Modeling Data** For the ordered pairs in the table, do the following.*

   (a) Plot the data. Could a line pass through all five points?

   (b) Sketch a line that models the data.

   (c) Determine an equation of the line. For data that are not exactly linear, answers may vary.

31.

x	0	1	2	3	4
y	4	2	0	−2	−4

32.

x	0	1	2	2	3
y	7	6	5	4	3

33.

x	−2	−1	0	1	2
y	4	1	0	−1	−4

34.

x	−2	−1	0	1	2
y	7	5	1	−3	−5

35.

x	−6	−4	−2	0	2
y	1	0	−1	−2	−3

APPLICATIONS

Exercises 39–46: Write an equation that models the described quantity. Specify what each variable represents.

39. A barrel contains 200 gallons of water and is being filled at a rate of 5 gallons per minute.

40. A barrel contains 40 gallons of gasoline and is being drained at a rate of 3 gallons per minute.

41. An athlete has run 5 miles and is jogging at 6 miles per hour.

42. A new car has 141 miles on its odometer and is traveling at 70 miles per hour.

43. A worker has already earned $200 and is being paid $8 per hour.

44. A gambler has lost $500 and is losing money at a rate of $150 per hour.

45. A carpenter has already shingled 5 garage roofs and is shingling roofs at a rate of 1 per day.

46. A hard drive has been spinning for 2 minutes and is spinning at 7200 revolutions per minute.

47. *Kilimanjaro Glacier* Mount Kilimanjaro is located in Tanzania, Africa, and has an elevation of 19,340 feet. In 1912, the glacier on this peak covered 5 acres. By 2002 this glacier had melted to only 1 acre. (*Source:* NBC News.)

   (a) Assume that this glacier melted at a constant rate each year. Find this *yearly* rate.

   (b) Use your answer in part (a) to write a linear equation that gives the acreage A of this glacier t years past 1912.

48. *World Population* In 1987, the world's population reached 5 billion people, and by 2012 the world's population reached 7 billion people.

(*Source:* U.S. Census Bureau.)

(a) Find the average yearly increase in the world's population from 1987 to 2012.

(b) Write a linear equation that estimates the world's population P in billions x years after 1987.

49. *Prison Population* The points in Figure 3.51 are (2003, 172), (2005, 188), (2007, 200), and (2009, 208), where the y-coordinates are in thousands.

(a) Use the first and last data points to determine a line that models the data. Write the equation in slope–intercept form.

(b) Use the line to estimate the population in 2013.

50. *Niagara Falls* The average flow of water over Niagara Falls is 212,000 cubic feet of water per second.

(a) Write an equation that gives the number of cubic feet of water F that flow over the falls in x seconds.

(b) How many cubic feet of water flow over Niagara Falls in 1 minute?

51. *Gas Mileage* The table shows the number of miles y traveled by a car on x gallons of gasoline.

x (gallons)	3	6	9	12
y (miles)	60	120	180	240

(a) Plot the data in the xy-plane. Label each axis.

(b) Sketch a line that models these data. (You may want to use a ruler.)

(c) Calculate the slope of the line. Interpret the slope.

(d) Find an equation of the line.

(e) How far could this car travel on 7 gallons of gasoline?

52. *Air Temperature* Generally, the air temperature becomes colder as the altitude above the ground increases. The table lists typical air temperatures x miles high when the ground temperature is $80°$ F.

x (miles)	0	1	2	3
y (°F)	80	61	42	23

(a) Plot the data in the xy-plane. Label each axis.

(b) Sketch a line that models these data. (You may want to use a ruler.)

(c) Calculate the slope of the line. Interpret the slope.

(d) Find the slope–intercept form of the line.

(e) Estimate the air temperature 5 miles high.

WRITING ABOUT MATHEMATICS

53. In Example 2 the gas mileage of an SUV is modeled with a linear equation. Explain why it is reasonable to use a linear equation to model this situation.

54. Explain the steps for finding the equation of a line that models data points in a table.

SECTION 3.7	**Checking Basic Concepts**

1. State whether the ordered pairs shown in the table are modeled exactly by $y = -5x + 10$.

x	-2	-1	0	1
y	20	15	10	5

2. State whether the linear model in the graph is exact or approximate. Find the equation of the line.

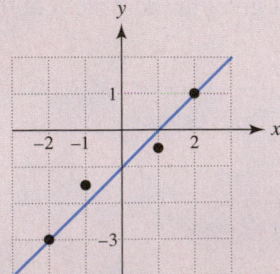

3. Find an equation, $y = mx + b$, that models the quantity y after x units of time.

(a) A quantity y is initially 50 pounds and increases at a rate of 10 pounds per day.

(b) A quantity y is initially $200°$ F and decreases at a rate of $2°$ F per minute.

4. The table contains ordered pairs.

(a) Plot the data.

(b) Sketch a line that models the data.

(c) Determine the equation of the line.

x	-2	0	2	4
y	2	1	0	-1

5. *Global Warming* Since 1945 the average annual recorded temperature on the Antarctic Peninsula has increased by $0.075°$ F per year.

(a) Write an equation that models the average temperature *increase T, x* years after 1945.

(b) Use your equation to calculate the temperature increase between 1945 and 2013.

CHAPTER 3   Summary

SECTION 3.1 ■ INTRODUCTION TO GRAPHING

The Rectangular Coordinate System (xy-plane)

Points   Plotted as (x, y) ordered pairs

Four Quadrants   The x- and y-axes divide the xy-plane into quadrants I, II, III, and IV.

NOTE: A point on an axis, such as $(1, 0)$, does not lie in a quadrant.

xy-plane   **Scatterplot**   **Line Graph**

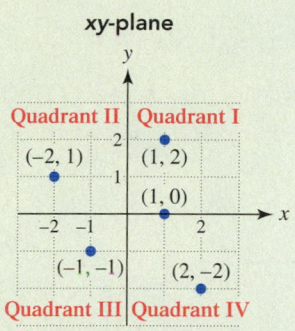   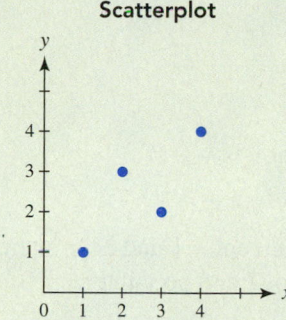   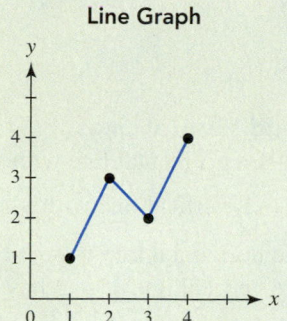

SECTION 3.2 ■ LINEAR EQUATIONS IN TWO VARIABLES

Equations in Two Variables   An equation with two variables and possibly some constants

Examples: $y = 3x + 7$ and $x + y = 100$

Solution to an Equation in Two Variables   The solution to an equation in two variables is an ordered pair that makes the equation a true statement.

Example: $(1, 2)$ is a solution to $2x + y = 4$ because $2(1) + 2 = 4$ is true.

Graphing a Linear Equation in Two Variables

Linear Equation $y = mx + b$ or $Ax + By = C$

Graphing   Plot at least three points and sketch a line passing through each one.

Example: $y = 2x - 1$

x	y
0	-1
1	1
2	3

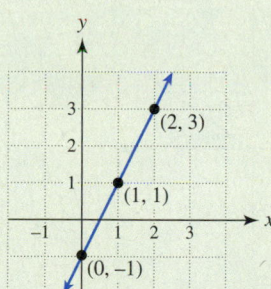

SECTION 3.3 ■ MORE GRAPHING OF LINES

Intercepts

x-Intercept   The x-coordinate of a point at which a graph intersects the x-axis; to find an x-intercept, let $y = 0$ in the equation and solve for x.

y-Intercept   The y-coordinate of a point at which a graph intersects the y-axis; to find a y-intercept, let $x = 0$ in the equation and solve for y.

Example: $x + 3y = 3$

x-intercept: Solve $x + 3(0) = 3$ to find the x-intercept of 3.

y-intercept: Solve $0 + 3y = 3$ to find the y-intercept of 1.

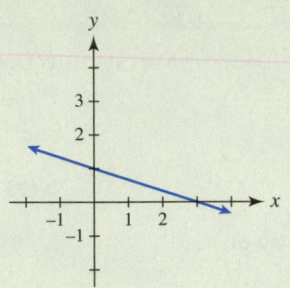

x-intercept: 3, y-intercept: 1

Horizontal and Vertical Lines

The equation of a horizontal line with y-intercept b is $y = b$.

The equation of a vertical line with x-intercept k is $x = k$.

Example: The horizontal line $y = -1$ has y-intercept -1 and no x-intercepts.
The vertical line $x = 2$ has x-intercept 2 and no y-intercepts.

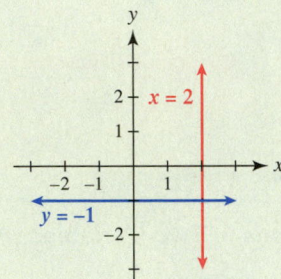

SECTION 3.4 ■ SLOPE AND RATES OF CHANGE

Slope   The ratio $\frac{\text{rise}}{\text{run}}$, or $\frac{\text{change in } y}{\text{change in } x}$, is the slope m of a line when run (change in x) is nonzero. A positive slope indicates that a line rises from left to right, and a negative slope indicates that a line falls from left to right.

Example: Slope $\frac{2}{3}$ indicates that a line rises 2 units for every 3 units of horizontal run.

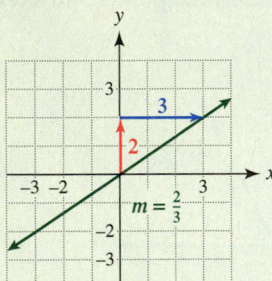

Calculating Slope   A line passing through (x_1, y_1) and (x_2, y_2) has slope

$$m = \frac{y_2 - y_1}{x_2 - x_1}, \text{ where } x_1 \neq x_2.$$

Example: The line through $(-2, 2)$ and $(3, 4)$ has slope

$$m = \frac{4 - 2}{3 - (-2)} = \frac{2}{5}.$$

Horizontal Line   Has zero slope

Vertical Line   Has undefined slope

Slope as a Rate of Change Slope measures the "tilt" or "angle" of a line. In applications, slope measures the rate of change in a quantity.

Example: The line shown in the graph has slope -2 and depicts an initial outside temperature of $6°F$. Slope -2 indicates that the temperature is *decreasing* at a rate of $2°F$ per hour.

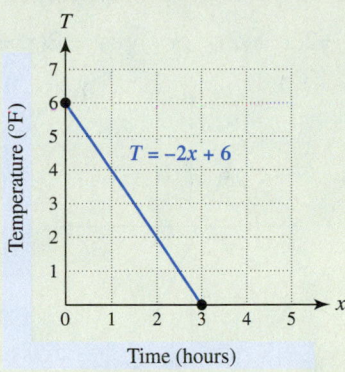

Time (hours)

SECTION 3.5 ■ SLOPE–INTERCEPT FORM

Slope–Intercept Form

For the line given by $y = mx + b$, the slope is m and the y-intercept is b.

Example: $y = -\frac{1}{2}x + 2$ has slope $-\frac{1}{2}$ and y-intercept 2, as shown in the graph.

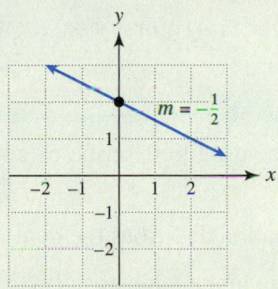

Parallel Lines Lines with the same slope are parallel; nonvertical parallel lines have the same slope. Two vertical lines are parallel.

Example: The equations $y = -2x + 1$ and $y = -2x$ determine two parallel lines because $m_1 = m_2 = -2$. See the graph shown below.

Perpendicular Lines

If two perpendicular lines have nonzero slopes m_1 and m_2, then $m_1 \cdot m_2 = -1$.

If two lines have nonzero slopes satisfying $m_1 \cdot m_2 = -1$, then they are perpendicular.

A vertical line and a horizontal line are perpendicular.

Example: The equations $y = -\frac{1}{2}x$ and $y = 2x - 2$ determine perpendicular lines because $m_1 \cdot m_2 = -\frac{1}{2} \cdot 2 = -1$. See the graph shown below.

Parallel Lines   **Perpendicular Lines**

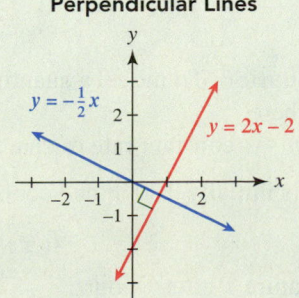

SECTION 3.6 ■ POINT–SLOPE FORM

Point–Slope Form An equation of the line passing through (x_1, y_1) with slope m is

$$y - y_1 = m(x - x_1) \quad \text{or} \quad y = m(x - x_1) + y_1.$$

Example: If $m = -2$ and $(x_1, y_1) = (-2, 3)$, then the point–slope form is either

$$y - 3 = -2(x + 2) \quad \text{or} \quad y = -2(x + 2) + 3.$$

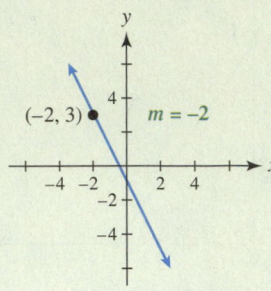

Example: To find an equation of a line passing through $(-2, 5)$ and $(4, 2)$, first find the slope.

$$m = \frac{2 - 5}{4 - (-2)} = \frac{-3}{6} = -\frac{1}{2}$$

Either $(-2, 5)$ or $(4, 2)$ may be used in the point–slope form. The point $(4, 2)$ results in the equation

$$y - 2 = -\frac{1}{2}(x - 4) \quad \text{or} \quad y = -\frac{1}{2}(x - 4) + 2.$$

SECTION 3.7 ■ INTRODUCTION TO MODELING

Mathematical Modeling Mathematics can be used to describe or approximate the behavior of real-world phenomena.

Exact Model The equation describes the data precisely without error.

Example: The equation $y = 3x$ models the data in the table exactly.

x	0	1	2	3
y	0	3	6	9

Approximate Model The modeling equation describes the data approximately. An approximate model occurs most often in applications.

Example: The line in the graph models the data approximately.

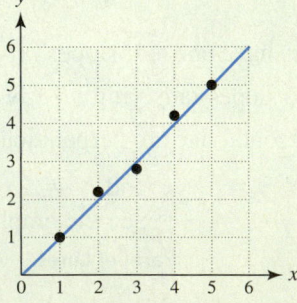

Modeling with a Linear Equation To model a quantity y that has a constant rate of change, use the equation $y = mx + b$, where

$$m = \text{(constant rate of change)} \quad \text{and} \quad b = \text{(initial amount)}.$$

Example: If the temperature is initially $100°\text{F}$ and cools at $5°\text{F}$ per hour, then

$$T = -5x + 100$$

models the temperature T after x hours.

CHAPTER 3   Review Exercises

SECTION 3.1

1. Identify the coordinates of each point in the graph. Identify the quadrant, if any, in which each point lies.

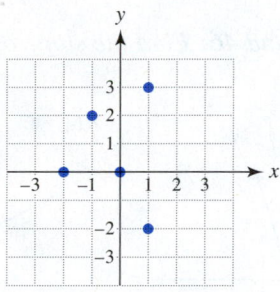

2. Make a scatterplot by plotting the following four points: $(-2, 3)$, $(-1, -1)$, $(0, 3)$, and $(2, -1)$.

Exercises 3 and 4: If possible, identify the quadrant in which each point is located.

3. (a) $(-4, 3)$ **(b)** $\left(\frac{1}{3}, -\frac{1}{2}\right)$

4. (a) $(0, 3.2)$ **(b)** $(-5, -1.7)$

Exercises 5 and 6: Use the table of xy-values to make a line graph.

5.

x	−2	−1	0	1	2
y	−3	2	−1	−2	3

6.

x	−10	−5	0	5	10
y	5	−10	10	−5	0

SECTION 3.2

Exercises 7–10: Determine whether the ordered pair is a solution for the given equation.

7. $y = x - 3$ $(6, 3)$

8. $y = 5 - 2x$ $(-2, 1)$

9. $3x - y = 3$ $(-1, 6)$

10. $\frac{1}{2}x + 2y = -8$ $(-4, -3)$

Exercises 11 and 12: Complete the table for the given equation.

11. $y = -3x$

x	−2	−1	0	1	2
y					

12. $2x + y = 5$

x					
y	−3	−1	0	1	3

Exercises 13–16: Use the given values of the variable to make a horizontal table of solutions for the equation.

13. $y = 3x + 2$ $x = -2, 0, 2, 4$

14. $y = 7 - x$ $x = 1, 2, 3, 4$

15. $y - 2x = 0$ $y = -1, 0, 1, 2$

16. $2y + x = 1$ $y = 1, 2, 3, 4$

Exercises 17–24: Graph the equation.

17. $y = 2x$ **18.** $y = x + 1$

19. $y = \frac{1}{2}x - 1$ **20.** $y = -3x + 2$

21. $x + y = 2$ **22.** $3x - 2y = 6$

23. $-4x + y = 8$ **24.** $2x + 3y = 12$

SECTION 3.3

Exercises 25 and 26: Identify the x- and y-intercepts.

25.

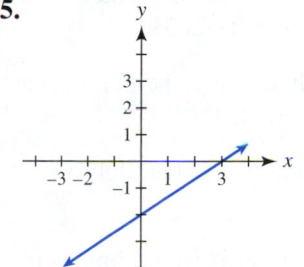

26.

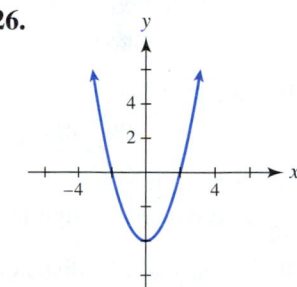

Exercises 27 and 28: Complete the table. Then determine the x- and y-intercepts for the graph of the equation.

27. $y = 2 - x$

x	−2	−1	0	1	2
y					

28. $x - 2y = 4$

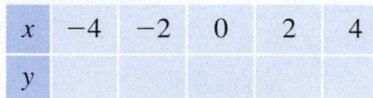

x	−4	−2	0	2	4
y					

Exercises 29–32: Find any intercepts for the graph of the equation and then graph the linear equation.

29. $2x - 3y = 6$

30. $5x - y = 5$

31. $0.1x - 0.2y = 0.4$

32. $\dfrac{x}{2} + \dfrac{y}{3} = 1$

Exercises 33 and 34: Write an equation for the line that passes through the points shown in the table.

33.

x	-2	-1	0	1	2
y	1	1	1	1	1

34.

x	3	3	3	3	3
y	-2	-1	0	1	2

35. Graph each equation.

   (a) $y = 1$ **(b)** $x = -3$

36. Write an equation for each line shown in the graph.

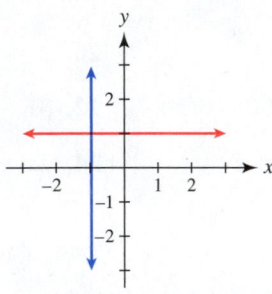

37. Write the equations of a horizontal line and a vertical line that pass through the point $(-2, 3)$.

38. Write the equation of a line that is perpendicular to $y = -\frac{1}{2}$ and passes through $(4, 1)$.

39. Write the equation of a line that is parallel to $y = 3$ and passes through $(-6, -5)$.

40. *Distance* The distance a driver is from home is illustrated in the graph.

   (a) Find the intercepts.

   (b) Interpret each intercept.

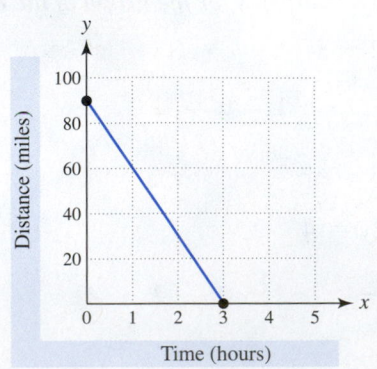

SECTION 3.4

Exercises 41–44: Find the slope, if possible, of the line passing through the two points.

41. $(2, 3), (4, 7)$ **42.** $(-3, 1), (2, -1)$

43. $(2, 1), (5, 1)$ **44.** $(-5, 6), (-5, 10)$

Exercises 45 and 46: Find the slope of the line shown in the graph.

45.

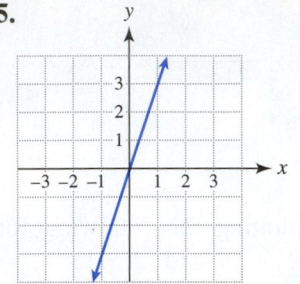

46.

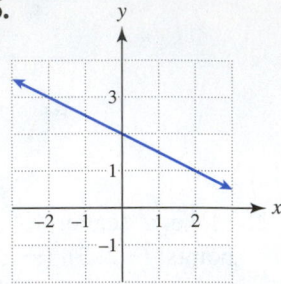

Exercises 47–50: Do the following.

   (a) Graph the linear equation.

   (b) What are the slope and y-intercept of the line?

47. $y = -2x$ **48.** $y = x - 1$

49. $x + 2y = 4$ **50.** $2x - 3y = -6$

Exercises 51–54: Sketch a line passing through the given point and having slope m.

51. $(0, -3), m = 2$ **52.** $(0, 1), m = -\frac{1}{2}$

53. $(-1, 1), m = -\frac{2}{3}$ **54.** $(2, 2), m = 1$

55. A line with slope $\frac{1}{2}$ passes through the first point shown in the table. Complete the table so that each point in the table lies on the line.

x	0	1	2	3
y	1			

56. If a line models the cost of buying x coffee drinks at $3.49 each, does the line have positive or negative slope?

SECTION 3.5

Exercises 57 and 58: Write the slope–intercept form for the line shown in the graph.

57. 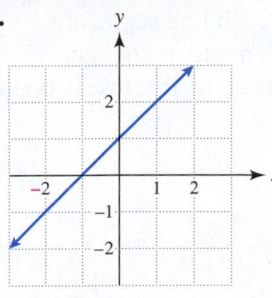   **58.**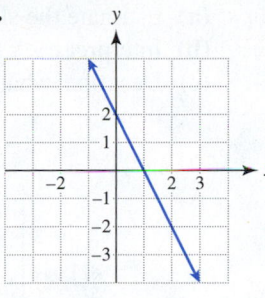

Exercises 59 and 60: Sketch a line with slope m and y-intercept b. Write its slope–intercept form.

59. $m = 2, b = -2$ **60.** $m = -\frac{3}{4}, b = 3$

Exercises 61–64: Do the following.

    (a) Write the equation in slope–intercept form.
    (b) Give the slope and y-intercept of the line.

61. $x + y = 3$ **62.** $-3x + 2y = -6$

63. $20x - 10y = 200$ **64.** $5x - 6y = 30$

Exercises 65–72: Graph the equation.

65. $y = \frac{1}{2}x + 1$ **66.** $y = 3x - 2$

67. $y = -\frac{1}{3}x$ **68.** $y = 3x$

69. $y = 2$ **70.** $y = -1$

71. $y = 4 - x$ **72.** $y = 2 - \frac{2}{3}x$

Exercises 73 and 74: All the points shown in the table lie on the same line. Find the slope–intercept form for the line.

73.

x	0	1	2
y	−5	0	5

74.

x	−1	0	1
y	2	0	−2

Exercises 75–78: Find the slope–intercept form for the line satisfying the given conditions.

75. Slope $-\frac{5}{6}$, y-intercept 2

76. Parallel to $y = -2x + 1$, passing through $(1, -5)$

77. Perpendicular to $y = -\frac{3}{2}x$, passing through $(3, 0)$

78. Perpendicular to $y = 5x - 3$, passing through the point $(0, -2)$

SECTION 3.6

Exercises 79 and 80: Determine whether the given point lies on the line.

79. $(-3, 1)$ $y - 1 = 2(x + 3)$

80. $(3, -8)$ $y = -3(x - 1) + 2$

Exercises 81–88: Find a point–slope form for the line that satisfies the conditions given. When two points are given, use the first point in the point–slope form.

81. Slope 5, passing through $(1, 2)$

82. Slope 20, passing through $(3, -5)$

83. Passing through $(-2, 1)$ and $(1, -1)$

84. Passing through $(20, -30)$ and $(40, 30)$

85. x-intercept 3, y-intercept -4

86. x-intercept $\frac{1}{2}$, y-intercept -1

87. Parallel to $y = 2x$, passing through $(5, 7)$

88. Perpendicular to $y - 4 = \frac{3}{2}(x + 1)$, passing through $(-1, 0)$

Exercises 89–92: Write the given point–slope form in slope–intercept form.

89. $y - 2 = 3(x + 1)$ **90.** $y - 9 = \frac{1}{3}(x - 6)$

91. $y = 2(x + 3) + 5$ **92.** $y = -\frac{1}{4}(x - 8) + 1$

SECTION 3.7

93. State whether the ordered pairs shown in the table are modeled exactly by $y = -x + 4$.

x	0	1	2	2
y	4	3	2	1

94. State whether the linear model shown in the graph is exact or approximate. Then find the equation of the line.

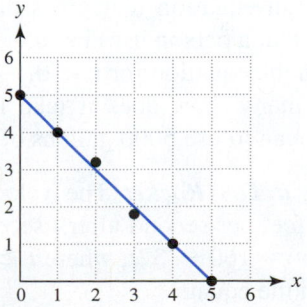

Exercises 95–98: Find an equation $y = mx + b$ that models y after x units of time.

95. *y* is initially 40 pounds and decreases at a rate of 2 pounds per minute.

96. *y* is initially 200 gallons and increases at 20 gallons per hour.

97. *y* is initially 50 and remains constant.

98. *y* is initially 20 feet *below* sea level and rises at 5 feet per second.

Exercises 99 and 100: For the ordered pairs in the table, do the following.

   (a) *Plot the data. Could a line pass through all five points?*

   (b) *Sketch a line that models the data.*

   (c) *Determine an equation of the line. For data that are not exactly linear, answers may vary.*

99.

x	0	1	2	3	4
y	10	6	2	−2	−6

100.

x	−4	−2	0	2	4
y	1	2.1	3	3.9	5

APPLICATIONS

101. *Graphing Real Data* The table contains real data on divorces *D* in millions during year *t*.
   (a) Make a line graph of the data. Label the axes.
   (b) Comment on any trends in the data.

t	1970	1980	1990	2000	2010
D	0.7	1.2	1.2	1.2	1.0

   Source: National Center for Health Statistics.

102. *Water Usage* The average American uses 100 gallons of water each day.
   (a) Write an equation that gives the gallons *G* of water that a person uses in *t* days.
   (b) Graph the equation for $t \geq 0$.
   (c) How many days does it take for the average American to use 5000 gallons of water?

103. *Modeling a Toy Rocket* The velocity *v* of a toy rocket in feet per second after *t* seconds of flight is given by $v = 160 - 32t$, where $t \geq 0$.
   (a) Graph the equation.
   (b) Interpret each intercept.

104. *Modeling* The accompanying line graph represents the insect population on 1 acre of land after *x* weeks. During this time a farmer sprayed pesticides on the land.
   (a) Estimate the slope of each line segment.
   (b) Interpret each slope as a rate of change.
   (c) Describe what happened to the insect population.

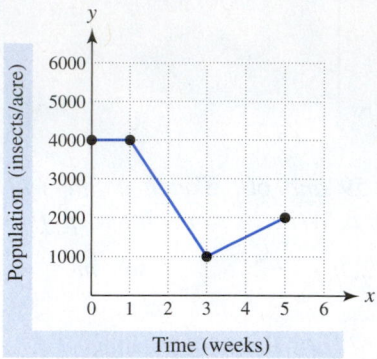

105. *Sketching a Graph* An athlete jogs 4 miles away from home at a constant rate of 8 miles per hour and then turns around and jogs back home at the same speed. Sketch a graph that shows the distance that the athlete is from home. Be sure to label each axis.

106. *Nursing Homes* In 1985, there were 19,100 nursing homes, and in 2010, there were 16,100. (*Source:* National Center for Health Statistics.)
   (a) Calculate the slope of the line passing through (1985, 19100) and (2010, 16100).
   (b) Interpret the slope as a rate of change.

107. *Rental Cars* The cost *C* in dollars for driving a rental car *x* miles is $C = 0.2x + 35$.
   (a) How much would it cost to rent the car but not drive it?
   (b) How much does it cost to drive the car one *additional* mile?
   (c) Determine the *C*-intercept for the graph of $C = 0.2x + 35$. What does it represent?
   (d) What is the slope of the graph of $C = 0.2x + 35$? What does it represent?

108. *Distance and Speed* A person is driving a car along a straight road. The graph at the top of the next page shows the distance *y* in miles that the driver is from home after *x* hours.
   (a) Is the person traveling toward or away from home? Why?
   (b) The graph passes through (1, 200) and (3, 100). Discuss the meaning of these points.
   (c) Find the slope–intercept form of the line. Interpret the slope as a rate of change.

(d) Use the graph to estimate the distance from home after 2 hours. Then check your answer by using your equation from part (c).

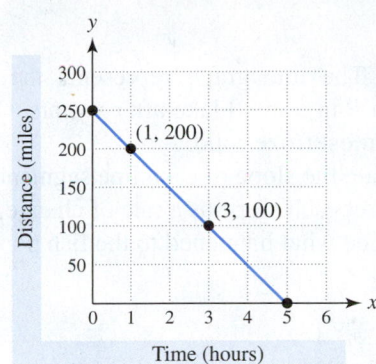

Time (hours)

109. *Arctic Glaciers* The arctic ice cap has been breaking up. As a result there has been an increase in the number of icebergs floating in the Arctic Ocean. The table gives an iceberg count I for various years t.
(*Source:* NBC News.)

t	1970	1980	2000
I	400	600	1000

(a) Make a scatterplot of the data.
(b) Find the slope–intercept form of a line that models the number of icebergs I in year t. Interpret the slope of this line.
(c) Is the line you found in part (b) an exact model for the data in the table?
(d) If this trend continued, what was the iceberg count in 2005?

110. *Gas Mileage* The table shows the number of miles y traveled by a car on x gallons of gasoline.

x	2	4	8	10
y	40	79	161	200

(a) Plot the data in the xy-plane. Be sure to label each axis.
(b) Sketch a line that models these data. (You may want to use a ruler.)
(c) Calculate the slope of the line. Interpret the slope of this line.
(d) Find the equation of the line. Is your line an *exact* model?
(e) How far could this car travel on 9 gallons of gasoline?

CHAPTER 3   Test

Step-by-step test solutions are found on the Chapter Test Prep Videos available in **MyMathLab** and on **You Tube** (search "RockswoldComboAlg" and click on "Channels").

1. Identify the coordinates of each point in the graph. State the quadrant, if any, in which each point lies.

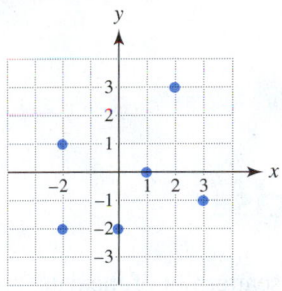

2. Make a scatterplot by plotting the four points $(0, 0)$, $(-2, -2)$, $(3, 0)$, and $(3, -2)$.

3. Complete the table for the equation $y = 2x - 4$. Then determine the x- and y-intercepts for the graph of the equation.

x	-2	-1	0	1	2
y					

4. Determine whether the ordered pair $(1, -3)$ is a solution for the equation $2x - y = 5$.

5. Sketch a line passing through the point $(2, 1)$ and having slope $-\frac{1}{2}$.

6. Find the x- and y-intercepts for the graph of the equation $5x - 3y = 15$.

Exercises 7–10: Graph the equation.

7. $y = 2$ **8.** $x = -3$

9. $y = -3x + 3$ **10.** $4x - 3y = 12$

11. Write the slope–intercept form for the line shown in the graph.

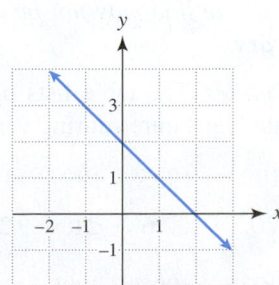

12. Write the equation $-4x + 2y = 1$ in slope–intercept form. Give the slope and the y-intercept.

13. Write the equation of a horizontal line and a vertical line passing through the point $(1, -5)$.

14. Find the slope of a line passing through the points $(-4, 3)$ and $(5, 1)$.

Exercises 15–18: Find the slope–intercept form for the line that satisfies the given conditions.

15. Slope $-\frac{4}{3}$, y-intercept -5

16. Parallel to $y = 3x - 1$, passing through $(2, -5)$

17. Perpendicular to $y = \frac{1}{3}x$, passing through $(1, 2)$

18. Passing through $(-4, 2)$ and $(2, -1)$

19. Write $y - 3 = \frac{1}{2}(x + 4)$ in slope–intercept form.

20. All of the points shown in the table lie on the same line. Find the slope–intercept form for the line.

x	-2	-1	0	1	2
y	-8	-5	-2	1	4

21. Write a point–slope form for a line that is parallel to $y = -3x$ and passes through $(-2, 7)$.

22. State whether the linear model shown is exact or approximate. Then find the equation of the line.

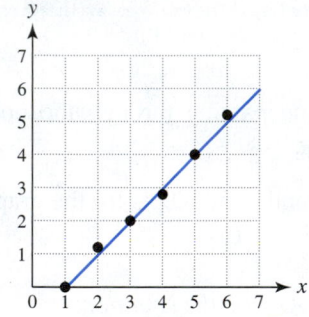

23. *Sketching a Graph* A cyclist rides a bicycle at 10 miles per hour for 2 hours and then at 8 miles per hour for 1 hour. Sketch a graph that shows the total distance d traveled after x hours. Be sure to label each axis.

24. *Modeling* The line graph represents the total fish population P in a small lake after x years. One winter the lake almost froze solid.
   (a) Estimate the slope of each line segment.
   (b) Interpret each slope as a rate of change.
   (c) Describe what happened to the fish population.

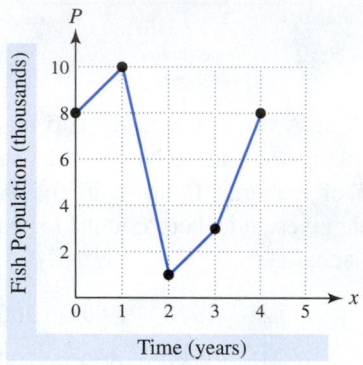

25. *Modeling Insects* Write an equation in slope–intercept form that models the number of insects N after x days if there are initially 2000 insects and the population increases at a rate of 100 insects per day.

CHAPTER 3   Extended and Discovery Exercises

Exercises 1 and 2: The table of real data can be modeled by a line. However, the line may not be an exact model, so answers may vary.

1. *Women in Politics* The table lists percentages P of women in state legislatures during year x.

x	1993	1995	1997	1999
P	20.5	20.6	21.6	22.4

x	2001	2003	2005	2007
P	22.4	22.4	22.7	23.5

Source: National Women's Political Caucus.

(a) Make a scatterplot of the data.
(b) Find a point–slope form of a line that models these data.
(c) Use your equation to estimate the percentage of women in state legislatures in 2010.

2. *U.S. Population* The population P of the United States in millions during year x is shown in the table.

x	1970	1980	1990	2000	2010
P	203	227	249	281	309

Source: U.S. Census Bureau.

(a) Make a scatterplot of the data.

(b) Find a point–slope form of a line that models these data.

(c) Use your equation to estimate what the U.S. population was in 2005.

CREATING GEOMETRIC SHAPES

Exercises 3–6: Many geometric shapes can be created with intersecting lines. For example, a triangle is formed when three lines intersect at three distinct points called vertices. Find equations of lines that satisfy the characteristics described. Sketching a graph may be helpful.

3. *Triangle* A triangle has vertices $(0, 0)$, $(2, 3)$, and $(3, 6)$.

   (a) Find slope–intercept forms for three lines that pass through each pair of points.

   (b) Graph the three lines. Is a triangle formed by the line segments connecting the three points?

4. *Parallelogram* A parallelogram has four sides, with opposite sides parallel. Three sides of a parallelogram are given by the equations $y = 2x + 2$, $y = 2x - 1$, and $y = -x - 2$.

   (a) If the fourth side of the parallelogram passes through the point $(-2, 3)$, find its equation.

   (b) Graph all four lines. Is a parallelogram formed?

5. *Rectangle* The two vertices $(0, 0)$ and $(4, 2)$ determine one side of a rectangle. The side parallel to this side passes through the point $(0, 3)$.

   (a) Find slope–intercept forms for the four lines that determine the sides of the rectangle.

   (b) Graph all four lines. Is a rectangle formed? (*Hint:* If you use a graphing calculator be sure to set a square window.)

6. *Square* Three vertices of a square are $(1, 2)$, $(4, 2)$, and $(4, 5)$.

   (a) Find the fourth vertex.

   (b) Find equations of lines that correspond to the four sides of the square.

   (c) Graph all four lines. Is a square formed?

CHAPTERS 1–3   Cumulative Review Exercises

Exercises 1 and 2: Classify the number as prime or composite. If a number is composite, write it as a product of prime numbers.

1. 40

2. 61

Exercises 3 and 4: Translate the phrase into an algebraic expression using the variable n.

3. Ten more than a number

4. A number squared decreased by 2

Exercises 5 and 6: Evaluate by hand and then simplify to lowest terms.

5. $\frac{3}{4} \div \frac{9}{8}$

6. $\frac{7}{10} - \frac{2}{15}$

Exercises 7 and 8: Evaluate the expression by hand.

7. $20 - 2 \cdot 3$

8. -3^2

Exercises 9 and 10: Classify the number as one or more of the following: natural number, whole number, integer, rational number, or irrational number.

9. $-\frac{4}{5}$

10. $\sqrt{3}$

Exercises 11 and 12: Simplify the expression.

11. $3 + 4x - 2 + 3x$

12. $2(x - 1) - (x + 2)$

Exercises 13 and 14: Solve the equation. Check your solution.

13. $3t - 5 = 1$

14. $2(x - 3) = -6 - x$

Exercises 15 and 16: Find the area of the figure shown.

15.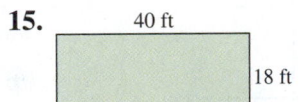

16.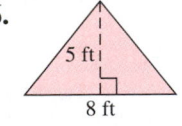

17. Complete the table. Then use the table to solve the equation $6 - 2x = 4$.

x	-2	-1	0	1	2
$6 - 2x$					

18. Translate the sentence "Twice a number increased by 2 equals the number decreased by 5" to an equation, using the variable n. Then solve the equation.

19. If $r = 10$ mph and $d = 80$ miles, use the formula $d = rt$ to find t.

20. Use an inequality to express the set of real numbers graphed.

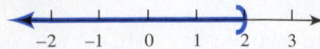

Exercises 21 and 22: Solve the inequality. Write the solution set in set-builder notation.

21. $3 - 6x < 3$ **22.** $2x \leq 1 - (2x - 1)$

23. Identify the coordinates of each point in the graph. State the quadrant, if any, in which each point lies.

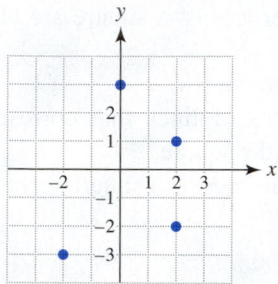

24. Complete the table for the equation $x + 2y = 4$.

x	-2	-1	0	1	2
y					

Exercises 25 and 26: Graph the equation.

25. $2x + 3y = -6$ **26.** $y = -\frac{3}{2}$

27. Determine the intercepts for the graph of the equation $-4x + 5y = 40$.

28. Write an equation for each line shown in the graph.

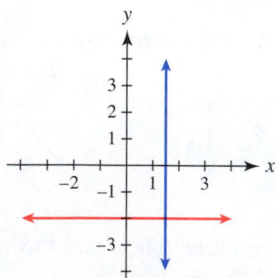

29. The table lists points located on a line. Write the slope–intercept form of the line.

x	-1	0	1	2
y	-6	-3	0	3

30. Write the equation $3x - 5y = 15$ in slope–intercept form. Graph the equation.

Exercises 31 and 32: Find the slope–intercept form for the line satisfying the given conditions.

31. Perpendicular to $3x - 2y = 6$, passing through the point $(0, -3)$

32. Passing through $(-1, 3)$ and $(2, -3)$

33. Let the initial value of y be 100 pounds, increasing at 5 pounds per hour. Find an equation in slope–intercept form that models y after x hours.

34. An insect population is initially 20,000 and is increasing at 5000 per day. Write an equation that gives the number of insects I after x days.

APPLICATIONS

35. *Pizzas* The table lists the cost C of buying x large pepperoni pizzas. Write an equation that relates C and x.

x	2	3	4
C	\$16	\$24	\$32

36. *U.S. Postal Service* In 2010, about $\frac{11}{20}$ of the mail consisted of first-class mail and periodicals. For every periodical there were 9 pieces of first-class mail. Estimate the fraction of the mail that was first-class mail. (*Source:* U.S. Postal Service.)

37. *Investment Money* A student invests two sums of money at 3% and 4% interest, receiving a total of \$110 in interest after 1 year. Twice as much money is invested at 4% as at 3%. Find the amount invested at each interest rate.

38. *Rental Cars* The cost C in dollars for driving a rental car x miles is $C = 0.3x + 25$.
   (a) How much does it cost to drive the car 200 miles?
   (b) How much does it cost to rent the car but not drive it?
   (c) How much does it cost to drive the car 1 *additional* mile?

4 Systems of Linear Equations in Two Variables

> We can do anything we want to do if we stick to it long enough.
>
> —HELEN KELLER

Source: National Center for Health Statistics.

Americans have been moving toward a more mobile lifestyle. In recent years, the percentage of U.S. households relying solely on mobile phone service has increased, while the percentage of households relying solely on landline phone service has decreased. The following figure shows that linear equations can be used to model the percentage P of households relying on each type of phone service during year x. What does the point of intersection represent?

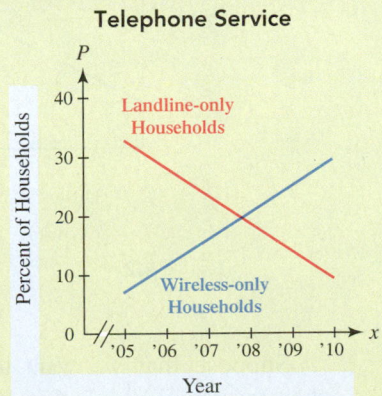

This graph illustrates a *system of linear equations*. If we use the graph to estimate both the year and the percentage at the intersection point, we are *solving* the system of linear equations graphically. In this chapter we discuss graphical, numerical, and symbolic methods for solving systems of linear equations.

4.1 Solving Systems of Linear Equations Graphically and Numerically

Basic Concepts • Solutions to Systems of Equations

A LOOK INTO MATH ▶ In business, linear equations are sometimes used to model supply and demand for a product. For example, if the price of a gourmet coffee drink is too high, the demand for the drink will decrease because consumers are interested in saving money. Similarly, if the price of the coffee drink is too low, supply will decrease because suppliers are interested in making money. To find an appropriate price for the coffee drink, a system of linear equations can be solved. In this section, we will solve systems of linear equations graphically and numerically.

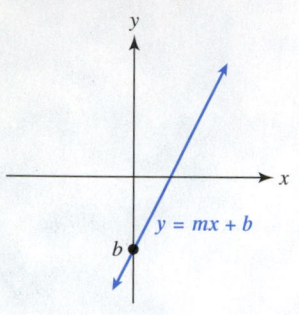

Figure 4.1

Basic Concepts

In Chapter 3 we showed that the graph of $y = mx + b$ is a line with slope m and y-intercept b, as illustrated in Figure 4.1. Each point on this line represents a solution to the equation $y = mx + b$. Because there are infinitely many points on a line, there are infinitely many solutions to this equation. However, many applications require that we find one particular solution to a linear equation. One way to find such a solution is to graph a second line in the same xy-plane and determine the point of intersection (if one exists).

NEW VOCABULARY

☐ Intersection-of-graphs method
☐ System of linear equations in two variables
☐ Solution to a system
☐ Inconsistent system
☐ Consistent system with independent equations
☐ Consistent system with dependent equations

▶ **REAL-WORLD CONNECTION** Consider the following application of a line. If renting a moving truck for one day costs $25 plus $0.50 per mile driven, then the equation $C = 0.5x + 25$ represents the cost C in dollars of driving the rental truck x miles. The graph of this line is shown in Figure 4.2(a) for $x \geq 0$.

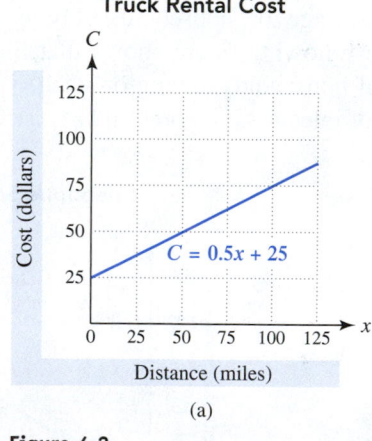
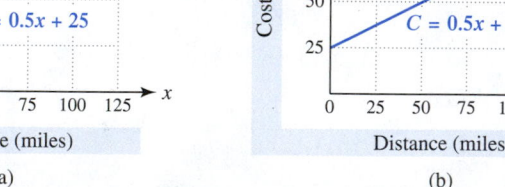

Figure 4.2

STUDY TIP

Try to find a consistent time and place to study your notes and do your homework. When the time comes to study for an exam, do so at your usual study time in your usual place rather than "pulling an all-nighter" in unfamiliar surroundings.

Suppose that we want to determine the number of miles that the truck is driven when the rental cost is $75. One way to solve this problem *graphically* is to graph both $C = 0.5x + 25$ and $C = 75$ in the same coordinate plane, as shown in Figure 4.2(b). The lines intersect at the point (**100**, **75**), which is a solution to $C = 0.5x + 25$ and to $C = 75$. That is, if the rental cost is **$75**, then the mileage must be **100** miles. This graphical technique for solving two equations is sometimes called the **intersection-of-graphs method**. To find a solution with this method, we locate a point where two graphs intersect.

EXAMPLE 1   ### Solving an equation graphically

The equation $P = 10x$ calculates an employee's pay for working x hours at $10 per hour. Use the intersection-of-graphs method to find the number of hours that the employee worked if the amount paid is $40.

Solution

Begin by graphing the equations $P = 10x$ and $P = 40$, as illustrated in Figure 4.3.

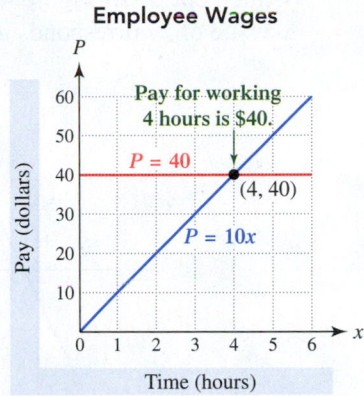

Figure 4.3

READING CHECK

• How is the intersection-of-graphs method used to find a solution to two equations?

The graphs intersect at the point $(4, 40)$. Since the x-coordinate represents the number of hours worked and the P-coordinate represents pay, the point $(\textbf{4}, \textbf{40})$ indicates that the employee must work **4** hours to earn **$40**.

▌ Now Try Exercise 67(a), (b)

Equations can be solved in more than one way. In Example 1 we determined graphically that $P = 10x$ is equal to 40 when $x = 4$. We could also solve this problem by making a table of values, as illustrated by Table 4.1. Note that, when $x = 4$, $P = \textbf{\$40}$. A table of values provides a *numerical solution*.

TABLE 4.1 Wages Earned at $10 per Hour

x (hours)	0	1	2	3	4	5	6
P (pay)	$0	$10	$20	$30	**$40**	$50	$60

NOTE: Although graphical and numerical methods are different, both methods should give the same solution. However, slight variations may occur because reading a graph precisely may be difficult, or a needed value may not appear in a table.

TECHNOLOGY NOTE

Intersection of Graphs and Table of Values
A graphing calculator can be used to find the intersection of the two graphs shown in Figure 4.3. It can also be used to create Table 4.1. The accompanying figures illustrate how a calculator can be used to determine that $y_1 = 10x$ equals $y_2 = 40$ when $x = 4$.

CALCULATOR HELP

To find a point of intersection, see Appendix A (page AP-6).

To make a table, see Appendix A (pages AP-2 and AP-3).

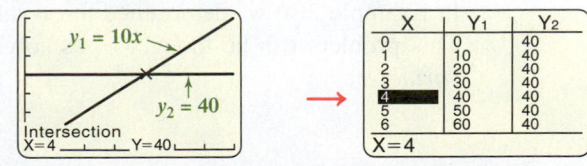

EXAMPLE 2 **Solving an equation graphically**

Use a graph to find the x-value when $y = 3$.
(a) $y = 2x - 1$ **(b)** $-3x + 2y = 12$

Solution

(a) Begin by graphing the equations $y = 2x - 1$ and $y = 3$. The graph of $y = 2x - 1$ is a line with slope 2 and y-intercept -1. The graph of $y = 3$ is a horizontal line with y-intercept 3. In Figure 4.4(a) the graphs intersect at the point $(2, 3)$. Therefore an x-value of **2** corresponds to a y-value of **3**.

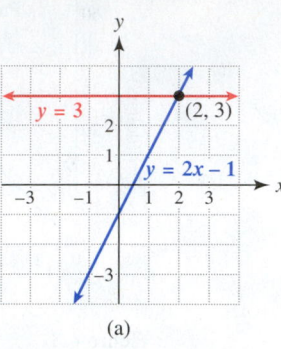 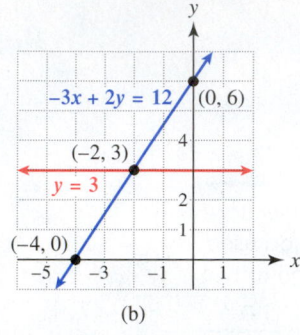

(a)   (b)

Figure 4.4

(b) One way to graph $-3x + 2y = 12$ is to write this equation in slope–intercept form.

$$-3x + 2y = 12 \qquad \text{Given equation}$$
$$2y = 3x + 12 \qquad \text{Add } 3x \text{ to each side.}$$
$$y = \frac{3}{2}x + 6 \qquad \text{Divide each side by 2.}$$

The line has slope $\frac{3}{2}$ and y-intercept 6. Its graph and the graph of $y = 3$ are shown in Figure 4.4(b). The graphs intersect at $(-2, 3)$. Therefore an x-value of -2 corresponds to a y-value of **3**.

Now Try Exercises 15, 17

MAKING CONNECTIONS

Different Ways to Graph the Same Line

In Example 2(b) the line $-3x + 2y = 12$ was graphed by finding its slope–intercept form. A second way to graph this line is to find the x- and y-intercepts of the line. If $y = 0$, then $x = -4$ makes this equation true, and if $x = 0$, then $y = 6$ makes this equation true. Thus the x-intercept is -4, and the y-intercept is 6. Note that the (slanted) line in Figure 4.4(b) passes through the points $(-4, 0)$ and $(0, 6)$, which could be used to graph the line.

Solutions to Systems of Equations

In Example 2(b) we determined the x-value when $y = 3$ in the equation $-3x + 2y = 12$. This problem can be thought of as solving the following *system of two equations in two variables.*

$$-3x + 2y = 12$$
$$y = 3$$

READING CHECK

• How do we express the solution to a system of equations in two variables?

The solution is the ordered pair $(-2, 3)$, which indicates that when $x = -2$ and $y = 3$, each equation is a true statement.

$$-3(-2) + 2(3) = 12 \checkmark \quad \text{A true statement}$$
$$3 = 3 \checkmark \quad \text{A true statement}$$

Suppose that the sum of two numbers is 10 and that their difference is 4. If we let x and y represent the two numbers, then the equations

$$x + y = 10 \quad \text{Sum is 10.}$$
$$x - y = 4 \quad \text{Difference is 4.}$$

describe this situation. Each equation is a linear equation in two variables, so we call these equations a **system of linear equations in two variables**. Its graph typically consists of two lines. A **solution to a system** of two equations is an ordered pair (x, y) that makes *both* equations true. If a single solution exists, the ordered pair gives the coordinates of a point where the two lines intersect.

NOTE: For two distinct lines, there can be no more than one intersection point. If such an intersection point exists, the ordered pair corresponding to it represents the only solution to the system of linear equations. In this case, we say the ordered pair is *the* solution to the system of equations.

TYPES OF EQUATIONS AND NUMBER OF SOLUTIONS When a system of two linear equations in two variables is graphed, exactly one of the following situations will result.

1. The two lines are parallel, as shown in Figure 4.5(a).
2. The two lines intersect exactly once, as shown in Figure 4.5(b).
3. The two lines are identical (they coincide), as shown in Figure 4.5(c).

READING CHECK

• How many solutions are possible for a system of linear equations in two variables?

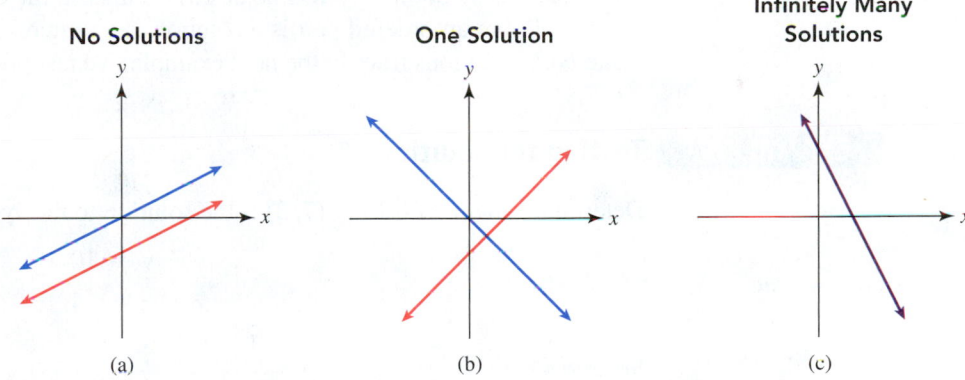

Figure 4.5

In the first situation, there are no solutions, and it is an **inconsistent system**. In the second situation, there is exactly one solution, and it is a **consistent system** with **independent equations**. In the third situation, there are infinitely many solutions, and it is a **consistent system** with **dependent equations**. This information is summarized in Table 4.2.

TABLE 4.2 Types of Systems of Equations

Type of Graph	Number of Solutions	Type of System	Type of Equations
Parallel Lines	0	Inconsistent	—
Intersecting Lines	1	Consistent	Independent
Identical Lines	Infinitely many	Consistent	Dependent

EXAMPLE 3 | **Identifying types of equations**

Graphs of two equations are shown. State the number of solutions to each system of equations. Then state whether the system is consistent or inconsistent. If it is consistent, state whether the equations are dependent or independent.

(a) 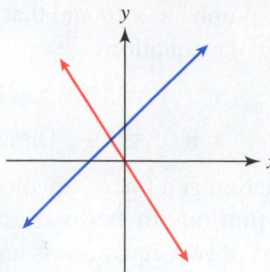 (b) 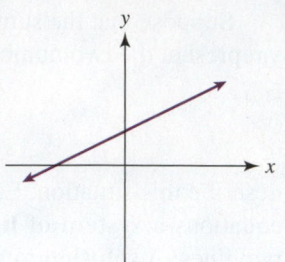 (c)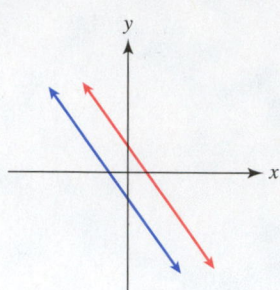

Solution

(a) The lines intersect at one point, so there is one solution. The system is consistent, and the equations are independent.

(b) There is only one line, which indicates that the graphs are identical, or coincide, so there are infinitely many solutions. The system is consistent and the equations are dependent.

(c) The lines are parallel, so there are no solutions. The system is inconsistent.

▮ **Now Try Exercises 19, 21, 23**

SYSTEMS WITH EXACTLY ONE SOLUTION For the remainder of this section, we focus on systems of linear equations in two variables that have exactly one solution. We will solve the systems by graphing and by making a table of values. Methods for solving two equations symbolically will be discussed later in the chapter.

Recall that an ordered pair is a solution to a system of two equations if its coordinates make *both* equations true. In the next example, we test possible solutions.

EXAMPLE 4 | **Testing for solutions**

Determine whether (4, 6) or (7, 3) is the solution to the system of equations

$$x + y = 10$$
$$x - y = \ \ 4.$$

Solution

To determine whether (4, 6) is the solution, substitute $x = 4$ and $y = 6$ in each equation. It must make *both* equations true.

$x + y = 10$	$x - y = 4$	Given equations
$4 + 6 \overset{?}{=} 10$	$4 - 6 \overset{?}{=} 4$	Let $x = 4, y = 6$.
$10 = 10$ (True) ✓	$-2 = 4$ (False) ✗	Second equation is false.

Because (4, 6) does not satisfy *both* equations, it is not the solution for the system of equations. Next let $x = 7$ and $y = 3$ to determine whether (7, 3) is the solution.

$x + y = 10$	$x - y = 4$	Given equations
$7 + 3 \overset{?}{=} 10$	$7 - 3 \overset{?}{=} 4$	Let $x = 7, y = 3$.
$10 = 10$ (True) ✓	$4 = 4$ (True) ✓	Both are true.

Because (7, 3) makes *both* equations true, it is the solution to the system of equations.

▮ **Now Try Exercise 25**

In the next example we find the solution to a system of linear equations graphically and numerically.

EXAMPLE 5 | **Solving a system graphically and numerically**

Solve the system of linear equations

$$x + 2y = 4$$
$$2x - y = 3$$

with a graph and with a table of values.

Solution

Graphically Begin by writing each equation in slope–intercept form.

$x + 2y = 4$	First equation		$2x - y = 3$	Second equation	
$2y = -x + 4$	Subtract x.		$-y = -2x + 3$	Subtract $2x$.	
$y = -\dfrac{1}{2}x + 2$	Divide by 2.		$y = 2x - 3$	Multiply by -1.	

The graphs of $y = -\frac{1}{2}x + 2$ and $y = 2x - 3$ are shown in Figure 4.6. The graphs intersect at the point $(2, 1)$, thus $(2, 1)$ is the solution.

Numerically Table 4.3 shows the equations $y = -\frac{1}{2}x + 2$ and $y = 2x - 3$ evaluated for various values of x. Note that when $x = 2$, both equations have a y-value of **1**. Thus $(2, 1)$ is the solution.

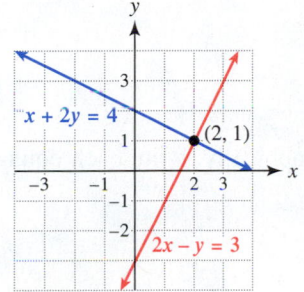

Figure 4.6

TABLE 4.3 A Numerical Solution

x	-1	0	1	2	3
$y = -\frac{1}{2}x + 2$	2.5	2	1.5	**1**	0.5
$y = 2x - 3$	-5	-3	-1	**1**	3

Find an x-value where the y-values are equal.

Now Try Exercise 45

EXAMPLE 6 | **Solving a system graphically**

Solve the system of equations graphically.

$$y = 2x$$
$$2x + y = 4$$

Solution

The equation $y = 2x$ can be written in slope–intercept form as $y = 2x + 0$. Its graph is a line passing through the origin with slope 2, as shown in Figure 4.7. The equation $2x + y = 4$ can be written in slope–intercept form as $y = -2x + 4$. Its graph is a line passing through the point $(0, 4)$ with slope -2. This line is also graphed in Figure 4.7. Because the intersection point is $(1, 2)$, the solution to the system of equations is the ordered pair $(1, 2)$.

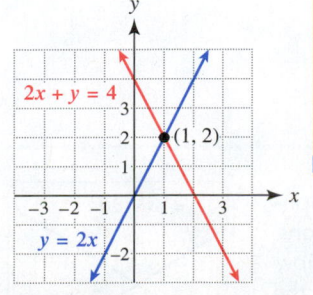

Figure 4.7

Now Try Exercise 59

A FOUR-STEP PROCESS FOR SOLVING APPLICATIONS In the next example, we use a four-step process to solve an application involving a system of linear equations. These steps are based on the four-step process discussed in Section 2.3, with Step 3 split into two parts to emphasize the importance of using the solution to the system of equations to determine the solution to the given problem.

EXAMPLE 7   **Traveling to watch sports**

In 2010, about 50 million Americans traveled to watch either football or basketball. About 10 million more people traveled to watch football than basketball. How many Americans traveled to watch each sport? (*Source: Sports Travel Magazine.*)

Solution

STEP 1:   *Identify each variable.*

x: millions of Americans who traveled to watch football
y: millions of Americans who traveled to watch basketball

STEP 2:   *Write a system of equations.* The total number of Americans who watched either sport is 50 million, so we know that $x + y = 50$. Because 10 million more people watched football than basketball, we also know that $x - y = 10$. Thus a system of equations representing this situation is

$$x + y = 50$$
$$x - y = 10.$$

STEP 3A: *Solve the system of equations.* To solve this system graphically, write each equation in slope–intercept form.

$$y = -x + 50$$
$$y = x - 10$$

Their graphs intersect at the point (30, 20), as shown in Figure 4.8.

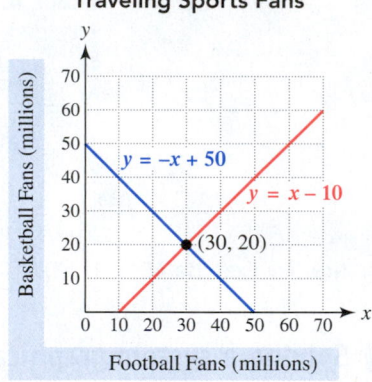

Traveling Sports Fans

Figure 4.8

STEP 3B: *Determine the solution to the problem.* The point (30, 20) corresponds to $x = 30$ and $y = 20$. Thus about 30 million Americans traveled to watch football and 20 million Americans traveled to watch basketball.

STEP 4:   *Check your solution.* Note that $30 + 20 = 50$ million Americans traveled to watch either football or basketball and that $30 - 20 = 10$ million more Americans watched football than basketball.

Now Try Exercise 69

TECHNOLOGY NOTE

CALCULATOR HELP

To find a point of intersection, see Appendix A (page AP-6).

Checking Solutions

The solution to the system in Example 7 can be checked with a graphing calculator, as shown in the accompanying figure where the graphs of $y_1 = -x + 50$ and $y_2 = x - 10$ intersect at the point (30, 20). However, a graphing calculator cannot read Example 7 and write down the system of equations. A human mind is needed for these tasks.

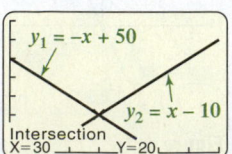

4.1 Putting It All Together

CONCEPT	EXPLANATION	EXAMPLE
System of Linear Equations in Two Variables	Can be written as $$Ax + By = C$$ $$Dx + Ey = F$$	$$x + y = 8$$ $$x - y = 2$$
Solution to a System of Equations	An ordered pair (x, y) that satisfies *both* equations	The solution to the preceding system is $(5, 3)$ because, when $x = 5$ and $y = 3$ are substituted, both equations are true. $$5 + 3 \stackrel{?}{=} 8 \checkmark \quad \text{True}$$ $$5 - 3 \stackrel{?}{=} 2 \checkmark \quad \text{True}$$
Inconsistent System	A system of linear equations in two variables that has no solutions is an inconsistent system. Graphing the system results in parallel lines.	**No Solutions**
Consistent System with Independent Equations	A system of linear equations in two variables that has 1 solution is a consistent system. Graphing the system results in intersecting lines.	**One Solution**
Consistent System with Dependent Equations	A system of linear equations in two variables that has infinitely many solutions is a consistent system. Graphing the system results in identical lines, or lines that coincide.	**Infinitely Many Solutions**

continued on next page

continued from previous page

CONCEPT	EXPLANATION	EXAMPLE
Graphical Solution to a System of Equations	Graph each equation. A point of intersection represents a solution.	The graphs of $y = -x + 8$ and $y = x - 2$ intersect at $(5, 3)$.
Numerical Solution to a System of Equations	Make a table for each equation. A solution occurs when one x-value gives the same y-values in both equations.	Make a table for $y = -x + 8$ and $y = x - 2$. When $x = 5$, $y = 3$ in both equations, so $(5, 3)$ is the solution.

For the numerical example:

x	4	5	6
$y = -x + 8$	4	3	2
$y = x - 2$	2	3	4

4.1 Exercises

MyMathLab

PRACTICE   WATCH   DOWNLOAD   READ   REVIEW

CONCEPTS AND VOCABULARY

1. A solution to a system of two equations in two variables is a(n) _____ pair.

2. A graphical technique for solving a system of two equations in two variables is the _____ method.

3. A system of linear equations can have _____, _____, or _____ solutions.

4. If a system of linear equations has at least one solution, then it is a(n) (consistent/inconsistent) system.

5. If a system of linear equations has no solutions, then it is a(n) (consistent/inconsistent) system.

6. If a system of linear equations has exactly one solution, then the equations are (dependent/independent).

7. If a system of linear equations has infinitely many solutions, then the equations are (dependent/independent).

8. To find a numerical solution to a system, start by creating a(n) _____ of values for the equations.

9. If a graphical method and a numerical method (table of values) are used to solve the same system of equations, then the two solutions should be (the same/different).

10. One way to graph a line is to write its equation in slope–intercept form. A second method is to find the x- and y-_____.

SOLVING SYSTEMS OF EQUATIONS

Exercises 11–18: Determine graphically the x-value when $y = 2$ in the given equation.

11. $y = 2x$

12. $y = \frac{1}{3}x$

13. $y = 4 - x$

14. $y = -2 - x$

15. $y = -\frac{1}{2}x + 1$

16. $y = 3x - 1$

17. $2x + y = 6$

18. $-3x + 4y = 11$

Exercises 19–24: The graphs of two equations are shown. State the number of solutions to each system of equations. Then state whether the system is consistent or inconsistent. If it is consistent, state whether the equations are dependent or independent.

19.

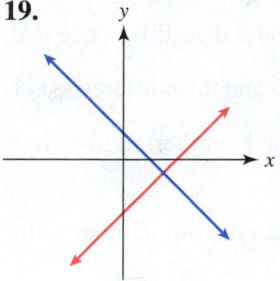

20.

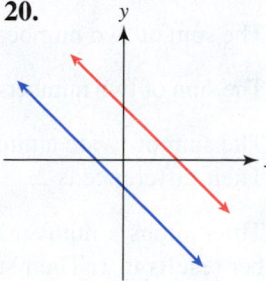

21.

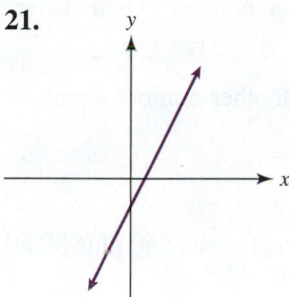

22.

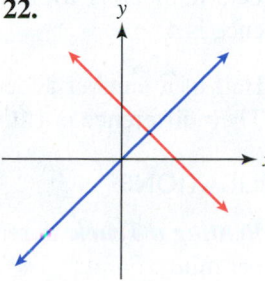

23.

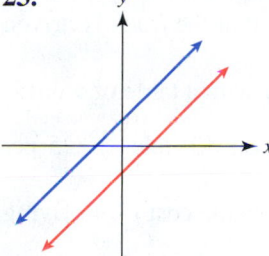

24.

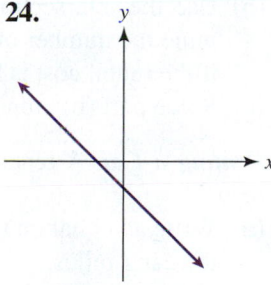

Exercises 25–30: Determine which ordered pair is a solution to the system of equations.

25. (0, 0), (1, 1)
$$x + y = 2$$
$$x - y = 0$$

26. (−1, 2), (1, −2)
$$2x + \ y = \ \ 0$$
$$x - 2y = -5$$

27. (−1, −1), (2, −3)
$$2x + 3y = -5$$
$$4x - 5y = \ 23$$

28. (2, −1), (−2, −2)
$$-x + 4y = -6$$
$$6x - 7y = \ 19$$

29. (2, 0), (−1, −3)
$$-5x + 5y = -10$$
$$4x + 9y = \ \ \ 8$$

30. $\left(\frac{1}{2}, \frac{3}{2}\right)$, $\left(\frac{3}{4}, \frac{5}{4}\right)$
$$x + y = 2$$
$$3x - y = 0$$

Exercises 31–36: The graphs of two equations are shown. Use the intersection-of-graphs method to identify the solution to both equations. Then check your answer.

31.

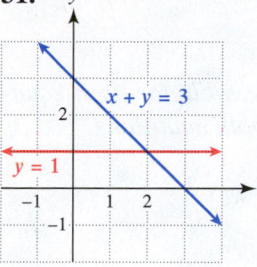

32.

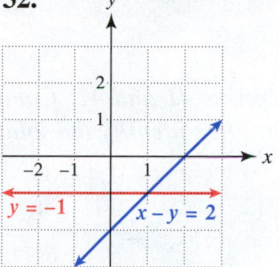

33.

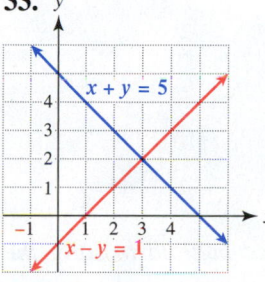

34.

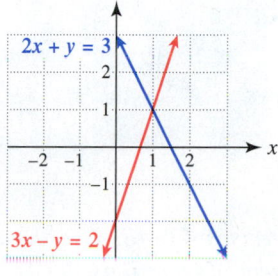

35.

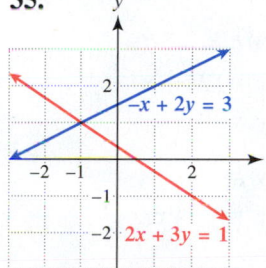

36.

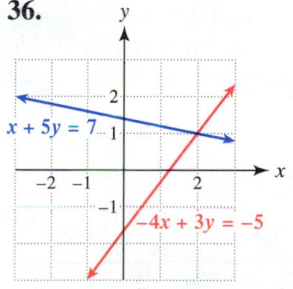

Exercises 37–40: A table for two equations is given. Identify the solution to both equations.

37.

x	1	2	3	4
$y = 2x$	2	4	6	8
$y = 4$	4	4	4	4

38.

x	2	3	4	5
$y = 6 - x$	4	3	2	1
$y = x - 2$	0	1	2	3

39.

x	1	2	3	4
$y = 4 - x$	3	2	1	0
$y = x - 2$	−1	0	1	2

40.

x	1	2	3	4
$y = 6 - 3x$	3	0	-3	-6
$y = 2 - x$	1	0	-1	-2

Exercises 41 and 42: Complete the table for each equation. Then identify the solution to both equations.

41.

x	0	1	2	3
$y = x + 2$				
$y = 4 - x$				

42.

x	-5	-4	-3	-2
$y = 2x + 1$				
$y = x - 1$				

Exercises 43–48: Use the specified method to solve the system of equations.

  (a) *Graphically*
  (b) *Numerically (table of values)*

43. $y = 2x + 3$
   $y = 1$

44. $y = 2 - x$
   $y = 0$

45. $y = 4 - x$
   $y = x - 2$

46. $y = 2x$
   $y = -\frac{1}{2}x$

47. $y = 3x$
   $y = x + 2$

48. $y = 2x - 3$
   $y = -x + 3$

Exercises 49–60: Solve the system of equations graphically.

49. $x + y = -3$
   $x - y = 1$

50. $x - y = 3$
   $2x + y = 3$

51. $2x - y = 3$
   $3x + y = 2$

52. $x + 2y = 6$
   $-x + 3y = 4$

53. $-4x + 2y = 0$
   $x - y = -1$

54. $4x - y = 2$
   $y = 2x$

55. $2x - y = 4$
   $x + 2y = 7$

56. $x - y = 2$
   $\frac{1}{2}x + y = 4$

57. $x = -y + 4$
   $x = 3y$

58. $x = 2y$
   $y = -\frac{1}{2}x$

59. $x + y = 3$
   $x = \frac{1}{2}y$

60. $2x - 4y = 8$
   $\frac{1}{2}x + y = -4$

*Exercises 61–66: **Number Problems** For each problem, complete each of the following.*

  (a) *Write a system of equations for the problem.*
  (b) *Find the unknown numbers by solving the system of equations graphically.*

61. The sum of two numbers is 4 and their difference is 0.

62. The sum of two numbers is −5 and their difference is 1.

63. The sum of twice a number and another number is 7. Their difference is 2.

64. Three times a number subtracted from another number results in 1. Their sum is 5.

65. One number is triple another number. Their difference is 4.

66. Half of a number added to another number equals 5. Their difference is 1.

APPLICATIONS

67. *Renting a Truck* A rental truck costs $50 plus $0.50 per mile.
  (a) Write an equation that gives the cost C of driving the truck x miles.
  (b) Use the intersection-of-graphs method to determine the number of miles that the truck is driven if the rental cost is $80.
  (c) Solve part (b) numerically with a table of values.

68. *Renting a Car* A rental car costs $25 plus $0.25 per mile.
  (a) Write an equation that gives the cost C of driving the car x miles.
  (b) Use the intersection-of-graphs method to determine the number of miles that the car is driven if the rental cost is $100.
  (c) Solve part (b) numerically with a table of values.

69. *Recorded Music* In 2009, rock and R&B music accounted for 42% of all music sales. Rock music sales were double the R&B music sales. (*Source:* Recording Industry Association of America.)
  (a) Let x be the percentage of sales due to rock music and let y be the percentage of music sales due to R&B music. Write a system of two equations that describes the given information.
  (b) Solve your system graphically.

70. *Sales of iPods* During the first and second quarters of 2010, about 32 million iPods were sold. The first quarter sales exceeded the second quarter sales by 10 million iPods. (*Source:* Apple.)

(a) Let x be the iPod sales in millions during the first quarter of 2010 and y be the iPod sales in millions during the second quarter of 2010. Write a system of two equations that describes the given information.

(b) Solve your system graphically.

71. *Dimensions of a Rectangle* A rectangle is 4 inches longer than it is wide. Its perimeter is 28 inches.

(a) Write a system of two equations in two variables that describes this information. Be sure to specify what each variable means.

(b) Solve your system graphically. Interpret your results.

72. *Dimensions of a Triangle* An isosceles triangle has a perimeter of 17 inches with its two shorter sides equal in length. The longest side measures 2 inches more than either of the shorter sides.

(a) Write a system of two equations in two variables that describes this information. Be sure to specify what each variable means.

(b) Solve your system graphically. Explain what your results mean.

WRITING ABOUT MATHEMATICS

73. Use the intersection-of-graphs method to help explain why you typically expect a linear system in two variables to have one solution.

74. Could a system of two linear equations in two variables have exactly two solutions? Explain your reasoning.

75. Give one disadvantage of using a table to solve a system of equations. Explain your answer.

76. Do the equations $y = 2x + 1$ and $y = 2x - 1$ have a common solution? Explain your answer.

4.2   Solving Systems of Linear Equations by Substitution

The Method of Substitution • Recognizing Other Types of Systems • Applications

A LOOK INTO MATH ▶

Suppose that a boy and a girl sent a total of 760 text messages in one month. With only this information, it is impossible to know how many messages were sent by each person. The boy may have sent 432 messages while the girl sent 328. However, 11 and 749 also total 760, as do many other possibilities. If we are told that the boy sent three times as many messages as the girl, a system of linear equations can be written and solved to find the answer. Refer to Example 4. In this section, we discuss a symbolic method for solving systems of linear equations.

The Method of Substitution

In Section 4.1 we solved systems of linear equations by using graphs and tables. A disadvantage of a graph is that reading the graph precisely can be difficult. A disadvantage of using a table is that locating the solution can be difficult when it is either a fraction or a large number. In this subsection we introduce the *method of substitution*, in which we solve systems of equations symbolically. The advantage of this method is that the *exact* solution can always be found (provided it exists).

NEW VOCABULARY

☐ Method of substitution

READING CHECK

• What is an advantage of using the method of substitution rather than using a graph or table to solve a system of linear equations?

STUDY TIP

Questions on exams do not always come in the order that they are presented in the text. When studying for an exam, choose review exercises randomly so that the topics are studied in the same random way that they may appear on an exam.

▶ **REAL-WORLD CONNECTION** Suppose that you and a friend earned $120 together. If x represents how much your friend earned and y represents how much you earned, then the equation $x + y = 120$ describes this situation. Now, if we also know that you earned twice as much as your friend, then we can include a second equation, $y = 2x$. The amount that each of you earned can now be determined by *substituting* $2x$ for y in the first equation.

$$x + y = 120 \qquad \text{First equation}$$
$$x + 2x = 120 \qquad \text{Substitute } 2x \text{ for } y.$$
$$3x = 120 \qquad \text{Combine like terms.}$$
$$x = 40 \qquad \text{Divide each side by 3.}$$

So your friend earned $40, and you earned twice as much, or $80.

This technique of substituting an expression for a variable and solving the resulting equation is called the **method of substitution**.

EXAMPLE 1 **Using the method of substitution**

Solve each system of equations.
(a) $2x + y = 10$ (b) $-2x + 3y = -8$
  $y = 3x$ $x = 3y + 1$

Solution
(a) From the second equation, substitute $3x$ for y in the first equation.

$$2x + y = 10 \qquad \text{First equation}$$
$$2x + 3x = 10 \qquad \text{Substitute } 3x \text{ for } y.$$
$$5x = 10 \qquad \text{Combine like terms.}$$
$$x = 2 \qquad \text{Divide each side by 5.}$$

The solution to this system is an *ordered pair*, so we must also find y. Because $y = 3x$ and $x = 2$, it follows that $y = 3(2) = 6$. The solution is $(2, 6)$. (Check it.)

(b) The second equation, $x = 3y + 1$, is solved for x. Substitute $(3y + 1)$ for x in the first equation. Be sure to include parentheses around the expression $3y + 1$ since this entire expression is to be multiplied by -2.

$$-2x + 3y = -8 \qquad \text{First equation}$$
$$-2(3y + 1) + 3y = -8 \qquad \text{Substitute } (3y + 1) \text{ for } x.$$
$$-6y - 2 + 3y = -8 \qquad \text{Distributive property}$$

Be sure to include parentheses

$$-3y - 2 = -8 \qquad \text{Combine like terms.}$$
$$-3y = -6 \qquad \text{Add 2 to each side.}$$
$$y = 2 \qquad \text{Divide each side by } -3.$$

To find x, substitute 2 for y in $x = 3y + 1$ to obtain $x = 3(2) + 1 = 7$. The solution is $(7, 2)$. (Check it.)

Now Try Exercises 5, 11

NOTE: When an expression contains two or more terms, it is usually best to place parentheses around it when substituting it for a single variable in an equation. In Example 1(b), the distributive property would not have been applied correctly without the parentheses.

READING CHECK

• When substituting an expression with two or more terms for a single variable, why is it important to use parentheses?

Sometimes it is necessary to solve for a variable before substitution can be used, as demonstrated in the next example.

> **EXAMPLE 2**   **Using the method of substitution**
>
> Solve each system of equations.
> | (a) | $x + y = 8$ | (b) | $3a - 2b = 2$ |
> |-----|-------------|-----|---------------|
> | | $2x - 3y = 6$ | | $a + 4b = 3$ |
>
> **Solution**
> (a) Neither equation is solved for a variable, but we can easily solve the first equation for y.
>
> $$x + y = 8 \qquad \text{First equation}$$
> $$y = 8 - x \qquad \text{Subtract } x \text{ from each side.}$$
>
> Now we can substitute $(8 - x)$ for y in the second equation.
>
> $$2x - 3y = 6 \qquad \text{Second equation}$$
> $$2x - 3(\mathbf{8 - x}) = 6 \qquad \text{Substitute } (8 - x) \text{ for } y.$$
> $$2x - 24 + 3x = 6 \qquad \text{Distributive property}$$
> $$5x = 30 \qquad \text{Combine like terms; add 24.}$$
> $$x = 6 \qquad \text{Divide each side by 5.}$$
>
> Because $y = 8 - x$ and $x = \mathbf{6}$, we know that $y = 8 - \mathbf{6} = \mathbf{2}$. The solution is $(\mathbf{6}, \mathbf{2})$.
> (b) Although we could solve either equation for either variable, solving the second equation for a is easiest because the coefficient of a is 1 and we can avoid fractions.
>
> $$a + 4b = 3 \qquad \text{Second equation}$$
> $$a = 3 - 4b \qquad \text{Subtract } 4b \text{ from each side.}$$
>
> Now substitute $(3 - 4b)$ for a in the first equation.
>
> $$3a - 2b = 2 \qquad \text{First equation}$$
> $$3(\mathbf{3 - 4b}) - 2b = 2 \qquad \text{Substitute } (3 - 4b) \text{ for } a.$$
> $$9 - 12b - 2b = 2 \qquad \text{Distributive property}$$
> $$-14b = -7 \qquad \text{Combine like terms; subtract 9.}$$
> $$b = \frac{1}{2} \qquad \text{Divide each side by } -14.$$
>
> To find a, substitute $b = \frac{1}{2}$ in $a = 3 - 4b$ to obtain $a = \mathbf{1}$. The solution is $\left(\mathbf{1}, \frac{1}{2}\right)$.
>
> **Now Try Exercises 17, 23**

NOTE: When a system of equations contains variables other than x and y, we will list them alphabetically in an ordered pair.

Recognizing Other Types of Systems

A system of linear equations typically has exactly one solution. However, in the next example we see how the method of substitution can be used on systems that have no solutions or infinitely many solutions.

EXAMPLE 3   ### Solving other types of systems

If possible, use substitution to solve the system of equations. Then use graphing to help explain the result.

(a) $3x + y = 4$ (b) $x + y = 2$
     $6x + 2y = 2$ $2x + 2y = 4$

Solution

(a) Solve the first equation for y to obtain $y = 4 - 3x$. Next substitute $(4 - 3x)$ for y in the second equation.

$$6x + 2y = 2 \qquad \text{Second equation}$$
$$6x + 2(\mathbf{4 - 3x}) = 2 \qquad \text{Substitute } (4 - 3x) \text{ for } y.$$
$$6x + 8 - 6x = 2 \qquad \text{Distributive property}$$
$$\mathbf{8 = 2 \ (False)} \qquad \text{Combine like terms.}$$

The equation $8 = 2$ is *always false*, which indicates that there are *no solutions*. One way to graph each equation is to write the equations in slope–intercept form.

$3x + y = 4$ First equation	$6x + 2y = 2$ Second equation
$y = -3x + 4$ Subtract 3x.	$y = -3x + 1$ Subtract 6x; divide by 2.

The graphs of these equations are parallel lines with slope -3, as shown in Figure 4.9(a). Because the lines *do not intersect*, there are *no solutions* to the system of equations.

(b) Solve the first equation for y to obtain $y = 2 - x$. Now substitute $(2 - x)$ for y in the second equation.

$$2x + 2y = 4 \qquad \text{Second equation}$$
$$2x + 2(\mathbf{2 - x}) = 4 \qquad \text{Substitute } (2 - x) \text{ for } y.$$
$$2x + 4 - 2x = 4 \qquad \text{Distributive property}$$
$$\mathbf{4 = 4 \ (True)} \qquad \text{Combine like terms.}$$

The equation $4 = 4$ is *always true*, which means that there are *infinitely many solutions*. One way to graph these equations is to write them in slope–intercept form.

$x + y = 2$ First equation	$2x + 2y = 4$ Second equation
$y = -x + 2$ Subtract x.	$y = -x + 2$ Subtract 2x; divide by 2.

Because the equations have the same slope–intercept form, their graphs are identical, resulting in a single line, as shown in Figure 4.9(b). *Every point* on this line *represents a solution* to the system of equations, so there are infinitely many solutions.

Now Try Exercises 33, 35

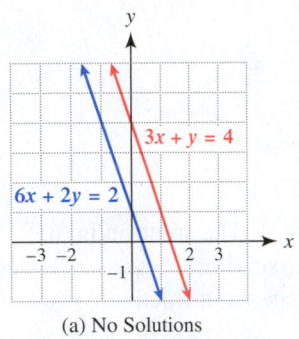

(a) No Solutions

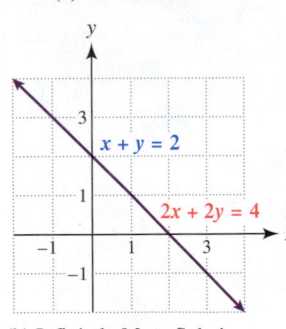

(b) Infinitely Many Solutions

Figure 4.9

Applications

In the next example, we use the method of substitution to solve a system of equations that represents the information presented in A Look Into Math at the beginning of this section.

EXAMPLE 4   ### Finding numbers of text messages

A boy and a girl sent a total of 760 text messages in one month. If the boy sent three times as many messages as the girl, how many text messages did each person send?

Solution

If x represents the number of text messages sent by the girl and y represents the number of text messages sent by the boy, then the equation $x + y = 760$ can be written because the total number of messages is 760. Also, the equation $y = 3x$ can be included because the boy sent three times as many messages as the girl. The system is

$$x + y = 760$$
$$y = 3x.$$

Substitute $3x$ for y in the first equation.

$x + y = 760$	First equation
$x + 3x = 760$	Substitute $3x$ for y.
$4x = 760$	Combine like terms.
$x = 190$	Divide each side by 4.

The girl sent 190 text messages and the boy sent 3 times as many, or $3(190) = 570$. This answer checks because $190 + 570 = 760$ and 570 is three times 190.

■ **Now Try Exercise 67**

The next two examples illustrate how the method of substitution can be used to solve applications. In these examples we apply the same four-step process that was used to solve application problems in Section 4.1.

EXAMPLE 5   **Determining pizza sales**

In 2009, combined sales of frozen and ready-to-eat pizza reached \$36.9 billion. Ready-to-eat pizza sales were 7.2 times frozen pizza sales. Find the amount of sales for each type of pizza. (*Source: Business Trend Analyst.*)

Solution

STEP 1:   *Identify each variable.* Clearly identify what each variable represents.

　　　　x: sales of ready-to-eat pizza, in billions of dollars
　　　　y: sales of frozen pizza, in billions of dollars

STEP 2:   *Write a system of equations.*

$x + y = 36.9$	Sales total \$36.9 billion.
$x = 7.2y$	Ready-to-eat pizza sales x are 7.2 times frozen pizza sales y.

STEP 3A: *Solve the system of linear equations.* Substitute $7.2y$ for x in the first equation.

$x + y = 36.9$	First equation
$7.2y + y = 36.9$	Substitute $7.2y$ for x.
$8.2y = 36.9$	Combine like terms.
$y = \dfrac{36.9}{8.2}$	Divide each side by 8.2.
$y = 4.5$	Simplify.

Because $x = 7.2y$, it follows that $x = 7.2(4.5) = 32.4$.

STEP 3B: *Determine the solution to the problem.* The solution $y = 4.5$ and $x = 32.4$ indicates that frozen pizza sales were \$4.5 billion and ready-to-eat pizza sales were \$32.4 billion in 2009.

STEP 4:   *Check the solution.* The sum of these sales was $32.4 + 4.5 = \$36.9$ billion, and ready-to-eat pizza sales were $\frac{32.4}{4.5} = 7.2$ times frozen pizza sales. The answer checks.

■ **Now Try Exercise 57**

EXAMPLE 6 | **Determining airplane speed and wind speed**

An airplane flies 2400 miles into (or against) the wind in 8 hours. The return trip takes 6 hours. Find the speed of the airplane with no wind and the speed of the wind.

Solution

STEP 1: *Identify each variable.*

x: the speed of the airplane without wind
y: the speed of the wind

STEP 2: *Write a system of equations.* The speed of the airplane against the wind is $\frac{2400}{8} = 300$ miles per hour, because it traveled 2400 miles in 8 hours. The wind slowed the plane, so $x - y = 300$. Similarly, the airplane flew $\frac{2400}{6} = 400$ miles per hour with the wind because it traveled 2400 miles in 6 hours. The wind made the plane fly faster, so $x + y = 400$.

$$x - y = 300 \quad \text{Speed against the wind}$$
$$x + y = 400 \quad \text{Speed with the wind}$$

STEP 3A: *Solve the system of linear equations.* Solve the first equation for x to obtain $x = y + 300$. Substitute $(y + 300)$ for x in the second equation.

$$x + y = 400 \quad \text{Second equation}$$
$$(y + 300) + y = 400 \quad \text{Substitute } (y + 300) \text{ for } x.$$
$$2y = 100 \quad \text{Combine like terms; subtract 300.}$$
$$y = 50 \quad \text{Divide each side by 2.}$$

CRITICAL THINKING

A boat travels 10 miles per hour upstream and 16 miles per hour downstream. How fast is the current?

Because $x = y + 300$, it follows that $x = 50 + 300 = 350$.

STEP 3B: *Determine the solution to the problem.* The solution $x = 350$ and $y = 50$ indicates that the airplane can fly 350 miles per hour with no wind, and the wind speed is 50 miles per hour.

STEP 4: *Check the solution.* The plane flies $350 - 50 = 300$ miles per hour into the wind, taking $\frac{2400}{300} = 8$ hours. The plane flies $350 + 50 = 400$ miles per hour with the wind, taking $\frac{2400}{400} = 6$ hours. The answers check.

Now Try Exercise 63

4.2 Putting It All Together

CONCEPT	EXPLANATION	EXAMPLE
Method of Substitution	Can be used to solve a system of equations Gives the exact solution, provided one exists **STEP 1:** Solve one equation for one variable. **STEP 2:** Substitute the result in the other equation and solve. **STEP 3:** Use the solution for the first variable to find the other variable. **STEP 4:** Check the solution.	$x + y = 5 \quad$ Sum is 5. $x - y = 1 \quad$ Difference is 1. **STEP 1:** Solve for x in the second equation. $x = y + 1$ **STEP 2:** Substitute $(y + 1)$ in the first equation for x. $(y + 1) + y = 5$ $2y = 4$ $y = 2$ **STEP 3:** $x = 2 + 1 = 3$ **STEP 4:** $3 + 2 = 5$ ✓ $3 - 2 = 1$ ✓ $(3, 2)$ checks.

4.2   Exercises

MyMathLab PRACTICE   WATCH   DOWNLOAD   READ   REVIEW

CONCEPTS AND VOCABULARY

1. One advantage of solving a linear system using the method of substitution rather than graphical or numerical methods is that the _____ solution can always be found (provided it exists).

2. When substituting an expression that contains two or more terms for a single variable in an equation, it is usually best to place _____ around it.

3. Suppose that the method of substitution results in the equation $1 = 1$. What does this indicate about the number of solutions to the system of equations?

4. Suppose that the method of substitution results in the equation $0 = 1$. What does this indicate about the number of solutions to the system of equations?

SOLVING SYSTEMS OF EQUATIONS

Exercises 5–32: Use the method of substitution to solve the system of linear equations.

5. $x + y = 9$
   $y = 2x$

6. $x + y = -12$
   $y = -3x$

7. $x + 2y = 4$
   $x = 2y$

8. $-x + 3y = -12$
   $x = 5y$

9. $2x + y = -2$
   $y = x + 1$

10. $-3x + y = -10$
    $y = x - 2$

11. $x + 3y = 3$
    $x = y + 3$

12. $x - 2y = -5$
    $x = 4 - y$

13. $3x + 2y = \frac{3}{2}$
    $y = 2x - 1$

14. $-3x + 5y = 4$
    $y = 2 - 3x$

15. $2x - 3y = -12$
    $x = 2 - \frac{1}{2}y$

16. $\frac{3}{4}x + \frac{1}{4}y = -\frac{7}{4}$
    $x = 1 - 2y$

17. $2x - 3y = -4$
    $3x - y = 1$

18. $\frac{1}{2}x - y = -1$
    $2x - \frac{1}{2}y = \frac{13}{2}$

19. $x - 5y = 26$
    $2x + 6y = -12$

20. $4x - 3y = -4$
    $x + 7y = -63$

21. $\frac{1}{2}y - z = 5$
    $y - 3z = 13$

22. $3y - 7z = -2$
    $5y - z = 2$

23. $10r - 20t = 20$
    $r + 60t = -29$

24. $-r + 10t = 22$
    $-10r + 5t = 30$

25. $3x + 2y = 9$
    $2x - 3y = -7$

26. $5x - 2y = -5$
    $2x - 5y = 19$

27. $2a - 3b = 6$
    $-5a + 4b = -8$

28. $-5a + 7b = -1$
    $3a + 2b = 13$

29. $-\frac{1}{2}x + 3y = 5$
    $2x - \frac{1}{2}y = 3$

30. $3x - \frac{1}{2}y = 2$
    $-\frac{1}{2}x + 5y = \frac{19}{2}$

31. $3a + 5b = 16$
    $-8a + 2b = 34$

32. $5a - 10b = 20$
    $10a + 5b = 15$

Exercises 33–50: Use the method of substitution to solve the system of linear equations. These systems may have no solutions, one solution, or infinitely many solutions.

33. $x + y = 9$
    $x + y = 7$

34. $x - y = 8$
    $x - y = 4$

35. $x - y = 4$
    $2x - 2y = 8$

36. $2x + y = 5$
    $4x + 2y = 10$

37. $x + y = 4$
    $x - y = 2$

38. $x - y = 3$
    $2x - y = 7$

39. $x - y = 7$
    $-x + y = -7$

40. $2u - v = 6$
    $-4u + 2v = -12$

41. $u - 2v = 5$
    $2u - 4v = -2$

42. $3r + 3t = 9$
    $2r + 2t = 4$

43. $2r + 3t = 1$
    $r - 3t = -5$

44. $5x - y = -1$
    $2x - 7y = 7$

45. $y = 5x$
    $y = -3x$

46. $a = b + 1$
    $a = b - 1$

47. $5a = 4 - b$
    $5a = 3 - b$

48. $3y = x$
    $3y = 2x$

49. $2x + 4y = 0$
    $3x + 6y = 5$

50. $-5x + 10y = 3$
    $\frac{1}{2}x - y = 1$

51. **Thinking Generally** If the method of substitution results in an equation that is always true, what can be said about the graphs of the two equations?

52. **Thinking Generally** If the method of substitution results in an equation that is always false, what can be said about the graphs of the two equations?

APPLICATIONS

53. *Rectangle* A rectangular garden is 10 feet longer than it is wide. Its perimeter is 72 feet.
(a) Let W be the width of the garden and L be the length. Write a system of linear equations whose solution gives the width and length of the garden.
(b) Use the method of substitution to solve the system. Check your answer.

54. *Isosceles Triangle* The measures of the two smaller angles in a triangle are equal and their sum equals the largest angle.
(a) Let x be the measure of each of the two smaller angles and y be the measure of the largest angle. Write a system of linear equations whose solution gives the measures of these angles.
(b) Use the method of substitution to solve the system. Check your answer.

55. *Complementary Angles* The smaller of two complementary angles is half the measure of the larger angle.
(a) Let x be the measure of the smaller angle and y be the measure of the larger angle. Write a system of linear equations whose solution gives the measures of these angles.
(b) Use the method of substitution to solve your system.
(c) Use graphing to solve the system.

56. *Supplementary Angles* The smaller of two supplementary angles is one-fourth the measure of the larger angle.
(a) Let x be the measure of the smaller angle and y be the measure of the larger angle. Write a system of linear equations whose solution gives the measures of these angles.
(b) Use the method of substitution to solve the system.

57. *Average Room Prices* In 2009, the average room price for a hotel chain in Seattle was $21 less than in 2008. The 2009 room price was 86% of the 2008 room price. (*Source:* Hotel Price Index.)
(a) Let x be the average room price in 2008 and y be this price in 2009. Write a system of linear equations whose solution gives the average room prices for each year.
(b) Use the method of substitution to solve the system.

58. *Ticket Prices* Two hundred tickets were sold for a baseball game, which amounted to $840. Student tickets cost $3, and adult tickets cost $5.
(a) Let x be the number of student tickets sold and y be the number of adult tickets sold. Write a system of linear equations whose solution gives the number of each type of ticket sold.
(b) Use the method of substitution to solve the system of equations.

59. *NBA Basketball Court* An official NBA basketball court is 44 feet longer than it is wide. If its perimeter is 288 feet, find its dimensions.

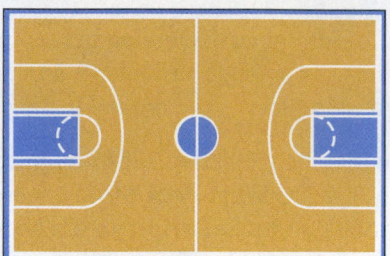

60. *Football Field* A U.S. football field is 139.5 feet longer than it is wide. If its perimeter is 921 feet, find its dimensions.

61. *Number Problem* The sum of two numbers is 70. The larger number is two more than three times the smaller number. Find the two numbers.

62. *Number Problem* The difference of two numbers is 12. The larger number is one less than twice the smaller number. Find the two numbers.

63. *Speed Problem* A tugboat goes 120 miles upstream in 15 hours. The return trip downstream takes 10 hours. Find the speed of the tugboat without a current and the speed of the current.

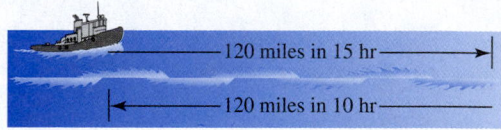

64. *Speed Problem* An airplane flies 1200 miles into the wind in 3 hours. The return trip takes 2 hours. Find the speed of the airplane without a wind and the speed of the wind.

65. *Mixture Problem* A chemist has 20% and 50% solutions of acid available. How many liters of each solution should be mixed to obtain 10 liters of a 40% acid solution?

66. *Mixture Problem* A mechanic needs a radiator to have a 40% antifreeze solution. The radiator currently is filled with 4 gallons of a 25% antifreeze solution. How much of the antifreeze mixture should be drained from the car if the mechanic replaces it with pure antifreeze?

67. *Great Lakes* Together, Lake Superior and Lake Michigan cover 54 thousand square miles. Lake Superior is approximately 10 thousand square miles larger than Lake Michigan. Find the size of each lake. (*Source:* National Oceanic and Atmospheric Administration.)

68. *Longest Rivers* The two longest rivers in the world are the Nile and the Amazon. Together, they are 8145 miles, with the Amazon being 145 miles shorter than the Nile. Find the length of each river. (*Source:* National Oceanic and Atmospheric Administration.)

WRITING ABOUT MATHEMATICS

69. State one advantage that the method of substitution has over the intersection-of-graphs method. Explain your answer.

70. When applying the method of substitution, how do you know that there are no solutions?

71. When applying the method of substitution, how do you know that there are infinitely many solutions?

72. When applying the intersection-of-graphs method, how do you know that there are no solutions?

SECTIONS 4.1 and 4.2

Checking Basic Concepts

1. Determine graphically the x-value in each equation when $y = 2$.

(a) $y = 1 - \frac{1}{2}x$ **(b)** $2x - 3y = 6$

2. Determine whether $(-1, 0)$ or $(4, 2)$ is a solution to the system

$$2x - 5y = -2$$
$$3x + 2y = 16.$$

3. Solve the system of equations graphically. Check your answer.

$$x - y = 1$$
$$2x + y = 5$$

4. If possible, use the method of substitution to solve each system of equations.

(a) $x + y = -1$
  $y = 2 - x$

(b) $4x - y = 5$
  $-x + y = -2$

(c) $x + 2y = 3$
  $-x - 2y = -3$

5. *Room Prices* A hotel rents single and double rooms for $150 and $200, respectively. The hotel receives $55,000 for renting 300 rooms.
(a) Let x be the number of single rooms rented and let y be the number of double rooms rented. Write a system of linear equations whose solution gives the values of x and y.
(b) Use the method of substitution to solve the system. Check your answer.

4.3 Solving Systems of Linear Equations by Elimination

The Elimination Method • Recognizing Other Types of Systems • Applications

A LOOK INTO MATH ▶ Anyone who has used Internet map services such as MapQuest or Google maps knows that there are often several ways to get from one location to another. Sometimes the route with the shortest physical distance is desirable. At other times, we want a route on faster roads with the least amount of driving time. In this section, we introduce a second symbolic method for solving systems of linear equations. With two symbolic "routes" to the same solution, you can choose the method that may work best for a particular system. This second method, called the *elimination method*, is very efficient for solving some types of systems of linear equations.

The Elimination Method

The **elimination method** is based on the addition property of equality. If

$$a = b \quad \text{and} \quad c = d,$$

then

$$a + c = b + d.$$

For example, if the sum of two numbers is 20 and their difference is 4, then the system of equations

$$x + y = 20$$
$$x - y = 4$$

describes these two numbers. By the addition property of equality, the sum of the left sides of these equations equals the sum of their right sides.

$$(x + y) + (x - y) = 20 + 4 \quad \text{Add left sides of these equations.}$$
$$\text{Add right sides of these equations.}$$
$$2x = 24 \quad \text{Combine terms.}$$

Note that the y-variable is eliminated when the left sides are added. The resulting equation, $2x = 24$, simplifies to $x = 12$. Thus the value of x in the solution to the system of equations is 12. The value of y can be found by substituting 12 for x in either of the given equations. Substituting 12 for x in the first equation, $x + y = 20$, results in $12 + y = 20$ or $y = 8$. The solution to the system of equations is $x = 12$, $y = 8$, which can be written as the ordered pair $(12, 8)$.

To organize the elimination method better, we can carry out the addition vertically.

$$\begin{aligned} x + y &= 20 \\ x - y &= 4 \\ \hline 2x + 0y &= 24 \qquad \text{Add left sides and add right sides.} \\ 2x &= 24 \qquad \text{Simplify.} \\ x &= 12 \qquad \text{Divide each side by 2.} \end{aligned}$$

Once the value of one variable is known, in this case $x = 12$, don't forget to find the value of the other variable by substituting this known value in either of the *given* equations. By substituting 12 for x in the second equation, we obtain $12 - y = 4$ or $y = 8$. The solution is the ordered pair $(12, 8)$.

EXAMPLE 1   **Applying the elimination method**

Solve each system of equations. Check each solution.
(a) $2x + y = 1$ (b) $-2a + b = -3$
    $3x - y = 9$ $2a + 3b = 7$

Solution
(a) Adding these two equations eliminates the y-variable.

$$\begin{aligned} 2x + y &= 1 \qquad && \text{First equation} \\ 3x - y &= 9 \qquad && \text{Second equation} \\ \hline 5x &= 10, \quad \text{or} \quad x = 2 \qquad && \text{Add and solve for } x. \end{aligned}$$

To find y, substitute **2** for x in either of the *given* equations. We will use $2x + y = 1$.

$$2(2) + y = 1 \qquad \text{Let } x = 2 \text{ in first equation.}$$
$$y = -3 \qquad \text{Subtract 4 from each side.}$$

The solution is the *ordered pair* $(2, -3)$, which can be checked by substituting 2 for x and -3 for y in each of the given equations.

$$2x + y = 1 \qquad\qquad 3x - y = 9 \qquad \text{Given equations}$$
$$2(2) + (-3) \overset{?}{=} 1 \qquad 3(2) - (-3) \overset{?}{=} 9 \qquad \text{Let } x = 2 \text{ and } y = -3.$$
$$4 - 3 \overset{?}{=} 1 \qquad\qquad 6 + 3 \overset{?}{=} 9 \qquad \text{Simplify.}$$
$$1 = 1 \checkmark \qquad\qquad 9 = 9 \checkmark \qquad \text{The solution checks.}$$

(b) Adding these two equations eliminates the a-variable.

$$-2a + b = -3 \qquad \text{First equation}$$
$$\underline{2a + 3b = 7} \qquad \text{Second equation}$$
$$4b = 4, \quad \text{or} \quad b = 1 \qquad \text{Add and solve for } b.$$

To find a, substitute **1** for b in either of the *given* equations. We will use $2a + 3b = 7$.

$$2a + 3(1) = 7 \qquad \text{Let } b = 1 \text{ in second equation.}$$
$$2a = 4 \qquad \text{Subtract 3 from each side.}$$
$$a = 2 \qquad \text{Divide each side by 2.}$$

The solution is the *ordered pair* $(2, 1)$, which can be checked by substituting 2 for a and 1 for b in each of the given equations.

$$-2a + b = -3 \qquad\qquad 2a + 3b = 7 \qquad \text{Given equations}$$
$$-2(2) + 1 \overset{?}{=} -3 \qquad 2(2) + 3(1) \overset{?}{=} 7 \qquad \text{Let } a = 2 \text{ and } b = 1.$$
$$-4 + 1 \overset{?}{=} -3 \qquad\qquad 4 + 3 \overset{?}{=} 7 \qquad \text{Simplify.}$$
$$-3 = -3 \checkmark \qquad\qquad 7 = 7 \checkmark \quad \text{The solution checks.}$$

▌ **Now Try Exercises 15, 17**

Adding two equations does not always eliminate a variable. For example, adding the following equations eliminates neither variable.

$$3x - 2y = 11 \qquad \text{First equation}$$
$$\underline{4x + y = 11} \qquad \text{Second equation}$$
$$7x - y = 22 \qquad \text{Add the equations.}$$

READING CHECK

• In using the elimination method, when is it necessary to apply the multiplication property of equality?

However, by the multiplication property of equality, we can multiply the second equation by 2. Then adding the equations eliminates the y-variable.

$$3x - 2y = 11 \qquad \text{First equation}$$
$$\underline{8x + 2y = 22} \qquad \text{Multiply the second equation by 2.}$$
$$11x = 33, \quad \text{or} \quad x = 3 \qquad \text{Add and solve for } x.$$

EXAMPLE 2 **Multiplying before applying elimination**

Solve each system of equations.
(a) $5x - y = -11$ **(b)** $3x + 2y = 1$
     $2x + 3y = -1$ $2x - 3y = 5$

Solution
(a) We multiply the first equation by 3 and then add to eliminate the y-variable.

$$15x - 3y = -33 \qquad \text{Multiply the first equation by 3.}$$
$$\underline{2x + 3y = -1} \qquad \text{Second equation}$$
$$17x = -34, \quad \text{or} \quad x = -2 \quad \text{Add and solve for } x.$$

We can find y by substituting -2 for x in the second equation, $2x + 3y = -1$.

$$2(-2) + 3y = -1 \qquad \text{Let } x = -2 \text{ in second equation.}$$
$$3y = 3 \qquad \text{Add 4 to each side.}$$
$$y = 1 \qquad \text{Divide each side by 3.}$$

The solution is $(-2, 1)$.

(b) We must apply the multiplication property to both equations. If we multiply the first equation, $3x + 2y = 1$, by 3, and the second equation, $2x - 3y = 5$, by 2, then the coefficients of the y-variables will be opposites. Adding eliminates the y-variable.

$$9x + 6y = 3 \qquad \text{Multiply the first equation by 3.}$$
$$\underline{4x - 6y = 10} \qquad \text{Multiply the second equation by 2.}$$
$$13x \qquad = 13, \quad \text{or} \quad x = 1 \qquad \text{Add and solve for } x.$$

To find y, substitute **1** for x in the first *given* equation, $3x + 2y = 1$.

$$3(1) + 2y = 1 \qquad \text{Let } x = 1 \text{ in first equation.}$$
$$2y = -2 \qquad \text{Subtract 3 from each side.}$$
$$y = -1 \qquad \text{Divide each side by 2.}$$

The solution is $(1, -1)$.

Now Try Exercises 29, 33

In practice, it is possible to eliminate *either* variable from a system of linear equations. It is often best to choose the variable that requires the least amount of computation to complete the elimination. In the next example, we solve a system of equations twice—first by using the multiplication property of equality to eliminate the x-variable and then by using it to eliminate the y-variable.

EXAMPLE 3 **Multiplying before applying elimination**

Solve the system of equations two times, first by eliminating x and then by eliminating y.

$$2y = -6 - 5x$$
$$2x = -5y + 6$$

Solution

It is best to write each equation in the standard form: $Ax + By = C$.

$$5x + 2y = -6 \qquad \text{First equation in standard form}$$
$$2x + 5y = 6 \qquad \text{Second equation in standard form}$$

Eliminate x If we multiply the first equation in standard form by -2 and the second equation in standard form by 5, then we can eliminate the x-variable by adding.

$$-10x - 4y = 12 \qquad \text{Multiply the first equation by } -2.$$
$$\underline{10x + 25y = 30} \qquad \text{Multiply the second equation by 5.}$$
$$21y = 42, \quad \text{or} \quad y = 2 \qquad \text{Add and solve for } y.$$

To find x, substitute **2** for y in the first *given* equation, $2y = -6 - 5x$.

$$2(2) = -6 - 5x \qquad \text{Let } y = 2 \text{ in first equation.}$$
$$10 = -5x \qquad \text{Add 6 to each side.}$$
$$-2 = x \qquad \text{Divide each side by } -5.$$

The solution is $(-2, 2)$.

Eliminate y If we multiply the first equation in standard form by -5 and the second equation in standard form by 2, then we can eliminate the y-variable by adding.

$$
\begin{aligned}
-25x - 10y &= 30 && \text{Multiply the first equation by } -5.\\
\underline{4x + 10y} &= \underline{12} && \text{Multiply the second equation by 2.}\\
-21x &= 42, \quad \text{or} \quad x = -2 && \text{Add and solve for } x.
\end{aligned}
$$

To find y, substitute -2 for x in the second given equation, $2x = -5y + 6$.

$$
\begin{aligned}
2(-2) &= -5y + 6 && \text{Let } x = -2 \text{ in second equation.}\\
-10 &= -5y && \text{Subtract 6 from each side.}\\
2 &= y && \text{Divide each side by } -5.
\end{aligned}
$$

The solution is $(-2, 2)$.

▌ **Now Try Exercise 31**

In the next example we use three different methods to solve a system of equations.

> **EXAMPLE 4**   **Solving a system with different methods**

Solve the system of equations symbolically, graphically, and numerically.

$$
\begin{aligned}
x + y &= 2\\
x - 3y &= 6
\end{aligned}
$$

Solution

Symbolic Solution Both the method of substitution and the elimination method are symbolic methods. The elimination method is used here. We can solve the system by multiplying the second equation by -1 and adding to eliminate the x-variable.

$$
\begin{aligned}
x + y &= 2 && \text{First equation}\\
\underline{-x + 3y} &= \underline{-6} && \text{Multiply the second equation by } -1.\\
4y &= -4, \quad \text{or} \quad y = -1 && \text{Add and solve for } y.
\end{aligned}
$$

We can find x by substituting -1 for y in the first equation, $x + y = 2$.

$$
\begin{aligned}
x + (-1) &= 2 && \text{Let } y = -1 \text{ in first equation.}\\
x &= 3 && \text{Add 1 to each side.}
\end{aligned}
$$

The solution is $(3, -1)$.

Graphical Solution For a graphical solution, we solve each equation for y to obtain the slope–intercept form.

$x + y = 2$	First equation		$x - 3y = 6$	Second equation
$y = -x + 2$	Subtract x.		$-3y = -x + 6$	Subtract x.
			$y = \dfrac{1}{3}x - 2$	Divide by -3.

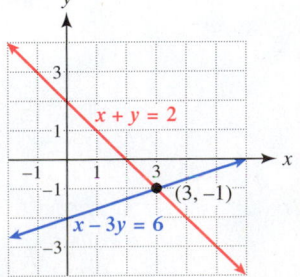

Figure 4.10

The graphs of $y = -x + 2$ and $y = \frac{1}{3}x - 2$ are shown in Figure 4.10. They intersect at $(3, -1)$. (These graphs could also be obtained by finding the x- and y-intercepts for each equation.)

Numerical Solution A numerical solution consists of a table of values, as shown in Table 4.4. Note that when $x = 3$, both y-values equal -1. Therefore the solution is $(3, -1)$.

TABLE 4.4

x	0	1	2	**3**	4
$y = -x + 2$	2	1	0	-1	-2
$y = \frac{1}{3}x - 2$	-2	$-\frac{5}{3}$	$-\frac{4}{3}$	-1	$-\frac{2}{3}$

▌ **Now Try Exercise 41**

Recognizing Other Types of Systems

In Section 4.1 we discussed how a system of linear equations can have no solutions, one solution, or infinitely many solutions. Elimination can also be used on systems that have no solutions or infinitely many solutions.

EXAMPLE 5 | **Solving other types of systems**

Solve each system of equations by using the elimination method. Then graph the system.
(a) $\quad x - 2y = \quad 4$ (b) $3x + 3y = 6$
$\quad -2x + 4y = -8$ $\quad\quad x + \quad y = 1$

Solution

(a) We multiply the first equation by 2 and then add, which eliminates both variables.

$$2x - 4y = \quad 8 \qquad \text{Multiply the first equation by 2.}$$
$$\underline{-2x + 4y = -8} \qquad \text{Second equation}$$
$$0 = \quad 0 \quad \textbf{(True)} \quad \text{Add.}$$

The equation $0 = 0$ is *always true*, which indicates that the system has *infinitely many solutions*. A graph of the two equations is shown in Figure 4.11. The two lines are identical so there actually is only one line, and *every point on this line represents a solution*. For example, $(0, -2)$ and $(4, 0)$ lie on the line and are both solutions.

(b) We multiply the second equation by -3 and add, eliminating both variables.

$$3x + 3y = \quad 6 \qquad \text{First equation}$$
$$\underline{-3x - 3y = -3} \qquad \text{Multiply the second equation by } -3.$$
$$0 = \quad 3 \quad \textbf{(False)} \quad \text{Add.}$$

The equation $0 = 3$ is *always false*, which indicates that the system has *no solutions*. A graph of the two equations is shown in Figure 4.12. Note that the two lines are parallel and thus do not intersect.

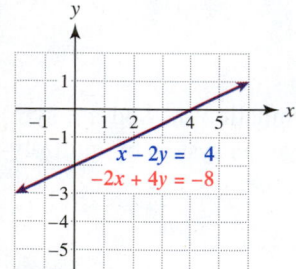

Figure 4.11

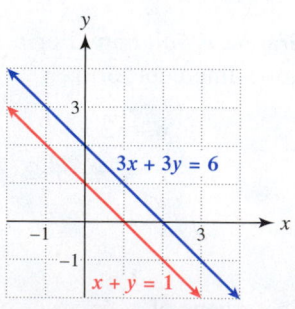

Figure 4.12

▌ **Now Try Exercises 51, 55**

Numerical and Graphical Solutions

The left-hand figure below shows a numerical solution for Example 5(a), where each equation is solved for y to obtain

$$y_1 = (4 - x)/(-2) \quad \text{and} \quad y_2 = (-8 + 2x)/4.$$

Note that $y_1 = y_2$ for each value of x, indicating that the graphs of y_1 and y_2 are the same line. Similarly, the right-hand figure shows a numerical solution for Example 5(b), where

$$y_1 = (6 - 3x)/3 \quad \text{and} \quad y_2 = 1 - x.$$

Note that $y_1 \neq y_2$ and the difference $y_1 - y_2$ is 1 for every value of x, indicating that the graphs of y_1 and y_2 are parallel lines that do not intersect.

X	Y1	Y2
0	-2	-2
1	-1.5	-1.5
2	-1	-1
3	-.5	-.5
4	0	0
5	.5	.5
6	1	1

Y1■(4−X)/(−2)

X	Y1	Y2
-1	3	2
0	2	1
1	1	0
2	0	-1
3	-1	-2
4	-2	-3
5	-3	-4

Y1■(6−3X)/3

Applications

▶ **REAL-WORLD CONNECTION** In the next two examples we use elimination to solve applications relating to new cancer cases and to burning calories during exercise.

EXAMPLE 6   ### Determining new cancer cases

In 2010, there were 1,530,000 new cancer cases. Men accounted for 50,000 more new cases than women. How many new cases of cancer were there for each gender? (*Source:* American Cancer Society.)

Solution

STEP 1: *Identify each variable.*

x: new cancer cases for men in 2010
y: new cancer cases for women in 2010

STEP 2: *Write a system of equations.*

$$x + y = 1{,}530{,}000 \quad \text{New cases totaled 1,530,000.}$$

$$x - y = 50{,}000 \quad \text{Men had 50,000 more cases than women.}$$

STEP 3A: *Solve the system of linear equations.* Add the two equations to eliminate the y-variable.

$$
\begin{array}{rcl}
x + y & = & 1{,}530{,}000 \\
x - y & = & 50{,}000 \\
\hline
2x & = & 1{,}580{,}000, \quad \text{or} \quad x = 790{,}000
\end{array}
$$

Substituting 790,000 for x in the first equation results in $790{,}000 + y = 1{,}530{,}000$ or $y = 740{,}000$. The solution is $x = 790{,}000$ and $y = 740{,}000$.

STEP 3B: *Determine the solution to the problem.* There were 790,000 new cases of cancer for men and 740,000 new cases for women in 2010.

STEP 4: *Check the solution.* The total number of cases was

$$790{,}000 + 740{,}000 = 1{,}530{,}000.$$

The number of new cases for men exceeded the number of new cases for women by

$$790{,}000 - 740{,}000 = 50{,}000.$$

The answer checks.

Now Try Exercise 57

EXAMPLE 7 **Burning calories during exercise**

During strenuous exercise, an athlete can burn 10 calories per minute on a rowing machine and 11.5 calories per minute on a stair climber. If an athlete uses both machines and burns 433 calories in a 40-minute workout, how many minutes does the athlete spend on each machine? (*Source: Runner's World.*)

Solution

STEP 1: *Identify each variable.*

x: number of minutes on a rowing machine
y: number of minutes on a stair climber

STEP 2: *Write a system of equations.* The total workout takes 40 minutes, so $x + y = 40$. The athlete burns $10x$ calories on the rowing machine and $11.5y$ calories on the stair climber. Because the total number of calories equals 433, it follows that $10x + 11.5y = 433$.

$$x + y = 40 \qquad \text{Workout is 40 minutes.}$$
$$10x + 11.5y = 433 \qquad \text{Total calories is 433.}$$

STEP 3A: *Solve the system of linear equations.* Multiply the first equation by -10 and add the two equations.

$$-10x - 10y = -400 \qquad \text{Multiply by } -10.$$
$$\underline{10x + 11.5y = 433} \qquad \text{Second equation}$$
$$1.5y = 33, \quad \text{or} \quad y = \frac{33}{1.5} = 22 \quad \text{Add and solve for } y.$$

Because $x + y = 40$ and $y = 22$, it follows that $x = 18$.

STEP 3B: *Determine the solution to the problem.* The athlete spends 18 minutes on the rowing machine and 22 minutes on the stair climber.

STEP 4: *Check your answer.* Because $18 + 22 = 40$, the athlete works out for 40 minutes. Also, the athlete burns $10(18) + 11.5(22) = 433$ calories. The answer checks.

Now Try Exercise 59

4.3 Putting It All Together

CONCEPT	EXPLANATION	EXAMPLE
Elimination Method	Is based on the addition property of equality If $a = b$ and $c = d$, then $$a + c = b + d.$$ May be used to solve systems of equations	$$\begin{aligned} x + y &= 5 \\ \underline{x - y} &= \underline{-1} \\ 2x &= 4, \quad \text{or} \quad x = 2 \quad \text{Add.} \end{aligned}$$ Because $x + y = 5$ and $x = 2$, it follows that $y = 3$. The solution is $(2, 3)$.
Other Types of Systems	Elimination can be used to recognize systems having **1.** no solutions **2.** infinitely many solutions.	**1.** $$\begin{aligned} -x - y &= -4 \\ \underline{x + y} &= \underline{2} \\ 0 &= -2 \quad \text{Add.} \end{aligned}$$ Because $0 = -2$ is always false, there are no solutions.

CONCEPT	EXPLANATION	EXAMPLE
Other Types of Systems (continued)		2. $\begin{aligned} x + y &= 4 \\ 2x + 2y &= 8 \end{aligned}$ Multiply the first equation by -2. $\begin{aligned} -2x - 2y &= -8 \\ \underline{2x + 2y} &= \underline{8} \\ 0 &= 0 \quad \text{Add.} \end{aligned}$ Because $0 = 0$ is always true, there are infinitely many solutions.

4.3 Exercises

MyMathLab

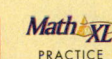

CONCEPTS AND VOCABULARY

1. Name two symbolic methods for solving a system of linear equations.

2. The elimination method is based on the _____ property of equality.

3. The addition property of equality states that if $a = b$ and $c = d$, then $a + c$ _____ $b + d$.

4. The multiplication property of equality states that if $a = b$, then ca _____ cb.

5. Suppose that the elimination method results in the equation $0 = 0$. What does this indicate about the number of solutions to the system of equations?

6. Suppose that the elimination method results in the equation $0 = 1$. What does this indicate about the number of solutions to the system of equations?

USING ELIMINATION

Exercises 7–14: If possible, use the given graph to solve the system of equations. Then use the elimination method to verify your answer.

7. $\begin{aligned} x - y &= 0 \\ x + y &= 2 \end{aligned}$

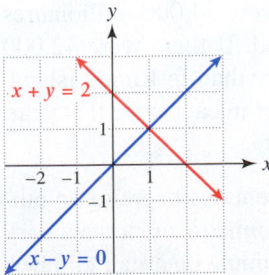

8. $\begin{aligned} x + y &= 6 \\ 2x - y &= 3 \end{aligned}$

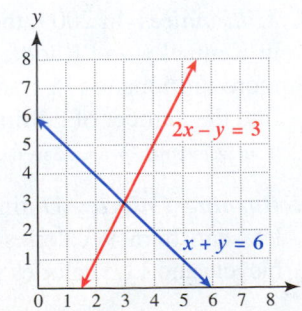

9. $\begin{aligned} 2x + 3y &= -1 \\ 2x - 3y &= -7 \end{aligned}$

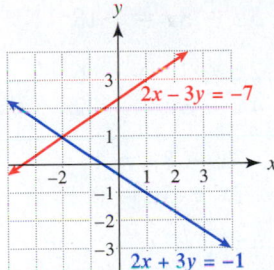

10. $\begin{aligned} -2x + y &= -3 \\ 4x - 3y &= 7 \end{aligned}$

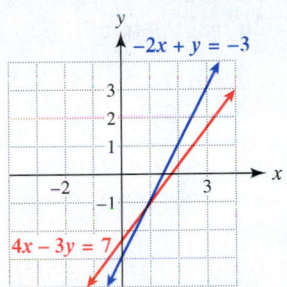

11. $\begin{aligned} x + y &= 3 \\ x + y &= -1 \end{aligned}$

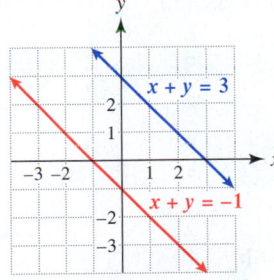

12. $\begin{aligned} 2x - y &= 4 \\ -2x + y &= -4 \end{aligned}$

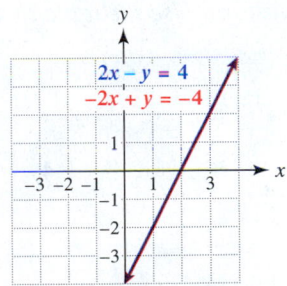

13. $\begin{aligned} 2x + 2y &= 6 \\ x + y &= 3 \end{aligned}$

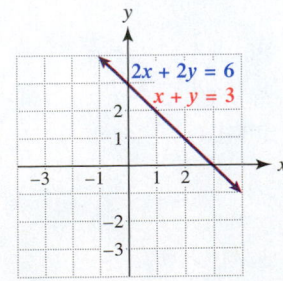

14. $\begin{aligned} -x + 3y &= 4 \\ x - 3y &= 3 \end{aligned}$

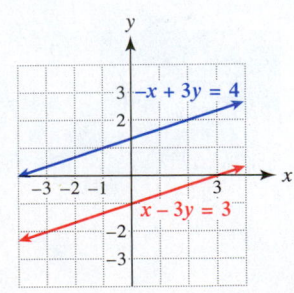

Exercises 15–36: Use the elimination method to solve the system of equations.

15. $x + y = 7$
$x - y = 5$

16. $x - y = 8$
$x + y = 4$

17. $-x + y = 5$
$x + y = 3$

18. $x - y = 10$
$-x - y = 20$

19. $2x + y = 8$
$3x - y = 2$

20. $-x + 2y = 3$
$x + 6y = 5$

21. $-2x + y = -3$
$2x - 4y = 0$

22. $2x + 6y = -5$
$7x - 6y = -4$

23. $\frac{1}{2}x - y = 3$
$\frac{3}{2}x + y = 5$

24. $x - \frac{1}{4}y = 4$
$-4x + \frac{1}{4}y = -9$

25. $a + 6b = 2$
$a + 3b = -1$

26. $3r - t = 7$
$2r - t = 2$

27. $5a - 6b = -2$
$5a + 5b = 9$

28. $-r + 2t = 0$
$3r + 2t = 8$

29. $3u + 2v = -16$
$2u + v = -9$

30. $5u - v = 0$
$3u + 3v = -18$

31. $5x - 7y = 5$
$-2x + 2y = -2$

32. $2x + 7y = 6$
$4x - 3y = -22$

33. $5x - 3y = 4$
$3x + 2y = 10$

34. $-3x - 8y = 1$
$2x + 5y = 0$

35. $-5x - 10y = -22$
$10x + 15y = 35$

36. $-15x + 4y = -20$
$5x + 7y = 90$

Exercises 37–40: A table of values is given for two linear equations. Use the table to solve this system.

37.

x	0	1	2	3	4
$y = -x + 5$	5	4	3	2	1
$y = 2x - 4$	-4	-2	0	2	4

38.

x	-3	-2	-1	0	1
$y = x + 1$	-2	-1	0	1	2
$y = -x - 3$	0	-1	-2	-3	-4

39.

x	-2	-1	0	1	2
$y = 3x + 1$	-5	-2	1	4	7
$y = -x + 1$	3	2	1	0	-1

40.

x	-2	-1	0	1	2
$y = 2x$	-4	-2	0	2	4
$y = -x$	2	1	0	-1	-2

USING MORE THAN ONE METHOD

Exercises 41–46: Solve the system of equations

(a) *symbolically,*
(b) *graphically, and*
(c) *numerically.*

41. $2x + y = 5$
$x - y = 1$

42. $-x + y = 2$
$3x + y = -2$

43. $2x + y = 5$
$x + y = 1$

44. $-x + y = 2$
$3x - y = -2$

45. $6x + 3y = 6$
$-2x + 2y = -2$

46. $-x + 2y = 5$
$2x + 2y = 8$

ELIMINATION AND OTHER TYPES OF SYSTEMS

Exercises 47–56: Use elimination to determine whether the system of equations has no solutions, one solution, or infinitely many solutions. Then graph the system.

47. $2x - 2y = 4$
$-x + y = -2$

48. $-2x + y = 4$
$4x - 2y = -8$

49. $x - y = 0$
$x + y = 0$

50. $x - y = 2$
$x + y = 2$

51. $x - y = 4$
$x - y = 1$

52. $-2x + 3y = 5$
$4x - 6y = 10$

53. $x - 3y = 2$
$-x + 3y = 4$

54. $6x + 9y = 18$
$4x + 6y = 12$

55. $4x - 8y = 24$
$6x - 12y = 36$

56. $x - y = 5$
$2x - y = 4$

APPLICATIONS

57. *Skin Cancer* In 2010, there were 68,000 new cases of skin cancer in the United States. Men represented 10,000 more cases than women. How many new cases of skin cancer were there for men and for women? (*Source:* American Cancer Society.)

58. *Millionaires* In 2009, there were 84,000 millionaires in Kentucky and Rhode Island. If there were 42,000 more millionaires in Kentucky than in Rhode Island, find the number of millionaires in each state that year. (*Source:* Internal Revenue Service.)

59. *Burning Calories* During strenuous exercise an athlete can burn 9 calories per minute on a stationary bicycle and 11.5 calories per minute on a stair climber.

In a 30-minute workout an athlete burns 300 calories. How many minutes does the athlete spend on each type of exercise equipment? (*Source: Runner's World.*)

60. *Distance Running* An athlete runs at 9 mph and then at 12 mph, covering 10 miles in 1 hour. How long does the athlete run at each speed?

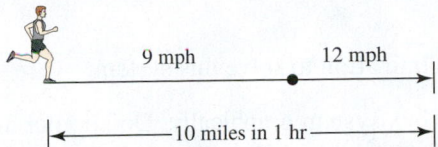

61. *River Current* A riverboat takes 8 hours to travel 64 miles downstream and 16 hours for the return trip. What is the speed of the current and the speed of the riverboat in still water?

62. *Airplane Speed* An airplane travels 3000 miles with the wind in 5 hours and takes 6 hours for the return trip into the wind. What is the speed of the wind and the speed of the airplane without any wind?

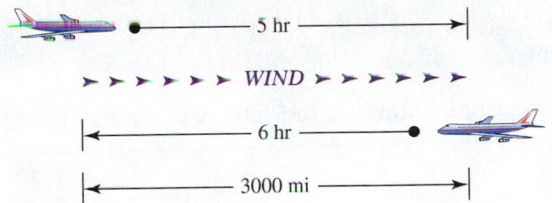

63. *Investments* A total of $5000 is invested at 3% and 5% annual interest. After 1 year the total interest equals $210. How much money is invested at each interest rate?

64. *Mixing Antifreeze* A car radiator holds 2 gallons of fluid and initially is empty. If a mixture of water and antifreeze contains 70% antifreeze and another mixture contains 15% antifreeze, how much of each should be combined to fill the radiator with a 50% antifreeze mixture?

65. *Number Problem* The sum of two integers is -17, and their difference is -69. Find the two integers.

66. *Supplementary Angles* The measures of two supplementary angles differ by 74°. Find the two angles.

67. *Picture Dimensions* The figure at the top of the next column shows a red graph that gives possible dimensions for a rectangular picture frame with perimeter 120 inches. The blue graph shows possible dimensions for a rectangular frame whose length L is twice its width W.

Picture Dimensions

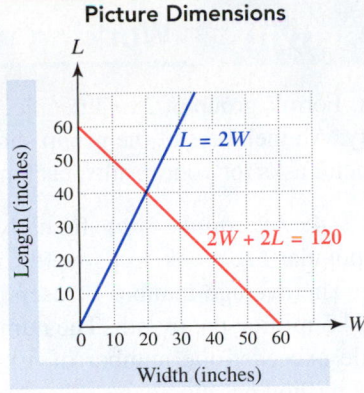

(a) Use the figure to determine the dimensions of a frame with a perimeter of 120 inches and a length that is twice the width.

(b) Solve this problem symbolically.

68. *Sales of DVDs and CDs* A company sells DVDs d and CDs c. The figure shows a red graph of $d + c = 2000$. The blue graph shows a revenue of $15,000 received from selling d DVDs at $12 each and c CDs at $6 each.

Digital Media

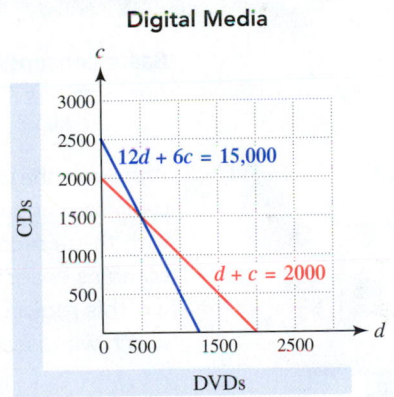

(a) If the total number of DVDs and CDs sold is 2000, determine how many of each were sold to obtain a revenue of $15,000.

(b) Solve this problem symbolically.

WRITING ABOUT MATHEMATICS

69. Suppose that a system of linear equations is solved symbolically, numerically, and graphically. How do the solutions from each method compare? Explain your answer.

70. When you are solving a system of linear equations by elimination, how can you recognize that the system has no solutions?

Group Activity   Working with Real Data

Directions: Form a group of 2 to 4 people. Select someone to record the group's responses for this activity. All members of the group should work cooperatively to answer the questions. If your instructor asks for your results, each member of the group should be prepared to respond.

Exercises 1–4: *Facebook Apps* In early 2011, the two most popular Facebook Apps were CityVille and FarmVille. The average number of users for the two apps was 74.5 million per month. The number of users of CityVille exceeded the number of users of Farm-Ville by 45 million per month. (*Source:* allfacebook.com)

1. Set up a system of equations whose solution gives the monthly number of users in millions for each app. Identify what each variable represents.

2. Use substitution to solve this system. Interpret the result.

3. Use elimination to solve this system.

4. Solve this system graphically. Do all your answers agree?

4.4   Systems of Linear Inequalities

Basic Concepts • Solutions to One Inequality • Solutions to Systems of Inequalities • Applications

A LOOK INTO MATH ▶

Although there is no *ideal* weight for a person, government agencies and insurance companies sometimes recommend a *range* of weights for various heights. *Inequalities* are used with these recommendations. One example is shown in Figure 4.13, where the blue region contains ordered pairs (w, h) that give recommended weight–height combinations. Describing this region mathematically requires an understanding of systems of linear inequalities, which we discuss in this section. (*Source:* Department of Agriculture.)

NEW VOCABULARY

☐ Test point
☐ Linear inequality in two variables
☐ System of linear inequalities in two variables

STUDY TIP

If you have not been studying with other students, consider getting together with classmates so that you can work on your math homework together.

Recommended Weight

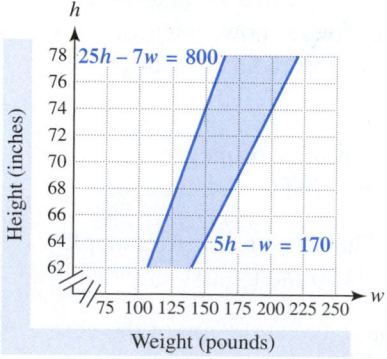

Figure 4.13

Basic Concepts

▶ **REAL-WORLD CONNECTION** Suppose that a college student works both at the library and at a department store. The library pays $10 per hour, and the department store pays $8 per hour. The equation $A = 10L + 8D$ calculates the amount of money earned from working

L hours at the library and *D* hours at the department store. If the cost of one college credit is $80, then solutions to the equation

$$10L + 8D = 80$$

are ordered pairs (*L*, *D*) that result in the student earning enough to pay for one credit. The equation's graph is the line shown in Figure 4.14(a). The point (4, 5) lies on this line, which indicates that, if the student works 4 hours at the library and 5 hours at the department store, then the pay is $80.

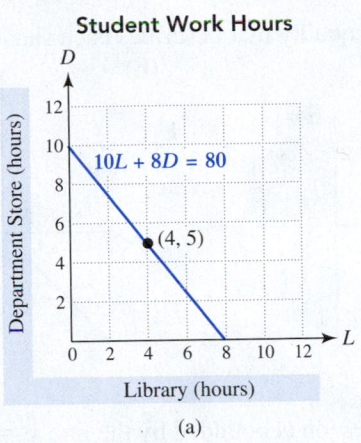

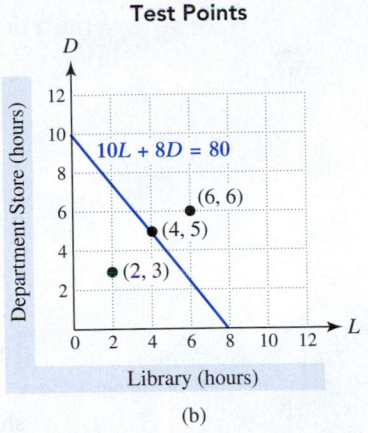

(a)   (b)

Figure 4.14

There are many situations in which the student can make more than $80. For example, the **test point** (6, 6) lies above the line in Figure 4.14(b), indicating that, if the student works 6 hours at both the library and the department store, then the pay is more than $80. In fact, any point *above* the line results in pay *greater than* $80. The region **above** the line is described by the inequality

$$10L + 8D > 80.$$

The test point (**6**, **6**) represents earnings of $108 and *satisfies* this inequality because

$$10(\mathbf{6}) + 8(\mathbf{6}) > 80$$

is a true statement. Similarly, any point *below* the line gives an ordered pair (*L*, *D*) that results in earnings *less than* $80. The point (**2**, **3**) in Figure 4.14(b) lies below the line and represents earnings of $44. That is,

$$10(\mathbf{2}) + 8(\mathbf{3}) < 80.$$

The region **below** the line is described by the inequality

$$10L + 8D < 80.$$

Solutions to One Inequality

Any linear equation in two variables can be written in standard form as

$$Ax + By = C,$$

where *A*, *B*, and *C* are constants. When the equals sign is replaced with $<$, $>$, $\leq$, or $\geq$, a **linear inequality in two variables** results. Examples of linear *equations* in two variables include

$$2x + 3y = 10 \quad \text{and} \quad y = \frac{1}{2}x - 5,$$

and so examples of linear *inequalities* in two variables include

$$2x + 3y < 10 \quad \text{and} \quad y \geq \frac{1}{2}x - 5.$$

A *solution* to a linear inequality in two variables is an ordered pair (x, y) that makes the inequality a true statement. The *solution set* is the set of all solutions to the inequality. The solution set to an inequality in two variables is typically a region in the xy-plane, which means that there are infinitely many solutions.

EXAMPLE 1

Writing a linear inequality

Write a linear inequality that describes each shaded region.

(a)

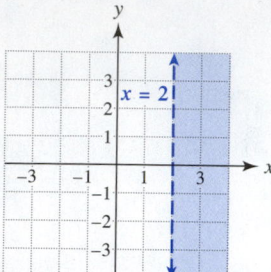

(b)

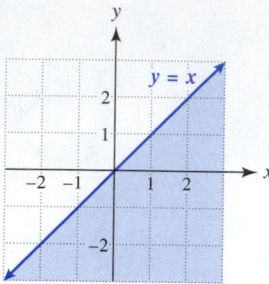

Solution

(a) The shaded region is bounded by the line $x = 2$. The *dashed* line indicates that the line is not included in the solution set. Only points with x-coordinates **greater than** 2 are shaded. Thus every point in the shaded region satisfies $x > 2$.

(b) The solution set includes all points that are on or **below** the line $y = x$. An inequality that describes this region is $y \leq x$, which can also be written as $-x + y \leq 0$.

Now Try Exercises 27, 29

TECHNOLOGY NOTE

Shading an Inequality
Graphing calculators can be used to shade a solution set to an inequality. The left-hand screen shows how to enter the equation from Example 1(b), and the right-hand screen shows the resulting graph.

CALCULATOR HELP

To shade an inquality, see Appendix A (pages AP-6 and AP-7).

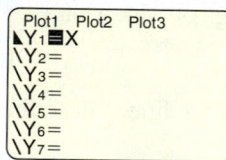

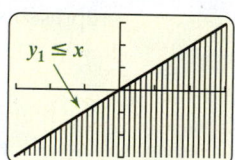

EXAMPLE 2

Graphing a linear inequality

Shade the solution set for each inequality.
(a) $y \leq 1$ (b) $x + y < 3$ (c) $-x + 2y \geq 2$

Solution

(a) First graph the horizontal line $y = 1$, where a solid line indicates that the line is included in the solution set. Next decide whether to shade above or below this line. The inequality $y \leq 1$ indicates that a solution (x, y) must have a y-coordinate less than or equal to 1. There are no restrictions on the x-coordinate. The shaded region **below** the line $y = 1$ in Figure 4.15(a) depicts all ordered pairs (x, y) satisfying $y \leq 1$. The horizontal line $y = 1$ is solid because it is included in the solution set.

(b) Graph the line $x + y = 3$, as shown in Figure 4.15(b). Because the inequality is $<$ and not $\leq$, the line is not included and is dashed rather than solid. To decide whether to shade above or below this dashed line, select a *test point* in either region. For example, the point $(0, 0)$ satisfies the inequality $x + y < 3$, because $0 + 0 < 3$ is a true statement. Therefore shade the region that contains $(0, 0)$, which is **below** the line.

(c) Graph $-x + 2y = 2$ as a solid line, as shown in Figure 4.15(c). To decide whether to shade above or below the solid line, use the test point $(0, 0)$ again. (Note that other test points can be used.) The point $(0, 0)$ does not satisfy the inequality $-x + 2y \geq 2$ because $-0 + 2(0) \geq 2$ is a false statement. Therefore shade the region that does not contain $(0, 0)$, which is **above** the line.

READING CHECK

• How is a test point used when shading a solution set for an inequality?

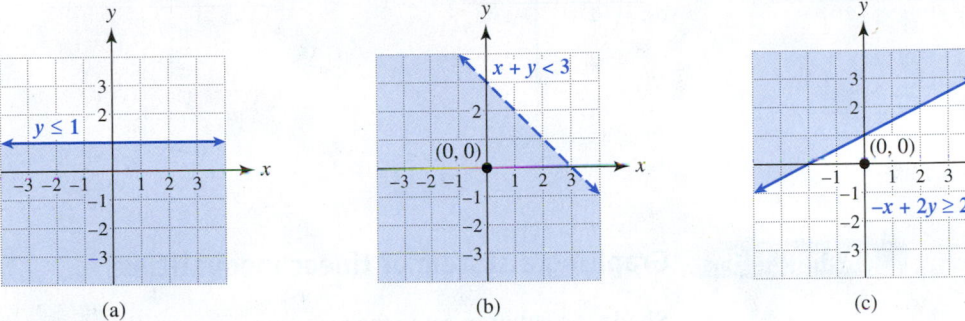

Figure 4.15

Now Try Exercises 35, 41, 43

NOTE: In both parts (b) and (c) of Example 2, the test point $(0, 0)$ was used to determine which region should be shaded. However, *any point that is not on* the solid or dashed line can be used as a test point. When the line does not pass through the origin, it is often convenient to use $(0, 0)$ as a test point because substituting 0 for both x and y in an inequality results in a very simple computation.

GRAPHING A LINEAR INEQUALITY IN TWO VARIABLES

1. Replace the inequality symbol with an equals sign and graph the resulting line. If the inequality is $<$ or $>$, use a dashed line, and if it is $\leq$ or $\geq$, use a solid line.
2. Pick a test point that does *not* lie on the line. Substitute this point in the given inequality. Determine whether the resulting statement is true or false.
3. If the statement is true, shade the region containing the test point. If the statement is false, shade the region that does not contain the test point.

READING CHECK

• How do you know whether to graph a solid line or a dashed line for an inequality?

Solutions to Systems of Inequalities

Sometimes a solution set must satisfy two inequalities in a **system of linear inequalities in two variables**. The point $(2, 1)$ satisfies both of the inequalities in the system of inequalities given by

$$x > 1$$
$$y < 2.$$

Figure 4.16(a) shows the solution set to $x > 1$ in blue, and Figure 4.16(b) shows the solution set to $y < 2$ in red. The two regions are shaded together in Figure 4.16(c) where the blue and red regions intersect to form a purple region. The solution set to the system of inequalities includes all points in the purple region, as shown in Figure 4.16(d). Points in this region satisfy *both* inequalities.

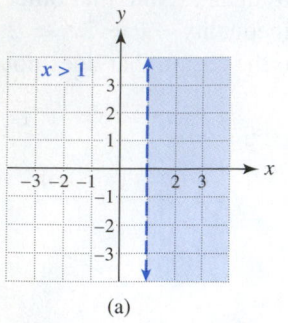
(a)

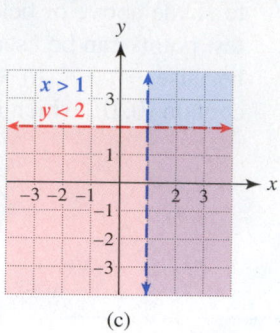
(b)

(c)

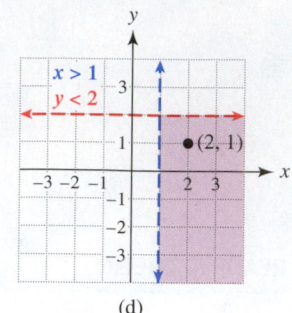
(d)

Figure 4.16

EXAMPLE 3 | **Graphing a system of linear inequalities**

Shade the solution set to the system of inequalities.

$$x > -1$$
$$x + y \le 1$$

Solution

Start by graphing the solution set to each inequality. The solution set to $x > -1$ is the blue region to the **right** of the dashed vertical line $x = -1$, as shown in Figure 4.17(a). The solution set to $x + y \le 1$ includes the solid line $x + y = 1$ and the red region that lies **below** it, as shown in Figure 4.17(b). For a point to satisfy the *system* of inequalities, it must satisfy *both* inequalities. Therefore the solution set is the *intersection* of the blue and red regions, shown as the purple region in Figure 4.17(c). Note that the test point $(0, 0)$, located in the shaded region, satisfies both inequalities.

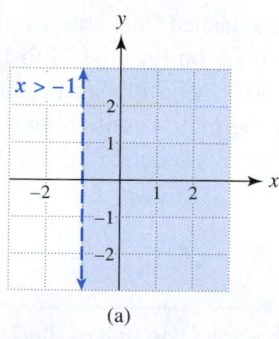
(a)

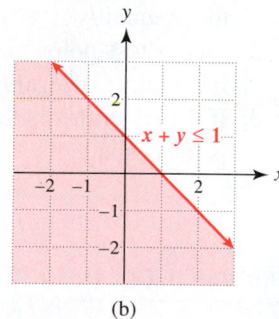
(b)

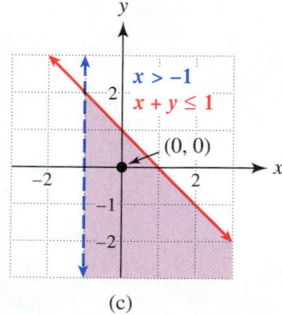
(c)

Figure 4.17

Now Try Exercise 67

EXAMPLE 4 | **Graphing a system of linear inequalities**

Shade the solution set to the system of inequalities.

$$x + 2y < -2$$
$$2x + y \ge 2$$

Solution

In this example we use test points to determine the solution set. We start by graphing the dashed line $x + 2y = -2$ and the solid line $2x + y = 2$, as shown in Figure 4.18(a). Note that these two lines divide the xy-plane into 4 regions, numbered **1**, **2**, **3**, and **4**. If we let $(0, 0)$ be a test point, it does not satisfy either inequality. Therefore we do not shade region **2**, which contains $(0, 0)$. However, there are still 3 possible regions. If we try the test point $(\mathbf{4}, \mathbf{-4})$ in region **4**, it satisfies both the given inequalities.

$$\mathbf{4} + 2(\mathbf{-4}) < -2 \quad \checkmark \quad \text{A true statement}$$

$$2(\mathbf{4}) + (\mathbf{-4}) \geq 2 \quad \checkmark \quad \text{A true statement}$$

Thus we shade region 4, as shown in Figure 4.18(b). Once a test point is found that makes both inequalities true, there is no need to check test points in other regions.

CRITICAL THINKING

Does the solution set in Figure 4.18 (b) include the point of intersection, $(2, -2)$? Explain your reasoning.

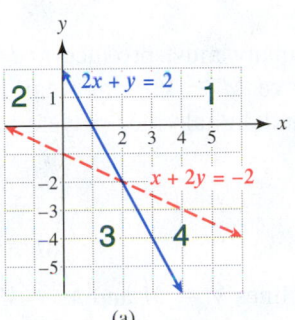

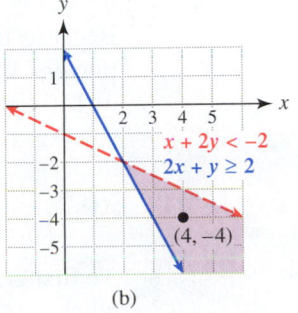

(a)   (b)

Figure 4.18

Now Try Exercise 63

GRAPHING A SYSTEM OF LINEAR INEQUALITIES IN TWO VARIABLES

1. For each inequality in the system, perform Step 1 from Graphing a Linear Inequality in Two Variables on page 279.
2. Pick a test point from one region and substitute it in the given inequalities.
3. If both of the resulting statements are true, shade the region containing the test point. If not, pick a test point from a different region and substitute it in the given inequalities. Repeat this step until the region to be shaded is found.

MAKING CONNECTIONS

Solving for y and shading inequalities

Another way to solve the system of inequalities in Example 4 that does not involve test points is to solve each inequality for y to obtain

$$y < -\frac{1}{2}x - 1$$

$$y \geq -2x + 2.$$

The solution set is the region in Figure 4.18(b) that lies **below** the line $y = -\frac{1}{2}x - 1$ and **above and including** the line $y = -2x + 2$.

Applications

▶ **REAL-WORLD CONNECTION** The next two examples illustrate applications involving systems of inequalities.

EXAMPLE 5 **Manufacturing MP3 players and digital video players**

A business manufactures MP3 players and digital video players. Because every digital video player contains an MP3 player, it must produce at least as many MP3 players as digital video players. In addition, the total number of MP3 players and digital video players produced each day cannot exceed 50 because of limited resources. Shade the region that shows the numbers of MP3 players M and digital video players V that can be produced within these restrictions. Label the horizontal axis M and the vertical axis V.

Solution

Because the company must produce *at least* as many MP3 players M as digital video players V, we have $M \geq V$, which can also be written as $V \leq M$. The total number of MP3 players and digital video players *cannot exceed* 50, so $M + V \leq 50$. To shade the solution set for

$$V \leq M$$

$$M + V \leq 50,$$

we first graph the lines $V = M$ and $M + V = 50$, as shown in Figure 4.19(a). Because the number of MP3 players and digital video players cannot be negative, the graph includes only quadrant I. These lines divide this quadrant into four regions, and we can determine the correct region to shade by selecting one test point from each region. The region containing the test point satisfying both inequalities is the one to be shaded. For example, the test point (**20**, **10**) with $M = 20$, $V = 10$ satisfies both inequalities.

$$\mathbf{10} \leq \mathbf{20} \ \checkmark \quad \text{A true statement; } V \leq M$$
$$\mathbf{20} + \mathbf{10} \leq \mathbf{50} \ \checkmark \quad \text{A true statement; } M + V \leq 50$$

The solution set is shaded in Figure 4.19(b).

NOTE: An alternative solution is to write the inequalities as $V \leq M$ and $V \leq -M + 50$. Then the solution set lies *below both lines*. This region is shaded in Figure 4.19(b).

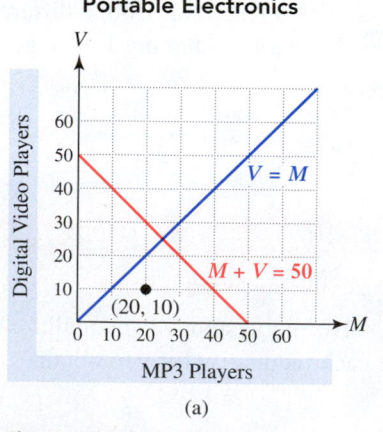

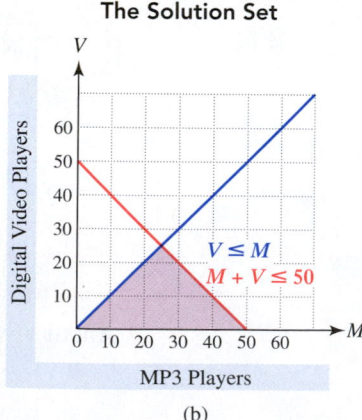

Figure 4.19

Now Try Exercise 73

The next example discusses the application from A Look into Math at the beginning of this section.

| EXAMPLE 6 | Finding weight–height combinations |

Figure 4.20 shows a shaded region containing recommended weights w for heights h.

Recommended Weight

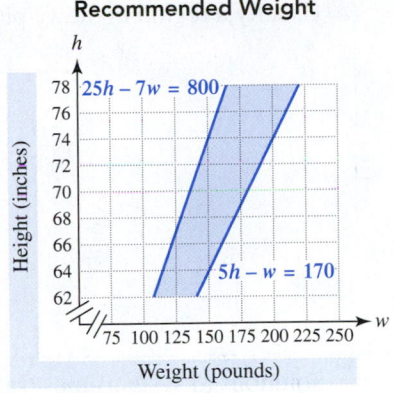

Figure 4.20

(a) What does this graph indicate about a 68-inch person who weighs 150 pounds?
(b) The shaded region in Figure 4.20 is determined by the following system of inequalities.

$$25h - 7w \leq 800$$
$$5h - w \geq 170$$

Verify that $h = 68$ and $w = 150$ satisfies the system of inequalities.
(c) What do ordered pairs (w, h) to the left of the shaded region indicate?

Solution

(a) The point $(150, 68)$ lies in the shaded region. Therefore someone who is 68 inches tall and weighs 150 pounds falls within the recommended guidelines.
(b) Both inequalities are satisfied by $h = 68$, $w = 150$.

$$25(68) - 7(150) = 650 \leq 800 \checkmark$$
$$5(68) - 150 = 190 \geq 170 \checkmark$$

(c) To the left of the shaded region are ordered pairs (w, h) that represent smaller weights and larger heights. These ordered pairs correspond to people who weigh less than recommended.

Now Try Exercise 77

4.4 Putting It All Together

CONCEPT	EXPLANATION	EXAMPLES
Linear Inequality in Two Variables	An inequality that can be written as $$Ax + By < C,$$ where $<$ can also be $\leq$, $>$, or $\geq$	$3x + y \geq 10$, $-x + 3y < 5$, $y \leq 5 - x$, and $x > 5$
Solution	A solution (x, y) makes the inequality a true statement.	The ordered pair $(0, 0)$ satisfies $$2x - y < 2,$$ so it is a solution to the inequality.

continued on next page

continued from previous page

CONCEPT	EXPLANATION	EXAMPLES
Solution Set	The set of all solutions Usually a region in the xy-plane	The solution set to $x + y > 2$ is all points above the line $x + y = 2$.
System of Linear Inequalities in Two Variables	Solutions to systems must satisfy both inequalities. The solution set usually includes infinitely many solutions.	The ordered pair $(0, 0)$ is a solution to $$x + y \leq 2$$ $$2x - y > -4,$$ because both inequalities are true when $x = 0$ and $y = 0$.

4.4 Exercises

MyMathLab Math XL PRACTICE WATCH DOWNLOAD READ REVIEW

CONCEPTS AND VOCABULARY

1. When the equals sign in $Ax + By = C$ is replaced with $<, >, \leq,$ or $\geq$, a linear _____ in two variables results.

2. A solution to a linear inequality in two variables is a(n) _____ that makes the inequality a true statement.

3. Describe the graph of the solution set to $y \leq k$ for some number k.

4. Describe the graph of the solution set to $x > k$ for some number k.

5. Describe the graph of the solution set to $y \geq x$.

6. When graphing the solution set to a linear inequality, one way to determine which region to shade is to use a(n) _____ point.

7. When graphing a linear inequality containing either $<$ or $>$, use a(n) _____ line.

8. When graphing a linear inequality containing either $\leq$ or $\geq$, use a(n) _____ line.

9. When graphing the linear inequality $Ax + By < C$, a first step is to graph the line _____.

10. A solution to a system of two inequalities must make (both inequalities/one inequality) true.

11. If two shaded regions represent the solution sets for two inequalities in a system, then the solution set for the system is where these two shaded regions _____.

12. If a test point is found that satisfies both inequalities in a system, do other test points still need to be checked?

SOLUTIONS TO LINEAR INEQUALITIES

Exercises 13–24: Determine whether the test point is a solution to the linear inequality.

13. $(3, 1)$, $x > 2$

14. $(-3, 4)$, $x \leq -3$

15. $(0, 0)$, $y \geq 2$

16. $(0, 0)$, $y < -3$

17. $(5, 4)$, $y \geq x$

18. $(-1, 2)$, $y < x$

19. $(3, 0)$, $y < x - 1$

20. $(0, 5)$, $y > 2x + 4$

21. $(-2, 6)$, $x + y \leq 4$

22. $(2, -4)$, $x - y \geq 7$

23. $(-1, -1)$, $2x + y \geq -1$

24. $(0, 1)$, $-x - 5y \geq -1$

Exercises 25–32: Write a linear inequality that describes the shaded region.

25.

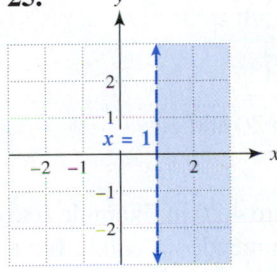

26.

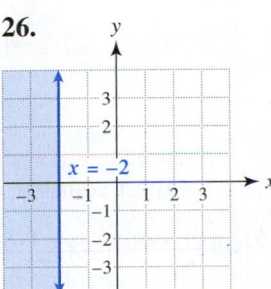

27.

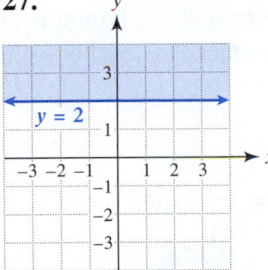

28.

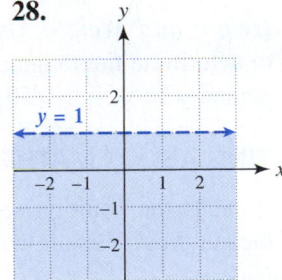

29.

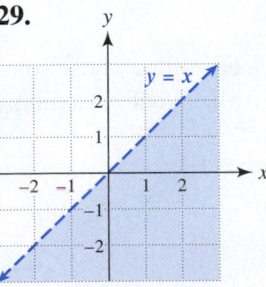

30.

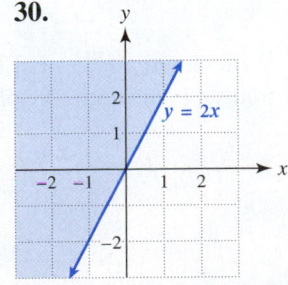

31.

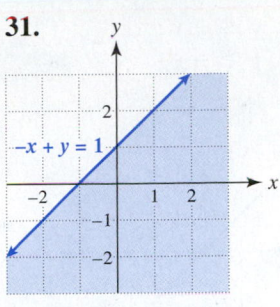

32.

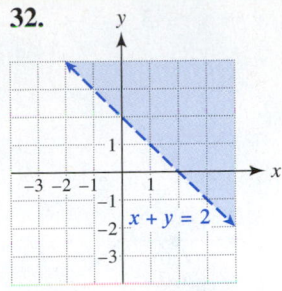

Exercises 33–44: Shade the solution set to the inequality.

33. $x \leq -1$

34. $x > 3$

35. $y < -2$

36. $y \geq 0$

37. $y > x$

38. $y \leq x$

39. $y \geq 3x$

40. $y < -2x$

41. $x + y \leq 1$

42. $x + y \geq -2$

43. $2x - y > 2$

44. $-x - y < 1$

Exercises 45–50: Determine if the test point is a solution to the system of linear inequalities.

45. $(3, 1)$
$$x - y < 3$$
$$x + y > 3$$

46. $(0, 0)$
$$x - 2y < 1$$
$$2x - y > -1$$

47. $(-2, 3)$
$$3x - 2y \geq 1$$
$$-x + 3y > 3$$

48. $(1, 2)$
$$2x - 2y < 5$$
$$x - y > -1$$

49. $(4, -2)$
$$x - 2y \geq 8$$
$$-2x - 5y > 0$$

50. $(-1, -2)$
$$x + y < 0$$
$$-2x - 3y \leq -1$$

Exercises 51–54: The graphs of two equations are shown with four test points labeled. Use these points to decide which region should be shaded to solve the given system of inequalities.

51. $\quad x \leq 2$
$$x + y \geq 2$$

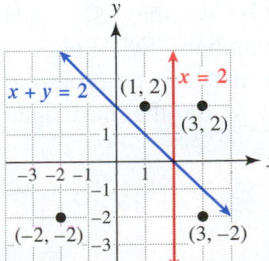

52. $\quad y \geq 1$
$$2x - y \geq 3$$

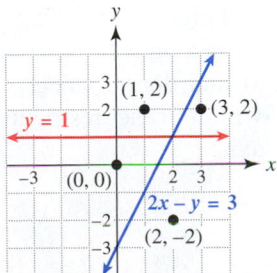

53. $x + y \leq 3$
$\quad\quad y \leq 2x$

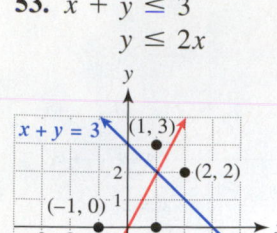

54. $y \leq \quad x$
$\quad\quad y \geq -x$

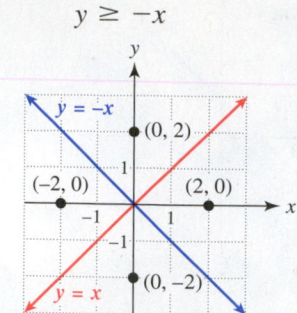

Exercises 55–72: Shade the solution set to the system of inequalities.

55. $x > 2$
$\quad\quad y < 3$

56. $x \leq -1$
$\quad\quad y \geq 3$

57. $x \leq -2$
$\quad\quad y < 2x$

58. $y > 2$
$\quad\quad y \geq -x$

59. $y \leq \quad x$
$\quad\quad y > -x$

60. $y \leq \quad \frac{1}{2}x$
$\quad\quad y \geq -2x$

61. $\quad x + y \leq 3$
$\quad\quad -x + y \leq 1$

62. $x + y > 2$
$\quad\quad x - y < 2$

63. $2x + y > -3$
$\quad\quad x + y \leq -1$

64. $-x + y \geq \quad 3$
$\quad\quad 2x - y \geq -2$

65. $2x + y \geq -3$
$\quad\quad x + y > -1$

66. $-x + y \geq \quad 2$
$\quad\quad 3x - y \geq -2$

67. $\quad\quad y > -2$
$\quad\quad x + 2y \leq -4$

68. $\quad x \geq 2$
$\quad\quad 3y < x - 3$

69. $\quad x + 2y > -4$
$\quad\quad 2x + \quad y \leq \quad 3$

70. $\quad x + 3y \geq 3$
$\quad\quad 3x - 2y \geq 6$

71. $3x + 4y \leq 12$
$\quad\quad 5x + 3y \geq 15$

72. $-2x + \quad y \geq \quad 6$
$\quad\quad\quad x - 2y \geq -4$

APPLICATIONS

73. *Manufacturing* (Refer to Example 5.) A business manufactures at least two MP3 players for each digital video player. The total number of MP3 players and digital video players must be less than 90. Shade the region that represents the number of MP3 players M and digital video players V that can be produced within these restrictions. Put V on the horizontal axis.

74. *Working on Two Projects* An employee is required to spend more time on project X than on project Y. The employee can work at most 40 hours on these two projects. Shade the region in the xy-plane that represents the number of hours that the employee can spend on each project.

75. *Maximum Heart Rate* When exercising, people often try to maintain target heart rates that are a percentage of their maximum heart rate. Maximum heart rate R is $R = 220 - A$, where A is the person's age and R is the heart rate in beats per minute.
   (a) Find R for a person 20 years old; 70 years old.
   (b) Sketch a graph of $R \leq 220 - A$. Assume that A is between 20 and 70 and put A on the horizontal axis of your graph.
   (c) Interpret your graph.

76. *Target Heart Rate* (Refer to the preceding exercise.) A target heart rate T that is half a person's maximum heart rate is given by $T = 110 - \frac{1}{2}A$, where A is a person's age.
   (a) What is T for a person 30 years old? 50 years old?
   (b) Sketch a graph of the system of inequalities.

$$T \geq 110 - \frac{1}{2}A$$

$$T \leq 220 - A$$

   Assume that A is between 20 and 60.
   (c) Interpret your graph.

77. *Height and Weight* Use Figure 4.20 in Example 6 to determine the range of recommended weights for a person who is 74 inches tall.

78. *Height and Weight* Use Figure 4.20 in Example 6 to determine the range of recommended heights for a person who weighs 150 pounds.

WRITING ABOUT MATHEMATICS

79. What is the solution set to the following system of inequalities? Explain your reasoning.

$$y > x$$

$$y < x - 1$$

80. Write down a system of linear inequalities whose solution set is the entire xy-plane. Explain your reasoning.

Checking Basic Concepts

1. Use elimination to solve the system of equations.

$$2x + 3y = 5$$
$$x - 7y = -6$$

2. Use elimination to solve each system of equations. How many solutions are there in each case?
   (a) $\quad x + y = -1$
   $\quad\quad\quad x - 2y = 2$
   (b) $\quad 5x - 6y = 4$
   $\quad\quad -5x + 6y = 1$
   (c) $\quad x - 3y = 0$
   $\quad\quad 2x - 6y = 0$

3. Solve the system of equations symbolically, graphically, and numerically. How many solutions are there?

$$-2x + y = 0$$
$$y = 2x$$

4. Shade the solution set to each inequality.
   (a) $y < -1$ **(b)** $x + y < 1$

5. Shade the solution set to the following system of inequalities.

$$x \le -1$$
$$-2x + y > -3$$

6. *Large U.S. Cities* The combined population of New York and Chicago was 11 million people in 2009. The population of New York exceeded the population of Chicago by 5 million people.
   (a) Let x be the population of New York and y be the population of Chicago. Write a system of equations whose solution gives the population of each city in 2009.
   (b) Solve the system of equations.

CHAPTER 4   Summary

SECTION 4.1 ■ SOLVING SYSTEMS OF LINEAR EQUATIONS GRAPHICALLY AND NUMERICALLY

System of Linear Equations

Solution	An ordered pair (x, y) that satisfies *both* equations
Solution Set	The set of all solutions
Graphical Solution	Graph each equation. A point of intersection is a solution. (Sometimes determining the exact answer when estimating from a graph may be difficult.)
Numerical Solution	Solve each equation for y and make a table for each equation. A solution occurs when two y-values are equal for a given x-value.

Example: The ordered pair (3, 1) is the solution to the following system.

$$x + y = 4 \quad\quad 3 + 1 = 4 \text{ is a true statement.}$$
$$x - y = 2 \quad\quad 3 - 1 = 2 \text{ is a true statement.}$$

A Graphical Solution
The point of intersection, (3, 1), is the solution to the system of equations.

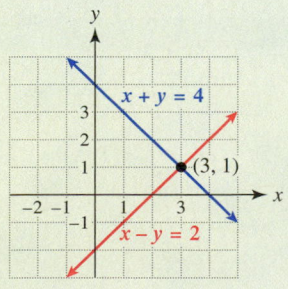

A Numerical Solution
The ordered pair (**3**, **1**) is the solution. When $x = 3$, both y-values equal 1.

x	1	2	**3**	3
$y = 4 - x$	3	2	**1**	0
$y = x - 2$	-1	0	**1**	2

Types of Systems of Linear Equations A system of linear equations can have no solutions, one solution, or infinitely many solutions.

No Solutions	One Solution	Infinitely Many Solutions
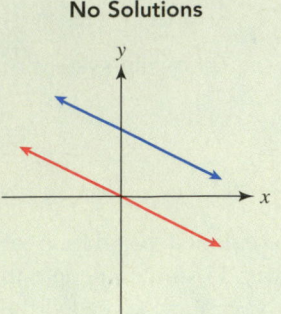	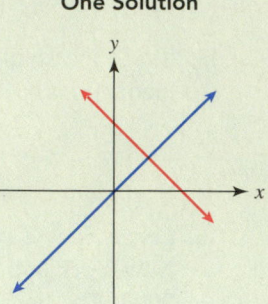	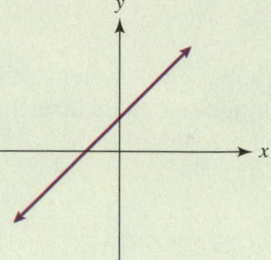
Inconsistent System	Consistent System Independent Equations	Consistent System Dependent Equations

SECTION 4.2 ■ SOLVING SYSTEMS OF LINEAR EQUATIONS BY SUBSTITUTION

Method of Substitution This method can be used to solve a system of equations symbolically and always gives the exact solution, provided one exists.

Example: $-2x + y = -3$
$x + y = 3$

STEP 1: Solve one of the equations for a convenient variable.

$$x + y = 3 \quad \text{becomes} \quad y = 3 - x.$$

STEP 2: Substitute this result in the other equation and then solve.

$$-2x + (3 - x) = -3 \quad \text{Substitute } (3 - x) \text{ for } y.$$
$$-3x = -6 \quad \text{Combine like terms; subtract 3.}$$
$$x = 2 \quad \text{Divide each side by } -3.$$

STEP 3: Find the value of the other variable. Because $y = 3 - x$ and $x = 2$, it follows that $y = 3 - 2 = 1$.

STEP 4: Check to determine that $(2, 1)$ is the solution.

$$-2(2) + (1) \stackrel{?}{=} -3 \;\checkmark \quad \text{A true statement}$$
$$2 + 1 \stackrel{?}{=} 3 \;\checkmark \quad \text{A true statement}$$

The solution $(2, 1)$ checks.

Recognizing Types of Systems

No solutions	The final equation is always false, such as $0 = 1$.
One solution	The final equation has one solution, such as $x = 1$.
Infinitely many solutions	The final equation is always true, such as $0 = 0$.

SECTION 4.3 ■ SOLVING SYSTEMS OF LINEAR EQUATIONS BY ELIMINATION

Method of Elimination This method can be used to solve a system of linear equations symbolically and always gives the exact solution, provided one exists.

Example: $x + 3y = 1$
$\underline{-x + y = 3}$
$4y = 4, \quad \text{or} \quad y = 1 \quad \text{Add and solve for } y.$

Substitute $y = 1$ in either of the given equations: $x + 3(1) = 1$ implies that $x = -2$, so $(-2, 1)$ is the solution.

NOTE: To eliminate a variable, it may be necessary to multiply one or both equations by a constant before adding.

Recognizing Types of Systems

No solutions	The final equation is always false, such as $0 = 1$.
One solution	The final equation has one solution, such as $x = 1$.
Infinitely many solutions	The final equation is always true, such as $0 = 0$.

SECTION 4.4 ■ SYSTEMS OF LINEAR INEQUALITIES

Graphing a Linear Inequality in Two Variables

1. Replace the inequality symbol with an equals sign and graph the resulting line. If the inequality is $<$ or $>$ use a dashed line, and if it is $\leq$ or $\geq$ use a solid line.
2. Pick a *test point* that does *not* lie on the line. Substitute this point in the given inequality. Determine whether the resulting statement is true or false.
3. If the statement is true, shade the region containing the test point. If the statement is false, shade the region that does not contain the test point.

Solving a System of Linear Inequalities in Two Variables

1. For each inequality in the system, perform Step 1 from Graphing a Linear Inequality in Two Variables.
2. Pick a test point from one region and substitute it in the given inequalities.
3. If both of the resulting statements are true, shade the region containing the test point. If not, pick a test point from a different region and substitute it in the given inequalities. Repeat this step until the region to be shaded is found.

Example:
$$x \leq 1$$
$$x + y \leq 2$$

Graph the lines $x = 1$ and $x + y = 2$. Then pick a test point, such as $(0, 0)$, and substitute it in each inequality.

$$0 \leq 1 \checkmark \quad \text{A true statement}$$
$$0 + 0 \leq 2 \checkmark \quad \text{A true statement}$$

Because $(0, 0)$ satisfies *both* inequalities, shade the region containing $(0, 0)$. See the graph.

NOTE: When shading the solution set to a *system* of inequalities, you may need to try more than one test point.

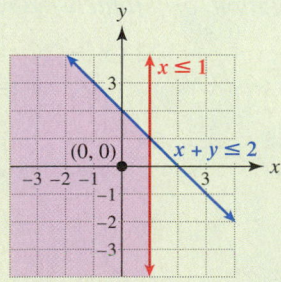

NOTE: An alternative way to determine the solution set is to shade the region that is both to the *left* of the line $x = 1$ (because $x \leq 1$) and *below* the line $y = -x + 2$ (because $y \leq -x + 2$).

CHAPTER 4 Review Exercises

SECTION 4.1

Exercises 1 and 2: Determine graphically the x-value for the equation when y = 3.

1. $y = 2x - 3$

2. $y = \frac{3}{2}x$

Exercises 3–6: The graphs of two equations are shown.

(a) State the number of solutions to the system of equations.

(b) Is the system consistent or inconsistent? If the system is consistent, state whether the equations are dependent or independent.

3.

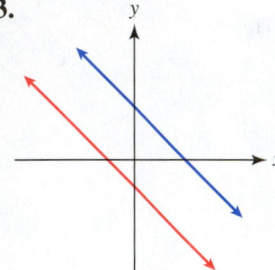

4.

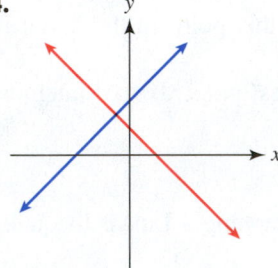

5.

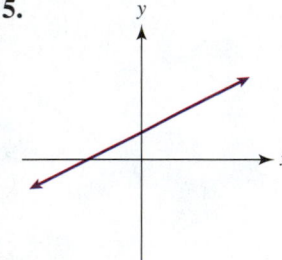

6.

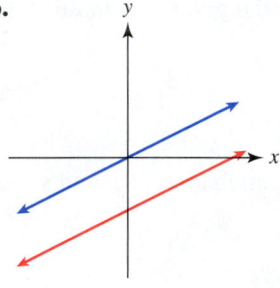

Exercises 7–10: Determine which ordered pair is a solution to the system of equations.

7. (0, 1), (1, 2)
$x + 2y = 5$
$x - y = -1$

8. (5, 2), (4, 0)
$2x - y = 8$
$x + 3y = 11$

9. (2, 2), (4, 3)
$\frac{1}{2}x = y - 1$
$2x = 3y - 1$

10. (2, −4), (−1, 2)
$5x - 2y = 18$
$y = -2x$

Exercises 11 and 12: The graphs for two equations are shown. Use the intersection-of-graphs method to identify the solution to both equations. Then check your result.

11.

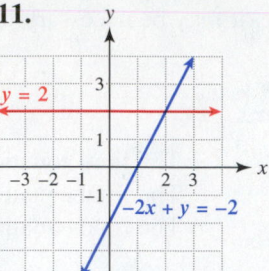

12.

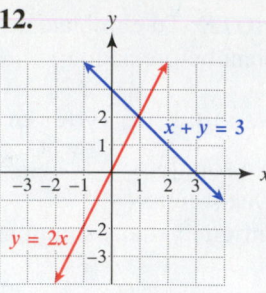

Exercises 13 and 14: A table for two equations is given. Identify the solution to both equations.

13.

x	1	2	3	4
$y = 3x$	3	6	9	12
$y = 6$	6	6	6	6

14.

x	−1	0	1	2
$y = 2x - 1$	−3	−1	1	3
$y = 2 - x$	3	2	1	0

Exercises 15–20: Solve the system of linear equations graphically.

15. $\phantom{x + {}}y = -3$
$x + y = 1$

16. $\phantom{x - {}}x = 1$
$x - y = -1$

17. $2x + y = 3$
$-x + y = 0$

18. $\phantom{2x + {}}y = 2x$
$2x + y = 4$

19. $x + 2y = 3$
$2x + y = 3$

20. $-3x - y = 7$
$2x + 3y = -7$

SECTION 4.2

Exercises 21–28: Use the method of substitution to solve the system of linear equations. These systems may have no solutions, one solution, or infinitely many solutions.

21. $x + y = 8$
$\phantom{x + {}}y = 3x$

22. $x - 2y = 22$
$\phantom{x - {}}y = -5x$

23. $x + 3y = 1$
$-2x + 2y = 6$

24. $3x - 2y = -4$
$2x - y = -4$

25. $x + y = 2$
$\phantom{x + {}}y = -x$

26. $x + y = -2$
$x + y = 3$

27. $-x + 2y = 2$
$x - 2y = -2$

28. $-x - y = -2$
$2x - y = 1$

SECTION 4.3

Exercises 29 and 30: Use the graph to solve the system of equations. Then use the elimination method to verify your answer.

29. $x + y = 3$
$x - y = 1$

30. $2x + 3y = 4$
$x - 2y = -5$

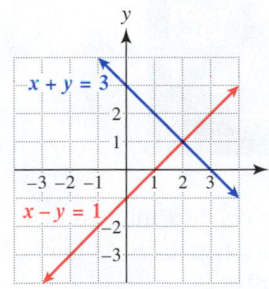

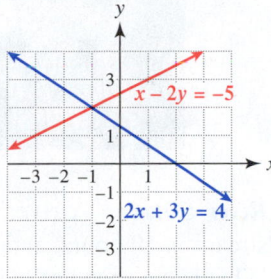

Exercises 31–38: Use the elimination method to solve the system of equations.

31. $x + y = 10$
$x - y = 12$

32. $2x - y = 2$
$3x + y = 3$

33. $-2x + 2y = -1$
$x - 3y = -3$

34. $2x - 5y = 0$
$2x + 4y = 9$

35. $2a + b = 3$
$-3a - 2b = -1$

36. $a - 3b = 2$
$3a + b = 26$

37. $5r + 3t = -1$
$-2r - 5t = -11$

38. $5r + 2t = 5$
$3r - 7t = 3$

Exercises 39 and 40: Solve the system of equations (a) symbolically, (b) graphically, and (c) numerically.

39. $3x + y = 6$
$x - y = -2$

40. $2x + y = 3$
$-x + 2y = -4$

Exercises 41–44: Use elimination to determine whether the system of equations has no solutions, one solution, or infinitely many solutions.

41. $x - y = 5$
$-x + y = -5$

42. $3x - 3y = 0$
$-x + y = 0$

43. $-2x + y = 3$
$2x - y = 3$

44. $-2x + y = 2$
$3x - y = 3$

SECTION 4.4

Exercises 45–48: Determine whether the test point is a solution to the linear inequality.

45. $(5, -3)$ $y \le 2$

46. $(-1, 3)$ $x > -1$

47. $(1, 2)$ $x + y < -2$

48. $(1, -4)$ $2x - 3y \ge 2$

Exercises 49 and 50: Write a linear inequality that describes the shaded region.

49.

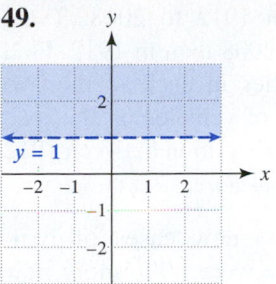

50.

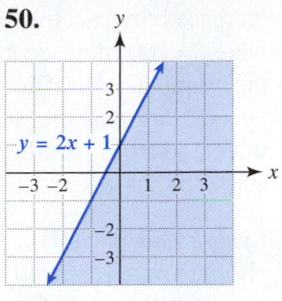

Exercises 51–56: Shade the solution set for the inequality.

51. $x \le 1$

52. $y > 2$

53. $y > 3x$

54. $x \ge 2y$

55. $y < x + 1$

56. $2x + y \ge -2$

Exercises 57 and 58: Determine whether the test point is a solution to the system of linear inequalities.

57. $(1, -2)$
$x - 2y > 3$
$2x + y < 3$

58. $(4, -3)$
$x - y \ge 1$
$4x + 3y \le 4$

Exercises 59 and 60: The graphs of two equations are shown with four test points labeled. Use these points to decide which region should be shaded to solve the system of inequalities.

59. $y \le 1$
$2x + y \ge -1$

60. $y \ge x$
$x + y \ge 2$

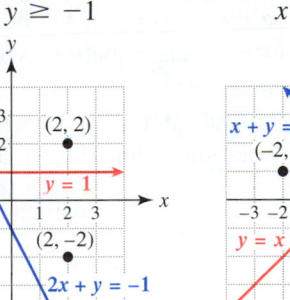

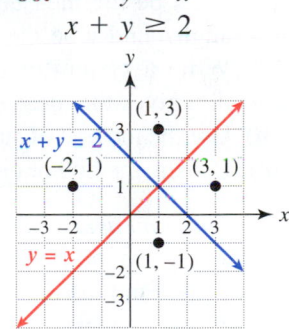

Exercises 61–66: Shade the solution set for the system of inequalities.

61. $x > -1$
$y < -2$

62. $y \le x$
$y \ge -2x$

63. $x + y \le 3$
$y \ge -x$

64. $2x + y < 3$
$y > x$

65. $\frac{1}{2}x + y \ge 2$
$x - 2y \le 0$

66. $2x - y < 3$
$4x + 2y > -6$

APPLICATIONS

67. *Traffic Fatalities* The number of traffic fatalities increased by 12 times from 1912 to 2008. There were 34,100 more deaths in 2008 than in 1912. Find the number of traffic fatalities in each of the two years. Note that the number of vehicles on the road increased from 1 million to 255 million between 1912 and 2008. (*Source:* Department of Health and Human Services.)

68. *Lymphoma* In 2010, 74,000 new cases of lymphoma were reported. There were 6000 more new cases for men than for women. How many new cases of lymphoma were there for men and for women? (*Source:* American Cancer Society.)

69. *Renting a Car* A rental car costs $40 plus $0.20 per mile that it is driven.
 (a) Write an equation that gives the cost C of driving the car x miles.
 (b) Use the intersection-of-graphs method to determine the number of miles that the car is driven if the rental cost is $90.
 (c) Solve part (b) numerically with a table of values.

70. *Supplementary Angles* The smaller of two supplementary angles is 30° less than the measure of the larger angle. Find each angle.

71. *Triangle* In an isosceles triangle, the measures of the two smaller angles are equal and their sum is 40° more than the larger angle.
 (a) Let x be the measure of each of the two smaller angles and y be the measure of the larger angle. Write a system of linear equations whose solution gives the measures of these angles.
 (b) Use the method of substitution to solve the system.
 (c) Use the method of elimination to solve the system.

72. *Garden Dimensions* A rectangular garden has 88 feet of fencing around it. The garden is 4 feet longer than it is wide. Find the dimensions of the garden.

73. *Room Prices* Ten rooms are rented at rates of $80 and $120 per night. The total collected for the 10 rooms is $920.
 (a) Write a system of linear equations whose solution gives the number of each type of room rented. Be sure to state what each variable represents.
 (b) Solve the system of equations.

74. *Mixture Problem* One type of candy sells for $2 per pound, and another type sells for $3 per pound. An order for 18 pounds of candy costs $47. How much of each type of candy was bought?

75. *Burning Calories* An athlete burns 9 calories per minute on a stationary bicycle and 11 calories per minute on a stair climber. In a 60-minute workout the athlete burns 590 calories. How many minutes does the athlete spend on each type of exercise equipment? (*Source:* Runner's World.)

76. *River Current* A riverboat travels 140 miles downstream in 10 hours, and the return trip takes 14 hours. What is the speed of the current?

77. *Garage Dimensions* The blue graph shown in the figure gives possible dimensions for a rectangular garage with perimeter 80 feet. The red graph shows possible dimensions for a garage that has width W two-thirds of its length L.

Garage Dimensions

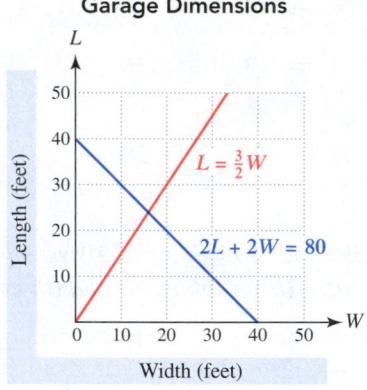

 (a) Use the graph to estimate the dimensions of a garage with perimeter 80 feet and width two-thirds its length.
 (b) Solve this problem symbolically.

78. *Wheels and Trailers* A business manufactures at least two wheels for each trailer it makes. The total number of trailers and wheels manufactured cannot exceed 30 per week. Shade the region that represents numbers of wheels W and trailers T that can be produced each week within these restrictions. Label the horizontal axis W and the vertical axis T.

79. *Target Heart Rate* A target heart rate T that is 70% of a person's maximum heart rate is approximated by $T = 150 - 0.7A$, where A is a person's age.
 (a) What is T for a person 20 years old? 60 years old?
 (b) Sketch a graph of $T \geq 150 - 0.7A$. Assume that A is between 20 and 60.
 (c) Interpret this graph.

CHAPTER 4   Test

1. Determine which ordered pair is a solution to the system of equations.

$$(3, -1), (1, 2)$$
$$3x + 2y = 7$$
$$2x - \;\; y = 0$$

2. The graphs for two equations are shown. Use the intersection-of-graphs method to identify the solution. Then check your solution.

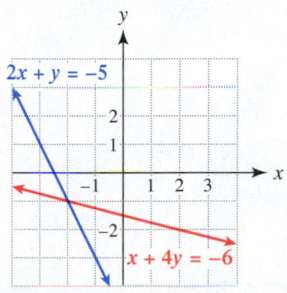

3. A table for two equations is given. Identify the solution to both equations.

x	-2	-1	0	1
$y = 2x$	-4	-2	0	2
$y = 3x + 1$	-5	-2	1	4

4. Solve the system of equations graphically.

$$x + 2y = 4$$
$$x + \;\; y = 1$$

5. Use the method of substitution to solve the system of linear equations.

$$3x + 2y = 9$$
$$y = 3x$$

6. Use the method of substitution to solve the system of linear equations. How many solutions are there? Is the system consistent or inconsistent?

(a) $x + 3y = 5$
     $3x - 2y = 4$

(b) $-x + \frac{1}{2}y = 12$
     $2x - \;\; y = -4$

Exercises 7 and 8: The graphs of two equations are shown.

(a) *State the number of solutions to the system of equations.*

(b) *Is the system consistent or inconsistent? If the system is consistent, state whether the equations are dependent or independent.*

7. 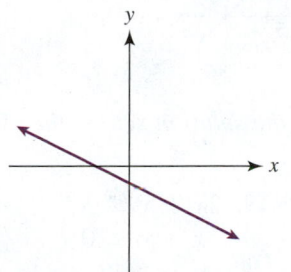   8.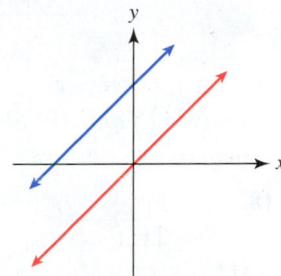

Exercises 9–12: Use the elimination method to solve the system of equations. Note that these systems may have no solutions, one solution, or infinitely many solutions.

9. $x + 2y = \;\;\; 5$
    $3x - 2y = -17$

10. $2x - 2y = 3$
     $-x + \;\; y = 5$

11. $x - 2y = \;\;\; 3$
     $-3x + 6y = -9$

12. $4x + 3y = \;\;\; 5$
    $3x - 2y = -9$

13. Write a linear inequality that describes the shaded region.

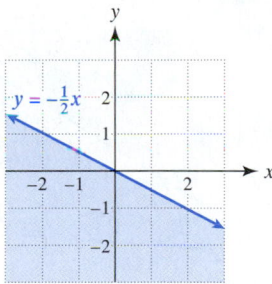

14. Determine whether the test point $(4, -3)$ is a solution to the system of linear inequalities.

$$2x + y > 3$$
$$x - y \geq 7$$

Exercises 15 and 16: Shade the solution set for the given inequality.

15. $x \leq 4$

16. $x + y > 2$

17. Determine which region —1, 2, 3, or 4— should be shaded to solve the system of inequalities.

$$x + y \le 5$$
$$x - y \ge 1$$

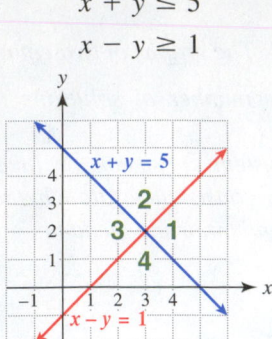

Exercises 18 and 19: Shade the solution set for the given system of inequalities.

18. $x > 2$
    $y < 2x$

19. $2x + y \le 3$
    $x - y \ge 0$

20. *IRS Collections* In 2009 and 2010, the IRS collected a total of $8 trillion in taxes. The IRS collected $0.6 trillion more in 2010 than in 2009. How much did the IRS collect in each year? (*Source*: Internal Revenue Service.)

21. *Mixture Problem* A chemist has 20% and 60% solutions of acid available. How many liters of each solution should be mixed to obtain 10 liters of a 30% acid solution?

22. *Jogging Speed* An athlete jogs at 6 miles per hour and at 9 miles per hour for a total time of 1 hour, covering a distance of 7 miles. How long does the athlete jog at each speed?

CHAPTER 4 Extended and Discovery Exercises

Exercises 1–4: ***Plant Growth*** *The following figure illustrates the relationships among forests, grasslands, and deserts, suggested by annual temperature T in degrees Fahrenheit and precipitation P in inches.*

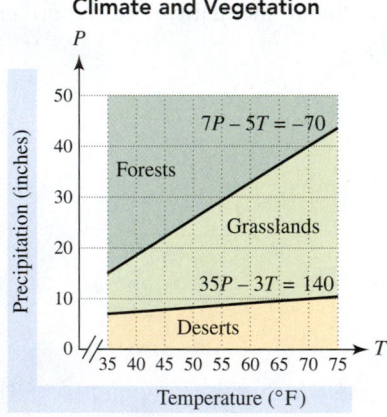

1. The equation of the line that separates grasslands and forests is

$$7P - 5T = -70.$$

Write an inequality that describes temperatures and amounts of precipitation that correspond to forested regions. (Include the line.)

2. The equation of the line that separates grasslands and deserts is

$$35P - 3T = 140.$$

Write an inequality that describes temperatures and amounts of precipitation that correspond to desert regions. (Include the line.)

3. Using the information from Exercises 1 and 2, write a system of inequalities that describes temperatures and amounts of precipitation that correspond to grassland regions.

4. Cheyenne, Wyoming, has an average annual temperature of about 50° F and an average annual precipitation of about 14 inches. Use the graph to predict the type of plant growth you might expect near Cheyenne. Then check to determine whether $T = 50$ and $P = 14$ satisfy the proper inequalities.

Exercises 5–8: ***Numerical Solutions*** *If a solution to a linear equation does not appear in a table of values, can you still find it? Use only the table to solve the equation. Then explain how you got your answer.*

5. $2x + 1 = 0$

x	-2	-1	0	1	2
$y = 2x + 1$	-3	-1	1	3	5

6. $4x + 3 = 5$

x	-2	-1	0	1	2
$y = 4x + 3$	-5	-1	3	7	11

7. $\frac{1}{2}x + 3 = 3.75$

x	-2	-1	0	1	2
$y = \frac{1}{2}x + 3$	2	2.5	3	3.5	4

8. $3x - 1 = 0$

x	-2	-1	0	1	2
$y = 3x - 1$	-7	-4	-1	2	5

CHAPTERS 1–4   Cumulative Review Exercises

1. Write 120 as a product of prime numbers.

2. Evaluate each expression by hand.
   (a) $2^3 \div \frac{5 + 7}{9 - 3}$ **(b)** $-\frac{2}{5} \cdot (5 - 25)$

3. Classify each number as rational or irrational.
   (a) -6.9 **(b)** $\sqrt{14}$

4. Insert $>$ or $<$ to make each statement true.
   (a) -5 _____ $|-5|$ **(b)** $|7|$ _____ $|-1|$

5. Simplify each expression.
   (a) $5x^2 - x^2$ **(b)** $3 - 2x + 7x - 5$

Exercises 6 and 7: Solve the equation.

6. $5(2x + 1) = 7 + x$ **7.** $1 - (x + 1) = x - 1$

8. Determine whether the equation $2(5x + 1) = 10x - 3$ has no solutions, one solution, or infinitely many solutions.

9. Find four consecutive integers whose sum is 50.

10. Find the area of a rectangle having a 36-inch length and a 1-foot width.

11. Solve the formula $W = 3x - 7y$ for x.

12. Solve the inequality $3 - (2x - 7) \le 8x$.

13. Use the table of xy-values to make a line graph.

x	-2	-1	0	1	2
y	-2	2	0	3	-1

14. Graph each equation.
   (a) $y = -2x + 2$ **(b)** $3y + 2x = 6$

15. Find the x- and y-intercepts for the graph of $4x - y = 8$.

16. Sketch a line with the given slope that passes through the given point.
   (a) $m = -2, (1, 1)$ **(b)** $m = \frac{4}{3}, (-3, 2)$

17. Write the slope–intercept form for each line.
   (a) 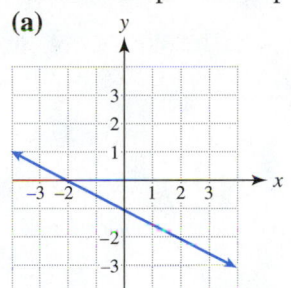   **(b)**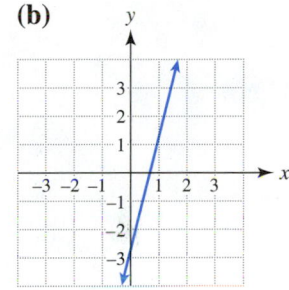

18. Find the slope–intercept form of the line passing through $(4, 2)$ and $(-4, 4)$.

19. Write the slope–intercept form of the line perpendicular to $2x - 6y = 7$, passing through $(1, 1)$.

20. Determine whether $(3, 1)$ or $(4, 4)$ is a solution to the system of equations.

$$3x - y = 8$$
$$2x + y = 12$$

Exercises 21–24: Solve the system of equations. Note that these systems may have no solutions, one solution, or infinitely many solutions.

21. $x - y = 4$
     $-2x - y = 1$

22. $2x + 3y = 4$
     $-4x - 6y = 7$

23. $3x - 4y = 8$
     $-15x + 20y = -40$

24. $7x + 2y = -3$
     $-5x - 3y = -1$

Exercises 25 and 26: Shade the solution set for the system of inequalities.

25. $y < x + 1$
     $x + y \le 3$

26. $3x + y \ge 6$
     $x - 3y \le 3$

27. *Rods to Feet* There are 16.5 feet in 1 rod. Write a formula that converts R rods to F feet.

28. *Temperature Change* Find the temperature change for a package of frozen carrots that is removed from a freezer at $-11°C$ and placed in water at $83°C$.

29. *Cost of a Digital Camera* A 7% sales tax on a digital camera amounts to $17.15. Find the cost of the digital camera.

30. *Tuition Increase* If tuition is currently $145 per credit and it is going to be increased by 9%, what will the new tuition per credit be?

31. *Gasoline Consumption* Write an equation in slope–intercept form that models the number of gallons G of gas in a truck's tank after x hours if the tank initially contains 30 gallons of gas and the truck uses 3 gallons every hour.

32. *Bank Loans* An individual has two loans totaling $2400. One loan charges 5% interest and the other charges 6% interest. If the interest for one year is $132, how much money is borrowed at each rate?

5

Polynomials and Exponents

If you want to do something, do it!

—PLAUTUS

Digital images were first sent between New York and London by cable in the early 1920s. Unfortunately, the transmission time was 3 hours and the quality was poor. Digital photography was developed further by NASA in the 1960s because ordinary pictures were subject to interference when transmitted through space. Today, digital pictures remain crystal clear even if they travel millions of miles. The following digital picture shows the planet Mars.

Source: NASA.

Whether they are taken with a webcam, with a smartphone, or by the Mars rover, digital images comprise tiny units called pixels, which are represented by numbers. As a result, mathematics plays an important role in digital images. In this chapter we illustrate some of the ways mathematics is used to describe digital pictures (see Example 4 and Exercise 80 in Section 5.4). We also discuss how mathematics is used to model things such as heart rate, computer sales, motion of the planets, and interest on money.

5.1 Rules for Exponents

Review of Bases and Exponents • Zero Exponents • The Product Rule • Power Rules

A LOOK INTO MATH ▶

Electronic devices such as tablet computers and smartphones store information as *bits*. A bit is either a 0 or a 1, and a string of 8 bits is called a *byte*. In the 1970s, IBM developed punch cards made out of paper that could hold up to 120 bits of information. Today, many computer hard drives can hold more than 1 terabyte of information; that's more than 8,000,000,000,000 bits! In mathematics, we often use *exponents* to express such large numbers. In this section, we discuss the rules for exponents.

Review of Bases and Exponents

The expression 5^3 is an exponential expression with *base* **5** and *exponent* **3**. Its value is

$$5 \cdot 5 \cdot 5 = 125.$$

In general, b^n is an exponential expression with base b and exponent n. If n is a natural number, it indicates the number of times the base b is to be multiplied with itself.

$$\underset{\text{Base}}{\overset{\text{Exponent}}{b^n}} = \underbrace{b \cdot b \cdot b \cdot \cdots \cdot b}_{n \text{ times}}$$

When evaluating expressions, evaluate exponents *before* performing addition, subtraction, multiplication, division, or negation.

STUDY TIP

Exponents occur throughout mathematics. Because exponents are so important, this section is essential for your success in mathematics. It takes practice, so set aside some extra time.

EVALUATING EXPRESSIONS

When evaluating expressions, use the following order of operations.

1. Evaluate exponents.
2. Perform negation.
3. Do multiplication and division from left to right.
4. Do addition and subtraction from left to right.

EXAMPLE 1 **Evaluating exponential expressions**

Evaluate each expression.

(a) $1 + \dfrac{2^4}{4}$ **(b)** $3\left(\dfrac{1}{3}\right)^2$ **(c)** -2^4 **(d)** $(-2)^4$

Solution
(a) Evaluate the exponent first.

$$1 + \frac{2^4}{4} = 1 + \frac{\overset{4 \text{ factors}}{\overbrace{2 \cdot 2 \cdot 2 \cdot 2}}}{4} = 1 + \frac{16}{4} = 1 + 4 = 5$$

(b) $3\left(\dfrac{1}{3}\right)^2 = 3\left(\underset{2 \text{ factors}}{\underbrace{\dfrac{1}{3} \cdot \dfrac{1}{3}}}\right) = 3 \cdot \dfrac{1}{9} = \dfrac{3}{9} = \dfrac{1}{3}$

(c) Because exponents are evaluated before negation is performed,

$$-2^4 = \overbrace{-(2 \cdot 2 \cdot 2 \cdot 2)}^{\text{4 factors}} = -16.$$

(d) $(-2)^4 = \overbrace{(-2)(-2)(-2)(-2)}^{\text{4 factors}} = 16$

Now Try Exercises 9, 11, 15, 17

NOTE: Parts (c) and (d) of Example 1 appear to be very similar. However, the negation sign is inside the parentheses in part (d), which means that the base for the exponential expression is -2. In part (c), no parentheses are used, indicating that the base of the exponential expression is 2.

READING CHECK

• Explain how to tell the difference between a negative number raised to a power and the opposite of a positive number raised to a power.

TECHNOLOGY NOTE

Evaluating Exponents
Exponents can often be evaluated on calculators by using the ^ key. The four expressions from Example 1 are evaluated with a calculator and the results are shown in the following two figures. When evaluating the last two expressions on your calculator, remember to use the negation key rather than the subtraction key.

CALCULATOR HELP

To evaluate exponents, see Appendix A (page AP-1).

```
1+2^4/4
               5
3(1/3)²▶Frac
             1/3
```

```
-2^4
             -16
(-2)^4
              16
```

Zero Exponents

So far we have discussed natural number exponents. What if an exponent is 0? What does 2^0 equal? To answer these questions, consider Table 5.1, which shows values for decreasing powers of 2. Note that each time the power of 2 decreases by 1, the resulting value is divided by 2. For this pattern to continue, we need to define 2^0 to be 1 because dividing 2 by 2 results in 1.

This discussion suggests that $2^0 = 1$, and is generalized as follows.

TABLE 5.1 Powers of 2

Power of 2	Value
2^3	8
2^2	4
2^1	2
2^0	?

ZERO EXPONENT

For any nonzero real number b,

$$b^0 = 1.$$

The expression 0^0 is undefined.

EXAMPLE 2 **Evaluating zero exponents**

Evaluate each expression. Assume that all variables represent nonzero numbers.

(a) 7^0 (b) $3\left(\frac{4}{9}\right)^0$ (c) $\left(\frac{x^2y^5}{3z}\right)^0$

Solution
(a) $7^0 = 1$
(b) $3\left(\frac{4}{9}\right)^0 = 3(1) = 3$. (Note that the exponent 0 does not apply to 3.)
(c) All variables are nonzero, so the expression inside the parentheses is also nonzero.
Thus $\left(\frac{x^2y^5}{3z}\right)^0 = 1$.

Now Try Exercises 13, 41, 67

The Product Rule

We can use a special rule to calculate products of exponential expressions *provided their bases are the same*. For example,

$$4^3 \cdot 4^2 = \underbrace{(4 \cdot 4 \cdot 4) \cdot (4 \cdot 4)}_{\text{5 factors}} = 4^5.$$

The expression $4^3 \cdot 4^2$ has $3 + 2 = 5$ factors of 4, so the result is $4^{3+2} = 4^5$. To multiply exponential expressions with the *same base*, add exponents and keep the base.

READING CHECK

• State the product rule in your own words.

THE PRODUCT RULE

For any real number a and natural numbers m and n,

$$a^m \cdot a^n = a^{m+n}.$$

NOTE: The product $2^4 \cdot 3^5$ cannot be simplified by using the product rule because the exponential expressions have different bases: **2** and **3**.

EXAMPLE 3 **Using the product rule**

Multiply and simplify.
(a) $2^3 \cdot 2^2$ (b) x^4x^5 (c) $2x^2 \cdot 5x^6$ (d) $x^3(2x + 3x^2)$

Solution
(a) $2^3 \cdot 2^2 = 2^{3+2} = 2^5 = 32$
(b) $x^4x^5 = x^{4+5} = x^9$
(c) Begin by applying the commutative property of multiplication to write the product in a more convenient order.

$$2x^2 \cdot 5x^6 = 2 \cdot 5 \cdot x^2 \cdot x^6 = 10x^{2+6} = 10x^8$$

(d) To simplify this expression, first apply the distributive property.

$$x^3(2x + 3x^2) = x^3 \cdot 2x + x^3 \cdot 3x^2 = 2x^4 + 3x^5$$

$$\overset{3+1}{} \qquad \overset{3+2}{}$$

Exponent is 1.

Now Try Exercises 21, 23, 27, 71

NOTE: If an exponent does not appear in an expression, it is assumed to be 1. For example, x can be written as x^1 and $(x + y)$ can be written as $(x + y)^1$.

EXAMPLE 4   **Applying the product rule**

Multiply and simplify.
(a) $x \cdot x^3$ **(b)** $(a + b)(a + b)^4$

Solution
(a) Begin by writing x as x^1. Then $x^1 \cdot x^3 = x^{1+3} = x^4$.
(b) First write $(a + b)$ as $(a + b)^1$. Then

$$(a + b)^1 \cdot (a + b)^4 = (a + b)^{1+4} = (a + b)^5.$$

Now Try Exercises 19, 63

Power Rules

How should $(4^3)^2$ be evaluated? To answer this question, consider how the product rule can be used in evaluating

$$(4^3)^2 = \underbrace{4^3 \cdot 4^3}_{\text{Product rule}} = 4^{\overbrace{3+3}^{3+3=3\cdot2}} = 4^6.$$

Similarly,

$$(a^5)^3 = \underbrace{a^5 \cdot a^5 \cdot a^5}_{\text{Product rule}} = a^{\overbrace{5+5+5}^{5+5+5=5\cdot3}} = a^{15}.$$

This discussion suggests that to raise a power to a power, we multiply the exponents.

> **RAISING A POWER TO A POWER**
>
> For any real number a and natural numbers m and n,
>
> $$(a^m)^n = a^{mn}.$$

EXAMPLE 5   **Raising a power to a power**

Simplify each expression.
(a) $(3^2)^4$ **(b)** $(a^3)^2$

Solution
(a) $(3^2)^4 = 3^{2\cdot4} = 3^8$ **(b)** $(a^3)^2 = a^{3\cdot2} = a^6$

Now Try Exercises 31, 33

To decide how to simplify the expression $(2x)^3$, consider

$$(2x)^3 = \underbrace{2x \cdot 2x \cdot 2x}_{\text{3 factors}} = \underbrace{(2 \cdot 2 \cdot 2)}_{\text{3 factors}} \cdot \underbrace{(x \cdot x \cdot x)}_{\text{3 factors}} = 2^3 x^3.$$

To raise a product to a power, we raise each factor to the power.

- State the rule for raising a product to a power in your own words.

RAISING A PRODUCT TO A POWER

For any real numbers a and b and natural number n,

$$(ab)^n = a^n b^n.$$

EXAMPLE 6 Raising a product to a power

Simplify each expression.
(a) $(3z)^2$ **(b)** $(-2x^2)^3$ **(c)** $4(x^2y^3)^5$ **(d)** $(-2^2a^5)^3$

Solution
(a) $(3z)^2 = 3^2z^2 = 9z^2$
(b) $(-2x^2)^3 = (-2)^3(x^2)^3 = -8x^6$
(c) $4(x^2y^3)^5 = 4(x^2)^5(y^3)^5 = 4x^{10}y^{15}$
(d) $(-2^2a^5)^3 = (-4a^5)^3 = (-4)^3(a^5)^3 = -64a^{15}$

Now Try Exercises 37, 39, 43, 45

The following equation illustrates another power rule.

$$\left(\frac{2}{3}\right)^4 = \underbrace{\frac{2}{3} \cdot \frac{2}{3} \cdot \frac{2}{3} \cdot \frac{2}{3}}_{4 \text{ factors}} = \frac{2 \cdot 2 \cdot 2 \cdot 2}{3 \cdot 3 \cdot 3 \cdot 3} = \frac{2^4}{3^4}$$

To raise a quotient to a power, raise both the numerator and the denominator to the power.

RAISING A QUOTIENT TO A POWER

For any real numbers a and b and natural number n,

$$\left(\frac{a}{b}\right)^n = \frac{a^n}{b^n}, \quad b \neq 0$$

EXAMPLE 7 Raising a quotient to a power

Simplify each expression.

(a) $\left(\dfrac{2}{3}\right)^3$ **(b)** $\left(\dfrac{a}{b}\right)^9$ **(c)** $\left(\dfrac{a+b}{5}\right)^2$

Solution

(a) $\left(\dfrac{2}{3}\right)^3 = \dfrac{2^3}{3^3} = \dfrac{8}{27}$ **(b)** $\left(\dfrac{a}{b}\right)^9 = \dfrac{a^9}{b^9}$

(c) Because the numerator is an expression with more than one term, we must place parentheses around it before raising it to the power 2.

$$\left(\frac{a+b}{5}\right)^2 = \frac{(a+b)^2}{5^2} = \frac{(a+b)^2}{25}$$

Now Try Exercises 51, 53, 55

Raising a Sum or Difference to a Power

Although there are power rules for products and quotients, there are not similar rules for sums and differences. In general, $(a + b)^n \neq a^n + b^n$ and $(a - b)^n \neq a^n - b^n$. For example, $(3 + 4)^2 = 7^2 = 49$ but $3^2 + 4^2 = 9 + 16 = 25$. Similarly, $(4 - 1)^3 = 3^3 = 27$ but $4^3 - 1^3 = 64 - 1 = 63$.

The five rules for exponents discussed in this section are summarized as follows.

RULES FOR EXPONENTS

The following rules hold for real numbers a and b, and natural numbers m and n.

Description	Rule	Example
Zero Exponent	$b^0 = 1$, for $b \neq 0$	$(-13)^0 = 1$
The Product Rule	$a^m \cdot a^n = a^{m+n}$	$5^4 \cdot 5^3 = 5^{4+3} = 5^7$
Power to a Power	$(a^m)^n = a^{m \cdot n}$	$(y^2)^5 = y^{2 \cdot 5} = y^{10}$
Product to a Power	$(ab)^n = a^n b^n$	$(pq)^7 = p^7 q^7$
Quotient to a Power	$\left(\dfrac{a}{b}\right)^n = \dfrac{a^n}{b^n}$, for $b \neq 0$	$\left(\dfrac{x}{y}\right)^3 = \dfrac{x^3}{y^3}$, for $y \neq 0$

Simplification of some expressions may require the application of more than one rule of exponents. This is demonstrated in the next example.

EXAMPLE 8   **Combining rules for exponents**

Simplify each expression.

(a) $(2a)^2(3a)^3$ **(b)** $\left(\dfrac{a^2 b^3}{c}\right)^4$ **(c)** $(2x^3 y)^2(-4x^2 y^3)^3$

Solution

(a) $(2a)^2(3a)^3 = 2^2 a^2 \cdot 3^3 a^3$ Raising a product to a power

$\qquad\qquad\quad = 4 \cdot 27 \cdot a^2 \cdot a^3$ Evaluate powers; commutative property

$\qquad\qquad\quad = 108a^5$ Product rule

(b) $\left(\dfrac{a^2 b^3}{c}\right)^4 = \dfrac{(a^2)^4 (b^3)^4}{c^4}$ Raising a quotient to a power; raising a product to a power

$\qquad\qquad\quad = \dfrac{a^8 b^{12}}{c^4}$ Raising a power to a power

(c) $(2x^3 y)^2(-4x^2 y^3)^3 = 2^2 (x^3)^2 y^2 (-4)^3 (x^2)^3 (y^3)^3$ Raising a product to a power

$\qquad\qquad\qquad\qquad = 4x^6 y^2 (-64) x^6 y^9$ Raising a power to a power

$\qquad\qquad\qquad\qquad = 4(-64) x^6 x^6 y^2 y^9$ Commutative property

$\qquad\qquad\qquad\qquad = -256 x^{12} y^{11}$ Product rule

Now Try Exercises 47, 49, 61

▶ **REAL-WORLD CONNECTION** Exponents occur frequently in calculations involving yearly percent increases, such as the increase in property value illustrated in the next example.

EXAMPLE 9 | **Calculating growth in property value**

If a parcel of property increases in value by about 11% each year for 20 years, then its value will double three times.
(a) Write an exponential expression that represents "doubling three times."
(b) If the property is initially worth $25,000, how much will it be worth after it doubles 3 times?

Solution
(a) Doubling three times is represented by 2^3.
(b) $2^3(25,000) = 8(25,000) = \$200,000$

Now Try Exercise 85

5.1   Putting It All Together

CONCEPT	EXPLANATION	EXAMPLES
Bases and Exponents	In the expression b^n, b is the base and n is the exponent. If n is a natural number, then $$b^n = \underbrace{b \cdot b \cdot \cdots \cdot b}_{n \text{ times}}.$$	2^3 has base 2 and exponent 3. $9^1 = 9$, $3^2 = 3 \cdot 3 = 9$, $4^3 = 4 \cdot 4 \cdot 4 = 64$, and $-6^2 = -(6 \cdot 6) = -36$
Zero Exponents	For any nonzero number b, $b^0 = 1$.	$5^0 = 1$, $x^0 = 1$, and $(xy^3)^0 = 1$
The Product Rule	For any real number a and natural numbers m and n, $$a^m \cdot a^n = a^{m+n}.$$	$2^4 \cdot 2^3 = 2^{4+3} = 2^7$, $x \cdot x^2 \cdot x^6 = x^{1+2+6} = x^9$, and $(x + 1) \cdot (x + 1)^2 = (x + 1)^3$
Raising a Power to a Power	For any real number a and natural numbers m and n, $$(a^m)^n = a^{mn}.$$	$(2^4)^2 = 2^{4 \cdot 2} = 2^8$, $(x^2)^5 = x^{2 \cdot 5} = x^{10}$, and $(a^4)^3 = a^{4 \cdot 3} = a^{12}$
Raising a Product to a Power	For any real numbers a and b and natural number n, $$(ab)^n = a^n b^n.$$	$(3x)^3 = 3^3 x^3 = 27x^3$, $(x^2 y)^4 = (x^2)^4 y^4 = x^8 y^4$, and $(-xy)^6 = (-x)^6 y^6 = x^6 y^6$
Raising a Quotient to a Power	For any real numbers a and b and natural number n, $$\left(\frac{a}{b}\right)^n = \frac{a^n}{b^n}. \quad b \neq 0$$	$\left(\dfrac{x}{y}\right)^5 = \dfrac{x^5}{y^5}$ and $\left(\dfrac{a^2 b}{d^3}\right)^4 = \dfrac{(a^2)^4 b^4}{(d^3)^4} = \dfrac{a^8 b^4}{d^{12}}$

5.1 Exercises

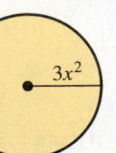

CONCEPTS AND VOCABULARY

1. In the expression b^n, b is the _____ and n is the _____.

2. The expression $b^0 =$ _____ for any nonzero number b.

3. $a^m \cdot a^n =$ _____

4. $(a^m)^n =$ _____

5. $(ab)^n =$ _____

6. $\left(\dfrac{a}{b}\right)^n =$ _____

PROPERTIES OF EXPONENTS

Exercises 7–18: Evaluate the expression.

7. 8^2

8. 4^3

9. $(-2)^3$

10. $(-3)^4$

11. -2^3

12. -3^4

13. 6^0

14. $(-0.5)^0$

15. $3 + \dfrac{4^2}{2}$

16. $6 - \left(\dfrac{-4}{2}\right)^2$

17. $4\left(\dfrac{1}{2}\right)^3$

18. $16\left(\dfrac{1}{4}\right)^2$

Exercises 19–74: Simplify the expression. Assume that all variables represent nonzero numbers.

19. $3 \cdot 3^2$

20. $5^3 \cdot 5^3$

21. $4^2 \cdot 4^6$

22. $10^4 \cdot 10^3$

23. $x^3 x^6$

24. $a^5 a^2$

25. $x^2 x^2 x^2$

26. $y^7 y^3 y^0$

27. $4x^2 \cdot 5x^5$

28. $-2y^6 \cdot 5y^2$

29. $3(-xy^3)(x^2y)$

30. $(a^2b^3)(-ab^2)$

31. $(2^3)^2$

32. $(10^3)^4$

33. $(n^3)^4$

34. $(z^7)^3$

35. $x(x^3)^2$

36. $(z^3)^2(5z^5)$

37. $(-7b)^2$

38. $(-4z)^3$

39. $(ab)^3$

40. $(xy)^8$

41. $(2x^2)^0$

42. $(3a^2)^4$

43. $(-4b^2)^3$

44. $(-3r^4t^3)^2$

45. $(x^2y^3)^7$

46. $(rt^2)^5$

47. $(y^3)^2(x^4y)^3$

48. $(ab^3)^2(ab)^3$

49. $(a^2b)^2(a^2b^2)^3$

50. $(x^3y)(x^2y^4)^2$

51. $\left(\dfrac{1}{3}\right)^3$

52. $\left(\dfrac{5}{2}\right)^2$

53. $\left(\dfrac{a}{b}\right)^5$

54. $\left(\dfrac{x}{2}\right)^4$

55. $\left(\dfrac{x-y}{3}\right)^3$

56. $\left(\dfrac{4}{x+y}\right)^2$

57. $\left(\dfrac{5}{a+b}\right)^2$

58. $\left(\dfrac{a-b}{2}\right)^3$

59. $\left(\dfrac{2x}{5}\right)^3$

60. $\left(\dfrac{3y}{2}\right)^4$

61. $\left(\dfrac{3x^2}{5y^4}\right)^3$

62. $\left(\dfrac{a^2b^3}{3}\right)^5$

63. $(x+y)(x+y)^3$

64. $(a-b)^2(a-b)$

65. $(a+b)^2(a+b)^3$

66. $(x-y)^5(x-y)^4$

67. $6(x^4y^6)^0$

68. $\left(\dfrac{xy}{z^2}\right)^0$

69. $a(a^2 + 2b^2)$

70. $x^3(3x - 5y^4)$

71. $3a^3(4a^2 + 2b)$

72. $2x^2(5 - 4y^3)$

73. $(r + t)(rt)$

74. $(x - y)(x^2y^3)$

75. **Thinking Generally** Students sometimes mistakenly apply the "rule" $a^m \cdot b^n \overset{?}{=} (ab)^{m+n}$. In general, this equation is *not true*. Find values for a, b, m, and n with $a \neq b$ and $m \neq n$ that will make this equation true.

76. **Thinking Generally** Students sometimes mistakenly apply the "rule" $(a + b)^n \overset{?}{=} a^n + b^n$. In general, this equation is *not true*. Find values for a, b, and n with $a \neq b$ that will make this equation true.

APPLICATIONS

Exercises 77–80: Write a simplified expression for the area of the given figure.

77.
$2x^2$
$5x^2$

78.
$2ab$
$2ab$

79.
$3x^2$

80.
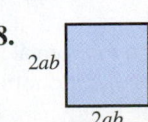
$4y^2$
$7y^3$

Exercises 81 and 82: Write a simplified expression for the volume of the given figure.

81.

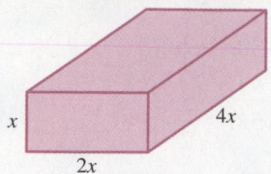

82.

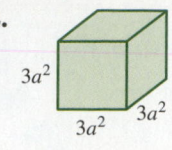

83. *Compound Interest* If P dollars are deposited in an account that pays 5% annual interest, then the amount of money in the account after 3 years is $P(1 + 0.05)^3$. Find the amount when $P = \$1000$.

84. *Compound Interest* If P dollars are deposited in an account that pays 9% annual interest, then the amount of money in the account after 4 years is $P(1 + 0.09)^4$. Find the amount when $P = \$500$.

85. *Investment Growth* If an investment increases in value by about 10% each year for 22 years, then its value will triple two times.

(a) Write an exponential expression that represents "tripling two times."

(b) If the investment has an initial value of $8000, how much will it be worth if it triples two times?

86. *Stock Value* If a stock decreases in value by about 23% each year for 9 years, then its value will be halved three times.

(a) Write an exponential expression that represents "halved three times."

(b) If the stock is initially worth $88 per share, how much will it be worth if it is halved three times?

WRITING ABOUT MATHEMATICS

87. Are the expressions $(4x)^2$ and $4x^2$ equal in value? Explain your answer.

88. Are the expressions $3^3 \cdot 2^3$ and 6^6 equal in value? Explain your answer.

5.2   Addition and Subtraction of Polynomials

Monomials and Polynomials • Addition of Polynomials • Subtraction of Polynomials • Evaluating Polynomial Expressions

A LOOK INTO MATH ▶

If you have ever exercised strenuously and then taken your pulse immediately afterward, you may have discovered that your pulse slowed quickly at first and then gradually leveled off. A typical scatterplot of this phenomenon is shown in Figure 5.1(a). These data points cannot be modeled accurately with a line, so a new expression, called a *polynomial*, is needed to model them. A graph of a polynomial that models these data is shown in Figure 5.1(b) and discussed in Exercise 71. (*Source:* V. Thomas, *Science and Sport.*)

NEW VOCABULARY

- ☐ Monomial
- ☐ Degree of a monomial
- ☐ Coefficient of a monomial
- ☐ Polynomial
- ☐ Polynomial in one variable
- ☐ Binomial
- ☐ Trinomial
- ☐ Degree of a polynomial
- ☐ Like terms

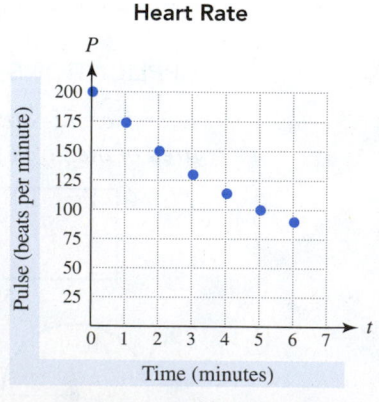

(a)

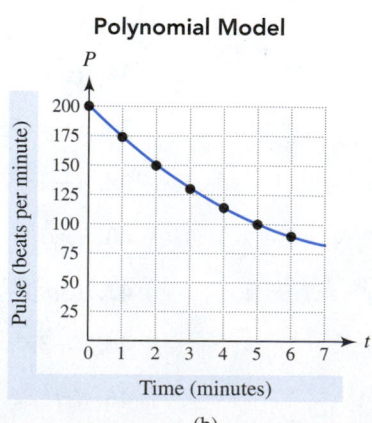

(b)

Figure 5.1 Heart Rate After Exercising

Monomials and Polynomials

A **monomial** is a number, a variable, or a product of numbers and variables raised to natural number powers. Examples of monomials include

$$-3, \quad xy^2, \quad 5a^2, \quad -z^3, \quad \text{and} \quad -\frac{1}{2}xy^3.$$

READING CHECK

• How do you determine the degree of a monomial?

A monomial may contain more than one variable, but monomials do not contain division by variables. For example, the expression $\frac{3}{z}$ is not a monomial. If an expression contains addition or subtraction signs, it is *not* a monomial.

The **degree of a monomial** is the sum of the exponents of the variables. If the monomial has only one variable, its degree is the exponent of that variable. Remember, when a variable does not have a written exponent, the exponent is implied to be 1. A nonzero number has degree 0, and the number 0 has *undefined* degree. The number in a monomial is called the **coefficient of the monomial**. Table 5.2 contains the degree and coefficient of several monomials.

TABLE 5.2 Properties of Monomials

Monomial	-5	$6a^3b$	$-xy$	$7y^3$
Degree	0	4	2	3
Coefficient	-5	6	-1	7

A **polynomial** is a monomial or the sum of two or more monomials. Each monomial is called a *term* of the polynomial. Addition or subtraction signs separate terms. The expression $2x^2 - 3x + 5$ is a **polynomial in one variable** with three terms. Examples of polynomials in one variable include

$$-2x, \quad 3x + 1, \quad 4y^2 - y + 7, \quad \text{and} \quad x^5 - 3x^3 + x - 7.$$

These polynomials have 1, 2, 3, and 4 terms, respectively. A polynomial with *two terms* is called a **binomial**, and a polynomial with *three terms* is called a **trinomial**.

A polynomial can have more than one variable, as in

$$x^2y^2, \quad 2xy^2 + 5x^2y - 1, \quad \text{and} \quad a^2 + 2ab + b^2.$$

READING CHECK

• How do you determine the degree of a polyomial?

Note that all variables in a polynomial are raised to natural number powers. The **degree of a polynomial** is the degree of the term (or monomial) with greatest degree.

EXAMPLE 1   **Identifying properties of polynomials**

Determine whether the expression is a polynomial. If it is, state how many terms and variables the polynomial contains and give its degree.

(a) $7x^2 - 3x + 1$ **(b)** $5x^3 - 3x^2y^3 + xy^2 - 2y^3$ **(c)** $4x^2 + \dfrac{5}{x+1}$

Solution

(a) The expression $7x^2 - 3x + 1$ is a polynomial with three terms and one variable. The first term $7x^2$ has degree 2 because the exponent on the variable is 2. The second term $-3x$ has degree 1 because the exponent on the variable is implied to be 1. The third term 1 has degree 0 because it is a nonzero number. The term with greatest degree is $7x^2$, so the polynomial has degree **2**.

(b) The expression $5x^3 - 3x^2y^3 + xy^2 - 2y^3$ is a polynomial with four terms and two variables, x and y. The first term has degree 3 because the exponent on the variable is 3. The second term has degree 5 because the *sum* of the exponents on the variables x and y is 5. Likewise, the third term has degree 3 and the fourth term has degree 3. The term with greatest degree is $-3x^2y^3$, so the polynomial has degree $2 + 3 = 5$.

(c) The expression $4x^2 + \frac{5}{x + 1}$ is not a polynomial because it contains division by the polynomial $x + 1$.

Now Try Exercises 21, 23, 27

Addition of Polynomials

Suppose that we have 2 identical rectangles with length l and width w, as illustrated in Figure 5.2. Then the area of one rectangle is lw and the total area is

$$lw + lw.$$

This area is equivalent to 2 times lw, which can be expressed as $2lw$, or

$$lw + lw = 2lw.$$

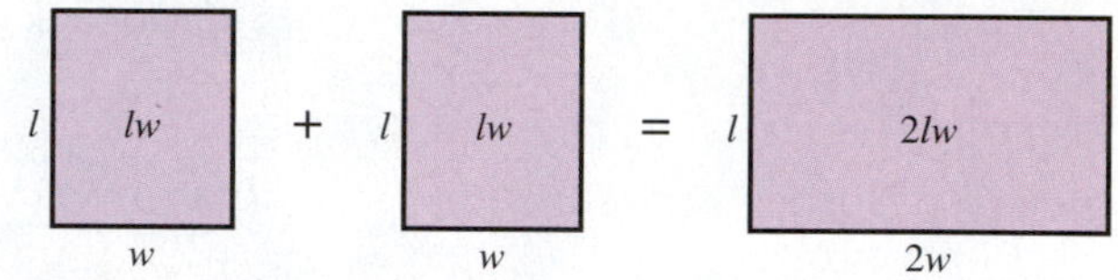

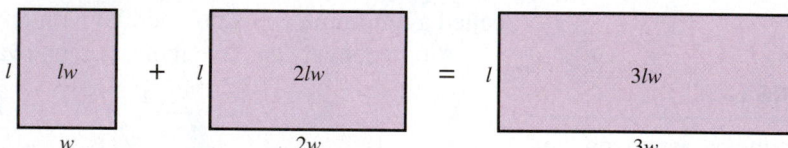

Figure 5.2 Adding $lw + lw$

If two monomials contain the same variables raised to the same powers, we call them **like terms**. We can add or subtract (combine) *like* terms but cannot combine *unlike* terms. The terms lw and $2lw$ are like terms and can be combined geometrically, as shown in Figure 5.3. If we joined one of the small rectangles with area lw and a larger rectangle with area $2lw$, then the total area is $3lw$.

Figure 5.3 Adding $lw + 2lw$

The *distributive property* justifies combining like terms.

$$1lw + 2lw = (1 + 2)lw = 3lw$$

The rectangles shown in Figure 5.4 have areas of ab and xy. Together, their area is the sum, $ab + xy$. However, because these monomials are unlike terms, they cannot be combined into one term.

Figure 5.4 Unlike terms: $ab + xy$

STUDY TIP

Do you want to know what material will be covered on your next exam? Often, the best place to look is on previously completed assignments and quizzes. If a topic is not discussed in class, is not found on the syllabus, and is not part of your assignments, then your time may be better spent studying other topics.

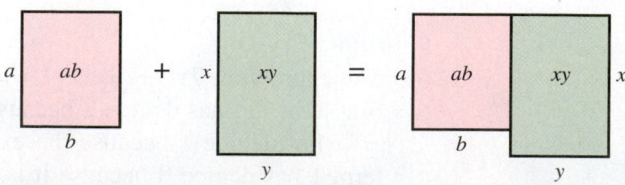

EXAMPLE 2   **Adding like terms**

State whether each pair of expressions contains like terms or unlike terms. If they are like terms, add them.

(a) $5x^2, -x^2$ **(b)** $7a^2b, 10ab^2$ **(c)** $4rt^2, \frac{1}{2}rt^2$

Solution
(a) The terms $5x^2$ and $-x^2$ have the same variable raised to the same power, so they are like terms. To add like terms, add their coefficients. Note that the coefficient of $-x^2$ is -1.

$$5x^2 + (-x^2) = (5 + (-1))x^2 \quad \text{Distributive property}$$
$$= 4x^2 \quad \text{Add.}$$

(b) The terms $7a^2b$ and $10ab^2$ have the same variables, but these variables are not raised to the same powers. They are unlike terms and cannot be added.
(c) The terms $4rt^2$ and $\frac{1}{2}rt^2$ have the same variables raised to the same powers, so they are like terms. We add them as follows.

$$4rt^2 + \frac{1}{2}rt^2 = \left(4 + \frac{1}{2}\right)rt^2 \quad \text{Distributive property}$$

$$= \frac{9}{2}rt^2 \quad \text{Add.}$$

Now Try Exercises 29, 31, 33

To add two polynomials, combine like terms, as illustrated in the next example.

EXAMPLE 3   **Adding polynomials**

Add by combining like terms.
(a) $(3x + 4) + (-4x + 2)$
(b) $(y^2 - 2y + 1) + (3y^2 + y + 11)$

Solution
(a) $(3x + 4) + (-4x + 2) = 3x + (-4x) + 4 + 2$
$$= (3 - 4)x + (4 + 2)$$
$$= -x + 6$$

(b) $(y^2 - 2y + 1) + (3y^2 + y + 11) = y^2 + 3y^2 - 2y + y + 1 + 11$
$$= (1 + 3)y^2 + (-2 + 1)y + (1 + 11)$$
$$= 4y^2 - y + 12$$

Now Try Exercises 37, 39

Recall that the commutative and associative properties of addition allow us to rearrange a sum in any order. For example, if we write each subtraction in $2x - 5 - 4x + 10$ as addition of the opposite, we have

$$2x - 5 - 4x + 10 = 2x + (-5) + (-4x) + 10,$$

and the terms can be rearranged as

$$2x + (-4x) + (-5) + 10 = 2x - 4x - 5 + 10.$$

If we pay attention to the sign in front of each term in a polynomial, the like terms can be combined without rearranging the terms, as demonstrated in the next example.

| EXAMPLE 4 | **Adding polynomials** |

Add $(x^3 - 3x^2 + 7x - 4) + (4x^3 - 5x + 9)$ by combining like terms.

Solution
Remove parentheses and identify like terms with their signs as shown.

$$x^3 - 3x^2 + 7x - 4 + 4x^3 - 5x + 9$$

When like terms (of the same color) are added, the resulting sum is

$$5x^3 - 3x^2 + 2x + 5.$$

Now Try Exercise 41

Polynomials can also be added vertically, as demonstrated in the next example.

| EXAMPLE 5 | **Adding polynomials vertically** |

Simplify $(3x^2 - 3x + 5) + (-x^2 + x - 6)$.

Solution
Write the polynomials in a vertical format and then add each column of like terms.

$$
\begin{array}{r}
3x^2 - 3x + 5 \\
\underline{-x^2 +\ \ x - 6} \\
2x^2 - 2x - 1
\end{array}
\quad \text{Add like terms in each column.}
$$

Regardless of the method used, the same answer should be obtained. However, adding vertically requires that *like terms be placed in the same column.*

Now Try Exercise 47

Subtraction of Polynomials

To subtract one integer from another, add the first integer and the *additive inverse* or *opposite* of the second integer. For example, $3 - 5$ is evaluated as follows.

$$3 - 5 = 3 + (-5) \quad \text{Add the opposite.}$$
$$= -2 \quad\quad\quad \text{Simplify.}$$

Similarly, to subtract one polynomial from another, add the first polynomial and the *opposite* of the second polynomial. To find the opposite of a polynomial, simply negate each term. Table 5.3 lists some polynomials and their opposites.

READING CHECK

• How do you subtract one polynomial from another?

CRITICAL THINKING

What is the result when a polynomial and its opposite are added?

TABLE 5.3 Opposites of Polynomials

Polynomial	Opposite
$2x - 4$	$-2x + 4$
$-x^2 - 2x + 9$	$x^2 + 2x - 9$
$6x^3 - 12$	$-6x^3 + 12$
$-3x^4 - 2x^2 - 8x + 3$	$3x^4 + 2x^2 + 8x - 3$

EXAMPLE 6   **Subtracting polynomials**

Simplify each expression.
(a) $(3x - 4) - (5x + 1)$
(b) $(5x^2 + 2x - 3) - (6x^2 - 7x + 9)$
(c) $(6x^3 + x^2) - (-3x^3 - 9)$

Solution
(a) To subtract $(5x + 1)$ from $(3x - 4)$, we add the opposite of $(5x + 1)$, or $(-5x - 1)$.

$$(3x - 4) - (5x + 1) = (3x - 4) + (-5x - 1)$$
$$= (3 - 5)x + (-4 - 1)$$
$$= -2x - 5$$

(b) The opposite of $(6x^2 - 7x + 9)$ is $(-6x^2 + 7x - 9)$.

$$(5x^2 + 2x - 3) - (6x^2 - 7x + 9) = (5x^2 + 2x - 3) + (-6x^2 + 7x - 9)$$
$$= (5 - 6)x^2 + (2 + 7)x + (-3 - 9)$$
$$= -x^2 + 9x - 12$$

(c) The opposite of $(-3x^3 - 9)$ is $(3x^3 + 9)$.

$$(6x^3 + x^2) - (-3x^3 - 9) = (6x^3 + x^2) + (3x^3 + 9)$$
$$= (6 + 3)x^3 + x^2 + 9$$
$$= 9x^3 + x^2 + 9$$

Now Try Exercises 57, 59, 61

NOTE: Some students prefer to subtract one polynomial from another by noting that a subtraction sign in front of parentheses changes the signs of all of the terms within the parentheses. For example, part (a) of the previous example could be worked as follows.

$$(3x - 4) - (5x + 1) = 3x - 4 - 5x - 1$$
$$= (3 - 5)x + (-4 - 1)$$
$$= -2x - 5$$

EXAMPLE 7   **Subtracting polynomials vertically**

Simplify $(5x^2 - 2x + 7) - (-3x^2 + 3)$.

Solution
To subtract one polynomial from another vertically, simply add the first polynomial and the opposite of the second polynomial. No x-term occurs in the second polynomial, so insert $0x$.

$$5x^2 - 2x + 7$$
$$\underline{3x^2 + 0x - 3} \quad \text{The opposite of } -3x^2 + 3 \text{ is } 3x^2 - 3 \text{ or } 3x^2 + 0x - 3.$$
$$8x^2 - 2x + 4 \quad \text{Add like terms in each column.}$$

Now Try Exercise 69

Evaluating Polynomial Expressions

Frequently, monomials and polynomials represent formulas that may be evaluated. We illustrate such applications in the next two examples.

EXAMPLE 8

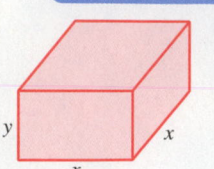

Figure 5.5

Writing and evaluating a monomial

Write the monomial that represents the volume of the box having a square bottom, as shown in Figure 5.5. Find the volume of the box if $x = 3$ feet and $y = 2$ feet.

Solution

The volume of a box is found by multiplying the length, width, and height together. Because the length and width are both x and the height is y, the monomial xxy represents the volume of the box. This can be written x^2y. To calculate the volume, let $x = 3$ and $y = 2$ in the monomial x^2y.

$$x^2y = 3^2 \cdot 2 = 9 \cdot 2 = 18 \text{ cubic feet}$$

Now Try Exercise 73

EXAMPLE 9

Modeling sales of personal computers

Worldwide sales of personal computers have increased dramatically in recent years, as illustrated in Figure 5.6. The polynomial

$$0.7868x^2 + 16.72x + 122.58$$

approximates the number of computers sold in millions, where $x = 0$ corresponds to 2000, $x = 1$ to 2001, and so on. Estimate the number of personal computers sold in 2008 by using both the graph and the polynomial. (*Source:* International Data Corporation.)

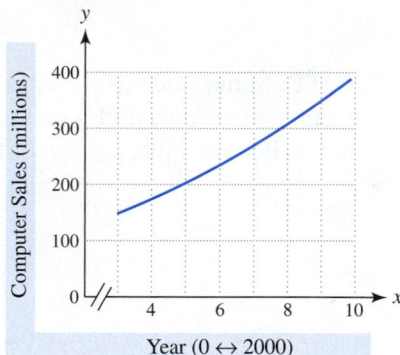

Figure 5.6

Solution

From the graph shown in Figure 5.7, it appears that personal computer sales were slightly more than 300 million, or about 310 million, in 2008.

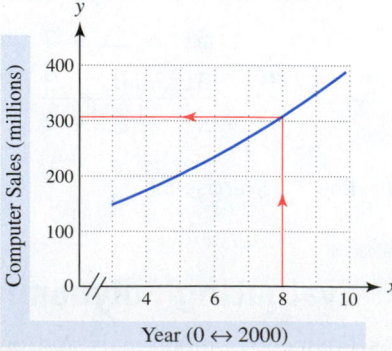

Figure 5.7

The year 2008 corresponds to $x = 8$ in the given polynomial, so substitute **8** for x and evaluate the resulting expression.

$$0.7868x^2 + 16.72x + 122.58 = 0.7868(8)^2 + 16.72(8) + 122.58$$
$$\approx 307 \text{ million}$$

The graph and the polynomial give similiar results.

Now Try Exercise 71

5.2 Putting It All Together

CONCEPT	EXPLANATION	EXAMPLES
Monomial	A number, variable, or product of numbers and variables raised to natural number powers The degree is the sum of the exponents. The coefficient is the number in a monomial.	$4x^2y$ Degree: 3; coefficient: 4 $-6x^2$ Degree: 2; coefficient: -6 $-a^4$ Degree: 4; coefficient: -1 x Degree: 1; coefficient: 1 -8 Degree: 0; coefficient: -8
Polynomial	A monomial or the sum of two or more monomials	$4x^2 + 8xy^2 + 3y^2$ Trinomial $-9x^4 + 100$ Binomial $-3x^2y^3$ Monomial
Like Terms	Monomials containing the same variables raised to the same powers	$10x$ and $-2x$, $4x^2$ and $3x^2$ $5ab^2$ and $-ab^2$, $5z$ and $\frac{1}{2}z$
Addition of Polynomials	To add polynomials, combine like terms.	$(x^2 + 3x + 1) + (2x^2 - 2x + 7)$ $= (1 + 2)x^2 + (3 - 2)x + (1 + 7)$ $= 3x^2 + x + 8$ $3xy + 5xy = (3 + 5)xy = 8xy$
Opposite of a Polynomial	To obtain the opposite of a polynomial, negate each term.	*Polynomial* *Opposite* $-2x^2 + x - 6$ $2x^2 - x + 6$ $a^2 - b^2$ $-a^2 + b^2$ $-3x - 18$ $3x + 18$
Subtraction of Polynomials	To subtract one polynomial from another, add the first polynomial and the opposite of the second polynomial.	$(x^2 + 3x) - (2x^2 - 5x)$ $= (x^2 + 3x) + (-2x^2 + 5x)$ $= (1 - 2)x^2 + (3 + 5)x$ $= -x^2 + 8x$
Evaluating a Polynomial	To evaluate a polynomial in x, substitute a value for x in the expression and simplify.	To evaluate the polynomial $3x^2 - 2x + 1$ for $x = 2$, substitute 2 for x and simplify. $3(2)^2 - 2(2) + 1 = 9$

5.2 Exercises

MyMathLab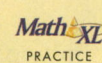
PRACTICE   WATCH   DOWNLOAD   READ   REVIEW

CONCEPTS AND VOCABULARY

1. A(n) _____ is a number, a variable, or a product of numbers and variables raised to a natural number power.

2. A(n) _____ is a monomial or a sum of monomials.

3. The _____ of a monomial is the sum of the exponents of the variables.

4. The _____ of a polynomial is the degree of the term with the greatest degree.

5. A polynomial with two terms is called a(n) _____.

6. A polynomial with three terms is called a(n) _____.

7. Two monomials with the same variables raised to the same powers are _____ terms.

8. To add two polynomials, combine _____ terms.

9. To subtract two polynomials, add the first polynomial to the _____ of the second polynomial.

10. Polynomials can be added horizontally or _____.

PROPERTIES OF POLYNOMIALS

Exercises 11–18: Identify the degree and coefficient of the monomial.

11. $3x^2$

12. y

13. $-ab$

14. $-2xy$

15. $-5rt$

16. $8x^2y^5$

17. 6

18. $-\frac{1}{2}$

Exercises 19–28: Determine whether the expression is a polynomial. If it is, state how many terms and variables the polynomial contains. Then state its degree.

19. $-x$

20. $7z$

21. $4x^2 - 5x + 9$

22. $x^3 - 9$

23. $x + \dfrac{1}{x}$

24. $\dfrac{5}{xy + 1}$

25. $3x^{-2}y^{-3}$

26. $5^2a^3b^4$

27. -2^3a^4bc

28. $-7y^{-1}z^{-3}$

Exercises 29–36: State whether the given pair of expressions are like terms. If they are like terms, add them.

29. $5x, -4x$

30. $x^2, 8x^2$

31. $x^3, -6x^3$

32. $4xy, -9xy$

33. $9x, -xy$

34. $5x^2y, -3xy^2$

35. ab, ba

36. $rt^2, -2t^2r$

ADDITION OF POLYNOMIALS

Exercises 37–46: Add the polynomials.

37. $(3x + 5) + (-4x + 4)$

38. $(-x + 5) + (2x - 5)$

39. $(3x^2 + 4x + 1) + (x^2 + 4x)$

40. $(-x^2 - x) + (2x^2 + 3x - 1)$

41. $(y^3 + 3y^2 - 5) + (3y^3 + 4y - 4)$

42. $(4z^4 + z^2 - 10) + (-z^4 + 4z - 5)$

43. $(-xy + 5) + (5xy - 4)$

44. $(2a^2 + b^2) + (3a^2 - 5b^2)$

45. $(a^3b^2 + a^2b^3) + (a^2b^3 - a^3b^2)$

46. $(a^2 + ab + b^2) + (a^2 - ab + b^2)$

Exercises 47–50: Add the polynomials vertically.

47. $4x^2 - 2x + 1$
    $\underline{5x^2 + 3x - 7}$

48. $8x^2 + 3x + 5$
    $\underline{-x^2 - 3x - 9}$

49. $-x^2 +\ \ x$
    $\underline{2x^2 - 8x - 1}$

50. $a^3 - 3a^2b + 3ab^2 - b^3$
    $\underline{a^3 + 3a^2b + 3ab^2 + b^3}$

SUBTRACTION OF POLYNOMIALS

Exercises 51–56: Write the opposite of the polynomial.

51. $5x^2$

52. $17x + 12$

53. $3a^2 - a + 4$

54. $-b^3 + 3b$

55. $-2t^2 - 3t + 4$

56. $7t^2 + t - 10$

Exercises 57–66: Subtract the polynomials.

57. $(3x + 1) - (-x + 3)$

58. $(-2x + 5) - (x + 7)$

59. $(-x^2 + 6x) - (2x^2 + x - 2)$

60. $(2y^2 + 3y - 2) - (y^2 - y)$

61. $(z^3 - 2z^2 - z) - (4z^2 + 5z + 1)$

62. $(3z^4 - z) - (-z^4 + 4z^2 - 5)$

63. $(4xy + x^2y^2) - (xy - x^2y^2)$

64. $(a^2 + b^2) - (-a^2 + b^2)$

65. $(ab^2) - (ab^2 + a^3b)$

66. $(x^2 + 3xy + 4y^2) - (x^2 - xy + 4y^2)$

Exercises 67–70: Subtract the polynomials vertically.

67. $(x^2 + 2x - 3) - (2x^2 + 7x + 1)$

68. $(5x^2 - 9x - 1) - (x^2 - x + 3)$

69. $(3x^3 - 2x) - (5x^3 + 4x + 2)$

70. $(a^2 + 3ab + 2b^2) - (a^2 - 3ab + 2b^2)$

APPLICATIONS

71. *Exercise and Heart Rate* The polynomial given by $1.6t^2 - 28t + 200$ calculates the heart rate shown in Figure 5.1(b) in A Look Into Math for this section, where t represents the elapsed time in minutes since exercise stopped.
   (a) What is the heart rate when the athlete first stops exercising?
   (b) What is the heart rate after 5 minutes?
   (c) Describe what happens to the heart rate after exercise stops.

72. *Cellular Phone Subscribers* In the early years of cellular phone technology—from 1986 through 1991—the number of subscribers in millions could be modeled by the polynomial $0.163x^2 - 0.146x + 0.205$, where $x = 1$ corresponds to 1986, $x = 2$ to 1987, and so on. The graph illustrates this growth.

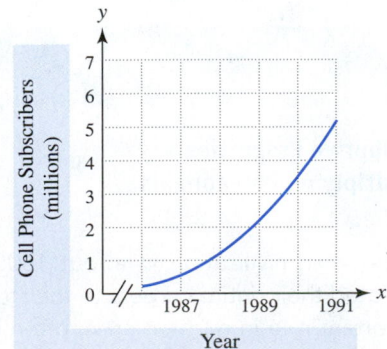

   (a) Use the graph to estimate the number of cellular phone subscribers in 1990.

   (b) Use the polynomial to estimate the number of cellular phone subscribers in 1990.
   (c) Do your answers from parts (a) and (b) agree?

73. *Areas of Squares* Write a monomial that equals the sum of the areas of the squares. Then calculate this sum for $z = 10$ inches.

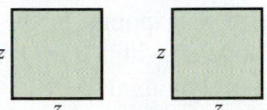

74. *Areas of Rectangles* Find a monomial that equals the sum of the areas of the three rectangles. Find this sum for $a = 5$ yards and $b = 3$ yards.

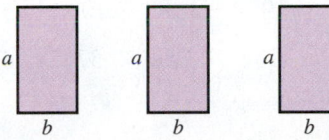

75. *Area of a Figure* Find a polynomial that equals the area of the figure. Calculate its area for $x = 6$ feet.

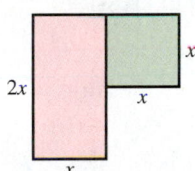

76. *Area of a Rectangle* Write a polynomial that gives the area of the rectangle. Calculate its area for $x = 3$ feet.

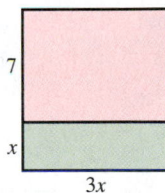

77. *Areas of Circles* Write a polynomial that gives the sum of the areas of two circles, one with radius x and the other with radius y. Find this sum for $x = 2$ feet and $y = 3$ feet. Leave your answer in terms of π.

78. *Squares and Circles* Write a polynomial that gives the sum of the areas of a square having sides of length x and a circle having diameter x. Approximate this sum to the nearest hundredth of a square foot for $x = 6$ feet.

79. *World Population* The table lists actual and projected world population P in billions for selected years t.

t	1974	1987	1999	2012
P	4	5	6	7

Source: U.S. Census Bureau.

(a) Find the slope of each line segment connecting consecutive data points in the table. Can these data be modeled with a line? Explain.

(b) Does the polynomial $0.077t - 148$ give good estimates for the world population in year t? Explain how you decided.

80. *Price of a Stamp* The table lists the price P of a first-class postage stamp for selected years t.

t	1963	1975	1987	2002	2007	2011
P	5¢	13¢	25¢	37¢	41¢	44¢

Source: U.S. Postal Service.

(a) Does the polynomial $0.835t - 1635$ model the data in the table exactly?

(b) Does it give approximations that are within 1.5¢ of the actual values?

WRITING ABOUT MATHEMATICS

81. Explain what the terms monomial, binomial, trinomial, and polynomial mean. Give an example of each.

82. Explain how to determine the degree of a polynomial having one variable. Give an example.

83. Explain how to obtain the opposite of a polynomial. Give an example.

84. Explain how to subtract two polynomials. Give an example.

SECTIONS 5.1 and 5.2

Checking Basic Concepts

1. Evaluate each expression.
(a) -5^2 **(b)** $3^2 - 2^3$

2. Simplify each expression.
(a) $10^3 \cdot 10^5$ **(b)** $(3x^2)(-4x^5)$
(c) $(a^3 b)^2$ **(d)** $\left(\dfrac{x}{z^3}\right)^4$

3. Simplify each expression.
(a) $(4y^3)^0$ **(b)** $(x^3)^2(3x^4)^2$
(c) $2a^2(5a^3 - 7)$

4. State the number of terms and variables in the polynomial $5x^3y - 2x^2y + 5$. What is its degree?

5. A box has a rectangular bottom twice as long as it is wide.
(a) If the bottom has width w and the box has height h, write a monomial that gives the volume of the box.
(b) Find the volume of the box for $w = 12$ inches and $h = 10$ inches.

6. Simplify each expression.
(a) $(2a^2 + 3a - 1) + (a^2 - 3a + 7)$
(b) $(4z^3 + 5z) - (2z^3 - 2z + 8)$
(c) $(x^2 + 2xy + y^2) - (x^2 - 2xy + y^2)$

5.3 Multiplication of Polynomials

**Multiplying Monomials • Review of the Distributive Properties •
Multiplying Monomials and Polynomials • Multiplying Polynomials**

A LOOK INTO MATH ▶

The study of polynomials dates back to Babylonian civilization in about 1800–1600 B.C. Many eighteenth-century mathematicians devoted their entire careers to the study of polynomials. Today, polynomials still play an important role in mathematics, often being used to approximate unknown quantities. In this section we discuss the basics of multiplying polynomials. (*Source: Historical Topics for the Mathematics Classroom, Thirty-first Yearbook, NCTM.*)

Multiplying Monomials

A monomial is a number, a variable, or a product of numbers and variables raised to natural number powers. To multiply monomials, we often use the product rule for exponents.

EXAMPLE 1 Multiplying monomials

Multiply.
(a) $-5x^2 \cdot 4x^3$ **(b)** $(7xy^4)(x^3y^2)$

Solution
(a) $-5x^2 \cdot 4x^3 = (-5)(4)x^2x^3$ Commutative property
$\qquad\qquad\quad = -20x^{2+3}$ The product rule
$\qquad\qquad\quad = -20x^5$ Simplify.

(b) $(7xy^4)(x^3y^2) = 7xx^3y^4y^2$ Commutative property
$\qquad\qquad\qquad = 7x^{1+3}y^{4+2}$ The product rule
$\qquad\qquad\qquad = 7x^4y^6$ Simplify.

Now Try Exercises 9, 13

READING CHECK

• Which rule for exponents is commonly used to multiply monomials?

Review of the Distributive Properties

Distributive Property
$3(x + 2) = 3x + 6$

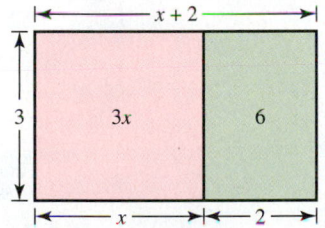

Figure 5.8 Area: $3x + 6$

Distributive properties are used frequently for multiplying monomials and polynomials. For all real numbers a, b, and c,

$$a(b + c) = ab + ac \quad \text{and} \quad a(b - c) = ab - ac.$$

The first distributive property above can be visualized geometrically. For example,

$$3(x + 2) = 3x + 6$$

is illustrated in Figure 5.8. The dimensions of the large rectangle are 3 by $x + 2$, and its area is $3(x + 2)$. The areas of the two small rectangles, $3x$ and 6, equal the area of the large rectangle. Therefore $3(x + 2) = 3x + 6$.

In the next example we use the distributive properties to multiply expressions.

EXAMPLE 2 Using distributive properties

Multiply.
(a) $2(3x + 4)$ **(b)** $(3x^2 + 4)5$ **(c)** $-x(3x - 6)$

Solution

(a) $2(3x + 4) = 2 \cdot 3x + 2 \cdot 4 = 6x + 8$

(b) $(3x^2 + 4)5 = 3x^2 \cdot 5 + 4 \cdot 5 = 15x^2 + 20$

(c) $-x(3x - 6) = -x \cdot 3x + x \cdot 6 = -3x^2 + 6x$

Now Try Exercises 15, 19, 21

READING CHECK

• What properties are commonly used to multiply a monomial and a polynomial?

Multiplying Monomials and Polynomials

A monomial consists of one term, whereas a polynomial consists of one or more terms separated by $+$ or $-$ signs. To multiply a monomial by a polynomial, we apply the distributive properties and the product rule.

EXAMPLE 3 **Multiplying monomials and polynomials**

Multiply.
(a) $9x(2x^2 - 3)$ (b) $(5x - 8)x^2$
(c) $-7(2x^2 - 4x + 6)$ (d) $4x^3(x^4 + 9x^2 - 8)$

Solution

(a) $\qquad 9x(2x^2 - 3) = 9x \cdot 2x^2 - 9x \cdot 3 \qquad$ Distributive property
$\qquad\qquad\qquad\qquad = 18x^3 - 27x \qquad$ The product rule

(b) $\qquad (5x - 8)x^2 = 5x \cdot x^2 - 8 \cdot x^2 \qquad$ Distributive property
$\qquad\qquad\qquad\quad = 5x^3 - 8x^2 \qquad$ The product rule

(c) $-7(2x^2 - 4x + 6) = -7 \cdot 2x^2 + 7 \cdot 4x - 7 \cdot 6 \qquad$ Distributive property
$\qquad\qquad\qquad\qquad = -14x^2 + 28x - 42 \qquad$ Simplify.

(d) $4x^3(x^4 + 9x^2 - 8) = 4x^3 \cdot x^4 + 4x^3 \cdot 9x^2 - 4x^3 \cdot 8 \qquad$ Distributive property
$\qquad\qquad\qquad\qquad = 4x^7 + 36x^5 - 32x^3 \qquad$ The product rule

Now Try Exercises 23, 25, 27, 29

We can also multiply monomials and polynomials that contain more than one variable.

EXAMPLE 4 **Multiplying monomials and polynomials**

Multiply.
(a) $2xy(7x^2y^3 - 1)$ (b) $-ab(a^2 - b^2)$

Solution

(a) $2xy(7x^2y^3 - 1) = 2xy \cdot 7x^2y^3 - 2xy \cdot 1 \qquad$ Distributive property
$\qquad\qquad\qquad = 14xx^2yy^3 - 2xy \qquad$ Commutative property
$\qquad\qquad\qquad = 14x^3y^4 - 2xy \qquad$ The product rule

(b) $\quad -ab(a^2 - b^2) = -ab \cdot a^2 + ab \cdot b^2 \qquad$ Distributive property
$\qquad\qquad\qquad = -aa^2b + abb^2 \qquad$ Commutative property
$\qquad\qquad\qquad = -a^3b + ab^3 \qquad$ The product rule

Now Try Exercises 31, 35

Multiplying Polynomials

Monomials, binomials, and trinomials are examples of polynomials. Recall that a monomial has one term, a binomial has two terms, and a trinomial has three terms. In the next example we multiply two binomials, using both geometric and symbolic techniques.

EXAMPLE 5   **Multiplying binomials**

Multiply $(x + 4)(x + 2)$
(a) geometrically and   **(b)** symbolically.

Solution

(a) To multiply $(x + 4)(x + 2)$ geometrically, draw a rectangle $x + 4$ long and $x + 2$ wide, as shown in Figure 5.9(a). The area of this rectangle equals length times width, or $(x + 4)(x + 2)$. The large rectangle can be divided into four smaller rectangles, which have areas of x^2, $4x$, $2x$, and 8, as shown in Figure 5.9(b). Thus

$$(x + 4)(x + 2) = x^2 + 4x + 2x + 8$$
$$= x^2 + 6x + 8.$$

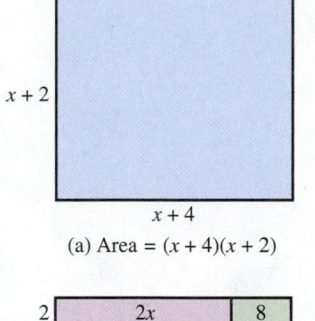

(a) Area = $(x + 4)(x + 2)$

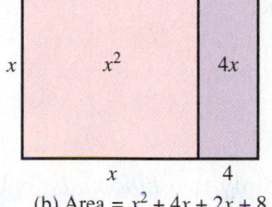

(b) Area = $x^2 + 4x + 2x + 8$

Figure 5.9

(b) To multiply $(x + 4)(x + 2)$ symbolically, apply the distributive property two times.

$$(x + 4)(x + 2) = (x + 4)(x) + (x + 4)(2)$$
$$= x \cdot x + 4 \cdot x + x \cdot 2 + 4 \cdot 2$$
$$= x^2 + 4x + 2x + 8$$
$$= x^2 + 6x + 8$$

Now Try Exercise 39

The distributive properties used in part (b) of the previous example show that if we want to multiply $(x + 4)$ by $(x + 2)$, we should multiply every term in $x + 4$ by every term in $x + 2$.

$$(x + 4)(x + 2) = x^2 + 2x + 4x + 8$$
$$= x^2 + 6x + 8$$

NOTE: This process of multiplying *binomials* is sometimes called *FOIL*. This acronym may be used to remind us to multiply the first terms (F), outside terms (O), inside terms (I), and last terms (L). The *FOIL* process is a shortcut for the process used in Example 5(b).

Multiply the *First terms* to obtain x^2. $(x + 4)(x + 2)$

Multiply the *Outside terms* to obtain $2x$. $(x + 4)(x + 2)$

Multiply the *Inside terms* to obtain $4x$. $(x + 4)(x + 2)$

Multiply the *Last terms* to obtain 8. $(x + 4)(x + 2)$

READING CHECK

• What kind of polynomials can be multiplied using the *FOIL* method?

The following statement summarizes how to multiply two polynomials in general.

MULTIPLYING POLYNOMIALS

The product of two polynomials may be found by multiplying every term in the first polynomial by every term in the second polynomial and then combining like terms.

EXAMPLE 6 **Multiplying binomials**

Multiply. Draw arrows to show how each term is found.
(a) $(3x + 2)(x + 1)$ **(b)** $(1 - x)(1 + 2x)$ **(c)** $(4x - 3)(x^2 - 2x)$

Solution

(a) $(3x + 2)(x + 1) = 3x \cdot x + 3x \cdot 1 + 2 \cdot x + 2 \cdot 1$
$$= 3x^2 + 3x + 2x + 2$$
$$= 3x^2 + 5x + 2$$

(b) $(1 - x)(1 + 2x) = 1 \cdot 1 + 1 \cdot 2x - x \cdot 1 - x \cdot 2x$
$$= 1 + 2x - x - 2x^2$$
$$= 1 + x - 2x^2$$

(c) $(4x - 3)(x^2 - 2x) = 4x \cdot x^2 - 4x \cdot 2x - 3 \cdot x^2 + 3 \cdot 2x$
$$= 4x^3 - 8x^2 - 3x^2 + 6x$$
$$= 4x^3 - 11x^2 + 6x$$

Now Try Exercises 51, 53, 59

The *FOIL* process may be helpful for remembering how to multiply two *binomials*, but it cannot be used for every product of polynomials. In the next example, the general process for multiplying polynomials is used to find products of binomials and trinomials.

EXAMPLE 7 **Multiplying polynomials**

Multiply.
(a) $(2x + 3)(x^2 + x - 1)$ **(b)** $(a - b)(a^2 + ab + b^2)$
(c) $(x^4 + 2x^2 - 5)(x^2 + 1)$

Solution
(a) Multiply every term in $(2x + 3)$ by every term in $(x^2 + x - 1)$.

$(2x + 3)(x^2 + x - 1) = 2x \cdot x^2 + 2x \cdot x - 2x \cdot 1 + 3 \cdot x^2 + 3 \cdot x - 3 \cdot 1$
$$= 2x^3 + 2x^2 - 2x + 3x^2 + 3x - 3$$
$$= 2x^3 + 5x^2 + x - 3$$

(b) $(a - b)(a^2 + ab + b^2) = a \cdot a^2 + a \cdot ab + a \cdot b^2 - b \cdot a^2 - b \cdot ab - b \cdot b^2$
$$= a^3 + a^2b + ab^2 - a^2b - ab^2 - b^3$$
$$= a^3 - b^3$$

(c) $(x^4 + 2x^2 - 5)(x^2 + 1) = x^4 \cdot x^2 + x^4 \cdot 1 + 2x^2 \cdot x^2 + 2x^2 \cdot 1 - 5 \cdot x^2 - 5 \cdot 1$
$$= x^6 + x^4 + 2x^4 + 2x^2 - 5x^2 - 5$$
$$= x^6 + 3x^4 - 3x^2 - 5$$

Now Try Exercises 63, 67, 69

STUDY TIP

Even if you know exactly how to do a math problem correctly, a simple computational error will often cause you to get an incorrect answer. Be sure to take your time on simple calculations.

Polynomials can be multiplied vertically in a manner similar to multiplication of real numbers. For example, multiplication of 123 times 12 is performed as follows.

$$
\begin{array}{r}
1\ 2\ 3 \\
\times\ 1\ 2 \\
\hline
2\ 4\ 6 \\
1\ 2\ 3 \\
\hline
1\ 4\ 7\ 6
\end{array}
$$

A similar method can be used to multiply polynomials vertically.

EXAMPLE 8   **Multiplying polynomials vertically**

Multiply $2x^2 - 4x + 1$ and $x + 3$ vertically.

Solution

Write the polynomials vertically. Then multiply every term in the first polynomial by each term in the second polynomial. Arrange the results so that *like terms are in the same column*.

$$
\begin{array}{r}
2x^2 - 4x + 1 \\
x + 3 \\
\hline
6x^2 - 12x + 3 \qquad \text{Multiply top polynomial by 3.} \\
2x^3 - 4x^2 + x \qquad \text{Multiply top polynomial by } x. \\
\hline
2x^3 + 2x^2 - 11x + 3 \qquad \text{Add each column.}
\end{array}
$$

Now Try Exercise 71

MAKING CONNECTIONS

Vertical and Horizontal Formats

Whether you decide to add, subtract, or multiply polynomials vertically or horizontally, remember that the same answer is obtained either way.

EXAMPLE 9   **Finding the volume of a box**

A box has a width 3 inches less than its height and a length 4 inches more than its height.
(a) If h represents the height of the box, write a polynomial that represents the volume of the box.
(b) Use this polynomial to calculate the volume of the box if $h = 10$ inches.

Solution
(a) If h is the height, then $h - 3$ is the width and $h + 4$ is the length, as illustrated in Figure 5.10. Its volume equals the product of these three expressions.

$$h(h - 3)(h + 4) = (h^2 - 3h)(h + 4)$$

$$= h^2 \cdot h + h^2 \cdot 4 - 3h \cdot h - 3h \cdot 4$$
$$= h^3 + 4h^2 - 3h^2 - 12h$$
$$= h^3 + h^2 - 12h$$

(b) If $h = 10$, then the volume is

$$10^3 + 10^2 - 12(10) = 1000 + 100 - 120 = 980 \text{ cubic inches.}$$

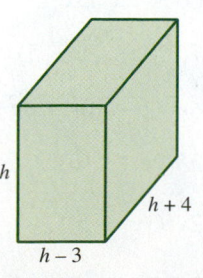

Figure 5.10

Now Try Exercise 79

5.3   Putting It All Together

CONCEPT	EXPLANATION	EXAMPLES
Distributive Properties	For all real numbers a, b, and c, $a(b + c) = ab + ac$ and $a(b - c) = ab - ac$.	$5(x + 3) = 5x + 15$, $3(x - 6) = 3x - 18$, and $-2x(3 - 5x^3) = -6x + 10x^4$
Multiplying Polynomials	The product of two polynomials may be found by multiplying every term in the first polynomial by every term in the second polynomial and then combining like terms.	$3x(5x^2 + 2x - 7)$ $= 3x \cdot 5x^2 + 3x \cdot 2x - 3x \cdot 7$ $= 15x^3 + 6x^2 - 21x$ $(x + 2)(7x - 3)$ $= x \cdot 7x - x \cdot 3 + 2 \cdot 7x - 2 \cdot 3$ $= 7x^2 - 3x + 14x - 6$ $= 7x^2 + 11x - 6$

5.3   Exercises

CONCEPTS AND VOCABULARY

1. The equation $x^2 \cdot x^3 = x^5$ illustrates what rule of exponents?

2. The equation $3(x - 2) = 3x - 6$ illustrates what property?

3. The product of two polynomials may be found by multiplying every _____ in the first polynomial by every _____ in the second polynomial and then combining like terms.

4. Polynomials can be multiplied horizontally or _____.

MULTIPLICATION OF MONOMIALS

Exercises 5–14: Multiply.

5. $x^2 \cdot x^5$ 6. $-a \cdot a^5$

7. $-3a \cdot 4a$ 8. $7x \cdot 5x$

9. $4x^3 \cdot 5x^2$ 10. $6b^6 \cdot 3b^5$

11. $xy^2 \cdot 4xy$ 12. $3ab \cdot ab^2$

13. $(-3xy^2)(4x^2y)$ 14. $(-r^2t^2)(-r^3t)$

MULTIPLICATION OF MONOMIALS AND POLYNOMIALS

Exercises 15–36: Multiply and simplify the expression.

15. $3(x + 4)$ 16. $-7(4x - 1)$

17. $-5(9x + 1)$ 18. $10(1 - 6x)$

19. $(4 - z)z$ 20. $3z(1 - 5z)$

21. $-y(5 + 3y)$ 22. $(2y - 8)2y$

23. $3x(5x^2 - 4)$ 24. $-6x(2x^3 + 1)$

25. $(6x - 6)x^2$ 26. $(1 - 2x^2)3x^2$

27. $-8(4t^2 + t + 1)$ 28. $7(3t^2 - 2t - 5)$

29. $n^2(-5n^2 + n - 2)$ 30. $6n^3(2 - 4n + n^2)$

31. $xy(x + y)$ 32. $ab(2a - 3b)$

33. $x^2(x^2y - xy^2)$ 34. $2y^2(xy - 5)$

35. $-ab(a^3 - 2b^3)$ 36. $5rt(r^2 + 2rt + t^2)$

MULTIPLICATION OF POLYNOMIALS

Exercises 37–42: (Refer to Example 5.) Multiply the given expression (a) geometrically and (b) symbolically.

37. $x(x + 3)$ **38.** $2x(x + 5)$

39. $(x + 2)(x + 2)$ **40.** $(x + 1)(x + 3)$

41. $(x + 3)(x + 6)$ **42.** $(x + 5)(x + 2)$

Exercises 43–70: Multiply and simplify the expression.

43. $(x + 3)(x + 5)$ **44.** $(x - 4)(x - 7)$

45. $(x - 8)(x - 9)$ **46.** $(x + 10)(x + 10)$

47. $(3z - 2)(2z - 5)$ **48.** $(z + 6)(2z - 1)$

49. $(8b - 1)(8b + 1)$ **50.** $(3t + 2)(3t - 2)$

51. $(10y + 7)(y - 1)$ **52.** $(y + 6)(2y + 7)$

53. $(5 - 3a)(1 - 2a)$ **54.** $(4 - a)(5 + 3a)$

55. $(1 - 3x)(1 + 3x)$ **56.** $(10 - x)(5 - 2x)$

57. $(x - 1)(x^2 + 1)$ **58.** $(x + 2)(x^2 - x)$

59. $(x^2 + 4)(4x - 3)$ **60.** $(3x^2 - 1)(3x^2 + 1)$

61. $(2n + 1)(n^2 + 3)$ **62.** $(2 - n^2)(1 + n^2)$

63. $(m + 1)(m^2 + 3m + 1)$

64. $(m - 2)(m^2 - m + 5)$

65. $(3x - 2)(2x^2 - x + 4)$

66. $(5x + 4)(x^2 - 3x + 2)$

67. $(x + 1)(x^2 - x + 1)$

68. $(x - 2)(x^2 + 4x + 4)$

69. $(4b^2 + 3b + 7)(b^2 + 3)$

70. $(-3a^2 - 2a + 1)(3a^2 - 3)$

Exercises 71–76: Multiply the polynomials vertically.

71. $(x + 2)(x^2 - 3x + 1)$

72. $(2y - 3)(3y^2 - 2y - 2)$

73. $(a - 2)(a^2 + 2a + 4)$

74. $(b - 3)(b^2 + 3b + 9)$

75. $(3x^2 - x + 1)(2x^2 + 1)$

76. $(2x^2 - 3x - 5)(2x^2 + 3)$

77. Thinking Generally If a polynomial with m terms and a polynomial with n terms are multiplied, how many terms are there in the product before like terms are combined?

78. Thinking Generally When a polynomial with m terms is multiplied by a second polynomial, the product contains k terms before like terms are combined. How many terms does the second polynomial contain?

APPLICATIONS

79. *Volume of a Box* (Refer to Example 9.) A box has a width 4 inches less than its height and a length 2 inches more than its height.
   (a) If h is the height of the box, write a polynomial that represents the volume of the box.
   (b) Use this polynomial to calculate the volume for $h = 25$ inches.

80. *Surface Area of a Box* Use the drawing of the box to write a polynomial that represents each of the following.

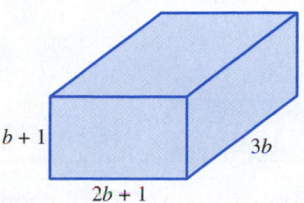

   (a) The area of its bottom
   (b) The area of its front
   (c) The area of its right side
   (d) The total area of its six sides

81. *Perimeter of a Pen* A rectangular pen for a pet has a perimeter of 100 feet. If one side of the pen has length x, then its area is given by $x(50 - x)$.
   (a) Multiply this expression.
   (b) Evaluate the expression obtained in part (a) for $x = 25$.

82. *Rectangular Garden* A rectangular garden has a perimeter of 500 feet.
   (a) If one side of the garden has length x, then write a polynomial expression that gives its area. Multiply this expression completely.
   (b) Evaluate the expression for $x = 50$ and interpret your answer.

83. *Surface Area of a Cube* Write a polynomial that represents the total area of the six sides of the cube having edges with length $x + 1$.

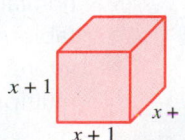

84. *Surface Area of a Sphere* The surface area of a sphere with radius r is $4\pi r^2$. Write a polynomial that gives the surface area of a sphere with radius $x + 2$. Leave your answer in terms of π.

85. *Toy Rocket* A toy rocket is shot straight up into the air. Its height h in feet above the ground after t seconds is represented by the expression $t(64 - 16t)$.
(a) Multiply this expression.
(b) Evaluate the expression obtained in part (a) and the given expression for $t = 2$.
(c) Are your answers in part (b) the same? Should they be the same?

86. *Toy Rocket on the Moon* (Refer to the preceding exercise.) If the same toy rocket were flown on the moon, then its height h in feet after t seconds would be $t\left(64 - \frac{5}{2}t\right)$.
(a) Multiply this expression.
(b) Evaluate the expression obtained in part (a) and the given expression for $t = 2$. Did the rocket go higher on the moon?

WRITING ABOUT MATHEMATICS

87. Explain how the acronym FOIL relates to multiplying two binomials, such as $x + 3$ and $2x + 1$.

88. Does the FOIL method work for multiplying a binomial and a trinomial? Explain.

89. Explain in words how to multiply any two polynomials. Give an example.

90. Give two properties of real numbers that are used for multiplying $3x(5x^2 - 3x + 2)$. Explain your answer.

Group Activity   Working with Real Data

Directions: Form a group of 2 to 4 people. Select someone to record the group's responses for this activity. All members of the group should work cooperatively to answer the questions. If your instructor asks for your results, each member of the group should be prepared to respond.

Biology Some types of worms have a remarkable ability to live without moisture. The following table from one study shows the number of worms W surviving after x days without moisture.

x (days)	0	20	40	80	120	160
W (worms)	50	48	45	36	20	3

Source: D. Brown and P. Rothery, *Models in Biology*.

(a) Use the equation $W = -0.0014x^2 - 0.076x + 50$ to find W for each x-value in the table.
(b) Discuss how well this equation approximates the data.
(c) Use this equation to estimate the number of worms on day 60 and on day 180. Which answer is most accurate? Explain.

5.4   Special Products

Product of a Sum and Difference • Squaring Binomials • Cubing Binomials

A LOOK INTO MATH ▶ Polynomials are often used to approximate real-world phenomena in applications. Polynomials have played an important role in the development of everyday products such as tablet computers, cell phones, and automobiles. Even digital images in computers and interest calculations at a bank make use of polynomials. In this section we discuss how to multiply some special types of binomials.

Product of a Sum and Difference

Products of the form $(a + b)(a - b)$ occur frequently in mathematics. Other examples include

$$(x + y)(x - y) \quad \text{and} \quad (2r + 3t)(2r - 3t).$$

These products can always be multiplied by using the techniques discussed in Section 5.3. However, there is a faster way to multiply these special products.

$$(a + b)(a - b) = a \cdot a - a \cdot b + b \cdot a - b \cdot b$$
$$= a^2 - ab + ba - b^2$$
$$= a^2 - b^2$$

In words, the product of a sum of two numbers and their difference equals the difference of their squares. We generalize this method as follows.

READING CHECK

- Explain in words how you can find the product of the sum of two numbers and their difference.

PRODUCT OF A SUM AND DIFFERENCE

For any real numbers a and b,
$$(a + b)(a - b) = a^2 - b^2.$$

STUDY TIP

In mathematics, there are often several correct ways to perform a particular process. If your instructor does not require you to use a specified method, choose the one that works best for you.

EXAMPLE 1 **Finding products of sums and differences**

Multiply.
(a) $(x + y)(x - y)$ **(b)** $(z - 2)(z + 2)$
(c) $(2r + 3t)(2r - 3t)$ **(d)** $(5m^2 - 4n^2)(5m^2 + 4n^2)$

Solution
(a) If we let $a = x$ and $b = y$, then we can apply the rule
$$(a + b)(a - b) = a^2 - b^2.$$

Thus
$$(x + y)(x - y) = (x)^2 - (y)^2$$
$$= x^2 - y^2.$$

(b) Because the expressions $(z + 2)(z - 2)$ and $(z - 2)(z + 2)$ are equal by the commutative property, we can apply the formula for the product of a sum and difference.
$$(z - 2)(z + 2) = (z)^2 - (2)^2$$
$$= z^2 - 4$$

(c) Let $a = 2r$ and $b = 3t$. Then the product can be evaluated as follows.
$$(2r + 3t)(2r - 3t) = (2r)^2 - (3t)^2$$
$$= 4r^2 - 9t^2$$

(d) $(5m^2 - 4n^2)(5m^2 + 4n^2) = (5m^2)^2 - (4n^2)^2 = 25m^4 - 16n^4$

Now Try Exercises 7, 13, 17

The next example demonstrates how the product of a sum and difference can be used to find some products of numbers mentally.

EXAMPLE 2 **Finding a product**

Use the product of a sum and difference to find $22 \cdot 18$.

Solution

Because $22 = 20 + 2$ and $18 = 20 - 2$, rewrite and evaluate $22 \cdot 18$ as follows.

$$\begin{aligned} 22 \cdot 18 &= (20 + 2)(20 - 2) && \text{Rewrite 22 as 20 + 2 and 18 as 20 − 2.} \\ &= 20^2 - 2^2 && \text{Product of a sum and difference} \\ &= 400 - 4 && \text{Evaluate exponents.} \\ &= 396 && \text{Subtract.} \end{aligned}$$

Now Try Exercise 21

Squaring Binomials

Area $= (a + b)^2$

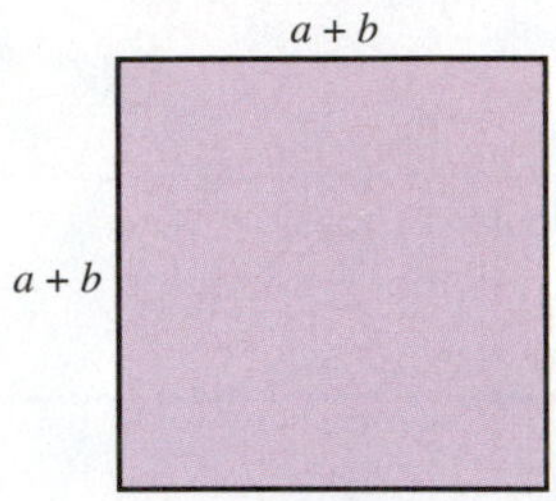

Figure 5.11

Because each side of the square shown in Figure 5.11 has length $(a + b)$, its area equals

$$(a + b)(a + b),$$

which can be written as $(a + b)^2$. We can multiply this expression as follows.

$$\begin{aligned} (a + b)^2 &= (a + b)(a + b) \\ &= a^2 + ab + ba + b^2 \\ &= a^2 + 2ab + b^2 \end{aligned}$$

This result is illustrated geometrically in Figure 5.12, where the area of the large square is $(a + b)^2$. This area can also be found by adding the areas of the four small rectangles.

$$a^2 + ab + ba + b^2 = a^2 + 2ab + b^2$$

$(a + b)^2 = a^2 + 2ab + b^2$

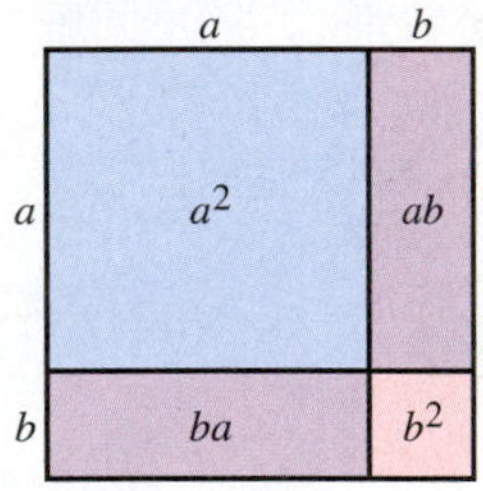

Figure 5.12

The geometric and symbolic results are the same. Note that to obtain the middle term, $2ab$, we can multiply the two terms, a and b, in the binomial and *double* the result.

A similar product that is also the square of a binomial can be calculated as

$$\begin{aligned} (a - b)^2 &= (a - b)(a - b) \\ &= a^2 - ab - ba + b^2 \\ &= a^2 - 2ab + b^2. \end{aligned}$$

These results are summarized as follows.

READING CHECK

- Explain in words how you can square a binomial.

SQUARING A BINOMIAL

For any real numbers a and b,

$$(a + b)^2 = a^2 + 2ab + b^2 \quad \text{and}$$
$$(a - b)^2 = a^2 - 2ab + b^2.$$

That is, the square of a binomial equals the square of the first term, plus (or minus) twice the product of the two terms, plus the square of the last term.

NOTE: $(a + b)^2 \neq a^2 + b^2$. Do not forget the middle term when squaring a binomial.

EXAMPLE 3 **Squaring a binomial**

Multiply.

(a) $(x + 3)^2$ **(b)** $(2x - 5)^2$ **(c)** $(1 - 5y)^2$ **(d)** $(7a^2 + 3b)^2$

Solution

(a) Let $a = x$ and $b = 3$ in the formula $(a + b)^2 = a^2 + 2ab + b^2$.

$$(x + 3)^2 = (x)^2 + 2(x)(3) + (3)^2$$
$$= x^2 + 6x + 9$$

(b) Apply the formula $(a - b)^2 = a^2 - 2ab + b^2$ with $a = 2x$ and $b = 5$.

$$(2x - 5)^2 = (2x)^2 - 2(2x)(5) + (5)^2$$
$$= 4x^2 - 20x + 25$$

(c) $(1 - 5y)^2 = (1)^2 - 2(1)(5y) + (5y)^2 = 1 - 10y + 25y^2$

(d) $(7a^2 + 3b)^2 = (7a^2)^2 + 2(7a^2)(3b) + (3b)^2 = 49a^4 + 42a^2b + 9b^2$

▌ **Now Try Exercises 27, 29, 35, 39**

MAKING CONNECTIONS

Multiplying Binomials and Special Products

If you forget these special products, you can still multiply polynomials by using earlier techniques. For example, the binomial in Example 3(b) can be multiplied as

$$(2x - 5)^2 = (2x - 5)(2x - 5)$$
$$= 2x \cdot 2x - 2x \cdot 5 - 5 \cdot 2x + 5 \cdot 5$$
$$= 4x^2 - 10x - 10x + 25$$
$$= 4x^2 - 20x + 25.$$

Figure 5.13 Digital Picture

▶ **REAL-WORLD CONNECTION** NASA first developed digital pictures because they were easy to transmit through space and because they provided clear images. A digital image from outer space is shown in Figure 5.13.

Today, digital cameras are readily available, and the Internet uses digital images exclusively. The next example shows how polynomials relate to digital pictures.

EXAMPLE 4   **Calculating the size of a digital picture**

A digital picture comprises tiny square units called *pixels*. Shading individual pixels creates a picture. A simplified version of a digital picture of the letter T is shown in Figure 5.14. This picture includes an image of the letter T that measures 3 pixels by 3 pixels and a 1-pixel border.

(a) If a square digital image measures x pixels by x pixels, then a picture that includes the image and a 1-pixel border will measure $x + 2$ pixels by $x + 2$ pixels. Find a polynomial that gives the total number of pixels in the picture, including the border.

(b) Let $x = 3$ and evaluate the polynomial. Does it agree with Figure 5.14?

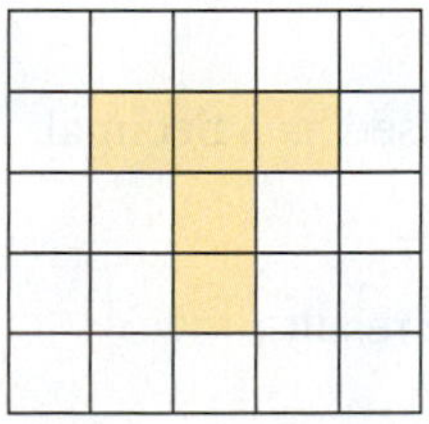

Figure 5.14

Solution

(a) The total number of pixels equals $(x + 2)$ times $(x + 2)$, or $(x + 2)^2$.

$$(x + 2)^2 = x^2 + 4x + 4$$

(b) For $x = 3$, the polynomial evaluates to $3^2 + 4 \cdot 3 + 4 = 25$, the total number of pixels. This result agrees with Figure 5.14, which has a total of $5 \cdot 5 = 25$ pixels with a 3 pixel by 3 pixel image of the letter T inside.

▌ **Now Try Exercise 79**

Cubing Binomials

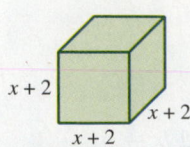

Figure 5.15
Volume = $(x + 2)^3$

To calculate the volume of the cube shown in Figure 5.15, we find the product of its length, width, and height. Because all sides have the same measure, its volume is $(x + 2)^3$. That is, the volume equals the *cube* of $x + 2$.

To multiply the expression $(x + 2)^3$, we proceed as follows.

$$(x + 2)^3 = (x + 2)(x + 2)^2 \qquad a^3 = a \cdot a^2$$

$$= (x + 2)(x^2 + 4x + 4) \qquad \text{Square the binomial.}$$

$$= x \cdot x^2 + x \cdot 4x + x \cdot 4 + 2 \cdot x^2 + 2 \cdot 4x + 2 \cdot 4 \qquad \text{Multiply the polynomials.}$$

$$= x^3 + 4x^2 + 4x + 2x^2 + 8x + 8 \qquad \text{Simplify terms.}$$

$$= x^3 + 6x^2 + 12x + 8 \qquad \text{Combine like terms.}$$

EXAMPLE 5 | **Cubing a binomial**

Multiply $(2z - 3)^3$.

Solution

$$(2z - 3)^3 = (2z - 3)(2z - 3)^2 \qquad a^3 = a \cdot a^2$$

$$= (2z - 3)(4z^2 - 12z + 9) \qquad \text{Square the binomial.}$$

$$= 8z^3 - 24z^2 + 18z - 12z^2 + 36z - 27 \qquad \text{Multiply the polynomials.}$$

$$= 8z^3 - 36z^2 + 54z - 27 \qquad \text{Combine like terms.}$$

NOTE: $(2z - 3)^3 \neq (2z)^3 - (3)^3 = 8z^3 - 27$.

Now Try Exercise 47

CRITICAL THINKING

Suppose that a student is convinced that the expressions

$$(x + y)^3 \quad \text{and} \quad x^3 + y^3$$

are equal. How could you convince the student otherwise?

EXAMPLE 6 | **Calculating interest**

If a savings account pays x percent annual interest, where x is expressed as a decimal, then after 3 years a sum of money will grow by a factor of $(1 + x)^3$.
(a) Multiply this expression.
(b) Evaluate the expression for $x = 0.10$ (or 10%), and interpret the result.

Solution
(a) $(1 + x)^3 = (1 + x)(1 + x)^2 \qquad a^3 = a \cdot a^2$

$$= (1 + x)(1 + 2x + x^2) \qquad \text{Square the binomial.}$$

$$= 1 + 2x + x^2 + x + 2x^2 + x^3 \qquad \text{Multiply the polynomials.}$$

$$= 1 + 3x + 3x^2 + x^3 \qquad \text{Combine like terms.}$$

(b) Let $x = \mathbf{0.1}$ in the expression $1 + 3x + 3x^2 + x^3$.

$$1 + 3(\mathbf{0.1}) + 3(\mathbf{0.1})^2 + (\mathbf{0.1})^3 = 1.331$$

The sum of money will increase by a factor of 1.331. For example, $1000 deposited in this account will grow to $1331 in 3 years.

Now Try Exercise 75

5.4   Putting It All Together

CONCEPT	EXPLANATION	EXAMPLES
Product of a Sum and Difference	For any real numbers x and y, $$(x + y)(x - y) = x^2 - y^2.$$	$(x + 6)(x - 6) = x^2 - 36,$ $(2x - 3)(2x + 3) = 4x^2 - 9,$ and $(x^2 + y^2)(x^2 - y^2) = x^4 - y^4$
Squaring a Binomial	For all real numbers x and y, $$(x + y)^2 = x^2 + 2xy + y^2 \quad \text{and}$$ $$(x - y)^2 = x^2 - 2xy + y^2.$$	$(x + 4)^2 = x^2 + 8x + 16,$ $(5x - 2)^2 = 25x^2 - 20x + 4,$ and $(x^2 + y^2)^2 = x^4 + 2x^2y^2 + y^4$
Cubing a Binomial	Multiply the binomial by its square.	$(x + 3)^3$ $= (x + 3)(x + 3)^2$ $= (x + 3)(x^2 + 6x + 9)$ $= x^3 + 6x^2 + 9x + 3x^2 + 18x + 27$ $= x^3 + 9x^2 + 27x + 27$

5.4   Exercises

MyMathLab   Math XL PRACTICE   WATCH   DOWNLOAD   READ   REVIEW

CONCEPTS AND VOCABULARY

1. $(a + b)(a - b) = $ _____

2. $(a + b)^2 = $ _____

3. $(a - b)^2 = $ _____

4. $(a + b)^3 = (a + b) \cdot $ _____

5. (True or False?) The two expressions $(x + y)^2$ and $x^2 + y^2$ are equal for all real numbers x and y.

6. (True or False?) The two expressions $(r - t)^2$ and $r^2 - t^2$ are equal for all real numbers r and t.

PRODUCT OF A SUM AND DIFFERENCE

Exercises 7–20: Multiply.

7. $(x - 3)(x + 3)$ **8.** $(x + 6)(x - 6)$

9. $(4x - 1)(4x + 1)$ **10.** $(10x + 3)(10x - 3)$

11. $(1 + 2a)(1 - 2a)$ **12.** $(4 - 9b)(4 + 9b)$

13. $(2x + 3y)(2x - 3y)$ **14.** $(5r - 6t)(5r + 6t)$

15. $(ab - 5)(ab + 5)$ **16.** $(2xy + 7)(2xy - 7)$

17. $(a^2 - b^2)(a^2 + b^2)$ **18.** $(3x^2 + y^2)(3x^2 - y^2)$

19. $(x^3 - y^3)(x^3 + y^3)$ **20.** $(2a^4 + b^4)(2a^4 - b^4)$

Exercises 21–26: (Refer to Example 2.) Use the product of a sum and a difference to evaluate the expression.

21. $101 \cdot 99$ **22.** $52 \cdot 48$

23. $23 \cdot 17$ **24.** $29 \cdot 31$

25. $90 \cdot 110$ **26.** $38 \cdot 42$

SQUARING BINOMIALS

Exercises 27–40: Multiply.

27. $(a - 2)^2$

28. $(x - 7)^2$

29. $(2x + 3)^2$

30. $(7x - 2)^2$

31. $(3b + 5)^2$

32. $(7t + 10)^2$

33. $\left(\frac{3}{4}a - 4\right)^2$

34. $\left(\frac{1}{5}a + 1\right)^2$

35. $(1 - b)^2$

36. $(1 - 4a)^2$

37. $(5 + y^3)^2$

38. $(9 - 5x^2)^2$

39. $(a^2 + b)^2$

40. $(x^3 - y^3)^2$

CUBING BINOMIALS

Exercises 41–48: Multiply.

41. $(a + 1)^3$

42. $(b + 4)^3$

43. $(x - 2)^3$

44. $(y - 7)^3$

45. $(2x + 1)^3$

46. $(4z + 3)^3$

47. $(6u - 1)^3$

48. $(5v + 3)^3$

MULTIPLICATION OF POLYNOMIALS

Exercises 49–66: Multiply, using any appropriate method.

49. $4(5x + 9)$

50. $(2x + 1)(3x - 5)$

51. $(x - 5)(x + 7)$

52. $(x + 10)(x + 10)$

53. $(3x - 5)^2$

54. $(x - 3)(x + 9)$

55. $(5x + 3)(5x + 4)$

56. $-x^3(x^2 - x + 1)$

57. $(4b - 5)(4b + 5)$

58. $(x + 5)^3$

59. $-5x(4x^2 - 7x + 2)$

60. $(4x^2 - 5)(4x^2 + 5)$

61. $(4 - a)^3$

62. $2x(x - 3)^3$

63. $x(x + 3)^2$

64. $(x - 1)^2(x + 1)$

65. $(x + 2)(x - 2)(x + 1)(x - 1)$

66. $(x - y)(x + y)(x^2 + y^2)$

67. Thinking Generally Multiply $(a^n + b^n)(a^n - b^n)$.

68. Thinking Generally Multiply $(a^n + b^n)^2$.

APPLICATIONS

Exercises 69–72: Do each part and verify that your answers are the same.

    (a) *Find the area of the large square by multiplying its length and width.*

    (b) *Find the sum of the areas of the smaller rectangles inside the large square.*

69.
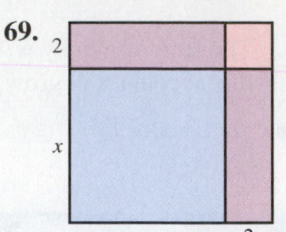

70.
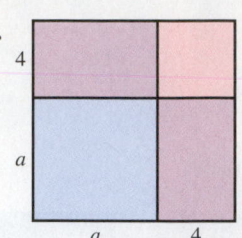

71.
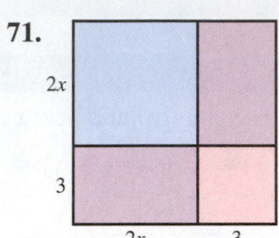

72.
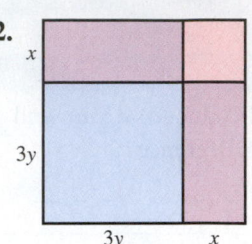

Exercises 73 and 74: Find a polynomial that represents the following.

    (a) *The outside surface area given by the six sides of the cube*

    (b) *The volume of the cube*

73.
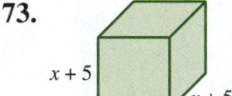

74.
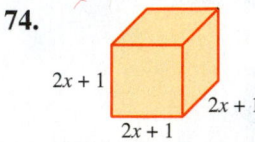

75. *Compound Interest* (Refer to Example 6.) If a sum of money is deposited in a savings account that is paying x percent annual interest (expressed as a decimal), then this sum of money increases by a factor of $(1 + x)^2$ after 2 years.
(a) Multiply this expression.
(b) Evaluate the polynomial expression found in part (a) for an annual interest rate of 10%, or $x = 0.10$, and interpret the answer.

76. *Compound Interest* If a sum of money is deposited in a savings account that is paying x percent annual interest, then this sum of money increases by a factor of $\left(1 + \frac{1}{100}x\right)^3$ after 3 years.
(a) Multiply this expression.
(b) Evaluate the polynomial expression in part (a) for an annual interest rate of 8%, or $x = 8$, and interpret the answer.

77. *Probability* If there is an x percent chance of rain on each of two consecutive days, then the expression $(1 - x)^2$ gives the percent chance that neither day will have rain. Assume that all percentages are expressed as decimals.

(a) Multiply this expression.

(b) Evaluate the polynomial expression in part (a) for a 50% chance of rain, or $x = 0.50$, and interpret the answer.

78. *Probability* If there is an x percent chance of rolling a 6 with one die, then the expression $(1 - x)^3$ gives the percent chance of not rolling a 6 with three dice. Assume that all percentages are expressed as decimals or fractions.

(a) Multiply this expression.

(b) Evaluate the polynomial expression found in part (a) for a 16.6% chance of rolling a 6, or $x = \frac{1}{6}$, and interpret the answer.

79. *Swimming Pool* A square swimming pool has an 8-foot-wide sidewalk around it.

(a) If the sides of the pool have length z, as shown in the accompanying figure, find a polynomial that gives the area of the sidewalk.

(b) Evaluate the polynomial in part (a) for $z = 60$ and interpret the answer.

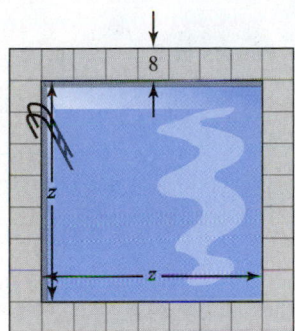

80. *Digital Picture* (Refer to Example 4.) Suppose that a digital picture, including its border, is $x + 2$ pixels by $x + 2$ pixels and that the actual image inside the border is $x - 2$ pixels by $x - 2$ pixels, as shown in the following figure.

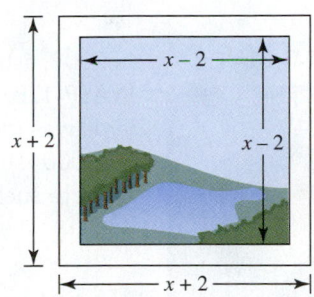

(a) Find a polynomial that gives the number of pixels in the border.

(b) Evaluate the polynomial in part (a) for $x = 5$.

(c) Sketch a digital picture of the letter H with $x = 5$. Does the picture agree with the answer in part (b)?

(d) Digital pictures typically have large values for x. If a picture has $x = 500$, find the total number of pixels in its border.

WRITING ABOUT MATHEMATICS

81. Explain why $(a + b)^2$ does not equal $a^2 + b^2$ in general for real numbers a and b.

82. Explain how to find the cube of a binomial.

SECTIONS 5.3 and 5.4   **Checking Basic Concepts**

1. Multiply each expression.

(a) $(-3xy^4)(5x^2y)$ (b) $-x(6 - 4x)$

(c) $3ab(a^2 - 2ab + b^2)$

2. Multiply each expression.

(a) $(x + 3)(4x - 3)$

(b) $(x^2 - 1)(2x^2 + 2)$

(c) $(x + y)(x^2 - xy + y^2)$

3. Multiply each expression.

(a) $(5x + 2)(5x - 2)$ (b) $(x + 3)^2$

(c) $(2 - 7x)^2$ (d) $(t + 2)^3$

4. Complete each part and verify that your answers are the same.

(a) Find the area of the large square by squaring the length of one of its sides.

(b) Find the sum of the areas of the smaller rectangles inside the large square.

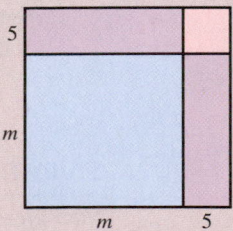

5.5 Integer Exponents and the Quotient Rule

Negative Integers as Exponents • The Quotient Rule • Other Rules for Exponents • Scientific Notation

A LOOK INTO MATH ▶

In 2009, astronomers discovered a large planet that orbits a distant star. The planet, named WASP-17b, is about 5,880,000,000,000,000 miles from Earth. Also in 2009, the H1N1 virus was identified in a worldwide influenza pandemic. A typical flu virus measures about 0.00000468 inch across. In this section, we will discuss how integer exponents can be used to write such numbers in *scientific notation*. (*Source: Scientific American.*)

Negative Integers as Exponents

So far we have defined exponents that are whole numbers. For example,

$$5^0 = 1 \quad \text{and} \quad 2^3 = 2 \cdot 2 \cdot 2 = 8.$$

What if an exponent is a negative integer? To answer this question, consider Table 5.4, which shows values for decreasing powers of 2. Note that each time the exponent on 2 decreases by 1, the resulting value is divided by 2.

TABLE 5.4 Powers of 2

Power of 2	Value
2^1	2
2^0	1
2^{-1}	$\frac{1}{2} = \frac{1}{2^1}$
2^{-2}	$\frac{1}{4} = \frac{1}{2^2}$

Decrease exponent by 1 — Divide by 2 (applied between each successive row)

Table 5.4 shows that $2^{-1} = \frac{1}{2^1}$ and $2^{-2} = \frac{1}{2^2}$. In other words, if the exponent on 2 is negative, then the expression is equal to the reciprocal of the corresponding expression with a positive exponent on 2. This discussion suggests the following definition for negative integer exponents.

READING CHECK

• How is a negative integer power on a base related to the corresponding positive integer power on that base?

NEW VOCABULARY

☐ Scientific notation

NEGATIVE INTEGER EXPONENTS

Let a be a nonzero real number and n be a positive integer. Then

$$a^{-n} = \frac{1}{a^n}.$$

That is, a^{-n} is the reciprocal of a^n.

STUDY TIP

Mathematics often builds on concepts that have already been studied. Try to get in the regular habit of reviewing topics from earlier parts of the text.

| EXAMPLE 1 | **Evaluating negative exponents** |

Simplify each expression.

(a) 2^{-3} **(b)** 7^{-1} **(c)** x^{-2} **(d)** $(x + y)^{-8}$

Solution

(a) Because $a^{-n} = \dfrac{1}{a^n}$, $2^{-3} = \dfrac{1}{2^3} = \dfrac{1}{2 \cdot 2 \cdot 2} = \dfrac{1}{8}$.

(b) $7^{-1} = \dfrac{1}{7^1} = \dfrac{1}{7}$

(c) $x^{-2} = \dfrac{1}{x^2}$

(d) $(x + y)^{-8} = \dfrac{1}{(x + y)^8}$

Now Try Exercises 7, 19, 25(b)

TECHNOLOGY NOTE

Negative Exponents

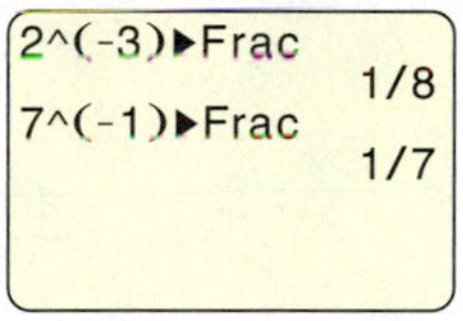

CALCULATOR HELP

To use the fraction feature (Frac), see Appendix A (pages AP-1 and AP-2).

Calculators can be used to evaluate negative exponents. The figure shows how a graphing calculator evaluates the expressions in parts (a) and (b) of Example 1.

```
2^(-3)▶Frac
              1/8
7^(-1)▶Frac
              1/7
```

The rules for exponents discussed in this chapter so far also apply to expressions having negative exponents. For example, we can apply the product rule, $a^m \cdot a^n = a^{m+n}$, as follows.

— **Add**

$$2^{-3} \cdot 2^2 = 2^{-3+2} = 2^{-1} = \dfrac{1}{2}$$

We can check this result by evaluating the expression without using the product rule.

$$2^{-3} \cdot 2^2 = \dfrac{1}{2^3} \cdot 2^2 = \dfrac{1}{8} \cdot 4 = \dfrac{4}{8} = \dfrac{1}{2}$$

| EXAMPLE 2 | **Using the product rule with negative exponents** |

Evaluate each expression.

(a) $5^2 \cdot 5^{-4}$ **(b)** $3^{-2} \cdot 3^{-1}$

Solution

— **Add**

(a) $5^2 \cdot 5^{-4} = 5^{2+(-4)} = 5^{-2} = \dfrac{1}{5^2} = \dfrac{1}{25}$

(b) $3^{-2} \cdot 3^{-1} = 3^{-2+(-1)} = 3^{-3} = \dfrac{1}{3^3} = \dfrac{1}{27}$

Now Try Exercise 9

EXAMPLE 3 **Using the rules of exponents**

Simplify the expression. Write the answer using positive exponents.

(a) $x^2 \cdot x^{-5}$ (b) $(y^3)^{-4}$ (c) $(rt)^{-5}$ (d) $(ab)^{-3}(a^{-2}b)^3$

Solution

(a) Using the product rule, $a^m \cdot a^n = a^{m+n}$, gives

$$x^2 \cdot x^{-5} = x^{2+(-5)} = x^{-3} = \frac{1}{x^3}.$$

(b) Using the power rule, $(a^m)^n = a^{mn}$, gives

$$(y^3)^{-4} = y^{3(-4)} = y^{-12} = \frac{1}{y^{12}}.$$

(c) Using the power rule, $(ab)^n = a^n b^n$, gives

$$(rt)^{-5} = r^{-5}t^{-5} = \frac{1}{r^5} \cdot \frac{1}{t^5} = \frac{1}{r^5 t^5}.$$

This expression could also be simplified as follows.

$$(rt)^{-5} = \frac{1}{(rt)^5} = \frac{1}{r^5 t^5}$$

(d) $(ab)^{-3}(a^{-2}b)^3 = a^{-3}b^{-3}a^{-6}b^3$

$$= a^{-3+(-6)}b^{-3+3}$$

$$= a^{-9}b^0$$

$$= \frac{1}{a^9} \cdot 1$$

$$= \frac{1}{a^9}$$

Now Try Exercises 21, 27, 29(a)

The Quotient Rule

Consider the division problem

$$\frac{3^4}{3^2} = \frac{3 \cdot 3 \cdot 3 \cdot 3}{3 \cdot 3} = \frac{3}{3} \cdot \frac{3}{3} \cdot 3 \cdot 3 = 1 \cdot 1 \cdot 3^2 = 3^2.$$

Because there are two more 3s in the numerator than in the denominator, the result is

$$\underset{\text{Subtract}}{3^{4-2}} = 3^2.$$

That is, to divide exponential expressions having the *same base*, subtract the exponent of the denominator from the exponent of the numerator and keep the same base. This rule is called the *quotient rule*, which we express in symbols as follows.

THE QUOTIENT RULE

For any nonzero number a and integers m and n,

$$\frac{a^m}{a^n} = a^{m-n}.$$

EXAMPLE 4   **Using the quotient rule**

Simplify each expression. Write the answer using positive exponents.

(a) $\dfrac{4^3}{4^5}$ **(b)** $\dfrac{6a^7}{3a^4}$ **(c)** $\dfrac{xy^7}{x^2y^5}$

Solution

Subtract

(a) $\dfrac{4^3}{4^5} = 4^{3-5} = 4^{-2} = \dfrac{1}{4^2} = \dfrac{1}{16}$

(b) $\dfrac{6a^7}{3a^4} = \dfrac{6}{3} \cdot \dfrac{a^7}{a^4} = 2a^{7-4} = 2a^3$

(c) $\dfrac{xy^7}{x^2y^5} = \dfrac{x^1}{x^2} \cdot \dfrac{y^7}{y^5} = x^{1-2}y^{7-5} = x^{-1}y^2 = \dfrac{y^2}{x}$

Now Try Exercises 13(b), 31(b), 33(a)

MAKING CONNECTIONS

The Quotient Rule and Simplifying Quotients

Some quotients can be simplified mentally. Because

$$\frac{x^5}{x^3} = \frac{x \cdot x \cdot x \cdot x \cdot x}{x \cdot x \cdot x},$$

the quotient $\dfrac{x^5}{x^3}$ has five factors of x in the numerator and three factors of x in the denominator. There are two more factors of x in the numerator than in the denominator, $5 - 3 = 2$, so this expression simplifies to x^2. Similarly,

$$\frac{x^3}{x^5} = \frac{x \cdot x \cdot x}{x \cdot x \cdot x \cdot x \cdot x}$$

has two more factors of x in the denominator than in the numerator. This quotient $\dfrac{x^3}{x^5}$ simplifies to $\dfrac{1}{x^2}$. Use this technique to simplify the expressions

$$\frac{z^7}{z^4}, \quad \frac{a^5}{a^8}, \quad \text{and} \quad \frac{x^6y^2}{x^3y^7}.$$

Other Rules for Exponents

Other rules can be used to simplify expressions with negative exponents.

QUOTIENTS AND NEGATIVE EXPONENTS

The following three rules hold for any nonzero real numbers a and b and positive integers m and n.

1. $\dfrac{1}{a^{-n}} = a^n$ **2.** $\dfrac{a^{-n}}{b^{-m}} = \dfrac{b^m}{a^n}$ **3.** $\left(\dfrac{a}{b}\right)^{-n} = \left(\dfrac{b}{a}\right)^n$

We demonstrate the validity of these rules as follows.

1. $\dfrac{1}{a^{-n}} = \dfrac{1}{\dfrac{1}{a^n}} = 1 \cdot \dfrac{a^n}{1} = a^n$

2. $\dfrac{a^{-n}}{b^{-m}} = \dfrac{\dfrac{1}{a^n}}{\dfrac{1}{b^m}} = \dfrac{1}{a^n} \cdot \dfrac{b^m}{1} = \dfrac{b^m}{a^n}$

3. $\left(\dfrac{a}{b}\right)^{-n} = \dfrac{a^{-n}}{b^{-n}} = \dfrac{\dfrac{1}{a^n}}{\dfrac{1}{b^n}} = \dfrac{1}{a^n} \cdot \dfrac{b^n}{1} = \dfrac{b^n}{a^n} = \left(\dfrac{b}{a}\right)^{n}$

EXAMPLE 5 **Working with quotients and negative exponents**

Simplify each expression. Write the answer using positive exponents.

(a) $\dfrac{1}{2^{-5}}$ **(b)** $\dfrac{3^{-3}}{4^{-2}}$ **(c)** $\dfrac{5x^{-4}y^2}{10x^2y^{-4}}$ **(d)** $\left(\dfrac{2}{z^2}\right)^{-4}$

Solution

(a) $\dfrac{1}{2^{-5}} = 2^5 = 2 \cdot 2 \cdot 2 \cdot 2 \cdot 2 = 32$

(b) $\dfrac{3^{-3}}{4^{-2}} = \dfrac{4^2}{3^3} = \dfrac{16}{27}$

(c) $\dfrac{5x^{-4}y^2}{10x^2y^{-4}} = \dfrac{y^2y^4}{2x^2x^4} = \dfrac{y^6}{2x^6}$

(d) $\left(\dfrac{2}{z^2}\right)^{-4} = \left(\dfrac{z^2}{2}\right)^4 = \dfrac{z^8}{2^4} = \dfrac{z^8}{16}$

Now Try Exercises 15(b), 17, 37, 47

The rules for natural number exponents that are summarized in Section 5.1 on page 303 also hold for integer exponents. Additional rules for integer exponents are summarized as follows.

RULES FOR INTEGER EXPONENTS

The following rules hold for nonzero real numbers a and b, and positive integers m and n.

Description	Rule	Example
Negative Exponents (1)	$a^{-n} = \dfrac{1}{a^n}$	$9^{-2} = \dfrac{1}{9^2} = \dfrac{1}{81}$
The Quotient Rule	$\dfrac{a^m}{a^n} = a^{m-n}$	$\dfrac{2^3}{2^{-2}} = 2^{3-(-2)} = 2^5$
Negative Exponents (2)	$\dfrac{1}{a^{-n}} = a^n$	$\dfrac{1}{7^{-5}} = 7^5$
Negative Exponents (3)	$\dfrac{a^{-n}}{b^{-m}} = \dfrac{b^m}{a^n}$	$\dfrac{4^{-3}}{2^{-5}} = \dfrac{2^5}{4^3}$
Negative Exponents (4)	$\left(\dfrac{a}{b}\right)^{-n} = \left(\dfrac{b}{a}\right)^{n}$	$\left(\dfrac{1}{5}\right)^{-2} = \left(\dfrac{5}{1}\right)^2 = 25$

READING CHECK

- What kinds of numbers are often expressed in scientific notation?

Scientific Notation

Powers of 10 are important because they are used in science to express numbers that are either very small or very large in absolute value. Table 5.5 lists some powers of 10. Note that if the power of 10 decreases by 1, the result decreases by a factor of $\frac{1}{10}$, or equivalently, the decimal point is moved one place to the left. Table 5.6 shows the names of some important powers of 10.

TECHNOLOGY NOTE

Powers of 10
Calculators make use of scientific notation, as illustrated in the accompanying figure. The letter E denotes a power of 10. That is,

$2.5\text{E}13 = 2.5 \times 10^{13}$ and
$5\text{E}{-6} = 5 \times 10^{-6}$.

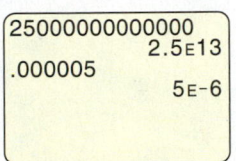

Note: The calculator has been set in *scientific mode*.

TABLE 5.5 Powers of 10

Power of 10	Value
10^3	1000
10^2	100
10^1	10
10^0	1
10^{-1}	$\frac{1}{10} = 0.1$
10^{-2}	$\frac{1}{100} = 0.01$
10^{-3}	$\frac{1}{1000} = 0.001$

TABLE 5.6 Important Powers of 10

Power of 10	Name
10^{12}	Trillion
10^9	Billion
10^6	Million
10^3	Thousand
10^{-1}	Tenth
10^{-2}	Hundredth
10^{-3}	Thousandth
10^{-6}	Millionth

Recall that numbers written in decimal notation are sometimes said to be in *standard form*. Decimal numbers that are either very large or very small in absolute value can be expressed in *scientific notation*.

REAL-WORLD CONNECTION As mentioned in A Look Into Math for this section, the distance to the planet WASP-17b is about 5,880,000,000,000,000 miles. This distance can be written in scientific notation as 5.88×10^{15} because

$$5{,}880{,}000{,}000{,}000{,}000 = 5.88 \times 10^{15}.$$

15 decimal places

The 10^{15} indicates that the decimal point in 5.88 should be moved **15** places to the **right**.

A typical virus is about 0.00000468 inch in diameter, which can be written in scientific notation as 4.68×10^{-6} because

$$0.00000468 = 4.68 \times 10^{-6}.$$

6 decimal places

The 10^{-6} indicates that the decimal point in 4.68 should be moved **6** places to the **left**.

The following definition provides a more complete explanation of scientific notation.

SCIENTIFIC NOTATION

A real number a is in **scientific notation** when a is written in the form $b \times 10^n$, where $1 \le |b| < 10$ and n is an integer.

EXAMPLE 6 **Converting scientific notation to standard form**

Write each number in standard form.
(a) 5.23×10^4 **(b)** 8.1×10^{-3} **(c)** 6×10^{-2}

Solution

(a) The positive exponent 4 indicates that the decimal point in 5.23 is to be moved **4** places to the **right**.

$$5.23 \times 10^4 = 5.\underset{1\ 2\ 3\ 4}{2\ 3\ 0\ 0.} = 52{,}300$$

(b) The negative exponent -3 indicates that the decimal point in 8.1 is to be moved **3** places to the **left**.

$$8.1 \times 10^{-3} = 0.\underset{1\ 2\ 3}{0\ 0\ 8.}1 = 0.0081$$

(c) $6 \times 10^{-2} = 0.\underset{1\ 2}{0\ 6.} = 0.06$

Now Try Exercises 61, 63

The following steps can be used for writing a positive number a in scientific notation.

WRITING A POSITIVE NUMBER IN SCIENTIFIC NOTATION

For a positive, rational number a expressed as a decimal, if $1 \le a < 10$, then $a = a \times 10^0$. Otherwise, use the following process to write a in scientific notation.

1. Move the decimal point in a until it becomes a number b such that $1 \le b < 10$.
2. Let the positive integer n be the number of places the decimal point was moved.
3. Write a in scientific notation as follows.
   - If $a \ge 10$, then $a = b \times 10^n$.
   - If $a < 1$, then $a = b \times 10^{-n}$.

NOTE: The scientific notation for a negative number a is the opposite of the scientific notation of $|a|$. For example, $450 = 4.5 \times 10^2$ and $-450 = -4.5 \times 10^2$.

EXAMPLE 7 **Writing a number in scientific notation**

Write each number in scientific notation.
(a) 308,000,000   (U.S. population in 2010)
(b) 0.001   (Approximate time in seconds for sound to travel one foot)

Solution

(a) Move the assumed decimal point in 308,000,000 eight places to obtain 3.08.

$$3.\underset{1\ 2\ 3\ 4\ 5\ 6\ 7\ 8}{0\ 8\ 0\ 0\ 0\ 0\ 0\ 0.}$$

Since $308{,}000{,}000 \ge 10$, the scientific notation is 3.08×10^8.

(b) Move the decimal point in 0.001 three places to obtain 1.

$$0.\underset{1\ 2\ 3}{0\,0\,1.}$$

Since $0.001 < 1$, the scientific notation is 1×10^{-3}.

Now Try Exercises 75, 79

Numbers in scientific notation can be multiplied by applying properties of real numbers and properties of exponents.

$$
\begin{aligned}
(6 \times 10^4) \cdot (3 \times 10^3) &= (6 \cdot 3) \times (10^4 \cdot 10^3) && \text{Properties of real numbers} \\
&= 18 \times 10^7 && \text{Product rule} \\
&= 1.8 \times 10^8 && \text{Scientific notation}
\end{aligned}
$$

CALCULATOR HELP

To display numbers in scientific notation, see Appendix A (page AP-2).

```
(6*10^4)(3*10^3)
              1.8E8
(6*10^4)/(3*10^3
)
              2E1
```

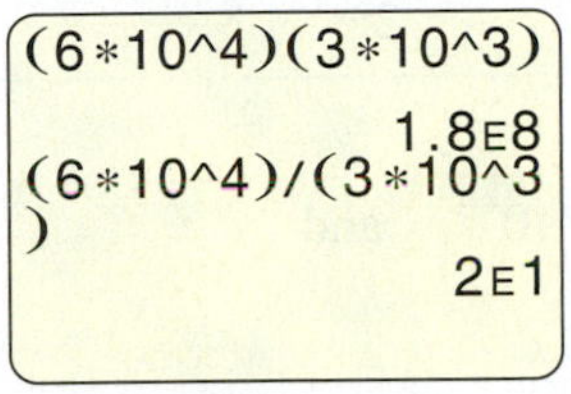

Figure 5.16

Division can also be performed with scientific notation.

$$
\begin{aligned}
\frac{6 \times 10^4}{3 \times 10^3} &= \frac{6}{3} \times \frac{10^4}{10^3} && \text{Property of fractions} \\
&= 2 \times 10^1 && \text{Quotient rule}
\end{aligned}
$$

These results are supported in Figure 5.16, where the calculator is in scientific mode.

In the next example we show how to use scientific notation in an application.

EXAMPLE 8 **Analyzing the cost of Internet advertising**

In 2009, a total of $\$2.38 \times 10^{10}$ was spent on Internet advertising in the United States. At that time the population of the United States was 3.05×10^8. Determine how much was spent per person on Internet advertising. (*Source: New York Times.*)

CRITICAL THINKING

Estimate the number of seconds that you have been alive. Write your answer in scientific notation.

Solution

To determine the amount spent per person, divide $\$2.38 \times 10^{10}$ by 3.05×10^8.

$$\frac{2.38 \times 10^{10}}{3.05 \times 10^8} = \frac{2.38}{3.05} \times 10^{10-8} \approx 0.78 \times 10^2 = 78$$

In 2009, about $78 per person was spent on Internet advertising.

Now Try Exercise 97

5.5 Putting It All Together

For the rules for integer exponents in this table, assume that a and b are nonzero real numbers and that m and n are integers.

CONCEPT	EXPLANATION	EXAMPLES
Negative Integer Exponents	$a^{-n} = \dfrac{1}{a^n}$	$2^{-4} = \dfrac{1}{2^4} = \dfrac{1}{16}, \quad a^{-8} = \dfrac{1}{a^8}, \quad$ and $(xy)^{-2} = \dfrac{1}{(xy)^2} = \dfrac{1}{x^2 y^2}$

continued on next page

continued from previous page

CONCEPT	EXPLANATION	EXAMPLES
Quotient Rule	$\dfrac{a^m}{a^n} = a^{m-n}$	$\dfrac{7^2}{7^4} = 7^{2-4} = 7^{-2} = \dfrac{1}{7^2} = \dfrac{1}{49}$ and $\dfrac{x^6}{x^3} = x^{6-3} = x^3$
Quotients and Negative Integer Exponents	1. $\dfrac{1}{a^{-n}} = a^n$ 2. $\dfrac{a^{-n}}{b^{-m}} = \dfrac{b^m}{a^n}$ 3. $\left(\dfrac{a}{b}\right)^{-n} = \left(\dfrac{b}{a}\right)^n$	1. $\dfrac{1}{5^{-2}} = 5^2 = 25$ 2. $\dfrac{x^{-4}}{y^{-2}} = \dfrac{y^2}{x^4}$ 3. $\left(\dfrac{2}{3}\right)^{-3} = \left(\dfrac{3}{2}\right)^3 = \dfrac{3^3}{2^3} = \dfrac{27}{8}$
Scientific Notation	Write a as $b \times 10^n$, where $1 \le |b| < 10$ and n is an integer.	$23{,}500 = 2.35 \times 10^4,$ $0.0056 = 5.6 \times 10^{-3},$ and $1000 = 1 \times 10^3$

5.5 Exercises

 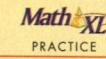

CONCEPTS AND VOCABULARY

Exercises 1–5: Complete the given rule for integer exponents m and n, where a and b are nonzero real numbers.

1. $a^{-n} = $ ____

2. $\dfrac{1}{a^{-n}} = $ ____

3. $\dfrac{a^m}{a^n} = $ ____

4. $\dfrac{a^{-n}}{b^{-m}} = $ ____

5. $\left(\dfrac{a}{b}\right)^{-n} = $ ____

6. To write a positive number a in scientific notation as $b \times 10^n$, the number b must satisfy ____.

NEGATIVE EXPONENTS

Exercises 7–18: Simplify the expression.

7. (a) 4^{-1} (b) $\left(\dfrac{1}{3}\right)^{-2}$

8. (a) 6^{-2} (b) 2.5^{-1}

9. (a) $2^3 \cdot 2^{-2}$ (b) $10^{-1} \cdot 10^{-2}$

10. (a) $3^{-4} \cdot 3^2$ (b) $10^4 \cdot 10^{-2}$

11. (a) $3^{-2} \cdot 3^{-1} \cdot 3^{-1}$ (b) $(2^3)^{-1}$

12. (a) $2^{-3} \cdot 2^5 \cdot 2^{-4}$ (b) $(3^{-2})^{-2}$

13. (a) $(3^2 4^3)^{-1}$ (b) $\dfrac{4^5}{4^2}$

14. (a) $(2^{-2} 3^2)^{-2}$ (b) $\dfrac{5^5}{5^3}$

15. (a) $\dfrac{1^9}{1^7}$ (b) $\dfrac{1}{4^{-3}}$

16. (a) $\dfrac{-6^4}{6}$ (b) $\dfrac{1}{6^{-2}}$

17. (a) $\dfrac{5^{-2}}{5^{-4}}$ (b) $\left(\dfrac{2}{7}\right)^{-2}$

18. (a) $\dfrac{7^{-3}}{7^{-1}}$ (b) $\left(\dfrac{3}{4}\right)^{-3}$

Exercises 19–50: Simplify the expression. Write the answer using positive exponents.

19. (a) x^{-1} **(b)** a^{-4}

20. (a) y^{-2} **(b)** z^{-7}

21. (a) $x^{-2} \cdot x^{-1} \cdot x$ **(b)** $a^{-5} \cdot a^{-2} \cdot a^{-1}$

22. (a) $y^{-3} \cdot y^4 \cdot y^{-5}$ **(b)** $b^5 \cdot b^{-3} \cdot b^{-6}$

23. (a) $x^2 y^{-3} x^{-5} y^6$ **(b)** $(xy)^{-3}$

24. (a) $a^{-2} b^{-6} b^3 a^{-1}$ **(b)** $(ab)^{-1}$

25. (a) $(2t)^{-4}$ **(b)** $(x+1)^{-7}$

26. (a) $(8c)^{-2}$ **(b)** $(a+b)^{-9}$

27. (a) $(a^{-2})^{-4}$ **(b)** $(rt^3)^{-2}$

28. (a) $(4x^3)^{-3}$ **(b)** $(xy^{-3})^{-2}$

29. (a) $(ab)^2(a^2)^{-3}$ **(b)** $\dfrac{x^4}{x^2}$

30. (a) $(x^3)^{-2}(xy)^4 y^{-5}$ **(b)** $\dfrac{y^9}{y^5}$

31. (a) $\dfrac{a^{10}}{a^{-3}}$ **(b)** $\dfrac{4z}{2z^4}$

32. (a) $\dfrac{b^5}{b^{-2}}$ **(b)** $\dfrac{12x^2}{24x^7}$

33. (a) $\dfrac{-4xy^5}{6x^3y^2}$ **(b)** $\dfrac{x^{-4}}{x^{-1}}$

34. (a) $\dfrac{12a^6b^2}{8ab^3}$ **(b)** $\dfrac{y^{-2}}{y^{-7}}$

35. (a) $\dfrac{10b^{-4}}{5b^{-5}}$ **(b)** $\left(\dfrac{a}{b}\right)^3$

36. (a) $\dfrac{8a^{-2}}{2a^{-3}}$ **(b)** $\left(\dfrac{2x}{y}\right)^5$

37. (a) $\dfrac{6x^2 y^{-4}}{18x^{-5}y^4}$ **(b)** $\dfrac{16a^{-3}b^{-5}}{4a^{-8}b}$

38. (a) $\dfrac{m^2 n^4}{3m^{-5}n^4}$ **(b)** $\dfrac{7x^{-3}y^{-5}}{x^{-3}y^{-2}}$

39. (a) $\dfrac{1}{y^{-5}}$ **(b)** $\dfrac{4}{2t^{-3}}$

40. (a) $\dfrac{1}{z^{-6}}$ **(b)** $\dfrac{5}{10b^{-5}}$

41. (a) $\dfrac{3a^4}{(2a^{-2})^3}$ **(b)** $\dfrac{(2b^5)^{-3}}{4b^{-6}}$

42. (a) $\dfrac{(2x^4)^{-2}}{5x^{-2}}$ **(b)** $\dfrac{2y^5}{(3y^{-4})^{-2}}$

43. (a) $\dfrac{1}{(xy)^{-2}}$ **(b)** $\dfrac{1}{(a^2 b)^{-3}}$

44. (a) $\dfrac{1}{(ab)^{-1}}$ **(b)** $\dfrac{1}{(rt^4)^{-2}}$

45. (a) $\dfrac{(3m^4 n)^{-2}}{(2mn^{-2})^3}$ **(b)** $\dfrac{(-4x^4 y)^2}{(xy^{-5})^{-3}}$

46. (a) $\dfrac{(x^4 y^2)^2}{(-2x^2 y^{-2})^3}$ **(b)** $\dfrac{(m^2 n^{-6})^{-2}}{(4m^2 n^{-4})^{-3}}$

47. (a) $\left(\dfrac{a}{b}\right)^{-2}$ **(b)** $\left(\dfrac{u}{4v}\right)^{-1}$

48. (a) $\left(\dfrac{2x}{y}\right)^{-3}$ **(b)** $\left(\dfrac{5u}{3v}\right)^{-2}$

49. (a) $\left(\dfrac{3a^4 b}{2ab^{-2}}\right)^{-2}$ **(b)** $\left(\dfrac{4m^4 n}{5m^{-3}n^2}\right)^2$

50. (a) $\left(\dfrac{2x^4 y^2}{3x^3 y^{-3}}\right)^3$ **(b)** $\left(\dfrac{a^{-5}b^2}{2ab^{-2}}\right)^{-2}$

51. Thinking Generally For positive integers m and n show that $\dfrac{a^n}{a^m} = \dfrac{1}{a^{m-n}}$.

52. Thinking Generally For positive integers m and n show that $\dfrac{a^{-n}}{a^{-m}} = a^{m-n}$.

SCIENTIFIC NOTATION

Exercises 53–58: (Refer to Table 5.6.) Write the value of the power of 10 in words.

53. 10^3 **54.** 10^6

55. 10^9 **56.** 10^{-1}

57. 10^{-2} **58.** 10^{-6}

Exercises 59–70: Write the expression in standard form.

59. 2×10^3 **60.** 5×10^2

61. 4.5×10^4 **62.** 7.1×10^6

63. 8×10^{-3} **64.** 9×10^{-1}

65. 4.56×10^{-4} **66.** 9.4×10^{-2}

67. 3.9×10^7 **68.** 5.27×10^6

69. -5×10^5 **70.** -9.5×10^3

Exercises 71–82: Write the number in scientific notation.

71. 2000

72. 11,000

73. 567

74. 9300

75. 12,000,000

76. 600,000

77. 0.004

78. 0.0008

79. 0.000895

80. 0.0123

81. −0.05

82. −0.934

Exercises 83–90: Evaluate the expression. Write the answer in standard form.

83. $(5 \times 10^3)(3 \times 10^2)$

84. $(2.1 \times 10^2)(2 \times 10^4)$

85. $(-3 \times 10^{-3})(5 \times 10^2)$

86. $(4 \times 10^2)(1 \times 10^3)(5 \times 10^{-4})$

87. $\dfrac{4 \times 10^5}{2 \times 10^2}$

88. $\dfrac{9 \times 10^2}{3 \times 10^6}$

89. $\dfrac{8 \times 10^{-6}}{4 \times 10^{-3}}$

90. $\dfrac{6.3 \times 10^2}{2 \times 10^{-3}}$

APPLICATIONS

91. *Light-year* The distance that light travels in 1 year is called a *light-year*. Light travels at 1.86×10^5 miles per second, and there are about 3.15×10^7 seconds in 1 year.
(a) Estimate the number of miles in 1 light-year.
(b) Except for the sun, Alpha Centauri is the nearest star, and its distance is 4.27 light-years from Earth. Estimate its distance in miles. Write your answer in scientific notation.

92. *Milky Way* It takes 2×10^8 years for the sun to make one orbit around the Milky Way galaxy. Write this number in standard form.

93. *Speed of the Sun* (Refer to the two previous exercises.) Assume that the sun's orbit in the Milky Way galaxy is circular with a diameter of 10^5 light-years. Estimate how many miles the sun travels in 1 year.

94. *Distance to the Moon* The moon is about 240,000 miles from Earth.
(a) Write this number in scientific notation.
(b) If a rocket traveled at 4×10^4 miles per hour, how long would it take for it to reach the moon?

95. **Online Exploration** In 1997, the creators of the Internet search engine BackRub renamed it Google. This new name is a play on the word *googol*, which is a very large number. Look up a googol and write it in scientific notation.

96. **Online Exploration** An astronomical unit (AU) is based on the distance from Earth to the sun. Look up the distance in kilometers from Earth to the sun.
(a) Write an astronomical unit in standard form to the nearest million kilometers.
(b) Convert your rounded answer from part (a) to scientific notation.

97. *Gross Domestic Product* The gross domestic product (GDP) is the total national output of goods and services valued at market prices *within* the United States. The GDP of the United States in 2005 was $12,460,000,000,000. (*Source:* Bureau of Economic Analysis.)
(a) Write this number in scientific notation.
(b) In 2005, the U.S. population was 2.98×10^8. On average, how many dollars of goods and services were produced by each individual?

98. *Average Family Net Worth* A family refers to a group of two or more people related by birth, marriage, or adoption who reside together. In 2000, the average family net worth was $280,000, and there were about 7.2×10^7 families. Calculate the total family net worth in the United States in 2000. (*Source:* U.S. Census Bureau.)

WRITING ABOUT MATHEMATICS

99. Explain what a negative exponent is and how it is different from a positive exponent. Give an example.

100. Explain why scientific notation is helpful for writing some numbers.

Group Activity   Working with Real Data

Directions: Form a group of 2 to 4 people. Select a person to record the group's responses for this activity. All members of the group should work cooperatively to answer the questions. If your instructor asks for your results, each member of the group should be prepared to respond.

Water in a Lake East Battle Lake in Minnesota covers an area of about 1950 acres, or 8.5×10^7 square feet, and its average depth is about 3.2×10^1 feet.
(a) Estimate the cubic feet of water in the lake. (*Hint:* volume = area $\times$ average depth.)
(b) One cubic foot of water equals about 7.5 gallons. How many gallons of water are in this lake?

(c) The population of the United States is about 3.1×10^8, and the average American uses about 1.5×10^2 gallons of water per day. Could this lake supply the American population with water for 1 day?

5.6   Division of Polynomials

Division by a Monomial • Division by a Polynomial

A LOOK INTO MATH ▶ The study of polynomials has occupied the minds of mathematicians for centuries. During the sixteenth century, Girolamo Cardano and other Italian mathematicians discovered how to solve higher degree polynomial equations. In this section we demonstrate how to divide polynomials. Division is often needed to factor polynomials and to solve polynomial equations. (*Source:* H. Eves, *An Introduction to the History of Mathematics.*)

Girolamo Cardano (1501–1576)

Division by a Monomial

To add two fractions with like denominators, we use the property

$$\frac{a}{d} + \frac{b}{d} = \frac{a + b}{d}.$$

For example, $\frac{1}{7} + \frac{3}{7} = \frac{1 + 3}{7} = \frac{4}{7}$.

To divide a polynomial by a monomial, we use the same property, only in reverse. That is,

$$\frac{a + b}{d} = \frac{a}{d} + \frac{b}{d}.$$

Note that each term in the numerator is divided by the monomial in the denominator. The next example shows how to divide a polynomial by a monomial.

EXAMPLE 1   **Dividing a polynomial by a monomial**

Divide.

(a) $\dfrac{a^5 + a^3}{a^2}$ **(b)** $\dfrac{5x^4 + 10x}{10x}$ **(c)** $\dfrac{3y^2 + 2y - 12}{6y}$

Solution

(a) $\dfrac{a^5 + a^3}{a^2} = \dfrac{a^5}{a^2} + \dfrac{a^3}{a^2} = a^{5-2} + a^{3-2} = a^3 + a$

(b) $\dfrac{5x^4 + 10x}{10x} = \dfrac{5x^4}{10x} + \dfrac{10x}{10x} = \dfrac{x^3}{2} + 1$

(c) $\dfrac{3y^2 + 2y - 12}{6y} = \dfrac{3y^2}{6y} + \dfrac{2y}{6y} - \dfrac{12}{6y} = \dfrac{y}{2} + \dfrac{1}{3} - \dfrac{2}{y}$

Now Try Exercises 17, 19, 21

MAKING CONNECTIONS

Division and Simplification

A common mistake made when dividing expressions is to "cancel" incorrectly. Note in Example 1(b) that

$$\frac{5x^4 + 10x}{10x} \neq 5x^4 + \frac{10x}{10x}.$$

The monomial must be divided into *every* term in the numerator.

When dividing two natural numbers, we can check our work by multiplying. For example, $\frac{10}{5} = 2$, and we can check this result by finding the product $5 \cdot 2 = 10$. Similarly, to check

$$\frac{a^5 + a^3}{a^2} = a^3 + a$$

in Example 1(a) we can multiply a^2 and $a^3 + a$.

$$a^2(a^3 + a) = a^2 \cdot a^3 + a^2 \cdot a \quad \text{Distributive property}$$
$$= a^5 + a^3 \checkmark \quad \text{It checks.}$$

EXAMPLE 2 | **Dividing and checking**

Divide the expression $\frac{8x^3 - 4x^2 + 6x}{2x^2}$ and then check the result.

Solution

Be sure to divide $2x^2$ into *every* term in the numerator.

$$\frac{8x^3 - 4x^2 + 6x}{2x^2} = \frac{8x^3}{2x^2} - \frac{4x^2}{2x^2} + \frac{6x}{2x^2} = 4x - 2 + \frac{3}{x}$$

Check:

$$2x^2\left(4x - 2 + \frac{3}{x}\right) = 2x^2 \cdot 4x - 2x^2 \cdot 2 + 2x^2 \cdot \frac{3}{x}$$
$$= 8x^3 - 4x^2 + 6x \checkmark$$

Now Try Exercise 23

EXAMPLE 3 **Finding the length of a rectangle**

The rectangle in Figure 5.17 has an area $A = x^2 + 2x$ and width x. Write an expression for its length l in terms of x.

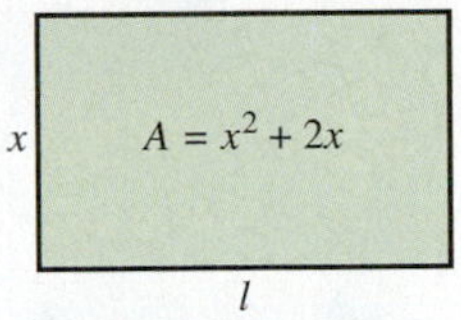

Figure 5.17

STUDY TIP

Do you have enough time to study your notes and complete your assignments? One way to manage your time is to make a list of your time commitments and determine the amount of time that each activity requires. Remember to include time for eating, sleeping, and relaxing!

Solution

The area A of a rectangle equals length l times width w, or $A = lw$. Solving for l gives

$$l = \frac{A}{w}.$$

Thus to find the length of the given rectangle, divide the area by the width.

$$l = \frac{x^2 + 2x}{x} = \frac{x^2}{x} + \frac{2x}{x} = x + 2$$

The length of the rectangle is $x + 2$. The answer checks because $x(x + 2) = x^2 + 2x$.

Now Try Exercise 49

Division by a Polynomial

To understand division by a polynomial better, we first need to review some terminology related to long division of natural numbers. To compute $271 \div 4$, we complete long division as follows.

$$\begin{array}{rl} \text{Quotient} \longrightarrow & 67 \text{ R } 3 \longleftarrow \text{Remainder} \\ \text{Divisor} \longrightarrow & 4\overline{)271} \longleftarrow \text{Dividend} \\ & \underline{24} \\ & 31 \\ & \underline{28} \\ & 3 \end{array}$$

To check this result, we find the product of the quotient and divisor and then add the remainder. Because $67 \cdot 4 + 3 = 271$, the answer checks. The quotient and remainder can also be expressed as $67\frac{3}{4}$. Division of polynomials is similar to long division of natural numbers.

EXAMPLE 4 **Dividing polynomials**

Divide $\frac{6x^2 + 13x + 3}{3x + 2}$ and check.

Solution

Begin by dividing the first term of $3x + 2$ into the first term of $6x^2 + 13x + 3$. That is, divide $3x$ into $6x^2$ to obtain $2x$. Then find the product of $2x$ and $3x + 2$, or $6x^2 + 4x$, place it below $6x^2 + 13x$, and subtract. Bring down the 3.

$$\begin{array}{rl} & 2x \\ 3x + 2 & \overline{)6x^2 + 13x + 3} \\ & \underline{6x^2 + 4x} \\ & \phantom{6x^2 +{}} 9x + 3 \end{array}$$

$\frac{6x^2}{3x} = 2x$

$2x(3x + 2) = 6x^2 + 4x$

Subtract: $13x - 4x = 9x$. Bring down the 3.

In the next step, divide $3x$ into the first term of $9x + 3$ to obtain 3. Then find the product of 3 and $3x + 2$, or $9x + 6$, place it below $9x + 3$, and subtract.

$$
\begin{array}{r}
2x + 3 \\
3x + 2\overline{)6x^2 + 13x + 3} \\
6x^2 + 4x \\
\hline
9x + 3 \\
9x + 6 \\
\hline
-3
\end{array}
\qquad
\begin{array}{l}
\dfrac{9x}{3x} = 3 \\
\\
\\
3(3x + 2) = 9x + 6 \\
\text{Subtract: } 3 - 6 = -3.
\end{array}
$$

The quotient is $2x + 3$ with remainder -3. This result can also be written as

$$
2x + 3 + \frac{-3}{3x + 2}, \qquad \text{Quotient} + \frac{\text{Remainder}}{\text{Divisor}}
$$

in the same manner that 67 R 3 was written as $67\frac{3}{4}$.

Check polynomial division by adding the remainder to the product of the divisor and the quotient. That is,

$$
\textbf{(Divisor)(Quotient)} + \textbf{Remainder} = \textbf{Dividend}.
$$

For this example, the equation becomes

$$
\begin{aligned}
(3x + 2)(2x + 3) + (-3) &= 3x \cdot 2x + 3x \cdot 3 + 2 \cdot 2x + 2 \cdot 3 - 3 \\
&= 6x^2 + 9x + 4x + 6 - 3 \\
&= 6x^2 + 13x + 3. \; \checkmark \quad \text{It checks.}
\end{aligned}
$$

Now Try Exercise 27

READING CHECK

- How can you check a polynomial division problem?

EXAMPLE 5 **Dividing polynomials having a missing term**

Simplify $(3x^3 + 2x - 4) \div (x - 2)$.

Solution
Because the dividend does not have an x^2-term, insert $0x^2$ as a "place holder." Then begin by dividing x into $3x^3$ to obtain $3x^2$.

$$
\begin{array}{r}
3x^2 \\
x - 2\overline{)3x^3 + 0x^2 + 2x - 4} \\
3x^3 - 6x^2 \\
\hline
6x^2 + 2x
\end{array}
\qquad
\begin{array}{l}
\dfrac{3x^3}{x} = 3x^2 \\
\\
3x^2(x - 2) = 3x^3 - 6x^2 \\
\text{Subtract } 0x^2 - (-6x^2) = 6x^2. \text{ Bring down } 2x.
\end{array}
$$

In the next step, divide x into $6x^2$.

$$
\begin{array}{r}
3x^2 + 6x \\
x - 2\overline{)3x^3 + 0x^2 + 2x - 4} \\
3x^3 - 6x^2 \\
\hline
6x^2 + 2x \\
6x^2 - 12x \\
\hline
14x - 4
\end{array}
\qquad
\begin{array}{l}
\dfrac{6x^2}{x} = 6x \\
\\
\\
6x(x - 2) = 6x^2 - 12x \\
\text{Subtract: } 2x - (-12x) = 14x. \text{ Bring down } -4.
\end{array}
$$

Now divide x into $\mathbf{14x}$.

$$\begin{array}{r} 3x^2 + 6x + \mathbf{14} \\ x - 2\overline{)3x^3 + 0x^2 + 2x - 4} \\ \underline{3x^3 - 6x^2} \\ 6x^2 + 2x \\ \underline{6x^2 - 12x} \\ \mathbf{14x} - 4 \\ \underline{14x - 28} \\ 24 \end{array}$$

$$\frac{\mathbf{14x}}{x} = 14$$

$14(x - 2) = 14x - 28$

Subtract: $-4 - (-28) = 24$.

The quotient is $3x^2 + 6x + 14$ with remainder 24. This result can also be written as

$$3x^2 + 6x + 14 + \frac{24}{x - 2}.$$

■ Now Try Exercise 37

EXAMPLE 6 **Dividing when the divisor is not linear**

Divide $x^3 - 3x^2 + 3x + 2$ by $x^2 + 1$.

Solution
Begin by writing $x^2 + 1$ as $x^2 + 0x + 1$.

$$\begin{array}{r} x - 3 \\ x^2 + 0x + 1\overline{)x^3 - 3x^2 + 3x + 2} \\ \underline{x^3 + 0x^2 + x} \\ -3x^2 + 2x + 2 \\ \underline{-3x^2 + 0x - 3} \\ 2x + 5 \end{array}$$

The quotient is $x - 3$ with remainder $2x + 5$. This result can also be written as

$$x - 3 + \frac{2x + 5}{x^2 + 1}.$$

■ Now Try Exercise 41

5.6 Putting It All Together

CONCEPT	EXPLANATION	EXAMPLES
Division by a Monomial	Use the property $$\frac{a + b}{d} = \frac{a}{d} + \frac{b}{d}.$$ Be sure to divide the denominator into every term in the numerator.	$\dfrac{2x^3 + 4x}{2x^2} = \dfrac{2x^3}{2x^2} + \dfrac{4x}{2x^2} = x + \dfrac{2}{x}$ and $\dfrac{a^2 - 2a}{4a} = \dfrac{a^2}{4a} - \dfrac{2a}{4a} = \dfrac{a}{4} - \dfrac{1}{2}$

continued on next page

continued from previous page

CONCEPT	EXPLANATION	EXAMPLES
Division by a Polynomial	Is done similarly to the way long division of natural numbers is performed If either the divisor or the dividend is missing a term, be sure to insert as a "place holder" the missing term with coefficient 0.	Divide $x^2 + 3x + 3$ by $x + 1$. $$\begin{array}{r} x + 2 \\ x + 1 \overline{)\, x^2 + 3x + 3} \\ \underline{x^2 + x} \\ 2x + 3 \\ \underline{2x + 2} \\ 1 \end{array}$$ The quotient is $x + 2$ with remainder 1, which can be expressed as $$x + 2 + \frac{1}{x + 1}.$$
Checking a Result	Dividend = (Divisor)(Quotient) + Remainder	When $x^2 + 3x + 3$ is divided by $x + 1$, the quotient is $x + 2$ with remainder 1. Thus $$(x + 1)(x + 2) + 1 = x^2 + 3x + 3,$$ and the answer checks.

5.6   Exercises

MyMathLab    PRACTICE    WATCH   DOWNLOAD   READ   REVIEW

CONCEPTS AND VOCABULARY

1. $\frac{a + b}{d} = $ _____

2. $\frac{a + b - c}{d} = $ _____

3. When dividing a polynomial by a monomial, the monomial must be divided into every _____ of the polynomial.

4. (True or False?) The expressions $\frac{5x^2 + 2x}{2x}$ and $5x^2 + 1$ are equal.

5. (True or False?) The expressions $\frac{5x^2 + 2x}{2x}$ and $\frac{5x^2}{2x}$ are equal.

6. Because $\frac{37}{9} = 4$ with remainder 1, it follows that $37 = $ _____ · _____ + _____.

7. Because $2x^3 - x + 5$ divided by $x + 1$ equals $2x^2 - 2x + 1$ with remainder 4, it follows that $2x^3 - x + 5 = $ _____ · _____ + _____.

8. When dividing $2x^3 + 3x - 1$ by $x - 1$, insert _____ into the dividend as a "place holder" for the missing x^2-term.

DIVISION BY A MONOMIAL

Exercises 9–16: Divide and check.

9. $\frac{6x^2}{3x}$

10. $\frac{-5x^2}{10x^4}$

11. $\frac{z^4 + z^3}{z}$

12. $\frac{t^3 - t}{t}$

13. $\frac{a^5 - 6a^3}{2a^3}$

14. $\frac{b^4 - 4b}{4b^2}$

15. $\frac{y + 6y^2}{3y^3}$

16. $\frac{8z^2 - z}{4z^2}$

Exercises 17–26: Divide.

17. $\frac{4x - 7x^4}{x^2}$

18. $\frac{1 + 6x^4}{3x^3}$

19. $\frac{6y^2 + 3y}{3y^3}$

20. $\frac{5z^2 - 10z^3}{5z^4}$

21. $\frac{9x^4 - 3x + 6}{3x}$

22. $\frac{y^3 - 4y + 6}{y}$

23. $\frac{12y^4 - 3y^2 + 6y}{3y^2}$

24. $\frac{2x^2 - 6x + 9}{12x}$

25. $\frac{15m^4 - 10m^3 + 20m^2}{5m^2}$

26. $\frac{n^8 - 8n^6 + 4n^4}{2n^5}$

Exercises 27–34: Divide and check.

27. $\dfrac{2x^2 - 3x + 1}{x - 2}$

28. $\dfrac{4x^2 - x + 3}{x + 2}$

29. $\dfrac{x^2 + 2x + 1}{x + 1}$

30. $\dfrac{4x^2 - 4x + 1}{2x - 1}$

31. $\dfrac{x^3 - x^2 + x - 2}{x - 1}$

32. $\dfrac{2x^3 + 3x^2 + 3x - 1}{2x + 1}$

33. $\dfrac{x^3 + x^2 - 7x + 2}{x - 2}$

34. $\dfrac{x^3 + x^2 - 2x + 12}{x + 3}$

Exercises 35–46: Divide.

35. $\dfrac{4x^3 - 3x^2 + 7x + 3}{4x + 1}$

36. $\dfrac{10x^3 - x^2 - 17x - 7}{5x + 2}$

37. $\dfrac{x^3 - x + 2}{x - 2}$

38. $\dfrac{6x^3 + 8x^2 + 4}{3x + 4}$

39. $(3x^3 + 2) \div (x - 1)$

40. $(-3x^3 + 8x^2 + x) \div (3x + 4)$

41. $(x^3 + 3x^2 + 1) \div (x^2 + 1)$

42. $(x^4 - x^3 + x^2 - x + 1) \div (x^2 - 1)$

43. $\dfrac{x^3 + 1}{x^2 - x + 1}$

44. $\dfrac{4x^3 + 3x + 2}{2x^2 - x + 1}$

45. $\dfrac{x^3 + 8}{x + 2}$

46. $\dfrac{x^4 - 16}{x - 2}$

47. **Thinking Generally** If the quotient in a polynomial division problem is an integer, what must be true about the degrees of the dividend and divisor?

48. **Thinking Generally** If the quotient in a polynomial division problem is a polynomial of degree 1, what must be true about the degrees of the dividend and divisor?

APPLICATIONS

Exercises 49 and 50: **Area of a Rectangle** *The area of a rectangle and its width are given. Find an expression for the length l.*

49.

50.

51. *Volume of a Box* The volume V of a box is $2x^3 + 4x^2$, and the area of its bottom is $2x^2$. Find the height of the box in terms of x. Make a possible sketch of the box, and label the length of each side.

52. *Area of a Triangle* A triangle has height h and area $A = 2h^2 - 4h$. Find its base b in terms of h. Make a possible sketch of the triangle, and label the height and base. (*Hint:* $A = \frac{1}{2}bh$.)

WRITING ABOUT MATHEMATICS

53. Suppose that one polynomial is divided into another polynomial and the remainder is 0. What does the product of the divisor and quotient equal? Explain.

54. A student simplifies the expression $\dfrac{4x^3 - 1}{4x^2}$ to $x - 1$. Explain the student's error.

SECTIONS 5.5 and 5.6

Checking Basic Concepts

1. Simplify each expression. Write the result with positive exponents.

   (a) 9^{-2} (b) $\dfrac{3x^{-3}}{6x^4}$ (c) $(4ab^{-4})^{-2}$

2. Simplify each expression. Write the result with positive exponents.

   (a) $\dfrac{1}{z^{-5}}$ (b) $\dfrac{x^{-3}}{y^{-6}}$ (c) $\left(\dfrac{3}{x^2}\right)^{-3}$

3. Write each number in scientific notation.

   (a) 45,000 (b) 0.000234 (c) 0.01

4. Write each expression in standard form.

   (a) 4.71×10^4 (b) 6×10^{-3}

5. Simplify $\dfrac{25a^4 - 15a^3}{5a^3}$.

6. Divide $3x^2 - x - 4$ by $x - 1$. State the quotient and remainder.

7. Divide $x^4 + 2x^3 - 2x^2 - 5x - 2$ by $x^2 - 3$. State the quotient and remainder.

8. *Distance to the Sun* The distance to the sun is approximately 93 million miles.

   (a) Write this distance in scientific notation.

   (b) Light travels at 1.86×10^5 miles per second. How long does it take for the sun's light to reach Earth?

CHAPTER 5   Summary

SECTION 5.1 ■ RULES FOR EXPONENTS

Bases and Exponents The expression b^n has base b and exponent n and equals the expression $\underbrace{b \cdot b \cdot b \cdots \cdot b}_{n \text{ times}}$, when n is a natural number.

Example: 2^3 has base 2 and exponent 3 and equals $2 \cdot 2 \cdot 2 = 8$.

Evaluating Expressions When evaluating expressions, evaluate exponents *before* performing addition, subtraction, multiplication, division, or negation. In general, operations within parentheses should be evaluated *before* using the order of operations.

1. Evaluate exponents.
2. Perform negation.
3. Do multiplication and division from left to right.
4. Do addition and subtraction from left to right.

Example: $-3^2 + 3 \cdot 4 = -9 + 3 \cdot 4 = -9 + 12 = 3$

Zero Exponents For any nonzero number b, $b^0 = 1$. Note that 0^0 is undefined.

Examples: $5^0 = 1$ and $\left(\dfrac{x}{y}\right)^0 = 1$, where x and y are nonzero.

Product Rule For any real number a and natural numbers m and n,

$$a^m \cdot a^n = a^{m+n}.$$

Examples: $3^4 \cdot 3^2 = 3^6$ and $x^3 x^2 x^4 = x^9$

Power Rules For any real numbers a and b and natural numbers m and n,

$$(a^m)^n = a^{mn}, \quad (ab)^n = a^n b^n, \quad \text{and} \quad \left(\frac{a}{b}\right)^n = \frac{a^n}{b^n}, \quad b \neq 0.$$

Examples: $(x^2)^3 = x^6$, $(3x)^4 = 3^4 x^4 = 81x^4$,   and $\left(\dfrac{2}{y}\right)^3 = \dfrac{2^3}{y^3} = \dfrac{8}{y^3}$

SECTION 5.2 ■ ADDITION AND SUBTRACTION OF POLYNOMIALS

Terms Related to Polynomials

Monomial	A number, variable, or product of numbers and variables raised to natural number powers
Degree of a Monomial	Sum of the exponents of the variables
Coefficient of a Monomial	The number in a monomial
	Example: The monomial $-3x^2y^3$ has degree 5 and coefficient -3.
Polynomial	A monomial or the sum of two or more monomials
Term of a Polynomial	Each monomial is a term of the polynomial.
Binomial	A polynomial with two terms
Trinomial	A polynomial with three terms
Degree of a Polynomial	The degree of the term with highest degree

Opposite of a Polynomial	The opposite is found by negating each term.
	Example: $2x^3 - 4x + 5$ is a trinomial with degree 3. Its opposite is $-2x^3 + 4x - 5$.
Like Terms	Two monomials with the same variables raised to the same powers
	Examples: $3xy^2$ and $-xy^2$ are like terms.
	$5x^3$ and $3x^3$ are like terms.
	$5x^2$ and $5x$ are *unlike* terms.

Addition of Polynomials Combine like terms, using the distributive property.

Example: $(2x^2 - 4x) + (-x^2 - x) = (2 - 1)x^2 + (-4 - 1)x$
$$= x^2 - 5x$$

Subtraction of Polynomials Add the first polynomial to the opposite of the second polynomial.

Example: $(4x^4 - 5x) - (7x^4 + 6x) = (4x^4 - 5x) + (-7x^4 - 6x)$
$$= (4 - 7)x^4 + (-5 - 6)x$$
$$= -3x^4 - 11x$$

SECTION 5.3 ■ MULTIPLICATION OF POLYNOMIALS

Multiplication of Monomials Use the commutative property and the product rule.

Examples: $-2x^3 \cdot 3x^2 = -2 \cdot 3 \cdot x^3 \cdot x^2 = -6x^5$
$$(2xy^2)(3x^2y^3) = 2 \cdot 3 \cdot x \cdot x^2 \cdot y^2 \cdot y^3 = 6x^3y^5$$

Assumed exponent of 1

Distributive Properties

$$a(b + c) = ab + ac \quad \text{and} \quad a(b - c) = ab - ac$$

Examples: $4x(3x + 6) = 4x \cdot 3x + 4x \cdot 6 = 12x^2 + 24x$

$$ab(a^2 - b^2) = ab \cdot a^2 - ab \cdot b^2 = a^3b - ab^3$$

Multiplication of Monomials and Polynomials Apply the distributive properties. Be sure to multiply every term in the polynomial by the monomial.

Example: $-2x^2(4x^2 - 5x - 3) = -8x^4 + 10x^3 + 6x^2$

Multiplication of Polynomials The product of two polynomials may be found by multiplying every term in the first polynomial by every term in the second polynomial. Be sure to combine like terms.

Examples: $(x + 3)(2x - 5) = 2x^2 - 5x + 6x - 15$
$$= 2x^2 + x - 15$$

$$(2x + 1)(x^2 - 5x + 2) = 2x^3 - 10x^2 + 4x + x^2 - 5x + 2$$
$$= 2x^3 - 9x^2 - x + 2$$

Product of a Sum and Difference

$$(a + b)(a - b) = a^2 - b^2$$

Examples: $(x + 4)(x - 4) = x^2 - 16$

$(2r - 3t)(2r + 3t) = (2r)^2 - (3t)^2 = 4r^2 - 9t^2$

Squaring Binomials

$$(a + b)^2 = a^2 + 2ab + b^2 \quad \text{and} \quad (a - b)^2 = a^2 - 2ab + b^2$$

Examples: $(2x + 1)^2 = (2x)^2 + 2(2x)1 + 1^2 = 4x^2 + 4x + 1$

$(z^2 - 2)^2 = (z^2)^2 - 2z^2(2) + 2^2 = z^4 - 4z^2 + 4$

Cubing Binomials To multiply $(a + b)^3$, write it as $(a + b)(a + b)^2$.

Example:
$$
\begin{aligned}
(x + 4)^3 &= (x + 4)(x + 4)^2 \\
&= (x + 4)(x^2 + 8x + 16) && \text{Square the binomial.} \\
&= x^3 + 8x^2 + 16x + 4x^2 + 32x + 64 && \text{Distributive property} \\
&= x^3 + 12x^2 + 48x + 64 && \text{Combine like terms.}
\end{aligned}
$$

Negative Integers as Exponents For any nonzero real number a and positive integer n,

$$a^{-n} = \frac{1}{a^n}.$$

Examples: $5^{-2} = \frac{1}{5^2} \quad \text{and} \quad x^{-4} = \frac{1}{x^4}$

The Quotient Rule For any nonzero real number a and integers m and n,

$$\frac{a^m}{a^n} = a^{m-n}.$$

Examples: $\dfrac{6^4}{6^2} = 6^{4-2} = 6^2 = 36 \quad \text{and} \quad \dfrac{xy^3}{x^4y^2} = x^{1-4}y^{3-2} = x^{-3}y^1 = \dfrac{y}{x^3}$

Other Rules For any nonzero real numbers a and b and positive integers m and n,

$$\frac{1}{a^{-n}} = a^n, \quad \frac{a^{-n}}{b^{-m}} = \frac{b^m}{a^n}, \quad \text{and} \quad \left(\frac{a}{b}\right)^{-n} = \left(\frac{b}{a}\right)^n.$$

Examples: $\dfrac{1}{4^{-3}} = 4^3, \quad \dfrac{x^{-3}}{y^{-2}} = \dfrac{y^2}{x^3}, \quad \text{and} \quad \left(\dfrac{4}{5}\right)^{-2} = \left(\dfrac{5}{4}\right)^2$

Scientific Notation A real number a written as $b \times 10^n$, where $1 \le |b| < 10$ and n is an integer

Examples: $2.34 \times 10^3 = 2340$ Move the decimal point 3 places to the right.

$2.34 \times 10^{-3} = 0.00234$ Move the decimal point 3 places to the left.

Division of a Polynomial by a Monomial Divide the monomial into *every* term of the polynomial.

Example: $\dfrac{5x^3 - 10x^2 + 15x}{5x} = \dfrac{5x^3}{5x} - \dfrac{10x^2}{5x} + \dfrac{15x}{5x} = x^2 - 2x + 3$

Division of a Polynomial by a Polynomial Division of polynomials is performed similarly to long division of natural numbers.

Example: Divide $2x^3 + 4x^2 - 3x + 1$ by $x + 1$.

$$
\begin{array}{r}
2x^2 + 2x - 5 \\
x + 1\overline{)2x^3 + 4x^2 - 3x + 1} \\
\underline{2x^3 + 2x^2} \\
2x^2 - 3x \\
\underline{2x^2 + 2x} \\
-5x + 1 \\
\underline{-5x - 5} \\
6
\end{array}
$$

The quotient is $2x^2 + 2x - 5$ with remainder 6, which can be written as

$$2x^2 + 2x - 5 + \frac{6}{x + 1}.$$

CHAPTER 5 Review Exercises

SECTION 5.1

Exercises 1–6: Evaluate the expression.

1. 5^3

2. -3^4

3. $4(-2)^0$

4. $3 + 3^2 - 3^0$

5. $\dfrac{-5^2}{5}$

6. $\left(\dfrac{-5}{5}\right)^2$

Exercises 7–24: Simplify the expression.

7. $6^2 \cdot 6^3$

8. $10^5 \cdot 10^7$

9. $z^4 \cdot z^5$

10. $y^2 \cdot y \cdot y^3$

11. $5x^2 \cdot 6x^7$

12. $(ab^3)(a^3b)$

13. $(2^5)^2$

14. $(m^4)^5$

15. $(ab)^3$

16. $(x^2y^3)^4$

17. $(xy)^3(x^2y^4)^2$

18. $(a^2b^9)^0$

19. $(r - t)^4(r - t)^5$

20. $(a + b)^2(a + b)^4$

21. $\left(\dfrac{3}{x - y}\right)^2$

22. $\left(\dfrac{x + y}{2}\right)^3$

23. $2x^2(3x - 5)$

24. $3x(4x + x^3)$

SECTION 5.2

Exercises 25 and 26: Identify the degree and coefficient of the monomial.

25. $6x^7$

26. $-x^2y^3$

Exercises 27–30: Determine whether the expression is a polynomial. If it is, state how many terms and variables the polynomial contains. Then state its degree.

27. $8y$

28. $8x^3 - 3x^2 + x - 5$

29. $a^2 + 2ab + b^2$

30. $\dfrac{1}{xy}$

31. Add the polynomials vertically.
$$
\begin{array}{r}
3x^2 + 4x + 8 \\
\underline{2x^2 - 5x - 5}
\end{array}
$$

32. Write the opposite of $6x^2 - 3x - 7$.

Exercises 33–40: Simplify.

33. $(4x - 3) + (-x + 7)$

34. $(3x^2 - 1) - (5x^2 + 12)$

35. $(x^2 + 5x + 6) - (3x^2 - 4x + 1)$

36. $(x^2 + 3x - 5) + (2x^2 - 5x - 1)$

37. $(a^3 + 4a^2) + (a^3 - 5a^2 + 7a)$

38. $(4x^3 - 2x + 6) - (4x^3 - 6)$

39. $(xy + y^2) + (4y^2 - 4xy)$

40. $(7x^2 + 2xy + y^2) - (7x^2 - 2xy + y^2)$

SECTION 5.3

Exercises 41–54: Multiply and simplify.

41. $-x^2 \cdot x^3$ **42.** $-(r^2 t^3)(rt)$

43. $-3(2t - 5)$ **44.** $2y(1 - 6y)$

45. $6x^3(3x^2 + 5x)$ **46.** $-x(x^2 - 2x + 9)$

47. $-ab(a^2 - 2ab + b^2)$ **48.** $(a - 2)(a + 5)$

49. $(8x - 3)(x + 2)$ **50.** $(2x - 1)(1 - x)$

51. $(y^2 + 1)(2y + 1)$ **52.** $(y^2 - 1)(2y^2 + 1)$

53. $(z + 1)(z^2 - z + 1)$

54. $(4z - 3)(z^2 - 3z + 1)$

Exercises 55 and 56: Multiply the expression
   (a) *geometrically and*
   (b) *symbolically.*

55. $z(z + 1)$ **56.** $2x(x + 2)$

SECTION 5.4

Exercises 57–72: Multiply.

57. $(z + 2)(z - 2)$ **58.** $(5z - 9)(5z + 9)$

59. $(1 - 3y)(1 + 3y)$ **60.** $(5x + 4y)(5x - 4y)$

61. $(rt + 1)(rt - 1)$

62. $(2m^2 - n^2)(2m^2 + n^2)$

63. $(x + 1)^2$ **64.** $(4x + 3)^2$

65. $(y - 3)^2$ **66.** $(2y - 5)^2$

67. $(4 + a)^2$ **68.** $(4 - a)^2$

69. $(x^2 + y^2)^2$ **70.** $(xy - 2)^2$

71. $(z + 5)^3$ **72.** $(2z - 1)^3$

Exercises 73 and 74: Use the product of a sum and a difference to evaluate the expression.

73. $59 \cdot 61$ **74.** $22 \cdot 18$

SECTION 5.5

Exercises 75–82: Simplify the expression.

75. 9^{-1} **76.** 3^{-2}

77. $4^3 \cdot 4^{-2}$ **78.** $10^{-6} \cdot 10^3$

79. $\dfrac{1}{6^{-2}}$ **80.** $\dfrac{5^7}{5^9}$

81. $(3^{-1} 2^2)^{-2}$ **82.** $(2^{-4} 5^3)^0$

Exercises 83–98: Simplify the expression. Write the answer using positive exponents.

83. z^{-2} **84.** y^{-4}

85. $a^{-4} \cdot a^2$ **86.** $x^2 \cdot x^{-5} \cdot x$

87. $(2t)^{-2}$ **88.** $(ab^2)^{-3}$

89. $(xy)^{-2}(x^{-2}y)^{-1}$ **90.** $\dfrac{x^6}{x^2}$

91. $\dfrac{4x}{2x^4}$ **92.** $\dfrac{20x^5y^3}{30xy^6}$

93. $\left(\dfrac{a}{b}\right)^5$ **94.** $\dfrac{4}{t^{-4}}$

95. $\dfrac{(3m^3n)^{-2}}{(2m^2n^{-3})^3}$ **96.** $\left(\dfrac{x^{-4}y^2}{3xy^{-3}}\right)^{-2}$

97. $\left(\dfrac{x}{y}\right)^{-2}$ **98.** $\left(\dfrac{3u}{2v}\right)^{-1}$

Exercises 99–102: Write the expression in standard form.

99. 6×10^2 **100.** 5.24×10^4

101. 3.7×10^{-3} **102.** 6.234×10^{-2}

Exercises 103–106: Write the number in scientific notation.

103. 10,000   **104.** 56,100,000

105. 0.000054   **106.** 0.001

Exercises 107 and 108: Evaluate the expression. Write the result in standard form.

107. $(4 \times 10^2)(6 \times 10^4)$ **108.** $\dfrac{8 \times 10^3}{4 \times 10^4}$

SECTION 5.6

Exercises 109–116: Divide and check.

109. $\dfrac{5x^2 + 3x}{3x}$ **110.** $\dfrac{6b^4 - 4b^2 + 2}{2b^2}$

111. $\dfrac{3x^2 - x + 2}{x - 1}$ **112.** $\dfrac{9x^2 - 6x - 2}{3x + 2}$

113. $\dfrac{4x^3 - 11x^2 - 7x - 1}{4x + 1}$

114. $\dfrac{2x^3 - x^2 - 1}{2x - 1}$ **115.** $\dfrac{x^3 - x^2 - x + 1}{x^2 + 1}$

116. $\dfrac{x^4 + 3x^3 + 8x^2 + 7x + 5}{x^2 + x + 1}$

APPLICATIONS

117. *Heart Rate* An athlete starts running and continues for 10 seconds. The polynomial $\frac{1}{2}t^2 + 60$ calculates the heart rate of the athlete in beats per minute t seconds after beginning the run, where $t \le 10$.
(a) What is the athlete's heart rate when the athlete first starts to run?
(b) What is the athlete's heart rate after 10 seconds?
(c) What happens to the athlete's heart rate while the athlete is running?

118. *Areas of Rectangles* Find a monomial equal to the sum of the areas of the rectangles. Calculate this sum for $x = 3$ feet and $y = 4$ feet.

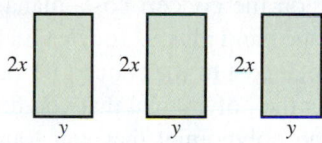

119. *Area of a Rectangle* Write a polynomial that gives the area of the rectangle. Calculate its area for $z = 6$ inches.

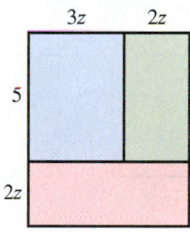

120. *Area of a Square* Find the area of the square whose sides have length $x^2 y$.

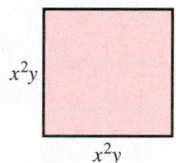

121. *Compound Interest* If P dollars are deposited in an account that pays 6% annual interest, then the amount of money after 3 years is given by $P(1 + 0.06)^3$. Find this amount when $P = \$700$.

122. *Volume of a Sphere* The expression for the volume of a sphere with radius r is $\frac{4}{3}\pi r^3$. Find a polynomial that gives the volume of a sphere with radius $x + 2$. Leave your answer in terms of π.

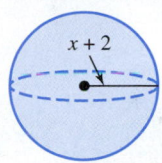

123. *Height Reached by a Baseball* A baseball is hit straight up. Its height h in feet above the ground after t seconds is given by $t(96 - 16t)$.
(a) Multiply this expression.
(b) Evaluate both the expression in part (a) and the given expression for $t = 2$. Interpret the result.

124. *Rectangular Building* A rectangular building has a perimeter of 1200 feet.
(a) If one side of the building has length L, write a polynomial expression that gives its area. (Be sure to multiply your expression.)
(b) Evaluate the expression in part (a) for $L = 50$ and interpret the answer.

125. *Geometry* Complete each part and verify that your answers are equal.
(a) Find the area of the large square by multiplying its length and width.
(b) Find the sum of the areas of the smaller rectangles inside the large square.

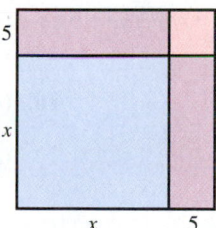

126. *Digital Picture* A digital picture, including its border, is $x + 4$ pixels by $x + 4$ pixels, and the actual picture inside the border is $x - 4$ pixels by $x - 4$ pixels.

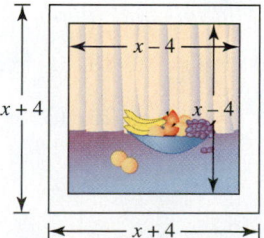

(a) Find a polynomial that gives the number of pixels in the border.
(b) Let $x = 100$ and evaluate the polynomial.

127. *Federal Debt* In 1990, the federal debt held by the public was \$2.19 trillion, and the population of the United States was 249 million. Use scientific notation to approximate the national debt per person. (*Source:* U.S. Department of the Treasury.)

128. *Alcohol Consumption* In 2007, about 239 million people in the United States were age 14 or older. They consumed, on average, 2.31 gallons of alcohol per person. Use scientific notation to estimate the total number of gallons of alcohol consumed by this age group. (*Source:* Department of Health and Human Services.)

CHAPTER 5   Test  Step-by-step test solutions are found on the Chapter Test Prep Videos available in **MyMathLab** and on **You Tube** (search "RockswoldComboAlg" and click on "Channels").

1. Simplify each expression.
 (a) -5^0 (b) -9^2

2. Evaluate each expression by hand.
 (a) $-4^2 + 10$ (b) 8^{-2}
 (c) $\dfrac{1}{2^{-3}}$ (d) $-3x^0$

3. State how many terms and variables the polynomial $5x^2 - 3xy - 7y^3$ contains. Then state its degree.

4. Write the opposite of $-x^3 + 4x - 8$.

Exercises 5–8: Simplify.

5. $(-3x + 4) + (7x + 2)$

6. $(y^3 - 2y + 6) - (4y^3 + 5)$

7. $(5x^2 - x + 3) - (4x^2 - 2x + 10)$

8. $(a^3 + 5ab) + (3a^3 - 3ab)$

Exercises 9–16: Write the given expression with positive exponents.

9. $6y^4 \cdot 4y^7$ **10.** $(a^2b^3)^2(ab^2)$

11. $x^7 \cdot x^{-3}$ **12.** $(a^{-1}b^2)^{-3}$

13. $ab(a^2 - b^2)$ **14.** $\left(\dfrac{3a^2}{2b^{-3}}\right)^{-2}$

15. $\dfrac{12xy^4}{6x^2y}$ **16.** $\left(\dfrac{2}{a + b}\right)^4$

Exercises 17–22: Multiply and simplify.

17. $3x^2(4x^3 - 6x + 1)$ **18.** $(z - 3)(2z + 4)$

19. $(7y^2 - 3)(7y^2 + 3)$ **20.** $(3x - 2)^2$

21. $(m + 3)^3$

22. $(y + 2)(y^2 - 2y + 3)$

23. Evaluate $78 \cdot 82$ using the product of a sum and a difference.

24. Write 6.1×10^{-3} in standard form.

25. Write 5410 in scientific notation.

Exercises 26 and 27: Divide.

26. $\dfrac{9x^3 - 6x^2 + 3x}{3x^2}$ **27.** $\dfrac{x^3 + x^2 - x + 1}{x + 2}$

28. *Concert Tickets* Tickets for a concert are sold for \$20 each.
 (a) Write a polynomial that gives the revenue from selling t tickets.
 (b) Putting on the concert costs management \$2000 to hire the band plus \$2 for each ticket sold. What is the total cost of the concert if t tickets are sold?
 (c) Subtract the polynomial that you found in part (b) from the polynomial that you found in part (a). What does this polynomial represent?

29. *Areas of Rectangles* Find a polynomial representing the sum of the areas of two identical rectangles that have width $2x$ and length $3x$. Calculate this sum for $x = 10$ feet.

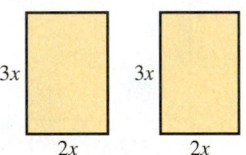

30. *Volume of a Box* Write a polynomial that represents the volume of the box. Be sure to multiply your answer completely.

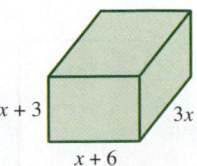

31. *Height Reached by a Golf Ball* When a golf ball is hit into the air, its height in feet above the ground after t seconds is given by $t(88 - 16t)$.
 (a) Multiply this expression.
 (b) Evaluate the expression in part (a) for $t = 3$. Interpret the result.

CHAPTER 5   Extended and Discovery Exercises

*Exercises 1–6: **Arithmetic and Scientific Notation** The product $(4 \times 10^3) \times (2 \times 10^2)$ can be evaluated as*

$$(4 \times 2) \times (10^3 \times 10^2) = 8 \times 10^5,$$

and the quotient $(4 \times 10^3) \div (2 \times 10^2)$ can be evaluated as

$$\frac{4 \times 10^3}{2 \times 10^2} = \frac{4}{2} \times \frac{10^3}{10^2} = 2 \times 10^1.$$

How would you evaluate $(4 \times 10^3) + (2 \times 10^2)$? How would you evaluate $(4 \times 10^3) - (2 \times 10^2)$? Make a conjecture as to how numbers in scientific notation should be added and subtracted. Try your method on these problems and then check your answers with a calculator set in scientific mode. Does your method work?

1. $(4 \times 10^3) + (3 \times 10^3)$

2. $(5 \times 10^{-2}) - (2 \times 10^{-2})$

3. $(1.2 \times 10^4) - (3 \times 10^3)$

4. $(2 \times 10^2) + (6 \times 10^1)$

5. $(2 \times 10^{-1}) + (4 \times 10^{-2})$

6. $(2 \times 10^{-3}) - (5 \times 10^{-2})$

*Exercises 7 and 8: **Constructing a Box** A box is constructed from a rectangular piece of metal by cutting squares from the corners and folding up the sides. The square, cutout corners are x inches by x inches.*

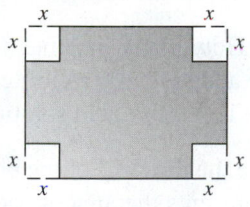 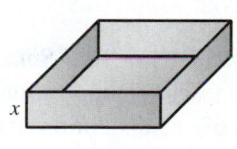

7. Suppose that the dimensions of the metal piece are 20 inches by 30 inches.
   (a) Write a polynomial that gives the volume of the box.
   (b) Find the volume of the box for $x = 4$ inches.

8. Suppose that the metal piece is square with sides of length 25 inches.
   (a) Write a polynomial expression that gives the outside surface area of the box. (Assume that the box does not have a top.)
   (b) Find this area for $x = 3$ inches.

*Exercises 9–12: **Calculators and Polynomials** A graphing calculator can be used to help determine whether two polynomial expressions in one variable are equal. For example, suppose that a student believes that $(x + 2)^2$ and $x^2 + 4$ are equal. Then the first two calculator tables shown demonstrate that the two expressions are not equal except for $x = 0$.*

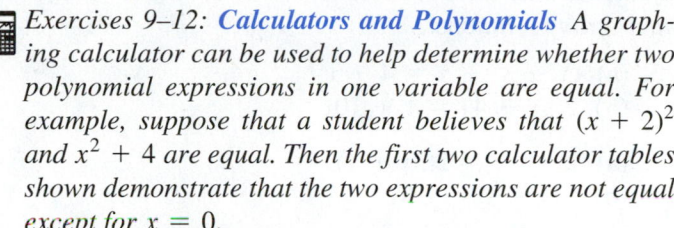

The next two calculator tables support the fact that $(x + 1)^2$ and $x^2 + 2x + 1$ are equal for all x.

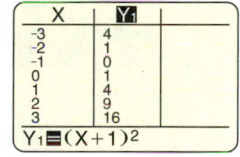 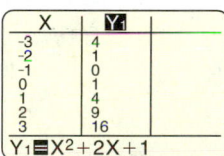

Use a graphing calculator to determine whether the first expression is equal to the second expression. If the expressions are not equal, multiply the first expression and simplify it.

9. $3x(4 - 5x)$, $12x - 5x$

10. $(x - 1)^2$, $x^2 - 1$

11. $(x - 1)(x^2 + x + 1)$, $x^3 - 1$

12. $(x - 2)^3$, $x^3 - 8$

CHAPTERS 1–5   Cumulative Review Exercises

Exercises 1 and 2: Evaluate each expression by hand.

1. (a) $18 - 2 \cdot 5$ (b) $42 \div 7 + 2$

2. (a) $21 - (-8)$ (b) $-\frac{7}{3} \div \left(-\frac{14}{9}\right)$

Exercises 3 and 4: Solve the equation. Note that these equations may have no solutions, one solution, or infinitely many solutions.

3. (a) $(x - 3) + x = 4 + x$
   (b) $2(5x - 4) = 1 + 10x$

4. (a) $2 + 6x = 2(3x + 1)$
   (b) $11x - 9 = -31$

5. Find the average speed of a car that travels 306 miles in 4 hours 30 minutes.

6. Write each value as a fraction in lowest terms.
   (a) 42% (b) 0.076

7. Graph the equation $4x - 5y = 20$.

8. Sketch a line with slope $-\frac{2}{3}$ that passes through the point $(1, 1)$.

9. Write the slope–intercept form for the line shown.

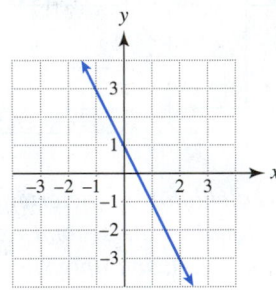

10. Find the *x*-and *y*-intercepts for the graph of the equation $2y = 3x - 6$.

Exercises 11 and 12: Write the slope–intercept form of a line that satisfies the given information.

11. Parallel to $3x - 6y = 7$, passing through $(2, -3)$

12. Passing through $(-2, -5)$ and $(1, 4)$

Exercises 13–16: Solve the system of equations. Note that these systems may have no solutions, one solution, or infinitely many solutions.

13. $4x + 3y = -6$
    $8x + 6y = 12$

14. $x - 3y = 5$
    $3x + y = 5$

15. $x + 4y = -8$
    $-3x - 12y = 24$

16. $x - 5y = 30$
    $2x + y = -6$

Exercises 17 and 18: Shade the solution set for the system of inequalities.

17. $x + y < 3$
    $y \geq x + 2$

18. $x - 2y > 4$
    $3x + y < 6$

19. Simplify the expression.
    (a) $3x^2 \cdot 5x^3$ (b) $(x^3 y)^2 (x^4 y^5)$

20. Simplify.
    (a) $(5x^2 - 3x + 4) - (3x^2 - 2x + 1)$
    (b) $(7a^3 - 4a^2 - 5) + (5a^3 + 4a^2 + a)$

21. Multiply and simplify.
    (a) $(2x + 3)(x - 7)$ (b) $(y + 3)(y^2 - 3y - 1)$
    (c) $(4x + 7)(4x - 7)$ (d) $(5a + 3)^2$

22. Simplify the expression. Write the answer using positive exponents.
    (a) $x^{-5} \cdot x^3 \cdot x$ (b) $\left(\dfrac{2}{x^3}\right)^{-3}$
    (c) $\dfrac{3x^2 y^{-1}}{6x^{-2} y}$ (d) $(xy^{-2})^3 (x^{-2} y)^{-2}$

23. Write 24,000,000,000 in scientific notation.

24. Write 4.71×10^{-7} in standard form.

25. Divide.
    (a) $\dfrac{8x^3 - 2x}{2x}$ (b) $\dfrac{2x^2 + x - 14}{x + 3}$

26. *Price Decrease* If the price of a computer is reduced from \$1200 to \$900, find the percent change.

27. *Mixing an Acid Solution* How many milliliters of a 3% acid solution should be added to 400 milliliters of a 6% acid solution to dilute it to a 5% acid solution?

28. *Surface Area of a Box* Use the drawing of the box to write a polynomial that represents the area of each of the following.

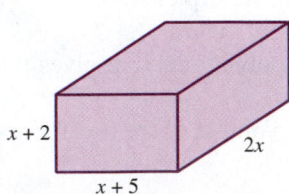

   (a) The bottom
   (b) The front
   (c) The right side
   (d) All six sides

6 Factoring Polynomials and Solving Equations

What we see depends mainly on what we look for.

—JOHN LUBBOCK

Most cars are designed so that their exteriors are curved and smooth. This characteristic is especially true for solar cars. In fact, a side view of a solar car often resembles the cross section of an airplane wing, as illustrated in the accompanying figure. This design reduces drag from air resistance and increases fuel efficiency.

Views of a Solar Car and a Standard Car

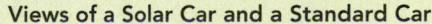
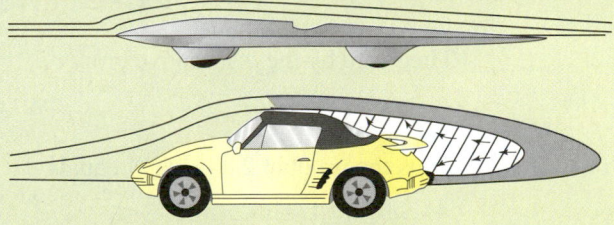

Mathematics plays an important role in the design of cars. Third-degree polynomials called *cubic splines* are used extensively by engineers to obtain a smooth shape for new cars. Although the topic of cubic splines is covered in advanced mathematics and engineering courses, in this chapter we introduce many concepts necessary for understanding polynomials and polynomial equations.

Source: R. Burden and J. Faires. *Numerical Analysis.*

6.1 Introduction to Factoring

Basic Concepts • Common Factors • Factoring by Grouping

A LOOK INTO MATH ▶ Polynomials are frequently used in applications to approximate data and to model such things as changes in the weather, sales of a new video game, and the growth of young children. As a result, scientists and mathematicians commonly solve equations involving polynomials. One way to solve these equations is to use *factoring*. In this section we introduce two basic methods of factoring polynomials.

NEW VOCABULARY

☐ Factoring a polynomial
☐ Greatest common factor (GCF)

Basic Concepts

When two or more numbers are multiplied, each number is called a factor. For example, in the equation $4 \cdot 6 = 24$, the numbers 4 and 6 are factors and we say that 24 can be *factored* into the product $4 \cdot 6$. Other ways to factor 24 include

$$2 \cdot 12, \quad 8 \cdot 3, \quad 2 \cdot 2 \cdot 6, \quad \text{and} \quad 2 \cdot 2 \cdot 2 \cdot 3.$$

When we factor a positive number, we reverse the multiplication process and write the number as a product of two or more smaller numbers. Similarly, we **factor a polynomial** by writing the polynomial as a product of two or more *lower degree* polynomials. For example, possible ways to factor the polynomial $12x^5 + 6x^4$ include

$$2x(6x^4 + 3x^3), \quad x^2(12x^3 + 6x^2), \quad \text{and} \quad 6x^4(2x + 1).$$

The second of these factorizations shows that the **fifth**-degree polynomial $12x^5 + 6x^4$ can be written as the product of the **second**-degree polynomial x^2 and the **third**-degree polynomial $12x^3 + 6x^2$, where each factor has a lower degree than the given polynomial. Similar statements can be made about the first and third factorizations.

Common Factors

When factoring a polynomial, we first look for factors that are common to each term. By applying the distributive property we can often write a polynomial as a product. For example, each term in the polynomial $8x^2 + 6x$ has a factor of $2x$ because

$$8x^2 = 2x \cdot 4x \quad \text{and} \quad 6x = 2x \cdot 3.$$

Therefore by the distributive property,

$$8x^2 + 6x = 2x(4x + 3).$$

Thus the product $2x(4x + 3)$ equals $8x^2 + 6x$. We check this result by multiplying.

$$2x(4x + 3) = 2x \cdot 4x + 2x \cdot 3$$
$$= 8x^2 + 6x \checkmark \qquad \text{It checks.}$$

This factorization is shown visually in Figure 6.1, where possible dimensions for a rectangle with an area of $8x^2 + 6x$ are $2x$ by $4x + 3$.

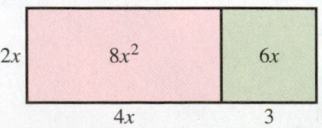

Figure 6.1 $8x^2 + 6x = 2x(4x + 3)$

CALCULATOR HELP

To make a table of values, see Appendix A (pages AP-2 and AP-3).

The expressions $8x^2 + 6x$ and $2x(4x + 3)$ are equal for all values of x. Figure 6.2 illustrates this fact with a partial table of values for each expression.

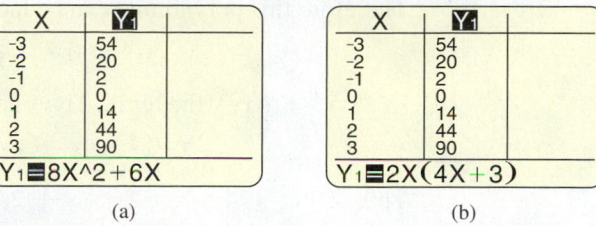

(a)   (b)

Figure 6.2 $8x^2 + 6x = 2x(4x + 3)$

EXAMPLE 1 | **Factoring an expression**

Factor the expression and sketch a rectangle that illustrates the factorization.
(a) $10x + 6$ **(b)** $6x^2 + 15x$

Solution

(a) Each term in the polynomial $10x + 6$ has a factor of 2 because

$$10x = \mathbf{2 \cdot 5x} \quad \text{and} \quad 6 = \mathbf{2 \cdot 3}.$$

By the distributive property, $10x + 6 = \mathbf{2(5x + 3)}$. This factorization is illustrated visually in Figure 6.3(a).

(b) Each term in the polynomial $6x^2 + 15x$ has a factor of $3x$ because

$$6x^2 = \mathbf{3x \cdot 2x} \quad \text{and} \quad 15x = \mathbf{3x \cdot 5}.$$

By the distributive property, $6x^2 + 15x = \mathbf{3x(2x + 5)}$. This factorization is illustrated visually in Figure 6.3(b).

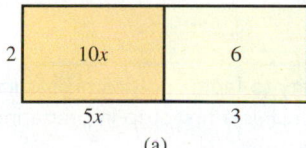   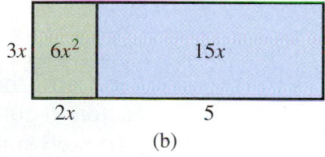

(a)   (b)

Figure 6.3

Now Try Exercises 9, 11

EXAMPLE 2 | **Finding common factors**

Factor.
(a) $15x^2 + 10x$ **(b)** $6y^3 - 2y^2$ **(c)** $3z^3 + 9z^2 - 6z$ **(d)** $2x^2y^2 + 4xy^3$

Solution

(a) In the expression $15x^2 + 10x$, the terms $15x^2$ and $10x$ contain a common factor of $5x$.

$$15x^2 = \mathbf{5x \cdot 3x} \quad \text{and} \quad 10x = \mathbf{5x \cdot 2}$$

Therefore this polynomial can be factored as

$$15x^2 + 10x = \mathbf{5x(3x + 2)}.$$

(b) In the expression $6y^3 - 2y^2$, the terms $6y^3$ and $2y^2$ contain a common factor of $2y^2$.

$$6y^3 = \mathbf{2y^2 \cdot 3y} \quad \text{and} \quad 2y^2 = \mathbf{2y^2 \cdot 1}$$

Therefore this polynomial can be factored as

$$6y^3 - 2y^2 = \mathbf{2y^2(3y - 1)}.$$

(c) In $3z^3 + 9z^2 - 6z$, the terms $3z^3$, $9z^2$, and $6z$ contain a common factor of $3z$.

$$3z^3 = \mathbf{3z} \cdot z^2, \quad 9z^2 = \mathbf{3z} \cdot \mathbf{3z}, \quad \text{and} \quad 6z = \mathbf{3z} \cdot \mathbf{2}$$

Therefore this polynomial can be factored as

$$3z^3 + 9z^2 - 6z = \mathbf{3z}(z^2 + \mathbf{3z} - \mathbf{2}).$$

(d) In $2x^2y^2 + 4xy^3$, the terms $2x^2y^2$ and $4xy^3$ contain a common factor of $2xy^2$.

$$2x^2y^2 = \mathbf{2xy^2} \cdot \mathbf{x} \quad \text{and} \quad 4xy^3 = \mathbf{2xy^2} \cdot \mathbf{2y}$$

Thus $2x^2y^2 + 4xy^3 = \mathbf{2xy^2}(x + \mathbf{2y})$.

Now Try Exercises 15, 17, 19, 21

In most situations we factor out the *greatest common factor* (GCF). For example, the polynomial $12b^3 + 8b^2$ has a common factor of $2b$. We could factor this polynomial as

$$12b^3 + 8b^2 = 2b(6b^2 + 4b).$$

However, we can factor out $4b^2$ instead.

$$12b^3 + 8b^2 = 4b^2(3b + 2)$$

Because $4b^2$ is the common factor with the greatest (integer) coefficient and highest degree, we say that $4b^2$ is the **greatest common factor** of $12b^3 + 8b^2$. In Examples 1 and 2 we factored out the greatest common factor for each expression.

When finding the greatest common factor for a polynomial, it is often helpful first to completely factor each term of the polynomial by writing its coefficient as the product of prime numbers and writing any powers of variables as repeated multiplication. For example, the complete factorization of $24x^4$ is $2 \cdot 2 \cdot 2 \cdot 3 \cdot x \cdot x \cdot x \cdot x$, and the complete factorization of $18xy^3$ is $2 \cdot 3 \cdot 3 \cdot x \cdot y \cdot y \cdot y$. The next example demonstrates this process.

READING CHECK

• How do you know if a common factor is actually the greatest common factor?

STUDY TIP

In general, the ability to factor polynomials is important in solving many types of equations. Factoring out the GCF is a first step in preparing a polynomial for other factoring techniques discussed in this chapter.

EXAMPLE 3 **Finding the greatest common factor**

Find the greatest common factor for each expression. Then factor the expression.
(a) $9x^2 + 6x$ **(b)** $4z^4 + 8z^2$ **(c)** $8a^2b^3 - 16a^3b^2$

Solution
(a) Because

$$9x^2 = \mathbf{3} \cdot 3 \cdot \mathbf{x} \cdot x \quad \text{and} \quad 6x = \mathbf{3} \cdot 2 \cdot \mathbf{x},$$

the terms have common factors of $\mathbf{3}$ and $\mathbf{x}$. The GCF is the product of these two factors, or $3 \cdot x = \mathbf{3x}$. Thus the expression $9x^2 + 6x$ can be factored as $\mathbf{3x}(3x + 2)$. Note that the term $3x$ inside the parentheses is the product of the (black) factors of $9x^2$ that were not part of the greatest common factor, and the term 2 inside the parentheses is the product of the (black) factors of $6x$ that were not part of the greatest common factor.
(b) Because

$$4z^4 = \mathbf{2} \cdot \mathbf{2} \cdot \mathbf{z} \cdot \mathbf{z} \cdot z \cdot z \quad \text{and}$$
$$8z^2 = \mathbf{2} \cdot \mathbf{2} \cdot 2 \cdot \mathbf{z} \cdot \mathbf{z},$$

the terms have common factors of $\mathbf{2}$, $\mathbf{2}$, $\mathbf{z}$, and $\mathbf{z}$. The GCF is the product of these four factors, or $2 \cdot 2 \cdot z \cdot z = \mathbf{4z^2}$. Thus $4z^4 + 8z^2$ can be factored as $\mathbf{4z^2}(z^2 + 2)$.

(c) Because

$$8a^2b^3 = 2 \cdot 2 \cdot 2 \cdot a \cdot a \cdot b \cdot b \cdot b \quad \text{and}$$
$$16a^3b^2 = 2 \cdot 2 \cdot 2 \cdot 2 \cdot a \cdot a \cdot a \cdot b \cdot b,$$

the terms have common factors of **2, 2, 2,** *a*, *a*, *b*, and *b*. The GCF is the product of these seven factors, or $2 \cdot 2 \cdot 2 \cdot a \cdot a \cdot b \cdot b = 8a^2b^2$. Here $8a^2b^3 - 16a^3b^2$ can be factored as $8a^2b^2(b - 2a)$.

▍ **Now Try Exercises 23, 25, 37**

NOTE: With practice, you may find that you can determine the GCF mentally without factoring each term as was done in Example 3.

MAKING CONNECTIONS

Checking Common Factors with Multiplication

When factoring we can check our work by multiplying. For example, if we are uncertain whether the equation

$$6y^3 - 2y^2 = 2y^2(3y - 1)$$

is correct, we can apply the distributive property to the right side of the above equation to obtain

$$2y^2(3y - 1) = 2y^2 \cdot 3y - 2y^2 \cdot 1$$
$$= 6y^3 - 2y^2. \checkmark \qquad \text{It checks.}$$

In the next example, we factor an expression that occurs in a scientific application.

EXAMPLE 4 | **Modeling the flight of a golf ball**

If a golf ball is hit upward at 66 feet per second (45 miles per hour), then its height in feet after *t* seconds is approximated by $66t - 16t^2$. Factor this expression.

Solution
The GCF for $66t$ and $16t^2$ is $2t$ because

$$66t = 2t \cdot 33 \quad \text{and} \quad 16t^2 = 2t \cdot 8t.$$

Therefore this polynomial can be factored as

$$66t - 16t^2 = 2t(33 - 8t).$$

▍ **Now Try Exercise 81**

Factoring by Grouping

Factoring by grouping is a technique that makes use of the associative and distributive properties. The next example illustrates one step in this factoring technique.

EXAMPLE 5 | **Factoring out binomials**

Factor.
(a) $5x(x + 3) + 6(x + 3)$ **(b)** $x^2(2x - 5) - 4x(2x - 5)$

Solution
(a) Each term in the expression $5x(x + 3) + 6(x + 3)$ contains the binomial $(x + 3)$. Therefore the distributive property can be used to factor this expression.

$$5x(x + 3) + 6(x + 3) = (5x + 6)(x + 3)$$

(b) Each term in $x^2(2x - 5) - 4x(2x - 5)$ contains the binomial $(2x - 5)$. Therefore the distributive property can be used to factor this expression.

$$x^2(2x - 5) - 4x(2x - 5) = (x^2 - 4x)(2x - 5)$$
$$= x(x - 4)(2x - 5)$$

Now Try Exercises 41, 45

Now consider the polynomial

$$x^3 + x^2 + 2x + 2.$$

We can factor this polynomial by first grouping it into two binomials.

$(x^3 + x^2) + (2x + 2)$ Associative property

$x^2(x + 1) + 2(x + 1)$ Factor out common factors.

$(x^2 + 2)(x + 1)$ Factor out $(x + 1)$.

When factoring by grouping, we factor out a common factor more than once.

The first step in factoring a four-term polynomial by grouping requires the use of the associative property to write the polynomial as the *sum* of two binomials. However, this property must be applied carefully to avoid sign errors. The next two examples illustrate that the middle arithmetic symbol ($+$ or $-$) in a four-term polynomial determines how the associative property is applied.

READING CHECK

• What factoring technique is sometimes used to factor four-term polynomials?

EXAMPLE 6   **Factoring by grouping when the middle symbol is $(+)$**

Factor each polynomial.
(a) $2x^3 - 4x^2 + 3x - 6$ **(b)** $3x + 3y + ax + ay$

Solution
(a) Use the associative property to write the polynomial as the *sum* of two binomials.

$$2x^3 - 4x^2 + 3x - 6 = (2x^3 - 4x^2) + (3x - 6)$$ Associative property
$$= 2x^2(x - 2) + 3(x - 2)$$ Factor out common factors.
$$= (2x^2 + 3)(x - 2)$$ Factor out $(x - 2)$.

(b) Group the polynomial into the *sum* of two binomials.

$$3x + 3y + ax + ay = (3x + 3y) + (ax + ay)$$ Associative property
$$= 3(x + y) + a(x + y)$$ Factor out common factors.
$$= (3 + a)(x + y)$$ Factor out $(x + y)$.

Now Try Exercises 47, 65

EXAMPLE 7   **Factoring by grouping when the middle symbol is $(-)$**

Factor each polynomial.
(a) $3y^3 - y^2 - 9y + 3$ **(b)** $z^3 + 4z^2 - 5z - 20$

Solution

(a) Begin by changing the middle subtraction to addition by adding the opposite of $9y$. Then apply the associative property to write the result as the *sum* of two binomials.

$$3y^3 - y^2 - 9y + 3 = 3y^3 - y^2 + (-9y) + 3 \qquad \text{Add the opposite of } 9y.$$
$$= (3y^3 - y^2) + (-9y + 3) \qquad \text{Associative property}$$
$$= y^2(3y - 1) - 3(3y - 1) \qquad \text{Factor out common factors.}$$
$$= (y^2 - 3)(3y - 1) \qquad \text{Factor out } (3y - 1).$$

Note that in the third step, -3 was factored from the second binomial.

(b) Begin by changing the middle subtraction to addition by adding the opposite of $5z$. Then apply the associative property to write the result as the *sum* of two binomials.

$$z^3 + 4z^2 - 5z - 20 = z^3 + 4z^2 + (-5z) - 20 \qquad \text{Add the opposite of } 5z.$$
$$= (z^3 + 4z^2) + (-5z - 20) \qquad \text{Associative property}$$
$$= z^2(z + 4) - 5(z + 4) \qquad \text{Factor out common factors.}$$
$$= (z^2 - 5)(z + 4) \qquad \text{Factor out } (z + 4).$$

▌ **Now Try Exercises 55, 61**

When factoring some polynomials, it may be necessary to factor out the greatest common factor before completing other factoring techniques such as factoring by grouping. In the next example the GCF is factored out before grouping is applied.

EXAMPLE 8   **Factoring out the GCF before grouping**

Completely factor each polynomial.

(a) $6x^3 - 12x^2 - 3x + 6$
(b) $2x^5 - 8x^4 + 6x^3 - 24x^2$

Solution

(a) The GCF of $6x^3 - 12x^2 - 3x + 6$ is 3, so factor out **3** before factoring the remaining polynomial by grouping.

$$6x^3 - 12x^2 - 3x + 6 = 3(2x^3 - 4x^2 - x + 2) \qquad \text{Factor out the GCF.}$$
$$= 3(2x^3 - 4x^2 + (-x) + 2) \qquad \text{Add the opposite of } x.$$
$$= 3\big((2x^3 - 4x^2) + (-x + 2)\big) \qquad \text{Associative property}$$
$$= 3\big(2x^2(x - 2) - 1(x - 2)\big) \qquad \text{Factor out common factors.}$$
$$= 3(2x^2 - 1)(x - 2) \qquad \text{Factor out } (x - 2).$$

(b) The GCF of $2x^5 - 8x^4 + 6x^3 - 24x^2$ is $2x^2$, so factor out **$2x^2$** before factoring the remaining polynomial by grouping.

$$2x^5 - 8x^4 + 6x^3 - 24x^2 = 2x^2(x^3 - 4x^2 + 3x - 12) \qquad \text{Factor out the GCF.}$$
$$= 2x^2\big((x^3 - 4x^2) + (3x - 12)\big) \qquad \text{Associative property}$$
$$= 2x^2\big(x^2(x - 4) + 3(x - 4)\big) \qquad \text{Factor out common factors.}$$
$$= 2x^2(x^2 + 3)(x - 4) \qquad \text{Factor out } (x - 4).$$

▌ **Now Try Exercises 67, 71**

6.1 Putting It All Together

CONCEPT	EXPLANATION	EXAMPLES
Common Factor	Factor out a monomial common to each term in a polynomial.	$6z^2 - 6z = 6z(z - 1)$ $4y^3 - 6y^2 = 2y^2(2y - 3)$ $5x^3 - 10x^2 + 15x = 5x(x^2 - 2x + 3)$ $2a^3b^3 - 4a^2b^3 = 2a^2b^3(a - 2)$
Greatest Common Factor (GCF)	The common factor with the greatest (integer) coefficient and highest degree	The GCF of $10x^4 + 15x^2$ is $5x^2$. Common factors include 1, 5, x, $5x$, x^2, and $5x^2$. However, $5x^2$ is *the greatest common factor*.
Factoring by Grouping	Factoring by grouping is a method that can be used to factor some *four-term* polynomials into a product of two binomials. It makes use of the associative and distributive properties.	$2x^3 + 3x^2 + 2x + 3$ $\quad = (2x^3 + 3x^2) + (2x + 3)$ $\quad = x^2(2x + 3) + 1(2x + 3)$ $\quad = (x^2 + 1)(2x + 3)$ $4x^3 - 24x^2 - 3x + 18$ $\quad = 4x^3 - 24x^2 + (-3x) + 18$ $\quad = (4x^3 - 24x^2) + (-3x + 18)$ $\quad = 4x^2(x - 6) - 3(x - 6)$ $\quad = (4x^2 - 3)(x - 6)$

6.1 Exercises

MyMathLab
PRACTICE   WATCH   DOWNLOAD   READ   REVIEW

CONCEPTS AND VOCABULARY

1. To _____ a polynomial, write it as a product of two or more lower degree polynomials.

2. A common factor in the expression $ab + ac$ is _____.

3. When factoring, we can check our work by _____.

4. When finding the GCF for a polynomial, it is often helpful to completely factor each term by writing its coefficient as the product of _____ numbers and writing any powers of variables as repeated _____.

5. The _____ of a polynomial is the common factor with the greatest (integer) coefficient and highest degree.

6. Factoring by _____ can be used to factor some four-term polynomials into a product of two binomials by using the associative and distributive properties.

7. Identify four common factors of $2x^2 + 4x$.

8. Identify the greatest common factor (GCF) of the expression $2x^2 + 4x$.

COMMON FACTORS

Exercises 9–14: Factor the expression. Then make a sketch of a rectangle that illustrates this factorization.

9. $2x + 4$ **10.** $6 + 3x$

11. $z^2 + 4z$ **12.** $a^2 + 5a$

13. $3y^2 + 12y$ **14.** $2z^2 + 10z$

Exercises 15–22: Factor the expression.

15. $3x^2 + 9x$ **16.** $10y^2 + 2y$

17. $4y^3 - 2y^2$

18. $6x^4 + 9x^2$

19. $2z^3 + 8z^2 - 4z$

20. $5x^4 - 15x^3 - 10x^2$

21. $6x^2y - 3xy^2$

22. $7x^3y^3 + 14x^2y^2$

Exercises 23–40: Identify the greatest common factor. Then factor the expression.

23. $6x - 18x^2$

24. $16x^2 - 24x^3$

25. $8y^3 - 12y^2$

26. $12y^3 - 8y^2 + 4y$

27. $6z^3 + 3z^2 + 9z$

28. $16z^3 - 24z^2 - 36z$

29. $x^4 - 5x^3 - 4x^2$

30. $2x^4 + 8x^2$

31. $5y^5 + 10y^4 - 15y^3 + 10y^2$

32. $7y^4 - 14y^3 - 21y^2 + 7y$

33. $xy + xz$

34. $ab - bc$

35. $ab^2 - a^2b$

36. $4x^2y + 6xy^2$

37. $5x^2y^4 + 10x^3y^3$

38. $3r^3t^3 - 6r^4t^2$

39. $a^2b + ab^2 + ab$

40. $6ab^2 - 9ab + 12b^2$

FACTORING BY GROUPING

Exercises 41–46: Factor.

41. $x(x + 1) - 2(x + 1)$

42. $5x(3x - 2) + 2(3x - 2)$

43. $(z + 5)z + (z + 5)4$

44. $3y^2(y - 2) + 5(y - 2)$

45. $4x^3(x - 5) - 2x(x - 5)$

46. $8x^2(x + 3) + (x + 3)$

Exercises 47–66: Factor by grouping.

47. $x^3 + 2x^2 + 3x + 6$

48. $x^3 + 6x^2 + x + 6$

49. $2y^3 + y^2 + 2y + 1$

50. $4y^3 + 10y^2 + 2y + 5$

51. $2z^3 - 6z^2 + 5z - 15$

52. $15z^3 - 5z^2 + 6z - 2$

53. $4t^3 - 20t^2 + 3t - 15$

54. $4t^3 - 12t^2 + 3t - 9$

55. $9r^3 + 6r^2 - 6r - 4$

56. $3r^3 + 12r^2 - 2r - 8$

57. $7x^3 + 21x^2 - 2x - 6$

58. $6x^3 + 3x^2 - 10x - 5$

59. $2y^3 - 7y^2 - 4y + 14$

60. $y^3 - 5y^2 - 3y + 15$

61. $z^3 - 4z^2 - 7z + 28$

62. $12z^3 - 18z^2 - 10z + 15$

63. $2x^4 - 3x^3 + 4x - 6$

64. $x^4 + x^3 + 5x + 5$

65. $ax + bx + ay + by$

66. $ax - bx + ay - by$

Exercises 67–78: Completely factor the polynomial.

67. $3x^3 + 6x^2 + 3x + 6$

68. $5x^3 - 5x^2 + 5x - 5$

69. $6y^4 - 24y^3 - 2y^2 + 8y$

70. $6x^4 - 12x^3 + 3x^2 - 6x$

71. $x^5 + 2x^4 - 3x^3 - 6x^2$

72. $y^6 + 3y^5 - 2y^4 - 6y^3$

73. $4x^5 + 2x^4 - 12x^3 - 6x^2$

74. $18y^5 + 27y^4 + 12y^3 + 18y^2$

75. $x^3y + x^2y^2 - 2x^2y - 2xy^2$

76. $6x^3y - 3x^2y^2 + 18x^2y - 9xy^2$

77. $2x^3y^3 - 2x^4y^2 + 4x^2y^3 - 4x^3y^2$

78. $4x^2y^6 - 4x^2y^5 - 8xy^7 + 8xy^6$

79. **Thinking Generally** Factor a from the polynomial expression $ax^2 + bx + c$.

80. **Thinking Generally** Factor c from the polynomial expression $ax^2 + bx + c$.

APPLICATIONS

81. *Flight of a Golf Ball* The height of a golf ball in feet after t seconds is given by $80t - 16t^2$.
   (a) Identify the greatest common factor.
   (b) Factor this expression.

82. *Flight of a Golf Ball* Repeat the previous exercise if the height of a golf ball in feet after t seconds is given by $128t - 16t^2$.

*Exercises 83–86: **Geometry** Use the information in the figure to write a polynomial expression that represents the area of the shaded region. Then factor the expression.*

83.

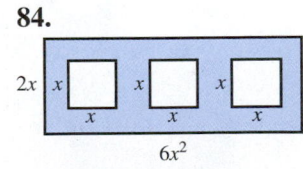

84.

85.

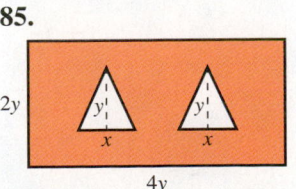

86.

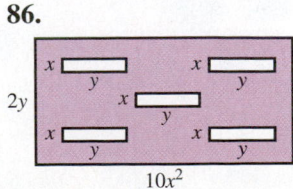

87. *Volume of a Box* A box is constructed by cutting out square corners of a rectangular piece of cardboard and folding up the sides. If the cutout corners have sides with length x, then the volume of the box is given by the polynomial $4x^3 - 60x^2 + 200x$.

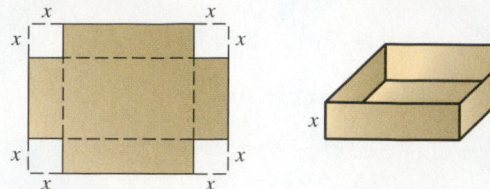

(a) Find the volume of the box when $x = 3$ inches.
(b) Factor out the greatest common factor for this expression.

88. *Volume of a Box* (Refer to the preceding exercise.) A box is constructed from a square piece of metal that is 20 inches on a side.
  (a) If the square corners of length x are cut out, write a polynomial that gives the volume of the box.
  (b) Evaluate the polynomial when $x = 4$ inches.
  (c) Factor out the greatest common factor for this polynomial expression.

WRITING ABOUT MATHEMATICS

89. Use an example to explain the difference between a common factor and the greatest common factor.

90. Use an example to explain how to factor a polynomial by grouping. What two properties of real numbers did you use?

6.2 Factoring Trinomials I ($x^2 + bx + c$)

Review of the FOIL Method • Factoring Trinomials with Leading Coefficient 1

A LOOK INTO MATH ▶

When astronauts are in training, they don't need to leave Earth's atmosphere to experience zero gravity. A specially modified jet following a precise curved path can provide about 30 seconds of weightlessness. The curved path can be modeled by a second degree polynomial with three terms (a trinomial). Specific information about the flight can be obtained by solving equations involving this trinomial. In this section, we discuss techniques for factoring trinomials with a *leading coefficient* of 1. (*Source:* NASA.)

Review of the FOIL Method

NEW VOCABULARY

☐ Leading coefficient
☐ Prime polynomial

Recall that a *trinomial* is a polynomial that has three terms. We begin by reviewing products of binomials that result in trinomials. For example, we multiply the binomials $(x + 2)$ and $(x + 3)$ as follows.

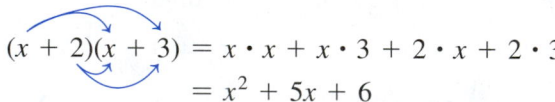

$$(x + 2)(x + 3) = x \cdot x + x \cdot 3 + 2 \cdot x + 2 \cdot 3$$
$$= x^2 + 5x + 6$$

Note that the first term, x^2, in the trinomial results from multiplying the *first* terms of each binomial. The middle term, $5x$, results from adding the product of the *outside* terms and the product of the *inside* terms. Finally, the last term, 6, results from multiplying the *last* terms of each binomial. We discussed this method of multiplying binomials, called FOIL, in Section 5.3 and illustrate it as follows.

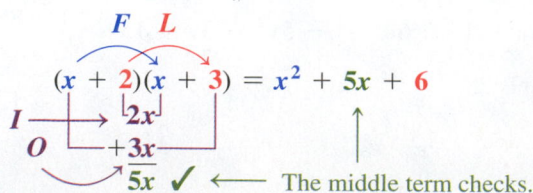

The middle term checks.

Factoring Trinomials with Leading Coefficient 1

Any trinomial of degree 2 in the variable x can be written in *standard form* as $ax^2 + bx + c$, where a, b, and c are constants. The constant a is called the **leading coefficient** of the trinomial. In this section we focus on trinomials where $a = 1$ and b and c are integers.

Recall that the binomials $(x + m)$ and $(x + n)$ are multiplied as follows.

$$(x + m)(x + n) = x^2 + nx + mx + mn$$
$$= x^2 + (m + n)x + mn$$

Note that the coefficient of the x-term is the sum of m and n and that the constant (or third) term is the product of m and n. Thus to factor a trinomial in the form $x^2 + bx + c$, we start by finding two numbers, m and n, such that when they are multiplied $m \cdot n = c$ and when they are added $m + n = b$. In the next example, we find integer pairs that have a specified product and sum.

| **EXAMPLE 1** | **Finding integer pairs with a given product and sum** |

For each of the following, find an integer pair that has the given product and sum.
(a) Product: 18; Sum: 11   **(b)** Product: –20; Sum: 1

Solution

(a) For two integers to have a product of (positive) 18, the integers must have the same sign. Because the specified sum is positive and any sum of two negative integers is negative, the two integers must be positive. Table 6.1 lists positive integer factor pairs for **18** along with the corresponding sum for each pair.

TABLE 6.1 Factor Pairs for 18

Factors	1, 18	**2, 9**	3, 6
Sum	19	**11**	9

From the table, we see that the integers **2** and **9** have a product of **18** and a sum of **11**.

(b) For two integers to have a negative product, they must have unlike signs. Integer factor pairs for -20 and the corresponding sum for each pair are listed in Table 6.2.

TABLE 6.2 Factor Pairs for -20

Factors	1, -20	2, -10	4, -5	$-1, 20$	$-2, 10$	**$-4, 5$**
Sum	-19	-8	-1	19	8	**1**

Here we see that the integers -4 and **5** have a product of -20 and a sum of **1**.

Now Try Exercises 5, 9

STUDY TIP

Spend extra time practicing the process illustrated in Example 1. Pay special attention to factor pairs in which at least one of the integers is negative.

The ability to find integer pairs as demonstrated in Example 1 is essential for factoring trinomials of the form $x^2 + bx + c$. To illustrate the factoring process, we will find an integer pair that can be used to factor $x^2 + 6x + 8$.

Standard Form	*Example*
$x^2 + bx + c$	$x^2 + 6x + 8$
$m \cdot n = c$	$m \cdot n = 8$
$m + n = b$	$m + n = 6$

TABLE 6.3
Factor Pairs for 8

Factors	1, 8	2, 4
Sum	9	6

To determine possible values for m and n, we list factor pairs for 8 and search for a pair whose sum is 6, as in Table 6.3.

Because $2 \cdot 4 = 8$ and $2 + 4 = 6$, we let $m = 2$ and $n = 4$. We then factor the given trinomial as

$$x^2 + 6x + 8 = (x + 2)(x + 4).$$

Note that, if you can find this factor pair mentally, making a table is not necessary. We check the result by multiplying the two binomials.

$$(x + 2)\ (x + 4) = x^2 + 6x + 8$$

$2x$

$+4x$

$6x$ ✔ ⟵ The middle term checks.

FACTORING $x^2 + bx + c$

To factor the trinomial $x^2 + bx + c$, find two numbers m and n that satisfy

$$m \cdot n = c \quad \text{and} \quad m + n = b.$$

Then $x^2 + bx + c = (x + m)(x + n)$.

EXAMPLE 2 **Factoring a trinomial having only positive coefficients**

Factor each trinomial.
(a) $x^2 + 7x + 12$ **(b)** $x^2 + 13x + 30$ **(c)** $z^2 + 9z + 20$

Solution
(a) To factor $x^2 + 7x + 12$ we need to find a factor pair for 12 whose sum is 7. To do so we make Table 6.4.

TABLE 6.4 Factor Pairs for 12

Factors	1, 12	2, 6	3, 4
Sum	13	8	7

The required factor pair is **3** and **4** because $3 \cdot 4 = 12$ and $3 + 4 = 7$. Therefore the given trinomial can be factored as

$$x^2 + 7x + 12 = (x + 3)(x + 4).$$

(b) To factor $x^2 + 13x + 30$ we need to find a factor pair for 30 whose sum is 13. The required pair is **3** and **10**. Thus

$$x^2 + 13x + 30 = (x + 3)(x + 10).$$

(c) To factor $z^2 + 9z + 20$ we need to find a factor pair for 20 whose sum is 9. The required pair is 4 and 5. Thus

$$z^2 + 9z + 20 = (z + 4)(z + 5).$$

Now Try Exercises 19, 21, 23

In the next example, the coefficients of the middle terms are negative.

EXAMPLE 3 **Factoring trinomials having a negative middle coefficient**

Factor each trinomial.
(a) $x^2 - 7x + 10$ (b) $x^2 - 8x + 15$ (c) $y^2 - 9y + 18$

Solution
(a) To factor $x^2 - 7x + 10$ we need to find a factor pair for 10 whose sum equals -7. To have a positive product *and* a negative sum, *both* numbers must be negative, as shown in Table 6.5.

TABLE 6.5 **Factor Pairs for 10**

Factors	$-1, -10$	$-2, -5$
Sum	-11	-7

The required pair is -2 and -5 because $-2 \cdot (-5) = 10$ and $-2 + (-5) = -7$. Therefore the given trinomial can be factored as

$$x^2 - 7x + 10 = (x - 2)(x - 5).$$

(b) To factor $x^2 - 8x + 15$ we need to find a factor pair for 15 whose sum is -8. The required pair is -3 and -5. Thus

$$x^2 - 8x + 15 = (x - 3)(x - 5).$$

(c) To factor $y^2 - 9y + 18$ we need to find a factor pair for 18 whose sum is -9. The required pair is -3 and -6. Thus

$$y^2 - 9y + 18 = (y - 3)(y - 6).$$

Now Try Exercises 27, 29, 31

In Examples 2 and 3 the coefficient of the last term was always positive. In the next example, this coefficient is negative and the coefficient of the middle term is either positive or negative.

EXAMPLE 4 **Factoring trinomials having a negative constant term**

Factor each trinomial.
(a) $x^2 - 3x - 4$ (b) $x^2 + 7x - 8$ (c) $t^2 - 2t - 24$

Solution
(a) To factor $x^2 - 3x - 4$ we need to find a factor pair for -4 whose sum is -3. To have a negative product one factor must be positive and the other factor must be negative, as shown in Table 6.6.

TABLE 6.6 **Factor Pairs for -4**

Factors	$-1, 4$	$1, -4$	$-2, 2$
Sum	3	-3	0

The required pair is 1 and -4 because $1 \cdot (-4) = -4$ and $1 + (-4) = -3$. Therefore the given trinomial can be factored as

$$x^2 - 3x - 4 = (x + 1)(x - 4),$$

which can be checked by multiplying $(x + 1)(x - 4)$.

(b) To factor $x^2 + 7x - 8$ we need to find a factor pair for -8 whose sum is 7. The required pair is -1 and **8**. Thus

$$x^2 + 7x - 8 = (x - 1)(x + 8).$$

(c) To factor $t^2 - 2t - 24$ we need to find a factor pair for -24 whose sum is -2. The required pair is -6 and 4. Thus

$$t^2 - 2t - 24 = (t - 6)(t + 4).$$

▌ **Now Try Exercises 37, 53**

READING CHECK

• What is a prime polynomial?

A polynomial with integer coefficients that cannot be factored by using integer coefficients is called a **prime polynomial**. The next example illustrates that some trinomials of the form $x^2 + bx + c$ cannot be factored into the product of two binomials.

EXAMPLE 5 ▌ **Discovering that a trinomial is prime**

Factor each trinomial, if possible.
(a) $x^2 + 9x + 12$ **(b)** $x^2 + 5x - 4$

Solution

(a) To factor $x^2 + 9x + 12$, we need to find a factor pair for 12 whose sum is 9. Table 6.7 reveals that no such factor pair exists.

TABLE 6.7 Factor Pairs for 12

Factors	1, 12	2, 6	3, 4
Sum	13	8	7

The trinomial $x^2 + 9x + 12$ is prime.

(b) At first glance it may appear that the required factor pair is 4 and 1 because $4 \cdot 1 = 4$ and $4 + 1 = 5$. However, it is important to pay close attention to the signs of the coefficients. To factor $x^2 + 5x - 4$, we need to find a factor pair for -4 whose sum is 5. No such factor pair exists. The trinomial $x^2 + 5x - 4$ is prime.

▌ **Now Try Exercises 17, 41**

READING CHECK

• How can you use the signs of the coefficients in a trinomial to determine the signs in the binomial factors?

MAKING CONNECTIONS

The Signs in the Binomial Factors

If a trinomial of the form $x^2 + bx + c$ can be factored, the signs of the coefficients in the trinomial can be used to determine the signs in the binomial factors. If c is positive, the binomial factors must have the same signs. If c is negative, the binomial factors must have opposite signs. If b and c represent positive numbers, this can be summarized as follows.

Form of the Trinomial	*Signs in the Binomial Factors*
$x^2 + bx + c$	$(\ +\ )(\ +\ )$
$x^2 - bx + c$	$(\ -\ )(\ -\ )$
$x^2 + bx - c$	$(\ -\ )(\ +\ )$
$x^2 - bx - c$	$(\ -\ )(\ +\ )$

When factoring some trinomials, it may be necessary to factor out the greatest common factor before attempting to factor the trinomial into the product of two binomials. The next example illustrates this process.

EXAMPLE 6 | **Factoring out the GCF before factoring further**

Factor each trinomial completely.
(a) $7x^2 + 35x + 42$ **(b)** $2x^4 - 4x^3 - 6x^2$

Solution
(a) Because the GCF of $7x^2 + 35x + 42$ is 7, factor out **7** first.

$$7x^2 + 35x + 42 = \mathbf{7}(x^2 + 5x + 6)$$

Now, to factor $x^2 + 5x + 6$, we need to find a factor pair for 6 whose sum is 5. The required pair is 2 and 3. Thus

$$7x^2 + 35x + 42 = 7(x + 2)(x + 3).$$

(b) Because the GCF of $2x^4 - 4x^3 - 6x^2$ is $2x^2$, factor out $\mathbf{2x^2}$ first.

$$2x^4 - 4x^3 - 6x^2 = \mathbf{2x^2}(x^2 - 2x - 3)$$

Now, to factor $x^2 - 2x - 3$, we need to find a factor pair for -3 whose sum is -2. The required pair is -3 and 1. Thus

$$2x^4 - 4x^3 - 6x^2 = 2x^2(x - 3)(x + 1).$$

CRITICAL THINKING

A cube has a surface area of $6x^2 + 24x + 24$. What is the length of each side?

■ **Now Try Exercises 59, 71**

EXAMPLE 7 | **Finding the dimensions of a rectangle**

Find one possibility for the dimensions of a rectangle that has an area of $x^2 + 6x + 5$.

Solution
The area of a rectangle equals length times width. If we can factor $x^2 + 6x + 5$, then the factors can represent its length and width. Because

$$x^2 + 6x + 5 = (x + 1)(x + 5),$$

one possibility for the rectangle's dimensions is width $x + 1$ and length $x + 5$, as illustrated in Figure 6.4.

■ **Now Try Exercise 87**

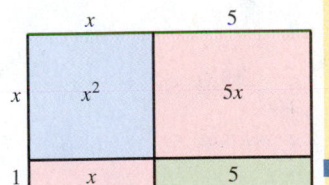

Figure 6.4 Area $= x^2 + 6x + 5$

6.2   Putting It All Together

CONCEPT	EXPLANATION	EXAMPLES
Factoring Trinomials of the Form $$x^2 + bx + c$$	Find two numbers m and n that satisfy $mn = c$ and $m + n = b$. Then $$x^2 + bx + c = (x + m)(x + n).$$	$x^2 + 9x + 20 = (x + 4)(x + 5)$ because $4 \cdot 5 = 20$ and $4 + 5 = 9$. $x^2 + x - 6 = (x - 2)(x + 3)$ because $-2 \cdot 3 = -6$ and $-2 + 3 = 1$. $x^2 - 8x + 12 = (x - 6)(x - 2)$ because $-6 \cdot (-2) = 12$ and $-6 + (-2) = -8$.

6.2   Exercises

MyMathLab

CONCEPTS AND VOCABULARY

1. In the trinomial $x^2 + bx + c$, the leading coefficient is _____.

2. Multiply $(x + m)(x + n)$. What is the coefficient of the x-term? What is the constant term?

3. To factor $x^2 + bx + c$, start by finding two numbers m and n that satisfy _____ $= c$ and _____ $= b$.

4. A trinomial with integer coefficients that cannot be factored using integer coefficients is _____.

PRODUCTS AND SUMS

Exercises 5–12: Find the integer pair that has the given product and sum.

5. Product: 28   Sum: 11

6. Product: 35   Sum: 12

7. Product: -30 Sum: -7

8. Product: -15 Sum: -2

9. Product: -50 Sum: 5

10. Product: -100 Sum: 21

11. Product: 28   Sum: -11

12. Product: 80   Sum: -42

FACTORING TRINOMIALS

Exercises 13–58: Factor the trinomial. If the trinomial cannot be factored, write "prime."

13. $x^2 + 3x + 2$ 14. $x^2 + 5x + 4$

15. $y^2 + 4y + 4$ 16. $y^2 + 8y + 7$

17. $z^2 + 3z + 7$ 18. $z^2 + 4z + 5$

19. $x^2 + 8x + 15$ 20. $x^2 + 9x + 14$

21. $m^2 + 13m + 36$ 22. $m^2 + 15m + 36$

23. $n^2 + 20n + 100$ 24. $n^2 + 52n + 100$

25. $x^2 - 6x + 5$ 26. $x^2 - 6x + 8$

27. $y^2 - 7y + 12$ 28. $y^2 - 12y + 27$

29. $z^2 - 13z + 40$ 30. $z^2 - 15z + 54$

31. $a^2 - 16a + 63$ 32. $a^2 - 82a + 81$

33. $y^2 - 6y + 10$ 34. $y^2 - 2y + 3$

35. $b^2 - 30b + 125$ 36. $b^2 - 19b + 90$

37. $x^2 + 13x - 90$ 38. $x^2 + 15x - 100$

39. $m^2 + 4m - 45$ 40. $m^2 + 4m - 60$

41. $a^2 + 16a - 63$ 42. $a^2 + 13a - 42$

43. $n^2 + 10n - 200$ 44. $n^2 + 2n - 120$

45. $x^2 + 22x - 23$ 46. $x^2 + 18x - 19$

47. $a^2 + 4a - 32$ 48. $a^2 + 9a - 36$

49. $b^2 - b - 20$ 50. $b^2 - b - 12$

51. $m^2 - 14m - 22$ 52. $m^2 - 11m - 24$

53. $x^2 - x - 72$ 54. $x^2 - 2x - 80$

55. $y^2 - 15y - 34$ 56. $y^2 - 10y - 39$

57. $z^2 - 5z - 66$ 58. $z^2 - 6z - 55$

Exercises 59–74: Factor the trinomial completely.

59. $5x^2 - 10x - 40$ 60. $2x^2 + 8x - 10$

61. $-3m^2 - 9m + 12$ 62. $-4n^2 + 20n - 24$

63. $y^3 - 7y^2 + 10y$ 64. $z^3 + 9z^2 + 20z$

65. $-x^3 - 2x^2 + 15x$ 66. $-y^3 + 9y^2 - 14y$

67. $3a^3 + 21a^2 + 18a$ 68. $5b^3 - 5b^2 - 60b$

69. $-2x^3 + 6x^2 - 8x$ 70. $-4y^3 + 20y^2 - 32y$

71. $2m^4 - 10m^3 - 28m^2$ 72. $6n^4 - 18n^3 + 12n^2$

73. $-3x^4 + 3x^3 + 6x^2$ 74. $-5y^4 + 25y^3 - 30y^2$

Exercises 75–78: Factor the trinomial. (Hint: Write the expression in standard form.)

75. $5 + 6x + x^2$ 76. $8 + 6x + x^2$

77. $3 - 4x + x^2$ 78. $10 - 7x + x^2$

Exercises 79–82: Factor the trinomial. (Hint: Write $(m - x)(n + x)$ and find m and n.)

79. $12 + 4x - x^2$ 80. $28 + 3x - x^2$

81. $32 - 4x - x^2$ 82. $40 - 3x - x^2$

83. Thinking Generally Factor the trinomial expression $x^2 + (k + 1)x + k$.

84. Thinking Generally Factor the trinomial expression $x^2 + (k - 2)x - 2k$.

GEOMETRY

85. A square has an area of $x^2 + 2x + 1$. Find the length of a side. Make a sketch of the square.

86. A square has an area of $x^2 + 6x + 9$. Find the length of a side. Make a sketch of the square.

87. A rectangle has an area of $x^2 + 3x + 2$. Find one possibility for its width and length. Make a sketch of the rectangle.

88. A rectangle has an area of $x^2 + 9x + 8$. Find one possibility for its width and length. Make a sketch of the rectangle.

89. A cube has a surface area of $6x^2 + 12x + 6$. Find the length of a side. (*Hint:* First factor out the GCF.)

90. A cube has a surface area of $6x^2 + 36x + 54$. Find the length of a side.

91. Write a polynomial in factored form that represents the total area of the figure.

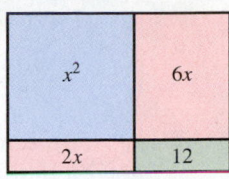

92. Write a polynomial in factored form that represents the total area of the figure.

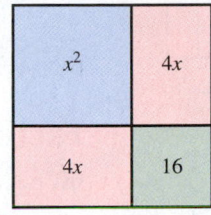

WRITING ABOUT MATHEMATICS

93. Explain how to determine whether a trinomial has been factored correctly. Give an example.

94. Factoring $x^2 + bx + c$ involves finding two integers m and n such that $mn = c$ and $m + n = b$. Is it better first to determine values of m and n so that the product is c or the sum is b? Explain your reasoning.

SECTIONS 6.1 and 6.2   **Checking Basic Concepts**

1. What is the greatest common factor for the expression $8x^3 - 12x^2 + 24x$?

2. Factor $12z^3 - 18z^2$.

3. Factor each expression completely.
   (a) $6y(y - 2) + 5(y - 2)$
   (b) $2x^3 + x^2 + 10x + 5$
   (c) $4z^3 - 12z^2 + 4z - 12$

4. Factor each trinomial completely, if possible.
   (a) $x^2 + 6x + 8$ **(b)** $x^2 - x - 42$

   (c) $a^2 + 3a - 5$ **(d)** $4a^3 + 20a^2 + 24a$

5. Write a polynomial in factored form that represents the total area of the figure.

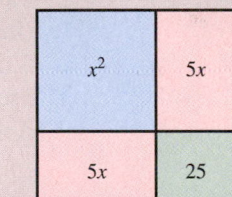

6.3 Factoring Trinomials II ($ax^2 + bx + c$)

Factoring Trinomials by Grouping • Factoring with FOIL in Reverse

A LOOK INTO MATH ▶

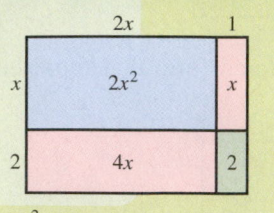

$2x^2 + 5x + 2 = (x + 2)(2x + 1)$

Figure 6.5

The sum of the areas of the four small rectangles shown in Figure 6.5 is $2x^2 + 5x + 2$. Note that the length of the large rectangle is $2x + 1$ and that its width is $x + 2$. One way to determine these dimensions is to factor $2x^2 + 5x + 2$. However, this trinomial has a leading coefficient of 2. Thus far we have factored only trinomials that have a leading coefficient of 1. In this section we discuss two methods used to factor trinomials in the form $ax^2 + bx + c$, where $a \neq 1$. In Example 1(a), we demonstrate how to factor the trinomial $2x^2 + 5x + 2$ as $(x + 2)(2x + 1)$.

Factoring Trinomials by Grouping

To factor the polynomial given by $x^2 + 6x + 5$ we find two numbers, m and n, such that $mn = 5$ and $m + n = 6$. For this trinomial we let $m = 1$ and $n = 5$, which gives

$$x^2 + 6x + 5 = (x + 1)(x + 5).$$

To factor the polynomial $2x^2 + 7x + 3$, which has a leading coefficient of 2, we must find two numbers m and n such that $mn = 2 \cdot 3 = 6$ and $m + n = 7$. One solution is $m = 1$ and $n = 6$. Using grouping, we can now factor this trinomial by writing $7x$ as the sum $1x + 6x$.

$$
\begin{aligned}
2x^2 + 7x + 3 &= 2x^2 + \overbrace{x + 6x}^{7x} + 3 && \text{Write } 7x \text{ as } x + 6x.\\
&= (2x^2 + x) + (6x + 3) && \text{Associative property}\\
&= x(2x + 1) + 3(2x + 1) && \text{Factor out common factors.}\\
&= (x + 3)(2x + 1) && \text{Factor out } (2x + 1).
\end{aligned}
$$

This technique of factoring trinomials by grouping is summarized as follows.

FACTORING $ax^2 + bx + c$ BY GROUPING

To factor $ax^2 + bx + c$ perform the following steps. (Assume that a, b, and c are integers and have no common factors.)

1. Find two numbers, m and n, such that $mn = ac$ and $m + n = b$.
2. Write the trinomial as $ax^2 + mx + nx + c$.
3. Use grouping to factor this expression into two binomials.

STUDY TIP

Be sure that you understand that the values of m and n are used in different ways, depending on whether or not the leading coefficient is 1.

MAKING CONNECTIONS

The Values of m and n

The values of m and n found when factoring $ax^2 + bx + c$, where $a \neq 1$, are not used in the same way as the values of m and n found when factoring $x^2 + bx + c$.

Factoring $x^2 + bx + c$

$mn = c$ and $m + n = b$

$x^2 + bx + c = (x + m)(x + n)$

Factoring $ax^2 + bx + c$, where $a \neq 1$

$mn = ac$ and $m + n = b$

$ax^2 + bx + c = ax^2 + mx + nx + c$

Factor the resulting four-term polynomial by grouping.

EXAMPLE 1   **Factoring $ax^2 + bx + c$ by grouping**

Factor each trinomial.
(a) $2x^2 + 5x + 2$ **(b)** $3z^2 + z - 2$ **(c)** $10t^2 - 11t + 3$

Solution
(a) To factor $2x^2 + 5x + 2$ we need to find m and n that satisfy $mn = 2 \cdot 2 = 4$ and $m + n = 5$. Table 6.8 shows that two such numbers are $m = 1$ and $n = 4$.

$$\overset{\lceil \quad 5x \quad \rceil}{2x^2 + 5x + 2 = 2x^2 + x + 4x + 2} \qquad \text{Write } 5x \text{ as } x + 4x.$$

$$= (2x^2 + x) + (4x + 2) \qquad \text{Associative property}$$

$$= x(2x + 1) + 2(2x + 1) \qquad \text{Factor out common factors.}$$

$$= (x + 2)(2x + 1) \qquad \text{Factor out } (2x + 1).$$

TABLE 6.8 **Factor Pairs for 4**

Factors	1, 4	2, 2
Sum	5	4

(b) To factor $3z^2 + 1z - 2$ we need to find m and n that satisfy $mn = 3 \cdot (-2) = -6$ and $m + n = 1$. Two such numbers are $m = 3$ and $n = -2$.

$$3z^2 + z - 2 = 3z^2 + 3z - 2z - 2 \qquad \text{Write } z \text{ as } 3z - 2z.$$

$$= (3z^2 + 3z) + (-2z - 2) \qquad \text{Associative property}$$

$$= 3z(z + 1) - 2(z + 1) \qquad \text{Factor out common factors.}$$

$$= (3z - 2)(z + 1) \qquad \text{Factor out } (z + 1).$$

(c) To factor $10t^2 - 11t + 3$ we need to find m and n that satisfy $mn = 10 \cdot 3 = 30$ and $m + n = -11$. Two such numbers are $m = -5$ and $n = -6$.

$$10t^2 - 11t + 3 = 10t^2 - 5t - 6t + 3 \qquad \text{Write } -11t \text{ as } -5t - 6t.$$

$$= (10t^2 - 5t) + (-6t + 3) \qquad \text{Associative property}$$

$$= 5t(2t - 1) - 3(2t - 1) \qquad \text{Factor out common factors.}$$

$$= (5t - 3)(2t - 1) \qquad \text{Factor out } (2t - 1).$$

Now Try Exercises 15, 25, 37

MAKING CONNECTIONS

Different Ways to Factor by Grouping

In Example 1(c) we could have written $-11t$ as $-6t - 5t$, rather than $-5t - 6t$. Then the factoring could have been written as

$$10t^2 - 11t + 3 = 10t^2 - 6t - 5t + 3$$

$$= (10t^2 - 6t) + (-5t + 3)$$

$$= 2t(5t - 3) - 1(5t - 3)$$

$$= (2t - 1)(5t - 3),$$

which gives the same result.

In the previous section we showed that some trinomials of the form $x^2 + bx + c$ are prime and cannot be factored into the product of two binomials with integer coefficients. The next example illustrates that some trinomials of the form $ax^2 + bx + c$, with $a \neq 1$, may also be prime.

EXAMPLE 2   **Discovering that a trinomial is prime**

Factor each trinomial.
(a) $3x^2 + 5x + 4$ **(b)** $2x^2 - 8x - 3$

Solution

(a) To factor $3x^2 + 5x + 4$, we need to find integers m and n such that $mn = 3 \cdot 4 = 12$ and $m + n = 5$. Because the middle term is positive, we consider only positive factors of 12. Table 6.9 reveals that no such integers exist.

TABLE 6.9 **Factor Pairs for 12**

Factors	1, 12	2, 6	3, 4	
Sum	13	8	7	← No sum equals 5.

The trinomial $3x^2 + 5x + 4$ is prime.

(b) To factor $2x^2 - 8x - 3$, we must find integers m and n such that $mn = 2 \cdot (-3) = -6$ and $m + n = -8$. Table 6.10 reveals that no such integers exist.

TABLE 6.10 **Factor Pairs for −6**

Factors	−1, 6	−2, 3	2, −3	1, −6	
Sum	5	1	−1	−5	← No sum equals −8.

The trinomial $2x^2 - 8x - 3$ is prime.

Now Try Exercises 17, 27

Although some trinomials may look as though they can be factored using the process discussed in Example 1, it is important to remember to first factor out the greatest common factor whenever possible. In the next example we factor out the GCF before factoring the trinomial further.

EXAMPLE 3 **Factoring out the GCF before factoring further**

Factor each trinomial completely.
(a) $15x^2 - 50x - 40$ **(b)** $4x^3 - 22x^2 + 30x$

Solution

(a) Because the GCF of $15x^2 - 50x - 40$ is 5, factor out 5 before factoring the remaining trinomial.

$$15x^2 - 50x - 40 = 5(3x^2 - 10x - 8)$$

To factor $3x^2 - 10x - 8$, we need numbers m and n such that $mn = 3 \cdot (-8) = -24$ and $m + n = -10$. Two such numbers are **−12** and **2**.

$$
\begin{aligned}
15x^2 - 50x - 40 &= 5(3x^2 \mathbf{- 12}x + \mathbf{2}x - 8) && \text{Write } -10x \text{ as } -12x + 2x. \\
&= 5\big((3x^2 - 12x) + (2x - 8)\big) && \text{Associative property} \\
&= 5\big(3x(x - 4) + 2(x - 4)\big) && \text{Factor out common factors.} \\
&= 5(3x + 2)(x - 4) && \text{Factor out } (x - 4).
\end{aligned}
$$

(b) Because the GCF of $4x^3 - 22x^2 + 30x$ is $2x$, factor out $2x$ before factoring the remaining trinomial.

$$4x^3 - 22x^2 + 30x = 2x(2x^2 - 11x + 15)$$

To factor the trinomial $2x^2 - 11x + 15$, we need to find numbers m and n such that $mn = 2 \cdot 15 = 30$ and $m + n = -11$. Two such numbers are -6 and -5.

$$
\begin{aligned}
4x^3 - 22x^2 + 30x &= 2x(2x^2 - \mathbf{6x} - \mathbf{5x} + 15) && \text{Write } -11x \text{ as } -6x - 5x. \\
&= 2x\big((2x^2 - 6x) + (-5x + 15)\big) && \text{Associative property} \\
&= 2x\big(2x(x - 3) - 5(x - 3)\big) && \text{Factor out common factors.} \\
&= 2x(2x - 5)(x - 3) && \text{Factor out } (x - 3).
\end{aligned}
$$

Now Try Exercises 55, 59

Factoring with FOIL in Reverse

Rather than factoring a trinomial by grouping, we can use FOIL in reverse to determine its binomial factors. For example, the factors of $2x^2 + 5x + 2$ are two binomials:

$$2x^2 + 5x + 2 = (\underline{\quad} + \underline{\quad})(\underline{\quad} + \underline{\quad}),$$

where the expressions to be placed in the four blanks are yet to be found. By the FOIL method, we know that the product of the first terms of these two binomials is $2x^2$. Because $2x^2 = 2x \cdot x$, we can write

$$2x^2 + 5x + 2 = (\underline{2x} + \underline{\quad})(\underline{x} + \underline{\quad}).$$

By FOIL, the product of the last terms in these binomials must be 2. Because $2 = 1 \cdot 2$, we could put 1 and 2 in the blanks. However, we must be sure to place them correctly so that the product of the outside terms plus the product of the inside terms is $5x$.

$$
(2x + 1)(x + 2) = 2x^2 + 5x + 2
$$

$1x$

$+ 4x$

$\overline{5x}$ ✓ ⟵ Middle term checks.

If we had interchanged the 1 and 2, we would have obtained an incorrect result.

$$
(2x + 2)(x + 1) = 2x^2 + 4x + 2
$$

$2x$

$+2x$

$\overline{4x}$ ✗ ⟵ Middle term is *not* $5x$.

MAKING CONNECTIONS

The Signs in the Binomial Factors

Let a, b, and c represent positive integers. If a trinomial of the form $ax^2 + bx + c$ can be factored, the signs in the binomial factors can be summarized as follows.

Form of the Trinomial	*Signs in the Binomial Factors*
$ax^2 + bx + c$	$(\ +\ )(\ +\ )$
$ax^2 - bx + c$	$(\ -\ )(\ -\ )$
$ax^2 + bx - c$	$(\ -\ )(\ +\ )$
$ax^2 - bx - c$	$(\ -\ )(\ +\ )$

In the next example we factor expressions of the form $ax^2 + bx + c$, where $a \neq 1$. We use FOIL in reverse and *trial and error* to find the correct factors.

EXAMPLE 4   **Factoring the form $ax^2 + bx + c$**

Factor each trinomial.
(a) $3x^2 + 5x + 2$ (b) $6x^2 + 7x - 3$ (c) $6 + 4y^2 - 11y$

Solution
(a) To factor $3x^2 + 5x + 2$, we start by finding factors of $3x^2$, which include **$3x$** and **x**, so we write

$$3x^2 + 5x + 2 = (\mathbf{3x} + \underline{\quad})(x + \underline{\quad}).$$

The factors of the last term 2 are **1** and **2**. (Because the middle term is positive, we do not consider -1 and -2.) If we place values of 1 and 2 in the binomial as follows, the middle term becomes $7x$ rather than $5x$.

$$(\mathbf{3x} + \mathbf{1})\,(x + \mathbf{2}) = 3x^2 + 7x + 2$$

$$\begin{array}{l} \underline{1x} \\ \underline{+6x} \\ 7x \;\, ✗ \end{array} \longleftarrow \;\text{Middle term is \textit{not} } 5x.$$

By reversing the positions of 1 and 2, we obtain a correct factorization.

$$(\mathbf{3x} + \mathbf{2})\,(x + \mathbf{1}) = 3x^2 + 5x + 2$$

$$\begin{array}{l} \underline{2x} \\ \underline{+3x} \\ 5x \;\, ✓ \end{array} \longleftarrow \;\text{Middle term checks.}$$

(b) To factor $6x^2 + 7x - 3$, we start by finding factors of $6x^2$, which include $2x$ and $3x$ or $6x$ and x. Factors of the last term, -3, include -1 and 3 or 1 and -3. The following two factorizations give incorrect results.

$$(2x + 1)\,(3x - 3) = 6x^2 - 3x - 3 \qquad\qquad (6x + 1)\,(x - 3) = 6x^2 - 17x - 3$$

$$\begin{array}{l} \underline{3x} \\ \underline{-6x} \\ -3x \;\, ✗ \end{array} \longleftarrow \text{Middle term is \textit{not} } 7x. \qquad \begin{array}{l} \underline{1x} \\ \underline{-18x} \\ -17x \;\, ✗ \end{array} \longleftarrow \text{Middle term is \textit{not} } 7x.$$

To obtain a middle term of $7x$ we use the following arrangement.

$$(3x - 1)\quad(2x + 3) = 6x^2 + 7x - 3$$

$$\begin{array}{l} \underline{-2x} \\ \underline{+9x} \\ 7x \;\, ✓ \end{array} \longleftarrow \;\text{Middle term checks.}$$

To find the correct factorization we often need to try more than one arrangement.

(c) To factor $6 + 4y^2 - 11y$, we start by writing the trinomial in standard form as $4y^2 - 11y + 6$. Then we find possible factors of the first term, $4y^2$, which include $2y$ and $2y$ or $4y$ and y. Factors of the last term, 6, include either -1 and -6 or -2 and -3. (Because the middle term is negative, we do not use the positive factors of 1 and 6 or 2 and 3.) To obtain a middle term of $-11y$ we use the following arrangement.

$$(4y - 3)\quad(y - 2) = 4y^2 - 11y + 6$$

$$\begin{array}{l} \underline{-3y} \\ \underline{-8y} \\ -11y \;\, ✓ \end{array} \longleftarrow \;\text{Middle term checks.}$$

Now Try Exercises 19, 33, 69

READING CHECK

• What is a good first step when factoring a trinomial with a negative leading coefficient?

In the next example we demonstrate how to factor trinomials having a negative leading coefficient. This task is sometimes accomplished by first factoring out -1.

| EXAMPLE 5 | **Factoring trinomials having a negative leading coefficient** |

Factor each trinomial.
(a) $-6x^2 + 17x - 5$ **(b)** $1 - x - 2x^2$

Solution
(a) One way to factor $-6x^2 + 17x - 5$ is to start by factoring out -1. Then we can apply FOIL in reverse.

Opposite of $-6x^2 + 17x - 5$

$$
\begin{aligned}
-6x^2 + 17x - 5 &= -1(6x^2 - 17x + 5) && \text{Factor out } -1. \\
&= -1(3x - 1)(2x - 5) && \text{Factor the trinomial.} \\
&= -(3x - 1)(2x - 5) && \text{Rewrite.}
\end{aligned}
$$

(b) To factor the trinomial $1 - x - 2x^2$, write it in standard form, factor out -1, and then apply FOIL in reverse.

$$
\begin{aligned}
1 - x - 2x^2 &= -2x^2 - x + 1 && \text{Standard form} \\
&= -1(2x^2 + x - 1) && \text{Factor out } -1. \\
&= -(2x - 1)(x + 1) && \text{Factor the trinomial.}
\end{aligned}
$$

Now Try Exercises 71, 75

6.3   Putting It All Together

CONCEPT	EXPLANATION	EXAMPLES
Factoring Trinomials by Grouping	To factor $ax^2 + bx + c$, find two numbers, m and n, such that $mn = ac$ and $m + n = b$. Then write $$ax^2 + mx + nx + c.$$ Use grouping to factor this expression into two binomials.	For $3x^2 + 10x + 8$ let $m = 6$ and $n = 4$ because $mn = 24$ and $m + n = 10$. $$\begin{aligned} 3x^2 + 10x + 8 &= 3x^2 + 6x + 4x + 8 \\ &= (3x^2 + 6x) + (4x + 8) \\ &= 3x(x + 2) + 4(x + 2) \\ &= (3x + 4)(x + 2) \end{aligned}$$
Factoring Trinomials by Using FOIL in Reverse	To factor $ax^2 + bx + c$, find factors of ax^2 and of c. Choose and arrange these factors in two binomials so that the middle term is bx.	For $3x^2 + 10x + 8$ the factors of $3x^2$ are $3x$ and x. The positive factors of 8 are either 2 and 4, or 1 and 8. A middle term of $10x$ can be obtained as follows. $(3x + 4)(x + 2) = 3x^2 + 10x + 8$ $4x$ $+6x$ $\overline{10x}$ ✓ ← Middle term checks.

continued on next page

continued from previous page

CONCEPT	EXPLANATION	EXAMPLES
Choosing Signs When Factoring $ax^2 + bx + c$	For $ax^2 + bx + c$ with $a > 0$, **1.** If $c > 0$ and $b > 0$, use $(\_ + \_)(\_ + \_)$. **2.** If $c > 0$ and $b < 0$, use $(\_ - \_)(\_ - \_)$. **3.** If $c < 0$, use $(\_ - \_)(\_ + \_)$ or $(\_ + \_)(\_ - \_)$.	**1.** $2x^2 + 7x + 3 = (2x + 1)(x + 3)$ $(c = 3, b = 7)$ **2.** $2x^2 - 7x + 3 = (2x - 1)(x - 3)$ $(c = 3, b = -7)$ **3.** $2x^2 + 5x - 3 = (2x - 1)(x + 3)$ $(c = -3, b = 5)$ $2x^2 - 5x - 3 = (2x + 1)(x - 3)$ $(c = -3, b = -5)$

6.3   Exercises

CONCEPTS AND VOCABULARY

1. To factor the polynomial $ax^2 + bx + c$ by grouping, you first find two numbers, m and n, such that $mn = $ _____ and $m + n = $ _____.

2. To factor the polynomial $ax^2 + bx + c$ with FOIL in reverse, you first find possible factors for _____ and for _____.

Exercises 3–6: Insert the symbol $+$ or $-$ in each binomial factor to make the equation true.

3. $3x^2 + 5x + 2 = (3x\_\_2)(x\_\_1)$.

4. $3x^2 - x - 2 = (3x\_\_2)(x\_\_1)$.

5. $3x^2 - 5x + 2 = (3x\_\_2)(x\_\_1)$.

6. $3x^2 + x - 2 = (3x\_\_2)(x\_\_1)$.

Exercises 7–14: Fill in the blank in each binomial factor to make the equation true.

7. $4x^2 + 11x + 6 = (4x + \_\_)(\_\_ + 2)$.

8. $4x^2 - 5x - 6 = (\_\_ - 2)(\_\_ + 3)$.

9. $4x^2 + 4x - 3 = (2x - \_\_)(2x + \_\_)$.

10. $4x^2 - 8x + 3 = (2x - \_\_)(\_\_ - 3)$.

11. $6x^2 + 11x - 7 = (\_\_ + 7)(2x - \_\_)$.

12. $6x^2 - 31x + 28 = (\_\_ - 7)(\_\_ - 4)$.

13. $6x^2 - x - 15 = (3x - \_\_)(2x + \_\_)$.

14. $6x^2 - 53x + 40 = (\_\_ - 5)(x - \_\_)$.

FACTORING TRINOMIALS

Exercises 15–54: Factor the trinomial. If the trinomial cannot be factored, write "prime."

15. $2x^2 + 7x + 3$

16. $2x^2 + 3x + 1$

17. $3y^2 + 2y + 4$

18. $2y^2 + 5y + 1$

19. $3x^2 + 4x + 1$

20. $3x^2 + 10x + 3$

21. $6x^2 + 11x + 3$

22. $6x^2 + 17x + 5$

23. $5x^2 - 11x + 2$

24. $7x^2 - 8x + 1$

25. $2y^2 - 7y + 5$

26. $2y^2 - 11y + 12$

27. $3m^2 - 11m - 6$

28. $5m^2 - 7m - 2$

29. $7z^2 - 37z + 10$

30. $3z^2 - 11z + 6$

31. $3t^2 - 7t - 6$

32. $8t^2 - 6t - 9$

33. $15r^2 + r - 6$

34. $12r^2 + r - 6$

35. $24m^2 - 23m - 12$

36. $24m^2 + 29m - 4$

37. $25x^2 + 5x - 2$

38. $30x^2 + 7x - 2$

39. $6x^2 + 11x - 2$

40. $12x^2 + 28x - 5$

41. $15y^2 - 7y + 2$

42. $14y^2 - 5y + 1$

43. $21n^2 + 4n - 1$

44. $21n^2 + 10n + 1$

45. $14y^2 + 23y + 3$

46. $28y^2 + 25y + 3$

47. $28z^2 - 25z + 3$

48. $15z^2 - 19z + 6$

49. $30x^2 - 29x + 6$

50. $50x^2 - 55x + 12$

51. $20a^2 + 18a - 5$ **52.** $40a^2 + 21a - 2$

53. $18t^2 + 23t - 6$ **54.** $33t^2 + 7t - 10$

Exercises 55–64: Factor the trinomial completely.

55. $12a^2 + 12a - 9$ **56.** $21b^2 - 14b - 56$

57. $12y^3 - 11y^2 + 2y$ **58.** $10z^3 + 19z^2 + 6z$

59. $24x^3 - 30x^2 + 9x$ **60.** $8y^3 - 16y^2 + 6y$

61. $8x^4 - 6x^3 + 2x^2$ **62.** $10y^3 + 15y^2 - 5y$

63. $28x^4 + 56x^3 + 21x^2$ **64.** $20y^4 + 42y^3 - 20y^2$

65. Thinking Generally Factor the trinomial expression $3x^2 + (3k + 1)x + k$.

66. Thinking Generally Factor the trinomial expression $3x^2 + (3k - 2)x - 2k$.

Exercises 67–76: Factor the trinomial.

67. $2 + 15x + 7x^2$ **68.** $3 + 16x + 5x^2$

69. $2 - 5x + 2x^2$ **70.** $5 - 6x + x^2$

71. $3 - 2x - 8x^2$ **72.** $5 - 3x - 2x^2$

73. $-2x^2 - 7x + 15$ **74.** $-5x^2 - 19x + 4$

75. $-5x^2 + 14x + 3$ **76.** $-6x^2 + 17x + 14$

77. A rectangle has an area of $6x^2 + 7x + 2$. Find possible dimensions for this rectangle. Make a sketch of the rectangle.

78. A rectangle has an area of $2x^2 + 5x + 3$. Find possible dimensions for the rectangle. Make a sketch of the rectangle.

79. Write a polynomial in factored form that represents the total area of the figure.

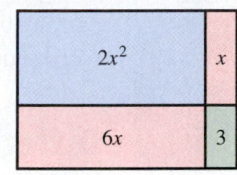

80. Write a polynomial in factored form that represents the total area of the figure.

WRITING ABOUT MATHEMATICS

81. Explain how the sign of the third term in the trinomial $ax^2 + bx + c$ affects how it is factored.

82. Explain the steps to be used to factor $ax^2 + bx + c$ by grouping.

Group Activity   Working with Real Data

Directions: Form a group of 2 to 4 people. Select someone to record the group's responses for this activity. All members of the group should work cooperatively to answer the questions. If your instructor asks for your results, each member of the group should be prepared to respond.

AIDS Cases From 1993 to 2009 the cumulative number N of AIDS cases in thousands can be approximated by $N = -2x^2 + 76x + 430$, where $x = 0$ corresponds to the year 1993.

Year	1993	1997	2001	2005	2009
Cases	422	677	844	984	1109

Source: U.S. Department of Health and Human Services.

(a) Use the equation to find N for each year in the table.
(b) Discuss how well this equation approximates the data.
(c) Rewrite the equation with the right side completely factored.
(d) Use your equation from part (c) to find N for each year in the table. Do your answers agree with those found in part (a)?

6.4 Special Types of Factoring

Difference of Two Squares • Perfect Square Trinomials • Sum and Difference of Two Cubes

A LOOK INTO MATH ▶ When children are first learning to read, they often spend a great deal of time "sounding out" the words that they encounter. In time, some words become easily recognizable as *sight words* that can be read more quickly. In mathematics, we can learn to recognize certain types of polynomials that can be factored more quickly using special factoring techniques.

NEW VOCABULARY

☐ Perfect square trinomial

STUDY TIP

The special types of factoring discussed in this section are helpful for factoring some (but not all) types of polynomials. Don't forget to review factoring techniques that were discussed earlier in this chapter.

Difference of Two Squares

In Section 5.4 we showed that $(a - b)(a + b) = a^2 - b^2$. If we rewrite this equation as

$$a^2 - b^2 = (a - b)(a + b),$$

then we can use it to factor a difference of two squares. For example, to factor the expression $x^2 - 25$, we first write it in the form $a^2 - b^2$, where $a = x$ and $b = 5$. Doing so results in expressing $x^2 - 25$ as $x^2 - 5^2$, and the equation

$$a^2 - b^2 = (a - b)(a + b)$$

becomes

$$x^2 - 5^2 = (x - 5)(x + 5).$$

This discussion suggests the following rule for factoring a difference of two squares.

DIFFERENCE OF TWO SQUARES

For any real numbers a and b,

$$a^2 - b^2 = (a - b)(a + b).$$

In the next example we apply this method to other expressions.

EXAMPLE 1 **Factoring the difference of two squares**

Factor each difference of two squares.
(a) $x^2 - 36$ **(b)** $4x^2 - 9$ **(c)** $100 - 16t^2$ **(d)** $49y^2 - 64z^2$

Solution
(a) Write $x^2 - 36$ as $x^2 - 6^2$ and then substitute x for a and 6 for b. The equation

$$a^2 - b^2 = (a - b)(a + b)$$

becomes

$$x^2 - 6^2 = (x - 6)(x + 6).$$

(b) The expression $4x^2 - 9$ can be written as $(2x)^2 - 3^2$. Thus

$$4x^2 - 9 = (2x - 3)(2x + 3).$$

(c) The expression $100 - 16t^2$ can be written as $(10)^2 - (4t)^2$. Thus

$$100 - 16t^2 = (10 - 4t)(10 + 4t).$$

(d) The expression $49y^2 - 64z^2$ can be written as $(7y)^2 - (8z)^2$. Thus

$$49y^2 - 64z^2 = (7y - 8z)(7y + 8z).$$

Now Try Exercises 17, 19, 25, 29

READING CHECK

• Is it possible to factor a *sum* of two squares by using real numbers?

MAKING CONNECTIONS

Sum of Squares versus Difference of Squares

The sum of two squares, $a^2 + b^2$, cannot be factored by using real numbers. However, the difference of two squares, $a^2 - b^2$, can be factored. For example, $x^2 + 4$ cannot be factored, but $x^2 - 4$ can be factored as $(x - 2)(x + 2)$.

Perfect Square Trinomials

In Section 5.4 we also showed how to compute $(a + b)^2$ and $(a - b)^2$ as

$$(a + b)^2 = a^2 + 2ab + b^2 \quad \text{and}$$
$$(a - b)^2 = a^2 - 2ab + b^2.$$

The expressions $a^2 + 2ab + b^2$ and $a^2 - 2ab + b^2$ are called **perfect square trinomials.** If we can recognize a perfect square trinomial, we can use the following formulas to factor it.

PERFECT SQUARE TRINOMIALS

For any real numbers a and b,

$$a^2 + 2ab + b^2 = (a + b)^2 \quad \text{and}$$
$$a^2 - 2ab + b^2 = (a - b)^2.$$

READING CHECK

• How can you recognize a perfect square trinomial?

When factoring a trinomial as a perfect square trinomial, we must first verify that the middle term is correct. For example, to factor $x^2 + 14x + 49$ we start by rewriting the expression as $x^2 + 14x + 7^2$. In order to factor this trinomial as a perfect square trinomial, the middle term $14x$ must be equal to twice the product of x and 7.

$$x^2 + 14x + 7^2$$
$$\downarrow$$
$$2(x)(7) \qquad \checkmark \quad \text{The middle term checks.}$$

When $a = x$ and $b = 7$, the equation $a^2 + 2ab + b^2 = (a + b)^2$ allows us to factor the given polynomial as $x^2 + 14x + 49 = (x + 7)^2$.

EXAMPLE 2   **Factoring perfect square trinomials**

If possible, factor each trinomial as a perfect square trinomial.
(a) $x^2 + 10x + 25$ **(b)** $4x^2 - 4x + 1$ **(c)** $9z^2 + 18z + 4$ **(d)** $x^2 - 4xy + 4y^2$

Solution
(a) Start by writing $x^2 + 10x + 25$ as $x^2 + 10x + 5^2$ and then check the middle term.

$$x^2 + 10x + 5^2$$
$$\downarrow$$
$$2(x)(5) \qquad \checkmark \quad \text{The middle term checks.}$$

When $a = x$ and $b = 5$, the equation $a^2 + 2ab + b^2 = (a + b)^2$ allows us to factor the given polynomial as $x^2 + 10x + 25 = (x + 5)^2$.

(b) The polynomial $4x^2 - 4x + 1$ can be written as $(2x)^2 - 4x + 1^2$.

$$(2x)^2 - 4x + 1^2$$
$$\downarrow$$
$$\longrightarrow 2(2x)(1) \longleftarrow \checkmark \quad \text{The middle term checks.}$$

When $a = 2x$ and $b = 1$, the equation $a^2 - 2ab + b^2 = (a - b)^2$ allows us to factor the given polynomial as $4x^2 - 4x + 1 = (2x - 1)^2$.

(c) Write $9z^2 + 18z + 4$ as $(3z)^2 + 18z + 2^2$ and then check the middle term.

$$(3z)^2 + 18z + 2^2$$
$$\downarrow$$
$$\longrightarrow 2(3z)(2) \longleftarrow \text{✗} \quad \text{The middle term is } 12z \text{ and does not equal } 18z.$$

The expression $9z^2 + 18z + 4$ cannot be factored as a perfect square trinomial.

(d) The polynomial $x^2 - 4xy + 4y^2$ can be written as $x^2 - 4xy + (2y)^2$.

$$x^2 - 4xy + (2y)^2$$
$$\downarrow$$
$$\longrightarrow 2(x)(2y) \longleftarrow \checkmark \quad \text{The middle term checks.}$$

When $a = x$ and $b = 2y$, the equation $a^2 - 2ab + b^2 = (a - b)^2$ allows us to factor the given polynomial as $x^2 - 4xy + 4y^2 = (x - 2y)^2$.

▌ **Now Try Exercises 37, 41, 43, 51**

MAKING CONNECTIONS

Special Factoring and General Techniques

If you do not recognize a polynomial as the difference of two squares or a perfect square trinomial, you can still factor the polynomial by using the methods discussed in earlier sections.

Sum and Difference of Two Cubes

The sum or difference of two cubes may be factored—a result of the two equations

$$(a + b)(a^2 - ab + b^2) = a^3 + b^3 \quad \text{and}$$
$$(a - b)(a^2 + ab + b^2) = a^3 - b^3.$$

These equations can be verified by multiplying the left side to obtain the right side. For example, multiplying the polynomials on the left side of the first equation results in

$$(a + b)(a^2 - ab + b^2) = a \cdot a^2 - a \cdot ab + a \cdot b^2 + b \cdot a^2 - b \cdot ab + b \cdot b^2$$
$$= a^3 - a^2b + ab^2 + a^2b - ab^2 + b^3$$
$$= a^3 + b^3.$$

SUM AND DIFFERENCE OF TWO CUBES

For any real numbers a and b,

Opposite Signs

$$a^3 + b^3 = (a + b)(a^2 - ab + b^2) \quad \text{and}$$
$$a^3 - b^3 = (a - b)(a^2 + ab + b^2).$$

Opposite Signs

Any binomial whose terms can be expressed as cubes can be factored as a sum or difference of cubes. We demonstrate this method in the next example.

EXAMPLE 3 **Factoring the sum and difference of two cubes**

Factor each polynomial.
(a) $z^3 + 8$ (b) $x^3 - 27$ (c) $8x^3 - 1$

Solution
(a) To factor $z^3 + 8$ we let $a^3 = z^3$ and $b^3 = 2^3$ so that $a = z$ and $b = 2$. Then

$$a^3 + b^3 = (a + b)(a^2 - ab + b^2)$$

becomes

$$z^3 + 2^3 = (z + 2)(z^2 - z \cdot 2 + 2^2)$$
$$= (z + 2)(z^2 - 2z + 4).$$

(b) To factor $x^3 - 27$ we let $a^3 = x^3$ and $b^3 = 3^3$ so that $a = x$ and $b = 3$. Then

$$a^3 - b^3 = (a - b)(a^2 + ab + b^2)$$

becomes

$$x^3 - 3^3 = (x - 3)(x^2 + x \cdot 3 + 3^2)$$
$$= (x - 3)(x^2 + 3x + 9).$$

(c) To factor $8x^3 - 1$ we let $a^3 = (2x)^3$ and $b^3 = 1^3$ so that $a = 2x$ and $b = 1$. Be sure to write $2x$ in **parentheses** when substituting for a in the term a^2.

$$(2x)^3 - 1^3 = (2x - 1)\left((2x)^2 + 2x \cdot 1 + 1^2\right)$$
$$= (2x - 1)(4x^2 + 2x + 1)$$

Now Try Exercises 57, 65

In this section we have discussed special methods for factoring polynomials that can be identified as the difference of two squares, perfect square trinomials, the sum of two cubes, or the difference of two cubes. The next example demonstrates how to recognize and factor such polynomials.

EXAMPLE 4 **Recognizing polynomials that can be factored with special methods**

Factor each polynomial.
(a) $8x^3 + 27$ (b) $4y^2 - 20y + 25$ (c) $9z^2 - 64$

Solution
(a) Because this polynomial has only two terms, it cannot be factored as a perfect square trinomial. Since it is not a difference, it cannot be factored as the difference of two squares. We will try to factor the polynomial as the sum of two cubes. To factor $8x^3 + 27$, we note that $8x^3 = (2x)^3$ and $27 = 3^3$. Then

$$8x^3 + 27 = (2x)^3 + 3^3$$
$$= (2x + 3)\left((2x)^2 - 2x \cdot 3 + 3^2\right)$$
$$= (2x + 3)(4x^2 - 6x + 9).$$

(b) Because this polynomial has three terms, we will try to factor it as a perfect square trinomial. Begin by writing $4y^2 - 20y + 25$ as $(2y)^2 - 20y + 5^2$ and then verify that the middle term checks.

$$(2y)^2 - 20y + 5^2$$

$$\longrightarrow 2(2y)(5) \longleftarrow \checkmark \quad \text{The middle term checks.}$$

When $a = 2y$ and $b = 5$, the equation $a^2 - 2ab + b^2 = (a - b)^2$ allows us to factor the given polynomial as $4y^2 - 20y + 25 = (2y - 5)^2$.

(c) Because this polynomial is a difference of two terms that appear to be square terms, we will try to factor the polynomial as the difference of two squares. The expression $9z^2 - 64$ can be written as $(3z)^2 - 8^2$. Thus

$$9z^2 - 64 = (3z - 8)(3z + 8).$$

▌ **Now Try Exercises 31, 47, 67**

When using the special factoring methods discussed in this section, it is important to remember to first factor out the greatest common factor whenever possible. In the next example we factor out the GCF before factoring further.

EXAMPLE 5 **Factoring out the GCF before factoring further**

Factor each polynomial completely.
(a) $27x^3 + 72x^2 + 48x$ **(b)** $18a^3 - 8ab^2$

Solution
(a) Because the GCF of $27x^3 + 72x^2 + 48x$ is $3x$, factor out $3x$ before factoring the remaining trinomial.

$$27x^3 + 72x^2 + 48x = 3x(9x^2 + 24x + 16)$$

The expression $9x^2 + 24x + 16$ is a perfect square trinomial that can be written as $(3x)^2 + 24x + 4^2$.

$$(3x)^2 + 24x + 4^2$$

$$\longrightarrow 2(3x)(4) \longleftarrow \checkmark \quad \text{The middle term checks.}$$

Thus $9x^2 + 24x + 16 = (3x + 4)^2$. As a result, the given polynomial can be factored as

$$27x^3 + 72x^2 + 48x = 3x(3x + 4)^2.$$

(b) Because the GCF of $18a^3 - 8ab^2$ is $2a$, factor out $2a$ before factoring the remaining polynomial.

$$18a^3 - 8ab^2 = 2a(9a^2 - 4b^2)$$

The expression $9a^2 - 4b^2$ is the difference of two squares and can be written as $(3a)^2 - (2b)^2$. Thus

$$18a^3 - 8ab^2 = 2a(3a - 2b)(3a + 2b).$$

▌ **Now Try Exercises 71, 79**

6.4   Putting It All Together

FACTORING	EXPLANATION	EXAMPLES
Difference of Two Squares	$a^2 - b^2 = (a - b)(a + b)$ **NOTE:** The *sum* of two squares, $a^2 + b^2$, cannot be factored by using real numbers.	$y^2 - 49 = (y - 7)(y + 7)$ $81 - z^2 = (9 - z)(9 + z)$ $4r^2 - 25t^2 = (2r - 5t)(2r + 5t)$ $16a^2 + b^2$ cannot be factored.
Perfect Square Trinomial	$a^2 + 2ab + b^2 = (a + b)^2$ $a^2 - 2ab + b^2 = (a - b)^2$ Be sure to verify that the given middle term equals $2ab$ before factoring.	$m^2 + 2m + 1 = (m + 1)^2$ $25y^2 - 30y + 9 = (5y - 3)^2$ $36r^2 + 12rt + t^2 = (6r + t)^2$ $x^2 + 5x + 4$ is *not* a perfect square trinomial because $\qquad 2ab = 2 \cdot x \cdot 2 = 4x \neq 5x.$
Sum and Difference of Two Cubes	$a^3 + b^3 = (a + b)(a^2 - ab + b^2)$ $a^3 - b^3 = (a - b)(a^2 + ab + b^2)$	$y^3 + 27 = (y + 3)(y^2 - y \cdot 3 + 3^2)$ $\qquad = (y + 3)(y^2 - 3y + 9)$ $27r^3 - 64t^3$ $\quad = (3r - 4t)\big((3r)^2 + 3r \cdot 4t + (4t)^2\big)$ $\quad = (3r - 4t)(9r^2 + 12rt + 16t^2)$

6.4   Exercises

MyMathLab   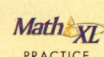 Math XL PRACTICE   WATCH   DOWNLOAD   READ   REVIEW

CONCEPTS AND VOCABULARY

1. $a^2 - b^2 =$ _____

2. If the expression $36x^2 - 49y^2$ is written in the form $a^2 - b^2$, then $a =$ _____ and $b =$ _____.

3. (True or False?) The expression $a^2 + b^2$ can be factored by using real numbers.

4. (True or False?) Using only integer coefficients, the expression $3x^2 - 5y^2$ can be factored as a difference of two squares.

5. $a^2 + 2ab + b^2 =$ _____

6. $a^2 - 2ab + b^2 =$ _____

7. $x^2 +$ _____ $+ 9$ is a perfect square trinomial.

8. $4r^2 -$ _____ $+ 25t^2$ is a perfect square trinomial.

9. $a^3 + b^3 =$ _____

10. $a^3 - b^3 =$ _____

11. If the expression $8x^3 + 27y^3$ is written in the form $a^3 + b^3$, then $a =$ _____ and $b =$ _____.

12. If the expression $x^3 - 1$ is written in the form $a^3 - b^3$, then $a =$ _____ and $b =$ _____.

13. $y^3 - 8 = (y\_\_2)(y^2\_\_2y + 4)$

14. $64z^3 + 27 = (4z\_\_3)(16z^2\_\_12z + 9)$

FACTORING THE DIFFERENCE OF TWO SQUARES

Exercises 15–32: Factor.

15. $x^2 - 1$

16. $x^2 - 16$

17. $z^2 - 100$

18. $z^2 - 81$

19. $4y^2 - 1$

20. $9y^2 - 16$

21. $36z^2 - 25$

22. $49z^2 - 64$

23. $9 - x^2$

24. $25 - x^2$

25. $1 - 9y^2$

26. $49 - 16y^2$

27. $4a^2 - 9b^2$

28. $16a^2 - b^2$

29. $36m^2 - 25n^2$

30. $49m^2 - 100n^2$

31. $81r^2 - 49t^2$

32. $625r^2 - 121t^2$

FACTORING PERFECT SQUARE TRINOMIALS

Exercises 33–54: Factor as a perfect square trinomial whenever possible.

33. $x^2 + 8x + 16$

34. $x^2 + 4x + 4$

35. $z^2 + 12z + 25$

36. $z^2 - 18z + 36$

37. $x^2 - 6x + 9$

38. $x^2 - 10x + 25$

39. $9y^2 + 6y + 1$

40. $16y^2 + 8y + 1$

41. $4z^2 - 4z + 1$

42. $25z^2 - 12z + 1$

43. $9t^2 + 16t + 4$

44. $4t^2 + 12t + 9$

45. $9x^2 + 30x + 25$

46. $25x^2 + 60x + 36$

47. $4a^2 - 36a + 81$

48. $9a^2 - 60a + 100$

49. $x^2 + 2xy + y^2$

50. $x^2 - 6xy + 9y^2$

51. $r^2 - 10rt + 25t^2$

52. $15r^2 + 10rt + t^2$

53. $4y^2 - 10yz + 9z^2$

54. $25y^2 - 20yz + 4z^2$

FACTORING SUMS AND DIFFERENCES OF TWO CUBES

Exercises 55–68: Factor.

55. $z^3 + 1$

56. $z^3 + 8$

57. $x^3 + 64$

58. $x^3 + 125$

59. $y^3 - 8$

60. $y^3 - 27$

61. $n^3 - 1$

62. $n^3 - 64$

63. $8x^3 + 1$

64. $27x^3 - 1$

65. $m^3 - 64n^3$

66. $m^3 + 8n^3$

67. $8x^3 + 125y^3$

68. $27x^3 + 64y^3$

GENERAL FACTORING USING SPECIAL METHODS

Exercises 69–86: Factor the expression completely.

69. $4x^2 - 16$

70. $12x^2 - 60x + 75$

71. $2y^2 - 28y + 98$

72. $y^3 - 9y$

73. $5z^3 + 40$

74. $4z^3 + 36z^2 + 100z$

75. $x^3y - xy^3$

76. $8m^3 - 8$

77. $2m^3 - 10m^2 + 18m$

78. $2a^3b - 18ab^3$

79. $700x^4 - 63x^2y^2$

80. $135r^3 - 5t^3$

81. $16a^3 + 2b^3$

82. $192x^2y^2 - 3y^4$

83. $4b^4 + 24b^3 + 36b^2$

84. $2y^4 + 24y^3 + 72y^2$

85. $500r^3 - 32t^3$

86. $8r^3 - 64t^3$

GEOMETRY

87. A square has an area of $4x^2 + 12x + 9$. Find the length of a side. Make a sketch of the square.

88. A square has an area of $9x^2 + 30x + 25$. Find the length of a side. Make a sketch of the square.

WRITING ABOUT MATHEMATICS

89. Explain how factoring $x^3 + y^3$ is different from factoring $x^3 - y^3$.

90. Using the techniques discussed in this section, can you factor the expression $4x^2 + 9y^2$ into two binomials? Explain your reasoning.

Checking Basic Concepts

1. Factor each trinomial.
(a) $2x^2 - 5x - 12$ **(b)** $6x^2 + 17x - 14$

2. Factor completely when possible.
(a) $3y^2 + 4y - 2$ **(b)** $6y^3 - 10y^2 - 4y$

3. Write a polynomial in factored form that represents the total area of the figure.

$3x^2$	$2x$
$9x$	6

4. Factor each polynomial.
(a) $z^2 - 64$ **(b)** $9r^2 - 4t^2$

5. Factor each trinomial.
(a) $x^2 + 12x + 36$ **(b)** $9a^2 - 12ab + 4b^2$

6. Factor.
(a) $m^3 - 27$ **(b)** $125n^3 + 27$

7. Factor completely.
(a) $16x^2 - 4$ **(b)** $3y^4 + 24y$

6.5 | Summary of Factoring

Guidelines for Factoring Polynomials • Factoring Polynomials

A LOOK INTO MATH ▶ So far in this chapter we have discussed several useful techniques for factoring polynomials. But in most factoring problems, the specific method that should be used is not stated. Instead, we must look carefully at each factoring problem and decide which approach is best. In this section we discuss general guidelines that can be used to factor polynomials.

Guidelines for Factoring Polynomials

The following guidelines can be used to factor polynomials in general.

FACTORING POLYNOMIALS

STEP 1: Factor out the greatest common factor (GCF), if possible.

STEP 2: **A.** If the polynomial has *four terms*, try factoring by grouping.

B. If the polynomial is a *binomial*, try one of the following.

1. $a^2 - b^2 = (a - b)(a + b)$ Difference of two squares

2. $a^3 - b^3 = (a - b)(a^2 + ab + b^2)$ Difference of two cubes

3. $a^3 + b^3 = (a + b)(a^2 - ab + b^2)$ Sum of two cubes

C. If the polynomial is a *trinomial*, check for a perfect square.

1. $a^2 + 2ab + b^2 = (a + b)^2$ Perfect square trinomial

2. $a^2 - 2ab + b^2 = (a - b)^2$ Perfect square trinomial

Otherwise, try to factor the trinomial by grouping or apply FOIL in reverse, as described in Sections 6.2 and 6.3.

STEP 3: Check to make sure that the polynomial is *completely* factored.

READING CHECK

• In your own words, rewrite the general guidelines for factoring polynomials.

NOTE: Always perform Step 1 first. Factoring out the greatest common factor usually makes it easier to factor the resulting polynomial. After a polynomial has been factored, remember to perform Step 3 so that you are sure the given polynomial is completely factored.

Factoring Polynomials

In the first example, we apply Step 1 to a polynomial with a common factor.

EXAMPLE 1 **Factoring out a common factor**

Factor $5x^3 - 10x^2 + 15x$.

Solution

STEP 1: The greatest common factor is **$5x$**.

$$5x^3 - 10x^2 + 15x = 5x(x^2 - 2x + 3)$$

STEP 2C: The trinomial $x^2 - 2x + 3$ is prime and cannot be factored further.

STEP 3: The completely factored polynomial is $5x(x^2 - 2x + 3)$.

Now Try Exercise 11

When factoring polynomials completely, it is often necessary to apply more than one factoring technique. In several of the next examples we factor polynomials that require more than one method of factoring.

EXAMPLE 2 **Factoring a difference of squares**

Factor $3x^4 - 48x^2$.

Solution

STEP 1: The greatest common factor is **$3x^2$**.

$$3x^4 - 48x^2 = 3x^2(x^2 - 16)$$

STEP 2B: The binomial $x^2 - 16$ can be factored as a difference of two squares.

$$3x^2(x^2 - 16) = 3x^2(x - 4)(x + 4)$$

STEP 3: The completely factored polynomial is $3x^2(x - 4)(x + 4)$.

Now Try Exercise 39

MAKING CONNECTIONS

Factoring Polynomials

We can often determine how a polynomial should be factored by considering the number of terms in the polynomial. This is summarized as follows.

Type of Polynomial	*Factoring Technique*
4-term Polynomial	Grouping
Trinomial	Perfect square trinomial
	FOIL in reverse or grouping
Binomial	Difference of two squares
	Sum or difference of two cubes

EXAMPLE 3   **Factoring a perfect square trinomial**

Factor $36y^3 - 24y^2 + 4y$.

Solution

STEP 1: The greatest common factor is **4y**.

$$36y^3 - 24y^2 + 4y = 4y(9y^2 - 6y + 1)$$

STEP 2C: We can factor $9y^2 - 6y + 1$ as a perfect square trinomial.

$$4y(9y^2 - 6y + 1) = 4y(3y - 1)(3y - 1)$$

STEP 3: The completely factored polynomial is $4y(3y - 1)^2$.

▌ **Now Try Exercise 41**

EXAMPLE 4   **Factoring a sum of cubes**

Factor $27z^3 + 64$.

Solution

STEP 1: There are no common factors.
STEP 2B: The binomial $27z^3 + 64$ can be written as $(3z)^3 + 4^3$ and can be factored as a sum of two cubes.

$$27z^3 + 64 = (3z + 4)\left((3z)^2 - 3z \cdot 4 + 4^2\right)$$
$$= (3z + 4)(9z^2 - 12z + 16)$$

NOTE: The trinomial $9z^2 - 12z + 16$ is prime and cannot be factored further.

STEP 3: The completely factored polynomial is $(3z + 4)(9z^2 - 12z + 16)$.

▌ **Now Try Exercise 31**

EXAMPLE 5   **Factoring a trinomial**

Factor $14x^4 + 7x^3 - 42x^2$.

Solution

STEP 1: The greatest common factor is $7x^2$.

$$14x^4 + 7x^3 - 42x^2 = 7x^2(2x^2 + x - 6)$$

STEP 2C: We can factor $2x^2 + x - 6$ using FOIL in reverse.

$$7x^2(2x^2 + x - 6) = 7x^2(2x - 3)(x + 2)$$

STEP 3: The completely factored polynomial is $7x^2(2x - 3)(x + 2)$.

▌ **Now Try Exercise 35**

EXAMPLE 6   **Factoring by grouping**

Factor $15x^3 + 10x^2 - 60x - 40$.

Solution

STEP 1: The greatest common factor is 5.

$$15x^3 + 10x^2 - 60x - 40 = 5(3x^3 + 2x^2 - 12x - 8)$$

STEP 2A: Because the resulting polynomial has four terms, we apply grouping.

$$5(3x^3 + 2x^2 - 12x - 8) = 5\big((3x^3 + 2x^2) + (-12x - 8)\big) \quad \text{Associative property}$$
$$= 5\big((x^2)(3x + 2) - 4(3x + 2)\big) \quad \text{Factor out common factors.}$$
$$= 5(x^2 - 4)(3x + 2) \quad \text{Factor out } (3x + 2).$$

STEP 2B: The binomial $x^2 - 4$ can now be factored as a difference of two squares.

$$5(x^2 - 4)(3x + 2) = 5(x - 2)(x + 2)(3x + 2)$$

STEP 3: The completely factored polynomial is $5(x - 2)(x + 2)(3x + 2)$.

Now Try Exercise 33

EXAMPLE 7   **Factoring a polynomial having two variables**

Factor $18x^3y - 8xy^3$.

Solution

STEP 1: The greatest common factor is $2xy$.

$$18x^3y - 8xy^3 = 2xy(9x^2 - 4y^2)$$

STEP 2B: The binomial $9x^2 - 4y^2$ can be written as $(3x)^2 - (2y)^2$ and can be factored as a difference of two squares.

$$9x^2 - 4y^2 = (3x - 2y)(3x + 2y)$$

STEP 3: The completely factored polynomial is $2xy(3x - 2y)(3x + 2y)$.

Now Try Exercise 55

EXAMPLE 8   **Applying several techniques**

Factor $3x^5 - 3x^3 - 24x^2 + 24$.

Solution

STEP 1: The greatest common factor is 3.

$$3x^5 - 3x^3 - 24x^2 + 24 = 3(x^5 - x^3 - 8x^2 + 8)$$

STEP 2A: The resulting four-term polynomial can be factored by grouping.

$$3(x^5 - x^3 - 8x^2 + 8) = 3\big((x^5 - x^3) + (-8x^2 + 8)\big) \quad \text{Associative property}$$
$$= 3\big(x^3(x^2 - 1) - 8(x^2 - 1)\big) \quad \text{Factor out common factors.}$$
$$= 3(x^3 - 8)(x^2 - 1) \quad \text{Factor out } x^2 - 1.$$

STEP 2B: Both binomials in this expression can be factored further. The binomial $x^3 - 8$ can be factored as a difference of two cubes, and the binomial $x^2 - 1$ can be factored as a difference of two squares.

$$3(x^3 - 8)(x^2 - 1) = 3(x - 2)(x^2 + 2x + 4)(x - 1)(x + 1)$$

NOTE: The trinomial $x^2 + 2x + 4$ is prime and cannot be factored further.

STEP 3: The completely factored polynomial is $3(x - 2)(x^2 + 2x + 4)(x - 1)(x + 1)$.

Now Try Exercise 45

6.5   Putting It All Together

CONCEPT	EXPLANATION	EXAMPLES
Greatest Common Factor	Factor out the greatest common factor, or monomial, that occurs in each term.	$2x^2 - 4x + 10 = 2(x^2 - 2x + 5)$ $3x^3 + 6x = 3x(x^2 + 2)$ $7xy - x^2y = xy(7 - x)$
Factoring by Grouping	Use the associative and distributive properties to factor a polynomial with four terms.	$x^3 - 3x^2 + 2x - 6 = (x^3 - 3x^2) + (2x - 6)$ $\qquad = x^2(x - 3) + 2(x - 3)$ $\qquad = (x^2 + 2)(x - 3)$
Factoring Binomials	Use the difference of two squares, the difference of two cubes, or the sum of two cubes.	$9x^2 - 4 = (3x - 2)(3x + 2)$ $x^3 - 27 = (x - 3)(x^2 + 3x + 9)$ $x^3 + 27 = (x + 3)(x^2 - 3x + 9)$
Factoring Trinomials	Use FOIL in reverse or grouping.	$x^2 + 5x - 6 = (x + 6)(x - 1)$ Check middle term: $-x + 6x = 5x$. ✓ $4x^2 + 4x - 3 = (4x^2 - 2x) + (6x - 3)$ $\qquad = 2x(2x - 1) + 3(2x - 1)$ $\qquad = (2x + 3)(2x - 1)$

6.5   Exercises

MyMathLab

PRACTICE   WATCH   DOWNLOAD   READ   REVIEW

CONCEPTS AND VOCABULARY

1. What do the letters GCF mean?

2. A good first step for factoring polynomials is to factor out the _____.

3. If a polynomial has four terms, what factoring method might be appropriate?

4. If a polynomial is a binomial, we can try to factor it as a difference of two _____, a difference of two _____, or a sum of two _____.

5. Can $x^2 + 1$ be factored? Explain.

6. Can $x^3 + 1$ be factored? Explain.

7. If a polynomial is a trinomial, we can try to factor it as a perfect _____ trinomial. Otherwise, try factoring it by _____ or apply _____ in reverse.

8. The last step for factoring is to be sure that the polynomial is _____ factored.

WARM UP

Exercises 9–24: Factor completely, if possible.

9. $4x - 2$

10. $x^2 + 3x$

11. $2y^2 - 4y + 4$

12. $5y^2 - 25y + 10$

13. $z^2 - 4$

14. $9z^2 - 25$

15. $a^3 + 8$

16. $8a^3 - 1$

17. $4b^2 - 12b + 9$

18. $b^2 + 4b + 4$

19. $m^2 + 9$

20. $4m^2 + 49$

21. $x^3 - x^2 + 5x - 5$

22. $3x^3 + 6x^2 + x + 2$

23. $y^2 - 5y + 4$

24. $y^2 - 3y - 10$

GENERAL FACTORING

Exercises 25–62: Factor completely.

25. $x^3 + 4x^2 - 9x - 36$

26. $6x^2 - 19x + 15$

27. $8a^3 - 64$ **28.** $ab^2 - 4a$

29. $12x^4 - 18x^3 + 4x^2 - 6x$

30. $3x^2y + 24xy + 48y$

31. $54t^4 + 16t$ **32.** $3t^3 + 18t^2 - 48t$

33. $2r^3 + 6r^2 - 2r - 6$

34. $3r^4 + 3r^3 - 24r - 24$ **35.** $6z^4 - 21z^3 - 45z^2$

36. $3x^4y + 24xy^4$ **37.** $12b^4 - 10b^3 + 2b^2$

38. $6a^4b + 4a^3b + 18a^2b + 12ab$

39. $6y^2z - 24z^3$ **40.** $6y^3z - 48z^4$

41. $3x^2y - 30xy + 75y$ **42.** $8x^3 + y^3$

43. $27m^3 - 8n^3$ **44.** $45m^3 - 69m^2 + 12m$

45. $3x^5 - 12x^3 - 3x^2 + 12$

46. $8x^3 - 8$ **47.** $5a^2 - 27a - 18$

48. $2a^2 - 6ab + 3a - 9b$

49. $3rt^2 + 33rt + 90r$ **50.** $9t^2 + 24t + 16$

51. $9b^3 + 6b^2 + 12b + 8$ **52.** $5b^3 - 55b^2 - 60b$

53. $6n^3 + 2n^2 - 10n$ **54.** $7n^4 + 28n^3 - 63n^2$

55. $4x^2 - 36y^2$ **56.** $64x^2 - 25y^2$

57. $2a^3 - 16a^2 + 32a$ **58.** $24a^3 + 72a^2 + 54a$

59. $32xy^3 + 4x$ **60.** $24x^3 - 4x^2 - 160x$

61. $8b^4 + 24b^3 - 2b^2 - 6b$

62. $3z^3 - 6z^2 - 27z + 54$

GEOMETRY

63. *Dimensions of a Square* If three identical squares have a total area of $27x^2 + 18x + 3$, find the length of one side of one of the squares.

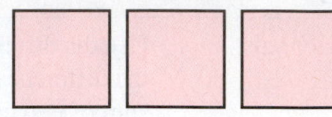

64. *Dimensions of a Cube* If three identical cubes have a total volume of $3x^3 + 18x^2 + 36x + 24$, find the length of one side of one of the cubes.

WRITING ABOUT MATHEMATICS

65. Explain how the number of terms in a polynomial can help determine what method should be used to factor it.

66. Describe a method for determining whether a polynomial has been factored correctly.

6.6   Solving Equations by Factoring I (Quadratics)

The Zero-Product Property • Solving Quadratic Equations • Applications

A LOOK INTO MATH ▶

If a golf ball is hit upward at 132 feet per second, or 90 miles per hour, then its height h in feet above the ground after t seconds is given by $h = 132t - 16t^2$. The expression $132t - 16t^2$ is an example of a *quadratic polynomial*. To determine the elapsed time between when the ball is hit and when it strikes the ground (or when $h = 0$), we solve the *quadratic equation*

$$132t - 16t^2 = 0.$$

One method for solving this equation is by factoring. (See Example 4.) In this section we discuss how to use factoring to solve a variety of equations.

NEW VOCABULARY

☐ Zero-product property
☐ Zeros (of a polynomial)
☐ Quadratic polynomial
☐ Quadratic equation
☐ Standard form

The Zero-Product Property

To solve equations we often use the **zero-product property**, which states that if the product of two numbers (or expressions) is 0, then at least one of the numbers (or expressions) must equal 0.

ZERO-PRODUCT PROPERTY

For all real numbers a and b, if $ab = 0$, then $a = 0$ or $b = 0$ (or both).

NOTE: The zero-product property works only for 0. If $ab = 1$, then it does *not* follow that $a = 1$ or $b = 1$. For example, $a = \frac{1}{3}$ and $b = 3$ satisfy the equation $ab = 1$.

After factoring an expression, we can use the zero-product property to solve an equation. The left side of the equation

$$3t^2 - 9t = 0$$

STUDY TIP

The zero-product property is used extensively in mathematics for solving many types of equations. Be sure that you learn how to apply this important property correctly.

may be factored to obtain

$$3t(t - 3) = 0.$$

Note that the product of $3t$ and $t - 3$ is 0. By the zero-product property, either

$$3t = 0 \quad \text{or} \quad t - 3 = 0.$$

Solving each equation for t results in

$$t = 0 \quad \text{or} \quad t = 3.$$

These values can be checked by substituting them into the given equation $3t^2 - 9t = 0$.

$$3(0)^2 - 9(0) = 0 \checkmark \quad \text{Let } t = 0. \text{ It checks.}$$
$$3(3)^2 - 9(3) = 0 \checkmark \quad \text{Let } t = 3. \text{ It checks.}$$

READING CHECK

• How can you tell if a number is a zero of a polynomial?

The t-values of 0 and 3 are called **zeros** of the polynomial $3t^2 - 9t$, because when either is substituted in this polynomial, the result is 0.

EXAMPLE 1   **Applying the zero-product property**

Solve each equation.
(a) $x(x - 1) = 0$ (b) $2z^2 = 0$
(c) $(t + 3)(t + 2) = 0$ (d) $x(x - 2)(2x + 1) = 0$

Solution
(a) By the zero-product property, $x(x - 1) = 0$ when $x = 0$ or $x - 1 = 0$. The solutions are 0 and 1.
(b) $2z^2 = 2 \cdot z \cdot z$ and $2 \neq 0$, so $2z^2 = 0$ when $z = 0$.
(c) $(t + 3)(t + 2) = 0$ implies that $t + 3 = 0$ or $t + 2 = 0$. The solutions to the given equation are -3 and -2.
(d) We can apply the zero-product property to $x(x - 2)(2x + 1) = 0$. Thus $x = 0$, or $x - 2 = 0$, or $2x + 1 = 0$. The solutions are $-\frac{1}{2}$, 0, and 2.

Now Try Exercises 13, 15, 19, 23

Solving Quadratic Equations

Any **quadratic polynomial** in the variable x can be written as $ax^2 + bx + c$ with $a \neq 0$. Any **quadratic equation** in the variable x can be written as $ax^2 + bx + c = 0$ with $a \neq 0$. This form of quadratic equation is called the **standard form** of a quadratic equation. Table 6.11 on the next page shows examples of quadratic polynomials along with related quadratic equations that can be expressed in standard form.

TABLE 6.11 Quadratic Polynomials and Equations

Quadratic Polynomial	Quadratic Equation	Standard Form
$x^2 - 2x + 1$	$x^2 - 2x + 1 = 9$	$x^2 - 2x - 8 = 0$
$3x^2 + 7x$	$3x^2 + 7x = 4$	$3x^2 + 7x - 4 = 0$
$x^2 - 9$	$x^2 - 9 = 2$	$x^2 - 11 = 0$
$6x - x^2 + 2$	$6x - x^2 + 2 = 1$	$-x^2 + 6x + 1 = 0$

READING CHECK

• What does it mean for a polynomial to be written so that it contains descending powers of x?

NOTE: When a quadratic polynomial in the variable x is in standard form and we read it from left to right, the terms contain *descending powers* of x. In other words, the first term contains x^2, the second term contains x, and the third term is a constant (the exponent on x is 0).

To solve a quadratic equation we often use factoring and the zero-product property. This method is summarized by the following steps. Although it is not necessary to label each step in the solution to a quadratic equation, it is important to keep these steps in mind.

SOLVING QUADRATIC EQUATIONS

To solve a quadratic equation by factoring, follow these steps.

STEP 1: If necessary, write the equation in standard form as $ax^2 + bx + c = 0$.

STEP 2: Factor the left side of the equation using any method.

STEP 3: Apply the zero-product property.

STEP 4: Solve each of the resulting equations. Check any solutions.

EXAMPLE 2 **Solving equations by factoring**

Solve each quadratic equation. Check your answers.
(a) $x^2 + 2x = 0$ (b) $y^2 = 16$ (c) $z^2 - 3z + 2 = 0$ (d) $2x^2 = 5 - 9x$

Solution
(a) Because $x^2 + 2x = 0$ is in standard form, we begin by factoring out the GCF, x.

$$x^2 + 2x = 0 \qquad \text{Given equation}$$
$$x(x + 2) = 0 \qquad \text{Factor out } x. \text{ (Step 2)}$$
$$x = 0 \quad \text{or} \quad x + 2 = 0 \qquad \text{Zero-product property (Step 3)}$$
$$x = 0 \quad \text{or} \quad x = -2 \qquad \text{Solve for } x. \text{ (Step 4)}$$

To check these values, substitute -2 and 0 for x in the given equation.

$$(-2)^2 + 2(-2) \stackrel{?}{=} 0 \qquad (0)^2 + 2(0) \stackrel{?}{=} 0 \qquad \text{Substitute } -2 \text{ and } 0.$$
$$0 = 0 \checkmark \qquad\qquad 0 = 0 \checkmark \qquad \text{Both answers check.}$$

Therefore the solutions are -2 and 0.

(b) To write $y^2 = 16$ in standard form we begin by subtracting 16 from each side to obtain 0 on the right side.

$$y^2 = 16 \qquad \text{Given equation}$$
$$y^2 - 16 = 0 \qquad \text{Subtract 16. (Step 1)}$$
$$(y - 4)(y + 4) = 0 \qquad \text{Difference of squares (Step 2)}$$
$$y - 4 = 0 \quad \text{or} \quad y + 4 = 0 \qquad \text{Zero-product property (Step 3)}$$
$$y = 4 \quad \text{or} \quad y = -4 \qquad \text{Solve for } y. \text{ (Step 4)}$$

To check these values, substitute -4 and 4 for y in the given equation.

$$(-4)^2 \overset{?}{=} 16 \qquad (4)^2 \overset{?}{=} 16 \qquad \text{Substitute } -4 \text{ and } 4.$$
$$16 = 16 \checkmark \qquad 16 = 16 \checkmark \qquad \text{Both answers check.}$$

The solutions are -4 and 4.

(c) We begin by factoring the left side of the equation, $z^2 - 3z + 2$.

$$
\begin{array}{ll}
z^2 - 3z + 2 = 0 & \text{Given equation} \\
(z - 1)(z - 2) = 0 & \text{Factor. (Step 2)} \\
z - 1 = 0 \quad \text{or} \quad z - 2 = 0 & \text{Zero-product property (Step 3)} \\
z = 1 \quad \text{or} \quad z = 2 & \text{Solve for } z. \text{ (Step 4)}
\end{array}
$$

To check these values, substitute 1 and 2 for z in the given equation.

$$1^2 - 3(1) + 2 \overset{?}{=} 0 \qquad 2^2 - 3(2) + 2 \overset{?}{=} 0 \qquad \text{Substitute } 1 \text{ and } 2.$$
$$0 = 0 \checkmark \qquad\qquad 0 = 0 \checkmark \qquad \text{Both answers check.}$$

The solutions are 1 and 2.

(d) We write $2x^2 = 5 - 9x$ in standard form by adding -5 and $9x$ to each side.

$$
\begin{array}{ll}
2x^2 = 5 - 9x & \text{Given equation} \\
2x^2 + 9x - 5 = 0 & \text{Add } -5 \text{ and } 9x. \text{ (Step 1)} \\
(2x - 1)(x + 5) = 0 & \text{Factor. (Step 2)} \\
2x - 1 = 0 \quad \text{or} \quad x + 5 = 0 & \text{Zero-product property (Step 3)} \\
x = \dfrac{1}{2} \quad \text{or} \quad x = -5 & \text{Solve for } x. \text{ (Step 4)}
\end{array}
$$

To check these values, substitute -5 and $\frac{1}{2}$ for x in the given equation.

$$2(-5)^2 \overset{?}{=} 5 - 9(-5) \qquad 2\left(\frac{1}{2}\right)^2 \overset{?}{=} 5 - 9\left(\frac{1}{2}\right) \qquad \text{Substitute } -5 \text{ and } \tfrac{1}{2}.$$

$$50 = 50 \checkmark \qquad\qquad \frac{1}{2} = \frac{1}{2} \checkmark \qquad \text{Both answers check.}$$

The solutions are -5 and $\frac{1}{2}$.

Now Try Exercises 27, 35, 43, 49

EXAMPLE 3 **Solving an equation by factoring**

Solve $6x^2 - x = 12$.

Solution
We *cannot* solve the equation $6x^2 - x = 12$ by factoring out the common factor of x in $6x^2 - x$ and setting each factor equal to 12. Instead we apply the zero-product property by first writing the given equation in the standard form as $ax^2 + bx + c = 0$.

$$
\begin{array}{ll}
6x^2 - x = 12 & \text{Given equation} \\
6x^2 - x - 12 = 0 & \text{Subtract 12. (Step 1)} \\
(2x - 3)(3x + 4) = 0 & \text{Factor. (Step 2)} \\
2x - 3 = 0 \quad \text{or} \quad 3x + 4 = 0 & \text{Zero-product property (Step 3)} \\
x = \dfrac{3}{2} \quad \text{or} \quad x = -\dfrac{4}{3} & \text{Solve for } x. \text{ (Step 4)}
\end{array}
$$

The solutions are $-\frac{4}{3}$ and $\frac{3}{2}$.

Now Try Exercise 51

MAKING CONNECTIONS

Equations and Expressions

The words "equation" and "expression" occur frequently in mathematics. However, they are *not* interchangeable. We often want to *solve* equations to find the values of the variable that make the equation true. We *factor* and *simplify* expressions. An equation is a statement that two expressions are equal. For example, $2x^2 - 3 = 5x + 1$ is an equation where $2x^2 - 3$ and $5x + 1$ are each expressions.

Applications

▶ **REAL-WORLD CONNECTION** To solve application problems we often need to solve equations. The next example illustrates how to solve the application presented in A Look Into Math at the beginning of this section.

EXAMPLE 4 **Modeling the flight of a golf ball**

If a golf ball is hit upward at 132 feet per second, or 90 miles per hour, then its height h in feet after t seconds is $h = 132t - 16t^2$. Find the time when the golf ball strikes the ground.

Solution
The golf ball strikes the ground when its height is 0.

$$132t - 16t^2 = 0 \qquad \text{Let } h = 0.$$
$$4t(33 - 4t) = 0 \qquad \text{Factor out } 4t.$$
$$4t = 0 \quad \text{or} \quad 33 - 4t = 0 \qquad \text{Zero-product property}$$
$$t = 0 \quad \text{or} \quad -4t = -33 \qquad \text{Divide by 4; subtract 33.}$$
$$t = 0 \quad \text{or} \quad t = \frac{33}{4} \qquad \text{Solve for } t.$$

The ball strikes the ground after $\frac{33}{4} = 8.25$ seconds. The solution of 0 is not used in this problem because it corresponds to the time when the ball is hit from ground level.

▍ **Now Try Exercise 67(a)**

▶ **REAL-WORLD CONNECTION** When you try to stop a car, the faster you are driving, the further the stopping distance. In fact, if you drive twice as fast, the braking distance will be about four times as much, and if you drive three times faster, the braking distance will be about nine times as much.

EXAMPLE 5 **Modeling braking distance**

The braking distance D in feet required to stop a car traveling at x miles per hour on dry, level pavement can be approximated by $D = \frac{1}{11}x^2$. (*Source:* L. Haefner.)
(a) Calculate the braking distance for a car traveling 70 miles per hour.
(b) If the braking distance is 44 feet, calculate the speed of the car.
(c) If you have a calculator available, use it to solve part (b) numerically with a table of values.

Solution

(a) If $x = 70$, then $D = \frac{1}{11}(70)^2 = \frac{4900}{11} \approx 445$ feet.

(b) *Symbolic Solution* Let $D = 44$ in the given equation and solve.

$$\frac{1}{11}x^2 = 44 \qquad \text{Let } D = 44.$$

$$\frac{1}{11}x^2 - 44 = 0 \qquad \text{Subtract 44.}$$

$$x^2 - 484 = 0 \qquad \text{Multiply by 11.}$$

$$(x - 22)(x + 22) = 0 \qquad \text{Difference of two squares } (22^2 = 484)$$

$$x - 22 = 0 \quad \text{or} \quad x + 22 = 0 \qquad \text{Zero-product property}$$

$$x = 22 \quad \text{or} \quad x = -22 \qquad \text{Solve for } x.$$

The car is traveling at (approximately) 22 miles per hour. (Note that $x = -22$ has no physical meaning in this problem.)

(c) *Numerical Solution* Let $Y_1 = X^2/11$ and make a table of values. Scroll through the table, as shown in Figure 6.6, to where $x = 22$ when $y_1 = 44$. Thus the solution is 22 miles per hour.

X	Y₁
20	36.364
21	40.091
22	44
23	48.091
24	52.364
25	56.818
26	61.455

X=22

Figure 6.6

Now Try Exercise 69

▶ **REAL-WORLD CONNECTION** Quadratic equations sometimes are used in applications involving rectangular shapes. The next application involves finding the dimensions of a digital photograph made up of tiny "dots" of color called *pixels*.

EXAMPLE 6   **Finding the dimensions of a digital photograph**

A small digital photograph is 20 pixels longer than it is wide, as illustrated in Figure 6.7. It has a total of 2400 pixels. Find the dimensions of this photograph.

Dimensions of a Photograph

x

2400 pixels

$x + 20$

Figure 6.7

Solution

From Figure 6.7 the rectangular photograph has an area of 2400 pixels.

$$x(x + 20) = 2400 \qquad \text{Area = width} \times \text{length}$$

$$x^2 + 20x = 2400 \qquad \text{Distributive property}$$

$$x^2 + 20x - 2400 = 0 \qquad \text{Subtract 2400.}$$

$$(x - 40)(x + 60) = 0 \qquad \text{Factor.}$$

$$x - 40 = 0 \quad \text{or} \quad x + 60 = 0 \qquad \text{Zero-product property}$$

$$x = 40 \quad \text{or} \quad x = -60 \qquad \text{Solve for } x.$$

The only valid solution is 40. Thus the dimensions of the photograph are 40 pixels by $40 + 20 = 60$ pixels.

CRITICAL THINKING

Are the two solutions to $x(2x + 1) = 1$ found by letting $x = 1$ or $2x + 1 = 1$? Explain your reasoning. What are the solutions to this equation?

Now Try Exercise 73

6.6   Putting It All Together

CONCEPT	EXPLANATION	EXAMPLES
Zero-Product Property	If the product of two or more expressions is 0, then at least one of the expressions must equal 0.	$ab = 0$ implies that $a = 0$ or $b = 0$. $x(x + 1) = 0$ implies that $x = 0$ or $x + 1 = 0$. $z(z - 1)(z + 2) = 0$ implies that $z = 0$ or $z - 1 = 0$ or $z + 2 = 0$.
Solving Quadratic Equations by Factoring	1. Write the equation as $\quad ax^2 + bx + c = 0$. 2. Factor the left side of this equation. 3. Apply the zero-product property. 4. Solve each resulting equation. Check any solutions.	$2x^2 + 11x = 6$ $2x^2 + 11x - 6 = 0$ Step 1 $(2x - 1)(x + 6) = 0$ Step 2 $2x - 1 = 0$ or $x + 6 = 0$ Step 3 $x = \dfrac{1}{2}$ or $x = -6$ Step 4

6.6   Exercises

MyMathLab    PRACTICE   WATCH   DOWNLOAD   READ   REVIEW

CONCEPTS AND VOCABULARY

1. If $ab = 0$, then either $a = $ _____ or $b = $ _____.

2. Can the zero-product property be used to state that if $(x - 1)(x - 2) = 3$, then either $x - 1 = 3$ or $x - 2 = 3$? Explain your answer.

3. If $2x(x + 6) = 0$, then either _____ or _____.

4. What is a good first step when you are solving the equation $4x^2 + 1 = 4x$ by factoring?

5. What is the next step when you are solving the equation $(x + 5)(x - 4) = 0$?

6. Factoring is an important method for _____ equations.

7. Because $2(4) - 8 = 0$, the value 4 is called a(n) _____ of the polynomial $2x - 8$.

8. What is the zero of the polynomial $3x - 6$?

9. Any quadratic equation in the variable x can be written in standard form as _____.

10. Standard form for $x^2 + 1 = 6x$ is _____.

11. When written in standard form and read from left to right, a quadratic polynomial in the variable x contains _____ powers of x.

12. For the constant term in a quadratic polynomial in the variable x, the exponent on x is _____.

ZERO-PRODUCT PROPERTY

Exercises 13–24: Solve the equation.

13. $x^2 = 0$

14. $5m^2 = 0$

15. $2x(x + 8) = 0$

16. $x(x + 10) = 0$

17. $(y - 1)(y - 2) = 0$

18. $(y + 4)(y - 3) = 0$

19. $(2z - 1)(4z - 3) = 0$

20. $(6z + 5)(z - 7) = 0$

21. $(1 - 3n)(3 - 7n) = 0$

22. $(5 - n)(5 + n) = 0$

23. $x(x - 5)(x - 8) = 0$

24. $x(x + 1)(x - 6) = 0$

SOLVING QUADRATIC EQUATIONS

Exercises 25–60: Solve and check.

25. $x^2 - x = 0$

26. $2x^2 + 4x = 0$

27. $z^2 - 5z = 0$ **28.** $6z^2 - 3z = 0$

29. $10y^2 + 15y = 0$ **30.** $2y^2 + 3y = 0$

31. $x^2 - 1 = 0$ **32.** $x^2 - 9 = 0$

33. $4n^2 - 1 = 0$ **34.** $9n^2 - 4 = 0$

35. $z^2 + 3z + 2 = 0$ **36.** $z^2 - 2z - 3 = 0$

37. $x^2 - 12x + 35 = 0$ **38.** $x^2 - x - 20 = 0$

39. $2b^2 + 3b - 2 = 0$ **40.** $3b^2 + b - 2 = 0$

41. $6y^2 + 19y + 10 = 0$ **42.** $4y^2 - 25y - 21 = 0$

43. $x^2 = 25$ **44.** $x^2 = 81$

45. $t^2 = 5t$ **46.** $10t^2 = -5t$

47. $3m^2 = -9m$ **48.** $4m^2 = 9$

49. $x^2 = 5x + 6$ **50.** $2x^2 + 3x = 14$

51. $12z^2 = 5 - 4z$ **52.** $12z^2 + 11z = 15$

53. $t(t + 1) = 2$ **54.** $t(t - 7) = -12$

55. $x(2x + 5) = 3$ **56.** $x(3x + 2) = 5$

57. $12x^2 + 12x = -3$ **58.** $18x^2 + 2 = 12x$

59. $30y^2 + 50y + 20 = 0$ **60.** $30y^2 - 25y + 5 = 0$

GEOMETRY

61. *Dimensions of a Square* A square has an area of 144 square feet. Find the length of a side.

62. *Dimensions of a Cube* A cube has a surface area of 96 square feet. Find the length of a side.

63. *Radius of a Circle* The numerical difference between the area and the circumference of a circle is 8π. Find the radius of the circle. (*Hint:* First factor out π in your equation.)

64. *Dimensions of a Rectangle* A rectangle is 5 feet longer than it is wide and has an area of 126 square feet. What are its dimensions?

Exercises 65 and 66: The Pythagorean Theorem Suppose that a right triangle has legs a and b and hypotenuse c, as illustrated in the figure. Then these values satisfy $a^2 + b^2 = c^2$.

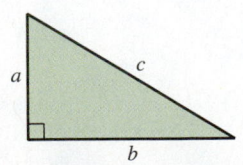

Use the Pythagorean theorem to find the value of x in the figure.

65.

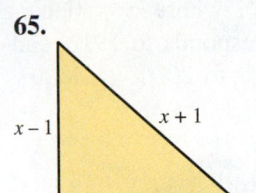

66.

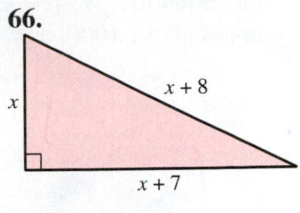

APPLICATIONS

67. *Flight of a Golf Ball* (Refer to Example 4.) The height h in feet of a golf ball after t seconds is given by $h = 96t - 16t^2$.
   (a) How long does it take for the golf ball to hit the ground?
   (b) Make a table of h for $t = 0, 1, 2, \ldots, 6$. After how many seconds does the golf ball reach its maximum height?

68. *Flight of a Baseball* The height h in feet of a baseball after t seconds is given by $h = -16t^2 + 88t + 3$. At what values of t is the height of the baseball 75 feet?

69. *Braking Distance* (Refer to Example 5.) The braking distance D in feet required to stop a car traveling x miles per hour on dry, level pavement can be approximated by $D = \frac{1}{11}x^2$.
   (a) Calculate the braking distance for 30 miles per hour and 60 miles per hour. How do your answers compare?
   (b) If the braking distance is 33 feet, estimate the speed of the car.
   (c) If you have a calculator, use it to solve part (b) numerically. Do your answers agree?

70. *Braking Distance* The braking distance D in feet required to stop a car traveling x miles per hour on wet, level pavement is approximated by $D = \frac{1}{9}x^2$.
   (a) Calculate the braking distance for 36 miles per hour and 72 miles per hour. How do your answers compare?
   (b) If the braking distance is 49 feet, estimate the speed of the car.
   (c) If you have a calculator, use it to solve part (b) numerically. Do your answers agree?

71. *Women in the Workforce* The number of women W in the workforce in millions can be estimated by the equation $W = \frac{19}{3125}x^2 + \frac{11}{2}$, where $x = 0$ corresponds to 1900, $x = 10$ corresponds to 1910, and so on until $x = 100$ corresponds to 2000. (*Source:* U.S. Census Bureau.)

(a) How many women were in the workforce in 1930 and in 2000?

(b) Use a table of values to estimate the year in which 45 million women were in the workforce.

72. *HIV/AIDS* Federal funding F in billions of dollars for HIV/AIDS research from 1981 to 2006 can be modeled by

$$F = 0.0316x^2 + 0.015x - 0.046,$$

where $x = 1$ corresponds to 1981, $x = 2$ corresponds to 1982, and so on until $x = 26$ corresponds to 2006. (*Source:* Kaiser Family Foundation.)

(a) Estimate federal funding for HIV/AIDS research in 2001.

(b) Use a table of values to estimate the year in which federal funding for HIV/AIDS research reached $18.5 billion.

73. *Digital Photographs* (Refer to Example 6.) A digital photograph is 10 pixels longer than it is wide and has a total area of 2000 pixels. Find the dimensions of this rectangular photograph.

74. *Dimensions of a Building* The rectangular floor of a shed has a length 4 feet longer than its width, and its area is 140 square feet. Let x be the width of the floor.
(a) Write a quadratic equation whose solution gives the width of the floor.
(b) Solve this equation.

WRITING ABOUT MATHEMATICS

75. List four steps for solving a quadratic equation by factoring.

76. Explain why factoring is important.

SECTIONS 6.5 and 6.6   Checking Basic Concepts

1. Factor out the greatest common factor.
(a) $9a^2 - 18a + 27$ (b) $7xy^2 + 28x$

2. Factor completely.
(a) $6z^4 - 28z^3 + 16z^2$ (b) $2r^2t^2 - 18r^2$

3. Factor completely.
(a) $36x^3 - 48x^2 + 16x$
(b) $24b^3 - 81$

4. Solve each quadratic equation.
(a) $4y^2 - 6y = 0$ (b) $5z^2 + 2z = 3$

5. Solve $x^2 + 2x - 3 = 0$ symbolically and numerically with a table of values.

6. If a golf ball is hit upward at 60 miles per hour, then its height h in feet after t seconds is given by $h = 88t - 16t^2$. Use factoring to determine when the golf ball strikes the ground.

6.7 Solving Equations by Factoring II (Higher Degree)

Polynomials with Common Factors • Special Types of Polynomials

A LOOK INTO MATH ▶

In this section we discuss factoring polynomials having higher degree. Polynomials of degree 2 or higher are often used in applications. For example, the polynomial

$$0.0013x^3 - 0.085x^2 + 1.6x + 12$$

models natural gas consumption in the United States in trillions of cubic feet, where $x = 0$ corresponds to 1960, $x = 10$ corresponds to 1970, and so on until $x = 40$ corresponds to 2000, as shown in Figure 6.8. (This trend did not continue after 2000.) In Exercises 79–82 we discuss further the consumption of natural gas. (*Source:* Department of Energy.)

Natural Gas Consumption

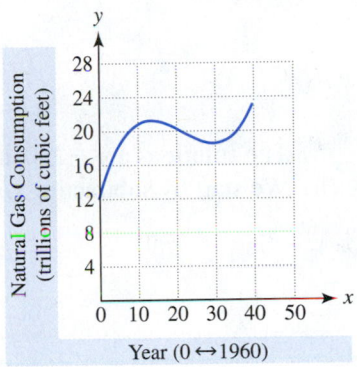

Figure 6.8

READING CHECK

• What is a good first step in factoring a polynomial?

Polynomials with Common Factors

The first step in factoring a polynomial is to factor out the greatest common factor (GCF). Once this is accomplished, the resulting expression should be factored further by using any of the factoring methods discussed earlier in this chapter.

EXAMPLE 1 **Factoring trinomials with common factors**

Factor each trinomial completely.
(a) $-4x^2 + 28x - 40$ **(b)** $10x^3 + 28x^2 - 6x$

Solution
(a) Each term in $-4x^2 + 28x - 40$ has a factor of -4, so we start by factoring out $\mathbf{-4}$. (In this example, we factor out the *opposite* of the GCF to obtain a positive leading coefficient in the resulting expression.)

$$-4x^2 + 28x - 40 = \mathbf{-4}(x^2 - 7x + 10) \quad \text{Factor out } -4.$$
$$= -4(x - 5)(x - 2) \quad \text{Factor the trinomial.}$$

(b) We start by factoring out the GCF for $10x^3 + 28x^2 - 6x$, which is $\mathbf{2x}$.

$$10x^3 + 28x^2 - 6x = \mathbf{2x}(5x^2 + 14x - 3) \quad \text{Factor out } 2x.$$
$$= 2x(5x - 1)(x + 3) \quad \text{Factor the trinomial.}$$

Now Try Exercises 13, 21

Many equations involving higher degree polynomials can be solved using the four-step process discussed in Section 6.6. It is important to remember that the zero-product property applies to *all* factors that contain the variable. In the next example we apply the zero-product property to three factors, resulting in three solutions.

EXAMPLE 2 **Solving polynomial equations**

Solve each equation.
(a) $x^3 - x^2 - 6x = 0$ (b) $4x^4 + 10x^3 = 6x^2$

Solution
(a) We start by factoring out the GCF, which is x.

$$x^3 - x^2 - 6x = 0 \qquad \text{Given equation}$$
$$x(x^2 - x - 6) = 0 \qquad \text{Factor out } x.$$
$$x(x - 3)(x + 2) = 0 \qquad \text{Factor the trinomial.}$$
$$x = 0 \quad \text{or} \quad x - 3 = 0 \quad \text{or} \quad x + 2 = 0 \qquad \text{Zero-product property}$$
$$x = 0 \quad \text{or} \quad x = 3 \quad \text{or} \quad x = -2 \qquad \text{Solve for } x.$$

The solutions are $-2, 0$, and 3.

(b) We start by subtracting $6x^2$ from each side to obtain 0 on one side of the equation.

$$4x^4 + 10x^3 = 6x^2 \qquad \text{Given equation}$$
$$4x^4 + 10x^3 - 6x^2 = 0 \qquad \text{Subtract } 6x^2.$$
$$2x^2(2x^2 + 5x - 3) = 0 \qquad \text{Factor out the GCF, } 2x^2.$$
$$2x^2(2x - 1)(x + 3) = 0 \qquad \text{Factor the trinomial.}$$
$$2x \cdot x = 0 \quad \text{or} \quad 2x - 1 = 0 \quad \text{or} \quad x + 3 = 0 \qquad \text{Zero-product property}$$
$$x = 0 \quad \text{or} \quad x = \frac{1}{2} \quad \text{or} \quad x = -3 \qquad \text{Solve for } x.$$

The solutions are $-3, 0$, and $\frac{1}{2}$.

Now Try Exercises 59, 65

STUDY TIP

If you are having difficulty with your studies, you may be able to find help at the student support services office on your campus.

▶ **REAL-WORLD CONNECTION** The corners of a square piece of metal are cut out to form a box, as shown in Figure 6.9. This square piece of metal has sides with length 10 inches, and the cutout corners are squares with length x. The outside surface area A of this box, including the bottom and the sides but *not* the top, is $A = 100 - 4x^2$. (See Critical Thinking in the margin.)

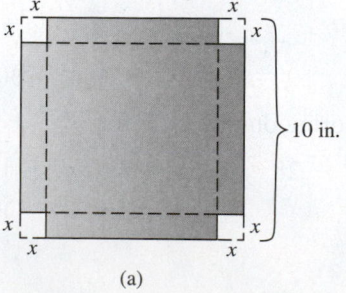

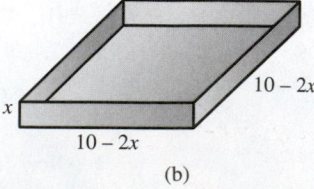

(a)   (b)

Figure 6.9

EXAMPLE 3   **Finding a dimension of a box**

Find the value of x in Figure 6.9 if the outside surface area A is 84 square inches.

Solution

Let $A = 84$ in the equation $A = 100 - 4x^2$ and solve for x.

$$
\begin{aligned}
100 - 4x^2 &= 84 && \text{Let } A = 84.\\
16 - 4x^2 &= 0 && \text{Subtract 84.}\\
4(4 - x^2) &= 0 && \text{Factor out 4.}\\
4(2 - x)(2 + x) &= 0 && \text{Difference of two squares}\\
2 - x = 0 \quad\text{or}\quad 2 + x &= 0 && \text{Zero-product property}\\
x = 2 \quad\text{or}\quad x &= -2 && \text{Solve for } x.
\end{aligned}
$$

If squares measuring 2 inches on a side are cut out of each corner, the surface area of the box is 84 square inches. (Note that $x = -2$ has no physical meaning in this problem.)

Now Try Exercise 77

Special Types of Polynomials

Some types of polynomials of higher degree can be factored by using methods that were presented earlier in this chapter.

EXAMPLE 4   **Factoring higher degree polynomials**

Factor each polynomial completely.

(a) $x^4 - 16$ **(b)** $y^4 + 5y^2 + 4$ **(c)** $x^4 + 2x^2y^2 + y^4$ **(d)** $r^4 - t^4$

Solution

(a) We view this polynomial as the difference of two squares, where $x^4 = (x^2)^2$ and $16 = 4^2$. Then we factor twice.

$$
\begin{aligned}
x^4 - 16 &= (x^2)^2 - 4^2 && \text{Rewrite.}\\
&= (x^2 - 4)(x^2 + 4) && \text{Difference of two squares}\\
&= (x - 2)(x + 2)(x^2 + 4) && \text{Difference of two squares}
\end{aligned}
$$

Note that $x^2 + 4$ does not factor.

(b) Because the trinomial $a^2 + 5a + 4$ factors as $(a + 1)(a + 4)$, we let $a = y^2$ and then factor the given trinomial.

$$
\begin{aligned}
y^4 + 5y^2 + 4 &= (y^2)^2 + 5(y^2) + 4\\
&= (y^2 + 1)(y^2 + 4)
\end{aligned}
$$

Note that neither $y^2 + 1$ nor $y^2 + 4$ can be factored further.

(c) Because the perfect square trinomial $a^2 + 2ab + b^2$ factors as $(a + b)^2$, we let $a = x^2$ and $b = y^2$ and then factor the given trinomial.

$$
\begin{aligned}
x^4 + 2x^2y^2 + y^4 &= (x^2)^2 + 2x^2y^2 + (y^2)^2\\
&= (x^2 + y^2)^2
\end{aligned}
$$

(d) Because the difference of squares $a^2 - b^2$ factors as $(a - b)(a + b)$, we let $a = r^2$ and $b = t^2$ and then factor the given binomial.

$$
\begin{aligned}
r^4 - t^4 &= (r^2)^2 - (t^2)^2\\
&= (r^2 - t^2)(r^2 + t^2)\\
&= (r - t)(r + t)(r^2 + t^2)
\end{aligned}
$$

Note that $r^2 + t^2$ cannot be factored.

Now Try Exercises 29, 37, 43, 45

EXAMPLE 5 **Solving an equation**

Solve $x^5 - 81x = 0$.

Solution

We start by factoring out the common factor of x.

$$x^5 - 81x = 0 \qquad \text{Given equation}$$

$$x(x^4 - 81) = 0 \qquad \text{Factor out } x.$$

$$x(x^2 - 9)(x^2 + 9) = 0 \qquad \text{Difference of two squares}$$

$$x(x - 3)(x + 3)(x^2 + 9) = 0 \qquad \text{Difference of two squares}$$

$$x = 0 \ \text{ or } \ x - 3 = 0 \ \text{ or } \ x + 3 = 0 \ \text{ or } \ x^2 + 9 = 0 \qquad \text{Zero-product property}$$

$$x = 0 \quad \text{ or } \quad x = 3 \quad \text{ or } \quad x = -3 \quad \text{Solve for } x.$$

Note that $x^2 + 9 = 0$ has no real-number solutions because the square of a number plus 9 is never 0. The solutions are $-3, 0,$ and 3.

▌ **Now Try Exercise 63**

6.7 Putting It All Together

CONCEPT	EXPLANATION	EXAMPLES
Common Factors	A first step when factoring polynomials is to factor out the GCF. Once this is done, factor the resulting expression, if possible.	$x^3 - 4x^2 - 5x = x(x^2 - 4x - 5)$ $\qquad\qquad = x(x - 5)(x + 1)$ $4x^4 - 16x^2 = 4x^2(x^2 - 4)$ $\qquad\qquad = 4x^2(x - 2)(x + 2)$
Factoring Higher Degree Polynomials	Some higher degree polynomials can be factored by using the same methods that we use to factor quadratic polynomials.	$x^4 + 6x^2 + 5 = (x^2 + 5)(x^2 + 1)$ $4y^4 - 25 = (2y^2 - 5)(2y^2 + 5)$ $2z^4 - 32 = 2(z^4 - 16)$ $\qquad\quad = 2(z^2 - 4)(z^2 + 4)$ $\qquad\quad = 2(z - 2)(z + 2)(z^2 + 4)$
Solving Equations by Factoring	Use factoring and the zero-product property to solve polynomial equations.	$3x^3 - 12x^2 + 9x = 0$ $3x(x^2 - 4x + 3) = 0$ $3x(x - 3)(x - 1) = 0$ $3x = 0 \ \text{ or } \ x - 3 = 0 \ \text{ or } \ x - 1 = 0$ $x = 0 \ \text{ or } \ x = 3 \ \text{ or } \ x = 1$ The solutions are 0, 1, and 3.

6.7 Exercises

CONCEPTS AND VOCABULARY

1. When you are factoring polynomials, a good first step is to factor out the _____.

2. When you are solving an equation by factoring, the _____ property is used.

3. The zero-product property applies to all _____ containing the variable.

4. Because $a^2 - 2ab + b^2 = (a - b)^2$, it follows that $x^4 - 2x^2y^2 + y^4 =$ _____.

5. Because $x^2 + 3x + 2 = (x + 1)(x + 2)$, it follows that $z^4 + 3z^2 + 2 =$ _____.

6. If $x^4 - 1$ is factored as $(x^2 - 1)(x^2 + 1)$, is it factored *completely*? Explain.

7. If $x^3 - x^2 + x - 1$ is factored as $(x - 1)(x^2 + 1)$, is it factored *completely*? Explain.

8. When you are solving the equation $x^3 = x$, what is a good first step?

9. When you are solving $x^4 - x^2 = 0$, what is a good first step?

10. How many real-number solutions does the equation $(x - 6)(x^2 + 4) = 0$ have?

FACTORING POLYNOMIALS

Exercises 11–48: Factor the polynomial completely.

11. $5x^2 - 5x - 30$
12. $3x^2 - 15x + 12$
13. $-4y^2 - 32y - 48$
14. $-7y^2 + 14y + 21$
15. $-20z^2 - 110z - 50$
16. $-12z^2 - 54z + 30$
17. $60 - 64t - 28t^2$
18. $18 - 45t - 27t^2$
19. $r^3 - r$
20. $r^3 + 2r^2 - 3r$
21. $3x^3 + 3x^2 - 18x$
22. $6x^3 - 26x^2 - 20x$
23. $72z^3 + 12z^2 - 24z$
24. $6z^3 - 4z^2 - 42z$
25. $x^4 - 4x^2$
26. $4x^4 - 36x^2$
27. $t^4 + t^3 - 2t^2$
28. $t^4 + 5t^3 - 24t^2$
29. $x^4 - 5x^2 + 6$
30. $x^4 - 3x^2 - 10$
31. $2x^4 + 7x^2 + 3$
32. $3x^4 - 8x^2 + 5$
33. $y^4 + 6y^2 + 9$
34. $y^4 - 10y^2 + 25$
35. $x^4 - 9$
36. $x^4 - 25$
37. $x^4 - 81$
38. $4x^4 - 64$
39. $z^5 + 2z^4 + z^3$
40. $6z^5 - 47z^4 + 35z^3$
41. $2x^2 + xy - y^2$
42. $2x^2 + 5xy + 2y^2$
43. $a^4 - 2a^2b^2 + b^4$
44. $a^3 + 2a^2b + ab^2$
45. $x^3 - xy^2$
46. $2x^2y - 2y^3$
47. $4x^3 + 4x^2y + xy^2$
48. $x^2y - 6xy^2 + 9y^3$

SOLVING EQUATIONS

Exercises 49–54: Do the following.

49. (a) Factor $x^3 - 4x$.
    (b) Solve $x^3 - 4x = 0$.

50. (a) Factor $4x^3 - 16x$.
    (b) Solve $4x^3 - 16x = 0$.

51. (a) Factor $2y^3 - 6y^2 - 36y$.
    (b) Solve $2y^3 - 6y^2 - 36y = 0$.

52. (a) Factor $z^4 - 13z^2 + 36$.
    (b) Solve $z^4 - 13z^2 + 36 = 0$.

53. (a) Factor $x^3 - x^2 + 4x - 4$.
    (b) Solve $x^3 - x^2 + 4x - 4 = 0$.

54. (a) Factor $y^4 - 8y^2 + 16$.
    (b) Solve $y^4 - 8y^2 + 16 = 0$.

Exercises 55–76: Solve.

55. $3x^2 + 33x + 72 = 0$
56. $4x^2 - 16x - 20 = 0$
57. $25x^2 = 50x + 75$
58. $10x^2 = 20x + 80$
59. $y^3 - 3y^2 - 4y = 0$
60. $y^3 - 3y^2 + 2y = 0$
61. $3z^3 + 6z^2 = 72z$
62. $4z^3 = 4z^2 + 24z$
63. $x^4 - 36x^2 = 0$
64. $4x^4 = 100x^2$
65. $r^4 + 6r^3 = 7r^2$
66. $r^4 + 30r^2 = 11r^3$
67. $x^4 - 13x^2 = -36$
68. $x^4 - 17x^2 + 16 = 0$
69. $x^4 + 1 = 2x^2$
70. $x^4 - 8x^2 + 16 = 0$
71. $a^4 = 81$
72. $b^3 = -8$
73. $x^3 - 2x^2 - x + 2 = 0$
74. $x^3 - x^2 + 4x - 4 = 0$
75. $x^3 - 5x^2 + x - 5 = 0$
76. $3x^3 + 2x^2 - 27x - 18 = 0$

APPLICATIONS

77. *Dimensions of a Box* (Refer to Example 3.) A box is made from a rectangular piece of metal with length 20 inches and width 15 inches. The box has no top.

    (a) What are the limitations on the size of x? Explain your answer.

    (b) Write an expression that gives the outside surface area of the box. (*Hint:* Consider the size of the metal sheet and how much was cut out.)

    (c) If the outside surface area of the box is 275 square inches, find x.

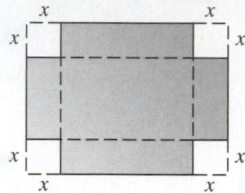 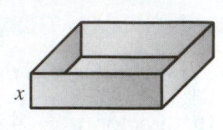

78. *Dimensions of a Box* Refer to the previous exercise.

    (a) Find a polynomial that gives the volume of the box for a given x.

    (b) Factor your polynomial completely.

    (c) What are the zeros of your polynomial? What do they represent in this problem?

Exercises 79–82: U.S. Natural Gas Consumption (Refer to A Look Into Math for this section.) *The polynomial*

$$0.0013x^3 - 0.085x^2 + 1.6x + 12$$

models natural gas consumption in trillions of cubic feet, where $x = 0$ corresponds to 1960, $x = 1$ corresponds to 1961, and so on until $x = 40$ corresponds to 2000.

79. How much natural gas was consumed in 1990?

80. In which year between 1970 and 1990 was natural gas consumption about 20.4 trillion cubic feet?

81. Explain any difficulties encountered when you try to solve the equation

$$0.0013x^3 - 0.085x^2 + 1.6x + 12 = 23.2.$$

82. How might you solve the equation in Exercise 81 without factoring? If you were to find the solution to this equation, what would it represent?

WRITING ABOUT MATHEMATICS

83. Compare factoring the polynomial $x^2 + 6x + 5$ with factoring the polynomial $z^4 + 6z^2 + 5$.

84. Suppose that a polynomial can be factored. Explain how its factors can be used to find the zeros of the polynomial. Give an example.

SECTION 6.7 — Checking Basic Concepts

1. Factor the trinomial completely.
    (a) $3x^2 - 6x - 24$ **(b)** $-10y^2 + 5y + 5$

2. Factor the binomial completely.
    (a) $z^4 - 25$ **(b)** $7t^4 - 7$

3. Factor.
    (a) $x^4 - 8x^2 + 16$ **(b)** $2y^3 + 17y^2 - 30y$

4. Solve $t^4 + t^3 = 12t^2$.

5. Solve $x^3 - 3x^2 + 2x - 6 = 0$.

CHAPTER 6   Summary

SECTION 6.1 ■ INTRODUCTION TO FACTORING

Terms Related to Factoring Polynomials

Factoring	Writing a polynomial as a product, usually of lower degree polynomials
Common Factor	An expression that is a factor of each term in a polynomial

    Example: Some common factors of $4x^4 + 8x^2$ are $2x$, x^2, and $4x^2$.

Greatest Common Factor (GCF)	The common factor with the greatest (integer) coefficient and the highest degree

    Example: The GCF of $4x^4 + 8x^2$ is $4x^2$.

$$4x^4 + 8x^2 = 4x^2(x^2 + 2)$$

Factoring by Grouping   Used to factor a four-term polynomial into two binomials

Example: $x^3 + 5x^2 + 3x + 15 = x^2(x + 5) + 3(x + 5)$
$$= (x^2 + 3)(x + 5)$$

SECTION 6.2 ■ FACTORING TRINOMIALS I ($x^2 + bx + c$)

Review of the FOIL Method A method used for multiplying two binomials

First Terms $(2x + 3)(5x + 4)$: $2x \cdot 5x = 10x^2$

Outside Terms $(2x + 3)(5x + 4)$: $2x \cdot 4 = 8x$

Inside Terms $(2x + 3)(5x + 4)$: $3 \cdot 5x = 15x$

Last Terms $(2x + 3)(5x + 4)$: $3 \cdot 4 = 12$

The product is the sum of these four terms:

$$(2x + 3)(5x + 4) = 10x^2 + 8x + 15x + 12 = 10x^2 + 23x + 12.$$

Factoring Trinomials Having a Leading Coefficient of 1 To factor the trinomial $x^2 + bx + c$, find two numbers, m and n, that satisfy
$$m \cdot n = c \quad \text{and} \quad m + n = b.$$

Then $x^2 + bx + c = (x + m)(x + n)$.

Example: Because $-3 \cdot 5 = -15$ and $-3 + 5 = 2$,
$$x^2 + 2x - 15 = (x - 3)(x + 5).$$

SECTION 6.3 ■ FACTORING TRINOMIALS II ($ax^2 + bx + c$)

Factoring Trinomials by Grouping To factor $ax^2 + bx + c$, perform the following steps. (Assume that a, b, and c are integers and have no factor in common.)

1. Find two numbers, m and n, such that $mn = ac$ and $m + n = b$.
2. Write the trinomial as $ax^2 + mx + nx + c$.
3. Use grouping to factor this expression into two binomials.

Example: To factor $3x^2 + 10x - 8$, find two numbers whose product is -24 and whose sum is 10. These two numbers are $m = 12$ and $n = -2$, so write $10x$ as $12x - 2x$.

$$3x^2 + 10x - 8 = 3x^2 + 12x - 2x - 8 \qquad \text{Write } 10x \text{ as } 12x - 2x.$$
$$= (3x^2 + 12x) + (-2x - 8) \qquad \text{Associative property}$$
$$= 3x(x + 4) - 2(x + 4) \qquad \text{Factor out common factors.}$$
$$= (3x - 2)(x + 4) \qquad \text{Factor out } (x + 4).$$

Factoring with FOIL in Reverse Use trial and error and FOIL in reverse to find the factors of a trinomial.

Example: To factor $3x^2 + 10x - 8$, first find factors of $3x^2$.
$$(3x + \underline{\quad})(x + \underline{\quad})$$

Then place factors of -8 so that the resulting middle term is $10x$.

$$(3x + -2) \cdot (x + 4)$$
$$\llcorner -2x \lrcorner$$
$$\underline{\quad + 12x \quad}$$
$$\overline{10x} \ \checkmark \ \leftarrow \text{Middle term checks.}$$

Difference of Two Squares

$$a^2 - b^2 = (a - b)(a + b)$$

Examples: $x^2 - 16 = (x - 4)(x + 4)$ $(a = x, b = 4)$

$\quad\quad\quad 4r^2 - 9t^2 = (2r - 3t)(2r + 3t)$ $(a = 2r, b = 3t)$

Perfect Square Trinomials

$$a^2 + 2ab + b^2 = (a + b)^2 \quad \text{and} \quad a^2 - 2ab + b^2 = (a - b)^2$$

Examples: $4x^2 + 4x + 1 = (2x)^2 + 2(2x)1 + 1^2 = (2x + 1)^2$ $(a = 2x, b = 1)$

$\quad\quad\quad x^2 - 10x + 25 = x^2 - 2 \cdot x \cdot 5 + 5^2 = (x - 5)^2$ $(a = x, b = 5)$

Sums and Differences of Two Cubes

$$a^3 + b^3 = (a + b)(a^2 - ab + b^2) \quad \text{and} \quad a^3 - b^3 = (a - b)(a^2 + ab + b^2)$$

Examples: $x^3 + 8 = (x + 2)(x^2 - 2x + 4)$ $(a = x, b = 2)$

$\quad\quad\quad 27x^3 - 1 = (3x - 1)(9x^2 + 3x + 1)$ $(a = 3x, b = 1)$

Guidelines for Factoring Polynomials The following guidelines can be used to factor polynomials in general.

STEP 1: Factor out the greatest common factor, if possible.

STEP 2: **A.** If the polynomial has *four terms*, try factoring by grouping.
  B. If the polynomial is a *binomial,* try one of the following.

1. $a^2 - b^2 = (a - b)(a + b)$	Difference of two squares
2. $a^3 - b^3 = (a - b)(a^2 + ab + b^2)$	Difference of two cubes
3. $a^3 + b^3 = (a + b)(a^2 - ab + b^2)$	Sum of two cubes

  C. If the polynomial is a *trinomial*, check for a perfect square.

1. $a^2 + 2ab + b^2 = (a + b)^2$	Perfect square trinomial
2. $a^2 - 2ab + b^2 = (a - b)^2$	Perfect square trinomial

  Otherwise, try to factor the trinomial by grouping or apply FOIL in reverse, as described in Sections 6.2 and 6.3.

STEP 3: Check to make sure that the polynomial is *completely* factored.

Examples: $12x^3 - 12x^2 + 3x = 3x(4x^2 - 4x + 1) = 3x(2x - 1)^2$ Steps 1, 2C, and 3

$\quad\quad\quad 9x^3 - 6x^2 + 18x - 12 = 3(3x^3 - 2x^2 + 6x - 4) = 3(x^2 + 2)(3x - 2)$ Steps 1, 2A, and 3

$\quad\quad\quad 16x^3 - 100x = 4x(4x^2 - 25) = 4x(2x - 5)(2x + 5)$ Steps 1, 2B, and 3

Zero-Product Property

For any real numbers a and b, if $ab = 0$, then $a = 0$ or $b = 0$ (or both).
The zero-product property is used to solve equations.

Examples: $xy = 0$ implies that $x = 0$ or $y = 0$.

$\quad\quad\quad (x + 5)(x - 3) = 0$ implies $x + 5 = 0$ or $x - 3 = 0$.

Zero of a Polynomial
A number a is a zero of a polynomial if the result is 0 when a is substituted in that polynomial.

Example: The number -2 is a zero of $x^2 - 4$ because $(-2)^2 - 4 = 0$.

Solving Quadratic Equations by Factoring To solve a quadratic equation by factoring, follow these steps.

STEP 1: If necessary, use algebra to write the equation as $ax^2 + bx + c = 0$.

STEP 2: Factor the left side of the equation using any method.

STEP 3: Apply the zero-product property.

STEP 4: Solve each of the resulting equations. Check any solutions.

Example:

$$x^2 + 7x = 8 \quad \text{Given equation}$$
$$x^2 + 7x - 8 = 0 \quad \text{Step 1}$$
$$(x + 8)(x - 1) = 0 \quad \text{Step 2}$$
$$x + 8 = 0 \quad \text{or} \quad x - 1 = 0 \quad \text{Step 3}$$
$$x = -8 \quad \text{or} \quad x = 1 \quad \text{Step 4}$$

SECTION 6.7 ■ SOLVING EQUATIONS BY FACTORING II (HIGHER DEGREE)

Factoring Polynomials of Higher Degree The distributive property and the techniques for factoring quadratic polynomials can also be applied to polynomials of higher degree.

Examples: $10r^3 + 15r = 5r(2r^2 + 3)$ (To check, multiply the right side.)

Because $2x^2 + x - 1 = (x + 1)(2x - 1)$,

it follows that $2z^4 + z^2 - 1 = (z^2 + 1)(2z^2 - 1)$.

Solving Equations by Factoring Use algebra to obtain 0 on one side of the equation. Factor the other side and apply the zero-product property.

Example:

$$x^3 = 4x \quad \text{Given equation}$$
$$x^3 - 4x = 0 \quad \text{Subtract } 4x.$$
$$x(x^2 - 4) = 0 \quad \text{Factor out the GCF, } x.$$
$$x(x - 2)(x + 2) = 0 \quad \text{Difference of two squares}$$
$$x = 0 \quad \text{or} \quad x - 2 = 0 \quad \text{or} \quad x + 2 = 0 \quad \text{Zero-product property}$$
$$x = 0 \quad \text{or} \quad x = 2 \quad \text{or} \quad x = -2 \quad \text{Solve for } x.$$

CHAPTER 6   Review Exercises

SECTION 6.1

Exercises 1–6: Identify the greatest common factor for the expression and then factor the expression.

1. $5x - 15$

2. $y^2 + 2y$

3. $8z^3 - 4z^2$

4. $6x^4 + 3x^3 - 12x^2$

5. $9xy + 15yz^2$

6. $a^2b^3 + a^3b^2$

Exercises 7–14: Use grouping to factor the given polynomial completely.

7. $x(x + 2) - 3(x + 2)$

8. $y^2(x - 5) + 3y(x - 5)$

9. $z^3 - 2z^2 + 5z - 10$

10. $t^3 + t^2 + 8t + 8$

11. $x^3 - 3x^2 + 6x - 18$

12. $ax + bx - ay - by$

13. $x^5 + 3x^4 - 2x^3 - 6x^2$

14. $2y^4 + 6y^3 + 2y^2 + 6y$

SECTION 6.2

Exercises 15–18: Find an integer pair that has the given product and sum.

15. Product: 20   Sum: 9

16. Product: −21   Sum: 4

17. Product: 36   Sum: −13

18. Product: −100   Sum: −21

Exercises 19–32: Factor the trinomial completely. If the trinomial cannot be factored, write "prime."

19. $x^2 - x - 12$ **20.** $x^2 + 10x + 24$

21. $x^2 + 6x - 16$ **22.** $x^2 - x - 42$

23. $x^2 + 4x - 6$ **24.** $x^2 - 5x + 8$

25. $x^2 + 2x - 3$ **26.** $x^2 + 22x + 120$

27. $2x^3 + 6x^2 - 20x$ **28.** $x^4 - 3x^3 - 28x^2$

29. $10 - 7x + x^2$ **30.** $24 + 2x - x^2$

31. $-2x^2 - 4x + 30$ **32.** $-x^3 - 9x^2 + 10x$

SECTION 6.3

Exercises 33–44: Factor the trinomial completely. If the trinomial cannot be factored, write "prime."

33. $9x^2 + 3x - 2$ **34.** $2x^2 + 3x - 5$

35. $3x^2 + 14x + 15$ **36.** $35x^2 - 2x - 1$

37. $3x^2 + 4x - 5$ **38.** $4x^2 - 12x - 5$

39. $24x^2 - 7x - 5$ **40.** $4x^2 + 33x - 27$

41. $12x^3 + 48x^2 + 21x$ **42.** $8x^4 + 14x^3 - 30x^2$

43. $12 - 5x - 2x^2$ **44.** $1 + 3x - 10x^2$

SECTION 6.4

Exercises 45–58: Factor completely.

45. $z^2 - 4$ **46.** $9z^2 - 64$

47. $36 - y^2$ **48.** $100a^2 - 81b^2$

49. $x^2 + 14x + 49$ **50.** $x^2 - 10x + 25$

51. $4x^2 - 12x + 9$ **52.** $9x^2 + 48x + 64$

53. $8t^3 - 1$ **54.** $27r^3 + 8t^3$

55. $2x^3 - 50x$ **56.** $24x^3 + 81$

57. $2x^3 + 28x^2 + 98x$ **58.** $2x^4 - 128x$

SECTION 6.5

Exercises 59–68: Factor completely.

59. $12x - 8$ **60.** $6x^3 + 9x^2$

61. $9y^2 - 6y + 6$ **62.** $yz^2 - 9y$

63. $x^4 + 7x^3 - 4x^2 - 28x$ **64.** $12x^3 + 36x^2 + 27x$

65. $3ab^3 - 24a$ **66.** $5x^3 + 20x$

67. $24x^3 - 6xy^2$ **68.** $x^3y + 27y$

SECTION 6.6

Exercises 69–80: Solve the equation.

69. $mn = 0$ **70.** $y^2 = 0$

71. $(4x - 3)(x + 9) = 0$

72. $(1 - 4x)(6 + 5x) = 0$

73. $z(z - 1)(z - 2) = 0$ **74.** $z^2 - 7z = 0$

75. $y^2 - 64 = 0$ **76.** $y^2 + 9y + 14 = 0$

77. $x^2 = x + 6$ **78.** $10x^2 + 11x = 6$

79. $t(t - 14) = 72$ **80.** $t(2t - 1) = 10$

SECTION 6.7

Exercises 81–90: Factor completely.

81. $5x^2 - 15x - 50$ **82.** $-3x^2 - 6x + 45$

83. $y^3 - 4y$ **84.** $3y^3 + 6y^2 - 9y$

85. $2z^4 + 14z^3 + 20z^2$ **86.** $8z^4 - 32z^2$

87. $x^4 - 6x^2 + 9$ **88.** $2x^4 - 15x^2 - 27$

89. $a^2 + 10ab + 25b^2$ **90.** $x^3 - xy^2$

Exercises 91–98: Solve.

91. $16x^2 - 72x - 40 = 0$

92. $2x^3 - 11x^2 + 15x = 0$ **93.** $t^3 = 25t$

94. $t^4 - 7t^3 + 12t^2 = 0$ **95.** $z^4 + 16 = 8z^2$

96. $z^4 - 256 = 0$ **97.** $y^3 = -64$

98. $y^3 - y^2 - y + 1 = 0$

APPLICATIONS

99. A square has area $9x^2 + 42x + 49$. Find the length of a side. Make a sketch of the square.

100. A rectangle has area $x^2 + 6x + 5$. Find possible dimensions for the rectangle. Make a sketch of the rectangle.

101. A cube has surface area $6x^2 + 12x + 6$. Find the length of a side.

102. Write a polynomial in factored form that represents the total area of the rectangle.

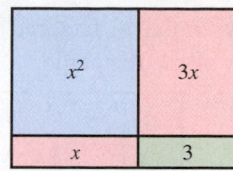

103. Write a polynomial in factored form that represents the total area of the rectangle.

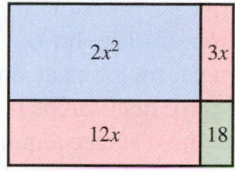

104. *Radius of a Circle* The area and the circumference of a circle are numerically equal. Find the radius of the circle.

105. *Dimensions of a Shed* The floor of a rectangular shed is 7 feet longer than it is wide and has an area of 120 square feet. What are its dimensions?

106. *Flight of a Ball* A ball is hit upward. Its height h in feet after t seconds is $h = -16t^2 + 80t + 4$. At what times is the ball 100 feet in the air?

107. *Stopping Distance* The distance D in feet that it takes to stop a car traveling x miles per hour on wet, level pavement can be approximated by $D = \frac{1}{9}x^2 + \frac{11}{3}x$.
  (a) Estimate the distance required for the car to stop when it is traveling 45 miles per hour.
  (b) If the stopping distance is 80 feet, what is the speed of the car?
  (c) If you have a calculator, use it to solve part (b) numerically with a table of values. Do your answers agree?

108. *Revenue* A company makes tops for the boxes of pickup trucks. The total revenue R in dollars from selling the tops for p dollars each is given by $R = p(200 - p)$, where $p \le 200$.

  (a) Find R when $p = \$100$.
  (b) Find p when $R = \$7500$.
  (c) If you have a calculator, use it to solve part (b) numerically with a table of values. Do your answers agree?

109. *Airline Passengers* The number N of worldwide airline passengers in millions from 1950 to 2000 is approximated by

$$N = 0.68y^2 + 3.8y + 24,$$

where $y = 0$ corresponds to 1950, $y = 1$ corresponds to 1951, and so on until $y = 50$ corresponds to 2000.
  (a) Estimate the number of airline passengers in 1970.
  (b) Use a table of values to estimate the year in which the number of airline passengers reached 544 million.

110. *Digital Photographs* A digital photograph is 30 pixels longer than it is wide and has a total area of 4000 pixels. Find the dimensions of this photograph.

111. *Dimensions of a Box* A box is made from a rectangular piece of metal with length 50 inches and width 40 inches by cutting out square corners of length x and folding up the sides.
  (a) Write an expression that gives the surface area of the inside of the box.
  (b) If the surface area of the box is 1900 square inches, find x.

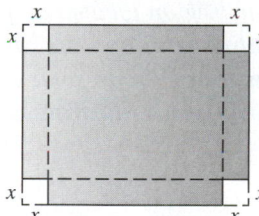 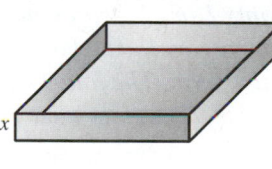

112. *Dimensions of a Cube* If two identical cubes have a total surface area of $12x^2 + 48x + 48$, find the length of one side of one of the cubes.

CHAPTER 6   **Test**  Step-by-step test solutions are found on the Chapter Test Prep Videos available in *MyMathLab* and on **YouTube** (search "RockswoldComboAlg" and click on "Channels").

Exercises 1 and 2: Identify the greatest common factor for the expression. Then factor the expression.

1. $4x^2y - 20xy^2 + 12xy$ **2.** $9a^3b^2 + 3a^2b^2$

Exercises 3 and 4: Factor by grouping.

3. $ay + by + az + bz$ **4.** $3x^3 + x^2 - 15x - 5$

Exercises 5–10: Factor the trinomial completely. If the trinomial cannot be factored, write "prime."

5. $y^2 + 4y - 12$

6. $4x^2 + 20x + 25$

7. $4z^2 - 19z + 12$

8. $21 - 17t + 2t^2$

9. $x^2 + 7x - 10$

10. $3y^2 + 4y + 2$

Exercises 11–16: Factor completely.

11. $6x^3 + 3x^2 - 3x$

12. $2z^4 - 12z^2 - 54$

13. $36y^3 - 100y$

14. $7x^4 + 56x$

15. $16a^4 + 24a^3 + 9a^2$

16. $2b^4 - 32$

Exercises 17–22: Solve the equation.

17. $x^2 - 16 = 0$

18. $y^2 = y + 20$

19. $9z^2 + 16 = 24z$

20. $x(x - 5) = 66$

21. $y^3 = 9y$

22. $x^4 - 5x^2 + 4 = 0$

23. A square has area $9x^2 + 30x + 25$. Find the length of a side in terms of x.

24. Write a polynomial in factored form that represents the total area of the rectangle.

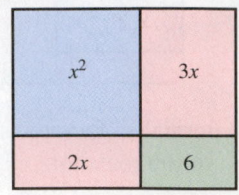

25. *Braking Distance* The braking distance D in feet required for a car traveling at x miles per hour to stop on dry, level pavement can be modeled by $D = \frac{1}{11}x^2$.
  (a) Calculate the distance required for the car to stop when it is traveling 55 miles per hour.
  (b) If the braking distance is 99 feet, estimate the speed of the car.

26. *Flight of a Ball* A ball is hit upward. Its height h in feet after t seconds is given by $h = -16t^2 + 48t + 4$. At what times is the ball 36 feet in the air?

CHAPTER 6   Extended and Discovery Exercises

FACTORING TRINOMIALS VISUALLY

Exercises 1–6: A special grid similar to the xy-plane can be used to factor trinomials visually. The grid has four quadrants, where the area of any region located in quadrants I or III represents a positive term and the area of any region located in quadrants II or IV represents a negative term. The following factoring grids illustrate how to factor trinomials visually.

$x^2 + 10x + 24$
factors as
$(x + 4)(x + 6)$

$x^2 - 4x - 21$
factors as
$(x - 7)(x + 3)$

Factor the following trinomials visually.

1. $x^2 + 5x + 6$

2. $x^2 + 9x + 20$

3. $x^2 - 11x + 30$

4. $x^2 - 3x - 10$

5. $x^2 + 4x - 12$

6. $x^2 - 8x + 16$

THE DIFFERENCE OF TWO SQUARES

Exercises 7–12: Difference of Two Squares The difference of two squares can be factored by using

$$a^2 - b^2 = (a - b)(a + b).$$

This equation can also be used in some situations where an expression may not appear to be the difference of two squares. For example, because $(\sqrt{3})^2 = 3$, $x^2 - 3$ can be written and then factored as

$$x^2 - 3 = x^2 - (\sqrt{3})^2$$
$$= (x - \sqrt{3})(x + \sqrt{3}).$$

Use this concept to factor the following expressions as the difference of two squares.

7. $x^2 - 5$

8. $y^2 - 7$

9. $3z^2 - 25$

10. $7t^2 - 2$

11. $x - 4$ for $x \geq 0$ (*Hint:* $(\sqrt{x})^2 = x$.)

12. $x - 7$ for $x \geq 0$

Exercises 13–18: Solving Equations (Refer to Exercises 7–12.) Solve the equation by factoring it as the difference of two squares.

13. $x^2 - 3 = 0$ **14.** $y^2 - 7 = 0$

15. $3x^2 - 25 = 0$ **16.** $7x^2 - 11 = 0$

17. $x^4 - 9 = 0$ **18.** $x^4 - 25 = 0$

CHAPTERS 1–6   Cumulative Review Exercises

Exercises 1 and 2: Evaluate by hand and then simplify to lowest terms.

1. $\frac{3}{5} \cdot \frac{15}{21}$ **2.** $\frac{4}{5} - \frac{1}{10}$

Exercises 3 and 4: Evaluate by hand.

3. $26 - 3 \cdot 6 \div 2$ **4.** $-2^2 + \frac{3+2}{8+2}$

5. Complete the table. Then use the table to solve the equation $2x + 3 = 5$.

x	-2	-1	0	1	2
$2x + 3$					

6. Translate the sentence "Triple a number decreased by 5 equals the number decreased by 7" into an equation using the variable n. Then solve the equation.

7. Convert 5.7% to fraction and decimal notation.

8. Solve $P = 2W + 2L$ for W.

9. Solve $5 - 3z < -1$.

10. Make a scatterplot having the following five points: $(-2, 3)$, $(-1, 2)$, $(0, -1)$, $(1, 1)$, and $(2, 2)$.

Exercises 11 and 12: Graph the given equation. Determine any intercepts.

11. $y = 3x - 2$ **12.** $y = -2$

Exercises 13 and 14: Find the slope–intercept form for the line satisfying the given conditions.

13. Perpendicular to $2x - 3y = -6$ and passing through the point $(1, 2)$

14. Passing through the points $(-2, 1)$ and $(1, 5)$

15. Identify the x-intercept and the y-intercept. Then write the slope–intercept form of the line.

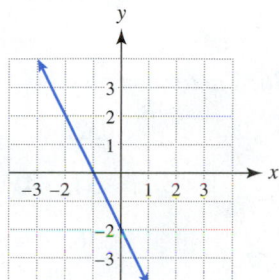

16. The graphs of two equations are shown. Use the graphs to identify the solution to the system of equations. Then check your answer.

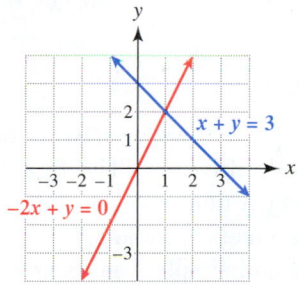

Exercises 17 and 18: Solve the system of equations.

17. $y = -1$
       $2x + y = 1$

18. $5x + y = -5$
       $-x + 2y = 12$

Exercises 19 and 20: Shade the solution set to the system of inequalities.

19. $x > -1$
       $y < x$

20. $2x - y \le 4$
       $x + 2y \ge 2$

Exercises 21–24: Simplify the expression.

21. -2^4

22. $(xy)^0$

23. $(xy)^4(x^3y^{-4})^2$

24. $7x^3(-2x^2 + 3x)$

Exercises 25–28: Simplify and write the expression using positive exponents.

25. $a^{-4} \cdot a^2$

26. $(2t^3)^{-2}$

27. $(xy)^{-3}(x^{-1}y^2)^{-1}$

28. $\left(\dfrac{2x}{y^{-2}}\right)^5$

Exercises 29 and 30: Divide and check.

29. $\dfrac{6x^3 + 12x^2}{3x}$

30. $\dfrac{3x^3 - x + 1}{x^2 + 1}$

Exercises 31–36: Factor completely.

31. $x^2 + 3x - 28$

32. $6y^2 + y - 12$

33. $25x^2 - 4y^2$

34. $64x^2 - 16x + 1$

35. $27t^3 - 8$

36. $-4x^2 + 4x + 24$

Exercises 37–40: Solve the equation.

37. $y^4 = 25y^2$

38. $8z^2 + 8z - 16 = 0$

39. $4z^3 = 49z$

40. $x^4 - 18x^2 + 81 = 0$

APPLICATIONS

41. *Shoveling the Driveway* Two people are shoveling snow from a driveway. The first person shovels 10 square feet per minute, while the second person shovels 8 square feet per minute.
 (a) Write and simplify an expression that gives the total square feet that the two people shovel in x minutes.
 (b) How many minutes would it take for them to clear a driveway with an area of 900 square feet?

42. *Running Distance* A person runs the following distances over three days: $1\frac{3}{4}$ miles, $2\frac{1}{2}$ miles, and $2\frac{2}{3}$ miles. How far does the person run altogether?

43. *Burning Calories* An athlete can burn 12 calories per minute while cross-country skiing and 9 calories per minute while running at 5 miles per hour. If the athlete burns 615 calories in 60 minutes, how long is spent on each activity?

44. *Renting a Car* A rental car costs $20 plus $0.25 per mile driven.
 (a) Write an equation that gives the cost C of driving the car x miles.
 (b) Determine the number of miles that the car is driven if the rental cost is $100.

45. *Triangle* In an isosceles triangle, the measures of the two smaller angles are equal and their sum is $20°$ more than the larger angle.
 (a) Let x be the measure of one of the two smaller angles and y be the measure of the larger angle. Write a system of linear equations whose solution gives the measures of these angles.
 (b) Solve your system.

46. *Areas of Rectangles* Find a monomial equal to the sum of the areas of the rectangles. Calculate this sum for $x = 2$ yards and $y = 3$ yards.

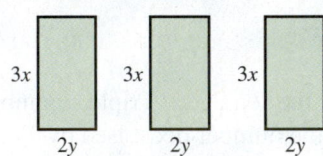

47. *Area of a Square* A square has area $x^2 + 12x + 36$. Find the length of a side.

48. *Flight of a Golf Ball* A golf ball is hit upward. Its height h in feet after t seconds is given by the formula $h = 64t - 16t^2$.
 (a) How long does it take for the ball to hit the ground?
 (b) At what times is the ball 48 feet in the air?

7 Rational Expressions

There is nothing wrong with making mistakes. Just don't respond with encores.

—ANONYMOUS

One of the most significant problems facing the U.S. transportation system is chronic highway congestion. According to a newly released "Highway Statistics," Americans drove about 3 trillion miles in 2009. Our ability to keep traffic moving smoothly and safely is key to keeping our economy strong, and traffic congestion costs motorists more than $87 billion annually in wasted time and fuel. A 2007 study found that, collectively, Americans spend as many as 4.2 billion hours stuck in traffic each year.

If the amount of traffic doubles on a highway, the time spent waiting in traffic may more than double. At times, only a slight increase in the traffic rate can result in a dramatic increase in the time spent waiting. (See Example 6 in Section 7.1.) Mathematics can describe this effect by using rational expressions. In this chapter we introduce rational expressions and some of their applications.

Source: Randy James, "America: Still Stuck in Traffic." Time.com, July 9, 2009; U.S. Department of Transportation, 2011.

7.1 Introduction to Rational Expressions

Basic Concepts • Simplifying Rational Expressions • Applications

A LOOK INTO MATH ▶

Have you ever been moving smoothly in traffic, only to come to a sudden halt? Mathematics shows that in certain conditions, if the number of cars on a road increases even slightly, then the movement of traffic can slow dramatically. To understand why this occurs, we will consider how *rational expressions* can be used to model traffic flow. (See Example 6.)

Basic Concepts

Recall that a *rational number* is any number that can be expressed as a ratio of two integers $\frac{p}{q}$, where $q \neq 0$. In this chapter, we discuss *rational expressions*, which can be written as the ratio of two polynomials. Because examples of polynomials include

$$3, \quad 2x, \quad x^2 + 4, \quad \text{and} \quad x^3 - 1,$$

it follows that examples of rational expressions include

$$\frac{3}{2x}, \quad \frac{2x}{x^2 + 4}, \quad \frac{x^2 + 4}{3}, \quad \text{and} \quad \frac{x^3 - 1}{x^2 + 4}.$$

NEW VOCABULARY

☐ Rational expression
☐ Lowest terms
☐ Vertical asymptote
☐ Probability

RATIONAL EXPRESSION

A **rational expression** can be written as $\frac{P}{Q}$, where P and Q are polynomials. A rational expression is defined whenever $Q \neq 0$.

We can evaluate polynomials for different values of a variable. For example, for $x = 2$ the polynomial $x^2 - 3x + 1$ evaluates to

$$(2)^2 - 3(2) + 1 = -1.$$

Rational expressions can be evaluated similarly.

EXAMPLE 1 **Evaluating rational expressions**

If possible, evaluate each expression for the given value of the variable.

(a) $\dfrac{1}{x + 1}$ $x = 2$ (b) $\dfrac{y^2}{2y - 1}$ $y = -4$

(c) $\dfrac{5z + 8}{z^2 - 2z + 1}$ $z = 1$ (d) $\dfrac{2 - x}{x - 2}$ $x = -3$

Solution

(a) If $x = 2$, then $\frac{1}{x + 1} = \frac{1}{2 + 1} = \frac{1}{3}$.

(b) If $y = -4$, then $\frac{y^2}{2y - 1} = \frac{(-4)^2}{2(-4) - 1} = -\frac{16}{9}$.

(c) If $z = 1$, then $\frac{5z + 8}{z^2 - 2z + 1} = \frac{5(1) + 8}{1^2 - 2(1) + 1}$, or $\frac{13}{0}$, which is undefined because division by 0 is not possible.

(d) If $x = -3$, then $\frac{2 - x}{x - 2} = \frac{2 - (-3)}{-3 - 2} = \frac{5}{-5} = -1$.

Now Try Exercises 7, 11, 13, 17

UNDEFINED EXPRESSIONS Division by 0 is undefined. As a result, rational expressions are different from polynomials because they are *undefined* whenever their *denominators are 0*. For example, the expression in Example 1(c) is undefined when $z = 1$.

EXAMPLE 2 **Determining when a rational expression is undefined**

Find all values of the variable for which each expression is undefined.

(a) $\dfrac{1}{x}$ (b) $\dfrac{4t}{t-3}$ (c) $\dfrac{1-6r}{r^2-4}$ (d) $\dfrac{4}{x^2+1}$

Solution

(a) A rational expression is undefined when its denominator is 0. Thus $\frac{1}{x}$ is undefined when $x = 0$.

(b) The expression $\frac{4t}{t-3}$ is undefined when its denominator, $t-3$, is 0, or when $t = 3$.

(c) The expression $\frac{1-6r}{r^2-4}$ is undefined when its denominator, r^2-4, is 0. Here

$$r^2 - 4 = (r-2)(r+2) = 0$$

implies that the denominator is 0 when $r = -2$ or $r = 2$.

(d) In the expression $\frac{4}{x^2+1}$ the denominator, x^2+1, is never 0 because any real number squared plus 1 is always greater than or equal to 1. Thus this rational expression is defined for all real numbers x.

Now Try Exercises 25, 27, 31, 33

Simplifying Rational Expressions

In Chapter 1, we used the *basic principle of fractions*,

$$\frac{a \cdot c}{b \cdot c} = \frac{a}{b}.$$

For example, this basic principle allows us to simplify the fraction $\frac{8}{12}$ as

$$\frac{8}{12} = \frac{2 \cdot 4}{3 \cdot 4} = \frac{2}{3}.$$

EXAMPLE 3 **Simplifying fractions**

Simplify each fraction by applying the basic principle of fractions.

(a) $\dfrac{5}{10}$ (b) $-\dfrac{36}{48}$

Solution

(a) $\dfrac{5}{10} = \dfrac{1 \cdot 5}{2 \cdot 5} = \dfrac{1}{2}$ (b) $-\dfrac{36}{48} = -\dfrac{3 \cdot 12}{4 \cdot 12} = -\dfrac{3}{4}$

Now Try Exercises 39, 43

We can also apply this basic principle to rational expressions. For example,

$$\frac{x(x-1)}{4(x-1)} = \frac{x}{4},$$

provided that $x \neq 1$.

NOTE: The simplification at the bottom of the previous page is not valid when $x = 1$ because the expression is undefined for this x-value. When simplifying a rational expression, *we assume that values of the variable that make the rational expression undefined are excluded*, unless stated otherwise.

BASIC PRINCIPLE OF RATIONAL EXPRESSIONS

The following property can be used to simplify rational expressions, where P, Q, and R are polynomials.

$$\frac{P \cdot R}{Q \cdot R} = \frac{P}{Q} \qquad Q \text{ and } R \text{ are nonzero.}$$

NOTE: $\frac{P \cdot R}{Q \cdot R} = \frac{P}{Q} \cdot \frac{R}{R} = \frac{P}{Q} \cdot \mathbf{1} = \frac{P}{Q}$, provided that $Q \neq 0$ and $R \neq 0$.

Like fractions, rational expressions can be written in *lowest terms*. For example, the rational expression $\frac{x^2 - 1}{x^2 + 2x + 1}$ can be written in lowest terms by factoring the numerator and the denominator and then applying the basic principle of rational expressions.

READING CHECK

• How do you know when a rational expression is written in lowest terms?

$$\frac{x^2 - 1}{x^2 + 2x + 1} = \frac{(x - 1)(x + 1)}{(x + 1)(x + 1)} \qquad \text{Factor the numerator and the denominator.}$$

$$= \frac{x - 1}{x + 1} \qquad \text{Apply } \frac{PR}{QR} = \frac{P}{Q} \text{ with } R = x + 1.$$

Because the basic principle of rational expressions cannot be applied further to $\frac{x - 1}{x + 1}$, we say that this expression is written in **lowest terms**.

EXAMPLE 4   **Simplifying rational expressions**

Simplify each expression.

(a) $\dfrac{8y}{4y^2}$ **(b)** $\dfrac{2x + 6}{3x + 9}$ **(c)** $\dfrac{(z + 1)(z - 5)}{(z - 5)(z + 3)}$ **(d)** $\dfrac{x^2 - 9}{2x^2 + 7x + 3}$

Solution
(a) Factor out the greatest common factor, $4y$, in the numerator and the denominator.

$$\frac{8y}{4y^2} = \frac{2 \cdot \mathbf{4y}}{y \cdot \mathbf{4y}} = \frac{2}{y} \qquad \text{Apply } \frac{PR}{QR} = \frac{P}{Q} \text{ with } R = 4y.$$

(b) Start by factoring the numerator and denominator.

$$\frac{2x + 6}{3x + 9} = \frac{2(x + 3)}{3(x + 3)} = \frac{2}{3} \qquad \text{Apply } \frac{PR}{QR} = \frac{P}{Q} \text{ with } R = x + 3.$$

(c) The commutative property allows us to write $\frac{PR}{QR}$ as $\frac{PR}{RQ}$.

$$\frac{(z + 1)(z - 5)}{(z - 5)(z + 3)} = \frac{z + 1}{z + 3} \qquad \text{Apply } \frac{PR}{RQ} = \frac{P}{Q} \text{ with } R = z - 5.$$

(d) Start by factoring the numerator and the denominator.

$$\frac{x^2 - 9}{2x^2 + 7x + 3} = \frac{(x - 3)(x + 3)}{(2x + 1)(x + 3)} = \frac{x - 3}{2x + 1} \qquad \text{Apply } \frac{PR}{QR} = \frac{P}{Q} \text{ with } R = x + 3.$$

Now Try Exercises 51, 55, 61, 79

Expressions and Equations

Expressions and equations are different concepts. An expression does not contain an equals sign, whereas an equation is a statement that two expressions are equal and *always* contains an equals sign. For example,

$$\frac{x}{x+4} \quad \text{and} \quad \frac{2}{x}$$

are two rational *expressions*, and

$$\frac{x}{x+4} = \frac{2}{x} \qquad \begin{array}{l}\text{An equation must contain}\\ \text{an equals sign.}\end{array}$$

is a rational *equation*. In this section we evaluate and simplify rational expressions. Later, we will *solve* rational equations by finding *x*-values that make the equations true.

READING CHECK

• How do rational expressions and rational equations differ?

A negative sign can be placed in a fraction in a number of ways. For example,

$$-\frac{5}{7} = \frac{-5}{7} = \frac{5}{-7}$$

illustrates three fractions that are equal. This property can also be applied to rational expressions, as demonstrated in the next example.

EXAMPLE 5 **Distributing a negative sign**

Simplify each expression.

(a) $\dfrac{-x-6}{2x+12}$ (b) $\dfrac{10-z}{z-10}$ (c) $-\dfrac{5-x}{x-5}$

Solution

(a) Factor -1 out of the numerator and 2 out of the denominator.

$$\frac{-x-6}{2x+12} = \frac{-1(x+6)}{2(x+6)} = -\frac{1}{2}$$

(b) Factor -1 out of the numerator.

$$\frac{10-z}{z-10} = \frac{-1(-10+z)}{z-10} = \frac{-1(z-10)}{z-10} = -1$$

(c) Rewrite the expression with the negative sign in the numerator and then apply the distributive property. Be sure to include parentheses around the numerator.

$$-\frac{5-x}{x-5} = \frac{-(5-x)}{x-5} = \frac{-5+x}{x-5} = \frac{x-5}{x-5} = 1$$

The same answer can be obtained by distributing the negative sign in the denominator.

$$-\frac{5-x}{x-5} = \frac{5-x}{-(x-5)} = \frac{5-x}{-x+5} = \frac{5-x}{5-x} = 1$$

Now Try Exercises 63, 67, 71

NOTE: The result for Example 5(b) becomes more obvious if we substitute a number for z. For example, if we let $z = 6$, then

$$\frac{10 - z}{z - 10} = \frac{10 - 6}{6 - 10} = \frac{4}{-4} = -1.$$

MAKING CONNECTIONS

Negative Signs and Rational Expressions

In general, $(b - a)$ equals $-1(a - b)$. Thus if $a \neq b$, then $\frac{b - a}{a - b} = -1$.

Applications

▶ **REAL-WORLD CONNECTION** The next example is based on the discussion in A Look Into Math for this section and illustrates modeling traffic flow with a rational expression.

EXAMPLE 6 **Modeling traffic flow**

Suppose that 10 cars per minute can pass through a construction zone. If traffic arrives *randomly* at an average rate of x cars per minute, the average time T in minutes spent waiting in line and passing through the construction zone is given by

$$T = \frac{1}{10 - x},$$

where $x < 10$. (*Source:* N. Garber and L. Hoel, *Traffic and Highway Engineering.*)

(a) Complete Table 7.1 by finding T for each value of x.

TABLE 7.1 Waiting in Traffic

x (cars/minute)	5	7	9	9.5	9.9	9.99
T (minutes)						

(b) Interpret the results.

Solution

(a) When $x = 5$ cars per minute, then $T = \frac{1}{10 - 5} = \frac{1}{5}$ minute. Other values are found similarly and are shown in Table 7.2.

TABLE 7.2 Waiting in Traffic

x (cars/minute)	5	7	9	9.5	9.9	9.99
T (minutes)	$\frac{1}{5}$	$\frac{1}{3}$	1	2	10	100

(b) As the average traffic rate increases from 9 cars per minute to 9.9 cars per minute, the time needed to pass through the construction zone increases from 1 minute to 10 minutes. As x nears 10 cars per minute, a small increase in x increases the waiting time dramatically.

Now Try Exercise 99

This nonlinear effect for traffic congestion in Example 6 is shown in Figure 7.1, where points from Table 7.2 have been plotted and a curve passing through them has been sketched. A vertical dashed line was also sketched at $x = 10$. This dashed line is called a

vertical asymptote and indicates that the rational expression is undefined at this value of x. Near the left side of the vertical asymptote, the waiting time T increases dramatically for small increases in x. The graph of T does not intersect or cross this vertical asymptote.

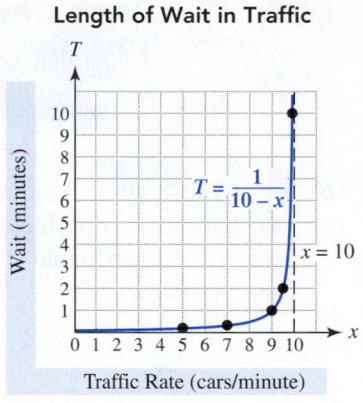

Figure 7.1

TECHNOLOGY NOTE

Making Tables

Table 7.2 can also be created with a graphing calculator by using the Ask feature, as illustrated in the following displays.

CALCULATOR HELP

To make a table of values, see Appendix A (pages AP-2 and AP-3).

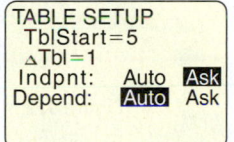

▶ **REAL-WORLD CONNECTION** When a new species of animal is introduced into an area that it did not previously inhabit, its population may grow quickly at first, and then level off over time. Rational expressions can be used to model such situations, as demonstrated in the next example.

EXAMPLE 7   **Modeling a fish population**

Suppose that a small fish species is introduced into a pond that had not previously held this type of fish, and that its population P in thousands is modeled by

$$P = \frac{3x + 1}{x + 4},$$

where $x \geq 0$ represents time in months.

(a) Complete Table 7.3 by finding P for each value of x. Round to 3 decimal places.

TABLE 7.3 Fish Population

x (months)	0	6	12	36	72
P (thousands)					

(b) How many fish were initially introduced into the pond?
(c) Interpret the results shown in your completed table.

Solution

(a) When $x = 0$, the population is $P = \frac{3(0) + 1}{0 + 4} = \frac{1}{4} = 0.25$ thousand fish. The other values are found similarly and are shown in Table 7.4.

TABLE 7.4 Fish Population

x (months)	0	6	12	36	72
P (thousands)	0.25	1.9	2.313	2.725	2.855

(b) Table 7.4 shows that initially (when $x = 0$) there were 0.25 thousand, or 250 fish.

(c) The fish population increased quickly at first but then leveled off. This population growth is shown graphically in Figure 7.2, where the population appears to be leveling off at 3 thousand fish.

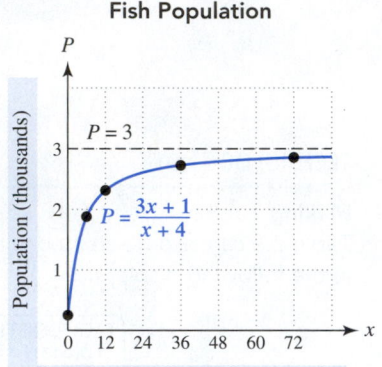

Fish Population

Figure 7.2

Now Try Exercise 101

AN APPLICATION INVOLVING PROBABILITY (OPTIONAL) If 10 marbles, one blue and nine red, are placed in a jar, then the *probability*, or *likelihood,* of picking the blue marble at random is 1 *chance* in 10, or $\frac{1}{10}$. The probability of drawing a red marble at random is 9 chances in 10, or $\frac{9}{10}$. **Probability** is a real number from 0 to 1. A probability of 0, or 0%, indicates that an event is impossible, whereas a probability of 1, or 100%, indicates that an event is certain. Rational expressions are often used to describe probability.

EXAMPLE 8   **Calculating probability**

Suppose that n balls, numbered 1 to n, are placed in a container and only one ball has the winning number.

(a) What is the probability of drawing the winning ball at random?

(b) Calculate this probability for $n = 100$, 1000, and 10,000.

(c) What happens to the probability of drawing the winning ball as the number of balls increases?

Solution

(a) There is 1 chance in n of drawing the winning ball, so the probability is $\frac{1}{n}$.

(b) For $n = 100$, 1000, and 10,000, the probabilities are $\frac{1}{100}$, $\frac{1}{1000}$, and $\frac{1}{10,000}$.

(c) As the number of balls increases, the probability of picking the winning ball decreases.

Now Try Exercise 105

7.1 Putting It All Together

CONCEPT	EXPLANATION	EXAMPLES
Rational Expression	An expression of the form $\frac{P}{Q}$, where P and Q are polynomials with $Q \neq 0$	$\frac{1}{x}$, $\frac{x-3}{2x^2-1}$, $\frac{2x+9}{5x}$, and $\frac{x^2+3x-5}{1}$
Undefined Rational Expressions	A rational expression is undefined for any value of the variable that makes the *denominator* equal to 0.	$\frac{1}{x-3}$ is undefined when $x = 3$. $\frac{5y}{y^2-1}$ is undefined when $y = 1$ or when $y = -1$.
Basic Principle of Rational Expressions	Factor the numerator and the denominator completely. Then apply $$\frac{P \cdot R}{Q \cdot R} = \frac{P}{Q}.$$	$\frac{4xy^2}{6xy^3} = \frac{2(2xy^2)}{3y(2xy^2)} = \frac{2}{3y}$ $\frac{4x(x-4)}{(x+1)(x-4)} = \frac{4x}{x+1}$

7.1 Exercises

 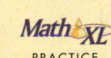

CONCEPTS AND VOCABULARY

1. A rational expression can be written as _____, where P and Q are _____ with $Q \neq 0$.

2. Is $\frac{x}{2x^2+1}$ a rational expression? Why or why not?

3. A rational expression is undefined whenever the _____ is equal to 0.

4. The rational expression $\frac{1}{x-a}$ is undefined whenever $x =$ _____.

5. The basic principle of fractions states that a fraction can be simplified by using $\frac{a \cdot c}{b \cdot c} =$ _____.

6. The basic principle of rational expressions can be used to simplify _____ expressions.

EVALUATING RATIONAL EXPRESSIONS

Exercises 7–20: If possible, evaluate the expression for the given value of the variable.

7. $\frac{3}{x}$; $x = -7$

8. $\frac{3}{x+3}$; $x = 0$

9. $-\frac{x}{x-5}$; $x = -4$

10. $-\frac{4x}{5x+1}$; $x = 1$

11. $\frac{y+1}{y^2}$; $y = -2$

12. $\frac{3y-1}{y^2+1}$; $y = -1$

13. $\frac{7z}{z^2-4}$; $z = -2$

14. $\frac{5}{z^2-3z+2}$; $z = -1$

15. $\frac{5}{3t+6}$; $t = -2$

16. $\frac{4t}{2t+5}$; $t = -\frac{5}{2}$

17. $\frac{4-x}{x-4}$; $x = -2$

18. $\frac{x-7}{7-x}$; $x = 4$

19. $-\frac{6-x}{x-6}$; $x = 0$

20. $\frac{8-2x}{2x-8}$; $x = -5$

Exercises 21–24: Complete the table for the given expression. If a value is undefined, place a dash in the table.

21.

x	-2	-1	0	1	2
$\frac{x}{x+1}$					

22.

x	-2	-1	0	1	2
$\frac{2x}{3x-1}$					

23.

x	-2	-1	0	1	2
$\dfrac{3x}{2x^2 + 1}$					

24.

x	-2	-1	0	1	2
$\dfrac{2x - 1}{x^2 - 1}$					

Exercises 25–38: Find any values of the variable that make the expression undefined.

25. $-\dfrac{8}{x}$

26. $\dfrac{7}{x + 1}$

27. $\dfrac{4}{z - 3}$

28. $\dfrac{7 - z}{z - 7}$

29. $\dfrac{4y}{5y + 4}$

30. $\dfrac{3 + y}{3y - 7}$

31. $\dfrac{5t + 2}{t^2 + 1}$

32. $\dfrac{8t}{t^2 + 25}$

33. $\dfrac{8x}{x^2 - 25}$

34. $\dfrac{x + 4}{x^2 - 36}$

35. $\dfrac{x^2 + 3x + 2}{x^2 + 5x + 6}$

36. $\dfrac{2x - 1}{x^2 - 7x + 10}$

37. $\dfrac{8z^2 + z + 1}{2z^2 - 7z + 5}$

38. $\dfrac{4n^2 + 17n - 15}{3n^2 - 8n + 4}$

SIMPLIFYING RATIONAL EXPRESSIONS

Exercises 39–46: Simplify the fraction to lowest terms.

39. $\frac{12}{18}$

40. $\frac{24}{32}$

41. $\frac{24}{48}$

42. $-\frac{22}{33}$

43. $-\frac{6}{15}$

44. $\frac{8}{22}$

45. $-\frac{25}{75}$

46. $-\frac{36}{42}$

Exercises 47–50: First simplify the fraction in part (a), then simplify the rational expression in part (b).

47. (a) $\dfrac{8}{16}$ **(b)** $\dfrac{x + 2}{2x + 4}$

48. (a) $\dfrac{6}{9}$ **(b)** $\dfrac{4x + 12}{6x + 18}$

49. (a) $\dfrac{7 - 3}{3 - 7}$ **(b)** $\dfrac{7 - x}{x - 7}$

50. (a) $-\dfrac{8 - 5}{5 - 8}$ **(b)** $-\dfrac{x - 5}{5 - x}$

Exercises 51–88: Simplify the expression.

51. $\dfrac{5x^4}{10x^6}$

52. $\dfrac{6y^2}{9y}$

53. $\dfrac{8xy^3}{6x^2y^2}$

54. $\dfrac{36x^2y^5}{6x^5y}$

55. $\dfrac{x + 4}{2x + 8}$

56. $\dfrac{5x - 10}{x - 2}$

57. $\dfrac{3z - 9}{5z - 15}$

58. $\dfrac{4z + 8}{10 + 5z}$

59. $\dfrac{(x + 1)(x - 1)}{(x + 6)(x - 1)}$

60. $\dfrac{(2x + 1)(x + 9)}{(4x - 3)(x + 9)}$

61. $\dfrac{(5y + 3)(2y - 1)}{(2y - 1)(y + 2)}$

62. $\dfrac{(4y - 1)(5y + 7)}{(5y + 7)(1 - 4y)}$

63. $\dfrac{x - 7}{7 - x}$

64. $\dfrac{5 - x}{x - 5}$

65. $\dfrac{a - b}{b - a}$

66. $\dfrac{2t - 3r}{3r - 2t}$

67. $\dfrac{-6 - x}{18 + 3x}$

68. $\dfrac{-2x - 6}{x + 3}$

69. $\dfrac{x + 1}{-2x - 2}$

70. $\dfrac{3x + 21}{-7 - x}$

71. $-\dfrac{9 - x}{x - 9}$

72. $-\dfrac{4 - 2x}{x - 2}$

73. $\dfrac{(3x + 5)(x - 1)}{(3x - 5)(1 - x)}$

74. $\dfrac{(2 - x)(x - 2)}{(x - 2)(2 - x)}$

75. $\dfrac{n^2 - n}{n^2 - 5n}$

76. $\dfrac{3n^2 - 4n}{n^2 + 4n}$

77. $\dfrac{x^2 - 3x}{6x - 18}$

78. $\dfrac{4x^2 + 16x}{5x^2 + 20x}$

79. $\dfrac{z^2 - 3z + 2}{z^2 - 4z + 3}$

80. $\dfrac{z^2 - 3z - 10}{z^2 - 2z - 8}$

81. $\dfrac{2x^2 + 7x - 4}{6x^2 + x - 2}$

82. $\dfrac{5x^2 + 3x - 2}{5x^2 + 13x - 6}$

83. $\dfrac{x - 3}{3x^2 - 11x + 6}$

84. $\dfrac{2x - 1}{4x^2 + 6x - 4}$

85. $-\dfrac{a - 9}{9 - a}$

86. $-\dfrac{-b - 6}{b + 6}$

87. $\dfrac{-2x - 1}{4x + 2}$

88. $\dfrac{4x + 3}{-8x - 6}$

Exercises 89 and 90: **Thinking Generally** *Complete the statement involving a rational expression.*

89. The expression $\frac{x-a}{a-x}$ simplifies to _____.

90. The expression $-\frac{x-a}{x+a}$ simplifies to $\frac{?}{x+a}$.

Exercises 91–98: Refer to Making Connections on page 423.
   (a) *Decide whether you are given an expression or an equation.*
   (b) *If you are given an expression, simplify it. If you are given an equation, solve it.*

91. $x + 1 = 7$

92. $x^2 - 4 = 0$

93. $\dfrac{x}{x(x+1)}$

94. $\dfrac{x-2}{(x-2)(x-8)}$

95. $\dfrac{x^2-4}{x+2}$

96. $\dfrac{x-4}{8-2x}$

97. $\dfrac{x}{2(1+3)} = 1$

98. $\dfrac{x}{2} + 2 = \dfrac{8}{2}$

APPLICATIONS

99. *Modeling Traffic Flow* (Refer to Example 6.) Five vehicles per minute can pass through a construction zone. If the traffic arrives randomly at an average rate of x vehicles per minute, the average time T in minutes spent waiting in line and passing through the construction zone is given by $T = \frac{1}{5-x}$ for $x < 5$. (*Source:* N. Garber.)
   (a) Evaluate T for $x = 3$ and interpret the result.
   (b) Complete the table and interpret the results.

x	2	4	4.5	4.9	4.99
T					

100. *Standing in Line* A worker at a poolside store can serve 20 customers per hour. If children arrive randomly at an average rate of x per hour, then the average number of children N waiting in line is given by $N = \frac{x^2}{400 - 20x}$ for $x < 20$. (*Source:* N. Garber.)
   (a) Complete the table.

x	5	10	18	19	20
N					

   (b) Compare the number of children waiting in line if the average rate increases from 18 to 19 children per hour.

101. *Frog Population* Suppose that a frog species is introduced into a wetland area and its population in hundreds is modeled by

$$P = \frac{7x + 3}{x + 6},$$

where $x \geq 0$ is time in months.
   (a) Complete the table by finding P for each given value of x. Round to 2 decimal places.

x (months)	0	12	36	72
P (hundreds)				

   (b) What was the initial frog population?
   (c) Interpret the results in your completed table.

102. *Insect Population* Suppose that an insect population in thousands per acre is modeled by

$$P = \frac{5x + 2}{x + 1},$$

where $x \geq 0$ is time in months.
   (a) Complete the table by finding P for each given value of x. Round to 3 decimal places.

x (months)	0	12	36	60
P (thousands)				

   (b) What was the initial insect population?
   (c) Interpret the results in your completed table.

103. *Probability* Suppose that a coin is flipped. What is the probability that a head appears?

104. *Probability* A die shows the numbers 1, 2, 3, 4, 5, and 6. If each number has an equal chance of appearing on any given roll, what is the probability that a 2 or 4 appears?

105. *Probability* (Refer to Example 8.) Suppose that there are n balls in a container and that three balls have a winning number. If a ball is drawn randomly, do each of the following.
   (a) Write a rational expression that gives the probability of drawing a winning ball.
   (b) Write a rational expression that gives the probability of not drawing a winning ball. Evaluate your expression for $n = 100$ and interpret the result.

106. *Surface Area of a Cylinder* If a cylindrical container has a volume of π cubic feet, then its surface area S in square feet (excluding the top and bottom) is given by $S = \frac{2\pi}{r}$, where r is the radius of the cylinder.

(a) Calculate S when $r = \frac{1}{2}$ foot.
(b) What happens to this surface area when r becomes large? Sketch this situation.
(c) What happens to the surface area when r becomes small (nearly 0)? Sketch this situation.

107. *Distance and Time* A car is traveling at 60 miles per hour.
(a) How long does it take the car to travel 360 miles?
(b) Write a rational expression that gives the time that it takes the car to travel M miles.

108. *Distance and Time* A bicyclist rides uphill at 10 miles per hour for 5 miles and then rides downhill at 20 miles per hour for 5 miles. What is the bicyclist's average speed? (*Hint:* Average speed equals distance divided by time.)

Exercises 109 and 110: Traffic Flow (Refer to Example 6.) The figure shows a graph of the waiting time T in minutes at a construction zone when cars are arriving randomly at an average rate of x cars per minute.

(a) *Give the equation of the vertical asymptote.*
(b) *Explain how the graph relates to traffic flow.*

109.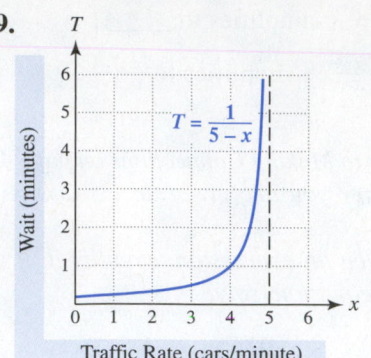

Traffic Rate (cars/minute)

110.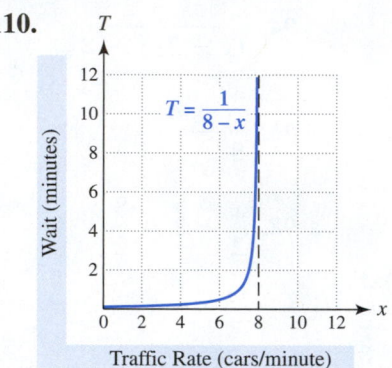

Traffic Rate (cars/minute)

WRITING ABOUT MATHEMATICS

111. What is a rational expression? When is a rational expression undefined?

112. Does the rational expression $\frac{5x + 2}{10x + 4}$ equal $\frac{5x}{10x} + \frac{2}{4}$? Explain your answer.

Group Activity   Working with Real Data

Directions: Form a group of 2 to 4 people. Select someone to record the group's responses for this activity. All members of the group should work cooperatively to answer the questions. If your instructor asks for your results, each member of the group should be prepared to respond.

Students Per Computer In the early years of personal computers, school districts could not afford to buy a computer for every student. As the price of computers decreased, more and more school districts were able to move toward this goal. The following table lists numbers of students per computer during these early years.

Year	1983	1985	1987	1989
Students/Computer	125	50	32	22

Year	1991	1993	1995	1997
Students/Computer	18	14	10	6

Source: Quality Education Data, Inc.

(a) Make a scatterplot of the data. Would a straight line model the data accurately? Explain.
(b) Discuss how well the formula

$$S = \frac{125}{1 + 0.7(x - 1983)}, \quad x \geq 1983$$

models these data, where S represents the students per computer and x represents the year.
(c) In what year does the formula suggest that there were about 17 students per computer?

7.2 Multiplication and Division of Rational Expressions

Review of Multiplication and Division of Fractions • Multiplication of Rational Expressions • Division of Rational Expressions

A LOOK INTO MATH ▶

Stopping distance for a car can vary depending on the road conditions. If the road is slippery, more distance is needed to stop. Also, cars that are traveling downhill require additional stopping distance. Rational expressions are frequently used by highway engineers to estimate the stopping distance of a car on slippery surfaces or on hills. (See Example 4.)

In previous chapters we reviewed how to add, subtract, multiply, and divide real numbers and polynomials. In this section we show how to multiply and divide rational expressions; in the next section we discuss addition and subtraction of rational expressions.

Review of Multiplication and Division of Fractions

To multiply two fractions we use the property

$$\frac{a}{b} \cdot \frac{c}{d} = \frac{ac}{bd}.$$

In the next example we review multiplication of fractions. (For a full review of fractions, see Section 1.2.)

EXAMPLE 1 **Multiplying fractions**

Multiply and simplify your answers to lowest terms.

(a) $\frac{3}{7} \cdot \frac{4}{5}$ (b) $2 \cdot \frac{3}{4}$ (c) $\frac{4}{21} \cdot \frac{7}{8}$

Solution

(a) $\frac{3}{7} \cdot \frac{4}{5} = \frac{12}{35}$ (b) $2 \cdot \frac{3}{4} = \frac{2}{1} \cdot \frac{3}{4} = \frac{6}{4} = \frac{3}{2}$

(c) $\frac{4}{21} \cdot \frac{7}{8} = \frac{7 \cdot 4}{21 \cdot 8} = \frac{7}{21} \cdot \frac{4}{8} = \frac{1}{3} \cdot \frac{1}{2} = \frac{1}{6}$

Now Try Exercises 5, 7, 9

To divide two fractions we change the division problem to a multiplication problem by using the property

$$\frac{a}{b} \div \frac{c}{d} = \frac{a}{b} \cdot \frac{d}{c}.$$

EXAMPLE 2 **Dividing fractions**

Divide and simplify your answers to lowest terms.

(a) $\frac{1}{3} \div \frac{5}{7}$ (b) $\frac{4}{5} \div 8$ (c) $\frac{8}{9} \div \frac{10}{3}$

Solution

(a) $\frac{1}{3} \div \frac{5}{7} = \frac{1}{3} \cdot \frac{7}{5} = \frac{7}{15}$

(b) $\frac{4}{5} \div 8 = \frac{4}{5} \cdot \frac{1}{8} = \frac{4}{40} = \frac{1}{10}$

(c) $\frac{8}{9} \div \frac{10}{3} = \frac{8}{9} \cdot \frac{3}{10} = \frac{24}{90} = \frac{4 \cdot 6}{15 \cdot 6} = \frac{4}{15}$

Now Try Exercises 13, 15, 17

Multiplication of Rational Expressions

Multiplying rational expressions is similar to multiplying fractions.

PRODUCTS OF RATIONAL EXPRESSIONS

To multiply two rational expressions, multiply the numerators and multiply the denominators. That is,

$$\frac{A}{B} \cdot \frac{C}{D} = \frac{AC}{BD},$$

where B and D are nonzero.

READING CHECK

- How do we multiply rational expressions?

EXAMPLE 3 **Multiplying rational expressions**

Multiply and simplify to lowest terms. Leave your answers in factored form.

(a) $\dfrac{3}{x} \cdot \dfrac{2x - 5}{x - 1}$ (b) $\dfrac{x - 1}{4x} \cdot \dfrac{x + 3}{x - 1}$

(c) $\dfrac{x^2 - 4}{x + 3} \cdot \dfrac{x + 3}{x + 2}$ (d) $\dfrac{4}{x^2 + 3x + 2} \cdot \dfrac{x^2 + 2x + 1}{8}$

Solution

(a) $\dfrac{3}{x} \cdot \dfrac{2x - 5}{x - 1} = \dfrac{3(2x - 5)}{x(x - 1)}$ Multiply the numerators and the denominators.

(b) $\dfrac{x - 1}{4x} \cdot \dfrac{x + 3}{x - 1} = \dfrac{(x - 1)(x + 3)}{4x(x - 1)}$ Multiply the numerators and the denominators.

$= \dfrac{x + 3}{4x}$ Simplify.

(c) $\dfrac{x^2 - 4}{x + 3} \cdot \dfrac{x + 3}{x + 2} = \dfrac{(x - 2)(x + 2)}{x + 3} \cdot \dfrac{x + 3}{x + 2}$ Factor.

$= \dfrac{(x - 2)(x + 2)(x + 3)}{(x + 3)(x + 2)}$ Multiply the numerators and the denominators.

$= x - 2$ Simplify.

(d) $\dfrac{4}{x^2 + 3x + 2} \cdot \dfrac{x^2 + 2x + 1}{8} = \dfrac{4(x^2 + 2x + 1)}{8(x^2 + 3x + 2)}$ Multiply the numerators and the denominators.

$= \dfrac{4(x + 1)(x + 1)}{8(x + 2)(x + 1)}$ Factor.

$= \dfrac{x + 1}{2(x + 2)}$ Simplify.

STUDY TIP

Have you been to your instructor's office lately? Visit with your instructor about your progress in the class to be sure that your work is complete and your grades are up-to-date.

▪ **Now Try Exercises 29, 31, 41, 45**

▶ **REAL-WORLD CONNECTION** The next example illustrates a rational expression that is used in highway design.

EXAMPLE 4 **Estimating stopping distance**

If a car is traveling at 60 miles per hour on a slippery road, then its stopping distance D in feet can be calculated by

$$D = \frac{3600}{30} \cdot \frac{1}{x},$$

where x is the coefficient of friction between the tires and the road and $0 < x \le 1$. The more slippery the road is, the smaller the value of x. (*Source:* L. Haefner, *Introduction to Transportation Systems.*)
(a) Multiply and simplify the formula for D.
(b) Compare the stopping distance on an icy road with $x = 0.1$ to the stopping distance on dry pavement with $x = 0.4$.

Solution
(a) Because

$$\frac{3600}{30} \cdot \frac{1}{x} = \frac{3600}{30x} = \frac{120 \cdot 30}{x \cdot 30} = \frac{120}{x},$$

it follows that $D = \frac{120}{x}$.
(b) When $x = \mathbf{0.1}$, $D = \frac{120}{0.1} = 1200$ feet, and when $x = \mathbf{0.4}$, $D = \frac{120}{0.4} = 300$ feet. A car traveling at 60 miles per hour on an icy road requires a stopping distance that is 4 times that of a car traveling at the same speed on dry pavement.

Now Try Exercise 73

Division of Rational Expressions

Dividing rational expressions is similar to dividing fractions.

QUOTIENTS OF RATIONAL EXPRESSIONS

READING CHECK

• How do we divide rational expressions?

To divide two rational expressions, multiply by the reciprocal of the divisor. That is,

$$\frac{A}{B} \div \frac{C}{D} = \frac{A}{B} \cdot \frac{D}{C},$$

where B, C, and D are nonzero.

EXAMPLE 5 **Dividing rational expressions**

Divide and simplify to lowest terms.
(a) $\frac{5}{2x} \div \frac{10}{x - 4}$ (b) $\frac{x^2 - 9}{x^2 + 4} \div (x - 3)$ (c) $\frac{x^2 - x}{x^2 - x - 2} \div \frac{x}{x - 2}$

Solution
(a)
$$\frac{5}{2x} \div \frac{10}{x - 4} = \frac{5}{2x} \cdot \frac{x - 4}{10}$$
Multiply by the reciprocal of the divisor.

$$= \frac{5(x - 4)}{20x}$$
Multiply the numerators and the denominators.

$$= \frac{x - 4}{4x}$$
Simplify. Note that $\frac{5}{20} = \frac{1}{4}$.

(b) $\dfrac{x^2 - 9}{x^2 + 4} \div (x - 3) = \dfrac{x^2 - 9}{x^2 + 4} \cdot \dfrac{1}{x - 3}$ Multiply by the reciprocal of the divisor.

$= \dfrac{x^2 - 9}{(x^2 + 4)(x - 3)}$ Multiply the numerators and the denominators.

$= \dfrac{(x + 3)(x - 3)}{(x^2 + 4)(x - 3)}$ Factor the numerator.

$= \dfrac{x + 3}{x^2 + 4}$ Simplify.

(c) $\dfrac{x^2 - x}{x^2 - x - 2} \div \dfrac{x}{x - 2} = \dfrac{x^2 - x}{x^2 - x - 2} \cdot \dfrac{x - 2}{x}$ Multiply by the reciprocal of the divisor.

$= \dfrac{(x^2 - x)(x - 2)}{(x^2 - x - 2)x}$ Multiply the numerators and the denominators.

$= \dfrac{x(x - 1)(x - 2)}{x(x + 1)(x - 2)}$ Factor the numerator and the denominator.

$= \dfrac{x - 1}{x + 1}$ Simplify.

▮ **Now Try Exercises 49, 57, 65**

7.2 Putting It All Together

CONCEPT	EXPLANATION	EXAMPLES
Multiplication of Rational Expressions	If A, B, C, and D represent rational expressions where B and D are non-zero, multiply the numerators and multiply the denominators: $$\frac{A}{B} \cdot \frac{C}{D} = \frac{AC}{BD}.$$ Then simplify the result to lowest terms.	$$\frac{4x}{x - 1} \cdot \frac{x - 1}{x + 1} = \frac{4x(x - 1)}{(x - 1)(x + 1)}$$ $$= \frac{4x}{x + 1}$$
Division of Rational Expressions	If A, B, C, and D represent rational expressions where B, C, and D are nonzero, multiply the first expression by the reciprocal of the second expression: $$\frac{A}{B} \div \frac{C}{D} = \frac{A}{B} \cdot \frac{D}{C}.$$ Then simplify the result to lowest terms.	$$\frac{x + 1}{x - 3} \div \frac{x + 1}{x - 5} = \frac{x + 1}{x - 3} \cdot \frac{x - 5}{x + 1}$$ $$= \frac{(x + 1)(x - 5)}{(x - 3)(x + 1)}$$ $$= \frac{x - 5}{x - 3}$$

7.2 Exercises

MyMathLab  PRACTICE WATCH DOWNLOAD READ REVIEW

CONCEPTS AND VOCABULARY

1. To multiply rational expressions, multiply the _____ and multiply the _____.

2. To divide two rational expressions, multiply the first expression by the _____ of the second expression.

3. $\dfrac{A}{B} \cdot \dfrac{C}{D} =$ _____

4. $\dfrac{A}{B} \div \dfrac{C}{D} =$ _____

REVIEW OF FRACTIONS

Exercises 5–20: Multiply or divide as indicated. Simplify to lowest terms.

5. $\dfrac{1}{2} \cdot \dfrac{4}{5}$

6. $\dfrac{6}{7} \cdot \dfrac{7}{18}$

7. $\dfrac{3}{7} \cdot 4$

8. $5 \cdot \dfrac{4}{5}$

9. $\dfrac{5}{4} \cdot \dfrac{8}{15}$

10. $\dfrac{3}{10} \cdot \dfrac{5}{9}$

11. $\dfrac{1}{3} \cdot \dfrac{2}{3} \cdot \dfrac{9}{11}$

12. $\dfrac{2}{5} \cdot \dfrac{10}{11} \cdot \dfrac{1}{4}$

13. $\dfrac{2}{3} \div \dfrac{1}{6}$

14. $\dfrac{5}{7} \div \dfrac{5}{8}$

15. $\dfrac{8}{9} \div \dfrac{5}{3}$

16. $\dfrac{7}{3} \div 6$

17. $8 \div \dfrac{4}{5}$

18. $\dfrac{1}{2} \div \dfrac{5}{4} \div \dfrac{2}{5}$

19. $\dfrac{4}{5} \div \dfrac{2}{3} \div \dfrac{1}{2}$

20. $\dfrac{7}{20} \div \dfrac{14}{5}$

MULTIPLYING RATIONAL EXPRESSIONS

Exercises 21–28: Simplify the expression.

21. $\dfrac{x+5}{x+5}$

22. $\dfrac{2x-3}{2x-3}$

23. $\dfrac{(z+1)(z+2)}{(z+4)(z+2)}$

24. $\dfrac{(2z-7)(z+5)}{(3z+5)(z+5)}$

25. $\dfrac{8y(y+7)}{12y(y+7)}$

26. $\dfrac{6(y+1)}{12(y+1)}$

27. $\dfrac{x(x+2)(x+3)}{x(x-2)(x+3)}$

28. $\dfrac{2(x+1)(x-1)}{4(x+1)(x-1)}$

Exercises 29–48: Multiply and simplify to lowest terms. Leave your answers in factored form.

29. $\dfrac{8}{x} \cdot \dfrac{x+1}{x}$

30. $\dfrac{7}{2x} \cdot \dfrac{x}{x-1}$

31. $\dfrac{8+x}{x} \cdot \dfrac{x-3}{x+8}$

32. $\dfrac{5x^2+x}{2x-1} \cdot \dfrac{1}{x}$

33. $\dfrac{z+3}{z+4} \cdot \dfrac{z+4}{z-7}$

34. $\dfrac{2z+1}{3z} \cdot \dfrac{3z}{z+2}$

35. $\dfrac{5x+1}{3x+2} \cdot \dfrac{3x+2}{5x+1}$

36. $\dfrac{x+1}{x+3} \cdot \dfrac{x+3}{x+1}$

37. $\dfrac{(t+1)^2}{t+2} \cdot \dfrac{(t+2)^2}{t+1}$

38. $\dfrac{(t-1)^2}{(t+5)^2} \cdot \dfrac{t+5}{t-1}$

39. $\dfrac{x^2}{x^2+4} \cdot \dfrac{x+4}{x}$

40. $\dfrac{x-1}{x^2} \cdot \dfrac{x^2}{x^2+1}$

41. $\dfrac{z^2-1}{z^2-4} \cdot \dfrac{z-2}{z+1}$

42. $\dfrac{z^2-9}{z-5} \cdot \dfrac{z-5}{z+3}$

43. $\dfrac{y^2-2y}{y^2-1} \cdot \dfrac{y+1}{y-2}$

44. $\dfrac{y^2-4y}{y+1} \cdot \dfrac{y+1}{y-4}$

45. $\dfrac{2x^2-x-3}{3x^2-8x-3} \cdot \dfrac{3x+1}{2x-3}$

46. $\dfrac{6x^2+11x-2}{3x^2+11x-4} \cdot \dfrac{3x-1}{6x-1}$

47. $\dfrac{(x-3)^3}{x^2-2x+1} \cdot \dfrac{x-1}{(x-3)^2}$

48. $\dfrac{x^2+4x+4}{x^2-2x+1} \cdot \dfrac{(x-1)^2}{(x+2)^2}$

DIVIDING RATIONAL EXPRESSIONS

Exercises 49–70: Divide and simplify to lowest terms. Leave your answers in factored form.

49. $\dfrac{2}{x} \div \dfrac{2x+3}{x}$

50. $\dfrac{6}{2x} \div \dfrac{x+2}{2x}$

51. $\dfrac{x-2}{3x} \div \dfrac{2-x}{6x}$

52. $\dfrac{x+1}{2x-1} \div \dfrac{x+1}{x}$

53. $\dfrac{z+2}{z+1} \div \dfrac{z+2}{z-1}$

54. $\dfrac{z+7}{z-4} \div \dfrac{z+7}{z-4}$

55. $\dfrac{3y+4}{2y+1} \div \dfrac{3y+4}{y+2}$

56. $\dfrac{y+5}{y-2} \div \dfrac{y}{y+3}$

57. $\dfrac{t^2-1}{t^2+1} \div \dfrac{t+1}{4}$

58. $\dfrac{4}{2t^3} \div \dfrac{8}{t^2}$

59. $\dfrac{y^2-9}{y^2-25} \div \dfrac{y+3}{y+5}$

60. $\dfrac{y+1}{y-4} \div \dfrac{y^2-1}{y^2-16}$

61. $\dfrac{2x^2-4x}{2x-1} \div \dfrac{x-2}{2x-1}$

62. $\dfrac{x-4}{x^2+x} \div \dfrac{5}{x+1}$

63. $\dfrac{2z^2-5z-3}{z^2+z-20} \div \dfrac{z-3}{z-4}$

64. $\dfrac{z^2+12z+27}{z^2-5z-14} \div \dfrac{z+3}{z+2}$

65. $\dfrac{t^2 - 1}{t^2 + 5t - 6} \div (t + 1)$

66. $\dfrac{t^2 - 2t - 3}{t^2 - 5t - 6} \div (t - 3)$

67. $\dfrac{a - b}{a + b} \div \dfrac{a - b}{2a + 3b}$

68. $\dfrac{x^3 - y^3}{x^2 - y^2} \div \dfrac{x^2 + xy + y^2}{x - y}$

69. $\dfrac{x - y}{x^2 + 2xy + y^2} \div \dfrac{1}{(x + y)^2}$

70. $\dfrac{a^2 - b^2}{4a^2 - 9b^2} \div \dfrac{a - b}{2a + 3b}$

71. Thinking Generally Simplify $\dfrac{a - b}{b - c} \cdot \dfrac{c - b}{b - a}$.

72. Thinking Generally Simplify $\dfrac{a - b}{b - c} \div \dfrac{b - a}{a - b}$.

APPLICATIONS

73. *Stopping on Slippery Roads* (Refer to Example 4.) If a car is traveling at 30 miles per hour on a slippery road, then its stopping distance D in feet can be calculated by

$$D = \dfrac{900}{30} \cdot \dfrac{1}{x},$$

where x is the coefficient of friction between the tires and the road and $0 < x \le 1$. (*Source:* L. Haefner.)
(a) Multiply and simplify the formula for D.
(b) Compare the stopping distance on an icy road with $x = 0.1$ and on dry pavement with $x = 0.4$.

74. *Stopping on Hills* If a car is traveling at 50 miles per hour on a hill with wet pavement, then its stopping distance D is given by

$$D = \dfrac{2500}{30} \cdot \dfrac{1}{x + 0.3},$$

where x equals the slope of the hill. (*Source:* L. Haefner.)
(a) Multiply and simplify the formula for D.
(b) Compare the stopping distance for an uphill slope of $x = 0.1$ to a downhill slope of $x = -0.1$.

75. *Probability* Suppose that one jar holds n balls and that a second jar holds $n + 1$ balls. Each jar contains one winning ball.
(a) The probability, or chance, of drawing the winning ball from the first jar and *not* drawing it from the second jar is

$$\dfrac{1}{n} \cdot \dfrac{n}{n + 1}.$$

Simplify this expression.
(b) Find this probability for $n = 99$.

76. *U.S. AIDS Cases* The cumulative number of AIDS cases C in the United States from 1982 to 1994 can be modeled by $C = 3200x^2 + 1586$, and the cumulative number of AIDS deaths D from 1982 to 1994 can be modeled by $D = 1900x^2 + 619$. In these equations $x = 0$ corresponds to 1982, $x = 1$ corresponds to 1983, and so on until $x = 12$ corresponds to 1994. (*Source:* U.S. Department of Health.)
(a) Write the rational expression $\frac{D}{C}$ in terms of x.
(b) Evaluate your expression for $x = 4, 7$, and 10. Round your answers to the nearest thousandth. Interpret the results.
(c) Explain what the rational expression $\frac{D}{C}$ represents.

WRITING ABOUT MATHEMATICS

77. Explain how to multiply two rational expressions.

78. Explain how to divide two rational expressions.

SECTIONS 7.1 and 7.2	**Checking Basic Concepts**

1. If possible, evaluate the expression $\dfrac{3}{x^2 - 1}$ for $x = -1$ and $x = 3$.

2. Simplify to lowest terms.
(a) $\dfrac{6x^3y^2}{15x^2y^3}$ **(b)** $\dfrac{5x - 15}{x - 3}$ **(c)** $\dfrac{x^2 - x - 6}{x^2 + x - 12}$

3. Multiply and simplify to lowest terms.
(a) $\dfrac{4}{3x} \cdot \dfrac{2x}{6}$ **(b)** $\dfrac{2x + 4}{x^2 - 1} \cdot \dfrac{x + 1}{x + 2}$

4. Divide and simplify to lowest terms.
(a) $\dfrac{7}{3z^2} \div \dfrac{14}{5z^3}$ **(b)** $\dfrac{x^2 + x}{x - 3} \div \dfrac{x}{x - 3}$

5. *Waiting in Line* Customers are waiting in line at a department store. They arrive randomly at an average rate of x per minute. If the clerk can wait on 2 customers per minute, then the average time in minutes spent waiting in line is given by $T = \dfrac{1}{2 - x}$ for $x < 2$. (*Source:* N. Garber, *Traffic and Highway Engineering*.)
(a) Complete the table.

x	0.5	1.0	1.5	1.9
T				

(b) What happens to the waiting time as x increases but remains less than 2?

7.3 Addition and Subtraction with Like Denominators

Review of Addition and Subtraction of Fractions ● Rational Expressions with Like Denominators

A LOOK INTO MATH ▶

When companies manufacture a large number of items, *quality control* is important. For example, suppose that a company makes computer flash drives. Because it is not practical to check every flash drive to make sure that it works properly, inspectors often check a random sample. By using mathematics and rational expressions, this technique helps determine the likelihood that all the flash drives are good. (See Example 6.)

In this section we discuss methods for adding and subtracting rational expressions with like denominators. These methods are similar to the ones used to add and subtract fractions with like denominators.

Review of Addition and Subtraction of Fractions

In Section 1.2 we demonstrated how the property

$$\frac{a}{c} + \frac{b}{c} = \frac{a+b}{c}$$

can be used to add fractions with like denominators. For example,

$$\frac{3}{7} + \frac{2}{7} = \frac{3+2}{7} = \frac{5}{7} \quad \text{Add the numerators.}$$

To subtract two fractions with like denominators, the property

$$\frac{a}{c} - \frac{b}{c} = \frac{a-b}{c}$$

is used. For example,

$$\frac{2}{5} - \frac{4}{5} = \frac{2-4}{5} = -\frac{2}{5} \quad \text{Subtract the numerators.}$$

STUDY TIP

By this time in the semester, it is likely that you know some of your classmates. Have you started or joined a study group? Be sure not to miss the opportunity to study math with your classmates.

EXAMPLE 1 **Adding and subtracting fractions with like denominators**

Simplify each expression to lowest terms.

(a) $\frac{3}{8} + \frac{4}{8}$ **(b)** $\frac{5}{9} + \frac{1}{9}$ **(c)** $\frac{12}{5} - \frac{7}{5}$ **(d)** $\frac{23}{20} - \frac{13}{20}$

Solution

(a) $\frac{3}{8} + \frac{4}{8} = \frac{3+4}{8} = \frac{7}{8}$

(b) $\frac{5}{9} + \frac{1}{9} = \frac{5+1}{9} = \frac{6}{9} = \frac{2}{3}$

(c) $\frac{12}{5} - \frac{7}{5} = \frac{12-7}{5} = \frac{5}{5} = 1$

(d) $\frac{23}{20} - \frac{13}{20} = \frac{23-13}{20} = \frac{10}{20} = \frac{1}{2}$

Now Try Exercises 7, 9, 11, 13

CALCULATOR HELP

To express a result as a fraction on a calculator, see Appendix A (pages AP-1 and AP-2).

TECHNOLOGY NOTE

Arithmetic of Fractions
Many calculators have the capability to perform addition and subtraction of fractions, as illustrated in the following figures. Compare these results with those from Example 1.

```
(3/8)+(4/8)►Frac
              7/8
(5/9)+(1/9)►Frac
              2/3
```

```
(12/5)-(7/5)►Fra
c
              1
(23/20)-(13/20)►
Frac
              1/2
```

Rational Expressions with Like Denominators

Addition and subtraction of rational expressions with like denominators are similar to addition and subtraction of fractions. The following property can be used to add two rational expressions with like denominators.

> ## SUMS OF RATIONAL EXPRESSIONS
>
> To add two rational expressions with like denominators, add their numerators. Keep the same denominator.
>
> $$\frac{A}{C} + \frac{B}{C} = \frac{A+B}{C} \qquad C \text{ is nonzero.}$$

When we add rational expressions with like denominators, we add the numerators. Then we combine like terms and simplify the resulting expression by applying the basic principle of rational expressions. For example, we can add $\frac{2x}{x+1}$ and $\frac{1-x}{x+1}$ as follows.

$$\frac{2x}{x+1} + \frac{1-x}{x+1} = \frac{2x+1-x}{x+1} \qquad \text{Add the numerators.}$$

$$= \frac{2x-x+1}{x+1} \qquad \text{Commutative property}$$

$$= \frac{x+1}{x+1} \qquad \text{Combine like terms.}$$

$$= 1 \qquad \text{Simplify.}$$

It is important to understand that the expressions $\frac{2x}{x+1} + \frac{1-x}{x+1}$ and 1 are *equivalent expressions*. That is, they are equal for *every* value of x except -1, for which the first expression is undefined.

In the next example, we add rational expressions with like denominators and simplify the result to lowest terms.

EXAMPLE 2 **Adding rational expressions with like denominators**

Add and simplify to lowest terms.

(a) $\dfrac{3}{b} + \dfrac{2}{b}$

(b) $\dfrac{z}{z+2} + \dfrac{2}{z+2}$

(c) $\dfrac{x-1}{x^2+x} + \dfrac{1}{x^2+x}$

(d) $\dfrac{t^2+t}{t-1} + \dfrac{1-3t}{t-1}$

Solution

(a) $\dfrac{3}{b} + \dfrac{2}{b} = \dfrac{3+2}{b} = \dfrac{5}{b}$ Add the numerators.

(b) $\dfrac{z}{z+2} + \dfrac{2}{z+2} = \dfrac{z+2}{z+2}$ Add the numerators.

$= 1$ Simplify.

(c) $\dfrac{x-1}{x^2+x} + \dfrac{1}{x^2+x} = \dfrac{x-1+1}{x^2+x}$ Add the numerators.

$= \dfrac{x}{x(x+1)}$ Factor the denominator.

$= \dfrac{1}{x+1}$ Simplify.

(d) $\dfrac{t^2+t}{t-1} + \dfrac{1-3t}{t-1} = \dfrac{t^2+t+1-3t}{t-1}$ Add the numerators.

$= \dfrac{t^2-2t+1}{t-1}$ Combine like terms.

$= \dfrac{(t-1)(t-1)}{t-1}$ Factor the numerator.

$= t-1$ Simplify.

▌ **Now Try Exercises 19, 25, 33, 35**

EXAMPLE 3   **Adding rational expressions with two variables**

Add and simplify to lowest terms.

(a) $\dfrac{4}{xy} + \dfrac{5}{xy}$ (b) $\dfrac{a}{a^2-b^2} + \dfrac{b}{a^2-b^2}$ (c) $\dfrac{1}{x-y} + \dfrac{-1}{y-x}$

Solution

(a) $\dfrac{4}{xy} + \dfrac{5}{xy} = \dfrac{4+5}{xy} = \dfrac{9}{xy}$ Add the numerators.

(b) $\dfrac{a}{a^2-b^2} + \dfrac{b}{a^2-b^2} = \dfrac{a+b}{a^2-b^2}$ Add the numerators.

$= \dfrac{a+b}{(a-b)(a+b)}$ Factor the denominator.

$= \dfrac{1}{a-b}$ Simplify.

(c) First write $\frac{1}{x-y} + \frac{-1}{y-x}$ with a common denominator. Note that if we multiply the second term by 1, written in the form $\frac{-1}{-1}$, it becomes

$$\dfrac{-1}{y-x} \cdot \dfrac{-1}{-1} = \dfrac{(-1)(-1)}{(y-x)(-1)} = \dfrac{1}{-y+x} = \dfrac{1}{x-y}.$$

Thus the given sum can be simplified as follows.

$$\dfrac{1}{x-y} + \dfrac{-1}{y-x} = \dfrac{1}{x-y} + \dfrac{1}{x-y}$$ Rewrite the second term.

$$= \dfrac{2}{x-y}$$ Add the numerators.

▌ **Now Try Exercises 51, 53, 55**

Next we consider subtraction of rational expressions with like denominators.

> ### DIFFERENCES OF RATIONAL EXPRESSIONS
>
> To subtract two rational expressions with like denominators, subtract their numerators. Keep the same denominator.
>
> $$\frac{A}{C} - \frac{B}{C} = \frac{A - B}{C} \qquad C \text{ is nonzero.}$$

Subtraction of rational expressions with like denominators is similar to addition except that instead of adding numerators, we subtract them. For example, the expressions $\frac{3x}{x-4}$ and $\frac{2x}{x-4}$ have like denominators and can be subtracted as follows.

READING CHECK

• How do we subtract rational expressions with like denominators?

$$\frac{3x}{x-4} - \frac{2x}{x-4} = \frac{3x - 2x}{x-4} \qquad \text{Subtract the numerators.}$$

$$= \frac{x}{x-4} \qquad \text{Combine like terms.}$$

In the next example, we subtract rational expressions with like denominators and simplify the result to lowest terms.

EXAMPLE 4 **Subtracting rational expressions with like denominators**

Subtract and simplify to lowest terms.

(a) $\dfrac{a+1}{a} - \dfrac{1}{a}$ (b) $\dfrac{2y}{3y-1} - \dfrac{3y}{3y-1}$ (c) $\dfrac{1+x}{2x^2+5x-3} - \dfrac{-2}{2x^2+5x-3}$

Solution

(a) $\dfrac{a+1}{a} - \dfrac{1}{a} = \dfrac{a+1-1}{a} = \dfrac{a}{a} = 1$

(b) $\dfrac{2y}{3y-1} - \dfrac{3y}{3y-1} = \dfrac{2y-3y}{3y-1} = \dfrac{-y}{3y-1}$ or $-\dfrac{y}{3y-1}$

(c) $\dfrac{1+x}{2x^2+5x-3} - \dfrac{-2}{2x^2+5x-3} = \dfrac{1+x-(-2)}{2x^2+5x-3}$

$$= \frac{x+3}{(2x-1)(x+3)}$$

$$= \frac{1}{2x-1}$$

Now Try Exercises 21, 27, 67

If the numerator of the second fraction in a difference has more than one term, it is important to put parentheses around the second numerator.

$$\frac{x+6}{2x+1} - \frac{3-x}{2x+1} = \frac{x+6-(3-x)}{2x+1} \qquad \text{Subtract the numerators; insert parentheses.}$$

$$= \frac{x+6-3-(-x)}{2x+1} \qquad \text{Distributive property}$$

$$= \frac{x+6-3+x}{2x+1} \qquad \text{Double negative property}$$

$$= \frac{2x+3}{2x+1} \qquad \text{Combine like terms.}$$

NOTE: If parentheses were not inserted in the previous calculation, the numerator would be

$$x + 6 - 3 - x = 3,$$

which would give an incorrect result.

EXAMPLE 5   **Subtracting rational expressions with like denominators**

Subtract and simplify to lowest terms.

(a) $\dfrac{2x}{x+1} - \dfrac{x-1}{x+1}$ (b) $\dfrac{x+y}{3y} - \dfrac{x-y}{3y}$

Solution

(a) $\dfrac{2x}{x+1} - \dfrac{x-1}{x+1} = \dfrac{2x - (x-1)}{x+1}$ Subtract the numerators.

$\qquad\qquad\qquad = \dfrac{2x - x + 1}{x+1}$ Distributive property

$\qquad\qquad\qquad = \dfrac{x+1}{x+1}$ Simplify the numerator.

$\qquad\qquad\qquad = 1$ Simplify.

(b) $\dfrac{x+y}{3y} - \dfrac{x-y}{3y} = \dfrac{x+y - (x-y)}{3y}$ Subtract the numerators.

$\qquad\qquad\qquad = \dfrac{x+y - x + y}{3y}$ Distributive property

$\qquad\qquad\qquad = \dfrac{2y}{3y}$ Simplify the numerator.

$\qquad\qquad\qquad = \dfrac{2}{3}$ Simplify.

Now Try Exercises 31, 59

▶ **REAL-WORLD CONNECTION** A Look Into Math for this section discusses how applications involving quality control use rational expressions. The next example illustrates one way this can occur.

EXAMPLE 6   **Analyzing quality control**

A container holds a mixture of 8-GB and 16-GB computer flash drives. In this container, there is a total of n flash drives, including 2 defective 8-GB flash drives and 4 defective 16-GB flash drives. If a flash drive is picked at random by a quality control inspector, then the probability, or chance, of one of the defective flash drives being chosen is given by the expression $\frac{2}{n} + \frac{4}{n}$.

(a) Simplify this expression.   (b) Interpret the result.

Solution

(a) Because the denominators are the same, we simply add the numerators.

$$\frac{2}{n} + \frac{4}{n} = \frac{2+4}{n} = \frac{6}{n} \qquad \text{Add the numerators.}$$

(b) There are 6 chances in n that a defective flash drive is chosen.

Now Try Exercise 73

7.3   Putting It All Together

CONCEPT	EXPLANATION	EXAMPLES
Addition of Rational Expressions	For polynomials A, B, and C, where C is nonzero, $$\frac{A}{C} + \frac{B}{C} = \frac{A+B}{C}.$$	$$\frac{x}{x+1} + \frac{1-x}{x+1} = \frac{x+1-x}{x+1} = \frac{1}{x+1}$$ $$\frac{2x}{x^2-1} + \frac{x}{x^2-1} = \frac{2x+x}{x^2-1} = \frac{3x}{x^2-1}$$
Subtraction of Rational Expressions	For polynomials A, B, and C, where C is nonzero, $$\frac{A}{C} - \frac{B}{C} = \frac{A-B}{C}.$$ If B consists of more than one term, put parentheses around B and apply the distributive property.	$$\frac{2x}{x^2-4} - \frac{x+2}{x^2-4} = \frac{2x-(x+2)}{x^2-4}$$ $$= \frac{2x-x-2}{x^2-4}$$ $$= \frac{x-2}{(x+2)(x-2)}$$ $$= \frac{1}{x+2}$$

7.3   Exercises

MyMathLab 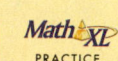 PRACTICE   WATCH   DOWNLOAD   READ   REVIEW

CONCEPTS AND VOCABULARY

1. When adding two rational expressions with like denominators, add their _____. The _____ do not change.

2. When subtracting two rational expressions with like denominators, subtract their _____. The _____ do not change.

3. $\dfrac{A}{C} + \dfrac{B}{C} =$ _____

4. $\dfrac{A}{C} - \dfrac{B}{C} =$ _____

ADDITION AND SUBTRACTION OF FRACTIONS

Exercises 5–18: Simplify to lowest terms.

5. $\dfrac{1}{2} + \dfrac{1}{2}$

6. $\dfrac{3}{7} + \dfrac{2}{7}$

7. $\dfrac{4}{5} + \dfrac{2}{5}$

8. $\dfrac{3}{11} + \dfrac{5}{11}$

9. $\dfrac{1}{6} + \dfrac{5}{6}$

10. $\dfrac{3}{10} + \dfrac{5}{10}$

11. $\dfrac{4}{7} - \dfrac{1}{7}$

12. $\dfrac{5}{13} - \dfrac{7}{13}$

13. $\dfrac{7}{8} - \dfrac{3}{8}$

14. $\dfrac{9}{16} - \dfrac{5}{16}$

15. $\dfrac{11}{12} - \dfrac{5}{12}$

16. $\dfrac{7}{24} - \dfrac{3}{24}$

17. $\dfrac{7}{15} + \dfrac{4}{15} - \dfrac{1}{15}$

18. $\dfrac{11}{36} - \dfrac{5}{36} + \dfrac{1}{36}$

ADDITION AND SUBTRACTION OF RATIONAL EXPRESSIONS

Exercises 19–70: Simplify to lowest terms.

19. $\dfrac{2}{x} + \dfrac{1}{x}$

20. $\dfrac{9}{x} - \dfrac{7}{x}$

21. $\dfrac{7+2x}{4x} - \dfrac{7}{4x}$

22. $\dfrac{x-1}{5x} + \dfrac{2x+1}{5x}$

23. $\dfrac{y+3}{y-3} + \dfrac{2y-12}{y-3}$

24. $\dfrac{5-y}{y+2} + \dfrac{3y-1}{y+2}$

25. $\dfrac{x}{x-3} + \dfrac{-3}{x-3}$

26. $\dfrac{2x}{2x+1} + \dfrac{1}{2x+1}$

27. $\dfrac{5z}{4z+3} - \dfrac{z}{4z+3}$

28. $\dfrac{z}{2z+1} - \dfrac{1-z}{2z+1}$

29. $\dfrac{t+5}{t+6} + \dfrac{t+7}{t+6}$

30. $\dfrac{4t-13}{t-4} + \dfrac{1-t}{t-4}$

31. $\dfrac{5x}{2x+3} - \dfrac{3x-3}{2x+3}$

32. $\dfrac{x}{5-x} - \dfrac{2x-5}{5-x}$

33. $\dfrac{x-4}{x^2-x} + \dfrac{4}{x^2-x}$

34. $\dfrac{2x-2}{4x^2-1} + \dfrac{1}{4x^2-1}$

35. $\dfrac{z^2 - 1}{z - 2} + \dfrac{3 - 3z}{z - 2}$

36. $\dfrac{x^2 + 2}{x + 1} + \dfrac{3x}{x + 1}$

37. $\dfrac{x^2 + 4x - 1}{4x + 2} - \dfrac{x^2 - 4x - 5}{4x + 2}$

38. $\dfrac{2x^2 - x + 5}{x^2 - 9} - \dfrac{x^2 - x + 14}{x^2 - 9}$

39. $\dfrac{3y}{5} + \dfrac{2y - 5}{5}$

40. $\dfrac{3y - 22}{11} + \dfrac{8y}{11}$

41. $\dfrac{x + y}{4} + \dfrac{x - y}{4}$

42. $\dfrac{x + y}{4} - \dfrac{x - y}{4}$

43. $\dfrac{z^2 + 4}{z - 2} - \dfrac{4z}{z - 2}$

44. $\dfrac{z^2 + 2z}{z + 1} + \dfrac{1}{z + 1}$

45. $\dfrac{2x^2 - 5x}{2x + 1} - \dfrac{3}{2x + 1}$

46. $\dfrac{2x^2}{x + 2} + \dfrac{9x + 10}{x + 2}$

47. $\dfrac{3n}{2n^2 - n + 5} + \dfrac{4n}{2n^2 - n + 5}$

48. $\dfrac{n}{n^2 + n + 1} - \dfrac{1}{n^2 + n + 1}$

49. $\dfrac{1}{x + 3} + \dfrac{2}{x + 3} + \dfrac{3}{x + 3}$

50. $\dfrac{x}{2x - 5} - \dfrac{1}{2x - 5} + \dfrac{2x + 1}{2x - 5}$

51. $\dfrac{8}{ab} + \dfrac{1}{ab}$

52. $\dfrac{6}{xy} + \dfrac{9}{xy}$

53. $\dfrac{x}{(x + y)^2} + \dfrac{y}{(x + y)^2}$

54. $\dfrac{x - 2y}{x^2 - y^2} + \dfrac{y}{x^2 - y^2}$

55. $\dfrac{5}{x - y} + \dfrac{-5}{y - x}$

56. $\dfrac{-4}{y - x} + \dfrac{4}{x - y}$

57. $\dfrac{8}{a - b} + \dfrac{8}{b - a}$

58. $\dfrac{6}{x - y} + \dfrac{6}{y - x}$

59. $\dfrac{a + b}{4a} - \dfrac{a - b}{4a}$

60. $\dfrac{x - y}{5x} - \dfrac{x + 9y}{5x}$

61. $\dfrac{x}{x + y} + \dfrac{y}{x + y}$

62. $\dfrac{x}{x - y} - \dfrac{y}{x - y}$

63. $\dfrac{a^2}{a + b} - \dfrac{b^2}{a + b}$

64. $\dfrac{a^2}{a + b} + \dfrac{2ab + b^2}{a + b}$

65. $\dfrac{2x - 5}{2x^2 + 5x + 2} + \dfrac{6}{2x^2 + 5x + 2}$

66. $\dfrac{4x}{3x^2 + 5x - 2} + \dfrac{2 - 3x}{3x^2 + 5x - 2}$

67. $\dfrac{3x + 7}{3x^2 - 2x - 5} - \dfrac{2x + 6}{3x^2 - 2x - 5}$

68. $\dfrac{5x - 4}{2x^2 - 7x + 6} - \dfrac{3x - 1}{2x^2 - 7x + 6}$

69. $\dfrac{4x^2}{2x + 3y} - \dfrac{9y^2}{2x + 3y}$

70. $\dfrac{x^3}{x^2 + xy + y^2} - \dfrac{y^3}{x^2 + xy + y^2}$

71. **Thinking Generally** If $\frac{2}{3 + x} + \frac{3}{3 + x}$ equals $\frac{5}{10}$, what must be true about x?

72. **Thinking Generally** If $\frac{8}{6 + x} - \frac{4}{3 + y}$ equals 0, what must be true about x and y?

APPLICATIONS

73. *Quality Control* (Refer to Example 6.) A container holds a total of $n + 1$ batteries. In this container, there are 6 defective AA batteries, 5 defective C batteries, and 3 defective D batteries. If a battery is chosen at random by a quality control inspector, the probability, or chance, of one of the defective batteries being chosen is

$$\dfrac{6}{n + 1} + \dfrac{5}{n + 1} + \dfrac{3}{n + 1}.$$

(a) Simplify this expression.
(b) Evaluate the simplified expression for $n = 99$ and interpret the result.

74. *Intensity of a Light Bulb* The farther a person is from a light bulb, the less intense its light. The equation $I = \frac{19}{4d^2}$ approximates the light intensity from a 60-watt light bulb at a distance of d meters, where I is measured in watts per square meter. (*Source:* R. Weidner.)

(a) Find I for $d = 2$ meters and interpret the result.
(b) The intensity of light from a 100-watt light bulb is about $I = \frac{32}{4d^2}$. Find an expression for the sum of the intensities of light from a 100-watt bulb and a 60-watt bulb.

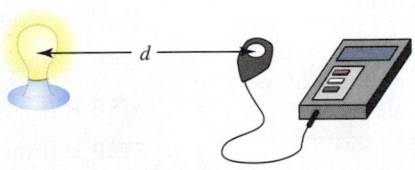

WRITING ABOUT MATHEMATICS

75. Explain how to add two rational expressions with like denominators. Give an example.

76. Explain how to subtract two rational expressions with like denominators. Give an example.

7.4   Addition and Subtraction with Unlike Denominators

Finding Least Common Multiples • Review of Fractions with Unlike Denominators • Rational Expressions with Unlike Denominators

A LOOK INTO MATH ▶   The sum and difference of rational expressions frequently occur in the design of electrical devices, such as smart phones and HD televisions. (See Example 7 and the discussion preceding it.) In Section 7.3, we added and subtracted rational expressions with like denominators. Although the denominators of rational expressions are often unlike, rational expressions can still be added and subtracted after a common denominator is found. One way to find the least common denominator for a sum or difference of rational expressions is to find the least common multiple of the denominators. In the following subsection we show how to find the least common multiple.

Finding Least Common Multiples

▶ **REAL-WORLD CONNECTION**   Two friends work part-time at a store. The first person works every sixth day, and the second person works every eighth day. If they both work today, how many days will pass before they work on the same day again?

We can answer this question by listing the days that each person works.

First person: 6,   12,   18,   **24**,   30,   36,   42,   **48**,   54
Second person: 8,   16,   **24**,   32,   40,   **48**,   56,   64

After **24** days, the two friends work on the same day. The next time is after **48** days. The numbers 24 and 48 are *common multiples* of 6 and 8. (Find another.) However, 24 is the *least* common multiple (LCM) of 6 and 8 because it is the smallest common multiple.

Another way to find the least common multiple of 6 and 8 is to factor each number into prime numbers.

$$6 = 2 \cdot \mathbf{3} \quad \text{and} \quad 8 = \mathbf{2 \cdot 2 \cdot 2}$$

To find the least common multiple, first list each factor the greatest number of times that it occurs in either factorization. Then find the product of these numbers. For this example, the factor 2 occurs three times in the factorization of 8 and only once in the factorization of 6, so list 2 three times. The factor 3 appears only once in the factorization of 6, so list it once:

$$\mathbf{2, \ 2, \ 2, \ 3}.$$

The least common multiple is their product: $\mathbf{2 \cdot 2 \cdot 2 \cdot 3} = 24$.

This same procedure can also be used to find the least common multiple of two or more polynomials.

FINDING THE LEAST COMMON MULTIPLE

The least common multiple (LCM) of two or more polynomials can be found as follows.

STEP 1: Factor each polynomial completely.

STEP 2: List each factor the greatest number of times that it occurs in any factorization.

STEP 3: Find the product of this list of factors. The result is the LCM.

EXAMPLE 1   **Finding least common multiples**

Find the least common multiple of each pair of expressions.
(a) $2x, \ 5x^2$ **(b)** $x^2 - x, \ x - 1$
(c) $x + 2, \ x - 3$ **(d)** $x^2 + 2x + 1, \ x^2 + 3x + 2$

Solution

(a) **STEP 1:** Factor $2x$ and $5x^2$ completely.

$$2x = \mathbf{2} \cdot \boldsymbol{x} \quad \text{and} \quad 5x^2 = \mathbf{5} \cdot \boldsymbol{x} \cdot \boldsymbol{x}$$

STEP 2: In either factorization, the factor 2 occurs at most once, the factor 5 occurs at most once, and the factor x occurs at most twice. The list of factors is $\mathbf{2}, \mathbf{5}, \boldsymbol{x}, \boldsymbol{x}$.

STEP 3: The LCM equals the product

$$\mathbf{2} \cdot \mathbf{5} \cdot \boldsymbol{x} \cdot \boldsymbol{x} = 10x^2.$$

(b) **STEP 1:** Factor $x^2 - x$ and $x - 1$ completely. Note that $x - 1$ cannot be factored.

$$x^2 - x = \boldsymbol{x}(\boldsymbol{x} - \mathbf{1}) \quad \text{and} \quad x - 1 = \boldsymbol{x} - \mathbf{1}$$

STEP 2: Both factors, x and $x - 1$, occur at most once in either factorization. The list of factors is $\boldsymbol{x}, (\boldsymbol{x} - \mathbf{1})$.

STEP 3: The LCM is the product $\boldsymbol{x}(\boldsymbol{x} - \mathbf{1})$, or $x^2 - x$.

(c) **STEP 1:** Neither $x + 2$ nor $x - 3$ can be factored.

STEP 2: The list of factors is $(x + 2), (x - 3)$.

STEP 3: The LCM is the product $(x + 2)(x - 3)$, or $x^2 - x - 6$.

(d) **STEP 1:** Factor $x^2 + 2x + 1$ and $x^2 + 3x + 2$ completely.

$$x^2 + 2x + 1 = (\boldsymbol{x} + \mathbf{1})(\boldsymbol{x} + \mathbf{1}) \quad \text{and} \quad x^2 + 3x + 2 = (\boldsymbol{x} + \mathbf{1})(\boldsymbol{x} + \mathbf{2})$$

STEP 2: In either factorization, the factor $(x + 1)$ occurs at most twice and the factor $(x + 2)$ occurs at most once. The list is $(\boldsymbol{x} + \mathbf{1}), (\boldsymbol{x} + \mathbf{1}), (\boldsymbol{x} + \mathbf{2})$.

STEP 3: The LCM is the product $(\boldsymbol{x} + \mathbf{1})^2(\boldsymbol{x} + \mathbf{2})$.

Now Try Exercises 15, 19, 27, 29

USING A STEP DIAGRAM TO FIND THE LCM The least common multiple for two polynomials can be found using a step diagram similar to that used in Section 1.2 for finding the least common multiple of two numbers. The next example illustrates this process.

EXAMPLE 2   **Using a step diagram to find the LCM**

Use a step diagram to find the LCM of $2x^3 - 8x^2$ and $2x^3 - 2x^2 - 24x$.

Solution

Start by factoring $2x^3 - 8x^2$ and $2x^3 - 2x^2 - 24x$ completely.

$$2x^3 - 8x^2 = 2x^2(x - 4) \quad \text{and} \quad 2x^3 - 2x^2 - 24x = 2x(x + 3)(x - 4)$$

Write the factored form of each expression on the top step of the diagram. We find the expressions in each of the following steps of the diagram by factoring out *any* common factor from the expressions in the previous step. The process continues until no common factor can be found, as shown in Figure 7.3.

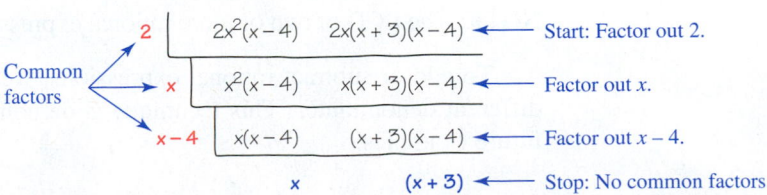

Figure 7.3 Factor Step Diagram for Finding the LCM

We find the LCM by multiplying the expressions along the side and at the bottom of the diagram. The LCM is $\mathbf{2} \cdot \boldsymbol{x} \cdot (\boldsymbol{x} - \mathbf{4}) \cdot \boldsymbol{x} \cdot (\boldsymbol{x} + \mathbf{3}) = 2x^2(x - 4)(x + 3)$.

Now Try Exercise 31

Review of Fractions with Unlike Denominators

Before we can find the sum $\frac{1}{2} + \frac{1}{3}$ by hand, we need to rewrite these fractions by using their least common denominator. The *least common denominator* (LCD) of $\frac{1}{2}$ and $\frac{1}{3}$ corresponds to the *least common multiple* of the denominators 2 and 3, which is 6. As a result, we can rewrite these fractions as

$$\frac{1}{2} \cdot \frac{3}{3} = \frac{3}{6} \quad \text{and} \quad \frac{1}{3} \cdot \frac{2}{2} = \frac{2}{6}.$$

Their sum is $\frac{1}{2} + \frac{1}{3} = \frac{3}{6} + \frac{2}{6} = \frac{5}{6}$.

EXAMPLE 3 **Adding and subtracting fractions with unlike denominators**

Simplify each expression.

(a) $\dfrac{3}{10} + \dfrac{4}{15}$ (b) $\dfrac{7}{8} - \dfrac{1}{6}$

Solution

(a) The LCD for $\frac{3}{10}$ and $\frac{4}{15}$ equals the LCM of 10 and 15, which is 30. We rewrite these fractions as

$$\frac{3}{10} \cdot \frac{3}{3} = \frac{9}{30} \quad \text{and} \quad \frac{4}{15} \cdot \frac{2}{2} = \frac{8}{30}.$$

Their sum is

$$\frac{3}{10} + \frac{4}{15} = \frac{9}{30} + \frac{8}{30} = \frac{17}{30}.$$

(b) The LCD for $\frac{7}{8}$ and $\frac{1}{6}$ is the LCM of 8 and 6, which is 24. We rewrite these fractions as

$$\frac{7}{8} \cdot \frac{3}{3} = \frac{21}{24} \quad \text{and} \quad \frac{1}{6} \cdot \frac{4}{4} = \frac{4}{24}.$$

Their difference is

$$\frac{7}{8} - \frac{1}{6} = \frac{21}{24} - \frac{4}{24} = \frac{17}{24}.$$

Now Try Exercises 45, 47

Rational Expressions with Unlike Denominators

The first step in adding or subtracting rational expressions with unlike denominators is to rewrite each expression by using the least common denominator. Then the sum or difference can be found by using the techniques discussed in Section 7.3.

NOTE: The LCD of two or more rational expressions equals the LCM of their denominators.

To add or subtract rational expressions, we often rewrite a rational expression with a different denominator. This technique is demonstrated in the next example and is used in future examples.

EXAMPLE 4 **Rewriting rational expressions**

Rewrite each rational expression so it has the given denominator D.

(a) $\dfrac{3}{2x}$, $D = 8x^2$ (b) $\dfrac{1}{x + 1}$, $D = x^2 - 1$

Solution

(a) We need to write $\frac{3}{2x}$ so that it is equivalent to $\frac{?}{8x^2}$.

$$\frac{3}{2x} \overset{\times\,4x}{=} \frac{?}{8x^2}$$
$$\times\,4x$$

Because $8x^2 = 2x \cdot \mathbf{4x}$, we can multiply $\frac{3}{2x}$ by 1 in the form $\frac{4x}{4x}$ as follows.

$$\frac{3}{2x} \cdot \frac{\mathbf{4x}}{\mathbf{4x}} = \frac{12x}{8x^2} \qquad \text{Multiply rational expressions.}$$

(b) We must write $\frac{1}{x+1}$ so that it is equivalent to $\frac{?}{x^2-1}$.

$$\frac{1}{x+1} \overset{\times\,(x-1)}{=} \frac{?}{x^2-1}$$
$$\times\,(x-1)$$

Because $x^2 - 1 = (x + 1)(\mathbf{x - 1})$, we can multiply $\frac{1}{x+1}$ by 1 in the form $\frac{x-1}{x-1}$ as follows.

$$\frac{1}{x+1} \cdot \frac{\mathbf{x - 1}}{\mathbf{x - 1}} = \frac{x - 1}{x^2 - 1} \qquad \text{Multiply rational expressions.}$$

▌ **Now Try Exercises 37, 39**

EXAMPLE 5 **Adding rational expressions with unlike denominators**

Find each sum and leave your answer in factored form.

(a) $\dfrac{5}{8y} + \dfrac{7}{4y^2}$ **(b)** $\dfrac{1}{x-1} + \dfrac{1}{x+1}$ **(c)** $\dfrac{x}{x^2 + 2x + 1} + \dfrac{1}{x+1}$

Solution

(a) First find the LCM of $8y$ and $4y^2$.

$$8y = \mathbf{2 \cdot 2 \cdot 2} \cdot y \quad \text{and} \quad 4y^2 = \mathbf{2 \cdot 2} \cdot \mathbf{y \cdot y}$$

Thus the LCM is $\mathbf{2 \cdot 2 \cdot 2 \cdot y \cdot y} = 8y^2$. Now, because

$$8y^2 = 8y \cdot \mathbf{y} \quad \text{and} \quad 8y^2 = 4y^2 \cdot \mathbf{2},$$

we multiply the first expression by $\frac{y}{y}$ and the second expression by $\frac{2}{2}$.

$$\frac{5}{8y} + \frac{7}{4y^2} = \frac{5}{8y} \cdot \frac{\mathbf{y}}{\mathbf{y}} + \frac{7}{4y^2} \cdot \frac{\mathbf{2}}{\mathbf{2}} \qquad \text{Rewrite by using the LCD.}$$

$$= \frac{5y}{8y^2} + \frac{14}{8y^2} \qquad \text{Multiply rational expressions.}$$

$$= \frac{5y + 14}{8y^2} \qquad \text{Add the numerators.}$$

(b) The LCM for $x - 1$ and $x + 1$ is their product, $(x - 1)(x + 1)$.

$$\frac{1}{x - 1} + \frac{1}{x + 1} = \frac{1}{x - 1} \cdot \frac{x + 1}{x + 1} + \frac{1}{x + 1} \cdot \frac{x - 1}{x - 1} \qquad \text{Rewrite by using the LCD.}$$

$$= \frac{x + 1}{(x - 1)(x + 1)} + \frac{x - 1}{(x + 1)(x - 1)} \qquad \text{Multiply rational expressions.}$$

$$= \frac{x + 1 + x - 1}{(x - 1)(x + 1)} \qquad \text{Add the numerators.}$$

$$= \frac{2x}{(x - 1)(x + 1)} \qquad \text{Simplify the numerator.}$$

(c) First find the LCM for $x^2 + 2x + 1$ and $x + 1$. Because

$$x^2 + 2x + 1 = (x + 1)(x + 1),$$

their LCM is $(x + 1)(x + 1) = (x + 1)^2$.

$$\frac{x}{x^2 + 2x + 1} + \frac{1}{x + 1} = \frac{x}{(x + 1)^2} + \frac{1}{x + 1} \cdot \frac{x + 1}{x + 1} \qquad \text{Rewrite by using the LCD.}$$

$$= \frac{x}{(x + 1)^2} + \frac{x + 1}{(x + 1)^2} \qquad \text{Multiply rational expressions.}$$

$$= \frac{2x + 1}{(x + 1)^2} \qquad \text{Add the numerators.}$$

Now Try Exercises 53, 65, 71

Subtraction of rational expressions is performed in a manner similar to addition and is illustrated in the next example.

EXAMPLE 6 **Subtracting rational expressions with unlike denominators**

Simplify each expression. Write your answer in lowest terms and leave it in factored form.

(a) $\dfrac{5}{z} - \dfrac{z}{z - 1}$ \qquad **(b)** $\dfrac{5}{x + 1} - \dfrac{1}{x^2 - 1}$

(c) $\dfrac{x}{x^2 - 2x} - \dfrac{1}{x^2 + 2x}$ \qquad **(d)** $\dfrac{3}{x - 1} - \dfrac{3}{x^2 - x} + \dfrac{5}{x}$

Solution
(a) The LCD is $z(z - 1)$.

$$\frac{5}{z} - \frac{z}{z - 1} = \frac{5}{z} \cdot \frac{z - 1}{z - 1} - \frac{z}{z - 1} \cdot \frac{z}{z} \qquad \text{Rewrite by using the LCD.}$$

$$= \frac{5(z - 1)}{z(z - 1)} - \frac{z^2}{z(z - 1)} \qquad \text{Multiply rational expressions.}$$

$$= \frac{5(z - 1) - z^2}{z(z - 1)} \qquad \text{Subtract the numerators.}$$

$$= \frac{-z^2 + 5z - 5}{z(z - 1)} \qquad \text{Simplify the numerator.}$$

(b) The LCD is $(x - 1)(x + 1)$. Note that $x^2 - 1 = (x - 1)(x + 1)$.

$$\frac{5}{x + 1} - \frac{1}{x^2 - 1} = \frac{5}{x + 1} \cdot \frac{x - 1}{x - 1} - \frac{1}{(x - 1)(x + 1)} \qquad \text{Rewrite by using the LCD.}$$

$$= \frac{5(x - 1) - 1}{(x - 1)(x + 1)} \qquad \text{Subtract the numerators.}$$

$$= \frac{5x - 5 - 1}{(x - 1)(x + 1)} \qquad \text{Distributive property}$$

$$= \frac{5x - 6}{(x - 1)(x + 1)} \qquad \text{Simplify the numerator.}$$

(c) Start by factoring each denominator. Because

$$x^2 - 2x = x(x - 2) \quad \text{and} \quad x^2 + 2x = x(x + 2),$$

the LCD is $x(x - 2)(x + 2)$.

$$\frac{x}{x^2 - 2x} - \frac{1}{x^2 + 2x} = \frac{x}{x(x - 2)} \cdot \frac{x + 2}{x + 2} - \frac{1}{x(x + 2)} \cdot \frac{x - 2}{x - 2} \qquad \text{Rewrite by using the LCD.}$$

$$= \frac{x(x + 2)}{x(x - 2)(x + 2)} - \frac{x - 2}{x(x - 2)(x + 2)} \qquad \text{Multiply rational expressions.}$$

$$= \frac{x(x + 2) - (x - 2)}{x(x - 2)(x + 2)} \qquad \text{Subtract the numerators.}$$

$$= \frac{x^2 + 2x - x + 2}{x(x - 2)(x + 2)} \qquad \text{Distributive property}$$

$$= \frac{x^2 + x + 2}{x(x - 2)(x + 2)} \qquad \text{Simplify the numerator.}$$

(d) The given expression contains three rational expressions. Begin by finding the LCM of the three denominators: $x - 1$, $x^2 - x$, and x. Because $x^2 - x = x(x - 1)$, the LCM is $x(x - 1)$.

$$\frac{3}{x - 1} - \frac{3}{x^2 - x} + \frac{5}{x} = \frac{3}{x - 1} \cdot \frac{x}{x} - \frac{3}{x(x - 1)} + \frac{5}{x} \cdot \frac{x - 1}{x - 1} \qquad \text{Rewrite by using the LCD.}$$

$$= \frac{3x}{x(x - 1)} - \frac{3}{x(x - 1)} + \frac{5(x - 1)}{x(x - 1)} \qquad \text{Multiply rational expressions.}$$

$$= \frac{3x - 3 + 5x - 5}{x(x - 1)} \qquad \text{Combine the expressions.}$$

$$= \frac{8x - 8}{x(x - 1)} \qquad \text{Simplify the numerator.}$$

$$= \frac{8(x - 1)}{x(x - 1)} \qquad \text{Factor the numerator.}$$

$$= \frac{8}{x} \qquad \text{Simplify to lowest terms.}$$

▌ **Now Try Exercises 63, 77, 79, 81**

▶ **REAL-WORLD CONNECTION** Sums of rational expressions occur in applications involving electricity. The flow of electricity through a wire can be compared to the flow of water through a hose. Voltage is the force "pushing" the electricity and corresponds to water pressure in a hose. Resistance is the opposition to the flow of electricity and corresponds to the diameter of a hose. More resistance results in less flow of electricity. An ordinary light bulb is an example of a resistor in an electrical circuit. Resistance is often measured in units called *ohms*. For example, a standard 60-watt light bulb has a resistance of about 200 ohms.

Suppose that two light bulbs are wired in parallel so that electricity can flow through either light bulb, as illustrated in Figure 7.4. If the individual resistances of the light bulbs are R and S, then the total resistance of the circuit is found by adding the expression

$$\frac{1}{R} + \frac{1}{S},$$

and then taking the *reciprocal*. (*Source:* R. Weidner and R. Sells, *Elementary Classical Physics*, vol. 2.)

Parallel Wiring

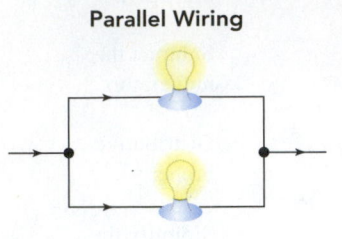

Figure 7.4

| EXAMPLE 7 | **Modeling electrical resistance** |

Add $\frac{1}{R} + \frac{1}{S}$, and then find the reciprocal of the result.

Solution
The LCD for $\frac{1}{R}$ and $\frac{1}{S}$ is **RS**.

$$\frac{1}{R} + \frac{1}{S} = \frac{1}{R} \cdot \frac{S}{S} + \frac{1}{S} \cdot \frac{R}{R} \qquad \text{Rewrite by using the LCD.}$$

$$= \frac{S}{RS} + \frac{R}{RS} \qquad \text{Multiply rational expressions.}$$

$$= \frac{S + R}{RS} \qquad \text{Add the numerators.}$$

CRITICAL THINKING

Find the reciprocal of the sum $x + \frac{1}{x}$.

In general, the reciprocal of $\frac{a}{b}$ is $\frac{b}{a}$, so the reciprocal of $\frac{S + R}{RS}$ is

$$\frac{RS}{S + R}.$$

This final expression can be used to find the total resistance of the circuit.

▌ **Now Try Exercise 101**

7.4 Putting It All Together

CONCEPT	EXPLANATION	EXAMPLES
Least Common Multiple (LCM) of Polynomials	1. Factor each polynomial completely. 2. List each factor the greatest number of times that it occurs in any factorization. 3. The LCM is the product of this list.	For $x^2 - 2x$ and $x^2 - 4x + 4$, 1. $x^2 - 2x = x(x - 2)$ $\quad x^2 - 4x + 4 = (x - 2)(x - 2)$ 2. $x, (x - 2), (x - 2)$ 3. LCM $= x(x - 2)(x - 2)$.
Least Common Denominator (LCD)	The LCD of two or more rational expressions equals the least common multiple (LCM) of their denominators.	The LCD of $\frac{2}{x^2 - 2x}$ and $\frac{3}{x^2 - 4x + 4}$ is $x(x - 2)(x - 2)$ because the LCM of $x^2 - 2x$ and $x^2 - 4x + 4$ is $x(x - 2)(x - 2)$, as shown above.

CONCEPT	EXPLANATION	EXAMPLES
Addition and Subtraction of Rational Expressions with Unlike Denominators	First rewrite each expression by using the LCD. Then add or subtract the expressions. Finally, write your answer in lowest terms. It is often helpful to leave the result in factored form.	The LCM of x and $x + 2$ is $x(x + 2)$. $$\frac{2}{x} + \frac{5}{x+2} = \frac{2}{x} \cdot \frac{x+2}{x+2} + \frac{5}{x+2} \cdot \frac{x}{x}$$ $$= \frac{2x+4}{x(x+2)} + \frac{5x}{x(x+2)}$$ $$= \frac{7x+4}{x(x+2)}$$

7.4 Exercises

MyMathLab  Math XL PRACTICE   WATCH   DOWNLOAD   READ   REVIEW

CONCEPTS AND VOCABULARY

1. Give a common multiple of 6 and 9 that is not the *least* common multiple.

2. The LCM of x and y is _____.

3. To rewrite $\frac{3}{4}$ with denominator 12, multiply $\frac{3}{4}$ by 1 written as the fraction _____.

4. To rewrite $\frac{4}{x-1}$ with denominator $x^2 - 1$, multiply $\frac{4}{x-1}$ by 1 written as the rational expression _____.

LEAST COMMON MULTIPLES

Exercises 5–12: Find the least common multiple.

5. 4, 6 **6.** 6, 9

7. 2, 3 **8.** 5, 4

9. 10, 15 **10.** 8, 12

11. 24, 36 **12.** 32, 40

Exercises 13–32: Find the least common multiple. Leave your answer in factored form.

13. $4x, 6x$ **14.** $6x, 9x$

15. $5x, 10x^2$ **16.** $4x^2, 12x$

17. $x, x + 1$ **18.** $4x, x - 1$

19. $2x + 1, x + 3$ **20.** $5x + 3, x + 9$

21. $x^2 - x, x^2 + x$ **22.** $x^2 + 2x, x^2$

23. $(x - 8)^2, (x - 8)(x + 1)$

24. $(2x - 1)^3, (2x - 1)(x + 3)$

25. $4x^2 - 1, 2x + 1$

26. $x^2 + 4x + 3, x + 3$

27. $x^2 - 1, x + 1$

28. $x^2 - 4, x - 2$

29. $2x^2 + 7x + 6, x^2 + 5x + 6$

30. $x^2 - 3x + 2, x^2 + 2x - 3$

31. $3y^2 + 6y, 3y^3 + 3y^2 - 6y$

32. $y^2 + 3y + 2, y^4 - 4y^2$

ADDITION AND SUBTRACTION OF RATIONAL EXPRESSIONS

Exercises 33–44: Rewrite the rational expression so it has the given denominator D. For example, if the fraction $\frac{1}{4}$ is written with denominator $D = 8$, it becomes $\frac{2}{8}$.

33. $\frac{1}{3}, D = 9$ **34.** $\frac{3}{4}, D = 24$

35. $\frac{5}{7}, D = 21$ **36.** $\frac{4}{5}, D = 30$

37. $\frac{1}{4x}, D = 8x^3$ **38.** $\frac{5}{3x}, D = 9x^2$

39. $\frac{1}{x+2}, D = x^2 - 4$ **40.** $\frac{3}{x-3}, D = x^2 - 9$

41. $\frac{1}{x+1}, D = x^2 + x$ **42.** $\frac{3}{x-3}, D = x^2 - 3x$

43. $\frac{2x}{x+1}, D = x^2 + 2x + 1$

44. $\frac{x}{2x-1}, D = 2x^2 + 11x - 6$

Exercises 45–100: Simplify the expression. Write your answer in lowest terms and leave it in factored form.

45. $\frac{4}{5} + \frac{1}{2}$

46. $\frac{3}{8} + \frac{1}{4}$

47. $\frac{5}{9} - \frac{1}{3}$

48. $\frac{7}{10} - \frac{3}{15}$

49. $\frac{4}{25} + \frac{2}{5}$

50. $\frac{7}{9} - \frac{1}{3}$

51. $\frac{1}{5} + \frac{3}{4} - \frac{1}{2}$

52. $\frac{6}{7} - \frac{8}{9} + \frac{2}{3}$

53. $\frac{1}{3x} + \frac{3}{4x}$

54. $\frac{4}{2x^2} + \frac{7}{3x}$

55. $\frac{5}{z^2} - \frac{7}{z^3}$

56. $\frac{8}{z} - \frac{3}{2z}$

57. $\frac{1}{x} - \frac{1}{y}$

58. $\frac{1}{xy} - \frac{4}{y}$

59. $\frac{a}{b} + \frac{b}{a}$

60. $\frac{3}{x} - \frac{4}{y}$

61. $\frac{1}{2x + 4} + \frac{3}{x + 2}$

62. $\frac{1}{5x - 10} - \frac{x}{x - 2}$

63. $\frac{2}{t - 2} - \frac{1}{t}$

64. $\frac{7}{2t} + \frac{1}{t + 5}$

65. $\frac{5}{n - 1} + \frac{n}{n + 1}$

66. $\frac{4n}{3n - 2} + \frac{n}{n + 1}$

67. $\frac{3}{x - 3} + \frac{6}{3 - x}$

68. $\frac{x}{x - 8} + \frac{x}{8 - x}$

69. $\frac{1}{5k - 1} + \frac{1}{1 - 5k}$

70. $\frac{4}{4 - 3k} + \frac{3k}{3k - 4}$

71. $\frac{2x}{(x - 1)^2} + \frac{4}{x - 1}$

72. $\frac{5}{(x + 5)} - \frac{x}{(x + 5)^2}$

73. $\frac{2y}{y(2y - 1)} + \frac{1}{2y - 1}$

74. $\frac{5y}{y(y + 1)} - \frac{5}{y + 1}$

75. $\frac{1}{x + 2} - \frac{1}{x^2 + 2x}$

76. $\frac{1}{x - 3} - \frac{2}{x^2 - 3x}$

77. $\frac{3}{x - 2} - \frac{1}{x^2 - 4}$

78. $\frac{x}{9 - x^2} - \frac{1}{3 - x}$

79. $\frac{2}{x^2 - 3x} - \frac{1}{x^2 + 3x}$

80. $\frac{3}{x^2 + 4x} - \frac{2}{x^2 - 4x}$

81. $\frac{1}{x - 2} - \frac{1}{x + 2} + \frac{1}{x}$

82. $\frac{1}{x^2} - \frac{2}{x} + \frac{2}{x - 1}$

83. $\frac{x}{x^2 + 4x + 4} + \frac{1}{x + 2}$

84. $\frac{1}{x^2 - 3x - 4} - \frac{1}{x + 1}$

85. $\frac{x}{(x + 1)(x + 2)} - \frac{1}{(x + 2)(x + 3)}$

86. $\frac{2x}{(x - 1)(x - 2)} - \frac{5}{x - 2}$

87. $\frac{1}{a + b} - \frac{1}{a - b}$

88. $\frac{x}{x^2 - y^2} - \frac{1}{x + y}$

89. $\frac{r}{r - t} + \frac{t}{t - r} - 1$

90. $\frac{1}{x} + \frac{2}{x^2 - 2x} + \frac{5}{x - 2}$

91. $\frac{1}{2a} + \frac{1}{3a} + \frac{1}{4a}$

92. $\frac{1}{b} + \frac{1}{b + 1} + \frac{1}{b + 2}$

93. $\frac{2}{x - y} + \frac{3}{y - x} + \frac{1}{x - y}$

94. $\frac{a}{a - b} + \frac{b}{b - a} + \frac{3}{b}$

95. $\frac{3}{x - 3} - \frac{3}{x^2 - 3x} - \frac{6}{x(x - 3)}$

96. $\frac{3}{2a - 4} + \frac{5}{2a} - \frac{3}{a^2 - 2a}$

97. $x + \frac{1}{x - 1} - \frac{1}{x + 1}$

98. $\frac{x}{x^2 - 4} - \frac{1}{x^2 + 4x + 4}$

99. $\frac{2x + 1}{x - 1} - \frac{3}{x + 1} + \frac{x}{x - 1}$

100. $\frac{1}{x - 3} - \frac{2}{x + 3} + \frac{x}{x^2 - 9}$

APPLICATIONS

101. *Electricity* In Example 7, we demonstrated that the expressions

$$\frac{1}{R} + \frac{1}{S} \quad \text{and} \quad \frac{S + R}{RS}$$

are equivalent. Evaluate both expressions by using $R = 120$ and $S = 200$. Are your answers the same?

102. *Intensity of a Light Bulb* The formula $I = \frac{32}{4d^2}$ approximates the intensity of light from a 100-watt light bulb at a distance of d meters, where I is in watts per square meter. For light from a 40-watt bulb the equation for intensity becomes $I = \frac{16}{5d^2}$. (*Source:* R. Weidner.)

(a) Find an expression for the sum of the intensities of light from the two light bulbs.

(b) Find the combined intensity of their light at $d = 5$ meters.

103. *Photography* A lens in a camera has a focal length, which is important for focusing the camera. If an object is at a distance D from a lens that has a focal length F, then to be in focus, the distance S between the lens and the film should satisfy the equation

$$\frac{1}{S} = \frac{1}{F} - \frac{1}{D},$$

as illustrated in the figure. Write the difference $\frac{1}{F} - \frac{1}{D}$ as one term.

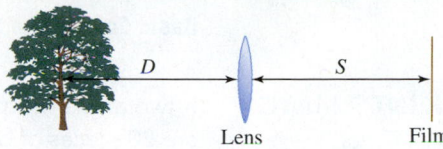

104. *Geometry* Find the sum of the areas of the two rectangles shown in the figure. Write your answer in factored form.

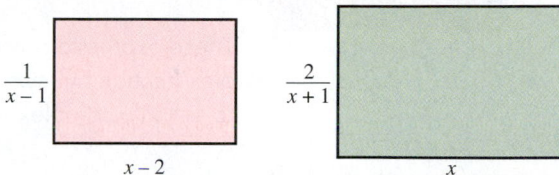

WRITING ABOUT MATHEMATICS

105. Explain how to find the least common multiple of two polynomials.

106. Explain how to subtract two rational expressions with unlike denominators.

SECTIONS 7.3 and 7.4

Checking Basic Concepts

1. Simplify each expression.

(a) $\dfrac{x}{x + 2} + \dfrac{2}{x + 2}$ (b) $\dfrac{2}{3x} - \dfrac{x}{3x}$

(c) $\dfrac{z^2 + z}{z + 2} + \dfrac{z}{z + 2}$

2. Find the least common multiple of each pair of expressions.

(a) $3x,\ 5x$ (b) $4x,\ x^2 + x$

(c) $x + 1,\ x - 1$

3. Simplify each expression.

(a) $\dfrac{1}{x + 1} + \dfrac{5}{x}$ (b) $\dfrac{5}{x - 3} + \dfrac{1}{3 - x}$

(c) $\dfrac{-4}{4x + 2} - \dfrac{x + 2}{2x + 1}$

4. Simplify the expression $\dfrac{a}{a - b} - \dfrac{b}{a + b}$.

7.5 Complex Fractions

Basic Concepts • Simplifying Complex Fractions

A LOOK INTO MATH ▶

If two and a half pizzas are cut so that each piece equals one-eighth of a pizza, then there are 20 pieces of pizza. The quotient two and a half divided by one-eighth can be written as

$$\frac{2\frac{1}{2}}{\frac{1}{8}} \quad \text{or} \quad \frac{2 + \frac{1}{2}}{\frac{1}{8}}.$$

These expressions are examples of *complex* fractions. Typically, we want to simplify a complex fraction by rewriting it as a standard fraction in the form $\frac{a}{b}$. In this section we show how to simplify complex fractions.

NEW VOCABULARY

☐ Complex fraction
☐ Basic complex fraction

Basic Concepts

A **complex fraction** is a rational expression that contains fractions in its numerator, denominator, or both. Examples of complex fractions include

$$\frac{1 + \frac{1}{2}}{1 - \frac{1}{3}}, \quad \frac{5z}{\frac{7}{z} - \frac{4}{2z}}, \quad \text{and} \quad \frac{\frac{2x}{3} + \frac{1}{x}}{x - \frac{1}{x + 1}}. \quad \text{Complex fractions}$$

The expression $\frac{a}{b} \div \frac{c}{d}$ can be written as a **basic complex fraction**, where both the numerator and denominator of the complex fraction are single fractions.

$$\frac{\dfrac{a}{b}}{\dfrac{c}{d}}. \quad \text{Basic complex fraction}$$

Because

$$\frac{a}{b} \div \frac{c}{d} = \frac{a}{b} \cdot \frac{d}{c},$$

we can simplify basic complex fractions by multiplying the numerator and the reciprocal of the denominator. This strategy is summarized as follows.

READING CHECK

• How do we simplify a basic complex fraction?

SIMPLIFYING BASIC COMPLEX FRACTIONS

For any real numbers a, b, c, and d,

$$\frac{\dfrac{a}{b}}{\dfrac{c}{d}} = \frac{a}{b} \cdot \frac{d}{c},$$

where b, c, and d are nonzero.

Simplifying Complex Fractions

There are two methods of simplifying a complex fraction. The first is to simplify both the numerator and the denominator and then divide the resulting two fractions. The second is to multiply the numerator and denominator by the least common denominator of all fractions within the complex fraction.

STUDY TIP

Two methods for simplifying complex fractions are discussed in this section. If your instructor does not require you to use a specific method, choose the one that works best for you.

SIMPLIFYING THE NUMERATOR AND DENOMINATOR The following steps outline Method I for simplifying complex fractions.

SIMPLIFYING COMPLEX FRACTIONS (METHOD I)

To simplify a complex fraction, perform the following steps.

STEP 1: Write the numerator as a single fraction; write the denominator as a single fraction.

STEP 2: Divide the denominator into the numerator by multiplying the numerator and the reciprocal of the denominator.

STEP 3: Simplify the result to lowest terms.

READING CHECK

• What is the first step when using Method I to simplify complex fractions?

Sometimes a complex fraction already has both its numerator and denominator written as single fractions. As a result, we can start with Step 2. This situation is shown in Example 1.

EXAMPLE 1 **Simplifying basic complex fractions**

Simplify each basic complex fraction.

(a) $\dfrac{\frac{3}{4}}{\frac{9}{8}}$ (b) $\dfrac{3\frac{3}{4}}{1\frac{1}{2}}$ (c) $\dfrac{\frac{x}{4}}{\frac{3}{2y}}$ (d) $\dfrac{\frac{(x-1)^2}{4}}{\frac{x-1}{8}}$

Solution
(a) We can skip Step 1. To simplify $\frac{3}{4} \div \frac{9}{8}$, multiply $\frac{3}{4}$ by the reciprocal of $\frac{9}{8}$, or $\frac{8}{9}$.

$$\dfrac{\frac{3}{4}}{\frac{9}{8}} = \frac{3}{4} \cdot \frac{8}{9} \qquad \text{Multiply by the reciprocal of } \tfrac{9}{8} \text{ (Step 2).}$$

$$= \frac{24}{36} \qquad \text{Multiply the fractions.}$$

$$= \frac{2}{3} \qquad \text{Simplify (Step 3).}$$

(b) Start by writing $3\frac{3}{4}$ and $1\frac{1}{2}$ as improper fractions.

$$\frac{3\frac{3}{4}}{1\frac{1}{2}} = \frac{\frac{15}{4}}{\frac{3}{2}}$$ Write as improper fractions (Step 1).

$$= \frac{15}{4} \cdot \frac{2}{3}$$ Multiply by the reciprocal of $\frac{3}{2}$ (Step 2).

$$= \frac{5}{2}$$ Multiply and simplify (Step 3).

(c) We can skip Step 1. To simplify $\frac{x}{4} \div \frac{3}{2y}$, multiply $\frac{x}{4}$ by the reciprocal of $\frac{3}{2y}$, or $\frac{2y}{3}$.

$$\frac{\frac{x}{4}}{\frac{3}{2y}} = \frac{x}{4} \cdot \frac{2y}{3}$$ Multiply by the reciprocal of $\frac{3}{2y}$ (Step 2).

$$= \frac{2xy}{12}$$ Multiply the fractions.

$$= \frac{xy}{6}$$ Simplify (Step 3).

(d) We can skip Step 1. Multiply $\frac{(x-1)^2}{4}$ by the reciprocal of $\frac{x-1}{8}$, or $\frac{8}{x-1}$.

$$\frac{\frac{(x-1)^2}{4}}{\frac{x-1}{8}} = \frac{(x-1)^2}{4} \cdot \frac{8}{x-1}$$ Multiply by the reciprocal of $\frac{x-1}{8}$ (Step 2).

$$= \frac{8(x-1)(x-1)}{4(x-1)}$$ Multiply the fractions.

$$= 2(x-1)$$ Simplify (Step 3).

Now Try Exercises 13, 15, 19, 29

EXAMPLE 2 **Simplifying complex fractions**

Simplify. Write your answer in lowest terms.

(a) $\dfrac{\frac{1}{a} + \frac{1}{b}}{\frac{1}{a} - \frac{1}{b}}$ **(b)** $\dfrac{x - \frac{4}{x}}{x + \frac{4}{x}}$ **(c)** $\dfrac{\frac{1}{x} + \frac{2}{x-1}}{\frac{2}{x} - \frac{1}{x-1}}$ **(d)** $\dfrac{\frac{1}{x} + \frac{1}{y}}{\frac{1}{x^2} - \frac{1}{y^2}}$

Solution

(a) In Step 1, write the numerator as one fraction by using the LCD, ab.

$$\frac{1}{a} + \frac{1}{b} = \frac{b}{ab} + \frac{a}{ab} = \frac{b+a}{ab}$$

Write the denominator as one fraction by using the LCD, ab.

$$\frac{1}{a} - \frac{1}{b} = \frac{b}{ab} - \frac{a}{ab} = \frac{b-a}{ab}$$

Finally, use these results to simplify the given complex fraction.

$$\frac{\dfrac{1}{a} + \dfrac{1}{b}}{\dfrac{1}{a} - \dfrac{1}{b}} = \frac{\dfrac{b + a}{ab}}{\dfrac{b - a}{ab}}$$ Simplify the numerator and the denominator (Step 1).

$$= \frac{b + a}{ab} \cdot \frac{ab}{b - a}$$ Multiply by the reciprocal of $\frac{b - a}{ab}$ (Step 2).

$$= \frac{ab(b + a)}{ab(b - a)}$$ Multiply.

$$= \frac{b + a}{b - a}$$ Simplify (Step 3).

(b) The LCD for both the numerator and the denominator is x.

$$\frac{x - \dfrac{4}{x}}{x + \dfrac{4}{x}} = \frac{\dfrac{x^2}{x} - \dfrac{4}{x}}{\dfrac{x^2}{x} + \dfrac{4}{x}}$$ Write by using the LCD (Step 1).

$$= \frac{\dfrac{x^2 - 4}{x}}{\dfrac{x^2 + 4}{x}}$$ Combine terms.

$$= \frac{x^2 - 4}{x} \cdot \frac{x}{x^2 + 4}$$ Multiply by the reciprocal of $\frac{x^2 + 4}{x}$ (Step 2).

$$= \frac{x(x^2 - 4)}{x(x^2 + 4)}$$ Multiply.

$$= \frac{x^2 - 4}{x^2 + 4}$$ Simplify (Step 3).

(c) The LCD for the numerator and the denominator is $x(x - 1)$.

$$\frac{\dfrac{1}{x} + \dfrac{2}{x - 1}}{\dfrac{2}{x} - \dfrac{1}{x - 1}} = \frac{\dfrac{x - 1}{x(x - 1)} + \dfrac{2x}{x(x - 1)}}{\dfrac{2(x - 1)}{x(x - 1)} - \dfrac{x}{x(x - 1)}}$$ Write by using the LCD (Step 1).

$$= \frac{\dfrac{x - 1 + 2x}{x(x - 1)}}{\dfrac{2(x - 1) - x}{x(x - 1)}}$$ Combine terms.

$$= \frac{\dfrac{3x - 1}{x(x - 1)}}{\dfrac{x - 2}{x(x - 1)}}$$ Simplify.

$$= \frac{3x - 1}{x(x - 1)} \cdot \frac{x(x - 1)}{x - 2}$$ Multiply by the reciprocal of $\frac{x - 2}{x(x - 1)}$ (Step 2).

$$= \frac{3x - 1}{x - 2}$$ Multiply and simplify (Step 3).

(d) The LCD for the numerator is xy, and the LCD for the denominator is x^2y^2.

$$\dfrac{\dfrac{1}{x} + \dfrac{1}{y}}{\dfrac{1}{x^2} - \dfrac{1}{y^2}} = \dfrac{\dfrac{y}{xy} + \dfrac{x}{xy}}{\dfrac{y^2}{x^2y^2} - \dfrac{x^2}{x^2y^2}} \qquad \text{Write by using the LCD (Step 1).}$$

$$= \dfrac{\dfrac{y + x}{xy}}{\dfrac{y^2 - x^2}{x^2y^2}} \qquad \text{Combine terms.}$$

$$= \dfrac{y + x}{xy} \cdot \dfrac{x^2y^2}{y^2 - x^2} \qquad \begin{array}{l}\text{Multiply by the reciprocal} \\ \text{of } \frac{y^2 - x^2}{x^2y^2} \text{ (Step 2).}\end{array}$$

$$= \dfrac{x^2y^2(y + x)}{xy(y^2 - x^2)} \qquad \text{Multiply.}$$

$$= \dfrac{x^2y^2(y + x)}{xy(y - x)(y + x)} \qquad \text{Factor the denominator (Step 3).}$$

$$= \dfrac{xy}{y - x} \qquad \text{Simplify.}$$

Now Try Exercises 31, 37, 41, 45.

MULTIPLYING BY THE LCD In Method II for simplifying a complex fraction, we multiply the numerator and denominator of the complex fraction by the least common denominator of all the fractions within the complex fraction.

The following steps outline Method II for simplifying complex fractions.

SIMPLIFYING COMPLEX FRACTIONS (METHOD II)

To simplify a complex fraction, perform the following steps.

STEP 1: Find the LCD of all fractions within the complex fraction.

STEP 2: Multiply the numerator and the denominator of the complex fraction by the LCD.

STEP 3: Simplify the result to lowest terms.

READING CHECK

• When using Method II, by what do we multiply the numerator and denominator of the complex fraction?

We can use this method to simplify the complex fraction in Example 2(a). Because the LCD for the numerator and the denominator is ab, we multiply the complex fraction by 1, expressed in the form $\frac{ab}{ab}$. This method is equivalent to multiplying the numerator and the denominator by ab.

$$\frac{\dfrac{1}{a} + \dfrac{1}{b}}{\dfrac{1}{a} - \dfrac{1}{b}} = \frac{\left(\dfrac{1}{a} + \dfrac{1}{b}\right)ab}{\left(\dfrac{1}{a} - \dfrac{1}{b}\right)ab}$$ Multiply by $\frac{ab}{ab} = 1$.

$$= \frac{\dfrac{ab}{a} + \dfrac{ab}{b}}{\dfrac{ab}{a} - \dfrac{ab}{b}}$$ Distributive property

$$= \frac{b + a}{b - a}$$ Simplify the fractions.

EXAMPLE 3 **Simplifying complex fractions**

Simplify.

(a) $\dfrac{2z}{\dfrac{4}{z} + \dfrac{3}{z}}$ **(b)** $\dfrac{\dfrac{1}{x - 3}}{\dfrac{1}{x} + \dfrac{3}{x - 3}}$ **(c)** $\dfrac{\dfrac{1}{a} - \dfrac{1}{b}}{\dfrac{1}{2b^2} - \dfrac{1}{2a^2}}$

Solution

(a) The LCD for all fractions within the complex fraction is z, so multiply the numerator and denominator of the complex fraction by z (Step 1).

$$\frac{2z}{\dfrac{4}{z} + \dfrac{3}{z}} = \frac{(2z)z}{\left(\dfrac{4}{z} + \dfrac{3}{z}\right)z}$$ Multiply by $\frac{z}{z} = 1$ (Step 2).

$$= \frac{2z^2}{\dfrac{4z}{z} + \dfrac{3z}{z}}$$ Distributive property

$$= \frac{2z^2}{4 + 3}$$ Simplify the fractions (Step 3).

$$= \frac{2z^2}{7}$$ Add.

(b) The LCD for all fractions within the complex fraction is the product $x(x - 3)$ (Step 1).

$$\frac{\dfrac{1}{x - 3}}{\dfrac{1}{x} + \dfrac{3}{x - 3}} = \frac{\dfrac{1}{x - 3}}{\left(\dfrac{1}{x} + \dfrac{3}{x - 3}\right)} \cdot \frac{x(x - 3)}{x(x - 3)}$$ Multiply by $\frac{x(x - 3)}{x(x - 3)} = 1$ (Step 2).

$$= \frac{\dfrac{x(x - 3)}{x - 3}}{\dfrac{x(x - 3)}{x} + \dfrac{3x(x - 3)}{x - 3}}$$ Distributive property

$$= \frac{x}{(x - 3) + 3x}$$ Simplify the fractions (Step 3).

$$= \frac{x}{4x - 3}$$ Simplify the denominator.

(c) The LCD for the numerator *and* the denominator is $2a^2b^2$ (Step 1).

$$\frac{\dfrac{1}{a} - \dfrac{1}{b}}{\dfrac{1}{2b^2} - \dfrac{1}{2a^2}} = \frac{\dfrac{1}{a} - \dfrac{1}{b}}{\dfrac{1}{2b^2} - \dfrac{1}{2a^2}} \cdot \frac{2a^2b^2}{2a^2b^2} \qquad \text{Multiply by } \frac{2a^2b^2}{2a^2b^2} = 1 \text{ (Step 2).}$$

$$= \frac{\dfrac{2a^2b^2}{a} - \dfrac{2a^2b^2}{b}}{\dfrac{2a^2b^2}{2b^2} - \dfrac{2a^2b^2}{2a^2}} \qquad \text{Distributive property}$$

$$= \frac{2ab^2 - 2a^2b}{a^2 - b^2} \qquad \text{Simplify the fractions (Step 3).}$$

$$= \frac{2ab(b - a)}{(a - b)(a + b)} \qquad \text{Factor.}$$

$$= -\frac{2ab}{a + b} \qquad \text{Simplify.}$$

▮ **Now Try Exercises 33, 35, 49**

CRITICAL THINKING

Are the expressions $\dfrac{\dfrac{a}{b}}{\dfrac{a}{b} + 1}$ and $\dfrac{1}{1 + 1}$ equal? Explain.

Are the expressions $\dfrac{\dfrac{a}{b} + 1}{\dfrac{a}{b}}$ and $1 + \dfrac{b}{a}$ equal? Explain.

7.5 Putting It All Together

CONCEPT	EXPLANATION	EXAMPLES
Complex Fraction	A rational expression that contains fractions in its numerator, denominator, or both	$\dfrac{3 + \dfrac{1}{x + 1}}{3 - \dfrac{1}{x + 1}}$ and $\dfrac{\dfrac{x}{y} - \dfrac{y}{x}}{\dfrac{x}{y} + \dfrac{y}{x}}$
Simplifying Basic Complex Fractions	When b, c, and d are nonzero, $$\frac{\dfrac{a}{b}}{\dfrac{c}{d}} = \frac{a}{b} \cdot \frac{d}{c}.$$	$\dfrac{\dfrac{2}{x}}{\dfrac{4}{x - 1}} = \dfrac{2}{x} \cdot \dfrac{x - 1}{4} = \dfrac{x - 1}{2x}$
Method I: Simplifying the Numerator and Denominator	Combine the terms in the numerator, combine the terms in the denominator, and then multiply the numerator by the reciprocal of the denominator.	$\dfrac{\dfrac{1}{x} + \dfrac{3}{x}}{\dfrac{5}{y} - \dfrac{4}{y}} = \dfrac{\dfrac{4}{x}}{\dfrac{1}{y}} = \dfrac{4}{x} \cdot \dfrac{y}{1} = \dfrac{4y}{x}$
Method II: Multiplying the Numerator and Denominator by the LCD	Multiply the numerator *and* the denominator by the LCD of *all* fractions within the expression.	$\dfrac{\dfrac{2}{x} + \dfrac{1}{y}}{\dfrac{4}{y} - \dfrac{1}{x}} = \dfrac{\left(\dfrac{2}{x} + \dfrac{1}{y}\right)xy}{\left(\dfrac{4}{y} - \dfrac{1}{x}\right)xy}$ $= \dfrac{2y + x}{4x - y}$ Note that the LCD is xy.

7.5 Exercises

MyMathLab 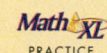 Math XL PRACTICE WATCH DOWNLOAD READ REVIEW

CONCEPTS AND VOCABULARY

1. $\dfrac{\frac{1}{2}}{\frac{3}{4}} = $ _____

2. $\dfrac{\frac{a}{b}}{\frac{c}{d}} = $ _____

3. A complex fraction is a rational expression that contains _____ in its numerator, denominator, or both.

4. Write the expression $\frac{x}{2} \div \frac{1}{x-1}$ as a complex fraction.

5. Write the expression $\frac{a}{b} \div \frac{c}{d}$ as a complex fraction.

6. What is the LCD for $\frac{1}{x+2}$ and $\frac{1}{x}$?

SIMPLIFYING COMPLEX FRACTIONS

Exercises 7–12: For the complex fraction, determine the LCD of all the fractions appearing within the expression.

7. $\dfrac{\frac{x}{5} - \frac{1}{6}}{\frac{2}{15} - 3x}$

8. $\dfrac{\frac{1}{2} - \frac{1}{x}}{\frac{1}{2} + \frac{1}{x}}$

9. $\dfrac{\frac{2}{x+1} - x}{\frac{2}{x-1} + x}$

10. $\dfrac{\frac{1}{4x} - \frac{4}{x}}{\frac{1}{2x} + \frac{1}{3x}}$

11. $\dfrac{\frac{1}{2x-1} - \frac{1}{2x+1}}{\frac{x+1}{x}}$

12. $\dfrac{\frac{1}{4x^2} - \frac{1}{2x^3}}{\frac{1}{x-1} + \frac{1}{x-1}}$

Exercises 13–52: Simplify the complex fraction.

13. $\dfrac{\frac{2}{3}}{\frac{5}{6}}$

14. $\dfrac{\frac{8}{9}}{\frac{5}{4}}$

15. $\dfrac{2\frac{1}{2}}{1\frac{3}{4}}$

16. $\dfrac{1\frac{2}{3}}{3\frac{1}{2}}$

17. $\dfrac{1\frac{1}{2}}{2\frac{1}{3}}$

18. $\dfrac{2\frac{1}{5}}{2\frac{1}{10}}$

19. $\dfrac{\frac{r}{t}}{\frac{2r}{t}}$

20. $\dfrac{\frac{8}{p}}{\frac{4}{p}}$

21. $\dfrac{\frac{6}{x}}{\frac{2}{y}}$

22. $\dfrac{\frac{3}{14x}}{\frac{6}{7x}}$

23. $\dfrac{\frac{6}{m-2}}{\frac{2}{m-2}}$

24. $\dfrac{\frac{3}{n+1}}{\frac{6}{n+1}}$

25. $\dfrac{\frac{p+1}{p}}{\frac{p+2}{p}}$

26. $\dfrac{\frac{2p}{2p+5}}{\frac{1}{4p+10}}$

27. $\dfrac{\frac{5}{z^2-1}}{\frac{z}{z^2-1}}$

28. $\dfrac{\frac{z}{z-2}}{\frac{z}{z-2}}$

29. $\dfrac{\frac{y}{y^2-9}}{\frac{1}{y+3}}$

30. $\dfrac{\frac{2y}{2y-1}}{\frac{1}{4y^2-1}}$

31. $\dfrac{x - \frac{1}{x}}{x + \frac{1}{x}}$

32. $\dfrac{4 - \frac{1}{x}}{4 + \frac{1}{x}}$

33. $\dfrac{x}{\frac{2}{x} + \frac{1}{x}}$

34. $\dfrac{5x}{1 + \frac{1}{x}}$

35. $\dfrac{\frac{3}{x+1}}{\frac{4}{x+1} - \frac{1}{x+1}}$

36. $\dfrac{\frac{5}{2x-3} - \frac{4}{2x-3}}{\frac{7}{2x-3} + \frac{8}{2x-3}}$

37. $\dfrac{\frac{1}{m^2n} + \frac{1}{mn^2}}{\frac{1}{m^2n} - \frac{1}{mn^2}}$

38. $\dfrac{\frac{3}{x-1} - \frac{2}{x}}{\frac{3}{x-1} + \frac{2}{x}}$

39. $\dfrac{\frac{1}{2x} + \frac{1}{y}}{\frac{1}{y} - \frac{1}{2x}}$

40. $\dfrac{\frac{3}{x} - \frac{2}{y}}{\frac{3}{x} + \frac{2}{y}}$

41. $\dfrac{\dfrac{1}{ab}+\dfrac{1}{a}}{\dfrac{1}{ab}-\dfrac{1}{b}}$

42. $\dfrac{\dfrac{1}{a}+\dfrac{2}{3b}}{\dfrac{1}{a}-\dfrac{5}{2b}}$

43. $\dfrac{\dfrac{2}{q}-\dfrac{1}{q+1}}{\dfrac{1}{q+1}}$

44. $\dfrac{\dfrac{5}{p}+\dfrac{4}{p-5}}{\dfrac{5}{p}-\dfrac{5}{p-5}}$

45. $\dfrac{\dfrac{1}{x+1}+\dfrac{1}{x+2}}{\dfrac{1}{x+1}-\dfrac{1}{x+2}}$

46. $\dfrac{\dfrac{1}{x-3}-\dfrac{1}{x+3}}{1-\dfrac{1}{x^2-9}}$

47. $\dfrac{\dfrac{1}{2x-1}-\dfrac{1}{2x+1}}{\dfrac{x+1}{x}}$

48. $\dfrac{\dfrac{1}{4x^2}-\dfrac{1}{x^3}}{\dfrac{1}{x-1}+\dfrac{1}{x-1}}$

49. $\dfrac{\dfrac{1}{ab^2}-\dfrac{1}{a^2b}}{\dfrac{1}{b}-\dfrac{1}{a}}$

50. $\dfrac{\dfrac{1}{x^2}-\dfrac{1}{y^2}}{\dfrac{1}{x}-\dfrac{1}{y}}$

51. $\dfrac{1}{a^{-1}+b^{-1}}$

52. $\dfrac{a^2-b^2}{a^{-2}-b^{-2}}$

APPLICATIONS

53. *Annuity* If P dollars are deposited every 2 weeks in an account paying an annual interest rate r expressed as a decimal, then the amount in the account after 2 years can be approximated by

$$\left(P\left(1+\dfrac{r}{26}\right)^{52}-P\right)\div\dfrac{r}{26}.$$

Write this expression as a complex fraction.

54. *Annuity* (Continuation of the preceding exercise) Use a calculator to evaluate the expression when $r = 0.026$ (2.6%) and $P = \$100$. Interpret the result.

55. *Resistance in Electricity* Light bulbs are often wired so that electricity can flow through either bulb, as illustrated in the accompanying figure.

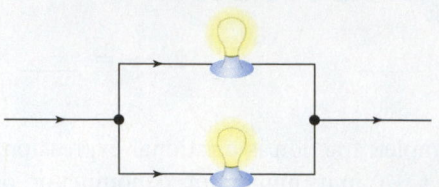

In this way, if one bulb burns out, the other bulb still works. If two light bulbs have resistances T and S, their combined resistance R is

$$R=\dfrac{1}{\dfrac{1}{T}+\dfrac{1}{S}}.$$

Simplify this formula.

56. *Resistance in Electricity* (Refer to the preceding exercise.) Evaluate the formula

$$R=\dfrac{1}{\dfrac{1}{T}+\dfrac{1}{S}}$$

when $T = 100$ and $S = 200$.

WRITING ABOUT MATHEMATICS

57. A student simplifies a complex fraction as shown below. Explain the student's mistake and how you would simplify the complex fraction correctly.

$$\dfrac{\dfrac{1}{x}+1}{\dfrac{1}{x}}\overset{?}{=}\dfrac{\dfrac{1}{x}}{\dfrac{1}{x}}+1=2$$

58. Explain one method for simplifying a complex fraction.

7.6   Rational Equations and Formulas

Solving Rational Equations • Rational Expressions and Equations • Graphical and Numerical Solutions • Solving a Formula for a Variable • Applications

A LOOK INTO MATH ▶ In Section 7.1 we demonstrated that if cars enter a construction zone at random at an average rate of x cars per minute and if 10 cars per minute can pass through the zone, then the average time T that a driver spends waiting in line and passing through the construction zone is

$$T = \frac{1}{10 - x},$$

where T is in minutes and $x < 10$. If the highway department wants to limit the average wait for a car to $\frac{1}{2}$ minute, then mathematics can be used to determine the corresponding value of x. (*Source:* N. Garber and L. Hoel, *Traffic and Highway Engineering.*)

NEW VOCABULARY

☐ Rational equation
☐ Basic rational equation
☐ Extraneous solutions

Solving Rational Equations

▶ **REAL-WORLD CONNECTION** If an equation contains one or more rational expressions, it is called a **rational equation**. Rational equations occur in mathematics whenever a rational expression is set equal to a constant. For example, the rational expression

$$\frac{1}{10 - x} \qquad \text{Rational expression}$$

from A Look Into Math can be used to estimate the average time that drivers wait to get through a construction site. If the wait is $\frac{1}{2}$ minute, we determine x by solving the *rational equation*

$$\frac{1}{10 - x} = \frac{1}{2}. \qquad \text{Rational equation}$$

To solve this equation, we multiply each side by the LCD: $2(10 - x)$.

$$\frac{2(10 - x)}{10 - x} = \frac{2(10 - x)}{2} \qquad \text{Multiply by the LCD.}$$

$$2 = 10 - x \qquad \text{Simplify.}$$

$$x = 8 \qquad \text{Add } x; \text{ subtract 2.}$$

The average wait is $\frac{1}{2}$ minute when cars arrive randomly at an average rate of 8 cars per minute. In general, if a rational equation is in the form

$$\frac{a}{b} = \frac{c}{d},$$

we can multiply each side of this equation by the common denominator bd to obtain

$$\frac{a(bd)}{b} = \frac{c(bd)}{d},$$

which simplifies to $ad = cb$. This technique can be used to solve some types of **basic rational equations**, which have a single rational expression on each side of the equals sign.

> **SOLVING BASIC RATIONAL EQUATIONS**
>
> The equations
>
> $$\frac{a}{b} = \frac{c}{d} \quad \text{and} \quad ad = bc$$
>
> are equivalent, provided that b and d are nonzero. Note that converting the first equation to the second equation is sometimes called *cross multiplying*.

READING CHECK

• Why do we cross multiply?

EXAMPLE 1 | **Solving rational equations**

Solve each equation.

(a) $\dfrac{5}{3} = \dfrac{4}{x}$ (b) $\dfrac{x+1}{5} = \dfrac{3x}{2}$ (c) $x = \dfrac{4}{3x-4}$ (d) $\dfrac{1}{x} + \dfrac{2}{x} = \dfrac{3}{7}$

Solution

(a)

$$\frac{5}{3} = \frac{4}{x} \qquad \text{Given equation}$$

$$5x = 12 \qquad \text{Cross multiply.}$$

$$x = \frac{12}{5} \qquad \text{Divide by 5.}$$

The solution is $\frac{12}{5}$.

(b)

$$\frac{x+1}{5} = \frac{3x}{2} \qquad \text{Given equation}$$

$$2(x+1) = 15x \qquad \text{Cross multiply.}$$

$$2x + 2 = 15x \qquad \text{Distributive property}$$

$$2 = 13x \qquad \text{Subtract } 2x.$$

$$\frac{2}{13} = x \qquad \text{Divide by 13.}$$

The solution is $\frac{2}{13}$.

(c)

$$\frac{x}{1} = \frac{4}{3x-4} \qquad \text{Write } x \text{ as } \frac{x}{1}.$$

$$x(3x-4) = 4 \cdot 1 \qquad \text{Cross multiply.}$$

$$3x^2 - 4x = 4 \qquad \text{Distributive property}$$

$$3x^2 - 4x - 4 = 0 \qquad \text{Subtract 4.}$$

$$(3x+2)(x-2) = 0 \qquad \text{Factor.}$$

$$3x + 2 = 0 \quad \text{or} \quad x - 2 = 0 \qquad \text{Zero-product property}$$

$$x = -\frac{2}{3} \quad \text{or} \quad x = 2 \qquad \text{Solve each equation.}$$

The solutions are $-\frac{2}{3}$ and 2.

(d)

$$\frac{1}{x} + \frac{2}{x} = \frac{3}{7} \qquad \text{Given equation}$$

$$\frac{3}{x} = \frac{3}{7} \qquad \text{Add the rational expressions.}$$

$$21 = 3x \qquad \text{Cross multiply.}$$

$$7 = x \qquad \text{Divide by 3.}$$

The solution is 7.

Now Try Exercises 9, 15, 31, 41

Another technique for solving rational equations is to multiply each side by the least common denominator. Unlike cross multiplying, this technique can *always* be used.

EXAMPLE 2 **Multiplying by the LCD**

Solve each equation. Check your answer.

(a) $\dfrac{1}{x-1} - \dfrac{1}{x} = \dfrac{1}{9x}$

(b) $\dfrac{1}{x-1} + \dfrac{1}{x+1} = \dfrac{12}{x^2-1}$

Solution

(a) Start by multiplying each term by the LCD, $9x(x-1)$.

$\dfrac{1}{x-1} - \dfrac{1}{x} = \dfrac{1}{9x}$	Given equation
$\dfrac{9x(x-1)}{x-1} - \dfrac{9x(x-1)}{x} = \dfrac{9x(x-1)}{9x}$	Multiply each term by the LCD.
$9x - 9(x-1) = x - 1$	Simplify each rational expression.
$9 = x - 1$	Distributive property; simplify.
$10 = x$	Add 1.

Check:

$\dfrac{1}{10-1} - \dfrac{1}{10} \overset{?}{=} \dfrac{1}{9(10)}$	Substitute 10 for x.
$\dfrac{10}{90} - \dfrac{9}{90} \overset{?}{=} \dfrac{1}{90}$	The LCD is 90.
$\dfrac{1}{90} = \dfrac{1}{90}$ ✓	The answer checks.

(b) Start by multiplying each term by the LCD, $x^2 - 1 = (x-1)(x+1)$.

$\dfrac{1}{x-1} + \dfrac{1}{x+1} = \dfrac{12}{x^2-1}$	Given equation
$\dfrac{(x-1)(x+1)}{x-1} + \dfrac{(x-1)(x+1)}{x+1} = \dfrac{12(x-1)(x+1)}{(x-1)(x+1)}$	Multiply each term by the LCD.
$(x+1) + (x-1) = 12$	Simplify.
$2x = 12$	Combine like terms.
$x = 6$	Divide by 2.

Check:

$\dfrac{1}{6-1} + \dfrac{1}{6+1} \overset{?}{=} \dfrac{12}{6^2-1}$	Substitute 6 for x.
$\dfrac{1}{5} + \dfrac{1}{7} \overset{?}{=} \dfrac{12}{35}$	Simplify.
$\dfrac{7}{35} + \dfrac{5}{35} \overset{?}{=} \dfrac{12}{35}$	The LCD is 35.
$\dfrac{12}{35} = \dfrac{12}{35}$ ✓	The answer checks.

Now Try Exercises 49, 55

In Example 2 we checked our answers. Although the answer checked in both cases, it may not always check. When multiplying a rational equation by the LCD, it is possible to obtain **extraneous solutions** that do not satisfy the *given* equation. This situation is demonstrated in Example 3. Before solving Example 3, we present a step-by-step strategy for solving rational equations.

STEPS FOR SOLVING A RATIONAL EQUATION

STEP 1: Find the LCD of the terms in the equation.

STEP 2: Multiply each side of the equation by the LCD.

STEP 3: Simplify each term.

STEP 4: Solve the resulting equation.

STEP 5: Check each answer in the *given* equation. Any value that makes a denominator equal 0 should be rejected because it is an extraneous solution.

READING CHECK

• What is the first step in solving a rational equation?

NOTE: If the rational equation is in the form $\frac{a}{b} = \frac{c}{d}$, or "fraction equals fraction," cross multiplying to obtain $ad = bc$ may be helpful. However, be sure to check your answers.

EXAMPLE 3   **Solving an equation having an extraneous solution**

If possible, solve $\dfrac{1}{x-2} + \dfrac{1}{x+2} = \dfrac{4}{x^2-4}$.

Solution
The LCD is $x^2 - 4 = (x-2)(x+2)$ (Step 1).

$$\frac{1}{x-2} + \frac{1}{x+2} = \frac{4}{x^2-4} \qquad \text{Given equation}$$

$$\frac{(x-2)(x+2)}{x-2} + \frac{(x-2)(x+2)}{x+2} = \frac{4(x^2-4)}{x^2-4} \qquad \text{Multiply by the LCD (Step 2).}$$

$$(x+2) + (x-2) = 4 \qquad \text{Simplify each term (Step 3).}$$

$$2x = 4 \qquad \text{Combine like terms (Step 4).}$$

$$x = 2 \qquad \text{Divide by 2.}$$

Check: $\dfrac{1}{\underline{2-2}} + \dfrac{1}{2+2} \overset{?}{=} \dfrac{4}{\underline{2^2-4}}$ ✗   Substitute 2 for x (Step 5).

Undefined terms

Note that terms on both sides of the equation are undefined because it is not possible to divide by 0. Therefore 2 is not a solution; rather, it is an *extraneous solution*. That is, there are no solutions to the *given* equation.

Now Try Exercise 53

Rational Expressions and Equations

Rational expressions and rational equations are not the same concepts. For example,

$$\frac{3}{x} \quad \text{and} \quad \frac{x}{x^2-x}$$

are both rational expressions. Neither one contains an equals sign. We often simplify or evaluate rational expressions. By factoring the denominator and applying the basic principle of rational expressions, we can simplify the second expression.

$$\frac{x}{x^2 - x} = \frac{x}{x(x - 1)} \qquad \text{Factor the denominator.}$$

$$= \frac{1}{x - 1} \qquad \text{Simplify.}$$

Thus the expressions $\frac{x}{x^2 - x}$ and $\frac{1}{x - 1}$ are equal for every value of x (except 0 and 1). Expressions can also be evaluated. For example, if we replace x with 3, the expression $\frac{3}{x}$ equals $\frac{3}{3}$, or 1.

When two expressions are set equal to each other, an equation is formed that is typically true for only a limited number of x-values. For example, $\frac{3}{x} = \frac{1}{x - 1}$ is a rational equation that is true for only one x-value.

$$\frac{3}{x} = \frac{1}{x - 1} \qquad \text{Given equation}$$

$$3(x - 1) = x \cdot 1 \qquad \text{Cross multiply.}$$

$$3x - 3 = x \qquad \text{Distributive property}$$

$$2x = 3 \qquad \text{Add 3; subtract } x.$$

$$x = \frac{3}{2} \qquad \text{Divide each side by 2.}$$

The only solution is $\frac{3}{2}$. (Check this solution.) If we replace x with any value other than $\frac{3}{2}$, the equation $\frac{3}{x} = \frac{1}{x - 1}$ is a false statement.

> **EXAMPLE 4** **Identifying expressions and equations**
>
> Determine whether you are given an expression or an equation. If it is an expression, simplify it and then evaluate it for $x = 5$. If it is an equation, solve it.
>
> **(a)** $\dfrac{x - 1}{x + 1} = \dfrac{x}{x + 3}$ **(b)** $\dfrac{x^2 - 2x}{x - 1} + \dfrac{1}{x - 1}$

Solution

(a) There is an equals sign, so it is an equation that can be solved.

$$\frac{x - 1}{x + 1} = \frac{x}{x + 3} \qquad \text{Given equation}$$

$$(x - 1)(x + 3) = x(x + 1) \qquad \text{Cross multiply.}$$

$$x^2 + 2x - 3 = x^2 + x \qquad \text{Multiply.}$$

$$2x - 3 = x \qquad \text{Subtract } x^2.$$

$$x = 3 \qquad \text{Subtract } x; \text{ add 3.}$$

$$\textit{Check:} \qquad \frac{3 - 1}{3 + 1} \overset{?}{=} \frac{3}{3 + 3} \qquad \text{Replace } x \text{ with 3 in given equation.}$$

$$\frac{1}{2} = \frac{1}{2} \ \checkmark \qquad \text{The answer checks.}$$

The solution to the equation is 3.

STUDY TIP

Be sure to look for an equals sign before you work on any problem that involves rational expressions. If the problem has an equals sign, it is an equation. If it does not, it is an expression. Refer to Making Connections on this page.

(b) There is no equals sign, so it is an expression that can be simplified. The common denominator is $x - 1$, so add the numerators.

$$\frac{x^2 - 2x}{x - 1} + \frac{1}{x - 1} = \frac{x^2 - 2x + 1}{x - 1} \qquad \text{Add the numerators.}$$

$$= \frac{(x - 1)(x - 1)}{x - 1} \qquad \text{Factor the numerator.}$$

$$= x - 1 \qquad \text{Simplify.}$$

The given expression simplifies to $x - 1$. When x equals **5**, the expression evaluates to **5** $- 1$, or 4.

▎**Now Try Exercises 63, 67**

MAKING CONNECTIONS

Expressions versus Equations

1. If a problem does not contain an equals sign, you are probably adding, subtracting, multiplying, dividing, or otherwise simplifying an expression. Your answer will be an expression, *not* a value for x.
2. If the problem has an equals sign, it is an equation to be solved. Your answer will be a value (or values) for x that makes the equation a true statement. Be sure to check all your answers.

Graphical and Numerical Solutions

Like other types of equations, rational equations can also be solved graphically and numerically. Graphs of rational expressions are not lines. They are typically curves that can be graphed by plotting several points and then sketching a graph or by using a graphing calculator.

EXAMPLE 5 **Solving a rational equation graphically and numerically**

Solve $\frac{2}{x} = x + 1$ graphically and numerically.

Solution

Graphical Solution To solve $\frac{2}{x} = x + 1$, graph $y_1 = \frac{2}{x}$ and $y_2 = x + 1$. The graph of y_2 is a line with slope 1 and y-intercept 1. To graph y_1 make a table of values, as shown in Table 7.5. Then plot the points and connect them with a smooth curve. Note that y_1 is undefined for $x = 0$. Generally, it is a good idea to plot at least three points on each side of an asymptote. These points were plotted and the curves sketched in Figure 7.5. Note that the graphs of y_1 and y_2 intersect at $(-2, -1)$ and $(1, 2)$. The solutions to the equation are -2 and 1. Check these solutions.

TABLE 7.5

x	$\frac{2}{x}$
-3	$-\frac{2}{3}$
-2	-1
-1	-2
0	—
1	2
2	1
3	$\frac{2}{3}$

$x = 0$ is an asymptote ⟶

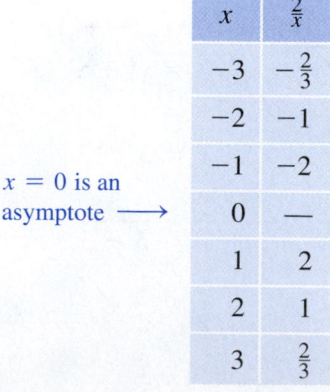
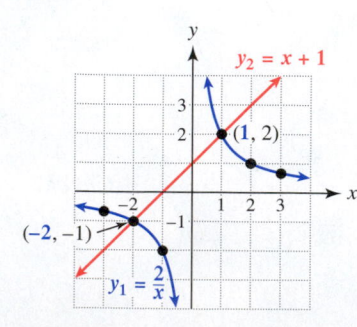

Figure 7.5

Numerical Solution Make a table of values for $y_1 = \frac{2}{x}$ and $y_2 = x + 1$, as shown in Table 7.6. Note that $\frac{2}{x} = x + 1$ when $x = -2$ or $x = 1$.

CRITICAL THINKING

If you solve Example 5 symbolically, what will be the result? Verify your answer.

TABLE 7.6

Solutions

x	-3	-2	-1	0	1	2	3
$y_1 = \frac{2}{x}$	$-\frac{2}{3}$	-1	-2	—	2	1	$\frac{2}{3}$
$y_2 = x + 1$	-2	-1	0	1	2	3	4

Now Try Exercise 75

A graphing calculator can be used to create Figure 7.5 and Table 7.6, as illustrated in Figure 7.6.

CALCULATOR HELP

To make a graph or table, see Appendix A (pages AP-2, AP-3, and AP-5).

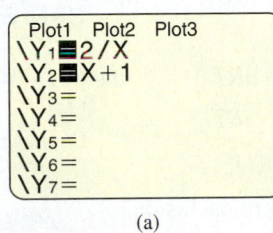
(a)

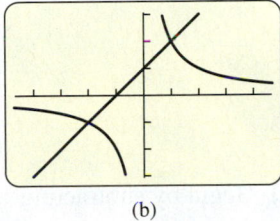
(b)

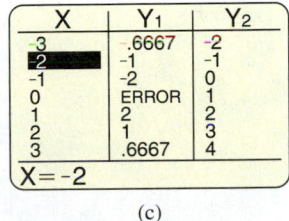
(c)

Figure 7.6

Solving a Formula for a Variable

▶ **REAL-WORLD CONNECTION** Formulas in applications often involve both the use of rational expressions and the need to solve an equation for a variable. This idea is illustrated in the next two examples.

EXAMPLE 6 **Finding time in a distance problem**

If a person travels at a speed, or rate, r for time t, then the distance d traveled is $d = rt$.
(a) How far does a person travel in 2 hours when traveling at 60 miles per hour?
(b) Solve the formula $d = rt$ for t.
(c) How long does it take a person to go 250 miles when traveling at 40 miles per hour?

Solution
(a) A person traveling at **60** miles per hour for **2** hours travels

$$d = rt = 60 \cdot 2 = 120 \text{ miles.}$$

(b) To solve $d = rt$ for t, divide each side by r.

$$d = rt \quad \text{Given equation}$$

$$\frac{d}{r} = t \quad \text{Divide each side by } r.$$

(c) At **40** miles per hour, a person can travel **250** miles in

$$t = \frac{d}{r} = \frac{250}{40} = \frac{25}{4} = 6.25 \text{ hours.}$$

Note that 6.25 hours equals 6 hours and 15 minutes.

Now Try Exercise 87

EXAMPLE 7   **Solving a formula for a variable**

Solve each equation for the specified variable.

(a) $A = \frac{bh}{2}$ for b (b) $P = \frac{nRT}{V}$ for V (c) $S = 2\pi rh + \pi r^2$ for h

Solution

(a) First, multiply each side by 2.

$$A = \frac{bh}{2}$$ Given equation

$$2A = bh$$ Multiply each side by 2.

$$\frac{2A}{h} = b$$ Divide each side by h.

(b) Begin by multiplying each side by V.

$$P = \frac{nRT}{V}$$ Given equation

$$PV = nRT$$ Multiply each side by V.

$$V = \frac{nRT}{P}$$ Divide each side by P.

(c) Begin by subtracting πr^2 from each side.

$$S = 2\pi rh + \pi r^2$$ Given equation

$$S - \pi r^2 = 2\pi rh$$ Subtract πr^2 from each side.

$$\frac{S - \pi r^2}{2\pi r} = h$$ Divide each side by $2\pi r$.

■ **Now Try Exercises 89, 91, 97**

Applications

▶ **REAL-WORLD CONNECTION** Rational equations sometimes occur in time and rate problems, as demonstrated in the next two examples.

EXAMPLE 8   **Mowing a lawn**

Two people are mowing a large lawn. One person has a riding mower, and the other person has a push mower. The person with the riding mower can cut the lawn alone in 4 hours, and the person with the push mower can cut the lawn alone in 9 hours. How long does it take for them, working together, to cut the lawn?

Solution

The first person can cut the entire lawn in 4 hours, so this person can cut $\frac{1}{4}$ of the lawn in 1 hour, $\frac{2}{4}$ of the lawn in 2 hours, and in general, $\frac{t}{4}$ of the lawn in t hours. The second person can cut the lawn in 9 hours, so (using similar reasoning) this person can cut $\frac{t}{9}$ of the lawn in t hours. Together they can cut

$$\frac{t}{4} + \frac{t}{9}$$

of the lawn in t hours. The job is complete when the fraction of the lawn cut reaches **1**. To find out how long this task takes, solve the equation

$$\frac{t}{4} + \frac{t}{9} = 1.$$

Begin by multiplying each side by the LCD, or **36**.

CRITICAL THINKING

If one person can mow a lawn in *x* hours and another person can mow it in *y* hours, how long does it take them, working together, to mow the lawn?

$$\frac{t}{4} + \frac{t}{9} = 1 \qquad \text{Equation to be solved}$$

$$\frac{36t}{4} + \frac{36t}{9} = 36(1) \qquad \text{Multiply each term by 36.}$$

$$9t + 4t = 36 \qquad \text{Simplify.}$$

$$13t = 36 \qquad \text{Combine like terms.}$$

$$t = \frac{36}{13} \qquad \text{Divide each side by 13.}$$

Working together they can cut the lawn in $\frac{36}{13} \approx 2.8$ hours.

Now Try Exercise 103

EXAMPLE 9   **Solving a distance problem**

Suppose that the winner of a 600-mile car race finishes 12 minutes ahead of the second-place finisher. If the winner averages 5 miles per hour faster than the second racer, find the average speed of each racer.

Solution
Here we apply the four-step method for solving an application problem.

STEP 1: *Identify any variables.*

$$x: \qquad \text{speed of slower car in miles per hour}$$
$$x + 5: \qquad \text{speed of faster car in miles per hour}$$

STEP 2: *Write an equation.* To determine the time required for each car to finish the race, we use the equation $t = \frac{d}{r}$. Because the race is 600 miles, the time for the slower car is $\frac{600}{x}$ and the time for the faster car is $\frac{600}{x + 5}$. The difference between these times is 12 minutes, or $\frac{12}{60} = \frac{1}{5}$ hour. Thus to determine *x*, we solve

$$\frac{600}{x} - \frac{600}{x + 5} = \frac{1}{5}.$$

NOTE: Speeds are in miles per *hour*, so time must be in hours, not minutes.

STEP 3: *Solve the equation.* Start by multiplying by the LCD, **5x(x + 5)**.

$$\frac{600}{x} - \frac{600}{x + 5} = \frac{1}{5} \qquad \text{Equation to be solved}$$

$$\frac{600 \cdot 5x(x + 5)}{x} - \frac{600 \cdot 5x(x + 5)}{x + 5} = \frac{1 \cdot 5x(x + 5)}{5} \qquad \text{Multiply each term by the LCD.}$$

$$3000(x + 5) - 3000x = x(x + 5) \qquad \text{Simplify.}$$

$$3000x + 15{,}000 - 3000x = x^2 + 5x \qquad \text{Distributive property}$$

$$15{,}000 = x^2 + 5x \qquad \text{Combine like terms.}$$

$$x^2 + 5x - 15{,}000 = 0 \qquad \text{Rewrite the equation.}$$

$$(x - 120)(x + 125) = 0 \qquad \text{Factor.}$$

$$x - 120 = 0 \quad \text{or} \quad x + 125 = 0 \qquad \text{Zero-product property}$$

$$x = 120 \quad \text{or} \quad x = -125 \qquad \text{Solve.}$$

The slower car travels at 120 miles per hour, and the faster car travels 5 miles per hour faster, or 125 miles per hour. (The solution -125 has no meaning in this problem.)

STEP 4: *Check your answer.* The slower car travels 600 miles at 120 miles per hour, which takes $\frac{600}{120} = 5$ hours. The faster car travels 600 miles at 125 miles per hour, which requires $\frac{600}{125} = 4.8$ hours. The difference between their times is $5 - 4.8 = 0.2$ hour, or $0.2 \times 60 = 12$ minutes, so the answer checks.

■ Now Try Exercise 105

CALCULATOR HELP

To make a table, see Appendix A (pages AP-2 and AP-3).

TECHNOLOGY NOTE

Solving Rational Equations Numerically

Tables can be used to find solutions to rational equations. The following displays show the positive solution of 120 from Example 9. Note the use of parentheses for entering the formula for Y_1.

7.6 Putting It All Together

CONCEPT	EXPLANATION	EXAMPLES
Basic Rational Equations	To solve an equation of the form $$\frac{a}{b} = \frac{c}{d}, \quad b \neq 0, d \neq 0,$$ cross multiply to obtain $ad = bc$.	$\frac{\mathbf{3}}{\mathbf{6}} = \frac{\mathbf{5}}{\mathbf{2x}}$ is equivalent to $\mathbf{6x = 30}$. $\frac{x}{2} = \frac{8}{x}$ is equivalent to $x^2 = 16$. (Cross multiplication works only for equations with *one* rational expression on each side: "fraction equals fraction.")
Rational Equations	To solve a rational equation, use the following steps. **STEP 1:** Find the LCD. **STEP 2:** Multiply each side of the equation by the LCD. **STEP 3:** Simplify each term. **STEP 4:** Solve the resulting equation. **STEP 5:** Check each possible solution.	Solve $\frac{1}{x} - \frac{2}{3x} = \frac{5}{6}$. **STEP 1:** The LCD is $6x$. **STEP 2:** $\frac{1(6x)}{x} - \frac{2(6x)}{3x} = \frac{5(6x)}{6}$ **STEP 3:** $6 - 4 = 5x$ **STEP 4:** $\frac{2}{5} = x$ **STEP 5:** $\frac{1}{\frac{2}{5}} - \frac{2}{3\left(\frac{2}{5}\right)} = \frac{5}{6}$ ✓ It checks.

7.6   Exercises

CONCEPTS AND VOCABULARY

1. If an equation contains one or more rational expressions, it is called a(n) _____ equation.

2. Give an example of a rational expression and an example of a rational equation.

3. The equation $\frac{a}{b} = \frac{c}{d}$ is equivalent to _____, provided that _____ and _____ are nonzero.

4. Are the equations $\frac{2}{x-1} = 5$ and $5(x-1) = 2$ equivalent provided that $x \neq 1$?

5. One way to solve the equation $\frac{5}{3x} + \frac{3}{4x} = 1$ is to multiply each side by the LCD, which is _____.

6. To solve the equation $T = \frac{R}{SV}$ for V, multiply each side by the variable _____ and then divide each side by the variable _____.

SOLVING RATIONAL EQUATIONS

Exercises 7–62: Solve and check your answer.

7. $\frac{x}{2} = \frac{3}{4}$

8. $\frac{2x}{3} = \frac{2}{5}$

9. $\frac{3}{z} = \frac{6}{5}$

10. $\frac{2}{7} = \frac{1}{z}$

11. $\frac{3y}{4} = \frac{7y}{2}$

12. $\frac{y}{6} = \frac{5y}{3}$

13. $\frac{2}{3} = \frac{1}{2x+1}$

14. $\frac{1}{x+4} = \frac{3}{5}$

15. $\frac{5}{2x} = \frac{8}{x+2}$

16. $\frac{1}{x-1} = \frac{5}{3x}$

17. $\frac{1}{z-1} = \frac{2}{z+1}$

18. $\frac{4}{z+3} = \frac{2}{z-2}$

19. $\frac{3}{n+5} = \frac{2}{n-5}$

20. $\frac{4}{3n+2} = \frac{1}{n-1}$

21. $\frac{m}{m-1} = \frac{5}{4}$

22. $\frac{5m}{2m-1} = \frac{3}{2}$

23. $\frac{5x}{5-x} = \frac{1}{3}$

24. $\frac{x+2}{3x} = \frac{4}{3}$

25. $\frac{6}{5-2x} = 2$

26. $\frac{x+1}{x} = 6$

27. $\frac{2x}{2x+1} = \frac{-1}{2x+1}$

28. $\frac{x}{x-4} = \frac{4}{x-4}$

29. $\frac{1}{1-x} = \frac{3}{1+x}$

30. $\frac{2x}{1-2x} = \frac{1}{2}$

31. $\frac{1}{z+2} = -z$

32. $\frac{1}{z-2} = \frac{z}{3}$

33. $\frac{-1}{2x+5} = \frac{x}{3}$

34. $\frac{x}{2} = \frac{1}{3x+5}$

35. $\frac{x}{2} + \frac{x}{4} = 3$

36. $\frac{x}{4} - \frac{x}{3} = 1$

37. $\frac{3x}{4} - \frac{x}{2} = 1$

38. $\frac{2x}{3} + \frac{x}{3} = 6$

39. $\frac{4}{t+1} + \frac{1}{t+1} = -1$

40. $\frac{2}{t-5} - \frac{5}{t-5} = 3$

41. $\frac{1}{x} + \frac{2}{x} = \frac{1}{2}$

42. $\frac{1}{2x} - \frac{2}{x} = -3$

43. $\frac{2}{x-1} + 1 = \frac{4}{x^2-1}$

44. $\frac{1}{x} + 2 = \frac{1}{x^2+x}$

45. $\frac{1}{x+2} = \frac{4}{4-x^2} - 1$

46. $\frac{1}{x-3} + 1 = \frac{6}{x^2-9}$

47. $\frac{5}{4z} - \frac{2}{3z} = 1$

48. $\frac{3}{z+1} - \frac{1}{z+1} = 2$

49. $\frac{4}{y-1} + \frac{1}{y} = \frac{6}{5}$

50. $\frac{6}{y+1} + \frac{6}{y} = 5$

51. $\frac{1}{2x} - \frac{1}{x+3} = 0$

52. $\frac{2}{x} - \frac{6}{2x-1} = -1$

53. $\frac{1}{x-1} + \frac{1}{x+1} = \frac{2}{x^2-1}$

54. $\frac{1}{2x+1} + \frac{1}{2x-1} = \frac{2}{4x^2-1}$

55. $\frac{1}{x-2} + \frac{1}{x+2} = \frac{6}{x^2-4}$

56. $\frac{2}{x+3} - \frac{1}{x-3} = \frac{1}{x^2-9}$

57. $\frac{1}{p+1} + \frac{1}{p+2} = \frac{1}{p^2+3p+2}$

58. $\frac{1}{p-1} - \frac{1}{p+3} = \frac{1}{p^2+2p-3}$

59. $\dfrac{1}{x-2} + \dfrac{3}{2x-4} = \dfrac{6}{3x-6}$

60. $\dfrac{4}{x+1} - \dfrac{4}{2x+2} = \dfrac{1}{(x+1)^2}$

61. $\dfrac{1}{r^2-r-2} + \dfrac{2}{r^2-2r} = \dfrac{1}{r^2+r}$

62. $\dfrac{3}{r^2-1} + \dfrac{1}{r^2+r} = \dfrac{3}{r^2-r}$

Exercises 63–70: **Expressions and Equations** *(Refer to Example 4.) Determine whether you are given an expression or an equation. If it is an expression, simplify (if possible) and evaluate it for $x = 2$. If it is an equation, solve it.*

63. $\dfrac{1}{x} - \dfrac{1-x}{x}$

64. $\dfrac{1}{x} - x = 0$

65. $\dfrac{1}{2x} - \dfrac{1}{4x} = \dfrac{1}{8}$

66. $\dfrac{1}{2x} - \dfrac{1}{4x}$

67. $\dfrac{x+1}{x-1} = \dfrac{2x-3}{2x-5}$

68. $\dfrac{2x-1}{4x+1} = \dfrac{x+1}{2x-1}$

69. $\dfrac{4x+4}{x+2} + \dfrac{x^2}{x+2}$

70. $\dfrac{x^2-2}{x-2} - \dfrac{x}{x-2}$

GRAPHICAL AND NUMERICAL SOLUTIONS

Exercises 71–74: Use the graph to solve the given equation. Check your answers.

71. $\dfrac{1}{x} = 4x$

72. $\dfrac{x}{x-1} = x$

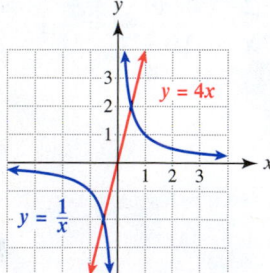

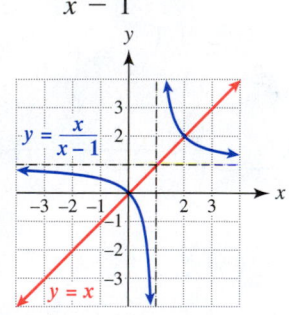

73. $\dfrac{x-1}{x} = -2x$

74. $\dfrac{3}{x^2-1} = 1$

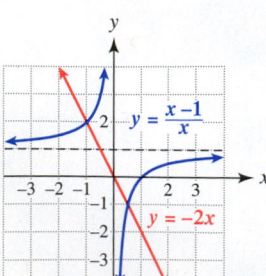

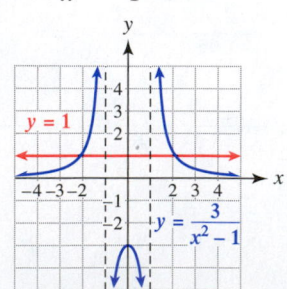

Exercises 75–82: (Refer to Example 5.) Solve the equation
(a) *graphically and*
(b) *numerically.*

75. $\dfrac{3}{x} = x + 2$

76. $-\dfrac{2}{x} = 1 - x$

77. $\dfrac{3x}{2} = \dfrac{1}{2}x - 1$

78. $\dfrac{x}{3} = 2 - \dfrac{2}{3}x$

79. $\dfrac{3}{x-1} = 3$

80. $\dfrac{2}{x+5} = 1$

81. $\dfrac{4}{x^2} = 1$

82. $\dfrac{-18}{x^2} = -2$

Exercises 83–86: Solve the rational equation graphically to the nearest thousandth.

83. $\dfrac{1}{\pi x - 2} + \dfrac{\sqrt{2}}{2.1x} = 1.3$

84. $\dfrac{6}{x^2-4} + \dfrac{\pi}{2x} = \dfrac{3}{2}$

85. $\dfrac{1}{x^3} - \dfrac{\pi}{4} = 0$

86. $\dfrac{1}{2\pi x^2} - \dfrac{x}{2} = 1$

SOLVING AN EQUATION FOR A VARIABLE

Exercises 87–98: (Refer to Examples 6 and 7.) Solve the equation for the specified variable.

87. $m = \dfrac{F}{a}$ for a

88. $m = \dfrac{2K}{v^2}$ for K

89. $I = \dfrac{V}{R+r}$ for r

90. $\dfrac{1}{T} = \dfrac{r}{R-r}$ for R

91. $h = \dfrac{2A}{b}$ for b

92. $h = \dfrac{2A}{b_1+b_2}$ for b_1

93. $\dfrac{3}{k} = \dfrac{z}{z+5}$ for z

94. $\dfrac{5}{r} = \dfrac{t+r}{t}$ for t

95. $T = \dfrac{ab}{a+b}$ for b

96. $A = \dfrac{2b}{a-b}$ for b

97. $\dfrac{3}{k} = \dfrac{1}{x} - \dfrac{2}{y}$ for x

98. $\dfrac{1}{R} = \dfrac{1}{R_1} + \dfrac{1}{R_2}$ for R_1

APPLICATIONS

99. **Waiting in Traffic** (Refer to A Look Into Math for this section.) Solve the equation

$$\dfrac{1}{10-x} = 1$$

to determine the traffic rate x in cars per minute corresponding to an average waiting time of 1 minute.

100. *Waiting in Line* At a post office, customers arrive at random at an average rate of x people per minute. The clerk can wait on 4 customers per minute. The average time T in minutes spent waiting in line is given by

$$T = \frac{1}{4 - x},$$

where $x < 4$.
(a) Evaluate T for $x = 3, 3.9$, and 3.99. What happens to the waiting time as the arrival rate nears 4 people per minute?
(b) Find x when the waiting time is 5 minutes.

101. *Shoveling a Sidewalk* It takes an older employee 4 hours to shovel the snow from a sidewalk, but a younger employee can shovel the same sidewalk in 3 hours. How long will it take them to clear the walk if they work together?

102. *Pumping Water* One pump can empty a pool in 5 days, whereas a second pump can empty the pool in 7 days. How long will it take the two pumps, working together, to empty the pool?

103. *Painting a House* One painter can paint a house in 8 days, yet a more experienced painter can paint the house in 4 days. How long will it take the two painters, working together, to paint the house?

104. *Highway Curves* To make a highway curve safe, highway engineers often bank it, as shown in the figure. If a curve is designed for a speed of 50 miles per hour and is banked with positive slope m, then a minimum radius R in feet for the curve is given by

$$R = \frac{2500}{15m + 2}.$$

(*Source:* N. Garber and L. Hoel, *Traffic and Highway Engineering.*)

(a) Find R for $m = 0.1$. Interpret the result.
(b) If $R = 500$, find m. Interpret the result.

105. *Bicycle Race* The winner of a 6-mile bicycle race finishes 2 minutes ahead of a teammate and travels, on average, 2 miles per hour faster than the teammate. Find the average speed of each racer.

106. *Freeway Travel* Two drivers travel 150 miles on a freeway and then stop at a wayside rest area. The first driver travels 5 miles per hour faster and arrives $\frac{1}{7}$ hour ahead of the second. Find the average speed of each car.

107. *Braking Distance* If a car is traveling *downhill* at 30 miles per hour on wet pavement, then the braking distance B in feet for this car is given by

$$B = \frac{30}{0.3 + m},$$

where $m < 0$ is the slope of the hill. (*Source:* L. Haefner, *Introduction to Transportation Systems.*)
(a) Find the braking distance for $m = -0.05$ and interpret the result.
(b) Find m if the braking distance is 150 feet.

108. *Slippery Roads* If a car is traveling at 30 miles per hour on a level road, then its braking distance in feet is $\frac{30}{x}$, where x is the coefficient of friction between the road and the tires. The variable x is positive and satisfies $x \le 1$. The closer the value of x is to 0, the more slippery the road is. (*Source:* L. Haefner.)
(a) Evaluate the expression for $x = 1, 0.5$, and 0.1. Interpret the results.
(b) Find x for a braking distance of 150 feet.

109. *River Current* A boat can travel 36 miles upstream in the same time that it can travel 54 miles downstream. If the speed of the current is 3 miles per hour, find the speed of the boat without a current.

110. *Airplane Speed* An airplane can travel 380 miles into the wind in the same time that it can travel 420 miles with the wind. If the wind speed is 10 miles per hour, find the speed of the airplane without any wind.

111. *Airplane Speed* An airplane can travel 450 miles into the wind in the same time that it can travel 750 miles with the wind. If the wind speed is 50 miles per hour, find the speed of the airplane without any wind.

112. *River Current* A boat can travel 114 miles upstream in the same time that it can travel 186 miles downstream. If the speed of the current is 6 miles per hour, find the speed of the boat without a current.

113. *Running and Jogging* An athlete runs 10 miles and then jogs home. The trip home takes 1 hour longer than it took to run that distance. If the athlete runs 5 miles per hour faster than she jogs, what are her average running and jogging speeds?

114. *Speed Limit* A person drives 390 miles on a stretch of road. Half the distance is driven traveling 5 miles per hour below the speed limit, and half the distance is driven traveling 5 miles per hour above the speed limit. If the time spent traveling at the slower speed exceeds the time spent traveling at the faster speed by 24 minutes, find the speed limit.

WRITING ABOUT MATHEMATICS

115. Do all rational equations have solutions? Explain.

116. Why is it important to check your answer when solving rational equations? Explain.

SECTIONS 7.5 and 7.6   Checking Basic Concepts

1. Simplify each complex fraction.

(a) $\dfrac{\dfrac{x}{3}}{\dfrac{2x}{5}}$ **(b)** $\dfrac{\dfrac{2}{2x} - \dfrac{1}{3x}}{6x}$

(c) $\dfrac{\dfrac{1}{a} - \dfrac{1}{b}}{\dfrac{1}{a} + \dfrac{1}{b}}$ **(d)** $\dfrac{\dfrac{1}{r^2} - \dfrac{1}{t^2}}{\dfrac{2}{r} - \dfrac{2}{t}}$

2. Solve each equation. Check your answer.

(a) $\dfrac{1}{2x} = \dfrac{3}{x + 1}$ **(b)** $\dfrac{x}{2x + 3} = \dfrac{4}{5}$

(c) $\dfrac{1}{2x} + \dfrac{3}{2x} = 1$ **(d)** $\dfrac{3}{x + 1} - \dfrac{2}{x} = -2$

(e) $\dfrac{1}{x - 1} = \dfrac{2}{x^2 - 1} - \dfrac{1}{2}$

3. Solve each equation for the specified variable.

(a) $\dfrac{ax}{2} - 3y = b$ for x

(b) $\dfrac{1}{2m - 1} = \dfrac{k}{m}$ for m

4. *Braking Distance* If a car is traveling *uphill* at 60 miles per hour on wet pavement, then the braking distance D in feet for this car is given by

$$D = \frac{120}{0.3 + m},$$

where $m > 0$ is the slope of the hill. (*Source:* L. Haefner, *Introduction to Transportation Systems.*)

(a) Find D for $m = 0.1$ and interpret the result.

(b) Find the slope of the road if D is 200 feet. Interpret the result.

7.7   Proportions and Variation

Proportions • Direct Variation • Inverse Variation • Analyzing Data • Joint Variation

A LOOK INTO MATH ▶   Proportions are frequently used to solve applications. The following are a few examples.

- If someone earns $120 per day, then that person can earn $600 in 5 days.
- If a car goes 280 miles on 10 gallons of gas, then it can go 560 miles on 20 gallons of gas.
- If a person walks 1 mile in 20 minutes, then that person can walk $\frac{1}{2}$ mile in 10 minutes.

In this section we discuss several applications of proportions.

NEW VOCABULARY

- ☐ Ratio
- ☐ Proportion
- ☐ Directly proportional (varies directly)
- ☐ Constant of proportionality/ variation
- ☐ Inversely proportional (varies inversely)
- ☐ Varies jointly

Proportions

▶ **REAL-WORLD CONNECTION** A **ratio** is a comparison of two quantities. For example, a math class might have 7 boys for every 8 girls. Thus the boy–girl ratio in this class is *7 to 8*, or $\frac{7}{8}$. In mathematics, ratios are typically expressed as fractions.

Ratios and proportions are sometimes used to find how much space remains for music on a compact disc (CD). A 700-megabyte CD can store about 80 minutes of music. Suppose that some music has been recorded on the CD and that 256 megabytes are still available. Using ratios and proportions, we can find how many more minutes of music could be recorded. A **proportion** is a statement that two *ratios* are equal.

Let x represent the number of minutes available on a CD. Then **80** minutes are to **700** megabytes as **x** minutes are to **256** megabytes. By setting the *ratios* $\frac{80}{700}$ and $\frac{x}{256}$ equal to each other, we obtain the *proportion*

$$\frac{80}{700} = \frac{x}{256}. \qquad \frac{\text{Minutes}}{\text{Megabytes}} = \frac{\text{Minutes}}{\text{Megabytes}}$$

Solving this equation for x gives

$$700x = 80(256) \qquad \text{Cross multiply.}$$

$$x = \frac{80 \cdot 256}{700} \approx 29.3 \text{ minutes.} \qquad \text{Divide by 700.}$$

About 29 minutes are available to record on the CD.

STUDY TIP

Look back at your progress so far this semester. Are there parts of your study process that need some adjustment? Are your notes and assignments organized? Are you spending enough time on homework and practice problems?

MAKING CONNECTIONS

Proportions and Fractional Parts

We could have solved the preceding problem by noting that the fraction of the CD still available for recording music is $\frac{256}{700}$. So $\frac{256}{700}$ of 80 minutes is

$$\frac{256}{700} \cdot 80 \approx 29.3 \text{ minutes.}$$

EXAMPLE 1 ## Calculating the water content in snow

Six inches of light, fluffy snow are equivalent to about half an inch of rain in terms of water content. If 15 inches of this type of snow fall, estimate the water content.

Solution

Let x be the equivalent amount of rain. Then **6** inches of snow are to $\frac{1}{2}$ inch of rain as **15** inches of snow are to **x** inches of rain, which can be written as the proportion

$$\frac{6}{\frac{1}{2}} = \frac{15}{x}. \qquad \frac{\text{Snow}}{\text{Rain}} = \frac{\text{Snow}}{\text{Rain}}$$

Solving this equation gives

$$6x = \frac{15}{2} \quad \text{or} \quad x = \frac{15}{12} = 1.25.$$

Thus 15 inches of light, fluffy snow are equivalent to about 1.25 inches of rain.

Now Try Exercise 65

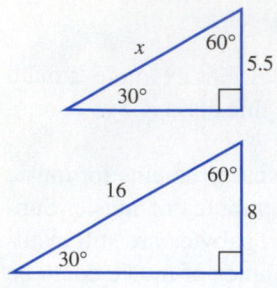

Figure 7.7

Proportions frequently occur in geometry when we work with similar figures. Two triangles are similar if the measures of their corresponding angles are equal. Corresponding sides of similar triangles are proportional. Figure 7.7 shows two right triangles that are similar because each has angles of 30°, 60°, and 90°.

We can find the length of side x by using proportions. Side x is to 16 as 5.5 is to 8, which can be written as the proportion

$$\frac{x}{16} = \frac{5.5}{8}. \qquad \frac{\text{Hypotenuse}}{\text{Hypotenuse}} = \frac{\text{Shorter leg}}{\text{Shorter leg}}$$

Solving yields the equation

$$8x = 5.5(16) \qquad \text{Cross multiply.}$$
$$x = 11. \qquad \text{Divide each side by 8.}$$

NOTE: Proportions can be set up in different ways and still produce the correct result. For example, we could say that x is to 5.5 in the smaller triangle as 16 is to 8 in the larger triangle.

$$\frac{x}{5.5} = \frac{16}{8} \qquad \frac{\text{Hypotenuse}}{\text{Shorter leg}} = \frac{\text{Hypotenuse}}{\text{Shorter leg}}$$

Solving, we obtain $8x = 5.5(16)$, or $x = 11$, which is the same answer.

EXAMPLE 2 **Calculating the height of a tree**

A 6-foot-tall person casts a 4-foot-long shadow. If a nearby tree casts a 36-foot-long shadow, estimate the height of the tree. See Figure 7.8.

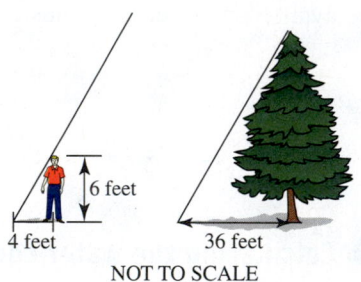

NOT TO SCALE

Figure 7.8

Solution

The triangles shown in Figure 7.9 represent the situation shown in Figure 7.8. These triangles are similar because the measures of the corresponding angles are equal. Therefore their sides are proportional. Let h be the height of the tree.

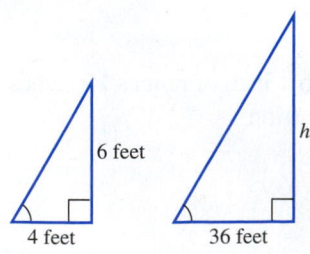

Figure 7.9

$$\frac{h}{6} = \frac{36}{4} \qquad \frac{\text{Height}}{\text{Height}} = \frac{\text{Shadow length}}{\text{Shadow length}}$$
$$4h = 6(36) \qquad \text{Cross multiply.}$$
$$h = \frac{6(36)}{4} \qquad \text{Divide each side by 4.}$$
$$h = 54 \qquad \text{Simplify.}$$

The tree is 54 feet tall.

Now Try Exercise 56

Direct Variation

▶ **REAL-WORLD CONNECTION** If your wage is $12 per hour, then your total pay P is proportional to the number of hours H that you work, which can be represented by the equation

$$\frac{P}{H} = \frac{12}{1}, \quad \frac{\text{Pay}}{\text{Hours}} = \frac{\text{Pay}}{\text{Hours}}$$

or, equivalently,

$$P = 12H.$$

We say that your pay P is *directly proportional* to the number of hours H that you work. The *constant of proportionality* is 12.

DIRECT VARIATION

Let x and y denote two quantities. Then y is **directly proportional** to x, or y **varies directly** with x, if there is a nonzero number k such that

$$y = kx.$$

The number k is called the **constant of proportionality**, or the **constant of variation**.

The following 4-step process is helpful when solving variation applications. This process is used in Examples 3 and 4 and can also be used for other types of variation problems.

SOLVING A VARIATION APPLICATION

When solving a variation problem, the following steps can be used.

STEP 1: Write the general equation for the type of variation problem that you are solving.

STEP 2: Substitute given values in this equation so the constant of variation k is the only unknown value in the equation. Solve for k.

STEP 3: Substitute the value of k in the general equation in Step 1.

STEP 4: Use this equation to find the requested quantity.

EXAMPLE 3 | **Solving a direct variation problem**

Let y be directly proportional to x, or vary directly with x. Suppose $y = 7$ when $x = 5$. Find y when $x = 11$.

Solution

STEP 1: The general equation for direct variation is $y = kx$.

STEP 2: Substitute **7** for y and **5** for x in $y = k\boldsymbol{x}$. Solve for k.

$$7 = k(\boldsymbol{5}) \quad \text{Let } y = 7 \text{ and } x = 5.$$

$$\frac{7}{5} = k \quad \text{Divide each side by 5.}$$

STEP 3: Replace k with $\frac{7}{5}$ in the equation $y = kx$ to obtain $y = \frac{7}{5}x$.

STEP 4: To find y when $x = 11$, let $x = \boldsymbol{11}$ in $y = \frac{7}{5}\boldsymbol{x}$.

$$y = \frac{7}{5}(\boldsymbol{11}) = \frac{77}{5} = 15.4$$

Now Try Exercise 33

EXAMPLE 4 **Solving a direct variation application**

The amount of weight that a beam of wood can support varies directly with its width. A beam that is 2.5 inches wide can support 800 pounds. How much weight can a similar beam support if its width is 3.2 inches?

Solution

STEP 1: The general equation for direct variation is $y = kx$, where y is the weight and x is the width.

STEP 2: Substitute **800** for y and **2.5** for x in $y = kx$. Solve for k.

$$\mathbf{800} = k(\mathbf{2.5}) \quad \text{Let } y = 800 \text{ and } x = 2.5.$$

$$\frac{\mathbf{800}}{\mathbf{2.5}} = k \quad \text{Divide each side by 2.5.}$$

$$k = 320 \quad \text{Simplify; rewrite the equation.}$$

STEP 3: Replace k with **320** in $y = kx$ to obtain $y = \mathbf{320}x$.

STEP 4: To find the weight y when $x = 3.2$, substitute **3.2** for x in $y = 320x$.

$$y = 320(\mathbf{3.2}) = 1024 \text{ pounds}$$

Now Try Exercise 67

The graph of $y = kx$ is a line passing through the origin, as illustrated in Figure 7.10. Sometimes data in a scatterplot indicate that two quantities are directly proportional. The constant of proportionality k corresponds to the slope (which may be negative) of the line passing through the points in the scatterplot.

READING CHECK

• How does the constant of proportionality for data that represent direct variation compare to the slope of a line that passes through the data?

Direct Variation

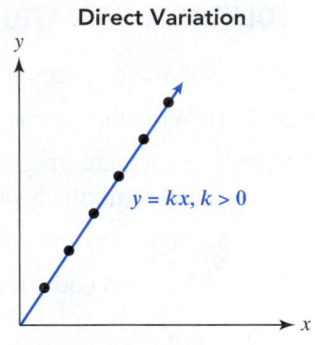

Figure 7.10

EXAMPLE 5 **Modeling college tuition**

Table 7.7 lists the tuition for taking various numbers of credits.
(a) A scatterplot of the data is shown in Figure 7.11. Could the data be modeled using a line?

Table 7.7

Credits	Tuition
3	$189
5	$315
8	$504
11	$693
17	$1071

Tuition

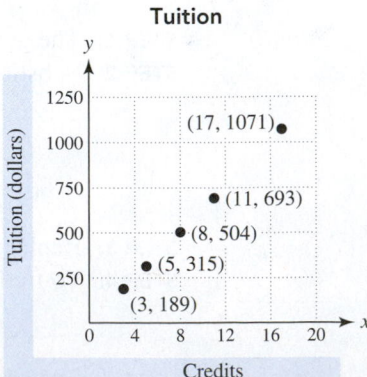

Figure 7.11

(b) Explain why tuition is directly proportional to the number of credits taken.

(c) Find the constant of proportionality. Interpret your result.

(d) Predict the cost of taking 16 credits.

Solution

(a) The data are linear and suggest a line passing through the origin.

(b) Because the data can be modeled by a line passing through the origin, tuition is directly proportional to the number of credits taken. Hence doubling the credits will double the tuition and tripling the credits will triple the tuition.

(c) The slope of the line equals the constant of proportionality k. If we use the first and last data points $(3, 189)$ and $(17, 1071)$, the slope is

$$k = \frac{1071 - 189}{17 - 3} = 63.$$

That is, tuition is $63 per credit. If we graph the line $y = 63x$, it models the data, as shown in Figure 7.12. This graph can also be created with a graphing calculator.

Tuition

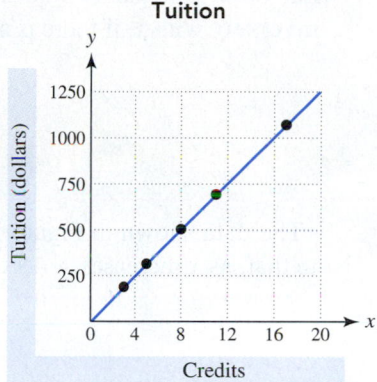

Figure 7.12

(d) If y represents tuition and x represents the credits taken, **16** credits would cost

$$y = 63(16) = \$1008.$$

Now Try Exercise 73

MAKING CONNECTIONS

Ratios and the Constant of Proportionality

The constant of proportionality in Example 5 can also be found by calculating the ratios $\frac{y}{x}$, where y is the tuition and x is the credits taken. Note that each ratio in the table is 63 because the equation $y = 63x$ is equivalent to the equation $\frac{y}{x} = 63$.

x	3	5	8	11	17
y	189	315	504	693	1071
$\frac{y}{x}$	63	63	63	63	63

Inverse Variation

▶ **REAL-WORLD CONNECTION** When two quantities vary inversely, an increase in one quantity results in a decrease in the second quantity. For example, at 30 miles per hour a car travels 120 miles in 4 hours, whereas at 60 miles per hour the car travels 120 miles in 2 hours.

Doubling the speed (or rate) decreases the travel time by half. Distance equals rate times time, so $d = rt$. Thus

$$120 = rt, \quad \text{or equivalently,} \quad t = \frac{120}{r}.$$

We say that the time t to travel 120 miles is *inversely proportional* to the speed or rate r. The constant of proportionality or constant of variation is 120.

READING CHECK

• When two quantities are inversely proportional, how does an increase in one quantity affect the other quantity?

INVERSE VARIATION

Let x and y denote two quantities. Then y is **inversely proportional** to x, or y **varies inversely** with x, if there is a nonzero number k such that

$$y = \frac{k}{x}.$$

The data shown in Figure 7.13 represent inverse variation and are modeled by $y = \frac{k}{x}$. Note that, as x increases, y decreases. We assume k is positive.

Inverse Variation

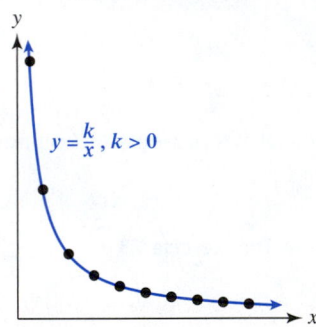

$$y = \frac{k}{x}, k > 0$$

Figure 7.13

EXAMPLE 6 | **Solving an inverse variation problem**

Let y be inversely proportional to x, or vary inversely with x. Suppose $y = 5$ when $x = 6$. Find y when $x = 21$.

Solution
STEP 1: The general equation for inverse variation is $y = \frac{k}{x}$.
STEP 2: Because $y = 5$ when $x = 6$, substitute **5** for y and **6** for x in $y = \frac{k}{x}$. Solve for k.

$$5 = \frac{k}{6} \quad \text{Let } y = 5 \text{ and } x = 6.$$

$$30 = k \quad \text{Multiply each side by 6.}$$

STEP 3: Replace k with **30** in the equation $y = \frac{k}{x}$ to obtain $y = \frac{30}{x}$.
STEP 4: To find y, let $x = \textbf{21}$. Then $y = \frac{30}{21} = \frac{10}{7}$.

Now Try Exercise 39

▶ **REAL-WORLD CONNECTION** A wrench is commonly used to loosen a nut on a bolt. See Figure 7.14. If the nut is difficult to loosen, a wrench with a longer handle is often helpful.

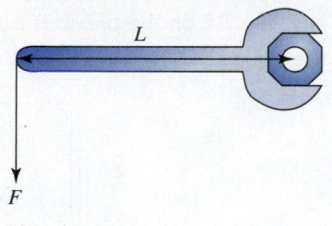

Figure 7.14

EXAMPLE 7   **Illustrating inverse variation with a wrench**

Table 7.8 lists the force F necessary to loosen a particular nut with wrenches of different lengths L.

Table 7.8

L (inches)	6	8	12	18	24
F (pounds)	12	9	6	4	3

(a) Make a scatterplot of the data and discuss the graph. Are the data linear?
(b) Explain why the force F is inversely proportional to the handle length L. Find k so that $F = \frac{k}{L}$ models the data.
(c) Predict the force needed to loosen the nut with a 15-inch wrench.

Solution
(a) The scatterplot shown in Figure 7.15 reveals that the data are nonlinear. As the length L of the wrench increases, the force F necessary to loosen the nut decreases.

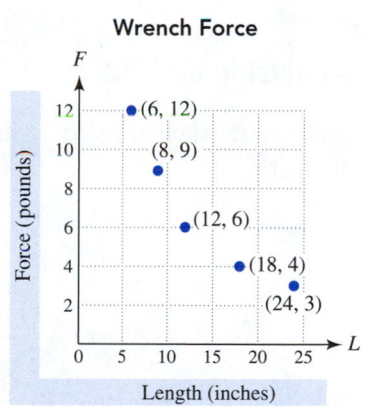

Figure 7.15

(b) If F is inversely proportional to L, then $F = \frac{k}{L}$, or $FL = k$. That is, the product of F and L equals the constant of proportionality k. In Table 7.8, the product of F and L always equals 72 for each data point. Thus F is inversely proportional to L with constant of proportionality $k = 72$, so $F = \frac{72}{L}$.
(c) If $L = \mathbf{15}$, then $F = \frac{72}{15} = 4.8$. A wrench with a 15-inch handle requires a force of 4.8 pounds to loosen the nut.

Now Try Exercise 75

TECHNOLOGY NOTE

Scatterplots and Graphs

A graphing calculator can be used to create scatterplots and graphs. A scatterplot of the data in Table 7.8 on the previous page is shown in the first figure. In the second figure, the data and the equation $y = \frac{72}{x}$ are graphed. Note that each tick mark represents 5 units.

CALCULATOR HELP

To make a scatterplot, see Appendix A (pages AP-3 and AP-4).

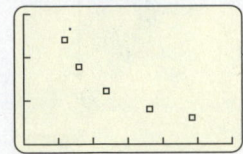 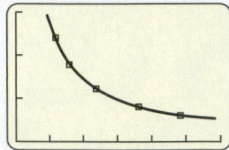

Analyzing Data

So far in this section, we have discussed *direct* and *inverse* variation. Table 7.9 gives a summary of these two types of variation.

TABLE 7.9 **Direct and Inverse Variation**

Type of Variation	Equation	Constant of Variation
y varies *directly* with x	$y = kx$	$k = \frac{y}{x}$
y varies *inversely* with x	$y = \frac{k}{x}$	$k = xy$

READING CHECK

• How can the constant of variation be used to determine whether data are directly or inversely proportional?

The last column in Table 7.9 shows that a set of data represents direct variation when the quotients $\frac{y}{x}$ equal a constant, and it represents inverse variation when the products xy equal a constant. In the next example, we determine if tables of data represent direct variation, inverse variation, or neither.

EXAMPLE 8 **Analyzing data**

Determine whether the data in each table represent direct variation, inverse variation, or neither.

(a)

x	4	5	10	20
y	40	32	16	8

(b)

x	2	5	9	11
y	18	45	81	99

(c)

x	2	4	6	8
y	8	20	30	56

Solution

(a) As x increases, y decreases. Because $xy = 160$ for each data point in the table, the equation $y = \frac{160}{x}$ models the data. The data represent inverse variation.

(b) Because $\frac{y}{x} = 9$ for each data point in the table, the equation $y = 9x$ models the data in the table. These data represent direct variation.

(c) Neither the products xy nor the ratios $\frac{y}{x}$ are constant for the data in the table. Therefore these data represent neither direct variation nor inverse variation.

Now Try Exercises 51(a), 53(a), 55(a)

Joint Variation

▶ **REAL-WORLD CONNECTION** In many applications a quantity depends on more than one variable. In *joint variation* a quantity varies with the product of more than one variable. For example, the formula for the area A of a rectangle is given by

$$A = WL,$$

where W and L are the width and length, respectively. Thus the area of a rectangle varies jointly with the width and length.

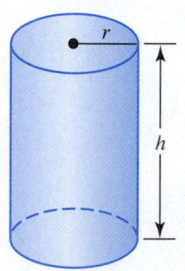

Figure 7.16

> ### JOINT VARIATION
>
> Let x, y, and z denote three quantities. Then z **varies jointly** with x and y if there is a nonzero number k such that
>
> $$z = kxy.$$

Sometimes joint variation can involve a power of a variable. For example, the volume V of a cylinder is given by $V = \pi r^2 h$, where r is its radius and h is its height, as illustrated in Figure 7.16. In this case we say that the volume varies jointly with the height and the *square* of the radius. The constant of variation is $k = \pi$.

EXAMPLE 9 **Finding the strength of a rectangular beam**

The strength S of a rectangular beam varies jointly with its width w and the square of its thickness t. See Figure 7.17. If a beam 3 inches wide and 5 inches thick supports 750 pounds, how much can a similar beam 2 inches wide and 6 inches thick support?

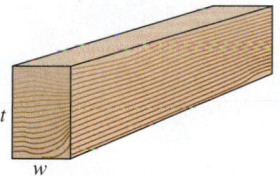

Figure 7.17

Solution
STEP 1: The strength of a beam is modeled by $S = kwt^2$, where k is a constant of variation.
STEP 2: We can find k by substituting $S = $ **750**, $w = $ **3**, and $t = $ **5** in the formula.

$$750 = k \cdot 3 \cdot 5^2 \quad \text{Substitute in } S = kwt^2.$$

$$k = \frac{750}{3 \cdot 5^2} \quad \text{Solve for } k; \text{ rewrite.}$$

$$= 10 \quad \text{Simplify.}$$

STEP 3: The equation $S = \mathbf{10}wt^2$ models the strength of this type of beam.
STEP 4: When $w = $ **2** and $t = $ **6**, the beam can support

$$S = 10 \cdot 2 \cdot 6^2 = 720 \text{ pounds.}$$

▌ **Now Try Exercise 83**

CRITICAL THINKING

Compare the increased strength of a beam if the width doubles and if the thickness doubles. What happens to the strength of a beam if both the width and thickness triple?

7.7 Putting It All Together

CONCEPT	EXPLANATION	EXAMPLES
Proportion	A statement that two ratios are equal	$\dfrac{8}{17} = \dfrac{49}{x}$ and $\dfrac{x}{6} = \dfrac{3}{14}$
Direct Variation	Two quantities x and y vary according to the equation $y = kx$, where k is a nonzero constant. The constant of proportionality (or variation) is k.	$y = 4x$ or $\dfrac{y}{x} = 4$ <table><tr><td>x</td><td>1</td><td>2</td><td>4</td></tr><tr><td>y</td><td>4</td><td>8</td><td>16</td></tr></table> Note that if x doubles, then y also doubles.
Inverse Variation	Two quantities x and y vary according to the equation $y = \frac{k}{x}$, where k is a nonzero constant. The constant of proportionality (or variation) is k.	$y = \dfrac{3}{x}$ or $xy = 3$ <table><tr><td>x</td><td>1</td><td>3</td><td>6</td></tr><tr><td>y</td><td>3</td><td>1</td><td>$\frac{1}{2}$</td></tr></table> Note that if x doubles from 3 to 6, then y decreases by half.
Joint Variation	Three quantities x, y, and z vary according to the equation $z = kxy$, where k is a nonzero constant.	The area A of a triangle varies jointly with b and h according to the equation $A = \frac{1}{2}bh$, where b is its base and h is its height. The constant of variation is $k = \frac{1}{2}$.

7.7 Exercises

MyMathLab   MathXL PRACTICE   WATCH   DOWNLOAD   READ   REVIEW

CONCEPTS AND VOCABULARY

1. What is a proportion?

2. If 5 is to 6 as x is to 7, write a proportion that allows you to find x.

3. Suppose that y is directly proportional to x. If x doubles, what happens to y?

4. Suppose that y is inversely proportional to x. If x doubles, what happens to y?

5. If y varies directly with x, then $\frac{y}{x}$ equals a(n) _____.

6. If y varies inversely with x, then xy equals a(n) _____.

7. Would the food bill B generally vary directly or inversely with the number of people N being fed? Explain your reasoning.

8. Would the time T needed to paint a building vary directly or inversely with the number of painters N working on the job? Explain your reasoning.

9. If xy equals a constant for every data point (x, y) in a table, then the data represent _____ variation.

10. If $\frac{y}{x}$ equals a constant for every data point (x, y) in a table, then the data represent _____ variation.

PROPORTIONS

Exercises 11–22: Solve the proportion.

11. $\dfrac{x}{24} = \dfrac{5}{8}$ **12.** $\dfrac{x}{5} = \dfrac{3}{7}$

13. $\dfrac{14}{x} = \dfrac{2}{3}$ **14.** $\dfrac{4}{9} = \dfrac{9}{x}$

15. $\dfrac{3}{16} = \dfrac{h}{256}$ **16.** $\dfrac{20}{a} = \dfrac{15}{4}$

17. $\dfrac{3}{4} = \dfrac{2x}{7}$ **18.** $\dfrac{7}{3z} = \dfrac{5}{4}$

19. $\dfrac{x}{6} = \dfrac{8}{3x}$ **20.** $\dfrac{4}{x} = \dfrac{4x}{9}$

21. $\dfrac{x}{7} = \dfrac{7}{4x}$ **22.** $\dfrac{2}{3x} = \dfrac{27x}{8}$

23. Thinking Generally Solve $\dfrac{a}{b} = \dfrac{c}{d}$ for b.

24. Thinking Generally Solve $\dfrac{a+b}{c^2} = \dfrac{1}{2}$ for b.

Exercises 25–32: Do the following.
   (a) *Write a proportion that models the situation.*
   (b) *Solve the proportion for x.*

25. 5 is to 8 as 9 is to x.

26. x is to 11 as 7 is to 4.

27. A triangle has sides of 4, 7, and 10. In a similar triangle the shortest side is 8 and the longest side is x.

28. A rectangle has sides of 5 and 12. In a similar rectangle the longer side is 10 and the shorter side is x.

29. If you earn $98 in 7 hours, then you can earn x dollars in 11 hours.

30. If 14 gallons of gasoline contain 1.4 gallons of ethanol, then 22 gallons of gasoline contain x gallons of ethanol.

31. If 3 MP3 players can hold 750 songs, then 7 similar MP3 players can hold x songs.

32. If a gas pump fills a 25-gallon tank in 6 minutes, it can fill a 14-gallon tank in x minutes.

VARIATION

Exercises 33–38: ***Direct Variation*** *Suppose that y is directly proportional to x.*
   (a) *Use the given information to find the constant of proportionality k.*
   (b) *Then use $y = kx$ to find y for $x = 6$.*

33. $y = 4$ when $x = 2$ **34.** $y = 5$ when $x = 10$

35. $y = 3$ when $x = 2$ **36.** $y = 11$ when $x = 55$

37. $y = -60$ when $x = 8$

38. $y = -17$ when $x = 68$

Exercises 39–44: ***Inverse Variation*** *Suppose that y is inversely proportional to x.*
   (a) *Use the given information to find the constant of proportionality k.*
   (b) *Then use $y = \frac{k}{x}$ to find y for $x = 8$.*

39. $y = 6$ when $x = 4$ **40.** $y = 2$ when $x = 24$

41. $y = 80$ when $x = \frac{1}{2}$ **42.** $y = \frac{1}{4}$ when $x = 32$

43. $y = 20$ when $x = 20$ **44.** $y = \frac{8}{3}$ when $x = 12$

Exercises 45–50: ***Joint Variation*** *Let z vary jointly with x and y.*
   (a) *Find the constant of variation k.*
   (b) *Use $z = kxy$ to find z when $x = 5$ and $y = 7$.*

45. $z = 6$ when $x = 3$ and $y = 8$

46. $z = 135$ when $x = 2.5$ and $y = 9$

47. $z = 5775$ when $x = 25$ and $y = 21$

48. $z = 1530$ when $x = 22.5$ and $y = 4$

49. $z = 25$ when $x = \frac{1}{2}$ and $y = 5$

50. $z = 12$ when $x = \frac{1}{4}$ and $y = 12$

ANALYZING DATA

Exercises 51–56: (Refer to Example 8.)
   (a) *Determine whether the data represent direct variation, inverse variation, or neither.*
   (b) *If the data represent either direct or inverse variation, find an equation that models the data.*
   (c) *Graph the equation and the data when possible.*

51.

x	2	3	4	5
y	3	4.5	6	7.5

52.

x	10	20	30	40
y	12	6	5	4

53.

x	3	6	9	12
y	12	6	4	3

54.

x	2	6	10	14
y	105	35	21	15

55.

x	4	6	12	20
y	10	20	30	40

56.

x	1	5	9	15
y	6	30	54	90

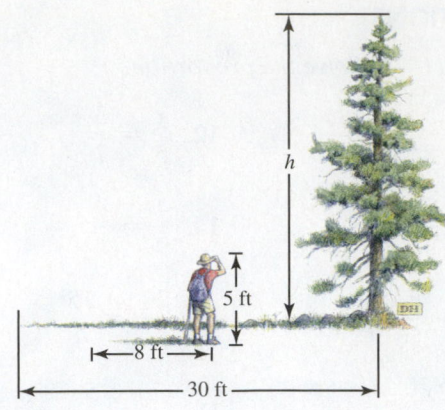

Exercises 57–62: Use the graph to determine whether the data represent direct variation, inverse variation, or neither. Find the constant of variation whenever possible.

57.
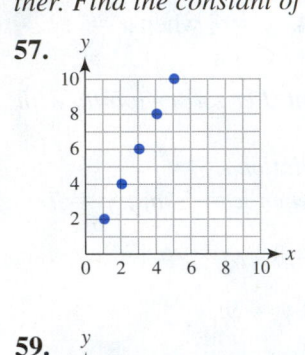

58.
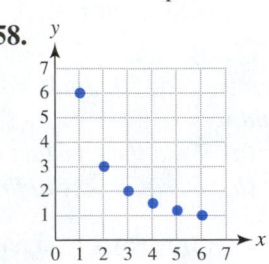

59.
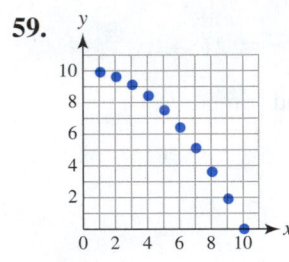

60.
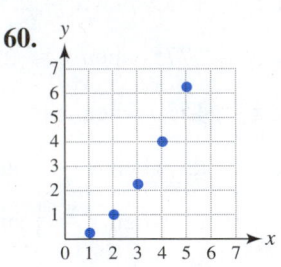

61.
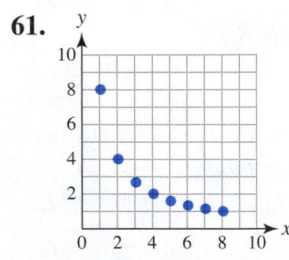

62.
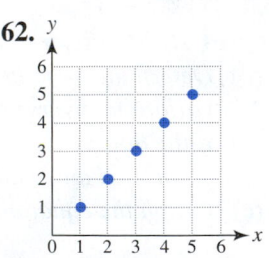

APPLICATIONS

63. *Recording Music* A 750-megabyte CD can record 85 minutes of music. How many minutes can be recorded on 420 megabytes?

64. *Height of a Tree* (Refer to Example 2.) A 5-foot-tall person casts an 8-foot-long shadow, and a nearby tree casts a 30-foot-long shadow. See the figure at the top of the next column. Estimate the height of the tree.

65. *Water Content in Snow* (Refer to Example 1.) Eight inches of heavy, wet snow are equivalent to an inch of rain. Estimate the water content in 13 inches of heavy, wet snow.

66. *Wages* If a person working for an hourly wage earns $143 in 13 hours, how much will that person earn in 15 hours?

67. *Strength of a Beam* (Refer to Example 4.) The strength of a metal beam varies directly with its width. A beam that is 6.2 inches wide can support 2800 pounds. How much weight can a similar beam support if it is 4.7 inches wide?

68. *Strength of a Beam* The strength of a wood beam varies inversely with its length. A beam that is 33 feet long can support 1200 pounds. How much weight can a similar beam support if it is 23 feet long?

69. *Making Fudge* If $2\frac{2}{3}$ cups of sugar can make 14 pieces of fudge, how much sugar is needed to make 49 pieces of fudge?

70. *Making Coffee* If 6 tablespoons of coffee grounds make 10 cups of coffee, how many tablespoons of coffee grounds are needed to make 35 cups of coffee?

71. *Rolling Resistance of Cars* If you were to try to push a car, you would experience *rolling resistance*. This resistance equals the force necessary to keep the car moving slowly in neutral gear. The following table shows the rolling resistance R for passenger cars of different gross weights W. (*Source:* N. Garber and L. Hoel, *Traffic and Highway Engineering.*)

W (pounds)	2000	2500	3000	3500
R (pounds)	24	30	36	42

(a) Do the data represent direct or inverse variation? Explain.

(b) Find an equation that models the data. Graph the equation with the data.

(c) Estimate the rolling resistance of a 3200-pound car.

72. *Transportation Costs* The use of a particular toll bridge varies inversely according to the toll. If the toll is $0.50, then 8000 vehicles use the bridge. Estimate the number of users if the toll is $0.80. (*Source:* N. Garber.)

73. *Flow of Water* The gallons of water G flowing in 1 minute through a hose with a cross-sectional area A are shown in the table.

A (square inch)	0.2	0.3	0.4	0.5
G (gallons)	5.4	8.1	10.8	13.5

(a) Do the data represent direct or inverse variation? Explain.

(b) Find an equation that models the data. Graph the equation with the data.

(c) Interpret the constant of variation k.

74. *Hooke's Law* The table shows the distance D that a spring stretches when a weight W is hung on it.

W (pounds)	2	6	9	15
D (inches)	1.6	4.8	7.2	12

(a) Do the data represent direct or inverse variation? Explain.

(b) Find an equation that models the data.

(c) How far will the spring stretch if an 11-pound weight is hung on it?

75. *Tightening Lug Nuts* (Refer to Example 7.) When a tire is mounted on a car, the lug nuts should not be over-tightened. The table at the top of the next column shows the maximum force used with wrenches of different lengths.

L (inches)	8	10	16
F (pounds)	150	120	75

Source: Tires Plus.

(a) Model the data, using the equation $F = \frac{k}{L}$.

(b) How much force should be used with a wrench 15 inches long?

76. *Ozone and UV Radiation* Ozone in the atmosphere filters out approximately 90% of the harmful ultraviolet (UV) rays from the sun. Depletion of the ozone layer increases the amount of UV radiation reaching Earth's surface. An increase in UV radiation is associated with skin cancer. The following graph shows the percentage increase y in UV radiation for a decrease in the ozone layer of x percent. (*Source:* R. Turner, D. Pearce, and I. Bateman, *Environmental Economics.*)

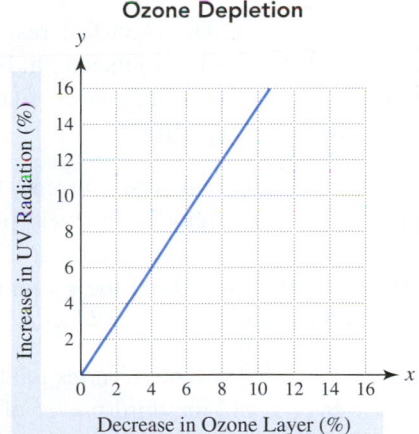

Ozone Depletion

(a) Does this graph represent direct or inverse variation?

(b) Find an equation for the line in the graph.

(c) Estimate the percentage increase in UV radiation if the ozone layer decreases by 5%.

77. *Air Temperature and Altitude* In the first 6 miles of Earth's atmosphere, air cools as the altitude increases. The following graph shows the temperature change y in degrees Fahrenheit at an altitude of x miles. (*Source:* A. Miller and R. Anthes, *Meteorology.*)

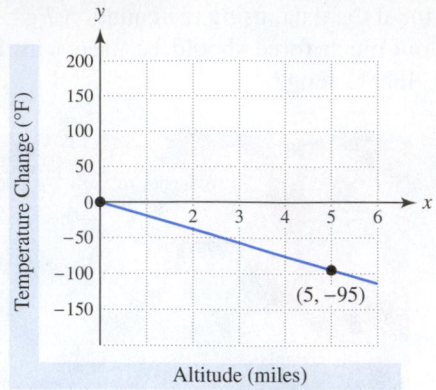

(a) Does this graph represent direct variation or inverse variation?
(b) Find an equation that models the data in the graph.
(c) Is the constant of proportionality k positive or negative? Interpret k.
(d) Find the change in air temperature 2.5 miles high.

78. *Cost of Tuition* (Refer to Example 5.) The cost of tuition is directly proportional to the number of credits taken. If 6 credits cost \$483, find the cost of 13 credits. What does the constant of proportionality represent?

79. *Electrical Resistance* The electrical resistance of a wire is directly proportional to its length. If a 30-foot-long wire has a resistance of 3 ohms, find the resistance of an 18-foot-long wire.

80. *Resistance and Current* The current that flows through an electrical circuit is inversely proportional to the resistance. When the resistance R is 180 ohms, the current I is 0.6 amp. Find the current when the resistance is 54 ohms.

81. *Joint Variation* The variable z varies jointly with the second power of x and the third power of y. Write a formula for z if $z = 31.9$ when $x = 2$ and $y = 2.5$.

82. *Wind Power* The electrical power generated by a windmill varies jointly with the square of the diameter of the area swept out by the blades and the cube of the wind velocity. If a windmill with a 10-foot diameter and a 16-mile-per-hour wind generates 15,392 watts, how much power would be generated if the blades swept out an area 12 feet in diameter and the wind speed was 15 miles per hour?

83. *Strength of a Beam* (Refer to Example 9.) If a wood beam 5 inches wide and 3 inches thick supports 300 pounds, how much can a similar beam 5 inches wide and 2 inches thick support?

84. *Carpeting* The cost of carpet for a rectangular room varies jointly with its width and length. If a room 11 feet wide and 14 feet long costs \$539 to carpet, find the cost to carpet a room 17 feet by 19 feet. Interpret the constant of variation k.

85. *Weight on the Moon* The weight of a person on the moon is directly proportional to the weight of the person on Earth. If a 175-pound person weighs 28 pounds on the moon, how much will a 220-pound person weigh on the moon?

86. *Weight Near Earth* The weight W of a person near Earth is inversely proportional to the square of the person's distance d from the *center* of Earth. If a person weighs 200 pounds when $d = 4000$ miles, how much does the same person weigh when $d = 7000$ miles? (*Note:* The radius of Earth is about 4000 miles.)

87. *Ohm's Law* The voltage V in an electrical circuit varies jointly with the amperage I and resistance R. If $V = 220$ when $I = 10$ and $R = 22$, find V when $I = 15$ and $R = 50$.

88. *Revenue* The revenue R from selling x items at price p varies jointly with x and p. If $R = \$24{,}000$ when $x = 3000$ and $p = \$8$, find the number of items x sold when $R = \$30{,}000$ and $p = \$6$.

WRITING ABOUT MATHEMATICS

89. Explain what it means for a quantity y to be directly proportional to a quantity x.

90. Explain what it means for a quantity y to be inversely proportional to a quantity x.

SECTION
7.7

Checking Basic Concepts

1. Solve each proportion.

   (a) $\dfrac{x}{9} = \dfrac{2}{5}$ (b) $\dfrac{4}{3} = \dfrac{5}{b}$

2. Write a proportion that models each situation. Then solve it.
   (a) 4 is to 6 as 8 is to x.
   (b) If 2 compact discs can record 148 minutes of music, then 5 compact discs can record x minutes of music.

3. Suppose that y is inversely proportional to x. If $y = 4$ when $x = 15$, find the constant of proportionality k. Find y when $x = 10$.

4. Decide whether the data in the table represent direct or inverse variation. Explain your reasoning. Find the constant of variation.

   (a)

x	2	4	6	8
y	3	6	9	12

   (b)

x	2	4	6	8
y	12	6	4	3

5. *Wages* If a person working for an hourly wage earns \$272 in 17 hours, how much will the person earn in 10 hours?

CHAPTER 7   Summary

SECTION 7.1 ■ INTRODUCTION TO RATIONAL EXPRESSIONS

Rational Expression
A rational expression can be written in the form $\dfrac{P}{Q}$, where P and Q are polynomials, and is defined whenever $Q \neq 0$.

Example: $\dfrac{x^2}{x - 5}$ is a rational expression that is defined for all real numbers except $x = 5$.

Undefined Rational Expression A rational expression is undefined for any value of the variable that makes the denominator equal to 0.

Examples: $\dfrac{1}{x - 5}$ is undefined when $x = 5$.

$\dfrac{3y}{y^2 - 4}$ is undefined when $y = -2$ or when $y = 2$.

Simplifying a Rational Expression To simplify a rational expression, factor the numerator and the denominator. Then apply the basic principle of rational expressions,

$$\frac{PR}{QR} = \frac{P}{Q}. \quad Q \text{ and } R \text{ nonzero}$$

Examples: $\dfrac{x^2 - 4}{x^2 - 3x + 2} = \dfrac{(x + 2)(x - 2)}{(x - 1)(x - 2)} = \dfrac{x + 2}{x - 1}, \quad \dfrac{x - 3}{3 - x} = -\dfrac{x - 3}{x - 3} = -1$

SECTION 7.2 ■ MULTIPLICATION AND DIVISION OF RATIONAL EXPRESSIONS

Multiplying Rational Expressions
To multiply two rational expressions, multiply the numerators and multiply the denominators.

$$\frac{A}{B} \cdot \frac{C}{D} = \frac{AC}{BD} \quad B \text{ and } D \text{ nonzero}$$

Example: $\dfrac{3}{x - 1} \cdot \dfrac{4}{x + 1} = \dfrac{12}{(x - 1)(x + 1)}$

Dividing Rational Expressions To divide two rational expressions, multiply by the reciprocal of the divisor.

$$\frac{A}{B} \div \frac{C}{D} = \frac{A}{B} \cdot \frac{D}{C} = \frac{AD}{BC} \qquad B, C, \text{ and } D \text{ nonzero}$$

Example: $\dfrac{x+2}{x^2+3x} \div \dfrac{x+2}{x} = \dfrac{x+2}{x(x+3)} \cdot \dfrac{x}{x+2} = \dfrac{x(x+2)}{x(x+3)(x+2)} = \dfrac{1}{x+3}$

SECTION 7.3 ■ ADDITION AND SUBTRACTION WITH LIKE DENOMINATORS

Addition of Rational Expressions Having Like Denominators To add two rational expressions having like denominators, add their numerators. Keep the same denominator.

$$\frac{A}{C} + \frac{B}{C} = \frac{A+B}{C} \qquad C \text{ nonzero}$$

Example: $\dfrac{5x}{x+4} + \dfrac{1}{x+4} = \dfrac{5x+1}{x+4}$

Subtraction of Rational Expressions Having Like Denominators To subtract two rational expressions having like denominators, subtract their numerators. Keep the same denominator.

$$\frac{A}{C} - \frac{B}{C} = \frac{A-B}{C} \qquad C \text{ nonzero}$$

Example: $\dfrac{6}{2x+1} - \dfrac{2}{2x+1} = \dfrac{4}{2x+1}$

SECTION 7.4 ■ ADDITION AND SUBTRACTION WITH UNLIKE DENOMINATORS

Finding Least Common Multiples The least common multiple (LCM) of two or more polynomials can be found as follows.

STEP 1: Factor each polynomial completely.

STEP 2: List each factor the greatest number of times that it occurs in any factorization.

STEP 3: Find the product of this list of factors. It is the LCM.

Example: $4x^2(x+1) = 2 \cdot 2 \cdot x \cdot x \cdot (x+1)$
$\qquad\qquad 2x(x^2-1) = 2 \cdot x \cdot (x+1) \cdot (x-1)$

Listing each factor the greatest number of times and multiplying gives the LCM.

$$2 \cdot 2 \cdot x \cdot x \cdot (x+1)(x-1) = 4x^2(x^2-1)$$

Finding the Least Common Denominator The least common denominator (LCD) is the least common multiple (LCM) of the denominators.

Example: From the preceding example, the LCD for $\dfrac{1}{4x^2(x+1)}$ and $\dfrac{1}{2x(x^2-1)}$ is $4x^2(x^2-1)$.

Addition and Subtraction of Rational Expressions Having Unlike Denominators First write each rational expression by using the LCD. Then add or subtract the resulting rational expressions. Finally, write your answer in lowest terms.

Example: $\dfrac{1}{x-1} - \dfrac{1}{x} = \dfrac{x}{x(x-1)} - \dfrac{x-1}{x(x-1)} = \dfrac{x-(x-1)}{x(x-1)} = \dfrac{1}{x(x-1)}$
$\qquad\qquad$ Note that the LCD is $x(x-1)$.

SECTION 7.5 ■ COMPLEX FRACTIONS

Complex Fractions A complex fraction is a rational expression that contains fractions in its numerator, denominator, or both. The following equation can be used to simplify basic complex fractions.

$$\frac{\dfrac{a}{b}}{\dfrac{c}{d}} = \frac{a}{b} \cdot \frac{d}{c} \qquad b, c, \text{ and } d \text{ nonzero}$$

Example:
$$\frac{\dfrac{x}{3}}{\dfrac{x}{x-1}} = \frac{x}{3} \cdot \frac{x-1}{x} = \frac{x(x-1)}{3x} = \frac{x-1}{3}$$

Simplifying Complex Fractions

Method I   Combine terms in the numerator, combine terms in the denominator, and then multiply the numerator by the reciprocal of the denominator.

Method II Multiply the numerator and denominator by the LCD of all fractions within the expression and simplify the resulting expression.

Example: *Method I*
$$\frac{\dfrac{1}{a} - \dfrac{1}{b}}{\dfrac{1}{a} + \dfrac{1}{b}} = \frac{\dfrac{b-a}{ab}}{\dfrac{b+a}{ab}} = \frac{b-a}{ab} \cdot \frac{ab}{b+a} = \frac{b-a}{b+a}$$

Method II   The least common denominator is ab.

$$\frac{\dfrac{1}{a} - \dfrac{1}{b}}{\dfrac{1}{a} + \dfrac{1}{b}} = \frac{\left(\dfrac{1}{a} - \dfrac{1}{b}\right)ab}{\left(\dfrac{1}{a} + \dfrac{1}{b}\right)ab} = \frac{\dfrac{ab}{a} - \dfrac{ab}{b}}{\dfrac{ab}{a} + \dfrac{ab}{b}} = \frac{b-a}{b+a}$$

SECTION 7.6 ■ RATIONAL EQUATIONS AND FORMULAS

Solving Rational Equations One way to solve the equation $\frac{a}{b} = \frac{c}{d}$ is to cross multiply to obtain $ad = bc$. (Check each answer.) A general way to solve rational equations is to follow these steps.

STEP 1: Find the LCD of the terms in the equation.

STEP 2: Multiply each side of the equation by the LCD.

STEP 3: Simplify each term.

STEP 4: Solve the resulting equation.

STEP 5: Check each answer in the *given* equation. Reject any value that makes a denominator equal 0.

Examples: $\frac{1}{2x} = \frac{2}{x+3}$ implies that $x + 3 = 4x$, or $x = 1$. This answer checks.

To solve $\frac{5}{x} - \frac{1}{3x} = \frac{7}{3}$ multiply each term by the LCD, $3x$:

$$\frac{5(3x)}{x} - \frac{3x}{3x} = \frac{7(3x)}{3},$$

which simplifies to $15 - 1 = 7x$, or $x = 2$. This answer checks.

Solving for a Variable   Many formulas contain more than one variable. To solve for a particular variable, use the rules of algebra to isolate the variable.

Example: To solve $S = \frac{2\pi}{r}$ for r, multiply each side by r to obtain $Sr = 2\pi$ and then divide each side by S to obtain $r = \frac{2\pi}{S}$.

SECTION 7.7 ■ PROPORTIONS AND VARIATION

Proportions   A proportion is a statement that two ratios are equal.

Example: $\frac{5}{x} = \frac{4}{7}$

Similar Triangles   Two triangles are similar if the measures of their corresponding angles are equal. Corresponding sides of similar triangles are proportional.

Example: The following triangles are similar.

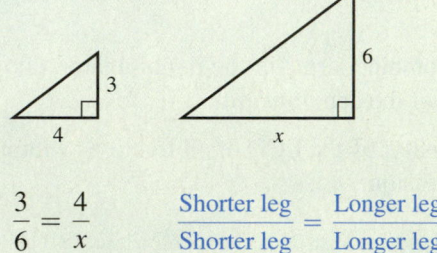

$$\frac{3}{6} = \frac{4}{x} \qquad \frac{\text{Shorter leg}}{\text{Shorter leg}} = \frac{\text{Longer leg}}{\text{Longer leg}}$$

By solving this proportion, we see that $x = 8$.

Direct Variation   A quantity y is *directly proportional* to a quantity x, or y *varies directly* with x, if there is a nonzero constant k such that $y = kx$. The number k is called the *constant of proportionality* or the *constant of variation*.

Example: If y varies directly with x, then the ratios $\frac{y}{x}$ always equal k. The following data satisfy $\frac{y}{x} = 4$, so the constant of variation is 4. Thus $y = 4x$.

x	1	2	3	4
y	4	8	12	16

Inverse Variation   A quantity y is *inversely proportional* to a quantity x, or y *varies inversely* with x, if there is a nonzero constant k such that $y = \frac{k}{x}$.

Example: If y varies inversely with x, then the products xy always equal k. The following data satisfy $xy = 12$, so the constant of variation is 12. Thus $y = \frac{12}{x}$.

x	1	2	4	6
y	12	6	3	2

Joint Variation   The quantity z *varies jointly* with x and y if $z = kxy$, $k \neq 0$.

Example: The area A of a rectangle varies jointly with the width W and length L because $A = LW$. Note that $k = 1$ in this example.

CHAPTER 7 Review Exercises

SECTION 7.1

Exercises 1–4: If possible, evaluate the expression for the given value of x.

1. $\dfrac{3}{x-3}$ $x=-2$

2. $\dfrac{4x}{5-x^2}$ $x=3$

3. $\dfrac{-x}{7-x}$ $x=7$

4. $\dfrac{4x}{x^2-3x+2}$ $x=2$

5. Complete the table for the rational expression. If a value is undefined, place a dash in the table.

x	-2	-1	0	1	2
$\dfrac{3x}{x-1}$					

6. Find the x-values that make $\dfrac{8}{x^2-4}$ undefined.

Exercises 7–12: Simplify to lowest terms.

7. $\dfrac{25x^3y^4}{15x^5y}$

8. $\dfrac{x^2-36}{x+6}$

9. $\dfrac{x-9}{9-x}$

10. $\dfrac{x^2-5x}{5x}$

11. $\dfrac{2x^2+5x-3}{2x^2+x-1}$

12. $\dfrac{3x^2+10x-8}{3x^2+x-2}$

Exercises 13 and 14: Do the following.
 (a) Decide whether you are given an expression or an equation.
 (b) If you are given an expression, simplify it. If you are given an equation, solve it.

13. $\dfrac{x+1}{(x-3)(x+1)}$

14. $\dfrac{x}{3(4-1)}=2$

SECTION 7.2

Exercises 15–18: Multiply and write in lowest terms.

15. $\dfrac{x-3}{x+1}\cdot\dfrac{2x+2}{x-3}$

16. $\dfrac{2x+5}{(x+5)(x-1)}\cdot\dfrac{x-1}{2x+5}$

17. $\dfrac{z+3}{z-4}\cdot\dfrac{z-4}{(z+3)^2}$

18. $\dfrac{x^2}{x^2-4}\cdot\dfrac{x+2}{x}$

Exercises 19–24: Divide and write in lowest terms.

19. $\dfrac{x+1}{2x}\div\dfrac{3x+3}{5x}$

20. $\dfrac{4}{x^3}\div\dfrac{x+1}{2x^2}$

21. $\dfrac{x-5}{x+2}\div\dfrac{2x-10}{x+2}$

22. $\dfrac{x^2-6x+5}{x^2-25}\div\dfrac{x-1}{x+5}$

23. $\dfrac{x^2-y^2}{x+y}\div\dfrac{x-y}{x+y}$

24. $\dfrac{a^3-b^3}{a+b}\div\dfrac{a-b}{2a+2b}$

SECTION 7.3

Exercises 25–32: Add or subtract and write in lowest terms.

25. $\dfrac{2}{x+10}+\dfrac{8}{x+10}$

26. $\dfrac{9}{x-1}-\dfrac{8}{x-1}$

27. $\dfrac{x+2y}{2x}+\dfrac{x-2y}{2x}$

28. $\dfrac{x}{x+3}+\dfrac{3}{x+3}$

29. $\dfrac{x}{x^2-1}-\dfrac{1}{x^2-1}$

30. $\dfrac{2x}{x^2-25}+\dfrac{10}{x^2-25}$

31. $\dfrac{3}{xy}-\dfrac{1}{xy}$

32. $\dfrac{x+y}{2y}+\dfrac{x-y}{2y}$

SECTION 7.4

Exercises 33–38: Find the least common multiple for the expressions. Leave your answer in factored form.

33. $3x,\ 5x$

34. $5x^2,\ 10x$

35. $x,\ x-5$

36. $10x^2,\ x^2-x$

37. $x^2-1,\ (x+1)^2$

38. $x^2-4x,\ x^2-16$

Exercises 39–44: Rewrite the rational expression by using the given denominator D.

39. $\dfrac{3}{8},D=24$

40. $\dfrac{4}{3x},D=12x$

41. $\dfrac{3x}{x-2},D=x^2-4$

42. $\dfrac{2}{x+1},D=x^2+x$

43. $\dfrac{3}{5x},D=5x^2-5x$

44. $\dfrac{2x}{2x-3},D=2x^2+x-6$

Exercises 45–56: Simplify the expression.

45. $\dfrac{5}{8} + \dfrac{1}{6}$

46. $\dfrac{3}{4x} + \dfrac{1}{x}$

47. $\dfrac{5}{9x} - \dfrac{2}{3x}$

48. $\dfrac{7}{x-1} - \dfrac{3}{x}$

49. $\dfrac{1}{x+1} + \dfrac{1}{x-1}$

50. $\dfrac{4}{3x^2} - \dfrac{3}{2x}$

51. $\dfrac{1+x}{3x} - \dfrac{3}{2x}$

52. $\dfrac{x}{x^2-1} - \dfrac{1}{x-1}$

53. $\dfrac{2}{x-y} - \dfrac{3}{x+y}$

54. $\dfrac{2}{x} - \dfrac{1}{2x} + \dfrac{2}{3x}$

55. $\dfrac{3}{2y} + \dfrac{1}{2x}$

56. $\dfrac{x}{y-x} + \dfrac{y}{x-y}$

SECTION 7.5

Exercises 57–66: Simplify the complex fraction.

57. $\dfrac{\frac{3}{4}}{\frac{7}{11}}$

58. $\dfrac{\frac{x}{5}}{\frac{2x}{7}}$

59. $\dfrac{\frac{m}{n}}{\frac{2m}{n^2}}$

60. $\dfrac{\frac{3}{p-1}}{\frac{1}{p+1}}$

61. $\dfrac{\frac{3}{m-1}}{\frac{2m-2}{m+1}}$

62. $\dfrac{\frac{2}{2n+1}}{\frac{8}{2n-1}}$

63. $\dfrac{\frac{1}{2x} - \frac{1}{3x}}{\frac{2}{3x} - \frac{1}{6x}}$

64. $\dfrac{\frac{2}{xy} - \frac{1}{y}}{\frac{2}{xy} + \frac{1}{y}}$

65. $\dfrac{\frac{1}{x} - \frac{1}{x+1}}{\frac{x}{x+1}}$

66. $\dfrac{\frac{2}{x-1} - \frac{1}{x+1}}{\frac{1}{x^2-1}}$

SECTION 7.6

Exercises 67–72: Solve and check your answer.

67. $\dfrac{x}{5} = \dfrac{4}{7}$

68. $\dfrac{4}{x} = \dfrac{3}{2}$

69. $\dfrac{3}{z+1} = \dfrac{1}{2z}$

70. $\dfrac{x+2}{x} = \dfrac{3}{5}$

71. $\dfrac{1}{x+1} = \dfrac{2}{x-2}$

72. $\dfrac{x}{3} = \dfrac{-1}{x+4}$

Exercises 73–86: If possible, solve. Check your answer.

73. $\dfrac{1}{5x} + \dfrac{3}{5x} = \dfrac{1}{5}$

74. $\dfrac{1}{x-1} + \dfrac{2x}{x-1} = 1$

75. $\dfrac{1}{x} + \dfrac{2}{3x} = \dfrac{1}{3}$

76. $\dfrac{1}{x+3} + \dfrac{2x}{x+3} = \dfrac{3}{2}$

77. $\dfrac{5}{x} - \dfrac{3}{x+1} = \dfrac{1}{2}$

78. $\dfrac{1}{x-1} - \dfrac{1}{x+1} = \dfrac{1}{4}$

79. $\dfrac{4}{p} - \dfrac{5}{p+2} = 0$

80. $\dfrac{1}{x-3} - \dfrac{1}{x+3} = \dfrac{1}{x^2-9}$

81. $\dfrac{1}{x+1} = \dfrac{-x}{x+1}$

82. $\dfrac{2}{x} = \dfrac{2}{x^2+x} - 4$

83. $\dfrac{2}{x^2-2x} + \dfrac{1}{x^2-4} = \dfrac{1}{x^2+2x}$

84. $\dfrac{3}{x^2-3x} - \dfrac{1}{x^2-9} = \dfrac{1}{x^2+3x}$

85. $\dfrac{1}{x^2} - \dfrac{5}{x^2+4x} = \dfrac{1}{x^2+4x}$

86. $\dfrac{5}{x^2-1} - \dfrac{1}{x^2+2x+1} = \dfrac{3}{x^2-1}$

Exercises 87 and 88: Do the following.
(a) Decide whether you are given an expression or an equation.
(b) If you are given an expression, simplify it. If you are given an equation, solve it.

87. $\dfrac{4}{x} - x = 0$

88. $\dfrac{x^2}{x-3} - \dfrac{9}{x-3}$

Exercises 89 and 90: Solve for the specified variable.

89. $\dfrac{1}{a} + \dfrac{2}{b} = \dfrac{3}{c}$ for b

90. $y = \dfrac{x}{x-1}$ for x

SECTION 7.7

Exercises 91 and 92: Solve the proportion.

91. $\dfrac{x}{6} = \dfrac{1}{3}$

92. $\dfrac{5}{x} = \dfrac{7}{3}$

Exercises 93 and 94: **Proportions** *Do the following.*

  (a) *Write a proportion that models the situation.*

  (b) *Solve the proportion for x.*

93. A rectangle has sides of 6 and 13. In a similar rectangle the longer side is 20 and the shorter side is x.

94. If you earn \$341 in 11 hours, then you can earn x dollars in 8 hours.

Exercises 95 and 96: **Direct Variation** *Suppose that y is directly proportional to x.*

  (a) *Use the given information to find the constant of proportionality k.*

  (b) *Then use $y = kx$ to find y for $x = 5$.*

95. $y = 8$ when $x = 2$

96. $y = 21$ when $x = 7$

Exercises 97 and 98: **Inverse Variation** *Suppose that y is inversely proportional to x.*

  (a) *Use the given information to find the constant of proportionality k.*

  (b) *Then use $y = \frac{k}{x}$ to find y for $x = 5$.*

97. $y = 2.5$ when $x = 4$

98. $y = 7$ when $x = 3$

99. **Joint Variation** Suppose that z varies jointly with x and y. If $z = 483$ when $x = 23$ and $y = 7$, find the constant of variation k.

100. **Joint Variation** Suppose that z varies jointly with x and the square of y. If $z = 891$ when $x = 22$ and $y = 3$, find z when $x = 10$ and $y = 4$.

Exercises 101 and 102: Do the following.

  (a) *Determine whether the data represent direct or inverse variation.*

  (b) *Find an equation that models the data.*

  (c) *Graph the data and your equation.*

101.

x	2	3	4	5
y	30	20	15	12

102.

x	2	4	6	8
y	6	12	18	24

Exercises 103 and 104: Use the graph to determine whether the data represent direct or inverse variation. Find the constant of variation.

103.

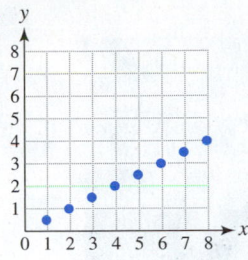

104.

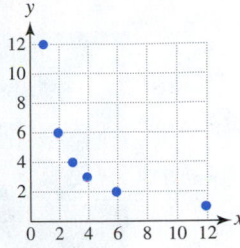

APPLICATIONS

105. **Modeling Traffic Flow** Fifteen vehicles per minute can pass through an intersection. If vehicles arrive randomly at an average rate of x per minute, the average waiting time T in minutes is given by $T = \frac{1}{15 - x}$ for $x < 15$. (Source: N. Garber, *Traffic and Highway Engineering.*)

  (a) Find T when $x = 10$ and interpret the result.

  (b) Complete the table.

x	5	10	13	14	14.9
T					

  (c) What happens to the waiting time as the traffic rate x approaches 15 vehicles per minute?

106. **Distance and Time** A car traveled at 50 miles per hour for 150 miles and then traveled at 75 miles per hour for 150 miles. What was the car's average speed?

107. **Emptying a Swimming Pool** A large pump can empty a swimming pool in 100 hours, whereas a small pump can empty the pool in 160 hours. How long will it take to empty the pool if both pumps are used?

108. **Running** Two athletes run 10 miles. One of the athletes runs 2 miles per hour faster and finishes 10 minutes ahead of the other athlete. Find the average speed of each athlete.

109. **River Current** A boat can travel 16 miles upstream in the same time that it can travel 48 miles downstream. If the speed of the current is 4 miles per hour, find the boat's speed.

110. **Height of a Tree** A 5-foot-tall person has a 6-foot-long shadow, and a nearby tree has a 40-foot-long shadow. Estimate the tree's height.

111. *Transportation Costs* Use of a toll road varies inversely with the toll. If the toll is $0.25, then 400 vehicles use the road. Estimate the number of users for a toll of $0.50.

112. *Cost of Carpet* The cost of carpet is directly proportional to the amount of carpet purchased. If 17 square yards cost $612, find the cost of 13 square yards.

113. *Tightening a Bolt* The torque exerted on a nut by a wrench is inversely proportional to the length of the wrench's handle. Suppose that a 12-inch wrench can be used to tighten a nut by using 30 pounds of force. How much force is necessary to tighten the same nut by using a 10-inch wrench?

114. *Water Content in Snow* Twenty inches of extremely dry, powdery snow are equivalent to an inch of rain. Estimate the water content in 32 inches of this type of snow.

115. *Polar Plunge* When a person swims in extremely cold water, the water removes body heat 25 times faster than air at the same temperature. To be safe, a person has between 30 and 90 seconds to get out of

the water. How long could a person safely remain in air at the same temperature?

116. *Strength of a Beam* The strength of a wood beam varies inversely with its length. A beam that is 18 feet long can support 900 pounds. How much weight can a similar beam support if its length is 21 feet?

117. *Wind Power* The electric power generated by a windmill varies jointly with the square of the diameter of the area swept out by the blades and the cube of the wind velocity. If a windmill with 6-foot-diameter blades and a 20-mile-per-hour wind generates 10,823 watts, how much power would be generated if the blades were 10 feet in diameter and the wind speed were 12 miles per hour?

118. *Strength of a Beam* The strength of a beam varies jointly with its width w and the square of its thickness t. If a beam 8 inches wide and 5 inches thick supports 650 pounds, how much can a similar beam 6 inches wide and 6 inches thick support?

CHAPTER 7   Test

Step-by-step test solutions are found on the Chapter Test Prep Videos available in **MyMathLab** and on **YouTube** (search "RockswoldComboAlg" and click on "Channels").

1. Evaluate the expression $\frac{3x}{2x-1}$ for $x = 3$.

2. Find any x-value that makes $\frac{x-1}{x+2}$ undefined.

Exercises 3 and 4: Simplify the expression.

3. $\dfrac{x^2 - 25}{x - 5}$

4. $\dfrac{3x^2 - 15x}{3x}$

Exercises 5–12: Simplify the expression. Write your answer in lowest terms.

5. $\dfrac{x-2}{x+4} \cdot \dfrac{3x+12}{x-2}$

6. $\dfrac{z+1}{z+3} \cdot \dfrac{2z+6}{z+1}$

7. $\dfrac{x+1}{5x} \div \dfrac{2x+2}{x-1}$

8. $\dfrac{2}{x^2} \div \dfrac{x+3}{3x}$

9. $\dfrac{x}{x+4} + \dfrac{3x+1}{x+4}$

10. $\dfrac{4t+1}{2t-3} - \dfrac{3t-6}{2t-3}$

11. Find the least common multiple for

$$6x^2 \quad \text{and} \quad 3x^2 - 3x.$$

12. Rewrite the rational expression $\frac{4}{7x}$ by using the denominator $7x^2 - 7x$.

Exercises 13 and 14: Simplify the expression. Write your answer in lowest terms.

13. $\dfrac{1}{y^2 + y} - \dfrac{y-1}{y^2 - y}$

14. $\dfrac{1}{xy} + \dfrac{x}{y} - \dfrac{1}{y^2}$

Exercises 15 and 16: Simplify the complex fraction.

15. $\dfrac{\dfrac{a}{3b}}{\dfrac{5a}{b^2}}$

16. $\dfrac{1 + \dfrac{1}{p - 1}}{1 - \dfrac{1}{p - 1}}$

Exercises 17–24: Solve the equation and check your answer.

17. $\dfrac{2}{7} = \dfrac{5}{x}$

18. $\dfrac{x + 3}{2x} = 1$

19. $\dfrac{1}{2x} + \dfrac{2}{5x} = \dfrac{9}{10}$

20. $\dfrac{1}{x - 1} + \dfrac{2}{x + 2} = \dfrac{3}{2}$

21. $\dfrac{1}{x^2 - 1} - \dfrac{4}{x + 1} = \dfrac{3}{x - 1}$

22. $\dfrac{1}{x^2 - 4x} + \dfrac{2}{x^2 - 16} = \dfrac{2}{x^2 + 4x}$

23. $\dfrac{x}{2x - 1} = \dfrac{1 - x}{2x - 1}$

24. $\dfrac{x}{x - 5} + \dfrac{x}{x + 5} = \dfrac{10x}{x^2 - 25}$

Exercises 25 and 26: Solve the equation for the specified variable.

25. $y = \dfrac{2}{3x - 5}$ for x

26. $\dfrac{a + b}{ab} = 1$ for b

27. Suppose that y is directly proportional to x.
  (a) If $y = 14$ when $x = 4$, find k so that $y = kx$.
  (b) Use $y = kx$ to find y for $x = 6$.

28. Use the table to determine whether y varies directly or inversely with x. Find the constant of variation.

x	2	4	8	16
y	16	8	4	2

29. *Emptying a Swimming Pool* It takes a large pump 40 hours to empty a swimming pool, whereas a small pump can empty the pool in 60 hours. How long will it take to empty the pool if both pumps are used?

30. *Height of a Building* A 5-foot-tall post has a 4-foot-long shadow, and a nearby building has a 54-foot-long shadow. Estimate the height of the building.

31. *Standing in Line* A department store clerk can wait on 30 customers per hour. If people arrive randomly at an average rate of x per hour, then the average number of customers N waiting in line is given by

$$N = \frac{x^2}{900 - 30x},$$

for $x < 30$. Evaluate the expression for $x = 24$ and interpret your result.

CHAPTER 7   Extended and Discovery Exercises

1. *Graph of a Rational Function* A car wash can clean 15 cars per hour. If cars arrive randomly at an average rate of x per hour, the average number N of cars waiting in line is given by

$$N = \frac{x^2}{225 - 15x},$$

where $x < 15$. (*Source:* N. Garber and L. Hoel, *Traffic and Highway Engineering.*)
  (a) Complete the table.

x	3	9	12	13	14
N					

  (b) For what value of x is N undefined?
  (c) Plot the points from the table. Then graph $x = 15$ as a vertical, dashed line.
  (d) Sketch a graph of N that passes through these points. Do not allow your graph to cross the vertical, dashed line, called an asymptote.
  (e) Use the graph to explain why a small increase in x can sometimes lead to a long wait.
  (f) Explain what happens over a long period of time if the arrival rate x exceeds 15 cars per hour. (*Hint:* The formula is not valid for $x \geq 15$.)

Exercises 2–6: **Graphing Rational Functions** *Complete the following.*

(a) *Use the given equation to complete the table of values for y.*

x	−4	−3	−2	−1	0	1	2	3	4
y									

(b) *Determine any x-value that will make the expression undefined.*

(c) *Sketch a dashed, vertical line (asymptote) in the xy-plane at any undefined values of x.*

(d) *Plot the points from the table.*

(e) *Sketch a graph of the equation. Do not let your graph cross the vertical, dashed line.*

2. $y = \dfrac{1}{x - 1}$

3. $y = \dfrac{1}{x + 1}$

4. $y = \dfrac{4}{x^2 + 1}$

5. $y = \dfrac{x}{x + 1}$

6. $y = \dfrac{x}{x - 1}$

CHAPTERS 1–7   Cumulative Review Exercises

1. Evaluate $\pi r^2 h$ when $r = 2$ and $h = 6$.

2. Translate the phrase "two less than twice a number" into an algebraic expression using the variable x.

Exercises 3 and 4: Evaluate and simplify to lowest terms.

3. $\frac{1}{2} \div \frac{5}{4}$

4. $\frac{5}{8} + \frac{1}{8}$

Exercises 5 and 6: Simplify the expression.

5. $-2 + 7x + 4 - 5x$

6. $-4(4 - y) + (5 - 3y)$

7. Solve $-2x + 11 = 13$ and check the solution.

8. Solve $-3x + 1 \geq x$.

Exercises 9 and 10: Graph the equation. Determine any intercepts.

9. $2x - 3y = 6$

10. $x = 1$

11. Sketch a line with slope $m = 3$ passing through $(-2, -1)$. Write its slope–intercept form.

12. The table lists points located on a line. Write the slope–intercept form of the line.

x	−2	−1	0	1
y	−5	−3	−1	1

Exercises 13 and 14: Find the slope–intercept form for the line satisfying the given conditions.

13. Parallel to $y = -\frac{2}{3}x + 1$ passing through $(2, -1)$

14. Passing through $(-1, 2)$ and $(2, 4)$

15. Determine which ordered pair is a solution to the system of equations: $(2, -6)$ or $(1, -2)$.

$$4x + \ y = 2$$
$$x - 4y = 9$$

16. Solve the system of equations.

$$-3r - t = \ \ 2$$
$$2r + t = -4$$

Exercises 17 and 18: Determine if the system of linear equations has no solutions, one solution, or infinitely many solutions.

17.
$$2x - y = \ \ 5$$
$$-2x + y = -5$$

18.
$$4x - 6y = 12$$
$$-6x + 9y = 18$$

Exercises 19–22: Simplify the expression.

19. $3z^2 \cdot 5z^6$

20. $(ab)^3$

21. $(2y - 3)(5y + 2)$

22. $(x^2 - y^2)^2$

Exercises 23 and 24: Simplify and write the expression using positive exponents.

23. $(3x^2)^{-3}$

24. $\dfrac{4x^2}{2x^4}$

25. Write 0.00123 in scientific notation.

26. Divide and check: $\dfrac{2x^2 - x + 3}{x - 1}$.

Exercises 27–30: Factor completely.

27. $6 + 13x - 5x^2$

28. $9z^2 - 4$

29. $t^2 + 16t + 64$

30. $x^3 - 16x$

Exercises 31 and 32: Solve the equation.

31. $y^2 + 5y - 14 = 0$ **32.** $x^3 = 4x$

Exercises 33 and 34: Simplify.

33. $\dfrac{3x}{4y} \cdot \dfrac{y}{9x^2}$ **34.** $\dfrac{x}{x^2 - 4} \div \dfrac{2x}{x - 2}$

35. Simplify $\dfrac{1 + \frac{2}{x}}{1 - \frac{2}{x}}$.

36. Solve $z = 3x - 2y$ for x.

Exercises 37 and 38: Solve.

37. $\dfrac{4}{3x} - \dfrac{3}{4x} = 1$ **38.** $\dfrac{1}{x - 1} + \dfrac{2}{x + 2} = \dfrac{3}{2}$

39. Suppose that y is directly proportional to x and that $y = 7$ when $x = 14$. Find y when $x = 11$.

40. Determine whether the data in the table represent direct or inverse variation. Find an equation that models the data.

x	1	2	4	10
y	20	10	5	2

APPLICATIONS

41. *Shoveling the Driveway* Two people are shoveling snow from a driveway. The first person shovels 12 square feet per minute, while the second person shovels 9 square feet per minute.
   (a) Write a simplified expression that gives the total square feet that the two people shovel in x minutes.
   (b) How long would it take them to clear a driveway with 1890 square feet?

42. *Burning Calories* An athlete can burn 10 calories per minute while running and 4 calories per minute while walking. If the athlete burns 450 calories in 60 minutes, how long is spent on each activity?

43. *Triangle* In an isosceles triangle, the measures of the two smaller angles are equal and their sum is 32° more than the largest angle.
   (a) Let x be the measure of one of the two smaller angles and y be the measure of the largest angle. Write a system of linear equations whose solution gives the measures of these angles.
   (b) Solve your system.

44. *Strength of a Beam* The strength of a wood beam varies inversely with its length. A beam that is 10 feet long can support 1100 pounds. How much weight can a similar beam support if it is 22 feet long?

8 Introduction to Functions

"Our competitive advantage is our *math skills*, which is probably not something you would expect of a media company."

—MAX LEVCHIN, CEO OF SLIDE

(Slide is the number one company for writing Facebook applications.)

Every day millions of people create trillions of bytes of information. The only way we can make sense out of these data and determine what is occurring within society is to use mathematics. One of the most important mathematical concepts used to discover trends and patterns is that of a *function*. A function typically receives an input (or question), performs a computation, and gives the output (or answer).

Functions have been used in science and engineering for centuries to answer questions related to things like eclipses, communication, and transportation. However, today functions are also being used to describe human behavior and to design social networks. (See Section 8.1, Exercise 75.) In fact, you may have noticed that new features available on Twitter and Facebook are sometimes referred to as applications or *functions*. People are creating thousands of new functions every day. *Math skills are essential* for writing successful applications and functions.

8.1   Functions and Their Representations

Basic Concepts • Representations of a Function • Definition of a Function • Identifying a Function • Graphing Calculators (Optional)

A LOOK INTO MATH ▶

In earlier chapters we showed how to use numbers to describe data. For example, instead of simply saying that there are *a lot* of people on Twitter, we might say that there are about 50 million tweets per day. A number helps explain what "a lot" means. We also showed that data can be summarized with formulas and graphs. Formulas and graphs are sometimes used to represent *functions*, which are essential in mathematics. In this section we introduce functions and their representations.

Basic Concepts

NEW VOCABULARY

☐ Function
☐ Function notation
☐ Input/Output
☐ Name of the function
☐ Dependent variable
☐ Independent variable
☐ Verbal representation
☐ Numerical representation
☐ Symbolic representation
☐ Graphical representation
☐ Diagrams/Diagrammatic
  representation
☐ Relation
☐ Domain/Range
☐ Nonlinear functions
☐ Vertical line test

▶ **REAL-WORLD CONNECTION** Functions are used to calculate many important quantities. For example, suppose that a person works for $7 per hour. Then we could use a function *named* f to calculate the amount of money the person earned after working x hours simply by multiplying the *input* x by 7. The result y is called the *output*. This concept is shown visually in the following diagram.

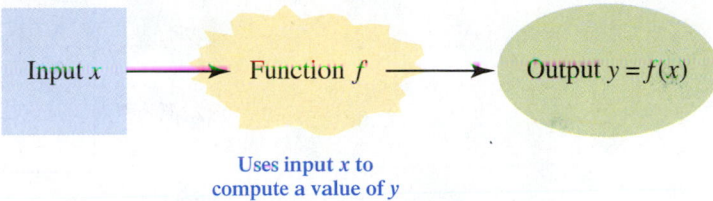

Uses input x to
compute a value of y

For each valid input x, a function computes *exactly one* output y, which may be represented by the ordered pair (x, y). If the input is 5 hours, f outputs $7 \cdot 5 = \$35$; if the input is 8 hours, f outputs $7 \cdot 8 = \$56$. These results can be represented by the ordered pairs $(5, 35)$ and $(8, 56)$. Sometimes an input may not be valid. For example, if $x = -3$, there is no reasonable output because a person cannot work -3 hours.

We say that y *is a function of x* because the output y is determined by and *depends* on the input x. As a result, y is called the *dependent variable* and x is the *independent variable*. To emphasize that y is a function of x, we use the notation $y = f(x)$. The symbol $f(x)$ does not represent multiplication of a variable f and a variable x. The notation $y = f(x)$ is called *function notation*, is read "y equals f of x," and means that function f with input x produces output y. For example, if $x = 3$ hours, $y = f(3) = \$21$.

FUNCTION NOTATION

The notation $y = f(x)$ is called **function notation**. The **input** is x, the **output** is y, and the **name of the function** is f.

$$\overset{\text{Name}}{\underset{\underset{\text{Output} \quad \text{Input}}{}}{y = f(x)}}$$

The variable y is called the **dependent variable** and the variable x is called the **independent variable**. The expression $f(4) = 28$ is read "f of 4 equals 28" and indicates that f outputs 28 when the input is 4. A function computes *exactly one* output for each valid input. The letters f, g, and h are often used to denote names of functions.

NOTE: Functions can be given *meaningful* names and variables. For example, function f could have been defined by $P(h) = 7h$, where function P calculates the pay after working h hours for $7 per hour.

▶ **REAL-WORLD CONNECTION** Functions can be used to compute a variety of quantities. For example, suppose that a boy has a sister who is exactly 5 years older than he is. If the age of the boy is x, then a function g can calculate the age of his sister by adding 5 to x. Thus $g(\mathbf{4}) = \mathbf{4} + 5 = \mathbf{9}$, $g(\mathbf{10}) = \mathbf{10} + 5 = \mathbf{15}$, and in general $g(\mathbf{x}) = \mathbf{x} + 5$. That is, function g adds 5 to input x to obtain the output $y = g(x)$.

Functions can be represented by an input–output machine, as illustrated in Figure 8.1. This machine represents function g and receives input $x = 4$, adds 5 to this value, and then outputs $g(4) = 4 + 5 = 9$.

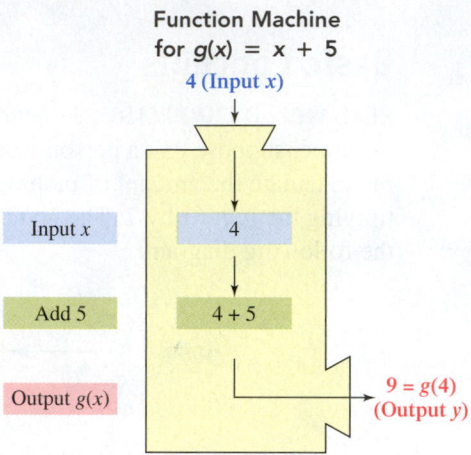

Figure 8.1

Representations of a Function

▶ **REAL-WORLD CONNECTION** A function f forms a relation between inputs x and outputs y that can be represented verbally, numerically, symbolically, and graphically. Functions can also be represented with diagrams. We begin by considering a function f that converts yards to feet.

VERBAL REPRESENTATION (WORDS) To convert x yards to y feet we multiply x by 3. Therefore, if function f computes the number of feet in x yards, a **verbal representation** of f is "Multiply the input x in yards by 3 to obtain the output y in feet."

TABLE 8.1

x (yards)	y (feet)
1	3
2	6
3	9
4	12
5	15
6	18
7	21

NUMERICAL REPRESENTATION (TABLE OF VALUES) A function f that converts yards to feet is shown in Table 8.1, where $y = f(x)$.

A *table of values* is called a **numerical representation** of a function. Many times it is impossible to list all valid inputs x in a table. On the one hand, if a table does not contain every x-input, it is a *partial* numerical representation. On the other hand, a *complete* numerical representation includes *all* valid inputs. Table 8.1 is a partial numerical representation of f because many valid inputs, such as $x = 10$ or $x = 5.3$, are not shown in it. Note that for each valid input x there is exactly one output y. *For a function, inputs are not listed more than once in a table.*

SYMBOLIC REPRESENTATION (FORMULA) A *formula* provides a **symbolic representation** of a function. The computation performed by f to convert x yards to y feet is expressed by $y = 3x$. A formula for f is $f(x) = 3x$, where $y = f(x)$. We say that function f is *defined by* or *given by* $f(x) = 3x$. Thus $f(\mathbf{2}) = 3 \cdot \mathbf{2} = \mathbf{6}$.

GRAPHICAL REPRESENTATION (GRAPH) A **graphical representation**, or **graph**, visually associates an x-input with a y-output. The ordered pairs

$$(\mathbf{1}, \mathbf{3}), (\mathbf{2}, \mathbf{6}), (\mathbf{3}, \mathbf{9}), (\mathbf{4}, \mathbf{12}), (\mathbf{5}, \mathbf{15}), (\mathbf{6}, \mathbf{18}), \text{ and } (\mathbf{7}, \mathbf{21})$$

from Table 8.1 are plotted in Figure 8.2(a). This scatterplot suggests a line for the graph f. For each real number x there is exactly one real number y determined by $y = 3x$. If we restrict inputs to $x \geq 0$ and plot all ordered pairs $(x, 3x)$, then a line with no breaks will appear, as shown in Figure 8.2(b).

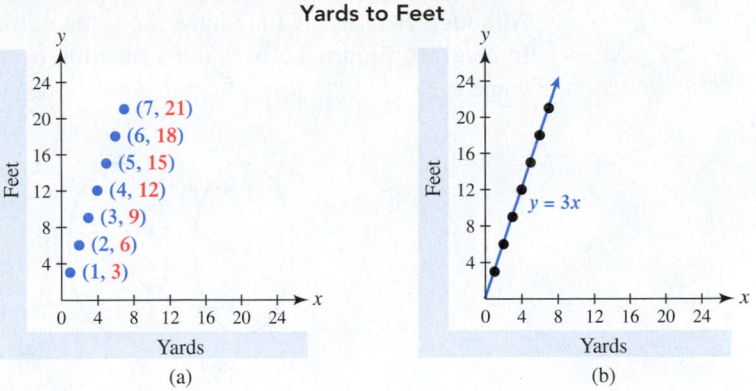

Figure 8.2

Because $f(1) = 3$, it follows that the point $(1, 3)$ lies on the graph of f, as shown in Figure 8.3. Graphs can sometimes be used to define a function f. For example, because the point $(1, 3)$ lies on the graph of f in Figure 8.3, we can conclude that $f(1) = 3$. That is, each point on the graph of f defines an input–output pair for f.

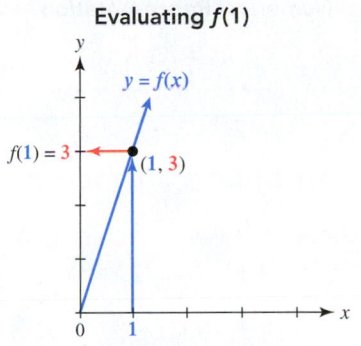

Figure 8.3

> **MAKING CONNECTIONS**
>
> **Functions, Points, and Graphs**
>
> If $f(a) = b$, then the point (a, b) lies on the graph of f. Conversely, if the point (a, b) lies on the graph of f, then $f(a) = b$. See Figure 8.4(a). Thus each point on the graph of f can be written in the form $(a, f(a))$. See Figure 8.4(b).

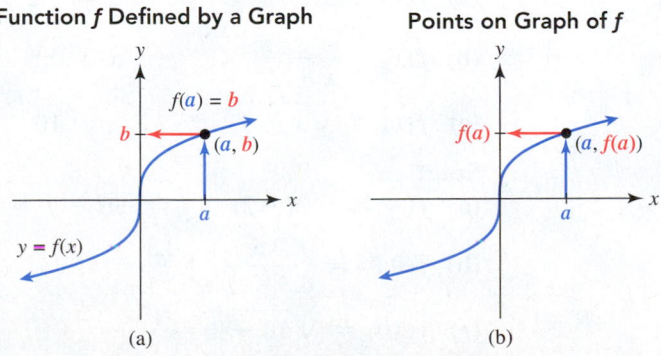

Figure 8.4

Yards to Feet

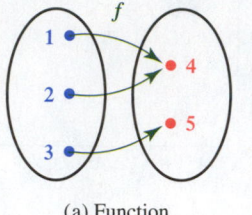

Function

Figure 8.5

DIAGRAMMATIC REPRESENTATION (DIAGRAM) Functions may be represented by **diagrams**. Figure 8.5 is a diagram of a function where an arrow is used to identify the output y associated with input x. For example, an arrow is drawn from input **2** to output **6**, which is written in function notation as $f(\mathbf{2}) = \mathbf{6}$. That is, **2** yards are equivalent to **6** feet.

Figure 8.6(a) shows a (different) function f even though $f(1) = 4$ and $f(2) = 4$. Although two inputs for f have the same output, each valid input has exactly one output. In contrast, Figure 8.6(b) is *not* a function because input 2 results in two different outputs, 5 and 6.

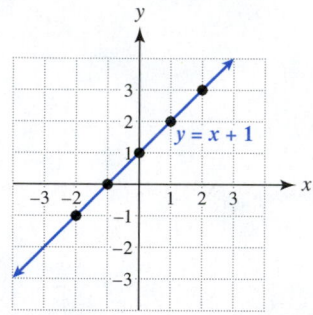

(a) Function   (b) Not a Function

Figure 8.6

MAKING CONNECTIONS

Four Representations of a Function

Symbolic Representation $f(x) = x + 1$

Numerical Representation

x	y
-2	-1
-1	0
0	1
1	2
2	3

Graphical Representation

Verbal Representation f adds 1 to an input x to produce an output y.

STUDY TIP

Be sure that you understand what verbal, numerical, graphical, and symbolic representations are.

EXAMPLE 1 **Evaluating symbolic representations (formulas)**

Evaluate each function f at the given value of x.
(a) $f(x) = 3x - 7$ $\qquad\qquad$ $x = -2$

(b) $f(x) = \dfrac{x}{x + 2}$ $\qquad\qquad$ $x = 0.5$

(c) $f(x) = \sqrt{x - 1}$ $\qquad\qquad$ $x = 10$

Solution
(a) $f(\mathbf{-2}) = 3(\mathbf{-2}) - 7 = -6 - 7 = \mathbf{-13}$

(b) $f(\mathbf{0.5}) = \dfrac{\mathbf{0.5}}{\mathbf{0.5} + 2} = \dfrac{0.5}{2.5} = \mathbf{0.2}$

(c) $f(\mathbf{10}) = \sqrt{\mathbf{10} - 1} = \sqrt{9} = \mathbf{3}$

Now Try Exercises 21, 23, 31

▶ **REAL-WORLD CONNECTION** In the next example we calculate sales tax by evaluating different representations of a function.

EXAMPLE 2 Calculating sales tax

Let a function f compute a sales tax of 7% on a purchase of x dollars. Use the given representation to evaluate $f(2)$.

(a) **Verbal Representation** Multiply a purchase of x dollars by 0.07 to obtain a sales tax of y dollars.

(b) **Numerical Representation (partial)** Shown in Table 8.2

(c) **Symbolic Representation** $f(x) = 0.07x$

(d) **Graphical Representation** Shown in Figure 8.7

(e) **Diagrammatic Representation** Shown in Figure 8.8

Sales Tax of 7%

TABLE 8.2

x	$f(x)$
$1.00	$0.07
$2.00	$0.14
$3.00	$0.21
$4.00	$0.28

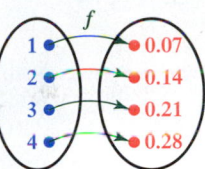

Figure 8.7   Figure 8.8

Solution

(a) **Verbal** Multiply the input 2 by 0.07 to obtain 0.14. The sales tax on a $2.00 purchase is $0.14.

(b) **Numerical** From Table 8.2, $f(2) = \$0.14$.

(c) **Symbolic** Because $f(x) = 0.07x$, $f(2) = 0.07(2) = 0.14$, or $0.14.

(d) **Graphical** To evaluate $f(2)$ with a graph, first find **2** on the x-axis in Figure 8.9. Then move vertically upward until you reach the graph of f. The point on the graph may be estimated as $(2, 0.14)$, meaning that $f(2) = 0.14$. Note that it may not be possible to find the exact answer from a graph. For example, one might estimate $f(2)$ to be 0.13 or 0.15 instead of 0.14.

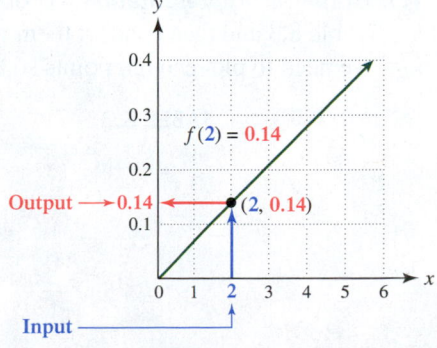

Figure 8.9

(e) **Diagrammatic** In Figure 8.8, follow the arrow from **2** to **0.14**. Thus $f(2) = 0.14$.

▌ **Now Try Exercises 25, 33, 53, 59, 61**

▶ **REAL-WORLD CONNECTION** There are many examples of functions. To give more meaning to a function, sometimes we change both its name and its input variable. For instance, if we know the radius r of a circle, we can calculate its circumference by using $C(r) = 2\pi r$. The next example illustrates how functions are used in physical therapy.

EXAMPLE 3   **Computing crutch length**

People who sustain leg injuries often require crutches. A proper crutch length can be estimated without using trial and error. The function L, given by $L(t) = 0.72t + 2$, outputs an appropriate crutch length L in inches for a person t inches tall. (*Source: Journal of the American Physical Therapy Association.*)
(a) Find $L(60)$ and interpret the result.
(b) If one person is 70 inches tall and another person is 71 inches tall, what should be the difference in their crutch lengths?

Solution
(a) $L(\mathbf{60}) = 0.72(\mathbf{60}) + 2 = \mathbf{45.2}$. Thus a person 60 inches tall needs crutches that are about 45.2 inches long.
(b) From the formula $L(t) = 0.72t + 2$, we can see that each 1-inch increase in t results in a 0.72-inch increase in $L(t)$. For example,

$$L(71) - L(70) = 53.12 - 52.4 = 0.72.$$

Now Try Exercise 75

In the next example we find a formula and then sketch a graph of a function.

EXAMPLE 4   **Finding representations of a function**

Let function f square the input x and then subtract 1 to obtain the output y.
(a) Write a formula, or symbolic representation, for f.
(b) Make a table of values, or numerical representation, for f. Use $x = -2, -1, 0, 1, 2$.
(c) Sketch a graph, or graphical representation, of f.

Solution
(a) **Symbolic Representation** If we square x and then subtract 1, we obtain $x^2 - 1$. Thus a formula for f is $f(x) = x^2 - 1$.
(b) **Numerical Representation** Make a table of values for $f(x)$, as shown in Table 8.3. For example,

$$f(\mathbf{-2}) = (\mathbf{-2})^2 - 1 = 4 - 1 = \mathbf{3}.$$

(c) **Graphical Representation** To obtain a graph of $f(x) = x^2 - 1$, plot the points from Table 8.3 and then connect them with a smooth curve, as shown in Figure 8.10. Note that we need to plot enough points so that we can determine the overall shape of the graph.

READING CHECK

Give a verbal, numerical, symbolic, and graphical representation of a function that calculates the number of days in a given number of weeks. Choose meaningful variables.

TABLE 8.3

x	$f(x)$
-2	**3**
-1	0
0	-1
1	0
2	3

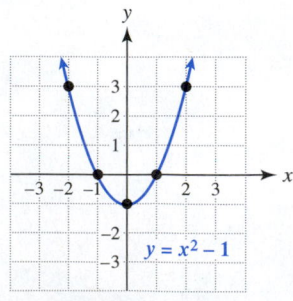

Figure 8.10

Now Try Exercise 63

Definition of a Function

A function is a fundamental concept in mathematics. Its definition should allow for all representations of a function. *A function receives an input x and produces exactly one output y,* which can be expressed as an ordered pair:

$$(x, y).$$

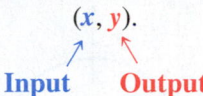

Input   Output

A **relation** is a set of ordered pairs, and a function is a special type of relation.

FUNCTION

A **function** f is a set of ordered pairs (x, y) where each x-value corresponds to exactly one y-value.

The **domain** of f is the set of all x-values, and the **range** of f is the set of all y-values. For example, a function f that converts 1, 2, 3, and 4 yards to feet could be expressed as

$$f = \{(1, 3), (2, 6), (3, 9), (4, 12)\}.$$

The domain of f is $D = \{1, 2, 3, 4\}$, and the range of f is $R = \{3, 6, 9, 12\}$.

MAKING CONNECTIONS

Relations and Functions

A relation can be thought of as a set of input–output pairs. A function is a special type of relation whereby each input results in exactly one output.

▶ **REAL-WORLD CONNECTION** In the next example, we see how education can improve a person's chances for earning a higher income.

EXAMPLE 5 **Computing average income**

A function f computes the average individual income in dollars in relation to educational attainment. This function is defined by $f(N) = 21{,}484$, $f(H) = 31{,}286$, $f(B) = 57{,}181$, and $f(M) = 70{,}181$, where N denotes no diploma, H a high school diploma, B a bachelor's degree, and M a master's degree. (*Source: 2010 Statistical Abstract.*)

(a) Write f as a set of ordered pairs.
(b) Give the domain and range of f.
(c) Discuss the relationship between education and income.

Solution

(a) $f = \{(N, 21484), (H, 31286), (B, 57181), (M, 70181)\}$
(b) The domain of function f is given by $D = \{N, H, B, M\}$, and the range of function f is given by $R = \{21484, 31286, 57181, 70181\}$.
(c) Education pays—the greater the educational attainment, the greater are annual earnings.

Now Try Exercise 101

EXAMPLE 6 **Finding the domain and range graphically**

Use the graphs of f shown in Figures 8.11 and 8.12 to find each function's domain and range.

(a)

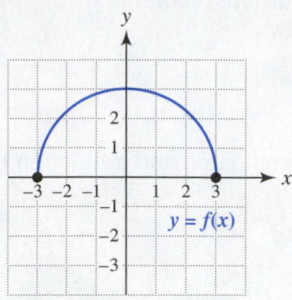

Figure 8.11

(b)

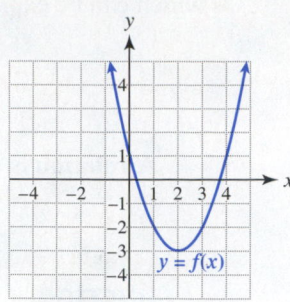

Figure 8.12

READING CHECK

Use the graph in
Figure 8.12 to evaluate $f(3)$.

Solution

(a) The domain is the set of all x-values that correspond to points on the graph of f. Figure 8.13 shows that the domain D includes all x-values satisfying $-3 \leq x \leq 3$. (Recall that the symbol $\leq$ is read "*less than or equal to.*") Because the graph is a semi-circle with no breaks, the domain includes all real numbers between and including -3 and 3. The range R is the set of y-values that correspond to points on the graph of f. Thus R includes all y-values satisfying $0 \leq y \leq 3$.

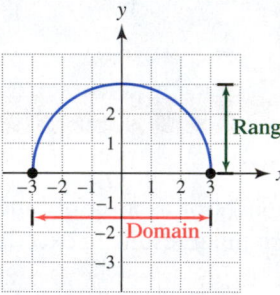

Figure 8.13

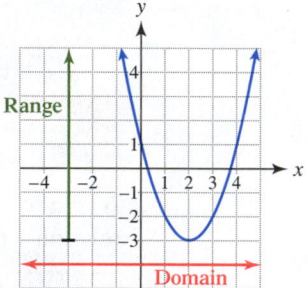

Figure 8.14

(b) The arrows on the ends of the graph in Figure 8.12 indicate that the graph extends indefinitely left and right, as well as upward. Thus D includes **all real numbers**. See Figure 8.14. The smallest y-value on the graph is $y = -3$, which occurs when $x = 2$. Thus the range R is $y \geq -3$. (Recall that the symbol $\geq$ is read "*greater than or equal to.*")

Now Try Exercises 77, 81

CRITICAL THINKING

Suppose that a car travels at 50 miles per hour to a city that is 250 miles away. Sketch a graph of a function f that gives the distance y traveled after x hours. Identify the domain and range of f.

The domain of a function is the set of all valid inputs. To determine the domain of a function from a formula, we must find x-values for which the formula is defined. To do this, we must determine if we can substitute any real number in the formula for $f(x)$. If we can, then the domain of f is *all real numbers*. However, there are situations in which we must limit the domain of f. For example, the domain must often be limited when there is either division or a square root in the formula for f. When division occurs, we must be careful to avoid values of the variable that result in division by 0, which is undefined. When a square root occurs, we must be careful to avoid values of the variable that result in the square root of a negative number, which is not a real number. This concept is demonstrated in the next example.

EXAMPLE 7 **Finding the domain of a function**

Use $f(x)$ to find the domain of f.

(a) $f(x) = 5x$ **(b)** $f(x) = \dfrac{1}{x-2}$ **(c)** $f(x) = \sqrt{x}$

Solution
(a) Because we can always multiply a real number x by 5, $f(x) = 5x$ is defined for all real numbers. Thus the domain of f includes all real numbers.
(b) Because we cannot divide by 0, input $x = 2$ is not valid for $f(x) = \dfrac{1}{x-2}$. The expression for $f(x)$ is defined for all other values of x. Thus the domain of f includes all real numbers except 2, or $x \neq 2$.
(c) Because square roots of negative numbers are not real numbers, the inputs for $f(x) = \sqrt{x}$ cannot be negative. Thus the domain of f includes all nonnegative numbers, or $x \geq 0$.

▮ **Now Try Exercises 87, 91, 95**

Symbolic, numerical, and graphical representations of three common functions are shown in Figure 8.15. Note that their graphs are not lines. For this reason they are called **nonlinear functions**. Use the graphs to find the domain and range of each function.

Absolute value: $f(x) = |x|$

x	-2	-1	0	1	2
$|x|$	2	1	0	1	2

Square: $f(x) = x^2$

x	-2	-1	0	1	2
x^2	4	1	0	1	4

Square root: $f(x) = \sqrt{x}$

x	0	1	4	9
$\sqrt{x}$	0	1	2	3

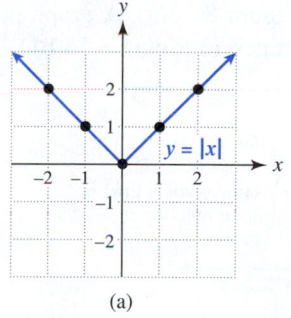

(a)

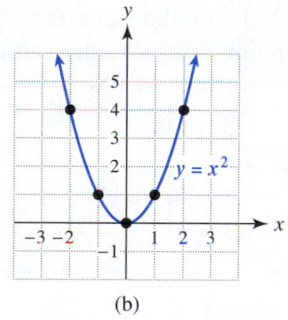

(b)

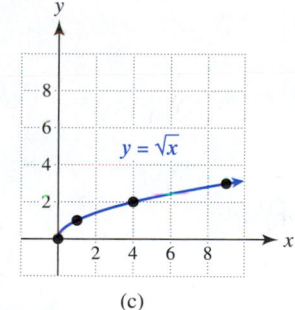

(c)

Figure 8.15

Identifying a Function

Recall that for a function each valid input x produces exactly one output y. In the next three examples we demonstrate techniques for identifying a function.

EXAMPLE 8 **Determining whether a set of ordered pairs is a function**

The set S of ordered pairs (x, y) represents the number of mergers and acquisitions y in 2010 for selected technology companies x.

$$S = \{(\text{IBM}, 12), (\text{HP}, 7), (\text{Oracle}, 5), (\text{Apple}, 5), (\text{Microsoft}, 0)\}$$

Determine if S is a function. (*Source:* cbinsights.)

Solution

The input x is the name of the technology company, and the output y is the number of mergers and acquisitions associated with that company. The set S *is* a function because each company x is associated with exactly one number y. Note that even though there were 5 mergers and acquisitions corresonding to both Oracle and Apple, S is nonetheless a function.

Now Try Exercise 123

EXAMPLE 9   **Determining whether a table of values represents a function**

TABLE 8.4

x	y
1	−4
2	8
3	2
1	5
4	−6

Determine whether Table 8.4 represents a function.

Solution

The table does not represent a function because input $x = 1$ produces two outputs: −4 and 5. That is, the following two ordered pairs both belong to this relation.

Same input x

$$(1, -4) \qquad (1, 5) \leftarrow \textbf{Not a function}$$

Different outputs y

Now Try Exercise 125

VERTICAL LINE TEST To determine whether a graph represents a function, we must be convinced that it is impossible for an input x to have two or more outputs y. If two distinct points have the *same x*-coordinate on a graph, then the graph cannot represent a function. For example, the ordered pairs $(-1, 1)$ and $(-1, -1)$ could not lie on the graph of a function because input -1 results in *two* outputs: 1 and −1. When the points $(-1, 1)$ and $(-1, -1)$ are plotted, they lie on the same vertical line, as shown in Figure 8.16(a). A graph passing through these points intersects the vertical line twice, as illustrated in Figure 8.16(b).

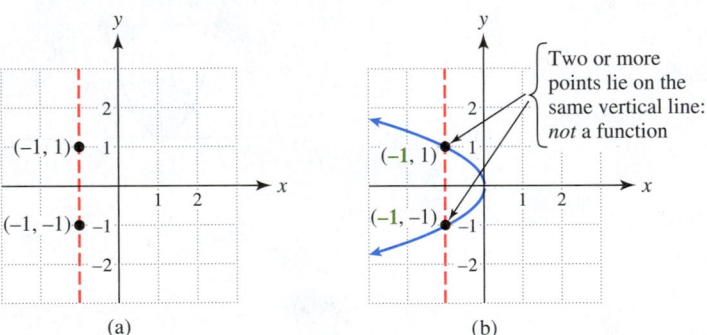

(a)   (b)

Figure 8.16

To determine whether a graph represents a function, visualize vertical lines moving across the *xy*-plane. If each vertical line intersects the graph *at most once*, then it is a graph of a function. This test is called the **vertical line test**. Note that the graph in Figure 8.16(b) fails the vertical line test and therefore does not represent a function.

READING CHECK

What is the vertical line test used for?

VERTICAL LINE TEST

If every vertical line intersects a graph at no more than one point, then the graph represents a function.

EXAMPLE 10   **Determining whether a graph represents a function**

Determine whether the graphs shown in Figure 8.17 represent functions.

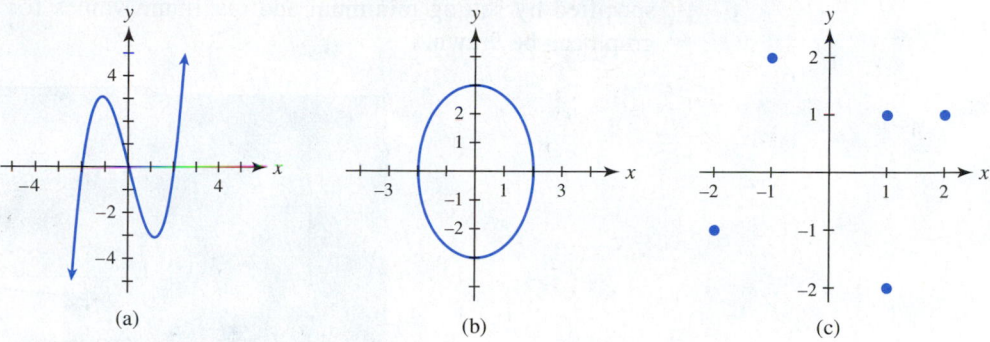

(a)   (b)   (c)

Figure 8.17

Solution

(a) Visualize vertical lines moving across the *xy*-plane from left to right. Any (red) vertical line will intersect the graph at most once, as depicted in Figure 8.18(a). Therefore the graph *does* represent a function.

Passes Vertical Line Test   **Fails Vertical Line Test**   **Fails Vertical Line Test**

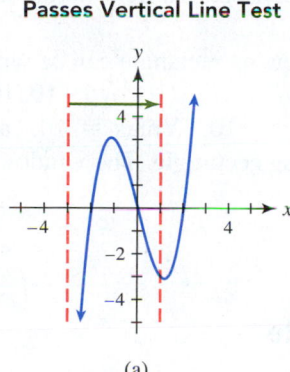   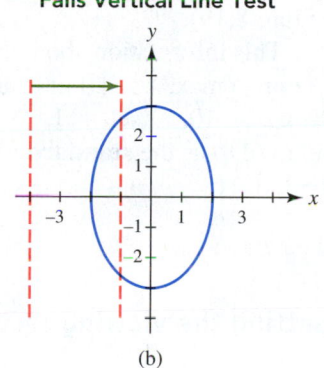   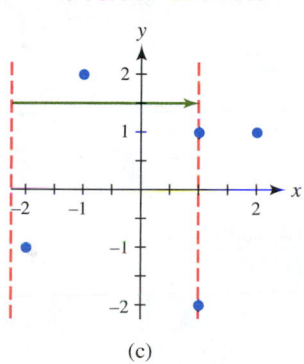

(a)   (b)   (c)

Figure 8.18

(b) Visualize vertical lines moving across the *xy*-plane from left to right. The graph *does not* represent a function because there exist (red) vertical lines that can intersect the graph twice. One such line is shown in Figure 8.18(b).

(c) Visualize vertical lines moving across the *xy*-plane from left to right. The graph is a scatterplot and *does not* represent a function because there exists one (red) vertical line that intersects two points: $(1, 1)$ and $(1, -2)$ with the same *x*-coordinate, as shown in Figure 8.18(c).

Now Try Exercises 111, 113, 119

Graphing Calculators (Optional)

Graphing calculators provide several features beyond those found on scientific calculators. Graphing calculators have additional keys that can be used to create tables, scatterplots, and graphs.

▶ **REAL-WORLD CONNECTION** The **viewing rectangle**, or **window**, on a graphing calculator is similar to the viewfinder in a camera. A camera cannot take a picture of an entire scene.

The camera must be centered on some object and can photograph only a portion of the available scenery. A camera can capture different views of the same scene by zooming in and out, as can graphing calculators. The *xy*-plane is infinite, but the calculator screen can show only a finite, rectangular region of the *xy*-plane. The viewing rectangle must be specified by setting minimum and maximum values for both the *x*- and *y*-axes before a graph can be drawn.

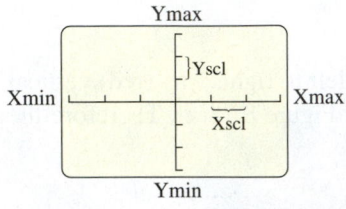

Figure 8.19

We use the following terminology regarding the size of a viewing rectangle. **Xmin** is the minimum *x*-value along the *x*-axis, and **Xmax** is the maximum *x*-value. Similarly, **Ymin** is the minimum *y*-value along the *y*-axis, and **Ymax** is the maximum *y*-value. Most graphs show an *x*-scale and a *y*-scale with tick marks on the respective axes. Sometimes the distance between consecutive tick marks is 1 unit, but at other times it might be 5 or 10 units. The distance represented by consecutive tick marks on the *x*-axis is called **Xscl**, and the distance represented by consecutive tick marks on the *y*-axis is called **Yscl** (see Figure 8.19).

This information about the viewing rectangle can be written as [Xmin, Xmax, Xscl] by [Ymin, Ymax, Yscl]. For example, [−10, 10, 1] by [−10, 10, 1] means that Xmin = −10, Xmax = 10, Xscl = 1, Ymin = −10, Ymax = 10, and Yscl = 1. This setting is referred to as the **standard viewing rectangle**. The window in Figure 8.19 is [−3, 3, 1] by [−3, 3, 1].

EXAMPLE 11   **Setting the viewing rectangle**

Show the viewing rectangle [−2, 3, 0.5] by [−100, 200, 50] on your calculator.

Solution
The window setting and viewing rectangle are displayed in Figure 8.20. Note that in Figure 8.20(b) there are 6 tick marks on the positive *x*-axis because its length is 3 units and the distance between consecutive tick marks is 0.5 unit.

CALCULATOR HELP

To set a viewing rectangle, see Appendix A (page AP-3).

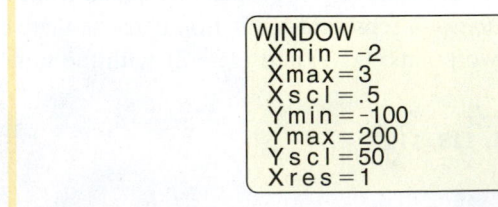

(a)   (b)

Figure 8.20

Now Try Exercise 133

SCATTERPLOTS Many graphing calculators have the capability to create scatterplots and line graphs. The next example illustrates how to make a scatterplot with a graphing calculator.

EXAMPLE 12 | **Making a scatterplot with a graphing calculator**

Plot the points $(-2, -2)$, $(-1, 3)$, $(1, 2)$, and $(2, -3)$ in $[-4, 4, 1]$ by $[-4, 4, 1]$.

Solution
We entered the points $(-2, -2)$, $(-1, 3)$, $(1, 2)$, and $(2, -3)$ shown in Figure 8.21(a), using the STAT EDIT feature. The variable L1 represents the list of x-values, and the variable L2 represents the list of y-values. In Figure 8.21(b) we set the graphing calculator to make a scatterplot with the STATPLOT feature, and in Figure 8.21(c) the points have been plotted. If you have a different model of calculator you may need to consult your owner's manual.

CALCULATOR HELP

To make a scatterplot, see Appendix A (pages AP-3 and AP-4).

$[-4, 4, 1]$ by $[-4, 4, 1]$

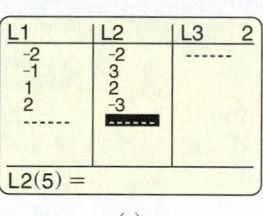
(a)

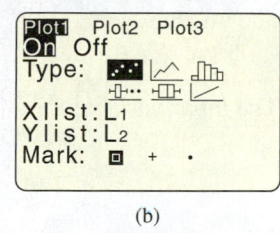
(b)

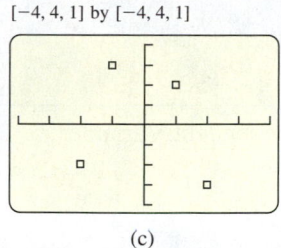
(c)

Figure 8.21

Now Try Exercise 137

GRAPHS AND TABLES We can use graphing calculators to create graphs and tables, usually more efficiently and reliably than with pencil-and-paper techniques. However, a graphing calculator uses the same techniques that we might use to sketch a graph. For example, one way to sketch a graph of $y = 2x - 1$ is first to make a table of values, as shown in Table 8.5.

We can plot these points in the xy-plane, as shown in Figure 8.22. Next we might connect the points, as shown in Figure 8.23.

TABLE 8.5

x	y
-1	-3
0	-1
1	1
2	3

Plotting Points

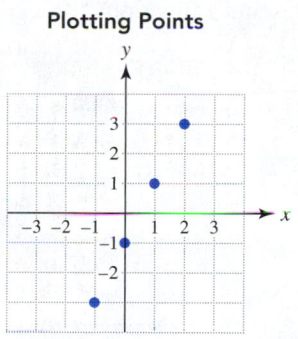

Figure 8.22

Graphing a Line

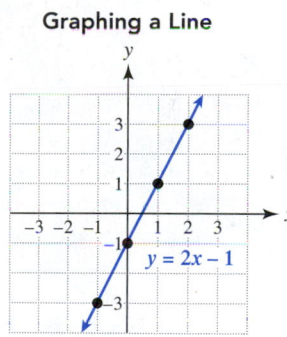

Figure 8.23

In a similar manner, a graphing calculator plots numerous points and connects them to make a graph. To create a similar graph with a graphing calculator, we enter the formula $Y_1 = 2X - 1$, set an appropriate viewing rectangle, and graph as shown in Figures 8.24 and 8.25. A table of values can also be generated as illustrated in Figure 8.26.

CALCULATOR HELP

To make a graph, see Appendix A (page AP-5). To make a table, see Appendix A (pages AP-2 and AP-3).

$[-10, 10, 1]$ by $[-10, 10, 1]$

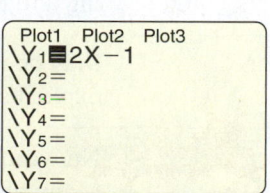

Figure 8.24

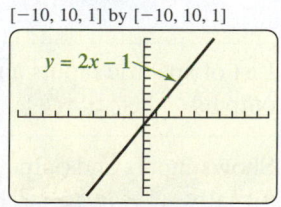

Figure 8.25

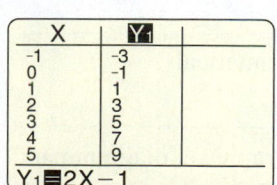

Figure 8.26

8.1 Putting It All Together

A function calculates exactly one output for each valid input and produces input–output ordered pairs in the form (x, y). A function typically computes something such as area, speed, or sales tax.

CONCEPT	EXPLANATION	EXAMPLES
Function	A set of ordered pairs (x, y), where each x-value corresponds to exactly one y-value	$f = \{(1, 3), (2, 3), (3, 1)\}$ $f(x) = 2x$ A graph of $y = x + 2$ A table of values for $y = 4x$
Independent Variable	The *input* variable for a function	*Function* *Independent Variable* $f(x) = 2x$ x $A(r) = \pi r^2$ r $V(s) = s^3$ s
Dependent Variable	The *output* variable of a function There is exactly one output for each valid input.	*Function* *Dependent Variable* $y = f(x)$ y $T = F(r)$ T $V = g(r)$ V
Domain and Range of a Function	The domain D is the set of all valid inputs. The range R is the set of all outputs.	For $S = \{(-1, 0), (3, 4), (5, 0)\}$, $D = \{-1, 3, 5\}$ and $R = \{0, 4\}$. For $f(x) = \frac{1}{x}$ the domain includes all real numbers except 0, or $x \neq 0$.
Vertical Line Test	If every vertical line intersects a graph at no more than one point, the graph represents a function.	This graph does *not* pass this test and thus does not represent a function. Two points lie on the same vertical line: *not* a function

A function can be represented verbally, symbolically, numerically, and graphically.

REPRESENTATION	EXPLANATION	COMMENTS
Verbal	Precise word description of what is computed	May be oral or written Must be stated *precisely*
Symbolic	Mathematical formula	Efficient and concise way of representing a function (e.g., $f(x) = 2x - 3$)
Numerical	List of specific inputs and their outputs	May be in the form of a table or an explicit set of ordered pairs
Graphical, diagrammatic	Shows inputs and outputs visually	No words, formulas, or tables Many types of graphs and diagrams are possible.

8.1 Exercises

MyMathLab   Math XL PRACTICE   WATCH   DOWNLOAD   READ   REVIEW

CONCEPTS AND VOCABULARY

1. The notation $y = f(x)$ is called _____ notation.

2. The notation $y = f(x)$ is read _____.

3. The notation $f(x) = x^2 + 1$ is a(n) _____ representation of a function.

4. A table of values is a(n) _____ representation of a function.

5. The set of valid inputs for a function is the _____.

6. The set of outputs for a function is the _____.

7. A function computes _____ output for each valid input.

8. (True or False?) The vertical line test is used to identify graphs of relations.

9. (True or False?) Four ways to represent functions are verbal, numerical, symbolic, and graphical.

10. If $f(3) = 4$, the point _____ is on the graph of f. If $(3, 6)$ is on the graph of f, then $f(\_\_\_\_) = \_\_\_\_$.

11. **Thinking Generally** If $f(a) = b$, the point _____ is on the graph of f.

12. **Thinking Generally** If (c, d) is on the graph of g, then $g(c) = \_\_\_\_$.

13. **Thinking Generally** If a is in the domain of f, then $f(a)$ represents how many outputs?

14. **Thinking Generally** If $f(x) = x$ for every x in the domain of f, then the domain and range of f are _____.

Exercises 15–20: Determine whether the phrase describes a function.

15. Calculating the square of a number

16. Calculating the low temperature for a day

17. Listing the students who passed a given math exam

18. Listing the children of parent x

19. Finding sales tax on a purchase

20. Naming the people in your class

REPRESENTING AND EVALUATING FUNCTIONS

Exercises 21–32: Evaluate $f(x)$ at the given values of x.

21. $f(x) = 4x - 2$ $x = -1, 0$

22. $f(x) = 5 - 3x$ $x = -4, 2$

23. $f(x) = \sqrt{x}$ $x = 0, \frac{9}{4}$

24. $f(x) = \sqrt[3]{x}$ $x = -1, 27$

25. $f(x) = x^2$ $x = -5, \frac{3}{2}$

26. $f(x) = x^3$ $x = -2, 0.1$

27. $f(x) = 3$ $x = -8, \frac{7}{3}$

28. $f(x) = 100$ $x = -\pi, \frac{1}{3}$

29. $f(x) = 5 - x^3$ $x = -2, 3$

30. $f(x) = x^2 + 5$ $x = -\frac{1}{2}, 6$

31. $f(x) = \dfrac{2}{x + 1}$ $x = -5, 4$

32. $f(x) = \dfrac{x}{x - 4}$ $x = -3, 1$

Exercises 33–38: Do the following.

    (a) *Write a formula for the function described.*
    (b) *Evaluate the function for input 10 and interpret the result.*

33. Function I computes the number of inches in x yards.

34. Function A computes the area of a circle with radius r.

35. Function M computes the number of miles in x feet.

36. Function C computes the circumference of a circle with radius r.

37. Function A computes the square feet in x acres. (*Hint:* There are 43,560 square feet in one acre.)

38. Function K computes the number of kilograms in x pounds. (*Hint:* There are about 2.2 pounds in one kilogram.)

Exercises 39–42: Write each function f as a set of ordered pairs. Give the domain and range of f.

39. $f(1) = 3, f(2) = -4, f(3) = 0$

40. $f(-1) = 4, f(0) = 6, f(1) = 4$

41. $f(a) = b, f(c) = d, f(e) = a, f(d) = b$

42. $f(a) = 7, f(b) = 7, f(c) = 7, f(d) = 7$

Exercises 43–52: Sketch a graph of f.

43. $f(x) = -x + 3$ 44. $f(x) = -2x + 1$

45. $f(x) = 2x$ 46. $f(x) = \frac{1}{2}x - 2$

47. $f(x) = 4 - x$ **48.** $f(x) = 6 - 3x$

49. $f(x) = x^2$ **50.** $f(x) = \sqrt{x}$

51. $f(x) = \sqrt{x + 1}$ **52.** $f(x) = \frac{1}{2}x^2 - 1$

Exercises 53–58: Use the graph of f to evaluate the given expressions.

53. $f(0)$ and $f(2)$ **54.** $f(-2)$ and $f(2)$

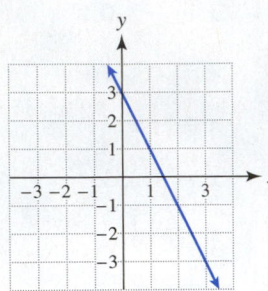
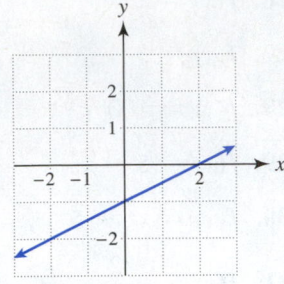

55. $f(-2)$ and $f(1)$ **56.** $f(-1)$ and $f(0)$

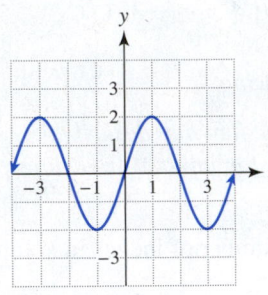
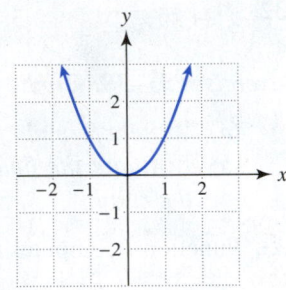

57. $f(1)$ and $f(2)$ **58.** $f(-1)$ and $f(4)$

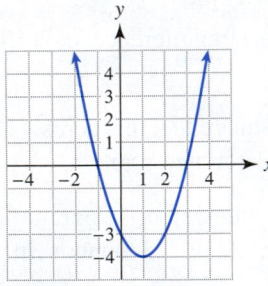
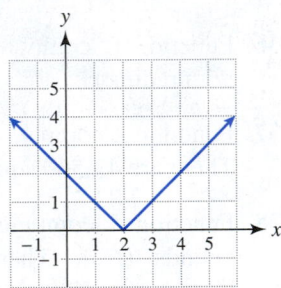

Exercises 59 and 60: Use the table to evaluate the given expressions.

59. $f(0)$ and $f(2)$

x	0	1	2	3	4
$f(x)$	5.5	4.3	3.7	2.5	1.9

60. $f(-10)$ and $f(5)$

x	-10	-5	0	5	10
$f(x)$	23	96	-45	-33	23

Exercises 61 and 62: Use the diagram to evaluate f(1990). Interpret your answer.

61. The function f computes average fuel efficiency of new U.S. passenger cars in miles per gallon during year x. (*Source:* Department of Transportation.)

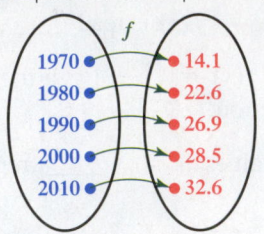

62. The function f computes average cost of tuition at public colleges and universities during academic year x. (*Source:* The College Board.)

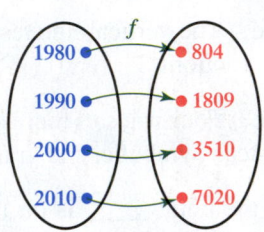

Exercises 63–66: Express the verbal representation for the function f numerically, symbolically, and graphically. Let $x = -3, -2, -1, \ldots, 3$ for the numerical representation (table), and let $-3 \leq x \leq 3$ for the graph.

63. Add 5 to the input x to obtain the output y.

64. Square the input x to obtain the output y.

65. Multiply the input x by 5 and then subtract 2 to obtain the output y.

66. Divide the input x by 2 and then add 3 to obtain the output y.

Exercises 67–72: Give a verbal representation for $f(x)$.

67. $f(x) = x - \frac{1}{2}$ **68.** $f(x) = \frac{3}{4}x$

69. $f(x) = \dfrac{x}{3}$ **70.** $f(x) = x^2 + 1$

71. $f(x) = \sqrt{x - 1}$ **72.** $f(x) = 1 - 3x$

73. *Cost of Driving* In 2010, the average cost of driving a new car in the United States was about 50 cents per mile. Symbolically, graphically, and numerically represent a function f that computes the cost in dollars of driving x miles. For the numerical representation (table) let $x = 10, 20, 30, \ldots, 70$. (*Source:* Associated Press.)

74. *Federal Income Taxes* In 2010, the lowest U.S. income tax rate was 10 percent. Symbolically, graphically, and numerically represent a function f that computes the tax on a taxable income of x dollars. For the numerical representation (table) let $x = 1000, 2000, 3000, \ldots, 7000$, and for the graphical representation let $0 \leq x \leq 10{,}000$. (*Source:* Internal Revenue Service.)

75. *Global Web Searches* The number of World Wide Web searches S in billions during year x can be approximated by $S(x) = 225x - 450{,}650$ from 2009 to 2012. Evaluate $S(2011)$ and interpret the result. (*Source:* RBC Capital Markets Corp.)

76. *Cost of Smartphones* The average cost difference D in dollars between smartphones and all other types of phones during year x can be approximated by $D(x) = -23.5x + 47{,}275$ from 2005 to 2009. Evaluate $D(2009)$ and interpret the result. (*Source:* Business Insider.)

IDENTIFYING DOMAINS AND RANGES

Exercises 77–84: Use the graph of f to identify its domain and range.

77.
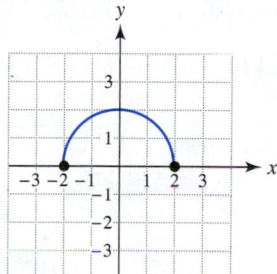

78.
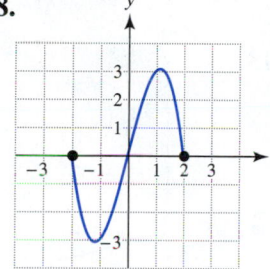

79.
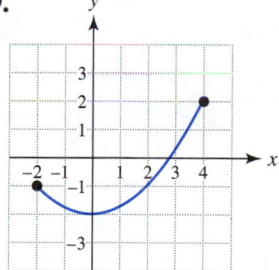

80.
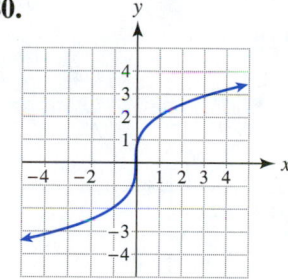

81.
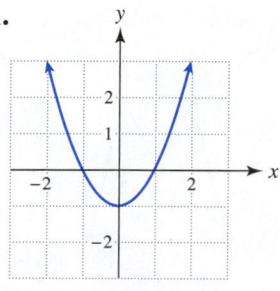

82.
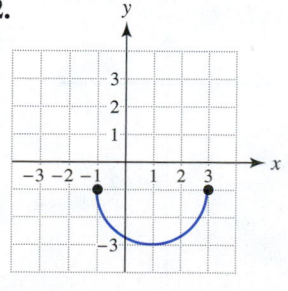

83.
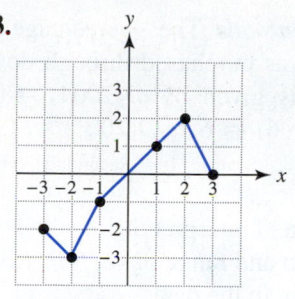

84.
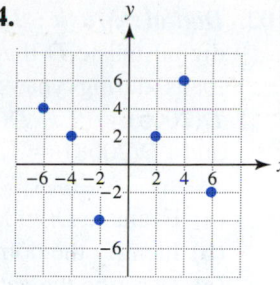

Exercises 85 and 86: Use the diagram to find the domain and range of f.

85.
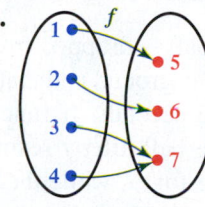

86.
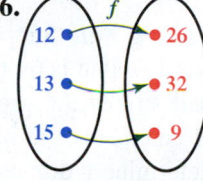

Exercises 87–100: Find the domain.

87. $f(x) = 10x$

88. $f(x) = 5 - x$

89. $f(x) = x^2 - 3$

90. $f(x) = \frac{1}{2}x^2$

91. $f(x) = \dfrac{3}{x - 5}$

92. $f(x) = \dfrac{x}{x + 1}$

93. $f(x) = \dfrac{2x}{x^2 + 1}$

94. $f(x) = \dfrac{6}{1 - x}$

95. $f(x) = \sqrt{x - 1}$

96. $f(x) = |x|$

97. $f(x) = |x - 5|$

98. $f(x) = \sqrt{2 - x}$

99. $f(x) = \dfrac{1}{x}$

100. $f(x) = 1 - 3x^2$

101. *Humpback Whales* The number of humpback whales W sighted in Maui's annual whale census for year x is given by $W(2005) = 649$, $W(2006) = 1265$, $W(2007) = 959$, $W(2008) = 1726$, and $W(2009) = 1010$. (*Source:* Pacific Whale Foundation.)
  (a) Evaluate $W(2008)$ and interpret the result.
  (b) Identify the domain and range of W.
  (c) Describe the pattern in the data.

102. *Digital Music Downloads* The percentage of digital music D that was purchased through downloads during year x is given by $D(2004) = 0.9$, $D(2005) = 5.7$, $D(2006) = 6.7$, $D(2007) = 11.2$, and $D(2008) = 12.8$. (*Source:* The Recording Industry Association of America.)

(a) Evaluate $D(2006)$ and interpret the result.
(b) Identify the domain and range of D.
(c) Describe the pattern in the data.

103. *Cost of Tuition* Suppose that a student can take from 1 to 20 credits at a college and that each credit costs $200. If function C calculates the cost of taking x credits, determine the domain and range of C.

104. *Falling Ball* Suppose that a ball is dropped from a window that is 64 feet above the ground and that the ball strikes the ground after 2 seconds. If function H calculates the height of the ball after t seconds, determine a domain and range for H, while the ball is falling.

IDENTIFYING A FUNCTION

Exercises 105–108: Determine whether the diagram could represent a function.

105.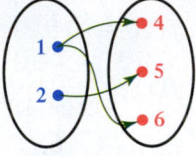

106.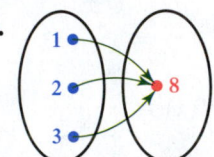

107.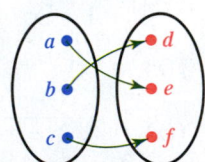

108.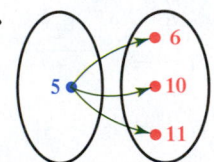

109. *Average Precipitation* The table lists the monthly average precipitation P in Las Vegas, Nevada, where $x = 1$ corresponds to January and $x = 12$ corresponds to December.

x (month)	1	2	3	4	5	6
P (inches)	0.5	0.4	0.4	0.2	0.2	0.1

x (month)	7	8	9	10	11	12
P (inches)	0.4	0.5	0.3	0.2	0.4	0.3

Source: J. Williams.

(a) Determine the value of P during May.
(b) Is P a function of x? Explain.
(c) If $P = 0.4$, find x.

110. *Wind Speeds* The table at the top of the next column lists the monthly average wind speed W in Louisville, Kentucky, where $x = 1$ corresponds to January and $x = 12$ corresponds to December.

x (month)	1	2	3	4	5	6
W (mph)	10.4	12.7	10.4	10.4	8.1	8.1

x (month)	7	8	9	10	11	12
W (mph)	6.9	6.9	6.9	8.1	9.2	9.2

Source: J. Williams.

(a) Determine the month with the highest average wind speed.
(b) Is W a function of x? Explain.
(c) If $W = 6.9$, find x.

Exercises 111–122: Determine whether the graph represents a function. If it does, identify the domain and range.

111.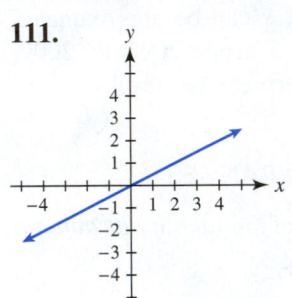

112.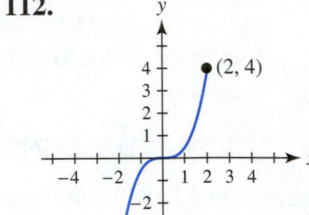

113.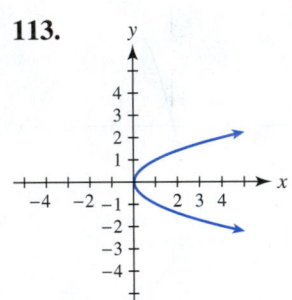

114.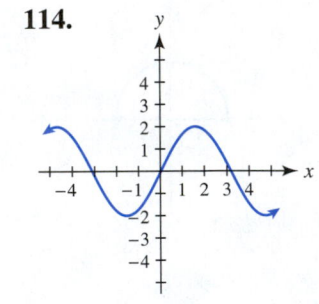

115.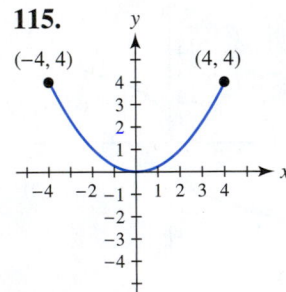

116.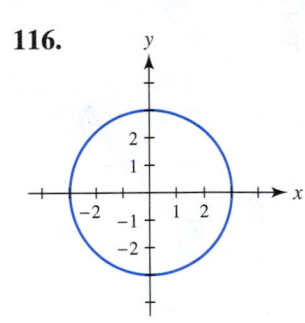

117.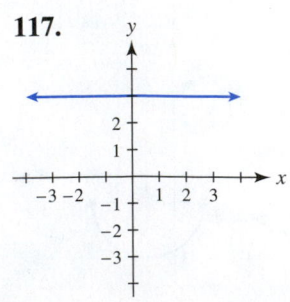

118.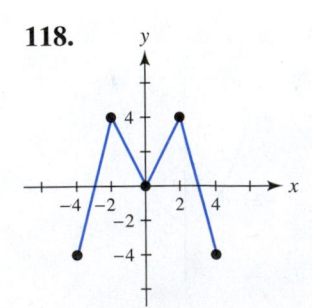

119.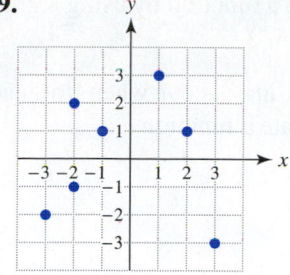

120.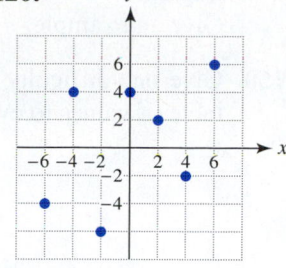

121.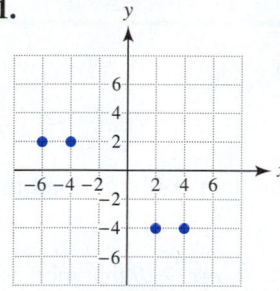

122.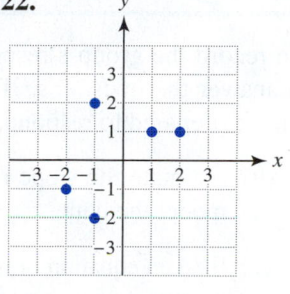

Exercises 123–126: Determine whether S is a function.

123. $S = \{(1, 2), (4, 5), (7, 8), (5, 4), (2, 2)\}$

124. $S = \{(4, 7), (-2, 1), (3, 8), (4, 9)\}$

125. *S* is given by the table.

x	5	10	5
y	2	1	0

126. *S* is given by the table.

x	−3	−2	−1
y	10	10	10

GRAPHICAL INTERPRETATION

Exercises 127 and 128: The graph represents the distance that a person is from home while walking on a straight path. The x-axis represents time and the y-axis represents distance. Interpret the graph.

127.   **128.**

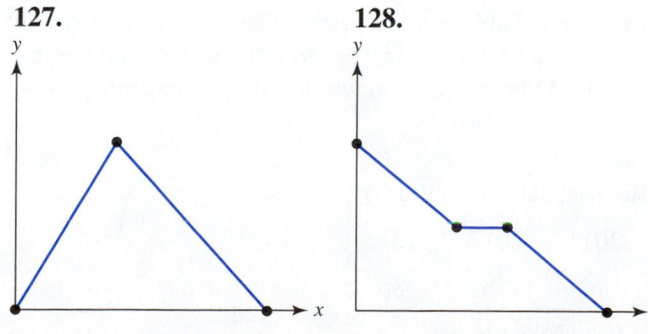

129. *Texting* The average 18- to 24-year-old person texts about 1500 messages per month. Sketch a graph that shows the total number of text messages sent over a period of 4 months. Assume that the same number of texts is sent each day. (*Source:* The Nielsen Company.)

130. *Computer Viruses* In 2000 there were about 50 thousand computer viruses. In 2010 there were about 1.6 million computer viruses. Sketch a graph of this increase from 2000 to 2010. Answers may vary. (*Source:* Symantec.)

GRAPHING CALCULATORS

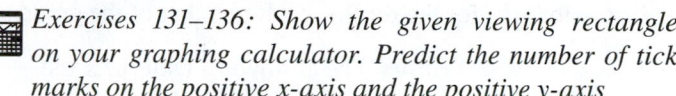 *Exercises 131–136: Show the given viewing rectangle on your graphing calculator. Predict the number of tick marks on the positive x-axis and the positive y-axis*

131. Standard viewing rectangle

132. [−12, 12, 2] by [−8, 8, 2]

133. [0, 100, 10] by [−50, 50, 10]

134. [−30, 30, 5] by [−20, 20, 5]

135. [1980, 1995, 1] by [12000, 16000, 1000]

136. [1900, 1990, 10] by [1700, 2800, 100]

Exercises 137–142: Use your calculator to make a scatterplot of the relation after determining an appropriate viewing rectangle.

137. $\{(4, 3), (-2, 1), (-3, -3), (5, -2)\}$

138. $\{(5, 5), (2, 0), (-2, 7), (2, -8), (-1, -5)\}$

139. $\{(20, 40), (-25, -15), (-20, 25), (15, -25)\}$

140. $\{(-13, 12), (3, 10), (-15, -4), (12, -9)\}$

141. $\{(100, -100), (50, 200), (-150, -140), (-30, 80)\}$

142. $\{(-125, 75), (45, 65), (-53, -67), (150, -80)\}$

Exercises 143–146: Make a table and graph of y = f(x). Let x = −3, −2, −1, …, 3 for your table and use the standard window for your graph.

143. $f(x) = \sqrt{x + 3}$ **144.** $f(x) = x^3 - \frac{1}{2}x^2$

145. $f(x) = \dfrac{5 - x}{5 + x}$ **146.** $f(x) = |2 - x| + \sqrt[3]{x}$

WRITING ABOUT MATHEMATICS

147. Give an example of a function. Identify the domain and range of your function.

148. Explain in your own words what a function is. How is a function different from other relations?

149. Explain how to evaluate a function by using a graph. Give an example.

150. Give one difficulty that may occur when you use a table of values to evaluate a function.

Group Activity   Working with Real Data

Directions: Form a group of 2 to 4 people. Select someone to record the group's responses for this activity. All members should work cooperatively to answer the questions. If your instructor asks for the results, each member of the group should be prepared to respond.

U.S. Craigslist Visitors The following table lists the average number of *unique* visitors to Craigslist for selected years.

Year	2006	2007	2008
Visitors	180,000	288,000	420,000

Year	2009	2010
Visitors	516,000	624,000

Source: Citi Investment Research and Analysis.

(a) Make a scatterplot of the data. Let *x* represent the number of years after 2006. Discuss any trend in numbers of visitors to Craigslist.

(b) Estimate the slope of a line that could be used to model the data.

(c) Find an equation of a line $y = mx + b$ that models the data.

(d) Interpret the slope as a rate of change.

(e) Use your results to estimate the number of *unique* visitors to Craigslist in 2012.

8.2   Linear Functions

**Basic Concepts • Representations of Linear Functions •
Modeling Data with Linear Functions • The Midpoint Formula (Optional)**

A LOOK INTO MATH ▶

Functions are frequently used to model, or describe, the real world. For example, people are becoming more energy conscious. As a result, there is an increase in the number of *green* buildings that are being constructed. Table 8.6 lists estimated U.S. sales of green building material. Because sales increase by $5 billion each year, a *linear function* can be used to model these data. (See Example 7.) In this section we discuss this important type of function.

NEW VOCABULARY

☐ Linear function
☐ Rate of change
☐ Constant function
☐ Midpoint

TABLE 8.6 Green Material Sales ($ billions)

Year	2010	2011	2012	2013
Sales	65	70	75	80

Source: Freedonia Group, Green Building Material.

Basic Concepts

▶ **REAL-WORLD CONNECTION** Suppose that the air conditioner is turned on when the temperature inside a house is $80°F$. The resulting temperatures are listed in Table 8.7 for various elapsed times. Note that for each 1-hour increase in elapsed time, the temperature decreases by $2°F$.

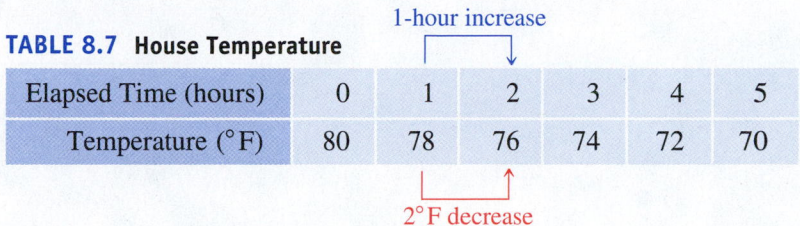

TABLE 8.7 **House Temperature**

Elapsed Time (hours)	0	1	2	3	4	5
Temperature (°F)	80	78	76	74	72	70

We want to determine a function f that models, or calculates, the house temperature after x hours. To do this, we will find numerical, graphical, verbal, and symbolic representations of f.

NUMERICAL REPRESENTATION (TABLE OF VALUES) We can think of Table 8.7 as a numerical representation (table of values) for the function f. A similar numerical representation that uses x and $f(x)$ is shown in Table 8.8.

TABLE 8.8 **Numerical Representation of $f(x)$**

x	0	1	2	3	4	5
$f(x)$	80	78	76	74	72	70

GRAPHICAL REPRESENTATION (GRAPH) To graph $y = f(x)$, we begin by plotting the points in Table 8.8, as shown in Figure 8.27. This scatterplot suggests that a line models these data, as shown in Figure 8.28. We call f a *linear function* because its graph is a *line*.

House Temperature

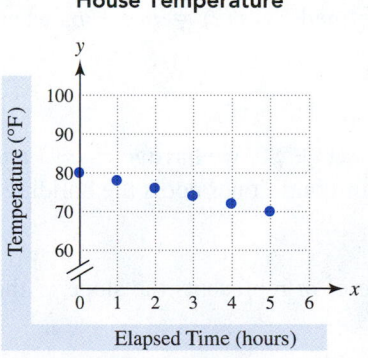

Figure 8.27 A Scatterplot

Graphical Representation of $f(x)$

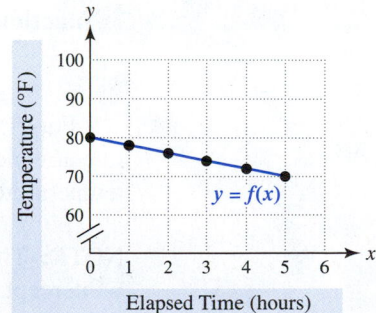

Figure 8.28 A Linear Function

Another graph of $y = f(x)$ with a different y-scale is shown in Figure 8.29 on the next page. Because the y-values always decrease by the same amount for each 1-hour increase on the x-axis, we say that function f has a *constant rate of change*. In this example, the constant rate of change is $-2°F$ per hour.

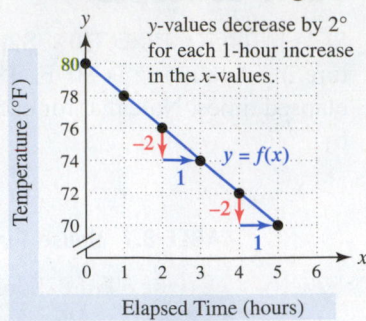

Figure 8.29

VERBAL REPRESENTATION (WORDS) Over a 5-hour period, the air conditioner lowers the initial temperature of 80°F by 2°F for each elapsed hour x. Thus a description of how to calculate the temperature is:

"Multiply x by -2°F and then add 80°F."   Verbal representation of $f(x)$

SYMBOLIC REPRESENTATION (FORMULA) Our verbal representation of $f(x)$ makes it straightforward for us to write a formula.

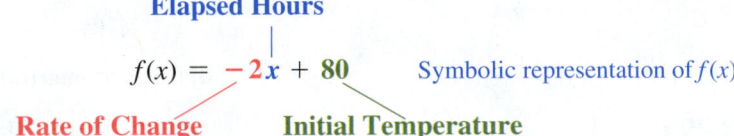

For example,

$$f(\mathbf{2.5}) = -2(\mathbf{2.5}) + 80 = \mathbf{75}$$

means that the temperature is **75**°F after the air conditioner has run for **2.5** hours. In this instance, it might be appropriate to *limit the domain* of f to x-values between 0 and 5, inclusive.

LINEAR FUNCTION

A function f defined by $f(x) = mx + b$, where m and b are constants, is a **linear function**.

For $f(x) = -2x + 80$, we have $m = -2$ and $b = 80$. The constant m represents the rate at which the air conditioner cools the building, and the constant b represents the initial temperature.

NOTE: The value of m represents the slope of the graph of $f(x) = mx + b$, and b is the y-intercept.

▶ **REAL-WORLD CONNECTION** In general, a linear function defined by $f(x) = mx + b$ changes by m units for each unit increase in x. This **rate of change** is an increase if $m > 0$ and a decrease if $m < 0$. For example, if new carpet costs $20 per square yard, then the linear function defined by $C(x) = 20x$ gives the cost of buying x square yards of carpet. The value of $m = 20$ gives the cost (rate of change) for each additional square yard of carpet. For function C, the value of b is 0 because it costs $0 to buy 0 square yards of carpet.

READING CHECK

Explain what a linear function is and what its graph looks like.

NOTE: If f is a linear function, then $f(0) = m(0) + b = b$. Thus b can be found by evaluating $f(x)$ at $x = 0$.

EXAMPLE 1 | **Identifying linear functions**

Determine whether f is a linear function. If f is a linear function, find values for m and b so that $f(x) = mx + b$.
(a) $f(x) = 4 - 3x$ **(b)** $f(x) = 8$ **(c)** $f(x) = 2x^2 + 8$

Solution
(a) Let $m = -3$ and $b = 4$. Then $f(x) = -3x + 4$, and f is a linear function.
(b) Let $m = 0$ and $b = 8$. Then $f(x) = 0x + 8$, and f is a linear function.
(c) Function f is not linear because its formula contains x^2. The formula for a linear function cannot contain an x with an exponent other than 1.

Now Try Exercises 11, 13, 15

EXAMPLE 2 | **Determining linear functions**

Use each table of values to determine whether $f(x)$ could represent a linear function. If f could be linear, write a formula for f in the form $f(x) = mx + b$.

(a)

x	0	1	2	3
$f(x)$	10	15	20	25

(b)

x	−2	0	2	4
$f(x)$	4	2	0	−2

(c)

x	0	1	2	3
$f(x)$	1	2	4	7

(d)

x	−2	0	3	5
$f(x)$	7	7	7	7

Solution
(a) For each unit increase in x, $f(x)$ increases by 5 units, so $f(x)$ could be linear with $m = 5$. Because $f(0) = 10$, $b = 10$. Thus $f(x) = 5x + 10$.
(b) For each 2-unit increase in x, $f(x)$ decreases by 2 units. Equivalently, each unit increase in x results in a 1-unit decrease in $f(x)$, so $f(x)$ could be linear with $m = -1$. Because $f(0) = 2$, $b = 2$. Thus $f(x) = -x + 2$.
(c) Each unit increase in x does not result in a constant change in $f(x)$. Thus $f(x)$ does *not* represent a linear function.
(d) For any change in x, $f(x)$ does *not* change, so $f(x)$ could be linear with $m = 0$. Because $f(0) = 7$, let $b = 7$. Thus $f(x) = 0x + 7$, or $f(x) = 7$. (When $m = 0$, we say that f is a *constant function*. See Example 8.)

Now Try Exercises 23, 25, 27, 31

Representations of Linear Functions

The graph of a linear function is a line. To graph a linear function f we can start by making a table of values and then plotting three or more points. We can then sketch the graph of f by drawing a line through these points, as demonstrated in the next example.

EXAMPLE 3   Graphing a linear function by hand

Sketch a graph of $f(x) = x - 1$. Use the graph to evaluate $f(-2)$.

Solution

Begin by making a table of values containing at least three points. Pick convenient values of x, such as $x = -1, 0, 1$.

$$f(-1) = -1 - 1 = -2$$
$$f(0) = 0 - 1 = -1$$
$$f(1) = 1 - 1 = 0$$

Display the results, as shown in Table 8.9.

Plot the points $(-1, -2), (0, -1)$, and $(1, 0)$. Sketch a line through these points to obtain the graph of f. A graph of a line results when *infinitely* many points are plotted, as shown in Figure 8.30.

To evaluate $f(-2)$, first find $x = -2$ on the x-axis. See Figure 8.31. Then move downward to the graph of f. By moving across to the y-axis, we see that the corresponding y-value is -3. Thus $f(-2) = -3$.

TABLE 8.9

x	y
-1	-2
0	-1
1	0

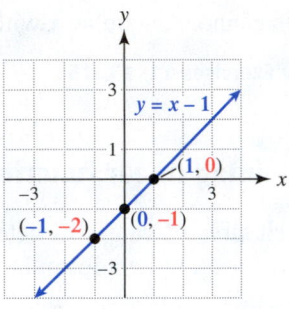

Figure 8.30

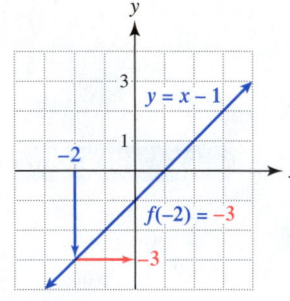

Figure 8.31

Now Try Exercises 39, 57

In the next example a graphing calculator is used to create a graph and table.

EXAMPLE 4   Using a graphing calculator

Give numerical and graphical representations of $f(x) = \frac{1}{2}x - 2$.

Solution

Numerical Representation To make a numerical representation, construct the table for $Y_1 = .5X - 2$, starting at $x = -3$ and incrementing by 1, as shown in Figure 8.32(a). (Other tables are possible.)

Graphical Representation Graph Y_1 in the standard viewing rectangle, as shown in Figure 8.32(b). (Other viewing rectangles may be used.)

CALCULATOR HELP

To make a table, see Appendix A (pages AP-2 and AP-3). To make a graph, see Appendix A (page AP-5).

$[-10, 10, 1]$ by $[-10, 10, 1]$

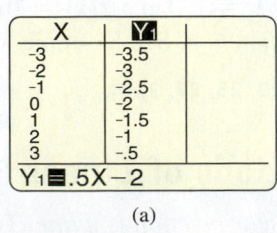

(a)

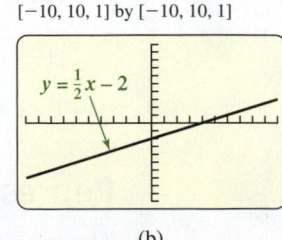

(b)

Figure 8.32

Now Try Exercise 75

CRITICAL THINKING

Two points determine a line. Why is it a good idea to plot at least three points when graphing a linear function by hand?

EXAMPLE 5   ## Representing a linear function

A linear function is given by $f(x) = -3x + 2$.
(a) Give a verbal representation of f.
(b) Make a numerical representation (table) of f by letting $x = -1, 0, 1$.
(c) Plot the points listed in the table from part (b). Then sketch a graph of $y = f(x)$.

Solution

(a) *Verbal Representation* Multiply the input x by -3 and add 2 to obtain the output.

(b) *Numerical Representation* Evaluate the formula $f(x) = -3x + 2$ at $x = -1, 0, 1$, which results in Table 8.10. Note that $f(-1) = 5$, $f(0) = 2$, and $f(1) = -1$.

(c) *Graphical Representation* To make a graph of f by hand without a graphing calculator, plot the points $(-1, 5)$, $(0, 2)$, and $(1, -1)$ from Table 8.10. Then draw a line passing through these points, as shown in Figure 8.33.

TABLE 8.10

x	$f(x)$
-1	5
0	2
1	-1

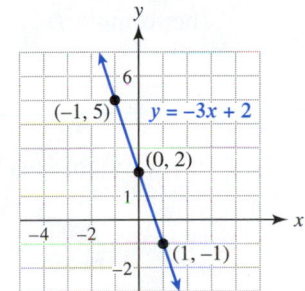

Figure 8.33

▌ **Now Try Exercise 71**

NOTE: To graph $y = -3x + 2$ in Example 5, we could also graph a line with slope -3 and y-intercept 2.

MAKING CONNECTIONS

Mathematics in Newspapers

Think of the mathematics that you see in newspapers or in online publications. Often, percentages are described *verbally*, numbers are displayed in *tables*, and data are shown in *graphs*. Seldom are *formulas* given, which is an important reason to study verbal, numerical, and graphical representations.

Modeling Data with Linear Functions

▶ **REAL-WORLD CONNECTION** A distinguishing feature of a linear function is that when the input x increases by 1 unit, the output $f(x) = mx + b$ always changes by an amount equal to m. For example, the percentage of wireless households during year x from 2005 to 2010 can be modeled by the linear function

$$f(x) = 4x - 8013,$$

where x is the year. The value of $m = 4$ indicates that the percentage of wireless households has increased, on average, by 4% per year. (*Source: National Center for Health Statistics.*)

The following are other examples of quantities that are modeled by linear functions. Try to determine the value of the constant m.

- The wages earned by an individual working x hours at $8 per hour
- The distance traveled by a jet airliner in x hours if its speed is 500 miles per hour
- The cost of tuition and fees when registering for x credits if each credit costs $200 and the fees are fixed at $300

When we are modeling data with a linear function defined by $f(x) = mx + b$, the following concepts are helpful to determine m and b.

MODELING DATA WITH A LINEAR FUNCTION

The formula $f(x) = mx + b$ may be interpreted as follows.

$$f(x) \quad = \quad mx \quad + \quad b$$

(New amount) = (Change) + (Fixed amount)

When x represents time, *change* equals (rate of change) $\times$ (time).

$$f(x) \quad = \quad m \quad \times \quad x \quad + \quad b$$

(Future amount) = (Rate of change) $\times$ (Time) + (Initial amount)

▶ **REAL-WORLD CONNECTION** These concepts are applied in the next three examples.

EXAMPLE 6 **Modeling growth of bamboo**

Bamboo is gaining popularity as a *green* building material because of its fast-growing, regenerative characteristics. Under ideal conditions, some species of bamboo grow at an astonishing 2 inches per hour. Suppose a bamboo plant is initially 6 inches tall. (*Source: Cali Bamboo.*)
(a) Find a function H that models the plant's height in inches under ideal conditions after t hours.
(b) Find $H(3)$ and interpret the result.

Solution
(a) The initial height is **6** inches and the rate of change is **2** inches per hour.

$$H(t) \quad = \quad 2 \quad \times \quad t \quad + \quad 6,$$

(Future height) = (Rate of change) $\times$ (Time) + (Initial height)

or $H(t) = 2t + 6$.
(b) $H(3) = 2(3) + 6 = 12$. After **3** hours the bamboo plant is **12** inches tall.

Now Try Exercise 117

EXAMPLE 7   **Modeling demand for building green**

Table 8.11 lists estimated sales of green building material in billions of dollars. (Refer to A Look Into Math at the beginning of this section.)

TABLE 8.11 Green Material Sales ($ billions)

Year	2010	2011	2012	2013
Sales	65	70	75	80

Source: Freedonia Group, Green Building Material.

(a) Make a scatterplot of the data and sketch the graph of a function f that models these data. Let x represent years after 2010. That is, let $x = 0$ correspond to 2010, $x = 1$ to 2011, and so on.
(b) What were the sales in 2010? What was the annual increase in sales each year?
(c) Find a formula for $f(x)$.
(d) Use your formula to estimate sales in 2014.

Solution
(a) In Figure 8.34 the scatterplot suggests that a linear function models the data. A line has been sketched with the data.

READING CHECK

How can you determine whether data in a table can be modeled by a linear function?

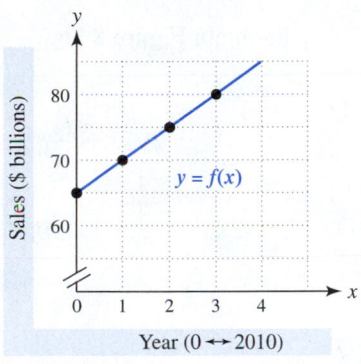

Green Building Material Sales

Figure 8.34 A Linear Model

(b) From Table 8.11, sales for green material were $65 billion in 2010, with sales increasing at a *constant rate* of $5 billion per year.
(c) From part (b) initial sales ($x = 0$) were $**65** billion, and sales increased by $**5** billion per year. Thus

$$f(x) \quad = \quad \mathbf{5} \quad \times \quad x \quad + \quad \mathbf{65},$$

(Future sales) = (Rate of change in sales) × (Time) + (Initial sales)

or $f(x) = 5x + 65$.
(d) Because $x = 4$ corresponds to 2014, evaluate $f(4)$.

$$f(\mathbf{4}) = 5(\mathbf{4}) + 65 = \mathbf{85}$$

This model estimates sales of green building material to be $85 billion in 2014.

Now Try Exercise 119

In the next example, we consider a simple function that models the speed of a car.

EXAMPLE 8 **Modeling with a constant function**

A car travels on a freeway with its speed recorded at regular intervals, as listed in Table 8.12.

TABLE 8.12 Speed of a Car

Elapsed Time (hours)	0	1	2	3	4
Speed (miles per hour)	70	70	70	70	70

(a) Discuss the speed of the car during this time interval.
(b) Find a formula for a function f that models these data.
(c) Sketch a graph of f together with the data.

Solution
(a) The speed of the car appears to be constant at 70 miles per hour.
(b) Because the speed is constant, the rate of change is 0. Thus

$$f(x) \quad = \quad 0x \quad + \quad 70$$

(Future speed) = (Change in speed) + (Initial speed)

and $f(x) = 70$. We call f a *constant function*.
(c) Because $y = f(x)$, graph $y = 70$ with the data points

$$(0, 70), (1, 70), (2, 70), (3, 70), \text{ and } (4, 70)$$

to obtain Figure 8.35.

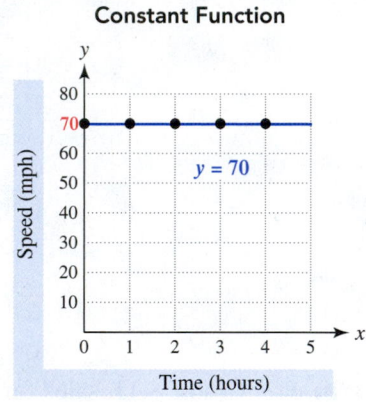

Constant Function

$y = 70$

Figure 8.35 Speed of a Car

Now Try Exercise 113

The function defined by $f(x) = 70$ is an example of a *constant function*. A **constant function** *is a linear function* with $m = 0$ and can be written as $f(x) = b$. Regardless of the input, a constant function always outputs the same value, b. Its graph is a horizontal line. Its domain is all real numbers and its range is $R = \{b\}$.

CRITICAL THINKING

Find a formula for a function D that calculates the *distance* traveled by the car in Example 8 after x hours. What is the rate of change for $D(x)$?

▶ **REAL-WORLD CONNECTION** The following are three applications of constant functions.

• A thermostat calculates a constant function regardless of the weather outside by maintaining a set temperature.
• A cruise control in a car calculates a constant function by maintaining a fixed speed, regardless of the type of road or terrain.
• A constant function calculates the 1250-foot height of the Empire State Building.

The Midpoint Formula (Optional)

▶ **REAL-WORLD CONNECTION** A common way to make estimations is to average data items. For example, in 2000 the average tuition and fees at public two-year colleges were about $1700, and in 2010 they were about $2700. (*Source:* The College Board.) To estimate tuition and fees in 2005, we could average the 2000 and 2010 amounts.

$$\frac{1700 + 2700}{2} = \$2200 \quad \text{Finding the average}$$

This technique predicts that tuition and fees were $2200 in 2005 and is referred to as finding the *midpoint*.

MIDPOINT FORMULA ON THE REAL NUMBER LINE The **midpoint** of a line segment is the unique point on the line segment that is an equal distance from the endpoints. For example, in Figure 8.36 the midpoint M of -3 and 5 on the real number line is 1.

Figure 8.36

We can calculate the value of M as follows.

$$M = \frac{x_1 + x_2}{2} = \frac{-3 + 5}{2} = 1$$

Average the x-values to find the midpoint.

MIDPOINT FORMULA IN THE xy-PLANE The midpoint of a line segment in the xy-plane can be found in a similar way. Figure 8.37(a) shows the midpoint on the line segment connecting the points (x_1, y_1) and (x_2, y_2). The x-coordinate of M is equal to the average of x_1 and x_2, and the y-coordinate of M is equal to the average of y_1 and y_2. For example, the line segment with endpoints $(-2, 1)$ and $(4, -3)$ is shown in Figure 8.37(b). The coordinates of the midpoint are

$$M = \left(\frac{-2 + 4}{2}, \frac{1 + (-3)}{2} \right) = (1, -1).$$

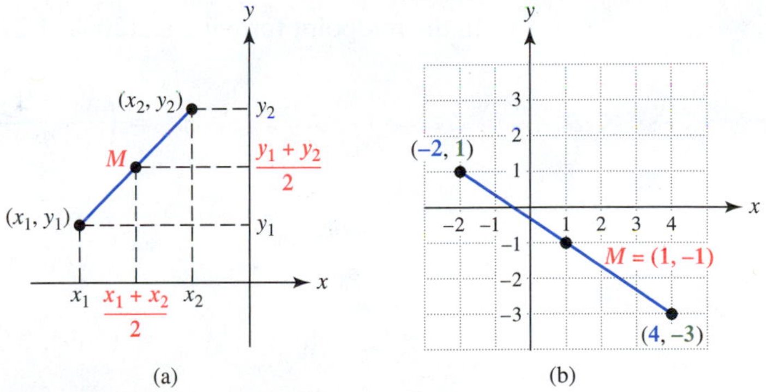

(a)   (b)

Figure 8.37

This discussion is summarized as follows.

MIDPOINT FORMULA IN THE xy-PLANE

The midpoint of the line segment with endpoints (x_1, y_1) and (x_2, y_2) in the xy-plane is

$$\left(\frac{x_1 + x_2}{2}, \frac{y_1 + y_2}{2} \right).$$

EXAMPLE 9 **Finding the midpoint**

Find the midpoint of the line segment connecting the points $(-3, -2)$ and $(4, 1)$.

Solution

In the midpoint formula let $(-3, -2)$ be (x_1, y_1) and $(4, 1)$ be (x_2, y_2).

$$M = \left(\frac{x_1 + x_2}{2}, \frac{y_1 + y_2}{2} \right) \qquad \text{Midpoint formula}$$

$$= \left(\frac{-3 + 4}{2}, \frac{-2 + 1}{2} \right) \qquad \text{Substitute.}$$

$$= \left(\frac{1}{2}, -\frac{1}{2} \right) \qquad \text{Simplify.}$$

The midpoint of the line segment is $\left(\frac{1}{2}, -\frac{1}{2} \right)$.

▌ **Now Try Exercise 93**

▶ **REAL-WORLD CONNECTION** In the next example we use the midpoint formula to estimate the divorce rate in the United States in 2005.

EXAMPLE 10 **Estimating the U.S. divorce rate**

The divorce rate per 1000 people in 2000 was 4.2, and in 2010 it was 3.4. (*Source:* Statistical Abstract of the United States.)
(a) Use the midpoint formula to estimate the divorce rate in 2005.
(b) Could the midpoint formula be used to estimate the divorce rate in 2003? Explain.

Solution

(a) In the midpoint formula, let $(2000, 4.2)$ be (x_1, y_1) and let $(2010, 3.4)$ be (x_2, y_2).

$$M = \left(\frac{x_1 + x_2}{2}, \frac{y_1 + y_2}{2} \right) \qquad \text{Midpoint formula}$$

$$= \left(\frac{2000 + 2010}{2}, \frac{4.2 + 3.4}{2} \right) \qquad \text{Substitute.}$$

$$= (2005, 3.8) \qquad \text{Simplify.}$$

The midpoint formula estimates that the divorce rate was **3.8** per 1000 people in **2005**. (Note that the actual rate was 3.6.)
(b) No, the midpoint formula can only be used to estimate data that are exactly halfway between two given data points. Because the year 2003 is not exactly halfway between 2000 and 2010, the midpoint formula cannot be used.

▌ **Now Try Exercise 107**

NOTE: An estimate obtained from the midpoint formula is equal to an estimate obtained from a linear function whose graph passes through the endpoints of the line segment. This fact is illustrated in the next example.

EXAMPLE 11   **Relating midpoints to linear functions**

The graph of a linear function f shown in Figure 8.38 passes through the points $(-1, 3)$ and $(2, -3)$.

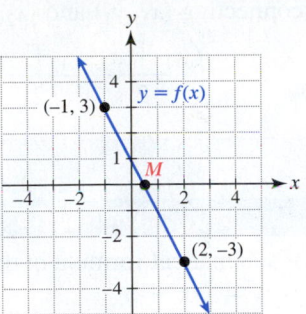

Figure 8.38

(a) Find a formula for $f(x)$.

(b) Evaluate $f\left(\frac{1}{2}\right)$. Does your answer agree with the graph?

(c) Find the midpoint M of the line segment connecting the points $(-1, 3)$ and $(2, -3)$. Comment on your result.

Solution

(a) The graph of f is a line that passes through $(-1, 3)$ and $(2, -3)$. The slope m of the line is

$$m = \frac{-3 - 3}{2 - (-1)} = -\frac{6}{3} = -2,$$

and from the graph, the y-intercept is **1**. Thus $f(x) = -2x + 1$.

(b) $f\left(\frac{1}{2}\right) = -2\left(\frac{1}{2}\right) + 1 = \mathbf{0}$. Yes, they agree because the point $\left(\frac{1}{2}, 0\right)$ lies on the graph of $y = f(x)$ in Figure 8.38.

(c) The midpoint of the line segment connecting $(-1, 3)$ and $(2, -3)$ is

$$M = \left(\frac{-1 + 2}{2}, \frac{3 + (-3)}{2}\right) = \left(\frac{1}{2}, 0\right).$$

Finding the midpoint $M = \left(\frac{1}{2}, 0\right)$ of the line segment with endpoints $(-1, 3)$ and $(2, -3)$ is equivalent to evaluating the linear function f, whose graph passes through $(-1, 3)$ and $(2, -3)$, at $x = \frac{1}{2}$.

Now Try Exercise 103

8.2   Putting It All Together

CONCEPT	EXPLANATION	EXAMPLES
Linear Function	Can be represented by $f(x) = mx + b$. Its graph is a line with slope m and y-intercept b.	$f(x) = 2x - 6$, $m = 2$ and $b = -6$ $f(x) = 10$, $m = 0$ and $b = 10$
Constant Function	Can be represented by $f(x) = b$. Its graph is a horizontal line.	$f(x) = -7$, $b = -7$ $f(x) = 22$, $b = 22$
Rate of Change for a Linear Function	The output of a linear function changes by a constant amount for each unit increase in the input.	$f(x) = -3x + 8$ decreases 3 units for each unit increase in x. $f(x) = 5$ neither increases nor decreases. The rate of change is 0.

continued on next page

continued from previous page

CONCEPT	EXPLANATION	EXAMPLES
Midpoint Formula	The midpoint of the line segment connecting (x_1, y_1) and (x_2, y_2) is $$\left(\frac{x_1 + x_2}{2}, \frac{y_1 + y_2}{2}\right).$$	The midpoint of the line segment connecting $(-2, 3)$ and $(4, 5)$ is $$\left(\frac{-2 + 4}{2}, \frac{3 + 5}{2}\right) = (1, 4).$$

REPRESENTATION	COMMENTS	EXAMPLE
Symbolic	Mathematical formula in the form $f(x) = mx + b$	$f(x) = 2x + 1$, where $m = 2$ and $b = 1$
Verbal	Multiply the input x by m and add b.	Multiply the input x by 2 and then add 1 to obtain the output.
Numerical (table of values)	For each unit increase in x in the table, the output of $f(x) = mx + b$ changes by an amount equal to m.	 1-unit increase | x	0	1	2
$f(x)$	1	3	5	 2-unit increase
Graphical	The graph of a linear function is a line. Plot at least 3 points and then sketch the line. If $f(x) = mx + b$, then the graph of f has slope m and y-intercept b.	*[graph showing line $y = 2x + 1$]*

8.2 Exercises

CONCEPTS AND VOCABULARY

1. The formula for a linear function is $f(x) = $ _____.

2. The formula for a constant function is $f(x) = $ _____.

3. The graph of a linear function is a(n) _____.

4. The graph of a constant function is a(n) _____ line.

5. If $f(x) = 7x + 5$, each time x increases by 1 unit, $f(x)$ increases by _____ units.

6. If $f(x) = 5$, each time x increases by 1 unit, $f(x)$ increases by _____ units.

7. (True or False?) Every constant function is a linear function.

8. (True or False?) Every linear function is a constant function.

9. If $C(x) = 2x$ calculates the cost in dollars of buying x square feet of carpet, what does 2 represent in the formula? Interpret the fact that the point $(10, 20)$ lies on the graph of C.

10. If $G(x) = 100 - 4x$ calculates the number of gallons of water in a tank after x minutes, what does -4 represent in the formula? Interpret the fact that the point $(5, 80)$ lies on the graph of G.

IDENTIFYING LINEAR FUNCTIONS

Exercises 11–18: Determine whether f is a linear function. If f is linear, give values for m and b so that f may be expressed as $f(x) = mx + b$.

11. $f(x) = \dfrac{1}{2}x - 6$

12. $f(x) = x$

13. $f(x) = \dfrac{5}{2} - x^2$

14. $f(x) = \sqrt{x} + 3$

15. $f(x) = -9$

16. $f(x) = 1.5 - 7.3x$

17. $f(x) = -9x$

18. $f(x) = \dfrac{1}{x}$

Exercises 19–22: Determine whether the graph represents a linear function.

19.   **20.**

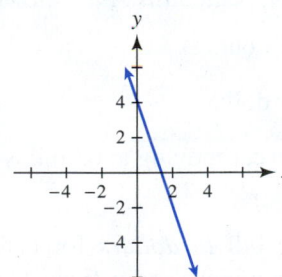 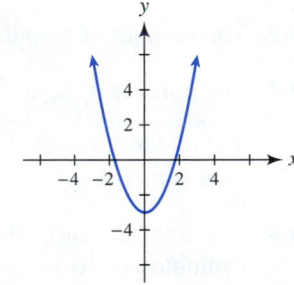

21.   **22.**

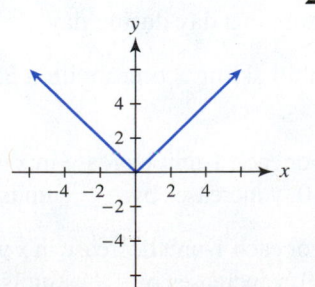 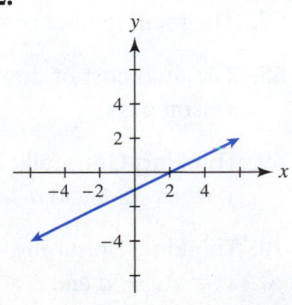

Exercises 23–32: (Refer to Example 2.) Use the table to determine whether $f(x)$ could represent a linear function. If it could, write $f(x)$ in the form $f(x) = mx + b$.

23.

x	0	1	2	3
$f(x)$	-6	-3	0	3

24.

x	0	2	4	6
$f(x)$	-2	2	6	10

25.

x	-2	0	2	4
$f(x)$	6	3	0	-3

26.

x	0	3	6	9
$f(x)$	8	4	2	1

27.

x	-2	-1	0	1
$f(x)$	-5	0	20	40

28.

x	-2	-1	0	1
$f(x)$	6	3	0	-3

29.

x	0	2	3	4
$f(x)$	0	4	6	8

30.

x	1	2	3	4
$f(x)$	0	1	3	7

31.

x	-1	0	1	2
$f(x)$	-4	-4	-4	-4

32.

x	2	5	6	8
$f(x)$	5	5	5	5

EVALUATING LINEAR FUNCTIONS

Exercises 33–38: Evaluate $f(x)$ at the given values of x.

33. $f(x) = 4x$ $\qquad\qquad$ $x = -4, 5$

34. $f(x) = -2x + 1$ $\qquad$ $x = -2, 3$

35. $f(x) = 5 - x$ $\qquad\quad$ $x = -\dfrac{2}{3}, 3$

36. $f(x) = \dfrac{1}{2}x - \dfrac{1}{4}$ $\qquad$ $x = 0, \dfrac{1}{2}$

37. $f(x) = -22$ $x = -\frac{3}{4}, 13$

38. $f(x) = 9x - 7$ $x = -1.2, 2.8$

Exercises 39–44: Use the graph of f to evaluate the given expressions.

39. $f(-1)$ and $f(0)$ **40.** $f(-2)$ and $f(2)$

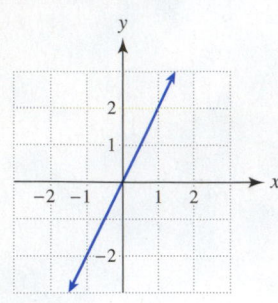
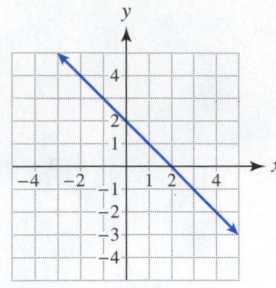

41. $f(-2)$ and $f(4)$ **42.** $f(0)$ and $f(3)$

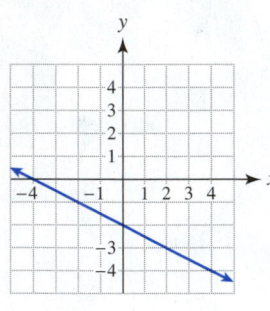
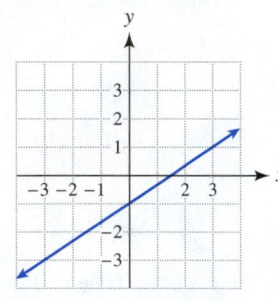

43. $f(-3)$ and $f(1)$ **44.** $f(1.5)$ and $f(0.5\pi)$

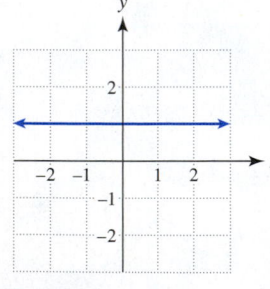
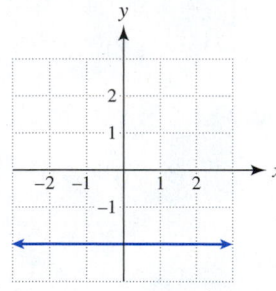

Exercises 45–48: Use the verbal description to write a formula for $f(x)$. Then evaluate $f(3)$.

45. Multiply the input by 6.

46. Multiply the input by -3 and add 7.

47. Divide the input by 6 and subtract $\frac{1}{2}$.

48. Output 8.7 for every input.

REPRESENTING LINEAR FUNCTIONS

Exercises 49–52: Match $f(x)$ with its graph (a.–d.) at the top of the next column.

49. $f(x) = 3x$ **50.** $f(x) = -2x$

51. $f(x) = x - 2$ **52.** $f(x) = 2x + 1$

a.   **b.**

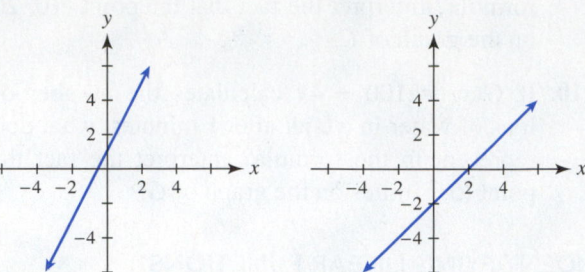

c.   **d.**

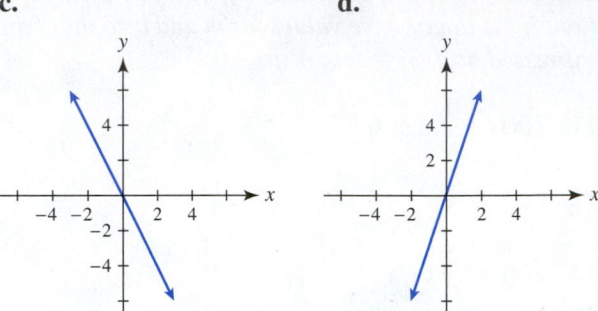

Exercises 53–62: Sketch a graph of $y = f(x)$.

53. $f(x) = 2$ **54.** $f(x) = -1$

55. $f(x) = -\frac{1}{2}x$ **56.** $f(x) = 2x$

57. $f(x) = x + 1$ **58.** $f(x) = x - 2$

59. $f(x) = 3x - 3$ **60.** $f(x) = -2x + 1$

61. $f(x) = 3 - x$ **62.** $f(x) = \frac{1}{4}x + 2$

Exercises 63–68: Write a symbolic representation (formula) for a linear function f that calculates the following.

63. The number of pounds in x ounces

64. The number of dimes in x dollars

65. The distance traveled by a car moving at 65 miles per hour for t hours

66. The long-distance phone bill *in dollars* for calling t minutes at 10 cents per minute and a fixed fee of $4.95

67. The total number of hours in a day during day x

68. The total cost of downhill skiing x times with a $500 season pass

69. **Thinking Generally** For each 1-unit increase in x with $y = ax + b$ and $a > 0$, y increases by _____ units.

70. **Thinking Generally** For each 1-unit decrease in x with $y = cx + d$ and $c < 0$, y increases by _____ units.

Exercises 71–74: Do the following.

  (a) Give a verbal representation of f.
  (b) Make a numerical representation (table) of f for x = −2, 0, 2.
  (c) Plot the points listed in the table from part (b), then sketch a graph of f.

71. $f(x) = -2x + 1$ **72.** $f(x) = 1 - x$

73. $f(x) = \frac{1}{2}x - 1$ **74.** $f(x) = \frac{3}{4}x$

 Exercises 75–78: Do the following.

  (a) Make a numerical representation (table) of f for x = −3, −2, −1, ..., 3.
  (b) Graph f in the window [−6, 6, 1] by [−4, 4, 1].

75. $f(x) = \frac{1}{3}x + \sqrt{2}$ **76.** $f(x) = -\frac{2}{3}x - \sqrt{3}$

77. $f(x) = \dfrac{x + 2}{5}$ **78.** $f(x) = \dfrac{2 - 3x}{7}$

MODELING

Exercises 79–82: Match the situation with the graph (a.–d.) that models it best, where x-values represent time from 2000 to 2010.

79. The cost of college tuition

80. The cost of 1 gigabyte of computer memory

81. The distance between Chicago and Denver

82. The total distance traveled by a satellite that is orbiting Earth if the satellite was launched in 2000

  a. 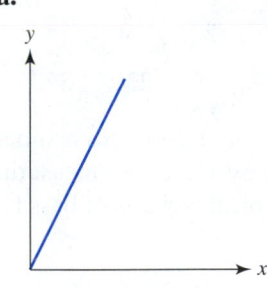   **b.**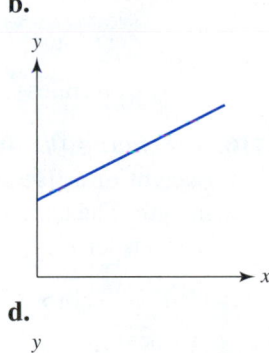

  c. 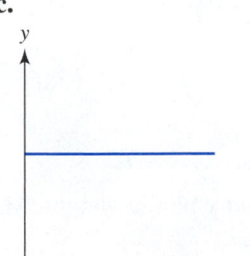   **d.**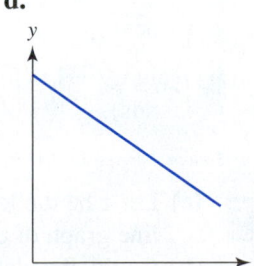

83. Online Exploration Look up the fuel efficiency *E* in miles per gallon for one of your favorite cars. (Answers will vary.)
  (a) Find a function *G* that calculates the number of gallons required to travel *x* miles.

(b) If the cost of gasoline is $3 per gallon, find function *C* that calculates the cost of fuel to travel *x* miles.

84. Online Exploration Suppose that you would like to drive to Miami for spring break (if it is possible) in the car that you chose in Exercise 83. Calculate the gallons of gasoline needed for the trip.

MIDPOINT FORMULA

Exercises 85–92: Find the midpoint of the line segment shown.

85.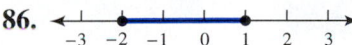

86.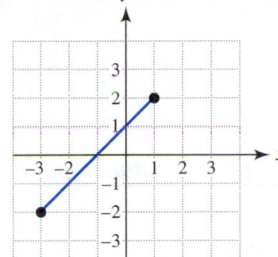

87. 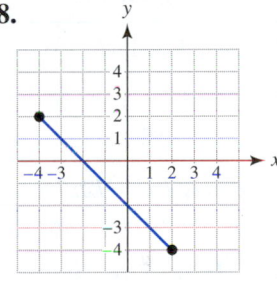   **88.**

89. 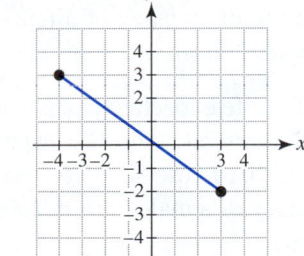   **90.**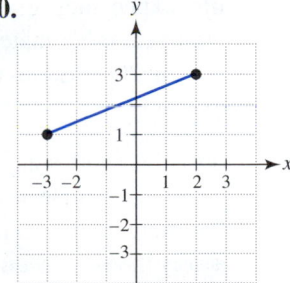

91. 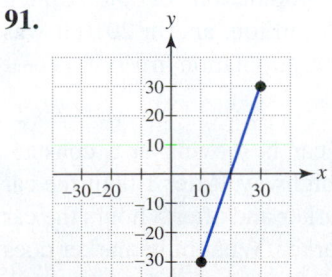   **92.**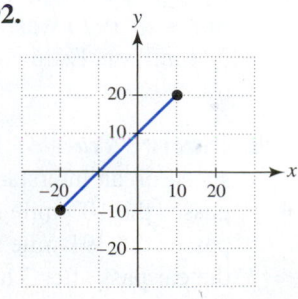

Exercises 93–102: Find the midpoint of the line segment connecting the given points.

93. $(-9, -3), (-7, 1)$ **94.** $(7, -2), (-5, 8)$

95. $\left(\frac{1}{2}, \frac{1}{3}\right), \left(-\frac{5}{2}, -\frac{2}{3}\right)$ **96.** $\left(-\frac{3}{5}, -\frac{1}{4}\right), \left(\frac{1}{10}, \frac{1}{2}\right)$

97. $(-0.3, 0.1), (0.7, 0.4)$

98. $(0.8, -0.4), (0.9, -0.1)$

99. $(2000, 5), (2010, 13)$ **100.** $(2005, 9), (2011, 3)$

101. Thinking Generally $(a, -b), (3a, 5b)$

102. Thinking Generally $(-a, b), (a, -b)$

Exercises 103–106: (Refer to Example 11.) The graph of a linear function f passes through the two given points.

   (a) *Find a formula for f(x). Determine f(2).*
   (b) *Determine f(2) by finding the midpoint of the line segment connecting the given points.*
   (c) *Compare your answers for parts (a) and (b).*

103. $(0, 5), (4, -3)$

104. $(0, 2), (4, 10)$

105. $(-3, -1), (7, 3)$

106. $(-1, 3), (5, -5)$

Exercises 107–112: Use the midpoint formula to make the requested estimation.

107. U.S. Life Expectancy The life expectancy of a female born in 1990 was 78.8 years, and the life expectancy of a female born in 2010 rose to 80.8 years. Estimate the life expectancy of a female born in 2000. (*Source:* Centers for Disease Control and Prevention.)

108. U.S. Life Expectancy The life expectancy of a male born in 1990 was 71.8 years, and the life expectancy of a male born in 2010 rose to 75.6 years. Estimate the life expectancy of a male born in 2000. (*Source:* Centers for Disease Control and Prevention.)

109. U.S. Population The population of the United States in 1970 was 205 million, and in 2010 it was 308 million. Estimate the population in 1990. (*Source:* U.S. Census Bureau.)

110. Distance Traveled A car is moving at a constant speed on an interstate highway. After 1 hour the car passes the 103-mile marker and after 5 hours the car passes the 391-mile marker. What mile marker does the car pass after 3 hours?

111. U.S. Median Income In 1999 the median family income was \$40,700, and in 2009 it was \$49,800. Estimate the median family income in 2004.

112. Estimating Fish Populations In 2008 there were approximately 3200 large-mouth bass in a lake. This number increased to 3800 in 2012. Estimate the number of large-mouth bass in the lake in 2010.

APPLICATIONS

113. Thermostat Let $y = f(x)$ describe the temperature y of a room that is kept at $70°\,F$ for x hours.
   (a) Represent f symbolically and graphically over a 24-hour period for $0 \le x \le 24$.
   (b) Construct a table of f for $x = 0, 4, 8, 12, \ldots, 24$.
   (c) What type of function is f?

114. Cruise Control Let $y = f(x)$ describe the speed y of an automobile after x minutes if the cruise control is set at 60 miles per hour.
   (a) Represent f symbolically and graphically over a 15-minute period for $0 \le x \le 15$.
   (b) Construct a table of f for $x = 0, 1, 2, \ldots, 6$.
   (c) What type of function is f?

115. Distance A car is initially 50 miles south of the Minnesota–Iowa border, traveling south on Interstate 35. Distances D between the car and the border are recorded in the table for various elapsed times t. Find a linear function D that models these data.

t (hours)	0	2	3	5
D (miles)	50	170	230	350

116. Estimating the Weight of a Bass Sometimes the weight of a fish can be estimated by measuring its length. The table lists typical weights of bass having various lengths.

Length (inches)	12	14	16	18	20	22
Weight (pounds)	1.0	1.7	2.5	3.6	5.0	6.6

Source: Minnesota Department of Natural Resources.

   (a) Let x be the length and y be the weight. Make a line graph of the data.
   (b) Could the data be modeled accurately with a linear function? Explain your answer.

117. Texting In 2010, the average American under age 18 sent approximately 93 texts per day, whereas the average adult over age 65 sent approximately 1 text per day.

(a) Find a formula for a function K that calculates the number of texts sent in x days by the average person under age 18.

(b) Find a formula for a function A that calculates the number of texts sent in x days by the average person over age 65.

(c) Evaluate $K(365)$ and $A(365)$. Interpret your results.

118. *Rain Forests* Rain forests are forests that grow in regions receiving more than 70 inches of rain per year. The world is losing about 49 million acres of rain forest each year. (*Source: New York Times Almanac.*)

(a) Find a linear function f that calculates the change in the acres of rain forest in millions in x years.

(b) Evaluate $f(7)$ and interpret the result.

119. *Car Sales* The table shows the number of U.S. Toyota vehicles sold in millions for past years.

Year	2000	2001	2002	2003	2004
Vehicles	1.6	1.7	1.8	1.9	2.0

Source: Autodata.

(a) What were the sales in 2000?

(b) What was the annual increase in sales?

(c) Find a linear function f that models these data. Let $x = 0$ correspond to 2000, $x = 1$ to 2001, and so on.

(d) Use f to estimate sales in 2006.

120. *Tuition and Fees* Suppose tuition costs $300 per credit and that student fees are fixed at $100.

(a) Find a formula for a linear function T that models the cost of tuition and fees for x credits.

(b) Evaluate $T(16)$ and interpret the result.

121. *Skype Users* The number of Skype users S in millions x years after 2006 can be modeled by the formula $S(x) = 110x + 123$.

(a) How many users were there in 2010?

(b) What does the number 123 indicate in the formula?

(c) What does the number 110 indicate in the formula?

122. *Temperature and Volume* If a sample of a gas such as helium is heated, it will expand. The formula $V(T) = 0.147T + 40$ calculates the volume V in cubic inches of a sample of gas at temperature T in degrees Celsius.

(a) Evaluate $V(0)$ and interpret the result.

(b) If the temperature increases by $10°C$, by how much does the volume increase?

(c) What is the volume of the gas when the temperature is $100°C$?

123. *Temperature and Volume* (Refer to the preceding exercise.) A sample of gas at $0°C$ has a volume V of 137 cubic centimeters, which increases in volume

by 0.5 cubic centimeter for every $1°C$ increase in temperature T.

(a) Write a formula $V(T) = aT + b$ that gives the volume of the gas at temperature T.

(b) Find the volume of the gas when $T = 50°C$.

124. *Cost* To make a music video it costs $750 to rent a studio plus $5 for each copy produced.

(a) Write a formula $C(x) = ax + b$ that calculates the cost of producing x videos.

(b) Find the cost of producing 2500 videos.

125. *Weight Lifting* Lifting weights can increase a person's muscle mass. Each additional pound of muscle burns an extra 40 calories per day. Write a linear function that models the number of calories burned each day by x pounds of muscle. By burning an extra 3500 calories a person can lose 1 pound of fat. How many pounds of muscle are needed to burn 1 pound of fat in 30 days? (*Source: Runner's World.*)

126. *Wireless Households* The percentage P of wireless households x years after 2005 can be modeled by the formula $P(x) = 4x + 7$, where $0 \le x \le 7$.

(a) Evaluate $P(0)$ and $P(3)$. Interpret your results.

(b) Explain the meaning of 4 and 7 in the formula.

127. *Mobile Data Penetration* The table lists the percentage P of people with cell phones who also subscribed to a data package during year x. (For example, with a data package one can surf the Web and check email.)

Year	2007	2008	2009
Percentage	55%	59%	63%

(a) What was this percentage in 2007?

(b) By how much did this percentage change each year?

(c) Write a function P that models these data. Let x be years after 2007.

(d) Estimate this percentage in 2010.

128. *Wal-Mart Sales* The table shows Wal-Mart's share as a percentage of overall U.S. retail sales for past years. (This percentage excludes restaurants and motor vehicles.)

Year	1998	1999	2000	2001	2002
Share (%)	6	6.5	7	7.5	8

Source: Commerce Department, Wal-Mart.

(a) What was Wal-Mart's share in 1998?

(b) By how much (percent) did Wal-Mart's share increase each year?

(c) Find a linear function f that models these data. Let $x = 0$ correspond to 1998.

(d) Use f to estimate Wal-Mart's share in 2005.

WRITING ABOUT MATHEMATICS

129. Explain how you can determine whether a function is linear by using its
(a) symbolic representation,
(b) graphical representation, and
(c) numerical representation.

130. Describe one way to determine whether data can be modeled by a linear function.

SECTIONS 8.1 and 8.2   **Checking Basic Concepts**

1. Find a formula and sketch a graph for a function that squares the input x and then subtracts 1.

2. Use the graph of f to do the following.
(a) Find the domain and range of f.
(b) Evaluate $f(0)$ and $f(2)$.
(c) Is f a linear function? Explain.

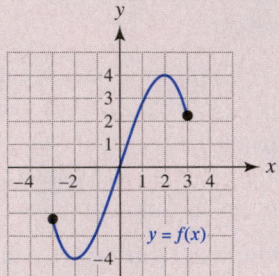

3. Determine whether f is a linear function.
(a) $f(x) = 4x - 2$
(b) $f(x) = 2\sqrt{x} - 5$
(c) $f(x) = -7$
(d) $f(x) = 9 - 2x + 5x$

4. Graph $f(x) = 4 - 3x$. Evaluate $f(-2)$.

5. Find a formula for a linear function that models the data.

x	0	1	2	3	4
$f(x)$	-1	$-\frac{1}{2}$	0	$\frac{1}{2}$	1

6. The median age in the United States from 1970 to 2010 can be approximated by
$$f(x) = 0.225x + 27.7,$$
where $x = 0$ corresponds to 1970, $x = 1$ to 1971, and so on.
(a) Evaluate $f(20)$ and interpret the result.
(b) Interpret the numbers 0.225 and 27.7.

7. Find the midpoint of the line segment connecting the points $(-3, 4)$ and $(5, -6)$.

8.3   Compound Inequalities

Basic Concepts • Symbolic Solutions and Number Lines • Numerical and Graphical Solutions • Interval Notation

A LOOK INTO MATH ▶

A person weighing 143 pounds and needing to purchase a life vest for white-water rafting is not likely to find one designed exactly for this weight. Life vests are manufactured to support a range of body weights. A vest approved for weights between 100 and 160 pounds might be appropriate for this person. In other words, if a person's weight is w, this life vest is safe if $w \geq 100$ *and* $w \leq 160$. This example illustrates the concept of a *compound inequality*.

Basic Concepts

A **compound inequality** consists of two inequalities joined by the words *and* or *or*. The following are two examples of compound inequalities.

1. $2x \geq -3$ **and** $2x < 5$

First compound inequality

2. $x + 2 \geq 3$ **or** $x - 1 < -5$

Second compound inequality

NEW VOCABULARY

☐ Compound inequality
☐ Intersection
☐ Three-part inequality
☐ Union
☐ Interval notation
☐ Infinity
☐ Negative infinity

If a compound inequality contains the word *and*, a solution must satisfy *both* inequalities. For example, 1 is a solution to the first compound inequality because

$$2(1) \geq -3 \quad \text{and} \quad 2(1) < 5 \qquad \text{First compound inequality}$$
$$\text{True} \qquad\qquad \text{True} \qquad\qquad \text{with } x = 1$$

are *both* true statements.

If a compound inequality contains the word *or*, a solution must satisfy *at least one* of the two inequalities. Thus 5 is a solution to the second compound inequality, because the first statement is true.

$$5 + 2 \geq 3 \quad \text{or} \quad 5 - 1 < -5 \qquad \text{Second compound inequality}$$
$$\text{True} \qquad\qquad \text{False} \qquad\qquad \text{with } x = 5$$

Note that 5 does not need to satisfy both statements for this compound inequality to be true.

EXAMPLE 1

Determining solutions to compound inequalities

Determine whether the given *x*-values are solutions to the compound inequalities.
(a) $x + 1 < 9$ and $2x - 1 > 8$ $\qquad x = 5, -5$
(b) $5 - 2x \leq -4$ or $5 - 2x \geq 4$ $\qquad x = 2, -3$

Solution
(a) Substitute $x = 5$ in the compound inequality $x + 1 < 9$ and $2x - 1 > 8$.

$$5 + 1 < 9 \quad \text{and} \quad 2(5) - 1 > 8$$
$$\text{True} \qquad\qquad \text{True}$$

Both inequalities are true, so 5 is a solution. Now substitute $x = -5$.

$$-5 + 1 < 9 \quad \text{and} \quad 2(-5) - 1 > 8$$
$$\text{True} \qquad\qquad \text{False}$$

To be a solution both inequalities must be true, so -5 is *not* a solution.
(b) Substitute $x = 2$ into the compound inequality $5 - 2x \leq -4$ or $5 - 2x \geq 4$.

$$5 - 2(2) \leq -4 \quad \text{or} \quad 5 - 2(2) \geq 4$$
$$\text{False} \qquad\qquad \text{False}$$

Neither inequality is true, so 2 is *not* a solution. Now substitute $x = -3$.

$$5 - 2(-3) \leq -4 \quad \text{or} \quad 5 - 2(-3) \geq 4$$
$$\text{False} \qquad\qquad \text{True}$$

At least one of the two inequalities is true, so -3 is a solution.

Now Try Exercises 7, 9

Symbolic Solutions and Number Lines

We can use a number line to graph solutions to compound inequalities, such as

$$x \leq 6 \quad \text{and} \quad x > -4.$$

The solution set for $x \leq 6$ is shaded to the left of 6, with a bracket placed at $x = 6$, as shown in Figure 8.39 on the next page. The solution set for $x > -4$ can be shown by shading a different number line to the right of -4 and placing a left parenthesis at -4. Because the inequalities are connected by *and*, the solution set consists of all numbers that are shaded on *both* number lines. The final number line represents the intersection of the

STUDY TIP

To review set-builder notation, refer to page 142.

two solution sets. That is, the solution set includes real numbers where the graphs "overlap." For any two sets A and B, the **intersection** of A and B, denoted $A \cap B$, is defined by

$$A \cap B = \{x \mid x \text{ is an element of } A \text{ and an element of } B\}.$$

Graphing a Compound Inequality

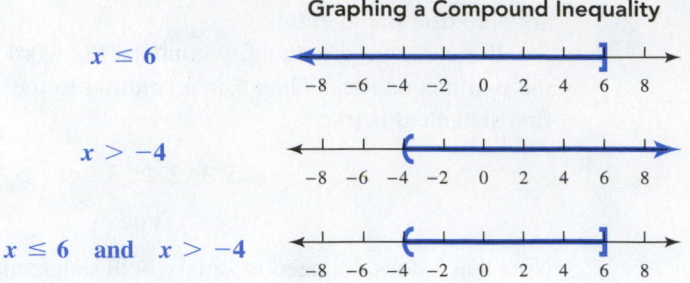

Figure 8.39

NOTE: A bracket, either [or], is used when an inequality contains $\leq$ or $\geq$. A parenthesis, either (or), is used when an inequality contains $<$ or $>$. This notation makes clear whether an endpoint is included in the inequality.

EXAMPLE 2 | **Solving a compound inequality containing "and"**

Solve $2x + 4 > 8$ and $5 - x < 9$. Graph the solution set.

Solution
First solve each linear inequality separately.

$$2x + 4 > 8 \quad \text{and} \quad 5 - x < 9$$
$$2x > 4 \quad \text{and} \quad -x < 4$$
$$x > 2 \quad \text{and} \quad x > -4$$

Graph $x > 2$ and $x > -4$ on two different number lines. On a third number line, shade solutions that appear on both of the first two number lines. As shown in Figure 8.40, the solution set is $\{x \mid x > 2\}$.

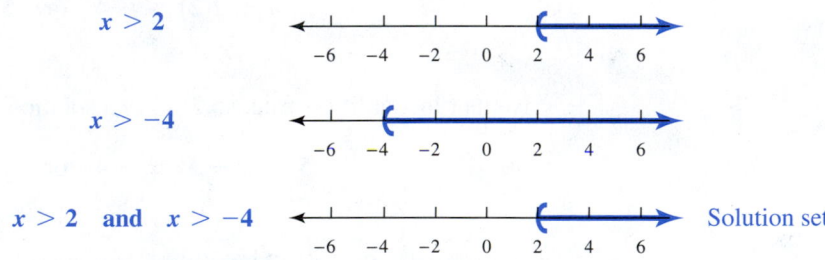

Figure 8.40
Now Try Exercise 43

Sometimes a compound inequality containing the word *and* can be combined into a three-part inequality. For example, rather than writing

$$x > 5 \quad \text{and} \quad x \leq 10,$$

we could write the **three-part inequality**

$$5 < x \leq 10.$$

This three-part inequality is represented by the number line shown in Figure 8.41.

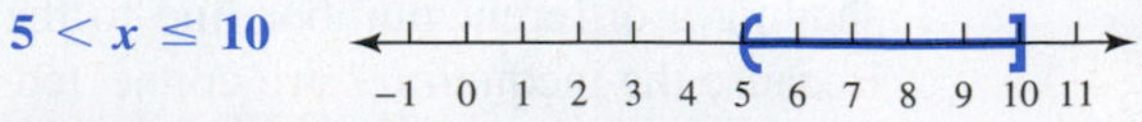

READING CHECK

What is a three-part inequality?

$5 < x \leq 10$![number line from -1 to 11]

Figure 8.41

EXAMPLE 3   **Solving three-part inequalities**

Solve each inequality. Graph each solution set. Write the solution set in set-builder notation. (To review set-builder notation, see page 142.)

(a) $4 < t + 2 \le 8$ **(b)** $-3 \le 3z \le 6$ **(c)** $-\dfrac{5}{2} < \dfrac{1 - m}{2} < 4$

Solution

(a) To solve a three-part inequality, isolate the variable by applying properties of inequalities to each part of the inequality.

$$4 < t + 2 \le 8 \qquad \text{Given three-part inequality}$$
$$4 - \mathbf{2} < t + 2 - \mathbf{2} \le 8 - \mathbf{2} \qquad \text{Subtract 2 from each part.}$$
$$2 < t \le 6 \qquad \text{Simplify each part.}$$

The solution set is $\{t \mid 2 < t \le 6\}$. See Figure 8.42.

$2 < t \le 6$

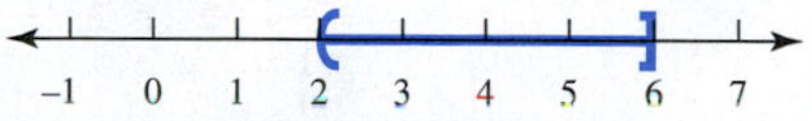

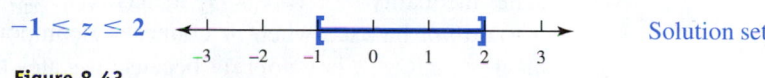

Solution set

Figure 8.42

CRITICAL THINKING

Graph the following inequalities and discuss your results.
1. $x < 2$ and $x > 5$
2. $x > 2$ or $x < 5$

(b) To simplify, divide each part by 3.

$$-3 \le 3z \le 6 \qquad \text{Given three-part inequality}$$
$$\frac{-3}{3} \le \frac{3z}{3} \le \frac{6}{3} \qquad \text{Divide each part by 3.}$$
$$-1 \le z \le 2 \qquad \text{Simplify each part.}$$

The solution set is $\{z \mid -1 \le z \le 2\}$. See Figure 8.43.

$-1 \le z \le 2$

Solution set

Figure 8.43

(c) Multiply each part by 2 to clear (eliminate) fractions.

$$-\frac{5}{2} < \frac{1 - m}{2} < 4 \qquad \text{Given three-part inequality}$$
$$2 \cdot \left(-\frac{5}{2}\right) < 2 \cdot \left(\frac{1 - m}{2}\right) < 2 \cdot 4 \qquad \text{Multiply each part by 2.}$$
$$-5 < 1 - m < 8 \qquad \text{Simplify each part.}$$
$$-5 - \mathbf{1} < 1 - m - \mathbf{1} < 8 - \mathbf{1} \qquad \text{Subtract 1 from each part.}$$
$$-6 < -m < 7 \qquad \text{Simplify each part.}$$
$$-\mathbf{1} \cdot (-6) > -\mathbf{1} \cdot (-m) > -\mathbf{1} \cdot 7 \qquad \text{Multiply each part by } -1;$$
$$\qquad \qquad \qquad \qquad \qquad \qquad \text{\textit{reverse} inequality symbols.}$$
$$6 > m > -7 \qquad \text{Simplify each part.}$$
$$-7 < m < 6 \qquad \text{Rewrite inequality.}$$

The solution set is $\{m \mid -7 < m < 6\}$. See Figure 8.44.

$-7 < m < 6$

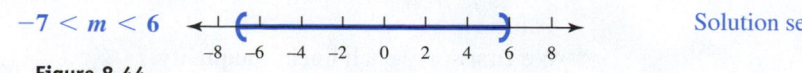

Solution set

Figure 8.44

STUDY TIP

When simplifying a three-part inequality, be sure to perform the same step on each of the three parts.

NOTE: Either $6 > m > -7$ or $-7 < m < 6$ is a correct way to write a three-part inequality. *However*, we usually write the smaller number on the left side and the larger number on the right side.

Now Try Exercises 59, 63, 79

▶ **REAL-WORLD CONNECTION** Three-part inequalities occur frequently in applications. In the next example we find altitudes at which the air temperature is within a certain range.

EXAMPLE 4 | **Solving a three-part inequality**

If the ground-level temperature is 80°F, the air temperature x miles above Earth's surface is cooler and can be modeled by $T(x) = 80 - 19x$. Find the altitudes at which the air temperature ranges from 42°F down to 23°F. (*Source:* A. Miller and R. Anthes, *Meteorology.*)

Solution
We write and solve the three-part inequality $23 \leq T(x) \leq 42$.

$$23 \leq 80 - 19x \leq 42 \qquad \text{Substitute for } T(x).$$

$$-57 \leq -19x \leq -38 \qquad \text{Subtract 80 from each part.}$$

$$\frac{-57}{-19} \geq x \geq \frac{-38}{-19} \qquad \text{Divide by } -19; \textit{ reverse} \text{ inequality symbols.}$$

$$3 \geq x \geq 2 \qquad \text{Simplify.}$$

$$2 \leq x \leq 3 \qquad \text{Rewrite inequality.}$$

The air temperature ranges from 42°F to 23°F for altitudes between 2 and 3 miles.

Now Try Exercise 113

> **MAKING CONNECTIONS**
>
> **Writing Three-Part Inequalities**
>
> The inequality $-2 < x < 1$ means that $x > -2$ *and* $x < 1$. A three-part inequality should *not* be used when *or* connects a compound inequality. Writing $x < -2$ or $x > 1$ as $1 < x < -2$ is **incorrect** because it states that x must be both greater than 1 *and* less than -2. It is impossible for any value of x to satisfy this statement.

We can also solve compound inequalities containing the word *or*. To write the solution to such an inequality we sometimes use *union* notation. For any two sets A and B, the **union** of A and B, denoted $A \cup B$, is defined by

$$A \cup B = \{x \mid x \text{ is an element of } A \textit{ or } \text{an element of } B\}.$$

If the solution to an inequality is $\{x \mid x < 1\}$ or $\{x \mid x \geq 3\}$, then it can also be written as

$$\{x \mid x < 1\} \cup \{x \mid x \geq 3\}.$$

That is, we can replace the word *or* with the $\cup$ symbol.

EXAMPLE 5 | **Solving a compound inequality containing "or"**

Solve $x + 2 < -1$ or $x + 2 > 1$. Graph the solution set.

Solution
We first solve each linear inequality.

$$x + 2 < -1 \quad \text{or} \quad x + 2 > 1 \qquad \text{Given compound inequality}$$

$$x < -3 \quad \text{or} \quad x > -1 \qquad \text{Subtract 2.}$$

We can graph the simplified inequalities on different number lines, as shown in Figure 8.45. A solution must satisfy at least one of the two inequalities. Thus the solution set for the compound inequality results from taking the *union* of the first two number lines. We can write the solution, using set-builder notation, as $\{x \mid x < -3\} \cup \{x \mid x > -1\}$ or as $\{x \mid x < -3 \text{ or } x > -1\}$.

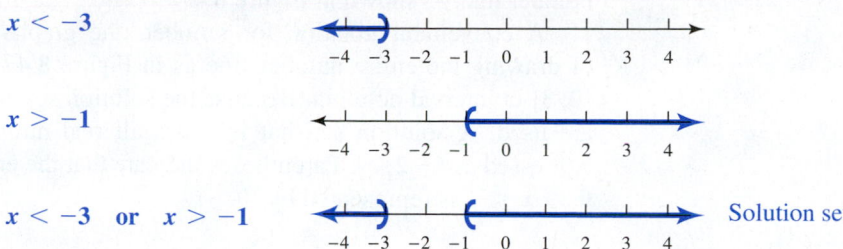

Figure 8.45

Now Try Exercise 47

CRITICAL THINKING

Carbon dioxide is emitted when human beings breathe. In one study of college students, the amount of carbon dioxide exhaled in grams per hour was measured during both lectures and exams. The average amount exhaled during lectures L satisfied $25.33 \le L \le 28.17$, whereas the average amount exhaled during exams E satisfied $36.58 \le E \le 40.92$. What do these results indicate? Explain. (*Source:* T. Wang, *ASHRAE Trans.*)

Numerical and Graphical Solutions

Compound inequalities can also be solved graphically and numerically, as illustrated in the next example.

> **EXAMPLE 6** **Estimating numbers of Internet users**
>
> The number of U.S. Internet users in millions during year x can be modeled by the formula $f(x) = 11.6(x - 2000) + 124$. Estimate the years when the number of users is expected to be between 240 and 275 million. (*Source:* The Nielsen Company.)

Solution

Numerical Solution Let $y_1 = 11.6(x - 2000) + 124$. Make a table of values, as shown in Figure 8.46(a). In 2010 the number of Internet users was 240 million, and in 2013 this number is about 275 million. Thus from 2010 to about 2013 the number of Internet users is expected to be between 240 million and 275 million.

Graphical Solution The graph of $y_1 = 11.6(x - 2000) + 124$ is shown between the graphs of $y_2 = 240$ and $y_3 = 275$ in Figures 8.46(b) and 8.46(c) from 2010 to about 2013, or when $2010 \le x \le 2013$.

CALCULATOR HELP

To find a point of intersection, see Appendix A (page AP-6).

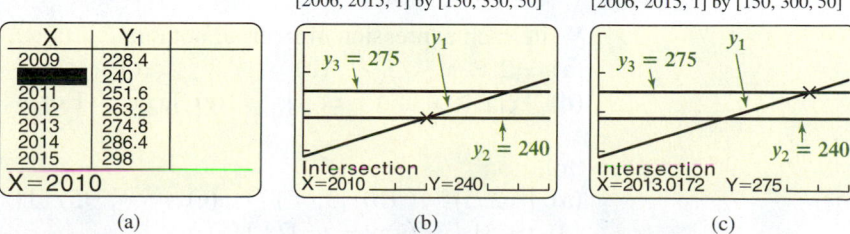

Figure 8.46

Now Try Exercise 107(a) and (b)

Interval Notation

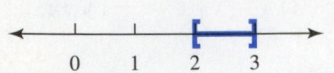

Figure 8.47

The solution set in Example 4 was $\{x \mid 2 \leq x \leq 3\}$. This solution set can be graphed on a number line, as shown in Figure 8.47.

A convenient notation for number line graphs is called **interval notation**. Instead of drawing the entire number line as in Figure 8.47, the solution set can be expressed as [2, 3] in interval notation. Because the solution set includes the endpoints 2 and 3, brackets are used. A solution set that includes all real numbers satisfying $-2 < x < 3$ can be expressed as $(-2, 3)$. Parentheses indicate that the endpoints are *not* included. The interval $0 \leq x < 4$ is represented by $[0, 4)$.

MAKING CONNECTIONS

Points and Intervals

The expression (1, 2) may represent a point in the *xy*-plane or the interval $1 < x < 2$. To alleviate confusion, phrases such as "the point (1, 2)" or "the interval (1, 2)" are used.

Table 8.13 provides examples of interval notation. The symbol ∞ refers to **infinity**, and it does not represent a real number. The interval $(5, \infty)$ represents $x > 5$, which has no maximum *x*-value, so ∞ is used for the right endpoint. The symbol $-\infty$ may be used similarly and denotes **negative infinity**. Real numbers are denoted $(-\infty, \infty)$.

TABLE 8.13 Interval Notation

Inequality	Interval Notation	Number Line Graph
$-1 < x < 3$	$(-1, 3)$	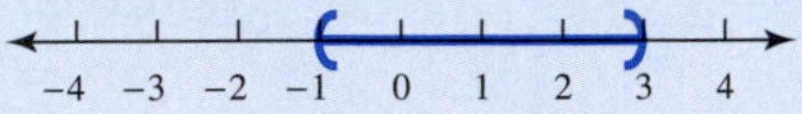
$-3 < x \leq 2$	$(-3, 2]$	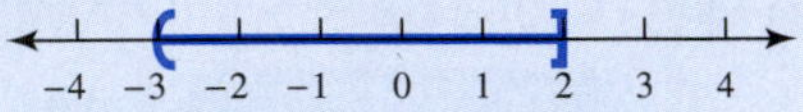
$-2 \leq x \leq 2$	$[-2, 2]$	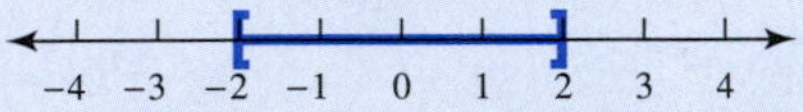
$x < -1$ or $x > 2$	$(-\infty, -1) \cup (2, \infty)$ ($\cup$ is the union symbol.)	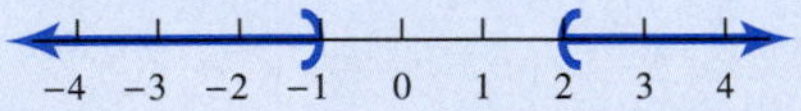
$x > -1$	$(-1, \infty)$	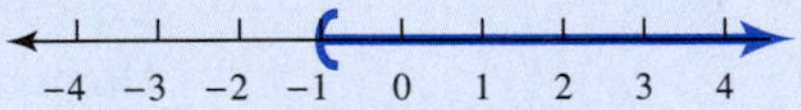
$x \leq 2$	$(-\infty, 2]$	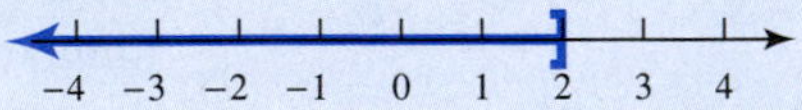

EXAMPLE 7   Writing inequalities in interval notation

Write each expression in interval notation.
(a) $-2 \leq x < 5$ **(b)** $x \geq 3$ **(c)** $x < -5$ or $x \geq 2$
(d) $\{x \mid x > 0 \text{ and } x \leq 3\}$ **(e)** $\{x \mid x \leq 1 \text{ or } x \geq 3\}$

Solution
(a) $[-2, 5)$ **(b)** $[3, \infty)$ **(c)** $(-\infty, -5) \cup [2, \infty)$
(d) $(0, 3]$ **(e)** $(-\infty, 1] \cup [3, \infty)$

Now Try Exercises 13, 17, 23, 27, 37

EXAMPLE 8 **Solving an inequality**

Solve $2x + 1 \leq -1$ or $2x + 1 \geq 3$. Write the solution set in interval notation.

Solution
First solve each inequality.

$$2x + 1 \leq -1 \quad \text{or} \quad 2x + 1 \geq 3 \qquad \text{Given compound inequality}$$

$$2x \leq -2 \quad \text{or} \qquad 2x \geq 2 \qquad \text{Subtract 1.}$$

$$x \leq -1 \quad \text{or} \qquad x \geq 1 \qquad \text{Divide by 2.}$$

The solution set may be written as $(-\infty, -1] \cup [1, \infty)$.

Now Try Exercise 55

8.3 Putting It All Together

CONCEPT	EXPLANATION	EXAMPLES
Compound Inequality	Two inequalities joined by *and* or *or*	$2x \geq 10$ and $x + 2 < 16$; $x < -1$ or $x > 2$
Three-Part Inequality	Can be used to write some types of compound inequalities involving *and*	$x > -2$ and $x \leq 3$ is equivalent to $-2 < x \leq 3$.
Interval Notation	Notation used to write sets of real numbers rather than using number lines or inequalities	$-2 \leq z \leq 4$ is equivalent to $[-2, 4]$. $x < 4$ is equivalent to $(-\infty, 4)$. $x \leq -2$ or $x > 0$ is equivalent to $(-\infty, -2] \cup (0, \infty)$.

TYPE OF INEQUALITY	METHOD TO SOLVE THE INEQUALITY
Solving a Compound Inequality Containing *and*	**STEP 1:** First solve each inequality individually. **STEP 2:** The solution set includes values that satisfy *both* inequalities from Step 1.
Solving a Compound Inequality Containing *or*	**STEP 1:** First solve each inequality individually. **STEP 2:** The solution set includes values that satisfy *at least one* of the inequalities from Step 1.
Solving a Three-Part Inequality	Work on all three parts at the same time. Be sure to perform the same step on each part. Continue until the variable is isolated in the middle part.

8.3 Exercises

CONCEPTS AND VOCABULARY

1. Give an example of a compound inequality containing the word *and*.

2. Give an example of a compound inequality containing the word *or*.

3. Is 1 a solution to $x > 3$ and $x \leq 5$?

4. Is 1 a solution to $x < 3$ or $x \geq 5$?

5. Is the compound inequality $x \geq -5$ and $x \leq 5$ equivalent to $-5 \leq x \leq 5$?

6. Name three ways to solve a compound inequality.

Exercises 7–12: Determine whether the given values of x are solutions to the compound inequality.

7. $x - 1 < 5$ and $2x > 3$ $x = 2, x = 6$

8. $2x + 1 \geq 4$ and $1 - x \leq 3$ $x = -2, x = 3$

9. $3x < -5$ or $2x \geq 3$ $x = 0, x = 3$

10. $x + 1 \leq -4$ or $x + 1 \geq 4$ $x = -5, x = 2$

11. $2 - x > -5$ and $2 - x \leq 4$ $x = -3, x = 0$

12. $x + 5 \geq 6$ or $3x \leq 3$ $x = -1, x = 1$

INTERVAL NOTATION

Exercises 13–38: Write the inequality in interval notation.

13. $2 \leq x \leq 10$

14. $-1 < x < 5$

15. $5 < x \leq 8$

16. $-\frac{1}{2} \leq x \leq \frac{5}{6}$

17. $x < 4$

18. $x \leq -3$

19. $x > -2$

20. $x \geq 6$

21. $x \geq -2$ and $x < 5$

22. $x \leq 6$ and $x \geq 2$

23. $x \leq 8$ and $x > -8$

24. $x \geq -4$ and $x < 3$

25. $x \geq 6$ or $x > 3$

26. $x \leq -4$ or $x < -3$

27. $x \leq -2$ or $x \geq 4$

28. $x \leq -1$ or $x > 6$

29. $x < 1$ or $x \geq 5$

30. $x < -3$ or $x > 3$

31.
![number line from -6 to 6, open circle at -2, closed bracket at 4]

32.
![number line from -6 to 6, closed bracket at 2, arrow right]

33.
![number line from -6 to 6, arrow left, open circle at -1]

34.
![number line from -6 to 6, closed bracket at -4 to closed bracket at 4]

35. $\{x \mid x < 4\}$

36. $\{x \mid -1 \leq x < 4\}$

37. $\{x \mid x < 1 \text{ or } x > 2\}$

38. $\{x \mid -\infty < x < \infty\}$

SYMBOLIC SOLUTIONS

Exercises 39–48: Solve the compound inequality. Graph the solution set on a number line.

39. $x \leq 3$ and $x \geq -1$

40. $x \geq 5$ and $x > 6$

41. $2x < 5$ and $2x > -4$

42. $2x + 1 < 3$ and $x - 1 \geq -5$

43. $x + 2 > 5$ and $3 - x < 10$

44. $x + 2 > 5$ or $3 - x < 10$

45. $x \leq -1$ or $x \geq 2$

46. $2x \leq -6$ or $x \geq 6$

47. $5 - x > 1$ or $x + 3 \geq -1$

48. $1 - 2x > 3$ or $2x - 4 \geq 4$

Exercises 49–58: Solve the compound inequality. Write your answer in interval notation.

49. $x - 3 \leq 4$ and $x + 5 \geq -1$

50. $2z \geq -10$ and $z < 8$

51. $3t - 1 > -1$ and $2t - \frac{1}{2} > 6$

52. $2(x + 1) < 8$ and $-2(x - 4) > -2$

53. $x - 4 \geq -3$ or $x - 4 \leq 3$

54. $1 - 3n \geq 6$ or $1 - 3n \leq -4$

55. $-x < 1$ or $5x + 1 < -10$

56. $7x - 6 > 0$ or $-\frac{1}{2}x \leq 6$

57. $1 - 7x < -48$ and $3x + 1 \leq -9$

58. $3x - 4 \leq 8$ or $4x - 1 \leq 13$

Exercises 59–80: Solve the three-part inequality. Write your answer in interval notation.

59. $-2 \leq t + 4 < 5$

60. $5 < t - 7 < 10$

61. $-\frac{5}{8} \leq y - \frac{3}{8} < 1$

62. $-\frac{1}{2} < y - \frac{3}{2} < \frac{1}{2}$

63. $-27 \le 3x \le 9$ **64.** $-4 < 2y < 22$

65. $\frac{1}{2} < -2y \le 8$ **66.** $-16 \le -4x \le 8$

67. $-4 < 5z + 1 \le 6$ **68.** $-3 \le 3z + 6 < 9$

69. $3 \le 4 - n \le 6$ **70.** $-1 < 3 - n \le 1$

71. $-1 < 2z - 1 < 3$ **72.** $2 \le 4z + 5 \le 6$

73. $-2 \le 5 - \frac{1}{3}m < 2$ **74.** $-\frac{3}{2} < 4 - 2m < \frac{7}{2}$

75. $100 \le 10(5x - 2) \le 200$

76. $-15 < 5(x - 1990) < 30$

77. $-3 < \dfrac{3z + 1}{4} < 1$ **78.** $-3 < \dfrac{z - 1}{2} < 5$

79. $-\dfrac{5}{2} \le \dfrac{2 - m}{4} \le \dfrac{1}{2}$

80. $\dfrac{4}{5} \le \dfrac{4 - 2m}{10} \le 2$

NUMERICAL AND GRAPHICAL SOLUTIONS

Exercises 81–84: Use the table to solve the three-part inequality. Write your answer in interval notation.

81. $-3 \le 3x \le 6$ **82.** $-5 \le 2x - 1 \le 1$

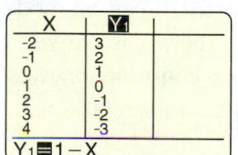

83. $-1 < 1 - x < 2$ **84.** $-2 \le -2x < 4$

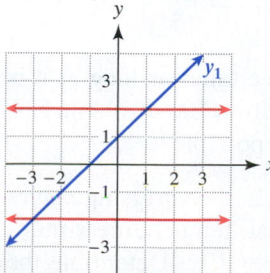 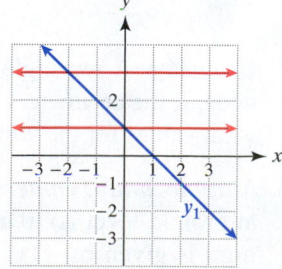

Exercises 85–88: Use the graph to solve the compound inequality. Write your answer in interval notation.

85. $-2 \le y_1 \le 2$ **86.** $1 \le y_1 < 3$

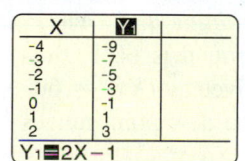

87. $y_1 < -2$ or $y_1 > 2$ **88.** $y_1 \le -2$ or $y_1 \ge 4$

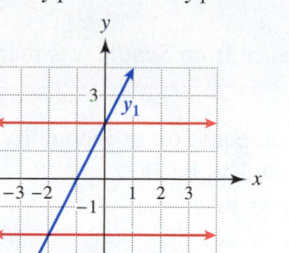 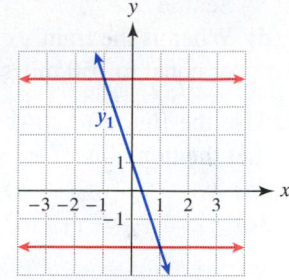

89. *Distance* The function f, shown in the figure, gives the distance y in miles between a car and Omaha, Nebraska, after x hours, where $0 \le x \le 6$.

(a) Is the car moving toward or away from Omaha? Explain.

(b) Determine the times when the car is 100 miles or 200 miles from Omaha.

(c) When is the car from 100 to 200 miles from Omaha?

(d) When is the car's distance from Omaha greater than or equal to 200 miles?

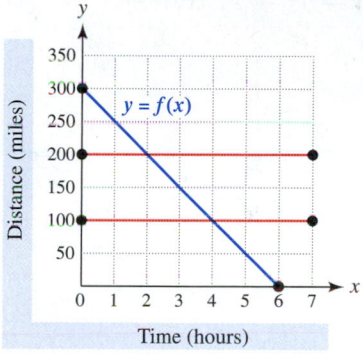

90. *Distance* The function g, shown in the figure, gives the distance y in miles between a train and Seattle after x hours, where $0 \le x \le 5$.

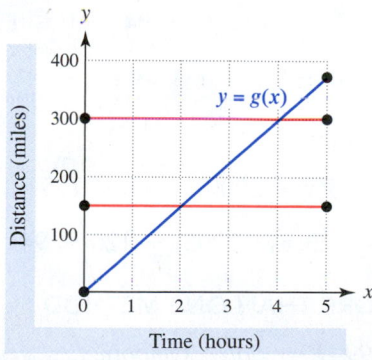

(a) Is the train moving toward or away from Seattle? Explain.

(b) Determine the times when the train is 150 miles or 300 miles from Seattle.

(continued on next page)

(c) When is the train from 150 to 300 miles from Seattle?

(d) When is the train's distance from Seattle less than or equal to 150 miles?

91. Use the figure to solve each equation or inequality. Let the domains of $y_1, y_2,$ and y_3 be $0 \leq x \leq 8$.

(a) $y_1 = y_2$ **(b)** $y_2 = y_3$
(c) $y_1 \leq y_2 \leq y_3$ **(d)** $y_2 < y_1$

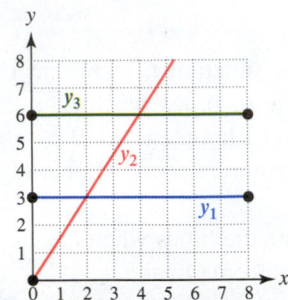

92. Use the figure to solve each equation or inequality. Let the domains of $y_1, y_2,$ and y_3 be $0 \leq x \leq 5$.

(a) $y_1 = y_2$ **(b)** $y_2 = y_3$
(c) $y_1 \leq y_2 \leq y_3$ **(d)** $y_2 < y_3$

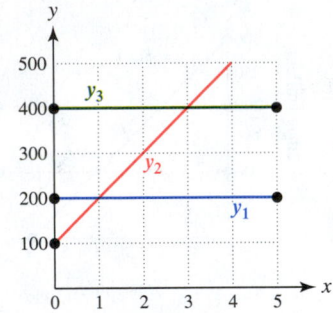

Exercises 93–98: Solve numerically or graphically. Write your answer in interval notation.

93. $-2 \leq 2x - 4 \leq 4$ **94.** $-1 \leq 1 - x \leq 3$

95. $x + 1 < -1$ or $x + 1 > 1$

96. $2x - 1 < -3$ or $2x - 1 > 5$

97. $95 \leq 25(x - 2000) + 45 \leq 295$

98. $42 \leq -13(x - 2005) + 120 \leq 94$

USING MORE THAN ONE METHOD

Exercises 99–104: Solve symbolically, graphically, and numerically. Write the solution set in interval notation.

99. $4 \leq 5x - 1 \leq 14$ **100.** $-4 < 2x < 4$

101. $4 - x \geq 1$ or $4 - x < 3$

102. $x + 3 \geq -2$ or $x + 3 \leq 1$

103. $2x + 1 < 3$ or $2x + 1 \geq 7$

104. $3 - x \leq 4$ or $3 - x > 8$

105. **Thinking Generally** Solve $c < x + b \leq d$ for x.

106. **Thinking Generally** Solve $c \leq ax + b \leq d$ for x, if $a < 0$.

APPLICATIONS

107. *Online Betting* Global online betting losses in billions can be modeled by $L(x) = 2.5x - 5000$, where x is a year from 2006 to 2011. Use each method to estimate when losses ranged from \$15 billion to \$20 billion. (*Source:* Christiansen Capital Advisors.)
(a) Numerical
(b) Graphical
(c) Symbolic

108. *College Tuition* From 1980 to 2000, college tuition and fees at private colleges could be modeled by the linear function $f(x) = 575(x - 1980) + 3600$. Use each method to estimate when the average tuition and fees ranged from \$8200 to \$10,500. (*Source:* The College Board.)
(a) Numerical
(b) Graphical
(c) Symbolic

109. *Altitude and Dew Point* If the dew point D on the ground is $60°F$, then the dew point x miles high is given by $D(x) = 60 - 5.8x$. Find the altitudes where the dew point ranges from $57.1°F$ to $51.3°F$. (*Source:* A. Miller.)

110. *Cigarette Consumption* Worldwide cigarette consumption in trillions from 1950 to 2010 can be modeled by $C(x) = 0.09x - 173.8$, where x is the year. Estimate the years when cigarette consumption was between 5.3 and 6.2 trillion. (*Source:* Department of Agriculture.)

111. *Geometry* For what values of x is the perimeter of the rectangle from 40 to 60 feet?

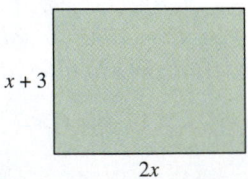

112. *Geometry* A rectangle is three times as long as it is wide. If the perimeter ranges from 100 to 160 inches, what values for the width are possible?

113. *Altitude and Temperature* If the air temperature at ground level is $70°F$, the air temperature x miles high is given by $T(x) = 70 - 19x$. Determine the altitudes at which the air temperature is from $41.5°F$ to $22.5°F$. (*Source:* A. Miller and R. Anthes, *Meteorology.*)

114. *Distance* A car's distance in miles from a rest stop after x hours is given by $f(x) = 70x + 50$.
(a) Make a table for $f(x)$ for $x = 4, 5, 6, \ldots, 10$ and use the table to solve the inequality $470 \le f(x) \le 680$. Explain your result.
(b) Solve the inequality in part (a) symbolically.

115. *Medicare Costs* In 2000 Medicare cost taxpayers $250 billion, and in 2010 it cost $500 billion. (*Source:* Department of Health and Human Services.)
(a) Find a linear function M that models these data x years after 2000.
(b) Estimate when Medicare costs were from $300 billion to $400 billion.

116. *Temperature Conversion* Water freezes at 32° F, or 0° C, and boils at 212° F, or 100° C.
(a) Find a linear function $C(F)$ that converts Fahrenheit temperature to Celsius temperature.

(b) The greatest temperature ranges on Earth are recorded in Siberia, where temperature has varied from about –70° C to 35° C. Find this temperature range in Fahrenheit.

WRITING ABOUT MATHEMATICS

117. Suppose that the solution set for a compound inequality can be written as $x < -3$ or $x > 2$. A student writes it as $2 < x < -3$. Is the student's three-part inequality correct? Explain your answer.

118. How can you determine whether an x-value is a solution to a compound inequality containing the word *and*? Give an example. Repeat the question for a compound inequality containing the word *or*.

8.4 Other Functions and Their Properties

Expressing Domain and Range in Interval Notation • Absolute Value Function • Polynomial Functions • Rational Functions (Optional) • Operations on Functions

A LOOK INTO MATH ▶

Many quantities in applications cannot be modeled with linear functions and equations. If data points do not lie on a line, we say that the data are *nonlinear*. For example, a scatterplot of the *cumulative* number of AIDS deaths from 1981 through 2007 is nonlinear, as shown in Figure 8.48. To model such data, we often use *nonlinear functions*, whose graphs are *not* a line. Because scatterplots of nonlinear data can have a variety of shapes, mathematicians use many different types of nonlinear functions, such as polynomial functions, which we discuss in this section. See Exercise 121. (*Source:* U.S. Department of Health and Human Services.)

NEW VOCABULARY

☐ Absolute value function
☐ Polynomial function of one variable
☐ Linear function
☐ Quadratic function
☐ Cubic function
☐ Rational function

U.S. AIDS Deaths

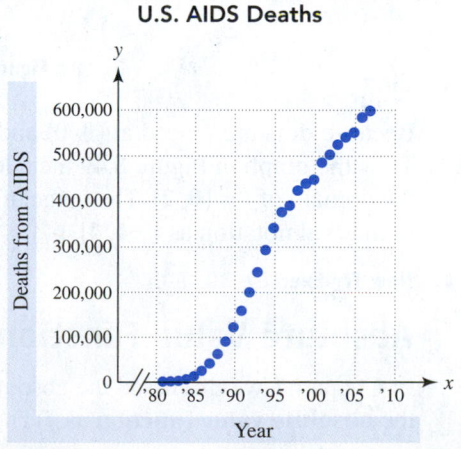

Figure 8.48

Expressing Domain and Range in Interval Notation

The set of all valid inputs for a function is called the *domain*, and the set of all outputs from a function is called the *range*. For example, all real numbers are valid inputs for $f(x) = x^2$. Rather than writing "the set of all real numbers" for the domain of f, we can use *interval notation* to express the domain as $(-\infty, \infty)$. The symbol ∞ represents infinity and is not a real number. Because $x^2 \geq 0$ for every real number x, the output from $f(x) = x^2$ is never negative. Therefore the range of f is $[0, \infty)$, which denotes all nonnegative real numbers, or $x \geq 0$. Note that 0 is in the range of f because $f(0) = 0$, and a bracket "[" is used to indicate that 0 is included in the range of f.

> **EXAMPLE 1** Writing domains in interval notation
>
> Write the domain for each function in interval notation.
> **(a)** $f(x) = 4x$ **(b)** $g(t) = \sqrt{t - 1}$ **(c)** $h(v) = \dfrac{1}{v + 3}$
>
> **Solution**
> **(a)** The expression $4x$ is defined for all real numbers x. Thus the domain of f is $(-\infty, \infty)$.
> **(b)** The square root $\sqrt{t - 1}$ is defined only when $t - 1$ is *not* negative. Thus the domain of g includes all real numbers satisfying $t - 1 \geq 0$ or $t \geq 1$. In interval notation this inequality is written as $[1, \infty)$.
> **(c)** The expression $\frac{1}{v + 3}$ is defined except when $v + 3 = 0$ or $v = -3$. Thus the domain of h includes all real numbers except -3 and can be written as $(-\infty, -3) \cup (-3, \infty)$. Parentheses are used because -3 is not included in the domain of h.
>
> **Now Try Exercises 13, 21, 25**

In the next example, we determine the domain and range of a function from its graph. Note that dots placed at each end of a graph indicate that the endpoints are included.

> **EXAMPLE 2** Writing the domain and range in interval notation
>
> Use the graph of f in Figure 8.49 to write its domain and range in interval notation.

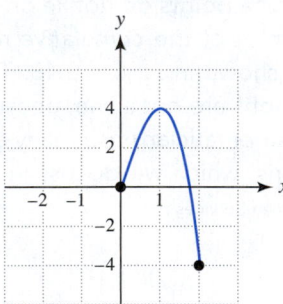

Figure 8.49

> **Solution**
> Because dots are placed at $(\mathbf{0}, 0)$ and $(\mathbf{2}, -4)$, the endpoints are included in the graph of f. Thus the graph in Figure 8.49 includes x-values from $x = \mathbf{0}$ to $x = \mathbf{2}$. In interval notation, the domain of f is $[\mathbf{0}, \mathbf{2}]$. The range of f includes y-values from -4 to 4 and can be expressed in interval notation as $[-4, 4]$.
>
> **Now Try Exercise 39**

Absolute Value Function

In Chapter 1 we discussed the absolute value of a number. We can define a function called the **absolute value function** as $f(x) = |x|$. We evaluate f as follows.

$$f(11) = |11| = 11, \quad f(-4) = |-4| = 4, \quad \text{and} \quad f(-\pi) = |-\pi| = \pi$$

To graph $y = |x|$ we begin by making a table of values, as shown in Table 8.14. Next we plot these points and sketch the graph, as shown in Figure 8.50. Note that the graph is V-shaped and never lies below the x-axis because the absolute value of a number cannot be negative.

TABLE 8.14

x	$|x|$
-2	2
-1	1
0	0
1	1
2	2

Absolute Value Function

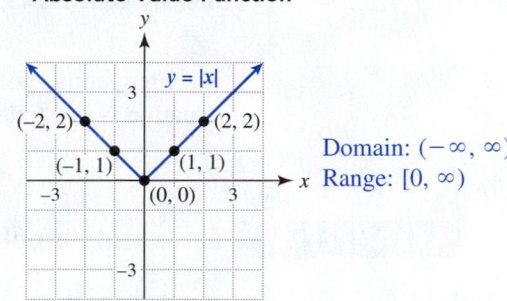

Domain: $(-\infty, \infty)$
Range: $[0, \infty)$

Figure 8.50

Because the input for $f(x) = |x|$ is any real number, the domain of f is all real numbers, or $(-\infty, \infty)$. The graph of the absolute value function shows that the output y (range) is any real number greater than or equal to 0. That is, the range is $[0, \infty)$.

Polynomial Functions

In Chapter 5 we introduced polynomials and defined their degrees. The following expressions are examples of polynomials of one variable.

$$1 - 5x, \quad 3t^2 - 5t + 1, \quad \text{and} \quad z^3 + 5$$

(The exponents on variables in polynomials must be nonnegative integers.) Recall that the *degree* of a polynomial of one variable equals the largest exponent on the variable. Thus the degree of $1 - 5x$ is 1, the degree of $3t^2 - 5t + 1$ is 2, and the degree of $z^3 + 5$ is 3.
    The equations

$$f(x) = 1 - 5x, \quad g(t) = 3t^2 - 5t + 1, \quad \text{and} \quad h(z) = z^3 + 5$$

define three **polynomial functions of one variable**. Function f is a **linear function** because it has degree 1, function g is a **quadratic function** because it has degree 2, and function h is a **cubic function** because it has degree 3.

NOTE: The domain of *every* polynomial function is *all* real numbers.

EXAMPLE 3   **Identifying polynomial functions**

Determine whether $f(x)$ represents a polynomial function. If possible, identify the type of polynomial function and its degree.
(a) $f(x) = 5x^3 - x + 10$
(b) $f(x) = x^{-2.5} + 1$
(c) $f(x) = 1 - 2x$
(d) $f(x) = \dfrac{3}{x - 1}$

Solution
(a) The expression $5x^3 - x + 10$ is a cubic polynomial, so $f(x)$ represents a cubic polynomial function. It has degree 3.

(b) $f(x) = x^{-2.5} + 1$ does not represent a polynomial function because the variables in a polynomial must have *nonnegative integer* exponents.

(c) $f(x) = 1 - 2x$ represents a polynomial function that is linear. It has degree 1.

(d) $f(x) = \dfrac{3}{x-1}$ does not represent a polynomial function because $\dfrac{3}{x-1}$ is not a polynomial.

Now Try Exercises 43, 45, 47, 51

Frequently, polynomials represent functions or formulas that can be evaluated. This situation is illustrated in the next two examples.

EXAMPLE 4 **Evaluating a polynomial function graphically and symbolically**

A graph of $f(x) = 4x - x^3$ is shown in Figure 8.51, where $y = f(x)$. Evaluate $f(-1)$ graphically and check your result symbolically.

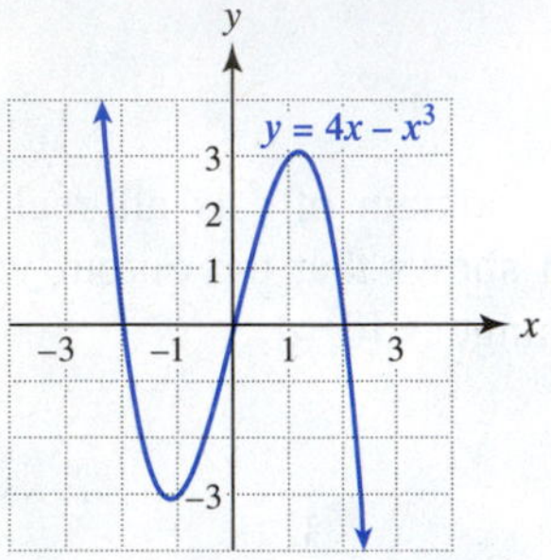

Solution

Graphical Evaluation To evaluate $f(-1)$ graphically, find -1 on the x-axis and move down until the graph of f is reached. Then move horizontally to the y-axis, as shown in Figure 8.52. Thus when $x = -1$, $y = -3$ and $f(-1) = -3$.

Symbolic Evaluation When $x = -1$, evaluation of $f(x) = 4x - x^3$ is performed as follows.

$$f(-1) = 4(-1) - (-1)^3 = -4 - (-1) = -3$$

Evaluating $f(-1) = -3$

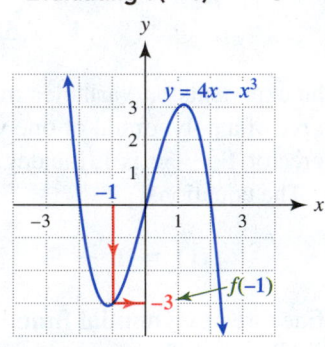

Figure 8.52

Figure 8.51

Now Try Exercises 59, 73

EXAMPLE 5 **Evaluating a polynomial function symbolically**

Evaluate $f(x)$ at the given value of x.

(a) $f(x) = -3x^4 - 2$, $x = 2$ **(b)** $f(x) = -2x^3 - 4x^2 + 5$, $x = -3$

Solution

(a) Be sure to evaluate exponents before multiplying.

$$f(2) = -3(2)^4 - 2 = -3 \cdot 16 - 2 = -50$$

(b) $f(-3) = -2(-3)^3 - 4(-3)^2 + 5 = -2(-27) - 4(9) + 5 = 23$

Now Try Exercises 61, 63

▶ **REAL-WORLD CONNECTION** A well-conditioned athlete's heart rate can reach 200 beats per minute during strenuous physical activity. Upon quitting, a typical heart rate decreases rapidly at first and then more gradually after a few minutes, as illustrated in the next example.

EXAMPLE 6   **Modeling heart rate of an athlete**

Let $P(t) = 1.875t^2 - 30t + 200$ model an athlete's heart rate (or pulse P) in beats per minute (bpm) t minutes after strenuous exercise has stopped, where $0 \leq t \leq 8$. (*Source: V. Thomas, Science and Sport.*)

(a) What is the initial heart rate when the athlete stops exercising?
(b) What is the heart rate after 8 minutes?
(c) A graph of P is shown in Figure 8.53. Interpret this graph.

Athlete's Heart Rate

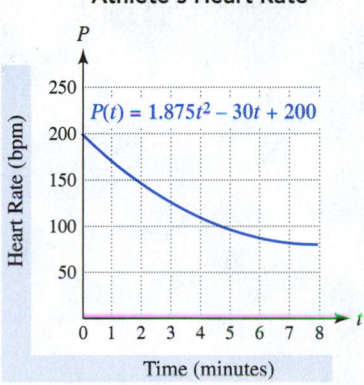

Figure 8.53

Solution
(a) To find the initial heart rate, evaluate $P(t)$ at $t = \mathbf{0}$, or

$$P(\mathbf{0}) = 1.875(\mathbf{0})^2 - 30(\mathbf{0}) + 200 = \mathbf{200}.$$

When the athlete stops exercising, the heart rate is **200** beats per minute. (This result agrees with the graph.)
(b) $P(\mathbf{8}) = 1.875(\mathbf{8})^2 - 30(\mathbf{8}) + 200 = 80$ beats per minute.
(c) The heart rate does not drop at a constant rate; rather, it drops rapidly at first and then gradually begins to level off.

▍ **Now Try Exercise 115**

Rational Functions (Optional)

A rational expression is formed when a polynomial is divided by a polynomial. For example, the expressions

$$\frac{2x - 1}{x}, \quad \frac{5}{x^2 - 1}, \quad \text{and} \quad \frac{2x - 5}{x^2 - 9}$$

are rational expressions. Rational expressions can be used to define *rational functions*.

RATIONAL FUNCTIONS

Let $p(x)$ and $q(x)$ be polynomials. Then a **rational function** is given by

$$f(x) = \frac{p(x)}{q(x)}.$$

The domain of f includes all x-values such that $q(x) \neq 0$.

From this definition, it follows that

$$f(x) = \frac{2x - 1}{x}, \quad g(x) = \frac{5}{x^2 + 1}, \quad \text{and} \quad h(x) = \frac{2x - 5}{x^2 - 9}$$

define rational functions. The domain of f includes all real numbers except 0, the domain of g includes all real numbers because $x^2 + 1 \neq 0$ for any x-value, and the domain of h includes all real numbers except ± 3.

Formulas for linear and polynomial functions are *defined* for all x-values. However, formulas for a rational function are *undefined* for x-values that make the denominator equal to 0. (Division by 0 is undefined.) For example, if $f(x) = \frac{1}{x-2}$, then $f(2) = \frac{1}{2-2} = \frac{1}{0}$ is undefined because the denominator equals 0. A graph of $y = f(x)$ is shown in Figure 8.54. The graph does not cross the dashed vertical line $x = 2$, because $f(x)$ is *undefined* at $x = 2$. The red vertical dashed line $x = 2$ is called a *vertical asymptote*, and is used as an aid for sketching a graph of f. It is not actually part of the graph of the function.

READING CHECK

How can you determine the domain of a rational function?

A Rational Function

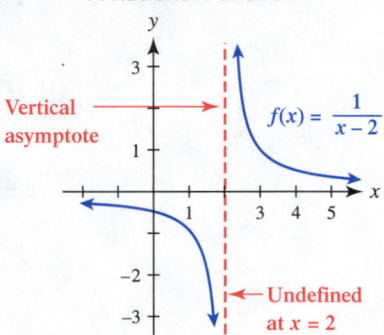

Figure 8.54

EXAMPLE 7   **Identifying the domains of rational functions**

Write the domain of each function in interval notation.

(a) $f(x) = \dfrac{1}{x + 2}$ **(b)** $g(x) = \dfrac{2x}{x^2 - 3x + 2}$ **(c)** $h(t) = \dfrac{4}{t^3 - t}$

Solution
(a) The domain of f includes all x-values except when the denominator equals 0.

$$x + 2 = 0 \qquad \text{Set the denominator equal to 0.}$$
$$x = -2 \qquad \text{Subtract 2.}$$

Thus $f(-2)$ is undefined and -2 must be excluded from the domain of f. In interval notation the domain of f is $(-\infty, -2) \cup (-2, \infty)$.

(b) The domain of g includes all real numbers except when $x^2 - 3x + 2 = 0$.

$$x^2 - 3x + 2 = 0 \qquad \text{Set the denominator equal to 0.}$$
$$(x - 1)(x - 2) = 0 \qquad \text{Factor.}$$
$$x = 1 \quad \text{or} \quad x = 2 \qquad \text{Zero-product property}$$

Because $g(1)$ and $g(2)$ are both undefined, 1 and 2 must be excluded from the domain of g. In interval notation the domain of g is $(-\infty, 1) \cup (1, 2) \cup (2, \infty)$.

(c) The domain of h includes all real numbers except when $t^3 - t = 0$.

$$t^3 - t = 0 \qquad \text{Set the denominator equal to 0.}$$
$$t(t^2 - 1) = 0 \qquad \text{Factor out } t.$$
$$t(t - 1)(t + 1) = 0 \qquad \text{Difference of squares}$$
$$t = 0 \quad \text{or} \quad t = 1 \quad \text{or} \quad t = -1 \qquad \text{Zero-product property}$$

In interval notation the domain of h is $(-\infty, -1) \cup (-1, 0) \cup (0, 1) \cup (1, \infty)$.

Now Try Exercises 27, 33, 35

To graph a rational function by hand, we usually start by making a table of values, as demonstrated in the next example. Because the graphs of rational functions are typically nonlinear, it is a good idea to plot at least 3 points on each side of an x-value where the formula is undefined—that is, where the denominator equals 0.

EXAMPLE 8 | **Graphing a rational function**

Graph $f(x) = \dfrac{1}{x}$. State the domain of f.

Solution
Make a table of values for $f(x) = \frac{1}{x}$, as shown in Table 8.15. Notice that $x = 0$ is not in the domain of f, and a dash can be used to denote this undefined value. The domain of f is all real numbers such that $x \neq 0$. Start by picking three x-values on each side of 0.

TABLE 8.15

x	-2	-1	$-\frac{1}{2}$	0	$\frac{1}{2}$	1	2
$\frac{1}{x}$	$-\frac{1}{2}$	-1	-2	—	2	1	$\frac{1}{2}$

↑————— **Undefined**

Plot the points shown in Table 8.15 and then connect the points with a smooth curve, as shown in Figure 8.55. Because $f(0)$ is undefined, the graph of $f(x) = \frac{1}{x}$ does not cross the line $x = 0$ (the y-axis). The line $x = 0$ is a vertical asymptote.

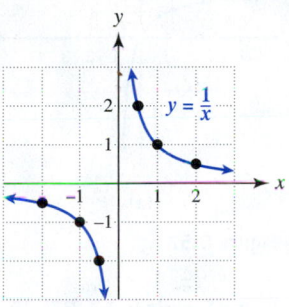

Figure 8.55

| Now Try Exercise 89

In the next example we evaluate a rational function in three ways.

EXAMPLE 9 | **Evaluating a rational function**

Use Table 8.16, the formula for $f(x)$, and Figure 8.56 to evaluate $f(-1)$, $f(1)$, and $f(2)$.

(a) TABLE 8.16

x	$f(x)$
-3	$\frac{3}{2}$
-2	$\frac{4}{3}$
-1	1
0	0
1	—
2	4
3	3

(b) $f(x) = \dfrac{2x}{x - 1}$

(c)

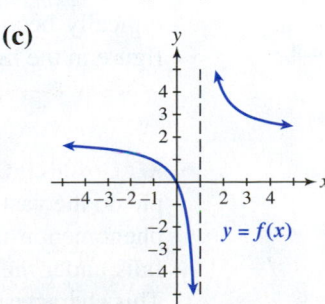

Figure 8.56

**TABLE 8.16
(repeated)**

x	$f(x)$
-3	$\frac{3}{2}$
-2	$\frac{4}{3}$
-1	1
0	0
1	$-$
2	4
3	3

Solution

(a) *Numerical Evaluation* Table 8.16 (repeated in the margin) shows that

$$f(-1) = 1, \quad f(1) \text{ is undefined}, \quad \text{and} \quad f(2) = 4.$$

(b) *Symbolic Evaluation* Let $f(x) = \dfrac{2x}{x - 1}$.

$$f(-1) = \frac{2(-1)}{-1 - 1} = 1$$

$$f(1) = \frac{2(1)}{1 - 1} = \frac{2}{0}, \text{ which is undefined. Input 1 is } not \text{ in the domain of } f.$$

$$f(2) = \frac{2(2)}{2 - 1} = 4$$

(c) *Graphical Evaluation* To evaluate $f(-1)$ graphically, find $x = -1$ on the x-axis and move upward to the graph of f. The y-value is **1** at the point of intersection, so $f(-1) = 1$, as shown in Figure 8.57(a). In Figure 8.57(b) the red, dashed vertical line $x = 1$ is a vertical asymptote. Because the graph of f does not intersect this line, $f(1)$ is undefined. Figure 8.57(c) reveals that $f(2) = 4$.

Evaluating a Rational Function Graphically

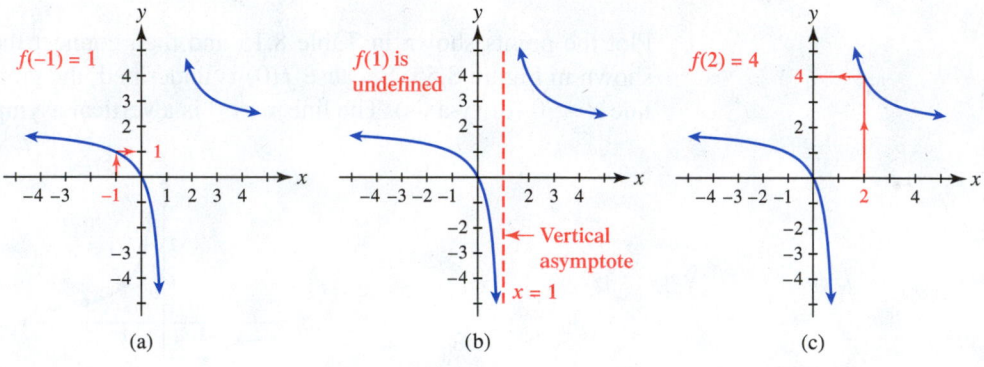

(a)   (b)   (c)

Figure 8.57

■ **Now Try Exercises 67, 77, 79**

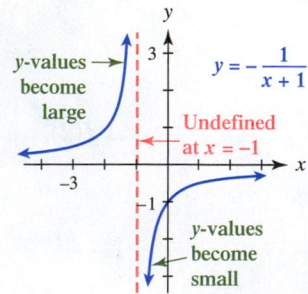

MAKING CONNECTIONS

Asymptotes and Graphs of Rational Functions

A *vertical asymptote* often occurs at x-values in the graph of a rational function $y = f(x)$ where the denominator equals 0. A vertical asymptote can be used as an aid to sketch the graph of a rational function. However, the graph of a rational function *never* crosses a vertical asymptote, so a vertical asymptote is *not* part of the graph of f.

On either side of a vertical asymptote, the y-values on the graph of a rational function typically become either very large (approach ∞) or very small (approach $-\infty$). See the figure in the margin.

▶ **REAL-WORLD CONNECTION** You may have noticed that a relatively small percentage of people do the vast majority of postings on social networks, such as Facebook and Twitter. This phenomenon is called *participation inequality*. That is, a vast majority of the population falls under the category of "lurkers," who are on the network but are not posting material. This characteristic of a social network can be modeled approximately by a rational function, as illustrated in the next example. (*Source:* Wu, Michael, *The Economics of 90-9-1.*)

EXAMPLE 10 **Modeling social network participation**

The rational function given by

$$f(x) = \frac{100}{101 - x}, \quad 5 \le x \le 100,$$

models participation inequality in a social network. In this formula, $f(x)$ outputs the percentage of the postings done by the least active (bottom) x percent of the population.

(a) Evaluate $f(95)$. Interpret your answer.

(b) A graph of $y = f(x)$ is shown in Figure 8.58. Interpret the graph.

Participation in a Social Network

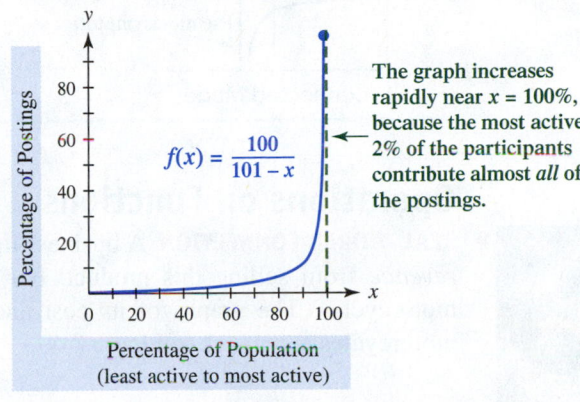

Figure 8.58 Participation Inequality

(c) Solve the rational equation $\frac{100}{101 - x} = 9$. Interpret your answer.

Solution

(a) $f(\mathbf{95}) = \dfrac{100}{101 - \mathbf{95}} = \dfrac{100}{6} \approx 16.7\%$ Let $x = 95$.

This means that the least active 95% of the population contributes only 16.7% of the postings, so the most active 5% of the population is responsible for the remaining 83.3% of the postings.

CRITICAL THINKING

Suppose that a social network had *participation equality*, in which every member contributed an equal number of postings. Sketch a graph like Figure 8.58 that describes this social network.

(b) The graph shows participation inequality visually. The graph remains at a relatively low percentage until $x = 90\%$. This means that the bottom 90% of the population does very few postings. For $x \ge 90\%$ the graph rises rapidly because the top 10% contributes a vast majority of the postings.

(c) To solve this equation, we begin by multiplying each side by $(\mathbf{101 - x})$.

$$\frac{100}{101 - x} = 9 \qquad\qquad \text{Given equation}$$

$$(\mathbf{101 - x}) \cdot \frac{100}{\mathbf{101 - x}} = 9(\mathbf{101 - x}) \qquad \text{Multiply by } (101 - x).$$

$$100 = 9(101 - x) \qquad\qquad \text{Simplify left side.}$$

$$100 = 909 - 9x \qquad\qquad \text{Distributive property}$$

$$\mathbf{9x} + 100 - \mathbf{100} = 909 - \mathbf{100} + \mathbf{9x} - 9x \qquad \text{Add } 9x. \text{ Subtract } 100.$$

$$9x = 809 \qquad\qquad \text{Simplify.}$$

$$x = \frac{809}{9} \approx 90\% \qquad\qquad \text{Simplify.}$$

This result indicates that the least active 90% of the population contributes only 9% of the postings.

Now Try Exercise 117

TECHNOLOGY NOTE

Asymptotes, Dot Mode, and Decimal Windows
When a rational function is graphed on a graphing calculator in connected mode, pseudo-asymptotes often occur because the calculator is simply connecting dots to draw a graph. The accompanying figures show the graph of $y = \frac{2}{x-2}$ in connected mode, in dot mode, and with a *decimal*, or *friendly*, window. In dot mode, pixels in the calculator screen are not connected. With dot mode (and sometimes with a decimal window) pseudo-asymptotes do not appear. To learn more about these features, consult your owner's manual.

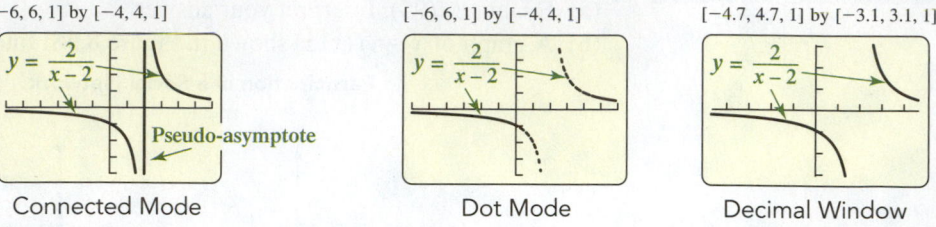

| Connected Mode | Dot Mode | Decimal Window |

CALCULATOR HELP

To set a calculator in dot mode or to set a decimal window, see Appendix A (page AP-11).

Operations on Functions

▶ **REAL-WORLD CONNECTION** A business incurs a *cost* to make its product and then it receives *revenue* from selling this product. For example, suppose a small business reconditions motorcycles. The graphs of its cost and of its revenue for reconditioning and selling x motorcycles are shown in Figure 8.59.

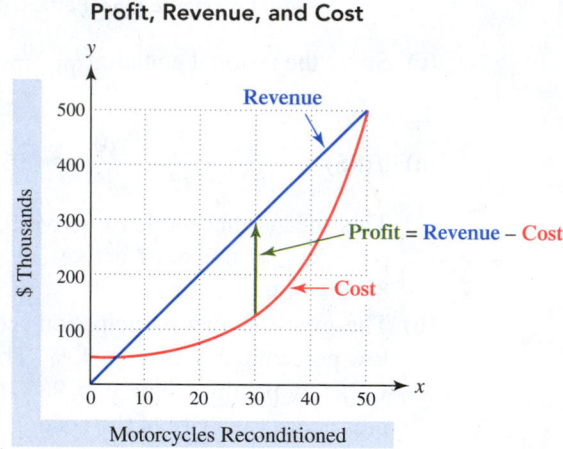

Figure 8.59

In general, *profit equals revenue minus cost*. In Figure 8.59, **profit** is shown visually as the length of the vertical green arrow between the graphs of **revenue** and **cost**. For any x-value, the distance by which revenue is above cost is called the profit for reconditioning and selling x motorcycles. *Maximum profit* for the company occurs at the x-value where the *length of the vertical green arrow is greatest*.

If we let $C(x)$, $R(x)$, and $P(x)$ be functions that calculate the cost, revenue, and profit, respectively, for reconditioning and selling x motorcycles, then

$$P(x) = R(x) - C(x).$$

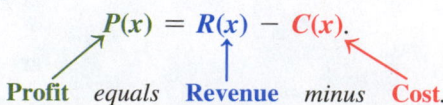

Profit *equals* **Revenue** *minus* **Cost**.

This example helps explain why we subtract functions in the real world. Functions can be added, multiplied, and divided in a similar manner.

Given two functions f and g, we define the sum $f + g$, difference $f - g$, product fg, and quotient $\frac{f}{g}$, as follows.

OPERATIONS ON FUNCTIONS

If $f(x)$ and $g(x)$ are both defined, then the sum, difference, product, and quotient of two functions f and g are defined by

$$(f + g)(x) = f(x) + g(x) \qquad \text{Sum}$$
$$(f - g)(x) = f(x) - g(x) \qquad \text{Difference}$$
$$(fg)(x) = f(x) \cdot g(x) \qquad \text{Product}$$
$$\left(\frac{f}{g}\right)(x) = \frac{f(x)}{g(x)}, \text{ where } g(x) \neq 0. \quad \text{Quotient}$$

EXAMPLE 11 **Performing arithmetic on functions**

Use $f(x) = x^2$ and $g(x) = 2x - 4$ to evaluate each of the following.

(a) $(f + g)(3)$ **(b)** $(fg)(-1)$ **(c)** $\left(\dfrac{f}{g}\right)(0)$ **(d)** $(f/g)(2)$

Solution
(a) $(f + g)(3) = f(3) + g(3) = 3^2 + (2 \cdot 3 - 4) = 9 + 2 = 11$
(b) $(fg)(-1) = f(-1) \cdot g(-1) = (-1)^2 \cdot (2 \cdot (-1) - 4) = 1 \cdot (-6) = -6$
(c) $\left(\dfrac{f}{g}\right)(0) = \dfrac{f(0)}{g(0)} = \dfrac{0^2}{2 \cdot 0 - 4} = \dfrac{0}{-4} = 0$

(d) Note that $(f/g)(2)$ is equivalent to $\left(\dfrac{f}{g}\right)(2)$.

$$(f/g)(2) = \frac{f(2)}{g(2)} = \frac{2^2}{2 \cdot 2 - 4} = \frac{4}{0},$$

which is not possible because division by 0 is undefined. Thus $(f/g)(2)$ is *undefined*.

Now Try Exercise 101

In the next example, we find the sum, difference, product, and quotient of two functions for a general x.

EXAMPLE 12 **Performing arithmetic on functions**

Use $f(x) = 4x - 5$ and $g(x) = 3x + 1$ to evaluate each of the following.

(a) $(f + g)(x)$ **(b)** $(f - g)(x)$ **(c)** $(fg)(x)$ **(d)** $\left(\dfrac{f}{g}\right)(x)$

Solution
(a) $(f + g)(x) = f(x) + g(x) = (4x - 5) + (3x + 1) = 7x - 4$
(b) $(f - g)(x) = f(x) - g(x) = (4x - 5) - (3x + 1) = x - 6$
(c) $(fg)(x) = f(x) \cdot g(x) = (4x - 5)(3x + 1) = 12x^2 - 11x - 5$

(d) $\left(\dfrac{f}{g}\right)(x) = \dfrac{f(x)}{g(x)} = \dfrac{4x - 5}{3x + 1}$

Now Try Exercise 105

8.4 Putting It All Together

CONCEPT	COMMENTS	EXAMPLES
Writing Domain and Range in Interval Notation	Interval notation can be used to specify the domain and range of a function.	If $f(x) = x^2 + 1$, the domain of f is $(-\infty, \infty)$, and the range of f is $[1, \infty)$.
Absolute Value Function	Defined by $$f(x) = |x|$$ and has a V-shaped graph	$f(-5) = |-5| = 5$ $f(0) = |0| = 0$ $f(4) = |4| = 4$
Polynomial Function of One Variable	Can be defined by a polynomial; its degree equals the largest exponent of the variable.	Because $x^3 - 4x^2 + 6$ is a polynomial with degree 3, $$f(x) = x^3 - 4x^2 + 6$$ defines a polynomial function of degree 3 and is called a cubic function.
Rational Function	A rational function can be written as $$f(x) = \frac{p(x)}{q(x)},$$ where $p(x)$ and $q(x)$ are polynomials. Note that $q(x) \neq 0$.	Because $2x - 3$ and $x + 1$ are polynomials, $$f(x) = \frac{2x - 3}{x + 1}$$ defines a rational function. Because $f(x)$ is undefined at $x = -1$, the domain of f is $(-\infty, -1) \cup (-1, \infty)$.
Operations on Functions	$(f + g)(x) = f(x) + g(x)$ Sum $(f - g)(x) = f(x) - g(x)$ Difference $(fg)(x) = f(x)g(x)$ Product $\left(\dfrac{f}{g}\right)(x) = \dfrac{f(x)}{g(x)}, g(x) \neq 0$ Quotient	Let $f(x) = x^2$ and $g(x) = 1 - x^2$. $\begin{aligned}(f + g)(x) &= f(x) + g(x) \\ &= x^2 + (1 - x^2) \\ &= 1\end{aligned}$ $\begin{aligned}(f - g)(x) &= f(x) - g(x) \\ &= x^2 - (1 - x^2) \\ &= 2x^2 - 1\end{aligned}$ $\begin{aligned}(fg)(x) &= f(x)g(x) \\ &= x^2(1 - x^2) \\ &= x^2 - x^4\end{aligned}$ $\begin{aligned}\left(\frac{f}{g}\right)(x) &= \frac{f(x)}{g(x)} \\ &= \frac{x^2}{1 - x^2}, x \neq -1, x \neq 1\end{aligned}$

8.4 Exercises

MyMathLab PRACTICE WATCH DOWNLOAD READ REVIEW

CONCEPTS AND VOCABULARY

1. The set of all valid inputs for a function is called its _____.

2. The set of all outputs for a function is called its _____.

3. The set of all real numbers can be written in interval notation as _____.

4. If the domain of a function includes all real numbers except 5, then its domain can be written in interval notation as _____.

5. The graph of the _____ function is V-shaped.

6. The degree of a polynomial of one variable equals the largest _____ of the variable.

7. A quadratic function has degree _____.

8. If a function is linear, then its degree is _____.

9. If $f(x) = \frac{x}{2x+1}$, then f is a(n) _____ function.

10. If $f(x) = \frac{x}{2x+1}$, then the domain of f includes all real numbers except _____.

11. Which of the following expressions (a.–d.) is not a rational function?

    a. $f(x) = \frac{1}{x}$ b. $f(x) = x^2 + 1$

    c. $f(x) = \sqrt{x}$ d. $f(x) = \frac{2x^2}{x-1}$

12. Which (a.–d.) is the domain of $f(x) = \frac{2x}{2x-1}$?

    a. $\{x \mid x \neq \frac{1}{2}\}$ b. $\{x \mid x \neq 1\}$

    c. $\{x \mid x \neq 0\}$ d. $\{x \mid x = 1\}$

DOMAIN AND RANGE

Exercises 13–24: Write the domain and the range of the function in interval notation. (Hint: You may want to consider the graph of the function.)

13. $f(x) = -2x$ 14. $f(x) = -\frac{1}{4}x + 1$

15. $g(t) = \frac{2}{3}t - 3$ 16. $g(t) = 9t$

17. $h(z) = z^2 + 2$ 18. $h(z) = z^2 - 1$

19. $f(z) = -z^2$ 20. $f(z) = -\frac{1}{4}z^2$

21. $g(x) = \sqrt{x+1}$ 22. $g(x) = \sqrt{x-2}$

23. $h(x) = |x - 1|$ 24. $h(x) = |2x|$

Exercises 25–36: Write the domain of the rational function in interval notation.

25. $f(x) = \frac{1}{x-1}$ 26. $f(x) = \frac{6}{x}$

27. $f(x) = \frac{x}{6-3x}$ 28. $f(x) = \frac{3x}{2x-4}$

29. $g(t) = \frac{2}{t^2-4}$ 30. $g(t) = \frac{5}{1-t^2}$

31. $g(t) = \frac{5t}{t^2-2t}$ 32. $g(t) = \frac{-t}{2t^2-3t}$

33. $h(z) = \frac{2-z}{z^3-1}$ 34. $h(z) = \frac{z+1}{z^3-z^2}$

35. $f(x) = \frac{4}{x^2-2x-3}$

36. $f(x) = \frac{1}{x^2+4x-5}$

Exercises 37–42: A graph of a function is shown. Write the domain and range of the function in interval notation.

37.

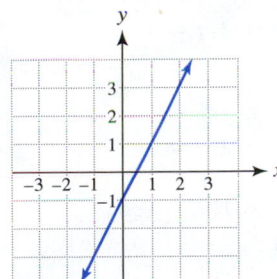

38.

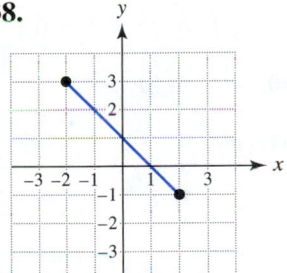

39.

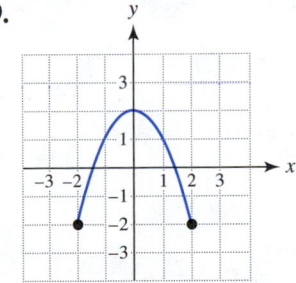

40.

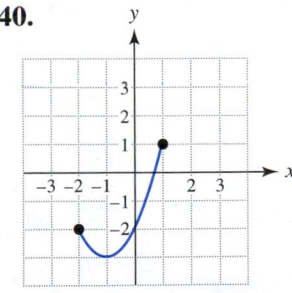

41.

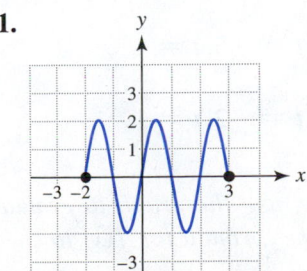

42.

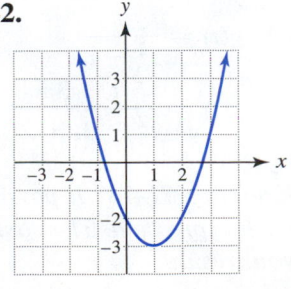

IDENTIFYING POLYNOMIAL FUNCTIONS

Exercises 43–54: Determine whether $f(x)$ represents a polynomial function. If possible, identify the degree and type of polynomial function.

43. $f(x) = 5x - 11$

44. $f(x) = 9 - x$

45. $f(x) = x^3$

46. $f(x) = x^2 + 3$

47. $f(x) = \dfrac{6}{x + 5}$

48. $f(x) = |x|$

49. $f(x) = 1 + 2x - x^2$

50. $f(x) = \frac{1}{4}x^3 - x$

51. $f(x) = 5x^{-2}$

52. $f(x) = x^2 + x^{-1}$

53. $f(x) = x^4 + 2x^2$

54. $f(x) = x^5 - 3x^3$

EVALUATING FUNCTIONS

Exercises 55–70: If possible, evaluate $g(t)$ for the given values of t.

55. $g(t) = |4t|$ \quad $t = 3, t = 0$

56. $g(t) = |t + 12|$ \quad $t = 18, t = -15$

57. $g(t) = |t - 2|$ \quad $t = 1, t = -\frac{3}{4}$

58. $g(t) = |2t + 1|$ \quad $t = 2, t = -\frac{1}{2}$

59. $g(t) = t^2 - t - 6$ \quad $t = 3, t = -3$

60. $g(t) = 3t^2 - 2t$ \quad $t = -2, t = 4$

61. $g(t) = -2t^3 + t$ \quad $t = 2, t = -2$

62. $g(t) = \frac{1}{3}t^3$ \quad $t = 1, t = -3$

63. $g(t) = t^2 - 2t - 6$ \quad $t = 0, t = -3$

64. $g(t) = 2t^3 - t^2 + 4$ \quad $t = 2, t = -1$

65. $g(t) = \dfrac{1}{t}$ \quad $t = 11, t = -7$

66. $g(t) = \dfrac{2}{3 - t}$ \quad $t = 10, t = 3$

67. $g(t) = -\dfrac{t}{t + 1}$ \quad $t = 5, t = -1$

68. $g(t) = -\dfrac{2 - t}{4t}$ \quad $t = 4, t = -1$

69. $g(t) = \dfrac{t^2}{t^2 - t}$ \quad $t = -5, t = 1$

70. $g(t) = \dfrac{t - 3}{t^2 - 3t + 2}$ \quad $t = -2, t = 1$

Exercises 71–78: If possible, use the graph to evaluate each expression. Then use the formula for $f(x)$ to check your results.

71. $f(0)$ and $f(1)$

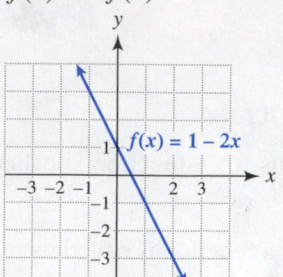

72. $f(-1)$ and $f(2)$

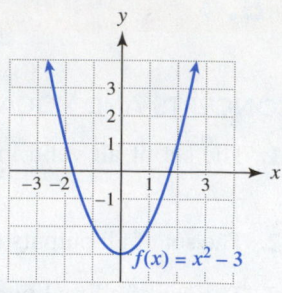

73. $f(-1)$ and $f(2)$

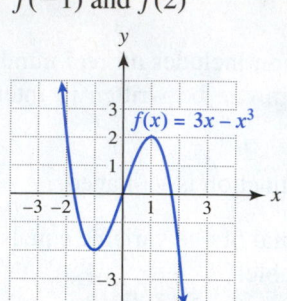

74. $f(0)$ and $f(-2)$

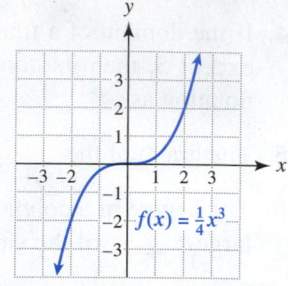

75. $f(-2)$ and $f(2)$

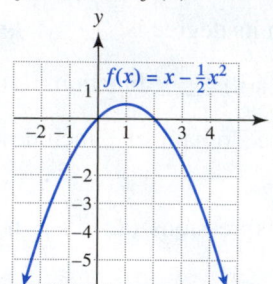

76. $f(-1)$ and $f(0)$

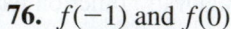
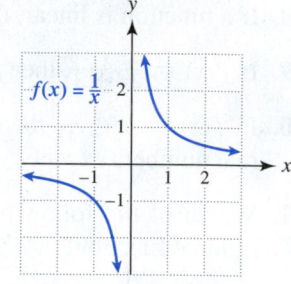

77. $f(-3)$ and $f(-1)$

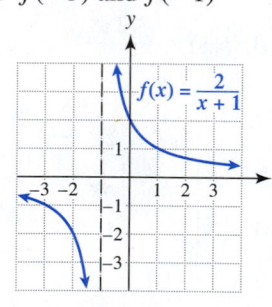

78. $f(0)$ and $f(1)$

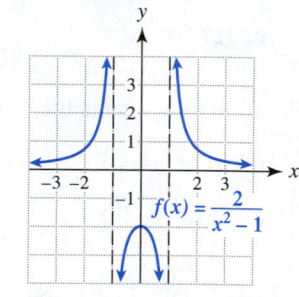

Exercises 79 and 80: Complete the table. Then evaluate $f(1)$.

79.

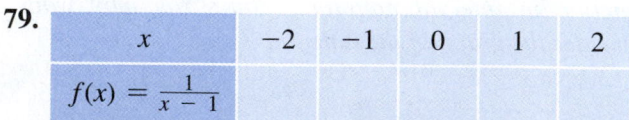

x	-2	-1	0	1	2
$f(x) = \dfrac{1}{x - 1}$					

80.

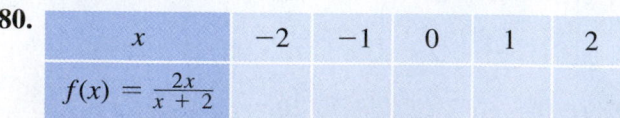

x	-2	-1	0	1	2
$f(x) = \dfrac{2x}{x + 2}$					

Exercises 81–100: Graph $y = f(x)$.

81. $f(x) = |2x|$

82. $f(x) = \left|\frac{1}{2}x\right|$

83. $f(x) = |x + 2|$ **84.** $f(x) = |x - 2|$

85. $f(x) = 1 - 2x$ **86.** $f(x) = \frac{1}{2}x + 1$

87. $f(x) = \frac{1}{2}x^2$ **88.** $f(x) = x^2 - 2$

89. $f(x) = \dfrac{1}{x - 1}$ **90.** $f(x) = \dfrac{1}{x + 1}$

91. $f(x) = \dfrac{1}{2x}$ **92.** $f(x) = \dfrac{2}{x}$

93. $f(x) = \dfrac{1}{x + 2}$ **94.** $f(x) = \dfrac{1}{x - 2}$

 **95.** $f(x) = \dfrac{4}{x^2 + 1}$ **96.** $f(x) = \dfrac{6}{x^2 + 2}$

97. $f(x) = \dfrac{3}{2x - 3}$ **98.** $f(x) = \dfrac{1}{3x + 2}$

 99. $f(x) = \dfrac{1}{x^2 - 1}$ **100.** $f(x) = \dfrac{4}{4 - x^2}$

OPERATIONS ON FUNCTIONS

Exercises 101–104: Use f(x) and g(x) to evaluate each of the following.

(a) $(f + g)(3)$ (b) $(f - g)(-2)$
(c) $(fg)(5)$ (d) $(f/g)(0)$

101. $f(x) = 5x,\ g(x) = x + 1$

102. $f(x) = x^2 + 2,\ g(x) = -2x$

103. $f(x) = 2x - 1,\ g(x) = 4x^2$

104. $f(x) = x^2 - 1,\ g(x) = x + 2$

Exercises 105–108: Use f(x) and g(x) to find each of the following.

(a) $(f + g)(x)$ (b) $(f - g)(x)$
(c) $(fg)(x)$ (d) $(f/g)(x)$

105. $f(x) = x + 1,\ g(x) = x + 2$

106. $f(x) = -3x,\ g(x) = x - 1$

107. $f(x) = 1 - x,\ g(x) = x^2$

108. $f(x) = x^2 + 4,\ g(x) = 6x$

109. Thinking Generally If $f(x) = x^2 - 2x$, then it follows that $f(a) = $ _____.

110. Thinking Generally If $f(x) = 2x - 1$, then it follows that $f(a + 2) = $ _____.

APPLICATIONS

Exercises 111–114: Graphical Interpretation Match the physical situation with the graph (a.–d.) of the rational function in the next column that models it best.

111. A population of fish that increases and then levels off

112. An insect population that dies out

113. The length of a ticket line as the rate at which people arrive in line increases

114. The wind speed during a day that is initially calm, becomes windy, and then is calm again

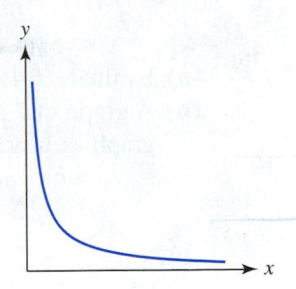

a.   b.

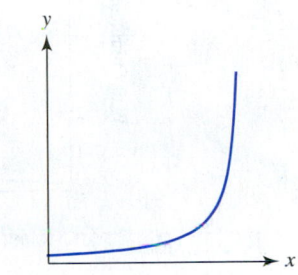

c.   d.

115. *Heart Rate of an Athlete* The following table lists the heart rate of an athlete running a 100-meter race. The race lasts 10 seconds.

Time (seconds)	0	2	4	6	8	10
Heart Rate (bpm)	90	100	113	127	143	160

(a) Does $P(t) = 0.2t^2 + 5t + 90$ model the data in the table exactly? Explain.
(b) Does P provide a reasonable model for the athlete's heart rate?
(c) Does $P(12)$ have significance in this situation? What should be the domain of P?

116. *Heart Rate of an Athlete* The following table lists an athlete's heart rate after the athlete finishes exercising strenuously.

Time (minutes)	0	2	4	6
Heart Rate (bpm)	180	137	107	90

(a) Does $P(t) = \frac{5}{3}t^2 - 25t + 180$ model the data in the table exactly? Explain.
(b) Does P provide a reasonable model for the athlete's heart rate?
(c) Does $P(12)$ have significance in this situation? What should be the domain of P?

117. *Time Spent in Line* If a parking lot attendant can wait on 5 vehicles per minute and vehicles are leaving the lot randomly at an average rate of x vehicles per minute, then the average time T in minutes spent waiting in line *and* paying the attendant is given by

$$T(x) = \frac{1}{5 - x},$$

where $x < 5$. (*Source:* N. Garber.)

(a) Evaluate $T(4)$ and interpret the result.

(b) A graph of T is shown in the figure. Interpret the graph as x increases from 0 to 5. Does this result agree with your intuition?

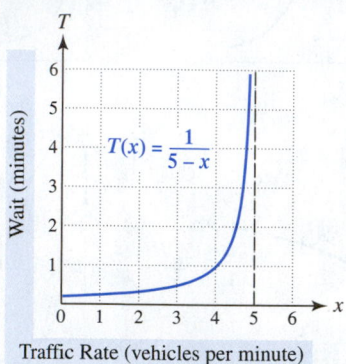

Traffic Rate (vehicles per minute)

(c) Find x if the waiting time is 3 minutes.

118. *People Waiting in Line* At a post office, workers can wait on 50 people per hour. If people arrive randomly at an average rate of x per hour, then the average number of people N waiting in line is given by

$$N(x) = \frac{x^2}{2500 - 50x},$$

where $x < 50$. (*Source:* N. Garber.)

(a) Evaluate $N(30)$ and interpret the result.

(b) A graph of N is shown in the figure. Interpret the graph as x increases from 0 to 50. Does this result agree with your intuition?

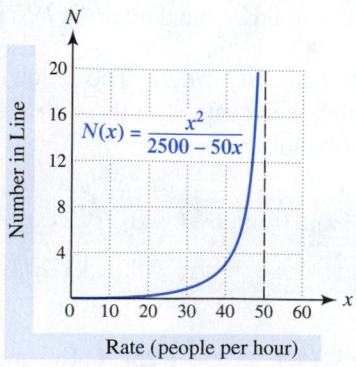

Rate (people per hour)

(c) Find x if $N = 8$.

119. *Uphill Highway Grade* The *grade* x of a hill is a measure of its steepness and corresponds to the slope of the road. For example, if a road rises 10 feet for every 100 feet of horizontal distance, it has an uphill grade of $x = \frac{10}{100}$, or 10%, as illustrated in the figure.

The braking distance for a car traveling 30 miles per hour on a wet, *uphill* grade x is given by

$$D(x) = \frac{900}{10.5 + 30x}.$$

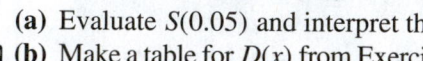

(*Source:* N. Garber.)

(a) Evaluate $D(0.05)$ and interpret the result.

(b) If the braking distance for this car is 60 feet, find the uphill grade x.

120. *Downhill Highway Grade* (See Exercise 119.) The braking distance for a car traveling 30 miles per hour on a wet, *downhill* grade x is given by

$$S(x) = \frac{900}{10.5 - 30x}.$$

(a) Evaluate $S(0.05)$ and interpret the result.

(b) Make a table for $D(x)$ from Exercise 119 and $S(x)$, starting at $x = 0$ and incrementing by 0.05.

(c) How do the braking distances for uphill and downhill grades compare? Does this result agree with your driving experience?

121. *U.S. AIDS Deaths* The following scatterplot shows the cumulative number of reported AIDS deaths. The data may be modeled x years after 1980 by $f(x) = 2.4x^2 - 14x + 23$, where the output is in thousands of deaths.

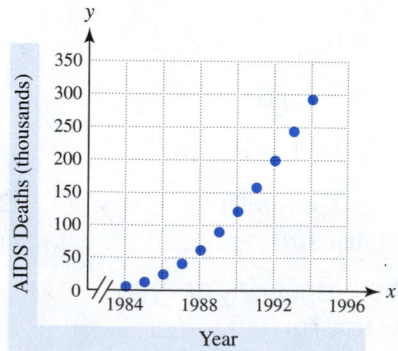

Year

(a) Use $f(x)$ to estimate the cumulative total of AIDS deaths in 1990. Compare it with the actual value of 121.6 thousand.

(b) In 1997 the cumulative number of AIDS deaths was 390 thousand. What estimate does $f(x)$ give? Discuss your result.

122. ***A PC for All?*** Worldwide sales of computers have climbed as prices have continued to drop. The function $f(x) = 0.29x^2 + 8x + 19$ models the number of personal computers sold in millions during year x, where $x = 0$ corresponds to 1990, $x = 1$ corresponds to 1991, and so on until $x = 25$ corresponds to 2015. Estimate the number of personal computers sold in 2010, using both the graph and the polynomial. (*Source:* eTForcasts.)

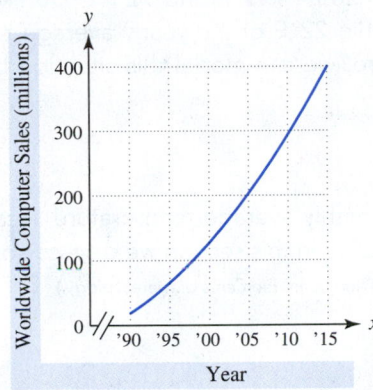

Exercises 123 and 124: ***Remembering What You Learn*** After a test students often forget what they learned. The rational function

$$R(x) = \frac{100}{1.2x + 1}, \quad 0 \le x \le 5,$$

gives an estimate of the percentage of the material a student remembers x days after a test.

123. Evaluate $R(1)$ and $R(3)$. Interpret your results.

124. If a student takes notes in class, these percentages increase by 30% for $1 \le x \le 5$. Write another function $N(x)$ that models this result. Evaluate $N(3)$.

125. ***Profit*** A company makes and sells notebook computers. The company's cost function in thousands of dollars is $C(x) = 0.3x + 100$, and the revenue function in thousands of dollars is $R(x) = 0.75x$, where x is the number of notebook computers.
(a) Evaluate and interpret $C(100)$.
(b) Interpret the y-intercepts on the graphs of C and R.
(c) Give the profit function $P(x)$.
(d) How many computers need to be sold to make a profit?

126. ***Profit*** A company makes and sells sailboats. The company's cost function in thousands of dollars is $C(x) = 2x + 20$, and the revenue function in thousands of dollars is $R(x) = 4x$, where x is the number of sailboats.
(a) Evaluate and interpret $C(5)$.
(b) Interpret the y-intercepts on the graphs of C and R.
(c) Give the profit function $P(x)$.
(d) How many sailboats need to be sold to break even?

WRITING ABOUT MATHEMATICS

127. Name two functions. Give their formulas, sketch their graphs, and state their domains and ranges.

128. Explain the difference between the domain and the range of a function.

SECTIONS 8.3 and 8.4	**Checking Basic Concepts**

1. (a) Is 3 a solution to the compound inequality $x + 2 < 4$ or $2x - 1 \ge 3$?
(b) Is 3 a solution to the compound inequality $x + 2 < 4$ and $2x - 1 \ge 3$?

2. Solve the following compound inequalities. Write your answer in interval notation.
(a) $-5 \le 2x + 1 \le 3$
(b) $1 - x \le -2$ or $1 - x \ge 2$
(c) $-2 < \dfrac{4 - 3x}{2} \le 6$

3. Write the domain of each function in interval notation.
(a) $f(x) = x^2$
(b) $g(t) = \dfrac{1}{t - 1}$
(c) $h(z) = \sqrt{z}$

4. Use the graph of f to do the following.
(a) Write the domain and range of f in interval notation.
(b) Evaluate $f(0)$ and $f(-2)$.

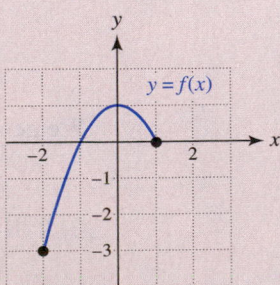

5. Graph $f(x) = |x - 3|$.

8.5 Absolute Value Equations and Inequalities

Absolute Value Equations • **Absolute Value Inequalities**

A LOOK INTO MATH ▶

Monthly average temperatures can vary greatly from one month to another, whereas yearly average temperatures remain fairly constant from one year to the next. In Boston, Massachusetts, the yearly average temperature is 50°F, but monthly average temperatures can vary from 28°F to 72°F. Because **50**°F − 28°F = **22**°F and 72°F − **50**°F = **22**°F, the monthly average temperatures are always within 22°F of the yearly average temperature. If *T* represents a monthly average temperature, we can model this situation by using the *absolute value inequality*

$$|T - 50| \le 22.$$

The absolute value is necessary because a monthly average temperature *T* can be either greater than or less than 50°F by as much as 22°F. In this section we discuss absolute value equations and inequalities. (*Source:* A. Miller and J. Thompson, *Elements of Meteorology.*)

NEW VOCABULARY

☐ Absolute value equation
☐ Absolute value inequality

Absolute Value Equations

An equation that contains an absolute value is an **absolute value equation**. Examples are

$$|x| = 2, \quad |2x - 1| = 5, \quad \text{and} \quad |5 - 3x| - 3 = 1.$$

Consider the absolute value equation $|x| = 2$. This equation has *two* solutions, −2 and 2, because $|-2| = 2$ and $|2| = 2$. We can also demonstrate this result with a table of values or a graph. In Table 8.17 $|x| = 2$ when $x = -2$ or $x = 2$. In Figure 8.60 the graph of $y_1 = |x|$ intersects the graph of $y_2 = 2$ at the points $(-2, 2)$ and $(2, 2)$. The *x*-values at these points of intersection correspond to the solutions −2 and 2.

TABLE 8.17

x	$|x|$
−2	2
−1	1
0	0
1	1
2	2

Solutions

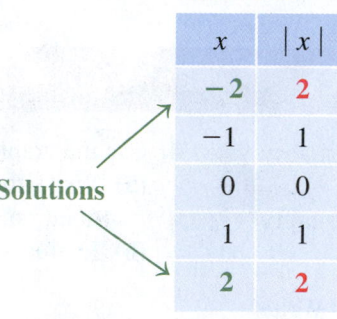

Solving |x| = 2 Graphically

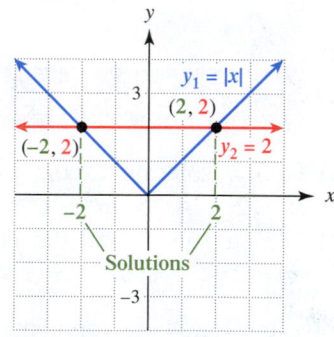

Figure 8.60

We generalize this discussion in the following manner.

SOLVING $|x| = k$

1. If $k > 0$, then $|x| = k$ is equivalent to $x = k$ or $x = -k$.

2. If $k = 0$, then $|x| = k$ is equivalent to $x = 0$.

3. If $k < 0$, then $|x| = k$ has no solutions.

EXAMPLE 1   **Solving absolute value equations**

Solve each equation.
(a) $|x| = 20$ **(b)** $|x| = -5$

Solution
(a) The solutions are -20 and 20.   **(b)** There are no solutions because $|x|$ is never negative.

▪ **Now Try Exercises 23, 25**

EXAMPLE 2   **Solving an absolute value equation**

Solve $|2x - 5| = 3$ symbolically.

Solution
If $|2x - 5| = 3$, then either $2x - 5 = 3$ or $2x - 5 = -3$. Solve each equation separately.

$$2x - 5 = 3 \quad \text{or} \quad 2x - 5 = -3 \quad \text{Equations to be solved}$$
$$2x = 8 \quad \text{or} \quad 2x = 2 \quad \text{Add 5.}$$
$$x = 4 \quad \text{or} \quad x = 1 \quad \text{Divide by 2.}$$

The solutions are **1** and **4**.

▪ **Now Try Exercise 31**

A table of values can be used to solve the equation $|2x - 5| = 3$ from Example 2. Table 8.18 shows that $|2x - 5| = 3$ when $x = 1$ or $x = 4$.

Numerical Solution

TABLE 8.18 $|2x - 5| = 3$

x	0	1	2	3	4	5	6
$|2x - 5|$	5	3	1	1	3	5	7

Solutions are **1** and **4**.

Graphical Solution

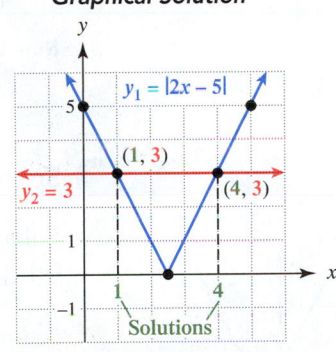

Figure 8.61

The equation from Example 2 can also be solved by graphing $y_1 = |2x - 5|$ and $y_2 = 3$. To graph y_1, first plot some of the points from Table 8.18. Its graph is V-shaped, as shown in Figure 8.61. Note that the x-coordinate of the "point" or vertex of the V can be found by solving the equation $2x - 5 = 0$ to obtain $\frac{5}{2}$. The graph of y_1 intersects the graph of y_2 at the points $(1, 3)$ and $(4, 3)$, giving the solutions **1** and **4**, so the graphical solutions agree with the numerical and symbolic solutions.

This discussion leads to the following result.

ABSOLUTE VALUE EQUATIONS

If $k > 0$, then
$$|ax + b| = k$$
is equivalent to
$$ax + b = k \quad \text{or} \quad ax + b = -k.$$

EXAMPLE 3 **Solving absolute value equations**

Solve.

(a) $|5 - x| - 2 = 8$ (b) $\left|\frac{1}{2}(x - 6)\right| = \frac{3}{4}$

Solution

(a) Start by adding 2 to each side to obtain $|5 - x| = 10$. This new equation is satisfied by the solution from either of the following equations.

$$5 - x = 10 \quad \text{or} \quad 5 - x = -10 \qquad \text{Equations to be solved}$$
$$-x = 5 \quad \text{or} \quad -x = -15 \qquad \text{Subtract 5.}$$
$$x = -5 \quad \text{or} \quad x = 15 \qquad \text{Multiply by } -1.$$

The solutions are -5 and 15.

(b) This equation is satisfied by the solution from either of the following equations.

$$\frac{1}{2}(x - 6) = \frac{3}{4} \quad \text{or} \quad \frac{1}{2}(x - 6) = -\frac{3}{4} \qquad \text{Equations to be solved}$$
$$4 \cdot \frac{1}{2}(x - 6) = 4 \cdot \frac{3}{4} \quad \text{or} \quad 4 \cdot \frac{1}{2}(x - 6) = 4\left(-\frac{3}{4}\right) \qquad \text{Multiply by 4 to clear fractions.}$$
$$2(x - 6) = 3 \quad \text{or} \quad 2(x - 6) = -3 \qquad \text{Simplify.}$$
$$2x - 12 = 3 \quad \text{or} \quad 2x - 12 = -3 \qquad \text{Distributive property}$$
$$2x = 15 \quad \text{or} \quad 2x = 9 \qquad \text{Add 12.}$$
$$x = \frac{15}{2} \quad \text{or} \quad x = \frac{9}{2} \qquad \text{Divide by 2.}$$

The solutions are $\frac{9}{2}$ and $\frac{15}{2}$.

Now Try Exercises 33, 37

EXAMPLE 4 **Solving absolute value equations with no solutions and one solution**

Solve.

(a) $|2x - 1| = -2$ (b) $|4 - 2x| = 0$

Solution

(a) Because an absolute value is never negative, there are no solutions. Figure 8.62 shows that the graph of $y_1 = |2x - 1|$ never intersects the graph of $y_2 = -2$.

(b) If $|y| = 0$, then $y = 0$. Thus $|4 - 2x| = 0$ when $4 - 2x = 0$ or when $x = 2$. The only solution is 2.

Now Try Exercises 35, 39

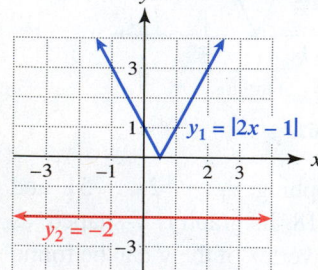

Figure 8.62

Sometimes an equation can have an absolute value on each side. An example would be $|2x| = |x - 3|$. In this situation either $2x = x - 3$ (the two expressions are equal) or $2x = -(x - 3)$ (the two expressions are opposites).

These concepts are summarized as follows.

SOLVING $|ax + b| = |cx + d|$

Let a, b, c, and d be constants. Then $|ax + b| = |cx + d|$ is equivalent to

$$ax + b = cx + d \quad \text{or} \quad ax + b = -(cx + d).$$

EXAMPLE 5   **Solving an absolute value equation**

Solve $|2x| = |x - 3|$.

Solution
Solve the following equations.

$$2x = x - 3 \quad \text{or} \quad 2x = -(x - 3)$$
$$x = -3 \quad \text{or} \quad 2x = -x + 3$$
$$3x = 3$$
$$x = 1$$

The solutions are -3 and 1.

Now Try Exercise 43

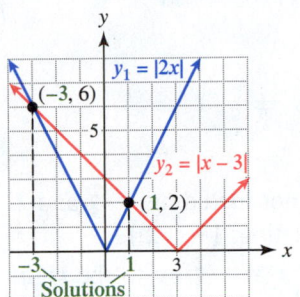

Figure 8.63

Example 5 is solved graphically in Figure 8.63. The graphs of $y_1 = |2x|$ and $y_2 = |x - 3|$ are V-shaped and intersect at $(-3, 6)$ and $(1, 2)$. The solutions are -3 and 1.

Absolute Value Inequalities

We can solve absolute value inequalities graphically. For example, to solve $|x| < 3$, let $y_1 = |x|$ and $y_2 = 3$ (see Figure 8.64). Their graphs intersect at $(-3, 3)$ and $(3, 3)$. The graph of y_1 is *below* the graph of y_2 for x-values *between*, but not including, $x = -3$ and $x = 3$. The solution set for $|x| < 3$ is $\{x \mid -3 < x < 3\}$ and is shaded on the x-axis.

Other absolute value inequalities can be solved graphically in a similar way. In Figure 8.65 the solutions to $|2x - 1| = 3$ are -1 and 2. The V-shaped graph of $y_1 = |2x - 1|$ is below the horizontal line $y_2 = 3$ when $-1 < x < 2$. Thus $|2x - 1| < 3$ whenever $-1 < x < 2$. The solution set is shaded on the x-axis.

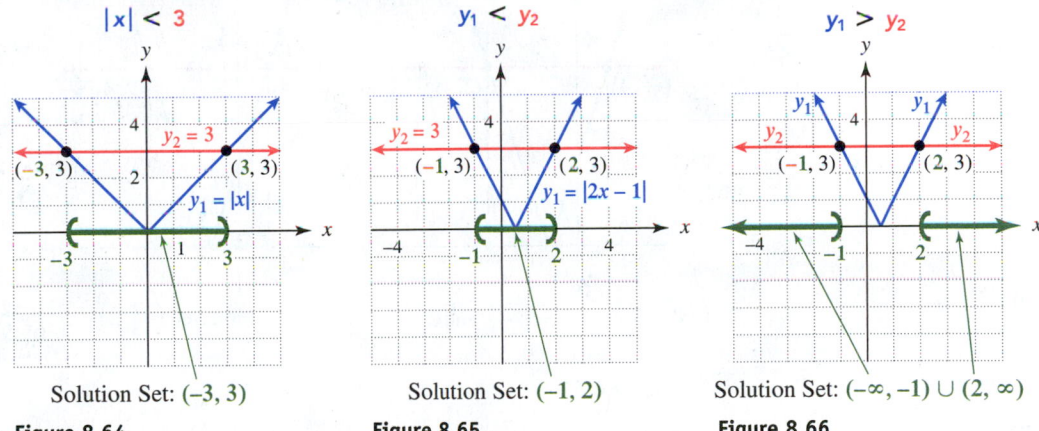

Solution Set: $(-3, 3)$

Figure 8.64

Solution Set: $(-1, 2)$

Figure 8.65

Solution Set: $(-\infty, -1) \cup (2, \infty)$

Figure 8.66

In Figure 8.66 the V-shaped graph of $y_1 = |2x - 1|$ is above the horizontal line $y_2 = 3$ both to the **left of $x = -1$** and to the **right of $x = 2$**. That is, $|2x - 1| > 3$ whenever $x < -1$ or $x > 2$. The solution set is shaded on the x-axis.

ABSOLUTE VALUE INEQUALITIES

Let the solutions to $|ax + b| = k$ be c and d, where $c < d$ and $k > 0$.

1. $|ax + b| < k$ is equivalent to $c < x < d$.
2. $|ax + b| > k$ is equivalent to $x < c$ or $x > d$.

Similar statements can be made for inequalities involving $\leq$ or $\geq$.

EXAMPLE 6 **Solving absolute value equations and inequalities**

Solve each absolute value equation and inequality.
(a) $|2 - 3x| = 4$ (b) $|2 - 3x| < 4$ (c) $|2 - 3x| > 4$

Solution
(a) The given equation is equivalent to the following equations.

$$2 - 3x = 4 \quad \text{or} \quad 2 - 3x = -4 \qquad \text{Equations to be solved}$$

$$-3x = 2 \quad \text{or} \quad -3x = -6 \qquad \text{Subtract 2.}$$

$$x = -\frac{2}{3} \quad \text{or} \quad x = 2 \qquad \text{Divide by } -3.$$

The solutions are $-\frac{2}{3}$ and **2**.

(b) Solutions to $|2 - 3x| < 4$ include x-values **between**, but not including, $-\frac{2}{3}$ and **2**. Thus the solution set is $\left\{ x \mid -\frac{2}{3} < x < 2 \right\}$, or in interval notation, $\left(-\frac{2}{3}, 2 \right)$.

(c) Solutions to $|2 - 3x| > 4$ include x-values to the **left** of $x = -\frac{2}{3}$ **or** to the **right** of $x = 2$. Thus the solution set is $\left\{ x \mid x < -\frac{2}{3} \text{ or } x > 2 \right\}$, or in interval notation, $\left(-\infty, -\frac{2}{3} \right) \cup (2, \infty)$.

Now Try Exercise 51

STUDY TIP

Be sure you understand how to write the solution to Example 6(c).

VISUALIZING SOLUTIONS Figure 8.67(a) can be used to visualize the solution to $|2 - 3x| = 4$ in Example 6(a). Similarly, Figures 8.67(b) and 8.67(c) can be used to visualize the solutions to $|2 - 3x| < 4$ and $|2 - 3x| > 4$ in parts (b) and (c) of Example 6.

Visualizing Solutions to Equations and Inequalities

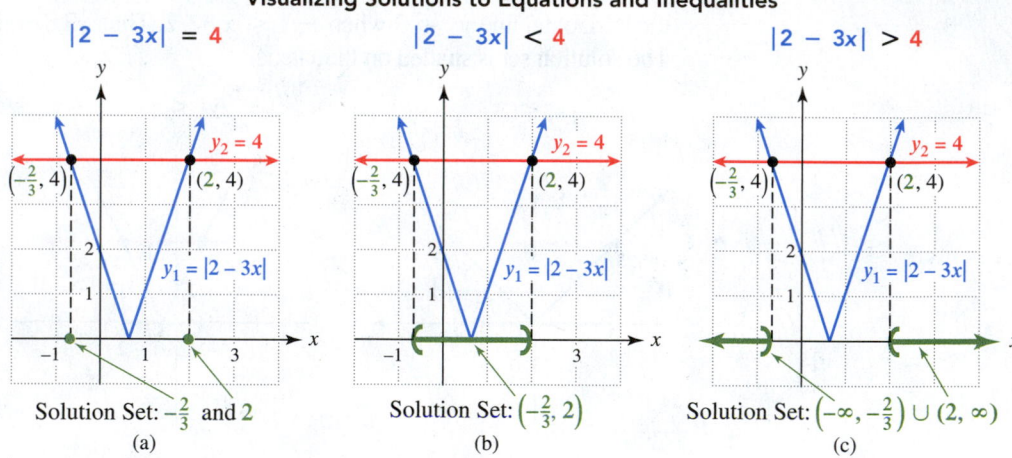

Figure 8.67

EXAMPLE 7 **Solving an absolute value inequality**

Solve $\left| \frac{2x - 5}{3} \right| > 3$. Write the solution set in interval notation.

Solution
Start by solving $\left| \frac{2x - 5}{3} \right| = 3$ as follows.

$$\frac{2x - 5}{3} = 3 \quad \text{or} \quad \frac{2x - 5}{3} = -3 \qquad \text{Equations to be solved}$$

$$2x - 5 = 9 \quad \text{or} \quad 2x - 5 = -9 \qquad \text{Multiply by 3.}$$

$$2x = 14 \quad \text{or} \quad 2x = -4 \qquad \text{Add 5.}$$

$$x = 7 \quad \text{or} \quad x = -2 \qquad \text{Divide by 2.}$$

Because the inequality symbol is $>$, the solution set is $x < -2$ or $x > 7$, or in interval notation, $(-\infty, -2) \cup (7, \infty)$.

Now Try Exercise 79

The results from Examples 6 and 7 can be generalized as follows.

CRITICAL THINKING

How many solutions are there to $|2x + 1| < 4$? How many solutions are there to $|2x + 1| < -4$?

INEQUALITIES AND ABSOLUTE VALUES

If $k > 0$ and $y = f(x)$, then

$$|y| < k \text{ is equivalent to } -k < y < k \text{ and}$$
$$|y| > k \text{ is equivalent to } y < -k \text{ or } y > k.$$

Similar statements can be made for inequalities involving $\leq$ and $\geq$.

In the next example, we use the fact that $-k \leq y \leq k$ is equivalent to $|y| \leq k$.

EXAMPLE 8 | **Analyzing error**

An engineer is designing a circular cover for a container. The diameter d of the cover is to be 4.25 inches and must be accurate to within 0.01 inch. Write an absolute value inequality that gives acceptable values for d.

Solution
The diameter d must satisfy $4.24 \leq d \leq 4.26$. Subtracting 4.25 from each part gives

$$-0.01 \leq d - 4.25 \leq 0.01,$$

CALCULATOR HELP

To graph an absolute value, see Appendix A (page AP-8).

which is equivalent to $|d - 4.25| \leq 0.01$. The "distance" or difference between 4.25 and the diameter is less than or equal to 0.01.

Now Try Exercise 121

EXAMPLE 9 | **Modeling temperature in Boston**

In A Look Into Math at the beginning of this section we discussed how the inequality $|T - 50| \leq 22$ models the range for the monthly average temperatures T in Boston.
(a) Solve this inequality and interpret the result.
(b) Give graphical support for part (a).

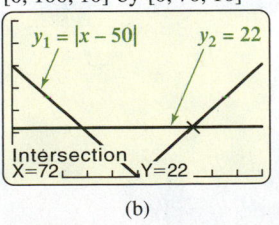

[0, 100, 10] by [0, 70, 10]

(a)

[0, 100, 10] by [0, 70, 10]

(b)

Solution
(a) *Symbolic Solution* Start by solving $|T - 50| = 22$.

$$T - 50 = 22 \quad \text{or} \quad T - 50 = -22 \qquad \text{Equations to be solved}$$
$$T = 72 \quad \text{or} \qquad\qquad T = 28 \qquad \text{Add 50 to each side.}$$

Thus the solution set to $|T - 50| \leq 22$ is $\{T \mid 28 \leq T \leq 72\}$. Monthly average temperatures in Boston vary from $28°F$ to $72°F$.

(b) *Graphical Solution* The graphs of $y_1 = |x - 50|$ and $y_2 = 22$ intersect at $(28, 22)$ and $(72, 22)$, as shown in Figures 8.68(a) and (b). The V-shaped graph of y_1 intersects the horizontal graph of y_2, or is below it, when $28 \leq x \leq 72$. Thus the solution set is $\{T \mid 28 \leq T \leq 72\}$. This result agrees with the symbolic result.

Figure 8.68

Now Try Exercise 113

Sometimes the solution set to an absolute value inequality can be either empty or the set of all real numbers. These two situations are illustrated in the next example.

EXAMPLE 10 **Solving absolute value inequalities**

Solve if possible.
(a) $|2x - 5| > -1$ (b) $|5x - 1| + 3 \le 2$

Solution
(a) Because the absolute value of an expression cannot be negative, $|2x - 5|$ is greater than -1 for every x-value. The solution set is all real numbers, or $(-\infty, \infty)$.
(b) Subtracting 3 from each side results in $|5x - 1| \le -1$. Because the absolute value is always greater than or equal to 0, no x-values satisfy this inequality. There are no solutions.

Now Try Exercises 75, 77

8.5 Putting It All Together

PROBLEM	SYMBOLIC SOLUTION	GRAPHICAL SOLUTION
$|ax + b| = k, k > 0$	Solve the equations $$ax + b = k$$ and $$ax + b = -k.$$	Graph $y_1 = |ax + b|$ and $y_2 = k$. Find the x-values of the two points of intersection.
$|ax + b| < k, k > 0$	If the solutions to $$|ax + b| = k$$ are c and d, $c < d$, then the solutions to $$|ax + b| < k$$ satisfy $$c < x < d.$$	Graph $y_1 = |ax + b|$ and $y_2 = k$. Find the x-values of the two points of intersection. The solutions are between these x-values on the number line, where the graph of y_1 lies below the graph of y_2.
$|ax + b| > k, k > 0$	If the solutions to $$|ax + b| = k$$ are c and d, $c < d$, then the solutions to $$|ax + b| > k$$ satisfy $$x < c \quad \text{or} \quad x > d.$$	Graph $y_1 = |ax + b|$ and $y_2 = k$. Find the x-values of the two points of intersection. The solutions are "outside" (left or right of) these x-values on the number line, where the graph of y_1 is above the graph of y_2.

8.5   Exercises

MyMathLab   Math XL PRACTICE   WATCH   DOWNLOAD   READ   REVIEW

CONCEPTS AND VOCABULARY

1. Give an example of an absolute value equation.

2. Give an example of an absolute value inequality.

3. Is -3 a solution to $|x| = 3$?

4. Is -4 a solution to $|x| > 3$?

5. Is the solution set to $|x| = 5$ equal to $\{-5, 5\}$?

6. Is $|x| < 3$ equivalent to $x < -3$ or $x > 3$? Explain.

7. How many times does the graph of $y = |2x - 1|$ intersect the graph of $y = 5$?

8. How many times does the graph of $y = |2x - 1|$ intersect the graph of $y = -5$?

Exercises 9–14: Determine whether the given values of x are solutions to the absolute value equation or inequality.

9. $|2x - 5| = 1$ $x = -3, x = 3$

10. $|5 - 6x| = 1$ $x = 1, x = 0$

11. $|7 - 4x| \le 5$ $x = -2, x = 2$

12. $|2 + x| < 2$ $x = -4, x = -1$

13. $|7x + 4| > -1$ $x = -\frac{4}{7}, x = 2$

14. $|12x + 3| \ge 3$ $x = -\frac{1}{4}, x = 2$

Exercises 15 and 16: Use the graph to solve the equation.

15. $|x - 2| = 2$ 16. $|2x + 1| = 3$

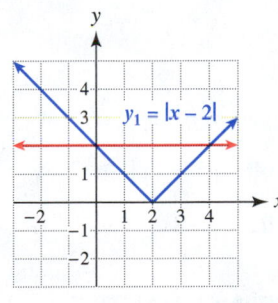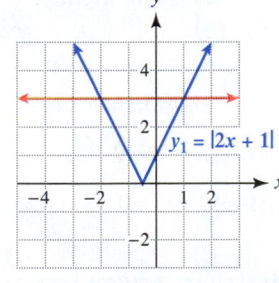

Exercises 17–22: Solve each equation or inequality.

17. $|x| = 3$ 18. $|x| = 5$

19. $|x| < 3$ 20. $|x| < 5$

21. $|x| > 3$ 22. $|x| > 5$

SYMBOLIC SOLUTIONS

Exercises 23–48: Solve the absolute value equation.

23. $|x| = 7$ 24. $|x| = 4$

25. $|x| = -6$ 26. $|x| = 0$

27. $|4x| = 9$ 28. $|-3x| = 7$

29. $|-2x| - 6 = 2$ 30. $|5x| + 1 = 5$

31. $|2x + 1| = 11$ 32. $|1 - 3x| = 4$

33. $|-2x + 3| + 3 = 4$

34. $|6x + 2| - 2 = 6$

35. $|3 - 4x| = 0$ 36. $|5x - 3| = -1$

37. $\left|\frac{1}{2}x - 1\right| = 5$ 38. $\left|6 - \frac{3}{4}x\right| = 3$

39. $|2x - 6| = -7$ 40. $\left|1 - \frac{2}{3}x\right| = 0$

41. $\left|\frac{2}{3}z - 1\right| - 3 = 8$ 42. $|1 - 2z| + 5 = 10$

43. $|z - 1| = |2z|$ 44. $|2z + 3| = |2 - z|$

45. $|3t + 1| = |2t - 4|$

46. $\left|\frac{1}{2}t - 1\right| = \left|3 - \frac{3}{2}t\right|$ 47. $\left|\frac{1}{4}x\right| = \left|3 + \frac{1}{4}x\right|$

48. $|2x - 1| = |2x + 2|$

Exercises 49–52: Solve each equation or inequality.

49. (a) $|2x| = 8$ (b) $|2x| < 8$
    (c) $|2x| > 8$

50. (a) $|3x - 9| = 6$
    (b) $|3x - 9| \le 6$
    (c) $|3x - 9| \ge 6$

51. (a) $|5 - 4x| = 3$
    (b) $|5 - 4x| \le 3$
    (c) $|5 - 4x| \ge 3$

52. (a) $\left|\frac{x - 5}{2}\right| = 2$ (b) $\left|\frac{x - 5}{2}\right| < 2$

    (c) $\left|\frac{x - 5}{2}\right| > 2$

Exercises 53–84: Solve the absolute value inequality. Write your answer in interval notation.

53. $|x| \le 3$ 54. $|x| < 2$

55. $|k| > 4$ 56. $|k| \ge 5$

57. $|t| \le -3$ **58.** $|t| < -1$

59. $|z| > 0$ **60.** $|2z| \ge 0$

61. $|2x| > 7$ **62.** $|-12x| < 30$

63. $|-4x + 4| < 16$ **64.** $|-5x - 8| > 2$

65. $2|x + 5| \ge 8$ **66.** $-3|x - 1| \ge -9$

67. $|8 - 6x| - 1 \le 2$ **68.** $4 - \left|\dfrac{2x}{3}\right| < -7$

69. $5 + \left|\dfrac{2 - x}{3}\right| \le 9$ **70.** $\left|\dfrac{x + 3}{5}\right| \le 12$

71. $|2x - 1| \le -3$ **72.** $|x + 6| \ge -5$

73. $|x + 1| - 1 > -3$ **74.** $-2|1 - 7x| \ge 2$

75. $|2z - 4| + 2 \le 1$ **76.** $|4 - z| \le 0$

77. $|3z - 1| > -3$ **78.** $|2z| \ge -2$

79. $\left|\dfrac{2 - t}{3}\right| \ge 5$ **80.** $\left|\dfrac{2t + 3}{5}\right| \ge 7$

81. $|t - 1| \le 0.1$

82. $|t - 2| \le 0.01$

83. $|b - 10| > 0.5$

84. $|b - 25| \ge 1$

NUMERICAL AND GRAPHICAL SOLUTIONS

Exercises 85 and 86: Use the table of $y = |ax + b|$ to solve each equation or inequality. Write your answers in interval notation for parts (b) and (c).

85. **(a)** $y = 2$ **(b)** $y < 2$ **(c)** $y > 2$

x	-2	-1	0	1	2	3	4
y	3	2	1	0	1	2	3

86. **(a)** $y = 6$ **(b)** $y \le 6$ **(c)** $y \ge 6$

x	-12	-6	0	6	12	18	24
y	9	6	3	0	3	6	9

Exercises 87 and 88: Use the graph of y_1 at the top of the next column to solve each equation or inequality. Write your answers in interval notation for parts (b) and (c).

87. **(a)** $|2x + 1| = 1$

     (b) $|2x + 1| \le 1$

     (c) $|2x + 1| \ge 1$

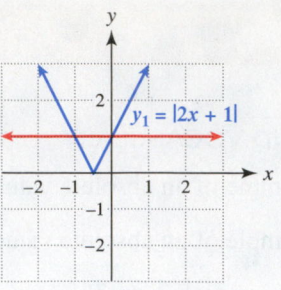

88. **(a)** $|x - 1| = 3$ **(b)** $|x - 1| < 3$

     (c) $|x - 1| > 3$

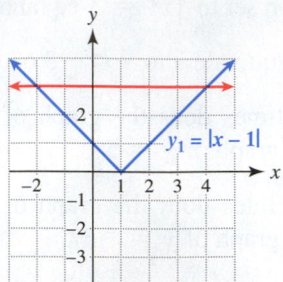

Exercises 89–98: Solve the inequality graphically. Write your answer in interval notation.

89. $|x| \ge 1$ **90.** $|x| < 2$

91. $|x - 1| \le 3$ **92.** $|x + 5| \ge 2$

93. $|4 - 2x| > 2$ **94.** $|1.5x - 3| \ge 6$

95. $|10 - 3x| < 4$ **96.** $|7 - 4x| \le 2.5$

97. $|8.1 - x| > -2$ **98.** $\left|\dfrac{5x - 9}{2}\right| \le -1$

USING MORE THAN ONE METHOD

Exercises 99–102: Solve the absolute value inequality

     (a) symbolically,
     (b) graphically, and
     (c) numerically.

Write your answer in set-builder notation.

99. $|3x| \le 9$ **100.** $|5 - x| \ge 3$

101. $|2x - 5| > 1$ **102.** $|-8 - 4x| < 6$

WRITING ABSOLUTE VALUE INEQUALITIES

Exercises 103–110: Write each compound inequality as an absolute value inequality. Do not simplify Exercises 107–110.

103. $-4 \le x \le 4$ **104.** $-0.1 < y < 0.1$

105. $y < -2$ or $y > 2$ **106.** $-0.1 \le x \le 0.1$

107. $-0.3 \le 2x + 1 \le 0.3$

108. $4x < -5$ or $4x > 5$ **109.** $\pi x \le -7$ or $\pi x \ge 7$

110. $-0.9 \le x - \sqrt{2} \le 0.9$

111. Thinking Generally If $a \ne 0$ and $k > 0$, then the graph of $y = |ax + b|$ intersects the graph of $y = k$ at _____ points.

112. Thinking Generally If a and k are positive, then the solution set to $|ax + b| < k$ is _____.

APPLICATIONS

Exercises 113–116: Average Temperatures (Refer to Example 9.) The given inequality models the range for the monthly average temperatures T in degrees Fahrenheit at the location specified.

(a) Solve the inequality.
(b) Give a possible interpretation of the inequality.

113. $|T - 43| \le 24$, Marquette, Michigan

114. $|T - 62| \le 19$, Memphis, Tennessee

115. $|T - 10| \le 36$, Chesterfield, Canada

116. $|T - 61.5| \le 12.5$, Buenos Aires, Argentina

117. *Highest Elevations* The table lists the highest elevation in each continent.

Continent	Elevation (feet)
Asia	29,028
S. America	22,834
N. America	20,320
Africa	19,340
Europe	18,510
Antarctica	16,066
Australia	7,310

Source: National Geographic.

(a) Calculate the average A of these elevations.
(b) Which continents have their highest elevations within 1000 feet of A?
(c) Which continents have their highest elevations within 5000 feet of A?
(d) Let E be an elevation. Write an absolute value inequality that says that E is within 5000 feet of A.

118. *Distance* Suppose that two cars, both traveling at a constant speed of 60 miles per hour, approach each other on a straight highway.

(a) If the two cars are initially 4 miles apart, sketch a graph of the distance between the two cars after x minutes, where $0 \le x \le 4$. (*Hint:* 60 miles per hour = 1 mile per minute.)
(b) Write an absolute value equation whose solution gives the times when the cars are 2 miles apart.
(c) Solve your equation from part (b).

119. *Error in Measurements* (Refer to Example 8.) The maximum error in the diameter of the can is restricted to 0.002 inch, so an acceptable diameter d must satisfy the absolute value inequality

$$|d - 2.5| \le 0.002.$$

Solve this inequality for d and interpret the result.

120. *Error in Measurements* Suppose that a person can operate a stopwatch accurately to within 0.02 second. If Byron Dyce's time in the 800-meter race is recorded as 105.30 seconds, write an absolute value inequality that gives the possible values for the actual time t.

121. *Error in Measurements* A circular lid is being designed for a container. The diameter d of the lid is to be 3.8 inches and must be accurate to within 0.03 inch. Write an absolute value inequality that gives acceptable values for d.

122. *Manufacturing a Tire* An engineer is designing a tire for a truck. The diameter d of the tire is to be 36 inches and the *circumference* must be accurate to within 0.1 inch. Write an absolute value inequality that gives acceptable values for d.

123. *Relative Error* If a quantity is measured to be x and the true value is t, then the relative error in the measurement is $\left|\frac{x - t}{t}\right|$. If the true measurement is $t = 20$ and you want the relative error to be less than 0.05 (5%), what values for x are possible?

124. *Relative Error* (Refer to the preceding exercise.) The volume V of a box is 50 cubic inches. How accurately must you measure the volume of the box for the relative error to be less than 3%?

WRITING ABOUT MATHEMATICS

125. If $a \neq 0$, how many solutions are there to the equation $|ax + b| = k$ when
     (a) $k > 0$, **(b)** $k = 0$, and **(c)** $k < 0$?
     Explain each answer.

126. Suppose that you know two solutions to the equation $|ax + b| = k$. How can you use these solutions to solve the inequalities $|ax + b| < k$ and $|ax + b| > k$? Give an example.

SECTION 8.5   **Checking Basic Concepts**

Write answers in interval notation whenever possible.

1. Solve $\left|\frac{3}{4}x - 1\right| - 3 = 5$.

2. Solve the absolute value equation and inequalities.
   (a) $|3x - 6| = 8$

   (b) $|3x - 6| < 8$

   (c) $|3x - 6| > 8$

3. Solve the inequality $|-2(3 - x)| < 6$, and then solve $|-2(3 - x)| \geq 6$.

4. Use the graph to solve the equation and inequalities.

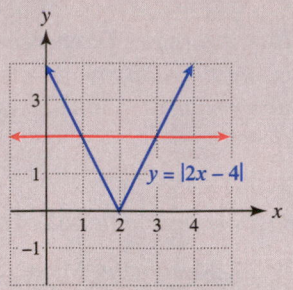

   (a) $|2x - 4| = 2$ **(b)** $|2x - 4| \leq 2$

   (c) $|2x - 4| \geq 2$

CHAPTER 8   Summary

SECTION 8.1 ■ FUNCTIONS AND THEIR REPRESENTATIONS

Function A function is a set of ordered pairs (x, y) where each x-value corresponds to exactly one y-value. A function takes a valid input x and computes exactly one output y, forming the ordered pair (x, y).

Domain and Range of a Function The domain D is the set of all valid inputs, or x-values, and the range R is the set of all outputs, or y-values.

Examples: $f = \{(1, 2), (2, 3), (3, 3)\}$ has $D = \{1, 2, 3\}$ and $R = \{2, 3\}$.

$f(x) = x^2$ has domain all real numbers and range $y \geq 0$. (See the graph below.)

Function Notation $y = f(x)$ and is read "y equals f of x."

Example: $f(x) = \frac{2x}{x - 1}$ implies that $f(3) = \frac{2 \cdot 3}{3 - 1} = \frac{6}{2} = 3$. Thus the point $(3, 3)$ is on the graph of f.

Function Representations A function can be represented symbolically, numerically, graphically, or verbally.

Symbolic Representation (Formula) $f(x) = x^2$

Numerical Representation (Table)

x	y
-2	4
-1	1
0	0
1	1
2	4

Graphical Representation (Graph)

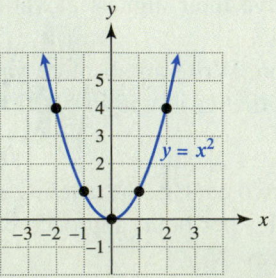

Verbal Representation (Words) f computes the square of the input x.

Vertical Line Test If every vertical line intersects a graph at most once, then the graph represents a function.

SECTION 8.2 ■ LINEAR FUNCTIONS

Linear Function A linear function can be represented by $f(x) = mx + b$. Its graph is a (straight) line, where m is the slope and b is the y-intercept. For each unit increase in x, $f(x)$ changes by an amount equal to m.

Example: $f(x) = 2x - 1$ represents a linear function with $m = 2$ and $b = -1$.

<table>
<tr><td align="center">*Numerical Representation*</td><td align="center">*Graphical Representation*</td></tr>
<tr><td>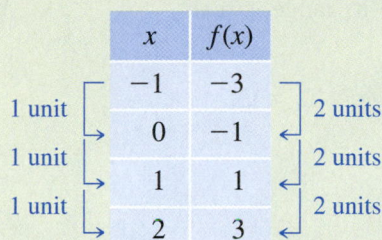</td><td>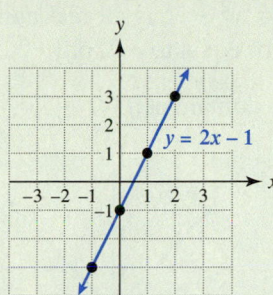</td></tr>
</table>

Each 1-unit increase in x results in a 2-unit increase in $f(x)$; thus $m = 2$.

NOTE: A numerical representation is a table of values of $f(x)$.

Modeling Data with Linear Functions When data have a constant rate of change, they can be modeled by $f(x) = mx + b$. The constant m represents the *rate of change*, and the constant b represents the *initial amount* or the value when $x = 0$. That is,

$$f(x) = (\textbf{Rate of change})x + (\textbf{Initial amount}).$$

Example: In the following table, the y-values decrease by 3 units for each 1-unit increase in x. When $x = 0, y = 4$. Thus the data are modeled by $f(x) = -3x + 4$.

x	-2	-1	0	1	2
y	10	7	4	1	-2

Midpoint Formula The midpoint of the line segment connecting (x_1, y_1) and (x_2, y_2) is

$$\left(\frac{x_1 + x_2}{2}, \frac{y_1 + y_2}{2} \right).$$

Example: The midpoint of the line segment connecting $(-5, 8)$ and $(9, 4)$ is

$$\left(\frac{-5 + 9}{2}, \frac{8 + 4}{2} \right) = (2, 6).$$

SECTION 8.3 ■ COMPOUND INEQUALITIES

Compound Inequality Two inequalities connected by *and* or *or*.

Examples: For $x + 1 < 3$ *or* $x + 1 > 6$, a solution satisfies *at least* one of the inequalities.

For $2x + 1 < 3$ *and* $1 - x > 6$, a solution satisfies *both* inequalities.

Three-Part Inequality A compound inequality in the form $x > a$ *and* $x < b$ can be written as the three-part inequality $a < x < b$.

Example: $1 \le x < 7$ means $x \ge 1$ *and* $x < 7$.

Interval Notation   Can be used to identify intervals on the real number line

Examples: $-2 < x \leq 3$ is equivalent to $(-2, 3]$.

$x < 5$ is equivalent to $(-\infty, 5)$.

All real numbers are denoted $(-\infty, \infty)$.

SECTION 8.4 ■ OTHER FUNCTIONS AND THEIR PROPERTIES

Domain and Range in Interval Notation   The domain and range of a function can often be expressed in interval notation.

Example: The domain of $f(x) = x^2 - 2$ is all real numbers, or $(-\infty, \infty)$, and its range is real numbers greater than or equal to -2, or $[-2, \infty)$.

Absolute Value Function   The domain of $f(x) = |x|$ is $(-\infty, \infty)$, and the range is $[0, \infty)$.

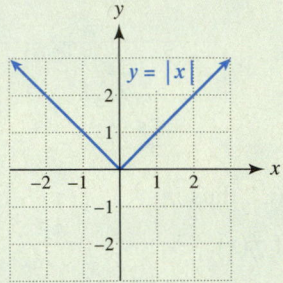

Polynomial Functions   The degree of a polynomial function (of one variable) equals the largest exponent of a variable. The graphs of polynomial functions with degree greater than 1 are not lines. The domain of a polynomial function is $(-\infty, \infty)$.

Examples: $f(x) = 4x - 1$ defines a linear function with degree 1.

$g(x) = 4x^2 + x - 4$ defines a quadratic function with degree 2.

$h(x) = x^3 + 0.7x - 1$ defines a cubic function with degree 3.

Rational Functions   If $f(x) = \frac{p(x)}{q(x)}$, where $p(x)$ and $q(x)$ are polynomials, f is a rational function. The domain of a rational function includes all real numbers, except x-values that make the denominator 0.

Examples: $f(x) = \frac{1}{x}$ has domain $(-\infty, 0) \cup (0, \infty)$, or $x \neq 0$.

$g(x) = \frac{x}{x^2 - 9}$ has domain $(-\infty, -3) \cup (-3, 3) \cup (3, \infty)$, or $x \neq -3, x \neq 3$.

Operations on Functions   If $f(x)$ and $g(x)$ are both defined, then the sum, difference, product, and quotient of two functions f and g are defined by

$$(f + g)(x) = f(x) + g(x) \qquad \text{Sum}$$

$$(f - g)(x) = f(x) - g(x) \qquad \text{Difference}$$

$$(fg)(x) = f(x) \cdot g(x) \qquad \text{Product}$$

$$\left(\frac{f}{g}\right)(x) = \frac{f(x)}{g(x)}, \text{ where } g(x) \neq 0. \quad \text{Quotient}$$

Example: Let $f(x) = x^2 - 1$ and $g(x) = x^2 + 1$.

$$(f + g)(x) = f(x) + g(x) = (x^2 - 1) + (x^2 + 1) = 2x^2$$

$$(f - g)(x) = f(x) - g(x) = (x^2 - 1) - (x^2 + 1) = -2$$

$$(fg)(x) = f(x) \cdot g(x) = (x^2 - 1)(x^2 + 1) = x^4 - 1$$

$$\left(\frac{f}{g}\right)(x) = \frac{f(x)}{g(x)} = \frac{x^2 - 1}{x^2 + 1}$$

SECTION 8.5 ■ ABSOLUTE VALUE EQUATIONS AND INEQUALITIES

Absolute Value Equations The graph of $y = |ax + b|, a \neq 0$, is V-shaped and intersects the horizontal line $y = k$ twice if $k > 0$. In this case there are two solutions to the equation $|ax + b| = k$ determined by $ax + b = k$ or $ax + b = -k$.

Example: The equation $|2x - 1| = 5$ has two solutions.

Symbolic Solution

$$2x - 1 = 5 \quad \text{or} \quad 2x - 1 = -5$$
$$2x = 6 \quad \text{or} \quad 2x = -4 \qquad \text{Add 1.}$$
$$x = 3 \quad \text{or} \quad x = -2 \qquad \text{Divide by 2.}$$

Graphical Solution

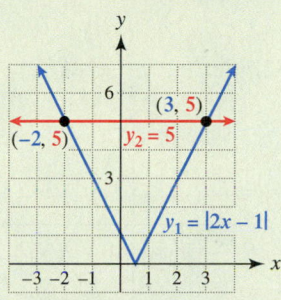

Numerical Solution

x	-3	-2	-1	0	1	2	3
$|2x - 1|$	7	5	3	1	1	3	5

The solutions are -2 and **3**.

The solutions are -2 and **3**.

Absolute Value Inequalities If the solutions to $|ax + b| = k$ are c and d with $c < d$, then the solution set for $|ax + b| < k$ is $\{x \mid c < x < d\}$, and the solution set for $|ax + b| > k$ is $\{x \mid x < c \text{ or } x > d\}$.

Example: The solutions to the equation $|2x - 1| = 5$ are -2 and 3, so the solution set for $|2x - 1| < 5$ is $\{x \mid -2 < x < 3\}$, and the solution set for $|2x - 1| > 5$ is $\{x \mid x < -2 \text{ or } x > 3\}$.

If $k > 0$ and $y = f(x)$, then

$$|y| < k \text{ is equivalent to } -k < y < k \text{ and}$$
$$|y| > k \text{ is equivalent to } y < -k \text{ or } y > k.$$

Examples: $|3 - x| < 5$ is equivalent to $-5 < 3 - x < 5$ and

$$|3 - x| > 5 \text{ is equivalent to } 3 - x < -5 \text{ or } 3 - x > 5.$$

CHAPTER 8 Review Exercises

SECTION 8.1

Exercises 1–4: Evaluate $f(x)$ for the given values of x.

1. $f(x) = 3x - 1$ $\qquad x = -2, \frac{1}{3}$

2. $f(x) = 5 - 3x^2$ $\qquad x = -3, 1$

3. $f(x) = \sqrt{x} - 2$ $\qquad x = 0, 9$

4. $f(x) = 5$ $\qquad x = -5, \frac{7}{5}$

Exercises 5 and 6: Do the following.

  (a) *Write a symbolic representation (formula) for the function described.*

  (b) *Evaluate the function for input 5 and interpret the result.*

5. Function P computes the number of pints in q quarts.

6. Function f computes 3 less than 4 times a number x.

7. If $f(3) = -2$, then the point _____ lies on the graph of f.

8. If $(4, -6)$ lies on the graph of f, then $f(\_\_\_) = \_\_\_$.

Exercises 9–12: Sketch a graph of f.

9. $f(x) = -2x$ $\qquad\qquad$ 10. $f(x) = \frac{1}{2}x - \frac{3}{2}$

11. $f(x) = x^2 - 1$ $\qquad\qquad$ 12. $f(x) = \sqrt{x + 1}$

Exercises 13 and 14: Use the graph of f to evaluate the given expressions.

13. $f(0)$ and $f(-3)$

14. $f(-2)$ and $f(1)$

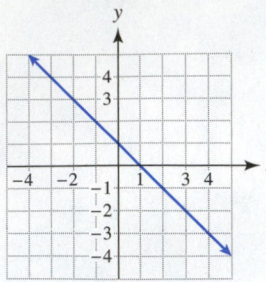
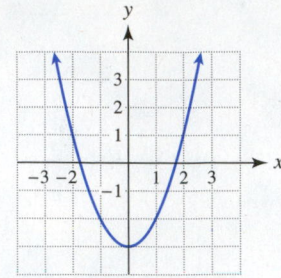

15. Use the table to evaluate $f(-1)$ and $f(3)$.

x	-1	1	3	5
$f(x)$	7	3	-1	-5

16. A function f is represented verbally by "Multiply the input x by 3 and then subtract 2." Give numerical, symbolic, and graphical representations for f. Let $x = -3, -2, -1, \ldots, 3$ in the table of values, and let $-3 \le x \le 3$ for the graph.

Exercises 17 and 18: Use the graph of f to estimate its domain and range.

17.

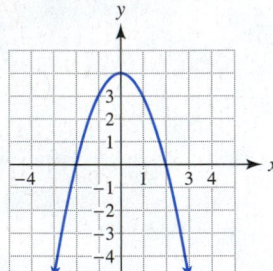

18.

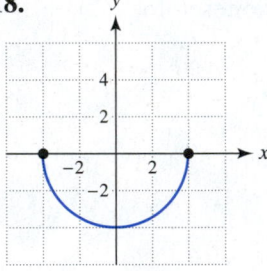

Exercises 19 and 20: Does the graph represent a function?

19.

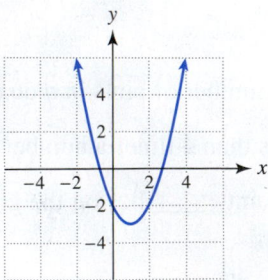

20.

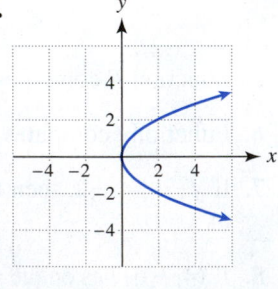

Exercises 21 and 22: Find the domain and range of S. Then state whether S defines a function.

21. $S = \{(-3, 4), (-1, 4), (2, 3), (4, -1)\}$

22. $S = \{(-1, 5), (0, 3), (1, -2), (-1, 2), (2, 4)\}$

Exercises 23–30: Find the domain.

23. $f(x) = -3x + 7$

24. $f(x) = \sqrt{x}$

25. $f(x) = \dfrac{3}{x}$

26. $f(x) = x^2 + 2$

27. $f(x) = \sqrt{5 - x}$

28. $f(x) = \dfrac{x}{x + 2}$

29. $f(x) = |2x + 1|$

30. $f(x) = x^3$

SECTION 8.2

Exercises 31 and 32: Does the graph represent a linear function?

31.

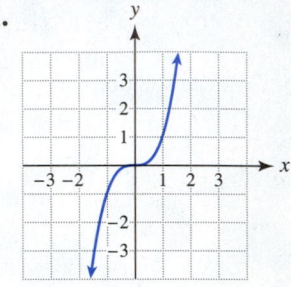

32.

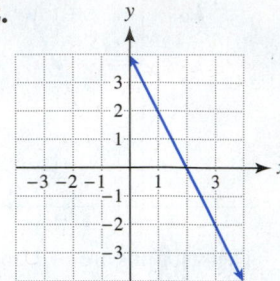

Exercises 33–36: Determine whether f is a linear function. If f is linear, give values for m and b so that f may be expressed as $f(x) = mx + b$.

33. $f(x) = -4x + 5$

34. $f(x) = 7 - x$

35. $f(x) = \sqrt{x}$

36. $f(x) = 6$

Exercises 37 and 38: Use the table to determine whether $f(x)$ could represent a linear function. If it could, write the formula for f in the form $f(x) = mx + b$.

37.

x	0	2	4	6
$f(x)$	-3	0	3	6

38.

x	-1	0	1	2
$f(x)$	-5	0	10	15

39. Evaluate $f(x) = \frac{1}{2}x + 3$ at $x = -4$.

40. Use the graph to evaluate $f(-2)$ and $f(1)$.

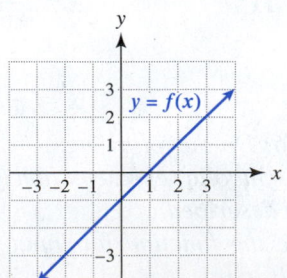

Exercises 41–44: Sketch a graph of $y = f(x)$.

41. $f(x) = x + 1$

42. $f(x) = 1 - 2x$

43. $f(x) = -\frac{1}{3}x$

44. $f(x) = -1$

45. Write a symbolic representation (formula) for a linear function H that calculates the number of hours in x days. Evaluate $H(2)$ and interpret the result.

46. Let $f(x) = \sqrt{x + 2} - x^2$.

(a) Make a numerical representation (table) for the function f with $x = 1, 2, 3, \ldots, 7$.
(b) Graph f in the standard window. What is the domain of f?

Exercises 47 and 48: Find the midpoint of the line segment connecting the given points.

47. $(-5, 3), (6, -9)$

48. $\left(\frac{2}{3}, -\frac{3}{4}\right), \left(\frac{1}{6}, \frac{3}{2}\right)$

SECTION 8.3

Use interval notation whenever possible for the remaining exercises.

Exercises 49–52: Solve the compound inequality. Graph the solution set on a number line.

49. $x + 1 \leq 3$ and $x + 1 \geq -1$

50. $2x + 7 < 5$ and $-2x \geq 6$

51. $5x - 1 \leq 3$ or $1 - x < -1$

52. $3x + 1 > -1$ or $3x + 1 < 10$

53. Use the table to solve $-2 \leq 2x + 2 \leq 4$.

x	-3	-2	-1	0	1	2	3
$2x + 2$	-4	-2	0	2	4	6	8

54. Use the following figure to solve each equation or inequality.
(a) $y_1 = y_2$ (b) $y_2 = y_3$
(c) $y_1 \leq y_2 \leq y_3$ (d) $y_2 < y_3$

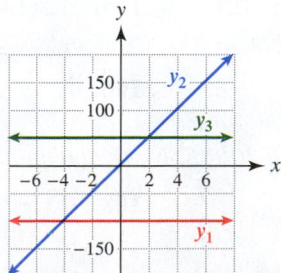

55. The graphs of y_1 and y_2 are shown at the top of the next column. Solve each equation or inequality.

(a) $y_1 = y_2$ (b) $y_1 < y_2$
(c) $y_1 > y_2$

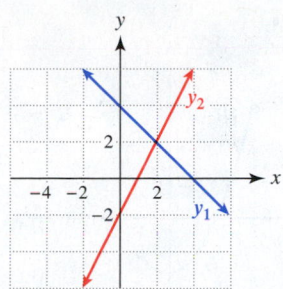

56. The graphs of three linear functions f, g, and h are shown in the following figure. Solve each equation or inequality.
(a) $f(x) = g(x)$ (b) $g(x) = h(x)$
(c) $f(x) < g(x) < h(x)$

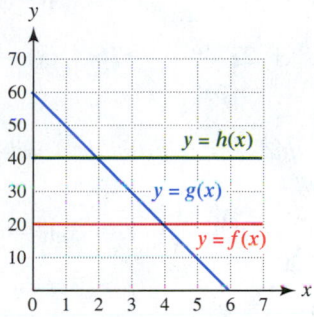

Exercises 57–62: Write the given inequality in interval notation.

57. $-3 \leq x \leq \frac{2}{3}$

58. $-6 < x \leq 45$

59. $x < \frac{7}{2}$

60. $x \geq 1.8$

61. $x > -3$ and $x < 4$

62. $x < 4$ or $x > 10$

Exercises 63–68: Solve the three-part inequality. Write the solution set in interval notation.

63. $-4 < x + 1 < 6$

64. $20 \leq 2x + 4 \leq 60$

65. $-3 < 4 - \frac{1}{3}x < 7$

66. $2 \leq \frac{1}{2}x - 2 \leq 12$

67. $-3 \leq \dfrac{4 - 5x}{3} - 2 < 3$

68. $30 \leq \dfrac{2x - 6}{5} - 4 < 50$

SECTION 8.4

Exercises 69 and 70: Write the domain and the range of the function.

69. $f(t) = \frac{1}{2}t^2$

70. $f(x) = |x + 2|$

71. Write the domain of $f(x) = \frac{x + 1}{2x - 8}$.

72. Write the domain and range of the function shown in the graph.

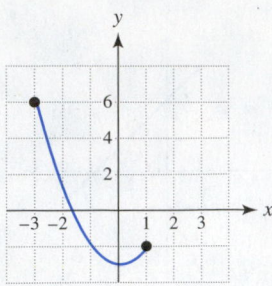

Exercises 73–76: Determine whether $f(x)$ represents a polynomial function. If possible, identify the degree and type of polynomial function.

73. $f(x) = 1 + 2x - 3x^2$

74. $f(x) = 5 + 7x$

75. $f(x) = x^3 + 2x$ **76.** $f(x) = |2x - 1|$

Exercises 77 and 78: If possible, evaluate $g(t)$ for the given values of t.

77. $g(t) = |1 - 4t|$ $t = 3, t = -\frac{1}{4}$

78. $g(t) = \dfrac{4}{4 - t^2}$ $t = 3, t = -2$

Exercises 79–82: Graph $y = f(x)$.

79. $f(x) = |x + 3|$ **80.** $f(x) = x^2 + 1$

81. $f(x) = \dfrac{1}{x}$ **82.** $f(x) = -3x$

83. Use $f(x) = 2x^2 - 3x$ and $g(x) = 2x - 3$ to find each of the following.
  (a) $(f + g)(3)$ **(b)** $(fg)(3)$

84. Use $f(x) = x^2 - 1$ and $g(x) = x - 1$ to find each of the following.
  (a) $(f - g)(x)$ **(b)** $(f/g)(x)$

SECTION 8.5

Exercises 85–88: Determine whether the given values of x are solutions to the absolute value equation or inequality.

85. $|12x - 24| = 24$ $x = -3; x = 2$

86. $|5 - 3x| > 3$ $x = \frac{4}{3}; x = 0$

87. $|3x - 6| \le 6$ $x = -3; x = 4$

88. $|2 + 3x| + 4 < 11$ $x = -3; x = \frac{2}{3}$

89. Use the table at the top of the next column to solve each equation or inequality.
  (a) $y_1 = 2$ **(b)** $y_1 < 2$ **(c)** $y_1 > 2$

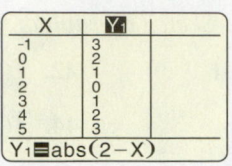

90. Use the graph of $y = |2x + 2|$ to solve each equation or inequality.
  (a) $|2x + 2| = 4$ **(b)** $|2x + 2| \le 4$
  (c) $|2x + 2| \ge 4$

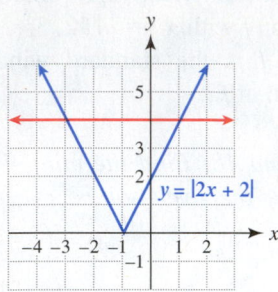

Exercises 91–96: Solve the absolute value equation.

91. $|x| = 22$ **92.** $|2x - 9| = 7$

93. $\left|4 - \frac{1}{2}x\right| = 17$ **94.** $\frac{1}{3}|3x - 1| + 1 = 9$

95. $|2x - 5| = |5 - 3x|$

96. $|-3 + 3x| = |-2x + 6|$

Exercises 97 and 98: Solve each absolute value equation or inequality.

97. (a) $|x + 1| = 7$ **(b)** $|x + 1| \le 7$
    (c) $|x + 1| \ge 7$

98. (a) $|1 - 2x| = 6$ **(b)** $|1 - 2x| \le 6$
    (c) $|1 - 2x| \ge 6$

Exercises 99–106: Solve the absolute value inequality.

99. $|x| > 3$ **100.** $|-5x| < 20$

101. $|4x - 2| \le 14$ **102.** $\left|1 - \frac{4}{5}x\right| \ge 3$

103. $|t - 4.5| \le 0.1$ **104.** $-2|13t - 5| \ge -4$

105. $|5 - 4x| > -5$ **106.** $|2t - 3| \le 0$

Exercises 107 and 108: Solve the inequality graphically.

107. $|2x| \ge 3$ **108.** $\left|\frac{1}{2}x - 1\right| \le 2$

Exercises 109 and 110: Write each compound inequality as an absolute value inequality.

109. $-0.05 \le x \le 0.05$

110. $5x - 1 < -4$ or $5x - 1 > 4$

APPLICATIONS

111. *Age at First Marriage* The median age at the first marriage for men from 1890 to 1960 can be modeled by $f(x) = -0.0492x + 119.1$, where x is the year. (*Source*: National Center of Health Statistics.)
  (a) Find the median age in 1910.
  (b) Graph f in [1885, 1965, 10] by [22, 26, 1]. What happened to the median age?
  (c) What is the slope of the graph of f? Interpret the slope as a rate of change.

112. *Marriages* From 2002 to 2008 the number of U.S. marriages in millions could be modeled by the formula $f(x) = 2.2$, where x is the year.
  (a) Estimate the number of marriages in 2006.
  (b) What information does f give about the number of marriages from 2002 to 2008?

113. *Fat Grams* A cup of milk contains 8 grams of fat.
  (a) Give a formula for $f(x)$ that calculates the number of fat grams in x cups of milk.
  (b) What is the slope of the graph of f?
  (c) Interpret the slope as a rate of change.

114. *Birth Rate* The U.S. birth rate per 1000 people for selected years is shown in the table.

Year	1950	1970	1990	2010
Birth Rate	24.1	18.4	16.7	13.5

Source: U.S. Census Bureau.

  (a) Make a scatterplot of the data.
  (b) Model the data with $f(x) = mx + b$, where x is the year. Answers may vary.
  (c) Use f to estimate the birth rate in 2000.

115. *Unhealthy Air Quality* The Environmental Protection Agency (EPA) monitors air quality in U.S. cities. The function f gives the annual number of days with unhealthy air quality in Los Angeles, California, for selected years.

x	1995	1999	2000	2003	2007
$f(x)$	113	56	87	88	100

Source: Environmental Protection Agency.

  (a) Find $f(1995)$ and interpret your result.
  (b) Identify the domain and range of f.
  (c) Discuss the trend of air pollution in Los Angeles.

116. *Temperature Scales* The table at the top of the next column shows equivalent temperatures in degrees Celsius and degrees Fahrenheit.

°C	−40	0	15	35	100
°F	−40	32	59	95	212

  (a) Plot the data. Let the x-axis correspond to the Celsius temperature and the y-axis correspond to the Fahrenheit temperature. What type of relation exists between the data?
  (b) Find $f(x) = mx + b$ so that f receives the Celsius temperature x as input and outputs the corresponding Fahrenheit temperature. Interpret the slope of the graph of f.
  (c) If the temperature is 20°C, what is the equivalent temperature in degrees Fahrenheit?

117. *Distance Between Bicyclists* The following graph shows the distance between two bicyclists traveling toward each other along a straight road after x hours.
  (a) After how long did the bicycle riders meet?
  (b) When were they 20 miles apart?
  (c) Find the times when they were less than 20 miles apart.
  (d) Estimate the sum of the speeds of the bicyclists.

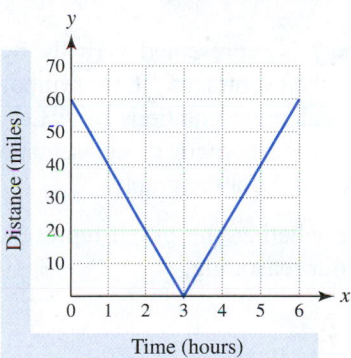

118. *Violent Crimes in the U.S.* The number of violent crimes reported has dropped from 10 million in 1992 to 5 million in 2008.
  (a) Find a linear function $f(x) = mx + b$ that models the data x years after 1992.
  (b) Use $f(x)$ to estimate the number of violent crimes in 2005.

119. *Average Precipitation* The average rainfall in Houston, Texas, is 3.9 inches per month. Each month's average A is within 1.7 inches of 3.9 inches. (*Source*: J. Williams, *The Weather Almanac 1995*.)
  (a) Write an absolute value inequality that models this situation.
  (b) Solve the inequality.

120. *Relative Error* If a quantity is measured to be T and the actual value is A, then the relative error in this measurement is $\left|\frac{T - A}{A}\right|$. If $A = 35$ and the relative error is to be less than 0.08 (8%), what values for T are possible?

CHAPTER 8   Test

 Step-by-step test solutions are found on the Chapter Test Prep Videos available in **MyMathLab** and on **You Tube** (search "RockswoldComboAlg" and click on "Channels").

1. Evaluate $f(4)$ if $f(x) = 3x^2 - \sqrt{x}$. Give a point on the graph of f.

2. Write a symbolic representation (formula) for a function C that calculates the cost of buying x pounds of candy at \$4 per pound. Evaluate $C(5)$ and interpret your result.

3. Sketch a graph of f.
   (a) $f(x) = -2x + 1$ (b) $f(x) = x^2 + 1$
   (c) $f(x) = \sqrt{x} + 3$ (d) $f(x) = |x + 1|$

4. Use the graph of f to evaluate $f(-3)$ and $f(0)$. Determine the domain and range of f.

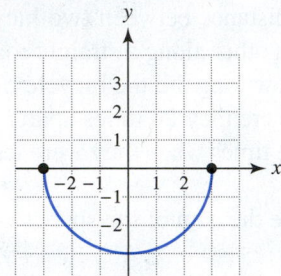

5. A function f is represented verbally by "Square the input x and then subtract 5." Give symbolic, numerical, and graphical representations of f. Let $x = -3, -2, -1, \ldots, 3$ in the numerical representation (table) and let $-3 \le x \le 3$ for the graph.

6. Determine whether the graph represents a function. Explain your reasoning.

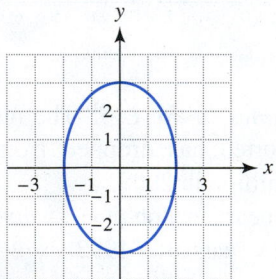

7. Find the domain of function f.
   (a) $f = \{(-2, 3), (-1, 5), (0, 3), (5, 7)\}$
   (b) $f(x) = \frac{3}{4}x - 5$
   (c) $f(x) = \sqrt{x + 4}$
   (d) $f(x) = 2x^2 - 1$
   (e) $f(x) = \frac{3x}{5 - x}$

8. Determine if $f(x) = 6 - 8x$ is a linear function. If it is, write it in the form $f(x) = mx + b$.

9. Graph the solution set to $2x + 6 < 2$ and $-3x \ge 3$ on a number line.

10. Use the table at the top of the next column to solve the compound inequality $-3x < -3$ or $-3x > 6$. Use interval notation.

x	-3	-2	-1	0	1	2	3
$-3x$	9	6	3	0	-3	-6	-9

11. Use the following figure to solve each equation or inequality. Write your answers for parts (c) and (d) in interval notation.
    (a) $y_1 = y_2$ (b) $y_2 = y_3$
    (c) $y_1 \le y_2 \le y_3$ (d) $y_2 < y_3$

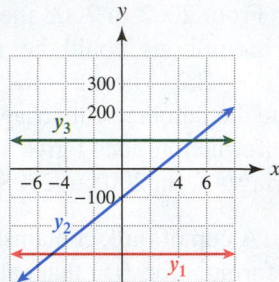

12. Solve $-2 < 2 + \frac{1}{2}x < 2$ and write the solution set in interval notation.

13. Solve the equation $|2 - \frac{1}{3}x| = 6$.

14. Solve each inequality. Write your answer in interval notation.
    (a) $|x| \le 5$ (b) $|x| > 0$

15. Determine whether $f(x) = 1 - 2x + x^3$ represents a polynomial function. If possible, identify the degree and type of polynomial function.

16. Evaluate $h(t) = -\frac{4t}{5 - t}$ at $t = -2$. Write the domain of h in interval notation.

17. Let $f(x) = x^2 + 1$ and $g(x) = 2x$. Find each of the following.
    (a) $(f - g)(-2)$ (b) $(fg)(x)$

18. *Drinking Fluids and Exercise* To determine the number of ounces of fluid that a person should drink in a day, divide his or her weight in pounds by 2 and then add 0.4 ounce for every minute of exercise.
    (a) Write a function that gives the fluid requirements for a person weighing 150 pounds and exercising x minutes a day.
    (b) If a 150-pound runner needs 89 ounces of fluid each day, determine the runner's daily minutes of exercise.

19. *Heart Rate of an Athlete* The table at the top of the next page lists the heart rate or pulse of an athlete running a 400-meter race. The race lasts 50 seconds.

Time(seconds)	0	20	30	50
Heart Rate (bpm)	100	134	150	180

(a) Does $P(t) = -\frac{1}{300}t^2 + \frac{53}{30}t + 100$ model the data in the table exactly? Explain.

(b) Does $P(60)$ have significance in this situation? What should be the domain of P?

20. *Time Spent in Line* Suppose that parking lot attendants can wait on 25 vehicles per minute and vehicles are arriving at the lot randomly at an average rate of x vehicles per minute. Then the average time T in minutes spent waiting in line *and* paying the attendant is given by

$$T(x) = \frac{1}{25 - x},$$

where $x < 25$. (*Source:* N. Garber.)

(a) Graph T in [0, 25, 5] by [0, 2, 0.5]. Identify any vertical asymptotes.

(b) If the wait is 1 minute, how many vehicles are arriving on average?

CHAPTER 8   Extended and Discovery Exercises

1. *Developing a Model* Two identical cylindrical tanks, A and B, each contain 100 gallons of water. Tank A has a pump that begins removing water at a constant rate of 8 gallons per minute. Tank B has a plug removed from its bottom and water begins to flow out—faster at first and then more slowly.

(a) Assuming that the tanks become empty at the same time, sketch a graph that models the amount of water in each tank. Explain your graphs.

(b) Which tank is half empty first? Explain.

2. *Modeling Real Data* Per capita personal incomes in the United States are listed in the following table.

Year	2002	2003	2004
Income	$31,461	$32,271	$33,881

Year	2005	2006	2007
Income	$35,424	$37,698	$39,458

Source: Department of Commerce.

(a) Make a scatterplot of the data.

(b) Find a function f that models the data. Explain your reasoning.

(c) Use f to estimate per capita income in 2000.

3. *Weight of a Small Fish* The graph shows a function f that models the weight in milligrams of a small fish, *Lebistes reticulatus*, during the first 14 weeks of its life. (*Source:* D. Brown and P. Rothery, *Models in Biology.*)

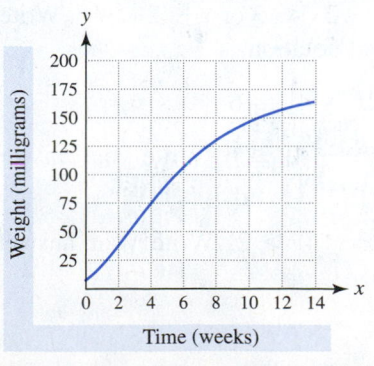

(a) Estimate the weight of the fish when it hatches, at 6 weeks, and at 12 weeks.

(b) If (x_1, y_1) and (x_2, y_2) are points on the graph of a function, the ***average rate of change of f from*** x_1 ***to*** x_2 is given by $\frac{y_2 - y_1}{x_2 - x_1}$. Approximate the average rates of change of f from hatching to 6 weeks and from 6 weeks to 12 weeks.

(c) Interpret these rates of change.

(d) During which time period does the fish gain weight the fastest?

4. *Recording Music* A compact disc (CD) can hold approximately 700 million bytes. One million bytes is commonly referred to as a *megabyte* (MB). Recording music requires an enormous amount of memory. The accompanying table lists the megabytes x needed to record y seconds (sec) of music.

x (MB)	0.129	0.231	0.415	0.491
y (sec)	6.010	10.74	19.27	22.83

x (MB)	0.667	1.030	1.160	1.260
y (sec)	31.00	49.00	55.25	60.18

Source: Gateway 2000 System CD.

(a) Make a scatterplot of the data.

(b) What relationship seems to exist between x and y? Why does this relationship seem reasonable?

(c) Find the slope–intercept form of a line that models the data. Interpret the slope of this line as a rate of change. Answers may vary.

(d) Check your answer in part (c) by graphing the line and data in the same graph.

(e) Write a linear equation whose solution gives the megabytes needed to record 120 seconds of music.

(f) Solve the equation in part (e) graphically or symbolically.

CHAPTERS 1–8 Cumulative Review Exercises

1. Write 120 as a product of prime numbers.

2. Translate the sentence "Triple a number decreased by 4 equals the number" into an equation using the variable n. Solve the equation.

Exercises 3 and 4: Simplify completely.

3. $\frac{1}{4} \div \frac{3}{4} - \frac{1}{2}$
   4. $12 - 3^2 \div 3 \cdot 2$

5. Solve $-3(3 - x) - 6 = 2x$.

6. Convert 0.075 to a percentage.

7. Solve $A = \frac{1}{3}(2a - b)$ for a.

8. Solve $5 - 3t < 1 - t$. Write the solution set in set-builder notation.

Exercises 9 and 10: Graph the equation.

9. $3x - 2y = -6$
   10. $y = |2x - 4|$

11. Write the slope–intercept form of a line that passes through the points $(-4, 4)$ and $(2, 1)$.

12. Solve the system of equations. Write your answer as an ordered pair.

$$2x + 3y = 5$$
$$3x - 2y = 1$$

Exercises 13 and 14: Shade the solution set for the given inequality.

13. $x \geq -1$
    14. $x - y \geq 2$

15. Shade the solution set for the system of inequalities.

$$x - y \geq 1$$
$$2x + y \leq 0$$

16. Simplify $(3x^2 + 2x - 4) - (4x^2 + 5)$.

Exercises 17–20: Simplify and use positive exponents.

17. $\dfrac{x^{-4}}{x^{-3}}$
    18. $(3b^{-3})(2b^4)$

19. $3(2t)^3$
    20. $\left(\dfrac{2x^3}{x^2y^{-1}}\right)^{-2}$

Exercises 21–24: Multiply the expression.

21. $2x^2(x^3 - 4x^2 - 5)$

22. $(5x + 1)(2x - 7)$

23. $(y - 3)(y + 3)$

24. $(x - 4y)^2$

25. Write 2.5×10^4 in standard form.

26. Write 0.028 in scientific notation.

Exercises 27 and 28: Divide.

27. $\dfrac{6x^3 - 4x^2 + 8x}{2x}$

28. $(3x^3 + 2x^2 + 1) \div (x - 1)$

Exercises 29–34: Factor completely.

29. $10x^2y^3 - 15x^3y^2$
    30. $x^3 + 3x^2 - x - 3$

31. $2z^2 + z - 3$
    32. $16x^2 - 25$

33. $a^3 - 8$
    34. $z^4 + 7z^2 + 6$

Exercises 35–38: Solve the equation.

35. $x(x + 5) = 0$
    36. $4x^2 = 0$

37. $2x^2 + 5x = 3$
    38. $x^3 = x$

39. Simplify $\dfrac{x^2 + 4x + 4}{x + 2}$ to lowest terms.

40. If possible, evaluate $\dfrac{x^2 + 1}{x - 1}$ for $x = -2$ and $x = 1$.

Exercises 41 and 42: Simplify to lowest terms.

41. $\dfrac{x^2 - 3x + 2}{x + 2} \div \dfrac{x - 1}{2x + 4}$

42. $\dfrac{1}{x + 3} + \dfrac{2}{x + 1}$

43. Solve $\frac{x + 2}{x - 1} = \frac{2}{3}$.

44. Suppose that y is directly proportional to x and that $y = 5$ when $x = 10$. Find y when $x = 20$.

45. Evaluate $f(x) = x^2 - 4x$ for $x = -2$.

46. Graph $f(x) = x^2 - 2$ and identify the domain and range of f. Write your answer in interval notation.

47. Solve $x - 2 < 3$ or $x - 2 > 6$. Write your answer in interval notation.

48. Solve $|2x - 4| = 6$.

49. Solve $|2x - 4| \leq 6$. Write your answer in interval notation.

50. Solve $|x - 4| > 2$. Write your answer in interval notation.

APPLICATIONS

51. *Modeling Motion* The table lists distance d in miles traveled by a car for various elapsed times t in hours. Find an equation that models these data.

t (hours)	2	3	4	6
d (miles)	144	216	288	432

52. *Rainfall* At noon 2 inches of rain had fallen. For the next 6 hours rain fell at $\frac{1}{4}$ inch per hour.
  (a) Find an equation in the form $y = mx + b$ that calculates the number of inches of rain that fell x hours past noon.
  (b) What is the slope of the graph of f?
  (c) Interpret the slope as a rate of change.
  (d) How much rain had fallen by 4 P.M.?

53. *Interest* A total of $5000 is deposited in two accounts paying 5% and 7% annual interest. If total interest received at the end of the year is $308, determine how much is invested at each interest rate.

54. *Height of a Building* An 8-foot-tall stop sign casts a 5-foot-long shadow, while a nearby building casts a 65-foot-long shadow. Find the height of the building.

55. *Shoveling the Driveway* Two people are shoveling snow from a driveway. The first person shovels 9 square feet per minute, while the second person shovels 11 square feet per minute.
  (a) Write a simplified expression that gives the total number of square feet the two people can shovel in x minutes.
  (b) How many minutes would it take for them to clear a driveway with 1000 square feet?

56. *Cost of a Television* A 5% sales tax on a television amounts to $82.50. Find the cost of the television.

9 Systems of Linear Equations

> The essence of mathematics is not to make simple things complicated, but to make complicated things simple.
>
> —STANLEY GUDDER

In 1940, a physicist named John Atanasoff at Iowa State University needed to solve 29 equations with 29 variables simultaneously. This task was too difficult to do by hand, so he and a graduate student invented the first fully electronic digital computer. Thus the desire to solve a mathematical problem led to one of the most important inventions of the twentieth century. Today people can solve thousands of equations with thousands of variables. Solutions to such equations have resulted in better airplanes, cars, electronic devices, weather forecasts, and medical equipment.

Equations are even used in biology. The following table contains the weight W, neck size N, and chest size C for three black bears. Suppose that park rangers find a bear with a neck size of 22 inches and a chest size of 38 inches. Can they use the data in the table to estimate the bear's weight? Using systems of linear equations, they *can* estimate the bear's weight. See Example 8 and Exercise 61 in Section 9.2.

Black Bear Measurements

W (pounds)	N (inches)	C (inches)
80	16	26
344	28	45
416	31	54
?	22	38

Sources: A. Tucker, *Fundamentals of Computing*; M. Triola, *Elementary Statistics*; Minitab, Inc.

9.1 Systems of Linear Equations in Three Variables

**Basic Concepts • Solving Linear Systems with Substitution and Elimination •
Modeling Data • Systems of Equations with No Solutions •
Systems of Equations with Infinitely Many Solutions**

A LOOK INTO MATH ▶ In the Chapter 9 opening discussion on page 590, we saw how systems of linear equations are used to predict the weight of black bears. In this section we will see how systems of linear equations are used to predict antelope populations in Wyoming. (See Examples 3 and 7.) The same mathematical concept, systems of linear equations, is used to model a variety of real-world situations. This is an amazing power of mathematics: *one simple but profound concept solves many complicated problems.*

Basic Concepts

NEW VOCABULARY

☐ Linear system in three
variables
☐ Ordered triple

When we solve a linear system in two variables, we can express a solution as an ordered pair (x, y). A linear equation in two variables can be represented graphically by a line. A system of two linear equations with a unique solution can be represented graphically by two lines intersecting at a point, as shown in Figure 9.1.

When solving **linear systems in three variables,** we often use the variables x, y, and z. A solution is expressed as an **ordered triple** (x, y, z), rather than an ordered pair (x, y). For example, if the ordered triple $(1, 2, 3)$ is a solution, $x = 1$, $y = 2$, and $z = 3$ satisfy each equation. A linear equation in three variables can be represented by a flat plane in space. If the solution is unique, we can represent a linear system of three equations in three variables graphically by three planes intersecting at a single point, as illustrated in Figure 9.2.

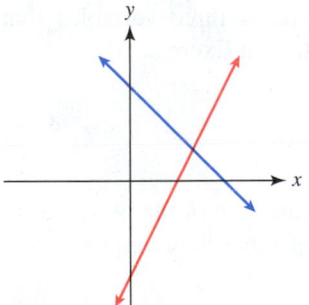

Figure 9.1

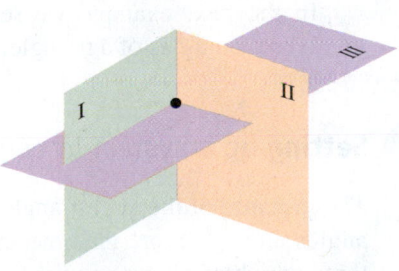

Figure 9.2

NOTE: The three planes in Figure 9.2 all intersect at right angles. In general, three planes can intersect at a point even if they are not at right angles to each other.

CRITICAL THINKING

Each of the following depictions of three planes in space represents a system of three linear equations in three variables. How many solutions are there in each case? Explain.

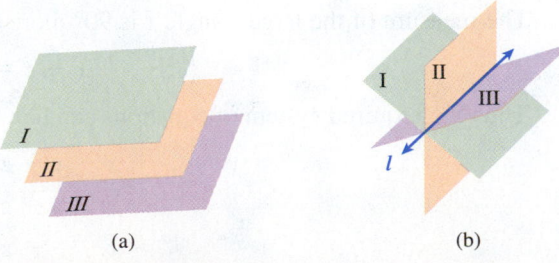

(a)   (b)

The next example shows how to check whether an ordered triple is a solution to a system of linear equations in three variables.

EXAMPLE 1 Checking for solutions to a system of three equations

Determine whether $(4, 2, -1)$ or $(-1, 0, 3)$ is a solution to the system.

$$
\begin{aligned}
2x - 3y + z &= 1 \\
x - 2y + 2z &= 5 \\
2y + z &= 3
\end{aligned}
$$

Solution

To check $(4, 2, -1)$, substitute $x = 4$, $y = 2$, and $z = -1$ in each equation.

$$
\begin{aligned}
2(4) - 3(2) + (-1) &\stackrel{?}{=} 1 \quad \checkmark \quad \text{True} \\
4 - 2(2) + 2(-1) &\stackrel{?}{=} 5 \quad \text{✗} \quad \text{False} \\
2(2) + (-1) &\stackrel{?}{=} 3 \quad \checkmark \quad \text{True}
\end{aligned}
$$

The ordered triple $(4, 2, -1)$ *does not satisfy all three equations, so it is not a solution.* Next, substitute $x = -1$, $y = 0$, and $z = 3$ in each equation.

$$
\begin{aligned}
2(-1) - 3(0) + 3 &\stackrel{?}{=} 1 \quad \checkmark \quad \text{True} \\
-1 - 2(0) + 2(3) &\stackrel{?}{=} 5 \quad \checkmark \quad \text{True} \\
2(0) + 3 &\stackrel{?}{=} 3 \quad \checkmark \quad \text{True}
\end{aligned}
$$

The ordered triple $(-1, 0, 3)$ *satisfies all three equations, so it is a solution.*

Now Try Exercise 9

In the next example we set up a system of three equations in three variables that involves the angles of a triangle. You are asked to solve this system in Exercise 51.

EXAMPLE 2 Setting up a system of equations

The measure of the largest angle in a triangle is 40° greater than the sum of the two smaller angles and 90° more than the smallest angle. Set up a system of three linear equations in three variables whose solution gives the measure of each angle.

Solution

Let x, y, and z be the measures of the three angles from largest to smallest. Because the sum of the measures of the angles in a triangle equals 180°, we have

$$x + y + z = 180.$$

The measure of the largest angle x is 40° greater than the sum of the measures of the two smaller angles $y + z$, so

$$x - (y + z) = 40 \quad \text{or} \quad x - y - z = 40.$$

The measure of the largest angle x is 90° more than the measure of the smallest angle z, so

$$x - z = 90.$$

Thus the required system of equations can be written as follows.

$$
\begin{aligned}
x + y + z &= 180 \\
x - y - z &= 40 \\
x - z &= 90
\end{aligned}
$$

Now Try Exercise 49(a)

▶ **REAL-WORLD CONNECTION**   In the next example we show how a linear system involving three equations and three variables can be used to model a real-world situation. We solve this system of equations in Example 7.

EXAMPLE 3   ## Modeling real data with a linear system

The Bureau of Land Management studies antelope populations in Wyoming. It monitors the number of adult antelope, the number of fawns each spring, and the severity of the winter. The first two columns of Table 9.1 contain counts of fawns and adults for three representative winters. The third column shows the severity of each winter. The severity of the winter is measured from 1 to 5, with 1 being mild and 5 being severe.

TABLE 9.1

Fawns (F)	Adults (A)	Winter (W)
405	870	3
414	848	2
272	684	5
?	750	4

We want to use the data in the first three rows of the table to estimate the number of fawns F in the fourth row when the number of adults is 750 and the severity of the winter is 4. To do so, we use the formula

$$F = a + bA + cW,$$

where a, b, and c are constants. Write a system of linear equations whose solution gives appropriate values for a, b, and c.

Solution
From the first row in the table, when $F = 405$, $A = 870$, and $W = 3$, the formula

$$F = a + bA + cW$$

becomes

$$405 = a + b(870) + c(3).$$

Similarly, $F = 414$, $A = 848$, and $W = 2$ gives

$$414 = a + b(848) + c(2),$$

and $F = 272$, $A = 684$, and $W = 5$ yields

$$272 = a + b(684) + c(5).$$

To find values for a, b, and c we can solve the following system of linear equations.

$$405 = a + 870b + 3c$$
$$414 = a + 848b + 2c$$
$$272 = a + 684b + 5c$$

We can also write these equations as a linear system in the following form.

$$a + 870b + 3c = 405$$
$$a + 848b + 2c = 414$$
$$a + 684b + 5c = 272$$

Finding values for a, b, and c will allow us to use the formula $F = a + bA + cW$ to predict the number of fawns F when the number of adults A is 750 and the severity of the winter W is 4 (see Example 7).

▌ **Now Try Exercise 53(a)**

NOTE: Linear systems of two equations can have no solutions, one solution, or infinitely many solutions. The same is true for larger linear systems. In the following subsection we focus on linear systems having one solution.

Solving Linear Systems with Substitution and Elimination

When solving systems of linear equations with more than two variables, we usually use both substitution and elimination. However, in the next example we use only substitution to solve a particular type of linear system in three variables.

EXAMPLE 4 | **Using substitution to solve a linear system of equations**

Solve the following system.

$$2x - y + z = 7$$
$$3y - z = 1$$
$$z = 2$$

Solution
The last equation gives us the value of z immediately. We can substitute $z = 2$ in the second equation and determine y.

$$3y - z = 1 \quad \text{Second equation}$$
$$3y - 2 = 1 \quad \text{Substitute } z = 2.$$
$$3y = 3 \quad \text{Add 2 to each side.}$$
$$y = 1 \quad \text{Divide each side by 3.}$$

Knowing that $y = 1$ and $z = 2$ allows us to find x by using the first equation.

$$2x - y + z = 7 \quad \text{First equation}$$
$$2x - 1 + 2 = 7 \quad \text{Let } y = 1 \text{ and } z = 2.$$
$$2x = 6 \quad \text{Simplify and subtract 1.}$$
$$x = 3 \quad \text{Divide each side by 2.}$$

Thus $x = 3$, $y = 1$, and $z = 2$, and the solution is $(3, 1, 2)$.

Now Try Exercise 13

In the next example we use elimination and substitution to solve a system of linear equations. This four-step method is summarized by the following.

STUDY TIP

Be sure to follow this four-step method. It will help you get the correct answer *consistently*.

SOLVING A LINEAR SYSTEM IN THREE VARIABLES

STEP 1: Eliminate one variable, such as x, from two of the equations.

STEP 2: Use the two resulting equations in two variables to eliminate one of the variables, such as y. Solve for the remaining variable, z.

STEP 3: Substitute z in one of the two equations from Step 2. Solve for the unknown variable y.

STEP 4: Substitute values for y and z in one of the given equations and find x. The solution is (x, y, z).

EXAMPLE 5 **Solving a linear system in three variables**

Solve the following system.

$$\begin{aligned} x - y + 2z &= 6 \\ 2x + y - 2z &= -3 \\ -x - 2y + 3z &= 7 \end{aligned}$$

Solution

STEP 1: We begin by eliminating the variable x from the second and third equations. To eliminate x from the second equation, we multiply the first equation by -2 and then add it to the second equation. To eliminate x from the third equation, we add the first and third equations.

$-2x + 2y - 4z = -12$	First equation times -2	$x - y + 2z = 6$	First equation
$\underline{2x + y - 2z = -3}$	Second equation	$\underline{-x - 2y + 3z = 7}$	Third equation
$3y - 6z = -15$	Add.	$-3y + 5z = 13$	Add.

STEP 2: Take the two resulting equations from Step 1 and eliminate either variable. Here we add the two equations to eliminate the variable y.

$$\begin{aligned} 3y - 6z &= -15 \\ \underline{-3y + 5z} &= \underline{13} \\ -z &= -2 \quad \text{Add the equations.} \\ z &= 2 \quad \text{Multiply by } -1. \end{aligned}$$

STEP 3: Now we can use substitution to find the values of x and y. We let $z = 2$ in either equation used in Step 2 to find y.

$$\begin{aligned} 3y - 6z &= -15 \quad \text{From Step 2} \\ 3y - 6(2) &= -15 \quad \text{Substitute } z = 2. \\ 3y - 12 &= -15 \quad \text{Multiply.} \\ 3y &= -3 \quad \text{Add 12.} \\ y &= -1 \quad \text{Divide by 3.} \end{aligned}$$

STEP 4: Finally, we substitute $y = -1$ and $z = 2$ in one of the given equations to find x.

$$\begin{aligned} x - y + 2z &= 6 \quad \text{First given equation} \\ x - (-1) + 2(2) &= 6 \quad \text{Let } y = -1 \text{ and } z = 2. \\ x + 1 + 4 &= 6 \quad \text{Simplify.} \\ x &= 1 \quad \text{Subtract 5.} \end{aligned}$$

The solution is $(1, -1, 2)$. Check this solution.

Now Try Exercise 25

▶ **REAL-WORLD CONNECTION** In the next example we solve a system of linear equations to determine the number of tickets sold at a play.

EXAMPLE 6 **Finding the number of tickets sold**

One thousand tickets were sold for a play, which generated $3800 in revenue. The prices of the tickets were $3 for children, $4 for students, and $5 for adults. There were 100 fewer student tickets sold than adult tickets. Find the number of each type of ticket sold.

Solution

Let x be the number of tickets sold to children, y be the number of tickets sold to students, and z be the number of tickets sold to adults. The total number of tickets sold was 1000, so

$$x + y + z = 1000. \quad (1)$$

Each child's ticket cost \$3, so the revenue generated from selling x tickets was $3x$. Similarly, the revenue generated from students was $4y$, and the revenue from adults was $5z$. Total ticket sales were \$3800, so

$$3x + 4y + 5z = 3800. \quad (2)$$

The equation $z - y = 100$, or

$$y - z = -100, \quad (3)$$

must also be satisfied, because 100 fewer tickets were sold to students than to adults.

To find the number of each type of ticket sold, we need to solve the following system of linear equations.

$$
\begin{aligned}
x + y + z &= 1000 \\
3x + 4y + 5z &= 3800 \\
y - z &= -100
\end{aligned}
$$

STEP 1: We begin by eliminating the variable x from the second equation. To do so, we multiply the first equation by -3 and add the second equation.

$$
\begin{array}{ll}
-3x - 3y - 3z = -3000 & \text{First given equation times } -3 \\
\underline{3x + 4y + 5z = 3800} & \text{Second equation} \\
y + 2z = 800 & \text{Add.}
\end{array}
$$

STEP 2: We then multiply the resulting equation from Step 1 by -1 and add the third given equation to eliminate y.

$$
\begin{array}{ll}
-y - 2z = -800 & \text{Equation from Step 1 times } -1 \\
\underline{y - z = -100} & \text{Third given equation} \\
-3z = -900 & \text{Add the equations.} \\
z = 300 & \text{Divide by } -3.
\end{array}
$$

STEP 3: To find y we can substitute $z = \mathbf{300}$ in one of the equations from Step 2.

$$
\begin{array}{ll}
y - z = -100 & \text{Step 2 equation} \\
y - \mathbf{300} = -100 & \text{Let } z = 300. \\
y = \mathbf{200} & \text{Add 300.}
\end{array}
$$

STEP 4: Finally, we substitute $y = \mathbf{200}$ and $z = \mathbf{300}$ in the first given equation.

$$
\begin{array}{ll}
x + \mathbf{y} + \mathbf{z} = 1000 & \text{First given equation} \\
x + \mathbf{200} + \mathbf{300} = 1000 & \text{Let } y = 200 \text{ and } z = 300. \\
x = \mathbf{500} & \text{Subtract 500.}
\end{array}
$$

Thus **500** tickets were sold to children, **200** to students, and **300** to adults.

Now Try Exercise 47

Modeling Data

In the next example we solve the system of equations that we discussed in Example 3.

> **EXAMPLE 7**

Predicting fawns in the spring

Solve the following linear system for a, b, and c.

$$a + 870b + 3c = 405$$
$$a + 848b + 2c = 414$$
$$a + 684b + 5c = 272$$

Then use $F = a + bA + cW$ to predict the number of fawns when there are 750 adults and the severity of the winter is 4.

Solution

STEP 1: We begin by eliminating the variable a from the second and third equations. To do so, we add the second and third equations to the first equation times -1.

$\begin{aligned} -a - 870b - 3c &= -405 \\ a + 848b + 2c &= 414 \\ \hline -22b - c &= 9 \end{aligned}$	$\begin{aligned} -a - 870b - 3c &= -405 \quad \text{First times } -1 \\ a + 684b + 5c &= 272 \quad \text{Second/third equation} \\ \hline -186b + 2c &= -133 \quad \text{Add.} \end{aligned}$

STEP 2: We use the two resulting equations from Step 1 to eliminate c. To do so we multiply the equation $-22b - c = 9$ by 2 and add it to the other equation.

$$
\begin{aligned}
-44b - 2c &= 18 \quad (-22b - c = 9) \text{ times 2} \\
-186b + 2c &= -133 \\
\hline
-230b &= -115 \quad \text{Add the equations.} \\
b &= 0.5 \quad \text{Divide by} -230.
\end{aligned}
$$

STEP 3: To find c we substitute $b = 0.5$ in either equation used in Step 2.

$$
\begin{aligned}
-44b - 2c &= 18 \quad \text{First equation in Step 2} \\
-44(0.5) - 2c &= 18 \quad \text{Let } b = 0.5. \\
-22 - 2c &= 18 \quad \text{Multiply.} \\
-2c &= 40 \quad \text{Add 22.} \\
c &= -20 \quad \text{Divide by } -2.
\end{aligned}
$$

STEP 4: Finally, we substitute $b = 0.5$ and $c = -20$ in one of the given equations to find a.

$$
\begin{aligned}
a + 870b + 3c &= 405 \quad \text{First given equation} \\
a + 870(0.5) + 3(-20) &= 405 \quad \text{Let } b = 0.5 \text{ and } c = -20. \\
a + 435 - 60 &= 405 \quad \text{Multiply.} \\
a &= 30 \quad \text{Solve for } a.
\end{aligned}
$$

CRITICAL THINKING

Give reasons why the coefficient for A is positive and the coefficient for W is negative in the formula

$$F = 30 + 0.5A - 20W.$$

The solution is $a = 30$, $b = 0.5$, and $c = -20$. Thus we may write

$$F = a + bA + cW$$
$$= 30 + 0.5A - 20W. \quad \text{Modeling equation}$$

If there are 750 adults and the winter has a severity of 4, this model predicts

$$F = 30 + 0.5(750) - 20(4)$$
$$= 325 \text{ fawns.}$$

Now Try Exercise 53

Systems of Equations with No Solutions

It is possible for a system of three linear equations in three variables to be inconsistent and have no solutions. If we apply substitution and elimination to this type of system, we arrive at a contradiction. This case is demonstrated in the next example.

EXAMPLE 8 **Recognizing an inconsistent system**

Solve the system, if possible.

$$\begin{aligned} x + y + z &= 4 \\ -x + y + z &= 2 \\ y + z &= 1 \end{aligned}$$

Solution

STEP 1: If we add the first two equations, we can eliminate x. The variable x is already eliminated from the third equation.

$$\begin{array}{ll} x + y + z = 4 & \text{First equation} \\ \underline{-x + y + z = 2} & \text{Second equation} \\ 2y + 2z = 6 & \text{Add.} \end{array}$$

STEP 2: If we multiply the third *given* equation by -2 and add it to the resulting equation in Step 1, we arrive at a contradiction.

$$\begin{array}{ll} -2y - 2z = -2 & \text{Third equation times } -2 \\ \underline{2y + 2z = 6} & \text{Equation from Step 1} \\ \mathbf{0} = \mathbf{4} \; ✗ & \text{Add. (Contradiction)} \end{array}$$

Because $\mathbf{0 = 4}$ is a contradiction, there are no solutions to the given system of equations.

Now Try Exercise 31

Systems of Equations with Infinitely Many Solutions

It is possible for a system of linear equations in three variables to have infinitely many solutions. If we apply substitution and elimination to this type of system, we arrive at an identity. This case is demonstrated in the next example.

EXAMPLE 9 **Solving a system with infinitely many solutions**

Solve the system.

$$\begin{aligned} x + y + z &= 2 \\ x - y + z &= 4 \\ 3x - y + 3z &= 10 \end{aligned}$$

Solution

STEP 1: To eliminate y from the second equation, add the first equation to the second. To eliminate y from the third equation, add the first equation to the third equation.

$$\begin{array}{ll} x + y + z = 2 & \text{First equation} \\ \underline{x - y + z = 4} & \text{Second equation} \\ 2x + 2z = 6 & \text{Add.} \end{array} \qquad \begin{array}{ll} x + y + z = 2 & \text{First equation} \\ \underline{3x - y + 3z = 10} & \text{Third equation} \\ 4x + 4z = 12 & \text{Add.} \end{array}$$

STEP 2: If we multiply the first resulting equation in Step 1 by -2 and add it to the second resulting equation in Step 1, we arrive at an identity.

$$
\begin{array}{rl}
-4x - 4z = -12 & \quad (2x + 2z = 6) \text{ times } -2 \\
\underline{4x + 4z = 12} & \quad \text{Second equation from Step 1} \\
0 = 0 & \quad \text{Add. (Identity)}
\end{array}
$$

The variable x can be written in terms of z by solving $2x + 2z = 6$ for x.

$$
\begin{array}{rl}
2x + 2z = 6 & \quad \text{Equation from Step 1} \\
2x = 6 - 2z & \quad \text{Subtract } 2z. \\
x = 3 - z & \quad \text{Divide by 2.}
\end{array}
$$

STEP 3: To find y in terms of z, substitute $3 - z$ for x in the first *given* equation.

$$
\begin{array}{rl}
x + y + z = 2 & \quad \text{First equation} \\
(3 - z) + y + z = 2 & \quad \text{Let } x = 3 - z. \\
y = -1 & \quad \text{Solve for } y.
\end{array}
$$

All solutions have the form $(3 - z, -1, z)$, where z can be any real number. For example, if $z = 1$, then $(2, -1, 1)$ is one of infinitely many solutions to the system of equations. Note that in this particular system, y must always equal -1.

Now Try Exercise 33

> **READING CHECK**
>
> What is the possible number of solutions to a system of linear equations in three variables?

| **9.1** | **Putting It All Together** |

In this section we discussed how to solve a system of three linear equations in three variables. Systems of linear equations can have no solutions, one solution, or infinitely many solutions.

CONCEPT	EXPLANATION/EXAMPLE
System of Linear Equations in Three Variables	The following is a system of three linear equations in three variables. Each equation is linear. $$\begin{array}{rl} x - 2y + z = 0 \\ -x + y + z = 4 \\ -y + 4z = 10 \end{array}$$
Solution to a Linear System in Three Variables	The solution to a linear system in three variables is an ordered triple, expressed as (x, y, z). The solution to the preceding system is $(1, 2, 3)$ because substituting $x = 1$, $y = 2$, and $z = 3$ in each equation results in a true statement. We can check solutions this way. $$\begin{array}{rl} (1) - 2(2) + (3) = 0 \ \checkmark & \quad \text{True} \\ -(1) + (2) + (3) = 4 \ \checkmark & \quad \text{True} \\ -(2) + 4(3) = 10 \ \checkmark & \quad \text{True} \end{array}$$

continued on next page

continued from previous page

CONCEPT	EXPLANATION
Solving a Linear System with Substitution and Elimination	**STEP 1:** Eliminate one variable, such as x, from two of the equations.
	STEP 2: Use the two resulting equations in two variables to eliminate one of the variables, such as y. Solve for the remaining variable z.
	STEP 3: Substitute z in one of the two equations from Step 2. Solve for the unknown variable y.
	STEP 4: Substitute values for y and z in one of the given equations and find x. The solution is (x, y, z).

9.1 Exercises

MyMathLab

CONCEPTS AND VOCABULARY

1. Can a system of three linear equations in three variables have exactly two solutions? Explain.

2. Give an example of a system of three linear equations in three variables.

3. Does the ordered triple $(1, 2, 3)$ satisfy the equation $x + y + z = 6$?

4. Does $(3, 4)$ represent a solution to the equation $x + y + z = 7$? Explain.

5. To solve uniquely for two variables, how many equations do you usually need?

6. To solve uniquely for three variables, how many equations do you usually need?

7. If a contradiction occurs while solving a system of linear equations, then there is/are _____ solution(s).

8. If an identity occurs while solving a linear system, how many solutions are there?

SOLVING LINEAR SYSTEMS

Exercises 9–12: Determine which ordered triple is a solution to the linear system.

9. $(1, 2, 3), (0, 2, 4)$
$$x + y + z = 6$$
$$x - y - z = -4$$
$$-x - y + z = 0$$

10. $(-1, 0, 2), (0, 4, 4)$
$$2x + y - 3z = -8$$
$$x - 3y + 2z = -4$$
$$3x - 2y + z = -4$$

11. $(1, 0, 3), (-1, 1, 2)$
$$3x - 2y + z = -3$$
$$-x + 3y - 2z = 0$$
$$x + 4y + 2z = 7$$

12. $\left(\frac{1}{2}, \frac{3}{2}, -\frac{1}{2}\right), (-1, 0, -2)$
$$x + 3y - 4z = 7$$
$$-x + 5y + 3z = \frac{11}{2}$$
$$3x - 2y - 7z = 2$$

Exercises 13–18: (Refer to Example 4.) Use substitution to solve the system of linear equations. Check your solution.

13. $x + y - z = 1$
$2y + z = -1$
$z = 1$

14. $2x + y - 3z = 1$
$y + 4z = 0$
$z = -1$

15. $-x - 3y + z = -2$
$2y + 3z = 3$
$z = 2$

16. $3x + 2y - 3z = -4$
$-y + 2z = 4$
$z = 0$

17. $a - b + 2c = 3$
$-3b + c = 4$
$c = -2$

18. $5a + 2b - 3c = 10$
$5b - 2c = -4$
$c = 3$

Exercises 19–44: Solve the system, if possible.

19. $x + y - z = 11$
$-x + 2y + 3z = -1$
$2z = 4$

20. $x + 2y - 3z = -7$
$-2x + y + z = -1$
$3z = 9$

21. $x + y - z = -2$
$-x + z = 1$
$y + 2z = 3$

22. $x + y - 3z = 11$
$-2x + y + 2z = 1$
$-3y + 3z = -21$

23. $x + y - 2z = -7$
$y + z = -1$
$-y + 3z = 9$

24. $2x + 3y + z = 5$
$y + 2z = 4$
$-2y + z = 2$

25. $x + 2y + 2z = 1$
$x + y + z = 0$
$-x - 2y + 3z = -11$

26. $x + y - z = 0$
$x - 3y + z = -2$
$x - y + 3z = 8$

27. $x + y + z = 5$
$y + z = 6$
$x + z = 3$

28. $x + y + z = 0$
$x - y - z = 6$
$-x + y - z = 4$

29. $x + 2y + 3z = 24$
$-x + y + 2z = 1$
$x + y - 2z = 9$

30. $5x - 15y + z = 22$
$-10x + 12y - 2z = -8$
$4x - 2y - 3z = 9$

31. $x + y + z = 2$
$x - y + z = 1$
$x + z = 3$

32. $4x - y + 3z = 3$
$2x + y + z = 2$
$x - y + z = 1$

33. $x + y + z = 6$
$x - y + z = 2$
$-x + 5y - z = 6$

34. $x - y + z = 3$
$2x - y + z = 2$
$-x - y + z = 5$

35. $2x + y + z = 3$
$2x - y - z = 9$
$x + y - z = 0$

36. $x + 3y + z = -8$
$x - 2y = 11$
$2y - z = -16$

37. $2x + 6y - 2z = 47$
$2x + y + 3z = -28$
$-x + y + z = -\frac{7}{2}$

38. $x + y + 2z = 23$
$3x - y + 3z = 8$
$2x + 2y + z = 13$

39. $x + 3y - 4z = \frac{13}{2}$
$-2x + 3y - z = \frac{1}{2}$
$3x + z = 4$

40. $x - 2y + z = \frac{9}{2}$
$4x - y + 3z = 9$
$x + 2y = -\frac{3}{2}$

41. $2x - 2y + z = 4$
$x - y + z = 1$
$x - y + z = 3$

42. $x - 2y + 3z = 1$
$x + 3y + z = 2$
$3x + 4y + 5z = 7$

43. $x + y + z = 5$
$x - y + z = 3$
$2x + y + 2z = 9$

44. $x + y - 2z = 0$
$x + 2y - 4z = -1$
$y - 2z = -1$

Exercises 45 and 46: **Thinking Generally** *Solve the system of linear equations, if $a \neq 0$.*

45. $x + y + z = a$
$x + y + z = 2a$
$-x + y + z = 0$

46. $x + y + z = a$
$-x - y - z = -a$
$y - z = 0$

APPLICATIONS

47. *Finding Costs* The accompanying table shows the costs of purchasing different combinations of hamburgers, fries, and soft drinks.

Hamburgers	Fries	Soft Drinks	Total Cost
1	2	4	$10
1	4	6	$15
0	3	2	$6

(a) Let x be the cost of a hamburger, y the cost of fries, and z the cost of a soft drink. Write a system of three linear equations that represents the data in the table.
(b) Solve the system and interpret your answer.

48. *Cost of CDs* The accompanying table shows the total cost of purchasing combinations of differently priced CDs. The types of CDs are labeled A, B, and C.

A	B	C	Total Cost
1	1	1	$37
3	2	1	$69
1	1	4	$82

(a) Let x be the cost of a CD of type A, y the cost of a CD of type B, and z the cost of a CD of type C. Write a system of three linear equations that represents the data in the table.
(b) Solve the system and interpret your answer.

49. *Geometry* The largest angle in a triangle is 55° more than the smallest angle. The sum of the measures of the two smaller angles is 10° more than the measure of the largest angle.
   (a) Let x, y, and z be the measures of the three angles from largest to smallest. Write a system of three linear equations whose solution gives the measure of each angle.
   (b) Solve the system.
   (c) Check your solution.

50. *Geometry* The perimeter of a triangle is 90 inches. The longest side is 20 inches longer than the shortest side and 10 inches longer than the remaining side.
   (a) Let x, y, and z be the lengths of the three sides from largest to smallest. Write a system of three linear equations whose solution gives the lengths of each side.
   (b) Solve the system.
   (c) Check your solution.

51. *Geometry* (Refer to Example 2.) Solve the system to find the measure of each angle.

$$\begin{aligned} x + y + z &= 180 \\ x - y - z &= 40 \\ x \quad\quad - z &= 90 \end{aligned}$$

52. *Loan Mixture* A student takes out a total of $5000 in three loans: one subsidized, one unsubsidized, and one from the parents of the student. The subsidized loan is $200 more than the combined total of the unsubsidized and parent loans. The unsubsidized loan is twice the amount of the parent loan. Find the amount of each loan.

53. *Predicting Fawns* (Refer to Examples 3 and 7.) The accompanying table shows counts for fawns and adult deer and the severity of the winter. These data may be modeled by the equation $F = a + bA + cW$.

Fawns (F)	Adults (A)	Winter (W)
525	600	4
365	400	2
805	900	5
?	500	3

   (a) Use the first three rows to write a system of three linear equations in three variables whose solution gives values for a, b, and c.
   (b) Solve the system.
   (c) Predict the number of fawns when there are 500 adults and the winter has severity 3.

54. *Business Production* A business has three machines that manufacture containers. Together they make 100 containers per day, whereas the two fastest machines can make 80 containers per day. The fastest machine makes 34 more containers per day than the slowest machine.
   (a) Let x, y, and z be the number of containers that the machines make from fastest to slowest. Write a system of three equations whose solution gives the number of containers that each machine can make.
   (b) Solve the system.

55. *Mixture Problem* One type of lawn fertilizer consists of a mixture of nitrogen, N, phosphorus, P, and potassium, K. An 80-pound sample contains 8 more pounds of nitrogen and phosphorus than potassium. There is 9 times as much potassium as phosphorus.
   (a) Write a system of three equations whose solution gives the amount of nitrogen, phosphorus, and potassium in this sample.
   (b) Solve the system.

56. *Predicting Home Prices* Selling prices of homes can depend on several factors such as size and age. The accompanying table shows the selling price for three homes. In this table, price P is given in thousands of dollars, age A in years, and home size S in thousands of square feet. These data may be modeled by the equation $P = a + bA + cS$.

Price (P)	Age (A)	Size (S)
190	20	2
320	5	3
50	40	1
?	10	2.5

   (a) Write a system of linear equations whose solution gives a, b, and c.
   (b) Solve the system.
   (c) Predict the price of a home that is 10 years old and has 2500 square feet.

57. *Investment Mixture* A sum of $30,000 was invested in three mutual funds. In one year, the first fund grew by 4%, the second by 5%, and the third by 7.5%. Total earnings were $1775. The amount invested in the third fund was $2000 less than the combined amount invested in the other two funds. Use a linear system of equations to determine the amount invested in each fund.

58. *Football Tickets* A total of 2500 tickets to a football game were sold. Prices were $2 for children, $3 for students, and $5 for adults. Twice as many tickets were sold to students as to children, and ticket revenues were $7250. Use a system of linear equations to determine how many of each type of ticket were sold.

WRITING ABOUT MATHEMATICS

59. In Chapter 4 we solved problems with two variables; to obtain a unique solution we needed two linear equations. In this section we solved problems with three variables; to obtain a unique solution we needed three equations. Try to generalize these results. In the design of aircraft, problems commonly involve 100,000 variables. How many equations are required to solve such problems? Can such problems be solved by hand? If not, how are such problems solved? Explain your answers.

60. In Exercise 56 the price of a home was estimated by its age and size. What other factors might affect the price of a home? Explain how these factors might affect the number of variables and equations in the linear system.

Group Activity   Working with Real Data

Directions: Form a group of 2 to 4 people. Select someone to record the group's responses for this activity. All of the members of the group should work cooperatively to answer the questions. If your instructor asks for your results, each member of the group should be prepared to respond.

CEO Salaries In 2010, hourly wages for the top three CEOs (chief executive officers) of U.S. corporations were calculated based on their working 14 hours per day for 365 days. Together the three CEOs made $75,000 per hour. The first CEO earned $3,000 more per hour than the hourly wage of the second CEO, and the first CEO earned $6,000 more per hour than the third CEO. (*Sources:* Forbes.com; Department of Labor.)

(a) Let x, y, and z be the hourly wages in *thousands* of dollars for the top CEOs from greatest to least. Write a system of equations whose solution gives these hourly wages.

(b) Solve the system. Interpret the answer.

(c) The average private sector American worker earned $22.78 per hour in 2010. How many hours did the average private sector American work to earn an amount equal to one hour of work by the first CEO in 2010?

9.2 Matrix Solutions of Linear Systems

Representing Systems of Linear Equations with Matrices •
Matrices and Social Networks • Gauss–Jordan Elimination •
Using Technology to Solve Systems of Linear Equations (Optional)

A LOOK INTO MATH ▶ Suppose that the size of a bear's head and its overall length are known. Can its weight be estimated from these variables? Can a bear's weight be estimated if its neck size and chest size are known? In this section we show that systems of linear equations can be used to make such estimates.

Representing Systems of Linear Equations with Matrices

NEW VOCABULARY

- ☐ Matrix
- ☐ Element
- ☐ Dimension of a matrix
- ☐ Square matrix
- ☐ Augmented matrix
- ☐ Main diagonal
- ☐ Reduced row–echelon form
- ☐ Gauss–Jordan elimination
- ☐ Matrix row transformations

Arrays of numbers are used frequently in many different real-world situations. Spreadsheets often make use of arrays. A **matrix** is a rectangular array of numbers. Each number in a matrix is called an **element**. The following are examples of *matrices* (plural of matrix), with their dimensions written below them.

Matrices and Their Dimensions

$$\begin{bmatrix} 2 & 0 \\ 3 & 1 \end{bmatrix} \quad \begin{bmatrix} -1.2 & 5 & 0 \\ 1 & 0 & 1 \\ 4 & -5 & 7 \end{bmatrix} \quad \begin{bmatrix} 3 & -6 & 0 & \sqrt{3} \\ 1 & 4 & 0 & 9 \\ -3 & 1 & 1 & 18 \\ -10 & -4 & 5 & -1 \end{bmatrix} \quad \begin{bmatrix} 4 & 2 \\ 0 & 1 \\ 1 & 0 \end{bmatrix} \quad \begin{bmatrix} 1 & 5 & -1 \\ 3 & 4 & 2 \end{bmatrix}$$

$$\boxed{2 \times 2} \qquad \boxed{3 \times 3} \qquad \qquad \boxed{4 \times 4} \qquad \qquad \boxed{3 \times 2} \qquad \boxed{2 \times 3}$$

Rows × Columns

▶ **REAL-WORLD CONNECTION** The dimension of a matrix is stated much like the dimensions of a rectangular room. We might say that a room is m feet long and n feet wide. Similarly, the **dimension of a matrix** is $m \times n$ (m by n) if it has m rows and n columns. For example, the last matrix in the preceding group has a dimension of 2×3 because it has 2 rows and 3 columns. If the number of rows equals the number of columns, the matrix is a **square matrix**. The first three matrices in that group are square matrices.

Matrices can be used to represent a system of linear equations. For example, if we have the system of equations

$$\begin{aligned} 3x - y + 2z &= 7 \\ x - 2y + z &= 0 \\ 2x + 5y - 7z &= -9, \end{aligned}$$

we can represent the system with the following **augmented matrix**. Note how the coefficients of the variables are placed in the matrix. A vertical line is positioned in the matrix where the equals signs occur in the system. The rows and columns are labeled, and the elements of the **main diagonal** of the augmented matrix are circled. The matrix has dimension 3×4.

Augmented Matrix

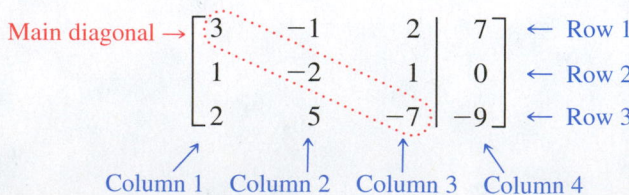

$$\text{Main diagonal} \rightarrow \begin{bmatrix} 3 & -1 & 2 & 7 \\ 1 & -2 & 1 & 0 \\ 2 & 5 & -7 & -9 \end{bmatrix} \begin{matrix} \leftarrow \text{Row 1} \\ \leftarrow \text{Row 2} \\ \leftarrow \text{Row 3} \end{matrix}$$

Column 1 Column 2 Column 3 Column 4

EXAMPLE 1 **Representing a linear system**

Represent each linear system with an augmented matrix. State the dimension of the matrix.

(a) $x - 2y = 9$
$6x + 7y = 16$

(b) $x - 3y + 7z = 4$
$2x + 5y - z = 15$
$2x + y = 8$

Solution

(a) This system can be represented by the following 2×3 augmented matrix.

$$\left[\begin{array}{rr|r} 1 & -2 & 9 \\ 6 & 7 & 16 \end{array}\right]$$

(b) This system can be represented by the following 3×4 augmented matrix.

$$\left[\begin{array}{rrr|r} 1 & -3 & 7 & 4 \\ 2 & 5 & -1 & 15 \\ 2 & 1 & 0 & 8 \end{array}\right]$$

Now Try Exercises 11, 13

Matrices and Social Networks

Today there are several different types of online social networks, such as Facebook and Twitter. Mathematics is essential to the success of these social networks. Consider Figure 9.3, which represents a simple social network of four people.

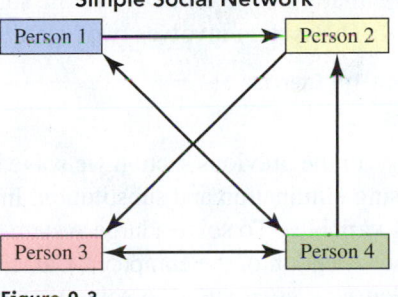

Simple Social Network

Figure 9.3

In Figure 9.3 the arrow from Person 1 to Person 2 indicates that Person 1 likes Person 2, but Person 2 does not like Person 1 because there is no arrow from Person 2 to Person 1. On the other hand, Person 3 and Person 4 like each other because there is a double arrow between them. In the next example we can see how a matrix can be used to represent this simple social network.

EXAMPLE 2 **Representing social networks**

Use a matrix to model the social network shown in Figure 9.3.

Solution

To model a social network with four people, we will use a square 4×4 matrix. Because Person 1 likes Person 2, we put a **1** in row 1 column 2. Because Person 1 also likes Person 4, we put a **1** in row 1 column 4. Continuing in the same manner, we arrive at the following matrix.

A Social Network

$$\left[\begin{array}{rrrr} 0 & 1 & 0 & 1 \\ 0 & 0 & 1 & 0 \\ 0 & 0 & 0 & 1 \\ 1 & 1 & 1 & 0 \end{array}\right]$$

Now Try Exercise 21

Gauss–Jordan Elimination

A convenient matrix form for representing a system of linear equations is **reduced row–echelon form**. The following augmented matrices are examples of reduced row–echelon form. Note that there are 1s on the main diagonal with 0s above and below the 1s.

$$\left[\begin{array}{cc|c} 1 & 0 & 3 \\ 0 & 1 & -2 \end{array}\right] \qquad \left[\begin{array}{ccc|c} 1 & 0 & 0 & 3 \\ 0 & 1 & 0 & 1 \\ 0 & 0 & 1 & -1 \end{array}\right] \qquad \left[\begin{array}{ccc|c} 1 & 0 & 0 & 8 \\ 0 & 1 & 0 & 2 \\ 0 & 0 & 1 & 3 \end{array}\right]$$

If an augmented matrix representing a linear system is in reduced row–echelon form, we can usually determine the solution easily.

EXAMPLE 3 **Determining a solution from reduced row–echelon form**

Each matrix represents a system of linear equations. Find the solution.

(a) $\left[\begin{array}{ccc|c} 1 & 0 & 0 & 2 \\ 0 & 1 & 0 & -3 \\ 0 & 0 & 1 & 5 \end{array}\right]$ (b) $\left[\begin{array}{cc|c} 1 & 0 & 10 \\ 0 & 1 & -4 \end{array}\right]$

Solution
(a) The top row represents $1x + 0y + 0z = 2$ or $x = 2$. The second and third rows tell us that $y = -3$ and $z = 5$. The solution is $(2, -3, 5)$.
(b) The system involves two equations in two variables. The solution is $(10, -4)$.

Now Try Exercise 19

In the previous section we solved systems of three linear equations in three variables by using elimination and substitution. In real life, systems of equations often contain thousands of variables. To solve a large system of equations, we need an efficient method. Long before the invention of the computer, Carl Friedrich Gauss (1777–1855) developed a method called *Gaussian elimination* to solve systems of linear equations. Even though it was developed more than 150 years ago, it is still used today in modern computers and calculators.

We will use a numerical method called **Gauss–Jordan elimination** to solve a linear system. It makes use of the following **matrix row transformations.**

MATRIX ROW TRANSFORMATIONS

For any augmented matrix representing a system of linear equations, the following row transformations result in an equivalent system of linear equations.

1. Any two rows may be interchanged.
2. The elements of any row may be multiplied by a nonzero constant.
3. Any row may be changed by adding to (or subtracting from) its elements a nonzero multiple of the corresponding elements of another row.

Gauss–Jordan elimination can be used to transform an augmented matrix into reduced row–echelon form. Its objective is to use these matrix row transformations to obtain a matrix that has the following reduced row–echelon form, where (a, b) represents the solution.

$$\left[\begin{array}{cc|c} \mathbf{1} & 0 & a \\ 0 & \mathbf{1} & b \end{array}\right]$$

This method is illustrated in the next example.

EXAMPLE 4   **Transforming a matrix into reduced row–echelon form**

Use Gauss–Jordan elimination to transform the augmented matrix of the linear system into reduced row–echelon form. Find the solution.

$$x + y = 5$$
$$-x + y = 1$$

Solution
Both the linear system and the augmented matrix are shown.

Linear System   **Augmented Matrix**

$$x + y = 5$$
$$-x + y = 1$$
$$\begin{bmatrix} 1 & 1 & | & 5 \\ -1 & 1 & | & 1 \end{bmatrix}$$

First, we want to obtain a 0 in the second row, where the -1 is highlighted. To do so, we add row 1 to row 2 and place the result in row 2. This step is denoted $R_2 + R_1$ and eliminates the x-variable from the second equation.

$$x + y = 5$$
$$2y = 6 \qquad R_2 + R_1 \rightarrow \begin{bmatrix} 1 & 1 & | & 5 \\ 0 & 2 & | & 6 \end{bmatrix}$$

To obtain a 1 where the 2 in the second row is located, we multiply the second row by $\frac{1}{2}$, denoted $\frac{1}{2}R_2$.

$$x + y = 5$$
$$y = 3 \qquad \frac{1}{2}R_2 \rightarrow \begin{bmatrix} 1 & 1 & | & 5 \\ 0 & 1 & | & 3 \end{bmatrix}$$

Next, we need to obtain a 0 where the 1 is highlighted. We do so by subtracting row 2 from row 1 and placing the result in row 1, denoted $R_1 - R_2$.

$$x = 2 \qquad R_1 - R_2 \rightarrow \begin{bmatrix} 1 & 0 & | & 2 \\ 0 & 1 & | & 3 \end{bmatrix}$$
$$y = 3$$

This matrix is in reduced row–echelon form. The solution is $(2, 3)$.

Now Try Exercise 23

In the next example, we use Gauss–Jordan elimination to solve a system with three linear equations and three variables. To do so we transform the matrix into the following reduced row–echelon form, where (a, b, c) represents the solution.

$$\begin{bmatrix} 1 & 0 & 0 & | & a \\ 0 & 1 & 0 & | & b \\ 0 & 0 & 1 & | & c \end{bmatrix}$$

EXAMPLE 5   **Transforming a matrix into reduced row–echelon form**

Use Gauss–Jordan elimination to transform the augmented matrix of the linear system into reduced row–echelon form. Find the solution.

$$x + y + 2z = 1$$
$$-x + z = -2$$
$$2x + y + 5z = -1$$

Solution

The linear system and the augmented matrix are both shown.

Linear System	Augmented Matrix

$$\begin{aligned} x + y + 2z &= 1 \\ -x + \quad\;\; z &= -2 \\ 2x + y + 5z &= -1 \end{aligned} \qquad \left[\begin{array}{ccc|c} 1 & 1 & 2 & 1 \\ -1 & 0 & 1 & -2 \\ 2 & 1 & 5 & -1 \end{array}\right]$$

First, we want to put 0s in the second and third rows, where the -1 and 2 are high-lighted. To obtain a 0 in the first position of the second row we add row 1 to row 2 and place the result in row 2, denoted $R_2 + R_1$. To obtain a 0 in the first position of the third row we subtract 2 times row 1 from row 3 and place the result in row 3, denoted $R_3 - 2R_1$. Row 1 does not change. These steps eliminate the x-variable from the second and third equations.

$$\begin{aligned} x + y + 2z &= 1 \\ y + 3z &= -1 \\ -y + \;\; z &= -3 \end{aligned} \qquad \begin{array}{l} \\ R_2 + R_1 \rightarrow \\ R_3 - 2R_1 \rightarrow \end{array} \left[\begin{array}{ccc|c} 1 & 1 & 2 & 1 \\ 0 & 1 & 3 & -1 \\ 0 & -1 & 1 & -3 \end{array}\right]$$

To eliminate the y-variable in row 1, we subtract row 2 from row 1. To eliminate the y-variable from row 3, we add row 2 to row 3.

$$\begin{aligned} x \quad\;\; - z &= 2 \\ y + 3z &= -1 \\ 4z &= -4 \end{aligned} \qquad \begin{array}{l} R_1 - R_2 \rightarrow \\ \\ R_3 + R_2 \rightarrow \end{array} \left[\begin{array}{ccc|c} 1 & 0 & -1 & 2 \\ 0 & 1 & 3 & -1 \\ 0 & 0 & 4 & -4 \end{array}\right]$$

To obtain a 1 in row 3, where the highlighted 4 is located, we multiply row 3 by $\frac{1}{4}$.

$$\begin{aligned} x \quad\;\; - z &= 2 \\ y + 3z &= -1 \\ z &= -1 \end{aligned} \qquad \begin{array}{l} \\ \\ \frac{1}{4}R_3 \rightarrow \end{array} \left[\begin{array}{ccc|c} 1 & 0 & -1 & 2 \\ 0 & 1 & 3 & -1 \\ 0 & 0 & 1 & -1 \end{array}\right]$$

For the matrix to be in reduced row–echelon form, we need 0s in the highlighted loca-tions. We first add row 3 to row 1 and then subtract 3 times row 3 from row 2.

$$\begin{aligned} x &= 1 \\ y &= 2 \\ z &= -1 \end{aligned} \qquad \begin{array}{l} R_1 + R_3 \rightarrow \\ R_2 - 3R_3 \rightarrow \end{array} \left[\begin{array}{ccc|c} 1 & 0 & 0 & 1 \\ 0 & 1 & 0 & 2 \\ 0 & 0 & 1 & -1 \end{array}\right]$$

This matrix is now in reduced row–echelon form. The solution is $(1, 2, -1)$.

Now Try Exercise 33

CRITICAL THINKING

An *inconsistent* system of linear equations has no solutions, and a system of *dependent* linear equations has infinitely many solutions. Suppose that an augmented matrix row reduces to either of the following matrices. Explain what each matrix indicates about the given system of linear equations.

$$\left[\begin{array}{ccc|c} 1 & 0 & 0 & 2 \\ 0 & 1 & 0 & 3 \\ 0 & 0 & 0 & 1 \end{array}\right] \qquad \left[\begin{array}{ccc|c} 1 & 0 & 1 & 2 \\ 0 & 1 & 2 & 3 \\ 0 & 0 & 0 & 0 \end{array}\right]$$

In the next example we find the amounts invested in three mutual funds.

EXAMPLE 6   **Determining investment amounts in mutual funds**

A total of \$8000 was invested in three funds that grew at a rate of 5%, 10%, and 20% over 1 year. After 1 year, the combined value of the three funds had grown by \$1200. Five times as much money was invested at 20% as at 10%. Find the amount invested in each fund.

Solution

Let x be the amount invested at 5%, y be the amount invested at 10%, and z be the amount invested at 20%. The total amount invested was \$8000, so

$$x + y + z = 8000.$$

The growth in the first mutual fund, paying 5% of x, is given by $0.05x$. Similarly, the growths in the other mutual funds are given by $0.10y$ and $0.20z$. As the total growth was \$1200, we can write

$$0.05x + 0.10y + 0.20z = 1200.$$

Multiplying each side of this equation by 20 to eliminate decimals results in

$$x + 2y + 4z = 24,000.$$

Five times as much was invested at 20% as at 10%, so $z = 5y$, or $5y - z = 0$.

These three equations can be written as a system of linear equations and as an augmented matrix.

Linear System   **Augmented Matrix**

$$\begin{aligned} x + y + z &= 8{,}000 \\ x + 2y + 4z &= 24{,}000 \\ 5y - z &= \phantom{24{,}00}0 \end{aligned} \qquad \left[\begin{array}{ccc|c} 1 & 1 & 1 & 8{,}000 \\ 1 & 2 & 4 & 24{,}000 \\ 0 & 5 & -1 & 0 \end{array}\right]$$

A 0 can be obtained in the highlighted position by subtracting row 1 from row 2.

$$\begin{aligned} x + y + z &= 8{,}000 \\ y + 3z &= 16{,}000 \\ 5y - z &= \phantom{16{,}00}0 \end{aligned} \qquad R_2 - R_1 \rightarrow \left[\begin{array}{ccc|c} 1 & 1 & 1 & 8{,}000 \\ 0 & 1 & 3 & 16{,}000 \\ 0 & 5 & -1 & 0 \end{array}\right]$$

Zeros can be obtained in the highlighted positions by subtracting row 2 from row 1 and by subtracting 5 times row 2 from row 3.

$$\begin{aligned} x \phantom{{}+ y} - 2z &= -8{,}000 \\ y + 3z &= 16{,}000 \\ -16z &= -80{,}000 \end{aligned} \qquad \begin{aligned} R_1 - R_2 \rightarrow \\ \\ R_3 - 5R_2 \rightarrow \end{aligned} \left[\begin{array}{ccc|c} 1 & 0 & -2 & -8{,}000 \\ 0 & 1 & 3 & 16{,}000 \\ 0 & 0 & -16 & -80{,}000 \end{array}\right]$$

To obtain a 1 in the highlighted position, multiply row 3 by $-\frac{1}{16}$.

$$\begin{aligned} x \phantom{{}+ y} - 2z &= -8{,}000 \\ y + 3z &= 16{,}000 \\ z &= 5{,}000 \end{aligned} \qquad -\frac{1}{16}R_3 \rightarrow \left[\begin{array}{ccc|c} 1 & 0 & -2 & -8{,}000 \\ 0 & 1 & 3 & 16{,}000 \\ 0 & 0 & 1 & 5{,}000 \end{array}\right]$$

To obtain a 0 in each of the highlighted positions, add twice row 3 to row 1 and subtract three times row 3 from row 2.

$$\begin{aligned} x &= 2{,}000 \\ y &= 1{,}000 \\ z &= 5{,}000 \end{aligned} \qquad \begin{aligned} R_1 + 2R_3 \rightarrow \\ R_2 - 3R_3 \rightarrow \\ \\ \end{aligned} \left[\begin{array}{ccc|c} 1 & 0 & 0 & 2{,}000 \\ 0 & 1 & 0 & 1{,}000 \\ 0 & 0 & 1 & 5{,}000 \end{array}\right]$$

Thus \$2000 was invested at 5%, \$1000 at 10%, and \$5000 at 20%.

Now Try Exercise 65

Using Technology to Solve Systems of Linear Equations (Optional)

▶ **REAL-WORLD CONNECTION** Examples 5 and 6 involve a lot of arithmetic. Trying to solve a large system of equations by hand is an enormous—if not impossible—task. In the real world, people use technology to solve large systems. In the next example we solve the linear systems from Examples 4 and 5 with a graphing calculator.

EXAMPLE 7 Using technology

Use a graphing calculator to solve the following systems of equations.

(a) $x + y = 5$ (b) $x + y + 2z = 1$
$-x + y = 1$ $-x + z = -2$
 $2x + y + 5z = -1$

Solution

(a) Enter the 2×3 augmented matrix from Example 4 in a graphing calculator, as shown in Figure 9.4(a). Then transform the matrix into reduced row–echelon form (rref), as shown in Figure 9.4(b). The solution is (2, 3).

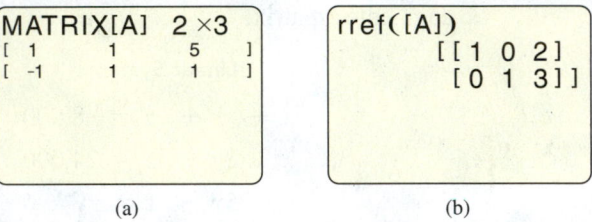

(a)　　　　　　　　　(b)

Figure 9.4

(b) Enter the 3×4 augmented matrix from Example 5, as shown in Figure 9.5(a). (The fourth column of A can be seen by scrolling right.) Then transform the matrix into reduced row–echelon form (rref), as shown in Figure 9.5(b). The solution is $(1, 2, -1)$.

CALCULATOR HELP

To enter a matrix and put it in reduced row–echelon form, see Appendix A (pages AP-8 and AP-9).

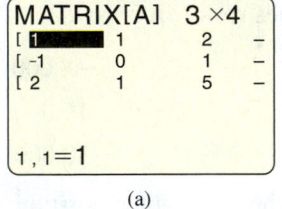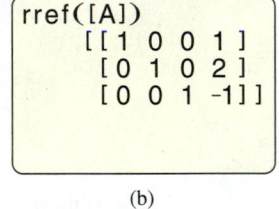

(a)　　　　　　　　　(b)

Figure 9.5

▮ **Now Try Exercises 41, 43**

▶ **REAL-WORLD CONNECTION** In the next example we use technology to solve a type of application that was first presented in the chapter introduction.

EXAMPLE 8 Modeling the weight of male bears

The data shown in Table 9.2 give the weight W, head length H, and overall length L of three bears. These data can be modeled with the equation $W = a + bH + cL$, where a, b, and c are constants that we need to determine. (*Sources:* M. Triola, *Elementary Statistics*; Minitab, Inc.)

TABLE 9.2

W (pounds)	H (inches)	L (inches)
362	16	72
300	14	68
147	11	52
?	13	65

(a) Set up a system of equations whose solution gives values for constants a, b, and c.
(b) Solve the system.
(c) Predict the weight of a bear with $H = 13$ inches and $L = 65$ inches.

Solution
(a) Substitute each row of Table 9.2 in the equation $W = a + bH + cL$.

$$362 = a + b(16) + c(72)$$
$$300 = a + b(14) + c(68)$$
$$147 = a + b(11) + c(52)$$

Rewrite this system as

$$a + 16b + 72c = 362$$
$$a + 14b + 68c = 300$$
$$a + 11b + 52c = 147$$

and represent it as the augmented matrix

$$A = \begin{bmatrix} 1 & 16 & 72 & | & 362 \\ 1 & 14 & 68 & | & 300 \\ 1 & 11 & 52 & | & 147 \end{bmatrix}.$$

(b) Enter A and put it in reduced row–echelon form, as shown in Figures 9.6(a) and (b), respectively. The solution is $a = -374$, $b = 19$, and $c = 6$.

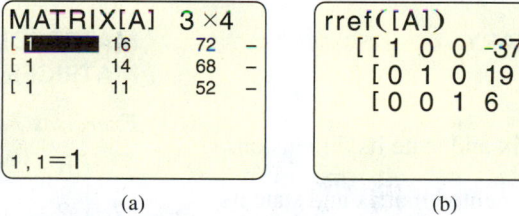

(a)   (b)

Figure 9.6

(c) For $W = a + bH + cL$, use

$$W = -374 + 19H + 6L$$

to predict the weight of a bear with head length $H = 13$ and overall length $L = 65$.

$$W = -374 + 19(13) + 6(65) = 263 \text{ pounds}$$

Now Try Exercise 61

9.2   Putting It All Together

A matrix is a rectangular array of numbers. An augmented matrix may be used to represent any system of linear equations. One common method for solving a system of linear equations is Gauss–Jordan elimination. Matrix row operations may be used to transform an augmented matrix to reduced row–echelon form.

CONCEPT	EXPLANATION/EXAMPLE
Augmented Matrix	A linear system can be represented by an augmented matrix. The following matrix has dimension 3×4. **Linear System** $\begin{aligned} x + 2y - z &= 6 \\ -2x + y - z &= 7 \\ 2x + 3z &= -11 \end{aligned}$ **Augmented Matrix** $\begin{bmatrix} 1 & 2 & -1 & 6 \\ -2 & 1 & -1 & 7 \\ 2 & 0 & 3 & -11 \end{bmatrix}$
Reduced Row–Echelon Form	The following augmented matrix is in reduced row–echelon form, which results from transforming the preceding system to reduced row–echelon form. There are 1s along the main diagonal and 0s elsewhere in the first three columns. The solution to the linear system is $(-1, 2, -3)$. $\begin{bmatrix} 1 & 0 & 0 & -1 \\ 0 & 1 & 0 & 2 \\ 0 & 0 & 1 & -3 \end{bmatrix}$

9.2   Exercises

 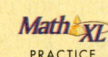

CONCEPTS AND VOCABULARY

1. What is a matrix?

2. Give an example of a matrix and state its dimension.

3. Give an example of an augmented matrix and state its dimension.

4. If an augmented matrix is used to solve a system of three linear equations in three variables, what will be its dimension?

5. Give an example of a matrix that is in reduced row–echelon form.

6. Identify the elements on the main diagonal in the augmented matrix.

$$\begin{bmatrix} 4 & -6 & -1 & 3 \\ 6 & 2 & -2 & 9 \\ 7 & 5 & -3 & 1 \end{bmatrix}$$

MATRIX DIMENSIONS AND AUGMENTED MATRICES

Exercises 7–10: State the dimension of the matrix.

7. $\begin{bmatrix} 3 & -3 & 7 \\ 2 & 6 & -2 \\ 4 & 2 & 5 \end{bmatrix}$

8. $\begin{bmatrix} -2 & 3 & 0 \\ 1 & -8 & 4 \end{bmatrix}$

9. $\begin{bmatrix} 1 & 7 \\ 0 & 2 \\ 2 & -5 \end{bmatrix}$

10. $\begin{bmatrix} 4 & 2 & -3 & -1 \\ 4 & -3 & 2 & -7 \\ 14 & 6 & 4 & 0 \end{bmatrix}$

Exercises 11–14: Represent the linear system as an augmented matrix.

11. $\begin{aligned} x - 3y &= 1 \\ -x + 3y &= -1 \end{aligned}$

12. $\begin{aligned} 4x + 2y &= -5 \\ 5x + 8y &= 2 \end{aligned}$

13.
$$2x - y + 2z = -4$$
$$x - 2y \quad\quad = 2$$
$$-x + y - 2z = -6$$

14.
$$3x - 2y + z = 5$$
$$-x \quad\quad + 2z = -4$$
$$x - 2y + z = -1$$

31.
$$x + y + z = 6$$
$$2y - z = 1$$
$$y + z = 5$$

32.
$$x + y + z = 3$$
$$x + y - z = 2$$
$$y + z = 2$$

33.
$$x + 2y + 3z = 6$$
$$-x + 3y + 4z = 0$$
$$x + y - 2z = -6$$

Exercises 15–20: Write the system of linear equations that the augmented matrix represents. Use the variables x, y, and z.

15. $\begin{bmatrix} 1 & 2 & | & -6 \\ 5 & -1 & | & 4 \end{bmatrix}$

16. $\begin{bmatrix} 1 & -5 & | & 7 \\ 0 & -3 & | & 6 \end{bmatrix}$

34.
$$2x - 4y + 2z = 10$$
$$-x + 3y - 4z = -19$$
$$2x - y - 6z = -28$$

17. $\begin{bmatrix} 1 & -1 & 2 & | & 6 \\ 2 & 1 & -2 & | & 1 \\ -1 & 2 & -1 & | & 3 \end{bmatrix}$

18. $\begin{bmatrix} 3 & -1 & 2 & | & -1 \\ 2 & -2 & 2 & | & 4 \\ 1 & 7 & -2 & | & 2 \end{bmatrix}$

35.
$$x + y + z = 0$$
$$2x + y + 2z = -1$$
$$x + y \quad\quad = 0$$

36.
$$x + y - 2z = 5$$
$$x + 2y - 2z = 4$$
$$-x - y + z = -4$$

19. $\begin{bmatrix} 1 & 0 & 0 & | & 4 \\ 0 & 1 & 0 & | & -2 \\ 0 & 0 & 1 & | & 7 \end{bmatrix}$

20. $\begin{bmatrix} 1 & 0 & 0 & | & 6 \\ 0 & 1 & 0 & | & -2 \\ 0 & 0 & 1 & | & 4 \end{bmatrix}$

37.
$$x + y + z = 3$$
$$-x \quad\quad - z = -2$$
$$x + y + 2z = 4$$

38.
$$x + 2y - z = 3$$
$$-x - y + z = 0$$
$$x + 2y \quad\quad = 5$$

39.
$$x + 2y + z = 3$$
$$2x + y - z = -6$$
$$-x - y + 2z = 5$$

40.
$$x + y + z = -3$$
$$x - y - z = -1$$
$$-2x + y + 4z = 4$$

SOCIAL NETWORKS

Exercises 21 and 22: (Refer to Example 2.) Use a 4 × 4 matrix to represent the social network.

21.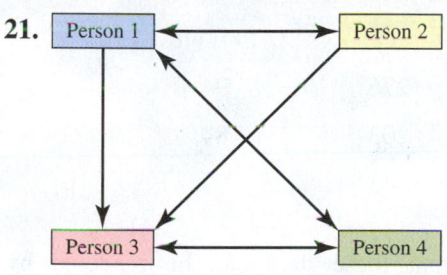

22.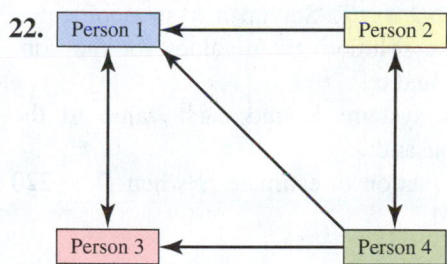

📱 *Exercises 41–50:* **Technology** *Use a graphing calculator to solve the system of linear equations.*

41.
$$x + 4y = 13$$
$$5x - 3y = -50$$

42.
$$9x - 11y = 7$$
$$5x + 6y = 16$$

43.
$$2x - y + 3z = 9$$
$$-4x + 5y + 2z = 12$$
$$2x \quad\quad + 7z = 23$$

44.
$$3x - 2y + 4z = 29$$
$$2x + 3y - 7z = -14$$
$$5x - y + 11z = 59$$

GAUSS–JORDAN ELIMINATION

Exercises 23–40: Use Gauss–Jordan elimination to find the solution. Write the solution as an ordered pair or ordered triple and check the solution.

45.
$$6x + 2y + z = 4$$
$$-2x + 4y + z = -3$$
$$2x - 8y \quad\quad = -2$$

46.
$$-x - 9y + 2z = -28.5$$
$$2x - y + 4z = -17$$
$$x - y + 8z = -9$$

23.
$$x + y = 4$$
$$x + 3y = 10$$

24.
$$x - 3y = -7$$
$$2x + y = 0$$

25.
$$2x + 3y = 3$$
$$-2x + 2y = 7$$

26.
$$x + 3y = -14$$
$$2x + 5y = -24$$

47.
$$4x + 3y + 12z = -9.25$$
$$15y + 8z = -4.75 + x$$
$$7z = -5.5 - 6y$$

27.
$$x - y = 5$$
$$x + 3y = -1$$

28.
$$x + 4y = 1$$
$$3x - 2y = 10$$

48.
$$5x + 4y = 13.3 + z$$
$$7y + 9z = 16.9 - x$$
$$x - 3y + 4z = -4.1$$

29.
$$4x - 8y = -10$$
$$x + y = 2$$

30.
$$x - 7y = -16$$
$$4x + 10y = 50$$

49.
$$1.2x - 0.9y + 2.7z = 5.37$$
$$3.1x - 5.1y + 7.2z = 14.81$$
$$1.8y + 6.38 = 3.6z - 0.2x$$

50.
$$11x + 13y - 17z = 380$$
$$5x - 14y - 19z = 24$$
$$-21y + 46z = -676 + 7x$$

Exercises 51–56: (Refer to the Critical Thinking box in this section.) Row-reduce the matrix associated with the given system to determine whether the system of linear equations is inconsistent or has dependent equations.

51. $x + 2y = 4$
$-2x - 4y = -8$

52. $x - 5y = 4$
$-2x + 10y = 8$

53. $x + y + z = 3$
$x + y - z = 1$
$x + y = 3$

54. $x + y + z = 5$
$x - y - z = 8$
$2x + 2y + 2z = 6$

55. $x + 2y + 3z = 14$
$2x - 3y - 2z = -10$
$3x - y + z = 4$

56. $x + 2y + 3z = 6$
$-x + 3y + 4z = 6$
$5y + 7z = 12$

Exercises 57 and 58: **Thinking Generally** *The matrix represents a linear system of equations. Solve the system, if a and b are nonzero constants.*

57. $\begin{bmatrix} a & 0 & 0 & | & 1 \\ 0 & b & 0 & | & 1 \\ 0 & 0 & ab & | & 2 \end{bmatrix}$

58. $\begin{bmatrix} a & 0 & 0 & | & 1 \\ 0 & b & 0 & | & 2 \\ 0 & 0 & 0 & | & a \end{bmatrix}$

APPLICATIONS

59. *Weight of a Bear* Use the results of Example 8 to estimate the weight of a bear with a head length of 12 inches and an overall length of 60 inches.

 60. *Garbage and Household Size* A larger household produces more garbage, on average, than a smaller household. If we know the amount of metal M and plastic P waste produced each week, we can estimate the household size H from $H = a + bM + cP$. The table contains representative data for three households. (*Source:* M. Triola, *Elementary Statistics.*)

H (people)	M (pounds)	P (pounds)
3	2.00	1.40
2	1.50	0.65
6	4.00	3.40

(a) Set up a system of equations whose solution gives values for the constants a, b, and c.
(b) Solve this system.
(c) Predict the size of a household that produces 3 pounds of metal waste and 2 pounds of plastic waste each week.

 61. *Weight of a Bear* (Refer to Example 8.) Head length and overall length are not the only variables that can be used to estimate the weight of a bear. The data in the accompanying table list the weight W, neck size N, and chest size C of three bears. These data

can be modeled by $W = a + bN + cC$. (*Sources:* M. Triola, *Elementary Statistics;* Minitab, Inc.)

W (pounds)	N (inches)	C (inches)
80	16	26
344	28	45
416	31	54

(a) Set up a system of equations whose solution gives values for the constants a, b, and c.
(b) Solve this system. Round each value to the nearest tenth.
(c) Predict the weight of a bear with neck size $N = 22$ inches and chest size $C = 38$ inches.

62. *Old Faithful Geyser* In Yellowstone National Park, Old Faithful Geyser has been a favorite attraction for decades. Although this geyser erupts about every 80 minutes, this time interval varies, as do the duration and height of the eruptions. The accompanying table shows the height H, duration D, and time interval T for three representative eruptions. (*Source:* National Park Service.)

H (feet)	D (seconds)	T (minutes)
160	276	94
125	203	84
140	245	79

(a) Assume that these data can be modeled by $H = a + bD + cT$. Set up a system of equations whose solution gives values for the constants a, b, and c.
(b) Solve this system. Round each value to the nearest thousandth.
(c) Use this equation to estimate H when $D = 220$ and $T = 81$.

63. *Jogging Speeds* A runner in preparation for a marathon jogs at 5, 6, and 8 miles per hour. The runner travels a total distance of 12.5 miles in 2 hours and jogs the same length of time at 5 miles per hour and at 8 miles per hour. How long does the runner jog at each speed?

64. *Mixture Problem* Three types of candy that cost $2, $3, and $4 per pound are to be mixed to produce a 5-pound bag of candy that costs $14.50. If there are to be equal amounts of the $3-per-pound candy and the $4-per-pound candy, how much of each type of candy should be included in the mixture?

65. *Interest and Investments* (Refer to Example 6.) A total of $3000 is invested at 3%, 4%, and 6% annual interest. The interest earned after 1 year equals $145. The amount invested at 6% is triple the amount invested at 3%. Find the amount invested at each rate.

66. *Geometry* The measure of the largest angle in a triangle is twice the measure of the smallest angle. The remaining angle is 10° less than the largest angle. Find the measure of each angle.

67. **Online Exploration** Go to your Facebook profile and select three or four of your friends plus yourself. Answers will vary.
  (a) Draw a social network that shows the friendships between each pair of people. Use a double arrow to represent this. (See Figure 9.3.)
  (b) Use a matrix to represent your social network.

68. **Online Exploration** Go online and look up a diagram of the molecular structure of water (H_2O).
  (a) Draw double arrows wherever the hydrogen and oxygen are connected.
  (b) Use a matrix to represent the structure of water. Let the first row represent oxygen.

WRITING ABOUT MATHEMATICS

69. Explain what the dimension of a matrix means. What is the difference between a matrix that has dimension 3×4 and one that has dimension 4×3?

70. Discuss the advantages of using technology to transform an augmented matrix to reduced row–echelon form. Are there any disadvantages? Explain.

SECTIONS 9.1 and 9.2   **Checking Basic Concepts**

1. Determine which ordered triple is the solution to the system of equations: $(5, -4, 0)$ or $(1, 3, -1)$.
$$x - y + 7z = -9$$
$$2x - 2y + 5z = -9$$
$$-x + 3y - 2z = 10$$

2. Solve the system of equations by using elimination and substitution.
$$x - y + z = 2$$
$$2x - 3y + z = -1$$
$$-x + y + z = 4$$

3. Use an augmented matrix to represent the system of equations. Solve the system by using
  (a) Gauss–Jordan elimination and
  (b) technology.
$$x + 2y + z = 1$$
$$x + y + z = -1$$
$$y + z = 1$$

4. A total of $1500 is invested at 1%, 2%, and 4% annual interest. The interest after 1 year equals $46. The amount invested at 2% is double the amount invested at 1%. Find the amount invested at each rate.

9.3   Determinants

Calculation of Determinants • Area of Regions • Cramer's Rule

A LOOK INTO MATH ▶

Surveyors commonly calculate the areas of parcels of land. To do so, they frequently divide the land into triangular regions. When the coordinates of the vertices of a triangle are known, determinants may be used to find the area of the triangle. *A determinant is a real number that can be calculated for any square matrix.* In this section we use determinants to find areas and to solve systems of linear equations.

Calculation of Determinants

NEW VOCABULARY

☐ Determinant
☐ Expansion by minors
☐ Minors
☐ Cramer's rule

▶ **REAL-WORLD CONNECTION** The concept of determinants originated with the Japanese mathematician Seki Kowa (1642–1708), who used them to solve systems of linear equations. Later, Gottfried Leibniz (1646–1716) formally described determinants and also used them to solve systems of linear equations. (*Source: Historical Topics for the Mathematical Classroom, NCTM.*)

We begin by defining a determinant of a 2×2 matrix.

DETERMINANT OF A 2×2 MATRIX

The **determinant** of

$$A = \begin{bmatrix} a & b \\ c & d \end{bmatrix}$$

is a *real number* defined by

$$\det A = ad - cb.$$

EXAMPLE 1   **Calculating determinants**

Find det A for each 2×2 matrix.

(a) $A = \begin{bmatrix} 1 & 2 \\ 3 & 4 \end{bmatrix}$ **(b)** $A = \begin{bmatrix} -1 & -3 \\ 2 & -8 \end{bmatrix}$

Solution
(a) The determinant is calculated as follows.

$$\det A = \det \begin{bmatrix} 1 & 2 \\ 3 & 4 \end{bmatrix} = (1)(4) - (3)(2) = -2$$

(b) Similarly,

$$\det A = \det \begin{bmatrix} -1 & -3 \\ 2 & -8 \end{bmatrix} = (-1)(-8) - (2)(-3) = 14.$$

Now Try Exercise 5

READING CHECK

How do you calculate the determinant of a 2×2 matrix?

We can use determinants of 2×2 matrices to find determinants of 3×3 matrices. This method is called **expansion of a determinant by minors**.

DETERMINANT OF A 3 × 3 MATRIX

$$\det A = \det \begin{bmatrix} a_1 & b_1 & c_1 \\ a_2 & b_2 & c_2 \\ a_3 & b_3 & c_3 \end{bmatrix}$$

$$= a_1 \cdot \det \begin{bmatrix} b_2 & c_2 \\ b_3 & c_3 \end{bmatrix} - a_2 \cdot \det \begin{bmatrix} b_1 & c_1 \\ b_3 & c_3 \end{bmatrix} + a_3 \cdot \det \begin{bmatrix} b_1 & c_1 \\ b_2 & c_2 \end{bmatrix}$$

The 2 × 2 matrices in this equation are called **minors**.

EXAMPLE 2 **Calculating 3 × 3 determinants**

Evaluate det A.

(a) $A = \begin{bmatrix} 2 & 1 & -1 \\ -1 & 3 & 2 \\ 4 & -3 & -5 \end{bmatrix}$ (b) $A = \begin{bmatrix} 5 & -2 & 4 \\ 0 & 2 & 1 \\ -1 & 4 & -4 \end{bmatrix}$

Solution
(a) We evaluate the determinant as follows.

$$\det \begin{bmatrix} \mathbf{2} & 1 & -1 \\ \mathbf{-1} & 3 & 2 \\ \mathbf{4} & -3 & -5 \end{bmatrix} = \mathbf{2} \cdot \det \begin{bmatrix} 3 & 2 \\ -3 & -5 \end{bmatrix} - (\mathbf{-1}) \cdot \det \begin{bmatrix} 1 & -1 \\ -3 & -5 \end{bmatrix}$$

$$+ \mathbf{4} \cdot \det \begin{bmatrix} 1 & -1 \\ 3 & 2 \end{bmatrix}$$

$$= 2(-9) + 1(-8) + 4(5)$$

$$= -6$$

(b) We evaluate the determinant as follows.

$$\det \begin{bmatrix} \mathbf{5} & -2 & 4 \\ \mathbf{0} & 2 & 1 \\ \mathbf{-1} & 4 & -4 \end{bmatrix} = \mathbf{5} \cdot \det \begin{bmatrix} 2 & 1 \\ 4 & -4 \end{bmatrix} - (\mathbf{0}) \cdot \det \begin{bmatrix} -2 & 4 \\ 4 & -4 \end{bmatrix}$$

$$+ (\mathbf{-1}) \cdot \det \begin{bmatrix} -2 & 4 \\ 2 & 1 \end{bmatrix}$$

$$= 5(-12) - 0(-8) + (-1)(-10)$$

$$= -50$$

Now Try Exercises 13, 15

Many graphing calculators can evaluate the determinant of a matrix, as illustrated in the next example, where we evaluate the determinants from Example 2.

EXAMPLE 3 **Using technology to find determinants**

Find each determinant of A, using a graphing calculator.

(a) $A = \begin{bmatrix} 2 & 1 & -1 \\ -1 & 3 & 2 \\ 4 & -3 & -5 \end{bmatrix}$ (b) $A = \begin{bmatrix} 5 & -2 & 4 \\ 0 & 2 & 1 \\ -1 & 4 & -4 \end{bmatrix}$

Solution

(a) Begin by entering the matrix and then evaluate the determinant, as shown in Figure 9.7. The result is det $A = -6$, which agrees with our earlier calculation.

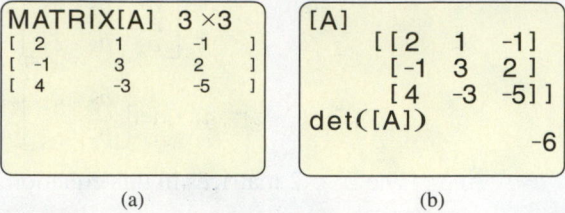

(a)   (b)

Figure 9.7

CALCULATOR HELP

To find a determinant, see Appendix A (pages AP-9 and AP-10).

(b) The determinant of A evaluates to -50 (see Figure 9.8).

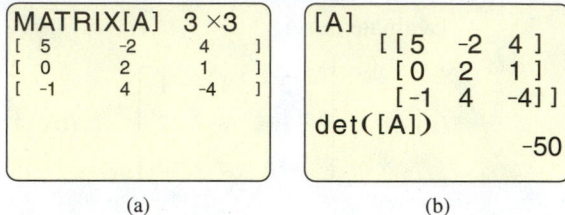

(a)   (b)

Figure 9.8

▌ **Now Try Exercise 21**

Area of Regions

CRITICAL THINKING

Suppose that you are given three distinct points in the xy-plane and $D = 0$. What must be true about the three points?

▶ **REAL-WORLD CONNECTION** A determinant may be used to find the area of a triangle. For example, if a triangle has vertices (a_1, a_2), (b_1, b_2), and (c_1, c_2), its area equals the absolute value of D, where

$$D = \frac{1}{2}\det\begin{bmatrix} a_1 & b_1 & c_1 \\ a_2 & b_2 & c_2 \\ 1 & 1 & 1 \end{bmatrix}.$$

If the vertices are entered in the columns of D counterclockwise as they appear in the xy-plane, D will be positive. (*Source:* W. Taylor, *The Geometry of Computer Graphics.*)

EXAMPLE 4 ▌ **Computing the area of a triangular parcel of land**

A triangular parcel of land is shown in Figure 9.9. If all units are miles, find the area of the parcel of land by using a determinant.

Solution

The vertices of the triangular parcel of land are $(2, 2)$, $(5, 4)$, and $(3, 8)$. The area of the triangle is

$$D = \frac{1}{2}\det\begin{bmatrix} 2 & 5 & 3 \\ 2 & 4 & 8 \\ 1 & 1 & 1 \end{bmatrix} = \frac{1}{2}\cdot 16 = 8.$$

The area of the triangle is 8 square miles.

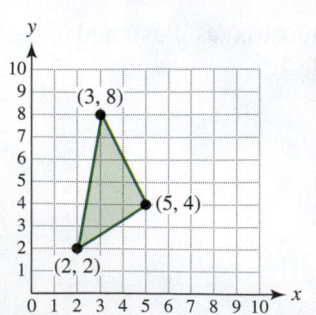

Figure 9.9

▌ **Now Try Exercise 27**

Cramer's Rule

▶ **REAL-WORLD CONNECTION** Determinants were developed by Gabriel Cramer (1704–1752). His work, published in 1750, provided a method called **Cramer's rule** for solving systems of linear equations.

CRAMER'S RULE FOR LINEAR SYSTEMS IN TWO VARIABLES

The solution to the system of linear equations

$$a_1x + b_1y = c_1$$
$$a_2x + b_2y = c_2$$

is given by $x = \frac{E}{D}$ and $y = \frac{F}{D}$, where

$$E = \det\begin{bmatrix} c_1 & b_1 \\ c_2 & b_2 \end{bmatrix}, \quad F = \det\begin{bmatrix} a_1 & c_1 \\ a_2 & c_2 \end{bmatrix}, \quad \text{and} \quad D = \det\begin{bmatrix} a_1 & b_1 \\ a_2 & b_2 \end{bmatrix} \neq 0.$$

NOTE: If $D = 0$, the system has either no solutions or infinitely many solutions.

EXAMPLE 5   **Using Cramer's rule**

Use Cramer's rule to solve the following linear systems.
(a) $3x - 4y = 18$ **(b)** $-4x + 9y = -24$
  $7x + 5y = -1$ $6x + 17y = -25$

Solution

(a) $E = \det\begin{bmatrix} c_1 & b_1 \\ c_2 & b_2 \end{bmatrix} = \det\begin{bmatrix} 18 & -4 \\ -1 & 5 \end{bmatrix} = (18)(5) - (-1)(-4) = \mathbf{86}$

$F = \det\begin{bmatrix} a_1 & c_1 \\ a_2 & c_2 \end{bmatrix} = \det\begin{bmatrix} 3 & 18 \\ 7 & -1 \end{bmatrix} = (3)(-1) - (7)(18) = \mathbf{-129}$

$D = \det\begin{bmatrix} a_1 & b_1 \\ a_2 & b_2 \end{bmatrix} = \det\begin{bmatrix} 3 & -4 \\ 7 & 5 \end{bmatrix} = (3)(5) - (7)(-4) = \mathbf{43}$

Because $x = \frac{E}{D} = \frac{86}{43} = 2$ and $y = \frac{F}{D} = \frac{-129}{43} = -3$, the solution is $(2, -3)$.

(b) $E = \det\begin{bmatrix} c_1 & b_1 \\ c_2 & b_2 \end{bmatrix} = \det\begin{bmatrix} -24 & 9 \\ -25 & 17 \end{bmatrix} = (-24)(17) - (-25)(9) = \mathbf{-183}$

$F = \det\begin{bmatrix} a_1 & c_1 \\ a_2 & c_2 \end{bmatrix} = \det\begin{bmatrix} -4 & -24 \\ 6 & -25 \end{bmatrix} = (-4)(-25) - (6)(-24) = \mathbf{244}$

$D = \det\begin{bmatrix} a_1 & b_1 \\ a_2 & b_2 \end{bmatrix} = \det\begin{bmatrix} -4 & 9 \\ 6 & 17 \end{bmatrix} = (-4)(17) - (6)(9) = \mathbf{-122}$

Because $x = \frac{E}{D} = \frac{-183}{-122} = 1.5$ and $y = \frac{F}{D} = \frac{244}{-122} = -2$, the solution is $(1.5, -2)$.

Now Try Exercises 33, 35

Cramer's rule can be applied to systems that have any number of linear equations. Cramer's rule for three linear equations is discussed in the Extended and Discovery Exercises at the end of this chapter.

9.3 **Putting It All Together**

CONCEPT	EXPLANATION
Determinant of a 2×2 Matrix	The determinant of a 2×2 matrix A is given by $$\det A = \det \begin{bmatrix} a & b \\ c & d \end{bmatrix} = ad - cb.$$
Determinant of a 3×3 Matrix	The determinant of a 3×3 matrix A is given by $$\det A = \det \begin{bmatrix} a_1 & b_1 & c_1 \\ a_2 & b_2 & c_2 \\ a_3 & b_3 & c_3 \end{bmatrix}$$ $$= a_1 \cdot \det \begin{bmatrix} b_2 & c_2 \\ b_3 & c_3 \end{bmatrix} - a_2 \cdot \det \begin{bmatrix} b_1 & c_1 \\ b_3 & c_3 \end{bmatrix}$$ $$+ a_3 \cdot \det \begin{bmatrix} b_1 & c_1 \\ b_2 & c_2 \end{bmatrix}.$$
Area of a Triangle	If a triangle has vertices (a_1, a_2), (b_1, b_2), and (c_1, c_2), its area equals the absolute value of D, where $$D = \frac{1}{2} \det \begin{bmatrix} a_1 & b_1 & c_1 \\ a_2 & b_2 & c_2 \\ 1 & 1 & 1 \end{bmatrix}.$$
Cramer's Rule for Linear Systems in Two Variables	The solution to the linear system $$a_1 x + b_1 y = c_1$$ $$a_2 x + b_2 y = c_2$$ is given by $x = \frac{E}{D}$ and $y = \frac{F}{D}$, where $$E = \det \begin{bmatrix} c_1 & b_1 \\ c_2 & b_2 \end{bmatrix}, \quad F = \det \begin{bmatrix} a_1 & c_1 \\ a_2 & c_2 \end{bmatrix}, \quad \text{and}$$ $$D = \det \begin{bmatrix} a_1 & b_1 \\ a_2 & b_2 \end{bmatrix} \neq 0.$$ **NOTE:** If $D = 0$, then the system has either no solutions or infinitely many solutions.

9.3 **Exercises**

PRACTICE   WATCH   DOWNLOAD   READ   REVIEW

CONCEPTS AND VOCABULARY

1. We can find the determinant of a(n) _____ matrix.

2. If we find a determinant, the answer is a(n) _____.

3. Cramer's rule can be used to solve a(n) _____.

4. If the first column of a matrix is all 0s, then its determinant equals _____.

CALCULATING DETERMINANTS

Exercises 5–20: Evaluate det A by hand where A is the given matrix.

5. $\begin{bmatrix} 1 & -2 \\ 3 & -8 \end{bmatrix}$

6. $\begin{bmatrix} 5 & -1 \\ 3 & 7 \end{bmatrix}$

7. $\begin{bmatrix} -3 & 7 \\ 8 & -1 \end{bmatrix}$

8. $\begin{bmatrix} 0 & -7 \\ -3 & 1 \end{bmatrix}$

9. $\begin{bmatrix} 23 & 4 \\ 6 & -13 \end{bmatrix}$

10. $\begin{bmatrix} 44 & -51 \\ -9 & 32 \end{bmatrix}$

11. $\begin{bmatrix} 1 & -1 & 2 \\ 0 & 1 & -3 \\ 0 & -4 & 7 \end{bmatrix}$

12. $\begin{bmatrix} 2 & -1 & -5 \\ -1 & 4 & -2 \\ 0 & 1 & 4 \end{bmatrix}$

13. $\begin{bmatrix} 2 & -1 & 0 \\ 1 & -2 & 6 \\ 0 & 1 & 8 \end{bmatrix}$

14. $\begin{bmatrix} 0 & 1 & -4 \\ 3 & -6 & 10 \\ 4 & -2 & 7 \end{bmatrix}$

15. $\begin{bmatrix} -1 & 3 & 5 \\ 3 & -3 & 5 \\ 2 & -3 & 7 \end{bmatrix}$

16. $\begin{bmatrix} 6 & -1 & 9 \\ 7 & 0 & -3 \\ 2 & 5 & -1 \end{bmatrix}$

17. $\begin{bmatrix} 5 & 0 & 0 \\ 0 & -2 & 0 \\ 0 & 0 & 5 \end{bmatrix}$

18. $\begin{bmatrix} 1 & 2 & 3 \\ 2 & 4 & 6 \\ 3 & 6 & 9 \end{bmatrix}$

19. $\begin{bmatrix} 0 & 2 & -3 \\ 0 & 3 & -9 \\ 0 & 5 & 9 \end{bmatrix}$

20. $\begin{bmatrix} 3 & -1 & 2 \\ 0 & 5 & 7 \\ 0 & 0 & -1 \end{bmatrix}$

Exercises 21–24: Use technology to calculate det A, where A is the given matrix.

21. $\begin{bmatrix} 2 & -5 & 13 \\ 10 & 15 & -10 \\ 17 & -19 & 22 \end{bmatrix}$

22. $\begin{bmatrix} 1.6 & 3.1 & 5.7 \\ 2.1 & 6.7 & 8.1 \\ -0.4 & -0.8 & -3.1 \end{bmatrix}$

23. $\begin{bmatrix} 17 & 0 & 4 \\ -9 & 14 & 1.5 \\ 13 & 67 & -11 \end{bmatrix}$

24. $\begin{bmatrix} 121 & 45 & -56 \\ -45 & 87 & 32 \\ -14 & -34 & 67 \end{bmatrix}$

Exercises 25 and 26: **Thinking Generally** *Find det A, if a, b, and c are nonzero constants.*

25. $A = \begin{bmatrix} a & 0 & 0 \\ 0 & b & 0 \\ 0 & 0 & c \end{bmatrix}$

26. $A = \begin{bmatrix} 0 & 0 & 0 \\ a & b & 0 \\ 0 & 0 & c \end{bmatrix}$

CALCULATING AREA

Exercises 27–32: Find the area of the figure by using a determinant. Assume that units are feet.

27.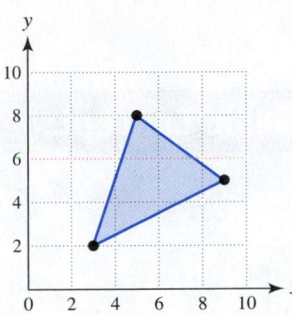

28.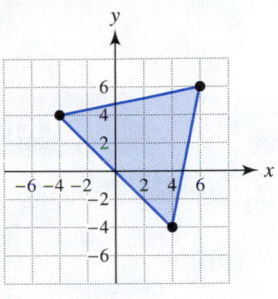

29.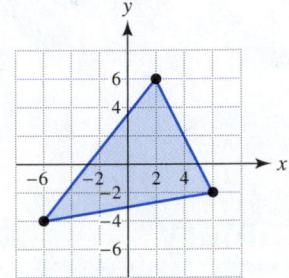

30.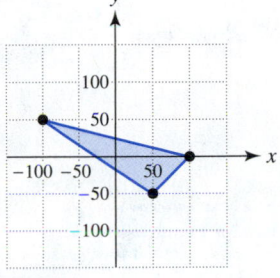

31.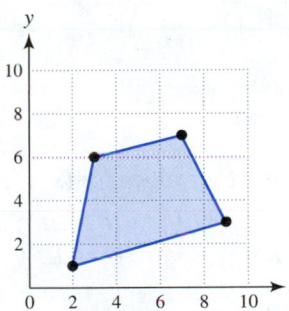

32.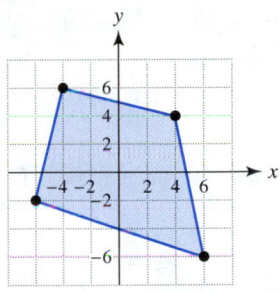

CRAMER'S RULE

Exercises 33–38: Solve the system of equations by using Cramer's rule.

33. $5x + 3y = 4$
    $6x - 4y = 20$

34. $-5x + 4y = -5$
    $4x + 4y = -32$

35. $7x - 5y = -3$
    $-4x + 6y = -8$

36. $-4x - 9y = -17$
    $8x + 4y = 9$

37. $8x = 3y - 61$
    $-x = 4y - 23$

38. $15y = -188 - 22x$
    $23y = -173 - 16x$

WRITING ABOUT MATHEMATICS

39. Suppose that one row of a 3×3 matrix A is all 0s. What is the value of det A? Give an example and explain your answer.

40. Explain how to evaluate a determinant of a 2×2 matrix. Can you find the determinant of a 2×3 matrix? Explain your answer.

SECTION 9.3   **Checking Basic Concepts**

1. Evaluate det A.

(a) $A = \begin{bmatrix} -3 & 4 \\ -2 & 3 \end{bmatrix}$

(b) $A = \begin{bmatrix} 1 & -2 & 3 \\ 5 & 1 & 1 \\ 0 & 2 & -1 \end{bmatrix}$

2. Use Cramer's rule to solve the system of linear equations.

$$2x - y = -14$$
$$3x - 4y = -36$$

3. Find the area of a triangle that has the vertices $(-1, 2)$, $(5, 6)$, and $(2, -3)$.

CHAPTER 9   Summary

SECTION 9.1 ■ SYSTEMS OF LINEAR EQUATIONS IN THREE VARIABLES

Solution to a System of Linear Equations in Three Variables An ordered triple (x, y, z) that satisfies all three equations

Example:
$$x - y + 2z = 3$$
$$2x - y + z = 5$$
$$x + y + z = 6$$

The solution is $(3, 2, 1)$ because these values for (x, y, z) satisfy all three equations.

$$3 - 2 + 2(1) = 3 \checkmark \quad \text{True}$$
$$2(3) - 2 + 1 = 5 \checkmark \quad \text{True}$$
$$3 + 2 + 1 = 6 \checkmark \quad \text{True}$$

Elimination and Substitution Systems of linear equations in three variables can be solved by elimination and substitution, using the following steps.

STEP 1: Eliminate one variable, such as x, from two of the given equations.

STEP 2: Use the two resulting equations in two variables to eliminate one of the variables, such as y. Solve for the remaining variable z.

STEP 3: Substitute z in one of the two equations from Step 2. Solve for the unknown variable y.

STEP 4: Substitute values for y and z in one of the given equations. Then find x. The solution is the ordered triple (x, y, z).

SECTION 9.2 ■ MATRIX SOLUTIONS OF LINEAR SYSTEMS

Matrix A rectangular array of numbers is a matrix. If a matrix has m rows and n columns, it has dimension $m \times n$.

Example: Matrix $A = \begin{bmatrix} 3 & -1 & 7 \\ 0 & 6 & -2 \end{bmatrix}$ has dimension 2×3.

Augmented Matrix   Any linear system can be represented with an augmented matrix.

Linear System

$4x - 3y = 5$
$x + 2y = 4$

Augmented Matrix

$$\begin{bmatrix} 4 & -3 & 5 \\ 1 & 2 & 4 \end{bmatrix}$$

Gauss–Jordan Elimination   A numerical method that uses matrix row transformations to tranform a matrix into reduced row–echelon form

Example: The matrix $\begin{bmatrix} 4 & -3 & 5 \\ 1 & 2 & 4 \end{bmatrix}$ reduces to $\begin{bmatrix} 1 & 0 & 2 \\ 0 & 1 & 1 \end{bmatrix}$.

The solution to the system is $(2, 1)$.

SECTION 9.3 ■ DETERMINANTS

Determinant for a 2 × 2 Matrix   A determinant is a *real number*. The determinant of a 2 × 2 matrix is

$$\det A = \det\begin{bmatrix} a & b \\ c & d \end{bmatrix} = ad - cb.$$

Example: $\det\begin{bmatrix} 2 & 3 \\ 4 & 5 \end{bmatrix} = (2)(5) - (4)(3) = -2$

Determinant for a 3 × 3 Matrix

$$\det A = \det\begin{bmatrix} a_1 & b_1 & c_1 \\ a_2 & b_2 & c_2 \\ a_3 & b_3 & c_3 \end{bmatrix}$$

$$= a_1 \cdot \det\begin{bmatrix} b_2 & c_2 \\ b_3 & c_3 \end{bmatrix} - a_2 \cdot \det\begin{bmatrix} b_1 & c_1 \\ b_3 & c_3 \end{bmatrix} + a_3 \cdot \det\begin{bmatrix} b_1 & c_1 \\ b_2 & c_2 \end{bmatrix}$$

Example: $\det\begin{bmatrix} 2 & 3 & 2 \\ 3 & 7 & -3 \\ 0 & 0 & -1 \end{bmatrix} = 2\det\begin{bmatrix} 7 & -3 \\ 0 & -1 \end{bmatrix} - 3\det\begin{bmatrix} 3 & 2 \\ 0 & -1 \end{bmatrix} + 0$

$$= 2(-7) - 3(-3) = -5$$

Cramer's rule uses determinants to solve linear systems of equations. Determinants can also be used to find areas of triangles. See Putting It All Together for Section 9.3.

CHAPTER 9   Review Exercises

SECTION 9.1

1. Is $(3, -4, 5)$ a solution for $x + y + z = 4$?

2. Decide whether either ordered triple is a solution: $(1, -1, 2)$ or $(1, 0, 5)$.

$$2x - 3y + z = 7$$
$$-x - y + 3z = 6$$
$$3x - 2y + z = 7$$

Exercises 3–8: Use elimination and substitution to solve the system of linear equations, if possible.

3. $\quad x - y - 2z = -11$
$\quad -x + 2y + 3z = 16$
$\quad\quad\quad\quad\quad 3z = 6$

4. $\quad x + y \quad\quad = 4$
$\quad -2x + y + 3z = -2$
$\quad\quad x - 2y + 5z = -26$

5. $2x - y \qquad = -5$
$\qquad x + 2y + z = \quad 7$
$\quad -2x + \quad y + z = \quad 7$

6. $2x + 3y + \quad z = 6$
$\quad -x + 2y + 2z = 3$
$\qquad x + \quad y + 2z = 4$

7. $x - \quad y + 3z = 2$
$2x + \quad y + 4z = 3$
$\quad x + 2y + \quad z = 5$

8. $x - y + 3z = 3$
$x + y - \quad z = 1$
$x \qquad + \quad z = 2$

SECTION 9.2

Exercises 9–12: Write the system of linear equations as an augmented matrix. Then use Gauss–Jordan elimination to solve the system, writing the solution as an ordered triple. Check your solution.

9. $x + \quad y + z = -6$
$\quad x + 2y + z = -8$
$\qquad y + z = -5$

10. $x + y + z = -3$
$\quad -x + y \qquad = \quad 5$
$\qquad y + z = -1$

11. $\quad x + 2y - \quad z = \quad 1$
$\quad -x + \quad y - 2z = \quad 5$
$\qquad 2y + \quad z = 10$

12. $\quad 2x + 2y - 2z = -14$
$\quad -2x - 3y + 2z = \quad 12$
$\qquad x + \quad y - 4z = -22$

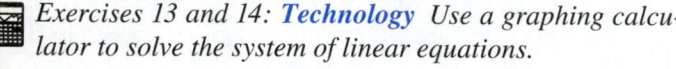 *Exercises 13 and 14:* **Technology** *Use a graphing calculator to solve the system of linear equations.*

13. $\quad 3x - 2y + 6z = -17$
$\quad -2x - \quad y + 5z = \quad 20$
$\qquad 4y + 7z = \quad 30$

14. $19x - 13y - 7z = \quad 7.4$
$22x + 33y - 8z = 110.5$
$10x - 56y + 9z = \quad 23.7$

SECTION 9.3

Exercises 15–18: Evaluate det A, if A is the given matrix.

15. $\begin{bmatrix} 6 & -5 \\ -4 & 2 \end{bmatrix}$

16. $\begin{bmatrix} 0 & -6 \\ 5 & 9 \end{bmatrix}$

17. $\begin{bmatrix} 3 & -5 & -3 \\ 1 & 4 & 7 \\ 0 & -3 & 1 \end{bmatrix}$

18. $\begin{bmatrix} -2 & -1 & -7 \\ 2 & 1 & -3 \\ 3 & -5 & 8 \end{bmatrix}$

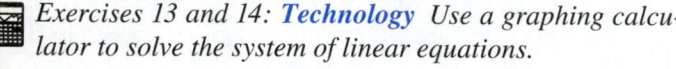 *Exercises 19 and 20: Use technology to calculate det A, where A is the given matrix.*

19. $\begin{bmatrix} 22 & -45 & 3 \\ 15 & -12 & -93 \\ 5 & 81 & -21 \end{bmatrix}$

20. $\begin{bmatrix} 0.5 & -7.3 & 9.6 \\ 0.1 & 3.1 & 9.2 \\ -0.5 & -1.9 & 5.4 \end{bmatrix}$

Exercises 21 and 22: Use a determinant to find the area of the triangle. Assume that the units are feet.

21.

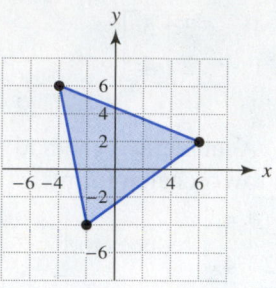

22.

Exercises 23–26: Use Cramer's rule to solve the system.

23. $7x + 6y = \quad 8$
$\quad 5x - 8y = 18$

24. $-2x + 5y = \quad 25$
$\quad 3x + 4y = -3$

25. $3x - 6y = 1.5$
$7x - 5y = 8$

26. $-5x + 4y = -47$
$\quad 6x - 7y = \quad 63$

APPLICATIONS

27. *Pedestrian Fatalities* Forty-seven percent of pedestrian fatalities occur on Friday and Saturday nights. The combined total of pedestrian fatalities in 1994 and 2004 was 10,130. There were 848 more fatalities in 1994 than in 2004. Find the number of pedestrian fatalities during each year. (*Source:* National Highway Traffic Safety Administration.)

28. *Tickets* Tickets for a football game cost $8 and $12. If 480 tickets were sold for total receipts of $4620, how many of each type of ticket were sold?

29. *Determining Costs* The accompanying table shows the costs for purchasing different combinations of malts, cones, and ice cream bars.

Malts	Cones	Bars	Total Cost
1	3	5	$14
1	2	4	$11
0	1	3	$5

(a) Let m be the cost of a malt, c the cost of a cone, and b the cost of an ice cream bar. Write a system of equations that represents the data in the table.

(b) Solve this system.

30. *Geometry* The largest angle in a triangle is 20° more than the sum of the two smaller angles. The measure of the largest angle is 85° more than the smallest angle. Find the measure of each angle in the triangle.

31. *Mixture Problem* Three types of candy that cost $1.50, $2.00, and $2.50 per pound are to be mixed to produce 12 pounds of candy worth $26.00. If there are to be 2 pounds more of the $2.50 candy than the $2.00 candy, how much of each type of candy should be used in the mixture?

32. *Estimating the Chest Size of a Bear* The accompanying table shows the chest size C, weight W, and overall length L of three bears. These data can be modeled with the formula $C = a + bW + cL$.

(*Sources:* M. Triola, *Elementary Statistics*; Minitab, Inc.)

C (inches)	W (pounds)	L (inches)
40	202	63
50	365	70
55	446	77
?	300	68

(a) Set up a system of linear equations whose solution gives values for the constants a, b, and c.

(b) Solve this system. Round each value to the nearest thousandth.

(c) Predict the chest size of a bear weighing 300 pounds and having a length of 68 inches.

CHAPTER 9   Test  CHAPTER Test Prep VIDEOS   Step-by-step test solutions are found on the Chapter Test Prep Videos available in **MyMathLab** and on **YouTube** (search "RockswoldComboAlg" and click on "Channels").

1. Can a system of three equations and three variables have exactly three solutions? Explain.

2. Determine which ordered triple is a solution to the linear system.

$$(-4, 3, 3), (1, -2, -2)$$
$$x - 3y + 4z = -1$$
$$2x + y - 3z = 6$$
$$x - y + z = 1$$

Exercises 3–6: Use substitution and elimination to solve the system, if possible.

3.
$$x + 3y = 2$$
$$-2x + y + z = 5$$
$$y + z = -3$$

4.
$$x + y - z = 1$$
$$2x - 3y + z = 0$$
$$x - 4y + 2z = 2$$

5.
$$x - y + 2z = 5$$
$$-x + y - 3z = -8$$
$$x - 2y + 2z = 3$$

6.
$$x - y + z = 1$$
$$x + y - z = 1$$
$$3x + y - z = 3$$

7. Consider the system of linear equations.

$$2x - 4y = -10$$
$$-3x - 2y = 7$$

(a) Write the system of linear equations as an augmented matrix.

(b) Use Gauss–Jordan elimination to solve the system, writing the solution as an ordered pair.

8. Consider the system of linear equations.

$$x + y + z = 2$$
$$x - y - z = 3$$
$$2x + 2y + z = 6$$

(a) Write the system as an augmented matrix.

(b) Use Gauss–Jordan elimination to solve the system, writing the solution as an ordered triple.

9. Evaluate det A if $A = \begin{bmatrix} -1 & 2 \\ -5 & 4 \end{bmatrix}$.

10. Evaluate det A if $A = \begin{bmatrix} 3 & 2 & -1 \\ 6 & 2 & -6 \\ 0 & 8 & -3 \end{bmatrix}$.

11. Solve the system of equations with Cramer's rule.

$$5x - 3y = 7$$
$$-4x + 2y = 11$$

12. *Jogging Speeds* A runner in preparation for a race jogs at 6, 7, and 9 miles per hour and travels a total of 7.1 miles in 1 hour. The runner jogs 12 minutes longer at 6 miles per hour than at 9 miles per hour. How many minutes did the runner jog at each speed?

13. *Geometry* The largest angle in a triangle is 50° more than the smallest angle. The sum of the measures of the smaller two angles is 10° more than the largest angle. Find the measure of each angle.

14. *Weight of a Bear* The data in the accompanying table list the weight W, neck size N, and chest size C of three bears. These data can be modeled by $W = a + bN + cC$.

W (pounds)	N (inches)	C (inches)
168	20	25
270	24	40
405	30	50

(a) Set up a system of equations whose solution gives values for the constants a, b, and c.

(b) Solve this system.

(c) Predict the weight W of a bear with neck size $N = 26$ and chest size $C = 44$.

CHAPTER 9 Extended and Discovery Exercises

CRAMER'S RULE

Exercises 1–6: Cramer's rule can be applied to systems of three equations in three variables. For the system of equations

$$a_1 x + b_1 y + c_1 z = d_1$$
$$a_2 x + b_2 y + c_2 z = d_2$$
$$a_3 x + b_3 y + c_3 z = d_3,$$

the solution can be written as follows.

$$D = \det \begin{bmatrix} a_1 & b_1 & c_1 \\ a_2 & b_2 & c_2 \\ a_3 & b_3 & c_3 \end{bmatrix}, \quad E = \det \begin{bmatrix} d_1 & b_1 & c_1 \\ d_2 & b_2 & c_2 \\ d_3 & b_3 & c_3 \end{bmatrix}$$

$$F = \det \begin{bmatrix} a_1 & d_1 & c_1 \\ a_2 & d_2 & c_2 \\ a_3 & d_3 & c_3 \end{bmatrix}, \quad G = \det \begin{bmatrix} a_1 & b_1 & d_1 \\ a_2 & b_2 & d_2 \\ a_3 & b_3 & d_3 \end{bmatrix}$$

If $D \neq 0$, a unique solution exists and is given by

$$x = \frac{E}{D}, \quad y = \frac{F}{D}, \quad z = \frac{G}{D}.$$

Use Cramer's rule to solve the equations.

1. $x + y + z = 6$
$2x + y + 2z = 9$
$\quad y + 3z = 9$

2. $\quad y + z = 1$
$2x - y - z = -1$
$x + y - z = 3$

3. $x + \quad z = 2$
$x + y \quad = 0$
$\quad y + 2z = 1$

4. $x + y + 2z = 1$
$-x - 2y - 3z = -2$
$\quad y - 3z = 5$

5. $x + \quad 2z = 7$
$-x + y + z = 5$
$2x - y + 2z = 6$

6. $x + 2y + 3z = -1$
$2x - 3y - z = 12$
$x + 4y - 2z = -12$

MATRICES AND ROAD MAPS

*Exercises 7–12: **Adjacency Matrix** A matrix A can be used to represent a map showing distances between cities. Let a_{ij} denote the number in row i and column j of a matrix A. Now consider the following map illustrating freeway distances in miles between four cities. Each city has been assigned a number. For example, there is a direct route from Denver, Colorado (city 1), to Colorado Springs, Colorado (city 2), of approximately 60 miles. Therefore $a_{12} = $ **60** in the accompanying matrix A. (Note that a_{12} is the number in row 1 and column 2.) The distance from Colorado Springs to Denver is also 60 miles, so $a_{21} = $ **60**. As there is no direct freeway connection between Las Vegas, Nevada (city 4), and Colorado Springs (city 2), we let $a_{24} = a_{42} = *$. The matrix A is called an **adjacency matrix**. (Source: S. Baase, Computer Algorithms: Introduction to Design and Analysis.)*

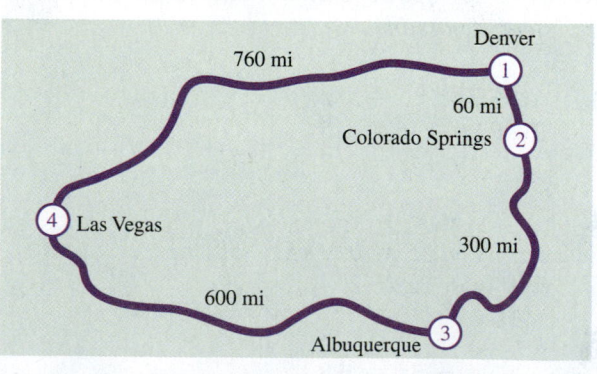

$$A = \begin{bmatrix} 0 & 60 & * & 760 \\ 60 & 0 & 300 & * \\ * & 300 & 0 & 600 \\ 760 & * & 600 & 0 \end{bmatrix}$$

7. Explain how to use A to find the freeway distance from Denver to Las Vegas.

8. Explain how to use A to find the freeway distance from Denver to Albuquerque.

9. If a map shows 20 cities, what would be the dimension of the adjacency matrix? How many elements would there be in this matrix?

10. Why are there only zeros on the main diagonal of A?

11. What does $a_{14} + a_{41}$ equal?

12. What does $a_{11} + a_{44}$ equal?

Exercises 13 and 14: (Refer to Exercises 7–12.) Determine an adjacency matrix A for the road map.

13.

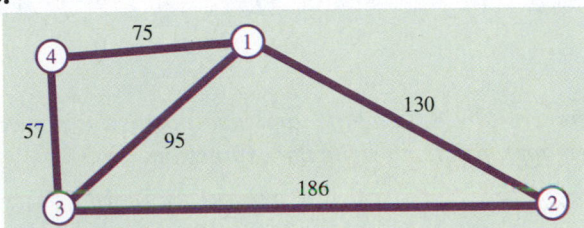

14.

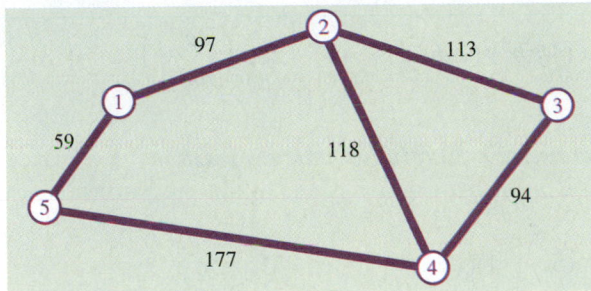

Exercises 15 and 16: (Refer to Exercises 7–12.) Sketch a road map represented by the adjacency matrix A. Is your answer unique?

15. $A = \begin{bmatrix} 0 & 30 & 20 & 5 \\ 30 & 0 & 15 & * \\ 20 & 15 & 0 & 25 \\ 5 & * & 25 & 0 \end{bmatrix}$

16. $A = \begin{bmatrix} 0 & 5 & * & 13 & 20 \\ 5 & 0 & 5 & * & * \\ * & 5 & 0 & 13 & * \\ 13 & * & 13 & 0 & 10 \\ 20 & * & * & 10 & 0 \end{bmatrix}$

SOLVING AN EQUATION IN FOUR VARIABLES

17. *Weight of a Bear* In Section 9.2 we estimated the weight of a bear by using two variables. We may be able to make more accurate estimates by using four variables. The accompanying table shows the weight W, neck size N, overall length L, and chest size C for four bears. (*Sources:* M. Triola, *Elementary Statistics*; Minitab, Inc.)

W (pounds)	N (inches)	L (inches)	C (inches)
125	19	57.5	32
316	26	65	42
436	30	72	48
514	30.5	75	54
?	24	63	39

(a) We can model the data in the table with the equation $W = a + bN + cL + dC$, where a, b, c, and d are constants. To do so, represent a system of linear equations by a 4×5 augmented matrix whose solution gives values for a, b, c, and d.

(b) Solve the system with a graphing calculator. Round each value to the nearest thousandth.

(c) Predict the weight of a bear with $N = 24$, $L = 63$, and $C = 39$. Interpret the result.

CHAPTERS 1–9 Cumulative Review Exercises

1. Write 360 as a product of prime numbers.

2. Translate the sentence "Double a number increased by 7 equals the number decreased by 2" into an equation by using the variable n. Then solve the equation.

Exercises 3 and 4: Simplify to lowest terms.

3. $\frac{2}{3} + \frac{4}{7} \cdot \frac{21}{28}$

4. $\frac{3}{5} \div \frac{6}{5} - \frac{2}{3}$

Exercises 5 and 6: Evaluate by hand.

5. $30 - 4 \div 2 \cdot 6$

6. $\dfrac{3^2 - 2^3}{20 - 5 \cdot 2}$

7. Solve $2(x + 1) - 6x = x - 4$.

8. Solve $4 - 3x = -2$ graphically. Check your answer.

9. Convert 124% to fraction and decimal notation.

10. If $A = 30$ square miles and $h = 10$ miles, use the formula $A = \frac{1}{2}bh$ to find b.

11. Solve $6t - 1 < 3 - t$.

12. Make a scatterplot with the points $(-1, 2)$, $(1, -2)$, $(0, 3)$, $(-2, 0)$, and $(2, 3)$.

Exercises 13 and 14: Graph the equation and determine any intercepts.

13. $-3x + 4y = 12$

14. $x = -2$

15. Use the graph to identify the x-intercept and the y-intercept. Then write the slope–intercept form of the line.

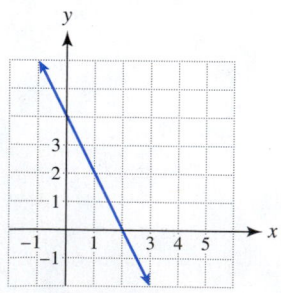

16. Write the slope–intercept form of a line that passes through the point $(-1, 3)$ with slope $m = -2$.

Exercises 17 and 18: Find the slope–intercept form of the line satisfying the given conditions.

17. Passing through the points $(-3, 5)$ and $(2, 8)$

18. Perpendicular to $x + 2y = 5$, passing through the point $(-1, 1)$

19. An insect population P is initially 4000 and increases by 500 insects per day. Write an equation in slope–intercept form that calculates P after x days.

20. Solve the system of equations. Write your answer as an ordered pair.

$$-2a + b = -5$$
$$4a - 3b = 0$$

Exercises 21 and 22: Shade the solution set.

21. $y \geq -1$

22. $x - y < 3$
 $2x + y \geq 6$

Exercises 23–28: Simplify and use positive exponents, when appropriate, to write the expression.

23. $5 - 3^4$

24. $(8t^{-3})(3t^2)(t^5)$

25. $\dfrac{2^{-4}}{4^{-2}}$

26. $(2t^3)^{-2}$

27. $(4a^2b^3)^2(2ab)^{-3}$

28. $\left(\dfrac{2a^{-1}}{ab^{-2}}\right)^{-3}$

Exercises 29–32: Multiply the expression.

29. $2a^2(a^2 - 2a + 3)$

30. $(5x + 1)(x - 7)$

31. $(2x + 3y)^2$

32. $(a + b)(a - b)$

Exercises 33 and 34: Divide.

33. $\dfrac{4x^3 - 8x^2 + 6x}{2x}$

34. $(x^4 - 9x^3 + 23x^2 - 17x + 11) \div (x - 5)$

Exercises 35–42: Factor completely.

35. $10ab^2 - 25a^3b^5$

36. $y^3 - 3y^2 + 2y - 6$

37. $6z^2 + 7z - 3$

38. $4z^2 - 9$

39. $4y^2 - 20y + 25$

40. $a^3 - 27$

41. $4z^4 - 17z^2 + 15$

42. $2a^3b + a^2b^2 - ab^3$

Exercises 43 and 44: Solve the equation.

43. $(x - 1)(x + 2) = 0$ **44.** $6y^2 - 7y = 3$

Exercises 45 and 46: Simplify to lowest terms.

45. $\dfrac{x^2 - 16}{x + 4}$ **46.** $\dfrac{2x^2 - 11x - 6}{6x^2 - 5x - 4}$

Exercises 47 and 48: Simplify and write in lowest terms.

47. $\dfrac{x^2 - 3x + 2}{x + 7} \div \dfrac{x - 2}{2x + 14}$

48. $\dfrac{x}{2x + 3} + \dfrac{x + 3}{2x + 3}$

Exercises 49 and 50: Solve the equation.

49. $\dfrac{x + 2}{5} = \dfrac{x}{4}$ **50.** $\dfrac{1}{3x} + \dfrac{5}{2x} = 2$

51. Suppose that y is inversely proportional to x and that $y = 25$ when $x = 4$. Find y for $x = 10$.

52. Evaluate $f(x) = 1 - 4x$ for $x = -3$.

53. Graph $f(x) = x^2 - 2$ and identify the domain and range of f. Write your answer in interval notation.

54. Use the graph to evaluate $f(0)$ and $f(-2)$.

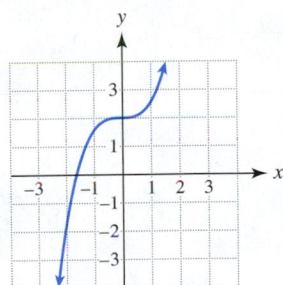

Exercises 55 and 56: Solve the inequality symbolically. Write the solution set in interval notation.

55. $\dfrac{4x - 9}{6} > \dfrac{1}{2}$

56. $\dfrac{2}{3}z - 2 \le \dfrac{1}{4}z - (2z + 2)$

Exercises 57 and 58: Solve the compound inequality. Graph the solution set on a number line.

57. $x + 2 > 1$ and $2x - 1 \le 9$

58. $4x + 7 < 1$ or $3x + 2 \ge 11$

Exercises 59 and 60: Solve the three-part inequality. Write the solution set in interval notation.

59. $-7 \le 2x - 3 \le 5$ **60.** $-8 \le -\dfrac{1}{2}x - 3 \le 5$

61. Use the graph to solve the equation and inequalities.
   (a) $y_1 = 2$ **(b)** $y_1 \le 2$
   (c) $y_1 \ge 2$

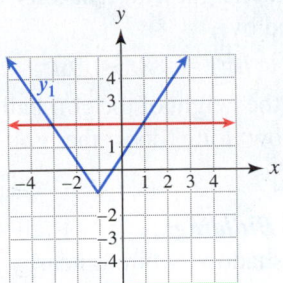

62. Solve the absolute value equation $\left|\dfrac{2}{3}x - 4\right| = 8$.

Exercises 63 and 64: Solve the absolute value inequality. Write the solution set in interval notation.

63. $|3x + 5| > 13$ **64.** $-3|2t - 11| \ge -9$

65. Use elimination and substitution to solve the system of linear equations.

$$2x + 3y - z = 3$$
$$3x - y + 4z = 10$$
$$2x + y - 2z = -1$$

66. Write the system of linear equations as an augmented matrix. Then use Gauss–Jordan elimination to solve the system. Write the solution as an ordered triple.

$$x + y - z = 4$$
$$-x - y - z = 0$$
$$x - 2y + z = -9$$

APPLICATIONS

67. *Graphical Model* An athlete rides a bicycle on a straight road away from home at 20 miles per hour for 1.5 hours. The athlete then turns around and rides home at 15 miles per hour. Sketch a graph that depicts the athlete's distance d from home after t hours.

68. *Ticket Prices* The price of admission to a baseball game is $15 for children and $25 for adults. If a group of 8 people pays $170 to enter the game, find the number of children and the number of adults in the group.

69. *Working Together* One person can shovel the snow from a sidewalk in 2 hours and another person can shovel the same sidewalk in 1.5 hours. How long will it take for them to shovel the snow from the sidewalk if they work together?

70. *Flight of a Golf Ball* If a golf ball is hit upward with a velocity of 88 feet per second (60 miles per hour), then its height h in feet after t seconds can be approximated by

$$h(t) = 88t - 16t^2.$$

(a) What is the height of the golf ball after 2 seconds?

(b) After how long does the golf ball strike the ground?

71. *Height of a Building* A 7-foot-tall stop sign casts a 4-foot-long shadow, while a nearby building casts a 35-foot-long shadow. Find the height of the building.

72. *Determining Costs* The accompanying table shows the costs for purchasing different combinations of burgers, fries, and malts.

Burgers	Fries	Malts	Total Cost
4	3	4	$23
1	2	1	$7
3	1	2	$13

(a) Let b be the cost of a burger, f be the cost of an order of fries, and m be the cost of a malt. Write a system of three linear equations that represents the data in the table.

(b) Solve this system and find the price of each item.

10

Radical Expressions and Functions

> Bear in mind that the wonderful things you learn in schools are the work of many generations, produced by enthusiastic effort and infinite labor in every country.
>
> —ALBERT EINSTEIN

Throughout history, people have created (or discovered) new numbers. Often these new numbers were met with resistance and regarded as being imaginary or unreal. The number 0 was not invented at the same time as the natural numbers. There was no Roman numeral for 0, which is one reason why our calendar started with A.D. 1 and, as a result, the twenty-first century began in 2001. No doubt there were skeptics during the time of the Roman Empire who questioned why anyone needed a number to represent nothing. Negative numbers also met strong resistance. After all, how could anyone possibly have −6 apples?

In this chapter we describe a new number system called *complex numbers*, which involve square roots of negative numbers. The Italian mathematician Cardano (1501–1576) was one of the first mathematicians to work with complex numbers and called them useless. René Descartes (1596–1650) originated the term *imaginary number*, which is associated with complex numbers. Today complex numbers are used in many applications, such as electricity, fiber optics, and the design of airplanes. We are privileged to study in a period of days what took people centuries to discover.

Source: Historical Topics for the Mathematics Classroom, Thirty-first Yearbook, NCTM.

10.1 Radical Expressions and Functions

Radical Notation • The Square Root Function • The Cube Root Function

A LOOK INTO MATH ▶ Suppose you start a summer painting business, and initially you are the only employee. Because business is good, you hire another employee and productivity goes up, more than enough to pay for the new employee. As time goes on, you hire more employees. Eventually, productivity starts to level off because there are too many employees to keep busy. This situation is common in business and can be modeled by using a function involving a square root. (See Example 12.)

Radical Notation

SQUARE ROOTS Recall the definition of the square root of a number a.

NEW VOCABULARY

☐ Radical sign
☐ Radicand
☐ Radical expression
☐ nth root
☐ Index
☐ Odd root
☐ Even root
☐ Principal nth root
☐ Square root function
☐ Cube root function

SQUARE ROOT

The number b is a *square root* of a if $b^2 = a$.

EXAMPLE 1 **Finding square roots**

Find the square roots of 100.

Solution

The square roots of 100 are **10** *and* **−10** because $\mathbf{10}^2 = 100$ and $(\mathbf{-10})^2 = 100$.

Now Try Exercise 1

Every positive number a has two square roots: one positive and one negative. Recall that the *positive square root* is called the *principal square root* and is denoted $\sqrt{a}$. The *negative square root* is denoted $-\sqrt{a}$. To identify both square roots, we write $\pm\sqrt{a}$. The symbol $\pm$ is read "plus or minus." The symbol $\sqrt{}$ is called the **radical sign**. The expression under the radical sign is called the **radicand**, and an expression containing a radical sign is called a **radical expression**. Examples of radical expressions include

$$\sqrt{6}, \quad 5 + \sqrt{x+1}, \quad \text{and} \quad \sqrt{\frac{3x}{2x-1}}.$$

MAKING CONNECTIONS

Expressions and Equations

Expressions and equations are *different* mathematical concepts. An expression does not contain an equals sign, whereas an equation *always* contains an equals sign. *An equation is a statement that two expressions are equal.* For example,

$$\sqrt{x+1} \quad \text{and} \quad \sqrt{5-x} \qquad \text{Expressions}$$

are two different expressions, and

$$\sqrt{x+1} = \sqrt{5-x} \qquad \text{Equation}$$

is an equation. We often *solve an equation*, but we *do not solve an expression*. Instead, we *simplify* and *evaluate* expressions.

In the next example we show how to find the principal square root of an expression.

EXAMPLE 2   **Finding principal square roots**

Evaluate each square root.

(a) $\sqrt{25}$ **(b)** $\sqrt{0.49}$ **(c)** $\sqrt{\frac{4}{9}}$ **(d)** $\sqrt{c^2}, c > 0$

Solution
(a) Because $5 \cdot 5 = 25$, the principal, or *positive*, square root of 25 is $\sqrt{25} = 5$.
(b) Because $(0.7)(0.7) = 0.49$, the principal square root of 0.49 is $\sqrt{0.49} = 0.7$.
(c) Because $\frac{2}{3} \cdot \frac{2}{3} = \frac{4}{9}$, the principal square root of $\frac{4}{9}$ is $\sqrt{\frac{4}{9}} = \frac{2}{3}$.
(d) The principal square root of c^2 is $\sqrt{c^2} = c$, as c is positive.

▌ **Now Try Exercises 15, 17, 19, 21**

The square roots of many real numbers, such as $\sqrt{17}$, $\sqrt{1.2}$, and $\sqrt{\frac{5}{7}}$, cannot be conveniently evaluated (or approximated) by hand. In these cases we sometimes use a calculator to give a decimal *approximation*, as demonstrated in the next example.

EXAMPLE 3   **Approximating a square root**

Approximate $\sqrt{17}$ to the nearest thousandth.

Solution
Figure 10.1 shows that $\sqrt{17} \approx 4.123$, rounded to the nearest thousandth. This result means that $4.123 \times 4.123 \approx 17$.

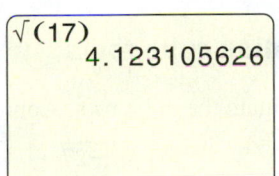

```
√(17)
          4.123105626
```

Figure 10.1

▌ **Now Try Exercise 39**

SQUARE ROOTS OF NEGATIVE NUMBERS The square root of a *negative* number is *not* a real number. For example, $\sqrt{-4} \neq 2$ because $2 \cdot 2 \neq -4$ and $\sqrt{-4} \neq -2$ because $(-2)(-2) \neq -4$. (Later in this chapter, we will use the complex numbers to identify square roots of negative numbers.)

READING CHECK

Are square roots of negative numbers real numbers? Explain.

▶ **REAL-WORLD CONNECTION** Have you ever noticed that if you climb up a hill or a tower you can see farther to the horizon? This phenomenon can be described by a formula containing a square root, as demonstrated in the next example.

EXAMPLE 4 **Seeing the horizon**

A formula for calculating the distance d in miles that one can see to the horizon on a clear day is approximated by $d = 1.22\sqrt{x}$, where x is the elevation, in feet, of a person.
(a) Approximate how far a 6-foot-tall person can see to the horizon.
(b) Approximate how far a person can see from an 8000-foot mountain.

Solution
(a) Let $x = \mathbf{6}$ in the formula $d = 1.22\sqrt{x}$.

$$d = 1.22\sqrt{\mathbf{6}} \approx 3, \text{ or about 3 miles.}$$

(b) If $x = \mathbf{8000}$, then $d = 1.22\sqrt{\mathbf{8000}} \approx 109$, or about 109 miles.

Now Try Exercise 107

CUBE ROOTS The cube root of a number a is denoted $\sqrt[3]{a}$.

CALCULATOR HELP

To calculate a cube root, see Appendix A (page AP-1).

CUBE ROOT

The number b is a *cube root* of a if $b^3 = a$.

Although the square root of a negative number is *not* a real number, the cube root of a negative number is a negative real number. *Every real number has one real cube root.*

EXAMPLE 5 **Finding cube roots**

Evaluate the cube root. Approximate your answer to the nearest hundredth when appropriate.
(a) $\sqrt[3]{8}$ **(b)** $\sqrt[3]{-27}$ **(c)** $\sqrt[3]{\frac{1}{64}}$ **(d)** $\sqrt[3]{d^6}$ **(e)** $\sqrt[3]{16}$

Solution
(a) $\sqrt[3]{8} = 2$ because $2^3 = 2 \cdot 2 \cdot 2 = 8$.
(b) $\sqrt[3]{-27} = -3$ because $(-3)^3 = (-3)(-3)(-3) = -27$.
(c) $\sqrt[3]{\frac{1}{64}} = \frac{1}{4}$ because $\left(\frac{1}{4}\right)^3 = \frac{1}{4} \cdot \frac{1}{4} \cdot \frac{1}{4} = \frac{1}{64}$.
(d) $\sqrt[3]{d^6} = d^2$ because $(d^2)^3 = d^2 \cdot d^2 \cdot d^2 = d^{2+2+2} = d^6$.
(e) $\sqrt[3]{16}$ is not an integer. Figure 10.2 shows that $\sqrt[3]{16} \approx 2.52$.

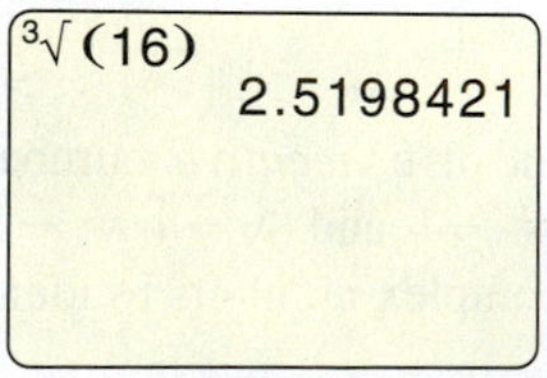

Now Try Exercises 23, 25, 27, 41

```
³√(16)
          2.5198421
```

Figure 10.2

NOTE: $\sqrt[3]{-b} = -\sqrt[3]{b}$ for any real number b. That is, the cube root of a negative is the negative of the cube root. For example, $\sqrt[3]{-8} = -\sqrt[3]{8} = -2$.

Table 10.1 illustrates how to evaluate both square roots and cube roots. If the radical expression is undefined, a dash is used.

TABLE 10.1 Radicals

Expression	$\sqrt{64}$	$-\sqrt{64}$	$\sqrt{-64}$	$\sqrt[3]{64}$	$-\sqrt[3]{64}$	$\sqrt[3]{-64}$
Evaluated	8	−8	—	4	−4	−4

*N*th **ROOTS** We can generalize square roots and cube roots to include *n*th roots of a number *a*. The number *b* is an **nth root** of *a* if $b^n = a$, where *n* is a positive integer. For example, $2^5 = 32$, so the 5th root of 32 is 2 and can be written as $\sqrt[5]{32} = 2$.

THE NOTATION $\sqrt[n]{a}$

The equation $\sqrt[n]{a} = b$ means that $b^n = a$, where *n* is a natural number called the **index**. If *n* is odd, we are finding an **odd root** and if *n* is even, we are finding an **even root**.

1. If $a > 0$, then $\sqrt[n]{a}$ is a positive number. $\sqrt[4]{16} = 2$ and is positive.
2. If $a < 0$ and
   (a) *n* is odd, then $\sqrt[n]{a}$ is a negative number. $\sqrt[3]{-8} = -2$ and is negative.
   (b) *n* is even, then $\sqrt[n]{a}$ is *not* a real number. $\sqrt[4]{-8}$ is undefined.

If $a > 0$ and *n* is even, then *a* has two real *n*th roots: one positive and one negative. The positive root is denoted $\sqrt[n]{a}$ and called the **principal *n*th root** of *a*. For example, $(-3)^4 = 81$ *and* $3^4 = 81$, but $\sqrt[4]{81} = 3$ in the same way *principal square roots* are calculated.

EXAMPLE 6   ### Finding *n*th roots

Find each root, if possible.

(a) $\sqrt[4]{16}$ **(b)** $\sqrt[5]{-32}$ **(c)** $\sqrt[4]{-81}$ **(d)** $-\sqrt[4]{81}$

Solution
(a) $\sqrt[4]{16} = 2$ because $2^4 = 2 \cdot 2 \cdot 2 \cdot 2 = 16$.
(b) $\sqrt[5]{-32} = -2$ because $(-2)^5 = (-2)(-2)(-2)(-2)(-2) = -32$.
(c) An *even* root of a *negative* number is *not* a real number.
(d) $-\sqrt[4]{81} = -3$ because $\sqrt[4]{81} = 3$.

Now Try Exercises 33, 35, 37

ABSOLUTE VALUE Consider the calculations

$$\sqrt{3^2} = \sqrt{9} = 3, \quad \sqrt{(-4)^2} = \sqrt{16} = 4, \quad \text{and} \quad \sqrt{(-6)^2} = \sqrt{36} = 6.$$

In general, the expression $\sqrt{x^2}$ equals $|x|$. Graphical support is shown in Figure 10.3, where the graphs of $Y_1 = \sqrt{(X^2)}$ and $Y_2 = abs(X)$ appear to be identical.

CRITICAL THINKING

Evaluate $\sqrt[6]{(-2)^6}$ and $\sqrt[3]{(-2)^3}$. Now simplify $\sqrt[n]{x^n}$ when *n* is even and when *n* is odd.

[−6, 6, 1] by [−4, 4, 1]   [−6, 6, 1] by [−4, 4, 1]

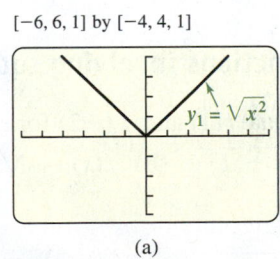
(a)

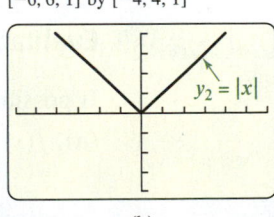
(b)

Figure 10.3

THE EXPRESSION $\sqrt{x^2}$

For every real number *x*, $\sqrt{x^2} = |x|$.

EXAMPLE 7 **Simplifying expressions**

Write each expression in terms of an absolute value.

(a) $\sqrt{(-3)^2}$ **(b)** $\sqrt{(x+1)^2}$ **(c)** $\sqrt{z^2 - 4z + 4}$

Solution

(a) $\sqrt{x^2} = |x|$, so $\sqrt{(-3)^2} = |-3|$

(b) $\sqrt{(x+1)^2} = |x+1|$

(c) $\sqrt{z^2 - 4z + 4} = \sqrt{(z-2)^2} = |z-2|$

Now Try Exercises 45, 49, 51

The Square Root Function

The **square root function** is given by $f(x) = \sqrt{x}$. The domain of the square root function is all nonnegative real numbers because we have *not* defined the square root of a negative number. Table 10.2 lists three points that lie on the graph of $f(x) = \sqrt{x}$. In Figure 10.4 these points are plotted and the graph of $y = \sqrt{x}$ has been sketched. The graph does not appear to the left of the origin because $f(x) = \sqrt{x}$ is undefined for negative inputs.

TABLE 10.2

x	$\sqrt{x}$
0	0
1	1
4	2

Square Root Function

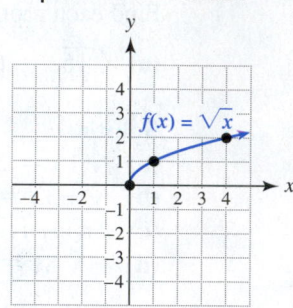

Figure 10.4

TECHNOLOGY NOTE

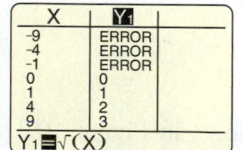

Square Roots of Negative Numbers

If a table of values for $y_1 = \sqrt{x}$ includes both negative and positive values for x, then many calculators give error messages when x is negative, as shown in the figure in the margin.

EXAMPLE 8 **Evaluating functions involving square roots**

If possible, evaluate $f(1)$ and $f(-2)$ for each $f(x)$.

(a) $f(x) = \sqrt{2x - 1}$ **(b)** $f(x) = \sqrt{4 - x^2}$

Solution

(a) $f(1) = \sqrt{2(1) - 1} = \sqrt{1} = 1$

$f(-2) = \sqrt{2(-2) - 1} = \sqrt{-5}$, which does not equal a real number.

(b) $f(1) = \sqrt{4 - (1)^2} = \sqrt{3}$

$f(-2) = \sqrt{4 - (-2)^2} = \sqrt{0} = 0$

Now Try Exercises 61, 63

▶ **REAL-WORLD CONNECTION** A good punter can kick a football so that the ball has a long *hang time*. Hang time is the length of time that the ball is in the air, and a long hang time gives the kicking team time to run down the field and stop the punt return. By using a function involving a square root, we can estimate hang time.

EXAMPLE 9 **Calculating hang time**

If a football is kicked x feet high, then the time T in seconds that the ball is in the air is given by the function

$$T(x) = \frac{1}{2}\sqrt{x}.$$

(a) Find the hang time if the ball is kicked 50 feet into the air.
(b) Does the hang time double if the ball is kicked 100 feet in the air?

Solution
(a) The hang time is $T(50) = \frac{1}{2}\sqrt{50} \approx 3.5$ seconds.
(b) The hang time is $T(100) = \frac{1}{2}\sqrt{100} = 5$ seconds. The time does *not* double when the height doubles.

▮ **Now Try Exercise 105**

CRITICAL THINKING

How high would a football have to be kicked to have twice the hang time of a football kicked 50 feet into the air?

EXAMPLE 10 **Finding the domain of a square root function**

Let $f(x) = \sqrt{x - 1}$.
(a) Find the domain of f. Write your answer in interval notation.
(b) Graph $y = f(x)$ and compare it to the graph of $y = \sqrt{x}$.

Solution
(a) For $f(x)$ to be defined, $x - 1$ cannot be negative. Thus valid inputs for x must satisfy

$$x - 1 \geq 0 \quad \text{or} \quad x \geq 1.$$

The domain is $[1, \infty)$.
(b) Table 10.3 lists points that lie on the graph of $y = \sqrt{x - 1}$. Note in Figure 10.5 that the graph appears only when $x \geq 1$. This graph is similar to $y = \sqrt{x}$ (see Figure 10.4) except that it is shifted one unit to the right.

TABLE 10.3

x	$\sqrt{x - 1}$
1	0
2	1
5	2

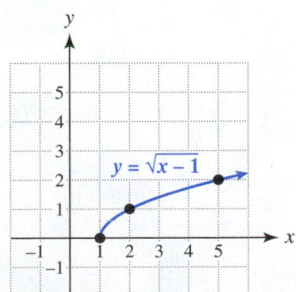

Figure 10.5

▮ **Now Try Exercises 75, 89**

EXAMPLE 11 ### Finding the domains of square root functions

Find the domain of each function. Write your answer in interval notation.

(a) $f(x) = \sqrt{4 - 2x}$ **(b)** $g(x) = \sqrt{x^2 + 1}$

Solution

(a) To determine when $f(x) = \sqrt{4 - 2x}$ is defined, we must solve $4 - 2x \geq 0$.

$$4 - 2x \geq 0 \qquad \text{Inequality to be solved}$$
$$4 \geq 2x \qquad \text{Add } 2x \text{ to each side.}$$
$$2 \geq x \qquad \text{Divide each side by 2.}$$

The domain is $(-\infty, 2]$.

(b) Regardless of the value of x in $g(x) = \sqrt{x^2 + 1}$, the expression $x^2 + 1$ is always positive because $x^2 \geq 0$. Thus $g(x)$ is defined for all real numbers, and its domain is $(-\infty, \infty)$.

Now Try Exercises 83, 85

MAKING CONNECTIONS

Domains of Functions and Their Graphs

In Example 11, the domains of f and g were found *symbolically*. Notice that the graph of f does not appear to the right of $x = 2$ because the domain of f is $(-\infty, 2]$, whereas the graph of g appears for all values of x because the domain of g is $(-\infty, \infty)$.

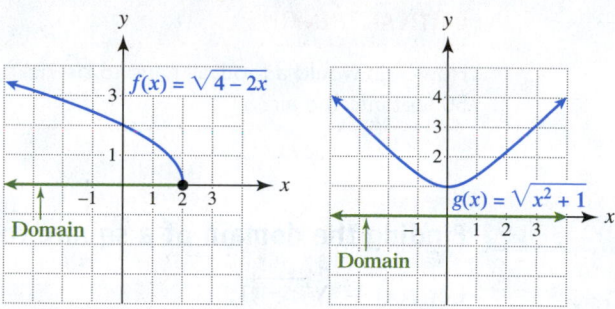

▶ **REAL-WORLD CONNECTION** (Refer to A Look Into Math at the beginning of this section.) In the next example, we analyze the benefit of adding employees to a small business.

EXAMPLE 12 ### Increasing productivity with more employees

The function $R(x) = 108\sqrt{x}$ gives the total revenue per year in thousands of dollars generated by a small business having x employees. A graph of $y = R(x)$ is shown in Figure 10.6.

Revenue as a Function of Employees

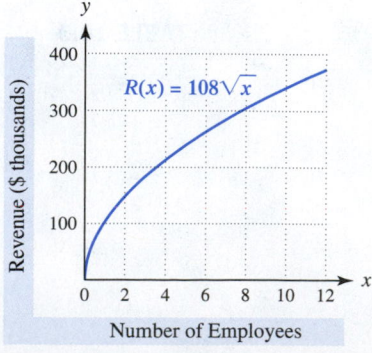

Figure 10.6

(a) Approximate values for $R(4)$, $R(8)$, and $R(12)$ by using both the formula and the graph.

(b) Evaluate $R(12) - R(8)$ and $R(8) - R(4)$ using the formula. Interpret your answer.

Solution

(a) $R(4) = 108\sqrt{4} = \$216$ thousand, $R(8) = 108\sqrt{8} \approx \$305$ thousand, and finally $R(12) = 108\sqrt{12} \approx \$374$ thousand.

We can graphically evaluate $R(\mathbf{4}) \approx \$\mathbf{215}$ thousand, $R(\mathbf{8}) \approx \$\mathbf{305}$ thousand, and $R(\mathbf{12}) \approx \$\mathbf{375}$ thousand, as shown in Figure 10.7.

Revenue as a Function of Employees

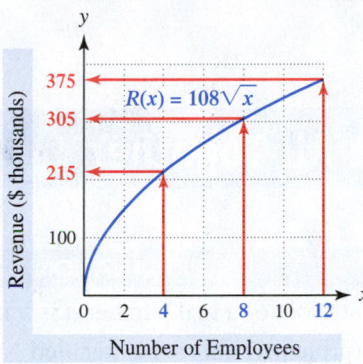

Figure 10.7

(b) $R(8) - R(4) \approx 305 - 216 = \$89$ thousand

$R(12) - R(8) \approx 374 - 305 = \$69$ thousand

There is more revenue gained from 4 to 8 employees than there is from 8 to 12 employees. Because the graph of $R(x) = 108\sqrt{x}$ starts to level off, there is a limited benefit to adding employees.

Now Try Exercise 109

Cube Root Function

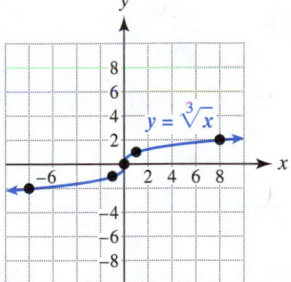

Figure 10.8

The Cube Root Function

We can define the **cube root function** by $f(x) = \sqrt[3]{x}$. Cube roots are defined for both positive and negative numbers, so *the domain of the cube root function includes all real numbers*. Table 10.4 lists points that lie on the graph of the cube root function. Figure 10.8 shows a graph of $y = \sqrt[3]{x}$.

TABLE 10.4 Cube Root Function

x	-27	-8	-1	0	1	8	27
$\sqrt[3]{x}$	-3	-2	-1	0	1	2	3

MAKING CONNECTIONS

Domains of the Square Root and Cube Root Functions

$f(x) = \sqrt{x}$ equals a real number for any nonnegative x. Thus $D = [0, \infty)$.

$f(x) = \sqrt[3]{x}$ equals a real number for any x. Thus $D = (-\infty, \infty)$.

Examples: $\sqrt{4} = 2$ but $\sqrt{-4}$ is *not* a real number. $\sqrt[3]{8} = 2$ and $\sqrt[3]{-8} = -2$.

EXAMPLE 13   **Evaluating functions involving cube roots**

Evaluate $f(1)$ and $f(-3)$ for each $f(x)$.

(a) $f(x) = \sqrt[3]{x^2 - 1}$ **(b)** $f(x) = \sqrt[3]{2 - x^2}$

Solution

(a) $f(\mathbf{1}) = \sqrt[3]{1^2 - 1} = \sqrt[3]{0} = 0$; $f(\mathbf{-3}) = \sqrt[3]{(-3)^2 - 1} = \sqrt[3]{8} = 2$

(b) $f(\mathbf{1}) = \sqrt[3]{2 - 1^2} = \sqrt[3]{1} = 1$; $f(\mathbf{-3}) = \sqrt[3]{2 - (-3)^2} = \sqrt[3]{-7}$ or $-\sqrt[3]{7}$

Now Try Exercises 65, 69

10.1   Putting It All Together

CONCEPT	EXPLANATION	EXAMPLES
nth Root of a Real Number	An nth root of a real number a is b if $b^n = a$, and the (principal) nth root is denoted $\sqrt[n]{a}$. If $a < 0$ and n is even, $\sqrt[n]{a}$ is not a real number.	The square roots of 25 are 5 and -5. The principal square root is $\sqrt{25} = 5$. $\sqrt[3]{-125} = -5$ because $(-5)^3 = -125$. $\sqrt[4]{-9}$ is not a real number.
Square Root and Cube Root Functions	$$f(x) = \sqrt{x} \quad \text{and} \quad g(x) = \sqrt[3]{x}$$ The cube root function g is defined for all inputs, whereas the square root function f is defined only for nonnegative inputs. **Square Root Function** **Cube Root Function** $y = \sqrt{x}$ Domain: $[0, \infty)$ $y = \sqrt[3]{x}$ Domain: $(-\infty, \infty)$	$f(64) = \sqrt{64} = 8$ $f(-64) = \sqrt{-64}$ is *not* a real number. $g(64) = \sqrt[3]{64} = 4$ $g(-64) = \sqrt[3]{-64} = -4$

10.1   Exercises

CONCEPTS AND VOCABULARY

1. What are the square roots of 9?

2. What is the principal square root of 9?

3. What is the cube root of 8?

4. Does every real number have a cube root?

5. If $b^n = a$ and $b > 0$, then $\sqrt[n]{a} =$ _____.

6. What is $\sqrt{x^2}$ equal to?

7. Evaluate $\sqrt{-25}$, if possible.

8. Evaluate $\sqrt[3]{-27}$, if possible.

9. Which of the following (a.–d.) equals -4?

   a. $\sqrt{16}$ **b.** $\sqrt{-16}$

   c. $-\sqrt[3]{16}$ **d.** $\sqrt[3]{-64}$

10. Which of the following (a.–d.) equals $|2x+1|$?

   a. $\sqrt{(2x+1)^2}$ **b.** $\sqrt[3]{(2x+1)^3}$

   c. $\left(\sqrt{2x+1}\right)^2$ **d.** $\sqrt{(2x)^2+1}$

11. Sketch a graph of the square root function.

12. Sketch a graph of the cube root function.

13. What is the domain of the square root function?

14. What is the domain of the cube root function?

RADICAL EXPRESSIONS

Exercises 15–38: Evaluate the expression by hand, if possible. Variables represent any real number.

15. $\sqrt{9}$ **16.** $\sqrt{121}$

17. $\sqrt{0.36}$ **18.** $\sqrt{0.64}$

19. $\sqrt{\frac{16}{25}}$ **20.** $\sqrt{\frac{9}{49}}$

21. $\sqrt{x^2}, x > 0$ **22.** $\sqrt{(x-1)^2}, x > 1$

23. $\sqrt[3]{27}$ **24.** $\sqrt[3]{64}$

25. $\sqrt[3]{-64}$ **26.** $-\sqrt[3]{-1}$

27. $\sqrt[3]{\frac{8}{27}}$ **28.** $\sqrt[3]{-\frac{1}{125}}$

29. $-\sqrt[3]{x^9}$ **30.** $\sqrt[3]{(x+1)^6}$

31. $\sqrt[3]{(2x)^6}$ **32.** $\sqrt[3]{27x^3}$

33. $\sqrt[4]{81}$ **34.** $\sqrt[5]{-1}$

35. $\sqrt[5]{-243}$ **36.** $\sqrt[4]{625}$

37. $\sqrt[4]{-16}$ **38.** $\sqrt[6]{-64}$

Exercises 39–44: Approximate to the nearest hundredth.

39. $-\sqrt{5}$ **40.** $\sqrt{11}$

41. $\sqrt[3]{5}$ **42.** $\sqrt[3]{-13}$

43. $\sqrt[5]{-7}$ **44.** $\sqrt[4]{6}$

Exercises 45–58: Simplify the expression. Assume that all variables are real numbers.

45. $\sqrt{(-4)^2}$ **46.** $\sqrt{9^2}$

47. $\sqrt{y^2}$ **48.** $\sqrt{z^4}$

49. $\sqrt{(x-5)^2}$ **50.** $\sqrt{(2x-1)^2}$

51. $\sqrt{x^2-2x+1}$ **52.** $\sqrt{4x^2+4x+1}$

53. $\sqrt[4]{y^4}$ **54.** $\sqrt[4]{x^8z^4}$

55. $\sqrt[4]{x^{12}}$ **56.** $\sqrt[6]{x^6}$

57. $\sqrt[5]{x^5}$ **58.** $\sqrt[5]{32(x+4)^5}$

SQUARE AND CUBE ROOT FUNCTIONS

Exercises 59–74: If possible, evaluate the function at the given value(s) of the variable.

59. $f(x) = \sqrt{x-1}$ $x = 10, 0$

60. $f(x) = \sqrt{4-3x}$ $x = -4, 1$

61. $f(x) = \sqrt{3-3x}$ $x = -1, 5$

62. $f(x) = \sqrt{x-5}$ $x = -1, 5$

63. $f(x) = \sqrt{x^2-x}$ $x = -4, 3$

64. $f(x) = \sqrt{2x^2-3}$ $x = -1, 2$

65. $f(x) = \sqrt[3]{x^2-8}$ $x = -3, 4$

66. $f(x) = \sqrt[3]{2x^2}$ $x = -2, 2$

67. $f(x) = \sqrt[3]{x-9}$ $x = 1, 10$

68. $f(x) = \sqrt[3]{5x-2}$ $x = -5, 2$

69. $f(x) = \sqrt[3]{3-x^2}$ $x = -2, 3$

70. $f(x) = \sqrt[3]{-1-x^2}$ $x = 0, 3$

71. $T(h) = \frac{1}{2}\sqrt{h}$ $h = 64$

72. $L(k) = 2\sqrt{k+2}$ $k = 23$

73. $f(x) = \sqrt{x+5} + \sqrt{x}$ $x = 4$

74. $f(x) = \dfrac{\sqrt{x-5} - \sqrt{x}}{2}$ $x = 9$

Exercises 75–88: Find the domain of f. Write your answer in interval notation.

75. $f(x) = \sqrt{x+2}$ **76.** $f(x) = \sqrt{x-1}$

77. $f(x) = \sqrt{x-2}$ **78.** $f(x) = \sqrt{x+1}$

79. $f(x) = \sqrt{2x-4}$ **80.** $f(x) = \sqrt{4x+2}$

81. $f(x) = \sqrt{1-x}$ **82.** $f(x) = \sqrt{6-3x}$

83. $f(x) = \sqrt{8-5x}$ **84.** $f(x) = \sqrt{3-2x}$

85. $f(x) = \sqrt{3x^2+4}$ **86.** $f(x) = \sqrt{1+2x^2}$

87. $f(x) = \dfrac{1}{\sqrt{2x+1}}$ **88.** $f(x) = \dfrac{1}{\sqrt{x-1}}$

Exercises 89–94: Graph the equation. Compare the graph to either $y = \sqrt{x}$ or $y = \sqrt[3]{x}$.

89. $y = \sqrt{x} + 2$

90. $y = \sqrt{x} - 1$

91. $y = \sqrt{x + 2}$

92. $y = \sqrt[3]{x} + 2$

93. $y = \sqrt[3]{x + 2}$

94. $y = \sqrt[3]{x} - 1$

REPRESENTATIONS OF ROOT FUNCTIONS

Exercises 95–102: Give symbolic, numerical, and graphical representations for the function f.

95. Function *f* takes the square root of *x* and then adds 1 to the result.

96. Function *f* takes the square root of *x* and then subtracts 2 from the result.

97. Function *f* takes the square root of the quantity three times *x*.

98. Function *f* takes the square root of the quantity *x* plus 1.

99. Function *f* takes the cube root of *x* and then multiplies the result by 2.

100. Function *f* takes the cube root of the quantity 4 times *x*.

101. Function *f* takes the cube root of the quantity *x* minus 1.

102. Function *f* takes the cube root of *x* and then adds 1.

AREA OF A TRIANGLE

*Exercises 103 and 104: **Heron's Formula** Suppose the lengths of the sides of a triangle are a, b, and c as illustrated in the figure.*

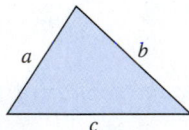

*If the **semiperimeter** (half of the perimeter) of the triangle is $s = \frac{1}{2}(a + b + c)$, then the area of the triangle is*

$$A = \sqrt{s(s - a)(s - b)(s - c)}.$$

Find the area A of the triangle with the given sides.

103. $a = 3, b = 4, c = 5$

104. $a = 5, b = 9, c = 10$

APPLICATIONS

105. *Jumping* (Refer to Example 9.) If a person jumps 4 feet off the ground, estimate how long the person is in the air.

106. *Hang Time* (Refer to Example 9.) Find the hang time for a golf ball hit 80 feet into the air.

107. *Distance to the Horizon* (Refer to Example 4.) Use the formula $d = 1.22\sqrt{x}$ to estimate how many miles a person can see from a jet airliner at 10,000 feet.

108. *Distance to the Horizon* (Refer to Example 4.) Use the formula $d = 1.22\sqrt{x}$ to estimate how many miles a 5-foot-tall person can see standing on the deck of a ship that is 50 feet above the ocean.

109. *Increasing Productivity* (Refer to Example 12.) Use $R(x) = 108\sqrt{x}$ to evaluate $R(16) - R(15)$. If the salary for the sixteenth employee is $25,000, is it a good decision to hire the sixteenth employee?

110. *Increasing Productivity* (Refer to Example 12.) Use $R(x) = 108\sqrt{x}$ to evaluate $R(2) - R(1)$. If the salary for the second employee is $25,000, is it a good decision to hire the second employee?

111. *Productivity of Workers* If workers are given more equipment, or *physical capital*, they are often more productive. For example, a carpenter definitely needs a hammer to be productive, but probably does not need 20 hammers. There is a leveling off in a worker's productivity as more is spent on equipment. The function $P(x) = 400\sqrt{x} + 8000$ approximates the worth of the goods produced by a typical U.S. worker in dollars when *x* dollars are spent on equipment per worker.
(a) Evaluate $P(25,000)$ and interpret the result.
(b) Sketch a graph of $y = P(x)$.
(c) Use the graph from (b) and the formula to evaluate $P(50,000) - P(25,000)$ and also to evaluate $P(75,000) - P(50,000)$. Interpret the result.

112. *Design of Open Channels* To protect cities from flooding during heavy rains, open channels are sometimes constructed to handle runoff. The rate R at which water flows through the channel is modeled by $R = k\sqrt{m}$, where m is the slope of the channel and k is a constant determined by the shape of the channel. (*Source:* N. Garber and L. Hoel, *Traffic and Highway Design.*)

(a) Suppose that a channel has a slope of $m = 0.01$ (or 1%) and a runoff rate of $R = 340$ cubic feet per second (cfs). Find k.

(b) If the slope of the channel increases to $m = 0.04$ (or 4%), what happens to R? Be specific.

WRITING ABOUT MATHEMATICS

113. Try to calculate $\sqrt{-7}$, $\sqrt[4]{-56}$, and $\sqrt[6]{-10}$ with a calculator. Describe what happens when you evaluate an even root of a negative number. Does the same difficulty occur when you evaluate an odd root of a negative number? Try to evaluate $\sqrt[3]{-7}$, $\sqrt[5]{-56}$, and $\sqrt[7]{-10}$. Explain.

114. Explain the difference between a root and a positive integer power of a number. Give examples.

10.2 Rational Exponents

Basic Concepts • Properties of Rational Exponents

A LOOK INTO MATH ▶ In today's society, people are used to having a lot of choices when it comes to online videos. In fact, if the average online video has 1,000,000 viewers, then about 200,000 of them, or 20%, have abandoned the video in 10 seconds or less. This average viewer abandonment rate can be modeled by a new function A, given by

$$A(x) = 7.3x^{7/16},$$

where A computes the percentage of viewers who abandon an online video after x seconds. But what does the exponent $\frac{7}{16}$ mean? To answer this question we need to discuss rational exponents and their properties. See Example 3. (*Source:* Business Insider.)

Basic Concepts

When m and n are integers, the product rule states that $a^m a^n = a^{m+n}$. This rule can be extended to include exponents that are fractions. For example,

$$4^{1/2} \cdot 4^{1/2} = 4^{1/2 + 1/2} = 4^1 = 4.$$

That is, if we multiply the expression $4^{1/2}$ by itself, the result is 4. Because we also know that $\sqrt{4} \cdot \sqrt{4} = 4$, this discussion suggests that $4^{1/2} = \sqrt{4}$ and leads to the following definition.

THE EXPRESSION $a^{1/n}$

If n is an integer greater than 1 and a is a real number, then
$$a^{1/n} = \sqrt[n]{a}.$$

$7^{1/5} = \sqrt[5]{7}$

NOTE: If $a < 0$ and n is an even positive integer, then $a^{1/n}$ is not a real number.

In the next two examples, we show how to interpret rational exponents.

EXAMPLE 1

Interpreting rational exponents

Write each expression in radical notation. Then evaluate the expression and round to the nearest hundredth when appropriate.

(a) $36^{1/2}$ (b) $23^{1/5}$ (c) $x^{1/3}$ (d) $(5x)^{1/2}$

Solution

(a) The exponent $\frac{1}{2}$ indicates a square root. Thus $36^{1/2} = \sqrt{36}$, which evaluates to 6.

(b) The exponent $\frac{1}{5}$ indicates a fifth root. Thus $23^{1/5} = \sqrt[5]{23}$, which is not an integer. Figure 10.9 shows this expression approximated in both exponential and radical notation. In either case $23^{1/5} \approx 1.87$.

(c) The exponent $\frac{1}{3}$ indicates a cube root, so $x^{1/3} = \sqrt[3]{x}$.

(d) The exponent $\frac{1}{2}$ indicates a square root, so $(5x)^{1/2} = \sqrt{5x}$.

Now Try Exercises 37, 45, 59, 63

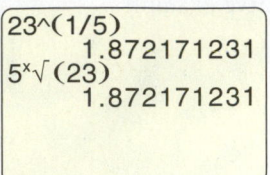

```
23^(1/5)
        1.872171231
5ˣ√(23)
        1.872171231
```

Figure 10.9

Suppose that we want to define the expression $8^{2/3}$. On the one hand, using properties of exponents we have

$$8^{1/3} \cdot 8^{1/3} = 8^{1/3 + 1/3} = 8^{2/3}.$$

On the other hand, we have

$$8^{1/3} \cdot 8^{1/3} = \sqrt[3]{8} \cdot \sqrt[3]{8} = 2 \cdot 2 = 4.$$

Thus $8^{2/3} = 4$, and that value is obtained whether we interpret $8^{2/3}$ as either

$$8^{2/3} = (8^{1/3})^2 = (\sqrt[3]{8})^2 = 2^2 = 4$$

or

$$8^{2/3} = (8^2)^{1/3} = \sqrt[3]{8^2} = \sqrt[3]{64} = 4.$$

This result suggests the following definition.

> ### THE EXPRESSION $a^{m/n}$
>
> **STUDY TIP**
>
> Be sure that you understand how rational exponents relate to radical notation.
>
> If m and n are positive integers with $\frac{m}{n}$ in lowest terms, then
>
> $$a^{m/n} = \sqrt[n]{a^m} = \left(\sqrt[n]{a}\right)^m. \qquad 2^{3/4} = \sqrt[4]{2^3} = \left(\sqrt[4]{2}\right)^3$$
>
> **NOTE:** If $a < 0$ and n is an even integer, then $a^{m/n}$ is *not* a real number.

The exponent $\frac{m}{n}$ indicates that we either take the *n*th root and then calculate the *m*th power of the result or calculate the *m*th power and then take the *n*th root. For example, $4^{3/2}$ means that we can either take the **square root** of 4 and then **cube** the result or we can **cube** 4 and then take the **square root** of the result. In either case the result is the same: $4^{3/2} = 8$. This concept is illustrated in the next example.

EXAMPLE 2

Interpreting rational exponents

Write each expression in radical notation. Evaluate the expression by hand when possible.

(a) $(-27)^{2/3}$ (b) $12^{3/5}$

Solution

(a) The exponent $\frac{2}{3}$ indicates either that we take the **cube root** of -27 and then **square** it or that we **square** -27 and then take the **cube root**. Thus

$$(-27)^{2/3} = (\sqrt[3]{-27})^2 = (-3)^2 = 9$$

or

$$(-27)^{2/3} = \sqrt[3]{(-27)^2} = \sqrt[3]{729} = 9.$$

Same result

(b) The exponent $\frac{3}{5}$ indicates either that we take the **fifth root** of 12 and then **cube** it or that we **cube** 12 and then take the **fifth root**. Thus

$$12^{3/5} = \left(\sqrt[5]{12}\right)^3 \quad \text{or} \quad 12^{3/5} = \sqrt[5]{12^3}.$$

This result cannot be evaluated by hand.

Now Try Exercises 47, 51

TECHNOLOGY NOTE

Rational Exponents

When evaluating expressions with rational (fractional) exponents, be sure to put parentheses around the fraction. For example, most calculators will evaluate 8^(2/3) and 8^2/3 differently. The accompanying figure shows that evaluating $8^{2/3}$ correctly as 8^(2/3) results in 4 but evaluating $8^{2/3}$ incorrectly as 8^2/3 results in $\frac{8^2}{3} = 21.\overline{3}$.

Correct ⟶

Incorrect ⟶

```
8^(2/3)
                    4
8^2/3
         21.33333333
```

▶ **REAL-WORLD CONNECTION** In the next example, we evaluate the function A presented in A Look Into Math at the beginning of this section. This function calculates the average viewer abandonment rate for online videos.

EXAMPLE 3 | **Calculating abandonment rates for online videos**

The function A, given by $A(x) = 7.3x^{7/16}$, computes the percentage of viewers who abandon an online video after x seconds.

(a) Find $A(10)$ and $A(90)$. Interpret your answers.

(b) Use radical notation to write $A(x)$.

Solution

(a) To approximate these expressions we will use a calculator, as shown in Figure 10.10.

$$A(\mathbf{10}) = 7.3(\mathbf{10})^{7/16} \approx \mathbf{20\%}$$
$$A(\mathbf{90}) = 7.3(\mathbf{90})^{7/16} \approx \mathbf{52\%}$$

After **10** seconds about **20%** of the viewers have abandoned the online video, and after **90** seconds over half, or **52%**, of the viewers have abandoned the online video.

(b) The exponent $\frac{7}{16}$ means we raise x to the **seventh** power and then take the **sixteenth** root. That is, $A(x) = 7.3\sqrt[16]{x^7}$.

```
7.3*10^(7/16)
         19.99046333
7.3*90^(7/16)
         52.27619364
```

Figure 10.10

Now Try Exercise 99

From properties of exponents we know that $a^{-n} = \frac{1}{a^n}$, where n is a positive integer. We now define this property for negative rational exponents.

READING CHECK

Explain how to evaluate a number with a rational exponent.

THE EXPRESSION $a^{-m/n}$

If m and n are positive integers with $\frac{m}{n}$ in lowest terms, then

$$a^{-m/n} = \frac{1}{a^{m/n}}, \qquad a \neq 0.$$

$$2^{-3/4} = \frac{1}{2^{3/4}}$$

Examples of changing rational exponents to radical expressions are shown in Table 10.5.

TABLE 10.5 Converting Forms

Expression	$a^{3/4}$	$5^{1/7}$	$x^{-5/3}$
Radical Form	$\sqrt[4]{a^3}$	$\sqrt[7]{5}$	$\dfrac{1}{\sqrt[3]{x^5}}$

EXAMPLE 4 | **Interpreting negative rational exponents**

Write each expression in radical notation and then evaluate.
(a) $64^{-1/3}$ (b) $81^{-3/4}$

Solution

(a) $64^{-1/3} = \dfrac{1}{64^{1/3}} = \dfrac{1}{\sqrt[3]{64}} = \dfrac{1}{4}.$

(b) $81^{-3/4} = \dfrac{1}{81^{3/4}} = \dfrac{1}{(\sqrt[4]{81})^3} = \dfrac{1}{3^3} = \dfrac{1}{27}.$

Now Try Exercises 53, 55

EXAMPLE 5 | **Converting to rational exponents**

Use rational exponents to write each radical expression.
(a) $\sqrt[5]{x^4}$ (b) $\dfrac{1}{\sqrt{b^5}}$ (c) $\sqrt[3]{(x-1)^2}$ (d) $\sqrt[3]{a^2+b^2}$

Solution
(a) $\sqrt[5]{x^4} = x^{4/5}$

(b) $\dfrac{1}{\sqrt{b^5}} = b^{-5/2}$

(c) $\sqrt[3]{(x-1)^2} = (x-1)^{2/3}$

(d) $\sqrt[3]{a^2+b^2} = (a^2+b^2)^{1/3}$ Note that $\sqrt[3]{a^2+b^2} \neq a^{2/3} + b^{2/3}$.

Now Try Exercises 27, 29, 33, 35

▶ **REAL-WORLD CONNECTION** In the next example, we use a formula from biology that involves a rational exponent.

| EXAMPLE 6 | **Analyzing stepping frequency** |

When smaller (four-legged) animals walk, they tend to take faster, shorter steps, whereas larger animals tend to take slower, longer steps. If an animal is h feet high at the shoulder, then the number N of steps per second that the animal takes *while walking* can be estimated by $N(h) = 1.6h^{-1/2}$. (*Source:* C. Pennycuick, *Newton Rules Biology*.)

(a) Use radical notation to write $N(h)$.

(b) Estimate the stepping frequency for an adult African elephant that is 10 feet high at the shoulders and a baby elephant that is only 33 inches high.

Solution

(a) We can rewrite the formula as $N(h) = \dfrac{1.6}{\sqrt{h}}$.

(b) We can evaluate N as follows. Note that 33 inches is $\dfrac{33}{12} = 2.75$ feet.

$$N(10) = \frac{1.6}{\sqrt{10}} \approx 0.51 \quad \text{and} \quad N(2.75) = \frac{1.6}{\sqrt{2.75}} \approx 0.96$$

While walking, the adult elephant takes about 1 step every 2 seconds, and the baby elephant takes about 1 step every second.

Now Try Exercise 101

Properties of Rational Exponents

Any rational number can be written as a ratio of two integers. That is, if p is a rational number, then $p = \frac{m}{n}$, where m and n are integers. Properties for integer exponents also apply to rational exponents, with one exception. If n is even in the expression $a^{m/n}$ and $\frac{m}{n}$ is written in lowest terms, then a must be nonnegative for the result to be a real number.

For example, the expression $(-8)^{3/2}$ is not a real number because -8 is negative and 2 is even. That is, $(-8)^{3/2} = (\sqrt{-8})^3$, which is undefined because the square root of a negative number is not a real number.

PROPERTIES OF EXPONENTS

Let p and q be rational numbers written in lowest terms. For all real numbers a and b for which the expressions are real numbers, the following properties hold.

1. $a^p \cdot a^q = a^{p+q}$ Product rule for exponents $3^{1/7} \cdot 3^{2/7} = 3^{3/7}$

2. $a^{-p} = \dfrac{1}{a^p}, \dfrac{1}{a^{-p}} = a^p$ Negative exponents $7^{-1/4} = \dfrac{1}{7^{1/4}}, \dfrac{1}{7^{-1/4}} = 7^{1/4}$

3. $\left(\dfrac{a}{b}\right)^{-p} = \left(\dfrac{b}{a}\right)^{p}$ Negative exponents for quotients $\left(\dfrac{3}{7}\right)^{-1/3} = \left(\dfrac{7}{3}\right)^{1/3}$

4. $\dfrac{a^p}{a^q} = a^{p-q}$ Quotient rule for exponents $\dfrac{5^{4/7}}{5^{1/7}} = 5^{3/7}$

5. $(a^p)^q = a^{pq}$ Power rule for exponents $(2^{1/3})^{1/2} = 2^{1/6}$

6. $(ab)^p = a^p b^p$ Power rule for products $(5x)^{2/3} = 5^{2/3} x^{2/3}$

7. $\left(\dfrac{a}{b}\right)^{p} = \dfrac{a^p}{b^p}$ Power rule for quotients $\left(\dfrac{5}{3}\right)^{3/4} = \dfrac{5^{3/4}}{3^{3/4}}$

In the next two examples, we apply these properties.

EXAMPLE 7 **Applying properties of exponents**

Write each expression using rational exponents and simplify. Write the answer with a positive exponent. Assume that all variables are positive numbers.

(a) $\sqrt{x} \cdot \sqrt[3]{x}$ **(b)** $\sqrt[3]{27x^2}$ **(c)** $\dfrac{\sqrt[4]{16x}}{\sqrt[3]{x}}$ **(d)** $\left(\dfrac{x^2}{81}\right)^{-1/2}$

Solution

(a) $\sqrt{x} \cdot \sqrt[3]{x} = x^{1/2} \cdot x^{1/3}$ Use rational exponents.

$= x^{1/2+1/3}$ Product rule for exponents

$= x^{5/6}$ Simplify: $\frac{1}{2} + \frac{1}{3} = \frac{3}{6} + \frac{2}{6} = \frac{5}{6}$.

(b) $\sqrt[3]{27x^2} = (27x^2)^{1/3}$ Use rational exponents.

$= 27^{1/3}(x^2)^{1/3}$ Power rule for products

$= 3x^{2/3}$ Simplify; power rule for exponents.

(c) $\dfrac{\sqrt[4]{16x}}{\sqrt[3]{x}} = \dfrac{(16x)^{1/4}}{x^{1/3}}$ Use rational exponents.

$= \dfrac{16^{1/4}x^{1/4}}{x^{1/3}}$ Power rule for products

$= 16^{1/4}x^{1/4-1/3}$ Quotient rule for exponents

$= 2x^{-1/12}$ Simplify: $\frac{1}{4} - \frac{1}{3} = \frac{3}{12} - \frac{4}{12} = -\frac{1}{12}$.

$= \dfrac{2}{x^{1/12}}$ Negative exponents

(d) $\left(\dfrac{x^2}{81}\right)^{-1/2} = \left(\dfrac{81}{x^2}\right)^{1/2}$ Negative exponents for quotients

$= \dfrac{(81)^{1/2}}{(x^2)^{1/2}}$ Power rule for quotients

$= \dfrac{9}{x}$ Simplify; power rule for exponents.

Now Try Exercises 77, 83, 91, 97

EXAMPLE 8 **Applying properties of exponents**

Write each expression with positive rational exponents and simplify, if possible.

(a) $\sqrt[3]{\sqrt{x+1}}$ **(b)** $\sqrt[5]{c^{15}}$ **(c)** $\dfrac{y^{-1/2}}{x^{-1/3}}$ **(d)** $\sqrt{x}(\sqrt{x}-1)$

Solution

(a) $\sqrt[3]{\sqrt{x+1}} = \left((x+1)^{1/2}\right)^{1/3} = (x+1)^{1/6}$

(b) $\sqrt[5]{c^{15}} = c^{15/5} = c^3$

(c) $\dfrac{y^{-1/2}}{x^{-1/3}} = \dfrac{x^{1/3}}{y^{1/2}}$

(d) $\sqrt{x}(\sqrt{x}-1) = x^{1/2}(x^{1/2}-1) = x^{1/2}x^{1/2} - x^{1/2} = x - x^{1/2}$

Now Try Exercises 85, 89, 95

10.2   Putting It All Together

CONCEPT	EXPLANATION	EXAMPLES
Rational Exponents	If m and n are positive integers with $\frac{m}{n}$ in lowest terms, $$a^{m/n} = \sqrt[n]{a^m} = \left(\sqrt[n]{a}\right)^m.$$ If $a < 0$ and n is even, $a^{m/n}$ is not a real number.	$8^{4/3} = \left(\sqrt[3]{8}\right)^4 = 2^4 = 16$ $(-27)^{3/4} = \left(\sqrt[4]{-27}\right)^3$ is *not* a real number.
Properties of Exponents	Let p and q be rational numbers. 1. $a^p \cdot a^q = a^{p+q}$ 2. $a^{-p} = \dfrac{1}{a^p}, \dfrac{1}{a^{-p}} = a^p$ 3. $\left(\dfrac{a}{b}\right)^{-p} = \left(\dfrac{b}{a}\right)^p$ 4. $\dfrac{a^p}{a^q} = a^{p-q}$ 5. $(a^p)^q = a^{pq}$ 6. $(ab)^p = a^p b^p$ 7. $\left(\dfrac{a}{b}\right)^p = \dfrac{a^p}{b^p}$	1. $2^{1/3} \cdot 2^{2/3} = 2^{1/3+2/3} = 2^1 = 2$ 2. $2^{-1/2} = \dfrac{1}{2^{1/2}}, \dfrac{1}{3^{-1/4}} = 3^{1/4}$ 3. $\left(\dfrac{3}{4}\right)^{-4/5} = \left(\dfrac{4}{3}\right)^{4/5}$ 4. $\dfrac{7^{2/3}}{7^{1/3}} = 7^{2/3-1/3} = 7^{1/3}$ 5. $\left(8^{2/3}\right)^{1/2} = 8^{(2/3)\cdot(1/2)} = 8^{1/3} = 2$ 6. $(2x)^{1/3} = 2^{1/3}x^{1/3}$ 7. $\left(\dfrac{x}{y}\right)^{1/6} = \dfrac{x^{1/6}}{y^{1/6}}$

10.2   Exercises

MyMathLab | 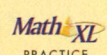 Math XL PRACTICE | WATCH | DOWNLOAD | READ | REVIEW

CONCEPTS AND VOCABULARY

1. Simplify $4^{1/2}$.

2. Simplify $8^{1/3}$.

3. Simplify $4^{-1/2}$.

4. Simplify $8^{-1/3}$.

5. Use a rational exponent to write $\sqrt{x}$.

6. Use a rational exponent to write $\sqrt[3]{a^4}$.

7. Write $a^{1/n}$ in radical notation.

8. Write $a^{m/n}$ in radical notation.

Exercises 9–16: Match the given expression to the expression (a.–h.) that it equals.

9. $\sqrt{x^3}$

10. $\sqrt[3]{8^2}$

11. $25^{-1/2}$

12. $27^{-2/3}$

13. $x^{1/5}$

14. $(x^{1/3})^2$

15. $\sqrt{x} \cdot \sqrt[3]{x}$

16. $\dfrac{x^{1/2}}{x^{1/3}}$

a. $\sqrt[6]{x^5}$

b. $\sqrt[6]{x}$

c. $x^{3/2}$

d. $\sqrt[5]{x}$

e. $\sqrt[3]{x^2}$

f. 4

g. $\dfrac{1}{5}$

h. $\dfrac{1}{9}$

CONVERTING FORMS

Exercises 17–26: Use radical notation to write each expression.

17. $7^{1/2}$

18. $10^{1/2}$

19. $a^{1/3}$

20. $y^{1/4}$

21. $x^{5/6}$

22. $x^{3/2}$

23. $(x + 5)^{1/2}$

24. $(x + y)^{2/3}$

25. $b^{-2/3}$

26. $b^{-3/4}$

Exercise 27–36: Use rational exponents to write each expression.

27. $\sqrt{t}$

28. $\sqrt[3]{t}$

29. $\sqrt[3]{(x + 1)}$

30. $\sqrt{y - 3}$

31. $\dfrac{1}{\sqrt{x + 1}}$

32. $\dfrac{1}{\sqrt[3]{x - 3}}$

33. $\sqrt{a^2 - b^2}$

34. $\sqrt[3]{a^3 + b^3}$

35. $\dfrac{1}{\sqrt[3]{x^7}}$

36. $\dfrac{1}{\sqrt[3]{y^4}}$

RATIONAL EXPONENTS

Exercises 37–44: Approximate to the nearest hundredth.

37. $16^{1/5}$

38. $7^{1/4}$

39. $5^{1/3}$

40. $11^{1/2}$

41. $9^{3/5}$

42. $13^{5/4}$

43. $4^{-3/7}$

44. $2^{-3/4}$

Exercises 45–64: Write each expression in radical notation. Evaluate the expression by hand when possible.

45. $9^{1/2}$

46. $100^{1/2}$

47. $8^{1/3}$

48. $(-8)^{1/3}$

49. $\left(\frac{4}{9}\right)^{1/2}$

50. $\left(\frac{64}{27}\right)^{1/3}$

51. $(-8)^{2/3}$

52. $(16)^{3/2}$

53. $\left(\frac{1}{8}\right)^{-1/3}$

54. $\left(\frac{1}{81}\right)^{-1/4}$

55. $16^{-3/4}$

56. $(32)^{-2/5}$

57. $(4^{1/2})^{-3}$

58. $(8^{1/3})^{-2}$

59. $z^{1/4}$

60. $b^{1/3}$

61. $y^{-2/5}$

62. $z^{-3/4}$

63. $(3x)^{1/3}$

64. $(5x)^{1/2}$

Exercises 65–72: Use a positive rational exponent to write the expression.

65. $\sqrt{y}$

66. $\sqrt{z + 1}$

67. $\sqrt{x} \cdot \sqrt{x}$

68. $\sqrt[3]{x} \cdot \sqrt[3]{x}$

69. $\sqrt[3]{8x^2}$

70. $\sqrt[3]{27z}$

71. $\dfrac{\sqrt{49x}}{\sqrt[3]{x^2}}$

72. $\dfrac{\sqrt[3]{8x}}{\sqrt[4]{x^3}}$

Exercises 73 and 74: **Thinking Generally** *Simplify the expression. Assume that a and b are positive integers.*

73. $(b^{1/a}b^{2/a})^a$

74. $b^{(b-1)/b} \cdot \sqrt[b]{b}$

Exercises 75–98: Use positive rational exponents to simplify the expression. Assume that all variables are positive.

75. $(x^2)^{3/2}$

76. $(y^4)^{1/2}$

77. $\sqrt[3]{x^3 y^6}$

78. $\sqrt{16x^4}$

79. $\sqrt{y^3} \cdot \sqrt[3]{y^2}$

80. $\left(\dfrac{x^6}{81}\right)^{1/4}$

81. $\left(\dfrac{x^6}{27}\right)^{2/3}$

82. $\left(\dfrac{1}{x^8}\right)^{-1/4}$

83. $\left(\dfrac{x^2}{y^6}\right)^{-1/2}$

84. $\dfrac{\sqrt{x}}{\sqrt[3]{27x^6}}$

85. $\sqrt{\sqrt{y}}$

86. $\sqrt{\sqrt[3]{(3x)^2}}$

87. $(a^{-1/2})^{4/3}$

88. $(x^{-3/2})^{2/3}$

89. $\dfrac{(k^{1/2})^{-3}}{(k^2)^{1/4}}$

90. $\dfrac{(b^{3/4})^4}{(b^{4/5})^{-5}}$

91. $\sqrt{b} \cdot \sqrt[4]{b}$

92. $\sqrt[3]{t} \cdot \sqrt[5]{t}$

93. $p^{1/2}\left(p^{3/2} + p^{1/2}\right)$

94. $d^{3/4}(d^{1/4} - d^{-1/4})$

95. $\sqrt[3]{x}\left(\sqrt{x} - \sqrt[3]{x^2}\right)$

96. $\frac{1}{2}\sqrt{x}\left(\sqrt{x} + \sqrt[4]{x^2}\right)$

97. $\dfrac{\sqrt[3]{27x}}{\sqrt{x}}$

98. $\dfrac{\sqrt[5]{32x^2}}{\sqrt[3]{8x}}$

APPLICATIONS

99. *Online Videos* (Refer to Example 3.) The function given by $A(x) = 7.3x^{7/16}$ computes the percentage of viewers who abandon an online video after x seconds for $0 \le x \le 120$.
(a) Make a table of values for $x = 0, 20, 40, 60,$ and 80. Round values to the nearest percent.
(b) Interpret the table in terms of how people watch online videos.

100. *Online Videos* (Refer to Example 3.) Suppose that the function given by $A(x) = 7.3x^{4/3}$ computes the percentage of viewers who abandon an online video after x seconds. Evaluate $A(10)$. Is this function realistic?

101. *Animal Stepping Frequency* (Refer to Example 6.) Use the formula $N(h) = 1.6h^{-1/2}$ to estimate the stepping frequency of a dog that is 2.5 feet high at the shoulders. (*Source:* C. Pennycuick.)

102. *Animal Pulse Rate* According to one model, an animal's heart rate varies according to its weight. The formula $N(w) = 885w^{-1/2}$ gives an estimate for the average number N of beats per minute for an animal that weighs w pounds. Use the formula to estimate the heart rate for a horse that weighs 800 pounds. (*Source:* C. Pennycuick.)

103. *Bird Wings* Heavier birds tend to have larger wings than lighter birds do. For some birds the relationship between the surface area A of the bird's wings in square inches and its weight W in pounds can be modeled by $A = 100\sqrt[3]{W^2}$. (*Source:* C. Pennycuick, *Newton Rules Biology.*)

(a) Find the area of the wings when the weight is 8 pounds.

(b) Write this formula with rational exponents.

104. *Planet Orbits and Distance* Johannes Kepler (1571–1630) discovered a relationship between a planet's distance D from the sun and the time T it takes to orbit the sun. This formula is $T = \sqrt{D^3}$, where T is in Earth years and $D = 1$ corresponds to the distance between Earth and the sun, or 93,000,000 miles.

(a) The planet Neptune is 30 times farther from the sun than Earth ($D = 30$). Estimate the number of years required for Neptune to orbit the sun.

(b) Write this formula with rational exponents.

105. *Baby's Head Size* If a female baby's head circumference is 32.5 centimeters at birth, then the function given by $H(x) = 35.2x^{3/40}$ gives the infant's head circumference at x months for $1 \le x \le 36$.

(a) Evaluate $H(12)$ and $H(24)$.

(b) Does an infant's head circumference increase the most in the first year or the second year?

106. *Musical Tones* One octave on a piano contains 12 keys (including both the black and white keys). The frequency of each successive key increases by a factor of $2^{1/12}$. For example, middle C is two keys below the first D above it. Therefore the frequency of this D is

$$2^{1/12} \cdot 2^{1/12} = 2^{1/6} \approx 1.12$$

times the frequency of middle C.

(a) If two tones are one octave apart, how do their frequencies compare?

(b) The A tone below middle C on a piano has a frequency of 220 cycles per second. Middle C is 3 keys above this A note. Estimate the frequency of middle C.

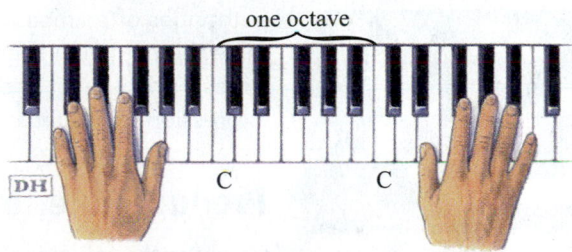

one octave

107. **Online Exploration** Go online and look up the shoulder height of three animals of different sizes. Use the formula in Exercise 101 to calculate their stepping frequencies while walking.

108. **Online Exploration** Go online and look up the weight of three animals of different sizes. Use the formula in Exercise 102 to calculate their heart rates.

WRITING ABOUT MATHEMATICS

109. Explain the meaning of the rational exponent in $x^{3/5}$.

110. Explain the meaning of each expression: $\sqrt[3]{x^5}$ and $(\sqrt[3]{x})^5$. Are these two expressions equal?

Checking Basic Concepts

1. Find the following.
  (a) The square roots of 49
  (b) The principal square root of 49

2. Evaluate.
  (a) $\sqrt[3]{-8}$ **(b)** $-\sqrt[4]{81}$

3. Write the expression in radical notation.
  (a) $x^{3/2}$ **(b)** $x^{2/3}$ **(c)** $x^{-2/5}$

4. Simplify $\sqrt{(x-1)^2}$ for any real number x.

5. Evaluate each function at the given value of x.
  (a) $f(x) = \sqrt{x}$, $x = 9$
  (b) $g(x) = \sqrt[3]{x}$, $x = 125$
  (c) $h(x) = x^{7/12}$, $x = \frac{12}{7}$

10.3   Simplifying Radical Expressions

Product Rule for Radical Expressions • Quotient Rule for Radical Expressions

A LOOK INTO MATH ▶

When a car stops suddenly, it often leaves skid marks. If the car is involved in an accident, authorities often measure the length of the skid marks M and use this information to determine the minimum speed S of the car. One formula that is sometimes used when the road surface is both wet and level is given by $S = \sqrt{9M}$. In Example 4 we simplify this radical expression and estimate the speed of a car based on the length of its skid marks.

Product Rule for Radical Expressions

Consider the following examples of multiplying radical expressions. The equations

NEW VOCABULARY

☐ Perfect nth power
☐ Perfect square
☐ Perfect cube

$$\sqrt{4} \cdot \sqrt{25} = 2 \cdot 5 = 10 \quad \text{and} \quad \sqrt{4 \cdot 25} = \sqrt{100} = 10$$

imply that

$$\sqrt{4} \cdot \sqrt{25} = \sqrt{4 \cdot 25} \quad \text{(see Figure 10.11(a))}.$$

Similarly, the equations

$$\sqrt[3]{8} \cdot \sqrt[3]{27} = 2 \cdot 3 = 6 \quad \text{and} \quad \sqrt[3]{8 \cdot 27} = \sqrt[3]{216} = 6$$

imply that

$$\sqrt[3]{8} \cdot \sqrt[3]{27} = \sqrt[3]{8 \cdot 27} \quad \text{(see Figure 10.11(b))}.$$

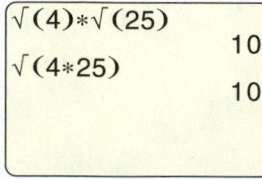

(a)

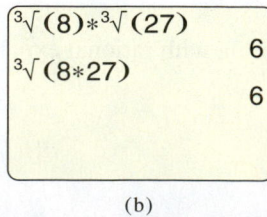

(b)

Figure 10.11

These examples suggest that *the product of the roots is equal to the root of the product.*

PRODUCT RULE FOR RADICAL EXPRESSIONS

Let a and b be real numbers, where $\sqrt[n]{a}$ and $\sqrt[n]{b}$ are both defined. Then

$$\sqrt[n]{a} \cdot \sqrt[n]{b} = \sqrt[n]{a \cdot b}. \qquad\qquad \sqrt[4]{2} \cdot \sqrt[4]{8} = \sqrt[4]{2 \cdot 8}$$

NOTE: The product rule works only when the radicals have the *same index*. For example, the product $\sqrt{2} \cdot \sqrt[3]{4}$ cannot be simplified because the indexes are 2 and 3. (However, by using rational exponents, we can simplify this product. See Example 6(b).)

We apply the product rule in the next two examples.

EXAMPLE 1   Multiplying radical expressions

Multiply each radical expression.

(a) $\sqrt{5} \cdot \sqrt{20}$ **(b)** $\sqrt[3]{-3} \cdot \sqrt[3]{9}$ **(c)** $\sqrt[4]{\frac{1}{3}} \cdot \sqrt[4]{\frac{1}{9}} \cdot \sqrt[4]{\frac{1}{3}}$

Solution
(a) $\sqrt{5} \cdot \sqrt{20} = \sqrt{5 \cdot 20} = \sqrt{100} = 10$
(b) $\sqrt[3]{-3} \cdot \sqrt[3]{9} = \sqrt[3]{-3 \cdot 9} = \sqrt[3]{-27} = -3$
(c) The product rule can also be applied to three or more factors. Thus

$$\sqrt[4]{\frac{1}{3}} \cdot \sqrt[4]{\frac{1}{9}} \cdot \sqrt[4]{\frac{1}{3}} = \sqrt[4]{\frac{1}{3} \cdot \frac{1}{9} \cdot \frac{1}{3}} = \sqrt[4]{\frac{1}{81}} = \frac{1}{3},$$

because $\frac{1}{3} \cdot \frac{1}{3} \cdot \frac{1}{3} \cdot \frac{1}{3} = \frac{1}{81}$.

Now Try Exercises 13, 15, 21

EXAMPLE 2   Multiplying radical expressions containing variables

Multiply each radical expression. Assume that all variables are positive.

(a) $\sqrt{x} \cdot \sqrt{x^3}$ **(b)** $\sqrt[3]{2a} \cdot \sqrt[3]{5a}$ **(c)** $\sqrt{11} \cdot \sqrt{xy}$ **(d)** $\sqrt[5]{\frac{2x}{y}} \cdot \sqrt[5]{\frac{16y}{x}}$

Solution
(a) $\quad \sqrt{x} \cdot \sqrt{x^3} = \sqrt{x \cdot x^3} = \sqrt{x^4} = x^2$
(b) $\quad \sqrt[3]{2a} \cdot \sqrt[3]{5a} = \sqrt[3]{2a \cdot 5a} = \sqrt[3]{10a^2}$
(c) $\quad \sqrt{11} \cdot \sqrt{xy} = \sqrt{11xy}$
(d) $\sqrt[5]{\frac{2x}{y}} \cdot \sqrt[5]{\frac{16y}{x}} = \sqrt[5]{\frac{2x}{y} \cdot \frac{16y}{x}}$ Product rule

$$= \sqrt[5]{\frac{32xy}{xy}} \qquad \text{Multiply fractions.}$$

$$= \sqrt[5]{32} \qquad \text{Simplify.}$$

$$= 2 \qquad 2^5 = 32$$

Now Try Exercises 23, 51, 57, 61

An integer a is a **perfect nth power** if there exists an integer b such that $b^n = a$. Thus 36 is a **perfect square** because $6^2 = 36$. Similarly, 8 is a **perfect cube** because $2^3 = 8$, and 81 is a *perfect fourth power* because $3^4 = 81$. Table 10.6 lists examples of perfect squares and perfect cubes.

TABLE 10.6

Perfect Squares	1	4	9	16	25	36
Perfect Cubes	1	8	27	64	125	216

To simplify a square root, we sometimes need to recognize the *largest* perfect square factor of the radicand. For example, 50 has several factors.

$$1, \ 2, \ 5, \ 10, \ 25, \ 50 \quad \text{Factors of 50}$$

From Table 10.6, the perfect square factors of 50 are **1** and **25**. The *largest* perfect square factor of 50 is **25**.

The product rule for radicals can be used to simplify radical expressions. For example, because the largest perfect square factor of 50 is 25, the expression $\sqrt{50}$ can be simplified as

$$\sqrt{50} = \sqrt{25 \cdot 2} = \sqrt{25} \cdot \sqrt{2} = 5\sqrt{2}.$$

This procedure is generalized as follows.

READING CHECK

Explain how to simplify $\sqrt{40}$.

SIMPLIFYING RADICALS (nth ROOTS)

STEP 1: Determine the largest perfect nth power factor of the radicand.

STEP 2: Use the product rule to factor out and simplify this perfect nth power.

EXAMPLE 3 | **Simplifying radical expressions**

Simplify each expression.

(a) $\sqrt{300}$ (b) $\sqrt[3]{16}$ (c) $\sqrt{54}$ (d) $\sqrt[4]{512}$

Solution

(a) First note that $300 = 100 \cdot 3$ and that 100 is the largest perfect square factor of 300.

$$\sqrt{300} = \sqrt{100 \cdot 3} = \sqrt{100} \cdot \sqrt{3} = 10\sqrt{3}$$

(b) The largest perfect cube factor of 16 is 8. (See Table 10.6.) Thus

$$\sqrt[3]{16} = \sqrt[3]{8} \cdot \sqrt[3]{2} = 2\sqrt[3]{2}.$$

(c) $\sqrt{54} = \sqrt{9} \cdot \sqrt{6} = 3\sqrt{6}$

(d) $\sqrt[4]{512} = \sqrt[4]{256} \cdot \sqrt[4]{2} = 4\sqrt[4]{2}$ because $4^4 = 256$.

Now Try Exercises 73, 75, 77, 79

▶ **REAL-WORLD CONNECTION** In A Look Into Math at the beginning of this section, we discussed a formula for estimating the minimum speed of a car by using its skid marks. In the next example, we simplify and use this formula.

EXAMPLE 4 **Calculating minimum speed**

A car is in an accident and leaves skid marks on wet, level pavement that are 121 feet long. The formula $S = \sqrt{9M}$ calculates the minimum speed S in miles per hour for skid marks M feet long.
(a) Simplify this formula.
(b) Determine the minimum speed of the car.

Solution
(a) Because 9 is a perfect square, we can factor it out.

$$\sqrt{9M} = \sqrt{9} \cdot \sqrt{M} = 3\sqrt{M}$$

Thus $S = 3\sqrt{M}$.
(b) $S = 3\sqrt{121} = 3 \cdot 11 = 33$ miles per hour.

▋ **Now Try Exercise 111**

NOTE: To simplify a cube root of a negative number we factor out the negative of the largest perfect cube factor. For example, $-16 = -8 \cdot 2$, so $\sqrt[3]{-16} = \sqrt[3]{-8} \cdot \sqrt[3]{2} = -2\sqrt[3]{2}$. This procedure can be used with any odd root of a negative number. (See Example 5(c).)

EXAMPLE 5 **Simplifying radical expressions**

Simplify each expression. Assume that all variables are positive.
(a) $\sqrt{25x^4}$ **(b)** $\sqrt{32n^3}$ **(c)** $\sqrt[3]{-16x^3y^5}$ **(d)** $\sqrt[3]{2a} \cdot \sqrt[3]{4a^2b}$

Solution
(a) $\sqrt{25x^4} = 5x^2$ Perfect square: $(5x^2)^2 = 25x^4$

(b) $\sqrt{32n^3} = \sqrt{(16n^2)2n}$ $16n^2$ is the largest perfect square factor.

            $= \sqrt{16n^2} \cdot \sqrt{2n}$ Product rule

            $= 4n\sqrt{2n}$ $(4n)^2 = 16n^2$

(c) $\sqrt[3]{-16x^3y^5} = \sqrt[3]{(-8x^3y^3)2y^2}$ $8x^3y^3$ is the largest perfect cube factor.

            $= \sqrt[3]{-8x^3y^3} \cdot \sqrt[3]{2y^2}$ Product rule

            $= -2xy\sqrt[3]{2y^2}$ $(-2xy)^3 = -8x^3y^3$

(d) $\sqrt[3]{2a} \cdot \sqrt[3]{4a^2b} = \sqrt[3]{(2a)(4a^2b)}$ Product rule

            $= \sqrt[3]{(8a^3)b}$ $8a^3$ is the largest perfect cube factor.

            $= \sqrt[3]{8a^3} \cdot \sqrt[3]{b}$ Product rule

            $= 2a\sqrt[3]{b}$ $(2a)^3 = 8a^3$

▋ **Now Try Exercises 45, 85, 89, 91**

The product rule for radical expressions cannot be used if the radicals do not have the same indexes. In this case we use rational exponents, as illustrated in the next example.

EXAMPLE 6 **Multiplying radicals with different indexes**

Simplify each expression. Write your answer in radical notation.

(a) $\sqrt{5} \cdot \sqrt[4]{5}$ (b) $\sqrt{2} \cdot \sqrt[3]{4}$ (c) $\sqrt[3]{x} \cdot \sqrt[4]{x}$

Solution

(a) Because $\sqrt{5} = 5^{1/2}$ and $\sqrt[4]{5} = 5^{1/4}$,

$$\sqrt{5} \cdot \sqrt[4]{5} = 5^{1/2} \cdot 5^{1/4} = 5^{1/2+1/4} = 5^{3/4}.$$

In radical notation, $5^{3/4} = \sqrt[4]{5^3} = \sqrt[4]{125}$.

(b) Because $\sqrt[3]{4} = \sqrt[3]{2^2} = 2^{2/3}$,

$$\sqrt{2} \cdot \sqrt[3]{4} = 2^{1/2} \cdot 2^{2/3} = 2^{1/2+2/3} = 2^{7/6}.$$

In radical notation, $2^{7/6} = \sqrt[6]{2^7} = \sqrt[6]{2^6 \cdot 2^1} = \sqrt[6]{2^6} \cdot \sqrt[6]{2} = 2\sqrt[6]{2}$.

(c) $\sqrt[3]{x} \cdot \sqrt[4]{x} = x^{1/3} \cdot x^{1/4} = x^{7/12} = \sqrt[12]{x^7}$

Now Try Exercises 101, 103, 107

Quotient Rule for Radical Expressions

Consider the following examples of dividing radical expressions. The equations

$$\sqrt{\frac{4}{9}} = \sqrt{\frac{2}{3} \cdot \frac{2}{3}} = \frac{2}{3} \quad \text{and} \quad \frac{\sqrt{4}}{\sqrt{9}} = \frac{2}{3}$$

imply that

$$\sqrt{\frac{4}{9}} = \frac{\sqrt{4}}{\sqrt{9}} \quad \text{(see Figure 10.12).}$$

These examples suggest that *the root of a quotient is equal to the quotient of the roots.*

Figure 10.12

```
√(4/9)▶Frac
              2/3
√(4)/√(9)▶Frac
              2/3
```

CALCULATOR HELP

To use the Frac feature, see Appendix A (pages AP-1 and AP-2).

QUOTIENT RULE FOR RADICAL EXPRESSIONS

Let a and b be real numbers, where $\sqrt[n]{a}$ and $\sqrt[n]{b}$ are both defined and $b \neq 0$. Then

$$\sqrt[n]{\frac{a}{b}} = \frac{\sqrt[n]{a}}{\sqrt[n]{b}}.$$

$$\sqrt[3]{\frac{23}{5}} = \frac{\sqrt[3]{23}}{\sqrt[3]{5}}$$

EXAMPLE 7 **Simplifying quotients**

Simplify each radical expression. Assume that all variables are positive.

(a) $\sqrt[3]{\dfrac{5}{8}}$ (b) $\sqrt[4]{\dfrac{x}{16}}$ (c) $\sqrt{\dfrac{16}{y^2}}$

Solution

(a) $\sqrt[3]{\dfrac{5}{8}} = \dfrac{\sqrt[3]{5}}{\sqrt[3]{8}} = \dfrac{\sqrt[3]{5}}{2}$ (b) $\sqrt[4]{\dfrac{x}{16}} = \dfrac{\sqrt[4]{x}}{\sqrt[4]{16}} = \dfrac{\sqrt[4]{x}}{2}$

(c) $\sqrt{\dfrac{16}{y^2}} = \dfrac{\sqrt{16}}{\sqrt{y^2}} = \dfrac{4}{y}$ because $y > 0$.

Now Try Exercises 25, 27, 29

EXAMPLE 8 **Simplifying quotients**

Simplify each radical expression. Assume that all variables are positive.

(a) $\dfrac{\sqrt{40}}{\sqrt{10}}$ **(b)** $\dfrac{\sqrt[3]{2}}{\sqrt[3]{16}}$ **(c)** $\dfrac{\sqrt{x^2 y}}{\sqrt{y}}$

Solution

(a) $\dfrac{\sqrt{40}}{\sqrt{10}} = \sqrt{\dfrac{40}{10}} = \sqrt{4} = 2$

(b) $\dfrac{\sqrt[3]{2}}{\sqrt[3]{16}} = \sqrt[3]{\dfrac{2}{16}} = \sqrt[3]{\dfrac{1}{8}} = \dfrac{1}{2}$ because $\dfrac{1}{2} \cdot \dfrac{1}{2} \cdot \dfrac{1}{2} = \dfrac{1}{8}$.

(c) $\dfrac{\sqrt{x^2 y}}{\sqrt{y}} = \sqrt{\dfrac{x^2 y}{y}} = \sqrt{x^2} = x$ because $x > 0$.

Now Try Exercises 33, 39, 41

MAKING CONNECTIONS

Rules for Radical Expressions and Rational Exponents

The rules for radical expressions are a result of the properties of rational exponents.

$$\sqrt[n]{a \cdot b} = \sqrt[n]{a} \cdot \sqrt[n]{b} \quad \text{is equivalent to} \quad (a \cdot b)^{1/n} = a^{1/n} \cdot b^{1/n}.$$

$$\sqrt[n]{\dfrac{a}{b}} = \dfrac{\sqrt[n]{a}}{\sqrt[n]{b}} \quad \text{is equivalent to} \quad \left(\dfrac{a}{b}\right)^{1/n} = \dfrac{a^{1/n}}{b^{1/n}}.$$

READING CHECK

Use rational exponents to explain why $\sqrt[3]{x} \cdot \sqrt[3]{x} \cdot \sqrt[3]{x} = x$.

EXAMPLE 9 **Simplifying radical expressions**

Simplify each radical expression. Assume that all variables are positive.

(a) $\sqrt[4]{\dfrac{16x^3}{y^4}}$ **(b)** $\sqrt{\dfrac{5a^2}{8}} \cdot \sqrt{\dfrac{5a^3}{2}}$

Solution

(a) To simplify this expression, we first use the quotient rule for radical expressions and then apply the product rule for radical expressions.

$$\sqrt[4]{\dfrac{16x^3}{y^4}} = \dfrac{\sqrt[4]{16x^3}}{\sqrt[4]{y^4}} \qquad \text{Quotient rule}$$

$$= \dfrac{\sqrt[4]{16}\,\sqrt[4]{x^3}}{\sqrt[4]{y^4}} \qquad \text{Product rule}$$

$$= \dfrac{2\,\sqrt[4]{x^3}}{y} \qquad \text{Evaluate 4th roots.}$$

(b) To simplify this expression, we use both the product and quotient rules.

$$\sqrt{\frac{5a^2}{8}} \cdot \sqrt{\frac{5a^3}{2}} = \sqrt{\frac{5a^2 \cdot 5a^3}{8 \cdot 2}} \qquad \text{Product rule}$$

$$= \sqrt{\frac{25a^5}{16}} \qquad \text{Multiply.}$$

$$= \frac{\sqrt{25a^5}}{\sqrt{16}} \qquad \text{Quotient rule}$$

$$= \frac{\sqrt{25a^4 \cdot a}}{\sqrt{16}} \qquad \text{Factor out largest perfect square.}$$

$$= \frac{\sqrt{25a^4} \cdot \sqrt{a}}{\sqrt{16}} \qquad \text{Product rule}$$

$$= \frac{5a^2\sqrt{a}}{4} \qquad (5a^2)^2 = 25a^4$$

Now Try Exercises 95, 97

EXAMPLE 10 **Simplifying products and quotients of roots**

Simplify each expression. Assume all radicands are positive.

(a) $\sqrt{x-3} \cdot \sqrt{x+3}$ **(b)** $\dfrac{\sqrt[3]{x^2 + 3x + 2}}{\sqrt[3]{x+1}}$

Solution

(a) Start by applying the product rule for radical expressions.

$$\sqrt{x-3} \cdot \sqrt{x+3} = \sqrt{(x-3)(x+3)} \qquad \text{Product rule}$$

$$= \sqrt{x^2 - 9} \qquad \text{Multiply binomials.}$$

NOTE: The expression $\sqrt{x^2 - 9}$ does *not* simplify further. It is important to realize that

$$\sqrt{x^2 - 9} \neq \sqrt{x^2} - \sqrt{9} = x - 3.$$

For example, $\sqrt{5^2 - 3^2} = \sqrt{16} = 4$, but $\sqrt{5^2} - \sqrt{3^2} = 5 - 3 = 2$.

(b) Start by applying the quotient rule for radical expressions.

$$\frac{\sqrt[3]{x^2 + 3x + 2}}{\sqrt[3]{x+1}} = \sqrt[3]{\frac{x^2 + 3x + 2}{x+1}} \qquad \text{Quotient rule}$$

$$= \sqrt[3]{\frac{(x+1)(x+2)}{x+1}} \qquad \text{Factor trinomial.}$$

$$= \sqrt[3]{x+2} \qquad \text{Simplify quotient.}$$

Now Try Exercises 63, 67

10.3 Putting It All Together

CONCEPT	EXPLANATION	EXAMPLES
Product Rule for Radical Expressions	Let a and b be real numbers, where $\sqrt[n]{a}$ and $\sqrt[n]{b}$ are both defined. Then $$\sqrt[n]{a} \cdot \sqrt[n]{b} = \sqrt[n]{a \cdot b}.$$	$$\sqrt{2} \cdot \sqrt{32} = \sqrt{64} = 8$$ $$\sqrt{500} = \sqrt{100} \cdot \sqrt{5} = 10\sqrt{5}$$
Simplifying Radicals (nth roots)	**STEP 1:** Find the largest perfect nth power factor of the radicand. **STEP 2:** Factor out and simplify this perfect nth power.	$$\sqrt{12} = \sqrt{4 \cdot 3} = \sqrt{4} \cdot \sqrt{3} = 2\sqrt{3}$$ $$\sqrt[3]{81} = \sqrt[3]{27 \cdot 3} = \sqrt[3]{27} \cdot \sqrt[3]{3} = 3\sqrt[3]{3}$$ $$\sqrt[4]{x^5} = \sqrt[4]{x^4 \cdot x} = \sqrt[4]{x^4} \cdot \sqrt[4]{x} = x\sqrt[4]{x},$$ provided $x \geq 0$.
Quotient Rule for Radical Expressions	Let a and b be real numbers, where $\sqrt[n]{a}$ and $\sqrt[n]{b}$ are both defined and $b \neq 0$. Then $$\sqrt[n]{\frac{a}{b}} = \frac{\sqrt[n]{a}}{\sqrt[n]{b}}.$$	$$\frac{\sqrt{60}}{\sqrt{15}} = \sqrt{\frac{60}{15}} = \sqrt{4} = 2$$ $$\sqrt[3]{\frac{x^2}{-27}} = \frac{\sqrt[3]{x^2}}{\sqrt[3]{-27}} = \frac{\sqrt[3]{x^2}}{-3} = -\frac{\sqrt[3]{x^2}}{3}$$

10.3 Exercises

MyMathLab   Math XL PRACTICE 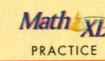 WATCH   DOWNLOAD   READ   REVIEW

CONCEPTS AND VOCABULARY

1. Does $\sqrt{2} \cdot \sqrt{3}$ equal $\sqrt{6}$?

2. Does $\sqrt{5} \cdot \sqrt[3]{5}$ equal 5?

3. $\sqrt[3]{a} \cdot \sqrt[3]{b} =$ _____

4. $\dfrac{\sqrt{a}}{\sqrt{b}} = \sqrt{?}$

5. $\dfrac{\sqrt[n]{a}}{\sqrt[n]{b}} = \sqrt[n]{?}$

6. $\dfrac{\sqrt{3}}{\sqrt{27}} =$ _____

7. Does $\sqrt{50} = \sqrt{25} + \sqrt{25}$?

8. Does $\sqrt{50} = 5\sqrt{2}$?

9. Is $\sqrt[3]{3}$ equal to 1? Explain.

10. Is 64 a perfect cube? Explain.

MULTIPLYING AND DIVIDING

Exercises 11–62: Simplify the expression. Assume that all variables are positive.

11. $\sqrt{3} \cdot \sqrt{3}$

12. $\sqrt{2} \cdot \sqrt{18}$

13. $\sqrt{2} \cdot \sqrt{50}$

14. $\sqrt[3]{-2} \cdot \sqrt[3]{-4}$

15. $\sqrt[3]{4} \cdot \sqrt[3]{16}$

16. $\sqrt[3]{x} \cdot \sqrt[3]{x^2}$

17. $\sqrt{\dfrac{9}{25}}$

18. $\sqrt[3]{\dfrac{x}{8}}$

19. $\sqrt{\dfrac{1}{2}} \cdot \sqrt{\dfrac{1}{8}}$

20. $\sqrt{\dfrac{5}{3}} \cdot \sqrt{\dfrac{1}{3}}$

21. $\sqrt[3]{\dfrac{2}{3}} \cdot \sqrt[3]{\dfrac{4}{3}} \cdot \sqrt[3]{\dfrac{1}{3}}$

22. $\sqrt[4]{\dfrac{8}{3}} \cdot \sqrt[4]{\dfrac{4}{9}} \cdot \sqrt[4]{\dfrac{8}{3}}$

23. $\sqrt{x^3} \cdot \sqrt{x^3}$

24. $\sqrt{z} \cdot \sqrt{z^7}$

25. $\sqrt[3]{\dfrac{7}{27}}$

26. $\sqrt[3]{\dfrac{9}{64}}$

27. $\sqrt[4]{\dfrac{x}{81}}$

28. $\sqrt[4]{\dfrac{16x}{81}}$

29. $\sqrt{\dfrac{9}{z^2}}$

30. $\sqrt{\dfrac{x^2}{81}}$

31. $\sqrt{\dfrac{x}{2}} \cdot \sqrt{\dfrac{x}{8}}$

32. $\sqrt{\dfrac{4}{y}} \cdot \sqrt{\dfrac{y}{5}}$

33. $\dfrac{\sqrt{45}}{\sqrt{5}}$

34. $\dfrac{\sqrt{7}}{\sqrt{28}}$

35. $\sqrt[3]{-4} \cdot \sqrt[3]{-16}$

36. $\sqrt[3]{9} \cdot \sqrt[3]{3}$

37. $\sqrt[4]{9} \cdot \sqrt[4]{9}$

38. $\sqrt[5]{16} \cdot \sqrt[5]{-2}$

39. $\dfrac{\sqrt[5]{64}}{\sqrt[5]{-2}}$

40. $\dfrac{\sqrt[4]{324}}{\sqrt[4]{4}}$

41. $\dfrac{\sqrt{a^2 b}}{\sqrt{b}}$

42. $\dfrac{\sqrt{4xy^2}}{\sqrt{x}}$

43. $\dfrac{\sqrt[3]{54}}{\sqrt[3]{2}}$

44. $\dfrac{\sqrt[3]{x^3 y^7}}{\sqrt[3]{y^4}}$

45. $\sqrt{4x^4}$

46. $\sqrt[3]{-8y^3}$

47. $\sqrt[3]{-5a^6}$

48. $\sqrt{9x^2y}$

49. $\sqrt[4]{16x^4y}$

50. $\sqrt[3]{8xy^3}$

51. $\sqrt{3x} \cdot \sqrt{12x}$

52. $\sqrt{6x^5} \cdot \sqrt{6x}$

53. $\sqrt[3]{8x^6y^3z^9}$

54. $\sqrt{16x^4y^6}$

55. $\sqrt[4]{\frac{3}{4}} \cdot \sqrt[4]{\frac{27}{4}}$

56. $\sqrt[5]{\frac{4}{-9}} \cdot \sqrt[5]{\frac{8}{-27}}$

57. $\sqrt[3]{12} \cdot \sqrt[3]{ab}$

58. $\sqrt{5x} \cdot \sqrt{5z}$

59. $\sqrt[4]{25z} \cdot \sqrt[4]{25z}$

60. $\sqrt[5]{3z^2} \cdot \sqrt[5]{7z}$

61. $\sqrt[5]{\frac{7a}{b^2}} \cdot \sqrt[5]{\frac{b^2}{7a^6}}$

62. $\sqrt[3]{\frac{8m}{n}} \cdot \sqrt[3]{\frac{n^4}{m^2}}$

Exercises 63–68: Use properties of polynomials to simplify the expression. Assume all radicands are positive.

63. $\sqrt{x+4} \cdot \sqrt{x-4}$

64. $\sqrt[3]{x-1} \cdot \sqrt[3]{x^2+x+1}$

65. $\sqrt[3]{a+1} \cdot \sqrt[3]{a^2-a+1}$

66. $\sqrt{b-1} \cdot \sqrt{b+1}$

67. $\dfrac{\sqrt{x^2+2x+1}}{\sqrt{x+1}}$

68. $\dfrac{\sqrt{x^2-4x+4}}{\sqrt{x-2}}$

Exercises 69–74: Complete the equation.

69. $\sqrt{500} = \underline{\quad} \sqrt{5}$ **70.** $\sqrt{28} = \underline{\quad} \sqrt{7}$

71. $\sqrt{8} = \underline{\quad} \sqrt{2}$ **72.** $\sqrt{99} = \underline{\quad} \sqrt{11}$

73. $\sqrt{45} = \underline{\quad} \sqrt{5}$ **74.** $\sqrt{243} = \underline{\quad} \sqrt{3}$

Exercises 75–98: Simplify the radical expression by factoring out the largest perfect nth power. Assume that all variables are positive.

75. $\sqrt{200}$

76. $\sqrt{72}$

77. $\sqrt[3]{81}$

78. $\sqrt[3]{256}$

79. $\sqrt[4]{64}$

80. $\sqrt[5]{27 \cdot 81}$

81. $\sqrt[5]{-64}$

82. $\sqrt[3]{-81}$

83. $\sqrt{b^5}$

84. $\sqrt{t^3}$

85. $\sqrt{8n^3}$

86. $\sqrt{32a^2}$

87. $\sqrt{12a^2b^5}$

88. $\sqrt{20a^3b^2}$

89. $\sqrt[3]{-125x^4y^5}$

90. $\sqrt[3]{-81a^5b^2}$

91. $\sqrt[3]{5t} \cdot \sqrt[3]{125t}$

92. $\sqrt[4]{4bc^3} \cdot \sqrt[4]{64ab^3c^2}$

93. $\sqrt[4]{\frac{9t^5}{r^8}} \cdot \sqrt[4]{\frac{9r}{5t}}$

94. $\sqrt[5]{\frac{4t^6}{r}} \cdot \sqrt[5]{\frac{8t}{r^6}}$

95. $\sqrt[3]{\frac{27x^2}{y^3}}$

96. $\sqrt[4]{\frac{32x^8}{z^4}}$

97. $\sqrt{\frac{7a^2}{27}} \cdot \sqrt{\frac{7a}{3}}$

98. $\sqrt{\frac{8a}{125}} \cdot \sqrt{\frac{2a}{5}}$

Exercises 99 and 100: **Thinking Generally** *Let m and n be positive integers and assume each expression exists.*

99. Simplify $\left(\sqrt[mn]{a^m b^m}\right)^n$ so that it equals ab.

100. Simplify $\left(\sqrt[mn]{a^n b^m}\right) \cdot \left(\sqrt[mn]{a^m b^m}\right)$ so that it equals $\sqrt[m]{ab} \cdot \sqrt[n]{ab}$.

Exercises 101–110: Simplify the expression. Let all variables be positive and write your answer in radical notation.

101. $\sqrt{3} \cdot \sqrt[3]{3}$

102. $\sqrt{5} \cdot \sqrt[3]{5}$

103. $\sqrt[4]{8} \cdot \sqrt[3]{4}$

104. $\sqrt[5]{16} \cdot \sqrt{2}$

105. $\sqrt[4]{27} \cdot \sqrt[3]{9} \cdot \sqrt{3}$

106. $\sqrt[5]{16} \cdot \sqrt[3]{16}$

107. $\sqrt[4]{x^3} \cdot \sqrt[3]{x}$

108. $\sqrt[4]{x^3} \cdot \sqrt{x}$

109. $\sqrt[4]{rt} \cdot \sqrt[3]{r^2t}$

110. $\sqrt[3]{a^3b^2} \cdot \sqrt{a^2b}$

APPLICATIONS

111. *Minimum Speed* (Refer to Example 4.) If the pavement is dry cement, then $S = \sqrt{25M}$ is an estimate for a car's speed S in miles per hour when it leaves skid marks M feet long.
   (a) Simplify this formula.
   (b) Determine the minimum speed of the car if the skid marks are 100 feet.

112. *Minimum Speed* (Refer to Example 4.) If the road surface is gravel, then $S = \sqrt{15M}$ is an estimate for a car's speed S in miles per hour when it leaves skid marks M feet long in the gravel. Determine the minimum speed of the car if the skid marks are 30 feet.

WRITING ABOUT MATHEMATICS

113. Explain what it means for a positive integer to be a *perfect square* or a *perfect cube*. List the positive integers that are perfect squares and less than 101. List the positive integers that are perfect cubes and less than 220.

114. Explain how the product and quotient rules for radical expressions are the result of properties of rational exponents.

Group Activity   Working with Real Data

Directions: Form a group of 2 to 4 people. Select someone to record the group's responses for this activity. All members of the group should work cooperatively to answer the questions. If your instructor asks for your results, each member of the group should be prepared to respond.

Designing a Paper Cup A paper drinking cup is being designed in the shape shown in the accompanying figure. The amount of paper needed to manufacture the cup is determined by the surface area S of the cup, which is given by

$$S = \pi r \sqrt{r^2 + h^2},$$

where r is the radius and h is the height.

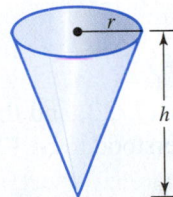

(a) Approximate S to the nearest hundredth when $r = 1.5$ inches and $h = 4$ inches.

(b) Could the formula for S be simplified as follows?

$$\pi r \sqrt{r^2 + h^2} \stackrel{?}{=} \pi r (\sqrt{r^2} + \sqrt{h^2})$$
$$\stackrel{?}{=} \pi r (r + h)$$

Try evaluating this second formula with $r = 1.5$ inches and $h = 4$ inches. Is your answer the same as in part (a)? Explain.

(c) Discuss why evaluating real-world formulas correctly is important.

(d) In general, does $\sqrt{a + b}$ equal $\sqrt{a} + \sqrt{b}$? Justify your answer by completing the following table. Approximate answers to the nearest hundredth when appropriate.

a	b	$\sqrt{a + b}$	$\sqrt{a} + \sqrt{b}$
0	4		
4	0		
5	4		
9	7		
4	16		
25	100		

10.4   Operations on Radical Expressions

Addition and Subtraction • Multiplication • Rationalizing the Denominator

A LOOK INTO MATH ▶

Our productivity as a nation has increased significantly during the past 70 years because of new technology. For example, if a typical business bought $10,000 in equipment per worker in 2005, it would have generated about $48,000 per worker in revenue. This same investment by a business in 1935 would have generated only $23,000 in revenue (adjusted to 2005 dollars). This situation can be described by the following two functions.

$$N(x) = 400\sqrt{x} + 8000 \quad \text{and} \quad O(x) = 195\sqrt{x} + 3500$$

2005 technology (new)   1935 technology (old)

What does the difference of these two functions represent? In this section we learn how to subtract expressions that contain radicals. See Example 7.

NEW VOCABULARY

☐ Like radicals
☐ Rationalizing the denominator
☐ Conjugate

Addition and Subtraction

We can add $2x^2$ and $5x^2$ to obtain $7x^2$ because they are *like* terms. That is,

$$2x^2 + 5x^2 = (2 + 5)x^2 = 7x^2. \quad \text{Like terms}$$

We can also add and subtract *like radicals*. **Like radicals** *have the same index and the same radicand*. For example, we can add $3\sqrt{2}$ and $5\sqrt{2}$ because they are like radicals.

$$3\sqrt{2} + 5\sqrt{2} = (3 + 5)\sqrt{2} = 8\sqrt{2} \quad \text{Like radicals}$$

EXAMPLE 1 | **Adding like radicals**

If possible, add the expressions and simplify.

(a) $10\sqrt{11} + 4\sqrt{11}$ **(b)** $5\sqrt[3]{6} + \sqrt[3]{6}$

(c) $4 + 5\sqrt{3}$ **(d)** $\sqrt{7} + \sqrt{11}$

Solution

(a) These terms are like radicals because they have the same index, 2, and the same radicand, 11.

$$10\sqrt{11} + 4\sqrt{11} = (10 + 4)\sqrt{11} = 14\sqrt{11}$$

(b) These terms are like radicals because they have the same index, 3, and the same radicand, 6. Note that the coefficient on the second term is understood to be 1.

$$5\sqrt[3]{6} + 1\sqrt[3]{6} = (5 + 1)\sqrt[3]{6} = 6\sqrt[3]{6}$$

(c) The expression $4 + 5\sqrt{3}$ can be written as $4\sqrt{1} + 5\sqrt{3}$. These terms cannot be added because they are not like radicals.

NOTE: $4 + 5\sqrt{3} \neq 9\sqrt{3}$ because multiplication is performed before addition.

(d) The expression $\sqrt{7} + \sqrt{11}$ contains unlike radicals that *cannot* be added.

Now Try Exercises 19, 21, 23, 25

Sometimes two radical expressions that are not alike can be added by changing them to like radicals. For example, $\sqrt{20}$ and $\sqrt{5}$ are unlike radicals. However,

because $\sqrt{20} = \sqrt{4 \cdot 5} = \sqrt{4} \cdot \sqrt{5} = 2\sqrt{5}$,

we have $\sqrt{20} + \sqrt{5} = 2\sqrt{5} + 1\sqrt{5} = 3\sqrt{5}$.

We cannot combine $x + x^2$ because they are unlike terms. Similarly, we cannot combine $\sqrt{2} + \sqrt{5}$ because they are unlike radicals. When combining radicals, the first step is to see if we can write pairs of terms as like radicals, as demonstrated in the next example.

EXAMPLE 2 | **Finding like radicals**

Write each pair of terms as like radicals, if possible.

(a) $\sqrt{45}, \sqrt{20}$ **(b)** $\sqrt{27}, \sqrt{5}$ **(c)** $5\sqrt[3]{16}, 4\sqrt[3]{54}$

Solution

(a) The expressions $\sqrt{45}$ and $\sqrt{20}$ are unlike radicals. However, they can be changed to like radicals, as follows.

$$\sqrt{45} = \sqrt{9 \cdot 5} = \sqrt{9} \cdot \sqrt{5} = 3\sqrt{5} \quad \text{and}$$

$$\sqrt{20} = \sqrt{4 \cdot 5} = \sqrt{4} \cdot \sqrt{5} = 2\sqrt{5}$$

The expressions $3\sqrt{5}$ and $2\sqrt{5}$ are like radicals.

READING CHECK

What are like radicals?

(b) Because $\sqrt{27} = \sqrt{9 \cdot 3} = \sqrt{9} \cdot \sqrt{3} = 3\sqrt{3}$, the given expressions $\sqrt{27}$ and $\sqrt{5}$ are unlike radicals and cannot be written as like radicals.

(c) $5\sqrt[3]{16} = 5\sqrt[3]{8 \cdot 2} = 5\sqrt[3]{8} \cdot \sqrt[3]{2} = 5 \cdot 2 \cdot \sqrt[3]{2} = 10\sqrt[3]{2}$ and

$4\sqrt[3]{54} = 4\sqrt[3]{27 \cdot 2} = 4\sqrt[3]{27} \cdot \sqrt[3]{2} = 4 \cdot 3 \cdot \sqrt[3]{2} = 12\sqrt[3]{2}$

The expressions $10\sqrt[3]{2}$ and $12\sqrt[3]{2}$ are like radicals.

Now Try Exercises 9, 11, 13

We use these techniques to add radical expressions in the next two examples.

EXAMPLE 3 Adding radical expressions

Add the expressions and simplify.
(a) $\sqrt{12} + 7\sqrt{3}$ **(b)** $\sqrt[3]{16} + \sqrt[3]{2}$ **(c)** $3\sqrt{2} + \sqrt{8} + \sqrt{18}$

Solution
(a)
$$\sqrt{12} + 7\sqrt{3} = \sqrt{4 \cdot 3} + 7\sqrt{3}$$
$$= \sqrt{4} \cdot \sqrt{3} + 7\sqrt{3}$$
$$= 2\sqrt{3} + 7\sqrt{3}$$
$$= 9\sqrt{3}$$

(b)
$$\sqrt[3]{16} + \sqrt[3]{2} = \sqrt[3]{8 \cdot 2} + \sqrt[3]{2}$$
$$= \sqrt[3]{8} \cdot \sqrt[3]{2} + \sqrt[3]{2}$$
$$= 2\sqrt[3]{2} + 1\sqrt[3]{2}$$
$$= 3\sqrt[3]{2}$$

(c)
$$3\sqrt{2} + \sqrt{8} + \sqrt{18} = 3\sqrt{2} + \sqrt{4 \cdot 2} + \sqrt{9 \cdot 2}$$
$$= 3\sqrt{2} + \sqrt{4} \cdot \sqrt{2} + \sqrt{9} \cdot \sqrt{2}$$
$$= 3\sqrt{2} + 2\sqrt{2} + 3\sqrt{2}$$
$$= 8\sqrt{2}$$

Now Try Exercises 29, 31, 43

EXAMPLE 4 Adding radical expressions

Add the expressions and simplify. Assume that all variables are positive.
(a) $\sqrt[4]{32} + 3\sqrt[4]{2}$ **(b)** $-2\sqrt{4x} + \sqrt{x}$ **(c)** $3\sqrt{3k} + 5\sqrt{12k} + 9\sqrt{48k}$

Solution
(a) Because $\sqrt[4]{32} = \sqrt[4]{16 \cdot 2} = \sqrt[4]{16} \cdot \sqrt[4]{2} = 2\sqrt[4]{2}$, we can add and simplify as follows.
$$\sqrt[4]{32} + 3\sqrt[4]{2} = 2\sqrt[4]{2} + 3\sqrt[4]{2} = 5\sqrt[4]{2}$$

(b) Note that $\sqrt{4x} = \sqrt{4} \cdot \sqrt{x} = 2\sqrt{x}$.
$$-2\sqrt{4x} + \sqrt{x} = -2(2\sqrt{x}) + \sqrt{x} = -4\sqrt{x} + 1\sqrt{x} = -3\sqrt{x}$$

(c) Note that $\sqrt{12k} = \sqrt{4} \cdot \sqrt{3k} = 2\sqrt{3k}$ and that $\sqrt{48k} = \sqrt{16} \cdot \sqrt{3k} = 4\sqrt{3k}$.
$$3\sqrt{3k} + 5\sqrt{12k} + 9\sqrt{48k} = 3\sqrt{3k} + 5(2\sqrt{3k}) + 9(4\sqrt{3k})$$
$$= (3 + 10 + 36)\sqrt{3k}$$
$$= 49\sqrt{3k}$$

Now Try Exercises 45, 47, 49

Subtraction of radical expressions is similar to addition, as illustrated in the next three examples.

EXAMPLE 5 **Subtracting like radicals**

Simplify the expressions.

(a) $5\sqrt{7} - 3\sqrt{7}$ (b) $8\sqrt[3]{5} - 3\sqrt[3]{5} + \sqrt[3]{11}$ (c) $5\sqrt{z} + \sqrt[3]{z} - 2\sqrt{z}$

Solution

(a) $\quad\quad\quad 5\sqrt{7} - 3\sqrt{7} = (5 - 3)\sqrt{7} = 2\sqrt{7}$

(b) $8\sqrt[3]{5} - 3\sqrt[3]{5} + \sqrt[3]{11} = (8 - 3)\sqrt[3]{5} + \sqrt[3]{11} = 5\sqrt[3]{5} + \sqrt[3]{11}$

(c) $\quad 5\sqrt{z} + \sqrt[3]{z} - 2\sqrt{z} = 5\sqrt{z} - 2\sqrt{z} + \sqrt[3]{z}$ Commutative property

$\quad\quad\quad\quad\quad\quad\quad = (5 - 2)\sqrt{z} + \sqrt[3]{z}$ Distributive property

$\quad\quad\quad\quad\quad\quad\quad = 3\sqrt{z} + \sqrt[3]{z}$ Subtract.

NOTE: We cannot combine $3\sqrt{z} + \sqrt[3]{z}$ because their indexes are different. That is, one term is a square root and the other is a cube root.

Now Try Exercises 33, 35, 39

EXAMPLE 6 **Subtracting radical expressions**

Subtract and simplify. Assume that all variables are positive.

(a) $3\sqrt[3]{xy^2} - 2\sqrt[3]{xy^2}$ (b) $\sqrt{16x^3} - \sqrt{x^3}$ (c) $\sqrt[3]{\dfrac{5x}{27}} - \dfrac{\sqrt[3]{5x}}{6}$

Solution

(a) $3\sqrt[3]{xy^2} - 2\sqrt[3]{xy^2} = (3 - 2)\sqrt[3]{xy^2} = \sqrt[3]{xy^2}$

(b) $\quad \sqrt{16x^3} - \sqrt{x^3} = \sqrt{16x^2} \cdot \sqrt{x} - \sqrt{x^2} \cdot \sqrt{x}$ Factor out perfect squares.

$\quad\quad\quad\quad\quad\quad\quad = 4x\sqrt{x} - x\sqrt{x}$ Simplify.

$\quad\quad\quad\quad\quad\quad\quad = (4x - x)\sqrt{x}$ Distributive property

$\quad\quad\quad\quad\quad\quad\quad = 3x\sqrt{x}$ Subtract.

(c) $\quad \sqrt[3]{\dfrac{5x}{27}} - \dfrac{\sqrt[3]{5x}}{6} = \dfrac{\sqrt[3]{5x}}{\sqrt[3]{27}} - \dfrac{\sqrt[3]{5x}}{6}$ Quotient rule for radical expressions

$\quad\quad\quad\quad\quad\quad\quad = \dfrac{\sqrt[3]{5x}}{3} - \dfrac{\sqrt[3]{5x}}{6}$ Evaluate $\sqrt[3]{27} = 3$.

$\quad\quad\quad\quad\quad\quad\quad = \dfrac{2\sqrt[3]{5x}}{6} - \dfrac{\sqrt[3]{5x}}{6}$ Find a common denominator.

$\quad\quad\quad\quad\quad\quad\quad = \dfrac{2\sqrt[3]{5x} - \sqrt[3]{5x}}{6}$ Subtract numerators.

$\quad\quad\quad\quad\quad\quad\quad = \dfrac{\sqrt[3]{5x}}{6}$ Simplify.

Now Try Exercises 55, 63, 65

▶ **REAL-WORLD CONNECTION** In A Look Into Math at the beginning of this section, we discussed how functions with radicals could describe an increase in productivity due to technology. In the next example, we calculate the difference of these two functions and interpret the result.

EXAMPLE 7   **Increasing productivity with new technology**

The functions given by

$$N(x) = 400\sqrt{x} + 8000 \quad \text{and} \quad O(x) = 195\sqrt{x} + 3500$$

2005 technology (new) **1935 technology (old)**

approximate the increase in revenue resulting from investing x dollars in equipment (per worker).
(a) Find their difference $D(x) = N(x) - O(x)$. Simplify your answer.
(b) Evaluate $D(40,000)$ and interpret the result.

Solution
(a) $D(x) = N(x) - O(x)$

$= (400\sqrt{x} + 8000) - (195\sqrt{x} + 3500)$	Substitute.
$= 400\sqrt{x} - 195\sqrt{x} + 8000 - 3500$	Rewrite terms.
$= (400 - 195)\sqrt{x} + 4500$	Distributive property
$= 205\sqrt{x} + 4500$	Simplify.

The difference is given by $D(x) = 205\sqrt{x} + 4500$.
(b) $D(40,000) = 205\sqrt{40,000} + 4500 = \$45,500$. An investment of $40,000 per worker resulted in $45,500 *more* revenue per worker in 2005 than in 1935 (adjusted to 2005 dollars).

Now Try Exercise 81

EXAMPLE 8   **Subtracting radical expressions**

Subtract and simplify. Assume that all variables are positive.

(a) $\dfrac{5\sqrt{2}}{3} - \dfrac{2\sqrt{2}}{4}$ **(b)** $\sqrt[4]{81a^5b^6} - \sqrt[4]{16ab^2}$ **(c)** $3\sqrt[3]{\dfrac{n^5}{27}} - 2\sqrt[3]{n^2}$

Solution

(a)

$\dfrac{5\sqrt{2}}{3} - \dfrac{2\sqrt{2}}{4} = \dfrac{5\sqrt{2}}{3} \cdot \dfrac{4}{4} - \dfrac{2\sqrt{2}}{4} \cdot \dfrac{3}{3}$	LCD is 12.
$= \dfrac{20\sqrt{2}}{12} - \dfrac{6\sqrt{2}}{12}$	Multiply fractions.
$= \dfrac{14\sqrt{2}}{12}$	Subtract numerators.
$= \dfrac{7\sqrt{2}}{6}$	Simplify.

(b)

$\sqrt[4]{81a^5b^6} - \sqrt[4]{16ab^2} = \sqrt[4]{81a^4b^4} \cdot \sqrt[4]{ab^2} - \sqrt[4]{16} \cdot \sqrt[4]{ab^2}$	Factor out perfect powers.
$= 3ab\sqrt[4]{ab^2} - 2\sqrt[4]{ab^2}$	Simplify.
$= (3ab - 2)\sqrt[4]{ab^2}$	Distributive property

(c) $3\sqrt[3]{\dfrac{n^5}{27}} - 2\sqrt[3]{n^2} = 3\sqrt[3]{\dfrac{n^3}{27}} \cdot \sqrt[3]{n^2} - 2\sqrt[3]{n^2}$ Factor out perfect cube.

$\qquad\qquad = \dfrac{3\sqrt[3]{n^3}}{\sqrt[3]{27}} \cdot \sqrt[3]{n^2} - 2\sqrt[3]{n^2}$ Quotient rule

$\qquad\qquad = n\sqrt[3]{n^2} - 2\sqrt[3]{n^2}$ Simplify: $\sqrt[3]{n^3} = n$ and $\sqrt[3]{27} = 3$.

$\qquad\qquad = (n - 2)\sqrt[3]{n^2}$ Distributive property

Now Try Exercises 69, 77, 79

Radicals often occur in geometry. In the next example, we find the perimeter of a triangle by adding radical expressions.

EXAMPLE 9 **Finding the perimeter of a triangle**

Find the *exact* perimeter of the right triangle shown in Figure 10.13. Then approximate your answer to the nearest hundredth of a foot.

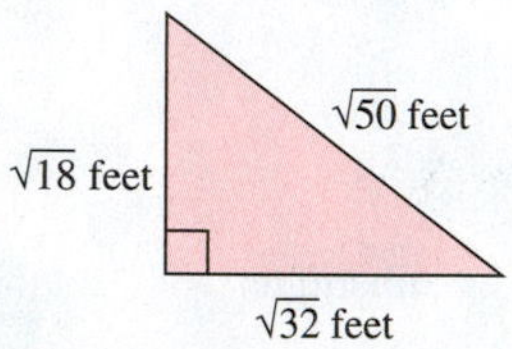

$\sqrt{50}$ feet

$\sqrt{18}$ feet

$\sqrt{32}$ feet

Figure 10.13

Solution
The sum of the lengths of the sides of the triangle is

$$\sqrt{18} + \sqrt{32} + \sqrt{50} = 3\sqrt{2} + 4\sqrt{2} + 5\sqrt{2} = 12\sqrt{2}.$$

The perimeter is $12\sqrt{2} \approx 16.97$ feet.

Now Try Exercise 125

Multiplication

Some types of radical expressions can be multiplied like binomials. For example, because

$$(x + 1)(x + 2) = x^2 + 3x + 2,$$

we have $\quad (\sqrt{x} + 1)(\sqrt{x} + 2) = (\sqrt{x})^2 + 2\sqrt{x} + 1\sqrt{x} + 2 = x + 3\sqrt{x} + 2,$

provided that $x \geq 0$. The next example demonstrates this technique.

EXAMPLE 10 **Multiplying radical expressions**

Multiply and simplify.
(a) $(\sqrt{b} - 4)(\sqrt{b} + 5)$ **(b)** $(4 + \sqrt{3})(4 - \sqrt{3})$

Solution
(a) This expression can be multiplied and then simplified.

$$(\sqrt{b} - 4)(\sqrt{b} + 5) = \sqrt{b} \cdot \sqrt{b} + 5\sqrt{b} - 4\sqrt{b} - 4 \cdot 5$$

$$= b + \sqrt{b} - 20$$

Compare this product with $(b - 4)(b + 5) = b^2 + b - 20$.

(b) This expression is in the form $(a + b)(a - b)$, which equals $a^2 - b^2$.

$$(4 + \sqrt{3})(4 - \sqrt{3}) = (4)^2 - (\sqrt{3})^2 \quad a = 4, b = \sqrt{3}$$

$$= 16 - 3$$

$$= 13$$

Now Try Exercises 85, 87

NOTE: Example 10(b) illustrates a special case for multiplying radicals. In general,

$$(\sqrt{a} + \sqrt{b})(\sqrt{a} - \sqrt{b}) = (\sqrt{a})^2 - (\sqrt{b})^2 = a - b,$$

provided a and b are nonnegative.

Rationalizing the Denominator

In mathematics it is common to write expressions without radicals in the denominator. Quotients containing radical expressions in the numerator or denominator can appear to be different but actually be equal. For example, $\frac{1}{\sqrt{3}}$ and $\frac{\sqrt{3}}{3}$ represent the same real number even though they look as though they are unequal. To show that they are equal, we multiply the first quotient by 1 in the form $\frac{\sqrt{3}}{\sqrt{3}}$.

$$\frac{1}{\sqrt{3}} \cdot \frac{\sqrt{3}}{\sqrt{3}} = \frac{1 \cdot \sqrt{3}}{\sqrt{3} \cdot \sqrt{3}} = \frac{\sqrt{3}}{3}$$

If the denominator of a quotient contains only one term with one square root, then we can rationalize the denominator by multiplying the numerator and denominator by this square root. For example, the denominator of $\frac{1}{\sqrt{3}}$ contains one term, which is $\sqrt{3}$. Therefore, we multiplied $\frac{1}{\sqrt{3}}$ by $\frac{\sqrt{3}}{\sqrt{3}}$ to rationalize the denominator.

NOTE: $\sqrt{b} \cdot \sqrt{b} = \sqrt{b^2} = b$ for any *positive* number b.

One way to *standardize* radical expressions is to remove any radical expressions from the denominator. This process is called **rationalizing the denominator**. Exercise 133 suggests one reason why people rationalized denominators before calculators were invented. The next example demonstrates how to rationalize the denominator of several quotients.

EXAMPLE 11

Rationalizing the denominator

Rationalize each denominator. Assume that all variables are positive.

(a) $\dfrac{1}{\sqrt{2}}$ **(b)** $\dfrac{3}{5\sqrt{3}}$ **(c)** $\sqrt{\dfrac{x}{24}}$ **(d)** $\dfrac{xy}{\sqrt{y^3}}$

Solution

(a) We start by multiplying this expression by 1 in the form $\frac{\sqrt{2}}{\sqrt{2}}$.

$$\frac{1}{\sqrt{2}} \cdot \frac{\sqrt{2}}{\sqrt{2}} = \frac{\sqrt{2}}{\sqrt{4}} = \frac{\sqrt{2}}{2}$$

Note that the expression $\frac{\sqrt{2}}{2}$ does not have a radical in the denominator.

(b) We multiply $\dfrac{3}{5\sqrt{3}}$ by 1 in the form $\dfrac{\sqrt{3}}{\sqrt{3}}$.

$$\frac{3}{5\sqrt{3}} \cdot \frac{\sqrt{3}}{\sqrt{3}} = \frac{3\sqrt{3}}{5\sqrt{9}} = \frac{3\sqrt{3}}{5 \cdot 3} = \frac{\sqrt{3}}{5}$$

(c) Because $\sqrt{24} = \sqrt{4} \cdot \sqrt{6} = 2\sqrt{6}$, we start by simplifying $\sqrt{\dfrac{x}{24}}$.

$$\sqrt{\frac{x}{24}} = \frac{\sqrt{x}}{\sqrt{24}} = \frac{\sqrt{x}}{2\sqrt{6}}$$

To rationalize the denominator we multiply this expression by 1 in the form $\dfrac{\sqrt{6}}{\sqrt{6}}$.

$$\frac{\sqrt{x}}{2\sqrt{6}} = \frac{\sqrt{x}}{2\sqrt{6}} \cdot \frac{\sqrt{6}}{\sqrt{6}} = \frac{\sqrt{6x}}{12}$$

(d) Because $\sqrt{y^3} = \sqrt{y^2} \cdot \sqrt{y} = y\sqrt{y}$, we start by simplifying $\dfrac{xy}{\sqrt{y^3}}$.

$$\frac{xy}{\sqrt{y^3}} = \frac{xy}{y\sqrt{y}} = \frac{x}{\sqrt{y}}$$

To rationalize the denominator we multiply by 1 in the form $\dfrac{\sqrt{y}}{\sqrt{y}}$.

$$\frac{x}{\sqrt{y}} \cdot \frac{\sqrt{y}}{\sqrt{y}} = \frac{x\sqrt{y}}{y}$$

Now Try Exercises 97, 101, 103, 105

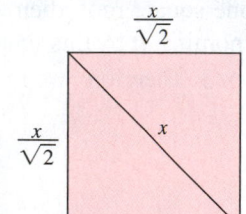

Figure 10.14

EXAMPLE 12

Rationalizing a denominator in geometry

A square with a diagonal of length x has sides of length $\dfrac{x}{\sqrt{2}}$. Find the perimeter and rationalize the denominator. See Figure 10.14.

Solution
The perimeter is 4 times the length of one side.

$$4 \cdot \frac{x}{\sqrt{2}} = \frac{4x}{\sqrt{2}}$$

To rationalize the denominator we multiply the ratio by 1 in the form $\dfrac{\sqrt{2}}{\sqrt{2}}$.

$$\frac{4x}{\sqrt{2}} \cdot \frac{\sqrt{2}}{\sqrt{2}} = \frac{4x\sqrt{2}}{2} = 2x\sqrt{2}$$

The perimeter is $2x\sqrt{2}$.

Now Try Exercise 127

When the denominator is either a sum or difference containing a square root, we rationalize the denominator by multiplying the numerator and denominator by the *conjugate* of the denominator. In this case, the **conjugate** of the denominator is found by changing a $+$ sign to a $-$ sign or vice versa. Table 10.7 lists examples of conjugates.

TABLE 10.7

Expression	$1 + \sqrt{2}$	$\sqrt{3} - 2$	$\sqrt{x} + 7$	$\sqrt{a} - \sqrt{b}$
Conjugate	$1 - \sqrt{2}$	$\sqrt{3} + 2$	$\sqrt{x} - 7$	$\sqrt{a} + \sqrt{b}$

In the next two examples we use this method to rationalize a denominator.

EXAMPLE 13 **Using a conjugate to rationalize the denominator**

Rationalize the denominator of $\dfrac{1}{1 + \sqrt{2}}$.

Solution
From Table 10.7, the conjugate of $1 + \sqrt{2}$ is $1 - \sqrt{2}$.

$$\frac{1}{1 + \sqrt{2}} = \frac{1}{1 + \sqrt{2}} \cdot \frac{(1 - \sqrt{2})}{(1 - \sqrt{2})}$$ Multiply numerator and denominator by the conjugate.

$$= \frac{1 - \sqrt{2}}{(1)^2 - (\sqrt{2})^2}$$ $(a + b)(a - b) = a^2 - b^2$

$$= \frac{1 - \sqrt{2}}{1 - 2}$$ Simplify.

$$= \frac{1 - \sqrt{2}}{-1}$$ Subtract.

$$= \frac{1}{-1} - \frac{\sqrt{2}}{-1}$$ $\dfrac{a - b}{c} = \dfrac{a}{c} - \dfrac{b}{c}$

$$= -1 + \sqrt{2}$$ Simplify.

READING CHECK

How do you rationalize a denominator?

Now Try Exercise 107

EXAMPLE 14 **Rationalizing the denominator**

Rationalize the denominator.

(a) $\dfrac{3 + \sqrt{5}}{2 - \sqrt{5}}$ **(b)** $\dfrac{\sqrt{x}}{\sqrt{x} - 2}$

Solution
(a) The conjugate of the denominator is $2 + \sqrt{5}$.

$$\frac{3 + \sqrt{5}}{2 - \sqrt{5}} = \frac{(3 + \sqrt{5})}{(2 - \sqrt{5})} \cdot \frac{(2 + \sqrt{5})}{(2 + \sqrt{5})}$$ Multiply by 1.

$$= \frac{6 + 3\sqrt{5} + 2\sqrt{5} + (\sqrt{5})^2}{(2)^2 - (\sqrt{5})^2}$$ Multiply.

$$= \frac{11 + 5\sqrt{5}}{4 - 5}$$ Combine terms.

$$= -11 - 5\sqrt{5}$$ Simplify.

(b) The conjugate of the denominator is $\sqrt{x} + 2$.

$$\frac{\sqrt{x}}{\sqrt{x} - 2} = \frac{\sqrt{x}}{(\sqrt{x} - 2)} \cdot \frac{(\sqrt{x} + 2)}{(\sqrt{x} + 2)}$$ Multiply by 1.

$$= \frac{x + 2\sqrt{x}}{x - 4}$$ Multiply.

Now Try Exercises 111, 115

EXAMPLE 15 **Rationalizing a denominator having a cube root**

Rationalize the denominator of $\dfrac{5}{\sqrt[3]{x}}$.

Solution

The expression $\dfrac{5}{\sqrt[3]{x}}$ is equal to $\dfrac{5}{x^{1/3}}$. To rationalize the denominator, $x^{1/3}$, we can multiply it by $x^{2/3}$ because $x^{1/3} \cdot x^{2/3} = x^{1/3+2/3} = x^1$.

$$\dfrac{5}{x^{1/3}} = \dfrac{5}{x^{1/3}} \cdot \dfrac{x^{2/3}}{x^{2/3}} \qquad \text{Multiply by 1.}$$

$$= \dfrac{5x^{2/3}}{x^{1/3+2/3}} \qquad \text{Product rule}$$

$$= \dfrac{5\sqrt[3]{x^2}}{x} \qquad \text{Add; write in radical notation.}$$

Now Try Exercise 121

10.4 Putting It All Together

CONCEPT	EXPLANATION	EXAMPLES
Like Radicals	Like radicals have the same index and the same radicand.	$7\sqrt{5}$ and $3\sqrt{5}$ are like radicals. $5\sqrt[3]{ab}$ and $\sqrt[3]{ab}$ are like radicals. $\sqrt[3]{5}$ and $\sqrt[3]{4}$ are unlike radicals. $\sqrt[3]{7}$ and $\sqrt{7}$ are unlike radicals.
Adding and Subtracting Radical Expressions	Combine like radicals when adding or subtracting. We cannot combine unlike radicals such as $\sqrt{2}$ and $\sqrt{5}$. But sometimes we can rewrite radicals and then combine.	$6\sqrt{13} + \sqrt{13} = (6+1)\sqrt{13} = 7\sqrt{13}$ $\begin{aligned}\sqrt{40} - \sqrt{10} &= \sqrt{4} \cdot \sqrt{10} - \sqrt{10} \\ &= 2\sqrt{10} - \sqrt{10} \\ &= \sqrt{10}\end{aligned}$
Multiplying Radical Expressions	Radical expressions can sometimes be multiplied like binomials.	$(\sqrt{a} - 5)(\sqrt{a} + 5) = a - 25$ and $(\sqrt{x} - 3)(\sqrt{x} + 1) = x - 2\sqrt{x} - 3$
Conjugate	The conjugate is found by changing a $+$ sign to a $-$ sign or vice versa.	*Expression* *Conjugate* $\sqrt{x} + 7$ $\sqrt{x} - 7$ $\sqrt{a} - 2\sqrt{b}$ $\sqrt{a} + 2\sqrt{b}$
Rationalizing a Denominator Having One Term	Write the quotient without a radical expression in the denominator.	To rationalize $\dfrac{5}{\sqrt{7}}$, multiply the expression by 1 in the form $\dfrac{\sqrt{7}}{\sqrt{7}}$. $\dfrac{5}{\sqrt{7}} \cdot \dfrac{\sqrt{7}}{\sqrt{7}} = \dfrac{5\sqrt{7}}{\sqrt{49}} = \dfrac{5\sqrt{7}}{7}$
Rationalizing a Denominator Having Two Terms	Multiply the numerator and denominator by the conjugate of the denominator.	$\dfrac{1}{2 - \sqrt{3}} = \dfrac{1}{2 - \sqrt{3}} \cdot \dfrac{(2 + \sqrt{3})}{(2 + \sqrt{3})}$ $= \dfrac{2 + \sqrt{3}}{4 - 3} = 2 + \sqrt{3}$

10.4 Exercises

CONCEPTS AND VOCABULARY

1. $\sqrt{a} + \sqrt{a} = $ _____

2. $\sqrt[3]{b} + \sqrt[3]{b} + \sqrt[3]{b} = $ _____

3. You cannot simplify $\sqrt[3]{4} + \sqrt[3]{7}$ because they are not _____ radicals.

4. Can you simplify $4\sqrt{15} - 3\sqrt{15}$? Explain.

5. Does $6 + 3\sqrt{5}$ equal $9\sqrt{5}$?

6. To rationalize the denominator of $\frac{2}{\sqrt{7}}$, multiply this expression by _____.

7. What is the conjugate of $\sqrt{t} - 5$?

8. To rationalize the denominator of $\frac{1}{5 - \sqrt{2}}$, multiply this expression by _____.

LIKE RADICALS

Exercises 9–18: (Refer to Example 2.) Write the terms as like radicals, if possible. Assume that all variables are positive.

9. $\sqrt{12}, \sqrt{24}$

10. $\sqrt{18}, \sqrt{27}$

11. $\sqrt{7}, \sqrt{28}, \sqrt{63}$

12. $\sqrt{200}, \sqrt{300}, \sqrt{500}$

13. $\sqrt[3]{16}, \sqrt[3]{-54}$

14. $\sqrt[3]{80}, \sqrt[3]{10}$

15. $\sqrt{x^2y}, \sqrt{4y^2}$

16. $\sqrt{x^5y^3}, \sqrt{9xy}$

17. $\sqrt[3]{8xy}, \sqrt[3]{x^4y^4}$

18. $\sqrt[3]{64x^4}, \sqrt[3]{-8x}$

ADDITION AND SUBTRACTION OF RADICALS

Exercises 19–80: If possible, simplify the expression. Assume that all variables are positive.

19. $2\sqrt{3} + 7\sqrt{3}$

20. $8\sqrt{7} + 2\sqrt{7}$

21. $4\sqrt[3]{5} + 2\sqrt[3]{5}$

22. $\sqrt[3]{13} + 3\sqrt[3]{13}$

23. $7 + 4\sqrt{7}$

24. $8 - 4\sqrt{3}$

25. $2\sqrt{3} + 3\sqrt{2}$

26. $\sqrt{6} + \sqrt{17}$

27. $\sqrt{3} + \sqrt[3]{3}$

28. $\sqrt{6} + \sqrt[4]{6}$

29. $\sqrt[3]{16} + 3\sqrt[3]{2}$

30. $\sqrt[3]{24} + \sqrt[3]{81}$

31. $\sqrt{2} + \sqrt{18} + \sqrt{32}$

32. $2\sqrt{3} + \sqrt{12} + \sqrt{27}$

33. $11\sqrt{11} - 5\sqrt{11}$

34. $9\sqrt{5} + \sqrt{2} - \sqrt{5}$

35. $\sqrt{x} + \sqrt{x} - \sqrt{y}$

36. $\sqrt{xy^2} - \sqrt{x}$

37. $\sqrt[3]{z} + \sqrt[3]{z}$

38. $\sqrt[3]{y} - \sqrt[3]{y}$

39. $2\sqrt[3]{6} - 7\sqrt[3]{6}$

40. $18\sqrt[3]{3} + 3\sqrt[3]{3}$

41. $\sqrt[3]{y^6} - \sqrt[3]{y^3}$

42. $2\sqrt{20} + 7\sqrt{5} + 3\sqrt{2}$

43. $3\sqrt{28} + 3\sqrt{7}$

44. $9\sqrt{18} - 2\sqrt{8}$

45. $\sqrt[4]{48} + 4\sqrt[4]{3}$

46. $\sqrt[4]{32} + \sqrt[4]{16}$

47. $\sqrt{9x} + \sqrt{16x}$

48. $-3\sqrt{x} + 5\sqrt{x}$

49. $3\sqrt{2k} + \sqrt{8k} + \sqrt{18k}$

50. $3\sqrt{k} + 2\sqrt{4k} + \sqrt{9k}$

51. $\sqrt{44} - 4\sqrt{11}$

52. $\sqrt[4]{5} + 2\sqrt[4]{5}$

53. $2\sqrt[3]{16} + \sqrt[3]{2} - \sqrt{2}$

54. $5\sqrt[3]{x} - 3\sqrt[3]{x}$

55. $\sqrt[3]{xy} - 2\sqrt[3]{xy}$

56. $3\sqrt{x^3} - \sqrt{x}$

57. $\sqrt{4x + 8} + \sqrt{x + 2}$

58. $\sqrt{2a + 1} + \sqrt{8a + 4}$

59. $\sqrt{9x + 18} - \sqrt{4x + 8}$

60. $\sqrt{25x - 25} - \sqrt{4x - 4}$

61. $\sqrt{x^3 + x^2} - \sqrt{x + 1}, x \geq 0$

62. $\sqrt{x^3 - x^2} - \sqrt{4x - 4}, x \geq 1$

63. $\sqrt{25x^3} - \sqrt{x^3}$

64. $\sqrt{36x^5} - \sqrt{25x^5}$

65. $\sqrt[3]{\frac{7x}{8}} - \frac{\sqrt[3]{7x}}{3}$

66. $\sqrt[3]{\frac{8x^2}{27}} - \sqrt[3]{\frac{x^2}{8}}$

67. $\frac{4\sqrt{3}}{3} + \frac{\sqrt{3}}{6}$

68. $\frac{8\sqrt{5}}{7} + \frac{4\sqrt{5}}{2}$

69. $\frac{15\sqrt{8}}{4} - \frac{2\sqrt{2}}{5}$

70. $\frac{23\sqrt{11}}{2} - \frac{\sqrt{44}}{8}$

71. $2\sqrt[4]{64} - \sqrt[4]{324} + \sqrt[4]{4}$

72. $2\sqrt[3]{16} - 5\sqrt[3]{54} + 10\sqrt[3]{2}$

73. $5\sqrt[4]{x^5} - \sqrt[4]{x}$

74. $20\sqrt[3]{b^4} - 4\sqrt[3]{b}$

75. $\sqrt{64x^3} - \sqrt{x} + 3\sqrt{x}$

76. $2\sqrt{3z} + 3\sqrt{12z} + 3\sqrt{48z}$

77. $\sqrt[4]{81a^5b^5} - \sqrt[4]{ab}$

78. $\sqrt[4]{xy^5} - \sqrt[4]{x^5y}$

79. $5\sqrt[3]{\frac{n^4}{125}} - 2\sqrt[3]{n}$

80. $\sqrt[3]{\frac{8x}{27}} - \frac{2\sqrt[3]{x}}{3}$

OPERATIONS ON FUNCTIONS

Exercises 81–84: Find $(f + g)(x)$ and $(f - g)(x)$.

81. $f(x) = 5\sqrt{x} - 2$, $g(x) = -2\sqrt{x} + 3$

82. $f(x) = 3\sqrt{4x} - 4$, $g(x) = 5\sqrt{x - 1}$

83. $f(x) = \sqrt[3]{8x} + 1$, $g(x) = 2\sqrt[3]{x} - 1$

84. $f(x) = \sqrt[3]{27x}$, $g(x) = \sqrt[3]{64x}$

MULTIPLYING BINOMIALS CONTAINING RADICALS

Exercises 85–96: Multiply and simplify.

85. $(\sqrt{x} - 3)(\sqrt{x} + 2)$ **86.** $(2\sqrt{x} + 1)(\sqrt{x} + 4)$

87. $(3 + \sqrt{7})(3 - \sqrt{7})$ **88.** $(5 - \sqrt{5})(5 + \sqrt{5})$

89. $(11 - \sqrt{2})(11 + \sqrt{2})$

90. $(6 + \sqrt{3})(6 - \sqrt{3})$

91. $(\sqrt{x} + 8)(\sqrt{x} - 8)$

92. $(\sqrt{ab} - 3)(\sqrt{ab} + 3)$

93. $(\sqrt{ab} - \sqrt{c})(\sqrt{ab} + \sqrt{c})$

94. $(\sqrt{2x} + \sqrt{3y})(\sqrt{2x} - \sqrt{3y})$

95. $(\sqrt{x} - 7)(\sqrt{x} + 8)$

96. $(\sqrt{ab} - 1)(\sqrt{ab} - 2)$

RATIONALIZING THE DENOMINATOR

Exercises 97–124: Rationalize the denominator.

97. $\frac{1}{\sqrt{7}}$ **98.** $\frac{1}{\sqrt{23}}$

99. $\frac{4}{\sqrt{3}}$ **100.** $\frac{8}{\sqrt{2}}$

101. $\frac{5}{3\sqrt{5}}$ **102.** $\frac{6}{11\sqrt{3}}$

103. $\sqrt{\frac{b}{12}}$ **104.** $\sqrt{\frac{5b}{72}}$

105. $\frac{rt}{2\sqrt{r^3}}$ **106.** $\frac{m^2 n}{2\sqrt{m^5}}$

107. $\frac{1}{3 - \sqrt{2}}$ **108.** $\frac{1}{\sqrt{3} - 2}$

109. $\frac{\sqrt{2}}{\sqrt{5} + 2}$ **110.** $\frac{\sqrt{3}}{\sqrt{3} + 2}$

111. $\frac{\sqrt{7} - 2}{\sqrt{7} + 2}$ **112.** $\frac{\sqrt{3} - 1}{\sqrt{3} + 1}$

113. $\frac{1}{\sqrt{7} - \sqrt{6}}$ **114.** $\frac{1}{\sqrt{8} - \sqrt{7}}$

115. $\frac{\sqrt{z}}{\sqrt{z} - 3}$ **116.** $\frac{2\sqrt{z}}{2 - \sqrt{z}}$

117. $\frac{\sqrt{a} + \sqrt{b}}{\sqrt{a} - \sqrt{b}}$ **118.** $\frac{\sqrt{x} - 2\sqrt{y}}{\sqrt{x} + 2\sqrt{y}}$

119. $\frac{1}{\sqrt{x + 1} - \sqrt{x}}$ **120.** $\frac{1}{\sqrt{a + 1} + \sqrt{a}}$

121. $\frac{3}{\sqrt[3]{x}}$ **122.** $\frac{6}{5\sqrt[3]{x}}$

123. $\frac{1}{\sqrt[3]{x^2}}$ **124.** $\frac{2}{\sqrt[3]{(x - 2)^2}}$

GEOMETRY

125. *Perimeter* (Refer to Example 9.) Find the exact perimeter of the right triangle. Then approximate your answer.

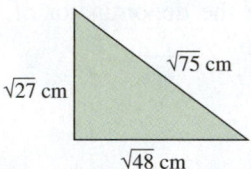

126. *Perimeter* Find the exact perimeter of the rectangle. Then approximate your answer.

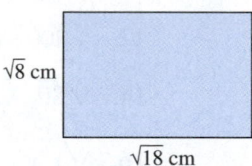

127. *Geometry* (Refer to Example 12.) A square has a diagonal with length $\sqrt{3}$. Find the perimeter and rationalize the denominator.

128. *Geometry* A rectangle has sides of length $\sqrt{8}$ and $\sqrt{32}$. Find the perimeter of the rectangle.

129. *Geometry* (Refer to Example 12.) A square has a diagonal that is 60 feet long. Find the exact perimeter of the rectangle and simplify your answer.

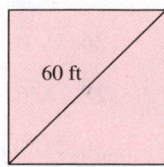

130. *Geometry* A square has a diagonal that is 10 feet long. Find the exact perimeter of the rectangle and simplify your answer.

131. *Geometry* A square has an area of x square feet. Find the length of the diagonal of the square.

132. *Geometry* A square has an area of 16 square feet. Find the length of the diagonal of the square.

WRITING ABOUT MATHEMATICS

133. Suppose that a student knows that $\sqrt{3} \approx 1.73205$ and does not have a calculator. Which of the following expressions would be easier to evaluate by hand? Why?

$$\frac{1}{\sqrt{3}} \quad \text{or} \quad \frac{\sqrt{3}}{3}$$

134. A student simplifies an expression *incorrectly*:

$$\sqrt{8} + \sqrt[3]{16} \overset{?}{=} \sqrt{4 \cdot 2} + \sqrt[3]{8 \cdot 2}$$
$$\overset{?}{=} \sqrt{4} \cdot \sqrt{2} + \sqrt[3]{8} \cdot \sqrt[3]{2}$$
$$\overset{?}{=} 2\sqrt{2} + 2\sqrt[3]{2}$$
$$\overset{?}{=} 4\sqrt{4}$$
$$\overset{?}{=} 8.$$

Explain any errors that the student made. What would you do differently?

SECTIONS 10.3 and 10.4 **Checking Basic Concepts**

1. Simplify each expression. Assume that all variables are positive.

(a) $(64^{-3/2})^{1/3}$ (b) $\sqrt{5} \cdot \sqrt{20}$

(c) $\sqrt[3]{-8x^4y}$ (d) $\sqrt{\dfrac{4b}{5}} \cdot \sqrt{\dfrac{4b^3}{5}}$

2. Simplify $\sqrt[3]{7} \cdot \sqrt{7}$

3. Simplify each expression.

(a) $\sqrt{3} \cdot \sqrt{12}$ (b) $\dfrac{\sqrt[3]{81}}{\sqrt[3]{3}}$

(c) $\sqrt{36x^6}, x > 0$

4. Simplify each expression.

(a) $5\sqrt{6} + 2\sqrt{6} + \sqrt{7}$

(b) $8\sqrt[3]{x} - 3\sqrt[3]{x}$

(c) $\sqrt{9x} - \sqrt{4x}$

5. Simplify each expression.

(a) $\sqrt[3]{xy^4} - \sqrt[3]{x^4y}$

(b) $(4 - \sqrt{2})(4 + \sqrt{2})$

6. Rationalize the denominator of $\dfrac{6}{2\sqrt{6}}$.

7. Rationalize the denominator of $\dfrac{2}{\sqrt{5} - 1}$.

10.5   More Radical Functions

Root Functions • Power Functions • Modeling with Power Functions (Optional)

A LOOK INTO MATH ▶ Radical expressions often occur in biology. In Example 5, we discuss how heavier birds tend to have larger wings. This relationship between weight and wing size can be modeled by a radical function. Another application of radical functions in biology is discussed in Example 3, where a function is used to predict the surface area of a person, given their weight.

Root Functions

NEW VOCABULARY

☐ Root function
☐ Power function

Earlier in this chapter we discussed the square root and cube root functions. There are also functions for higher roots. Figures 10.15–10.18 on the next page show graphs of four common root functions. Note that the higher the root, the more slowly the graph increases for $x \geq 1$. Also, note that the domains for functions of even roots include only the nonnegative real numbers, whereas the domains of functions of odd roots include all real numbers.

Square Root Function

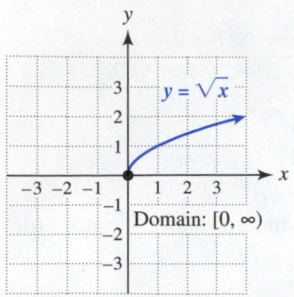

Figure 10.15

Cube Root Function

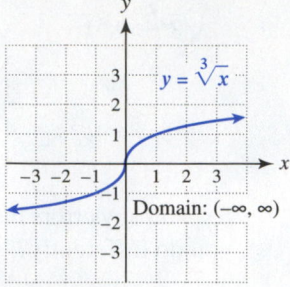

Figure 10.16

Fourth Root Function

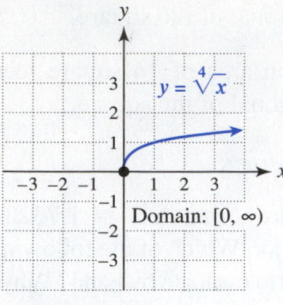

Figure 10.17

Fifth Root Function

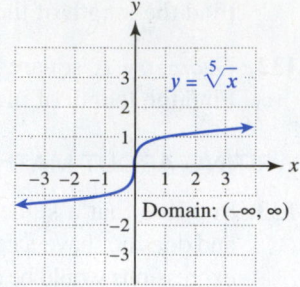

Figure 10.18

EXAMPLE 1 | **Evaluating root functions**

If possible, evaluate each root function f at the given x-values.

(a) $f(x) = \sqrt{x + 1}, \quad x = -5, x = 3$

(b) $f(x) = \sqrt[3]{1 - x}, \quad x = -7, x = 28$

(c) $f(x) = \sqrt[4]{x - 7}, \quad x = 5, x = 88$

(d) $f(x) = \sqrt[5]{x}, \qquad x = -1, x = 32$

Solution

(a) $f(-5) = \sqrt{-5 + 1} = \sqrt{-4}$, which is *not* a real number.

$\quad f(3) = \sqrt{3 + 1} = \sqrt{4} = 2$ because $2^2 = 4$.

(b) $f(-7) = \sqrt[3]{1 - (-7)} = \sqrt[3]{8} = 2$ because $2^3 = 8$.

$\quad f(28) = \sqrt[3]{1 - 28} = \sqrt[3]{-27} = -3$ because $(-3)^3 = -27$.

(c) $\quad f(5) = \sqrt[4]{5 - 7} = \sqrt[4]{-2}$, which is not a real number.

$\quad f(88) = \sqrt[4]{88 - 7} = \sqrt[4]{81} = 3$ because $3^4 = 81$.

(d) $f(-1) = \sqrt[5]{-1} = -1$ because $(-1)^5 = -1$.

$\quad f(32) = \sqrt[5]{32} = 2$ because $2^5 = 32$.

Now Try Exercises 9, 11, 13, 15

Power Functions

STUDY TIP

Power functions are important in many areas of study. Be sure to become familiar with them.

Power functions are a generalization of root functions. Examples of power functions include

$$f(x) = x^{1/2}, \ g(x) = x^{2/3}, \quad \text{and} \quad h(x) = x^{-3/5}.$$

The exponent for a power function is frequently a rational number p, written in lowest terms. That is, we usually write $f(x) = x^{1/2}$, rather than $f(x) = x^{3/6}$. The power function $f(x) = x^{m/n}$ can also be written as $f(x) = \sqrt[n]{x^m}$. A power function f is defined for all real numbers when n is odd and defined for only nonnegative numbers when n is even.

POWER FUNCTION

If a function f can be represented by

$$f(x) = x^p,$$

where p is a rational number, then f is a **power function**. If $p = \frac{1}{n}$, where $n \geq 2$ is an integer, then f is also a **root function**, which is given by

$$f(x) = \sqrt[n]{x}.$$

EXAMPLE 2   **Evaluating power functions**

If possible, evaluate $f(x)$ at the given value of x.
(a) $f(x) = x^{0.75}$ at $x = 16$ **(b)** $f(x) = x^{1/4}$ at $x = -81$

Solution
(a) $0.75 = \frac{3}{4}$, so $f(x) = x^{3/4}$. Thus $f(\mathbf{16}) = \mathbf{16}^{3/4} = (16^{1/4})^3 = 2^3 = 8$.

NOTE: $16^{1/4} = \sqrt[4]{16} = 2$.

(b) $f(-81) = (-81)^{1/4} = \sqrt[4]{-81}$, which is undefined. There is no real number a such that $a^4 = -81$ because a^4 is never negative.

Now Try Exercises 23, 27

▶ **REAL-WORLD CONNECTION** The surface area of the skin covering the human body is influenced by both the height and weight of a person. A taller person tends to have a larger surface area, as does a heavier person. See Extended and Discovery Exercise 2 at the end of this chapter. In the next example, we use a power function from biology to model this situation.

EXAMPLE 3   **Modeling surface area of the human body**

The surface area of a person who is 66 inches tall and weighs w pounds can be estimated by $S(w) = 327w^{0.425}$, where S is in square inches. (*Source:* H. Lancaster, *Quantitative Methods in Biological and Medical Sciences.*)
(a) Find S if this person weighs 130 pounds.
(b) If the person gains 20 pounds, by how much does the person's surface area increase?
(c) A graph of $y = S(w)$ is shown in Figure 10.19. Suppose that a person's weight is more than 50 pounds and it doubles. Use the graph to determine if the person's surface area also doubles.

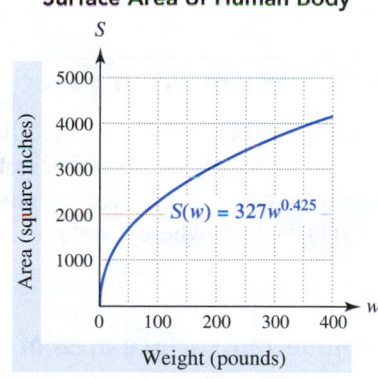

Surface Area of Human Body

$S(w) = 327w^{0.425}$

Area (square inches) vs. Weight (pounds)

Figure 10.19

Solution
(a) $S(130) = 327(130)^{0.425} \approx 2588$ square inches
(b) Because $S(150) = 327(150)^{0.425} \approx 2750$ square inches, the surface area of the person increases by about $2750 - 2588 = 162$ square inches.
(c) For $w \geq 50$ the graph increases more slowly. This means that the person's surface area (S-values) will not double even if his or her weight doubles (w-values). For example, $A(100) \approx 2300$ square inches and $A(200) \approx 3100$ square inches. The surface area did not double.

Now Try Exercise 51

In the next example, we investigate the graph of $y = x^p$ for different values of p.

EXAMPLE 4 **Graphing power functions**

The graphs of three power functions,

$$f(x) = x^{1/3}, \quad g(x) = x^{0.75}, \quad \text{and} \quad h(x) = x^{1.4},$$

are shown in Figure 10.20, where the two decimal exponents have been written as fractions. Discuss how the value of p affects the graph of $y = x^p$ when $x > 1$ and when $0 < x < 1$.

Power Functions

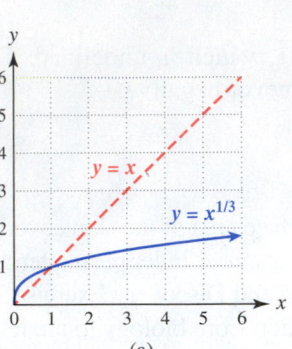
(a)

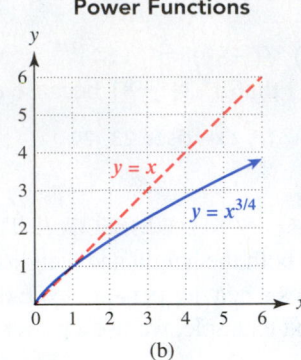
(b)

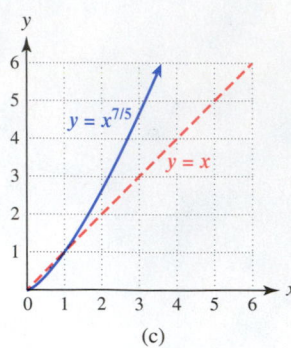
(c)

Figure 10.20

Solution
The graphs of these power functions have the following characteristics.
1. If $x > 1$, then $x^{1/3} < x^{3/4} < x^{7/5}$. Larger exponents result in graphs that increase (rise) *faster* for $x > 1$.
2. If $0 < x < 1$, then $x^{7/5} < x^{3/4} < x^{1/3}$. Smaller exponents result in graphs that have larger y-values for $0 < x < 1$.
3. All graphs pass through the points $(0, 0)$ and $(1, 1)$.

Now Try Exercise 35

Modeling with Power Functions (Optional)

▶ **REAL-WORLD CONNECTION** Allometry is the study of the relative sizes of different characteristics of an organism. For example, the weight of a bird is related to the surface area of its wings; heavier birds tend to have larger wings. Allometric relations are often modeled with $f(x) = kx^p$, where k and p are constants. (*Source:* C. Pennycuick, *Newton Rules Biology.*)

EXAMPLE 5 **Modeling surface area of wings**

Weights w of various birds and the corresponding surface area of their wings A are shown in Table 10.8.

TABLE 10.8 Weight and Wing Size

w (kilograms)	0.5	2.0	3.5	5.0
A (square meters)	0.069	0.175	0.254	0.325

(a) Make a scatterplot of the data. Discuss any trends in the data.
(b) Biologists modeled the data with $A(w) = kw^{2/3}$, where k is a constant. Find k.
(c) Graph A and the data in the same viewing rectangle.
(d) Estimate the area of the wings of a 3-kilogram bird.

CALCULATOR HELP

To make a scatterplot, see
Appendix A (pages AP-3 and AP-4).

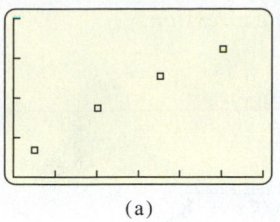

[0, 6, 1] by [0, 0.4, 0.1]

(a)

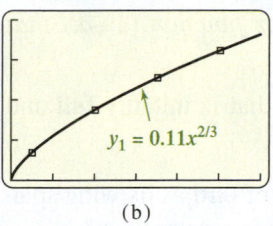

[0, 6, 1] by [0, 0.4, 0.1]

$y_1 = 0.11x^{2/3}$

(b)

Figure 10.21

Solution

(a) A scatterplot of the data is shown in Figure 10.21(a). As the weight of a bird increases so does the surface area of its wings.

(b) To determine k, substitute one of the data points into $A(w)$. We use $(2.0, 0.175)$.

$$A(w) = kw^{2/3} \quad \text{Given formula}$$

$$0.175 = k(2)^{2/3} \quad \text{Let } w = 2 \text{ and } A(w) = 0.175$$
$$\text{(any data point could be used).}$$

$$k = \frac{0.175}{2^{2/3}} \quad \text{Solve for } k; \text{ rewrite equation.}$$

$$k \approx 0.11 \quad \text{Approximate } k.$$

Thus $A(w) = 0.11w^{2/3}$.

(c) The data and graph of $Y_1 = 0.11X^{\wedge}(2/3)$ are shown in Figure 10.21(b). Note that the graph appears to pass through each data point.

(d) $A(3) = 0.11(3)^{2/3} \approx 0.23$ square meter

▮ **Now Try Exercise 59**

10.5   Putting It All Together

FUNCTION	EXPLANATION
Root	$f(x) = \sqrt[3]{x}, \quad g(x) = \sqrt[4]{x}, \quad \text{and} \quad h(x) = \sqrt[5]{x}$ Odd root functions are defined for all real numbers, and even root functions are defined for nonnegative real numbers.
Power	$f(x) = x^p$, where p is a rational number, is a power function. If $p = \frac{1}{n}$, where $n \geq 2$ is an integer, f is also a root function, given by $f(x) = \sqrt[n]{x}$. ***Examples:*** $f(x) = x^{5/3}$ Power function $g(x) = x^{1/4} = \sqrt[4]{x}$ Root and power function

---

10.5   Exercises

MyMathLab 

CONCEPTS AND VOCABULARY

1. Sketch a graph of $y = x^{1/2}$.

2. Sketch a graph of $y = x^{1/3}$.

3. What is the domain of $f(x) = x^{1/2}$?

4. What is the domain of $f(x) = x^{1/3}$?

5. Give a symbolic representation for a power function.

6. Give a symbolic representation for a root function.

7. What is the domain of $f(x) = \sqrt[4]{x}$?

8. What is the domain of $f(x) = \sqrt[5]{x}$?

ROOT FUNCTIONS

Exercises 9–16: If possible, evaluate each root function f at the given x-values. When the result is not an integer, approximate it to the nearest hundredth.

9. $f(x) = \sqrt{x^2 - 1}, \quad x = -2, x = 0$

10. $f(x) = \sqrt{2x - 5}, \quad x = 0, x = 3$

11. $f(x) = \sqrt[4]{1 - x}$, $x = -5$, $x = 2$

12. $f(x) = \sqrt[4]{x + 1}$, $x = -15$, $x = 15$

13. $f(x) = \sqrt[5]{4 - 3x}$, $x = -3$, $x = 1$

14. $f(x) = \sqrt[5]{1 - x^2}$, $x = 3$, $x = 1$

15. $f(x) = \sqrt[3]{1 - x}$, $x = -5$, $x = 2$

16. $f(x) = \sqrt[3]{(x + 1)^2}$, $x = -2$, $x = 7$

POWER FUNCTIONS

Exercises 17–22: Use radical notation to write f(x).

17. $f(x) = x^{1/2}$ **18.** $f(x) = x^{1/3}$

19. $f(x) = x^{2/3}$ **20.** $f(x) = x^{3/4}$

21. $f(x) = x^{-1/5}$ **22.** $f(x) = x^{-2/5}$

Exercises 23–30: If possible, evaluate f(x) at the given values of x. When appropriate, approximate the answer to the nearest hundredth.

23. $f(x) = x^{5/2}$ $x = 4, x = 5$

24. $f(x) = x^{-3/4}$ $x = 1, x = 3$

25. $f(x) = x^{-7/5}$ $x = -32, x = 10$

26. $f(x) = x^{4/3}$ $x = -8, x = 27$

27. $f(x) = x^{1/4}$ $x = 256, x = -10$

28. $f(x) = x^{3/4}$ $x = 16, x = -1$

29. $f(x) = x^{2/5}$ $x = 32, x = -32$

30. $f(x) = x^{5/6}$ $x = -5, x = 64$

Exercises 31–34: Graph $y = f(x)$. Write the domain of f in interval notation.

31. $f(x) = x^{1/4}$ **32.** $f(x) = x^{1/5}$

33. $f(x) = x^{2/3}$ **34.** $f(x) = (x + 1)^{1/4}$

Exercises 35–38: Graph f and g in the window [0, 6, 1] by [0, 6, 1]. Which function is greater when $x > 1$?

35. $f(x) = x^{1/5}, g(x) = x^{1/3}$

36. $f(x) = x^{4/5}, g(x) = x^{5/4}$

37. $f(x) = x^{1.2}, g(x) = x^{0.45}$

38. $f(x) = x^{-1.4}, g(x) = x^{1.4}$

Exercises 39 and 40: **Thinking Generally** *Let $0 < q < p$, where p and q are rational numbers.*

39. For $x > 1$, is $x^p > x^q$ or $x^p < x^q$?

40. For $0 < x < 1$, is $x^p > x^q$ or $x^p < x^q$?

Exercises 41 and 42: **Operations on Functions** *For the given f(x) and g(x), evaluate each expression.*

(a) $(f + g)(2)$ **(b)** $(f - g)(x)$
(c) $(fg)(x)$ **(d)** $(f/g)(x)$

41. $f(x) = \sqrt{8x}, g(x) = \sqrt{2x}$

42. $f(x) = \sqrt{9x + 18}, g(x) = \sqrt{4x + 8}$

Exercises 43–46: **Graphical Interpretation** *Match the situation with the graph of the power function (a.–d.) that models it best.*

43. Amount of water in a barrel that is initially full and has a hole near the bottom

44. The average weight of a type of bird as its wing span increases (*Hint:* The exponent is greater than 1.)

45. Money made after x hours by a person who works for a fixed hourly wage

46. Rapid growth of an insect population that eventually slows down

a. 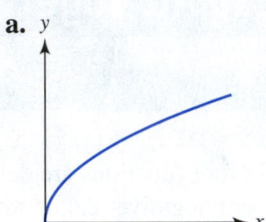   **b.**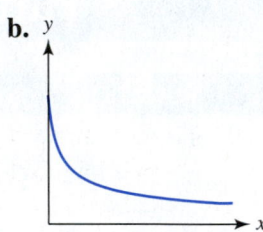

c. 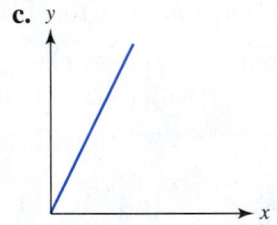   **d.**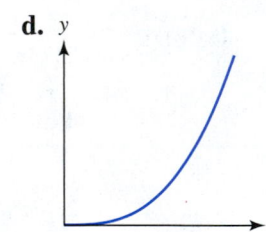

Exercises 47–50 **Graphical Recognition** *Match the equation with its graph (a.–d.). Do not use a calculator.*

47. $y = x^{7/4}$ **48.** $y = x^{4/3}$

49. $y = x^{1/3}$ **50.** $y = x^{1/2}$

a. 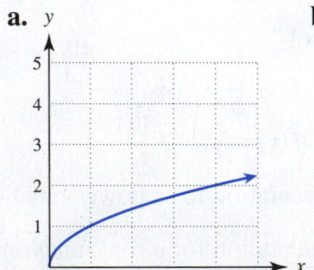   **b.**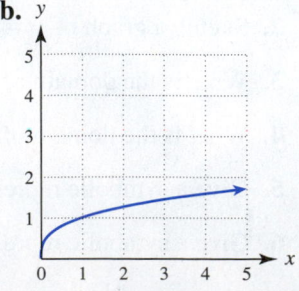

c. 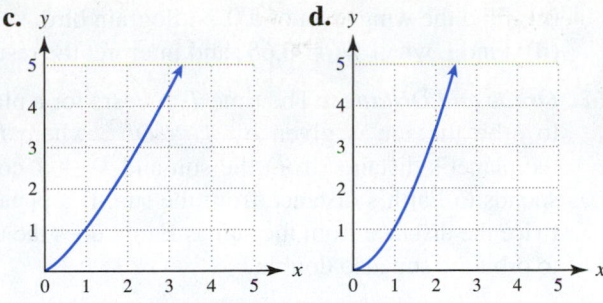 **d.**

APPLICATIONS

Exercises 51 and 52: **Surface Area** *(Refer to Example 3.) The surface area of the skin of a person who is 70 inches tall and weighs w pounds can be estimated by* $S(w) = 342w^{0.425}$, *where S is in square inches. Evaluate each of the following.*

51. $S(150)$ **52.** $S(200)$

Exercises 53–56 **Learning Curves** *The graph shows the percentage P of times that individuals completed a simple task correctly after practicing it x times.*

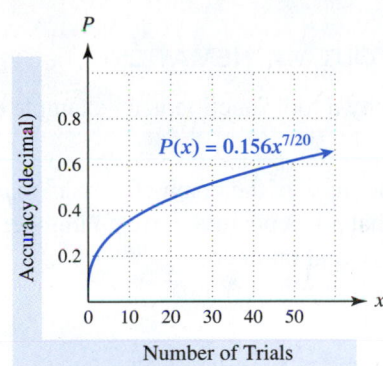

Number of Trials

53. If a group of people does 10 practice trials each, what percentage of the time is the task done correctly?

54. How many trials need to be performed to reach 50% accuracy?

55. If a person doubles the number of practice trials, does the accuracy always double? Explain.

56. What happens to a person's accuracy after a large number of trials?

57. *Aging More Slowly* In his theory of relativity, Albert Einstein showed that if a person travels at nearly the speed of light, then time slows down significantly. Suppose that there are twins; one remains on Earth and the other leaves in a very fast spaceship having velocity v. If the twin on Earth ages T_0 years, then according to Einstein the twin in the spaceship ages T years, where

$$T(v) = T_0\sqrt{1 - (v/c)^2}.$$

In this formula c represents the speed of light, which is 186,000 miles per second.
(a) Evaluate T when $v = 0.8c$ (eight-tenths the speed of light) and $T_0 = 10$ years. (*Hint:* Simplify $\frac{v}{c}$ without using 186,000 miles per second.)
(b) Interpret your result.

58. *Increasing Your Weight* (Refer to Exercise 57.) Albert Einstein also showed that the weight (mass) of an object increases when traveling near the speed of light. If a person's weight on Earth is W_0, then the same person's weight W in a spaceship traveling at velocity v is

$$W(v) = \frac{W_0}{\sqrt{1 - (v/c)^2}}.$$

(a) Evaluate W when $v = 0.6c$ (six-tenths the speed of light) and $W_0 = 220$ pounds (100 kilograms).
(b) Interpret your result.

59. *Modeling Wing Size* (Refer to Example 5.) The surface area of wings for a species of bird with weight w is shown in the table.

w (kilograms)	2	3	4
A (square meters)	0.254	0.333	0.403

(a) Make a scatterplot of the data.
(b) The data can be modeled with $A(w) = kw^{2/3}$. Find k.
(c) Graph A and the data in the same viewing rectangle. Does the graph of A pass through the data points?
(d) Estimate the area of the wings of a 2.5-kilogram bird.

60. *Pulse Rate in Animals* The following table lists typical pulse rates R in beats per minute (bpm) for animals with various weights W in pounds.
(*Source:* C. Pennycuick.)

W (pounds)	20	150	500	1500
R (beats per minute)	198	72	40	23

(a) Describe what happens to the pulse rate as the weight of the animal increases.
(b) Plot the data in [0, 1600, 400] by [0, 220, 20].
(c) If $R = kW^{-1/2}$, find k.
(d) Find R when $W = 700$, and interpret the result.

61. *Modeling Wing Span* (Refer to Example 5.) Biologists have found that the weight W of a bird and the length L of its wing span are related by $L = kW^{1/3}$, where k is a constant. The following table lists L and W for one species of bird. (*Source:* C. Pennycuick.)

W (kilograms)	0.1	0.4	0.8	1.1
L (meters)	0.422	0.670	0.844	0.938

(a) Use the data to approximate the value of k.

(b) Graph L and the data in the same viewing rectangle. What happens to L as W increases?

(c) Find the wing span of a 0.7-kilogram bird.

(d) Find L when $W = 0.65$, and interpret the result.

62. *Orbits and Distances* The time T in years for a planet to orbit the sun is given by $T = D^{3/2}$, where D is the planet's distance from the sun and $D = 1$ corresponds to Earth's distance from the sun. If a planet is twice the distance from the sun as Earth, does the time to orbit the sun also double?

WRITING ABOUT MATHEMATICS

63. Explain why a root function is an example of a power function.

64. Discuss the shape of the graph of $y = x^p$ as p increases. Assume that p is a positive rational number and that x is a positive real number.

Group Activity   Working with Real Data

Directions: Form a group of 2 to 4 people. Select someone to record the group's responses for this activity. All members of the group should work cooperatively to answer the questions. If your instructor asks for your results, each member of the group should be prepared to respond.

Simple Pendulum Gravity is responsible for an object falling toward Earth. The farther the object falls, the faster it is moving when it hits the ground. For each second that an object falls, its speed increases by a constant amount, called the *acceleration due to gravity*, denoted g. One way to calculate the value of g is to use a simple pendulum. See the accompanying figure.

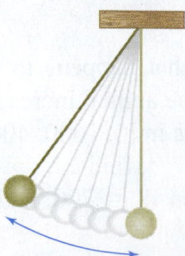

The time T for a pendulum to swing back and forth once is called its *period* and is given by

$$T = 2\pi\sqrt{\frac{L}{g}},$$

where L equals the length of the pendulum. The table lists the periods of pendulums with different lengths.

L (feet)	0.5	1.0	1.5
T (seconds)	0.78	1.11	1.36

(a) Solve the formula for g.

(b) Use the table to determine the value of g. (*Note:* The units for g are feet per second per second.)

(c) Interpret your result.

10.6 Equations Involving Radical Expressions

Solving Radical Equations • The Distance Formula • Solving the Equation $x^n = k$

A LOOK INTO MATH ▶

In a fast-paced society, people do not spend time watching videos that they find boring. For example, earlier in this chapter we learned how the function

$$A(x) = 7.3\sqrt[16]{x^7}$$

computes the percentage of viewers who abandon an online video after x seconds. If we want to find the number of seconds that it takes for **50%** of viewers to abandon a typical online video, we could *solve* the *radical equation*

$$7.3\sqrt[16]{x^7} = 50$$

(See Exercise 119.) In this section we discuss how to solve radical equations and equations with rational exponents.

Solving Radical Equations

NEW VOCABULARY

☐ Extraneous solutions
☐ Pythagorean theorem
☐ Distance

Many times, equations contain either radical expressions or rational exponents. Examples include

$$\sqrt{x} = 6, \quad 5x^{1/2} = 1, \quad \text{and} \quad \sqrt[3]{x-1} = 3. \quad \text{Radical equations}$$

One strategy for solving an equation containing a square root is to isolate the square root (if necessary) and then square each side of the equation. This technique is an example of the *power rule for solving equations.*

POWER RULE FOR SOLVING EQUATIONS

If each side of an equation is raised to the same positive integer power, then any solutions to the given equation are among the solutions to the new equation. That is, the solutions to the equation $a = b$ are among the solutions to $a^n = b^n$.

NOTE: After applying the power rule, the new equation can have solutions that do not satisfy the given equation, which are sometimes called **extraneous solutions**. We must *always* check our answers in the *given* equation after applying the power rule.

READING CHECK

When you apply the power rule to solve an equation, what must you do with your solutions? Why?

We illustrate the power rule in the next example.

EXAMPLE 1 **Solving a radical equation symbolically**

Solve $\sqrt{2x - 1} = 3$. Check your solution.

Solution

Begin by squaring each side of the equation. That is, apply the power rule with $n = 2$.

$$\sqrt{2x - 1} = 3 \qquad \text{Given equation}$$
$$(\sqrt{2x - 1})^2 = 3^2 \qquad \text{Square each side.}$$
$$2x - 1 = 9 \qquad \text{Simplify.}$$
$$2x = 10 \qquad \text{Add 1.}$$
$$x = 5 \qquad \text{Divide by 2.}$$

To check this answer we substitute $x = 5$ in the given equation.

$$\sqrt{2(5) - 1} \stackrel{?}{=} 3$$
$$3 = 3 \checkmark \quad \text{It checks.}$$

Now Try Exercise 21

NOTE: To simplify $(\sqrt{2x - 1})^2$ in Example 1, we used the fact that

$$(\sqrt{a})^2 = \sqrt{a} \cdot \sqrt{a} = a.$$

The following steps can be used to solve a radical equation.

SOLVING A RADICAL EQUATION

STEP 1: Isolate a radical term on one side of the equation.

STEP 2: Apply the power rule by raising each side of the equation to the power equal to the index of the isolated radical term.

STEP 3: Solve the equation. If it still contains a radical, repeat Steps 1 and 2.

STEP 4: Check your answers by substituting each result in the *given* equation.

NOTE: In earlier sections, we *simplified expressions*. In this section, we *solve equations*. Equations contain equals signs and when we solve an equation, we try to find values of the variable that make the equation a true statement.

In the next example, we apply these steps to a radical equation.

EXAMPLE 2 **Isolating the radical term**

Solve $\sqrt{4 - x} + 5 = 8$.

Solution

STEP 1: To isolate the radical term, we subtract 5 from each side of the equation.

$$\sqrt{4 - x} + 5 = 8 \qquad \text{Given equation}$$
$$\sqrt{4 - x} = 3 \qquad \text{Subtract 5.}$$

STEP 2: The isolated term involves a square root, so we must **square** each side.

$$(\sqrt{4-x})^2 = (3)^2 \qquad \text{Square each side.}$$

STEP 3: Next we solve the resulting equation. (It is not necessary to repeat Steps 1 and 2 because the resulting equation does not contain any radical expressions.)

$$4 - x = 9 \qquad \text{Simplify.}$$
$$-x = 5 \qquad \text{Subtract 4.}$$
$$x = -5 \qquad \text{Multiply by } -1.$$

STEP 4: To check this answer we substitute $x = -5$ in the given equation.

$$\sqrt{4 - (-5)} + 5 \overset{?}{=} 8$$
$$\sqrt{9} + 5 \overset{?}{=} 8$$
$$8 = 8 \checkmark \quad \text{It checks.}$$

Now Try Exercise 23

The next example shows that we must check our answers to identify extraneous solutions when squaring each side of an equation.

EXAMPLE 3   **Solving a radical equation**

Solve $\sqrt{3x + 3} = 2x - 1$. Check your results and then solve the equation graphically.

Solution
Symbolic Solution Begin by squaring each side of the equation.

$$\sqrt{3x + 3} = 2x - 1 \qquad \text{Given equation}$$
$$(\sqrt{3x + 3})^2 = (2x - 1)^2 \qquad \text{Square each side.}$$
$$3x + 3 = 4x^2 - 4x + 1 \qquad \text{Multiply.}$$
$$0 = 4x^2 - 7x - 2 \qquad \text{Subtract } 3x + 3.$$
$$0 = (4x + 1)(x - 2) \qquad \text{Factor.}$$
$$x = -\frac{1}{4} \quad \text{or} \quad x = 2 \qquad \text{Solve for } x.$$

To check these values substitute $x = -\frac{1}{4}$ and $x = 2$ in the given equation.

$$\sqrt{3\left(-\frac{1}{4}\right) + 3} \overset{?}{=} 2\left(-\frac{1}{4}\right) - 1$$
$$\sqrt{2.25} \overset{?}{=} -1.5$$
$$1.5 \neq -1.5 \; ✗ \qquad \text{It does not check.}$$

Thus $-\frac{1}{4}$ is an *extraneous solution*. Next substitute $x = 2$ in the given equation.

$$\sqrt{3 \cdot 2 + 3} \overset{?}{=} 2 \cdot 2 - 1$$
$$\sqrt{9} \overset{?}{=} 3$$
$$3 = 3 \checkmark \qquad \text{It checks.}$$

The only solution is 2. Next we check with a graphical solution.

[−5, 5, 1] by [−5, 5, 1]

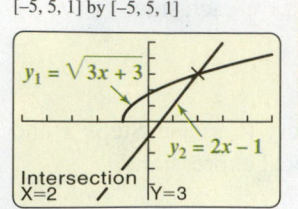

Figure 10.22

CALCULATOR HELP

To find a point of intersection, see Appendix A (page AP-6).

Graphical Solution The solution **2** to the equation $\sqrt{3x + 3} = 2x - 1$ is supported graphically in Figure 10.22, where the graphs of $y_1 = \sqrt{3x + 3}$ and $y_2 = 2x - 1$ intersect at the point (**2**, 3). *Note that the graphical solution does not give an extraneous solution.*

Now Try Exercises 27, 79(a), (b)

Example 3 demonstrates that *checking a solution is essential when you are squaring each side of an equation.* Squaring may introduce extraneous solutions, which are solutions to the resulting equation but are not solutions to the given equation.

CRITICAL THINKING

Will a numerical solution give extraneous solutions?

When an equation contains two or more terms with square roots, it may be necessary to square each side of the equation more than once. In these situations, isolate one of the square roots and then square each side of the equation. If a radical term remains after simplifying, repeat these steps. We apply this technique in the next example.

EXAMPLE 4 **Squaring twice**

Solve $\sqrt{2x} - 1 = \sqrt{x + 1}$.

Solution
Begin by squaring each side of the equation.

$$\sqrt{2x} - 1 = \sqrt{x + 1} \qquad \text{Given equation}$$

$$(\sqrt{2x} - 1)^2 = (\sqrt{x + 1})^2 \qquad \text{Square each side.}$$

$$(\sqrt{2x})^2 - 2(\sqrt{2x})(1) + 1^2 = x + 1 \qquad (a - b)^2 = a^2 - 2ab + b^2$$

$$2x - 2\sqrt{2x} + 1 = x + 1 \qquad \text{Simplify.}$$

$$2x - 2\sqrt{2x} = x \qquad \text{Subtract 1.}$$

$$x = 2\sqrt{2x} \qquad \text{Subtract } x \text{ and add } 2\sqrt{2x}.$$

$$x^2 = 4(2x) \qquad \text{Square each side again.}$$

$$x^2 - 8x = 0 \qquad \text{Subtract } 8x.$$

$$x(x - 8) = 0 \qquad \text{Factor.}$$

$$x = 0 \quad \text{or} \quad x = 8 \qquad \text{Solve.}$$

To check these answers substitute $x = \mathbf{0}$ and $x = \mathbf{8}$ in the given equation.

$$\sqrt{2 \cdot \mathbf{0}} - 1 \overset{?}{=} \sqrt{\mathbf{0} + 1} \qquad\qquad \sqrt{2 \cdot \mathbf{8}} - 1 \overset{?}{=} \sqrt{\mathbf{8} + 1}$$

$$-1 \neq 1 \ \textbf{✗} \quad \text{It does not check.} \qquad\qquad 3 = 3 \ \textbf{✓} \quad \text{It checks.}$$

The only solution is 8.

Now Try Exercise 43

In the next example we apply the power rule to an equation that contains a cube root.

EXAMPLE 5 **Solving an equation containing a cube root**

Solve $\sqrt[3]{4x - 7} = 4$.

Solution

STEP 1: The cube root term is already isolated, so we proceed to Step 2.

STEP 2: Because the index is 3, we **cube** each side of the equation.

$$\sqrt[3]{4x - 7} = 4 \qquad \text{Given equation}$$

$$(\sqrt[3]{4x - 7})^3 = (4)^3 \qquad \text{Cube each side.}$$

STEP 3: We solve the resulting equation.

$$4x - 7 = 64 \qquad \text{Simplify.}$$

$$4x = 71 \qquad \text{Add 7 to each side.}$$

$$x = \frac{71}{4} \qquad \text{Divide each side by 4.}$$

STEP 4: To check this answer we substitute $x = \frac{71}{4}$ in the given equation.

$$\sqrt[3]{4\left(\tfrac{71}{4}\right) - 7} \stackrel{?}{=} 4$$

$$\sqrt[3]{64} \stackrel{?}{=} 4$$

$$4 = 4 \checkmark \quad \text{It checks.}$$

Now Try Exercise 31

▶ **REAL-WORLD CONNECTION** Larger birds tend to have larger wings. This relationship can sometimes be modeled by $A(W) = 100\sqrt[3]{W^2}$, where W is the weight in pounds of the bird and A is the area of the wings in square inches. In the next example, we use the formula to determine the weight of a bird that has wings with a surface area of 600 square inches. (*Source:* C. Pennycuick, *Newton Rules Biology.*)

EXAMPLE 6 **Finding the weight of a bird**

Solve the equation $600 = 100\sqrt[3]{W^2}$ to determine the weight in pounds of a bird having wings with an area of 600 square inches.

Solution

Begin by dividing each side of the equation $600 = 100\sqrt[3]{W^2}$ by 100 to isolate the radical term on the right side of the equation.

$$\frac{600}{100} = \sqrt[3]{W^2} \qquad \text{Divide each side by 100.}$$

$$(6)^3 = (\sqrt[3]{W^2})^3 \qquad \text{Cube each side.}$$

$$216 = W^2 \qquad \text{Simplify.}$$

$$\sqrt{216} = W \qquad \text{Take principal square root, } W > 0.$$

$$W \approx 14.7 \qquad \text{Approximate.}$$

The weight of the bird is approximately 14.7 pounds.

Now Try Exercise 117

TECHNOLOGY NOTE

Graphing Radical Expressions
The equation in Example 6 can be solved graphically. Sometimes it is more convenient to use rational exponents than radical notation. Thus $y = 100\sqrt[3]{W^2}$ can be entered as $Y_1 = 100X^{\wedge}(2/3)$. (Be sure to include parentheses around the 2/3.) The accompanying figure shows y_1 intersecting the line $y_2 = 600$ near the point (14.7, 600), which supports our symbolic result.

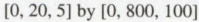

[0, 20, 5] by [0, 800, 100]

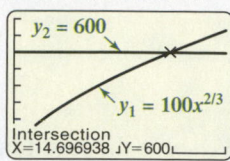

In the next example we solve an equation that would be difficult to solve symbolically, but an *approximate* solution can be found graphically.

EXAMPLE 7 **Solving an equation with rational exponents graphically**

Solve $x^{2/3} = 3 - x^2$ graphically.

Solution
Graph $Y_1 = X^{\wedge}(2/3)$ and $Y_2 = 3 - X^{\wedge}2$. The graphs intersect near $(-1.34, 1.21)$ and $(1.34, 1.21)$, as shown in Figure 10.23. Thus the solutions are given by $x \approx \pm 1.34$.

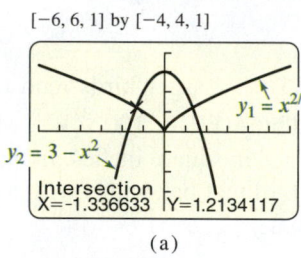

| $[-6, 6, 1]$ by $[-4, 4, 1]$ | $[-6, 6, 1]$ by $[-4, 4, 1]$ |

(a)   (b)

Figure 10.23

Now Try Exercise 73

$a^2 + b^2 = c^2$

The Distance Formula

▶ **REAL-WORLD CONNECTION** One of the most famous theorems in mathematics is the **Pythagorean theorem**. It states that if a right triangle has legs a and b with hypotenuse c (see Figure 10.24), then

Figure 10.24

$$a^2 + b^2 = c^2. \quad \text{Pythagorean Theorem}$$

For example, if the legs of a right triangle are $a = 3$ and $b = 4$, the hypotenuse is $c = 5$ because $3^2 + 4^2 = 5^2$. Also, if the sides of a triangle satisfy $a^2 + b^2 = c^2$, it is a right triangle.

EXAMPLE 8 **Applying the Pythagorean theorem**

A rectangular television screen has a width of 20 inches and a height of 15 inches. Find the diagonal of the television. Why is it called a 25-inch television?

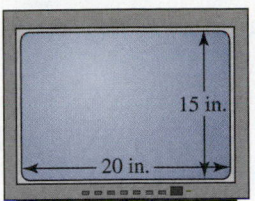

Figure 10.25

Solution

In Figure 10.25, let $a = 20$ and $b = 15$. Then the diagonal of the television corresponds to the hypotenuse of a right triangle with legs of 20 inches and 15 inches.

$$c^2 = a^2 + b^2 \qquad \text{Pythagorean theorem}$$
$$c = \sqrt{a^2 + b^2} \qquad \text{Take the principal square root, } c > 0.$$
$$c = \sqrt{20^2 + 15^2} \qquad \text{Substitute } a = 20 \text{ and } b = 15.$$
$$c = 25 \qquad \text{Simplify.}$$

A 25-inch television has a diagonal of 25 inches.

▌ **Now Try Exercise 125**

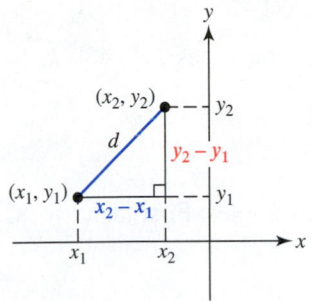

Figure 10.26

The Pythagorean theorem can be used to determine the distance between two points. Suppose that a line segment has endpoints (x_1, y_1) and (x_2, y_2), as illustrated in Figure 10.26. The lengths of the legs of the right triangle are $x_2 - x_1$ and $y_2 - y_1$. The distance d is the hypotenuse of a right triangle. Applying the Pythagorean theorem, we have

$$d^2 = (x_2 - x_1)^2 + (y_2 - y_1)^2.$$

Distance is nonnegative, so we let d be the principal square root and obtain

$$d = \sqrt{(x_2 - x_1)^2 + (y_2 - y_1)^2}.$$

DISTANCE FORMULA

The **distance** d between the points (x_1, y_1) and (x_2, y_2) in the xy-plane is

$$d = \sqrt{(x_2 - x_1)^2 + (y_2 - y_1)^2}.$$

EXAMPLE 9   **Finding distance between points**

Find the distance between the points $(-2, 3)$ and $(1, -4)$.

Solution

Start by letting $(x_1, y_1) = (-2, 3)$ and $(x_2, y_2) = (1, -4)$. Then substitute these values into the distance formula.

$$d = \sqrt{(x_2 - x_1)^2 + (y_2 - y_1)^2} \qquad \text{Distance formula}$$
$$= \sqrt{(1 - (-2))^2 + (-4 - 3)^2} \qquad \text{Substitute.}$$
$$= \sqrt{9 + 49} \qquad \text{Simplify.}$$
$$= \sqrt{58} \qquad \text{Add.}$$
$$\approx 7.62 \qquad \text{Approximate.}$$

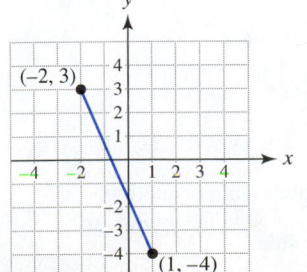

Figure 10.27

The distance between the points, as shown in Figure 10.27, is exactly $\sqrt{58}$ units, or about 7.62 units. Note that we would obtain the same result if we let $(x_1, y_1) = (1, -4)$ and $(x_2, y_2) = (-2, 3)$.

▌ **Now Try Exercise 109**

NOTE: In Example 9, $\sqrt{9+49} \neq \sqrt{9} + \sqrt{49} = 3 + 7 = 10$. In general, for any a and b, $\sqrt{a^2 + b^2} \neq a + b$.

Solving the Equation $x^n = k$

The equation $x^n = k$, where n is a *positive integer*, can be solved by taking the nth root of each side of the equation. The following technique allows us to find all *real* solutions to this equation.

SOLVING THE EQUATION $x^n = k$

Take the nth root of each side of $x^n = k$ to obtain $\sqrt[n]{x^n} = \sqrt[n]{k}$.

1. If n is odd, then $\sqrt[n]{x^n} = x$ and the equation becomes $x = \sqrt[n]{k}$.
2. If n is *even* and $k > 0$, then $\sqrt[n]{x^n} = |x|$ and the equation becomes $|x| = \sqrt[n]{k}$.
   (If $k < 0$, there are no real solutions.)

To understand this technique better, consider the following examples. First let $x^3 = 8$, so that n is odd. Taking the cube root of each side gives

$$\sqrt[3]{x^3} = \sqrt[3]{8}, \text{ which is equivalent to } x = \sqrt[3]{8}, \text{ or } x = 2.$$

Next let $x^2 = 4$, so that n is even. Taking the square root of each side gives

$$\sqrt{x^2} = \sqrt{4}, \text{ which is equivalent to } |x| = \sqrt{4}, \text{ or } |x| = 2.$$

The solutions to $|x| = 2$ are -2 or 2, which can be written as ± 2.

EXAMPLE 10   **Solving the equation $x^n = k$**

Solve each equation.
(a) $x^3 = -64$ (b) $x^2 = 12$ (c) $2(x - 1)^4 = 32$

Solution
(a) Taking the cube root of each side of $x^3 = -64$ gives

$$\sqrt[3]{x^3} = \sqrt[3]{-64} \text{ or } x = -4.$$

(b) Taking the square root of each side of $x^2 = 12$ gives

$$\sqrt{x^2} = \sqrt{12} \text{ or } |x| = \sqrt{12}.$$

The equation $|x| = \sqrt{12}$ is equivalent to $x = \pm\sqrt{12}$.
(c) First divide each side of $2(x - 1)^4 = 32$ by 2 to isolate the power of $x - 1$.

$$(x - 1)^4 = 16 \qquad \text{Divide each side by 2.}$$

$$\sqrt[4]{(x - 1)^4} = \sqrt[4]{16} \qquad \text{Take the 4th root of each side.}$$

$$|x - 1| = 2 \qquad \text{Simplify: } \sqrt[4]{y^4} = |y|.$$

$$x - 1 = -2 \quad \text{or} \quad x - 1 = 2 \qquad \text{Solve the absolute value equation.}$$

$$x = -1 \quad \text{or} \quad x = 3 \qquad \text{Add 1 to each side.}$$

Now Try Exercises 47, 55, 67

▶ **REAL-WORLD CONNECTION** In some parts of the United States, wind power is used to generate electricity. Suppose that the diameter of the circular path created by the blades of a wind-powered generator is 8 feet. Then the wattage W generated by a wind velocity of v miles per hour is modeled by

$$W(v) = 2.4v^3.$$

If the wind blows at 10 miles per hour, the generator can produce about

$$W(10) = 2.4 \cdot 10^3 = 2400 \text{ watts.}$$

(*Source: Conquering the Sciences, Sharp Electronics.*)

EXAMPLE 11   **Modeling a wind generator**

The formula $W(v) = 2.4v^3$ is used to calculate the watts generated when there is a wind velocity of v miles per hour.
(a) Find a function f that calculates the wind velocity when W watts are being produced.
(b) If the wattage doubles, has the wind velocity also doubled? Explain.

Solution
(a) Given W we need a formula to find v, so solve $W = 2.4v^3$ for v.

$$W = 2.4v^3 \qquad \text{Given formula}$$

$$\frac{W}{2.4} = v^3 \qquad \text{Divide by 2.4.}$$

$$\sqrt[3]{\frac{W}{2.4}} = \sqrt[3]{v^3} \qquad \text{Take the cube root of each side.}$$

$$v = \sqrt[3]{\frac{W}{2.4}} \qquad \text{Simplify and rewrite equation.}$$

Thus $f(W) = \sqrt[3]{\frac{W}{2.4}}$.
(b) Suppose that the power generated is 1000 watts. Then the wind speed is

$$f(1000) = \sqrt[3]{\frac{1000}{2.4}} \approx 7.5 \text{ miles per hour.}$$

If the power doubles to 2000 watts, then the wind speed is

$$f(2000) = \sqrt[3]{\frac{2000}{2.4}} \approx 9.4 \text{ miles per hour.}$$

Thus for the wattage to double, the wind speed does not need to double.

▌ **Now Try Exercise 133**

10.6   Putting It All Together

CONCEPT	DESCRIPTION	EXAMPLE
Power Rule for Solving Equations	If each side of an equation is raised to the same positive integer power, any solutions to the given equation are among the solutions to the new equation.	$\sqrt{2x} = x$ $2x = x^2$ Square each side. $x^2 - 2x = 0$ Rewrite equation. $x = 0$ or $x = 2$ Factor and solve. *Be sure to check any solutions.*

continued on next page

continued from previous page

CONCEPT	DESCRIPTION	EXAMPLE
Pythagorean Theorem	If c is the hypotenuse of a right triangle and a and b are its legs, then $a^2 + b^2 = c^2$.	If the sides of the right triangle are $a = 5$, $b = 12$, and $c = 13$, then they satisfy $a^2 + b^2 = c^2$ or $5^2 + 12^2 = 13^2$.
Distance Formula	The distance d between the points (x_1, y_1) and (x_2, y_2) is $$d = \sqrt{(x_2 - x_1)^2 + (y_2 - y_1)^2}.$$	The distance between the points $(2, 3)$ and $(-3, 4)$ is $$d = \sqrt{(-3 - 2)^2 + (4 - 3)^2}$$ $$= \sqrt{(-5)^2 + (1)^2} = \sqrt{26}.$$
Solving the Equation $x^n = k$, Where n Is a Positive Integer	Take the nth root of each side to obtain $\sqrt[n]{x^n} = \sqrt[n]{k}$. Then, 1. $x = \sqrt[n]{k}$, if n is odd. 2. $x = \pm\sqrt[n]{k}$, if n is even and $k \geq 0$.	1. n odd: If $x^5 = 32$, then $x = \sqrt[5]{32} = 2$. 2. n even: If $x^4 = 81$, then $x = \pm\sqrt[4]{81} = \pm 3$.

Equations involving radical expressions can be solved symbolically, numerically, and graphically. All symbolic solutions must be checked; *numerical and graphical methods do not give extraneous solutions.* These concepts are illustrated for the equation $\sqrt{x + 2} = x$, where $y_1 = \sqrt{x + 2}$ and $y_2 = x$.

Symbolic Solution

$$\sqrt{x + 2} = x$$
$$x + 2 = x^2$$
$$x^2 - x - 2 = 0$$
$$(x - 2)(x + 1) = 0$$
$$x = 2 \quad \text{or} \quad x = -1$$
Check: $\sqrt{2 + 2} = 2$ ✓
$$\sqrt{-1 + 2} \neq -1$$ ✗

The only solution is 2.

Numerical Solution

X	Y₁	Y₂
-2	0	-2
-1	1	-1
0	1.4142	0
1	1.7321	1
2	2	2
3	2.2361	3
4	2.4495	4

X=2

$y_1 = y_2$ when $x = 2$, so 2 is a solution. Note that -1 is *not* a solution.

Graphical Solution

[-4.7, 4.7, 1] by [-3.1, 3.1, 1]

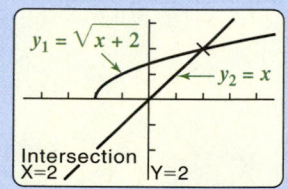

The graphs intersect at $(2, 2)$. Note that there is no point of intersection when $x = -1$.

10.6 Exercises

MyMathLab   Math XL PRACTICE    WATCH   DOWNLOAD   READ   REVIEW

CONCEPTS AND VOCABULARY

1. What is a good first step for solving $\sqrt{4x - 1} = 5$?

2. What is a good first step for solving $\sqrt[3]{x + 1} = 6$?

3. Can an equation involving rational exponents have more than one solution?

4. When you square each side of an equation to solve for an unknown, what must you do with any answers?

5. What is the Pythagorean theorem used for?

6. If the legs of a right triangle are 3 and 4, what is the length of the hypotenuse?

7. What formula can you use to find the distance d between two points?

8. Write the equation $\sqrt{x} + \sqrt[4]{x^3} = 2$ with rational exponents.

SIMPLIFYING RADICALS

Exercises 9–16: Simplify. Assume radicands of square roots are positive.

9. $\sqrt{2} \cdot \sqrt{2}$ **10.** $\sqrt{5} \cdot \sqrt{5}$

11. $\sqrt{x} \cdot \sqrt{x}$ **12.** $\sqrt{2x} \cdot \sqrt{2x}$

13. $(\sqrt{2x+1})^2$ **14.** $(\sqrt{7x})^2$

15. $(\sqrt[3]{5x^2})^3$ **16.** $(\sqrt[3]{2x-5})^3$

SYMBOLIC SOLUTIONS

Exercises 17–46: Solve the equation symbolically. Check your results.

17. $\sqrt{x} = 8$ **18.** $\sqrt{3z} = 6$

19. $\sqrt[4]{x} = 3$ **20.** $\sqrt[3]{x} - 4 = 2$

21. $\sqrt{2t+4} = 4$ **22.** $\sqrt{y+4} = 3$

23. $\sqrt{x+1} - 3 = 4$ **24.** $\sqrt{2x+5} + 2 = 5$

25. $2\sqrt{x-2} + 1 = 5$ **26.** $-\sqrt{x+7} - 1 = -7$

27. $\sqrt{x+6} = x$ **28.** $\sqrt{z+6} = z$

29. $\sqrt[3]{x} = 3$ **30.** $\sqrt[3]{x+10} = 4$

31. $\sqrt[3]{2z-4} = -2$ **32.** $\sqrt[3]{z-1} = -3$

33. $\sqrt[4]{t+1} = 2$ **34.** $\sqrt[4]{5t} = 5$

35. $\sqrt{5z-1} = \sqrt{z+1}$ **36.** $y = \sqrt{y+1} + 1$

37. $\sqrt{1-x} = 1 - x$ **38.** $\sqrt[3]{4x} = x$

39. $\sqrt{b^2-4} = b - 2$ **40.** $\sqrt{b^2-2b+1} = b$

41. $\sqrt{1-2x} = x + 7$ **42.** $\sqrt{4-y} = y - 2$

43. $\sqrt{x} = \sqrt{x-5} + 1$ **44.** $\sqrt{x-1} = \sqrt{x+4} - 1$

45. $\sqrt{2t-2} + \sqrt{t} = 7$ **46.** $\sqrt{x+1} - \sqrt{x-6} = 1$

SOLVING THE EQUATION $x^n = k$

Exercises 47–68: (Refer to Example 10.) Solve.

47. $x^2 = 49$ **48.** $x^2 = 9$

49. $2z^2 = 200$ **50.** $3z^2 = 48$

51. $(t+1)^2 = 16$ **52.** $(t-5)^2 = 81$

53. $(4-2x)^2 = 100$ **54.** $(3x-6)^2 = 25$

55. $b^3 = 64$ **56.** $a^3 = 1000$

57. $2t^3 = -128$ **58.** $3t^3 = -81$

59. $(x+1)^3 = 8$ **60.** $(4-x)^3 = -1$

61. $(2-5z)^3 = -125$ **62.** $(2x+4)^3 = 125$

63. $x^4 = 16$ **64.** $x^4 = 7$

65. $x^5 = 12$ **66.** $x^5 = -32$

67. $2(x+2)^4 = 162$ **68.** $\frac{1}{2}(x-1)^5 = 16$

GRAPHICAL SOLUTIONS

Exercises 69–78: Solve graphically. Approximate solutions to the nearest hundredth when appropriate.

69. $\sqrt[3]{x+5} = 2$ **70.** $\sqrt[3]{x} + \sqrt{x} = 3.43$

71. $\sqrt{2x-3} = \sqrt{x} - \frac{1}{2}$ **72.** $x^{4/3} - 1 = 2$

73. $x^{5/3} = 2 - 3x^2$ **74.** $x^{3/2} = \sqrt{x+2} - 2$

75. $z^{1/3} - 1 = 2 - z$ **76.** $z^{3/2} - 2z^{1/2} - 1 = 0$

77. $\sqrt{y+2} + \sqrt{3y+2} = 2$

78. $\sqrt{x+1} - \sqrt{x-1} = 4$

USING MORE THAN ONE METHOD

Exercises 79–82: Solve the equation
   (**a**) *symbolically,*
   (**b**) *graphically, and*
   (**c**) *numerically.*

79. $2\sqrt{x} = 8$ **80.** $\sqrt[3]{5-x} = 2$

81. $\sqrt{6z-2} = 8$ **82.** $\sqrt{y+4} = \frac{y}{3}$

SOLVING AN EQUATION FOR A VARIABLE

Exercises 83–86: Solve for the indicated variable.

83. $T = 2\pi\sqrt{\dfrac{L}{32}}$ for L

84. $Z = \sqrt{L^2 + R^2}$ for R

85. $r = \sqrt{\dfrac{A}{\pi}}$ for A

86. $F = \dfrac{1}{2\pi\sqrt{LC}}$ for C

PYTHAGOREAN THEOREM

Exercises 87–94: If the sides of a triangle are a, b, and c and they satisfy $a^2 + b^2 = c^2$, the triangle is a right triangle. Determine whether the triangle with the given sides is a right triangle.

87. $a = 6$ $b = 8$ $c = 10$

88. $a = 5$ $b = 12$ $c = 13$

89. $a = \sqrt{5}$ $b = \sqrt{9}$ $c = \sqrt{14}$

90. $a = 4$ $b = 5$ $c = 7$

91. $a = 7$ $b = 24$ $c = 25$

92. $a = 1$ $b = \sqrt{3}$ $c = 2$

93. $a = 8$ $b = 8$ $c = 16$

94. $a = 11$ $b = 60$ $c = 61$

Exercises 95–98: Find the length of the missing side in the right triangle.

95.
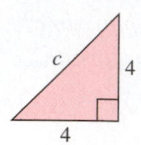

96.
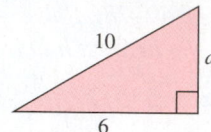

97.
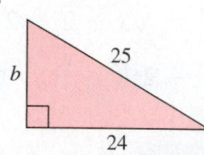

98.
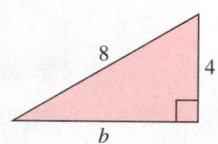

Exercises 99–104: A right triangle has legs a and b with hypotenuse c. Find the length of the missing side.

99. $a = 3, b = 4$

100. $a = 4, b = 7$

101. $a = \sqrt{3}, c = 8$

102. $a = \sqrt{6}, c = \sqrt{10}$

103. $b = 48, c = 50$

104. $b = 10, c = 26$

DISTANCE FORMULA

Exercises 105–108: Find the length of the line segment.

105.
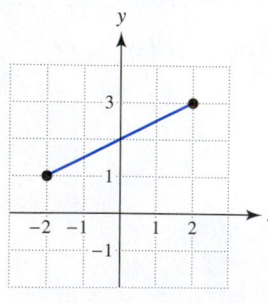

106.
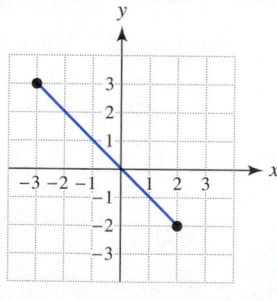

107.
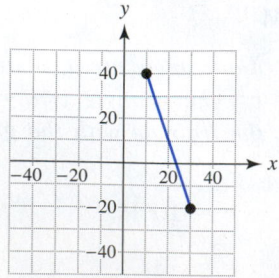

108.
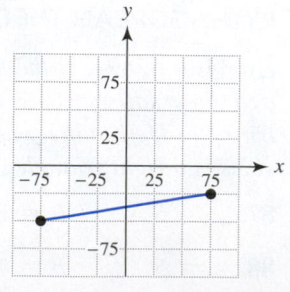

Exercises 109–112: Find the distance between the points.

109. $(-1, 2), (4, 10)$ **110.** $(5, -40), (-6, 20)$

111. $(0, -3), (4, 0)$ **112.** $(3, 9), (-4, 2)$

Exercises 113–116: Find x if the distance between the points is d. Assume that $x \geq 0$.

113. $(x, 3), (0, 6)$ $d = 5$

114. $(x, -1), (6, 11)$ $d = 13$

115. $(x, -5), (62, 6)$ $d = 61$

116. $(x, 3), (12, -4)$ $d = 25$

APPLICATIONS

Exercises 117 and 118: **Weight of a Bird** *(Refer to Example 6.) Estimate the weight W of a bird having wings of area A. Let $A = 100\sqrt[3]{W^2}$.*

117. $A = 400$ square inches

118. $A = 1000$ square inches

Exercises 119 and 120: **Abandonment Rate** *(Refer to A Look Into Math for this section.) Use $A(x) = 7.3\sqrt[16]{x^7}$ to find the average number of seconds x that it takes for A percent of the viewers to abandon an online movie.*

119. $A = 50$ **120.** $A = 20$

Exercises 121–124: **Distance to the Horizon** *Because of Earth's curvature, a person can see a limited distance to the horizon. The higher the location of the person, the farther that person can see. The distance D in miles to the horizon can be estimated by $D(h) = 1.22\sqrt{h}$, where h is the height of the person above the ground in feet.*

121. Find D for a 6-foot-tall person standing on level ground.

122. Find D for a person on top of Mount Everest with a height of 29,028 feet.

123. What height allows a person to see 20 miles?

124. How high does a plane need to fly for the pilot to be able to see 100 miles?

125. *Diagonal of a Television* (Refer to Example 8.) A rectangular television screen is 11.4 inches by 15.2 inches. Find the diagonal of the television screen.

126. *Dimensions of a Television* The height of a television with a 13-inch diagonal is $\frac{3}{4}$ of its width. Find the width and height of the television set.

127. *DVD and Picture Dimensions* If the picture shown on a television set is h units high and w units wide, the *aspect ratio* of the picture is $\frac{w}{h}$ (see the accompanying figure). Digital video discs support the newer aspect ratio of $\frac{16}{9}$ rather than the older ratio of $\frac{4}{3}$. If the width of a picture with an aspect ratio of $\frac{16}{9}$ is 29 inches, approximate the height and diagonal of the rectangular picture. (**Source:** J. Taylor, *DVD Demystified.*)

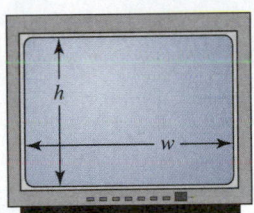

128. *Flood Control* The spillway capacity of a dam is important in flood control. Spillway capacity Q in cubic feet of water per second flowing over the spillway depends on the width W and the depth D of the spillway, as illustrated in the accompanying figure. If W and D are measured in feet, capacity can be modeled by $Q = 3.32WD^{3/2}$. (**Source:** D. Callas, Project Director, *Snapshots of Applications in Mathematics.*)
  (a) Find the capacity of a spillway with $W = 20$ feet and $D = 5$ feet.
  (b) A spillway with a width of 30 feet is to have a capacity of $Q = 2690$ cubic feet per second. Estimate to the nearest foot the appropriate depth of the spillway.

129. *Sky Diving* When sky divers initially fall from an airplane, their velocity v in miles per hour after free falling d feet can be approximated by $v = \frac{60}{11}\sqrt{d}$. (Because of air resistance, they will eventually reach a terminal velocity.) How far do sky divers need to fall to attain the following velocities? (These values for d represent minimum distances.)
  (a) 60 miles per hour   **(b)** 100 miles per hour

130. *Guy Wire* A guy wire attached to the top of a 30-foot-long pole is anchored 10 feet from the base of the pole, as illustrated in the figure. Find the length of the guy wire to the nearest tenth of a foot.

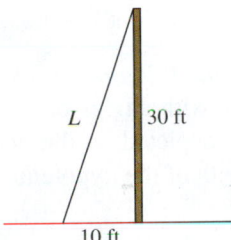

131. *Skid Marks* Vehicles involved in accidents often leave skid marks, which can be used to determine how fast a vehicle was traveling. To determine this speed, officials often use a test vehicle to compare skid marks on the same section of road. Suppose that a vehicle in a crash left skid marks D feet long and that a test vehicle traveling at v miles per hour leaves skid marks d feet long. Then the speed V of the vehicle involved in the crash is given by

$$V = v\sqrt{\dfrac{D}{d}}.$$

(**Source:** N. Garber and L. Hoel, *Traffic and Highway Engineering.*)
  (a) Find V if $v = 30$ mph, $D = 285$ feet, and $d = 178$ feet. Interpret your result.
  (b) A test vehicle traveling at 45 mph leaves skid marks 255 feet long. How long would the skid marks be for a vehicle traveling 60 miles per hour?

132. *Highway Curves* If a circular highway curve without any banking has a radius of R feet, the speed limit L in miles per hour for the curve is $L = 1.5\sqrt{R}$. (*Source:* N. Garber.)
  (a) Find the speed limit for a curve having a radius of 400 feet.
  (b) If the radius of a curve doubles, what happens to the speed limit?
  (c) A curve with a 40-mile-per-hour speed limit is being designed. What should be its radius?

133. *Wind Power* (Refer to Example 11.) If a wind-powered generator has blades that create a circular path with a diameter of 10 feet, then the wattage W generated by a wind velocity of v miles per hour is modeled by $W(v) = 3.8v^3$.
  (a) If the wind velocity doubles, what happens to the wattage generated?
  (b) Solve $W = 3.8v^3$ for v.
  (c) If the wind generator is producing 30,400 watts, find the wind speed.

134. *Height and Weight* Suppose that the weight of a person is directly proportional to the cube of the person's height. If one person weighs twice as much as a (similarly proportioned) second person, by what factor is the heavier person's height greater than the shorter person's height?

135. *45°–45° Right Triangle* Suppose that the legs of a right triangle with angles of 45° and 45° both have length a, as depicted in the accompanying figure. Find the length of the hypotenuse.

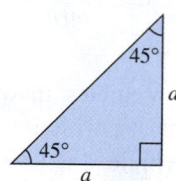

136. *30°–60° Right Triangle* In a right triangle with angles of 30° and 60°, the shortest side is half the length of the hypotenuse (see the accompanying figure). If the hypotenuse has length c, find the length of the other two sides, a and b, in terms of c.

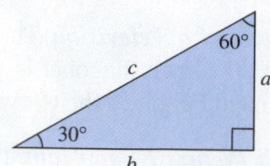

WRITING ABOUT MATHEMATICS

137. A student solves an equation *incorrectly* as follows.

$$\sqrt{3 - x} = \sqrt{x} - 1$$
$$(\sqrt{3 - x})^2 \overset{?}{=} (\sqrt{x})^2 - (1)^2$$
$$3 - x \overset{?}{=} x - 1$$
$$-2x \overset{?}{=} -4$$
$$x \overset{?}{=} 2$$

  (a) How could you convince the student that the answer is wrong?
  (b) Discuss where any errors were made.

138. When each side of an equation is squared, you must check your results. Explain why.

Checking Basic Concepts

  1. Sketch a graph of each function and then evaluate $f(-1)$, if possible.
  (a) $f(x) = \sqrt{x}$
  (b) $f(x) = \sqrt[3]{x}$
  (c) $f(x) = \sqrt{x^2}$

  2. Evaluate $f(x) = 0.2x^{2/3}$ when $x = 64$.

  3. Find the domain of $f(x) = \sqrt{x - 4}$. Write your answer in interval notation.

  4. Solve each equation. Check your answers.
  (a) $\sqrt{2x - 4} = 2$
  (b) $\sqrt[3]{x - 1} = 3$
  (c) $\sqrt{3x} = 1 + \sqrt{x + 1}$

  5. Find the distance between $(-3, 5)$ and $(2, -7)$.

  6. A 16-inch diagonal television set has a rectangular picture with a width of 12.8 inches. Find the height of the picture.

  7. Solve $(x + 1)^4 = 16$.

10.7 Complex Numbers

Basic Concepts • Addition, Subtraction, and Multiplication • Powers of i • Complex Conjugates and Division

A LOOK INTO MATH ▶

Mathematics is both applied and theoretical. A common misconception is that abstract or theoretical mathematics is unimportant in today's world. Many new ideas with great practical importance were first developed as abstract concepts with no particular application in mind. For example, complex numbers, which are related to square roots of negative numbers, started as an abstract concept to solve equations. Today, complex numbers are used in many sophisticated applications, such as the design of electrical circuits, ships, and airplanes. The *fractal image* shown to the left would not have been discovered without complex numbers.

NEW VOCABULARY

☐ Imaginary unit
☐ Complex number
☐ Standard form
☐ Real part
☐ Imaginary part
☐ Imaginary number
☐ Pure imaginary number
☐ Complex conjugate

Basic Concepts

A graph of $y = x^2 + 1$ is shown in Figure 10.28. There are no x-intercepts, so the quadratic equation $x^2 + 1 = 0$ has no real number solutions.

If we try to solve $x^2 + 1 = 0$ by subtracting 1 from each side, the result is $x^2 = -1$. Because $x^2 \geq 0$ for any real number x, there are no real solutions. However, mathematicians have invented (or discovered) solutions.

$$x^2 = -1$$
$$x = \pm \sqrt{-1} \quad \text{Solve for } x.$$

We now define a number called the **imaginary unit**, denoted i.

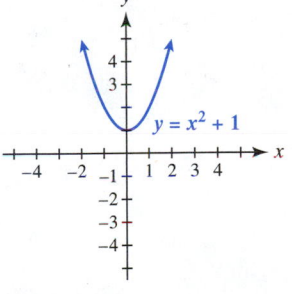

Figure 10.28

> **PROPERTIES OF THE IMAGINARY UNIT i**
>
> $$i = \sqrt{-1} \quad \text{and} \quad i^2 = -1$$

By creating the number i, the solutions to the equation $x^2 + 1 = 0$ are i and $-i$. Using the real numbers and the imaginary unit i, we can define a new set of numbers called the *complex numbers*. A **complex number** can be written in **standard form**, as $a + bi$, where a and b are real numbers. The **real part** is a and the **imaginary part** is b. Every real number a is also a complex number because it can be written $a + 0i$. A complex number $a + bi$ with $b \neq 0$ is an **imaginary number**. A complex number $a + bi$ with $a = 0$ and $b \neq 0$ is sometimes called a **pure imaginary number**. Examples of pure imaginary numbers include $4i$ and $-2i$. Table 10.9 lists several complex numbers with their real and imaginary parts.

TABLE 10.9

Complex Number: $a + bi$	$-3 + 2i$	5	$-3i$	$-1 + 7i$	$-5 - 2i$	$4 + 6i$
Real Part: a	-3	5	0	-1	-5	4
Imaginary Part: b	2	0	-3	7	-2	6

Figure 10.29 on the next page shows how different sets of numbers are related. Note that *the set of complex numbers contains the set of real numbers.*

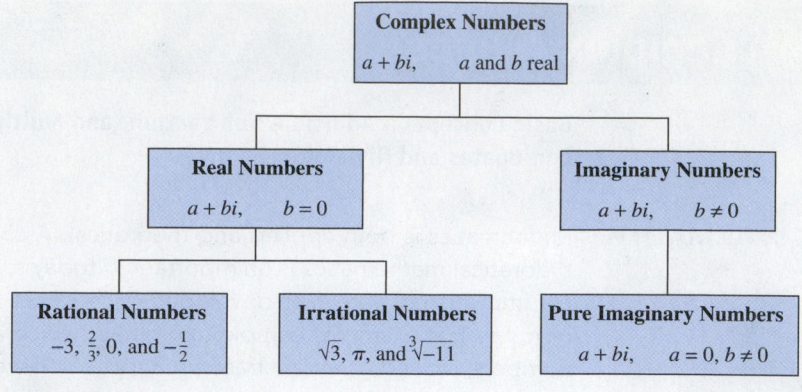

Figure 10.29

Using the imaginary unit i, we may write the square root of a negative number as a complex number. For example, $\sqrt{-2} = i\sqrt{2}$, and $\sqrt{-4} = i\sqrt{4} = 2i$. This method is summarized as follows.

CALCULATOR HELP

To set your calculator in $a + bi$ mode, see Appendix A (page AP-7).

THE EXPRESSION $\sqrt{-a}$

If $a > 0$, then $\sqrt{-a} = i\sqrt{a}$.

NOTE: Although it is standard for a complex number to be expressed as $a + bi$, we often write $\sqrt{2}i$ as $i\sqrt{2}$ so that it is clear that i is not under the square root. Similarly $\frac{1}{2}\sqrt{2}i$ is sometimes written as $\frac{1}{2}i\sqrt{2}$ or $\frac{i\sqrt{2}}{2}$.

EXAMPLE 1 | **Writing the square root of a negative number**

Write each square root using the imaginary unit i.

(a) $\sqrt{-25}$ **(b)** $\sqrt{-7}$ **(c)** $\sqrt{-20}$

Solution
(a) $\sqrt{-25} = i\sqrt{25} = 5i$
(b) $\sqrt{-7} = i\sqrt{7}$
(c) $\sqrt{-20} = i\sqrt{20} = i\sqrt{4}\sqrt{5} = 2i\sqrt{5}$

Now Try Exercises 13, 15, 19

Addition, Subtraction, and Multiplication

Addition can be defined for complex numbers in a manner similar to how we add binomials. For example,

$$(-3 + 2x) + (2 - x) = (-3 + 2) + (2x - x) = -1 + x.$$

ADDITION AND SUBTRACTION To add the complex numbers $(-3 + 2i)$ and $(2 - i)$, add the real parts and then add the imaginary parts.

$$(-3 + 2i) + (2 - i) = (-3 + 2) + (2i - i)$$
$$= (-3 + 2) + (2 - 1)i$$
$$= -1 + i$$

This same process works for subtraction.

$$(6 - 3i) - (2 + 5i) = (6 - 2) + (-3i - 5i)$$
$$= (6 - 2) + (-3 - 5)i$$
$$= 4 - 8i$$

This method is summarized as follows.

SUM OR DIFFERENCE OF COMPLEX NUMBERS

Let $a + bi$ and $c + di$ be two complex numbers. Then

$$(a + bi) + (c + di) = (a + c) + (b + d)i \quad \text{Sum}$$

and
$$(a + bi) - (c + di) = (a - c) + (b - d)i. \quad \text{Difference}$$

READING CHECK

How do you add and subtract complex numbers?

EXAMPLE 2

Adding and subtracting complex numbers

Write each sum or difference in standard form.
(a) $(-7 + 2i) + (3 - 4i)$ **(b)** $3i - (5 - i)$

Solution
(a) $(-7 + 2i) + (3 - 4i) = (-7 + 3) + (2 - 4)i = -4 - 2i$
(b) $3i - (5 - i) = 3i - 5 + i = -5 + (3 + 1)i = -5 + 4i$

■ **Now Try Exercises 23, 29**

TECHNOLOGY NOTE

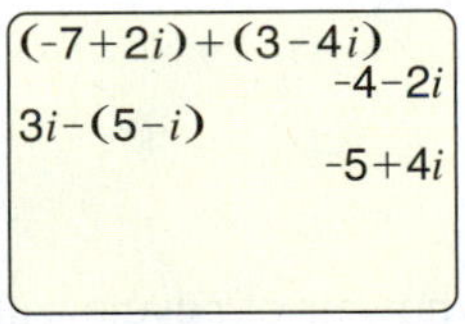

CALCULATOR HELP

To access the imaginary unit i, see Appendix A (page AP-7).

Complex Numbers
Many calculators can perform arithmetic with complex numbers. The figure shows a calculator display for the results in Example 2.

```
(-7+2i)+(3-4i)
                -4-2i
3i-(5-i)
                -5+4i
```

MULTIPLICATION We multiply two complex numbers in the same way that we multiply binomials and then we apply the property $i^2 = -1$. For example,

$$(2 - 3x)(1 + 4x) = 2 + 8x - 3x - 12x^2 = 2 + 5x - 12x^2.$$

In the next example we find the product of $2 - 3i$ and $1 + 4i$ in a similar manner.

EXAMPLE 3 **Multiplying complex numbers**

Write each product in standard form.
(a) $(2 - 3i)(1 + 4i)$ **(b)** $(5 - 2i)(5 + 2i)$

Solution
(a) Multiply the complex numbers like binomials.

$$
\begin{aligned}
(2 - 3i)(1 + 4i) &= (2)(1) + (2)(4i) - (3i)(1) - (3i)(4i) \\
&= 2 + 8i - 3i - 12i^2 \\
&= 2 + 5i - 12(-1) \qquad i^2 = -1 \\
&= 14 + 5i
\end{aligned}
$$

(b)
$$
\begin{aligned}
(5 - 2i)(5 + 2i) &= (5)(5) + (5)(2i) - (2i)(5) - (2i)(2i) \\
&= 25 + 10i - 10i - 4i^2 \\
&= 25 - 4(-1) \qquad i^2 = -1 \\
&= 29
\end{aligned}
$$

These results are supported in Figure 10.30.

Now Try Exercises 31, 35

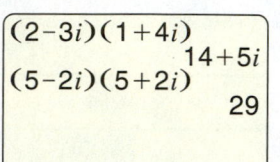

```
(2-3i)(1+4i)
              14+5i
(5-2i)(5+2i)
                 29
```

Figure 10.30

Powers of i

An interesting pattern appears when powers of i are calculated.

$$
\begin{aligned}
i^1 &= i \\
i^2 &= -1 \\
i^3 &= i^2 \cdot i = -1 \cdot i = -i \\
i^4 &= i^2 \cdot i^2 = (-1)(-1) = 1 \\
i^5 &= i^4 \cdot i = (1)i = i \\
i^6 &= i^4 \cdot i^2 = (1)(-1) = -1 \\
i^7 &= i^4 \cdot i^3 = (1)(-i) = -i \\
i^8 &= i^4 \cdot i^4 = (1)(1) = 1
\end{aligned}
$$

The powers of i cycle with the pattern i, -1, $-i$, and 1. These examples suggest the following method for calculating powers of i.

POWERS OF i

The value of i^n can be found by dividing n (a positive integer) by 4. If the remainder is r, then

$$
i^n = i^r.
$$

Note that $i^0 = 1$, $i^1 = i$, $i^2 = -1$, and $i^3 = -i$.

<table>
<tr><td>**EXAMPLE 4**</td><td>**Calculating powers of *i***</td></tr>
</table>

Evaluate each expression.

(a) i^9 **(b)** i^{19} **(c)** i^{40}

Solution

(a) When 9 is divided by 4, the result is 2 with remainder **1**. Thus $i^9 = i^1 = i$.

(b) When 19 is divided by 4, the result is 4 with remainder **3**. Thus $i^{19} = i^3 = -i$.

(c) When 40 is divided by 4, the result is 10 with remainder **0**. Thus $i^{40} = i^0 = 1$.

Now Try Exercises 49, 51, 55

Complex Conjugates and Division

The **complex conjugate** of $a + bi$ is $a - bi$. To find the conjugate, we change the sign of the imaginary part b. Table 10.10 contains some complex numbers and their conjugates.

TABLE 10.10   Complex Conjugates

Number	$2 + 5i$	$6 - 3i$	$-2 + 7i$	$-1 - i$	5	$-5i$
Conjugate	$2 - 5i$	$6 + 3i$	$-2 - 7i$	$-1 + i$	5	$5i$

<table>
<tr><td>**EXAMPLE 5**</td><td>**Finding complex conjugates**</td></tr>
</table>

Find the complex conjugate of each number.

(a) $5 + 3i$ **(b)** $-2 - i$ **(c)** $-4i$ **(d)** -6

Solution

(a) The conjugate is found by changing the imaginary part of $5 + 3i$ to its opposite. The conjugate is $5 - 3i$.

(b) The conjugate is $-2 + i$.

(c) $-4i$ can be written as $0 - 4i$. The conjugate is $0 + 4i$, or $4i$.

(d) -6 can be written as $-6 + 0i$. The conjugate is $-6 - 0i$, or -6. The conjugate of a real number is simply the same real number.

Now Try Exercises 57, 59, 61, 63

NOTE: The product of a complex number and its conjugate is a real number. That is,

$$(a + bi)(a - bi) = a^2 + b^2.$$

For example, $(3 + 4i)(3 - 4i) = 3^2 + 4^2 = 25$, which is a real number.

This property of complex conjugates is used to divide two complex numbers. To convert the quotient $\frac{2 + 3i}{3 - i}$ into standard form $a + bi$, we multiply the numerator and the denominator by the complex conjugate of the *denominator*, which is $3 + i$. The next example illustrates this method.

EXAMPLE 6   **Dividing complex numbers**

Write each quotient in standard form.

(a) $\dfrac{2 + 3i}{3 - i}$ (b) $\dfrac{4}{2i}$

Solution

(a) Multiply the numerator and denominator by **3 + i**.

$$\frac{2 + 3i}{3 - i} = \frac{(2 + 3i)(3 + i)}{(3 - i)(3 + i)} \qquad \text{Multiply by 1.}$$

$$= \frac{2(3) + (2)(i) + (3i)(3) + (3i)(i)}{(3)(3) + (3)(i) - (i)(3) - (i)(i)} \qquad \text{Multiply.}$$

$$= \frac{6 + 2i + 9i + 3i^2}{9 + 3i - 3i - i^2} \qquad \text{Simplify.}$$

$$= \frac{6 + 11i + 3(-1)}{9 - (-1)} \qquad i^2 = -1$$

$$= \frac{3 + 11i}{10} \qquad \text{Simplify.}$$

$$= \frac{3}{10} + \frac{11}{10}i \qquad \frac{a + bi}{c} = \frac{a}{c} + \frac{b}{c}i$$

(b) Multiply the numerator and denominator by **− 2i**.

$$\frac{4}{2i} = \frac{(4)(-2i)}{(2i)(-2i)} \qquad \text{Multiply by 1.}$$

$$= \frac{-8i}{-4i^2} \qquad \text{Simplify.}$$

$$= \frac{-8i}{-4(-1)} \qquad i^2 = -1$$

$$= \frac{-8i}{4} \qquad \text{Simplify.}$$

$$= -2i \qquad \text{Divide.}$$

```
(2+3i)/(3-i)▶Fra
c
        3/10+11/10i
4/(2i)
             -2i
```

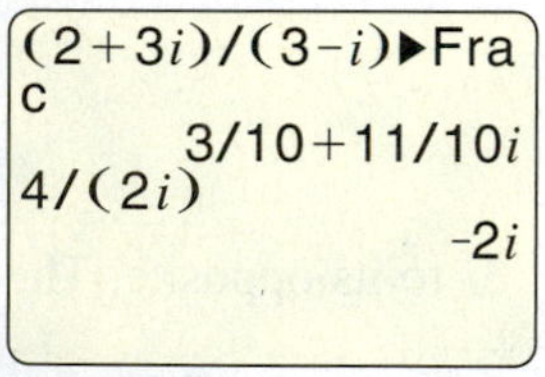

Figure 10.31

These results are supported in Figure 10.31.

Now Try Exercises 71, 73

10.7   Putting It All Together

CONCEPT	EXPLANATION	EXAMPLES
Complex Numbers	A complex number can be expressed as $a + bi$, where a and b are real numbers. The imaginary unit i satisfies $i = \sqrt{-1}$ and $i^2 = -1$. As a result, we can write $\sqrt{-a} = i\sqrt{a}$ if $a > 0$.	The real part of $5 - 3i$ is 5 and the imaginary part is -3. $\sqrt{-13} = i\sqrt{13}$ and $\sqrt{-9} = 3i$

CONCEPT	EXPLANATION	EXAMPLES
Addition, Subtraction, and Multiplication	To add (subtract) complex numbers, add (subtract) the real parts and then add (subtract) the imaginary parts.	$(3 + 6i) + (-1 + 2i)$ Sum $= (3 + -1) + (6 + 2)i$ $= 2 + 8i$ $(2 - 5i) - (1 + 4i)$ Difference $= (2 - 1) + (-5 - 4)i$ $= 1 - 9i$
	Multiply complex numbers in a manner similar to the way *FOIL* is used to multiply binomials. Then apply the property $i^2 = -1$.	$(-1 + 2i)(3 + i)$ Product $= (-1)(3) + (-1)(i) + (2i)(3) + (2i)(i)$ $= -3 - i + 6i + 2i^2$ $= -3 + 5i + 2(-1)$ $= -5 + 5i$
Complex Conjugates	The conjugate of $a + bi$ is $a - bi$.	The conjugate of $3 - 5i$ is $3 + 5i$. The conjugate of $2i$ is $-2i$.
Division	To simplify a quotient, multiply the numerator and denominator by the complex conjugate of the *denominator*. Then simplify the expression and write it in standard form as $a + bi$.	$\dfrac{10}{1 + 2i} = \dfrac{10(1 - 2i)}{(1 + 2i)(1 - 2i)}$ Quotient $= \dfrac{10 - 20i}{5}$ $= 2 - 4i$

10.7 Exercises

MyMathLab   Math XL PRACTICE  WATCH DOWNLOAD READ REVIEW

CONCEPTS AND VOCABULARY

1. Give an example of a complex number that is not a real number.

2. Can you give an example of a real number that is not a complex number? Explain.

3. $\sqrt{-1} =$ _____   4. $i^2 =$ _____

5. $\sqrt{-a} =$ _____, if $a > 0$.

6. The complex conjugate of $10 + 7i$ is _____.

7. The standard form for a complex number is _____.

8. Write $\dfrac{2 + 4i}{2}$ in standard form.

9. The real part of $4 - 5i$ is _____.

10. The imaginary part of $4 - 5i$ is _____.

11. The imaginary part of -7 is _____.

12. The number $7i$ is a(n) _____ imaginary number.

COMPLEX NUMBERS

Exercises 13–22: Use i to write the expression.

13. $\sqrt{-5}$ 14. $\sqrt{-21}$

15. $\sqrt{-100}$ 16. $\sqrt{-49}$

17. $\sqrt{-144}$ 18. $\sqrt{-64}$

19. $\sqrt{-12}$ 20. $\sqrt{-8}$

21. $\sqrt{-18}$ 22. $\sqrt{-48}$

Exercises 23–46: Write the expression in standard form.

23. $(5 + 3i) + (-2 - 3i)$

24. $(1 - i) + (5 - 7i)$

25. $2i + (-8 + 5i)$

26. $-3i + 5i$

27. $(2 - 7i) - (1 + 2i)$

28. $(1 + 8i) - (3 + 9i)$ **29.** $5i - (10 - 2i)$

30. $-3(4 - 3i)$

31. $(3 + 2i)(-1 + 5i)$

32. $(1 + i) - (1 - i)$ **33.** $4(5 - 3i)$

34. $(1 + 2i)(-6 - i)$ **35.** $(5 + 4i)(5 - 4i)$

36. $(3 + 5i)(3 - 5i)$ **37.** $(-4i)(5i)$

38. $(-6i)(-4i)$

39. $3i + (2 - 3i) - (1 - 5i)$

40. $4 - (5 - 7i) + (3 + 7i)$

41. $(2 + i)^2$ **42.** $(-1 + 2i)^2$

43. $2i(-3 + i)$ **44.** $5i(1 - 9i)$

45. $i(1 + i)^2$ **46.** $2i(1 - i)^2$

Exercises 47 and 48: **Thinking Generally** *Write in standard form.*

47. $(a + 3bi)(a - 3bi)$ **48.** $(a + bi) - (a - bi)$

Exercises 49–56: (Refer to Example 4.) Simplify.

49. i^{11} **50.** i^{50}

51. i^{21} **52.** i^{103}

53. i^{58} **54.** i^{61}

55. i^{64} **56.** i^{28}

Exercises 57–64: Write the complex conjugate.

57. $3 + 4i$ **58.** $1 - 4i$

59. $-6i$ **60.** -10

61. $5 - 4i$ **62.** $7 + 2i$

63. -1 **64.** $19i$

Exercises 65–78: Write the expression in standard form.

65. $\dfrac{2}{1 + i}$ **66.** $\dfrac{-6}{2 - i}$

67. $\dfrac{3i}{5 - 2i}$ **68.** $\dfrac{-8}{2i}$

69. $\dfrac{8 + 9i}{5 + 2i}$ **70.** $\dfrac{3 - 2i}{1 + 4i}$

71. $\dfrac{5 + 7i}{1 - i}$ **72.** $\dfrac{-7 + 4i}{3 - 2i}$

73. $\dfrac{2 - i}{i}$ **74.** $\dfrac{3 + 2i}{-i}$

75. $\dfrac{1}{i} + \dfrac{1}{2i}$ **76.** $\dfrac{3}{4i} + \dfrac{2}{i}$

77. $\dfrac{1}{-1 + i} - \dfrac{2}{i}$ **78.** $-\dfrac{3}{2i} - \dfrac{2}{1 + i}$

APPLICATIONS

Exercises 79 and 80: **Corrosion in Airplanes** *Corrosion in the metal surface of an airplane can be difficult to detect visually. One test used to locate it involves passing an alternating current through a small area on the plane's surface. If the current varies from one region to another, it may indicate that corrosion is occurring. The impedance Z (or opposition to the flow of electricity) of the metal is related to the voltage V and current I by the equation $Z = \frac{V}{I}$, where Z, V, and I are complex numbers. Calculate Z for the given values of V and I.* (**Source:** Society for Industrial and Applied Mathematics.)

79. $V = 40 + 70i, I = 2 + 3i$

80. $V = 10 + 20i, I = 3 + 7i$

81. **Online Exploration** We graph real numbers on a number line. Can we graph complex numbers on a number line? Go online and find out how to graph complex numbers.

82. **Online Exploration** Go online and find out how to calculate the absolute value of a complex number. If a complex number is graphed, what does the absolute value tell us?

WRITING ABOUT MATHEMATICS

83. A student multiplies $(2 + 3i)(4 - 5i)$ *incorrectly* to obtain $8 - 15i$. What is the student's mistake?

84. A student divides the ratio $\frac{6 - 10i}{3 + 2i}$ *incorrectly* to obtain $2 - 5i$. What is the student's mistake?

SECTION 10.7   Checking Basic Concepts

1. Use i to write each expression.

(a) $\sqrt{-64}$

(b) $\sqrt{-17}$

2. Simplify each expression.

(a) $(2 - 3i) + (1 - i)$

(b) $4i - (2 + i)$

(c) $(3 - 2i)(1 + i)$

(d) $\dfrac{3}{2 - 2i}$

CHAPTER 10   Summary

SECTION 10.1 ■ RADICAL EXPRESSIONS AND FUNCTIONS

Radicals and Radical Notation

Square Root b is a square root of a if $b^2 = a$.

Principal Square Root $\sqrt{a} = b$ if $b^2 = a$ and $b \geq 0$.

> **Examples:** $\sqrt{16} = 4$, $-\sqrt{9} = -3$,
>
> and $\pm\sqrt{36} = \pm 6$

Cube Root b is a cube root of a if $b^3 = a$.

> **Examples:** $\sqrt[3]{27} = 3$, $\sqrt[3]{-8} = -2$

nth Root b is an nth root of a if $b^n = a$.

> **Example:** $\sqrt[4]{16} = 2$ because $2^4 = 16$.

> **NOTE:** An *even* root of a *negative* number is not a real number. Also, $\sqrt[n]{a}$ denotes the *principal* nth root.

Absolute Value   The expressions $|x|$ and $\sqrt{x^2}$ are equivalent.

> **Example:** $\sqrt{(x + y)^2} = |x + y|$

The Square Root Function   The square root function is denoted $f(x) = \sqrt{x}$. Its domain is $\{x \mid x \geq 0\}$ and its graph is shown in the figure.

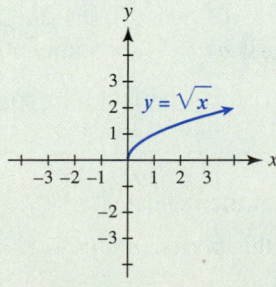

The Cube Root Function The cube root function is denoted $f(x) = \sqrt[3]{x}$. Its domain is all real numbers and its graph is shown in the figure.

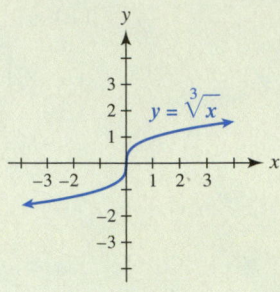

The Expression $a^{1/n}$

$a^{1/n} = \sqrt[n]{a}$ if n is an integer greater than 1.

Examples: $5^{1/2} = \sqrt{5}$ and $64^{1/3} = \sqrt[3]{64} = 4$

The Expression $a^{m/n}$

$a^{m/n} = \sqrt[n]{a^m}$ or $a^{m/n} = (\sqrt[n]{a})^m$

Examples: $8^{2/3} = \sqrt[3]{8^2} = \sqrt[3]{64} = 4$ and
$$8^{2/3} = (\sqrt[3]{8})^2 = (2)^2 = 4$$

Properties of Exponents

Product Rule

$a^p a^q = a^{p+q}$

Negative Exponents

$a^{-p} = \dfrac{1}{a^p}, \dfrac{1}{a^{-p}} = a^p$

Negative Exponents for Quotients

$\left(\dfrac{a}{b}\right)^{-p} = \left(\dfrac{b}{a}\right)^p$

Quotient Rule for Exponents

$\dfrac{a^p}{a^q} = a^{p-q}$

Power Rule for Exponents

$(a^p)^q = a^{pq}$

Power Rule for Products

$(ab)^p = a^p b^p$

Power Rule for Quotients

$\left(\dfrac{a}{b}\right)^p = \dfrac{a^p}{b^p}$

Product Rule for Radical Expressions Provided each expression is defined,
$$\sqrt[n]{a} \cdot \sqrt[n]{b} = \sqrt[n]{a \cdot b}.$$

Example: $\sqrt[3]{3} \cdot \sqrt[3]{9} = \sqrt[3]{27} = 3$

Perfect nth Power An integer a is a perfect nth power if $b^n = a$ for some integer b.

Examples: 25 is a perfect square, 8 is a perfect cube, and 16 is a perfect fourth power.

Simplifying Radicals (nth roots)

STEP 1: Determine the largest perfect nth power factor of the radicand.

STEP 2: Use the product rule to factor out and simplify this perfect nth power.

Examples: $\sqrt{32} = \sqrt{16 \cdot 2} = \sqrt{16} \cdot \sqrt{2} = 4\sqrt{2}$
$$\sqrt[3]{32} = \sqrt[3]{8 \cdot 4} = \sqrt[3]{8} \cdot \sqrt[3]{4} = 2\sqrt[3]{4}$$

Quotient Rule for Radical Expressions Provided each expression is defined,

$$\sqrt[n]{\frac{a}{b}} = \frac{\sqrt[n]{a}}{\sqrt[n]{b}}.$$

Example: $\dfrac{\sqrt[3]{24}}{\sqrt[3]{3}} = \sqrt[3]{\dfrac{24}{3}} = \sqrt[3]{8} = 2$

SECTION 10.4 ■ OPERATIONS ON RADICAL EXPRESSIONS

Addition and Subtraction Combine like radicals.

Examples: $2\sqrt[3]{4} + 3\sqrt[3]{4} = 5\sqrt[3]{4}$ and $\sqrt{5} - 2\sqrt{5} = -\sqrt{5}$

Multiplication Sometimes radical expressions can be multiplied like binomials.

Examples: $(4 - \sqrt{2})(2 + \sqrt{2}) = 8 + 4\sqrt{2} - 2\sqrt{2} - 2 = 6 + 2\sqrt{2}$

$(5 - \sqrt{3})(5 + \sqrt{3}) = (5)^2 - (\sqrt{3})^2 = 25 - 3 = 22$ because

$(a - b)(a + b) = a^2 - b^2.$

Rationalizing the Denominator One technique is to multiply the numerator and denominator by the conjugate of the denominator if the denominator is a binomial containing one or more square roots.

Examples: $\dfrac{1}{4 + \sqrt{2}} = \dfrac{1}{(4 + \sqrt{2})} \cdot \dfrac{(4 - \sqrt{2})}{(4 - \sqrt{2})} = \dfrac{4 - \sqrt{2}}{(4)^2 - (\sqrt{2})^2} = \dfrac{4 - \sqrt{2}}{14}$

$\dfrac{4}{\sqrt{7}} = \dfrac{4}{\sqrt{7}} \cdot \dfrac{\sqrt{7}}{\sqrt{7}} = \dfrac{4\sqrt{7}}{7}$

SECTION 10.5 ■ MORE RADICAL FUNCTIONS

Root Functions $f(x) = x^{1/n} = \sqrt[n]{x}$

Odd root functions are defined for all real numbers, and even root functions are defined for nonnegative real numbers.

Examples: $f(x) = x^{1/3} = \sqrt[3]{x}$, $g(x) = x^{1/4} = \sqrt[4]{x}$, and $h(x) = x^{1/5} = \sqrt[5]{x}$

Power Functions If a function can be defined by $f(x) = x^p$, where p is a rational number, then it is a power function.

Examples: $f(x) = x^{4/5}$ and $g(x) = x^{2.3}$

SECTION 10.6 ■ EQUATIONS INVOLVING RADICAL EXPRESSIONS

Power Rule for Solving Radical Equations The solutions to $a = b$ are among the solutions to $a^n = b^n$, where n is a positive integer.

Example: The solutions to the equation $\sqrt{3x + 3} = 2x - 1$ are among the solutions to the equation $3x + 3 = (2x - 1)^2.$

Solving Radical Equations

STEP 1: Isolate a radical term on one side of the equation.

STEP 2: Apply the power rule by raising each side of the equation to the power equal to the index of the isolated radical term.

STEP 3: Solve the equation. If it still contains a radical, repeat Steps 1 and 2.

STEP 4: Check your answers by substituting each result in the *given* equation.

Example: To isolate the radical in $\sqrt{x + 1} + 4 = 6$ subtract 4 from each side to obtain $\sqrt{x + 1} = 2$. Next square each side, which gives $x + 1 = 4$ or $x = 3$. Checking verifies that 3 is a solution.

Pythagorean Theorem If a right triangle has legs a and b with hypotenuse c, then

$$a^2 + b^2 = c^2.$$

Example: If a right triangle has legs 8 and 15, then the hypotenuse equals

$$c = \sqrt{8^2 + 15^2} = \sqrt{289} = 17.$$

The Distance Formula The distance d between (x_1, y_1) and (x_2, y_2) is

$$d = \sqrt{(x_2 - x_1)^2 + (y_2 - y_1)^2}.$$

Example: The distance between $(-1, 3)$ and $(4, 5)$ is

$$d = \sqrt{\left(4 - (-1)\right)^2 + (5 - 3)^2} = \sqrt{25 + 4} = \sqrt{29}.$$

Solving the Equation $x^n = k$ Let n be a positive integer.
Take the nth root of each side of $x^n = k$ to obtain $\sqrt[n]{x^n} = \sqrt[n]{k}$.

1. If n is *odd*, then $\sqrt[n]{x^n} = x$ and the equation becomes $x = \sqrt[n]{k}$.
2. If n is *even* and $k \geq 0$, then $\sqrt[n]{x^n} = |x|$ and the equation becomes $|x| = \sqrt[n]{k}$.

Examples: $x^3 = -27$ implies that $x = \sqrt[3]{-27} = -3$.

$x^4 = 81$ implies that $|x| = \sqrt[4]{81} = 3$ or $x = \pm 3$.

SECTION 10.7 ■ COMPLEX NUMBERS

Complex Numbers

Imaginary Unit	$i = \sqrt{-1}$ and $i^2 = -1$
Standard Form	$a + bi$, where a and b are real numbers
	Examples: $4 + 3i$, $5 - 6i$, 8, and $-2i$
Real Part	The real part of $a + bi$ is a.
	Example: The real part of $3 - 2i$ is 3.
Imaginary Part	The imaginary part of $a + bi$ is b.
	Example: The imaginary part of $2 - i$ is -1.
Arithmetic Operations	Arithmetic operations are similar to arithmetic operations on binomials.

Examples: $(2 + 2i) + (3 - i) = 5 + i$ *Sum*

$(1 - i) - (1 - 2i) = i$ *Difference*

$(1 - i)(1 + i) = 1^2 - i^2 = 1 - (-1) = 2$ *Product*

$\dfrac{2}{1 - i} = \dfrac{2}{1 - i} \cdot \dfrac{1 + i}{1 + i} = \dfrac{2 + 2i}{2} = 1 + i$ *Quotient*

Powers of i   The value of i^n equals i^r, where r is the remainder when n is divided by 4. Note that $i^0 = 1, i^1 = i, i^2 = -1$, and $i^3 = -i$.

Example: $i^{21} = i^1 = i$ because when 21 is divided by 4 the remainder is 1.

CHAPTER 10   Review Exercises

SECTION 10.1

Exercises 1–12: Simplify the expression.

1. $\sqrt{4}$

2. $\sqrt{36}$

3. $\sqrt{9x^2}$

4. $\sqrt{(x-1)^2}$

5. $\sqrt[3]{-64}$

6. $\sqrt[3]{-125}$

7. $\sqrt[3]{x^6}$

8. $\sqrt[3]{27x^3}$

9. $\sqrt[4]{16}$

10. $\sqrt[5]{-1}$

11. $\sqrt[4]{x^8}$

12. $\sqrt[5]{(x+1)^5}$

SECTION 10.2

Exercises 13–16: Write the expression in radical notation.

13. $14^{1/2}$

14. $(-5)^{1/3}$

15. $\left(\dfrac{x}{y}\right)^{3/2}$

16. $(xy)^{-2/3}$

Exercises 17–20: Evaluate the expression.

17. $(-27)^{2/3}$

18. $16^{1/4}$

19. $16^{3/2}$

20. $81^{3/4}$

Exercises 21–24: Simplify the expression. Assume that all variables are positive.

21. $(z^3)^{2/3}$

22. $(x^2y^4)^{1/2}$

23. $\left(\dfrac{x^2}{y^6}\right)^{3/2}$

24. $\left(\dfrac{x^3}{y^6}\right)^{-1/3}$

SECTION 10.3

Exercises 25–40: Simplify the expression. Assume that all variables are positive.

25. $\sqrt{2} \cdot \sqrt{32}$

26. $\sqrt[3]{-4} \cdot \sqrt[3]{2}$

27. $\sqrt[3]{x^4} \cdot \sqrt[3]{x^2}$

28. $\dfrac{\sqrt{80}}{\sqrt{20}}$

29. $\sqrt[3]{-\dfrac{x}{8}}$

30. $\sqrt{\dfrac{1}{3}} \cdot \sqrt{\dfrac{1}{3}}$

31. $\sqrt{48}$

32. $\sqrt{54}$

33. $\sqrt[3]{\dfrac{3}{x}} \cdot \sqrt[3]{\dfrac{9}{x^2}}$

34. $\sqrt{32a^3b^2}$

35. $\sqrt{3xy} \cdot \sqrt{27xy}$

36. $\sqrt[3]{-25z^2} \cdot \sqrt[3]{-5z^2}$

37. $\sqrt{x^2 + 2x + 1}$

38. $\sqrt[4]{\dfrac{2a^2}{b}} \cdot \sqrt[4]{\dfrac{8a^3}{b^3}}$

39. $2\sqrt{x} \cdot \sqrt[3]{x}$

40. $\sqrt[3]{rt} \cdot \sqrt[4]{r^2t^4}$

SECTION 10.4

Exercises 41–50: Simplify the expression. Assume that all variables are positive.

41. $3\sqrt{3} + \sqrt{3}$

42. $\sqrt[3]{x} + 2\sqrt[3]{x}$

43. $3\sqrt[3]{5} - 6\sqrt[3]{5}$

44. $\sqrt[4]{y} - 2\sqrt[4]{y}$

45. $2\sqrt{12} + 7\sqrt{3}$

46. $3\sqrt{18} - 2\sqrt{2}$

47. $7\sqrt[3]{16} - \sqrt[3]{2}$

48. $\sqrt{4x+4} + \sqrt{x+1}$

49. $\sqrt{4x^3} - \sqrt{x}$

50. $\sqrt[3]{ab^4} + 2\sqrt[3]{a^4b}$

Exercises 51–56: Multiply and simplify.

51. $(1 + \sqrt{2})(3 + \sqrt{2})$

52. $(7 - \sqrt{5})(1 + \sqrt{3})$

53. $(3 + \sqrt{6})(3 - \sqrt{6})$

54. $(10 - \sqrt{5})(10 + \sqrt{5})$

55. $(\sqrt{a} + \sqrt{2b})(\sqrt{a} - \sqrt{2b})$

56. $(\sqrt{xy} - 1)(\sqrt{xy} + 2)$

Exercises 57–62: Rationalize the denominator.

57. $\dfrac{4}{\sqrt{5}}$

58. $\dfrac{r}{2\sqrt{t}}$

59. $\dfrac{1}{\sqrt{2}+3}$

60. $\dfrac{2}{5-\sqrt{7}}$

61. $\dfrac{1}{\sqrt{8}-\sqrt{7}}$

62. $\dfrac{\sqrt{a}-\sqrt{b}}{\sqrt{a}+\sqrt{b}}$

SECTION 10.5

Exercises 63 and 64: Graph the equation.

63. $y = \sqrt[4]{x}$

64. $y = \sqrt[3]{x}$

Exercises 65 and 66: Write f(x) in radical notation and evaluate f(4).

65. $f(x) = x^{1/2}$

66. $f(x) = x^{2/7}$

Exercises 67 and 68: Graph the equation. Compare the graph to either $y = \sqrt{x}$ or $y = \sqrt[3]{x}$.

67. $y = \sqrt{x} - 2$

68. $y = \sqrt[3]{x} - 1$

Exercises 69–72: Find the domain of f. Write your answer in interval notation.

69. $f(x) = \sqrt{x - 1}$

70. $f(x) = \sqrt{6 - 2x}$

71. $f(x) = \sqrt{x^2 + 1}$

72. $f(x) = \dfrac{1}{\sqrt{x + 2}}$

SECTION 10.6

Exercises 73–78: Solve. Check your answer.

73. $\sqrt{x + 2} = x$

74. $\sqrt{2x - 1} = \sqrt{x + 3}$

75. $\sqrt[3]{x - 1} = 2$

76. $\sqrt[3]{3x} = 3$

77. $\sqrt{2x} = x - 4$

78. $\sqrt{x + 1} = \sqrt{x + 2}$

 Exercises 79 and 80: Solve graphically. Approximate solutions to the nearest hundredth when appropriate.

79. $\sqrt[3]{2x - 1} = 2$

80. $x^{2/3} = 3 - x$

Exercises 81 and 82: A right triangle has legs a and b with hypotenuse c. Find the length of the missing side.

81. $a = 4, b = 7$

82. $a = 5, c = 8$

Exercises 83 and 84: Find the exact distance between the given points.

83. $(-2, 3), (2, -2)$

84. $(2, -3), (-4, 1)$

Exercises 85–94: Solve.

85. $x^2 = 121$

86. $2z^2 = 32$

87. $(x - 1)^2 = 16$

88. $x^3 = 64$

89. $(x - 1)^3 = 8$

90. $(2x - 1)^3 = 27$

91. $x^4 = 256$

92. $x^5 = -1$

93. $(x - 3)^5 = -32$

94. $3(x + 1)^4 = 3$

SECTION 10.7

Exercises 95–100: Write the complex expression in standard form.

95. $(1 - 2i) + (-3 + 2i)$

96. $(1 + 3i) - (3 - i)$

97. $(1 - i)(2 + 3i)$

98. $\dfrac{3 + i}{1 - i}$

99. $\dfrac{i(4 + i)}{2 - 3i}$

100. $(1 - i)^2(1 + i)$

APPLICATIONS

101. *Hang Time* A football is punted and has a hang time T of 4.6 seconds. Use the formula $T(h) = \frac{1}{2}\sqrt{h}$ to estimate to the nearest foot the height h to which it was kicked.

102. *Baseball Diamond* The four bases of a baseball diamond form a square that is 90 feet on a side. Find the distance from home plate to second base.

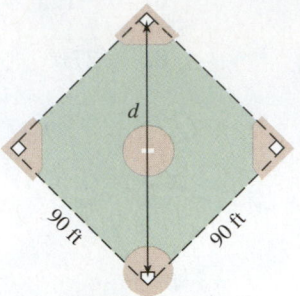

103. *Falling Time* The time T in seconds for an object to fall from a height of h feet is given by $T = \frac{1}{4}\sqrt{h}$. If a person steps off a 10-foot-high board into a swimming pool, how long is the person in the air?

104. *Geometry* A cube has sides of length $\sqrt{5}$.
   (a) Find the area of one side of the cube.
   (b) Find the volume of the cube.
   (c) Find the length of the diagonal of one of the sides.
   (d) Find the distance from A to B in the figure.

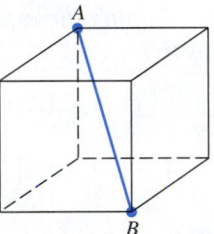

105. *Pendulum* The time required for a pendulum to swing back and forth is called its *period* (see the figure). The period T of a pendulum does not depend on its weight, only on its length L and gravity. It is given by $T = 2\pi\sqrt{\frac{L}{32.2}}$, where T is in seconds and L is in feet. Estimate the length of a pendulum with a period of 1 second.

106. *Pendulum* (Refer to Exercise 105.) If a pendulum were on the moon, its period could be calculated by $T = 2\pi\sqrt{\frac{L}{5.1}}$. Estimate the length of a pendulum with a period of 1 second on the moon. Compare your answer to that for Exercise 105.

107. *Population Growth* In 1790 the population of the United States was 4 million, and by 2000 it had grown to 281 million. The *average* annual percentage growth in the population r (expressed as a decimal) can be determined by the polynomial equation $281 = 4(1 + r)^{210}$. Solve this equation for r and interpret the result.

108. *Highway Curves* If a circular highway curve is banked with a slope of $m = \frac{1}{10}$ (see the accompanying figure) and has a radius of R feet, then the speed limit L in miles per hour for the curve is given by

$$L = \sqrt{3.75R}.$$

(*Source:* N. Garber and L. Hoel, *Traffic and Highway Engineering.*)

(a) Find the speed limit if R is 500 feet.
(b) With no banking, the speed limit is given by $L = 1.5\sqrt{R}$. Find the speed limit for a curve with no banking and a radius of 500 feet. How does banking affect the speed limit? Does this result agree with your intuition?

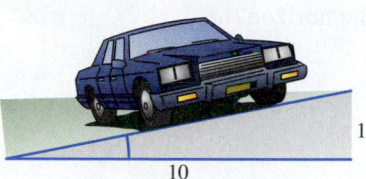

109. *Geometry* Find the length of a side of a square if the square has an area of 7 square feet.

CHAPTER 10   Test

 Step-by-step test solutions are found on the Chapter Test Prep Videos available in **MyMathLab** and on YouTube (search "RockswoldComboAlg" and click on "Channels").

Exercises 1–6: Simplify the expression.

1. $\sqrt[3]{-27}$

2. $\sqrt{(z + 1)^2}$

3. $\sqrt{25x^4}$

4. $\sqrt[3]{8z^6}$

5. $\sqrt[4]{16x^4y^5}$; $x > 0$, $y > 0$

6. $(\sqrt{3} - \sqrt{2})(\sqrt{3} + \sqrt{2})$

Exercises 7 and 8: Write the expression in radical notation.

7. $7^{2/5}$

8. $\left(\frac{x}{y}\right)^{-2/3}$

Exercises 9 and 10: Evaluate the expression by hand.

9. $(-8)^{4/3}$

10. $36^{-3/2}$

Exercises 11 and 12: Use rational exponents to write the expression.

11. $\sqrt[3]{x^4}$

12. $\sqrt{x} \cdot \sqrt[5]{x}$

13. Find the domain of $f(x) = \sqrt{4 - x}$. Write your answer in interval notation.

14. Sketch a graph of $y = \sqrt{x + 3}$.

Exercises 15–22: Simplify the expression. Assume that all variables are positive.

15. $(2z^{1/2})^3$

16. $\left(\frac{y^2}{z^3}\right)^{-1/3}$

17. $\sqrt{3} \cdot \sqrt{27}$

18. $\frac{\sqrt{y^3}}{\sqrt{4y}}$

19. $7\sqrt{7} - 3\sqrt{7} + \sqrt{5}$

20. $7\sqrt[3]{x} - \sqrt[3]{x}$

21. $4\sqrt{18} + \sqrt{8}$

22. $\frac{\sqrt[3]{32}}{\sqrt[3]{4}}$

23. Solve each equation.
(a) $\sqrt{x - 2} = 5$ **(b)** $\sqrt[3]{x + 1} = 2$
(c) $(x - 1)^3 = 8$ **(d)** $\sqrt{2x + 2} = x - 11$

24. Rationalize the denominator.
(a) $\frac{2}{3\sqrt{7}}$ **(b)** $\frac{1}{1 + \sqrt{5}}$

25. Solve $\sqrt{3x} - x + 1 = \sqrt[3]{x - 1}$ graphically. Round solutions to the nearest hundredth.

26. One leg of a right triangle has length 7 and the hypotenuse has length 13. Find the length of the third side.

27. Find the distance between $(-3, 5)$ and $(-1, 7)$.

Exercises 28–31: Write the complex expression in standard form.

28. $(-5 + i) + (7 - 20i)$

29. $(3i) - (6 - 5i)$

30. $\left(\frac{1}{2} - i\right)\left(\frac{1}{2} + i\right)$ **31.** $\frac{2i}{5 + 2i}$

32. *Distance to the Horizon* A formula for calculating the distance d in miles that one can see to the horizon on a clear day is approximated by the formula $d = 1.22\sqrt{x}$, where x is the elevation, in feet, of a person.

(a) If a person climbs a cliff and stops at an elevation of 200 feet, how far can the person see to the horizon?

(b) How high should the person climb to see 25 miles to the horizon?

33. *Wing Span of a Bird* The wing span L of a bird with weight W can sometimes be modeled by $L = 27.4W^{1/3}$, where L is in inches and W is in pounds. Use this formula to estimate the weight of a bird that has a wing span of 30 inches. (*Source:* C. Pennycuick, Newton Rules Biology.)

CHAPTER 10 Extended and Discovery Exercises

1. *Modeling Wood in a Tree* Forestry services estimate the volume of timber in a given area of forest. To make such estimates, scientists have developed formulas to find the amount of wood contained in a tree with height h in feet and diameter d in inches. One study concluded that the volume V of wood in cubic feet in a tree is given by $V = kh^{1.12}d^{1.98}$, where k is a constant. Note that the diameter is measured 4.5 feet above the ground. (*Source:* B. Ryan, B. Joiner, and T. Ryan, *Minitab Handbook.*)

(a) A tree with an 11-inch diameter and a 47-foot height has a volume of 11.4 cubic feet. Approximate the constant k.

(b) Estimate the volume of wood in the same type of tree with $d = 20$ inches and $h = 105$ feet.

2. *Area of Skin* The surface area of the skin covering the human body is a function of more than one variable. Both height and weight influence the surface area of a person's body. Hence a taller person tends to have a larger surface area, as does a heavier person. A formula to determine the area of a person's skin in square inches is $S = 15.7w^{0.425}h^{0.725}$, where w is weight in pounds and h is height in inches. (*Source:* H. Lancaster, *Quantitative Methods in Biological and Medical Sciences.*)

(a) Use S to estimate the area of a person's skin who is 65 inches tall and weighs 154 pounds.

(b) If a person's weight doubles, what happens to the area of the person's skin? Explain.

(c) If a person's height doubles, what happens to the area of the person's skin? Explain.

3. *Minimizing Cost* A natural gas line running along a river is to be connected from point A to a cabin on the other bank located at point D, as illustrated in the figure. The width of the river is 500 feet, and the distance from point A to point C is 1000 feet. The cost of running the pipe along the shoreline is \$30 per foot, and the cost of running it underwater is \$50 per foot. The cost of connecting the gas line from A to D is to be minimized.

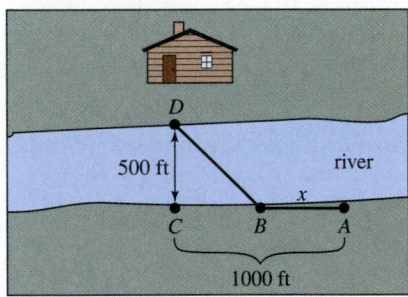

(a) Write an expression that gives the cost of running the line from A to B if the distance between these points is x feet.

(b) Find the distance from B to D in terms of x.

(c) Write an expression that gives the cost of running the line from B to D.

(d) Use your answer from parts (a) and (c) to write an expression that gives the cost of running the line from A to B to D.

(e) Graph your expression from part (d) in the window [0, 1000, 100] by [40000, 60000, 5000] to determine the value of x that minimizes the cost of the line going from A to D. What is the minimum cost?

CHAPTERS 1–10   Cumulative Review Exercises

1. Evaluate $S = 4\pi r^2$ when $r = 3$.

2. Identify the domain and range of the relation given by $S = \{(-1, 2), (0, 4), (1, 2)\}$.

3. Simplify each expression. Write the result using positive exponents.

   (a) $\left(\dfrac{ab^2}{b^{-1}}\right)^{-3}$ (b) $\dfrac{(x^2y)^3}{x^2(y^2)^{-3}}$

   (c) $(rt)^2(r^2t)^3$

4. Write 0.00043 in scientific notation.

5. Find $f(3)$ if $f(x) = \dfrac{x}{x-2}$. What is the domain of f?

6. Find the domain of $f(x) = \sqrt[3]{2x}$.

Exercises 7 and 8: Graph f by hand.

7. $f(x) = 1 - 2x$

8. $f(x) = \sqrt{x+1}$

9. Sketch a graph of $f(x) = x^2 - 4$.
   (a) Identify the domain and range.
   (b) Evaluate $f(-2)$.
   (c) Identify the x-intercepts.
   (d) Solve the equation $f(x) = 0$.

10. Find the slope–intercept form of the line that passes through $(-1, 2)$ and is perpendicular to $y = -2x$.

11. Let f be a linear function. Find a formula for $f(x)$.

x	-2	-1	1	2
$f(x)$	7	4	-2	-5

12. Sketch a graph of a line passing through $(-1, 1)$ with slope $-\frac{1}{2}$.

13. Solve $5x - (3 - x) = \frac{1}{2}x$.

14. Solve $2x - 5 \le 4 - x$.

15. Solve $|x - 2| \le 3$.

16. Solve $-1 \le 1 - 2x \le 6$.

17. Solve each system symbolically, if possible.
    (a) $2x - y = 4$ (b) $3x - 4y = 2$
        $x + y = 8$ $x - \frac{4}{3}y = 1$

18. Shade the solution set in the xy-plane.
    $$x + 2y \le 2$$
    $$-x + 3y \ge 3$$

19. Solve the system.
    $$x + 2y - z = 6$$
    $$x - 3y + z = -2$$
    $$x + y + z = 6$$

20. Calculate $\det \begin{bmatrix} 4 & 2 & -1 \\ 2 & 1 & 0 \\ 0 & -2 & 1 \end{bmatrix}$.

Exercises 21–24: Multiply the expression.

21. $4x(4 - x^3)$ 22. $(x - 4)(x + 4)$

23. $(5x + 3)(x - 2)$ 24. $(4x + 9)^2$

Exercises 25–30: Factor the expression.

25. $9x^2 - 16$ 26. $x^2 - 4x + 4$

27. $15x^3 - 9x^2$ 28. $12x^2 - 5x - 3$

29. $r^3 - 1$ 30. $x^3 - 3x^2 + 5x - 15$

Exercises 31 and 32: Solve each equation.

31. $x^2 - 3x + 2 = 0$ 32. $x^3 = 4x$

Exercises 33 and 34: Simplify the expression.

33. $\dfrac{x^2 + 3x + 2}{x - 3} \div \dfrac{x + 1}{2x - 6}$

34. $\dfrac{2}{x - 1} + \dfrac{5}{x}$

Exercises 35–42: Simplify the expression. Assume all variables are positive.

35. $\sqrt{36x^2}$ 36. $\sqrt[3]{64}$

37. $16^{-3/2}$ 38. $\sqrt[4]{625}$

39. $\sqrt{2x} \cdot \sqrt{8x}$ 40. $\sqrt{x} \cdot \sqrt[4]{x}$

41. $\dfrac{\sqrt[3]{16x^4}}{\sqrt[3]{2x}}$ 42. $4\sqrt{12x} - 2\sqrt{3x}$

43. Multiply $(2x + \sqrt{3})(x - \sqrt{3})$.

44. Find the domain of $f(x) = \sqrt{1 - x}$.

45. Graph $f(x) = 3\sqrt[3]{x}$.

46. Find the distance between $(-2, 3)$ and $(1, 2)$.

47. Write $(1 - i)(2 + 3i)$ in standard form.

48. The lengths of the legs of a right triangle are 5 and 12. What is the length of the hypotenuse?

Exercises 49–52: Solve symbolically.

49. $2\sqrt{x + 3} = x$ **50.** $\sqrt[3]{x - 1} = 3$

51. $\sqrt{x + 4} = 2\sqrt{x + 5}$ **52.** $\frac{1}{3}x^4 = 27$

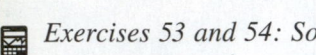

Exercises 53 and 54: Solve graphically and approximate your answer to the nearest hundredth.

53. $\sqrt[3]{x^2 - 2} + x = \sqrt{x}$ **54.** $x - \sqrt[3]{x} = \sqrt{x + 2}$

APPLICATIONS

55. *Calculating Water Flow* The gallons G of water in a tank after t minutes are given by $G(t) = 300 - 15t$. Interpret the y-intercept and the slope of the graph of G as a rate of change.

56. *Distance* A person jogs away from home at 8 miles per hour for 1 hour, rests in a park for 2 hours, and then walks back home at 4 miles per hour. Sketch a graph that models the distance y that the person is from home after x hours.

57. *Investment* Suppose $2000 is deposited in two accounts paying 5% and 4% annual interest. If the total interest after 1 year is $93, how much is invested at each rate?

58. *Geometry* A rectangular box with a square base has a volume of 256 cubic inches. If its height is 4 inches less than the length of an edge of the base, find the dimensions of the box.

59. *Height of a Building* A 5-foot 3-inch person casts a shadow that is 7.5 feet while a nearby building casts a 32-foot shadow. Find the height of the building.

60. *Angles in a Triangle* The measure of the largest angle in a triangle is 20° less than the sum of the measures of the two smaller angles. The sum of the measures of the two larger angles is 90° greater than the measure of the smaller angle. Find the measure of each angle.

11 Quadratic Functions and Equations

> There is no branch of mathematics, however abstract, which may not some day be applied to the real world.
>
> —NIKOLAI LOBACHEVSKY

W hat size television should you buy? Should you buy a 32-inch screen or a 50-inch screen? According to a home entertainment article in *Money* magazine, the answer depends on how far you sit from your television. The farther you sit from your television, the larger it should be. For example, if you sit only 6 feet from the screen, then a 32-inch television would be adequate; if you sit 10 feet from the screen, then a 50-inch television is more appropriate.

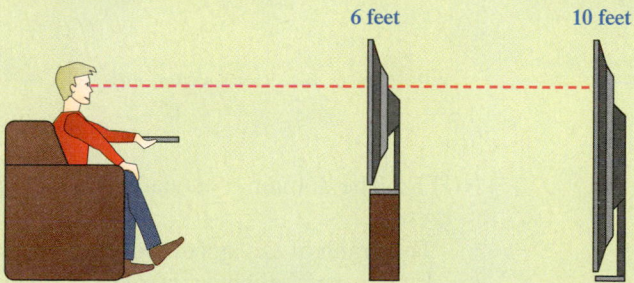

We can use a *quadratic function* to calculate the size S of the television screen needed for a person who sits x feet from the screen. You might want to use S to determine the size of screen that is recommended for you. (See Exercises 97 and 98 in Section 11.1.) Quadratic functions are a special type of polynomial function that occur frequently in applications involving economics, road construction, falling objects, geometry, and modeling real-world data. In this chapter we discuss quadratic functions and equations.

Source: Money, January 2007, p. 107; hdguru.com.

11.1   Quadratic Functions and Their Graphs

Graphs of Quadratic Functions • Min–Max Applications • Basic Transformations of $y = ax^2$ • Transformations of $y = ax^2 + bx + c$ (Optional)

A LOOK INTO MATH ▶

Suppose that a hotel is considering giving a group discount on room rates. The regular price is $80, but for each room rented the price decreases by $2. On the one hand, if the hotel rents one room, it makes only $78. On the other hand, if the hotel rents 40 rooms, the rooms are all free and the hotel makes nothing. Is there an optimal number of rooms between 1 and 40 that should be rented to maximize the revenue from the group?

In Figure 11.1 the hotel's revenue is graphed. From the graph it is apparent that "peak" revenue occurs when 20 rooms are rented. Obviously, this graph is not described by a linear function; rather, in Example 7 we see that it is described by a *quadratic function*.

NEW VOCABULARY

☐ Quadratic function
☐ Vertex
☐ Axis of symmetry
☐ Reflection
☐ Vertical shift

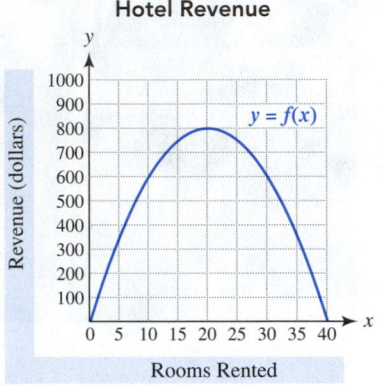

Figure 11.1

Graphs of Quadratic Functions

In Section 8.4 we discussed how a quadratic function could be represented by a polynomial of degree 2. We now give an alternative definition of a quadratic function.

> **QUADRATIC FUNCTION**
>
> A **quadratic function** can be written in the form
> $$f(x) = ax^2 + bx + c,$$
> where a, b, and c are constants with $a \neq 0$.

NOTE: The domain of a quadratic function is all real numbers.

READING CHECK

How can you identify the vertex and the axis of symmetry on the graph of a parabola?

The graph of *any* quadratic function is a *parabola*. Recall that a parabola is a ∪-shaped graph that opens either upward or downward. The graph of the simple quadratic function $y = x^2$ is a parabola that opens upward, with its *vertex* located at the origin, as shown in Figure 11.2(a). The **vertex** is the *lowest* point on the graph of a parabola that opens upward and the *highest* point on the graph of a parabola that opens downward. A parabola opening downward is shown in Figure 11.2(b). Its vertex is the point $(0, 2)$ and is the highest point on the graph. If we were to fold the xy-plane along the y-axis, or the line $x = 0$, the left and right sides of the graph would match. That is, the graph is symmetric with respect to the line $x = 0$. In this case the line $x = 0$ is the **axis of symmetry** for the graph. Figure 11.2(c) shows a parabola that opens upward with vertex $(2, -1)$ and axis of symmetry $x = 2$.

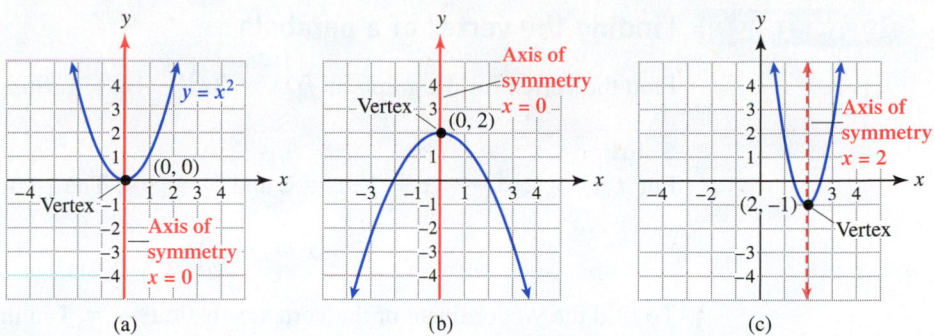

Figure 11.2

EXAMPLE 1   **Identifying the vertex and the axis of symmetry**

Use the graph of the quadratic function to identify the vertex, the axis of symmetry, and whether the parabola opens upward or downward.

(a)

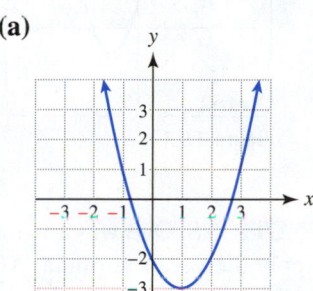

Figure 11.3

(b)

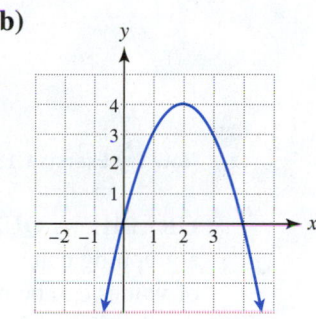

Figure 11.4

Solution

(a) The vertex is the lowest point on the graph shown in Figure 11.3, and its coordinates are $(1, -3)$. The axis of symmetry is the vertical line passing through the vertex, so its equation is $x = 1$. The parabola opens upward.

(b) The vertex is the highest point on the graph shown in Figure 11.4, and its coordinates are $(2, 4)$. The axis of symmetry is the vertical line passing through the vertex, so its equation is $x = 2$. The parabola opens downward.

Now Try Exercises 29, 31

THE VERTEX FORMULA In order to graph a parabola by hand, it is helpful to know the location of the vertex. The following formula can be used to find the coordinates of the vertex for *any* parabola. This formula is derived by *completing the square*, a technique discussed later in the next section.

READING CHECK

How do you use the vertex formula to find the coordinates of the vertex?

VERTEX FORMULA

The x-coordinate of the vertex of the graph of $y = ax^2 + bx + c$, $a \neq 0$, is given by

$$x = -\frac{b}{2a}.$$

To find the y-coordinate of the vertex, substitute this x-value in $y = ax^2 + bx + c$.

NOTE: The equation of the axis of symmetry for $f(x) = ax^2 + bx + c$ is $x = -\frac{b}{2a}$, and the vertex is the point $\left(-\frac{b}{2a}, f\left(-\frac{b}{2a}\right)\right)$.

EXAMPLE 2 **Finding the vertex of a parabola**

Find the vertex for the graph of $f(x) = 2x^2 - 4x + 1$. Support your answer graphically.

Solution

For $f(x) = 2x^2 - 4x + 1$, $a = 2$ and $b = -4$. The x-coordinate of the vertex is

$$x = -\frac{b}{2a} = -\frac{(-4)}{2(2)} = 1. \qquad \text{\textit{x}-coordinate of vertex}$$

To find the y-coordinate of the vertex, substitute $x = 1$ in the given formula.

$$f(1) = 2(1)^2 - 4(1) + 1 = -1 \qquad \text{\textit{y}-coordinate of vertex}$$

Thus the vertex is located at $(1, -1)$, which is supported by Figure 11.5.

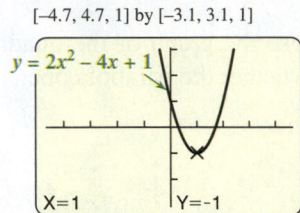

[−4.7, 4.7, 1] by [−3.1, 3.1, 1]

$y = 2x^2 - 4x + 1$

X=1   Y=−1

Figure 11.5

■ **Now Try Exercise 15**

GRAPHING QUADRATIC FUNCTIONS One way to graph a quadratic function without a graphing calculator is to first apply the vertex formula. After the vertex is located, a table of values can be made to locate points on either side of the vertex. This technique is used in the next example to graph three quadratic functions.

EXAMPLE 3 **Graphing quadratic functions**

Identify the vertex and axis of symmetry on the graph of $y = f(x)$. Graph $y = f(x)$.

(a) $f(x) = x^2 - 1$ **(b)** $f(x) = -(x + 1)^2$ **(c)** $f(x) = x^2 + 4x + 3$

Solution

(a) Begin by applying the vertex formula with $a = 1$ and $b = 0$ to locate the vertex.

$$x = -\frac{b}{2a} = -\frac{0}{2(1)} = 0 \qquad \text{\textit{x}-coordinate of vertex}$$

The y-coordinate of the vertex is found by evaluating $f(0)$.

$$y = f(0) = 0^2 - 1 = -1 \qquad \text{\textit{y}-coordinate of vertex}$$

Thus the coordinates of the vertex are $(0, -1)$. Table 11.1 is made by finding points on either side of the vertex. Plotting these points and connecting a smooth ∪-shaped graph results in Figure 11.6. The axis of symmetry is the vertical line passing through the vertex, so its equation is $x = 0$.

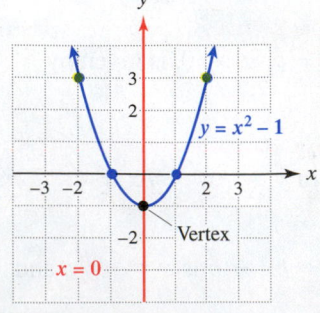

Figure 11.6

TABLE 11.1

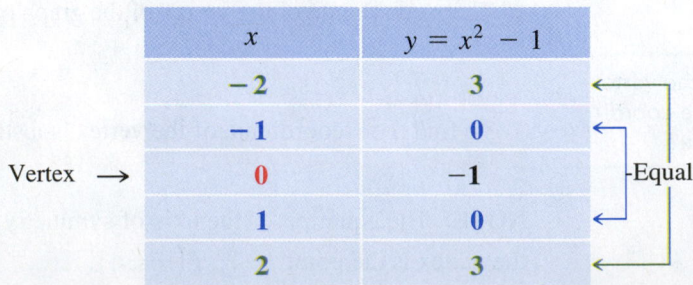

	x	$y = x^2 - 1$	
	-2	3	
	-1	0	
Vertex →	0	-1	Equal
	1	0	
	2	3	

(b) Before we can apply the vertex formula, we need to determine values for a and b by multiplying the expression for $f(x) = -(x + 1)^2$.

$$-(x + 1)^2 = -(x^2 + 2x + 1) \qquad \text{Multiply.}$$
$$= -x^2 - 2x - 1 \qquad \text{Distribute the negative sign.}$$

Substitute $a = -1$ and $b = -2$ into the vertex formula.

$$x = -\frac{b}{2a} = -\frac{(-2)}{2(-1)} = -1 \qquad \text{x-coordinate of vertex}$$

The y-coordinate of the vertex is found by evaluating $f(-1)$.

$$y = f(-1) = -(-1 + 1)^2 = 0 \qquad \text{y-coordinate of vertex}$$

Thus the coordinates of the vertex are $(-1, 0)$. Table 11.2 is made by finding points on either side of the vertex. Plotting these points and connecting a smooth ∩-shaped graph results in Figure 11.7. The axis of symmetry is the vertical line passing through the vertex, so its equation is $x = -1$.

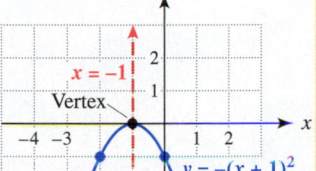

Figure 11.7

TABLE 11.2

	x	$y = -(x + 1)^2$	
	-3	-4	
	-2	-1	
Vertex →	-1	0	Equal
	0	-1	
	1	-4	

(c) The graph of $f(x) = x^2 + 4x + 3$ can be found in a manner similar to that used for the graphs in (a) and (b). See Table 11.3 and Figure 11.8. With $a = 1$ and $b = 4$, the vertex formula can be used to show that the vertex is located at $(-2, -1)$. The equation of the axis of symmetry is $x = -2$.

$y = x^2 + 4x + 3$

$x = -2$

Vertex

Figure 11.8

TABLE 11.3

	x	$y = x^2 + 4x + 3$	
	-5	8	
	-4	3	
	-3	0	
Vertex →	-2	-1	Equal
	-1	0	
	0	3	
	1	8	

Now Try Exercises 35, 41, 43

STUDY TIP

The vertex formula is important to memorize because it is often used when graphing parabolas.

Min–Max Applications

▶ **REAL-WORLD CONNECTION** Have you ever noticed that the first piece of pizza tastes really great and gives a person a lot of satisfaction? The second piece may be almost as good. After a few more pieces, there is often a point where the satisfaction starts to level off, and it can even go down as a person starts to eat too much. Eventually, if a person overeats, he or she might regret eating any pizza at all.

In Figure 11.9 a parabola models the total satisfaction received from eating x slices of pizza. Notice that the satisfaction level increases and then decreases. Although maximum satisfaction would vary with the individual, it always occurs at the vertex. (This is a seemingly simple model, but it is nonetheless used frequently in economics to describe consumer satisfaction from buying x identical items.)

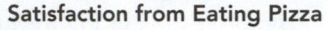

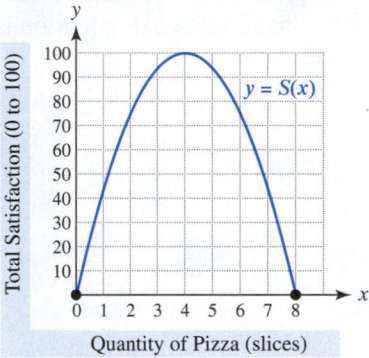

Figure 11.9

| EXAMPLE 4 | **Finding maximum satisfaction from pizza** |

The graph of $S(x) = -6.25x^2 + 50x$ is shown in Figure 11.9. Use the vertex formula to determine the number of slices of pizza that give maximum satisfaction. Does this maximum agree with what is shown in the graph of $y = S(x)$?

Solution
The parabola in Figure 11.9 opens downward, so the greatest y-value occurs at the vertex. To find the x-coordinate of the vertex, we can apply the vertex formula. Substitute $a = \mathbf{-6.25}$ and $b = \mathbf{50}$ into the vertex formula.

$$x = -\frac{b}{2a} = -\frac{\mathbf{50}}{2(\mathbf{-6.25})} = \mathbf{4} \qquad \text{\textit{x}-coordinate of vertex}$$

Maximum satisfaction on a scale of 1 to 100 occurs when **4** slices of pizza are eaten, which agrees with Figure 11.9. (This number obviously varies from individual to individual.)

Now Try Exercise 81

READING CHECK

How do you determine if a parabola has a maximum y-value or a minimum y-value?

FINDING MIN–MAX The graph in Figure 11.9 is a parabola that opens downward and whose maximum y-value of 100 occurs at the vertex. When a quadratic function f is used to model real data, the y-coordinate of the vertex represents either a maximum value of $f(x)$ or a minimum value of $f(x)$. For example, Figure 11.10(a) shows a parabola that opens upward. The minimum y-value on this graph is **1** and occurs at the vertex $(2, \mathbf{1})$. Similarly, Figure 11.10(b) shows a parabola that opens downward. The maximum y-value on this graph is **3** and occurs at the vertex $(-3, \mathbf{3})$.

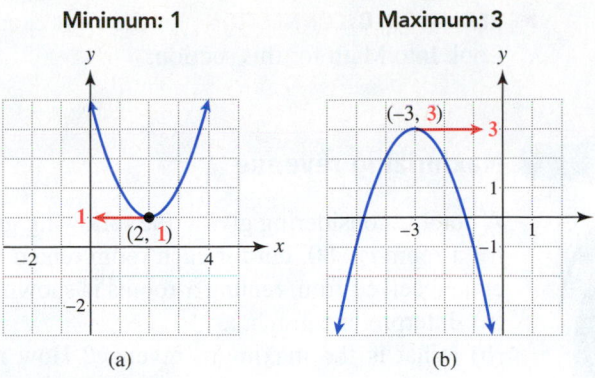

Figure 11.10

EXAMPLE 5 **Finding a minimum *y*-value**

Find the minimum *y*-value on the graph of $f(x) = x^2 - 4x + 3$.

Solution

To locate the minimum *y*-value on a parabola, first apply the vertex formula with $a = 1$ and $b = -4$ to find the *x*-coordinate of the vertex.

$$x = -\frac{b}{2a} = -\frac{(-4)}{2(1)} = 2 \qquad \text{x-coordinate of vertex}$$

The minimum *y*-value on the graph is found by evaluating $f(2)$.

$$y = f(2) = 2^2 - 4(2) + 3 = -1 \qquad \text{y-coordinate of vertex}$$

Thus the minimum *y*-coordinate on the graph of $y = f(x)$ is -1. This result is supported by Figure 11.11.

Now Try Exercise 53

Locating a Minimum y-Value

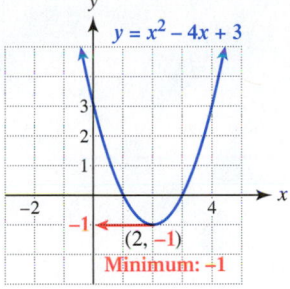

Figure 11.11

▶ **REAL-WORLD CONNECTION** In the next example, we demonstrate finding a maximum height reached by a baseball.

EXAMPLE 6 **Finding maximum height**

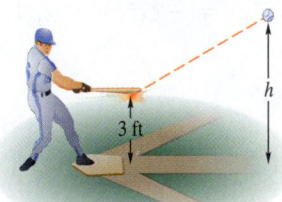

A baseball is hit into the air, and its height *h* in feet after *t* seconds can be calculated by $h(t) = -16t^2 + 96t + 3$.

(a) What is the height of the baseball when it is hit?

(b) Determine the maximum height of the baseball.

Solution

(a) The baseball is hit when $t = 0$, so $h(0) = -16(0)^2 + 96(0) + 3 = 3$ feet.

(b) The graph of *h* opens downward because $a = -16 < 0$. Thus the maximum height of the baseball occurs at the vertex. To find the vertex, we apply the vertex formula with $a = -16$ and $b = 96$ because $h(t) = -16t^2 + 96t + 3$.

$$t = -\frac{b}{2a} = -\frac{96}{2(-16)} = 3 \text{ seconds}$$

The maximum height of the baseball occurs at $t = 3$ seconds and is

$$h(3) = -16(3)^2 + 96(3) + 3 = 147 \text{ feet}.$$

Now Try Exercise 87

▶ **REAL-WORLD CONNECTION** In the next example, we answer the question presented in A Look Into Math for this section.

EXAMPLE 7   Maximizing revenue

A hotel is considering giving the following group discount on room rates. The regular price for a room is $80, but for each room rented the price decreases by $2. A graph of the revenue received from renting x rooms is shown in Figure 11.12.

(a) Interpret the graph.
(b) What is the maximum revenue? How many rooms should be rented to receive the maximum revenue?
(c) Write a formula for $f(x)$ whose graph is shown in Figure 11.12.
(d) Use $f(x)$ to determine symbolically the maximum revenue and the number of rooms that should be rented.

Hotel Revenue

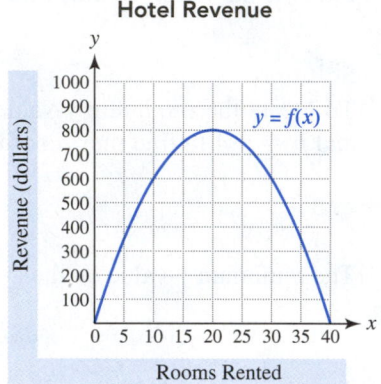

Figure 11.12

Solution

(a) The revenue increases at first, reaches a maximum (which corresponds to the vertex), and then decreases.

(b) In Figure 11.12 the vertex is (**20**, **800**). Thus the maximum revenue of **$800** occurs when **20** rooms are rented.

(c) If x rooms are rented, the price for each room is $80 - 2x$. The revenue equals the number of rooms rented times the price of each room. Thus $f(x) = x(80 - 2x)$.

(d) First, multiply $x(80 - 2x)$ to obtain $80x - 2x^2$ and then let $f(x) = -2x^2 + 80x$. The x-coordinate of the vertex is

$$x = -\frac{b}{2a} = -\frac{80}{2(-2)} = \mathbf{20}.$$

The y-coordinate is $f(\mathbf{20}) = -2(\mathbf{20})^2 + 80(\mathbf{20}) = \mathbf{800}$. These calculations verify our results in part (b).

▌ **Now Try Exercise 91**

TECHNOLOGY NOTE

Locating a Vertex
Graphing calculators can locate a vertex with the MAXIMUM or MINIMUM utility. The maximum in Example 7 is found in the figure.

[0, 50, 10] by [0, 1000, 100]

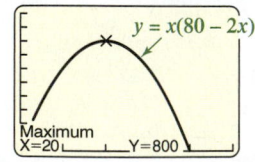

CALCULATOR HELP

To find a minimum or maximum, see Appendix A (page AP-11).

Basic Transformations of $y = ax^2$

THE GRAPH OF $y = ax^2$, $a > 0$ First, graph $y_1 = \frac{1}{2}x^2$, $y_2 = x^2$, and $y_3 = 2x^2$, as shown in Figure 11.13(a). Note that $a = \frac{1}{2}$, $a = \mathbf{1}$, and $a = \mathbf{2}$, respectively, and that as a increases, the resulting parabola becomes narrower. The graph of $y_1 = \frac{1}{2}x^2$ is wider than the graph of $y_2 = x^2$, and the graph of $y_3 = 2x^2$ is narrower than the graph of $y_2 = x^2$. In general, the graph of $y = ax^2$ is wider than the graph of $y = x^2$ when $0 < a < 1$ and narrower than the graph of $y = x^2$ when $a > 1$. When $a > 0$, the graph of $y = ax^2$ opens upward and never lies *below* the x-axis.

THE GRAPH OF $y = ax^2$, $a < 0$ When $a < 0$, the graph of $y = ax^2$ never lies *above* the x-axis because, for any input x, the product $ax^2 \le 0$. The graphs of $y_4 = -\frac{1}{2}x^2$, $y_5 = -x^2$, and $y_6 = -2x^2$ are shown in Figure 11.13(b) and open downward. The graph of $y_4 = -\frac{1}{2}x^2$ is wider than the graph of $y_5 = -x^2$ and the graph of $y_6 = -2x^2$ is narrower than the graph of $y_5 = -x^2$.

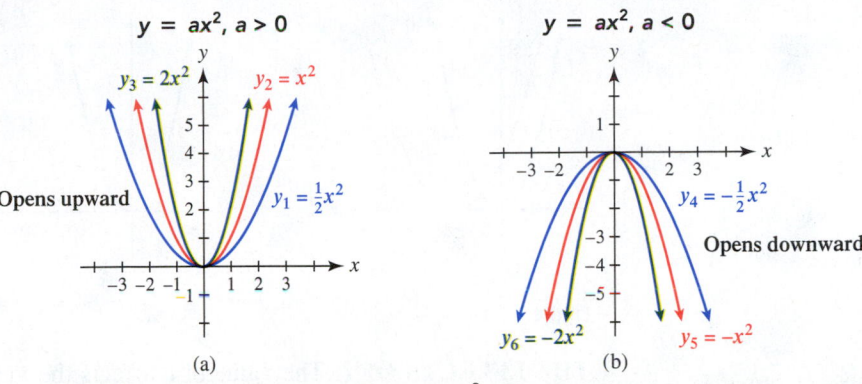

Figure 11.13 The Effect of a on $y = ax^2$

THE GRAPH OF $y = ax^2$

The graph of $y = ax^2$ is a parabola with the following characteristics.

1. The vertex is $(0, 0)$, and the axis of symmetry is given by $x = 0$.
2. It opens upward if $a > 0$ and opens downward if $a < 0$.
3. It is wider than the graph of $y = x^2$, if $0 < |a| < 1$. It is narrower than the graph of $y = x^2$, if $|a| > 1$.

Reflection Across the x-Axis

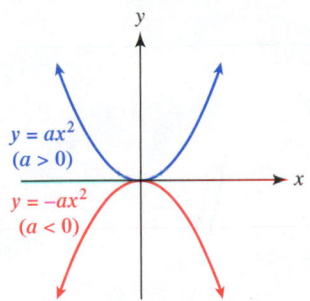

$y = ax^2$
$(a > 0)$

$y = -ax^2$
$(a < 0)$

Figure 11.14

REFLECTIONS OF $y = ax^2$ In Figure 11.13 the graph of $y_1 = \frac{1}{2}x^2$ can be *transformed* into the graph of $y_4 = -\frac{1}{2}x^2$ by *reflecting* it across the x-axis. The graph of y_4 is a **reflection** of the graph of y_1 across the x-axis. In general, *the graph of $y = -ax^2$ is a reflection of the graph of $y = ax^2$ across the x-axis*, as shown in Figure 11.14. That is, if we folded the xy-plane along the x-axis the two graphs would match.

EXAMPLE 8   **Graphing $y = ax^2$**

Compare the graph of $g(x) = -3x^2$ to the graph of $f(x) = x^2$. Then graph both functions on the same coordinate axes.

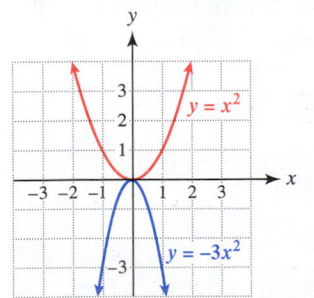

Figure 11.15

Solution
Both graphs are parabolas. However, the graph of g opens downward and is narrower than the graph of f. These graphs are shown in Figure 11.15.

Now Try Exercise 67

Transformations of $y = ax^2 + bx + c$ (Optional)

The graph of a quadratic function is a parabola, and any quadratic function f can be written as $f(x) = ax^2 + bx + c$, where a, b, and c are constants with $a \ne 0$. As a result, the values of a, b, and c determine both the shape and position of the parabola in the xy-plane. In this subsection, we summarize some of the effects that these constants have on the graph of f.

THE EFFECTS OF a The effects of a on the graph of $y = ax^2$ were already discussed, and these effects can be generalized to include the graph of $y = ax^2 + bx + c$.
1. **Width:** The graph of $f(x) = ax^2 + bx + c$ is wider than the graph of $y = x^2$ if $0 < |a| < 1$ and narrower if $|a| > 1$. See Figures 11.16(a) and (b).
2. **Opening:** The graph of $f(x) = ax^2 + bx + c$ opens upward if $a > 0$ and downward if $a < 0$. See Figure 11.16(c).

The Effects of a on $y = ax^2 + bx + c$

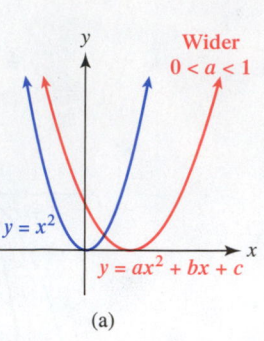 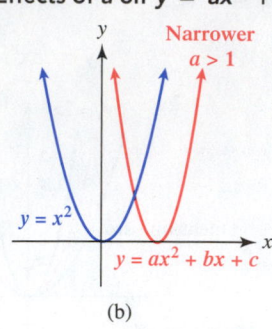 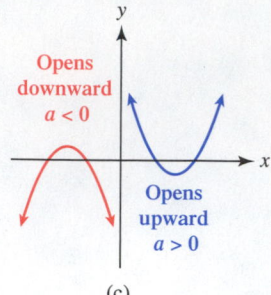

(a)   (b)   (c)

Figure 11.16

THE EFFECTS OF c The value of c affects the vertical placement of the parabola in the xy-plane. This placement is often called **vertical shift**.
1. **y-Intercept:** Because $f(0) = a(0)^2 + b(0) + c = c$, it follows that the y-intercept for the graph of $y = f(x)$ is c. See Figure 11.17(a).
2. **Vertical Shift:** The graph of $f(x) = ax^2 + bx + c$ is shifted vertically c units compared to the graph of $y = ax^2 + bx$. If $c < 0$, the shift is downward; if $c > 0$, the shift is upward. See Figures 11.17(b) and (c). The parabolas in both figures have identical shapes.

The Effects of c on $y = ax^2 + bx + c$

The Combined Effects of a and b

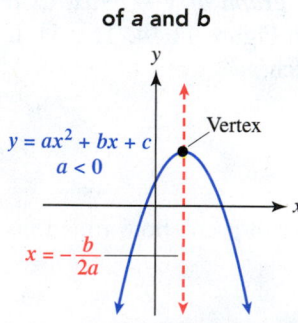

Figure 11.18

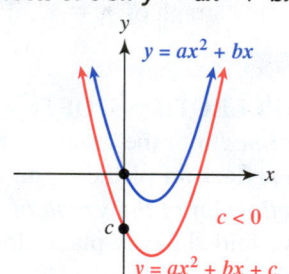 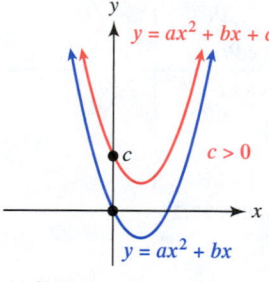

(a) y-intercept: c (b) Shifted downward: $c < 0$ (c) Shifted upward: $c > 0$

Figure 11.17

THE COMBINED EFFECTS OF a AND b The combined values of a and b determine the x-coordinate of the vertex and the equation of the axis of symmetry.
1. **Vertex:** The x-coordinate of the vertex is $-\frac{b}{2a}$.
2. **Axis of Symmetry:** The axis of symmetry is given by $x = -\frac{b}{2a}$.

Figure 11.18 illustrates these concepts.

EXAMPLE 9   **Analyzing the graph of $y = ax^2 + bx + c$**

Let $f(x) = -\frac{1}{2}x^2 + x + \frac{3}{2}$.
(a) Does the graph of f open upward or downward? Is this graph wider or narrower than the graph of $y = x^2$?
(b) Find the axis of symmetry and the vertex.
(c) Find the y-intercept and any x-intercepts.
(d) Sketch a graph of f.

Solution

(a) If $f(x) = -\frac{1}{2}x^2 + x + \frac{3}{2}$, then $a = -\frac{1}{2}$, $b = 1$, and $c = \frac{3}{2}$. Because $a = -\frac{1}{2} < 0$, the parabola opens downward. Also, because $0 < |a| < 1$, the graph is wider than the graph of $y = x^2$.

(b) The axis of symmetry is $x = -\dfrac{b}{2a} = -\dfrac{1}{2\left(-\frac{1}{2}\right)} = 1$, or $x = 1$. Because

$$f(1) = -\frac{1}{2}(1)^2 + (1) + \frac{3}{2} = -\frac{1}{2} + 1 + \frac{3}{2} = 2,$$

the vertex is $(1, 2)$.

(c) The y-intercept equals c, or $\frac{3}{2}$. To find x-intercepts we let $y = 0$ and solve for x.

$$-\frac{1}{2}x^2 + x + \frac{3}{2} = 0 \qquad \text{Equation to be solved}$$

$$x^2 - 2x - 3 = 0 \qquad \text{Multiply by –2; clear fractions.}$$

$$(x + 1)(x - 3) = 0 \qquad \text{Factor.}$$

$$x + 1 = 0 \quad \text{or} \quad x - 3 = 0 \qquad \text{Zero-product property}$$

$$x = -1 \quad \text{or} \quad x = 3 \qquad \text{Solve.}$$

The x-intercepts are -1 and 3.

(d) Start by plotting the vertex and intercepts, as shown in Figure 11.19. Then sketch a smooth, $\cap$-shaped graph that connects these points.

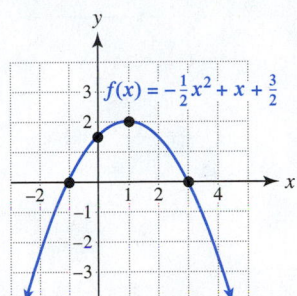

Figure 11.19

Now Try Exercise 75

11.1 Putting It All Together

CONCEPT	EXPLANATION	EXAMPLES
Quadratic Function	Can be written as $f(x) = ax^2 + bx + c$, $a \neq 0$	$f(x) = x^2 + x - 2$ and $g(x) = -2x^2 + 4$ $(b = 0)$
Vertex of a Parabola	The x-coordinate of the vertex for the function $f(x) = ax^2 + bx + c$ with $a \neq 0$ is given by $$x = -\frac{b}{2a}.$$ The y-coordinate of the vertex is found by substituting this x-value in the equation. Hence the vertex is $\left(-\frac{b}{2a}, f\left(-\frac{b}{2a}\right)\right)$.	If $f(x) = -2x^2 + 8x - 7$, then $$x = -\frac{8}{2(-2)} = 2$$ and $$f(2) = -2(2)^2 + 8(2) - 7 = 1.$$ The vertex is $(2, 1)$. The graph of f opens downward because $a < 0$.
Graph of a Quadratic Function	Its graph is a parabola that opens upward if $a > 0$ and downward if $a < 0$. The value of $|a|$ affects the width of the parabola. The vertex can be used to determine the maximum or minimum output of a quadratic function.	The graph of $y = -\frac{1}{4}x^2$ opens downward and is wider than the graph of $y = x^2$, as shown in the figure. Each graph has its vertex at $(0, 0)$.

continued on next page

continued from previous page

The following gives symbolic, numerical, and graphical representations for the quadratic function f that *squares the input x and then subtracts* 1.

Symbolic Representation

$$f(x) = x^2 - 1$$

Numerical Representation

x	$f(x) = x^2 - 1$
-2	3
-1	0
0	-1
1	0
2	3

Vertex →

Equal

Graphical Representation

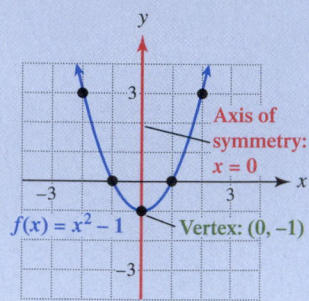

Axis of symmetry: $x = 0$

$f(x) = x^2 - 1$ Vertex: $(0, -1)$

11.1 Exercises

CONCEPTS AND VOCABULARY

1. The graph of a quadratic function is called a(n) _____.

2. If a parabola opens upward, what is the lowest point on the parabola called?

3. If a parabola is symmetric with respect to the y-axis, the y-axis is called the _____.

4. The vertex on the graph of $y = x^2$ is _____.

5. Sketch a parabola that opens downward with a vertex of $(1, 2)$.

6. If $y = ax^2 + bx + c$, the x-coordinate of the vertex is given by $x = $ _____.

7. Compared to the graph of $y = x^2$, the graph of $y = 2x^2$ is (wider/narrower).

8. The graph of $y = -x^2$ is similar to the graph of $y = x^2$ except that it is _____ across the x-axis.

9. Any quadratic function can be written in the form $f(x) = $ _____.

10. If a parabola opens downward, the point with the largest y-value is called the _____.

11. (True or False?) The axis of symmetry for the graph of $y = ax^2 + bx + c$ is given by $x = -\frac{b}{2a}$.

12. (True or False?) If the vertex of a parabola is located at (a, b), then the axis of symmetry is given by $x = b$.

13. (True or False?) If a parabola opens downward and its vertex is (a, b), then the minimum y-value on the parabola is b.

14. (True or False?) The graph of $y = -ax^2$ with $a > 0$ opens downward.

VERTEX FORMULA

Exercises 15–24: Find the vertex of the parabola.

15. $f(x) = x^2 - 4x - 2$

16. $f(x) = 2x^2 + 6x - 3$

17. $f(x) = -\frac{1}{3}x^2 - 2x + 1$

18. $f(x) = 5 - 4x + x^2$

19. $f(x) = 3 - 2x^2$

20. $f(x) = \frac{1}{4}x^2 - 3x - 2$

21. $f(x) = -0.3x^2 + 0.6x + 1.1$

22. $f(x) = 25 - 10x + 20x^2$

23. $f(x) = 6x - x^2$

24. $f(x) = x - \frac{1}{2}x^2$

GRAPHS OF QUADRATIC FUNCTIONS

Exercises 25–28: Use the given graph of f to evaluate the expressions.

25. $f(-2)$ and $f(0)$ **26.** $f(-2)$ and $f(2)$

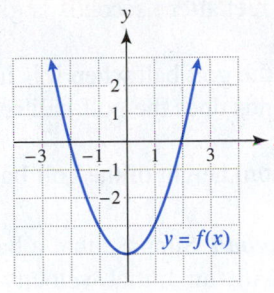   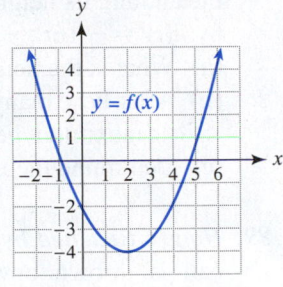

27. $f(-3)$ and $f(1)$ **28.** $f(-1)$ and $f(2)$

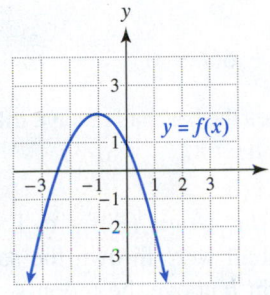   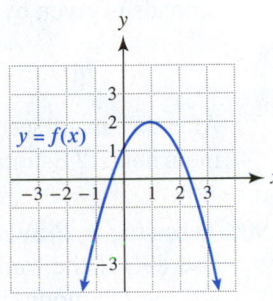

Exercises 29–32: Identify the vertex, axis of symmetry, and whether the parabola opens upward or downward.

29.   **30.**

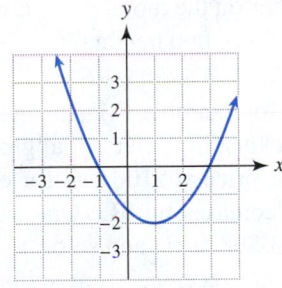   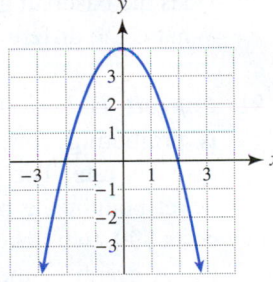

31.   **32.**

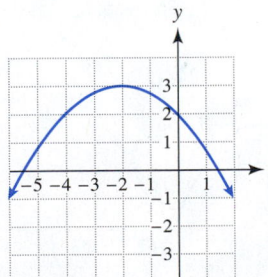   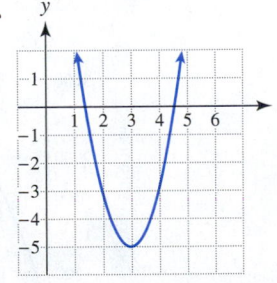

Exercises 33–52: Do the following for the given f(x).
(a) *Identify the vertex and axis of symmetry on the graph of y = f(x).*
(b) *Graph y = f(x).*
(c) *Evaluate f(−2) and f(3).*

33. $f(x) = \frac{1}{2}x^2$ **34.** $f(x) = -3x^2$

35. $f(x) = x^2 - 2$ **36.** $f(x) = x^2 - 1$

37. $f(x) = -3x^2 + 1$ **38.** $f(x) = \frac{1}{2}x^2 + 2$

39. $f(x) = (x - 1)^2$ **40.** $f(x) = (x + 2)^2$

41. $f(x) = -(x + 2)^2$ **42.** $f(x) = -(x - 1)^2$

43. $f(x) = x^2 + x - 2$ **44.** $f(x) = x^2 - 2x + 2$

45. $f(x) = 2x^2 - 3$ **46.** $f(x) = 1 - 2x^2$

47. $f(x) = 2x - x^2$ **48.** $f(x) = x^2 + 2x - 8$

49. $f(x) = -2x^2 + 4x - 1$

50. $f(x) = -\frac{1}{2}x^2 + 2x - 3$

51. $f(x) = \frac{1}{4}x^2 - x + 5$ **52.** $f(x) = 3 - 6x - 4x^2$

MIN–MAX

Exercises 53–58: Find the minimum y-value on the graph of y = f(x).

53. $f(x) = x^2 + 2x - 1$ **54.** $f(x) = x^2 + 6x + 2$

55. $f(x) = x^2 - 5x$ **56.** $f(x) = x^2 - 3x$

57. $f(x) = 2x^2 + 2x - 3$ **58.** $f(x) = 3x^2 - 3x + 7$

Exercises 59–64: Find the maximum y-value on the graph of y = f(x).

59. $f(x) = -x^2 + 2x + 5$

60. $f(x) = -x^2 + 4x - 3$

61. $f(x) = 4x - x^2$

62. $f(x) = 6x - x^2$

63. $f(x) = -2x^2 + x - 5$

64. $f(x) = -5x^2 + 15x - 2$

65. *Numbers* Find two positive numbers whose sum is 20 and whose product is maximum.

66. **Thinking Generally** Find two positive numbers whose sum is k and whose product is maximum.

TRANSFORMATIONS OF GRAPHS

Exercises 67–74: Graph f(x). Compare the graph of y = f(x) to the graph of y = x^2.

67. $f(x) = -x^2$ **68.** $f(x) = -2x^2$

69. $f(x) = 2x^2$ **70.** $f(x) = 3x^2$

71. $f(x) = \frac{1}{4}x^2$ **72.** $f(x) = \frac{1}{2}x^2$

73. $f(x) = -\frac{1}{2}x^2$ **74.** $f(x) = -\frac{3}{2}x^2$

Exercises 75–80: (Refer to Example 9.) Use the given $f(x)$ to complete the following.

(a) *Does the graph of f open upward or downward? Is this graph wider, narrower, or the same as the graph of $y = x^2$?*

(b) *Find the axis of symmetry and the vertex.*

(c) *Find the y-intercept and any x-intercepts.*

(d) *Sketch a graph of f.*

75. $f(x) = \frac{1}{2}x^2 + x - \frac{3}{2}$ **76.** $f(x) = -x^2 + 4x + 5$

77. $f(x) = 2x - x^2$ **78.** $f(x) = x - 2x^2$

79. $f(x) = 2x^2 + 2x - 4$ **80.** $f(x) = \frac{1}{2}x^2 - \frac{1}{2}x - 1$

APPLICATIONS

Exercises 81 and 82: Eating Pizza (Refer to Example 4.) Let $S(x)$ denote the satisfaction, on a scale from 0 to 100, from eating x pieces of pizza. Find the number of pieces that gives the maximum satisfaction.

81. $S(x) = -\frac{100}{9}x^2 + \frac{200}{3}x$

82. $S(x) = -16x^2 + 80x$

Exercises 83–86: Quadratic Models Match the physical situation with the graph (a.–d.) that models it best.

83. The height y of a stone thrown from ground level after x seconds

84. The number of people attending a popular movie x weeks after its opening

85. The temperature after x hours in a house when the furnace quits and then a repair person fixes it

86. U.S. population from 1800 to the present

a.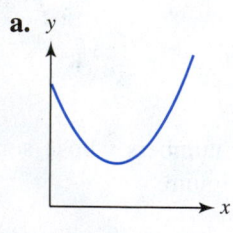

b.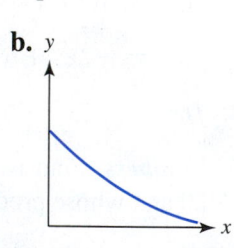

c.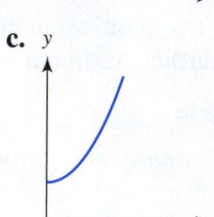

d.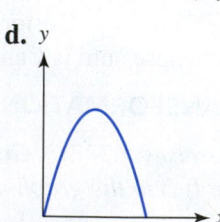

87. *Height Reached by a Baseball* (Refer to Example 6.) A baseball is hit into the air, and its height h in feet after t seconds is given by $h(t) = -16t^2 + 64t + 2$.

(a) What is the height of the baseball when it is hit?

(b) After how many seconds does the baseball reach its maximum height?

(c) Determine the maximum height of the baseball.

88. *Height Reached by a Golf Ball* A golf ball is hit into the air, and its height h in feet after t seconds is given by $h(t) = -16t^2 + 128t$.

(a) What is the height of the golf ball when it is hit?

(b) After how many seconds does the golf ball reach its maximum height?

(c) Determine the maximum height of the golf ball.

89. *Height Reached by a Baseball* Suppose that a baseball is thrown upward with an initial velocity of 66 feet per second (45 miles per hour) and it is released 6 feet above the ground. Its height h after t seconds is given by

$$h(t) = -16t^2 + 66t + 6.$$

After how many seconds does the baseball reach a maximum height? Estimate this height.

90. *Throwing a Baseball on the Moon* (Refer to Exercise 89.) If the same baseball were thrown the same way on the moon, its height h above the moon's surface after t seconds would be

$$h(t) = -2.55t^2 + 66t + 6.$$

Does the baseball go higher on the moon or on Earth? What is the difference in these two heights?

91. *Concert Tickets* (Refer to Example 7.) An agency is promoting concert tickets by offering a group-discount rate. The regular price is $100 and for each ticket bought the price decreases by $1. (One ticket costs $99, two tickets cost $98 *each*, and so on.)

(a) A graph of the revenue received from selling x tickets is shown in the figure. Interpret the graph.

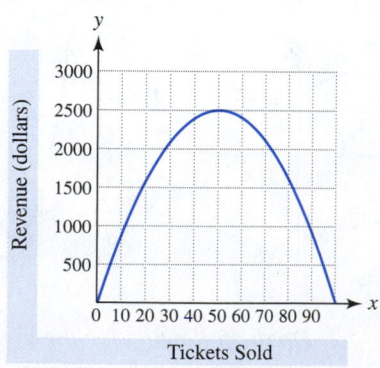

(b) What is the maximum revenue? How many tickets should be sold to maximize revenue?

(c) Write a formula for $y = f(x)$ whose graph is shown in the figure.

(d) Use $f(x)$ to determine symbolically the maximum revenue and the number of tickets that should be sold to maximize revenue.

92. *Maximizing Revenue* The regular price for a round-trip ticket to Las Vegas, Nevada, charged by an airline charter company is $300. For a group rate the company will reduce the price of each ticket by $1.50 for every passenger on the flight.

(a) Write a formula $f(x)$ that gives the revenue from selling x tickets.

(b) Determine how many tickets should be sold to maximize the revenue. What is the maximum revenue?

93. *Monthly Facebook Visitors* The number of *unique* monthly Facebook visitors in millions x years after 2006 can be modeled by

$$V(x) = 10.75x^2 - 24x + 35.$$

(a) Evaluate $V(1), V(2), V(3)$, and $V(4)$. Interpret your answer.

(b) Explain why numbers of Facebook visitors can not be modeled by a linear function.

94. *Cell Phone Complexity* Consumers often enjoy having certain features on their cell phones, such as e-mail and the ability to surf the Web. However, as phones become more and more complicated to operate, the benefits from the additional complexity start to decrease. The following parabola models this general situation. Interpret why a parabola is appropriate to model this consumer experience.

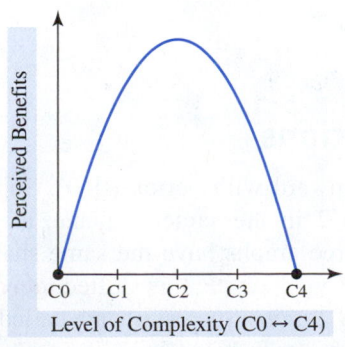

95. *Maximizing Area* A farmer is fencing a rectangular area for cattle and uses a straight portion of a river as one side of the rectangle, as illustrated in the figure. Note that there is no fence along the river. If the farmer has 1200 feet of fence, find the dimensions for the rectangular area that give a maximum area for the cattle.

River

x

96. *Maximizing Area* A rectangular pen being constructed for a pet requires 60 feet of fence.

(a) Write a formula $f(x)$ that gives the area of the pen if one side of the pen has length x.

(b) Find the dimensions of the pen that give the largest area. What is the largest area?

Exercises 97 and 98: Large-Screen Televisions (Refer to the introduction to this chapter.) Use the formula

$$S(x) = -0.227x^2 + 8.155x - 8.8,$$

where $6 \leq x \leq 16$, to estimate the recommended screen size in inches when viewers sit x feet from the screen.

97. $x = 8$ feet

98. $x = 12$ feet

99. *Carbon Emissions* Past and future carbon emissions in billions of metric tons during year x can be modeled by

$$C(x) = \frac{1}{300}x^2 - \frac{199}{15}x + \frac{39,619}{3},$$

where $1990 \leq x \leq 2020$. (*Source:* U.S. Department of Energy.)

(a) Evaluate $C(1990)$. Interpret your answer.

(b) Find the expected *increase* in carbon emissions from 1990 to 2020.

100. *Seedling Growth* In a study of the effect of temperature on the growth of melon seedlings, the seedlings were grown at different temperatures, and their heights were measured after a fixed period of time. The findings of this study can be modeled by

$$f(x) = -0.095x^2 + 5.4x - 52.2,$$

where x is the temperature in degrees Celsius and the output $f(x)$ gives the resulting average height in centimeters. (*Source:* R. Pearl, "The growth of *Cucumis melo* seedlings at different temperatures.")

 (a) Graph f in [20, 40, 5] by [0, 30, 5].
(b) Estimate graphically the temperature that resulted in the greatest height for the melon seedlings.
(c) Solve part (b) symbolically.

WRITING ABOUT MATHEMATICS

101. If $f(x) = ax^2 + bx + c$, explain how the values of a and c affect the graph of f.

102. Suppose that a quantity Q is modeled by the formula $Q(x) = ax^2 + bx + c$ with $a < 0$. Explain how to find the x-value that maximizes $Q(x)$. How do you find the maximum value of $Q(x)$?

11.2 Parabolas and Modeling

Vertical and Horizontal Translations • Vertex Form • Modeling with Quadratic Functions (Optional)

A LOOK INTO MATH ▶

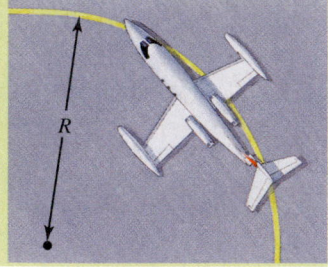

NEW VOCABULARY

☐ Translations
☐ Vertex form
☐ Completing the square

A taxiway used by an airplane to exit a runway often contains curves. A curve that is too sharp for the speed of the plane is a safety hazard. The scatterplot shown in Figure 11.20 gives an appropriate radius R of a curve designed for an airplane taxiing at x miles per hour. The data are nonlinear because they do not lie on a line. In this section we explain how a quadratic function may be used to model such data. First, we discuss translations of parabolas. (*Source:* FAA.)

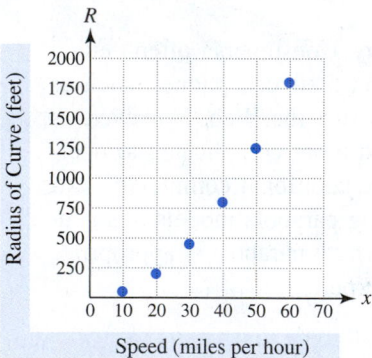

Figure 11.20

Vertical and Horizontal Translations

The graph of $y = x^2$ is a parabola opening upward with vertex $(0, 0)$. Suppose that we graph $y_1 = x^2$, $y_2 = x^2 + 1$, and $y_3 = x^2 - 2$ in the same xy-plane, as calculated in Table 11.4 and shown in Figure 11.21. All three graphs have the same shape. However, compared to the graph of $y_1 = x^2$, the graph of $y_2 = x^2 \mathbf{+ 1}$ is shifted ***upward* 1** unit and the graph of $y_3 = x^2 \mathbf{- 2}$ is shifted ***downward* 2** units. Such shifts are called **translations** because they do not change the shape of a graph—only its position.

Vertical Shifts of $y = x^2$

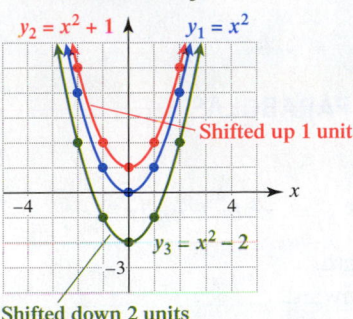

Figure 11.21

TABLE 11.4

x	$y_2 = x^2 + 1$	$y_1 = x^2$	$y_3 = x^2 - 2$
-2	5	4	2
-1	2	1	-1
0	1	0	-2
1	2	1	-1
2	5	4	2

The x-values do NOT change.   Add 1 to find the y-values.   Subtract 2 to find the y-values.

Next, suppose that we graph $y_1 = x^2$ and $y_2 = (x - 1)^2$ in the same xy-plane. Compare Tables 11.5 and 11.6. Note that the y-values are equal when the x-value for y_2 is 1 unit *larger* than the x-value for y_1. For example, $y_1 = 4$ when $x = -2$ and $y_2 = 4$ when $x = -1$. Thus the graph of $y_2 = (x - 1)^2$ has the same shape as the graph of $y_1 = x^2$ except that it is translated *horizontally to the right* 1 unit, as illustrated in Figure 11.22.

Horizontal Shift of $y = x^2$

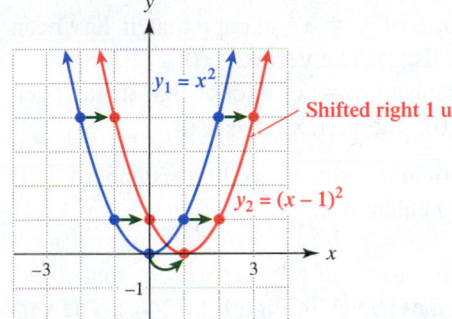

Figure 11.22

TABLE 11.5

x	$y_1 = x^2$
-2	4
-1	1
0	0
1	1
2	4

TABLE 11.6

x	$y_2 = (x - 1)^2$
-1	4
0	1
1	0
2	1
3	4

Add 1 to find the x-values.   The y-values do NOT change.

The graphs $y_1 = x^2$ and $y_2 = (x + 2)^2$ are shown in Figure 11.23. Note that Tables 11.7 and 11.8 show their y-values to be equal when the x-value for y_2 is 2 units *smaller* than the x-value for y_1. As a result, the graph of $y_2 = (x + 2)^2$ has the same shape as the graph of $y_1 = x^2$ except that it is translated *horizontally to the left* 2 units.

Horizontal Shift of $y = x^2$

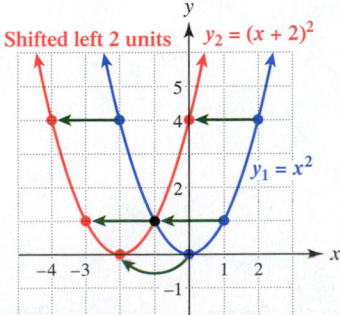

Figure 11.23

TABLE 11.7

x	$y_1 = x^2$
-2	4
-1	1
0	0
1	1
2	4

TABLE 11.8

x	$y_2 = (x + 2)^2$
-4	4
-3	1
-2	0
-1	1
0	4

Subtract 2 to find the x-values.   The y-values do NOT change.

These results are summarized as follows.

VERTICAL AND HORIZONTAL TRANSLATIONS OF PARABOLAS

Let h and k be positive numbers.

To graph	*shift the graph of $y = x^2$ by k units*
$y = x^2 + k$	upward.
$y = x^2 - k$	downward.

To graph	*shift the graph of $y = x^2$ by h units*
$y = (x - h)^2$	right.
$y = (x + h)^2$	left.

EXAMPLE 1 **Translating the graph $y = x^2$**

Sketch the graph of the equation and identify the vertex.

(a) $y = x^2 + 2$ **(b)** $y = (x + 3)^2$ **(c)** $y = (x - 2)^2 - 3$

Solution

(a) The graph of $y = x^2 + 2$ is similar to the graph of $y = x^2$ except that it has been translated *upward* 2 units, as shown in Figure 11.24(a). The vertex is $(0, 2)$.

(b) The graph of $y = (x + 3)^2$ is similar to the graph of $y = x^2$ except that it has been translated *left* 3 units, as shown in Figure 11.24(b). The vertex is $(-3, 0)$.

NOTE: If you are thinking that the graph should be shifted right (instead of left) 3 units, try graphing $y = (x + 3)^2$ on a graphing calculator.

(c) The graph of $y = (x - 2)^2 - 3$ is similar to the graph of $y = x^2$ except that it has been translated downward 3 units *and* right 2 units, as shown in Figure 11.24(c). The vertex is $(2, -3)$.

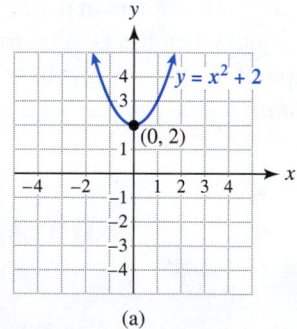

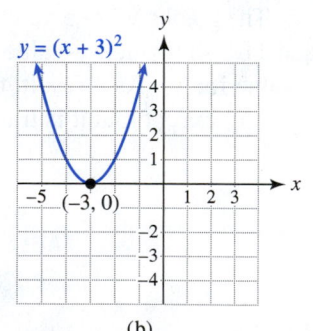

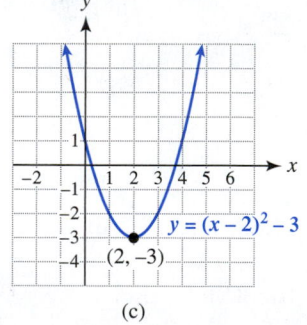

(a)   (b)   (c)

Figure 11.24

Now Try Exercises 15, 19, 27

Vertex Form

The graphs of $y = 2x^2$ and $y = 2x^2 - 12x + 20$ have *exactly* the same shape, as illustrated in Figures 11.25 and 11.26. However, the vertex for $y = 2x^2$ is $(0, 0)$, whereas the vertex for $y = 2x^2 - 12x + 20$ is $(3, 2)$.

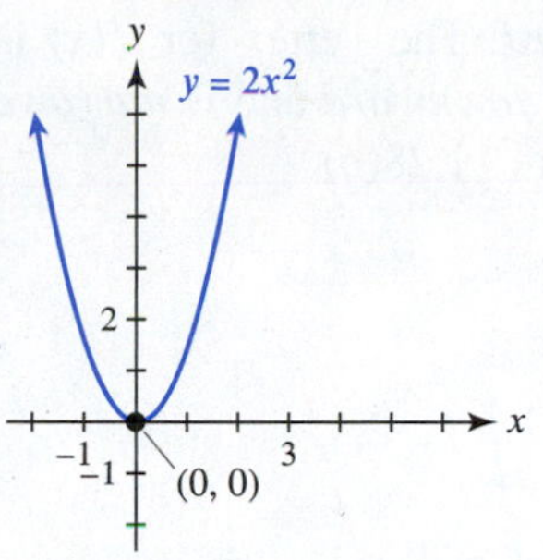

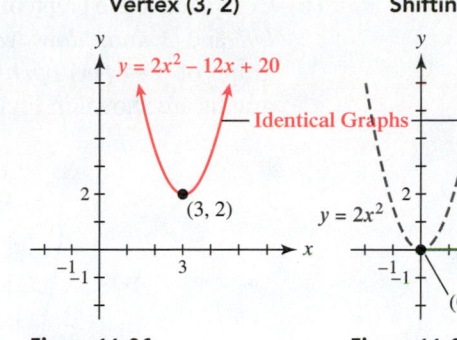

Figure 11.25

Figure 11.26

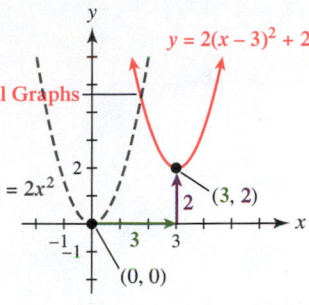

Figure 11.27

When the graph of $y = 2x^2$ (in Figure 11.25) is shifted **3** units right and **2** units upward by graphing the equation

$$y = 2(x - 3)^2 + 2, \quad \text{Vertex form}$$

Shift 2 units upward

Shift 3 units right

we get the graph in Figure 11.27, which is identical to the graph of $y = 2x^2 - 12x + 20$ in Figure 11.26. Thus $y = 2(x - 3)^2 + 2$ and $y = 2x^2 - 12x + 20$ are equivalent equations for the same parabola.

In general, the equation for any parabola can be written as either $y = ax^2 + bx + c$ or $y = a(x - h)^2 + k$, where the vertex is (h, k). The second form is sometimes called *vertex form*.

CRITICAL THINKING

Expand $y = 2(x - 3)^2 + 2$ and combine like terms. What does it equal?

READING CHECK

Give the vertex form. Explain it with an example.

VERTEX FORM

The **vertex form** of the equation of a parabola with vertex (h, k) is

$$y = a(x - h)^2 + k,$$

where $a \neq 0$ is a constant. If $a > 0$, the parabola opens upward; if $a < 0$, the parabola opens downward.

NOTE: Vertex form is sometimes called **standard form for a parabola with a vertical axis**.

In the next three examples, we demonstrate graphing parabolas in vertex form, finding their equations, and writing vertex forms of equations.

EXAMPLE 2   **Graphing parabolas in vertex form**

Compare the graph of $y = f(x)$ to the graph of $y = x^2$. Then sketch a graph of $y = f(x)$ and $y = x^2$ in the same xy-plane.

(a) $f(x) = \frac{1}{2}(x - 5)^2 + 2$ **(b)** $f(x) = -3(x + 5)^2 - 3$

Solution

(a) Compared to the graph of $y = x^2$, the graph of $y = f(x)$ is *translated 5 units right* and *2 units upward*. The vertex for $f(x)$ is $(5, 2)$, whereas the vertex of $y = x^2$ is $(0, 0)$. Because $a = \frac{1}{2}$, the graph of $y = f(x)$ *opens upward* and is *wider* than the graph of $y = x^2$. These graphs are shown in Figure 11.28(a) on the next page.

(b) Compared to the graph of $y = x^2$, the graph of $y = -3(x + 5)^2 - 3$ is *translated* 5 *units left* and 3 *units downward*. The vertex for $f(x)$ is $(-5, -3)$. Because $a = -3$, the graph of $y = f(x)$ *opens downward* and is *narrower* than the graph of $y = x^2$. These graphs are shown in Figure 11.28(b).

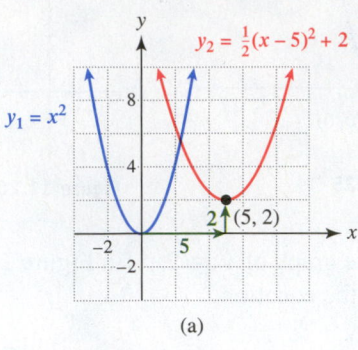
(a)

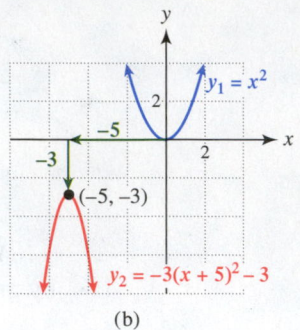
(b)

Figure 11.28

Now Try Exercises 35, 37

EXAMPLE 3 **Finding equations of parabolas**

Write the vertex form of a parabola with $a = 2$ and vertex $(-2, 1)$. Then express this equation in the form $y = ax^2 + bx + c$.

Solution
The vertex form of a parabola is given by $y = a(x - h)^2 + k$, where the vertex is (h, k). For $a = \mathbf{2}, h = \mathbf{-2}$, and $k = \mathbf{1}$, the equation becomes

$$y = \mathbf{2}(x - (\mathbf{-2}))^2 + \mathbf{1} \quad \text{or} \quad y = 2(x + 2)^2 + 1.$$

To write $y = \mathbf{2(x + 2)^2 + 1}$ in the form $y = ax^2 + bx + c$, do the following.

$$
\begin{aligned}
y &= 2(x^2 + 4x + 4) + 1 && \text{Multiply } (x + 2)^2 \text{ as } x^2 + 4x + 4.\\
&= 2x^2 + 8x + 8 + 1 && \text{Distributive property}\\
&= 2x^2 + 8x + 9 && \text{Add.}
\end{aligned}
$$

The equivalent equation is $y = \mathbf{2x^2 + 8x + 9}$.

Now Try Exercise 41

In the next example, the vertex formula is used to write the equation of a parabola in vertex form.

EXAMPLE 4 **Using the vertex formula to write vertex form**

Find the vertex on the graph of $y = 3x^2 + 6x + 1$. Write this equation in vertex form.

Solution
We can use the vertex formula, $x = -\frac{b}{2a}$, to find the x-coordinate of the vertex with $a = \mathbf{3}$ and $b = \mathbf{6}$.

$$x = -\frac{b}{2a} = -\frac{\mathbf{6}}{2(\mathbf{3})} = \mathbf{-1} \quad x\text{-coordinate of vertex}$$

To find the y-coordinate, let $x = -1$ in $y = 3x^2 + 6x + 1$.

$$y = 3(-1)^2 + 6(-1) + 1 = -2 \quad \text{Let } x = -1.$$

The vertex is $(-1, -2)$. We can now find the vertex form with $a = 3$, $h = -1$, and $k = -2$.

$$y = a(x - h)^2 + k \qquad \text{Vertex form}$$
$$y = 3(x - (-1))^2 - 2 \qquad \text{Substitute.}$$
$$y = 3(x + 1)^2 - 2 \qquad \text{Simplify.}$$

The equations $y = 3x^2 + 6x + 1$ and $y = 3(x + 1)^2 - 2$ are equivalent, and their graphs represent the same parabola with vertex $(-1, -2)$.

Now Try Exercise 53

Completing the Square

$$x^2 + 4x + 4 = (x + 2)^2$$

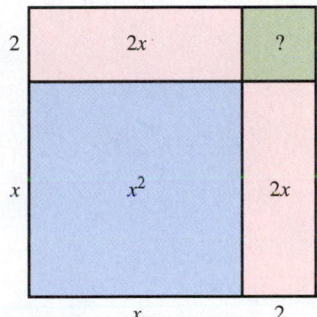

Figure 11.29

COMPLETING THE SQUARE TO FIND THE VERTEX The vertex of a parabola can be found by using a technique called **completing the square**. To complete the square for the expression $x^2 + 4x$, consider Figure 11.29.

Note that the blue and pink areas sum to

$$x^2 + 2x + 2x = x^2 + 4x.$$

The area of the small green square must be $2 \cdot 2 = 4$. Thus to *complete the square* for $x^2 + 4x$, we add 4. That is,

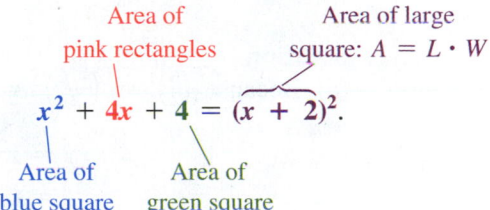

In the next example, we use the above result to complete the square and find the vertex on the graph of $y = x^2 + 4x$.

EXAMPLE 5 ▶ **Writing vertex form by completing the square**

Write the equation $y = x^2 + 4x$ in vertex form by completing the square. Identify the vertex.

Solution

As discussed, we must add 4 to $x^2 + 4x$ to complete the square. Because we are given an equation, we will add 4 *and* subtract 4 from the right side of the equation in order to keep the equation "balanced."

$$y = x^2 + 4x. \qquad \text{Given equation}$$
$$y = x^2 + 4x + 4 - 4 \qquad \text{Add and subtract 4 on the right.}$$
$$y = (x^2 + 4x + 4) - 4 \qquad \text{Associative property}$$
$$y = (x + 2)^2 - 4 \qquad \text{Perfect square trinomial}$$

The vertex form is $y = (x + 2)^2 - 4$, and the vertex is $(-2, -4)$.

Now Try Exercise 61

NOTE: Adding 4 and subtracting 4 from the right side of the equation in Example 5 is equivalent to adding 0 to the right side, which does not change the equation.

In general, to complete the square for $x^2 + bx$ we must add $\left(\frac{b}{2}\right)^2$, as illustrated in Figure 11.30. This technique is shown in Example 6.

READING CHECK

Use the figure to *complete the square* for $x^2 + 8x$.

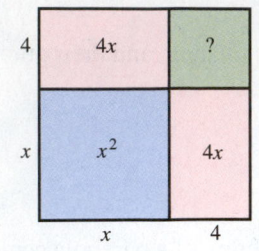

Completing the Square

$$x^2 + bx + \left(\tfrac{b}{2}\right)^2 = \left(x + \tfrac{b}{2}\right)^2$$

Figure 11.30

EXAMPLE 6 **Writing vertex form**

Write each equation in vertex form. Identify the vertex.
(a) $y = x^2 - 6x - 1$ **(b)** $y = x^2 + 3x + 4$ **(c)** $y = 2x^2 + 4x - 1$

Solution
(a) Because $\left(\frac{b}{2}\right)^2 = \left(\frac{-6}{2}\right)^2 = 9$, add *and* subtract 9 on the right side.

$y = x^2 - 6x - 1$	Given equation
$= (x^2 - 6x + 9) - 9 - 1$	Add and subtract 9.
$= (x - 3)^2 - 10$	Perfect square trinomial

The vertex is $(3, -10)$.
(b) Because $\left(\frac{b}{2}\right)^2 = \left(\frac{3}{2}\right)^2 = \frac{9}{4}$, add *and* subtract $\frac{9}{4}$ on the right side.

$y = x^2 + 3x + 4$	Given equation
$= \left(x^2 + 3x + \dfrac{9}{4}\right) - \dfrac{9}{4} + 4$	Add and subtract $\frac{9}{4}$.
$= \left(x + \dfrac{3}{2}\right)^2 + \dfrac{7}{4}$	Perfect square trinomial

The vertex is $\left(-\frac{3}{2}, \frac{7}{4}\right)$.
(c) This equation is slightly different because the leading coefficient is 2 rather than 1. Start by factoring 2 from the first two terms on the right side.

$y = 2x^2 + 4x - 1$	Given equation
$= 2(x^2 + 2x) - 1$	Factor out 2.
$= 2(x^2 + 2x + 1 - 1) - 1$	$\left(\frac{b}{2}\right)^2 = \left(\frac{2}{2}\right)^2 = 1$
$= 2(x^2 + 2x + 1) - 2 - 1$	Distributive property: $2 \cdot (-1)$
$= 2(x + 1)^2 - 3$	Perfect square trinomial

The vertex is $(-1, -3)$.

Now Try Exercises 65, 69, 73

Modeling with Quadratic Functions (Optional)

▶ **REAL-WORLD CONNECTION** In A Look Into Math for this section, we discussed airport taxiway curves designed for airplanes. The data previously shown in Figure 11.20 are listed in Table 11.9.

TABLE 11.9 Safe Taxiway Speed

x (mph)	10	20	30	40	50	60
R (ft)	50	200	450	800	1250	1800

Source: Federal Aviation Administration.

A second scatterplot of the data is shown in Figure 11.31. The data may be modeled by $R(x) = ax^2$ for some value a. To illustrate this relation, graph R for different values of a. In Figures 11.32–11.34, R has been graphed for $a = 2, -1$, and $\frac{1}{2}$, respectively. When $a > 0$ the parabola opens upward, and when $a < 0$ the parabola opens downward. Larger values of $|a|$ make a parabola narrower, whereas smaller values of $|a|$ make the parabola wider. Through trial and error, $a = \frac{1}{2}$ gives a good fit to the data, so $R(x) = \frac{1}{2}x^2$ models the data.

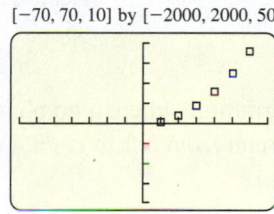

$[-70, 70, 10]$ by $[-2000, 2000, 500]$

Figure 11.31

$[-70, 70, 10]$ by $[-2000, 2000, 500]$

$y = 2x^2$

Figure 11.32 $a = 2$

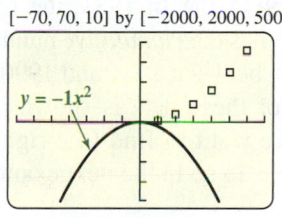

$[-70, 70, 10]$ by $[-2000, 2000, 500]$

$y = -1x^2$

Figure 11.33 $a = -1$

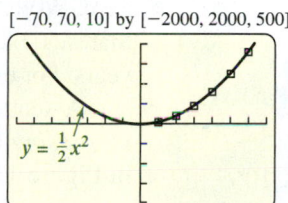

$[-70, 70, 10]$ by $[-2000, 2000, 500]$

$y = \frac{1}{2}x^2$

Figure 11.34 $a = \frac{1}{2}$

CALCULATOR HELP

To make a scatterplot, see Appendix A (pages AP-3 and AP-4).

This value of a can also be found *symbolically*, as demonstrated in the next example.

EXAMPLE 7 **Modeling safe taxiway speed**

Find a value for the constant a so that $R(x) = ax^2$ models the data in Table 11.9. Check your result by making a table of values for $R(x)$.

Solution
From Table 11.9, when $x = \mathbf{10}$ miles per hour, the curve radius is $\mathbf{50}$ feet. Therefore

$$R(\mathbf{10}) = \mathbf{50} \quad \text{or} \quad a(\mathbf{10})^2 = \mathbf{50}. \qquad R(x) = ax^2$$

Solving for a gives

$$a = \frac{\mathbf{50}}{\mathbf{10}^2} = \frac{1}{2}.$$

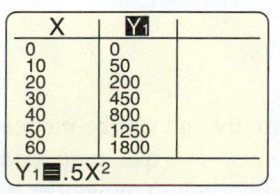

Figure 11.35

To be sure that $R(x) = \frac{1}{2}x^2$ is correct, make a table, as shown in Figure 11.35. Its values agree with those in Table 11.9.

Now Try Exercise 87

INCREASING AND DECREASING The concept of increasing and decreasing is frequently used when modeling data with functions. Suppose that the graph of the equation $y = x^2$ shown in Figure 11.36 represents a valley. If we walk from *left to right*, the valley "goes down" and then "goes up." Mathematically, we say that the graph of $y = x^2$ is *decreasing* when $x \leq 0$ and *increasing* when $x \geq 0$. In Figure 11.33, the graph of $y = -1x^2$ increases when $x \leq 0$ and decreases when $x \geq 0$, and in Figure 11.24(c) on page 730 the graph decreases when $x \leq 2$ and increases when $x \geq 2$.

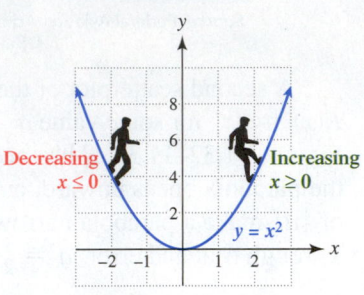

Figure 11.36

NOTE: When determining where a graph is increasing and where it is decreasing, we must "walk" along the graph *from left to right*. (We read English from left to right, which might help you remember.)

TABLE 11.10

Year	U.S. AIDS Cases
1981	425
1984	11,106
1987	71,414
1990	199,608
1993	417,835
1996	609,933

Source: Department of Health and Human Services.

▶ **REAL-WORLD CONNECTION** In 1981, the first cases of AIDS were reported in the United States. Table 11.10 lists the *cumulative* number of AIDS cases in the United States for various years. For example, between 1981 and 1990, a total of 199,608 AIDS cases were reported.

A scatterplot of these data is shown in Figure 11.37. To model these nonlinear and *increasing* data, we want to find (the right half of) a parabola with the shape illustrated in Figure 11.38. We do so in the next example.

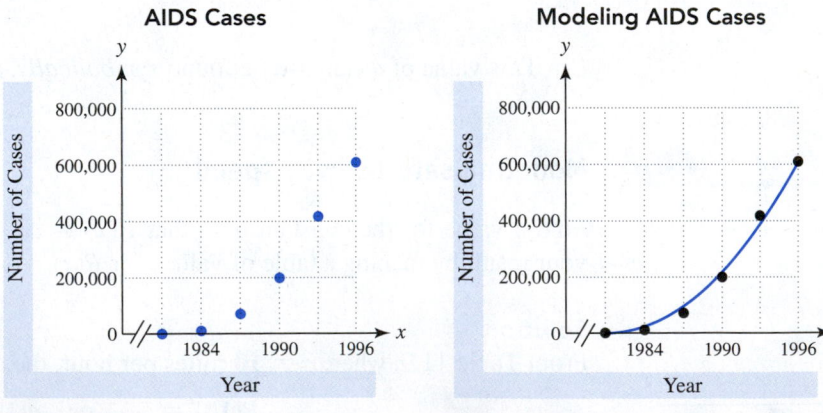

Figure 11.37　　　　　　　**Figure 11.38**

NOTE: From 1981 to 1996, the number of AIDS cases grew rapidly and can be modeled by a quadratic function. After 1996, the growth in AIDS cases slowed so a quadratic model is much less accurate. When modeling data with functions, it is important to remember that data can change character over time. As a result, the same modeling function may not be appropriate for every time period.

EXAMPLE 8 | Modeling AIDS cases

Use the data in Table 11.10 to complete the following.
(a) Make a scatterplot of the data in [1980, 1997, 2] by [−10000, 800000, 100000].
(b) The lowest data point in Table 11.10 is (1981, 425). Let this point be the vertex of a parabola that opens upward. Graph $y = a(x − 1981)^2 + 425$ together with the data by first letting $a = 1000$.
(c) Use trial and error to adjust the value of a until the graph models the data.
(d) Use your final equation to estimate the number of AIDS cases in 1992. Compare it to the known value of 338,786.

Solution

(a) A scatterplot of the data is shown in Figure 11.39.

[1980, 1997, 2] by
[−10000, 800000, 100000]

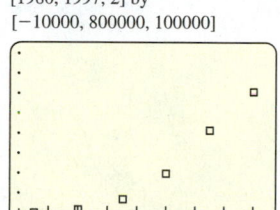

Figure 11.39

[1980, 1997, 2] by
[−10000, 800000, 100000]

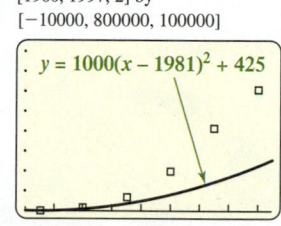

$y = 1000(x − 1981)^2 + 425$

Figure 11.40 $a = 1000$

[1980, 1997, 2] by
[−10000, 800000, 100000]

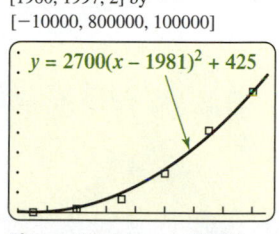

$y = 2700(x − 1981)^2 + 425$

Figure 11.41 $a = 2700$

(b) A graph of $y = 1000(x − 1981)^2 + 425$ is shown in Figure 11.40. To have a better fit of the data, a larger value for a is needed.
(c) Figure 11.41 shows the effect of adjusting the value of a to **2700**. This value provides a reasonably good fit. (Note that you may decide on a slightly different value for a.)
(d) If $a = $ **2700**, the modeling equation becomes

$$y = 2700(x − 1981)^2 + 425.$$

To estimate the number of AIDS cases in 1992, substitute $x = $ **1992** to obtain

$$y = 2700(\mathbf{1992} − 1981)^2 + 425 = \mathbf{327{,}125}.$$

This number is about 12,000 less than the known value of **338,786**.

Now Try Exercise 91

11.2   Putting It All Together

CONCEPT	EXPLANATION	EXAMPLES
Translations of Parabolas	Compared to the graph $y = x^2$, the graph of $y = x^2 + k$ is shifted vertically k units and the graph of $y = (x − h)^2$ is shifted horizontally h units.	Compared to the graph of $y = x^2$, the graph of $y = x^2 − 4$ is shifted *downward* 4 units. Compared to the graph of $y = x^2$, the graph of $y = (x − 4)^2$ is shifted *right* 4 units and the graph of $y = (x + 4)^2$ is shifted *left* 4 units.
Vertex Form	The vertex form of the equation of a parabola with vertex (h, k) is $$y = a(x − h)^2 + k,$$ where $a \neq 0$ is a constant. If $a > 0$, the parabola opens upward; if $a < 0$, the parabola opens downward.	The graph of $y = 3(x + 2)^2 − 7$ has a vertex of $(−2, −7)$ and opens upward because $3 > 0$.

continued on next page

continued from previous page

CONCEPT	EXPLANATION	EXAMPLES
Completing the Square Method	To complete the square to obtain the vertex form, add *and* subtract $\left(\frac{b}{2}\right)^2$ on the right side of the equation $y = x^2 + bx + c$. Then factor the perfect square trinomial.	If $y = x^2 + 10x - 3$, then add *and* subtract $\left(\frac{b}{2}\right)^2 = \left(\frac{10}{2}\right)^2 = 25$ on the right side of this equation. $$y = (x^2 + 10x + 25) - 25 - 3$$ $$= (x + 5)^2 - 28$$ The vertex is $(-5, -28)$.

11.2 Exercises

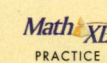

CONCEPTS AND VOCABULARY

1. Compared to the graph of $y = x^2$, the graph of $y =$ _____ is shifted upward 2 units.

2. Compared to the graph of $y = x^2$, the graph of $y =$ _____ is shifted to the right 2 units.

3. The vertex of $y = (x - 1)^2 + 2$ is _____.

4. The vertex of $y = (x + 1)^2 - 2$ is _____.

5. A quadratic function f may be written either in the form _____ or _____.

6. The vertex form of a parabola is given by _____ and its vertex is _____.

7. The graph of the equation $y = -x^2$ is a parabola that opens _____.

8. The x-coordinate of the vertex of $y = ax^2 + bx + c$ is $x =$ _____.

9. Compared to the graph of $y = x^2$, the graph of $y = x^2 + k$ with $k > 0$ is shifted k units _____.
a. upward   **b.** downward
c. left   **d.** right

10. Compared to the graph of $y = x^2$, the graph of $y = (x - k)^2$ with $k > 0$ is shifted k units _____.
a. upward   **b.** downward
c. left   **d.** right

TABLES AND TRANSLATIONS

Exercises 11–14: Complete the table for each translation of $y = x^2$. State what the translation does.

11.

x	-2	-1	0	1	2
$y = x^2$					
$y = x^2 - 3$					

12.

x	-2	-1	0	1	2
$y = x^2$					
$y = x^2 + 5$					

13.

x					
$y = x^2$	4	1	0	1	4

x					
$y = (x - 3)^2$	4	1	0	1	4

14.

x					
$y = x^2$	16	4	0	4	16

x					
$y = (x + 4)^2$	16	4	0	4	16

GRAPHS OF PARABOLAS

Exercises 15–34: Do the following.
 (a) Sketch a graph of the equation.
 (b) Identify the vertex.
 (c) Compare the graph of $y = f(x)$ to the graph of $y = x^2$. (State any transformations used.)

15. $f(x) = x^2 - 4$ **16.** $f(x) = x^2 - 1$

17. $f(x) = 2x^2 + 1$ **18.** $f(x) = \frac{1}{2}x^2 + 1$

19. $f(x) = (x - 3)^2$ **20.** $f(x) = (x + 1)^2$

21. $f(x) = -x^2$ **22.** $f(x) = -(x + 2)^2$

23. $f(x) = 2 - x^2$ **24.** $f(x) = (x - 1)^2$

25. $f(x) = (x + 2)^2$ **26.** $f(x) = (x - 2)^2 - 3$

27. $f(x) = (x + 1)^2 - 2$ **28.** $f(x) = (x - 3)^2 + 1$

29. $f(x) = (x - 1)^2 + 2$

30. $f(x) = \frac{1}{2}(x + 3)^2 - 3$

31. $f(x) = 2(x - 5)^2 - 4$

32. $f(x) = -3(x + 4)^2 + 5$

33. $f(x) = -\frac{1}{2}(x + 3)^2 + 1$

34. $f(x) = 2(x - 5)^2 + 10$

Exercises 35–38: Compare the graph of $y = f(x)$ to the graph of $y = x^2$. Then sketch a graph of $y = f(x)$ and $y = x^2$ in the same xy-plane.

35. $f(x) = \frac{1}{2}(x - 1)^2 - 2$

36. $f(x) = 2(x + 2)^2 - 1$

37. $f(x) = -2(x + 1)^2 + 3$

38. $f(x) = -\frac{1}{2}(x - 2)^2 + 2$

 Exercises 39 and 40: Graph the equation in a window that shows the vertex and all intercepts.

39. $f(x) = -0.4x^2 + 6x - 10$

40. $f(x) = 3x^2 - 40x + 50$

VERTEX FORM

Exercises 41–44: (Refer to Example 3.) Write the vertex form of a parabola that satisfies the conditions given. Then write the equation in the form $y = ax^2 + bx + c$.

41. Vertex $(3, 4)$ and $a = 3$

42. Vertex $(-1, 3)$ and $a = -5$

43. Vertex $(5, -2)$ and $a = -\frac{1}{2}$

44. Vertex $(-2, -6)$ and $a = \frac{3}{4}$

Exercises 45–48: Write the vertex form of a parabola that satisfies the conditions given. Assume that $a = \pm 1$.

45. Opens upward, vertex $(1, 2)$

46. Opens downward, vertex $(-1, -2)$

47. Opens downward, vertex $(0, -3)$

48. Opens upward, vertex $(5, -4)$

Exercises 49–52: Write the vertex form of the parabola shown in the graph. Assume that $a = \pm 1$.

49. 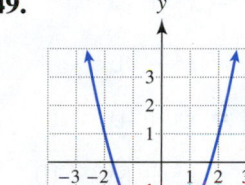   **50.**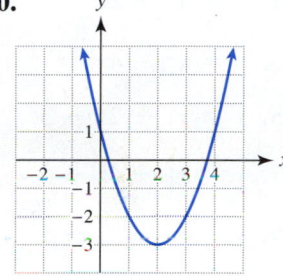

51. 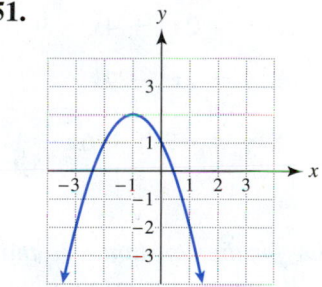   **52.**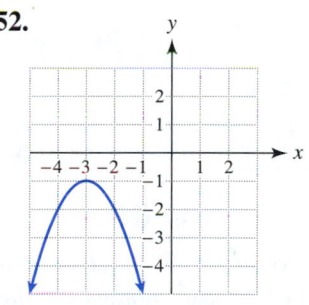

Exercises 53–58: (Refer to Example 4.) Do the following.
(a) Find the vertex on the graph of the equation.
(b) Write the equation in vertex form.

53. $y = 4x^2 - 8x + 5$

54. $y = 3x^2 - 12x + 15$

55. $y = -x^2 - 2x - 3$

56. $y = -x^2 - 4x - 5$

57. $y = -2x^2 - 4x + 1$

58. $y = 2x^2 - 16x + 27$

COMPLETING THE SQUARE

Exercises 59 and 60: Use the given figure to determine what number should be added to the expression to complete the square.

59. $x^2 + 2x$ **60.** $x^2 + 6x$

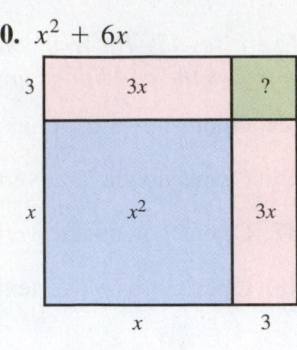

Exercises 61–78: (Refer to Example 6.) Write the equation in vertex form. Identify the vertex.

61. $y = x^2 + 2x$ **62.** $y = x^2 + 6x$

63. $y = x^2 - 4x$ **64.** $y = x^2 - 8x$

65. $y = x^2 + 2x - 3$ **66.** $y = x^2 + 4x + 1$

67. $y = x^2 - 4x + 5$ **68.** $y = x^2 - 8x + 10$

69. $y = x^2 + 3x - 2$ **70.** $y = x^2 + 5x - 4$

71. $y = x^2 - 7x + 1$ **72.** $y = x^2 - 3x + 5$

73. $y = 3x^2 + 6x - 1$ **74.** $y = 2x^2 + 4x - 9$

75. $y = 2x^2 - 3x$ **76.** $y = 3x^2 - 7x$

77. $y = -2x^2 - 8x + 5$ **78.** $y = -3x^2 + 6x + 1$

MODELING DATA

Exercises 79–82: Find a value for the constant a so that $f(x) = ax^2$ models the data. If you are uncertain about your value for a, check it by making a table of values.

79.

x	1	2	3
y	2	8	18

80.

x	−2	0	2
y	6	0	6

81.

x	2	4	6	8
y	1.2	4.8	10.8	19.2

82.

x	5	10	15	20
y	17.5	70	157.5	280

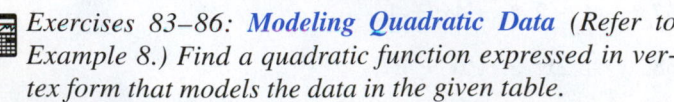 *Exercises 83–86: Modeling Quadratic Data (Refer to Example 8.) Find a quadratic function expressed in vertex form that models the data in the given table.*

83.

x	1	2	3	4
y	−3	−1	5	15

84.

x	−2	−1	0	1	2
y	5	2	−7	−22	−43

85.

x	1980	1990	2000	2010
y	6	55	210	450

86.

x	1990	1995	2000	2005
y	10	60	205	470

87. *Braking Distance* The table lists approximate braking distances D in feet for cars traveling at x miles per hour on dry, level pavement.

x	12	24	36	48
D	12	48	108	192

(a) Make a scatterplot of the data.
(b) Find a function given by $D(x) = ax^2$ that models these data.

88. *Health Care Costs* The table lists approximate *annual* percent increases in the cost of health insurance premiums between 1992 and 2000.

Year	1992	1994	1996	1998	2000
Increase	11%	4%	1%	4%	11%

Source: Kaiser Family Foundation.

(a) Describe what happened to health care costs from 1992 to 2000.
(b) Would a linear function model these data? Explain.
(c) What type of function might model these data? Explain.
(d) What might be a good choice for the vertex? Explain.
(e) Find a quadratic function C expressed in vertex form that models the data. (Answers may vary.)
(f) Graph C and the data.

89. *Sub-Saharan Africa* The table lists actual and projected real gross domestic product (GDP) per capita for selected years in Sub-Saharan Africa.

Year	1980	1995	2010	2025
Real GDP	$1000	$600	$1000	$2200

Source: IMF WEO, Standard Chartered Research.

(a) Describe what happens to the real GDP.
(b) Would a linear function model these data? Explain.
(c) What type of function might model these data? Explain.

(d) What might be a good choice for the vertex? Explain.
(e) Find a quadratic function C expressed in vertex form that models the data. (Answers may vary.)
(f) Graph C and the data.

90. *Tax Theory and Modeling* If a government wants to generate enough revenue from income taxes, then it is important to assess the correct tax rates. On the one hand, if tax rates are too low, then enough revenue may not be generated. On the other hand, if taxes are too high, people may work less and not earn as much money. Again, enough revenue may not be generated. The following parabolic graph illustrates this phenomenon.

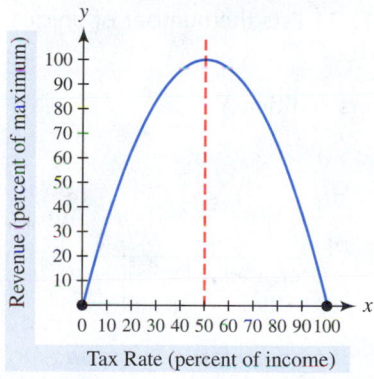

(a) The points $(0, 0)$ and $(100, 0)$ are on this graph. Explain why.
(b) Explain why a linear model is not suitable.
(c) According to *this* graph, what rate maximizes revenue?

91. *U.S. AIDS Deaths* (Refer to Example 8.) The table lists cumulative numbers of AIDS deaths *in thousands* for selected years.

Year	1982	1986	1990	1994
Deaths (thousands)	1	25	123	300

Source: Department of Health and Human Services.

(a) Determine $f(x) = a(x - h)^2 + k$, so that f models these data.
(b) Estimate the number of AIDS deaths in 1992 and compare it to the actual value of 202 thousand.

92. *Head Start Enrollment* The table lists numbers of students *in thousands* enrolled in Head Start for selected years.

Year	1966	1980	1995
Students (thousands)	733	376	750

(a) Determine $f(x) = a(x - h)^2 + k$, so that f models these data.
(b) Estimate Head Start enrollment in 1990 and compare it to the actual value of 541 thousand.

INCREASING AND DECREASING

Exercises 93–96: Use the graph of $y = f(x)$ to determine the intervals where f is increasing and where f is decreasing.

93.

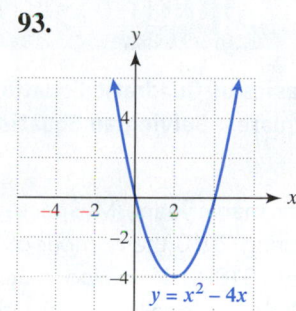

94.

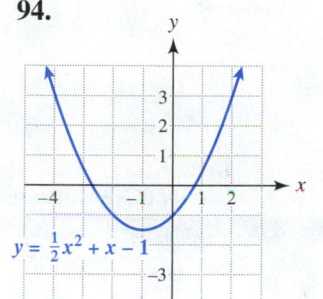

95.

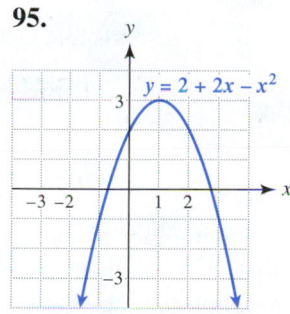

96.

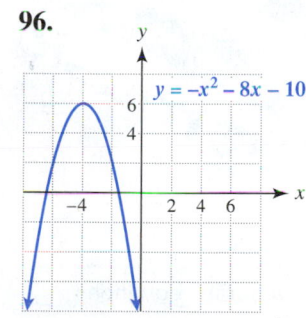

Exercises 97–102: Determine the intervals where $y = f(x)$ is increasing and where it is decreasing.

97. $y = 3x^2 - 4x + 1$ **98.** $y = 2x^2 - 5x - 2$

99. $y = -x^2 - 3x$ **100.** $y = -2x^2 + 3x - 4$

101. $y = 5 - x - 4x^2$ **102.** $y = 3 + x + x^2$

WRITING ABOUT MATHEMATICS

103. Explain how to find the vertex of $y = x^2 + bx + c$ by completing the square.

104. If $f(x) = a(x - h)^2 + k$, explain how the values of a, h, and k affect the graph of $y = f(x)$.

SECTIONS 11.1 and 11.2

Checking Basic Concepts

1. Graph each quadratic function. Identify the vertex and axis of symmetry.
   (a) $f(x) = x^2 - 2$
   (b) $f(x) = x^2 - 2x - 2$

2. Compare the graph of $y_1 = 2x^2$ to the graph of $y_2 = -\frac{1}{2}x^2$.

3. Find the maximum y-value on the graph of the equation $y = -3x^2 + 12x - 5$.

4. Sketch a graph of $y = f(x)$. Compare this graph to the graph of $y = x^2$.
   (a) $f(x) = (x - 1)^2 + 2$
   (b) $f(x) = -(x + 3)^2$

5. Write the vertex form for each equation.
   (a) $y = x^2 + 14x - 7$
   (b) $y = 4x^2 + 8x - 2$

11.3 Quadratic Equations

Basics of Quadratic Equations • The Square Root Property • Completing the Square • Solving an Equation for a Variable • Applications of Quadratic Equations

A LOOK INTO MATH ▶

NEW VOCABULARY

☐ Quadratic equation
☐ Square root property

For many years, MySpace was the largest social network in the United States. However, during 2010 the number of unique monthly visitors fell dramatically from 70 million in January 2010 to 45 million in January 2011. Table 11.11 lists the number of unique visitors V to MySpace in millions x months after January 2010.

TABLE 11.11 Unique MySpace Visitors (millions)

x (months after Jan. 2010)	0	6	12
V (unique monthly visitors)	70	63	45

Source: comScore.

Decreased 7 million   Decreased 18 million

The number of unique visitors to MySpace decreased faster in the second half of 2010 than it did in the first half, so a linear function will *not* model these data. Instead, these data can be modeled by the *quadratic function*

$$V(x) = -\frac{25}{144}x^2 + 70. \qquad \text{Quadratic function}$$

If we want to determine the month when MySpace had **56** million unique visitors, then we could solve the *quadratic equation*

$$-\frac{25}{144}x^2 + 70 = 56. \qquad \text{Quadratic equation}$$

In this section we learn how to solve this and other quadratic equations. See Exercise 125.

Basics of Quadratic Equations

Any quadratic function f can be represented by $f(x) = ax^2 + bx + c$ with $a \neq 0$. Examples of quadratic functions include

$$f(x) = 2x^2 - 1, \; g(x) = -\tfrac{1}{3}x^2 + 2x, \text{ and } h(x) = x^2 + 2x - 1. \qquad \text{Quadratic functions}$$

Quadratic functions can be used to write quadratic equations. Examples of quadratic equations include

$$2x^2 - 1 = 0, \quad -\tfrac{1}{3}x^2 + 2x = 0, \quad \text{and} \quad x^2 + 2x - 1 = 0. \quad \text{Quadratic equations}$$

READING CHECK

How can you identify a quadratic equation?

QUADRATIC EQUATION

A **quadratic equation** is an equation that can be written as

$$ax^2 + bx + c = 0,$$

where a, b, and c are constants with $a \neq 0$.

Solutions to the quadratic equation $ax^2 + bx + c = 0$ correspond to x-intercepts of the graph of $y = ax^2 + bx + c$. Because the graph of a quadratic function is either $\cup$-shaped or $\cap$-shaped, it can intersect the x-axis zero, one, or two times, as illustrated in Figure 11.42. Hence a quadratic equation can have zero, one, or two real solutions.

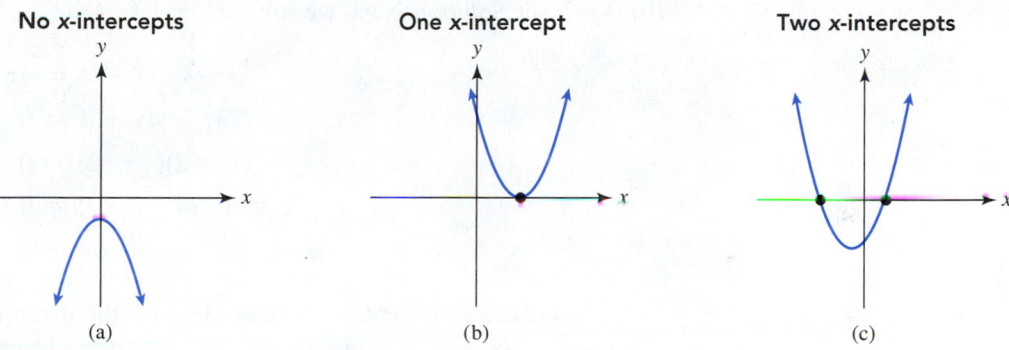

Figure 11.42

NOTE: Some quadratic equations may not have 0 on the right side of the equation, such as $4x^2 = 1$. To solve this quadratic equation graphically, we will sometimes rewrite it as $4x^2 - 1 = 0$ and graph $y = 4x^2 - 1$. Then solutions are x-intercepts, because $y = 0$ on the x-axis.

We have already solved quadratic equations by factoring, graphing, and constructing tables. In the next example we apply these three techniques to quadratic equations that have no real solutions, one real solution, and two real solutions.

EXAMPLE 1 **Solving quadratic equations**

Solve each quadratic equation. Support your results numerically and graphically.
(a) $2x^2 + 1 = 0$ (No real solutions)
(b) $x^2 + 4 = 4x$ (One real solution)
(c) $x^2 - 6x + 8 = 0$ (Two real solutions)

Solution
(a) *Symbolic Solution*

$$\begin{aligned}
2x^2 + 1 &= 0 && \text{Given equation} \\
2x^2 &= -1 && \text{Subtract 1 from each side.} \\
x^2 &= -\frac{1}{2} && \text{Divide each side by 2.}
\end{aligned}$$

This equation has no real-number solutions because $x^2 \geq 0$ for all real numbers x.

Numerical and Graphical Solution The points in Table 11.12 for $y = 2x^2 + 1$ are plotted in Figure 11.43 and connected with a parabolic graph. The graph of $y = 2x^2 + 1$ has no x-intercepts, indicating that there are no real solutions.

TABLE 11.12

x	y
-2	9
-1	3
0	1
1	3
2	9

For all x, $y \neq 0$, so there are no solutions.

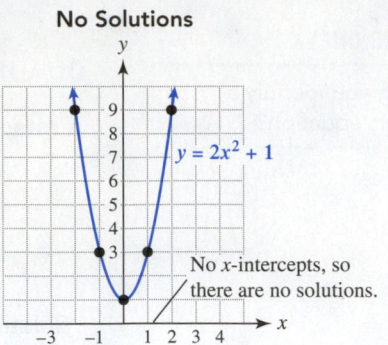

No Solutions

Figure 11.43

(b) *Symbolic Solution* Next, we solve $x^2 + 4 = 4x$.

$$x^2 + 4 = 4x \qquad \text{Given equation}$$
$$x^2 - 4x + 4 = 0 \qquad \text{Subtract } 4x \text{ from each side.}$$
$$(x - 2)(x - 2) = 0 \qquad \text{Factor.}$$
$$x - 2 = 0 \quad \text{or} \quad x - 2 = 0 \qquad \text{Zero-product property}$$
$$x = \mathbf{2} \qquad \text{There is one solution.}$$

Numerical and Graphical Solution Because the given quadratic equation is equivalent to $x^2 - 4x + 4 = 0$, we let $y = x^2 - 4x + 4$. The points in Table 11.13 are plotted in Figure 11.44 and connected with a parabolic graph. The graph of $y = x^2 - 4x + 4$ has one x-intercept, **2**. Note that in Table 11.13, $y = \mathbf{0}$ when $x = \mathbf{2}$, indicating that the equation has one solution.

TABLE 11.13

x	y
0	4
1	1
2	**0**
3	1
4	4

One solution: **2**

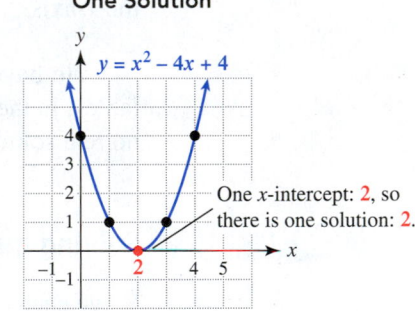

One Solution

Figure 11.44

(c) *Symbolic Solution* Next, we solve $x^2 - 6x + 8 = 0$.

$$x^2 - 6x + 8 = 0 \qquad \text{Given equation}$$
$$(x - 2)(x - 4) = 0 \qquad \text{Factor.}$$
$$x - 2 = 0 \quad \text{or} \quad x - 4 = 0 \qquad \text{Zero-product property}$$
$$x = \mathbf{2} \quad \text{or} \quad x = \mathbf{4} \qquad \text{There are two solutions.}$$

Numerical and Graphical Solution The points in Table 11.14 for $y = x^2 - 6x + 8$ are plotted in Figure 11.45 and connected with a parabolic graph. The graph of

$y = x^2 - 6x + 8$ has two x-intercepts, **2** and **4**, indicating two solutions. Note that in Table 11.14 $y = \mathbf{0}$ when $x = \mathbf{2}$ or $x = \mathbf{4}$.

TABLE 11.14

x	y
0	8
1	3
2	0
3	−1
4	0
5	3
6	8

Two solutions: **2** and **4**

Two Solutions

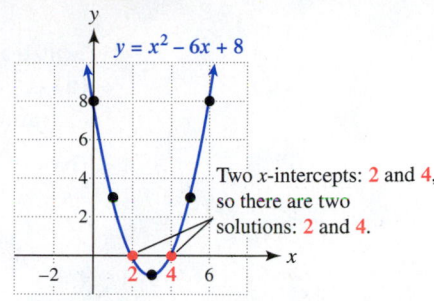

Two x-intercepts: **2** and **4**, so there are two solutions: **2** and **4**.

Figure 11.45

Now Try Exercises 29, 37, 39

READING CHECK

How many solutions can a quadratic equation have?

The Square Root Property

The **square root property** is used to solve quadratic equations that have no x-terms. The following is an example of the square root property.

$$x^2 = 25 \quad \text{is equivalent to} \quad x = \pm 5.$$

The equation $x = \pm 5$ (read "x equals plus or minus 5") indicates that either $x = 5$ or $x = -5$. Each value is a solution because $(5)^2 = 25$ and $(-5)^2 = 25$.

We can derive this result in general for $k \geq 0$.

$$x^2 = k \qquad \text{Given quadratic equation}$$
$$\sqrt{x^2} = \sqrt{k} \qquad \text{Take the square root of each side.}$$
$$|x| = \sqrt{k} \qquad \sqrt{x^2} = |x| \text{ for all } x.$$
$$x = \pm\sqrt{k} \qquad |x| = b \text{ implies } x = \pm b, b \geq 0.$$

This result is summarized by the *square root property*.

SQUARE ROOT PROPERTY

Let k be a nonnegative number. Then the solutions to the equation

$$x^2 = k$$

are given by $x = \pm\sqrt{k}$. If $k < 0$, then this equation has no real solutions.

Before applying the square root property in the next two examples, we review a quotient property of square roots. If a and b are positive numbers, then

$$\sqrt{\frac{a}{b}} = \frac{\sqrt{a}}{\sqrt{b}}.$$

For example, $\sqrt{\frac{25}{36}} = \frac{\sqrt{25}}{\sqrt{36}} = \frac{5}{6}.$

EXAMPLE 2   **Using the square root property**

Solve each equation.
(a) $x^2 = 7$ **(b)** $16x^2 - 9 = 0$ **(c)** $(x - 4)^2 = 25$

Solution
(a) $x^2 = 7$ is equivalent to $x = \pm\sqrt{7}$ by the square root property. The solutions are $\sqrt{7}$ and $-\sqrt{7}$.

(b)

$16x^2 - 9 = 0$	Given equation
$16x^2 = 9$	Add 9 to each side.
$x^2 = \dfrac{9}{16}$	Divide each side by 16.
$x = \pm\sqrt{\dfrac{9}{16}}$	Square root property
$x = \pm\dfrac{3}{4}$	Simplify.

The solutions are $\frac{3}{4}$ and $-\frac{3}{4}$.

(c)

$(x - 4)^2 = 25$	Given equation
$(x - 4) = \pm\sqrt{25}$	Square root property
$x - 4 = \pm 5$	Simplify.
$x = 4 \pm 5$	Add 4 to each side.
$x = 9 \quad \text{or} \quad x = -1$	Evaluate $4 + 5$ and $4 - 5$.

The solutions are 9 and -1.

READING CHECK

When do you use the square root property to solve an equation?

Now Try Exercises 51, 53, 57

▶ **REAL-WORLD CONNECTION** If an object is dropped from a height of h feet, its distance d above the ground after t seconds is given by

$$d(t) = h - 16t^2.$$

This formula can be used to estimate the time it takes for a falling object to hit the ground.

EXAMPLE 3   **Modeling a falling object**

A toy falls 30 feet from a window. How long does the toy take to hit the ground?

Solution
The height h of the window above the ground is 30 feet, so let $d(t) = 30 - 16t^2$. The toy strikes the ground when the distance d above the ground equals 0.

$30 - 16t^2 = 0$	Equation to solve for t
$-16t^2 = -30$	Subtract 30 from each side.
$t^2 = \dfrac{30}{16}$	Divide each side by -16.
$t = \pm\sqrt{\dfrac{30}{16}}$	Square root property
$t = \pm\dfrac{\sqrt{30}}{4}$	Simplify.

Time cannot be negative in this problem, so the appropriate solution is $t = \frac{\sqrt{30}}{4} \approx 1.4$. The toy hits the ground after about 1.4 seconds.

Now Try Exercise 115

Completing the Square

In Section 11.2 we used the *method of completing the square* to find the vertex of a parabola. This method can also be used to solve quadratic equations. Because

$$x^2 + bx + \left(\frac{b}{2}\right)^2 = \left(x + \frac{b}{2}\right)^2,$$

we can solve a quadratic equation in the form $x^2 + bx = d$, where b and d are constants, by adding $\left(\frac{b}{2}\right)^2$ to each side and then factoring the resulting perfect square trinomial.

In the equation $x^2 + 6x = 7$ we have $b = 6$, so we add $\left(\frac{6}{2}\right)^2 = 9$ to each side.

$$
\begin{array}{ll}
x^2 + 6x = 7 & \text{Given equation} \\
x^2 + 6x + 9 = 7 + 9 & \text{Add 9 to each side.} \\
(x + 3)^2 = 16 & \text{Perfect square trinomial} \\
x + 3 = \pm 4 & \text{Square root property} \\
x = -3 \pm 4 & \text{Add } -3 \text{ to each side.} \\
x = 1 \quad \text{or} \quad x = -7 & \text{Simplify } -3 + 4 \text{ and } -3 - 4.
\end{array}
$$

The solutions are 1 and -7. Note that after completing the square, the left side of the equation is a perfect square trinomial. We show how to create one in the next example.

READING CHECK

Use the figure to *complete the square* for $x^2 + 6x$.

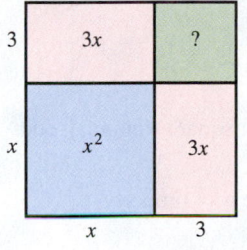

EXAMPLE 4 **Creating a perfect square trinomial**

Find the term that should be added to $x^2 - 10x$ to form a perfect square trinomial.

Solution
The coefficient of the x-term is -10, so we let $b = -10$. To complete the square we divide b by 2 and then square the result.

$$\left(\frac{b}{2}\right)^2 = \left(\frac{-10}{2}\right)^2 = 25$$

If we add 25 to $x^2 - 10x$, a perfect square trinomial is formed.

$$x^2 - 10x + 25 = (x - 5)^2$$

Now Try Exercise 67

EXAMPLE 5 **Completing the square when the leading coefficient is 1**

Solve the equation $x^2 - 4x + 2 = 0$.

Solution
Start by writing the equation in the form $x^2 + bx = d$.

$$
\begin{array}{ll}
x^2 - 4x + 2 = 0 & \text{Given equation} \\
x^2 - 4x = -2 & \text{Subtract 2.} \\
x^2 - 4x + 4 = -2 + 4 & \text{Add } \left(\frac{b}{2}\right)^2 = \left(\frac{-4}{2}\right)^2 = 4. \\
(x - 2)^2 = 2 & \text{Perfect square trinomial; add.} \\
x - 2 = \pm\sqrt{2} & \text{Square root property} \\
x = 2 \pm \sqrt{2} & \text{Add 2.}
\end{array}
$$

The solutions are $2 + \sqrt{2} \approx 3.41$ and $2 - \sqrt{2} \approx 0.59$.

Now Try Exercise 73

EXAMPLE 6 | **Completing the square when the leading coefficient is not 1**

Solve the equation $2x^2 + 7x - 5 = 0$.

Solution
Start by writing the equation in the form $x^2 + bx = d$. That is, add 5 to each side and then divide each side by 2 so that the leading coefficient of the x^2-term becomes 1.

$$2x^2 + 7x - 5 = 0 \qquad \text{Given equation}$$

$$2x^2 + 7x = 5 \qquad \text{Add 5 to each side.}$$

$$x^2 + \frac{7}{2}x = \frac{5}{2} \qquad \text{Divide each side by 2.}$$

$$x^2 + \frac{7}{2}x + \frac{49}{16} = \frac{5}{2} + \frac{49}{16} \qquad \text{Add } \left(\frac{b}{2}\right)^2 = \left(\frac{7}{4}\right)^2 = \frac{49}{16}.$$

$$\left(x + \frac{7}{4}\right)^2 = \frac{89}{16} \qquad \text{Perfect square trinomial; add.}$$

$$x + \frac{7}{4} = \pm \frac{\sqrt{89}}{4} \qquad \text{Square root property}$$

$$x = -\frac{7}{4} \pm \frac{\sqrt{89}}{4} \qquad \text{Add } -\frac{7}{4}.$$

$$x = \frac{-7 \pm \sqrt{89}}{4} \qquad \text{Combine fractions.}$$

The solutions are $\frac{-7 + \sqrt{89}}{4} \approx 0.61$ and $\frac{-7 - \sqrt{89}}{4} \approx -4.1$.

Now Try Exercise 81

CRITICAL THINKING

What happens if you try to solve

$$2x^2 - 13 = 1$$

by completing the square? What method could you use to solve this problem?

Solving an Equation for a Variable

We often need to solve an equation or formula for a variable. For example, the formula $V = \frac{1}{3}\pi r^2 h$ calculates the volume of the cone shown in Figure 11.46. Let's say that we know the volume V is 120 cubic inches and the height h is 15 inches. We can then find the radius of the cone by solving the equation for r.

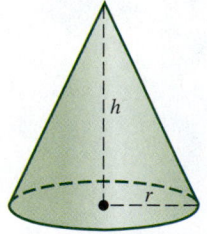

Figure 11.46

$$V = \frac{1}{3}\pi r^2 h \qquad \text{Solve the equation for } r.$$

$$3V = \pi r^2 h \qquad \text{Multiply by 3.}$$

$$\frac{3V}{\pi} = r^2 h \qquad \text{Divide by } \pi.$$

$$\frac{3V}{\pi h} = r^2 \qquad \text{Divide by } h.$$

$$r = \pm \sqrt{\frac{3V}{\pi h}} \qquad \text{Square root property; rewrite.}$$

Because $r \geq 0$, we use the positive or *principal square root*. Thus for $V = \mathbf{120}$ cubic inches and $h = \mathbf{15}$ inches,

$$r = \sqrt{\frac{3(\mathbf{120})}{\pi(\mathbf{15})}} = \sqrt{\frac{24}{\pi}} \approx 2.8 \text{ inches.}$$

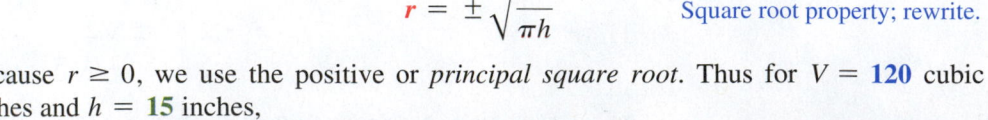

EXAMPLE 7 Solving equations for variables

Solve each equation for the specified variable.
(a) $s = -\frac{1}{2}gt^2 + h$ for t **(b)** $d^2 = x^2 + y^2$ for y

Solution
(a) Begin by subtracting h from each side of the equation.

$$s = -\frac{1}{2}gt^2 + h \qquad \text{Solve the equation for } t.$$

$$s - h = -\frac{1}{2}gt^2 \qquad \text{Subtract } h.$$

$$-2(s - h) = gt^2 \qquad \text{Multiply by } -2.$$

$$\frac{2h - 2s}{g} = t^2 \qquad \text{Divide by } g; \text{ simplify.}$$

$$t = \pm\sqrt{\frac{2h - 2s}{g}} \qquad \text{Square root property; rewrite.}$$

(b) Begin by subtracting x^2 from each side of the equation.

$$d^2 = x^2 + y^2 \qquad \text{Solve the equation for } y.$$

$$d^2 - x^2 = y^2 \qquad \text{Subtract } x^2.$$

$$y = \pm\sqrt{d^2 - x^2} \qquad \text{Square root property; rewrite.}$$

Now Try Exercises 105, 107

Applications of Quadratic Equations

▶ **REAL-WORLD CONNECTION** In Section 11.2 we modeled curves on airport taxiways by using $R(x) = \frac{1}{2}x^2$. In this formula x represented the airplane's speed in miles per hour, and R represented the radius of the curve in feet. This formula may be used to determine the speed limit for a curve with a radius of **650** feet by solving the *quadratic equation*

$$\frac{1}{2}x^2 = \mathbf{650}.$$

In the next example, we solve this quadratic equation. (*Source: FAA.*)

EXAMPLE 8 Finding a safe speed limit

Solve $\frac{1}{2}x^2 = 650$ and interpret any solutions.

Solution
Use the square root property to solve this problem.

$$\frac{1}{2}x^2 = \mathbf{650} \qquad \text{Given equation}$$

$$x^2 = 1300 \qquad \text{Multiply by 2.}$$

$$x = \pm\sqrt{1300} \qquad \text{Square root property}$$

The solutions are $\sqrt{1300} \approx 36$ and $-\sqrt{1300} \approx -36$. The solution of $x \approx 36$ indicates that a safe speed limit for a curve with a radius of 650 feet is about 36 miles per hour. (The negative solution has no physical meaning in this problem.)

Now Try Exercise 113

▶ **REAL-WORLD CONNECTION** From 1989 to 2009 there was a dramatic increase in compact disc and tape sales and then a dramatic decrease. Their global music sales S in billions of dollars x years after 1989 are modeled by

$$S(x) = -0.095x^2 + 1.85x + 6.$$

A graph of $S(x) = -0.095x^2 + 1.85x + 6$ is shown in Figure 11.47. (*Source:* RIAA, Bain Analysis.)

Global Music Sales from CDs and Tapes

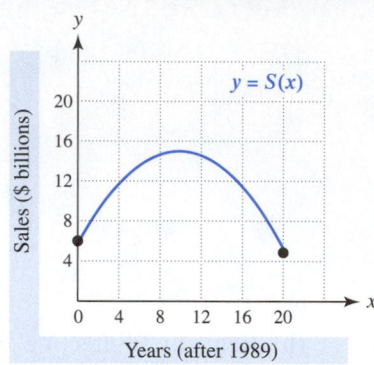

Figure 11.47

When were sales equal to $12 billion? Because the graph is parabolic, it appears that there are two answers to the question: about 4 years and about 15 years after 1989. In the next example, we use graphical and numerical methods to find these years. In Section 11.4 (Exercise 113), you are asked to answer this question symbolically.

EXAMPLE 9 | **Modeling global CD and tape sales**

Use each method to determine the years when global sales of CDs and tapes were $12 billion.
(a) Graphical   **(b)** Numerical

Solution
(a) *Graphical Solution* Graph $y_1 = -0.095x^2 + 1.85x + 6$ and $y_2 = 12$, as shown in Figures 11.48(a) and (b). Their graphs intersect near the points $(4.11, 12)$ and $(15.36, 12)$. Because x is years *after* 1989, sales were $12 billion dollars in about 1993 (1989 + 4) and about 2004 (1989 + 15).

Determining When Sales Were $12 Billion

CALCULATOR HELP

To find a point of intersection, see Appendix A (page AP-6).

[0, 20, 5] by [0, 20, 5]   [0, 20, 5] by [0, 20, 5]

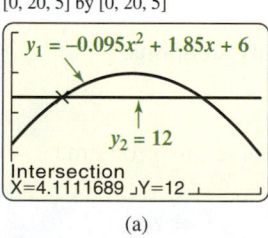

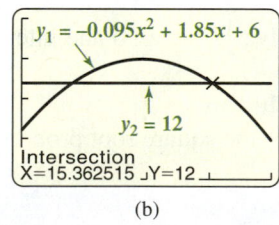

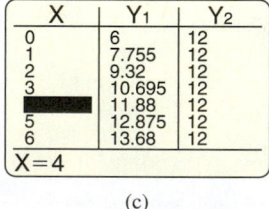

(a)   (b)   (c)

Figure 11.48

(b) *Numerical Solution* Make a table of $y_1 = -0.095x^2 + 1.85x + 6$ and $y_2 = 12$, as shown in Figure 11.48(c). When $x = 4$, y_1 is approximately 12. Thus sales were $12 billion around 1993. If we were to scroll down in the table further, we would see that sales also equal $12 billion when $x \approx 15$, or in 2004. The graphical and numerical solutions agree.

Now Try Exercise 117

11.3 Putting It All Together

Quadratic equations can have no real solutions, one real solution, or two real solutions. Symbolic techniques for solving quadratic equations include factoring, the square root property, and completing the square. We discussed factoring in Chapter 6, so the following table summarizes only the square root property and the method of completing the square.

TECHNIQUE	DESCRIPTION	EXAMPLES
Square Root Property	If $k \geq 0$, the solutions to the equation $x^2 = k$ are $\pm\sqrt{k}$.	$x^2 = 100$ is equivalent to $x = \pm 10$ and $x^2 = 13$ is equivalent to $x = \pm\sqrt{13}$. $x^2 = -2$ has no real solutions.
Method of Completing the Square	To solve an equation in the form $x^2 + bx = d$, add $\left(\frac{b}{2}\right)^2$ to each side of the equation. Factor the resulting perfect square trinomial and solve for x by applying the square root property.	To solve $x^2 + 8x = 3$, add $\left(\frac{8}{2}\right)^2 = \mathbf{16}$ to each side. $x^2 + 8x + \mathbf{16} = 3 + \mathbf{16}$ Add 16. $(x + 4)^2 = 19$ Perfect square trinomial $x + 4 = \pm\sqrt{19}$ Square root property $x = -4 \pm\sqrt{19}$ Add -4.

11.3 Exercises

MyMathLab   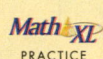 PRACTICE   WATCH   DOWNLOAD   READ   REVIEW

CONCEPTS AND VOCABULARY

1. Give an example of a quadratic equation. How many real solutions can a quadratic equation have?

2. Is a quadratic equation a linear equation or a nonlinear equation?

3. Name three symbolic methods that can be used to solve a quadratic equation.

4. Sketch a graph of a quadratic function that has two x-intercepts and opens downward.

5. Sketch a graph of a quadratic function that has no x-intercepts and opens upward.

6. If the graph of $y = ax^2 + bx + c$ intersects the x-axis twice, how many solutions does the equation $ax^2 + bx + c = 0$ have? Explain.

7. Solve $x^2 = 64$. What property did you use?

8. To solve $x^2 + bx = 6$ by completing the square, what value should be added to each side of the equation?

Exercises 9–16: Is the given equation quadratic?

9. $x^2 - 3x + 1 = 0$

10. $2x^2 - 3 = 0$

11. $3x + 1 = 0$

12. $x^3 - 3x^2 + x = 0$

13. $-3x^2 + x = 16$

14. $x^2 - 1 = 4x$

15. $x^2 = \sqrt{x} + 1$

16. $\frac{1}{x - 1} = 5$

SOLVING QUADRATIC EQUATIONS

Exercises 17–20: Approximate to the nearest hundredth.

17. (a) $1 \pm \sqrt{7}$ (b) $-2 \pm \sqrt{11}$

18. (a) $\pm\dfrac{\sqrt{3}}{2}$ (b) $\pm\dfrac{2\sqrt{5}}{7}$

19. (a) $\dfrac{3 \pm \sqrt{13}}{5}$ (b) $\dfrac{-5 \pm \sqrt{6}}{9}$

20. (a) $\dfrac{2}{5} \pm \dfrac{\sqrt{5}}{5}$ (b) $-\dfrac{3}{7} \pm \dfrac{\sqrt{3}}{7}$

Exercises 21–24: A graph of $y = ax^2 + bx + c$ is given. Use this graph to solve $ax^2 + bx + c = 0$, if possible.

21.

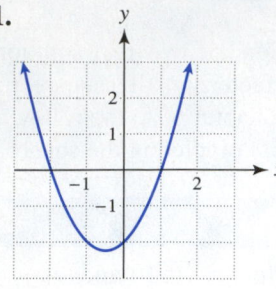

22.

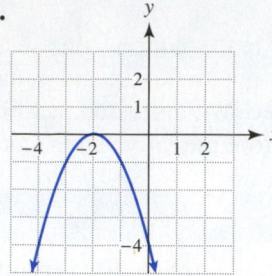

23.

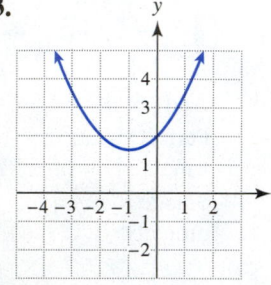

24.

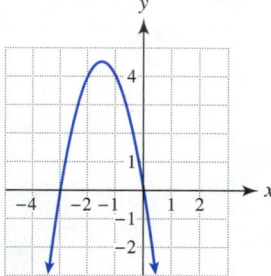

Exercises 25–28: A table of $y = ax^2 + bx + c$ is given. Use this table to solve $ax^2 + bx + c = 0$.

25.

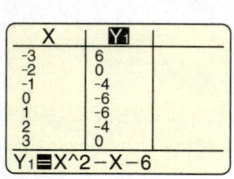

26.

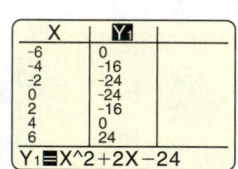

27.

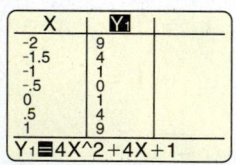

28.

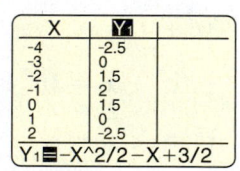

Exercises 29–40: Use each method to solve the equation.

(a) *Symbolic* **(b)** *Graphical* **(c)** *Numerical*

29. $x^2 - 4x - 5 = 0$

30. $x^2 - x - 6 = 0$

31. $x^2 + 2x = 3$

32. $x^2 + 4x = 5$

33. $x^2 = 9$

34. $x^2 = 4$

35. $4x^2 - 4x = 3$

36. $2x^2 + x = 1$

37. $x^2 + 2x = -1$

38. $4x^2 - 4x + 1 = 0$

39. $x^2 + 2 = 0$

40. $-4x^2 - 1 = 2$

Exercises 41–50: Solve by factoring.

41. $x^2 + 2x - 35 = 0$

42. $2x^2 - 7x + 3 = 0$

43. $6x^2 - x - 1 = 0$

44. $x^2 + 4x + 6 = -3x$

45. $4x^2 + 13x + 9 = x$

46. $9x^2 + 4 = 12x$

47. $25x^2 - 350 = 125x$

48. $20x^2 + 150 = 130x$

49. $2(5x^2 + 9) = 27x$

50. $15(3x^2 + x) = 10$

Exercises 51–62: Use the square root property to solve.

51. $x^2 = 144$

52. $4x^2 - 5 = 0$

53. $5x^2 - 64 = 0$

54. $3x^2 = 7$

55. $(x + 1)^2 = 25$

56. $(x + 4)^2 = 9$

57. $(x - 1)^2 = 64$

58. $(x - 3)^2 = 0$

59. $(2x - 1)^2 = 5$

60. $(5x + 3)^2 = 7$

61. $10(x - 5)^2 = 50$

62. $7(3x + 1)^2 = 14$

COMPLETING THE SQUARE

Exercises 63–66: To solve by completing the square, what value should you add to each side of the equation?

63. $x^2 + 4x = -3$

64. $x^2 - 6x = 4$

65. $x^2 - 5x = 4$

66. $x^2 + 3x = 1$

Exercises 67–70: (Refer to Example 4.) Find the term that should be added to the expression to form a perfect square trinomial. Write the resulting perfect square trinomial in factored form.

67. $x^2 - 8x$

68. $x^2 - 5x$

69. $x^2 + 9x$

70. $x^2 + x$

Exercises 71–86: Solve by completing the square.

71. $x^2 - 2x = 24$

72. $x^2 - 2x + \frac{1}{2} = 0$

73. $x^2 + 6x - 2 = 0$

74. $x^2 - 16x = 5$

75. $x^2 - 3x = 5$

76. $x^2 + 5x = 2$

77. $x^2 - 5x + 1 = 0$

78. $x^2 - 9x + 7 = 0$

79. $x^2 - 4 = 2x$

80. $x^2 + 1 = 7x$

81. $2x^2 - 3x = 4$

82. $3x^2 + 6x - 5 = 0$

83. $4x^2 - 8x - 7 = 0$

84. $25x^2 - 20x - 1 = 0$

85. $36x^2 + 18x + 1 = 0$ **86.** $12x^2 + 8x - 2 = 0$

Exercises 87–96: Solve by any method.

87. $3x^2 + 12x = 36$ **88.** $6x^2 + 9x = 27$

89. $x^2 + 4x = -2$ **90.** $x^2 + 6x + 3 = 0$

91. $3x^2 - 4 = 2$ **92.** $-2x^2 + 3 = 1$

93. $-6x^2 + 70 = 16x$

94. $-15x^2 + 25x + 10 = 0$

95. $-3x(x - 8) = 6$ **96.** $-2x(4 - x) = 8$

Exercises 97 and 98: **Thinking Generally** *Solve for x. Assume a and c are positive.*

97. $ax^2 - c = 0$ **98.** $ax^2 + bx = 0$

SOLVING EQUATIONS BY MORE THAN ONE METHOD

Exercises 99–104: Solve the quadratic equation

   (a) symbolically,
   (b) graphically, and
   (c) numerically.

99. $x^2 - 3x - 18 = 0$ **100.** $\frac{1}{2}x^2 + 2x - 6 = 0$

101. $x^2 - 8x + 15 = 0$ **102.** $2x^2 + 3 = 7x$

103. $4(x^2 + 35) = 48x$ **104.** $4x(2 - x) = -5$

SOLVING AN EQUATION FOR A VARIABLE

Exercises 105–112: Solve for the specified variable.

105. $x = y^2 - 1$ for y

106. $x = 9y^2$ for y **107.** $K = \frac{1}{2}mv^2$ for v

108. $c^2 = a^2 + b^2$ for b

109. $E = \dfrac{k}{r^2}$ for r **110.** $W = I^2R$ for I

111. $LC = \dfrac{1}{(2\pi f)^2}$ for f

112. $F = \dfrac{KmM}{r^2}$ for r

APPLICATIONS

113. *Safe Curve Speed* (Refer to Example 8.) Find a safe speed limit x for an airport taxiway curve with the given radius R by using $R = \frac{1}{2}x^2$.
   (a) $R = 450$ feet   **(b)** $R = 800$ feet

114. *Braking Distance* The braking distance y in feet that it takes for a car to stop on wet, level pavement can be estimated by $y = \frac{1}{9}x^2$, where x is the speed of the car in miles per hour. Find the speed associated with each braking distance. (*Source:* L. Haefner, *Introduction to Transportation Systems.*)
   (a) 25 feet   **(b)** 361 feet   **(c)** 784 feet

115. *Falling Object* (Refer to Example 3.) How long does it take for a toy to hit the ground if it is dropped out of a window 60 feet above the ground? Does it take twice as long as it takes to fall from a window 30 feet above the ground?

116. *Falling Object* If a metal ball is thrown *downward* with an initial velocity of 22 feet per second (15 mph) from a 100-foot water tower, its height h in feet above the ground after t seconds is modeled by

$$h(t) = -16t^2 - 22t + 100.$$

   (a) Determine symbolically when the height of the ball is 62 feet.

   (b) Support your result in part (a) either graphically or numerically.

   (c) If the ball is thrown *upward* at 22 feet per second, then its height is given by

$$h(t) = -16t^2 + 22t + 100.$$

Determine when the height of the ball is 80 feet.

Exercises 117 and 118: *Television Size* (*Refer to the introduction of this chapter.*) *The size S of the television screen recommended for a person who sits x feet from the screen ($6 \le x \le 15$) is given by*

$$S(x) = -0.227x^2 + 8.155x - 8.8.$$

If a person buys a television set with a size S screen, how far from the screen should the person sit?

117. $S = 42$ inches   **118.** $S = 50$ inches

119. *Distance* Two athletes start jogging at the same time. One jogs north at 6 miles per hour while the second jogs east at 8 miles per hour. After how long are the two athletes 20 miles apart?

120. *Geometry* A triangle has an area of 35 square inches, and its base is 3 inches more than its height. Find the base and height of the triangle.

121. *Construction* A rectangular plot of land has an area of 520 square feet and is 6 feet longer than it is wide.
(a) Write a quadratic equation in the form $ax^2 + bx + c = 0$, whose solution gives the width of the rectangular plot of land.
(b) Solve the equation.

122. *Modeling Motion* The height y in feet of a tennis ball after x seconds is shown in the graph. Estimate when the ball was 25 feet above the ground.

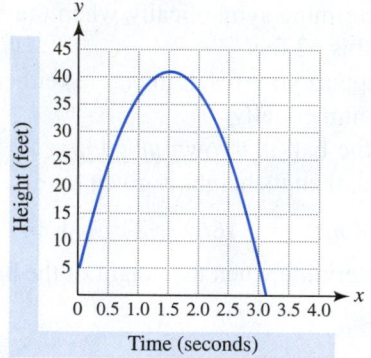

123. *Seedling Growth* (Refer to Exercise 100, Section 11.1.) The heights of melon seedlings grown at different temperatures are shown in the following graph. At what temperatures were the heights of the seedlings about 22 centimeters? (*Source:* R. Pearl, "The growth of *Cucumis melo* seedlings at different temperatures.")

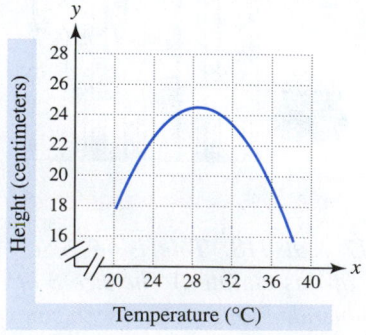

124. *U.S. Population* The three tables show the population of the United States in millions from 1800 through 2010 for selected years.

Year	1800	1820	1840	1860
Population	5	10	17	31

Year	1880	1900	1920	1940
Population	50	76	106	132

Year	1960	1980	2000	2010
Population	179	226	269	308

Source: U.S. Census Bureau.

(a) Without plotting the data, how do you know that the data are nonlinear?
(b) These data are modeled (approximately) by

$$f(x) = 0.0066(x - 1800)^2 + 5.$$

Find the vertex of the graph of f and interpret it.
(c) Estimate when the U.S. population reached 85 million.

125. *MySpace Visitors* (Refer to A Look Into Math at the beginning of this section.) The number of unique monthly visitors V in millions to MySpace x months after January 2010 can be modeled by

$$V(x) = -\frac{25}{144}x^2 + 70.$$

Determine the month when this number of visitors was 56 million.

126. *Federal Debt* The federal debt D in trillions of dollars held by foreign and international investors x years after 1970 can be modeled by

$$D(x) = \frac{1}{320}x^2 - \frac{3}{80}x,$$

where $15 \le x \le 40$.
(a) Evaluate $D(40)$ and interpret the result.
(b) Determine the year when the federal debt held by foreign and international investors first reached $500 billion ($0.5 trillion).

WRITING ABOUT MATHEMATICS

127. Suppose that you are asked to solve

$$ax^2 + bx + c = 0.$$

Explain how the graph of $y = ax^2 + bx + c$ can be used to find any real solutions to the equation.

128. Explain why a quadratic equation could not have more than two solutions. (*Hint:* Consider the graph of $y = ax^2 + bx + c$.)

Group Activity | Working With Real Data

Directions: Form a group of 2 to 4 people. Select someone to record the group's responses for this activity. All members of the group should work cooperatively to answer the questions. If your instructor asks for the results, each member of the group should be prepared to respond.

Personal Consumption Although there was an economic downturn in 2008, personal consumption has grown overall. Table 11.15 lists (approximate) personal consumption C in dollars for selected years x.

TABLE 11.15 Personal Consumption (dollars)

x	1959	1982	1990	1998	2009
C	\$300	\$2000	\$4000	\$6000	\$10,000

Source: BEA.

 (a) Make a scatterplot of these data. Use the window [1955, 2010, 10] by [0, 12000, 2000].

(b) Find a quadratic function given by

$$C(x) = a(x - 1959)^2 + k$$

that models the data. Graph C and the data. (Answers may vary.)

(c) Estimate when personal consumption was \$8000. Compare your answer to the actual value of 2005.

(d) Use your function C to predict consumption in 2020 to the nearest thousand dollars.

11.4 The Quadratic Formula

Solving Quadratic Equations • The Discriminant • Quadratic Equations Having Complex Solutions

A LOOK INTO MATH ▶

To model the stopping distance of a car, highway engineers compute two quantities. The first quantity is the *reaction distance*, which is the distance a car travels from the time a driver first recognizes a hazard until the brakes are applied. The second quantity is *braking distance*, which is the distance a car travels after a driver applies the brakes. *Stopping distance* equals the sum of the reaction distance and the braking distance. If a car is traveling x miles per hour, highway engineers estimate the reaction distance in feet as $\frac{11}{3}x$ and the braking distance in feet as $\frac{1}{9}x^2$. See Figure 11.49. (*Source:* L. Haefner, *Introduction to Transportation Systems.*)

NEW VOCABULARY

☐ Quadratic formula
☐ Discriminant

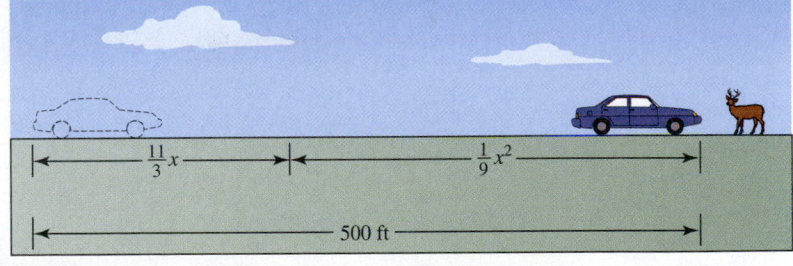

Figure 11.49 Stopping Distance

To estimate the total stopping distance d in feet, add the two expressions to obtain

$$d(x) = \frac{1}{9}x^2 + \frac{11}{3}x.$$

If a car's headlights don't illuminate the road beyond **500** feet, a safe nighttime speed limit x for the car can be determined by solving the quadratic equation

$$\underbrace{\frac{1}{9}x^2}_{\text{Braking Distance}} + \underbrace{\frac{11}{3}x}_{\text{Reaction Distance}} = \underbrace{\mathbf{500}}_{\text{Stopping Distance}}. \qquad \text{Quadratic equation}$$

In this section we learn how to solve this equation with the quadratic formula. (See Example 4 and Exercises 109–112.)

Solving Quadratic Equations

Thus far, we have solved quadratic equations by factoring, the square root property, and completing the square. In this subsection, we derive a formula that can be used to solve *any* quadratic equation. To do this, we solve the general quadratic equation $ax^2 + bx + c = 0$ for x by completing the square. The resulting formula is called the **quadratic formula**. We assume that $a > 0$ and derive this formula as follows.

$ax^2 + bx + c = 0$	Quadratic equation
$ax^2 + bx = -c$	Subtract c.
$x^2 + \dfrac{b}{a}x = -\dfrac{c}{a}$	Divide by a.
$x^2 + \dfrac{b}{a}x + \dfrac{b^2}{4a^2} = -\dfrac{c}{a} + \dfrac{b^2}{4a^2}$	Add $\left(\dfrac{b/a}{2}\right)^2 = \dfrac{b^2}{4a^2}$.
$\left(x + \dfrac{b}{2a}\right)^2 = -\dfrac{c}{a} + \dfrac{b^2}{4a^2}$	Perfect square trinomial
$\left(x + \dfrac{b}{2a}\right)^2 = -\dfrac{c \cdot 4a}{a \cdot 4a} + \dfrac{b^2}{4a^2}$	Multiply $-\dfrac{c}{a}$ by $\dfrac{4a}{4a}$.
$\left(x + \dfrac{b}{2a}\right)^2 = -\dfrac{4ac}{4a^2} + \dfrac{b^2}{4a^2}$	Simplify.
$\left(x + \dfrac{b}{2a}\right)^2 = \dfrac{-4ac + b^2}{4a^2}$	Add fractions.
$\left(x + \dfrac{b}{2a}\right)^2 = \dfrac{b^2 - 4ac}{4a^2}$	Rewrite.
$x + \dfrac{b}{2a} = \pm\sqrt{\dfrac{b^2 - 4ac}{4a^2}}$	Square root property
$x = -\dfrac{b}{2a} \pm \sqrt{\dfrac{b^2 - 4ac}{4a^2}}$	Add $-\dfrac{b}{2a}$.
$x = -\dfrac{b}{2a} \pm \dfrac{\sqrt{b^2 - 4ac}}{2a}$	Property of square roots
$x = \dfrac{-b \pm \sqrt{b^2 - 4ac}}{2a}$	Combine fractions.

QUADRATIC FORMULA

The solutions to $ax^2 + bx + c = 0$ with $a \neq 0$ are given by

$$x = \frac{-b \pm \sqrt{b^2 - 4ac}}{2a}.$$

NOTE: The quadratic formula can be used to solve *any* quadratic equation. It always "works."

EXAMPLE 1 **Solving a quadratic equation having two solutions**

Solve the equation $2x^2 - 3x - 1 = 0$. Support your results graphically.

Solution

Symbolic Solution Let $a = 2$, $b = -3$, and $c = -1$ in $2x^2 - 3x - 1 = 0$.

$$x = \frac{-b \pm \sqrt{b^2 - 4ac}}{2a} \qquad \text{Quadratic formula}$$

$$x = \frac{-(-3) \pm \sqrt{(-3)^2 - 4(2)(-1)}}{2(2)} \qquad \text{Substitute for } a, b, \text{ and } c.$$

$$x = \frac{3 \pm \sqrt{17}}{4} \qquad \text{Simplify.}$$

The solutions are $\frac{3 + \sqrt{17}}{4} \approx 1.78$ and $\frac{3 - \sqrt{17}}{4} \approx -0.28$.

Graphical Solution The graph of $y = 2x^2 - 3x - 1$ is shown in Figure 11.50. Note that the two x-intercepts correspond to the two solutions for $2x^2 - 3x - 1 = 0$. Estimating from this graph, we see that the solutions are approximately -0.25 and 1.75, which supports our symbolic solution. (You could also use a graphing calculator to find the x-intercepts.)

Two Solutions

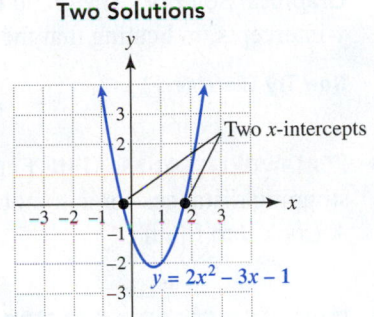

Figure 11.50

Now Try Exercise 9

CRITICAL THINKING

Use the equation and results from Example 1 to evaluate each expression mentally.

$$2\left(\frac{3 + \sqrt{17}}{4}\right)^2 - 3\left(\frac{3 + \sqrt{17}}{4}\right) - 1 \quad \text{and} \quad 2\left(\frac{3 - \sqrt{17}}{4}\right)^2 - 3\left(\frac{3 - \sqrt{17}}{4}\right) - 1$$

EXAMPLE 2   **Solving a quadratic equation having one solution**

Solve the equation $25x^2 + 20x + 4 = 0$. Support your result graphically.

Solution

Symbolic Solution Let $a = 25$, $b = 20$, and $c = 4$ in $25x^2 + 20x + 4 = 0$.

$$x = \frac{-b \pm \sqrt{b^2 - 4ac}}{2a} \qquad \text{Quadratic formula}$$

$$= \frac{-20 \pm \sqrt{20^2 - 4(25)(4)}}{2(25)} \qquad \text{Substitute for } a, b, \text{ and } c.$$

$$= \frac{-20 \pm \sqrt{0}}{50} \qquad \text{Simplify.}$$

$$= \frac{-20}{50} = -0.4 \qquad \sqrt{0} = 0$$

There is one solution, -0.4.

Graphical Solution The graph of $y = 25x^2 + 20x + 4$ is shown in Figure 11.51. Note that the one x-intercept, -0.4, corresponds to the solution to $25x^2 + 20x + 4 = 0$.

Now Try Exercise 11

One Solution

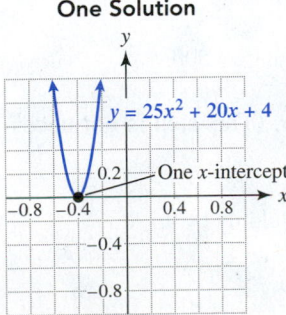

Figure 11.51

EXAMPLE 3 | **Recognizing a quadratic equation having no real solutions**

Solve the equation $5x^2 - x + 3 = 0$. Support your result graphically.

Solution

Symbolic Solution Let $a = 5$, $b = -1$, and $c = 3$ in $5x^2 - 1x + 3 = 0$.

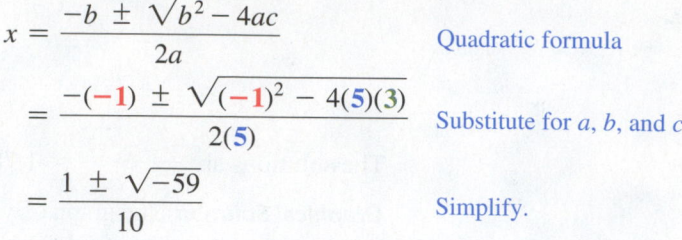

No Real Solutions

$y = 5x^2 - x + 3$

No x-intercepts

Figure 11.52

$$x = \frac{-b \pm \sqrt{b^2 - 4ac}}{2a} \qquad \text{Quadratic formula}$$

$$= \frac{-(-1) \pm \sqrt{(-1)^2 - 4(5)(3)}}{2(5)} \qquad \text{Substitute for } a, b, \text{ and } c.$$

$$= \frac{1 \pm \sqrt{-59}}{10} \qquad \text{Simplify.}$$

There are *no real solutions* to this equation because $\sqrt{-59}$ *is not a real number*. (Later in this section we discuss how to find complex solutions to quadratic equations like this one.)

Graphical Solution The graph of $y = 5x^2 - x + 3$ is shown in Figure 11.52. There are no x-intercepts, indicating that the equation $5x^2 - x + 3 = 0$ has no real solutions.

Now Try Exercise 13

▶ **REAL-WORLD CONNECTION** Earlier in this section we discussed how engineers estimate safe stopping distances for automobiles. In the next example we solve the equation presented in A Look Into Math.

EXAMPLE 4 | **Modeling stopping distance**

If a car's headlights do not illuminate the road beyond 500 feet, estimate a safe nighttime speed limit x for the car by solving $\frac{1}{9}x^2 + \frac{11}{3}x = 500$.

500 feet

Solution

Begin by subtracting 500 from each side of the given equation.

$$\frac{1}{9}x^2 + \frac{11}{3}x - 500 = 0 \qquad \text{Subtract 500.}$$

To eliminate fractions, multiply each side by the LCD, which is 9. (This step is not necessary, but it makes the problem easier to work.)

$$x^2 + 33x - 4500 = 0 \qquad \text{Multiply by 9.}$$

Now let $a = 1$, $b = 33$, and $c = -4500$ in the quadratic formula.

$$x = \frac{-b \pm \sqrt{b^2 - 4ac}}{2a} \qquad \text{Quadratic formula}$$

$$= \frac{-33 \pm \sqrt{33^2 - 4(1)(-4500)}}{2(1)}$$ Substitute for a, b, and c.

$$= \frac{-33 \pm \sqrt{19,089}}{2}$$ Simplify.

The solutions are

$$\frac{-33 + \sqrt{19,089}}{2} \approx 52.6 \quad \text{and} \quad \frac{-33 - \sqrt{19,089}}{2} \approx -85.6.$$

The negative solution has no physical meaning because negative speeds are not possible. The other solution is 52.6, so an appropriate speed limit might be 50 miles per hour.

Now Try Exercise 109

The Discriminant

The expression $b^2 - 4ac$ in the quadratic formula is called the **discriminant**. It provides information about the number of solutions to a quadratic equation.

READING CHECK

How can you use the discriminant to determine the number of solutions to a quadratic equation?

THE DISCRIMINANT AND QUADRATIC EQUATIONS

To determine the number of solutions to the quadratic equation $ax^2 + bx + c = 0$, evaluate the discriminant $b^2 - 4ac$.

1. If $b^2 - 4ac > 0$, there are two real solutions.
2. If $b^2 - 4ac = 0$, there is one real solution.
3. If $b^2 - 4ac < 0$, there are no real solutions; there are two complex solutions.

GRAPHS AND THE DISCRIMINANT The graph of $y = ax^2 + bx + c$ can be used to determine the sign of the discriminant, $b^2 - 4ac$. Figure 11.53 illustrates several possibilities. For example, in Figure 11.53(a) neither graph with $a > 0$ or $a < 0$ intersects the x-axis. Therefore these graphs indicate that the equation $ax^2 + bx + c = 0$ has no real solutions and that the discriminant must be negative: $b^2 - 4ac < 0$. Figures 11.53(b) and (c) can be done similarly.

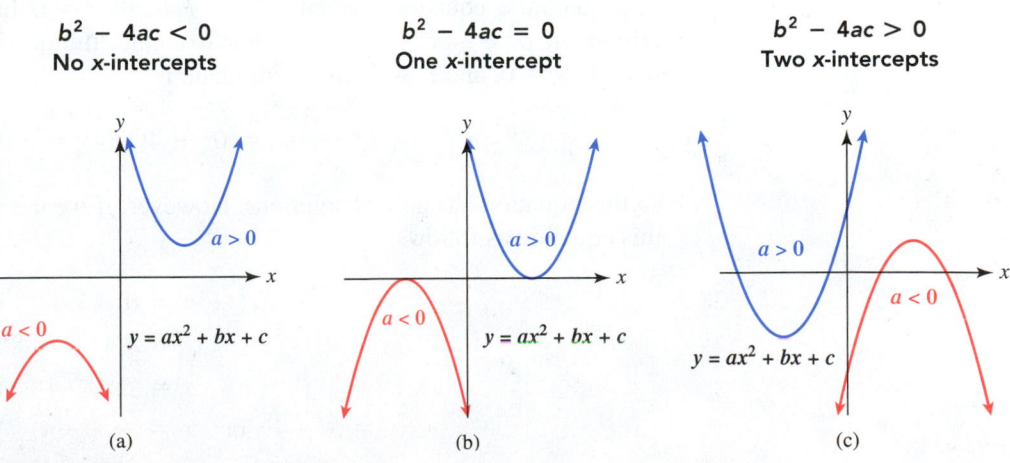

Figure 11.53

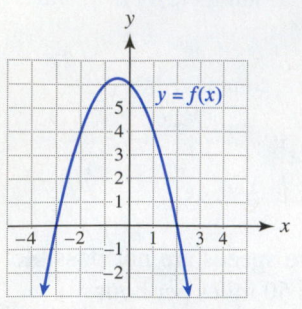

Figure 11.54

EXAMPLE 5 **Analyzing graphs of quadratic functions**

A graph of $f(x) = ax^2 + bx + c$ is shown in Figure 11.54.
(a) State whether $a > 0$ or $a < 0$.
(b) Solve the equation $ax^2 + bx + c = 0$.
(c) Determine whether the discriminant is positive, negative, or zero.

Solution
(a) The parabola opens downward, so $a < 0$.
(b) The graph of $f(x) = ax^2 + bx + c$ intersects the x-axis at -3 and 2. Therefore $f(-3) = 0$ and $f(2) = 0$. The solutions to $ax^2 + bx + c = 0$ are -3 and 2.
(c) There are two solutions, so the discriminant is positive.

Now Try Exercise 33

EXAMPLE 6 **Using the discriminant**

Use the discriminant to determine the number of solutions to $4x^2 + 25 = 20x$. Then solve the equation, using the quadratic formula.

Solution
Write the equation as $4x^2 - 20x + 25 = 0$ so that $a = 4$, $b = -20$, and $c = 25$. The discriminant evaluates to

$$b^2 - 4ac = (-20)^2 - 4(4)(25) = 0.$$

Thus there is **one real solution**.

$$x = \frac{-b \pm \sqrt{b^2 - 4ac}}{2a} \qquad \text{Quadratic formula}$$

$$= \frac{-(-20) \pm \sqrt{0}}{2(4)} \qquad \text{Substitute.}$$

$$= \frac{20}{8} = 2.5 \qquad \text{Simplify.}$$

The only solution is 2.5.

Now Try Exercise 39(a), (b)

Quadratic Equations Having Complex Solutions

The quadratic equation written as $ax^2 + bx + c = 0$ has no real solutions if the discriminant, $b^2 - 4ac$, is negative. For example, the quadratic equation $x^2 + 4 = 0$ has $a = 1$, $b = 0$, and $c = 4$. Its discriminant is

$$b^2 - 4ac = 0^2 - 4(1)(4) = -16 < 0,$$

so this equation has no real solutions. However, if we use complex numbers, we can solve this equation as follows.

$$x^2 + 4 = 0 \qquad \text{Given equation}$$

$$x^2 = -4 \qquad \text{Subtract 4.}$$

$$x = \pm\sqrt{-4} \qquad \text{Square root property}$$

$$x = \sqrt{-4} \quad \text{or} \quad x = -\sqrt{-4} \qquad \text{Meaning of } \pm$$

$$x = 2i \quad \text{or} \quad x = -2i \qquad \text{The expression } \sqrt{-a}$$

No x-intercepts

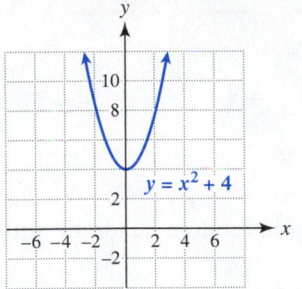

Figure 11.55

The solutions are $\pm 2i$. We check each solution to $x^2 + 4 = 0$ as follows.

$$(2i)^2 + 4 = (2)^2 i^2 + 4 = 4(-1) + 4 = 0 \checkmark \quad \text{It checks.}$$
$$(-2i)^2 + 4 = (-2)^2 i^2 + 4 = 4(-1) + 4 = 0 \checkmark \quad \text{It checks.}$$

The fact that the equation $x^2 + 4 = 0$ has only imaginary solutions is apparent from the graph of $y = x^2 + 4$, shown in Figure 11.55. This parabola does not intersect the x-axis, so the equation $x^2 + 4 = 0$ has no real solutions.

These results can be generalized as follows.

THE EQUATION $x^2 + k = 0$

If $k > 0$, the solutions to $x^2 + k = 0$ are given by $x = \pm i\sqrt{k}$.

NOTE: This result is a form of the *square root property* that includes complex solutions.

EXAMPLE 7   **Solving a quadratic equation having complex solutions**

Solve $x^2 + 5 = 0$.

Solution
The solutions are $\pm i\sqrt{5}$. That is, $x = i\sqrt{5}$ or $x = -i\sqrt{5}$.

Now Try Exercise 57

When $b \neq 0$, the preceding method cannot be used. Consider the quadratic equation $2x^2 + x + 3 = 0$, which has $a = 2$, $b = 1$, and $c = 3$. Its discriminant is negative.

$$b^2 - 4ac = 1^2 - 4(2)(3) = -23 < 0$$

This equation has two complex solutions, as demonstrated in the next example.

EXAMPLE 8   **Solving a quadratic equation having complex solutions**

Solve $2x^2 + x + 3 = 0$. Write your answer in standard form: $a + bi$.

Solution
Let $a = 2$, $b = 1$, and $c = 3$.

$$x = \frac{-b \pm \sqrt{b^2 - 4ac}}{2a} \qquad \text{Quadratic formula}$$

$$= \frac{-1 \pm \sqrt{1^2 - 4(2)(3)}}{2(2)} \qquad \text{Substitute for } a, b, \text{ and } c.$$

$$= \frac{-1 \pm \sqrt{-23}}{4} \qquad \text{Simplify.}$$

$$= \frac{-1 \pm i\sqrt{23}}{4} \qquad \sqrt{-23} = i\sqrt{23}$$

$$= -\frac{1}{4} \pm i\frac{\sqrt{23}}{4} \qquad \text{Property of fractions}$$

The solutions are $-\frac{1}{4} + i\frac{\sqrt{23}}{4}$ and $-\frac{1}{4} - i\frac{\sqrt{23}}{4}$.

Now Try Exercise 73

CALCULATOR HELP

To set your calculator in $a + bi$ mode or to access the imaginary unit i, see Appendix A (page AP-7).

CRITICAL THINKING

Use the results of Example 8 to evaluate each expression mentally.

$$2\left(-\tfrac{1}{4} + i\tfrac{\sqrt{23}}{4}\right)^2 + \left(-\tfrac{1}{4} + i\tfrac{\sqrt{23}}{4}\right) + 3 \quad \text{and} \quad 2\left(-\tfrac{1}{4} - i\tfrac{\sqrt{23}}{4}\right)^2 + \left(-\tfrac{1}{4} - i\tfrac{\sqrt{23}}{4}\right) + 3$$

Sometimes we can use properties of radicals to simplify a solution to a quadratic equation, as demonstrated in the next example.

EXAMPLE 9 **Solving a quadratic equation having complex solutions**

Solve $\tfrac{3}{4}x^2 + 1 = x$. Write your answer in standard form: $a + bi$.

Solution

Begin by subtracting x from each side of the equation and then multiply by 4 to clear fractions. The resulting equation is $3x^2 - 4x + 4 = 0$. Substitute $a = 3$, $b = -4$, and $c = 4$ in the quadratic formula.

$$x = \frac{-b \pm \sqrt{b^2 - 4ac}}{2a} \qquad \text{Quadratic formula}$$

$$= \frac{-(-4) \pm \sqrt{(-4)^2 - 4(3)(4)}}{2(3)} \qquad \text{Substitute.}$$

$$= \frac{4 \pm \sqrt{-32}}{6} \qquad \text{Simplify.}$$

$$= \frac{4 \pm 4i\sqrt{2}}{6} \qquad \sqrt{-32} = i\sqrt{32} = i\sqrt{16}\sqrt{2} = 4i\sqrt{2}$$

$$= \frac{2}{3} \pm \frac{2}{3}i\sqrt{2} \qquad \text{Property of fractions; simplify.}$$

▮ **Now Try Exercise 79**

In the next example, we use completing the square to obtain complex solutions.

EXAMPLE 10 **Completing the square to find complex solutions**

Solve $x(x + 2) = -2$ by completing the square.

Solution

After applying the distributive property, we obtain the equation $x^2 + 2x = -2$. Because $b = 2$, add $\left(\tfrac{b}{2}\right)^2 = \left(\tfrac{2}{2}\right)^2 = 1$ to each side of the equation.

$$x^2 + 2x = -2 \qquad \text{Equation to be solved}$$

$$x^2 + 2x + 1 = -2 + 1 \qquad \text{Add 1 to each side.}$$

$$(x + 1)^2 = -1 \qquad \text{Perfect square trinomial; add.}$$

$$x + 1 = \pm\sqrt{-1} \qquad \text{Square root property}$$

$$x + 1 = \pm i \qquad \sqrt{-1} = i, \text{ the imaginary unit}$$

$$x = -1 \pm i \qquad \text{Add } -1 \text{ to each side.}$$

The solutions are $-1 + i$ and $-1 - i$.

▮ **Now Try Exercise 91**

11.4 Putting It All Together

CONCEPT	EXPLANATION	EXAMPLES
Quadratic Formula	The quadratic formula can be used to solve *any* quadratic equation written as $ax^2 + bx + c = 0$. The solutions are given by $$x = \frac{-b \pm \sqrt{b^2 - 4ac}}{2a}.$$	For the equation $$2x^2 - 3x + 1 = 0$$ with $a = 2$, $b = -3$, and $c = 1$, the solutions are $$\frac{-(-3) \pm \sqrt{(-3)^2 - 4(2)(1)}}{2(2)} = \frac{3 \pm \sqrt{1}}{4} = 1, \frac{1}{2}.$$
The Discriminant	The expression $b^2 - 4ac$ is called the discriminant. **1.** $b^2 - 4ac > 0$ indicates two real solutions. **2.** $b^2 - 4ac = 0$ indicates one real solution. **3.** $b^2 - 4ac < 0$ indicates no real solutions; rather, there are two complex solutions.	For the equation $$x^2 + 4x - 1 = 0$$ with $a = 1$, $b = 4$, and $c = -1$, the discriminant is $$b^2 - 4ac = 4^2 - 4(1)(-1) = 20 > 0,$$ indicating two real solutions.
Quadratic Formula and Complex Solutions	If the discriminant is negative ($b^2 - 4ac < 0$), the two solutions are complex numbers that are not real numbers. If $k > 0$, the solutions to $x^2 + k = 0$ are given by $x = \pm i\sqrt{k}$.	Solve $2x^2 - x + 3 = 0$. $$x = \frac{-(-1) \pm \sqrt{(-1)^2 - 4(2)(3)}}{2(2)}$$ $$= \frac{1 \pm \sqrt{-23}}{4} = \frac{1}{4} \pm i\frac{\sqrt{23}}{4}$$ $x^2 + 7 = 0$ is equivalent to $x = \pm i\sqrt{7}$.

Three Ways to Solve $x^2 - x - 2 = 0$

Symbolic Solution

Solve $x^2 - x - 2 = 0$.

$$x = \frac{-(-1) \pm \sqrt{(-1)^2 - 4(1)(-2)}}{2(1)}$$

$$x = \frac{1 \pm 3}{2}$$

$$x = 2, -1$$

Solutions are -1 and 2.

Numerical Solution

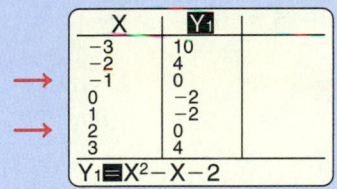

$x^2 - x - 2 = 0$ when $x = -1$ or 2.

Graphical Solution

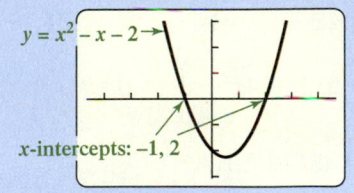

x-intercepts: -1, 2

The *x*-intercepts are -1 and 2.

11.4 Exercises

MyMathLab

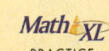 Math XL PRACTICE WATCH DOWNLOAD READ REVIEW

CONCEPTS AND VOCABULARY

1. What is the quadratic formula used for?

2. What basic algebraic technique is used to derive the quadratic formula?

3. Write the discriminant.

4. If the discriminant evaluates to 0, what does that indicate about the quadratic equation?

5. Name four symbolic techniques for solving a quadratic equation.

6. Does every quadratic equation have at least one real solution? Explain.

7. Solve $x^2 - k = 0$, if $k > 0$.

8. Solve $x^2 + k = 0$, if $k > 0$.

THE QUADRATIC FORMULA

Exercises 9–14: Use the quadratic formula to solve the equation. Support your result graphically. If there are no real solutions, say so.

9. $2x^2 + 11x - 6 = 0$ **10.** $x^2 + 2x - 24 = 0$

11. $-x^2 + 2x - 1 = 0$ **12.** $3x^2 - x + 1 = 0$

13. $-2x^2 + x - 1 = 0$ **14.** $-x^2 + 4x - 4 = 0$

Exercises 15–32: Solve by using the quadratic formula. If there are no real solutions, say so.

15. $x^2 - 6x - 16 = 0$ **16.** $2x^2 - 9x + 7 = 0$

17. $4x^2 - x - 1 = 0$ **18.** $-x^2 + 2x + 1 = 0$

19. $-3x^2 + 2x - 1 = 0$ **20.** $x^2 + x + 3 = 0$

21. $36x^2 - 36x + 9 = 0$ **22.** $4x^2 - 5.6x + 1.96 = 0$

23. $2x(x - 3) = 2$ **24.** $x(x + 1) + x = 5$

25. $(x - 1)(x + 1) + 2 = 4x$

26. $\frac{1}{2}(x - 6) = x^2 + 1$ **27.** $\frac{1}{2}x(x + 1) = 2x^2 - \frac{3}{2}$

28. $\frac{1}{2}x^2 - \frac{1}{4}x + \frac{1}{2} = x$ **29.** $2x(x - 1) = 7$

30. $3x(x - 4) = 4$ **31.** $-3x^2 + 10x - 5 = 0$

32. $-2x^2 + 4x - 1 = 0$

THE DISCRIMINANT

Exercises 33–38: A graph of $y = ax^2 + bx + c$ is shown.
  (a) State whether $a > 0$ or $a < 0$.
  (b) Solve $ax^2 + bx + c = 0$, if possible.
  (c) Determine whether the discriminant is positive, negative, or zero.

33.   **34.**

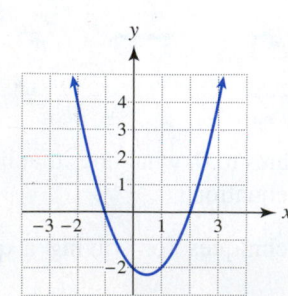 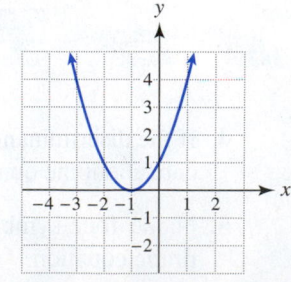

35.   **36.**

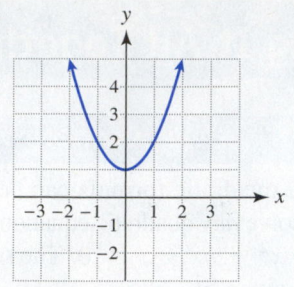 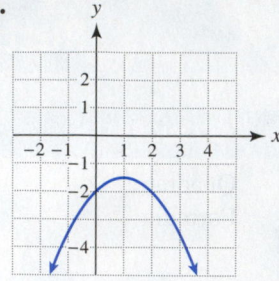

37.   **38.**

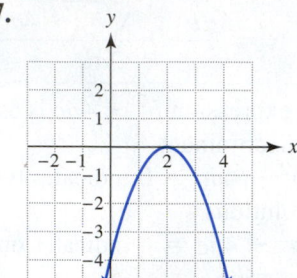 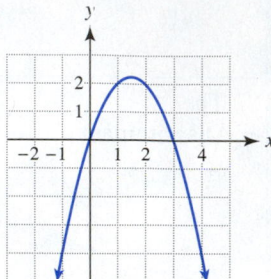

Exercises 39–46: Do the following for the given equation.
  (a) Evaluate the discriminant.
  (b) How many real solutions are there?
  (c) Support your answer for part (b) graphically.

39. $3x^2 + x - 2 = 0$ **40.** $5x^2 - 13x + 6 = 0$

41. $x^2 - 4x + 4 = 0$ **42.** $\frac{1}{4}x^2 + 4 = 2x$

43. $\frac{1}{2}x^2 + \frac{3}{2}x + 2 = 0$ **44.** $x - 3 = 2x^2$

45. $x(x + 3) = 3$

46. $(4x - 1)(x - 3) = -25$

Exercises 47–56: Use the quadratic formula to find any x-intercepts on the graph of the equation.

47. $y = x^2 - 2x - 1$ **48.** $y = x^2 + 3x + 1$

49. $y = -2x^2 - x + 3$ **50.** $y = -3x^2 - x + 4$

51. $y = x^2 + x + 5$ **52.** $y = 3x^2 - 2x + 5$

53. $y = x^2 + 9$ **54.** $y = x^2 + 11$

55. $y = 3x^2 + 4x - 2$ **56.** $y = 4x^2 - 2x - 3$

COMPLEX SOLUTIONS

Exercises 57–88: Solve the equation. Write complex solutions in standard form.

57. $x^2 + 9 = 0$ **58.** $x^2 + 16 = 0$

59. $x^2 + 80 = 0$ **60.** $x^2 + 20 = 0$

61. $x^2 + \frac{1}{4} = 0$ **62.** $x^2 + \frac{9}{4} = 0$

63. $16x^2 + 9 = 0$ **64.** $25x^2 + 36 = 0$

65. $x^2 = -6$ **66.** $x^2 = -75$

67. $x^2 - 3 = 0$ **68.** $x^2 - 8 = 0$

69. $x^2 + 2 = 0$ **70.** $x^2 + 4 = 0$

71. $x^2 - x + 2 = 0$ **72.** $x^2 + 2x + 3 = 0$

73. $2x^2 + 3x + 4 = 0$ **74.** $3x^2 - x = 1$

75. $x^2 + 1 = 4x$ **76.** $3x^2 + 2 = x$

77. $x(x + 1) = -2$ **78.** $x(x - 4) = -8$

79. $5x^2 + 2x + 4 = 0$ **80.** $7x^2 - 2x + 4 = 0$

81. $\frac{1}{2}x^2 + \frac{3}{4}x = -1$ **82.** $-\frac{1}{3}x^2 + x = 2$

83. $x(x + 2) = x - 4$ **84.** $x - 5 = 2x(2x + 1)$

85. $x(2x - 1) = 1 + x$ **86.** $2x = x(3 - 4x)$

87. $x^2 = x(1 - x) - 2$ **88.** $2x^2 = 2x(5 - x) - 8$

COMPLETING THE SQUARE

Exercises 89–94: Solve by completing the square.

89. $x^2 + 2x + 4 = 0$ **90.** $x^2 - 2x + 2 = 0$

91. $x(x + 4) = -5$ **92.** $x(8 - x) = 25$

93. $2x^2 - 4x + 6 = 0$ **94.** $2x^2 + 2x + 1 = 0$

YOU DECIDE THE METHOD

Exercises 95–108: Find exact solutions to the quadratic equation, using a method of your choice. Explain why you chose the method you did. Answers may vary.

95. $x^2 - 3x + 2 = 0$ **96.** $x^2 + 2x + 1 = 0$

97. $0.5x^2 - 1.75x = 1$ **98.** $\frac{3}{5}x^2 + \frac{9}{10}x = \frac{3}{5}$

99. $x^2 - 5x + 2 = 0$ **100.** $2x^2 - x - 4 = 0$

101. $2x^2 + x = -8$ **102.** $4x^2 = 2x - 3$

103. $4x^2 - 1 = 0$ **104.** $3x^2 = 9$

105. $3x^2 + 6 = 0$ **106.** $4x^2 + 7 = 0$

107. $9x^2 + 1 = 6x$ **108.** $10x^2 + 15x = 25$

APPLICATIONS

*Exercises 109–112: **Modeling Stopping Distance** (Refer to Example 4.) Use $d = \frac{1}{9}x^2 + \frac{11}{3}x$ to find a safe speed x for the following stopping distances d.*

109. 42 feet

110. 152 feet

111. 390 feet

112. 726 feet

113. *Global Music Sales* (Refer to Example 9, Section 11.3.) From 1989 to 2009, global music sales S of compact discs and tapes in billions of dollars can be modeled by $S(x) = -0.095x^2 + 1.85x + 6$, where x is years after 1989. Estimate symbolically when sales were $12 billion. (*Source: RIAA, Bain Analysis.*)

114. *Monthly Facebook Visitors* The number of *unique* monthly Facebook visitors in millions x years after 2006 can be modeled by

$$V(x) = 10.75x^2 - 24x + 35.$$

Assuming current trends continue, estimate symbolically when this number might reach 530 million.

115. *Groupon's Growth* Groupon negotiates coupons for items discounted 50–90% off by having thousands of subscribers. As a result, it experienced a dramatic increase in value from October 2010 to March 2011. The function

$$G(x) = 0.4x^2 + 1.8x + 6$$

approximates the company's value in billions of dollars x months after October 2010.

(a) Evaluate $G(0)$. Interpret the result.

(b) Determine when Groupon's value was about $15 billion.

116. *Foursquare Users* Foursquare provides a service that allows your friends to know your whereabouts by "checking in." From March 2010 to March 2011, it experienced amazing growth. The function

$$F(x) = \frac{1}{18}x^2 - \frac{1}{12}x + \frac{1}{2}$$

approximates Foursquare users in millions x months after March 2010.
(a) Evaluate $F(3)$. Interpret the result.
(b) Determine symbolically when Foursquare had 4.25 million users.

117. *U.S. AIDS Deaths* The cumulative numbers in thousands of AIDS deaths from 1984 through 1994 may be modeled by

$$f(x) = 2.39x^2 + 5.04x + 5.1,$$

where $x = 0$ corresponds to 1984, $x = 1$ corresponds to 1985, and so on until $x = 10$ corresponds to 1994. See the accompanying graph. Use the formula for $f(x)$ to estimate the year when the total number of AIDS deaths reached 200 thousand. Compare your result with that shown in the graph.

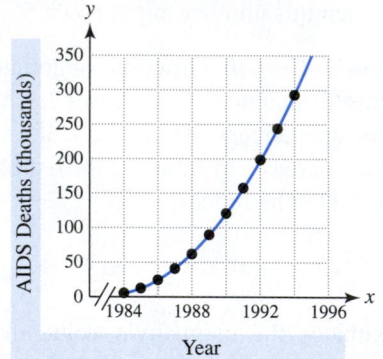

118. *Canoeing* A camper paddles a canoe 2 miles downstream in a river that has a 2-mile-per-hour current. To return to camp, the canoeist travels upstream on a different branch of the river. It is 4 miles long and has a 1-mile-per-hour current. The total trip (both ways) takes 3 hours. Find the average speed of the canoe in still water. (*Hint:* Time equals distance divided by rate.)

119. *Airplane Speed* A pilot flies 500 miles against a 20-mile-per-hour wind. On the next day, the pilot flies back home with a 10-mile-per-hour tail wind. The total trip (both ways) takes 4 hours. Find the speed of the airplane without a wind.

120. *Distance* Two cars leave an intersection, one traveling south and one traveling east, as shown in the figure at the top of the next column. After 1 hour the two cars are 50 miles apart and the car traveling east has traveled 10 miles farther than the car traveling south. How far did each car travel?

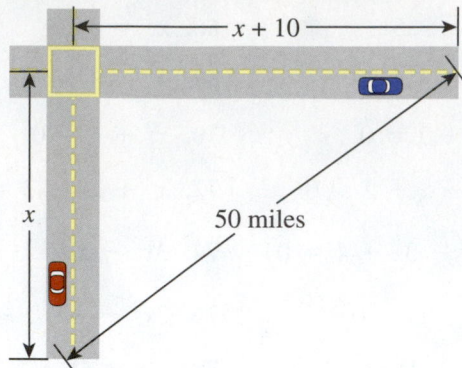

121. *Screen Dimensions* The width of a rectangular computer screen is 3 inches more than its height. If the area of the screen is 154 square inches, find its dimensions
(a) graphically,
(b) numerically, and
(c) symbolically.

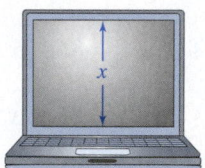

122. *Sidewalk Dimension* A rectangular flower garden in a park is 30 feet wide and 40 feet long. A sidewalk around the perimeter of the garden is being planned, as shown in the figure. The gardener has enough money to pour 624 square feet of cement sidewalk. Find the width of the sidewalk.

123. *Modeling Water Flow* When water runs out of a hole in a cylindrical container, the height of the water in the container can often be modeled by a quadratic function. The data in the table show the height y in centimeters of water at 30-second intervals in a metal can that has a small hole in it.

Time	0	30	60	90
Height	16	11.9	8.4	5.3

Time	120	150	180
Height	3.1	1.4	0.5

These data are modeled by

$$f(x) = 0.0004x^2 - 0.15x + 16.$$

(a) Explain why a linear function would not be appropriate for modeling these data.

(b) Use the table to estimate the time at which the height was 7 centimeters.

(c) Use the quadratic formula to solve part (b).

124. *Hospitals* The general trend in the number of hospitals in the United States from 1945 through 2000 is shown in the graph and can be modeled by

$$f(x) = -1.38x^2 + 84x + 5865,$$

where $x = 5$ corresponds to 1945, $x = 10$ corresponds to 1950, and so on until $x = 60$ represents 2000.

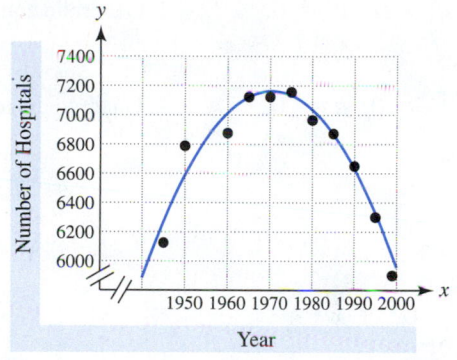

(a) Describe any trends in the numbers of hospitals from 1945 to 2000.

(b) What information does the vertex give?

(c) Use the formula for $f(x)$ to estimate the number of hospitals in 1970. Compare your result with that shown in the graph.

(d) Use the formula for $f(x)$ to estimate the year (or years) when there were 6300 hospitals. Compare your result with that shown in the graph.

WRITING ABOUT MATHEMATICS

125. Explain how the discriminant $b^2 - 4ac$ can be used to determine the number of solutions to a quadratic equation.

126. Let $f(x) = ax^2 + bx + c$ be a quadratic function. If you know the value of $b^2 - 4ac$, what information does this give you about the graph of f? Explain your answer.

SECTIONS 11.3 and 11.4 **Checking Basic Concepts**

1. Solve the quadratic equation $2x^2 - 7x + 3 = 0$ symbolically and graphically.

2. Use the square root property to solve $x^2 = 5$.

3. Complete the square to solve $x^2 - 4x + 1 = 0$.

4. Solve the equation $x^2 + y^2 = 1$ for y.

5. Use the quadratic formula to solve each equation.
 (a) $2x^2 = 3x + 1$
 (b) $9x^2 - 24x + 16 = 0$
 (c) $x^2 + x + 2 = 0$

6. Calculate the discriminant for each equation and give the number of *real* solutions.
 (a) $x^2 - 5x + 5 = 0$
 (b) $2x^2 - 5x + 4 = 0$
 (c) $49x^2 - 56x + 16 = 0$

7. Solve each equation.
 (a) $x^2 + 5 = 0$
 (b) $x^2 + x + 3 = 0$

11.5 Quadratic Inequalities

Basic Concepts • Graphical and Numerical Solutions • Symbolic Solutions

A LOOK INTO MATH ▶ Parabolas are frequently used in highway design. Sometimes it is necessary for engineers to determine where a highway is below a particular elevation. To do this, they may need to solve a *quadratic inequality*. (See Example 5.) In this section we discuss how to solve quadratic inequalities graphically, numerically, and symbolically.

Basic Concepts

If the equals sign in a quadratic equation is replaced with $>$, $\geq$, $<$, or $\leq$, a **quadratic inequality** results. Examples of quadratic inequalities include

$$x^2 + 4x - 3 < 0, \quad 5x^2 \geq 5, \quad \text{and} \quad 1 - z \leq z^2. \qquad \text{Quadratic inequalities}$$

Any quadratic equation can be written as

$$ax^2 + bx + c = 0, \quad a \neq 0, \qquad \text{Quadratic equation}$$

so any quadratic inequality can be written as

$$ax^2 + bx + c > 0, \quad a \neq 0, \qquad \text{Quadratic inequality}$$

where $>$ may be replaced with $\geq$, $<$, or $\leq$.

NEW VOCABULARY

☐ Quadratic inequality
☐ Test value

EXAMPLE 1   **Identifying a quadratic inequality**

Determine whether the inequality is quadratic.
(a) $5x + x^2 - x^3 \leq 0$ **(b)** $4 + 5x^2 > 4x^2 + x$

Solution
(a) The inequality $5x + x^2 - x^3 \leq 0$ is not quadratic because it has an x^3-term.
(b) Write the inequality as follows.

$$
\begin{aligned}
4 + 5x^2 &> 4x^2 + x && \text{Given inequality} \\
4 + 5x^2 - 4x^2 - x &> 0 && \text{Subtract } 4x^2 \text{ and } x. \\
4 + x^2 - x &> 0 && \text{Combine like terms.} \\
x^2 - x + 4 &> 0 && \text{Rewrite.}
\end{aligned}
$$

Because the inequality can be written in the form $ax^2 + bx + c > 0$ with $a = 1$, $b = -1$, and $c = 4$, it is a quadratic inequality.

Now Try Exercises 7, 11

Graphical and Numerical Solutions

Equality often is the boundary between *greater than* and *less than*, so a first step in solving an inequality is to determine the x-values where equality occurs. We begin by using this concept with graphical techniques.

A graph of $y = x^2 - x - 2$ has x-intercepts -1 and 2. See Figure 11.56. The solutions to $x^2 - x - 2 = 0$ are given by $x = -1$ or $x = 2$. Between the x-intercepts the graph dips **below** the x-axis and the y-values are **negative**. Thus solutions to $x^2 - x - 2 < 0$ satisfy $-1 < x < 2$. To support this result we select a **test value**. For example, 0 lies between -1 and 2. If we substitute $x = 0$ in $x^2 - x - 2 < 0$, it results in a true statement.

$$0^2 - 0 - 2 < 0 \checkmark \qquad \text{True}$$

Visualizing Quadratic Inequalities

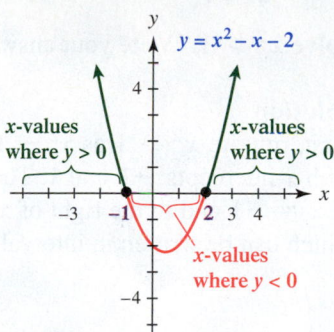

Figure 11.56

When $x < -1$ or $x > 2$, the graph lies **above** the x-axis and the y-values are **positive**. Thus the solutions to $x^2 - x - 2 > 0$ satisfy $x < -1$ or $x > 2$. For example, 3 is greater than 2 and -3 is less than -1. Therefore both 3 and -3 are solutions. We can verify this result by substituting **3** and **-3** as test values in $x^2 - x - 2 > 0$.

$$3^2 - 3 - 2 > 0 \quad ✓ \quad \text{True}$$
$$(-3)^2 - (-3) - 2 > 0 \quad ✓ \quad \text{True}$$

In the next three examples, we use these concepts to solve quadratic inequalities.

EXAMPLE 2 **Solving a quadratic inequality**

Make a table of values for $y = x^2 - 3x - 4$ and then sketch the graph. Use the table and graph to solve $x^2 - 3x - 4 \leq 0$. Write your answer in interval notation.

Solution
The points calculated for Table 11.16 are plotted in Figure 11.57 and connected with a smooth ∪-shaped graph.

Numerical Solution Table 11.16 shows that $x^2 - 3x - 4$ equals 0 when $x = -1$ or $x = 4$. Between these values, $x^2 - 3x - 4$ is negative, so the solution set to $x^2 - 3x - 4 \leq 0$ is given by $-1 \leq x \leq 4$ or, in interval notation, $[-1, 4]$.

Graphical Solution In Figure 11.57 the graph of $y = x^2 - 3x - 4$ shows that the x-intercepts are -1 and **4**. Between these values, the graph dips **below** the x-axis. Thus the solution set is $[-1, 4]$.

STUDY TIP

Learn how to use a parabola to help solve a quadratic inequality.

TABLE 11.16

x	$y = x^2 - 3x - 4$
-2	6
-1	0
0	-4
1	-6
2	-6
3	-4
4	0
5	6

x-intercepts

Less than or equal to 0

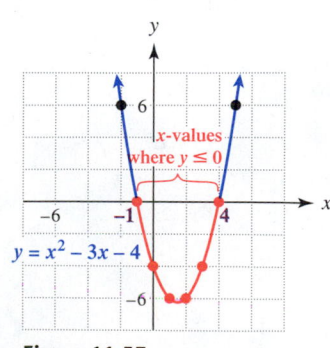

Solving $x^2 - 3x - 4 \leq 0$
x-Intercepts: $-1, 4$

Figure 11.57

Now Try Exercises 19, 27

| EXAMPLE 3 | **Solving a quadratic inequality** |

Solve $x^2 > 1$. Write your answer in interval notation.

Solution
First, rewrite $x^2 > 1$ as $x^2 - 1 > 0$. The graph of $y = x^2 - 1$ is shown in Figure 11.58 with x-intercepts -1 and 1. The graph lies **above** the x-axis and is shaded green to the left of $x = -1$ and to the right of $x = 1$. Thus the solution set is given by $x < -1$ or $x > 1$, which can be written in interval notation as $(-\infty, -1) \cup (1, \infty)$.

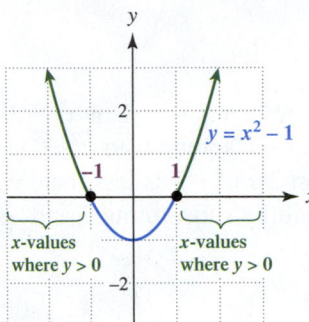

Solving $x^2 - 1 > 0$
x-Intercepts: $-1, 1$

Figure 11.58

■ **Now Try Exercise 49**

| EXAMPLE 4 | **Solving some special cases** |

Solve each of the inequalities graphically.
(a) $x^2 + 1 > 0$ **(b)** $x^2 + 1 < 0$ **(c)** $(x - 1)^2 \le 0$

Solution
(a) Because the graph of $y = x^2 + 1$, shown in Figure 11.59, is always above the x-axis, $x^2 + 1$ is always greater than 0. The solution set includes all real numbers, or $(-\infty, \infty)$.
(b) Because the graph of $y = x^2 + 1$, shown in Figure 11.59, never goes below the x-axis, $x^2 + 1$ is never less than 0. Thus there are no real solutions.
(c) Because the graph of $y = (x - 1)^2$, shown in Figure 11.60, never goes below the x-axis, $(x - 1)^2$ is never less than 0. When $x = 1$, $y = 0$, so **1** is the only solution to the inequality $(x - 1)^2 \le 0$.

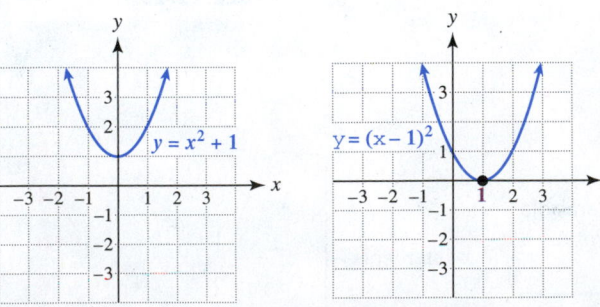

Figure 11.59   **Figure 11.60**

■ **Now Try Exercises 51, 55, 57**

The following chart summarizes visual solutions for several possibilities when solving quadratic equations and inequalities.

Visualizing Solutions to Quadratic Equations and Inequalities

Two x-Intercepts

Case 1: Upward Opening

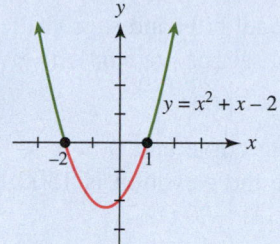

Equation/Inequality	Solutions
$x^2 + x - 2 = 0$	$x = -2, 1$
$x^2 + x - 2 < 0$	$-2 < x < 1$
$x^2 + x - 2 > 0$	$x < -2$ or $x > 1$

Case 2: Downward Opening

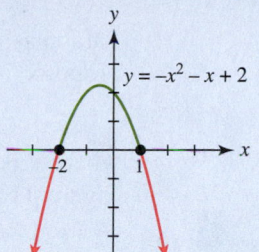

Equation/Inequality	Solutions
$-x^2 - x + 2 = 0$	$x = -2, 1$
$-x^2 - x + 2 < 0$	$x < -2$ or $x > 1$
$-x^2 - x + 2 > 0$	$-2 < x < 1$

One x-Intercept

Case 1: Upward Opening

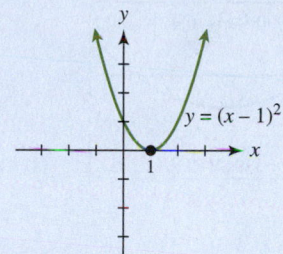

Equation/Inequality	Solutions
$(x - 1)^2 = 0$	$x = 1$
$(x - 1)^2 < 0$	no solutions
$(x - 1)^2 > 0$	$x < 1$ or $x > 1$

Case 2: Downward Opening

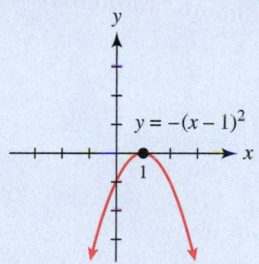

Equation/Inequality	Solutions
$-(x - 1)^2 = 0$	$x = 1$
$-(x - 1)^2 < 0$	$x < 1$ or $x > 1$
$-(x - 1)^2 > 0$	no solutions

No x-Intercepts

Case 1: Upward Opening

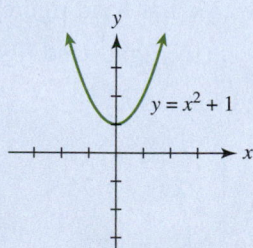

Equation/Inequality	Solutions
$x^2 + 1 = 0$	no solutions
$x^2 + 1 < 0$	no solutions
$x^2 + 1 > 0$	all real numbers

Case 2: Downward Opening

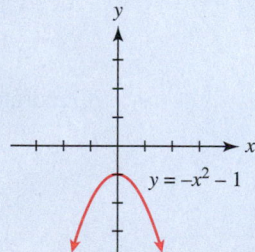

Equation/Inequality	Solutions
$-x^2 - 1 = 0$	no solutions
$-x^2 - 1 < 0$	all real numbers
$-x^2 - 1 > 0$	no solutions

▶ **REAL-WORLD CONNECTION** In the next example we show how quadratic inequalities are used in highway design, as discussed in A Look Into Math for this section.

EXAMPLE 5

Determining elevations on a sag curve

Parabolas are frequently used in highway design to model hills and sags (valleys) along a proposed route. Suppose that the elevation E in feet of a sag, or *sag curve*, is given by

$$E(x) = 0.00004x^2 - 0.4x + 2000,$$

where x is the horizontal distance in feet along the sag curve and $0 \leq x \leq 10{,}000$. See Figure 11.61. Estimate graphically the x-values where the elevation is 1500 feet or less. (*Source:* F. Mannering and W. Kilareski, *Principles of Highway Engineering and Traffic Analysis.*)

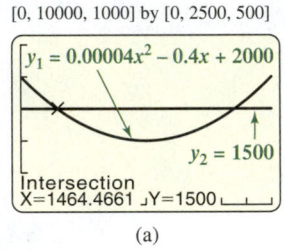

E

Figure 11.61

Solution

Graphical Solution We must solve the quadratic inequality

$$0.00004x^2 - 0.4x + 2000 \leq 1500.$$

Let $y_1 = 0.00004x^2 - 0.4x + 2000$ be the sag curve and $y_2 = 1500$ be a line with an elevation of 1500 feet. Their graphs intersect at $x \approx 1464$ and $x \approx 8536$, as shown in Figure 11.62. The elevation of the road is less than 1500 feet between these x-values. Therefore the elevation is 1500 feet or less when $1464 \leq x \leq 8536$ (approximately).

Determining Where the Elevation is 1500 Feet

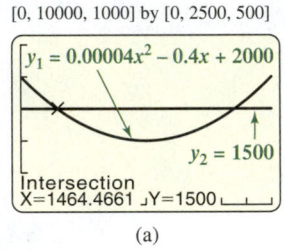

[0, 10000, 1000] by [0, 2500, 500]

$y_1 = 0.00004x^2 - 0.4x + 2000$

$y_2 = 1500$

Intersection
X=1464.4661 Y=1500

(a)

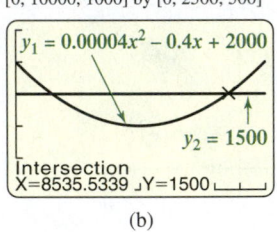

[0, 10000, 1000] by [0, 2500, 500]

$y_1 = 0.00004x^2 - 0.4x + 2000$

$y_2 = 1500$

Intersection
X=8535.5339 Y=1500

(b)

Figure 11.62

CALCULATOR HELP

To find a point of intersection, see Appendix A (page AP-6).

▌ **Now Try Exercise 63**

Symbolic Solutions

To solve a quadratic inequality symbolically, we first solve the corresponding equality. We can then write the solution to the inequality, using the following method.

SOLUTIONS TO QUADRATIC INEQUALITIES

Let $ax^2 + bx + c = 0$, $a > 0$, have two real solutions p and q, where $p < q$.

$ax^2 + bx + c < 0$ is equivalent to $p < x < q$ (see left-hand figure).

$ax^2 + bx + c > 0$ is equivalent to $x < p$ or $x > q$ (see right-hand figure).

Quadratic inequalities involving $\leq$ or $\geq$ can be solved similarly.

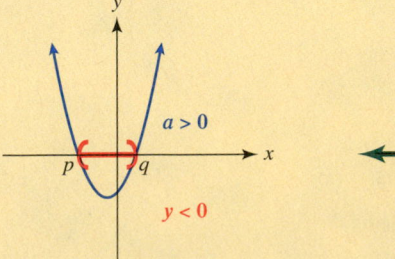

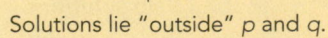

Solutions lie between p and q.   Solutions lie "outside" p and q.

READING CHECK

How can you distinguish a linear inequality from a quadratic inequality?

One way to handle the situation where $a < 0$ is to multiply each side of the inequality by -1, in which case we must be sure to *reverse* the inequality symbol. For example, the inequality $-2x^2 + 8 \leq 0$ has $a = -2$, which is negative. If we multiply each side of this inequality by -1, we obtain $2x^2 - 8 \geq 0$ and now $a = 2$, which is positive.

EXAMPLE 6   Solving quadratic inequalities

Solve each inequality symbolically. Write your answer in interval notation.
(a) $6x^2 - 7x - 5 \geq 0$ **(b)** $x(3 - x) > -18$

Solution
(a) Begin by solving $6x^2 - 7x - 5 = 0$.

$$6x^2 - 7x - 5 = 0 \qquad \text{Quadratic equation}$$
$$(2x + 1)(3x - 5) = 0 \qquad \text{Factor.}$$
$$2x + 1 = 0 \quad \text{or} \quad 3x - 5 = 0 \qquad \text{Zero-product property}$$
$$x = -\frac{1}{2} \quad \text{or} \quad x = \frac{5}{3} \qquad \text{Solve.}$$

Therefore solutions to $6x^2 - 7x - 5 \geq 0$ lie "**outside**" these two values and satisfy $x \leq -\frac{1}{2}$ or $x \geq \frac{5}{3}$. In interval notation the solution set is $\left(-\infty, -\frac{1}{2}\right] \cup \left[\frac{5}{3}, \infty\right)$.

(b) First, rewrite the inequality as follows.

$$x(3 - x) > -18 \qquad \text{Given inequality}$$
$$3x - x^2 > -18 \qquad \text{Distributive property}$$
$$3x - x^2 + 18 > 0 \qquad \text{Add 18.}$$
$$-x^2 + 3x + 18 > 0 \qquad \text{Rewrite.}$$
$$x^2 - 3x - 18 < 0 \qquad \text{Multiply by } -1 \text{ because } a < 0;$$
$$\text{reverse the inequality symbol.}$$

Next, solve $x^2 - 3x - 18 = 0$.

$$(x + 3)(x - 6) = 0 \qquad \text{Factor.}$$
$$x = -3 \quad \text{or} \quad x = 6 \qquad \text{Solve.}$$

Solutions to $x^2 - 3x - 18 < 0$ lie **between** these two values and satisfy $-3 < x < 6$. In interval notation the solution set is $(-3, 6)$.

Now Try Exercises 31, 37

CRITICAL THINKING

The graph of $y = -x^2 + x + 12$ is a parabola opening downward with x-intercepts -3 and 4. Solve each inequality.
(a) $-x^2 + x + 12 > 0$
(b) $-x^2 + x + 12 < 0$

EXAMPLE 7   Finding the dimensions of a building

A rectangular building needs to be 7 feet longer than it is wide, as illustrated in Figure 11.63. The area of the building must be at least 450 square feet. What widths x are possible for this building? Support your results with a table of values.

Figure 11.63

Solution
Symbolic Solution   If x is the width of the building, $x + 7$ is the length of the building and its area is $x(x + 7)$. The area must be at least 450 square feet, so the inequality $x(x + 7) \geq 450$ must be satisfied.

First solve the following quadratic equation.

$$x(x + 7) = 450 \quad \text{Quadratic equation}$$

$$x^2 + 7x = 450 \quad \text{Distributive property}$$

$$x^2 + 7x - 450 = 0 \quad \text{Subtract 450.}$$

$$x = \frac{-7 \pm \sqrt{7^2 - 4(1)(-450)}}{2(1)} \quad \text{Quadratic formula: } a = 1, \ b = 7, \text{ and } c = -450$$

$$= \frac{-7 \pm \sqrt{1849}}{2} \quad \text{Simplify.}$$

$$= \frac{-7 \pm 43}{2} \quad \sqrt{1849} = 43$$

$$= \mathbf{18, -25} \quad \text{Evaluate.}$$

Thus the solutions to $x(x + 7) \geq 450$ are $x \leq \mathbf{-25}$ or $x \geq \mathbf{18}$. The width is positive, so the building width must be 18 feet or more.

Numerical Solution A table of values is shown in Figure 11.64, where $y_1 = x(x + 7)$ equals 450 when $x = 18$. For $x \geq 18$ the area is *at least* 450 square feet.

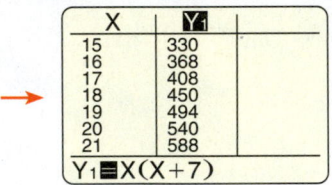

X	Y1
15	330
16	368
17	408
18	450
19	494
20	540
21	588

Y1⬛X(X+7)

Figure 11.64

Now Try Exercise 67

11.5 Putting It All Together

METHOD	EXPLANATION
Solving a Quadratic Inequality Symbolically	Let $ax^2 + bx + c = 0$, $a > 0$, have two real solutions p and q, where $p < q$. $ax^2 + bx + c < 0$ is equivalent to $p < x < q$. $ax^2 + bx + c > 0$ is equivalent to $x < p$ or $x > q$. *Examples:* The solutions to $x^2 - 3x + 2 = 0$ are 1 and 2. The solutions to $x^2 - 3x + 2 < 0$ satisfy $1 < x < 2$. The solutions to $x^2 - 3x + 2 > 0$ satisfy $x < 1$ or $x > 2$.
Solving a Quadratic Inequality Graphically	Given $ax^2 + bx + c < 0$ with $a > 0$, graph $y = ax^2 + bx + c$ and locate any x-intercepts. If there are two x-intercepts, then solutions correspond to x-values between the x-intercepts. Solutions to $ax^2 + bx + c > 0$ correspond to x-values "outside" the x-intercepts. See the box on page 771.
Solving a Quadratic Inequality Numerically	If a quadratic inequality is expressed as $ax^2 + bx + c < 0$ with $a > 0$, then we can solve $y = ax^2 + bx + c = 0$ with a table. If there are two solutions, then the solutions to the given inequality lie between these values. Solutions to $ax^2 + bx + c > 0$ lie before or after these values in the table.

11.5 Exercises

CONCEPTS AND VOCABULARY

1. How is a quadratic inequality different from a quadratic equation?

2. Do quadratic inequalities typically have two solutions? Explain.

3. Is 3 a solution to $x^2 < 7$?

4. Is 5 a solution to $x^2 \geq 25$?

5. The solutions to $x^2 - 2x - 8 = 0$ are -2 and 4. What are the solutions to $x^2 - 2x - 8 < 0$? Write your answer as an inequality.

6. The solutions to $x^2 + 2x - 3 = 0$ are -3 and 1. What are the solutions to $x^2 + 2x - 3 > 0$? Write your answer as an inequality.

QUADRATIC INEQUALITIES

Exercises 7–12: Is the inequality quadratic?

7. $x^2 + 4x + 5 < 0$

8. $x > x^3 - 5$

9. $x^2 > 19$

10. $x(x - 1) - 2 \geq 0$

11. $4x > 1 - x$

12. $2x(x^2 + 3) < 0$

Exercises 13–18: Is the given value of x a solution?

13. $2x^2 + x - 1 > 0$ $x = 3$

14. $x^2 - 3x + 2 \leq 0$ $x = 2$

15. $x^2 + 2 \leq 0$ $x = 0$

16. $2x(x - 3) \geq 0$ $x = 1$

17. $x^2 - 3x \leq 1$ $x = -3$

18. $4x^2 - 5x + 1 > 30$ $x = -2$

GRAPHICAL SOLUTIONS

Exercises 19–24: The graph of $y = ax^2 + bx + c$ is given. Solve each equation or inequality.
   (a) $ax^2 + bx + c = 0$
   (b) $ax^2 + bx + c < 0$
   (c) $ax^2 + bx + c > 0$

19.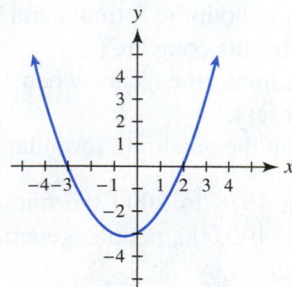

20.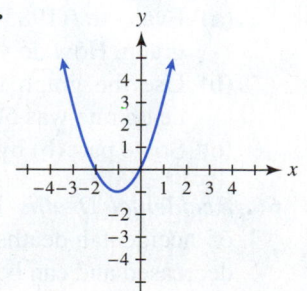

21.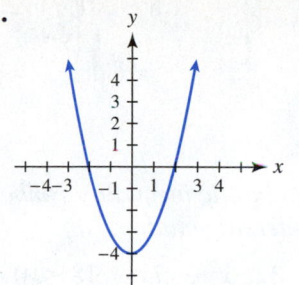

22.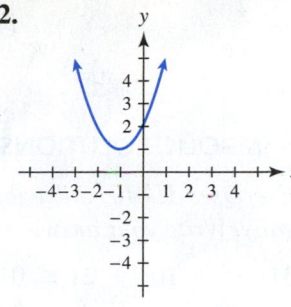

23.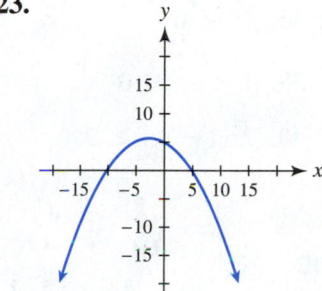

24.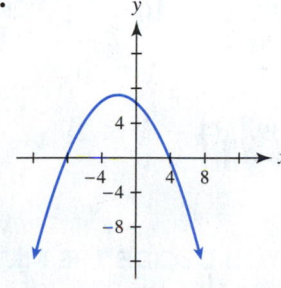

Exercises 25 and 26: Solve the inequality graphically to the nearest thousandth.

25. $\pi x^2 - \sqrt{3}x \leq \frac{3}{11}$

26. $\sqrt{5}x^2 - \pi^2 x \geq 10.3$

NUMERICAL SOLUTIONS

Exercises 27–30: A table for $y = ax^2 + bx + c$ is given. Solve each equation or inequality.
   (a) $ax^2 + bx + c = 0$
   (b) $ax^2 + bx + c < 0$
   (c) $ax^2 + bx + c > 0$

27. $y = x^2 - 4$

x	-3	-2	-1	0	1	2	3
y	5	0	-3	-4	-3	0	5

28. $y = x^2 - x - 2$

x	-3	-2	-1	0	1	2	3
y	10	4	0	-2	-2	0	4

29. $y = x^2 + 4x$

x	-5	-4	-3	-2	-1	0	1
y	5	0	-3	-4	-3	0	5

30. $y = -2x^2 - 2x + 1.5$

x	-2	-1.5	-1	-0.5	0	0.5	1
y	-2.5	0	1.5	2	1.5	0	-2.5

SYMBOLIC SOLUTIONS

Exercises 31–40: Solve the quadratic inequality symbolically. Write your answer in interval notation.

31. $x^2 + 10x + 21 \leq 0$ **32.** $x^2 - 7x - 18 < 0$

33. $3x^2 - 9x + 6 > 0$ **34.** $7x^2 + 34x - 5 \geq 0$

35. $x^2 < 10$ **36.** $x^2 \geq 64$

37. $x(6 - x) < 0$ **38.** $1 - x^2 \leq 0$

39. $x(4 - x) \leq 2$ **40.** $2x(1 - x) \geq 2$

YOU DECIDE THE METHOD

Exercises 41–44: Solve the equation in part (a). Use the results to solve the inequalities in parts (b) and (c).

41. (a) $x^2 - 4 = 0$ **42.** (a) $x^2 - 5 = 0$
　　(b) $x^2 - 4 < 0$ 　　(b) $x^2 - 5 \leq 0$
　　(c) $x^2 - 4 > 0$ 　　(c) $x^2 - 5 \geq 0$

43. (a) $x^2 + x - 1 = 0$
　　(b) $x^2 + x - 1 < 0$
　　(c) $x^2 + x - 1 > 0$

44. (a) $x^2 + 4x - 5 = 0$
　　(b) $x^2 + 4x - 5 \leq 0$
　　(c) $x^2 + 4x - 5 \geq 0$

Exercises 45–62: Solve the inequality by any method. Write your answer in interval notation when appropriate.

45. $x^2 + 4x + 3 < 0$ **46.** $x^2 + x - 2 \leq 0$

47. $2x^2 - x - 15 \geq 0$ **48.** $3x^2 - 3x - 6 > 0$

49. $2x^2 \leq 8$ **50.** $x^2 < 9$

51. $x^2 > -5$ **52.** $-x^2 \geq 1$

53. $-x^2 + 3x > 0$ **54.** $-8x^2 - 2x + 1 \leq 0$

55. $x^2 + 2 \leq 0$ **56.** $x^2 + 3 \geq -5$

57. $(x - 2)^2 \leq 0$ **58.** $(x + 2)^2 \leq 0$

59. $(x + 1)^2 > 0$ **60.** $(x - 3)^2 > 0$

61. $x(1 - x) \geq -2$ **62.** $x(x - 2) < 3$

APPLICATIONS

63. *Highway Design* (Refer to Example 5.) The elevation E of a sag curve, in feet, is given by

$$E(x) = 0.0000375x^2 - 0.175x + 1000,$$

where $0 \leq x \leq 4000$.

(a) Estimate graphically the x-values for which the elevation is 850 feet or less. (*Hint:* Use the window [0, 4000, 1000] by [500, 1200, 100].)

(b) For which x-values is the elevation 850 feet or more?

64. *Early Cellular Phone Use* The number of cellular subscribers in the United States in thousands from 1985 to 1991 can be modeled by

$$f(x) = 163x^2 - 146x + 205,$$

where x is the year and $x = 0$ corresponds to 1985, $x = 1$ to 1986, and so on. (*Source:* M. Paetsch, *Mobile Communication in the U.S. and Europe.*)

(a) Write a quadratic inequality whose solution set represents the years when there were 2 million subscribers or more.

(b) Solve this inequality.

65. *Heart Disease Death Rates* From 1960 to 2010, age-adjusted heart disease rates decreased dramatically. The number of deaths per 100,000 people can be modeled by

$$f(x) = -0.05107x^2 + 194.74x - 184{,}949,$$

where x is the year, as illustrated in the accompanying figure. (*Source:* Department of Health and Human Services.)

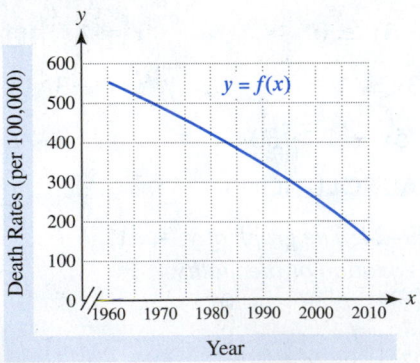

(a) Evaluate $f(1985)$, using both the formula and the graph. How do your results compare?

(b) Use the graph to estimate the years when this death rate was 500 or less.

(c) Solve part (b) by using the quadratic formula.

66. *Accidental Deaths* From 1910 to 2000 the number of accidental deaths per 100,000 people generally decreased and can be modeled by

$$f(x) = -0.001918x^2 + 6.93x - 6156,$$

where x is the year, as shown in the accompanying figure. Note that after 2000 these rates increased and are not modeled by $f(x)$. (*Source:* Department of Health and Human Services.)

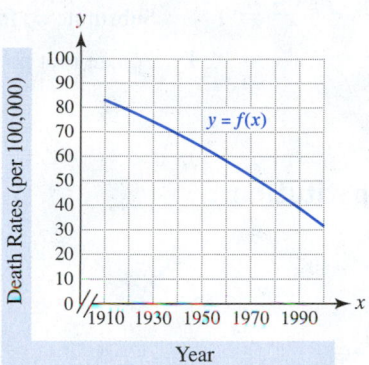

(a) Evaluate $f(1955)$, using both the formula and the graph. How do your results compare?
(b) Use the graph to estimate when this death rate was 60 or more.
(c) Solve part (b) by using the quadratic formula.

67. *Dimensions of a Pen*   A rectangular pen for a pet is 5 feet longer than it is wide. Give possible values for the width w of the pen if its area must be between 176 and 500 square feet, inclusively.

68. *Dimensions of a Cylinder*   The volume of a cylindrical can is given by $V = \pi r^2 h$, where r is its radius and h is its height. See the accompanying figure. If $h = 6$ inches and the volume of the can must be 50 cubic inches or more, estimate to the nearest tenth of an inch possible values for r.

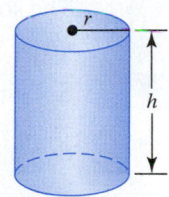

WRITING ABOUT MATHEMATICS

69. Consider the inequality $x^2 < 0$. Discuss the solutions to this inequality and explain your reasoning.

70. Explain how the graph of $y = ax^2 + bx + c$ can be used to solve the inequality

$$ax^2 + bx + c > 0$$

when $a < 0$. Assume that the x-intercepts of the graph are p and q with $p < q$.

11.6   Equations in Quadratic Form

Higher Degree Polynomial Equations • Equations Having Rational Exponents • Equations Having Complex Solutions

A LOOK INTO MATH ▶   Although many equations are *not* quadratic equations, they can sometimes be written in quadratic form. These equations are *reducible to quadratic form*. To express such an equation in quadratic form we often use substitution. In this section we discuss this process.

Higher Degree Polynomial Equations

Sometimes a fourth-degree polynomial can be factored like a quadratic trinomial, provided it does not have an x-term or an x^3-term. Let's consider the equation $x^4 - 5x^2 + 4 = 0$.

$x^4 - 5x^2 + 4 = 0$	Given equation
$(x^2)^2 - 5(x^2) + 4 = 0$	Properties of exponents

We use the substitution $u = x^2$.

$u^2 - 5u + 4 = 0$	Let $u = x^2$.
$(u - 4)(u - 1) = 0$	Factor.
$u - 4 = 0$ or $u - 1 = 0$	Zero-product property
$u = 4$ or $u = 1$	Solve each equation.

Because the given equation $x^4 - 5x^2 + 4 = 0$ uses the variable x, we must give the solutions in terms of x. We substitute x^2 for u in $u = 4$ and $u = 1$ and then solve to obtain the following four solutions.

$$
\begin{array}{lcll}
u = 4 & \text{or} & u = 1 & \text{Solutions in terms of } u \\
x^2 = 4 & \text{or} & x^2 = 1 & \text{Substitute } x^2 \text{ for } u. \\
x = \pm 2 & \text{or} & x = \pm 1 & \text{Square root property}
\end{array}
$$

The solutions are $-2, -1, 1,$ and 2.

EXAMPLE 1 **Solving a sixth-degree equation by substitution**

Solve $2x^6 + x^3 = 1$.

Solution
Start by subtracting 1 from each side.

$$
\begin{array}{ll}
2x^6 + x^3 - 1 = 0 & \text{Subtract 1.} \\
2(x^3)^2 + (x^3) - 1 = 0 & \text{Properties of exponents} \\
2u^2 + u - 1 = 0 & \text{Let } u = x^3. \\
(2u - 1)(u + 1) = 0 & \text{Factor.} \\
2u - 1 = 0 \quad \text{or} \quad u + 1 = 0 & \text{Zero-product property} \\
u = \dfrac{1}{2} \quad \text{or} \quad u = -1 & \text{Solve.}
\end{array}
$$

Now substitute x^3 for u, and solve for x to obtain the following two solutions.

$$
\begin{array}{ll}
x^3 = \dfrac{1}{2} \quad \text{or} \quad x^3 = -1 & \text{Substitute } x^3 \text{ for } u. \\[2mm]
x = \sqrt[3]{\dfrac{1}{2}} \quad \text{or} \quad x = -1 & \text{Take cube root of each side.}
\end{array}
$$

■ **Now Try Exercise 3**

Equations Having Rational Exponents

Equations that have negative exponents are sometimes reducible to quadratic form. Consider the following example, in which two methods are presented.

EXAMPLE 2 **Solving an equation having negative exponents**

Solve $-6m^{-2} + 13m^{-1} + 5 = 0$.

Solution
Method I Use the substitution $u = m^{-1} = \frac{1}{m}$ and $u^2 = m^{-2} = \frac{1}{m^2}$.

$$
\begin{array}{ll}
-6m^{-2} + 13m^{-1} + 5 = 0 & \text{Given equation} \\
-6u^2 + 13u + 5 = 0 & \text{Let } u = m^{-1} \text{ and } u^2 = m^{-2}. \\
6u^2 - 13u - 5 = 0 & \text{Multiply by } -1. \\
(2u - 5)(3u + 1) = 0 & \text{Factor.} \\
2u - 5 = 0 \quad \text{or} \quad 3u + 1 = 0 & \text{Zero-product property} \\
u = \dfrac{5}{2} \quad \text{or} \quad u = -\dfrac{1}{3} & \text{Solve for } u.
\end{array}
$$

Because $u = \frac{1}{m}$, $m = \frac{1}{u}$. Thus the solutions are given by $m = \frac{2}{5}$ or $m = -3$.

Method II Another way to solve this equation is to multiply each side by the LCD, m^2.

$$-6m^{-2} + 13m^{-1} + 5 = 0 \qquad \text{Given equation}$$
$$m^2(-6m^{-2} + 13m^{-1} + 5) = m^2 \cdot 0 \qquad \text{Multiply by } m^2.$$
$$-6m^2m^{-2} + 13m^2m^{-1} + 5m^2 = 0 \qquad \text{Distributive property}$$
$$-6 + 13m + 5m^2 = 0 \qquad \text{Add exponents.}$$
$$5m^2 + 13m - 6 = 0 \qquad \text{Rewrite the equation.}$$
$$(5m - 2)(m + 3) = 0 \qquad \text{Factor.}$$
$$5m - 2 = 0 \quad \text{or} \quad m + 3 = 0 \qquad \text{Zero-product property}$$
$$m = \frac{2}{5} \quad \text{or} \quad m = -3 \qquad \text{Solve.}$$

Now Try Exercise 5

In the next example we solve an equation having fractional exponents.

EXAMPLE 3 **Solving an equation having fractional exponents**

Solve $x^{2/3} - 2x^{1/3} - 8 = 0$.

Solution
Use the substitution $u = x^{1/3}$.

$$x^{2/3} - 2x^{1/3} - 8 = 0 \qquad \text{Given equation}$$
$$(x^{1/3})^2 - 2(x^{1/3}) - 8 = 0 \qquad \text{Properties of exponents}$$
$$u^2 - 2u - 8 = 0 \qquad \text{Let } u = x^{1/3}.$$
$$(u - 4)(u + 2) = 0 \qquad \text{Factor.}$$
$$u - 4 = 0 \quad \text{or} \quad u + 2 = 0 \qquad \text{Zero-product property}$$
$$u = 4 \quad \text{or} \quad u = -2 \qquad \text{Solve.}$$

Because $u = x^{1/3}$, $u^3 = (x^{1/3})^3 = x$. Thus $x = 4^3 = 64$ or $x = (-2)^3 = -8$. The solutions are -8 and 64.

Now Try Exercise 13

Equations Having Complex Solutions

Sometimes an equation that is reducible to quadratic form also has complex solutions. This situation is discussed in the next two examples.

EXAMPLE 4 **Solving a fourth-degree equation**

Find all complex solutions to $x^4 - 1 = 0$.

Solution

$$x^4 - 1 = 0 \qquad \text{Given equation}$$
$$(x^2)^2 - 1 = 0 \qquad \text{Properties of exponents}$$
$$u^2 - 1 = 0 \qquad \text{Let } u = x^2.$$
$$(u - 1)(u + 1) = 0 \qquad \text{Factor difference of squares.}$$
$$u - 1 = 0 \quad \text{or} \quad u + 1 = 0 \qquad \text{Zero-product property}$$
$$u = 1 \quad \text{or} \quad u = -1 \qquad \text{Solve for } u.$$

Now substitute x^2 for u, and solve for x.

$$u = 1 \quad \text{or} \quad u = -1 \qquad \text{Solutions in terms of } u$$

$$x^2 = 1 \quad \text{or} \quad x^2 = -1 \qquad \text{Let } x^2 = u.$$

$$x = \pm 1 \quad \text{or} \quad x = \pm i \qquad \text{Square root property}$$

There are four complex solutions: $-1, 1, -i,$ and i.

■ **Now Try Exercise 25**

EXAMPLE 5 **Solving a rational equation**

Find all complex solutions to $\dfrac{1}{x} + \dfrac{1}{x^2} = -1$.

Solution

This equation is a rational equation. However, if we multiply through by the LCD, x^2, we clear fractions and obtain a quadratic equation with complex solutions.

$$\frac{1}{x} + \frac{1}{x^2} = -1 \qquad \text{Given equation}$$

$$\frac{x^2}{x} + \frac{x^2}{x^2} = -1x^2 \qquad \text{Multiply each term by } x^2.$$

$$x + 1 = -x^2 \qquad \text{Simplify.}$$

$$x^2 + x + 1 = 0 \qquad \text{Add } x^2.$$

$$x = \frac{-1 \pm \sqrt{1^2 - 4(1)(1)}}{2(1)} \qquad \text{Quadratic formula}$$

$$x = \frac{-1 \pm i\sqrt{3}}{2} \qquad \sqrt{-3} = i\sqrt{3}$$

$$x = -\frac{1}{2} \pm \frac{i\sqrt{3}}{2} \qquad \frac{a \pm b}{c} = \frac{a}{c} \pm \frac{b}{c}$$

■ **Now Try Exercise 31**

11.6 Putting It All Together

EQUATION	SUBSTITUTION	EXAMPLES
Higher Degree Polynomial	Let $u = x^n$ for some integer n.	To solve $x^4 - 3x^2 - 4 = 0$, let $u = x^2$. This equation becomes $$u^2 - 3u - 4 = 0.$$
Rational Exponents	Pick a substitution that reduces the equation to quadratic form.	To solve $n^{-2} + 6n^{-1} + 9 = 0$, let $u = n^{-1}$. This equation becomes $$u^2 + 6u + 9 = 0.$$ To solve $6x^{2/5} - 5x^{1/5} - 4 = 0$, let $u = x^{1/5}$. This equation becomes $$6u^2 - 5u - 4 = 0.$$

EQUATION	SUBSTITUTION	EXAMPLES
Equations Having Complex Solutions	Both polynomial and rational equations can have complex solutions. Use the fact that if $a > 0$, then $\sqrt{-a} = i\sqrt{a}$.	$$1 + \frac{1}{x^2} = 0$$ $$x^2 \cdot \left(1 + \frac{1}{x^2}\right) = 0 \cdot x^2$$ $$x^2 + 1 = 0$$ $$x^2 = -1$$ $$x = \pm i$$

11.6 Exercises

MyMathLab   PRACTICE   WATCH   DOWNLOAD   READ   REVIEW

EQUATIONS REDUCIBLE TO QUADRATIC FORM

Exercises 1–6: Use the given substitution to solve the equation.

1. $x^4 - 7x^2 + 6 = 0$ $u = x^2$

2. $2k^4 - 7k^2 + 6 = 0$ $u = k^2$

3. $3z^6 + z^3 - 10 = 0$ $u = z^3$

4. $2x^6 + 17x^3 + 8 = 0$ $u = x^3$

5. $4n^{-2} + 17n^{-1} + 15 = 0$ $u = n^{-1}$

6. $m^{-2} + 24 = 10m^{-1}$ $u = m^{-1}$

Exercises 7–24: Solve. Find all real solutions.

7. $x^4 = 8x^2 + 9$ **8.** $3x^4 = 10x^2 + 8$

9. $3x^6 - 5x^3 - 2 = 0$ **10.** $6x^6 + 11x^3 + 4 = 0$

11. $2z^{-2} + 11z^{-1} = 40$ **12.** $z^{-2} - 10z^{-1} + 25 = 0$

13. $x^{2/3} - 2x^{1/3} + 1 = 0$ **14.** $3x^{2/3} + 18x^{1/3} = 48$

15. $x^{2/5} - 33x^{1/5} = -32$ **16.** $x^{2/5} - 80x^{1/5} = 81$

17. $x - 13\sqrt{x} + 36 = 0$ **18.** $x - 17\sqrt{x} + 16 = 0$

19. $z^{1/2} - 2z^{1/4} + 1 = 0$ **20.** $z^{1/2} - 4z^{1/4} + 4 = 0$

21. $(x + 1)^2 - 5(x + 1) - 14 = 0$

22. $2(x - 5)^2 + 5(x - 5) + 3 = 0$

23. $(x^2 - 1)^2 - 4 = 0$

24. $(x^2 - 9)^2 - 8(x^2 - 9) + 16 = 0$

EQUATIONS HAVING COMPLEX SOLUTIONS

Exercises 25–34: Find all complex solutions.

25. $x^4 - 16 = 0$ **26.** $\frac{1}{3}x^4 - 27 = 0$

27. $x^3 + x = 0$ **28.** $4x^3 + x = 0$

29. $x^4 - 2 = x^2$ **30.** $x^4 - 3 = 2x^2$

31. $\frac{1}{x} + \frac{1}{x^2} = -\frac{1}{2}$ **32.** $\frac{2}{x - 1} - \frac{1}{x} = -1$

33. $\frac{2}{x - 2} - \frac{1}{x} = -\frac{1}{2}$ **34.** $\frac{1}{x} - \frac{1}{x^2} = \frac{1}{2}$

WRITING ABOUT MATHEMATICS

35. Explain how to solve $ax^4 - bx^2 + c = 0$. Assume that the left side of the equation factors.

36. Explain what it means for an equation to be reducible to quadratic form.

SECTIONS 11.5 AND 11.6   Checking Basic Concepts

1. Solve the inequality $x^2 - x - 6 > 0$. Write your answer in interval notation.

2. Solve the inequality $3x^2 + 5x + 2 \le 0$. Write your answer in interval notation.

3. Solve $x^6 + 6x^3 - 16 = 0$.

4. Solve $x^{2/3} - 7x^{1/3} - 8 = 0$.

5. Find all complex solutions to $x^4 + 2x^2 + 1 = 0$.

CHAPTER 11   Summary

SECTION 11.1 ■ QUADRATIC FUNCTIONS AND THEIR GRAPHS

Quadratic Function Any quadratic function f can be written as

$$f(x) = ax^2 + bx + c \quad (a \neq 0).$$

Graph of a Quadratic Function Its graph is a parabola that is wider than the graph of $y = x^2$ if $0 < |a| < 1$, and narrower than the graph of $y = x^2$ if $|a| > 1$. The y-intercept is c.

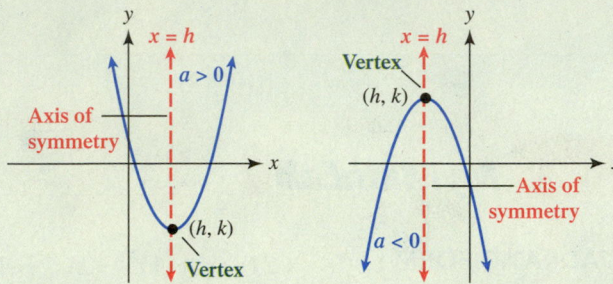

Axis of Symmetry The parabola is symmetric with respect to this vertical line. The axis of symmetry passes through the vertex. Its equation is $x = -\frac{b}{2a}$.

Vertex Formula The x-coordinate of the vertex is $-\frac{b}{2a}$.

Example: Let $y = x^2 - 4x + 1$ with $a = 1$ and $b = -4$.

$$x = -\frac{-4}{2(1)} = 2 \quad \text{and} \quad y = 2^2 - 4(2) + 1 = -3. \text{ The vertex is } (2, -3).$$

SECTION 11.2 ■ PARABOLAS AND MODELING

Vertical and Horizontal Translations Let h and k be positive numbers.

To graph	*shift the graph of $y = x^2$ by k units*
$y = x^2 + k$	upward.
$y = x^2 - k$	downward.

To graph	*shift the graph of $y = x^2$ by h units*
$y = (x - h)^2$	right.
$y = (x + h)^2$	left.

Example: Compared to $y = x^2$, the graph of $y = (x - 1)^2 + 2$ is translated right 1 unit and upward 2 units.

Vertex Form Any quadratic function can be expressed as $f(x) = a(x - h)^2 + k$. In this form the point (h, k) is the vertex. A quadratic function can be put in this form by completing the square or by applying the vertex formula.

Example:
$$y = x^2 + 10x - 4 \qquad \text{Given equation}$$
$$= (x^2 + 10x + 25) - 25 - 4 \qquad \left(\tfrac{b}{2}\right)^2 = \left(\tfrac{10}{2}\right)^2 = 25; \text{ complete the square.}$$
$$= (x + 5)^2 - 29 \qquad \text{Perfect square trinomial; add.}$$
The vertex is $(-5, -29)$.

SECTION 11.3 ■ QUADRATIC EQUATIONS

Quadratic Equations Any quadratic equation can be written as $ax^2 + bx + c = 0$ and can have no real solutions, one real solution, or two real solutions. These solutions correspond to the x-intercepts for the graph of $y = ax^2 + bx + c$.

Example:
$$x^2 + x - 2 = 0$$
$$(x + 2)(x - 1) = 0$$
$$x = -2 \quad \text{or} \quad x = 1$$

The solutions are -2 and 1.
See the graph to the right.

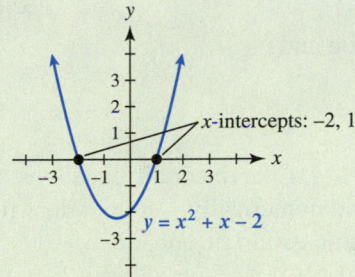

x-intercepts: $-2, 1$

$y = x^2 + x - 2$

Completing the Square Write the equation in the form $x^2 + bx = d$. Complete the square by adding $\left(\frac{b}{2}\right)^2$ to each side of the equation.

Example:
$$x^2 - 8x = 3$$
$$x^2 - 8x + 16 = 3 + 16 \qquad \text{Add } \left(\tfrac{-8}{2}\right)^2 = 16 \text{ to each side.}$$
$$(x - 4)^2 = 19 \qquad \text{Perfect square trinomial}$$
$$x - 4 = \pm\sqrt{19} \qquad \text{Square root property}$$
$$x = 4 \pm \sqrt{19} \qquad \text{Add 4 to each side.}$$

SECTION 11.4 ■ THE QUADRATIC FORMULA

The Quadratic Formula The solutions to $ax^2 + bx + c = 0$ $(a \neq 0)$ are given by

$$x = \frac{-b \pm \sqrt{b^2 - 4ac}}{2a}.$$

Example: Solve $2x^2 + 3x - 1 = 0$ by letting $a = 2$, $b = 3$, and $c = -1$.

$$x = \frac{-3 \pm \sqrt{3^2 - 4(2)(-1)}}{2(2)} = \frac{-3 \pm \sqrt{17}}{4} \approx 0.28, -1.78$$

The Discriminant The expression $b^2 - 4ac$ is called the discriminant. If $b^2 - 4ac > 0$, there are two real solutions; if $b^2 - 4ac = 0$, there is one real solution; and if $b^2 - 4ac < 0$, there are no real solutions—rather there are two complex solutions.

Example: For $2x^2 + 3x - 1 = 0$, the discriminant is

$$b^2 - 4ac = 3^2 - 4(2)(-1) = 17 > 0.$$

There are two real solutions to this quadratic equation, as shown in the previous example.

Quadratic Equations with Complex Solutions A quadratic equation sometimes has no real solutions.

Example: $x^2 + 4 = 0$
$$x^2 = -4 \qquad \text{Subtract 4 from each side.}$$
$$x = \pm 2i \qquad \text{Square root property; two complex solutions}$$

SECTION 11.5 ■ QUADRATIC INEQUALITIES

Quadratic Inequalities When the equals sign in a quadratic equation is replaced with $<, >, \leq,$ or $\geq$, a quadratic inequality results. For example,

$$3x^2 - x + 1 = 0$$

is a quadratic equation and

$$3x^2 - x + 1 > 0$$

is a quadratic inequality. Like quadratic equations, quadratic inequalities can be solved symbolically, graphically, and numerically. An important first step in solving a quadratic inequality is to solve the corresponding quadratic equation.

Examples: The solutions to $x^2 - 5x - 6 = 0$ are -1 and 6.

The solutions to $x^2 - 5x - 6 < 0$ satisfy $-1 < x < 6$.

The solutions to $x^2 - 5x - 6 > 0$ satisfy $x < -1$ or $x > 6$.

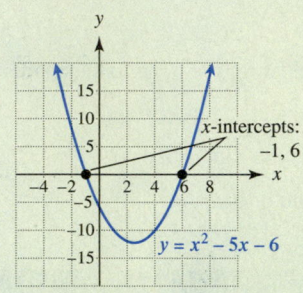

SECTION 11.6 ■ EQUATIONS IN QUADRATIC FORM

Equations Reducible to Quadratic Form An equation that is not quadratic, but can be put into quadratic form by using a substitution, is reducible to quadratic form.

Example: To solve $x^{2/3} - 2x^{1/3} - 15 = 0$, let $u = x^{1/3}$. This equation becomes

$$u^2 - 2u - 15 = 0.$$

Factoring results in $(u + 3)(u - 5) = 0$, so $u = -3$ or $u = 5$.
Because $u = x^{1/3}$, $x = u^3$ and $x = (-3)^3 = -27$ or $x = (5)^3 = 125$.

CHAPTER 11   Review Exercises

SECTION 11.1

Exercises 1 and 2: Identify the vertex, axis of symmetry, and whether the parabola opens upward or downward.

1.

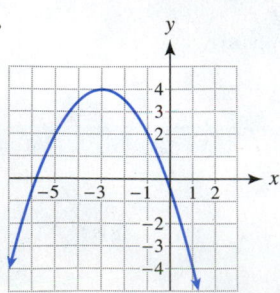

2.

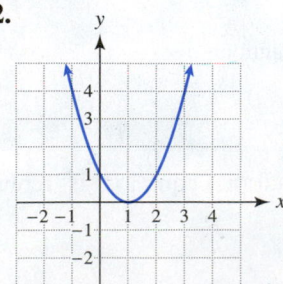

Exercises 3–6: Do the following.
  (a) Graph f.
  (b) Identify the vertex and axis of symmetry.
  (c) Evaluate f(x) at the given value of x.

3. $f(x) = x^2 - 2,$ $\qquad\qquad x = -1$

4. $f(x) = -x^2 + 4x - 3,$ $\qquad x = 3$

5. $f(x) = -\frac{1}{2}x^2 + x + \frac{3}{2},$ $\qquad x = -2$

6. $f(x) = 2x^2 + 8x + 5,$ $\qquad x = -3$

7. Find the minimum y-value located on the graph of $y = 2x^2 - 6x + 1.$

8. Find the maximum y-value located on the graph of $y = -3x^2 + 2x - 5$.

Exercises 9–12: Find the vertex of the parabola.

9. $f(x) = x^2 - 4x - 2$ 10. $f(x) = 5 - x^2$

11. $f(x) = -\frac{1}{4}x^2 + x + 1$ 12. $f(x) = 2 + 2x + x^2$

SECTION 11.2

Exercises 13–18: Do the following.
   (a) Graph f.
   (b) Compare the graph of f with the graph of $y = x^2$.

13. $f(x) = x^2 + 2$ 14. $f(x) = 3x^2$

15. $f(x) = (x - 2)^2$ 16. $f(x) = (x + 1)^2 - 3$

17. $f(x) = \frac{1}{2}(x + 1)^2 + 2$

18. $f(x) = 2(x - 1)^2 - 3$

19. Write the vertex form of a parabola with $a = -4$ and vertex $(2, -5)$.

20. Write the vertex form of a parabola that opens downward with vertex $(-4, 6)$. Assume that $a = \pm 1$.

Exercises 21–24: Write the equation in vertex form. Identify the vertex.

21. $y = x^2 + 4x - 7$ 22. $y = x^2 - 7x + 1$

23. $y = 2x^2 - 3x - 8$ 24. $y = 3x^2 + 6x - 2$

Exercises 25 and 26: Find a value for the constant a so that $f(x) = ax^2 - 1$ models the data.

25.
x	1	2	3
$f(x)$	2	11	26

26.
x	-1	0	1
$f(x)$	$-\frac{3}{4}$	-1	$-\frac{3}{4}$

Exercises 27 and 28: Write $f(x)$ in the form given by $f(x) = ax^2 + bx + c$. Identify the y-intercept on the graph of f.

27. $f(x) = -5(x - 3)^2 + 4$

28. $f(x) = 3(x + 2)^2 - 4$

SECTION 11.3

Exercises 29–32: Use the graph of $y = ax^2 + bx + c$ to solve $ax^2 + bx + c = 0$.

29. 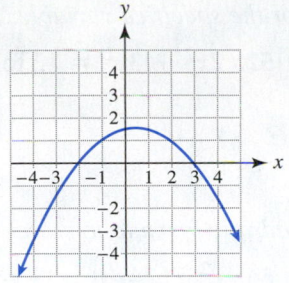   30.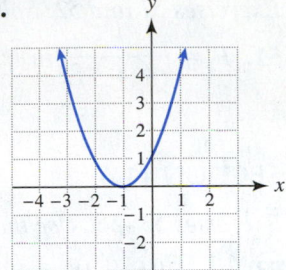

31. 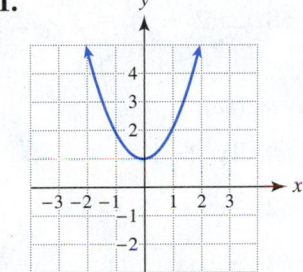   32.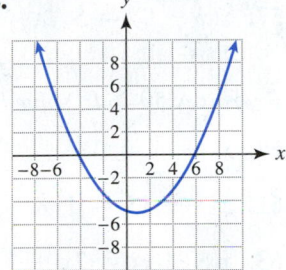

Exercises 33 and 34: A table of $y = ax^2 + bx + c$ is given. Solve $ax^2 + bx + c = 0$.

33.

X	Y1
-20	250
-15	100
-10	0
-5	-50
0	-50
5	0
10	100

Y1■X^2+5X−50

34.

X	Y1
-.75	2
-.5	0
-.25	-1
0	-1
.25	0
.5	2
.75	5

Y1■8X^2+2X−1

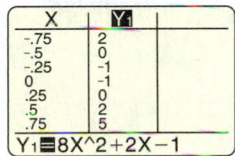 *Exercises 35–38: Solve the quadratic equation*
   (a) graphically and
   (b) numerically.

35. $x^2 - 5x - 50 = 0$ 36. $\frac{1}{2}x^2 + x - \frac{3}{2} = 0$

37. $\frac{1}{4}x^2 + \frac{1}{2}x = 2$ 38. $\frac{1}{2}x + \frac{3}{4} = \frac{1}{4}x^2$

Exercises 39–42: Solve by factoring.

39. $x^2 + x - 20 = 0$ 40. $x^2 + 11x + 24 = 0$

41. $15x^2 - 4x - 4 = 0$ 42. $7x^2 - 25x + 12 = 0$

Exercises 43–46: Use the square root property to solve.

43. $x^2 = 100$ 44. $3x^2 = \frac{1}{3}$

45. $4x^2 - 6 = 0$ 46. $5x^2 = x^2 - 4$

Exercises 47–50: Solve by completing the square.

47. $x^2 + 6x = -2$ 48. $x^2 - 4x = 6$

49. $x^2 - 2x - 5 = 0$ 50. $2x^2 + 6x - 1 = 0$

Exercises 51 and 52: Solve for the specified variable.

51. $F = \dfrac{k}{(R + r)^2}$ for R **52.** $2x^2 + 3y^2 = 12$ for y

SECTION 11.4

Exercises 53–58: Use the quadratic formula to solve.

53. $x^2 - 9x + 18 = 0$ **54.** $x^2 - 24x + 143 = 0$

55. $6x^2 + x = 1$ **56.** $5x^2 + 1 = 5x$

57. $x(x - 8) = 5$ **58.** $2x(2 - x) = 3 - 2x$

Exercises 59–64: Solve by any method.

59. $x^2 - 4 = 0$ **60.** $4x^2 - 1 = 0$

61. $2x^2 + 15 = 11x$ **62.** $2x^2 + 15 = 13x$

63. $x(5 - x) = 2x + 1$ **64.** $-2x(x - 1) = x - \frac{1}{2}$

Exercises 65–68: A graph of $y = ax^2 + bx + c$ is shown.
  (a) *State whether $a > 0$ or $a < 0$.*
  (b) *Solve $ax^2 + bx + c = 0$.*
  (c) *Determine whether the discriminant is positive, negative, or zero.*

65.
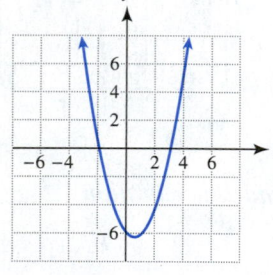

66.
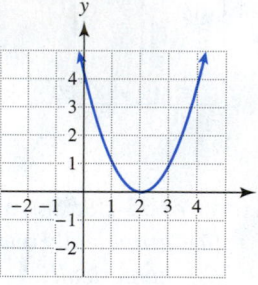

67.
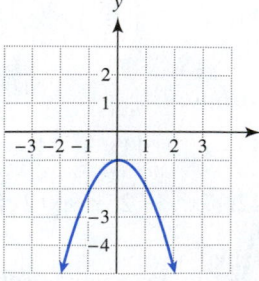

68.
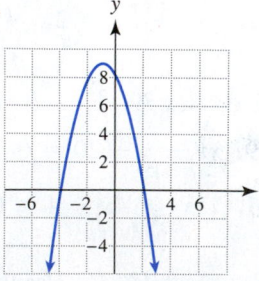

Exercises 69–72: Do the following for the given equation.
  (a) *Evaluate the discriminant.*
  (b) *How many real solutions are there?*
  (c) *Support your answer for part (b) graphically.*

69. $2x^2 - 3x + 1 = 0$ **70.** $7x^2 + 2x - 5 = 0$

71. $3x^2 + x + 2 = 0$

72. $4.41x^2 - 12.6x + 9 = 0$

Exercises 73–76: Solve. Write any complex solutions in standard form.

73. $x^2 + x + 5 = 0$ **74.** $2x^2 + 8 = 0$

75. $2x^2 = x - 1$ **76.** $7x^2 = 2x - 5$

SECTION 11.5

Exercises 77 and 78: The graph of $y = ax^2 + bx + c$ is shown. Solve each equation or inequality.
  (a) $ax^2 + bx + c = 0$
  (b) $ax^2 + bx + c < 0$
  (c) $ax^2 + bx + c > 0$

77.
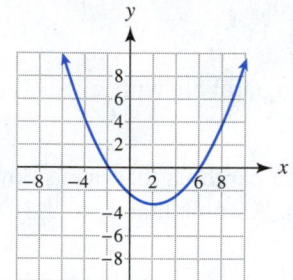

78.
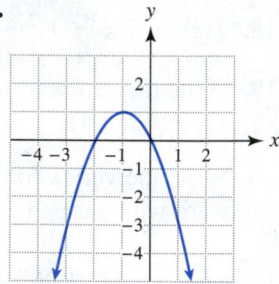

Exercises 79 and 80: A table of $y = ax^2 + bx + c$ is shown. Solve each equation or inequality.
  (a) $ax^2 + bx + c = 0$
  (b) $ax^2 + bx + c < 0$
  (c) $ax^2 + bx + c > 0$

79. $y = x^2 - 16$

x	-6	-4	-2	0	2	4	6
y	20	0	-12	-16	-12	0	20

80. $y = x^2 + x - 2$

x	-3	-2	-1	0	1	2	3
y	4	0	-2	-2	0	4	10

Exercises 81 and 82: Solve the equation in part (a). Use the results to solve the inequalities in parts (b) and (c).

81. **(a)** $x^2 - 2x - 3 = 0$
    (b) $x^2 - 2x - 3 < 0$
    (c) $x^2 - 2x - 3 > 0$

82. **(a)** $2x^2 - 7x - 15 = 0$
    (b) $2x^2 - 7x - 15 \leq 0$
    (c) $2x^2 - 7x - 15 \geq 0$

Exercises 83–88: Solve the quadratic inequality. Write your answer in interval notation.

83. $x^2 + 4x + 3 \le 0$ **84.** $5x^2 - 16x + 3 < 0$

85. $6x^2 - 13x + 2 > 0$ **86.** $x^2 \ge 5$

87. $(x - 1)^2 \ge 0$ **88.** $x^2 + 3 < 2$

SECTION 11.6

Exercises 89–92: Solve the equation.

89. $x^4 - 14x^2 + 45 = 0$

90. $2z^{-2} + z^{-1} - 28 = 0$

91. $x^{2/3} - 9x^{1/3} + 8 = 0$

92. $(x - 1)^2 + 2(x - 1) + 1 = 0$

Exercises 93 and 94: Find all complex solutions.

93. $4x^4 + 4x^2 + 1 = 0$ **94.** $\dfrac{1}{x - 2} - \dfrac{3}{x} = -1$

APPLICATIONS

95. *Construction* A rain gutter is being fabricated from a flat sheet of metal so that the cross section of the gutter is a rectangle, as shown in the accompanying figure. The width of the metal sheet is 12 inches.

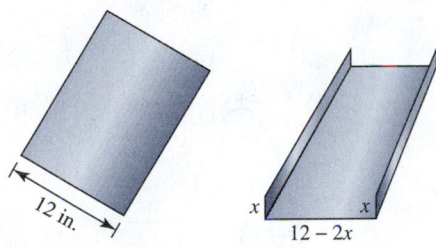

(a) Write a formula $f(x)$ that gives the area of the cross section.
(b) To hold the greatest amount of rainwater, the cross section should have maximum area. Find the dimensions that result in this maximum.

96. *Height of a Stone* Suppose that a stone is thrown upward with an initial velocity of 44 feet per second (30 miles per hour) and is released 4 feet above the ground. Its height h in feet after t seconds is given by

$$h(t) = -16t^2 + 44t + 4.$$

(a) When does the stone reach a height of 32 feet?
(b) After how many seconds does the stone reach maximum height? Estimate this height.

97. *Maximizing Revenue* Suppose that hotel rooms cost $90 per night. However, for a group rate the management is considering reducing the cost of a room by $3 for every room rented.
(a) Write a formula $f(x)$ that gives the revenue from renting x rooms at the group rate.
(b) Graph f in [0, 30, 5] by [0, 800, 100].
(c) How many rooms should be rented to receive revenue of $600?
(d) How many rooms should be rented to maximize revenue?

98. *Numbers* The product of two numbers is 143. One number is 2 more than the other.
(a) Write a quadratic equation whose solution gives the smaller number x.
(b) Solve the equation.

99. *Braking Distance* On dry pavement a safe braking distance d in feet for a car traveling x miles per hour is $d = \frac{x^2}{12}$. For each distance d, find x. (*Source:* F. Mannering, *Principles of Highway Engineering and Traffic Control.*)
(a) $d = 144$ feet **(b)** $d = 300$ feet

100. *U.S. Energy Consumption* From 1950 to 1970 per capita consumption of energy in millions of Btu can be modeled by $f(x) = \frac{1}{4}(x - 1950)^2 + 220$, where x is the year. (*Source:* Department of Energy.)
(a) Find and interpret the vertex.
(b) Graph f in [1950, 1970, 5] by [200, 350, 25]. What happened to energy consumption during this time period?
(c) Use f to predict the consumption in 2010. Actual consumption was 321 million Btu. Did f provide a good model for 2010? Explain.

101. *Screens* A square computer screen has an area of 123 square inches. Approximate its dimensions to the nearest tenth of an inch.

102. *Flying a Kite* A kite is being flown, as illustrated in the accompanying figure. If 130 feet of string have been let out, find the value of x.

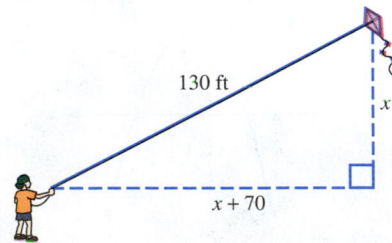

103. *Area* A uniform strip of grass is to be planted around a rectangular swimming pool, as illustrated in the accompanying figure on the next page. The swimming pool is 30 feet wide and 50 feet long. If there is only enough grass seed to cover 250 square feet, estimate the width x that the strip of grass should be.

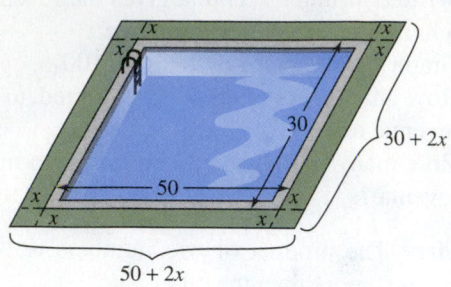

104. *Dimensions of a Cone* The volume V of a cone is given by $V = \frac{1}{3}\pi r^2 h$, where r is its base radius and h is its height. See the accompanying figure. If $h = 20$ inches and the volume of the cone must be between 750 and 1700 cubic inches, inclusively, estimate, to the nearest tenth of an inch, possible values for r.

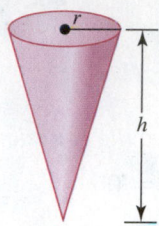

CHAPTER 11 Test

CHAPTER Test Prep VIDEOS — Step-by-step test solutions are found on the Chapter Test Prep Videos available in *MyMathLab* and on YouTube (search "RockswoldComboAlg" and click on "Channels").

1. Find the vertex and axis of symmetry for the graph of $f(x) = -\frac{1}{2}x^2 + x + 1$. Evaluate $f(-2)$.

2. Find the minimum y-value located on the graph of $y = x^2 + 3x - 5$.

3. Find the exact value for the constant a so that $f(x) = ax^2 + 2$ models the data in the table.

x	-2	0	2	4
$f(x)$	0	2	0	-6

4. Compare the graph of $y = f(x)$ to the graph of $y = x^2$. Then graph $y = f(x)$.
   (a) $f(x) = (x - 1)^2$ **(b)** $f(x) = x^2 - 2$
   (c) $f(x) = \frac{1}{2}(x - 3)^2 + 2$

5. Write $y = x^2 - 6x + 2$ in vertex form. Identify the vertex and axis of symmetry.

6. Use the graph of $f(x) = ax^2 + bx + c$ to solve $ax^2 + bx + c = 0$. Then evaluate $f(1)$.

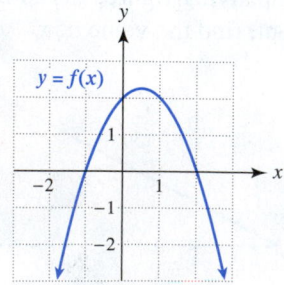

Exercises 7 and 8: Solve the quadratic equation.

7. $3x^2 + 11x - 4 = 0$ 8. $2x^2 = 2 - 6x^2$

9. Solve $x^2 - 8x = 1$ by completing the square.

10. Solve $x(-2x + 3) = -1$ by using the quadratic formula.

11. Solve $9x^2 - 16 = 0$.

12. Solve $F = \dfrac{Gm^2}{r^2}$ for m.

13. A graph of $y = ax^2 + bx + c$ is shown.
    (a) State whether $a > 0$ or $a < 0$.
    (b) Solve $ax^2 + bx + c = 0$.
    (c) Determine whether the discriminant is positive, negative, or zero.

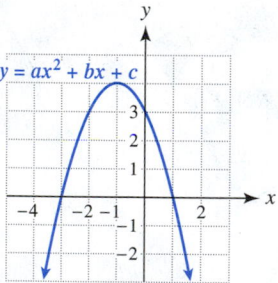

14. Complete the following for $-3x^2 + 4x - 5 = 0$.
    (a) Evaluate the discriminant.
    (b) How many real solutions are there?
    (c) Support your answer for part (b) graphically.

Exercises 15 and 16: The graph of $y = ax^2 + bx + c$ is shown. Solve each equation or inequality.
    (a) $ax^2 + bx + c = 0$
    (b) $ax^2 + bx + c < 0$
    (c) $ax^2 + bx + c > 0$

15.

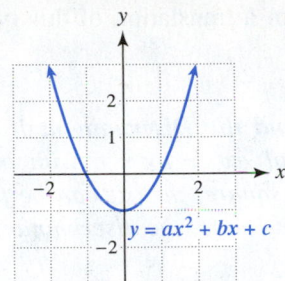

16.

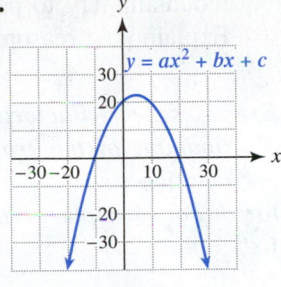

17. Solve the quadratic equation in part (a). Use the result to solve the inequalities in parts (b) and (c) and write your answer in interval notation.
(a) $8x^2 - 2x - 3 = 0$
(b) $8x^2 - 2x - 3 \le 0$
(c) $8x^2 - 2x - 3 \ge 0$

18. Solve $x^2 + 2x \le 0$.

19. Solve $x^6 - 3x^3 + 2 = 0$. Find all real solutions.

20. Solve $2x^2 + 4x + 3 = 0$. Write all complex solutions in standard form.

21. Solve $\sqrt{2} - \pi x^2 = 2.12x - 0.5\pi$ graphically. Round your answers to the nearest hundredth.

22. *Braking Distance* On wet pavement a safe braking distance d in feet for a car traveling x miles per hour is $d = \frac{x^2}{9}$. What speed corresponds to a braking distance of 250 feet? (*Source:* F. Mannering, *Principles of Highway Engineering and Traffic Control.*)

23. *Construction* A fence is being constructed along a 20-foot building, as shown in the accompanying figure. No fencing is used along the building.
(a) If 200 feet of fence are available, find a formula $f(x)$ using only the variable x that gives the area enclosed.
(b) What value of x gives the greatest area?

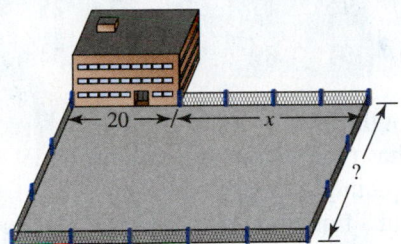

24. *Height of a Stone* Suppose that a stone is thrown upward with an initial velocity of 88 feet per second (60 miles per hour) and is released 8 feet above the ground. Its height h in feet after t seconds is given by

$$h(t) = -16t^2 + 88t + 8.$$

(a) Graph h in [0, 6, 1] by [0, 150, 50].
(b) When does the stone strike the ground?
(c) After how many seconds does the stone reach maximum height? Estimate this height.

CHAPTER 11   Extended and Discovery Exercises

MODELING DATA WITH A QUADRATIC FUNCTION

1. *Survival Rate of Birds* The survival rate of sparrowhawks varies according to their age. The following table summarizes the results of one study by listing the age in years and the percentage of birds that survived the previous year. For example, 52% of sparrowhawks that reached age 6 lived to be 7 years old. (*Source:* D. Brown and P. Rothery, *Models in Biology.*)

Age	1	2	3	4	5
Percent (%)	45	60	71	67	67

Age	6	7	8	9
Percent (%)	61	52	30	25

(a) Try to explain the relationship between age and the likelihood of surviving the next year.

(b) Make a scatterplot of the data. What type of function might model the data? Explain.
(c) Graph each function. Which of the following functions models the data better?

$$f_1(x) = -3.57x + 71.1$$
$$f_2(x) = -2.07x^2 + 17.1x + 33$$

(d) Use one of these functions to estimate the likelihood of a 5.5-year-old sparrowhawk surviving for 1 more year.

2. *Photosynthesis and Temperature* Photosynthesis is the process by which plants turn sunlight into energy. At very cold temperatures photosynthesis may halt even though the sun is shining. In one study the efficiency of photosynthesis for an Antarctic species of grass was investigated. The following table lists results for various temperatures. The temperature x is in degrees Celsius, and the efficiency y is given as a percent. The purpose of the research was to determine the temperature at which photosynthesis is most efficient. (*Source*: D. Brown.)

x (°C)	−1.5	0	2.5	5	7	10	12
y (%)	33	46	55	80	87	93	95

x (°C)	15	17	20	22	25	27	30
y(%)	91	89	77	72	54	46	34

(a) Plot the data.
(b) What type of function might model these data? Explain your reasoning.
(c) Find a function f that models the data.
(d) Use f to estimate the temperature at which photosynthesis is most efficient in this type of grass.

TRANSLATIONS OF PARABOLAS IN COMPUTER GRAPHICS

Exercises 3 and 4: In older video games with two-dimensional graphics, the background is often translated to give the illusion that a character in the game is moving. The simple scene on the left shows a mountain and an airplane. To make it appear that the airplane is flying, the mountain can be translated to the left, as shown in the figure on the right. (Reference: C. Pokorny and C. Gerald, *Computer Graphics*.)

3. *Video Games* Suppose that the mountain in the figure on the left is modeled by $f(x) = -0.4x^2 + 4$ and that the airplane is located at the point $(1, 5)$.
   (a) Graph f in $[-4, 4, 1]$ by $[0, 6, 1]$, where the units are kilometers. Plot the point $(1, 5)$ to show the location of the airplane.
   (b) Assume that the airplane is moving horizontally to the right at 0.2 kilometer per second. To give a video game player the illusion that the airplane is moving, graph the image of the mountain and the position of the airplane after 10 seconds.

4. *Video Games* (Refer to Exercise 3.) Discuss how you could create the illusion of the airplane moving to the left and gaining altitude as it passes over the mountain. Try to perform a translation of this type. Explain your reasoning.

*Exercises 5–8: **Factoring and the Discriminant** If the discriminant of the trinomial $ax^2 + bx + c$ with integer coefficients is a perfect square, then it can be factored. For example, on the one hand, the discriminant of $6x^2 + x - 2$ is*

$$1^2 - 4(6)(-2) = 49,$$

which is a perfect square ($7^2 = 49$), so we can factor the trinomial as

$$6x^2 + x - 2 = (2x - 1)(3x + 2).$$

On the other hand, the discriminant for $x^2 + x - 1$ is

$$1^2 - 4(1)(-1) = 5,$$

which is not a perfect square, so we cannot factor this trinomial by using integers as coefficients. Similarly, if the discriminant is negative, the trinomial cannot be factored by using integer coefficients. Use the discriminant to predict whether the trinomial can be factored. Then test your prediction.

5. $10x^2 - x - 3$ 6. $4x^2 - 3x - 6$

7. $3x^2 + 2x - 2$ 8. $2x^2 + x + 3$

*Exercises 9–14: **Polynomial Inequalities** The solution set for a polynomial inequality can be found by first determining the boundary numbers. For example, to solve $f(x) = x^3 - 4x > 0$ begin by solving $x^3 - 4x = 0$. The solutions (boundary numbers) are $-2, 0,$ and 2. The function $f(x) = x^3 - 4x$ is either only positive or only negative on intervals between consecutive zeros. To determine the solution set, we can evaluate test values for each interval as shown below.*

Interval	Test Value	$f(x) = x^3 - 4x$
$(-\infty, -2)$	$x = -3$	$f(-3) = -15 < 0$
$(-2, 0)$	$x = -1$	$f(-1) = 3 > 0$
$(0, 2)$	$x = 1$	$f(1) = -3 < 0$
$(2, \infty)$	$x = 3$	$f(3) = 15 > 0$

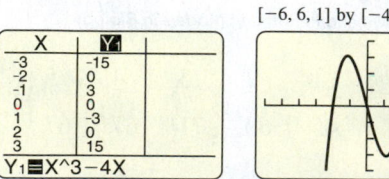

$[-6, 6, 1]$ by $[-4, 4, 1]$

We can see that $f(x) > 0$ for $(-2, 0) \cup (2, \infty)$. These results are also supported graphically in the figure on the right above, where the graph of f is above the x-axis when $-2 < x < 0$ or when $x > 2$.

Use these concepts to solve the polynomial inequality.

9. $x^3 - x^2 - 6x > 0$

10. $x^3 - 3x^2 + 2x < 0$

11. $x^3 - 7x^2 + 14x \leq 8$

12. $9x - x^3 \geq 0$

13. $x^4 - 5x^2 + 4 > 0$

14. $1 < x^4$

Exercises 15–20: Rational Inequalities Rational inequalities can be solved using many of the same techniques that are used to solve other types of inequalities. However, there is one important difference. For a rational inequality, the boundary between greater than and less than can be either an x-value where equality occurs or an x-value where a rational expression is undefined. For example, consider the inequality $f(x) = \frac{2 - x}{2x} > 0$. The solution to the equation $\frac{2 - x}{2x} = 0$ is 2. The rational expression $\frac{2 - x}{2x}$ is undefined when $x = 0$. Therefore we select test values on the intervals $(-\infty, 0)$, $(0, 2)$, and $(2, \infty)$. The table in the next column reveals that $f(x) > 0$ for $(0, 2)$.

Interval	Test Value	$f(x) = \dfrac{2 - x}{2x}$
$(-\infty, 0)$	$x = -0.5$	$f(-0.5) = -2.5 < 0$
$(0, 2)$	$x = 1$	$f(1) = 0.5 > 0$
$(2, \infty)$	$x = 2.5$	$f(2.5) = -0.1 < 0$

$[-4.7, 4.7, 1]$ by $[-3.1, 3.1, 1]$

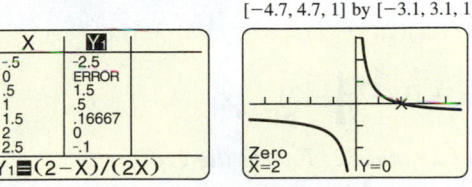

Note that $f(x)$ changes from negative to positive at $x = 0$, where $f(x)$ is undefined. These results are supported graphically in the figure on the right above.

Solve the rational inequality.

15. $\dfrac{3 - x}{3x} \geq 0$

16. $\dfrac{x - 2}{x + 2} > 0$

17. $\dfrac{3 - 2x}{1 + x} < 3$

18. $\dfrac{x + 1}{4 - 2x} \geq 1$

19. $\dfrac{5}{x^2 - 4} < 0$

20. $\dfrac{x}{x^2 - 1} \geq 0$

CHAPTERS 1–11 Cumulative Review Exercises

1. Evaluate $F = \dfrac{5}{z^2 + 1}$ when $z = -2$.

2. Classify each number as one or more of the following: natural number, whole number, integer, rational number, or irrational number: $0.\overline{4}$, $\sqrt{7}$, 0, -5, $\sqrt[3]{8}$, $-\frac{4}{3}$.

3. Simplify each expression. Write the result using positive exponents.

(a) $\left(\dfrac{x^2 y^6}{x^{-3}}\right)^2$

(b) $\dfrac{(xy^{-3})^2}{x(y^{-2})^{-1}}$

(c) $(a^2 b)^2 (ab^3)^{-4}$

4. Write 9,290,000 in scientific notation.

5. Find $f(-2)$, if $f(x) = \sqrt{2 - x}$. What is the domain of f?

6. If $f(2) = 5$, then what point lies on the graph of f?

Exercises 7 and 8: Graph f by hand.

7. $f(x) = x^2 + 2x$

8. $f(x) = |2x - 4|$

9. Find the slope–intercept form of the line that passes through $(4, -1)$ and is parallel to the line passing through $(0, 1)$ and $(-2, 4)$.

10. Find the equation of a vertical line that passes through $(-3, 4)$.

11. Solve $2x - 3(x + 2) = 6$.

Exercises 12–14: Solve the inequality. Write your answer in interval notation.

12. $7 - x > 3x$

13. $|3x - 2| \leq 1$

14. $-4 \leq 1 - x < 2$

15. Solve the system.

$$-x - 4y = -3$$
$$5x + y = -4$$

16. Shade the solution set in the xy-plane.

$$3x + y \leq 3$$
$$x - 3y \leq 3$$

17. Solve the system.

$$x + y - z = 3$$
$$x - y + z = 1$$
$$2x - y - z = 1$$

Exercises 18–20: Multiply the expression.

18. $(3x - 2)(2x + 7)$ **19.** $3xy(x^2 + y^2)$

20. $(\sqrt{x} + 3)(\sqrt{x} - 3)$

Exercises 21 and 22: Factor the expression.

21. $x^3 - x^2 - 2x$ **22.** $4x^2 - 25$

Exercises 23 and 24: Solve each equation.

23. $x^2 - 3 = 0$ **24.** $x^2 + 1 = 2x$

Exercises 25 and 26: Simplify the expression.

25. $\dfrac{(x + 3)^2}{x + 2} \cdot \dfrac{x + 2}{2x + 6}$ **26.** $\dfrac{1}{x + 2} - \dfrac{1}{x}$

Exercises 27–30: Simplify the expression. Assume all variables are positive.

27. $\sqrt{16x^6}$ **28.** $16^{-3/2}$

29. $\dfrac{\sqrt[3]{81x}}{\sqrt[3]{3x}}$ **30.** $\sqrt{8x} + \sqrt{2x}$

31. Graph $f(x) = \sqrt{4x}$.

32. Find the distance between $(-1, 2)$ and $(4, 3)$.

33. Write $\dfrac{3 - i}{2 + i}$ in standard form.

34. Solve $3\sqrt{x + 1} = 2x$.

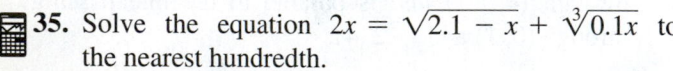**35.** Solve the equation $2x = \sqrt{2.1 - x} + \sqrt[3]{0.1x}$ to the nearest hundredth.

36. Sketch a graph of $f(x) = x^2 - 2x + 3$.
  (a) Find the vertex.
  (b) Evaluate $f(-1)$.
  (c) What is the axis of symmetry?
  (d) Where is f increasing?

37. Write $f(x) = 2x^2 - 4x - 1$ in vertex form.

38. Compare the graph of $f(x) = 4(x + 1)^2 - 2$ to the graph of $y = x^2$.

39. Solve $x^2 + 6x = 2$ by completing the square.

40. Solve $2x^2 - 3x = 1$ by using the quadratic formula.

41. Solve $x(4 - x) = 3$.

42. The graph of $y = ax^2 + bx + c$ is shown in the next column. Solve each equation or inequality.

  (a) $ax^2 + bx + c = 0$
  (b) $ax^2 + bx + c \le 0$

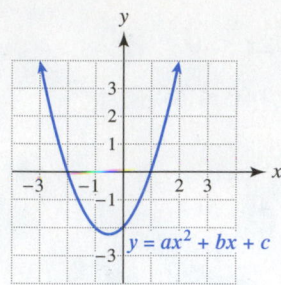

43. Solve $x^2 - 3x + 2 > 0$.

44. Solve $x^4 - 256 = 0$. Find all complex solutions.

Exercises 45–52: **Thinking Generally** *Match the graph (a.–h.) with its equation. Assume that a, b, and c are positive constants.*

45. $y = ax - b$ **46.** $y = b$

47. $y = -ax^2 + c$ **48.** $y = \dfrac{a}{x}$

49. $y = ax^3$ **50.** $y = |ax + b|$

51. $y = a\sqrt{x}$ **52.** $y = a\sqrt[3]{x}$

a.
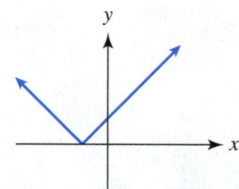

b.
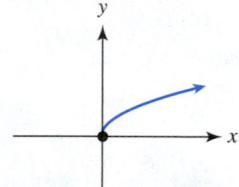

c.
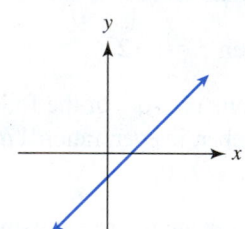

d.
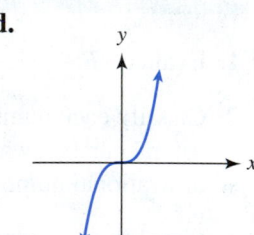

e.
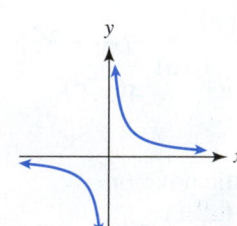

f.
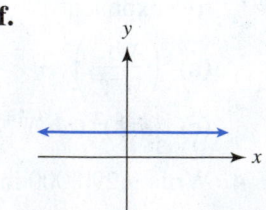

g.
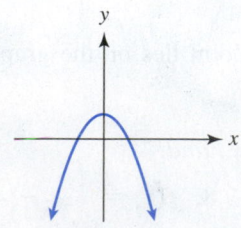

h.
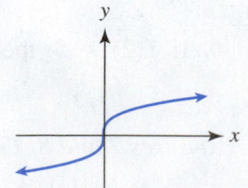

APPLICATIONS

53. *Calculating Water Flow* Water is being pumped out of a tank. The gallons G of water in the tank after t minutes is shown in the figure, where $y = G(t)$.
 (a) Evaluate $G(0)$. Interpret your answer.
 (b) What is the t-intercept? Interpret your answer.
 (c) What is the slope of the graph of G? Interpret your answer.
 (d) Find a formula for $G(t)$.

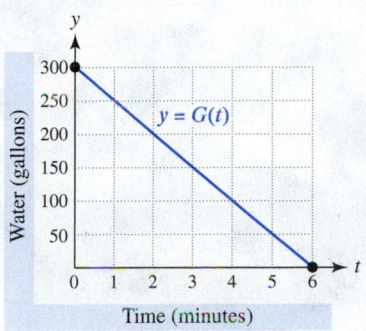

54. *Investment* Suppose $4000 is deposited in three accounts paying 4%, 5%, and 6% annual interest. The amount invested at 6% is $1000 more than the amount invested at 5%. The interest after 1 year is $216. How much is invested at each rate?

55. *Maximizing Area* There are 490 feet of fence available to surround the perimeter of a rectangular garden. On one side, there is a 10-foot gate that requires no fencing. What dimensions for the garden give the largest area?

56. *Height of a Tree* A 6-foot-tall person casts a shadow that is 10 feet long while a nearby tree casts a 55-foot shadow. Find the height of the tree.

12 Exponential and Logarithmic Functions

Time is money.

—BENJAMIN FRANKLIN

If you deposit money into a savings account, you might *earn* an interest rate of 0.18%, whereas if you charge purchases on your credit card, you might *pay* an interest rate of 18%. Do interest rates and decimal points really matter? To find out, consider the following.

If $1000 is deposited in a savings account at 0.18% interest per year for 15 years, the final amount would be about $1027.37. However, if the interest rate were a typical credit card rate of 18%, the final amount would be about $14,584.37. Interest rates and decimal points clearly matter! After *doing the math*, it becomes obvious that the best financial decision is to pay off high-interest credit cards first, before putting money into a savings account that pays only 0.18%.

In this chapter we will learn ways to calculate interest on investments by studying a new type of function called an *exponential function*. (See Section 12.2.) Exponential functions are frequently used in business and finance, but they also have many other applications.

12.1 Composite and Inverse Functions

Composition of Functions • One-to-One Functions • Inverse Functions • Tables and Graphs of Inverse Functions

A LOOK INTO MATH ▶

Suppose that you walk into a classroom, turn on the lights, and sit down at your desk. How could you undo or reverse these actions? You would stand up from the desk, turn off the lights, and walk out of the classroom. Note that you must not only perform the "inverse" of each action, but you also must do them in the *reverse order*. In mathematics, we undo an arithmetic operation by performing its inverse operation. For example, the inverse operation of addition is subtraction, and the inverse operation of multiplication is division. In this section, we explore these concepts further by discussing inverse functions.

NEW VOCABULARY

☐ Composite function
☐ Composition
☐ One-to-one correspondence
☐ One-to-one
☐ Horizontal line test
☐ Inverse function

Composition of Functions

▶ **REAL-WORLD CONNECTION** Many tasks in life are performed in *sequence*, such as putting on your socks and then your shoes. These types of situations also occur in mathematics. For example, suppose that we want to calculate the number of ounces in 3 tons. Because there are 2000 pounds in one ton, we might first multiply **3** by 2000 to obtain **6000** pounds. There are 16 ounces in a pound, so we could multiply **6000** by 16 to obtain **96,000** ounces. This particular calculation involves a *sequence* of calculations that can be represented by the diagram shown in Figure 12.1.

Converting Tons to Ounces

P multiplies 3, the input for *P*, by 2000.

O multiplies 6000, the output from *P*, by 16.

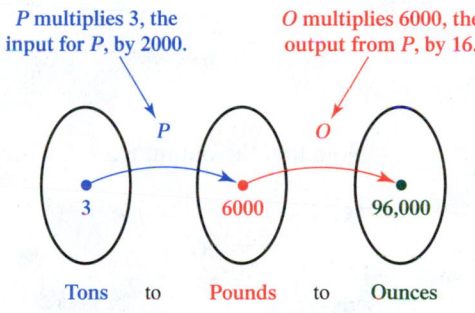

| Tons | to | Pounds | to | Ounces |

Figure 12.1

The results shown in Figure 12.1 can be calculated by using function notation. Suppose that we let $P(x) = 2000x$ convert x tons to **P** pounds and also let $O(x) = 16x$ convert x pounds to **O** ounces. We can calculate the number of ounces in 3 tons by performing the *composition of O and P*. This method can be expressed symbolically as follows.

Composition of *O* and *P*

Start by evaluating the innermost function, $P(3)$.

$$(O \circ P)(3) = O\big(P(3)\big)$$

The output from *P* becomes the input for *O*.

$$= O(2000 \cdot 3) \quad \text{Let } x = 3 \text{ in } P(x) = 2000x.$$

$$= O(6000) \quad \text{Simplify.}$$

$$= 16(6000) \quad \text{Let } x = 6000 \text{ in } O(x) = 16x.$$

$$= 96,000. \quad \text{Multiply.}$$

P first multiplies 3 by 2000 and then *O* multiplies 6000 by 16 to get 96,000. This is written as $(O \circ P)(3)$.

COMPOSITION OF FUNCTIONS

If f and g are functions, then the **composite function** $g \circ f$, or **composition** of g and f, is defined by

$$(g \circ f)(x) = g\big(f(x)\big).$$

NOTE: We read $g\big(f(x)\big)$ as "g of f of x."

READING CHECK

Explain how to compute
$(f \circ g)(x)$ and $(g \circ f)(x)$.

The compositions $g \circ f$ and $f \circ g$ represent evaluating functions f and g in different orders. When evaluating $g \circ f$, function f is evaluated first followed by function g, whereas for $f \circ g$ function g is evaluated first followed by function f. Note that in general $(g \circ f)(x) \neq (f \circ g)(x)$. That is, *the order in which functions are applied makes a difference,* in the same way that putting on your socks and then your shoes is quite different from putting on your shoes and then your socks. See Example 2(a) and (b).

EXAMPLE 1   **Finding composite functions**

Evaluate $(g \circ f)(2)$ and then find a formula for $(g \circ f)(x)$.

(a) $f(x) = x^3$, $g(x) = 3x - 2$ (b) $f(x) = 5x$, $g(x) = x^2 - 3x + 1$

(c) $f(x) = \sqrt{2x}$, $g(x) = \dfrac{1}{x - 1}$

Solution

(a) $(g \circ f)(2) = g\big(f(2)\big)$ Composition of functions

$\qquad\qquad = g(8)$ $f(2) = 2^3 = 8$

$\qquad\qquad = 22$ $g(8) = 3(8) - 2 = 22$

Note that the output from f, which is $f(2)$, becomes the input for g.

$(g \circ f)(x) = g\big(f(x)\big)$ Composition of functions

$\qquad\qquad = g(x^3)$ $f(x) = x^3$

$\qquad\qquad = 3x^3 - 2$ Replace x with x^3 in $g(x) = 3x - 2$.

(b) $(g \circ f)(2) = g\big(f(2)\big)$ Composition of functions

$\qquad\qquad = g(10)$ $f(2) = 5(2) = 10$

$\qquad\qquad = 71$ $g(10) = 10^2 - 3(10) + 1 = 71$

$(g \circ f)(x) = g\big(f(x)\big)$ Composition of functions

$\qquad\qquad = g(5x)$ $f(x) = 5x$

$\qquad\qquad = (5x)^2 - 3(5x) + 1$ Replace x with $5x$ in $g(x) = x^2 - 3x + 1$.

$\qquad\qquad = 25x^2 - 15x + 1$ Simplify.

(c) $(g \circ f)(2) = g\big(f(2)\big)$ Composition of functions

$\qquad\qquad = g(2)$ $f(2) = \sqrt{2(2)} = 2$

$\qquad\qquad = 1$ $g(2) = \dfrac{1}{2 - 1} = 1$

$(g \circ f)(x) = g\big(f(x)\big)$ Composition of functions

$\qquad\qquad = g\big(\sqrt{2x}\big)$ $f(x) = \sqrt{2x}$

$\qquad\qquad = \dfrac{1}{\sqrt{2x} - 1}$ Replace x with $\sqrt{2x}$ in $g(x) = \dfrac{1}{x - 1}$.

Now Try Exercises 13, 15, 19

Composite functions can also be evaluated numerically and graphically, as demonstrated in the next two examples.

EXAMPLE 2 **Evaluating composite functions with tables**

Use Tables 12.1 and 12.2 to evaluate each expression.
(a) $(f \circ g)(2)$ **(b)** $(g \circ f)(2)$ **(c)** $(f \circ f)(0)$

TABLE 12.1

x	0	1	2	3
$f(x)$	3	2	0	1

TABLE 12.2

x	0	1	2	3
$g(x)$	1	3	2	0

Solution
(a) $(f \circ g)(2) = f\big(g(2)\big)$ Composition of functions
$= f(2)$ $g(2) = 2$
$= 0$ $f(2) = 0$
(b) $(g \circ f)(2) = g\big(f(2)\big)$ Composition of functions
$= g(0)$ $f(2) = 0$
$= 1$ $g(0) = 1$

NOTE: From parts (a) and (b), we see that $(f \circ g)(2) \neq (g \circ f)(2)$.

(c) $(f \circ f)(0) = f\big(f(0)\big)$ Composition of functions
$= f(3)$ $f(0) = 3$
$= 1$ $f(3) = 1$

Now Try Exercises 23, 25

EXAMPLE 3 **Evaluating composite functions graphically**

Use Figure 12.2 in the margin to evaluate $(g \circ f)(2)$.

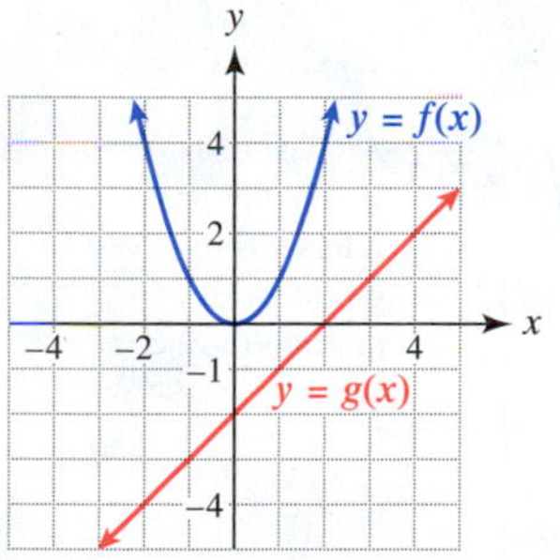

Solution
Because $(g \circ f)(2) = g\big(f(2)\big)$, start by using Figure 12.2 to evaluate $f(2)$. Figure 12.3(a) shows that $f(2) = 4$, which becomes the input for g. Figure 12.3(b) reveals that $g(4) = 2$.

$$(g \circ f)(2) = g\big(f(2)\big) = g(4) = 2$$

$f(2) = 4$ $g(4) = 2$

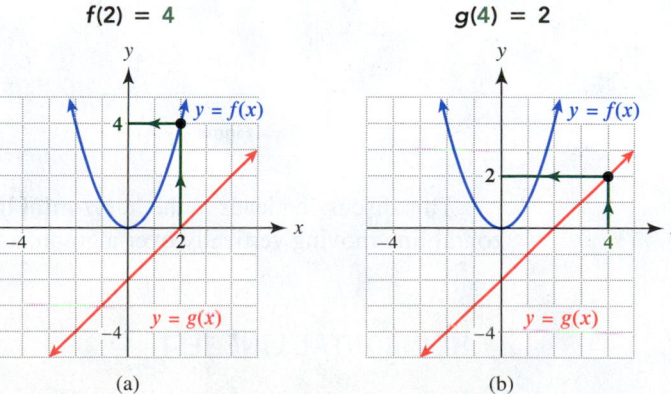

(a) (b)

Figure 12.3

Figure 12.2

Now Try Exercise 29

One-to-One Functions

▶ **REAL-WORLD CONNECTION** One-to-one correspondences occur in real life. For example, in a typical city there is *usually* a **one-to-one correspondence** between homes and addresses. Each home has one address, and each address corresponds to one home.

If we change the input for a function, does the output also change? Do *different inputs* always result in *different outputs* for every function? The answer is no. For example, if $f(x) = x^2 + 1$, then the inputs -2 and 2 result in the *same* output, 5. That is, $f(-2) = 5$ and $f(2) = 5$. However, for $g(x) = 2x$, *different inputs* always result in *different outputs*. For function g, there is a one-to-one correspondence between inputs and outputs. Thus we say that g is a *one-to-one function*, whereas f is not.

ONE-TO-ONE FUNCTION

A function f is **one-to-one** if, for any c and d in the domain of f,

$$c \neq d \quad \text{implies that} \quad f(c) \neq f(d).$$

That is, different inputs always result in different outputs.

READING CHECK

How can you determine if a function is one-to-one?

One way to determine whether a function f is one-to-one is to look at its graph. Suppose that a function has two *different inputs* that result in the *same output*. Then there must be two points on its graph that have the same y-value but different x-values. For example, if $f(x) = 2x^2$, then $f(-1) = 2$ and $f(1) = 2$. Thus the points $(-1, 2)$ and $(1, 2)$ both lie on the graph of f, as shown in Figure 12.4(a). Two points with different x-values and the same y-value determine a horizontal line, as shown in Figure 12.4(b). This horizontal line intersects the graph of f more than once, indicating that different inputs do *not* always have different outputs. Thus $f(x) = 2x^2$ is *not* one-to-one.

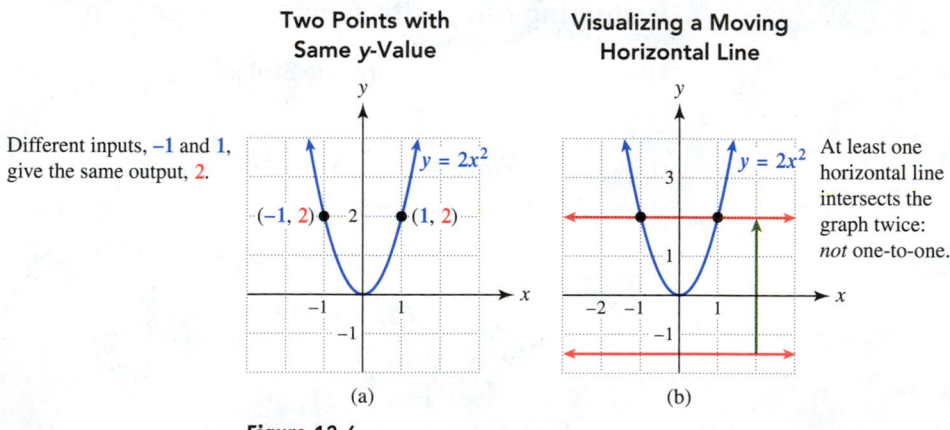

Figure 12.4

This discussion leads to the **horizontal line test**, which is applied by visualizing a horizontal line moving vertically over a graph, as shown in Figure 12.4(b).

HORIZONTAL LINE TEST

If every horizontal line intersects the graph of a function f at most once, then f is a one-to-one function.

We apply the horizontal line test in the next example.

EXAMPLE 4   **Using the horizontal line test**

Determine whether each graph in Figure 12.5 represents a one-to-one function.

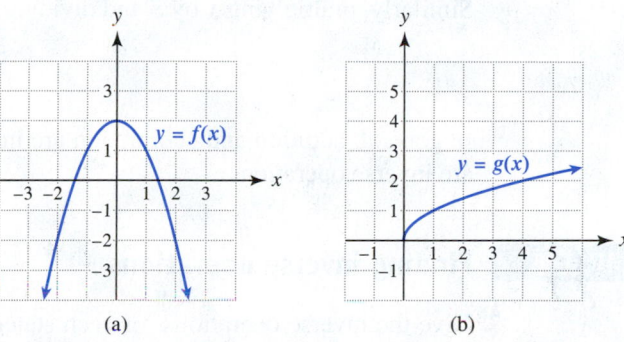

(a)   (b)

Figure 12.5

Solution

Figure 12.6(a) shows one of many horizontal lines that intersect the graph of $y = f(x)$ twice. Therefore function f is *not* one-to-one.

Two Points of Intersection:
f Is Not One-to-One   At Most One Point of
Intersection: g Is One-to-One

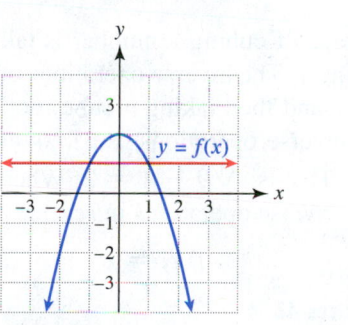   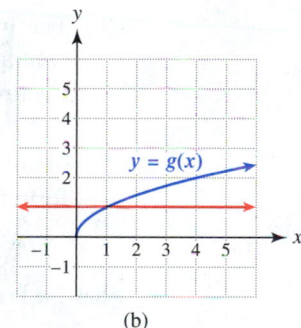

(a)   (b)

Figure 12.6

Figure 12.6(b) suggests that every horizontal line will intersect the graph of $y = g(x)$ *at most* once. Therefore function g is one-to-one.

Now Try Exercises 37, 39

MAKING CONNECTIONS

STUDY TIP

Be sure that you understand when to use the vertical line test and when to use the horizontal line test.

Vertical and Horizontal Line Tests

The **vertical line test** is used to identify **functions**, whereas the **horizontal line test** is used to identify **one-to-one** functions. For example, consider the graph of $f(x) = x^2$. A vertical line never intersects the graph more than once, so f is a function. A horizontal line can intersect the graph twice, so f *is not* a one-to-one function.

Function; Not One-to-One

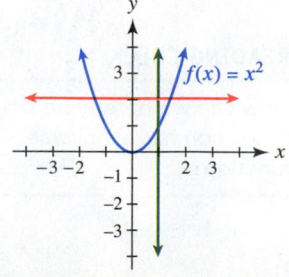

Inverse Functions

▶ **REAL-WORLD CONNECTION** Turning on a light and turning off a light are inverse operations from ordinary life. Inverse operations undo each other. In mathematics, adding **5** to x and subtracting **5** from x are inverse operations because

$$x + 5 - 5 = x.$$

Similarly, multiplying x by **5** and dividing x by **5** are inverse operations because

$$\frac{5x}{5} = x.$$

In general, addition and subtraction are inverse operations, and multiplication and division are inverse operations.

EXAMPLE 5 | **Finding inverse operations**

Give the inverse operations for each statement. Then write a function f for the given statement and a function g for its inverse operations.
(a) Divide x by 3.
(b) Cube x and then add 1 to the result.

Solution
(a) The inverse of dividing x by 3 is to *multiply x by 3*. Thus

$$f(x) = \frac{x}{3} \quad \text{and} \quad g(x) = 3x.$$

(b) The inverse of cubing a number is taking a cube root, and the inverse of adding 1 is subtracting 1. The inverse operations of "cubing a number and then adding 1" are "subtracting 1 and then taking a cube root." For example, 2 cubed plus 1 is $2^3 + 1 = 9$. For the inverse operations, we *first* subtract 1 from 9 and then take the cube root to obtain 2. That is, $\sqrt[3]{9 - 1} = 2$. When there is more than one operation, we must perform the inverse operations in *reverse order*. Thus

$$f(x) = x^3 + 1 \quad \text{and} \quad g(x) = \sqrt[3]{x - 1}.$$

Now Try Exercises 43, 49

Functions f and g in each part of Example 5 are examples of *inverse functions*. Note that in part (a), if $f(x) = \frac{x}{3}$ and $g(x) = 3x$, then

$$f(\mathbf{15}) = \mathbf{5} \quad \text{and} \quad g(\mathbf{5}) = \mathbf{15}.$$

In general, *if f and g are inverse functions, $f(\mathbf{a}) = \mathbf{b}$ implies $g(\mathbf{b}) = \mathbf{a}$.* Thus

$$(g \circ f)(\mathbf{a}) = g\big(f(a)\big) = g(\mathbf{b}) = \mathbf{a} \qquad \text{f and g are inverse functions.}$$

for any $\mathbf{a}$ in the domain of f, whenever g and f are inverse functions. The composition of a function with its inverse leaves the input unchanged.

READING CHECK

How can we determine if a function has an inverse function?

INVERSE FUNCTIONS

Let f be a one-to-one function. Then f^{-1} is the **inverse function** of f if

$$(f^{-1} \circ f)(x) = f^{-1}\big(f(x)\big) = x, \quad \text{for every x in the domain of f,} \quad \text{and}$$
$$(f \circ f^{-1})(x) = f\big(f^{-1}(x)\big) = x, \quad \text{for every x in the domain of f^{-1}.}$$

NOTE: In the expression $f^{-1}(x)$, the -1 is *not* an exponent. That is, $f^{-1}(x) \neq \frac{1}{f(x)}$. Rather, if $f(x) = \frac{x}{3}$, then $f^{-1}(x) = 3x$, and if $f(x) = x^3 + 1$, then $f^{-1}(x) = \sqrt[3]{x - 1}$.

EXAMPLE 6   **Verifying inverses**

Verify that $f^{-1}(x) = 3x$ if $f(x) = \frac{x}{3}$.

Solution
We must show that $(f^{-1} \circ f)(x) = x$ and that $(f \circ f^{-1})(x) = x$.

$$(f^{-1} \circ f)(x) = f^{-1}\big(f(x)\big) \qquad \text{Composition of functions}$$

$$= f^{-1}\left(\frac{x}{3}\right) \qquad f(x) = \frac{x}{3}$$

$$= 3\left(\frac{x}{3}\right) \qquad f^{-1}(x) = 3x$$

$$= x \checkmark \qquad \text{Simplify; it checks.}$$

$$(f \circ f^{-1})(x) = f\big(f^{-1}(x)\big) \qquad \text{Composition of functions}$$

$$= f(3x) \qquad f^{-1}(x) = 3x$$

$$= \frac{3x}{3} \qquad f(x) = \frac{x}{3}$$

$$= x \checkmark \qquad \text{Simplify; it checks.}$$

Now Try Exercise 51

The definition of inverse functions states that *f must be a one-to-one function*. To understand why, consider Figure 12.7(a), where a one-to-one function *f* is represented by a diagram. To find f^{-1}, the arrows are simply reversed. For example, $f(1) = 3$ implies that $f^{-1}(3) = 1$, so the arrow from **1** to **3** for *f* must be redrawn in Figure 12.7(b), from **3** to **1** for f^{-1}.

One-to-One, So Inverse Exists   **Inverse Is Also One-to-One**

No two arrows go to the same output.

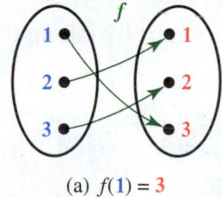

Reverse each arrow to find f^{-1}.

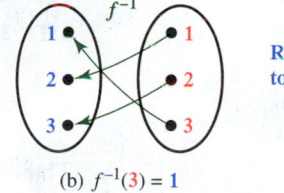

(a) $f(1) = 3$ (b) $f^{-1}(3) = 1$

Figure 12.7

In Figure 12.8(a), function *g* is *not* one-to-one because two different arrows, from **2** and **3**, both go to **1**. In Figure 12.8(b), the arrows are reversed in an attempt to draw the inverse *function*. However, **two** arrows now go from input **1** to outputs **2** and **3**. This means that *one input has two outputs*, which is not possible if g^{-1} is a function.

g Is Not One-to-One   **g Has No Inverse Function**

Two inputs give the same output: not one-to-one.

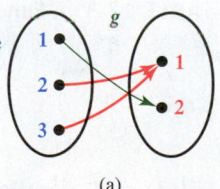

One input gives two outputs: not a function.

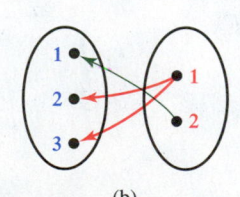

(a)   (b)

Figure 12.8

The following steps can be used to find the inverse of a function symbolically. Note that for a function f to have an inverse function, f must be one-to-one.

FINDING AN INVERSE FUNCTION

To find f^{-1} for a one-to-one function f, perform the following steps.

STEP 1: Let $y = f(x)$.

STEP 2: Interchange x and y.

STEP 3: Solve the formula for y. The resulting formula is $y = f^{-1}(x)$.

We apply these steps in the next example.

EXAMPLE 7 **Finding an inverse function**

Find the inverse of each one-to-one function.
(a) $f(x) = 3x - 7$
(b) $g(x) = (x + 2)^3$

Solution
(a) **STEP 1:** Let $y = 3x - 7$.

STEP 2: Interchange x and y to get $x = 3y - 7$.

STEP 3: To solve for y start by adding 7 to each side.

$$x + 7 = 3y \qquad \text{Add 7 to each side.}$$

$$\frac{x + 7}{3} = y \qquad \text{Divide each side by 3.}$$

Thus $f^{-1}(x) = \frac{x + 7}{3}$ or $f^{-1}(x) = \frac{1}{3}x + \frac{7}{3}$.

(b) **STEP 1:** Let $y = (x + 2)^3$.

STEP 2: Interchange x and y to get $x = (y + 2)^3$.

STEP 3: To solve for y start by taking the cube root of each side.

$$\sqrt[3]{x} = y + 2 \qquad \text{Take the cube root of each side.}$$

$$\sqrt[3]{x} - 2 = y \qquad \text{Subtract 2 from each side.}$$

Thus $g^{-1}(x) = \sqrt[3]{x} - 2$.

Now Try Exercises 63, 71

Tables and Graphs of Inverse Functions

TABLES OF INVERSE FUNCTIONS Inverse functions can be represented with tables. Table 12.3 shows a table of values for a function f.

TABLE 12.3 A Function f

x	1	2	3	4	5
$f(x)$	3	6	9	12	15

Because $f(1) = 3$, $f^{-1}(3) = 1$. Similarly, $f(2) = 6$ implies that $f^{-1}(6) = 2$ and so on. Table 12.4 lists values for $f^{-1}(x)$.

TABLE 12.4 The Inverse Function f^{-1}

x	3	6	9	12	15
$f^{-1}(x)$	1	2	3	4	5

Interchange values

Note that the domain of f is {**1, 2, 3, 4, 5**} and that the range of f is {**3, 6, 9, 12, 15**}, whereas the domain of f^{-1} is {**3, 6, 9, 12, 15**} and the range of f^{-1} is {**1, 2, 3, 4, 5**}. *The domain of f is the range of f^{-1}, and the range of f is the domain of f^{-1}.* This statement is true in general for a function and its inverse.

GRAPHS OF INVERSE FUNCTIONS If $f(a) = b$, then the point (a, b) lies on the graph of f. This statement also means that $f^{-1}(b) = a$ and that the point (b, a) lies on the graph of f^{-1}. These points are shown in Figure 12.9(a) with a solid green line segment connecting them. The line $y = x$ is perpendicular to this line segment and divides it into two equal parts. As a result, the graph of f^{-1} can be sketched from the graph of f by reflecting the graph of f across the line $y = x$. See Figure 12.9(b).

Reflections Across the Line y = x

Point (b, a) is a reflection of point (a, b) across $y = x$.

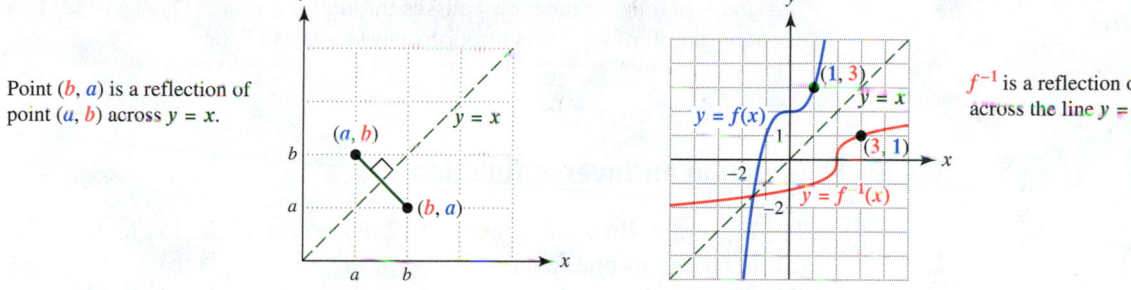

f^{-1} is a reflection of f across the line $y = x$.

Figure 12.9

The relationship between the graph of a function and the graph of its inverse function is summarized as follows.

GRAPHS OF FUNCTIONS AND THEIR INVERSES

The graph of f^{-1} is a reflection of the graph of f across the line $y = x$.

EXAMPLE 8   **Graphing an inverse function**

The graph of $y = f(x)$ is shown in Figure 12.10. Sketch a graph of $y = f^{-1}(x)$.

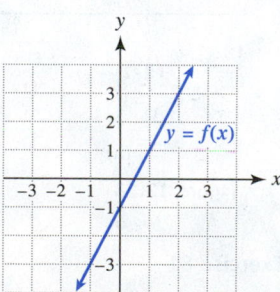

Figure 12.10

Solution

The graph of $y = f^{-1}(x)$ is the reflection of the graph of $y = f(x)$ across the line $y = x$ and is shown in Figure 12.11. Note that the reflection of a line is another line.

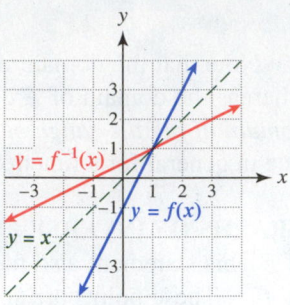

Figure 12.11

■ **Now Try Exercise 83**

CRITICAL THINKING

The graph of a linear function f passes through the points (1, 2) and (2, 1). What two points does the graph of f^{-1} pass through? Find $f(x)$ and $f^{-1}(x)$.

EXAMPLE 9 **Graphing an inverse function**

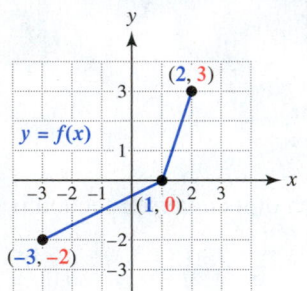

Figure 12.12

The line graph shown in Figure 12.12 represents a function f.
(a) Is f a one-to-one function?
(b) Sketch a graph of $y = f^{-1}(x)$.

Solution
(a) Every horizontal line intersects the graph of f at most once. By the horizontal line test, the graph represents a one-to-one function.
(b) The points $(-3, -2)$, $(1, 0)$, and $(2, 3)$ lie on the graph of f. It follows that the points $(-2, -3)$, $(0, 1)$, and $(3, 2)$ lie on the graph of f^{-1}. Plot these three points and then connect them with line segments, as shown in Figure 12.13(a). Note that the graph of $y = f^{-1}(x)$ is a reflection of the graph of $y = f(x)$ across the line $y = x$, as shown in Figure 12.13(b).

READING CHECK

If we are given the graph of $y = f(x)$, where f is one-to-one, how can we sketch the graph of $y = f^{-1}(x)$?

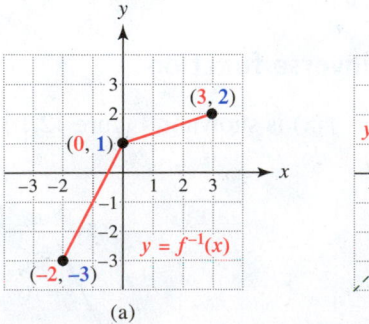
(a)

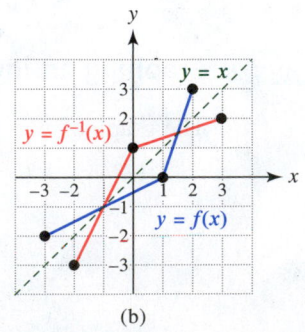
(b)

Figure 12.13

■ **Now Try Exercise 81**

12.1   Putting It All Together

CONCEPT	EXPLANATION	EXAMPLES
Composite Functions	The composite of g and f is given by $$(g \circ f)(x) = g(f(x)),$$ and represents a *new* function whose name is $g \circ f$.	If $f(x) = 1 - 4x$ and $g(x) = x^3$, then $$(g \circ f)(x) = g(f(x))$$ $$= g(1 - 4x)$$ $$= (1 - 4x)^3.$$
One-to-One Functions	Function f is one-to-one if different inputs always give different outputs.	$f(x) = x^2$ is not one-to-one because $f(-4) = f(4) = 16$, whereas $g(x) = x + 1$ is one-to-one because every input has a unique output.
Horizontal Line Test	This test is used to determine whether a function is one-to-one from its graph: If every horizontal line intersects the graph of a function f at most once, then f is a one-to-one function.	$f(x) = x^2$ is not one-to-one because a horizontal line can intersect its graph more than once. **Not One-to-One** 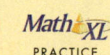
Inverse Functions	f^{-1} will undo the operations performed by f. That is, $$(f^{-1} \circ f)(x) = x \quad \text{and}$$ $$(f \circ f^{-1})(x) = x.$$	If $f(x) = x^3$, then $f^{-1}(x) = \sqrt[3]{x}$ because cubing a number x and then taking its cube root results in the number x.

12.1   Exercises

MyMathLab | Math XL PRACTICE | WATCH | DOWNLOAD | READ | REVIEW

CONCEPTS AND VOCABULARY

1. $(g \circ f)(7) = $ _____

2. $(f \circ g)(x) = $ _____

3. Does $(f \circ g)(x)$ always equal $(g \circ f)(x)$?

4. If a function f is one-to-one, then different _____ always result in different _____.

5. If $f(3) = 5$ and $f(7) = 5$, could f be one-to-one?

6. If every horizontal line intersects the graph of f at most once, then f is _____.

7. The inverse operation of subtracting 10 is _____.

8. $(f^{-1} \circ f)(7) = $ _____.

9. If $f(6) = 8$, then $f^{-1}($_____$) = $ _____.

10. If $f^{-1}(y) = x$, then $f($_____$) = $ _____.

11. For f to have an inverse function, f must be _____.

12. The graph of f^{-1} is a(n) _____ of the graph of f across the line _____.

COMPOSITE FUNCTIONS

Exercises 13–22: For the given f(x) and g(x), find each of the following.

   (a) $(g \circ f)(-2)$ *(b)* $(f \circ g)(4)$
   (c) $(g \circ f)(x)$ *(d)* $(f \circ g)(x)$

13. $f(x) = x^2$ $g(x) = x + 3$

14. $f(x) = 4x^2$ $g(x) = 5x$

15. $f(x) = 2x$ $g(x) = x^3 - 1$

16. $f(x) = 3x + 1$ $g(x) = x^2 + 4x$

17. $f(x) = \frac{1}{2}x$ $g(x) = |x - 2|$

18. $f(x) = 6x$ $g(x) = \dfrac{2}{x - 5}$

19. $f(x) = \dfrac{1}{x}$ $g(x) = 3 - 5x$

20. $f(x) = \sqrt{x + 3}$ $g(x) = x^3 - 3$

21. $f(x) = 2x$ $g(x) = 4x^2 - 2x + 5$

22. $f(x) = 9x - \dfrac{1}{3x}$ $g(x) = \dfrac{x}{3}$

Exercises 23–28: Evaluate each expression numerically.

x	-2	-1	0	1	2
$f(x)$	2	1	0	-1	-2

x	-2	-1	0	1	2
$g(x)$	0	1	-1	2	-2

23. (a) $(f \circ g)(0)$ (b) $(g \circ f)(-1)$

24. (a) $(f \circ g)(1)$ (b) $(g \circ f)(-2)$

25. (a) $(f \circ f)(-1)$ (b) $(g \circ g)(0)$

26. (a) $(g \circ g)(2)$ (b) $(f \circ f)(1)$

27. (a) $(f^{-1} \circ g)(-2)$ (b) $(g^{-1} \circ f)(2)$

28. (a) $(f \circ g^{-1})(1)$ (b) $(g \circ f^{-1})(-2)$

Exercises 29 and 30: Evaluate each expression graphically.

29. (a) $(f \circ g)(0)$

    (b) $(g \circ f)(1)$

    (c) $(f \circ f)(-1)$

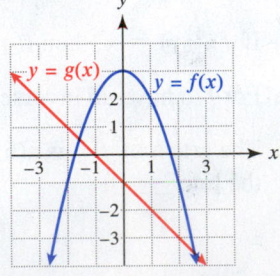

30. (a) $(f \circ g)(1)$

    (b) $(g \circ f)(-2)$

    (c) $(g \circ g)(-2)$

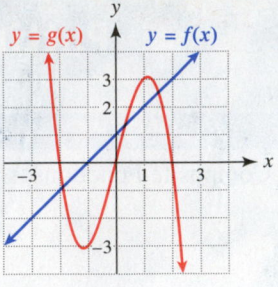

Exercises 31–36: Show that f is not a one-to-one function by finding two inputs that result in the same output. Answers may vary.

31. $f(x) = 5x^2$ **32.** $f(x) = 4 - x^2$

33. $f(x) = x^4 + 100$ **34.** $f(x) = \dfrac{x^2}{x^2 + 1}$

35. $f(x) = x^4 - 3x^2$ **36.** $f(x) = \sqrt{x^2 - 1}$

Exercises 37–42: Use the horizontal line test to determine whether the graph represents a one-to-one function.

37.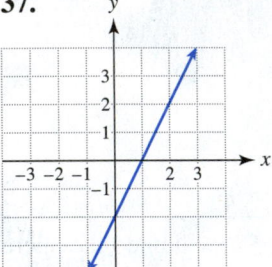

38.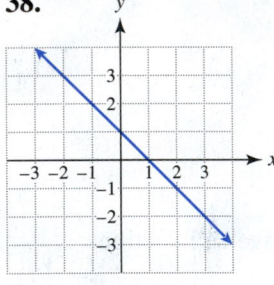

39.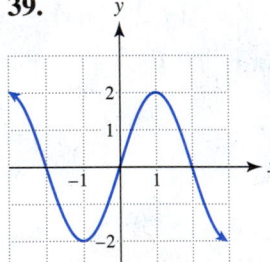

40.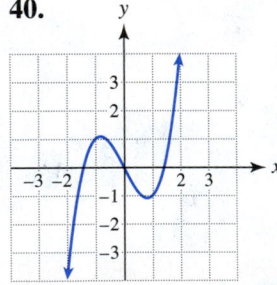

41.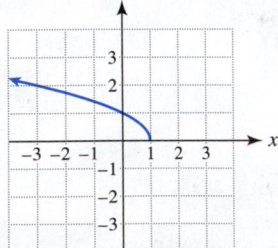

42.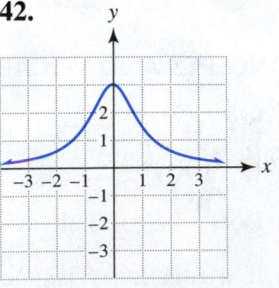

Exercises 43–50: (Refer to Example 5.) Give the inverse operation for the statement. Then write a function f for the given statement and a function g for its inverse.

43. Multiply x by 7.

44. Subtract 10 from x.

45. Add 5 to x and then divide the result by 2.

46. Multiply x by 6 and then add 8 to the result.

47. Multiply x by $\frac{1}{2}$ and then subtract 3 from the result.

48. Divide x by 10 and then add 20 to the result.

49. Cube the sum of x and 5.

50. Take the cube root of x and then subtract 2.

Exercises 51–58: (Refer to Example 6.) Verify that f(x) and $f^{-1}(x)$ are indeed inverse functions.

51. $f(x) = 4x$ $f^{-1}(x) = \dfrac{x}{4}$

52. $f(x) = \dfrac{2x}{3}$ $f^{-1}(x) = \dfrac{3x}{2}$

53. $f(x) = 3x + 5$ $f^{-1}(x) = \dfrac{x-5}{3}$

54. $f(x) = x + 7$ $f^{-1}(x) = x - 7$

55. $f(x) = x^3$ $f^{-1}(x) = \sqrt[3]{x}$

56. $f(x) = \sqrt[3]{x-4}$ $f^{-1}(x) = x^3 + 4$

57. $f(x) = \dfrac{1}{x}$ $f^{-1}(x) = \dfrac{1}{x}$

58. $f(x) = \dfrac{x+7}{7}$ $f^{-1}(x) = 7x - 7$

Exercises 59–74: (Refer to Example 7.) Find $f^{-1}(x)$.

59. $f(x) = 12x$ **60.** $f(x) = \frac{3}{4}x$

61. $f(x) = x + 8$ **62.** $f(x) = x - 3$

63. $f(x) = 5x - 2$ **64.** $f(x) = 3x + 4$

65. $f(x) = -\frac{1}{2}x + 1$ **66.** $f(x) = \frac{3}{4}x - \frac{1}{4}$

67. $f(x) = 8 - x$ **68.** $f(x) = 5 - x$

69. $f(x) = \dfrac{x+1}{2}$ **70.** $f(x) = \dfrac{3-x}{5}$

71. $f(x) = \sqrt[3]{2x}$ **72.** $f(x) = \sqrt[3]{x+4}$

73. $f(x) = x^3 - 8$ **74.** $f(x) = (x-5)^3$

Exercises 75–78: Use the table to make a table of values for $f^{-1}(x)$. State the domain and range for f and for f^{-1}.

75.

x	0	1	2	3	4
$f(x)$	0	5	10	15	20

76.

x	−4	−2	0	2	4
$f(x)$	1	2	3	4	5

77.

x	−5	0	5	10	15
$f(x)$	4	2	0	−2	−4

78.

x	0	2	4	6	8
$f(x)$	8	6	4	2	0

Exercises 79–82: (Refer to Example 9.) Use the graph of y = f(x) to sketch a graph of y = $f^{-1}(x)$.

79.
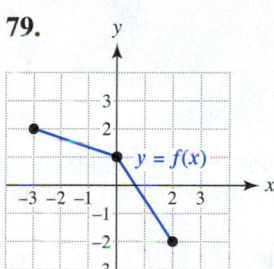

80.
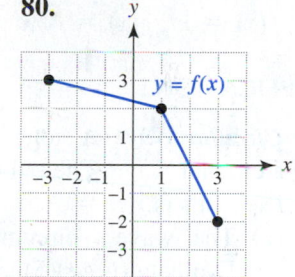

81.
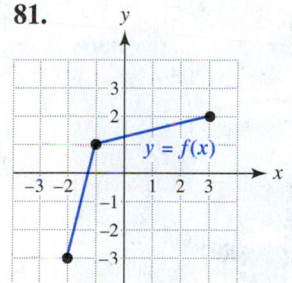

82.
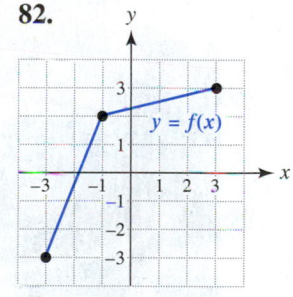

Exercises 83–86: (Refer to Example 8.) Use the graph of y = f(x) to sketch a graph of $f^{-1}(x)$. Include the graph of f and the line y = x in your graph.

83.
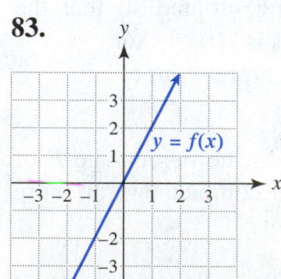

84.
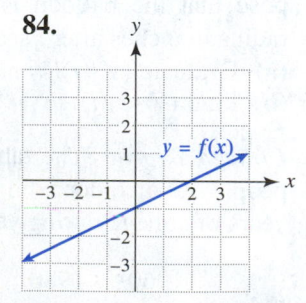

85.

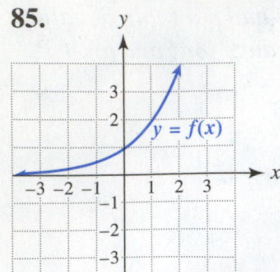

86.

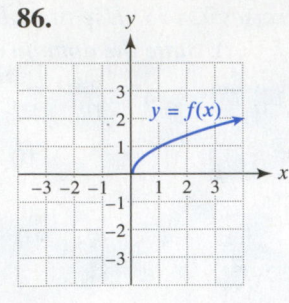

OPERATIONS ON FUNCTIONS

Exercises 87–90: Find each of the following for the given f(x) and g(x).

   (a) $(fg)(2)$ **(b)** $(f - g)(x)$ **(c)** $(f \circ g)(x)$

87. $f(x) = x^2 - 2, g(x) = x^2 + 2$

88. $f(x) = 2x^2, g(x) = 2x - 1$

89. $f(x) = \dfrac{1}{x}, g(x) = \dfrac{2}{x}$

90. $f(x) = x^3, g(x) = \sqrt[3]{x}$

APPLICATIONS

91. *Circular Wave* A stone is dropped in a lake, creating a circular wave. The radius r of the wave in feet after t seconds is $r(t) = 2t$.
  (a) The wave's circumference C is $C(r) = 2\pi r$. Evaluate $(C \circ r)(5)$ and interpret your result.
  (b) Find $(C \circ r)(t)$.

92. *Volume of a Balloon* The volume V of a spherical balloon with radius r is given by $V(r) = \frac{4}{3}\pi r^3$. Suppose that the balloon is being inflated so that the radius in inches after t seconds is $r(t) = \sqrt[3]{t}$.
  (a) Evaluate $(V \circ r)(3)$ and interpret your result.
  (b) Find $(V \circ r)(t)$.

93. *College Degree* The table lists the percentage P of people 25 or older who have completed 4 or more years of college during year x.

x	1960	1980	2000	2010
$P(x)$	8	16	27	29

Source: U.S. Census Bureau.

  (a) Evaluate $P(1980)$ and interpret the results.
  (b) Make a table for $P^{-1}(x)$.
  (c) Evaluate $P^{-1}(16)$.

94. *Skin Cancer and Ozone* Ozone in the stratosphere filters out most of the harmful ultraviolet (UV) rays from the sun. However, depletion of the ozone layer affects this protection. The formula $U(x) = 1.5x$ calculates the percent increase in UV radiation for an x percent decrease in the thickness of the ozone layer. The formula $C(x) = 3.5x$ calculates the percent increase in skin cancer cases when the UV radiation increases by x percent. (*Source:* R. Turner, D. Pierce, and I. Bateman, *Environmental Economics.*)

  (a) Evaluate $U(2)$ and $C(3)$ and interpret each result.
  (b) Find $(C \circ U)(2)$ and interpret the result.
  (c) Find $(C \circ U)(x)$. What does it calculate?

95. *Temperature and Mosquitoes* Temperature can affect the number of mosquitoes observed on a summer night. Graphs of two functions, T and M, are shown. Function T calculates the temperature on a summer evening h hours past midnight, and M calculates the number of mosquitoes observed per 100 square feet when the outside temperature is T.

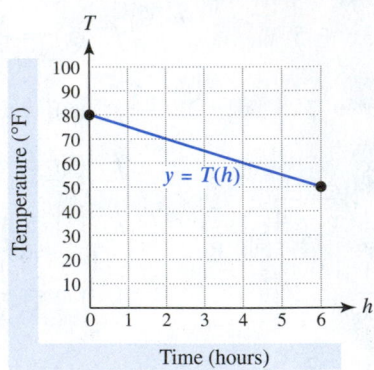

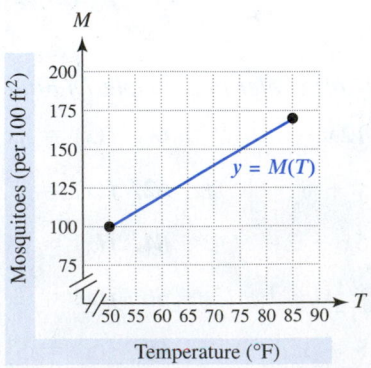

  (a) Find $T(1)$ and $M(75)$.
  (b) Evaluate $(M \circ T)(1)$ and interpret your result.
  (c) What does $(M \circ T)(h)$ calculate?
  (d) Find equations for the lines in each graph.
  (e) Use your answers from part (d) to write a formula for $(M \circ T)(h)$.

96. *High School Grades* The table lists the percentage P of college freshmen with a high school grade average of A or A– during year x.

x	1970	1980	1990	2000	2010
$P(x)$	20	26	29	43	46

Source: Department of Education.

(a) Evaluate $P(1970)$ and interpret the results.
(b) Make a table for $P^{-1}(x)$.
(c) Evaluate $P^{-1}(43)$.

97. *Temperature* The function given by $f(x) = \frac{9}{5}x + 32$ converts x degrees Celsius to an equivalent temperature in degrees Fahrenheit.
(a) Is f a one-to-one function? Why or why not?
(b) Find $f^{-1}(x)$ and interpret what it calculates.

98. *Feet and Yards* The function given by $f(x) = 3x$ converts x yards to feet.
(a) Is f a one-to-one function? Why or why not?
(b) Find $f^{-1}(x)$ and interpret what it calculates.

99. *Quarts and Gallons* Write a function f that converts x gallons to quarts. Then find $f^{-1}(x)$ and interpret what it computes.

100. *One-to-One Function* The table lists monthly average wind speeds at Hilo, Hawaii, in miles per hour from July through December, where x is the month.

x	July	Aug	Sept	Oct	Nov	Dec
$f(x)$	7	7	7	7	7	7

(a) Is function f one-to-one? Explain.
(b) Does f^{-1} exist?

(c) What happens if you try to make a table for f^{-1}?
(d) Could f be one-to-one if it were computed at a different location? What would have to be true about the monthly average wind speeds?

101. **Online Exploration** Go online and find how many cups there are in one gallon and how many teaspoons there are in one cup.
(a) Write a function C that converts x gallons to cups.
(b) Write a function T that converts x cups to teaspoons.
(c) Write the composite function $(T \circ C)(x)$, where x represents gallons.
(d) Evaluate $(T \circ C)(3)$ and interpret the result.

102. **Online Exploration** Go online and find the number of yuan (Chinese currency) in 1 dollar and the number of rupees (Indian currency) in 1 yuan. Answers may vary over time.
(a) Write a function Y that converts x dollars to yuan.
(b) Write a function R that converts x yuan to rupees.
(c) Write the composite function $(R \circ Y)(x)$, where x represents dollars.
(d) Evaluate $(R \circ Y)(100)$ and interpret the result.

WRITING ABOUT MATHEMATICS

103. Explain the difference between $(g \circ f)(2)$ and $(f \circ g)(2)$. Are they always equal? If f and g are inverse functions, evaluate $(g \circ f)(2)$ and $(f \circ g)(2)$.

104. Explain what it means for a function to be one-to-one.

12.2 Exponential Functions

Basic Concepts • Graphs of Exponential Functions • Percent Change and Exponential Functions • Compound Interest • Models Involving Exponential Functions • The Natural Exponential Function

A LOOK INTO MATH ▶ Suppose that we deposit $100 into a savings account that pays 2% annual interest. If neither the money nor the interest is withdrawn, then after 1 year the account balance will be $102 and after 2 years the account balance will be $104.04. The extra 4 cents in interest in the second year occurs because we get not only 2% interest on the original $100 but also 2% interest on the $2 interest earned during the first year. In general, when an amount A experiences a constant percent change over each fixed time period, such as a year or a month, then the growth (or possibly decay) in A can be described by an exponential function. In this section we discuss exponential functions and many of their applications, including how to calculate interest.

Basic Concepts

▶ **REAL-WORLD CONNECTION** Suppose that an insect population (per acre) doubles each week. Table 12.5 shows the size of the population after x weeks. Note that, as the population of insects becomes larger, the *increase* in population each week becomes greater. The population is increasing by 100%, or doubling, each week. When a quantity increases by a constant percentage (or constant factor) at regular intervals, its growth is *exponential*.

TABLE 12.5 Insect Population

Week	0	1	2	3	4	5
Population	100	200	400	800	1600	3200

$\times 2 \quad \times 2 \quad \times 2 \quad \times 2 \quad \times 2$

Doubles each week

We can model the data in Table 12.5 by using the exponential function

$$f(x) = 100(2)^x.$$

For example,

$$f(0) = 100(2)^0 = 100 \cdot 1 = 100,$$
$$f(1) = 100(2)^1 = 100 \cdot 2 = 200,$$
$$f(2) = 100(2)^2 = 100 \cdot 4 = 400,$$

and so on. Note that the exponential function f has a *variable as an exponent*.

EXPONENTIAL FUNCTION

A function represented by

$$f(x) = Ca^x, \quad a > 0 \quad \text{and} \quad a \neq 1,$$

is an **exponential function with base a and coefficient C**. (Unless stated otherwise, we assume that $C > 0$.)

In the formula $f(x) = Ca^x$, a is called the **growth factor** when $a > 1$ and the **decay factor** when $0 < a < 1$. For an exponential function, each time x increases by 1 unit, $f(x)$ increases by a factor of a when $a > 1$ and decreases by a factor of a when $0 < a < 1$. Moreover, because

$$f(0) = Ca^0 = C(1) = C,$$

the value of C equals the value of $f(x)$ when $x = 0$. That is, C is the y-intercept. If x represents time, C represents the initial value of f when time equals 0. Figure 12.14 illustrates **exponential growth** and **exponential decay** for $x > 0$.

Exponential Growth for Positive x

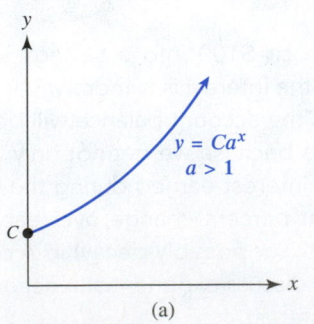

(a)

Exponential Decay for Positive x

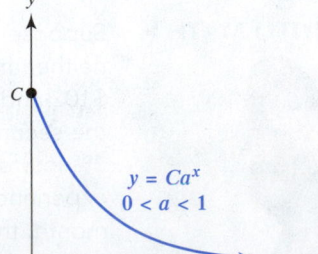

(b)

Figure 12.14

The set of valid inputs (domain) for an exponential function includes all real numbers. The set of corresponding outputs (range) includes all positive real numbers.

When evaluating an exponential function, we evaluate exponents *before* we multiply. For example, if $f(x) = 4(3)^x$, then

$$f(2) = 4(3)^2 = 4(9) = 36.$$

Evaluate exponents first.

The next example illustrates how exponential functions are evaluated.

EXAMPLE 1   **Evaluating exponential functions**

Evaluate $f(x)$ for the given value of x.
(a) $f(x) = 10(3)^x$ $x = 2$ (b) $f(x) = 5\left(\frac{1}{2}\right)^x$ $x = 3$
(c) $f(x) = \frac{1}{3}(2)^x$ $x = -1$

Solution
(a) $f(2) = 10(3)^2 = 10 \cdot 9 = 90$
(b) $f(3) = 5\left(\frac{1}{2}\right)^3 = 5 \cdot \frac{1}{8} = \frac{5}{8}$
(c) $f(-1) = \frac{1}{3}(2)^{-1} = \frac{1}{3} \cdot \frac{1}{2} = \frac{1}{6}$

NOTE: $f(x)$ is also defined for all *negative* inputs.

Now Try Exercises 11, 13, 15

MAKING CONNECTIONS

The Expressions a^{-x} and $\left(\frac{1}{a}\right)^x$

Using properties of exponents, we can write 2^{-x} as

$$2^{-x} = \frac{1}{2^x} = \left(\frac{1}{2}\right)^x.$$

In general, the expressions a^{-x} and $\left(\frac{1}{a}\right)^x$ are equal for all positive values of a.

Graphs of Exponential Functions

We can graph $f(x) = 2^x$ by first evaluating some points, as in Table 12.6. If we plot these points and sketch the graph, we obtain Figure 12.15.

TABLE 12.6
$f(x) = 2^x$

x	2^x
-2	$\frac{1}{4}$
-1	$\frac{1}{2}$
0	1
1	2
2	4

Graphing $f(x) = 2^x$

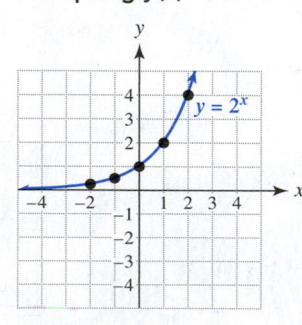

Figure 12.15

- This graph is always above the x-axis.
- The graph passes through the point $(0, 1)$.
- Negative x-values give y-values between 0 and 1.
- Positive x-values give y-values greater than 1.

In Figure 12.16, we investigate the graph of $y = a^x$ for values of a that are greater than 1 by graphing $y = 1.3^x$, $y = 1.7^x$, and $y = 2.5^x$. Graphs of $y = a^x$ where a is between 0 and 1 are shown in Figure 12.17.

Graphing $y = a^x$
for $a > 1$

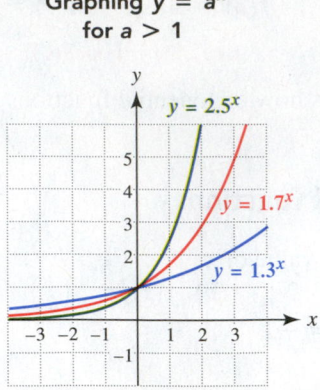

Figure 12.16

Graphing $y = a^x$
for $0 < a < 1$

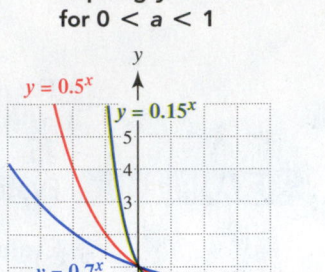

Figure 12.17

- As x increases, the y-values increase.
- Larger values of a ($a > 1$) result in y-values that increase more rapidly.
- The graphs pass through $(0, 1)$ because $C = 1$.
- The graphs are always above the x-axis.

- As x increases, the y-values decrease.
- Smaller values of a ($0 < a < 1$) result in y-values that decrease more rapidly.
- The graphs pass through $(0, 1)$ because $C = 1$.
- The graphs are always above the x-axis.

In the next example we show the dramatic difference between the outputs of linear and exponential functions.

EXAMPLE 2   **Comparing exponential and linear functions**

Compare $f(x) = 3^x$ and $g(x) = 3x$ graphically and numerically for $x \geq 0$.

Solution
Graphical Comparison   The graphs of $y_1 = 3^x$ and $y_2 = 3x$ are shown in Figure 12.18. For large values of x, the graph of the exponential function y_1 increases much faster than the graph of the linear function y_2.
Numerical Comparison   The tables of $y_1 = 3^x$ and $y_2 = 3x$ are shown in Figure 12.19. For large values of x, the values for y_1 increase much faster than the values for y_2.

Comparing Exponential and Linear Growth

Exponential growth ———   ┌── Linear growth

[0, 5, 1] by [0, 120, 20]

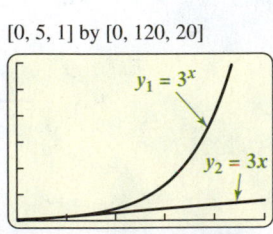

Figure 12.18

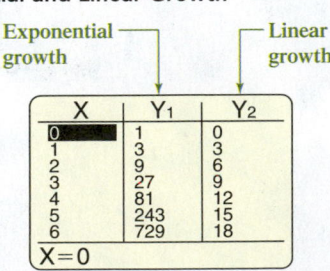

Figure 12.19

Now Try Exercise 89

NOTE: The results of Example 2 are true in general: for large enough inputs, exponential functions with $a > 1$ grow far faster than any linear function.

MAKING CONNECTIONS

Exponential and Polynomial Functions

The function $f(x) = 2^x$ is an exponential function. The base **2** is a constant and the exponent x is a variable, so $f(3) = 2^3 = 8$.

The function $g(x) = x^2$ is a quadratic (polynomial) function. The base x is a variable and the exponent **2** is a constant, so $g(3) = 3^2 = 9$.

The table clearly shows that the exponential function grows much faster than the quadratic function for large values of x.

x	0	2	4	6	8	10	12	
2^x	1	4	16	64	256	1024	4096	← Exponential growth
x^2	0	4	16	36	64	100	144	← Quadratic growth

In the next example, we determine whether a function is linear or exponential.

EXAMPLE 3   **Finding linear and exponential functions**

For each table, determine whether f is a linear function or an exponential function. Find a formula for f.

(a)

x	0	1	2	3	4
$f(x)$	16	8	4	2	1

(b)

x	0	1	2	3	4
$f(x)$	5	7	9	11	13

(c)

x	0	1	2	3	4
$f(x)$	1	3	9	27	81

Solution

(a) Each time x increases by 1 unit, $f(x)$ decreases by a factor of $\frac{1}{2}$. Therefore f is an exponential function with a *decay factor* of $\frac{1}{2}$. Because $f(0) = 16$, $C = $ **16**, so $f(x) = $ **16** $\left(\frac{1}{2}\right)^x$. This formula can also be written as $f(x) = 16(2)^{-x}$.

(b) Each time x increases by 1 unit, $f(x)$ increases by **2** units. Therefore f is a linear function, and the slope of its graph equals **2**. The y-intercept is **5**, so $f(x) = $ **2**$x + $ **5**.

(c) Each time x increases by 1 unit, $f(x)$ increases by a factor of **3**. Therefore f is an exponential function with a *growth factor* of **3**. Because $f(0) = 1$, it follows that $C = $ **1** and $f(x) = $ **1**$(3)^x$, or $f(x) = 3^x$.

Now Try Exercises 45, 47, 49

READING CHECK

How can you distinguish between a linear function and an exponential function?

MAKING CONNECTIONS

Linear and Exponential Functions

For a *linear function*, given by $f(x) = ax + b$, each time x increases by 1 unit y increases (or decreases) *by a units*, where a equals the slope of the graph of f.

For an *exponential function*, given by $f(x) = Ca^x$, each time x increases by 1 unit y increases *by a factor of a* when $a > 1$ and decreases by a factor of a when $0 < a < 1$. The constant a equals either the growth factor or the decay factor.

Percent Change and Exponential Functions

PERCENT CHANGE When an amount A changes to a new amount B, then the **percent change** is calculated by

$$\frac{B - A}{A} \times 100. \qquad \text{Percent change formula}$$

We multiply the ratio by 100 to change decimal form to percent form.

EXAMPLE 4 | **Finding percent change**

Complete the following.
(a) Find the percent change if an account balance increases from $500 to $1000.
(b) Find the percent change if an account balance decreases from $1000 to $500.

Solution
(a) Let $A = 500$ and $B = 1000$.

$$\frac{1000 - 500}{500} \times 100 = \frac{500}{500} \times 100 \qquad \frac{B - A}{A} \times 100$$

$$= 1 \times 100 \qquad \text{Simplify.}$$

$$= 100\% \qquad \text{Multiply.}$$

The percent change (increase) is 100%; that is, the account balance doubles.

(b) Let $A = 1000$ and $B = 500$.

$$\frac{500 - 1000}{1000} \times 100 = -\frac{500}{1000} \times 100 \qquad \frac{B - A}{A} \times 100$$

$$= -\frac{1}{2} \times 100 \qquad \text{Simplify.}$$

$$\approx -50\% \qquad \text{Multiply.}$$

The percent change (decrease) is -50%.

Now Try Exercise 51

NOTE: In Example 4, the account did *not* increase by 100% and then decrease by 100% to return the account to its initial value of $500. In part (b) the initial amount is $A = \$1000$ and needs to decrease only by a factor of $\frac{1}{2}$, or 50%, to decrease to $500.

PERCENT CHANGE AND GROWTH FACTOR Suppose a savings account A increases from $200 to $600. The percent change, $R\%$, equals

$$\frac{600 - 200}{200} \times 100 = 2 \times 100 = 200\%.$$

Note that the account balance *tripled* from $200 to 600, but the percent change is 200%, *not* 300%. Also, the *increase* in the account balance A is $400 because $\$600 - \$200 = \$400$. This $400 increase equals 200% of $200. If we let r represent the percent change as a decimal, then

$$\underbrace{200\% \text{ of } \$200}_{R\% \text{ of } A} = \underbrace{2.00 \times \$200}_{rA} = \underbrace{\$400.}_{\text{Increase in } A} \qquad R\% = 200\%; r = 2.00$$

In general, if the *percent increase* in an account balance A is given by r in decimal form, then the *amount of increase* in A is equal to rA and the new balance after the increase is $A + rA$. For example, if $200 increases by 200%, then $r = 2.00$, $A = \$200$, and

$$rA = 2.00(200) = \$400, \qquad \text{Increase in account balance}$$

so the balance increases by $400. The new balance for the account is

$$A + rA = \$200 + \$400 = \$600.$$

Initial amount + Increase = Final amount

If we factor out A in the expression $A + rA$, we get

$$A + rA = A(1 + r) \qquad \text{Factor out } A.$$

Initial amount + Increase in A = Initial amount $\times$ Growth factor

Thus if an account increases from $200 to $600, the increase is $400 and the percent increase is 200%. In addition, the account balance increased by a growth factor equal to

$$a = 1 + r = 1 + 2.00 = 3, \qquad \text{Growth factor is } a = 3.$$

which means that the account balance tripled.

EXAMPLE 5 **Analyzing the decrease in an account balance**

An account that contains $2000 decreases in value by 20%.
(a) Find the decrease in value of the account.
(b) Find the final value of the account.
(c) By what factor a did the account decrease?

Solution
(a) Let $A = 2000$ and $r = -0.20$ (-20% in decimal form). The decrease is

$$rA = -0.20(2000) = -400. \quad \text{Decrease in } A \text{ is } rA.$$

The account decreased in value by $400.

(b) The final value of the account is

$$A + rA = 2000 + (-400) = \$1600.$$

Initial amount + Decrease = New amount

(c) The account decreased by a factor of

$$a = 1 + r = 1 + (-0.20) = 0.80. \quad \text{Decay factor is } a = 0.80.$$

The account value decreased to 80% of its original value because 80% of $2000 is $1600.

Now Try Exercise 55

NOTE: In general, a positive amount A cannot *decrease* by more than 100% because a 100% decrease would reduce A to 0. However, percent *increases* can be more than 100%.

PERCENT CHANGE AND EXPONENTIAL FUNCTIONS An exponential function results when an *initial value* C is multiplied by a *growth* (or *decay*) *factor* a for each unit increase in x. For example, if the population P of a city is 100,000 people, and the city is growing at 5% per year, then $C = 100,000$ and the growth factor is

$$a = 1 + r = 1 + 0.05 = 1.05.$$

STUDY TIP

Be sure that you understand how a percent change relates to the growth factor of an exponential function.

Thus the exponential function

$$P(x) = 100,000(1.05)^x \qquad \text{$P(x) = Ca^x$ with $C = 100,000$ and $a = 1.05$.}$$

models the city's population after x years. For example,

$$P(6) = 100,000(1.05)^6 \approx 134,000$$

indicates that after 6 years the city's population has grown to about 134,000 people.
These concepts are summarized by the following.

PERCENT CHANGE AND EXPONENTIAL FUNCTIONS

Suppose that an amount A increases or decreases by $R\%$ (or by r expressed in decimal form) for each unit increase in x.

1. If $r > 0$, the *growth factor* is $a = 1 + r$ and $a > 1$.
2. If $-1 < r < 0$, the *decay factor* is $a = 1 + r$ and $0 < a < 1$.
3. If the initial amount is C, the amount A after x-unit increases is given by

$$A(x) = Ca^x \quad \text{or equivalently,} \quad A(x) = C(1 + r)^x.$$

READING CHECK

If your income increases by 200%, what is the growth factor?

EXAMPLE 6   **Analyzing growth of bacteria**

Initially a laboratory culture contains 50,000 bacteria per milliliter and it is increasing in numbers by 20% per hour.
(a) Write the formula for an exponential function B that gives the number of bacteria per milliliter after x hours.
(b) Evaluate $B(3)$ and interpret your result.

Solution
(a) The initial value is $C = 50,000$ and the hourly percent increase is 20%, or $r = 0.20$ in decimal form. Thus the growth factor is

$$a = 1 + r = 1 + 0.20 = 1.20. \qquad \text{Growth factor = 1.20 with $r = 0.20$.}$$

Because the exponential function can be written as $B(x) = Ca^x$, it follows that

$$B(x) = 50,000(1.20)^x.$$

(b) $B(3) = 50,000(1.20)^3 = 50,000(1.728) = 86,400$. Thus after 3 hours, the culture contains 86,400 bacteria per milliliter.

Now Try Exercise 85

Compound Interest

▶ **REAL-WORLD CONNECTION** If $100 is deposited in a savings account paying 10% annual interest, the interest earned after 1 year equals $100 \times 0.10 = $10. The total amount of money in the account after 1 year is $100(1 + 0.10) = $110. Each year the money in the account increases by a growth factor of 1.10, so after x years there will be $100(1.10)^x$ dollars in the account. Thus **compound interest** is an example of exponential growth.

COMPOUND INTEREST

If P dollars is deposited in an account and if interest is paid at the end of each year with an annual rate of interest r, expressed in decimal form, then after t years the account will contain A dollars, where

$$A = P(1 + r)^t.$$

The growth factor is $(1 + r)$.

NOTE: The compound interest formula takes the form of an exponential function with

$$a = 1 + r.$$

EXAMPLE 7   ## Calculating compound interest

A 20-year-old worker deposits \$2000 in a retirement account that pays 6% annual interest at the end of each year. How much money will be in the account when the worker is 65 years old? What is the growth factor?

Solution
Here, $P = \mathbf{2000}$, $r = \mathbf{0.06}$, and $t = \mathbf{45}$. The amount in the account after 45 years is

$$A = \mathbf{2000}(1 + \mathbf{0.06})^{45} \approx \$27{,}529.22, \quad A = P(1 + r)^t$$

which is supported by Figure 12.20. Each year the amount of money in the account is multiplied by a factor of $(1 + 0.06)$, so the growth factor is 1.06.

```
2000(1+.06)^45
         27529.22165
```

Figure 12.20

Now Try Exercise 65

INTEREST PAID MORE THAN ONCE A YEAR Many times, interest is paid more than once a year. For example, suppose an account gives 8% annual interest that is paid every 3 months, or *quarterly*. It follows that $\frac{1}{4}$ of the annual 8% interest, or 2%, is paid every 3 months. The growth factor is $a = 1 + 0.02 = 1.02$ for each 3-month period. If the initial balance is \$1000, then after **5** years, the **quarterly** 2% interest has been paid $\mathbf{4} \cdot \mathbf{5} = 20$ times, and the account contains

$$\$1000(1.02)^{20} \approx \$1485.95.$$

In general, if P dollars are deposited in an account paying an *annual* interest rate r (in decimal form) and this interest rate is compounded or paid n times per year, then after t years the account contains A dollars, given by

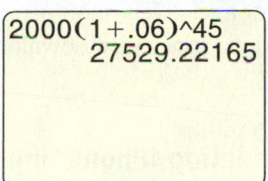

$$\text{Amount after } t \text{ years} \qquad \text{Annual interest (decimal)} \\ \text{Number of years} \\ A = P\left(1 + \frac{r}{n}\right)^{nt} \\ \text{Initial deposit} \qquad \text{Number of times interest is paid per year}$$

EXAMPLE 8   **Calculating compound interest**

Initially, $1500 is deposited in an account paying 6% annual interest, compounded monthly. What is the account balance after 5 years?

Solution

Let $P = \mathbf{1500}$, $r = \mathbf{0.06}$, $n = \mathbf{12}$, and $t = \mathbf{5}$. The balance after 5 years is

$$
\begin{aligned}
A &= P\left(1 + \frac{r}{n}\right)^{nt} && \text{Interest formula} \\
&= \mathbf{1500}\left(1 + \frac{\mathbf{0.06}}{\mathbf{12}}\right)^{(\mathbf{12 \cdot 5})} && \text{Substitute.} \\
&= 1500(1.005)^{60} && \text{Evaluate.} \\
&\approx \$2023.28. && \text{Approximate.}
\end{aligned}
$$

Now Try Exercise 73

NOTE: Generally, compounding interest more frequently results in more interest being paid. In Example 8, if the annual 6% interest had been paid only once a year, the growth factor would be larger, $a = 1.06$, but this 6% annual interest would have been paid only 5 times, once each year. The new balance would be $1500(1.06)^5 \approx \$2007.34$, rather than the $2023.28 that resulted from monthly compounding.

Models Involving Exponential Functions

▶ **REAL-WORLD CONNECTION** Apple's path to 2 billion iPhone application downloads occurred over a relatively short period of time. In the next example, we model this exponential growth of iPhone application downloads.

EXAMPLE 9   **Modeling iPhone application downloads**

In September 2008 about 100 million iPhone applications were downloaded, and the number of iPhone downloads was increasing at a rate of 28.4% per month. (*Source:* Apple Corp.)
(a) What was the monthly growth factor?
(b) Write an exponential function $P(x) = Ca^x$ that models the number of downloads in millions for the iPhone x months after September 2008.
(c) Evaluate $P(12)$ and interpret the result.

Solution

(a) Because $r = 0.284$, the monthly growth factor was $a = 1 + 0.284 = 1.284$.
(b) Let $C = 100$ and $a = 1.284$. Then $P(x) = 100(1.284)^x$.
(c) $P(12) = 100(1.284)^{12} \approx 2000$. In approximately 12 months, from September 2008 to September 2009, iPhone application downloads grew from 100 million to about 2000 million, or 2 billion.

Now Try Exercise 91

▶ **REAL-WORLD CONNECTION** Traffic flow at intersections can be modeled by exponential functions whenever traffic patterns occur randomly. In the next example we model traffic at an intersection by using an exponential function.

EXAMPLE 10 | **Modeling traffic flow**

On average, a particular intersection has 360 vehicles arriving randomly each hour. Traffic engineers use $f(x) = (0.905)^x$ to estimate the likelihood, or probability, that *no* vehicle will enter the intersection within an interval of x seconds. (*Source:* F. Mannering and W. Kilareski, *Principles of Highway Engineering and Traffic Analysis.*)

(a) Compute $f(5)$ and interpret the results.
(b) A graph of $y = f(x)$ is shown in Figure 12.21. Discuss this graph.
(c) Is this function an example of exponential growth or decay?

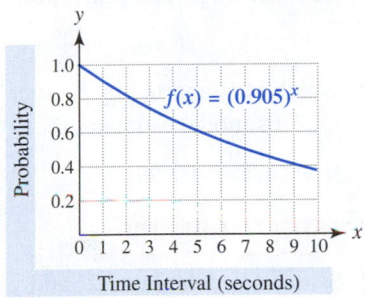

Figure 12.21

Solution
(a) The result $f(5) = (0.905)^5 \approx 0.61$ indicates that there is a 61% chance that no vehicle will enter the intersection during any particular 5-second interval.
(b) The graph decreases, which means that as the interval of time increases there is less chance (likelihood) that a car will *not* enter the intersection.
(c) Because the graph is decreasing and $a = 0.905 < 1$, this function is an example of exponential decay.

Now Try Exercise 97

The Natural Exponential Function

A special type of exponential function is called the *natural exponential function*, expressed as $f(x) = e^x$. The **base** e is a special number in mathematics similar to π. The number π is approximately 3.14, whereas the number e is approximately 2.72. The number e is named for the great Swiss mathematician Leonhard Euler (1707–1783). Most calculators have a special key that can be used to compute the natural exponential function.

NATURAL EXPONENTIAL FUNCTION

The function represented by

$$f(x) = e^x$$

is the **natural exponential function**, where $e \approx 2.71828$.

CALCULATOR HELP

To evaluate the natural exponential function, see Appendix A (page AP-8).

▶ **REAL-WORLD CONNECTION** The natural exponential function is frequently used to model **continuous growth**. For example, the fact that births and deaths occur throughout the year, not just at one time during the year, must be recognized when population growth is being modeled. If a population P is growing continuously at r percent per year, expressed as a decimal, we can model this population after x years by

$$P = Ce^{rx}, \quad \text{Continuous growth, } r > 0$$

where C is the initial population. To evaluate natural exponential functions, we use a calculator, as demonstrated in the next example.

| EXAMPLE 11 | Modeling population |

In 2011 Texas' population was 26 million people and was growing at a continuous rate of 2% per year. This population in millions x years after 2011 can be modeled by

$$f(x) = 26e^{0.02x}.$$

Estimate the population in 2015.

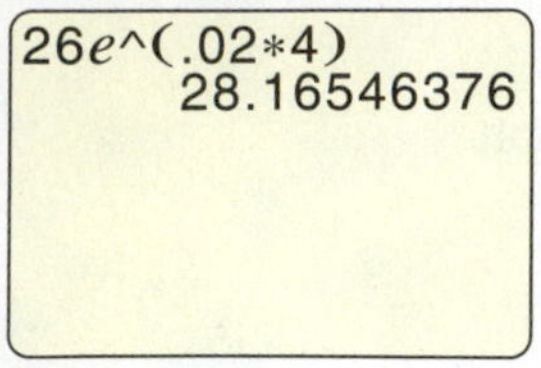

```
26e^(.02*4)
        28.16546376
```

Figure 12.22

Solution
Because 2015 is 4 years after 2011, we evaluate $f(4)$ to obtain

$$f(4) = 26e^{0.02(4)} \approx 28.2,$$

which is supported by Figure 12.22. (Be sure to include parentheses around the exponent of e.) This model estimates the population of Texas to be about 28.2 million in 2015.

Now Try Exercise 99

CRITICAL THINKING

Sketch a graph of $y = 2^x$ and $y = 3^x$ in the same xy-plane. Then use these two graphs to sketch a graph of $y = e^x$. How do these graphs compare?

12.2 Putting It All Together

TOPIC	EXPLANATION	EXAMPLE
Exponential Function	The variable is an exponent. $$f(x) = Ca^x$$ Initial value when $x = 0$ Growth factor (base) Growth: $a > 1$ Decay: $0 < a < 1$	$f(x) = 3(2)^x$ models exponential **growth**, and $g(x) = 2\left(\frac{1}{3}\right)^x$ models exponential **decay**.
Graphs of Exponential Functions	• Above the x-axis for all inputs • Increases for $a > 1$ • Decreases for $0 < a < 1$ • y-intercept is C.	**Exponential Functions:** $f(x) = Ca^x$ 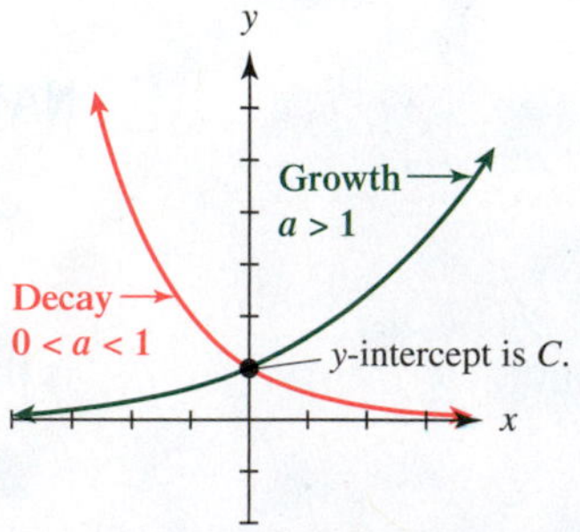 Growth → $a > 1$ Decay → $0 < a < 1$ y-intercept is C.
Percent Change	An amount changes from A to B: $$\text{Percent change} = \frac{B - A}{A} \times 100.$$	If \$200 increases to \$300, then $$\frac{300 - 200}{200} \times 100 = 50\%$$ is the percent change.

TOPIC	EXPLANATION	EXAMPLE
Constant Percent Change and Exponential Functions	Constant percent change (decimal) Time $$f(x) = C\underbrace{(1 + r)}^{x}$$ Initial amount Growth factor	If 3000 bacteria increase in number by 12% daily for x days, then $$f(x) = 3000(1.12)^x$$ models these numbers after x days.

Compound Interest	Compounded annually:
	
	$$A = P(1 + r)^t$$
	
	Compounded n times per year:
	
	$$A = P\left(1 + \frac{r}{n}\right)^{nt}$$
	
	P: Initial deposit
	A: Final amount
	r: Annual interest rate (decimal)
	n: Interest paid n times per year
	t: Number of years
		If \$1000 is deposited at 3% annual interest for 4 years, then
		
		$$A = \$1000(1.03)^4$$
		$$\approx 1125.51.$$
		
		If \$1000 is deposited at 3% annual interest compounded monthly for 4 years, then
		
		$$A = \$1000\left(1 + \frac{0.03}{12}\right)^{(12 \cdot 4)}$$
		$$\approx \$1127.33.$$

Natural Exponential Function	$$f(x) = e^x$$ Growth factor, or base, is $e \approx 2.72$.	Using a calculator, $$f(2) = e^2 \approx 7.39.$$

12.2 Exercises

MyMathLab

Math XL PRACTICE   WATCH   DOWNLOAD   READ   REVIEW

CONCEPTS AND VOCABULARY

1. Give a general formula for an exponential function f.

2. Sketch a graph of an exponential function that illustrates exponential decay.

3. Give the domain and range of an exponential function.

4. Evaluate the expressions 2^x and x^2 for $x = 5$.

5. Approximate e to the nearest thousandth.

6. Evaluate e^2 and π^2 using your calculator.

7. If a quantity y grows exponentially, then for each unit increase in x, y increases by a constant _____.

8. If $f(x) = 1.5^x$, what is the growth factor?

9. If a quantity increases from A to B, then the percent change equals _____.

10. If a quantity increases by 35% each year, then the growth factor a is _____.

EVALUATING EXPONENTIAL FUNCTIONS

Exercises 11–22: Evaluate the exponential function for the given values of x by hand when possible. Approximate answers to the nearest hundredth when appropriate.

11. $f(x) = 3^x$ $x = -2, x = 2$

12. $f(x) = 5^x$ $x = -1, x = 3$

13. $f(x) = 5(2^x)$ $x = 0, x = 5$

14. $f(x) = 3(7^x)$ $x = -2, x = 0$

15. $f(x) = \left(\frac{1}{2}\right)^x$ $x = -2, x = 3$

16. $f(x) = \left(\frac{1}{4}\right)^x$ $x = 0, x = 2$

17. $f(x) = 5(3)^{-x}$ $x = -1, x = 2$

18. $f(x) = 4\left(\frac{3}{7}\right)^x$ $x = 1, x = 4$

19. $f(x) = 1.8^x$ $x = -3, x = 1.5$

20. $f(x) = 0.91^x$ $x = 5.1, x = 10$

21. $f(x) = 3(0.6)^x$ $\qquad$ $x = -1, x = 2$

22. $f(x) = 5(4.5)^{-x}$ $\qquad$ $x = -2.1, x = 1.9$

Exercises 23 and 24: **Thinking Generally** *For the given exponential function, evaluate $f(0)$ and $f(-1)$.*

23. $f(x) = a^x$

24. $f(x) = (1 + r)^{2x}$

GRAPHS OF EXPONENTIAL FUNCTIONS

Exercises 25–28: Match the formula with its graph (a.–d.). Do not use a calculator.

25. $f(x) = 1.5^x$ $\qquad$ **26.** $f(x) = \frac{1}{4}(2^x)$

27. $f(x) = 4\left(\frac{1}{2}\right)^x$ $\qquad$ **28.** $f(x) = \left(\frac{1}{3}\right)^x$

a.
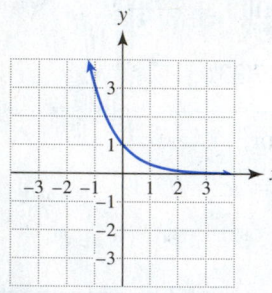

b.
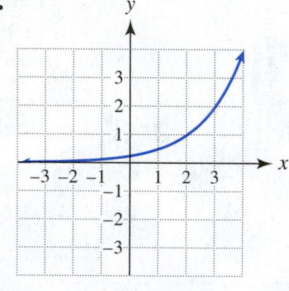

c.
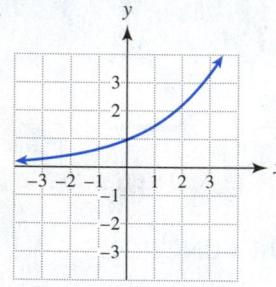

d.
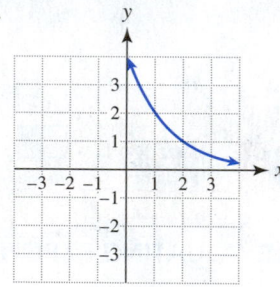

Exercises 29–32: Use the graph of $y = Ca^x$ to determine the constants C and a.

29.
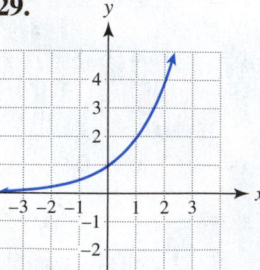

30.
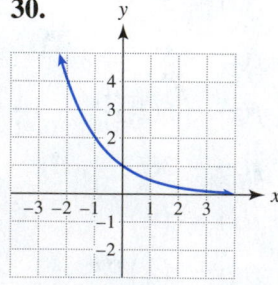

31.
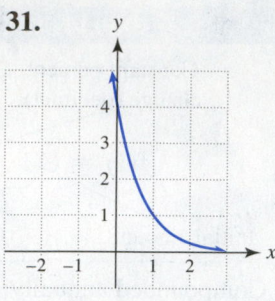

32.
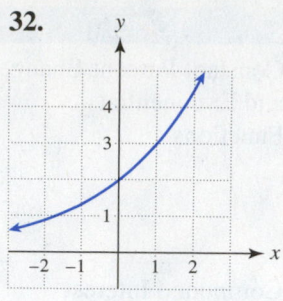

Exercises 33–44: Graph $y = f(x)$. State whether the graph depicts exponential growth or exponential decay.

33. $f(x) = 2^x$ $\qquad$ **34.** $f(x) = 3^x$

35. $f(x) = \left(\frac{1}{4}\right)^x$ $\qquad$ **36.** $f(x) = \left(\frac{1}{2}\right)^x$

37. $f(x) = 2^{-x}$ $\qquad$ **38.** $f(x) = 3^{-x}$

39. $f(x) = 3^x - 1$ $\qquad$ **40.** $f(x) = 2^x + 1$

41. $f(x) = 2^{x-1}$ $\qquad$ **42.** $f(x) = 2^{x+1}$

43. $f(x) = 4\left(\frac{1}{3}\right)^x$ $\qquad$ **44.** $f(x) = 3\left(\frac{1}{2}\right)^x$

LINEAR AND EXPONENTIAL GROWTH

Exercises 45–50: (Refer to Example 3.) A table for a function f is given.

   (a) *Determine whether function f represents exponential growth, exponential decay, or linear growth.*

   (b) *Find a formula for f.*

45.

x	0	1	2	3	4
$f(x)$	64	16	4	1	$\frac{1}{4}$

46.

x	0	1	2	3	4
$f(x)$	$\frac{1}{2}$	1	2	4	8

47.

x	0	1	2	3	4
$f(x)$	8	11	14	17	20

48.

x	-2	-1	0	1	2
$f(x)$	4	2	1	$\frac{1}{2}$	$\frac{1}{4}$

49.

x	-2	-1	0	1	2
$f(x)$	2.56	3.2	4	5	6.25

50.

x	-2	-1	0	1	2
$f(x)$	-6	-2	2	6	10

PERCENT CHANGE AND EXPONENTIAL FUNCTIONS

Exercises 51–54: For the given amounts A and B, find each of the following.

   (a) *The percent change if A changes to B*
   (b) *The percent change if B changes to A*

51. $A = \$200, B = \$400$

52. $A = \$1.50, B = \$1.00$

53. $A = 150, B = 30$

54. $A = 80, B = 200$

Exercises 55–60: An account contains A dollars and increases or decreases by R%. For each A and R, answer the following.

   (a) *Find the increase or decrease in the value of the account.*
   (b) *Find the final value of the account.*
   (c) *By what factor did the account value increase or decrease?*

55. $A = \$1000, R = 120\%$

56. $A = \$500, R = 230\%$

57. $A = \$650, R = 20\%$

58. $A = \$70, R = 35\%$

59. $A = \$800, R = -10\%$

60. $A = \$950, R = -60\%$

Exercises 61–64: For the given f(x), state the initial value C, the growth or decay factor a, and percent change R for each unit increase in x.

61. $f(x) = 9(1.07)^x$

62. $f(x) = 3(1.351)^x$

63. $f(x) = 1.5(0.45)^x$

64. $f(x) = 0.9^x$

COMPOUND INTEREST

Exercises 65–70: (Refer to Example 7.) If P dollars is deposited in an account paying R percent annual interest, approximate the amount in the account after x years.

65. $P = \$1500$ $R = 9\%$ $x = 10$ years

66. $P = \$1500$ $R = 15\%$ $x = 10$ years

67. $P = \$200$ $R = 20\%$ $x = 50$ years

68. $P = \$5000$ $R = 8.4\%$ $x = 7$ years

69. $P = \$560$ $R = 1.4\%$ $x = 25$ years

70. $P = \$750$ $R = 10\%$ $x = 13$ years

71. Thinking Generally Suppose that \$1000 is deposited in an account paying 8% annual interest for 10 years. If \$2000 had been deposited instead of \$1000, would there be twice the money in the account after 10 years? Explain.

72. Thinking Generally Suppose that \$500 is deposited in an account paying 5% annual interest for 10 years. If the interest rate had been 10% instead of 5%, would the total interest earned after 10 years be twice as much? Explain.

Exercises 73–76: (Refer to Example 8.) The amount P is deposited in an account giving R% annual interest compounded n times a year. Find the amount A in the account after t years.

73. $P = \$700, R = 4\%, n = 4, t = 3$

74. $P = \$550, R = 3\%, n = 2, t = 5$

75. $P = \$1200, R = 2.5\%, n = 12, t = 7$

76. $P = \$1500, R = 6.5\%, n = 365, t = 20$

THE NATURAL EXPONENTIAL FUNCTION

Exercises 77–80: Evaluate f(x) for the given value of x. Approximate answers to the nearest hundredth.

77. $f(x) = e^x$ $x = 1.2$

78. $f(x) = 2e^x$ $x = 2$

79. $f(x) = 1 - e^x$ $x = -2$

80. $f(x) = 4e^{-x}$ $x = 1.5$

Exercises 81–84: Graph f(x) in $[-4, 4, 1]$ by $[0, 8, 1]$. State whether the graph illustrates exponential growth or exponential decay.

81. $f(x) = e^{0.5x}$ **82.** $f(x) = e^x + 1$

83. $f(x) = 1.5e^{-0.32x}$ **84.** $f(x) = 2e^{-x} + 1$

APPLICATIONS

Exercises 85–88: **Exponential Models** *Write the formula for an exponential function, $f(x) = Ca^x$, that models the situation. Evaluate f(4).*

85. A sample of 5000 bacteria decreases in number by 25% per week.

86. A sample of 700 insects increases in number by 200% per day.

87. A sample of 50 birds increases in number by 10% per month.

88. A sample of 137 fish decreases in number by 2% per week.

89. *Salary Growth* Suppose your salary is $50,000 per year. Would you rather have a 20% raise each year or a $20 raise each year?

90. *Salary Growth* Suppose your salary is $35,000 per year. Would you rather have a 10% raise each year or a $4000 raise each year? Assume that you will keep this job for 10 years.

91. *Blood Alcohol* Suppose that a person's peak blood alcohol level is 0.07 (grams per 100 mL) and that this level decreases by 40% each hour. (*Source:* National Institutes of Health.)
  (a) What is the hourly decay factor?
  (b) Write an exponential function $B(x) = Ca^x$ that models the blood alcohol level after x hours.
  (c) Evaluate $B(2)$ and interpret the result.

92. *Apple's Revenue* From 2001 to 2010 Apple's revenue grew at an annual rate of 40%, from $1 billion to about $21 billion. (*Source:* Apple Corp.)
  (a) What is the annual growth factor?
  (b) Write an exponential function $R(x) = Ca^x$ that models the revenue in billions of dollars x years after 2001.
  (c) Evaluate $R(9)$ and interpret the result.

93. *Tweets per Month* From July 2008 to July 2010, the number of tweets T per month increased dramatically and could be modeled in millions by

$$T(x) = 0.5(1.242)^x,$$

where x represents months after July 2008. (*Source:* Silicon Alley.)
  (a) What is the monthly growth factor?
  (b) Interpret the 0.5 in the formula.
  (c) Evaluate $T(24)$ and interpret the result.

94. *Dating Artifacts* Radioactive carbon-14 is found in all living things and is used to date objects containing organic material. Suppose that an object initially contains C grams of carbon-14. After x years it will contain A grams, where

$$A = C(0.99988)^x.$$

  (a) Let $C = 10$ and graph A over a 20,000-year period. Is this function an example of exponential growth or decay?
  (b) If $C = 10$, how many grams are left after 5700 years? What fraction of the carbon-14 is left?

95. *E. coli Bacteria* A strain of bacteria that inhabits the intestines of animals is named *Escherichia coli* (*E. coli*). These bacteria are capable of rapid growth and can be dangerous to humans—particularly children. The table shows the results of one study of the growth of *E. coli* bacteria, where concentrations are listed in *thousands* of bacteria per milliliter.

t (minutes)	0	50	100
Concentration	500	1000	2000

t (minutes)	150	200	250
Concentration	4000	8000	16,000

Source: G. S. Stent, *Molecular Biology of Bacterial Viruses.*

  (a) Find C and a so that $f(t) = Ca^{t/50}$ models these data.
  (b) Use $f(t)$ to estimate the concentration of bacteria after 170 minutes.
  (c) Discuss the growth of this strain of bacteria over a 250-minute time period.

96. *Swimming Pool Maintenance* Chlorine is frequently used to disinfect swimming pools. The chlorine concentration should remain between 1.5 and 2.5 parts per million (ppm). After a warm, sunny day only 80% of the chlorine may remain in the water, with the other 20% dissipating into the air or combining with other chemicals in the water. (*Source:* D. Thomas, *Swimming Pool Operator's Handbook.*)
  (a) Let $f(x) = 3(0.8)^x$ model the concentration of chlorine in parts per million after x days. What is the initial concentration of chlorine in the pool?
  (b) If no more chlorine is added, estimate when the chlorine level drops below 1.6 parts per million.

97. *Modeling Traffic Flow* (Refer to Example 10.) Construct a table of $f(x) = (0.905)^x$, starting at $x = 0$ and incrementing by 10, until $x = 50$.
  (a) Evaluate $f(0)$ and interpret the result.
  (b) For a time interval of what length is there only a 5% chance that no cars will enter the intersection?

98. *Pros and Putts* The percentage of putts P from 3 feet to 25 feet made by professional golfers can be modeled by the exponential function

$$P(x) = 99(0.872)^{x-3},$$

where x is the length of the putt.

(a) Find the percentage of putts made by professionals from 3 feet.
(b) Evaluate $P(8)$ and interpret the results.
(c) What is the decay factor? Interpret the decay factor.

Exercises 99–102: *Population Growth* (Refer to Example 11.) The population P in 2010 for a state is given along with R%, its annual percentage rate of continuous growth.

(a) *Write the formula $f(x) = Pe^{rx}$, where r is in decimal notation, that models the population in millions x years after 2010.*
(b) *Estimate the population in 2020.*

99. Nevada: $P = 2.7$ million, $R = 1.4\%$

100. North Carolina: $P = 9.4$ million, $R = 1.36\%$

101. California: $P = 38$ million, $R = 1.02\%$

102. Arizona: $P = 6.6$ million, $R = 1.44\%$

103. **Online Exploration** In many states, landlords are mandated to return a tenant's security deposit plus interest. Go online and look up this interest rate for your state. Answers may vary.
(a) If you are a tenant, calculate the interest you should receive after 1 year. (If you are not, assume the security deposit is $1200.)
(b) Write a function I that calculates the interest that you should receive after x years. (Does your landlord have to pay compound interest?)

104. **Online Exploration** Go online and determine how long it takes the fastest growing bacteria to double in number.
(a) Suppose that you start with a sample of 4 million such bacteria. Write an exponential function that gives the number N of bacteria in millions after x minutes.
(b) How many bacteria are there after 2 hours?

WRITING ABOUT MATHEMATICS

105. A student evaluates $f(x) = 4(2)^x$ at $x = 3$ and obtains 512. Did the student evaluate the function correctly? What was the student's error?

106. For a set of data, how can you distinguish between linear growth and exponential growth? Give an example of each type of data.

SECTIONS 12.1 AND 12.2   **Checking Basic Concepts**

1. If $f(x) = 2x^2 + 5x - 1$ and $g(x) = x + 1$, find each expression.
  (a) $(g \circ f)(1)$ **(b)** $(f \circ g)(x)$

2. Sketch a graph of $f(x) = x^2 - 1$.
  (a) Is f a one-to-one function? Explain.
  (b) Does f have an inverse function?

3. If $f(x) = 4x - 3$, find $f^{-1}(x)$.

4. Evaluate $f(-2)$ if $f(x) = 3(2)^x$.

5. Sketch a graph of $f(x) = \left(\frac{1}{3}\right)^x$.

6. Use the graph of $y = Ca^x$ to determine the constants C and a.

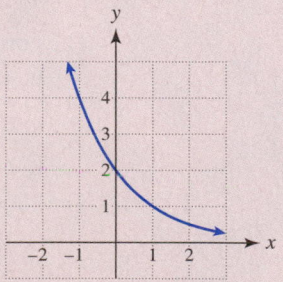

12.3 Logarithmic Functions

The Common Logarithmic Function ● **The Inverse of the Common Logarithmic Function** ● **Logarithms with Other Bases**

A LOOK INTO MATH ▶

NEW VOCABULARY

☐ Common logarithm of a positive number x
☐ Common logarithmic function
☐ Natural logarithm
☐ Logarithm with base a of a positive number x
☐ Logarithmic function with base a

Logarithmic functions are used in many applications. For example, if one airplane weighs twice as much as another, does the heavier airplane typically need a runway that is twice as long? Using a logarithmic function, we can answer this question. (See Example 10.) In this section we discuss logarithmic functions and several of their applications.

The Common Logarithmic Function

In applications, measurements can vary greatly in size. Table 12.7 lists some examples of objects, with the approximate distances in meters across each.

TABLE 12.7 Sizes of Objects

Object	Distance (meters)
Atom	10^{-9}
Protozoan	10^{-4}
Small Asteroid	10^{2}
Earth	10^{7}
Universe	10^{26}

Source: C. Ronan, *The Natural History of the Universe.*

Each distance is listed in the form 10^{k} for some k. The value of k distinguishes one measurement from another. The *common logarithmic function* or *base-10 logarithmic function*, denoted *log* or *log_{10}*, outputs k *if* the input x can be written as 10^{k} for some real number k. For example, $\log 10^{-9} = -9$, $\log 10^{2} = 2$, and $\log 10^{1.43} = 1.43$. For any real number k, $\log 10^{k} = k$. Some values for $\log x$ are given in Table 12.8.

TABLE 12.8 Common Logarithms of Powers of 10   Write x as a power of 10.

x	10^{-4}	10^{-3}	10^{-2}	10^{-1}	10^{0}	10^{1}	10^{2}	10^{3}	10^{4}
$\log x$	-4	-3	-2	-1	0	1	2	3	4

A common logarithm is the *exponent* on base 10.

We use this information to define the common logarithm.

COMMON LOGARITHM

The **common logarithm of a positive number x**, denoted $\log x$, is calculated as follows. If x is written as $x = 10^{k}$, then

$$\log x = k,$$

where k is a real number. That is, $\log 10^{k} = k$.
The function given by

$$f(x) = \log x$$

is called the **common logarithmic function**.

NOTE: Previously, we have always used one letter, such as f or g, to name a function. The common logarithm is the *first* function for which we use *three* letters, log, to name it. Thus $f(x)$, $g(x)$, and $\log(x)$ all represent functions. Generally, $\log(x)$ is written without parentheses as $\log x$.

The following steps can be used to calculate $\log x$.

> **STEP 1:** Is x positive? If not, then $\log x$ is *undefined*.
> **STEP 2:** Write x as 10^k for some real number k. If this is not possible, then use a calculator to approximate $\log x$.
> **STEP 3:** If $x = 10^k$, then $\log x = k$. That is, $\log 10^k = k$.

In the next two examples, we evaluate $\log x$ for various values of x.

EXAMPLE 1   **Calculating log x**

Evaluate each expression, if possible.
(a) $\log(-4)$ **(b)** $\log 1000$

Solution
(a) STEP 1: *Is x positive?* No. Because $x = -4$ is negative, $\log x$ is undefined.

> **NOTE:** Because 10^k is always positive, there is no k such that $10^k = -4$. The common logarithm of *any* negative number is undefined.

(b) STEP 1: *Is x positive?* Yes, because $x = 1000$.
STEP 2: *Write x as 10^k for some real number k.* Because $1000 = 10^3$, we can write $x = 10^3$.
STEP 3: *If $x = 10^k$, then $\log x = k$.* Because $x = 10^3$, it follows that

$$\log x = \log 1000 = \log 10^3 = 3.$$

■ **Now Try Exercises 27, 31**

EXAMPLE 2   **Evaluating common logarithms**

Evaluate each common logarithm.
(a) $\log 100$ **(b)** $\log \frac{1}{10}$ **(c)** $\log \sqrt{1000}$ **(d)** $\log 45$

Solution
(a) $x = 100 = 10^2$, so $\log x = \log 100 = \log 10^2 = 2$
(b) $x = \frac{1}{10} = 10^{-1}$, so $\log x = \log \frac{1}{10} = \log 10^{-1} = -1$
(c) $\log \sqrt{1000} = \log(1000)^{1/2} = \log(10^3)^{1/2} = \log 10^{3/2} = \frac{3}{2}$
(d) How to write $x = 45$ as a power of 10 is not obvious. However, we can use a calculator to determine that $\log 45 \approx 1.6532$. Thus $x \approx 10^{\wedge}(1.6532) \approx 45$.

Figure 12.23 supports these answers.

CALCULATOR HELP

To evaluate the common logarithmic function, see Appendix A (page AP-8).

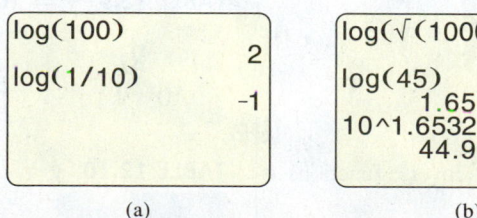

(a)   (b)

Figure 12.23

■ **Now Try Exercises 17, 23, 29, 43**

NOTE: To find log x, we ask ourselves, "What exponent should 10 be raised to in order to get x?" This *exponent* equals log x.

THE GRAPH OF THE COMMON LOGARITHM The four points $(10^{-1}, 1)$, $(10^0, 0)$, $(10^{0.5}, 0.5)$, and $(10^1, 1)$ are on the graph of $y = \log x$. See Table 12.8 at the beginning of this section. Plotting these points, as shown in Figure 12.24(a), and sketching the graph of $y = \log x$ results in Figure 12.24(b). Note some important features of this graph.

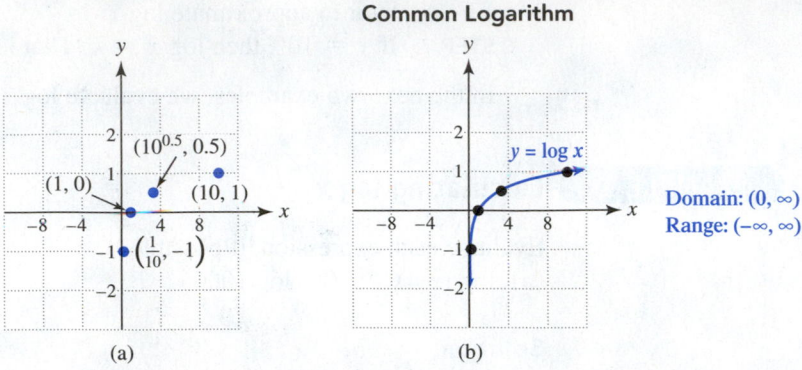

Figure 12.24

- The graph of the common logarithm increases very slowly for large values of x. For example, x must be 100 for log x to reach 2 and x must be 1000 for log x to reach 3.
- The graph passes through the point $(1, 0)$. Thus log $1 = 0$.
- The graph does not exist for negative values of x. The *domain* of log x includes only positive numbers. The *range* of log x includes all real numbers.
- When $0 < x < 1$, log x outputs negative values. The y-axis is a vertical asymptote, so as x approaches 0, log x approaches $-\infty$.

MAKING CONNECTIONS

The Common Logarithmic Function and the Square Root Function

Much like the square root function, the common logarithmic function does not have an easy-to-evaluate formula. We can calculate $\sqrt{4} = 2$ and $\sqrt{100} = 10$ mentally, but for $\sqrt{2}$ we use a calculator. Similarly, we can mentally calculate log $1000 = \log 10^3 = 3$, whereas we use a calculator for log 45. The notation log x is understood as $\log_{10} x$ in the same way that $\sqrt{x}$ is understood to be $\sqrt[2]{x}$. Another similarity between the square root and common logarithmic functions is that their domains do not include negative numbers. If only real numbers are allowed as outputs, both $\sqrt{-3}$ and log (-3) are undefined expressions.

The Inverse of the Common Logarithmic Function

The graph of $y = \log x$ shown in Figure 12.24(b) is a one-to-one function because it passes the horizontal line test. Thus the common logarithmic function has an inverse function. To determine this inverse function for log x, consider Tables 12.9 and 12.10.

TABLE 12.9 $y = 10^x$

x	-2	-1	0	1	2
10^x	10^{-2}	10^{-1}	10^0	10^1	10^2

(1)

Inverse functions —

TABLE 12.10 $y = \log x$

x	10^{-2}	10^{-1}	10^0	10^1	10^2
$\log x$	-2	-1	0	1	2

(2)

If we start with the input **2** for 10^x, we compute $\mathbf{10^2}$, as shown by (1) in Table 12.9. If we use $\mathbf{10^2}$ as the input for log x, we compute **2**, as shown by (2) in Table 12.10. That is,

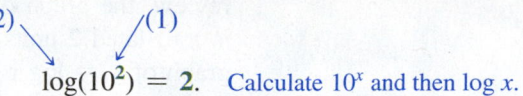

$$\log(10^2) = 2. \quad \text{Calculate } 10^x \text{ and then } \log x.$$

Next, suppose we find the logarithm first and then compute a power of 10. If we start with the input $\mathbf{10^2}$, we compute **2**, as shown by (2) in Table 12.10. If we use **2** as the input for 10^x, we compute $\mathbf{10^2}$, as shown by (1) in Table 12.9. That is,

$$10^{\log(10^2)} = 10^2. \quad \text{Calculate } \log x \text{ and then } 10^x.$$

In general, if $10^a = b$, then $\log b = a$ and if $\log b = a$, then $10^a = b$. Thus the *inverse function* of $f(x) = \log x$ is $f^{-1}(x) = 10^x$. Note that composition of these two functions satisfies the definition of an inverse function.

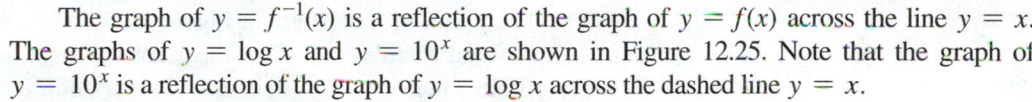

$$
\begin{aligned}
(f \circ f^{-1})(x) &= f\big(f^{-1}(x)\big) \quad \text{and} \quad (f^{-1} \circ f)(x) = f^{-1}\big(f(x)\big) \\
&= f(10^x) \qquad\qquad\qquad\qquad\quad = f^{-1}(\log x) \\
&= \log 10^x \qquad\qquad\qquad\qquad\;\; = 10^{\log x} \\
&= x \qquad\qquad\qquad\qquad\qquad\quad = x
\end{aligned}
$$

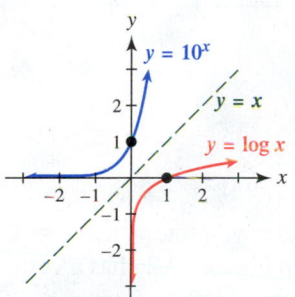

Figure 12.25

The graph of $y = f^{-1}(x)$ is a reflection of the graph of $y = f(x)$ across the line $y = x$. The graphs of $y = \log x$ and $y = 10^x$ are shown in Figure 12.25. Note that the graph of $y = 10^x$ is a reflection of the graph of $y = \log x$ across the dashed line $y = x$.

These inverse properties of log x and 10^x are summarized as follows.

INVERSE PROPERTIES OF THE COMMON LOGARITHM

The following properties hold for common logarithms.

$$\log 10^x = x, \qquad \text{for any real number } x$$
$$10^{\log x} = x, \qquad \text{for any positive real number } x$$

EXAMPLE 3   Applying inverse properties

Use inverse properties to simplify each expression.
(a) $\log 10^\pi$ **(b)** $\log 10^{x^2+1}$ **(c)** $10^{\log 7}$ **(d)** $10^{\log 3x}, x > 0$

Solution
(a) Because $\log 10^x = x$ for any real number x, $\log 10^\pi = \boldsymbol{\pi}$.
(b) $\log 10^{x^2+1} = \boldsymbol{x^2 + 1}$
(c) Because $10^{\log x} = x$ for any positive real number x, $10^{\log 7} = \boldsymbol{7}$.
(d) $10^{\log 3x} = \boldsymbol{3x}$, provided x is a *positive* number.

■ **Now Try Exercises 25, 35, 37, 41**

EXAMPLE 4   Graphing logarithmic functions

Graph each function f and compare its graph to $y = \log x$.
(a) $f(x) = \log(x - 2)$ **(b)** $f(x) = \log(x) + 1$

Solution

(a) We can use our knowledge of translations to sketch the graph of $y = \log(x - 2)$. To review, the graph of $y = (x - 2)^2$ is similar to the graph of $y = x^2$, except that it is translated 2 units to the *right*. Thus the graph of $y = \log(x - 2)$ is similar to the graph of $y = \log x$ (see Figure 12.25) except that it is translated 2 units to the right, as shown in Figure 12.26. The graph of $y = \log x$ passes through $(1, 0)$, so the graph of $y = \log(x - 2)$ passes through $(3, 0)$. Also, instead of the y-axis being the vertical asymptote, the line $x = 2$ is the vertical asymptote.

Translated 2 Units Right

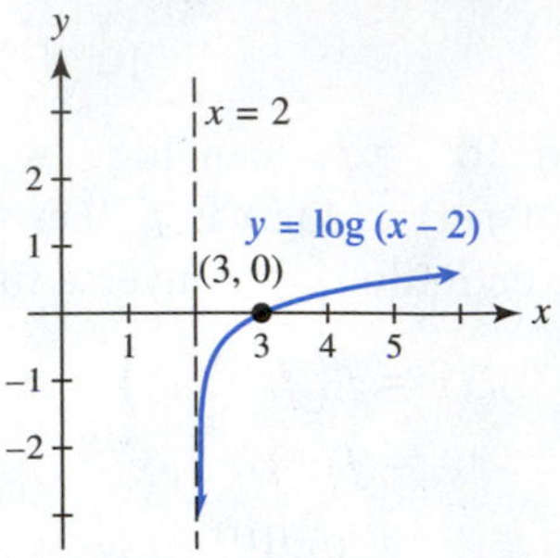

Translated 1 Unit Upward

Figure 12.26

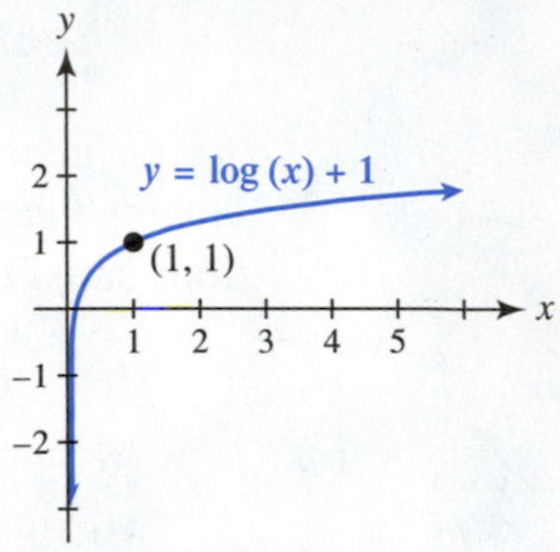

Figure 12.27

NOTE: The graph of $y = \log x$ in Figure 12.25 has a vertical asymptote when $x = 0$ and is undefined when $x < 0$. Thus the graph of $\log(x - 2)$ in Figure 12.26 has a vertical asymptote when $x - 2 = 0$, or $x = 2$. It is undefined when $x - 2 < 0$, or $x < 2$.

(b) The graph of $y = \log(x) + 1$ is similar to the graph of $y = \log x$, except that it is translated 1 unit *upward*. This graph is shown in Figure 12.27. Note that the graph of $y = \log(x) + 1$ passes through the point $(1, 1)$.

▌ **Now Try Exercises 47, 49**

▶ **REAL-WORLD CONNECTION** Logarithms are used to model quantities that vary greatly in intensity. For example, the human ear is extremely sensitive and able to detect intensities on the eardrum ranging from 10^{-16} watts per square centimeter (w/cm^2) to 10^{-4} w/cm^2, which is usually painful. The next example illustrates modeling sound with logarithms.

EXAMPLE 5 **Modeling sound levels**

Sound levels in decibels (dB) can be computed by $f(x) = 160 + 10 \log x$, where x is the intensity of the sound in watts per square centimeter. Ordinary conversation has an intensity of 10^{-10} w/cm^2. What decibel level is this? (*Source:* R. Weidner and R. Sells, *Elementary Classical Physics*, Vol. 2.)

CRITICAL THINKING

If the sound level increases by 10 dB, by what factor does the intensity x increase?

Solution

To find the decibel level for ordinary conversation, evaluate $f(10^{-10})$.

$$f(10^{-10}) = 160 + 10 \log(10^{-10}) \quad \text{Substitute } x = 10^{-10}.$$
$$= 160 + 10(-10) \quad \text{Evaluate } \log(10^{-10}).$$
$$= 60 \quad \text{Simplify.}$$

Ordinary conversation corresponds to 60 dB.

▌ **Now Try Exercise 105**

Logarithms with Other Bases

Common logarithms are base-10 logarithms, but we can define logarithms having other bases. For example, base-2 logarithms are frequently used in computer science. Some values for the base-2 logarithmic function, denoted $f(x) = \log_2 x$, are shown in Table 12.11. If x can be expressed as $x = 2^k$ for some real number k, then $\log_2 x = \log_2 2^k = k$.

READING CHECK

What does $\log_2 x$ mean?

Write x as a power of 2.

TABLE 12.11 Base-2 Logarithms of Powers of 2

x	2^{-3}	2^{-2}	2^{-1}	2^0	2^1	2^2	2^3
$\log_2 x$	-3	-2	-1	0	1	2	3

A base-2 logarithm is the *exponent* on base 2.

NOTE: A base-2 logarithm is an *exponent* on base 2.

Logarithms with other bases are evaluated in the next three examples.

EXAMPLE 6 **Evaluating base-2 logarithms**

Simplify each logarithm.
(a) $\log_2 8$ **(b)** $\log_2 \frac{1}{4}$

Solution
(a) The logarithmic expression $\log_2 8$ represents the exponent on base 2 that gives 8. Because $8 = 2^3$, $\log_2 8 = \log_2 2^3 = \mathbf{3}$.
(b) Because $\frac{1}{4} = \frac{1}{2^2} = 2^{-2}$, $\log_2 \frac{1}{4} = \log_2 2^{-2} = \mathbf{-2}$.

Now Try Exercises 81, 83

Some values of base-e logarithms are shown in Table 12.12. A base-e logarithm is referred to as a **natural logarithm** and is denoted either $\log_e x$ or $\ln x$. Natural logarithms are used in mathematics, science, economics, electronics, and communications.

CALCULATOR HELP

To evaluate the natural logarithmic function, see Appendix A (page AP-8).

TABLE 12.12 Natural Logarithms of Powers of e

x	e^{-3}	e^{-2}	e^{-1}	e^0	e^1	e^2	e^3
$\ln x$	-3	-2	-1	0	1	2	3

NOTE: A natural logarithm is an *exponent* on base e.

To evaluate natural logarithms we usually use a calculator.

EXAMPLE 7 **Evaluating natural logarithms**

Approximate to the nearest hundredth.
(a) $\ln 10$ **(b)** $\ln \frac{1}{2}$

```
ln(10)
        2.302585093
ln(1/2)
       -.6931471806
```

Figure 12.28

Solution
(a) Figure 12.28 shows that $\ln 10 \approx 2.30$.
(b) Figure 12.28 shows that $\ln \frac{1}{2} \approx -0.69$.

Now Try Exercises 61, 63

We now define base-*a* logarithms.

> ### BASE-*a* LOGARITHMS
>
> The **logarithm with base *a* of a positive number *x***, denoted $\log_a x$, is calculated as follows. If *x* is written as $x = a^k$, then
>
> $$\log_a x = k,$$
>
> where $a > 0$, $a \neq 1$, and *k* is a real number. That is, $\log_a a^k = k$. The function given by
>
> $$f(x) = \log_a x$$
>
> is called the **logarithmic function with base *a***.

Remember that *a logarithm is an exponent*. The expression $\log_a x$ equals the exponent *k* such that $a^k = x$. The graph of $y = \log_a x$ with $a > 1$ is shown in Figure 12.29. Note that the graph passes through the point $(1, 0)$. Thus $\log_a 1 = 0$.

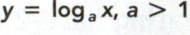

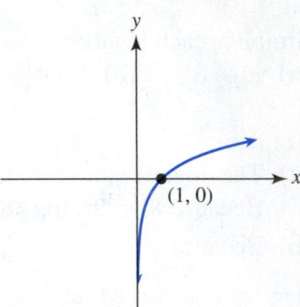

Figure 12.29

CRITICAL THINKING

Explain why $\log_a 1 = 0$ for any positive base *a*, $a \neq 1$.

NOTE: The natural logarithm, $\ln x$, is a base-*a* logarithm with $a = e$. That is, $\ln x = \log_e x$.

EXAMPLE 8 **Evaluating base-*a* logarithms**

Simplify each logarithm.
(a) $\log_5 25$ **(b)** $\log_4 \frac{1}{64}$ **(c)** $\log_7 1$ **(d)** $\log_3 9^{-1}$

Solution
(a) $25 = 5^2$, so $\log_5 25 = \log_5 5^2 = \mathbf{2}$.
(b) $\frac{1}{64} = \frac{1}{4^3} = 4^{-3}$, so $\log_4 \frac{1}{64} = \log_4 4^{-3} = \mathbf{-3}$.
(c) $1 = 7^0$, so $\log_7 1 = \log_7 7^0 = \mathbf{0}$. (The logarithm of 1 is 0, regardless of the base.)
(d) $9^{-1} = (3^2)^{-1} = 3^{-2}$, so $\log_3 9^{-1} = \log_3 3^{-2} = \mathbf{-2}$.

Now Try Exercises 55, 75, 85, 87

The graph of $y = \log_a x$ in Figure 12.29 passes the horizontal line test, so it is a one-to-one function and it has an inverse. If we let $f(x) = \log_a x$, then $f^{-1}(x) = a^x$. This statement is a generalization of the fact that $f(x) = \log x$ and $f^{-1}(x) = 10^x$ represent inverse functions. The graphs of $y = \log_a x$ and $y = a^x$ with $a > 1$ are shown in Figure 12.30. Note that the graph of $y = a^x$ is a reflection of the graph of $y = \log_a x$ across the line $y = x$.

Inverse Functions

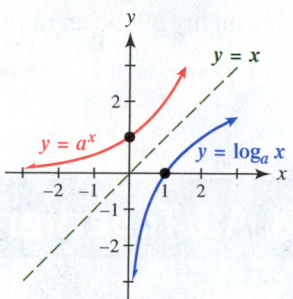

Figure 12.30 $a > 1$

These inverse properties are summarized by the following.

INVERSE PROPERTIES OF BASE-a LOGARITHMS

The following properties hold for logarithms with base a.

$$\log_a a^x = x, \qquad \text{for any real number } x$$
$$a^{\log_a x} = x, \qquad \text{for any positive real number } x$$

NOTE: The inverse function of $f(x) = \ln x$ is $f^{-1}(x) = e^x$.
The inverse function of $f(x) = \log x$ is $f^{-1}(x) = 10^x$.

EXAMPLE 9   Applying inverse properties

Simplify each expression.
(a) $\ln e^{0.5x}$ **(b)** $e^{\ln 4}$ **(c)** $2^{\log_2 7x}$ **(d)** $10^{\log(9x-3)}$

Solution
(a) $\ln e^{0.5x} = 0.5x$ because $\ln e^k = k$ for all k.
(b) $e^{\ln 4} = 4$ because $e^{\ln k} = k$ for all positive k.
(c) $2^{\log_2 7x} = 7x$ for $x > 0$ because $a^{\log_a k} = k$ for all positive k.
(d) $10^{\log(9x-3)} = 9x - 3$ for $x > \frac{1}{3}$ because $10^{\log k} = k$ for all positive k.

Now Try Exercises 39, 57, 59, 89

▶ **REAL-WORLD CONNECTION** Logarithms occur in many applications. One application is runway length for airplanes, as discussed in A Look Into Math and in the next example.

EXAMPLE 10   Calculating runway length

There is a mathematical relationship between an airplane's weight x and the runway length required at takeoff. For certain types of airplanes, the minimum runway length L in thousands of feet may be modeled by $L(x) = 1.3 \ln x$, where x is in thousands of pounds.
(*Source:* L. Haefner, *Introduction to Transportation Systems.*)
(a) Estimate the runway length needed for an airplane weighing 10,000 pounds.
(b) Does a 20,000-pound airplane need twice the runway length that a 10,000-pound airplane needs? Explain.

Solution

(a) Because $L(x) = 1.3 \ln x$, it follows that $L(10) = 1.3 \ln(10) \approx 3$. An airplane weighing 10,000 pounds requires a runway (at least) 3000 feet long.

(b) Because $L(20) = 1.3 \ln(20) \approx 3.9$, a 20,000-pound airplane does not need twice the runway length needed by a 10,000-pound airplane. Rather, the heavier airplane needs roughly 3900 feet of runway, or only an extra 900 feet.

Now Try Exercise 107

12.3 Putting It All Together

Common logarithms are base-10 logarithms. If a positive number x is written as $x = 10^k$, then $\log x = k$. The value of $\log x$ represents the exponent on the base 10 that gives x.

CONCEPT	DESCRIPTION	EXAMPLES
Base-a Logarithms	*Definition:* $\log_a x = k$ means $x = a^k$, where $a > 0$ and $a \neq 1$. *Domain:* all *positive* real numbers *Range:* all real numbers *Graph:* $a > 1$ (shown to the right for $\log x$ and $\ln x$); passes through $(1, 0)$; vertical asymptote: y-axis *Common Logarithm:* base-10 logarithm and denoted $\log x$ *Natural Logarithm:* base-e logarithm, where $e \approx 2.718$, and denoted $\ln x$	$\log 1000 = \log 10^3 = 3$, $\log_2 16 = \log_2 2^4 = 4$, and $\log_3 \dfrac{1}{81} = \log_3 3^{-4} = -4$ (graph of $y = \ln x$ and $y = \log x$ passing through $(1,0)$)
Inverse Properties	The following properties hold for base-a logarithms. $\log_a a^x = x$, for any real number x $a^{\log_a x} = x$, for any positive number x	$\log 10^{7.48} = 7.48$ $2^{\log_2 63} = 63$ $10^{\log 23} = 23$ $\ln e^4 = 4$

12.3 Exercises

CONCEPTS AND VOCABULARY

1. What is the base of the common logarithm?

2. What is the base of the natural logarithm?

3. What are the domain and range of $\log x$?

4. What are the domain and range of 10^x?

5. $\log 10^k = $ _____

6. $\ln e^k = $ _____

7. If $\log x = k$, then $10^k = $ _____.

8. If $x > 0$, then $10^{\log x} =$ _____.

9. What does k equal if $10^k = 5$?

10. What does k equal if $e^k = 5$?

11. $\log 1 =$ _____

12. $\log(-1)$ is _____.

Exercises 13–16: Complete the table.

13.

x	10^{-5}	10^0	$10^{0.5}$	$10^{2.2}$
$\log x$				

14.

x	10^{-6}	10^{-1}	$10^{\frac{5}{7}}$	10^{π}
$\log x$				

15.

x	e^{-6}	e^{-1}	$e^{\frac{5}{7}}$	e^{π}
$\ln x$				

16.

x	2^{-5}	2^0	$2^{0.5}$	$2^{2.2}$
$\log_2 x$				

EVALUATING AND GRAPHING COMMON LOGARITHMS

Exercises 17–42: Simplify the expression, if possible.

17. $\log 10^5$

18. $\log 10$

19. $\log 10^{-4}$

20. $\log 10^{-1}$

21. $\log 1$

22. $\log \sqrt[3]{100}$

23. $\log \frac{1}{100}$

24. $\log \frac{1}{10}$

25. $\log 10^{4.7}$

26. $\log 10^{2x+4}$

27. $\log 10{,}000$

28. $\log 100{,}000$

29. $\log \sqrt{10}$

30. $\log 1{,}000{,}000$

31. $\log(-23)$

32. $\log(-8)$

33. $\log 0.001$

34. $\log 0.0001$

35. $10^{\log 2}$

36. $10^{\log 7.5}$

37. $10^{\log x^2}$

38. $10^{\log |x|}$

39. $10^{\log 5}$

40. $\log 10^{\frac{3}{4}}$

41. $\log 10^{(2x-7)}$

42. $\log 10^{(8-4x)}$

Exercises 43–46: Evaluate the common logarithm, using a calculator. Round values to the nearest thousandth.

43. $\log 25$

44. $\log 0.501$

45. $\log 1.45$

46. $\log \frac{1}{35}$

Exercises 47–54: Graph $y = f(x)$. Compare the graph to the graph of $y = \log x$.

47. $f(x) = \log(x) - 1$

48. $f(x) = \log(x) + 2$

49. $f(x) = \log(x + 1)$

50. $f(x) = \log(x + 2)$

51. $f(x) = \log(x - 1)$

52. $f(x) = \log(x - 3)$

53. $f(x) = 2\log x$

54. $f(x) = -\log x$

EVALUATING AND GRAPHING NATURAL LOGARITHMS

Exercises 55–60: Simplify the expression.

55. $\ln 1$

56. $\ln e^2$

57. $\ln e^{-5x}$

58. $e^{\ln 2x}$

59. $e^{\ln x^2}$

60. $\ln e^{7x}$

Exercises 61–64: Evaluate the natural logarithm, using a calculator. Round values to the nearest thousandth.

61. $\ln 7$

62. $\ln 126$

63. $\ln \frac{4}{7}$

64. $\ln 0.67$

Exercises 65–68: Graph f in $[-4, 4, 1]$ by $[-4, 4, 1]$. Compare this graph to the graph of $y = \ln x$. Identify the domain of f.

65. $f(x) = \ln|x|$

66. $f(x) = \ln(x) - 2$

67. $f(x) = \ln(x + 2)$

68. $f(x) = 2\ln x$

EVALUATING AND GRAPHING BASE-a LOGARITHMS

Exercises 69–94: Simplify the expression, if possible.

69. $\log_5 5^{6x}$

70. $\log_3 3^3$

71. $\log_2 \sqrt{\frac{1}{8}}$

72. $\log_2 2^6$

73. $\log_2 2^8$

74. $\log_2 2^{-5}$

75. $\log_2 \sqrt{8}$

76. $\log_2 \sqrt{32}$

77. $\log_2 \sqrt[3]{\frac{1}{4}}$

78. $\log_2 \frac{1}{64}$

79. $\log_2 -8$

80. $\log_2 -7$

81. $\log_2 4$

82. $\log_2 \frac{1}{32}$

83. $\log_2 \frac{1}{16}$

84. $\log_3 27$

85. $\log_3 \frac{1}{9}$　　　　**86.** $\log_4 16$

87. $\log_5 5^{-2}$　　　　**88.** $\log_8 64$

89. $5^{\log_5 17}$　　　　**90.** $9^{\log_9 73}$

91. $4^{\log_4 (2x)^2}$　　　　**92.** $b^{\log_b (x-1)}$

93. $5^{\log_5 0.6z}$　　　　**94.** $7^{\log_7 (x-9)}$

Exercises 95 and 96: Complete the table.

95.

x	$\frac{1}{4}$	$\frac{1}{2}$	1	$\sqrt{2}$	64
$\log_2 x$					

96.

x	$\frac{1}{7}$	1	$\sqrt{7}$	7	49
$\log_7 x$					

Exercises 97–100: Graph $y = f(x)$.

97. $f(x) = \log_2 x$　　　　**98.** $f(x) = \log_2 (x - 2)$

99. $f(x) = 2 + \log_2 x$　　　　**100.** $f(x) = \log_2 (x + 2)$

Exercises 101–104: Without using a calculator match $f(x)$ with its graph (a.–d.).

101. $f(x) = \log x$　　　　**102.** $f(x) = \log_3 x$

103. $f(x) = \log_3 (x) + 2$　　　　**104.** $f(x) = \log (x + 1)$

a.

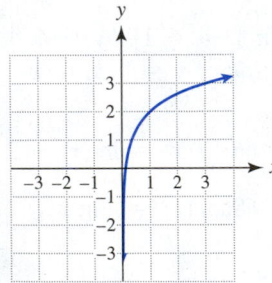

b.

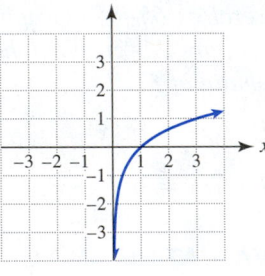

c.

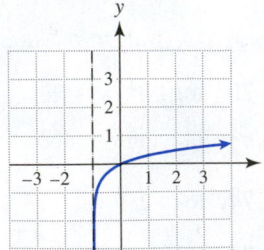

d.

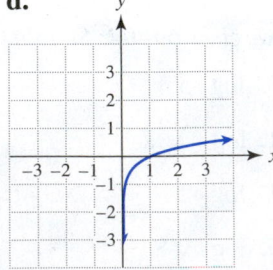

APPLICATIONS

105. *Modeling Sound* (Refer to Example 5.) At professional football games in domed stadiums the decibel level may reach 120. The eardrum usually experiences pain when the intensity of the sound reaches 10^{-4} watts per square centimeter. How many decibels does this quantity represent? Is the noise at a football game likely to hurt some people's eardrums?

106. *Hurricanes* The barometric air pressure P in inches of mercury at a distance of d miles from the eye of a severe hurricane can sometimes be modeled by the formula $P(d) = 0.48 \ln (d + 1) + 27$. Average air pressure is about 30 inches of mercury. (*Source:* A. Miller and R. Anthes, *Meteorology.*)

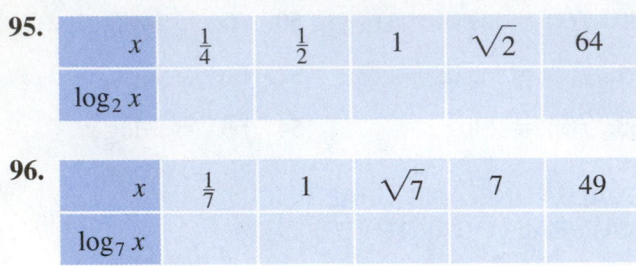

(a) Evaluate $P(0)$ and $P(50)$. Interpret the results.

(b) A graph of $y = P(d)$ is shown in the figure. Describe how the air pressure changes as the distance from the eye of the hurricane increases.

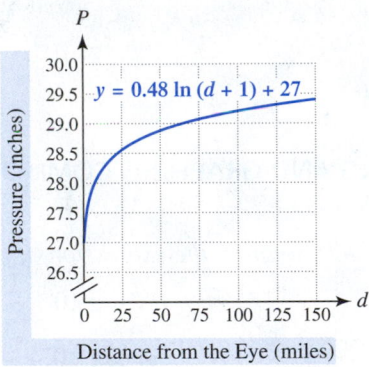

Distance from the Eye (miles)

(c) Is the eye of the hurricane a low-pressure area or a high-pressure area?

107. *Runway Length* (Refer to Example 10.)

(a) A graph of $L(x) = 1.3 \ln x$ is shown in the figure, where $y = L(x)$. As the weight of the plane increases, what can be said about the length of the runway required?

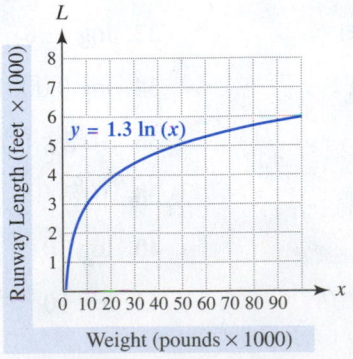

Weight (pounds × 1000)

(b) Evaluate $L(50)$ and interpret the result.

108. *Growth in Salary* Suppose that a person's salary is initially $40,000 and could be determined by either $f(x)$ or $g(x)$, where x represents the number of years of experience.
a. $f(x) = 40,000(1.1)^x$
b. $g(x) = 40,000 \log(10 + x)$
Would most people prefer that their salaries increase exponentially or logarithmically? Explain.

109. *Magnitude of a Star* The first stellar brightness scale was developed 2000 years ago by two Greek astronomers, Hipparchus and Ptolemy. The brightest star in the sky was given a magnitude of 1, and the faintest star was given a magnitude of 6. In 1856 this scale was described mathematically by the formula

$$M = 6 - 2.5 \log \frac{I}{I_0},$$

where M is the magnitude of a star with an intensity of I and I_0 is the intensity of the faintest star seen in the sky. (*Source:* M. Zeilik, *Introductory Astronomy and Astrophysics.*)
(a) Find M if $I = 10$ and $I_0 = 1$.
(b) What is the magnitude of a star that is 100 times more intense than the faintest star?
(c) If the intensity of a star increases by a factor of 10, what happens to its magnitude?

110. *Population of Urban Regions* Although less industrialized urban regions of the world are experiencing exponential population growth, industrialized urban regions are experiencing logarithmic population growth. Population in less industrialized urban regions can be modeled by

$$f(x) = 0.338(1.035)^x,$$

whereas the population in industrialized urban regions can be modeled by

$$g(x) = 0.36 + 0.15 \ln(x + 1).$$

In these formulas the output is in billions of people and x is in years, where $x = 0$ corresponds to 1950, $x = 10$ corresponds to 1960, and so on until $x = 80$ corresponds to 2030. (*Source:* D. Meadows, *Beyond The Limits.*)
(a) Evaluate $f(50)$ and $g(50)$. Interpret the results.
(b) Graph f and g in [0, 80, 10] by [0, 5, 1]. Compare the two graphs.
(c) If x increases from 20 to 40, by what factor does $f(x)$ increase? By what factor does $g(x)$ increase?

111. *Earthquakes* The Richter scale is used to determine the intensity of earthquakes, which corresponds to the amount of energy released. If an earthquake has an intensity of x, its *magnitude*, as computed by the Richter scale, is given by $R(x) = \log \frac{x}{I_0}$, where I_0 is the intensity of a small, measurable earthquake.
(a) On July 26, 1963, an earthquake in Yugoslavia had a magnitude of 6.0 on the Richter scale, and on August 19, 1977, an earthquake in Indonesia measured 8.0. Find the intensity x for each of these earthquakes if $I_0 = 1$.
(b) How many times more intense was the Indonesian earthquake than the Yugoslavian earthquake?

112. *Path Loss for Cellular Phones* For cellular phones to work throughout a country, large numbers of cellular towers are necessary. How well the signal is propagated throughout a region depends on the location of these towers. One quick way to estimate the strength of a signal at x kilometers is to use the formula

$$D(x) = -121 - 36 \log x.$$

This formula computes the decrease in the signal, using decibels, so it is always negative. For example, $D(1) = -121$ means that at a distance of 1 kilometer the signal has decreased in strength by 121 decibels.
(*Source:* C. Smith, *Practical Cellular & PCS Design.*)

(a) Evaluate $D(3)$ and interpret the result.
(b) Graph D in [1, 10, 1] by [−160, −120, 10].
(c) What happens to the signal as x increases?

WRITING ABOUT MATHEMATICS

113. Explain what $\log_a x$ means and give an example.

114. How would you explain to a student that $\log_a 1 \neq 1$?

Group Activity   Working With Real Data

Directions: Form a group of 2 to 4 people. Select someone to record the group's responses for this activity. All members of the group should work cooperatively to answer the questions. If your instructor asks for your results, each member of the group should be prepared to respond.

Greenhouse Gases Carbon dioxide (CO_2) is a greenhouse gas in the atmosphere that may raise average temperatures on Earth. The burning of fossil fuels could be responsible for the increased levels of carbon dioxide. If current trends continue, future concentrations of atmospheric carbon dioxide in parts per million (ppm) could reach the levels shown in the accompanying table. The CO_2 concentration in the year 2000 was greater than it had been at any time in the previous 160,000 years.

Year	2000	2050	2100	2150	2200
CO_2 (ppm)	364	467	600	769	987

Source: R. Turner, *Environmental Economics.*

(a) Let x be in years, where $x = 0$ corresponds to 2000, $x = 1$ to 2001, and so on. Find values for C and a so that $f(x) = Ca^x$ models the data.

(b) Graph f and the data in the same viewing rectangle.

(c) Use $f(x)$ to estimate graphically the year when the carbon dioxide concentration will be double the preindustrial level of 280 ppm.

12.4   Properties of Logarithms

Basic Properties • Change of Base Formula

A LOOK INTO MATH ▶ The discovery of logarithms by John Napier (1550–1617) played an important role in the history of science. Logarithms were instrumental in allowing Johannes Kepler (1571–1630) to calculate the positions of the planet Mars, which led to his discovery of the laws of planetary motion. Kepler's laws were used by Isaac Newton (1642–1727) to discover the universal laws of gravity. Although calculators and computers have made tables of logarithms obsolete, applications involving logarithms still play an important role in modern-day computation. One reason for their continued importance is that *logarithms possess several important properties.*

Basic Properties

In this subsection we discuss three important properties of logarithms. The first property is the product rule for logarithms.

NEW VOCABULARY

☐ Change of base formula

PRODUCT RULE FOR LOGARITHMS

For positive numbers m, n, and $a \neq 1$,

$$\log_a mn = \log_a m + \log_a n.$$

This product rule may be verified by using properties of exponents and the fact that $\log_a a^k = k$ for any real number k. Here, we verify the product property for logarithms. The two other properties presented later can be verified in a similar manner.

If m and n are positive numbers, we can write $m = a^c$ and $n = a^d$ for some real numbers c and d.

$$\log_a mn = \log_a(a^c a^d) = \log_a(a^{c+d}) = c + d \quad \text{and}$$

$$\log_a m + \log_a n = \log_a a^c + \log_a a^d = c + d$$

Thus $\log_a mn = \log_a m + \log_a n$.

This property is illustrated in Figure 12.31, which shows that

$$\log 10 = \log(2 \cdot 5) = \log 2 + \log 5.$$

In the next two examples, we demonstrate various operations involving logarithms.

```
log(10)
                1
log(2)+log(5)
                1
```

Figure 12.31

EXAMPLE 1 **Writing logarithms as sums**

Write each expression as a sum of logarithms. Assume that x is positive.
(a) $\log 21$ **(b)** $\ln 5x$ **(c)** $\log x^3$

Solution
(a) $\log 21 = \log(3 \cdot 7) = \log 3 + \log 7$
(b) $\ln 5x = \ln(5 \cdot x) = \ln 5 + \ln x$
(c) $\log x^3 = \log(x \cdot x \cdot x) = \log x + \log x + \log x$

Now Try Exercises 13, 15, 17

EXAMPLE 2 **Combining logarithms**

Write each expression as one logarithm. Assume that x and y are positive.
(a) $\log 5 + \log 6$ **(b)** $\ln x + \ln xy$ **(c)** $\log 2x + \log 5x$

Solution
(a) $\log 5 + \log 6 = \log(5 \cdot 6) = \log 30$
(b) $\ln x + \ln xy = \ln(x \cdot xy) = \ln x^2 y$
(c) $\log 2x + \log 5x = \log(2x \cdot 5x) = \log 10x^2$

Now Try Exercises 25, 27, 29

The second property is the quotient rule for logarithms.

QUOTIENT RULE FOR LOGARITHMS

For positive numbers m, n, and $a \neq 1$,

$$\log_a \frac{m}{n} = \log_a m - \log_a n.$$

```
log(10)
                1
log(20)-log(2)
                1
```

Figure 12.32

This property is illustrated in Figure 12.32, which shows that

$$\log 10 = \log \frac{20}{2} = \log 20 - \log 2.$$

EXAMPLE 3 **Writing logarithms as differences**

Write each expression as a difference of two logarithms. Assume that variables are positive.

(a) $\log \dfrac{3}{2}$ (b) $\ln \dfrac{3x}{y}$ (c) $\log_5 \dfrac{x}{z^4}$

Solution

(a) $\log \dfrac{3}{2} = \log 3 - \log 2$ (b) $\ln \dfrac{3x}{y} = \ln 3x - \ln y$

(c) $\log_5 \dfrac{x}{z^4} = \log_5 x - \log_5 z^4$

Now Try Exercises 19, 21, 23

NOTE: $\log_a (m + n) \neq \log_a m + \log_a n$; $\log_a (m - n) \neq \log_a m - \log_a n$;

$$\log_a (mn) \neq \log_a m \cdot \log_a n; \quad \log_a \left(\dfrac{m}{n}\right) \neq \dfrac{\log_a m}{\log_a n}$$

EXAMPLE 4 **Combining logarithms**

Write each expression as one term. Assume that x is positive.
(a) $\log 50 - \log 25$ (b) $\ln x^3 - \ln x$ (c) $\log 15x - \log 5x$

Solution

(a) $\log 50 - \log 25 = \log \dfrac{50}{25} = \log 2$

(b) $\ln x^3 - \ln x = \ln \dfrac{x^3}{x} = \ln x^2$

(c) $\log 15x - \log 5x = \log \dfrac{15x}{5x} = \log 3$

Now Try Exercises 33, 35, 37

The third property is the power rule for logarithms. To illustrate this rule we use

$$\log x^3 = \log (x \cdot x \cdot x) = \log x + \log x + \log x = 3 \log x.$$

Thus $\log x^3 = 3 \log x$. This example is generalized in the following rule.

POWER RULE FOR LOGARITHMS

For positive numbers m and $a \neq 1$ and any real number r,

$$\log_a (m^r) = r \log_a m.$$

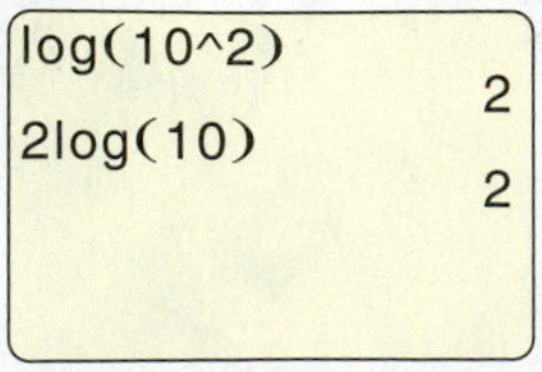

```
log(10^2)
              2
2log(10)
              2
```

Figure 12.33

We use a calculator and the equation

$$\log 10^2 = 2 \log 10$$

to illustrate this property. Figure 12.33 shows the result. We apply the power rule in the next example.

EXAMPLE 5   **Applying the power rule**

Rewrite each expression, using the power rule.
(a) $\log 5^6$ **(b)** $\ln (0.55)^{x-1}$ **(c)** $\log_5 8^{kx}$

Solution
(a) $\log 5^6 = \mathbf{6} \log 5$
(b) $\ln (0.55)^{x-1} = (\mathbf{x - 1}) \ln (0.55)$
(c) $\log_5 8^{kx} = \mathbf{kx} \log_5 8$

▮ **Now Try Exercises 39, 41, 47**

Sometimes we use more than one property to simplify an expression. We assume that all variables are positive in the next two examples.

EXAMPLE 6   **Combining logarithms**

Write each expression as the logarithm of a single expression.
(a) $3 \log x + \log x^2$ **(b)** $2 \ln x - \ln \sqrt{x}$

Solution
(a) $3 \log x + \log x^2 = \log x^3 + \log x^2$ Power rule

$\qquad\qquad\qquad\quad = \log (x^3 \cdot x^2)$ Product rule

$\qquad\qquad\qquad\quad = \log x^5$ Properties of exponents

(b) $2 \ln x - \ln \sqrt{x} = 2 \ln x - \ln x^{1/2}$ $\sqrt{x} = x^{1/2}$

$\qquad\qquad\qquad\quad = \ln x^2 - \ln x^{1/2}$ Power rule

$\qquad\qquad\qquad\quad = \ln \dfrac{x^2}{x^{1/2}}$ Quotient rule

$\qquad\qquad\qquad\quad = \ln x^{3/2}$ Properties of exponents

▮ **Now Try Exercises 49, 57**

EXAMPLE 7   **Expanding logarithms**

Write each expression in terms of logarithms of x, y, and z.

(a) $\log \dfrac{x^2 y^3}{\sqrt{z}}$ **(b)** $\ln \sqrt[3]{\dfrac{xy}{z}}$

Solution
(a) $\log \dfrac{x^2 y^3}{\sqrt{z}} = \log x^2 y^3 - \log \sqrt{z}$ Quotient rule

$\qquad\qquad\quad = \log x^2 + \log y^3 - \log z^{1/2}$ Product rule; $\sqrt{z} = z^{1/2}$

$\qquad\qquad\quad = 2 \log x + 3 \log y - \tfrac{1}{2} \log z$ Power rule

(b) $\ln \sqrt[3]{\dfrac{xy}{z}} = \ln \left(\dfrac{xy}{z}\right)^{1/3}$ $\sqrt[3]{m} = m^{1/3}$

$\qquad\qquad\quad = \tfrac{1}{3} \ln \dfrac{xy}{z}$ Power rule

$\qquad\qquad\quad = \tfrac{1}{3} (\ln xy - \ln z)$ Quotient rule

$\qquad\qquad\quad = \tfrac{1}{3} (\ln x + \ln y - \ln z)$ Product rule

$\qquad\qquad\quad = \tfrac{1}{3} \ln x + \tfrac{1}{3} \ln y - \tfrac{1}{3} \ln z$ Distributive property

▮ **Now Try Exercises 63, 65**

EXAMPLE 8 | **Applying properties of logarithms**

Using only properties of logarithms and the approximations $\ln 2 \approx 0.7$, $\ln 3 \approx 1.1$, and $\ln 5 \approx 1.6$, find an approximation for each expression.

(a) $\ln 8$ **(b)** $\ln 15$ **(c)** $\ln \dfrac{10}{3}$

Solution

(a) $\ln 8 = \ln 2^3 = 3 \ln 2 \approx 3(0.7) = 2.1$

(b) $\ln 15 = \ln (3 \cdot 5) = \ln 3 + \ln 5 \approx 1.1 + 1.6 = 2.7$

(c) $\ln \dfrac{10}{3} = \ln \left(\dfrac{2 \cdot 5}{3} \right) = \ln 2 + \ln 5 - \ln 3 \approx 0.7 + 1.6 - 1.1 = 1.2$

Now Try Exercises 77, 79, 83

Change of Base Formula

▶ **REAL-WORLD CONNECTION** Most calculators only have keys to evaluate common and natural logarithms. Occasionally, it is necessary to evaluate a logarithmic function with a base other than 10 or e. In these situations we use the following **change of base formula**, which we illustrate in the next example.

CHANGE OF BASE FORMULA

Let x and $a \neq 1$ be positive real numbers. Then

$$\log_a x = \frac{\log x}{\log a} \quad \text{or} \quad \log_a x = \frac{\ln x}{\ln a}.$$

EXAMPLE 9 | **Change of base formula**

Approximate $\log_2 14$ to the nearest thousandth.

Solution

Using the change of base formula,

$$\log_2 14 = \frac{\log 14}{\log 2} \approx 3.807 \quad \text{or} \quad \log_2 14 = \frac{\ln 14}{\ln 2} \approx 3.807.$$

Figure 12.34 supports these results.

```
log(14)/log(2)
        3.807354922
ln(14)/ln(2)
        3.807354922
```

Figure 12.34

Now Try Exercise 89

12.4 Putting It All Together

The following table summarizes some important properties for base-a logarithms. Common and natural logarithms satisfy the same properties.

CONCEPTS	DESCRIPTION	EXAMPLES
Properties of Logarithms	The following properties hold for positive numbers m, n, and $a \neq 1$ and for any real number r.	
1. Product Rule	1. $\log_a mn = \log_a m + \log_a n$	1. $\log(10 \cdot 2) = \log 10 + \log 2$
2. Quotient Rule	2. $\log_a \frac{m}{n} = \log_a m - \log_a n$	2. $\log \frac{45}{6} = \log 45 - \log 6$
3. Power Rule	3. $\log_a (m^r) = r \log_a m$	3. $\ln x^6 = 6 \ln x$
Change of Base Formula	Let x and $a \neq 1$ be positive numbers. Then $$\log_a x = \frac{\log x}{\log a} \quad \text{and} \quad \log_a x = \frac{\ln x}{\ln a}.$$	The expression $\log_3 6$ is equivalent to either $$\frac{\log 6}{\log 3} \quad \text{or} \quad \frac{\ln 6}{\ln 3}.$$

12.4 Exercises

MyMathLab

NOTE: Assume that variables are positive and that expressions are defined in this exercise set.

CONCEPTS AND VOCABULARY

1. $\log 12 = \log 3 + \log (\underline{\quad})$

2. $\ln 5 = \ln 20 - \ln (\underline{\quad})$

3. $\log 8 = (\underline{\quad}) \log 2$

4. $\log mn = \underline{\quad}$

5. $\log \frac{m}{n} = \underline{\quad}$

6. $\log (m^r) = \underline{\quad}$

7. Does $\log x + \log y$ equal $\log(x + y)$?

8. Does $\log x - \log y$ equal $\log\left(\frac{x}{y}\right)$?

9. Does $\log(xy)$ equal $(\log x)(\log y)$?

10. Does $\log\left(\frac{x}{y}\right)$ equal $\frac{\log x}{\log y}$?

11. Give the change of base formula.

12. $\log_a 1 = \underline{\quad}$ and $\log_a a = \underline{\quad}$.

BASIC PROPERTIES OF LOGARITHMS

Exercises 13–18: Write the expression as a sum of two or more logarithms.

13. $\ln(15)$

14. $\log(77)$

15. $\log xy$

16. $\ln 10z$

17. $\log y^2$

18. $\log x^2 y$

Exercises 19–24: Write the expression as a difference of two logarithms.

19. $\log \frac{7}{3}$

20. $\ln \frac{11}{13}$

21. $\ln \frac{x}{y}$

22. $\log \frac{2x}{z}$

23. $\log_2 \frac{45}{x}$

24. $\log_7 \frac{5x}{4z}$

Exercises 25–32: Write the expression as one logarithm.

25. $\log 45 + \log 5$

26. $\log 30 - \log 10$

27. $\ln x + \ln y$

28. $\ln m + \ln n - \ln n$

29. $\ln 7x^2 + \ln 2x$

30. $\ln x + \ln y - \ln z$

31. $\ln x + \ln y^2 - \ln y$

32. $\ln \sqrt{z} - \ln z^3 + \ln y^3$

Exercises 33–38: Write the expression as one term. Evaluate if possible.

33. $\log 20 - \log 4$ **34.** $\log 900 - \log 9$

35. $\ln x^4 - \ln x^2$ **36.** $\ln 9x^2 - \ln 3x$

37. $\log 300x - \log 3x$ **38.** $\log 18x^2 - \log 2x^2$

Exercises 39–48: Rewrite using the power rule.

39. $\log 3^6$ **40.** $\log x^7$

41. $\ln 2^x$ **42.** $\ln (0.77)^{x+1}$

43. $\log_2 5^{1/4}$ **44.** $\log_3 \sqrt{x}$

45. $\log_4 \sqrt[3]{z}$ **46.** $\log_7 3^\pi$

47. $\log x^{y-1}$ **48.** $\ln a^{2b}$

Exercises 49–60: Use properties of logarithms to write the expression as the logarithm of a single expression.

49. $4 \log z - \log z^3$ **50.** $2 \log_5 y + \log_5 x$

51. $\log x + 2 \log x + 2 \log y$

52. $\log x^2 + 3 \log z - 5 \log y$

53. $\log x - 2 \log \sqrt{x}$ **54.** $\ln y^2 - 6 \ln \sqrt[3]{y}$

55. $\ln 2^{x+1} - \ln 2$ **56.** $\ln 8^{1/2} + \ln 2^{1/2}$

57. $\ln \sqrt[3]{x} + \ln \sqrt{x}$ **58.** $2 \log_3 \sqrt{x} - 3 \log_3 x$

59. $2 \log_a (x + 1) - \log_a (x^2 - 1)$

60. $\log_b (x^2 - 9) - \log_b (x - 3)$

Exercises 61–72: Use properties of logarithms to write the expression in terms of logarithms of x, y, and z.

61. $\log xy^2$ **62.** $\log \dfrac{x^2}{y^3}$

63. $\ln \dfrac{x^4 y}{z}$ **64.** $\ln \dfrac{\sqrt{x}}{y}$

65. $\log \dfrac{\sqrt[3]{z}}{\sqrt{y}}$ **66.** $\log \sqrt{\dfrac{x}{y}}$

67. $\log (x^4 y^3)$ **68.** $\log (x^2 y^4 z^3)$

69. $\ln \dfrac{1}{y} - \ln \dfrac{1}{x}$ **70.** $\ln \dfrac{1}{xy}$

71. $\log_4 \sqrt{\dfrac{x^3 y}{z^2}}$ **72.** $\log_3 \left(\dfrac{x^2 \sqrt{z}}{y^3} \right)$

Exercises 73–76: Graph f and g in the window $[-6, 6, 1]$ by $[-4, 4, 1]$. If the two graphs appear to be identical, prove that they are, using properties of logarithms.

73. $f(x) = \log x^3$, $g(x) = 3 \log x$

74. $f(x) = \ln x + \ln 3$, $g(x) = \ln 3x$

75. $f(x) = \ln (x + 5)$, $g(x) = \ln x + \ln 5$

76. $f(x) = \log (x - 2)$, $g(x) = \log x - \log 2$

Exercises 77–86: (Refer to Example 8.) Using only properties of logarithms and the approximations $\log 2 \approx 0.3$, $\log 5 \approx 0.7$, and $\log 13 \approx 1.1$, find an approximation for the expression.

77. $\log 16$ **78.** $\log 125$

79. $\log 65$ **80.** $\log 26$

81. $\log 130$ **82.** $\log 100$

83. $\log \dfrac{5}{2}$ **84.** $\log \dfrac{26}{5}$

85. $\log \dfrac{1}{13}$ **86.** $\log \dfrac{1}{65}$

CHANGE OF BASE FORMULA

Exercises 87–92: Use the change of base formula to approximate each expression to the nearest hundredth.

87. $\log_3 5$ **88.** $\log_5 12$

89. $\log_2 25$ **90.** $\log_7 8$

91. $\log_9 102$ **92.** $\log_6 293$

APPLICATIONS

93. *Modeling Sound* (See Example 5, Section 12.3.) The formula $f(x) = 10 \log (10^{16} x)$ can be used to calculate the decibel level of a sound with an intensity x. Use properties of logarithms to simplify this formula to $f(x) = 160 + 10 \log x$.

94. *Cellular Phone Technology* A formula used to calculate the strength of a signal for a cellular phone is

$$L = 110.7 - 19.1 \log h + 55 \log d,$$

where h is the height of the cellular phone tower and d is the distance the phone is from the tower. Use properties of logarithms to write an expression for L that contains only one logarithm. (*Source:* C. Smith, *Practical Cellular & PCS Design.*)

WRITING ABOUT MATHEMATICS

95. State the three basic properties of logarithms and give an example of each.

96. A student insists that $\log (x - y)$ is equal to the expression $\log x - \log y$. How could you convince the student otherwise?

SECTIONS
12.3 and 12.4
Checking Basic Concepts

1. Simplify each expression by hand.
   (a) $\log 10^4$ (b) $\ln e^x$
   (c) $\log_2 \frac{1}{8}$ (d) $\log_5 \sqrt{5}$

2. Sketch a graph of $f(x) = \log x$.
   (a) What are the domain and range of f?
   (b) Evaluate $f(1)$.
   (c) Can the common logarithm of a positive number be negative? Explain.
   (d) Can the common logarithm of a negative number be positive? Explain.

3. Write the expression in terms of logarithms of x, y, and z. Assume that variables are positive.
   (a) $\log xy$ (b) $\ln \dfrac{x}{yz}$
   (c) $\ln x^2$ (d) $\log \dfrac{x^2 y^3}{\sqrt{z}}$

4. Write as the logarithm of a single expression.
   (a) $\log x + \log y$
   (b) $\ln 2x - 3 \ln y$
   (c) $2 \log_2 x + 3 \log_2 y - \log_2 z$

12.5 Exponential and Logarithmic Equations

Exponential Equations and Models • Logarithmic Equations and Models

A LOOK INTO MATH ▶

Wind power capacity in the world grew 31% in 2009. With the problems concerning nuclear power, it is expected that wind power will continue to grow at exponential rates. For example, from 2006 to 2010, China's wind power capacity almost doubled each year. In fact, in 2010 China's capacity to produce wind power exceeded the capacity of the United States. In order to model and answer questions about this growth, we will use exponential and logarithmic equations. See Exercise 93. (*Sources: Business and Economy, Clean Energy, Wind Energy, 2010.*)

Exponential Equations and Models

To solve the equation $10 + x = 100$, we subtract 10 from each side because addition and subtraction are inverse operations.

$$10 + x - 10 = 100 - 10 \qquad \text{Subtract 10 from each side.}$$
$$x = 90$$

To solve the equation $10x = 100$, we divide each side by 10 because multiplication and division are inverse operations.

$$\frac{10x}{10} = \frac{100}{10} \qquad \text{Divide each side by 10.}$$
$$x = 10$$

Now suppose that we want to solve the *exponential equation*

The exponent is a variable

$$10^x = 100. \qquad \text{Exponential equation}$$

What is new about this type of equation is that the variable x is an *exponent*. The inverse operation of 10^x is $\log x$. Rather than subtracting 10 from each side or dividing each side by 10, we take the base-10 logarithm of each side. Doing so results in

$$\log 10^x = \log 100. \quad \text{Take base-10 logarithm of each side.}$$

Because $\log 10^x = x$ for all real numbers x, the equation becomes

$$x = \log 100 \quad \text{or, equivalently,} \quad x = 2.$$

These concepts are applied in the next example.

EXAMPLE 1 **Solving exponential equations**

Solve and approximate to the nearest hundredth.
(a) $10^x = 150$ **(b)** $e^x = 40$ **(c)** $2^x = 50$ **(d)** $0.9^x = 0.5$

Solution
(a)
$$10^x = 150 \qquad \text{Given equation}$$
$$\log 10^x = \log 150 \qquad \text{Take the common logarithm of each side.}$$
$$x = \log 150 \approx 2.18 \qquad \text{Inverse property: } \log 10^k = k \text{ for all } k$$

(b) The inverse operation of e^x is $\ln x$, so we take the natural logarithm of each side.
$$e^x = 40 \qquad \text{Given equation}$$
$$\ln e^x = \ln 40 \qquad \text{Take the natural logarithm of each side.}$$
$$x = \ln 40 \approx 3.69 \qquad \text{Inverse property: } \ln e^k = k \text{ for all } k$$

(c) The inverse operation of 2^x is $\log_2 x$. Calculators do not usually have a base-2 logarithm key, so we take the common logarithm of each side and then apply the power rule.
$$2^x = 50 \qquad \text{Given equation}$$
$$\log 2^x = \log 50 \qquad \text{Take the common logarithm of each side.}$$
$$x \log 2 = \log 50 \qquad \text{Power rule: } \log(m^r) = r \log m$$
$$x = \frac{\log 50}{\log 2} \approx 5.64 \qquad \text{Divide by } \log 2 \text{ and approximate.}$$

(d) This time we begin by taking the natural logarithm of each side.
$$0.9^x = 0.5 \qquad \text{Given equation}$$
$$\ln 0.9^x = \ln 0.5 \qquad \text{Take the natural logarithm of each side.}$$
$$x \ln 0.9 = \ln 0.5 \qquad \text{Power rule: } \ln(m^r) = r \ln m$$
$$x = \frac{\ln 0.5}{\ln 0.9} \approx 6.58 \qquad \text{Divide by } \ln 0.9 \text{ and approximate.}$$

Now Try Exercises 13, 15, 25, 27

MAKING CONNECTIONS

Logarithms of Quotients and Quotients of Logarithms
The solution in Example 1(c) is $\frac{\log 50}{\log 2}$. Note that
$$\frac{\log 50}{\log 2} \neq \log 50 - \log 2.$$

However, $\log 50 - \log 2 = \log \frac{50}{2} = \log 25$ by the quotient rule for logarithms, as shown in the figure.

```
log(50)/log(2)
          5.64385619
log(50)−log(2)
          1.397940009
log(25)
          1.397940009
```

The next two examples illustrate methods for solving exponential equations.

EXAMPLE 2 | **Solving exponential equations**

Solve each equation and approximate to the nearest hundredth.

(a) $2e^x - 1 = 5$ (b) $3^{x-5} = 15$ (c) $e^{2x} = e^{x+5}$ (d) $3^{2x} = 2^{x+3}$

Solution

(a) Begin by solving for e^x.

$2e^x - 1 = 5$	Given equation
$2e^x = 6$	Add 1 to each side.
$e^x = 3$	Divide each side by 2.
$\ln e^x = \ln 3$	Take the natural logarithm.
$x = \ln 3 \approx 1.10$	Inverse property: $\ln e^k = k$

(b) Start by taking the common logarithm of each side. (We could also take the natural logarithm of each side.)

$3^{x-5} = 15$	Given equation
$\log 3^{x-5} = \log 15$	Take the common logarithm of each side.
$(x - 5) \log 3 = \log 15$	Power rule for logarithms
$x - 5 = \dfrac{\log 15}{\log 3}$	Divide by $\log 3$.
$x = \dfrac{\log 15}{\log 3} + 5 \approx 7.46$	Add 5 to each side and approximate.

(c) For $e^{2x} = e^{x+5}$, the bases are equal, so the exponents must also be equal. To verify this assertion, take the natural logarithm of each side.

$e^{2x} = e^{x+5}$	Given equation
$\ln e^{2x} = \ln e^{x+5}$	Take the natural logarithm.
$2x = x + 5$	Inverse property: $\ln e^k = k$
$x = 5$	Subtract x.

(d) For $3^{2x} = 2^{x+3}$, the bases are not equal. However, we can still solve the equation by taking the common logarithm of each side. A logarithm of any base could be used.

$3^{2x} = 2^{x+3}$	Given equation
$\log 3^{2x} = \log 2^{x+3}$	Take the common logarithm.
$2x \log 3 = (x + 3) \log 2$	Power rule for logarithms
$2x \log 3 = x \log 2 + 3 \log 2$	Distributive property
$2x \log 3 - x \log 2 = 3 \log 2$	Subtract $x \log 2$.
$x(2 \log 3 - \log 2) = 3 \log 2$	Factor out x.
$x = \dfrac{3 \log 2}{2 \log 3 - \log 2}$	Divide by $2 \log 3 - \log 2$.
$x \approx 1.38$	Approximate.

Now Try Exercises 29, 31, 35, 43

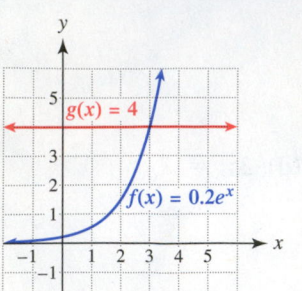

Figure 12.35

EXAMPLE 3 **Solving an exponential equation**

Graphs for $f(x) = 0.2e^x$ and $g(x) = 4$ are shown in Figure 12.35.
(a) Use the graphs to estimate the solution to the equation $f(x) = g(x)$.
(b) Check your estimate by solving the equation symbolically.

Solution
(a) The graphs intersect near the point $(3, 4)$. Therefore the solution is given by $x \approx 3$.
(b) We must solve the equation $0.2e^x = 4$.

$0.2e^x = 4$	Given equation
$e^x = 20$	Divide each side by 0.2.
$\ln e^x = \ln 20$	Take the natural logarithm of each side.
$x = \ln 20$	Inverse property: $\ln e^k = k$
$x \approx 2.996$	Approximate.

NOTE: The graphical estimate did not give the *exact* solution of $\ln 20$.

▌ **Now Try Exercise 45**

▶ **REAL-WORLD CONNECTION** As discussed in Section 12.2, if \$1000 is deposited in a savings account paying 10% annual interest at the end of each year, the amount A in the account after x years is given by

$$A(x) = 1000(1.1)^x.$$

After 10 years there will be

$$A(\mathbf{10}) = 1000(1.1)^{\mathbf{10}} \approx \$2593.74$$

in the account. To calculate how long it will take for \$4000 to accrue in the account, we need to solve the exponential equation

$$1000(1.1)^x = 4000.$$

We do so in the next example.

EXAMPLE 4 **Solving an exponential equation**

Solve $1000(1.1)^x = 4000$ symbolically. Give graphical support for your answer.

Solution
Symbolic Solution Begin by dividing each side of the equation by 1000.

$1000(1.1)^x = 4000$	Given equation
$1.1^x = 4$	Divide by 1000.
$\log 1.1^x = \log 4$	Take the common logarithm of each side.
$x \log 1.1 = \log 4$	Power rule for logarithms
$x = \dfrac{\log 4}{\log 1.1} \approx 14.5$	Divide by log 1.1 and approximate.

Interest is paid at the end of the year, so it will take 15 years for \$1000 earning 10% annual interest to grow to (at least) \$4000.

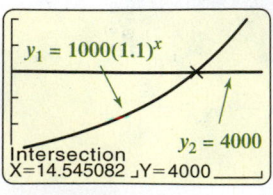

Figure 12.36

Graphical Solution Graphical support is shown in Figure 12.36, where the graphs of $y_1 = 1000(1.1)^x$ and $y_2 = 4000$ intersect when $x \approx 14.5$.

▌ **Now Try Exercise 97**

▶ **REAL-WORLD CONNECTION** In the next example, we model the life span of a robin with an exponential function.

EXAMPLE 5

Modeling the life span of a robin

The life spans of 129 robins were monitored over a 4-year period in one study. The formula $f(x) = 10^{-0.42x}$ can be used to calculate the percentage of robins remaining after x years. For example, $f(1) \approx 0.38$ means that after 1 year 38% of the robins were still alive. (*Source:* D. Lack, *The Life Span of a Robin*.)

(a) Evaluate $f(2)$ and interpret the result.
(b) Determine when 5% of the robins remained.

Solution

(a) $f(2) = 10^{-0.42(2)} \approx 0.145$. After **2** years about **14.5%** of the robins were still alive.
(b) Use 5% = 0.05 and solve the following equation.

$$10^{-0.42x} = 0.05 \qquad \text{Equation to solve}$$

$$\log 10^{-0.42x} = \log 0.05 \qquad \text{Take the common logarithm of each side.}$$

$$-0.42x = \log 0.05 \qquad \text{Inverse property: } \log 10^k = k$$

$$x = \frac{\log 0.05}{-0.42} \approx 3.1 \qquad \text{Divide by } -0.42.$$

After about 3 years only 5% of the robins were still alive.

Now Try Exercise 103

Logarithmic Equations and Models

To solve an exponential equation we use logarithms. To solve a logarithmic equation we *exponentiate* each side of the equation. To do so we use the fact that if $x = y$, then $a^x = a^y$ for any positive base a. For example, to solve

$$\log x = 3 \qquad \text{Logarithmic equation}$$

we exponentiate each side of the equation, using base 10.

$$10^{\log x} = 10^3 \qquad \text{If } a = b, \text{ then } 10^a = 10^b.$$

Because $10^{\log x} = x$ for all positive x, it follows that

$$x = 10^3. \qquad \text{Inverse property}$$

To solve logarithmic equations, we frequently use the inverse property

$$a^{\log_a x} = x.$$

Bases must be equal.

Examples of this inverse property include

$$e^{\ln 2k} = 2k, \qquad 2^{\log_2 x} = x, \qquad \text{and} \qquad 10^{\log (x+5)} = x + 5.$$

Both are base-e.   Both are base-2.   Both are base-10.

EXAMPLE 6

Solving logarithmic equations

Solve and approximate solutions to the nearest hundredth when appropriate.

(a) $2 \log x = 4$ (b) $\ln 3x = 5.5$ (c) $\log_2 (x + 4) = 7$

Solution

(a)

$2 \log x = 4$	Given equation
$\log x = 2$	Divide each side by 2.
$10^{\log x} = 10^2$	Exponentiate each side, using base 10.
$x = 100$	Inverse property: $10^{\log k} = k$

(b)

$\ln 3x = 5.5$	Given equation
$e^{\ln 3x} = e^{5.5}$	Exponentiate each side, using base e.
$3x = e^{5.5}$	Inverse property: $e^{\ln k} = k$
$x = \dfrac{e^{5.5}}{3} \approx 81.56$	Divide each side by 3 and approximate.

(c)

$\log_2 (x + 4) = 7$	Given equation
$2^{\log_2 (x+4)} = 2^7$	Exponentiate each side, using base 2.
$x + 4 = 2^7$	Inverse property: $2^{\log_2 k} = k$
$x = 2^7 - 4$	Subtract 4 from each side.
$x = 124$	Simplify.

Now Try Exercises 61, 63, 69

Because the domain of any logarithmic function includes only positive numbers, it is important to check answers, as emphasized in the next example.

EXAMPLE 7 **Solving a logarithmic equation**

Solve $\log (x + 2) + \log (x - 2) = \log 5$. Check any answers.

Solution
Start by applying the product rule for logarithms.

$\log (x + 2) + \log (x - 2) = \log 5$	Given equation
$\log ((x + 2)(x - 2)) = \log 5$	Product rule
$\log (x^2 - 4) = \log 5$	Multiply.
$10^{\log (x^2 - 4)} = 10^{\log 5}$	Exponentiate using base 10.
$x^2 - 4 = 5$	Inverse properties
$x^2 = 9$	Add 4.
$x = \pm 3$	Square root property

Check each answer.

$$\log (\mathbf{3} + 2) + \log (\mathbf{3} - 2) \stackrel{?}{=} \log 5 \qquad \log (\mathbf{-3} + 2) + \log (\mathbf{-3} - 2) \stackrel{?}{=} \log 5$$

$$\log 5 + \log 1 \stackrel{?}{=} \log 5 \qquad\qquad \log (-1) + \log (-5) \neq \log 5 \ \textcolor{red}{\textbf{✗}}$$

$$\log 5 + 0 \stackrel{?}{=} \log 5 \qquad\qquad\qquad\qquad \underleftarrow{\hspace{1cm}} \text{Undefined} \underrightarrow{\hspace{1cm}}$$

$$\log 5 = \log 5 \ \textcolor{blue}{✓}$$

Although 3 is a solution, -3 is not, because both $\log (-1)$ and $\log (-5)$ are undefined expressions. Be sure to check your answers.

Now Try Exercise 77

EXAMPLE 8   Modeling bird populations

Near New Guinea there is a relationship between the number of different species of birds and the size of an island. Larger islands tend to have a greater variety of birds. Table 12.13 lists the number of species of birds y found on islands with an area of x square kilometers.

TABLE 12.13   Bird Species on Islands

x (km^2)	0.1	1	10	100	1000
y (species)	10	15	20	25	30

Source: B. Freedman, *Environmental Ecology.*

(a) Find values for the constants a and b so that $y = a + b \log x$ models the data.
(b) Predict the number of bird species on an island of 4000 square kilometers.

Solution
(a) Because **log 1 = 0**, substitute $x = 1$ and $y = $ **15** in the equation to find a.

$$15 = a + b \, \mathbf{log \, 1} \qquad y = a + b \log x$$
$$15 = a + b \cdot \mathbf{0} \qquad \log 1 = 0$$
$$15 = a \qquad \text{Simplify.}$$

Thus $y = 15 + b \log x$. To find b substitute $x = 10$ and $y = $ **20**.

$$\mathbf{20} = 15 + b \, \mathbf{log \, 10} \qquad y = 15 + b \log x$$
$$20 = 15 + b \cdot \mathbf{1} \qquad \log 10 = 1$$
$$5 = b \qquad \text{Simplify.}$$

The data in Table 12.13 are modeled by $y = \mathbf{15 + 5 \log x}$. This result is supported by Figure 12.37.

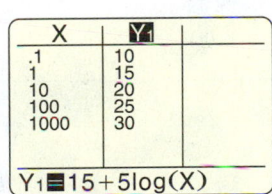

X	Y1
.1	10
1	15
10	20
100	25
1000	30

$Y_1 \blacksquare 15 + 5\log(X)$

Figure 12.37

(b) To predict the number of species on an island of 4000 square kilometers, let $x = $ **4000** and find y.

$$y = 15 + 5 \log \mathbf{4000} \approx 33$$

The model estimates about 33 different species of birds on this island.

Now Try Exercise 105

EXAMPLE 9   Modeling runway length

For some types of airplanes with weight x, the minimum runway length L required at take-off is modeled by

$$L(x) = 3 \log x.$$

In this equation L is measured in thousands of feet and x is measured in thousands of pounds. Estimate the weight of the heaviest airplane that can take off from a runway 5100 feet long. (*Source:* L. Haefner, *Introduction to Transportation Systems.*)

CRITICAL THINKING

In Example 10, Section 12.3, we used the formula $L(x) = 1.3 \ln x$ to model runway length. Are $L(x) = 1.3 \ln x$ and $L(x) = 3 \log x$ equivalent formulas? Explain.

Solution

Runway length is measured in thousands of feet, so we must solve the equation $L(x) = 5.1$.

$3 \log x = 5.1$	$L(x) = 5.1$
$\log x = 1.7$	Divide each side by 3.
$10^{\log x} = 10^{1.7}$	Exponentiate each side, using base 10.
$x = 10^{1.7}$	Inverse property: $10^{\log k} = k$
$x \approx 50.1$	Approximate.

The largest airplane that can take off from this runway weighs about 50,000 pounds.

▮ **Now Try Exercise 101**

12.5   Putting It All Together

TYPE OF EQUATION	PROCEDURE	EXAMPLE
Exponential	Begin by solving for the exponential expression a^x. Then take a logarithm of each side.	$4e^x + 1 = 9$ — Given equation $e^x = 2$ — Solve for e^x. $\ln e^x = \ln 2$ — Take the natural logarithm. $x = \ln 2$ — Inverse property: $\ln e^k = k$
Logarithmic	Begin by solving for the logarithm in the equation. Then exponentiate each side of the equation, using the same base as the logarithm.	$\frac{1}{3} \log 2x = 1$ — Given equation $\log 2x = 3$ — Multiply by 3. $10^{\log 2x} = 10^3$ — Exponentiate using base 10. $2x = 1000$ — Inverse property: $10^{\log k} = k$ $x = 500$ — Divide by 2.

12.5   Exercises

CONCEPTS AND VOCABULARY

1. To solve $x - 5 = 50$, what should be done?

2. To solve $5x = 50$, what should be done?

3. To solve $10^x = 50$, what should be done?

4. To solve $\log x = 5$, what should be done?

5. $\log 10^x = $ _____

6. $10^{\log x} = $ _____

7. $\ln e^{2x} = $ _____

8. $e^{\ln (x+7)} = $ _____

9. Does $\frac{\log 5}{\log 4}$ equal $\log \frac{5}{4}$? Explain.

10. Does $\frac{\log 5}{\log 4}$ equal $\log 5 - \log 4$? Explain.

11. How many solutions are there to $\log x = k$, where k is any real number?

12. How many solutions are there to $10^x = k$, where k is a positive number?

EXPONENTIAL EQUATIONS

Exercises 13–44: Solve the given exponential equation. Approximate answers to the nearest hundredth when appropriate.

13. $10^x = 1000$

14. $10^x = 0.01$

15. $2^x = 64$

16. $3^x = 27$

17. $2^{x-3} = 8$

18. $3^{2x} = 81$

19. $4^x + 3 = 259$

20. $3(5^{2x}) = 300$

21. $10^{0.4x} = 124$

22. $0.75^x = 0.25$

23. $e^{-x} = 1$

24. $0.5^{-5x} = 5$

25. $e^x = 25$

26. $e^x = 0.4$

27. $0.4^x = 2$

28. $0.7^x = 0.3$

29. $e^x - 1 = 6$

30. $2e^{4x} = 15$

31. $2(10)^{x+2} = 35$

32. $10^{3x} + 10 = 1500$

33. $3.1^{2x} - 4 = 16$

34. $5.4^{x-1} = 85$

35. $e^{3x} = e^{2x-1}$

36. $e^{x^2} = e^{3x-2}$

37. $5^{4x} = 5^{x^2-5}$

38. $2^{4x} = 2^{x+3}$

39. $e^{2x} \cdot e^x = 10$

40. $10^{x-2} \cdot 10^x = 1000$

41. $e^x = 2^{x+2}$

42. $2^{2x} = 3^{x-1}$

43. $4^{0.5x} = 5^{x+2}$

44. $3^{2x} = 7^{x+1}$

Exercises 45–48: (Refer to Example 3.) The symbolic and graphical representations of f and g are given.

(a) *Use the graph to solve* $f(x) = g(x)$.

(b) *Solve* $f(x) = g(x)$ *symbolically.*

45. $f(x) = 0.2(10^x)$,
   $g(x) = 2$

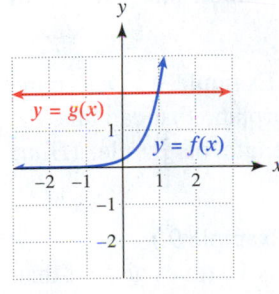

46. $f(x) = e^x$,
   $g(x) = 7.4$

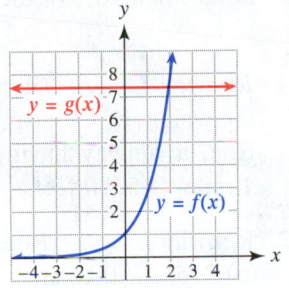

47. $f(x) = 2^{-x}$,
   $g(x) = 4$

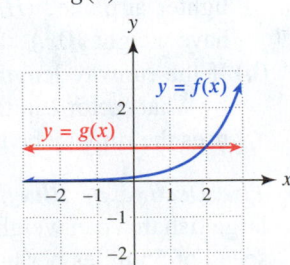

48. $f(x) = 0.1(3^x)$,
   $g(x) = 0.9$

Exercises 49–56: Solve the equation symbolically. Give graphical or numerical support. Approximate answers to the nearest hundredth when appropriate.

49. $10^x = 0.1$

50. $2(10^x) = 2000$

51. $4e^x + 5 = 9$

52. $e^x + 6 = 36$

53. $4^x = 1024$

54. $3^x = 729$

55. $(0.55)^x + 0.55 = 2$

56. $5(0.9)^x = 3$

Exercises 57–60: The given equation cannot be solved symbolically. Find any solutions graphically to the nearest hundredth.

57. $e^x - x = 2$

58. $x \log x = 1$

59. $\ln x = e^{-x}$

60. $10^x - 2 = \log(x + 2)$

LOGARITHMIC EQUATIONS

Exercises 61–82: Solve the given equation. Approximate answers to the nearest hundredth when appropriate.

61. $\log x = 2$

62. $\log x = 0.01$

63. $\ln x = 5$

64. $2 \ln x = 4$

65. $\log 2x = 7$

66. $6 \ln 4x = 12$

67. $\log_2 x = 4$

68. $\log_2 x = 32$

69. $\log_2 5x = 2.3$

70. $2 \log_3 4x = 10$

71. $2 \log x + 5 = 7.8$

72. $\ln(x - 1) = 3.3$

73. $5 \ln(2x + 1) = 55$

74. $5 - \log(x + 3) = 2.6$

75. $\log x^2 = \log x$

76. $\ln x^2 = \ln(3x - 2)$

77. $\ln x + \ln(x + 1) = \ln 30$

78. $\log(x - 1) + \log(2x + 1) = \log 14$

79. $\log_3 3x - \log_3(x + 2) = \log_3 2$

80. $\log_4(x^2 - 1) - \log_4(x - 1) = \log_4 6$

81. $\log_2(x - 1) + \log_2(x + 1) = 3$

82. $\log_4(x^2 + 2x + 1) - \log_4(x + 1) = 2$

Exercises 83–88: Solve the equation symbolically. Give graphical or numerical support. Approximate answers to the nearest hundredth.

83. $\log x = 1.6$

84. $\ln x = 2$

85. $\ln(x + 1) = 1$

86. $2\log(2x + 3) = 8$

87. $17 - 6\log_3 x = 5$

88. $4\log_2 x + 7 = 12$

Exercises 89–92: Two functions, f and g, are given.

(a) *Use the graph to solve $f(x) = g(x)$.*

(b) *Solve $f(x) = g(x)$ symbolically.*

89. $f(x) = \ln x$,
$g(x) = 0.7$

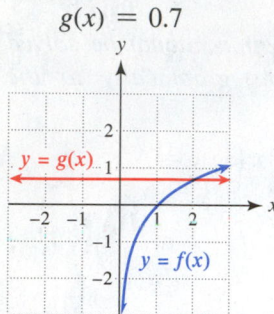

90. $f(x) = \log_2 x$,
$g(x) = 1.6$

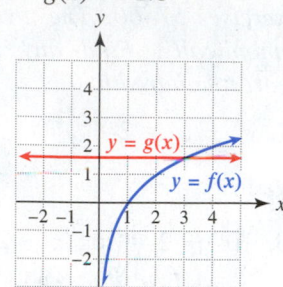

91. $f(x) = 5\log 2x$,
$g(x) = 3$

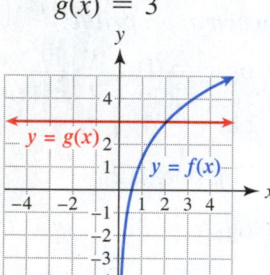

92. $f(x) = 2\ln(x) - 3$,
$g(x) = 0.9$

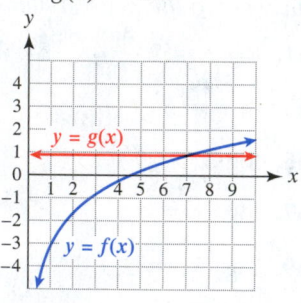

APPLICATIONS

93. *China's Wind Power*   China overtook the United States in 2010 for generating the most wind power. China's wind power capacity W in gigawatts x years after 2005 can be modeled by $W(x) = 1.73(10^{0.276x})$.

(a) Evaluate $W(5)$ and interpret the result.

(b) Determine when China generated 22 gigawatts.

94. *Risk of Down Syndrome*   The likelihood of a live birth having Down syndrome varies with the age of the mother. The function

$$D(x) = 0.0000816(10^{0.102x})$$

gives the percentage of live births that have Down syndrome when the mother is x years old.

(a) Evaluate $D(35)$ and interpret the result.

(b) Determine the mother's age when this percentage reaches 4%.

95. *Bacteria Growth*   A sample of 5 million bacteria increases by 15% each hour. To the nearest hour, how long does it take for the sample to reach 20 million?

96. *Bacteria Growth*   A sample of 3 million bacteria increases by 125% each day. To the nearest day, how long does it take for the bacteria sample to reach 173 million?

97. *Growth of a Mutual Fund*   (Refer to Example 4.) An investor deposits $2000 in a mutual fund that returns 15% at the end of 1 year. Determine the length of time required for the investment to triple its value if the annual rate of return remains the same.

98. *Savings Account*   (Refer to Example 4.) If a savings account pays 6% annual interest at the end of each year, how many years will it take for the account to double in value?

99. *Liver Transplants*   In the United States the gap between available organs for liver transplants and people who need them has widened. The number of individuals waiting for liver transplants can be modeled by

$$f(x) = 2339(1.24)^{x-1988},$$

where x is the year. (*Source:* United Network for Organ Sharing.)

(a) Evaluate $f(1994)$ and interpret the result.

(b) Determine when the number of individuals waiting for liver transplants was 20,000.

100. *Life Span of a Robin*   (Refer to Example 5.) Determine when 50% of the robins in the study were still alive.

101. *Runway Length*   (Refer to Example 9.) Determine the weight of the heaviest airplane that can take off from a runway having a length of $\frac{3}{4}$ mile. (*Hint:* 1 mile = 5280 feet.)

102. *Runway Length*   (Refer to Example 9.)

(a) Suppose that an airplane is 10 times heavier than a second airplane. How much longer should the runway be for the heavier airplane than for the lighter airplane? (*Hint:* Let the heavier airplane have weight $10x$.)

(b) If the runway length is increased by 3000 feet, by what factor can the weight of an airplane that uses the runway be increased?

103. *The Decline of Bluefin Tuna*   Bluefin tuna are large fish that can weigh 1500 pounds and swim at a speed of 55 miles per hour. They are used for sushi, and a prime fish can be worth more than $50,000. As a result, the number of western Atlantic bluefin tuna has declined dramatically. Their numbers in thousands between 1974 and 1991 can be modeled by $f(x) = 230(10^{-0.055x})$, where x is the number of years after 1974. See the accompanying graph.

(*Source*: B. Freedman.)

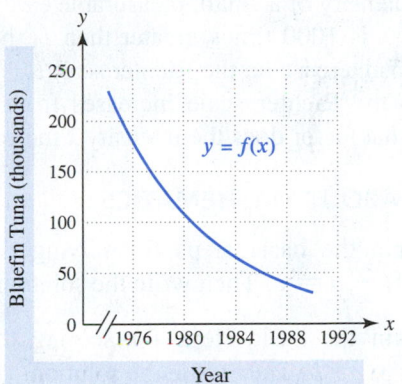

(a) Evaluate $f(1)$ and interpret the result.
(b) Use the graph to estimate the year in which bluefin tuna numbered 115 thousand.
(c) Solve part (b) symbolically.

104. *Insect Populations* (Refer to Example 8.) The table lists numbers of species of insects y found on islands having areas of x square miles.

x (square miles)	1	2	4
y (species)	1000	1500	2000

x (square miles)	8	16	32
y (species)	2500	3000	3500

(a) Find values for the constants a and b so that $y = a + b \log_2 x$ models the data.
(b) Construct a table for y and verify that your equation models the data.
(c) Estimate the number of species of insects on an island having an area of 12 square miles.

Exercises 105 and 106: (Refer to Example 8.) Find values for a and b so that $y = a + b \log x$ models the data.

105.

x	0.1	1	10	100
y	22	25	28	31

106.

x	0.01	1	100	1000
y	-10	-2	6	10

107. *Calories Consumed and Land Ownership* In developing countries there is a relationship between the amount of land a person owns and the average number of calories that person consumes daily. This relationship is modeled by $f(x) = 645 \log(x + 1) + 1925$,

where x is the amount of land owned in acres and $0 \le x \le 4$. (*Source*: D. Grigg, *The World Food Problem*.)

(a) Estimate graphically the number of acres owned by a person consuming 2200 calories per day.
(b) Solve part (a) symbolically.

108. *Population of Industrialized Urban Regions* The number of people living in industrialized urban regions throughout the world has not grown exponentially. Instead it has grown logarithmically and is modeled by

$$f(x) = 0.36 + 0.15 \ln(x - 1949).$$

In this formula the output is billions of people and the input x is the year, where $1950 \le x \le 2030$.

(*Source*: D. Meadows, *Beyond The Limits*.)

(a) Determine graphically when this population may reach 1 billion.
(b) Solve part (a) symbolically.

109. *Fertilizer Use* Between 1950 and 1980, the use of chemical fertilizers increased worldwide. The table lists worldwide average use y in kilograms per acre of cropland during year x. (*Source*: D. Grigg, *The World Food Problem*.)

x	1950	1963	1972	1979
y	5.0	11.3	22.0	31.2

(a) Are the data linear or nonlinear? Explain.
(b) The equation $y = 5(1.06)^{(x-1950)}$ may be used to model the data. The growth factor is 1.06. What does this growth factor indicate about fertilizer use during this time period?
(c) Estimate the year when fertilizer use was 15 kilograms per acre of cropland.

110. *Greenhouse Gases* If current trends continue, concentrations of atmospheric carbon dioxide (CO_2) in parts per million (ppm) are expected to increase. This increase in concentration of CO_2 has been accelerated by burning fossil fuels and deforestation. The exponential equation $y = 364(1.005)^x$ may be used to model CO_2 in parts per million x years after 2000. Estimate the year when the CO_2 concentration could be double the preindustrial level of 280 parts per million. (*Source*: R. Turner, *Environmental Economics*.)

111. *Modeling Sound* The formula

$$f(x) = 160 + 10 \log x$$

is used to calculate the decibel level of a sound with intensity x measured in watts per square centimeter. The noise level at a basketball game can reach 100 decibels. Find the intensity x of this sound.

112. *Loudness of a Sound* (Refer to Exercise 111.)
   (a) Show that, if the intensity of a sound increases by a factor of 10 from x to $10x$, the decibel level increases by 10 decibels. *Hint:* Show that

$$160 + 10 \log 10x = 170 + 10 \log x.$$

   (b) Find the increase in decibels if the intensity x increases by a factor of 1000.
   (c) Find the increase in the intensity x if the decibel level increases by 20.

113. *Hurricanes* (Refer to Exercise 106, Section 12.3.) The barometric air pressure in inches of mercury at a distance of x miles from the eye of a severe hurricane is given by $f(x) = 0.48 \ln(x + 1) + 27$. How far from the eye is the pressure 28 inches of mercury?
   (*Source:* A. Miller and R. Anthes, *Meteorology.*)

114. *Earthquakes* The Richter scale is used to determine the intensity of earthquakes, which corresponds to the amount of energy released. If an earthquake has an intensity of x, its magnitude, as computed by the Richter scale, is given by $R(x) = \log \frac{x}{I_0}$, where I_0 is the intensity of a small, measurable earthquake.
   (a) If x is 1000 times greater than I_0, how large is this increase on the Richter scale?
   (b) If the Richter scale increases from 5 to 8, by what factor does the intensity x increase?

WRITING ABOUT MATHEMATICS

115. Explain the basic steps for solving the equation $a(10^x) - b = c$. Then write the solution.

116. Explain the basic steps for solving the equation $a \log 3x = b$. Then write the solution.

SECTION 12.5

Checking Basic Concepts

1. Solve the equation. Approximate answers to the nearest hundredth when appropriate.
   (a) $2(10^x) = 40$ **(b)** $2^{3x} + 3 = 150$
   (c) $\ln x = 4.1$ **(d)** $4 \log 2x = 12$

2. Solve $\log(x + 4) + \log(x - 4) = \log 48$. Check the answers.

3. If \$500 is deposited in a savings account that pays 3% annual interest at the end of each year, the amount of money A in the account after x years is given by $A = 500(1.03)^x$. Estimate the number of years required for this amount to reach \$900.

CHAPTER 12   Summary

SECTION 12.1 ■ COMPOSITE AND INVERSE FUNCTIONS

Composition of Functions If f and g are functions, then the composite function $g \circ f$, or composition of g and f, is defined by $(g \circ f)(x) = g(f(x))$.

Example: If $f(x) = x - 5$ and $g(x) = 2x^2 + 4x - 6$, then $(g \circ f)(x)$ is

$$g(f(x)) = g(x - 5)$$
$$= 2(x - 5)^2 + 4(x - 5) - 6.$$

One-to-One Function A function f is one-to-one if, for any c and d in the domain of f,

$$c \neq d \quad \text{implies that} \quad f(c) \neq f(d).$$

That is, different inputs always result in different outputs.

Example: $f(x) = x^2 + 4$ is *not* one-to-one because $f(-3) = f(3) = 13$.   Different inputs; same output

Horizontal Line Test If every horizontal line intersects the graph of a function f at most once, then f is a one-to-one function.

Examples:

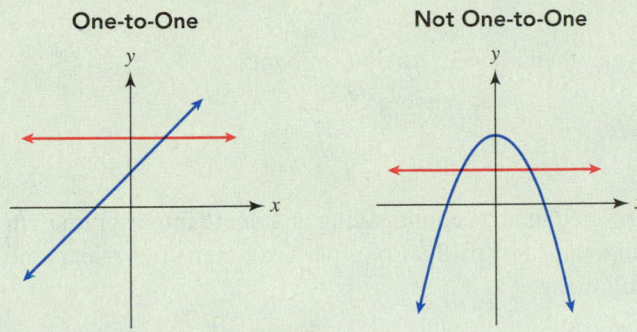

Inverse Functions If f is one-to-one, then f has an inverse function, denoted f^{-1}, that satisfies $(f^{-1} \circ f)(x) = x$ and $(f \circ f^{-1})(x) = x$.

Example: $f(x) = 7x$ and $f^{-1}(x) = \frac{x}{7}$ are inverse functions.

SECTION 12.2 ■ EXPONENTIAL FUNCTIONS

Exponential Function An exponential function is defined by $f(x) = Ca^x$, where $a > 0$, $C > 0$, and $a \neq 1$. Its domain (set of valid inputs) is all real numbers and its range (outputs) is all positive real numbers. The base is a.

Example: $f(x) = e^x$ is the natural exponential function and $e \approx 2.71828$.

Exponential Growth and Decay When $a > 1$, the graph of $f(x) = Ca^x$ models exponential growth, and when $0 < a < 1$, it models exponential decay. The base a represents either the growth factor or the decay factor. The constant C equals the y-intercept.

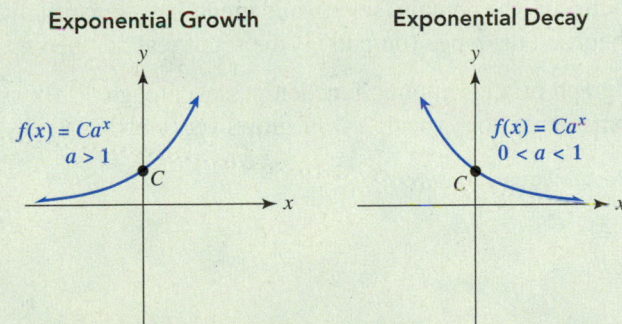

Example: $f(x) = 1.5(2)^x$ is an exponential function with $a = 2$ and $C = 1.5$. It models exponential growth because $a > 1$. The growth factor is 2 because for each unit increase in x, the output from $f(x)$ increases by a *factor* of 2.

Percent Change If an amount A changes to a new amount B, then the percent change is given by

$$\frac{B - A}{A} \times 100.$$

Example: If \$500 increases to \$700, the percent change is

$$\frac{700 - 500}{500} \times 100 = 40\%.$$

Percent Change and Growth Factor If an initial amount C experiences a constant percent change r (in decimal form) for each x-unit increase in time, then the new amount A is

$$A = C(1 + r)^x,$$

where the growth factor is $a = 1 + r$.

Example: If \$800 increases by 4% each year, then after 5 years the amount is

$$A = 800(1 + 0.04)^5 \approx \$973.32.$$

The growth factor is $a = 1.04$.

Compound Interest If P dollars are deposited in an account paying an *annual* interest rate r (in decimal form) and this interest rate is compounded or paid n times per year, then after t years the account contains A dollars given by the following.

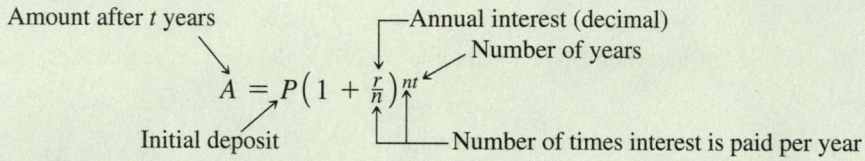

Example: If \$1200 is deposited at 6% annual interest, compounded quarterly, then after 8 years the account contains

$$A = 1200\left(1 + \frac{0.06}{4}\right)^{(4 \cdot 8)} \approx \$1932.39.$$

SECTION 12.3 ■ LOGARITHMIC FUNCTIONS

Base-a Logarithms The logarithm with base a of a positive number x is denoted $\log_a x$. If $\log_a x = b$, then $x = a^b$. That is, $\log_a x$ represents the exponent on base a that results in x.

Examples: $\log_2 16 = 4$ because $16 = 2^4$ and $\log 100 = 2$ because $100 = 10^2$.

Domain and Range of Logarithmic Functions The domain (set of valid inputs) of a logarithmic function is the set of all positive real numbers and the range (outputs) is the set of real numbers.

Graph of a Logarithmic Function The graph of a logarithmic function passes through $(1, 0)$, as illustrated in the following graph. As x becomes large, $\log_a x$ with $a > 1$ grows very slowly.

Base-a Logarithm

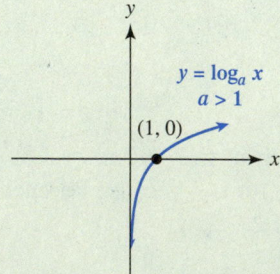

SECTION 12.4 ■ PROPERTIES OF LOGARITHMS

Basic Properties Logarithms have several important properties. For positive numbers m, n, and $a \neq 1$ and any real number r,

1. $\log_a mn = \log_a m + \log_a n$.
2. $\log_a \frac{m}{n} = \log_a m - \log_a n$.
3. $\log_a (m^r) = r \log_a m$.

Examples: **1.** $\log 5 + \log 20 = \log (5 \cdot 20) = \log 100 = 2$

2. $\log 100 - \log 5 = \log \frac{100}{5} = \log 20 \approx 1.301$

3. $\ln 2^6 = 6 \ln 2 \approx 4.159$

NOTE: $\log_a 1 = 0$ for any valid base a. Thus $\log 1 = 0$ and $\ln 1 = 0$.

Inverse Properties The following inverse properties are important for solving exponential and logarithmic equations.

1. $\log_a a^x = x$, for any real number x

2. $a^{\log_a x} = x$, for any positive number x

Examples: **1.** $\log_2 2^\pi = \pi$ **2.** $10^{\log 2.5} = 2.5$

3. $e^{\ln 5} = 5$ **4.** $\log 10^7 = 7$

NOTE: If $f(x) = \log x$, then $f^{-1}(x) = 10^x$, and if $g(x) = \ln x$, then $g^{-1}(x) = e^x$.

SECTION 12.5 ■ EXPONENTIAL AND LOGARITHMIC EQUATIONS

Solving Equations When solving an exponential equation, we usually take a logarithm of each side. When solving a logarithmic equation, we usually exponentiate each side.

Examples: **1.**

$2(5)^x = 22$	Exponential equation
$5^x = 11$	Divide by 2.
$\log 5^x = \log 11$	Take the common logarithm.
$x \log 5 = \log 11$	Power rule
$x = \dfrac{\log 11}{\log 5}$	Divide by log 5.

2.

$\log 2x = 2$	Logarithmic equation
$10^{\log 2x} = 10^2$	Exponentiate each side.
$2x = 100$	Inverse properties
$x = 50$	Divide by 2.

CHAPTER 12   Review Exercises

SECTION 12.1

Exercises 1 and 2: Find the following.

(a) $(g \circ f)(-2)$ (b) $(f \circ g)(x)$

1. $f(x) = 2x^2 - 4x$, $g(x) = 5x + 1$

2. $f(x) = \sqrt[3]{x - 6}$, $g(x) = 4x^3$

3. Use the two tables to evaluate each expression.

(a) $(f \circ g)(2)$ (b) $(g \circ f)(1)$

x	0	1	2	3
$f(x)$	3	2	1	0

x	0	1	2	3
$g(x)$	1	2	3	0

4. Use the graph to evaluate each expression.

(a) $(f \circ g)(-1)$

(b) $(g \circ f)(2)$

(c) $(f \circ f)(1)$

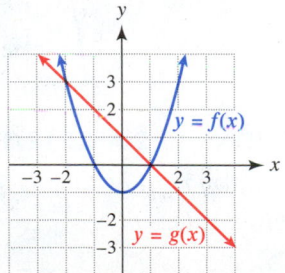

Exercises 5 and 6: Show that f is not one-to-one by find-ing two inputs that result in the same output. Answers may vary.

5. $f(x) = \dfrac{4}{1 + x^2}$ **6.** $f(x) = x^2 - 2x + 1$

Exercises 7 and 8: Use the horizontal line test to deter-mine whether the graph represents a one-to-one function.

7.

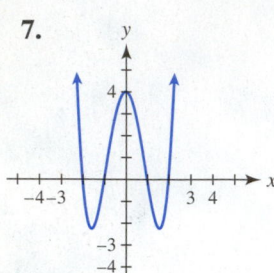

8.

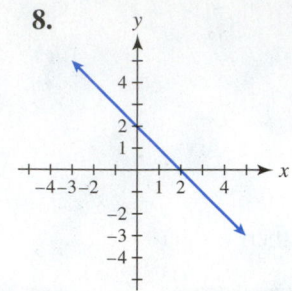

Exercises 9 and 10: Verify that f(x) and $f^{-1}(x)$ are indeed inverse functions.

9. $f(x) = 2x - 9$ $f^{-1}(x) = \dfrac{x + 9}{2}$

10. $f(x) = x^3 + 1$ $f^{-1}(x) = \sqrt[3]{x - 1}$

Exercises 11–14: Find $f^{-1}(x)$.

11. $f(x) = 5x$ **12.** $f(x) = x - 11$

13. $f(x) = 2x + 7$ **14.** $f(x) = \dfrac{4}{x}$

15. Use the table to make a table of values for $f^{-1}(x)$. What are the domain and range for f^{-1}?

x	0	1	2	3
$f(x)$	10	8	7	3

16. Use the graph of $y = f(x)$ to sketch a graph of $y = f^{-1}(x)$. Include the graph of f and the line $y = x$.

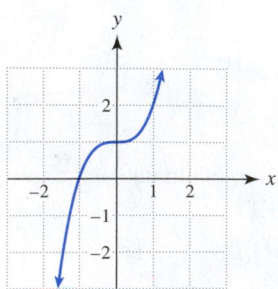

SECTIONS 12.2 AND 12.3

Exercises 17–20: Evaluate the exponential function for the given values of x.

17. $f(x) = 6^x$ $x = -1, \quad x = 2$

18. $f(x) = 5(2^{-x})$ $x = 0, \quad x = 3$

19. $f(x) = \left(\frac{1}{3}\right)^x$ $x = -1, \quad x = 4$

20. $f(x) = 3\left(\frac{1}{6}\right)^x$ $x = 0, \quad x = 1$

Exercises 21–24: Graph f. State whether the graph illus-trates exponential growth, exponential decay, or loga-rithmic growth.

21. $f(x) = 2^x$ **22.** $f(x) = \left(\frac{1}{2}\right)^x$

23. $f(x) = \ln(x + 1)$ **24.** $f(x) = 3^{-x}$

Exercises 25 and 26: A table for a function f is given.

   (a) *Determine whether f represents linear or expo-nential growth.*

   (b) *Find a formula for f.*

25.

x	0	1	2	3	4
$f(x)$	5	10	20	40	80

26.

x	0	1	2	3	4
$f(x)$	5	10	15	20	25

27. Use the graph of $y = Ca^x$ to find C and a.

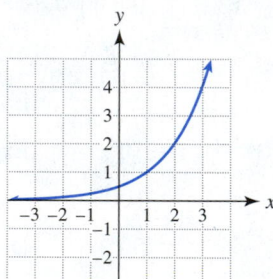

28. Use the graph of $y = k \log_2 x$ to find k.

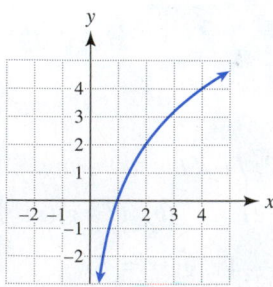

29. Find the percent change if \$150 decreases to \$120.

30. Find the growth factor if \$1500 increases by 7% each year.

Exercises 31 and 32: For the given amounts A and B, find each of the following. Round values to the nearest hundredth of a percent when appropriate.

   (a) *The percent change if A changes to B*
   (b) *The percent change if B changes to A*

31. $A = \$600$, $B = \$1200$

32. $A = \$2.20$, $B = \$1.00$

Exercises 33 and 34: An account contains A dollars and increases or decreases by R%. For each A and R, answer the following.

   (a) *Find the increase or decrease in value of the account.*
   (b) *Find the final value of the account.*
   (c) *By what factor did the account value increase or decrease?*

33. $A = \$500$, $R = \$210\%$

34. $A = \$700$, $R = -25\%$

Exercises 35 and 36: **Exponential Models** *Write the formula for an exponential function, $f(x) = Ca^x$, that models the situation. Evaluate $f(2)$.*

35. A city's population of 20,000 decreases in number by 5% per year.

36. A sample of 1500 insects increases in number by 300% per day.

Exercises 37 and 38: If P dollars is deposited in an account that pays R percent annual interest at the end of each year, approximate the amount in the account after x years.

37. $P = \$1200$ $R = 10\%$ $x = 9$ years

38. $P = \$900$ $R = 18\%$ $x = 40$ years

Exercises 39–42: Evaluate $f(x)$ for the given value of x. Approximate answers to the nearest hundredth.

39. $f(x) = 2e^x - 1$ $x = 5.3$

40. $f(x) = 0.85^x$ $x = 2.1$

41. $f(x) = 2 \log x$ $x = 55$

42. $f(x) = \ln (2x + 3)$ $x = 23$

Exercises 43–46: Evaluate the logarithm by hand.

43. $\log 0.001$

44. $\log \sqrt{10,000}$

45. $\ln e^{-4}$

46. $\log_4 16$

Exercises 47–50: Approximate to the nearest thousandth.

47. $\log 65$

48. $\ln 0.85$

49. $\ln 120$

50. $\log \frac{2}{5}$

Exercises 51–54: Simplify, using inverse properties.

51. $10^{\log 7}$

52. $\log_2 2^{5/9}$

53. $\ln e^{6-x}$

54. $e^{2 \ln x}$

SECTION 12.4

Exercises 55–60: Write the expression by using sums and differences of logarithms of x, y, and z.

55. $\ln xy$

56. $\log \dfrac{x}{y}$

57. $\ln (x^2 y^3)$

58. $\log \dfrac{\sqrt{x}}{z^3}$

59. $\log_2 \dfrac{x^2 y}{z}$

60. $\log_3 \sqrt[3]{\dfrac{x}{y}}$

Exercises 61–64: Write as the logarithm of one expression.

61. $\log 45 + \log 5 - \log 3$

62. $\log_4 2x + \log_4 5x$

63. $2 \ln x - 3 \ln y$

64. $\log x^4 - \log x^3 + \log y$

Exercises 65–68: Rewrite, using the power rule.

65. $\log 6^3$

66. $\ln x^2$

67. $\log_2 5^{2x}$

68. $\log_4 (0.6)^{x+1}$

SECTION 12.5

Exercises 69–78: Solve the given equation. Approximate answers to the nearest hundredth when appropriate.

69. $10^x = 100$

70. $2^{2x} = 256$

71. $3e^x + 1 = 28$

72. $0.85^x = 0.2$

73. $5 \ln x = 4$

74. $\ln 2x = 5$

75. $2 \log x = 80$

76. $3 \log x - 5 = 1$

77. $2^{x+4} = 3^x$

78. $\ln (2x + 1) + \ln (x - 5) = \ln 13$

Exercises 79 and 80: Do the following.

    (a) Solve $f(x) = g(x)$ graphically.

    (b) Solve $f(x) = g(x)$ symbolically.

79. $f(x) = \frac{1}{2}(2^x)$,
$g(x) = 4$

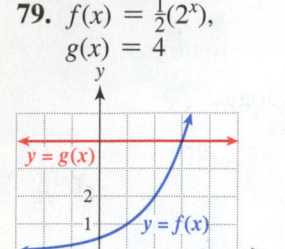

80. $f(x) = \log_2 2x$,
$g(x) = 3$

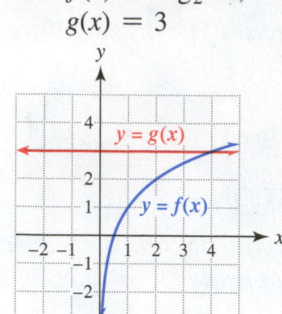

APPLICATIONS

81. *Surface Area of a Balloon* The surface area S of a spherical balloon with radius r is given by the formula $S(r) = 4\pi r^2$. Suppose that the balloon is being inflated so that its radius in inches after t seconds is $r(t) = \sqrt{2t}$.

    (a) Evaluate $(S \circ r)(8)$ and interpret your result.

    (b) Find $(S \circ r)(t)$.

82. *Sales Tax* Suppose that $f(x) = 0.08x$ calculates the sales tax in dollars on an item that costs x dollars.

    (a) Is f a one-to-one function? Why?

    (b) Find a formula for f^{-1} and interpret what it calculates.

83. *Growth of a Mutual Fund* An investor deposits $1500 in a mutual fund that returns 11% annually. Determine the time required for the investment to double in value.

84. *Modeling Data* Find values for the constants a and b so that $y = a + b \log x$ models these data.

x	0.1	1	10	100	1000
y	50	100	150	200	250

85. *Modeling Data* Find values for the constants C and a so that $y = Ca^x$ models these data.

x	0	1	2	3	4
y	3	6	12	24	48

86. *Earthquakes* The Richter scale, used to determine the magnitude of earthquakes, is based on the formula $R(x) = \log \frac{x}{I_0}$, where x is the measured intensity. Let $I_0 = 1$. Find the intensity x for an earthquake with $R = 7$.

87. *Modeling Population* In 2000 the population of Nevada was 2 million and growing continuously at an annual rate of 5.1%. The population of Nevada in millions x years after 2000 can be modeled by

$$f(x) = 2e^{0.051x}.$$

    (a) Graph f in the window $[0, 10, 2]$ by $[0, 4, 1]$. Does this function represent exponential growth or decay?

    (b) Estimate the population of Nevada in 2010.

    (c) Estimate the year when the population was 3 million.

88. *Modeling Bacteria* A colony of bacteria can be modeled by $N(t) = 1000e^{0.0014t}$, where N is measured in bacteria per milliliter and t is in minutes.

    (a) Evaluate $N(0)$ and interpret the result.

    (b) Estimate how long it takes for N to double.

89. *Modeling Wind Speed* Wind speeds vary at different heights above the ground. For a particular day, $f(x) = 1.2 \ln(x) + 5$ computes the wind speed in meters per second x meters above the ground, where $x \geq 1$. (*Source*: A. Miller and R. Anthes, *Meteorology*.)

    (a) Find the wind speed at a height of 5 meters.

    (b) Estimate the height at which the wind speed is 8 meters per second.

1. If $f(x) = 4x^3 - 5x$ and $g(x) = x + 7$, evaluate $(g \circ f)(1)$ and $(f \circ g)(x)$.

2. Use the graph to evaluate each expression.
   (a) $(f \circ g)(-1)$ (b) $(g \circ f)(1)$

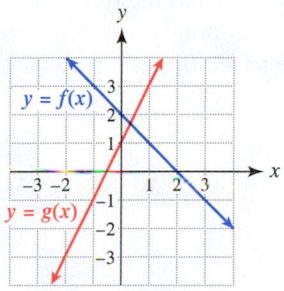

3. Explain why $f(x) = x^2 - 25$ is not a one-to-one function.

4. If $f(x) = 5 - 2x$, find $f^{-1}(x)$.

5. Use the graph of $y = f(x)$ to sketch a graph of $y = f^{-1}(x)$. Include the graph of f and the line $y = x$ in your graph.

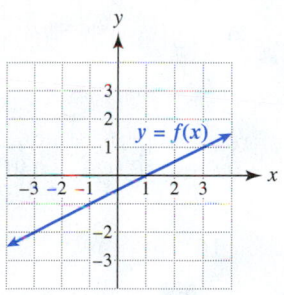

6. Use the table to write a table of values for $f^{-1}(x)$. What are the domain and range of f^{-1}?

x	1	2	3	4
$f(x)$	8	6	4	2

7. Evaluate $f(x) = 3\left(\frac{1}{4}\right)^x$ at $x = 2$.

8. Graph $f(x) = 1.5^{-x}$. State whether the graph of f illustrates exponential growth, exponential decay, or logarithmic growth.

Exercises 9 and 10: A table for a function f is given.

  (a) *Determine whether f represents linear or exponential growth.*
  (b) *Find a formula for f(x).*

9.
x	-2	-1	0	1	2
$f(x)$	0.75	1.5	3	6	12

10.
x	-2	-1	0	1	2
$f(x)$	-4	-2.5	-1	0.5	2

11. Use the graph of $y = Ca^x$ to find C and a.

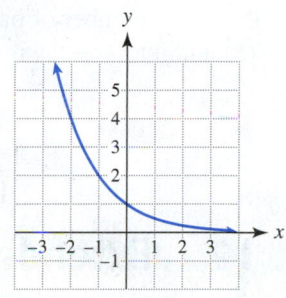

12. Find the percent change if \$600 increases to \$900.

13. If \$1000 increases by 5% per year, what is the growth factor?

14. If \$750 is deposited in an account paying 7% annual interest at the end of each year, approximate the amount in the account after 5 years.

15. Let $f(x) = 1.5 \ln(x - 5)$. Approximate $f(21)$ to the nearest hundredth.

16. Evaluate $\log \sqrt{10}$ by hand.

17. Approximate $\log_2 43$ to the nearest thousandth.

18. Graph $f(x) = \log(x - 2)$. Compare this graph to the graph of $y = \log x$.

19. Write $\log \dfrac{x^3 y^2}{\sqrt{z}}$, using sums and differences of logarithms of x, y, and z.

20. Write $4 \ln x - 5 \ln y + \ln z$ as one logarithm.

21. Rewrite $\log 7^{2x}$, using the power rule.

22. Simplify $\ln e^{1-3x}$, using inverse properties.

Exercises 23–26: Solve the given equation. Approximate answers to the nearest hundredth when appropriate.

23. $2e^x = 50$ 24. $3(10)^x - 7 = 143$

25. $5 \log x = 9$ **26.** $3 \ln 5x = 27$

27. *Modeling Data* Find values for constants a and b so that $y = a + b \log x$ models the data.

x	0.01	0.1	1	10	100
y	−1	2	5	8	11

28. *Modeling Bacteria Growth* A sample of bacteria is growing at a rate of 9% per hour and can be modeled by $f(x) = 4(1.09)^x$, where the input x represents elapsed time in hours and the output $f(x)$ is in millions of bacteria.
   (a) What was the initial number of bacteria?
   (b) Evaluate $f(5)$ and interpret the result.
   (c) Does this function represent exponential growth or exponential decay?
   (d) Estimate the elapsed time when there were 8 million bacteria.

29. A mutual fund account containing $5000 decreases by 2% per year.
   (a) Write a function A that gives the amount in the account after x years.
   (b) Evaluate $A(3)$ and interpret the result.
   (c) Estimate the number of years for the account to decrease to $4500.

CHAPTER 12   Extended and Discovery Exercises

Exercises 1–4: Radioactive Carbon Dating While an animal is alive, it breathes both carbon dioxide and oxygen. Because a small portion of normal atmospheric carbon dioxide is made up of radioactive carbon-14, a fixed percentage of the animal's body is composed of carbon-14. When the animal dies, it quits breathing and the carbon-14 disintegrates without being replaced. One method used to determine when an animal died is to estimate the percentage of carbon-14 remaining in its bones. The **half-life** of carbon-14 is 5730 years. That is, half the original amount of carbon-14 in bones of a fossil will remain after 5730 years. The percentage P, in decimal form, of carbon-14 remaining after x years is modeled by $P(x) = a^x$. (Because of nuclear testing, radioactive carbon dating is no longer accurate for animals that died after 1950.)

1. Find the value of a. (*Hint:* $P(5730) = 0.5$.)

2. Calculate the percentage of carbon-14 that remains after 10,000 years.

3. Estimate the age of a fossil with $P = 0.9$.

4. Estimate the age of a fossil with $P = 0.01$.

Exercises 5–8: Modeling Blood Flow in Animals For medical reasons, dyes are injected into the bloodstream to determine the health of internal organs. In one study that involved animals, the dye BSP was injected to assess blood flow in the liver. The results are listed in the accompanying table, where x represents the elapsed time in minutes and y is the concentration of the dye in the bloodstream in milligrams per milliliter (mg/mL). Scientists modeled the data with $f(x) = 0.133(0.878(0.73^x) + 0.122(0.92^x))$.

x (minutes)	1	2	3	4
y (mg/mL)	0.102	0.077	0.057	0.045

x (minutes)	5	7	9	13
y (mg/mL)	0.036	0.023	0.015	0.008

x (minutes)	16	19	22
y (mg/mL)	0.005	0.004	0.003

Source: F. Harrison, "The measurement of liver blood flow in conscious calves."

5. Graph f together with the data. Comment on the fit.

6. Determine the y-intercept and interpret the result.

7. What happens to the concentration of the dye after a long period of time? Explain.

8. Estimate graphically the time at which the concentration of the dye reached 40% of its initial amount. Would you want to solve this problem symbolically? Explain.

Exercises 9 and 10: Acid Rain Air pollutants frequently cause acid rain. An index of acidity is pH, which measures the concentration of the hydrogen ions in a solution, and ranges from 1 to 14. Pure water is neutral and has a pH of 7, acid solutions have a pH less than 7, and alkaline solutions have a pH greater than 7. The pH of a substance can be computed by $f(x) = -\log x$, where x represents the hydrogen ion concentration in moles per liter. Pure water exposed to normal carbon dioxide in the atmosphere has a pH of 5.6. If the pH of a lake drops below this level, it is indicative of an acidic lake. (Source: G. Howells, Acid Rain and Acid Water.)

9. In rural areas of Europe, rainwater typically has a hydrogen ion concentration of $x = 10^{-4.7}$. Find its pH. What effect might this rain have on a lake with a pH of 5.6?

10. Seawater has a pH of 8.2. Compared to seawater, how many times greater is the hydrogen ion concentration in rainwater from rural Europe?

Exercises 11 and 12: *Investment Account* If x dollars are deposited every 2 weeks (26 times per year) in an account paying an annual interest rate r, expressed in decimal form, the amount A in the account after n years can be approximated by the formula

$$A = x \left[\frac{(1 + r/26)^{26n} - 1}{(r/26)} \right].$$

11. If \$100 is deposited every 2 weeks in an account paying 9% interest, approximate the amount in the account after 10 years.

12. Suppose that your retirement account pays 12% annual interest. Determine how much you should deposit in this account every 2 weeks, in order to have one million dollars at age 65.

CHAPTERS 1–12   Cumulative Review Exercises

1. Write the number 0.000429 in scientific notation.

2. Classify each real number as one or more of the following: natural number, whole number, integer, rational number, or irrational number.

$$-\frac{11}{7}, -3, 0, \sqrt{6}, \pi, 5.\overline{18}$$

3. Select the formula that models the data best.

x	-2	-1	0	1	2
y	-7	-5	-3	-1	1

   a. $y = 3x + 1$ **b.** $y = x - 3$ **c.** $y = 2x - 3$

4. State whether the equation illustrates an identity, commutative, associative, or distributive property.

   $$(5 - y) + 9 = 9 + (5 - y)$$

Exercises 5–8: *Simplify the expression. Write the result with positive exponents.*

5. $\left(\dfrac{1}{d^2} \right)^{-2}$

6. $\left(\dfrac{8a^2}{2b^3} \right)^{-3}$

7. $\dfrac{(2x^{-2}y^3)^2}{xy^{-2}}$

8. $\dfrac{x^{-3}y}{4x^2y^{-3}}$

9. Use the graph to express the equation of the line in slope–intercept form.

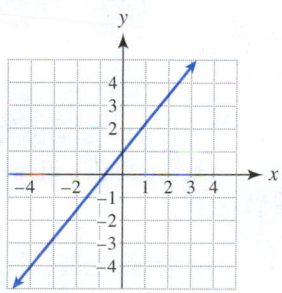

10. Find the domain of $f(x) = \dfrac{10}{x + 3}$.

11. Use the table to write the formula for $f(x) = ax + b$.

x	-2	-1	0	1	2
$f(x)$	-11	-7	-3	1	5

12. Write the equation of the vertical line passing through the point $(4, 7)$.

13. Calculate the slope of the line passing through the points $(4, -1)$ and $(2, -3)$.

14. Sketch the graph of a line passing through the point $(-1, -2)$ with slope $m = 3$.

Exercises 15 and 16: Write the slope–intercept form for a line satisfying the given conditions.

15. Perpendicular to $y = -\frac{1}{7}x - 8$, passing through $(1, 1)$

16. Parallel to $y = 3x - 1$, passing through $(0, 5)$

17. Use the graph to solve the equation $y_1 = y_2$.

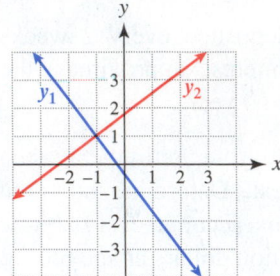

18. Use the table to solve the inequality $y < -4$, where y represents a linear function. Write the solution set as an inequality.

x	-2	-1	0	1	2
y	-24	-14	-4	6	16

Exercises 19–24: Solve the equation or inequality. Write the solutions to the inequalities in interval notation.

19. $\frac{2}{3}(x - 3) + 8 = -6$

20. $\frac{1}{3}z + 6 < \frac{1}{4}z - (5z - 6)$

21. $\left(\frac{t + 2}{3}\right) - 10 = \frac{1}{3}t - (5t + 8)$

22. $-10 \le -\frac{3}{5}x - 4 < -1$

23. $-2|t - 4| \ge -12$ **24.** $\left|\frac{1}{2}x - 5\right| = 3$

25. Shade the solution set in the xy-plane.
$$x + y > 3$$
$$2x - y \ge 3$$

26. Evaluate $\det A$ if $A = \begin{bmatrix} -1 & -2 \\ 3 & 4 \end{bmatrix}$.

Exercises 27–30: Solve the system of equations, if possible. Write the solution as an ordered pair or ordered triple where appropriate.

27. $4x - 3y = 1$
    $5x + 2y = 7$

28. $2x - 3y = -2$
    $-6x + 9y = 5$

29. $2x - y + 3z = -2$
    $x + 5y - 2z = -8$
    $-3x - y - 3z = 6$

30. $x + y - z = -1$
    $-x - y - z = -1$
    $x - 2y + z = 1$

31. Maximize the objective function R, subject to the given constraints.
$$R = 2x + 5y$$
$$3x + y \le 12$$
$$x + 3y \le 12$$
$$x \ge 0, y \ge 0$$

32. Find the area of the triangle by using a determinant. Assume that units are inches.

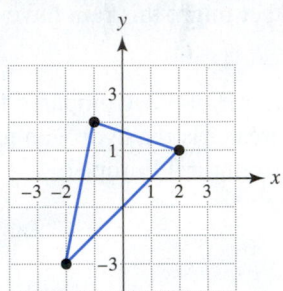

Exercises 33–36: Factor completely.

33. $2x^3 - 4x^2 + 2x$ **34.** $4a^2 - 25b^2$

35. $8t^3 - 27$ **36.** $4a^3 - 2a^2 + 10a - 5$

Exercises 37–40: Solve the equation.

37. $6x^2 - 7x - 10 = 0$ **38.** $9x^2 = 4$

39. $x^4 - 2x^3 = 15x^2$ **40.** $5x - 10x^2 = 0$

Exercises 41 and 42: Simplify the expression.

41. $\dfrac{x^2 + 5x + 6}{x^2 - 9} \cdot \dfrac{x - 3}{x + 2}$

42. $\dfrac{x^2 - 2x - 8}{x^2 + x - 12} \div \dfrac{(x - 4)^2}{x^2 - 16}$

Exercises 43 and 44: Solve the rational equation. Check your result.

43. $\dfrac{2}{x + 2} - \dfrac{1}{x - 2} = \dfrac{-3}{x^2 - 4}$

44. $\dfrac{3y}{y^2 + y - 2} = \dfrac{1}{y - 1} - 2$

45. Solve the equation for J.
$$P = \frac{J + 2z}{J}$$

46. Simplify the complex fraction.

$$\frac{\dfrac{3}{x^2} + x}{x - \dfrac{3}{x^2}}$$

47. Suppose that y varies directly with x. If $y = 15$ when $x = 3$, find y when x is 8.

48. Divide $(3x^3 - 2x - 15)$ by $(x - 2)$.

Exercises 49–54: Simplify the expression. Assume that all variables are positive.

49. $\left(\dfrac{x^6}{y^9}\right)^{2/3}$

50. $\sqrt[3]{-x^4} \cdot \sqrt[3]{-x^5}$

51. $\sqrt{5ab} \cdot \sqrt{20ab}$

52. $2\sqrt{24} - \sqrt{54}$

53. $\sqrt[3]{a^5 b^4} + 3\sqrt[3]{a^5 b}$

54. $\left(5 + \sqrt{5}\right)\left(5 - \sqrt{5}\right)$

55. Rationalize the denominator.

$$\frac{2}{5 - \sqrt{3}}$$

56. Find the domain of f. Write your answer in interval notation.

$$f(x) = \frac{3}{\sqrt{x - 4}}$$

Exercises 57 and 58: Solve. Check your answer.

57. $2(x + 1)^2 = 50$

58. $\sqrt{x + 6} = x$

Exercises 59 and 60: Write in standard form.

59. $(-2 + 3i) - (-5 - 2i)$

60. $\dfrac{3 - i}{1 + 3i}$

61. Find the vertex on the graph of the function given by $f(x) = 3x^2 - 12x + 13$.

62. Find the maximum y-value on the parabola determined by $y = -2x^2 + 6x - 1$.

63. Compare the graph of $f(x) = (x - 3)^2 + 2$ to the graph of $y = x^2$.

64. Write the equation $y = x^2 + 6x - 2$ in vertex form and identify the vertex.

Exercises 65–68: Solve the quadratic equation by using the method of your choice.

65. $x^2 - 13x + 40 = 0$

66. $2d^2 - 5 = d$

67. $z^2 - 4z = -2$

68. $x^4 - 10x^2 + 24 = 0$

69. A graph of $y = ax^2 + bx + c$ is shown.
 (a) Solve $ax^2 + bx + c = 0$.
 (b) State whether $a > 0$ or $a < 0$.
 (c) Determine whether the discriminant is positive, negative, or zero.

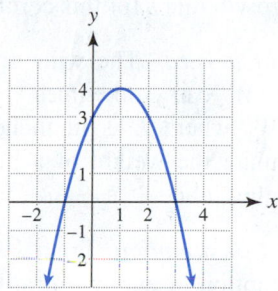

70. Solve $x^2 + 5x - 14 \geq 0$. Write your answer in interval notation.

71. For $f(x) = x^2 - 2$ and $g(x) = 2x + 1$, find each of the following.
 (a) $(f \circ g)(1)$ (b) $(g \circ f)(x)$

72. Show that f is not one-to-one by finding two inputs that result in the same output. Answers may vary.

$$f(x) = x^2 + x - 6$$

73. Find $f^{-1}(x)$ for $f(x) = \frac{3}{x}$.

74. If \$800 is deposited in an account that pays 7.5% annual interest at the end of each year, approximate the amount in the account after 15 years.

Exercises 75 and 76: Evaluate without a calculator.

75. $\log_3 81$

76. $e^{\ln 2x}$

77. Write the expression by using sums and differences of logarithms of x and y. Assume x and y are positive.

$$\log \frac{\sqrt{x}}{y^2}$$

78. Write $2 \ln x + \ln 5x$ as the logarithm of a single expression.

Exercises 79 and 80: Solve the equation. Approximate answers to the nearest hundredth.

79. $6 \log x - 2 = 9$

80. $2^{3x} = 17$

APPLICATIONS

81. *Population Growth*   The population P of a community with an annual percentage growth rate r (expressed as a decimal) after t years is given by $P = P_0(1 + r)^t$, where P_0 represents the initial population of the community. If a community having an initial population of $P_0 = 12{,}000$ grew to a population of $P = 14{,}600$ in $t = 5$ years, find the annual percentage growth rate r for this community.

82. *Wing Span of a Bird*   The wing span L of a bird with weight W can sometimes be modeled by $L = 27.4\sqrt[3]{W}$, where L is in inches and W is in pounds. Estimate the weight of a bird with a wing span of 36 inches. (*Source:* C. Pennycuick, *Newton Rules Biology.*)

83. *U.S. Energy Consumption*   From 1950 to 1970, per capita consumption of energy in millions of Btu can be modeled by $f(x) = 0.25x^2 - 975x + 950{,}845$, where x is the year. (*Source:* Department of Energy.)
   (a) During what year was per capita energy consumption at its lowest?
   (b) Find this minimum value.

84. *Braking Distance*   On dry, level pavement a safe braking distance d in feet for a car traveling x miles per hour is $d = \frac{x^2}{12}$. What speed corresponds to a braking distance of 350 feet? (*Source:* F. Mannering, *Principles of Highway Engineering and Traffic Control.*)

85. *Investing for Retirement*   A college student invests $8000 in an account that pays interest annually. If the student would like this investment to be worth $1,000,000 in 45 years, what annual interest rate would the account need to pay?

86. *Modeling Wind Speed*   Wind speeds are usually measured at heights from 5 to 10 meters above the ground. For a particular day, $f(x) = 1.4 \ln(x) + 7$ computes the wind speed in meters per second x meters above the ground, where $x \geq 1$. (*Source:* A. Miller and R. Anthes, *Meteorology.*)
   (a) Find the wind speed at a height of 8 meters.
   (b) Estimate the height at which the wind speed is 10 meters per second.

13

Conic Sections

The art of asking the right questions in mathematics is more important than the art of solving them.

—GEORG CANTOR

Throughout history people have been fascinated with the universe around them and compelled to understand its mysteries. Conic sections, which include parabolas, circles, ellipses, and hyperbolas, have played an important role in gaining this understanding. Although conic sections were described and named by the Greek astronomer Apollonius in 200 B.C., it was not until much later that they were used to model motion in the universe. In the sixteenth century Tycho Brahe, the greatest observational astronomer of the age, recorded precise data on planetary movement in the sky. Using Brahe's data, in 1619 Johannes Kepler determined that planets move in elliptical orbits around the sun. In 1686 Newton used Kepler's work to show that elliptical orbits are the result of his famous theory of gravitation. We now know that all celestial objects—including planets, comets, asteroids, and satellites—travel in paths described by conic sections.

Today scientists search the sky for information about the universe with enormous radio telescopes in the shape of parabolic dishes. The Hubble telescope also makes use of a parabolic mirror. As a result, our understanding of the universe has changed dramatically in recent years.

Conic sections have had a profound influence on people's understanding of their world and the cosmos. In this chapter we introduce you to these age-old curves.

Source: Historical Topics for the Mathematics Classroom, Thirty-first Yearbook, NCTM.

13.1 Parabolas and Circles

Types of Conic Sections • Parabolas with Horizontal Axes of Symmetry •
Equations of Circles

A LOOK INTO MATH ▶

In this section we discuss two types of conic sections: parabolas and circles. Recall that we discussed parabolas with vertical axes of symmetry in Chapter 11. In this section we discuss parabolas with horizontal axes of symmetry, but first we introduce the three basic types of conic sections. The Hubble telescope travels in a path described by a conic section called an ellipse, as do all planets and satellites.

Types of Conic Sections

Conic sections are named after the different ways that a plane can intersect a cone. The three basic curves are parabolas, ellipses, and hyperbolas. A circle is a special case of an ellipse. Figure 13.1 shows the three types of conic sections along with an example of the graph associated with each.

NEW VOCABULARY

☐ Conic sections
☐ Circle
☐ Radius
☐ Center
☐ Standard equation
  of a circle

Types of Conic Sections

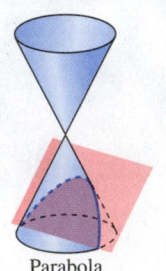

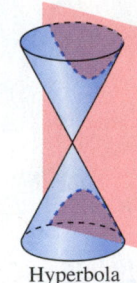

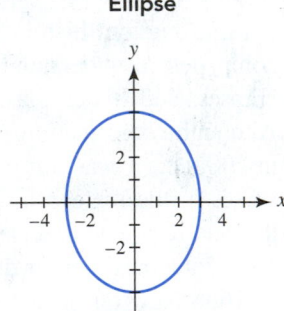

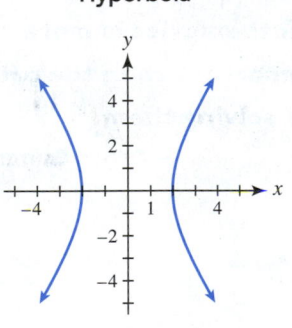

Figure 13.1

READING CHECK

State the types of conic sections.

Parabolas with Horizontal Axes of Symmetry

Recall that the *vertex form of a parabola* with a vertical axis of symmetry is

$$y = a(x - h)^2 + k, \quad \text{Vertex form}$$

where (h, k) is the vertex. If $a > 0$, the parabola opens upward; if $a < 0$, the parabola opens downward, as shown in Figure 13.2. The preceding equation can also be expressed in the form

$$y = ax^2 + bx + c. \quad \text{General form}$$

In this form the x-coordinate of the vertex is $x = -\frac{b}{2a}$.

Parabolas with a Vertical Axis of Symmetry

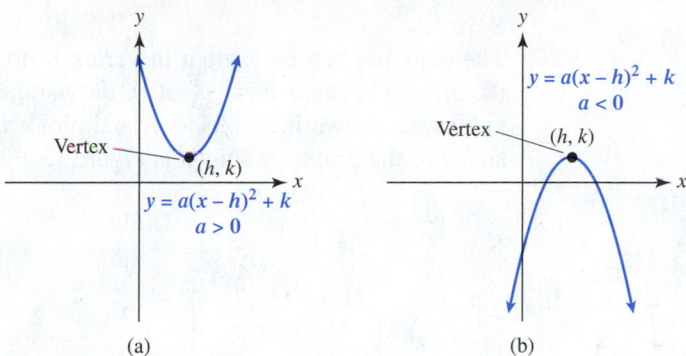

Figure 13.2 Functions

Interchanging the roles of x and y (and also h and k) gives equations for parabolas that open to the right or the left. In this case, their axes of symmetry are horizontal.

PARABOLAS WITH HORIZONTAL AXES OF SYMMETRY

The graph of $x = a(y - k)^2 + h$ is a parabola that opens to the right if $a > 0$ and to the left if $a < 0$. The vertex of the parabola is located at (h, k).

The graph of $x = ay^2 + by + c$ is a parabola opening to the right if $a > 0$ and to the left if $a < 0$. The y-coordinate of its vertex is given by $y = -\frac{b}{2a}$.

READING CHECK

Give a general equation for a parabola with a horizontal axis of symmetry.

These parabolas are illustrated in Figure 13.3.

Parabolas with a Horizontal Axis of Symmetry

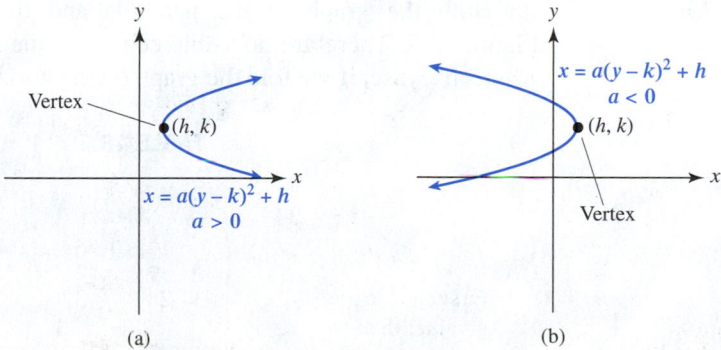

Figure 13.3 Not Functions

NOTE: If the variable x is squared in an equation for a parabola, then its axis of symmetry is vertical. If the variable y is squared instead, then its axis of symmetry is horizontal.

EXAMPLE 1   ### Graphing a parabola

Graph $x = -\frac{1}{2}y^2$. Find its vertex and axis of symmetry.

Solution

The equation can be written in vertex form because $x = -\frac{1}{2}(y - \mathbf{0})^2 + \mathbf{0}$. The vertex is $(\mathbf{0}, \mathbf{0})$, and because $a = -\frac{1}{2} < 0$, the parabola opens to the left. We can make a table of values, as shown in Table 13.1, and plot a few points to help determine the location and shape of the graph, as shown in Figure 13.4. Its axis of symmetry is the x-axis, or $y = 0$.

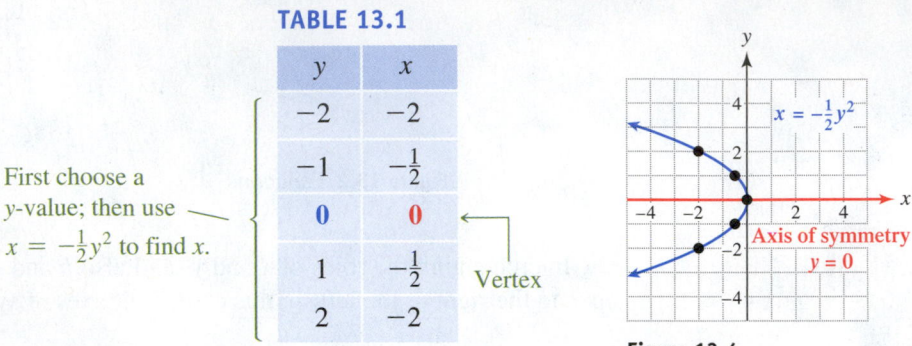

TABLE 13.1

First choose a y-value; then use $x = -\frac{1}{2}y^2$ to find x.

y	x
-2	-2
-1	$-\frac{1}{2}$
$\mathbf{0}$	$\mathbf{0}$ ← Vertex
1	$-\frac{1}{2}$
2	-2

Figure 13.4

Now Try Exercise 15

EXAMPLE 2   ### Graphing a parabola

Graph $x = (y - 3)^2 + 2$. Find its vertex and axis of symmetry.

Solution

Because $h = 2$ and $k = 3$ in the equation $x = a(y - k)^2 + h$, the vertex is $(\mathbf{2}, \mathbf{3})$, and because $a = 1 > 0$, the parabola opens to the right. This parabola has the same shape as $y = x^2$, except that it opens to the right rather than upward. To graph this parabola we can make a table of values and plot a few points, as shown in Table 13.2 and Figure 13.5.

NOTE: Sometimes, finding the x- and y-intercepts of the parabola is helpful when you are graphing. To find the x-intercept let $y = \mathbf{0}$ in $x = (y - 3)^2 + 2$. The x-intercept is $x = (\mathbf{0} - 3)^2 + 2 = 11$. To find any y-intercepts let $x = \mathbf{0}$ in $x = (y - 3)^2 + 2$. Here $\mathbf{0} = (y - 3)^2 + 2$ means that $(y - 3)^2 = -2$, which has no real solutions, and thus this parabola has no y-intercepts.

Both the graph of the parabola and the points from Table 13.2 are shown in Figure 13.5. There are no y-intercepts and the x-intercept is 11. The axis of symmetry is $y = 3$ because, if we fold the graph on the horizontal line $y = 3$, the two sides match.

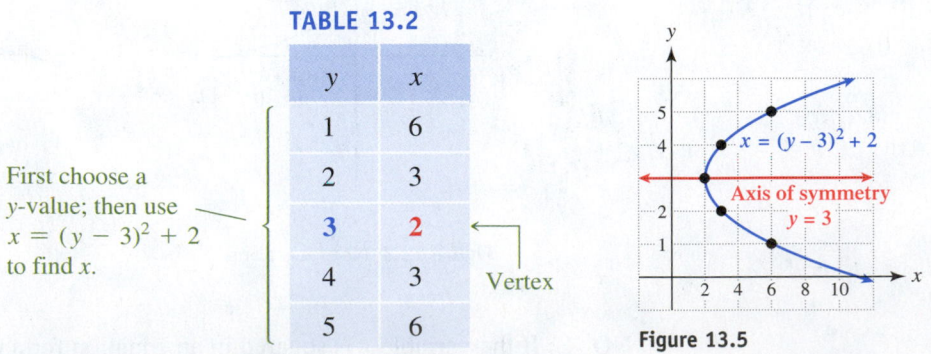

TABLE 13.2

First choose a y-value; then use $x = (y - 3)^2 + 2$ to find x.

y	x
1	6
2	3
$\mathbf{3}$	$\mathbf{2}$ ← Vertex
4	3
5	6

Figure 13.5

Now Try Exercise 23

EXAMPLE 3   **Graphing a parabola and finding its vertex**

Identify the vertex and then graph each parabola.
(a) $x = -y^2 + 1$ **(b)** $x = y^2 - 2y - 1$

Solution

(a) If we rewrite $x = -y^2 + 1$ as $x = -(y - 0)^2 + 1$, then $h = 1$ and $k = 0$, so the vertex is $(1, 0)$. By letting $y = 0$ in $x = -y^2 + 1$, we find that the x-intercept is $x = -0^2 + 1 = 1$. Similarly, we let $x = 0$ in $x = -y^2 + 1$ to find the y-intercepts. The equation $0 = -y^2 + 1$ has solutions -1 and 1.

The parabola opens to the left because $a = -1 < 0$. Additional points given in Table 13.3 help in graphing the parabola shown in Figure 13.6.

TABLE 13.3

y	x
-2	-3
-1	0
0	**1**
1	0
2	-3

Vertex is $(\mathbf{1}, \mathbf{0})$, so choose y-values → on each side of $\mathbf{0}$.

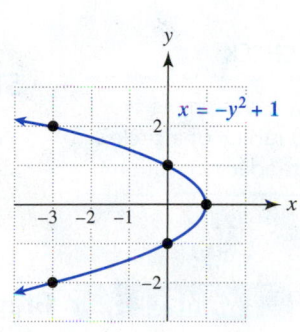

Figure 13.6

(b) The y-coordinate of the vertex for the graph of $x = y^2 - 2y - 1$ is given by

$$y = -\frac{b}{2a} = -\frac{-2}{2(1)} = \mathbf{1}.$$

To find the x-coordinate of the vertex, substitute $y = \mathbf{1}$ into the given equation.

$$x = (\mathbf{1})^2 - 2(\mathbf{1}) - 1 = \mathbf{-2}$$

The vertex is $(\mathbf{-2}, \mathbf{1})$. The parabola opens to the right because $a = 1 > 0$. Additional points given in Table 13.4 help in graphing the parabola shown in Figure 13.7. Note that the y-intercepts do not have integer values and that the quadratic formula could be used to find approximations for these values. The x-intercept is -1.

TABLE 13.4

y	x
-1	2
0	-1
1	**-2**
2	-1
3	2

Vertex is $(\mathbf{-2}, \mathbf{1})$, so choose y-values → on each side of $\mathbf{1}$.

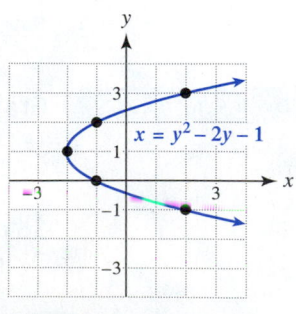

Figure 13.7

Now Try Exercises 17, 39

Equations of Circles

A **circle** consists of the set of points in a plane that are the same distance from a fixed point. The fixed distance is called the **radius**, and the fixed point is called the **center**. In Figure 13.8 all points lying on the circle are a distance of 2 units from the center (2, 1). Therefore the radius of the circle equals 2.

We can find the equation of the circle shown in Figure 13.8 by using the distance formula. If a point (x, y) lies on the graph of a circle, its distance from the center $(2, 1)$ is 2, and

$$\sqrt{(x - 2)^2 + (y - 1)^2} = 2. \quad d = \sqrt{(x_2 - x_1)^2 + (y_2 - y_1)^2}$$

Squaring each side gives

$$(x - 2)^2 + (y - 1)^2 = 2^2.$$

This equation represents the standard equation for a circle with center $(2, 1)$ and radius 2.

Circle

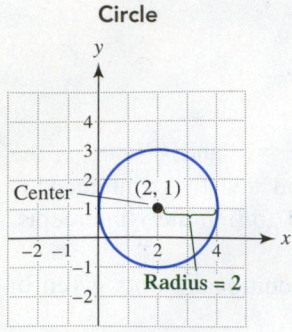

Figure 13.8

READING CHECK

How can you determine the center and radius of a circle given its standard equation?

STANDARD EQUATION OF A CIRCLE

The **standard equation of a circle** with center (h, k) and radius r is

$$(x - h)^2 + (y - k)^2 = r^2.$$

EXAMPLE 4 **Graphing a circle**

Graph $x^2 + y^2 = 9$. Find the radius and center.

Solution
The equation $x^2 + y^2 = 9$ can be written in standard form as

$$(x - 0)^2 + (y - 0)^2 = 3^2.$$

Therefore the center is $(0, 0)$ and the radius is 3. Its graph is shown in Figure 13.9.

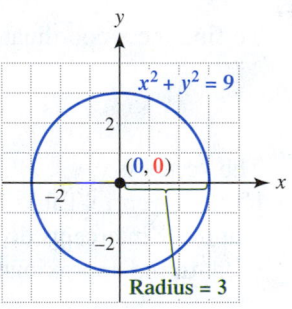

Figure 13.9

Now Try Exercise 65

EXAMPLE 5 **Graphing a circle**

Graph $(x + 1)^2 + (y - 3)^2 = 4$. Find the radius and center.

Solution
Write the equation as

$$(x - (-1))^2 + (y - 3)^2 = 2^2.$$

The center is $(-1, 3)$, and the radius is 2. The circle's graph is shown in Figure 13.10.

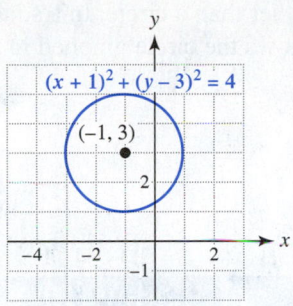

Figure 13.10

■ **Now Try Exercise 69**

In the next example we use the *method of completing the square* to find the center and radius of a circle. (To review completing the square, refer to Sections 11.2 and 11.3.)

EXAMPLE 6   **Finding the center of a circle**

Find the center and radius of the circle given by $x^2 + 4x + y^2 - 6y = 5$.

Solution
Begin by writing the equation as

$$(x^2 + 4x + \underline{\quad}) + (y^2 - 6y + \underline{\quad}) = 5.$$

To complete the square, add $\left(\frac{4}{2}\right)^2 = 4$ and $\left(\frac{-6}{2}\right)^2 = 9$ to each side of the equation.

$$(x^2 + 4x + 4) + (y^2 - 6y + 9) = 5 + 4 + 9$$

Factoring each perfect square trinomial yields

$$(x + 2)^2 + (y - 3)^2 = 18.$$

The center is $(-2, 3)$, and because $18 = \left(\sqrt{18}\right)^2$, the radius is $\sqrt{18}$, or $3\sqrt{2}$.

■ **Now Try Exercise 71**

CRITICAL THINKING

Does the following equation represent a circle? If so, give its center and radius.

$$x^2 + y^2 + 10y = -32$$

TECHNOLOGY NOTE

Graphing Circles
The graph of a circle does not represent a function. One way to graph a circle with a graphing calculator is to solve the equation for y and obtain two equations. One equation gives the upper half of the circle, and the other equation gives the lower half.

For example, to graph $x^2 + y^2 = 4$ begin by solving for y.

$$y^2 = 4 - x^2 \qquad \text{Subtract } x^2.$$

$$y = \pm\sqrt{4 - x^2} \qquad \text{Square root property}$$

Then graph $Y_1 = \sqrt{(4 - X^2)}$ and $Y_2 = -\sqrt{(4 - X^2)}$. The graph of y_1 is the upper half of the circle, and the graph of y_2 is the lower half of the circle, as shown in Figure 13.11.

$[-4.7, 4.7, 1]$ by $[-3.1, 3.1, 1]$ $[-4.7, 4.7, 1]$ by $[-3.1, 3.1, 1]$ $[-4.7, 4.7, 1]$ by $[-3.1, 3.1, 1]$

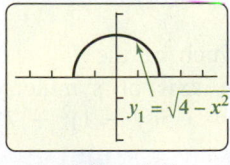

(a) Upper half   (b) Lower half   (c) Both halves

Figure 13.11

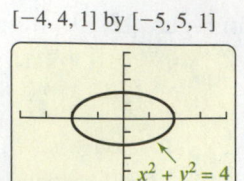

[−4, 4, 1] by [−5, 5, 1]

Figure 13.12

NOTE: If a circle is not graphed in a *square viewing rectangle*, it will appear to be an oval rather than a circle. In a square viewing rectangle a circle will appear circular. Figure 13.12 shows the circle graphed in a viewing rectangle that is not square.

13.1 Putting It All Together

CONCEPT	EXPLANATION	EXAMPLE
Parabola with Horizontal Axis	Vertex form: $x = a(y - k)^2 + h$ If $a > 0$, it opens to the right. If $a < 0$, it opens to the left. The vertex is (h, k). See Figure 13.3. These parabolas may also be expressed as $x = ay^2 + by + c$, where the y-coordinate of the vertex is $y = -\frac{b}{2a}$.	$x = 2(y - 1)^2 + 4$ opens to the right and its vertex is $(4, 1)$. Axis of symmetry $y = 1$, $(4, 1)$, $x = 2(y - 1)^2 + 4$
Standard Equation of a Circle	$(x - h)^2 + (y - k)^2 = r^2$. The radius is r and the center is (h, k).	$(x + 2)^2 + (y - 1)^2 = 16$ has center $(-2, 1)$ and radius 4.

13.1 Exercises

CONCEPTS AND VOCABULARY

1. Name the three general types of conic sections.

2. What is the difference between the parabolas given by $y = ax^2 + bx + c$ and $x = ay^2 + by + c$?

3. If a parabola has a horizontal axis of symmetry, does it represent a function?

4. Sketch a graph of a parabola with a horizontal axis of symmetry.

5. If a parabola has two y-intercepts, does it represent a function? Why or why not?

6. If $x = a(y - k)^2 + h$, what is the vertex?

7. The graph of $x = -y^2$ opens to the _____.

8. The graph of $x = 2y^2 + y - 1$ opens to the _____.

9. The graph of $(x - h)^2 + (y - k)^2 = r^2$ is a(n) _____ with center _____.

10. The graph of $x^2 + y^2 = r^2$ is a circle with center _____ and radius _____.

11. Which of the following (a. – d.) are the coordinates of the vertex for the parabola given by $x = 4(y + 2)^2 - 3$?
    a. $(-3, 4)$ b. $(3, -2)$
    c. $(-3, 2)$ d. $(-3, -2)$

12. Which of the following (a.–d.) is the equation of the axis of symmetry for the parabola given by $x = -5(y - 1)^2 + 7$?
    a. $y = 1$ b. $x = -5$
    c. $x = 7$ d. $y = 7$

13. Which of the following (a.–d.) is the center and radius of the circle given by $x^2 + (y - 2)^2 = 9$?
    a. $(0, -2), r = 3$ b. $(2, 0), r = 3$
    c. $(0, 2), r = 3$ d. $(0, 2), r = 9$

14. Which of the following (a.–d.) is the y-coordinate of the vertex of the parabola given by $x = y^2 - 2y + 3$?
    a. -2 b. 1
    c. 0 d. 3

PARABOLAS

Exercises 15–40: Graph the parabola. Find the vertex and axis of symmetry.

15. $x = y^2$

16. $x = -y^2$

17. $x = y^2 + 1$

18. $x = y^2 - 1$

19. $y = x^2 - 1$

20. $y = 2x^2$

21. $x = 2y^2$

22. $x = \frac{1}{4}y^2$

23. $x = (y - 1)^2 + 2$

24. $x = (y - 2)^2 + 1$

25. $y = (x + 2)^2 + 1$

26. $y = (x - 4)^2 + 5$

27. $y = -2(x + 2)^2$

28. $y = (x - 1)^2 - 2$

29. $x = \frac{1}{2}(y + 1)^2 - 3$

30. $x = -2(y + 3)^2 + 1$

31. $x = -3(y - 1)^2$

32. $x = \frac{1}{4}(y + 2)^2 - 3$

33. $y = 2x^2 - x + 1$

34. $y = -x^2 + 2x + 2$

35. $x = -2y^2 + 3y + 2$

36. $x = \frac{1}{2}y^2 + y - 1$

37. $x = 3y^2 + y$

38. $x = -\frac{3}{2}y^2 - 2y + 1$

39. $x = y^2 + 2y + 1$

40. $x = y^2 - 3y - 4$

Exercises 41–44: Use the graph to determine the equation of the parabola. (Hint: Either $a = 1$ or $a = -1$.)

41.
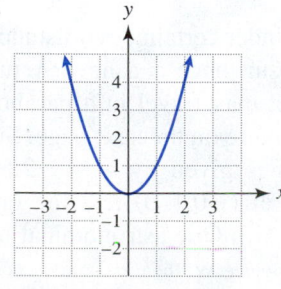

42.
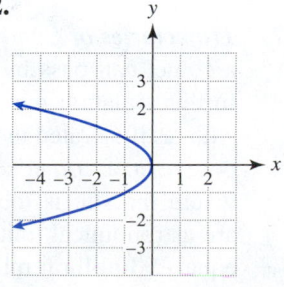

43.
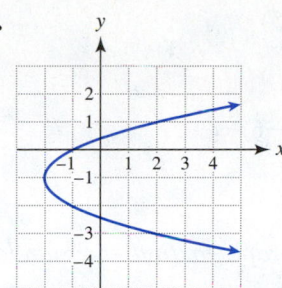

44.
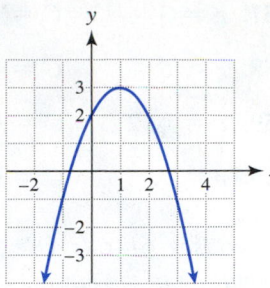

Exercises 45–48: Determine the direction that the parabola opens if it satisfies the given conditions.

45. Passing through $(2, 0)$, $(-2, 0)$, and $(0, -2)$

46. Passing through $(0, -3)$, $(0, 2)$, and $(1, 1)$

47. Vertex $(1, 2)$ passing through the point $(-1, -2)$ with a vertical axis

48. Vertex $(-1, 3)$ passing through the point $(0, 0)$ with a horizontal axis

49. What x-values are possible for the graph of the equation $x = 2y^2$?

50. What y-values are possible for the graph of the equation $x = 2y^2$?

51. **Thinking Generally** How many y-intercepts does a parabola given by
$$x = a(y - k)^2 + h$$
have if $a > 0$ and $h < 0$?

52. **Thinking Generally** Does the graph of the equation $x = ay^2 + by + c$ always have a y-intercept? Explain.

53. What is the x-intercept for the graph of the equation given by
$$x = 3y^2 - y + 1?$$

54. What are the y-intercepts for the graph of the equation given by
$$x = y^2 - 3y + 2?$$

CIRCLES

Exercises 55–60: Write the standard equation of the circle with the given radius r and center C.

55. $r = 1$ $C = (0, 0)$

56. $r = 4$ $C = (2, 3)$

57. $r = 3$ $C = (-1, 5)$

58. $r = 5$ $C = (5, -3)$

59. $r = \sqrt{2}$ $C = (-4, -6)$

60. $r = \sqrt{6}$ $C = (0, 4)$

Exercises 61–64: Use the graph to find the standard equation of the circle.

61.
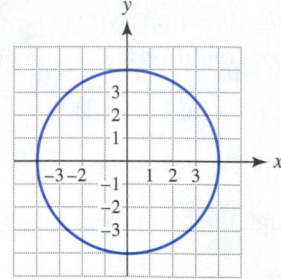

62.
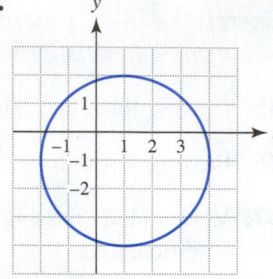

63.
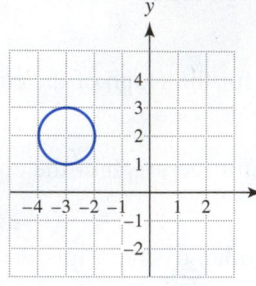

64.
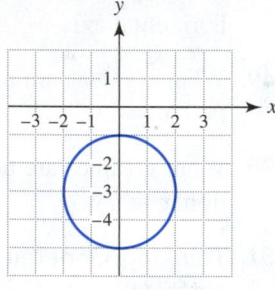

Exercises 65–74: Find the radius and center of the circle. Then graph the circle.

65. $x^2 + y^2 = 4$ **66.** $x^2 + y^2 = 1$

67. $(x - 1)^2 + (y - 3)^2 = 9$

68. $(x + 2)^2 + (y + 1)^2 = 4$

69. $(x + 5)^2 + (y - 5)^2 = 25$

70. $(x - 4)^2 + (y + 3)^2 = 16$

71. $x^2 + 6x + y^2 - 2y = -1$

72. $x^2 + y^2 + 12y + 32 = 0$

73. $x^2 + 6x + y^2 - 2y + 3 = 0$

74. $x^2 - 4x + y^2 + 4y = -3$

APPLICATIONS

75. *Radio Telescopes* The Parks radio telescope has the shape of a parabolic dish, as depicted in the figure at the top of the next column. A cross section of this telescope can be modeled by $x = \frac{32}{11,025}y^2$, where $-105 \le y \le 105$; the units are feet. (*Source:* J. Mar, *Structure Technology for Large Radar and Telescope Systems.*)

🖩 **(a)** Graph the cross-sectional shape of the dish in $[-40, 40, 10]$ by $[-120, 120, 20]$.

     (b) Find the depth d of the dish.

76. *Train Tracks* To make a curve safer for trains, parabolic curves are sometimes used instead of circular curves. See the accompanying figures. (*Source:* F. Mannering and W. Kilareski, *Principles of Highway Engineering and Traffic Analysis.*)

     (a) Suppose that a curve must pass through the points $(-1, 0)$, $(0, 3)$, and $(0, -3)$, where the units are kilometers. Find an equation for the train tracks in the form

$$x = a(y - h)^2 + k.$$

     (b) Find another point that lies on the train tracks.

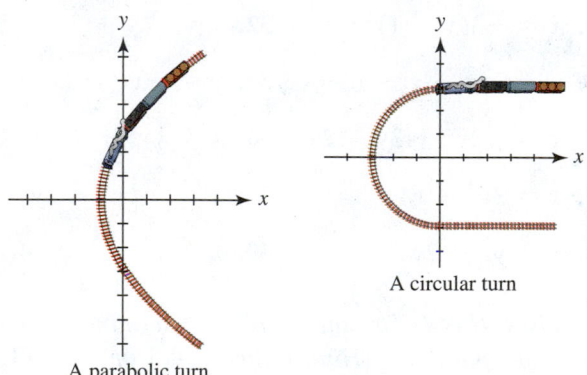

A parabolic turn   A circular turn

77. *Trajectories of Comets* Under certain circumstances, a comet can pass by the sun once and never return. In this situation the comet may travel in a parabolic path, as illustrated in the figure on the next page. Suppose that a comet's path is given by $x = -2.5y^2$, where the sun is located at $(-0.1, 0)$ and the units are astronomical units (A.U.). One astronomical unit equals 93 million miles. (*Source:* W. Thomson, *Introduction to Space Dynamics.*)

     (a) Plot a point for the sun's location and then graph the path of the comet.

     (b) Find the distance from the sun to the comet when the comet is located at $(-2.5, 1)$.

y

Parabolic trajectory

Comet   Sun

x

78. *Speed of a Comet*   (Continuation of Exercise 77.) The velocity V in meters per second of a comet traveling in a parabolic trajectory around the sun is given by $V = \frac{k}{\sqrt{D}}$, where D is the comet's distance from the sun in meters and $k = 1.15 \times 10^{10}$.

(a) How does the velocity of the comet change as its distance from the sun changes?

(b) Calculate the velocity of the comet when it is closest to the sun. (*Hint:* 1 mile $\approx$ 1609 meters.)

WRITING ABOUT MATHEMATICS

79. Suppose that you are given the equation
$$x = a(y - k)^2 + h.$$

(a) Explain how you can determine the direction that the parabola opens.

(b) Explain how to find the axis of symmetry and the vertex.

(c) If the points $(0, 4)$ and $(0, -2)$ lie on the graph of x, what is the axis of symmetry?

(d) Generalize part (c) if $(0, y_1)$ and $(0, y_2)$ lie on the graph of x.

80. Suppose that you are given the vertex of a parabola. Can you determine the axis of symmetry? Explain.

Group Activity   Working With Real Data

Directions: Form a group of 2 to 4 people. Select someone to record the group's responses for this activity. All members of the group should work cooperatively to answer the questions. If your instructor asks for your results, each member of the group should be prepared to respond.

Radio Telescope   The U.S. Naval Research Laboratory designed a giant radio telescope weighing 3450 tons. Its parabolic dish has a diameter of 300 feet and a depth of 44 feet, as shown in the accompanying figure. (*Source:* J. Mar, *Structure Technology for Large Radio and Radar Telescope Systems.*)

(a) Determine an equation of the form $x = ay^2$, $a > 0$, that models a cross section of the dish.

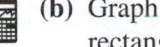 (b) Graph your equation in an appropriate viewing rectangle.

13.2   Ellipses and Hyperbolas

Equations of Ellipses • Equations of Hyperbolas

A LOOK INTO MATH ▶   Celestial objects travel in paths or trajectories determined by conic sections. For this reason, conic sections have been studied for centuries. In modern times, physicists have learned that subatomic particles can also travel in trajectories determined by conic sections. Recall that the three main types of conic sections are parabolas, ellipses, and hyperbolas and that circles are a special type of ellipse. In Section 11.2 and Section 13.1, we discussed parabolas and circles. In this section we focus on ellipses and hyperbolas and some of their applications.

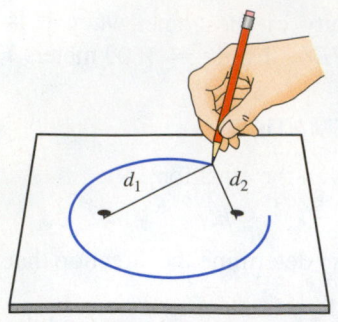

Figure 13.13

Equations of Ellipses

One method used to sketch an ellipse is to tie the ends of a string to two nails driven into a flat board. If a pencil is placed against the string anywhere between the nails, as shown in Figure 13.13, and is used to draw a curve, the resulting curve is an ellipse. The sum of the distances d_1 and d_2 between the pencil and each of the nails is always fixed by the length of the string. The location of the nails corresponds to the *foci* of the ellipse. An **ellipse** is the set of points in a plane the sum of whose distances from two fixed points is constant. Each fixed point is called a **focus** (plural, foci) of the ellipse.

CRITICAL THINKING

What happens to the shape of the ellipse shown in Figure 13.13 as the nails are moved farther apart? What happens to its shape as the nails are moved closer together? When would a circle be formed?

NEW VOCABULARY

- ☐ Ellipse
- ☐ Focus of an ellipse
- ☐ Major/minor axis
- ☐ Vertices of an ellipse
- ☐ Center of an ellipse
- ☐ Hyperbola
- ☐ Focus of a hyperbola
- ☐ Branches
- ☐ Vertices of a hyperbola
- ☐ Transverse axis
- ☐ Fundamental rectangle
- ☐ Asymptotes

In Figure 13.14 the **major axis** and the **minor axis** are labeled for each ellipse. The major axis is the longer of the two axes. Figure 13.14(a) shows an ellipse with a *horizontal* major axis, and Figure 13.14(b) shows an ellipse with a *vertical* major axis. The **vertices**, V_1 and V_2, of each ellipse are located at the endpoints of the major axis, and the **center of the ellipse** is the midpoint of the major axis (or the intersection of the major and minor axes).

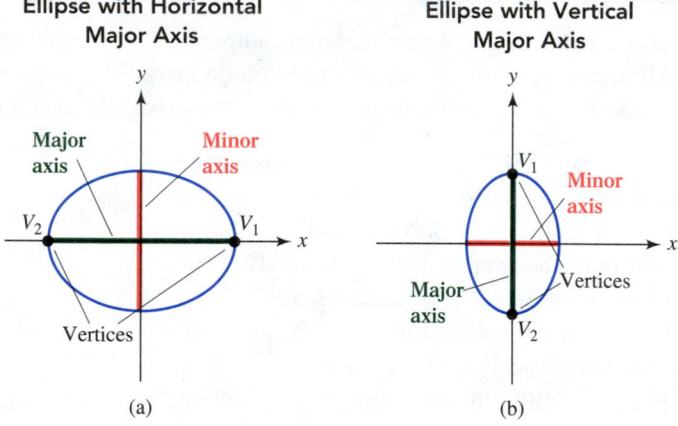

Figure 13.14 Not Functions

A vertical line can intersect the graph of an ellipse twice, so an ellipse cannot be represented by a function. However, some ellipses can be represented by the following equations.

READING CHECK

How can you determine from the standard equation of an ellipse if the major axis is horizontal or vertical?

STANDARD EQUATIONS FOR ELLIPSES CENTERED AT (0, 0)

The ellipse with center at the origin, *horizontal* major axis, and equation

$$\frac{x^2}{a^2} + \frac{y^2}{b^2} = 1, \qquad a > b > 0,$$

has vertices $(\pm a, 0)$ and endpoints of the minor axis $(0, \pm b)$.

The ellipse with center at the origin, *vertical* major axis, and equation

$$\frac{x^2}{b^2} + \frac{y^2}{a^2} = 1, \qquad a > b > 0,$$

has vertices $(0, \pm a)$ and endpoints of the minor axis $(\pm b, 0)$.

Figure 13.15(a) shows an ellipse having a horizontal major axis; Figure 13.15(b) shows one having a vertical major axis. The coordinates of the vertices V_1 and V_2 and endpoints of the minor axis U_1 and U_2 are labeled.

CRITICAL THINKING

Suppose that $a = b$ for an ellipse centered at (0, 0). What can be said about the ellipse? Explain.

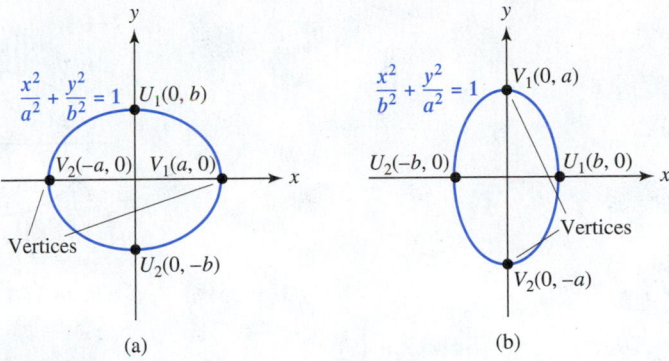

(a)   (b)

Figure 13.15

In the next example we show how to sketch graphs of ellipses.

EXAMPLE 1   **Sketching ellipses**

Sketch a graph of each ellipse. Label the vertices and endpoints of the minor axes.

(a) $\dfrac{x^2}{25} + \dfrac{y^2}{4} = 1$ (b) $9x^2 + 4y^2 = 36$

Solution

(a) The equation $\dfrac{x^2}{25} + \dfrac{y^2}{4} = 1$ describes an ellipse with $a^2 = 25$ and $b^2 = 4$. (When you are deciding whether 25 or 4 represents a^2, let a^2 be the larger of the two numbers.) Thus $a = 5$ and $b = 2$, so the ellipse has a horizontal major axis with vertices $(\pm 5, 0)$ and the endpoints of the minor axis are $(0, \pm 2)$. Plot these four points and then sketch the ellipse, as shown in Figure 13.16(a).

(b) To write $9x^2 + 4y^2 = 36$ as a standard equation, divide each side by 36.

$$9x^2 + 4y^2 = 36 \quad \text{Given equation}$$

$$\frac{9x^2}{36} + \frac{4y^2}{36} = \frac{36}{36} \quad \text{Divide each side by 36.}$$

$$\frac{x^2}{4} + \frac{y^2}{9} = 1 \quad \text{Simplify.}$$

This ellipse has a vertical major axis with $a = 3$ and $b = 2$. The vertices are $(0, \pm 3)$, and the endpoints of the minor axis are $(\pm 2, 0)$, as shown in Figure 13.16(b).

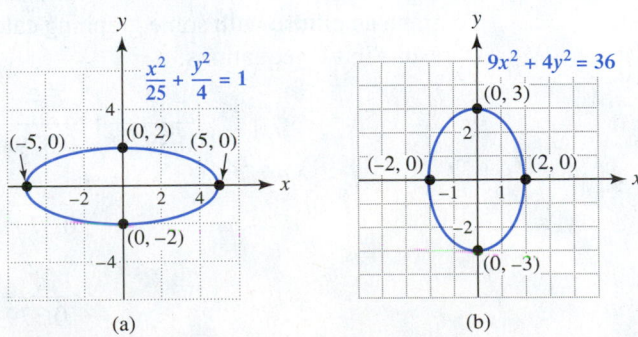

(a)   (b)

Figure 13.16

Now Try Exercises 15, 21

Finding the standard equation of an ellipse

Use the graph in Figure 13.17 to determine the standard equation of the ellipse.

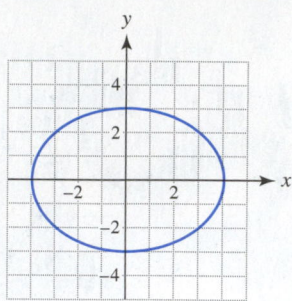

Figure 13.17

Solution

The ellipse is centered at $(0, 0)$ with a horizontal major axis. The length of the major axis is 8, so $a = 4$. The length of the minor axis is 6, so $b = 3$. Thus $a^2 = 16$ and $b^2 = 9$, and the standard equation of the ellipse is

$$\frac{x^2}{16} + \frac{y^2}{9} = 1.$$

Now Try Exercise 25

▶ **REAL-WORLD CONNECTION** Planets travel around the sun in elliptical orbits. Astronomers have measured the values of a and b for each planet. Using this information, we can find the equation of a planet's orbit, as illustrated in the next example.

Modeling the orbit of Mercury

The planet Mercury has one of the least circular orbits of the eight major planets. For Mercury, $a = 0.387$ and $b = 0.379$. The units are astronomical units (A.U.), where 1 A.U. equals 93 million miles—the distance between Earth and the sun. Graph $\frac{x^2}{a^2} + \frac{y^2}{b^2} = 1$ to model the orbit of Mercury in $[-0.6, 0.6, 0.1]$ by $[-0.4, 0.4, 0.1]$. Then plot the sun at the point $(0.08, 0)$. (*Source:* M. Zeilik, *Introductory Astronomy and Astrophysics.*)

Solution

The orbit of Mercury is given by

$$\frac{x^2}{0.387^2} + \frac{y^2}{0.379^2} = 1. \qquad\qquad \frac{x^2}{a^2} + \frac{y^2}{b^2} = 1$$

To graph an ellipse with some graphing calculators, we must solve the equation for y. Doing so results in two equations.

$$\frac{x^2}{0.387^2} + \frac{y^2}{0.379^2} = 1$$

$$\frac{y^2}{0.379^2} = 1 - \frac{x^2}{0.387^2} \qquad\qquad \text{Subtract } \tfrac{x^2}{0.387^2}.$$

$$\frac{y}{0.379} = \pm\sqrt{1 - \frac{x^2}{0.387^2}} \qquad\qquad \text{Square root property}$$

$$y = \pm 0.379\sqrt{1 - \frac{x^2}{0.387^2}} \qquad\qquad \text{Multiply by 0.379.}$$

The orbit of Mercury can be found by graphing these equations. See Figures 13.18(a) and (b). The point (0.08, 0) represents the position of the sun in Figure 13.18(b).

CRITICAL THINKING

Use Figure 13.18 and the information in Example 3 to estimate the minimum and maximum distances that Mercury is from the sun.

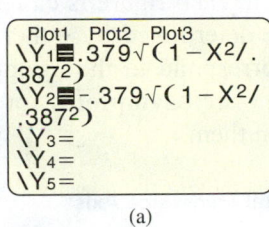

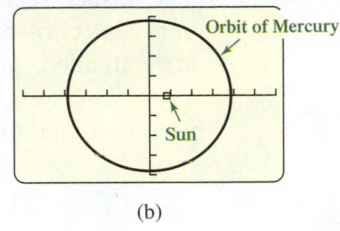

(a)

(b)

Figure 13.18

■ **Now Try Exercise 47**

Equations of Hyperbolas

The third type of conic section is the **hyperbola**, which is the set of points in a plane the difference of whose distances from two fixed points is constant. Each fixed point is called a **focus** of the hyperbola. Figure 13.19 shows a hyperbola whose equation is

$$\frac{x^2}{4} - \frac{y^2}{9} = 1.$$

This hyperbola is centered at the origin and has two **branches**, a *left branch* and a *right branch*. The **vertices** are $(-2, 0)$ and $(2, 0)$, and the line segment connecting the vertices is called the **transverse axis**. (The transverse axis is not part of the hyperbola.)

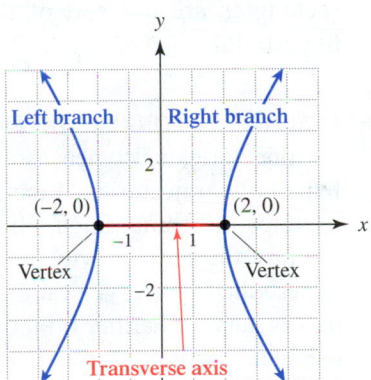

Figure 13.19

By the vertical line test, a hyperbola cannot be represented by a function, but many hyperbolas can be described by the following equations.

READING CHECK

How can you determine from the standard equation of a hyperbola if the transverse axis is horizontal or vertical?

STANDARD EQUATIONS FOR HYPERBOLAS CENTERED AT (0, 0)

The hyperbola with center at the origin, *horizontal* transverse axis, and equation

$$\frac{x^2}{a^2} - \frac{y^2}{b^2} = 1$$

has vertices $(\pm a, 0)$.

The hyperbola with center at the origin, *vertical* transverse axis, and equation

$$\frac{y^2}{a^2} - \frac{x^2}{b^2} = 1$$

has vertices $(0, \pm a)$.

Hyperbolas, along with the coordinates of their vertices, are shown in Figure 13.20. The two parts of the hyperbola in Figure 13.20(a) are the *left branch* and *right branch*, whereas in Figure 13.20(b) the hyperbola has an *upper branch* and a *lower branch*. The dashed rectangle in each figure is called the **fundamental rectangle**, and its four vertices (corners) are determined by either $(\pm a, \pm b)$ or $(\pm b, \pm a)$. If its diagonals are extended, they correspond to the asymptotes of the hyperbola. The dashed lines $y = \pm\frac{b}{a}x$ and $y = \pm\frac{a}{b}x$ are **asymptotes** for the hyperbolas, respectively, and may be used as an aid to graph them.

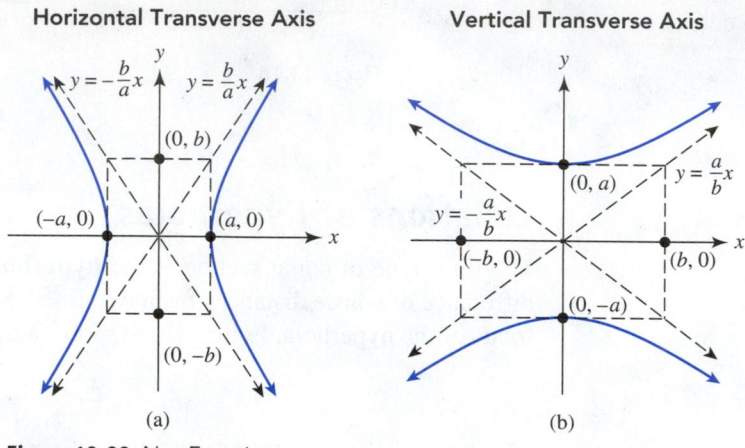

Figure 13.20 Not Functions

NOTE: A hyperbola consists of two solid curves, or branches. The dashed lines and rectangles are not part of the actual graph but are used as an aid for sketching the hyperbola.

▶ **REAL-WORLD CONNECTION** One interpretation of an asymptote of a hyperbola can be based on trajectories of comets as they approach the sun. Comets travel in parabolic, elliptic, or hyperbolic trajectories. If the speed of a comet is too slow, the gravitational pull of the sun captures the comet in an elliptic orbit (see Figure 13.21(a)). If the speed of the comet is too fast, the sun's gravity is too weak to capture the comet and the comet passes by it in a hyperbolic trajectory. Near the sun the gravitational pull is stronger, and the comet's trajectory is curved. Farther from the sun, the gravitational pull becomes weaker, and the comet eventually returns to a straight-line trajectory determined by the *asymptote* of the hyperbola (see Figure 13.21(b)). Finally, if the speed is neither too slow nor too fast, the comet will travel in a parabolic path (see Figure 13.21(c)).

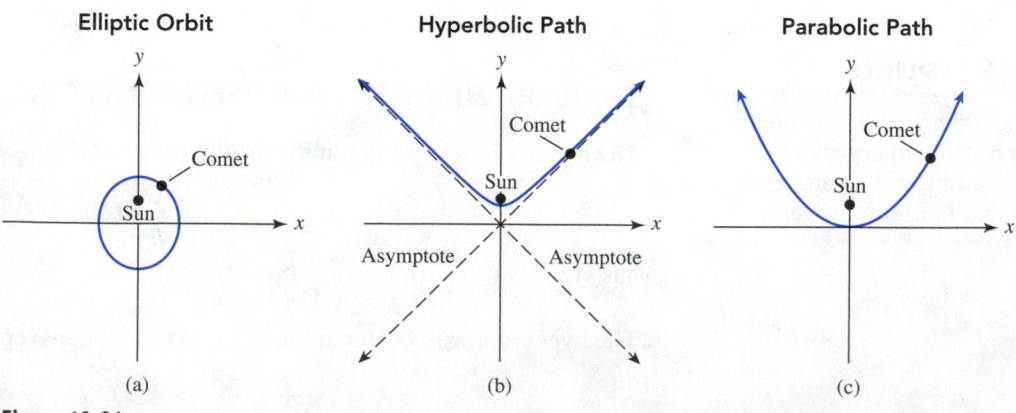

Figure 13.21

EXAMPLE 4   ### Sketching a hyperbola

Sketch a graph of $\frac{y^2}{4} - \frac{x^2}{9} = 1$. Label the vertices and show the asymptotes.

Solution

The equation is in standard form with $a^2 = 4$ and $b^2 = 9$, so $a = 2$ and $b = 3$. It has a vertical transverse axis with vertices $(0, -2)$ and $(0, 2)$. The vertices (corners) of the fundamental rectangle are $(\pm 3, \pm 2)$, that is, $(3, 2)$, $(3, -2)$, $(-3, 2)$, and $(-3, -2)$. The asymptotes are the diagonals of this rectangle and are given by $y = \pm \frac{a}{b}x$, or $y = \pm \frac{2}{3}x$. Figure 13.22 shows the hyperbola and these features.

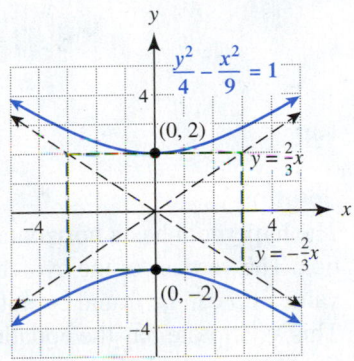

Figure 13.22

■ **Now Try Exercise 29**

TECHNOLOGY NOTE

Graphing a Hyperbola

The graph of a hyperbola does not represent a function. One way to graph a hyperbola with a graphing calculator is to solve the equation for y and obtain two equations. One equation gives the upper (or right) half of the hyperbola and the other equation gives the lower (or left) half. For example, to graph $\frac{y^2}{4} - \frac{x^2}{8} = 1$, begin by solving for y.

$$\frac{y^2}{4} = 1 + \frac{x^2}{8} \qquad \text{Add } \tfrac{x^2}{8}.$$

$$y^2 = 4\left(1 + \frac{x^2}{8}\right) \qquad \text{Multiply by 4.}$$

$$y = \pm 2\sqrt{1 + \frac{x^2}{8}} \qquad \text{Square root property}$$

Graph $Y_1 = 2\sqrt{(1 + X^2/8)}$ and $Y_2 = -2\sqrt{(1 + X^2/8)}$, which give the upper and lower branches, respectively. See Figure 13.23.

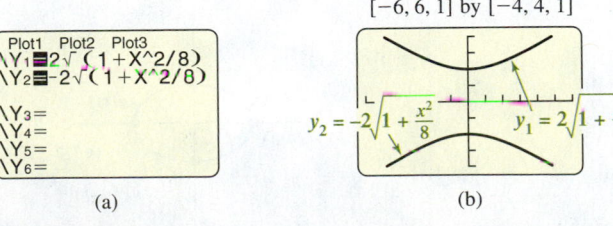

(a)   (b)

Figure 13.23

EXAMPLE 5 | **Determining the standard equation of a hyperbola**

Use the graph in Figure 13.24 to determine the standard equation of the hyperbola.

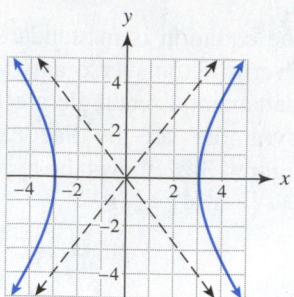

Figure 13.24

Solution

The hyperbola has a horizontal transverse axis, so the x^2-term must come first in the equation. The vertices of the hyperbola are $(\pm 3, 0)$, which indicates that $a = 3$ and $a^2 = 9$. The value of b can be found by noting that one of the asymptotes passes through the point $(3, 4)$. This asymptote has the equation $y = \frac{b}{a}x$ or $y = \frac{4}{3}x$, so let $b = 4$ and $b^2 = 16$. The equation of the hyperbola is

$$\frac{x^2}{9} - \frac{y^2}{16} = 1.$$

Now Try Exercise 41

13.2 Putting It All Together

CONCEPT	DESCRIPTION	
Ellipses Centered at $(0, 0)$ with $a > b > 0$	**_Horizontal Major Axis_** Vertices: $(a, 0)$ and $(-a, 0)$ Endpoints of minor axis: $(0, b)$ and $(0, -b)$ $$\frac{x^2}{a^2} + \frac{y^2}{b^2} = 1$$ 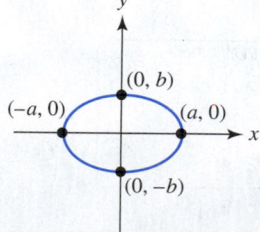	**_Vertical Major Axis_** Vertices: $(0, a)$ and $(0, -a)$ Endpoints of minor axis: $(-b, 0)$ and $(b, 0)$ $$\frac{x^2}{b^2} + \frac{y^2}{a^2} = 1$$ 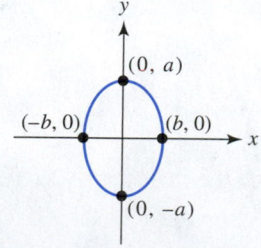

CONCEPT	DESCRIPTION	
Hyperbolas Centered at $(0, 0)$ with $a > 0$ and $b > 0$	**Horizontal Transverse Axis** Vertices: $(a, 0)$ and $(-a, 0)$ Asymptotes: $y = \pm\dfrac{b}{a}x$ $$\dfrac{x^2}{a^2} - \dfrac{y^2}{b^2} = 1$$ 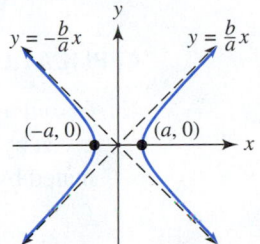	**Vertical Transverse Axis** Vertices: $(0, a)$ and $(0, -a)$ Asymptotes: $y = \pm\dfrac{a}{b}x$ $$\dfrac{y^2}{a^2} - \dfrac{x^2}{b^2} = 1$$ 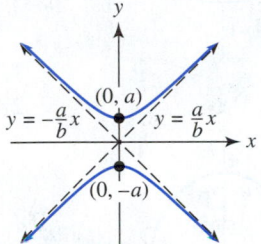

13.2 Exercises

MyMathLab   Math XL PRACTICE   WATCH   DOWNLOAD   READ   REVIEW

CONCEPTS AND VOCABULARY

1. Sketch an ellipse with a horizontal major axis.

2. Sketch a hyperbola with a vertical transverse axis.

3. The ellipse whose standard equation is $\frac{x^2}{a^2} + \frac{y^2}{b^2} = 1$, $a > b > 0$, has a(n) _____ major axis.

4. The ellipse whose standard equation is $\frac{x^2}{b^2} + \frac{y^2}{a^2} = 1$, $a > b > 0$, has a(n) _____ major axis.

5. What is the maximum number of times that a line can intersect an ellipse?

6. What is the maximum number of times that a parabola can intersect an ellipse?

7. The hyperbola whose equation is $\frac{x^2}{a^2} - \frac{y^2}{b^2} = 1$ has _____ and _____ branches.

8. The hyperbola whose equation is $\frac{y^2}{a^2} - \frac{x^2}{b^2} = 1$ has _____ and _____ branches.

9. How are the asymptotes of a hyperbola related to the fundamental rectangle?

10. Could an ellipse be centered at the origin and have vertices $(4, 0)$ and $(0, -5)$?

11. Which of the following (a.–d.) represents the coordinates of the vertices on the ellipse $\frac{x^2}{4} + \frac{y^2}{9} = 1$?
    a. $(0, \pm 3)$ b. $(\pm 2, 0)$
    c. $(\pm 4, 0)$ d. $(0, \pm 9)$

12. Which of the following (a.–d.) represents the coordinates of the vertices on the hyperbola $\frac{x^2}{4} - \frac{y^2}{9} = 1$?
    a. $(0, \pm 3)$ b. $(\pm 2, 0)$
    c. $(\pm 4, 0)$ d. $(0, \pm 9)$

ELLIPSES

Exercises 13–24: Graph the ellipse. Label the vertices and endpoints of the minor axis.

13. $\dfrac{x^2}{9} + \dfrac{y^2}{25} = 1$ 14. $\dfrac{y^2}{9} + \dfrac{x^2}{25} = 1$

15. $\dfrac{x^2}{9} + \dfrac{y^2}{4} = 1$ 16. $\dfrac{x^2}{3} + \dfrac{y^2}{9} = 1$

17. $x^2 + \dfrac{y^2}{4} = 1$ 18. $\dfrac{x^2}{9} + y^2 = 1$

19. $\dfrac{y^2}{5} + \dfrac{x^2}{7} = 1$ 20. $\dfrac{y^2}{11} + \dfrac{x^2}{6} = 1$

21. $36x^2 + 4y^2 = 144$ 22. $25x^2 + 16y^2 = 400$

23. $6y^2 + 7x^2 = 42$ 24. $9x^2 + 5y^2 = 45$

Exercises 25–28: Use the graph to determine the standard equation of the ellipse.

25.

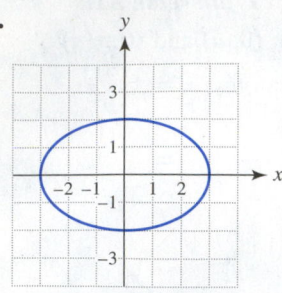

26.

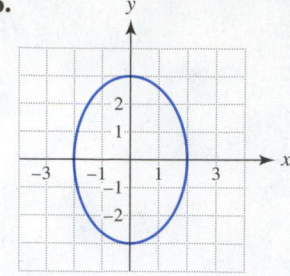

27.

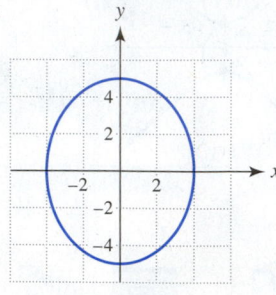

28.

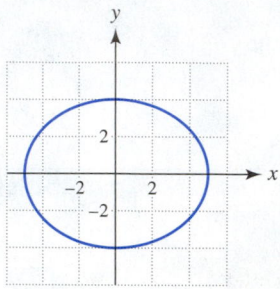

HYPERBOLAS

Exercises 29–40: Graph the hyperbola. Show the asymptotes and vertices.

29. $\dfrac{x^2}{4} - \dfrac{y^2}{9} = 1$

30. $\dfrac{y^2}{4} - \dfrac{x^2}{9} = 1$

31. $\dfrac{x^2}{25} - \dfrac{y^2}{16} = 1$

32. $\dfrac{y^2}{25} - \dfrac{x^2}{16} = 1$

33. $x^2 - y^2 = 1$

34. $y^2 - x^2 = 1$

35. $\dfrac{x^2}{3} - \dfrac{y^2}{4} = 1$

36. $\dfrac{y^2}{5} - \dfrac{x^2}{8} = 1$

37. $9y^2 - 4x^2 = 36$

38. $36x^2 - 25y^2 = 900$

39. $16x^2 - 4y^2 = 64$

40. $y^2 - 9x^2 = 9$

Exercises 41–44: Use the graph to determine the standard equation of the hyperbola.

41.

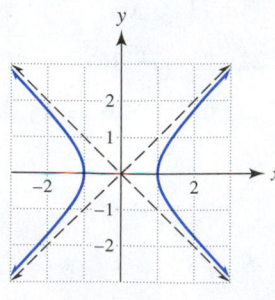

42.

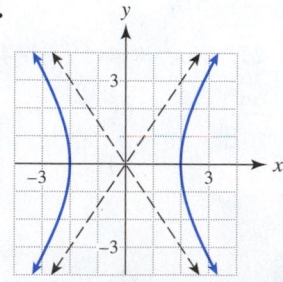

43.

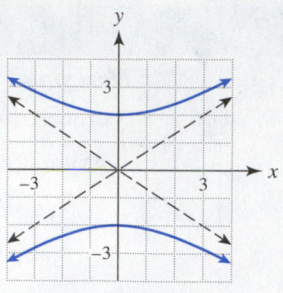

44.

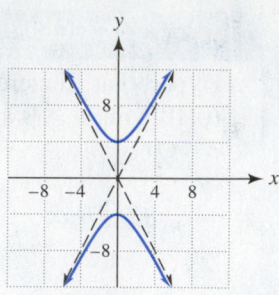

APPLICATIONS

45. *Geometry of an Ellipse* The area inside an ellipse is given by $A = \pi ab$, and its perimeter can be approximated by

$$P = 2\pi \sqrt{\dfrac{a^2 + b^2}{2}}.$$

Approximate A and P to the nearest hundredth for each ellipse.

(a) $\dfrac{x^2}{16} + \dfrac{y^2}{25} = 1$ **(b)** $\dfrac{x^2}{7} + \dfrac{y^2}{2} = 1$

46. *Geometry of an Ellipse* (Refer to Exercise 45.) If $a = b$ in the equation for an ellipse, the ellipse becomes a circle. Let $a = b$ in the formulas for the area and perimeter of an ellipse. Do the equations simplify to the area and perimeter for a circle? Explain.

47. *Planet Orbit* (Refer to Example 3.) Pluto's orbit is less circular than that of any of the eight planets. For Pluto (now a dwarf planet), $a = 39.44$ A.U. and $b = 38.20$. A.U.

    (a) Graph the elliptic orbit of Pluto in the window $[-60, 60, 10]$ by $[-40, 40, 10]$. Plot the point $(9.81, 0)$ to show the position of the sun. Assume that the major axis is horizontal.

    (b) Use the information in Exercise 45 to determine how far Pluto travels in one orbit around the sun and approximate the area inside its orbit.

48. *Halley's Comet* (Refer to Example 3.) The famous Halley's comet travels in an elliptical orbit with $a = 17.95$ and $b = 4.44$ and passes by Earth roughly every 76 years. The most recent pass by Earth was in February 1986. (*Source:* M. Zeilik.)

    (a) Graph the orbit of Halley's comet in $[-21, 21, 5]$ by $[-14, 14, 5]$. Assume that the major axis is horizontal and that all units are in astronomical units. Plot a point at $(17.39, 0)$ to represent the position of the sun.

(b) Use the formula in Exercise 45 to estimate how many miles Halley's comet travels in one orbit around the sun.

(c) Estimate the average speed of Halley's comet in miles per hour.

49. *Satellite Orbit* The orbit of Explorer VII and the outline of Earth's surface are shown in the accompanying figure. This orbit is described by

$$\frac{x^2}{4464^2} + \frac{y^2}{4462^2} = 1,$$

and the surface of Earth is described by

$$\frac{(x - 164)^2}{3960^2} + \frac{y^2}{3960^2} = 1.$$

Find the maximum and minimum heights of the satellite above Earth's surface if all units are miles. (*Source:* W. Thomson, *Introduction to Space Dynamics.*)

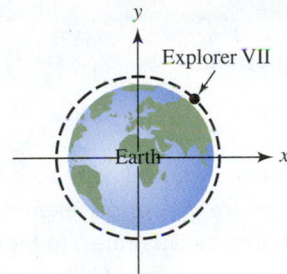

50. *Weight Machines* Elliptic shapes are used rather than circular shapes in some weight machines. Suppose that the ellipse shown in the accompanying figure is represented by the equation

$$\frac{x^2}{16} + \frac{y^2}{100} = 1,$$

where the units are inches. Find r_1 and r_2.

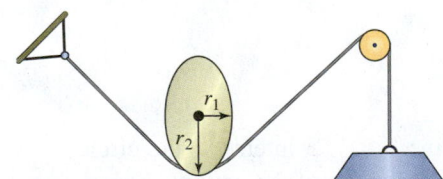

51. *Arch Bridge* The arch under a bridge is designed as the upper half of an ellipse, as illustrated in the accompanying figure. Its equation is modeled by

$$400x^2 + 10,000y^2 = 4,000,000,$$

where the units are feet. Find the height and width of the arch.

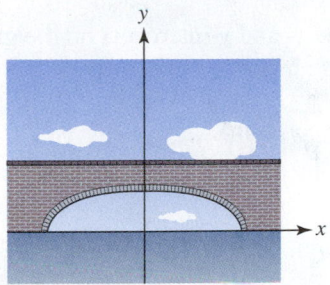

52. **Thinking Generally** Suppose that the population y of a country can be modeled by the upper right branch of the hyperbola

$$\frac{x^2}{a^2} - \frac{y^2}{b^2} = 1,$$

where x represents time in years. What happens to the population after a long period of time?

53. **Online Exploration** Go online and find one application of ellipses. Write a short paragraph about your findings.

54. **Online Exploration** Go online and find one application of hyperbolas. Write a short paragraph about your findings.

WRITING ABOUT MATHEMATICS

55. Explain how the values of a and b affect the graph of $\frac{x^2}{a^2} + \frac{y^2}{b^2} = 1$. Assume that $a > b > 0$.

56. Explain how the values of a and b affect the graph of $\frac{x^2}{a^2} - \frac{y^2}{b^2} = 1$. Assume that a and b are positive.

Checking Basic Concepts

1. Graph the parabola $x = (y - 2)^2 + 1$. Find the vertex and axis of symmetry.

2. Find the equation of the circle with center $(1, -2)$ and radius 2. Graph the circle.

3. Find the x- and y-intercepts on the graph of
$$\frac{x^2}{4} + \frac{y^2}{9} = 1.$$

4. Graph the following. Label any vertices and state the type of conic section that it represents.

   (a) $x = y^2$

   (b) $\dfrac{x^2}{16} + \dfrac{y^2}{25} = 1$

   (c) $\dfrac{x^2}{4} - \dfrac{y^2}{9} = 1$

   (d) $(x - 1)^2 + (y + 2)^2 = 9$

13.3 Nonlinear Systems of Equations and Inequalities

Basic Concepts • Solving Nonlinear Systems of Equations • Solving Nonlinear Systems of Inequalities

A LOOK INTO MATH ▶

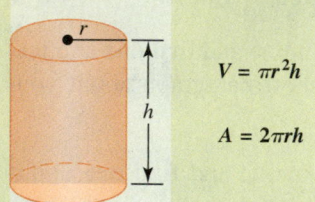

$V = \pi r^2 h$

$A = 2\pi rh$

Figure 13.25 Cylindrical Container

To describe characteristics of curved objects we often need *nonlinear equations*. The equations of the conic sections discussed in this chapter are but a few examples of nonlinear equations. For instance, cylinders have a curved shape, as illustrated in Figure 13.25. If the radius of a cylinder is denoted r and its height h, then its volume V is given by the nonlinear equation $V = \pi r^2 h$ and its side area A is given by the nonlinear equation $A = \mathbf{2\pi rh}$.

If we want to manufacture a cylindrical container that holds **35** cubic inches and whose side area is **50** square inches, we need to solve the following **nonlinear system of equations**. (This system is solved in Example 4.)

$$\pi r^2 h = 35$$
$$2\pi rh = 50$$

In this section we solve nonlinear systems of equations and inequalities.

NEW VOCABULARY

☐ Nonlinear system of equations
☐ Method of substitution
☐ Nonlinear system of inequalities

Basic Concepts

One way to locate the points at which the line $y = 2x$ intersects the circle $x^2 + y^2 = 5$ is to graph both equations (see Figure 13.26 at the top of the next page).

The equation describing the circle is nonlinear. Another way to locate the points of intersection is symbolically, by solving the nonlinear system of equations.

$$y = 2x \quad \text{Line}$$
$$x^2 + y^2 = 5 \quad \text{Circle}$$

Linear systems of equations can have no solutions, one solution, or infinitely many solutions. It is possible for a nonlinear system of equations to have *any number* of solutions. Figure 13.26 shows that this nonlinear system of equations has two solutions: $(-1, -2)$ and $(1, 2)$.

Two Solutions

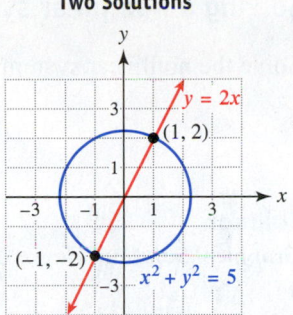

Figure 13.26

Solving Nonlinear Systems of Equations

Nonlinear systems of equations in two variables can sometimes be solved graphically, numerically, and symbolically. One symbolic technique is the **method of substitution**, which we demonstrate in the next example.

EXAMPLE 1   **Solving a nonlinear system of equations symbolically**

Solve the following system of equations symbolically. Check any solutions and compare them with Figure 13.26.

$$y = 2x$$
$$x^2 + y^2 = 5$$

Solution
Substitute $2x$ for y in the second equation and solve for x.

$x^2 + (2x)^2 = 5$	Let $y = 2x$ in the second equation.
$x^2 + 4x^2 = 5$	Properties of exponents
$5x^2 = 5$	Combine like terms.
$x^2 = 1$	Divide by 5.
$x = \pm 1$	Square root property

To determine corresponding y-values, substitute $x = \pm 1$ in $y = 2x$; the solutions are $(1, 2)$ and $(-1, -2)$. To check $(1, 2)$, substitute $x = 1$ and $y = 2$ in the given equations.

$$2 \overset{?}{=} 2(1) \checkmark \qquad \text{True}$$
$$(1)^2 + (2)^2 \overset{?}{=} 5 \checkmark \qquad \text{True}$$

To check $(-1, -2)$, substitute $x = -1$ and $y = -2$ in the given equations.

$$-2 \overset{?}{=} 2(-1) \checkmark \qquad \text{True}$$
$$(-1)^2 + (-2)^2 \overset{?}{=} 5 \checkmark \qquad \text{True}$$

The solutions check and agree with Figure 13.26.

Now Try Exercise 13

In the next example we solve a nonlinear system of equations graphically and symbolically. The symbolic solution gives the exact solutions.

EXAMPLE 2 **Solving a nonlinear system of equations**

Solve the nonlinear system of equations graphically and symbolically.

$$x^2 - y = 2$$
$$x^2 + y = 4$$

Solution

Graphical Solution Begin by solving each equation for y.

$$y = x^2 - 2$$
$$y = 4 - x^2$$

Graph $y_1 = x^2 - 2$ and $y_2 = 4 - x^2$. The solutions are approximately $(-1.73, 1)$ and $(1.73, 1)$, as shown in Figure 13.27. The graphs consist of two parabolas intersecting at two points.

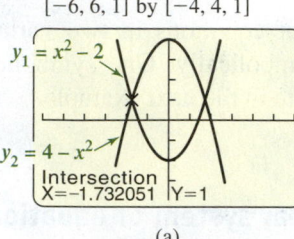

(a)   (b)

Figure 13.27

Symbolic Solution Solving the first equation for y gives $y = x^2 - 2$. Substitute this expression for y in the second equation and solve for x.

$$x^2 + \mathbf{y} = 4 \qquad \text{Second equation}$$
$$x^2 + (\mathbf{x^2 - 2}) = 4 \qquad \text{Substitute } y = x^2 - 2.$$
$$2x^2 = 6 \qquad \text{Combine like terms; add 2.}$$
$$x^2 = 3 \qquad \text{Divide by 2.}$$
$$x = \pm\sqrt{3} \qquad \text{Square root property}$$

To determine y, substitute $x = \pm\sqrt{3}$ in $y = x^2 - 2$.

$$y = (\sqrt{3})^2 - 2 = 3 - 2 = \mathbf{1}$$
$$y = (-\sqrt{3})^2 - 2 = 3 - 2 = \mathbf{1}$$

The *exact* solutions are $(\sqrt{3}, \mathbf{1})$ and $(-\sqrt{3}, \mathbf{1})$.

Now Try Exercise 29(a), (b)

EXAMPLE 3 **Solving a nonlinear system of equations**

Solve the nonlinear system of equations symbolically and graphically.

$$x^2 - y^2 = 3$$
$$x^2 + y^2 = 5$$

Solution

Symbolic Solution Instead of using substitution on this nonlinear system of equations, we use elimination. Note that, if we add the two equations, the y-variable will be eliminated.

$$
\begin{array}{ll}
x^2 - y^2 = 3 & \text{First equation} \\
\underline{x^2 + y^2 = 5} & \text{Second equation} \\
2x^2 \quad\quad = 8 & \text{Add equations.}
\end{array}
$$

Solving gives $x^2 = 4$, or $x = \pm 2$. To determine y, substitute **4** for x^2 in $x^2 + y^2 = 5$.

$$4 + y^2 = 5 \quad \text{or} \quad y^2 = 1$$

Because $y^2 = 1$, $y = \pm 1$. There are four solutions: $(2, 1)$, $(2, -1)$, $(-2, 1)$, and $(-2, -1)$.

Graphical Solution The graph of the first equation is a hyperbola, and the graph of the second is a circle with radius $\sqrt{5}$. The four points of intersection (solutions) are $(\pm 2, \pm 1)$, as shown in Figure 13.28.

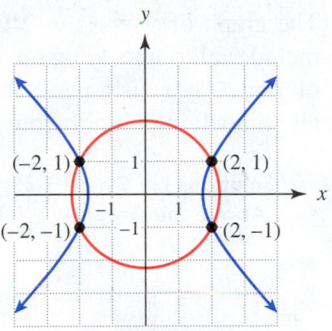

Figure 13.28

> **Now Try Exercises 23, 25**

In the next example we solve the system of equations presented in A Look Into Math for this section.

EXAMPLE 4

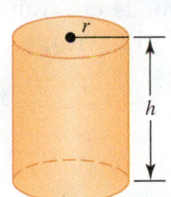

Modeling the dimensions of a can

Find the dimensions of a can having a volume V of 35 cubic inches and a side area A of 50 square inches by solving the following nonlinear system of equations symbolically. (See the figure in the margin.)

$$\pi r^2 h = 35$$
$$2\pi r h = 50$$

Solution

We can find r by solving each equation for h and then setting them equal. This eliminates the variable h.

$$\frac{50}{2\pi r} = \frac{35}{\pi r^2} \qquad h = \frac{50}{2\pi r} \text{ and } h = \frac{35}{\pi r^2}$$

$$50\pi r^2 = 70\pi r \qquad \text{Eliminate fractions (cross multiply).}$$

$$50\pi r^2 - 70\pi r = 0 \qquad \text{Subtract } 70\pi r.$$

$$10\pi r(5r - 7) = 0 \qquad \text{Factor out } 10\pi r.$$

$$10\pi r = 0 \quad \text{or} \quad 5r - 7 = 0 \qquad \text{Zero-product property}$$

$$r = 0 \quad \text{or} \quad r = \tfrac{7}{5} = 1.4 \qquad \text{Solve.}$$

Because $h = \frac{50}{2\pi r}$, $r = 0$ is not possible, but we can find h by substituting **1.4** for r in the formula.

$$h = \frac{50}{2\pi (1.4)} \approx 5.68$$

A can having a volume of 35 cubic inches and a side area of 50 square inches has a radius of 1.4 inches and a height of about 5.68 inches.

READING CHECK

How many solutions can a nonlinear system of equations have?

> **Now Try Exercise 51(b)**

Solving Nonlinear Systems of Inequalities

In Chapter 4 we solved systems of *linear* inequalities. A **nonlinear system of inequalities** in two variables can be solved similarly by using graphical techniques. For example, consider the nonlinear system of inequalities

$$y \geq x^2 - 2$$
$$y \leq 4 - x^2.$$

The graph of $y = x^2 - 2$ is a parabola opening upward. The solution set to $y \geq x^2 - 2$ includes all points lying on or above this parabola. See Figure 13.29(a). Similarly, the graph of $y = 4 - x^2$ is a parabola opening downward. The solution set to $y \leq 4 - x^2$ includes all points lying on or below this parabola. See Figure 13.29(b).

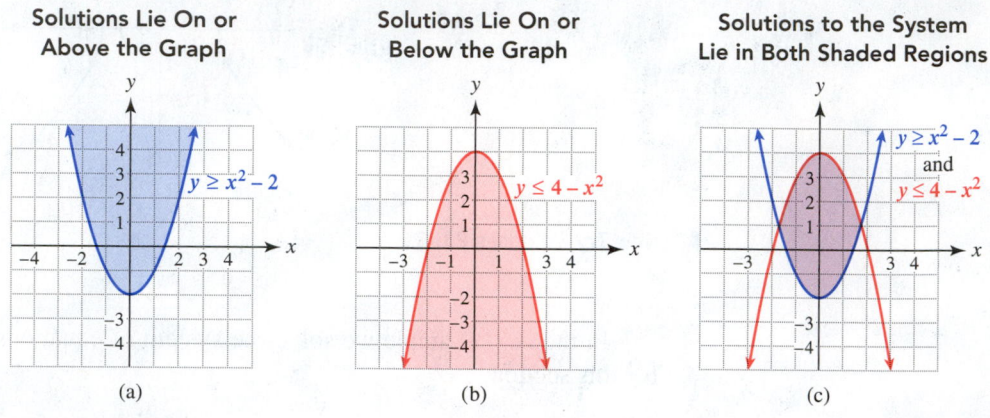

(a)   (b)   (c)

Figure 13.29

The solution set for this nonlinear *system* of inequalities includes all points (x, y) in *both* shaded regions. The *intersection* of the shaded regions is shown in Figure 13.29(c).

EXAMPLE 5   Solving a nonlinear system of inequalities graphically

Shade the solution set for the system of inequalities.

$$\frac{x^2}{4} + \frac{y^2}{9} < 1$$
$$y > 1$$

Solution
The solutions to $\frac{x^2}{4} + \frac{y^2}{9} < 1$ lie *inside* the ellipse $\frac{x^2}{4} + \frac{y^2}{9} = 1$. See Figure 13.30(a). Solutions to $y > 1$ lie above the line $y = 1$, as shown in Figure 13.30(b). The intersection of these two regions is shown in Figure 13.30(c). Any point in this region is a solution.

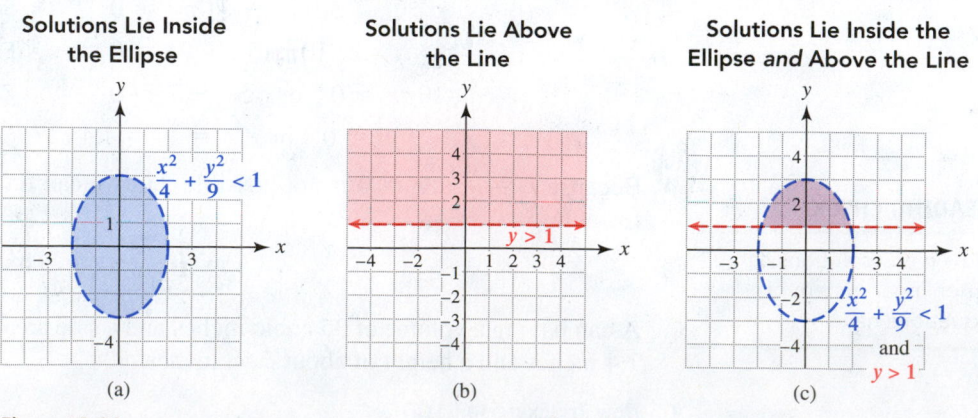

(a)   (b)   (c)

Figure 13.30

For example, the point (0, 2) lies in the shaded region and is a solution to the system. Note that a dashed curve and a dashed line are used when equality is not included.

■ **Now Try Exercise 39**

EXAMPLE 6 **Solving a nonlinear system of inequalities graphically**

Shade the solution set for the following system of inequalities.

$$x^2 + y \leq 4$$
$$-x + y \geq 2$$

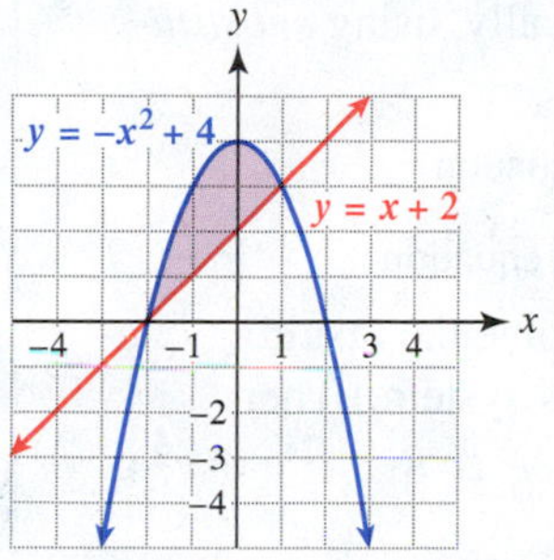

Solution

The solutions to $x^2 + y \leq 4$ lie *on* or *below* the parabola $y = -x^2 + 4$, and the solutions to $-x + y \geq 2$ lie *on* or *above* the line $y = x + 2$. The appropriate shaded region is shown in Figure 13.31. Both the parabola and the line are solid because equality is included in both inequalities.

■ **Now Try Exercise 37**

Figure 13.31

In the next example we use a graphing calculator to shade a region that lies above both graphs, using the "$Y_1 =$" menu. This feature allows us to shade either above or below the graph of a function.

EXAMPLE 7 **Solving a system of inequalities with a graphing calculator**

Shade the solution set for the following system of inequalities.

$$y \geq x^2 - 2$$
$$y \geq -1 - x$$

Solution

Enter $y_1 = x^2 - 2$ and $y_2 = -1 - x$, as shown in Figure 13.32(a). Note that the option to shade above the graphs of Y_1 and Y_2 was selected to the left of Y_1 and Y_2. Then the two inequalities were graphed in Figure 13.32(b). The solution set corresponds to the region where there are both vertical and horizontal lines.

CALCULATOR HELP

To shade the solution set to a system of inequalities, see Appendix A (pages AP-6 and AP-7).

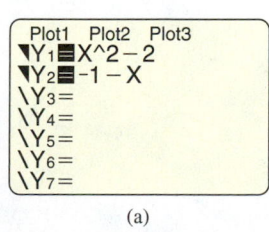

[−4.7, 4.7, 1] by [−3.1, 3.1, 1]

(a)   (b)

Figure 13.32

■ **Now Try Exercise 49**

13.3   Putting It All Together

Unlike a linear system of equations, a nonlinear system of equations can have *any number of solutions*. Nonlinear systems of inequalities involving two variables usually have infinitely many solutions, which can be represented by a shaded region in the *xy*-plane.

CONCEPT	EXPLANATION
Nonlinear Systems of Equations in Two Variables	To solve the following system of equations symbolically, using *substitution*, solve the first equation for y to get $y = 5 - x$. $$x + y = 5 \quad \text{First equation}$$ $$x^2 - y = 1 \quad \text{Second equation}$$ Substitute $5 - x$ for y in the second equation and solve the resulting quadratic equation. (*Elimination* can also be used on this system.) Then $$x^2 - (5 - x) = 1 \quad \text{or} \quad x^2 + x - 6 = 0,$$ and factoring can be used to determine that the solutions are given by $$x = -3 \quad \text{or} \quad x = 2.$$ Thus $y = 5 - (-3) = 8$ or $y = 5 - 2 = 3$. The solutions are $(-3, 8)$ and $(2, 3)$. Graphical support is shown in the accompanying figure. 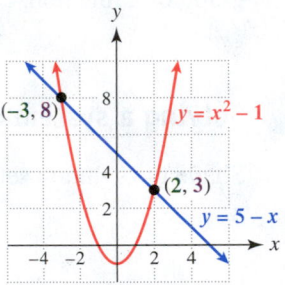
Nonlinear Systems of Inequalities in Two Variables	To solve the following system of inequalities graphically, solve each inequality for y. $$x + y \leq 5 \quad \text{or} \quad y \leq 5 - x \quad \text{First inequality}$$ $$x^2 - y \leq 1 \quad \text{or} \quad y \geq x^2 - 1 \quad \text{Second inequality}$$ The solutions lie on or above the parabola and on or below the line, as shown in the figure. 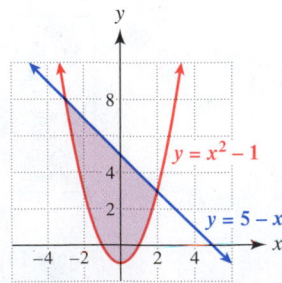

13.3 Exercises

CONCEPTS AND VOCABULARY

1. How many solutions can a nonlinear system of equations have?

2. If a nonlinear system of equations has two equations, how many equations does a solution have to satisfy?

3. Determine visually the number of solutions to the following system of equations. Explain your reasoning.

$$y = x$$
$$x^2 + y^2 = 4$$

4. Describe the solution set to $x^2 + y^2 \leq 1$.

5. Does $(-2, -1)$ satisfy $5x^2 - 2y^2 > 18$?

6. Does $(3, 4)$ satisfy $x^2 - 2y \geq 4$?

7. Sketch a parabola and ellipse with four points of intersection.

8. Sketch a line and a hyperbola with two points of intersection.

NONLINEAR SYSTEMS OF EQUATIONS

Exercises 9–12: Use the graph to estimate all solutions to the system of equations. Check each solution.

9. $x^2 + y^2 = 10$
$\quad\ y = 3x$

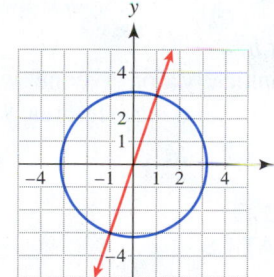

10. $x^2 + 3y^2 = 16$
$\quad\ y = -x$

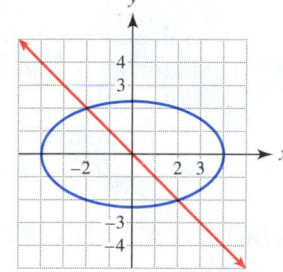

11. $y^2 - x^2 = 1$
$\quad x^2 + 3y^2 = 3$

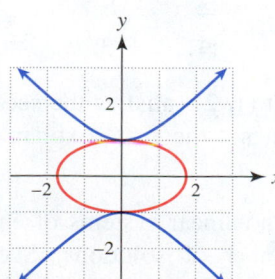

12. $\quad\ y = 1 - x^2$
$\quad x^2 + y^2 = 1$

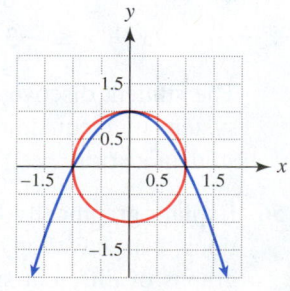

Exercises 13–24: Solve the system of equations symbolically. Check your solutions.

13. $\quad\ y = 2x$
$\quad x^2 + y^2 = 45$

14. $y = x$
$\quad y^2 = 3 - 2x^2$

15. $x + y = 1$
$\quad x^2 - y^2 = 3$

16. $y - x = -1$
$\quad\quad y = 2x^2$

17. $y - x^2 = 0$
$\quad x^2 + y^2 = 6$

18. $x^2 - y^2 = 4$
$\quad x^2 + y^2 = 4$

19. $3x^2 + 2y^2 = 5$
$\quad\ x - y = -2$

20. $\quad x^2 + 2y^2 = 9$
$\quad 2x^2 - y^2 = -2$

21. $x^2 + y^2 = 4$
$\quad x^2 - 9y^2 = 9$

22. $\quad x^2 + y^2 = 15$
$\quad 9x^2 + 4y^2 = 36$

23. $\quad x^2 + y^2 = 10$
$\quad 2x^2 - y^2 = 17$

24. $2x^2 + 3y^2 = 5$
$\quad 3x^2 - 4y^2 = -1$

Exercises 25–28: Solve the system of equations graphically. Check your solutions.

25. $\quad\quad y = x^2 - 3$
$\quad 2x^2 - y = 1 - 3x$

26. $x + y = 2$
$\quad x - y^2 = 3$

27. $y - x = -4$
$\quad x - y^2 = -2$

28. $xy = 1$
$\quad\ y = x$

Exercises 29–32: Solve the system of equations
  (a) symbolically,
  (b) graphically, and
  (c) numerically.

29. $\quad\quad y = -2x$
$\quad x^2 + y = 3$

30. $4x - y = 0$
$\quad x^3 - y = 0$

31. $\quad xy = 1$
$\quad x - y = 0$

32. $x^2 + y^2 = 4$
$\quad y - x = 2$

NONLINEAR SYSTEMS OF INEQUALITIES

Exercises 33–36: Shade the solution set in the xy-plane.

33. $y \geq x^2$

34. $y \leq x^2 - 1$

35. $\dfrac{x^2}{4} + \dfrac{y^2}{9} > 1$

36. $x^2 + y^2 \leq 1$

Exercises 37–44: Shade the solution set in the xy-plane. Then use the graph to select one solution.

37. $y > x^2 + 1$
$\quad y < 3$

38. $y > x^2$
$\quad y < x + 2$

39. $x^2 + y^2 \le 1$
$ y < x$

40. $y > x^2 - 2$
$ y \le 2 - x^2$

41. $ x^2 + y^2 \le 1$
$(x - 2)^2 + y^2 \le 1$

42. $ x^2 - y \ge 2$
$(x + 1)^2 + y^2 \le 4$

43. $x^2 - y^2 \le 4$
$ x^2 + y^2 \le 9$

44. $3x + 2y < 6$
$ x^2 + y^2 \le 16$

Exercises 45 and 46: Match the inequality or system of inequalities with its graph (a. or b.).

45. $y \le \dfrac{1}{2}x^2$

46. $y \ge x^2 + 1$
$ y \le 5$

a.

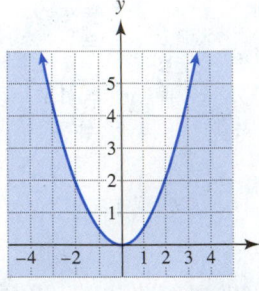

b.

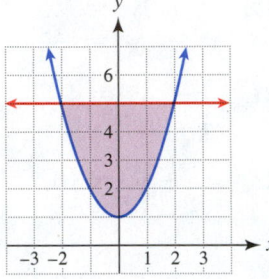

Exercises 47 and 48: Use the graph to write the inequality or system of inequalities.

47.

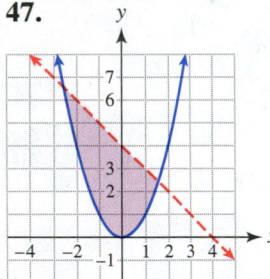

48.

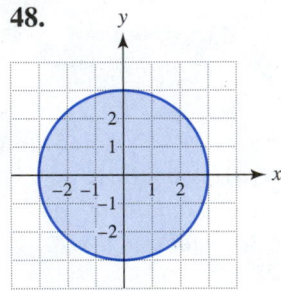

 Exercises 49 and 50: Use a graphing calculator to shade the solution set to the system of inequalities.

49. $y \ge x^2 - 1$
$ y \ge 2 - x$

50. $y \le 4 - x^2$
$ y \le x^2 - 4$

APPLICATIONS

51. *Dimensions of a Can* (Refer to Example 4.) Find the dimensions of a cylindrical container with a volume of 40 cubic inches and a side area of 50 square inches **(a)** graphically and **(b)** symbolically.

52. *Dimensions of a Can* (Refer to Example 4.) Is it possible to design an aluminum can with volume of 60 cubic inches and side area of 60 square inches? If so, find the dimensions of the can.

53. *Area and Perimeter* The area of a room is 143 feet, and its perimeter is 48 feet. Let x be the width and y be the length of the room. See the accompanying figure.
(a) Write a nonlinear system of equations that models this situation.
(b) Solve the system.

54. *Dimensions of a Cone* The volume V of a cone is given by $V = \frac{1}{3}\pi r^2 h$, and the surface area S of its side is given by $S = \pi r\sqrt{r^2 + h^2}$, where h is the height and r is the radius of the base (see the accompanying figure).

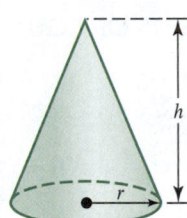

(a) Solve each equation for h.
(b) Estimate r and h graphically for a cone with volume V of 34 cubic feet and surface area S of 52 square feet.

WRITING ABOUT MATHEMATICS

55. A student *incorrectly* changes the following system of inequalities

$$\begin{array}{c} x^2 - y \ge 6 \\ 2x - y \le -3 \end{array} \quad \text{to} \quad \begin{array}{c} y \ge x^2 - 6 \\ y \le 2x + 3. \end{array}$$

The student discovers that $(1, 2)$ satisfies the second system of inequalities but not the first. Explain the student's error.

56. Explain graphically how nonlinear systems of equations can have any number of solutions. Sketch graphs of different systems with zero, one, two, and three solutions.

SECTION 13.3 Checking Basic Concepts

1. Solve the following system of equations symbolically and graphically.

$$x^2 - y = 2x$$
$$2x - y = 3$$

2. Determine visually the number of solutions to the following system of equations.

$$y = x^2 - 4$$
$$y = x$$

3. The solution set for a system of inequalities is shown in the accompanying figure.
   (a) Find one ordered pair (x, y) that is a solution and one that is not.

(b) Write the system of inequalities represented by the graph.

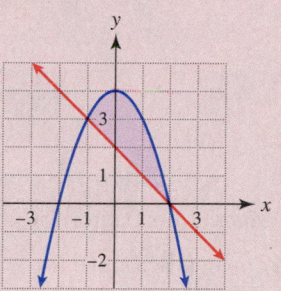

4. Shade the solution set for the following system of inequalities.

$$x^2 + y^2 \leq 4$$
$$y < 1$$

CHAPTER 13   Summary

SECTION 13.1 ■ PARABOLAS AND CIRCLES

Parabolas   There are three basic types of conic sections: parabolas, ellipses, and hyperbolas. A parabola can have a vertical or a horizontal axis. Two forms of an equation for a parabola with a *vertical axis* are

$$y = ax^2 + bx + c \quad \text{and} \quad y = a(x - h)^2 + k.$$

If $a > 0$ the parabola opens upward, and if $a < 0$ it opens downward (see the figure on the left). The vertex is located at (h, k). Two forms of an equation for a parabola with a *horizontal axis* are

$$x = ay^2 + by + c \quad \text{and} \quad x = a(y - k)^2 + h.$$

If $a > 0$ the parabola opens to the right, and if $a < 0$ it opens to the left (see the figure on the right). The vertex is located at (h, k).

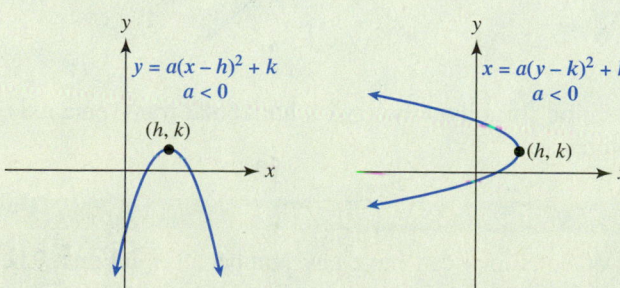

Example: $x = -2(y - 4)^2 + 1$ has vertex $(1, 4)$, has axis of symmetry $y = 4$, and opens to the left.

Circles The standard equation for a circle with center (h, k) and radius r is

$$(x - h)^2 + (y - k)^2 = r^2.$$

Example: $(x + 2)^2 + (y - 3)^2 = 36$ has center $(-2, 3)$ and radius 6.

SECTION 13.2 ■ ELLIPSES AND HYPERBOLAS

Ellipses The standard equation for an ellipse centered at the origin with a *horizontal major axis* is $\frac{x^2}{a^2} + \frac{y^2}{b^2} = 1$, $a > b > 0$, and the vertices are $(\pm a, 0)$, as shown in the figure on the left. The standard equation for an ellipse centered at the origin with a *vertical major axis* is $\frac{x^2}{b^2} + \frac{y^2}{a^2} = 1$, $a > b > 0$, and the vertices are $(0, \pm a)$, as shown in the figure on the right. Circles are a special type of ellipse, with the major and minor axes having equal lengths.

Horizontal Major Axis   **Vertical Major Axis**

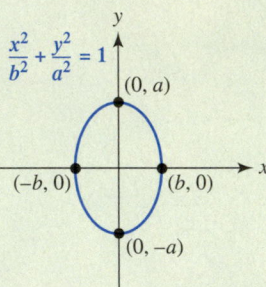

Example: $\frac{x^2}{9} + \frac{y^2}{4} = 1$ is the standard equation for an ellipse with vertices $(\pm 3, 0)$, centered at $(0, 0)$.

Hyperbolas The standard equation for a hyperbola centered at the origin with a *horizontal transverse axis* is $\frac{x^2}{a^2} - \frac{y^2}{b^2} = 1$, the asymptotes are given by $y = \pm \frac{b}{a}x$, and the vertices are $(\pm a, 0)$, as shown in the figure on the left. The standard equation for a hyperbola centered at the origin with a *vertical transverse axis* is $\frac{y^2}{a^2} - \frac{x^2}{b^2} = 1$, the asymptotes are $y = \pm \frac{a}{b}x$, and the vertices are $(0, \pm a)$, as shown in the figure on the right.

Horizontal Transverse Axis   **Vertical Transverse Axis**

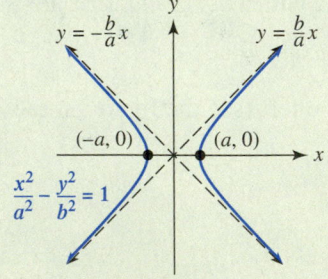 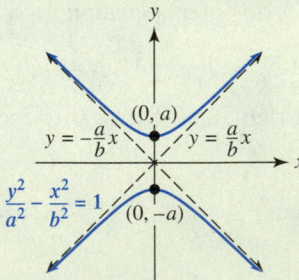

Example: $\frac{x^2}{9} - \frac{y^2}{4} = 1$ is the standard equation for a hyperbola with horizontal transverse axis, vertices $(\pm 3, 0)$, center $(0, 0)$, and asymptotes $y = \pm \frac{2}{3}x$.

SECTION 13.3 ■ NONLINEAR SYSTEMS OF EQUATIONS AND INEQUALITIES

Nonlinear Systems Nonlinear systems of equations can have any number of solutions. The methods of substitution and elimination can often be used to solve a nonlinear system of equations symbolically. Nonlinear systems can also be solved graphically. The solution set for a nonlinear system of two inequalities in two variables is typically a region in the *xy*-plane. A solution is an ordered pair (x, y) that satisfies both inequalities.

Example: Solve $y \geq x^2 - 2$

$$y \leq 4 - \tfrac{1}{2}x^2.$$

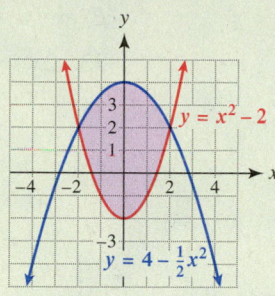

CHAPTER 13   Review Exercises

SECTION 13.1

Exercises 1–6: Graph the parabola. Find the vertex and axis of symmetry.

1. $x = 2y^2$

2. $x = -(y + 1)^2$

3. $x = -2(y - 2)^2$

4. $x = (y + 2)^2 - 1$

5. $x = -3y^2 + 1$

6. $x = \tfrac{1}{2}y^2 + y - 3$

7. Use the graph to determine the equation of the parabola.

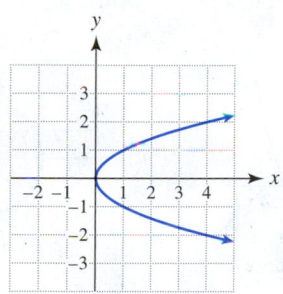

8. Use the graph to find the equation of the circle.

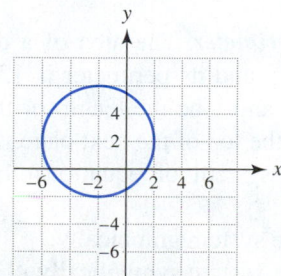

9. Write the equation of the circle with radius 1 and center $(0, 0)$.

10. Write the equation of the circle with radius 4 and center $(2, -3)$.

Exercises 11–14: Find the radius and center of the circle. Then graph the circle.

11. $x^2 + y^2 = 25$

12. $(x - 2)^2 + y^2 = 9$

13. $(x + 3)^2 + (y - 1)^2 = 5$

14. $x^2 - 2x + y^2 + 2y = 7$

SECTION 13.2

Exercises 15–18: Graph the ellipse. Label the vertices and endpoints of the minor axis.

15. $\dfrac{x^2}{4} + \dfrac{y^2}{25} = 1$

16. $x^2 + \dfrac{y^2}{4} = 1$

17. $25x^2 + 20y^2 = 500$

18. $4x^2 + 9y^2 = 36$

19. Use the graph to determine the equation of the ellipse.

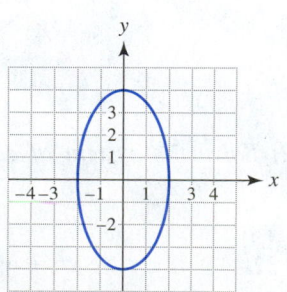

20. Use the graph to find the equation of the hyperbola.

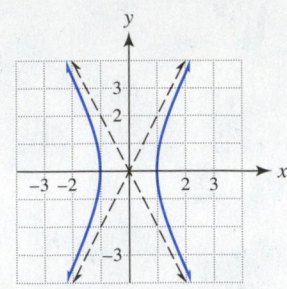

Exercises 21–24: Graph the hyperbola. Include the asymptotes in your graph.

21. $\dfrac{x^2}{9} - \dfrac{y^2}{4} = 1$ **22.** $\dfrac{y^2}{25} - \dfrac{x^2}{16} = 1$

23. $y^2 - x^2 = 1$ **24.** $25x^2 - 16y^2 = 400$

SECTION 13.3

Exercises 25–28: Use the graph to estimate all solutions to the system of equations. Check each solution.

25. $x^2 + y^2 = 9$
    $x \ + y = 3$

26. $xy = 2$
    $y = 2x$

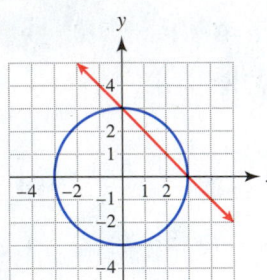 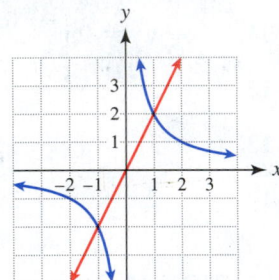

27. $x^2 - y = x$
    $y = x$

28. $x^2 + y^2 = 5$
    $x^2 - y^2 = 3$

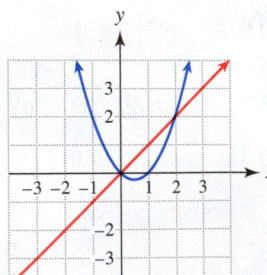 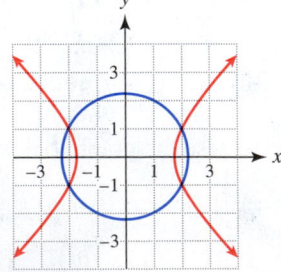

Exercises 29–32: Solve the system of equations. Check your solutions.

29. $y \ = x$
    $x^2 + y^2 = 32$

30. $x \ - y \ = 4$
    $x^2 + y^2 = 16$

31. $y = x^2$
    $2x^2 + y = 3$

32. $y = x^2 + 1$
    $2x^2 - y = 3x - 3$

Exercises 33 and 34: Solve the system graphically.

33. $2x \ - y = 4$
    $x^2 + y = 4$

34. $x^2 + y \ = 0$
    $x^2 + y^2 = 2$

Exercises 35 and 36: Solve the system of equations
    (a) *symbolically,*
    (b) *graphically, and*
    (c) *numerically.*

35. $y = x$
    $x^2 + 2y = 8$

36. $y = x^3$
    $x^2 - y = 0$

Exercises 37–44: Shade the solution set in the xy-plane.

37. $y \geq 2x^2$ **38.** $y < 2x - 3$

39. $y < -x^2$ **40.** $\dfrac{x^2}{9} + \dfrac{y^2}{16} \leq 1$

41. $y - x^2 \geq 1$
    $y \leq 2$

42. $x^2 + \ y \leq 4$
    $3x + 2y \geq 6$

43. $y > x^2$
    $y < 4 - x^2$

44. $\dfrac{x^2}{4} + \dfrac{y^2}{9} > 1$
    $x^2 + y^2 < 16$

Exercises 45 and 46: Use the graph to write the system of inequalities.

45. 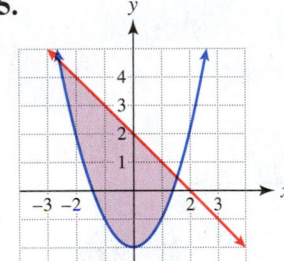   **46.**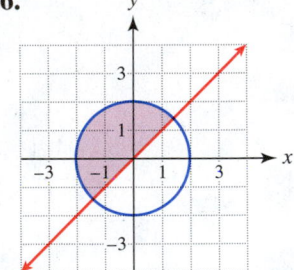

APPLICATIONS

47. *Area and Perimeter* The area of a desktop is 1000 square inches, and its perimeter is 130 inches. Let x be the width and y be the length of the desktop. See the figure at the top of the next page.
    (a) Write a system of equations that models this situation.
    (b) Solve the system graphically.
    (c) Solve the system symbolically.

Graphically find the dimensions of a can with $A = 80$ square inches and $V = 35$ cubic inches. Is your answer unique?

51. *Geometry of an Ellipse* The area inside an ellipse is given by $A = \pi ab$, and its perimeter P can be approximated by

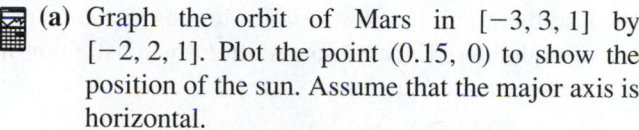

$$P = 2\pi\sqrt{\frac{a^2 + b^2}{2}}.$$

48. *Numbers* The product of two positive numbers is 60, and their difference is 7. Let x be the smaller number and y be the larger number.

(a) Graph $\dfrac{x^2}{5} + \dfrac{y^2}{12} = 1$.

(b) Estimate the ellipse's area and perimeter.

(a) Write a system of equations whose solution gives the two numbers.

(b) Solve the system graphically.

(c) Solve the system symbolically.

49. *Dimensions of a Container* The volume of a cylindrical container is $V = \pi r^2 h$, and its surface area, *excluding* the top and bottom, is $A = 2\pi rh$. Find the dimensions of a container with $A = 100$ square feet and $V = 50$ cubic feet. Is your answer unique?

 50. *Dimensions of a Container* The volume of a cylindrical container is $V = \pi r^2 h$, and its surface area, *including* the top and bottom, is $A = 2\pi rh + 2\pi r^2$.

52. *Orbit of Mars* Mars has an elliptical orbit that is nearly circular, with $a = 1.524$ and $b = 1.517$, where the units are astronomical units (1 A.U. equals 93 million miles). (*Source:* M. Zeilik.)

(a) Graph the orbit of Mars in $[-3, 3, 1]$ by $[-2, 2, 1]$. Plot the point $(0.15, 0)$ to show the position of the sun. Assume that the major axis is horizontal.

(b) Use the information in Exercise 51 to estimate how far Mars travels in one orbit around the sun. Approximate the area inside its orbit.

Chapter 13   Test

Step-by-step test solutions are found on the Chapter Test Prep Videos available in **MyMathLab** and on You**Tube** (search "RockswoldComboAlg" and click on "Channels").

1. Graph the parabola $y = -(x - 1)^2 + 2$. Find the vertex and axis of symmetry.

2. Graph the parabola $x = (y - 4)^2 - 2$. Find the vertex and axis of symmetry.

3. Use the graph to find the equation of the parabola.

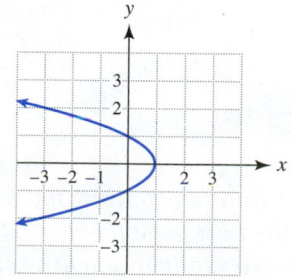

4. Use the graph to find the equation of the circle.

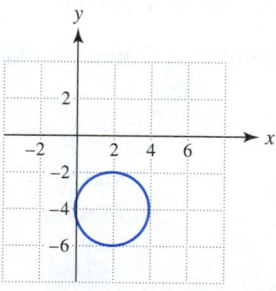

5. Write the equation of the circle with radius 10 and center $(-5, 2)$.

6. Find the radius and center of the circle given by

$$x^2 + 4x + y^2 - 6y = 3.$$

Then graph the circle.

7. Graph the ellipse $\frac{x^2}{16} + \frac{y^2}{49} = 1$. Label the vertices and endpoints of the minor axis.

8. Use the graph to determine the standard equation of the ellipse.

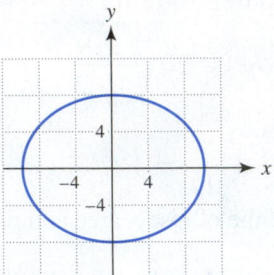

9. Graph the hyperbola $4x^2 - 9y^2 = 36$. Include the asymptotes in your graph.

10. Use the graph to estimate all solutions to the system of equations. Check each solution by substitution in the system of equations.

$$x^2 + y^2 = 16$$
$$x - y = 4$$

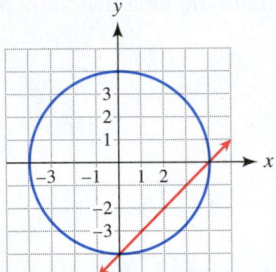

11. Solve the system of equations symbolically.

$$x - y = 3$$
$$x^2 + y^2 = 17$$

12. Solve the system of equations graphically.

$$2x^2 - y = 4$$
$$x^2 + y = 8$$

13. Shade the solution set in the *xy*-plane.

$$3x + y > 6$$
$$x^2 + y^2 < 25$$

14. Use the graph to write the system of inequalities.

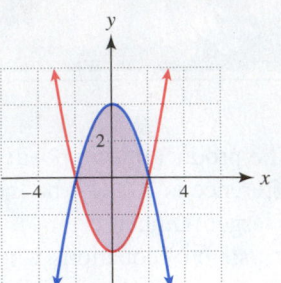

15. *Area and Perimeter* The area of a rectangular swimming pool is 5000 square feet, and its perimeter is 300 feet. Let *x* be the width and *y* be the length of the pool.
(a) Write a nonlinear system of equations that models this situation.
(b) Solve the system.

16. *Dimensions of a Box* The volume of a rectangular box with a square bottom and open top is $V = x^2y$, and its surface area is $A = x^2 + 4xy$, where *x* represents its width and length and *y* represents its height. Estimate graphically the dimensions of a box with $V = 1183$ cubic inches and $A = 702$ square inches. Is your answer unique? (*Hint:* Substitute appropriate values for *V* and *A*, and then solve each equation for *y*.)

17. *Orbit of Uranus* The planet Uranus has an elliptical orbit that is nearly circular, with $a = 19.18$ and $b = 19.16$, where the units are astronomical units (1 A.U. equals 93 million miles). (*Source:* M. Zeilik.)
(a) Graph the orbit of Uranus in $[-30, 30, 10]$ by $[-20, 20, 10]$. Plot the point $(0.9, 0)$ to show the position of the sun. Assume that the major axis is horizontal.
(b) Find the minimum distance between Uranus and the sun.

CHAPTER 13 Extended and Discovery Exercises

*Exercises 1–2: Foci of Parabolas The focus of a parabola is a point that has special significance. When a parabola is rotated about its axis, it sweeps out a shape called a **paraboloid**, as illustrated in the top figure on the following page. Paraboloids have an important reflective property. When incoming rays of light from the sun or distant stars strike the surface of a parabo-loid, each ray is reflected toward the focus, as shown in the figure on the next page labeled "Reflective Property." If the rays are sunlight, intense heat is produced, which can be used to generate solar heat. Radio signals from distant space also concentrate at the focus, and scientists can measure these signals by placing a receiver there.*

The same reflective property of a paraboloid can be used in reverse. If a light source is placed at the focus, the light is reflected straight ahead, as depicted in the figure labeled "Headlight." Searchlights, flashlights, and car headlights make use of this reflective property.

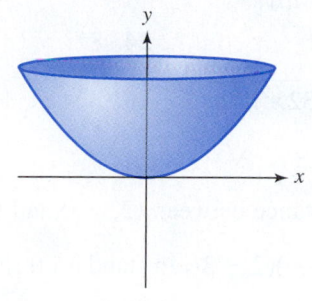

Paraboloid

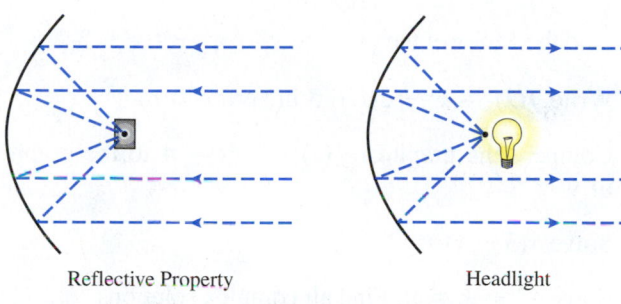

Reflective Property   Headlight

The focus is always located inside a parabola, on its axis of symmetry. If the distance between the vertex and the focus is $|p|$, the following equations can be used to locate the focus. Note that the value of p may be either positive or negative.

EQUATION OF A PARABOLA WITH VERTEX (0, 0)

Vertical Axis

The parabola with a focus at $(0, p)$ has the equation
$$x^2 = 4py.$$

Horizontal Axis

The parabola with a focus at $(p, 0)$ has the equation
$$y^2 = 4px.$$

1. Graph each parabola. Label the vertex and focus.
   (a) $x^2 = 4y$ (b) $y^2 = -8x$ (c) $x = 2y^2$

2. The reflective property of paraboloids is used in satellite dishes and radio telescopes. The U.S. Naval Research Laboratory designed a giant radio telescope weighing 3450 tons. Its parabolic dish has a diameter of 300 feet and a depth of 44 feet, as shown in the figures in the next column. (*Source:* J. Mar, Structure Technology for Large Radio and Radar Telescope Systems.)
   (a) Determine an equation in the form $y = ax^2$ that describes a cross section of this dish.
   (b) If the receiver is located at the focus, how far should it be from the vertex?

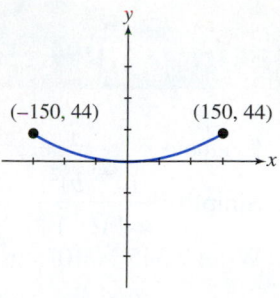

Radio Telescope

Exercises 3 and 4: *Translations of Ellipses and Hyperbolas* Ellipses and hyperbolas can be translated so that they are centered at a point (h, k), rather than at the origin. These techniques are the same as those used for parabolas and circles. To translate a conic section so that it is centered at (h, k) rather than $(0, 0)$, replace x with $(x - h)$ and replace y with $(y - k)$. For example, to center $\frac{x^2}{9} + \frac{y^2}{4} = 1$ at $(-1, 2)$, change its equation to $\frac{(x + 1)^2}{9} + \frac{(y - 2)^2}{4} = 1$. See the accompanying figures.

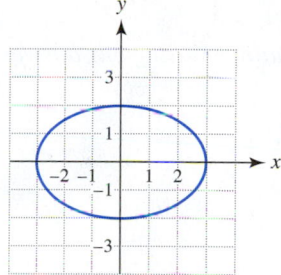

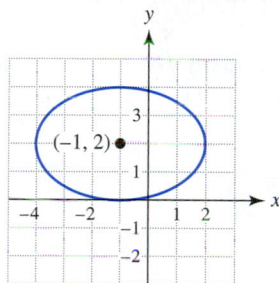

Ellipse centered at $(0, 0)$ Ellipse centered at $(-1, 2)$

3. Graph each conic section. For hyperbolas give the equations of the asymptotes.
   (a) $\dfrac{(x - 3)^2}{25} + \dfrac{(y - 1)^2}{9} = 1$
   (b) $\dfrac{(x + 1)^2}{4} + \dfrac{(y + 2)^2}{16} = 1$
   (c) $\dfrac{(x + 1)^2}{4} - \dfrac{(y - 3)^2}{9} = 1$
   (d) $\dfrac{(y - 4)^2}{16} - \dfrac{(x + 1)^2}{4} = 1$

4. Write the equation of an ellipse having the following properties.
   (a) Horizontal major axis of length 8, minor axis of length 4, and centered at $(-3, 5)$.
   (b) Vertical major axis of length 10, minor axis of length 6, and centered at $(2, -3)$.

5. Determine the center of the conic section.
   (a) $9x^2 - 18x + 4y^2 + 24y + 9 = 0$
   (b) $25x^2 + 150x - 16y^2 + 32y - 191 = 0$

CHAPTERS 1–13   Cumulative Review Exercises

1. Evaluate $K = x^2 + y^2$ when $x = 4$ and $y = -3$.

2. Simplify $\dfrac{(a^{-2} b)^2}{a^{-1} (b^3)^{-2}}$.

3. Write 7.345×10^{-3} in standard (decimal) notation.

4. Find $f(-4)$, if $f(x) = \dfrac{x}{x-4}$. State the domain of f.

5. If the point $(3, 4)$ is on the graph of $y = f(x)$, then $f(\underline{\quad}) = \underline{\quad}$.

Exercises 6 and 7: Graph f by hand.

6. $f(x) = x^2 - x$

7. $f(x) = 1 - 2x$

8. Find the slope–intercept form of the line that passes through the point $(2, -2)$ and is perpendicular to the line given by $y = -\frac{2}{3}x + 1$.

9. Solve $2(1 - x) - 4x = x$.

Exercises 10–12: Solve the inequality. Write your answer in interval notation.

10. $-5 \le 1 - 2x < 3$

11. $x^2 - 4 \le 0$

12. $|1 - x| \ge 2$

13. Solve the system.
$$-2x + y = 1$$
$$5x - y = 2$$

14. Shade the solution set in the xy-plane.
$$x + y \le 4$$
$$x - y \ge 2$$

Exercises 15 and 16: Multiply the expression.

15. $(2x - 1)(x + 5)$

16. $xy(2x - 3y^2 + 1)$

Exercises 17 and 18: Factor the expression.

17. $6x^2 - 13x - 5$

18. $x^3 - 4x$

Exercises 19 and 20: Solve the equation.

19. $x^2 + 3x + 2 = 0$

20. $x^2 + 1 = -3x$

Exercises 21 and 22: Simplify the expression.

21. $\dfrac{x-2}{x+2} \div \dfrac{2x-4}{3x+6}$

22. $\dfrac{1}{x+1} + \dfrac{1}{x-1}$

Exercises 23–26: Simplify the expression. Assume all variables are positive.

23. $\sqrt{8x^2}$

24. $8^{2/3}$

25. $\sqrt[3]{2x} \cdot \sqrt[3]{32x^2}$

26. $3\sqrt{3x} + \sqrt{12x}$

27. Graph $f(x) = -\sqrt{x}$.

28. Find the distance between $(2, -3)$ and $(-2, 0)$.

29. Write $(2 + 3i)(2 - 3i)$ in standard form.

30. Solve $\sqrt{x + 2} = x$.

31. Find the vertex of the graph of $y = x^2 - 6x + 3$.

32. Write $f(x) = x^2 - 2x + 3$ in vertex form.

33. Compare the graph of $f(x) = \sqrt{x - 4}$ to the graph of $y = \sqrt{x}$.

34. Solve $x(3 - x) = 2$.

35. Solve $x^3 + x = 0$. Find all complex solutions.

36. Simplify by hand.
   (a) $\log 10{,}000$
   (b) $\log_2 8$
   (c) $\log_3 3^x$
   (d) $e^{\ln 6}$
   (e) $\log 2 + \log 50$
   (f) $\log_2 24 - \log_2 3$

37. If $f(x) = x^2 + 1$ and $g(x) = 2x$, find the following.
   (a) $(f \circ g)(2)$
   (b) $(g \circ f)(x)$

38. If $f(x) = 2 - 3x$, find $f^{-1}(x)$.

39. If \$1000 is deposited in an account that pays 5% annual interest, approximate the amount in the account after 6 years.

40. Write $\log \dfrac{x^2 \sqrt{y}}{z^3}$ in terms of logarithms of x, y, and z.

41. Solve $2e^x - 1 = 17$.

42. Solve $3 + \log 4x = 5$.

43. Graph each conic section.
   (a) $x = (y - 1)^2$
   (b) $(x - 1)^2 + (y + 1)^2 = 4$
   (c) $\dfrac{x^2}{4} + \dfrac{y^2}{25} = 1$
   (d) $4x^2 - 9y^2 = 36$

44. Solve the nonlinear system of equations.
$$x^2 + y^2 = 1$$
$$x^2 + 9y^2 = 9$$

45. Shade the solution set in the xy-plane.

$$x^2 + y^2 \leq 4$$
$$x^2 - y \leq 2$$

APPLICATIONS

46. *Calculating Distance* The distance D in miles that a driver of a car is from home after x hours is given by $D(x) = 400 - 50x$.
  (a) Evaluate $D(0)$. Interpret your answer.
  (b) What is the x-intercept on the graph of D? Interpret your answer.
  (c) What is the slope of the graph of D? Interpret your answer.

47. *Investment* Suppose $2000 is deposited in three accounts paying 5%, 6%, and 7% annual interest. The amount invested at 6% is $500 more than the amount invested at 5%. The interest after 1 year is $120. How much is invested at each rate?

48. *Area* There are 1200 feet of fence available to surround the perimeter of a rectangular garden. What dimensions for the garden give the largest area?

49. *Population* The population of a city in millions after x years is given by $P = 2e^{0.02x}$. How long will it take for the population to double?

50. *Dimensions of a Can* Find the radius of a cylindrical can with a volume V of 60 cubic inches and a side area S of 50 square inches. (*Hint: $V = \pi r^2 h$ and $S = 2\pi rh$.*)

14 Sequences and Series

> Go deep enough into anything and you will find mathematics.
>
> —DEAN SCHLICTER

In this final chapter we present sequences and series, which are essential topics because they are used to model and approximate important quantities. Complicated population growth can be modeled with sequences, and accurate approximations for numbers such as π and e are made with series. For example, in about 2000 B.C. the Babylonians thought that π equaled 3.125, whereas at the same time the Chinese thought π equaled 3. By 1700, mathematicians were able to calculate π to 100 decimal places due to the discovery of series. Series are also essential to the solutions of many modern applied mathematics problems.

Although you may not always recognize the impact of mathematics on everyday life, its influence is nonetheless profound. Mathematics is the *language of technology*—it allows experiences to be quantified. In the preceding chapters we showed numerous examples of mathematics being used to model the real world. Computers, DVD players, cars, highway design, weather, hurricanes, electricity, social networks, government data, cellular phones, medicine, ecology, business, sports, and psychology represent only some of the applications of mathematics. In fact, if a subject is studied in enough detail, mathematics usually appears in one form or another. Although predicting what the future may bring is difficult, one thing *is* certain—mathematics will continue to play an important role in both theoretical research and new technology.

Source: Mathforum.org, Drexel University 1994–2007.

14.1 Sequences

Basic Concepts • Representations of Sequences • Models and Applications

A LOOK INTO MATH ▶ Sequences are *ordered lists*. For example, names listed alphabetically represent a sequence. Figure 14.1 shows an insect population in thousands per acre over a 6-year period. Listing populations by year is another example of a sequence. In mathematics, a sequence is a function for which valid inputs must be natural numbers. For example, we can use a function f to define this sequence by letting $f(1)$ represent the insect population after 1 year, $f(2)$ represent the insect population after 2 years, and in general let $f(n)$ represent the population after n years. In this section we discuss sequences.

NEW VOCABULARY

☐ Terms of a sequence
☐ Finite sequence
☐ Infinite sequence
☐ *n*th term
☐ General term

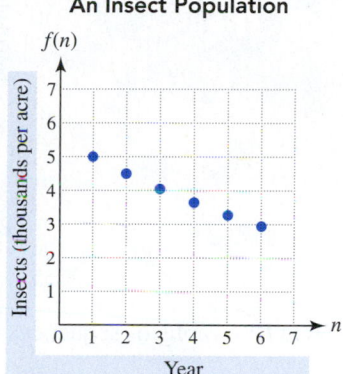

Figure 14.1

Basic Concepts

▶ **REAL-WORLD CONNECTION** Suppose that an individual's starting salary is \$40,000 per year and that the person's salary is increased by 10% each year. This situation is modeled by the formula

$$f(n) = 40{,}000(1.10)^n. \quad \text{Symbolic representation}$$

We do not allow the input n to be any real number, but rather limit n to a *natural number* because the individual's salary is constant throughout a particular year. The first five *terms of the sequence* are

$$f(1), f(2), f(3), f(4), f(5).$$

They can be computed as follows.

$$
\begin{aligned}
f(1) &= 40{,}000(1.10)^1 = \textbf{44,000} \\
f(2) &= 40{,}000(1.10)^2 = \textbf{48,400} \\
f(3) &= 40{,}000(1.10)^3 = \textbf{53,240} \quad \text{The first five terms of a sequence} \\
f(4) &= 40{,}000(1.10)^4 = \textbf{58,564} \\
f(5) &= 40{,}000(1.10)^5 \approx \textbf{64,420}
\end{aligned}
$$

This sequence is represented *numerically* in Table 14.1 on the next page.

TABLE 14.1 Numerical Representation

n	1	2	3	4	5
$f(n)$	44,000	48,400	53,240	58,564	64,420

A *graphical* representation results when each data point in Table 14.1 is plotted, as illustrated in Figure 14.2.

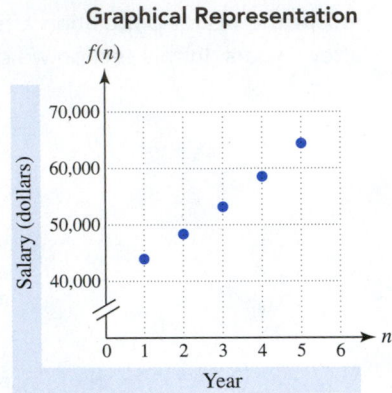

Graphical Representation

Figure 14.2

NOTE: Graphs of sequences are scatterplots.

The preceding sequence is an example of a *finite sequence* of numbers. The even natural numbers,

$$2, 4, 6, 8, 10, 12, 14, \ldots,$$ Even natural numbers

are an example of an *infinite sequence* represented by $f(n) = 2n$, where n is a natural number. The three dots, or periods (called an *ellipsis*), indicate that the pattern continues indefinitely.

SEQUENCES

A **finite sequence** is a function whose domain is $D = \{1, 2, 3, \ldots, n\}$ for some fixed natural number n.

An **infinite sequence** is a function whose domain is the set of natural numbers.

READING CHECK

How are functions and sequences related?

Because sequences are functions, many of the concepts discussed in previous chapters apply to sequences. Instead of letting y represent the output, however, the convention is to write $a_n = f(n)$, where n is a natural number in the domain of the sequence. The *terms* of a sequence are

$$\underset{}{} a_1, a_2, a_3, \ldots, a_n, \ldots .$$

— nth term

Sequence

The first term is $a_1 = f(1)$, the second term is $a_2 = f(2)$, and so on. The **nth term**, or **general term**, of a sequence is $a_n = f(n)$.

EXAMPLE 1   **Computing terms of a sequence**

Write the first four terms of each sequence for $n = 1, 2, 3,$ and 4.

(a) $f(n) = 2n - 1$ (b) $f(n) = 3(-2)^n$ (c) $f(n) = \dfrac{n}{n + 1}$

Solution

(a) For $a_n = f(n) = 2n - 1$, we write the first four terms as follows.

$$a_1 = f(1) = 2(1) - 1 = 1$$
$$a_2 = f(2) = 2(2) - 1 = 3$$
$$a_3 = f(3) = 2(3) - 1 = 5$$
$$a_4 = f(4) = 2(4) - 1 = 7$$

The first four terms are **1**, **3**, **5**, and **7**.

(b) For $a_n = f(n) = 3(-2)^n$, we write the first four terms as follows.

$$a_1 = f(1) = 3(-2)^1 = -6$$
$$a_2 = f(2) = 3(-2)^2 = 12$$
$$a_3 = f(3) = 3(-2)^3 = -24$$
$$a_4 = f(4) = 3(-2)^4 = 48$$

The first four terms are -6, **12**, -24, and **48**.

(c) For $a_n = f(n) = \dfrac{n}{n + 1}$, we write the first four terms as follows.

$$a_1 = f(1) = \frac{1}{1 + 1} = \frac{1}{2}$$
$$a_2 = f(2) = \frac{2}{2 + 1} = \frac{2}{3}$$
$$a_3 = f(3) = \frac{3}{3 + 1} = \frac{3}{4}$$
$$a_4 = f(4) = \frac{4}{4 + 1} = \frac{4}{5}$$

The first four terms are $\frac{1}{2}, \frac{2}{3}, \frac{3}{4},$ and $\frac{4}{5}$. Note that, although the input to a sequence is a natural number, the output need not be a natural number.

Now Try Exercises 9, 11, 13

TECHNOLOGY NOTE

Generating Sequences
Many graphing calculators can generate sequences if you change the MODE from function (Func) to sequence (Seq). In Figures 14.3 and 14.4 the sequences from Example 1 are generated. On some calculators the sequence utility is found in the LIST OPS menus. The expression

$$\text{seq}\,(2n - 1, n, 1, 4)$$

represents terms 1 through 4 of the sequence $f(n) = 2n - 1$ with the variable n.

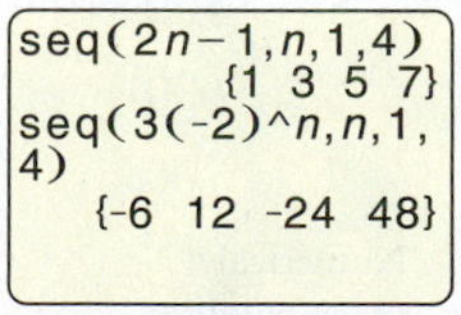

```
seq(2n-1,n,1,4)
          {1 3 5 7}
seq(3(-2)^n,n,1,
4)
     {-6 12 -24 48}
```

Figure 14.3

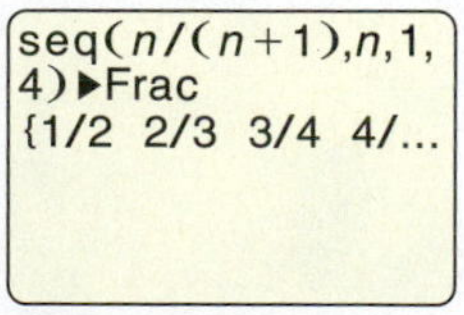

```
seq(n/(n+1),n,1,
4)▶Frac
{1/2 2/3 3/4 4/...
```

Figure 14.4

Representations of Sequences

Because sequences are functions, they can be represented symbolically, graphically, and numerically. The next two examples illustrate such representations.

EXAMPLE 2 **Using a graphical representation**

Use Figure 14.5 to write the terms of the sequence.

A Finite Sequence

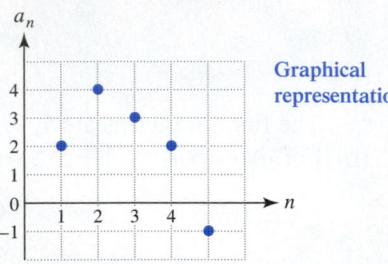

Graphical representation

Figure 14.5

Solution
The points $(1, 2)$, $(2, 4)$, $(3, 3)$, $(4, 2)$, and $(5, -1)$ are shown in the graph. The terms of the sequence are $2, 4, 3, 2,$ and -1.

Now Try Exercise 29

EXAMPLE 3 **Representing a sequence**

In 2011 the average person in the United States used 100 gallons of water at home each day. Give symbolic, numerical, and graphical representations for a sequence that models the total amount of water used over a 7-day period. (*Source:* U.S. Geological Survey.)

Solution
Symbolic Representation Let

$$a_n = 100n \text{ for } n = 1, 2, 3, \ldots, 7. \quad \text{Symbolic representation}$$

Numerical Representation Table 14.2 contains the sequence.

Graphical Representation Plot the points $(1, 100)$, $(2, 200)$, $(3, 300)$, $(4, 400)$, $(5, 500)$, $(6, 600)$, and $(7, 700)$, as shown in Figure 14.6.

TABLE 14.2

n	a_n
1	100
2	200
3	300
4	400
5	500
6	600
7	700

Numerical representation

Personal Water Usage

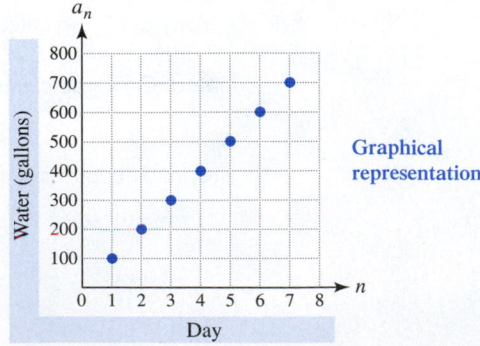

Graphical representation

Figure 14.6

Now Try Exercise 39

Models and Applications

▶ **REAL-WORLD CONNECTION** A population model for a species of insect with a life span of 1 year can be described with a sequence. Suppose that each adult female insect produces, on average, r female offspring that survive to reproduce the following year. Let a_n represent the female insect population at the beginning of year n. Then the number of female insects is given by

$$a_n = Cr^{n-1},$$

where C is the initial population of female insects. (*Source:* D. Brown and P. Rothery, *Models in Biology.*)

EXAMPLE 4 **Modeling numbers of insects**

Suppose that the initial population of adult female insects is 500 per acre and that $r = 1.04$. Then the average number of female insects per acre at the beginning of year n is described by

$$a_n = 500(1.04)^{n-1}.$$

Represent the female insect population numerically and graphically for 7 years. Round values to the nearest whole number. Discuss the results. By what percent is the population increasing each year?

Solution

Numerical Representation Table 14.3 contains *approximations* for the first 7 terms of the sequence. The insect population increases from 500 to about 633 insects per acre during this time period.

TABLE 14.3 An Insect Population (per acre)

n	1	2	3	4	5	6	7
a_n	500	520	541	562	585	608	633

Graphical Representation Plot the points (1, 500), (2, 520), (3, 541), (4, 562), (5, 585), (6, 608), and (7, 633), as shown in Figure 14.7. These results indicate that the insect population gradually increases. Because the growth factor is 1.04, the population is increasing by 4% each year.

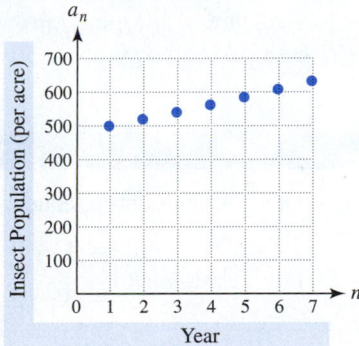

An Insect Population

Figure 14.7

Now Try Exercise 45

CRITICAL THINKING

Explain how the value of *r* in Example 4 affects the population of female insects over time. Assume that $r > 0$.

TECHNOLOGY NOTE

Graphs and Tables of Sequences
In the sequence mode, many graphing calculators are capable of representing sequences graphically and numerically. Figure 14.8(a) shows how to enter the sequence from Example 4 to produce the table of values shown in Figure 14.8(b).

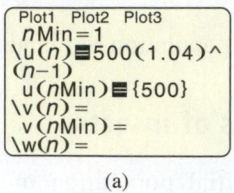
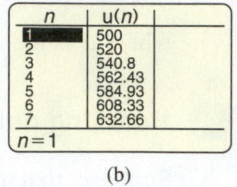

(a)   (b)

Figure 14.8

Figures 14.9(a) and (b) show the set-up for graphing the sequence from Example 4 to produce the graph shown in Figure 14.9(c).

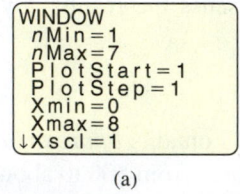
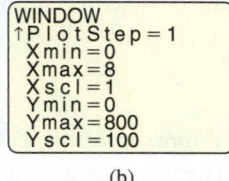
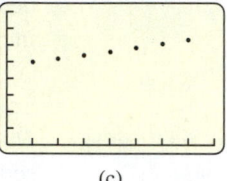

(a)   (b)   (c)

Figure 14.9

14.1   Putting It All Together

An infinite sequence is a function whose domain is the set of natural numbers. A finite sequence has the domain $D = \{1, 2, 3, \ldots, n\}$ for some fixed natural number n. Graphs of sequences are scatterplots, *not* continuous lines and curves. Sequences are functions that may be represented symbolically, numerically, and graphically.

REPRESENTATION	EXAMPLE
Symbolic	$a_n = n - 3$ represents a sequence. The first four terms of the sequence are $-2, -1, 0$, and 1: $a_1 = \mathbf{1} - 3 = -2, \qquad a_2 = \mathbf{2} - 3 = -1,$ $a_3 = \mathbf{3} - 3 = 0, \qquad a_4 = \mathbf{4} - 3 = 1.$
Numerical	A numerical representation for $a_n = n - 3$ with $n = 1, 2, 3$, and 4 is shown in the table.

n	1	2	3	4
a_n	-2	-1	0	1

REPRESENTATION	EXAMPLE
Graphical	For a graphical representation of the first four terms of $a_n = n - 3$, the points $(1, -2)$, $(2, -1)$, $(3, 0)$, and $(4, 1)$ from the previous table are plotted. 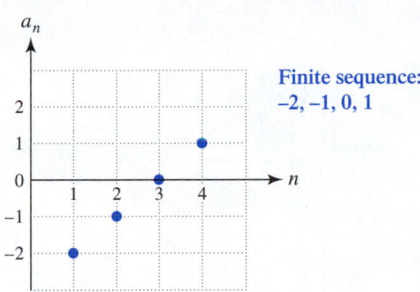

14.1 Exercises

CONCEPTS AND VOCABULARY

1. Give an example of a finite sequence.

2. Give an example of an infinite sequence.

3. An infinite sequence is a(n) _____ whose domain is the set of _____.

4. An ordered list is a(n) _____.

5. The third term in the sequence $4, -5, 6, -7, 8$ is _____.

6. The graph of a sequence is not a continuous graph but rather a(n) _____.

7. If $f(n)$ represents a sequence, the second term of the sequence is given by _____.

8. If a_n represents a sequence, the fourth term of the sequence is given by _____.

EVALUATING AND REPRESENTING SEQUENCES

Exercises 9–16: Write the first four terms of the sequence for $n = 1, 2, 3,$ and 4.

9. $f(n) = n^2$

10. $f(n) = 3n + 4$

11. $f(n) = \dfrac{1}{n + 5}$

12. $f(n) = 3^n$

13. $f(n) = 5\left(\frac{1}{2}\right)^n$

14. $f(n) = n^2 + 2n$

15. $f(n) = 9$

16. $f(n) = (-1)^n$

Exercises 17–24: Write the first three terms of the sequence for $n = 1, 2,$ and 3.

17. $a_n = n^3$

18. $a_n = 5 - n$

19. $a_n = \dfrac{4n}{3 + n}$

20. $a_n = 3^{-n}$

21. $a_n = 2n^2 + n - 1$

22. $a_n = n^4 - 1$

23. $a_n = -2$

24. $a_n = n^n$

Exercises 25 and 26: **Thinking Generally** *Let b and c be fixed numbers (constants). Find a_1 and a_2.*

25. $a_n = bn + c$

26. $a_n = \dfrac{n + b}{n - c}$, $c \neq 1$ and $c \neq 2$

Exercises 27 and 28: Use the numerical representation to evaluate $\frac{1}{2}(a_1 + a_4)$.

27.

n	1	2	3	4	5
a_n	10	8	6	4	2

28.

n	1	2	3	4	5
a_n	−5	0	10	30	60

Exercises 29–32: Write the terms of the sequence.

29. a_n

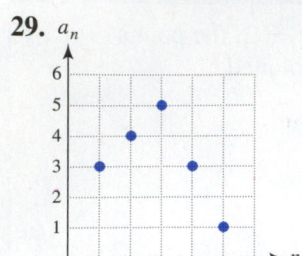

30. a_n

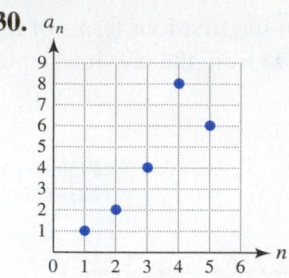

31. a_n

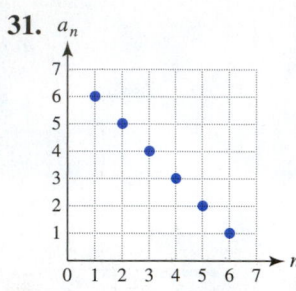

32. a_n

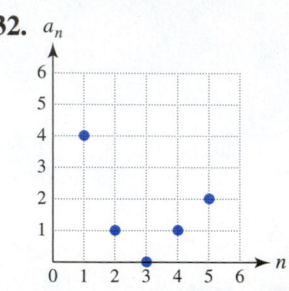

Exercises 33–38: Represent the first seven terms of the sequence numerically and graphically.

33. $a_n = n + 1$

34. $a_n = \frac{1}{2}n - \frac{1}{2}$

35. $a_n = n^2 - n$

36. $a_n = \frac{1}{2}n^2$

37. $a_n = 2^n$

38. $a_n = 2(0.5)^n$

APPLICATIONS

39. *Solid Waste* On average, each U.S. resident in 2000 generated about 30 pounds of solid waste per week. Give symbolic, numerical, and graphical representations for a sequence that models the total amount of waste produced over a 7-week period. (*Source:* Environmental Protection Agency.)

40. *Carbon Dioxide Emitters* Because people burn fossil fuels, the United States emits about 5.8 billion metric tons of carbon dioxide per year. (A metric ton is about 2200 pounds.) Give symbolic, numerical, and graphical representations for a sequence that models the total amount of carbon dioxide emitted in billions of metric tons in the United States during a 5-year period. (*Source:* Energy Information Administration.)

41. *Geometry* The lengths of the sides of a sequence of squares are given by 1, 2, 3, and 4, as shown in the figure at the top of the next column. Write sequences that give
(a) the areas of the squares and

(b) the perimeters of the squares.

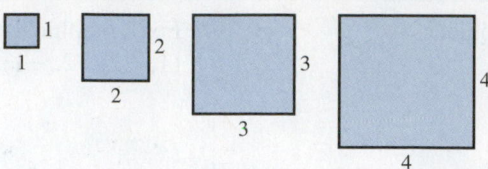

42. *Salaries* An individual's starting salary is $50,000, and the individual receives an increase of 8% per year. Give symbolic, numerical, and graphical representations for this person's salary over 5 years.

43. *Depreciation* Automobiles usually depreciate in value over time. Often, a newer automobile may be worth only 80% of its previous year's value. Suppose that a car is worth $25,000 new.
(a) How much is it worth after 1 year? After 2 years?
(b) Write a formula for a sequence that gives the car's value after n years.
(c) Make a table that shows how much the car was worth each year during the first 7 years.

44. *Falling Object* The distance d that an object falls during *consecutive* seconds is shown in the table. For example, during the third second (from $n = 2$ to $n = 3$) an object falls a distance of 80 feet.

n (seconds)	1	2	3	4	5
d (feet)	16	48	80	112	144

(a) Find values for c and b so that $d = cn + b$ models these data.
(b) How far does an object fall during the sixth second?

45. *Modeling Insect Populations* (Refer to Example 4.) Suppose that the initial population of insects is 2048 per acre and that $r = 0.5$. Use a sequence to represent the insect population over a 7-year period
(a) symbolically,
(b) numerically, and
(c) graphically.

46. *Auditorium Seating* An auditorium has 50 seats in the first row, 55 seats in the second row, 60 seats in the third row, and so on.
(a) Make a table that shows the number of seats in the first seven rows.
(b) Write a formula that gives the number of seats in row n.
(c) How many seats are there in row 23?
(d) Graph the number of seats in each row for $n = 1, 2, 3, \ldots, 10$.

47. Compare the graph of the function $f(x) = 2x + 1$, where x is a real number, with the graph of the sequence $f(n) = 2n + 1$, where n is a natural number.

48. Explain what a sequence is. Describe the difference between a finite and an infinite sequence.

14.2 Arithmetic and Geometric Sequences

Representations of Arithmetic Sequences • Representations of Geometric Sequences • Applications and Models

A LOOK INTO MATH ▶ Indoor air pollution has become more hazardous as people spend 80% to 90% of their time in tightly sealed, energy-efficient buildings, which often lack proper ventilation. Many contaminants such as tobacco smoke, formaldehyde, radon, lead, and carbon monoxide are often allowed to increase to unsafe levels. One way to alleviate this problem is to use efficient ventilation systems. Mathematics plays an important role in determining the proper amount of ventilation. In this section we use sequences to model ventilation in classrooms. (See Example 7.) Before implementing this model we discuss the basic concepts relating to two special types of sequences.

Representations of Arithmetic Sequences

If a sequence is defined by a linear function, it is an *arithmetic sequence*. For example,

$$f(n) = 2n - 3$$

represents an arithmetic sequence because $f(x) = 2x - 3$ defines a linear function. The first five terms of this sequence, $-1, 1, 3, 5, 7$, are shown in Table 14.4.

Each time n increases by **1**, the next term is **2** more than the previous term. We say that the *common difference* of this arithmetic sequence is $d = 2$. That is, the difference between successive terms equals 2. When the points associated with these terms are graphed, they lie on the line $y = 2x - 3$, as illustrated in Figure 14.10. Arithmetic sequences are represented by linear functions, so their graphical representations consist of collinear points (points that lie on a line). The slope m of the line equals the common difference d.

NEW VOCABULARY

☐ Arithmetic sequence
☐ Common difference
☐ Geometric sequence
☐ Common ratio

TABLE 14.4

	n	$f(n)$	
+1	1	−1	+2
+1	2	1	+2
+1	3	3	+2
+1	4	5	+2
	5	7	

Common difference is 2.

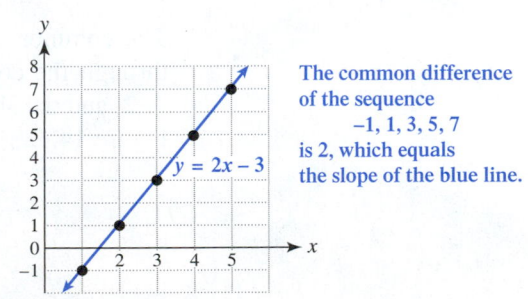

The common difference of the sequence $-1, 1, 3, 5, 7$ is 2, which equals the slope of the blue line.

Figure 14.10

ARITHMETIC SEQUENCE

An **arithmetic sequence** is a linear function given by $a_n = dn + c$ whose domain is the set of natural numbers. The value of d is called the **common difference**.

NOTE: If each term after the first term is obtained by adding a fixed number to the previous term, then the sequence is an arithmetic sequence. The fixed number is the *common difference*. For example, 1, 6, 11, 16, 21, . . . is an arithmetic sequence because each term (after the first) is found by adding the common difference of **5** to the previous term. That is, $6 = 1 + \mathbf{5}$, $11 = 6 + \mathbf{5}$, $16 = 11 + \mathbf{5}$, and so on.

EXAMPLE 1 **Recognizing arithmetic sequences**

Determine whether f is an arithmetic sequence. If it is, identify the common difference d.
(a) $f(n) = 2 - 3n$

(b)

n	$f(n)$
1	10
2	5
3	0
4	−5
5	−10

(c) A graph of f is shown in Figure 14.11.

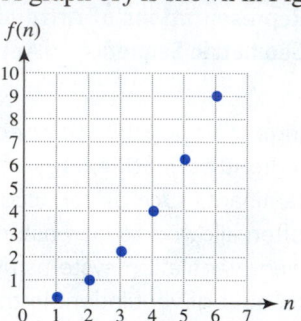

Figure 14.11

Solution
(a) This sequence is arithmetic because $f(x) = -3x + 2$ defines a linear function. The common difference is $d = -3$.
(b) The table reveals that each term is found by adding -5 to the previous term. This represents an arithmetic sequence with common difference $d = -5$.
(c) The sequence shown in Figure 14.11 is not an arithmetic sequence because the points are not collinear. That is, the points do not lie on a line and there is no common difference.

Now Try Exercises 11, 17, 25

MAKING CONNECTIONS

Common Difference and Slope

The common difference d of an arithmetic sequence equals the slope of the line passing through the collinear points. For example, if $a_n = -\mathbf{2}n + 4$, the common difference is $-\mathbf{2}$, and the slope of the line passing through the points on the graph of a_n is also $-\mathbf{2}$ (see Figure 14.12).

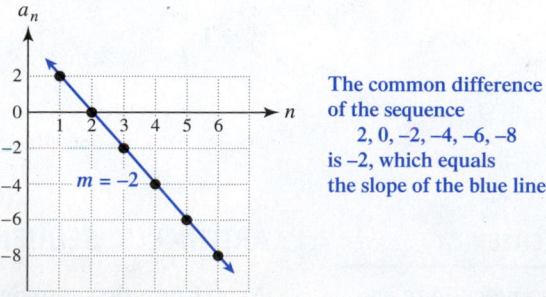

The common difference of the sequence
2, 0, –2, –4, –6, –8
is –2, which equals the slope of the blue line.

Figure 14.12

EXAMPLE 2 **Finding symbolic representations**

Find the general term a_n for each arithmetic sequence.
(a) $a_1 = 3$ and $d = 4$ **(b)** $a_1 = 3$ and $a_4 = 12$

Solution
(a) Let $a_n = dn + c$. For $d = 4$, we write $a_n = 4n + c$, and to find c we use $a_1 = 3$.

$$a_1 = 4(1) + c = 3 \quad \text{or} \quad c = -1$$

Thus $a_n = 4n - 1$.
(b) Because $a_1 = 3$ and $a_4 = 12$, the common difference d equals the slope of the line passing through the points $(1, 3)$ and $(4, 12)$, or

$$d = \frac{12 - 3}{4 - 1} = 3.$$

Therefore $a_n = 3n + c$. To find c we use $a_1 = 3$ and obtain

$$a_1 = 3(1) + c = 3 \quad \text{or} \quad c = 0.$$

Thus $a_n = 3n$.

Now Try Exercises 29, 31

Consider the arithmetic sequence

$$1, 5, 9, 13, 17, 21, 25, 29, \ldots. \qquad \text{Common difference is 4.}$$

The common difference is $d = 4$, and the first term is $a_1 = 1$. To find the second term we add d to the first term. To find the third term we add $2d$ to the first term, and to find the fourth term we add $3d$ to the first term. That is,

$$
\left.
\begin{aligned}
a_1 &= 1, \\
a_2 &= a_1 + 1d = 1 + 1 \cdot 4 = 5, \\
a_3 &= a_1 + 2d = 1 + 2 \cdot 4 = 9, \\
a_4 &= a_1 + 3d = 1 + 3 \cdot 4 = 13,
\end{aligned}
\right\} \quad \text{First four terms}
$$

and, in general, a_n is determined by

$$a_n = a_1 + (n - 1)d = 1 + (n - 1)4.$$

This result suggests the following formula.

READING CHECK

Explain how to write the general term of an arithmetic sequence, given the first term and the common difference.

GENERAL TERM OF AN ARITHMETIC SEQUENCE

The nth term a_n of an arithmetic sequence is given by

$$a_n = a_1 + (n - 1)d,$$

where a_1 is the first term and d is the common difference.

EXAMPLE 3 **Finding terms of an arithmetic sequence**

If $a_1 = 5$ and $d = 3$, find a_{54}.

Solution
To find a_{54}, apply the formula $a_n = a_1 + (n - 1)d$ with $a_1 = 5$, $n = 54$, and $d = 3$.

$$a_{54} = 5 + (54 - 1)3 = 164$$

Now Try Exercise 35

Representations of Geometric Sequences

If a sequence is defined by an exponential function, it is a *geometric sequence*. For example,

$$f(n) = 3(2)^{n-1}$$

represents a geometric sequence because $f(x) = 3(2)^{x-1}$ defines an exponential function. The first five terms of this sequence are shown in Table 14.5.

Geometric Sequence:
3, 6, 12, 24, 48

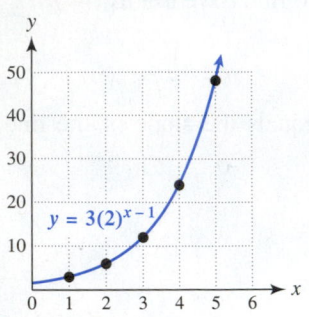

Figure 14.13

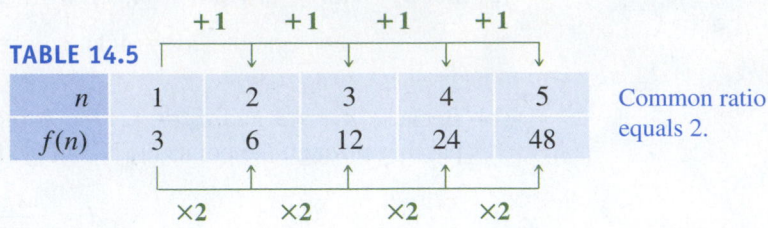

TABLE 14.5

n	1	2	3	4	5
$f(n)$	3	6	12	24	48

Common ratio equals 2.

Successive terms are found by multiplying the previous term by 2. We say that the *common ratio* of this geometric sequence equals 2. Note that the ratios of successive terms are $\frac{6}{3}, \frac{12}{6}, \frac{24}{12}$, and $\frac{48}{24}$, and that they all equal the common ratio 2. When the points associated with the terms in Table 14.5 are graphed, they do *not* lie on a line. Rather, they lie on the exponential curve $y = 3(2)^{x-1}$, as shown in Figure 14.13. A geometric sequence with a positive common ratio is an exponential function whose domain is the set of natural numbers. Its terms reflect either *exponential growth* or *exponential decay*.

GEOMETRIC SEQUENCE

A **geometric sequence** is given by $a_n = a_1(r)^{n-1}$, where n is a natural number and $r \neq 0$ or 1. The value of r is called the **common ratio**, and a_1 is the first term of the sequence.

NOTE: If each term after the first term is obtained by multiplying the previous term by a fixed number, the sequence is a geometric sequence. The fixed number can be either positive or negative, but it cannot be 0 or 1.

EXAMPLE 4 **Recognizing geometric sequences**

Determine whether f is a geometric sequence. If it is, identify the common ratio.

(a) $f(n) = 2(0.9)^{n-1}$ **(c)** A graph of f is shown in Figure 14.14.

(b)

n	$f(n)$
1	8
2	4
3	2
4	1
5	$\frac{1}{2}$

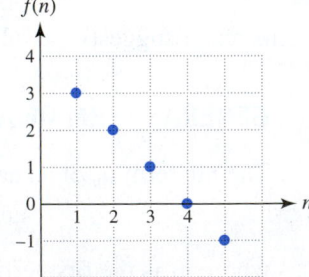

Figure 14.14

Solution

(a) This sequence is geometric because $f(x) = 2(0.9)^{x-1}$ defines an exponential function. The common ratio is $r = 0.9$.

(b) The table shows that each successive term is half the previous term. This sequence represents a geometric sequence with a common ratio of $r = \frac{1}{2}$.

(c) The sequence shown in Figure 14.14 is not a geometric sequence because the points are collinear. There is no common ratio.

Now Try Exercises 41, 45, 53

MAKING CONNECTIONS

Common Ratios and Growth or Decay Factors

If the common ratio r of a geometric sequence is positive, then r equals either the growth factor or the decay factor for an exponential function.

EXAMPLE 5 **Finding symbolic representations**

Find a general term a_n for each geometric sequence.
(a) $a_1 = \frac{1}{2}$ and $r = 5$
(b) $a_1 = 2$, $a_3 = 18$, and $r < 0$.

Solution
(a) Let $a_n = a_1(r)^{n-1}$. Because $a_1 = \frac{1}{2}$ and $r = 5$, we can write $a_n = \frac{1}{2}(5)^{n-1}$.
(b) $a_1 = 2$ and $a_3 = 18$, so

$$a_3 = a_1(r)^{3-1}$$
$$18 = 2(r)^2.$$

This equation simplifies to

$$r^2 = 9 \quad \text{or} \quad r = \pm 3.$$

It is specified that $r < 0$, so $r = -3$ and $a_n = 2(-3)^{n-1}$. Note that the common ratio r can be *negative*.

Now Try Exercises 55, 59

CRITICAL THINKING

If we are given a_1 and a_5, can we determine the common ratio of a geometric series? Explain.

EXAMPLE 6 **Finding a term of a geometric sequence**

If $a_1 = 5$ and $r = 3$, find a_{10}.

Solution
To find a_{10}, apply the formula $a_n = a_1(r)^{n-1}$ with $a_1 = 5$, $r = 3$, and $n = 10$.

$$a_{10} = 5(3)^{10-1} = 5(3)^9 = 98,415$$

Now Try Exercise 61

Applications and Models

▶ **REAL-WORLD CONNECTION** Sequences are frequently used to describe a variety of situations. In the next example, we use a sequence to model classroom ventilation.

EXAMPLE 7 **Modeling classroom ventilation**

Ventilation is an effective means for removing indoor air pollutants. According to the American Society of Heating, Refrigerating, and Air-Conditioning Engineers (ASHRAE), a classroom should have a ventilation rate of 900 cubic feet per hour per person.
(a) Write a sequence that gives the hourly ventilation necessary for 1, 2, 3, 4, and 5 people in a classroom. Is this sequence arithmetic, geometric, or neither?

(b) Write the general term for this sequence. Why is it reasonable to limit the domain to natural numbers?

(c) Find a_{30} and interpret the result.

Solution

(a) One person requires 900 cubic feet of air circulated per hour, two people require 1800, three people 2700, and so on. The first five terms of this sequence are

$$900, 1800, 2700, 3600, 4500.$$

This sequence is arithmetic, with a common difference of 900.

(b) The nth term equals $900n$, so we let $a_n = 900n$. Because we cannot have a fraction of a person, limiting the domain to the natural numbers is reasonable.

(c) The result $a_{30} = 900(30) = \textbf{27,000}$ indicates that a classroom with **30** people should have a ventilation rate of **27,000** cubic feet per hour.

■ **Now Try Exercise 65**

▶ **REAL-WORLD CONNECTION** Chlorine is frequently added to the water to disinfect swimming pools. The chlorine concentration should remain between 1.5 and 2.5 parts per million (ppm). On a warm, sunny day 30% of the chlorine may dissipate from the water. In the next example we use a sequence to model the amount of chlorine in a pool at the beginning of each day. (*Source:* D. Thomas, *Swimming Pool Operator's Handbook.*)

EXAMPLE 8   ## Modeling chlorine in a swimming pool

A swimming pool on a warm, sunny day begins with a high chlorine content of 4 parts per million. (Assume that each day 30% of the chlorine dissipates.)

(a) Write a sequence that models the amount of chlorine in the pool at the beginning of the first 3 days, assuming that no additional chlorine is added and that the days are warm and sunny. Is this sequence arithmetic, geometric, or neither?

(b) Write the general term for this sequence.

(c) At the beginning of what day does the chlorine first drop below 1.5 parts per million?

Solution

(a) Because 30% of the chlorine dissipates, 70% remains in the water at the beginning of the next day. If the concentration at the beginning of the first day is 4 parts per million, then at the beginning of the second day it is

$$4 \cdot 0.70 = 2.8 \text{ parts per million,}$$

and at the start of the third day it is

$$2.8 \cdot 0.70 = 1.96 \text{ parts per million.}$$

The first three terms are 4, 2.8, 1.96. Successive terms are found by multiplying the previous term by 0.7. Thus the sequence is geometric, with common ratio 0.7.

(b) The initial amount is $a_1 = \textbf{4}$ and the common ratio is $r = \textbf{0.7}$, so the sequence can be represented by $a_n = \textbf{4}(\textbf{0.7})^{n-1}$.

(c) The table shown in Figure 14.15 reveals that $a_4 = 4(0.7)^{4-1} \approx 1.372 < 1.5$. Thus, at the beginning of the fourth day, the chlorine level in the swimming pool drops below the recommended minimum of 1.5 parts per million.

n	$u(n)$
1	4
2	2.8
3	1.96
4	1.372
5	.9604
6	.67228
7	.4706

$u(n) \blacksquare 4(.7)^{\wedge}(n-1)$

Figure 14.15

■ **Now Try Exercise 67**

14.2 Putting It All Together

In this section we discussed two types of sequences: arithmetic and geometric. Arithmetic sequences are linear functions, and geometric sequences with *positive r* are exponential functions. The inputs for both are limited to the natural numbers. The graph of an arithmetic sequence consists of points that lie on a line, whereas the graph of a geometric sequence (with a positive r) consists of points that lie on an exponential curve.

SEQUENCE	FORMULA	EXAMPLE
Arithmetic	$a_n = dn + c$ or $a_n = a_1 + (n - 1)d$, where d is the common difference and a_1 is the first term.	If $a_n = 5n + 2$, then the common difference is $d = 5$ and the terms of the sequence are $7, 12, 17, 22, 27, 32, 37, \ldots .$ Each term after the first is found by adding 5 to the previous term. The general term can be written as $a_n = 7 + 5(n - 1).$
Geometric	$a_n = a_1(r)^{n-1},$ where r is the common ratio ($r \neq 0, r \neq 1$) and a_1 is the first term.	If $a_n = 4(-2)^{n-1}$, then the common ratio is $r = -2$ and the first term is $a_1 = 4$. The terms of the sequence are $4, -8, 16, -32, 64, -128, 256, \ldots .$ Each term after the first is found by multiplying the previous term by -2.

14.2 Exercises

MyMathLab   Math XL
 PRACTICE   WATCH   DOWNLOAD   READ   REVIEW

CONCEPTS AND VOCABULARY

1. An arithmetic sequence is a(n) _linear/exponential_ function.

2. A geometric sequence with $r > 0$ is a(n) _linear/exponential_ function.

3. Give an example of an arithmetic sequence. State the common difference.

4. Give an example of a geometric sequence. State the common ratio.

5. To find successive terms in an arithmetic sequence, _____ the common difference to the _____ term.

6. To find successive terms in a geometric sequence, _____ the previous term by the _____.

7. Find the next term in the arithmetic sequence given by 3, 7, 11, 15. What is the common difference?

8. Find the next term in the geometric sequence given by 2, -4, 8, -16. What is the common ratio?

9. Write the general term a_n for a geometric sequence, using a_1 and r.

10. Write the general term a_n for an arithmetic sequence, using a_1 and d.

ARITHMETIC SEQUENCES

Exercises 11–28: (Refer to Example 1.) Determine whether f is an arithmetic sequence. Identify the common difference when possible.

11. $f(n) = 10n - 5$

12. $f(n) = -3n - 5$

13. $f(n) = 6 - n$

14. $f(n) = 6 + \frac{1}{2}n$

15. $f(n) = n^3 + 1$

16. $f(n) = 5\left(\frac{1}{3}\right)^{n-1}$

17.

n	1	2	3	4
$f(n)$	3	6	9	12

18.

n	1	2	3	4
$f(n)$	-7	-5	-3	-1

19.

n	1	2	3	4
$f(n)$	10	7	4	1

20.

n	1	2	3	4
$f(n)$	1	2	4	8

21.

n	1	2	3	4
$f(n)$	-4	0	8	12

22.

n	1	2	3	4
$f(n)$	1	2.5	4	5.5

23. $f(n)$

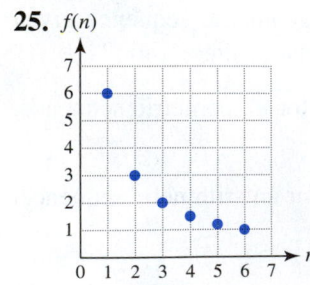

24. $f(n)$

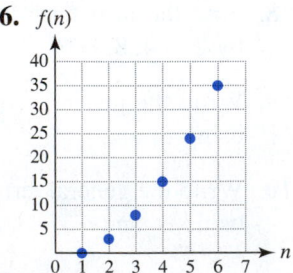

25. $f(n)$

26. $f(n)$

27. $f(n)$

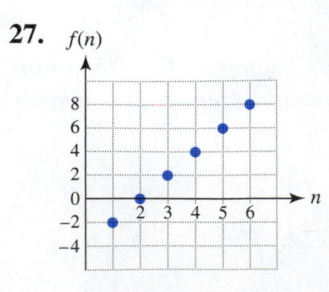

28. $f(n)$

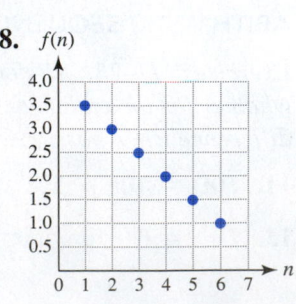

Exercises 29–34: (Refer to Example 2.) Find the general term a_n for the arithmetic sequence.

29. $a_1 = 7$ and $d = -2$

30. $a_1 = 5$ and $a_2 = 9$

31. $a_1 = -2$ and $a_3 = 6$

32. $a_2 = 7$ and $a_3 = 10$

33. $a_8 = 16$ and $a_{12} = 8$

34. $a_3 = 7$ and $d = -5$

Exercises 35–38: (Refer to Example 3.)

35. If $a_1 = -3$ and $d = 2$, find a_{32}.

36. If $a_1 = 2$ and $d = -3$, find a_{19}.

37. If $a_1 = -3$ and $a_2 = 0$, find a_9.

38. If $a_3 = -3$ and $d = 4$, find a_{62}.

GEOMETRIC SEQUENCES

Exercises 39–54: (Refer to Example 4.) Determine whether f is a geometric sequence. Identify the common ratio when possible.

39. $f(n) = 3^n$ **40.** $f(n) = 2(4)^n$

41. $f(n) = \frac{2}{3}(0.8)^{n-1}$ **42.** $f(n) = 7 - 3n$

43. $f(n) = 2(n-1)^2$ **44.** $f(n) = 2\left(-\frac{3}{4}\right)^{n-1}$

45.

n	1	2	3	4
$f(n)$	2	4	8	16

46.

n	1	2	3	4
$f(n)$	-6	3	-1.5	0.75

47.

n	1	2	3	4
$f(n)$	1	4	9	16

48.

n	1	2	3	4
$f(n)$	7	4	-1	-8

49.

n	1	2	3	4
$f(n)$	2	8	32	128

50.

n	1	2	3	4
$f(n)$	1	$\frac{1}{2}$	$\frac{1}{4}$	$\frac{1}{8}$

51. $f(n)$

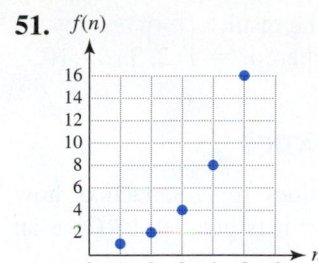

52. $f(n)$

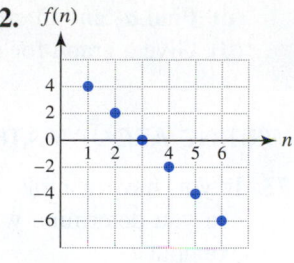

53. $f(n)$

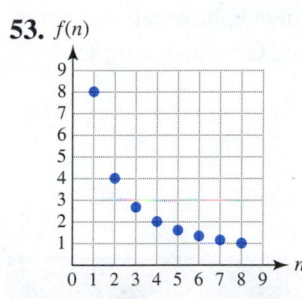

54. $f(n)$

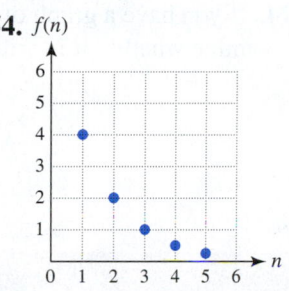

Exercises 55–60: (Refer to Example 5.) Find the general term a_n for the geometric sequence.

55. $a_1 = 1.5$ and $r = 4$

56. $a_1 = 3$ and $r = \frac{1}{4}$

57. $a_1 = -3$ and $a_2 = 6$

58. $a_1 = 2$ and $a_4 = 54$

59. $a_1 = 1$, $a_3 = 16$, and $r > 0$

60. $a_2 = 3$, $a_4 = 12$, and $r < 0$

Exercises 61–64: (Refer to Example 6.)

61. If $a_1 = 2$ and $r = 3$, find a_8.

62. If $a_1 = 4$ and $a_2 = 2$, find a_9.

63. If $a_1 = -1$ and $a_2 = 3$, find a_6.

64. If $a_3 = 5$ and $r = -3$, find a_7.

APPLICATIONS

65. *Room Ventilation* (Refer to Example 7.) In areas such as bars and lounges that sometimes allow smoking, the ventilation rate should be 3000 cubic feet per hour per person. (*Source:* ASHRAE.)
   (a) Write a sequence that gives the hourly ventilation necessary for 1, 2, 3, 4, and 5 people in a barroom. Is this sequence arithmetic, geometric, or neither?
   (b) Write the general term for this sequence.
   (c) Find a_{20} and interpret the result.
   (d) Give a graphical representation for eight terms of this sequence, using $n = 1, 2, 3, \ldots, 8$. Are the points collinear?

66. *Salary* Suppose that an employee receives a $2000 raise each year and that the sequence a_n models the employee's salary after n years. Is this sequence arithmetic, geometric, or neither? Explain.

67. *Chlorine in Swimming Pools* (Refer to Example 8.) Suppose that the water in a swimming pool initially has a chlorine content of 3 parts per million and that 20% of the chlorine dissipates each day.
   (a) If no additional chlorine is added, write the general term for a sequence that gives the chlorine concentration at the beginning of each day.
   (b) Give a graphical representation for this sequence, using $n = 1, 2, 3, \ldots, 8$. Are the points collinear? Is this sequence arithmetic, geometric, or neither?

68. *Salary* Suppose that an employee receives a 7% increase in salary each year and that the sequence a_n models the employee's salary after n years. Is this sequence arithmetic, geometric, or neither? Explain.

69. *Bouncing Ball* A tennis ball bounces back to 85% of the height from which it was dropped and then to 85% of the height of each successive bounce.
   (a) Write the general term for a sequence a_n that gives the maximum height of the ball on the nth bounce. Let $a_1 = 5$ feet.
   (b) Is the sequence arithmetic or geometric? Explain.
   (c) Find a_8 and interpret the result.

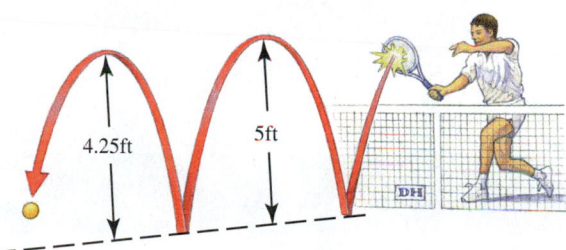

70. *Falling Object* The total distance D_n that an object falls in n seconds is shown in the table. Is the sequence arithmetic, geometric, or neither? Explain your reasoning.

n (seconds)	1	2	3	4	5
D_n (feet)	16	64	144	256	400

71. *Theater Seating* A theater has 40 seats in the first row, 42 seats in the second row, 44 seats in the third row, and so on.
   (a) Can the number of seats in each row be modeled by an arithmetic or geometric sequence? Explain.

(b) Write the general term for a sequence a_n that gives the number of seats in row n.

(c) How many seats are there in row 20?

72. *Appreciation of Lake Property* A certain type of lake property in northern Minnesota is increasing in value by 15% per year. Let the sequence a_n give the value of this type of lake property at the beginning of year n.

(a) Is a_n arithmetic, geometric, or neither? Explain your reasoning.

(b) Write the general term a_n for this sequence if $a_1 = \$100,000$.

(c) Find a_7 and interpret the result.

(d) Give a graph for a_n, where $n = 1, 2, 3, \ldots, 10$.

WRITING ABOUT MATHEMATICS

73. If you have a table of values for a sequence, how can you determine whether it is geometric? Give an example.

74. If you have a graph of a sequence, how can you determine whether it is arithmetic? Give an example.

SECTIONS 14.1 and 14.2   Checking Basic Concepts

1. Write the first four terms of the sequence defined by $a_n = \dfrac{n}{n+4}$.

2. Represent the sequence $a_n = n + 1$ numerically and graphically for $n = 1, 2, 3, 4, 5$.

3. Use the table to determine whether the sequence is arithmetic or geometric. Write the general term for the sequence.

(a)

n	1	2	3	4	5
a_n	−2	1	4	7	10

(b)

n	1	2	3	4	5
a_n	3	−6	12	−24	48

4. Find the general term a_n for an arithmetic sequence with $a_1 = 5$ and $d = 2$.

5. Determine the general term a_n for a geometric sequence with $a_1 = 5$ and $r = 2$.

14.3   Series

Basic Concepts • Arithmetic Series • Geometric Series • Summation Notation

A LOOK INTO MATH ▶ Although the terms *sequence* and *series* are sometimes used interchangeably in everyday life, they represent different mathematical concepts. In mathematics a sequence is a function whose domain is the set of natural numbers, whereas a series is a summation of the terms in a sequence. Series have played a central role in the development of modern mathematics. Today series are often used to approximate functions that are too complicated to have formulas. Series are also used to calculate accurate approximations for π and e.

Basic Concepts

Suppose that a person has a starting salary of \$30,000 per year and receives a \$2000 raise each year. Then the *sequence*

$$30{,}000, \ 32{,}000, \ 34{,}000, \ 36{,}000, \ 38{,}000$$

$$\underbrace{\hspace{6cm}}_{\text{Sequence}}$$

lists these salaries over a 5-year period. The total amount earned is given by the *series*

$$30{,}000 + 32{,}000 + 34{,}000 + 36{,}000 + 38{,}000,$$

Series

whose sum is $170,000. We now define the concept of a series.

NEW VOCABULARY

- ☐ Finite series
- ☐ Arithmetic series
- ☐ Sum of the first n terms of an arithmetic series
- ☐ Geometric series
- ☐ Sum of the first n terms of a geometric series
- ☐ Annuity
- ☐ Summation notation
- ☐ Index of summation
- ☐ Lower limit
- ☐ Upper limit

FINITE SERIES

A **finite series** is an expression of the form

$$a_1 + a_2 + a_3 + a_4 + a_n.$$

EXAMPLE 1   MySpace U.S. advertising revenues

Table 14.6 presents a sequence a_n that gives MySpace U.S. advertising revenues in millions of dollars n years after 2006. For example, $a_3 = 450$ indicates that in 2009, advertising revenues were $450 million.

TABLE 14.6   MySpace Advertising

n	1	2	3	4	5
a_n	480	595	450	290	190

Source: eMarketer (via WSJ), March 2011.

(a) Write a series whose sum represents the total U.S. advertising revenues for MySpace from 2007 to 2011.

(b) Interpret $a_1 + a_2 + a_3 + \cdots + a_5$.

Solution

(a) The required series and sum are given by

$$480 + 595 + 450 + 290 + 190 = 2005.$$

Revenues from 2007 to 2011 were $2005 million, or $2.005 billion.

(b) This series represents total revenues for 5 years from 2007 to 2011.

Now Try Exercise 41

MAKING CONNECTIONS

Sequences and Series

A *sequence* is an *ordered list*; a *series* is the *sum of the terms of a sequence*. For example, the even integers from 2 to 20 are represented by the sequence

$$2, 4, 6, 8, 10, 12, 14, 16, 18, 20. \qquad \text{Sequence}$$

The corresponding series is

$$2 + 4 + 6 + 8 + 10 + 12 + 14 + 16 + 18 + 20, \qquad \text{Series}$$

which sums to 110.

READING CHECK

How do you distinguish between a sequence and a series?

Arithmetic Series

Summing the terms of an arithmetic sequence results in an **arithmetic series**. For example, $a_n = 2n - 1$ for $n = 1, 2, 3, \ldots, 7$ defines the arithmetic sequence

$$\underbrace{1, 3, 5, 7, 9, 11, 13.}_{\textbf{Arithmetic sequence}}$$

The corresponding arithmetic *series* is

$$\underbrace{1 + 3 + 5 + 7 + 9 + 11 + 13,}_{\textbf{Arithmetic series}}$$

whose sum is 49. The following formula gives the sum of the first n terms of an arithmetic sequence. Note that the sum of the first n terms of a sequence is a finite series.

SUM OF THE FIRST n TERMS OF AN ARITHMETIC SEQUENCE

The **sum of the first n terms of an arithmetic sequence**, denoted S_n, is found by averaging the first and nth terms and then multiplying by n. That is,

$$S_n = a_1 + a_2 + a_3 + \cdots + a_n = n\left(\frac{a_1 + a_n}{2}\right).$$

READING CHECK

How do you find the sum of an arithmetic series?

The series $1 + 3 + 5 + 7 + 9 + 11 + 13$ consists of **7** terms, where the first term is **1** and the last term is **13**. Substituting in the formula gives

$$S_7 = 7\left(\frac{\textbf{1 + 13}}{2}\right) = 49,$$

which agrees with the sum obtained by adding the 7 terms.

Because $a_n = a_1 + (n - 1)d$ for an arithmetic sequence, S_n can also be written

$$S_n = n\left(\frac{a_1 + a_n}{2}\right)$$

$$= \frac{n}{2}(a_1 + a_n)$$

$$= \frac{n}{2}(a_1 + a_1 + (n - 1)d)$$

$$= \frac{n}{2}(2a_1 + (n - 1)d).$$

EXAMPLE 2 **Finding the sum of the terms of an arithmetic sequence**

Suppose that a person has a starting annual salary of \$30,000 and receives a \$1500 raise each year. Calculate the total amount earned after 10 years.

Solution
The sequence that gives the salary during year n is given by

$$a_n = 30{,}000 + 1500(n - 1).$$

One way to calculate the sum of the first 10 terms, denoted S_{10}, is to find a_1 and a_{10}.

$$a_1 = 30{,}000 + 1500(1 - 1) = 30{,}000$$

$$a_{10} = 30{,}000 + 1500(10 - 1) = 43{,}500$$

Thus the total amount earned during this 10-year period is

$$S_{10} = 10\left(\frac{a_1 + a_{10}}{2}\right)$$

$$= 10\left(\frac{30{,}000 + 43{,}500}{2}\right)$$

$$= \$367{,}500.$$

This sum can also be found with the second formula by letting $d = 1500$.

$$S_n = \frac{n}{2}(2a_1 + (n - 1)d)$$

$$= \frac{10}{2}(2 \cdot 30{,}000 + (10 - 1)1500)$$

$$= 5(60{,}000 + 9 \cdot 1500)$$

$$= \$367{,}500.$$

Now Try Exercise 49

EXAMPLE 3 **Finding the sum of the terms of an arithmetic sequence**

Find the sum of the series $2 + 4 + 6 + \cdots + 100$.

Solution

The first term of this series is $a_1 = 2$, and the common difference is $d = 2$. This series represents the even numbers from 2 to 100, so the number of terms is $n = 50$. Using the formula

$$S_n = n\left(\frac{a_1 + a_n}{2}\right) \qquad \text{Sum formula}$$

for the sum of an arithmetic series, we obtain

$$S_{50} = 50\left(\frac{2 + 100}{2}\right) = 2550.$$

Now Try Exercise 13

TECHNOLOGY NOTE

Sum of a Series
The "seq(" utility, found on some calculators under the LIST OPS menus, generates a sequence. The "sum(" utility, found on some calculators under the LIST MATH menus, calculates the sum of the sequence inside the parentheses. To verify the result in Example 2, let $a_n = 30{,}000 + 1500(n - 1)$. The value 367,500 for S_{10} is shown in Figure 14.16(a). The result found in Example 3 is shown in Figure 14.16(b).

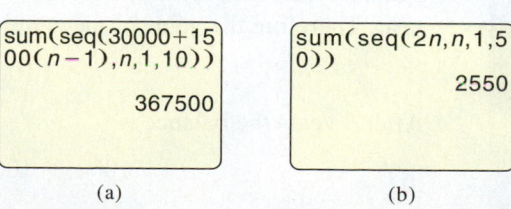

```
sum(seq(30000+15
00(n-1),n,1,10))
            367500
```
(a)

```
sum(seq(2n,n,1,5
0))
              2550
```
(b)

Figure 14.16

Geometric Series

A **geometric series** is the sum of the terms of a geometric sequence. For example,

$$\underbrace{1, 2, 4, 8, 16, 32}_{\text{Geometric sequence}}$$

is a geometric sequence with $a_1 = 1$ and $r = 2$. Then

$$\underbrace{1 + 2 + 4 + 8 + 16 + 32}_{\text{Geometric series}}$$

is a geometric series. We can use the following formula to sum a finite geometric series.

READING CHECK

How do you find the sum of a geometric series?

SUM OF THE FIRST n TERMS OF A GEOMETRIC SEQUENCE

If its first term is a_1 and its common ratio is r, then the **sum of the first n terms of a geometric sequence** is given by

$$S_n = a_1\left(\frac{1 - r^n}{1 - r}\right),$$

provided $r \neq 1$.

EXAMPLE 4 **Finding the sum of the terms of a geometric sequence**

Find the sum of each series.
(a) $1 + 2 + 4 + 8 + 16 + 32 + 64 + 128 + 256$

(b) $\frac{1}{2} - \frac{1}{4} + \frac{1}{8} - \frac{1}{16} + \frac{1}{32}$

Solution
(a) This series is geometric, with $n = \mathbf{9}$, $a_1 = \mathbf{1}$, and $r = \mathbf{2}$, so

$$S_9 = \mathbf{1}\left(\frac{1 - \mathbf{2}^{\mathbf{9}}}{1 - \mathbf{2}}\right) = 511. \qquad\qquad S_n = a_1\left(\frac{1 - r^n}{1 - r}\right)$$

(b) This series is geometric, with $n = \mathbf{5}$, $a_1 = \frac{1}{2} = \mathbf{0.5}$, and $r = -\frac{1}{2} = \mathbf{-0.5}$, so

$$S_5 = \mathbf{0.5}\left(\frac{1 - (\mathbf{-0.5})^5}{1 - (\mathbf{-0.5})}\right) = \frac{11}{32} = 0.34375.$$

Now Try Exercises 17, 19

▶ **REAL-WORLD CONNECTION** A sum of money from which regular payments are made is called an **annuity**. An annuity may be purchased with a lump sum deposit or by deposits made at various intervals. Suppose that $1000 is deposited at the end of each year in an annuity account that pays an annual interest rate I expressed as a decimal. At the end of the first year the account contains $1000. At the end of the second year $1000 is deposited again. In addition, the first deposit of $1000 would have received interest during the second year. Therefore the value of the annuity after 2 years is

$$1000 + 1000(1 + I).$$

After 3 years the balance is

$$1000 + 1000(1 + I) + 1000(1 + I)^2,$$

and after n years this amount is given by

$$1000 + 1000(1 + I) + 1000(1 + I)^2 + \cdots + 1000(1 + I)^{n-1}.$$

This series is a geometric series with its first term $a_1 = 1000$ and the common ratio $r = (1 + I)$. The sum of the first n terms is given by

$$S_n = a_1\left(\frac{1 - (1 + I)^n}{1 - (1 + I)}\right) = a_1\left(\frac{(1 + I)^n - 1}{I}\right).$$

EXAMPLE 5 **Finding the future value of an annuity**

Suppose that a 20-year-old worker deposits $1000 into an annuity account at the end of each year. If the interest rate is 5%, find the future value of the annuity when the worker is 65 years old.

Solution
Let $a_1 = 1000$, $I = 0.05$, and $n = 45$. The future value of the annuity is

$$S_n = a_1\left(\frac{(1 + I)^n - 1}{I}\right)$$

$$= 1000\left(\frac{(1 + 0.05)^{45} - 1}{0.05}\right)$$

$$\approx \$159{,}700$$

Now Try Exercise 23

Summation Notation

Summation notation is used to write series efficiently. The symbol Σ, the uppercase Greek letter *sigma*, is used to indicate a sum.

SUMMATION NOTATION

$$\sum_{k=1}^{n} a_k = a_1 + a_2 + a_3 + \cdots + a_n$$

The letter k is called the **index of summation**. The numbers 1 and n represent the subscripts of the first and last terms in the series. They are called the **lower limit** and **upper limit** of the summation, respectively.

EXAMPLE 6 **Using summation notation**

Find each sum.

(a) $\displaystyle\sum_{k=1}^{5} k^2$ **(b)** $\displaystyle\sum_{k=1}^{4} 5$ **(c)** $\displaystyle\sum_{k=3}^{6} (2k - 5)$

Solution

(a) $\displaystyle\sum_{k=1}^{5} k^2 = 1^2 + 2^2 + 3^2 + 4^2 + 5^2 = 55$

(b) $\displaystyle\sum_{k=1}^{4} 5 = 5 + 5 + 5 + 5 = 20$

(c) $\displaystyle\sum_{k=3}^{6} (2k - 5) = \underbrace{(2(3) - 5)}_{k=3} + \underbrace{(2(4) - 5)}_{k=4} + \underbrace{(2(5) - 5)}_{k=5} + \underbrace{(2(6) - 5)}_{k=6}$

$$= 1 + 3 + 5 + 7 = 16$$

Now Try Exercises 27, 29, 31

Summation notation is used frequently in statistics. The next example demonstrates how averages can be expressed in summation notation.

EXAMPLE 7 **Applying summation notation**

Express the average of the n numbers $x_1, x_2, x_3, \ldots, x_n$ in summation notation.

Solution
The average of n numbers can be written as

$$\frac{x_1 + x_2 + x_3 + \cdots + x_n}{n}.$$

This expression is equivalent to $\frac{1}{n}\left(\sum_{k=1}^{n} x_k\right)$.

Now Try Exercise 35

NOTE: $\sum_{k=1}^{n} x_k$ is equivalent to $\displaystyle\sum_{k=1}^{n} x_k$.

▶ **REAL-WORLD CONNECTION** Series are used to model air filtration.

EXAMPLE 8 **Modeling air filtration**

Suppose that an air filter removes 90% of the impurities entering it.
(a) Find a series that represents the amount of impurities removed by a sequence of n air filters. Express this answer in summation notation.
(b) How many air filters would be necessary to remove 99.99% of the impurities?

Solution

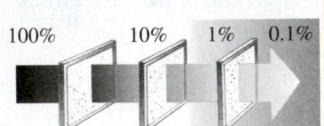

100%   10%   1%   0.1%

Figure 14.17 Impurities Passing Through Air Filters

(a) The first filter removes 90% of the impurities, so 10%, or 0.1, passes through it. Of the 0.1 that passes through the first filter, 90% is removed by the second filter, while 10% of 10%, or 0.01, passes through. Then, 10% of 0.01, or 0.001, passes through the third filter. Figure 14.17 depicts these results, from which we can establish a pattern. If we let 100%, or 1, represent the amount of impurities entering the first air filter, the amount removed by n filters equals

$$(0.9)(1) + (0.9)(0.1) + (0.9)(0.01) + (0.9)(0.001) + \cdots + (0.9)(0.1)^{n-1}.$$

In summation notation we write this series as $\sum_{k=1}^{n} 0.9(0.1)^{k-1}$.

(b) To remove 99.99%, or 0.9999, of the impurities requires 4 air filters, because

$$\sum_{k=1}^{4} 0.9(0.1)^{k-1} = (0.9)(1) + (0.9)(0.1) + (0.9)(0.01) + (0.9)(0.001)$$

$$= 0.9 + 0.09 + 0.009 + 0.0009$$

$$= 0.9999.$$

Now Try Exercise 43

14.3 Putting It All Together

A finite sequence is an ordered list such as

$$a_1, a_2, a_3, a_4, a_5, \ldots, a_n.$$

A finite series is the summation of the terms of a sequence and can be expressed as

$$a_1 + a_2 + a_3 + a_4 + a_5 + \cdots + a_n.$$

SERIES	DESCRIPTION	EXAMPLE
Finite Arithmetic	$a_1 + a_2 + a_3 + \cdots + a_n$, where $a_n = dn + c$ or $a_n = a_1 + (n-1)d$. The sum of the first n terms is $$S_n = n\left(\frac{a_1 + a_n}{2}\right) \quad \text{or}$$ $$S_n = \frac{n}{2}(2a_1 + (n-1)d),$$ where a_1 is the first term and d is the common difference.	The series $$4 + 7 + 10 + 13 + 16 + 19 + 22$$ is obtained from the sequence $$a_n = 3n + 1 \quad \text{or} \quad a_n = 4 + 3(n-1).$$ Its sum is $$S_7 = 7\left(\frac{4+22}{2}\right) = 91 \quad \text{or}$$ $$S_7 = \frac{7}{2}(2 \cdot 4 + (7-1)3) = 91.$$
Finite Geometric	$a_1 + a_2 + a_3 + \cdots + a_n$, where $a_n = a_1(r)^{n-1}$ for nonzero constants a_1 and r. The sum of the first n terms is $$S_n = a_1\left(\frac{1-r^n}{1-r}\right),$$ where a_1 is the first term and r is the common ratio ($r \neq 1$).	The series $$3 + 6 + 12 + 24 + 48 + 96$$ has $n = 6$, $a_1 = 3$, and $r = 2$. Its sum is $$S_6 = 3\left(\frac{1-2^6}{1-2}\right) = 189.$$

14.3 Exercises

MyMathLab

 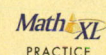

PRACTICE | WATCH | DOWNLOAD | READ | REVIEW

CONCEPTS AND VOCABULARY

1. The summation of the terms of a sequence is called a(n) _____.

2. Find the sum of the series $1 + 2 + 3 + 4$.

3. The series $1 + 3 + 5 + 7 + 9$ is an example of a(n) _____ series.

4. The series $1 + 3 + 9 + 27 + 81$ is an example of a(n) _____ series.

5. If $a_1 + a_2 + a_3 + \cdots + a_n$ is an arithmetic series, its sum is $S_n =$ _____.

6. If $a_1 + a_2 + a_3 + \cdots + a_n$ is a geometric series with the common ratio $r \neq 1$, its sum is $S_n =$ _____.

7. The symbol Σ is used to indicate a(n) _____.

8. Write $\sum_{k=1}^{4} a_k$ as a sum.

9. $\sum_{n=1}^{5} a_1 + (n-1)d$ is an example of a(n) _____ series.

10. $\sum_{n=1}^{4} a_1 r^{n-1}$ is an example of a(n) _____ series.

SUMS OF SERIES

Exercises 11–16: Find the sum of the arithmetic series by using a formula.

11. $3 + 5 + 7 + 9 + 11 + 13$

12. $7.5 + 6 + 4.5 + 3 + 1.5 + 0 + (-1.5)$

13. $1 + 2 + 3 + 4 + \cdots + 40$

14. $1 + 3 + 5 + 7 + \cdots + 99$

15. $-7 + (-4) + (-1) + 2 + 5$

16. $89 + 84 + 79 + 74 + 69 + 64 + 59 + 54$

Exercises 17–22: Find the sum of the geometric series by using a formula.

17. $3 + 9 + 27 + 81 + 243 + 729 + 2187$

18. $2 - 1 + \frac{1}{2} - \frac{1}{4} + \frac{1}{8} - \frac{1}{16} + \frac{1}{32}$

19. $1 - 2 + 4 - 8 + 16 - 32 + 64 - 128$

20. $2 + \frac{1}{2} + \frac{1}{8} + \frac{1}{32} + \frac{1}{128} + \frac{1}{512}$

21. $0.5 + 1.5 + 4.5 + 13.5 + 40.5 + 121.5$

22. $0.6 + 0.3 + 0.15 + 0.075 + 0.0375$

Exercises 23–26: Annuities (Refer to Example 5.) Find the future value of the annuity.

23. $a_1 = \$2000 \quad I = 0.08 \quad n = 20$

24. $a_1 = \$500 \quad I = 0.15 \quad n = 10$

25. $a_1 = \$10,000 \quad I = 0.11 \quad n = 5$

26. $a_1 = \$3000 \quad I = 0.19 \quad n = 45$

SUMMATION NOTATION

Exercises 27–34: Write the terms of the series and find their sum.

27. $\displaystyle\sum_{k=1}^{4} 2k$

28. $\displaystyle\sum_{k=1}^{6} (k-1)$

29. $\displaystyle\sum_{k=1}^{8} 4$

30. $\displaystyle\sum_{k=2}^{6} (5-2k)$

31. $\displaystyle\sum_{k=1}^{7} k^2$

32. $\displaystyle\sum_{k=1}^{4} 5(2)^{k-1}$

33. $\displaystyle\sum_{k=4}^{5} (k^2-k)$

34. $\displaystyle\sum_{k=1}^{4} \log k$

Exercises 35–38: Write in summation notation.

35. $1^4 + 2^4 + 3^4 + 4^4 + 5^4 + 6^4$

36. $1 + \frac{1}{5^1} + \frac{1}{5^2} + \frac{1}{5^3} + \frac{1}{5^4}$

37. $1 + \frac{1}{2^2} + \frac{1}{3^2} + \frac{1}{4^2} + \frac{1}{5^2}$

38. $1 + \frac{1}{10} + \frac{1}{100} + \frac{1}{1000} + \frac{1}{10,000}$

39. Verify that $\sum_{k=1}^{n} k = \frac{n(n+1)}{2}$ by using a formula for the sum of the first n terms of an arithmetic series.

40. Use Exercise 39 to find the sum of the series $\sum_{k=1}^{200} k$.

APPLICATIONS

41. *Prison Escapees* The following table lists the number of escapees from state prisons each year.

Year	1990	1991	1992
Escapees	8518	9921	10,706

Year	1993	1994	1995
Escapees	14,035	14,307	12,249

Source: Bureau of Justice Statistics.

(a) Write a series whose sum is the total number of escapees from 1990 to 1995.

(b) Find its sum.

42. *Captured Prison Escapees* (Refer to Exercise 41.) The table lists the number of escapees from state prisons who were captured, including inmates who may have escaped during a previous year.

Year	1990	1991	1992
Captured	9324	9586	10,031

Year	1993	1994	1995
Captured	12,872	13,346	12,166

Source: Bureau of Justice Statistics.

(a) Write a series whose sum is the total number of escapees captured from 1990 to 1995.

(b) Find its sum.

(c) Compare the number of escapees to the number captured during this time period.

43. *Air Filtration* (Refer to Example 8.) Suppose that an air filter removes 80% of the impurities entering it.

(a) Find a series that represents the amount of impurities removed by a sequence of n air filters. Express the answer in summation notation.

(b) How many filters would be necessary to remove 96% of the impurities?

44. *Air Filtration* Suppose that an air filter removes 70% of the impurities entering it.

(a) Find a series that represents the amount of impurities removed by a sequence of n air filters. Express the answer in summation notation.

(b) How many filters would be necessary to remove 97.3% of the impurities?

45. *Area* A sequence of smaller squares is formed by connecting the midpoints of the sides of a larger square as shown in the figure.

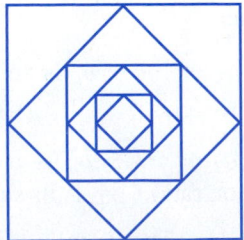

(a) If the area of the largest square is 1 square unit, determine the first five terms of a sequence that describes the area of each successive square.

(b) Use a formula to sum the areas of the first 10 squares.

46. *Perimeter* (Refer to Exercise 45.) Use a formula to find the sum of the perimeters of the first 10 squares.

47. *Stacking Logs* A stack of logs is made in layers, with one log less in each layer, as shown in the accompanying figure. If the top layer has 6 logs and the bottom layer has 14 logs, what is the total number of logs in the pile? Use a formula to find this sum.

48. *Stacking Logs* (Refer to Exercise 47.) Suppose that a stack of logs has 15 logs in the top layer and a total of 10 layers. How many logs are in the stack?

49. *Salaries* (Refer to Example 2.) Suppose that an individual's starting salary is $35,000 per year and that the individual receives a $2000 raise each year. Find the total amount earned over 20 years.

50. *Salaries* Suppose that an individual's starting salary is $35,000 per year and that the individual receives a 10% raise each year. Find the total amount earned over 20 years.

51. *Bouncing Ball* A tennis ball first bounces to 75% of the height from which it was dropped and then to 75% of the height of each successive bounce. If it is dropped from a height of 10 feet, find the distance it *falls* between the fourth and fifth bounce.

52. *Bouncing Ball* A tennis ball first bounces to 75% of the height from which it was dropped and then to 75% of the height of each successive bounce. If it is dropped from a height of 10 feet, find the *total* distance it travels before it reaches its fifth bounce. (*Hint:* Make a sketch.)

53. Online Exploration Go online and find a series that can be used to calculate π. Use the first five terms of your series to estimate π.

54. Online Exploration Go online and find a series that can be used to calculate the number e. Use the first five terms of your series to estimate e.

WRITING ABOUT MATHEMATICS

55. Discuss the difference between a sequence and a series. Give an example of each.

56. Suppose that an arithmetic series has $a_1 = 1$ and a common difference of $d = 2$, whereas a geometric series has $a_1 = 1$ and a common ratio of $r = 2$. Discuss how their sums compare as the number of terms n becomes large. (*Hint:* Calculate each sum for $n = 10$, 20, and 30.)

Group Activity Working With Real Data

Directions: Form a group of 2 to 4 people. Select someone to record the group's responses for this activity. All members of the group should work cooperatively to answer the questions. If your instructor asks for your results, each member of the group should be prepared to respond.

Depreciation For tax purposes, businesses frequently depreciate equipment. Two different methods of depreciation are called *straight-line depreciation* and *sum-of-the-years'-digits*. Suppose that a college student buys a $3000 computer to start a business that provides Internet services. This student estimates the life of the computer at 4 years, after which its value will be $200. The difference between $3000 and $200, or $2800, may be deducted from the student's taxable income over a 4-year period.

In straight-line depreciation, equal portions of $2800 are deducted each year over the 4 years. The sum-of-the-years'-digits method gives depreciation differently. For a computer having a useful life of 4 years, the sum of the years is computed by

$$1 + 2 + 3 + 4 = 10.$$

With this method, $\frac{4}{10}$ of $2800 is deducted the first year, $\frac{3}{10}$ the second year, and so on, until $\frac{1}{10}$ is deducted the fourth year. Both depreciation methods yield a total deduction of $2800 over the 4 years. (*Source:* Sharp Electronics Corporation, *Conquering the Sciences.*)

(a) Find an arithmetic sequence that gives the amount depreciated each year by each method.

(b) Write a series whose sum is the amount depreciated over 4 years by each method.

14.4 The Binomial Theorem

Pascal's Triangle • Factorial Notation and Binomial Coefficients • Using the Binomial Theorem

A LOOK INTO MATH ▶ In this section we demonstrate how to expand expressions of the form $(a + b)^n$, where n is a natural number. These expressions occur in statistics, finite mathematics, computer science, and calculus. The two methods that we discuss are Pascal's triangle and the binomial theorem.

Pascal's Triangle

Expanding $(a + b)^n$ for increasing values of n gives the following results.

Expanding Binomials

$$(a + b)^0 = 1$$
$$(a + b)^1 = 1a + 1b$$
$$(a + b)^2 = 1a^2 + 2ab + 1b^2$$
$$(a + b)^3 = 1a^3 + 3a^2b + 3ab^2 + 1b^3$$
$$(a + b)^4 = 1a^4 + 4a^3b + 6a^2b^2 + 4ab^3 + 1b^4$$
$$(a + b)^5 = 1a^5 + 5a^4b + 10a^3b^2 + 10a^2b^3 + 5ab^4 + 1b^5$$

NEW VOCABULARY

☐ Pascal's triangle
☐ Factorial notation
☐ Binomial coefficient
☐ Binomial theorem

Note that $(a + b)^1$ has two terms, starting with a and ending with b; $(a + b)^2$ has three terms, starting with a^2 and ending with b^2; and, in general, $(a + b)^n$ has $n + 1$ terms, starting with a^n and ending with b^n. From left to right, the exponent on a decreases by 1 each successive term, and the exponent on b increases by 1 each successive term.

The triangle formed by the red numbers is called **Pascal's triangle**. This triangle consists of 1s along the sides, and each element inside the triangle is the sum of the two numbers above it, as shown in Figure 14.18. Pascal's triangle is usually written without variables and can be extended to include as many rows as needed.

Pascal's Triangle

```
            1
          1   1
        1   2   1
      1   3   3   1
    1   4   6   4   1
  1   5  10  10   5   1
```

Figure 14.18

We can use this triangle to expand $(a + b)^n$, where n is a natural number. For example, the expression $(m + n)^4$ consists of five terms written as

$$(m + n)^4 = \_m^4 + \_m^3n^1 + \_m^2n^2 + \_m^1n^3 + \_n^4.$$

Because there are five terms, the coefficients can be found in the fifth row of Pascal's triangle, which is

$$1\quad 4\quad 6\quad 4\quad 1.$$

READING CHECK

Explain how to find the numbers in Pascal's triangle.

Thus

$$(m + n)^4 = \underline{1}\,m^4 + \underline{4}\,m^3n^1 + \underline{6}\,m^2n^2 + \underline{4}\,m^1n^3 + \underline{1}\,n^4$$

$$= m^4 + 4m^3n + 6m^2n^2 + 4mn^3 + n^4.$$

EXAMPLE 1   **Expanding a binomial**

Expand each binomial, using Pascal's triangle.
(a) $(x + 2)^5$ **(b)** $(2m - n)^3$

Solution

(a) To find the coefficients, use the sixth row (1 5 10 10 5 1) in Pascal's triangle.

$$(x + 2)^5 = \underline{1}\,x^5 + \underline{5}\,x^4 \cdot 2^1 + \underline{10}\,x^3 \cdot 2^2 + \underline{10}\,x^2 \cdot 2^3 + \underline{5}\,x^1 \cdot 2^4 + \underline{1}(2^5)$$

$$= x^5 + 10x^4 + 40x^3 + 80x^2 + 80x + 32$$

(b) To find the coefficients, use the fourth row (1 3 3 1) in Pascal's triangle.

$$(2m - n)^3 = \underline{1}(2m)^3 + \underline{3}(2m)^2(-n)^1 + \underline{3}(2m)^1(-n)^2 + \underline{1}(-n)^3$$

$$= 8m^3 - 12m^2n + 6mn^2 - n^3$$

Now Try Exercises 13, 15

Factorial Notation and Binomial Coefficients

An alternative to Pascal's triangle is the binomial theorem, which uses **factorial notation**.

---

n FACTORIAL ($n!$)

For any positive integer n,

$$n! = 1 \cdot 2 \cdot 3 \cdot \cdots \cdot n.$$

We also define $0! = 1$.

NOTE: Because multiplication is commutative, n factorial can also be defined as

$$n! = n \cdot (n - 1) \cdot (n - 2) \cdot \cdots \cdot 2 \cdot 1.$$

Examples include the following.

$$0! = 1$$
$$1! = 1$$
$$2! = 1 \cdot 2 = 2$$
$$3! = 1 \cdot 2 \cdot 3 = 6$$
$$4! = 1 \cdot 2 \cdot 3 \cdot 4 = 24$$
$$5! = 1 \cdot 2 \cdot 3 \cdot 4 \cdot 5 = 120$$

Figure 14.19 supports these results. On some calculators, factorial (!) can be accessed in the MATH PRB menus.

Factorials

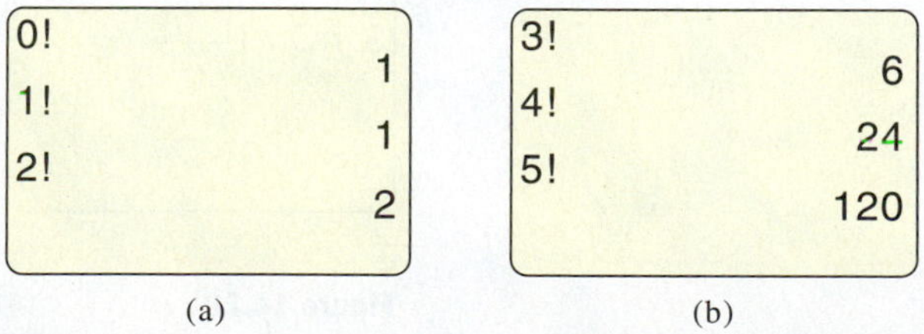

0!		3!	
1!	1	4!	6
2!	1	5!	24
	2		120

(a)   (b)

Figure 14.19

EXAMPLE 2   **Evaluating factorial expressions**

Simplify the expression.

(a) $\dfrac{5!}{3!2!}$ **(b)** $\dfrac{4!}{4!0!}$

Solution

(a) $\dfrac{5!}{3!2!} = \dfrac{1 \cdot 2 \cdot 3 \cdot 4 \cdot 5}{(1 \cdot 2 \cdot 3)(1 \cdot 2)} = \dfrac{120}{6 \cdot 2} = 10$ **(b)** $0! = 1$, so $\dfrac{4!}{4!0!} = \dfrac{4!}{4!(1)} = \dfrac{4!}{4!} = 1$

Now Try Exercises 23, 25

The expression $_nC_r$ represents a *binomial coefficient* that can be used to calculate the numbers in Pascal's triangle.

BINOMIAL COEFFICIENT $_nC_r$

For n and r nonnegative integers, $n \geq r$,

$$_nC_r = \dfrac{n!}{(n - r)!\, r!}$$

is a **binomial coefficient**.

Values of $_nC_r$ for $r = 0, 1, 2, \ldots, n$ correspond to the $n + 1$ numbers in row $n + 1$ of Pascal's triangle.

EXAMPLE 3   **Calculating $_nC_r$**

Calculate $_3C_r$ for $r = 0, 1, 2, 3$ by hand. Check your results on a calculator. Compare these numbers with the fourth row in Pascal's triangle.

Solution

$$_3C_0 = \dfrac{3!}{(3 - 0)!\, 0!} = \dfrac{6}{6 \cdot 1} = 1 \qquad _3C_1 = \dfrac{3!}{(3 - 1)!\, 1!} = \dfrac{6}{2 \cdot 1} = 3$$

$$_3C_2 = \dfrac{3!}{(3 - 2)!\, 2!} = \dfrac{6}{1 \cdot 2} = 3 \qquad _3C_3 = \dfrac{3!}{(3 - 3)!\, 3!} = \dfrac{6}{1 \cdot 6} = 1$$

These results are supported in Figure 14.20. The fourth row of Pascal's triangle is

1   3   3   1,

which agrees with the calculated values for $_3C_r$. On some calculators, the MATH PRB menus are used to calculate $_nC_r$.

Binomial Coefficients

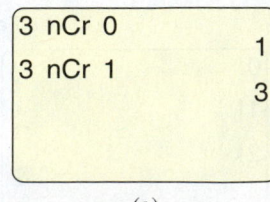

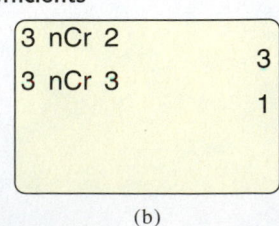

(a)   (b)

Figure 14.20

Now Try Exercises 27, 31

Using the Binomial Theorem

The binomial coefficients can be used to expand expressions of the form $(a + b)^n$. To do so, we use the **binomial theorem**.

BINOMIAL THEOREM

For any positive integer n and any numbers a and b,

$$(a + b)^n = {}_nC_0 a^n + {}_nC_1 a^{n-1}b^1 + \cdots + {}_nC_{n-1} a^1 b^{n-1} + {}_nC_n b^n.$$

Using the results of Example 3, we write

$$(a + b)^3 = {}_3C_0 a^3 + {}_3C_1 a^2 b^1 + {}_3C_2 a^1 b^2 + {}_3C_3 b^3$$
$$= 1a^3 + 3a^2 b + 3ab^2 + 1b^3$$
$$= a^3 + 3a^2 b + 3ab^2 + b^3.$$

EXAMPLE 4   **Expanding a binomial**

Use the binomial theorem to expand each expression.
(a) $(x + y)^5$ **(b)** $(3 - 2x)^4$

Solution
(a) The coefficients are calculated as follows.

$${}_5C_0 = \frac{5!}{(5 - 0)!\,0!} = 1, \quad {}_5C_1 = \frac{5!}{(5 - 1)!\,1!} = 5, \quad {}_5C_2 = \frac{5!}{(5 - 2)!\,2!} = 10$$

$${}_5C_3 = \frac{5!}{(5 - 3)!\,3!} = 10, \quad {}_5C_4 = \frac{5!}{(5 - 4)!\,4!} = 5, \quad {}_5C_5 = \frac{5!}{(5 - 5)!\,5!} = 1$$

Using the binomial theorem, we arrive at the following result.

$$(x + y)^5 = {}_5C_0 x^5 + {}_5C_1 x^4 y^1 + {}_5C_2 x^3 y^2 + {}_5C_3 x^2 y^3 + {}_5C_4 x^1 y^4 + {}_5C_5 y^5$$
$$= 1x^5 + 5x^4 y + 10x^3 y^2 + 10x^2 y^3 + 5xy^4 + 1y^5$$
$$= x^5 + 5x^4 y + 10x^3 y^2 + 10x^2 y^3 + 5xy^4 + y^5$$

(b) The coefficients are calculated as follows.

$${}_4C_0 = \frac{4!}{(4 - 0)!\,0!} = 1, \quad {}_4C_1 = \frac{4!}{(4 - 1)!\,1!} = 4, \quad {}_4C_2 = \frac{4!}{(4 - 2)!\,2!} = 6,$$

$${}_4C_3 = \frac{4!}{(4 - 3)!\,3!} = 4, \quad {}_4C_4 = \frac{4!}{(4 - 4)!\,4!} = 1$$

Using the binomial theorem with $a = 3$ and $b = (-2x)$, we arrive at the following result.

$$(3 - 2x)^4 = {}_4C_0 (3)^4 + {}_4C_1 (3)^3 (-2x) + {}_4C_2 (3)^2 (-2x)^2$$
$$+ {}_4C_3 (3)(-2x)^3 + {}_4C_4 (-2x)^4$$
$$= 1(81) + 4(27)(-2x) + 6(9)(4x^2) + 4(3)(-8x^3) + 1(16x^4)$$
$$= 81 - 216x + 216x^2 - 96x^3 + 16x^4$$

Now Try Exercises 39, 49

The binomial theorem gives *all* of the terms of $(a + b)^n$. However, we can find any individual term by noting that the $(r + 1)$st term in the binomial expansion for $(a + b)^n$ is given by the formula ${}_nC_r a^{n-r} b^r$, for $0 \le r \le n$. The next example shows how to use this formula to find the $(r + 1)$st term of $(a + b)^n$.

EXAMPLE 5 **Finding the kth term in a binomial expansion**

Find the third term of $(x - y)^5$.

Solution

In this example the $(r + 1)$st term is the *third* term in the expansion of $(x - y)^5$. That is, $r + 1 = 3$, or $r = 2$. Also, the exponent in the expression is $n = 5$. To get this binomial into the form $(a + b)^n$, we note that the first term in the binomial is $a = x$ and that the second term in the binomial is $b = -y$. Substituting the values for r, n, a, and b in the formula $_nC_r a^{n-r} b^r$ for the $(r + 1)$st term yields

$$_5C_2(x)^{5-2}(-y)^2 = 10x^3y^2.$$

The third term in the binomial expansion of $(x - y)^5$ is $10x^3y^2$.

Now Try Exercise 53

14.4 Putting It All Together

TOPIC	EXPLANATION	EXAMPLE
Pascal's Triangle	1 1 1 1 2 1 **1 3 3 1** 1 4 6 4 1 1 5 10 10 5 1	$(a + b)^3 = 1a^3 + 3a^2b + 3ab^2 + 1b^3$ (Row 4) To expand $(a + b)^n$, use row $n + 1$ in the triangle.
Factorial Notation	The expression $n!$ equals $1 \cdot 2 \cdot 3 \cdot \cdots \cdot n$.	$5! = 1 \cdot 2 \cdot 3 \cdot 4 \cdot 5 = 120$
Binomial Coefficient $_nC_r$	$_nC_r = \dfrac{n!}{(n - r)!\, r!}$	$_6C_4 = \dfrac{6!}{(6 - 4)!\, 4!} = \dfrac{6!}{2!\, 4!} = \dfrac{720}{2 \cdot 24} = 15$
Binomial Theorem	$(a + b)^n = {}_nC_0 a^n + {}_nC_1 a^{n-1}b^1 + \cdots$ $+ {}_nC_{n-1} a^1 b^{n-1} + {}_nC_n b^n$	$(a + b)^4 = {}_4C_0 a^4 + {}_4C_1 a^3b + {}_4C_2 a^2b^2$ $+ {}_4C_3 ab^3 + {}_4C_4 b^4$ $= 1a^4 + 4a^3b + 6a^2b^2 + 4ab^3 + 1b^4$ $= a^4 + 4a^3b + 6a^2b^2 + 4ab^3 + b^4$

14.4 Exercises

MyMathLab

CONCEPTS AND VOCABULARY

1. How many terms result from expanding $(a + b)^4$?

2. How many terms result from expanding $(a + b)^n$?

3. To find the coefficients for the expansion of $(a + b)^3$, what row of Pascal's triangle do you use?

4. Write down the first 5 rows of Pascal's triangle.

5. $4! = $ _____

6. $1 \cdot 2 \cdot 3 \cdot 4 \cdot 5 \cdot 6 = $ _____

7. $_nC_r = $ _____

8. $(a + b)^2 = $ _____

9. **Thinking Generally** $\frac{n!}{(n - 1)!} = $ _____

10. **Thinking Generally** $_nC_n = $ _____

USING PASCAL'S TRIANGLE

Exercises 11–18: Use Pascal's triangle to expand the given expression.

11. $(x + y)^3$
12. $(x + y)^4$

13. $(2x + 1)^4$
14. $(2x - 1)^4$

15. $(a - b)^5$
16. $(3x + 2y)^3$

17. $(x^2 + 1)^3$
18. $\left(\frac{1}{2} - x^2\right)^5$

FACTORIALS AND BINOMIAL COEFFICIENTS

Exercises 19–32: Evaluate the expression.

19. $3!$
20. $6!$

21. $\frac{4!}{3!}$
22. $\frac{6!}{3!}$

23. $\frac{2!}{0!}$
24. $\frac{5!}{1!}$

25. $\frac{5!}{2!3!}$
26. $\frac{6!}{4!2!}$

27. $_5C_4$
28. $_3C_1$

29. $_6C_5$
30. $_2C_2$

31. $_4C_0$
32. $_4C_3$

Exercises 33–38: Evaluate the binomial coefficient with a calculator.

33. $_{12}C_7$
34. $_{13}C_8$

35. $_9C_5$
36. $_{25}C_{14}$

37. $_{19}C_{11}$
38. $_{10}C_6$

THE BINOMIAL THEOREM

Exercises 39–50: Use the binomial theorem to expand the expression.

39. $(m + n)^3$
40. $(m + n)^5$

41. $(x - y)^4$
42. $(1 - 3x)^4$

43. $(2a + 1)^3$
44. $(x^2 - 1)^3$

45. $(x + 2)^5$
46. $(a - 3)^5$

47. $(3 + 2m)^4$
48. $(m - 3n)^3$

49. $(2x - y)^3$
50. $(2a + 3b)^4$

Exercises 51–56: The $(r + 1)$st term of the expression $(a + b)^n$, $0 \leq r \leq n$, is given by $_nC_r a^{n-r}b^r$. Find the specified term. Refer to Example 5.

51. The first term of $(a + b)^8$

52. The second term of $(a - b)^{10}$

53. The fourth term of $(x + y)^7$

54. The sixth term of $(a + b)^9$

55. The first term of $(2m + n)^9$

56. The eighth term of $(2a - b)^8$

WRITING ABOUT MATHEMATICS

57. Explain how to find the numbers in Pascal's triangle.

58. Compare the expansion of $(a + b)^n$ to the expansion of $(a - b)^n$. Give an example.

Checking Basic Concepts

1. Determine whether the series is arithmetic or geometric.
   (a) $\frac{1}{2} + \frac{1}{4} + \frac{1}{8} + \cdots + \frac{1}{256}$
   (b) $\frac{1}{2} + \frac{5}{2} + \frac{9}{2} + \frac{13}{2} + \frac{17}{2}$

2. Use a formula to find the sum of the arithmetic series

$$4 + 8 + 12 + \cdots + 48.$$

3. Use a formula to find the sum of the geometric series

$$1 - 2 + 4 - 8 + 16 - 32 + 64 - 128 + 256 - 512.$$

4. Use Pascal's triangle to expand $(x - y)^4$.

5. Use the binomial theorem to expand $(x + 2)^3$.

CHAPTER 14   Summary

SECTION 14.1 ■ SEQUENCES

Sequences An *infinite sequence* is a function whose domain is the natural numbers. A *finite sequence* is a function whose domain is $D = \{1, 2, 3, \ldots, n\}$ for some natural number n.

Example: $a_n = 2n$ is a symbolic representation of the even natural numbers. The first six terms, 2, 4, 6, 8, 10, 12, of this sequence are represented numerically and graphically in the table and figure.

Numerical Representation | Graphical Representation

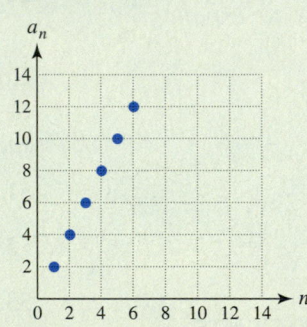

n	a_n
1	2
2	4
3	6
4	8
5	10
6	12

SECTION 14.2 ■ ARITHMETIC AND GEOMETRIC SEQUENCES

Two common types of sequences are arithmetic and geometric.

Arithmetic Sequence An arithmetic sequence is determined by a linear function of the form $f(n) = dn + c$ or $f(n) = a_1 + (n - 1)d$. Successive terms in an arithmetic sequence are found by adding the common difference d to the previous term.

Example: The sequence 1, 3, 5, 7, 9, 11, . . . is an arithmetic sequence with its first term $a_1 = 1$, common difference $d = 2$, and general term $a_n = 2n - 1$.

Geometric Sequence The general term for a geometric sequence is given by $f(n) = a_1 r^{n-1}$. Successive terms in a geometric sequence are found by multiplying the previous term by the common ratio r.

Example: The sequence 3, 6, 12, 24, 48, . . . is a geometric sequence with its first term $a_1 = 3$, common ratio $r = 2$, and general term $a_n = 3(2)^{n-1}$.

SECTION 14.3 ■ SERIES

Series A series results when the terms of a sequence are summed. The series associated with the sequence 2, 4, 6, 8, 10 is

$$2 + 4 + 6 + 8 + 10, \qquad \text{Series}$$

and its sum equals 30. An arithmetic series results when the terms of an arithmetic sequence are summed, and a geometric series results when the terms of a geometric sequence are summed. In this section, we discussed formulas for finding sums of arithmetic and geometric series. See Putting It All Together for Section 14.3.

Summation Notation Summation notation can be used to write series efficiently.

Example: $1^2 + 2^2 + 3^2 + 4^2 + 5^2 = \sum_{k=1}^{5} k^2.$

SECTION 14.4 ■ THE BINOMIAL THEOREM

Pascal's triangle may be used to find the coefficients for the expansion of $(a + b)^n$, where n is a natural number.

$$
\begin{array}{ccccccccccc}
 & & & & & 1 & & & & & \\
 & & & & 1 & & 1 & & & & \\
 & & & 1 & & 2 & & 1 & & & \\
 & & 1 & & 3 & & 3 & & 1 & & \\
 & 1 & & 4 & & 6 & & 4 & & 1 & \\
1 & & 5 & & 10 & & 10 & & 5 & & 1 \\
\end{array}
$$

Example: To expand $(x + y)^4$, use the fifth row of Pascal's triangle.

$$(x + y)^4 = \mathbf{1}x^4 + \mathbf{4}\,x^3y + \mathbf{6}\,x^2y^2 + \mathbf{4}\,xy^3 + \mathbf{1}y^4$$
$$= x^4 + 4x^3y + 6x^2y^2 + 4xy^3 + y^4$$

The binomial theorem can also be used to expand powers of binomials. See Putting It All Together for Section 14.4.

CHAPTER 14   Review Exercises

SECTION 14.1

Exercises 1–4: Write the first four terms of the sequence for n = 1, 2, 3, and 4.

1. $f(n) = n^3$

2. $f(n) = 5 - 2n$

3. $f(n) = \dfrac{2n}{n^2 + 1}$

4. $f(n) = (-2)^n$

Exercises 5–6: Write the terms of the sequence.

5. a_n

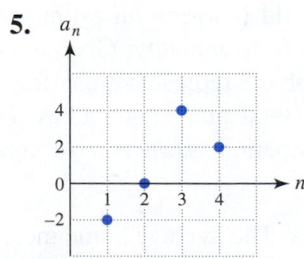

6. a_n

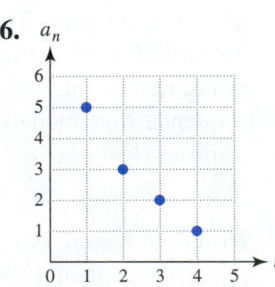

Exercises 7–10: Represent the first seven terms of the sequence numerically and graphically.

7. $a_n = 2n$

8. $a_n = n^2 - 4$

9. $a_n = 4\left(\frac{1}{2}\right)^n$

10. $a_n = \sqrt{n}$

SECTION 14.2

Exercises 11–18: Determine whether f is an arithmetic sequence. Identify the common difference when possible.

11. $f(n) = 5n - 1$

12. $f(n) = 4 - n^2$

13. $f(n) = 2^n$

14. $f(n) = 4 - \frac{1}{3}n$

15.

n	1	2	3	4
$f(n)$	20	17	14	11

16.

n	1	2	3	4
$f(n)$	-3	0	6	12

17. $f(n)$

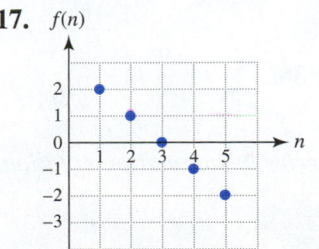

18. $f(n)$

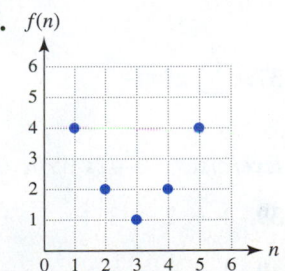

Exercises 19 and 20: Find the general term a_n for the arithmetic sequence.

19. $a_1 = -3$ and $d = 4$ **20.** $a_1 = 2$ and $a_2 = -3$

Exercises 21–28: Determine whether f is a geometric sequence. Identify the common ratio when possible.

21. $f(n) = 2(4)^n$ **22.** $f(n) = 2n^4$

23. $f(n) = 1 - 2n$ **24.** $f(n) = 5(0.7)^n$

25.

n	1	2	3	4
$f(n)$	5	4	3	1

26.

n	1	2	3	4
$f(n)$	27	−9	3	−1

27. 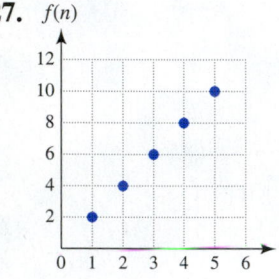   **28.**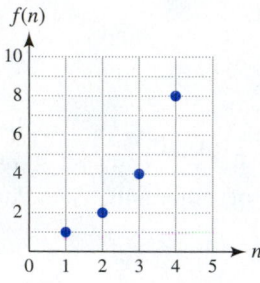

Exercises 29 and 30: Find the general term a_n for the geometric sequence.

29. $a_1 = 5$ and $r = 0.9$ **30.** $a_1 = 2$ and $a_2 = 8$

SECTION 14.3

Exercises 31–34: Find the sum, using a formula.

31. $4 + 9 + 14 + 19 + 24 + 29 + 34 + 39 + 44$

32. $4.5 + 3.0 + 1.5 + 0 - 1.5$

33. $1 - 4 + 16 - 64 + \cdots + 4096$

34. $1 + \frac{1}{2} + \frac{1}{4} + \frac{1}{8} + \frac{1}{16} + \cdots + \frac{1}{256}$

Exercises 35–38: Write the terms of the series.

35. $\sum_{k=1}^{5} 2k + 1$ **36.** $\sum_{k=1}^{4} \frac{1}{k+1}$

37. $\sum_{k=1}^{4} k^3$ **38.** $\sum_{k=2}^{7} (1 - k)$

Exercises 39–42: Write the series in summation notation.

39. $1 + 2 + 3 + \cdots + 20$

40. $1 + \frac{1}{2} + \frac{1}{3} + \cdots + \frac{1}{20}$

41. $\frac{1}{2} + \frac{2}{3} + \frac{3}{4} + \cdots + \frac{9}{10}$

42. $1^2 + 2^2 + 3^2 + 4^2 + 5^2 + 6^2 + 7^2$

SECTION 14.4

Exercises 43–46: Use Pascal's triangle to expand the given expression.

43. $(x + 4)^3$ **44.** $(2x + 1)^4$

45. $(x - y)^5$

46. $(a - 1)^6$

Exercises 47–50: Evaluate the expression.

47. $3!$ **48.** $\frac{5!}{3!2!}$

49. $_6C_3$ **50.** $_4C_3$

Exercises 51–54: Use the binomial theorem to expand the given expression.

51. $(m + 2)^4$ **52.** $(a + b)^5$

53. $(x - 3y)^4$ **54.** $(3x - 2)^3$

APPLICATIONS

55. *Salaries* An individual's starting salary is $45,000, and the individual receives a 10% raise each year. Give symbolic, numerical, and graphical representations for this person's salary over 7 years. What type of sequence is it?

56. *Salaries* An individual's starting salary is $45,000, and the individual receives an increase of $5000 each year. Give symbolic, numerical, and graphical representations for this person's salary over 7 years. What type of sequence is it?

57. *Rain Forests* Rain forests are defined as forests that grow in regions that receive more than 70 inches of rain each year. The world is losing an estimated 49 million acres of rain forests annually. Give symbolic, numerical, and graphical representations for a sequence that models the total number of acres (in millions) lost over a 7-year period. (*Source: New York Times Almanac, 1999.*)

58. *Home Mortgage Payments* The average home mortgage payment in 1996 was $1087 per month. Since then, mortgage payments have risen, on average, by 2.5% per year.
(**a**) Write a sequence a_n that models the average mortgage payment, where $n = 1$ corresponds to 1996, $n = 2$ to 1997, and so on.

(b) Is a_n arithmetic, geometric, or neither? Explain your reasoning.

(c) Find a_5 and interpret the result.

(d) Give a graphical representation for a_n, where $n = 1, 2, 3, \ldots, 10$.

| CHAPTER 14 | Test | CHAPTER Test Prep VIDEOS | Step-by-step test solutions are found on the Chapter Test Prep Videos available in **MyMathLab** and on **You Tube** (search "RockswoldComboAlg" and click on "Channels"). |

1. Write the first four terms of the given sequence for $n = 1, 2, 3,$ and 4.

$$f(n) = \frac{n^2}{n + 1}$$

2. Use the graph to write the terms of the sequence.

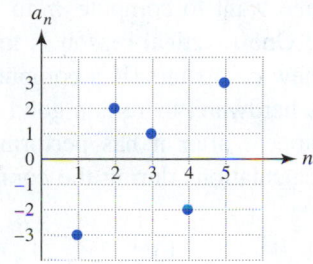

3. List the first seven terms of $a_n = n^2 - n$ in a table. Let $n = 1, 2, \ldots, 7$.

4. Expand the expression $(2x - 1)^4$.

Exercises 5 and 6: Determine whether the sequence is arithmetic or geometric. Identify either the common difference or the common ratio.

5. $f(n) = 7 - 3n$

6.

n	1	2	3	4
$f(n)$	-2	4	-8	16

7. Find the general term a_n for the arithmetic sequence if $a_1 = 2$ and $d = -3$.

8. Find the general term a_n for the geometric sequence if $a_1 = 2$ and $a_3 = 4.5$. Assume $r > 0$.

Exercises 9 and 10: Determine whether $a_n = f(n)$ is a geometric sequence. Identify the common ratio when possible.

9. $f(n) = -3(2.5)^n$ **10.**

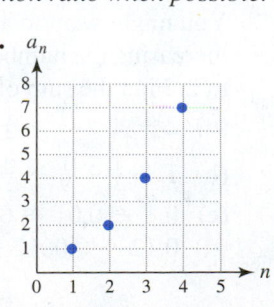

Exercises 11–12: Find the sum, using a formula.

11. $-1 + 2 + 5 + 8 + 11 + 14 + 17 + 20 + 23$

12. $1 - \frac{2}{3} + \frac{4}{9} - \frac{8}{27} + \frac{16}{81} - \frac{32}{243} + \frac{64}{729}$

13. Write the terms of the series $\sum_{k=2}^{7} 3k$.

14. Write the series $1^3 + 2^3 + 3^3 + \cdots + 60^3$ in summation notation.

15. Evaluate $\frac{7!}{4!\,3!}$.

16. Evaluate $_5C_3$.

17. *Auditorium Seating* An auditorium has 50 seats in the first row, 57 seats in the second row, 64 seats in the third row, and so on. Use a formula to find the total number of seats in the first 45 rows.

18. *Median Home Price* In 2011 the median price of a single-family home was \$180,000 and was decreasing at a rate of 4% per year. Give symbolic, numerical, and graphical representations for the median home price over a 5-year period, starting in 2011. What type of sequence is it?

19. *Tent Caterpillars* Large numbers of tent caterpillars can defoliate trees and ruin crops. After they mature, they spin a cocoon and develop into moths that lay eggs. Suppose that an initial population of 2000 tent caterpillars doubles every 5 days.

(a) Write a formula for a_n that models the number of tent caterpillars after $n - 1$ five-day time periods. (*Hint:* $a_1 = 2000$, $a_2 = 4000$, and $a_3 = 8000$.)

(b) Is a_n arithmetic, geometric, or neither? Explain your reasoning.

(c) Find a_6 and interpret the result.

(d) Give a graph for a_n, where $n = 1, 2, 3, 4, 5, 6$.

CHAPTER 14   Extended and Discovery Exercises

SEQUENCES AND SERIES

*Exercises 1 and 2: **Recursive Sequences** Some sequences are not defined by a formula for a_n. Instead they are defined recursively. With a **recursive formula** you must find terms a_1 through a_{n-1} before you can find a_n. For example, let*

$$a_1 = 2$$
$$a_n = a_{n-1} + 3, \quad \text{for } n \geq 2.$$

To find a_2, a_3, and a_4, we let $n = 2, 3, 4$.

$$a_2 = a_1 + 3 = 2 + 3 = 5$$
$$a_3 = a_2 + 3 = 5 + 3 = 8$$
$$a_4 = a_3 + 3 = 8 + 3 = 11$$

The first four terms of the sequence are 2, 5, 8, 11.

1. **Fibonacci Sequence** The Fibonacci sequence dates back to 1202 and is one of the most famous sequences in mathematics. It can be defined recursively as follows.

$$a_1 = 1, \qquad a_2 = 1$$
$$a_n = a_{n-1} + a_{n-2}, \quad \text{for } n \geq 3$$

Find the first 12 terms of this sequence.

2. **Insect Populations** Frequently the population of a particular insect does not continue to grow indefinitely. Instead, its population grows rapidly at first and then levels off because of competition for limited resources. In one study, the behavior of the winter moth was modeled with a sequence similar to the following, where a_n gives the population density in thousands per acre during year n.

$$a_1 = 1$$
$$a_n = 2.85a_{n-1} - 0.19a_{n-1}^2, \qquad n \geq 2$$

(**Source:** G. Varley and G. Gradwell, "Population models for the winter moth.")

(a) Make a table for $n = 1, 2, 3, \ldots, 7$. Describe what happens to the population density of the winter moth.

(b) Graph the sequence for $n = 1, 2, 3, \ldots, 20$. Discuss the graph.

NOTE: Many graphing calculators are capable of generating tables and graphs for a recursive sequence.

3. **Calculating π** The quest for an accurate estimation for π is a fascinating story covering thousands of years. Because π is an irrational number, it cannot be represented exactly by a fraction. Its decimal expansion neither repeats nor has a pattern. The ability

to compute π was essential to the development of societies because π appears in formulas used in construction, surveying, and geometry. In early historical records, π was given the value of 3. Later the Egyptians used a value of

$$\frac{256}{81} \approx 3.1605.$$

Not until the discovery of series was an exceedingly accurate decimal approximation of π possible. In 1989, π was computed to 1,073,740,000 digits, which required 100 hours of supercomputer time. Why would anyone want to compute π to so many decimal places? One practical reason is to test electrical circuits in new computers. If a computer has a small defect in its hardware, there is a good chance that an error will appear after it has performed trillions of arithmetic calculations during the computation of π. The series

$$\frac{\pi^4}{90} \approx \frac{1}{1^4} + \frac{1}{2^4} + \frac{1}{3^4} + \frac{1}{4^4} + \frac{1}{5^4} + \cdots + \frac{1}{n^4}$$

gives an estimate of π, where larger values of n give better approximations. (**Source:** P. Beckmann, *A History of Pi*.)

(a) Approximate π by finding the sum of the first four terms.

(b) Use a calculator to approximate π by summing the first 50 terms. Compare the result to the actual value of π.

4. **Infinite Series** The sum S of an infinite geometric series can be found if its common ratio r satisfies $|r| < 1$. It is given by

$$S = \frac{a_1}{1 - r}.$$

(If $|r| \geq 1$, this sum does not exist.) For example, the infinite geometric series

$$1 + \frac{1}{2} + \frac{1}{4} + \frac{1}{8} + \frac{1}{16} + \cdots$$

has $a_1 = 1$ and $r = \frac{1}{2}$. Therefore its sum S equals

$$S = \frac{1}{1 - \frac{1}{2}} = 2.$$

You might want to add terms of this series to see how increasing the number of terms results in a sum closer to 2. Find the sum of each infinite geometric series.

(a) $2 - 1 + \frac{1}{2} - \frac{1}{4} + \frac{1}{8} - \frac{1}{16} + \cdots$

(b) $1 + \frac{1}{3} + \frac{1}{9} + \frac{1}{27} + \frac{1}{81} + \cdots$

(c) $0.1 + 0.01 + 0.001 + 0.0001 + \cdots$

(d) $0.12 + 0.0012 + 0.000012$
$$+ 0.00000012 + \cdots$$

Cumulative Review Exercises

1. State whether the equation illustrates an identity, commutative, associative, or distributive property.

$$29(102) = 29(100) + 29(2)$$

2. Identify the domain and range of the relation given by $S = \{(-6, 5), (-2, 1), (0, 3), (2, 0)\}$.

Exercises 3–6: Simplify the expression. Write the result using positive exponents.

3. $\dfrac{x^{-2}y^3}{(3xy^{-2})^3}$

4. $\left(\dfrac{3b}{6a^2}\right)^{-4}$

5. $\left(\dfrac{1}{z^2}\right)^{-5}$

6. $\dfrac{8x^{-3}y^2}{4x^3y^{-1}}$

7. Find the domain of $f(x) = \dfrac{-5}{x-8}$.

8. Use the table to write the formula for $f(x) = ax + b$.

x	-2	-1	0	1	2
$f(x)$	5	3	1	-1	-3

9. Write the equation of the horizontal line that passes through the point $(2, 3)$.

10. Find the slope and the y-intercept of the graph of $f(x) = -3x + 5$.

Exercises 11 and 12: Write the slope-intercept form for a line satisfying the given conditions.

11. Perpendicular to $y = -\frac{2}{3}x - 4$, passing through $(1, 4)$

12. Parallel to $y = 2x - 7$, passing through $(5, 2)$

Exercises 13–18: Solve the equation or inequality. Write the solutions to inequalities in interval notation.

13. $\frac{2}{5}(x - 4) = -12$

14. $\frac{2}{5}z + \frac{1}{4}z > 2 - (z - 1)$

15. $-3|t - 5| \le -18$

16. $\left|4 + \frac{2}{3}x\right| = 6$

17. $\frac{1}{4}t - (2t + 5) + 6 = \dfrac{t + 3}{4}$

18. $-3 \le \frac{2}{3}x + 5 < 11$

19. Determine which of the following is a solution to the given system of equations.

$$(3, -2), (-1, 3)$$
$$3x + y = 7$$
$$-2x - 3y = 0$$

20. Shade the solution set in the xy-plane.

$$x - y < 4$$
$$x + 2y \ge 7$$

Exercises 21 and 22: Solve the system of equations. Write the solution as an ordered pair or ordered triple where appropriate.

21. $\begin{aligned} x - 2y &= 1 \\ -2x + 7y &= 4 \end{aligned}$

22. $\begin{aligned} x + y + z &= 5 \\ -2x - y + z &= -10 \\ x + 2y + 8z &= 1 \end{aligned}$

23. Maximize the objective function R subject to the given constraints.

$$R = 3x + 8y$$
$$x + 4y \le 10$$
$$4x + y \le 10$$
$$x \ge 0, y \ge 0$$

24. Evaluate $\det A$.

$$A = \begin{bmatrix} 4 & -3 \\ 3 & 2 \end{bmatrix}$$

Exercises 25 and 26: Multiply the expressions.

25. $2x^3(4x^4 - 3x^3 + 5)$

26. $(2z - 7)(3z + 4)$

Exercises 27 and 28: Factor completely.

27. $4x^2 - 9y^2$

28. $2a^3 - a^2 + 8a - 4$

Exercises 29 and 30: Solve the equation.

29. $4x^2 - x - 3 = 0$

30. $x^4 - 10x^3 = -24x^2$

Exercises 31 and 32: Simplify the expression.

31. $\dfrac{x^2 - 7x + 10}{x^2 - 25} \cdot \dfrac{x + 5}{x + 1}$

32. $\dfrac{x^2 + 7x + 12}{x^2 - 9} \div \dfrac{x^2 - 5x + 6}{(x - 3)^2}$

Exercises 33 and 34: Solve the rational equation. Check your result.

33. $\dfrac{2}{x+5} = \dfrac{-3}{x^2-25} + \dfrac{1}{x-5}$

34. $\dfrac{2y}{y^2-3y+2} = \dfrac{1}{y-2} + 2$

35. Solve the equation for W.

$$R = \dfrac{3C-2W}{5}$$

36. Simplify the complex fraction.

$$\dfrac{\dfrac{1}{x^2} + \dfrac{2}{x}}{\dfrac{1}{x^2} - \dfrac{4}{x}}$$

Exercises 37 and 38: Simplify the expression. Assume that all variables are positive.

37. $\sqrt[3]{x^4y^4} - 2\sqrt[3]{xy}$

38. $(4 + \sqrt{2})(4 - \sqrt{2})$

Exercises 39 and 40: Solve. Check your answers.

39. $8(x-3)^2 = 200$

40. $3\sqrt{2x+6} = 6x$

Exercises 41 and 42: Write the complex expression in standard form.

41. $(-3 + i)(-4 - 2i)$

42. $\dfrac{2-6i}{1+2i}$

43. Find the minimum y-value located on the graph of $y = 3x^2 + 8x + 5$.

44. Write the equation $y = 2x^2 + 8x + 17$ in vertex form and identify the vertex.

Exercises 45 and 46: Solve by using the method of your choice. Write any complex solutions in standard form.

45. $x^2 - 4x + 13 = 0$

46. $z^2 - 4z = 32$

47. A graph of $y = ax^2 + bx + c$ is shown at the top of the next column.
   (a) Solve $ax^2 + bx + c = 0$.
   (b) State whether $a > 0$ or $a < 0$.

(c) Determine whether the discriminant is positive, negative, or zero.

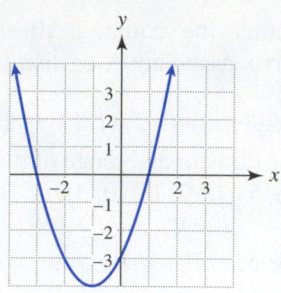

48. Solve the quadratic inequality. Write your answer in interval notation.

$$x^2 + 2x - 3 < 0$$

49. For $f(x) = x^2 + 1$ and $g(x) = 3x - 2$, find each of the following.
   (a) $(f \circ g)(-2)$
   (b) $(g \circ f)(x)$

50. Find $f^{-1}(x)$ for the one-to-one function

$$f(x) = \dfrac{3x+1}{2}.$$

51. Write $\ln(x^3\sqrt{y})$ by using sums and differences of logarithms of x and y.

52. Write $2 \log x - \log 4xy$ as one logarithm. Assume x and y are positive.

Exercises 53 and 54: Solve the equation. Approximate answers to the nearest hundredth.

53. $8 \log x + 3 = 17$

54. $4^{2x} = 5$

55. Graph the parabola $x = (y-3)^2 + 1$. Find the vertex and the axis of symmetry.

56. Find the center and the radius of the circle whose equation is $x^2 - 6x + y^2 + 2y = -6$.

Exercises 57 and 58: Graph the ellipse or hyperbola. Label the vertices and the endpoints of the minor axis on the ellipse. Show the asymptotes on the hyperbola.

57. $\dfrac{x^2}{4} + \dfrac{y^2}{9} = 1$

58. $\dfrac{x^2}{16} - \dfrac{y^2}{4} = 1$

Exercises 59 and 60: Use the graph to determine the equation of the ellipse or hyperbola.

59.

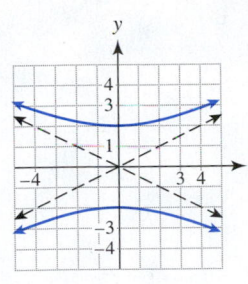

60.

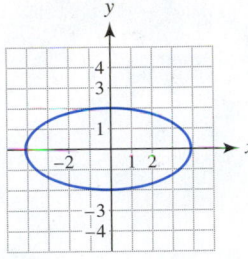

61. Solve the system of equations. Check your solutions.

$$y = x^2 + 1$$
$$x^2 + 2y = 5$$

62. Shade the solution set in the xy-plane.

$$y \geq x^2 - 2$$
$$y \leq -x$$

Exercises 63–66: Determine whether f is an arithmetic or a geometric sequence. If it is arithmetic, find the common difference. If it is geometric, find the common ratio.

63. $f(n) = 5 - 2n$ **64.** $f(n) = 3(0.2)^n$

65. $f(n) = 7(4)^n$ **66.** $f(n) = 6n + 1$

67. Find the general term a_n for the arithmetic sequence where $a_1 = 2$ and $a_2 = 5$.

68. Find the general term a_n for the geometric sequence where $a_1 = 4$ and $a_2 = 12$.

Exercises 69 and 70: Find the sum using a formula.

69. $3 + 7 + 11 + 15 + 19 + \cdots + 35$

70. $1 - 2 + 4 - 8 + 16 - \cdots + 1024$

Exercises 71 and 72: Expand the binomial expression.

71. $(2x + 3)^4$ **72.** $(2a - 5b)^3$

APPLICATIONS

73. *Radius of a Circle* If a circle has an area of A square units, its radius r is given by $r = \sqrt{\frac{A}{\pi}}$. Approximate the radius of a circle with an area of 14 square inches to the nearest hundredth of an inch.

74. *Distance from Work* Starting at a warehouse, a delivery truck driver travels down a straight highway for 3 hours at 40 miles per hour, stops and unloads the truck for 1 hour, and then returns to the warehouse at 60 miles per hour. Sketch a graph that shows the distance between the truck and the warehouse during this period of time.

75. *Exercise and Fluid Consumption* When a person exercises, the total amount of fluid he or she will need that day increases depending on the person's weight and the duration of the exercise. To determine the number of ounces of fluid needed, divide the person's weight by 2 and then add 0.4 ounces for every minute of exercise. (*Source: Runner's World.*)
   (a) Write a function that gives the fluid requirements for a person weighing 170 pounds who exercises for x minutes a day.
   (b) If an athlete who exercises for 90 minutes requires 130 ounces of fluid, find the athlete's weight.

76. *Airplane Speed* An airplane travels 1080 miles into the wind in 3 hours. The return trip with the wind takes 2.7 hours. Find the average speed of the airplane and the average wind speed.

77. *Size of a Tent* The length of a rectangular tent floor is 6 feet shorter than twice the width. If the area of the tent floor is 108 square feet, what are the dimensions of the tent?

78. *Working Together* Suppose that one person can weed a garden in 60 minutes and a second person can weed the same garden in 90 minutes. How long would it take these two people to weed the garden if they worked together?

79. *Numbers* The product of two positive numbers is 96. If the larger number is subtracted from 3 times the smaller number, the result is 12. Let x be the smaller number and let y be the larger number.
   (a) Write a system of equations for this situation.
   (b) What are the two numbers?

80. *Marching Band* A band is marching in a triangular formation so that 1 person is in the first row, 3 people are in the second row, 5 people are in the third row, and so on. Use a formula to find the total number of musicians in the marching band if the last row contains 23 people.

Appendix A
Using the Graphing Calculator

Overview of the Appendix

This appendix provides instruction for the TI-83, TI-83 Plus, and TI-84 Plus graphing calculators that may be used in conjunction with this textbook. It includes specific keystrokes needed to work several examples from the text. Students are advised to consult the *Graphing Calculator Guidebook* provided by the manufacturer.

Entering Mathematical Expressions

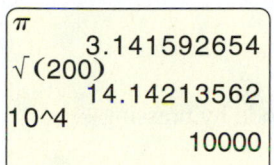

Figure A.1

EVALUATING π To evaluate π, use the following keystrokes, as shown in the first and second lines of Figure A.1. (Do *not* use 3.14 or $\frac{22}{7}$ for π.)

EVALUATING A SQUARE ROOT To evaluate a square root, such as $\sqrt{200}$, use the following keystrokes, as shown in the third and fourth lines of Figure A.1.

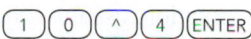

EVALUATING AN EXPONENTIAL EXPRESSION To evaluate an exponential expression, such as 10^4, use the following keystrokes, as shown in the last two lines of Figure A.1.

1 0 ^ 4 ENTER

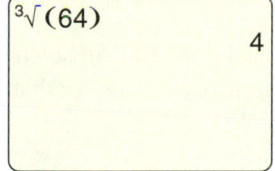

Figure A.2

EVALUATING A CUBE ROOT To evaluate a cube root, such as $\sqrt[3]{64}$, use the following keystrokes, as shown in Figure A.2.

MATH 4 6 4) ENTER

SUMMARY: ENTERING MATHEMATICAL EXPRESSIONS

To access the *number π*, use (2nd)(^[π]).

To evaluate a *square root*, use (2nd)(x²[√]).

To evaluate an *exponential expression*, use the (^) key. To square a number, the (x²) key can also be used.

To evaluate a *cube root*, use (MATH)(4).

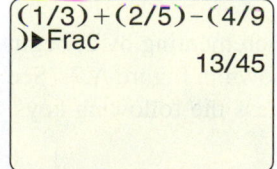

Figure A.3

Expressing Answers as Fractions

To evaluate $\frac{1}{3} + \frac{2}{5} - \frac{4}{9}$ in fraction form, use the following keystrokes, as shown in Figure A.3.

SUMMARY: EXPRESSING ANSWERS AS FRACTIONS

Enter the arithmetic expression. To access the "Frac" feature, use the keystrokes $\boxed{\text{MATH}}\,\boxed{1}$. Then press $\boxed{\text{ENTER}}$.

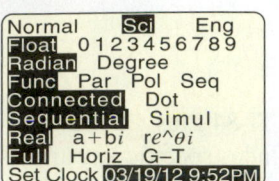

Figure A.4

Displaying Numbers in Scientific Notation

To display numbers in scientific notation, set the graphing calculator in scientific mode (Sci) by using the following keystrokes. See Figure A.4. (These keystrokes assume that the calculator is starting from normal mode.)

In scientific mode we can display the numbers 5432 and 0.00001234 in scientific notation, as shown in Figure A.5.

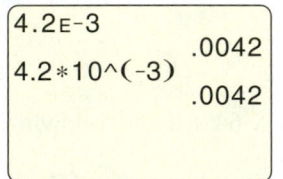

Figure A.5

SUMMARY: SETTING SCIENTIFIC MODE

If your calculator is in normal mode, it can be set in scientific mode by pressing

These keystrokes return the graphing calculator to the home screen.

Entering Numbers in Scientific Notation

Numbers can be entered in scientific notation. For example, to enter 4.2×10^{-3} in scientific notation, use the following keystrokes. (Be sure to use the negation key (−) rather than the subtraction key.)

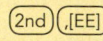

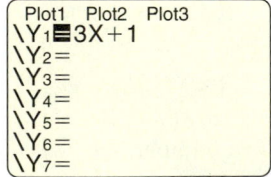

Figure A.6

This number can also be entered using the following keystrokes. See Figure A.6.

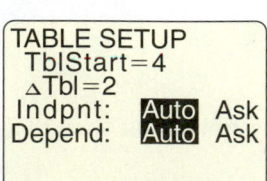

Figure A.7

SUMMARY: ENTERING NUMBERS IN SCIENTIFIC NOTATION

One way to enter a number in scientific notation is to use the keystrokes

$\boxed{\text{2nd}}\,\boxed{\text{,[EE]}}$

to access an exponent (EE) of 10.

Making a Table

To make a table of values for $y = 3x + 1$ starting at $x = 4$ and incrementing by 2, begin by pressing $\boxed{\text{Y=}}$ and then entering the formula $Y_1 = 3X + 1$, as shown in Figure A.7. (See Entering a Formula on page AP-4.) To set the table parameters, press the following keys. See Figure A.8.

Figure A.8

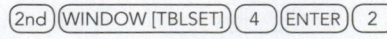

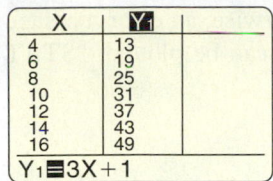

Figure A.9

These keystrokes specify a table that starts at $x = 4$ and increments the x-values by 2. Therefore, the values of Y_1 at $x = 4, 6, 8, \ldots$ appear in the table. To create this table, press the following keys.

$$\boxed{2\text{nd}}\ \boxed{\text{GRAPH [TABLE]}}$$

We can scroll through x- and y-values by using the arrow keys. See Figure A.9. Note that there is no first or last x-value in the table.

SUMMARY: MAKING A TABLE

1. Enter the formula for the equation using $\boxed{Y =}$.
2. Press $\boxed{2\text{nd}}\,\boxed{\text{WINDOW [TBLSET]}}$ to set the starting x-value and the increment between x-values appearing in the table.
3. Create the table by pressing $\boxed{2\text{nd}}\,\boxed{\text{GRAPH [TABLE]}}$.

Setting the Viewing Rectangle (Window)

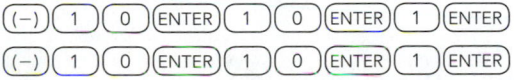

Figure A.10

There are at least two ways to set the standard viewing rectangle of $[-10, 10, 1]$ by $[-10, 10, 1]$. The first involves pressing $\boxed{\text{ZOOM}}$ followed by $\boxed{6}$. See Figure A.10. The second method for setting the standard viewing rectangle is to press $\boxed{\text{WINDOW}}$ and enter the following keystrokes. See Figure A.11.

$$\boxed{(-)}\ \boxed{1}\ \boxed{0}\ \boxed{\text{ENTER}}\ \boxed{1}\ \boxed{0}\ \boxed{\text{ENTER}}\ \boxed{1}\ \boxed{\text{ENTER}}$$
$$\boxed{(-)}\ \boxed{1}\ \boxed{0}\ \boxed{\text{ENTER}}\ \boxed{1}\ \boxed{0}\ \boxed{\text{ENTER}}\ \boxed{1}\ \boxed{\text{ENTER}}$$

```
WINDOW
 Xmin=-10
 Xmax=10
 Xscl=1
 Ymin=-10
 Ymax=10
 Yscl=1
 Xres=1
```

Figure A.11

(Be sure to use the negation key $(-)$ rather than the subtraction key.) Other viewing rectangles can be set in a similar manner by pressing $\boxed{\text{WINDOW}}$ and entering the appropriate values. To see the viewing rectangle, press $\boxed{\text{GRAPH}}$.

SUMMARY: SETTING THE VIEWING RECTANGLE

To set the standard viewing rectangle, press $\boxed{\text{ZOOM}}\ \boxed{6}$. To set any viewing rectangle, press $\boxed{\text{WINDOW}}$ and enter the necessary values. To see the viewing rectangle, press $\boxed{\text{GRAPH}}$.

NOTE: You do not need to change "Xres" from 1.

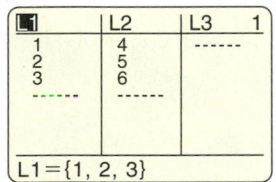

Figure A.12

Making a Scatterplot or a Line Graph

To make a scatterplot with the points $(-5, -5), (-2, 3), (1, -7)$, and $(4, 8)$, begin by following these steps.

1. Press $\boxed{\text{STAT}}$ followed by $\boxed{1}$.
2. If list L1 is not empty, use the arrow keys to place the cursor on L1, as shown in Figure A.12. Then press $\boxed{\text{CLEAR}}$ followed by $\boxed{\text{ENTER}}$. This deletes all elements in the list. Similarly, if L2 is not empty, clear the list.
3. Input each x-value into list L1 followed by $\boxed{\text{ENTER}}$. Input each y-value into list L2 followed by $\boxed{\text{ENTER}}$. See Figure A.13.

```
L1      L2      L3    1
-5      -5      ------
-2       3
 1      -7
 4       8
------  ------


L1(5)=
```

Figure A.13

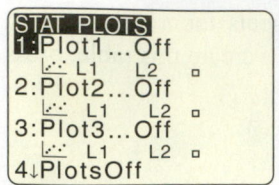

Figure A.14

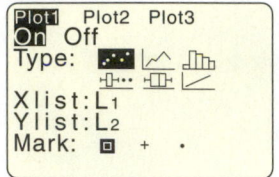

Figure A.15

$[-10, 10, 1]$ by $[-10, 10, 1]$

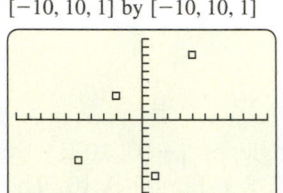

Figure A.16

It is essential that both lists have the same number of values—otherwise, an error message appears when a scatterplot is attempted. Before these four points can be plotted, "STAT PLOT" must be turned on. It is accessed by pressing

as shown in Figure A.14.

There are three possible "STAT PLOTS," numbered 1, 2, and 3. Any one of the three can be selected. The first plot can be selected by pressing (1). Next, place the cursor over "On" and press (ENTER) to turn "Plot1" on. There are six types of plots that can be selected. The first type is a *scatterplot* and the second type is a *line graph*, so place the cursor over the first type of plot and press (ENTER) to select a scatterplot. (To make the line graph, place the cursor over the second type of plot and press (ENTER).) The *x*-values are stored in list L1, so select L1 for "Xlist" by pressing (2nd)(1). Similarly, press (2nd)(2) for the "Ylist," since the *y*-values are stored in list L2. Finally, there are three styles of marks that can be used to show data points in the graph. We will usually use the first because it is largest and shows up the best. Make the screen appear as in Figure A.15. Before plotting the four data points, be sure to set an appropriate viewing rectangle. Then press (GRAPH). The data points appear as in Figure A.16.

REMARK 1: A fast way to set the viewing rectangle for any scatterplot is to select the "ZOOMSTAT" feature by pressing (ZOOM)(9). This feature automatically scales the viewing rectangle so that all data points are shown.

REMARK 2: If an equation has been entered into the (Y =) menu and selected, it will be graphed with the data. This feature is used frequently to model data.

SUMMARY: MAKING A SCATTERPLOT OR A LINE GRAPH

The following are basic steps necessary to make either a scatterplot or a line graph.

1. Use (STAT)(1) to access lists L1 and L2.
2. If list L1 is not empty, place the cursor on L1 and press (CLEAR)(ENTER). Repeat for list L2, if it is not empty.
3. Enter the *x*-values into list L1 and the *y*-values into list L2.
4. Use (2nd)(Y = [STAT PLOT]) to select appropriate parameters for the scatterplot or line graph.
5. Set an appropriate viewing rectangle. Press (GRAPH). Otherwise, press (ZOOM)(9). This feature automatically sets the viewing rectangle and plots the data.

NOTE: (ZOOM)(9) *cannot* be used to set a viewing rectangle for the graph of an equation.

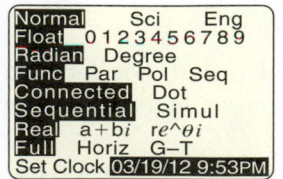

Figure A.17

Figure A.18

Entering a Formula

To enter a formula, press (Y =). For example, use the following keystrokes after "$Y_1 =$" to enter $y = x^2 - 4$. See Figure A.17.

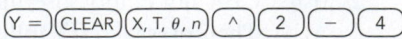

Note that there is a built-in key to enter the variable X. If "$Y_1 =$" does not appear after pressing (Y =), press (MODE) and make sure the calculator is set in *function mode*, denoted "Func". See Figure A.18.

SUMMARY: ENTERING A FORMULA

To enter a formula, press (Y =). To delete a formula, press (CLEAR).

[−10, 10, 1] by [−10, 10, 1]

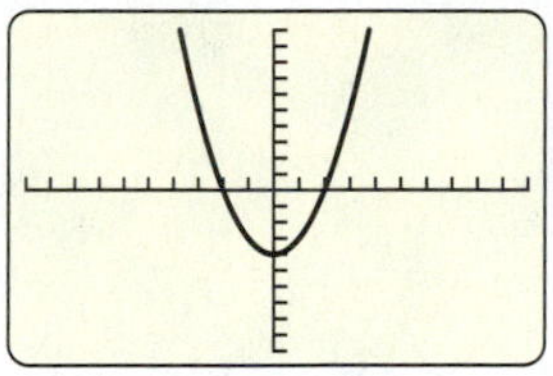

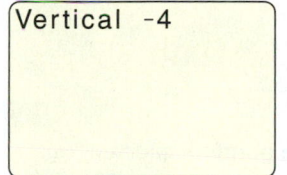

Figure A.19

Graphing an Equation

To graph an equation, such as $y = x^2 - 4$, start by pressing (Y =) and enter Y_1 = X^2 − 4. If there is an equation already entered, remove it by pressing (CLEAR). The equals signs in "Y_1 =" should be in reverse video (a dark rectangle surrounding a white equals sign), which indicates that the equation will be graphed. If the equals sign is not in reverse video, place the cursor over it and press (ENTER). Set an appropriate viewing rectangle and then press (GRAPH). The graph will appear in the specified viewing rectangle. See Figures A.17 and A.19.

SUMMARY: GRAPHING AN EQUATION

1. Use the (Y =) menu to enter the formula.
2. Use the (WINDOW) menu to set an appropriate viewing rectangle.
3. Press (GRAPH).

Figure A.20

Graphing a Vertical Line

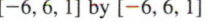

Set an appropriate window (or viewing rectangle). Then return to the home screen by pressing

(2nd)(MODE [QUIT]).

To graph a vertical line, such as $x = -4$, press

(2nd)(PRGM [DRAW])(4)((−))(4).

See Figure A.20. Pressing (ENTER) will make the vertical line appear, as shown in Figure A.21.

[−6, 6, 1] by [−6, 6, 1]

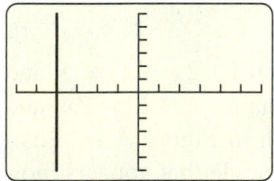

Figure A.21

SUMMARY: GRAPHING THE VERTICAL LINE $x = h$

1. Set an appropriate window by pressing (WINDOW).
2. Return to the home screen by pressing (2nd)(MODE [QUIT]).
3. Draw a vertical line by pressing (2nd)(PRGM [DRAW])(4)(h)(ENTER).

Figure A.22

Squaring a Viewing Rectangle

In a square viewing rectangle the graph of $y = x$ is a line that makes a 45° angle with the positive x-axis, a circle appears circular, and all sides of a square have the same length. An approximate square viewing rectangle can be set if the distance along the x-axis is 1.5 times the distance along the y-axis. Examples of viewing rectangles that are (approximately) square include

[−6, 6, 1] by [−4, 4, 1]   and   [−9, 9, 1] by [−6, 6, 1].

Square viewing rectangles can be set automatically by pressing either

(ZOOM)(4)   or   (ZOOM)(5).

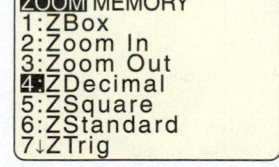

ZOOM 4 provides a *decimal window*, which is discussed on page AP-11. See Figure A.22.

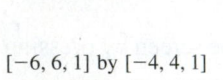

SUMMARY: SQUARING A VIEWING RECTANGLE

Either [ZOOM][4] or [ZOOM][5] may be used to produce a square viewing rectangle. An (approximately) square viewing rectangle has the form

$$[-1.5k, 1.5k, 1] \text{ by } [-k, k, 1],$$

where k is a positive number.

Locating a Point of Intersection

Figure A.23

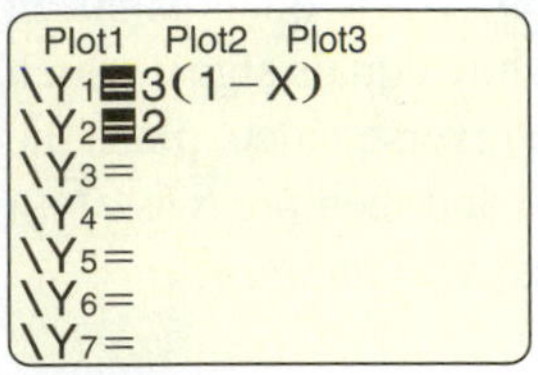

To find the point of intersection for the graphs of

$$y_1 = 3(1 - x) \quad \text{and} \quad y_2 = 2,$$

start by entering Y_1 and Y_2, as shown in Figure A.23. Set the window, and graph both equations. Then press the following keys to find the intersection point.

[2nd][TRACE [CALC]][5]

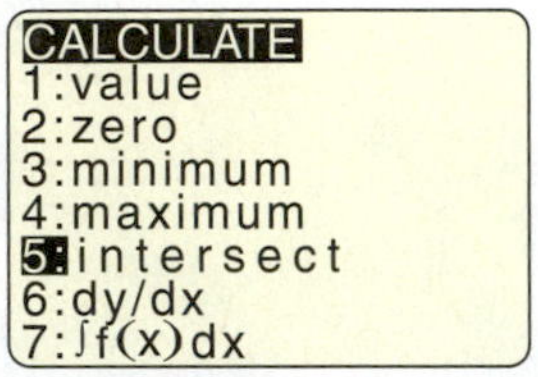

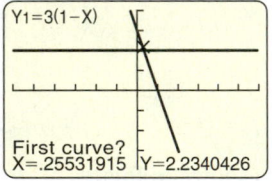

Figure A.24

See Figure A.24, where the "intersect" utility is being selected. The calculator prompts for the first curve, as shown in Figure A.25. Use the arrow keys to locate the cursor near the point of intersection and press [ENTER]. Repeat these steps for the second curve. Finally we are prompted for a guess. For each of the three prompts, place the free-moving cursor near the point of intersection and press [ENTER]. The approximate coordinates of the point of intersection will be shown.

[−6, 6, 1] by [−4, 4, 1]

Figure A.25

SUMMARY: FINDING A POINT OF INTERSECTION

1. Graph the two equations in an appropriate viewing rectangle.
2. Press [2nd][TRACE [CALC]][5].
3. Use the arrow keys to select an approximate location for the point of intersection. Press [ENTER] to make the three selections for "First curve?", "Second curve?", and "Guess?". (Note that if the cursor is near the point of intersection, you usually do not need to move the cursor for each selection. Just press [ENTER] three times.)

Shading Inequalities

To shade the solution set for one or more linear inequalities such as $2x + y \le 5$ and $-2x + y \ge 1$, begin by solving each inequality for y to obtain $y \le 5 - 2x$ and $y \ge 2x + 1$. Then let $Y_1 = 5 - 2X$ and $Y_2 = 2X + 1$, as shown in Figure A.26. Position the cursor to the left of Y_1 and press [ENTER] three times. The triangle that appears indicates that the calculator will shade the region below the graph of Y_1. Next locate the cursor to the left of Y_2 and press [ENTER] twice. This triangle indicates that the calculator will shade the region above the graph of Y_2. After setting the viewing rectangle to $[-15, 15, 5]$ by $[-10, 10, 5]$, press [GRAPH]. The result is shown in Figure A.27.

Figure A.26

[−15, 15, 5] by [−10, 10, 5]

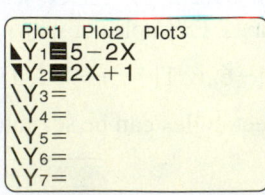

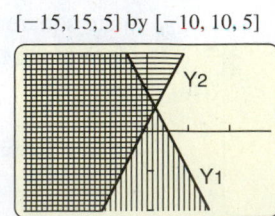

Figure A.27

SUMMARY: SHADING ONE OR MORE INEQUALITIES

1. Solve each inequality for y.
2. Enter each formula as Y_1 and Y_2 in the $\boxed{Y=}$ menu.
3. Locate the cursor to the left of Y_1 and press $\boxed{\text{ENTER}}$ two or three times to shade either above or below the graph of Y_1. Repeat for Y_2.
4. Set an appropriate viewing rectangle.
5. Press $\boxed{\text{GRAPH}}$.

NOTE: The "Shade" utility in the DRAW menu can also be used to shade the region *between* two graphs.

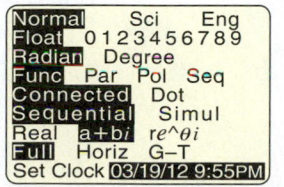

Figure A.28

Setting $a + bi$ Mode

To evaluate expressions containing square roots of negative numbers, such as $\sqrt{-25}$, set your calculator in $a + bi$ mode by using the following keystrokes.

See Figures A.28 and A.29.

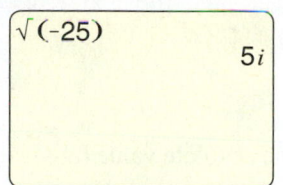

Figure A.29

SUMMARY: SETTING $a + bi$ MODE

1. Press $\boxed{\text{MODE}}$.
2. Move the cursor to the seventh line, highlight $a + bi$, and press $\boxed{\text{ENTER}}$.
3. Press $\boxed{\text{2nd}}$ $\boxed{\text{MODE [QUIT]}}$ and return to the home screen.

Evaluating Complex Arithmetic

Complex arithmetic can be performed much like other arithmetic expressions. This is done by entering

$$\boxed{\text{2nd}}\ \boxed{.\,[i]}$$

to obtain the imaginary unit i from the home screen. For example, to find the sum $(-2 + 3i) + (4 - 6i)$, perform the following keystrokes on the home screen.

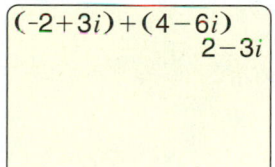

Figure A.30

The result is shown in Figure A.30. Other complex arithmetic operations are done similarly.

SUMMARY: EVALUATING COMPLEX ARITHMETIC

Enter a complex expression in the same way as you would any arithmetic expression. To obtain the complex number i, use $\boxed{\text{2nd}}$ $\boxed{.\,[i]}$.

Figure A.31

Other Mathematical Expressions

EVALUATING OTHER ROOTS To evaluate a fifth root, such as $\sqrt[5]{23}$, use the following keystrokes, as shown in Figure A.31.

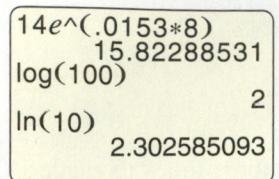

Figure A.32

EVALUATING THE NATURAL EXPONENTIAL FUNCTION To evaluate $14e^{0.0153(8)}$, use the following keystrokes, as shown in the first and second lines of Figure A.32.

EVALUATING THE COMMON LOGARITHMIC FUNCTION To evaluate $\log(100)$, use the following keystrokes, as shown in the third and fourth lines of Figure A.32.

EVALUATING THE NATURAL LOGARITHMIC FUNCTION To evaluate $\ln(10)$, use the following keystrokes, as shown in the last two lines of Figure A.32.

SUMMARY: OTHER MATHEMATICAL EXPRESSIONS

To evaluate a *kth root*, use ⬚k⬚ ⬚MATH⬚ ⬚5⬚.

To access the *natural exponential function*, use ⬚2nd⬚ ⬚LN [eˣ]⬚.

To access the *common logarithmic function*, use ⬚LOG⬚.

To access the *natural logarithmic function*, use ⬚LN⬚.

Accessing the Absolute Value

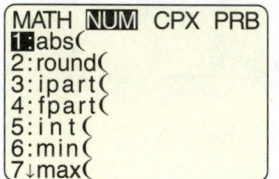

Figure A.33

To graph $y_1 = |x - 50|$, begin by entering $Y_1 = \text{abs}(X - 50)$. The absolute value (abs) is accessed by pressing

⬚MATH⬚ ⬚▷⬚ ⬚1⬚.

See Figure A.33.

SUMMARY: ACCESSING THE ABSOLUTE VALUE

1. Press ⬚MATH⬚.
2. Position the cursor over "NUM".
3. Press ⬚1⬚ to select the absolute value.

Entering the Elements of a Matrix

The elements of the augmented matrix A given by

$$A = \left[\begin{array}{ccc|c} 1 & 1 & 2 & 1 \\ -1 & 0 & 1 & -2 \\ 2 & 1 & 5 & -1 \end{array}\right]$$

can be entered by using the following keystrokes on the TI-83 Plus or TI-84 Plus to define a matrix A with dimension 3×4. (*Note:* On the TI-83 the matrix menu is found by pressing ⬚MATRX⬚.)

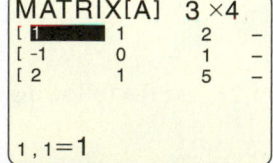

Figure A.34

⬚2nd⬚ ⬚x⁻¹ [MATRIX]⬚ ⬚▷⬚ ⬚▷⬚ ⬚1⬚ ⬚3⬚ ⬚ENTER⬚ ⬚4⬚ ⬚ENTER⬚

Input the 12 elements of the matrix A, row by row. Finish each entry by pressing ⬚ENTER⬚. See Figure A.34. After these elements have been entered, press

⬚2nd⬚ ⬚MODE [QUIT]⬚

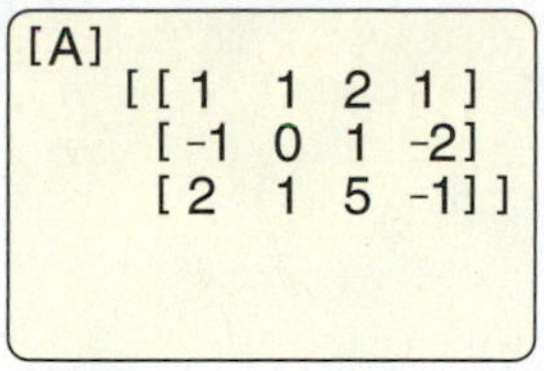

```
[A]
  [[ 1   1   2   1 ]
   [-1   0   1  -2]
   [ 2   1   5  -1]]
```

Figure A.35

to return to the home screen. To display the matrix A, press

2nd x^{-1} [MATRIX] 1 ENTER .

See Figure A.35.

Reduced Row–Echelon Form

```
rref([A])
  [[ 1   0   0   1 ]
   [ 0   1   0   2 ]
   [ 0   0   1  -1]]
```

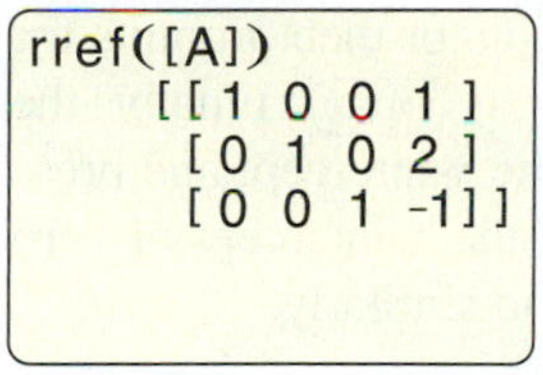

Figure A.36

To find the reduced row–echelon form of matrix A (entered above in Figure A.35), use the following keystrokes from the home screen on the TI-83 Plus or TI-84 Plus.

2nd x^{-1} [MATRIX] ▷ ALPHA APPS [B] 2nd x^{-1} [MATRIX] 1) ENTER

The resulting matrix is shown in Figure A.36. On the TI-83 graphing calculator, use the following keystrokes to find the reduced row–echelon form.

MATRX ▷ ALPHA MATRX [B] MATRX 1) ENTER

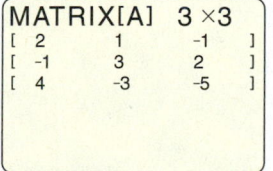

Figure A.37

Evaluating a Determinant

To evaluate the determinant of matrix A given by

$$A = \begin{bmatrix} 2 & 1 & -1 \\ -1 & 3 & 2 \\ 4 & -3 & -5 \end{bmatrix},$$

start by entering the 9 elements of the 3×3 matrix, as shown in Figure A.37. To compute det A, perform the following keystrokes from the home screen.

2nd x^{-1} [MATRIX] ▷ 1 2nd x^{-1} [MATRIX] 1) ENTER

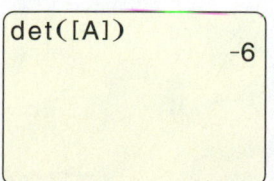

Figure A.38

The results are shown in Figure A.38.

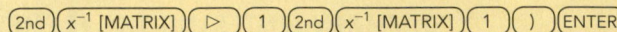

SUMMARY: EVALUATING A DETERMINANT OF A MATRIX

1. Enter the dimension and elements of the matrix A.
2. Return to the home screen by pressing (2nd)(MODE [QUIT])·
3. On the TI-83 Plus or TI-84 Plus, perform the following keystrokes.

(2nd)(x⁻¹ [MATRIX]) (▷) (1)(2nd)(x⁻¹ [MATRIX]) (1) ( )) (ENTER)

NOTE: On the TI-83, replace the keystrokes (2nd)(x⁻¹ [MATRIX]) with (MATRX).

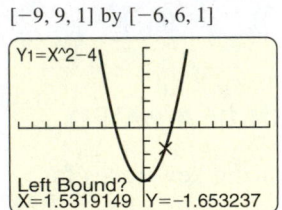

Figure A.39

Locating an *x*-Intercept or Zero

To locate an *x*-intercept or *zero* of $f(x) = x^2 - 4$, start by entering $Y_1 = X^2 - 4$ into the (Y =) menu. Set the viewing rectangle to $[-9, 9, 1]$ by $[-6, 6, 1]$ and graph Y_1. Afterwards, press the following keys to invoke the zero finder. See Figure A.39.

(2nd)(TRACE [CALC])(2)

    The graphing calculator prompts for a left bound. Use the arrow keys to set the cursor to the left of the *x*-intercept and press (ENTER). The graphing calculator then prompts for a right bound. Set the cursor to the right of the *x*-intercept and press (ENTER). Finally, the graphing calculator prompts for a guess. Set the cursor roughly at the *x*-intercept and press (ENTER). See Figures A.40–A.42. The calculator then approximates the *x*-intercept or zero automatically, as shown in Figure A.43. The zero of -2 can be found similarly.

$[-9, 9, 1]$ by $[-6, 6, 1]$ $[-9, 9, 1]$ by $[-6, 6, 1]$ $[-9, 9, 1]$ by $[-6, 6, 1]$ $[-9, 9, 1]$ by $[-6, 6, 1]$

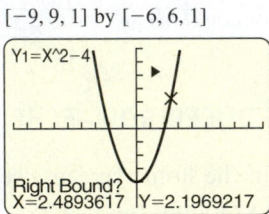   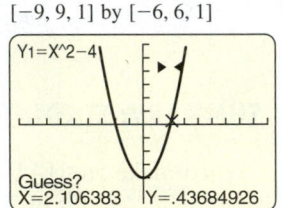   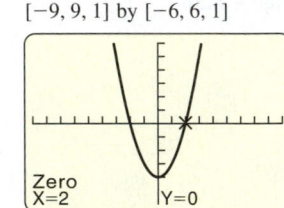

Figure A.40   **Figure A.41**   **Figure A.42**   **Figure A.43**

SUMMARY: LOCATING AN *x*-INTERCEPT OR ZERO

1. Graph the function in an appropriate viewing rectangle.
2. Press (2nd)(TRACE [CALC])(2).
3. Select the left and right bounds, followed by a guess. Press (ENTER) after each selection. The calculator then approximates the *x*-intercept or zero.

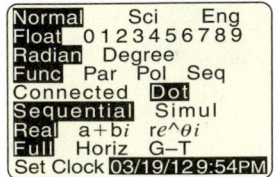

Figure A.44

Setting Connected or Dot Mode

To set your graphing calculator in dot mode, press (MODE), position the cursor over "Dot," and press (ENTER). See Figure A.44. Graphs will now appear in dot mode rather than connected mode.

SUMMARY: SETTING CONNECTED OR DOT MODE

1. Press (MODE).
2. Position the cursor over "Connected" or "Dot". Press (ENTER).

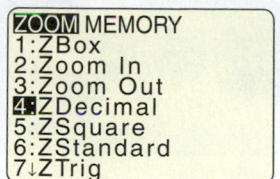

Figure A.45

Setting a Decimal Window

With a decimal window, the cursor stops on convenient x-values. In the decimal window $[-9.4, 9.4, 1]$ by $[-6.2, 6.2, 1]$ the cursor stops on x-values that are multiples of 0.2. If we reduce the viewing rectangle to $[-4.7, 4.7, 1]$ by $[-3.1, 3.1, 1]$, the cursor stops on x-values that are multiples of 0.1. To set this smaller window automatically, press (ZOOM)(4). See Figure A.45. Decimal windows are also useful when graphing rational functions with asymptotes in connected mode.

> ### SUMMARY: SETTING A DECIMAL WINDOW
>
> 1. Press (ZOOM)(4) to set the viewing rectangle $[-4.7, 4.7, 1]$ by $[-3.1, 3.1, 1]$.
> 2. A larger decimal window is $[-9.4, 9.4, 1]$ by $[-6.2, 6.2, 1]$.

Finding Maximum and Minimum Values

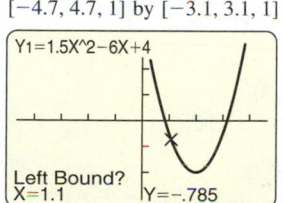

Figure A.46

To find a minimum y-value (or vertex) on the graph of $f(x) = 1.5x^2 - 6x + 4$, start by entering $Y_1 = 1.5X^\wedge 2 - 6X + 4$ from the (Y =) menu. Set the viewing rectangle and then perform the following keystrokes to find the minimum y-value.

$$\boxed{\text{2nd}}\;\boxed{\text{TRACE [CALC]}}\;\boxed{3}$$

See Figure A.46.

The calculator prompts for a left bound. Use the arrow keys to position the cursor left of the vertex and press (ENTER). Similarly, position the cursor to the right of the vertex for the right bound and press (ENTER). Finally, the graphing calculator asks for a guess between the left and right bounds. Place the cursor near the vertex and press (ENTER). See Figures A.47–A.49. The minimum value is shown in Figure A.50.

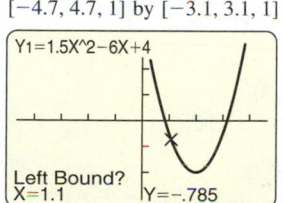

$[-4.7, 4.7, 1]$ by $[-3.1, 3.1, 1]$

Figure A.47

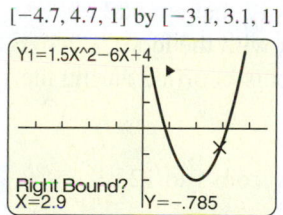

$[-4.7, 4.7, 1]$ by $[-3.1, 3.1, 1]$

Figure A.48

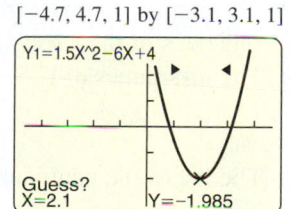

$[-4.7, 4.7, 1]$ by $[-3.1, 3.1, 1]$

Figure A.49

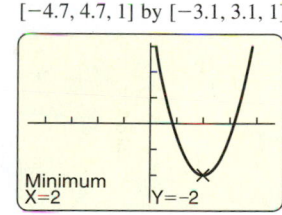

$[-4.7, 4.7, 1]$ by $[-3.1, 3.1, 1]$

Figure A.50

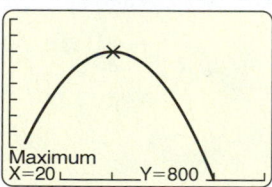

$[0, 50, 10]$ by $[0, 1000, 100]$

Figure A.51

A maximum of the function f on an interval can be found similarly, except enter

$$\boxed{\text{2nd}}\;\boxed{\text{TRACE [CALC]}}\;\boxed{4}.$$

The calculator prompts for left and right bounds, followed by a guess. Press (ENTER) after the cursor has been located appropriately for each prompt. The graphing calculator will display the maximum y-value. An example is shown in Figure A.51, where $f(x) = x(80 - 2x)$.

> ### SUMMARY: FINDING MAXIMUM AND MINIMUM VALUES
>
> 1. Graph the function in an appropriate viewing rectangle.
> 2. Press (2nd)(TRACE [CALC])(3) to find a minimum y-value.
> 3. Press (2nd)(TRACE [CALC])(4) to find a maximum y-value.
> 4. Use the arrow keys to locate the left and right x-bounds, followed by a guess. Press (ENTER) to select each position of the cursor.

Appendix B
Sets

Basic Terminology

A **set** is a collection of things, and the members of a set are called **elements**. A set can be described by listing its elements between braces. For example, the set W containing the *weekdays* is

$$W = \{\text{Monday, Tuesday, Wednesday, Thursday, Friday}\}.$$

This set has 5 elements. For example, Monday *is an element of W*, which is denoted

$$\text{Monday} \in W.$$

However, Sunday *is not an element of W*, which is denoted

$$\text{Sunday} \notin W.$$

If a set contains no elements, then it is called the **empty set** or **null set**. The empty set is denoted $\varnothing$, or $\{\ \}$. For example, the set Z that contains the names of U.S. states starting with the letter Z is the empty set. That is, $Z = \varnothing$ or $Z = \{\ \}$.

NOTE: Do *not* write the empty set as $\{\varnothing\}$.

EXAMPLE 1 | **Listing the elements of a set**

Use set notation to list the elements of each set S described.
(a) The natural numbers from 1 to 12 that are odd
(b) The days of the week that start with the letter T
(c) The last names of U.S. presidents in office during the 1990s

Solution
(a) The list of the natural numbers from 1 to 12 is

$$1, 2, 3, 4, 5, 6, 7, 8, 9, 10, 11, \text{ and } 12.$$

The set of odd natural numbers from this list is

$$S = \{1, 3, 5, 7, 9, 11\}.$$

(b) $S = \{\text{Tuesday, Thursday}\}$
(c) $S = \{\text{Bush, Clinton}\}$

Now Try Exercises 1, 3, 5

EXAMPLE 2 | **Determining the elements of sets**

Use $\in$ or $\notin$ to make each statement true.
(a) 5 _____ $\{1, 2, 3, 4, 5, 6\}$
(b) -2 _____ $\{-4, 0, 2, 4, 6\}$
(c) $\frac{1}{2}$ _____ $\{0, 0.5, 1.0, 1.5, 2.0\}$

Solution
(a) Because 5 is an element of $\{1, 2, 3, 4, 5, 6\}$, we write

$$5 \in \{1, 2, 3, 4, 5, 6\}.$$

(b) Because -2 is not an element of $\{-4, 0, 2, 4, 6\}$, we write

$$-2 \notin \{-4, 0, 2, 4, 6\}.$$

(c) Because $\frac{1}{2} = 0.5$, we write

$$\frac{1}{2} \in \{0, 0.5, 1.0, 1.5, 2.0\}.$$

Now Try Exercises 11, 13, 15

Universal Set

When discussing sets, we assume that there is a *universal set*. The **universal set** contains all elements under consideration. For example, if the universal set U is all days of the week, then the set S containing the days that start with the letter S is

$$S = \{\text{Sunday, Saturday}\}.$$

However, if the universal set U is only the weekdays, then the set S containing the days starting with the letter S is the *empty set*, or $S = \{\ \}$.

EXAMPLE 3 **Using different universal sets**

Determine the set O of odd integers that belong to each universal set U.
(a) $U = \{1, 2, 3, 4, 5, 6, 7, 8, 9, 10\}$
(b) $U = \{1, 6, 11, 16, 21, 26\}$
(c) $U = \{1, 2, 3, 4, \ldots\}$

Solution
(a) The odd integers in U are 1, 3, 5, 7, and 9, so

$$O = \{1, 3, 5, 7, 9\}.$$

(b) The odd integers in $U = \{\mathbf{1}, 6, \mathbf{11}, 16, \mathbf{21}, 26\}$ are $\mathbf{1}$, $\mathbf{11}$, and $\mathbf{21}$. Thus

$$O = \{1, 11, 21\}.$$

(c) The three dots in $\{1, 2, 3, 4, \ldots\}$ indicate that U contains all natural numbers. Thus the set O contains all odd natural numbers, or

$$O = \{1, 3, 5, 7, 9, \ldots\}.$$

Now Try Exercises 21, 23

Subsets

If every element in a set B is contained in a set A, then we say that B is a **subset** of A, denoted $B \subseteq A$. For example, if $A = \{1, 2, 3, 4\}$ and $B = \{2, 4\}$, then $B \subseteq A$ because every element in B belongs to A. However, A is not a subset of B, denoted $A \not\subseteq B$, because the elements 1 and 3 are in A but *not* in B. The symbol $\not\subseteq$ is read "is not a subset of."

If every element in set A is in set B and every element in set B is in set A, then A and B are **equal sets**, denoted $A = B$. Note that if $A \subseteq B$ and $B \subseteq A$, then $A = B$. Why?

EXAMPLE 4 **Determining subsets**

Let $A = \{a, b, c, d, e\}$, $B = \{b, c, d\}$, $C = \{b, e\}$, and $D = \{e, b\}$. Determine whether each statement is true or false.
(a) $A \subseteq B$ **(b)** $B \subseteq A$ **(c)** $C = D$ **(d)** $C \not\subseteq A$ **(e)** $\varnothing \subseteq B$ **(f)** $A \subseteq A$

Solution

(a) False; the elements a and e in *A* are not in *B*, so *A* is *not* a subset of *B*.

(b) True; every element in *B* is in *A*, so *B* is a subset of *A*.

(c) True; although the elements in *C* and *D* are listed in a different order, they contain exactly the same elements, so *C* and *D* are equal.

(d) False; every element in *C* is in *A*, so *C* *is* a subset of *A*.

(e) True; the empty set, or null set, is a subset of *every* set.

(f) True; every element in *A* is in *A*, so *A* is a subset of itself.

NOTE: Every set is a subset of itself.

■ **Now Try Exercises 29, 31, 33, 35**

Venn Diagrams

Venn diagrams are often used to depict relationships among sets. A large rectangle typically represents the universal set, and subsets of the universal set are represented by regions within the universal set. In Figure B.1 the universal set *U* is represented by everything inside the large rectangle. The set *A* is represented by the red circular region within this rectangle because *A* is a subset of *U*.

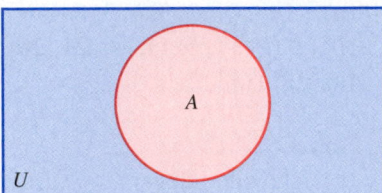

Figure B.1 $A \subseteq U$

The **complement** of a set *A*, denoted *A'*, is the set containing all elements in the universal set that are *not* in *A*. That is, if $a \notin A$, then $a \in A'$. For example, if

$$U = \{1, 2, 3, 4, 5, 6\} \quad \text{and} \quad A = \{1, 2, 3\},$$

then

$$A' = \{4, 5, 6\}$$

because the elements 4, 5, and 6 are found in *U* but not in *A*. This situation is illustrated by the Venn diagram in Figure B.2. The red region is *A*, and the blue region is *A'*. Together, the red and blue regions comprise the universal set *U*.

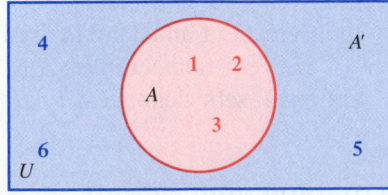

Figure B.2 *A* and *A'*

NOTE: Every element in *U* must be in either *A* or *A'*, but not both.

EXAMPLE 5   **Determining complements**

Let the universal set U be $U = \{$red, blue, yellow, green, black, white$\}$, and let two subsets of U be $A = \{$red, blue, yellow$\}$ and $B = \{$black, white$\}$. Find each of the following.
(a) A' **(b)** B' **(c)** U'

Solution

(a) The elements in U that are not in A are in $A' = \{$green, black, white$\}$.
(b) $B' = \{$red, blue, yellow, green$\}$
(c) Because the universal set U contains every element under consideration, the complement of U, or U', is empty. That is, $U' = \{\ \ \}$.

Now Try Exercises 73, 75, 81

Union and Intersection

Although we do not perform arithmetic operations, such as multiplication or division, on sets, we can find the *union* or *intersection* of two or more sets. The **union** of two sets A and B, denoted $A \cup B$ and read "A union B," is the set containing any element that can be found in *either* set A *or* set B. If an element is in both A and B, then this element is listed only once in $A \cup B$. For example, if

$$A = \{1, 2, 3, 4\} \quad \text{and} \quad B = \{3, 4, 5, 6\},$$

then

$$A \cup B = \{1, 2, 3, 4, 5, 6\}.$$

Note that elements 3 and 4 are in both A and B but are listed only once in $A \cup B$. A Venn diagram of this situation is shown in Figure B.3. The region that represents the union is shaded blue.

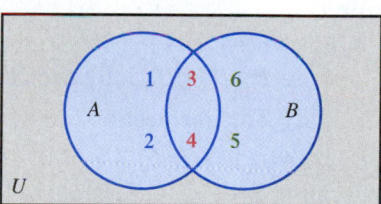

Figure B.3

The **intersection** of two sets A and B, denoted $A \cap B$ and read "A intersect B," is the set containing elements that can be found in *both* set A and set B. For example, if

$$A = \{1, 2, 3, 4\} \quad \text{and} \quad B = \{3, 4, 5, 6\},$$

then

$$A \cap B = \{3, 4\}.$$

Note that elements 3 and 4 belong to *both A and B*. This situation is illustrated by the Venn diagram in Figure B.4 on the next page. The purple region containing elements 3 and 4 represents the intersection of A and B.

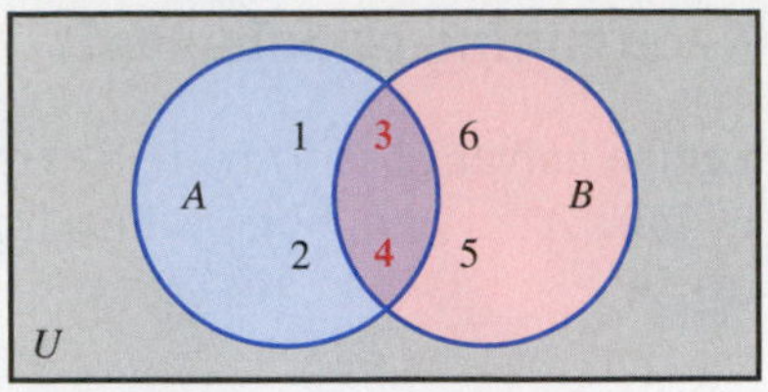

Figure B.4

EXAMPLE 6 | **Finding unions and intersections of sets**

Let $A = \{a, b, c, x, z\}$, $B = \{a, x, y\}$, and $C = \{c, x, y, z\}$. Find each set.
(a) $A \cup B$ **(b)** $B \cap C$ **(c)** $A \cup C$ **(d)** $A \cap B \cap C$

Solution
(a) The union contains the elements belonging to either *A or B* or both. Thus

$$A \cup B = \{a, b, c, x, y, z\}.$$

(b) The intersection contains the elements belonging to both *B and C*. Thus

$$B \cap C = \{x, y\}.$$

(c) $A \cup C = \{a, b, c, x, y, z\}$
(d) For an element to be in the intersection of three sets, the element must belong to each set. The only element belonging to *A and B and C* is x. Thus $A \cap B \cap C = \{x\}$.

Now Try Exercises 63, 65, 69, 79

EXAMPLE 7 | **Using set operations**

Let $U = \{1, 2, 3, 4, 5, 6, 7, 8\}$, $A = \{2, 4, 6, 8\}$, $B = \{1, 3, 5, 6\}$, and $C = \{5, 6, 7, 8\}$.
Find each set.
(a) $A \cap B$ **(b)** $B \cap B'$ **(c)** $B \cup C$ **(d)** $A \cup C'$

Solution
(a) $A \cap B = \{6\}$
(b) Because $B = \{1, 3, 5, 6\}$ and $B' = \{2, 4, 7, 8\}$, $B \cap B' = \{\ \}$.

     NOTE: The intersection of a set and its complement is always the empty set.
(c) $B \cup C = \{1, 3, 5, 6, 7, 8\}$
(d) Because $A = \{2, 4, 6, 8\}$ and $C' = \{1, 2, 3, 4\}$, $A \cup C' = \{1, 2, 3, 4, 6, 8\}$.

Now Try Exercises 67, 71, 77, 83

Finite and Infinite Sets

Sets can either have a finite number of elements or infinitely many elements. The elements of a **finite set** can be listed explicitly, whereas the elements of an **infinite set** cannot be listed because there are infinitely many elements. For example, the set of integers *S* from -3 to 3 is a finite set with 7 elements because

$$S = \{-3, -2, -1, 0, 1, 2, 3\}.$$

In contrast, the set of natural numbers,

$$N = \{1, 2, 3, 4, \ldots\},$$

is an infinite set because the list continues without end.

| EXAMPLE 8 | **Recognizing finite and infinite sets** |

Identify each set as finite or infinite.
(a) The set of integers
(b) The set of natural numbers between 1 and 5, inclusive
(c) The set of rational numbers between 1 and 5, inclusive

Solution
(a) There are infinitely many integers, so the set of integers is infinite.
(b) The set of natural numbers between 1 and 5, inclusive, is {1, 2, 3, 4, 5}. Because we can list each element of this set, it is a finite set.
(c) Because there are infinitely many rational numbers (fractions) between 1 and 5, this set is an infinite set.

Now Try Exercises 87, 89

B Exercises

BASIC CONCEPTS

Exercises 1–10: Use set notation to list the elements of the set described.

1. The natural numbers less than 8

2. The natural numbers greater than 4 and less than or equal to 10

3. The days of the week starting with the letter S

4. U.S. states whose names start with the letter A

5. The letters of the alphabet from A to G

6. The letters of the alphabet from D to M

7. Dogs that can fly under their own power

8. People having an income of $10 trillion in 2011

9. The even integers greater than −2 and less than 11

10. The odd integers less than 21 and greater than 12

Exercises 11–20: Insert $\in$ or $\notin$ to make the statement true.

11. 10 _____ {5, 10, 15, 20}

12. 7 _____ {1, 3, 5, 7, 9, 11}

13. 5 _____ {2, 4, 6, 8, 10}

14. −1 _____ {0, 1, 2, 3, 4, 5}

15. $\frac{1}{4}$ _____ {0, 0.25, 0.5, 0.75, 1}

16. $\frac{2}{5}$ _____ {0, 0.2, 0.4, 0.6, 0.8, 1}

17. $\frac{1}{3}$ _____ {0, 0.33, 0.67, 1}

18. $\frac{4}{2}$ _____ {1, 3, 4, 5, 7}

19. red _____ {blue, green, red, yellow}

20. M _____ {M, T, W, R, F}

Exercises 21–24: Determine the set E of even integers that belong to the given universal set U.

21. $U = \{1, 2, 3, 4, 5, 6, 7, 8, 9, 10\}$

22. $U = \{-3, -2, -1, 0, 1, 2, 3\}$

23. $U = \{1, 2, 3, 4, \ldots\}$

24. $U = \{1, 3, 5, 7, 9, 11, \ldots\}$

Exercises 25–28: Determine the set A of words starting with the letter A that belong to the given universal set U.

25. $U = \{$Apple, Orange, Pear, Apricot$\}$

26. $U = \{$Apple, Orange$\}$

27. $U = \{$Calculus, Algebra, Geometry$\}$

28. $U = \{$Calculus, Algebra, Geometry, Arithmetic$\}$

SUBSETS

Exercises 29–36: Let $A = \{1, 2, 3, 4\}$, $B = \{3, 4, 5, 6\}$, and $C = \{3, 6\}$. Determine whether the statement is true or false.

29. $A \subseteq B$

30. $B \subseteq A$

31. $C \subseteq B$

32. $\varnothing \subseteq C$

33. $C \subseteq C$

34. $A \nsubseteq B$

35. $B \nsubseteq C$

36. $C \subseteq A$

Exercises 37–44: Let $A = \{a, b, c, d, e\}$, $B = \{b, d\}$, and $C = \{a, c, e\}$. Determine whether the statement is true or false.

37. $B \subseteq A$

38. $C \subseteq A$

39. $B \nsubseteq C$

40. $C = B$

41. $C \subseteq \varnothing$

42. $C \nsubseteq B$

43. $A \subseteq A$

44. $C \nsubseteq C$

VENN DIAGRAMS

Exercises 45–52: Let A and B be two sets and U be the universal set. Sketch a Venn diagram and shade the region that illustrates the given set.

45. $A \cup B$

46. $A \cap B$

47. A'

48. B'

49. $A \cap U$

50. $B \cup U$

51. $(A \cap B)'$

52. $(A \cup B)'$

UNIONS, INTERSECTIONS, AND COMPLEMENTS

Exercises 53–62: Write the expression in terms of one set.

53. $\{1, 2, 3\} \cup \{3, 4\}$

54. $\{1, 2, 3\} \cup \{6\}$

55. $\{a, b, c\} \cap \{a, b, d\}$

56. $\{x, y, z\} \cap \{a, b, c\}$

57. $\{a, b, c\} \cup \{\ \}$

58. $\{a, b, c\} \cap \varnothing$

59. $\{a, b\} \cup \{c, b\} \cup \{d, a\}$

60. $\{1, 3, 5\} \cup \{4, 5\} \cup \{4, 5, 6\}$

61. $\{4, 5, 8, 9\} \cap \{3, 5, 8, 9\} \cap \{4, 5, 8\}$

62. $\{1, 2, 5\} \cap \{2, 1\} \cap \{1, 2, 3\}$

Exercises 63–86: Let

$$U = \{1, 2, 3, 4, 5, 6, 7, 8, 9, 10\},$$
$$A = \{1, 3, 5, 7, 9\}, B = \{2, 4, 6, 8, 10\},$$
$$C = \{3, 4, 5, 6, 7\}, \text{ and } D = \{4, 5, 6, 9\}.$$

Use set notation to list the elements in the given set.

63. $A \cup B$

64. $A \cap B$

65. $C \cap D$

66. $A \cup D$

67. $U \cap \varnothing$

68. $\varnothing \cup U$

69. $A \cup \varnothing$

70. $\varnothing \cap B$

71. $D \cap D'$

72. $A \cup A$

73. U'

74. A'

75. B'

76. D'

77. $A \cup D'$

78. $A' \cap B'$

79. $A \cap B \cap C$

80. $B \cup C \cup D$

81. C'

82. $D' \cup D$

83. $B \cap C$

84. $B \cap C \cap D$

85. $C \cup C'$

86. $D' \cap B \cap C$

FINITE AND INFINITE SETS

Exercises 87–92: Determine whether the given set is finite or infinite.

87. The set of whole numbers

88. The set of real numbers

89. The set of natural numbers less than 1000

90. The set of natural numbers greater than 4 and less than 20

91. The days of the week

92. The names of the students in your class

Appendix C
Linear Programming

Basic Concepts

NEW VOCABULARY

☐ Objective function
☐ Constraint
☐ Feasible solutions
☐ Linear programming problem
☐ Optimal solution
☐ Vertex

▶ **REAL-WORLD CONNECTION** Suppose that a small business sells candy for $3 per pound and freshly ground coffee for $5 per pound. All inventory is sold by the end of the day. The revenue R collected in dollars is given by

$$R = 3x + 5y,$$

where x is the pounds of candy sold and y is the pounds of coffee sold. For example, if the business sells **80** pounds of candy and **40** pounds of coffee during a day, then its revenue is

$$R = 3(\mathbf{80}) + 5(\mathbf{40}) = \$440.$$

The function $R = 3x + 5y$ is called an **objective function**.

Suppose also that the company cannot package more than 150 pounds of candy and coffee per day. Then the inequality

$$x + y \leq 150$$

represents a **constraint** on the objective function, which limits the company's revenue for any one day. A goal of this business might be to maximize the objective function

$$R = 3x + 5y, \quad \text{Revenue equation}$$

subject to the following constraints.

$$x + y \leq 150 \quad \text{Constraint on production}$$
$$x \geq 0, y \geq 0 \quad \text{Variables cannot be negative.}$$

Note that the constraints $x \geq 0$ and $y \geq 0$ are included because the number of pounds of candy or coffee cannot be negative. The problem that we have described is called a *linear programming problem*. Before learning how to solve a linear programming problem, we need to discuss the region of *feasible solutions*.

Region of Feasible Solutions

The constraints for a linear programming problem consist of linear inequalities. These inequalities are satisfied by some points in the *xy*-plane but not by others. The set of solutions to these constraints is called the **feasible solutions**. For example, the *region of feasible solutions* to the constraints for the business just described is shaded in Figure C.1.

Constraints on Sales

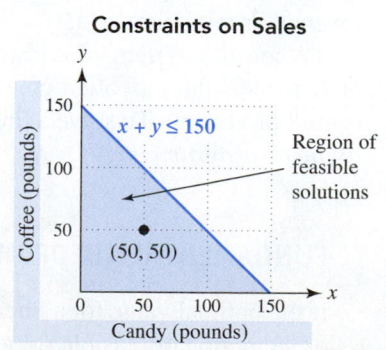

Figure C.1

The point (50, 50) lies in the shaded region and represents the business selling 50 pounds of candy and 50 pounds of coffee. In the next example, we shade the region of feasible solutions to a set of constraints.

EXAMPLE 1 ## Finding the region of feasible solutions

Shade the region of feasible solutions for the following constraints.

$$x + 2y \leq 30$$
$$2x + y \leq 30$$
$$x \geq 0, y \geq 0$$

Solution

The feasible solutions are the ordered pairs (x, y) that *satisfy all four* inequalities. They lie on or below the lines $x + 2y = 30$ and $2x + y = 30$, and on or above the line $y = 0$ and on or to the right of $x = 0$, as shown in Figure C.2. Note that the inequalities $x \geq 0$ and $y \geq 0$ restrict the region of feasible solutions to quadrant I and the x- and y-axes.

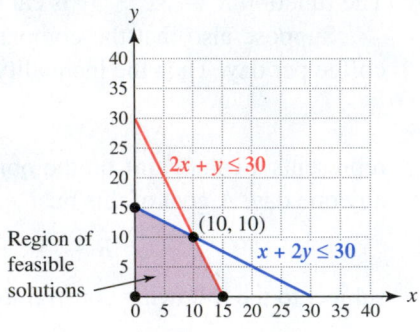

Figure C.2

Now Try Exercise 15

Solving Linear Programming Problems

A **linear programming problem** consists of an *objective function* and a system of linear inequalities called *constraints*. The solution set for the system of linear inequalities is called the *region of feasible solutions*. The objective function describes a quantity that is to be optimized. The **optimal value** for a linear programming problem often results in maximum revenue or minimum cost.

When the system of constraints has only two variables, the boundary of the region of feasible solutions often consists of line segments intersecting at points called *vertices* (plural of **vertex**). To solve a linear programming problem we use the *fundamental theorem of linear programming*.

FUNDAMENTAL THEOREM OF LINEAR PROGRAMMING

If the optimal value for a linear programming problem exists, then it occurs at a vertex of the region of feasible solutions.

The fundamental theorem of linear programming is used to solve the following linear programming problem. A justification of this theorem is given after Example 2.

EXAMPLE 2 **Maximizing an objective function**

Maximize the objective function $R = 2x + 3y$ subject to

$$x + 2y \leq 30$$
$$2x + y \leq 30$$
$$x \geq 0, y \geq 0.$$

Solution
The region of feasible solutions is shaded in Figure C.2 from Example 1. Note that the vertices on the boundary of feasible solutions are $(0, 0)$, $(15, 0)$, $(10, 10)$, and $(0, 15)$. To find the maximum value of R, substitute each vertex in the formula for R. The maximum value of R is **50** when $x = $ **10** and $y = $ **10**. See Table C.1.

TABLE C.1

Vertex	$R = 2x + 3y$	
$(0, 0)$	$2(0) + 3(0) = 0$	
$(15, 0)$	$2(15) + 3(0) = 30$	
$(\mathbf{10}, \mathbf{10})$	$2(\mathbf{10}) + 3(\mathbf{10}) = \mathbf{50}$	← **Maximum R is 50.**
$(0, 15)$	$2(0) + 3(15) = 45$	

Now Try Exercise 33

The following steps are helpful in solving linear programming word problems.

STEPS FOR SOLVING A LINEAR PROGRAMMING WORD PROBLEM

STEP 1: Read the problem carefully. Consider making a table.

STEP 2: Write the objective function and all the constraints.

STEP 3: Sketch a graph of the region of feasible solutions. Identify all vertices.

STEP 4: Evaluate the objective function at each vertex. A maximum (or a minimum) occurs at a vertex.

NOTE: If the region is unbounded, a maximum (or minimum) may not exist.

JUSTIFICATION OF THE FUNDAMENTAL THEOREM To better understand the fundamental theorem of linear programming, consider the following example. Suppose that we want to maximize profit $P = 3x + 7y$ subject to the following four constraints:

$$x \geq 1, \quad x \leq 5, \quad y \geq x, \quad \text{and} \quad y \leq 6.$$

The corresponding region of feasible solutions is shown in Figure C.3 on the next page.

Each value of P determines a unique line. For example, if $P = 70$, then the equation for P becomes $3x + 7y = 70$. The resulting line, shown in Figure C.4 on the next page, does not intersect the region of feasible solutions. Thus there are no values for x and y that lie in this region and result in a profit of 70. Figure C.4 also shows the lines that result from letting $P = 0, 10,$ and 30. If $P = 10$, then the line intersects the region of feasible solutions only

CRITICAL THINKING

What is the minimum value
for *P* subject to the given
constraints?

at the vertex (1, 1). This means that if $x = 1$ and $y = 1$, then $P = 3(1) + 7(1) = 10$. If
$P = 30$, then the line $3x + 7y = 30$ intersects the region of feasibility infinitely many times.
However, it appears that values greater than 30 are possible for *P*. In Figure C.5 lines are
drawn for $P = 57, 63$, and 70. Notice that there are no points of intersection for $P = 63$ or
$P = 70$, but there is one vertex in the region of feasible solutions at (5, 6) that gives $P = 57$.
Thus the maximum value of *P* is 57, and this maximum occurs at a vertex of the region of
feasible solutions. The fundamental theorem of linear programming generalizes this result.

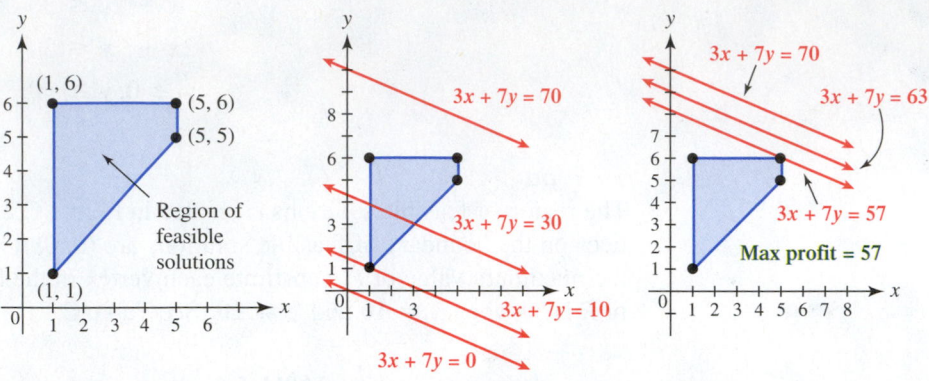

Figure C.3   Figure C.4   Figure C.5

EXAMPLE 3   **Minimizing the cost of vitamins**

A breeder is mixing two different vitamins, Brand X and Brand Y, into pet food.
Each serving of pet food should contain at least 60 units of vitamin A and 30 units of
vitamin C. Brand X costs 80 cents per ounce and Brand Y costs 50 cents per ounce. Each
ounce of Brand X contains 15 units of vitamin A and 10 units of vitamin C, whereas each
ounce of Brand Y contains 20 units of vitamin A and 5 units of vitamin C. Determine how
much of each brand of vitamin should be mixed to produce a minimum cost per serving.

Solution

STEP 1: Begin by listing the information, as illustrated in Table C.2.

TABLE C.2

Brand	Amount	Vitamin A	Vitamin C	Cost
X	x	15	10	80 cents
Y	y	20	5	50 cents
Minimum		60	30	

STEP 2: If *x* ounces of Brand X are purchased at 80 cents per ounce and if *y* ounces of
Brand Y are purchased at 50 cents per ounce, then the total cost *C* is given by
$C = 80x + 50y$. Because each ounce of Brand X contains 15 units of vitamin A
and each ounce of Brand Y contains 20 units of vitamin A, the total number of
units of vitamin A is $15x + 20y$. If each serving of pet food must contain at least
60 units of vitamin A, the constraint is $15x + 20y \geq 60$. Similarly, because each
serving requires at least 30 units of vitamin C, $10x + 5y \geq 30$. The linear programming problem then becomes the following.

$$
\begin{array}{lll}
\text{Minimize:} & C = 80x + 50y & \text{Cost (in cents)} \\
\text{Subject to:} & 15x + 20y \geq 60 & \text{Vitamin A constraint} \\
& 10x + 5y \geq 30 & \text{Vitamin C constraint} \\
& x \geq 0, y \geq 0 & \text{Amounts cannot be negative.}
\end{array}
$$

STEP 3: The region containing the feasible solutions is shown in Figure C.6.

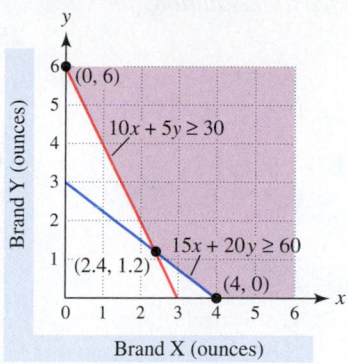

Figure C.6

NOTE: To determine the vertex (2.4, 1.2), solve the system of equations

$$15x + 20y = 60$$
$$10x + 5y = 30$$

by using elimination.

STEP 4: In Figure C.6, the vertices are (0, 6), (2.4, 1.2), and (4, 0). Evaluate the objective function $C = 80x + 50y$ at each vertex, as shown in Table C.3.

TABLE C.3

Vertex	$C = 80x + 50y$	
(0, 6)	$80(0) + 50(6) = 300$	
(2.4, 1.2)	$80(\mathbf{2.4}) + 50(\mathbf{1.2}) = \mathbf{252}$	←—Minimum cost (cents)
(4, 0)	$80(4) + 50(0) = 320$	

The minimum cost occurs when **2.4** ounces of Brand X and **1.2** ounces of Brand Y are mixed, at a cost of **$2.52** per serving.

▌ Now Try Exercise 47

C   Exercises

CONCEPTS AND VOCABULARY

1. A procedure used in business to optimize quantities such as cost and profit is called _____.

2. In linear programming, the function to be optimized is called the _____ function.

3. The region in the *xy*-plane that satisfies the constraints is called the region of _____.

4. In linear programming, constraints typically consist of a system of _____.

5. If the optimal value for a linear programming problem exists, then it occurs at a(n) _____ of the region of feasible solutions.

6. To find the optimal value in a linear programming problem, substitute each vertex in the _____ function.

REGIONS OF FEASIBLE SOLUTIONS

Exercises 7–20: Shade the region of feasible solutions for the following constraints.

7. $x + y \leq 3$
$x \geq 0, y \geq 0$

8. $2x + y \leq 4$
$x \geq 0, y \geq 0$

9. $4x + 3y \leq 12$
$x \geq 0, y \geq 0$

10. $5x + 3y \leq 15$
$x \geq 0, y \geq 0$

11. $x \leq 5$
$y \leq 2$
$x \geq 0, y \geq 0$

12. $x \leq 3$
$y \leq 4$
$x \geq 1, y \geq 1$

13. $x + y \leq 5$
$x + y \geq 2$
$x \geq 0, y \geq 0$

14. $2x + y \leq 6$
$x + y \geq 3$
$x \geq 0, y \geq 0$

15. $3x + 2y \leq 6$
$2x + 3y \leq 6$
$x \geq 0, y \geq 0$

16. $5x + 3y \leq 30$
$3x + 5y \leq 30$
$x \geq 0, y \geq 0$

17. $x + y \leq 3$
$x + 3y \geq 3$
$x \geq 0, y \geq 0$

18. $3x + y \geq 6$
$x + 2y \geq 6$
$x \geq 0, y \geq 0$

19. $x + 2y \geq 4$
$3x + 2y \geq 6$
$x \geq 0, y \geq 0$

20. $4x + 3y \geq 12$
$3x + 4y \geq 12$
$x \geq 0, y \geq 0$

LINEAR PROGRAMMING

Exercises 21–24: Find the maximum of R on the region of feasible solutions shown in the figure.

21. $R = 4x + 5y$

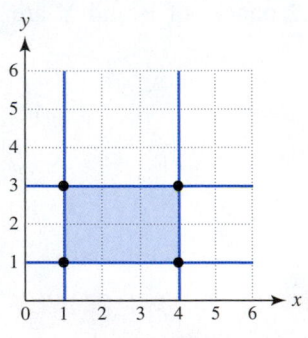

22. $R = 2x + 3y$

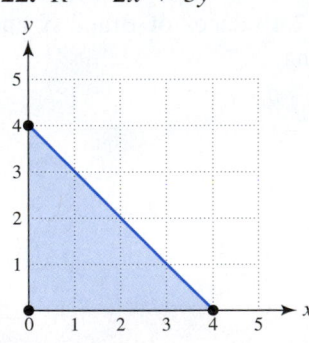

23. $R = x + 3y$

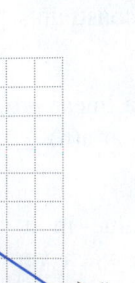

24. $R = 12x + 9y$

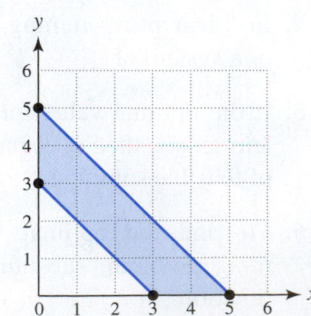

Exercises 25–28: Use the figures in Exercises 21–24 to complete Exercises 25–28, respectively, to minimize C.

25. $C = 2x + 3y$

26. $C = 3x + y$

27. $C = 5x + y$

28. $C = 2x + 7y$

Exercises 29–36: Maximize the objective function R, subject to the given constraints.

29. $R = 3x + 5y$
$x + y \leq 150$
$x \geq 0, y \geq 0$

30. $R = 6x + 5y$
$x + y \leq 8$
$x \geq 0, y \geq 0$

31. $R = 3x + 2y$
$2x + y \leq 6$
$x \geq 0, y \geq 0$

32. $R = x + 3y$
$x + 2y \leq 4$
$x \geq 0, y \geq 0$

33. $R = 12x + 9y$
$3x + y \leq 6$
$x + 3y \leq 6$
$x \geq 0, y \geq 0$

34. $R = 10x + 30y$
$x + 3y \leq 12$
$3x + y \leq 12$
$x \geq 0, y \geq 0$

35. $R = 4x + 5y$
$x + y \geq 2$
$x + 2y \leq 4$
$x \geq 0, y \geq 0$

36. $R = 3x + 7y$
$x + y \geq 1$
$3x + y \leq 3$
$x \geq 0, y \geq 0$

Exercises 37–42: Minimize the objective function C, subject to the given constraints.

37. $C = x + 2y$
$x \leq 3, y \leq 2$
$x \geq 0, y \geq 0$

38. $C = 3x + y$
$x \leq 5, y \leq 3$
$x \geq 1, y \geq 1$

39. $C = 8x + 15y$
$x + y \geq 4$
$x \geq 0, y \geq 0$

40. $C = x + 2y$
$3x + 4y \geq 12$
$x \geq 0, y \geq 0$

41. $C = 30x + 40y$
$2x + y \leq 6$
$x + y \geq 2$
$x \geq 0, y \geq 0$

42. $C = 50x + 70y$
$2x + 3y \leq 6$
$x + y \geq 1$
$x \geq 0, y \geq 0$

APPLICATIONS

Exercises 43–50: Solve the linear programming problem.

43. *Maximizing Revenue* A small business sells candy for $4 per pound and coffee for $6 per pound. The business can package and sell at most a total of 100 pounds of candy and coffee per day, but at least 20 pounds of candy must be sold each day. Determine how many pounds of candy and coffee need to be sold each day to maximize revenue.

44. *Maximizing Revenue* An organic foods store sells almonds for $10 per pound and flax seed for $5 per pound. The business can package and sell, at most, a total of 50 pounds of almonds and flax seed per day, but it must sell at least 15 pounds of flax seed per day. Determine the maximum daily revenue.

45. *Minimizing Cost* It costs a business $50 to make a graphing calculator and $20 to make a scientific calculator. Each week the company must produce at least 90 calculators. At least twice as many scientific calculators must be made as graphing calculators. Determine the minimum weekly cost.

46. *Minimizing Cost* It costs a business $20 to make one compact disc player and $10 to make one radio. Each week the company must make a combined total of at least 50 compact disc players and radios. At least as many compact disc players as radios must be manufactured. Determine how many compact disc players and radios should be made to minimize weekly costs.

47. *Vitamin Cost* A pet owner is mixing two different brands of vitamins, Brand X and Brand Y, into pet food. Brand X costs 90 cents per ounce and Brand Y costs 60 cents per ounce. Each serving is a mixture of the two brands and should contain at least 40 units of vitamin A and 30 units of vitamin C. Each ounce of Brand X contains 20 units of vitamin A and 10 units of vitamin C, whereas each ounce of Brand Y contains 10 units of vitamin A and 10 units of vitamin C. Determine how much of each brand of vitamin should be mixed to produce a minimum cost per serving.

48. *Pet Food Cost* A pet owner is buying two brands of food, X and Y, for his animals. Each serving of the mixture of the two foods should contain at least 60 grams of protein and 40 grams of fat. Brand X costs 75 cents per unit and Brand Y costs 50 cents per unit. Each unit of Brand X contains 20 grams of protein and 10 grams of fat, whereas each unit of Brand Y contains 10 grams of protein and 10 grams of fat. Determine how much of each brand should be bought to obtain a minimum cost per serving.

49. *Raising Animals* A breeder can raise no more than 50 hamsters and mice but no more than 20 hamsters. If she sells the hamsters for $15 each and the mice for $10 each, find the maximum revenue.

50. *Maximizing Profit* A business manufactures two parts, X and Y. Machines A and B are needed to make each part. To make part X, machine A is needed 3 hours and machine B is needed 1 hour. To make part Y, machine A is needed 1 hour and machine B is needed 2 hours. Machine A is available 60 hours per week and machine B is available 50 hours per week. The profit from part X is $300 and the profit from part Y is $250. How many parts of each type should be made to maximize weekly profit?

Appendix D
Synthetic Division

Basic Concepts

NEW VOCABULARY

☐ Synthetic division

A shortcut called **synthetic division** can be used to divide $x - k$, where k is a constant, into a polynomial. For example, to divide $x - 2$ into $3x^3 - 8x^2 + 7x - 6$, we do the following (with the equivalent long division shown at the right).

$$
\begin{array}{c}
\textbf{Synthetic Division} \\[4pt]
\underline{2}\; \begin{array}{rrrr} 3 & -8 & 7 & -6 \\ & 6 & -4 & 6 \\ \hline 3 & -2 & 3 & 0 \end{array} \\[6pt]
\end{array}
\qquad
\begin{array}{c}
\textbf{Long Division of Polynomials} \\[4pt]
\begin{array}{r}
3x^2 - 2x + 3 \\
x - 2\,\overline{)\,3x^3 - 8x^2 + 7x - 6} \\
\underline{3x^3 - 6x^2} \\
-2x^2 + 7x \\
\underline{-2x^2 + 4x} \\
3x - 6 \\
\underline{3x - 6} \\
0
\end{array}
\end{array}
$$

Add to find row 3.

Note that the blue numbers in the expression for long division correspond to the third row in synthetic division. The remainder is 0, which is the last number in the third row. The quotient is $3x^2 - 2x + 3$. Its coefficients are **3**, **−2**, and **3** and are located in the third row. To divide $x - 2$ into $3x^3 - 8x^2 + 7x - 6$ with synthetic division use the following steps.

STEP 1: In the top row write **2** (the value of k) on the left and then write the coefficients of the dividend $3x^3 - 8x^2 + 7x - 6$.

STEP 2: (a) Copy the leading coefficient **3** of $3x^3 - 8x^2 + 7x - 6$ into the third row and multiply it by **2** (the value of k). Write the result 6 in the second row below **−8**. Add **−8** and 6 in the second column to obtain the **−2** in the third row.

(b) Repeat the process by multiplying **−2** by **2** and place the result −4 below **7**. Then add **7** and −4 to obtain **3**.

(c) Multiply **3** by **2** and place the result 6 below the **−6**. Adding 6 and **−6** gives **0**.

STEP 3: The last number in the third row is **0**, which is the remainder. The other numbers in the third row are the coefficients of the quotient, which is $3x^2 - 2x + 3$.

EXAMPLE 1 **Performing synthetic division**

Use synthetic division to divide $x^4 - 5x^3 + 9x^2 - 10x + 3$ by $x - 3$.

Solution

Because the divisor is $x - 3$, the value of k is **3**.

$$
\underline{3}\; \begin{array}{rrrrr} 1 & -5 & 9 & -10 & 3 \\ & 3 & -6 & 9 & -3 \\ \hline 1 & -2 & 3 & -1 & 0 \end{array}
$$

The quotient is $x^3 - 2x^2 + 3x - 1$ and the remainder is 0. This result is expressed by

$$\frac{x^4 - 5x^3 + 9x^2 - 10x + 3}{x - 3} = x^3 - 2x^2 + 3x - 1 + \frac{0}{x - 3}, \quad \text{or}$$

$$\frac{x^4 - 5x^3 + 9x^2 - 10x + 3}{x - 3} = x^3 - 2x^2 + 3x - 1.$$

■ **Now Try Exercise 3**

EXAMPLE 2 **Performing synthetic division**

Use synthetic division to divide $2x^3 - x + 5$ by $x + 1$.

Solution
Write $2x^3 - x + 5$ as $2x^3 + 0x^2 - x + 5$. The divisor $x + 1$ can be written as

$$x + 1 = x - (-1),$$

so we let $k = -1$.

$$
\begin{array}{r|rrrr}
-1 & 2 & 0 & -1 & 5 \\
   &   & -2 & 2 & -1 \\
\hline
   & 2 & -2 & 1 & 4
\end{array}
$$

The remainder is 4, and the quotient is $2x^2 - 2x + 1$. This result can also be expressed as

$$\frac{2x^3 - x + 5}{x + 1} = 2x^2 - 2x + 1 + \frac{4}{x + 1}.$$

■ **Now Try Exercise 13**

D   Exercises

Exercises 1–16: Use synthetic division to divide.

1. $\dfrac{x^2 + 3x - 1}{x - 1}$

2. $\dfrac{2x^2 + x - 1}{x - 3}$

3. $(3x^2 - 22x + 7) \div (x - 7)$

4. $(5x^2 + 29x - 6) \div (x + 6)$

5. $\dfrac{x^3 + 7x^2 + 14x + 8}{x + 4}$

6. $\dfrac{2x^3 + 3x^2 + 2x + 4}{x + 1}$

7. $\dfrac{2x^3 + x^2 - 1}{x - 2}$

8. $\dfrac{x^3 + x - 2}{x + 3}$

9. $(x^3 - 2x^2 - 2x + 4) \div (x - 4)$

10. $(2x^4 + 3x^2 - 4) \div (x + 2)$

11. $(x^3 - 3x^2 - 8x - 10) \div (x - 5)$

12. $(x^4 - 1) \div (x + 1)$

13. $(2x^4 - x) \div (x + 2)$

14. $(2x^3 + x - 1) \div (x - 3)$

15. $(b^4 - 1) \div (b - 1)$

16. $(a^2 + a) \div (a + 2.5)$

Answers to Selected Exercises

1 Introduction to Algebra

SECTION 1.1 (pp. 9–11)

1. natural **2.** 0 **3.** 1 **4.** composite **5.** factors
6. formula **7.** variable **8.** equals sign **9.** sum
10. product **11.** quotient **12.** difference
13. Composite; $4 = 2 \cdot 2$ **15.** Neither **17.** Prime
19. Composite; $92 = 2 \cdot 2 \cdot 23$
21. Composite; $225 = 3 \cdot 3 \cdot 5 \cdot 5$ **23.** Prime
25. $6 = 2 \cdot 3$ **27.** $12 = 2 \cdot 2 \cdot 3$
29. $32 = 2 \cdot 2 \cdot 2 \cdot 2 \cdot 2$ **31.** $39 = 3 \cdot 13$
33. $294 = 2 \cdot 3 \cdot 7 \cdot 7$ **35.** $300 = 2 \cdot 2 \cdot 3 \cdot 5 \cdot 5$
37. Yes **39.** No **41.** Yes **43.** No **45.** 15 **47.** 5 **49.** 4
51. 18 **53.** 4 **55.** 22 **57.** 12 **59.** 9 **61.** 5 **63.** 28
65. 7 **67.** 5 **69.** 0 **71.** 3 **73.** $x + 5$; x is the number.
75. $3s$; s is the cost of soda. **77.** $n + 5$; n is the number.
79. $p - 200$; p is the population. **81.** $\frac{z}{6}$; z is the number.
83. st; s is the speed and t is the time.
85. $\frac{x + 7}{y}$; x is one number, y is the other number.
87. $P = 100D$
89.

Yards (y)	1	2	3	4	5	6	7
Feet (F)	3	6	9	12	15	18	21

$F = 3y$

91. $M = 3x$; 108 mi **93.** $B = 6D$ **95.** $C = 12x$
97. 198 square feet

SECTION 1.2 (pp. 23–26)

1. $\frac{3}{4}; \frac{1}{4}$ **2.** 11; 21 **3.** 0 **4.** lowest **5.** True **6.** False
7. $\frac{a}{b}$ **8.** multiply **9.** $\frac{1}{a}$ **10.** $\frac{1}{5}$ **11.** $\frac{ac}{bd}$ **12.** $\frac{ad}{bc}$ **13.** $\frac{a + c}{b}$
14. $\frac{a - c}{b}$ **15.** 4 **17.** 25 **19.** 10 **21.** $\frac{3}{5}$ **23.** $\frac{3}{5}$ **25.** $\frac{1}{2}$
27. $\frac{2}{5}$ **29.** $\frac{1}{3}$ **31.** $\frac{2}{5}$ **33.** $\frac{1}{4}$ **35.** $\frac{3}{20}$ **37.** 1 **39.** $\frac{3}{5}$
41. $\frac{12}{5}$ **43.** $\frac{3}{4}$ **45.** 1 **47.** $\frac{3a}{2b}$ **49.** $\frac{3}{16}$ **51.** 4 **53.** $\frac{1}{3}$
55. (a) $\frac{1}{5}$ (b) $\frac{1}{7}$ (c) $\frac{7}{4}$ (d) $\frac{8}{9}$
57. (a) 2 (b) 9 (c) $\frac{101}{12}$ (d) $\frac{17}{31}$
59. $\frac{3}{2}$ **61.** 6 **63.** 1 **65.** $\frac{4}{3}$ **67.** 12 **69.** $\frac{3}{10}$ **71.** $\frac{a}{2}$
73. 1 **75.** (a) $\frac{1}{2}$ (b) $\frac{1}{3}$ **77.** (a) $\frac{25}{29}$ (b) $\frac{11}{29}$ **79.** 45
81. 15 **83.** 24 **85.** 12 **87.** 24 **89.** $\frac{3}{6}, \frac{4}{6}$ **91.** $\frac{28}{36}, \frac{15}{36}$
93. $\frac{3}{48}, \frac{28}{48}$ **95.** $\frac{4}{12}, \frac{9}{12}, \frac{10}{12}$ **97.** $\frac{13}{16}$ **99.** $\frac{1}{6}$ **101.** $\frac{59}{70}$
103. $\frac{13}{36}$ **105.** $\frac{17}{600}$ **107.** $\frac{9}{8}$ **109.** $4\frac{3}{4}$ ft **111.** $32\frac{5}{16}$ inches
113. $\frac{5}{8}$ yd^2 **115.** $8\frac{1}{4}$ miles **117.** $\frac{1}{6250}$ **119.** $\frac{961}{1260}$

CHECKING BASIC CONCEPTS 1.1 & 1.2 (p. 26)

1. (a) Prime (b) Composite; $28 = 2 \cdot 2 \cdot 7$
(c) Neither (d) Composite; $180 = 2 \cdot 2 \cdot 3 \cdot 3 \cdot 5$
2. 2 **3.** 30 **4.** $x + 5$ **5.** $I = 12F$ **6.** (a) 3 (b) 8
7. (a) $\frac{5}{7}$ (b) $\frac{2}{3}$ **8.** $\frac{3}{4}$ **9.** (a) $\frac{1}{2}$ (b) $\frac{1}{4}$ (c) $\frac{2}{5}$ (d) $\frac{7}{12}$
10. $3\frac{1}{3}$ cups

SECTION 1.3 (pp. 32–33)

1. multiply **2.** 2 **3.** base; exponent **4.** 6^2 **5.** 8^3
6. 17; multiplication; addition **7.** 2; exponents; subtraction
8. 4; left; right **9.** False **10.** False **11.** 3^4 **13.** 2^5
15. $\left(\frac{1}{2}\right)^4$ **17.** a^5 **19.** $(x + 3)^2$ **21.** (a) 16 (b) 16
23. (a) 6 (b) 1 **25.** (a) 32 (b) 1000 **27.** (a) $\frac{4}{9}$ (b) $\frac{1}{32}$
29. 2^3 **31.** 5^2 **33.** 7^2 **35.** 10^3 **37.** $\left(\frac{1}{2}\right)^4$ **39.** $\left(\frac{2}{3}\right)^5$
41. 29 **43.** 4 **45.** 90 **47.** 3 **49.** 2 **51.** 19 **53.** 32
55. 125 **57.** 3 **59.** 4 **61.** 80 **63.** $\frac{49}{16}$ **65.** $2^3 - 8$; 0
67. $30 - 4 \cdot 3$; 18 **69.** $\frac{4^2}{2^3}$; 2 **71.** $\frac{40}{10} + 2$; 6
73. $100(2 + 3)$; 500 **75.** 536,870,912 bytes
77. (a) $k = 7$ (b) 32 **79.** (a) 8 yr (b) $80,000

SECTION 1.4 (pp. 42–44)

1. $-b$ **2.** natural **3.** rational **4.** real **5.** irrational
6. True **7.** True **8.** $0.\overline{27}$ **9.** 1; 4 **10.** 4 **11.** principal
12. not equal **13.** approximately equal **14.** 0 **15.** left
16. origin **17.** (a) -9 (b) 9 **19.** (a) $-\frac{2}{3}$ (b) $\frac{2}{3}$
21. (a) -8 (b) 8 **23.** (a) $-a$ (b) a **25.** 6 **27.** $\frac{1}{2}$
29. 0.25 **31.** 0.875 **33.** 1.5 **35.** 0.05 **37.** $0.\overline{6}$ **39.** $0.\overline{7}$
41. Natural, whole, integer, and rational
43. Natural, whole, integer, and rational
45. Whole, integer, and rational **47.** Rational **49.** 5
51. 7 **53.** 2.646 **55.** Rational **57.** Rational
59. Irrational **61.** Natural, integer, and rational
63. Natural, integer, and rational **65.** Rational
67.
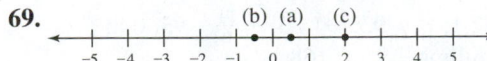

69.

71.

73.

75. 5.23 **77.** 8 **79.** 0 **81.** $\pi - 3$ **83.** $-b$ **85.** <
87. > **89.** > **91.** < **93.** > **95.** < **97.** > **99.** =
101. $-9, -2^3, -3, 0, 1$ **103.** $-2, -\frac{3}{2}, \frac{1}{3}, \sqrt{5}, \pi$
105. (a) 25.5 **(b)** Answers may vary.
(c) 25.45; answers may vary.

CHECKING BASIC CONCEPTS 1.3 & 1.4 (p. 44)

1. (a) 5^4 **(b)** 7^5 **2. (a)** 8 **(b)** 10,000 **(c)** $\frac{8}{27}$
3. (a) 4^3 **(b)** 2^6 **4. (a)** 26 **(b)** 9 **(c)** 2 **(d)** $\frac{1}{2}$ **(e)** 4
(f) 0 **5.** $5^3 \div 3$, or $\frac{5^3}{3}$ **6. (a)** 17 **(b)** $-a$
7. (a) 0.15 **(b)** 0.625
8. (a) Natural, integer, and rational **(b)** Integer and rational
(c) Irrational **(d)** Rational
9.

10. (a) 12 **(b)** 0 **11. (a)** < **(b)** < **(c)** <
12. $-7, -1.6, 0, \frac{1}{3}, \sqrt{3}, 3^2$

SECTION 1.5 (pp. 50–51)

1. sum **2.** zero **3.** True **4.** True **5.** absolute value
6. difference **7.** addition **8.** opposite; $(-b)$ **9.** addition
10. subtraction **11.** $-25; 0$ **13.** $\sqrt{21}; 0$ **15.** $-5.63; 0$
17. 4 **19.** 2 **21.** -3 **23.** 4 **25.** 2 **27.** -1 **29.** 10
31. -150 **33.** 1 **35.** -7 **37.** $\frac{1}{4}$ **39.** $-\frac{9}{14}$ **41.** -1.1
43. 34 **45.** -3 **47.** 7 **49.** $-\frac{1}{14}$ **51.** $\frac{1}{2}$ **53.** 2.9
55. -164 **57.** -9 **59.** 42 **61.** -5 **63.** 50 **65.** 8.8
67. $\frac{1}{2}$ **69.** -1 **71.** $2 + (-5); -3$ **73.** $-5 + 7; 2$
75. $-(2^3); -8$ **77.** $-6 - 7; -13$ **79.** $6 + (-10) - 5; -9$
81. Mt. Whitney: 14,497 ft; Death Valley: -282 ft; 14,779 ft
83. $230 **85.** 25 yards **87.** 64,868 feet

SECTION 1.6 (pp. 58–59)

1. product **2.** negative **3.** positive **4.** quotient **5.** $\frac{1}{a}$
6. reciprocal **7.** reciprocal, or multiplicative inverse
8. $-a$ **9.** $\frac{1}{b}$ **10.** positive **11.** negative **12.** 5; 8
13. -12 **15.** -18 **17.** 0 **19.** 60 **21.** $\frac{1}{4}$ **23.** -1
25. 200 **27.** -50 **29.** 120 **31.** $-\frac{3}{2}$ **33.** Negative
35. 25 **37.** -1 **39.** -16 **41.** 8 **43.** -40 **45.** -2
47. 10 **49.** -4 **51.** -32 **53.** 0 **55.** Undefined
57. $-\frac{1}{22}$ **59.** $\frac{4}{15}$ **61.** $-\frac{15}{16}$ **63.** Undefined **65.** -1
67. $-\frac{4}{3}$ **69.** 0.5 **71.** 0.1875 **73.** 3.5 **75.** $5.\overline{6}$
77. 1.4375 **79.** 0.875 **81.** $\frac{1}{4}$ **83.** $\frac{4}{25}$ **85.** $\frac{5}{8}$ **87.** $\frac{11}{16}$
89. $2.\overline{3}; \frac{7}{3}$ **91.** $2.1; \frac{21}{10}$ **93.** $-1.8\overline{3}; -\frac{11}{6}$ **95.** $0.8; \frac{4}{5}$
97. About 131 million **99.** 0.168

CHECKING BASIC CONCEPTS 1.5 & 1.6 (p. 60)

1. (a) 0 **(b)** -19 **2. (a)** $\frac{8}{9}$ **(b)** -3.2
3. (a) $-1 + 5; 4$ **(b)** $4 - (-3); 7$ **4.** 145°F
5. (a) 35 **(b)** $\frac{4}{15}$ **6. (a)** -9 **(b)** -32 **(c)** 25

7. (a) $-\frac{15}{2}$ **(b)** $\frac{15}{32}$ **8.** $-\frac{6}{7}$ **9. (a)** -5 **(b)** -5 **(c)** -5
(d) 5 **10. (a)** 0.6 **(b)** 3.875

SECTION 1.7 (pp. 68–71)

1. commutative; addition **2.** commutative; multiplication
3. associative; addition **4.** associative; multiplication
5. True **6.** False **7.** distributive **8.** distributive
9. identity; addition **10.** identity; multiplication
11. $-a$ **12.** $\frac{1}{a}$ **13.** $10 + (-6)$ **15.** $6 \cdot (-5)$
17. $10 + a$ **19.** $7b$ **21.** $1 + (2 + 3)$ **23.** $(2 \cdot 3) \cdot 4$
25. $a + (5 + c)$ **27.** $x \cdot (3 \cdot 4)$
29. $a + b + c = (a + b) + c$
$\qquad\qquad\quad = c + (a + b)$
$\qquad\qquad\quad = c + (b + a)$
$\qquad\qquad\quad = c + b + a$
31. 20 **33.** $ab - 8a$ **35.** $-4t + 4z$ **37.** $-5 + a$
39. $3a + 15$ **41.** $17 - a$
43. $a \cdot (b + c + d) = a \cdot ((b + c) + d)$
$\qquad\qquad\qquad\quad = a \cdot (b + c) + ad$
$\qquad\qquad\qquad\quad = ab + ac + ad$
45. $11x$ **47.** $-b$ **49.** $2a$ **51.** $-14w$
53. Commutative (multiplication) **55.** Associative (addition)
57. Distributive **59.** Distributive, commutative (multiplication) **61.** Distributive **63.** Associative (multiplication)
65. Distributive **67.** Identity (addition)
69. Identity (multiplication) **71.** Identity (multiplication)
73. Inverse (multiplication) **75.** Inverse (addition)
77. 30 **79.** 100 **81.** 178 **83.** 79 **85.** 90 **87.** 816
89. 1 **91.** $\frac{1}{6}$ **93. (a)** 410 **(b)** 9970 **(c)** -6300
(d) $-140,000$ **95. (a)** 19,000 **(b)** $-45,100$ **(c)** 60,000
(d) $-7,900,000$ **97. (a)** 1.256 **(b)** 0.96 **(c)** 0.0987
(d) -0.0056 **(e)** 120 **(f)** 457.8
99. Commutative (addition) **101.** 19.8 miles
103. (a) $13 \cdot (5 \cdot 2)$; multiplying by 10 is easy.
(b) Associative (multiplication)

SECTION 1.8 (pp. 77–78)

1. term **2.** coefficient **3.** factors; terms **4.** like **5.** like
6. distributive **7.** Yes; 91 **9.** Yes; -6 **11.** No
13. Yes; 1 **15.** No **17.** Like **19.** Like **21.** Unlike
23. Unlike **25.** Like **27.** Unlike **29.** Like **31.** $3x$
33. $14y$ **35.** $41a$ **37.** 0 **39.** Not possible
41. Not possible **43.** $3x^2$ **45.** $-y$ **47.** $3x + 2$
49. $-2z + \frac{1}{2}$ **51.** $11y$ **53.** $4z - 1$ **55.** $12y - 7z$
57. $-3x - 4$ **59.** $-6x - 1$ **61.** $-\frac{1}{3}x + \frac{2}{3}$
63. $\frac{2}{5}x + \frac{3}{5}y + \frac{1}{5}$ **65.** $0.4x^2$ **67.** $7x^2 - 7x$ **69.** $2b$
71. $7x^3 + 2y$ **73.** x **75.** 3 **77.** z **79.** $3x - 2$
81. $2z + 3$ **83.** $5x + 6x; 11x$ **85.** $x^2 + 2x^2; 3x^2$
87. $6x - 4x; 2x$ **89. (a)** $1570w$ **(b)** 65,940 square feet
91. (a) $50x$ **(b)** 2400 cubic feet **(c)** 24 minutes

CHECKING BASIC CONCEPTS 1.7 & 1.8 (p. 79)

1. (a) $18y$ **(b)** $x + 10$ **2.** $20y$ **3. (a)** $5 - x$
(b) $5x - 35$ **4.** Distributive **5.** 0 **6. (a)** 60 **(b)** 7
(c) 368 **7. (a)** Unlike **(b)** Like **8. (a)** $14z$ **(b)** $-3y + 3$

9. (a) $-3y - 3$ **(b)** $-14y$ **(c)** x **(d)** 35
10. $3x + 5x$; $8x$

CHAPTER 1 REVIEW (pp. 84–86)

1. Prime **2.** Composite; $27 = 3 \cdot 3 \cdot 3$
3. Composite; $108 = 2 \cdot 2 \cdot 3 \cdot 3 \cdot 3$
4. Composite; $91 = 7 \cdot 13$ **5.** Neither **6.** Neither
7. 3 **8.** 5 **9.** 12 **10.** 6 **11.** 7 **12.** 7 **13.** 8
14. 6 **15.** $3^2 + 5$ **16.** $2^3 \div (3 + 1)$
17. $3x$, where x is the number
18. $x - 4$, where x is the number **19.** 5 **20.** 6
21. (a) $\frac{5}{8}$ **(b)** $\frac{3}{4}$ **22. (a)** $\frac{3}{4}$ **(b)** $\frac{3}{5}$ **23.** $\frac{5}{8}$ **24.** $\frac{2}{9}$ **25.** $\frac{3}{5}$
26. $\frac{24}{23}$ **27.** $\frac{5}{2}$ **28.** 6 **29.** 2 **30.** $\frac{3x}{2y}$ **31.** $\frac{3}{35}$
32. (a) $\frac{1}{8}$ **(b)** 1 **(c)** $\frac{19}{5}$ **(d)** $\frac{2}{3}$ **33.** 9 **34.** $\frac{9}{14}$ **35.** 12
36. $\frac{1}{8}$ **37.** $\frac{x}{3}$ **38.** $\frac{20}{27y}$ **39.** 24 **40.** 42 **41.** $\frac{1}{3}$ **42.** $\frac{1}{2}$
43. $\frac{19}{24}$ **44.** $\frac{9}{22}$ **45.** $\frac{5}{12}$ **46.** $\frac{13}{18}$ **47.** 5^6 **48.** $\left(\frac{7}{6}\right)^3$
49. x^5 **50.** 3^4 **51.** $(x + 1)^2$ **52.** $(a - 5)^3$
53. (a) 64 **(b)** 49 **(c)** 8 **54.** 5 **55.** 25 **56.** 7 **57.** 3
58. 25 **59.** 1 **60.** 2 **61.** 0 **62.** 0 **63.** 10 **64.** 1 **65.** 5
66. 19 **67. (a)** 8 **(b)** 3 **68. (a)** $-\frac{3}{7}$ **(b)** $-\frac{2}{5}$
69. (a) 0.8 **(b)** 0.15 **70. (a)** $0.\overline{5}$ **(b)** $0.\overline{63}$
71. Whole, integer, and rational **72.** Rational
73. Integer and rational **74.** Irrational **75.** Irrational
76. Rational
77.

78. (a) 5 **(b)** π **(c)** 0 **79. (a)** $<$ **(b)** $>$ **(c)** $<$ **(d)** $>$
80. $-3, -\frac{2}{3}, \sqrt{3}, \pi - 1, 3$ **81.** 4 **82.** -3 **83.** 1
84. -5 **85.** 1 **86.** -2 **87.** -44 **88.** 40 **89.** $-\frac{11}{4}$
90. -7 **91.** $-\frac{2}{9}$ **92.** $-\frac{5}{4}$ **93.** $\frac{2}{7}$ **94.** $-\frac{4}{35}$ **95.** 4
96. $-\frac{3}{4}$ **97.** $3 + (-5)$; -2 **98.** $2 - (-4)$; 6 **99.** $0.\overline{7}$
100. 2.2 **101.** $\frac{3}{5}$ **102.** $\frac{3}{8}$ **103.** $16 + 3$ **104.** $-x \cdot 14$
105. $(-4 + 1) + 3$ **106.** $x \cdot (y \cdot 5)$ **107.** $5x + 60$
108. $-a + 3$ **109.** Identity (addition)
110. Identity (multiplication) **111.** Inverse (multiplication)
112. Inverse (addition) **113.** Commutative (multiplication)
114. Associative (addition) **115.** Distributive
116. Commutative (addition) **117.** Identity (multiplication)
118. Associative (multiplication) **119.** Distributive
120. Identity (addition) **121.** Inverse (addition)
122. Inverse (multiplication) **123.** 40 **124.** 301
125. 2475 **126.** 6580 **127.** 549.8 **128.** 43.56
129. Yes; 55 **130.** Yes; -1 **131.** No **132.** No
133. $-6x$ **134.** $15z$ **135.** $4x^2$ **136.** $3x + 1$
137. $\frac{1}{2}z + 2$ **138.** $x - 18$ **139.** $9x^2 - 6$ **140.** 0
141. 5 **142.** c **143.** $3y + 2$ **144.** $2x - 5$
145. (a) $7x$ **(b)** 420 square feet **(c)** 24 minutes
146. 16 square feet
147.

Gallons (G)	1	2	3	4	5	6
Pints (P)	8	16	24	32	40	48

$P = 8G$

148. $C = 0.05x$ **149.** $\frac{3}{20}$ **150.** \$200,000 **151.** $1\frac{3}{20}$ feet
152. $15\frac{3}{8}$ miles **153.** \$1101 **154.** $124°$F
155. About 129 million

CHAPTER 1 TEST (p. 87)

1. (a) Prime **(b)** Composite; $56 = 2 \cdot 2 \cdot 2 \cdot 7$ **2.** $\frac{15}{7}$
3. $4^2 - 3$; 13 **4.** $\frac{3}{4}$ **5.** $\frac{1}{30}$ **6.** 24 **7. (a)** $\frac{3}{4}$ **(b)** $\frac{16}{45}$
(c) $\frac{2}{7}$ **(d)** $\frac{15}{4}$ **(e)** $\frac{31}{36}$ **(f)** $\frac{2}{13}$ **8.** y^4 **9. (a)** 8 **(b)** 71
(c) -40 **(d)** 9 **10.** 0.35 **11. (a)** Integer and rational
(b) Irrational
12.

13. (a) $<$ **(b)** $>$ **14. (a)** -6 **(b)** 21 **15.** $\frac{3}{4}$
16. (a) Distributive **(b)** Associative (multiplication)
(c) Commutative (addition) **17.** 1734 **18. (a)** $12 - 4z$
(b) $15x - 6$ **(c)** $x - 19$ **19. (a)** $\frac{19}{12}x$ **(b)** $12\frac{2}{3}$ acres
20. $2\frac{3}{5}$ feet **21. (a)** $C = 13x$ **(b)** \$221
22. \$640

2 Linear Equations and Inequalities

SECTION 2.1 (PP. 97–99)

1. solution **2.** true **3.** false **4.** solution set **5.** solutions
6. equivalent **7.** $b + c$ **8.** subtraction **9.** bc
10. division **11.** equivalent **12.** given **13.** 22 **15.** 3
17. -5 **19.** 9 **21.** 17 **23.** $\frac{17}{10}$ **25.** 5.1 **27.** -3 **29.** $\frac{2}{3}$
31. a **33.** $\frac{1}{5}$ **35.** 6 **37.** 3 **39.** 0 **41.** 7 **43.** -6
45. 3 **47.** $\frac{5}{4}$ **49.** 7 **51.** -8.5 **53.** a
55. (a)

Hours (x)	0	1	2	3	4	5	6
Rainfall (R)	3	3.5	4	4.5	5	5.5	6

(b) $R = 0.5x + 3$ **(c)** 4.5 inches; yes **(d)** 4.125 inches
57. (a) $L = 300x$ **(b)** $870 = 300x$ **(c)** 2.9 **59.** 50 days
61. (a) $50°$ N **(b)** 5.7 **63.** \$25,000

SECTION 2.2 (pp. 108–110)

1. constant **2.** $ax + b = 0$ **3.** Exactly one
4. numerically **5.** Addition, multiplication **6.** LCD
7. None **8.** Infinitely many **9.** Yes; $a = 3, b = -7$
11. Yes; $a = \frac{1}{2}, b = 0$ **13.** No **15.** No **17.** Yes;
$a = 1.1, b = -0.9$ **19.** Yes; $a = 2, b = -6$ **21.** No
23.

x	-1	0	1	2	3
$x - 3$	-4	-3	-2	-1	0

2

25.

x	0	1	2	3	4
$-3x + 7$	7	4	1	-2	-5

2

27.

x	-2	-1	0	1	2
$4 - 2x$	8	6	4	2	0

-1

29. $\frac{3}{11}$ **31.** 23 **33.** 28 **35.** $-\frac{11}{5}$ **37.** $-\frac{7}{2}$ **39.** 3 **41.** $\frac{9}{4}$
43. $-\frac{5}{2}$ **45.** $\frac{10}{3}$ **47.** $\frac{9}{8}$ **49.** 1 **51.** -0.055 **53.** 8 **55.** $\frac{1}{3}$
57. 1 **59.** $-\frac{b}{a}$ **61.** No solutions **63.** One solution **65.** No
solutions **67.** Infinitely many solutions **69.** No solutions
71. (a)

Hours (x)	0	1	2	3	4
Distance (D)	4	12	20	28	36

(b) $D = 8x + 4$ **(c)** 28 miles; yes **(d)** 2.25 hours; the
bicyclist is 22 miles from home after 2 hours and 15 minutes.
73. 2009 **75.** 2004 **77.** 2008

CHECKING BASIC CONCEPTS 2.1 & 2.2 (p. 110)

1. (a) No **(b)** Yes

2.

x	3	3.5	4	4.5	5
$4x - 3$	9	11	13	15	17

4

3. (a) 18 **(b)** $\frac{1}{6}$ **(c)** $2.\overline{6}$ or $\frac{8}{3}$ **(d)** 3 **4. (a)** One solution
(b) Infinitely many solutions **(c)** No solutions
5. (a) $D = 300 - 75x$ **(b)** $0 = 300 - 75x$ **(c)** 4 hours

SECTION 2.3 (pp. 120–123)

1. Check your solution. **2.** + **3.** = **4.** $n + 1$ and
$n + 2$ **5.** $\frac{x}{100}$ **6.** 0.01 **7.** left **8.** right **9.** $\frac{B - A}{A} \cdot 100$
10. increase; decrease **11.** rt **12.** distance; time
13. $2 + x = 12$; 10 **15.** $\frac{x}{5} = x - 24$; 30
17. $\frac{x + 5}{2} = 7$; 9 **19.** $\frac{x}{2} = 17$; 34 **21.** 31, 32, 33
23. 34 **25.** 8 **27.** 21 **29.** 7 **31.** 210 lb **33.** 13
35. 40 **37.** 1812 million **39.** 37,000 **41.** 24
43. Length: 19 inches; width: 12 inches **45.** Facebook:
60 million; MySpace: 110 million **47.** 2003: 150,000;
2010: 98,000 **49.** $\frac{37}{100}$; 0.37 **51.** $\frac{37}{25}$; 1.48
53. $\frac{69}{1000}$; 0.069 **55.** $\frac{1}{2000}$; 0.0005 **57.** 45%
59. 180% **61.** 0.6% **63.** 40% **65.** 75% **67.** $83.\overline{3}\%$
69. About 53.3% **71.** $988 per credit **73.** About
35.8 million acres **75.** 25%; -20% **77.** $d = 8$ miles
79. $r = 20$ feet per second **81.** $t = 5$ hours
83. 60 mph **85.** $\frac{3}{8}$ hour **87.** 0.6 hr at 5 mph; 0.5 hr
at 8 mph **89.** 3.75 hours **91.** 30 ounces **93.** $1500
at 6%; $2500 at 5% **95.** 6 gallons **97.** 10 grams

SECTION 2.4 (pp. 133–135)

1. formula **2.** lw **3.** $\frac{1}{2}bh$ **4.** $\frac{1}{360}$ **5.** 360 **6.** 180
7. lwh **8.** $2lw + 2wh + 2lh$ **9.** $2\pi r$ **10.** πr^2
11. $\pi r^2 h$ **12.** $\frac{1}{2}(a + b)h$ **13.** 18 ft^2 **15.** 9 in^2
17. $16\pi \approx 50.3$ cm^2 **19.** 11 mm^2 **21.** 91 in^2
23. 105 ft^2 or $\frac{35}{3}$ yd^2 **25.** $8\pi \approx 25.1$ in. **27.** 4602 ft^2
29. 65° **31.** 81° **33.** 36° **35.** 36°, 36°, 108°
37. $C = 12\pi \approx 37.7$ in.; $A = 36\pi \approx 113.1$ in^2
39. $r = 1$ in.; $A = \pi \approx 3.14$ in^2
41. $V = 2640$ in^3; $S = 1208$ in^2
43. $V = 2$ ft^3; $S = \frac{32}{3}$ ft^2 **45.** 20π in^3 **47.** 600π in^3
49. (a) $\frac{45}{32}\pi \approx 4.4$ in^3 **(b)** About 2.4 fl oz
51. $y = -3x + 2$ **53.** $y = -\frac{4}{3}x + 4$ **55.** $w = \frac{A}{l}$
57. $h = \frac{V}{\pi r^2}$ **59.** $a = \frac{2A}{h} - b$ **61.** $w = \frac{V}{lh}$
63. $b = 2s - a - c$ **65.** $b = a - c$ **67.** $a = \frac{cd}{b - d}$
69. 15 in. **71.** 2.98 **73.** 2.24 **75.** 77°F **77.** -40°F
79. -5°C **81.** -20°C **83.** 11 mpg **85.** 2.4 mi

CHECKING BASIC CONCEPTS 2.3 & 2.4 (p. 135)

1. (a) $3x = 36$; 12 **(b)** $35 - x = 43$; -8
2. $-32, -31, -30$ **3.** 0.095 **4.** 125%
5. About 75,747 **6.** 6.5 hr **7.** $5000 at 6%; $3000
at 7% **8.** 18 gal **9.** 12 in.
10. $A = 9\pi \approx 28.3$ ft^2; $C = 6\pi \approx 18.8$ ft **11.** 30°
12. $l = \frac{A - \pi r^2}{\pi r}$

SECTION 2.5 (pp. 146–148)

1. $<, \leq, >, \geq$ **2.** greater than; less than **3.** solution
4. equivalent **5.** True **6.** True **7.** number line **8.** $>$
9. $<$ **10.** $>$

11. <number line: open at 1, arrow right>

13. <number line: closed at 0, arrow right>

15. <number line: closed at 1, arrow left>

17. <number line: bracket at -2, closed to right>

19. $x < 0$ **21.** $x \leq 3$ **23.** $x \geq 10$ **25.** $[6, \infty)$
27. $(-2, \infty)$ **29.** $(-\infty, 7]$ **31.** Yes **33.** No **35.** Yes
37. No **39.** $x > -2$ **41.** $x < 1$

43.

x	1	2	3	4	5
$-2x + 6$	4	2	0	-2	-4

$x \geq 3$

45.

x	-3	-2	-1	0	1
$5 - x$	8	7	6	5	4
$x + 7$	4	5	6	7	8

$x < -1$

47. $x > 3$; <number line>

49. $y \geq -2$; <number line>

51. $z > 8$; <number line>

53. $t \leq -5$; <number line>

55. $x < 5$; <number line>

57. $t \leq -2$; <number line>

59. $y > -\frac{3}{20}$; <number line>

61. $z \geq -\frac{14}{3}$; <number line>

63. $\{x | x > 1\}$ **65.** $\{x | x \geq -7\}$ **67.** $\{x | x < 6\}$
69. $x < 7$ **71.** $x \leq -\frac{4}{3}$ **73.** $x > -\frac{39}{2}$ **75.** $x \leq \frac{3}{2}$
77. $x > \frac{1}{2}$ **79.** $x \leq \frac{5}{4}$ **81.** $x > -\frac{8}{3}$ **83.** $x \leq -\frac{1}{3}$
85. $x < 5$ **87.** $x \leq -\frac{3}{2}$ **89.** $x > -\frac{9}{8}$ **91.** $x \leq \frac{4}{7}$
93. $x < \frac{21}{11}$ **95.** $x > 60$ **97.** $x \geq 21$ **99.** $x > 40,000$
101. $x \leq 70$ **103.** Less than 10 feet
105. 86 or more **107.** 4.5 hours **109.** 4 days
111. (a) $C = 1.5x + 2000$ **(b)** $R = 12x$
(c) $P = 10.5x - 2000$ **(d)** 191 or more compact discs
113. (a) After 48 minutes **(b)** After more than 48 minutes
115. Altitudes more than 4.5 miles

CHECKING BASIC CONCEPTS 2.5 (p. 149)

1.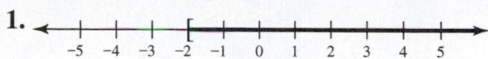

2. $x < 1$ **3.** Yes

4.

x	-2	-1	0	1	2
$5 - 2x$	9	7	5	3	1

; $x \geq -1$

5. (a) $x > 3$ **(b)** $x \geq -35$ **(c)** $x \leq -1$
6. $x \leq 12$ **7.** More than 13 inches

CHAPTER 2 REVIEW (pp. 152–155)

1. -6 **2.** 2 **3.** $\frac{9}{4}$ **4.** $-\frac{1}{2}$ **5.** 3 **6.** $-\frac{7}{3}$ **7.** -2.5 **8.** $-\frac{7}{2}$
9. Yes; $a = -4, b = 1$ **10.** No **11.** 2 **12.** 22 **13.** $\frac{27}{5}$
14. 7 **15.** -7 **16.** $-\frac{2}{3}$ **17.** $\frac{9}{4}$ **18.** 0 **19.** $\frac{7}{8}$ **20.** $\frac{11}{4}$
21. No solutions **22.** Infinitely many solutions
23. Infinitely many solutions **24.** One solution
25.

x	0.5	1.0	1.5	2.0	2.5
$-2x + 3$	2	1	0	-1	-2

1.5

26.

x	-2	-1	0	1	2
$-(x+1)+3$	4	3	2	1	0

0

27. $6x = 72$; 12 **28.** $x + 18 = -23$; -41
29. $2x - 5 = x + 4$; 9 **30.** $x + 4 = 3x$; 2 **31.** 9
32. $-52, -51, -50$ **33.** $\frac{17}{20}$; 0.85 **34.** $\frac{7}{125}$; 0.056
35. $\frac{3}{10,000}$; 0.0003 **36.** $\frac{171}{50}$; 3.42 **37.** 89% **38.** 0.5%
39. 230% **40.** 100% **41.** $d = 24$ mi **42.** $d = 3850$ ft
43. $r = 25$ yards per second **44.** $t = \frac{25}{3}$ hr **45.** 7.5 m^2
46. $36\pi \approx 113.1$ ft^2 **47.** 864 in^2, or 6 ft^2 **48.** 40 in.
49. $18\pi \approx 56.5$ ft **50.** $25\pi \approx 78.5$ in^2 **51.** 50°
52. 22.5° **53.** $625\pi \approx 1963.5$ in^3
54. 1620 in^2, or 11.25 ft^2 **55.** 174 in^2 **56.** About 60.6 ft^2
57. $y = 3x - 5$ **58.** $y = -x + 8$ **59.** $y = \frac{z}{2x}$
60. $b = 3S - a - c$ **61.** $b = \frac{12T - 4a}{3}$
62. $c = \frac{ab}{d - b}$ **63.** 2.82 **64.** 3.12 **65.** 59°F **66.** 45°C
67.

68.

69.

70.

71. $x < 3$ **72.** $x \geq -1$ **73.** Yes **74.** No
75.

x	0	1	2	3	4
$5 - x$	5	4	3	2	1

$x < 2$

76.

x	1	1.5	2	2.5	3
$2x - 5$	-3	-2	-1	0	1

$x \leq 2.5$

77. $x > 3$ **78.** $x \geq -5$ **79.** $x \leq -1$ **80.** $x < \frac{23}{3}$
81. $x \leq \frac{1}{9}$ **82.** $x \geq \frac{9}{4}$ **83.** $x < 50$ **84.** $x \leq 45,000$
85. $x \geq 16$ **86.** $x < 1995$ or $x \leq 1994$
87. (a)

Time	12:00	1:00	2:00	3:00	4:00	5:00
Rainfall (R)	2	2.75	3.5	4.25	5	5.75

(b) $R = \frac{3}{4}x + 2$ **(c)** $\frac{23}{4} = 5\frac{3}{4}$ in.; yes **(d)** $\frac{77}{16} = 4\frac{13}{16}$ in.
88. $2125
89. (a)

Hours (x)	1	2	3	4	5
Distance (D)	40	30	20	10	0

(b) $D = 50 - 10x$ **(c)** 20 miles; yes **(d)** 3 hours or less
or from noon to 3:00 P.M.
90. About 33.4% **91.** $\frac{1}{6}$ hr, or 10 min **92.** 33 in. by 23 in.
93. 50 mL **94.** $500 at 3%; $800 at 5% **95.** 25 inches or
less **96.** 74 or more **97.** 6 hr **98. (a)** $C = 85x + 150,000$
(b) $R = 225x$ **(c)** $P = 140x - 150,000$ **(d)** 1071 or
fewer

CHAPTER 2 TEST (pp. 155–156)

1. -6 **2.** $\frac{5}{2}$ **3.** $-\frac{16}{11}$ **4.** $\frac{25}{6}$ **5.** No solutions
6. Infinitely many solutions
7.

x	0	1	2	3	4
$6 - 2x$	6	4	2	0	-2

3

8. $x + (-7) = 6$; 13 **9.** $2x + 6 = x - 7$; -13
10. 111, 112, 113 **11.** $\frac{4}{125}$; 0.032 **12.** 34.5%
13. $r = 50$ ft/sec **14.** 7.5 in^2
15. $C = 30\pi \approx 94.2$ in.; $A = 225\pi \approx 706.9$ in^2
16. 30° **17.** $x = \frac{y - z}{3y}$ **18.** $x = \frac{20R - 4y}{5}$
19.
20. $x > 0$ **21.** $x \leq 6$ **22.** $x > -\frac{1}{7}$
23. (a) $S = 2x + 5$ **(b)** 21 in. **(c)** 17.5 in.
24. 2000 mL **25.** About 26.7%

CHAPTERS 1 AND 2 CUMULATIVE REVIEW (p. 157)

1. Composite; $3 \cdot 3 \cdot 5$ **2.** Prime **3.** $\frac{1}{2}$ **4.** $\frac{1}{9}$ **5.** $\frac{13}{24}$
6. $\frac{13}{15}$ **7.** Integer and rational **8.** Irrational **9.** 3 **10.** 10
11. 15 **12.** 9 **13.** $4x^3$ **14.** $5x + 3$ **15.** 14 **16.** -4
17. $\frac{1}{2}$ **18.** 1 **19.** Infinitely many **20.** No solutions
21. 29, 30, 31 **22.** 0.047 **23.** 17% **24.** 65 mph **25.** 20°
26. $25\pi \approx 78.5$ in^2 **27.** $x = \frac{a + 4}{3y}$
28. $x = 3A - y - z$
29. $x < 1$

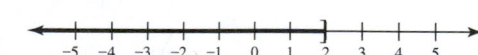

30. $x \leq 2$

31. $I = 36Y$ **32.** $618 **33.** 300 mL
34. 272 billion kilowatt-hours

3   Graphing Equations

SECTION 3.1 (pp. 165–168)

1. xy-plane **2.** origin **3.** $(0, 0)$ **4.** 4 **5.** False **6.** x; y
7. scatterplot **8.** line **9.** $(-2, -2), (-2, 2), (0, 0), (2, 2)$
11. $(-1, 0), (0, -3), (0, 2), (2, 0)$

13. (1, 3): I; (0, −3): None; (−2, 2): II

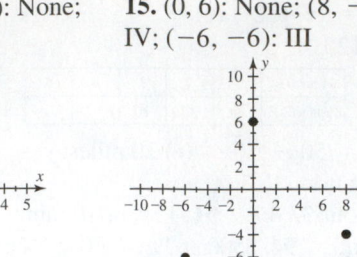

15. (0, 6): None; (8, −4): IV; (−6, −6): III

17. (a) I **(b)** III **19. (a)** None **(b)** I
21. (a) II **(b)** IV **23.** I and III

25.

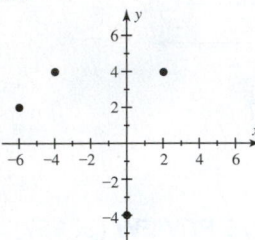

27.

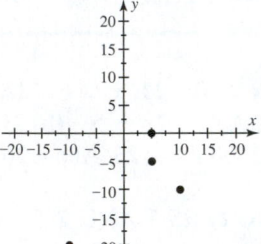

29.

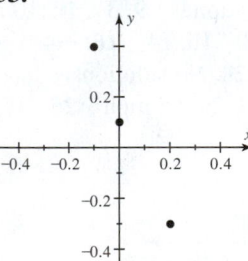

31.

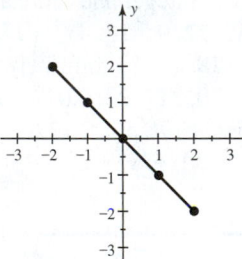

33.

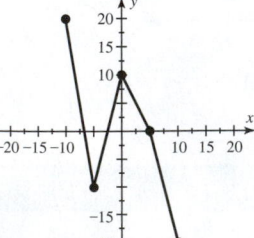

35.

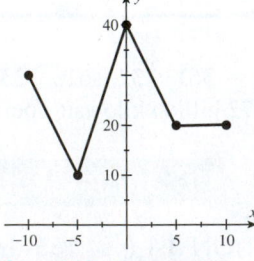

37.

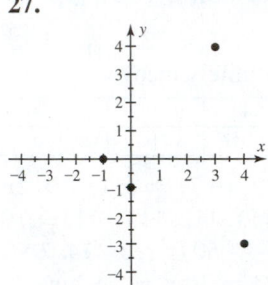

39.

41. (1960, 484), (1980, 632), (2000, 430), (2010, 325); in 1960 there were 484 billion cigarettes consumed in the United States (answers may vary slightly).

43. (a)

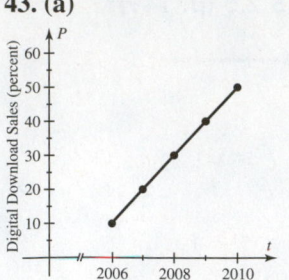

(b) The percentage of digital download sales increased.

45. (a)

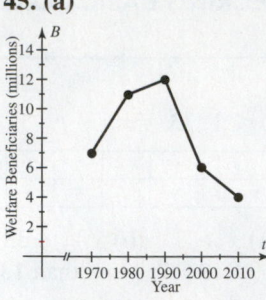

(b) The number of welfare beneficiaries increased and then decreased.

47. (a)

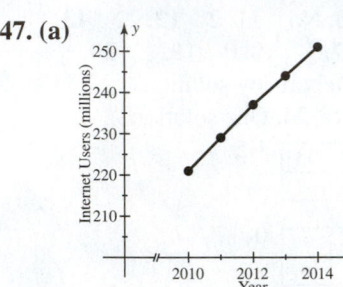

(b) The number of Internet users is increasing.

49. (a) The rate decreased. **(b)** 9 **(c)** About −70%

SECTION 3.2 (pp. 175–177)

1. True **2.** False **3.** ordered pair **4.** linear **5.** False
6. True **7.** graph **8.** line **9.** False **10.** False **11.** Yes
13. No **15.** No **17.** Yes **19.** No

21.

x	−2	−1	0	1	2
y	−8	−4	0	4	8

23.

x	6	3	0	−3	−9
y	−2	0	2	4	8

25.

x	y
−8	−4
−4	0
0	4
4	8
8	12

27.

x	y
−6	−4
0	−2
6	0
12	2
18	4

29.

x	−3	0	3	6
y	−9	0	9	18

31.

x	8	6	4	2
y	−2	0	2	4

33.

x	y
−8	−2
−4	0
0	2
4	4

35.

x	y
$-\frac{1}{2}$	−2
$-\frac{1}{4}$	−1
0	0
$\frac{1}{4}$	1

37. They must be multiples of 5.

39.

x	−1	0	1
y	2	0	−2

Table values may vary.

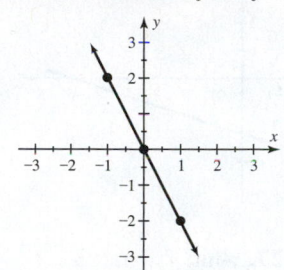

41.

x	0	1	2
y	3	2	1

Table values may vary.

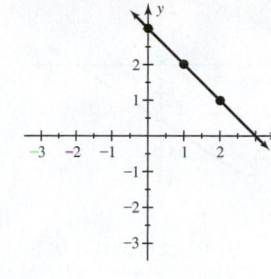

59.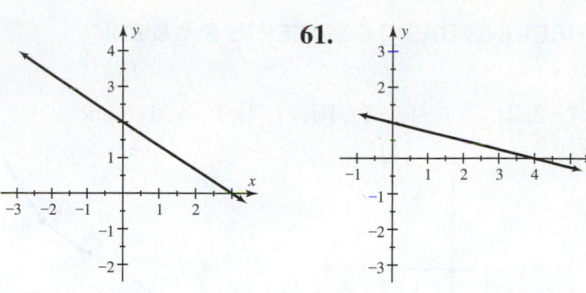

61.

63.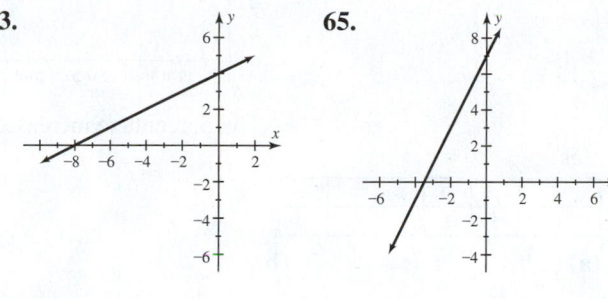

65.

43.

x	−2	0	2
y	3	2	1

Table values may vary.

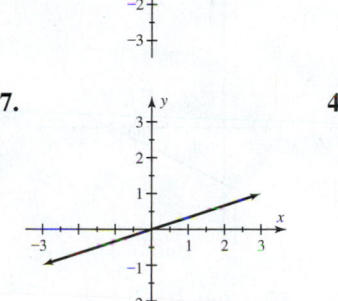

45.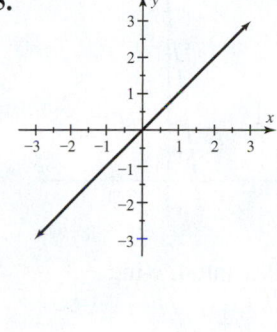

67.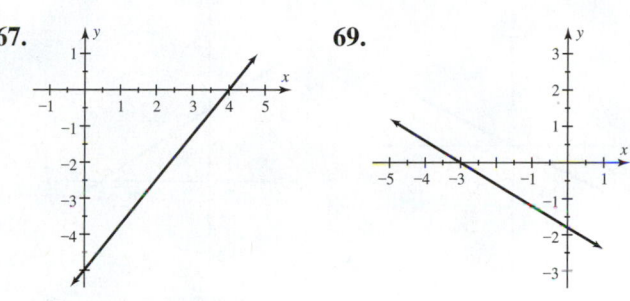

69.

47.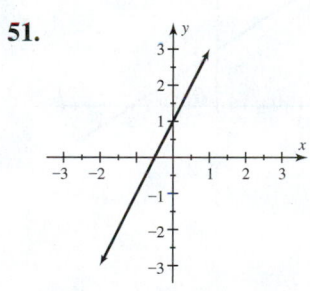

49.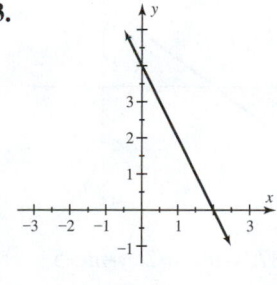

71. (a)

t	2010	2020	2030	2040	2050
P	13.2	15.0	16.7	18.5	20.3

(b) 2030

73. (a)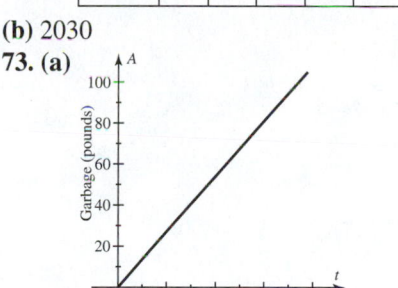

(b) About 37 days

75. (a) 10; 50

51.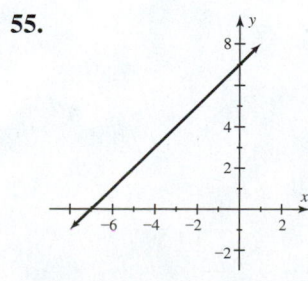

53.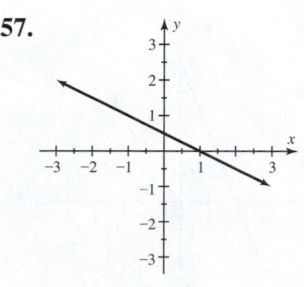

(b)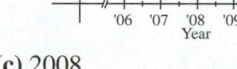

55.

57.

(c) 2008

CHECKING BASIC CONCEPTS 3.1 & 3.2 (p. 178)

1. $(-2, 2)$, II; $(-1, -2)$, III; $(1, 3)$, I; $(3, 0)$, none

2.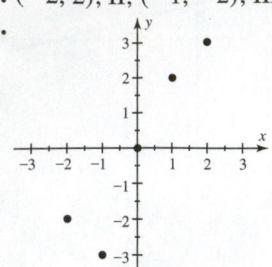

3.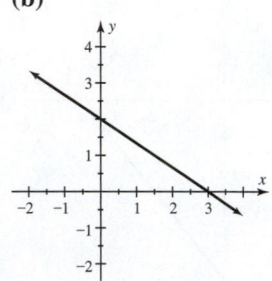

The percentage increased.

4. Yes

5.

x	-2	-1	0	1	2
y	5	3	1	-1	-3

6. (a) 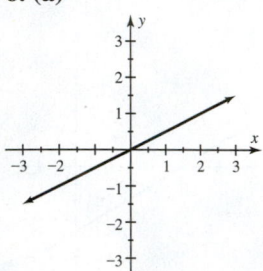 **(b)**

7. (a) In 1990, receipts were \$0.47 trillion; in 2010, receipts were \$2.17 trillion.

(b)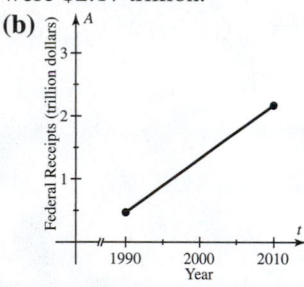

(c) 2006

SECTION 3.3 (pp. 186–189)

1. x-intercept **2.** 0 **3.** y-intercept **4.** 0 **5.** horizontal
6. $y = b$ **7.** vertical **8.** $x = k$ **9.** 3; -2
11. 0; 0 **13.** -2 and 2; 4 **15.** 1; 1

17.

x	-2	-1	0	1	2
y	0	1	2	3	4

-2; 2

19.

x	-4	-2	0	2	4
y	-6	-4	-2	0	2

2; -2

21. x-int: 3; y-int: -2 **23.** x-int: 6; y-int: -2

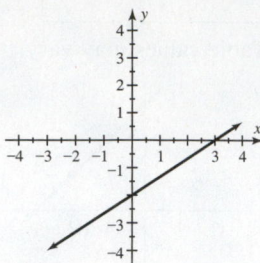   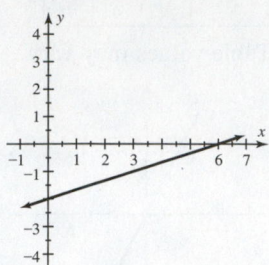

25. x-int: -1; y-int: 6   **27.** x-int: 7; y-int: 3

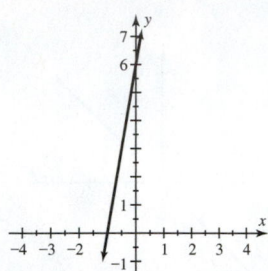   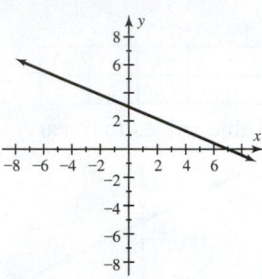

29. x-int: 4; y-int: -3 **31.** x-int: 4; y-int: -2

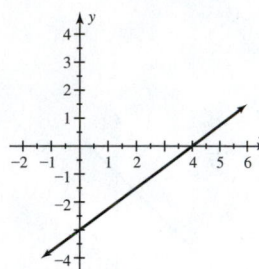   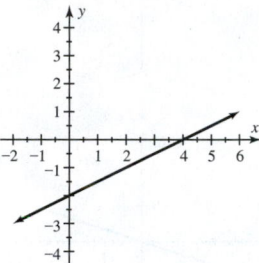

33. x-int: -4; y-int: 3   **35.** x-int: 3; y-int: 2

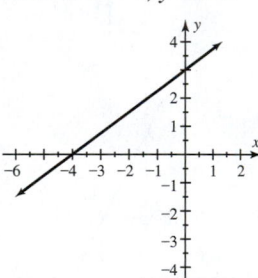   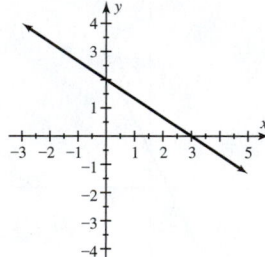

37. x-int: -2; y-int: 5

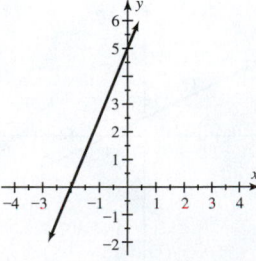

39. x-int: $\frac{C}{A}$; y-int: $\frac{C}{B}$ **41.** $y = 1$ **43.** $x = -6$

45. (a) 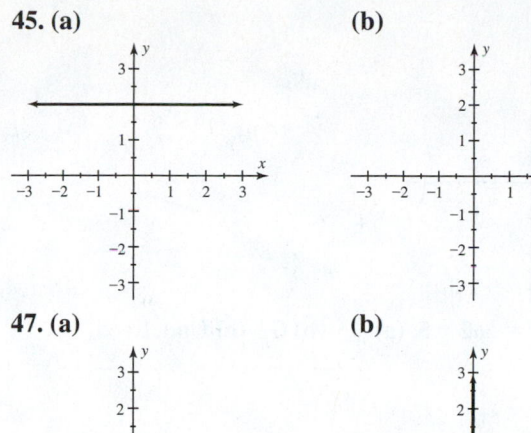 **(b)**

47. (a) 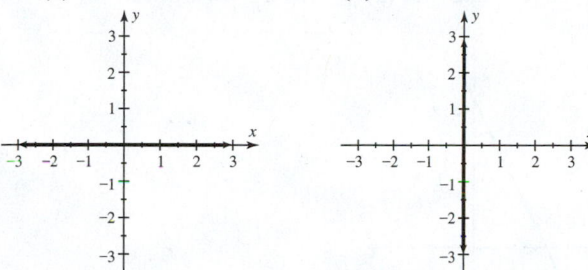 **(b)**

49. (a) 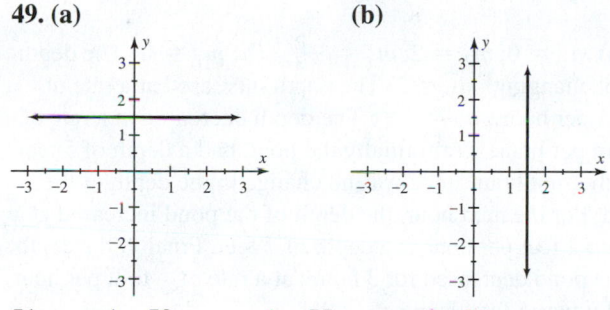 **(b)**

51. $y = 4$ **53.** $x = -1$ **55.** $y = -6$ **57.** $x = 5$
59. $y = 2; x = 1$ **61.** $y = -45; x = 20$
63. $y = 5; x = 0$ **65.** $x = -1$ **67.** $y = -\frac{5}{6}$
69. $x = 4$ **71.** $x = -\frac{2}{3}$ **73.** $y = 0$
75. (a) y-int: 200; x-int: 4   **(b)** The driver was initially
200 miles from home; the driver arrived home after 4 hours.
77. (a) v-int: 128; t-int: 4

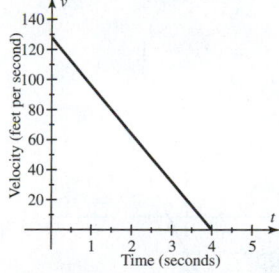

(b) The initial velocity was 128 ft/sec; the velocity after
4 seconds was 0.
79. (a) y-int: 2000; x-int: 4   **(b)** The pool initially contained
2000 gallons; the pool was empty after 4 hours.

SECTION 3.4 (pp. 198–203)

1. True   **2.** True   **3.** rise; run   **4.** horizontal   **5.** vertical
6. $\frac{y_2 - y_1}{x_2 - x_1}$ **7.** positive   **8.** negative   **9.** rate   **10.** Toward
11. Positive   **13.** Zero   **15.** Negative   **17.** Undefined
19. 0; the rise always equals 0.
21. 1; the graph rises 1 unit for each unit of run.
23. 2; the graph rises 2 units for each unit of run.

25. Undefined; the run always equals 0.
27. $-\frac{3}{2}$; the graph falls 3 units for each 2 units of run.
29. -1; the graph falls 1 unit for each unit of run.
31. 2;   **33.** Undefined;

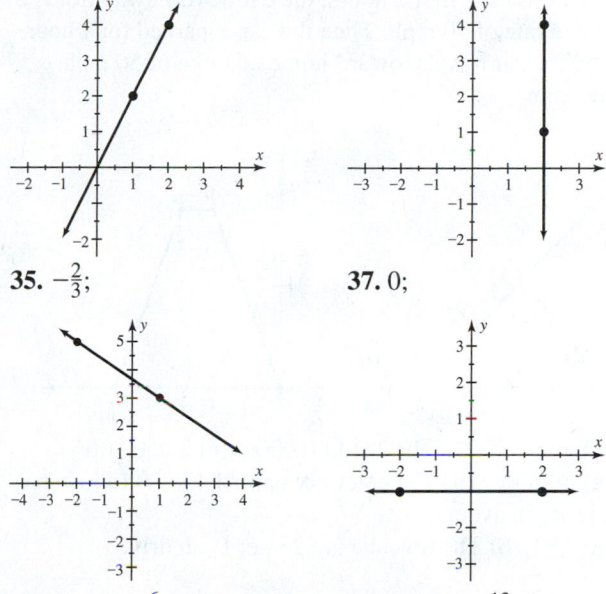

35. $-\frac{2}{3}$;   **37.** 0;

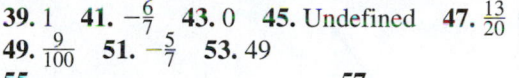

39. 1   **41.** $-\frac{6}{7}$ **43.** 0   **45.** Undefined   **47.** $\frac{13}{20}$
49. $\frac{9}{100}$ **51.** $-\frac{5}{7}$ **53.** 49
55.   **57.**

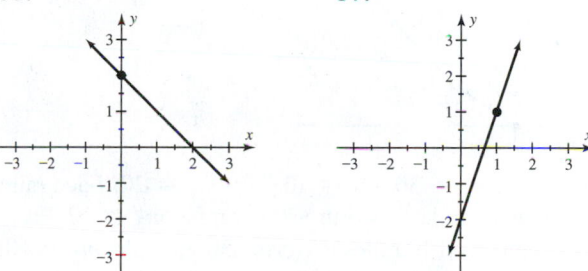

59.   **61.**

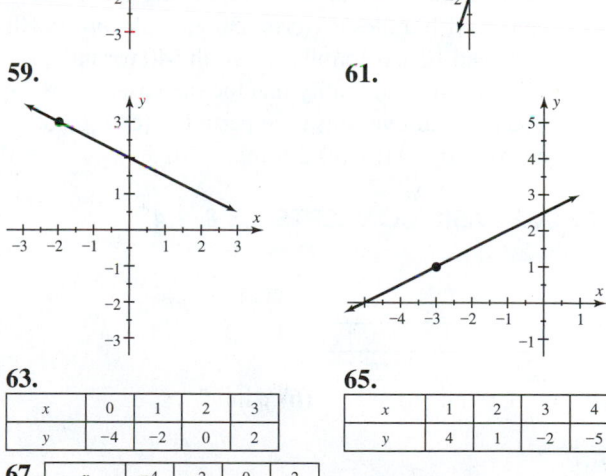

63.

x	0	1	2	3
y	-4	-2	0	2

65.

x	1	2	3	4
y	4	1	-2	-5

67.

x	-4	-2	0	2
y	0	3	6	9

69. c.   **71.** b.   **73. (a)** $m_1 = 1000; m_2 = -1000$
(b) $m_1 = 1000$: Water is being added to the pool at a rate
of 1000 gallons per hour. $m_2 = -1000$: Water is being
removed from the pool at a rate of 1000 gallons per hour.
(c) Initially the pool contained 2000 gallons of water. Over
the first 3 hours, water was pumped into the pool at a rate
of 1000 gallons per hour. For the next 2 hours, water was
pumped out of the pool at a rate of 1000 gallons per hour.

75. (a) $m_1 = 50$; $m_2 = 0$; $m_3 = -50$ (b) $m_1 = 50$: The car is moving away from home at a rate of 50 mph. $m_2 = 0$: The car is not moving. $m_3 = -50$: The car is moving toward home at a rate of 50 mph. (c) Initially the car is at home. Over the first 2 hours, the car travels away from home at a rate of 50 mph. Then the car is parked for 1 hour. Finally, the car travels toward home at a rate of 50 mph.
77. m^3/min
79.   **81.**

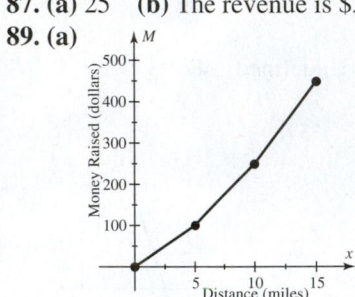

83. (a) 76,000,000 (b) 281,000,000 (c) 2,050,000/yr
85. (a) 40,800 (b) The celebrity gained 40,800 followers each year, on average.
87. (a) 25 (b) The revenue is $25 per flash drive.
89. (a)

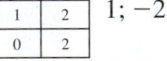

(b) $m_1 = 20$; $m_2 = 30$; $m_3 = 40$ (c) $m_1 = 20$: Each mile between 0 and 5 miles is worth $20 per mile. $m_2 = 30$: Each mile between 5 and 10 miles is worth $30 per mile. $m_3 = 40$: Each mile between 10 and 15 miles is worth $40 per mile.
91. (a) 1000 (b) Median family income increased on average by $1000 per year over this time period. (c) $56,000
93. (a) Toward (b) 10 ft (c) 2 ft/min (d) 5

CHECKING BASIC CONCEPTS 3.3 & 3.4 (pp. 203–204)

1. $-2; 3$
2.

x	-2	-1	0	1	2
y	-6	-4	-2	0	2

$1; -2$

3. (a) x-int: 6; y-int: -3 (b) y-int: 2

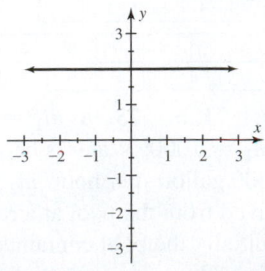

(c) x-int: -1

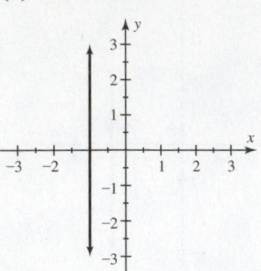

4. $y = 4$; $x = -2$ **5.** (a) $\frac{3}{4}$ (b) 0 (c) Undefined
6. $-\frac{1}{2}$
7.

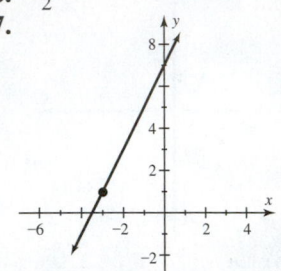

8. (a) $m_1 = 0$; $m_2 = 2$; $m_3 = -\frac{2}{3}$ (b) $m_1 = 0$: The depth is not changing. $m_2 = 2$: The depth increased at a rate of 2 feet per hour. $m_3 = -\frac{2}{3}$: The depth decreased at a rate of $\frac{2}{3}$ foot per hour. (c) Initially the pond had a depth of 5 feet. For the first hour, there was no change in the depth of the pond. For the next hour, the depth of the pond increased at a rate of 2 feet per hour to a depth of 7 feet. Finally, the depth of the pond decreased for 3 hours at a rate of $\frac{2}{3}$ foot per hour until it was 5 feet deep.

SECTION 3.5 (pp. 211–213)

1. $y = mx + b$ **2.** slope **3.** y-intercept **4.** origin
5. parallel **6.** slope **7.** perpendicular **8.** reciprocals
9. f. **10.** d. **11.** a. **12.** b. **13.** e. **14.** c.
15. $y = x - 1$ **17.** $y = -2x + 1$ **19.** $y = -2x$
21. $y = \frac{3}{4}x + 2$
23. $y = x + 2$ **25.** $y = 2x - 1$

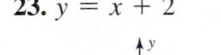
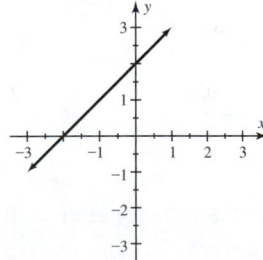
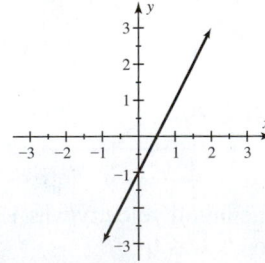

27. $y = -\frac{1}{2}x - 2$ **29.** $y = \frac{1}{3}x$

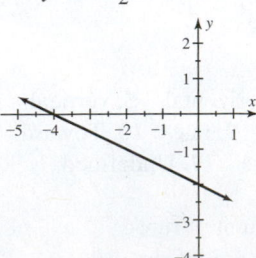
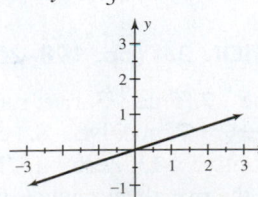

31. $y = 3$

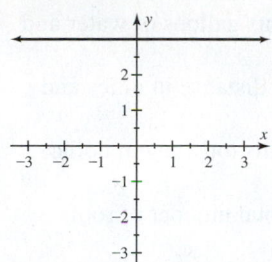

33. (a) $y = -x + 4$ **(b)** $-1; 4$
35. (a) $y = -2x + 4$ **(b)** $-2; 4$
37. (a) $y = \frac{1}{2}x + 2$ **(b)** $\frac{1}{2}; 2$
39. (a) $y = \frac{2}{3}x - 2$ **(b)** $\frac{2}{3}; -2$
41. (a) $y = \frac{1}{4}x + \frac{3}{2}$ **(b)** $\frac{1}{4}; \frac{3}{2}$
43. (a) $y = -\frac{1}{3}x + \frac{2}{3}$ **(b)** $-\frac{1}{3}; \frac{2}{3}$

45.   **47.**

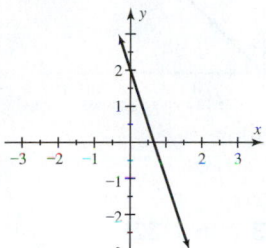   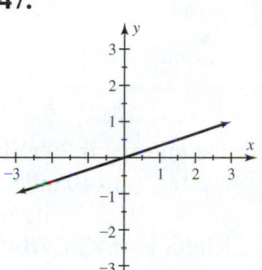

49.   **51.**

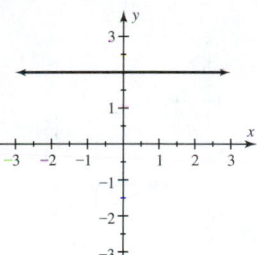   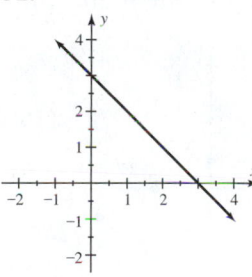

53.   **55.**

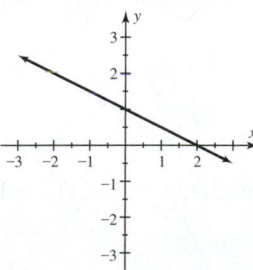   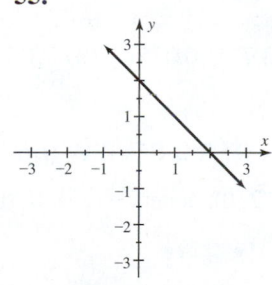

57. $y = 2x + 2$ **59.** $y = x - 2$ **61.** $y = \frac{4}{7}x + 3$
63. $y = 3x$ **65.** $y = -\frac{1}{2}x + \frac{5}{2}$ **67.** $y = 2x$
69. $y = \frac{1}{3}x + \frac{1}{3}$ **71. (a)** \$25   **(b)** 25 cents   **(c)** 25; the
fixed cost of renting the car   **(d)** 0.25; the cost per mile of
driving the car   **73. (a)** \$7.45   **(b)** $C = 0.07x + 3.95$
(c) 67 min   **75. (a)** $m = 0.35; b = 164.3$
(b) The fixed cost of owning the car for one month

SECTION 3.6 (pp. 221–224)

1. True   **2.** True   **3.** $y = mx + b$
4. $y - y_1 = m(x - x_1)$ or $y = m(x - x_1) + y_1$
5. 1; 3   **6.** distributive   **7.** Yes; every nonvertical line has
exactly one slope and one y-intercept.   **8.** No; it depends on
the point used.   **9.** No   **11.** Yes   **13.** $y - 3 = -2(x + 2)$
15. $y - 2 = \frac{3}{4}(x - 1)$ **17.** $y + 1 = -\frac{1}{2}(x - 3)$
19. $y - 1 = 4(x + 3)$ **21.** $y + 3 = \frac{1}{2}(x + 5)$
23. $y - 30 = 1.5(x - 2010)$ **25.** $y - 4 = \frac{7}{3}(x - 2)$
27. $y = \frac{3}{5}(x - 5)$ **29.** $y - 15 = 5(x - 2003)$
31. $y = 3x - 2$ **33.** $y = \frac{1}{3}x$ **35.** $y = \frac{2}{3}x + \frac{1}{12}$
37. $y = -2x + 9$ **39.** $y = \frac{3}{5}x - 2$ **41.** $y = -16x - 19$
43. $y = -2x + 5$ **45.** $y = -x + 1$ **47.** $y = -\frac{1}{9}x + \frac{1}{3}$
49. $y = 2x - 7$ **51.** $y = 2x - 15$ **53.** $y = -2x - 1$
55. $y = \frac{1}{2}x - \frac{5}{2}$ **57.** $-mx_1 + y_1$
59. (a) $-5°$F per hour   **(b)** $T = -5x + 45$; the temperature is decreasing at a rate of $5°$F per hour.
(c) 9; at 9 A.M. the temperature was $0°$F.
(d)

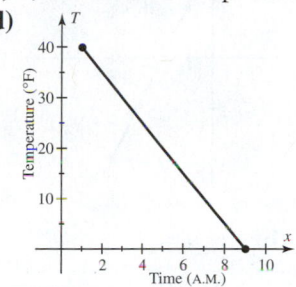

(e) At 4 A.M. the temperature was $25°$F.
61. (a) Entering; 300 gallons   **(b)** 100; initially the tank
contains 100 gallons.   **(c)** $y = 50x + 100$; the amount of
water is increasing at a rate of 50 gallons per minute.
(d) 8 minutes   **63. (a)** Toward   **(b)** After 1 hour the person
is 250 miles from home. After 4 hours the person is 100
miles from home.   **(c)** 50 mph   **(d)** $y = -50x + 300$; the
car is traveling toward home at 50 mph.
65. (a) In 2002, average home size was 2334 square feet.
(b) $y - 2334 = 34(x - 2002)$
(c) About 2500 square feet; 2504 square feet
(d) Home size increased, on average, by 34 square feet per year.
67. (a) $y - 731 = 515(x - 2006)$ **(b)** 1761
69. Cigarette consumption decreased on average by 10.33
billion cigarettes per year.
71. (a) $y = 2.7x - 5391.3$ **(b)** 41.1 million

CHECKING BASIC CONCEPTS 3.5 & 3.6 (pp. 224–225)

1. $y = -3x + 1$ **2.** $y = \frac{4}{5}x - 4; \frac{4}{5}; -4$
3.

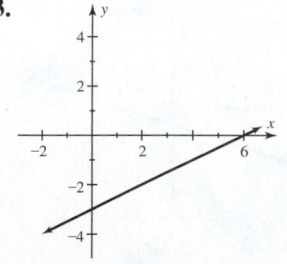

4. (a) $y = 3x - 2$ **(b)** $y = -\frac{3}{2}x$ **(c)** $y = -\frac{7}{3}x - \frac{5}{3}$
5. $y - 3 = -2(x + 1)$ **6.** $y = -2x + 1$
7. $y = 2x + 3$ **8. (a)** $y = -12x + 48$ **(b)** 12 mph
(c) 4 P.M. **(d)** 48 miles **9. (a)** 5 inches **(b)** 2 inches per
hour **(c)** 5; total inches of snow that fell before noon
(d) 2; the rate of snowfall was 2 inches per hour.

SECTION 3.7 (pp. 230–232)

1. linear **2.** exact **3.** approximate **4.** constant
5. constant rate of change **6.** initial amount **7.** f. **8.** d.
9. a. **10.** c. **11.** e. **12.** b. **13.** Yes **15.** No **17.** No
19. Exact; $y = 2x - 2$ **21.** Approximate; $y = 2x + 2$
23. Approximate; $y = 2$ **25.** $y = 5x + 40$
27. $y = -20x - 5$ **29.** $y = 8$
31. (a) Yes **(b)**

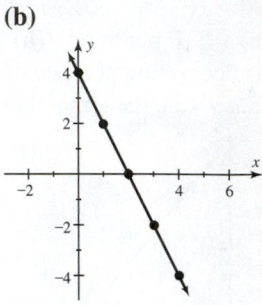

(c) $y = -2x + 4$
33. (a) No **(b)**

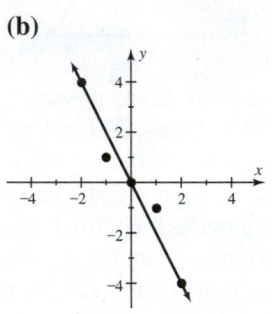

(c) $y = -2x$
35. (a) Yes **(b)**

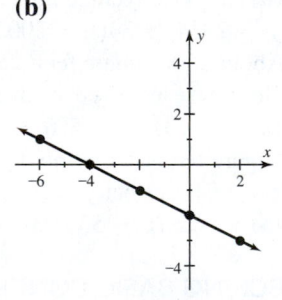

(c) $y = -\frac{1}{2}x - 2$
37. (a) No **(b)**

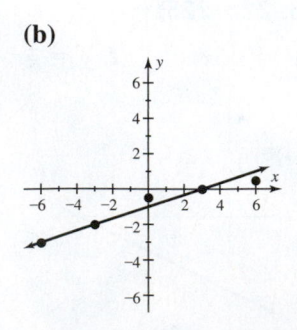

(c) $y = \frac{1}{3}x - 1$
39. $g = 5t + 200$, where g represents gallons of water and
t represents time in minutes.
41. $d = 6t + 5$, where d represents distance in miles and t
represents time in hours.
43. $p = 8t + 200$, where p represents total pay in dollars
and t represents time in hours.
45. $r = t + 5$, where r represents total number of roofs
shingled and t represents time in days.
47. (a) $-\frac{2}{45}$ acre per year **(b)** $A = -\frac{2}{45}t + 5$
49. (a) $y = 6x - 11,846$ **(b)** About 232,000
51. (a)   **(b)**

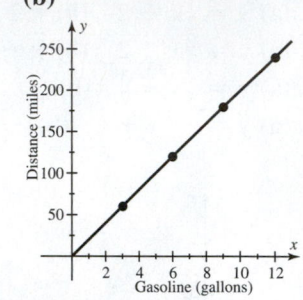

(c) 20; the mileage is 20 miles per gallon.
(d) $y = 20x$ **(e)** 140 miles

CHECKING BASIC CONCEPTS 3.7 (p. 232)

1. Yes **2.** Approximate; $y = x - 1$
3. (a) $y = 10x + 50$ **(b)** $y = -2x + 200$
4. (a)   **(b)**

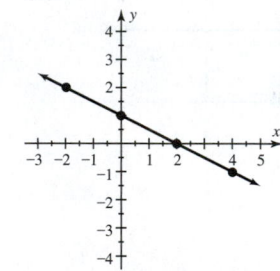

(c) $y = -\frac{1}{2}x + 1$
5. (a) $T = 0.075x$ **(b)** 5.1°F

CHAPTER 3 REVIEW (pp. 237–241)

1. $(-2, 0)$: none; $(-1, 2)$: II; $(0, 0)$: none; $(1, -2)$: IV; $(1, 3)$: I
2.

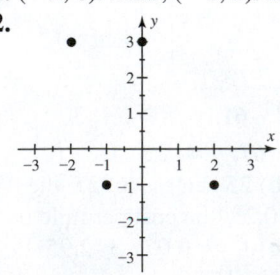

3. (a) II **(b)** IV **4. (a)** None **(b)** III

5.

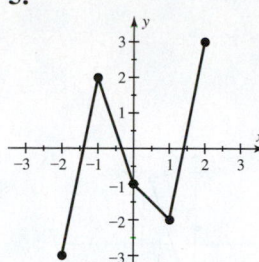

6.

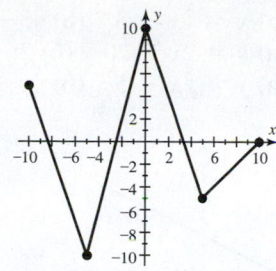

23.

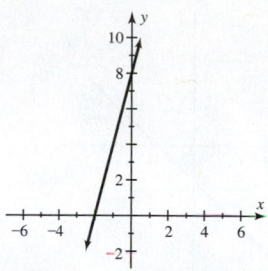

24.

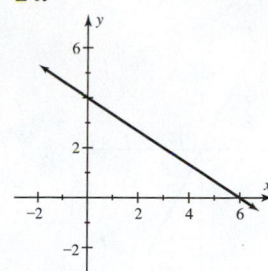

7. Yes   **8.** No   **9.** No   **10.** Yes

25. 3; −2   **26.** −2 and 2; −4

11.

x	−2	−1	0	1	2
y	6	3	0	−3	−6

12.

x	4	3	2.5	2	1
y	−3	−1	0	1	3

13.

x	−2	0	2	4
y	−4	2	8	14

14.

x	1	2	3	4
y	6	5	4	3

15.

x	−0.5	0	0.5	1
y	−1	0	1	2

16.

x	−1	−3	−5	−7
y	1	2	3	4

27.

x	−2	−1	0	1	2
y	4	3	2	1	0

2; 2

28.

x	−4	−2	0	2	4
y	−4	−3	−2	−1	0

4; −2

17.

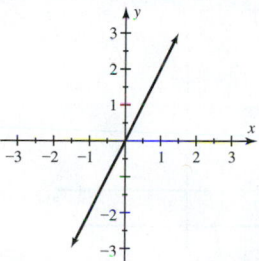

18.

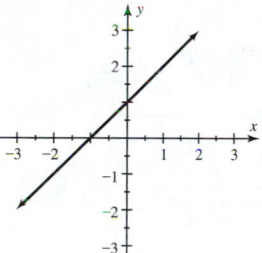

29. x-int: 3; y-int: −2

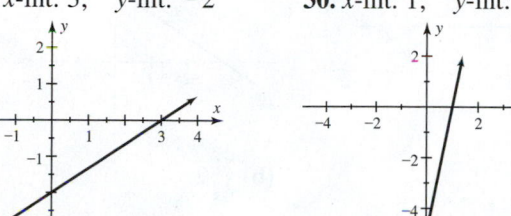

30. x-int: 1; y-int: −5

19.

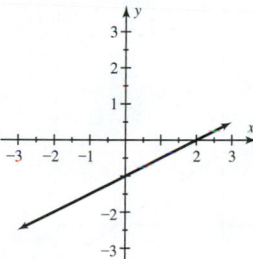

20.

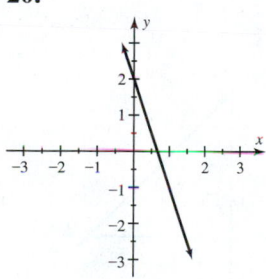

31. x-int: 4; y-int: −2

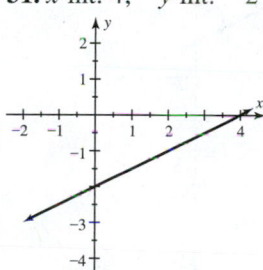

32. x-int: 2; y-int: 3

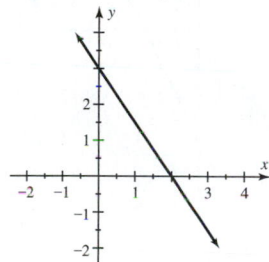

33. $y = 1$ **34.** $x = 3$

35. (a)

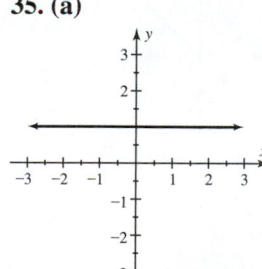

(b)

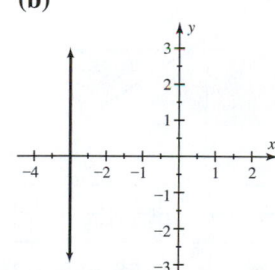

21.

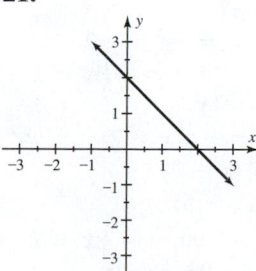

22.

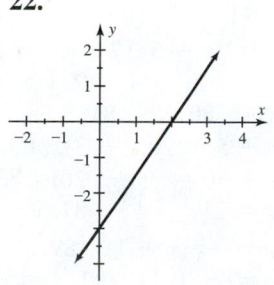

36. $x = -1$; $y = 1$ **37.** $y = 3$; $x = -2$ **38.** $x = 4$
39. $y = -5$ **40. (a)** y-int: 90; x-int: 3   **(b)** The driver is
initially 90 miles from home; the driver arrives home after
3 hours.   **41.** 2   **42.** $-\frac{2}{5}$ **43.** 0   **44.** Undefined   **45.** 3
46. $-\frac{1}{2}$

47. (a)

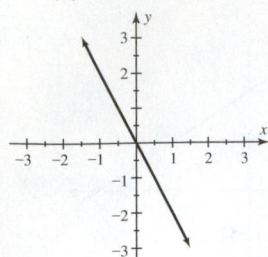

(b) $-2; 0$

48. (a)

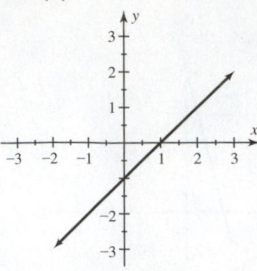

(b) $1; -1$

49. (a)

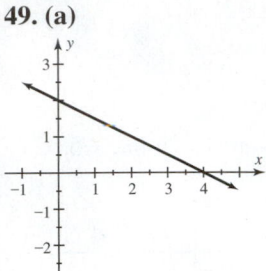

(b) $-\frac{1}{2}; 2$

50. (a)

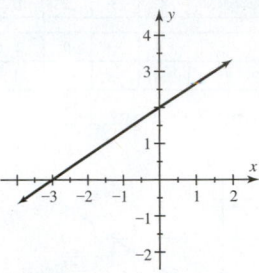

(b) $\frac{2}{3}; 2$

51.

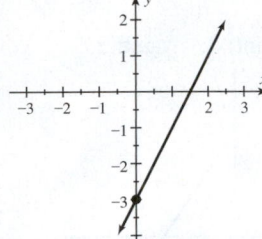

52.

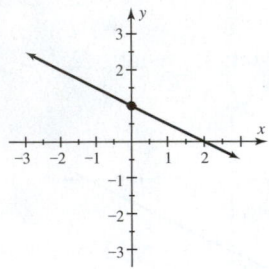

53.

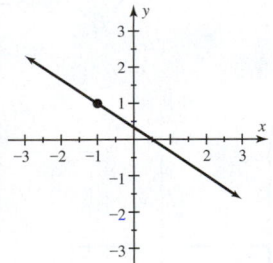

54.

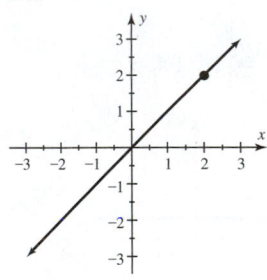

55.

x	0	1	2	3
y	1	$\frac{3}{2}$	2	$\frac{5}{2}$

56. Positive

57. $y = x + 1$ **58.** $y = -2x + 2$

59. $y = 2x - 2$

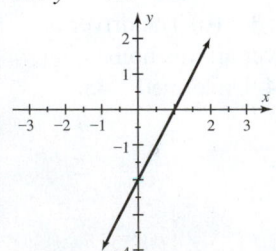

60. $y = -\frac{3}{4}x + 3$

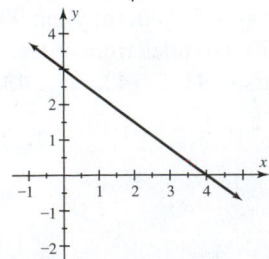

61. (a) $y = -x + 3$ **(b)** $-1; 3$
62. (a) $y = \frac{3}{2}x - 3$ **(b)** $\frac{3}{2}; -3$
63. (a) $y = 2x - 20$ **(b)** $2; -20$
64. (a) $y = \frac{5}{6}x - 5$ **(b)** $\frac{5}{6}; -5$

65.

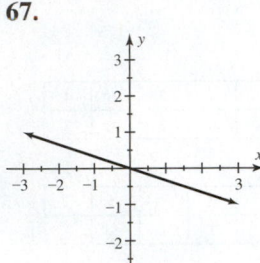

66.

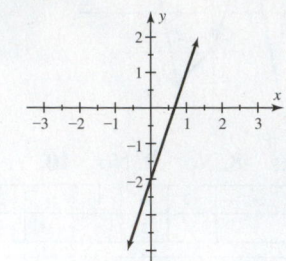

67.

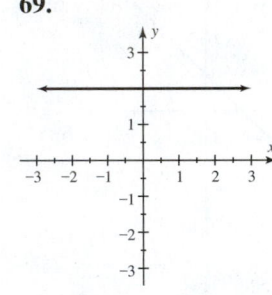

68.

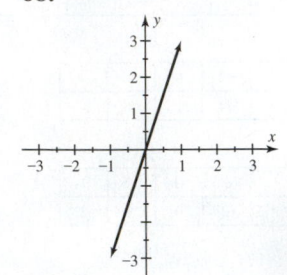

69.

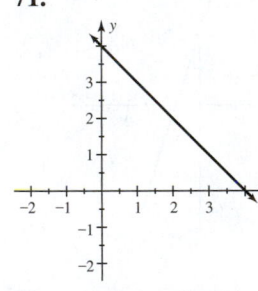

70.

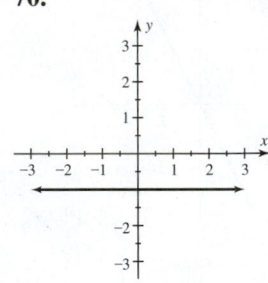

71.

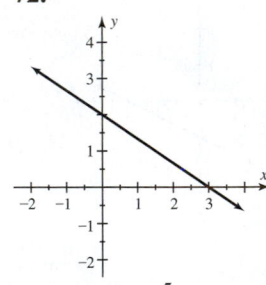

72.

73. $y = 5x - 5$ **74.** $y = -2x$ **75.** $y = -\frac{5}{6}x + 2$
76. $y = -2x - 3$ **77.** $y = \frac{2}{3}x - 2$ **78.** $y = -\frac{1}{5}x - 2$
79. Yes   **80.** No   **81.** $y - 2 = 5(x - 1)$
82. $y + 5 = 20(x - 3)$ **83.** $y - 1 = -\frac{2}{3}(x + 2)$
84. $y + 30 = 3(x - 20)$ **85.** $y = \frac{4}{3}(x - 3)$
86. $y = 2\left(x - \frac{1}{2}\right)$ **87.** $y - 7 = 2(x - 5)$
88. $y = -\frac{2}{3}(x + 1)$ **89.** $y = 3x + 5$ **90.** $y = \frac{1}{3}x + 7$
91. $y = 2x + 11$ **92.** $y = -\frac{1}{4}x + 3$ **93.** No
94. Approximate; $y = -x + 5$ **95.** $y = -2x + 40$
96. $y = 20x + 200$ **97.** $y = 50$ **98.** $y = 5x - 20$

99. (a) Yes   **(b)**

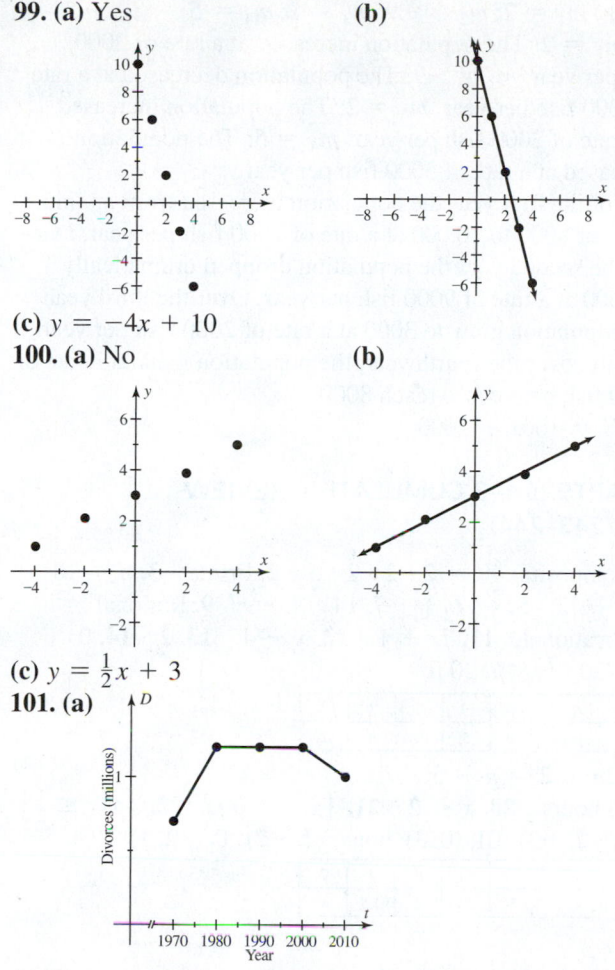

(c) $y = -4x + 10$
100. (a) No   **(b)**

(c) $y = \frac{1}{2}x + 3$
101. (a)

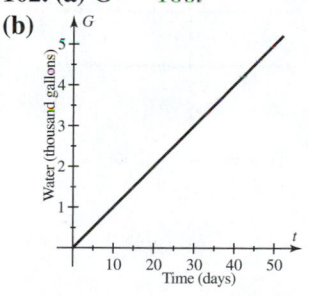

(b) The number of divorces increased significantly between 1970 and 1980, remained unchanged from 1980 to 2000, and then decreased.
102. (a) $G = 100t$
(b)

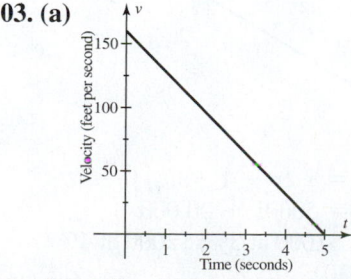

(c) 50 days
103. (a)

(b) v-int: 160; t-int: 5; the initial velocity was 160 ft/sec, and the velocity after 5 seconds was 0.

104. (a) $m_1 = 0$; $m_2 = -1500$; $m_3 = 500$ **(b)** $m_1 = 0$: The population remained unchanged. $m_2 = -1500$: The population decreased at a rate of 1500 insects per week. $m_3 = 500$: The population increased at a rate of 500 insects per week.   **(c)** For the first week the population did not change from its initial value of 4000. Over the next two weeks the population decreased at a rate of 1500 insects per week until it reached 1000. Finally, the population increased at a rate of 500 per week for two weeks, reaching 2000.
105.

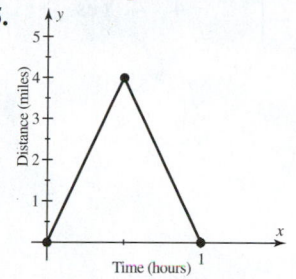

106. (a) -120 **(b)** The number of nursing homes decreased at an average rate of 120 per year.
107. (a) \$35   **(b)** 20¢   **(c)** 35; the fixed cost of renting the car   **(d)** 0.2; the cost for each mile driven
108. (a) Toward; the slope is negative.   **(b)** After 1 hour the car is 200 miles from home; after 3 hours the car is 100 miles from home.   **(c)** $y = -50x + 250$; the car is moving toward home at 50 mph.   **(d)** 150 miles; 150 miles
109. (a)

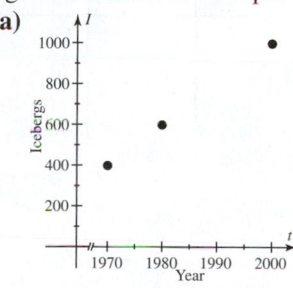

(b) $I = 20t - 39,000$; the number of icebergs increased at an average rate of 20 per year.   **(c)** Yes   **(d)** 1100 icebergs
110. (a)   **(b)**

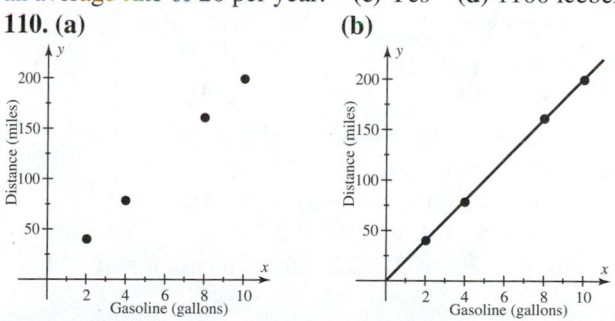

(c) 20; the mileage is 20 miles per gallon.
(d) $y = 20x$; no   **(e)** About 180 miles

CHAPTER 3 TEST (pp. 241–242)

1. $(-2, -2)$: III; $(-2, 1)$: II; $(0, -2)$: none; $(1, 0)$: none; $(2, 3)$: I; $(3, -1)$: IV

2.

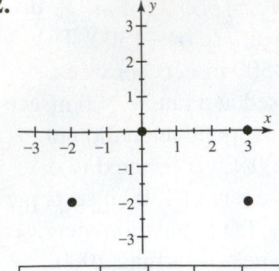

3.

x	-2	-1	0	1	2
y	-8	-6	-4	-2	0

2; -4   **4.** Yes

5.

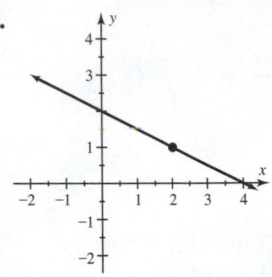

6. x-int: 3; y-int: -5

7.

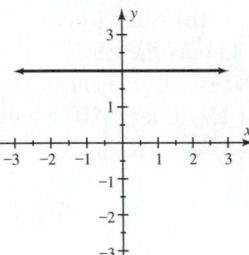

8.

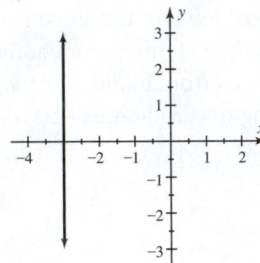

9.

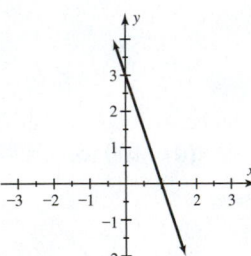

10.

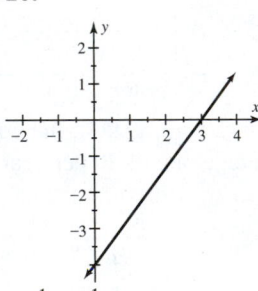

11. $y = -x + 2$ **12.** $y = 2x + \frac{1}{2}; 2; \frac{1}{2}$

13. $y = -5; x = 1$ **14.** $-\frac{2}{9}$ **15.** $y = -\frac{4}{3}x - 5$

16. $y = 3x - 11$ **17.** $y = -3x + 5$ **18.** $y = -\frac{1}{2}x$

19. $y = \frac{1}{2}x + 5$ **20.** $y = 3x - 2$

21. $y - 7 = -3(x + 2)$ **22.** Approximate; $y = x - 1$

23.

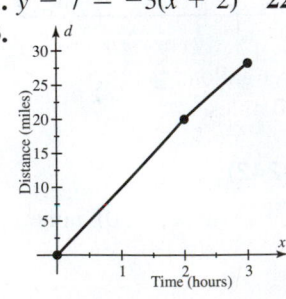

24. (a) $m_1 = 2; m_2 = -9; m_3 = 2; m_4 = 5$

(b) $m_1 = 2$: The population increased at a rate of 2000 fish per year. $m_2 = -9$: The population decreased at a rate of 9000 fish per year. $m_3 = 2$: The population increased at a rate of 2000 fish per year. $m_4 = 5$: The population increased at a rate of 5000 fish per year.

(c) For the first year the population increased from an initial value of 8000 to 10,000 at a rate of 2000 fish per year. During the second year the population dropped dramatically to 1000 at a rate of 9000 fish per year. Over the third year the population grew to 3000 at a rate of 2000 fish per year. Finally, over the fourth year, the population grew at a rate of 5000 fish per year to reach 8000.

25. $N = 100x + 2000$

CHAPTERS 1–3 CUMULATIVE REVIEW (pp. 243–244)

1. Composite; $40 = 2 \cdot 2 \cdot 2 \cdot 5$ **2.** Prime   **3.** $n + 10$

4. $n^2 - 2$ **5.** $\frac{2}{3}$ **6.** $\frac{17}{30}$ **7.** 14   **8.** -9   **9.** Rational

10. Irrational   **11.** $7x + 1$ **12.** $x - 4$ **13.** 2   **14.** 0

15. 720 ft^2 **16.** 20 ft^2

17.

x	-2	-1	0	1	2
y	10	8	6	4	2

1

18. $2n + 2 = n - 5; -7$

19. 8 hours   **20.** $x < 2$ **21.** $\{x \mid x > 0\}$ **22.** $\{x \mid x \le \frac{1}{2}\}$

23. $(-2, -3)$: III; $(0, 3)$: none; $(2, -2)$: IV; $(2, 1)$: I

24.

x	-2	-1	0	1	2
y	3	2.5	2	1.5	1

25.

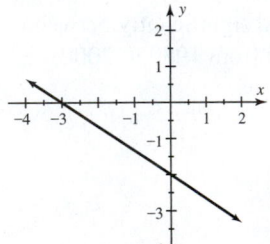

26.

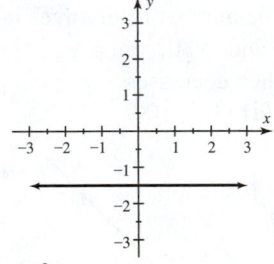

27. x-int: -10; y-int: 8   **28.** $x = \frac{3}{2}; y = -2$

29. $y = 3x - 3$

30. $y = \frac{3}{5}x - 3$;

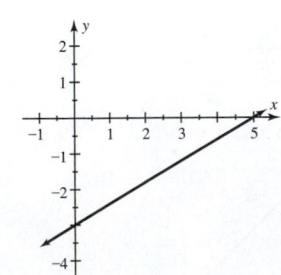

31. $y = -\frac{2}{3}x - 3$ **32.** $y = -2x + 1$

33. $y = 5x + 100$ **34.** $I = 5000x + 20,000$

35. $C = 8x$ **36.** $\frac{99}{200}$ **37.** \$1000 at 3%; \$2000 at 4%

38. (a) \$85   **(b)** \$25   **(c)** 30¢

4 Systems of Linear Equations in Two Variables

SECTION 4.1 (pp. 254–257)

1. ordered   **2.** intersection-of-graphs   **3.** no solutions; one solution; infinitely many   **4.** consistent   **5.** inconsistent
6. independent   **7.** dependent   **8.** table   **9.** the same
10. intercepts   **11.** 1   **13.** 2   **15.** −2   **17.** 2
19. One; consistent; independent   **21.** Infinitely many; consistent; dependent   **23.** None; inconsistent   **25.** (1, 1)
27. (2, −3)   **29.** (2, 0)   **31.** (2, 1)   **33.** (3, 2)
35. (−1, 1)   **37.** (2, 4)   **39.** (3, 1)
41. (1, 3);

x	0	1	2	3
y = x + 2	2	3	4	5
y = 4 − x	4	3	2	1

43. (a) (−1, 1)   **(b)** (−1, 1)   **45. (a)** (3, 1)   **(b)** (3, 1)
47. (a) (1, 3)   **(b)** (1, 3)   **49.** (−1, −2)   **51.** (1, −1)
53. (1, 2)   **55.** (3, 2)   **57.** (3, 1)   **59.** (1, 2)
61. (a) x + y = 4, x − y = 0   **(b)** 2, 2
63. (a) 2x + y = 7, x − y = 2   **(b)** 3, 1
65. (a) x = 3y, x − y = 4   **(b)** 6, 2
67. (a) C = 0.5x + 50   **(b)** 60 mi   **(c)** 60 mi
69. (a) x + y = 42; x = 2y   **(b)** (28, 14)
71. (a) x − y = 4; 2x + 2y = 28 where x is length, y is width   **(b)** (9, 5); the rectangle is 9 in. × 5 in.

SECTION 4.2 (pp. 263–265)

1. exact   **2.** parentheses   **3.** It has infinitely many solutions.
4. It has no solutions.   **5.** (3, 6)   **7.** (2, 1)   **9.** (−1, 0)
11. (3, 0)   **13.** $\left(\frac{1}{2}, 0\right)$ **15.** (0, 4)   **17.** (1, 2)
19. (6, −4)   **21.** (4, −3)   **23.** $\left(1, -\frac{1}{2}\right)$ **25.** (1, 3)
27. (0, −2)   **29.** (2, 2)   **31.** (−3, 5)   **33.** No solutions
35. Infinitely many   **37.** (3, 1)   **39.** Infinitely many
41. No solutions   **43.** $\left(-\frac{4}{3}, \frac{11}{9}\right)$ **45.** (0, 0)
47. No solutions   **49.** No solutions   **51.** They are a single line.
53. (a) L − W = 10, 2L + 2W = 72   **(b)** (23, 13)
55. (a) x = $\frac{1}{2}$y; x + y = 90   **(b)** (30, 60)   **(c)** (30, 60)
57. (a) x − y = 21; y = 0.86x   **(b)** (150, 129)
59. 94 ft × 50 ft   **61.** 17 and 53   **63.** 10 mph; 2 mph
65. 3.$\overline{3}$ L of 20% solution, 6.$\overline{6}$ L of 50% solution
67. Superior: 32,000 mi^2; Michigan: 22,000 mi^2

CHECKING BASIC CONCEPTS 4.1 & 4.2 (p. 265)

1. (a) −2   **(b)** 6   **2.** (4, 2)   **3.** (2, 1)   **4. (a)** No solutions
(b) (1, −1)   **(c)** Infinitely many solutions
5. (a) x + y = 300, 150x + 200y = 55,000   **(b)** (100, 200)

SECTION 4.3 (pp. 273–275)

1. Substitution; elimination   **2.** addition   **3.** =   **4.** =
5. It has infinitely many solutions.   **6.** It has no solutions.
7. (1, 1)   **9.** (−2, 1)   **11.** No solutions   **13.** Infinitely many   **15.** (6, 1)   **17.** (−1, 4)   **19.** (2, 4)   **21.** (2, 1)
23. (4, −1)   **25.** (−4, 1)   **27.** $\left(\frac{4}{5}, 1\right)$ **29.** (−2, −5)

31. (1, 0)   **33.** (2, 2)   **35.** $\left(\frac{4}{5}, \frac{9}{5}\right)$ **37.** (3, 2)   **39.** (0, 1)
41. (2, 1)   **43.** (4, −3)   **45.** (1, 0)
47. Infinitely many;   **49.** One;

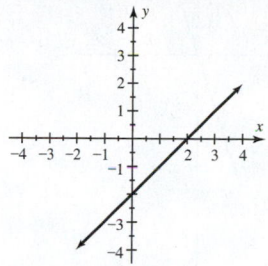
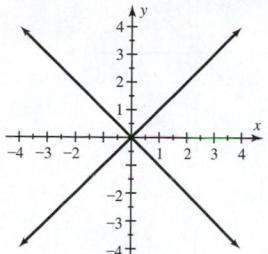

51. No solutions;   **53.** No solutions;

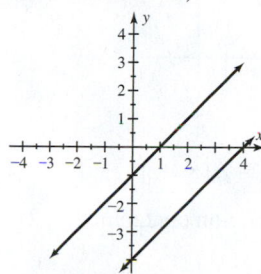
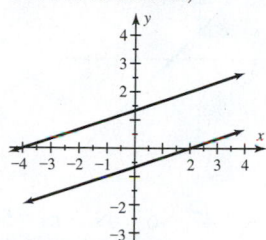

55. Infinitely many;

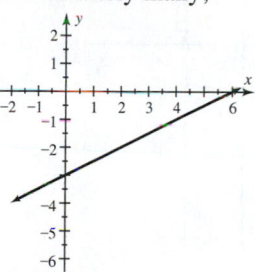

57. Men: 39,000; women: 29,000   **59.** Bicycle: 18 min; stair climber: 12 min   **61.** Current: 2 mph; boat: 6 mph
63. $2000 at 3%; $3000 at 5%   **65.** −43, 26
67. (a) 20 in. × 40 in.   **(b)** 20 in. × 40 in.

SECTION 4.4 (pp. 284–286)

1. inequality   **2.** ordered pair   **3.** All points below and including the line y = k   **4.** All points to the right of the line x = k.   **5.** All points above and including the line y = x   **6.** test   **7.** dashed   **8.** solid   **9.** Ax + By = C
10. both inequalities   **11.** intersect   **12.** No   **13.** Yes
15. No   **17.** No   **19.** Yes   **21.** Yes   **23.** No
25. x > 1   **27.** y ≥ 2   **29.** y < x   **31.** −x + y ≤ 1
33.   **35.**

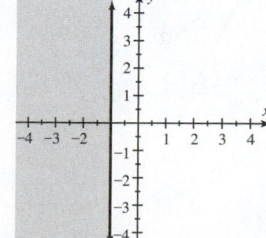
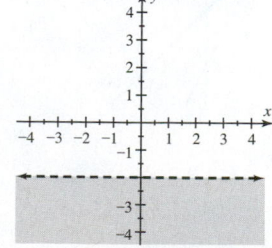

37.
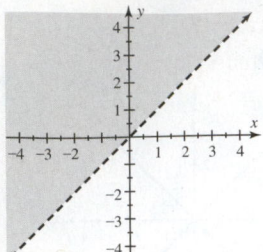

39.
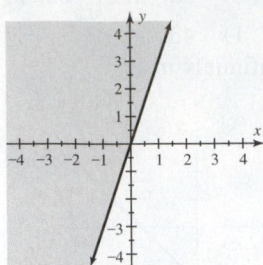

41.
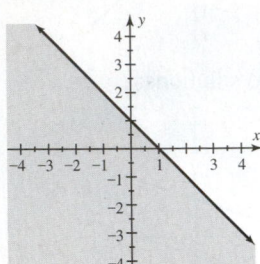

43.
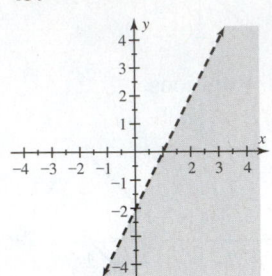

45. Yes   **47.** No   **49.** Yes   **51.** The region containing (1, 2)
53. The region containing (1, 0)

55.
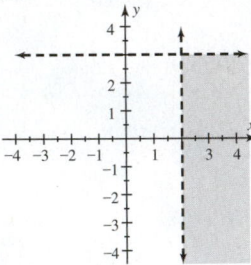

57.
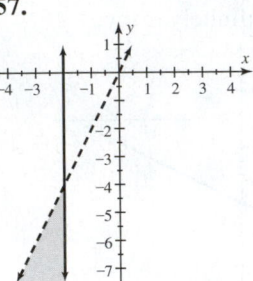

59.
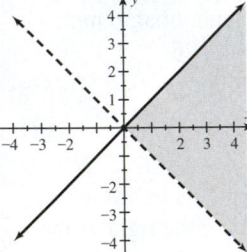

61.
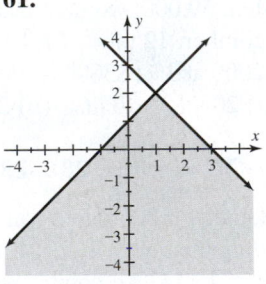

63.
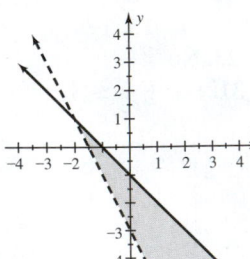

65.
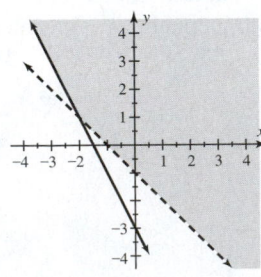

67.
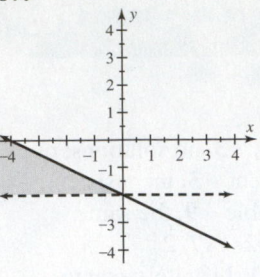

69.
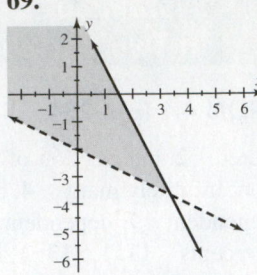

71.
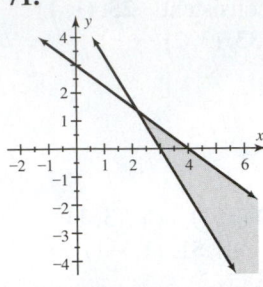

73.
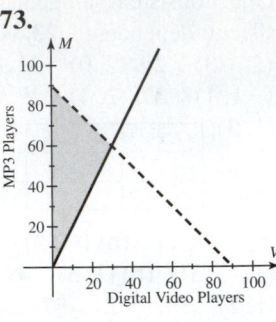

75. (a) 200 bpm; 150 bpm
(b)
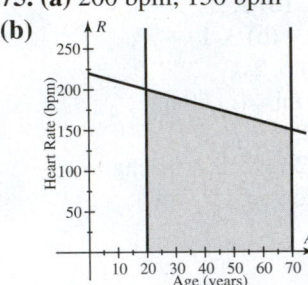

(c) Possible heart rates for ages 20 to 70   **77.** 150 to 200 lb

CHECKING BASIC CONCEPTS 4.3 & 4.4 (p. 287)

1. (1, 1)   **2. (a)** (0, −1); one   **(b)** No solutions
(c) Infinitely many   **3.** Infinitely many
4. (a)
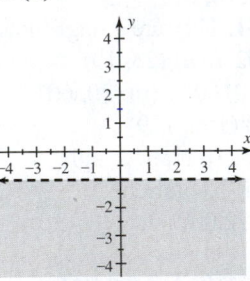

(b)
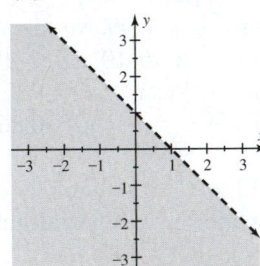

5.
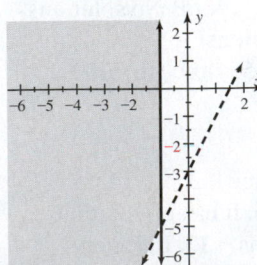

6. (a) $x + y = 11$, $x - y = 5$ **(b)** (8, 3)

CHAPTER 4 REVIEW (pp. 290–292)

1. 3 **2.** 2 **3. (a)** None **(b)** Inconsistent **4. (a)** One
(b) Consistent; independent **5. (a)** Infinitely many
(b) Consistent; dependent **6. (a)** None **(b)** Inconsistent
7. $(1, 2)$ **8.** $(5, 2)$ **9.** $(4, 3)$ **10.** $(2, -4)$ **11.** $(2, 2)$
12. $(1, 2)$ **13.** $(2, 6)$ **14.** $(1, 1)$ **15.** $(4, -3)$
16. $(1, 2)$ **17.** $(1, 1)$ **18.** $(1, 2)$ **19.** $(1, 1)$ **20.** $(-2, -1)$
21. $(2, 6)$ **22.** $(2, -10)$ **23.** $(-2, 1)$ **24.** $(-4, -4)$
25. No solutions **26.** No solutions **27.** Infinitely many
28. $(1, 1)$ **29.** $(2, 1)$ **30.** $(-1, 2)$ **31.** $(11, -1)$
32. $(1, 0)$ **33.** $\left(\frac{9}{4}, \frac{7}{4}\right)$ **34.** $\left(\frac{5}{2}, 1\right)$ **35.** $(5, -7)$
36. $(8, 2)$ **37.** $(-2, 3)$ **38.** $(1, 0)$ **39.** $(1, 3)$ **40.** $(2, -1)$
41. Infinitely many **42.** Infinitely many **43.** No solutions
44. One **45.** Yes **46.** No **47.** No **48.** Yes **49.** $y > 1$
50. $y \leq 2x + 1$

51.

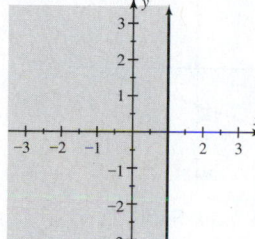

52.

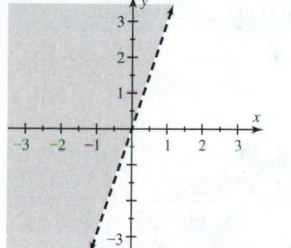

53.

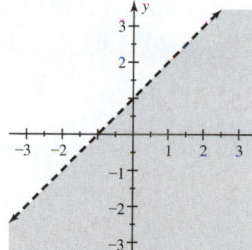

54.

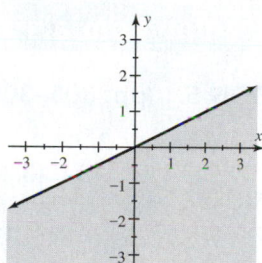

55.

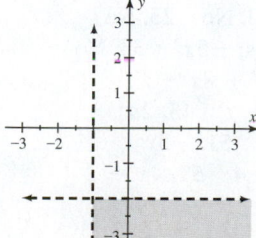

56.

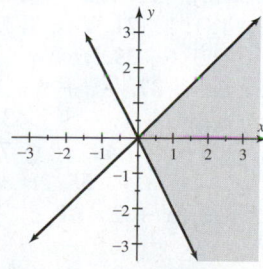

57. Yes **58.** No **59.** The region containing $(2, -2)$
60. The region containing $(1, 3)$

61.

62.

63.

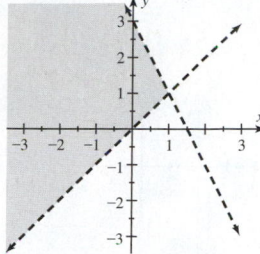

64.

65.

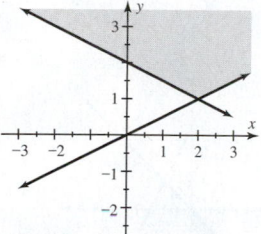

66.

67. 3100 deaths in 1912; 37,200 deaths in 2008
68. Men: 40,000 cases; women: 34,000 cases
69. (a) $C = 0.2x + 40$ **(b)** 250 mi **(c)** 250 mi
70. 75°, 105° **71. (a)** $2x + y = 180$, $2x - y = 40$
(b) $(55, 70)$ **(c)** $(55, 70)$ **72.** 20 ft $\times$ 24 ft
73. (a) $x + y = 10$, $80x + 120y = 920$; x is \$80 rooms,
y is \$120 rooms. **(b)** $(7, 3)$ **74.** 7 lb of \$2 candy; 11 lb of
\$3 candy **75.** Bicycle: 35 min; stair climber: 25 min
76. 2 mph **77. (a)** 16 ft $\times$ 24 ft (answers may vary
slightly) **(b)** 16 ft $\times$ 24 ft
78.

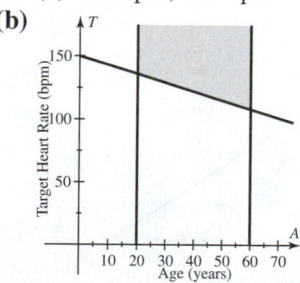

79. (a) 136 bpm; 108 bpm
(b)

(c) Target heart rates above 70% of the maximum heart rate
for ages 20 to 60

CHAPTER 4 TEST (pp. 293–294)

1. $(1, 2)$ **2.** $(-2, -1)$ **3.** $(-1, -2)$ **4.** $(-2, 3)$ **5.** $(1, 3)$
6. (a) $(2, 1)$; one; consistent **(b)** No solutions; zero;
inconsistent **7. (a)** Infinitely many **(b)** Consistent;
dependent **8. (a)** None **(b)** Inconsistent **9.** $(-3, 4)$
10. No solutions **11.** Infinitely many solutions

12. $(-1, 3)$ **13.** $y \leq -\frac{1}{2}x$ **14.** Yes
15.

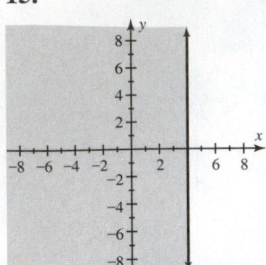

16.

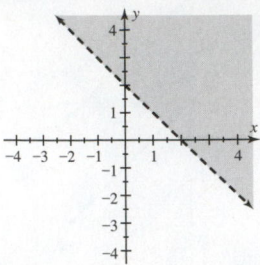

16. (a)

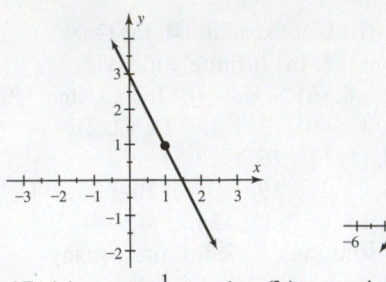

(b)

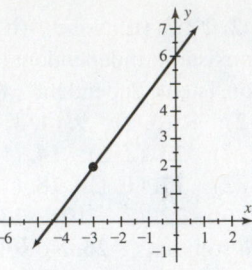

17. 4
18.

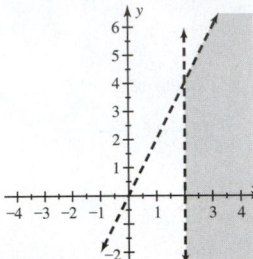

19.

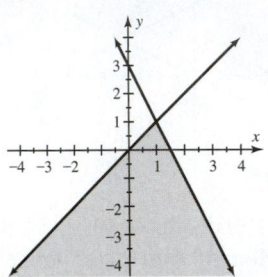

17. (a) $y = -\frac{1}{2}x - 1$ **(b)** $y = 4x - 3$
18. $y = -\frac{1}{4}x + 3$ **19.** $y = -3x + 4$ **20.** $(4, 4)$
21. $(1, -3)$ **22.** No solutions   **23.** Infinitely many
solutions   **24.** $(-1, 2)$
25.

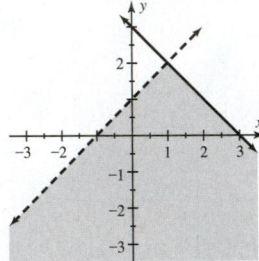

26.

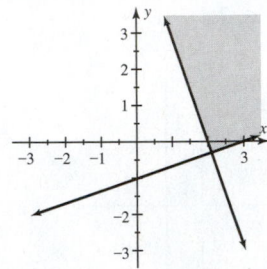

20. $3.7 trillion in 2009; $4.3 trillion in 2010
21. 7.5 L of 20% solution and 2.5 L of 60% solution
22. $\frac{2}{3}$ hr at 6 mph; $\frac{1}{3}$ hr at 9 mph

27. $F = 16.5R$ **28.** 94°C   **29.** $245   **30.** $158.05
31. $G = -3x + 30$ **32.** $1200 at 5%; $1200 at 6%

CHAPTERS 1–4 CUMULATIVE REVIEW (pp. 295–296)

1. $2 \cdot 2 \cdot 2 \cdot 3 \cdot 5$ **2. (a)** 4   **(b)** 8   **3. (a)** Rational
(b) Irrational   **4. (a)** $<$ **(b)** $>$ **5. (a)** $4x^2$ **(b)** $5x - 2$
6. $\frac{2}{9}$ **7.** $\frac{1}{2}$ **8.** No solutions   **9.** 11, 12, 13, 14
10. 432 in^2 or 3 ft^2 **11.** $x = \frac{W + 7y}{3}$ **12.** $x \geq 1$
13.

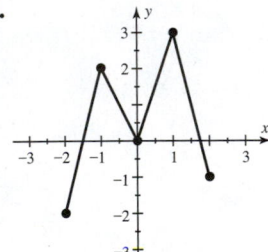

5 Polynomials and Exponents

SECTION 5.1 (pp. 305–306)

1. base; exponent   **2.** 1   **3.** a^{m+n} **4.** a^{mn} **5.** $a^n b^n$ **6.** $\frac{a^n}{b^n}$
7. 64   **9.** -8 **11.** -8 **13.** 1   **15.** 11   **17.** $\frac{1}{2}$
19. 3^3 or 27   **21.** 4^8 or 65,536   **23.** x^9 **25.** x^6 **27.** $20x^7$
29. $-3x^3y^4$ **31.** 2^6 or 64   **33.** n^{12} **35.** x^7 **37.** $49b^2$
39. a^3b^3 **41.** 1   **43.** $-64b^6$ **45.** $x^{14}y^{21}$ **47.** $x^{12}y^9$
49. $a^{10}b^8$ **51.** $\frac{1}{27}$ **53.** $\frac{a^5}{b^5}$ **55.** $\frac{(x - y)^3}{27}$ **57.** $\frac{25}{(a + b)^2}$
59. $\frac{8x^3}{125}$ **61.** $\frac{27x^6}{125y^{12}}$ **63.** $(x + y)^4$ **65.** $(a + b)^5$ **67.** 6
69. $a^3 + 2ab^2$ **71.** $12a^5 + 6a^3b$ **73.** $r^2t + rt^2$
75. $a = 3, b = 1, m = 1, n = 0$ (answers may vary)
77. $10x^4$ **79.** $9\pi x^4$ **81.** $8x^3$ **83.** $1157.63
85. (a) 3^2 **(b)** $72,000

14. (a)

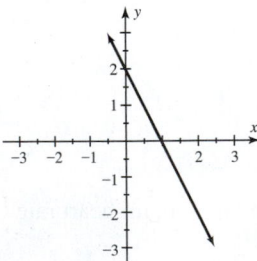

(b)

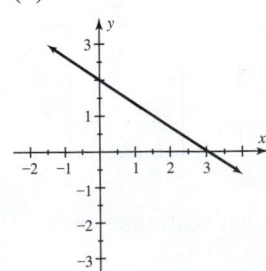

15. 2; -8

SECTION 5.2 (pp. 314–316)

1. monomial   **2.** polynomial   **3.** degree   **4.** degree
5. binomial   **6.** trinomial   **7.** like   **8.** like   **9.** opposite
10. vertically   **11.** 2; 3   **13.** 2; -1 **15.** 2; -5 **17.** 0; 6
19. Yes; 1; 1; 1   **21.** Yes; 3; 1; 2   **23.** No   **25.** No
27. Yes; 1; 3; 6   **29.** Yes; x **31.** Yes; $-5x^3$ **33.** No
35. Yes; $2ab$ **37.** $-x + 9$ **39.** $4x^2 + 8x + 1$
41. $4y^3 + 3y^2 + 4y - 9$ **43.** $4xy + 1$ **45.** $2a^2b^3$
47. $9x^2 + x - 6$ **49.** $x^2 - 7x - 1$ **51.** $-5x^2$
53. $-3a^2 + a - 4$ **55.** $2t^2 + 3t - 4$ **57.** $4x - 2$
59. $-3x^2 + 5x + 2$ **61.** $z^3 - 6z^2 - 6z - 1$
63. $3xy + 2x^2y^2$ **65.** $-a^3b$ **67.** $-x^2 - 5x - 4$
69. $-2x^3 - 6x - 2$ **71. (a)** 200 bpm   **(b)** 100 bpm

(c) It decreases quickly at first, then more slowly.
73. $2z^2$; 200 in^2 **75.** $2x^2 + x^2$ or $3x^2$; 108 ft^2
77. $\pi x^2 + \pi y^2$; 13π ft^2
79. (a) $m_1 \approx 0.077$; $m_2 \approx 0.083$; $m_3 \approx 0.077$. A line is
reasonable but not exact.   **(b)** For the given years, its estimates are reasonable.

CHECKING BASIC CONCEPTS 5.1 & 5.2 (p. 316)

1. (a) -25 **(b)** 1   **2. (a)** 10^8 **(b)** $-12x^7$ **(c)** a^6b^2
(d) $\frac{x^4}{z^{12}}$ **3. (a)** 1   **(b)** $9x^{14}$ **(c)** $10a^5 - 14a^2$
4. 3; 2; 4   **5. (a)** $2w^2h$ **(b)** 2880 in^3 **6. (a)** $3a^2 + 6$
(b) $2z^3 + 7z - 8$ **(c)** $4xy$

SECTION 5.3 (pp. 322–324)

1. The product rule   **2.** Distributive   **3.** term; term
4. vertically   **5.** x^7 **7.** $-12a^2$ **9.** $20x^5$ **11.** $4x^2y^3$
13. $-12x^3y^3$ **15.** $3x + 12$ **17.** $-45x - 5$ **19.** $4z - z^2$
21. $-5y - 3y^2$ **23.** $15x^3 - 12x$ **25.** $6x^3 - 6x^2$
27. $-32t^2 - 8t - 8$ **29.** $-5n^4 + n^3 - 2n^2$
31. $x^2y + xy^2$ **33.** $x^4y - x^3y^2$ **35.** $-a^4b + 2ab^4$
37. $x^2 + 3x$ **39.** $x^2 + 4x + 4$ **41.** $x^2 + 9x + 18$
43. $x^2 + 8x + 15$ **45.** $x^2 - 17x + 72$
47. $6z^2 - 19z + 10$ **49.** $64b^2 - 1$ **51.** $10y^2 - 3y - 7$
53. $5 - 13a + 6a^2$ **55.** $1 - 9x^2$ **57.** $x^3 - x^2 + x - 1$
59. $4x^3 - 3x^2 + 16x - 12$ **61.** $2n^3 + n^2 + 6n + 3$
63. $m^3 + 4m^2 + 4m + 1$ **65.** $6x^3 - 7x^2 + 14x - 8$
67. $x^3 + 1$ **69.** $4b^4 + 3b^3 + 19b^2 + 9b + 21$
71. $x^3 - x^2 - 5x + 2$ **73.** $a^3 - 8$
75. $6x^4 - 2x^3 + 5x^2 - x + 1$ **77.** $m \cdot n$
79. (a) $h^3 - 2h^2 - 8h$ **(b)** 14,175 in^3
81. (a) $50x - x^2$ **(b)** 625   **83.** $6x^2 + 12x + 6$
85. (a) $64t - 16t^2$ **(b)** 64; 64   **(c)** Yes; yes

SECTION 5.4 (pp. 329–331)

1. $a^2 - b^2$ **2.** $a^2 + 2ab + b^2$ **3.** $a^2 - 2ab + b^2$
4. $(a + b)^2$ **5.** False   **6.** False   **7.** $x^2 - 9$ **9.** $16x^2 - 1$
11. $1 - 4a^2$ **13.** $4x^2 - 9y^2$ **15.** $a^2b^2 - 25$
17. $a^4 - b^4$ **19.** $x^6 - y^6$ **21.** 9999   **23.** 391   **25.** 9900
27. $a^2 - 4a + 4$ **29.** $4x^2 + 12x + 9$
31. $9b^2 + 30b + 25$ **33.** $\frac{9}{16}a^2 - 6a + 16$
35. $1 - 2b + b^2$ **37.** $25 + 10y^3 + y^6$
39. $a^4 + 2a^2b + b^2$ **41.** $a^3 + 3a^2 + 3a + 1$
43. $x^3 - 6x^2 + 12x - 8$ **45.** $8x^3 + 12x^2 + 6x + 1$
47. $216u^3 - 108u^2 + 18u - 1$ **49.** $20x + 36$
51. $x^2 + 2x - 35$ **53.** $9x^2 - 30x + 25$
55. $25x^2 + 35x + 12$ **57.** $16b^2 - 25$
59. $-20x^3 + 35x^2 - 10x$ **61.** $64 - 48a + 12a^2 - a^3$
63. $x^3 + 6x^2 + 9x$ **65.** $x^4 - 5x^2 + 4$ **67.** $a^{2n} - b^{2n}$
69. (a) $x^2 + 4x + 4$ **(b)** $x^2 + 4x + 4$
71. (a) $4x^2 + 12x + 9$ **(b)** $4x^2 + 12x + 9$
73. (a) $6x^2 + 60x + 150$ **(b)** $x^3 + 15x^2 + 75x + 125$
75. (a) $1 + 2x + x^2$ **(b)** 1.21; the money increases by
1.21 times in 2 years if the interest rate is 10%.

77. (a) $1 - 2x + x^2$ **(b)** 0.25; if the chance of rain on
each day is 50%, then there is a 25% chance that it will not
rain on either day.
79. (a) $32z + 256$ **(b)** 2176; the area of an 8-foot-wide
sidewalk around a 60×60 foot pool is 2176 ft^2.

CHECKING BASIC CONCEPTS 5.3 & 5.4 (p. 331)

1. (a) $-15x^3y^5$ **(b)** $-6x + 4x^2$ **(c)** $3a^3b - 6a^2b^2 + 3ab^3$
2. (a) $4x^2 + 9x - 9$ **(b)** $2x^4 - 2$ **(c)** $x^3 + y^3$
3. (a) $25x^2 - 4$ **(b)** $x^2 + 6x + 9$ **(c)** $4 - 28x + 49x^2$
(d) $t^3 + 6t^2 + 12t + 8$ **4. (a)** $m^2 + 10m + 25$
(b) $m^2 + 10m + 25$

SECTION 5.5 (pp. 340–342)

1. $\frac{1}{a^n}$ **2.** a^n **3.** a^{m-n} **4.** $\frac{b^m}{a^n}$ **5.** $\left(\frac{b}{a}\right)^n$ **6.** $1 \le b < 10$
7. (a) $\frac{1}{4}$ **(b)** 9   **9. (a)** 2   **(b)** $\frac{1}{1000}$ **11. (a)** $\frac{1}{81}$ **(b)** $\frac{1}{8}$
13. (a) $\frac{1}{576}$ **(b)** 64   **15. (a)** 1   **(b)** 64   **17. (a)** 25   **(b)** $\frac{49}{4}$
19. (a) $\frac{1}{x}$ **(b)** $\frac{1}{a^4}$ **21. (a)** $\frac{1}{x^2}$ **(b)** $\frac{1}{a^8}$ **23. (a)** $\frac{y^3}{x^3}$ **(b)** $\frac{1}{x^3y^3}$
25. (a) $\frac{1}{16t^4}$ **(b)** $\frac{1}{(x+1)^7}$ **27. (a)** a^8 **(b)** $\frac{1}{r^2t^6}$
29. (a) $\frac{b^2}{a^4}$ **(b)** x^2 **31. (a)** a^{13} **(b)** $\frac{2}{z^3}$ **33. (a)** $-\frac{2y^3}{3x^2}$ **(b)** $\frac{1}{x^3}$
35. (a) $2b$ **(b)** $\frac{a^3}{b^3}$ **37. (a)** $\frac{x^7}{3y^8}$ **(b)** $\frac{4a^5}{b^6}$ **39. (a)** y^5 **(b)** $2t^3$
41. (a) $\frac{3a^{10}}{8}$ **(b)** $\frac{1}{32b^9}$ **43. (a)** x^2y^2 **(b)** a^6b^3
45. (a) $\frac{n^4}{72m^{11}}$ **(b)** $\frac{16x^{11}}{y^{13}}$ **47. (a)** $\frac{b^2}{a^2}$ **(b)** $\frac{4v}{u}$
49. (a) $\frac{4}{9a^6b^6}$ **(b)** $\frac{16m^{14}}{25n^2}$ **51.** $\frac{a^n}{a^m} = a^{n-m} = a^{-(m-n)} = \frac{1}{a^{m-n}}$
53. Thousand   **55.** Billion   **57.** Hundredth   **59.** 2000
61. 45,000   **63.** 0.008   **65.** 0.000456   **67.** 39,000,000
69. $-500,000$ **71.** 2×10^3 **73.** 5.67×10^2
75. 1.2×10^7 **77.** 4×10^{-3} **79.** 8.95×10^{-4}
81. -5×10^{-2} **83.** 1,500,000   **85.** -1.5 **87.** 2000
89. 0.002   **91. (a)** About 5.859×10^{12} mi
(b) About 2.5×10^{13} mi   **93.** About 9.2×10^9 mi
95. 1×10^{100} **97. (a)** 1.246×10^{13} **(b)** About $41,812

SECTION 5.6 (pp. 348–349)

1. $\frac{a}{d} + \frac{b}{d}$ **2.** $\frac{a}{d} + \frac{b}{d} - \frac{c}{d}$ **3.** term   **4.** False   **5.** False
6. 9; 4; 1   **7.** $x + 1$; $2x^2 - 2x + 1$; 4   **8.** $0x^2$ **9.** $2x$
11. $z^3 + z^2$ **13.** $\frac{a^2}{2} - 3$ **15.** $\frac{1}{3y^2} + \frac{2}{y}$ **17.** $\frac{4}{x} - 7x^2$
19. $\frac{2}{y} + \frac{1}{y^2}$ **21.** $3x^3 - 1 + \frac{2}{x}$ **23.** $4y^2 - 1 + \frac{2}{y}$
25. $3m^2 - 2m + 4$ **27.** $2x + 1 + \frac{3}{x-2}$ **29.** $x + 1$
31. $x^2 + 1 + \frac{-1}{x-1}$ **33.** $x^2 + 3x - 1$
35. $x^2 - x + 2 + \frac{1}{4x+1}$ **37.** $x^2 + 2x + 3 + \frac{8}{x-2}$
39. $3x^2 + 3x + 3 + \frac{5}{x-1}$ **41.** $x + 3 + \frac{-x-2}{x^2+1}$
43. $x + 1$ **45.** $x^2 - 2x + 4$ **47.** They are the same.
49. $4x$
51. $x + 2$;

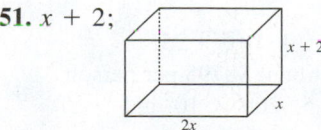

CHECKING BASIC CONCEPTS 5.5 & 5.6 (p. 349)

1. (a) $\frac{1}{81}$ (b) $\frac{1}{2x^7}$ (c) $\frac{b^8}{16a^2}$ **2.** (a) z^5 (b) $\frac{y^6}{x^3}$ (c) $\frac{x^6}{27}$

3. (a) 4.5×10^4 (b) 2.34×10^{-4} (c) 1×10^{-2}

4. (a) 47,100   (b) 0.006   **5.** $5a - 3$ **6.** $3x + 2$; R: -2

7. $x^2 + 2x + 1$; R: $x + 1$ **8.** (a) 9.3×10^7

(b) 500 sec (8 min 20 sec)

CHAPTER 5 REVIEW (pp. 353–356)

1. 125   **2.** -81 **3.** 4   **4.** 11   **5.** -5 **6.** 1   **7.** 6^5

8. 10^{12} **9.** z^9 **10.** y^6 **11.** $30x^9$ **12.** a^4b^4 **13.** 2^{10}

14. m^{20} **15.** a^3b^3 **16.** x^8y^{12} **17.** x^7y^{11} **18.** 1

19. $(r - t)^9$ **20.** $(a + b)^6$ **21.** $\frac{9}{(x - y)^2}$ **22.** $\frac{(x + y)^3}{8}$

23. $6x^3 - 10x^2$ **24.** $12x^2 + 3x^4$ **25.** 7; 6   **26.** 5; -1

27. Yes; 1; 1; 1   **28.** Yes; 4; 1; 3   **29.** Yes; 3; 2; 2   **30.** No

31. $5x^2 - x + 3$ **32.** $-6x^2 + 3x + 7$ **33.** $3x + 4$

34. $-2x^2 - 13$ **35.** $-2x^2 + 9x + 5$ **36.** $3x^2 - 2x - 6$

37. $2a^3 - a^2 + 7a$ **38.** $-2x + 12$ **39.** $5y^2 - 3xy$

40. $4xy$ **41.** $-x^5$ **42.** $-r^3t^4$ **43.** $-6t + 15$

44. $2y - 12y^2$ **45.** $18x^5 + 30x^4$ **46.** $-x^3 + 2x^2 - 9x$

47. $-a^3b + 2a^2b^2 - ab^3$ **48.** $a^2 + 3a - 10$

49. $8x^2 + 13x - 6$ **50.** $-2x^2 + 3x - 1$

51. $2y^3 + y^2 + 2y + 1$ **52.** $2y^4 - y^2 - 1$ **53.** $z^3 + 1$

54. $4z^3 - 15z^2 + 13z - 3$ **55.** $z^2 + z$ **56.** $2x^2 + 4x$

57. $z^2 - 4$ **58.** $25z^2 - 81$ **59.** $1 - 9y^2$

60. $25x^2 - 16y^2$ **61.** $r^2t^2 - 1$ **62.** $4m^4 - n^4$

63. $x^2 + 2x + 1$ **64.** $16x^2 + 24x + 9$

65. $y^2 - 6y + 9$ **66.** $4y^2 - 20y + 25$

67. $16 + 8a + a^2$ **68.** $16 - 8a + a^2$

69. $x^4 + 2x^2y^2 + y^4$ **70.** $x^2y^2 - 4xy + 4$

71. $z^3 + 15z^2 + 75z + 125$ **72.** $8z^3 - 12z^2 + 6z - 1$

73. 3599   **74.** 396   **75.** $\frac{1}{9}$ **76.** $\frac{1}{9}$ **77.** 4   **78.** $\frac{1}{1000}$ **79.** 36

80. $\frac{1}{25}$ **81.** $\frac{9}{16}$ **82.** 1   **83.** $\frac{1}{z^2}$ **84.** $\frac{1}{y^4}$ **85.** $\frac{1}{a^2}$ **86.** $\frac{1}{x^3}$

87. $\frac{1}{4t^2}$ **88.** $\frac{1}{a^3b^6}$ **89.** $\frac{1}{y^3}$ **90.** x^4 **91.** $\frac{2}{x^3}$ **92.** $\frac{2x^4}{3y^3}$ **93.** $\frac{a^5}{b^5}$

94. $4t^4$ **95.** $\frac{n^7}{72m^{12}}$ **96.** $\frac{9x^{10}}{y^{10}}$ **97.** $\frac{y^2}{x^2}$ **98.** $\frac{2v}{3u}$ **99.** 600

100. 52,400   **101.** 0.0037   **102.** 0.06234   **103.** 1×10^4

104. 5.61×10^7 **105.** 5.4×10^{-5} **106.** 1×10^{-3}

107. 24,000,000   **108.** 0.2   **109.** $\frac{5x}{3} + 1$

110. $3b^2 - 2 + \frac{1}{b^2}$ **111.** $3x + 2 + \frac{4}{x - 1}$

112. $3x - 4 + \frac{6}{3x + 2}$ **113.** $x^2 - 3x - 1$

114. $x^2 + \frac{-1}{2x - 1}$ **115.** $x - 1 + \frac{-2x + 2}{x^2 + 1}$

116. $x^2 + 2x + 5$ **117.** (a) 60 bpm   (b) 110 bpm

(c) It increases.   **118.** $6xy$; 72 ft^2

119. $10z^2 + 25z$; 510 in^2 **120.** x^4y^2 **121.** \$833.71

122. $\frac{4}{3}\pi x^3 + 8\pi x^2 + 16\pi x + \frac{32}{3}\pi$ **123.** (a) $96t - 16t^2$

(b) 128; after 2 seconds the ball is 128 ft high.

124. (a) $600L - L^2$ (b) 27,500; a rectangular building with a perimeter of 1200 ft and a side of length 50 ft has an area of 27,500 ft^2.

125. (a) $x^2 + 10x + 25$ (b) $x^2 + 10x + 25$

126. (a) $16x$ (b) 1600   **127.** About \$8795 per person

128. About 552,090,000 gal or 5.5209×10^8 gal

CHAPTER 5 TEST (p. 356)

1. (a) -1 (b) -81 **2.** (a) -6 (b) $\frac{1}{64}$ (c) 8   (d) -3

3. 3; 2; 3   **4.** $x^3 - 4x + 8$ **5.** $4x + 6$

6. $-3y^3 - 2y + 1$ **7.** $x^2 + x - 7$ **8.** $4a^3 + 2ab$

9. $24y^{11}$ **10.** a^5b^8 **11.** x^4 **12.** $\frac{a^3}{b^6}$ **13.** $a^3b - ab^3$

14. $\frac{4}{9a^4b^6}$ **15.** $\frac{2y^3}{x}$ **16.** $\frac{16}{(a + b)^4}$ **17.** $12x^5 - 18x^3 + 3x^2$

18. $2z^2 - 2z - 12$ **19.** $49y^4 - 9$ **20.** $9x^2 - 12x + 4$

21. $m^3 + 9m^2 + 27m + 27$ **22.** $y^3 - y + 6$ **23.** 6396

24. 0.0061   **25.** 5.41×10^3 **26.** $3x - 2 + \frac{1}{x}$

27. $x^2 - x + 1 + \frac{-1}{x + 2}$ **28.** (a) $20t$ (b) $2t + 2000$

(c) $18t - 2000$; profit from selling t tickets

29. $12x^2$; 1200 ft^2 **30.** $3x^3 + 27x^2 + 54x$

31. (a) $88t - 16t^2$ (b) 120; after 3 seconds the ball is 120 ft high.

CHAPTERS 1–5 CUMULATIVE REVIEW (p. 358)

1. (a) 8   (b) 8   **2.** (a) 29   (b) $\frac{3}{2}$ **3.** (a) 7   (b) No solutions

4. (a) Infinitely many solutions   (b) -2 **5.** 68 mph

6. (a) $\frac{21}{50}$ (b) $\frac{19}{250}$

7.

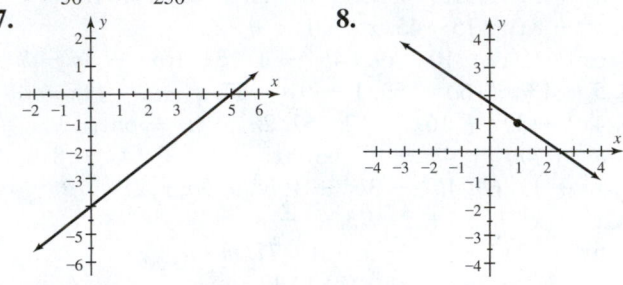

8.

9. $y = -2x + 1$ **10.** 2; -3 **11.** $y = \frac{1}{2}x - 4$

12. $y = 3x + 1$ **13.** No solutions   **14.** $(2, -1)$

15. Infinitely many solutions   **16.** $(0, -6)$

17.

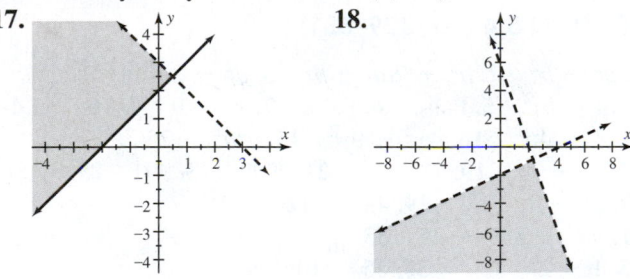

18.

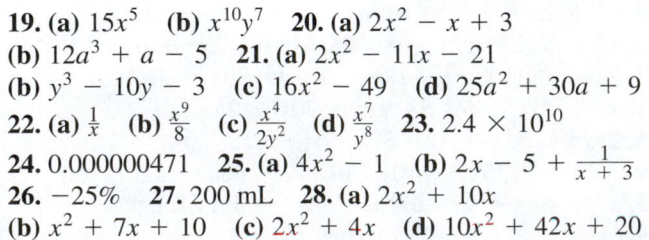

19. (a) $15x^5$ (b) $x^{10}y^7$ **20.** (a) $2x^2 - x + 3$

(b) $12a^3 + a - 5$ **21.** (a) $2x^2 - 11x - 21$

(b) $y^3 - 10y - 3$ (c) $16x^2 - 49$ (d) $25a^2 + 30a + 9$

22. (a) $\frac{1}{x}$ (b) $\frac{x^9}{8}$ (c) $\frac{x^4}{2y^2}$ (d) $\frac{x^7}{y^8}$ **23.** 2.4×10^{10}

24. 0.000000471   **25.** (a) $4x^2 - 1$ (b) $2x - 5 + \frac{1}{x + 3}$

26. -25% **27.** 200 mL   **28.** (a) $2x^2 + 10x$

(b) $x^2 + 7x + 10$ (c) $2x^2 + 4x$ (d) $10x^2 + 42x + 20$

6	**Factoring Polynomials and Solving Equations**

SECTION 6.1 (pp. 366–368)

1. factor   **2.** a **3.** multiplying   **4.** prime; multiplication
5. greatest common factor (GCF)   **6.** grouping
7. $1, 2, x, 2x$ **8.** $2x$
9. $2(x + 2)$;

11. $z(z + 4)$;

13. $3y(y + 4)$;

15. $3x(x + 3)$ **17.** $2y^2(2y - 1)$ **19.** $2z(z^2 + 4z - 2)$
21. $3xy(2x - y)$ **23.** $6x; 6x(1 - 3x)$
25. $4y^2; 4y^2(2y - 3)$ **27.** $3z; 3z(2z^2 + z + 3)$
29. $x^2; x^2(x^2 - 5x - 4)$
31. $5y^2; 5y^2(y^3 + 2y^2 - 3y + 2)$ **33.** $x; x(y + z)$
35. $ab; ab(b - a)$ **37.** $5x^2y^3; 5x^2y^3(y + 2x)$
39. $ab; ab(a + b + 1)$ **41.** $(x - 2)(x + 1)$
43. $(z + 4)(z + 5)$ **45.** $2x(2x^2 - 1)(x - 5)$
47. $(x^2 + 3)(x + 2)$ **49.** $(y^2 + 1)(2y + 1)$
51. $(2z^2 + 5)(z - 3)$ **53.** $(4t^2 + 3)(t - 5)$
55. $(3r^2 - 2)(3r + 2)$ **57.** $(7x^2 - 2)(x + 3)$
59. $(y^2 - 2)(2y - 7)$ **61.** $(z^2 - 7)(z - 4)$
63. $(x^3 + 2)(2x - 3)$ **65.** $(x + y)(a + b)$
67. $3(x^2 + 1)(x + 2)$ **69.** $2y(3y^2 - 1)(y - 4)$
71. $x^2(x^2 - 3)(x + 2)$ **73.** $2x^2(x^2 - 3)(2x + 1)$
75. $xy(x + y)(x - 2)$ **77.** $2x^2y^2(x + 2)(y - x)$
79. $a\left(x^2 + \frac{b}{a}x + \frac{c}{a}\right)$ **81. (a)** $16t$ **(b)** $16t(5 - t)$
83. $2x^2 - 4x; 2x(x - 2)$ **85.** $8y^2 - xy; y(8y - x)$
87. (a) 168 in^3 **(b)** $4x(x^2 - 15x + 50)$

SECTION 6.2 (pp. 374–375)

1. 1 **2.** $x^2 + (m + n)x + mn; m + n; mn$ **3.** $mn; m + n$
4. prime   **5.** $4, 7$ **7.** $3, -10$ **9.** $-5, 10$ **11.** $-7, -4$
13. $(x + 1)(x + 2)$ **15.** $(y + 2)(y + 2)$ **17.** Prime
19. $(x + 3)(x + 5)$ **21.** $(m + 4)(m + 9)$
23. $(n + 10)(n + 10)$ **25.** $(x - 1)(x - 5)$
27. $(y - 3)(y - 4)$ **29.** $(z - 5)(z - 8)$
31. $(a - 7)(a - 9)$ **33.** Prime   **35.** $(b - 5)(b - 25)$
37. $(x - 5)(x + 18)$ **39.** $(m - 5)(m + 9)$ **41.** Prime
43. $(n - 10)(n + 20)$ **45.** $(x - 1)(x + 23)$
47. $(a - 4)(a + 8)$ **49.** $(b + 4)(b - 5)$ **51.** Prime
53. $(x + 8)(x - 9)$ **55.** $(y + 2)(y - 17)$
57. $(z + 6)(z - 11)$ **59.** $5(x - 4)(x + 2)$
61. $-3(m + 4)(m - 1)$ **63.** $y(y - 2)(y - 5)$
65. $-x(x + 5)(x - 3)$ **67.** $3a(a + 1)(a + 6)$

69. $-2x(x^2 - 3x + 4)$ **71.** $2m^2(m - 7)(m + 2)$
73. $-3x^2(x + 1)(x - 2)$ **75.** $(x + 1)(x + 5)$
77. $(x - 1)(x - 3)$ **79.** $(6 - x)(2 + x)$
81. $(8 + x)(4 - x)$ **83.** $(x + 1)(x + k)$
85. $x + 1$;   **87.** $x + 1, x + 2$;

89. $x + 1$ **91.** $(x + 2)(x + 6)$

CHECKING BASIC CONCEPTS 6.1 & 6.2 (p. 375)

1. $4x$ **2.** $6z^2(2z - 3)$ **3. (a)** $(6y + 5)(y - 2)$
(b) $(x^2 + 5)(2x + 1)$ **(c)** $4(z^2 + 1)(z - 3)$
4. (a) $(x + 2)(x + 4)$ **(b)** $(x - 7)(x + 6)$ **(c)** Prime
(d) $4a(a + 2)(a + 3)$ **5.** $(x + 5)(x + 5)$

SECTION 6.3 (pp. 382–383)

1. $ac; b$ **2.** $ax^2; c$ **3.** $+; +$ **4.** $+; -$ **5.** $-; -$
6. $-; +$ **7.** $3; x$ **8.** $x; 4x$ **9.** $1; 3$ **10.** $1; 2x$ **11.** $3x; 1$
12. $6x; x$ **13.** $5; 3$ **14.** $6x; 8$ **15.** $(x + 3)(2x + 1)$
17. Prime   **19.** $(x + 1)(3x + 1)$ **21.** $(2x + 3)(3x + 1)$
23. $(x - 2)(5x - 1)$ **25.** $(y - 1)(2y - 5)$ **27.** Prime
29. $(z - 5)(7z - 2)$ **31.** $(t - 3)(3t + 2)$
33. $(3r + 2)(5r - 3)$ **35.** $(3m - 4)(8m + 3)$
37. $(5x - 1)(5x + 2)$ **39.** $(x + 2)(6x - 1)$ **41.** Prime
43. $(3n + 1)(7n - 1)$ **45.** $(2y + 3)(7y + 1)$
47. $(4z - 3)(7z - 1)$ **49.** $(3x - 2)(10x - 3)$ **51.** Prime
53. $(2t + 3)(9t - 2)$ **55.** $3(2a - 1)(2a + 3)$
57. $y(3y - 2)(4y - 1)$ **59.** $3x(4x - 3)(2x - 1)$
61. $2x^2(4x^2 - 3x + 1)$ **63.** $7x^2(2x + 1)(2x + 3)$
65. $(3x + 1)(x + k)$ **67.** $(7x + 1)(x + 2)$
69. $(2x - 1)(x - 2)$ **71.** $-(4x + 3)(2x - 1)$
73. $-(x + 5)(2x - 3)$ **75.** $-(x - 3)(5x + 1)$
77. $3x + 2$ by $2x + 1$;

79. $(2x + 1)(x + 3)$

SECTION 6.4 (pp. 389–390)

1. $(a - b)(a + b)$ **2.** $6x; 7y$ **3.** False   **4.** False
5. $(a + b)^2$ **6.** $(a - b)^2$ **7.** $6x$ **8.** $20rt$
9. $(a + b)(a^2 - ab + b^2)$ **10.** $(a - b)(a^2 + ab + b^2)$
11. $2x; 3y$ **12.** $x; 1$ **13.** $-; +$ **14.** $+; -$
15. $(x - 1)(x + 1)$ **17.** $(z - 10)(z + 10)$
19. $(2y - 1)(2y + 1)$ **21.** $(6z - 5)(6z + 5)$
23. $(3 - x)(3 + x)$ **25.** $(1 - 3y)(1 + 3y)$
27. $(2a - 3b)(2a + 3b)$ **29.** $(6m - 5n)(6m + 5n)$
31. $(9r - 7t)(9r + 7t)$ **33.** $(x + 4)^2$ **35.** Not possible
37. $(x - 3)^2$ **39.** $(3y + 1)^2$ **41.** $(2z - 1)^2$
43. Not possible   **45.** $(3x + 5)^2$ **47.** $(2a - 9)^2$
49. $(x + y)^2$ **51.** $(r - 5t)^2$ **53.** Not possible
55. $(z + 1)(z^2 - z + 1)$ **57.** $(x + 4)(x^2 - 4x + 16)$

59. $(y - 2)(y^2 + 2y + 4)$ **61.** $(n - 1)(n^2 + n + 1)$
63. $(2x + 1)(4x^2 - 2x + 1)$
65. $(m - 4n)(m^2 + 4mn + 16n^2)$
67. $(2x + 5y)(4x^2 - 10xy + 25y^2)$ **69.** $4(x - 2)(x + 2)$
71. $2(y - 7)^2$ **73.** $5(z + 2)(z^2 - 2z + 4)$
75. $xy(x - y)(x + y)$ **77.** $2m(m^2 - 5m + 9)$
79. $7x^2(10x - 3y)(10x + 3y)$
81. $2(2a + b)(4a^2 - 2ab + b^2)$ **83.** $4b^2(b + 3)^2$
85. $4(5r - 2t)(25r^2 + 10rt + 4t^2)$
87. $2x + 3$;

$2x$	$4x^2$	$6x$
3	$6x$	9
	$2x$	3

CHECKING BASIC CONCEPTS 6.3 & 6.4 (p. 391)

1. (a) $(x - 4)(2x + 3)$ **(b)** $(2x + 7)(3x - 2)$
2. (a) Prime   **(b)** $2y(3y + 1)(y - 2)$
3. $(3x + 2)(x + 3)$ **4. (a)** $(z - 8)(z + 8)$
(b) $(3r - 2t)(3r + 2t)$ **5. (a)** $(x + 6)^2$ **(b)** $(3a - 2b)^2$
6. (a) $(m - 3)(m^2 + 3m + 9)$
(b) $(5n + 3)(25n^2 - 15n + 9)$
7. (a) $4(2x - 1)(2x + 1)$ **(b)** $3y(y + 2)(y^2 - 2y + 4)$

SECTION 6.5 (pp. 395–396)

1. Greatest common factor   **2.** GCF   **3.** Grouping
4. squares; cubes; cubes   **5.** No; a sum of squares cannot
be factored.   **6.** Yes; a sum of cubes can be factored.
7. square; grouping; FOIL   **8.** completely
9. $2(2x - 1)$ **11.** $2(y^2 - 2y + 2)$ **13.** $(z - 2)(z + 2)$
15. $(a + 2)(a^2 - 2a + 4)$ **17.** $(2b - 3)^2$
19. Not possible   **21.** $(x^2 + 5)(x - 1)$
23. $(y - 4)(y - 1)$ **25.** $(x - 3)(x + 3)(x + 4)$
27. $8(a - 2)(a^2 + 2a + 4)$ **29.** $2x(3x^2 + 1)(2x - 3)$
31. $2t(3t + 2)(9t^2 - 6t + 4)$ **33.** $2(r - 1)(r + 1)(r + 3)$
35. $3z^2(2z + 3)(z - 5)$ **37.** $2b^2(3b - 1)(2b - 1)$
39. $6z(y - 2z)(y + 2z)$ **41.** $3y(x - 5)^2$
43. $(3m - 2n)(9m^2 + 6mn + 4n^2)$
45. $3(x - 2)(x + 2)(x - 1)(x^2 + x + 1)$
47. $(5a + 3)(a - 6)$ **49.** $3r(t + 5)(t + 6)$
51. $(3b^2 + 4)(3b + 2)$ **53.** $2n(3n^2 + n - 5)$
55. $4(x - 3y)(x + 3y)$ **57.** $2a(a - 4)^2$
59. $4x(2y + 1)(4y^2 - 2y + 1)$
61. $2b(2b - 1)(2b + 1)(b + 3)$ **63.** $3x + 1$

SECTION 6.6 (pp. 402–404)

1. 0; 0   **2.** No; one side of the equation must be zero.
3. $2x = 0$; $x + 6 = 0$ **4.** Subtract $4x$ from each side.
5. Apply the zero-product property.   **6.** solving   **7.** zero
8. 2   **9.** $ax^2 + bx + c = 0$ with $a \neq 0$
10. $x^2 - 6x + 1 = 0$ **11.** descending   **12.** 0   **13.** 0
15. $-8, 0$ **17.** $1, 2$ **19.** $\frac{1}{2}, \frac{3}{4}$ **21.** $\frac{1}{3}, \frac{3}{7}$ **23.** $0, 5, 8$
25. $0, 1$ **27.** $0, 5$ **29.** $-\frac{3}{2}, 0$ **31.** $-1, 1$ **33.** $-\frac{1}{2}, \frac{1}{2}$
35. $-2, -1$ **37.** $5, 7$ **39.** $-2, \frac{1}{2}$ **41.** $-\frac{5}{2}, -\frac{2}{3}$ **43.** $-5, 5$
45. $0, 5$ **47.** $-3, 0$ **49.** $-1, 6$ **51.** $-\frac{5}{6}, \frac{1}{2}$ **53.** $-2, 1$

55. $-3, \frac{1}{2}$ **57.** $-\frac{1}{2}$ **59.** $-1, -\frac{2}{3}$ **61.** 12 ft   **63.** 4   **65.** 4
67. (a) 6 sec

(b)

Time (t)	0	1	2	3	4	5	6
Height (h)	0	80	128	144	128	80	0

; 3 sec

69. (a) 81.8 ft; 327.3 ft; when the speed doubles, the
braking distance quadruples.   **(b)** About 19 mph
(c) About 19 mph; yes   **71. (a)** About 11 million;
about 66 million   **(b)** About 1981   **73.** 40 by 50 pixels

CHECKING BASIC CONCEPTS 6.5 & 6.6 (p. 404)

1. (a) $9(a^2 - 2a + 3)$ **(b)** $7x(y^2 + 4)$
2. (a) $2z^2(3z - 2)(z - 4)$ **(b)** $2r^2(t - 3)(t + 3)$
3. (a) $4x(3x - 2)^2$ **(b)** $3(2b - 3)(4b^2 + 6b + 9)$
4. (a) $0, \frac{3}{2}$ **(b)** $-1, \frac{3}{5}$ **5.** $-3, 1$ **6.** After $\frac{1}{2}$ sec

SECTION 6.7 (pp. 409–410)

1. GCF   **2.** zero-product   **3.** factors
4. $(x^2 - y^2)^2$ **5.** $(z^2 + 1)(z^2 + 2)$
6. No; $x^4 - 1 = (x - 1)(x + 1)(x^2 + 1)$
7. Yes; $x^2 + 1$ cannot be factored further.
8. Subtract x from each side.   **9.** Factor out x^2.   **10.** One
11. $5(x - 3)(x + 2)$ **13.** $-4(y + 2)(y + 6)$
15. $-10(z + 5)(2z + 1)$ **17.** $-4(t + 3)(7t - 5)$
19. $r(r - 1)(r + 1)$ **21.** $3x(x - 2)(x + 3)$
23. $12z(2z - 1)(3z + 2)$ **25.** $x^2(x - 2)(x + 2)$
27. $t^2(t - 1)(t + 2)$ **29.** $(x^2 - 3)(x^2 - 2)$
31. $(x^2 + 3)(2x^2 + 1)$ **33.** $(y^2 + 3)^2$
35. $(x^2 - 3)(x^2 + 3)$ **37.** $(x - 3)(x + 3)(x^2 + 9)$
39. $z^3(z + 1)^2$ **41.** $(x + y)(2x - y)$
43. $(a + b)^2(a - b)^2$ **45.** $x(x + y)(x - y)$
47. $x(2x + y)^2$ **49. (a)** $x(x - 2)(x + 2)$ **(b)** $-2, 0, 2$
51. (a) $2y(y - 6)(y + 3)$ **(b)** $-3, 0, 6$
53. (a) $(x^2 + 4)(x - 1)$ **(b)** 1   **55.** $-8, -3$ **57.** $-1, 3$
59. $-1, 0, 4$ **61.** $-6, 0, 4$ **63.** $-6, 0, 6$ **65.** $-7, 0, 1$
67. $-3, -2, 2, 3$ **69.** $-1, 1$ **71.** $-3, 3$ **73.** $-1, 1, 2$
75. 5   **77. (a)** $x < 7.5$ in. because the width is 15 in.
(b) $300 - 4x^2$ **(c)** 2.5 in.
79. 18.6 trillion ft^3 **81.** Factoring is very difficult (answers
may vary).

CHECKING BASIC CONCEPTS 6.7 (p. 410)

1. (a) $3(x - 4)(x + 2)$ **(b)** $-5(2y + 1)(y - 1)$
2. (a) $(z^2 - 5)(z^2 + 5)$ **(b)** $7(t - 1)(t + 1)(t^2 + 1)$
3. (a) $(x - 2)^2(x + 2)^2$ **(b)** $y(y + 10)(2y - 3)$
4. $-4, 0, 3$ **5.** 3

CHAPTER 6 REVIEW (pp. 413–415)

1. 5; $5(x - 3)$ **2.** y; $y(y + 2)$ **3.** $4z^2$; $4z^2(2z - 1)$
4. $3x^2$; $3x^2(2x^2 + x - 4)$ **5.** $3y$; $3y(3x + 5z^2)$
6. a^2b^2; $a^2b^2(b + a)$ **7.** $(x - 3)(x + 2)$
8. $y(y + 3)(x - 5)$ **9.** $(z^2 + 5)(z - 2)$
10. $(t^2 + 8)(t + 1)$ **11.** $(x^2 + 6)(x - 3)$
12. $(x - y)(a + b)$ **13.** $x^2(x^2 - 2)(x + 3)$
14. $2y(y^2 + 1)(y + 3)$ **15.** $4, 5$ **16.** $-3, 7$ **17.** $-9, -4$

18. $-25, 4$ **19.** $(x - 4)(x + 3)$ **20.** $(x + 4)(x + 6)$
21. $(x - 2)(x + 8)$ **22.** $(x - 7)(x + 6)$ **23.** Prime
24. Prime **25.** $(x - 1)(x + 3)$ **26.** $(x + 10)(x + 12)$
27. $2x(x - 2)(x + 5)$ **28.** $x^2(x + 4)(x - 7)$
29. $(2 - x)(5 - x)$ **30.** $(6 - x)(4 + x)$
31. $-2(x + 5)(x - 3)$ **32.** $-x(x + 10)(x - 1)$
33. $(3x - 1)(3x + 2)$ **34.** $(x - 1)(2x + 5)$
35. $(x + 3)(3x + 5)$ **36.** $(5x - 1)(7x + 1)$ **37.** Prime
38. Prime **39.** $(3x + 1)(8x - 5)$ **40.** $(x + 9)(4x - 3)$
41. $3x(2x + 7)(2x + 1)$ **42.** $2x^2(x + 3)(4x - 5)$
43. $(3 - 2x)(4 + x)$ **44.** $(1 - 2x)(1 + 5x)$
45. $(z - 2)(z + 2)$ **46.** $(3z - 8)(3z + 8)$
47. $(6 - y)(6 + y)$ **48.** $(10a - 9b)(10a + 9b)$
49. $(x + 7)^2$ **50.** $(x - 5)^2$ **51.** $(2x - 3)^2$
52. $(3x + 8)^2$ **53.** $(2t - 1)(4t^2 + 2t + 1)$
54. $(3r + 2t)(9r^2 - 6rt + 4t^2)$ **55.** $2x(x - 5)(x + 5)$
56. $3(2x + 3)(4x^2 - 6x + 9)$ **57.** $2x(x + 7)^2$
58. $2x(x - 4)(x^2 + 4x + 16)$ **59.** $4(3x - 2)$
60. $3x^2(2x + 3)$ **61.** $3(3y^2 - 2y + 2)$
62. $y(z - 3)(z + 3)$ **63.** $x(x - 2)(x + 2)(x + 7)$
64. $3x(2x + 3)^2$ **65.** $3a(b - 2)(b^2 + 2b + 4)$
66. $5x(x^2 + 4)$ **67.** $6x(2x - y)(2x + y)$
68. $y(x + 3)(x^2 - 3x + 9)$ **69.** $m = 0$ or $n = 0$
70. 0 **71.** $-9, \frac{3}{4}$ **72.** $-\frac{6}{5}, \frac{1}{4}$ **73.** $0, 1, 2$ **74.** $0, 7$
75. $-8, 8$ **76.** $-7, -2$ **77.** $-2, 3$ **78.** $-\frac{3}{2}, \frac{2}{5}$
79. $-4, 18$ **80.** $-2, \frac{5}{2}$ **81.** $5(x - 5)(x + 2)$
82. $-3(x - 3)(x + 5)$ **83.** $y(y - 2)(y + 2)$
84. $3y(y - 1)(y + 3)$ **85.** $2z^2(z + 2)(z + 5)$
86. $8z^2(z - 2)(z + 2)$ **87.** $(x^2 - 3)^2$
88. $(x - 3)(x + 3)(2x^2 + 3)$ **89.** $(a + 5b)^2$
90. $x(x + y)(x - y)$ **91.** $-\frac{1}{2}, 5$ **92.** $0, \frac{5}{2}, 3$ **93.** $-5, 0, 5$
94. $0, 3, 4$ **95.** $-2, 2$ **96.** $-4, 4$ **97.** -4 **98.** $-1, 1$
99. $3x + 7$; **100.** $x + 1$ by $x + 5$;

$3x$	$9x^2$	$21x$
7	$21x$	49
	$3x$	7

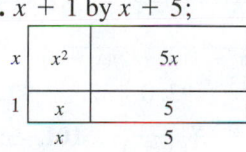

101. $x + 1$ **102.** $(x + 1)(x + 3)$ **103.** $(2x + 3)(x + 6)$
104. 2 **105.** 8 ft by 15 ft **106.** After 2 sec and 3 sec
107. (a) 390 ft **(b)** 15 mph **(c)** 15 mph; yes
108. (a) \$10,000 **(b)** \$50 or \$150 **(c)** \$50 or \$150; yes
109. (a) 372 million **(b)** 1975 **110.** 50 by 80 pixels
111. (a) $2000 - 4x^2$ **(b)** 5 in. **112.** $x + 2$

CHAPTER 6 TEST (pp. 415–416)

1. $4xy$; $4xy(x - 5y + 3)$ **2.** $3a^2b^2$; $3a^2b^2(3a + 1)$
3. $(y + z)(a + b)$ **4.** $(x^2 - 5)(3x + 1)$
5. $(y - 2)(y + 6)$ **6.** $(2x + 5)^2$ **7.** $(z - 4)(4z - 3)$
8. $(t - 7)(2t - 3)$ **9.** Prime **10.** Prime
11. $3x(x + 1)(2x - 1)$ **12.** $2(z - 3)(z + 3)(z^2 + 3)$
13. $4y(3y - 5)(3y + 5)$ **14.** $7x(x + 2)(x^2 - 2x + 4)$
15. $a^2(4a + 3)^2$ **16.** $2(b - 2)(b + 2)(b^2 + 4)$
17. $-4, 4$ **18.** $-4, 5$ **19.** $\frac{4}{3}$ **20.** $-6, 11$ **21.** $-3, 0, 3$
22. $-2, -1, 1, 2$ **23.** $3x + 5$ **24.** $(x + 2)(x + 3)$
25. (a) 275 ft **(b)** 33 mph **26.** 1 sec and 2 sec

CHAPTERS 1–6 CUMULATIVE REVIEW
(pp. 417–418)

1. $\frac{3}{7}$ **2.** $\frac{7}{10}$ **3.** 17 **4.** $-\frac{7}{2}$
5. 1;

x	-2	-1	0	1	2
$2x + 3$	-1	1	3	5	7

6. $3n - 5 = n - 7$; -1 **7.** $\frac{57}{1000}$; 0.057
8. $W = \frac{P - 2L}{2}$ **9.** $z > 2$
10.

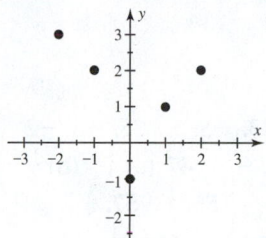

11. x-int: $\frac{2}{3}$; y-int: -2

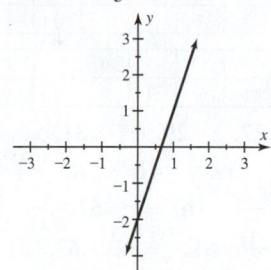

12. y-int: -2

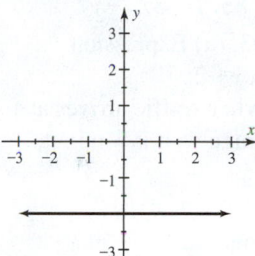

13. $y = -\frac{3}{2}x + \frac{7}{2}$ **14.** $y = \frac{4}{3}x + \frac{11}{3}$
15. -1; -2; $y = -2x - 2$ **16.** $(1, 2)$ **17.** $(1, -1)$
18. $(-2, 5)$
19.   **20.**

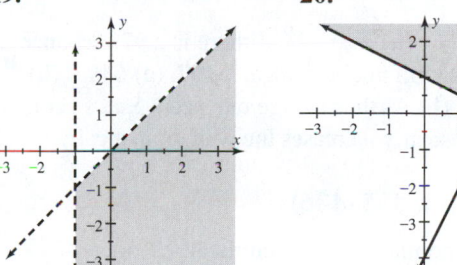

21. -16 **22.** 1 **23.** $\frac{x^{10}}{y^4}$ **24.** $-14x^5 + 21x^4$ **25.** $\frac{1}{a^2}$
26. $\frac{1}{4t^6}$ **27.** $\frac{1}{x^2y^5}$ **28.** $32x^5y^{10}$ **29.** $2x^2 + 4x$
30. $3x + \frac{-4x + 1}{x^2 + 1}$ **31.** $(x - 4)(x + 7)$
32. $(2y + 3)(3y - 4)$ **33.** $(5x - 2y)(5x + 2y)$
34. $(8x - 1)^2$ **35.** $(3t - 2)(9t^2 + 6t + 4)$
36. $-4(x - 3)(x + 2)$ **37.** $-5, 0, 5$ **38.** $-2, 1$
39. $-\frac{7}{2}, 0, \frac{7}{2}$ **40.** $-3, 3$ **41. (a)** $10x + 8x$ or $18x$
(b) 50 min **42.** $6\frac{11}{12}$ mi
43. 25 min skiing; 35 min running
44. (a) $C = 0.25x + 20$ **(b)** 320 mi
45. (a) $2x + y = 180$; $2x - y = 20$ **(b)** $(50, 80)$, the
angles are 50°, 50°, and 80°.
46. $18xy$; 108 yd^2 **47.** $x + 6$ **48. (a)** 4 sec **(b)** After
1 sec and 3 sec

7 Rational Expressions

SECTION 7.1 (pp. 427–430)

1. $\frac{P}{Q}$; polynomials **2.** Yes; both x and $2x^2 + 1$ are polynomials. **3.** denominator **4.** a **5.** $\frac{a}{b}$ **6.** rational **7.** $-\frac{3}{7}$
9. $-\frac{4}{9}$ **11.** $-\frac{1}{4}$ **13.** Undefined **15.** Undefined **17.** -1
19. 1

21.

x	-2	-1	0	1	2
$\frac{x}{x+1}$	2	—	0	$\frac{1}{2}$	$\frac{2}{3}$

23.

x	-2	-1	0	1	2
$\frac{3x}{2x^2+1}$	$-\frac{2}{3}$	-1	0	1	$\frac{2}{3}$

25. 0 **27.** 3 **29.** $-\frac{4}{5}$ **31.** None **33.** $-5, 5$ **35.** $-3, -2$
37. $1, \frac{5}{2}$ **39.** $\frac{2}{3}$ **41.** $\frac{1}{2}$ **43.** $-\frac{2}{5}$ **45.** $-\frac{1}{3}$ **47.** (a) $\frac{1}{2}$ (b) $\frac{1}{2}$
49. (a) -1 (b) -1 **51.** $\frac{1}{2x^2}$ **53.** $\frac{4y}{3x}$ **55.** $\frac{1}{2}$ **57.** $\frac{3}{5}$
59. $\frac{x+1}{x+6}$ **61.** $\frac{5y+3}{y+2}$ **63.** -1 **65.** -1 **67.** $-\frac{1}{3}$
69. $-\frac{1}{2}$ **71.** 1 **73.** $-\frac{3x+5}{3x-5}$ **75.** $\frac{n-1}{n-5}$ **77.** $\frac{x}{6}$
79. $\frac{z-2}{z-3}$ **81.** $\frac{x+4}{3x+2}$ **83.** $\frac{1}{3x-2}$ **85.** 1 **87.** $-\frac{1}{2}$
89. -1 **91.** (a) Equation (b) 6 **93.** (a) Expression
(b) $\frac{1}{x+1}$ **95.** (a) Expression (b) $x - 2$
97. (a) Equation (b) 8 **99.** (a) $\frac{1}{2}$; when traffic arrives at a
rate of 3 vehicles per minute, the average wait is $\frac{1}{2}$ minute.
(b)

x	2	4	4.5	4.9	4.99
T	$\frac{1}{3}$	1	2	10	100

As x nears 5 vehicles per minute, a small increase in x
increases the wait dramatically.
101. (a)

x	0	12	36	72
P	0.5	4.83	6.07	6.5

(b) 50 (c) The population increased quickly at first, but
then leveled off.
103. $\frac{1}{2}$ **105.** (a) $\frac{3}{n}$ (b) $\frac{n-3}{n}$; $\frac{97}{100}$; there is a 97% chance
that a winning ball will not be drawn. **107.** (a) 6 hr (b) $\frac{M}{60}$
109. (a) $x = 5$ (b) As the average rate nears 5 cars per minute, a small increase in x increases the wait dramatically.

SECTION 7.2 (pp. 435–436)

1. numerators; denomiators **2.** reciprocal **3.** $\frac{AC}{BD}$ **4.** $\frac{AD}{BC}$
5. $\frac{2}{5}$ **7.** $\frac{12}{7}$ **9.** $\frac{2}{3}$ **11.** $\frac{2}{11}$ **13.** 4 **15.** $\frac{8}{15}$ **17.** 10 **19.** $\frac{12}{5}$
21. 1 **23.** $\frac{z+1}{z+4}$ **25.** $\frac{2}{3}$ **27.** $\frac{x+2}{x-2}$ **29.** $\frac{8(x+1)}{x^2}$
31. $\frac{x-3}{x}$ **33.** $\frac{z+3}{z-7}$ **35.** 1 **37.** $(t+1)(t+2)$
39. $\frac{x(x+4)}{x^2+4}$ **41.** $\frac{z-1}{z+2}$ **43.** $\frac{y}{y-1}$ **45.** $\frac{x+1}{x-3}$ **47.** $\frac{x-3}{x-1}$
49. $\frac{2}{2x+3}$ **51.** -2 **53.** $\frac{z-1}{z+1}$ **55.** $\frac{y+2}{2y+1}$ **57.** $\frac{4(t-1)}{t^2+1}$
59. $\frac{y-3}{y-5}$ **61.** $2x$ **63.** $\frac{2z+1}{z+5}$ **65.** $\frac{1}{t+6}$ **67.** $\frac{2a+3b}{a+b}$
69. $x - y$ **71.** 1 **73.** (a) $D = \frac{30}{x}$ (b) 300 ft; 75 ft; stop-
ping distance on dry pavement is one-fourth as long.
75. (a) $\frac{1}{n+1}$ (b) $\frac{1}{100}$

CHECKING BASIC CONCEPTS 7.1 & 7.2 (p. 436)

1. Undefined; $\frac{3}{8}$ **2.** (a) $\frac{2x}{5y}$ (b) 5 (c) $\frac{x+2}{x+4}$
3. (a) $\frac{4}{9}$ (b) $\frac{2}{x-1}$ **4.** (a) $\frac{5z}{6}$ (b) $x + 1$
5. (a)

x	0.5	1.0	1.5	1.9
T	$\frac{2}{3}$	1	2	10

(b) As x nears 2 persons per minute, a small increase in x
increases the wait dramatically.

SECTION 7.3 (pp. 442–443)

1. numerators; denominators **2.** numerators; denominators
3. $\frac{A+B}{C}$ **4.** $\frac{A-B}{C}$ **5.** 1 **7.** $\frac{6}{5}$ **9.** 1 **11.** $\frac{3}{7}$ **13.** $\frac{1}{2}$
15. $\frac{1}{2}$ **17.** $\frac{2}{3}$ **19.** $\frac{3}{x}$ **21.** $\frac{1}{2}$ **23.** 3 **25.** 1 **27.** $\frac{4z}{4z+3}$
29. 2 **31.** 1 **33.** $\frac{1}{x-1}$ **35.** $z - 1$ **37.** 2 **39.** $y - 1$
41. $\frac{x}{2}$ **43.** $z - 2$ **45.** $x - 3$ **47.** $\frac{7n}{2n^2-n+5}$ **49.** $\frac{6}{x+3}$
51. $\frac{9}{ab}$ **53.** $\frac{1}{x+y}$ **55.** $\frac{10}{x-y}$ **57.** 0 **59.** $\frac{b}{2a}$ **61.** 1
63. $a - b$ **65.** $\frac{1}{x+2}$ **67.** $\frac{1}{3x-5}$ **69.** $2x - 3y$
71. It equals 7. **73.** (a) $\frac{14}{n+1}$ (b) $\frac{7}{50}$; when there are
100 batteries, there are 7 chances in 50 that a defective bat-
tery is chosen.

SECTION 7.4 (pp. 451–453)

1. Examples include 36 and 54 (answers may vary). **2.** xy
3. $\frac{3}{3}$ **4.** $\frac{x+1}{x+1}$ **5.** 12 **7.** 6 **9.** 30 **11.** 72 **13.** $12x$
15. $10x^2$ **17.** $x(x+1)$ **19.** $(2x+1)(x+3)$
21. $x(x-1)(x+1)$ **23.** $(x-8)^2(x+1)$
25. $(2x-1)(2x+1)$ **27.** $(x-1)(x+1)$
29. $(2x+3)(x+2)(x+3)$ **31.** $3y(y+2)(y-1)$
33. $\frac{3}{9}$ **35.** $\frac{15}{21}$ **37.** $\frac{2x^2}{8x^3}$ **39.** $\frac{x-2}{x^2-4}$ **41.** $\frac{x}{x^2+x}$
43. $\frac{2x^2+2x}{x^2+2x+1}$ **45.** $\frac{13}{10}$ **47.** $\frac{2}{9}$ **49.** $\frac{14}{25}$ **51.** $\frac{9}{20}$ **53.** $\frac{13}{12x}$
55. $\frac{5z-7}{z^3}$ **57.** $\frac{y-x}{xy}$ **59.** $\frac{a^2+b^2}{ab}$ **61.** $\frac{7}{2(x+2)}$
63. $\frac{t+2}{t(t-2)}$ **65.** $\frac{n^2+4n+5}{(n-1)(n+1)}$ **67.** $-\frac{3}{x-3}$ **69.** 0
71. $\frac{6x-4}{(x-1)^2}$ **73.** $\frac{3}{2y-1}$ **75.** $\frac{x-1}{x(x+2)}$ **77.** $\frac{3x+5}{(x-2)(x+2)}$
79. $\frac{x+9}{x(x-3)(x+3)}$ **81.** $\frac{x^2+4x-4}{x(x-2)(x+2)}$ **83.** $\frac{2x+2}{(x+2)^2}$
85. $\frac{x^2+2x-1}{(x+1)(x+2)(x+3)}$ **87.** $-\frac{2b}{(a+b)(a-b)}$ **89.** 0
91. $\frac{13}{12a}$ **93.** 0 **95.** $\frac{3}{x}$ **97.** $\frac{x^3-x+2}{(x-1)(x+1)}$
99. $\frac{3x^2+x+4}{(x-1)(x+1)}$ **101.** $\frac{1}{75}$; $\frac{1}{75}$; yes **103.** $\frac{D-F}{FD}$

CHECKING BASIC CONCEPTS 7.3 & 7.4 (p. 453)

1. (a) 1 (b) $\frac{2-x}{3x}$ (c) z **2.** (a) $15x$ (b) $4x(x+1)$
(c) $(x+1)(x-1)$ **3.** (a) $\frac{6x+5}{x(x+1)}$ (b) $\frac{4}{x-3}$
(c) $-\frac{x+4}{2x+1}$ **4.** $\frac{a^2+b^2}{(a-b)(a+b)}$

SECTION 7.5 (pp. 461–462)

1. $\frac{1}{2} \cdot \frac{4}{3} = \frac{2}{3}$ **2.** $\frac{a}{b} \cdot \frac{d}{c} = \frac{ad}{bc}$ **3.** fractions **4.** $\frac{\frac{x}{2}}{x-1}$
5. $\frac{\frac{a}{b}}{\frac{c}{d}}$ **6.** $x(x+2)$ **7.** 30 **9.** $(x-1)(x+1)$
11. $x(2x-1)(2x+1)$ **13.** $\frac{4}{5}$ **15.** $\frac{10}{7}$ **17.** $\frac{9}{14}$ **19.** $\frac{1}{2}$
21. $\frac{3y}{x}$ **23.** 3 **25.** $\frac{p+1}{p+2}$ **27.** $\frac{5}{z}$ **29.** $\frac{y}{y-3}$ **31.** $\frac{x^2-1}{x^2+1}$
33. $\frac{x^2}{3}$ **35.** 1 **37.** $\frac{n+m}{n-m}$ **39.** $\frac{2x+y}{2x-y}$ **41.** $\frac{1+b}{1-a}$
43. $\frac{q+2}{q}$ **45.** $2x + 3$ **47.** $\frac{2x}{(x+1)(2x-1)(2x+1)}$
49. $\frac{1}{ab}$ **51.** $\frac{ab}{a+b}$ **53.** $\frac{P\left(1+\frac{r}{26}\right)^{52}-P}{\frac{r}{26}}$ **55.** $R = \frac{ST}{S+T}$

SECTION 7.6 (pp. 473–476)

1. rational   **2.** $\frac{2x+5}{3x}$; $\frac{2x+5}{3x}=9$ (answers may vary)
3. $ad=bc$; b; d **4.** Yes   **5.** $12x$ **6.** V; T **7.** $\frac{3}{2}$ **9.** $\frac{5}{2}$
11. 0   **13.** $\frac{1}{4}$ **15.** $\frac{10}{11}$ **17.** 3   **19.** 25   **21.** 5   **23.** $\frac{5}{16}$
25. 1   **27.** No solutions $\left(\text{extraneous: } -\frac{1}{2}\right)$ **29.** $\frac{1}{2}$
31. -1 **33.** $-\frac{3}{2}, -1$ **35.** 4   **37.** 4   **39.** -6 **41.** 6
43. -3 (extraneous: 1)   **45.** 1 (extraneous: -2)   **47.** $\frac{7}{12}$
49. $\frac{1}{6}, 5$ **51.** 3   **53.** No solutions (extraneous: 1)   **55.** 3
57. No solutions (extraneous: -1)   **59.** No solutions
61. -2 **63.** Expression; 1; 1   **65.** Equation; 2
67. Equation; 4   **69.** Expression; $x+2$; 4   **71.** $-\frac{1}{2}; \frac{1}{2}$
73. $-1; \frac{1}{2}$ **75. (a)** $-3, 1$ **(b)** $-3, 1$ **77. (a)** -1 **(b)** -1
79. (a) 2   **(b)** 2   **81. (a)** $-2, 2$ **(b)** $-2, 2$ **83.** 0.300,
1.100   **85.** 1.084   **87.** $a=\frac{F}{m}$ **89.** $r=\frac{V}{I}-R$
91. $b=\frac{2A}{h}$ **93.** $z=\frac{15}{k-3}$ **95.** $b=\frac{aT}{a-T}$
97. $x=\frac{ky}{3y+2k}$ **99.** 9 cars per minute
101. $\frac{12}{7} \approx 1.7$ hours   **103.** $\frac{8}{3} \approx 2.7$ days   **105.** 20 mph;
18 mph   **107. (a)** 120; the braking distance is 120 feet when
the slope of the road is -0.05.   **(b)** -0.1 **109.** 15 mph
111. 200 mph   **113.** 10 mph running; 5 mph jogging

CHECKING BASIC CONCEPTS 7.5 & 7.6 (p. 476)

1. (a) $\frac{5}{6}$ **(b)** $\frac{1}{9x^2}$ **(c)** $\frac{b-a}{b+a}$ **(d)** $\frac{r+t}{2rt}$ **2. (a)** $\frac{1}{5}$ **(b)** -4
(c) 2   **(d)** $-2, \frac{1}{2}$ **(e)** -3 (extraneous: 1)
3. (a) $x=\frac{2(b+3y)}{a}$ **(b)** $m=\frac{k}{2k-1}$ **4. (a)** 300; when
the slope of the hill is 0.1, the braking distance is 300 feet.
(b) 0.3; the braking distance is 200 feet when the slope of
the road is 0.3.

SECTION 7.7 (pp. 486–490)

1. A statement that two ratios are equal   **2.** $\frac{5}{6}=\frac{x}{7}$
3. It doubles.   **4.** It is halved.   **5.** constant   **6.** constant
7. Directly; if the number being fed doubles, the bill will
double.   **8.** Inversely; doubling the number of painters will
halve the time.   **9.** inverse   **11.** 15   **13.** 21   **15.** 48
17. $\frac{21}{8}$ **19.** $-4, 4$ **21.** $-\frac{7}{2}, \frac{7}{2}$ **23.** $b=\frac{ad}{c}$ **25. (a)** $\frac{5}{8}=\frac{9}{x}$
(b) $\frac{72}{5}$ **27. (a)** $\frac{4}{8}=\frac{10}{x}$ **(b)** 20   **29. (a)** $\frac{98}{7}=\frac{x}{11}$ **(b)** \$154
31. (a) $\frac{3}{750}=\frac{7}{x}$ **(b)** 1750   **33. (a)** 2   **(b)** 12   **35. (a)** $\frac{3}{2}$
(b) 9   **37. (a)** $-\frac{15}{2}$ **(b)** -45 **39. (a)** 24   **(b)** 3
41. (a) 40   **(b)** 5   **43. (a)** 400   **(b)** 50   **45. (a)** $k=0.25$
(b) $z=8.75$ **47. (a)** $k=11$ **(b)** $z=385$
49. (a) $k=10$ **(b)** $y=350$
51. (a) Direct   **(b)** $y=\frac{3}{2}x$ **53. (a)** Inverse   **(b)** $y=\frac{36}{x}$
(c)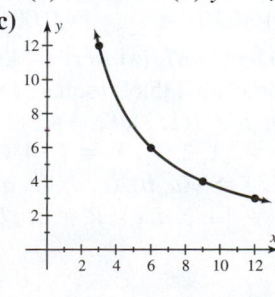

55. (a) Neither   **(b)** NA   **(c)** NA   **57.** Direct; 2
59. Neither   **61.** Inverse; 8   **63.** 47.6 minutes
65. 1.625 inches   **67.** About 2123 lb   **69.** $9\frac{1}{3}$ c
71. (a) Direct; the ratios $\frac{R}{W}$ always equal 0.012.
73. (a) Direct; the ratios $\frac{G}{A}$ always equal 27.
(b) $R=0.012W$
(b) $G=27A$

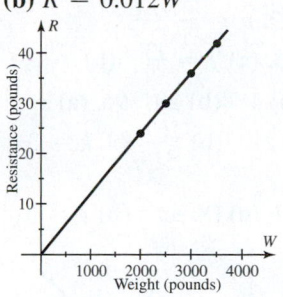

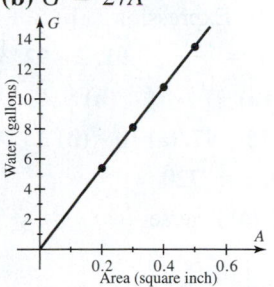

(c) 38.4 pounds

(c) For each square-inch increase in the cross-sectional area of the hose, the flow increases by 27 gallons per minute.

75. (a) $F=\frac{1200}{L}$ **(b)** 80 pounds
77. (a) Direct   **(b)** $y=-19x$ **(c)** Negative; for each
1-mile increase in altitude the temperature decreases by 19°F.
(d) 47.5°F decrease   **79.** 1.8 ohms   **81.** $z=0.5104x^2y^3$
83. About 133 pounds   **85.** About 35.2 pounds   **87.** 750

CHECKING BASIC CONCEPTS 7.7 (p. 491)

1. (a) $\frac{18}{5}$ **(b)** $\frac{15}{4}$ **2. (a)** $\frac{4}{6}=\frac{8}{x}$; 12   **(b)** $\frac{2}{148}=\frac{5}{x}$;
370 minutes   **3.** 60; 6   **4. (a)** Direct; the ratios $\frac{y}{x}$ always
equal $\frac{3}{2}; \frac{3}{2}$.   **(b)** Inverse; the products xy always equal
24; 24.   **5.** \$160

CHAPTER 7 REVIEW (pp. 495–498)

1. $-\frac{3}{5}$ **2.** -3 **3.** Undefined   **4.** Undefined
5.

x	-2	-1	0	1	2
$\frac{3x}{x-1}$	2	$\frac{3}{2}$	0	—	6

6. $-2, 2$ **7.** $\frac{5y^3}{3x^2}$ **8.** $x-6$ **9.** -1 **10.** $\frac{x-5}{5}$ **11.** $\frac{x+3}{x+1}$
12. $\frac{x+4}{x+1}$ **13. (a)** Expression   **(b)** $\frac{1}{x-3}$
14. (a) Equation   **(b)** 18   **15.** 2   **16.** $\frac{1}{x+5}$ **17.** $\frac{1}{z+3}$
18. $\frac{x}{x-2}$ **19.** $\frac{5}{6}$ **20.** $\frac{8}{x(x+1)}$ **21.** $\frac{1}{2}$ **22.** 1   **23.** $x+y$
24. $2(a^2+ab+b^2)$ **25.** $\frac{10}{x+10}$ **26.** $\frac{1}{x-1}$ **27.** 1
28. 1   **29.** $\frac{1}{x+1}$ **30.** $\frac{2}{x-5}$ **31.** $\frac{2}{xy}$ **32.** $\frac{x}{y}$ **33.** $15x$
34. $10x^2$ **35.** $x(x-5)$ **36.** $10x^2(x-1)$
37. $(x-1)(x+1)^2$ **38.** $x(x-4)(x+4)$ **39.** $\frac{9}{24}$
40. $\frac{16}{12x}$ **41.** $\frac{3x^2+6x}{x^2-4}$ **42.** $\frac{2x}{x^2+x}$ **43.** $\frac{3x-3}{5x^2-5x}$
44. $\frac{2x^2+4x}{2x^2+x-6}$ **45.** $\frac{19}{24}$ **46.** $\frac{7}{4x}$ **47.** $-\frac{1}{9x}$ **48.** $\frac{4x+3}{x(x-1)}$
49. $\frac{2x}{(x-1)(x+1)}$ **50.** $\frac{8-9x}{6x^2}$ **51.** $\frac{2x-7}{6x}$
52. $-\frac{1}{(x-1)(x+1)}$ **53.** $\frac{5y-x}{(x-y)(x+y)}$ **54.** $\frac{13}{6x}$ **55.** $\frac{3x+y}{2xy}$
56. -1 **57.** $\frac{33}{28}$ **58.** $\frac{7}{10}$ **59.** $\frac{n}{2}$ **60.** $\frac{3(p+1)}{p-1}$
61. $\frac{3(m+1)}{2(m-1)^2}$ **62.** $\frac{2n-1}{4(2n+1)}$ **63.** $\frac{1}{3}$ **64.** $\frac{2-x}{2+x}$ **65.** $\frac{1}{x^2}$

66. $x + 3$ **67.** $\frac{20}{7}$ **68.** $\frac{8}{3}$ **69.** $\frac{1}{5}$ **70.** -5 **71.** -4
72. $-3, -1$ **73.** 4 **74.** -2 **75.** 5 **76.** 7 **77.** $-2, 5$
78. $-3, 3$ **79.** 8 **80.** No solutions **81.** No solutions
(extraneous: -1) **82.** $-\frac{3}{2}$ (extraneous: 0) **83.** -3
84. -12 **85.** $\frac{4}{5}$ **86.** -3 **87.** (a) Equation (b) $-2, 2$
88. (a) Expression (b) $x + 3$ **89.** $b = \frac{2ac}{3a - c}$
90. $x = \frac{y}{y - 1}$ **91.** 2 **92.** $\frac{15}{7}$ **93.** (a) $\frac{6}{x} = \frac{13}{20}$ (b) $\frac{120}{13}$
94. (a) $\frac{341}{11} = \frac{x}{8}$ (b) \$248 **95.** (a) 4 (b) 20 **96.** (a) 3
(b) 15 **97.** (a) 10 (b) 2 **98.** (a) 21 (b) $\frac{21}{5}$ **99.** $k = 3$
100. $z = 720$

101. (a) Inverse (b) $y = \frac{60}{x}$ **102.** (a) Direct (b) $y = 3x$
(c) 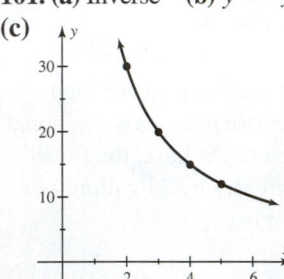   (c)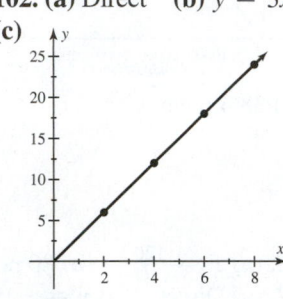

103. Direct; $\frac{1}{2}$ **104.** Inverse; 12 **105.** (a) $\frac{1}{5} = 0.2$; when
the average rate of arrival is 10 cars per minute, the average
wait is 0.2 minute, or 12 seconds.

(b)

x	5	10	13	14	14.9
T	$\frac{1}{10}$	$\frac{1}{5}$	$\frac{1}{2}$	1	10

(c) It increases dramatically. **106.** 60 mph
107. $\frac{800}{13} \approx 61.5$ hours **108.** 10 mph and 12 mph
109. 8 mph **110.** About 33.3 feet **111.** 200 vehicles
112. \$468 **113.** 36 pounds **114.** 1.6 inches **115.** 750 to
2250 seconds, or 12.5 to 37.5 minutes **116.** About 771 lb
117. 6493.8 watts **118.** 702 pounds

CHAPTER 7 TEST (pp. 498–499)

1. $\frac{9}{5}$ **2.** -2 **3.** $x + 5$ **4.** $x - 5$ **5.** 3 **6.** 2 **7.** $\frac{x - 1}{10x}$
8. $\frac{6}{x(x + 3)}$ **9.** $\frac{4x + 1}{x + 4}$ **10.** $\frac{t + 7}{2t - 3}$ **11.** $6x^2(x - 1)$
12. $\frac{4x - 4}{7x^2 - 7x}$ **13.** $-\frac{1}{y + 1}$ **14.** $\frac{x^2y - x + y}{xy^2}$ **15.** $\frac{b}{15}$
16. $\frac{p}{p - 2}$ **17.** $\frac{35}{2}$ **18.** 3 **19.** 1 **20.** $-1, 2$ **21.** $\frac{2}{7}$
22. -12 **23.** No solutions $\left(\text{extraneous: } \frac{1}{2}\right)$
24. 0 (extraneous: 5) **25.** $x = \frac{2 + 5y}{3y}$ **26.** $b = \frac{a}{a - 1}$
27. (a) $\frac{7}{2}$ (b) 21 **28.** Inversely; 32 **29.** 24 hours
30. 67.5 feet **31.** $\frac{16}{5} = 3.2$; when the arrival rate is 24
people per hour, there are about 3 people in line, on average.

CHAPTERS 1–7 CUMULATIVE REVIEW
(pp. 500–501)

1. $24\pi \approx 75.4$ **2.** $2x - 2$ **3.** $\frac{2}{5}$ **4.** $\frac{3}{4}$ **5.** $2x + 2$
6. $y - 11$ **7.** -1 **8.** $x \le \frac{1}{4}$

9. x-intercept: 3   **10.** x-intercept: 1
y-intercept: -2 y-intercept: none

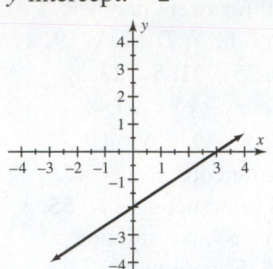 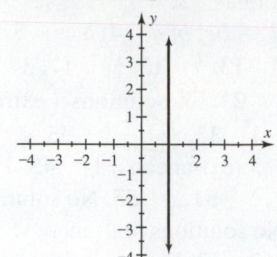

11. $y = 3x + 5$

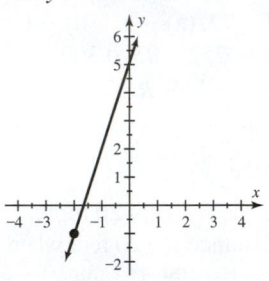

12. $y = 2x - 1$ **13.** $y = -\frac{2}{3}x + \frac{1}{3}$ **14.** $y = \frac{2}{3}x + \frac{8}{3}$
15. $(1, -2)$ **16.** $(2, -8)$ **17.** Infinitely many solutions
18. No solutions **19.** $15z^8$ **20.** a^3b^3 **21.** $10y^2 - 11y - 6$
22. $x^4 - 2x^2y^2 + y^4$ **23.** $\frac{1}{27x^6}$ **24.** $\frac{2}{x^2}$ **25.** 1.23×10^{-3}
26. $2x + 1 + \frac{4}{x - 1}$ **27.** $(3 - x)(2 + 5x)$
28. $(3z - 2)(3z + 2)$ **29.** $(t + 8)^2$ **30.** $x(x - 4)(x + 4)$
31. $-7, 2$ **32.** $-2, 0, 2$ **33.** $\frac{1}{12x}$ **34.** $\frac{1}{2x + 4}$ **35.** $\frac{x + 2}{x - 2}$
36. $x = \frac{z + 2y}{3}$ **37.** $\frac{7}{12}$ **38.** $-1, 2$ **39.** 5.5
40. Inverse; $y = \frac{20}{x}$ **41.** (a) $21x$ (b) 90 min
42. 35 min running, 25 min walking
43. (a) $2x + y = 180, 2x - y = 32$ (b) $(53, 74)$ or $53°$,
$53°, 74°$ **44.** About 500 lb

8 | Introduction to Functions

SECTION 8.1 (pp. 517–522)

1. function **2.** y equals f of x **3.** symbolic **4.** numerical
5. domain **6.** range **7.** one **8.** False **9.** True
10. $(3, 4)$; 3; 6 **11.** (a, b) **12.** d **13.** 1 **14.** equal
15. Yes **16.** Yes **17.** No **18.** No **19.** Yes **20.** No
21. -6; -2 **23.** 0; $\frac{3}{2}$ **25.** 25; $\frac{9}{4}$ **27.** 3; 3 **29.** 13; -22
31. $-\frac{1}{2}$; $\frac{2}{5}$ **33.** (a) $I(x) = 36x$ (b) $I(10) = 360$; There are
360 inches in 10 yards. **35.** (a) $M(x) = \frac{x}{5280}$
(b) $M(10) = \frac{10}{5280} \approx 0.0019$; There is $\frac{10}{5280}$ mile in
10 feet. **37.** (a) $A(x) = 43,560x$ (b) $A(10) = 435,600$;
There are 435,600 square feet in 10 acres.
39. $f = \{(1, 3), (2, -4), (3, 0)\}$;
$D = \{1, 2, 3\}$; $R = \{-4, 0, 3\}$
41. $f = \{(a, b), (c, d), (e, a), (d, b)\}$;
$D = \{a, c, d, e\}$; $R = \{a, b, d\}$

43.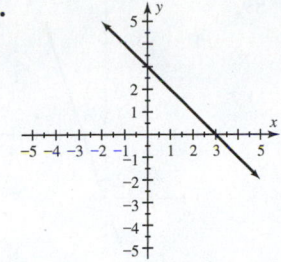

45.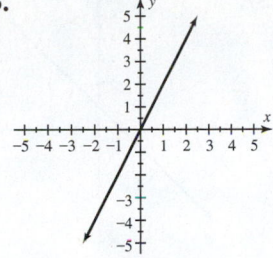

47.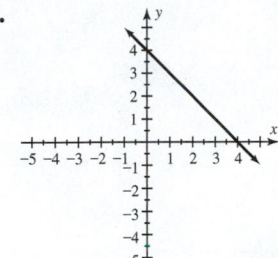

49.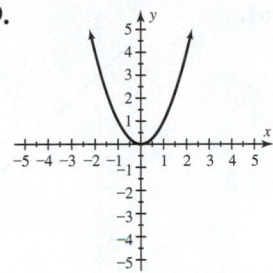

51.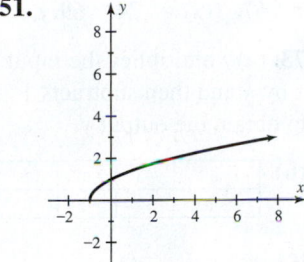

53. 3; −1   **55.** 0; 2   **57.** −4; −3   **59.** 5.5; 3.7

61. 26.9; in 1990 average fuel efficiency was **26.**9 mpg.

63. Numerical:

x	−3	−2	−1	0	1	2	3
y = f(x)	2	3	4	5	6	7	8

Graphical:

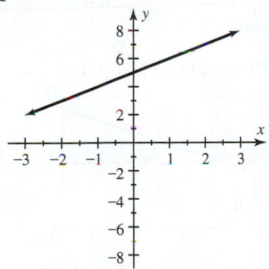

Symbolic: $y = x + 5$

65. Numerical:

x	−3	−2	−1	0	1	2	3
y = f(x)	−17	−12	−7	−2	3	8	13

Graphical:

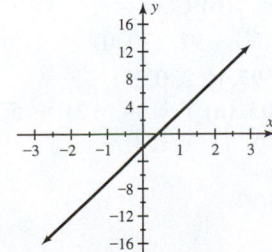

Symbolic: $y = 5x - 2$

67. Subtract $\frac{1}{2}$ from the input x to obtain the output y.

69. Divide the input x by 3 to obtain the output y.

71. Subtract 1 from the input x and then take the square root to obtain the output y.

73. $f(x) = 0.50x$

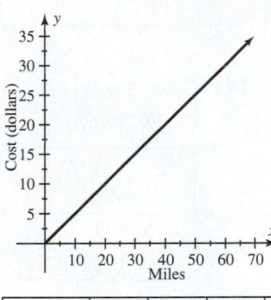

Miles	10	20	30	40	50	60	70
Cost	$5	$10	$15	$20	$25	$30	$35

75. 1825; in 2011 there were 1825 billion, or 1.825 trillion, World Wide Web searches.

77. D: $-2 \le x \le 2$; R: $0 \le y \le 2$

79. D: $-2 \le x \le 4$; R: $-2 \le y \le 2$

81. D: All real numbers; R: $y \ge -1$

83. D: $-3 \le x \le 3$; R: $-3 \le y \le 2$

85. $D = \{1, 2, 3, 4\}$; $R = \{5, 6, 7\}$

87. All real numbers   **89.** All real numbers   **91.** $x \ne 5$

93. All real numbers   **95.** $x \ge 1$ **97.** All real numbers

99. $x \ne 0$ **101. (a)** 1726; 1726 whales were sighted in 2008.   **(b)** $D = \{2005, 2006, 2007, 2008, 2009\}$, $R = \{649, 1265, 959, 1726, 1010\}$ **(c)** Increased every other year   **103.** $D = \{1, 2, 3, \ldots, 20\}$; $R = \{200, 400, 600, \ldots, 4000\}$ **105.** No   **107.** Yes

109. (a) 0.2   **(b)** Yes. Each month has one average amount of precipitation.   **(c)** 2, 3, 7, 11   **111.** Yes. D: All real numbers; R: All real numbers   **113.** No

115. Yes. D: $-4 \le x \le 4$; R: $0 \le y \le 4$

117. Yes. D: All real numbers; R: $y = 3$ **119.** No

121. It does. $D = \{-6, -4, 2, 4\}$; $R = \{-4, 2\}$ **123.** Yes

125. No   **127.** The person walks away from home, then turns around and walks back a little slower.

129.

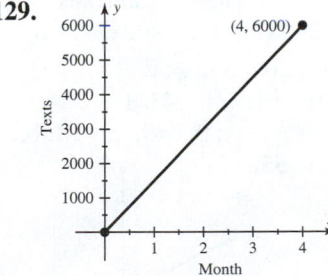

131. 10; 10

$[-10, 10, 1]$ by $[-10, 10, 1]$

133. 10; 5

$[0, 100, 10]$ by $[-50, 50, 10]$

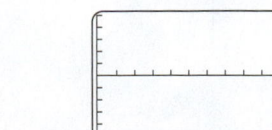

135. 16; 5

[1980, 1995, 1] by
[12000, 16000, 1000]

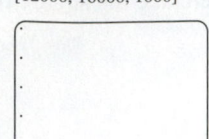

137.

[−6, 6, 1] by [−6, 6, 1]

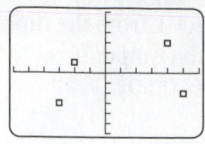

139. [−30, 30, 5] by
[−50, 50, 5]

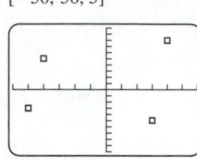

141. [−200, 200, 50] by
[−250, 250, 50]

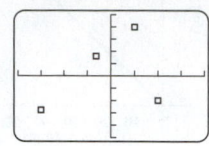

143. Numerical

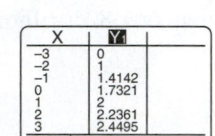

Graphical

[−10, 10, 1] by [−10, 10, 1]

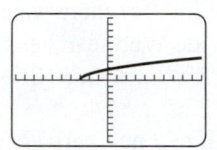

145. Numerical

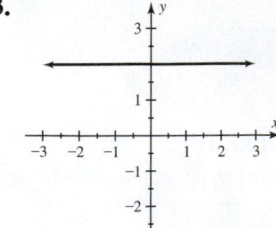

Graphical (Dot Mode)

[−10, 10, 1] by [−10, 10, 1]

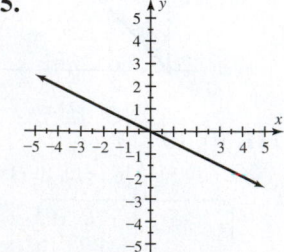

SECTION 8.2 (pp. 534–540)

1. $mx + b$ **2.** b **3.** line **4.** horizontal **5.** 7 **6.** 0
7. True **8.** False **9.** Carpet costs $2 per square foot.
Ten square feet of carpet costs $20. **10.** The rate at which
water is leaving the tank is 4 gallons per minute. After
5 minutes the tank contains 80 gallons of water.
11. Yes; $m = \frac{1}{2}, b = -6$ **13.** No
15. Yes; $m = 0, b = -9$ **17.** Yes; $m = -9, b = 0$
19. Yes **21.** No **23.** Yes; $f(x) = 3x - 6$ **25.** Yes;
$f(x) = -\frac{3}{2}x + 3$ **27.** No **29.** Yes; $f(x) = 2x$
31. Yes; $f(x) = -4$ **33.** $-16; 20$ **35.** $\frac{17}{3}; 2$
37. $-22; -22$ **39.** $-2; 0$ **41.** $-1; -4$ **43.** 1; 1
45. $f(x) = 6x; 18$ **47.** $f(x) = \frac{x}{6} - \frac{1}{2}; 0$ **49.** d. **51.** b.
53.

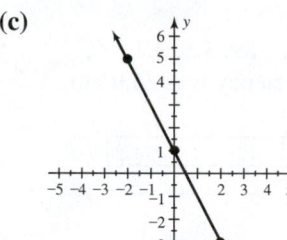

55.

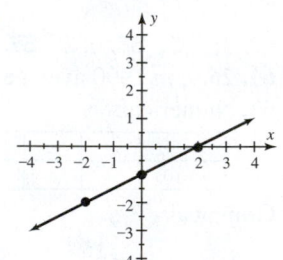

57.

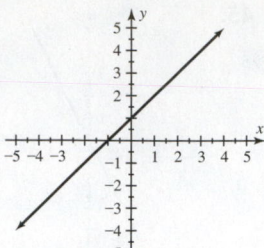

59.

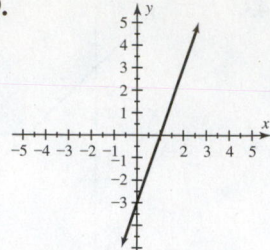

61.

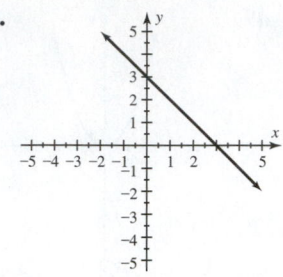

63. $f(x) = \frac{1}{16}x$ **65.** $f(t) = 65t$ **67.** $f(x) = 24$ **69.** a

71. (a) f multiplies the input
x by -2 and then adds 1 to
obtain the output y.

73. (a) f multiplies the input
x by $\frac{1}{2}$ and then subtracts 1
to obtain the output y.

(b)

x	−2	0	2
$y = f(x)$	5	1	−3

(b)

x	−2	0	2
$y = f(x)$	−2	−1	0

(c)

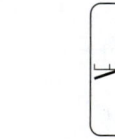

(c)

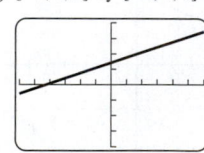

75. (a)

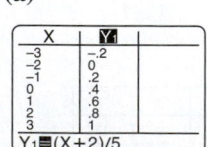

(b) [−6, 6, 1] by [−4, 4, 1]

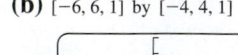

77. (a)

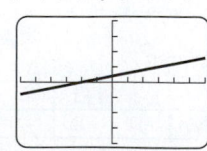

(b) [−6, 6, 1] by [−4, 4, 1]

79. b. **81.** c. **83. (a)** $G(x) = \frac{X}{E}$ **(b)** $C(x) = \frac{3x}{E}$
85. -1 **87.** $(-1, 0)$ **89.** $\left(-\frac{1}{2}, \frac{1}{2}\right)$ **91.** $(20, 0)$
93. $(-8, -1)$ **95.** $\left(-1, -\frac{1}{6}\right)$ **97.** $(0.2, 0.25)$
99. $(2005, 9)$ **101.** $(2a, 2b)$ **103. (a)** $f(x) = -2x + 5; 1$
(b) 1 **(c)** Equal **105. (a)** $f(x) = \frac{2}{5}x + \frac{1}{5}; 1$
(b) 1 **(c)** Equal **107.** 79.8 years
109. 256.5 million **111.** $45,250

113. (a) Symbolic: $f(x) = 70$
Graphical:

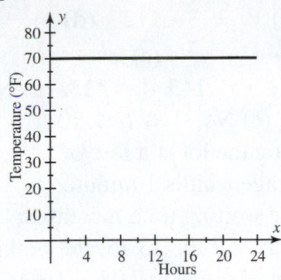

(b)

Hours	0	4	8	12	16	20	24
Temp. (°F)	70	70	70	70	70	70	70

(c) Constant
115. $D(t) = 60t + 50$ **117. (a)** $K(x) = 93x$
(b) $A(x) = x$ **(c)** 33,945; 365; On average, someone under 18 sends 33,945 texts in 1 year while someone over 65 sends 365 texts.
119. (a) 1.6 million   **(b)** 0.1 million
(c) $f(x) = 0.1x + 1.6$ **(d)** 2.2 million
121. (a) 563 million   **(b)** In 2006, there were about 123 million users.   **(c)** Users increased, on average, by 110 million per year.   **123. (a)** $V(T) = 0.5T + 137$
(b) 162 cm^3 **125.** $f(x) = 40x$; about 2.92 pounds
127. (a) 55%   **(b)** 4%   **(c)** $P(x) = 4x + 55$ **(d)** 67%

CHECKING BASIC CONCEPTS 8.1 & 8.2 (p. 540)

1. Symbolic: $f(x) = x^2 - 1$
Graphical:

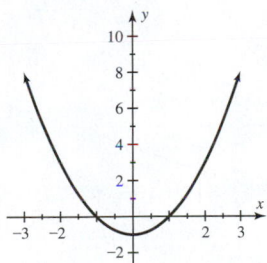

2. (a) D: $-3 \le x \le 3$; R: $-4 \le y \le 4$ **(b)** 0; 4
(c) No. The graph is not a line.
3. (a) Yes   **(b)** No   **(c)** Yes   **(d)** Yes
4. $f(-2) = 10$

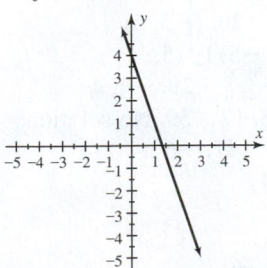

5. $f(x) = \frac{1}{2}x - 1$ **6. (a)** 32.2: In 1990 the median age was about 32 years.   **(b)** 0.225: The median age is increasing by 0.225 year each year. 27.7: In 1970 the median age was 27.7 years.   **7.** $(1, -1)$

SECTION 8.3 (pp. 548–551)

1. $x > 1$ and $x \le 7$ (answers may vary)
2. $x \le 3$ or $x > 5$ (answers may vary)   **3.** No   **4.** Yes
5. Yes   **6.** Numerically, graphically, symbolically
7. Yes, no   **8.** No, yes   **9.** No, yes   **10.** Yes, no
11. No, yes   **12.** Yes, yes   **13.** $[2, 10]$ **15.** $(5, 8]$
17. $(-\infty, 4)$ **19.** $(-2, \infty)$ **21.** $[-2, 5)$ **23.** $(-8, 8]$
25. $(3, \infty)$ **27.** $(-\infty, -2] \cup [4, \infty)$
29. $(-\infty, 1) \cup [5, \infty)$ **31.** $(-3, 5]$ **33.** $(-\infty, -2)$
35. $(-\infty, 4)$ **37.** $(-\infty, 1) \cup (2, \infty)$
39. $\{x \mid -1 \le x \le 3\}$

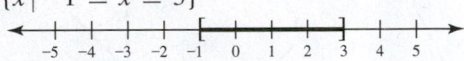

41. $\{x \mid -2 < x < 2.5\}$

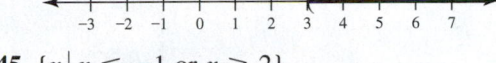

43. $\{x \mid x > 3\}$

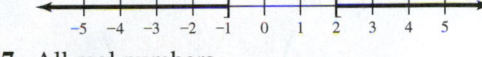

45. $\{x \mid x \le -1 \text{ or } x \ge 2\}$

47. All real numbers

49. $[-6, 7]$ **51.** $\left(\frac{13}{4}, \infty\right)$ **53.** $(-\infty, \infty)$
55. $\left(-\infty, -\frac{11}{5}\right) \cup (-1, \infty)$ **57.** No solutions
59. $[-6, 1)$ **61.** $\left[-\frac{1}{4}, \frac{11}{8}\right)$ **63.** $[-9, 3]$
65. $\left[-4, -\frac{1}{4}\right)$ **67.** $(-1, 1]$ **69.** $[-2, 1]$ **71.** $(0, 2)$
73. $(9, 21]$ **75.** $\left[\frac{12}{5}, \frac{22}{5}\right]$ **77.** $\left(-\frac{13}{3}, 1\right)$ **79.** $[0, 12]$
81. $[-1, 2]$ **83.** $(-1, 2)$ **85.** $[-3, 1]$
87. $(-\infty, -2) \cup (0, \infty)$
89. (a) Toward, because distance is decreasing
(b) 4 hours, 2 hours   **(c)** From 2 to 4 hours   **(d)** During the first 2 hours   **91. (a)** 2   **(b)** 4   **(c)** $\{x \mid 2 \le x \le 4\}$
(d) $\{x \mid 0 \le x < 2\}$ **93.** $[1, 4]$ **95.** $(-\infty, -2) \cup (0, \infty)$
97. $[2002, 2010]$ **99.** $[1, 3]$ **101.** $(-\infty, \infty)$
103. $(-\infty, 1) \cup [3, \infty)$ **105.** $(c - b, d - b]$
107. (a) From 2006 to 2008   **(b)** From 2006 to 2008
(c) From 2006 to 2008   **109.** From 0.5 to 1.5 miles
111. From $5.\overline{6}$ to 9 feet   **113.** From 1.5 to 2.5 miles
115. (a) $M(x) = 25x + 250$ **(b)** From 2002 to 2006

SECTION 8.4 (pp. 563–567)

1. domain   **2.** range   **3.** $(-\infty, \infty)$
4. $(-\infty, 5) \cup (5, \infty)$ **5.** absolute value
6. exponent   **7.** 2   **8.** 1   **9.** rational   **10.** $-\frac{1}{2}$ **11.** c.
12. a.   **13.** D: $(-\infty, \infty)$; R: $(-\infty, \infty)$
15. D: $(-\infty, \infty)$; R: $(-\infty, \infty)$
17. D: $(-\infty, \infty)$; R: $[2, \infty)$
19. D: $(-\infty, \infty)$; R: $(-\infty, 0]$
21. D: $[-1, \infty)$; R: $[0, \infty)$
23. D: $(-\infty, \infty)$; R: $[0, \infty)$
25. $(-\infty, 1) \cup (1, \infty)$ **27.** $(-\infty, 2) \cup (2, \infty)$
29. $(-\infty, -2) \cup (-2, 2) \cup (2, \infty)$
31. $(-\infty, 0) \cup (0, 2) \cup (2, \infty)$ **33.** $(-\infty, 1) \cup (1, \infty)$
35. $(-\infty, -1) \cup (-1, 3) \cup (3, \infty)$

37. $D: (-\infty, \infty)$; $R: (-\infty, \infty)$

39. $D: [-2, 2]$; $R: [-2, 2]$

41. $D: [-2, 3]$; $R: [-2, 2]$ **43.** Yes; 1; linear

45. Yes; 3; cubic   **47.** No   **49.** Yes; 2; quadratic

51. No   **53.** Yes; 4; fourth degree   **55.** 12; 0   **57.** 1; $\frac{11}{4}$

59. 0; 6   **61.** -14; 14   **63.** -6; 9   **65.** $\frac{1}{11}$; $-\frac{1}{7}$

67. $-\frac{5}{6}$; undefined   **69.** $\frac{5}{6}$; undefined   **71.** 1; -1

73. -2; -2 **75.** -4; 0   **77.** -1; undefined

79.

x	-2	-1	0	1	2
$f(x) = \frac{1}{x-1}$	$-\frac{1}{3}$	$-\frac{1}{2}$	-1	—	1

81.

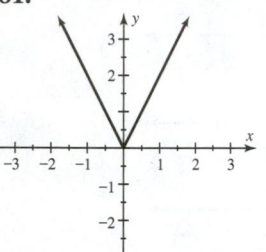

83.

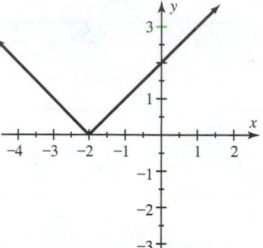

85.

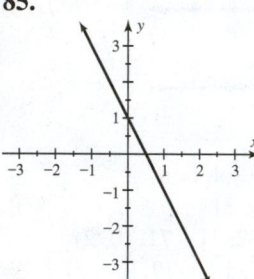

87.

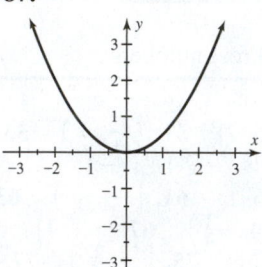

89.

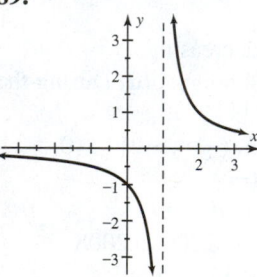

91.

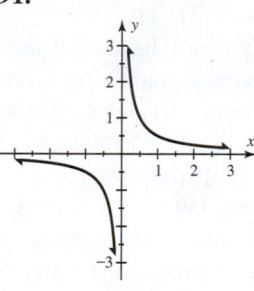

93.

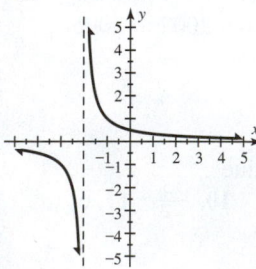

95.

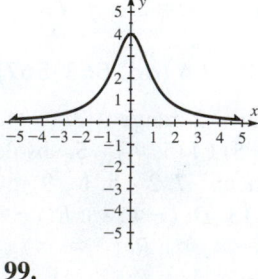

97.

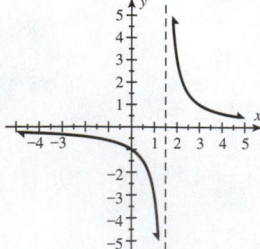

99.

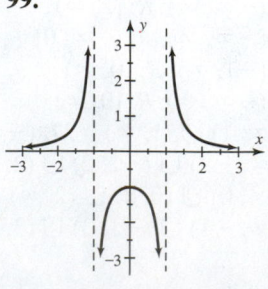

101. (a) 19   **(b)** -9 **(c)** 150   **(d)** 0

103. (a) 41   **(b)** -21 **(c)** 900   **(d)** Undefined

105. (a) $2x + 3$ **(b)** -1 **(c)** $x^2 + 3x + 2$ **(d)** $\frac{x+1}{x+2}$

107. (a) $x^2 - x + 1$ **(b)** $1 - x - x^2$ **(c)** $x^2 - x^3$

(d) $\frac{1-x}{x^2}$ **109.** $a^2 - 2a$ **111.** c.   **113.** d.   **115. (a)** No;

answers may vary.   **(b)** Yes   **(c)** No; $0 \le t \le 10$

117. (a) 1; when cars are leaving the lot at a rate of

4 vehicles per minute, the average wait is 1 minute.

(b) As more cars try to exit, the waiting time increases;

yes.   **(c)** $4.\overline{6}$ vehicles per minute   **119. (a)** 75; the braking

distance is 75 feet when the uphill grade is 0.05.   **(b)** 0.15

121. (a) 123 thousand, which is close to the actual value.

(b) 478.6 thousand, which is too high; AIDS deaths did not

continue to rise as rapidly as the model predicts.

123. About 45% and 22%; after 1 day (3 days) students

remember 45% (22%) of what they have learned.

125. (a) 130; it costs $130 thousand to make 100 notebook

computers.   **(b)** The y-intercept for C is 100. The company

has $100 thousand in fixed costs even if it makes 0

computers. The y-intercept for R is 0. If the company sells

0 computers, its revenue is $0.

(c) $P(x) = 0.45x - 100$ **(d)** 223 or more

CHECKING BASIC CONCEPTS 8.3 & 8.4 (p. 567)

1. (a) Yes   **(b)** No   **2. (a)** $[-3, 1]$

(b) $(-\infty, -1] \cup [3, \infty)$ **(c)** $\left[-\frac{8}{3}, \frac{8}{3}\right)$

3. (a) $(-\infty, \infty)$ **(b)** $(-\infty, 1) \cup (1, \infty)$ **(c)** $[0, \infty)$

4. (a) $D: [-2, 1]$; $R: [-3, 1]$ **(b)** 1; -3

5.

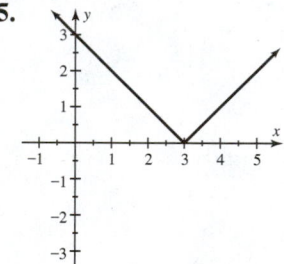

SECTION 8.5 (pp. 575–578)

1. $|3x + 2| = 6$ (answers may vary)

2. $|2x - 1| \le 17$ (answers may vary)   **3.** Yes   **4.** Yes

5. Yes   **6.** No, it is equivalent to $-3 < x < 3$.   **7.** 2

8. 0   **9.** No, yes   **10.** Yes, no   **11.** No, yes   **12.** No, yes

13. Yes, yes   **14.** No, yes   **15.** 0, 4   **16.** -2, 1

17. $-3, 3$ **18.** $-5, 5$ **19.** $(-3, 3)$ **20.** $(-5, 5)$

21. $(-\infty, -3) \cup (3, \infty)$ **22.** $(-\infty, -5) \cup (5, \infty)$

23. $-7, 7$ **25.** No solutions   **27.** $-\frac{9}{4}, \frac{9}{4}$ **29.** $-4, 4$

31. $-6, 5$ **33.** 1, 2   **35.** $\frac{3}{4}$ **37.** $-8, 12$ **39.** No solutions

41. $-15, 18$ **43.** $-1, \frac{1}{3}$ **45.** $-5, \frac{3}{5}$ **47.** -6

49. (a) $-4, 4$ **(b)** $\{x \mid -4 < x < 4\}$

(c) $\{x \mid x < -4 \text{ or } x > 4\}$

51. (a) $\frac{1}{2}, 2$ **(b)** $\{x \mid \frac{1}{2} \le x \le 2\}$

(c) $\{x \mid x \le \frac{1}{2} \text{ or } x \ge 2\}$ **53.** $[-3, 3]$

55. $(-\infty, -4) \cup (4, \infty)$ **57.** No solutions

59. $(-\infty, 0) \cup (0, \infty)$ **61.** $\left(-\infty, -\frac{7}{2}\right) \cup \left(\frac{7}{2}, \infty\right)$

63. $(-3, 5)$ **65.** $(-\infty, -9] \cup [-1, \infty)$ **67.** $\left[\frac{5}{6}, \frac{11}{6}\right]$

69. $[-10, 14]$ **71.** No solutions   **73.** $(-\infty, \infty)$
75. No solutions   **77.** $(-\infty, \infty)$
79. $(-\infty, -13] \cup [17, \infty)$ **81.** $[0.9, 1.1]$
83. $(-\infty, 9.5) \cup (10.5, \infty)$ **85. (a)** $-1, 3$ **(b)** $(-1, 3)$
(c) $(-\infty, -1) \cup (3, \infty)$ **87. (a)** $-1, 0$ **(b)** $[-1, 0]$
(c) $(-\infty, -1] \cup [0, \infty)$ **89.** $(-\infty, -1] \cup [1, \infty)$
91. $[-2, 4]$ **93.** $(-\infty, 1) \cup (3, \infty)$ **95.** $(2, 4.\overline{6})$
97. $(-\infty, \infty)$ **99.** $\{x \mid -3 \le x \le 3\}$
101. $\{x \mid x < 2 \text{ or } x > 3\}$ **103.** $|x| \le 4$ **105.** $|y| > 2$
107. $|2x + 1| \le 0.3$ **109.** $|\pi x| \ge 7$ **111.** two
113. (a) $\{T \mid 19 \le T \le 67\}$ **(b)** Monthly average
temperatures vary from $19°$ F to $67°$ F.
115. (a) $\{T \mid -26 \le T \le 46\}$ **(b)** Monthly average
temperatures vary from $-26°$ F to $46°$ F.   **117. (a)** About
19,058 feet   **(b)** Africa and Europe   **(c)** South America,
North America, Africa, Europe, and Antarctica
(d) $|E - A| \le 5000$ **119.** $\{d \mid 2.498 \le d \le 2.502\}$; the
diameter can vary from 2.498 to 2.502 inches.
121. $|d - 3.8| \le 0.03$
123. Values between 19 and 21, exclusive

CHECKING BASIC CONCEPTS 8.5 (p. 578)

1. $-\frac{28}{3}, 12$ **2. (a)** $-\frac{2}{3}, \frac{14}{3}$ **(b)** $\left(-\frac{2}{3}, \frac{14}{3}\right)$
(c) $\left(-\infty, -\frac{2}{3}\right) \cup \left(\frac{14}{3}, \infty\right)$ **3.** $(0, 6); (-\infty, 0] \cup [6, \infty)$
4. (a) $1, 3$ **(b)** $[1, 3]$ **(c)** $(-\infty, 1] \cup [3, \infty)$

CHAPTER 8 REVIEW (pp. 581–585)

1. $-7; 0$ **2.** $-22; 2$ **3.** $-2; 1$ **4.** $5; 5$
5. (a) $P(q) = 2q$ **(b)** $P(5) = 10$; there are
10 pints in 5 quarts.   **6. (a)** $f(x) = 4x - 3$
(b) $f(5) = 17$; three less than four times 5 is 17.
7. $(3, -2)$ **8.** $4; -6$
9.
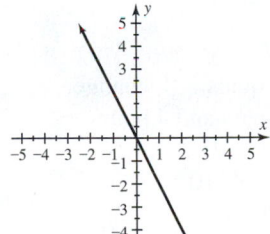
10.
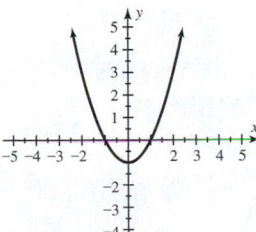
11.
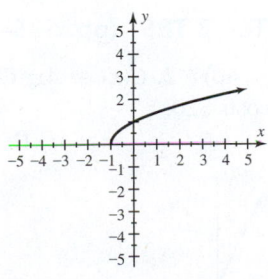
12.

13. $1; 4$ **14.** $1; -2$ **15.** $7; -1$

16. Numerical:

x	-3	-2	-1	0	1	2	3
$y = f(x)$	-11	-8	-5	-2	1	4	7

Symbolic: $f(x) = 3x - 2$
Graphical:

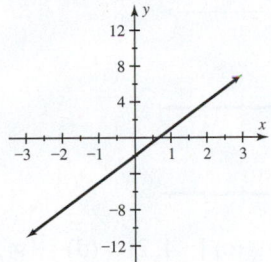

17. D: All real numbers; R: $y \le 4$ **18.** D: $-4 \le x \le 4$;
R: $-4 \le y \le 0$ **19.** Yes   **20.** No
21. $D = \{-3, -1, 2, 4\}; R = \{-1, 3, 4\}$; yes
22. $D = \{-1, 0, 1, 2\}; R = \{-2, 2, 3, 4, 5\}$; no
23. All real numbers   **24.** $x \ge 0$ **25.** $x \ne 0$
26. All real numbers   **27.** $x \le 5$ **28.** $x \ne -2$
29. All real numbers   **30.** All real numbers   **31.** No
32. Yes   **33.** Yes; $m = -4, b = 5$ **34.** Yes;
$m = -1, b = 7$ **35.** No   **36.** Yes; $m = 0, b = 6$
37. Yes; $f(x) = \frac{3}{2}x - 3$ **38.** No   **39.** 1
40. $f(-2) = -3; f(1) = 0$
41.
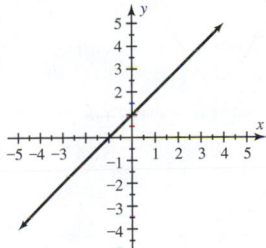
42.
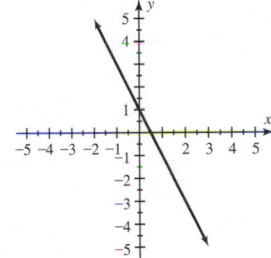

43.
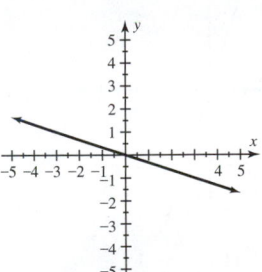
44.
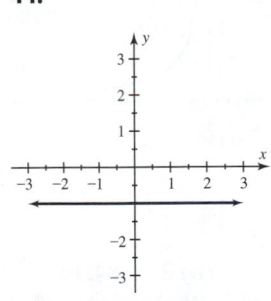

45. $H(x) = 24x; H(2) = 48$, there are 48 hours in 2 days.
46. (a)
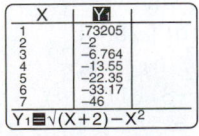
(b) Domain: $x \ge -2$

$[-10, 10, 1]$ by $[-10, 10, 1]$
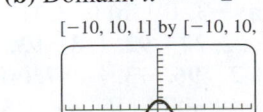

47. $(0.5, -3)$ **48.** $\left(\frac{5}{12}, \frac{3}{8}\right)$
49. $[-2, 2]$

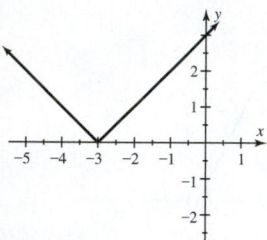

50. $(-\infty, -3]$

51. $\left(-\infty, \frac{4}{5}\right] \cup (2, \infty)$

52. $(-\infty, \infty)$

53. $[-2, 1]$ **54. (a)** -4 **(b)** 2 **(c)** $[-4, 2]$ **(d)** $(-\infty, 2)$
55. (a) 2 **(b)** $(2, \infty)$ **(c)** $(-\infty, 2)$ **56. (a)** 4 **(b)** 2
(c) $(2, 4)$ **57.** $\left[-3, \frac{2}{3}\right]$ **58.** $(-6, 45]$ **59.** $\left(-\infty, \frac{7}{2}\right)$
60. $[1.8, \infty)$ **61.** $(-3, 4)$ **62.** $(-\infty, 4) \cup (10, \infty)$
63. $(-5, 5)$ **64.** $[8, 28]$ **65.** $(-9, 21)$ **66.** $[8, 28]$
67. $\left(-\frac{11}{5}, \frac{7}{5}\right]$ **68.** $[88, 138)$
69. $D = (-\infty, \infty); R = [0, \infty)$
70. $D = (-\infty, \infty); R = [0, \infty)$
71. $(-\infty, 4) \cup (4, \infty)$ **72.** $D = [-3, 1]; R = [-3, 6]$
73. Yes; 2; quadratic   **74.** Yes; 1; linear   **75.** Yes; 3;
cubic   **76.** No   **77.** 11; 2   **78.** $-\frac{4}{5}$; undefined
79.   **80.**

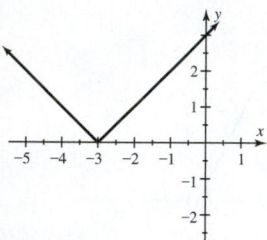
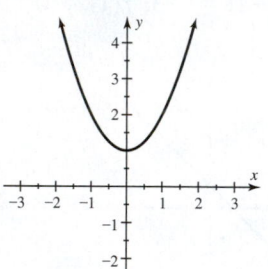

81.   **82.**

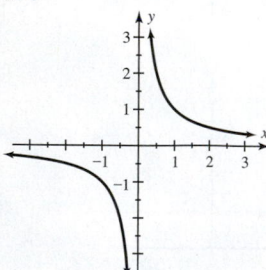
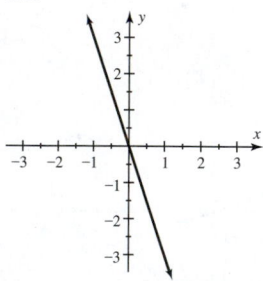

83. (a) 12 **(b)** 27 **84. (a)** $x^2 - x$ **(b)** $x + 1$
85. No, no   **86.** No, yes   **87.** No, yes   **88.** No, yes
89. (a) $0, 4$ **(b)** $(0, 4)$ **(c)** $(-\infty, 0) \cup (4, \infty)$
90. (a) $-3, 1$ **(b)** $[-3, 1]$ **(c)** $(-\infty, -3] \cup [1, \infty)$
91. $-22, 22$ **92.** $1, 8$ **93.** $-26, 42$ **94.** $-\frac{23}{3}, \frac{25}{3}$
95. $0, 2$ **96.** $-3, \frac{9}{5}$ **97. (a)** $-8, 6$ **(b)** $[-8, 6]$
(c) $(-\infty, -8] \cup [6, \infty)$ **98. (a)** $-\frac{5}{2}, \frac{7}{2}$ **(b)** $\left[-\frac{5}{2}, \frac{7}{2}\right]$
(c) $\left(-\infty, -\frac{5}{2}\right] \cup \left[\frac{7}{2}, \infty\right)$ **99.** $(-\infty, -3) \cup (3, \infty)$
100. $(-4, 4)$ **101.** $[-3, 4]$ **102.** $\left(-\infty, -\frac{5}{2}\right] \cup [5, \infty)$
103. $[4.4, 4.6]$ **104.** $\left[\frac{3}{13}, \frac{7}{13}\right]$ **105.** $(-\infty, \infty)$ **106.** $\frac{3}{2}$
107. $(-\infty, -1.5] \cup [1.5, \infty)$ **108.** $[-2, 6]$
109. $|x| \le 0.05$ **110.** $|5x - 1| > 4$
111. (a) About 25.1

(b) Decreased

[1885, 1965, 10] by [22, 26, 1]

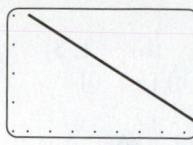

(c) -0.0492; the median age decreased by about 0.0492
year per year.
112. (a) 2.2 million   **(b)** The number of marriages each
year did not change.   **113. (a)** $f(x) = 8x$ **(b)** 8
(c) The total fat increases at the rate of 8 grams per cup.
114. (a)

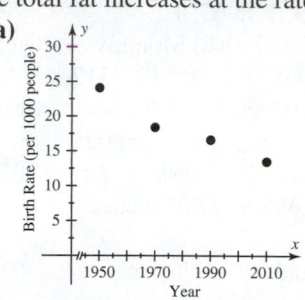

(b) $f(x) = -0.1675x + 350$ **(c)** About 15 per
1000 people (answers may vary)
115. (a) 113; in 1995, there were 113 unhealthy
days.   **(b)** $D = \{1995, 1999, 2000, 2003, 2007\}$
$R = \{56, 87, 88, 100, 113\}$ **(c)** It decreased and then
increased.
116. (a) Linear

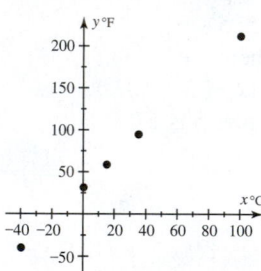

(b) $f(x) = \frac{9}{5}x + 32$; a 1°C change equals $\frac{9}{5}$° F change.
(c) 68°F   **117. (a)** 3 hours   **(b)** 2 hours and 4 hours
(c) Between 2 and 4 hours, exclusive   **(d)** 20 miles
per hour   **118. (a)** $f(x) = -0.3125x + 10$
(b) About 5.9 million   **119. (a)** $|A - 3.9| \le 1.7$
(b) $2.2 \le A \le 5.6$
120. Values between 32.2 and 37.8, exclusive

CHAPTER 8 TEST (pp. 586–587)

1. 46; (4, 46)   **2.** $C(x) = 4x$; $C(5) = 20$, 5 pounds of
candy costs $20.
3. (a)   **(b)**

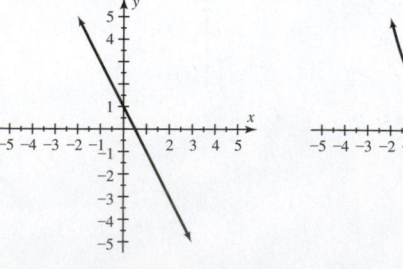
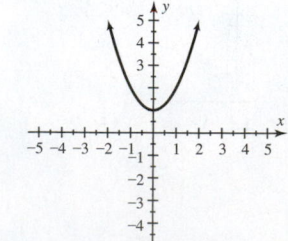

(c)

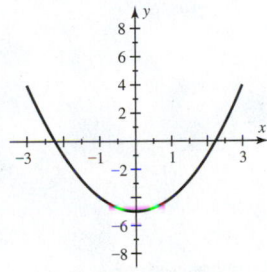

(d)

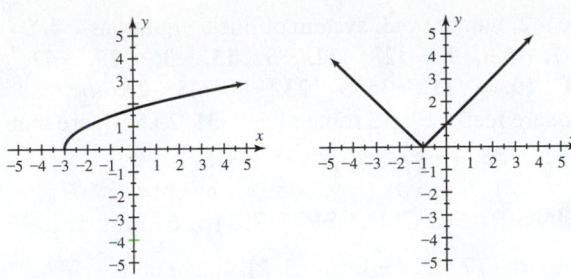

4. $0, -3$; D: $-3 \le x \le 3$; R: $-3 \le y \le 0$

5. Symbolic: $f(x) = x^2 - 5$

Numerical:

x	-3	-2	-1	0	1	2	3
$y = f(x)$	4	-1	-4	-5	-4	-1	4

Graphical:

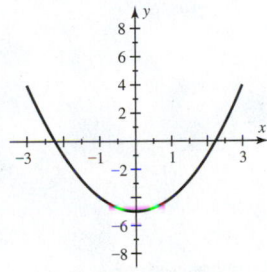

6. No, it fails the vertical line test.

7. (a) $D = \{-2, -1, 0, 5\}$ **(b)** All real numbers

(c) $x \ge -4$ **(d)** All real numbers **(e)** $x \ne 5$

8. It is; $f(x) = -8x + 6$

9.

10. $(-\infty, -2) \cup (1, \infty)$ **11. (a)** -5 **(b)** 5

(c) $[-5, 5]$ **(d)** $(-\infty, 5)$ **12.** $(-8, 0)$ **13.** $-12, 24$

14. (a) $[-5, 5]$ **(b)** $(-\infty, 0) \cup (0, \infty)$ **15.** Yes; 3;

cubic **16.** $\frac{8}{7}$; $(-\infty, 5) \cup (5, \infty)$ **17. (a)** 9

(b) $2x^3 + 2x$ **18. (a)** $f(x) = 0.4x + 75$ **(b)** 35 minutes

19. (a) Yes; $P(0) = 100$, $P(20) = 134$, $P(30) = 150$,

$P(50) = 180$ **(b)** Probably not because the race is over

after 50 seconds; $[0, 50]$. **20. (a)** $x = 25$ [0, 25, 5] by [0, 2, 0.5]

(b) 24 vehicles/min

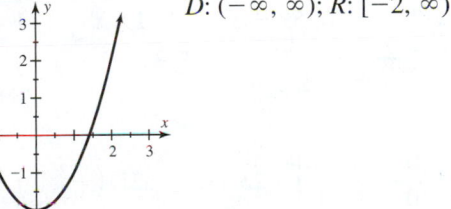

CHAPTERS 1–8 CUMULATIVE REVIEW
(pp. 588–589)

1. $2^3 \cdot 3 \cdot 5$ **2.** $3n - 4 = n$; 2 **3.** $-\frac{1}{6}$ **4.** 6 **5.** 15

6. 7.5% **7.** $a = \frac{3A + b}{2}$ **8.** $\{t \mid t > 2\}$

9.

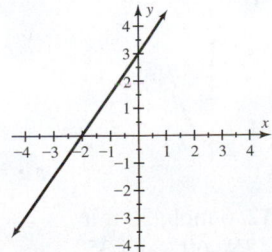

10.

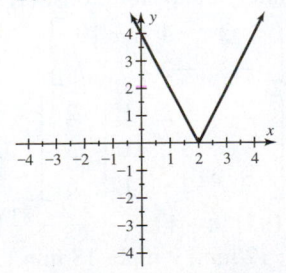

11. $y = -\frac{1}{2}x + 2$ **12.** $(1, 1)$

13.

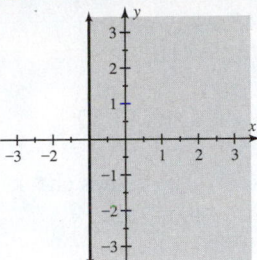

14.

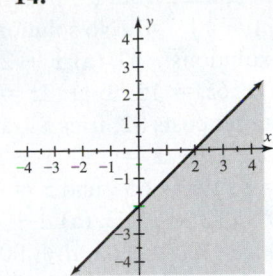

15.

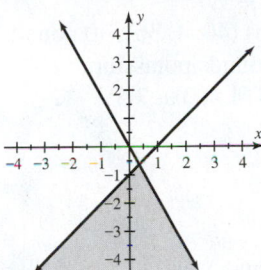

16. $-x^2 + 2x - 9$ **17.** $\frac{1}{x}$ **18.** $6b$ **19.** $24t^3$ **20.** $\frac{1}{4x^2y^2}$

21. $2x^5 - 8x^4 - 10x^2$ **22.** $10x^2 - 33x - 7$ **23.** $y^2 - 9$

24. $x^2 - 8xy + 16y^2$ **25.** 25,000 **26.** 2.8×10^{-2}

27. $3x^2 - 2x + 4$ **28.** $3x^2 + 5x + 5 + \frac{6}{x - 1}$

29. $5x^2y^2(2y - 3x)$ **30.** $(x - 1)(x + 1)(x + 3)$

31. $(z - 1)(2z + 3)$ **32.** $(4x - 5)(4x + 5)$

33. $(a - 2)(a^2 + 2a + 4)$ **34.** $(z^2 + 1)(z^2 + 6)$

35. $-5, 0$ **36.** 0 **37.** $-3, \frac{1}{2}$ **38.** $-1, 0, 1$ **39.** $x + 2$

40. $-\frac{5}{3}$; undefined **41.** $2x - 4$ **42.** $\frac{3x + 7}{(x + 1)(x + 3)}$

43. -8 **44.** 10 **45.** 12

46.

D: $(-\infty, \infty)$; R: $[-2, \infty)$

47. $(-\infty, 5) \cup (8, \infty)$ **48.** $-1, 5$ **49.** $[-1, 5]$

50. $(-\infty, 2) \cup (6, \infty)$ **51.** $d = 72t$

52. (a) $y = \frac{1}{4}x + 2$ **(b)** $\frac{1}{4}$ **(c)** Rain is falling at a rate

of $\frac{1}{4}$ inch per hour. **(d)** 3 inches **53.** $2100 at 5%, $2900

at 7% **54.** 104 feet **55. (a)** $20x$ **(b)** 50 min **56.** $1650

9 Systems of Linear Equations

SECTION 9.1 (pp. 600–603)

1. No; three planes cannot intersect at exactly 2 points.

2. $x + y + z = 5$, $2x - 3y + z = 7$, $x + 2y - 4z = 2$

(answers may vary) **3.** Yes **4.** No; a solution must be an

ordered triple. **5.** Two **6.** Three **7.** no **8.** Infinitely

many **9.** $(1, 2, 3)$ **11.** $(-1, 1, 2)$ **13.** $(3, -1, 1)$

15. $\left(\frac{17}{2}, -\frac{3}{2}, 2\right)$ **17.** $(5, -2, -2)$ **19.** $(11, 2, 2)$

21. $(1, -1, 2)$ **23.** $(0, -3, 2)$ **25.** $(-1, 3, -2)$

27. $(-1, 2, 4)$ **29.** $(8, 5, 2)$ **31.** No solutions
33. $(4 - z, 2, z)$ **35.** $(3, -3, 0)$ **37.** $\left(-\frac{3}{2}, 5, -10\right)$
39. $\left(\frac{3}{2}, 1, -\frac{1}{2}\right)$ **41.** No solutions   **43.** $(4 - z, 1, z)$
45. No solutions   **47. (a)** $x + 2y + 4z = 10$,
$x + 4y + 6z = 15, 3y + 2z = 6$ **(b)** $(2, 1, 1.5)$;
a hamburger costs \$2, fries \$1, and a soft drink \$1.50.
49. (a) $x + y + z = 180, x - z = 55, x - y - z = -10$
(b) $x = 85°, y = 65°$, and $z = 30°$ **(c)** These values check.
51. $110°, 50°, 20°$ **53. (a)** $a + 600b + 4c = 525$,
$a + 400b + 2c = 365, a + 900b + 5c = 805$
(b) $a = 5, b = 1, c = -20, F = 5 + A - 20W$
(c) 445 fawns   **55. (a)** $N + P + K = 80$,
$N + P - K = 8, 9P - K = 0$ **(b)** $(40, 4, 36)$; 40 pounds
nitrogen, 4 pounds phosphorus, 36 pounds potassium
57. \$7500 at 4%, \$8500 at 5%, and \$14,000 at 7.5%

SECTION 9.2 (pp. 612–615)

1. A rectangular array of numbers

2. $\begin{bmatrix} 2 & 1 & 3 \\ 0 & -4 & 2 \end{bmatrix}$ is 2×3 (answers may vary).

3. $\left[\begin{array}{cc|c} 1 & 3 & 10 \\ 2 & -6 & 4 \end{array}\right]$ is 2×3 (answers may vary).

4. 3×4 **5.** $\left[\begin{array}{cc|c} 1 & 0 & -3 \\ 0 & 1 & 5 \end{array}\right]$ (answers may vary)

6. $4, 2, -3$ **7.** 3×3 **9.** 3×2

11. $\left[\begin{array}{cc|c} 1 & -3 & 1 \\ -1 & 3 & -1 \end{array}\right]$ **13.** $\left[\begin{array}{ccc|c} 2 & -1 & 2 & -4 \\ 1 & -2 & 0 & 2 \\ -1 & 1 & -2 & -6 \end{array}\right]$

15. $x + 2y = -6, 5x - y = 4$ **17.** $x - y + 2z = 6$,
$2x + y - 2z = 1, -x + 2y - z = 3$
19. $x = 4, y = -2, z = 7$

21. $\begin{bmatrix} 0 & 1 & 1 & 1 \\ 1 & 0 & 1 & 0 \\ 0 & 0 & 0 & 1 \\ 1 & 0 & 1 & 0 \end{bmatrix}$ **23.** $(1, 3)$ **25.** $\left(-\frac{3}{2}, 2\right)$ **27.** $\left(\frac{7}{2}, -\frac{3}{2}\right)$

29. $\left(\frac{1}{2}, \frac{3}{2}\right)$ **31.** $(1, 2, 3)$ **33.** $(3, -3, 3)$ **35.** $(-1, 1, 0)$
37. $(1, 1, 1)$ **39.** $(-3, 2, 2)$ **41.** $(-7, 5)$ **43.** $(1, 2, 3)$
45. $(1, 0.5, -3)$ **47.** $(0.5, 0.25, -1)$ **49.** $(0.5, -0.2, 1.7)$
51. Dependent   **53.** Inconsistent   **55.** Dependent
57. $\left(\frac{1}{a}, \frac{1}{b}, \frac{2}{ab}\right)$ **59.** 214 pounds

61. (a) $a + 16b + 26c = 80$
$a + 28b + 45c = 344$
$a + 31b + 54c = 416$

(b) $a \approx -272.9, b \approx 19.8$, and $c \approx 1.4$ **(c)** About 216 lb
63. $\frac{1}{2}$ hour at 5 miles per hour, 1 hour at 6 miles per hour, and
$\frac{1}{2}$ hour at 8 miles per hour   **65.** \$500 at 3%, \$1000 at 4%,
and \$1500 at 6%   **67.** Answers may vary.

CHECKING BASIC CONCEPTS 9.1 & 9.2 (p. 615)

1. $(1, 3, -1)$ **2.** $(1, 2, 3)$ **3.** $(-2, 2, -1)$ **4.** \$200 at 1%;
\$400 at 2%; \$900 at 4%

SECTION 9.3 (pp. 620–622)

1. square   **2.** number   **3.** system of linear equations   **4.** 0
5. -2 **7.** -53 **9.** -323 **11.** -5 **13.** -36 **15.** -42
17. -50 **19.** 0   **21.** -3555 **23.** -7466.5 **25.** abc
27. 15 square feet   **29.** 52 square feet   **31.** 25.5 square feet
33. $(2, -2)$ **35.** $\left(-\frac{29}{11}, -\frac{34}{11}\right)$ **37.** $(-5, 7)$

CHECKING BASIC CONCEPTS 9.3 (p. 622)

1. (a) -1 **(b)** 17   **2.** $(-4, 6)$ **3.** 21 square units

CHAPTER 9 REVIEW (pp. 623–625)

1. Yes   **2.** $(1, -1, 2)$ **3.** $(-4, 3, 2)$ **4.** $(-1, 5, -3)$
5. $(-1, 3, 2)$ **6.** $(1, 1, 1)$ **7.** No solutions
8. $(2 - z, 2z - 1, z)$

9. $\left[\begin{array}{ccc|c} 1 & 1 & 1 & -6 \\ 1 & 2 & 1 & -8 \\ 0 & 1 & 1 & -5 \end{array}\right]$; $(-1, -2, -3)$

10. $\left[\begin{array}{ccc|c} 1 & 1 & 1 & -3 \\ -1 & 1 & 0 & 5 \\ 0 & 1 & 1 & -1 \end{array}\right]$; $(-2, 3, -4)$

11. $\left[\begin{array}{ccc|c} 1 & 2 & -1 & 1 \\ -1 & 1 & -2 & 5 \\ 0 & 2 & 1 & 10 \end{array}\right]$; $(-5, 4, 2)$

12. $\left[\begin{array}{ccc|c} 2 & 2 & -2 & -14 \\ -2 & -3 & 2 & 12 \\ 1 & 1 & -4 & -22 \end{array}\right]$; $(-4, 2, 5)$

13. $(-7, 4, 2)$ **14.** $(5.4, 2.1, 9.7)$ **15.** -8 **16.** 30   **17.** 89
18. 130   **19.** 181,845   **20.** 67.688   **21.** 46 square feet
22. 128 square feet   **23.** $(2, -1)$ **24.** $(-5, 3)$ **25.** $\left(\frac{3}{2}, \frac{1}{2}\right)$
26. $(7, -3)$ **27.** 5489 in 1994; 4641 in 2004
28. \$8 tickets: 285; \$12 tickets: 195
29. (a) $m + 3c + 5b = 14, m + 2c + 4b = 11$,
$c + 3b = 5$ **(b)** Malts: \$3; cones: \$2; bars: \$1
30. $100°, 65°$, and $15°$ **31.** 2 pounds of \$1.50 candy,
4 pounds of \$2 candy, 6 pounds of \$2.50 candy
32. (a) $a + 202b + 63c = 40, a + 365b + 70c = 50$,
$a + 446b + 77c = 55$ **(b)** $a \approx 27.134; b \approx 0.061$,
$c \approx 0.009$ **(c)** About 46 inches

CHAPTER 9 TEST (pp. 625–626)

1. No. Three planes cannot intersect at exactly three
points.   **2.** $(1, -2, -2)$ **3.** $(-4, 2, -5)$
4. No solutions   **5.** $(1, 2, 3)$
6. Infinitely many solutions: $(1, z, z)$

7. (a) $\left[\begin{array}{cc|c} 2 & -4 & -10 \\ -3 & -2 & 7 \end{array}\right]$ **(b)** $(-3, 1)$

8. (a) $\left[\begin{array}{ccc|c} 1 & 1 & 1 & 2 \\ 1 & -1 & -1 & 3 \\ 2 & 2 & 1 & 6 \end{array}\right]$ **(b)** $\left(\frac{5}{2}, \frac{3}{2}, -2\right)$

9. 6   **10.** 114   **11.** $\left(-\frac{47}{2}, -\frac{83}{2}\right)$ **12.** 6 mph: 30 min;
7 mph: 12 min; 9 mph: 18 min   **13.** $85°, 60°$, and $35°$

14. (a) $a + 20b + 25c = 168$; $a + 24b + 40c = 270$;
$a + 30b + 50c = 405$
(b) $a = -270$, $b = 20.1$, $c = 1.44$ **(c)** About 316 lb

CHAPTERS 1–9 CUMULATIVE REVIEW
(pp. 628–630)

1. $2^3 \cdot 3^2 \cdot 5$ **2.** $2n + 7 = n - 2$; -9 **3.** $\frac{23}{21}$
4. $-\frac{1}{6}$ **5.** 18   **6.** $\frac{1}{10}$ **7.** $\frac{6}{5}$ **8.** 2   **9.** $\frac{31}{25}$; 1.24
10. 6 miles   **11.** $t < \frac{4}{7}$
12.

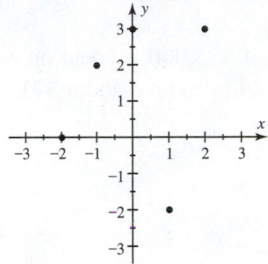

13. x-int: -4; y-int: 3

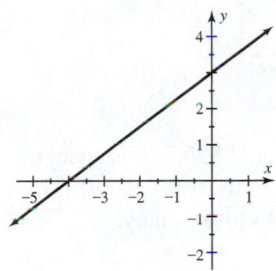

14. x-int: -2; y-int: none

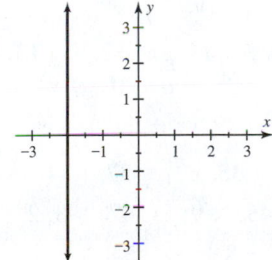

15. x-int: 2; y-int: 4; $y = -2x + 4$ **16.** $y = -2x + 1$
17. $y = \frac{3}{5}x + \frac{34}{5}$ **18.** $y = 2x + 3$
19. $P = 500x + 4000$ **20.** (7.5, 10)
21.   **22.**

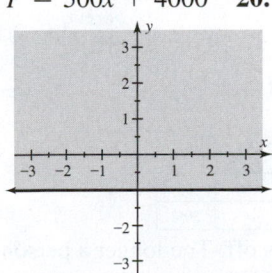
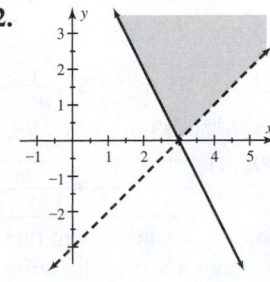

23. -76
24. $24t^4$ **25.** 1   **26.** $\frac{1}{4t^6}$ **27.** $2ab^3$ **28.** $\frac{a^6}{8b^6}$
29. $2a^4 - 4a^3 + 6a^2$ **30.** $5x^2 - 34x - 7$
31. $4x^2 + 12xy + 9y^2$ **32.** $a^2 - b^2$ **33.** $2x^2 - 4x + 3$
34. $x^3 - 4x^2 + 3x - 2 + \frac{1}{x - 5}$ **35.** $5ab^2(2 - 5a^2b^3)$
36. $(y - 3)(y^2 + 2)$ **37.** $(2z + 3)(3z - 1)$

38. $(2z - 3)(2z + 3)$ **39.** $(2y - 5)^2$
40. $(a - 3)(a^2 + 3a + 9)$ **41.** $(z^2 - 3)(4z^2 - 5)$
42. $ab(a + b)(2a - b)$ **43.** $-2, 1$ **44.** $-\frac{1}{3}, \frac{3}{2}$
45. $x - 4$ **46.** $\frac{x - 6}{3x - 4}$ **47.** $2(x - 1)$ or $2x - 2$
48. 1   **49.** 8   **50.** $\frac{17}{12}$ **51.** 10   **52.** 13
53. $D = (-\infty, \infty)$; $R = [-2, \infty)$

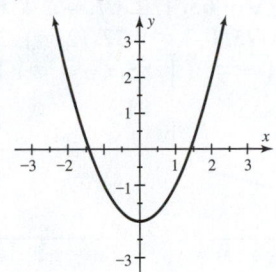

54. $2, -2$ **55.** $(3, \infty)$ **56.** $(-\infty, 0]$
57. $(-1, 5]$

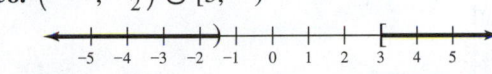

58. $\left(-\infty, -\frac{3}{2}\right) \cup [3, \infty)$

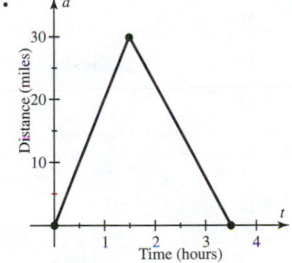

59. $[-2, 4]$ **60.** $[-16, 10]$ **61. (a)** $-3, 1$ **(b)** $[-3, 1]$
(c) $(-\infty, -3] \cup [1, \infty)$ **62.** $-6, 18$
63. $(-\infty, -6) \cup \left(\frac{8}{3}, \infty\right)$ **64.** $[4, 7]$ **65.** $(1, 1, 2)$

66. $\begin{bmatrix} 1 & 1 & -1 & \mid & 4 \\ -1 & -1 & -1 & \mid & 0 \\ 1 & -2 & 1 & \mid & -9 \end{bmatrix}$; $(-1, 3, -2)$

67.

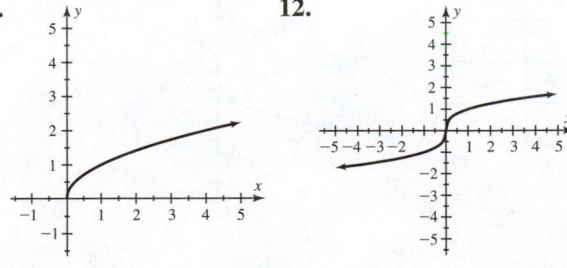

68. 3 children and 5 adults   **69.** $\frac{6}{7}$ hour   **70. (a)** 112 feet
(b) 5.5 seconds   **71.** 61.25 feet
72. (a) $4b + 3f + 4m = 23$; $b + 2f + m = 7$;
$3b + f + 2m = 13$ **(b)** Burger: \$2; fries: \$1; malt: \$3

10   Radical Expressions and Functions

SECTION 10.1 (pp. 640–643)

1. ± 3 **2.** 3   **3.** 2   **4.** Yes   **5.** b   **6.** $|x|$ **7.** Undefined   **8.** -3 **9.** d.
10. a.
11.   **12.**

13. $\{x|x \geq 0\}$ **14.** All real numbers **15.** 3 **17.** 0.6
19. $\frac{4}{5}$ **21.** x **23.** 3 **25.** -4 **27.** $\frac{2}{3}$ **29.** $-x^3$ **31.** $4x^2$ **33.** 3
35. -3 **37.** Not possible **39.** -2.24 **41.** 1.71
43. -1.48 **45.** 4 **47.** $|y|$ **49.** $|x - 5|$ **51.** $|x - 1|$ **53.** $|y|$
55. $|x^3|$ **57.** x **59.** 3, not possible **61.** $\sqrt{6}$, not possible
63. $\sqrt{20}$ or $2\sqrt{5}$, $\sqrt{6}$ **65.** 1, 2 **67.** -2, 1 **69.** -1, $\sqrt[3]{-6}$ or
$-\sqrt[3]{6}$ **71.** 4 **73.** 5 **75.** $[-2, \infty)$ **77.** $[2, \infty)$ **79.** $[2, \infty)$
81. $(-\infty, 1]$ **83.** $\left(-\infty, \frac{8}{5}\right]$ **85.** $(-\infty, \infty)$ **87.** $\left(-\frac{1}{2}, \infty\right)$

89.

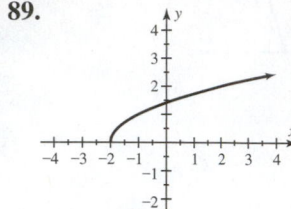

91.

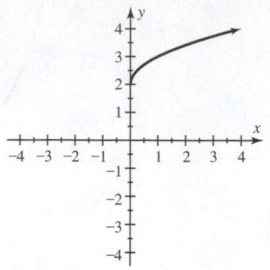

Shifted 2 units left   Shifted 2 units upward

93.

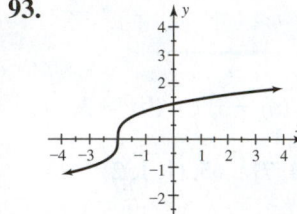

Shifted 2 units left

95.

x	$\sqrt{x+1}$
-1	—
0	1
1	2
4	3
9	4

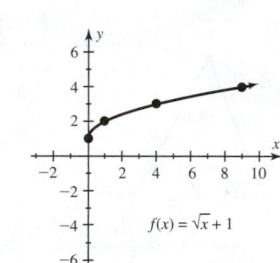

97.

x	$\sqrt{3x}$
-1	—
0	0
$\frac{1}{3}$	1
$\frac{4}{3}$	2
3	3

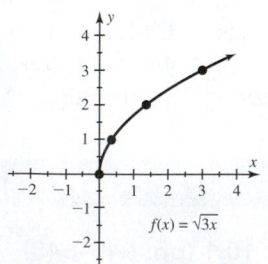

99.

x	$2\sqrt[3]{x}$
-8	-4
-1	-2
0	0
1	2
8	4

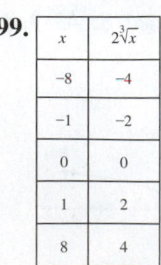

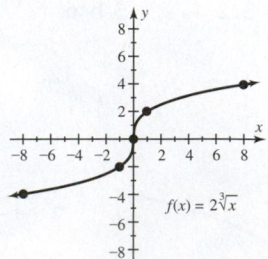

101.

x	$\sqrt[3]{x-1}$
-7	-2
0	-1
1	0
2	1
9	2

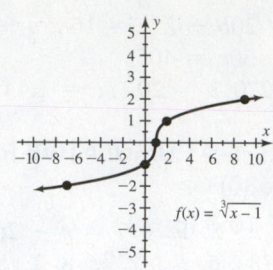

103. $A = 6$ **105.** 1 sec   **107.** 122 mi
109. About $14 thousand; no
111. (a) $P(25,000) = \$71,246$; if $25,000 is spent on equipment per worker, each worker will produce about $71,246 worth of goods.
(b)

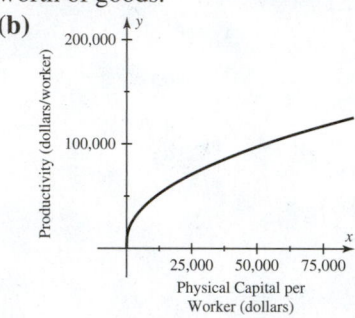

(c) $26,197; $20,102; an additional $25,000 is spent on equipment per worker, but productivity levels off. There is a point where the business starts to lose money.

SECTION 10.2 (pp. 649–651)

1. 2 **2.** 2 **3.** $\frac{1}{2}$ **4.** $\frac{1}{2}$ **5.** $x^{1/2}$ **6.** $a^{4/3}$ **7.** $\sqrt[n]{a}$
8. $\left(\sqrt[n]{a}\right)^m$ or $\sqrt[n]{a^m}$ **9.** c. **10.** f. **11.** g. **12.** h. **13.** d.
14. e. **15.** a. **16.** b. **17.** $\sqrt{7}$ **19.** $\sqrt[3]{a}$ **21.** $\sqrt[6]{x^5}$
23. $\sqrt{x+y}$ **25.** $\frac{1}{\sqrt[3]{b^2}}$ **27.** $t^{1/2}$ **29.** $(x+1)^{1/3}$
31. $(x+1)^{-1/2}$ **33.** $(a^2 - b^2)^{1/2}$ **35.** $x^{-7/3}$ **37.** 1.74
39. 1.71 **41.** 3.74 **43.** 0.55 **45.** $\sqrt{9}$; 3 **47.** $\sqrt[3]{8}$; 2
49. $\sqrt{\frac{4}{9}}$; $\frac{2}{3}$ **51.** $\sqrt[3]{(-8)^2}$ or $\left(\sqrt[3]{-8}\right)^2$; 4 **53.** $\sqrt[3]{8}$; 2
55. $\frac{1}{\sqrt[4]{16^3}}$ or $\frac{1}{\left(\sqrt[4]{16}\right)^3}$; $\frac{1}{8}$ **57.** $\frac{1}{\left(\sqrt{4}\right)^3}$; $\frac{1}{8}$ **59.** $\sqrt[4]{z}$
61. $\frac{1}{\sqrt[5]{y^2}}$ **63.** $\sqrt[3]{3x}$ **65.** $y^{1/2}$ **67.** x **69.** $2x^{2/3}$
71. $\frac{7}{x^{1/6}}$ **73.** b^3 **75.** x^3 **77.** xy^2 **79.** $y^{13/6}$ **81.** $\frac{x^4}{9}$
83. $\frac{y^3}{x}$ **85.** $y^{1/4}$ **87.** $\frac{1}{a^{2/3}}$ **89.** $\frac{1}{k^2}$
91. $b^{3/4}$ **93.** $p^2 + p$ **95.** $x^{5/6} - x$ **97.** $\frac{3}{x^{1/6}}$
99. (a)

x	0	20	40	60	80
$A(x)$	0	27%	37%	44%	50%

(b) The abandonment rate levels off. The longer a person watches a video, the more likely he or she will continue to watch. **101.** About 1 step/sec **103. (a)** 400 in^2
(b) $A = 100W^{2/3}$ **105. (a)** 42.4 cm; 44.7 cm **(b)** First year
107. Answers may vary.

CHECKING BASIC CONCEPTS 10.1 & 10.2 (p. 652)

1. (a) ± 7 **(b)** 7   **2. (a)** -2 **(b)** -3 **3. (a)** $\sqrt{x^3}$
or $\left(\sqrt{x}\right)^3$ **(b)** $\sqrt[3]{x^2}$ or $(\sqrt[3]{x})^2$ **(c)** $\frac{1}{\sqrt[5]{x^2}}$ or $\frac{1}{(\sqrt[5]{x})^2}$
4. $|x-1|$ **5. (a)** 3   **(b)** 5   **(c)** $\left(\frac{12}{7}\right)^{7/12}$ or about 1.37.

SECTION 10.3 (pp. 659–660)

1. Yes   **2.** No   **3.** $\sqrt[3]{ab}$ **4.** $\frac{a}{b}$ **5.** $\frac{a}{b}$ **6.** $\frac{1}{3}$ **7.** No
8. Yes   **9.** No, since $1^3 \neq 3$ **10.** Yes; $4^3 = 64$
11. 3   **13.** 10   **15.** 4   **17.** $\frac{3}{5}$ **19.** $\frac{1}{4}$ **21.** $\frac{2}{3}$ **23.** x^3
25. $\frac{\sqrt[3]{7}}{3}$ **27.** $\frac{\sqrt[4]{x}}{3}$ **29.** $\frac{3}{z}$ **31.** $\frac{x}{4}$ **33.** 3   **35.** 4   **37.** 3
39. -2 **41.** a **43.** 3   **45.** $2x^2$ **47.** $-a^2\sqrt[3]{5}$
49. $2x\sqrt[4]{y}$ **51.** $6x$ **53.** $2x^2yz^3$ **55.** $\frac{3}{2}$ **57.** $\sqrt[3]{12ab}$
59. $5\sqrt{z}$ **61.** $\frac{1}{a}$ **63.** $\sqrt{x^2-16}$ **65.** $\sqrt[3]{a^3+1}$
67. $\sqrt{x+1}$ **69.** 10   **71.** 2   **73.** 3   **75.** $10\sqrt{2}$ **77.** $3\sqrt[3]{3}$
79. $2\sqrt{2}$ **81.** $-2\sqrt[5]{2}$ **83.** $b^2\sqrt{b}$ **85.** $2n\sqrt{2n}$
87. $2ab^2\sqrt{3b}$ **89.** $-5xy\sqrt[3]{xy^2}$ **91.** $5\sqrt[3]{5t^2}$ **93.** $\frac{3t}{r\sqrt[4]{5r^3}}$
95. $\frac{3\sqrt[3]{x^2}}{y}$ **97.** $\frac{7a\sqrt{a}}{9}$
99. $\left(\sqrt[mn]{a^m b^m}\right)^n = (a^m b^m)^{n/mn}$
$= (a^m b^m)^{1/m}$
$= a^{m/m} b^{m/m}$
$= ab$

101. $\sqrt[6]{3^5}$ **103.** $2\sqrt[12]{2^5}$ **105.** $3\sqrt[12]{3^{11}}$ **107.** $x\sqrt[12]{x}$
109. $\sqrt[12]{r^{11}t^7}$ **111. (a)** $S = 5\sqrt{M}$
(b) 50 miles per hour

SECTION 10.4 (pp. 671–673)

1. $2\sqrt{a}$ **2.** $3\sqrt[3]{b}$ **3.** like
4. Yes; $4\sqrt{15} - 3\sqrt{15} = \sqrt{15}$
5. No   **6.** $\frac{\sqrt{7}}{\sqrt{7}}$ **7.** $\sqrt{t} + 5$ **8.** $\frac{5+\sqrt{2}}{5+\sqrt{2}}$

9. Not possible   **11.** $\sqrt{7}, 2\sqrt{7}, 3\sqrt{7}$ **13.** $2\sqrt[3]{2}, -3\sqrt[3]{2}$
15. Not possible   **17.** $2\sqrt[3]{xy}, xy\sqrt[3]{xy}$ **19.** $9\sqrt{3}$ **21.** $6\sqrt[3]{5}$
23. Not possible   **25.** Not possible   **27.** Not possible
29. $5\sqrt[3]{2}$ **31.** $8\sqrt{2}$ **33.** $6\sqrt{11}$ **35.** $2\sqrt{x} - \sqrt{y}$
37. $2\sqrt[3]{z}$ **39.** $-5\sqrt[3]{6}$ **41.** $y^2 - y$ **43.** $9\sqrt{7}$ **45.** $6\sqrt[4]{3}$
47. $7\sqrt{x}$ **49.** $8\sqrt{2k}$ **51.** $-2\sqrt{11}$ **53.** $5\sqrt[3]{2} - \sqrt{2}$
55. $-\sqrt[3]{xy}$ **57.** $3\sqrt{x+2}$ **59.** $\sqrt{x+2}$
61. $(x-1)\sqrt{x+1}$ **63.** $4x\sqrt{x}$ **65.** $\frac{\sqrt[3]{7x}}{6}$ **67.** $\frac{3\sqrt{3}}{2}$
69. $\frac{71\sqrt{2}}{10}$ **71.** $2\sqrt{2}$ **73.** $(5x-1)\sqrt[4]{x}$
75. $(8x+2)\sqrt{x}$ or $2\sqrt{x}(4x+1)$
77. $(3ab-1)\sqrt[4]{ab}$ **79.** $(n-2)\sqrt[3]{n}$
81. $(f+g)(x) = 3\sqrt{x} + 1; (f-g)(x) = 7\sqrt{x} - 5$
83. $(f+g)(x) = 4\sqrt[3]{x}; (f-g)(x) = 2$ **85.** $x - \sqrt{x} - 6$
87. 2   **89.** 119   **91.** $x - 64$ **93.** $ab - c$
95. $x + \sqrt{x} - 56$ **97.** $\frac{\sqrt{7}}{7}$
99. $\frac{4\sqrt{3}}{3}$ **101.** $\frac{\sqrt{5}}{3}$ **103.** $\frac{\sqrt{3b}}{6}$ **105.** $\frac{t\sqrt{r}}{2r}$
107. $\frac{3+\sqrt{2}}{7}$ **109.** $\sqrt{10} - 2\sqrt{2}$ **111.** $\frac{11-4\sqrt{7}}{3}$
113. $\sqrt{7} + \sqrt{6}$ **115.** $\frac{z + 3\sqrt{z}}{z-9}$ **117.** $\frac{a + 2\sqrt{ab} + b}{a-b}$

119. $\sqrt{x+1} + \sqrt{x}$ **121.** $\frac{3\sqrt[3]{x^2}}{x}$ **123.** $\frac{\sqrt[3]{x}}{x}$
125. $12\sqrt{3} \approx 20.8$ cm   **127.** $2\sqrt{6}$ **129.** $120\sqrt{2}$ ft
131. $\sqrt{2x}$ ft

CHECKING BASIC CONCEPTS 10.3 & 10.4 (p. 673)

1. (a) $\frac{1}{8}$ **(b)** 10   **(c)** $-2x\sqrt[3]{xy}$ **(d)** $\frac{4b^2}{5}$ **2.** $\sqrt[6]{7^5}$
3. (a) 6   **(b)** 3   **(c)** $6x^3$ **4. (a)** $7\sqrt{6} + \sqrt{7}$
(b) $5\sqrt[3]{x}$ **(c)** $\sqrt{x}$ **5. (a)** $(y-x)\sqrt[3]{xy}$
(b) 14   **6.** $\frac{\sqrt{6}}{2}$ **7.** $\frac{\sqrt{5}+1}{2}$

SECTION 10.5 (pp. 677–680)

1.   **2.**

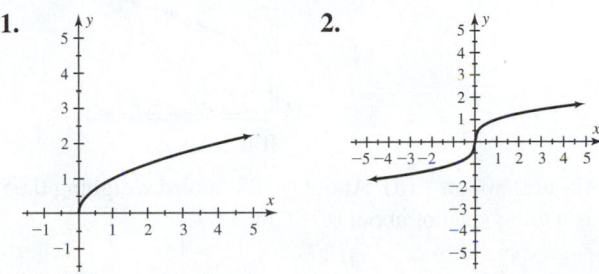

3. $\{x \mid x \geq 0\}$ **4.** All real numbers   **5.** $f(x) = x^p$,
p is rational.   **6.** $f(x) = \sqrt[n]{x}$, where n is an integer greater
than 1.   **7.** $\{x \mid x \geq 0\}$ **8.** All real numbers
9. $\sqrt{3} \approx 1.73$; undefined   **11.** $\sqrt[4]{6} \approx 1.57$; undefined
13. $\sqrt[5]{13} \approx 1.67$; 1   **15.** $\sqrt[3]{6} \approx 1.82$; -1 **17.** $f(x) = \sqrt{x}$
19. $f(x) = \sqrt[3]{x^2}$ **21.** $f(x) = \frac{1}{\sqrt[5]{x}}$ **23.** 32; 55.90
25. $-\frac{1}{128} \approx -0.01; 0.04$ **27.** 4; not possible   **29.** 4; 4
31. $[0, \infty)$ **33.** $(-\infty, \infty)$

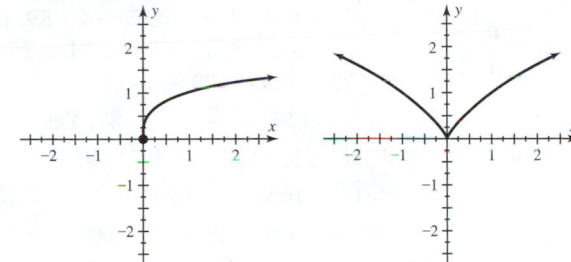

35. $g(x)$ is greater.   **37.** $f(x)$ is greater.
[0, 6, 1] by [0, 6, 1]   [0, 6, 1] by [0, 6, 1]

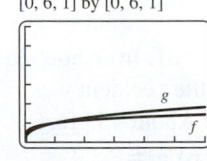

39. $x^p > x^q$ **41. (a)** 6   **(b)** $\sqrt{2x}$ **(c)** $4|x|$
(d) 2   **43.** b.   **45.** c.   **47.** d.   **49.** b.   **51.** 2877 in^2
53. About 35%   **55.** No, for $x \geq 10$ it is less than double.
57. (a) 6 yr   **(b)** The twin in the spaceship will be 4 years
younger than the twin on Earth.

59. (a) [0, 5, 1] by [0, 0.5, 0.1]   **(b)** $k \approx 0.16$

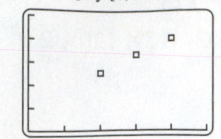

(c) [0, 5, 1] by [0, 0.5, 0.1]; yes   **(d)** 0.295 m²

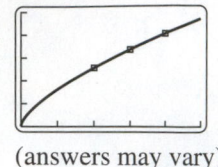

(answers may vary)

61. (a) $k \approx 0.91$ **(b)** [0, 1.5, 0.1] by [0, 1, 0.1]

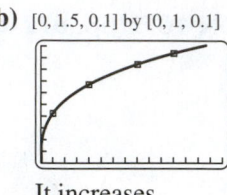

It increases.

(c) About 0.808 m   **(d)** About 0.788; a bird weighing 0.65 kg has a wing span of about 0.788 m.

SECTION 10.6 (pp. 690–694)

1. Square each side.   **2.** Cube each side.   **3.** Yes
4. Check them in the given equation.   **5.** Finding an unknown side of a right triangle (answers may vary)   **6.** 5
7. $d = \sqrt{(x_2 - x_1)^2 + (y_2 - y_1)^2}$ **8.** $x^{1/2} + x^{3/4} = 2$
9. 2   **11.** x **13.** $2x + 1$ **15.** $5x^2$ **17.** 64   **19.** 81   **21.** 6
23. 48   **25.** 6   **27.** 3   **29.** 27   **31.** -2 **33.** 15   **35.** $\frac{1}{2}$
37. 0, 1   **39.** 2   **41.** -4 **43.** 9   **45.** 9   **47.** ± 7
49. ± 10 **51.** $-5, 3$ **53.** $-3, 7$ **55.** 4   **57.** -4 **59.** 1
61. $\frac{7}{5}$ **63.** ± 2 **65.** $\sqrt[5]{12}$ **67.** $-5, 1$ **69.** 3   **71.** 1.88
73. $-1, 0.70$ **75.** 1.79   **77.** -0.47 **79.** (a)–(c) 16
81. (a)–(c) 11   **83.** $L = \frac{8T^2}{\pi^2}$ **85.** $A = \pi r^2$ **87.** Yes
89. Yes   **91.** Yes   **93.** No   **95.** $\sqrt{32} = 4\sqrt{2}$ **97.** 7
99. $c = 5$ **101.** $b = \sqrt{61}$ **103.** $a = 14$
105. $\sqrt{20} = 2\sqrt{5}$ **107.** $\sqrt{4000} = 20\sqrt{10}$ **109.** $\sqrt{89}$
111. 5   **113.** 4   **115.** 2, 122   **117.** $W = 8$ pounds
119. About 81 sec   **121.** About 3 miles   **123.** About 269 feet
125. 19 inches   **127.** $h \approx 16.3$ inches, $d \approx 33.3$ inches
129. (a) 121 feet   **(b)** About 336 feet   **131. (a)** About 38 miles per hour; the vehicle involved in the accident was traveling about 38 miles per hour.   **(b)** About 453 feet
133. (a) It increases by a factor of 8.   **(b)** $v = \sqrt[3]{\frac{W}{3.8}}$
(c) 20 mph   **135.** $a\sqrt{2}$

CHECKING BASIC CONCEPTS 10.5 & 10.6 (p. 694)

1. (a)   **(b)**

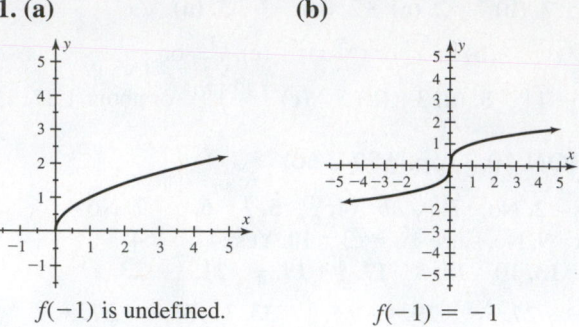

$f(-1)$ is undefined. $f(-1) = -1$

(c)

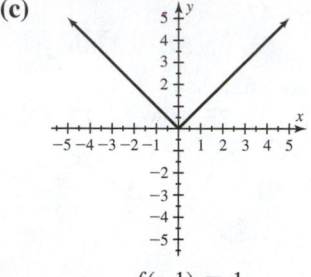

$f(-1) = 1$

2. 3.2   **3.** $[4, \infty)$ **4. (a)** 4   **(b)** 28   **(c)** 3   **5.** 13   **6.** 9.6 in.
7. $-3, 1$

SECTION 10.7 (pp. 701–702)

1. $2 + 3i$ (answers may vary)   **2.** No; any real number a can be written as $a + 0i$.   **3.** i **4.** -1 **5.** $i\sqrt{a}$
6. $10 - 7i$ **7.** $a + bi$ **8.** $1 + 2i$ **9.** 4   **10.** -5 **11.** 0
12. pure   **13.** $i\sqrt{5}$ **15.** $10i$ **17.** $12i$ **19.** $2i\sqrt{3}$
21. $3i\sqrt{2}$ **23.** 3   **25.** $-8 + 7i$ **27.** $1 - 9i$
29. $-10 + 7i$ **31.** $-13 + 13i$ **33.** $20 - 12i$ **35.** 41
37. 20   **39.** $1 + 5i$ **41.** $3 + 4i$ **43.** $-2 - 6i$ **45.** -2
47. $a^2 + 9b^2$ **49.** $-i$ **51.** i **53.** -1 **55.** 1   **57.** $3 - 4i$
59. $6i$ **61.** $5 + 4i$ **63.** -1 **65.** $1 - i$ **67.** $-\frac{6}{29} + \frac{15}{29}i$
69. $2 + i$ **71.** $-1 + 6i$ **73.** $-1 - 2i$ **75.** $-\frac{3}{2}i$
77. $-\frac{1}{2} + \frac{3}{2}i$ **79.** $\frac{290}{13} + \frac{20}{13}i$ **81.** They are graphed using a real axis and an imaginary axis.

CHECKING BASIC CONCEPTS 10.7 (p. 703)

1. (a) $8i$ **(b)** $i\sqrt{17}$ **2. (a)** $3 - 4i$ **(b)** $-2 + 3i$
(c) $5 + i$ **(d)** $\frac{3}{4} + \frac{3}{4}i$

CHAPTER 10 REVIEW (pp. 707–709)

1. 2   **2.** 6   **3.** $3|x|$ **4.** $|x - 1|$ **5.** -4 **6.** -5 **7.** x^2
8. $3x$ **9.** 2   **10.** -1 **11.** x^2 **12.** $x + 1$ **13.** $\sqrt{14}$
14. $\sqrt[3]{-5}$ **15.** $\left(\sqrt{\frac{x}{y}}\right)^3$ or $\sqrt{\left(\frac{x}{y}\right)^3}$ **16.** $\frac{1}{\sqrt[3]{(xy)^2}}$ or $\frac{1}{(\sqrt[3]{xy})^2}$
17. 9   **18.** 2   **19.** 64   **20.** 27   **21.** z^2 **22.** xy^2 **23.** $\frac{x^3}{y^9}$
24. $\frac{y^2}{x}$ **25.** 8   **26.** -2 **27.** x^2 **28.** 2   **29.** $-\frac{\sqrt[3]{x}}{2}$ **30.** $\frac{1}{3}$
31. $4\sqrt{3}$ **32.** $3\sqrt{6}$ **33.** $\frac{3}{x}$ **34.** $4ab\sqrt{2a}$ **35.** $9xy$
36. $5z\sqrt[3]{z}$ **37.** $x + 1$ **38.** $\frac{2a\sqrt[4]{a}}{b}$ **39.** $2\sqrt[6]{x^5}$
40. $\sqrt[6]{r^5 t^8}$ or $t\sqrt[6]{r^5 t^2}$ **41.** $4\sqrt{3}$ **42.** $3\sqrt[3]{x}$ **43.** $-3\sqrt[3]{5}$

44. $-\sqrt[4]{y}$ **45.** $11\sqrt{3}$ **46.** $7\sqrt{2}$ **47.** $13\sqrt[3]{2}$
48. $3\sqrt{x+1}$ **49.** $(2x-1)\sqrt{x}$ **50.** $(b+2a)\sqrt[3]{ab}$
51. $5+4\sqrt{2}$ **52.** $7+7\sqrt{3}-\sqrt{5}-\sqrt{15}$ **53.** 3
54. 95   **55.** $a-2b$ **56.** $xy+\sqrt{xy}-2$ **57.** $\frac{4\sqrt{5}}{5}$
58. $\frac{r\sqrt{t}}{2t}$ **59.** $\frac{3-\sqrt{2}}{7}$ **60.** $\frac{5+\sqrt{7}}{9}$
61. $\sqrt{8}+\sqrt{7}$ **62.** $\frac{a-2\sqrt{ab}+b}{a-b}$

63. 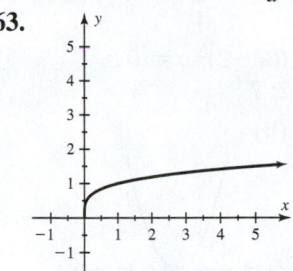   **64.**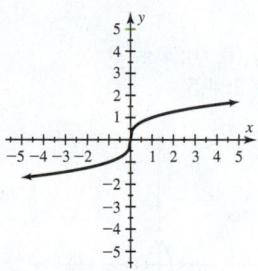

65. $f(x)=\sqrt{x}$; 2   **66.** $f(x)=\sqrt[7]{x^2}$; $\sqrt[7]{16}$
67. 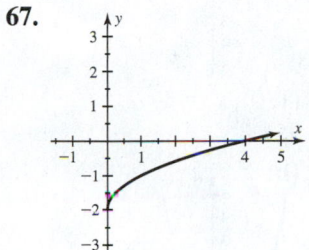   **68.**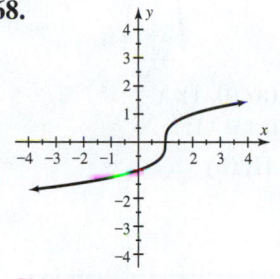

Shifted 2 units downward   Shifted 1 unit to the right

69. $[1,\infty)$ **70.** $(-\infty,3]$ **71.** $(-\infty,\infty)$ **72.** $(-2,\infty)$
73. 2   **74.** 4   **75.** 9   **76.** 9   **77.** 8   **78.** $\frac{1}{4}$ **79.** 4.5
80. 1.62   **81.** $c=\sqrt{65}$ **82.** $b=\sqrt{39}$ **83.** $\sqrt{41}$
84. $\sqrt{52}=2\sqrt{13}$ **85.** ± 11 **86.** ± 4 **87.** $-3,5$
88. 4   **89.** 3   **90.** 2   **91.** ± 4 **92.** -1 **93.** 1
94. $-2,0$ **95.** -2 **96.** $-2+4i$ **97.** $5+i$ **98.** $1+2i$
99. $-\frac{14}{13}+\frac{5}{13}i$ **100.** $2-2i$ **101.** About 85 feet
102. $\sqrt{16{,}200}=90\sqrt{2}\approx 127.3$ feet   **103.** About 0.79
second   **104. (a)** 5 square units   **(b)** $5\sqrt{5}$ cubic units
(c) $\sqrt{10}$ units   **(d)** $\sqrt{15}$ units   **105.** About 0.82 foot
106. About 0.13 foot; the length of the pendulum is shorter.
107. $r=\sqrt[210]{\frac{281}{4}}-1\approx 0.02$; from 1790 through 2000 the
average annual percent growth rate was about 2%.
108. (a) About 43 miles per hour   **(b)** About 34 miles per
hour; a steeper bank allows for a higher speed limit; yes
109. $\sqrt{7}\approx 2.65$ feet

CHAPTER 10 TEST (pp. 709–710)

1. -3 **2.** $|z+1|$ **3.** $5x^2$ **4.** $2z^2$ **5.** $2xy\sqrt[4]{y}$ **6.** 1
7. $\sqrt[5]{7^2}$ or $(\sqrt[5]{7})^2$ **8.** $\sqrt[3]{\left(\frac{y}{x}\right)^2}$ or $\left(\sqrt[3]{\frac{y}{x}}\right)^2$ **9.** 16
10. $\frac{1}{216}$ **11.** $x^{4/3}$ **12.** $x^{7/10}$ **13.** $(-\infty,4]$
14.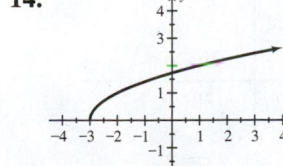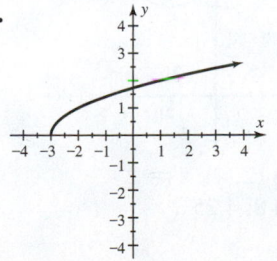

15. $8z^{3/2}$ **16.** $\frac{z}{y^{2/3}}$ **17.** 9   **18.** $\frac{y}{2}$ **19.** $4\sqrt{7}+\sqrt{5}$
20. $6\sqrt[3]{x}$ **21.** $14\sqrt{2}$ **22.** 2   **23. (a)** 27   **(b)** 7   **(c)** 3
(d) 17   **24. (a)** $\frac{2\sqrt{7}}{21}$ **(b)** $\frac{-1+\sqrt{5}}{4}$ **25.** 2.63
26. $\sqrt{120}\approx 10.95$ **27.** $\sqrt{8}=2\sqrt{2}$ **28.** $2-19i$
29. $-6+8i$ **30.** $\frac{5}{4}$ **31.** $\frac{4}{29}+\frac{10}{29}i$ **32. (a)** About 17.25 mi
(b) About 420 ft   **33.** 1.31 pounds

CHAPTERS 1–10 CUMULATIVE REVIEW
(pp. 711–712)

1. 36π **2.** $D=\{-1,0,1\}$, $R=\{2,4\}$ **3. (a)** $\frac{1}{a^3b^9}$
(b) x^4y^9 **(c)** r^8t^5 **4.** 4.3×10^{-4} **5.** $3; x\neq 2$
6. All real numbers
7. 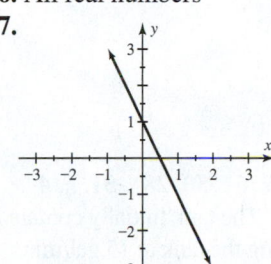   **8.**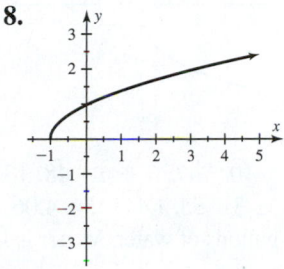

9.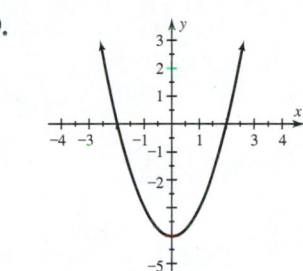

(a) D: all real numbers; R: $y\geq -4$ **(b)** 0   **(c)** $-2,2$
(d) $-2,2$ **10.** $y=\frac{1}{2}x+\frac{5}{2}$ **11.** $f(x)=1-3x$
12.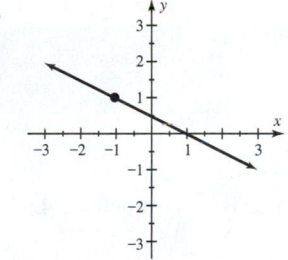

13. $\frac{6}{11}$ **14.** $(-\infty,3]$ **15.** $[-1,5]$ **16.** $\left[-\frac{5}{2},1\right]$
17. (a) $(4,4)$ **(b)** No solutions
18.

19. $(3, 2, 1)$ **20.** 4 **21.** $-4x^4 + 16x$ **22.** $x^2 - 16$
23. $5x^2 - 7x - 6$ **24.** $16x^2 + 72x + 81$
25. $(3x - 4)(3x + 4)$ **26.** $(x - 2)^2$ **27.** $3x^2(5x - 3)$
28. $(4x - 3)(3x + 1)$ **29.** $(r - 1)(r^2 + r + 1)$
30. $(x - 3)(x^2 + 5)$ **31.** $1, 2$ **32.** $-2, 0, 2$ **33.** $2x + 4$
34. $\frac{7x - 5}{x^2 - x}$ **35.** $6x$ **36.** 4 **37.** $\frac{1}{64}$ **38.** 5 **39.** $4x$
40. $x^{3/4}$ or $\sqrt[4]{x^3}$ **41.** $2x$ **42.** $6\sqrt{3x}$ **43.** $2x^2 - \sqrt{3}x - 3$
44. $(-\infty, 1]$
45.

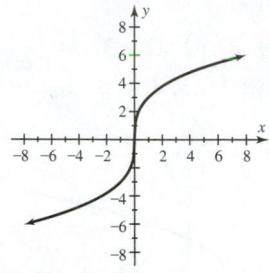

46. $\sqrt{10}$ **47.** $5 + i$ **48.** 13 **49.** 6 **50.** 28 **51.** $\frac{4}{9}, 4$
52. ± 3 **53.** 1.41 **54.** 4.06 **55.** The tank initially contains 300 gallons of water. Water is leaving the tank at 15 gal/min.
56.

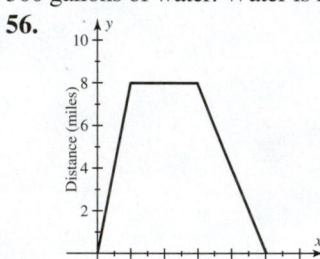

57. $1300 at 5%, $700 at 4%   **58.** 8 in. by 8 in. by 4 in.
59. 22.4 feet   **60.** $45°, 55°, 80°$

11 **Quadratic Functions and Equations**

SECTION 11.1 (pp. 724–728)

1. parabola   **2.** The vertex   **3.** axis of symmetry   **4.** $(0, 0)$
5.

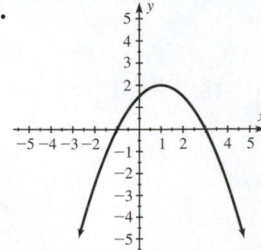

6. $-\frac{b}{2a}$ **7.** narrower   **8.** reflected   **9.** $ax^2 + bx + c$
with $a \neq 0$ **10.** vertex   **11.** True   **12.** False
13. False   **14.** True   **15.** $(2, -6)$ **17.** $(-3, 4)$
19. $(0, 3)$ **21.** $(1, 1.4)$ **23.** $(3, 9)$ **25.** $0, -4$
27. $-2, -2$ **29.** $(1, -2)$; $x = 1$; upward
31. $(-2, 3)$; $x = -2$; downward

33. (b)

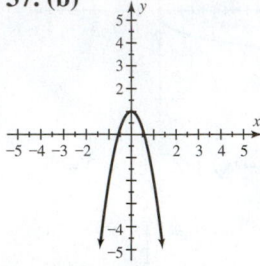

(a) $(0, 0)$; $x = 0$
(c) 2; 4.5

35. (b)

(a) $(0, -2)$; $x = 0$
(c) 2; 7

37. (b)

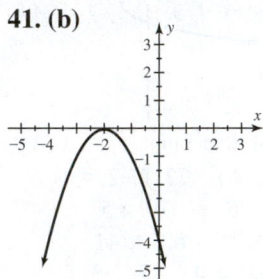

(a) $(0, 1)$; $x = 0$
(c) -11; -26

39. (b)

(a) $(1, 0)$; $x = 1$
(c) 9; 4

41. (b)

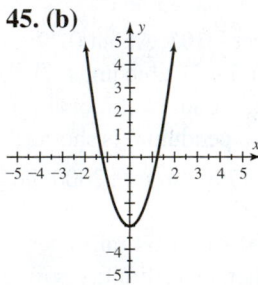

(a) $(-2, 0)$; $x = -2$
(c) 0; -25

43. (b)

(a) $(-0.5, -2.25)$; $x = -0.5$
(c) 0; 10

45. (b)

(a) $(0, -3)$; $x = 0$
(c) 5; 15

47. (b)

(a) $(1, 1)$; $x = 1$
(c) -8; -3

49. (b)

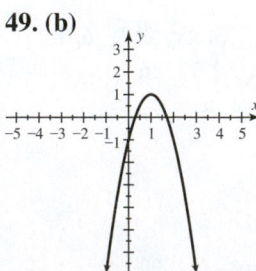

(a) $(1, 1)$; $x = 1$
(c) -17; -7

51. (b)

(a) $(2, 4)$; $x = 2$
(c) 8; 4.25

53. -2 **55.** $-\frac{25}{4}$ **57.** $-\frac{7}{2}$ **59.** 6   **61.** 4
63. $-\frac{39}{8}$ **65.** 10, 10
67.   **69.**

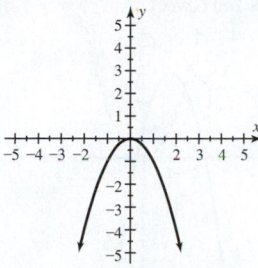   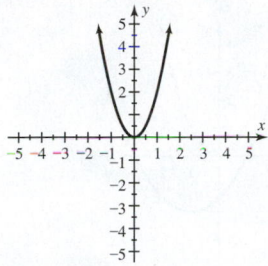

Reflected across the x-axis   Narrower

71.   **73.**

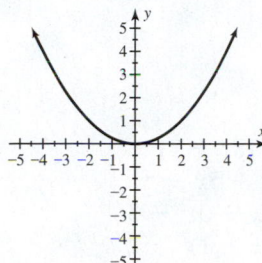   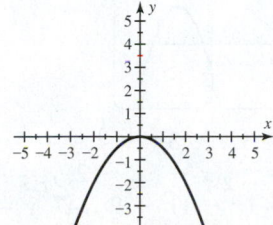

Wider   Reflected across the x-axis
                                  and wider

75. (a) Upward; wider   **77. (a)** Downward; the same
(b) $x = -1$; $(-1, -2)$ **(b)** $x = 1$; $(1, 1)$
(c) y-int: $-\frac{3}{2}$; x-int: $-3, 1$ **(c)** y-int: 0; x-int: 0, 2
(d)   **(d)**

79. (a) Upward; narrower **(b)** $x = -\frac{1}{2}$; $\left(-\frac{1}{2}, -\frac{9}{2}\right)$
(c) y-int: -4; x-int: $-2, 1$
(d)

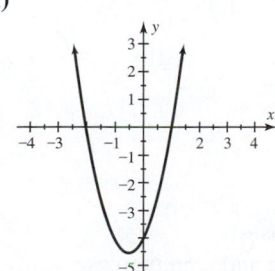

81. 3 slices   **83.** d.   **85.** a.   **87. (a)** 2 feet   **(b)** 2 seconds
(c) 66 feet   **89.** $\frac{66}{32} \approx 2$ seconds; about 74 feet

91. (a) The revenue increases at first up to 50 tickets and then it decreases.   **(b)** $2500; 50 **(c)** $f(x) = x(100 - x)$
(d) $2500; 50   **93. (a)** $V(1) = 21.75$, $V(2) = 30$, $V(3) = 59.75$, and $V(4) = 111$; in 2007 there were 21.75 million unique Facebook visitors in one month. Other values can be interpreted similarly.
(b) The increases between consecutive years are 8.25 million, 29.75 million, and 51.25 million. A linear function does not model the data because these three increases are not equal (or nearly equal).
95. 300 feet by 600 feet   **97.** 42 in.
99. (a) In 1990, emissions were 6 billion metric tons.
(b) 3 billion metric tons.

SECTION 11.2 (pp. 738–741)

1. $x^2 + 2$ **2.** $(x - 2)^2$ **3.** $(1, 2)$ **4.** $(-1, -2)$
5. $f(x) = ax^2 + bx + c$; $f(x) = a(x - h)^2 + k$
6. $y = a(x - h)^2 + k$; (h, k) **7.** downward   **8.** $-\frac{b}{2a}$
9. a.   **10.** d.

11.

x	-2	-1	0	1	2
$y = x^2$	4	1	0	1	4
$y = x^2 - 3$	1	-2	-3	-2	1

Shifted 3 units downward

13.

x	-2	-1	0	1	2
$y = x^2$	4	1	0	1	4

x	1	2	3	4	5
$y = (x - 3)^2$	4	1	0	1	4

Shifted 3 units right

15. (a)   **17. (a)**

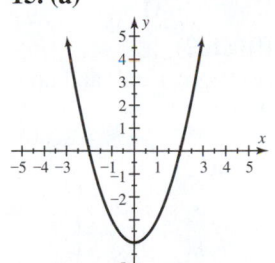   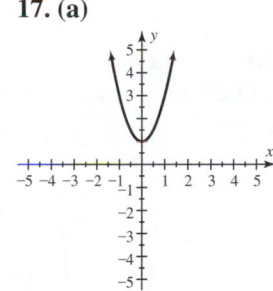

(b) $(0, -4)$ **(b)** $(0, 1)$
(c) Down 4 units   **(c)** Narrower and up 1 unit

19. (a)

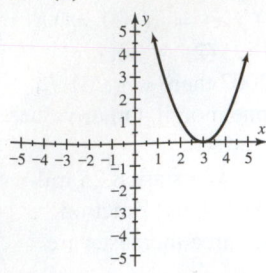

(b) $(3, 0)$
(c) Right 3 units

21. (a)

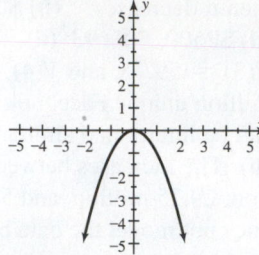

(b) $(0, 0)$
(c) Reflected across the
x-axis

23. (a)

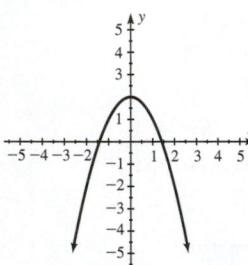

(b) $(0, 2)$
(c) Reflected across the
x-axis and up 2 units

25. (a)

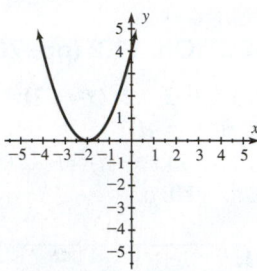

(b) $(-2, 0)$
(c) Left 2 units

27. (a)

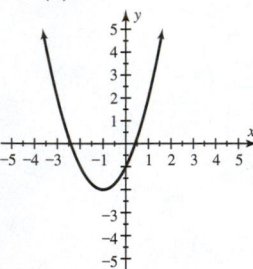

(b) $(-1, -2)$
(c) Left 1 unit and down
2 units

29. (a)

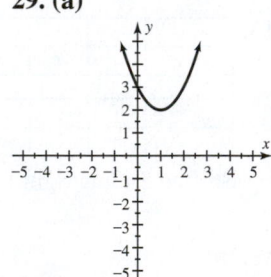

(b) $(1, 2)$
(c) Right 1 unit and up
2 units

31. (a)

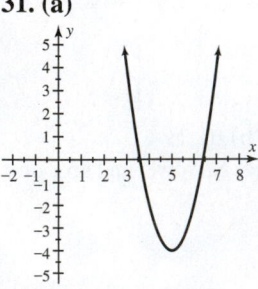

(b) $(5, -4)$
(c) Narrower, right 5 units
and down 4 units

33. (a)

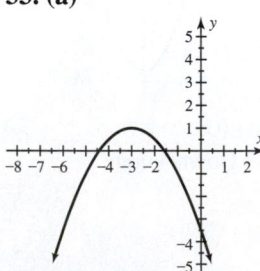

(b) $(-3, 1)$
(c) Wider, reflected
across the x-axis, left
3 units and up 1 unit

35. Translated 1 unit right,
2 units downward, and
is wider

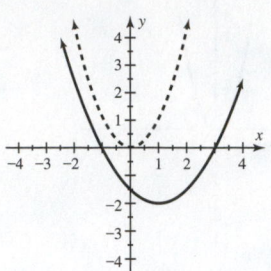

37. Translated 1 unit
left, 3 units upward,
opens downward, and
is narrower

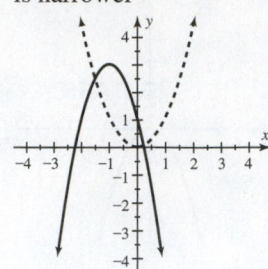

39. $[-20, 20, 2]$ by $[-20, 20, 2]$

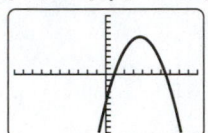

41. $y = 3(x - 3)^2 + 4; y = 3x^2 - 18x + 31$
43. $y = -\frac{1}{2}(x - 5)^2 - 2; y = -\frac{1}{2}x^2 + 5x - \frac{29}{2}$
45. $y = (x - 1)^2 + 2$ **47.** $y = -(x - 0)^2 - 3$
49. $y = (x - 0)^2 - 3$ **51.** $y = -(x + 1)^2 + 2$
53. (a) $(1, 1)$ **(b)** $y = 4(x - 1)^2 + 1$ **55. (a)** $(-1, -2)$
(b) $y = -(x + 1)^2 - 2$ **57. (a)** $(-1, 3)$
(b) $y = -2(x + 1)^2 + 3$ **59.** 1
61. $y = (x + 1)^2 - 1; (-1, -1)$
63. $y = (x - 2)^2 - 4; (2, -4)$
65. $y = (x + 1)^2 - 4; (-1, -4)$
67. $y = (x - 2)^2 + 1; (2, 1)$
69. $y = \left(x + \frac{3}{2}\right)^2 - \frac{17}{4}; \left(-\frac{3}{2}, -\frac{17}{4}\right)$
71. $y = \left(x - \frac{7}{2}\right)^2 - \frac{45}{4}; \left(\frac{7}{2}, -\frac{45}{4}\right)$
73. $y = 3(x + 1)^2 - 4; (-1, -4)$
75. $y = 2\left(x - \frac{3}{4}\right)^2 - \frac{9}{8}; \left(\frac{3}{4}, -\frac{9}{8}\right)$
77. $y = -2(x + 2)^2 + 13; (-2, 13)$ **79.** $a = 2$
81. $a = 0.3$ **83.** $y = 2(x - 1)^2 - 3$
85. $y = 0.5(x - 1980)^2 + 6$
87. (a)   **(b)** $D(x) = \frac{1}{12}x^2$

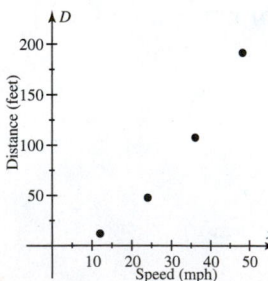

89. (a) Decreases and then increases
(b) No, because the data decrease and then increase
(c) Quadratic; it can model data that decrease and increase.
(d) $(1995, 600)$; it has the minimum y-value.
(e) $C(x) = 1.8(x - 1995)^2 + 600$
(f) $[1970, 2030, 10]$ by $[0, 2500, 500]$

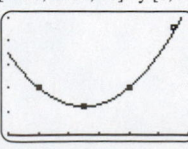

91. (a) $f(x) = 2(x - 1982)^2 + 1$ (answers may vary)
(b) 201 thousand (answers may vary)   **93.** incr: $x \geq 2$,
decr: $x \leq 2$ **95.** incr: $x \leq 1$, decr: $x \geq 1$
97. incr: $x \geq \frac{2}{3}$, decr: $x \leq \frac{2}{3}$ **99.** incr: $x \leq -\frac{3}{2}$,
decr: $x \geq -\frac{3}{2}$ **101.** incr: $x \leq -\frac{1}{8}$, decr: $x \geq -\frac{1}{8}$

CHECKING BASIC CONCEPTS 11.1 & 11.2 (p. 742)

1. (a)   **(b)**

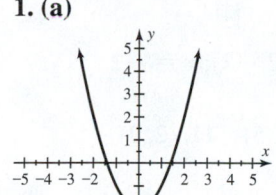

$(0, -2)$; $x = 0$ $(1, -3)$; $x = 1$

2. y_1 opens upward, whereas y_2 opens downward, y_1 is
narrower than y_2.   **3.** 7

4. (a)   **(b)**

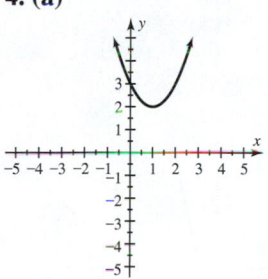
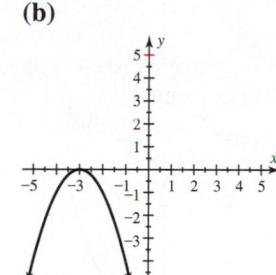

1 unit right, 2 units up   Reflected across the
                              x-axis, 3 units left

5. (a) $y = (x + 7)^2 - 56$ **(b)** $y = 4(x + 1)^2 - 6$

SECTION 11.3 (pp. 751–754)

1. $x^2 + 3x - 2 = 0$ (answers may vary); it can have 0, 1, or
2 solutions   **2.** Nonlinear   **3.** Factoring, square root property,
completing the square

4.   **5.**

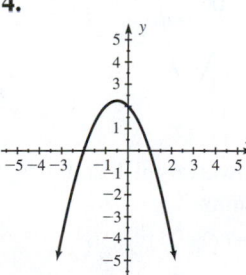

(answers may vary)   (answers may vary)

6. Two; the solutions are the x-intercepts.   **7.** ± 8; the
square root property (answers may vary)   **8.** $\left(\frac{b}{2}\right)^2$
9. Yes   **10.** Yes   **11.** No   **12.** No   **13.** Yes   **14.** Yes
15. No   **16.** No   **17. (a)** $3.65, -1.65$ **(b)** $1.32, -5.32$
19. (a) $1.32, -0.12$ **(b)** $-0.28, -0.83$ **21.** $-2, 1$
23. No real solutions   **25.** $-2, 3$ **27.** -0.5 **29.** $-1, 5$
31. $-3, 1$ **33.** $-3, 3$ **35.** $-\frac{1}{2}, \frac{3}{2}$ **37.** -1 **39.** No real
solutions   **41.** $-7, 5$ **43.** $-\frac{1}{3}, \frac{1}{2}$ **45.** $-\frac{3}{2}$ **47.** $-2, 7$

49. $\frac{6}{5}, \frac{3}{2}$ **51.** ± 12 **53.** $\pm\frac{8}{\sqrt{5}}$ or $\pm\frac{8\sqrt{5}}{5}$ **55.** $-6, 4$
57. $-7, 9$ **59.** $\frac{1 \pm \sqrt{5}}{2}$ **61.** $5 \pm \sqrt{5}$ **63.** 4   **65.** $\frac{25}{4}$
67. 16; $(x - 4)^2$ **69.** $\frac{81}{4}$; $\left(x + \frac{9}{2}\right)^2$ **71.** $-4, 6$
73. $-3 \pm \sqrt{11}$ **75.** $\frac{3 \pm \sqrt{29}}{2}$ **77.** $\frac{5 \pm \sqrt{21}}{2}$
79. $1 \pm \sqrt{5}$ **81.** $\frac{3 \pm \sqrt{41}}{4}$ **83.** $\frac{2 \pm \sqrt{11}}{2}$
85. $\frac{-3 \pm \sqrt{5}}{12}$ **87.** $-6, 2$ **89.** $-2 \pm \sqrt{2}$
91. $\pm\sqrt{2}$ **93.** $-5, \frac{7}{3}$ **95.** $4 \pm \sqrt{14}$ **97.** $\pm\sqrt{\frac{c}{a}}$
99. $-3, 6$ **101.** $3, 5$ **103.** $5, 7$
105. $y = \pm\sqrt{x + 1}$ **107.** $v = \pm\sqrt{\frac{2K}{m}}$
109. $r = \pm\sqrt{\frac{k}{E}}$ **111.** $f = \pm\frac{1}{2\pi\sqrt{LC}}$ **113. (a)** 30 miles
per hour   **(b)** 40 miles per hour   **115.** About 1.9
seconds; no   **117.** About 8 feet   **119.** 2 hours
121. (a) $x^2 + 6x - 520 = 0$ **(b)** -26 or 20;
20 feet   **123.** About $23°C$ and $34°C$ **125.** October 2010

SECTION 11.4 (pp. 763–767)

1. To solve quadratic equations that are written in the form
$ax^2 + bx + c = 0$ **2.** Completing the square
3. $b^2 - 4ac$ **4.** One solution   **5.** Factoring, square root
property, completing the square, and the quadratic
formula   **6.** No; not when $b^2 - 4ac < 0$ **7.** $\pm\sqrt{k}$
8. $\pm i \sqrt{k}$ **9.** $-6, \frac{1}{2}$ **11.** 1   **13.** No real solutions
15. $-2, 8$ **17.** $\frac{1 \pm \sqrt{17}}{8}$ **19.** No real solutions   **21.** $\frac{1}{2}$
23. $\frac{3 \pm \sqrt{13}}{2}$ **25.** $2 \pm \sqrt{3}$ **27.** $\frac{1 \pm \sqrt{37}}{6}$
29. $\frac{1 \pm \sqrt{15}}{2}$ **31.** $\frac{5 \pm \sqrt{10}}{3}$ **33. (a)** $a > 0$ **(b)** $-1, 2$
(c) Positive   **35. (a)** $a > 0$ **(b)** No real solutions
(c) Negative   **37. (a)** $a < 0$ **(b)** 2   **(c)** Zero   **39. (a)** 25
(b) 2   **41. (a)** 0   **(b)** 1   **43. (a)** $-\frac{7}{4}$ **(b)** 0   **45. (a)** 21
(b) 2   **47.** $1 \pm \sqrt{2}$ **49.** $-\frac{3}{2}, 1$ **51.** None   **53.** None
55. $\frac{-2 \pm \sqrt{10}}{3}$ **57.** $\pm 3i$ **59.** $\pm 4i\sqrt{5}$ **61.** $\pm\frac{1}{2}i$
63. $\pm\frac{3}{4}i$ **65.** $\pm i\sqrt{6}$ **67.** $\pm\sqrt{3}$ **69.** $\pm i\sqrt{2}$
71. $\frac{1}{2} \pm i\frac{\sqrt{7}}{2}$ **73.** $-\frac{3}{4} \pm i\frac{\sqrt{23}}{4}$ **75.** $2 \pm \sqrt{3}$
77. $-\frac{1}{2} \pm i\frac{\sqrt{7}}{2}$ **79.** $-\frac{1}{5} \pm i\frac{\sqrt{19}}{5}$ **81.** $-\frac{3}{4} \pm i\frac{\sqrt{23}}{4}$
83. $-\frac{1}{2} \pm i\frac{\sqrt{15}}{2}$ **85.** $\frac{1 \pm \sqrt{3}}{2}$ **87.** $\frac{1}{4} \pm i\frac{\sqrt{15}}{4}$
89. $-1 \pm i\sqrt{3}$ **91.** $-2 \pm i$ **93.** $1 \pm i\sqrt{2}$ **95.** 1, 2
97. $-\frac{1}{2}, 4$ **99.** $\frac{5 \pm \sqrt{17}}{2}$ **101.** $-\frac{1}{4} \pm \frac{3}{4}i\sqrt{7}$ **103.** $\pm\frac{1}{2}$
105. $\pm i\sqrt{2}$ **107.** $\frac{1}{3}$ **109.** 9 miles per hour
111. 45 miles per hour   **113.** About 1993 and 2004
($x \approx 4.11, 15.36$)   **115. (a)** 6; in Oct. 2010 Groupon's
value was $6 billion.   **(b)** Jan. 2011   **117.** $x \approx 8.04$,
or about 1992; this agrees with the graph.
119. $130 + 5\sqrt{634} \approx 256$ mph   **121. (a)–(c)** 11 in. by
14 in.   **123. (a)** The rate of change is not constant.
(b) 75 seconds (answers may vary)   **(c)** 75 seconds

CHECKING BASIC CONCEPTS 11.3 & 11.4 (p. 767)

1. $\frac{1}{2}, 3$ **2.** $\pm\sqrt{5}$ **3.** $2 \pm \sqrt{3}$ **4.** $y = \pm\sqrt{1 - x^2}$

5. (a) $\frac{3 \pm \sqrt{17}}{4}$ **(b)** $\frac{4}{3}$ **(c)** $-\frac{1}{2} \pm i\frac{\sqrt{7}}{2}$ **6. (a)** 5; two real
solutions **(b)** -7; no real solutions **(c)** 0; one real
solution **7. (a)** $\pm i\sqrt{5}$ **(b)** $-\frac{1}{2} \pm i\frac{\sqrt{11}}{2}$

SECTION 11.5 (pp. 775–777)

1. It has an inequality symbol rather than an equals sign.
2. No, they often have infinitely many. **3.** No **4.** Yes
5. $-2 < x < 4$ **6.** $x < -3$ or $x > 1$ **7.** Yes **9.** Yes
11. No **13.** Yes **15.** No **17.** No **19. (a)** $-3, 2$
(b) $-3 < x < 2$ **(c)** $x < -3$ or $x > 2$ **21. (a)** $-2, 2$
(b) $-2 < x < 2$ **(c)** $x < -2$ or $x > 2$ **23. (a)** $-10, 5$
(b) $x < -10$ or $x > 5$ **(c)** $-10 < x < 5$
25. $[-0.128, 0.679]$ **27. (a)** $-2, 2$ **(b)** $-2 < x < 2$
(c) $x < -2$ or $x > 2$ **29. (a)** $-4, 0$ **(b)** $-4 < x < 0$
(c) $x < -4$ or $x > 0$ **31.** $[-7, -3]$
33. $(-\infty, 1) \cup (2, \infty)$ **35.** $(-\sqrt{10}, \sqrt{10})$
37. $(-\infty, 0) \cup (6, \infty)$
39. $(-\infty, 2 - \sqrt{2}] \cup [2 + \sqrt{2}, \infty)$ **41. (a)** $-2, 2$
(b) $-2 < x < 2$ **(c)** $x < -2$ or $x > 2$ **43. (a)** $\frac{-1 \pm \sqrt{5}}{2}$
(b) $\frac{-1 - \sqrt{5}}{2} < x < \frac{-1 + \sqrt{5}}{2}$
(c) $x < \frac{-1 - \sqrt{5}}{2}$ or $x > \frac{-1 + \sqrt{5}}{2}$ **45.** $(-3, -1)$
47. $(-\infty, -2.5] \cup [3, \infty)$ **49.** $[-2, 2]$ **51.** $(-\infty, \infty)$
53. $(0, 3)$ **55.** No real solutions **57.** 2
59. $(-\infty, -1) \cup (-1, \infty)$ **61.** $[-1, 2]$ **63. (a)** From
1131 feet to 3535 feet (approximately) **(b)** Before 1131 feet
or after 3535 feet (approximately) **65. (a)** About 383; they
agree (approx.). **(b)** About 1969 or after **(c)** About 1969
or after **67.** From 11 feet to 20 feet

SECTION 11.6 (p. 781)

1. $\pm 1, \pm\sqrt{6}$ **3.** $-\sqrt[3]{2}, \sqrt[3]{\frac{5}{3}}$ **5.** $-\frac{4}{5}, -\frac{1}{3}$ **7.** $-3, 3$
9. $-\sqrt[3]{\frac{1}{3}}, \sqrt[3]{2}$ **11.** $-\frac{1}{8}, \frac{2}{5}$ **13.** 1 **15.** 1, $32^5 = 33{,}554{,}432$
17. 16, 81 **19.** 1 **21.** $-3, 6$ **23.** $-\sqrt{3}, \sqrt{3}$
25. $\pm 2, \pm 2i$ **27.** $0, \pm i$ **29.** $\pm\sqrt{2}, \pm i$
31. $-1 \pm i$ **33.** $\pm 2i$

CHECKING BASIC CONCEPTS 11.5 & 11.6 (p. 781)

1. $(-\infty, -2) \cup (3, \infty)$ **2.** $\left[-1, -\frac{2}{3}\right]$ **3.** $-2, \sqrt[3]{2}$
4. $-1, 8^3 = 512$ **5.** $\pm i$

CHAPTER 11 REVIEW (pp. 784–788)

1. $(-3, 4); x = -3$; downward **2.** $(1, 0); x = 1$; upward

3. (a)

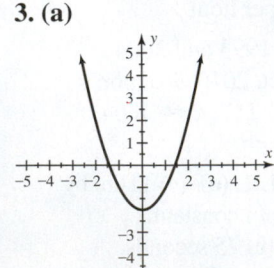

4. (a)

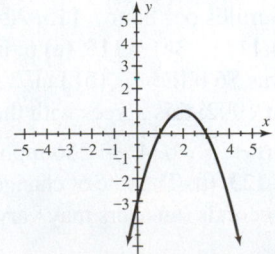

(b) $(0, -2); x = 0$ **(b)** $(2, 1); x = 2$
(c) -1 **(c)** 0

5. (a)

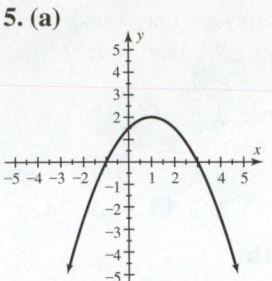

6. (a)

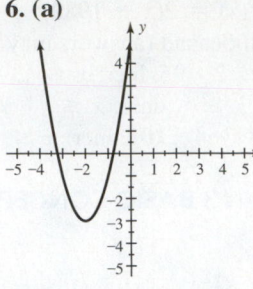

(b) $(1, 2); x = 1$ **(b)** $(-2, -3); x = -2$
(c) -2.5 **(c)** -1

7. $-\frac{7}{2}$ **8.** $-\frac{14}{3}$ **9.** $(2, -6)$ **10.** $(0, 5)$ **11.** $(2, 2)$
12. $(-1, 1)$

13. (a)

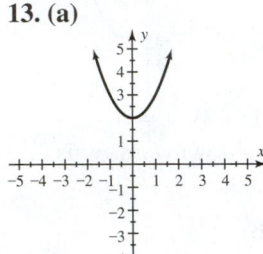

14. (a)

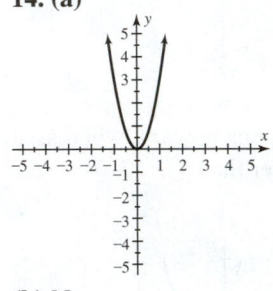

(b) Up 2 units **(b)** Narrower

15. (a)

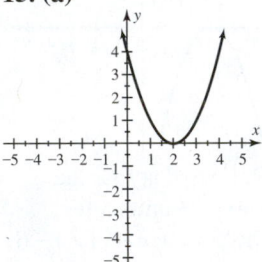

16. (a)

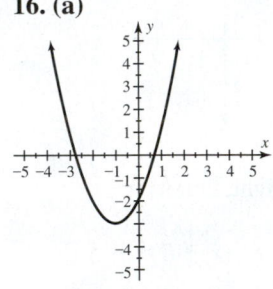

(b) Right 2 units **(b)** Left 1 unit, down 3 units

17. (a)

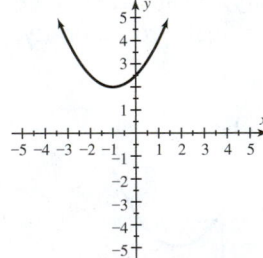

18. (a)

(b) Wider, left 1 unit, **(b)** Narrower, right 1 unit,
up 2 units down 3 units
19. $y = -4(x - 2)^2 - 5$ **20.** $y = -(x + 4)^2 + 6$
21. $y = (x + 2)^2 - 11; (-2, -11)$
22. $y = \left(x - \frac{7}{2}\right)^2 - \frac{45}{4}; \left(\frac{7}{2}, -\frac{45}{4}\right)$
23. $y = 2\left(x - \frac{3}{4}\right)^2 - \frac{73}{8}; \left(\frac{3}{4}, -\frac{73}{8}\right)$
24. $y = 3(x + 1)^2 - 5; (-1, -5)$ **25.** $a = 3$
26. $a = \frac{1}{4}$ **27.** $f(x) = -5x^2 + 30x - 41; -41$
28. $f(x) = 3x^2 + 12x + 8; 8$ **29.** $-2, 3$ **30.** -1
31. No real solutions **32.** $-4, 6$ **33.** $-10, 5$
34. $-0.5, 0.25$ **35.** $-5, 10$ **36.** $-3, 1$ **37.** $-4, 2$

38. $-1, 3$ **39.** $-5, 4$ **40.** $-8, -3$ **41.** $-\frac{2}{5}, \frac{2}{3}$

42. $\frac{4}{7}, 3$ **43.** ± 10 **44.** $\pm \frac{1}{3}$ **45.** $\pm \frac{\sqrt{6}}{2}$ **46.** No real

solutions **47.** $-3 \pm \sqrt{7}$ **48.** $2 \pm \sqrt{10}$

49. $1 \pm \sqrt{6}$ **50.** $\frac{-3 \pm \sqrt{11}}{2}$ **51.** $R = -r \pm \sqrt{\frac{k}{F}}$

52. $y = \pm \sqrt{\frac{12 - 2x^2}{3}}$ **53.** $3, 6$ **54.** $11, 13$ **55.** $-\frac{1}{2}, \frac{1}{3}$

56. $\frac{5 \pm \sqrt{5}}{10}$ **57.** $4 \pm \sqrt{21}$ **58.** $\frac{3 \pm \sqrt{3}}{2}$

59. ± 2 **60.** $\pm \frac{1}{2}$ **61.** $\frac{5}{2}, 3$ **62.** $\frac{3}{2}, 5$ **63.** $\frac{3 \pm \sqrt{5}}{2}$

64. $\frac{1 \pm \sqrt{5}}{4}$ **65. (a)** $a > 0$ **(b)** $-2, 3$ **(c)** Positive

66. (a) $a > 0$ **(b)** 2 **(c)** Zero **67. (a)** $a < 0$

(b) No real solutions **(c)** Negative **68. (a)** $a < 0$

(b) $-4, 2$ **(c)** Positive **69. (a)** 1 **(b)** 2 **70. (a)** 144

(b) 2 **71. (a)** -23 **(b)** 0 **72. (a)** 0 **(b)** 1

73. $-\frac{1}{2} \pm i \frac{\sqrt{19}}{2}$ **74.** $\pm 2i$ **75.** $\frac{1}{4} \pm i \frac{\sqrt{7}}{4}$ **76.** $\frac{1}{7} \pm i \frac{\sqrt{34}}{7}$

77. (a) $-2, 6$ **(b)** $-2 < x < 6$ **(c)** $x < -2$ or $x > 6$

78. (a) $-2, 0$ **(b)** $x < -2$ or $x > 0$ **(c)** $-2 < x < 0$

79. (a) $-4, 4$ **(b)** $-4 < x < 4$ **(c)** $x < -4$ or $x > 4$

80. (a) $-2, 1$ **(b)** $-2 < x < 1$ **(c)** $x < -2$ or $x > 1$

81. (a) $-1, 3$ **(b)** $-1 < x < 3$ **(c)** $x < -1$ or $x > 3$

82. (a) $-\frac{3}{2}, 5$ **(b)** $-\frac{3}{2} \le x \le 5$ **(c)** $x \le -\frac{3}{2}$ or $x \ge 5$

83. $[-3, -1]$ **84.** $\left(\frac{1}{5}, 3\right)$ **85.** $\left(-\infty, \frac{1}{6}\right) \cup (2, \infty)$

86. $(-\infty, -\sqrt{5}] \cup [\sqrt{5}, \infty)$ **87.** $(-\infty, \infty)$

88. No solutions **89.** $\pm \sqrt{5}, \pm 3$ **90.** $-\frac{1}{4}, \frac{2}{7}$

91. $1, 512$ **92.** 0 **93.** $\pm i \frac{\sqrt{2}}{2}$ **94.** $2 \pm i\sqrt{2}$

95. (a) $f(x) = x(12 - 2x)$ **(b)** 6 inches by 3 inches

96. (a) After 1 second and 1.75 seconds **(b)** 1.375 seconds;

34.25 feet **97. (a)** $f(x) = x(90 - 3x)$

(b) [0, 30, 5] by [0, 800, 100]

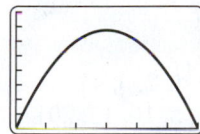

(c) 10 or 20 rooms **(d)** 15 rooms

98. (a) $x(x + 2) = 143$ **(b)** $x = -13$ or $x = 11$; the

numbers are -13 and -11 or 11 and 13.

99. (a) $\sqrt{1728} \approx 41.6$ miles per hour **(b)** 60 miles per hour

100. (a) $(1950, 220)$; in 1950, the per capita consumption

was at a low of 220 million Btu.

(b) [1950, 1970, 5] by [200, 350, 25]

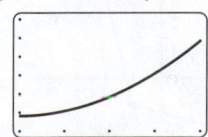

It increased.

(c) $f(2010) = 1120$; no; the trend represented by this

model did not continue after 1970. **101.** About 11.1 inches

by 11.1 inches **102.** 50 feet **103.** About 1.5 feet

104. About 6.0 to 9.0 inches

CHAPTER 11 TEST (pp. 788–789)

1. $\left(1, \frac{3}{2}\right); x = 1; -3$ **2.** $-\frac{29}{4}$ **3.** $a = -\frac{1}{2}$

4. (a) Same as $y = x^2$ **(b)** Same as $y = x^2$ except

except shifted 1 unit right shifted 2 units downward

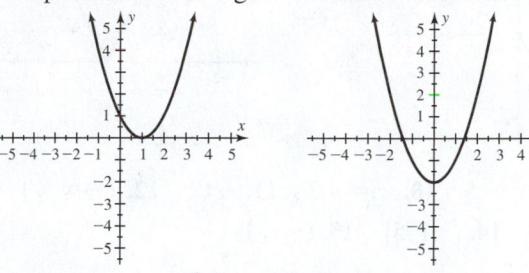

(c) Same as $y = x^2$ except it is wider, shifted right 3 units,

and shifted upward 2 units

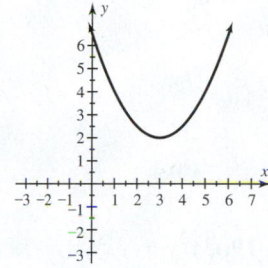

5. $y = (x - 3)^2 - 7; (3, -7); x = 3$ **6.** $-1, 2; 2$

7. $-4, \frac{1}{3}$ **8.** $-\frac{1}{2}, \frac{1}{2}$ **9.** $4 \pm \sqrt{17}$ **10.** $\frac{3 \pm \sqrt{17}}{4}$

11. $\pm \frac{4}{3}$ **12.** $m = \pm \sqrt{\frac{Fr^2}{G}}$ **13. (a)** $a < 0$ **(b)** $-3, 1$

(c) Positive **14. (a)** -44 **(b)** No real solutions

(c) The graph of $y = -3x^2 + 4x - 5$ does not intersect the

x-axis. **15. (a)** $-1, 1$ **(b)** $-1 < x < 1$

(c) $x < -1$ or $x > 1$ **16. (a)** $-10, 20$

(b) $x < -10$ or $x > 20$ **(c)** $-10 < x < 20$

17. (a) $-\frac{1}{2}, \frac{3}{4}$ **(b)** $\left[-\frac{1}{2}, \frac{3}{4}\right]$ **(c)** $\left(-\infty, -\frac{1}{2}\right] \cup \left[\frac{3}{4}, \infty\right)$

18. $[-2, 0]$ **19.** $\sqrt[3]{2}, 1$ **20.** $-1 \pm i \frac{\sqrt{2}}{2}$

21. $-1.37, 0.69$ **22.** $\sqrt{2250} \approx 47.4$ miles per hour

23. (a) $f(x) = (x + 20)(90 - x)$ **(b)** 35

24. (a) [0, 6, 1] by [0, 150, 50]

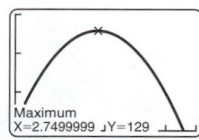

(b) After about 5.6 seconds

(c) 2.75 seconds; 129 feet

CHAPTERS 1–11 CUMULATIVE REVIEW (pp. 791–793)

1. 1 **2.** Natural: $\sqrt[3]{8}$; whole: $0, \sqrt[3]{8}$; integer: $0, -5, \sqrt[3]{8}$;

rational: $0.\overline{4}, 0, -5, \sqrt[3]{8}, -\frac{4}{3}$; irrational: $\sqrt{7}$ **3. (a)** $x^{10}y^{12}$

(b) $\frac{x}{y^8}$ **(c)** $\frac{1}{b^{10}}$ **4.** 9.29×10^6 **5.** $2; x \le 2$ **6.** $(2, 5)$

7.

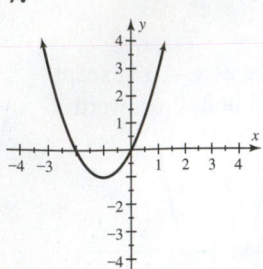

8.

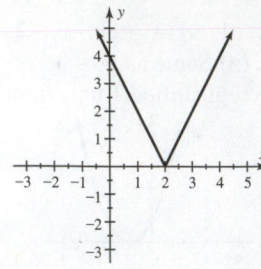

9. $y = -\frac{3}{2}x + 5$ **10.** $x = -3$ **11.** -12 **12.** $\left(-\infty, \frac{7}{4}\right)$
13. $\left[\frac{1}{3}, 1\right]$ **14.** $(-1, 5]$ **15.** $(-1, 1)$

16.

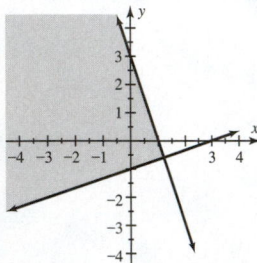

17. $(2, 2, 1)$ **18.** $6x^2 + 17x - 14$ **19.** $3x^3y + 3xy^3$
20. $x - 9$ **21.** $x(x + 1)(x - 2)$ **22.** $(2x + 5)(2x - 5)$
23. $\pm\sqrt{3}$ **24.** 1 **25.** $\frac{x+3}{2}$ **26.** $-\frac{2}{x(x+2)}$ **27.** $4x^3$
28. $\frac{1}{64}$ **29.** 3 **30.** $3\sqrt{2x}$

31.

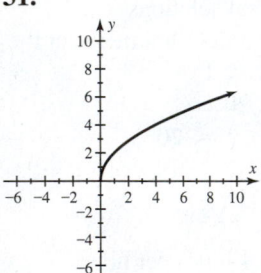

32. $\sqrt{26}$ **33.** $1 - i$ **34.** 3 **35.** 0.79

36.

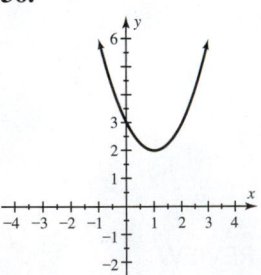

(a) $(1, 2)$ **(b)** 6 **(c)** $x = 1$ **(d)** $x \geq 1$
37. $f(x) = 2(x - 1)^2 - 3$ **38.** Shifted 1 unit left, 2 units
downward; narrower **39.** $-3 \pm \sqrt{11}$ **40.** $\frac{3 \pm \sqrt{17}}{4}$
41. 1, 3 **42. (a)** $-2, 1$ **(b)** $-2 \leq x \leq 1$
43. $x < 1$ or $x > 2$ **44.** $\pm 4, \pm 4i$ **45.** c. **46.** f. **47.** g.
48. e. **49.** d. **50.** a. **51.** b. **52.** h. **53. (a)** 300; initially,
the tank holds 300 gal. **(b)** 6; after 6 minutes the tank is empty.
(c) -50; water is pumped out at 50 gal/min.
(d) $G(t) = 300 - 50t$ **54.** $2200 at 6%, $1200 at 5%,
$600 at 4% **55.** 125 ft by 125 ft **56.** 33 feet

12 Exponential and Logarithmic Functions

SECTION 12.1 (pp. 805–809)

1. $g(f(7))$ **2.** $f(g(x))$ **3.** No **4.** inputs; outputs
5. No **6.** one-to-one **7.** adding 10 **8.** 7 **9.** 8; 6
10. $x; y$ **11.** one-to-one **12.** reflection; $y = x$
13. (a) 7 **(b)** 49 **(c)** $(g \circ f)(x) = x^2 + 3$
(d) $(f \circ g)(x) = (x + 3)^2$ **15. (a)** -65 **(b)** 126
(c) $(g \circ f)(x) = 8x^3 - 1$ **(d)** $(f \circ g)(x) = 2x^3 - 2$
17. (a) 3 **(b)** 1 **(c)** $(g \circ f)(x) = \left|\frac{1}{2}x - 2\right|$
(d) $(f \circ g)(x) = \frac{1}{2}|x - 2|$ **19. (a)** $\frac{11}{2}$ **(b)** $-\frac{1}{17}$
(c) $(g \circ f)(x) = 3 - \frac{5}{x}$ **(d)** $(f \circ g)(x) = \frac{1}{3 - 5x}$
21. (a) 77 **(b)** 122 **(c)** $(g \circ f)(x) = 16x^2 - 4x + 5$
(d) $(f \circ g)(x) = 8x^2 - 4x + 10$ **23. (a)** 1 **(b)** 2
25. (a) -1 **(b)** 1 **27. (a)** 0 **(b)** 2 **29. (a)** 2
(b) -3 **(c)** -1 **31.** $f(1) = f(-1) = 5$
33. $f(1) = f(-1) = 101$ **35.** $f(2) = f(-2) = 4$
37. Yes **39.** No **41.** Yes **43.** Divide x by 7;
$f(x) = 7x; g(x) = \frac{x}{7}$ **45.** Multiply x by 2 then subtract 5;
$f(x) = \frac{x + 5}{2}; g(x) = 2x - 5$ **47.** Add 3 to x and multiply
the result by 2; $f(x) = \frac{1}{2}x - 3; g(x) = 2(x + 3)$
49. Take the cube root of x and subtract 5;
$f(x) = (x + 5)^3; g(x) = \sqrt[3]{x} - 5$
51. $(f \circ f^{-1})(x) = 4\left(\frac{x}{4}\right) = x; (f^{-1} \circ f)(x) = \frac{4x}{4} = x$
53.–57. Show $(f \circ f^{-1})(x) = (f^{-1} \circ f)(x) = x$. See the
answer to Exercise 51 above. **59.** $f^{-1}(x) = \frac{x}{12}$
61. $f^{-1}(x) = x - 8$ **63.** $f^{-1}(x) = \frac{x + 2}{5}$
65. $f^{-1}(x) = -2(x - 1)$ **67.** $f^{-1}(x) = 8 - x$
69. $f^{-1}(x) = 2x - 1$ **71.** $f^{-1}(x) = \frac{x^3}{2}$
73. $f^{-1}(x) = \sqrt[3]{x + 8}$

75.

x	0	5	10	15	20
$f^{-1}(x)$	0	1	2	3	4

Domain of f = range of f^{-1} = $\{0, 1, 2, 3, 4\}$
Range of f = domain of f^{-1} = $\{0, 5, 10, 15, 20\}$

77.

x	4	2	0	-2	-4
$f^{-1}(x)$	-5	0	5	10	15

Domain of f = range of f^{-1} = $\{-5, 0, 5, 10, 15\}$
Range of f = domain of f^{-1} = $\{-4, -2, 0, 2, 4\}$

79.

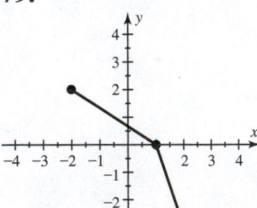

81.

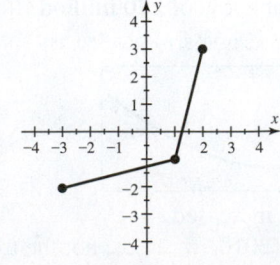

83.
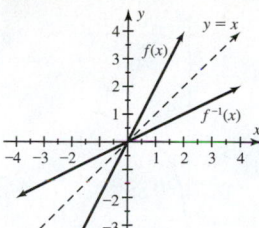

85.
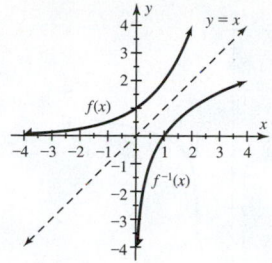

87. (a) 12   **(b)** −4   **(c)** $x^4 + 4x^2 + 2$ **89. (a)** $\frac{1}{2}$
(b) $-\frac{1}{x}$ **(c)** $\frac{x}{2}$ **91. (a)** 20π; after 5 seconds, the wave has a
circumference of $20\pi \approx 62.8$ feet.
(b) $(C \circ r)(t) = 4\pi t$ **93. (a)** 16; in 1980, 16% of people
25 or older had completed four or more years of college.

(b)

x	8	16	27	29
$P^{-1}(x)$	1960	1980	2000	2010

(c) 1980

95. (a) 75°; 150   **(b)** 150; one hour after midnight there are
150 mosquitoes per 100 square feet.   **(c)** The number of
mosquitoes per 100 square feet, h hours after midnight
(d) $T(h) = -5h + 80$; $M(T) = 2T$
(e) $(M \circ T)(h) = -10h + 160$
97. (a) Yes, different inputs result in different outputs.
(b) $f^{-1}(x) = \frac{5}{9}(x - 32)$ converts x degrees Fahrenheit to an
equivalent temperature in degrees Celsius.
99. $f(x) = 4x$; $f^{-1}(x) = \frac{x}{4}$ converts x quarts to gallons.
101. (a) $C(x) = 16x$ **(b)** $T(x) = 48x$
(c) $(T \circ C)(x) = 768x$ **(d)** 2304; there are 2304 tsp in 3 gal.

SECTION 12.2 (pp. 821–825)

1. $f(x) = Ca^x$
2.
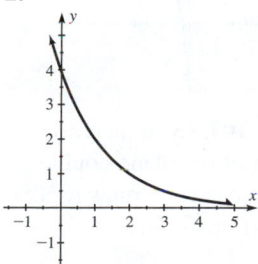

3. D: all real numbers; R: all positive real numbers
4. 32; 25   **5.** 2.718   **6.** 7.389; 9.870 (approximately)
7. factor   **8.** 1.5   **9.** $\frac{B-A}{A} \times 100$ **10.** 1.35
11. $\frac{1}{9}$; 9   **13.** 5; 160   **15.** 4; $\frac{1}{8}$ **17.** 15; $\frac{5}{9}$
19. 0.17; 2.41   **21.** 5; 1.08   **23.** 1; $\frac{1}{a}$ **25.** c.   **27.** d.
29. $C = 1, a = 2$ **31.** $C = 4, a = \frac{1}{4}$
33.
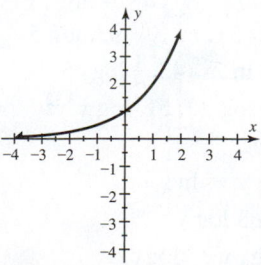

35.
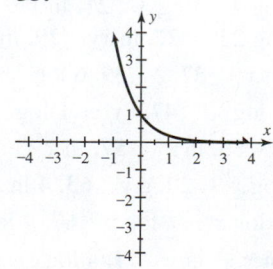

Growth   Decay

37.
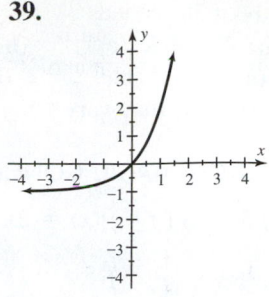

39.

Decay   Growth

41.
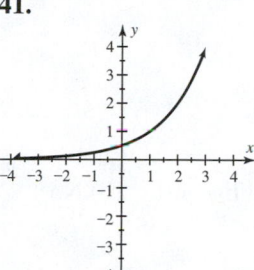

43.
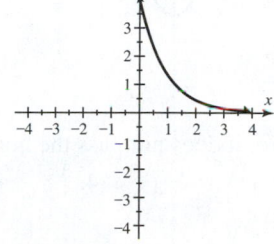

Growth   Decay

45. (a) Exponential decay   **(b)** $f(x) = 64\left(\frac{1}{4}\right)^x$
47. (a) Linear growth   **(b)** $f(x) = 3x + 8$
49. (a) Exponential growth   **(b)** $f(x) = 4(1.25)^x$
51. (a) 100%   **(b)** −50%
53. (a) −80%   **(b)** 400%
55. (a) $1200   **(b)** $2200   **(c)** 2.2
57. (a) $130   **(b)** $780   **(c)** 1.2
59. (a) −$80   **(b)** $720   **(c)** 0.9
61. $C = 9, a = 1.07, R = 7\%$
63. $C = 1.5, a = 0.45, R = -55\%$ **65.** $3551.05
67. $1,820,087.63   **69.** $792.75   **71.** Yes; this is equivalent to having two accounts, each containing $1000 initially.
73. $788.78   **75.** $1429.24   **77.** 3.32   **79.** 0.86

81. $[-4, 4, 1]$ by $[0, 8, 1]$
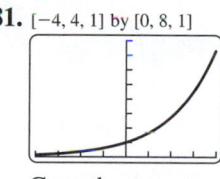

83. $[-4, 4, 1]$ by $[0, 8, 1]$
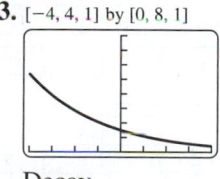

Growth   Decay

85. $f(x) = 5000(0.75)^x$; $f(4) \approx 1582$
87. $f(x) = 50(1.1)^x$; $f(4) \approx 73$ **89.** 20% is much better
91. (a) 0.6   **(b)** $B(x) = 0.07(0.6)^x$ **(c)** 0.0252; after 2 hours
blood alcohol is 0.0252 g/100 mL   **93. (a)** 1.242
(b) In July 2008 there were about 0.5 million tweets per month.
(c) About 90.8; after 24 months, there were 90.8 million
tweets per month.   **95. (a)** $C = 500, a = 2$ **(b)** About
5278 thousand per milliliter   **(c)** The growth is exponential
and doubles every 50 seconds.

97.
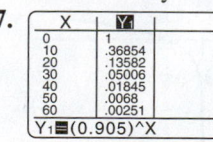

(a) 1; the probability that no vehicle will enter the intersection during a period of 0 seconds is 1 or 100%.

(b) About 30 seconds
99. (a) $f(x) = 2.7e^{0.014x}$ **(b)** 3.1 million
101. (a) $f(x) = 38e^{0.0102x}$ **(b)** 42 million
103. Answers may vary.

CHECKING BASIC CONCEPTS 12.1 & 12.2 (p. 825)

1. (a) 7 **(b)** $(f \circ g)(x) = 2x^2 + 9x + 6$
2.

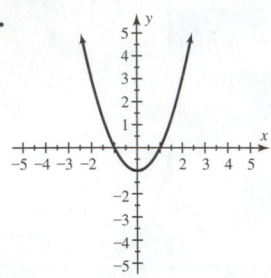

(a) No, it does not pass the horizontal line test. **(b)** No
3. $f^{-1}(x) = \frac{x+3}{4}$ **4.** $\frac{3}{4}$
5.

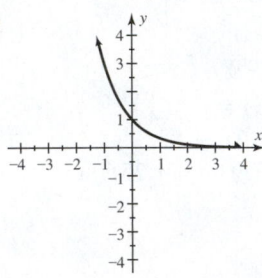

6. $C = 2; a = \frac{1}{2}$

SECTION 12.3 (pp. 834–837)

1. 10 **2.** e **3.** $D = \{x \mid x > 0\}$; R: all real numbers
4. D: all real numbers; $R = \{x \mid x > 0\}$ **5.** k **6.** k
7. x **8.** x **9.** $\log 5$ **10.** $\ln 5$ **11.** 0 **12.** undefined

13.

x	10^{-5}	10^0	$10^{0.5}$	$10^{2.2}$
$\log x$	-5	0	0.5	2.2

14.

x	10^{-6}	10^{-1}	$10^{5/7}$	10^π
$\log x$	-6	-1	5/7	π

15.

x	e^{-6}	e^{-1}	$e^{5/7}$	e^π
$\ln x$	-6	-1	5/7	π

16.

x	2^{-5}	2^0	$2^{0.5}$	$2^{2.2}$
$\log_2 x$	-5	0	0.5	2.2

17. 5 **19.** -4 **21.** 0 **23.** -2 **25.** 4.7 **27.** 4 **29.** $\frac{1}{2}$
31. Undefined **33.** -3 **35.** 2 **37.** $x^2, x \neq 0$ **39.** 5
41. $2x - 7$ **43.** 1.398 **45.** 0.161

47.   **49.**

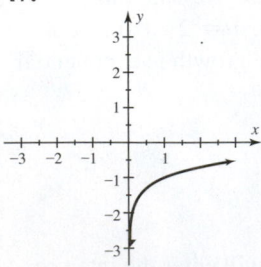   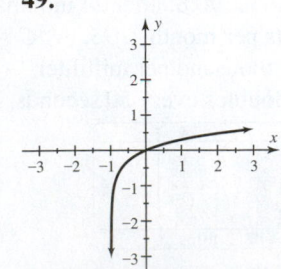

1 unit downward   1 unit to the left

51.

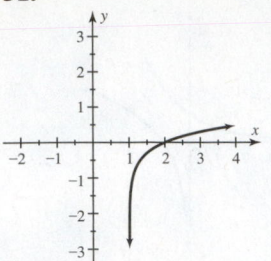

53.

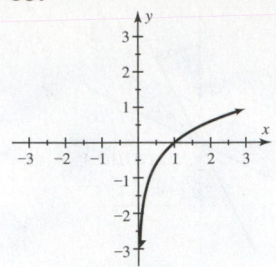

1 unit to the right   Increases faster

55. 0 **57.** $-5x$ **59.** $x^2, x \neq 0$ **61.** 1.946 **63.** -0.560
65. $[-4, 4, 1]$ by $[-4, 4, 1]$ **67.** $[-4, 4, 1]$ by $[-4, 4, 1]$

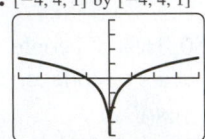   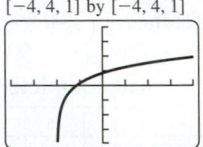

Reflected across the   2 units to the left;
y-axis together with $D = \{x \mid x > -2\}$
the graph of $y = \ln x$;
$D = \{x \mid x \neq 0\}$

69. $6x$ **71.** $-\frac{3}{2}$ **73.** 8 **75.** $\frac{3}{2}$ **77.** $-\frac{2}{3}$ **79.** Undefined
81. 2 **83.** -4 **85.** -2 **87.** -2 **89.** 17 **91.** $(2x)^2$
93. $0.6z, z > 0$
95.

x	1/4	1/2	1	$\sqrt{2}$	64
$\log_2 x$	-2	-1	0	1/2	6

97.   **99.**

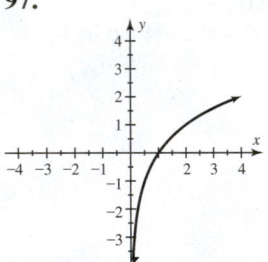

101. d. **103.** a. **105.** 120 dB; yes **107. (a)** It increases,
but it doesn't double when the weight of the plane doubles.
(b) About 5.086; a 50,000-pound airplane needs a runway 5086
feet long. **109. (a)** 3.5 **(b)** 1 **(c)** It decreases by 2.5.
111. (a) 10^6; 10^8 **(b)** 100 times

SECTION 12.4 (pp. 843–844)

1. 4 **2.** 4 **3.** 3 **4.** $\log m + \log n$ **5.** $\log m - \log n$
6. $r \log m$ **7.** No **8.** Yes **9.** No **10.** No
11. $\log_a x = \frac{\log x}{\log a}$ or $\log_a x = \frac{\ln x}{\ln a}$ **12.** 0; 1
13. $\ln 3 + \ln 5$ **15.** $\log x + \log y$ **17.** $\log y + \log y$
19. $\log 7 - \log 3$ **21.** $\ln x - \ln y$ **23.** $\log_2 45 - \log_2 x$
25. $\log 225$ **27.** $\ln xy$ **29.** $\ln 14x^3$ **31.** $\ln xy$ **33.** $\log 5$
35. $\ln x^2$ **37.** 2 **39.** $6 \log 3$ **41.** $x \ln 2$ **43.** $\frac{1}{4} \log_2 5$
45. $\frac{1}{3} \log_4 z$ **47.** $(y - 1)\log x$ **49.** $\log z$ **51.** $\log x^3 y^2$
53. 0 **55.** $\ln 2^x$ **57.** $\ln x^{5/6}$ **59.** $\log_a \frac{x+1}{x-1}$
61. $\log x + 2 \log y$ **63.** $4 \ln x + \ln y - \ln z$
65. $\frac{1}{3} \log z - \frac{1}{2} \log y$ **67.** $4 \log x + 3 \log y$
69. $\ln x - \ln y$ **71.** $\frac{3}{2} \log_4 x + \frac{1}{2} \log_4 y - \log_4 z$

73.

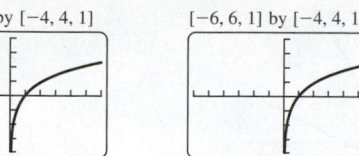

[−6, 6, 1] by [−4, 4, 1]   [−6, 6, 1] by [−4, 4, 1]

By the power rule, $\log x^3 = 3 \log x$

75.

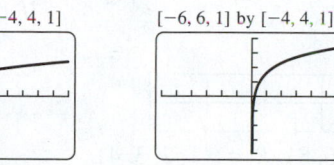

[−6, 6, 1] by [−4, 4, 1]   [−6, 6, 1] by [−4, 4, 1]

Not the same

77. 1.2   **79.** 1.8   **81.** 2.1   **83.** 0.4   **85.** −1.1
87. 1.46   **89.** 4.64   **91.** 2.10
93. $10 \log (10^{16}x) = 10(\log 10^{16} + \log x) =$
$10(16 + \log x) = 160 + 10 \log x$

CHECKING BASIC CONCEPTS 12.3 & 12.4 (p. 845)

1. (a) 4   (b) x (c) −3   (d) $\frac{1}{2}$
2.

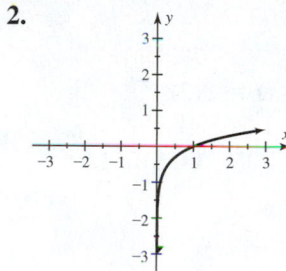

(a) $D = \{x \mid x > 0\}$; R: all real numbers   (b) 0
(c) Yes; for example, $\log \frac{1}{10} = -1$.   (d) No; negative numbers are not in the domain of $\log x$.   **3.** (a) $\log x + \log y$
(b) $\ln x - \ln y - \ln z$ (c) $2 \ln x$
(d) $2 \log x + 3 \log y - \frac{1}{2} \log z$ **4.** (a) $\log xy$
(b) $\ln \frac{2x}{y^3}$ (c) $\log_2 \frac{x^2 y^3}{z}$

SECTION 12.5 (pp. 852–856)

1. Add 5 to each side.   **2.** Divide each side by 5.
3. Take the common logarithm of each side.
4. Exponentiate each side using base 10.   **5.** x
6. $x, x > 0$ **7.** $2x$ **8.** $x + 7, x > -7$
9. No; $\log \frac{5}{4} = \log 5 - \log 4$
10. No; $\log 5 - \log 4 = \log \frac{5}{4}$ **11.** 1   **12.** 1   **13.** 3
15. 6   **17.** 6   **19.** 4   **21.** $\frac{\log 124}{0.4} \approx 5.23$ **23.** 0
25. $\ln 25 \approx 3.22$ **27.** $\frac{\ln 2}{\ln 0.4} \approx -0.76$ **29.** $\ln 7 \approx 1.95$
31. $\log \frac{35}{2} - 2 \approx -0.76$ **33.** $\frac{\log 20}{2 \log 3.1} \approx 1.32$ **35.** −1
37. −1, 5   **39.** $\frac{\ln 10}{3} \approx 0.77$ **41.** $\frac{2 \ln 2}{1 - \ln 2} \approx 4.52$
43. $\frac{2 \log 5}{0.5 \log 4 - \log 5} \approx -3.51$ **45.** (a) 1   (b) 1
47. (a) −2   (b) −2   **49.** −1   **51.** 0   **53.** 5
55. $\frac{\log 1.45}{\log 0.55} \approx -0.62$ **57.** −1.84, 1.15   **59.** 1.31
61. 100   **63.** $e^5 \approx 148.41$ **65.** 5,000,000   **67.** 16
69. $\frac{2^{2.3}}{5} \approx 0.98$ **71.** $10^{1.4} \approx 25.12$
73. $\frac{e^{11} - 1}{2} \approx 29,936.57$ **75.** 1   **77.** 5   **79.** 4   **81.** 3
83. $10^{1.6} \approx 39.81$ **85.** $e - 1 \approx 1.72$ **87.** 9

89. (a) 2   (b) $e^{0.7} \approx 2.01$ **91.** (a) 2   (b) $\frac{1}{2}(10^{0.6}) \approx 1.99$
93. (a) About 41.5; in 2010 China generated about 41.5 gigawatts.   (b) 2009   **95.** 10 hours   **97.** 8 years
99. (a) About 8503; in 1994 about 8503 people were waiting for liver transplants.   (b) In about 1998
101. About 20,893 pounds   **103.** (a) About 203; in 1975 there were about 203 thousand bluefin tuna.
(b) In 1979   (c) In 1979   **105.** $a = 25, b = 3$
107. (a) About 1.67 acres   (b) About 1.67 acres
109. (a) Nonlinear; they do not increase at a constant rate.   (b) Each year the amount of fertilizer increased by a factor of 1.06, or by 6%.   (c) In 1968
111. 10^{-6} watts/square centimeter   **113.** About 7 miles

CHECKING BASIC CONCEPTS 12.5 (p. 856)

1. (a) $\log 20 \approx 1.30$ (b) $\frac{\log 147}{3 \log 2} \approx 2.40$
(c) $e^{4.1} \approx 60.34$ (d) 500   **2.** 8   **3.** 20 years

CHAPTER 12 REVIEW (pp. 859–862)

1. (a) 81   (b) $(f \circ g)(x) = 50x^2 - 2$ **2.** (a) −32
(b) $(f \circ g)(x) = \sqrt[3]{4x^3 - 6}$ **3.** (a) 0   (b) 3   **4.** (a) 3
(b) −2   (c) −1   **5.** $f(1) = f(-1) = 2$
6. $f(0) = f(2) = 1$ **7.** No   **8.** Yes
9. $(f \circ f^{-1})(x) = 2\left(\frac{x + 9}{2}\right) - 9 = x$
$(f^{-1} \circ f)(x) = \frac{(2x - 9) + 9}{2} = x$
10. $(f \circ f^{-1}) = (\sqrt[3]{x - 1})^3 + 1 = x$
$(f^{-1} \circ f)(x) = \sqrt[3]{(x^3 + 1)} - 1 = x$
11. $f^{-1}(x) = \frac{x}{5}$ **12.** $f^{-1}(x) = x + 11$
13. $f^{-1}(x) = \frac{x - 7}{2}$ **14.** $f^{-1}(x) = \frac{4}{x}$
15.

x	10	8	7	3
$f^{-1}(x)$	0	1	2	3

$D = \{3, 7, 8, 10\}$; $R = \{0, 1, 2, 3\}$
16.

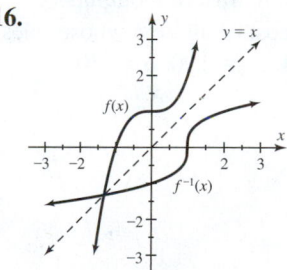

17. $\frac{1}{6}$; 36   **18.** 5; $\frac{5}{8}$ **19.** 3; $\frac{1}{81}$ **20.** 3; $\frac{1}{2}$

21.

Exponential growth

22.

Exponential decay

23.

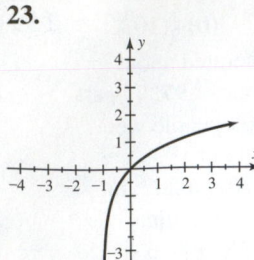

24.

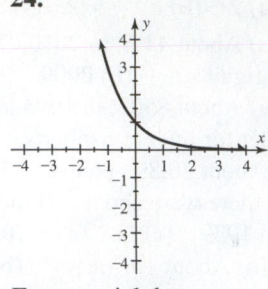

Logarithmic growth   Exponential decay

25. (a) Exponential growth   **(b)** $f(x) = 5(2)^x$
26. (a) Linear growth   **(b)** $f(x) = 5x + 5$
27. $C = \frac{1}{2}, a = 2$ **28.** $k = 2$ **29.** -20% **30.** 1.07
31. (a) 100% **(b)** -50% **32. (a)** $-54.\overline{54}\%$ **(b)** 120%
33. (a) $\$1050$ **(b)** $\$1550$ **(c)** 3.1 **34. (a)** $-\$175$
(b) $\$525$ **(c)** 0.75 **35.** $f(x) = 20{,}000(0.95)^x$;
$f(2) = 18{,}050$ **36.** $f(x) = 1500(4)^x$; $f(2) = 24{,}000$
37. $\$2829.54$ **38.** $\$675{,}340.51$ **39.** 399.67 **40.** 0.71
41. 3.48 **42.** 3.89 **43.** -3 **44.** 2 **45.** -4 **46.** 2
47. 1.813 **48.** -0.163 **49.** 4.787 **50.** -0.398 **51.** 7
52. $\frac{5}{9}$ **53.** $6 - x$ **54.** $x^2, x > 0$ **55.** $\ln x + \ln y$
56. $\log x - \log y$ **57.** $2 \ln x + 3 \ln y$
58. $\frac{1}{2} \log x - 3 \log z$ **59.** $2 \log_2 x + \log_2 y - \log_2 z$
60. $\frac{1}{3} \log_3 x - \frac{1}{3} \log_3 y$ **61.** $\log 75$ **62.** $\log_4 (10x^2)$
63. $\ln \frac{x^2}{y^3}$ **64.** $\log xy$ **65.** $3 \log 6$ **66.** $2 \ln x$
67. $2x \log_2 5$ **68.** $(x + 1) \log_4 0.6$ **69.** 2 **70.** 4
71. $\ln 9 \approx 2.20$ **72.** $\frac{\log 0.2}{\log 0.85} \approx 9.90$ **73.** $e^{0.8} \approx 2.23$
74. $\frac{1}{2} e^5 \approx 74.21$ **75.** 10^{40} **76.** 100
77. $\frac{4 \log 2}{\log 3 - \log 2} \approx 6.84$ **78.** 6 **79. (a)** 3 **(b)** 3
80. (a) 4 **(b)** 4 **81. (a)** 64π; after 8 seconds, the balloon
has a surface area of $64\pi \approx 201$ in^2.   **(b)** $(S \circ r)(t) = 8\pi t$
82. (a) Yes, different inputs result in different outputs.
(b) $f^{-1}(x) = \frac{x}{0.08}$ calculates the cost of an item whose sales
tax is x dollars.   **83.** 7 years   **84.** $a = 100, b = 50$
85. $C = 3, a = 2$ **86.** 10^7

87. (a) [0, 10, 2] by [0, 4, 1]

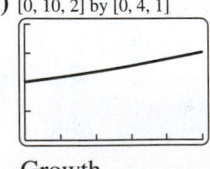

Growth

(b) About 3.3 million   **(c)** 2008   **88. (a)** 1000; there were
1000 bacteria/mL initially.   **(b)** About 495.11 minutes
89. (a) About 6.93 meters/second   **(b)** 12.18 meters

CHAPTER 12 TEST (pp. 863–864)

1. 6; $(f \circ g)(x) = 4(x + 7)^3 - 5(x + 7)$
2. (a) 3 **(b)** 3 **3.** $f(-5) = f(5) = 0$
(answers may vary)   **4.** $f^{-1}(x) = \frac{5 - x}{2}$

5.

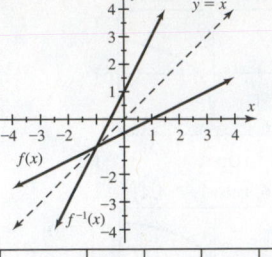

6.

x	8	6	4	2
$f^{-1}(x)$	1	2	3	4

$D = \{2, 4, 6, 8\}; R = \{1, 2, 3, 4\}$
7. $\frac{3}{16}$
8.

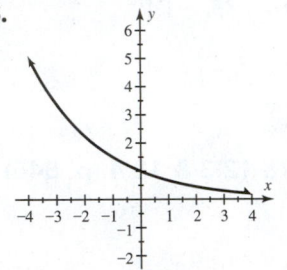

Exponential decay

9. (a) Exponential growth   **(b)** $f(x) = 3(2)^x$
10. (a) Linear growth   **(b)** $f(x) = 1.5x - 1$
11. $C = 1, a = \frac{1}{2}$ **12.** 50% **13.** 1.05 **14.** $\$1051.91$
15. 4.16 **16.** $\frac{1}{2}$ **17.** 5.426
18.

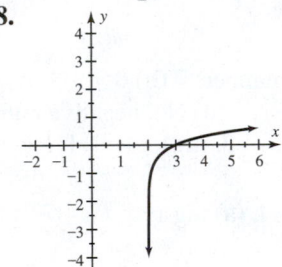

Shifted to the right 2 units
19. $3 \log x + 2 \log y - \frac{1}{2} \log z$ **20.** $\ln \frac{x^4 z}{y^5}$ **21.** $2x \log 7$
22. $1 - 3x$ **23.** $\ln 25 \approx 3.22$ **24.** $\log 50 \approx 1.70$
25. $10^{1.8} \approx 63.10$ **26.** $\frac{1}{5} e^9 \approx 1620.62$
27. $a = 5, b = 3$ **28. (a)** 4 million   **(b)** About 6.15; after
5 hours there were about 6.15 million bacteria.   **(c)** Growth
(d) After 8 hours
29. (a) $A(x) = 5000 (0.98)^x$ **(b)** $A(3) \approx 4705.96$; after
3 years the account contains $\$4705.96$.   **(c)** About 5

CHAPTERS 1–12 CUMULATIVE REVIEW (pp. 865–868)

1. 4.29×10^{-4} **2.** Natural: none; whole: 0; integer: $-3, 0$;
rational: $-\frac{11}{7}, -3, 0, 5.\overline{18}$; irrational: $\sqrt{6}, \pi$ **3.** c.
4. Commutative   **5.** d^4 **6.** $\frac{b^9}{64a^6}$ **7.** $\frac{4y^8}{x^5}$ **8.** $\frac{y^4}{4x^5}$
9. $y = \frac{5}{4}x + 1$ **10.** $\{x \mid x \neq -3\}$ **11.** $f(x) = 4x - 3$
12. $x = 4$ **13.** 1

14.

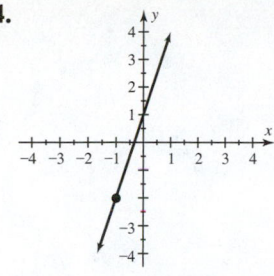

15. $y = 7x - 6$ **16.** $y = 3x + 5$ **17.** -1 **18.** $x < 0$
19. -18 **20.** $(-\infty, 0)$ **21.** $\frac{4}{15}$ **22.** $(-5, 10]$
23. $[-2, 10]$ **24.** $4, 16$
25.

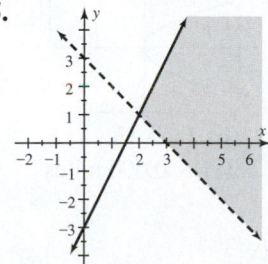

26. 2 **27.** $(1, 1)$ **28.** No solutions   **29.** $(-4, 0, 2)$
30. $(0, 0, 1)$ **31.** $R = 21$ **32.** 8 square inches
33. $2x(x - 1)^2$ **34.** $(2a - 5b)(2a + 5b)$
35. $(2t - 3)(4t^2 + 6t + 9)$ **36.** $(2a^2 + 5)(2a - 1)$
37. $-\frac{5}{6}, 2$ **38.** $-\frac{2}{3}, \frac{2}{3}$ **39.** $-3, 0, 5$ **40.** $0, \frac{1}{2}$ **41.** 1
42. $\frac{x + 2}{x - 3}$ **43.** 3 **44.** -3 **45.** $J = \frac{2z}{P - 1}$ **46.** $\frac{x^3 + 3}{x^3 - 3}$
47. 40 **48.** $3x^2 + 6x + 10 + \frac{5}{x - 2}$ **49.** $\frac{x^4}{y^6}$ **50.** x^3
51. $10ab$ **52.** $\sqrt{6}$ **53.** $(b + 3)a\sqrt[3]{a^2b}$ **54.** 20
55. $\frac{5 + \sqrt{3}}{11}$ **56.** $(4, \infty)$ **57.** $-6, 4$ **58.** 3
59. $3 + 5i$ **60.** $-i$ **61.** $(2, 1)$ **62.** $\frac{7}{2}$
63. Shifted right 3 units and up 2 units
64. $y = (x + 3)^2 - 11; (-3, -11)$ **65.** $5, 8$
66. $\frac{1 \pm \sqrt{41}}{4}$ **67.** $2 \pm \sqrt{2}$ **68.** $\pm 2, \pm \sqrt{6}$
69. (a) $-1, 3$ **(b)** $a < 0$ **(c)** Positive
70. $(-\infty, -7] \cup [2, \infty)$ **71. (a)** 7
(b) $(g \circ f)(x) = 2x^2 - 3$ **72.** $f(-4) = f(3) = 6$
73. $f^{-1}(x) = \frac{3}{x}$ **74.** \$2367.10   **75.** 4 **76.** $2x, x > 0$
77. $\frac{1}{2} \log x - 2 \log y$ **78.** $\ln (5x^3)$ **79.** $10^{11/6} \approx 68.13$
80. $\frac{\log 17}{3 \log 2} \approx 1.36$ **81.** 4% **82.** 2.27 pounds
83. (a) 1950   **(b)** 220 million Btu
84. $\sqrt{4200} \approx 64.8$ mph   **85.** About 11.3%
86. (a) 9.91 meters per second   **(b)** 8.52 meters

13 Conic Setctions

SECTION 13.1 (pp. 876–879)

1. Parabola, ellipse, hyperbola   **2.** The axes of symmetry
are vertical and horizontal, respectively.   **3.** No

4.

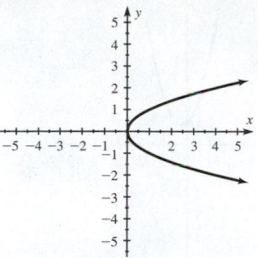

5. No; it does not pass the vertical line test.   **6.** (h, k)
7. left   **8.** right   **9.** circle; (h, k) **10.** $(0, 0); r$
11. d.   **12.** a.   **13.** c.   **14.** b.
15.

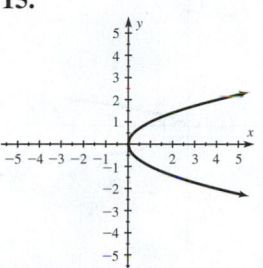

$(0, 0); y = 0$

17.

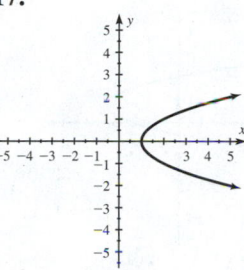

$(1, 0); y = 0$

19.

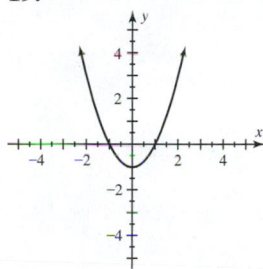

$(0, -1); x = 0$

21.

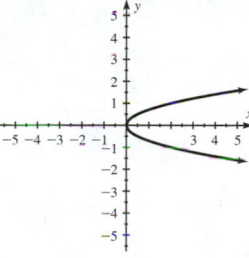

$(0, 0); y = 0$

23.

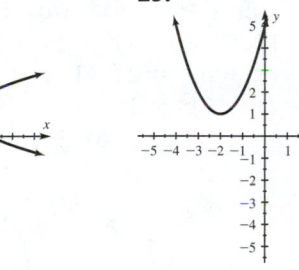

$(2, 1); y = 1$

25.

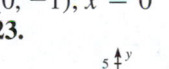

$(-2, 1); x = -2$

27.

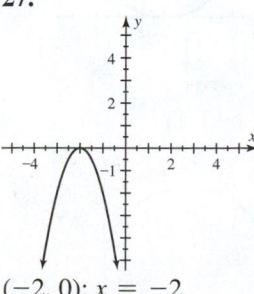

$(-2, 0); x = -2$

29.

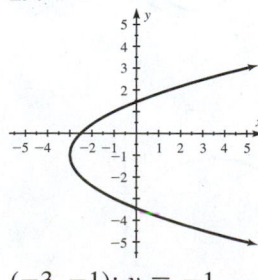

$(-3, -1); y = -1$

31.

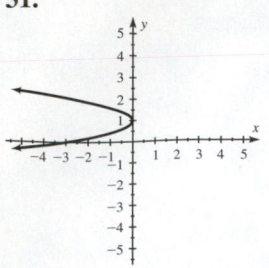

$(0, 1); y = 1$

33.

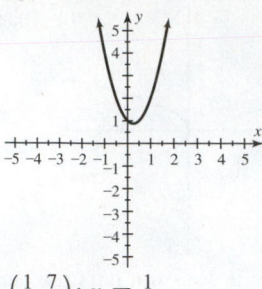

$\left(\frac{1}{4}, \frac{7}{8}\right); x = \frac{1}{4}$

35.

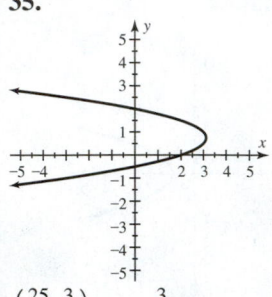

$\left(\frac{25}{8}, \frac{3}{4}\right); y = \frac{3}{4}$

37.

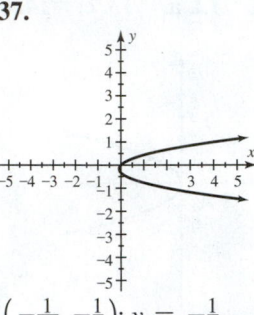

$\left(-\frac{1}{12}, -\frac{1}{6}\right); y = -\frac{1}{6}$

39.

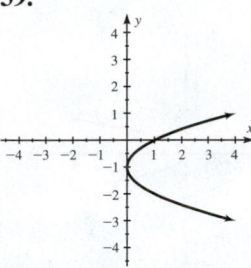

$(0, -1); y = -1$
41. $y = x^2$ **43.** $x = (y + 1)^2 - 2$ **45.** Upward
47. Downward **49.** $x \geq 0$ **51.** Two **53.** 1
55. $x^2 + y^2 = 1$ **57.** $(x + 1)^2 + (y - 5)^2 = 9$
59. $(x + 4)^2 + (y + 6)^2 = 2$ **61.** $x^2 + y^2 = 16$
63. $(x + 3)^2 + (y - 2)^2 = 1$
65. 2; (0, 0)

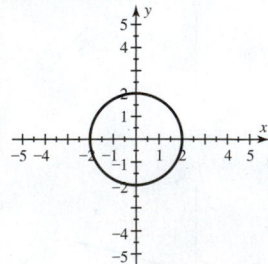

67. 3; (1, 3)

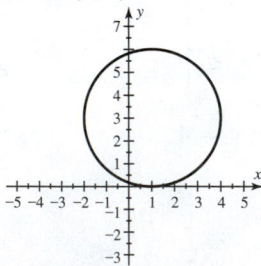

69. 5; (−5, 5)

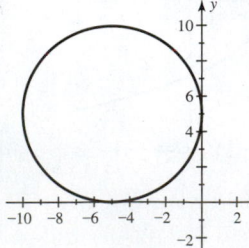

71. 3; (−3, 1)

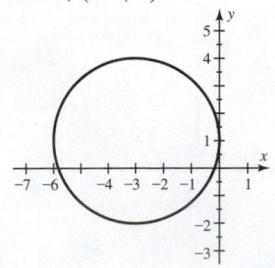

73. $\sqrt{7}$; (−3, 1)

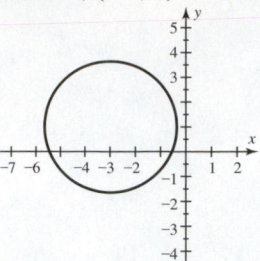

75. (a)

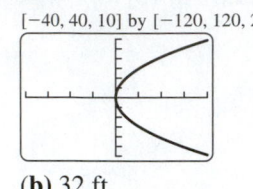

[−40, 40, 10] by [−120, 120, 20]

(b) 32 ft

77. (a)

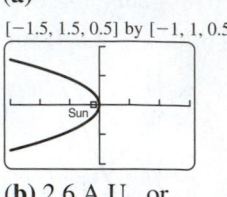

[−1.5, 1.5, 0.5] by [−1, 1, 0.5]

(b) 2.6 A.U., or
241,800,000 miles

SECTION 13.2 (pp. 887–889)

1.

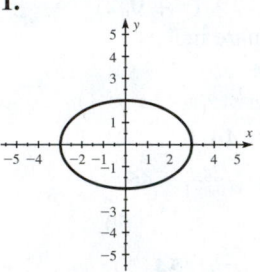

(answers may vary)

2.

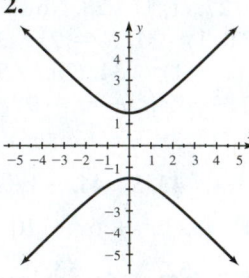

(answers may vary)
3. horizontal **4.** vertical **5.** 2 **6.** 4 **7.** left; right
8. lower; upper **9.** They are the diagonals extended.
10. No **11.** a. **12.** b.

13.

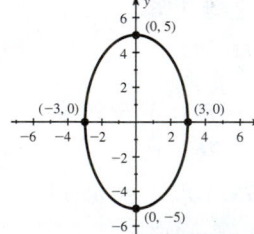

15.

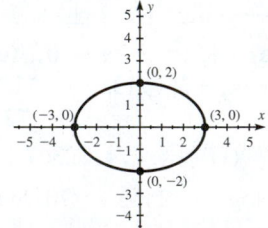

17.

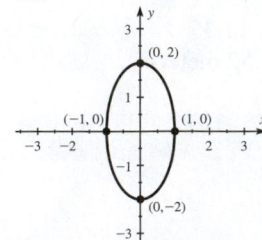

19.

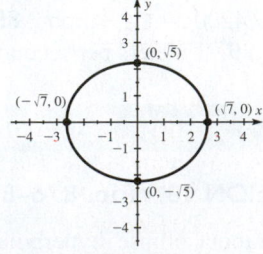

21.

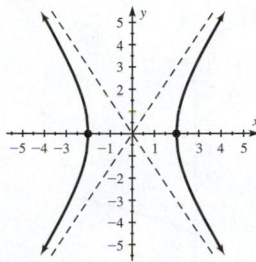

23.

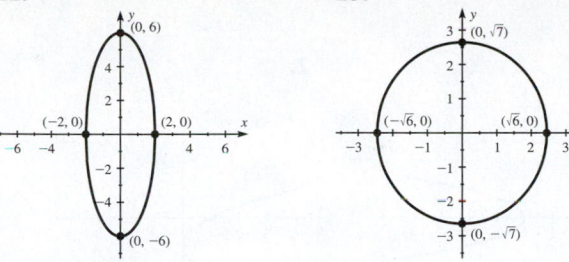

25. $\dfrac{x^2}{9} + \dfrac{y^2}{4} = 1$ **27.** $\dfrac{y^2}{25} + \dfrac{x^2}{16} = 1$

29.

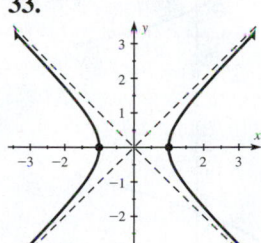

31.

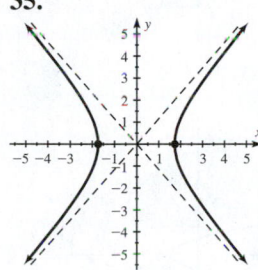

33.

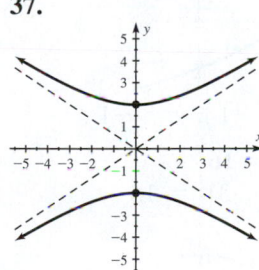

35.

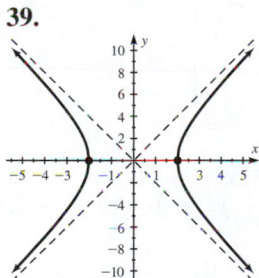

37.

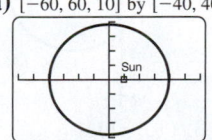

39.

41. $x^2 - y^2 = 1$ **43.** $\dfrac{y^2}{4} - \dfrac{x^2}{9} = 1$

45. (a) $A \approx 62.83$; $P \approx 28.45$
(b) $A \approx 11.75$; $P \approx 13.33$
47. (a) $[-60, 60, 10]$ by $[-40, 40, 10]$

(b) $P \approx 243.9$ A.U., or about 2.27×10^{10} miles;
$A \approx 4733$ square A.U., or about 4.09×10^{19} square miles
49. Maximum: 668 miles; minimum: 340 miles
51. Height: 20 feet; width: 200 feet
53. Answers may vary.

CHECKING BASIC CONCEPTS 13.1 & 13.2 (p. 890)

1.

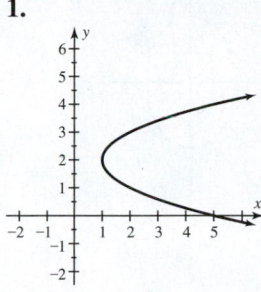

2. $(x - 1)^2 + (y + 2)^2 = 4$

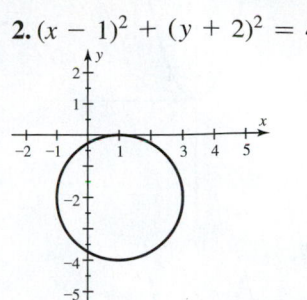

$(1, 2)$; $y = 2$
3. x-intercepts: ± 2; y-intercepts: ± 3
4. (a)

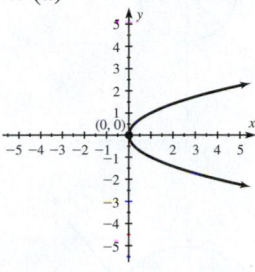

(b)

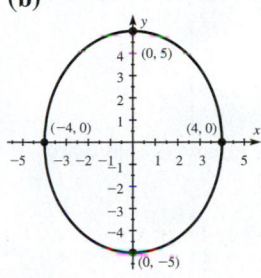

Parabola

Ellipse

(c)

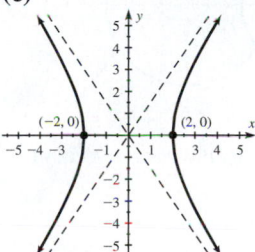

(d)

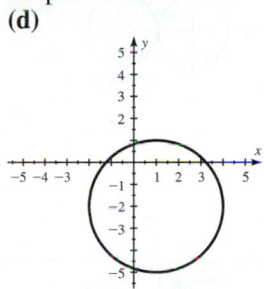

Hyperbola

Circle (and ellipse)

SECTION 13.3 (pp. 897–898)

1. Any number   **2.** Two   **3.** Two; the line intersects the circle twice.   **4.** All points inside and including a circle of radius 1 centered at the origin   **5.** No   **6.** No

7.

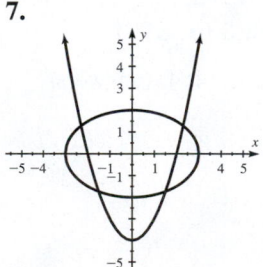

8.

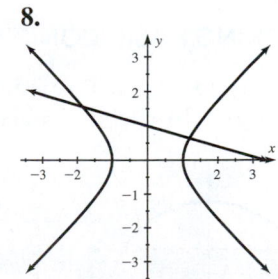

(answers may vary)   (answers may vary)
9. $(1, 3)$, $(-1, -3)$ **11.** $(0, -1)$, $(0, 1)$
13. $(3, 6)$, $(-3, -6)$ **15.** $(2, -1)$ **17.** $(-\sqrt{2}, 2)$, $(\sqrt{2}, 2)$
19. $\left(-\dfrac{3}{5}, \dfrac{7}{5}\right)$, $(-1, 1)$ **21.** No solutions   **23.** $(\pm 3, \pm 1)$
25. $(-1, -2)$, $(-2, 1)$ **27.** $(7, 3)$, $(2, -2)$
29. $(-1, 2)$, $(3, -6)$ **31.** $(-1, -1)$, $(1, 1)$

33.

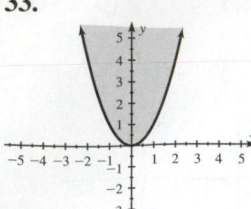

35.

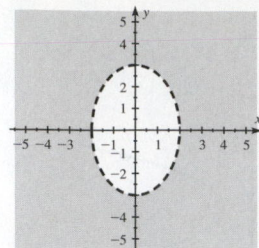

37.

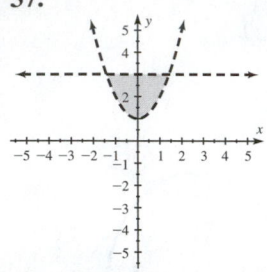

(0, 2), (answers may vary)

39.

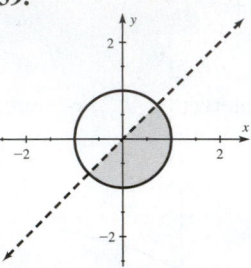

$\left(\frac{1}{2}, -\frac{1}{2}\right)$, (answers may vary)

41.

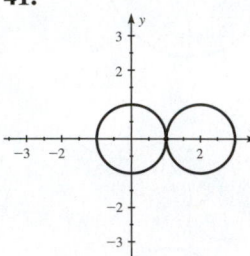

(1, 0)

43.

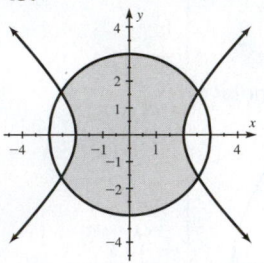

(0, 0), (answers may vary)

45. a. **47.** $y \geq x^2$; $y < 4 - x$

49. $[-10, 10, 1]$ by $[-10, 10, 1]$

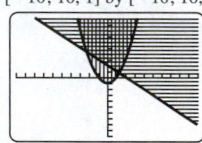

51. $r = 1.6$ inches; $h \approx 4.97$ inches

53. (a) $xy = 143, 2x + 2y = 48$
(b) $x = 11$ ft, $y = 13$ ft

CHECKING BASIC CONCEPTS 13.3 (p. 899)

1. (1, −1), (3, 3) **2.** Two **3.** (a) (0, 3), (4, 4) (answers
may vary) (b) $y \geq 2 - x$ and $y \leq 4 - x^2$

4.

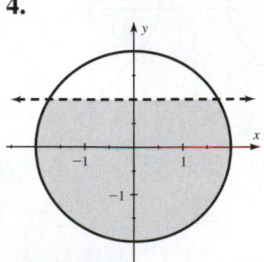

CHAPTER 13 REVIEW (pp. 901–903)

1.

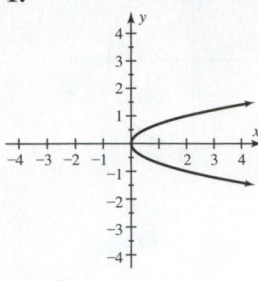

(0, 0); $y = 0$

2.

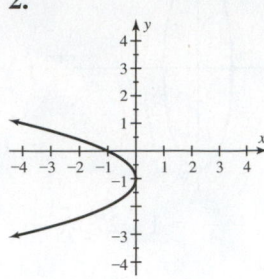

(0, −1); $y = -1$

3.

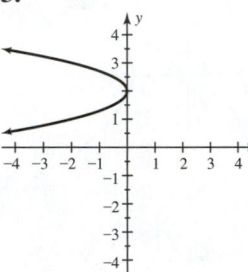

(0, 2); $y = 2$

4.

(−1, −2); $y = -2$

5.

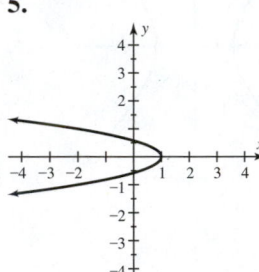

(1, 0); $y = 0$

6.

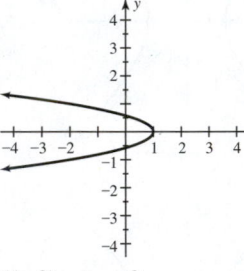

$\left(-\frac{7}{2}, -1\right)$; $y = -1$

7. $x = y^2$ **8.** $(x + 2)^2 + (y - 2)^2 = 16$
9. $x^2 + y^2 = 1$ **10.** $(x - 2)^2 + (y + 3)^2 = 16$

11. 5; (0, 0)

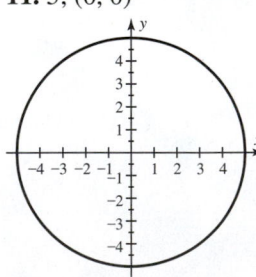

12. 3; (2, 0)

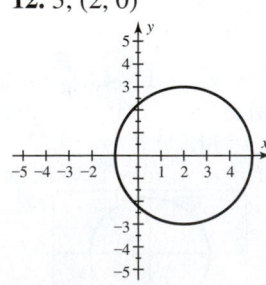

13. $\sqrt{5}$; (−3, 1)

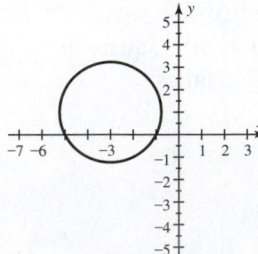

14. 3; (1, −1)

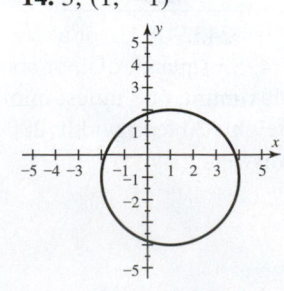

15.
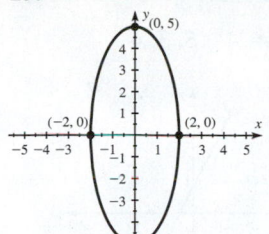

16.
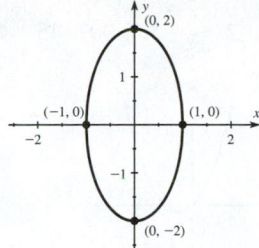

17.
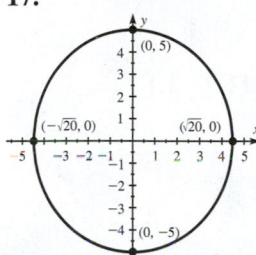

18.
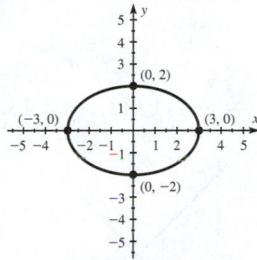

19. $\dfrac{y^2}{16} + \dfrac{x^2}{4} = 1$ **20.** $x^2 - \dfrac{y^2}{4} = 1$

21.
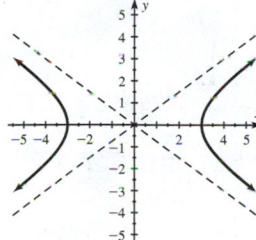

22.
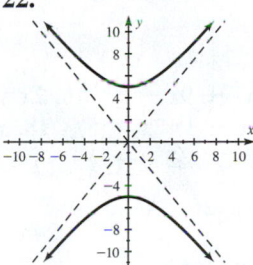

23.
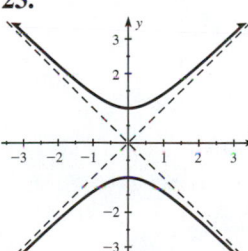

24.
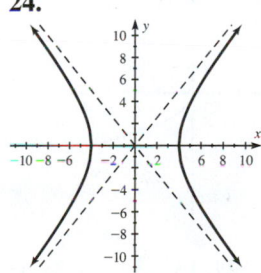

25. $(0, 3), (3, 0)$ **26.** $(-1, -2), (1, 2)$ **27.** $(0, 0), (2, 2)$

28. $(-2, -1), (-2, 1), (2, -1), (2, 1)$ **29.** $(-4, -4), (4, 4)$

30. $(0, -4), (4, 0)$ **31.** $(-1, 1), (1, 1)$ **32.** $(1, 2), (2, 5)$

33. $(-4, -12), (2, 0)$ **34.** $(-1, -1), (1, -1)$

35. $(2, 2), (-4, -4)$ **36.** $(1, 1), (0, 0)$

37.
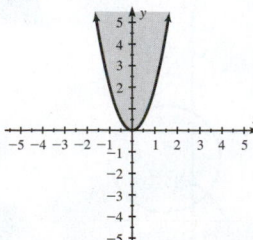

38.
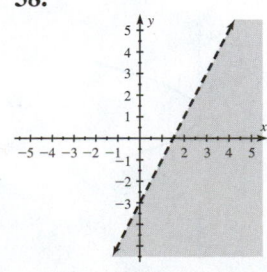

39.
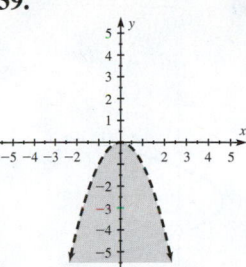

40.
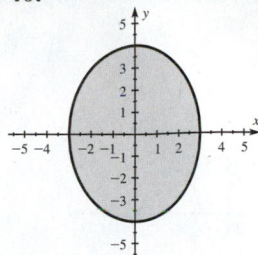

41.
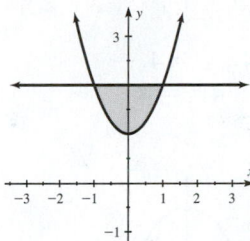

42.
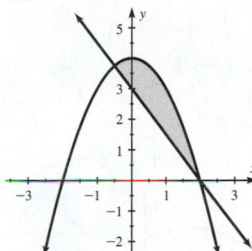

43.
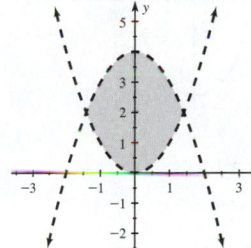

44.
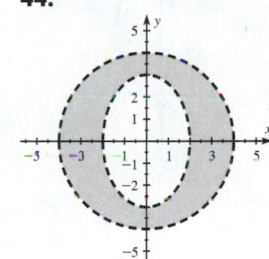

45. $y \geq x^2 - 2, \; y \leq 2 - x$ **46.** $y \geq x, \; x^2 + y^2 \leq 4$

47. (a) $xy = 1000, \; 2x + 2y = 130$

(b) $x = 25$ inches; $y = 40$ inches

(c) $x = 25$ inches; $y = 40$ inches

48. (a) $xy = 60, \; y - x = 7$ **(b)** $x = 5; y = 12$

(c) $x = 5; y = 12$ **49.** $r = 1$ foot, $h \approx 15.92$ feet; yes

50. Either $r \approx 0.94$ inches, $h \approx 12.60$ inches or $r \approx 3.00$ inches, $h \approx 1.23$ inches; no

51. (a) $[-7.5, 7.5, 1]$ by $[-5, 5, 1]$

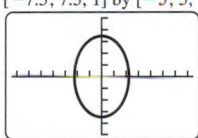

(b) $A \approx 24.33, \; P \approx 18.32$

52. (a) $[-3, 3, 1]$ by $[-2, 2, 1]$

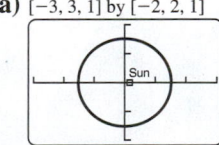

(b) $P \approx 9.55$ A.U., or about 8.9×10^8 miles; $A \approx 7.26$ square A.U., or about 6.3×10^{16} square miles

CHAPTER 13 TEST (pp. 903–904)

1.

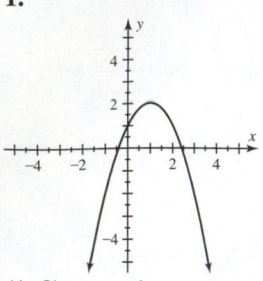

2.

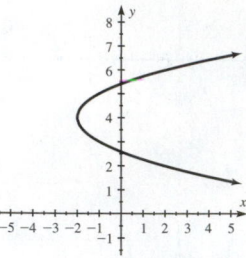

$(1, 2); x = 1$ $(-2, 4); y = 4$
3. $x = -y^2 + 1$ **4.** $(x - 2)^2 + (y + 4)^2 = 4$
5. $(x + 5)^2 + (y - 2)^2 = 100$
6. $r = 4$, center $= (-2, 3)$ **7.**

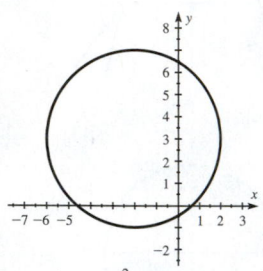

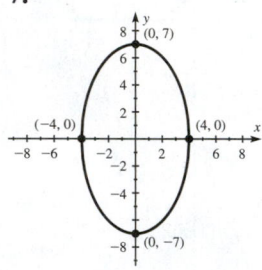

8. $\dfrac{x^2}{100} + \dfrac{y^2}{64} = 1$

9.

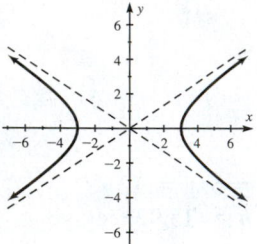

10. $(0, -4), (4, 0)$ **11.** $(-1, -4), (4, 1)$
12. $(-2, 4), (2, 4)$
13.

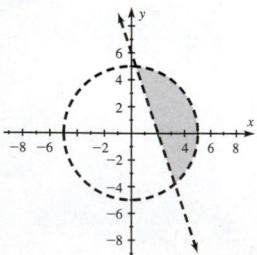

14. $y \le 4 - x^2, y \ge x^2 - 4$
15. (a) $xy = 5000, 2x + 2y = 300$
(b) 50 feet by 100 feet
16. Either $x \approx 22.08$ in., $y \approx 2.43$ in. or
$x \approx 7.29$ in., $y \approx 22.24$ in.; no

17. (a) $[-30, 30, 10]$ by $[-20, 20, 10]$

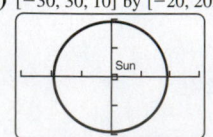

(b) 18.28 A.U., or about 1,700,040,000 miles

CHAPTERS 1–13 CUMULATIVE REVIEW (pp. 906–907)

1. 25   **2.** $\dfrac{b^8}{a^3}$ **3.** 0.007345   **4.** $\dfrac{1}{2}; x \ne 4$ **5.** 3; 4
6.   **7.**

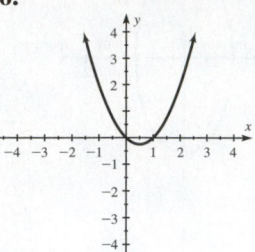

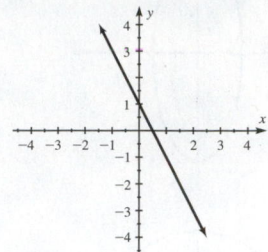

8. $y = \dfrac{3}{2}x - 5$ **9.** $\dfrac{2}{7}$ **10.** $(-1, 3]$ **11.** $[-2, 2]$
12. $(-\infty, -1] \cup [3, \infty)$ **13.** $(1, 3)$
14.

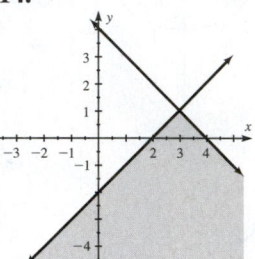

15. $2x^2 + 9x - 5$ **16.** $2x^2y - 3xy^3 + xy$
17. $(3x + 1)(2x - 5)$ **18.** $x(x + 2)(x - 2)$ **19.** $-2, -1$
20. $\dfrac{-3 \pm \sqrt{5}}{2}$ **21.** $\dfrac{3}{2}$ **22.** $\dfrac{2x}{x^2 - 1}$ **23.** $2x\sqrt{2}$ **24.** 4
25. $4x$ **26.** $5\sqrt{3x}$
27.

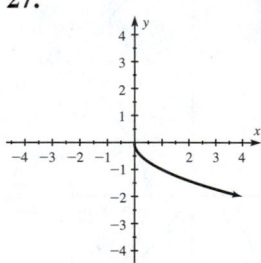

28. 5   **29.** 13   **30.** 2   **31.** $(3, -6)$
32. $f(x) = (x - 1)^2 + 2$ **33.** Shifted 4 units right
34. 1, 2   **35.** 0, $\pm i$ **36. (a)** 4   **(b)** 3   **(c)** x **(d)** 6   **(e)** 2
(f) 3   **37. (a)** 17   **(b)** $2x^2 + 2$ **38.** $\dfrac{2}{3} - \dfrac{1}{3}x$
39. \$1340.10   **40.** $2 \log x + \dfrac{1}{2} \log y - 3 \log z$ **41.** $\ln 9$
42. 25
43. (a)   **(b)**

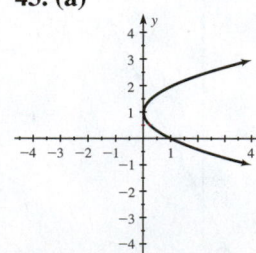

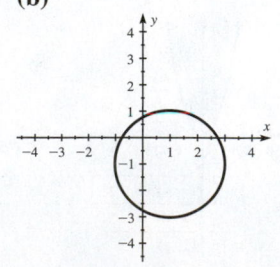

(c)

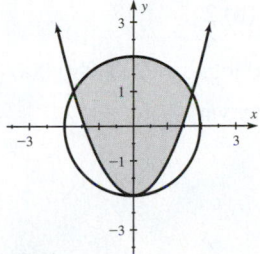

(d)

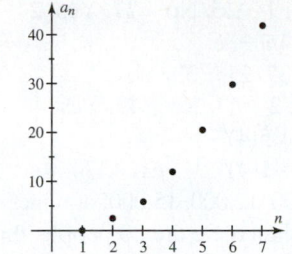

44. $(0, -1), (0, 1)$

45.

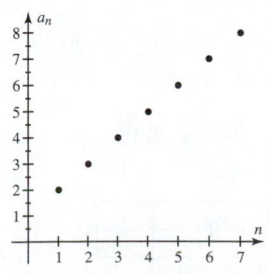

46. (a) 400; initially, the driver is 400 miles from home.
(b) 8; after 8 hours, the driver arrives at home.
(c) -50; the driver is traveling 50 mph toward home.
47. $500 at 5%, $1000 at 6%, $500 at 7%
48. 300 ft by 300 ft   **49.** $50 \ln 2 \approx 34.7$ yr
50. 2.4 in.

14 Sequences and Series

SECTION 14.1 (pp. 915–917)

1. 1, 2, 3, 4 (answers may vary)   **2.** 1, 3, 5, 7, … (answers
may vary)   **3.** function; natural numbers   **4.** sequence
5. 6   **6.** scatterplot   **7.** $f(2)$ **8.** a_4 **9.** 1, 4, 9, 16
11. $\frac{1}{6}, \frac{1}{7}, \frac{1}{8}, \frac{1}{9}$ **13.** $\frac{5}{2}, \frac{5}{4}, \frac{5}{8}, \frac{5}{16}$ **15.** 9, 9, 9, 9   **17.** 1, 8, 27
19. $1, \frac{8}{5}, 2$ **21.** 2, 9, 20   **23.** $-2, -2, -2$
25. $b + c$; $2b + c$ **27.** 7   **29.** 3, 4, 5, 3, 1
31. 6, 5, 4, 3, 2, 1
33.

n	1	2	3	4	5	6	7
a_n	2	3	4	5	6	7	8

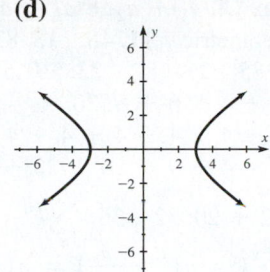

35.

n	1	2	3	4	5	6	7
a_n	0	2	6	12	20	30	42

37.

n	1	2	3	4	5	6	7
a_n	2	4	8	16	32	64	128

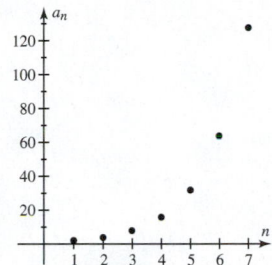

39. $a_n = 30n$ for $n = 1, 2, 3, \ldots, 7$

n	1	2	3	4	5	6	7
a_n	30	60	90	120	150	180	210

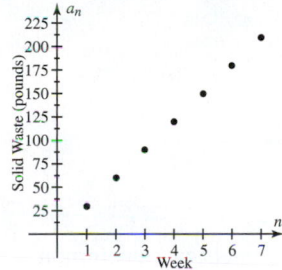

41. (a) 1, 4, 9, 16   **(b)** 4, 8, 12, 16
43. (a) $20,000; $16,000   **(b)** $a_n = 25,000(0.8)^n$
(c)

n	1	2	3	4	5	6	7
a_n	20,000	16,000	12,800	10,240	8192	6553.6	5242.9

45. (a) $a_n = 2048(0.5)^{n-1}$, for $n = 1, 2, 3, \ldots, 7$
(b)

n	1	2	3	4	5	6	7
a_n	2048	1024	512	256	128	64	32

(c)

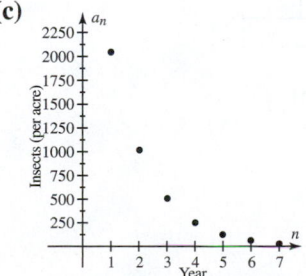

SECTION 14.2 (pp. 923–926)

1. linear   **2.** exponential   **3.** $a_n = 3n + 1$; 3 (answers
may vary)   **4.** $a_n = 5(2)^{n-1}$; 2 (answers may vary)
5. add; previous   **6.** multiply; common ratio   **7.** 19; 4
8. 32; -2 **9.** $a_n = a_1(r)^{n-1}$ **10.** $a_n = a_1 + (n - 1)d$
11. Yes; 10   **13.** Yes; -1 **15.** No   **17.** Yes; 3

19. Yes; -3 **21.** No   **23.** Yes; 1   **25.** No   **27.** Yes; 2
29. $a_n = -2n + 9$ **31.** $a_n = 4n - 6$
33. $a_n = -2n + 32$ **35.** 59   **37.** 21   **39.** Yes; 3
41. Yes; 0.8   **43.** No   **45.** Yes; 2   **47.** No   **49.** Yes; 4
51. Yes; 2   **53.** No   **55.** $a_n = 1.5(4)^{n-1}$
57. $a_n = -3(-2)^{n-1}$ **59.** $a_n = 1(4)^{n-1}$ **61.** 4374
63. 243   **65.** (a) 3000, 6000, 9000, 12,000, 15,000; arithmetic
(b) $a_n = 3000n$ **(c)** 60,000; when there are 20 people, the
ventilation rate should be 60,000 cubic feet per hour.
(d)

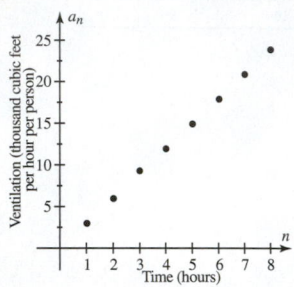

Yes
67. (a) $a_n = 3(0.8)^{n-1}$
(b) No; geometric

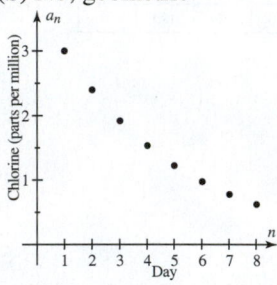

69. (a) $a_n = 5(0.85)^{n-1}$ **(b)** Geometric; the common
ratio is 0.85.   **(c)** About 1.6; on the 8th bounce the ball
reaches a maximum height of about 1.6 ft.
71. (a) Arithmetic; the common difference is 2.
(b) $a_n = 40 + 2(n - 1)$ or $a_n = 38 + 2n$ **(c)** 78

CHECKING BASIC CONCEPTS 14.1 & 14.2 (p. 926)

1. $\frac{1}{5}, \frac{1}{3}, \frac{3}{7}, \frac{1}{2}$

2.

n	1	2	3	4	5
a_n	2	3	4	5	6

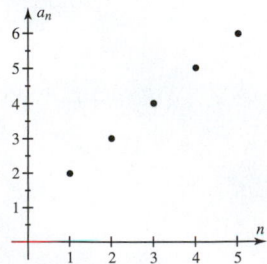

3. (a) Arithmetic; $a_n = 3n - 5$ **(b)** Geometric;
$a_n = 3(-2)^{n-1}$ **4.** $a_n = 2n + 3$ **5.** $a_n = 5(2)^{n-1}$

SECTION 14.3 (pp. 933–935)

1. series   **2.** 10   **3.** arithmetic   **4.** geometric
5. $n\left(\frac{a_1 + a_n}{2}\right)$ or $\frac{n}{2}(2a_1 + (n - 1)d)$ **6.** $a_1\left(\frac{1 - r^n}{1 - r}\right)$

7. sum   **8.** $a_1 + a_2 + a_3 + a_4$ **9.** arithmetic
10. geometric   **11.** 48   **13.** 820   **15.** -5 **17.** 3279
19. -85 **21.** 182   **23.** \$91,523.93   **25.** \$62,278.01
27. $2 + 4 + 6 + 8$; 20
29. $4 + 4 + 4 + 4 + 4 + 4 + 4 + 4$; 32
31. $1 + 4 + 9 + 16 + 25 + 36 + 49$; 140
33. $12 + 20$; 32   **35.** $\sum_{k=1}^{6} k^4$ **37.** $\sum_{k=1}^{5} \frac{1}{k^2}$
39. $\sum_{k=1}^{n} k = n\left(\frac{a_1 + a_n}{2}\right) = n\left(\frac{1 + n}{2}\right) = \frac{n(n + 1)}{2}$
41. (a) $8518 + 9921 + 10,706 + 14,035 + 14,307 + 12,249$
(b) 69,736   **43.** (a) $\sum_{k=1}^{n} 0.8(0.2)^{k-1}$ **(b)** 2
45. (a) $1, \frac{1}{2}, \frac{1}{4}, \frac{1}{8}, \frac{1}{16}$ **(b)** $\frac{1023}{512}$ **47.** 90 logs   **49.** \$1,080,000
51. About 3.16 feet   **53.** Answers may vary.

SECTION 14.4 (p. 941)

1. 5   **2.** $n + 1$ **3.** 4
4.
$$1$$
$$1 \quad 1$$
$$1 \quad 2 \quad 1$$
$$1 \quad 3 \quad 3 \quad 1$$
$$1 \quad 4 \quad 6 \quad 4 \quad 1$$
5. 24   **6.** $6! = 720$ **7.** $\frac{n!}{(n - r)! \, r!}$ **8.** $a^2 + 2ab + b^2$
9. n **10.** 1   **11.** $x^3 + 3x^2y + 3xy^2 + y^3$
13. $16x^4 + 32x^3 + 24x^2 + 8x + 1$
15. $a^5 - 5a^4b + 10a^3b^2 - 10a^2b^3 + 5ab^4 - b^5$
17. $x^6 + 3x^4 + 3x^2 + 1$ **19.** 6   **21.** 4   **23.** 2
25. 10   **27.** 5   **29.** 6   **31.** 1   **33.** 792   **35.** 126
37. 75,582   **39.** $m^3 + 3m^2n + 3mn^2 + n^3$
41. $x^4 - 4x^3y + 6x^2y^2 - 4xy^3 + y^4$
43. $8a^3 + 12a^2 + 6a + 1$
45. $x^5 + 10x^4 + 40x^3 + 80x^2 + 80x + 32$
47. $81 + 216m + 216m^2 + 96m^3 + 16m^4$
49. $8x^3 - 12x^2y + 6xy^2 - y^3$ **51.** a^8 **53.** $35x^4y^3$
55. $512m^9$

CHECKING BASIC CONCEPTS 14.3 & 14.4 (p. 942)

1. (a) Geometric   **(b)** Arithmetic   **2.** 312
3. -341 **4.** $x^4 - 4x^3y + 6x^2y^2 - 4xy^3 + y^4$
5. $x^3 + 6x^2 + 12x + 8$

CHAPTER 14 REVIEW (pp. 943–945)

1. 1, 8, 27, 64   **2.** 3, 1, -1, -3 **3.** $1, \frac{4}{5}, \frac{3}{5}, \frac{8}{17}$
4. $-2, 4, -8, 16$ **5.** $-2, 0, 4, 2$ **6.** 5, 3, 2, 1
7.

n	1	2	3	4	5	6	7
a_n	2	4	6	8	10	12	14

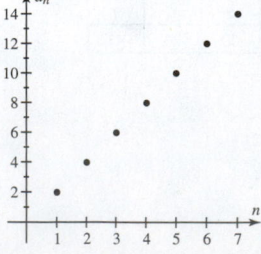

8.

n	1	2	3	4	5	6	7
a_n	-3	0	5	12	21	32	45

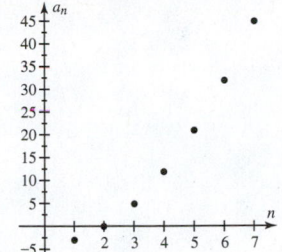

9.

n	1	2	3	4	5	6	7
a_n	2	1	0.5	0.25	0.125	0.0625	0.0313

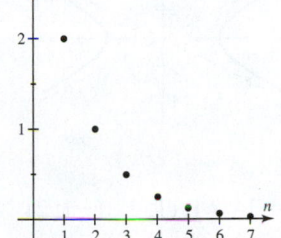

10.

n	1	2	3	4	5	6	7
a_n	1	1.4142	1.7321	2	2.2361	2.4495	2.6458

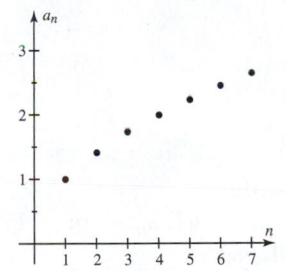

11. Yes; 5 **12.** No **13.** No **14.** Yes; $-\frac{1}{3}$ **15.** Yes; -3

16. No **17.** Yes; -1 **18.** No **19.** $a_n = 4n - 7$

20. $a_n = -5n + 7$ **21.** Yes; 4 **22.** No **23.** No

24. Yes; 0.7 **25.** No **26.** Yes; $-\frac{1}{3}$ **27.** No **28.** Yes; 2

29. $a_n = 5(0.9)^{n-1}$ **30.** $a_n = 2(4)^{n-1}$ **31.** 216 **32.** 7.5

33. 3277 **34.** $\frac{511}{256}$ **35.** $3 + 5 + 7 + 9 + 11$

36. $\frac{1}{2} + \frac{1}{3} + \frac{1}{4} + \frac{1}{5}$ **37.** $1 + 8 + 27 + 64$

38. $-1 + (-2) + (-3) + (-4) + (-5) + (-6)$

39. $\sum_{k=1}^{20} k$ **40.** $\sum_{k=1}^{20} \frac{1}{k}$ **41.** $\sum_{k=1}^{9} \frac{k}{k+1}$ **42.** $\sum_{k=1}^{7} k^2$

43. $x^3 + 12x^2 + 48x + 64$

44. $16x^4 + 32x^3 + 24x^2 + 8x + 1$

45. $x^5 - 5x^4y + 10x^3y^2 - 10x^2y^3 + 5xy^4 - y^5$

46. $a^6 - 6a^5 + 15a^4 - 20a^3 + 15a^2 - 6a + 1$

47. 6 **48.** 10 **49.** 20 **50.** 4

51. $m^4 + 8m^3 + 24m^2 + 32m + 16$

52. $a^5 + 5a^4b + 10a^3b^2 + 10a^2b^3 + 5ab^4 + b^5$

53. $x^4 - 12x^3y + 54x^2y^2 - 108xy^3 + 81y^4$

54. $27x^3 - 54x^2 + 36x - 8$

55. $a_n = 45,000(1.1)^{n-1}$ for $n = 1, 2, 3, \ldots, 7$; geometric

n	1	2	3	4	5	6	7
a_n	45,000	49,500	54,450	59,895	65,885	72,473	79,720

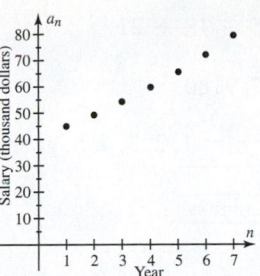

56. $a_n = 45,000 + 5000(n - 1)$ for $n = 1, 2, 3, \ldots, 7$; arithmetic

n	1	2	3	4	5	6	7
a_n	45,000	50,000	55,000	60,000	65,000	70,000	75,000

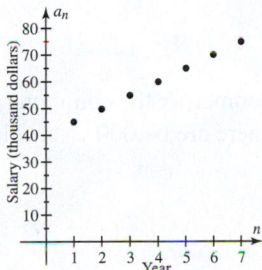

57. $a_n = 49n$ for $n = 1, 2, 3, \ldots, 7$

n	1	2	3	4	5	6	7
a_n	49	98	147	196	245	294	343

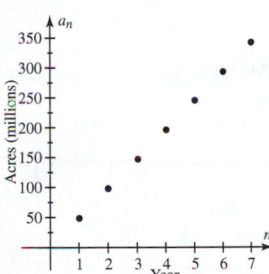

58. (a) $a_n = 1087(1.025)^{n-1}$ (b) Geometric; the common ratio is 1.025. (c) About 1200; the average mortgage payment in 2000 was about $1200 per month.

(d)

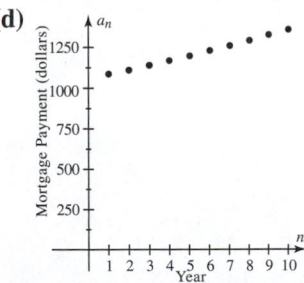

CHAPTER 14 TEST (p. 945)

1. $\frac{1}{2}, \frac{4}{3}, \frac{9}{4}, \frac{16}{5}$ **2.** $-3, 2, 1, -2, 3$

3.

n	1	2	3	4	5	6	7
a_n	0	2	6	12	20	30	42

4. $16x^4 - 32x^3 + 24x^2 - 8x + 1$

5. Arithmetic; -3 **6.** Geometric; -2

7. $a_n = 2 - 3(n - 1)$ or $a_n = 5 - 3n$

8. $a_n = 2(1.5)^{n-1}$ **9.** Yes; 2.5 **10.** No **11.** 99

12. $\frac{463}{729}$ **13.** $6 + 9 + 12 + 15 + 18 + 21$

14. $\sum_{k=1}^{60} k^3$ **15.** 35 **16.** 10 **17.** 9180

18. $a_n = 180,000(0.96)^{n-1}$ for $n = 1, 2, 3, 4, 5$; geometric

n	1	2	3	4	5
a_n	180,000	172,800	165,888	159,252	152,882

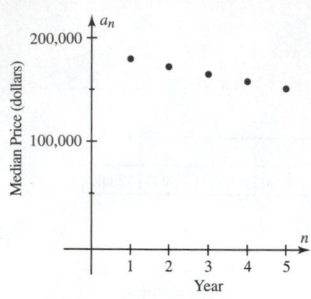

19. (a) $a_n = 2000(2)^{n-1}$ **(b)** Geometric; the common ratio is 2. **(c)** 64,000; after 25 days there are 64,000 caterpillars.
(d)

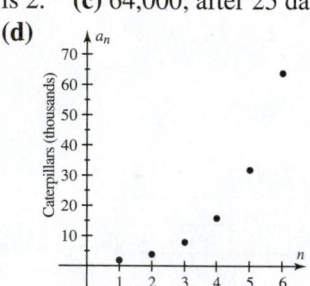

CHAPTERS 1–14 CUMULATIVE REVIEW (pp. 947–949)

1. Distributive **2.** $D = \{-6, -2, 0, 2\}$; $R = \{0, 1, 3, 5\}$
3. $\frac{y^9}{27x^5}$ **4.** $\frac{16a^8}{b^4}$ **5.** z^{10} **6.** $\frac{2y^3}{x^6}$ **7.** $D = \{x \mid x \neq 8\}$
8. $f(x) = -2x + 1$ **9.** $y = 3$ **10.** $-3; 5$
11. $y = \frac{3}{2}x + \frac{5}{2}$ **12.** $y = 2x - 8$ **13.** -26
14. $\left(\frac{20}{11}, \infty\right)$ **15.** $(-\infty, -1] \cup [11, \infty)$ **16.** $-15, 3$
17. $\frac{1}{8}$ **18.** $[-12, 9)$ **19.** $(3, -2)$
20.

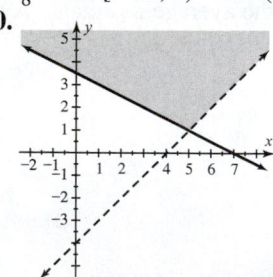

21. $(5, 2)$ **22.** $(3, 3, -1)$ **23.** $R = 22$ **24.** 1
25. $8x^7 - 6x^6 + 10x^3$ **26.** $6z^2 - 13z - 28$
27. $(2x - 3y)(2x + 3y)$ **28.** $(a^2 + 4)(2a - 1)$
29. $-\frac{3}{4}, 1$ **30.** $0, 4, 6$ **31.** $\frac{x - 2}{x + 1}$ **32.** $\frac{x + 4}{x - 2}$ **33.** 12
34. $\frac{1}{2}, 3$ **35.** $W = \frac{3C - 5R}{2}$ **36.** $\frac{1 + 2x}{1 - 4x}$
37. $(xy - 2)\sqrt[3]{xy}$ **38.** 14 **39.** $-2, 8$ **40.** $\frac{3}{2}$
41. $14 + 2i$ **42.** $-2 - 2i$ **43.** $-\frac{1}{3}$
44. $y = 2(x + 2)^2 + 9$; $(-2, 9)$ **45.** $2 \pm 3i$
46. $-4, 8$ **47. (a)** $-3, 1$ **(b)** $a > 0$ **(c)** Positive
48. $(-3, 1)$ **49. (a)** 65 **(b)** $(g \circ f)(x) = 3x^2 + 1$
50. $f^{-1}(x) = \frac{2x - 1}{3}$ **51.** $3 \ln x + \frac{1}{2} \ln y$ **52.** $\log \frac{x}{4y}$
53. 56.23 **54.** 0.58

55.

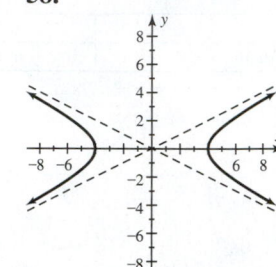

$(1, 3)$; $y = 3$
56. $(3, -1)$; 2
57.

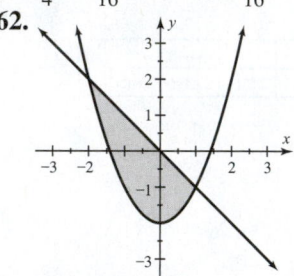

58.

59. $\frac{y^2}{4} - \frac{x^2}{16} = 1$ **60.** $\frac{x^2}{16} + \frac{y^2}{4} = 1$ **61.** $(1, 2), (-1, 2)$
62.

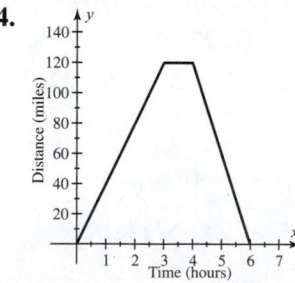

63. Arithmetic; -2 **64.** Geometric; 0.2
65. Geometric; 4 **66.** Arithmetic; 6 **67.** $a_n = 3n - 1$
68. $a_n = 4(3)^{n-1}$ **69.** 171 **70.** 683
71. $16x^4 + 96x^3 + 216x^2 + 216x + 81$
72. $8a^3 - 60a^2b + 150ab^2 - 125b^3$
73. $\sqrt{\frac{14}{\pi}} \approx 2.11$ inches
74.

75. (a) $f(x) = 0.4x + 85$ **(b)** 188 pounds **76.** Airplane: 380 mph; wind: 20 mph **77.** 9 feet by 12 feet **78.** 36 minutes **79. (a)** $xy = 96, 3x - y = 12$ **(b)** 8 and 12
80. 144

B   Sets

(pp. AP-17–AP-18)

1. $\{1, 2, 3, 4, 5, 6, 7\}$ **2.** $\{5, 6, 7, 8, 9, 10\}$ **3.** {Sunday, Saturday} **4.** {Alabama, Alaska, Arizona, Arkansas}
5. $\{A, B, C, D, E, F, G\}$ **6.** $\{D, E, F, G, H, I, J, K, L, M\}$
7. $\varnothing$ **8.** $\varnothing$ **9.** $\{0, 2, 4, 6, 8, 10\}$ **10.** $\{13, 15, 17, 19\}$

11. $\in$ **12.** $\in$ **13.** $\notin$ **14.** $\notin$ **15.** $\in$ **16.** $\in$ **17.** $\notin$
18. $\notin$ **19.** $\in$ **20.** $\in$ **21.** $E = \{2, 4, 6, 8, 10\}$
22. $E = \{-2, 0, 2\}$ **23.** $E = \{2, 4, 6, 8, 10, \ldots\}$
24. $E = \varnothing$ **25.** $A = \{\text{Apple, Apricot}\}$ **26.** $A = \{\text{Apple}\}$
27. $A = \{\text{Algebra}\}$ **28.** $A = \{\text{Algebra, Arithmetic}\}$
29. False   **31.** True   **33.** True   **35.** True   **37.** True
39. True   **41.** False   **43.** True

45. 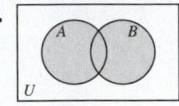   **47.**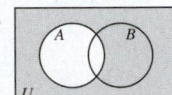

49. 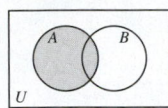   **51.**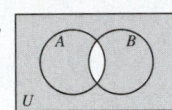

53. $\{1, 2, 3, 4\}$ **55.** $\{a, b\}$ **57.** $\{a, b, c\}$ **59.** $\{a, b, c, d\}$
61. $\{5, 8\}$ **63.** $\{1, 2, 3, 4, 5, 6, 7, 8, 9, 10\}$ **65.** $\{4, 5, 6\}$
67. $\varnothing$ **69.** $\{1, 3, 5, 7, 9\}$ **71.** $\varnothing$ **73.** $\varnothing$
75. $\{1, 3, 5, 7, 9\}$ **77.** $\{1, 2, 3, 5, 7, 8, 9, 10\}$ **79.** $\varnothing$
81. $\{1, 2, 8, 9, 10\}$ **83.** $\{4, 6\}$ **85.** $\{1, 2, 3, 4, 5, 6, 7, 8, 9, 10\}$
87. Infinite   **89.** Finite   **91.** Finite

C   Linear Programming

(pp. AP-23–AP-25)

1. linear programming   **2.** objective   **3.** feasible solutions
4. linear inequalities   **5.** vertex   **6.** objective

7.   **9.**
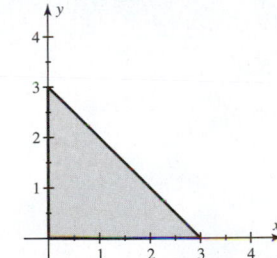   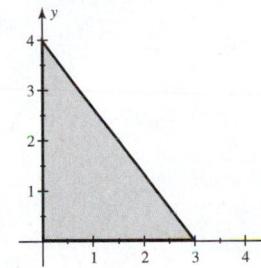

11.   **13.**
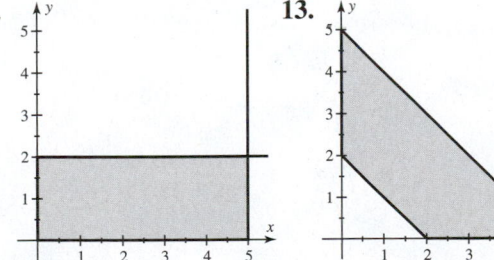

15.   **17.**
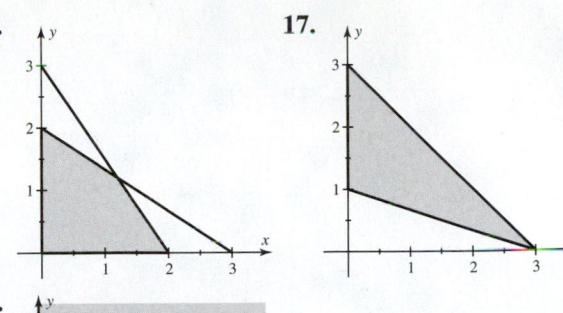

19.
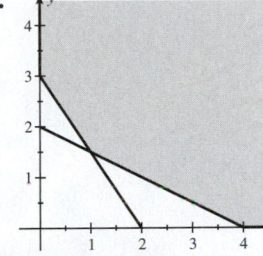

21. $R = 31$ **23.** $R = 15$ **25.** $C = 5$ **27.** $C = 2$
29. $R = 750$ **31.** $R = 12$ **33.** $R = 31.5$ **35.** $R = 16$
37. $C = 0$ **39.** $C = 32$ **41.** $C = 60$ **43.** 20 pounds of
candy, 80 pounds of coffee   **45.** \$1800   **47.** 1 ounce
Brand X, 2 ounces Brand Y   **49.** \$600

D   Synthetic Division

(p. AP-27)

1. $x + 4 + \dfrac{3}{x - 1}$

3. $3x - 1$

5. $x^2 + 3x + 2$

7. $2x^2 + 5x + 10 + \dfrac{19}{x - 2}$

9. $x^2 + 2x + 6 + \dfrac{28}{x - 4}$

11. $x^2 + 2x + 2$

13. $2x^3 - 4x^2 + 8x - 17 + \dfrac{34}{x + 2}$

15. $b^3 + b^2 + b + 1$

Glossary

absolute value A nonnegative number, written $|a|$, that is equal to the distance of a from the origin on the number line.

absolute value equation An equation that contains an absolute value.

absolute value function The function defined by $f(x) = |x|$.

absolute value inequality An inequality that contains an absolute value.

addends In an addition problem, the two numbers that are added.

addition property of equality If a, b, and c are real numbers, then $a = b$ is equivalent to $a + c = b + c$.

additive identity The number 0.

additive inverse (opposite) The additive inverse, or opposite, of a number a is $-a$.

adjacency matrix A matrix used to represent a map showing distances between cities or a social network.

algebraic expression An expression consisting of numbers, variables, operation symbols, such as $+$, $-$, $\times$, and $\div$, and grouping symbols, such as parentheses.

annuity A sum of money from which regular payments are made.

approximately equal The symbol $\approx$ indicates that two quantities are nearly equal.

arithmetic sequence A linear function given by $a_n = dn + c$ whose domain is the set of natural numbers.

arithmetic series The sum of the terms of an arithmetic sequence.

associative property for addition For any real numbers a, b, and c, $(a + b) + c = a + (b + c)$.

associative property for multiplication For any real numbers a, b, and c, $(a \cdot b) \cdot c = a \cdot (b \cdot c)$.

asymptotes of a hyperbola The two lines determined by the diagonals of the hyperbola's fundamental rectangle.

augmented matrix A matrix used to represent a system of linear equations; a vertical line is positioned in the matrix where the equals signs occur in the system of equations.

average The result of adding up the numbers of a set and then dividing the sum by the number of elements in the set.

axis of symmetry of a parabola The line passing through the vertex of the parabola that divides the parabola into two symmetric parts.

base The value of b in the expression b^n.

base e If the base of an exponential expression is e (approximately 2.72), then we say this expression has base e.

basic complex fraction A complex fraction where both the numerator and denominator are single fractions.

basic principle of fractions When simplifying fractions, the principle that states $\frac{a \cdot c}{b \cdot c} = \frac{a}{b}$.

basic rational equation A rational equation that has a single rational expression on each side of the equals sign.

binary operation An operation that requires two numbers to calculate an answer.

binomial A polynomial with two terms.

binomial coefficient The expression $_nC_r = n!/((n - r)!r!)$, where n and r are nonnegative integers, $n \geq r$, that can be used to calculate the numbers in Pascal's triangle.

binomial theorem A theorem that provides a formula to expand expressions of the form $(a + b)^n$.

braces { } used to enclose the elements of a set.

branches A hyperbola has two branches, a left branch and a right branch, or an upper branch and a lower branch.

byte A unit of computer memory, capable of storing one letter of the alphabet.

center of a circle The point that is a fixed distance from all the points on a circle.

center of an ellipse The midpoint of the major axis.

change of base formula A formula used to evaluate a logarithm of one base by using a logarithm of a different base and given by $\log_a x = \frac{\log x}{\log a}$ or $\log_a x = \frac{\ln x}{\ln a}$.

circle The set of points in a plane that are the same distance from a fixed point.

circumference The perimeter of a circle.

coefficient The numeric constant of a term.

coefficient of a monomial The number in a monomial.

common difference The value of d in an arithmetic sequence, $a_n = dn + c$.

common logarithmic function The function given by $f(x) = \log x$.

common logarithm of a positive number x Denoted $\log x$, it may be calculated as follows: If x is expressed as $x = 10^k$, then $\log x = k$, where k is a real number. That is, $\log 10^k = k$.

common ratio The value of r in a geometric sequence, $a_n = a_1(r)^{n-1}$.

commutative property for addition For any real numbers a and b, $a + b = b + a$.

commutative property for multiplication For any real numbers a and b, $a \cdot b = b \cdot a$.

complement The set containing all elements in the universal set that are not in A, denoted A'.

completing the square method An important technique in mathematics that involves adding a constant to a binomial so that a perfect square trinomial results.

complex conjugate The complex conjugate of $a + bi$ is $a - bi$.

complex fraction A rational expression that contains fractions in its numerator, denominator, or both.

complex number A complex number can be written in standard form as $a + bi$, where a and b are real numbers and i is the imaginary unit.

composite function If f and g are functions, then g of f, or the composition of g and f, is defined by $(g \circ f)(x) = g(f(x))$ and is read "g of f of x."

composite number A natural number greater than 1 that is not a prime number.

composition Replacing a variable with an algebraic expression in function notation is called composition of functions; when functions are applied to a variable in sequence.

compound inequality Two inequalities joined by the word *and* or the word *or*.

compound interest A type of interest paid at the end of each year using the formula $A = P(1 + r)^t$, where P is the amount deposited, r is the annual interest rate in decimal form, and t is the number of years for the account to contain A dollars.

conic section The curve formed by the intersection of a plane and a cone.

conjugate The conjugate of $a + b$ is $a - b$.

consistent system with dependent equations A system of linear equations with infinitely many solutions.

consistent system with independent equations A system of linear equations with exactly one solution.

constant function A linear function with $m = 0$ that can be written as $f(x) = b$.

constant of proportionality (constant of variation) In the equation $y = kx$, the nonzero number k.

constraint In linear programming, an inequality which limits the objective function.

continuous growth Growth in a quantity that is directly proportional to the amount present.

contradiction An equation that is always false regardless of the values of any variables.

Cramer's rule A method that uses determinants to solve linear systems of equations.

cube root The number b is a cube root of a if $b^3 = a$.

cube root function The function defined by $f(x) = \sqrt[3]{x}$.

cubic function A function f of degree 3 represented by $f(x) = ax^3 + bx^2 + cx + d$, where $a, b, c,$ and d are constants and $a \neq 0$.

decay factor The value of a in an exponential function, $f(x) = Ca^x$, when $0 < a < 1$.

degree A degree (°) is 1/360 of a revolution.

degree of a monomial The sum of the exponents of the variables.

degree of a polynomial The degree of the term (or monomial) with highest degree.

dependent equations Equations in a linear system that have infinitely many solutions.

dependent variable The variable that represents the output of a function.

determinant A real number associated with a square matrix.

diagrammatic representation A function represented by a diagram.

difference The answer to a subtraction problem.

difference of two cubes Expression in the form $a^3 - b^3$, which can be factored as $(a - b)(a^2 + ab + b^2)$.

difference of two squares Expression in the form $a^2 - b^2$, which can be factored as $(a - b)(a + b)$.

dimension of a matrix The size expressed in number of rows and columns. For example, if a matrix has m rows and n columns, its dimension is $m \times n$ (m by n).

directly proportional A quantity y is directly proportional to x if there is a nonzero number k such that $y = kx$.

discriminant The expression $b^2 - 4ac$ in the quadratic formula.

distance The distance d between the points (x_1, y_1) and (x_2, y_2) in the xy-plane is $d = \sqrt{(x_2 - x_1)^2 + (y_2 - y_1)^2}$.

distributive properties For any real numbers $a, b,$ and c, $a(b + c) = ab + ac$ and $a(b - c) = ab - ac$.

dividend In a division problem, the number being divided.

divisor In a division problem, the number being divided *into* the dividend.

domain The set of all x-values of the ordered pairs in a function.

element of a matrix Each number in a matrix.

elements of a set The members of a set.

elimination method A symbolic method used to solve a system of equations that is based on the property that if "equals are added to equals the results are equal."

ellipse The set of points in a plane the sum of whose distances from two fixed points is constant.

empty set (null set) A set that contains no elements.

equal sets If every element in a set A is in set B and every element in set B is in set A, then A and B are equal sets, denoted $A = B$.

equation A mathematical statement that two algebraic expressions are equal.

equation of a line Point-slope form and slope-intercept form are examples of an equation of a line.

equivalent equations Equations that have the same solution set.

even root The nth root, $\sqrt[n]{a}$, where n is even.

expansion of a determinant by minors A method of finding a 3×3 determinant by using determinants of 2×2 matrices.

exponent The value of n in the expression b^n.

exponential decay When $0 < a < 1$, the graph of $f(x) = Ca^x$ models exponential decay.

exponential equation An equation that has a variable as an exponent.

exponential expression An expression that has an exponent.

exponential function with base a and coefficient C A function represented by $f(x) = Ca^x$, where $a > 0, C > 0,$ and $a \neq 1$.

exponential growth When $a > 1$, the graph of $f(x) = Ca^x$ models exponential growth.

extraneous solution A solution that does not satisfy the given equation.

factorial notation $n! = 1 \cdot 2 \cdot 3 \cdot \cdots \cdot n$ for any positive integer.

factoring a polynomial The process of writing a polynomial as a product of lower degree polynomials.

factoring by grouping A technique that uses the associative and distributive properties by grouping four terms of a polynomial in such a way that the polynomial can be factored even though its greatest common factor is 1.

factors In a multiplication problem, the two numbers multiplied.

feasible solutions In linear programming, the set of solutions that satisfy the constraints.

finite sequence A function with domain $D = \{1, 2, 3, \ldots, n\}$ for some fixed natural number n.

finite series A series that contains a finite number of terms, and that can be expressed in the form $a_1 + a_2 + a_3 + \cdots + a_n$ for some n.

finite set A set whose elements can be listed explicitly.

focus (plural: **foci**) A fixed point used to determine the points that form a parabola, an ellipse, or a hyperbola.

FOIL A method for multiplying two binomials $(A + B)$ and $(C + D)$. Multiply **F**irst terms AC, **O**utside terms AD, **I**nside terms BC, and **L**ast terms BD; then combine like terms.

formula A special type of equation used to calculate one quantity from given values of other quantities.

function A set of ordered pairs (x, y), where each x-value corresponds to exactly one y-value.

function notation The notation $y = f(x)$, where the input x produces output y.

fundamental rectangle The rectangle of a hyperbola whose four vertices are determined by either $(\pm a, \pm b)$ or $(\pm b, \pm a)$, where $\frac{x^2}{a^2} - \frac{y^2}{b^2} = 1$ or $\frac{y^2}{a^2} - \frac{x^2}{b^2} = 1$.

Gauss–Jordan elimination A method used to solve a linear system in which matrix row transformations are applied to an augmented matrix.

general term (nth term) of a sequence a_n, where n is a natural number in the domain of a sequence $a_n = f(n)$.

geometric sequence An exponential function given by $a_n = a_1(r)^{n-1}$, where n is a natural number and $r \neq 0$ or 1.

geometric series The sum of the terms of a geometric sequence.

graphical representation A graph of a function.

graphical solution A solution to an equation obtained by graphing.

greater than If a real number b is located to the right of a real number a on the number line, we say that b is greater than a, and write $b > a$.

greater than or equal to If a real number a is greater than or equal to b, denoted $a \geq b$, then either $a > b$ or $a = b$ is true.

greatest common factor (GCF) of a polynomial The term with the highest degree and greatest coefficient that is a factor of all terms in the polynomial.

growth factor The value of a in the exponential function, $f(x) = Ca^x$, when $a > 1$.

half-life The time it takes for a radioactive sample to decay to half its original amount.

horizontal line test If every horizontal line intersects the graph of a function f at most once, then f is a one-to-one function.

hyperbola The set of points in a plane the difference of whose distances from two fixed points is constant.

identity An equation that is always true regardless of the values of any variables.

identity property of 1 If any number a is multiplied by 1, the result is a, that is, $a \cdot 1 = 1 \cdot a = a$.

identity property of 0 If 0 is added to any real number a, the result is a, that is, $a + 0 = 0 + a = a$.

imaginary number A complex number $a + bi$ with $b \neq 0$.

imaginary part The value of b in the complex number $a + bi$.

imaginary unit A number denoted i whose properties are $i = \sqrt{-1}$ and $i^2 = -1$.

improper fraction A fraction whose numerator is greater than or equal to its denominator in absolute value.

inconsistent system A system of linear equations that has no solution.

independent equations Equations in a linear system that have different graphs.

independent variable The variable that represents the input of a function.

index The value of n in the expression $\sqrt[n]{a}$.

index of summation The variable k in the expression $\sum_{k=1}^{n}$.

inequality When the equals sign in an equation is replaced with any one of the symbols $<$, $\leq$, $>$, or $\geq$, an inequality results.

infinite sequence A function whose domain is the set of natural numbers.

infinite set A set with infinitely many elements.

infinity Values that increase without bound.

input An element of the domain of a function.

integers A set of numbers including natural numbers, their opposites, and 0, or $\ldots, -3, -2, -1, 0, 1, 2, 3, \ldots$.

intercept form A linear equation in the form $x/a + y/b = 1$.

intersection The set containing elements that belong to *both A and B*, denoted $A \cap B$ and read "A intersect B."

intersection-of-graphs method A graphical technique for solving two equations.

interval notation A notation for number line graphs that eliminates the need to draw the entire line.

inverse function If f is a one-to-one function, then f^{-1} is the inverse function of f, if $(f^{-1} \circ f)(x) = f^{-1}(f(x)) = x$ for every x in the domain of f, and $(f \circ f^{-1})(x) = f(f^{-1}(x)) = x$ for every x in the domain of f^{-1}.

inversely proportional A quantity y is inversely proportional to x if there is a nonzero number k such that $y = k/x$.

irrational numbers Real numbers that cannot be expressed as fractions, such as π or $\sqrt{2}$.

joint variation A quantity z varies jointly with x and y if there is a nonzero number k such that $z = kxy$.

leading coefficient In a polynomial of one variable, the coefficient of the monomial with highest degree.

least common denominator (LCD) The common denominator with the fewest factors.

least common multiple (LCM) The smallest number that two or more numbers will divide into evenly.

less than If a real number a is located to the left of a real number b on the number line, we say that a is less than b and write $a < b$.

less than or equal to If a real number a is less than or equal to b, denoted $a \leq b$, then either $a < b$ or $a = b$ is true.

like radicals Radicals that have the same index and the same radicand.

like terms Two terms, or monomials, that contain the same variables raised to the same powers.

linear equation An equation that can be written in the form $ax + b = 0$, where $a \neq 0$.

linear equation in two variables An equation that can be written in the form $Ax + By = C$, where A, B, and C are fixed numbers and A and B are not both equal to 0.

linear function A function f represented by $f(x) = mx + b$, where m and b are constants.

linear inequality A linear inequality results whenever the equals sign in a linear equation is replaced with any one of the symbols $<$, $\leq$, $>$, or $\geq$.

linear inequality in two variables When the equals sign in a linear equation in two variables is replaced with $<$, $\leq$, $>$, or $\geq$, a linear inequality in two variables results.

linear polynomial A polynomial of degree 1 that can be written as $ax + b$, where $a \neq 0$.

linear programming problem A problem consisting of an objective function and a system of linear inequalities called constraints.

linear system in three variables A system of three equations in which each equation can be written in the form $ax + by + cz = d$; an ordered triple (x, y, z) is a solution to the system of equations if the values for x, y, and z make *all three* equations true.

line graph The resulting graph when consecutive data points in a scatterplot are connected with straight line segments.

logarithm with base a of a positive number x Denoted $\log_a x$, it may be calculated as follows: If x can be expressed as $x = a^k$, then $\log_a x = k$, where $a > 0$, $a \neq 1$, and k is a real number. That is, $\log_a a^k = k$.

logarithmic function with base a The function represented by $f(x) = \log_a x$.

logistic function A function used to model growth of a population.

lower limit In summation notation, the number representing the subscript of the first term of the series.

lowest terms A fraction is in lowest terms if its numerator and denominator have no factors in common.

main diagonal In an augmented matrix, the diagonal set of numbers from the upper left of the matrix to the lower right.

major axis The longer axis of an ellipse, which connects the vertices.

matrix A rectangular array of numbers.

matrix row transformations Operations performed on rows of an augmented matrix that result in an equivalent system of linear equations.

method of substitution A symbolic method for solving a system of equations in which one equation is solved for one of the variables and then the result is substituted into the other equation.

minor axis The shorter axis of an ellipse.

minors The 2×2 matrices that are used to find a determinant of a 3×3 matrix.

monomial A number, a variable, or a product of numbers and variables raised to natural number powers.

multiplication property of equality If a, b, and c are real numbers with $c \neq 0$, then $a = b$ is equivalent to $ac = bc$.

multiplicative identity The number 1.

multiplicative inverse (reciprocal) The multiplicative inverse of a nonzero number a is $1/a$.

name of function In the function given by $f(x)$, we call the function f.

natural exponential function The function represented by $f(x) = e^x$, where $e \approx 2.71828$.

natural logarithm The base-e logarithm, denoted either $\log_e x$ or $\ln x$.

natural numbers The set of (counting) numbers expressed as $1, 2, 3, 4, 5, 6, \ldots$.

negative infinity Values that decrease without bound.

negative reciprocals Slopes of two lines satisfy $m_1 = -\frac{1}{m_2}$, or $m_1 \cdot m_2 = -1$. These lines are perpendicular.

negative slope On a graph, the slope of a line that falls from left to right.

negative square root of a Denoted $-\sqrt{a}$.

nonlinear data If data points do not lie on a (straight) line, the data are nonlinear.

nonlinear function A function that is *not* a linear function; its graph is not a line.

nonlinear system of equations Two or more equations at least one of which is nonlinear.

nonlinear system of inequalities Two or more inequalities at least one of which is nonlinear.

nth root The number b is an nth root of a if $b^n = a$, where n is a positive integer.

nth term (general term) of a sequence See general term (nth term) of a sequence.

null set (empty set) A set that contains no elements.

numerical representation A table of values for a function.

numerical solution A solution often obtained by using a table of values.

objective function The given function to be optimized in a linear programming problem.

odd root The nth root, $\sqrt[n]{a}$, where n is odd.

one-to-one function A function f in which for any c and d in the domain of f, $c \neq d$ implies that $f(c) \neq f(d)$. That is, different inputs always result in different outputs.

opposite (additive inverse) The opposite, or additive inverse, of a number a is $-a$.

opposite of a polynomial The polynomial obtained by negating each term in a given polynomial.

optimal value In linear programming, the value that maximizes or minimizes the objective function.

ordered pair A pair of numbers written in parentheses (x, y), in which the order of the numbers is important.

ordered triple Can be expressed as (x, y, z), where x, y, and z are numbers and represent a solution to a linear system in three variables.

origin On the number line, the point associated with the real number 0; in the xy-plane, the point where the axes intersect, $(0, 0)$.

output An element of the range of a function.

parabola The $\cup$-shaped graph of a quadratic function that opens either upward or downward.

parallel lines Two or more lines in the same plane that never intersect; they have the same slope.

Pascal's triangle A triangle made up of numbers in which there are 1s along the sides and each element inside the triangle is the sum of the two numbers above it.

percent change If a quantity changes from x to y, then the percent change is $[(y - x)/x] \times 100$.

perfect cube An integer with an integer cube root.

perfect nth power The value of a if there exists an integer b such that $b^n = a$.

perfect square An integer with an integer square root.

perfect square trinomial A trinomial that can be factored as the square of a binomial, for example, $a^2 + 2ab + b^2 = (a + b)^2$.

perpendicular lines Two lines in a plane that intersect to form a right (90°) angle.

point–slope form The line with slope m passing through the point (x_1, y_1), given by the equation $y - y_1 = m(x - x_1)$ or, equivalently, $y = m(x - x_1) + y_1$.

polynomial The sum of one or more monomials.

polynomial functions in one variable Functions that are defined by a polynomial in one variable.

polynomials in one variable Polynomials that contain one variable.

positive slope On a graph, the slope of a line that rises from left to right.

power function A function that can be represented by $f(x) = x^p$, where p is a rational number.

prime factorization A number written as a product of prime numbers.

prime number A natural number greater than 1 that has *only* itself and 1 as natural number factors.

prime polynomial A polynomial with integer coefficients that cannot be factored by using integer coefficients.

principal *n*th root of *a* Denoted $\sqrt[n]{a}$.

principal square root The square root of a that is nonnegative, denoted $\sqrt{a}$.

probability A real number between 0 and 1, inclusive. A probability of 0 indicates that an event is impossible, whereas a probability of 1 indicates that an event is certain.

product The answer to a multiplication problem.

proportion A statement that two ratios are equal.

pure imaginary number A complex number $a + bi$ with $a = 0$ and $b \neq 0$.

Pythagorean theorem If a right triangle has legs a and b with hypotenuse c, then $a^2 + b^2 = c^2$.

quadrants The four regions determined by the xy-plane.

quadratic equation An equation that can be written in the form $ax^2 + bx + c = 0$, where a, b, and c are constants, with $a \neq 0$.

quadratic formula The solutions to the quadratic equation, $ax^2 + bx + c = 0$, $a \neq 0$, are $(-b \pm \sqrt{b^2 - 4ac})/(2a)$.

quadratic function A function f represented by the equation $f(x) = ax^2 + bx + c$, where a, b, and c are constants, with $a \neq 0$.

quadratic inequality If the equals sign in a quadratic equation is replaced with $>$, $\geq$, $<$, or $\leq$, a quadratic inequality results.

quadratic polynomial A polynomial of degree 2 that can be written as $ax^2 + bx + c$, with $a \neq 0$.

quotient The answer to a division problem.

radical expression An expression that contains a radical sign.

radical sign The symbol $\sqrt{\ }$ or $\sqrt[n]{\ }$ for some positive integer n.

radicand The expression under the radical sign.

radius The fixed distance between the center and any point on a circle.

range The set of all y-values of the ordered pairs in a function.

rate of change Slope can be interpreted as a rate of change. It indicates how fast the graph of a line is changing.

ratio A comparison of two quantities, expressed as a quotient.

rational equation An equation that contains one or more rational expressions.

rational expression A polynomial divided by a nonzero polynomial.

rational function A function defined by $f(x) = p(x)/q(x)$, where $p(x)$ and $q(x)$ are polynomials and the domain of f includes all x-values such that $q(x) \neq 0$.

rational number Any number that can be expressed as the ratio of two integers p/q, where $q \neq 0$; a fraction.

rationalizing the denominator The process of removing radicals from a denominator so that the denominator contains only rational numbers.

real numbers All rational and irrational numbers; any number that can be represented by decimal numbers.

real part The value of a in the complex number $a + bi$.

reciprocal (multiplicative inverse) The reciprocal of a nonzero number a is $1/a$.

rectangular coordinate system (*xy*-plane) The xy-plane used to plot points and graph data.

reduced row–echelon form A matrix form for representing a system of linear equations in which there are 1s on the main diagonal with 0s above and below each 1.

reflection If the point (x, y) is on the graph of a function, then $(x, -y)$ is on the graph of its reflection across the x-axis.

relation A set of ordered pairs.

rise The change in y between two points on a line, that is, $y_2 - y_1$.

root function In the power function $f(x) = x^p$, if $p = 1/n$, where $n \geq 2$ is an integer, then f is also a root function, which is given by $f(x) = \sqrt[n]{x}$.

run The change in x between two points on a line, that is, $x_2 - x_1$.

scatterplot A graph of distinct points plotted in the xy-plane.

scientific notation A real number a written as $b \times 10^n$, where $1 \leq |b| < 10$ and n is an integer.

set A collection of things.

set-builder notation Notation to describe a set of numbers without having to list all of the elements. For example, $\{x | x > 5\}$ is read as "the set of all real numbers x such that x is greater than 5."

slope The ratio of the change in y (rise) to the change in x (run) along a line. The slope m of a line passing through the points (x_1, y_1) and (x_2, y_2) is $m = (y_2 - y_1)/(x_2 - x_1)$, where $x_1 \neq x_2$.

slope–intercept form The line with slope m and y-intercept b is given by $y = mx + b$.

solution Each value of the variable that makes the equation true.

solution set The set of all solutions to an equation.

solution to a system In a system of two equations in two variables, an ordered pair, (x, y), that makes *both* equations true.

square matrix A matrix in which the number of rows and the number of columns are equal.

square root The number b is a square root of a number a if $b^2 = a$.

square root function The function given by $f(x) = \sqrt{x}$, where $x \geq 0$.

square root property If k is a nonnegative number, then the solutions to the equation $x^2 = k$ are given by $x = \pm \sqrt{k}$. If $k < 0$, then this equation has no real solutions.

standard equation of a circle The standard equation of a circle with center (h, k) and radius r is $(x - h)^2 + (y - k)^2 = r^2$.

standard form (of a linear equation in two variables) The form $Ax + By = C$, where A, B, and C are constants, with A and B not both 0.

standard form of a complex number $a + bi$, where a and b are real numbers.

standard form of a quadratic equation The equation given by $ax^2 + bx + c = 0$, where $a \neq 0$.

standard viewing rectangle of a graphing calculator Xmin $= -10$, Xmax $= 10$, Xscl $= 1$, Ymin $= -10$, Ymax $= 10$, and Yscl $= 1$, denoted $[-10, 10, 1]$ by $[-10, 10, 1]$.

subscript The symbol x_1 has a subscript of 1 and is read "x sub one" or "x one".

subset If every element in a set B is contained in a set A, then we say that B is a subset of A, denoted $B \subseteq A$.

sum The answer to an addition problem.

sum of the first n terms of an arithmetic sequence Denoted S_n, is found by averaging the first and nth terms and then multiplying by n.

sum of the first n terms of a geometric sequence Given by $S_n = a_1(1 - r^n)/(1 - r)$, if its first term is a_1 and its common ratio is r, provided $r \neq 1$.

sum of two cubes Expression in the form $a^3 + b^3$, which can be factored as $(a + b)(a^2 - ab + b^2)$.

summation notation Notation in which the uppercase Greek letter sigma represents the sum, for example,
$$\sum_{k=1}^{n} a_k = a_1 + a_2 + a_3 + \cdots + a_n.$$

symbolic representation Representing a function with a formula; for example, $f(x) = x^2 - 2x$.

symbolic solution A solution to an equation obtained by using properties of equations; the resulting solution set is exact.

synthetic division A shortcut that can be used to divide $x - k$, where k is a number, into a polynomial.

system of linear equations in two variables A system of equations in which each equation can be written as $Ax + By = C$.

system of linear inequalities in two variables Two or more linear inequalities to be solved at the same time, the solution to which must satisfy each inequality.

table of values An organized way to display the inputs and outputs of a function; a numerical representation.

term A number, a variable, or a product of numbers and variables raised to powers.

terms of a sequence $a_1, a_2, a_3, \ldots a_n, \ldots$ where the first term is $a_1 = f(1)$, the second term is $a_2 = f(2)$, and so on.

test point When graphing the solution set to an inequality, a point chosen to determine which region of the xy-plane to include in the solution set.

test value A real number chosen to determine the solution set to an inequality.

three-part inequality A compound inequality written in the form $a < x < b$, where $\leq$ may replace $<$.

translation The shifting of a graph upward, downward, to the right, or to the left in such a way that the shape of the graph stays the same.

transverse axis In a hyperbola, the line segment that connects the vertices.

trinomial A polynomial with three terms.

unary operation An operation that requires only one number.

undefined slope The slope of a line that is vertical.

union Denoted $A \cup B$ and read "A union B," it is the set containing any element that can be found in *either A or B*.

universal set A set that contains all elements under consideration.

upper limit In summation notation, the number representing the subscript of the last term of the series.

variable A symbol, such as x, y, or z, used to represent any unknown quantity.

varies directly A quantity y varies directly with x if there is a nonzero number k such that $y = kx$.

varies inversely A quantity y varies inversely with x if there is a nonzero number k such that $y = k/x$.

varies jointly A quantity z varies jointly with x and y if there is a nonzero number k such that $z = kxy$.

Venn diagrams Diagrams used to depict relationships between sets.

verbal representation A description, in words, of what a function computes.

vertex The lowest point on the graph of a parabola that opens upward or the highest point on the graph of a parabola that opens downward.

vertex form of a parabola The vertex form of a parabola with vertex (h, k) is $y = a(x - h)^2 + k$, where $a \neq 0$ is a constant.

vertical asymptote A vertical asymptote typically occurs in the graph of a rational function when the denominator of a rational expression equals 0 but the numerator does not equal 0; it can be represented by a vertical line in the graph of a rational function.

vertical line test If every vertical line intersects a graph at no more than one point, then the graph represents a function.

vertical shift A translation of a graph upward or downward.

vertices of an ellipse The endpoints of the major axis.

vertices of a hyperbola The endpoints of the transverse axis.

viewing rectangle (window) On a graphing calculator, the window that determines the x- and y-values shown in the graph.

whole numbers The set of numbers 0, 1, 2, 3, 4, 5, ...

x-axis The horizontal axis in the xy-plane.

x-coordinate The first value in an ordered pair.

x-intercept The x-coordinate of a point where a graph intersects the x-axis.

Xmax Regarding the viewing rectangle of a graphing calculator, Xmax is the maximum x-value along the x-axis.

Xmin Regarding the viewing rectangle of a graphing calculator, Xmin is the minimum x-value along the x-axis.

Xscl Regarding the viewing rectangle of a graphing calculator, the distance between consecutive tick marks on the x-axis.

xy-plane (rectangular coordinate system) The system used to plot points and graph data.

y-axis The vertical axis in the xy-plane.

y-coordinate The second value in an ordered pair.

y-intercept The y-coordinate of a point where a graph intersects the y-axis.

Ymax Regarding the viewing rectangle of a graphing calculator, Ymax is the maximum y-value along the y-axis.

Ymin Regarding the viewing rectangle of a graphing calculator, Ymin is the minimum y-value along the y-axis.

Yscl Regarding the viewing rectangle of a graphing calculator, the distance between consecutive tick marks on the y-axis.

zero of a polynomial An x-value that results in 0 when it is substituted into a polynomial; for example, the zeros of $x^2 - 4$ are 2 and -2.

zero-product property If the product of two numbers is 0, then at least one of the numbers must be 0, that is, $ab = 0$ implies $a = 0$ or $b = 0$ (or both).

zero slope The slope of a line that is horizontal.

Photo Credits

Bibliography

Baase, S. *Computer Algorithms: Introduction to Design and Analysis.* 2nd ed. Reading, Mass.: Addison-Wesley Publishing Company, 1988.

Beckmann, P. *A History of Pi.* New York: Barnes and Noble, Inc., 1993.

Brown, D., and P. Rothery. *Models in Biology: Mathematics, Statistics and Computing.* West Sussex, England: John Wiley and Sons Ltd, 1993.

Burden, R., and J. Faires. *Numerical Analysis.* 5th ed. Boston: PWS-KENT Publishing Company, 1993.

Callas, D. *Snapshots of Applications in Mathematics.* Delhi, New York: State University College of Technology, 1994.

Conquering the Sciences. Sharp Electronics Corporation, 1986.

Eves, H. *An Introduction to the History of Mathematics.* 5th ed. Philadelphia: Saunders College Publishing, 1983.

Freedman, B. *Environmental Ecology: The Ecological Effects of Pollution, Disturbance, and Other Stresses.* 2nd ed. San Diego: Academic Press, 1995.

Garber, N., and L. Hoel. *Traffic and Highway Engineering.* Boston, Mass.: PWS Publishing Co., 1997.

Greenspan, A. *The Economic Importance of Improving Math-Science Education,* Speech before the Committee on Education and the Workforce, U.S. House of Representatives, September 2000.

Grigg, D. *The World Food Problem.* Oxford: Blackwell Publishers, 1993.

Haefner, L. *Introduction to Transportation Systems.* New York: Holt, Rinehart and Winston, 1986.

Harrison, F., F. Hills, J. Paterson, and R. Saunders. "The measurement of liver blood flow in conscious calves." *Quarterly Journal of Experimental Physiology* 71: 235–247.

Historical Topics for the Mathematics Classroom, Thirty-first Yearbook. National Council of Teachers of Mathematics, 1969.

Howells, G. *Acid Rain and Acid Waters.* 2nd ed. New York: Ellis Horwood, 1995.

Kraljic, M. *The Greenhouse Effect.* New York: The H. W. Wilson Company, 1992.

Lack, D. *The Life of a Robin.* London: Collins, 1965.

Lancaster, H. *Quantitative Methods in Biological and Medical Sciences: A Historical Essay.* New York: Springer-Verlag, 1994.

Mannering, F., and W. Kilareski. *Principles of Highway Engineering and Traffic Analysis.* New York: John Wiley and Sons, 1990.

Mar, J., and H. Liebowitz. *Structure Technology for Large Radio and Radar Telescope Systems.* Cambridge, Mass.: The MIT Press, 1969.

Meadows, D. *Beyond the Limits.* Post Mills, Vermont: Chelsea Green Publishing Co., 1992.

Miller, A., and R. Anthes. *Meteorology.* 5th ed. Columbus, Ohio: Charles E. Merrill Publishing Company, 1985.

Miller, A., and J. Thompson. *Elements of Meteorology.* 2nd ed. Columbus, Ohio: Charles E. Merrill Publishing Company, 1975.

Paetsch, M. *Mobile Communications in the U.S. and Europe: Regulation, Technology, and Markets.* Norwood, Mass.: Artech House, Inc., 1993.

Pearl, R., T. Edwards, and J. Miner. "The growth of *Cucumis melo* seedlings at different temperatures." *J. Gen. Physiol.* 17: 687–700.

Pennycuick, C. *Newton Rules Biology.* New York: Oxford University Press, 1992.

Pokorny, C., and C. Gerald. *Computer Graphics: The Principles behind the Art and Science.* Irvine, Calif.: Franklin, Beedle, and Associates, 1989.

Ronan, C. *The Natural History of the Universe.* New York: MacMillan Publishing Company, 1991.

Ryan, B., B. Joiner, and T. Ryan. *Minitab Handbook.* Boston: Duxbury Press, 1985.

Smith, C. *Practical Cellular and PCS Design.* New York: McGraw-Hill, 1998.

Thomas, D. *Swimming Pool Operators Handbook.* National Swimming Pool Foundation of Washington, D.C., 1972.

Thomas, V. *Science and Sport.* London: Faber and Faber, 1970.

Thomson, W. *Introduction to Space Dynamics.* New York: John Wiley and Sons, 1961.

Triola, M. *Elementary Statistics.* 7th ed. Reading, Mass.: Addison-Wesley Publishing Company, 1998.

Tucker, A., A. Bernat, W. Bradley, R. Cupper, and G. Scragg. *Fundamentals of Computing I: Logic, Problem Solving, Programs, and Computers.* New York: McGraw-Hill, 1995.

Turner, R. K., D. Pierce, and I. Bateman. *Environmental Economics, An Elementary Approach.* Baltimore: The Johns Hopkins University Press, 1993.

Van Sickle, J. *GPS for Land Surveyors.* Chelsey, Mich.: Ann Arbor Press, 1996.

Varley, G., and G. Gradwell. "Population models for the winter moth." *Symposium of the Royal Entomological Society of London* 4: 132–142.

Wang, T. *ASHRAE Trans.* 81, Part 1 (1975): 32.

Weidner, R., and R. Sells. *Elementary Classical Physics,* Vol. 2. Boston: Allyn and Bacon, Inc., 1965.

Williams, J. *The Weather Almanac 1995.* New York: Vintage Books, 1994.

Wright, J. *The New York Times Almanac 1999.* New York: Penguin Group, 1998.

Zeilik, M., S. Gregory, and D. Smith. *Introductory Astronomy and Astrophysics.* 3rd ed. Philadelphia: Saunders College Publishers, 1992.

Index

Formulas from Geometry

Rectangle

$A = lw$

$P = 2l + 2w$

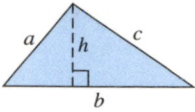

Triangle

$A = \frac{1}{2}bh$

$P = a + b + c$

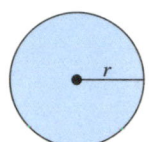

Circle

$A = \pi r^2$

$C = 2\pi r$

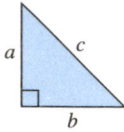

Pythagorean Theorem

$a^2 + b^2 = c^2$

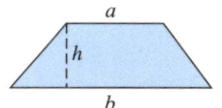

Trapezoid

$A = \frac{1}{2}(a + b)h$

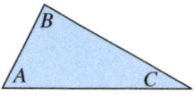

Sum of the Angles in a Triangle

$A + B + C = 180°$

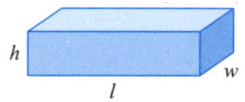

Rectangular Prism

$V = lwh$

$S = 2lw + 2lh + 2wh$

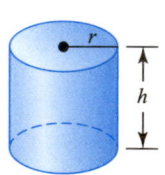

Circular Cylinder

$V = \pi r^2 h$

$S = 2\pi rh + 2\pi r^2$

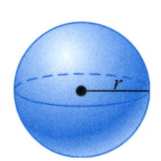

Sphere

$V = \frac{4}{3}\pi r^3$

$S = 4\pi r^2$

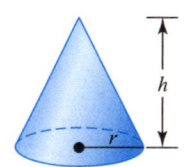

Cone

$V = \frac{1}{3}\pi r^2 h$

$S = \pi r\sqrt{r^2 + h^2} + \pi r^2$

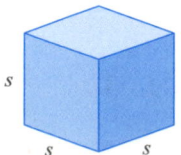

Cube

$V = s^3$

$S = 6s^2$

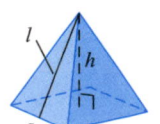

Square-Based Prism

$V = \frac{1}{3}s^2 h$

$S = s^2 + 2sl$